AP® Statistics Course Skills Al

The Practice of Statistics for the AP® Course, 7th Edition

The table below lists the skills that students are expected to develop in AP® Statistics. **Bold** references indicate the unit(s) of *TPS* 7e where the skill is explicitly addressed. The text provides frequent opportunities for students to practice each skill, and every example in the book cites the relevant skill(s) in the upper-right corner.

Skill Category 1	Skill Category 2	Skill Category 3	Skill Category 4
Selecting Statistical Methods	**Data Analysis**	**Using Probability and Simulation**	**Statistical Argumentation**
Select methods for collecting and/or analyzing data for statistical inference.	*Describe patterns, trends, associations, and relationships in data.*	*Explore random phenomena.*	*Develop an explanation or justify a conclusion using evidence from data, definitions, or statistical inference.*
1.A Identify the question to be answered or problem to be solved (*not assessed*). **Units 1–9**	**2.A** Describe data presented numerically or graphically. **Units 1 and 2**	**3.A** Determine relative frequencies, proportions, or probabilities using simulation or calculations. **Units 1, 4, 5**	**4.A** Make an appropriate claim or draw an appropriate conclusion. **Units 1–9**
1.B Identify key and relevant information to answer a question or solve a problem. **Units 1–9**	**2.B** Construct numerical or graphical representations of distributions. **Units 1 and 2**	**3.B** Determine parameters for probability distributions. **Units 4 and 5**	**4.B** Interpret statistical calculations and findings to assign meaning or assess a claim. **Units 1–9**
1.C Describe an appropriate method for gathering and representing data. **Gathering data: Unit 3** **Representing data: Units 1 and 2**	**2.C** Calculate summary statistics, relative positions of points within a distribution, correlation, and predicted response. **Units 1 and 2**	**3.C** Describe probability distributions. **Units 4 and 5**	
	2.D Compare distributions or relative positions of points within a distribution. **Units 1 and 2**		
INFERENCE			
1.D Identify an appropriate inference method for confidence intervals. **Units 6, 7, 9**		**3.D** Construct a confidence interval, provided the conditions for inference are met. **Units 6, 7, 9**	**4.C** Verify that inference procedures apply in a given situation. **Units 6–9**
1.E Identify an appropriate inference method for significance tests. **Units 6–9**		**3.E** Calculate a test statistic and find a *p*-value, provided the conditions for inference are met. **Units 6–9**	**4.D** Justify a claim based on a confidence interval. **Units 6, 7, 9**
1.F Identify null and alternative hypotheses. **Units 6–9**			**4.E** Justify a claim using a decision based on significance tests. **Units 6–9**

for the AP® Course

The Practice of Statistics

for the AP® Course

The Practice of Statistics

SEVENTH EDITION

Daren Starnes
Statistics Education Consultant

Josh Tabor
The Potter's School

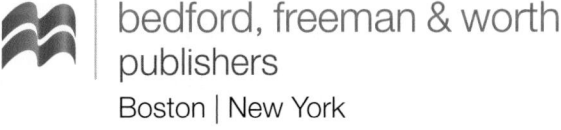

bedford, freeman & worth publishers
Boston | New York

Program Director, High School: Yolanda Cossio
Senior Program Manager: Raj Desai
Development Editor: Ann Heath
Editorial Assistant: Calyn Claire Liss
Director of High School Marketing: Janie Pierce-Bratcher
Senior Marketing Manager: Thomas Menna
Media Manager, High School: Lisa Samols
Senior Media Editor: Justin Perry
Senior Director, Content Management Enhancement: Tracey Kuehn
Executive Managing Editor: Michael Granger
Senior Manager, Publishing Services: Andrea Cava
Executive Content Project Manager: Vivien Weiss
Senior Workflow Project Manager: Paul Rohloff
Production Supervisor: Robert Cherry
Director of Design, Content Management: Diana Blume
Senior Design Services Manager: Natasha A. S. Wolfe
Interior Design: Maureen McCutcheon
Senior Cover Design Manager: John Callahan
Art Manager: Matthew McAdams
Illustrations: Lumina Datamatics, Inc.
Executive Permissions Editor: Robin Fadool
Director of Digital Production: Keri deManigold
Lead Media Project Manager: Jodi Isman
Composition: Lumina Datamatics, Inc.
Printing and Binding: Transcontinental

ISBN 978-1-319-40934-0

Library of Congress Control Number: 2023947318

Printed in Canada.
1 2 3 4 5 6 29 28 27 26 25 24

Acknowledgments and copyrights appear on the same page as the text and art selections they cover; these acknowledgments and copyrights constitute an extension of the copyright page.

Bedford, Freeman & Worth Publishers
120 Broadway, New York, NY 10271
bfwpub.com/catalog

In memory of Dan Yates

Dan Yates was a pioneer in the world of high school statistics education. With the launch of AP® Statistics in 1997, Dan quickly recognized the need to provide resources aimed specifically at high school students and teachers. He reconceived the successful college introductory statistics books by David Moore and George McCabe, and *The Practice of Statistics* was born. In the formative years of AP® Statistics, Dan provided invaluable support for countless educators who were new to teaching statistics at workshops, via email, and on the AP® Statistics electronic discussion board. Dan's generosity was exceeded only by his kindness. He was well liked and highly respected by all who knew him. Dan was a mentor and friend. We miss him dearly. But his spirit and influence will live on through the many contributions Dan made to our community.

Brief Contents

Contents

Appendices

Additional Online Content

About the Authors

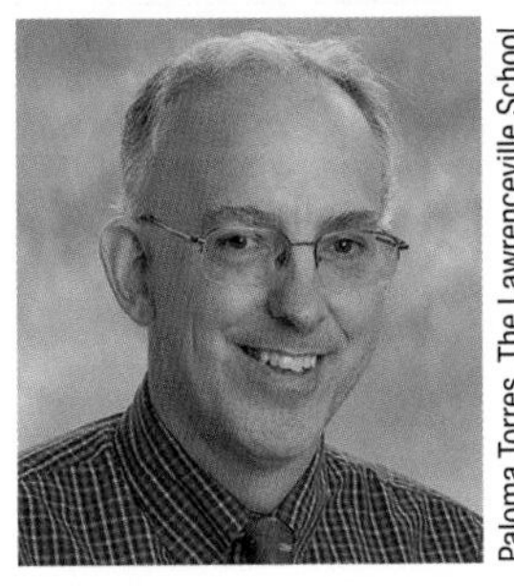

Paloma Torres, The Lawrenceville School

DAREN STARNES has taught a variety of statistics courses — including AP® Statistics, Introductory Statistics, and Mathematical Statistics — for 25 years. He earned his MA in Mathematics from the University of Michigan and his BS in Mathematics from the University of North Carolina at Charlotte. Daren has been a Reader, Table Leader, and Question Leader for the AP® Statistics exam for more than 20 years. As a College Board consultant since 1999, Daren has led hundreds of workshops for AP® Statistics teachers throughout the United States and overseas. He frequently presents in-person and online sessions about statistics teaching and learning for high school and college faculty. Daren is an active member of the National Council of Teachers of Mathematics (NCTM), the American Statistical Association (ASA), the American Mathematical Association of Two-Year Colleges (AMATYC), and the International Association for Statistical Education (IASE). He served on the ASA/NCTM Joint Committee on the Curriculum in Statistics and Probability for six years. While on the committee, he edited the *Guidelines for Assessment and Instruction in Statistics Education* (GAISE) pre-K–12 report. Daren is also coauthor of the popular on-level text *Statistics and Probability with Applications* (now in its fourth edition) and of the new college text *Introductory Statistics: A Student-Centered Approach*. Daren and his wife Judy enjoy traveling, rambling walks, jigsaw puzzles, and spending time with their three sons and seven grandchildren.

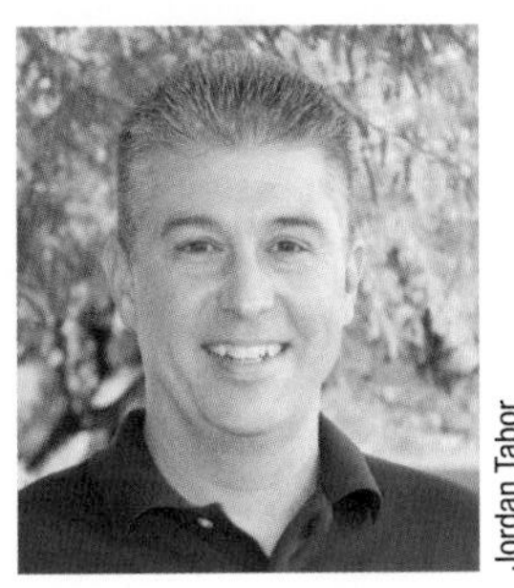

Jordan Tabor

JOSH TABOR has enjoyed teaching on-level and AP® Statistics to high school students for more than 26 years, most recently at The Potter's School. He received a BS in Mathematics from Biola University, in La Mirada, California. In recognition of his outstanding work as an educator, Josh was named one of the five finalists for Arizona Teacher of the Year in 2011. He is a past member of the AP® Statistics Development Committee (2005–2009) as well as an experienced Reader, Table Leader, Question Leader, and Exam Leader at the AP® Statistics Reading since 1999. In 2013, Josh was named to the SAT® Mathematics Development Committee. Each year, Josh leads one-week AP® Summer Institutes and one-day College Board workshops around the country. He also frequently speaks at local, national, and international conferences. In addition to teaching and speaking, Josh has authored articles in *The American Statistician, The Mathematics Teacher, STATS Magazine,* and *The Journal of Statistics Education.* Combining his love of statistics and love of sports, Josh teamed with Christine Franklin to write *Statistical Reasoning in Sports*, an innovative textbook for on-level statistics courses. Josh is also coauthor of the popular on-level text *Statistics and Probability with Applications*, Fourth Edition, and the new college text *Introductory Statistics: A Student-Centered Approach.* Outside of work, Josh enjoys gardening, traveling, and playing board games with his family.

Content Advisory Board and Program Resources Team

Cornell College

ANN CANNON, Cornell College, Mount Vernon, IA
Content Advisor, Accuracy Checker

Ann has served as Reader, Table Leader, Question Leader, and Assistant Chief Reader for the AP® Statistics exam for the past 17 years. She is also the 2017 recipient of the Mu Sigma Rho William D. Warde Statistics Education Award for lifetime devotion to the teaching of statistics. Ann has taught introductory statistics at the college level for 25 years and is very active in the Statistics Education Section of the American Statistical Association, currently serving as secretary/treasurer. She is coauthor of *STAT2: Modeling with Regression and ANOVA* and the new college text *Introductory Statistics: A Student-Centered Approach*.

Erica Chauvet

ERICA CHAUVET, Waynesburg University, PA
Content Advisor, Solutions, Assessments, Online Homework, Videos, Strive for a 5 Guide

Erica has more than 20 years of experience in teaching AP® Statistics, AP® Calculus, and college statistics. She has been an AP® Statistics Reader and Table Leader for the past 15 years. She also serves as the AP® Statistics Reading social director and famously hosts the annual 1.96-mile Prediction Fun Run. Since its inception in 2017, Erica has assisted as co-editor of the weekly *New York Times* feature: "What's Going On In This Graph?" Erica has worked as a writer, consultant, and reviewer of statistics and calculus material for the past 15 years.

Philip Savard

DOUG TYSON, Central York High School, York, PA
Content Advisor, Teacher's Edition Author, Videos

Doug has taught mathematics and statistics to high school and undergraduate students for more than 30 years. He has taught AP® Statistics for 15 years and has been active as an AP® Reader, Table Leader, and Question Leader for a decade. He serves as a Professional Development/Workshop Consultant for the College Board, running workshops and teacher training events in statistical education around the country. Doug may also be recognized as one of the AP® Daily Video presenters. He is the author of a College Board curriculum module on random sampling and random assignment, as well as of the Teacher's Edition for *Statistics and Probability with Applications*, Fourth Edition. He also serves on the NCTM/ASA Joint Committee on Curriculum in Statistics and Probability.

James Bush

JAMES BUSH, Waynesburg University, Waynesburg, PA
Content Advisor, Video Program Manager, Videos

James has taught introductory and advanced courses in statistics for more than 35 years. He is currently a Professor of Mathematics at Waynesburg University and is the recipient of the Lucas Hathaway Teaching Excellence Award. Active as an AP® Statistics Reader, Table Leader, and Rubric Team member, James also serves as a content consultant and conducts AP® Statistics preparation workshops for students and teachers.

Program Resources Team

To learn more about our team members go to: **bfwpub.com/TPS7eTeam**

Bob Amar, The Lovett School, Atlanta, GA	*Applets*
Beth Benzing, Strath Haven High School, Wallingford/Swarthmore School District, PA	*Activity Videos*
Bridget Matamoros, Guyer High School, Denton, TX	*Videos*
David Cardenas, New York City Museum School, New York, NY	*Videos*
David Wilcox, The Lawrenceville School, Lawrenceville, NJ	*Additional AP® Practice Exams for Units 1, 4, 6*
Dorothy Davis, Alpharetta High School, Fulton County, GA	*Videos*
Evan Quinter, University of Delaware, Newark, DE	*Videos*
Jeff Eicher, Jr., Eastern University, St. Davids, PA	*Videos*
Joanna Lu, Bergen County Technical High School, Teterboro, NJ	*Videos*
Joshua Sawyer, Elizabeth City – Pasquotank Public Schools, NC	*Videos*
Kathy Dickensheets, Pittsburgh Prep, Pittsburgh, PA	*Accuracy checker – assessment and online homework*
Leigh Nataro, Moravian Academy, Bethlehem, PA	*Tech Corner videos*
Nicholas Simonetti, Early College at Guilford, Greensboro, NC	*Online Homework*
Ricky Yip, Bonita High School, La Verne, CA	*Videos*
Sherri Sharbaffan, Central Park East High School, New York, NY	*Videos*
Thomas Rothery, Perry High School, Gilbert, AZ	*Online Homework*
Vicki Greenberg, The Lovett School, Atlanta, GA	*Videos*

Acknowledgments

First and foremost, we owe a tremendous debt of gratitude to David Moore and Dan Yates. Professor Moore reshaped the college introductory statistics course through publication of three pioneering texts: *Introduction to the Practice of Statistics (IPS)*, *The Basic Practice of Statistics (BPS)*, and *Statistics: Concepts and Controversies.* He was also one of the original architects of the AP® Statistics course. When the course first launched in the 1996–1997 school year, there were no textbooks written specifically for the high school student audience that were aligned to the AP® Statistics topic outline. Along came Dan Yates. His vision for such a text became reality with the publication of *The Practice of Statistics* (*TPS*) in 1998. More than a million students have used one of the first six editions of *TPS* for AP® Statistics! Dan also championed the importance of developing high-quality resources for AP® Statistics teachers, which were originally provided in a *Teachers' Resource Binder.* We stand on the shoulders of two giants in statistics education as we carry forward their visions in this and future editions.

The Practice of Statistics has continued to evolve, thanks largely to the support of our longtime editor and team captain, Ann Heath. Her keen eye for design is evident throughout the pages of the student and teacher's editions. More importantly, Ann's ability to oversee all of the complex pieces of this project while maintaining a good sense of humor is legendary. Ann has continually challenged everyone involved with *TPS* to innovate in ways that benefit AP® Statistics students and teachers. She is a good friend and an inspirational leader.

Teamwork is the secret sauce of *TPS*. We have been blessed to collaborate with many talented AP® Statistics teachers and introductory statistics college professors over the years we have been working on this project. We sincerely appreciate their willingness to give us candid feedback about early drafts of the student edition, and to assist with the development of an expanding cadre of resources for students and teachers.

On the seventh edition, we are especially grateful to the individuals who played lead roles in key components of the project. Ann Cannon did yeoman's work once again in reading, reviewing, and accuracy checking every line in the student edition. Her sage advice and willingness to ask tough questions were much appreciated throughout the writing of this edition of *TPS*. Doug Tyson took on the herculean task of producing the *Teacher's Edition* (TE). We know teachers will appreciate his careful thinking about effective pedagogy and the importance of engaging students with relevant context throughout the TE units. Doug also was a key contributor to the many videos that accompany *TPS* 7e.

Erica Chauvet wrote all of the solutions for *TPS* 7e exercises. Her thorough attention to matching the solutions to the details in the worked examples was exceeded only by her remarkable speed in completing this burdensome task. Erica also managed an overhaul of the online assessments in the Achieve platform, crafted prototype quizzes and tests, and revised the test bank. Kathy Dickensheets reviewed the solutions and other assessment items, offering many helpful suggestions.

James Bush is overseeing production of the vast collection of *TPS* 7e tutorial videos for students and teachers. We are thankful for James's expertise in video creation and for his willingness to pitch in wherever we need him. Bob Amar did amazing work developing the applets at www.stapplet.com, which are featured throughout the seventh edition.

Every member of the *TPS* 7e Content Advisory Board and Program Resources Team is an experienced teacher with significant involvement in the AP® Statistics program. In addition to the individuals mentioned earlier, we offer our heartfelt thanks to the following superstars for their tireless work and commitment to excellence: Beth Benzing, Bridget Matamoros, David Cardenas, David Wilcox, Dotty Davis, Evan Quinter, Jeff Eicher, Jr., Joanna Lu, Joshua Sawyer, Leigh Nataro, Nicholas Simonetti, Ricky Yip, Sherri Sharbaffan, Tom Rothery, and Vicki Greenberg.

Sincere gratitude also goes to everyone at Bedford, Freeman and Worth (BFW) involved in producing this new edition of *TPS*. Yolanda Cossio, Raj Desai, and Calyn Clare Liss joined the team for this edition and provided helpful guidance, leadership, and support. Vivien Weiss and Paul Rohloff oversaw the production process with their usual care and professionalism. Jill Hobbs kept us clear and consistent with her thoughtful copyediting. Natasha Wolfe, Maureen McCutcheon, Matt McAdams, and Joe Belbruno ensured that the design of the finished book and the art program exceeded our expectations. We are grateful to Lisa Samols and Justin Perry for their help in overseeing the development of the new Achieve program that supports the seventh edition. Special thanks go to Tommy Menna and all the dedicated people on the high school sales and marketing team at BFW who promote *TPS* 7e with enthusiasm. We also offer our thanks to Murugesh Rajkumar Namasivayam, Ranjani Saravanan, and the team at Lumina Datamatics for turning a complex manuscript into good-looking pages.

Thank you to all the reviewers over the past 20 years who have offered encouraging words and thoughtful suggestions for improvement in this and previous editions. And to the many outstanding statistics educators who have taken the time to share their questions and insights with us online, at conferences and workshops, at the AP® Reading, and in assorted other venues, we express our appreciation.

Daren Starnes and Josh Tabor

A final note from Daren: I feel extremely fortunate to have partnered with Josh Tabor in writing *TPS* 7e. He is a gifted teacher and talented author in his own right. Josh's willingness to once again oversee the revision of half of the units in this edition pays tribute to his unwavering commitment to excellence. He enjoys exploring new possibilities, which ensures that *TPS* will keep evolving in future editions. Josh is a good friend and trusted colleague.

My biggest thank you goes to my wife, Judy. She has made incredible sacrifices throughout my years as a textbook author. For Judy's unconditional love and support, I offer my utmost appreciation. She is truly my inspiration.

A final note from Josh: I have greatly enjoyed working with Daren Starnes on this edition of *TPS*. No one I know works harder or holds himself to a higher standard than he does. His wealth of experience and vision for this edition made him an excellent writing partner. For your friendship, encouragement, and support — thanks!

I especially want to thank the two most important people in my life. To my wife, Anne, your patience while I spent countless hours working on this project is greatly appreciated. I couldn't have survived without your consistent support and encouragement. To my daughter, Jordan, it's amazing how quickly you are growing up. I can't believe that you are already in my AP® Statistics class this year! I love you both very much.

To the Student

During the Covid-19 pandemic, researchers collected mountains of data. These data were used to assess the quality of diagnostic tests, to model the spread and impacts of the disease, and to test the effectiveness of vaccines and emerging therapeutics. The researchers relied on statistics to guide the collection, analysis, and interpretation of the data. Careful use of statistics enabled them to get clear answers to challenging questions in the face of uncertainty.

Statistics is also essential for tracking the economy, measuring changes in climate, gauging people's views about important social issues, and performing research in many other fields of study. Companies, governments, and other organizations regularly use statistics to make critical decisions that affect us all. In short, statistics matters!

The Practice of Statistics (*TPS*), Seventh Edition, is full of real-world data that illustrate the usefulness of statistics. Each set of data is presented along with some brief background to help you understand where the data come from. We deliberately chose contexts and data sets in the examples and exercises to pique your interest.

Statistical Thinking and You

The purpose of *TPS*, Seventh Edition, is to give you an enduring understanding of the big ideas of statistics and of the methods used in solving statistics problems. We want to help you develop the statistical thinking skills you will need to make informed, data-based decisions in your future studies, career, and daily life. AP® Statistics focuses on four course skill categories:

- *Selecting Statistical Methods:* Select methods for collecting and/or analyzing data for statistical inference.
- *Data Analysis:* Describe patterns, trends, associations, and relationships in data.
- *Using Probability and Simulation*: Explore random phenomena.
- *Statistical Argumentation:* Develop an explanation or justify a conclusion using evidence from data, definitions, or statistical inference.*

See the inside front cover for a full list of the course skills in each of these four categories.

The Practice of Statistics and AP® Statistics

The Practice of Statistics was the first book written specifically for the Advanced Placement (AP®) Statistics course. *TPS* 7e is organized to follow the AP® Statistics *Course and Exam Description* (CED) closely. Every learning objective and

*AP® *Statistics Course and Exam Description* (effective Fall 2020), the College Board.

essential knowledge statement in the CED course framework is covered thoroughly in the text.

Most importantly, *TPS* 7e is designed to prepare you for the AP® Statistics exam. The author team has more than 50 years combined experience teaching AP® Statistics and more than 45 years combined experience grading the AP® exam! Both of us have served on the AP® Reading leadership team for more than 15 years, helping finalize scoring guidelines for free-response questions. Including our Content Advisory Board and Program Resources Team (page xiii), we have very extensive knowledge of how the AP® Statistics exam is developed and scored.

The seventh edition of *TPS* will help you get ready for the AP® Statistics exam throughout the course by:

- **Using terms, notation, formulas, and tables consistent with those found in the CED and on the AP® Statistics exam.** Key terms are shown in bold in the text, defined in the Glossary, and cross-referenced in the Index. The formulas and notation used in *TPS* 7e are exactly the same as on the formula sheet and tables provided on both sections of the AP® Statistics exam, which can be found in the back of the book.
- **Following accepted conventions from AP® Statistics exam rubrics in model solutions.** Since the current CED was released in 2019, the scoring guidelines for free-response questions have been quite consistent. We carefully considered these guidelines when writing the solutions that appear in *TPS* 7e. For example, the four-step State–Plan–Do–Conclude process that we use to solve inference problems in Units 6–9 was inspired by AP® Statistics exam scoring rubrics.
- **Including specific AP® Statistics exam details and advice at point of use.**
    - We give numerous **AP® Exam Tips** as "on-the-spot" reminders of common mistakes and ways to avoid them. Specific **Formula Sheet** tips appear whenever we introduce a formula, so you know whether it is included on the AP® Statistics exam formula sheet. At the end of each section, we provide **AP® Daily Video** tips that alert you to videos available in AP® Classroom that review the content of the section.
    - The **Strategies for Success on the AP® Statistics Exam** appendix summarizes the structure and scoring of the exam, as well as our top tips for maximizing your exam score.
- **Providing about 1500 AP®-style exercises throughout the book.**
    - **Section Exercises** include paired odd- and even-numbered problems that test the same concept or skill from that section. *For Investigation* exercises ask you to apply the skills from the section in a new context or nonroutine way. Every section includes *Multiple Choice* exercises like those found on the AP® Statistics exam. *Recycle and Review* exercises at the end of each exercise set let you practice what you learned in previous sections.

- At the end of each Unit, or Unit Part, you will find a **FRAPPY!** — Free Response AP® Problem, Yay! Each FRAPPY! gives you the chance to answer an AP®-style free-response question based on the material in the unit. After you finish, you can view and critique two example solutions on the book's Student Site (bfwpub.com/tps7e). Then you can score your own response using a rubric provided by your teacher.
- **Unit Review Exercises** consist of free-response questions aligned to specific learning targets from the unit.
- The **AP® Statistics Practice Test** at the end of each unit (or part of a unit) will help you prepare for in-class exams. These tests include both multiple-choice and free-response questions, very much in the style of the AP® Statistics exam.
- The **Cumulative AP® Practice Tests** after Units 3, 5, 7, and 9 provide challenging, cumulative multiple-choice and free-response questions like those you might find on a midterm, final, or the AP® Statistics exam.

Learning with *The Practice of Statistics,* Seventh Edition

TPS 7e is designed to be easy to read and easy to use. This book is written by experienced high school AP® Statistics teachers, for high school students. We aimed to provide clear, concise explanations and to use a conversational approach that encourages you to read the book. We also tried to enhance both the visual appeal and the book's clear organization in the layout of the pages.

Be sure to take advantage of all that *TPS* 7e has to offer. You can gain a lot by reading the text, but you learn statistics best by *doing* statistics. To deepen your understanding of key concepts and skills, do the Activities, answer the Check Your Understanding questions along the way, and work lots of exercises. Use the answers at the back of the book to help identify strengths and weaknesses in your solutions.

The main ideas of statistics took a long time to discover and thus take some time to master. The basic principle of learning them is to be persistent.

Turn the page for a tour of the text. See how the book can help you realize success in the course and on the AP® Statistics exam.

Get the Most from This Program

Your AP® Statistics adventure begins here! This program was created with YOU in mind. It is packed with features and resources to help you learn effectively and to do well on the AP® Statistics exam.

Written to support your AP® Statistics journey.

Your book matches the course both in organization and in scope. Units, Parts, and Sections focus on the concepts and skills that are central to the **AP® STATISTICS COURSE.** Daily practice is built in to support you at every step.

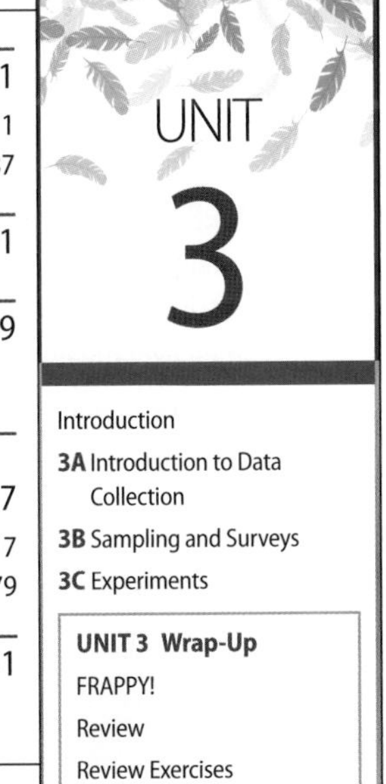

AP® EXAM TIP

Students often do not get full credit on the AP® Statistics exam because they use option (ii) with "calculator-speak" to show their work on normal distribution calculation questions — for example, normalcdf(−1000,6,6.84,1.55). This is *not* considered clear communication. To get full credit, be sure to carefully label each of the inputs in the calculator command if you use technology in Step 2: normalcdf(lower: −1000, upper: 6, mean: 6.84, SD: 1.55).

AP® EXAM TIP
Formula sheet

The regression line is defined as $\hat{y} = a + bx$ on the formula sheet provided on both sections of the AP® Statistics exam.

AP® EXAM TIP
AP® Daily Videos

Review the content of this section and get extra help by watching the AP® Daily Videos for Topics 1.1–1.3, which are available in AP® Classroom.

Pay attention to the Integrated **AP® EXAM TIPS** that support you in three ways:

- By providing advice from experts about how to be successful on the AP® Statistics exam
- By offering insights into the AP® Statistics formula sheet so you know where to focus your study
- By guiding you to the relevant AP® Daily Video that corresponds to each section in the text

Examples Are Your Friends

Study the examples for detailed instruction, with text and video guidance, for every concept and skill in the course.

Examples with paired exercises are the heart of the program.

EXAMPLES. Each example title is followed by the content and skills being taught and a "play button" link to a **Worked Example Video.** These videos were prepared by experienced AP® Statistics teachers to help you study, review, and get extra support when needed.

Example: Call me, maybe?

(b) Do these data support the claim that a majority of students prefer to communicate with their friends using text messaging? Justify your answer.

Relative frequency table

Preferred method	Relative frequency
In person	12/50 = 0.24 or 24%
Internet chat/IM	6/50 = 0.12 or 12%
Phone call	1/50 = 0.02 or 2%
Social media	6/50 = 0.12 or 12%
Text messaging	25/50 = 0.50 or 50%
Total	50/50 = 1.00 or 100%

EXAMPLE Call me, maybe?
Summarizing data with tables

Skills 2.A, 2.B

PROBLEM: Here are the data on preferred method of communicating with friends for all 50 students in the Census at School sample:

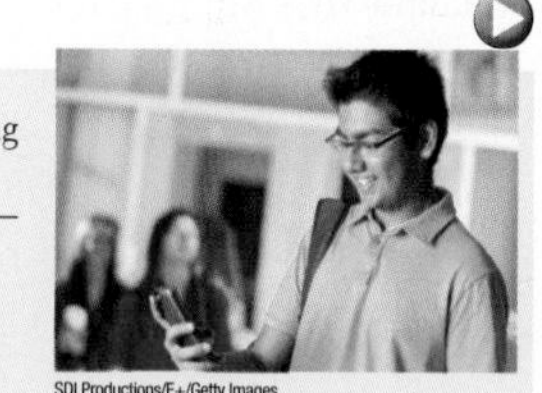
SDI Productions/E+/Getty Images

Internet chat/IM	In person	In person	Internet chat/IM	Phone call
Text messaging	Text messaging	Text messaging	Social media	Social media
Text messaging	Text messaging	Text messaging	In person	In person
Text messaging	Text messaging	Text messaging	In person	In person
Text messaging	Text messaging	Internet chat/IM	Social media	Text messaging
Text messaging	Text messaging	In person	Text messaging	Internet chat/IM
Text messaging	Social media	Text messaging	Social media	Text messaging
Text messaging	In person	Social media	In person	Text messaging
Text messaging	Text messaging	Text messaging	Internet chat/IM	In person
Text messaging	In person	Internet chat/IM	Text messaging	In person

(a) Make a frequency table and a relative frequency table to summarize the distribution of preferred communication method.
(b) Do these data support the claim that a majority of students prefer to communicate with their friends using text messaging? Justify your answer.

SOLUTION:

(a) Frequency table

Preferred method	Frequency
In person	12
Internet chat/IM	6
Phone call	1
Social media	6
Text messaging	25
Total	50

The frequency table shows the *number* of students who prefer each communication method. To create the frequency table, count how many students said "In person," how many said "Internet chat or instant messenger," and so on.

Relative frequency table

Preferred method	Relative frequency
In person	12/50 = 0.24 or 24%
Internet chat/IM	6/50 = 0.12 or 12%
Phone call	1/50 = 0.02 or 2%
Social media	6/50 = 0.12 or 12%
Text messaging	25/50 = 0.50 or 50%
Total	50/50 = 1.00 or 100%

The relative frequency table shows the *proportion* or *percentage* of students who prefer each communication method. Note that in statistics, a proportion is a value between 0 and 1 that is equivalent to a percentage.

(b) No. Exactly 50% of the students in the sample said that they prefer to communicate with their friends via text messaging, but a majority is more than half.

FOR PRACTICE, TRY EXERCISE 5

Note that the **SOLUTION** is presented in a special font to model the style, steps, and language you should use to earn full credit on the AP® Statistics exam.

THE VOICE OF THE TEACHER. Pay special attention to the carefully placed "teacher talk" boxes that guide you through the solution. These comments offer lots of good advice—as if your teacher is working directly with you to solve a problem.

PRACTICE! Take time to work the paired problem in the Section Exercises. Practicing a similar problem after studying the worked example helps you know whether or not you've "got it."

Exercises: Practice Makes Perfect!

Work the exercises assigned by your teacher. Watch the videos when you need extra help.

Time spent working exercises is key to success in this course.

SECTION EXERCISES ORGANIZED BY TOPIC
Use the handy headings to guide you to topics on which you need practice.

THE TPS "BACKWARD/FORWARD" NAVIGATION SYSTEM

Do you read the text first, or do you start with the assigned exercises? Either way, *TPS* is designed to support you.

Look for the icons that appear next to selected exercises. They will guide you

- Back to the **example** that models the exercise so that you can review the worked solution, if you are confused.
- To a short **video** that provides step-by-step instruction for solving the exercise.

Summarizing Data with Tables

5. **New York squirrels** How do squirrels in New York City's Central Park respond to humans? The Squirrel Census project enlisted volunteers to record data on all squirrels observed in Central Park in a recent year. One of the variables measures how the squirrels responded to humans. We selected a random sample of 75 of the more than 3000 squirrels observed. Here are the data on their reactions to humans:[7]

pg 6

Approach	Run from	Indifferent	Indifferent	Run from
Indifferent	Approach	Indifferent	Indifferent	Indifferent
Approa				
Indiffere				

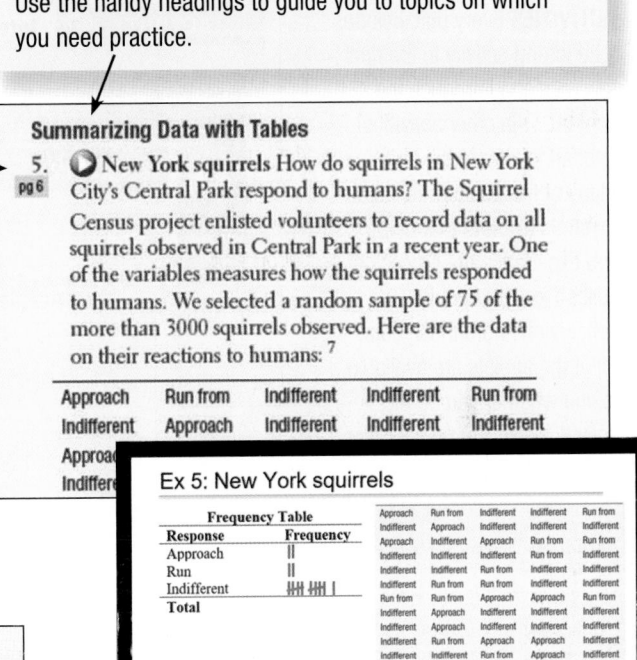

Ex 5: New York squirrels

Frequency Table

Response	Frequency
Approach	||
Run	||
Indifferent	~~||||~~ ~~||||~~ |
Total	

Approach	Run from	Indifferent	Indifferent	Run from
Indifferent	Approach	Indifferent	Indifferent	Indifferent
Approach	Indifferent	Approach	Run from	Run from
Indifferent	Indifferent	Indifferent	Run from	Indifferent
Indifferent	Indifferent	Run from	Indifferent	Indifferent
Indifferent	Run from	Run from	Indifferent	Indifferent
Run from	Run from	Approach	Approach	Run from
Indifferent	Approach	Indifferent	Indifferent	Indifferent
Indifferent	Approach	Indifferent	Indifferent	Indifferent
Indifferent	Run from	Approach	Approach	Indifferent
Indifferent	Indifferent	Run from	Approach	Indifferent
Indifferent	Approach	Indifferent	Run from	Approach
Run from	Approach	Run from	Indifferent	Indifferent
Indifferent	Approach	Indifferent	Run from	Indifferent
Indifferent	Indifferent	Approach	Approach	Indifferent

For Investigation *Apply the skills from the section in a new context or nonroutine way.*

38. **Population pyramids** The histograms show the age distribution of the population in Australia and Ethiopia in a recent ye[ar] ... tional databa[se] ... *population py[ramids]*

Multiple Choice *Select the best answer for each question.*

40. Here are the amounts of money (cents) in coins carried by 10 students in a statistics class: 50, 35, 0, 46, 86, 0, 5, 47, 23, 65. To ma[ke a stemplot of these data, you] would use stems

(A) 0, 2, 3, 4, 5, 6, ...

(B) 0, 1, 2, 3, 4, 5, ...

Recycle and Review *Practice what you learned in previous sections.*

46. **Marital Status (1B)** The side-by-side bar graph compares the marital status of U.S. adult residents (18 years old or older) in 1980 and 2020.[63] Compare the distributions of marital status for these two years.

Additional problem types in the Section Exercises let you practice solving many different types of questions, including:

- **For Investigation** exercises that assess multiple skills and concepts, and let you apply your knowledge in nonroutine ways.
- **AP®-style Multiple Choice** questions that give you regular practice for the AP® Statistics exam.
- **Recycle and Review** problems that refer back to concepts and skills learned in an earlier section, as noted in the references in parentheses (1B).

Learn Statistics by Doing Statistics

Experience statistical principles with simulations and technology tools.

ACTIVITIES Every unit includes a hands-on activity in the first few pages to introduce the content, with other activities appearing later in the unit. Many of the 30 activities use dynamic **applets** that help you experience the process of collecting data and drawing conclusions from those data. All of the applets are available to you whether you do the activity in class or on your own.

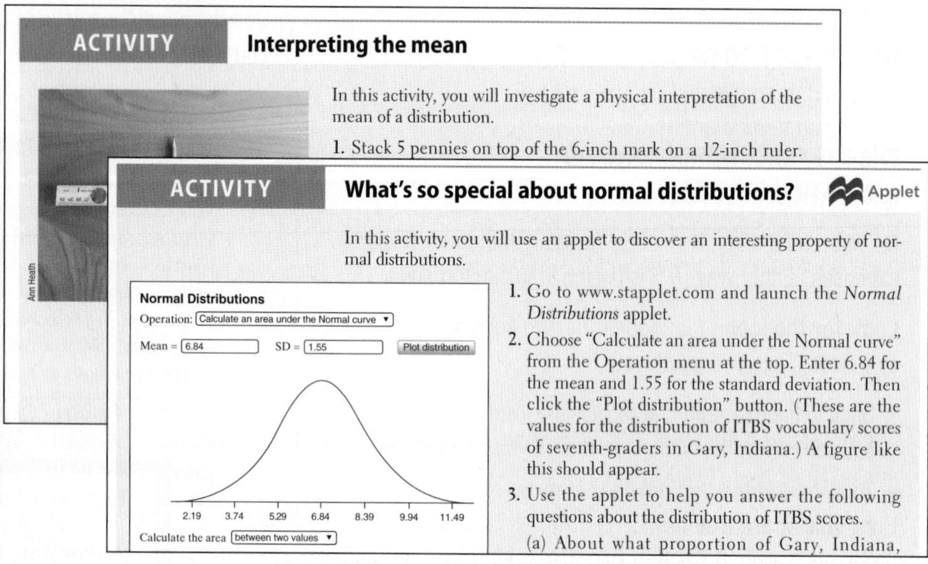

ACTIVITY **Interpreting the mean**

In this activity, you will investigate a physical interpretation of the mean of a distribution.

1. Stack 5 pennies on top of the 6-inch mark on a 12-inch ruler.

ACTIVITY **What's so special about normal distributions?** Applet

In this activity, you will use an applet to discover an interesting property of normal distributions.

1. Go to www.stapplet.com and launch the *Normal Distributions* applet.
2. Choose "Calculate an area under the Normal curve" from the Operation menu at the top. Enter 6.84 for the mean and 1.55 for the standard deviation. Then click the "Plot distribution" button. (These are the values for the distribution of ITBS vocabulary scores of seventh-graders in Gary, Indiana.) A figure like this should appear.
3. Use the applet to help you answer the following questions about the distribution of ITBS scores.
   (a) About what proportion of Gary, Indiana,

TECHNOLOGY Use technology as a tool for discovery and analysis. The 30 **Tech Corners,** placed strategically throughout the book at the optimal point of use, give step-by-step instructions for using the TI-83/84 calculator. Instructions for additional calculators are available on the book's website.

Supporting **Tech Corner Videos** are available to walk you through the keystrokes needed to perform each analysis.

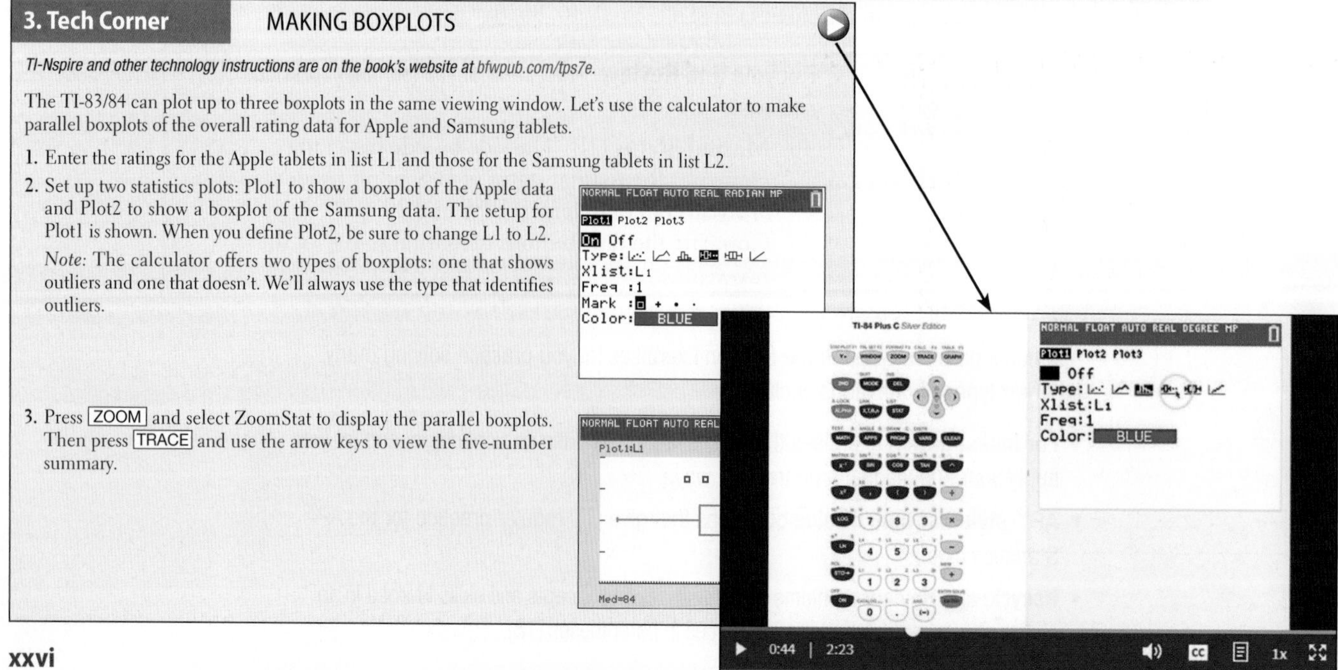

3. Tech Corner MAKING BOXPLOTS

TI-Nspire and other technology instructions are on the book's website at bfwpub.com/tps7e.

The TI-83/84 can plot up to three boxplots in the same viewing window. Let's use the calculator to make parallel boxplots of the overall rating data for Apple and Samsung tablets.

1. Enter the ratings for the Apple tablets in list L1 and those for the Samsung tablets in list L2.
2. Set up two statistics plots: Plot1 to show a boxplot of the Apple data and Plot2 to show a boxplot of the Samsung data. The setup for Plot1 is shown. When you define Plot2, be sure to change L1 to L2.
   Note: The calculator offers two types of boxplots: one that shows outliers and one that doesn't. We'll always use the type that identifies outliers.
3. Press ZOOM and select ZoomStat to display the parallel boxplots. Then press TRACE and use the arrow keys to view the five-number summary.

Achieve bfw publishers **Your Goals**

Access everything you need for success through the ACHIEVE platform.

Read the text, watch videos, and receive guided feedback on your homework.

EASY ACCESS TO RESOURCES WITH THE E-BOOK Interactive and mobile-ready, the e-book allows you to read and reference the text when you are online and offline. All offline highlights and notes sync when you connect to the internet. What's more, clicking on ▶ gives you instant access to each video while you are online.

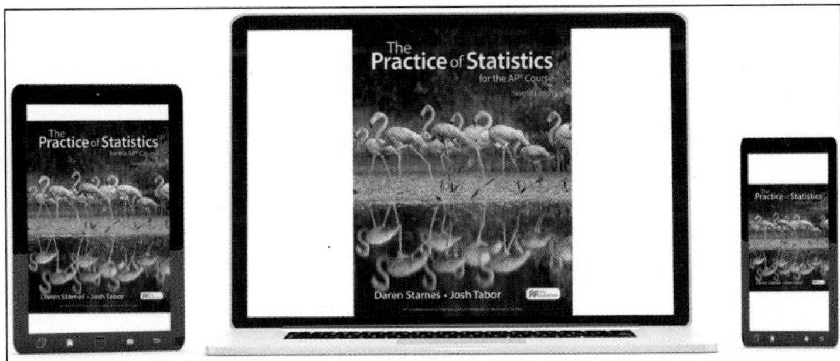

The **online homework** system helps you learn with targeted feedback based on common misconceptions, so you learn even if you get the answer wrong!

Hundreds of practice problems from the text are available, including the **Check Your Understanding** problems, selected **Section Exercises**, all **Unit/Part Review exercises,** and each **Unit/Part AP® Statistics Practice Test.**

Sprinkled throughout the text after key concepts and skills are presented, **Check Your Understanding** questions do just that. Working them in Achieve or on paper will help you stay on track.

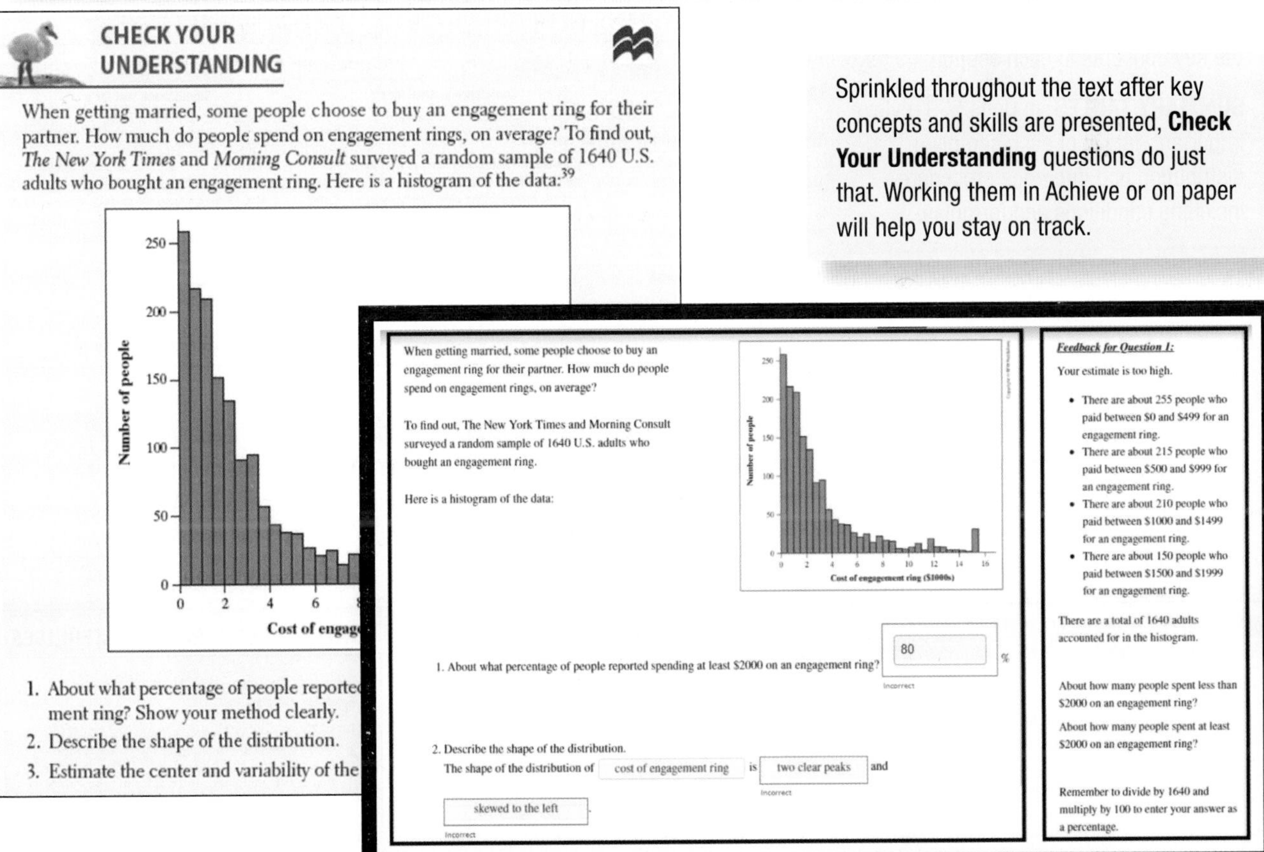

CHECK YOUR UNDERSTANDING

When getting married, some people choose to buy an engagement ring for their partner. How much do people spend on engagement rings, on average? To find out, *The New York Times* and *Morning Consult* surveyed a random sample of 1640 U.S. adults who bought an engagement ring. Here is a histogram of the data:[39]

1. About what percentage of people reporte[...] ment ring? Show your method clearly.
2. Describe the shape of the distribution.
3. Estimate the center and variability of the [...]

When getting married, some people choose to buy an engagement ring for their partner. How much do people spend on engagement rings, on average?

To find out, The New York Times and Morning Consult surveyed a random sample of 1640 U.S. adults who bought an engagement ring.

Here is a histogram of the data:

1. About what percentage of people reported spending at least \$2000 on an engagement ring? 80 %
Incorrect

2. Describe the shape of the distribution.
The shape of the distribution of cost of engagement ring is two clear peaks and skewed to the left.
Incorrect

Feedback for Question 1:

Your estimate is too high.

- There are about 255 people who paid between \$0 and \$499 for an engagement ring.
- There are about 215 people who paid between \$500 and \$999 for an engagement ring.
- There are about 210 people who paid between \$1000 and \$1499 for an engagement ring.
- There are about 150 people who paid between \$1500 and \$1999 for an engagement ring.

There are a total of 1640 adults accounted for in the histogram.

About how many people spent less than \$2000 on an engagement ring?

About how many people spent at least \$2000 on an engagement ring?

Remember to divide by 1640 and multiply by 100 to enter your answer as a percentage.

Review and Practice for Quizzes and Tests

Take advantage of the review tools included at the end of every Unit and Part.

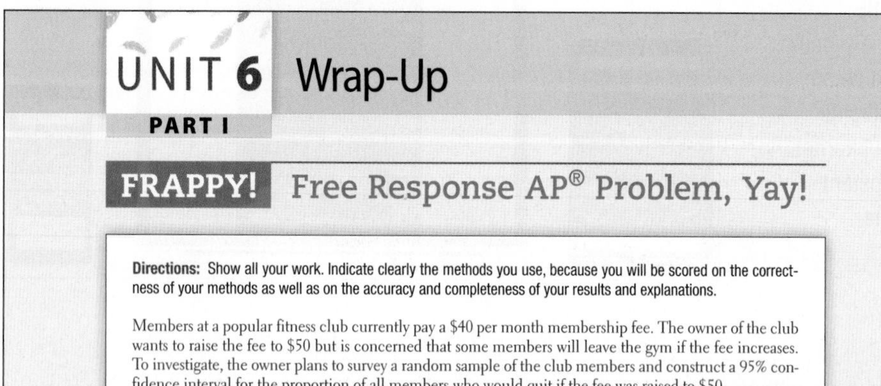

UNIT 6 Wrap-Up

PART I

FRAPPY! Free Response AP® Problem, Yay!

Directions: Show all your work. Indicate clearly the methods you use, because you will be scored on the correctness of your methods as well as on the accuracy and completeness of your results and explanations.

Members at a popular fitness club currently pay a $40 per month membership fee. The owner of the club wants to raise the fee to $50 but is concerned that some members will leave the gym if the fee increases. To investigate, the owner plans to survey a random sample of the club members and construct a 95% confidence interval for the proportion of all members who would quit if the fee was raised to $50.

DO THE FRAPPY! Learn how to answer FRQs successfully by working the FRAPPY!— the Free Response AP® Problem, Yay!—that begins each Unit/Part Wrap-up.

SNAPSHOT REVIEW Study the Unit Review, which gives a short summary of each section, to be sure you understand the key concepts in each section.

SUMMARY TABLES in Units 5–9 review important details of each sampling distribution and inference procedure, including conditions and formulas.

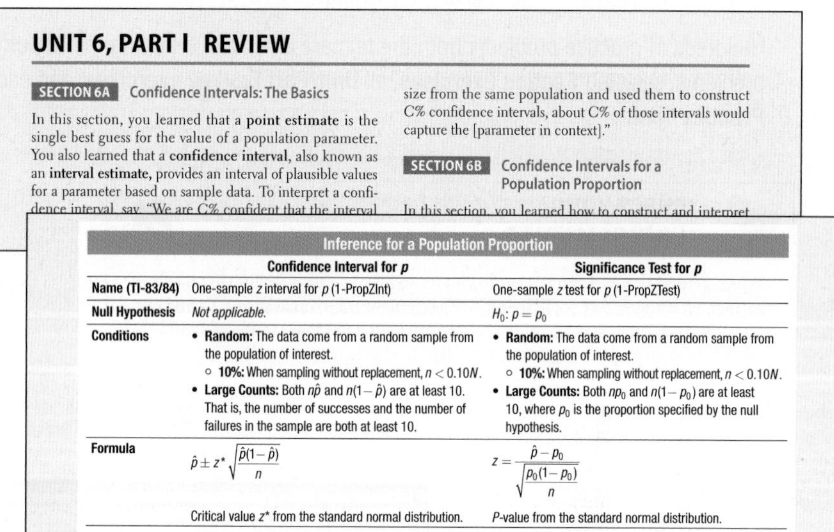

UNIT 6, PART I REVIEW

SECTION 6A Confidence Intervals: The Basics

In this section, you learned that a **point estimate** is the single best guess for the value of a population parameter. You also learned that a **confidence interval**, also known as an **interval estimate**, provides an interval of plausible values for a parameter based on sample data. To interpret a confidence interval, say, "We are C% confident that the interval

size from the same population and used them to construct C% confidence intervals, about C% of those intervals would capture the [parameter in context]."

SECTION 6B Confidence Intervals for a Population Proportion

In this section, you learned how to construct and interpret

Inference for a Population Proportion

	Confidence Interval for p	Significance Test for p
Name (TI-83/84)	One-sample z interval for p (1-PropZInt)	One-sample z test for p (1-PropZTest)
Null Hypothesis	*Not applicable.*	$H_0: p = p_0$
Conditions	• **Random:** The data come from a random sample from the population of interest. ○ **10%:** When sampling without replacement, $n < 0.10N$. • **Large Counts:** Both $n\hat{p}$ and $n(1-\hat{p})$ are at least 10. That is, the number of successes and the number of failures in the sample are both at least 10.	• **Random:** The data come from a random sample from the population of interest. ○ **10%:** When sampling without replacement, $n < 0.10N$. • **Large Counts:** Both np_0 and $n(1-p_0)$ are at least 10, where p_0 is the proportion specified by the null hypothesis.
Formula	$\hat{p} \pm z^* \sqrt{\frac{\hat{p}(1-\hat{p})}{n}}$ Critical value z^* from the standard normal distribution.	$z = \frac{\hat{p} - p_0}{\sqrt{\frac{p_0(1-p_0)}{n}}}$ P-value from the standard normal distribution.

Use the **WHAT DID YOU LEARN?** table to verify your mastery of each topic and to find help when needed. All of the individual **LEARNING TARGETS** are listed with references to the sections in which they are introduced, as well as related examples and relevant Unit Review Exercises. And, of course, watch the Unit Review Exercise Videos to get tips on solving these multifaceted problems.

What Did You Learn?

Learning Target	Section	Related Example on Page(s)	Relevant Unit Review Exercise(s)
Interpret a confidence interval in context.	6A	533	R1, R2
Use a confidence interval to make a decision about the value of a parameter.	6A	534	R1, R5
Interpret a confidence level in context.	6A	536	R1

UNIT 6, PART I REVIEW EXERCISES

These exercises are designed to help you review the important concepts and skills of the unit.

R1 **Sports fans (6A)** Are you a sports fan? That's the question the Gallup polling organization asked a random sample of 1527 U.S. adults.[37] Gallup reported that a 95% confidence interval for the proportion of all U.S. adults who are sports fans is 0.565 to 0.615.

(a) Interpret the confidence interval.

(b) Interpret the confidence level.

(c) Based on the interval, is there convincing evidence that a majority of U.S. adults are sports fans? Explain your answer.

... and for the AP® Statistics Exam

Consistent daily practice will help you avoid cramming before the AP® Statistics exam.

Each Unit/Part concludes with an **AP® STATISTICS PRACTICE TEST** that usually consists of 10 multiple-choice questions and 3 free-response questions.

UNIT 6, PART I AP® STATISTICS PRACTICE TEST

Section I: Multiple Choice *Select the best answer for each question.*

T1 A confidence interval for a proportion is 0.272 to 0.314. What are the point estimate and the margin of error for this interval?
(A) 0.293, 0.021
(B) 0.293, 0.042
(C) 0.293, 0.084
(D) 0.586, 0.021
(E) 0.586, 0.042

(A) 0.05
(B) 0.10
(C) 0.20
(D) 0.40
(E) 0.90

T5 Bags of a certain brand of tortilla chips are labeled as having a net weight of 14 ounces. A representative of a consumer advocacy group wishes to see if there

Section II: Free Response *Show all your work. Indicate clearly the methods you use, because you will be graded on the correctness of your methods as well as on the accuracy and completeness of your results and explanations.*

T11 The U.S. Forest Service is considering additional restrictions on the number of vehicles allowed to enter Yellowstone National Park. To assess public reaction, the service asks a random sample of 150 visitors if they favor the proposal. Of these, 89 say "Yes."
(a) Construct and interpret a 99% confidence interval for the proportion of all visitors to Yellowstone who favor the restrictions.
(b) Interpret the confidence level.

(a) Do these data provide convincing evidence for the Pew Research Center's claim?
(b) Which type of error, Type I or Type II, could you have made in part (a)? Explain your answer.

T13 A mattress manufacturer has designed a new production process that is intended to reduce the proportion of mattresses with defects in the stitching. Before the change, 15% of the mattresses had this type of defect. After the new production process is implemented, 100

Cumulative AP® Practice Test 1

Section I: Multiple Choice *Choose the best answer for each question.*

AP1 You look at real estate ads for houses in Sarasota, Florida. Many houses have prices from $200,000 to $400,000. The few houses on the water, however, have prices up to $15 million. Which of the following statements best describes the distribution of home prices in Sarasota?
(A) The distribution is most likely skewed to the left, and the mean is greater than the median.
(B) The distribution is most likely skewed to the left, and the mean is less than the median.
(C) The distribution is roughly symmetric with a few high outliers, and the mean is approximately equal to the median.
(D) The distribution is most likely skewed to the right, and the mean is greater than the median.
(E) The distribution is most likely skewed to the right, and the mean is less than the median.

AP4 For a biology project, you measure the weight in grams (g) and the tail length in millimeters (mm) of a group of mice. The equation of the least-squares line for predicting tail length from weight is

$$\text{predicted tail length} = 20 + 3 \times \text{weight}$$

Which of the following is *not* correct?
(A) A mouse's predicted tail length increases by about 3 mm for each additional gram of weight.
(B) The pred... 38 g is 1...
(C) The corr... positive.
(D) If you ha... ters inste... sion line...
(E) Mice tha... length 20...

Section II: Part A *Show all your work. Indicate clearly the methods you use, because you will be graded on the correctness of your methods as well as on the accuracy and completeness of your results and explanations.*

AP16 The manufacturer of exercise machines for fitness centers has designed two new elliptical machines that are meant to increase cardiovascular fitness. The two machines are being tested on 30 volunteers at a fitness center near the company's headquarters. The volunteers are randomly assigned to one of the machines and use it daily for two months. A measure of cardiovascular fitness is administered at the start of the experiment and again at the end. The following stemplot contains the differences (After − Before) in the two scores for subjects using the two machines. Note that greater differences indicate larger gains in fitness.

...ents a ... − Before) ...ores.

(a) Write a few sentences comparing the distributions of cardiovascular fitness gains from the two elliptical machines.
(b) Which machine should be chosen if the company wants to advertise it as achieving the highest overall gain in cardiovascular fitness? Explain your reasoning.
(c) Which machine should be chosen if the company wants to advertise it as achieving the most consistent gain in cardiovascular fitness? Explain your reasoning.
(d) What is the largest population to which we can apply the results of this study?

AP17 A national tire company wants to compare two different types of incentives for increasing sales: awarding cash bonuses or awarding noncash prizes such as vacations. The company has 60 retail sales districts of various sizes across the country and will randomly

Four **CUMULATIVE AP® PRACTICE TESTS** simulate the real exam. Placed after Units 3, 5, 7, and 9, these tests expand in coverage of content and skills as you work through the book. The last test models a full AP® Statistics exam.

Strategies for Success on the AP® Statistics Exam

ABOUT THE AP® STATISTICS EXAM

The AP® Statistics exam consists of two distinct sections: Multiple Choice and Free Response. Here are specific details about the structure and scoring of the exam, as of Fall 2023.

AP® STATISTICS EXAM STRUCTURE		
Section I: Multiple Choice *40 multiple choice questions, each with 5 answer choices*	90 minutes	50% of exam score
The composition of the Multiple Choice section is based on both content and skills. Here are the number of questions for each unit and skill category.		

Consult the **STRATEGIES FOR SUCCESS ON THE AP® STATISTICS EXAM** appendix for insight into the AP® Statistics exam and advice on how to take it with confidence.

UNIT

1

Exploring One-Variable Data

PART I

Data Analysis

Manuela Schewe-Behnisch/EyeEm/Getty Images

Introduction

We live in a world of *data*. Every day, the media report poll results, outcomes of medical studies, and analyses of data on everything from gasoline prices to elections to weather to the latest technology. These data are trying to tell us a story. To understand what the data are saying, we use **statistics.**

DEFINITION Statistics

Statistics is the science and art of collecting, analyzing, and drawing conclusions from data.

A solid understanding of statistics will help you make informed decisions based on data in your daily life. The following activity illustrates one of the many uses of statistics in the real world.

ACTIVITY Smelling Parkinson's

Joy Milne, a retired nurse from Scotland, noticed a "subtle musky odor" on her husband Les that she had never encountered before. At first, Joy thought the smell might be from Les's sweat after long hours of work. But when Les was diagnosed with Parkinson's disease 6 years later, Joy suspected the odor might be a result of the disease.

Scientists were intrigued by Joy's claim and designed an experiment to test her ability to "smell Parkinson's." Joy was presented with 12 different shirts, each worn by a different person, some of whom had Parkinson's disease and some of whom did not. The shirts were given to Joy in a random order, and she had to decide whether or not each shirt was worn by a patient with Parkinson's disease. Joy identified 11 of the 12 shirts correctly.

Although the researchers wanted to believe that Joy could detect Parkinson's disease by smell, it is possible that she was just a lucky guesser. You and your classmates will perform a simulation with cards or technology to determine which explanation is more believable.

1. Your teacher will hand each pair of students a set of 12 cards (shirts). On the back of some cards is "Parkinson's" and on the back of others is "No Parkinson's." Shuffle the cards thoroughly.
2. Decide who will guess first and have your partner act as the researcher. For each card, guess "Parkinson's" or "No Parkinson's." The researcher will not reveal whether each guess is right or wrong, but will record the number of correct guesses. Now switch roles and repeat the process.
3. Your teacher will draw and label a number line for a class dotplot. Plot the number of correct guesses you made in Step 2 on the graph.
4. Repeat Steps 1 and 2 until you have a total of at least 50 trials of the simulation for your class.
5. How often were 11 or more shirts correctly identified by chance alone? Based on this result, which seems more believable: Joy is just a lucky guesser, or Joy really can smell Parkinson's disease? Explain your reasoning.

The Smelling Parkinson's activity outlined the steps in the *statistical problem-solving process*: (1) Ask questions, (2) collect data, (3) analyze the data, and (4) interpret the results.[1] Researchers began by identifying the question to be answered: "Can Joy Milne smell Parkinson's disease?" To answer this question, the researchers designed and carried out an appropriate plan to collect data. The resulting data consisted of Joy's 11 correct and 1 incorrect shirt identifications. When analyzing the data, researchers had to consider the possibility that Joy was just guessing and correctly identified 11 of the 12 shirts by chance alone. After careful analysis, the researchers concluded that Joy Milne could actually smell Parkinson's disease. To make the case even stronger, the researchers later discovered that Joy had correctly identified all 12 shirts. Her one "mistake" was a person who was diagnosed with Parkinson's disease a few months later. That's pretty amazing!

In AP® Statistics, you will be asked to solve a variety of statistical problems. Your success will depend on steadily developing *skills* in four categories: selecting statistical methods, analyzing data, using probability and simulation, and making statistical arguments.[2] We'll emphasize both content and skills throughout the book.

SECTION 1A Statistics: The Language of Variation

LEARNING TARGETS *By the end of the section, you should be able to:*

- Identify the individuals and variables in a set of data, and classify the variables as categorical or quantitative.
- Make and interpret a frequency table or a relative frequency table for a distribution of data.

People's opinions vary on all kinds of issues. Data on physical characteristics of people, animals, and plants vary from one individual to another. Chance outcomes — like tosses of a coin or guesses about which shirts were worn by patients with Parkinson's disease — vary. Statistics is the language of variation.

This section starts by showing you how to organize and classify data. Then, you will learn how to represent data in tabular form.

Individuals and Variables

Every year, the U.S. Census Bureau collects data from more than 3.5 million randomly selected U.S. households as part of the American Community Survey (ACS). The table displays some data from 10 households included in the ACS in a recent year.[3]

Bruce Leighty/Photodisc/Getty Images

Household	Region	People in household	Time in dwelling (years)	Response mode	Household income ($)	Internet access?
1	South	3	20–29	Phone	272,000	Yes
2	Midwest	3	30+	Internet	54,600	Yes
5	South	1	10–19	Mail	49,900	Yes
6	South	2	2–4	Phone	297,000	Yes
7	Northeast	4	5–9	Internet	130,000	Yes
11	Midwest	3	2–4	Internet	82,000	Yes
14	Midwest	3	10–19	Internet	57,000	Yes
16	West	3	2–4	Mail	36,800	Yes
17	South	2	2–4	Mail	133,000	Yes
25	Midwest	1	20–29	Mail	80,000	No

Most data tables follow this format: each row describes an **individual,** and each column contains the values of a **variable.** Sometimes the individuals in a data set are called *cases* or *observational units*.

DEFINITION Individual, Variable

An **individual** is an object described in a set of data. Individuals can be people, animals, or things.

A **variable** is a characteristic that can take different values for different individuals.

For the ACS data set, the *individuals* are households. It is important to note that household is *not* a variable in this data set — the numbers in that column of the data table are just labels for the individuals. The *variables* recorded for each household are region, people in household, time in their current dwelling, survey response mode, household income, and whether the dwelling has internet access. Region, time in dwelling, response mode, and internet access status are **categorical variables.** People in household and household income (in dollars) are **quantitative variables.**

DEFINITION Categorical variable, Quantitative variable

A **categorical variable** takes values that are labels, which place each individual into a particular group, called a category.

A **quantitative variable** takes number values that are quantities — counts or measurements.

Not every variable that takes number values is quantitative. Zip code is one example. Although zip codes are numbers, they are neither counts nor measurements of anything. Instead, they are simply labels for a regional location, making zip code a categorical variable. Time in dwelling from the ACS data set is also a categorical variable because the values are recorded as intervals of time, such as 2–4 years. If time in dwelling had been recorded to the nearest year for each household, this variable would be quantitative.

The variable "year" is often treated as categorical. But it depends on how the data are being used. Consider a data set about cars, in which one of the variables recorded is model year. If we want to know what percentage of cars on the road are 2022 models, we treat year as categorical. If we want to know the average age of cars on the road, we would convert model year to age (in years) and treat this variable as quantitative.

EXAMPLE Census at school
Individuals and variables

Skill 2.A

PROBLEM: Census at School is an international project that collects data about primary and secondary school students using surveys. Since its launch in 2000, students from Australia, Canada, Ireland, Japan, New Zealand, South Africa, the United Kingdom, and the United States have taken part in the project. We selected a random sample of 50 U.S. high school students who completed the survey in a recent year. The table displays data on some of the survey questions for the first 10 students in the sample.

EF Volart/Moment/Getty Images

State	Birth month	Age (years)	Handed	Height (cm)	Home occupants	Allergies	Preferred communication
WI	11	17	Right	175	4	Yes	Internet chat/IM
IN	6	16	Right	175.5	5	No	In person
NY	6	17	Right	157	5	Yes	In person
NC	6	17	Right	169	3	No	Internet chat/IM
MA	6	18	Right	169	3	Yes	Phone call
MO	10	18	Right	170	5	No	Text messaging
PA	5	14	Right	170	6	No	Text messaging
IA	1	17	Left	176	2	No	Text messaging
NC	5	17	Right	175	5	No	Social media
CA	2	17	Right	158	8	Yes	Social media
⋮	⋮	⋮	⋮	⋮	⋮	⋮	⋮

(a) Identify the individuals and variables in this data set.
(b) Classify each variable as categorical or quantitative.

SOLUTION:

(a) Individuals: 50 randomly selected U.S. high school students who completed the Census at School survey. Variables: State, birth month, age (years), handedness, height (cm), number of home occupants, whether the student has allergies, preferred communication method.

We'll see in Unit 3 why choosing at random, as we did in this example, is a good idea.

(b) Categorical: State, birth month, handedness, whether the student has allergies, preferred communication method. Quantitative: Age (years), height (cm), and number of home occupants.

Note that birth month is categorical, even though the values listed are numbers.

FOR PRACTICE, TRY EXERCISE 1

AP® EXAM TIP

If you learn to distinguish categorical from quantitative variables now, it will pay big rewards later. You will be expected to analyze categorical and quantitative data appropriately on the AP® Statistics exam.

The proper method of data analysis depends on whether a variable is categorical or quantitative. For that reason, it is important to distinguish these two types of variables. Be sure to include any units of measurement for a quantitative variable (like centimeters for height). To make life simpler, we sometimes refer to *categorical data* or *quantitative data* instead of identifying the variable as categorical or quantitative.

Summarizing Data with Tables

A variable generally takes values that vary from one individual to another. That's why we call it a variable! The **distribution** of a variable describes the pattern of variation of the values.

DEFINITION Distribution

The **distribution** of a variable tells us what values the variable takes and how often it takes each value.

We can summarize a variable's distribution with a **frequency table** or a **relative frequency table.** To make either kind of table, start by tallying the number of times that the variable takes each value.

DEFINITION Frequency table, Relative frequency table

A **frequency table** shows the number of individuals having each value.

A **relative frequency table** shows the proportion or percentage of individuals having each value.

We can use a frequency table or a relative frequency table to help describe the distribution of a variable, as the following example illustrates.

EXAMPLE

Call me, maybe?
Summarizing data with tables

Skills 2.A, 2.B

SDI Productions/E+/Getty Images

PROBLEM: Here are the data on preferred method of communicating with friends for all 50 students in the Census at School sample:

Internet chat/IM	In person	In person	Internet chat/IM	Phone call
Text messaging	Text messaging	Text messaging	Social media	Social media
Text messaging	Text messaging	Text messaging	In person	In person
Text messaging	Text messaging	Text messaging	In person	In person
Text messaging	Text messaging	Internet chat/IM	Social media	Text messaging
Text messaging	Text messaging	In person	Text messaging	Internet chat/IM
Text messaging	Social media	Text messaging	Social media	Text messaging
Text messaging	In person	Social media	In person	Text messaging
Text messaging	Text messaging	Text messaging	Internet chat/IM	In person
Text messaging	In person	Internet chat/IM	Text messaging	In person

(a) Make a frequency table and a relative frequency table to summarize the distribution of preferred communication method.

(b) Do these data support the claim that a majority of students prefer to communicate with their friends using text messaging? Justify your answer.

SOLUTION:

(a) Frequency table

Preferred method	Frequency
In person	12
Internet chat/IM	6
Phone call	1
Social media	6
Text messaging	25
Total	50

The frequency table shows the *number* of students who prefer each communication method. To create the frequency table, count how many students said "In person," how many said "Internet chat or instant messenger," and so on.

Relative frequency table

Preferred method	Relative frequency
In person	12/50 = 0.24 or 24%
Internet chat/IM	6/50 = 0.12 or 12%
Phone call	1/50 = 0.02 or 2%
Social media	6/50 = 0.12 or 12%
Text messaging	25/50 = 0.50 or 50%
Total	50/50 = 1.00 or 100%

The relative frequency table shows the *proportion* or *percentage* of students who prefer each communication method. Note that in statistics, a proportion is a value between 0 and 1 that is equivalent to a percentage.

(b) No. Exactly 50% of the students in the sample said that they prefer to communicate with their friends via text messaging, but a majority is more than half.

FOR PRACTICE, TRY EXERCISE 5

Note that the frequencies and relative frequencies listed in these tables are not data. The frequency and relative frequency tables summarize the data by telling us how many, or what proportion or percentage of, students in the Census at School sample prefer each method of communicating with friends.

The same process can be used to summarize the distribution of a quantitative variable. However, it would be hard to make a frequency table or a relative frequency table for quantitative data that take many different values, like height (cm) in the Census at School data set. We'll look at a better option for summarizing quantitative variables in Section 1C.

FROM DATA ANALYSIS TO INFERENCE

Sometimes we're interested in drawing conclusions that go beyond the data at hand. That's the idea of *statistical inference*. In the Census at School survey, 12 of the 50 randomly selected students prefer communicating in person with their friends. That's 24% of the *sample*. Can we conclude that exactly 24% of the *population* of students who completed the online survey prefer communicating in person with their friends? No.

If another random sample of 50 students who completed the survey were selected, the percentage who prefer communicating in person with their friends would probably not be exactly 24%. Can we at least say that the actual population value is "close" to 24%? As you will learn in future units, that depends on what we mean by "close."

Our ability to do statistical inference is determined by how the data are produced. Unit 3 discusses the two main methods of collecting data — sampling and experiments — and the types of conclusions that can be drawn from each. As the Smelling Parkinson's activity at the beginning of this unit illustrates, the logic of inference rests on asking, "What are the chances?" *Probability*, the study of chance behavior, is the focus of Units 4 and 5. We'll introduce the most common methods of statistical inference in Units 6–9.

CHECK YOUR UNDERSTANDING

Malena is a car buff who wants to find out more about the cars that high school students drive. The principal of a local high school gives Malena permission to go to the student parking lot and record some data. Later, Malena does some internet research on each model of car in the parking lot and makes a spreadsheet that includes each car's license plate, model, year, number of stickers on the car, color, weight (in kilograms), whether it has a navigation system, and highway gas mileage.

1. Identify the individuals and variables in Malena's study.
2. Classify each variable as categorical or quantitative.

A professor suspects that most students in an 8 A.M. introductory statistics class did not complete the pre-class reading assignment. To find out, the professor gave the students a 10-question pop quiz about the assigned reading at the beginning of class. Here are the number of correct answers on the pop quiz for the 50 students in the class:

9	8	6	7	7	8	4	7	7	8	8	8	6	7	8	8	7	7	6	8	9	7	6	5	7
8	8	7	9	6	6	6	8	9	5	8	7	7	7	7	2	4	8	3	6	5	5	8	7	3

3. Make a frequency table and a relative frequency table to summarize the distribution of number of correct answers on the pop quiz.
4. What proportion of students got fewer than 7 correct answers on the pop quiz? Does this result support the professor's belief that most students did not complete the pre-class reading assignment? Justify your answer.

SECTION 1A Summary

- **Statistics** is the science and art of collecting, analyzing, and drawing conclusions from data.
- A data set contains information about a number of **individuals.** Individuals may be people, animals, or things. For each individual, the data give values for one or more **variables.** A variable is a characteristic that can take different values for different individuals.
- A **categorical variable** takes values that are labels, which place each individual into a particular group, called a category. A **quantitative variable** takes numerical values that count or measure some characteristic of each individual.
- The **distribution** of a variable describes what values the variable takes and how often it takes each value.
- To summarize the distribution of a variable, you can use a **frequency table** that shows the number of individuals having each value or a **relative frequency table** that shows the proportion or percentage of individuals having each value. You can use a frequency table or a relative frequency table to help describe the distribution of a variable.

AP® EXAM TIP
AP® Daily Videos
Review the content of this section and get extra help by watching the AP® Daily Videos for Topics 1.1–1.3, which are available in AP® Classroom.

SECTION 1A Exercises

The solutions to all exercises numbered in red can be found in the Solutions Appendix.

Individuals and Variables

1. **A class survey** Here is a small part of a data set that describes the students in a math class. The data come from anonymous responses to a questionnaire filled out on the second day of class.

pg 4

Grade level	Dominant hand	GPA	Children in family	Homework last night (min)	Android or iPhone?
9	L	2.3	3	0–14	iPhone
11	R	3.8	6	15–29	Android
10	R	3.1	2	15–29	Android
10	R	4.0	1	45–59	iPhone
10	R	3.4	4	0–14	iPhone
10	L	3.0	3	30–44	Android
9	R	3.9	2	15–29	iPhone
12	R	3.5	2	0–14	iPhone

(a) Identify the individuals and variables in this data set.

(b) Classify each variable as categorical or quantitative.

2. **Coaster craze** Many people like to ride roller coasters. Amusement parks try to increase attendance by building exciting new coasters. The following table displays data on several roller coasters from around the world.[4]

Roller coaster	Type	Height (ft)	Design	Speed (mph)	Duration (sec)
Copperhead Strike	Steel	82.0	Sit down	50.0	144
Eurostar	Steel	98.9	Inverted	50.2	140
Jungle Trailblazer	Wood	108.3	Sit down	54.1	150
Falcon	Steel	197.5	Wing	73.3	156
Olympia Looping	Steel	106.7	Sit down	49.7	105
Time Traveler	Steel	100.0	Sit down	50.3	117

(a) Identify the individuals and variables in this data set.

(b) Classify each variable as categorical or quantitative.

3. **Hit movies** *Avatar* was the top-earning movie based on box-office receipts worldwide as of December 2022. The following table displays data on several popular movies.[5] Identify the individuals and variables in this data set. Then classify each variable as categorical or quantitative.

Movie	Year	Rating	Time (min)	Genre	Box office ($)
Avatar	2009	PG-13	162	Action	2,910,284,102
Avengers: Endgame	2019	PG-13	181	Action	2,797,732,053
Titanic	1997	PG-13	194	Drama	2,207,986,545
Star Wars: The Force Awakens	2015	PG-13	136	Adventure	2,064,615,817
Avengers: Infinity War	2018	PG-13	156	Action	2,048,359,754
Spider Man: No Way Home	2021	PG-13	148	Action	1,910,041,582
Jurassic World	2015	PG-13	124	Action	1,669,963,641
The Lion King	2019	PG	118	Adventure	1,647,733,638
The Avengers	2012	PG-13	143	Action	1,515,100,211
Furious 7	2015	PG-13	137	Action	1,514,553,486
Top Gun: Maverick	2022	PG-13	131	Action	1,488,187,968
Frozen II	2019	PG	103	Adventure	1,444,735,491

4. **Skyscrapers** Here is some information about the tallest buildings in the world as of 2023.[6] Identify the individuals and variables in this data set. Then classify each variable as categorical or quantitative.

Building	Country	Height (m)	Floors	Use	Year completed
Burj Khalifa	United Arab Emirates	828.0	163	Mixed	2010
Merdeka 118	Malaysia	678.9	118	Mixed	2023
Shanghai Tower	China	632.0	128	Mixed	2015
Makkah Royal Clock Tower Hotel	Saudi Arabia	601.0	120	Hotel	2012
Ping An Finance Center	China	599.1	115	Mixed	2017
Lotte World Tower	South Korea	554.5	123	Mixed	2017
One World Trade Center	United States	541.3	94	Office	2014
Guangzhou CTF Finance Center	China	530.0	111	Mixed	2016
Tianjin CTF Finance Center	China	530.0	97	Mixed	2019
China Zun Tower	China	527.7	101	Mixed	2019

Summarizing Data with Tables

5. **New York squirrels** How do squirrels in New York City's Central Park respond to humans? The Squirrel Census project enlisted volunteers to record data on all squirrels observed in Central Park in a recent year. One of the variables measures how the squirrels responded to humans. We selected a random sample of 75 of the more than 3000 squirrels observed. Here are the data on their reactions to humans:[7]

pg 6

Approach	Run from	Indifferent	Indifferent	Run from
Indifferent	Approach	Indifferent	Indifferent	Indifferent
Approach	Indifferent	Approach	Run from	Run from
Indifferent	Indifferent	Indifferent	Run from	Indifferent
Indifferent	Indifferent	Run from	Indifferent	Indifferent
Indifferent	Run from	Run from	Indifferent	Indifferent
Run from	Run from	Approach	Approach	Run from
Indifferent	Approach	Indifferent	Indifferent	Indifferent
Indifferent	Approach	Indifferent	Indifferent	Indifferent
Indifferent	Run from	Approach	Approach	Indifferent
Indifferent	Indifferent	Run from	Approach	Indifferent
Indifferent	Approach	Indifferent	Run from	Approach
Run from	Approach	Run from	Indifferent	Indifferent
Indifferent	Approach	Indifferent	Run from	Indifferent
Indifferent	Indifferent	Approach	Approach	Indifferent

(a) Make a frequency table and a relative frequency table to summarize the distribution of response to humans.

(b) Were these Central Park squirrels more likely to approach humans, run from humans, or be indifferent to humans? Justify your answer.

6. **Disc dogs** Here is a list of the breeds of dogs that won the World Canine Disc Championships over a 45-year period.[8]

Whippet	Australian shepherd	Australian shepherd
Whippet	Border collie	Border collie
Whippet	Australian shepherd	Border collie
Mixed breed	Mixed breed	Australian shepherd
Mixed breed	Mixed breed	Border collie
Other purebred	Mixed breed	Border collie
Labrador retriever	Border collie	Other purebred
Mixed breed	Border collie	Border collie
Mixed breed	Australian shepherd	Border collie
Border collie	Border collie	Border collie
Mixed breed	Australian shepherd	Mixed breed
Mixed breed	Border collie	Australian shepherd
Labrador retriever	Mixed breed	Australian shepherd
Labrador retriever	Australian shepherd	Mixed breed
Mixed breed	Australian shepherd	Other purebred

(a) Make a frequency table and a relative frequency table to summarize the distribution of dog breed among World Canine Disc Championship winners.

(b) Which breeds of dog were the most frequent winners? Justify your answer.

7. **CubeSat missions** A CubeSat is a miniature satellite used for space research that can be easily deployed from a launch vehicle or from the International Space Station. More than 1000 CubeSats have been launched since specifications for these satellites were jointly developed by Cal Poly San Luis Obispo and Stanford University in 1999. The relative frequency table summarizes data on the type of mission for each CubeSat launched in November and December 2019.[9]

Mission type	Proportion
Science	0.18
Technology	0.28
Imaging	0.34
Communications	0.12
Education	0.08

(a) There were a total of 50 CubeSats launched during these two months. How many had a communications mission type?

(b) A report states that a majority of these CubeSat launches had an imaging mission type. Explain why this statement is incorrect.

Multiple Choice *Select the best answer for each question.*

Exercises 8 and 9 refer to the following setting. A realtor in Hilton Head, South Carolina, collects data on all of the homes sold in the town during 2023. The resulting data set lists each home's address, along with the following information: zip code, number of bedrooms, primary building material (wood, brick, etc.), square footage, whether it has a pool, age of home, and sales price.

8. The individuals in this data set are

(A) the Hilton Head realtor who collected the data.

(B) all Hilton Head homeowners who sold their homes during 2023.

(C) the addresses of all homes sold in Hilton Head during 2023.

(D) all homes sold in Hilton Head during 2023.

(E) all homes in Hilton Head during 2023.

9. This data set contains

(A) 7 variables, 2 of which are categorical.

(B) 7 variables, 3 of which are categorical.

(C) 7 variables, 4 of which are categorical.

(D) 8 variables, 3 of which are categorical.

(E) 8 variables, 4 of which are categorical.

SECTION 1B Displaying and Describing Categorical Data

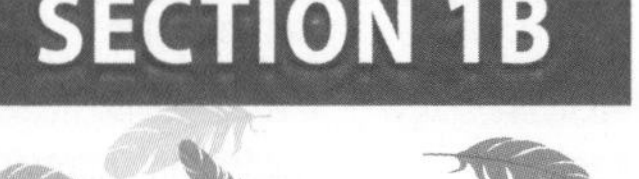

LEARNING TARGETS *By the end of the section, you should be able to:*

- Make and interpret bar graphs of categorical data.
- Compare distributions of categorical data.
- Identify what makes some graphs of categorical data misleading.

In this section, you will learn how to display and describe the distribution of a categorical variable. Sections 1C, 1D, 1E, and 1F focus on displaying and describing the distribution of a quantitative variable. Unit 2 examines relationships between two categorical or two quantitative variables. This process of exploratory data analysis is known as *descriptive statistics*.

Displaying Categorical Data: Bar Graphs

A frequency table or relative frequency table summarizes a variable's distribution with numbers. To display the distribution more clearly, use a graph. The most common way to display categorical data is with a **bar graph** (also known as a *bar chart*).

> **DEFINITION Bar graph**
>
> A **bar graph** shows each category as a bar. The heights of the bars show the category frequencies or relative frequencies.

Figure 1.1 shows a bar graph of the data on preferred communication method for the random sample of 50 U.S. high school students who completed the Census at School survey. Note that the percentages for each category come from the relative frequency table.

FIGURE 1.1 Bar graph of the distribution of preferred communication method for a random sample of 50 U.S. high school students.

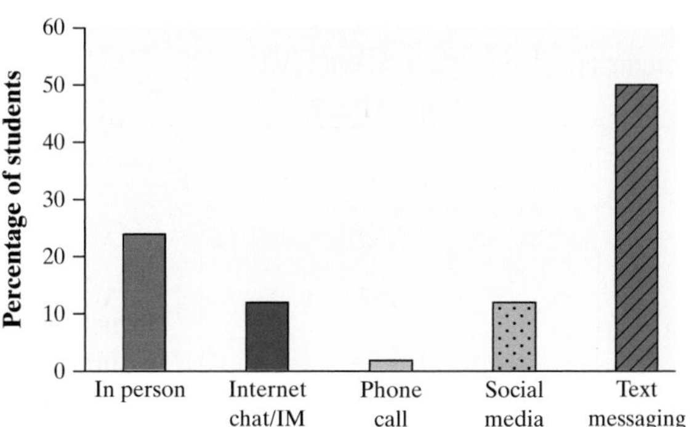

Relative frequency table

Preferred method	Relative frequency
In person	12/50 = 0.24 or 24%
Internet chat/IM	6/50 = 0.12 or 12%
Phone call	1/50 = 0.02 or 2%
Social media	6/50 = 0.12 or 12%
Text messaging	25/50 = 0.50 or 50%

It is fairly straightforward to make a bar graph by hand. Here's how you do it.

HOW TO MAKE A BAR GRAPH

1. **Draw and label the axes.** Put the name of the categorical variable under the horizontal axis. To the left of the vertical axis, indicate whether the graph shows the frequency (count) or relative frequency (percentage or proportion) of individuals in each category.
2. **"Scale" the axes.** Write the names of the categories at equally spaced intervals under the horizontal axis. On the vertical axis, start at 0 and place tick marks at equal intervals until you equal or exceed the largest frequency or relative frequency in any category.
3. **Draw bars** above the category names. Make the bars equal in width and leave gaps between them. Be sure that the height of each bar corresponds to the frequency or relative frequency of individuals in that category.

Making a graph is not an end in itself. The real purpose of a graph is to help us interpret the data. When you look at a graph, always ask, "What do I see?" In Figure 1.1, the bar graph reveals that the most preferred method of communicating with friends for these high school students is text messaging (50%). The next most preferred method is talking with their friends in person (24%). Social media and internet chat/instant messenger are somewhat less popular (12% each) methods of communicating with friends. By far the least preferred method of communication for these students is a phone call (2%).

EXAMPLE

What's on the radio?
Displaying categorical data: Bar graphs

Skills 2.A, 2.B

ACORN 1/Alamy Stock Photo

PROBLEM: Nielsen Audio, the rating service for radio audiences, places U.S. radio stations into categories that describe the kinds of programs they broadcast. The frequency table summarizes the distribution of station format in a recent year.[10] Make a frequency bar graph to display the data. Describe what you see.

Format	Number of stations	Format	Number of stations
Adult contemporary	1357	Religious	3837
All sports	669	Rock	1466
Classic hits	1140	Spanish language	1228
Country	2200	Variety	1257
News/talk/information	2002	Other formats	1769
Oldies	405	**Total**	**17,330**

SOLUTION:

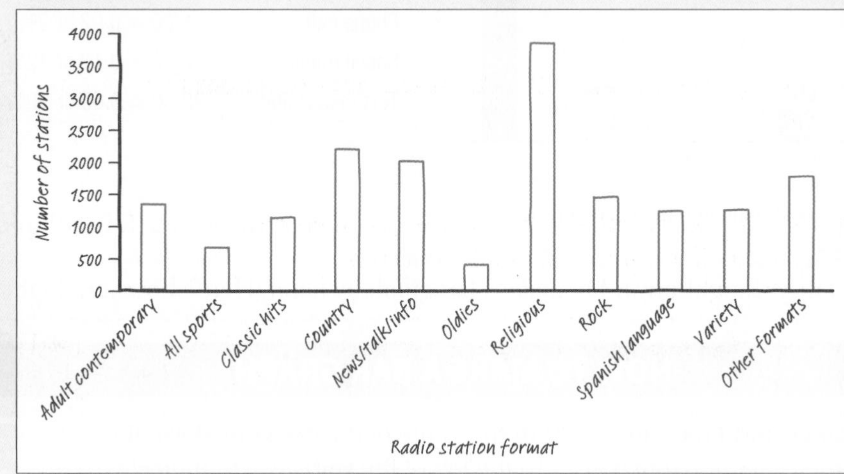

To make the bar graph:
1. **Draw and label the axes.**
2. **"Scale" the axes.** The largest frequency is 3837, so we choose a vertical scale from 0 to 4000, with tick marks 500 units apart.
3. **Draw bars** above the category names.

On U.S. radio stations, the most common formats are Religious (3837), Country (2200), and News/Talk/Information (2002), while the least common formats are Oldies (405) and All Sports (669). Moderately common formats offered by a similar number of stations include Classic Hits (1140), Spanish Language (1228), Variety (1257), Adult Contemporary (1357), and Rock (1466). There are 1769 stations with Other formats.

FOR PRACTICE, TRY EXERCISE 1

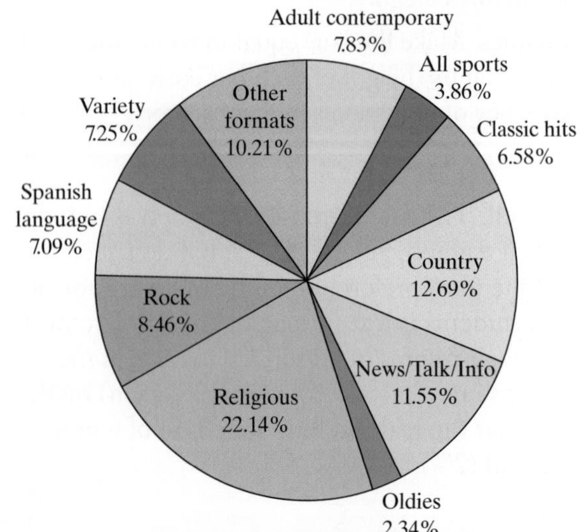

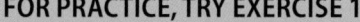

Radio station format

PIE CHARTS

Here is a **pie chart** of the radio station format data from the preceding example. Notice that this graph shows the *relative frequency* for each response category. For instance, the "Spanish language" slice makes up 7.09% of the graph because the relative frequency for this category is 1228/17,330 = 0.0709 = 7.09%.

DEFINITION Pie chart

A **pie chart** shows each category as a slice of the "pie." The areas of the slices are proportional to the category frequencies or relative frequencies.

Use a pie chart when you want to emphasize each category's relation to the whole. Each slice of the pie shows the count or percentage of individuals in that category. Pie charts are challenging to make by hand, but technology will do the job for you.

A bar graph or pie chart that displays a distribution of categorical data must include *all* individuals in the data set. This might require including an "Other" category, as in the radio station example. Bar graphs are more flexible than pie charts. Both types of graphs can display the distribution of a categorical variable, but a bar graph can also compare any set of quantities that are measured in the same units.

Comparing Distributions of Categorical Data

You can use a bar graph to display the distribution of a categorical variable. A **side-by-side bar graph** can be used to compare the distribution of a categorical variable in two or more groups.

DEFINITION Side-by-side bar graph

A **side-by-side bar graph** displays the distribution of one categorical variable in each of two or more groups.

It's a good idea to use relative frequencies when comparing data for multiple groups, especially if the groups have different sizes.

EXAMPLE

Screen time for teens and tweens
Comparing distributions of categorical data

Skill 2.D

PROBLEM: How much time do tweens (ages 8–12) and teens (ages 13–18) spend using digital devices each day? Researchers surveyed a random sample of more than 1300 U.S. 8- to 18-year-olds to find out. The side-by-side bar graph summarizes the data on daily screen time reported by the 560 tweens and 746 teens in the sample.[11]

Melanie Acevedo/DigitalVision/Getty Images

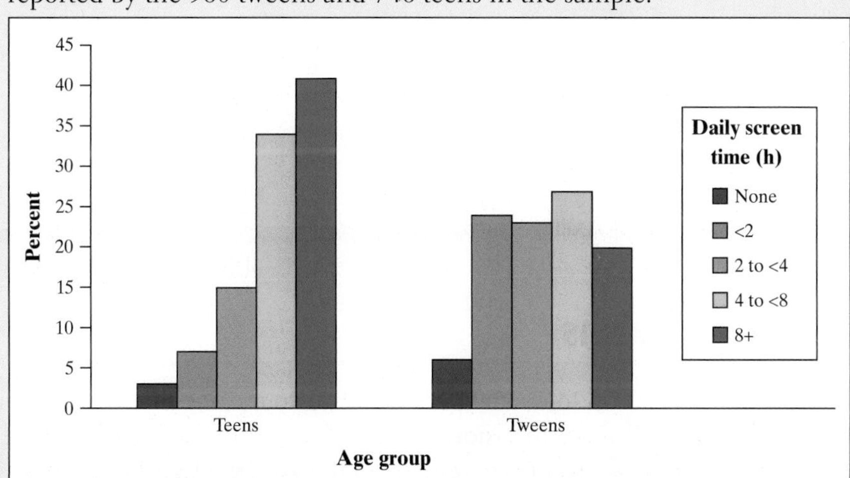

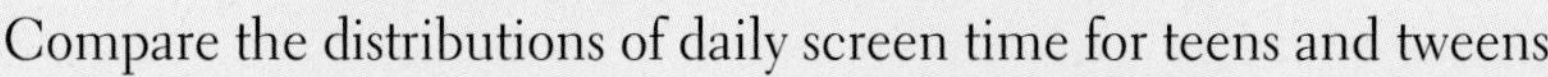
Compare the distributions of daily screen time for teens and tweens.

SOLUTION:

Tweens tend to report less daily screen time than teens. Tweens are more likely than teens to spend no time (6% versus 3%), less than 2 hours (24% versus 7%), and from 2 to less than 4 hours (23% versus 15%) on digital devices per day. Teens are more likely than tweens to spend from 4 to less than 8 hours (34% versus 27%) and 8 or more hours (41% versus 20%) on digital devices per day. More than half (53%) of tweens report less than 4 hours of screen time per day, while 75% of teens report 4 or more hours per day of screen time.

FOR PRACTICE, TRY EXERCISE 7

The following relative frequency table summarizes the data on daily screen time from the example by age group. As a result, the percentages in each row add to 100%. Note that we could have also compared the distributions with this relative frequency table.

		Daily time spent on digital devices				
		None	<2 hours	2 to <4 hours	4 to <8 hours	8+ hours
Age group	Teens	3%	7%	15%	34%	41%
	Tweens	6%	24%	23%	27%	20%

In the example, we grouped the bars by age group in the side-by-side bar graph. This arrangement clearly shows the two distributions of daily screen time — one for the teens and one for the tweens in the sample. It is also possible to group the bars by daily screen time, as in the following graph. This arrangement makes category-by-category comparisons for the daily time spent on digital devices easier — but now it's harder to see the individual distributions of daily screen time for the teens and for the tweens. Whichever way you organize the bars, be sure to include a key that describes what each color or type of shading on the side-by-side bar graph represents.

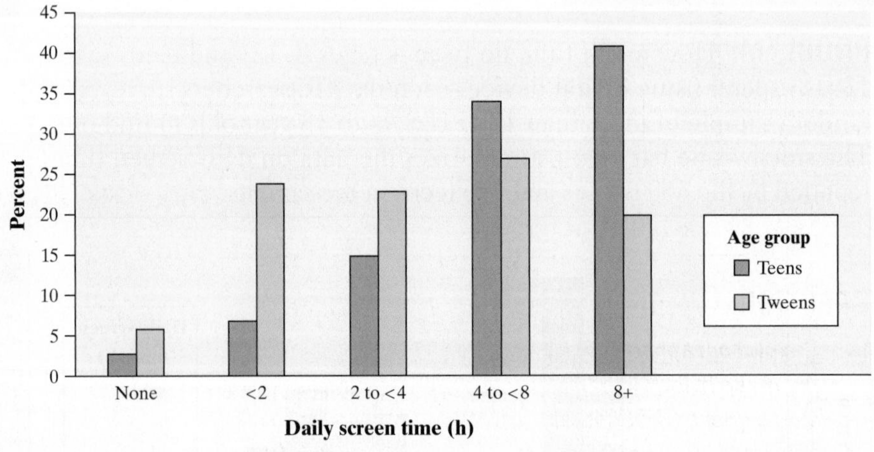

Misleading Graphs

Bar graphs can be a bit dull to look at. It is tempting to replace the bars with pictures or to use special three-dimensional (3D) effects to make the graphs seem more interesting. Don't do it! Our eyes react to the area of the bars as well as to their height. When all bars have the same width, the area (width × height) varies in proportion to the height, and our eyes receive the right impression about the quantities being compared.

EXAMPLE Who buys iMacs? Misleading graphs

Skill 2.A

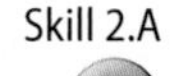

PROBLEM: When Apple, Inc., first introduced the iMac in 1998, the company wanted to know whether this new computer was expanding its market share. (The iMac has enjoyed great success ever since!) Was the iMac mainly being bought by previous Macintosh owners, or was it being purchased by first-time computer buyers and by previous PC users who were switching over? To find out, Apple hired a firm to conduct a survey of 500 randomly selected iMac customers. The firm categorized each customer as a new computer purchaser, a previous PC owner, or a previous Macintosh owner. The table summarizes the survey data.[12]

Previous ownership	Count	Percentage (%)
None	85	17.0
PC	60	12.0
Macintosh	355	71.0
Total	**500**	**100.0**

(a) A graph of the data that uses pictures instead of the more traditional bars is shown in figure (a). How is this pictograph misleading?

(b) A bar graph of the data is shown in figure (b). Explain why this graph could be considered deceptive.

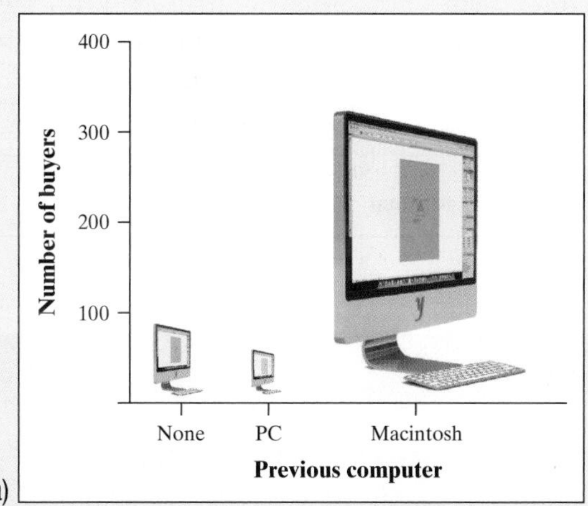

(a)

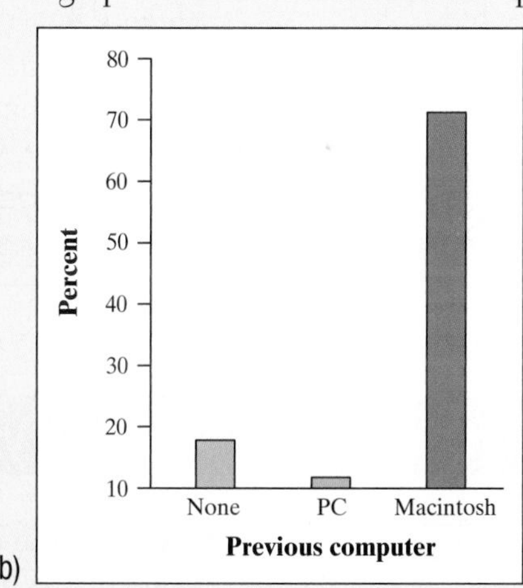

(b)

SOLUTION:

(a) The pictograph is misleading because the areas of the computers make it look like the number of iMac buyers who are former Mac owners (355) is at least 10 times as large as the number of buyers in either of the other two categories (None: 85, PC owner: 60), which isn't true.

In part (a), the *heights* of the images are correct. But the *areas* of the images are misleading. The Macintosh image is about 6 times as tall as the PC image, but its area is about 36 times as large!

(b) The bar graph is misleading because starting the vertical scale at 10 instead of 0 makes it look like the percentage of iMac buyers who previously owned a PC (12.0%) is less than half the percentage who are first-time computer buyers (17.0%), which isn't true.

FOR PRACTICE, TRY EXERCISE 11

There are two important lessons to be learned from this example: (1) **beware the pictograph** and (2) **watch those scales.**

CHECK YOUR UNDERSTANDING

The Pew Research Center asked a random sample of 1502 U.S. adults about their cell phone ownership. The frequency table summarizes their responses.[13]

Type of cell phone	Frequency
None	73
Cell phone, not smartphone	212
Smartphone	1217
Total	**1502**

1. Make a relative frequency bar graph to display the distribution of cell phone ownership for the 1502 people in the sample. Describe what you see.
2. The following graph displays the distribution of cell phone ownership for each of four age groups in the sample. Compare the distributions.

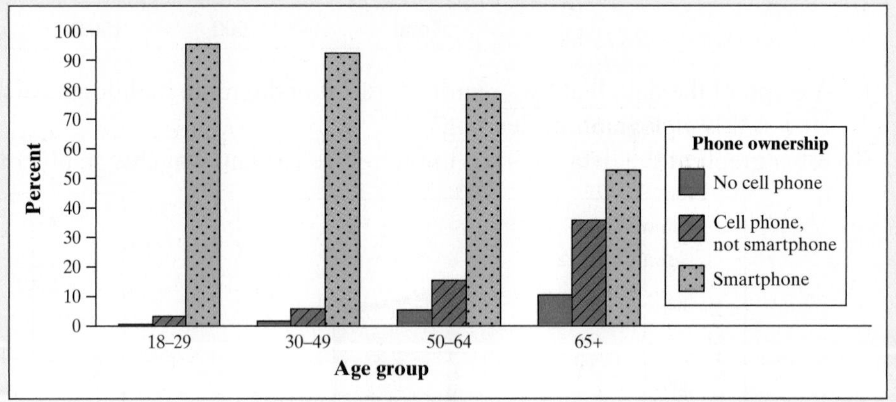

SECTION 1B Summary

> **AP® EXAM TIP**
> **AP® Daily Videos**
> Review the content of this section and get extra help by watching the AP® Daily Videos for Topic 1.4, which are available in AP® Classroom.

- A **bar graph** or **pie chart** can be used to display the distribution of a categorical variable. When examining any graph, ask yourself, "What do I see?"
- You can use a **side-by-side bar graph** to compare the distribution of one categorical variable for two or more groups. Be sure to use relative frequencies when comparing groups of different sizes.
- Beware of graphs that mislead the eye. Look at the scales to see if they have been distorted to create a particular impression. Avoid making graphs that replace the bars of a bar graph with pictures whose height and width both change.

SECTION 1B Exercises

Displaying Categorical Data: Bar Graphs

1. **Birth days** The frequency table summarizes data on the numbers of babies born on each day of a single week in the United States.[14]

pg 12

Day	Births
Sunday	7374
Monday	11,704
Tuesday	13,169
Wednesday	13,038
Thursday	13,013
Friday	12,664
Saturday	8459

Make a frequency bar graph to display the data. Describe what you see.

2. **Going up?** The website fivethirtyeight.com published an interesting article about the more than 75,000 elevators in New York City. The frequency table summarizes data on the number of elevators of each type at that time.[15]

Type	Count
Passenger elevator	66,602
Freight elevator	4140
Escalator	2663
Dumbwaiter	1143
Sidewalk elevator	943
Private elevator	252
Handicap lift	227
Manlift	73
Public elevator	45

Make a frequency bar graph to display the data. Describe what you see.

3. **Cool car colors** The popularity of colors for cars and light trucks changes over time. Silver passed green in 2000 to become the most popular color worldwide, then gave way to shades of white in 2007. The relative frequency table summarizes data on the colors of vehicles sold worldwide in 2020.[16]

Color	Percentage of vehicles
Black	19
Blue	7
Brown	3
Gray	15
Green	1
Red	5
Silver	9
White	38
Yellow	2
Other	??

(a) What percentage of vehicles would fall in the "Other" category?

(b) Make a bar graph to display the data. Describe what you see.

(c) Would it be appropriate to make a pie chart of these data? Explain your answer.

4. **Spam** Email spam is very annoying. The relative frequency table summarizes data on the most common types of spam.[17]

Type of spam	Percentage
Adult	19
Financial	20
Health	7
Internet	7
Leisure	6
Products	25
Scams	9
Other	??

(a) What percentage of spam would fall in the "Other" category?

(b) Make a bar graph to display the data. Describe what you see.

(c) Would it be appropriate to make a pie chart of these data? Explain your answer.

5. **Family origins** Here is a pie chart of U.S. Census Bureau data showing the origin of more than 62 million Hispanic people in the United States.[18] About what percentage of Hispanic people are of Mexican origin? Puerto Rican origin?

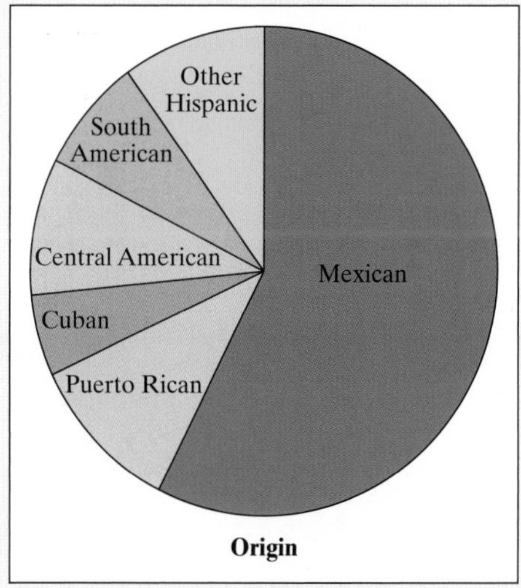

6. **Which major?** More than 2 million students earn bachelor's degrees in U.S. colleges and universities each year. The pie chart displays data on bachelor's degrees earned by field of study in a recent year.[19] About what percentage of students earned their bachelor's degrees in business? In health professions? In social science?

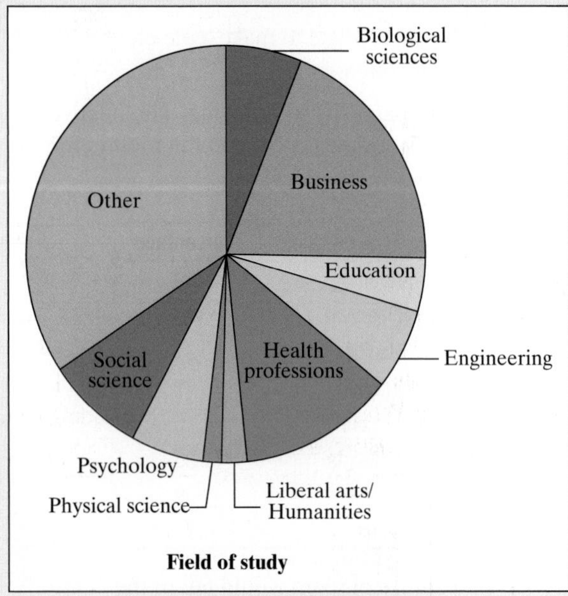

Comparing Distributions of Categorical Data

7. **Far from home** A survey asked first-year college students, "How many miles is this college from your permanent home?" Students selected from the following options: 5 or fewer, 6 to 10, 11 to 50, 51 to 100, 101 to 500, or more than 500. The side-by-side bar graph shows the percentage of students at public and private 4-year colleges who chose each option.[20] Compare the distributions of distance from home for students from private and public 4-year colleges who completed the survey.

pg 13

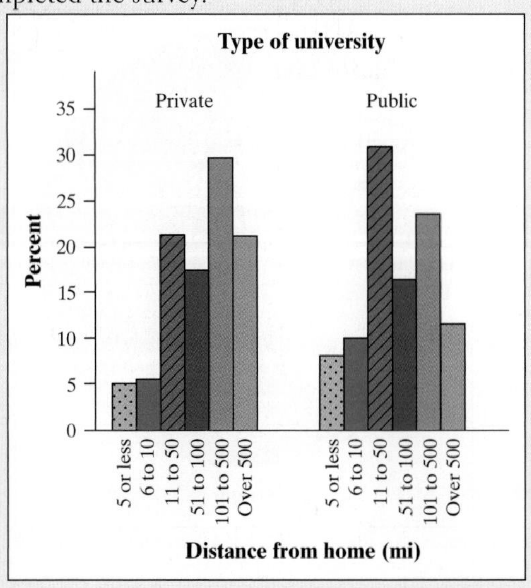

8. **More CubeSat missions** A CubeSat is a miniature satellite used for space research that can be easily deployed from a launch vehicle or from the International Space Station. More than 1000 CubeSats have been launched since specifications for these satellites were jointly developed by Cal Poly San Luis Obispo and Stanford University in 1999. The side-by-side bar graph displays data on the types of missions undertaken by CubeSats launched in 2017 and 2019. Compare the distributions of mission type for these two years.[21]

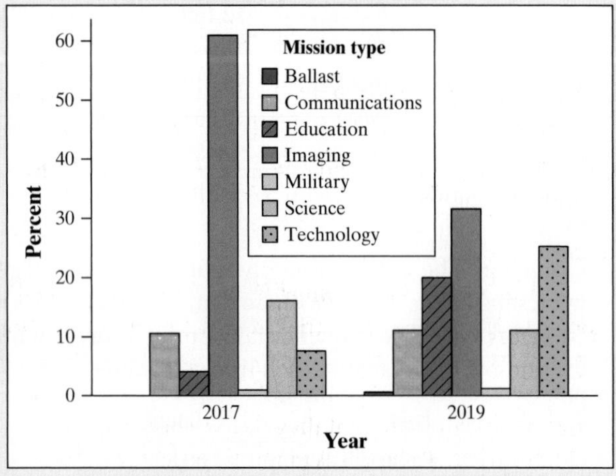

9. **Reading for pleasure** Researchers surveyed a random sample of more than 1300 U.S. tweens (ages 8–12) and teens (ages 13–18) about how often they read for pleasure. The following table summarizes the responses in each age group.[22]

		Age group	
		Teens	Tweens
How often they read for pleasure	Every day	21%	34%
	At least once a week	27%	29%
	At least once a month	14%	12%
	Less than once a month	20%	12%
	Never	18%	13%

(a) Make a side-by-side bar graph to compare the distributions of how often teens and tweens read for pleasure.

(b) Describe similarities and differences between the two distributions.

10. **Popular car colors** Favorite car colors may differ among countries in different parts of the world. The following table summarizes data on the most popular car colors in a recent year for the countries in North America and Asia.[23]

		Continent	
		North America	Asia
Color	White	27%	45%
	Gray	21%	13%
	Black	20%	18%
	Blue	10%	5%
	Silver	10%	7%
	Red	8%	4%
	Brown	2%	3%
	Green	1%	2%
	Yellow	0.5%	2%
	Other	0.5%	1%

(a) Make a side-by-side bar graph to compare the distributions of most popular car color in North America and Asia.

(b) Describe similarities and differences between the two distributions.

Misleading Graphs

11. **Who dislikes shopping?** Harris Interactive asked adults from several countries if they like or dislike shopping for clothes. The following pictograph displays the percentage of adults in the United States, Germany, and Spain who said that they dislike shopping for clothes. Explain how this graph is misleading.[24] pg 15

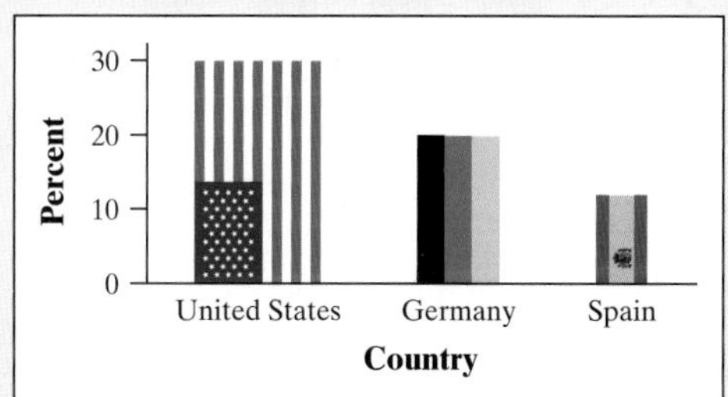

12. **Social media** The Pew Research Center surveyed a random sample of U.S. teens and adults about their use of social media. The following pictograph displays the percentage of people in various age groups who report using social media. Explain how this graph is misleading.

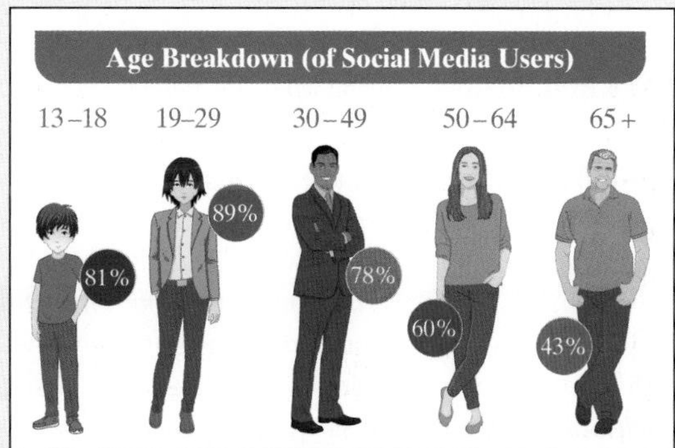

13. **Binge-watching** Do you "binge-watch" television series by viewing multiple episodes of a series at one sitting? A survey of 800 people who binge-watch were asked how many episodes is too many to watch in one viewing session. The results are displayed in the bar graph.[25] Explain how this graph is misleading.

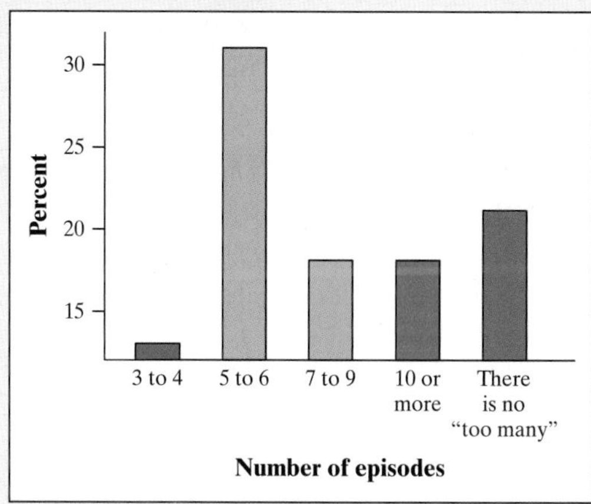

14. **Support the court?** A news network reported the results of a survey about a controversial court decision. The network initially posted on its website a bar graph of the data similar to the one shown here. Explain how this graph is misleading. (*Note:* When notified about the misleading nature of its graph, the network posted a corrected version.)

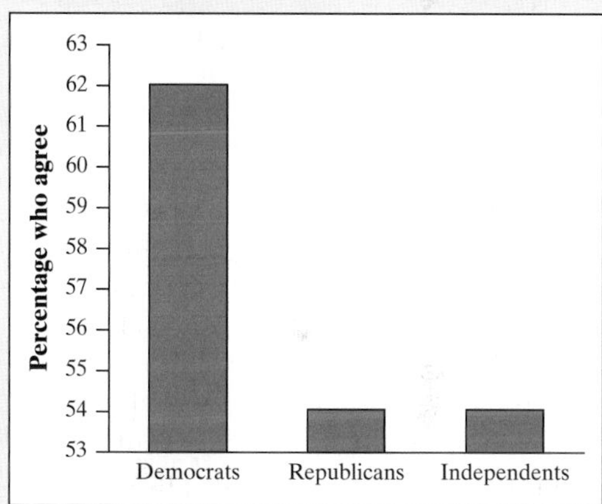

15. **Lotteries and education** A Gallup Poll asked respondents about their highest level of education and whether they had bought a state lottery ticket in the last 12 months.[26] Here is a bar graph of the data:

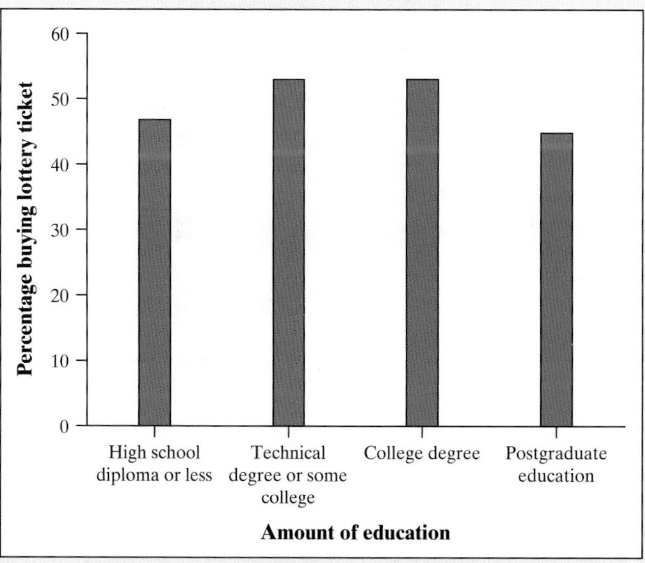

(a) Describe what this graph reveals about lottery ticket buying habits among the different education groups.

(b) Would it be appropriate to make a pie chart for this data set? Explain your answer.

For Investigation *Apply the skills from the section in a new context or nonroutine way.*

16. **Choropleth maps** A *choropleth map* is a graphical representation of data by geographic region in which values are depicted by color. For instance, the choropleth map presented here shows the percentage of people who were internet users in each country in 2020.[27] Write a thorough analysis of what the graph shows. Be sure to include a comparison of internet use in different regions of the world.

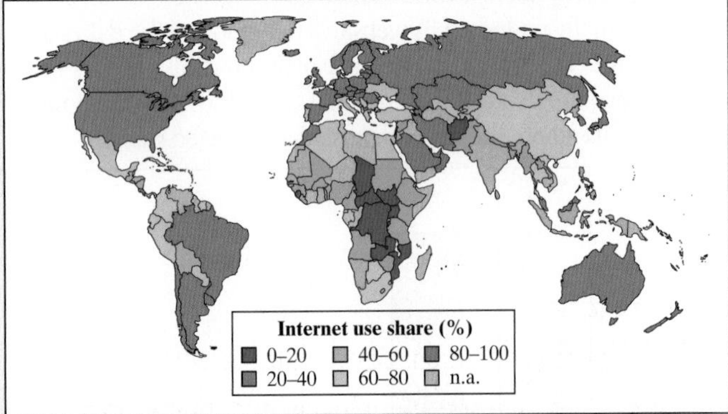

17. **Cow pie chart?** Methane is a greenhouse gas that may contribute to climate change. The modified pie chart shown here displays data on the sources of methane emissions in the United States.[28] Write a thorough analysis of what the graph shows. Be sure to include a comparison of methane emissions by beef cows, dairy cows, and pigs.

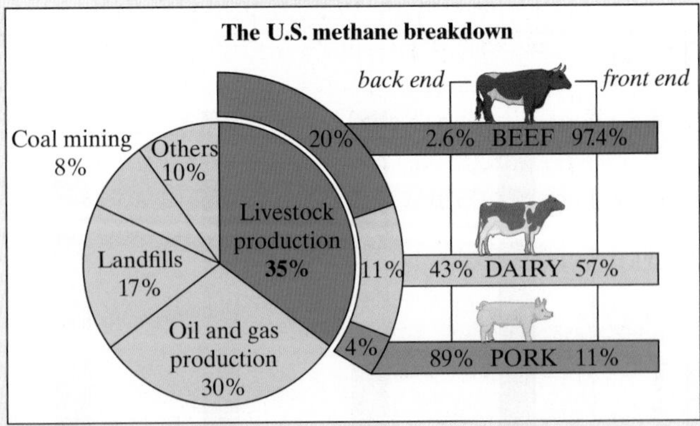

Multiple Choice *Select the best answer for each question.*

18. For which of the following would it be *inappropriate* to display the data with a single pie chart?

(A) The distribution of car colors for vehicles purchased in the last month

(B) The distribution of unemployment percentages for each of the 50 states

(C) The distribution of favorite sport for a sample of 30 middle school students

(D) The distribution of shoe type worn by shoppers at a local mall

(E) The distribution of presidential candidate preference for voters in a state

19. The following bar graph shows the distribution of favorite subject for a sample of 1000 students. What is the most serious problem with the graph?

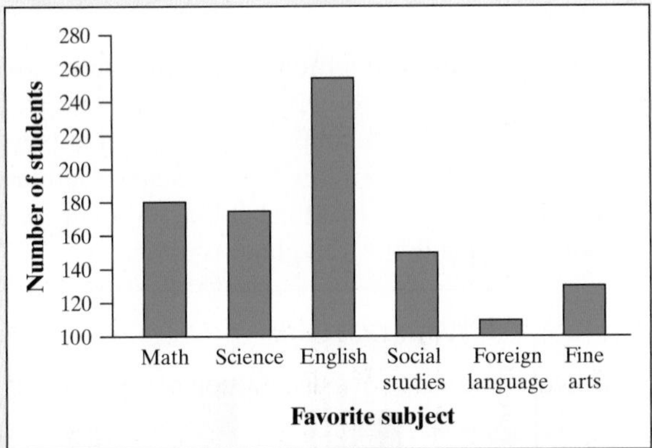

(A) The subjects are not listed in the correct order.

(B) This distribution should be displayed with a pie chart.

(C) The vertical axis should show the percentage of students.

(D) The vertical axis should start at 0 rather than 100.

(E) The foreign language bar should be broken up by language.

Recycle and Review *Practice what you learned in previous sections.*

20. **Hotels (1A)** A high school lacrosse team is planning to go to Buffalo for a three-day tournament. The tournament's sponsor provides a list of available hotels, along with some information about each hotel. The following table displays data about hotel options. Identify the individuals and variables in this data set. Classify each variable as categorical or quantitative.

Table for Exercise #20

Hotel	Pool	Exercise room?	Internet cost ($/day)	Restaurants	Distance to site (mi)	Room service?	Room rate ($/day)
Comfort Inn	Out	Y	0.00	1	8.2	Y	149
Fairfield Inn & Suites	In	Y	0.00	1	8.3	N	119
Baymont Inn & Suites	Out	Y	0.00	1	3.7	Y	60
Chase Suite Hotel	Out	N	15.00	0	1.5	N	139
Courtyard	In	Y	0.00	1	0.2	Dinner only	114
Hilton	In	Y	10.00	2	0.1	Y	156
Marriott	In	Y	9.95	2	0.0	Y	145

SECTION 1C Displaying Quantitative Data with Graphs

LEARNING TARGETS *By the end of the section, you should be able to:*

- Make and interpret dotplots of quantitative data.
- Describe the shape of a distribution of quantitative data.
- Describe the distribution of a quantitative variable.
- Compare distributions of quantitative data.
- Make and interpret stemplots of quantitative data.
- Make and interpret histograms of quantitative data.

As you learned in the previous section, bar graphs are typically used to display the distribution of a categorical variable. In this section, you will learn to make and interpret several types of graphs that can be used to display the distribution of a quantitative variable: *dotplots, stemplots,* and *histograms.*

There are two types of quantitative variables: *discrete* and *continuous*. Most **discrete variables** result from counting something, like the number of people in a household or the number of lottery tickets a person buys until they win the jackpot. Note that the number of possible values of a discrete variable can be finite or infinite. **Continuous variables** typically result from measuring something, like height (in inches) or time to run a 100-meter dash (in seconds). Note that there are infinitely many possible values for a continuous variable. Age is technically a continuous variable — a high school student's age might be 17.30162... years. But it is often treated as a discrete variable — for example, age = 17 years.

DEFINITION Discrete variable, Continuous variable

A quantitative variable that takes a countable set of possible values with gaps between them on the number line is a **discrete variable.**

A quantitative variable that can take any value in an interval on the number line is a **continuous variable.**

For a discrete variable, countable means that we can number the possible values of the quantitative variable using whole numbers. As an example, the number of lottery tickets a person buys until they win the jackpot can take any of the values 1, 2, 3, and so on.

Displaying Quantitative Data: Dotplots

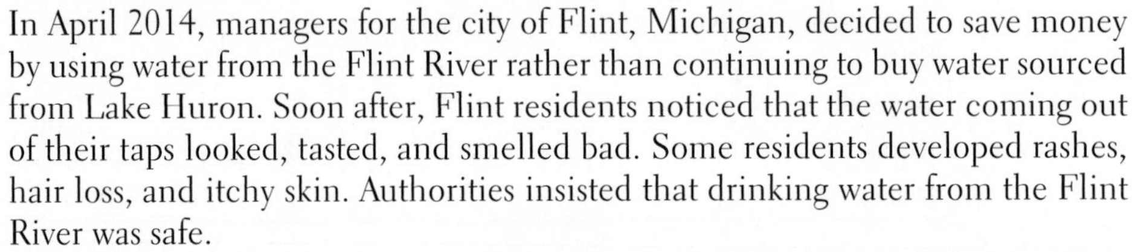

JIM WATSON/AFP/Getty Images

In April 2014, managers for the city of Flint, Michigan, decided to save money by using water from the Flint River rather than continuing to buy water sourced from Lake Huron. Soon after, Flint residents noticed that the water coming out of their taps looked, tasted, and smelled bad. Some residents developed rashes, hair loss, and itchy skin. Authorities insisted that drinking water from the Flint River was safe.

As part of its regular testing program, city officials measured lead levels in water samples collected from 71 randomly selected Flint dwellings between January and June 2015. Here are the data (in parts per billion, ppb).[29] The U.S. Environmental Protection Agency (EPA) requires action if a water system's lead level exceeds 15 parts per billion.

0	0	0	0	0	0	0	0	0	0	0	0
0	1	1	1	1	2	2	2	2	2	2	2
2	2	2	2	3	3	3	3	3	3	3	3
3	3	3	4	4	5	5	5	5	5	5	5
5	6	6	6	6	7	7	7	8	8	9	10
10	11	13	18	20	21	22	29	42	42	104	

Figure 1.2 is a **dotplot** of these data.

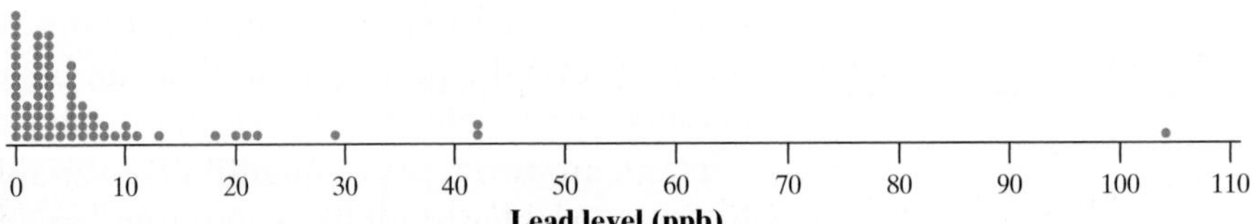

FIGURE 1.2 Dotplot of lead levels in water samples from 71 randomly selected Flint, Michigan, dwellings in January to June, 2015.

DEFINITION Dotplot

A **dotplot** shows each data value as a dot above its location on a number line.

A dotplot is the simplest graph for displaying the distribution of a quantitative variable. It is fairly easy to make a dotplot by hand for small sets of quantitative data.

HOW TO MAKE A DOTPLOT

1. **Draw and label the axis.** Draw a horizontal axis and put the name of the quantitative variable underneath it. Be sure to include units of measurement if appropriate.
2. **Scale the axis.** Find the smallest and largest values in the data set. Start the horizontal axis at a convenient number equal to or less than the smallest value and place tick marks at equal intervals until you equal or exceed the largest value.
3. **Plot the values.** Mark a dot above the location on the horizontal axis corresponding to each data value. Try to make all the dots the same size and space them out equally as you stack them.

Remember what we said in Section 1B: making a graph is not an end in itself. When you look at a graph, always ask, "What do I see?" The dotplot in Figure 1.2 shows that 8/71 = 0.113 = 11.3% of the Flint water samples had lead levels that exceeded 15 parts per billion. Does that mean Flint's water system required action based on the EPA guideline? We'll answer that question — and tell you the rest of the story about the Flint water crisis — in Section 1E.

EXAMPLE

Eating healthy
Displaying quantitative data: Dotplots

Skills 2.A, 2.B

PROBLEM: How healthy is plant-based yogurt? The *Nutrition Action Healthletter* provided data on calories, saturated fat, protein, calcium, and total sugars for many popular brands of yogurt. Here are the data on the amount of added sugar, in teaspoons (tsp), in single-serving containers for all of the plant-based yogurt brands:[30]

Clare Gainey/Alamy Stock Photo

0.0	1.0	1.0	2.5	0.0	0.0	1.0	1.5	2.0	2.0	3.0	3.5	2.5
3.0	5.0	3.0	3.5	3.5	3.5	2.0	4.0	3.5	2.0	3.5	1.5	2.0

(a) Make a dotplot of these data.

(b) The American Heart Association (AHA) recommends a maximum of 6 teaspoons of added sugar per day for women.[31] What percentage of these plant-based yogurt brands have less than half of the AHA's daily recommendation?

SOLUTION:

(a)

0.0 0.5 1.0 1.5 2.0 2.5 3.0 3.5 4.0 4.5 5.0

Amount of added sugar (tsp)

To make the dotplot:

1. **Draw and label the axis.** Be sure to include units along with the variable name.
2. **Scale the axis.** The smallest value is 0.0 and the largest value is 5.0. So we choose a scale from 0 to 5 with tick marks 0.5 unit apart.
3. **Plot the values.**

(b) Half of the AHA's recommended daily maximum would be 3 teaspoons of added sugar, and 15/26 = 0.577 = 57.7% of these plant-based yogurt brands have less than 3 teaspoons of added sugar.

FOR PRACTICE, TRY EXERCISE 3

Describing Shape

When you describe the shape of a dotplot or another graph of quantitative data, focus on the main features. Look for clear *peaks,* not for minor ups and downs in the graph. Look for *clusters* of values and obvious *gaps.* Decide if the distribution is **roughly symmetric, skewed to the left,** or **skewed to the right.**

DEFINITION Roughly symmetric, Skewed to the left, Skewed to the right

A distribution is **roughly symmetric** if the right side of the graph (containing the half of the observations with the largest values) is approximately a mirror image of the left side.

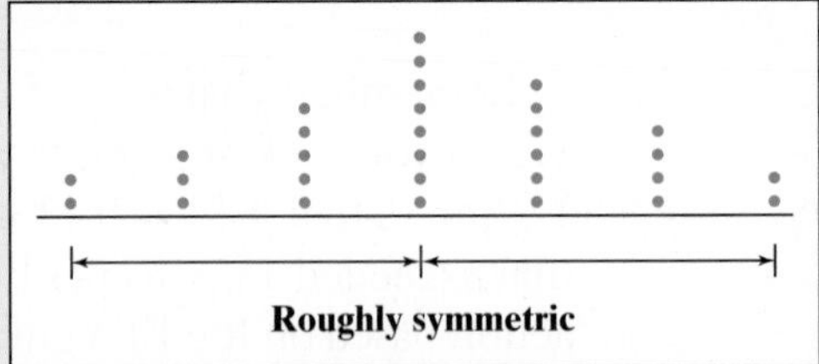

A distribution is **skewed to the left** if the left side of the graph is much longer than the right side.

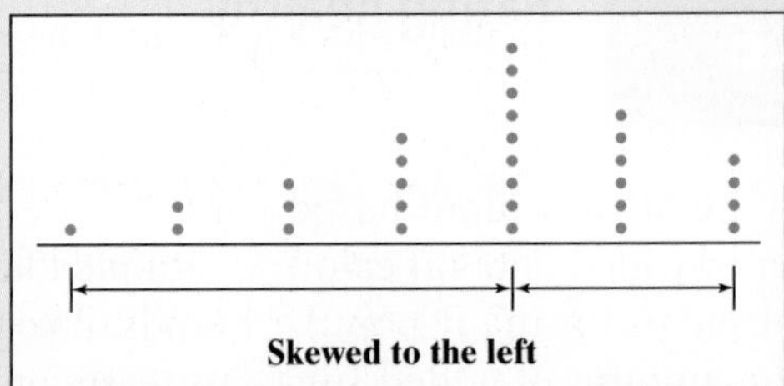

A distribution is **skewed to the right** if the right side of the graph is much longer than the left side.

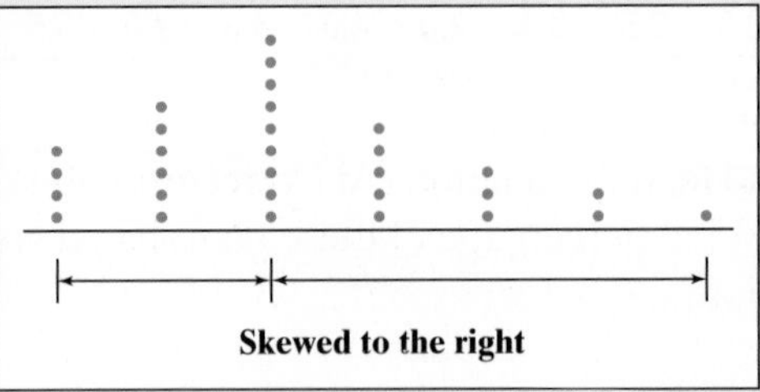

We could also describe a distribution with a long tail to the left as "skewed toward negative values" or "negatively skewed," and a distribution with a long right tail as "positively skewed." For ease, we sometimes say "left-skewed" instead of "skewed to the left" and "right-skewed" instead of "skewed to the right." **The direction of skewness is toward the long tail, not in the direction where most observations are clustered.** The drawing is a cute but corny way to help you keep this straight. To avoid danger, Mr. Starnes skis on the gentler slope — in the direction of the skewness.

EXAMPLE

Quiz scores and die rolls
Describing shape

Skill 2.A

PROBLEM: The dotplots display two different sets of quantitative data. Graph (a) shows the scores on a 20-point quiz for each of the 30 students in a statistics class. Graph (b) shows the results of 100 rolls of a six-sided die. Describe the shape of each distribution.

(a)

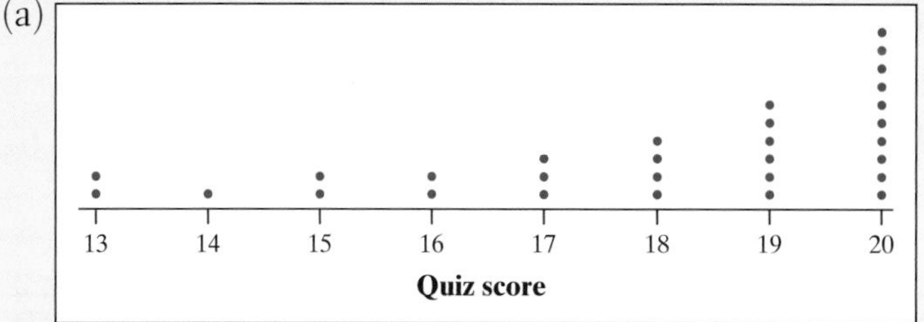

(b)

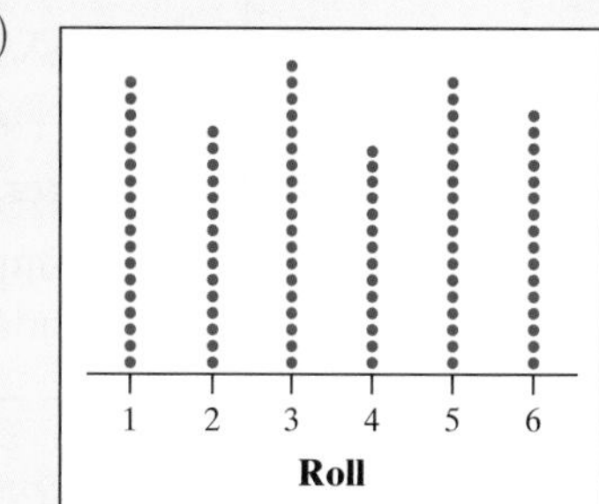

SOLUTION:

(a) The distribution of statistics quiz scores is skewed to the left, with a single peak at 20 (a perfect score).

(b) The distribution of die rolls is roughly symmetric. It has no clear peak.

FOR PRACTICE, TRY EXERCISE 7

We can describe the shape of the distribution in part (b) of the example as **approximately uniform** because the frequencies are about the same for all possible rolls.

DEFINITION Approximately uniform

A distribution in which the frequency (relative frequency) of each possible value is about the same is **approximately uniform.**

Some people refer to graphs with a single peak, like the dotplot of quiz scores in part (a) of the example, as *unimodal*. Figure 1.3 is a dotplot of the duration (in minutes) of 263 eruptions of the Old Faithful geyser in July 1995, when the Starnes family made its first trip to Yellowstone National Park. We describe this graph as *bimodal* because it has two clear peaks: one at about 2 minutes and one at about 4.5 minutes. Although we could continue the pattern with "trimodal" for three peaks, and so on, it's more common to refer to distributions with more than two clear peaks as *multimodal*. When you examine a graph of quantitative data, describe any pattern you see as clearly as you can.

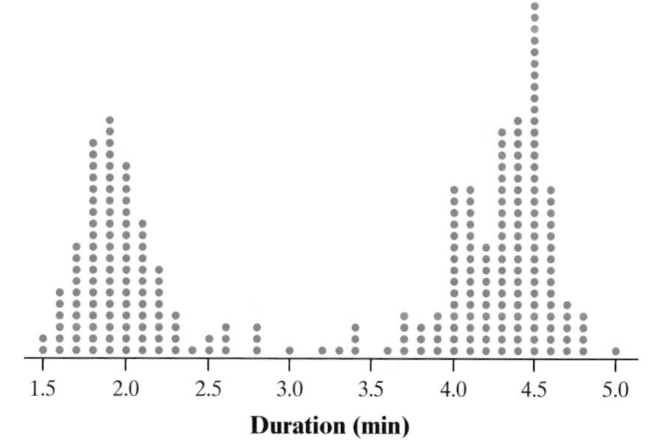

FIGURE 1.3 Dotplot displaying the duration (in minutes) of 263 eruptions of the Old Faithful geyser in July 1995. This distribution has two main clusters of data and two clear peaks—one near 2 minutes and the other near 4.5 minutes.

Describing Distributions of Quantitative Data

Here is a general strategy for describing the distribution of a quantitative variable.

HOW TO DESCRIBE THE DISTRIBUTION OF A QUANTITATIVE VARIABLE

In any graph, look for the *overall pattern* and for clear *departures* from that pattern.

- You can describe the overall pattern of a distribution by its **shape, center,** and **variability.**
- An important kind of departure is an **outlier,** an observation that falls outside the overall pattern.

Variability is sometimes referred to as *spread*. We prefer variability because students often think that spread refers only to the distance between the maximum and minimum values of a quantitative data set (the *range*).

We will discuss more formal ways to measure center and variability and to identify outliers in Section 1D. For now, just use the middle value in the ordered data set (what you may have learned as the *median* in previous math classes) when describing center and the *minimum* and *maximum* values when describing variability.

Let's practice with the dotplot of lead levels in water samples from 71 randomly selected Flint, Michigan, dwellings that you saw in Figure 1.2.

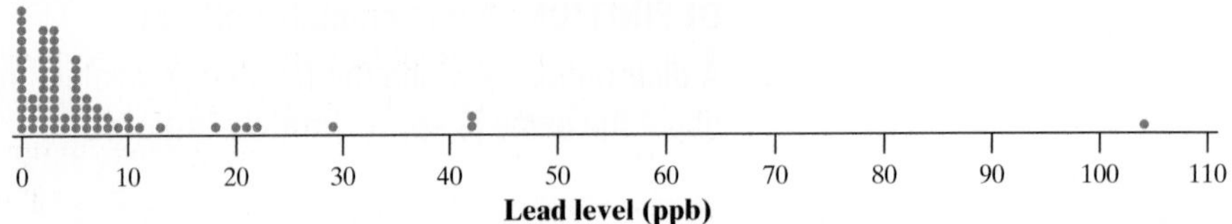

Shape: The distribution of lead level is skewed to the right, with a single peak at 0 ppb.

Outliers: There is one obvious outlier at 104 ppb. The two lead levels of 42 ppb may also be outliers.

Center: The middle value (median) is a lead level of 3 ppb.

Variability: The lead levels vary from 0 to 104 ppb.

When describing a distribution of quantitative data, don't forget: **S**tatistical **O**pinions **C**an **V**ary (Shape, Outliers, Center, Variability).

EXAMPLE

Eating healthy
Describing distributions of quantitative data

Skill 2.A

PROBLEM: How healthy is plant-based yogurt? Here is a dotplot of data on the amount of added sugar, in teaspoons (tsp), in single-serving containers of several plant-based yogurt brands:[32]

Sara Stathas/Alamy Stock Photo

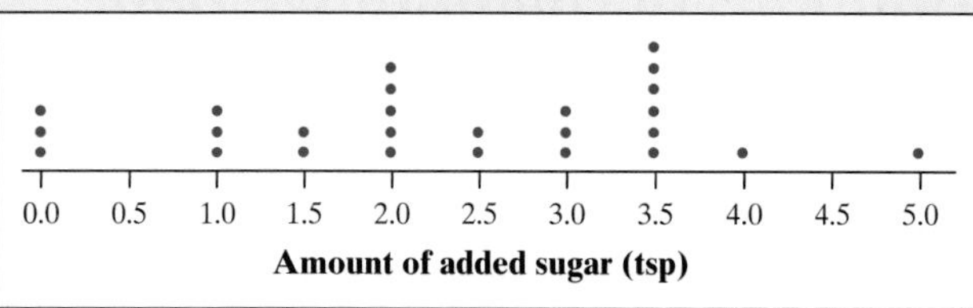

Describe the distribution.

SOLUTION:

Shape: The distribution of amount of added sugar for these plant-based yogurt brands is roughly symmetric, with two clear peaks at 2 tsp and 3.5 tsp of added sugar. There are small gaps at 0.5 tsp and 4.5 tsp.

Outliers: The three brands of yogurt with 0 tsp of added sugar and the one brand with 5 tsp of added sugar are possible outliers.

Center: The middle value is between 2 and 2.5 tsp of added sugar (median rating = 2.25 tsp).

Variability: The amount of added sugar varies from 0 to 5 tsp.

AP® EXAM TIP

Always be sure to include context when you are asked to describe a distribution. This means using the variable name (e.g., amount of added sugar), not just the units the variable is measured in (tsp).

FOR PRACTICE, TRY EXERCISE 9

CHECK YOUR UNDERSTANDING

Knoebels Amusement Park in Elysburg, Pennsylvania, has earned acclaim for being an affordable, family-friendly entertainment venue. Knoebels does not charge for general admission or parking, but it does charge customers for each ride they take. How much do the rides cost at Knoebels? The table shows the cost for each ride in a sample of 22 rides in a recent year.[33]

Name	Cost	Name	Cost
Merry Mixer	$1.50	Looper	$1.75
Italian Trapeze	$1.50	Flying Turns	$3.00
Satellite	$1.50	Flyer	$1.50
Galleon	$1.50	The Haunted Mansion	$1.75
Whipper	$1.25	StratosFear	$2.00
Skooters	$1.75	Twister	$2.50
Ribbit	$1.25	Cosmotron	$1.75
Roundup	$1.50	Paratrooper	$1.50
Paradrop	$1.25	Downdraft	$1.50
The Phoenix	$2.50	Rockin' Tug	$1.25
Gasoline Alley	$1.75	Sklooosh!	$1.75

1. Make a dotplot of the data.
2. Describe the distribution.

Comparing Distributions of Quantitative Data

Some of the most interesting statistics questions involve comparing two or more groups. Do high school graduates earn more, on average, than students who do not graduate from high school? Which of several popular diets leads to greater long-term weight loss? Just like when describing the distribution of a quantitative variable, you should always discuss shape, outliers, center, and variability whenever you compare distributions of quantitative data.

EXAMPLE

Household size in South Africa and the United Kingdom
Comparing distributions of quantitative data

Skill 2.D

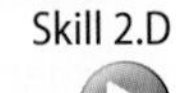

PROBLEM: How do the numbers of people living in households in the United Kingdom (U.K.) and South Africa compare? To help answer this question, we selected separate random samples of 50 households from each country.[34] Here are dotplots of the number of people in each household. Compare the distributions of household size for these two countries.

FrankvandenBergh/Getty Images

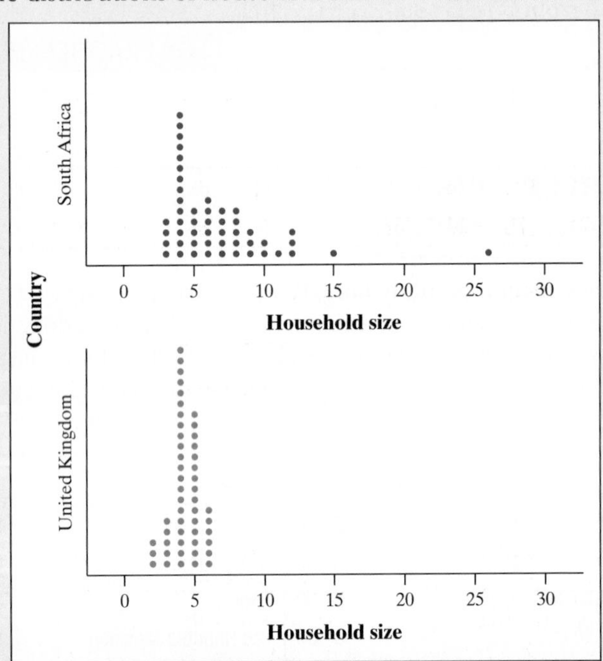

SOLUTION:

Shape: The distribution of household size for the U.K. sample is roughly symmetric, with a single peak at 4 people. The distribution of household size for the South Africa sample is skewed to the right, with a single peak at 4 people and a clear gap between 15 and 26 people.

Outliers: There are no apparent outliers in the U.K. dotplot. The two largest values in the South Africa dotplot—households with 15 and 26 people—appear to be outliers.

Center: Household sizes for the South Africa sample tend to be larger (median = 6 people) than for the U.K. sample (median = 4 people).

Variability: The household sizes for the South Africa sample vary more (from 3 to 26 people) than for the U.K. sample (from 2 to 6 people).

AP® EXAM TIP

When comparing distributions of quantitative data, it's not enough just to list values for the center and variability of each distribution. You have to *compare* these values explicitly, using words like "greater than," "less than," or "about the same as. And be sure to include context!

FOR PRACTICE, TRY EXERCISE 13

Notice that in the preceding example, we discussed the distributions of household size only for the two *samples* of 50 households. We might be interested in whether the sample data give us convincing evidence of a difference in the *population* distributions of household size for South Africa and the United Kingdom. We'll have to wait several units to decide whether we can reach such a conclusion, but our ability to make such an inference later will be helped by the fact that the households in our samples were chosen at random.

Displaying Quantitative Data: Stemplots

Another simple type of graph for displaying quantitative data is a **stemplot**, also called a *stem-and-leaf plot.*

DEFINITION Stemplot

A **stemplot** shows each data value separated into two parts: a *stem,* which consists of the leftmost digits, and a *leaf,* consisting of the final digit. The stems are ordered from least to greatest and arranged in a vertical column. The leaves are arranged in increasing order out from the appropriate stems.

Here are data on the resting pulse rates (beats per minute, bpm) of 19 middle school students:

71	104	76	88	78	71	68	86	70	90
74	76	69	68	88	96	68	82	120	

Figure 1.4 shows a stemplot of these data.

FIGURE 1.4 Stemplot of the resting pulse rates of 19 middle school students.

```
 6 | 8889
 7 | 0114668
 8 | 2688
 9 | 06
10 | 4
11 |
12 | 0
```

Key: 8|2 is a student whose resting pulse rate is 82 beats per minute.

According to the American Heart Association, a resting pulse rate greater than 100 bpm is considered high for this age group. We can see that $2/19 = 0.105 = 10.5\%$ of these students have high resting pulse rates by this standard. Also, the distribution of pulse rates for these 19 students is skewed to the right (toward the larger values).

Stemplots give us a quick picture of a distribution that includes the individual data values in the graph. It is fairly easy to make a stemplot by hand for small sets of quantitative data.

HOW TO MAKE A STEMPLOT

1. **Make stems.** Separate each data value into a stem (all but the final digit) and a leaf (the final digit). Write the stems in a vertical column from smallest to largest. Draw a vertical line at the right of this column. Do not skip any stems, even if there is no data value for a particular stem.
2. **Add leaves.** Write each leaf in the row to the right of its stem.
3. **Order the leaves.** Arrange the leaves in increasing order out from the stem.
4. **Add a key.** Provide a key that explains in context what the stems and leaves represent.

EXAMPLE

Preventing concussions
Displaying quantitative data: Stemplots

Skills 2.A, 2.B

PROBLEM: Many athletes (and their parents) worry about the risk of concussions when playing sports. A youth football coach plans to obtain specially made helmets for the players that are designed to reduce their chance of getting a concussion. Here are the measurements of head circumference (in inches) for the 30 players on the team:[35]

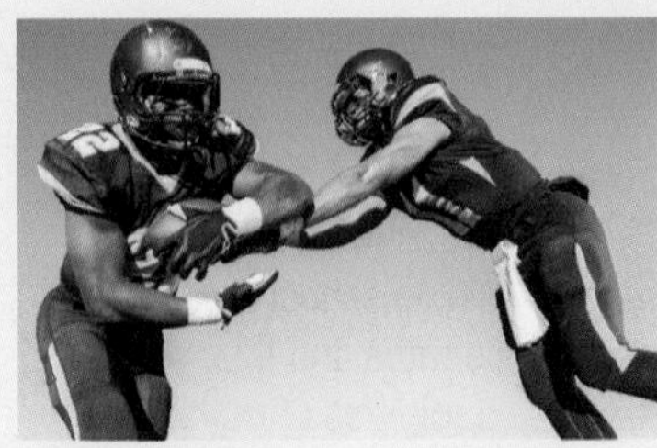

Pete Saloutos/AGE Fotostock

23.0	22.2	21.7	22.0	22.3	22.6	22.7	21.5	22.7	25.6
20.8	23.0	24.2	23.5	20.8	24.0	22.7	22.6	23.9	22.5
23.1	21.9	21.0	22.4	23.5	22.5	23.9	23.4	21.6	23.3

(a) Make a stemplot of these data.
(b) Describe the shape of the distribution. Are there any obvious outliers?

SOLUTION:

(a)

```
20 | 88
21 | 05679
22 | 02345566777
23 | 001345599
24 | 02
25 | 6
```

Key: 23|5 is a player with a head circumference of 23.5 inches.

To make the stemplot:

1. **Make stems.** The smallest head circumference is 20.8 inches and the largest is 25.6 inches. We use the first two digits as the stem and the final digit as the leaf. So we need stems from 20 to 25.
2. **Add leaves.** For the player with a head circumference of 23.0 inches, place a 0 on the 23 stem. For the player with a head circumference of 22.2 inches, place a 2 on the 22 stem. Continue in this way until you have added the data for all the players.
3. **Order the leaves.**
4. **Add a key.**

(b) The distribution of head circumference for the 30 players on the youth football team is roughly symmetric, with a single peak on the 22-inch stem. There are no obvious outliers.

FOR PRACTICE, TRY EXERCISE 17

We can get a better picture of the head circumference data by *splitting stems*. In Figure 1.5(a), the leaves from 0 to 9 are placed on the same stem. Figure 1.5(b) shows another stemplot of the same data. This time, leaves 0 through 4 are placed on one stem, while leaves 5 through 9 are placed on another stem. Now we can see the shape of the distribution more clearly — including the possible outlier at 25.6 inches.

FIGURE 1.5 Two stemplots showing the head circumference data. The graph in (b) improves on the graph in (a) by splitting stems.

(a)

```
20 | 88
21 | 05679
22 | 02345566777
23 | 001345599
24 | 02
25 | 6
```

Key: 23|5 is a player with a head circumference of 23.5 inches.

(b)

```
20 | 88
21 | 0
21 | 5679
22 | 0234
22 | 5566777
23 | 00134
23 | 5599
24 | 02
24 |
25 |
25 | 6
```

Be sure to include these stems even though they include no data.

Here are a few tips to consider before making a stemplot:

- There is no magic number of stems to use. Too few or too many stems will make it difficult to see the distribution's shape. Five stems is a good minimum.
- If you split stems, make sure that each stem is assigned an equal number of possible leaf digits.
- When the data have too many digits, you can get more flexibility by rounding or truncating the data.

COMPARING DISTRIBUTIONS WITH STEMPLOTS

You can use a *back-to-back stemplot* with common stems to compare the distribution of a quantitative variable in two groups. The leaves on each side are placed in order leading out from the common stem. For example, Figure 1.6 shows a back-to-back stemplot of the 19 middle school students' resting pulse rates and their pulse rates after 5 minutes of running.

FIGURE 1.6 Back-to-back stemplot of 19 middle school students' resting pulse rates and their pulse rates after 5 minutes of running.

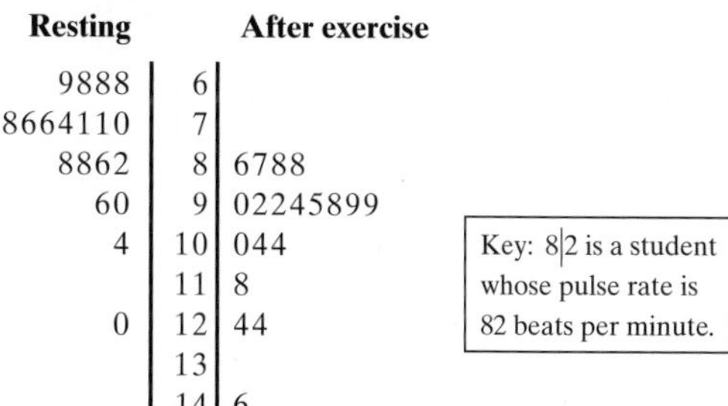

Resting
After exercise
9888 | 6 |
8664110 | 7 |
8862 | 8 | 6788
60 | 9 | 02245899
4 | 10 | 044
| 11 | 8
0 | 12 | 44
| 13 |
| 14 | 6
Key: 8|2 is a student whose pulse rate is 82 beats per minute.

Let's compare the distributions. As always, we discuss shape, outliers, center, and variability.

Shape: The distribution of resting pulse rates is skewed to the right, with a single peak on the 70s stem. The distribution of after-exercise pulse rates is also skewed to the right, with a single peak on the 90s stem.

Outliers: The student whose resting pulse rate was 120 bpm appears to be an outlier in that group. The student whose after-exercise pulse rate was 146 bpm appears to be an outlier in that group.

Center: The students' resting pulse rates tended to be lower (median = 76 bpm) than their after-exercise pulse rates (median = 98 bpm).

Variability: The after-exercise pulse rates vary more (from 86 bpm to 146 bpm) than the resting pulse rates (from 68 bpm to 120 bpm).

CHECK YOUR UNDERSTANDING

1. Which is healthier: dairy yogurt or plant-based yogurt? The *Nutrition Action Healthletter* provided data on calories, saturated fat, protein, calcium, and total sugars for many popular brands of yogurt.[36] Here are parallel dotplots that compare the calories in single-serving containers of 57 brands of dairy yogurt and 26 brands of plant-based yogurt. Describe similarities and differences in the two distributions.

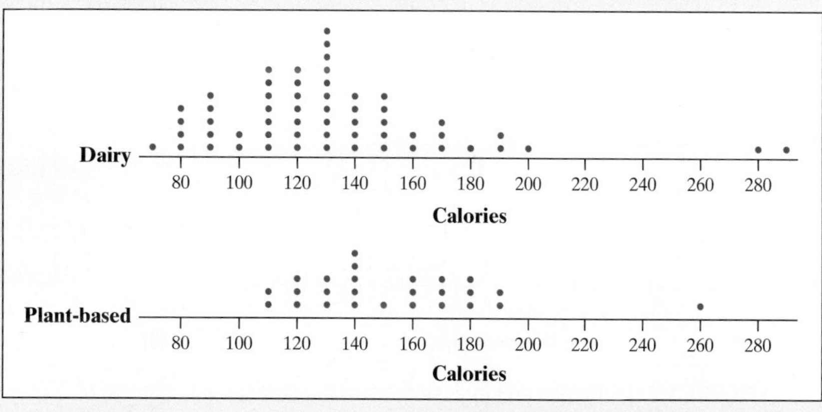

Here is a stemplot of the percentage of residents aged 65 and older in the 50 U.S. states:[37]

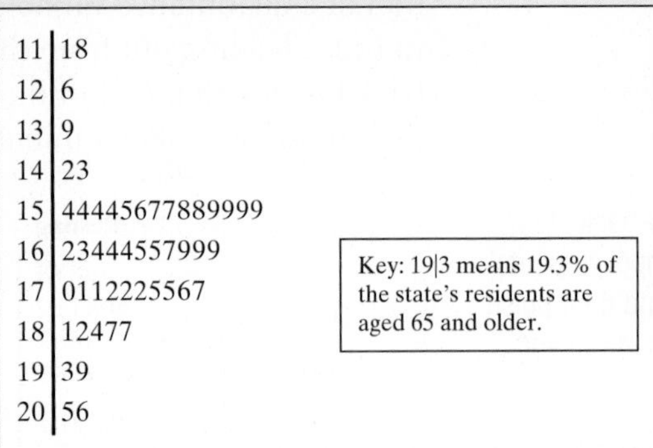

11 | 18
12 | 6
13 | 9
14 | 23
15 | 44445677889999
16 | 23444557999
17 | 0112225567
18 | 12477
19 | 39
20 | 56
Key: 19|3 means 19.3% of the state's residents are aged 65 and older.

2. The largest data value is for Maine. What percentage of Maine residents are 65 or older? (Florida has the second largest data value.)
3. Make another stemplot of the data by splitting stems. What does this new graph reveal that is not apparent from the original stemplot?

Displaying Quantitative Data: Histograms

You can use a dotplot or a stemplot to display quantitative data. Both of these types of graphs show every individual data value. However, for large data sets, this can make it difficult to see the overall pattern in the graph. We often get a cleaner picture of the distribution by grouping nearby values together. Doing so allows us to make a new type of graph: a **histogram.**

DEFINITION Histogram

A **histogram** shows each interval as a bar. The heights of the bars show the frequencies or relative frequencies of values in each interval.

Figure 1.7 shows a dotplot and a histogram of the durations (in minutes) of 263 eruptions of the Old Faithful geyser in July 1995. Notice how the histogram groups nearby values together into equal-width intervals.

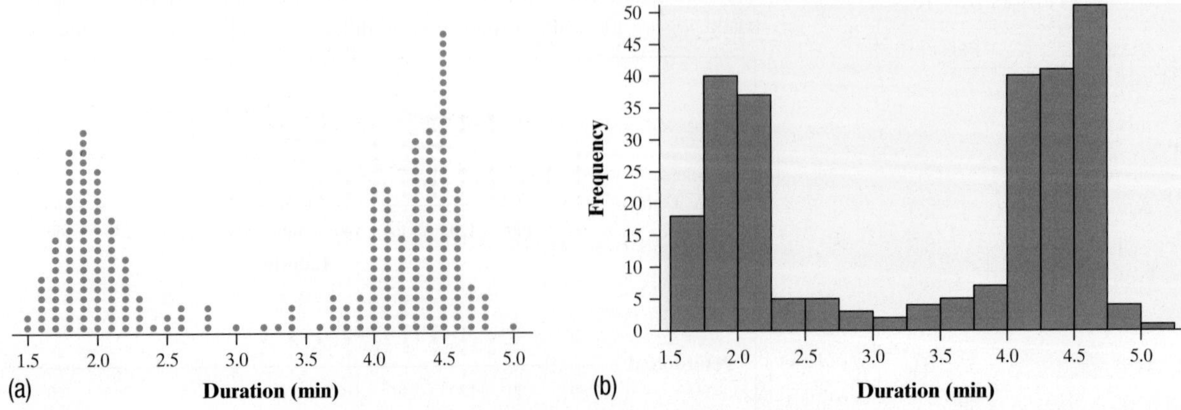

FIGURE 1.7 (a) Dotplot and (b) histogram of the duration (in minutes) of 263 eruptions of the Old Faithful geyser in July 1995.

You can make a histogram by hand, even for fairly large sets of quantitative data. For details on making histograms with technology, see the Tech Corner at the end of this section.

HOW TO MAKE A HISTOGRAM

1. **Choose equal-width intervals** that span the data. Five intervals is a good minimum.
2. **Make a table** that shows the frequency (count) or relative frequency (percentage or proportion) of individuals in each interval.
3. **Draw and label the axes.** Draw horizontal and vertical axes. Put the name of the quantitative variable under the horizontal axis. To the left of the vertical axis, indicate whether the graph shows the frequency (count) or relative frequency (percentage or proportion) of individuals in each interval.
4. **Scale the axes.** Place equally spaced tick marks at the smallest value in each interval along the horizontal axis until you equal or exceed the largest data value. On the vertical axis, start at 0 and place equally spaced tick marks until you equal or exceed the largest frequency or relative frequency in any interval.
5. **Draw bars** above the intervals. Make the bars equal in width and leave no gaps between them. Be sure that the height of each bar corresponds to the frequency or relative frequency of data values in that interval. An interval with no data values will appear as a bar of height 0 on the graph.

It is possible to choose intervals of unequal widths when making a histogram, but such graphs are beyond the scope of this book.

EXAMPLE

Foreign-born residents
Displaying quantitative data: Histograms

Skills 2.A, 2.B

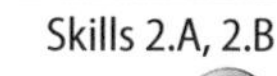

Sarah Reingewirtz/MediaNews Group/Los Angeles Daily News/Getty Images

PROBLEM: How does the percentage of foreign-born residents in each U.S. state compare to the rest of the country? The table presents the data for all 50 states in a recent year.[38]

State	Percentage	State	Percentage	State	Percentage
Alabama	3.6	Georgia	10.3	Maine	3.9
Alaska	8.0	Hawaii	19.3	Maryland	15.4
Arizona	13.4	Idaho	5.8	Massachusetts	17.3
Arkansas	5.1	Illinois	13.9	Michigan	7.0
California	26.7	Indiana	5.3	Minnesota	8.4
Colorado	9.5	Iowa	5.6	Mississippi	2.1
Connecticut	14.8	Kansas	7.2	Missouri	4.3
Delaware	10.0	Kentucky	4.4	Montana	2.3
Florida	21.1	Louisiana	4.2	Nebraska	7.4

State	Percentage	State	Percentage	State	Percentage
Nevada	19.8	Oklahoma	6.1	Utah	8.6
New Hampshire	6.4	Oregon	9.7	Vermont	4.7
New Jersey	23.4	Pennsylvania	7.0	Virginia	12.7
New Mexico	9.6	Rhode Island	13.7	Washington	14.9
New York	22.4	South Carolina	5.6	West Virginia	1.6
North Carolina	8.4	South Dakota	4.1	Wisconsin	5.1
North Dakota	4.1	Tennessee	5.5	Wyoming	3.1
Ohio	4.8	Texas	17.1		

(a) Make a frequency histogram to display the data.
(b) What proportion of states have less than 10% foreign-born residents?

SOLUTION:

(a)

Interval	Frequency
0 to <5	13
5 to <10	20
10 to <15	8
15 to <20	5
20 to <25	3
25 to <30	1

To make the histogram:

1. **Choose equal-width intervals** that span the data. The data vary from 1.6% to 26.7%. We choose intervals of width 5, beginning at 0: 0 to <5, 5 to <10, and so on. This choice results in more than the minimum of five intervals.
2. **Make a table.** Record the number of states in each interval when making a frequency histogram.
3. **Draw and label the axes.**
4. **Scale the axes.** The scale on the horizontal axis matches the intervals we chose in Step 1. The highest frequency in an interval is 20, so we scale the vertical axis from 0 to 20, placing tick marks every 5 units.
5. **Draw bars.**

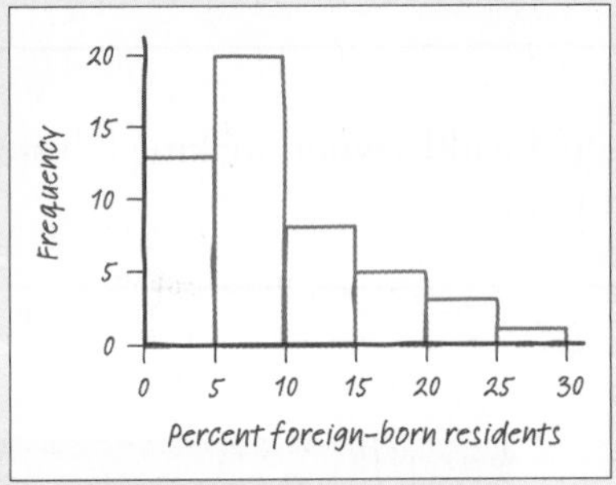

(b) $(13+20)/50 = 33/50 = 0.66$ of states have less than 10% foreign-born residents.

FOR PRACTICE, TRY EXERCISE 25

From the histogram, we can see that the distribution of the percentage of foreign-born residents in the 50 U.S. states is skewed to the right. The graph does not show any obvious outliers. How should we describe the center and variability of the distribution? Because a histogram does not show individual data values, we can only give *estimates* of the center and variability using the intervals on the graph. With 50 data values, the middle value (median) falls between the 25th and 26th values in the ordered data set, so the median percentage of foreign-born residents is in the 5% to < 10% interval. The data vary from at least 0% to at most 29.9% foreign-born residents. Using the raw data from the example, we can confirm that the median is 7.3% and that the data vary from 1.6% to 26.7%.

Figure 1.8 shows two different histograms of the foreign-born resident data. The one on the left (a) uses the intervals of width 5 from the example. The distribution has a single peak in the 5 to <10 interval. The one on the right (b) uses

intervals half as wide: 0 to <2.5, 2.5 to <5, and so on. Now we see the shape of the distribution more clearly. **The choice of intervals in a histogram can affect the appearance of a distribution.** Histograms with more intervals show more detail but may have a less clear overall pattern.

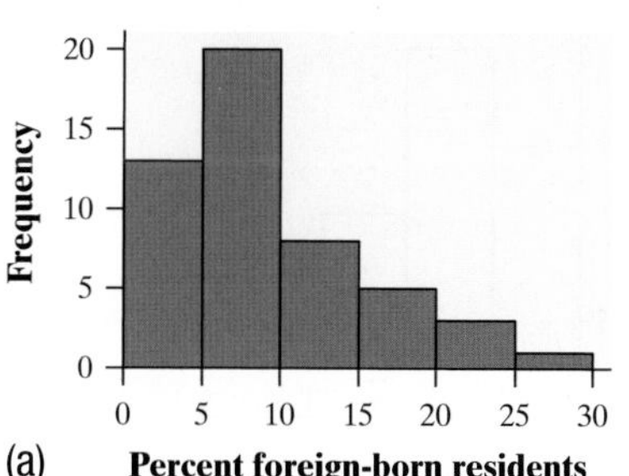

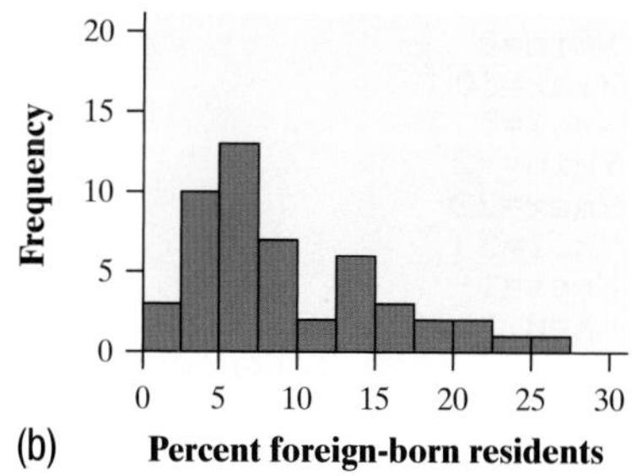

FIGURE 1.8 (a) Frequency histogram of the percentage of foreign-born residents in the 50 states with intervals of width 5, from the previous example. (b) Frequency histogram of the data with intervals of width 2.5.

1. Tech Corner MAKING HISTOGRAMS

TI-Nspire and other technology instructions are on the book's website at bfwpub.com/tps7e.

You can use technology to make a histogram. The technology's default choice of intervals is a good starting point, but you should adjust the intervals to fit with common sense. We'll illustrate using data on the percentage of foreign-born residents in the states from the preceding example.

1. Enter the percentage of foreign-born residents in each state in your statistics/list editor. Press STAT and choose Edit..., then type the values into list L1.

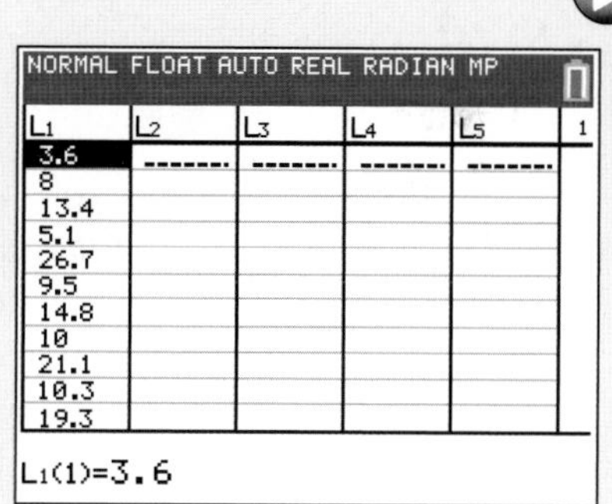

2. Set up a histogram in the statistics plots menu. Press 2nd Y= (STAT PLOT), press ENTER or 1 to go into Plot1, and adjust the settings as shown.

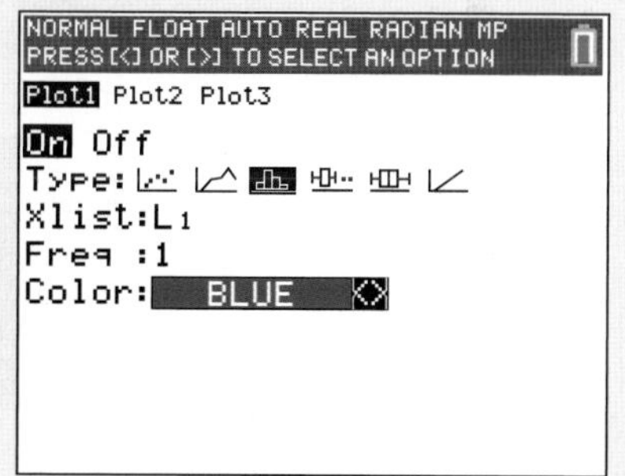

3. Use ZoomStat to let the calculator choose intervals and make a histogram. Press ZOOM and choose ZoomStat. Press TRACE and use the left and right arrow keys to examine the intervals.

Note the calculator's unusual choice of intervals.

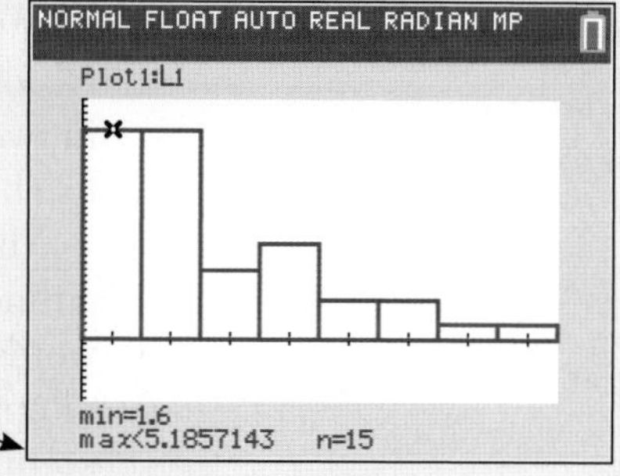

4. Adjust the intervals to match those in Figure 1.8(a), then graph the histogram. Press WINDOW and enter the values shown for Xmin, Xmax, Xscl, Ymin, Ymax, and Yscl. Then press GRAPH. Press TRACE and use the left and right arrow keys to examine the intervals.

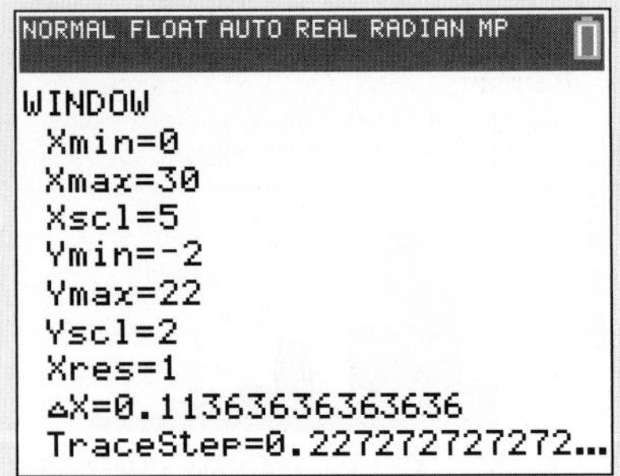

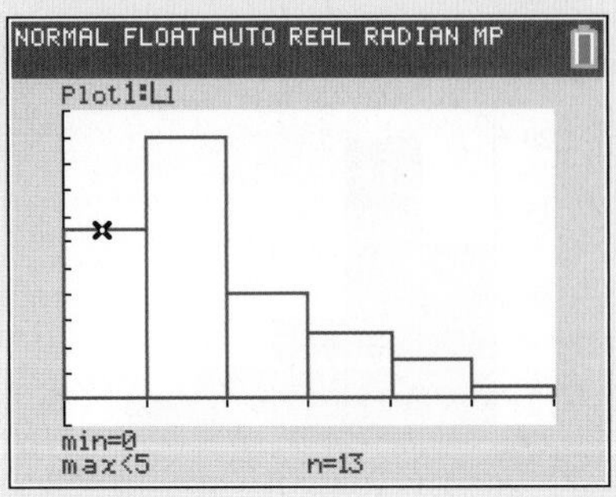

AP® EXAM TIP

If you're asked to make a graph on a free-response question, be sure to label and scale your axes. Don't just transfer what you see on a TI-83/84 calculator screen to your paper and expect to earn full credit.

5. See if you can modify the graph so that it matches the histogram in Figure 1.8(b).

Don't confuse histograms and bar graphs. Although histograms resemble bar graphs, their details and uses are different. A histogram displays the distribution of a *quantitative variable*. Its horizontal axis identifies intervals of values that the variable can take. A bar graph displays the distribution of a *categorical variable*. Its horizontal axis identifies the categories. Be sure to draw bar graphs with blank space between the bars to separate the categories. Draw histograms with no space between the bars for adjacent intervals. For comparison, here is one of each type of graph from earlier examples:

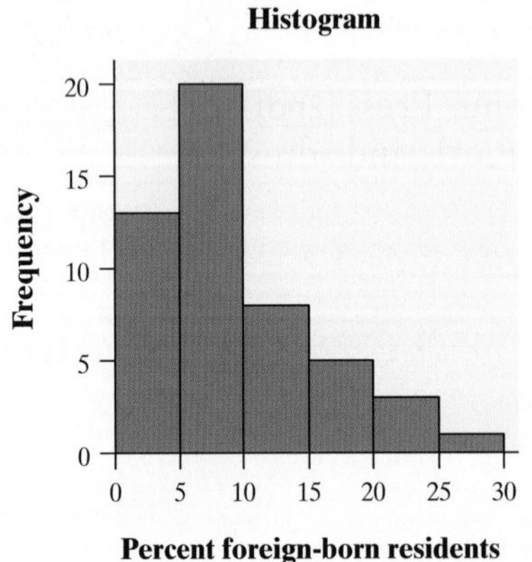

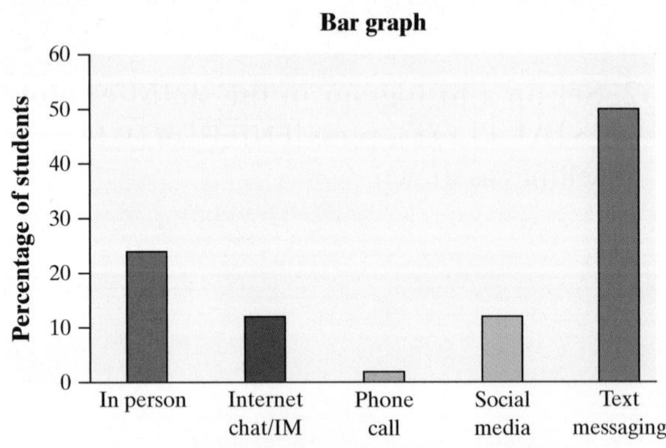

COMPARING DISTRIBUTIONS WITH HISTOGRAMS

Histograms can also be used to compare the distribution of a quantitative variable in two or more groups. Be sure to use the same intervals when making comparative histograms, so the graphs can be drawn using a common horizontal axis scale.

Mazie is interested in comparing the reading levels of a biology journal and an airline magazine. To do so, Mazie counts the number of letters in a sample of 400 words from an article in the journal, and then counts the number of letters in a sample of 100 words from an article in the airline magazine. Figure 1.9(a) displays comparative histograms of the data. This figure is misleading — it compares frequencies, but the two samples were of very different sizes (400 and 100).

Using the same data, we produced the histograms in Figure 1.9(b). By using relative frequencies, this figure makes a valid comparison of word lengths in the two samples. The moral of the story: **use relative frequencies (percentages or proportions) when comparing distributions, especially if the groups have different sizes.**

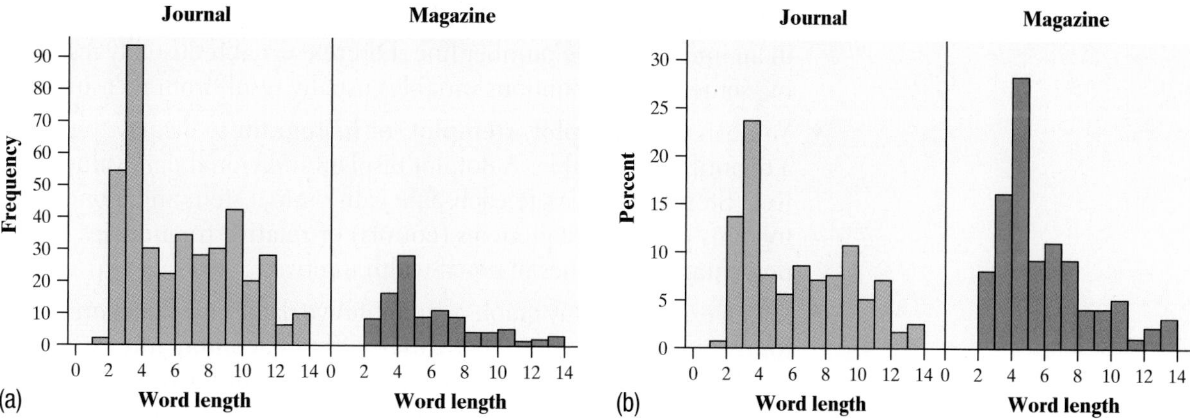

FIGURE 1.9 Two sets of histograms comparing word lengths in articles from a biology journal and from an airline magazine. In graph (a), the vertical scale uses frequencies. Graph (b) fixes the problem of different sample sizes by using percentages (relative frequencies) on the vertical scale.

When getting married, some people choose to buy an engagement ring for their partner. How much do people spend on engagement rings, on average? To find out, *The New York Times* and *Morning Consult* surveyed a random sample of 1640 U.S. adults who bought an engagement ring. Here is a histogram of the data:[39]

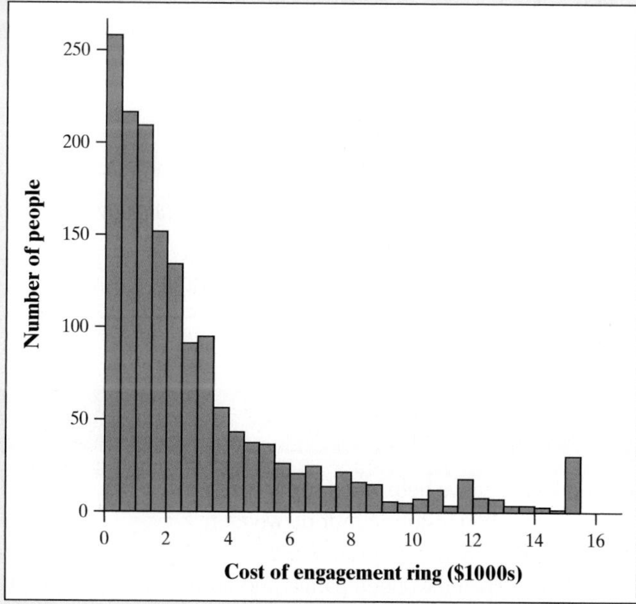

1. About what percentage of people reported spending at least $2000 on an engagement ring? Show your method clearly.
2. Describe the shape of the distribution.
3. Estimate the center and variability of the distribution.

SECTION 1C Summary

- There are two types of quantitative variables: discrete and continuous. A **discrete variable** can take a countable set of possible numeric values with gaps between them on the number line. A **continuous variable** can take any value in an interval on the number line. Discrete variables usually result from counting something; continuous variables usually result from measuring something.
- You can use a **dotplot, stemplot,** or **histogram** to display the distribution of a quantitative variable. A dotplot displays individual data values on a number line. Stemplots separate each data value into a stem and a one-digit leaf. Histograms plot the frequencies (counts) or relative frequencies (proportions or percentages) of values in equal-width intervals.
- When examining any graph of quantitative data, look for an *overall pattern* and for clear *departures* from that pattern. **Shape, center,** and **variability** describe the overall pattern of the distribution of a quantitative variable. **Outliers** are observations that lie outside the overall pattern of a distribution.
- Some distributions have simple shapes, such as **roughly symmetric, skewed to the left,** or **skewed to the right.** A distribution in which the frequency (relative frequency) of each possible value is about the same is **approximately uniform.**
- The number of peaks is another aspect of overall shape. So are distinct clusters and gaps. A single-peaked graph is sometimes called *unimodal*, and a double-peaked graph is sometimes called *bimodal*.
- Dotplots, back-to-back stemplots, and histograms can also be used to compare distributions of quantitative data. When comparing the distribution of a quantitative variable in two or more groups, be sure to compare shape, outliers, center, and variability.
- Histograms are for quantitative data; bar graphs are for categorical data. In both types of graphs, be sure to use relative frequencies when comparing data sets of different sizes.

AP® EXAM TIP
AP® Daily Videos

Review the content of this section and get extra help by watching the AP® Daily Videos for Topics 1.5 and 1.6, which are available in AP® Classroom.

1C Tech Corner

TI-Nspire and other technology instructions are on the book's website at bfwpub.com/tps7e.

1. Making histograms Page 35

SECTION 1C Exercises

1. **Protecting wood** How can we help wood surfaces resist weathering, especially when restoring historic wooden buildings? In a study that attempted to answer this question, researchers prepared wooden panels and then exposed them to the weather. Here are some of the variables the researchers recorded: type of wood, paint thickness, paint color, weathering time (1, 2, or 3 months), and number of blemishes. Classify each variable as categorical, quantitative (discrete), or quantitative (continuous).

2. **Social media** A social media company records data on each of its users for several variables: internet provider, age (in years), how many times they visited the site, total time spent on the site, country where they live, and how long since they created a member profile. Classify each variable as categorical, quantitative (discrete), or quantitative (continuous).

Displaying Quantitative Data: Dotplots

3. **Women's soccer** How good was the 2019 U.S. women's soccer team? With players like Carli Lloyd, Alex Morgan, and Megan Rapinoe, the team put on an impressive showing en route to winning the 2019 Women's World Cup. Here are data on the number of goals scored by the team in games played in the 2019 season:[40]

pg 23

1	1	2	2	1	5	6	3	5	3	13	3	2	2	2	2	2	2	3	4	3	2	1	3	6

(a) Make a dotplot of these data.

(b) In what proportion of games did the team score 4 or more goals?

4. **Fuel efficiency** The Environmental Protection Agency (EPA) is in charge of determining and reporting fuel economy ratings for cars. Here are the EPA estimates of highway gas mileage in miles per gallon (mpg) for a sample of 21 model-year 2022 midsize cars:[41]

25	30	27	31	38	26	28	40	25	28	30
31	30	30	34	30	31	31	32	48	31	

(a) Make a dotplot of these data.

(b) What percentage of the car models in the sample get more than 35 mpg on the highway?

5. **More women's soccer** The following dotplot shows the difference in the number of goals scored (U.S. women's team − Opponent) in each game from Exercise 3.

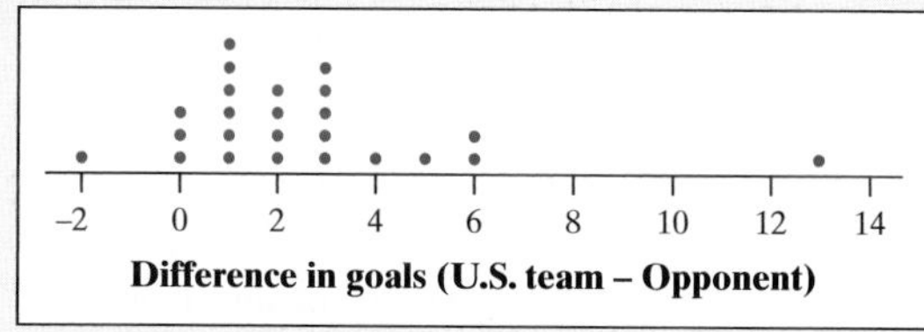

(a) Explain what the dot above −2 represents.

(b) What does the graph tell us about how well the team did in 2019? Be specific.

6. **Better fuel efficiency** The following dotplot shows the difference in EPA mileage ratings (Highway − City) for each of the 21 model-year 2022 midsize cars from Exercise 4.

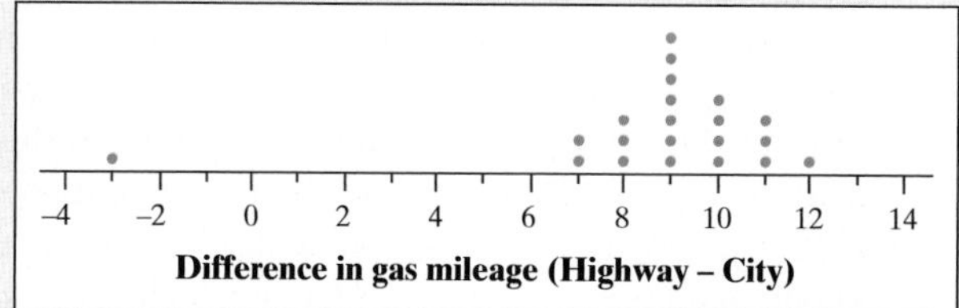

(a) The dot above −3 is for the Toyota Prius. Explain what this dot represents.

(b) What does the graph tell us about fuel economy in the city versus on the highway for these car models? Be specific.

Describing Shape

7. **Getting older and random digits**

pg 25

(a) How old is the oldest person you know? Prudential Insurance Company asked 400 people to place a blue sticker on a huge wall next to the age of the oldest person they have ever known. An image of the graph is shown here. Describe the shape of the distribution.

(b) The dotplot displays the results of using a random number generator to produce 100 digits between 0 and 9, inclusive. Describe the shape of the distribution.

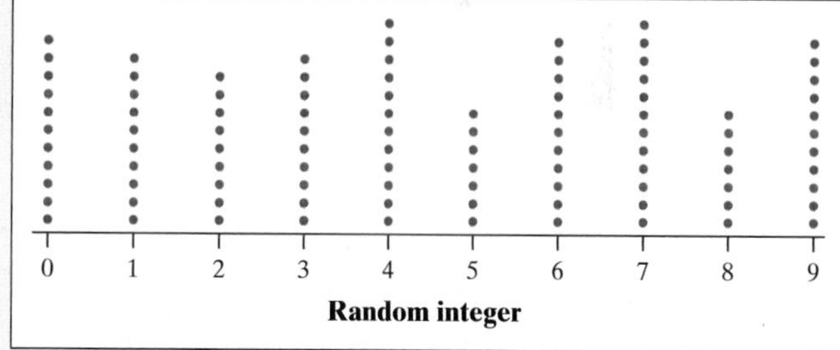

8. **Roll the dice and family ages**

(a) The dotplot shows the results of rolling a pair of fair, six-sided dice and finding the sum of the up-faces 100 times. Describe the shape of the distribution.

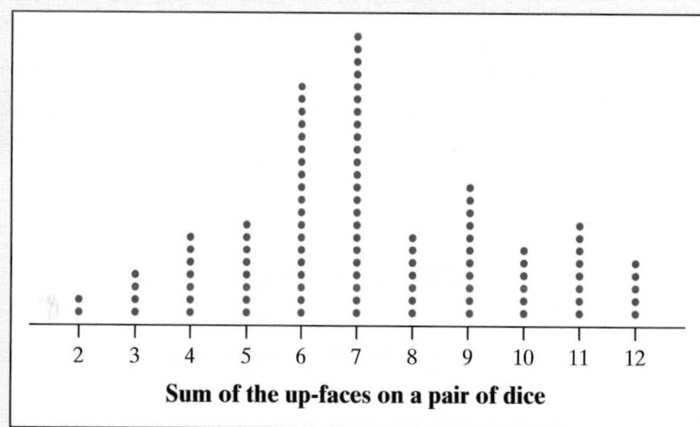

(b) Statistics instructor Paul Myers collected data on the ages (in years) of family members for each student in his class. The dotplot displays the data. Describe the shape of the distribution.

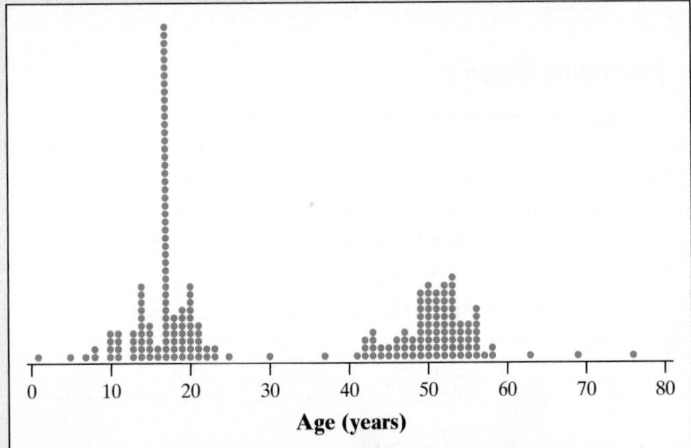

Describing Distributions of Quantitative Data

9. Feeling sleepy? A statistics professor asked how much sleep (in hours) students got on the night prior to their first exam. Here is a dotplot of the data from the 50 students in the class. Describe the distribution.

pg 26

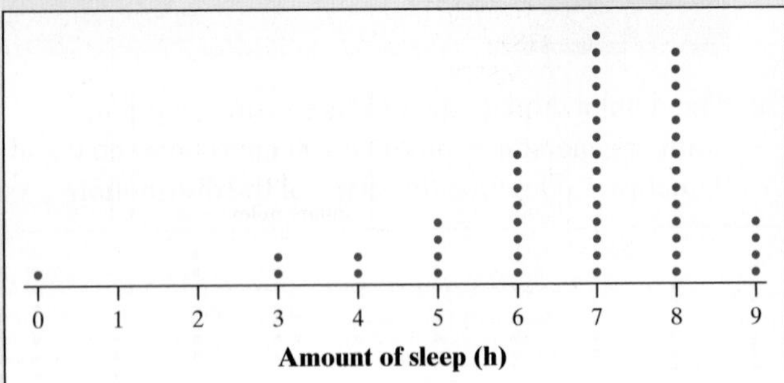

10. **Frozen pizza** *Consumer Reports* collected data on the number of calories per serving for 16 brands of frozen cheese pizza as part of its product reviews. Here is a dotplot of the data.[42] Describe the distribution.

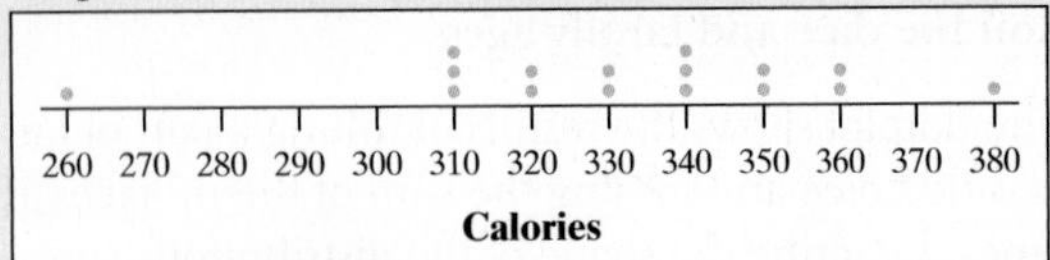

11. **Soccer distribution** Refer to Exercise 3.

(a) Describe the shape of the distribution. Are there any outliers?

(b) How many goals did the team score in a typical game that season? Explain your answer.

12. **Fuel distribution** Refer to Exercise 4.

(a) Describe the shape of the distribution. Are there any outliers?

(b) What is the typical fuel efficiency of the 21 cars in the sample? Explain your answer.

Comparing Distributions of Quantitative Data

13. **Calcium in yogurt** Which dairy yogurt is healthier: plain or flavored? The *Nutrition Action Healthletter* provided data on calories, saturated fat, protein, calcium, and total sugars for many popular brands of yogurt.[43] Here are parallel dotplots that display the data on calcium content as a percentage of the recommended daily value (%DV) of 1000 milligrams for several brands of plain and flavored dairy yogurt. Compare the distributions of calcium content in these two types of yogurt.

pg 28

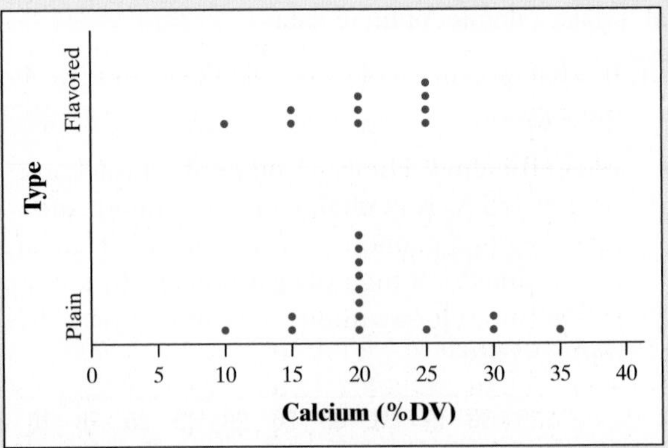

14. **Healthy streams** Nitrates are organic compounds that are a main ingredient in fertilizers. When fertilizers run off into streams, the nitrates can have a toxic effect on fish. An ecologist studying nitrate pollution in two streams measures nitrate concentrations at 42 places in Stony Brook and 42 places in Mill Brook. The parallel dotplots display the data. Compare the distributions of nitrate concentration in these two streams.

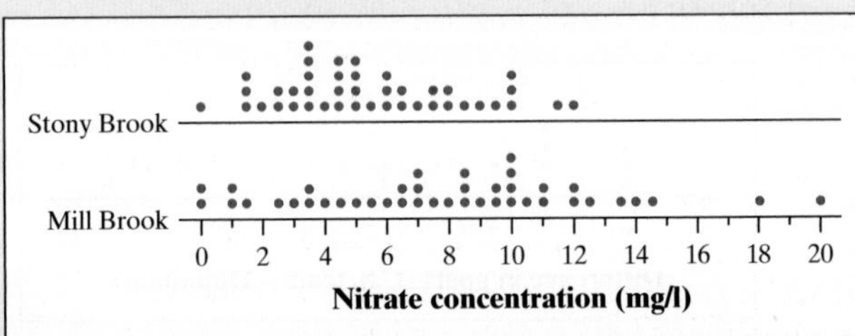

15. **Enhancing creativity** Do external rewards — like money, praise, fame, and grades — promote creativity? Researcher Teresa Amabile recruited 47 experienced creative writers who were college students and divided them at random into two groups. The students in one group were given a list of statements about external reasons (E) for writing, such as public recognition, making money, or pleasing their parents. Students in the other group were given a list of statements about internal reasons (I) for writing, such as expressing yourself and enjoying playing with words. Both groups were then instructed to write a poem about laughter. Each student's poem was rated separately by 12 different poets using a creativity scale.[44] These ratings were averaged to obtain an overall creativity score for each poem.

Parallel dotplots of the two groups' creativity scores are shown here.

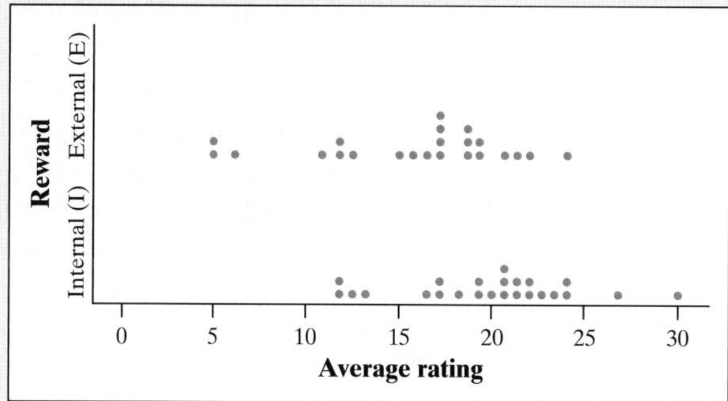

(a) Is the variability in creativity scores similar or different for the two groups? Justify your answer.

(b) Do the data suggest that external rewards promote creativity? Justify your answer.

16. **Healthy cereal?** Researchers collected data on 76 brands of cereal at a local supermarket.[45] For each brand, the sugar content (grams per serving) and the shelf in the store on which the cereal was located (1 = bottom, 2 = middle, 3 = top) were recorded. Here are parallel dotplots of the data:

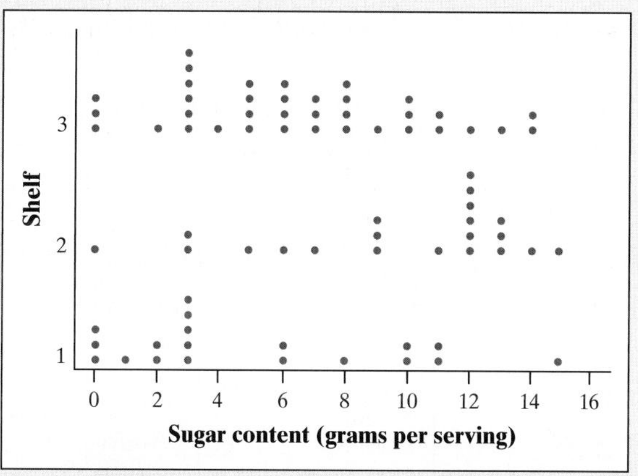

(a) Is the variability in sugar content of the cereals on the three shelves similar or different? Justify your answer.

(b) Critics claim that supermarkets tend to put sugary cereals on the middle shelf, where kids can better see them. Do the data from this study support this claim? Justify your answer.

Displaying Quantitative Data: Stemplots

17. **Snickers® are fun!** Here are the weights (in grams) of 17 Snickers Fun Size bars from a single bag:
pg 30

17.1	17.4	16.6	17.4	17.7	17.1	17.3	17.7	17.8
19.2	16.0	15.9	16.5	16.8	16.5	17.1	16.7	

(a) Make a stemplot of these data. What interesting feature does the graph reveal?

(b) The advertised weight of a Snickers Fun Size bar is 17 grams. What percentage of candy bars in this sample weigh less than advertised?

18. **Eat your beans!** Beans and other legumes are a great source of protein. The following data give the protein content of 30 different varieties of beans, in grams per 100 grams of cooked beans.[46]

7.5	8.2	8.9	9.3	7.1	8.3	8.7	9.5	8.2	9.1
9.0	9.0	9.7	9.2	8.9	8.1	9.0	7.8	8.0	7.8
7.0	7.5	13.5	8.3	6.8	10.6	8.3	7.6	7.7	8.1

(a) Make a stemplot of these data. What interesting feature does the graph reveal?

(b) What proportion of these bean varieties contain more than 9 grams of protein per 100 grams of cooked beans?

19. **South Carolina counties** Here is a stemplot of the areas of the 46 counties in South Carolina. Note that the data have been rounded to the nearest 10 square miles (mi²).

3	9999
4	0116689
5	01115566778
6	47899
7	01245579
8	0011
9	13
10	8
11	233
12	2

Key: 6|4 represents a county with an area of 635 to 644.99 square miles.

(a) What is the area of the largest South Carolina county?

(b) Describe the distribution of area for the 46 South Carolina counties.

20. **Shopping spree** The stemplot displays data on the amount spent by 50 shoppers at a grocery store. Note that the values have been rounded to the nearest dollar.

0	399
1	1345677889
2	000123455668888
3	25699
4	1345579
5	0359
6	1
7	0
8	366
9	3

Key: 9|3 = \$92.50 to \$93.49 spent

(a) What was the smallest amount spent by any of the shoppers?

(b) Describe the distribution of amount spent by these 50 shoppers.

21. **Arizona heat** Here is a stemplot of the high temperature readings (in degrees Fahrenheit) for Phoenix, Arizona, for each day in July in a recent year:[47]

8	4
8	
9	3
9	799
10	011223444
10	556667788999
11	0113
11	5

(a) Why did we split the stems?
(b) Give an appropriate key for this stemplot.
(c) Describe the shape of the distribution. Are there any outliers?

22. **Watch that caffeine!** The U.S. Food and Drug Administration (FDA) limits the amount of caffeine in a 12-ounce can of carbonated beverage to 72 milligrams. That translates to a maximum of 48 milligrams of caffeine per 8-ounce serving. Data on the caffeine content of popular soft drinks (in milligrams per 8-ounce serving) are displayed in the stemplot.

1	556
2	033344
2	55667778888899
3	113
3	55567778
4	33
4	77

(a) Why did we split the stems?
(b) Give an appropriate key for this graph.
(c) Describe the shape of the distribution. Are there any outliers?

23. **Acorns and oak trees** Of the many species of oak trees in the United States, 28 grow on the Atlantic Coast and 11 grow in California. The back-to-back stemplot displays data on the average volume of acorns (in cubic centimeters) for these 39 oak species. Compare the distributions of acorn size for the oak trees in these two regions.[48]

Atlantic Coast		California
998643	0	4
88864211111	1	06
50	2	06
6640	3	
8	4	1
	5	59
8	6	0
	7	1
1	8	
1	9	
5	10	
	11	
	12	
	13	
	14	
	15	
	16	
	17	1

Key: 2|6 = An oak species whose acorn volume is 2.6 cm³.

24. **Finch evolution** Biologists Peter and Rosemary Grant spent many years collecting data on finches in a remote part of the Galápagos Islands. Their research team caught and measured all of the birds in more than 20 generations of finches. The back-to-back stemplot shows the beak depths (in millimeters) of 89 finches captured the year before a drought occurred and 89 finches captured the year after the drought occurred.[49] Compare the distributions of beak depth in these two years.

Before drought		After drought
2	6	
8	6	
411	7	1
98	7	9
44420	8	044
9999977765555	8	778
444321111100000	9	0011123344
99999888887777655	9	5666666777789999
444433322111111000	10	000222223333334444
8766655555	10	5555556666677777778999
440	11	0000111134444
7	11	5667

KEY: 10|0 = 10.0 mm beak depth

Displaying Quantitative Data: Histograms

25. **Carbon dioxide emissions** Burning fuels in power plants and motor vehicles emits carbon dioxide (CO_2), which contributes to global warming. The table displays CO_2 emissions in metric tons per person from 48 countries with populations of at least 20 million in a recent year.[50]

pg 33

Country	CO_2	Country	CO_2	Country	CO_2
Algeria	4.0	Italy	5.6	South Africa	8.2
Argentina	4.0	Japan	8.7	Spain	5.4
Australia	16.3	Kenya	0.3	Sudan	0.5
Bangladesh	0.6	Korea, North	1.5	Tanzania	0.2
Brazil	2.2	Korea, South	11.9	Thailand	4.1
Canada	15.4	Malaysia	7.8	Turkey	4.9
China	7.1	Mexico	3.4	Ukraine	5.1
Colombia	2.0	Morocco	2.0	United Kingdom	5.5
Congo	0.6	Myanmar	0.5	United States	16.1
Egypt	2.5	Nepal	0.5	Uzbekistan	3.3
Ethiopia	0.2	Nigeria	0.7	Venezuela	4.1
France	5.0	Pakistan	1.2	Vietnam	2.6
Germany	8.4	Peru	1.7		
Ghana	0.5	Philippines	1.3		
India	1.9	Poland	8.5		
Indonesia	2.3	Romania	3.9		
Iran	9.4	Russia	11.5		
Iraq	5.6	Saudi Arabia	17.0		

(a) Make a histogram of the data using intervals of width 2, starting at 0.

(b) What proportion of countries had CO_2 emissions of at least 10 metric tons per person?

26. **Traveling to work** How long do people travel each day to get to work? The following table gives the average travel times to work (in minutes) in a recent year for workers in each state and the District of Columbia who are at least 16 years old and don't work at home.[51]

State	Travel time to work (min)	State	Travel time to work (min)	State	Travel time to work (min)
AL	24.9	KY	23.6	ND	17.3
AK	19.1	LA	25.7	OH	23.7
AZ	25.7	ME	24.2	OK	21.9
AR	21.7	MD	33.2	OR	23.9
CA	29.8	MA	30.2	PA	27.2
CO	25.8	MI	24.6	RI	25.2
CT	26.6	MN	23.7	SC	25.0
DE	26.3	MS	24.8	SD	17.2
DC	30.8	MO	23.9	TN	25.2
FL	27.8	MT	18.3	TX	26.6
GA	28.8	NE	18.8	UT	21.9
HI	27.5	NV	24.6	VT	23.3
ID	21.1	NH	27.5	VA	28.7
IL	29.2	NJ	32.2	WA	28.0
IN	23.8	NM	22.3	WV	25.9
IA	19.3	NY	33.6	WI	22.2
KS	19.4	NC	24.8	WY	17.9

(a) Make a histogram to display the travel time data using intervals of width 2 minutes, starting at 16 minutes.

(b) In what proportion of states plus the District of Columbia is the average travel time at least 20 minutes?

27. **Home runs** Here are the number of home runs hit by each of the 30 Major League Baseball teams in a single season. Make a histogram that effectively displays the distribution of number of home runs hit. Describe the distribution.

220	249	213	245	256	182	227	223	224	149
288	162	220	279	146	250	307	242	306	257
215	163	219	239	167	210	217	223	247	231

28. **Country music** Here are the lengths, in minutes, of 50 songs by country artist Dierks Bentley. Make a histogram that effectively displays the distribution of song length. Describe the distribution.

4.2	4.0	3.9	3.8	3.7	4.7
3.4	4.0	4.4	5.0	4.6	3.7
4.6	4.4	4.1	3.0	3.2	4.7
3.5	3.7	4.3	3.7	4.8	4.4
4.2	4.7	6.2	4.0	7.0	3.9
3.4	3.4	2.9	3.3	4.0	4.2
3.2	3.4	3.7	3.5	3.4	3.7
3.9	3.7	3.8	3.1	3.7	3.6
4.5	3.7				

29. **Returns on common stocks** The return on a stock is the change in its market price plus any dividend payments made. Return is usually expressed as a percentage of the beginning price. The figure shows a histogram of the distribution of monthly percent return for the U.S. stock market (total return on all common stocks) in 273 consecutive months.[52]

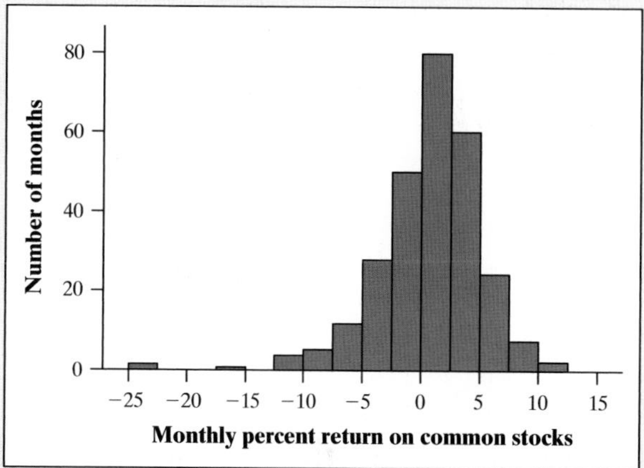

(a) A return less than zero means that stocks lost value in that month. About what percentage of all months had returns less than zero?

(b) Describe the shape of the distribution. Are there any outliers?

(c) Estimate the center and variability of the distribution.

30. **More healthy cereal?** Researchers collected data on calories per serving for 77 brands of breakfast cereal. The histogram displays the data.[53]

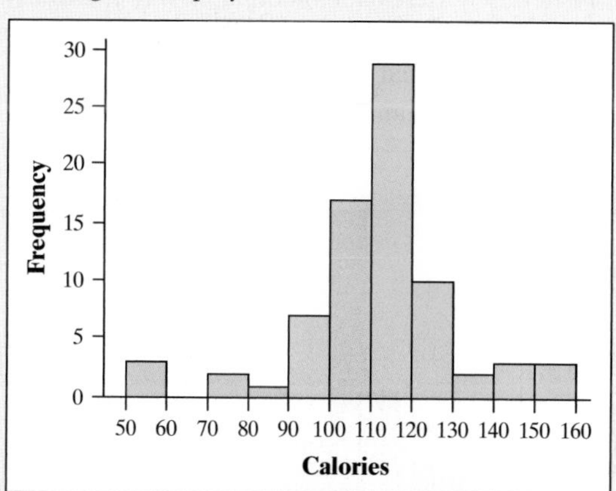

(a) About what percentage of the cereal brands have 130 or more calories per serving?

(b) Describe the shape of the distribution. Are there any outliers?

(c) Estimate the center and variability of the distribution.

31. **Value of a college degree** Is it true that students who earn an associate's degree or a bachelor's degree make more money than students who attend college but do not earn a degree? To find out, we selected a random sample of 500 U.S. residents aged 18 and older who had attended college from a recent Current Population Survey.[54] The educational attainment and annual income of each person were recorded. Here are relative frequency histograms of the income data for the 327 college graduates and the 173 nongraduates:

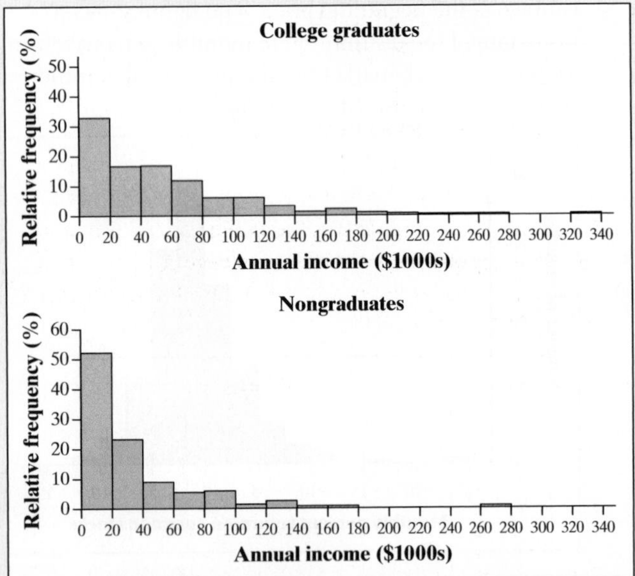

(a) Compare the distributions of annual income for the two groups.

(b) Would it be appropriate to use frequency histograms instead of relative frequency histograms in this setting? Explain why or why not.

32. **Irises** U.S. botanist Edgar Anderson collected data on 50 iris flowers from each of three different species: *Setosa, Versicolor,* and *Virginia.*[55] British statistician Ronald Fisher used these data to develop a way of classifying an individual iris based on measurements (in centimeters) of sepal length, sepal width, petal length, and petal width. Here are histograms of the data on sepal length:

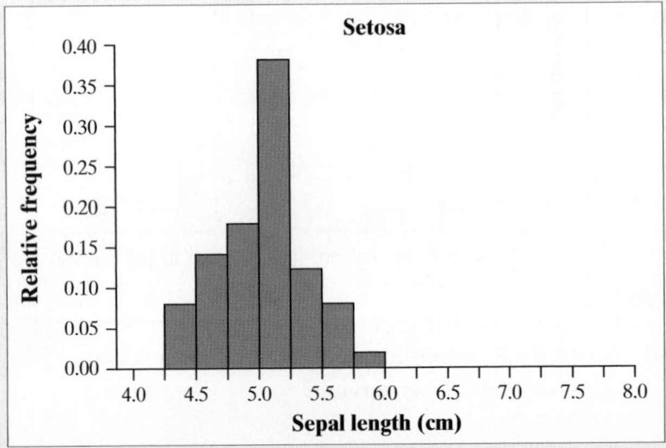

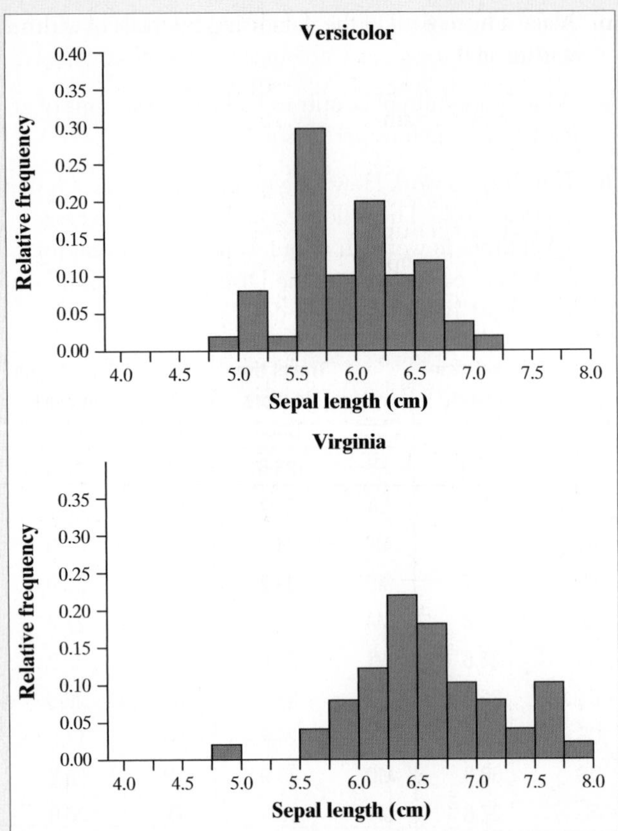

(a) Compare the distributions of sepal length for the three species.

(b) Would it be appropriate to use frequency histograms instead of relative frequency histograms in this setting? Explain why or why not.

33. **Body mass index** Researchers wanted to investigate the relationship between blood pressure and body mass index (BMI) in predominantly lean African populations. As part of the research, they collected data on BMI from a sample of 338 rural and 290 semi-urban women in Ghana. Here are comparative histograms of the data for the two groups:[56]

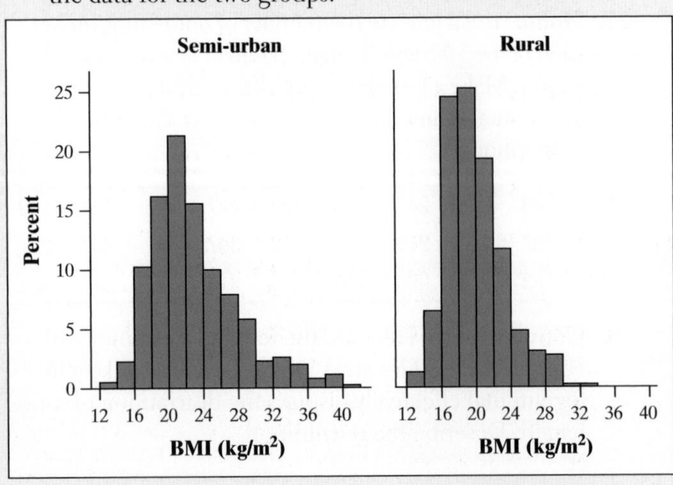

(a) Describe similarities and differences between the two distributions.

(b) According to the World Health Organization, a person with a BMI less than 18 is considered underweight. Compare the percentages of rural and semi-urban women in the sample who are underweight.

34. **Paper towels** In commercials for Bounty paper towels, the manufacturer claims that they are the "quicker picker-upper"—but are they also the stronger picker-upper? Two statistics students decided to investigate. They selected a random sample of 30 Bounty paper towels and a random sample of 30 generic paper towels and measured their strength when wet. To do this, the students uniformly soaked each paper towel with 4 ounces of water, held two opposite edges of the paper towel, and counted how many quarters each paper towel could hold until ripping. The data are displayed in the relative frequency histograms.[57]

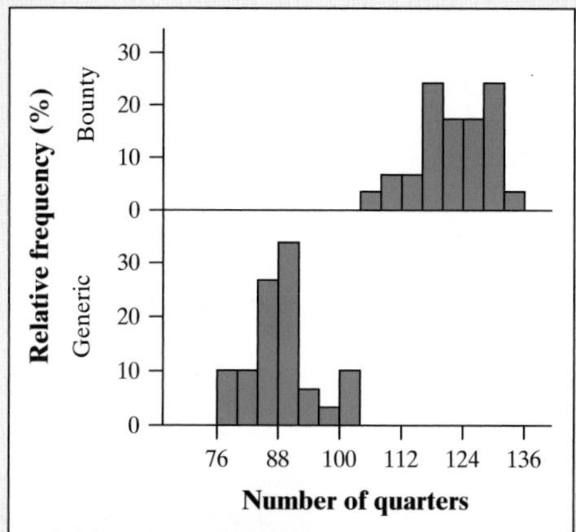

(a) Describe similarities and differences between the two distributions.

(b) What conclusion should the students make from their research? Explain your answer.

35. **Birth months** Imagine asking a random sample of 60 students from your school about their birth months. Draw a plausible (believable) graph of the distribution of birth month. Should you use a bar graph or a histogram to display the data?

36. **Die rolls** Imagine rolling a fair, six-sided die 60 times. Draw a plausible (believable) graph of the distribution of die rolls. Should you use a bar graph or a histogram to display the data?

37. **Seat belts** Each year, the National Highway Traffic Safety Administration (NHTSA) conducts an observational study of seat belt use in all 50 states. Trained observers station themselves at randomly selected locations along roadways in each state, and then record data on seat belt use by people in passing vehicles. Here is a histogram of the percentage of people who were wearing seat belts in each state during a recent year:[58]

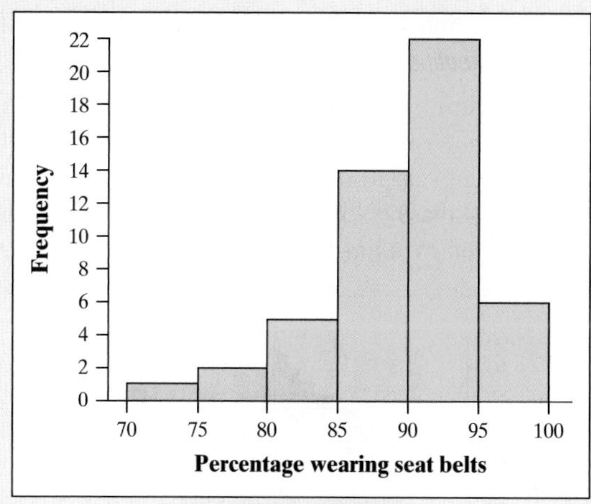

(a) In what proportion of states was at least 90% seat belt use observed?

(b) Describe the shape of the distribution.

Thirty-four states enforce a seat belt violation as a primary offense, while 16 states enforce a seat belt violation only as a secondary offense. The histograms show the percentage of people who were wearing seat belts in primary enforcement versus secondary enforcement states during the NHTSA study.

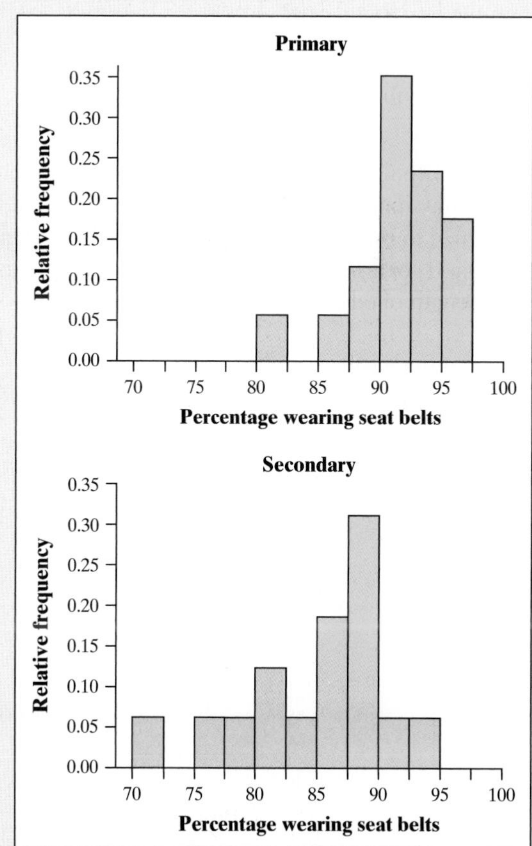

(c) Why did we use relative frequency instead of frequency when making these histograms?

(d) Compare the distributions of seat belt use in primary and secondary enforcement states.

For Investigation *Apply the skills from the section in a new context or nonroutine way.*

38. **Population pyramids** The histograms show the age distribution of the population in Australia and Ethiopia in a recent year from the U.S. Census Bureau's international database.[59] (This type of graph is referred to as a *population pyramid.*)

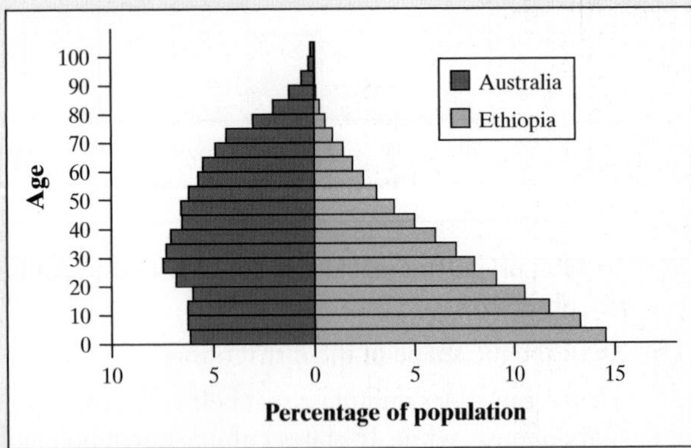

(a) The total population of Australia at this time was about 26 million. Ethiopia's population was about 111 million. Why did we use percentages rather than counts on the horizontal axis?

(b) What important differences do you see between the age distributions? Be specific.

39. **Vitamin C and teeth** Researchers performed an experiment with 60 guinea pigs to investigate the effect of vitamin C on tooth growth. Each animal was randomly assigned to receive one of three dose levels of vitamin C (0.5, 1, or 2 mg/day). The response variable was the length of odontoblast cells (responsible for tooth growth), in micrometers (μm). Here is a special type of dotplot that displays the data:[60]

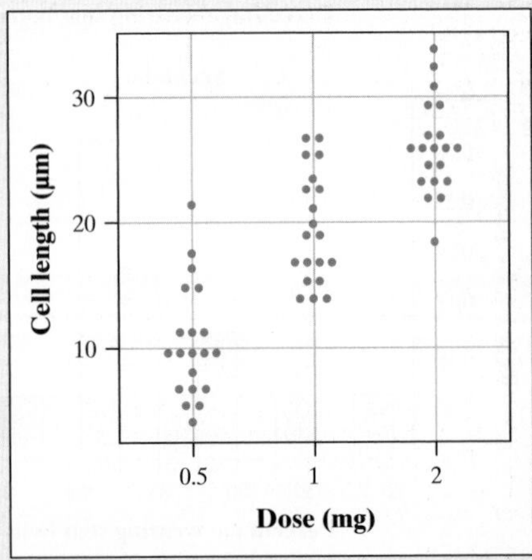

(a) How does this dotplot differ from the ones shown in this section?

(b) Do these data provide evidence that vitamin C helps teeth grow in guinea pigs similar to the ones used in this experiment? Justify your answer.

Multiple Choice *Select the best answer for each question.*

40. Here are the amounts of money (cents) in coins carried by 10 students in a statistics class: 50, 35, 0, 46, 86, 0, 5, 47, 23, 65. To make a stemplot of these data, you would use stems

(A) 0, 2, 3, 4, 5, 6, 8.

(B) 0, 1, 2, 3, 4, 5, 6, 7, 8.

(C) 0, 3, 5, 6, 7.

(D) 00, 10, 20, 30, 40, 50, 60, 70, 80, 90.

(E) None of these.

41. The histogram shows the heights (in inches) of 300 randomly selected high school students. Which of the following is the best description of the shape of the distribution of height?

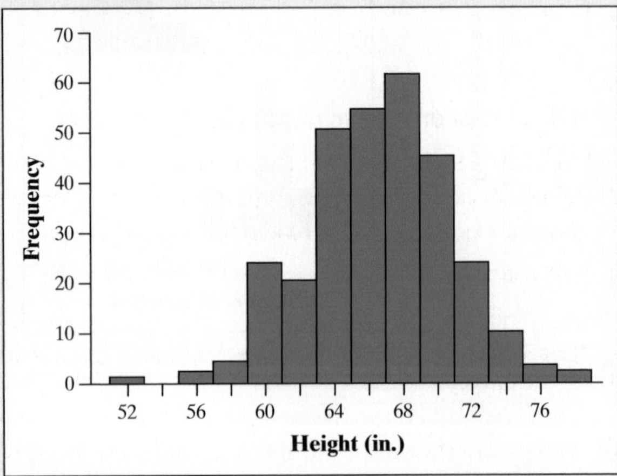

(A) Roughly symmetric and single-peaked (unimodal)

(B) Roughly symmetric and double-peaked (bimodal)

(C) Roughly symmetric and multi-peaked (multimodal)

(D) Skewed to the left

(E) Skewed to the right

42. The dotplot shows the tuition (to the nearest $1000) for the 63 largest colleges and universities in North Carolina in a recent year.[61]

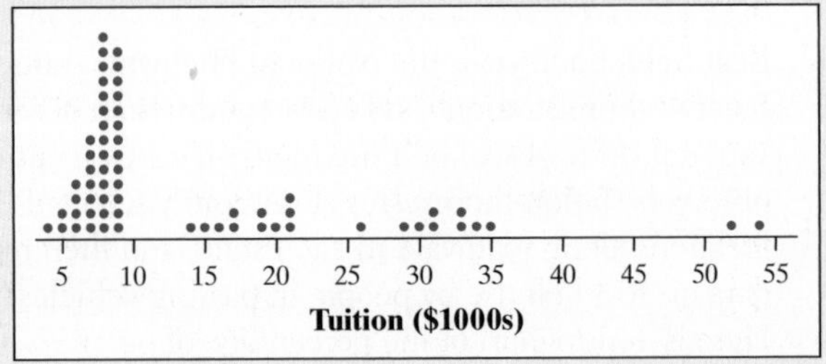

Which of the following statements about the dotplot is *not* correct?

(A) The center (median) of the distribution is $8000.

(B) There are more North Carolina colleges and universities with tuitions less than $10,000 than with tuitions greater than $10,000.

(C) The tuitions vary from about $4000 to about $54,000.

(D) There are two obvious outliers — institutions with tuitions greater than $50,000.

(E) The distribution of tuition is skewed to the right.

43. When comparing the distribution of a quantitative variable for two groups, it would be best to use relative frequency histograms rather than frequency histograms when

(A) the distributions have different shapes.

(B) the distributions have different amounts of variability.

(C) the distributions have different centers.

(D) the distributions have different numbers of data values.

(E) at least one of the distributions has outliers.

44. Which of the following is the best reason for choosing a stemplot rather than a histogram to display the distribution of a quantitative variable?

(A) Stemplots allow you to split stems; histograms don't.

(B) Stemplots allow you to see individual data values.

(C) Stemplots are better for displaying very large sets of data.

(D) Stemplots never require rounding of values.

(E) Stemplots make it easier to determine the shape of a distribution.

45. For their final project, two statistics students designed an experiment to investigate the effect of the interviewer on people's responses to the question, "How many hours a week do you work out?" Interviewer 1 wore jeans and a school T-shirt, while Interviewer 2 wore athletic attire with the school logo. The two students were each randomly assigned three 15-minute time slots during a 90-minute period to serve as the interviewer. They asked the question of people who passed by their location near campus. Here are parallel dotplots of the data obtained by each interviewer:[62]

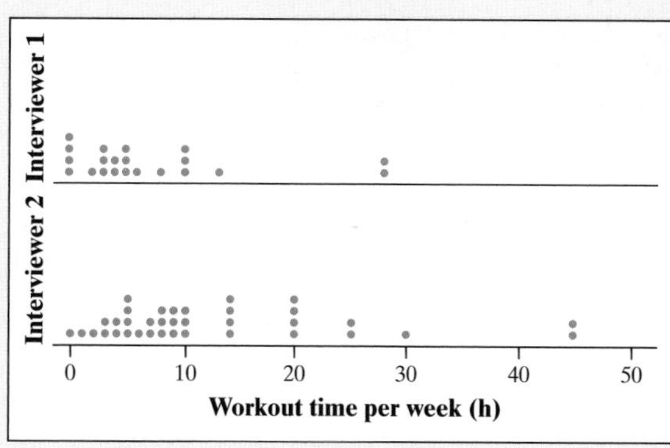

Which of the following statements about the graph is correct?

(A) For both interviewers, the distribution of workout time per week is roughly symmetric.

(B) There is a similar amount of variability in people's reported workout times for both interviewers.

(C) Interviewer 1's distribution of workout time per week has outliers; Interviewer 2's distribution of workout time per week has no outliers.

(D) People who responded to Interviewer 2 tended to report a higher amount of workout time per week than people who responded to Interviewer 1.

(E) For both interviewers, more than 75% of people reported working out 10 or less hours per week.

Recycle and Review *Practice what you learned in previous sections.*

46. **Marital Status (1B)** The side-by-side bar graph compares the marital status of U.S. adult residents (18 years old or older) in 1980 and 2020.[63] Compare the distributions of marital status for these two years.

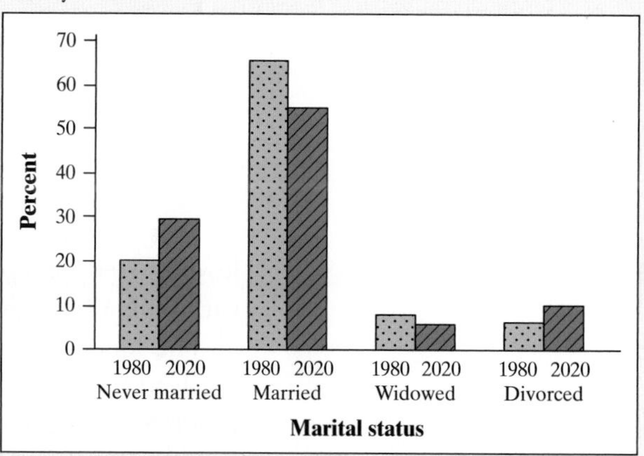

SECTION 1D Describing Quantitative Data with Numbers

LEARNING TARGETS *By the end of the section, you should be able to:*

- Find the median of a distribution of quantitative data.
- Calculate the mean of a distribution of quantitative data.
- Find the range of a distribution of quantitative data.
- Calculate and interpret the standard deviation of a distribution of quantitative data.
- Find the interquartile range (*IQR*) of a distribution of quantitative data.
- Choose appropriate measures of center and variability to summarize a distribution of quantitative data.
- Identify outliers in a distribution of quantitative data.
- Make and interpret boxplots of quantitative data.
- Use boxplots and summary statistics to compare distributions of quantitative data.

In Section 1C, you learned how to display quantitative data with graphs. You also explored how to use these graphs to describe and compare distributions of a quantitative variable. In this section, we'll focus on numerical summaries of quantitative data.

Let's return to a familiar context from the preceding section. Recall that city managers in Flint, Michigan, switched the city's water source from Lake Huron to the Flint River to save money. Here once again are the lead levels (in parts per billion, ppb) in 71 water samples collected from randomly selected Flint dwellings after the switch, along with a dotplot of the data.[64]

0	0	0	0	0	0	0	0	0	0	0	0
0	1	1	1	1	2	2	2	2	2	2	2
2	2	2	2	3	3	3	3	3	3	3	3
3	3	3	4	4	5	5	5	5	5	5	5
5	6	6	6	6	7	7	7	8	8	9	10
10	11	13	18	20	21	22	29	42	42	104	

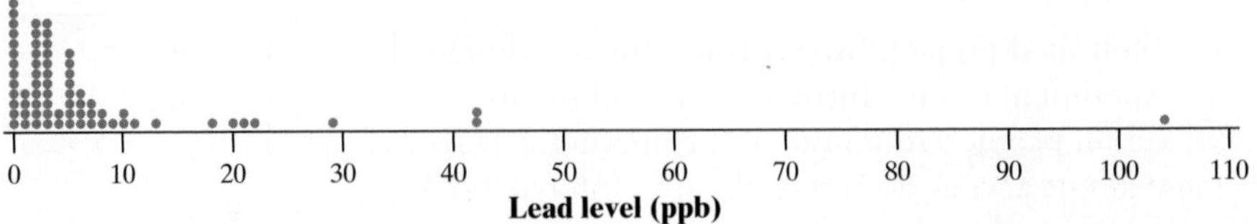

This distribution is right-skewed and single-peaked. The dwelling with a lead level of 104 ppb appears to be an outlier. How should we describe the center and variability of this distribution?

The *mode* of a distribution is the most frequently occurring data value. For the lead level data, the mode is 0 ppb. But that value isn't representative of how much lead a typical Flint dwelling has in its water. We want to report a value that is in the "center" of the distribution. The mode is often not a good measure of the center because it can fall anywhere in a distribution. Also, a distribution can have multiple modes, or no mode at all.

Measuring Center: The Median

In Section 1C, we advised you to simply use the "middle value" in an ordered quantitative data set to describe its center. That's the idea of the **median.**

DEFINITION Median

The **median** is the midpoint of a distribution—the number such that about half the observations are smaller and about half are larger. To find the median, arrange the data values from smallest to largest.

- If the number n of data values is odd, the median is the middle value in the ordered list.
- If the number n of data values is even, use the average of the two middle values in the ordered list as the median.

You can find the median by hand for small sets of data. For instance, here are data on the population density (in number of people per square kilometer) for all seven countries in Central America:[65]

Country	Belize	Costa Rica	El Salvador	Guatemala	Honduras	Nicaragua	Panama
Population density (people per km^2)	17	100	308	158	82	48	52

To find the median, start by sorting the data values from smallest to largest:

17	48	52	(82)	100	158	308

Because there are $n = 7$ data values (an odd number), the median is the middle value in the ordered list: 82.

Here is a dotplot of the population density data. You can confirm that the median is 82 by "counting inward" from the minimum and maximum values.

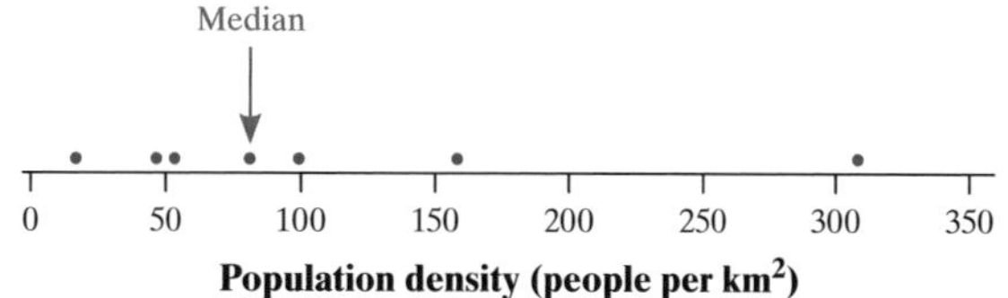

EXAMPLE

More chips please
Measuring center: The median

Skill 2.C

PROBLEM: Have you ever noticed that bags of chips seem to contain lots of air and not enough chips? A group of chip enthusiasts collected data on the percentage of air in a sample of 14 popular brands of chips. Here are their data:[66]

Ann Heath

Brand	Percent air	Brand	Percent air
Cape Cod	46	Popchips	45
Cheetos	59	Pringles	28
Doritos	48	Ruffles	50
Fritos	19	Stacy's Pita Chips	50
Kettle Brand	47	Sun Chips	41
Lays	41	Terra	49
Lays Baked	39	Tostitos Scoops	34

Find the median.

SOLUTION:

19 28 34 39 41 41 (45 | 46) 47 48 49 50 50 59

The median is $\frac{45+46}{2} = 45.5$ % air.

Sort the data values from smallest to largest. Because there are $n = 14$ data values (an even number), use the average of the middle two values in the ordered list as the median.

FOR PRACTICE, TRY EXERCISE 1

Here is a dotplot of the air in chips data from the preceding example with the median of 45.5 marked. How should we interpret the median? About half of the chip brands have less than 45.5% air, and about half have more.

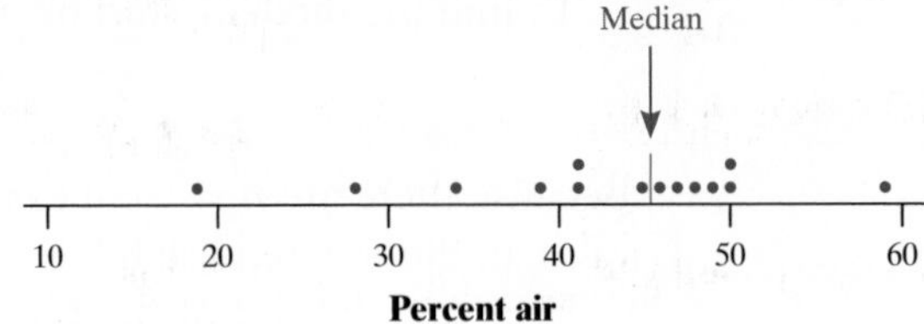

Measuring Center: The Mean

The most commonly used measure of center is the **mean.**

DEFINITION The mean

The **mean** of a distribution of quantitative data is the average of all the individual data values. To find the mean, add all the values and divide by the total number of data values.

If the n data values are $x_1, x_2, \ldots, x_n$, the *sample mean* $\bar{x}$ (pronounced "x-bar") is given by the formula

$$\bar{x} = \frac{\text{sum of data values}}{\text{number of data values}} = \frac{x_1 + x_2 + \ldots + x_n}{n} = \frac{\sum x_i}{n}$$

AP® EXAM TIP
Formula sheet

The formula for the sample mean $\bar{x}$ is included on the formula sheet provided on both sections of the AP® Statistics exam.

The $\sum$ (capital Greek letter sigma) in the formula is short for "add them all up." The subscripts on the observations x_i are just a way of keeping the n data values distinct. They do not necessarily indicate the order or any other special facts about the data.

EXAMPLE

More chips please
Measuring center: The mean

Skill 2.C

PROBLEM: Here are the data on percent air in the sample of 14 bags of chips from the preceding example, along with a dotplot:

Brand	Percent air	Brand	Percent air
Cape Cod	46	Popchips	45
Cheetos	59	Pringles	28
Doritos	48	Ruffles	50
Fritos	19	Stacy's Pita Chips	50
Kettle Brand	47	Sun Chips	41
Lays	41	Terra	49
Lays Baked	39	Tostitos Scoops	34

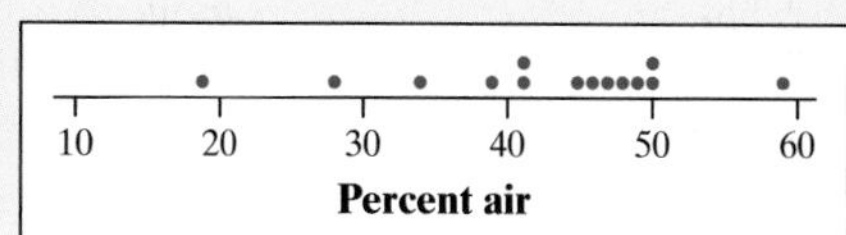

(a) Calculate the mean percent air in the bag for these 14 brands of chips.
(b) The bag of Fritos chips, with only 19% air, is a possible outlier. Calculate the mean percent air in the bag for the other 13 brands of chips. What do you notice?

SOLUTION:

(a) $\bar{x} = \frac{46+59+48+19+47+\ldots+34}{14} = \frac{596}{14} = 42.57\%$ air

$$\bar{x} = \frac{x_1 + x_2 + \ldots + x_n}{n} = \frac{\sum x_i}{n}$$

(b) $\bar{x} = \frac{46+59+48+47+\ldots+34}{13} = \frac{577}{13} = 44.38\%$ air

The bag of Fritos decreased the mean percent air by 1.81 percentage points.

FOR PRACTICE, TRY EXERCISE 3

The notation $\bar{x}$ refers to the mean of a *sample*. Most of the time, the data we encounter can be thought of as a sample from some larger population, like the 14 bags of chips in the example. When we need to refer to a *population mean*, we'll use the symbol μ (Greek letter mu, pronounced "mew"). If you have the entire population of data available, then you can calculate μ in just the way you'd expect: add the values of all the observations, and divide by the number of observations. For instance, the population mean density in all seven South American countries is

$$\mu = \frac{17+48+52+82+100+158+308}{7} = 109.286 \text{ people per square kilometer}$$

We call $\bar{x}$ a **statistic** and μ a **parameter.** Remember **s** and **p**: statistics come from **s**amples and **p**arameters come from **p**opulations. In later units, you will learn how to use sample statistics to make conclusions about population parameters (known as *inferential statistics*).

DEFINITION Statistic, Parameter

A **statistic** is a number that describes some characteristic of a sample.

A **parameter** is a number that describes some characteristic of a population.

PROPERTIES OF THE MEAN

The preceding example illustrates an important weakness of the mean as a measure of center: The mean is not **resistant** to extreme values, such as outliers. The bag of Fritos, with only 19% air, decreases the mean by 1.81 percentage points.

DEFINITION Resistant

A statistical measure is **resistant** if it is not affected much by extreme data values.

The median *is* a resistant measure of center. In the preceding example, the median percent air in all 14 bags of chips is 45.5. If we remove the possible outlier bag of Fritos, the median percent air in the remaining 13 bags is nearly the same (46).

Why is the mean so sensitive to extreme values? The following activity provides some insight.

ACTIVITY Interpreting the mean

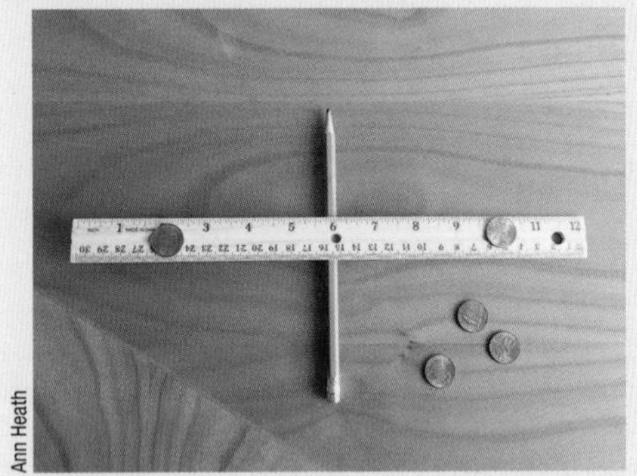

Ann Heath

In this activity, you will investigate a physical interpretation of the mean of a distribution.

1. Stack 5 pennies on top of the 6-inch mark on a 12-inch ruler. Place a pencil under the ruler to make a "seesaw" on a desk or table. Move the pencil until the ruler balances. What is the relationship between the location of the pencil and the mean of the five data values 6, 6, 6, 6, and 6?
2. Move one penny off the stack to the 8-inch mark on your ruler. Now move one other penny so that the ruler balances again without moving the pencil. Where did you put the other penny? What is the mean of the five data values represented by the pennies now?
3. Move one more penny off the stack to the 2-inch mark on your ruler. Now move both remaining pennies from the 6-inch mark so that the ruler still balances with the pencil in the same location. Is the mean of the data values still 6?
4. Discuss with your classmates: Why is the mean called the "balance point" of a distribution?

The activity gives a physical interpretation of the mean as the balance point of a distribution. For the data on percent air in each of 14 brands of chips, the dotplot balances at $\bar{x} = 42.57\%$.

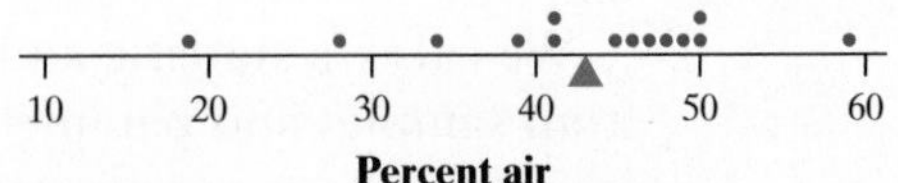

COMPARING THE MEAN AND MEDIAN

Which measure — the mean or the median — should we report as the center of a distribution? That depends on both the shape of the distribution and whether there are any outliers.

- **Shape:** Figure 1.10 shows the mean and median for dotplots with three different shapes. Notice how these two measures of center compare in each case. The mean is pulled in the direction of the long tail in a skewed distribution.

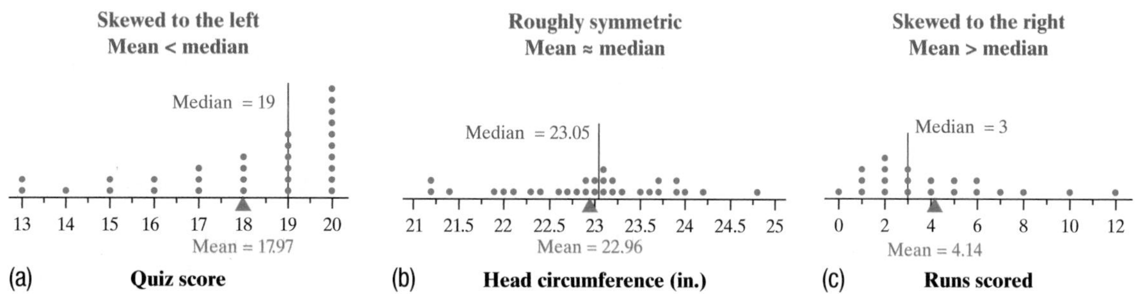

FIGURE 1.10 Dotplots that show the relationship between the mean and the median in distributions with different shapes: (a) scores on a 20-point quiz for each of the 30 students in a statistics class, (b) head circumference (in inches) for 30 players on a youth football team, and (c) runs scored by a softball team in 21 games played.

- **Outliers:** We noted earlier that the median is a resistant measure of center, but the mean is not. If we remove the two possible outliers (10 and 12) in Figure 1.10(c), the mean number of runs scored per game decreases from 4.14 to 3.42, but the median number of runs scored is still 3.

EFFECT OF SKEWNESS AND OUTLIERS ON MEASURES OF CENTER

- If a distribution of quantitative data is roughly symmetric and has no outliers, the mean and the median will be similar.
- If the distribution is strongly skewed, the mean will be pulled in the direction of the skewness but the median won't. For a right-skewed distribution, we expect the mean to be greater than the median. For a left-skewed distribution, we expect the mean to be less than the median.
- The median is resistant to outliers but the mean isn't.

The mean and the median measure center in different ways, and both are useful. In Major League Baseball (MLB), the distribution of player salaries is strongly skewed to the right. Most players earn close to the minimum salary, which was \$700,000 in 2022, while a few earn more than \$20 million. The median salary for MLB players in 2022 was about \$1.2 million — but the mean salary was about \$4.4 million. Max Scherzer, Mike Trout, Anthony Rendon, Gerrit Cole, Carlos Correa, Manny Machado, and several other highly paid superstars pulled the mean up but their huge paychecks did not affect the median. The median gives us a good idea of what a "typical" MLB salary is. If we want to know the total salary paid to MLB players in 2022, however, we would multiply the mean salary by the total number of players: (\$4.4 million)(975) ≈ \$4.3 billion!

Some students purchased pumpkins for a carving contest. Before the contest began, they weighed the pumpkins. The weights (in pounds) are shown here, along with a histogram of the data.

3.6	4.0	9.6	14.0	11.0
12.4	13.0	2.0	6.0	6.6
15.0	3.4	12.7	6.0	2.8
9.6	4.0	6.1	5.4	11.9
5.4	31.0	33.0		

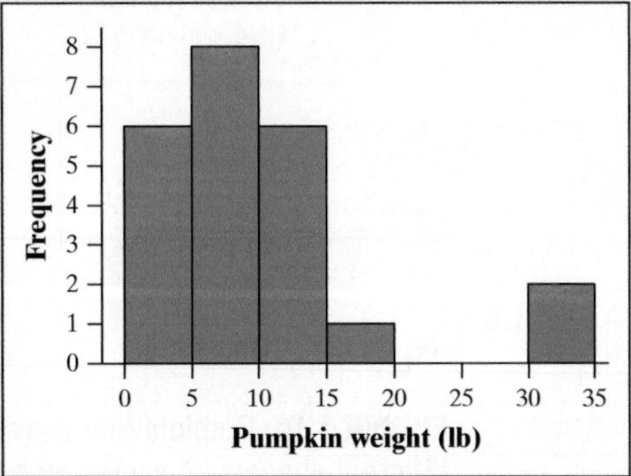

1. Find the median weight of the pumpkins.
2. Calculate the mean weight of the pumpkins.
3. Explain why the mean of the distribution would be larger than the median before doing any calculations.

Measuring Variability: The Range

Being able to describe the shape and center of a distribution of quantitative data is a great start. However, two distributions can have the same shape and center, but still look quite different.

Figure 1.11 shows comparative dotplots of the length (in millimeters, mm) of separate random samples of PVC pipe from two suppliers, A and B.[67] Both distributions are roughly symmetric and single-peaked (unimodal), with centers at about 600 mm, but the variability of these two distributions is quite different. The sample of pipes from Supplier A has a much more consistent length (less variability) than the sample from Supplier B.

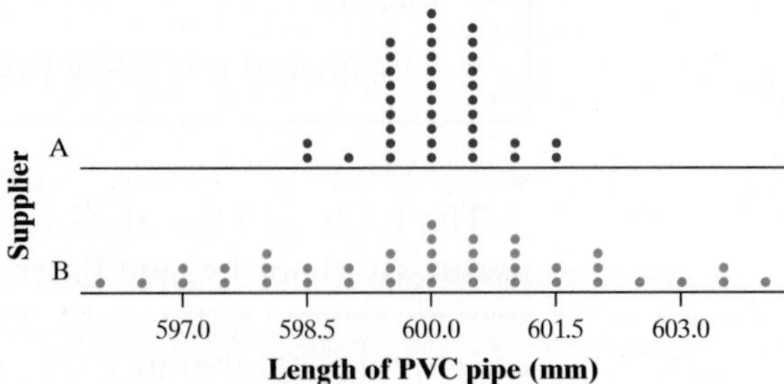

FIGURE 1.11 Comparative dotplots of the length of PVC pipes in separate random samples from Supplier A and Supplier B.

There are several ways to measure the variability of a distribution. The simplest is the **range**.

DEFINITION Range

The **range** of a distribution is the distance between the minimum value and the maximum value. That is,

$$\text{range} = \text{maximum} - \text{minimum}$$

Refer to the parallel dotplots of the length of PVC pipe from the two suppliers in Figure 1.11. For Supplier A, the range of the distribution is $601.5 - 598.5 = 3.0$ mm. That's quite a bit smaller than the range of Supplier B's distribution: $604.0 - 596.0 = 8.0$ mm.

Note that **the range of a data set is a single number**. In everyday language, people sometimes say things like "The data values range from 17 to 308." Be sure to use the term *range* correctly, now that you know its statistical definition.

EXAMPLE

More chips please
Measuring variability: The range

Skill 2.C

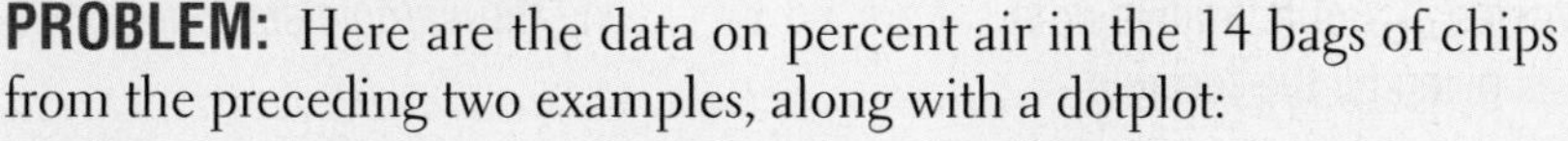

PROBLEM: Here are the data on percent air in the 14 bags of chips from the preceding two examples, along with a dotplot:

Brand	Percent air	Brand	Percent air
Cape Cod	46	Popchips	45
Cheetos	59	Pringles	28
Doritos	48	Ruffles	50
Fritos	19	Stacy's Pita Chips	50
Kettle Brand	47	Sun Chips	41
Lays	41	Terra	49
Lays Baked	39	Tostitos Scoops	34

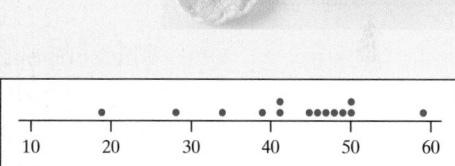

PackStock/Alamy Stock Photo

Find the range of the distribution.

SOLUTION:

range $= 59 - 19 = 40\%$ *air*

range = maximum − minimum

FOR PRACTICE, TRY EXERCISE 11

The range is *not* a resistant measure of variability. It depends on only the maximum and minimum values, which may be outliers. Look again at the data on the percent air in bags of chips. Without the possible outlier at 19%, the range of the distribution would decrease from 40% air to $59\% - 28\% = 31\%$ air.

The following graph illustrates another problem with using the range as a measure of variability. Many older movies were recorded on high-resolution film with a width of 70 mm, much larger than the standard 35-mm film used for modern movies. The parallel dotplots show the widths (in millimeters) of a sample of 11 strips of 70-mm movie film produced by each of two machines.[68] Both distributions are symmetric, centered at 70 mm, and have a range of $70.2 - 69.8 = 0.4$ mm. But the widths of the film strips made by Machine B clearly vary more from the center of 70 mm than the film strips made by Machine A.

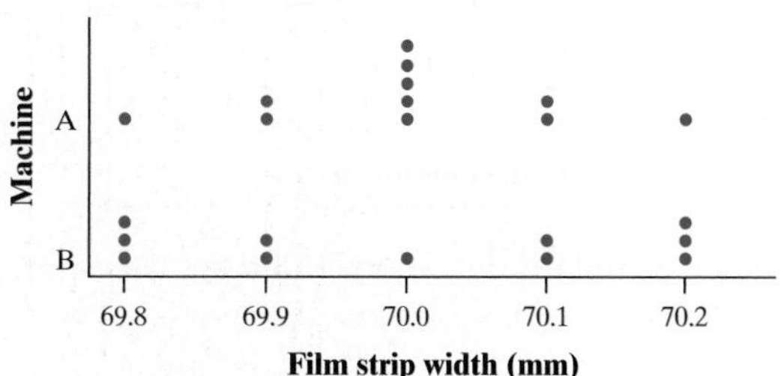

Measuring Variability: The Standard Deviation

To avoid the problem illustrated by the film strip example, we prefer a measure of variability that uses more than just the minimum and maximum values of a quantitative data set. If we summarize the center of a distribution with the mean, then we should use the **standard deviation** to describe the variation of data values around the mean.

DEFINITION Standard deviation

The **standard deviation** measures the typical distance of the values in a distribution from the mean. It is calculated by finding an "average" of the squared deviations of the individual data values from the mean, and then taking the square root.

If the values in a quantitative data set are $x_1, x_2, \ldots, x_n$, the *sample standard deviation s_x* is given by the formula

$$s_x = \sqrt{\frac{(x_1 - \bar{x})^2 + (x_2 - \bar{x})^2 + \ldots + (x_n - \bar{x})^2}{n-1}} = \sqrt{\frac{\sum(x_i - \bar{x})^2}{n-1}}$$

AP® EXAM TIP
Formula sheet

The formula for the sample standard deviation s_x is included on the formula sheet provided on both sections of the AP® Statistics exam.

The value obtained before taking the square root in the standard deviation calculation is called the *sample variance*, denoted by s_x^2. Because variance is measured in squared units, it is not a very helpful way to describe the variability of a distribution.

HOW TO CALCULATE THE SAMPLE STANDARD DEVIATION s_x

1. Find the mean of the distribution.
2. Calculate the *deviation* of each value from the mean:

$$\text{deviation} = \text{value} - \text{mean}$$

3. Square each of the deviations.
4. Add all of the squared deviations, then divide by $n - 1$ to get the sample variance s_x^2.
5. Take the square root to return to the original units.

EXAMPLE

How many friends?
Measuring variability: The standard deviation

Skill 2.C

PROBLEM: A random sample of 11 high school students are asked how many "close" friends they have. Here are their responses, along with a dotplot:

1	2	2	2	3	3	3	3	4	4	6

Ian Allenden/Alamy Stock Photo

Calculate the sample standard deviation. Interpret this value.

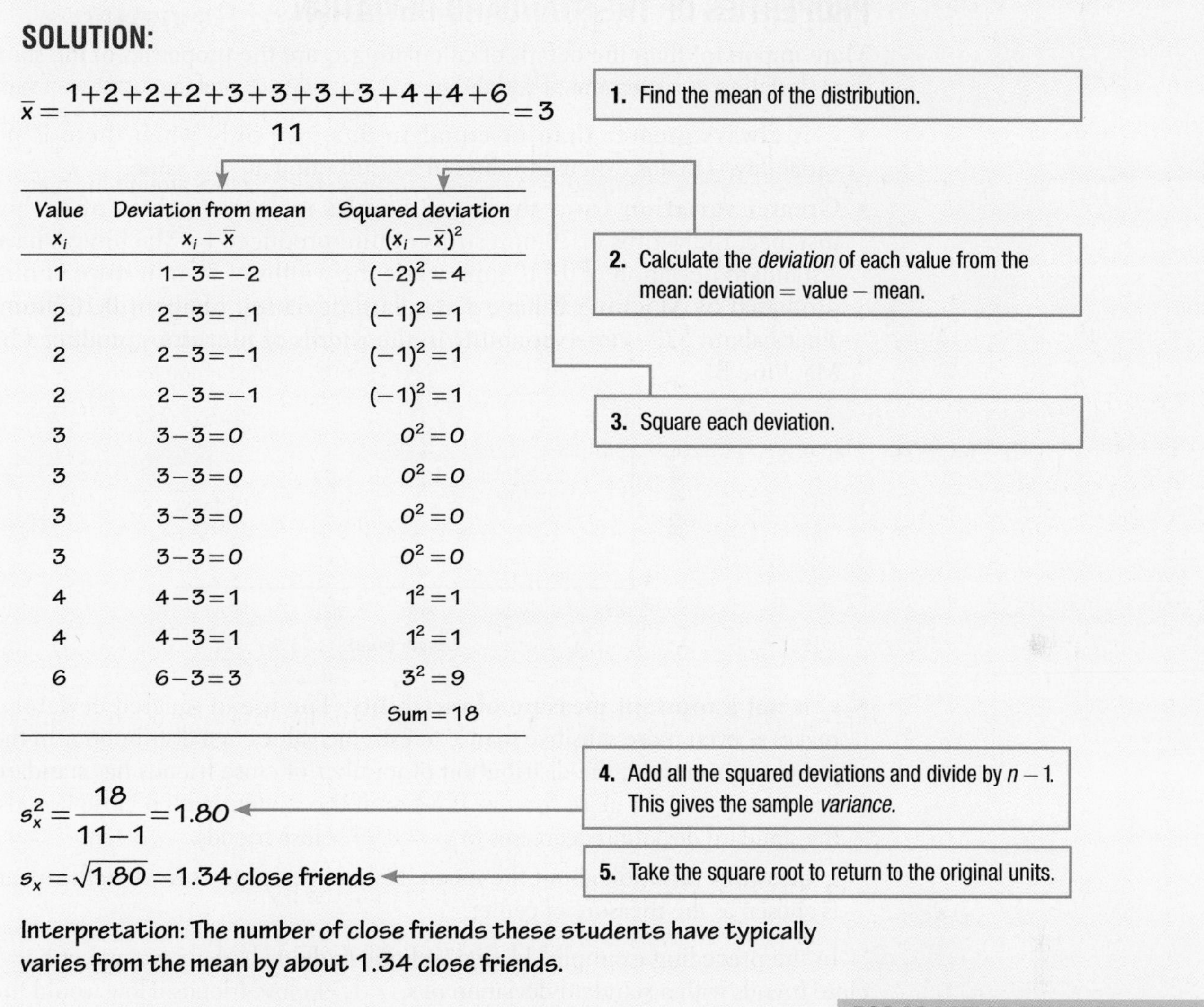

SOLUTION:

$$\bar{x} = \frac{1+2+2+2+3+3+3+3+4+4+6}{11} = 3$$

1. Find the mean of the distribution.

Value x_i	Deviation from mean $x_i - \bar{x}$	Squared deviation $(x_i - \bar{x})^2$
1	$1-3=-2$	$(-2)^2=4$
2	$2-3=-1$	$(-1)^2=1$
2	$2-3=-1$	$(-1)^2=1$
2	$2-3=-1$	$(-1)^2=1$
3	$3-3=0$	$0^2=0$
3	$3-3=0$	$0^2=0$
3	$3-3=0$	$0^2=0$
3	$3-3=0$	$0^2=0$
4	$4-3=1$	$1^2=1$
4	$4-3=1$	$1^2=1$
6	$6-3=3$	$3^2=9$
		Sum = 18

2. Calculate the *deviation* of each value from the mean: deviation = value − mean.

3. Square each deviation.

$$s_x^2 = \frac{18}{11-1} = 1.80$$

4. Add all the squared deviations and divide by $n-1$. This gives the sample *variance*.

$$s_x = \sqrt{1.80} = 1.34 \text{ close friends}$$

5. Take the square root to return to the original units.

Interpretation: The number of close friends these students have typically varies from the mean by about 1.34 close friends.

FOR PRACTICE, TRY EXERCISE 13

The notation s_x refers to the standard deviation of a *sample*. When we need to refer to the standard deviation of a population, we'll use the symbol σ (lowercase Greek letter sigma). We often use the sample statistic s_x to estimate the population parameter σ. The population standard deviation σ is calculated by dividing the sum of squared deviations from the population mean μ by the population size N (not N − 1) before taking the square root.

Think About It

WHY IS THE STANDARD DEVIATION CALCULATED IN SUCH A COMPLEX WAY? Add the deviations from the mean in the preceding example. You should get a sum of 0. Why? Because the mean is the balance point of the distribution. We square the deviations to avoid the positive and negative deviations balancing each other out and adding to 0. It might seem strange to "average" the squared deviations by dividing by $n-1$. We'll explain the reason for doing this in Unit 5. It's easier to understand why we take the square root: to return to the original units (close friends).

PROPERTIES OF THE STANDARD DEVIATION

More important than the details of calculating s_x are the properties of the standard deviation as a measure of variability:

- **s_x is always greater than or equal to 0.** $s_x = 0$ only when there is no variability — that is, when all values in a distribution are the same.
- **Greater variation from the mean results in larger values of s_x.** For instance, the widths of 70-mm strips of film produced by Machine A have a standard deviation of 0.110 mm, while the widths of 70-mm strips of film produced by Machine B have a standard deviation of about 0.167 mm. That's about 52% more variability in the widths of film strips produced by Machine B!

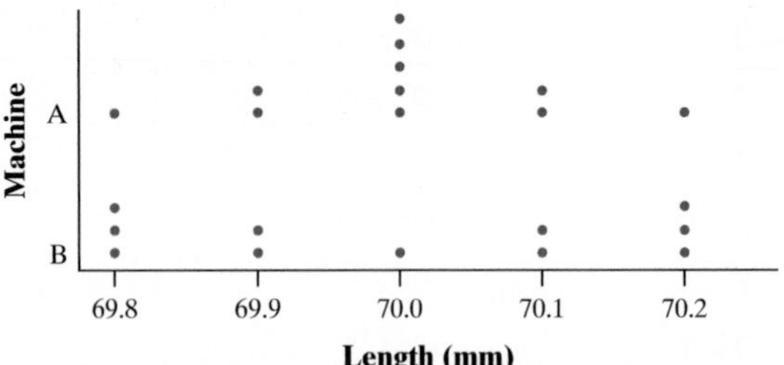

- **s_x is not a resistant measure of variability.** The use of squared deviations makes s_x even more sensitive than $\bar{x}$ to extreme values in a distribution. In the preceding example, the distribution of number of close friends has standard deviation $s_x = 1.34$ close friends. If we omit the student with 6 close friends, the standard deviation decreases to $s_x = 0.949$ close friends.
- **s_x measures variation about the mean.** It should be used only when the mean is chosen as the measure of center.

In the preceding example, 11 high school students had an average of $\bar{x} = 3$ close friends with a standard deviation of $s_x = 1.34$ close friends. How would the sample standard deviation be affected if a 12th high school student was added to the sample who had 3 close friends? The mean number of close friends in the sample would still be $\bar{x} = 3$. Because the standard deviation measures the typical distance of the values in a distribution from the mean, s_x would *decrease* because this 12th value is at a distance of 0 from the mean. In fact, the new standard deviation would be

$$s_x = \sqrt{\frac{\sum(x_i - \bar{x})^2}{n-1}} = \sqrt{\frac{18}{12-1}} = 1.28 \text{ close friends}$$

Measuring Variability: The Interquartile Range (*IQR*)

We can avoid the impact of extreme values on our measure of variability by focusing on the middle of the distribution. Here's the basic strategy: Order the data values from smallest to largest. Then find the **quartiles**, the values that divide the distribution into four groups of roughly equal size. The **first quartile Q_1** lies one-fourth of the way through the ordered list. The second quartile is the median, which is halfway through the list. The **third quartile Q_3** lies three-fourths of the way through the list. The first and third quartiles mark out the middle half of the distribution.

DEFINITION Quartiles, First quartile Q_1, Third quartile Q_3

The **quartiles** of a distribution divide an ordered data set into four groups having roughly the same number of values. To find the quartiles, arrange the data values left to right from smallest to largest and find the median.

The **first quartile Q_1** is the median of the data values that are to the left of the median in the ordered list.

The **third quartile Q_3** is the median of the data values that are to the right of the median in the ordered list.

For example, here are the amounts collected each hour by a charity at a local store:

\$19 \$26 \$25 \$37 \$31 \$28 \$22 \$22 \$29 \$34 \$39 \$31

The dotplot displays the data. Because there are 12 data values, the quartiles divide the distribution into 4 groups of 3 values each.

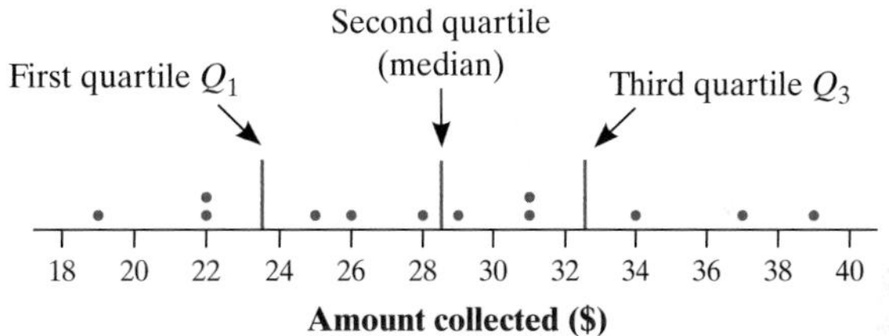

The quartiles can be used to help us describe the *position* of an individual data value in a distribution of quantitative data. We'll discuss ways to measure position/location in Section 1E.

The **interquartile range (*IQR*)** measures variability using the quartiles. It's simply the range of the middle half of the distribution.

DEFINITION Interquartile range (*IQR*)

The **interquartile range (*IQR*)** is the distance between the first and third quartiles of a distribution. In symbols:

$$IQR = Q_3 - Q_1$$

For the distribution of amount collected each hour at the charity store, you can check that $Q_1 = \$23.50$, Median $= \$28.50$, and $Q_3 = \$32.50$. The interquartile range of the distribution is

$$IQR = \$32.50 - \$23.50 = \$9.00$$

EXAMPLE

More chips please
Measuring variability: The interquartile range (*IQR*)

Skill 2.C

PROBLEM: Here again are the data on the percent air in a sample of 14 bags of chips.

Brand	Percent air	Brand	Percent air
Cape Cod	46	Popchips	45
Cheetos	59	Pringles	28
Doritos	48	Ruffles	50
Fritos	19	Stacy's Pita Chips	50
Kettle Brand	47	Sun Chips	41
Lays	41	Terra	49
Lays Baked	39	Tostitos Scoops	34

Find the interquartile range.

SOLUTION:

19 28 34 39 41 41 (45|46) 47 48 49 50 50 59

Median = 45.5

Sort the data values left to right from smallest to largest and find the median.

[19 28 34 (39) 41 41 45]|46 47 48 49 50 50 59

$Q_1 = 39$ Median = 45.5

Find the first quartile Q_1, which is the median of the data values to the left of the median in the ordered list.

19 28 34 (39) 41 41 45|[46 47 48 (49) 50 50 59]

$Q_1 = 39$ Median = 45.5 $Q_3 = 49$

Find the third quartile Q_3, which is the median of the data values to the right of the median in the ordered list.

IQR = 49 − 39 = 10% air

$IQR = Q_3 - Q_1$

FOR PRACTICE, TRY EXERCISE 19

How should we interpret the interquartile range from the example, *IQR* = 10% air? The middle half of the distribution of percent air in these 14 bags of chips has a range of 10%.

The quartiles and the interquartile range are *resistant* because they are not affected by a few extreme values. For the air in chips data, Q_3 would still be 49 and the *IQR* would still be 10 if the maximum were 69 rather than 59.

Be sure to leave out the median when you locate the quartiles. In the preceding example, the median was not one of the data values. For the earlier Close Friends example data set, we ignore the circled median of 3 when finding Q_1 and Q_3.

[1 2 (2) 2 3] (3) [3 3 (4) 4 6]

Q_1 Median Q_3

Choosing Summary Statistics

Which measure of center and variability should we use to summarize the distribution of a quantitative variable? That depends on the shape of the distribution and whether there are any outliers. As you learned earlier, the median is a resistant measure of center but the mean is not. Among measures of variability, the interquartile range (*IQR*) is resistant to extreme values but the range and standard deviation are not. For now, follow this advice:

- If a distribution of quantitative data is roughly symmetric with no outliers, the mean $\bar{x}$ and standard deviation s_x are the preferred measures of center and variability.
- If the distribution is clearly skewed or has outliers, use the median as the measure of center and the *IQR* as the measure of variability.

CHOOSING MEASURES OF CENTER AND VARIABILITY

The median and *IQR* are usually better choices than the mean and standard deviation for describing a skewed distribution or a distribution with outliers. Use $\bar{x}$ and s_x for roughly symmetric distributions that don't have outliers.

We recommend using the range to measure variability only as a last resort because it gives so little information about how the individual data values are distributed.

EXAMPLE

Lead in the water
Choosing summary statistics

Skill 4.B

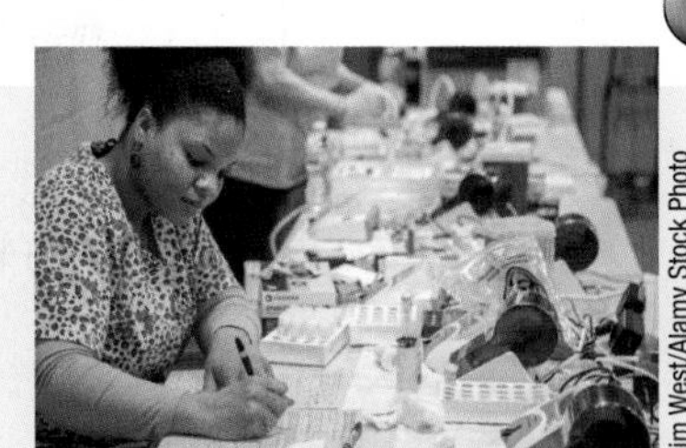
Jim West/Alamy Stock Photo

PROBLEM: Here once again is a dotplot of the lead levels (in parts per billion, ppb) in 71 water samples taken from randomly selected Flint, Michigan, dwellings after the city switched its water supply from Lake Huron to the Flint River. Summary statistics for the data set are also provided.[69]

0 10 20 30 40 50 60 70 80 90 100 110
Lead level (ppb)

n	Mean	SD	Min	Q_1	Med	Q_3	Max
71	7.31	14.347	0	2	3	7	104

Which measures of center and variability should we choose to summarize this distribution? Explain your answer.

SOLUTION:

The distribution of lead level is right-skewed and has an obvious outlier at 104 ppb. Consequently, we should choose measures of center and variability that are resistant: the median of 3 ppb and the interquartile range of $IQR = 7 - 2 = 5$ ppb.

FOR PRACTICE, TRY EXERCISE 23

For other statistical reasons, the mean and standard deviation will become the preferred measures of center and variability for the distribution of a quantitative variable in future sections. You'll see why when we get there!

Graphing calculators and computer software will calculate summary statistics for you. Using technology to perform calculations will allow you to focus on choosing the right methods and interpreting your results.

2. Tech Corner CALCULATING SUMMARY STATISTICS

TI-Nspire and other technology instructions are on the book's website at bfwpub.com/tps7e.

You can use technology to calculate measures of center and variability for a distribution of quantitative data. We'll illustrate using data on the percent air in a sample of 14 bags of chips from previous examples.

1. Type the values into list L1.
2. Press STAT, arrow over to the CALC menu, and choose 1-Var Stats.

   OS 2.55 or later: In the dialog box, press 2nd 1 (L1) and ENTER to specify L1 as the List. Leave FreqList blank. Arrow down to Calculate and press ENTER.

   Older OS: Press 2nd 1 (L1) and ENTER.

   Press the down arrow to see the rest of the one-variable statistics.

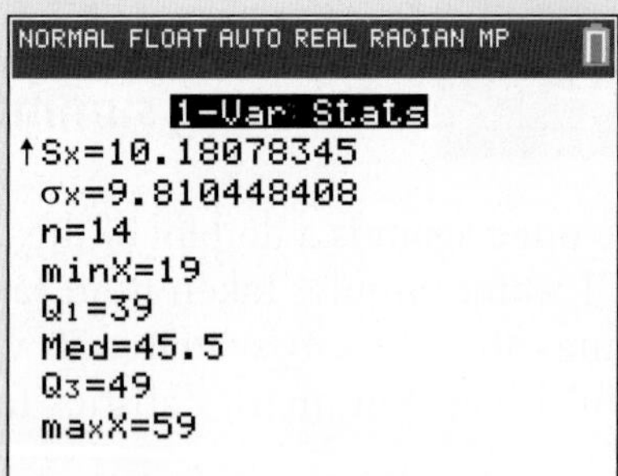

Note: The TI-83/84 does not give the range or interquartile range (*IQR*) directly when you calculate one-variable statistics. But the calculator provides the five-number summary, so you can calculate range = max − min and $IQR = Q_3 - Q_1$.

Some statistical software uses slightly different rules for calculating quartiles. We used Minitab statistical software to calculate summary statistics for the percent air in chips data. In the output, Minitab reports that $Q_1 = 37.75$ and $Q_3 = 49.25$. The TI-83/84 gives the first and third quartiles as $Q_1 = 39$ and $Q_3 = 49$, which match the values we calculated in an earlier example. Results from the various rules are usually close to each other, especially for large data sets.

Descriptive Statistics

Variable	N	Mean	StDev	Minimum	Q_1	Median	Q_3	Maximum
Percent air	14	42.57	10.18	19.00	37.75	45.50	49.25	59.00

CHECK YOUR UNDERSTANDING

Some students purchased pumpkins for a carving contest. Before the contest began, they weighed the pumpkins. The weights in pounds are shown here, along with a histogram of the data.

3.6	4.0	9.6	14.0	11.0	12.4	13.0	2.0	6.0	6.6	15.0	3.4
12.7	6.0	2.8	9.6	4.0	6.1	5.4	11.9	5.4	31.0	33.0	

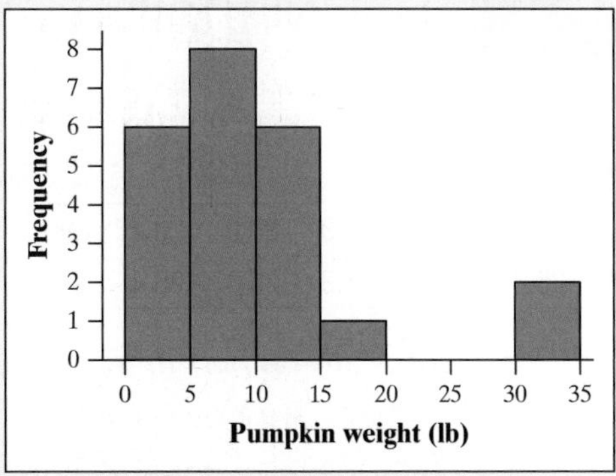

1. Explain why you cannot calculate the range exactly from the histogram. Then use the data to calculate the range of the distribution.
2. The mean and standard deviation of the distribution are 9.93 lb and 8.01 lb, respectively. Interpret the standard deviation.
3. Calculate the interquartile range of the distribution.
4. Which measures of center and variability would you choose to describe the distribution? Explain your answer.

Identifying Outliers

LeBron James emerged as a superstar in the National Basketball Association (NBA) during his rookie season (2003–2004). He maintained a consistent level of excellence over the first 16 years of his professional career, reaching the NBA Finals eight consecutive times and winning three NBA championships. The dotplot shows the average number of points per game that LeBron scored in each of these 16 seasons.[70]

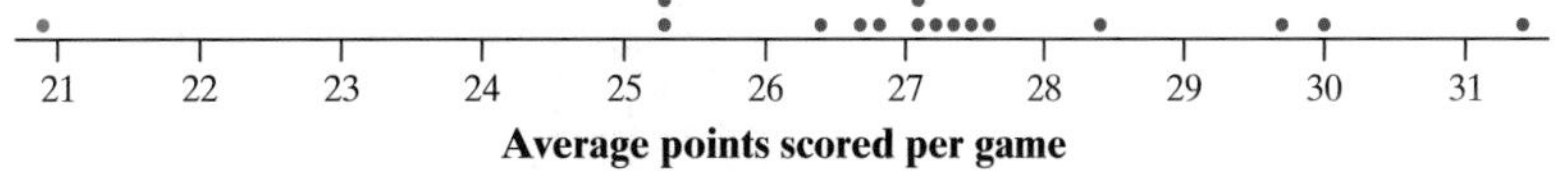

LeBron's 20.9 points per game average in his rookie season stands out (in red) from the rest of the distribution. Should this value be classified as an *outlier?*

The most common method for identifying outliers in a distribution of quantitative data uses the interquartile range (*IQR*). Besides being a resistant measure of variability, the *IQR* serves as a kind of "ruler" for determining how extreme an individual data value must be to be classified as an outlier.

HOW TO IDENTIFY OUTLIERS: THE $1.5 \times IQR$ RULE

Call an observation an outlier if it falls more than $1.5 \times IQR$ above the third quartile or more than $1.5 \times IQR$ below the first quartile. That is,

$$\text{Low outliers} < Q_1 - 1.5 \times IQR \quad \text{High outliers} > Q_3 + 1.5 \times IQR$$

EXAMPLE

An NBA legend
Identifying outliers

Skill 4.B

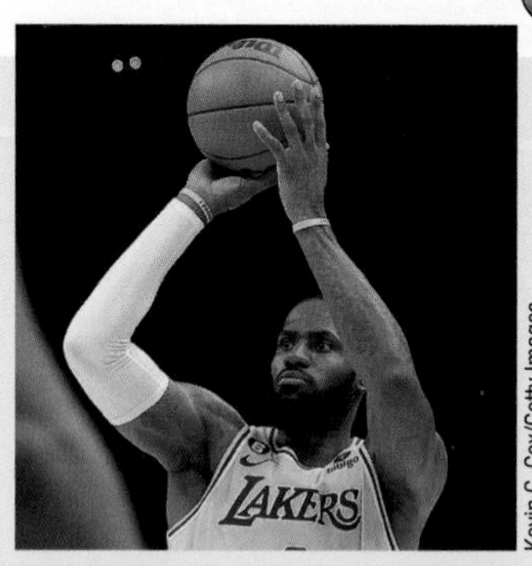

Kevin C. Cox/Getty Images

PROBLEM: Here are data on the average number of points per game that LeBron James scored in each of his first 16 NBA seasons:

20.9	27.2	31.4	27.3	30.0	28.4	29.7	26.7
27.1	26.8	27.1	25.3	25.3	26.4	27.5	27.4

Use the $1.5 \times IQR$ rule to identify any outliers in the distribution.

SOLUTION:

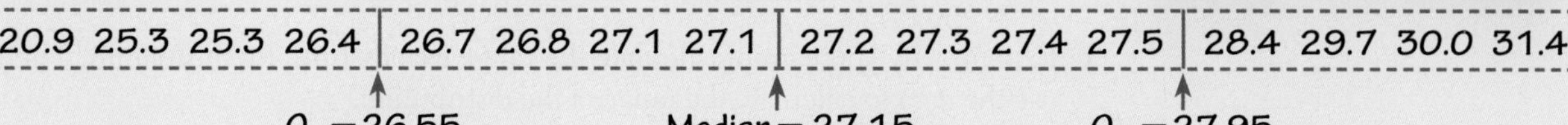
20.9 25.3 25.3 26.4 | 26.7 26.8 27.1 27.1 | 27.2 27.3 27.4 27.5 | 28.4 29.7 30.0 31.4

$Q_1 = 26.55$ Median $= 27.15$ $Q_3 = 27.95$

$IQR = 27.95 - 26.55 = 1.40$

Low outliers $< 26.55 - 1.5 \times 1.40 = 24.45$
High outliers $> 27.95 + 1.5 \times 1.40 = 30.05$

- Find the interquartile range: $IQR = Q_3 - Q_1$.
- Then calculate the upper and lower cutoff values for outliers:
  Low outliers $< Q_1 - 1.5 \times IQR$
  High outliers $> Q_3 + 1.5 \times IQR$

LeBron James's rookie-season average of 20.9 points per game is a low outlier because it is less than 24.45. The season when LeBron averaged 31.4 points per game is a high outlier because it is greater than 30.55.

FOR PRACTICE, TRY EXERCISE 27

There are other methods for determining outliers, such as "any value that is more than 2 standard deviations from the mean." Let's apply this $2 \times SD$ *rule* to the data from the preceding example. Here are summary statistics on the average number of points scored by LeBron James in each of his first 16 NBA seasons:

n	Mean	SD	Min	Q_1	Med	Q_3	Max
16	27.156	2.328	20.9	26.55	27.15	27.95	31.4

$$\text{Low outliers} < \text{Mean} - 2 \times \text{SD} = 27.156 - 2 \times 2.328 = 22.50$$

$$\text{High outliers} > \text{Mean} + 2 \times \text{SD} = 27.156 + 2 \times 2.328 = 31.812$$

By the $2 \times$ SD rule, the season in which LeBron James averaged 20.9 points per game is an outlier because it is less than 22.50. However, the season in which LeBron averaged 31.4 points per game is *not* an outlier by this rule because 31.4 is not greater than the upper cutoff of 31.812.

Unless otherwise indicated, we always use the $1.5 \times IQR$ rule in this book to identify outliers because it is based on statistics that are resistant to extreme data values,

> **AP® EXAM TIP**
>
> You may be asked to determine whether a quantitative data set has any outliers. Be prepared to state and use the $1.5 \times IQR$ rule and the $2 \times SD$ rule for identifying outliers.

unlike the mean and standard deviation. As we will explain in Section 1F, there is some justification for using the $2 \times SD$ rule in the special case of roughly symmetric, single-peaked, mound-shaped distributions called *normal distributions*.

It is important to identify outliers in a distribution for several reasons:

1. **They might be inaccurate data values.** Maybe someone recorded a value as 10.1 instead of 101. Perhaps a measuring device broke down. Or maybe someone gave a silly response, like the student in a class survey who claimed to study 30,000 minutes per night! Try to correct errors like these if possible. If you can't, give summary statistics with and without the outlier.
2. **They can indicate a remarkable occurrence.** For example, in a graph of career earnings of professional tennis players, Serena Williams is likely to be an outlier.
3. **They can heavily influence the values of some summary statistics,** such as the mean, range, and standard deviation.

Displaying Summary Statistics: Boxplots

You can use a dotplot, stemplot, or histogram to display the distribution of a quantitative variable. Another graphical option for quantitative data is a **boxplot** (sometimes called a *box-and-whisker* plot). A boxplot summarizes a distribution by displaying the location of five important values within the distribution, known as its **five-number summary.**

> **DEFINITION Five-number summary, Boxplot**
>
> The **five-number summary** of a distribution of quantitative data consists of the minimum, the first quartile Q_1, the median, the third quartile Q_3, and the maximum.
>
> A **boxplot** is a visual representation of the five-number summary.

Figure 1.12 illustrates the process of making a boxplot. The dotplot in Figure 1.12(a) shows LeBron James's average points per game for each of 16 seasons.

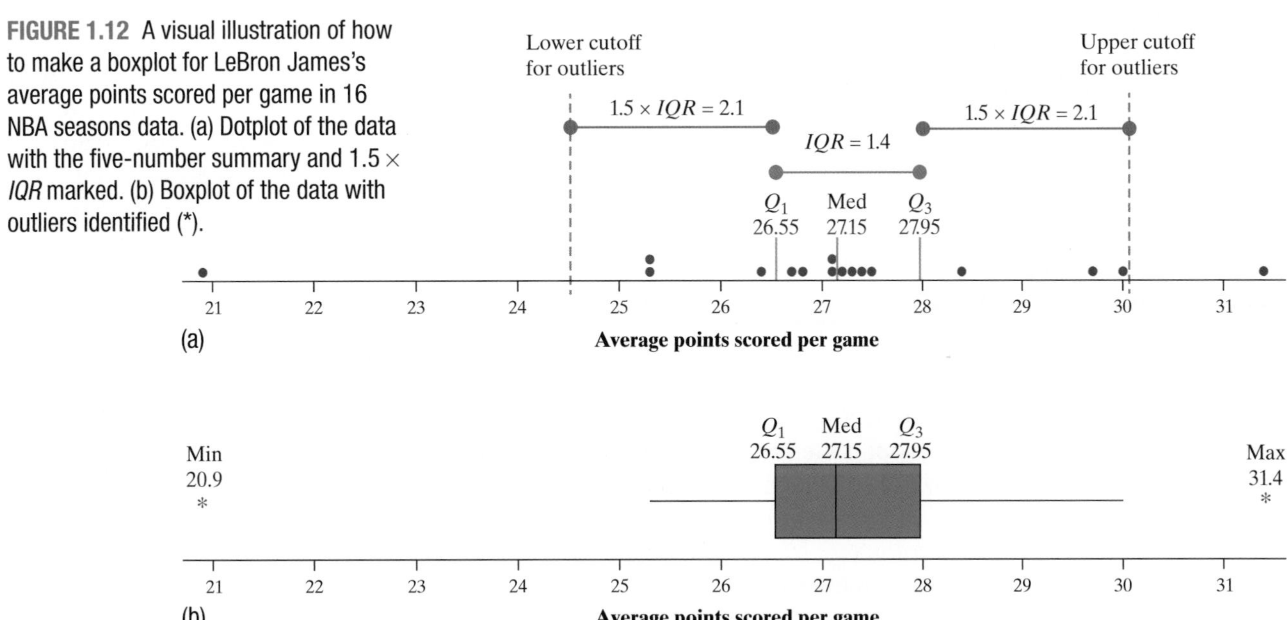

FIGURE 1.12 A visual illustration of how to make a boxplot for LeBron James's average points scored per game in 16 NBA seasons data. (a) Dotplot of the data with the five-number summary and $1.5 \times IQR$ marked. (b) Boxplot of the data with outliers identified (*).

We have labeled the first quartile, the median, and the third quartile and marked them with orange line segments. The process of testing for outliers with the $1.5 \times IQR$ rule is illustrated above the dotplot. Because the values of 20.9 and 31.4 are outliers, we mark these separately. To get the finished boxplot in Figure 1.12(b), we make a box spanning from Q_1 to Q_3 and then draw whiskers to the smallest and largest data values that are not outliers.

As you can see, it is fairly easy to make a boxplot by hand for small sets of quantitative data. Here's a summary of the steps.

HOW TO MAKE A BOXPLOT

1. **Find the five-number summary** for the distribution.
2. **Identify outliers** using the $1.5 \times IQR$ rule.
3. **Draw and label the axis.** Draw a horizontal axis and put the name of the quantitative variable underneath it, including units if applicable.
4. **Scale the axis.** Look at the minimum and maximum values in the data set. Start the horizontal axis at a convenient number less than or equal to the minimum and place tick marks at equal intervals until you equal or exceed the maximum.
5. **Draw a box** that spans from the first quartile (Q_1) to the third quartile (Q_3).
6. **Mark the median** with a vertical line segment that's the same height as the box.
7. **Mark any outliers** with a special symbol such as an asterisk (*).
8. **Draw whiskers** — lines that extend from the ends of the box to the smallest and largest data values that are *not* outliers.

We can see from the boxplot in Figure 1.12 that the middle 50% of LeBron James's season scoring averages fall between 26.55 and 27.95 points per game.

EXAMPLE

How big are the large fries?
Displaying summary statistics: Boxplots

Skills 2.A, 2.B

Michael Neelon/Alamy Stock Photo

PROBLEM: According to nutrition information provided by Burger King, the serving size for its large french fries is 173 grams.[71] Two young researchers wondered if the company is exaggerating the serving size. To find out, they went to several Burger King restaurants in their area and ordered a total of 15 large fries. The weights of the 15 orders (in grams) are shown here.[72]

178	176	173	172	179	165	179	181	186	184	181	180	183	183	187

(a) Make a boxplot to display the data.
(b) Does the graph in part (a) support the researchers' suspicion that Burger King is exaggerating the serving size of its large fries? Explain your reasoning.

SOLUTION:

(a) 165 172 173 (176) 178 179 179 (180) 181 181 183 (183) 184 186 187

Min Q_1 Med Q_3 Max

$IQR = Q_3 - Q_1 = 183 - 176 = 7$

Low outliers $< Q_1 - 1.5 \times IQR = 176 - 1.5 \times 7 = 165.5$

High outliers $> Q_3 + 1.5 \times IQR = 183 + 1.5 \times 7 = 193.5$

The order of large fries that weighed 165 grams is an outlier.

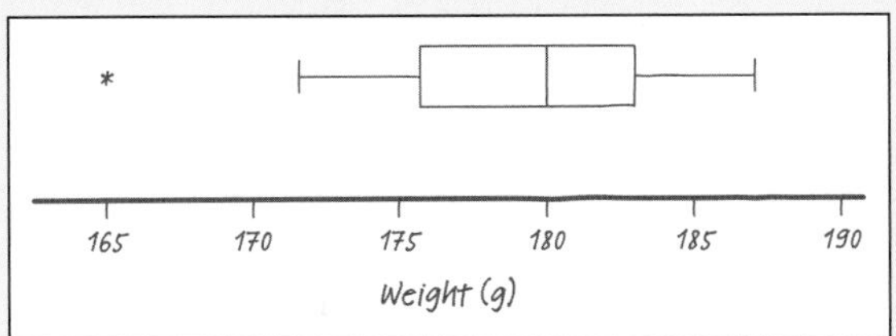

1. Find the five-number summary.
2. Identify outliers.
3. Draw and label the axis.
4. Scale the axis.
5. Draw a box from Q_1 to Q_3.
6. Mark the median.
7. Mark any outliers.
8. Draw whiskers.

(b) No. From the boxplot, $Q_1 = 176$, so at least 75% of the orders of large fries that the researchers bought from local Burger King restaurants weighed 176 grams or more. Only the outlier (165 grams) and one other order (172 grams) of large fries weighed less than the advertised 173-gram serving size.

FOR PRACTICE, TRY EXERCISE 31

Boxplots provide a quick summary of the center and variability of a distribution. The median is displayed as a vertical line segment in the central box, the interquartile range is the length of the box, and the range is the length of the entire plot, including outliers. *Note that some statistical software orients boxplots vertically.*

Boxplots don't give a complete picture of the shape of a distribution because they do not display each individual data value. From the boxplot in the example, we can see that the distribution of weight for orders of Burger King's large fries is skewed to the left because the left half of the distribution varies from 165 to 180 grams, while the right half of the distribution varies from 180 to 187 grams. But we can't identify peaks, gaps, or clusters from a boxplot. For instance, the following dotplot displays the duration, in minutes, of 263 eruptions of the Old Faithful geyser. The distribution of eruption durations is clearly bimodal (two-peaked). However, the boxplot of the data hides this important information about shape.

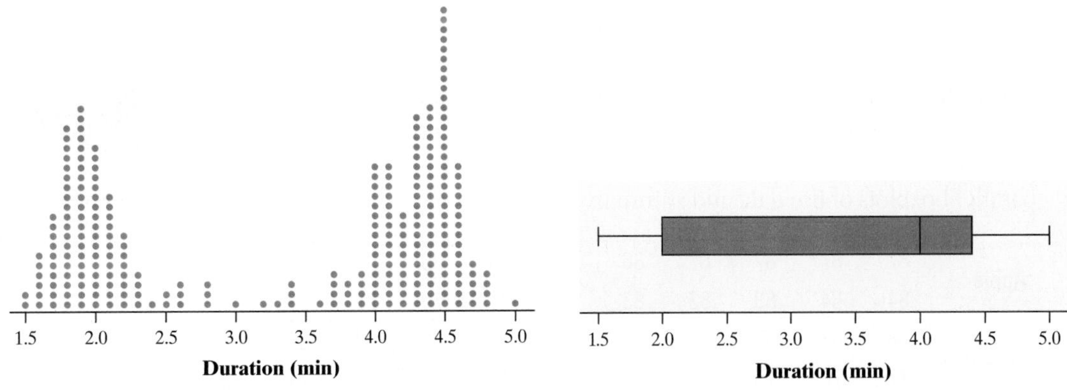

CHECK YOUR UNDERSTANDING

Some students purchased pumpkins for a carving contest. Before the contest began, they weighed the pumpkins. The weights in pounds are shown here, along with a histogram of the data.

3.6	4.0	9.6	14.0	11.0	12.4	13.0	2.0	6.0	6.6	15.0	3.4
12.7	6.0	2.8	9.6	4.0	6.1	5.4	11.9	5.4	31.0	33.0	

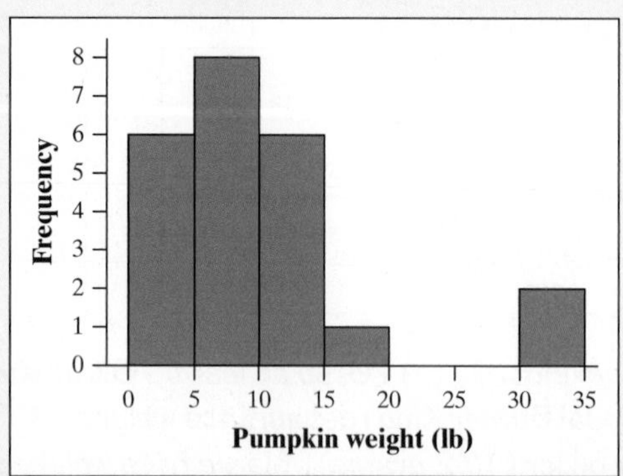

1. Identify any outliers in the distribution.
2. Make a boxplot to display the data.
3. Explain why the boxplot does not completely display the shape of the distribution.

Comparing Distributions with Boxplots and Summary Statistics

Boxplots are especially effective for comparing the distribution of a quantitative variable in two or more groups. Just remember to discuss shape, outliers, center, and variability as you did with comparative dotplots, stemplots, and histograms in Section 1C.

EXAMPLE

Which company makes better tablets?
Comparing distributions with boxplots and summary statistics

Skill 2.D

Scott Olson/Getty Images

PROBLEM: To help potential purchasers make informed decisions, *Consumer Reports* rated many tablet computers for performance and quality. Based on several variables, the magazine gave each tablet an overall rating, where higher scores indicate better ratings. The overall ratings of a sample of tablets produced by Apple and Samsung are given here, along with parallel boxplots of the data and summary statistics.[73]

Apple	87	87	87	87	86	86	86	86	84	84		
	84	84	83	83	83	83	81	79	76	73		
Samsung	88	87	87	86	86	86	86	86	84	84	83	83
	77	76	76	75	75	75	75	75	74	71	62	

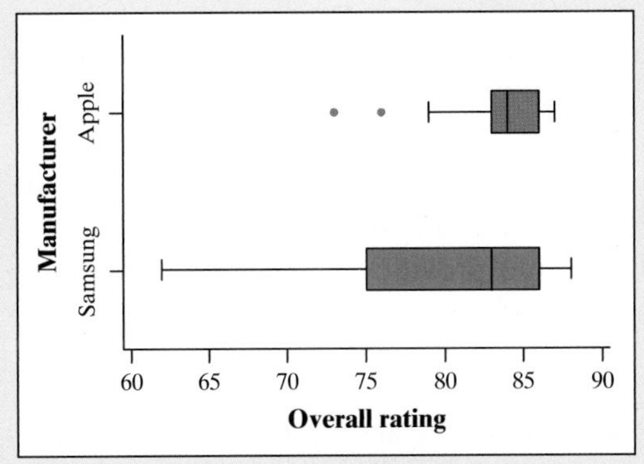

	$\bar{x}$	s_x	Min	Q_1	Median	Q_3	Max	*IQR*
Apple	83.45	3.762	73	83	84	86	87	3
Samsung	79.87	6.74	62	75	83	86	88	11

Compare the distributions of overall rating for Apple and Samsung.

SOLUTION:

Shape: Both distributions of overall rating are skewed to the left.

Outliers: There are two low outliers in the Apple tablet distribution with overall ratings of 73 and 76. The Samsung tablet distribution has no outliers.

Center: The Apple tablets had a slightly higher median overall rating (84) than the Samsung tablets (83). More importantly, about 75% of the Apple tablets had overall ratings that were greater than or equal to the median for the Samsung tablets.

Variability: There is much more variation in overall rating among the Samsung tablets than among the Apple tablets. The *IQR* for Samsung tablets (11) is almost four times larger than the *IQR* for Apple tablets (3).

Be sure to discuss shape, outliers, center, and variability, and to include context (variable name)!

Don't forget to explicitly *compare* measures of center and variability, using words like "greater than," "less than," or "about the same as." Due to the strong skewness and outliers, use the median and *IQR* instead of the mean and standard deviation when comparing center and variability.

FOR PRACTICE, TRY EXERCISE 35

AP® EXAM TIP

Use statistical terms carefully and correctly on the AP® Statistics exam. Don't say "mean" if you really mean "median." Range is a single number; so are Q_1, Q_3, and *IQR*. Avoid sloppy use of language, like "the outlier *skews* the mean" or "the median is in the middle of the *IQR*." Skewed is a shape and the *IQR* is a single number, not a region. If you misuse a term, expect to lose some credit.

Here's an activity that gives you a chance to use the skills you have learned in this unit: selecting statistical methods, data analysis, and making statistical arguments.

ACTIVITY

Team challenge: Did Mr. Starnes stack his class?

In this activity, you will work in a team of three or four students to resolve a dispute.

Mr. Starnes teaches AP® Statistics, but he also does the class scheduling for the high school. There are two AP® Statistics classes — one taught by Mr. Starnes and one taught by Ms. McGrail. The two teachers give the same first test to their classes and grade the test together. Mr. Starnes's students earned an average score that was 8 points higher than the average for Ms. McGrail's class. Ms. McGrail wonders whether Mr. Starnes might have "adjusted" the class rosters from the computer scheduling program. In other words, she thinks he might have "stacked" his class. He denies this, of course.

To help resolve the dispute, Mr. Starnes provides data on the cumulative grade point averages of the students in both classes from his computer. The following table displays the data.

McGrail	2.900	3.300	3.980	2.900	3.200	3.500	2.800	2.900	3.950
	3.100	2.850	2.900	3.245	3.000	3.000	2.800	2.900	3.200
Starnes	2.900	2.860	2.600	3.600	3.200	2.700	3.100	3.085	3.750
	3.400	3.338	3.560	3.800	3.200	3.100			

Based on these data, did Mr. Starnes stack his class? Give appropriate graphical and numerical evidence to support your conclusion.

You can use technology to make boxplots, as the following Tech Corner illustrates.

3. Tech Corner MAKING BOXPLOTS

TI-Nspire and other technology instructions are on the book's website at bfwpub.com/tps7e.

The TI-83/84 can plot up to three boxplots in the same viewing window. Let's use the calculator to make parallel boxplots of the overall rating data for Apple and Samsung tablets.

1. Enter the ratings for the Apple tablets in list L1 and those for the Samsung tablets in list L2.
2. Set up two statistics plots: Plot1 to show a boxplot of the Apple data and Plot2 to show a boxplot of the Samsung data. The setup for Plot1 is shown. When you define Plot2, be sure to change L1 to L2.

   Note: The calculator offers two types of boxplots: one that shows outliers and one that doesn't. We'll always use the type that identifies outliers.

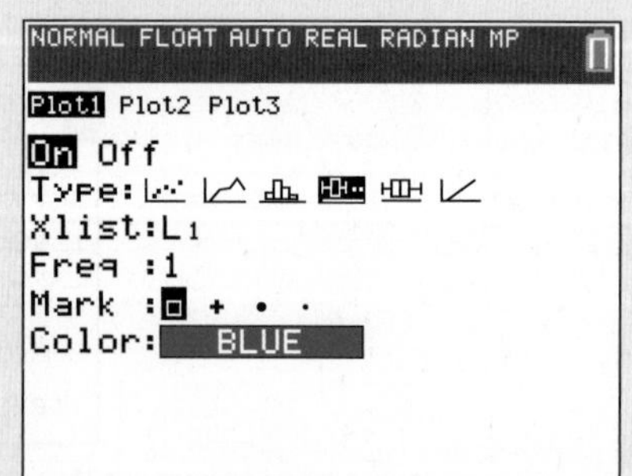

3. Press ZOOM and select ZoomStat to display the parallel boxplots. Then press TRACE and use the arrow keys to view the five-number summary.

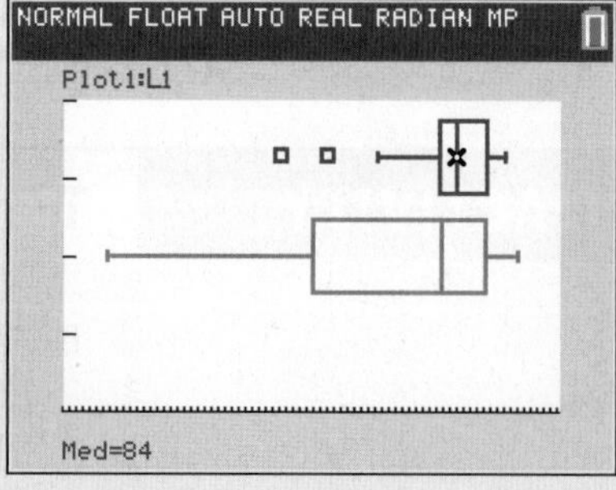

SECTION 1D Summary

- A numerical summary of a distribution of quantitative data should include measures of *center* and *variability*.
- The **mean** and the **median** measure the center of a distribution in different ways. The median is the midpoint of the distribution, the number such that about half the observations are smaller and half are larger. The mean is the average of the observations. In symbols, the sample mean is given by

$$\bar{x} = \frac{\sum x_i}{n}.$$

- A **statistic** is a number that describes a sample. A **parameter** is a number that describes a population. We often use statistics (like the sample mean $\bar{x}$) to estimate parameters (like the population mean μ).
- The simplest measure of variability for a distribution of quantitative data is the **range,** which is the distance from the minimum value to the maximum value.
- When you use the mean to describe the center of a distribution, use the **standard deviation** to measure variability. The standard deviation gives the typical distance of the values in a distribution from the mean. In symbols, the sample standard deviation is given by $s_x = \sqrt{\frac{\sum(x_i - \bar{x})^2}{n-1}}$. The value obtained before taking the square root is known as the *sample variance*, denoted by s_x^2. The standard deviation s_x is 0 when there is no variability and gets larger as variability from the mean increases.
- When you use the median to describe the center of a distribution, use the **interquartile range (*IQR*)** to describe the distribution's variability. The **first quartile Q_1** has about one-fourth of the individual data values at or below it, and the **third quartile Q_3** has about three-fourths of the individual data values at or below it. The interquartile range measures variability in the middle half of the distribution and is found by calculating $IQR = Q_3 - Q_1$.
- The median is a **resistant** measure of center because it is relatively unaffected by extreme values. The mean is not resistant. Among measures of variability, the *IQR* is resistant, but the range and standard deviation are not.
- The mean and standard deviation are good descriptions for roughly symmetric distributions with no outliers. The median and *IQR* are a better description for skewed distributions or distributions with outliers.
- The most common method of identifying outliers in a distribution of quantitative data is the **1.5 × *IQR* rule**. According to this rule, an individual data value is an outlier if it is less than $Q_1 - 1.5 \times IQR$ or greater than $Q_3 + 1.5 \times IQR$. Another method for identifying outliers is the 2 × SD rule, which says that any data value more than 2 standard deviations from the mean of the distribution is an outlier.
- The **five-number summary** of a distribution consists of the minimum, Q_1, the median, Q_3, and the maximum. **A boxplot** displays the five-number summary, marking outliers with a special symbol. The box shows the variability in the middle half of the distribution. The median is marked within the box. Lines extend from the box to the smallest and largest observations that are not outliers. Boxplots are helpful for comparing the center (median) and variability (range, *IQR*) of multiple distributions. Boxplots aren't as useful for identifying the shape of a distribution because they do not display peaks, clusters, gaps, and other interesting features.

AP® EXAM TIP
AP® Daily Videos

Review the content of this section and get extra help by watching the AP® Daily Videos for Topics 1.7–1.9, which are available in AP® Classroom.

1D Tech Corners

TI-Nspire and other technology instructions are on the book's website at bfwpub.com/tps7e.

SECTION 1D Exercises

Measuring Center: The Median

1. **Soccer stars** How good was the 2019 U.S. women's soccer team? With players like Carli Lloyd, Alex Morgan, and Megan Rapinoe, the team put on an impressive showing en route to winning the 2019 Women's World Cup. Here are data on the number of goals scored by the team in 24 games played in the 2019 season:[74]
pg 49

1	1	2	2	1	5	6	3	5	3	13	3
2	2	2	2	2	3	4	3	2	1	3	6

Find the median.

2. **How fuel efficient?** The EPA is in charge of determining and reporting fuel economy ratings for cars. Here are the EPA estimates of highway gas mileage in miles per gallon (mpg) for a sample of 21 model-year 2022 midsize cars:[75]

25	30	27	31	38	26	28	40	25	28	30
31	30	30	34	30	31	31	32	48	31	

Find the median.

Measuring Center: The Mean

3. **Goooaaaalll!** Refer to Exercise 1.
pg 51

(a) Calculate the mean number of goals scored by the U.S. women's soccer team in the 24 games it played during the 2019 season.

(b) The one game when the team scored 13 goals is a possible outlier in the distribution. Recalculate the mean number of goals scored by the team in the remaining 23 games. What do you notice?

4. **Average fuel efficiency** Refer to Exercise 2.

(a) Calculate the mean fuel efficiency in the sample of 21 model-year 2020 midsize cars.

(b) The Toyota Prius, with its 48 mpg fuel efficiency, is a possible outlier in the distribution. Recalculate the mean fuel efficiency for the other 20 car models. What do you notice?

5. **Electing the president** To become president of the United States, a candidate does not have to receive a majority of the popular vote, but that candidate does have to win a majority of the 538 votes that are cast in the Electoral College. Here is a dotplot of the number of electoral votes in 2024 for each of the 50 states and the District of Columbia:

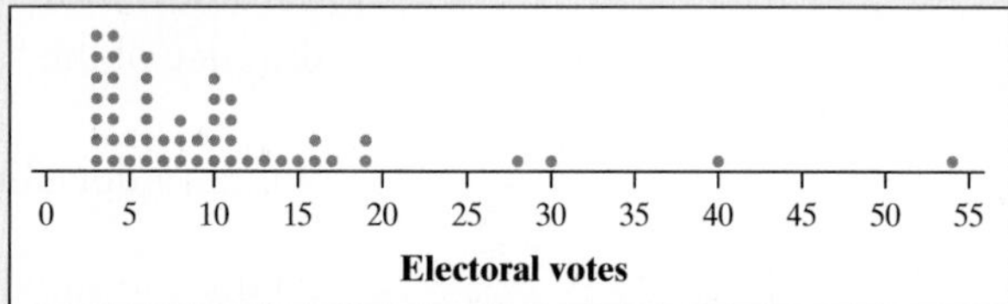

(a) Find and interpret the median.

(b) Without doing any calculations, explain how the mean and the median of this distribution compare.

(c) Is the value you found in part (a) a statistic or a parameter? Justify your answer.

6. **Birth rates in Africa** One of the important factors in determining population growth rates is the birth rate in a country. The dotplot shows the birth rates per 1000 individuals (rounded to the nearest whole number) for all 54 African nations in a recent year.[76]

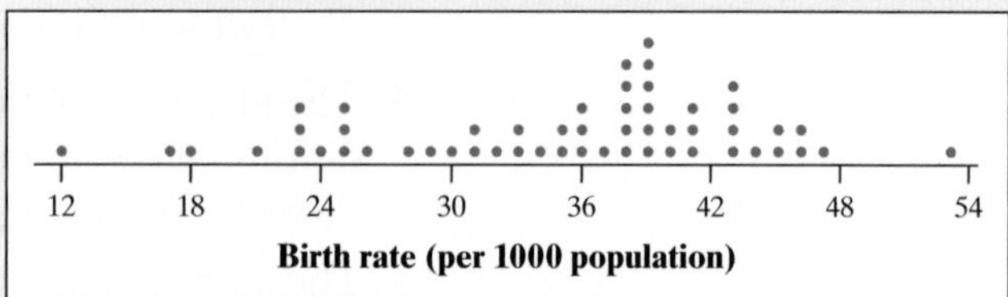

(a) Find and interpret the median.

(b) Without doing any calculations, explain how the mean and the median compare.

(c) Is the value you found in part (a) a statistic or a parameter? Justify your answer.

7. **College loans** In 2021, college student loan debt in the United States reached $1.71 trillion. At the time, there were 45.3 million U.S. borrowers with student loan debt. More than 3 million had loan debts greater than $100,000; about 900,000 had loan debts exceeding $200,000.[77]

(a) Find the mean amount of student loan debt among U.S. borrowers in 2021.

(b) Do you think the median amount of student loan debt was greater than or less than the mean amount? Justify your answer.

8. **House prices** The mean and median selling prices of existing single-family homes sold in the United States in the first 3 months of 2022 were \$428,700 and \$507,800.[78]

(a) Which of these numbers is the mean and which is the median? Explain your reasoning.

(b) What shape would you expect the distribution of home prices to have? Justify your answer.

9. **Teens and fruit** We all know that fruit is good for us. Here is a histogram of the number of servings of fruit per day that 74 seventeen-year-olds claimed they consumed in a study in Pennsylvania.[79] Find the mean and the median. Show your method clearly.

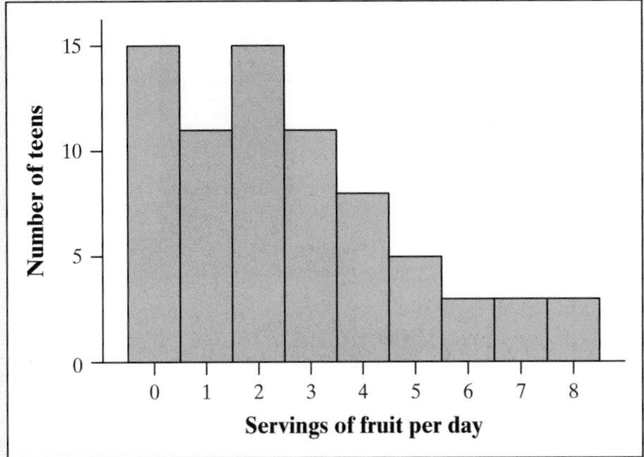

10. **Shakespeare** The following histogram shows the distribution of lengths of words used in Shakespeare's plays.[80] Find the mean and the median. Show your method clearly.

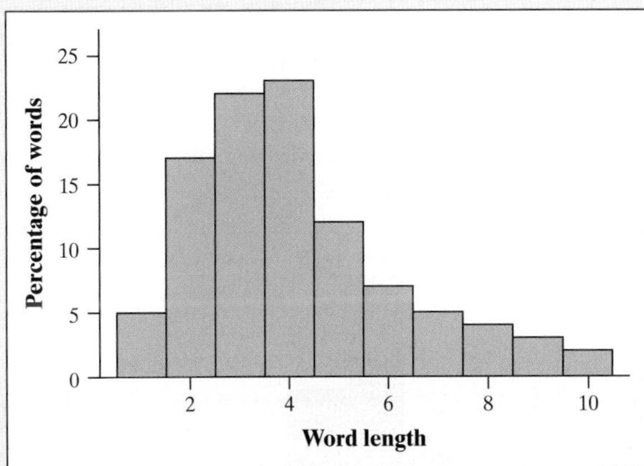

Measuring Variability: The Range

11. **Soccer range** Refer to Exercise 1.
pg 55

(a) Find the range of the distribution of goals scored by the U.S. women's soccer team in all 24 games played that season.

(b) The one game when the team scored 13 goals is a possible outlier in the distribution. Recalculate the range excluding this game. Explain why the result is so different than the answer to part (a).

12. **Mileage range** Refer to Exercise 2.

(a) Find the range of the distribution of highway gas mileage for the sample of 21 model-year 2022 midsize cars.

(b) The Toyota Prius, with its 48 mpg fuel efficiency, is a possible outlier in the distribution. Recalculate the range excluding this car model. Explain why the result is so different than the answer to part (a).

Measuring Variability: The Standard Deviation

13. **Foot length** Here are the foot lengths (in centimeters) for a random sample of seven 14-year-olds from the United Kingdom:
pg 56

25	22	20	25	24	24	28

Calculate the sample standard deviation. Interpret this value.

14. **Well rested?** A random sample of 6 students in a first-period statistics class was asked how much sleep (to the nearest hour) they got last night. Their responses were 6, 7, 7, 8, 10, and 10. Calculate the sample standard deviation. Interpret this value.

15. **Digital photos** How much storage space do digital photos use? Here is a dotplot of the file sizes (to the nearest tenth of a megabyte) for 18 randomly selected photos in Noemi's cloud storage:[81]

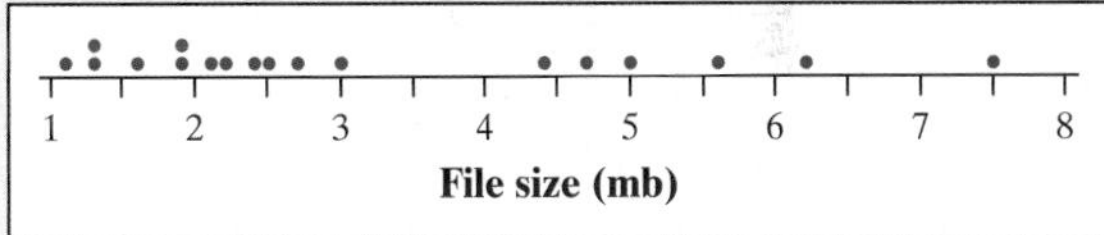

(a) The distribution of file size has a mean of $\bar{x} = 3.2$ megabytes and a standard deviation of $s_x = 1.9$ megabytes. Interpret the standard deviation.

(b) Suppose the music file that takes up 7.5 megabytes of storage space is replaced with another version of the file that takes up only 4 megabytes. How would this affect the mean and the standard deviation? Justify your answer.

16. **Healthy fast food?** Here is a dotplot of the amount of fat (to the nearest gram) in 12 different hamburgers served at a fast-food restaurant:

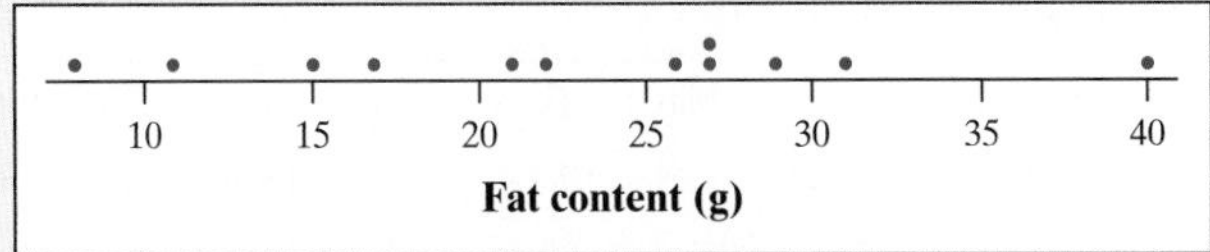

(a) The distribution of fat content has a mean of $\bar{x} = 22.83$ grams and a standard deviation of $s_x = 9.06$ grams. Interpret the standard deviation.

(b) Suppose the restaurant replaces the burger that has 22 grams of fat with a new burger that has 35 grams of fat. How would this affect the mean and the standard deviation? Justify your answer.

17. **Comparing SD** Which of the following distributions has a smaller standard deviation? Justify your answer.

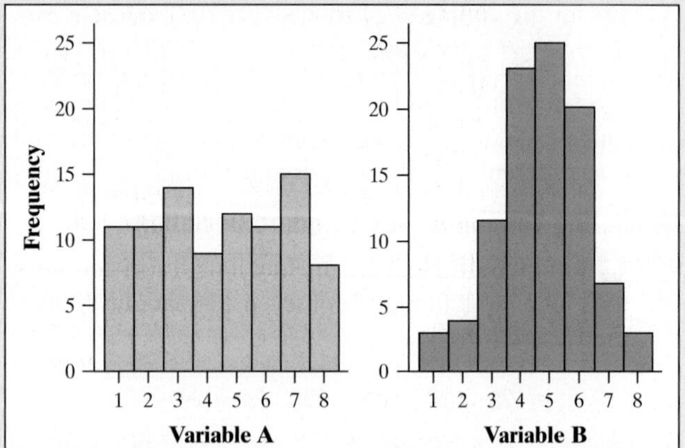

18. **Estimating SD** The dotplot shows the number of shots per game taken by National Hockey League (NHL) player Sidney Crosby in 81 regular season games in a recent season.[82] Is the standard deviation of this distribution closest to 2, 5, or 10? Explain your answer without doing any calculations.

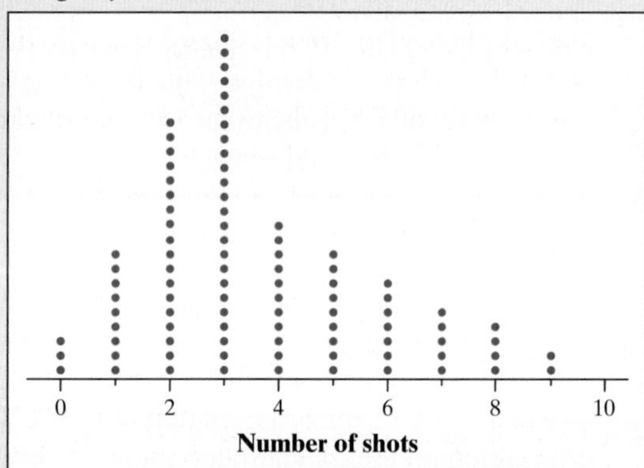

Measuring Variability: The Interquartile Range (*IQR*)

19. **Soccer *IQR*** Refer to Exercise 1. Find the interquartile range of the distribution.
pg 60

20. **Fuel efficiency *IQR*** Refer to Exercise 2. Find the interquartile range of the distribution.

21. **Shopping spree** The table displays summary statistics for data on the amount (in dollars) spent by a sample of 50 grocery shoppers.

$\bar{x}$	s_x	Min	Q_1	Med	Q_3	Max
34.70	21.70	3.11	19.27	27.86	45.40	93.34

(a) Find and interpret the interquartile range.

(b) What would you guess is the shape of the distribution based only on the summary statistics? Explain your reasoning.

22. **C-sections** A study in Switzerland examined the number of cesarean sections (surgical deliveries of babies) performed in a year by a sample of doctors. Here are summary statistics for the distribution:

$\bar{x}$	s_x	Min	Q_1	Med	Q_3	Max
19.1	10.126	5	10	18.5	29	33

(a) Find and interpret the interquartile range.

(b) What would you guess is the shape of the distribution based only on the summary statistics? Explain your reasoning.

Choosing Summary Statistics

23. **Cow's milk** Here are a histogram and summary statistics of data on the percentage of butterfat in milk from a random sample of 100 three-year-old Ayershire cows:[83]
pg 61

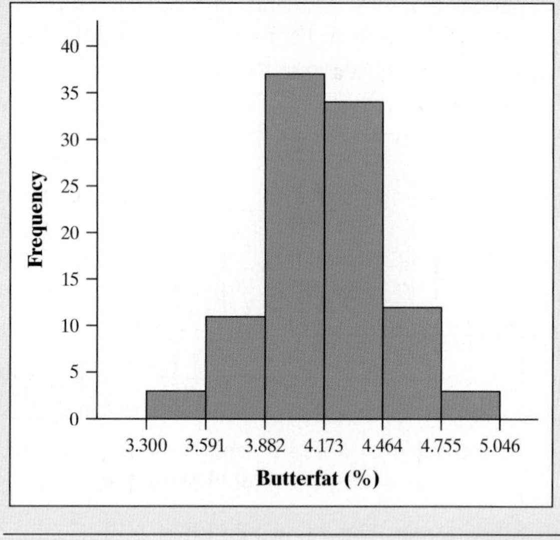

n	Mean	SD	Min	Q_1	Med	Q_3	Max
100	4.173	0.291	3.52	3.97	4.17	4.37	4.91

Which measures of center and variability should we choose to summarize this distribution? Explain your answer.

24. **Country songs** Here are a histogram and summary statistics of the lengths, in minutes, of 50 songs by country artist Dierks Bentley:

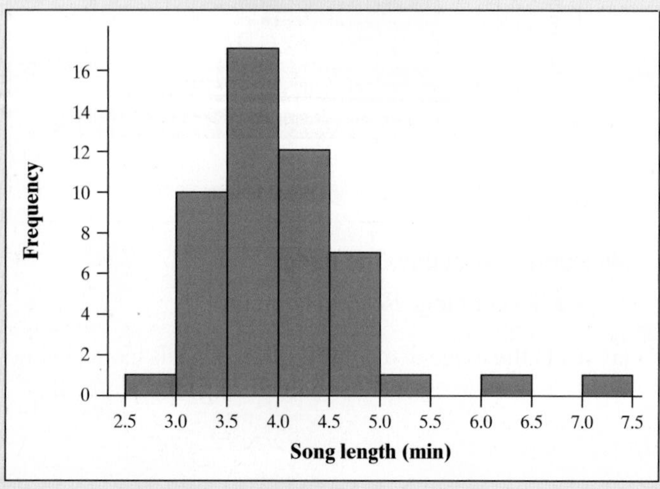

n	Mean	SD	Min	Q_1	Med	Q_3	Max
50	3.998	0.739	2.9	3.5	3.85	4.4	7

Which measures of center and variability should we choose to summarize this distribution? Explain your answer.

25. **SD contest** This is a standard deviation contest. You must choose four numbers from the whole numbers 0 to 10, with repeats allowed.

(a) Choose four numbers that have the smallest possible standard deviation.

(b) Choose four numbers that have the largest possible standard deviation.

(c) Is more than one choice possible in either part (a) or (b)? Explain your answer.

26. **What do they measure?** For each of the following summary statistics, decide (i) whether it could be used to measure center or variability and (ii) whether it is resistant.

(a) $\frac{Q_1 + Q_3}{2}$

(b) $\frac{\text{Max} - \text{Min}}{2}$

Identifying Outliers

27. **Soccer outliers** Refer to Exercise 1. Use the $1.5 \times IQR$ rule to identify any outliers in the distribution.
pg 64

28. **Fuel efficiency outliers** Refer to Exercise 2. Use the $1.5 \times IQR$ rule to identify any outliers in the distribution.

29. **Cow's milk outliers** Refer to Exercise 23. Determine whether the distribution has low or high outliers using

(a) The $1.5 \times IQR$ rule

(b) The $2 \times$ SD rule

30. **Country music outliers** Refer to Exercise 24. Determine whether the distribution has low or high outliers using

(a) The $1.5 \times IQR$ rule

(b) The $2 \times$ SD rule

Displaying Summary Statistics: Boxplots

31. **Daily texts** According to blogger Kenneth Burke, adults ages 18 to 24 send an average of 64 texts per day.[84] Dr. Williams suspected that this value was exaggerated, and collected data from a class of Introductory Statistics students on the number of texts they had sent in the past 24 hours. Here are the data:
pg 66

0	7	1	29	25	8	5	1	25	98	9	0
268	118	72	0	92	52	14	3	3	44	5	42

(a) Make a boxplot to display the data.

(b) Based on the boxplot in part (a), what percentage of Dr. Williams's students sent more than 64 texts in the previous day: less than 25%, between 25% and 50%, between 50% and 75%, or more than 75%? Do the data confirm Dr. Williams's suspicion? Explain your answer.

32. **Copper mining** In March 2021, a Canadian company found substantial deposits of copper at shallow depths in Arizona's Copper World region, which previously yielded about 440,000 tons of high-quality copper from 1874 to 1969. The company drilled test holes in several randomly selected locations at each of the old mine sites, and measured the amount of copper (as a percentage) in the rock extracted from each hole. Here are the data from the 28 test holes drilled at the Broad Top Butte site:[85]

0.00	0.75	1.43	0.43	0.19	0.52	0.00
0.00	0.20	1.38	0.00	0.28	0.71	0.30
0.30	0.44	0.59	0.91	0.33	0.39	0.00
0.70	0.37	0.38	0.67	0.00	0.38	0.30

(a) Make a boxplot to display the data.

(b) An economically viable copper mine has about 0.6% copper in its rock. Based on the boxplot in part (a), what percentage of the rock at Broad Top Butte should the company estimate is more than 0.6% copper: less than 25%, between 25% and 50%, between 50% and 75%, or more than 75%? Explain your answer.

33. **Electoral boxplot** Refer to Exercise 5. Here is a boxplot of the electoral vote data:

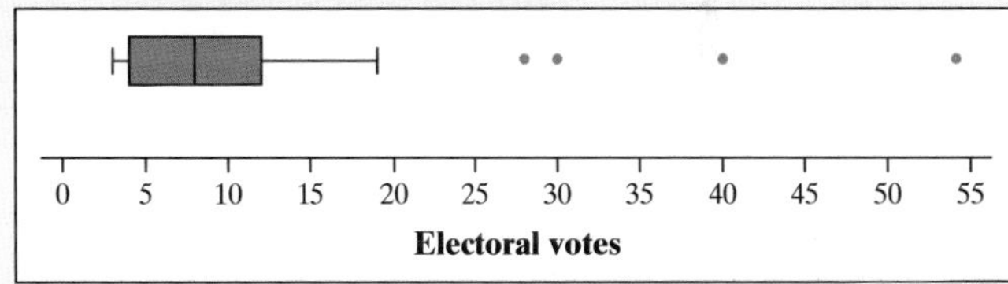

(a) Use the boxplot to estimate the IQR of the distribution.

(b) Identify an aspect of the distribution that the dotplot in Exercise 5 reveals that the boxplot does not.

34. **Birth rate boxplot** Refer to Exercise 6. Here is a boxplot of the birth rate data:

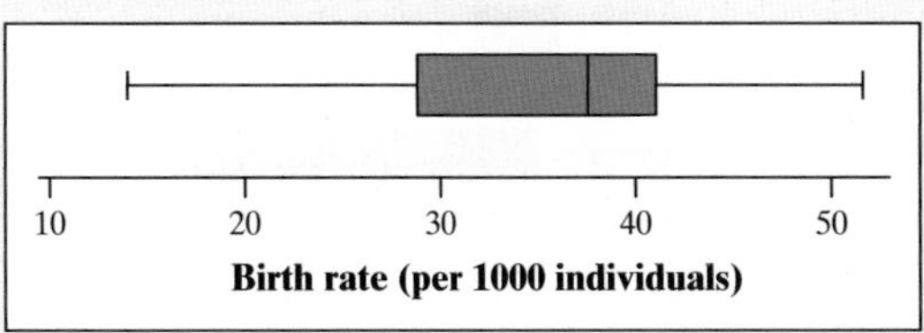

(a) Use the boxplot to estimate the IQR of the distribution.

(b) Identify an aspect of the distribution that the dotplot in Exercise 6 reveals that the boxplot does not.

Comparing Distributions with Boxplots and Summary Statistics

35. **Overthinking it?** Athletes often comment that they try not to "overthink it" when competing in their sport. Is it possible to "overthink it"? To investigate, researchers put some golfers to the test. They recruited 40 experienced golfers and allowed them some time to practice their putting. After practicing, they randomly assigned the golfers in equal numbers to two groups. Golfers in one group were asked to write a detailed description of their putting technique (which could lead to "overthinking it"). Golfers in the other group were asked to do an unrelated verbal task for the same amount of time. After completing their tasks, each golfer attempted putts from a fixed distance until they made three putts in a row. Here are parallel boxplots displaying the number of putts required for the golfers in each group to make three putts in a row, along with summary statistics.[86] Compare the distributions.

pg 68

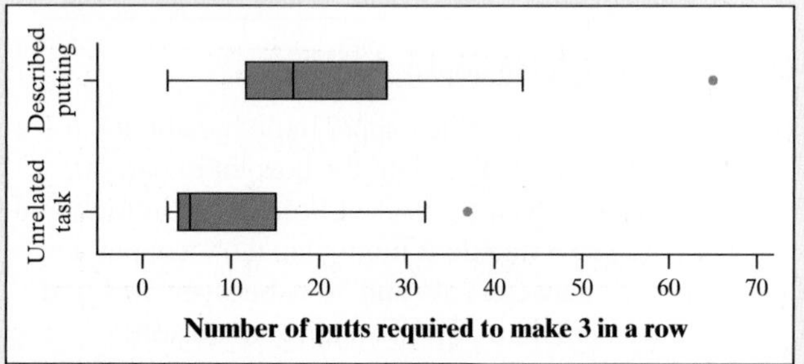

Group Name	n	Mean	SD	Min	Q_1	Med	Q_3	Max
Described putting	20	21.2	15.07	3	11.5	17	28	65
Unrelated task	20	10.6	9.832	3	4	5.5	15.5	37

36. **Who is the GOAT?** Which NBA player is the Greatest of All Time: LeBron James or Michael Jordan? Here are parallel boxplots displaying the average points scored per game by each player in the seasons they played through 2019, along with summary statistics.[87] Compare the distributions.

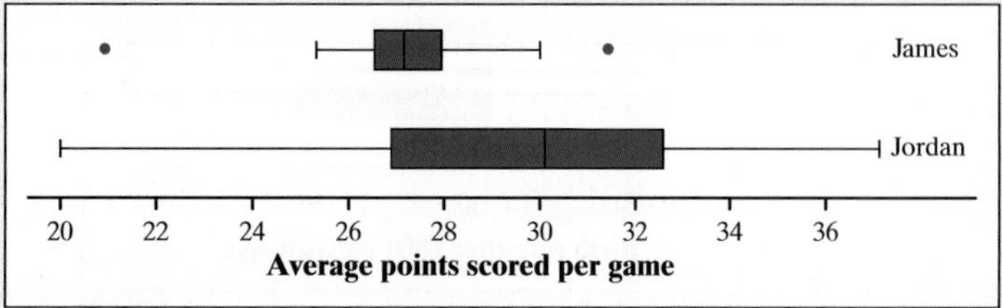

Player	$\bar{x}$	s_x	Min	Q_1	Med	Q_3	Max	IQR
James	27.16	2.33	20.90	26.55	27.15	27.95	31.40	1.40
Jordan	29.45	4.76	20.00	26.90	30.10	32.60	37.10	5.70

37. **Energetic refrigerators** *Consumer Reports* magazine rated different types of refrigerators, including those with bottom freezers, those with top freezers, and those with side freezers. One of the variables it measured was annual energy cost (in dollars). The following boxplots show the energy cost distributions for each of these types.[88]

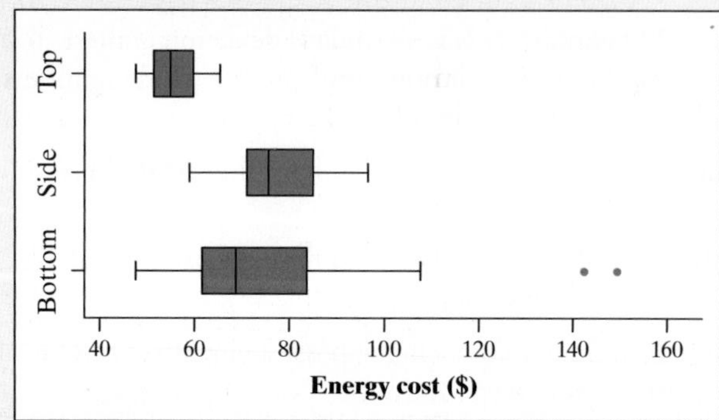

(a) From the boxplots, what can you say about the percentage of each type of freezer that costs more than $60 per year to operate?

(b) Compare the energy cost distributions for the three types of refrigerators.

38. **Weight-loss strategies** Which dieting strategy works best? Researchers used data from the U.K. television show *How to Lose Weight Well* to investigate. In each episode, six participants are assigned in pairs to adopt three dieting strategies: crashers, shape shifters, and life changers. The crashers follow extreme diets for 1–2 weeks, such as replacing meals with bone broth or baby food. The shape shifters follow moderately extreme diets for 4–6 weeks, like the blood sugar diet that involves consuming at most 800 calories per day. The life changers follow less-extreme diets for 12–16 weeks, like low-carbohydrate or Mediterranean diets. Here are boxplots that summarize the percentage of body weight lost by 137 pairs of participants in the TV show over a 5-year period:[89]

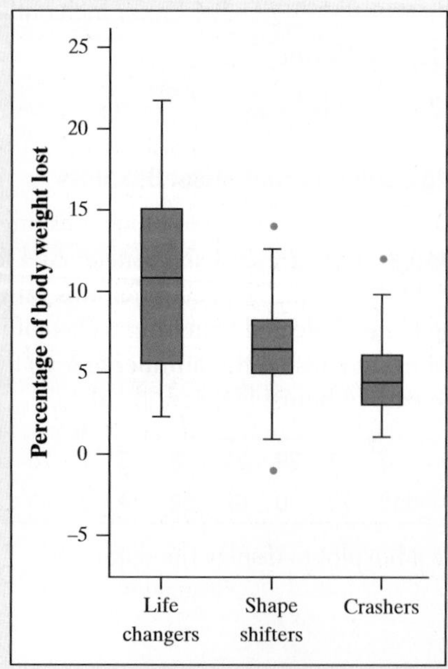

(a) From the boxplots, what can you say about the percentage of people in each of the three dieting groups who lost less than 5% of their body weight?

(b) Compare the distributions of percentage of body weight lost for the three dieting groups.

39. **On-site worker commutes** How long do on-site workers typically spend traveling to their workplace? The answer may depend on where they live. Here are the travel times (in minutes) of 20 randomly chosen on-site workers in New York state and 15 randomly chosen on-site workers in North Carolina:[90]

New York	10	30	5	25	40	20	10	15	30	20	15
	20	85	15	65	15	60	60	40	45		
North Carolina	30	20	10	40	25	20	10	60	15	40	5
	30	10	12	10							

(a) Make parallel boxplots to display the data.

(b) Do these data provide strong evidence that travel times to work differ for workers in these two states? Give appropriate evidence to support your answer.

40. **SSHA scores** Higher scores on the Survey of Study Habits and Attitudes (SSHA) indicate good study habits and attitudes toward learning. Here are scores for 18 first-year college students and 20 second-year college students:

First-year students	154	109	137	115	152	140	154	178	101	
	103	126	126	137	165	165	129	200	148	
Second-year students	108	140	114	91	180	115	126	92	169	146
	109	132	75	88	113	151	70	115	187	104

(a) Make parallel boxplots to compare the distributions.

(b) Do these data support the belief that first-year and second-year college students differ in their study habits and attitudes toward learning? Give appropriate evidence to support your answer.

For Investigation *Apply the skills from the section in a new context or nonroutine way.*

41. **Trimmed mean** Another measure of center for a quantitative data set is the *trimmed mean*. To calculate the trimmed mean, order the data set from lowest to highest, remove the same number of data values from each end, and calculate the mean of the remaining values. For example, to calculate the 10% trimmed mean, start by removing the smallest 10% and the largest 10% of values in the data set. For a data set with 50 values, you would remove 5 values from each "end" of the data set because 5/50 = 0.10 or 10%.

Researchers asked a random sample of 20 students from a large high school how many pairs of shoes they had. Here is a dotplot of the data:

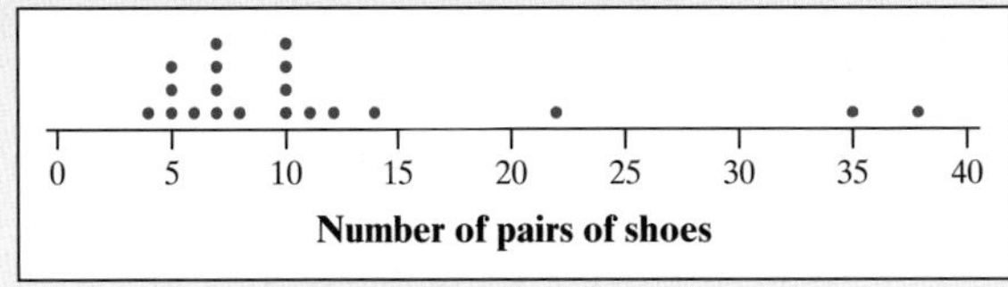

(a) Calculate the mean of the distribution.

(b) Calculate the 10% trimmed mean.

(c) Why is the trimmed mean a better summary of the center of this distribution than the mean?

42. **Measuring skewness** Here is a boxplot of the number of electoral votes in 2024 for each of the 50 states and the District of Columbia, along with summary statistics. You can see that the distribution is skewed to the right with four high outliers. How might we compute a numerical measure of skewness?

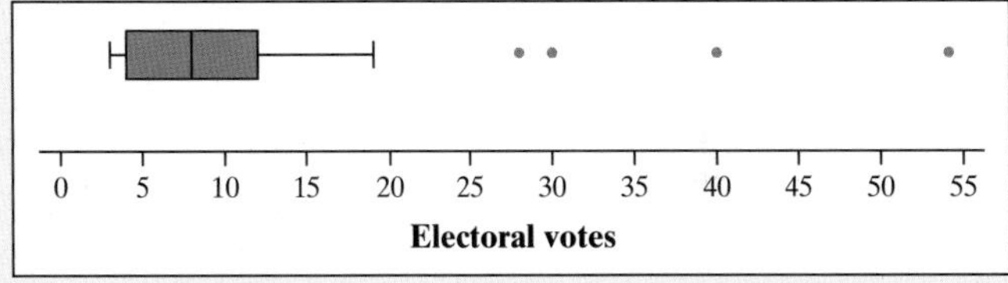

n	Mean	SD	Min	Q_1	Med	Q_3	Max
51	10.549	9.653	3	4	8	12	54

(a) One simple formula for calculating skewness is $\frac{\text{maximum} - \text{median}}{\text{median} - \text{minimum}}$. Compute this value for the electoral vote data. Explain why this formula should yield a value greater than 1 for a right-skewed distribution.

(b) Based solely on the summary statistics provided, define a formula for a different statistic that measures skewness. Compute the value of this statistic for the electoral vote data. What values of the statistic might indicate that a distribution is skewed to the right? Explain your reasoning.

Multiple Choice *Select the best answer for each question.*

43. If a distribution is strongly skewed to the right with no outliers, which of the following relationships is most likely correct?

(A) mean < median
(B) mean ≈ median
(C) mean = median
(D) mean > median
(E) We can't tell without examining the data.

44. The scores on a statistics test had a mean of 81 and a standard deviation of 9. One student was absent on the test day, and their score wasn't included in the calculation. If this student's score of 84 was added to the distribution of scores, what would happen to the mean and standard deviation?

(A) Mean will increase and standard deviation will increase.
(B) Mean will increase and standard deviation will decrease.
(C) Mean will increase and standard deviation will stay the same.
(D) Mean will decrease and standard deviation will increase.
(E) Mean will decrease and standard deviation will decrease.

45. The stemplot shows the number of home runs hit by each of the 30 Major League Baseball teams in a single season. Home run totals greater than what value should be considered outliers?

Stem	Leaves
09	15
10	3789
11	47
12	19
13	
14	89
15	34445
16	239
17	223
18	356
19	1
20	3
21	0
22	2

Key: 14|8 is a team with 148 home runs.

(A) 173
(B) 210
(C) 222
(D) 229
(E) 257

46. Which of the following boxplots best matches the distribution shown in the histogram?

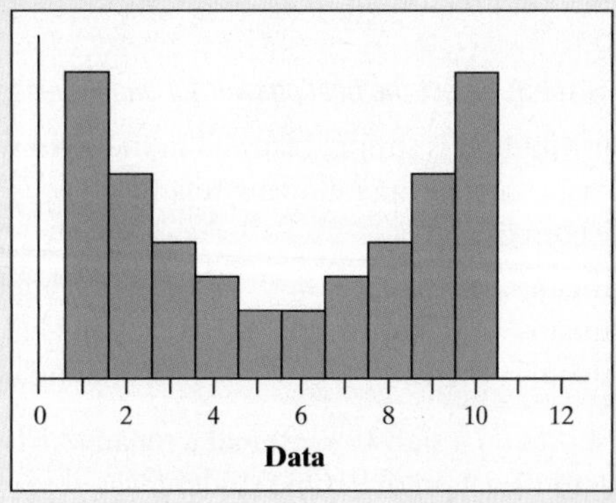

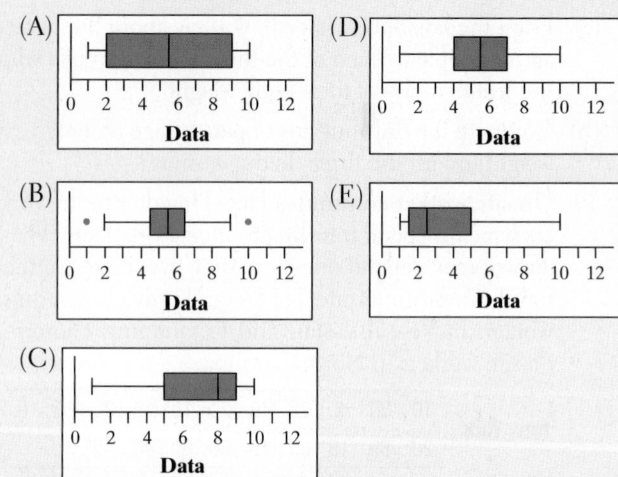

Recycle and Review *Practice what you learned in previous sections.*

47. **How tall are you? (1C)** We chose a random sample of 50 Canadian students who completed an online survey that included the question, "How tall are you without your shoes on? Answer to the nearest half-centimeter." Here are the students' responses:

166.5	170.0	178.0	163.0	150.5	169.0	173.0	169.0	171.0	166.0
190.0	183.0	178.0	161.0	171.0	170.0	191.0	168.5	178.5	173.0
175.0	160.5	166.0	164.0	163.0	174.0	160.0	174.0	182.0	167.0
166.0	170.0	170.0	181.0	171.5	160.0	178.0	157.0	165.0	187.0
168.0	157.5	145.5	156.0	182.0	168.5	177.0	162.5	160.5	185.5

Make an appropriate graph to display the data. Describe the shape, center, and variability of the distribution. Are there any outliers?

48. **Climate change insurance? (1A, 1B)** Qualtrics conducted a survey of 1070 randomly selected U.S. homeowners. Respondents were asked, "How much extra would you be willing to spend per year on insurance policies that focus on coverage for climate change–induced risks?" The bar graph summarizes the responses.[91]

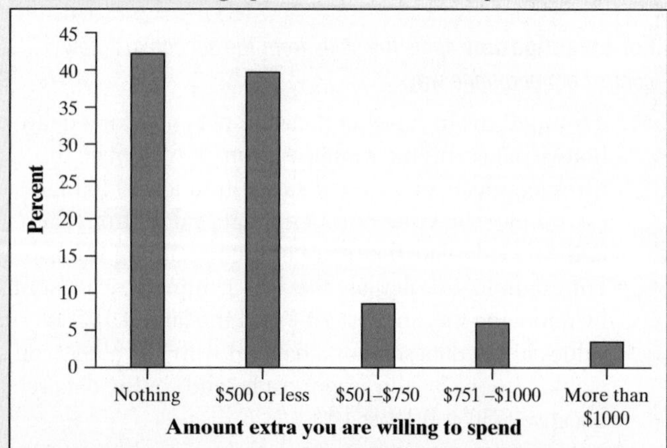

(a) Identify the individuals and the variable displayed in the bar graph. Is the variable categorical or quantitative? Justify your answer.
(b) Describe what you see.

UNIT 1 Wrap-Up

PART I

FRAPPY! Free Response AP® Problem, Yay!

Directions: Show all your work. Indicate clearly the methods you use, because you will be scored on the correctness of your methods as well as on the accuracy and completeness of your results and explanations.

A realtor collects data about the sales price (in \$1000s) of 43 homes that sold in their city in the past several months. Here are the data, sorted from lowest to highest sales price:[92]

107	133	135	141	147	165	165	170	173	175	182
190	191	195	199	205	210	215	220	226	228	242
249	251	255	255	260	265	265	275	275	280	285
289	295	301	310	315	350	365	365	397	503	

(a) Make a histogram of the data using intervals of width 50, starting at 100, with axes scaled as shown.

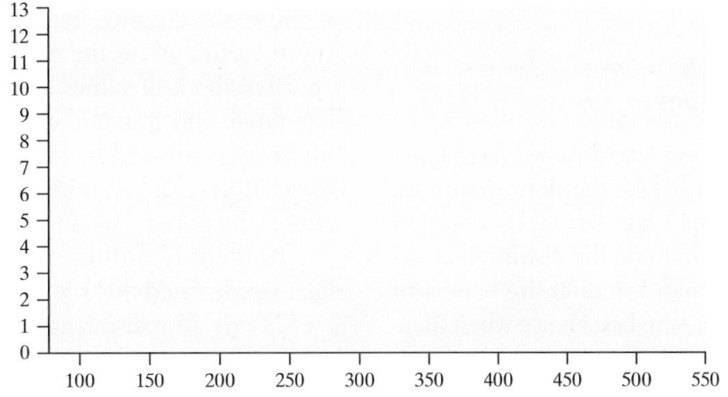

(b) Determine whether or not the maximum sales price of \$503,000 is an outlier in the distribution of sales price for these 43 homes. Show your method clearly.

Here are parallel boxplots of the sales prices for the homes in each of three regions of the city:

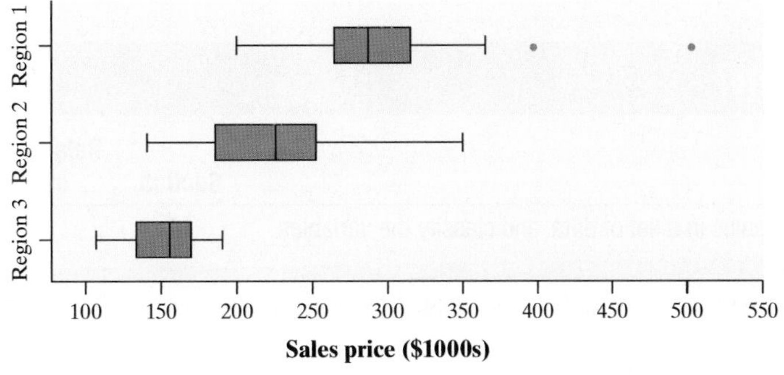

(c) Compare the distributions of sales price in the three regions of the city.

After you finish the FRAPPY!, you can view two example solutions on the book's website (**bfwpub.com/tps7e**). Determine whether you think each solution is "complete," "substantial," "developing," or "minimal." If the solution is not complete, what improvements would you suggest to the student who wrote it? Finally, your teacher will provide a scoring rubric. Score your response and note what, if anything, you would do differently to improve your own score.

UNIT 1, PART I REVIEW

SECTION 1A Statistics: The Language of Variation

In this brief section, you learned how to identify the **individuals** and **variables** in a data set. You also learned how to distinguish between **categorical variables** and **quantitative variables,** as well as how to summarize the **distribution** of a variable with a **frequency table** or **relative frequency table.**

SECTION 1B Displaying and Describing Categorical Data

In this section, you learned how to display the distribution of a categorical variable with a **bar graph** or a **pie chart,** and what to look for when describing these graphs. Next, you learned how to compare the distribution of a categorical variable in two or more groups using **side-by-side bar graphs.** Remember to properly label your graphs! Poor labeling is an easy way to lose credit on the AP® Statistics exam. You should also be able to recognize misleading graphs and be careful to avoid making misleading graphs yourself.

SECTION 1C Displaying Quantitative Data with Graphs

In this section, you learned how to make three different types of graphs for displaying quantitative data: **dotplots, stemplots,** and **histograms.** Each of the graphs has distinct benefits, but all of them are good tools for examining the distribution of a quantitative variable. Dotplots and stemplots are handy for smaller sets of data. Histograms are the best choice when there are a large number of data values. On the AP® Statistics exam, you will be expected to make, interpret, and describe each of these types of graphs.

When you are describing the distribution of a quantitative variable, you should look at a graph of the data to determine the overall pattern (**shape, center, variability**) and striking departures from that pattern (**outliers**). When comparing distributions of quantitative data, you should include explicit comparison words for center and variability such as "is greater than" or "is approximately the same as." When asked to compare distributions, a very common mistake on the AP® Statistics exam is describing the characteristics of each distribution separately without making these explicit comparisons.

SECTION 1D Describing Quantitative Data with Numbers

To measure the *center* of a distribution of quantitative data, you learned how to calculate the **median** and the **mean** of a distribution. You also learned that the median is a **resistant** measure of center, but the mean isn't resistant because it can be greatly affected by skewness or outliers.

To measure the *variability* of a distribution of quantitative data, you learned how to calculate the **range, standard deviation,** and **interquartile range (*IQR*).** The range measures the distance from the minimum value to the maximum value, while the standard deviation measures the typical distance of values in the distribution from the mean. The range and standard deviation are not resistant — both are heavily affected by extreme values. The interquartile range (*IQR*) is a resistant measure of variability because it ignores the upper 25% and lower 25% of the distribution.

To identify outliers in a distribution of quantitative data, you learned the **1.5 × *IQR* rule** and the less common 2 × *SD rule*. You also learned that **boxplots** are a great way to visually summarize a distribution of quantitative data. Boxplots are helpful for comparing the center (median) and the variability (range, *IQR*) for multiple distributions. Boxplots aren't as useful for identifying the shape of a distribution because they do not display peaks, clusters, gaps, and other interesting features.

What Did You Learn?

Learning Target	Section	Related Example on Page(s)	Relevant Review Exercise(s)
Identify the individuals and variables in a set of data, and classify the variables as categorical or quantitative.	1A	4	R1
Make and interpret a frequency table or a relative frequency table for a distribution of data.	1A	6	R2
Make and interpret bar graphs of categorical data.	1B	12	R2
Compare distributions of categorical data.	1B	13	R4
Identify what makes some graphs of categorical data misleading.	1B	15	R3
Make and interpret dotplots of quantitative data.	1C	23	R5

Learning Target	Section	Related Example on Page(s)	Relevant Review Exercise(s)
Describe the shape of a distribution of quantitative data.	1C	25	R6, R7
Describe the distribution of a quantitative variable.	1C	26	R6
Compare distributions of quantitative data.	1C	28	R8
Make and interpret stemplots of quantitative data.	1C	30	R6
Make and interpret histograms of quantitative data.	1C	33	R7
Find the median of a distribution of quantitative data.	1D	49	R5
Calculate the mean of a distribution of quantitative data.	1D	51	R6
Find the range of a distribution of quantitative data.	1D	55	R9
Calculate and interpret the standard deviation of a distribution of quantitative data.	1D	56	R9
Find the interquartile range (*IQR*) of a distribution of quantitative data.	1D	60	R7
Choose appropriate measures of center and variability to summarize a distribution of quantitative data.	1D	61	R9
Identify outliers in a distribution of quantitative data.	1D	64	R7, R9
Make and interpret boxplots of quantitative data.	1D	66	R7
Use boxplots and summary statistics to compare distributions of quantitative data.	1D	68	R10

UNIT 1, PART I REVIEW EXERCISES

These exercises are designed to help you review the important concepts and skills in Part I of this unit.

R1 **Who buys cars? (1A)** A car dealer keeps records on car buyers for future marketing purposes. The table gives information on the last four buyers.

Buyer's name	Zip code	First-time buyer?	Buyer's distance from dealer (mi)	Car model	Model year	Price
P. Smith	27514	Y	13	Fiesta	2023	$34,490
K. Ewing	27510	N	10	Mustang	2021	$27,155
L. Shipman	27516	N	2	Fusion	2020	$23,170
S. Reice	27243	Y	4	F-150	2022	$42,210

(a) Identify the individuals in this data set.

(b) What variables were measured? Classify each as categorical or quantitative.

R2 **I want candy! (1A, 1B)** Mr. Starnes bought some candy for his AP® Statistics class to eat on Halloween. He offered the students an assortment of Snickers®, Milky Way®, Butterfinger®, Twix®, and 3 Musketeers® candies. Each student was allowed to choose one option. Here are the data on the type of candy selected:

Twix	Snickers	Butterfinger
Butterfinger	Snickers	Snickers
3 Musketeers	Snickers	Snickers
Butterfinger	Twix	Twix
Twix	Twix	Twix
Snickers	Snickers	Twix
Snickers	Milky Way	Twix
Twix	Twix	Butterfinger
Milky Way	Butterfinger	3 Musketeers
Milky Way	Butterfinger	Butterfinger

(a) Summarize the data in a relative frequency table.

(b) Make a bar graph to display the data. Describe what you see.

R3 **Popular online sites (1B)** Common Sense Media surveyed a random sample of 1306 U.S. 8- to 18-year-olds in 2021. One question asked on the survey was "If you had to pick one online site you didn't want to live without, which would it be?" The figure shows the percentage of respondents who chose YouTube, Snapchat, TikTok, and Instagram.[93]

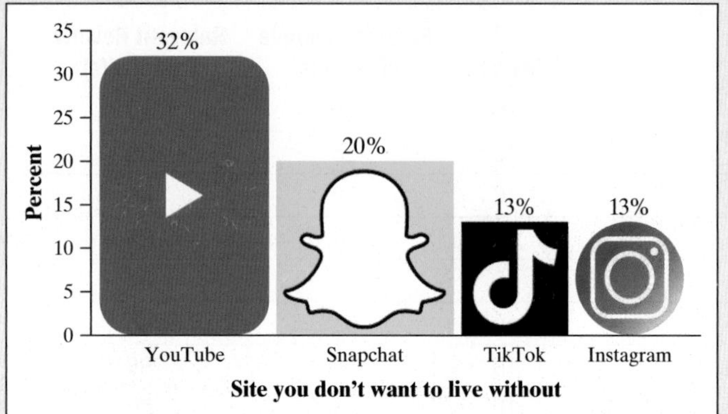

(a) Explain how the graph gives a misleading impression.

(b) Would it be appropriate to make a pie chart to display the data? Why or why not?

R4 **Success in college (1B)** A national survey asked 95,505 first-year college students about specific academic behaviors identified by college faculty as being important for student success. One question asked, "How often in the past year did you ask questions during class?" The figure is a side-by-side bar graph comparing the percentage who answered "frequently" by race/ethnic group and whether the respondent was a first-generation college student.[94] Describe what you see.

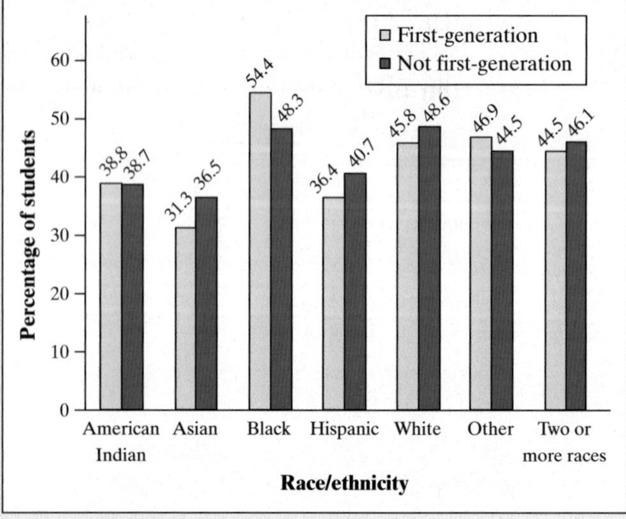

R5 **Music and memory (1C, 1D)** For a final project in their AP® Statistics class, two students studied the impact of different types of background music on students' ability to remember words from a list they were allowed to study for 5 minutes. Here are data on the number of words remembered by one group of students who listened to Beethoven's Fifth Symphony:[95]

11	12	23	15	14	15	14	15
10	14	15	9	11	13	25	11
13	13	12	20	17	23	11	12
12	11	20	20	12	12	19	13
15	10	14	11	7	17	13	18

(a) Make a dotplot to display the data.

(b) What proportion of this group of students remembered 20 or more words?

(c) Find the median of the distribution.

(d) Is the mean of the distribution less than, about the same as, or greater than the median? Explain how you know without performing any calculations.

R6 **Density of the earth (1C, 1D)** In 1798, the English scientist Henry Cavendish measured the density of the earth several times by careful work with a torsion balance. The variable recorded was the density of the earth as a multiple of the density of water. Here are Cavendish's 29 measurements:[96]

5.50	5.61	4.88	5.07	5.26	5.55	5.36	5.29	5.58	5.65
5.57	5.53	5.62	5.29	5.44	5.34	5.79	5.10	5.27	5.39
5.42	5.47	5.63	5.34	5.46	5.30	5.75	5.68	5.85	

(a) Make a stemplot of the data.

(b) Describe the distribution of density measurements.

(c) The currently accepted value for the density of the earth is 5.51 times the density of water. How does this value compare to the mean of the distribution of density measurements?

R7 **Eat your oatmeal (1C, 1D)** Researchers collected data on the calories in 39 different brands of single-serve oatmeal. Here are the data:[97]

100	110	130	130	140	150	150	150	150	160	160	160	160
170	170	170	170	170	180	190	190	200	200	210	210	210
210	220	220	230	230	240	240	250	270	280	300	310	350

(a) Make a histogram of the data. Describe the shape of the distribution.

(b) Find the interquartile range.

(c) Make a boxplot of the data.

(d) Compare the histogram from part (a) with the boxplot from part (c). Identify an aspect of the distribution that one graph reveals but the other does not.

R8 **High versus low incomes (1C)** Rich and poor households differ in ways that go beyond income. Here are histograms that compare the distributions of household size (number of people) for low-income and high-income households.[98] Low-income households had annual incomes less than \$15,000, and high-income households had annual incomes of at least \$100,000.

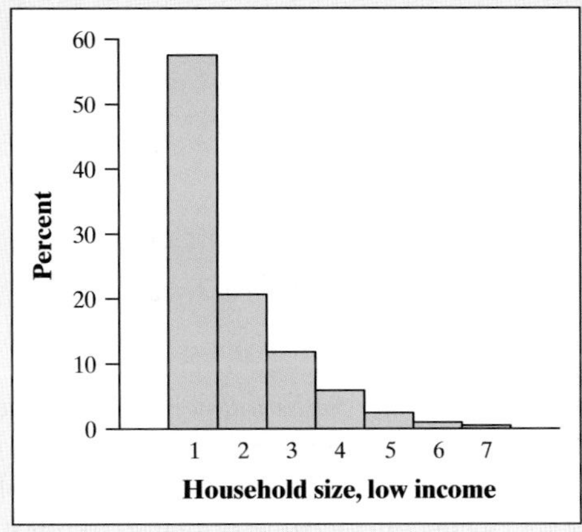

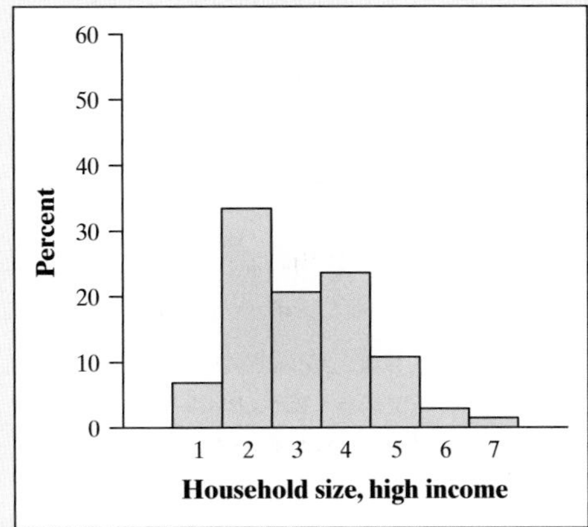

(a) About what percentage of each group of households consisted of four or more people?

(b) Describe the similarities and differences in these two distributions of household size.

Exercises R9 and R10 refer to the following setting. Do you like to eat tuna? Many people do. Unfortunately, some of the tuna that people eat may contain high levels of mercury. Exposure to mercury can be especially hazardous for pregnant women and small children. How much mercury is safe to consume? The U.S. Food and Drug Administration will act (for example, by removing the product from store shelves) if the mercury concentration in a 6-ounce can of tuna is 1.00 part per million (ppm) or higher.

What is the typical mercury concentration in cans of tuna sold in stores? A study conducted by Defenders of Wildlife set out to answer this question. Defenders collected a sample of 164 cans of tuna from stores across the United States. They sent the selected cans to a laboratory that is often used by the U.S. Environmental Protection Agency for mercury testing.[99]

R9 Mercury in tuna (1D) Here are a dotplot and summary statistics of the data on mercury concentration in the sampled cans (in parts per million, ppm):

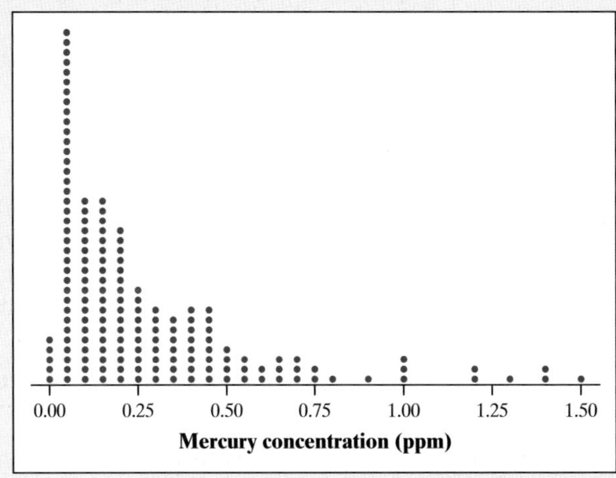

Variable	n	Mean	SD	Min	Q_1	Med	Q_3	Max
Mercury	164	0.285	0.300	0.012	0.071	0.180	0.380	1.500

(a) Find the range of the distribution.

(b) Interpret the standard deviation.

(c) Determine whether there are any outliers. Show your method clearly.

(d) Which measures of center and variability would you use to summarize the distribution? Justify your answer.

R10 More mercury in tuna (1D) Is there a difference in the mercury concentration of light tuna and albacore tuna? Use the parallel boxplots and the summary statistics to compare the two distributions.

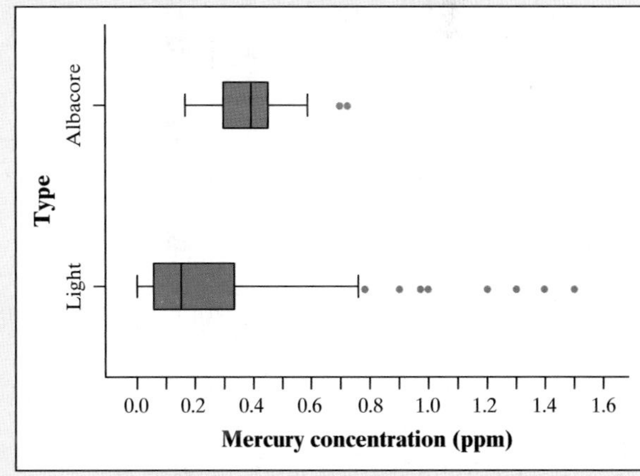

Type	n	Mean	SD	Min	Q_1	Med	Q_3	Max
Albacore	20	0.401	0.152	0.170	0.293	0.400	0.460	0.730
Light	144	0.269	0.312	0.012	0.059	0.160	0.347	1.500

UNIT 1, PART I AP® STATISTICS PRACTICE TEST

Section I: Multiple Choice *Select the best answer for each question.*

T1 An airline records data on several variables for each of its flights: model of plane, amount of fuel used, time in flight, number of passengers, and whether the flight arrived on time. The number and type of variables recorded are

(A) 1 categorical, 4 quantitative (1 discrete, 3 continuous).

(B) 1 categorical, 4 quantitative (2 discrete, 2 continuous).

(C) 2 categorical, 3 quantitative (1 discrete, 2 continuous).

(D) 2 categorical, 3 quantitative (2 discrete, 1 continuous).

(E) 3 categorical, 2 quantitative (1 discrete, 1 continuous).

T2 The bar graph summarizes responses of dog owners to the question, "Where in the car do you let your dog ride?" Which of the following statements is true?

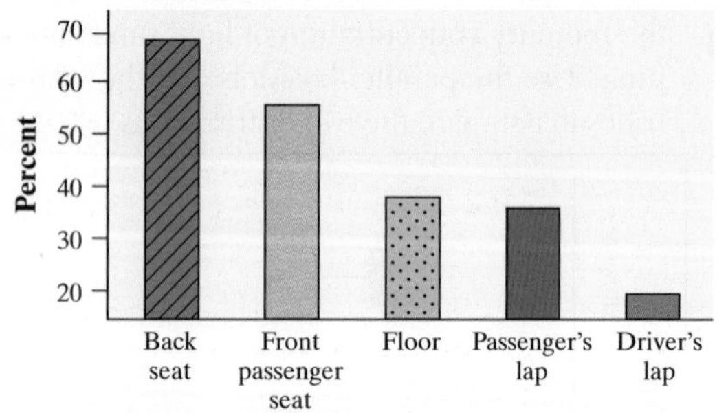

(A) A majority of owners do not allow their pets to ride in the front passenger seat.

(B) About twice as many pets are allowed to sit in the front passenger seat as in the passenger's lap.

(C) The vertical axis scale should start at 0.

(D) These data could also be presented in a pie chart.

(E) The distribution of where the dog rides is skewed to the right.

T3 Forty students took a statistics test worth 50 points. The dotplot displays the data. The third quartile is

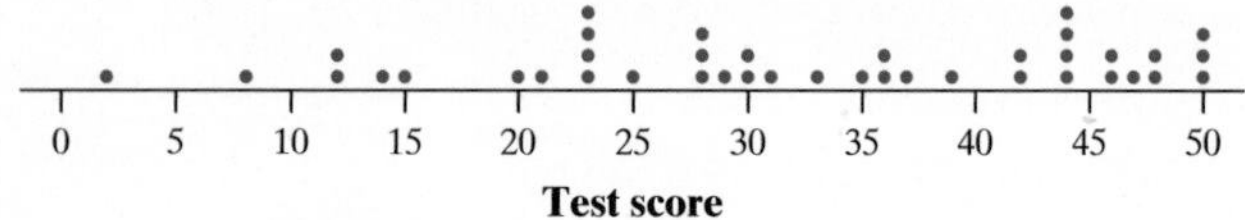

(A) 45.

(B) 44.

(C) 43.

(D) 32.

(E) 23.

Questions T4–T6 refer to the following setting. Realtors collect data so that they can serve their clients more effectively. In a recent week, data on the age of all homes sold in a particular area were collected and displayed in this histogram.

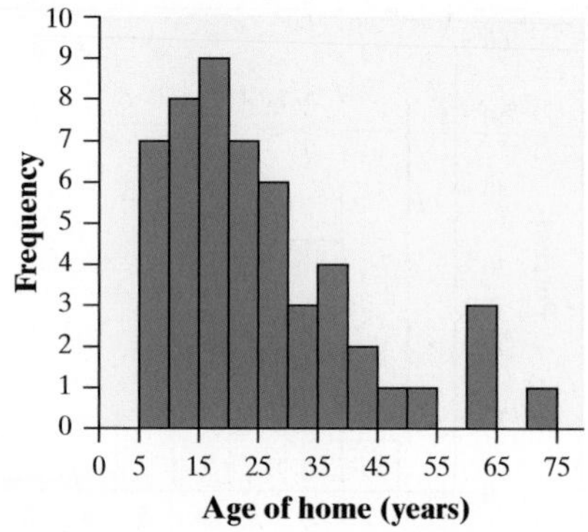

T4 Which of the following could be the median age?

(A) 19 years

(B) 24 years

(C) 29 years

(D) 34 years

(E) 39 years

T5 Which of the following is most likely true for this distribution?

(A) mean $>$ median, range $<$ *IQR*

(B) mean $<$ median, range $<$ *IQR*

(C) mean $>$ median, range $>$ *IQR*

(D) mean $<$ median, range $>$ *IQR*

(E) mean $=$ median, range $>$ *IQR*

T6 The standard deviation of the distribution of home age is about 16 years. Which of the following is a correct interpretation of this value?

(A) The age of all houses in the sample is within 16 years of the mean age.

(B) The gap between the youngest and oldest house is 16 years.

(C) The age of all the houses in the sample is 16 years from the mean age.

(D) The gap between the first quartile and the third quartile is 16 years.

(E) The age of the houses in the sample typically varies by about 16 years from the mean age.

T7 The mean salary of all first-year workers at a company is \$45,000. The mean salary of all workers who have been at the company for more than one year is \$51,000. What must be true about the mean salary of all workers at the company?

(A) It must be \$48,000.

(B) It must be larger than the median salary.

(C) It could be any number between \$45,000 and \$51,000.

(D) It must be larger than \$48,000.

(E) It cannot be larger than \$50,000.

T8 The back-to-back stemplot shows the lifetimes of several Brand X and Brand Y batteries.

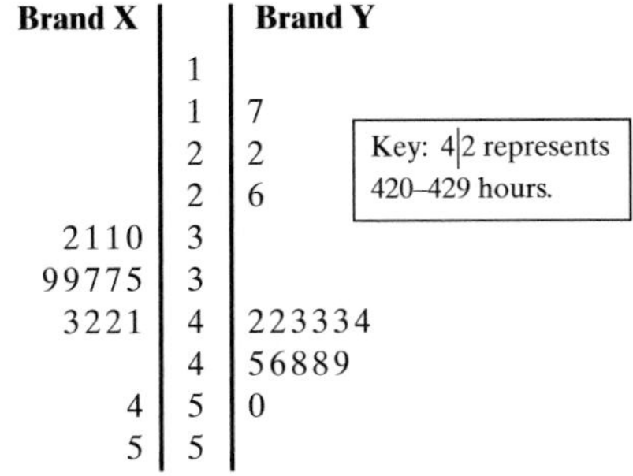

Brand X		Brand Y
	1	
	1	7
	2	2
	2	6
2110	3	
99775	3	
3221	4	223334
	4	56889
4	5	0
5	5	

Key: 4|2 represents 420–429 hours.

One reason that someone might prefer a Brand Y battery is

(A) a Brand Y battery had the longest lifetime.

(B) Brand Y batteries had the larger range of lifetimes.

(C) Brand X batteries had a longer minimum lifetime.

(D) Brand Y batteries had a longer median lifetime.

(E) most Brand X batteries had a lifetime of at least 300 hours.

T9 How has teen internet use changed over time? The Pew Research Center surveyed separate random samples of U.S. teens aged 13 to 17 in 2014–2015 and in 2022. The side-by-side bar graph summarizes the teens' responses to the question, "About how often do you use the internet, either on a computer or a cellphone?"[100]

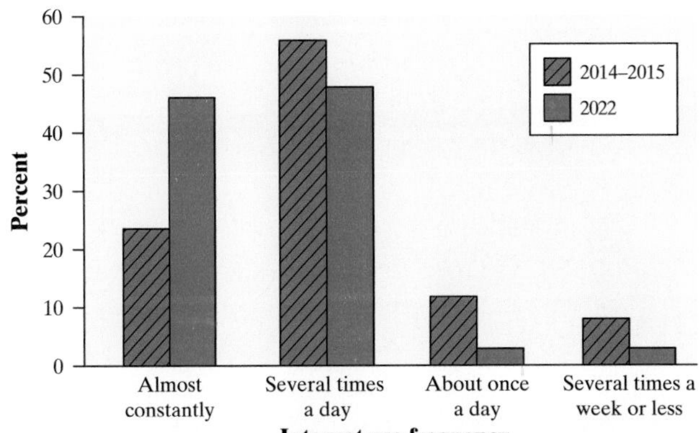

Which of the following is a correct statement?

(A) Teens reported using the internet more frequently in 2014–2015 than in 2022.

(B) A majority of teens said "almost constantly" in 2022.

(C) The percentage of teens who said "almost constantly" doubled from 2014–2015 to 2022.

(D) A majority of teens said "several times a day" in both 2014–2015 and 2022.

(E) The percentage of teens who said "about once a day" doubled from 2014–2015 to 2022.

T10 Researchers conducted an experiment to investigate the effect of a new weed killer to prevent weed growth in onion crops. They used two chemicals, the standard weed killer (S) and the new chemical (N), and tested both at high and low concentrations on 50 test plots. They then recorded the number of weeds that grew in each plot. Here are some boxplots of the results:

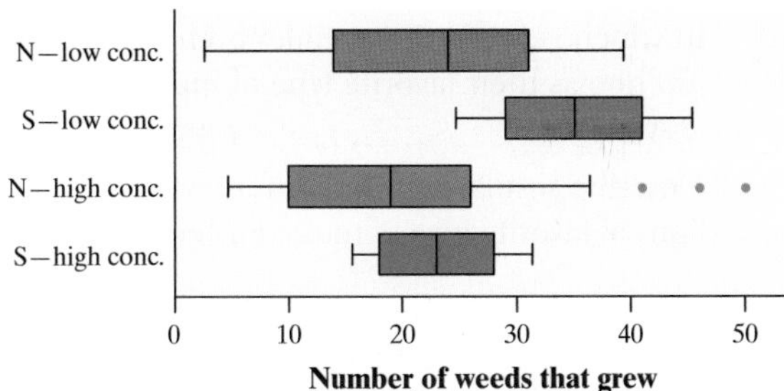

Which of the following is *not* a correct statement about the results of this experiment?

(A) At both high and low concentrations, the new chemical results in better weed control than the standard weed killer.

(B) For both chemicals, a smaller number of weeds typically grew at higher concentrations than at lower concentrations.

(C) The results for the standard weed killer are less variable than those for the new chemical.

(D) High and low concentrations of either chemical have approximately the same effects on weed growth.

(E) Some of the results for the low concentration of weed killer show a smaller number of weeds growing than some of the results for the high concentration.

Section II: Free Response *Show all your work. Indicate clearly the methods you use, because you will be graded on the correctness of your methods as well as on the accuracy and completeness of your results and explanations.*

T11 As part of an annual survey, high school students were asked to choose their favorite type of music from a list. The side-by-side bar graph compares the preferences for separate random samples of 250 students in consecutive years.

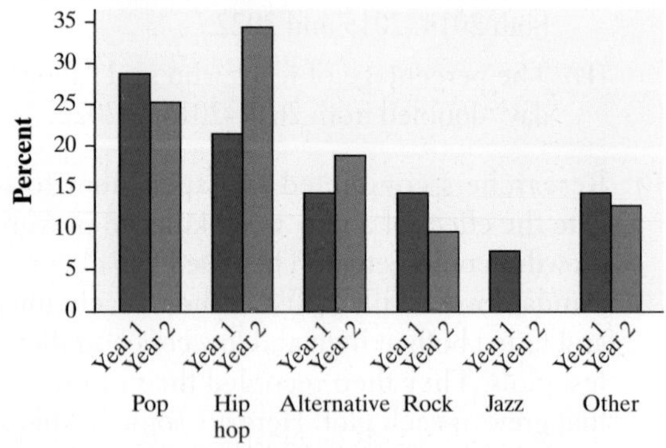

(a) In which year did more students choose either pop or hip hop as their favorite type of music? Explain your answer.

(b) Describe similarities and differences in the distributions of favorite type of music for the two years.

T12 A researcher is interested in how many contacts older adults have in their smartphones. Here are data on the number of contacts for a random sample of 30 adults older than age 65 with smartphones in a large city:

7	20	24	25	25	28	28	30	32	35
42	43	44	45	46	47	48	48	50	51
72	75	77	78	79	83	87	88	135	151

(a) Construct a histogram of these data.

(b) Determine whether this distribution has any outliers. Show your method clearly.

(c) Would it be better to use the mean and standard deviation or the median and *IQR* to describe the center and variability of this distribution? Why?

T13 Researchers suspect that athletes typically have a faster reaction time than non-athletes do. To test this theory, the researchers gave an online reflex test to separate random samples of 33 athletes and 30 non-athletes at one school. Here are parallel boxplots and summary statistics of the data on reaction times (in milliseconds) for the two groups of students:

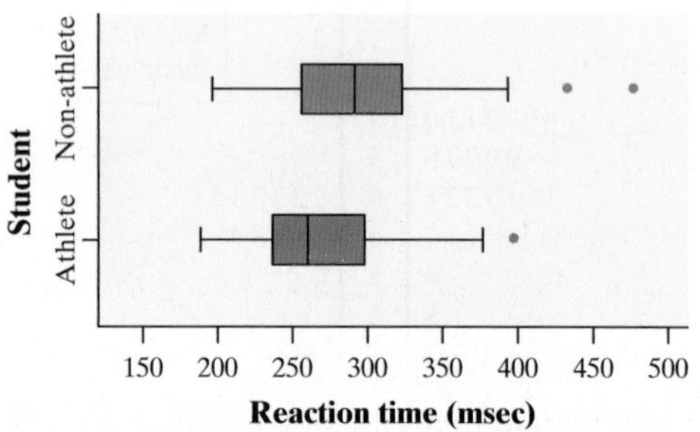

Student	n	Mean	StDev	Min	Q_1	Med	Q_3	Max
Non-athlete	30	297.3	65.9	197.0	255.0	292.0	325.0	478.0
Athlete	33	270.1	57.7	189.6	236.0	261.0	300.0	398.0

Compare the distribution of reaction time for athletes and non-athletes.

UNIT 1 Exploring One-Variable Data

PART II Modeling Distributions of Quantitative Data

Manuela Schewe-Behnisch/EyeEm/Getty Images

Introduction

Suppose Emilio earns 43 out of 50 points on a statistics test. Should he be satisfied or disappointed with this result? That depends on how Emilio's score compares with the scores of the other students who took the test. Section 1E focuses on describing the position of an individual data value in a distribution of quantitative data. We begin by discussing a familiar measure of position: *percentiles*. Next, we introduce a new type of graph that is useful for displaying percentiles. Then, we consider another way to describe an individual's position in a distribution that is based on the mean and standard deviation. Finally, we examine the effects of transforming data on the shape, center, and variability of a distribution of quantitative data.

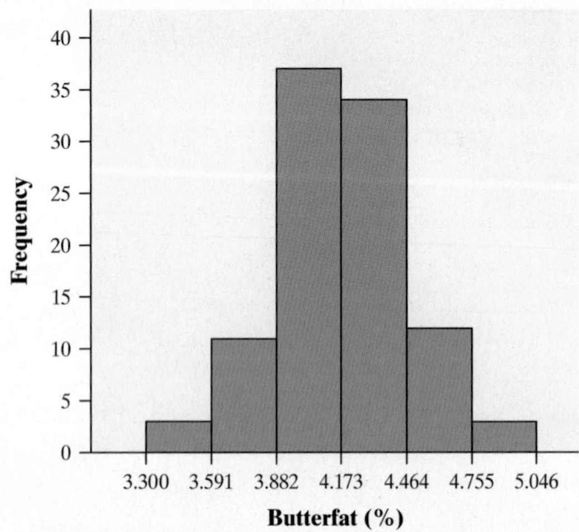

FIGURE 1.13 Histogram of the percentage of butterfat in milk for a random sample of 100 three-year-old Ayrshire cows. The distribution is roughly symmetric, single-peaked, and mound-shaped.

Figure 1.13 is a histogram of data on the percentage of butterfat in milk for a random sample of 100 three-year-old Ayrshire cows.[101] As this graph illustrates, repeated measurements of the same attribute for a large group of similar individuals often have distributions that are roughly symmetric, single-peaked, and mound-shaped. Section 1F shows you how to model quantitative data sets like this one with a *normal distribution*. You can use normal distributions to estimate the proportion or percentage of individuals with data values in a specified interval, as well as to describe an individual's position in the distribution.

SECTION 1E Describing Position and Transforming Data

LEARNING TARGETS *By the end of the section, you should be able to:*

- Calculate and interpret a percentile in a distribution of quantitative data.
- Calculate and interpret a standardized score (*z*-score) in a distribution of quantitative data.
- Use percentiles or standardized scores (*z*-scores) to compare the relative positions of individual values in distributions of quantitative data.
- Use a cumulative relative frequency graph to estimate percentiles and individual values in a distribution of quantitative data.
- Analyze the effect of adding, subtracting, multiplying by, or dividing by a constant on the shape, center, and variability of a distribution of quantitative data.

Here are the scores of all 25 students in Mr. Tabor's statistics class on a 50-point test:

35	18	37	38	42	41	25	37	36	32	12	**43**	31
29	32	48	44	45	38	40	45	38	38	40	22	

The score marked in red is Emilio's 43. How did Emilio perform on this test relative to the other students in the class?

Figure 1.14 displays a dotplot of the class's test scores, with Emilio's score marked in red. The distribution is skewed to the left and single-peaked with some possible low outliers. From the dotplot, we can see that Emilio's score is above the mean (balance point) of the distribution. We can also see that he did better on the test than most other students in the class.

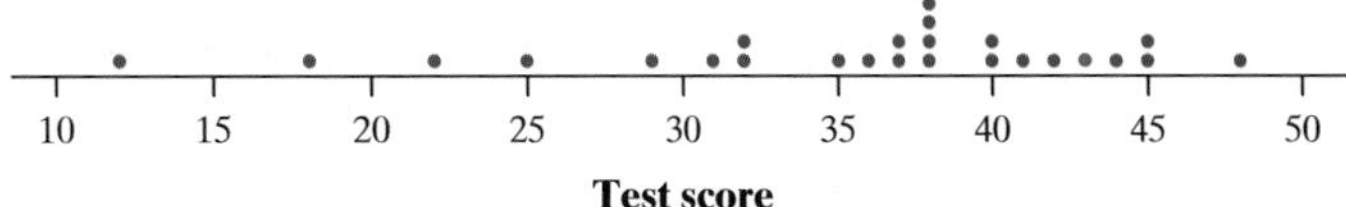

FIGURE 1.14 Dotplot of scores (out of 50 points) on Mr. Tabor's statistics test. Emilio's score of 43 is marked in red.

Measuring Position: Percentiles

One way to describe Emilio's position in the distribution of test scores is to calculate the **percentile** corresponding to a test score of 43. Recall that the three quartiles (Q_1, median, Q_3) divide a distribution of quantitative data into four groups of roughly equal size. The idea of a percentile is similar: the 99 percentiles divide a distribution into 100 roughly equal-size groups. This idea makes sense if a quantitative data set has a large number of values, but it isn't as helpful for smaller data sets.

DEFINITION Percentile

The *p*th **percentile** of a distribution is the value with *p*% of observations less than or equal to it.

Examining the dotplot, we see that four students in the class earned test scores greater than Emilio's 43. Because 21 of the 25 observations (84%) are less than or equal to 43, Emilio is at the 84th percentile in the class's test score distribution.

Here are a few important notes about percentiles:

- **Be careful with your language when describing percentiles.** Percentiles are specific locations in a distribution, so an observation isn't "in" the 84th percentile. Rather, it is "at" the 84th percentile.
- Our definition of percentile matches the one used in AP® Statistics. Some people define percentile as the percentage of values in a distribution that are *less than* an individual data value. Using this alternative definition, it is possible for an individual to fall at the 0th percentile. Of course, it is possible for an individual to be at the 100th percentile using our definition.
- Percentiles are usually reported as whole numbers. Consider a quantitative data set with 75 values. How should we report the percentile for the individual with 50 of the 75 values in the distribution less than or equal to their data value? Because $50/75 = 0.667$, we say that this individual is at the 66th percentile of the distribution. We can't say the individual is at the 67th percentile because only 66.7% of the values in the data set are less than or equal to this individual's data value.
- The median of a distribution is roughly the 50th percentile. The first quartile Q_1 is roughly the 25th percentile of a distribution because it separates the lowest 25% of values from the upper 75%. Likewise, the third quartile Q_3 is roughly the 75th percentile.

- **A high percentile is not always a good thing.** For example, a person whose cholesterol level is at the 99th percentile for their age group may need treatment for high cholesterol!

EXAMPLE

How much lead is in the water?
Measuring position: Percentiles

Skill 2.C

Jim West/Alamy Stock Photo

PROBLEM: Here once again is a dotplot of the lead levels (in parts per billion, ppb) in 71 water samples taken from randomly selected Flint, Michigan, dwellings after the city switched its water supply from Lake Huron to the Flint River.[102]

(a) Calculate the percentile for the water sample with a lead level of 9 ppb. Interpret this value.

(b) U.S. EPA regulations require action to be taken if the 90th percentile of lead level exceeds 15 ppb. Based on these data, should action have been taken?

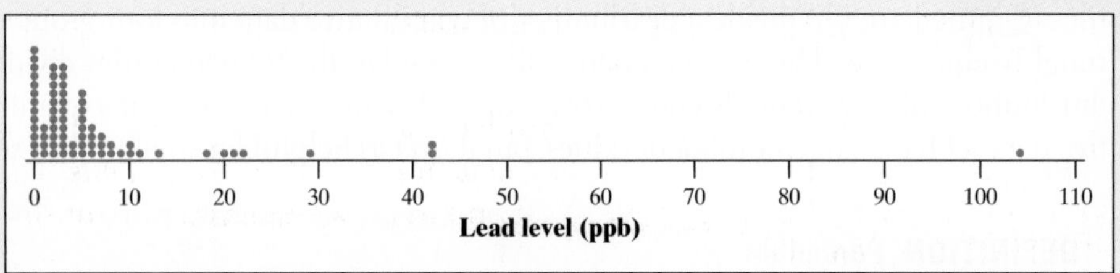

SOLUTION:

(a) 59/71 = 0.831, so this water sample is at the 83rd percentile in the distribution of lead level. About 83% of the Flint dwellings tested had water samples with lead levels less than or equal to 9 ppb.

It is easier to start by counting how many values are greater than 9 ppb!

(b) (0.90)(71) = 63.9. Because 64 of the values are less than or equal to 18, the 90th percentile is 18 ppb. This exceeds 15 ppb, so action should have been taken in Flint, Michigan, based on these data.

The 90th percentile is greater than or equal to 64 of the 71 lead levels.

FOR PRACTICE, TRY EXERCISE 1

City officials in Flint decided to omit two water samples from their analysis: one with 20 ppb of lead that came from a business, and one with 104 ppb of lead that came from a home using a water filter. With those two values excluded, you can check that only 6/69 = 0.087 = 8.7% of the remaining water samples had lead levels greater than 15 ppb. So authorities determined that no action was needed.

Flint residents worked with scientists from Virginia Tech University to retest the lead level of water in the city's dwellings. Their result: far more than 10% of the water samples had lead levels greater than 15 ppb. Authorities finally agreed that Flint's water contained dangerously high lead levels and switched back to buying water sourced from Lake Huron. By the end of 2016, the amount of lead in Flint's water was back to a safe level. However, several state and local officials have been charged with crimes for endangering the health of Flint residents.

Measuring Position: Standardized Scores (*z*-Scores)

A percentile is one way to describe an individual's position in a distribution of quantitative data. Another way is to give the **standardized score (*z*-score)** for the individual's data value.

DEFINITION Standardized score (*z*-score)

The **standardized score (*z*-score)** for an individual value in a distribution tells us how many standard deviations from the mean the value falls, and in what direction. To find the standardized score (*z*-score), compute

$$z = \frac{\text{value} - \text{mean}}{\text{standard deviation}}$$

Values larger than the mean have positive *z*-scores. Values smaller than the mean have negative *z*-scores. Notice that *standardized scores have no units.* That's because the units of measurement in the numerator and denominator of the *z*-score formula cancel each other.

Let's return to the data from Mr. Tabor's statistics test. Figure 1.15 shows a dotplot of the data, along with summary statistics. The relationship between the mean and the median is what you'd expect in this left-skewed distribution.

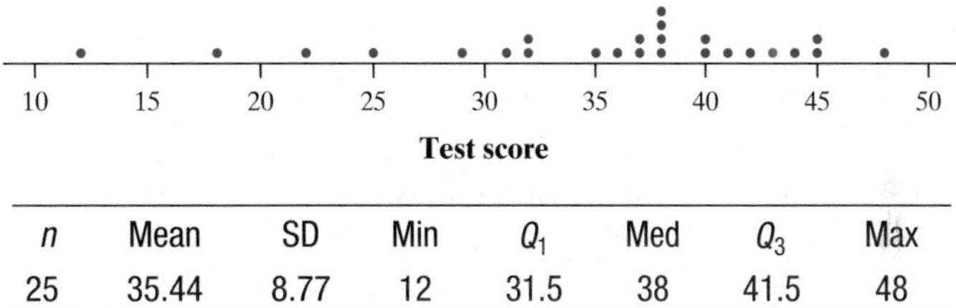

FIGURE 1.15 Dotplot and summary statistics of the scores on Mr. Tabor's statistics test. Emilio's score of 43 is marked in red on the dotplot.

n	Mean	SD	Min	Q_1	Med	Q_3	Max
25	35.44	8.77	12	31.5	38	41.5	48

Where does Emilio's 43 (marked in red on the dotplot) fall in the distribution? His standardized score (*z*-score) is

$$z = \frac{\text{value} - \text{mean}}{\text{standard deviation}} = \frac{43 - 35.44}{8.77} = 0.862$$

That is, Emilio's test score is 0.862 standard deviation above the class's mean score.

EXAMPLE

How much lead is in the water?
Measuring position: Standardized scores (*z*-scores)

Skill 3.A

Ryan Garza/ZUMA Press/Newscom

PROBLEM: Refer to the preceding example in this section about the Flint water crisis. Summary statistics for the data on lead levels (in ppb) in the water samples from 71 Flint dwellings are shown in the following table.

n	Mean	SD	Min	Q_1	Med	Q_3	Max
71	7.31	14.347	0	2	3	7	104

(a) Calculate and interpret the *z*-score for the water sample from LeeAnne Walters's home, which had a lead level of 104 ppb.
(b) The lead level in a different Flint home's water sample had a standardized score of −0.37. Find the lead level in that home's water sample.

SOLUTION:

(a) $z = \frac{104 - 7.31}{14.347} = 6.74$

$z = \frac{\text{value} - \text{mean}}{\text{SD}}$

The water sample from LeeAnne Walters's home had a lead level 6.74 standard deviations above the mean for all 71 dwellings tested. That's a pretty extreme outlier!

Be sure to interpret the magnitude (number of standard deviations) and direction (greater than or less than the mean) of a *z*-score in context.

(b) $-0.37 = \frac{\text{value} - 7.31}{14.347}$

$-0.37(14.347) + 7.31 = \text{value}$

$2.00 = \text{value}$

Be sure to show your work when finding the value that corresponds to a given *z*-score.

The home's water sample had a lead level of 2.00 ppb.

FOR PRACTICE, TRY EXERCISE 7

Can we use the standardized score (*z*-score) of an individual data value to find its corresponding percentile, or vice versa? The answer is "no" for most distributions. For roughly symmetric, single-peaked, mound-shaped distributions, the answer is "yes." We will discuss the connection between *z*-scores and percentiles for these *normal distributions* in Section 1F.

Comparing Relative Positions in Distributions of Quantitative Data

How can we compare the relative positions of individual data values in distributions of quantitative data? For instance, Shanice and Deiondre are both applying to nursing schools. Shanice takes the National League of Nursing (NLN) Pre-Admission Exam and earns a score of 111. Deiondre takes the Test of Essential Academic Skills (TEAS) Exam and earns a score of 88. Which person did better relative to the others taking their respective exams? We can't compare their scores directly (111 versus 88) because the two exams are scored on different scales. Shanice's 111 is at the 70th percentile of the NLN score distribution, while Deiondre's 88 is at the 80th percentile of the TEAS score distribution.[103] So Deiondre performed better relative to fellow exam-takers.

Percentiles are one option for comparing the relative positions of individuals in distributions of quantitative data. Standardized scores (*z*-scores) are another option if we know the mean and standard deviation of each distribution.

EXAMPLE

Growing like a beanstalk
Comparing relative positions in distributions of quantitative data

Skill 2.D

Jamie Grill/Tetra Images, LLC./Alamy Stock Photo

PROBLEM: Jordan (Mr. Tabor's daughter) was 55 inches tall at age 9. The distribution of height for 9-year-old girls has mean 52.5 inches and standard deviation 2.5 inches. Zayne (Mr. Starnes's grandson) was 58 inches tall at age 11. The distribution of height for 11-year-old boys has mean 56.5 inches and standard deviation 3.0 inches.[104] Who is taller relative to other children of their sex and age, Jordan or Zayne? Justify your answer.

SOLUTION:

Jordan: $\frac{55-52.5}{2.5}=1.0$ Zayne: $\frac{58-56.5}{3.0}=0.5$

> The standardized heights tell us where each child stands (pun intended!) in the distribution of height for their age group.

Jordan is 1 standard deviation above the mean height of 9-year-old girls, while Zayne is one-half standard deviation above the mean height of 11-year-old boys. So Jordan is taller relative to girls her age than Zayne is relative to boys his age.

FOR PRACTICE, TRY EXERCISE 11

CHECK YOUR UNDERSTANDING

The U.S. House of Representatives has 435 voting members. The number of representatives that each state has is based on its population when the national census is taken every 10 years. A dotplot of the number of representatives from each of the 50 states in 2024 is shown, along with summary statistics. The red point on the graph is for the state of Ohio, which has 15 representatives.

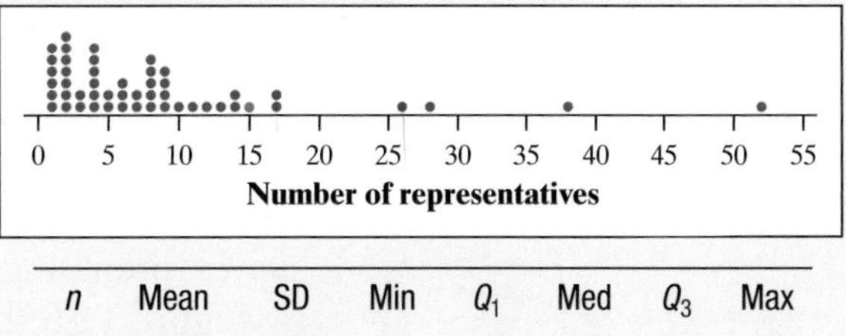

n	Mean	SD	Min	Q_1	Med	Q_3	Max
50	8.7	9.69	1	2	6	10	52

1. Calculate and interpret the percentile for Ohio.
2. South Carolina is at the 52nd percentile of the distribution. How many representatives did South Carolina have in 2024?
3. Calculate the standardized score (z-score) for Ohio. Interpret this value.

How many counties does each state have? The table displays summary statistics for these data.

n	Mean	SD	Min	Q_1	Med	Q_3	Max
50	62.82	46.421	3	24	63	88	254

4. Ohio has 88 counties. In which distribution—number of representatives or number of counties—is Ohio's relative position more unusual? Justify your answer.

Cumulative Relative Frequency Graphs

The following table summarizes data on the age at which each of the first 46 U.S. presidents first took office. The first column, Age, uses intervals with a width of 5 years starting at age 40. The table includes columns for frequency, relative frequency, *cumulative frequency*, and *cumulative relative frequency*. Note that age is a *continuous* variable. For instance, Theodore Roosevelt was the youngest to take office at 42 years, 322 days (about 42.882 years) and Joe Biden was the oldest to take office at 78 years, 61 days (about 78.167 years).

Age	Frequency	Relative frequency	Cumulative frequency	Cumulative relative frequency
40 to <45	2	2/46 = 0.0435 = 4.35%	2	2/46 = 0.0435 = 4.35%
45 to <50	6	6/46 = 0.1304 = 13.04%	8	8/46 = 0.1739 = 17.39%
50 to <55	10	10/46 = 0.2174 = 21.74%	18	18/46 = 0.3913 = 39.13%
55 to <60	15	15/46 = 0.3261 = 32.61%	33	33/46 = 0.7174 = 71.74%
60 to <65	8	8/46 = 0.1739 = 17.39%	41	41/46 = 0.8913 = 89.13%
65 to <70	3	3/46 = 0.0652 = 6.52%	44	44/46 = 0.9565 = 95.65%
70 to <75	1	1/46 = 0.0217 = 2.17%	45	45/46 = 0.9783 = 97.83%
75 to <80	1	1/46 = 0.0217 = 2.17%	46	46/46 = 1.00 = 100%

The 45 to <50 row of the table reveals that 8/46 = 0.1739 = 17.39% of U.S. presidents first took office by the time they turned 50. In other words, the 17th percentile of the distribution of inauguration age is 50.000 years. We can display the corresponding percentiles from the table in a **cumulative relative frequency graph.**

DEFINITION Cumulative relative frequency graph

A **cumulative relative frequency graph** plots a point corresponding to the percentile of a given value in a distribution of quantitative data. Consecutive points are then connected with a line segment to form the graph.

Some people refer to cumulative relative frequency graphs as *ogives* (pronounced "o-jives") or as *percentile plots*.

Figure 1.16 shows a cumulative relative frequency graph for the ages of the first 46 U.S. presidents when they first took office. Notice the following details about the graph:

- The leftmost point is plotted at a height of 0% at age = 40. This point tells us that none of the first 46 U.S. presidents took office by the time they turned 40.
- The next point to the right is plotted at a height of 4.35% at age = 45. This point tells us that 4.35% of these presidents took office by their 45th birthday.
- The graph grows very gradually at first because few presidents took office when they were in their 40s. Then the graph gets

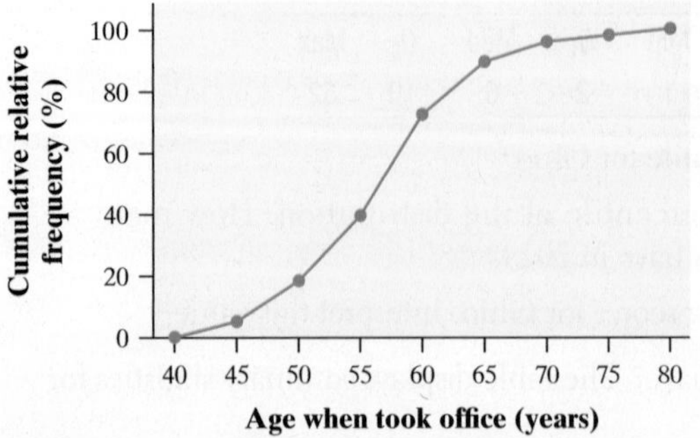

FIGURE 1.16 Cumulative relative frequency graph of the ages of U.S. presidents when they first took office.

very steep beginning at age 50 because most U.S. presidents were in their 50s when they took office. The rapid growth in the graph slows at age 60.

- The rightmost point on the graph is plotted directly above age = 80 and has cumulative relative frequency 100%. That's because 100% of these U.S. presidents took office by age 80.

A cumulative relative frequency graph can be used to estimate the position of an individual value in a distribution or to estimate a specified percentile of the distribution.

EXAMPLE

Ages of U.S. presidents
Cumulative relative frequency graphs

Skill 2.B

Jae C. Hong/AP Photo

PROBLEM: Use the graph in Figure 1.16 to help you answer each question.

(a) Was Barack Obama, who first took office at age 47 years, 169 days (about 47.463 years) unusually young? Explain your answer.

(b) Estimate the 65th percentile of the distribution.

SOLUTION:

(a)

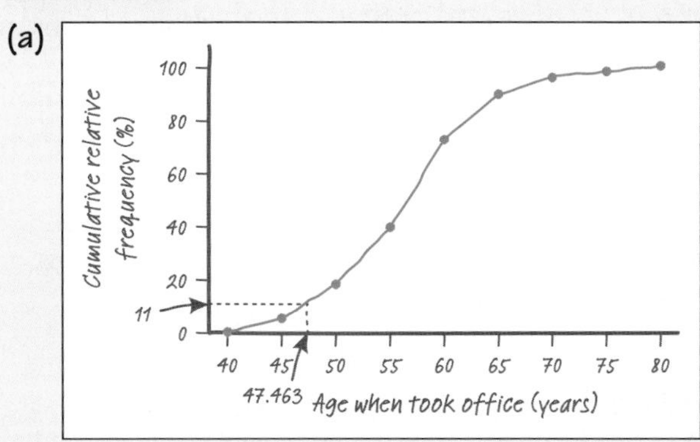

To find President Obama's percentile in the distribution, draw a vertical line up from his age (47.463) on the horizontal axis until it meets the graph. Then draw a horizontal line from this point to the vertical axis.

Barack Obama's age when he first took office places him at about the 11th percentile of the distribution. About 11% of the first 46 U.S. presidents took office by the time they were 47.463 years old. So Obama was fairly young, but not unusually young, when he took office.

(b)

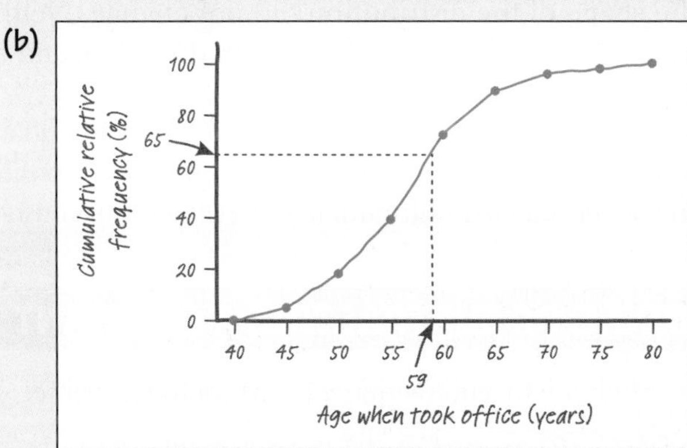

The 65th percentile of the distribution is the age with a cumulative relative frequency (percentile) of 65%. To find this value, draw a horizontal line across from the vertical axis at a height of 65% until it meets the graph. Then draw a vertical line from this point down to the horizontal axis.

The 65th percentile of the distribution is about 59 years old.

FOR PRACTICE, TRY EXERCISE 15

Transforming Data

It is sometimes useful to *transform data* when analyzing the distribution of a quantitative variable. We may want to change the units of measurement for a data set from kilograms to pounds (1 kg $\approx$ 2.2 lb), or from degrees Fahrenheit to degrees Celsius $\left[°\text{C} = \frac{5}{9}(°\text{F} - 32)\right]$. Or perhaps a measuring device is calibrated wrong, so we have to add a constant to each data value to get accurate measurements. What effect do these kinds of transformations—adding or subtracting; multiplying or dividing—have on the shape, center, and variability of a distribution of quantitative data?

EFFECT OF ADDING OR SUBTRACTING A CONSTANT

Recall that Mr. Tabor gave his class of 25 statistics students a test worth 50 points. Here is a dotplot of the students' scores along with summary statistics:

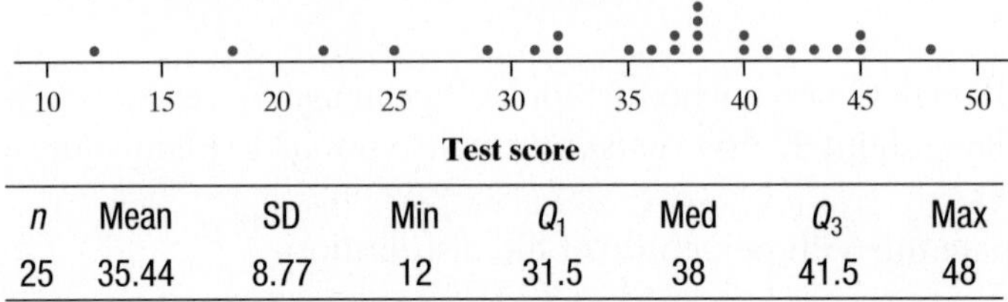

n	Mean	SD	Min	Q_1	Med	Q_3	Max
25	35.44	8.77	12	31.5	38	41.5	48

Suppose Mr. Tabor was nice and added 5 points to each student's test score. How would this affect the distribution of scores? Figure 1.17 shows graphs and numerical summaries for the original test scores and adjusted scores.

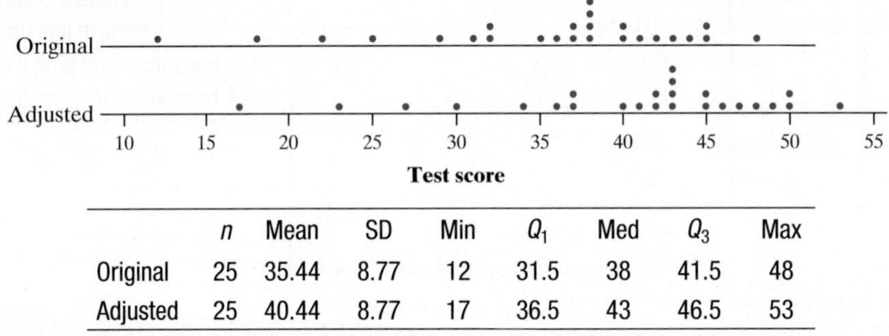

	n	Mean	SD	Min	Q_1	Med	Q_3	Max
Original	25	35.44	8.77	12	31.5	38	41.5	48
Adjusted	25	40.44	8.77	17	36.5	43	46.5	53

FIGURE 1.17 Dotplots and summary statistics for the original scores and adjusted scores (with 5 points added) on Mr. Tabor's statistics test.

From both the graph and summary statistics, we can see that the measures of center (mean and median) increased by 5 points, along with the minimum, Q_1, Q_3, and the maximum. The shape of the distribution did not change. Neither did the variability of the distribution—the range, the standard deviation, and the interquartile range (*IQR*) all stayed the same.

As this example shows, adding the same positive number to each value in a quantitative data set shifts the distribution to the right by that number. Subtracting a positive constant from each data value would shift the distribution to the left by that constant.

THE EFFECT OF ADDING OR SUBTRACTING A CONSTANT

Adding the same positive number a to (subtracting a from) each data value:

- Adds a to (subtracts a from) measures of center (mean, median).
- Does not change measures of variability (range, *IQR*, standard deviation).
- Does not change the shape of the distribution.

Does adding or subtracting a constant affect an individual's position in a distribution? Consider Emilio, who earned a score of 43 on Mr. Tabor's statistics test. Emilio is at the 84th percentile of the original test score distribution because 21 of the 25 observations (84%) are less than or equal to 43. After Mr. Tabor adds 5 points to each student's score, Emilio's 48 is still in the same position relative to other students in the adjusted test score distribution: at the 84th percentile. What about Emilio's standardized scores (z-scores) in the two distributions?

$$\text{Original score: } z = \frac{43 - 35.44}{8.77} = 0.862 \quad \text{Adjusted score: } z = \frac{48 - 40.44}{8.77} = 0.862$$

As you can see, transforming a quantitative data set by adding or subtracting a constant does not affect an individual's percentile or standardized score (z-score).

How about outliers? You can check that the minimum score of 12 is an outlier in the original test score distribution by using the $1.5 \times IQR$ rule (or the $2 \times SD$ rule). This same individual will still be an outlier in the adjusted test score distribution with a score of 17.

EFFECT OF MULTIPLYING OR DIVIDING BY A CONSTANT

Suppose that Mr. Tabor wants to convert his students' adjusted test scores to percentages. Because the test was worth 50 points, he multiplies each student's adjusted test score by 2 to get their score out of 100 points. (Note that, with the 5-point adjustment prior to doubling, one student ended up with a final score of 106!) Figure 1.18 shows graphs and numerical summaries of the adjusted scores and the doubled scores:

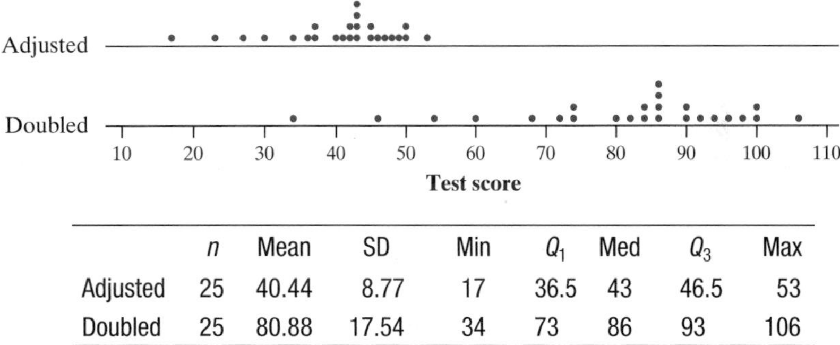

	n	Mean	SD	Min	Q_1	Med	Q_3	Max
Adjusted	25	40.44	8.77	17	36.5	43	46.5	53
Doubled	25	80.88	17.54	34	73	86	93	106

FIGURE 1.18 Dotplots and summary statistics for the adjusted scores (out of 50) and doubled scores (out of 100) on Mr. Tabor's statistics test.

From the graphs and summary statistics, we can see that the measures of center and variability have all doubled, just like the individual data values. So have the minimum, Q_1, Q_3, and the maximum. But the shape of the two distributions is the same. Multiplying or dividing each value in a quantitative data set by a positive constant stretches (or compresses) the distribution by that factor.

EFFECT OF MULTIPLYING OR DIVIDING BY A CONSTANT

Multiplying (or dividing) each data value by the same positive number b:

- Multiplies (divides) measures of center (mean, median) by b.
- Multiplies (divides) measures of variability (range, IQR, standard deviation) by b.
- Does not change the shape of the distribution.

It is not common to multiply (or divide) each data value by a *negative* number b. Doing so would multiply (or divide) the measures of center by b, but would multiply the measures of variability by the *absolute value of* b. We can't have a negative amount of variability! Multiplying or dividing by a negative number would also affect the shape of the distribution, as all values would be reflected over the y axis.

Note that we didn't include variance in the measures of variability in the summary box. Recall that the variance of a distribution is equal to the square of the standard deviation. Therefore, if you multiply each data value by b, the variance is multiplied by b^2.

PUTTING IT ALL TOGETHER: ADDING/SUBTRACTING AND MULTIPLYING/DIVIDING

What happens if we transform a quantitative data set by both adding or subtracting a constant and multiplying or dividing by a constant? We just use the facts about transforming data that we've already established and the order of operations.

EXAMPLE

Too cool at the cabin? Transforming data

Skill 2.C

gulfix/Getty Images

PROBLEM: During the winter months, temperatures at the Starnes family Colorado cabin can stay well below freezing (32°F, or 0°C) for weeks at a time. To prevent the pipes from freezing, Mrs. Starnes sets the thermostat at 50°F. She also buys a digital thermometer that records the indoor temperature each night at midnight. Unfortunately, the thermometer is programmed to measure the temperature in degrees Celsius. Here are a dotplot and numerical summaries of the midnight temperature readings for a 30-day period:

0 2 4 6 8 10 12 14 16

Indoor temperature reading (°C)

n	Mean	SD	Min	Q_1	Med	Q_3	Max
30	8.43	2.27	3.00	7.00	8.50	10.00	14.00

Suppose that Mrs. Starnes converts all of the temperature readings from degrees Celsius to degrees Fahrenheit by using the formula °F = (9/5)°C + 32.

(a) What shape would the distribution of indoor temperature reading in degrees Fahrenheit have?

(b) Find the mean indoor temperature reading in degrees Fahrenheit. Does the thermostat setting seem accurate?

(c) Calculate the standard deviation of the indoor temperature readings in degrees Fahrenheit.

SOLUTION:

(a) The same shape as the distribution of indoor temperature reading in degrees Celsius: roughly symmetric and single-peaked.

Neither multiplying/dividing by a (positive) constant nor adding/subtracting a constant changes the shape of a distribution of quantitative data.

(b) Mean = (9/5)(8.43) + 32 = 47.17°F. The thermostat doesn't seem to be very accurate. It is set at 50°F, but the mean indoor temperature reading over the 30-day period is about 47°F.

(c) SD = (9/5) (2.27) = 4.09°F

> Multiplying each observation by 9/5 multiplies the standard deviation by 9/5. However, adding 32 to each observation doesn't affect the variability.

FOR PRACTICE, TRY EXERCISE 19

Many other types of transformations—such as powers and logarithms—can be very useful when analyzing quantitative data. We have only studied what happens when you transform data by adding, subtracting, multiplying, or dividing by a constant.

CONNECTING TRANSFORMATIONS AND *z*-SCORES

What happens if we standardize *all* the values in a distribution of quantitative data? Here is a dotplot of the original test scores for the 25 students in Mr. Tabor's statistics class, along with summary statistics:

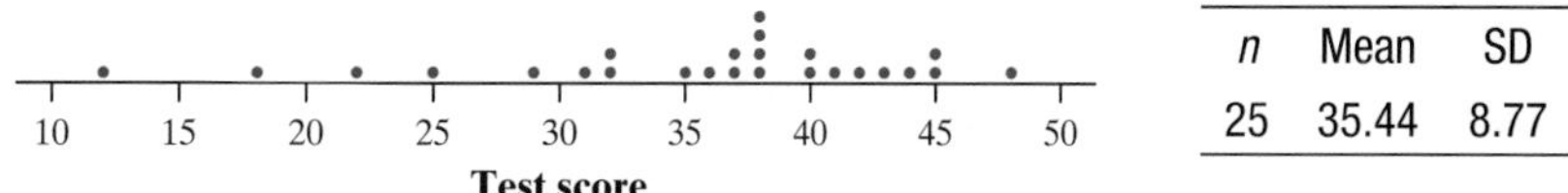

n	Mean	SD
25	35.44	8.77

We calculate the standardized score for each student using

$$z = \frac{\text{score} - 35.44}{8.77}$$

In other words, we subtract 35.44 from each student's test score and then divide by 8.77. What effect do these transformations have on the shape, center, and variability of the distribution?

Here is a dotplot of the class's *z*-scores, along with summary statistics. Let's describe the distribution.

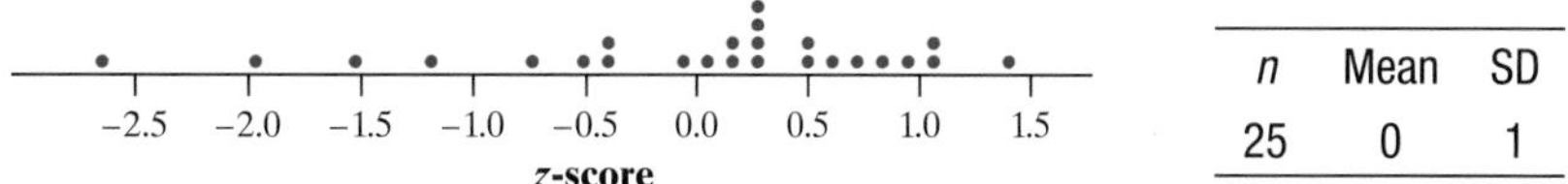

n	Mean	SD
25	0	1

- **Shape:** *The shape of the distribution of z-scores is the same as the shape of the original distribution of test scores*—skewed to the left. Neither subtracting a constant nor dividing by a constant changes the shape of the graph.
- **Center:** *The mean of the distribution of z-scores is 0.* Subtracting 35.44 from each test score would reduce the mean from 35.44 to 0. Dividing each of these new data values by 8.77 would divide the new mean of 0 by 8.77, which still yields a mean of 0.
- **Variability:** *The standard deviation of the distribution of z-scores is 1.* Subtracting 35.44 from each test score does not affect the standard deviation. However, dividing all of the resulting values by 8.77 would divide the original standard deviation of 8.77 by 8.77, yielding 1.

We would get the same results no matter what the original distribution looks like: the distribution of *z*-scores always has the same shape as the original distribution, a mean of 0, and a standard deviation of 1.

CHECK YOUR UNDERSTANDING

Questions 1 and 2 refer to the following setting. The cumulative relative frequency graph summarizes data on the length of phone calls made from the office at Gabalot High School last month.

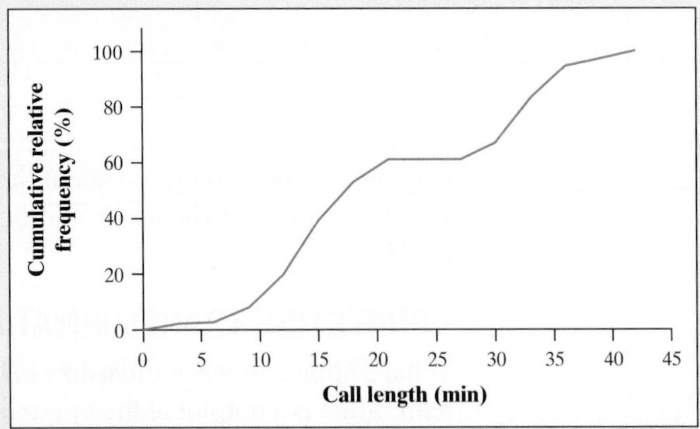

1. About what percentage of calls lasted less than or equal to 30 minutes? More than 30 minutes?
2. Estimate Q_1, Q_3, and the *IQR* of the distribution of phone call length.

Questions 3–5 refer to the following setting. Knoebels Amusement Park in Elysburg, Pennsylvania, has earned acclaim for being an affordable, family-friendly entertainment venue. Knoebels does not charge for general admission or parking, but it does charge customers for each ride they take. How much do the rides cost at Knoebels? The figure shows a dotplot of the cost for each of 22 rides in a recent year, along with summary statistics.[105]

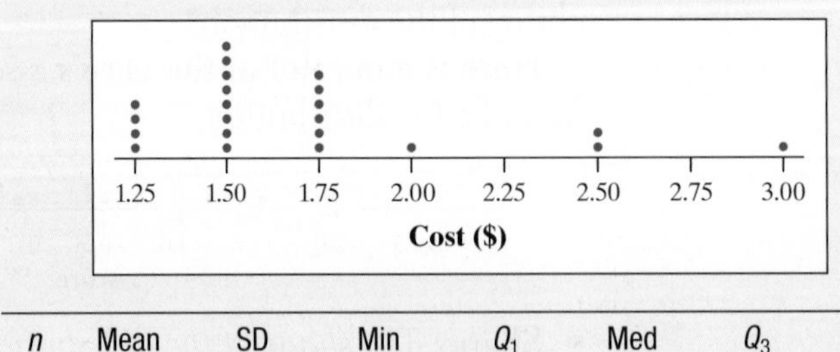

n	Mean	SD	Min	Q_1	Med	Q_3	Max
22	1.71	0.45	1.25	1.50	1.50	1.75	3.00

3. Suppose you convert the cost of the rides from dollars to cents ($1 = 100 cents). Describe the shape, mean, and standard deviation of the distribution of ride cost in cents.
4. Managers at Knoebels Amusement Park decide to increase the cost of each ride by 25 cents. How would the shape, mean, and standard deviation of the distribution in Question 3 be affected? Explain your answers.
5. Now suppose you convert the increased costs from Question 4 to *z*-scores. What would be the shape, mean, and standard deviation of this distribution?

SECTION 1E Summary

- Two ways of describing an individual value's position in a distribution of quantitative data are **percentiles** and **standardized scores (*z*-scores).**
  - An individual's percentile is the percentage of values in a distribution that are less than or equal to the individual's data value.
  - To standardize any data value, subtract the mean of the distribution and then divide the difference by the standard deviation. The resulting standardized score (*z*-score)

$$z = \frac{\text{value} - \text{mean}}{\text{standard deviation}}$$

    measures how many standard deviations the data value lies above or below the mean of the distribution.
- We can use percentiles and *z*-scores to compare the relative positions of individuals in one or more distributions of quantitative data.
- A **cumulative relative frequency graph** allows you to estimate the percentile for a specific value and the value corresponding to a given percentile in a distribution of quantitative data.
- It is necessary to **transform data** when changing units of measurement.
  - When you add a positive constant *a* to (subtract *a* from) all the values in a quantitative data set, measures of center increase (decrease) by *a*. Measures of variability do not change, nor does the shape of the distribution.
  - When you multiply (divide) all the values in a quantitative data set by a positive constant *b*, measures of center and variability are multiplied (divided) by *b*. However, the shape of the distribution does not change.
  - A common transformation is to standardize all the values in a distribution: for each value, subtract the mean of the distribution and then divide the difference by the standard deviation to get its *z*-score. The distribution of standardized scores (*z*-scores) has the same shape as the original distribution, a mean of 0, and a standard deviation of 1.

AP® EXAM TIP

AP® Daily Videos

Review the content of this section and get extra help by watching the AP® Daily Video for Topic 1.10: Video 1, which is available in AP® Classroom.

SECTION 1E Exercises

Measuring Position: Percentiles

1. **Shoes** How many pairs of shoes does a typical high school student have? To find out, researchers asked a random sample of 20 students from a large high school how many pairs of shoes they had. Here is a dotplot of the data:

pg 90

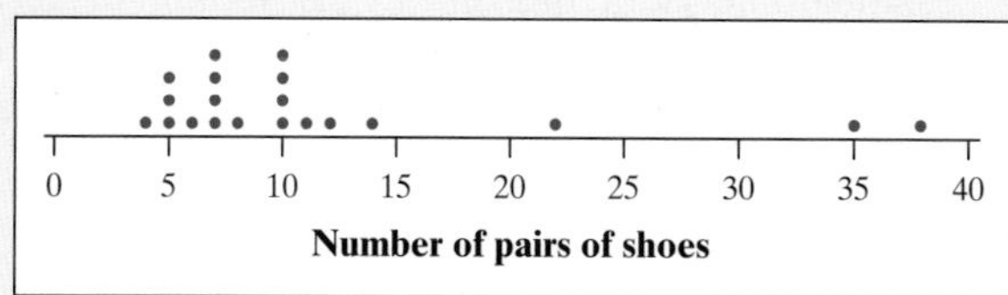

(a) Calculate the percentile for Jackie, who reported having 22 pairs of shoes. Interpret this value.

(b) Raul's reported number of pairs of shoes is at the 45th percentile of the distribution. How many pairs of shoes does Raul have?

2. **Seniors in the states** Here is a stemplot of the percent of residents aged 65 and older in the 50 states:[106]

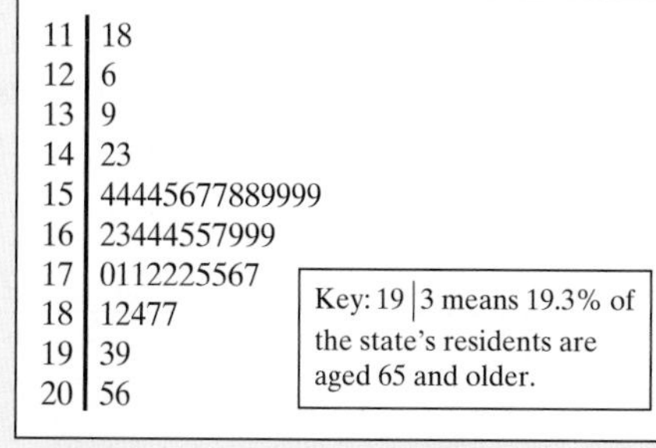

11 | 18
12 | 6
13 | 9
14 | 23
15 | 44445677889999
16 | 23444557999
17 | 0112225567
18 | 12477
19 | 39
20 | 56
Key: 19 | 3 means 19.3% of the state's residents are aged 65 and older.

(a) Calculate the percentile for California, where 14.3% of the residents are aged 65 and older. Interpret this value.

(b) Pennsylvania is at the 86th percentile of the distribution. What percentage of Pennsylvania's residents are aged 65 and older?

3. **Wear your helmet!** Many athletes (and their parents) worry about the risk of concussions when playing sports. A football coach plans to obtain specially made helmets for the players that are designed to reduce the chance of getting a concussion. Here are the measurements of head circumference (in inches) for the players on the team:[107]

23.0	22.2	21.7	22.0	22.3	22.6	22.7	21.5
22.7	25.6	20.8	23.0	24.2	23.5	20.8	24.0
22.7	22.6	23.9	22.5	23.1	21.9	21.0	22.4
23.5	22.5	23.9	23.4	21.6	23.3		

(a) Antawn, the team's starting quarterback, has a head circumference of 22.3 inches. What is Antawn's percentile?

(b) Find the head circumference of the player at the 90th percentile of the distribution.

4. **Unlocked phones** The "sold" listings on a popular auction website included 20 sales of used "unlocked" phones of one popular model. Here are the sales prices (in dollars):

450	415	495	300	325	430	370	400	325	400
235	330	304	415	355	405	449	355	425	299

(a) Find the percentile of the phone that sold for $330.

(b) What was the sales price of the phone at the 75th percentile of the distribution?

5. **Setting speed limits** According to the *Los Angeles Times*, speed limits on California highways are set at the 85th percentile of vehicle speeds on those stretches of road. Explain to someone who knows little statistics what that means.

6. **Blood pressure** Dinesh told a friend, "My doctor says my blood pressure is at the 90th percentile. That means I'm better off than about 90% of people like me." How should Dinesh's friend, who has taken statistics, respond to this statement?

Measuring Position: Standardized Scores (z-Scores)

7. **Foreign-born residents** Here are summary statistics for the percentage of foreign-born residents in each of the 50 U.S. states in a recent year:[108]

pg 91

n	Mean	SD	Min	Q_1	Med	Q_3	Max
50	9.494	6.214	1.6	4.8	7.3	13.7	26.7

(a) Calculate and interpret the z-score for Montana, which had 2.3% foreign-born residents.

(b) New York had a standardized score of 2.08. Find the percentage of foreign-born residents in New York.

8. **Household incomes** How do household incomes compare in different states? Here are summary statistics for the median household income in the 50 states in a recent year:[109]

n	Mean	SD	Min	Q_1	Med	Q_3	Max
50	67,680.70	11,469.23	44,836	60,000	67,122.50	77,320	93,236

(a) Calculate and interpret the z-score for North Carolina, with a median household income of $60,000.

(b) New Jersey had a standardized score of 1.52. Find New Jersey's median household income.

9. **Play ball!** Researchers recorded data on the number of wins for each of the 30 Major League Baseball (MLB) teams in the 2019 season. The mean number of wins was 81.[110]

(a) The Washington Nationals won 93 games (and the World Series!) in 2019. The team's standardized score is 0.75. Interpret this value.

(b) Find the standard deviation of the number of wins during the 2019 season.

10. **Stocks** The Dow Jones Industrial Average (DJIA) is a commonly used index of the overall strength of the U.S. stock market. Researchers recorded data on the daily change in the DJIA for each day that the market was open in 2019. The mean daily change was 20.94 points.

(a) The change in the DJIA on May 7, 2019, was −473.39 points. The standardized score for that day is −2.39. Interpret this value.

(b) Find the standard deviation of the daily change in the DJIA during 2019.

Comparing Relative Positions in Distributions of Quantitative Data

11. **SAT versus ACT** Some students who are applying for college take both the SAT and the ACT. One such student, Alejandra, scored 1280 on the SAT and 27 on the ACT. In the year when Alejandra took these tests, the distribution of SAT scores had a mean of 1059 and a standard deviation of 210. The distribution of ACT scores had a mean of 20.7 and a standard deviation of 5.9.[111] Which of Alejandra's two test scores was better, relatively speaking? Justify your answer.

pg 93

12. **Biles by miles?** Simone Biles won the gold medal in women's artistic gymnastics at the 2019 World Championships. Her overall score in the all-around

competition was 58.999. More than 40 years earlier, Romanian gymnast Nadia Comaneci took the world by storm with the first perfect 10. With an overall score of 79.275, Comaneci also won the all-around gold medal.[112] Because the scoring systems have changed, these two scores aren't directly comparable. In 2019, the 23 gymnasts who completed the all-around competition had a mean score of 54.719 points and a standard deviation of 1.800 points. In 1976, the top 24 gymnasts in the all-around competition had a mean score of 76.527 points and a standard deviation of 1.327 points. Which gymnast, Biles or Comaneci, had a better performance, relatively speaking? Justify your answer.

13. **Big or little?** Mrs. Munson wants to know how her son's height and weight compare with those of other boys his age. She uses an online calculator to determine that her son is at the 48th percentile for weight and the 76th percentile for height. Explain to Mrs. Munson what these values mean.

14. **Track star** Petra is a star thrower on the track team. At the final regular-season meet, Petra throws the discus 150 feet, 2 inches. That distance falls at the 80th percentile of Petra's distribution of discus throws during the season. At the league championship meet, Petra throws the discus an identical 150 feet, 2 inches. This result is at the 50th percentile of discus throws recorded by competitors at the meet. Explain how Petra's performances compare.

Cumulative Relative Frequency Graphs

15. **Run fast!** As part of a student project, high school students were asked to sprint 50 yards, and their times (in seconds) were recorded. A cumulative relative frequency graph of the sprint times is shown here.

pg 95

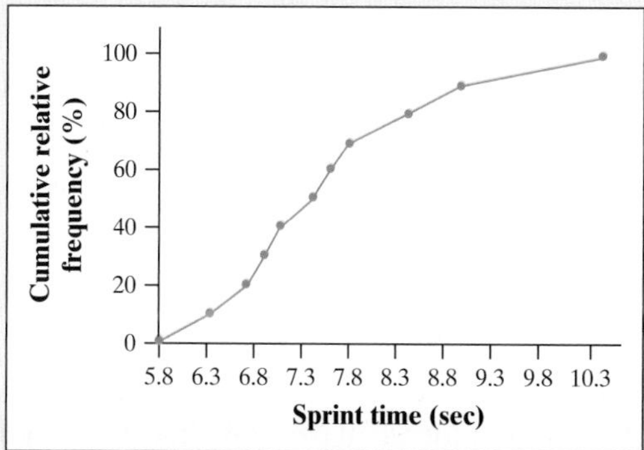

(a) One student ran the 50 yards in 8.05 seconds. Is a sprint time of 8.05 seconds unusually slow? Explain your answer.

(b) Estimate the 20th percentile of the distribution.

16. **Cumulative household incomes** The cumulative relative frequency graph describes the distribution of median household incomes in the 50 states in a recent year.[113]

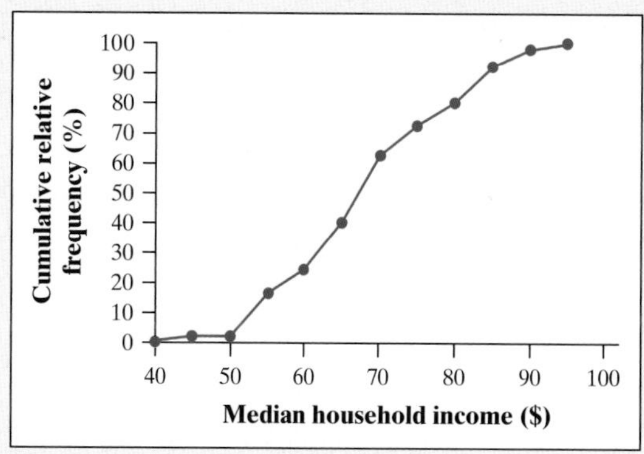

(a) The median household income in Illinois that year was $73,653. Is Illinois an unusually wealthy state? Explain your answer.

(b) Estimate the 35th percentile of the distribution.

17. **Light life** The cumulative relative frequency graph describes the lifetimes (in hours) of 200 light bulbs.[114]

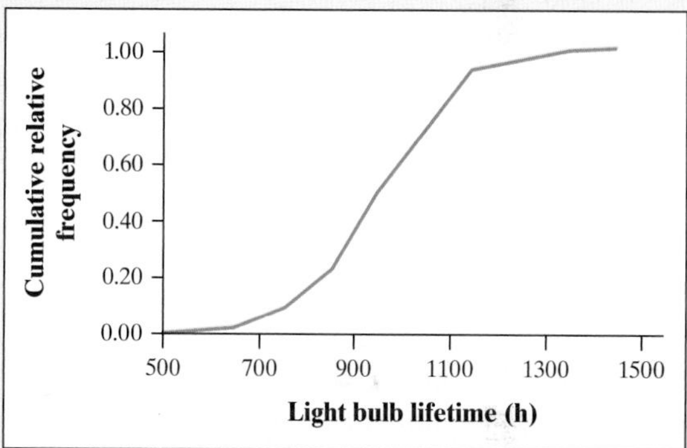

(a) Estimate the interquartile range (*IQR*) of this distribution. Show your method.

(b) Explain why the graph is less steep between 650 and 850 than it is between 850 and 1150.

18. **Shopping spree** The figure is a cumulative relative frequency graph of the amount spent by 50 consecutive grocery shoppers at a store.

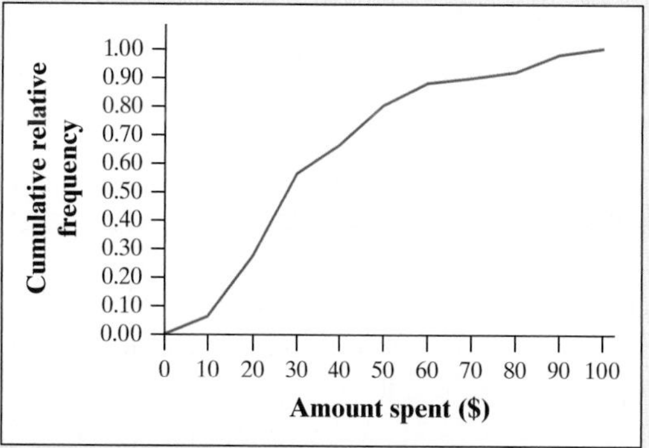

(a) Estimate the interquartile range (*IQR*) of this distribution. Show your method.

(b) Explain why the graph is steepest between $10 and $30.

Transforming Data

19. **Long jump** There were 40 athletes competing in the long jump at a major track meet. The meet official recorded the distance, to the nearest centimeter, of each athlete's best jump. Here are a dotplot and some numerical summaries of the data:

pg 98

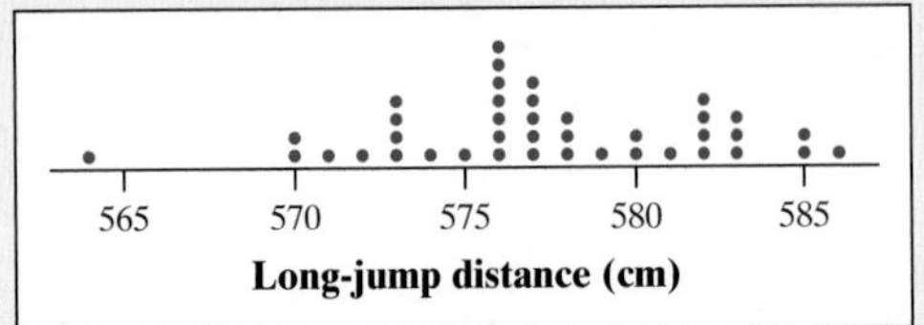

n	Mean	SD	Min	Q_1	Med	Q_3	Max
40	577.3	4.713	564	574.5	577	581.5	586

The meet official realized that they measured all the jumps from the back of the board instead of the front. Thus, they had to subtract 20 centimeters from each jump to get the correct measurement.

(a) What shape would the distribution of corrected long-jump distance have?

(b) Find the median of the distribution of corrected long-jump distance.

(c) Calculate the interquartile range (*IQR*) of the distribution of corrected long-jump distance.

20. **Step right up!** A dotplot of the distribution of height (in inches) for students in Mrs. Nataro's class is shown, along with some numerical summaries of the data.

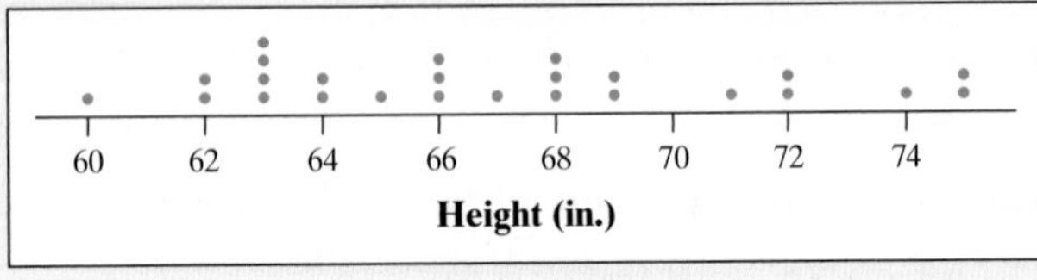

n	Mean	SD	Min	Q_1	Med	Q_3	Max
25	67	4.29	60	63	66	70	75

Suppose that Mrs. Nataro has the entire class stand on a 6-inch-high platform for a photo and then asks the students to measure the distance from the top of their heads to the ground.

(a) What shape would this distribution of adjusted height have?

(b) Find the median of the distribution of adjusted height.

(c) Calculate the interquartile range (*IQR*) of the distribution of adjusted height.

21. **Teacher raises** A school system employs teachers at salaries between $38,000 and $70,000. The teachers' union and school board are negotiating the form of next year's increase in the salary schedule. Suppose that every teacher is given a 5% raise. What effect will this raise have on each of the following characteristics of the resulting distribution of salary?

(a) Shape

(b) Center

(c) Variability

22. **Used cars, cheap!** A used-car dealership has 28 cars in inventory, with prices ranging from $11,500 to $25,000. For a Labor Day sale, the dealer reduces the price of each car by 10%. What effect will this reduction have on each of the following characteristics of the resulting distribution of price?

(a) Shape

(b) Center

(c) Variability

23. **Cool pool?** Coach Ferguson uses a thermometer to measure the temperature (in degrees Fahrenheit) at 20 different locations in the school swimming pool. An analysis of the data yields a mean of 77°F and a standard deviation of 3°F. The coach converts the temperature measurements to degrees Celsius using the formula $°C = \frac{5}{9}°F - \frac{160}{9}$.

(a) Find the mean temperature reading in degrees Celsius.

(b) Calculate the variance of the temperature readings in degrees Celsius.

24. **Acid rain?** Rainwater was collected in water receptacles at 30 different sites near an industrial complex, and the amount of acidity (pH level) was measured. The mean and standard deviation of the values are 4.60 and 2.10, respectively. When the pH meter was recalibrated back at the laboratory, it was found to be in error. The error can be corrected by adding 0.1 pH unit to all of the values and then multiplying the result by 1.2.

(a) Find the mean of the corrected distribution of acidity (pH).

(b) Calculate the variance of the corrected distribution of acidity (pH).

25. **Taxi!** In 2022, taxicabs in Los Angeles charged an initial fee of $2.85 plus $2.70 per mile. In equation form, fare = 2.85 + 2.70(miles). At the end of a month, a frequent LA taxi passenger collects all of their taxicab receipts and calculates some numerical summaries. The mean fare the passenger paid was $15.45 with a standard deviation of $10.20. Find the mean and standard deviation of the lengths of the passenger's cab rides in miles.

26. **Quiz scores** The scores on Ms. Martin's statistics quiz had a mean of 12 and a standard deviation of 3. Ms. Martin wants to transform the scores to have a mean of 75 and a standard deviation of 12. What transformations should she apply to each test score? Explain your reasoning.

For Investigation *Apply the skills from the section in a new context or nonroutine way.*

27. **Comparing relative frequency graphs** Nitrates are organic compounds that are a substantial component of agricultural fertilizers. When those fertilizers run off into streams, the nitrates can have a toxic effect on animals that live in those streams. An ecologist studying nitrate pollution in two streams collects data on nitrate concentrations at 42 places on Stony Brook and 42 places on Mill Brook. Here are cumulative relative frequency graphs of the data for each stream:

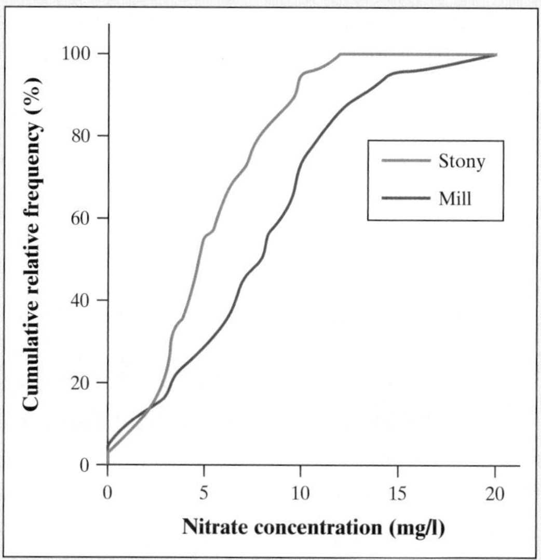

(a) Find and compare the median nitrate concentrations for the two streams.

(b) Which stream has more variability in its nitrate concentrations? Justify your answer with appropriate numerical evidence from the graph.

28. **Logging representatives** Here are data on the number of representatives in 2024 for each of the 50 U.S. states, along with a histogram. You can see that the distribution is skewed to the right with some high outliers.

1	1	1	1	1	1	2	2	2	2
2	2	2	3	3	4	4	4	4	4
4	5	5	6	6	6	7	7	8	8
8	8	8	9	9	9	9	10	11	12
13	14	14	15	17	17	26	28	38	52

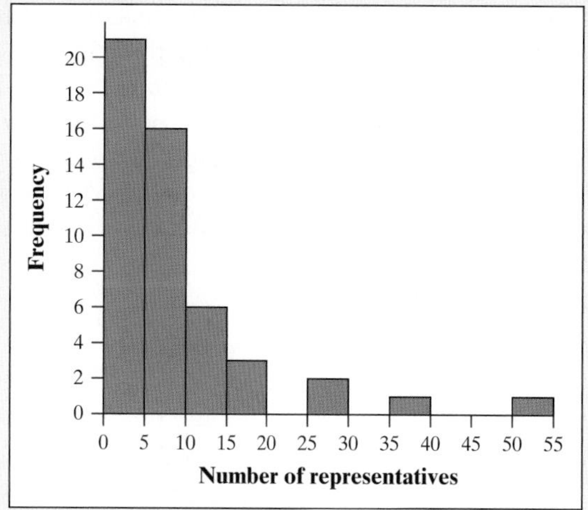

When the distribution of a quantitative variable is clearly skewed, transforming the data using powers, roots, or logarithms can sometimes yield a distribution that is roughly symmetric.

(a) Use technology to take the square root of the number of representatives in each state, and make a histogram of the transformed data. Is the resulting distribution roughly symmetric?

(b) Use technology to take the logarithm (base 10) of the number of representatives in each state, and make a histogram of the transformed data. Is the resulting distribution roughly symmetric?

Multiple Choice *Select the best answer for each question.*

29. Jorge's score on Exam 1 was at the 64th percentile of the scores for all students in a large statistics class. Jorge's score falls

(A) between the minimum and the first quartile.

(B) between the first quartile and the median.

(C) between the median and the third quartile.

(D) between the third quartile and the maximum.

(E) at the mean score for all students.

30. At a restaurant, Sam always tips the server \$2 plus 10% of the cost of the meal. If Sam's distribution of meal costs has a mean of \$9 and a standard deviation of \$3, what are the mean and standard deviation of Sam's distribution of tips?

(A) \$2.90, \$0.30

(B) \$2.90, \$2.30

(C) \$9.00, \$3.00

(D) \$11.00, \$2.00

(E) \$2.00, \$0.90

31. Scores on the Graduate Record Examinations (GRE) are widely used to help predict the performance of applicants to graduate schools. The scores on the GRE Quantitative Reasoning Test follow a bell-shaped distribution with mean = 154 and standard deviation = 9.6.[115] Wayne's standardized score on the GRE Quantitative Reasoning Test was −0.6. What was Wayne's actual score?

(A) 5.76

(B) 96.40

(C) 148.24

(D) 153.40

(E) 159.76

32. Georgina's average bowling score is 180 in a league where the average for all bowlers is 150 and the standard deviation is 20. River's average bowling score is 190 in a league where the average is 160 and the standard deviation is 15. Who ranks higher in their own league, Georgina or River?

(A) River, because 190 is higher than 180.

(B) River, because River's standardized score is higher than Georgina's.

(C) River and Georgina have the same rank in their leagues, because both are 30 pins above the mean.

(D) Georgina, because Georgina's standardized score is higher than River's.

(E) Georgina, because the standard deviation of bowling scores is higher in Georgina's league.

Exercises 33 and 34 refer to the following setting. The number of absences during the fall semester was recorded for each student in a large elementary school. These data were used to create the following cumulative relative frequency graph.

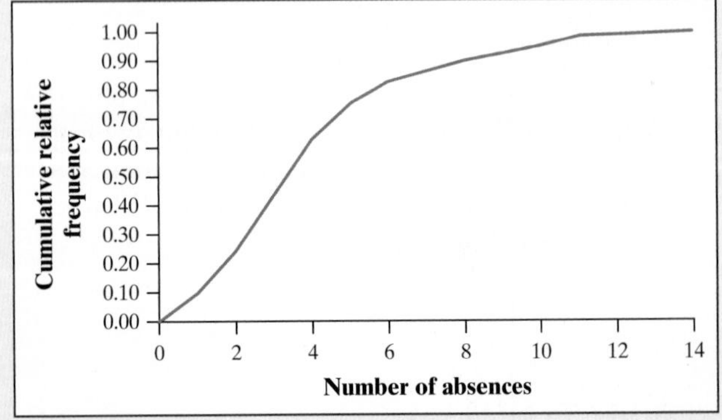

33. What is the interquartile range (*IQR*) for the distribution of absences?

(A) 1 (B) 2 (C) 3 (D) 5 (E) 14

34. If the distribution of absences was displayed in a histogram with intervals of width 2, what would be the best description of the histogram's shape?

(A) Symmetric

(B) Uniform

(C) Skewed left

(D) Skewed right

(E) Cannot be determined

Recycle and Review ***Practice what you learned in previous sections.***

Exercises 35 and 36 refer to the following setting. We used an online random sampling tool to select 50 Canadian students who completed a survey in a recent year.

35. **Lefties (1B)** Students were asked, "Are you right-handed, left-handed, or ambidextrous?" The responses (R = right-handed, L = left-handed, A = ambidextrous) are shown here:

R R R R R R R R R R R L R R
R R R R R R R R R R R R R A
R R R R A R R L R R R R L A
R R R R R R R R

(a) Make an appropriate graph to display these data.

(b) More than 10,000 Canadian high school students took the survey that year. What proportion of this population would you estimate is left-handed? Justify your answer.

36. **Travel time (1C)** The dotplot displays data on 50 students' responses to the question, "How long does it usually take you to travel to school?" Describe the distribution.

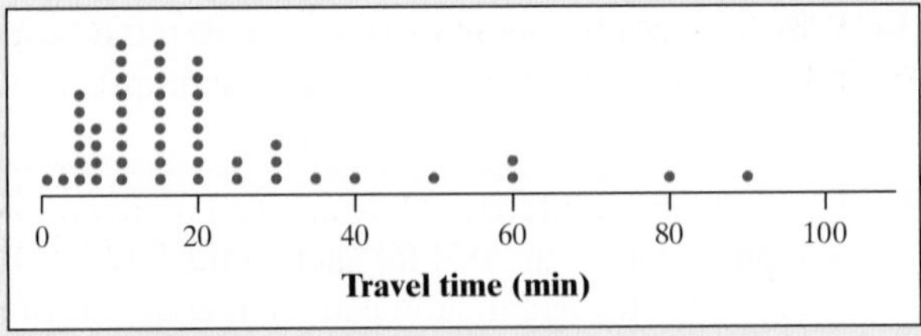

SECTION 1F Normal Distributions

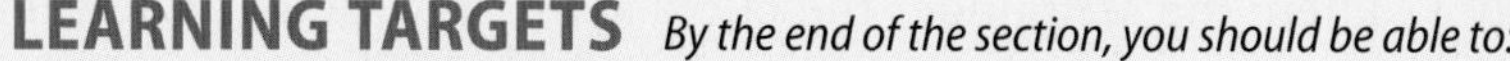

LEARNING TARGETS *By the end of the section, you should be able to:*

- Draw a normal curve to model the distribution of a quantitative variable.
- Use the empirical rule to estimate the proportion of values in a specified interval in a normal distribution.
- Determine whether a distribution of quantitative data is approximately normal using graphical and numerical evidence.
- Find the proportion of values in a specified interval in a normal distribution.
- Find the value that corresponds to a given percentile in a normal distribution.
- Calculate the mean or standard deviation of a normal distribution given the value of a percentile.

In Part I of this unit, we developed graphical and numerical tools for describing distributions of quantitative data. Our work gave us a clear strategy for exploring data on a single quantitative variable:

- Always plot your data: make a graph—usually a dotplot, stemplot, or histogram.
- Look for the overall pattern (shape, center, variability) and for striking departures such as outliers.
- Calculate numerical summaries to describe center and variability.

In this section, we add one more step to this strategy:

- When there's a regular overall pattern, use a simplified model to describe it.

Normal Curves

Figure 1.19 shows a histogram of the scores of all seventh-grade students in Gary, Indiana, on the vocabulary part of the Iowa Test of Basic Skills (ITBS).[116] The scores are grade-level equivalents, so a score of 6.3 indicates that the student's performance is typical for a student in the third month of grade 6.

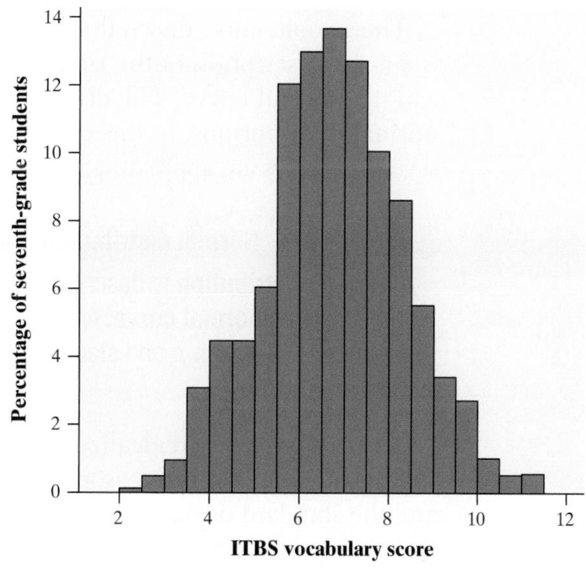

FIGURE 1.19 Histogram of the Iowa Test of Basic Skills (ITBS) vocabulary scores of all seventh-grade students in Gary, Indiana.

The distribution is roughly symmetric, and both tails fall off smoothly from a single center peak. There are no large gaps or obvious outliers.

Figure 1.20(a) and Figure 1.20(b) are copies of the histogram in Figure 1.19 with a smooth curve added. We adjusted the vertical scale of the graphs so that the total *area* of the bars in the histogram is 100% and the total area under the curve is exactly 1. Then, the area under the curve above a particular interval of values on the horizontal axis can be used to estimate the actual proportion or percentage of data values in that interval.

- In Figure 1.20(a), the area of the shaded bars represents the actual proportion of students with vocabulary scores less than 6.0. There are 287 such students, who make up the proportion 287/947 = 0.303 of all Gary seventh-graders.
- In Figure 1.20(b), the shaded area under the curve to the left of 6.0 estimates the proportion of students with scores less than 6.0. This area is 0.293, only 0.010 away from the actual proportion 0.303. The method for finding this area will be presented shortly. For now, note that the area under the curve gives a good estimate for the actual proportion of all seventh-grade students in Gary, Indiana, with ITBS vocabulary scores below sixth-grade level.

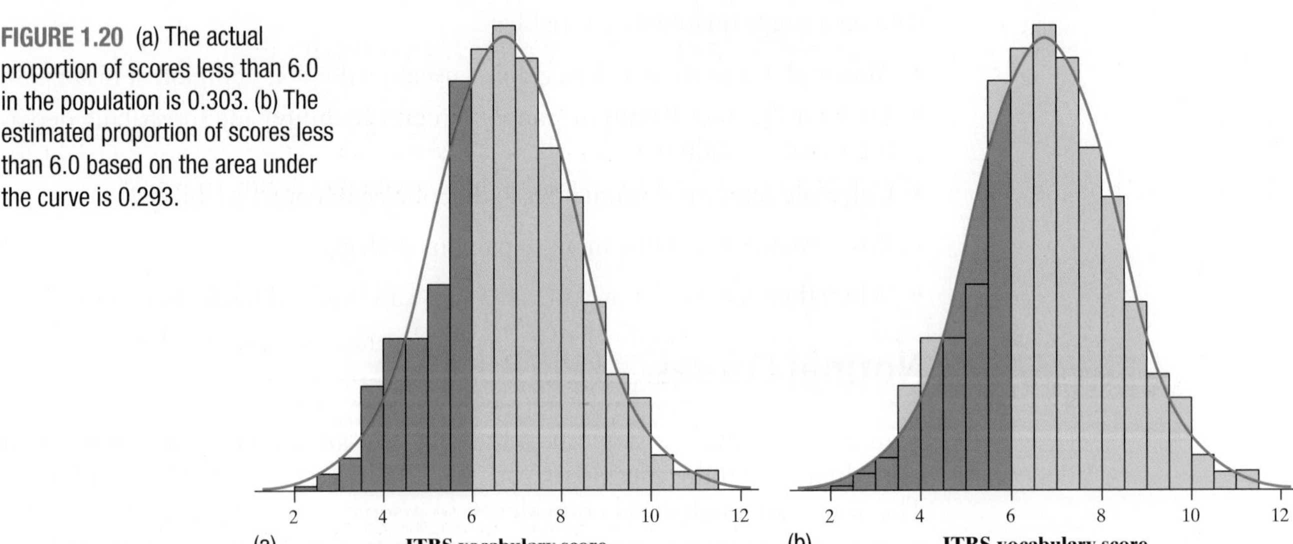

FIGURE 1.20 (a) The actual proportion of scores less than 6.0 in the population is 0.303. (b) The estimated proportion of scores less than 6.0 based on the area under the curve is 0.293.

The smooth curve drawn through the tops of the histogram bars in Figure 1.20 is a good description of the overall pattern of the ITBS score distribution. We call it a **normal curve.** The distributions described by normal curves are called **normal distributions.** In this case, the ITBS vocabulary scores of Gary, Indiana, seventh-graders are approximately normally distributed.

DEFINITION Normal distribution, Normal curve

A **normal distribution** is described by a symmetric, single-peaked, mound-shaped curve called a **normal curve.** Any normal distribution is completely specified by two parameters: its mean μ and standard deviation σ.

A normal curve is an idealized model for the population distribution of a quantitative variable. For this reason, we label the mean of a normal distribution as μ and the standard deviation of a normal distribution as σ. This is the same notation we used for the population mean and population standard deviation in Part I of this unit. In both cases, we refer to μ and σ as *parameters*.

Look at the two normal distributions in Figure 1.21. They illustrate several important facts:

- **Shape:** All normal distributions have the same overall shape: symmetric, single-peaked (unimodal), and mound-shaped (also called bell-shaped).
- **Center:** The mean μ is located at the balance point of the symmetric normal distribution and is the same as the median.
- **Variability:** The standard deviation σ measures the variability (width) of a normal distribution.

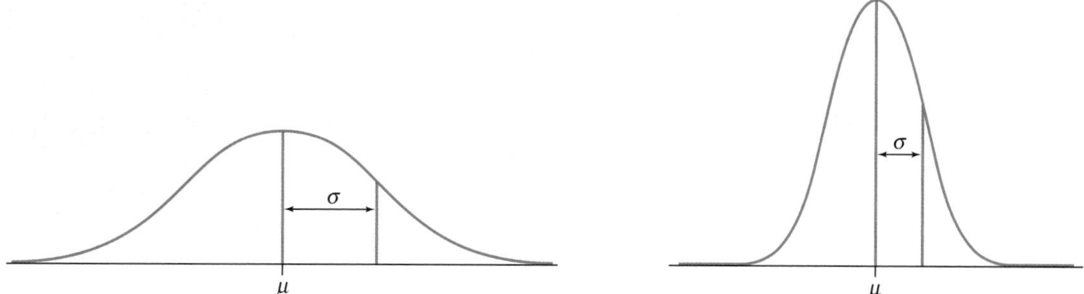

FIGURE 1.21 Two normal distributions, showing the mean μ and the standard deviation σ.

You can estimate σ by eye on a normal curve. Here's how: Imagine that you are skiing down a mountain that has the shape of a normal distribution. At first, you descend at an increasingly steep angle as you go out from the peak.

Fortunately, instead of skiing straight down the mountain, you find that the slope begins to get flatter rather than steeper as you go out and down:

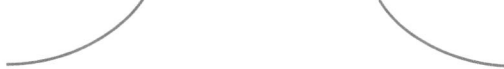

The points at which this change of curvature takes place are located at a distance σ on either side of the mean μ. (Calculus students know these as "inflection points.") You can feel the change in curvature as you run a pencil along a normal curve, which will allow you to estimate the standard deviation.

The distribution of ITBS vocabulary scores for seventh-grade students in Gary, Indiana, is modeled well by a normal distribution with mean $\mu = 6.84$ and standard deviation $\sigma = 1.55$. The following figure shows this distribution with the points 1, 2, and 3 standard deviations from the mean labeled on the horizontal axis.

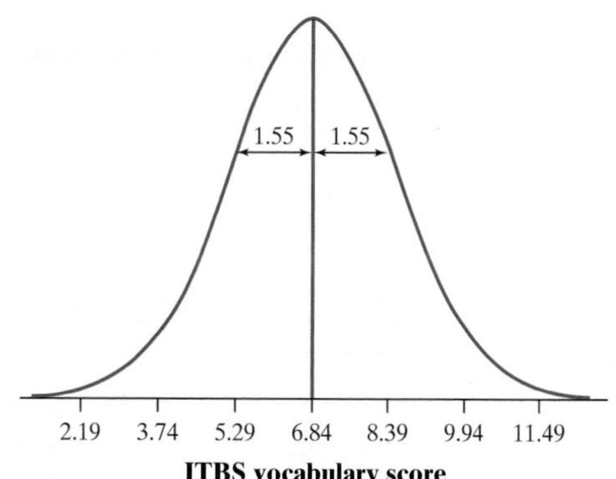

You will be asked to draw reasonably accurate normal curves throughout this course. The best way to learn is to practice.

EXAMPLE

Skill 2.D

Stop the car!
Normal curves

fstop123/Getty Images

PROBLEM: Many studies on automobile safety suggest that when automobile drivers make emergency stops, their stopping distances follow an approximately normal distribution. Suppose that for one model of car traveling at 62 mph under typical conditions on dry pavement, the mean stopping distance is $\mu = 155$ ft with a standard deviation of $\sigma = 3$ ft. Draw the normal curve that approximates the distribution of stopping distance. Label the mean and the points that are 1, 2, and 3 standard deviations from the mean.

SOLUTION:

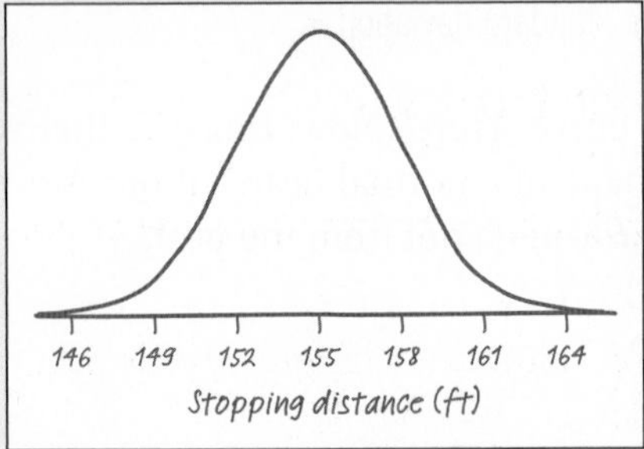

The mean (155) is at the balance point/midpoint of the normal distribution. The standard deviation (3) is the distance from the mean to the change-of-curvature points on either side. Label the mean and the points that are 1, 2, and 3 SDs from the mean:
1 SD: $155 - 1(3) = 152$ and $155 + 1(3) = 158$
2 SD: $155 - 2(3) = 149$ and $155 + 2(3) = 161$
3 SD: $155 - 3(3) = 146$ and $155 + 3(3) = 164$
Be sure to label the horizontal axis with the variable name!

FOR PRACTICE, TRY EXERCISE 1

Remember that μ and σ alone do not determine the appearance of most distributions. These are special properties of normal distributions.

Why are normal distributions important in statistics? Here are three reasons.

1. Normal distributions are good descriptions of some distributions of real data. Distributions that are often close to normal include
   - Scores on tests taken by many people (such as SAT or ACT exams).
   - Repeated careful measurements of the same quantity (like stopping distances of a car).
   - Characteristics of biological populations (such as lengths of crickets and yields of corn).
2. Normal distributions can approximate the results of many kinds of chance outcomes, such as the number of heads in many tosses of a fair coin.
3. Many of the inference methods described in Units 6–9 are based on normal distributions.

The Empirical Rule

Earlier, we saw that the distribution of Iowa Test of Basic Skills (ITBS) vocabulary scores for seventh-grade students in Gary, Indiana, is approximately normal with mean $\mu = 6.84$ and standard deviation $\sigma = 1.55$. How unusual is it for a Gary seventh-grader to get an ITBS score less than 3.74? The figure shows the normal curve for this distribution with the area of interest shaded. Note that the boundary value, 3.74, is exactly 2 standard deviations below the mean.

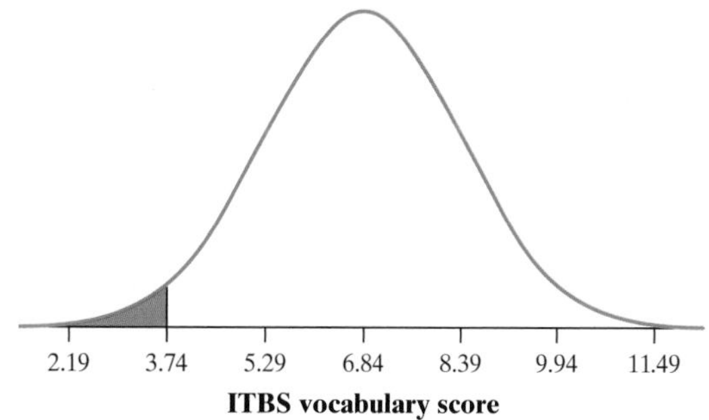

How can we estimate the shaded area, which represents the proportion of all Gary, Indiana, seventh-graders with ITBS vocabulary scores less than 3.74? The following activity reveals one way to do it.

ACTIVITY What's so special about normal distributions?

In this activity, you will use an applet to discover an interesting property of normal distributions.

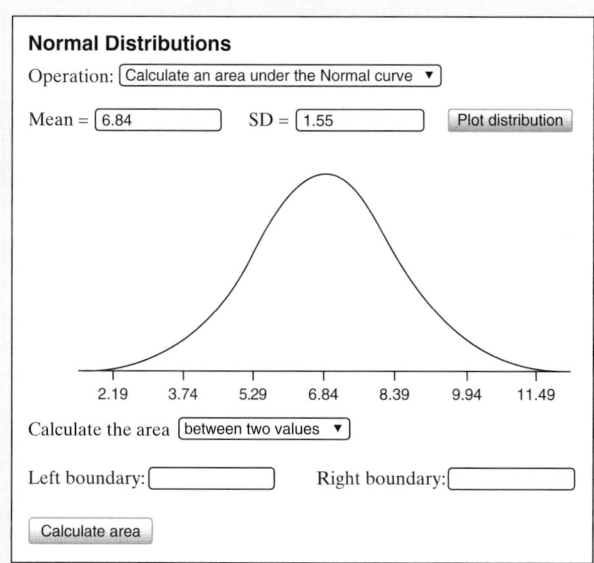

1. Go to www.stapplet.com and launch the *Normal Distributions* applet.
2. Choose "Calculate an area under the Normal curve" from the Operation menu at the top. Enter 6.84 for the mean and 1.55 for the standard deviation. Then click the "Plot distribution" button. (These are the values for the distribution of ITBS vocabulary scores of seventh-graders in Gary, Indiana.) A figure like this should appear.
3. Use the applet to help you answer the following questions about the distribution of ITBS scores.
   (a) About what proportion of Gary, Indiana, seventh-graders have ITBS vocabulary scores between 5.29 and 8.39? That is, what percentage of the area under the normal curve lies within 1 standard deviation of the mean?
   (b) About what proportion of Gary, Indiana, seventh-graders have ITBS vocabulary scores between 3.74 and 9.94? That is, what percentage of the area under the normal curve lies within 2 standard deviations of the mean?
   (c) About what proportion of Gary, Indiana, seventh-graders have ITBS vocabulary scores between 2.19 and 11.49? That is, what percentage of the area under the normal curve lies within 3 standard deviations of the mean?

4. The distribution of IQ scores in a population is approximately normal with mean $\mu = 100$ and standard deviation $\sigma = 15$. Adjust the applet to display this distribution. About what percentage of individuals in the population have IQ scores within 1, 2, and 3 standard deviations of the mean?
5. Adjust the applet to display a normal distribution with mean 0 and standard deviation 1. This is called the *standard normal distribution*, based on the fact that standardized scores (z-scores) have a mean of 0 and a standard deviation of 1. What percentage of the area under the standard normal curve lies within 1, 2, and 3 standard deviations of the mean?
6. Summarize by completing this sentence: "For any normal distribution, the area under the normal curve within 1, 2, and 3 standard deviations of the mean is about ___________%, ___________%, and ___________%."

As the activity reveals, although there are many normal distributions, they all obey the **empirical rule.** (*Empirical* means "learned from experience or by observation.")

DEFINITION The empirical rule

In a normal distribution with mean μ and standard deviation σ, the **empirical rule** states that

- About 68% of the values fall within 1σ of the mean μ.
- About 95% of the values fall within 2σ of the mean μ.
- About 99.7% of the values fall within 3σ of the mean μ.

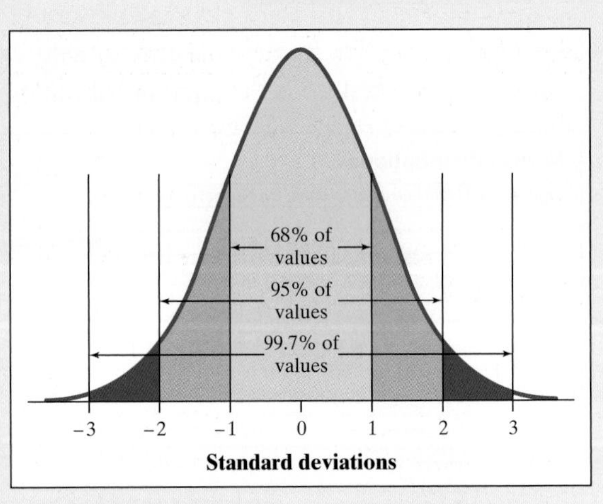

In Section 1D, we introduced the $2 \times$ SD rule for determining outliers: an outlier is any data value that is more than 2 standard deviations from the mean. This rule makes better sense when we are discussing a normal distribution, where less than 5% of the values would meet this criterion according to the empirical rule.

Some people refer to the empirical rule as the *68–95–99.7 rule*. By remembering these three numbers, you can quickly estimate the proportion or percentage of values in a specified interval in a normal distribution.

Earlier, we asked how unusual it would be for a Gary, Indiana, seventh-grader to get an ITBS vocabulary score less than 3.74. Figure 1.22 gives the answer in graphical form. By the empirical rule, about 95% of these students have ITBS vocabulary scores between 3.74 and 9.94, which means that about 5% of the students have scores less than 3.74 or greater than 9.94. Due to the symmetry of the normal distribution, about 5% / 2 = 2.5% of students have scores less than 3.74.

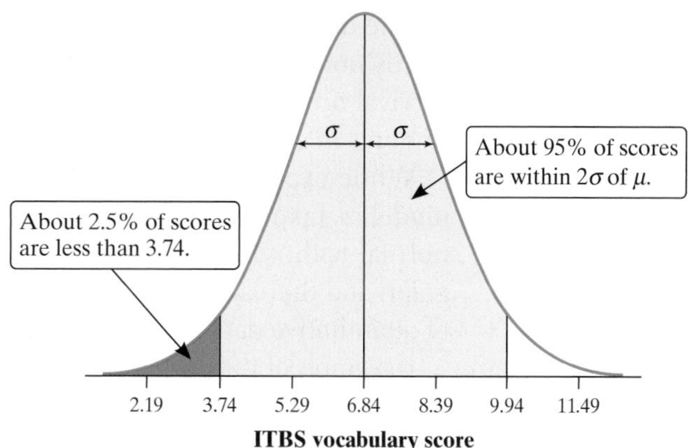

FIGURE 1.22 Using the empirical rule to estimate the percentage of Gary, Indiana, seventh-graders with ITBS vocabulary scores less than 3.74.

EXAMPLE

Stop the car!
The empirical rule

Skill 2.D

PROBLEM: Many studies on automobile safety suggest that when automobile drivers must make emergency stops, the stopping distances follow an approximately normal distribution. Suppose that for one model of car traveling at 62 mph under typical conditions on dry pavement, the mean stopping distance is $\mu = 155$ ft with a standard deviation of $\sigma = 3$ ft. About what percentage of cars of this model would take more than 158 feet to make an emergency stop?

Peter Cade/Stone/Getty Images

SOLUTION:

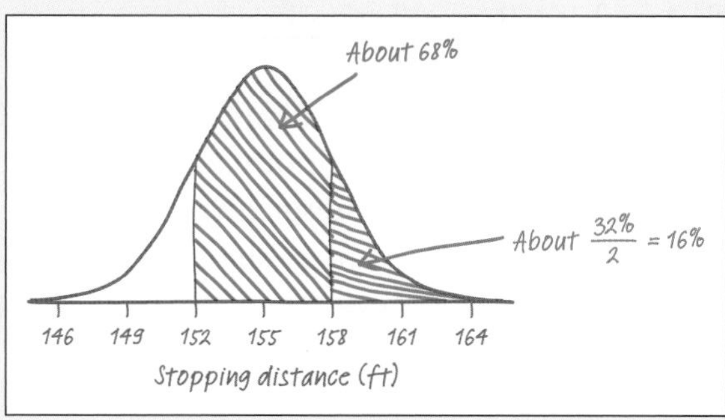

Start by sketching a normal curve and labeling the values 1, 2, and 3 standard deviations from the mean. Then shade the area of interest.

Use the empirical rule and the symmetry of the normal distribution to find the desired area.

About 16% of cars of this model would take more than 158 feet to make an emergency stop.

FOR PRACTICE, TRY EXERCISE 5

Note that the empirical rule applies *only* to normal distributions. It does not apply to skewed distributions or even to other symmetric distributions, such as a uniform distribution.

Assessing Normality

Normal distributions provide good models for some distributions of quantitative data. Examples include SAT and ACT test scores, the highway gas mileage of 2024 Corvette convertibles, weights of 9-ounce bags of potato chips, and heights of 3-year-old girls.

The distributions of other quantitative variables are skewed and therefore distinctly non-normal. Examples include single-family home prices in a certain city, survival times of patients with cancer after treatment, and number of siblings for students in a statistics class. (All of these distributions are right-skewed.)

While experience can suggest whether a normal distribution is a reasonable model in a particular case, it is risky to assume that a distribution is approximately normal without first analyzing the data. As in Part I of this unit, we start with a graph and then add numerical summaries to assess the normality of a distribution of quantitative data.

If a graph of the data is clearly skewed, has multiple peaks, or isn't bell-shaped, that's evidence the distribution is not normal. Figure 1.23 shows a dotplot and summary statistics of the number of siblings reported by each student in a statistics class. This distribution is skewed to the right, so it is not approximately normal. We can also see from the summary statistics that the minimum of the distribution is less than 2 standard deviations from the mean. In a normal distribution, about 2.5% of the values should fall more than 2 standard deviations below the mean.

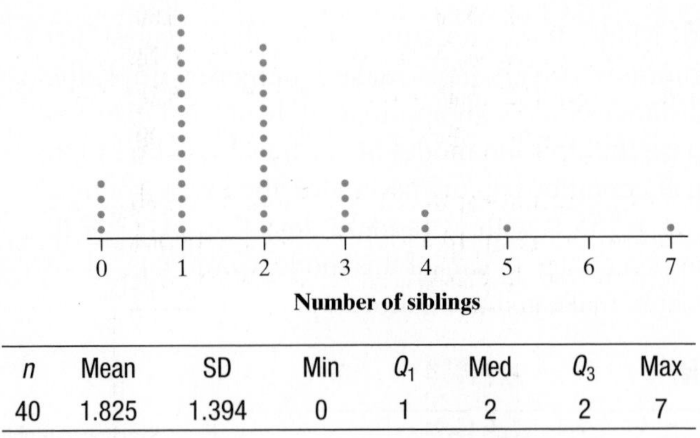

n	Mean	SD	Min	Q_1	Med	Q_3	Max
40	1.825	1.394	0	1	2	2	7

FIGURE 1.23 Dotplot and summary statistics of data on the number of siblings reported by students in a statistics class.

Even if a graph of the data looks roughly symmetric and bell-shaped, we shouldn't assume that the distribution is approximately normal. The empirical rule can give additional evidence in favor of or against normality.

Figure 1.24 shows a dotplot and summary statistics for data on calories per serving in 77 brands of breakfast cereal.[117] The graph is roughly symmetric, single-peaked (unimodal), and somewhat bell-shaped.

Let's count the number of data values within 1, 2, and 3 standard deviations of the mean:

Mean $\pm$ 1SD: $106.883 \pm 1(19.484)$ 87.399 to 126.367 63 out of $77 = 81.8\%$

Mean $\pm$ 2SD: $106.883 \pm 2(19.484)$ 67.915 to 145.851 71 out of $77 = 92.2\%$

Mean $\pm$ 3SD: $106.883 \pm 3(19.484)$ 48.431 to 165.335 77 out of $77 = 100\%$

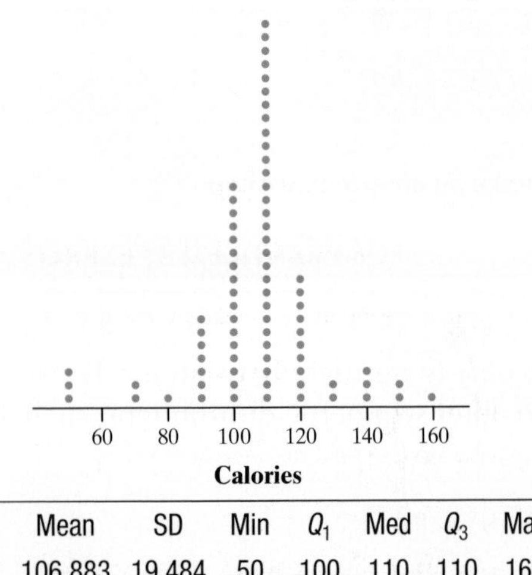

n	Mean	SD	Min	Q_1	Med	Q_3	Max
77	106.883	19.484	50	100	110	110	160

FIGURE 1.24 Dotplot and summary statistics for data on calories per serving in 77 different brands of breakfast cereal.

In a normal distribution, about 68% of the values fall within 1 standard deviation of the mean. For the cereal data, almost 82% of the brands had between 87.399 and 126.367 calories per serving. These two percentages are far apart—so this distribution of calories in breakfast cereals is not approximately normal.

How far do the percentages of values within 1, 2, and 3 standard deviations of the mean have to be from 68%, 95%, and 99.7% for us to say that a distribution of quantitative data is *not* approximately normal? Unfortunately, there is no hard-and-fast rule to make this decision, as it depends on the sample size and how close to normal you want to be.

EXAMPLE

Skill 2.D

Is the amount of cream in Oreo cookies normally distributed?

Assessing normality

Dan Anderson

PROBLEM: Student researchers collected data on the amount of cream (in grams) in a random sample of 45 Double Stuf Oreo cookies as part of their final project in a statistics class. Here are their data:[118]

6.37	6.44	6.44	6.45	6.48	6.49	6.49	6.52	6.58
6.59	6.62	6.62	6.63	6.67	6.67	6.68	6.69	6.71
6.72	6.73	6.73	6.73	6.73	6.74	6.74	6.76	6.80
6.80	6.83	6.85	6.85	6.85	6.86	6.87	6.88	6.90
6.91	6.92	6.92	6.92	6.94	7.01	7.04	7.09	7.15

A histogram and summary statistics for the data are shown. Is this distribution of amount of cream in Double Stuf Oreo cookies approximately normal? Give appropriate evidence to justify your answer.

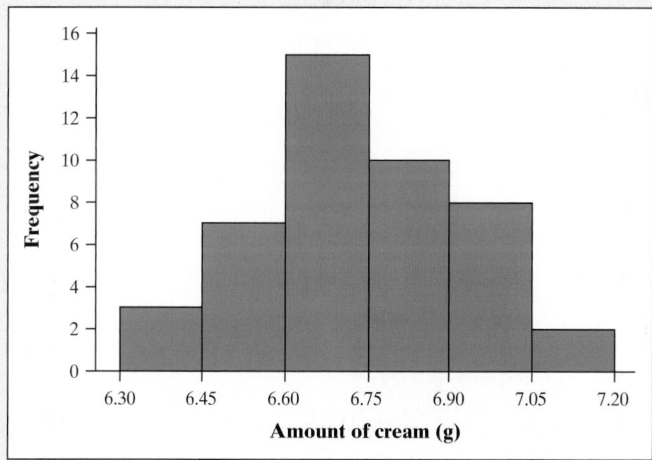

n	Mean	SD	Min	Q_1	Med	Q_3	Max
45	6.742	0.184	6.37	6.62	6.73	6.875	7.15

SOLUTION:

The histogram looks roughly symmetric, single-peaked, and somewhat mound-shaped. The percentages of values within 1, 2, and 3 standard deviations of the mean are

Mean $\pm$ 1SD: $6.742 \pm 1(0.184)$ 6.558 to 6.926 32 out of 45 = 71.1%

Mean $\pm$ 2SD: $6.742 \pm 2(0.184)$ 6.374 to 7.110 43 out of 45 = 95.6%

Mean $\pm$ 3SD: $6.742 \pm 3(0.184)$ 6.190 to 7.294 45 out of 45 = 100.0%

These percentages are close to the 68%, 95%, and 99.7% targets for a normal distribution. The graphical and numerical evidence suggests that this distribution of amount of cream in Oreo Double Stuf cookies is approximately normal.

FOR PRACTICE, TRY EXERCISE 9

AP® EXAM TIP

Never say that a distribution of quantitative data *is* normal. Real-world data always show at least slight departures from a normal distribution. The most you can say is that the distribution is "approximately normal."

The example shows that the distribution of amount of cream in the student researchers' random sample of 45 Double Stuf Oreo cookies is *approximately* normal. As you will learn in Unit 3, random sampling allows us to generalize results from a sample to the larger population. It is therefore reasonable to believe that the distribution of amount of cream in the population of Double Stuf Oreo cookies is *approximately* normal. Assessing the normality of a population distribution is an important part of several inference methods that you will encounter in later units.

CHECK YOUR UNDERSTANDING

Questions 1 and 2 refer to the following setting. An automaker has found that the life of its batteries varies from car to car according to an approximately normal distribution with mean $\mu = 48$ months and standard deviation $\sigma = 8$ months.

1. Draw the normal curve that approximates the distribution of battery life. Label the mean and the points that are 1, 2, and 3 standard deviations from the mean.
2. About what percentage of these car batteries would last more than 40 months?

Here are the ages of the first 46 U.S. presidents when they first took office:

42.88	43.65	46.42	46.85	47.46	48.28	49.29	49.33	50.50	51.02	51.09	51.96
51.96	52.05	52.30	54.24	54.41	54.54	55.24	55.47	55.96	56.03	56.29	56.67
57.18	57.65	57.75	57.89	57.97	58.56	58.62	58.85	59.54	60.93	61.07	61.27
61.34	61.97	62.27	64.18	64.61	65.86	68.06	69.96	70.60	78.17		

A histogram and summary statistics for these data are shown here:

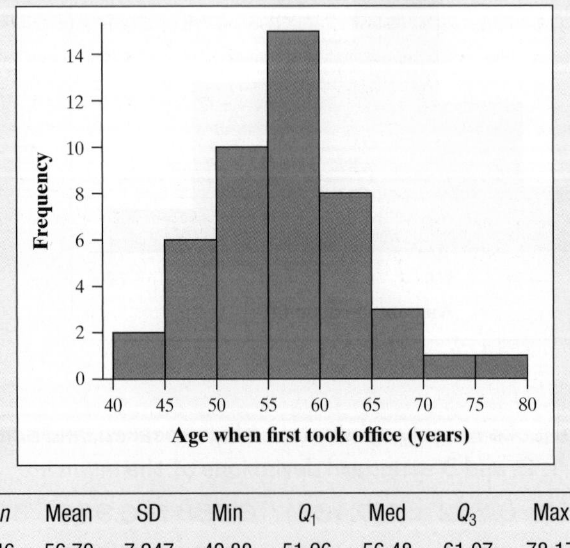

n	Mean	SD	Min	Q_1	Med	Q_3	Max
46	56.70	7.247	42.88	51.96	56.48	61.07	78.17

3. Is the distribution of age when U.S. presidents first took office approximately normal? Give appropriate evidence to justify your answer.

Finding Areas in a Normal Distribution

Let's return to the distribution of ITBS vocabulary scores among all Gary, Indiana, seventh-graders. Recall that this distribution is approximately normal with mean $\mu = 6.84$ and standard deviation $\sigma = 1.55$. What proportion of these seventh-graders have vocabulary scores that are below sixth-grade level (i.e., less than 6)? Figure 1.25 shows the normal curve with the area of interest shaded. We can't use the empirical rule to find this area because the boundary value of 6 is not exactly 1, 2, or 3 standard deviations from the mean.

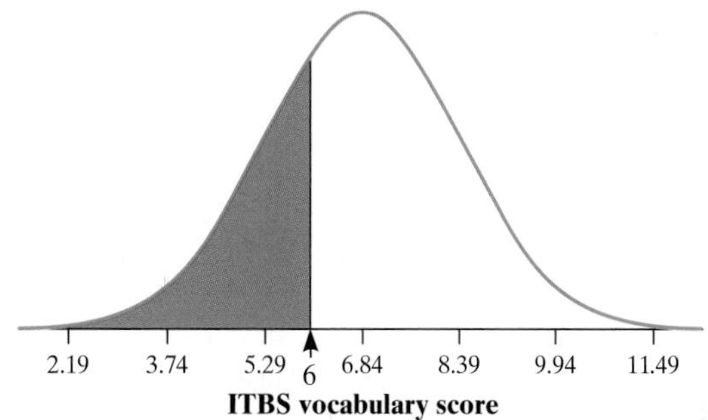

FIGURE 1.25 Normal curve we would use to estimate the proportion of Gary, Indiana, seventh-graders with ITBS vocabulary scores that are less than 6—that is, below sixth-grade level.

FINDING AREAS TO THE LEFT IN A NORMAL DISTRIBUTION

As the empirical rule suggests, all normal distributions are the same if we measure them in units of size σ from the mean μ. Changing to these units requires us to standardize each value, just as we did in Section 1E:

$$z = \frac{\text{value} - \text{mean}}{\text{standard deviation}} = \frac{x - \mu}{\sigma}$$

Recall that subtracting a constant and dividing by a constant don't change the shape of a distribution. If the quantitative variable we standardize has an approximately normal distribution, then so does the new variable z. This new distribution of standardized values can be modeled with a normal curve having mean $\mu = 0$ and standard deviation $\sigma = 1$. It is called the **standard normal distribution.**

DEFINITION Standard normal distribution

The **standard normal distribution** is the normal distribution with mean 0 and standard deviation 1.

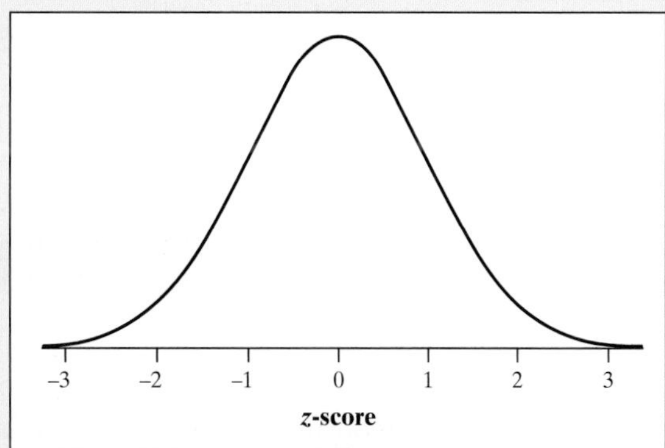

Because all normal distributions are the same when we standardize, we can find areas under any normal curve using the standard normal distribution. For the ITBS test score data, we want to find the area to the left of 6 in the normal distribution with mean 6.84 and standard deviation 1.55. See Figure 1.26(a). We start by standardizing the boundary value $x = 6$:

$$z = \frac{\text{value} - \text{mean}}{\text{standard deviation}} = \frac{6 - 6.84}{1.55} = -0.542$$

Figure 1.26(b) shows the standard normal distribution with the area to the left of $z = -0.542$ shaded. Notice that the shaded areas in the two graphs are the same.

After standardizing the boundary value, we can use technology or Table A in the back of the book to find the relevant area. We recommend using technology, for reasons that will become clear shortly.

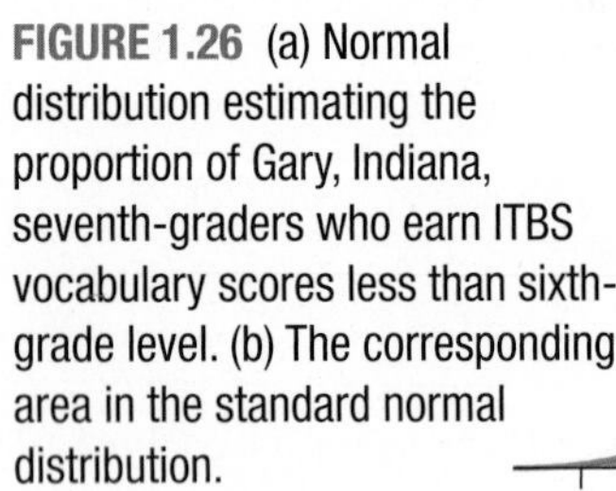

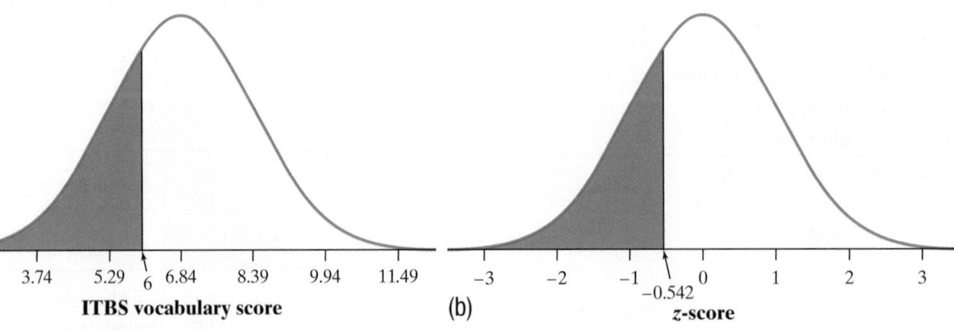

FIGURE 1.26 (a) Normal distribution estimating the proportion of Gary, Indiana, seventh-graders who earn ITBS vocabulary scores less than sixth-grade level. (b) The corresponding area in the standard normal distribution.

AP® EXAM TIP
Formula sheet
Table A is included with the formula sheet provided on both sections of the AP® Statistics exam. However, technology gives a more accurate answer that doesn't require rounding the z-score to two decimal places, as the following Tech Corner illustrates.

In Table A, the entry for each z-score is the area under the standard normal curve *to the left* of z. Note that Table A gives z-scores to only *two* decimal places. So we have to round the calculated z-score of -0.542 to -0.54 before using the table. To find the area to the left of $z = -0.54$, locate -0.5 in the left-hand column, then locate the remaining digit 4 as .04 in the top row. The entry to the right of -0.5 and under .04 is .2946. This is the area we seek. We estimate that about 29.46% of Gary, Indiana, seventh-graders' scores fall below the sixth-grade level on the ITBS vocabulary test.

z	.03	.04	.05
−0.6	.2643	.2611	.2578
−0.5	.2981	.2946	.2912
−0.4	.3336	.3300	.3264

4. Tech Corner — FINDING AREAS IN A NORMAL DISTRIBUTION

TI-Nspire and other technology instructions are on the book's website at bfwpub.com/tps7e.

You can use the normalcdf command on the TI-83/84 to find areas under a normal curve. The syntax is normalcdf(lower bound, upper bound, mean, standard deviation). Let's use this command to calculate the proportion of ITBS vocabulary scores in Gary, Indiana, that are less than 6. Note that we can do the area calculation using the standard normal distribution or the "unstandardized" normal distribution with mean 6.84 and standard deviation 1.55.

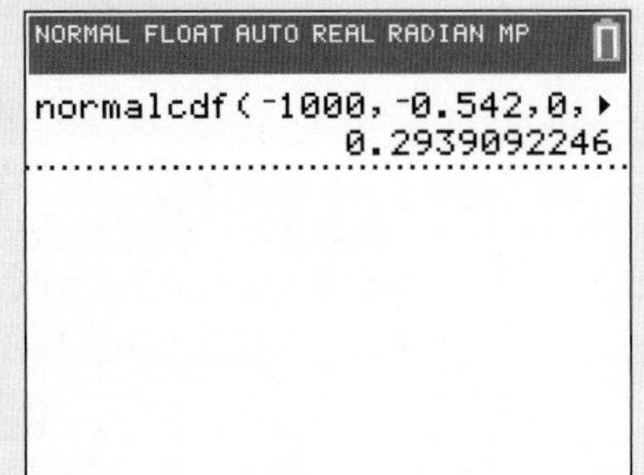

i. *Using the standard normal distribution:* What proportion of observations in a standard normal distribution are less than $z = -0.542$? Recall that the standard normal distribution has mean $\mu = 0$ and standard deviation $\sigma = 1$.

- Press 2nd VARS (Distr) and choose normalcdf(.

OS 2.55 or later: In the dialog box, enter these values: lower: −1000, upper: −0.542, μ:0, σ:1, choose Paste, and then press ENTER.

Older OS: Complete the command normalcdf(−1000, −0.542,0,1) and press ENTER.

Note: We chose −1000 as the lower bound because it's many, many standard deviations less than the mean.

ii. *Using the unstandardized normal distribution*: What proportion of observations in a normal distribution with mean $\mu = 6.84$ and standard deviation $\sigma = 1.55$ are less than 6?

- Press 2nd VARS (Distr) and choose normalcdf(.

  OS 2.55 or later: In the dialog box, enter these values: lower: −1000, upper: 6, μ: 6.84, σ: 1.55, choose Paste, and then press ENTER.

  Older OS: Complete the command normalcdf (−1000,6,6.84,1.55) and press ENTER.

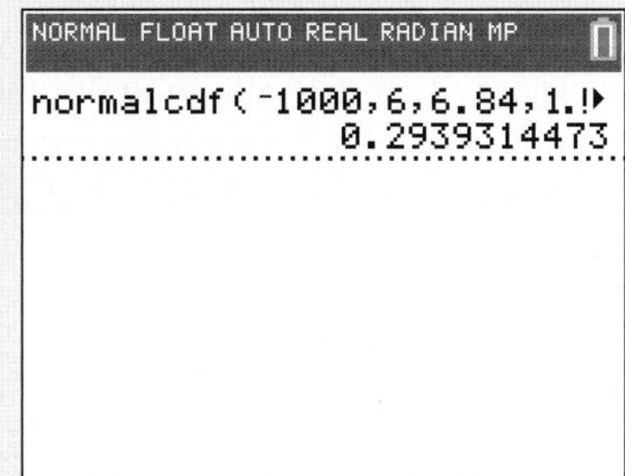

As the Tech Corner illustrates, it is possible to find the proportion of Gary, Indiana, seventh-graders with ITBS vocabulary scores less than 6 directly from the original (unstandardized) normal distribution using technology. *Check with your teacher to see if this method will be allowed in your class.*

HOW TO FIND AREAS IN ANY NORMAL DISTRIBUTION

Step 1: Draw a normal distribution with the horizontal axis labeled and scaled using the mean and standard deviation, the boundary value(s) clearly identified, and the area of interest shaded.

Step 2: Perform calculations—show your work! Do one of the following:

(i) Standardize the boundary value(s) and use technology or Table A to find the desired area under the standard normal curve; or

(ii) Use technology to find the desired area without standardizing. Label the calculator inputs.

Be sure to answer the question that was asked.

AP® EXAM TIP

Students often do not get full credit on the AP® Statistics exam because they use option (ii) with "calculator-speak" to show their work on normal distribution calculation questions—for example, normalcdf(−1000,6,6.84,1.55). This is *not* considered clear communication. To get full credit, be sure to carefully label each of the inputs in the calculator command if you use technology in Step 2: normalcdf(lower: −1000, upper: 6, mean: 6.84, SD: 1.55).

EXAMPLE

Stop the car!
Finding area to the left in a normal distribution

Skill 3.A

PROBLEM: Many studies on automobile safety suggest that when automobile drivers must make emergency stops, the stopping distances follow an approximately normal distribution. Suppose that for one model of car traveling at 62 mph under typical conditions on dry pavement, the mean stopping distance is $\mu = 155$ ft with a standard deviation of $\sigma = 3$ ft. Dontrelle is driving one of these cars at 62 mph and spots an accident 160 feet ahead that requires an emergency stop. Is Dontrelle likely to be able to stop the car in less than 160 feet? Justify your answer.

RuslanDashinsky/E+/Getty Images

SOLUTION:

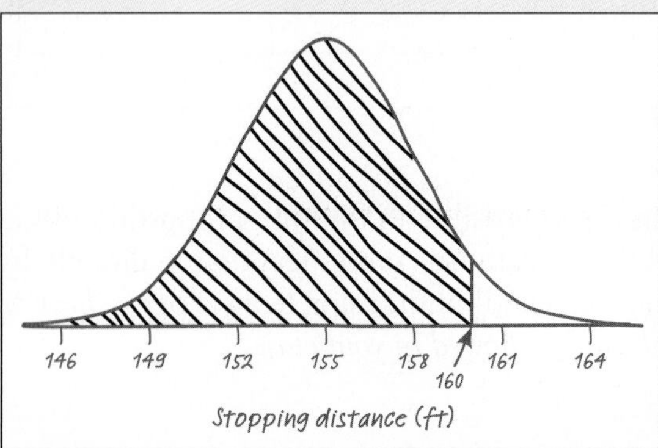

> **1. Draw a normal distribution.** Be sure to
> - Scale the horizontal axis.
> - Label the horizontal axis with the variable name, including units of measurement.
> - Clearly identify the boundary value(s).
> - Shade the area of interest.

(i) $z = \dfrac{160 - 155}{3} = 1.667$

Using technology: normalcdf(lower: −1000, upper: 1.667, mean: 0, SD: 1) = 0.9522

Using Table A: Area to the left of $z = 1.67$ is 0.9525.

(ii) normalcdf(lower: −1000, upper: 160, mean: 155, SD: 3) = 0.9522

> **2. Perform calculations—show your work!**
> i. Standardize the boundary value(s) and use technology or Table A to find the desired probability; or
> ii. Use technology to find the desired area without standardizing. Label the calculator inputs.

About 95% of cars of this model would be able to make an emergency stop in less than 160 feet. So Dontrelle is likely to be able to stop safely.

> Be sure to answer the question.

FOR PRACTICE, TRY EXERCISE 15

What percentage of cars of this model can make an emergency stop in *exactly* 160 feet? Because a point on the number line has no width, there is no area directly above the point 160.000... under the normal curve in the example. So the areas under the curve with values < 160 and values ≤ 160 are the same. According to the normal model, the percentage of cars of this model that can make an emergency stop in less than 160 feet is the same as the percentage that can make an emergency stop in less than or equal to 160 feet.

FINDING AREAS TO THE RIGHT IN A NORMAL DISTRIBUTION

Finding areas to the right of a boundary line in a normal distribution is very similar to finding areas to the left, especially when using technology. When using

Table A, the values in the table give areas to the left. Because the total area is 1 under the standard normal curve, to find the area to the right, simply subtract the area to the left from 1. The following example shows what we mean.

EXAMPLE

Can Nelly clear the trees?
Finding area to the right in a normal distribution

Skill 3.A

PROBLEM: Nelly Korda was one of the top golfers on the Ladies Professional Golf Association (LPGA) tour in 2022. When Nelly hits her driver, the distance the ball travels can be modeled by a normal distribution with mean 274 yards and standard deviation 8 yards.[119] On a specific hole, Korda would need to hit the ball at least 260 yards to have a clear second shot that avoids a large group of trees. Is she likely to have a clear second shot?

Mike Ehrmann/Getty Images

SOLUTION:

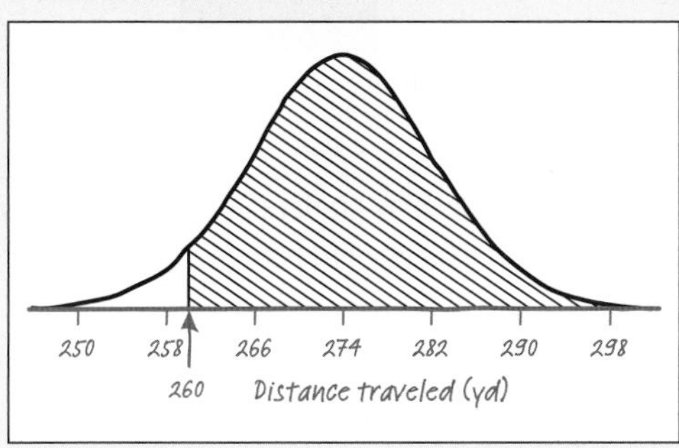

1. Draw a normal distribution.

(i) $z = \frac{260 - 274}{8} = -1.750$

2. Perform calculations—show your work!

Using technology: normalcdf(lower: −1.750, upper: 1000, mean: 0, SD: 1) = 0.9599

Using Table A: Area to the right of $z = -1.75$ is $1 - 0.0401 = 0.9599$.

(ii) normalcdf(lower: 260, upper: 1000, mean: 274, SD: 8) = 0.9599

About 96% of Nelly Korda's drives travel at least 290 yards. So she is likely to have a clear second shot.

Be sure to answer the question.

FOR PRACTICE, TRY EXERCISE 17

A common student mistake is to find the area to the left of a given value in a normal distribution and to report that area as the answer, regardless of whether the problem asks for the area to the left or to the right. This mistake can usually be prevented by comparing your answer to the shaded area of interest in the normal distribution. Think about whether the area should be closer to 0 or closer to 1. In the preceding example, for instance, it should be obvious that 0.0401 is *not* the correct area.

FINDING AREAS BETWEEN TWO VALUES IN A NORMAL DISTRIBUTION

Finding the area between two values in a normal distribution is very similar to finding the area to the left or to the right of a given value, especially when using technology. When using Table A, it's a little more complicated.

For instance, suppose we want to estimate the proportion of Gary, Indiana, seventh-graders with ITBS vocabulary scores between 6 and 9. Figure 1.27(a) shows the desired area under the normal curve with mean $\mu = 6.84$ and standard deviation $\sigma = 1.55$.

Option (i): If we standardize each boundary value, we get

$$z = \frac{6 - 6.84}{1.55} = -0.542 \quad \text{and} \quad z = \frac{9 - 6.84}{1.55} = 1.394$$

Figure 1.27(b) shows the corresponding area of interest in the standard normal distribution.

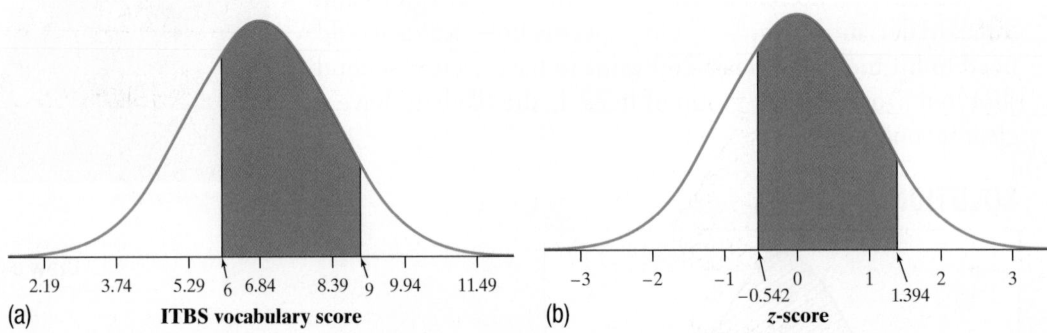

FIGURE 1.27 (a) Normal distribution approximating the proportion of seventh-graders in Gary, Indiana, with ITBS vocabulary scores between 6 and 9. (b) The corresponding area in the standard normal distribution.

Using technology: normalcdf(lower: −0.542, upper: 1.394, mean: 0, SD: 1) = 0.6244

Using Table A: The table makes this process a bit trickier because it shows only areas to the left of a given two-decimal-place z-score. The visual shows one way to think about the calculation.

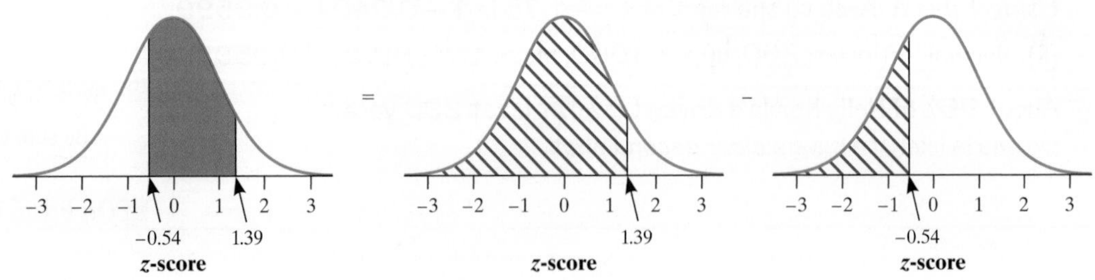

Area between $z = -0.54$ and $z = 1.39$

$$\begin{aligned} &= (\text{Area to the left of } z = 1.39) - (\text{Area to the left of } z = -0.54) \\ &= 0.9177 - 0.2946 \\ &= 0.6231 \end{aligned}$$

Option (ii): normalcdf(lower: 6, upper: 9, mean: 6.84, SD: 1.55) = 0.6243.

About 62% of Gary, Indiana, seventh-graders earned grade-equivalent ITBS vocabulary scores between 6 and 9.

(ii) invNorm(area: 0.25, mean: 94.5, SD: 4) = 91.802

About 75% of 3-year-old girls are taller than 91.80 centimeters.

Be sure to answer the question.

FOR PRACTICE, TRY EXERCISE 25

Another approach to finding the 25th percentile in the example is to use the interpretation of the z-score. A standardized score of $z = -0.674$ means we are looking for the value that is 0.674 standard deviation below the mean:

$$\text{mean} - 0.674(\text{SD}) = 94.5 - 0.674(4) = 91.804$$

So 75% of 3-year-old girls are taller than 91.804 centimeters.

Calculating the Mean or Standard Deviation of a Normal Distribution

You have seen how to find the value corresponding to a given percentile in a normal distribution with known mean and standard deviation. It is also possible to estimate parameters of a normal distribution—the mean or the standard deviation—using the value of one or more percentiles.

EXAMPLE

Get off your phone!
Calculating the mean or SD of a normal distribution

Skill 3.A

PROBLEM: According to a study by RescueTime, people spend an average of 1.25 minutes on their smartphone each time they pick it up.[122] If Noel's distribution of smartphone use is approximately normal with a mean of 1.25 minutes and a 97th percentile of 2 minutes, calculate its standard deviation.

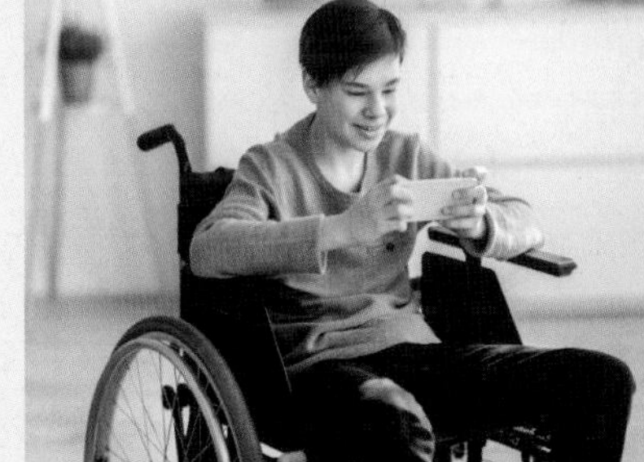

Prostock-studio/Alamy Stock Photo

SOLUTION:

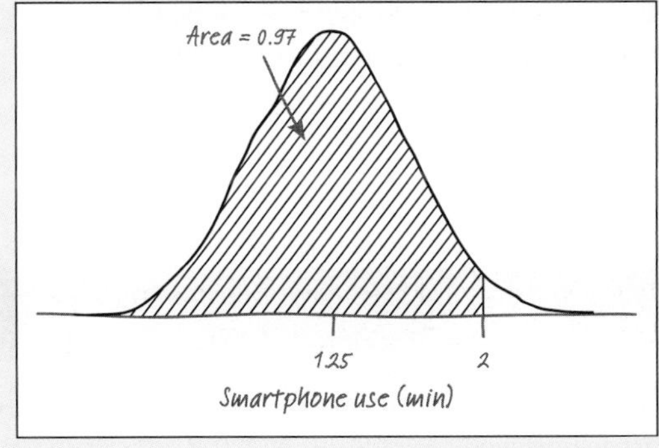

1. Draw a normal distribution.
Because we don't know the standard deviation, we can only label the mean and the boundary value, along with the corresponding area.

Using technology: invNorm(area: 0.97, mean: 0, SD: 1) = 1.881

Using Table A: 0.97 area to the left → $z = 1.88$

$$1.881 = \frac{2 - 1.25}{\sigma}$$

$$1.881\sigma = 0.75$$

$$\sigma = \frac{0.75}{1.881}$$

$$\sigma = 0.399 \text{ minute}$$

2. Perform calculations—show your work!
Because we are trying to find the standard deviation, we need to use the *z*-score formula and fill in the other three values. We know the boundary value is 2 and the mean is 1.25. The last value we need is the *z*-score for the area to the left of 0.97. Substitute these three values into the *z*-score formula and solve for σ.

FOR PRACTICE, TRY EXERCISE 31

The preceding example showed you how to find the standard deviation if you are given the mean of a normal distribution and the value of another percentile. If you are given the standard deviation and the value of a percentile, you can use a similar approach to find the mean of a normal distribution. But what if you don't know the mean or the standard deviation? With the values of two different percentiles in a normal distribution, you can solve a system of equations to find μ and σ.

CHECK YOUR UNDERSTANDING

High levels of cholesterol in the blood increase the risk of heart disease.

1. For teenagers, the distribution of blood cholesterol is approximately normal with mean $\mu = 151.6$ milligrams of cholesterol per deciliter of blood (mg/dl) and standard deviation $\sigma = 25$ mg/dl.[123] Find the 20th percentile of the distribution of blood cholesterol for teenagers.
2. For young adults, the distribution of blood cholesterol is approximately normal with mean $\mu = 157.5$ milligrams of cholesterol per deciliter of blood (mg/dl). About 8.9% of young adults have high cholesterol—that is, levels of 200 mg/dl or greater. Find the standard deviation of the distribution of blood cholesterol for young adults.

SECTION 1F Summary

- Some distributions of quantitative data can be modeled by symmetric, single-peaked, mound-shaped **normal curves.**
- Any **normal distribution** is completely specified by two numbers: its mean and its standard deviation. The mean μ is the center of the distribution and the standard deviation σ is the distance from μ to the change-of-curvature points on either side.

- An area under a normal curve above any interval of values on the horizontal axis estimates the proportion of values in the distribution that fall within that interval.
- The **empirical rule** describes what percentage of observations in any normal distribution fall within 1, 2, and 3 standard deviations of the mean: about 68%, 95%, and 99.7%, respectively.
- To assess normality for a distribution of quantitative data, we first observe the shape of a dotplot, stemplot, or histogram. Then we check how well the data fit the empirical rule.
- All normal distributions are the same when values are standardized. If a quantitative variable can be modeled by a normal distribution with mean μ and standard deviation σ, we can standardize using

$$z = \frac{\text{value} - \text{mean}}{\text{standard deviation}} = \frac{x - \mu}{\sigma}$$

  The standardized values can be modeled using the **standard normal distribution** with mean 0 and standard deviation 1.
- You can use technology or Table A in the back of the book to find areas or percentiles in any normal distribution. Table A gives percentiles for the standard normal distribution.
- To find the area in a normal distribution corresponding to a given interval of values:

  Step 1: Draw a normal distribution with the horizontal axis labeled and scaled using the mean and standard deviation, the boundary value(s) clearly identified, and the area of interest shaded.

  Step 2: Perform calculations—show your work! Do one of the following:

  (i) Standardize each boundary value and use technology or Table A to find the desired area under the standard normal curve; or

  (ii) Use technology to find the desired area without standardizing. Label the inputs you used for the calculator command.

  Be sure to answer the question that was asked.
- To find the value in a normal distribution corresponding to a given area (percentile):

  Step 1: Draw a normal distribution with the horizontal axis labeled and scaled using the mean and standard deviation, the area of interest shaded and labeled, and the unknown boundary value clearly marked.

> **AP® EXAM TIP**
> **AP® Daily Videos**
> Review the content of this section and get extra help by watching the AP® Daily Videos for Topic 1.10: Videos 2 and 3, which are available in AP® Classroom.

Step 2: Perform calculations—show your work! Do one of the following:

(i) Use technology or Table A to find the value of z with the indicated area under the standard normal curve, then "unstandardize" to transform back to the original distribution; or

(ii) Use technology to find the desired area without standardizing. Label the inputs you used for the calculator command.

Be sure to answer the question that was asked.

- You can find the mean or standard deviation of a normal distribution using one or more percentiles by solving for the missing value in the z-score formula.

1F Tech Corners

TI-Nspire and other technology instructions are on the book's website at bfwpub.com/tps7e.

4. Finding areas in a normal distribution — Page 118

5. Finding percentiles in a normal distribution — Page 125

SECTION 1F Exercises

Normal Curves

1. **Potato chips** The weights of 9-ounce bags of a particular brand of potato chips can be modeled by a normal distribution with mean $\mu = 9.12$ ounce and standard deviation $\sigma = 0.05$ ounce. Draw the normal curve that approximates the distribution of weight. Label the mean and the points that are 1, 2, and 3 standard deviations from the mean. pg 110

2. **Batter up!** In baseball, a player's batting average is the proportion of times the player gets a hit out of his total number of times at bat. The distribution of batting average in a recent season for Major League Baseball players with at least 100 plate appearances can be modeled by a normal distribution with mean $\mu = 0.261$ and standard deviation $\sigma = 0.034$. Draw the normal curve that approximates the distribution of batting average. Label the mean and the points that are 1, 2, and 3 standard deviations from the mean.

3. **Normal curve** Estimate the mean and standard deviation of the following normal distribution.

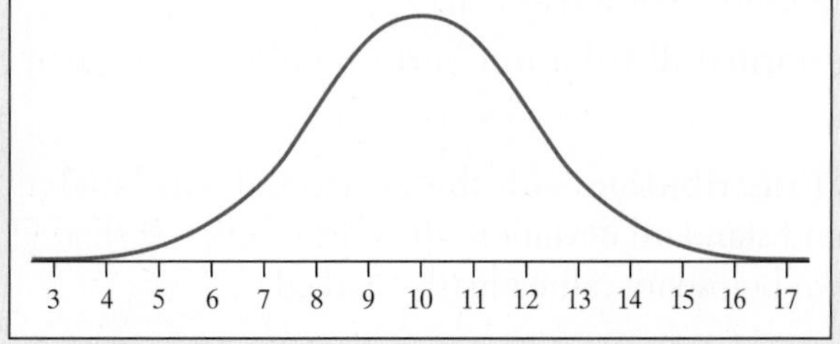

4. **Another normal curve** Estimate the mean and standard deviation of the following normal distribution.

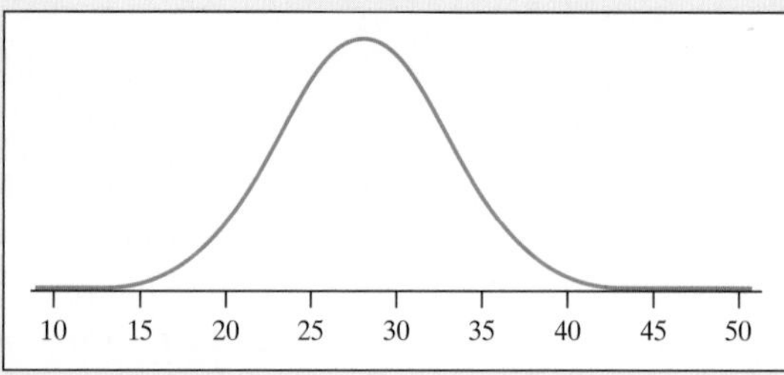

The Empirical Rule

5. **Empirical potato chips** Refer to Exercise 1. About what percentage of bags weigh less than 9.02 ounces? pg 113

6. **Empirical batter up!** Refer to Exercise 2. About what percentage of Major League Baseball players with 100 plate appearances had batting averages of 0.363 or higher?

7. **Black bears** The distribution of weight for male black bears in a large region is approximately normal with a mean of $\mu = 250$ pounds. About 99.7% of these bears have weights between 130 and 370 pounds. Find the standard deviation of the distribution.

8. **Oranges** Mandarin oranges from a certain grove have weights that follow an approximately normal distribution with mean $\mu = 3$ ounces. About 95% of the mandarin oranges in this grove have weights between 2 and 4 ounces. Find the standard deviation of the distribution.

Assessing Normality

9. **Refrigerators** *Consumer Reports* magazine collected data on the usable capacity (in cubic feet) of a sample of 36 side-by-side refrigerators. Here are the data:[124]

pg 115

12.9	13.7	14.1	14.2	14.5	14.5	14.6	14.7	15.1	15.2	15.3	15.3
15.3	15.3	15.5	15.6	15.6	15.8	16.0	16.0	16.2	16.2	16.3	16.4
16.5	16.6	16.6	16.6	16.8	17.0	17.0	17.2	17.4	17.4	17.9	18.4

A histogram of the data and summary statistics are shown here. Is the distribution of usable capacity approximately normal? Give appropriate evidence to justify your answer.

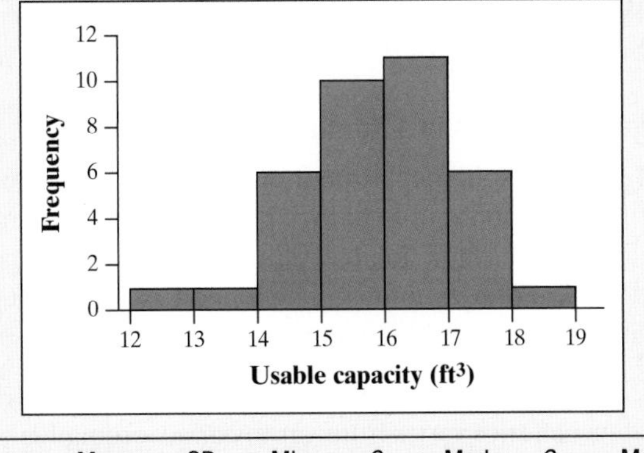

n	Mean	SD	Min	Q_1	Med	Q_3	Max
36	15.825	1.217	12.9	15.15	15.9	16.6	18.4

10. **Big sharks** Here are the lengths (in feet) of 44 great white sharks:[125]

18.7	12.3	18.6	16.4	15.7	18.3	14.6	15.8	14.9	17.6	12.1
16.4	16.7	17.8	16.2	12.6	17.8	13.8	12.2	15.2	14.7	12.4
13.2	15.8	14.3	16.6	9.4	18.2	13.2	13.6	15.3	16.1	13.5
19.1	16.2	22.8	16.8	13.6	13.2	15.7	19.7	18.7	13.2	16.8

A dotplot of the data and summary statistics are shown here. Is the distribution of shark length approximately normal? Give appropriate evidence to justify your answer.

Length (ft)

n	Mean	SD	Min	Q_1	Med	Q_3	Max
44	15.586	2.55	9.4	13.55	15.75	17.2	22.8

11. **Normal highway driving?** The dotplot shows the EPA highway gas mileage estimates in miles per gallon (mpg) for a random sample of 21 model-year 2022 midsize cars.[126] Explain why the distribution of highway gas mileage is not approximately normal.

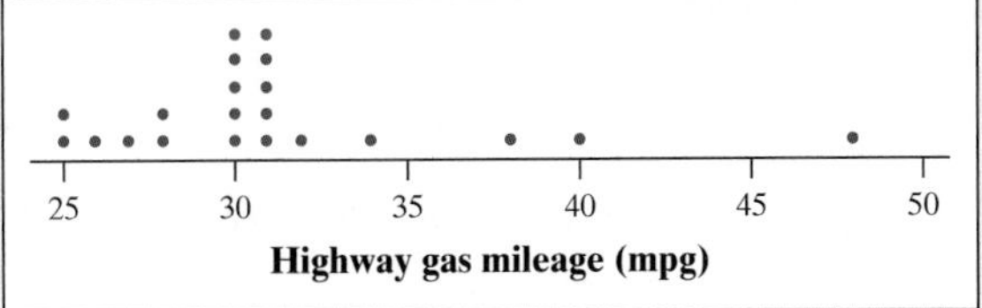

12. **Normal to be foreign born?** The histogram displays the percentage of foreign-born residents in each of the 50 states in a recent year.[127] Explain why the distribution of the percentage of foreign-born residents in the states is not approximately normal.

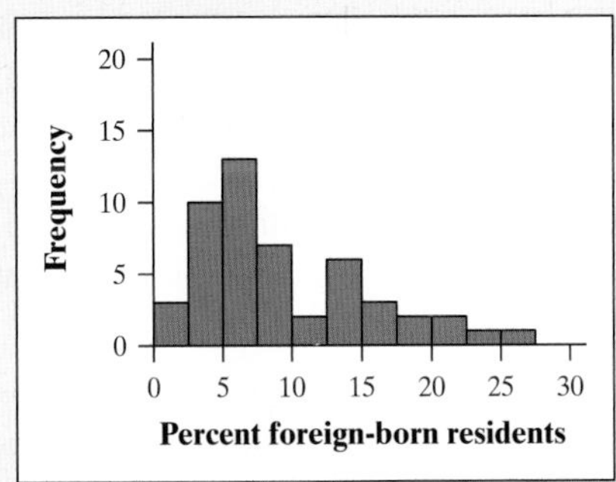

13. **Is tuition normal?** We collected data on the tuition charged by a sample of colleges and universities in Michigan. Here are some numerical summaries for the data:

Mean	SD	Min	Max
10,614	8049	1873	30,823

Based on the relationship between the mean, standard deviation, minimum, and maximum, is it reasonable to believe that the distribution of tuition in Michigan is approximately normal? Explain your answer.

14. **Is cholesterol normal?** The National Health and Nutrition Examination Survey (NHANES) is an ongoing research program conducted by the National Center for Health Statistics. Researchers selected a random sample of more than 7000 U.S. residents for the most recent survey and recorded lab measurements of their blood cholesterol levels.[128] Here are some numerical summaries of the data:

Mean	SD	Min	Q_1	Med	Q_3	Max
179.895	40.642	76	151	176	204	446

Based on the summary statistics, is it reasonable to believe that the distribution of cholesterol level is approximately normal? Explain your answer.

Finding Areas in a Normal Distribution

15. **More potato chips** The weights of 9-ounce bags of a particular brand of potato chips can be modeled by a normal distribution with mean $\mu = 9.12$ ounce and standard deviation $\sigma = 0.05$ ounce. About what percentage of 9-ounce bags of this brand of potato chips weigh less than the advertised 9 ounces? Is this likely to pose a problem for the company that produces these chips?

pg 120

16. **Get a hit!** The distribution of batting average in a recent season for Major League Baseball players with at least 100 plate appearances can be modeled by a normal distribution with mean $\mu = 0.261$ and standard deviation $\sigma = 0.034$. A player with a batting average below

0.200 is at risk of sitting on the bench during important games. About what percentage of players are at risk?

17. **Watch the salt!** A study investigated about 3000 meals ordered from Chipotle restaurants using the online site Grubhub. Researchers calculated the sodium content (in milligrams) for each order based on Chipotle's published nutrition information. The distribution of sodium content is approximately normal with mean 1993 mg and standard deviation 593 mg.[129] About what proportion of the meals ordered exceeded the recommended daily allowance of 2400 mg of sodium?
pg 121

18. **Blood pressure** According to a health information website, the distribution of adults' diastolic blood pressure (in millimeters of mercury, mmHg) can be modeled by a normal distribution with mean 70 mmHg and standard deviation 20 mmHg. A diastolic pressure greater than 100 mmHg for an adult is classified as very high blood pressure. About what proportion of adults have very high blood pressure according to this criterion?

19. **Limit the salt!** Refer to Exercise 17. About what percentage of the meals ordered at Chipotle contained between 1200 mg and 1800 mg of sodium?
pg 123

20. **High blood pressure?** Refer to Exercise 18. According to the same health information website, a diastolic blood pressure between 80 and 90 mmHg indicates borderline high blood pressure. About what percentage of adults have borderline high blood pressure?

21. **Standard normal areas** Find the proportion of values in a standard normal distribution that satisfies each of the following statements.

(a) $z > -1.66$

(b) $-1.66 < z < 2.85$

(c) $z > 3.90$

22. **More standard normal areas** Find the proportion of values in a standard normal distribution that satisfies each of the following statements.

(a) $z < -2.46$

(b) $0.89 < z < 2.46$

(c) $z < -4.02$

23. **Sudoku** Mrs. Starnes enjoys doing Sudoku puzzles. The time she takes to complete an easy puzzle can be modeled by a normal distribution with mean 5.3 minutes and standard deviation 0.9 minute.

(a) What proportion of the time does Mrs. Starnes finish an easy Sudoku puzzle in less than 3 minutes?

(b) How often does it take Mrs. Starnes more than 6 minutes to complete an easy puzzle?

(c) What percentage of easy Sudoku puzzles takes Mrs. Starnes between 6 and 8 minutes to complete?

24. **Hit an ace!** Professional tennis player Novak Djokovic hits the ball extremely hard. His first-serve speeds can be modeled by a normal distribution with mean 112 miles per hour (mph) and standard deviation 5 mph.

(a) How often does Djokovic hit his first serve faster than 120 mph?

(b) What percentage of Djokovic's first serves are slower than 100 mph?

(c) What proportion of Djokovic's first serves have speeds between 100 and 110 mph?

Finding Percentiles in a Normal Distribution

25. **Fire!** A fire department in a small county reports that its response time to fires is approximately normally distributed with a mean of 22 minutes and a standard deviation of 6.9 minutes. The longest 1% of response times take at least how many minutes?
pg 126

26. **Car batteries** An automaker has found that the life of its batteries varies from car to car according to an approximately normal distribution with mean $\mu = 48$ months and standard deviation $\sigma = 8$ months. Find the 30th percentile of the distribution.

27. **Quartiles** Find the 25th percentile (Q_1) and the 75th percentile (Q_3) of the standard normal distribution.

28. **Deciles** The deciles of any distribution are the values at the 10th, 20th, ..., 90th percentiles. The first and last deciles are the 10th and the 90th percentiles, respectively. What are the first and last deciles of the standard normal distribution?

29. **Birth weights** Researchers in Norway analyzed data on the birth weights of 400,000 newborns over a 6-year period. The distribution of birth weight is approximately normal with a mean of 3668 grams and a standard deviation of 511 grams.[130] Babies who weigh less than 2500 grams at birth are classified as "low birth weight."

(a) What proportion of newborns would be identified as low birth weight?

(b) Find the 80th percentile of the distribution of birth weight.

30. **Post office** A local post office weighs outgoing mail and finds that the weights of first-class letters are approximately normally distributed with a mean of 0.69 ounce and a standard deviation of 0.16 ounce.

(a) First-class letters weighing more than 1 ounce require extra postage. What proportion of first-class letters at this post office require extra postage?

(b) Forty percent of first-class letters weigh more than what amount?

Calculating the Mean or Standard Deviation of a Normal Distribution

31. NBA heights The distribution of height for National Basketball Association (NBA) players is approximately normal with a mean of 78.4 inches.[131] If 5.7% of players have heights greater than 84 inches, calculate the standard deviation of the distribution.

pg 127

32. **Helmet sizes** The army reports that the distribution of head circumference for its soldiers is approximately normal with a mean of 22.8 inches. If 23.4% of soldiers have head circumferences less than 22 inches, calculate the standard deviation of the distribution.

33. **Sub shop** The lengths of foot-long sub sandwiches at a local sub shop follow an approximately normal distribution with unknown mean μ and standard deviation 0.2 inch. If 20% of these sandwiches are shorter than 11.7 inches, find the mean length μ.

34. **Rally time** A tennis ball machine fires balls a distance that is approximately normally distributed. The mean distance μ is unknown and the standard deviation is 1.2 feet. If 5% of balls go farther than 70 feet, find μ.

Exercises 35 and 36 refer to the following setting. At some fast-food restaurants, customers who want a lid for their drinks get them from a large stack near the straws, napkins, and condiments. The lids are made with a small amount of flexibility so they can be stretched across the mouth of the cup and then snugly secured. When lids are too small or too large, customers can get very frustrated, especially if they end up spilling their drinks. At one particular restaurant, large drink cups require lids with a diameter of between 3.95 and 4.05 inches. The restaurant's lid supplier claims that the diameter of its large lids follows a normal distribution with mean 3.98 inches and standard deviation 0.02 inch. Assume that the supplier's claim is true.

35. **Put a lid on it!**

(a) What percentage of large lids are too small to fit?

(b) What percentage of large lids are too big to fit?

(c) Compare your answers to parts (a) and (b). Does it make sense for the lid manufacturer to try to make one of these values larger than the other? Why or why not?

36. **A better fit?** The supplier is considering two changes to reduce to 1% the percentage of its large-cup lids that are too small. One strategy is to adjust the mean diameter of its lids. Another option is to decrease the standard deviation of the lid diameters by altering the production process.

(a) If the standard deviation remains $\sigma = 0.02$ inch, at what value should the supplier set the mean diameter of its large-cup lids so that only 1% are too small to fit?

(b) If the mean diameter remains $\mu = 3.98$ inches, what value of the standard deviation will result in only 1% of lids that are too small to fit?

(c) Which of the two options in parts (a) and (b) do you think is preferable? Justify your answer. (Be sure to consider the effect of these changes on the percentage of lids that are too large to fit.)

For Investigation *Apply the skills from the section in a new context or nonroutine way.*

37. **Outliers** Researchers in Norway analyzed data on the birth weights of 400,000 newborns over a 6-year period. The distribution of birth weight is approximately normal with a mean of 3668 grams and a standard deviation of 511 grams.[132]

(a) What birth weights would be considered outliers according to the $1.5 \times IQR$ rule?

(b) Based on your answer to part (a), what proportion of Norwegian newborns would have birth weights that are considered outliers?

38. **Flight times** An airline flies the same route at the same time each day. The flight time can be modeled by a normal distribution with unknown mean and standard deviation. On 15% of days, the flight takes more than 60 minutes. On 3% of days, the flight lasts 75 minutes or more. Use this information to determine the mean and standard deviation of the flight time distribution.

39. **How close to 68%?** According to the empirical rule, about 68% of the values in a normal distribution should be within 1 standard deviation of the mean. For a quantitative data set with $n = 77$, how close should the percentage of values within 1 SD of the mean be to 68%? To find out, we used software to take 500 random samples of size $n = 77$ from a normally distributed population.[133] The following dotplot shows the proportion of values that were within 1 standard deviation of the mean in each simulated sample.

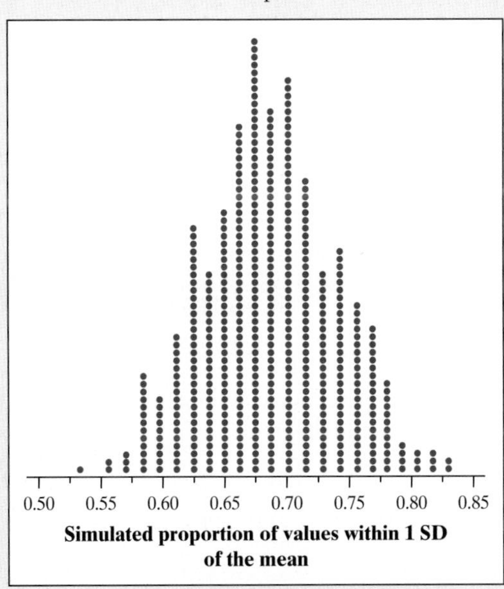

(a) In what percentage of the 500 simulated samples was the proportion of values within 1 standard deviation of the mean greater than or equal to 0.818?

(b) Based on your answer to part (a), would it be surprising to get a sample of 77 values from a normally distributed population in which 81.8% of the values are within 1 SD of the mean? Explain your answer.

(c) In the calories in cereal data set displayed in Figure 1.24 (page 114), 63/77 = 81.8% of the values were within 1 standard deviation of the mean. Based on this information and your answer to part (b), what should you conclude about whether the distribution of calories in cereal is approximately normally distributed? Explain your answer.

Multiple Choice *Select the best answer for each question.*

40. Which of the following graphs would be the least helpful in determining whether a distribution of quantitative data is approximately normal?

(A) A boxplot
(B) A dotplot
(C) A histogram
(D) A stemplot
(E) All of the graphs in choices (A)–(D) would be equally helpful.

Exercises 41–43 refer to the following setting. The weights of laboratory cockroaches can be modeled with a normal distribution having mean 80 grams and standard deviation 2 grams. The following figure shows the normal curve for this distribution of weight.

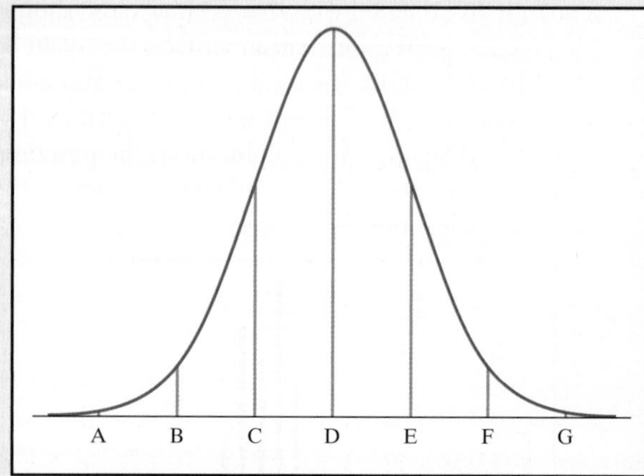

41. Point C on this normal distribution corresponds to

(A) 74 grams.
(B) 76 grams.
(C) 78 grams.
(D) 82 grams.
(E) 84 grams.

42. About what percentage of the cockroaches weigh between 74 and 86 grams?

(A) 47.5%
(B) 49.85%
(C) 68%
(D) 95%
(E) 99.7%

43. About what proportion of the cockroaches will have weights greater than 83 grams?

(A) 0.0228
(B) 0.0668
(C) 0.0772
(D) 0.1587
(E) 0.9332

44. A different species of cockroach has weights that are approximately normally distributed with a mean of 50 grams. After measuring the weights of many of these cockroaches, a lab assistant reports that 14% of the cockroaches weigh more than 55 grams. Based on this report, what is the approximate standard deviation of weight for this species of cockroach?

(A) 4.6

(B) 5.0

(C) 6.2

(D) 14.0

(E) Cannot determine without more information.

Recycle and Review *Practice what you learned in previous sections.*

45. **Making money (1E)** The parallel dotplots show the total family income of randomly chosen individuals from Indiana (38 individuals) and New Jersey (44 individuals). Means and standard deviations are given below the dotplots.

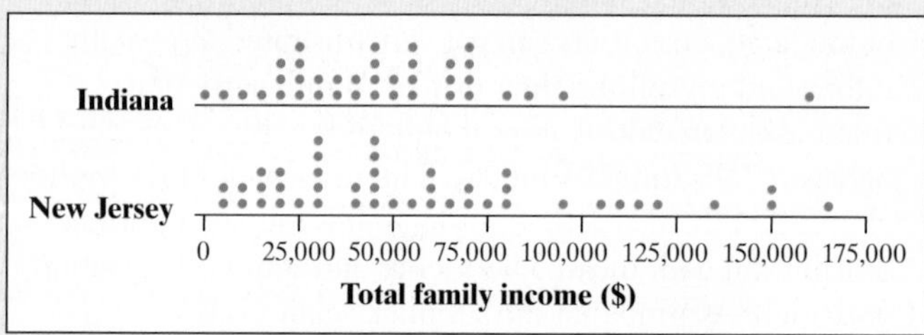

State	Mean	Standard deviation
Indiana	$47,400	$29,400
New Jersey	$58,100	$41,900

Consider individuals in each state with total family incomes of $95,000. Which individual has a higher income, relative to others in their state? Use percentiles and z-scores to support your answer.

46. **More money (1D)** Refer to Exercise 45.

(a) How do the ranges of the two distributions compare? Justify your answer.

(b) Explain why the standard deviation of the total family income in the New Jersey sample is so much larger than that for the Indiana sample.

UNIT 1 Wrap-Up

PART II

FRAPPY! Free Response AP® Problem, Yay!

Directions: Show all your work. Indicate clearly the methods you use, because you will be scored on the correctness of your methods as well as on the accuracy and completeness of your results and explanations.

The distribution of scores on a recent test closely followed a normal distribution with a mean of 22 points and a standard deviation of 4 points.

(a) What proportion of the students scored at least 25 points on this test?

(b) What is the 31st percentile of the distribution of test scores?

(c) The teacher wants to transform the test scores so that they have an approximately normal distribution with a mean of 80 points and a standard deviation of 10 points. To do this, she will use a formula in the form

$$\text{new score} = a + b(\text{old score})$$

Find the values of a and b that the teacher should use to transform the distribution of test scores.

(d) Before the test, the teacher gave a review assignment for homework. The maximum score on the assignment was 10 points. The distribution of scores on this assignment had a mean of 9.2 points and a standard deviation of 2.1 points. Would it be appropriate to use a normal distribution to calculate the proportion of students who scored below 7 points on this assignment? Explain your answer.

After you finish the FRAPPY!, you can view two example solutions on the book's website **(bfwpub.com/tps7e)**. Determine whether you think each solution is "complete," "substantial," "developing," or "minimal." If the solution is not complete, what improvements would you suggest to the student who wrote it? Finally, your teacher will provide a scoring rubric. Score your response and note what, if anything, you would do differently to improve your own score.

UNIT 1, PART II REVIEW

SECTION 1E Describing Position and Transforming Data

In this section, you learned two different ways to describe the *position* of individuals in a distribution of quantitative data: **percentiles** and **standardized scores (*z*-scores)**. Percentiles describe the position of an individual value in a distribution by measuring what percentage of the observations are less than or equal to that value. Standardized scores (*z*-scores) describe the position of an individual value in a distribution by measuring how many standard deviations the value is above or below the mean. To find the standardized score for a particular observation, transform the value by subtracting the mean and then dividing the difference by the standard deviation. Besides describing the position of an individual in a distribution, you can also use percentiles and *z*-scores to compare the relative positions of individual values in one or more distributions of quantitative data.

A **cumulative relative frequency graph** is a handy tool for displaying percentiles in the distribution of a quantitative variable. You can use it to estimate the percentile for a particular value of the variable or to estimate the value of the variable at a particular percentile.

You also learned to describe the effects of *transforming data* on the shape, center, and variability of a distribution of quantitative data. Adding a positive constant to (or subtracting it from) each value in a data set changes measures of center, but not the shape or variability of the distribution. Multiplying or dividing each value in a data set by a positive constant changes measures of center and variability, but not the shape of the distribution.

SECTION 1F Normal Distributions

In this section, you learned how a **normal curve** can be used to model some distributions of quantitative data. A **normal distribution** is symmetric, single-peaked, and mound-shaped with mean μ and standard deviation σ. For a quantitative variable that can be modeled by a normal distribution, the area under the normal curve above an interval of values on the horizontal axis estimates the proportion or percentage of data values in that interval. You learned how to draw a normal curve with the mean and the points that are 1, 2, and 3 standard deviations from the mean labeled.

For any distribution that is approximately normal, about 68% of the values will be within 1 standard deviation of the mean, about 95% of the values will be within 2 standard deviations of the mean, and about 99.7% of the values will be within 3 standard deviations of the mean. This handy result is known as the **empirical rule.** You learned how to determine whether a distribution of quantitative data is approximately normal using graphs (dotplots, stemplots, histograms) and the empirical rule.

When values in a distribution do not fall exactly 1, 2, or 3 standard deviations from the mean, you learned how to use technology or Table A to calculate the proportion of values in any specified interval under a normal curve. This includes areas to the left, areas to the right, and areas between two values in a normal distribution. You also learned how to determine the value of an individual that falls at a specified percentile in a normal distribution, and how to find the mean or standard deviation of a normal distribution given the value of a percentile.

On the AP® Statistics exam, it is extremely important that you clearly communicate your methods when answering free-response questions that involve a normal distribution. Start by drawing a normal distribution with the horizontal axis labeled and scaled using the mean and standard deviation, the boundary value(s) clearly identified, and the area of interest shaded. Then perform calculations, being sure to show your work. If you use technology without standardizing as your method, be sure to label the inputs of your calculator command.

What Did You Learn?

Learning Target	Section	Related Example on Page(s)	Relevant Chapter Review Exercise(s)
Calculate and interpret a percentile in a distribution of quantitative data.	1E	90	R1
Calculate and interpret a standardized score (*z*-score) in a distribution of quantitative data.	1E	91	R1
Use percentiles or standardized scores (*z*-scores) to compare the relative positions of individual values in distributions of quantitative data.	1E	93	R1
Use a cumulative relative frequency graph to estimate percentiles and individual values in a distribution of quantitative data.	1E	95	R2
Analyze the effect of adding, subtracting, multiplying by, or dividing by a constant on the shape, center, and variability of a distribution of quantitative data.	1E	98	R3
Draw a normal curve to model the distribution of a quantitative variable.	1F	110	R4
Use the empirical rule to estimate the proportion of values in a specified interval in a normal distribution.	1F	113	R4
Determine whether a distribution of quantitative data is approximately normal using graphical and numerical evidence.	1F	115	R5
Find the proportion of values in a specified interval in a normal distribution.	1F	120, 121, 123	R6, R7
Find the value that corresponds to a given percentile in a normal distribution.	1F	126	R6
Calculate the mean or standard deviation of a normal distribution given the value of a percentile.	1F	127	R7

T5 The average yearly snowfall in Chillyville is approximately normally distributed with a standard deviation of 4.83 inches. If the snowfall in Chillyville exceeds 60 inches in 15% of the years, what is the mean?

(A) 55 inches
(B) 55.34 inches
(C) 58.96 inches
(D) 65 inches
(E) The mean cannot be computed from the given information.

T6 How much oil the wells in a given field will ultimately produce is crucial information in deciding whether to drill more wells. Some descriptive statistics for the total amount of oil produced by 38 wells in one region, in thousands of barrels, are shown here.

Descriptive Statistics

n	Mean	StDev	Minimum	Q_1	Median	Q_3	Maximum
38	50.397	28.28	3.00	34.50	47.00	65.25	157.00

One well in the region had a standardized score of $z = -1.676$. How many barrels of oil did this well produce?

(A) 3
(B) 48.721
(C) 3000
(D) 48,721
(E) 97,794

T7 If the distribution of height in a population is approximately normally distributed, and the middle 99.7% of individuals have heights between 60 inches and 84 inches, what is the standard deviation of the distribution?

(A) 1 inch
(B) 3 inches
(C) 4 inches
(D) 6 inches
(E) 12 inches

T8 The distribution of the time it takes for different people to solve a certain crossword puzzle is skewed to the right with a mean of 30 minutes and a standard deviation of 15 minutes. The distribution of *z*-scores for those times is

(A) normally distributed with mean 30 and SD 15.
(B) skewed to the right with mean 30 and SD 15.
(C) normally distributed with mean 0 and SD 1.
(D) skewed to the right with mean 0 and SD 1.
(E) skewed to the right, but the mean and standard deviation cannot be determined without more information.

T9 In 1965, the mean price of a new car was \$2650 and the standard deviation was \$1000. In 2021, the mean price was \$41,000 and the standard deviation was \$9500. If a Ford Mustang cost \$2300 in 1965 and \$38,000 in 2021, in which year was it more expensive relative to other cars?

(A) 1965, because the standardized score is greater than in 2021
(B) 1965, because the standard deviation is smaller
(C) 2021, because the standardized score is greater than in 1965
(D) 2021, because \$38,000 is greater than \$2300
(E) The 1965 and 2021 Ford Mustangs are equally expensive relative to other cars in those years.

T10 A small company estimating its copying expenses finds that the number of copies made each day during the past year has mean 358 and standard deviation 34. Which of the following is a correct interpretation of the standard deviation?

(A) The number of copies made per day was always between 256 and 460.
(B) On about 95% of days, the number of copies made per day was between 290 and 426.
(C) The typical difference between the mean and the median number of copies made per day was about 34.
(D) The number of copies made each day typically varied by about 34 copies.
(E) The number of copies made each day typically varied from the mean by about 34 copies.

Section II: Free Response *Show all your work. Indicate clearly the methods you use, because you will be graded on the correctness of your methods as well as on the accuracy and completeness of your results and explanations.*

T11 According to the truecar.com website, the asking prices for cars of a certain model are approximately normally distributed with mean \$35,987 and standard deviation \$607.50.[135]

(a) According to the website, any asking price between \$34,772 and \$36,225 for this car model is considered "good" or "great." What proportion of cars have "good" or "great" asking prices?

(b) The manufacturer's suggested retail price (MSRP) for this car model is at the 98th percentile of the distribution of asking price. Find the MSRP.

Questions 12 and 13 refer to the following setting. For more than a century, doctors have been telling patients that normal body temperature is 98.6°F (37.0°C). This value dates back to a study done by Carl Wunderlich in the mid-1800s. More recently, researchers conducted a study to determine whether the "accepted" value for normal body temperature is accurate. They collected body temperatures (in degrees Celsius, °C) from 130 healthy individuals.[136] Here is a dotplot of the data along with summary statistics:

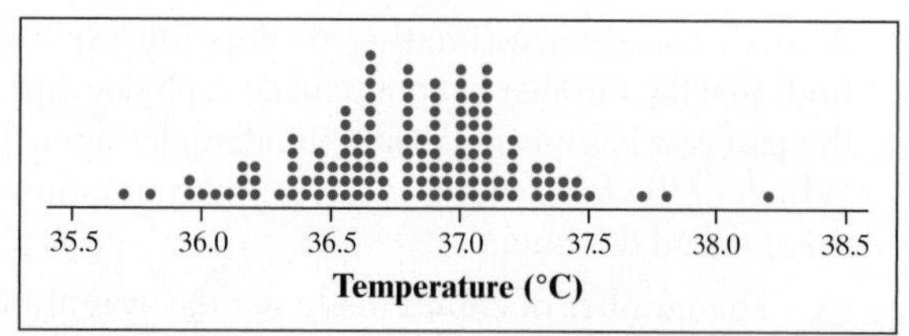

n	Mean	SD	Min	Q_1	Med	Q_3	Max
130	36.805	0.407	35.722	36.556	36.833	37.056	38.222

T12 One of the participants in the study, Mikaela, had a temperature of 37.5°C.

(a) Calculate and interpret Mikaela's standardized score.

(b) Find Mikaela's percentile.

(c) Another participant in the study, Jin-Yu, had a temperature of 36.3°C. Whose temperature was more unusual relative to the other individuals in this study: Mikaela's or Jin-Yu's? Justify your answer.

T13 (a) Do the data provide some evidence that the accepted value for normal body temperature of 37.0°C is incorrect? Explain your answer.

(b) Is the distribution of body temperature approximately normal? Give appropriate evidence to support your answer.

UNIT

2

Exploring Two-Variable Data

Introduction

Although we can gain many insights into the world around us by analyzing one variable at a time, investigating relationships between variables is central to what we do in statistics. When we understand the relationship between two variables, we can use the value of one variable to help us make predictions about the other variable. In this unit, we'll start by exploring relationships between two *categorical* variables, such as membership in an environmental club and snowmobile use by visitors to Yellowstone National Park. In Sections 2B–2D, we investigate relationships between two *quantitative* variables, such as the number of miles a used car has been driven and its price.

SECTION 2A Relationships Between Two Categorical Variables

LEARNING TARGETS *By the end of the section, you should be able to:*

- Identify the explanatory and response variables in a given setting.
- Calculate statistics for two categorical variables.
- Display the relationship between two categorical variables.
- Describe the relationship between two categorical variables.

In Section 1B, you learned techniques for displaying and describing the distribution of a single categorical variable. In this section, you'll learn how to perform calculations and make graphs to investigate the relationship between two categorical variables.

Explanatory and Response Variables

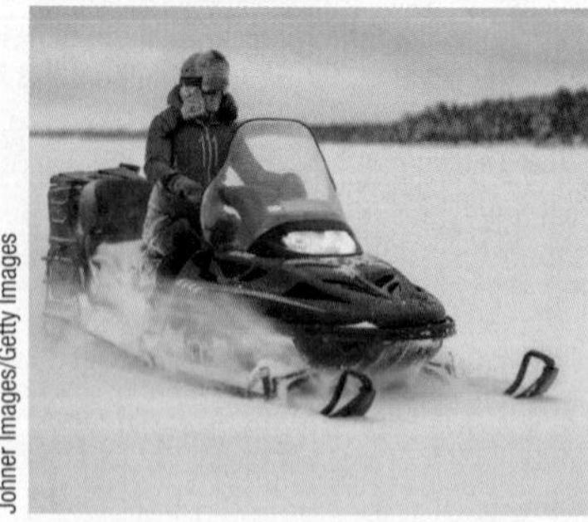
Johner Images/Getty Images

Yellowstone National Park staff surveyed a random sample of 1526 winter visitors to the park.[1] They asked each person whether they belonged to an environmental club (like the Sierra Club). Respondents were also asked if they owned, rented, or never had used a snowmobile. The data set looks like this:

Respondent	Environmental club membership	Snowmobile use
1	No	Own
2	No	Rent
3	Yes	Never
4	Yes	Rent
5	No	Never
⋮	⋮	⋮

In this data set, snowmobile use is the **response variable.** Environmental club membership is the **explanatory variable** because we anticipate that knowing whether or not an individual is a member of an environmental club will help predict their snowmobile use.

DEFINITION Response variable, Explanatory variable

A **response variable** measures an outcome of a study.
An **explanatory variable** may help predict or explain changes in a response variable.

You will often see explanatory variables called *independent variables* and response variables called *dependent variables*. Because the words *independent* and *dependent* have other meanings in statistics, we won't use them in this way.

It is easiest to identify explanatory and response variables when we actively change the values of one variable to see how it affects another variable. For instance, to study the effect of alcohol on body temperature, researchers gave several different amounts of alcohol to mice. Then they measured the change in each mouse's body temperature 15 minutes later. In this case, amount of alcohol is the explanatory variable, and change in body temperature is the response variable. When we don't specify the values of either variable before collecting the data, there may or may not be a clear explanatory variable.

EXAMPLE

Diamonds and color
Explanatory and response variables

Skill 2.B

PROBLEM: Identify the explanatory variable and response variable for the following relationships, if possible. Explain your reasoning.

(a) The weight (in carats) and the price (in dollars) for a sample of diamonds
(b) The hair color and eye color for a sample of students

stori/Deposit Photos

SOLUTION:

(a) Explanatory: weight; Response: price. The weight of a diamond helps explain how expensive it is.
(b) Either hair color or eye color could be the explanatory variable because each one could be used to predict the other.

FOR PRACTICE, TRY EXERCISE 1

In many studies, the goal is to show that changes in one or more explanatory variables actually *cause* changes in a response variable. However, other explanatory–response relationships don't involve direct causation. In the alcohol and mice study, alcohol actually *causes* a change in body temperature. In contrast, there is no cause-and-effect relationship between hair color and eye color.

Summarizing Data on Two Categorical Variables

It is common to summarize data on two categorical variables with a **two-way table** (sometimes called a *contingency* table). Here is the two-way table that summarizes the Yellowstone survey responses.

		Environmental club membership	
		No	Yes
	Never	445	212
Snowmobile use	Rent	497	77
	Own	279	16

DEFINITION Two-way table

A **two-way table** is a table of counts or relative frequencies that summarizes data on the relationship between two categorical variables for some group of individuals.

It's easier to grasp the information in a two-way table if row and column totals are included, like the one shown here.

		Environmental club membership		
		No	Yes	Total
Snowmobile use	Never	445	212	657
	Rent	497	77	574
	Own	279	16	295
	Total	1221	305	1526

Now we can quickly answer questions like these:

- What percentage of people in the sample are environmental club members?

$$\frac{305}{1526} = 0.200 = 20.0\%$$

- What proportion of people in the sample have never used a snowmobile?

$$\frac{657}{1526} = 0.431$$

These percentages and proportions are known as **marginal relative frequencies** because they are calculated using values in the margins of the two-way table. Note that a proportion is always a number between 0 and 1, whereas a percentage is a number between 0 and 100. To convert from a proportion to a percentage, multiply by 100.

DEFINITION Marginal relative frequency

A **marginal relative frequency** gives the percentage or proportion of individuals in a two-way table that have a specific value for one categorical variable. A marginal relative frequency is calculated by dividing a row or column total by the total for the entire two-way table.

We can compute marginal relative frequencies for the column totals to give the distribution of environmental club membership in the entire sample of 1526 park visitors:

$$\text{No: } \frac{1221}{1526} = 0.800 \text{ or } 80.0\% \quad \text{Yes: } \frac{305}{1526} = 0.200 \text{ or } 20.0\%$$

We could also compute marginal relative frequencies for the row totals to give the distribution of snowmobile use for all the individuals in the sample. Note that we could use a bar graph or a pie chart to display either of these distributions, which are sometimes called *marginal distributions*.

A marginal relative frequency tells you about only *one* of the variables in a two-way table. It won't help you answer questions like these, which involve values of *both* variables:

- What percentage of people in the sample are environmental club members and own snowmobiles?

$$\frac{16}{1526} = 0.010 = 1.0\%$$

- What proportion of people in the sample are not environmental club members and have never used snowmobiles?

$$\frac{445}{1526} = 0.292$$

These percentages or proportions are known as **joint relative frequencies** because they are found using the value where a row and a column come together. You can remember this by thinking about the joints in your body—they are where two bones come together.

DEFINITION Joint relative frequency

A **joint relative frequency** gives the percentage or proportion of individuals in a two-way table that have a specific value for one categorical variable and a specific value for another categorical variable. A joint relative frequency is calculated by dividing the value in one cell by the total for the entire two-way table.

Some two-way tables display marginal and joint relative frequencies instead of counts. For example, here are two different two-way tables for the Yellowstone survey data. The first table has frequencies (counts) in the cells. The second table divides the frequencies in each cell by the total number of individuals in the sample (1526) to give joint and marginal relative frequencies. *Note:* Relative frequency totals might be slightly off due to rounding.

		Environmental club membership		
		No	Yes	Total
Snowmobile use	Never	445	212	657
	Rent	497	77	574
	Own	279	16	295
	Total	1221	305	1526

		Environmental club membership		
		No	Yes	Total
Snowmobile use	Never	0.292	0.139	0.431
	Rent	0.326	0.050	0.376
	Own	0.183	0.010	0.193
	Total	0.800	0.200	1.000

Marginal and joint relative frequencies don't tell us much about the *relationship* between environmental club membership and snowmobile use for the people in the sample. However, we can use the two-way table of counts to investigate this relationship with questions like these:

- What percentage of environmental club members in the sample are snowmobile owners?

$$\frac{16}{305} = 0.052 = 5.2\%$$

- What proportion of snowmobile renters in the sample are not environmental club members?

$$\frac{497}{574} = 0.866$$

These percentages or proportions are known as **conditional relative frequencies** because they give the relative frequency when a certain condition is met (e.g., the person is not a member of an environmental club).

DEFINITION Conditional relative frequency

A **conditional relative frequency** gives the percentage or proportion of individuals that have a specific value for one categorical variable among a group of individuals that share the same value of another categorical variable (the condition). A conditional relative frequency is calculated by dividing the value in one cell of a two-way table by the total for the appropriate row or column.

EXAMPLE

A *Titanic* disaster
Summarizing data on two categorical variables

Skill 2.C

Blank Archives/Getty Images

PROBLEM: In 1912, the luxury liner *Titanic*, on its first voyage across the Atlantic, struck an iceberg and sank. Some passengers got off the ship in lifeboats, but many died. The two-way table gives information about adult passengers who survived and who died, by class of travel.[2]

		Class of travel			
		First	Second	Third	Total
Survival status	Survived	197	94	151	442
	Died	122	167	476	765
	Total	319	261	627	1207

(a) What proportion of adult passengers on the *Titanic* survived?
(b) What percentage of adult *Titanic* passengers traveled in third class and survived?
(c) What proportion of survivors were third-class passengers?
(d) What percentage of third-class passengers survived?

SOLUTION:

(a) $\frac{442}{1207} = 0.366$

This is an example of a marginal relative frequency.

(b) $\frac{151}{1207} = 0.125 = 12.5\%$

This is an example of a joint relative frequency.

(c) $\frac{151}{442} = 0.342$

(d) $\frac{151}{627} = 0.241 = 24.1\%$

Although the conditional relative frequencies asked for in parts (c) and (d) sound similar, they are different. The denominator for part (c) is the number of survivors and the denominator for part (d) is the number of third-class passengers.

FOR PRACTICE, TRY EXERCISE 3

Displaying the Relationship Between Two Categorical Variables

Let's return to the two-way table that summarizes data from the Yellowstone National Park survey of 1526 randomly selected winter visitors.

		Environmental club membership		
		No	Yes	Total
Snowmobile use	Never	445	212	657
	Rent	497	77	574
	Own	279	16	295
	Total	1221	305	1526

To explore the relationship between environmental club membership and snowmobile use, we start by calculating the relative frequency of Never, Rent, and Own for each value of the explanatory variable (No, Yes). The following table summarizes the distribution of snowmobile use for environmental club members and non-club members. Note that the percentages should add to 100% (up to the limits of rounding) for each of these distributions, which are sometimes called *conditional distributions*.

Snowmobile use	Non-club members	Environmental club members
Never	$\frac{445}{1221} = 0.364$ or 36.4%	$\frac{212}{305} = 0.695$ or 69.5%
Rent	$\frac{497}{1221} = 0.407$ or 40.7%	$\frac{77}{305} = 0.252$ or 25.2%
Own	$\frac{279}{1221} = 0.229$ or 22.9%	$\frac{16}{305} = 0.052$ or 5.2%

AP® EXAM TIP

When comparing groups of different sizes, be sure to use relative frequencies (percentages or proportions) instead of frequencies (counts) when analyzing data on two categorical variables. Comparing only the frequencies can be misleading, as in this setting. There are many more people who never use snowmobiles among the non–environmental club members in the sample (445) than among the environmental club members (212). However, the *percentage* of environmental club members who never use snowmobiles (69.5%) is much higher than the percentage of non-members who never use snowmobiles (36.4%). Finally, avoid making statements like "More club members never use snowmobiles" when you mean "A greater percentage of club members never use snowmobiles." It is important to use statistical language precisely.

In Section 1B, you learned how to use side-by-side bar graphs to compare distributions of a categorical variable for two or more groups. Figure 2.1(a) shows the side-by-side bar graph comparing the distribution of snowmobile use for environmental club members and non-club members. Figure 2.1(b) shows a **segmented bar graph** of the same data. Notice that the segmented bar graph can be constructed by stacking the bars in the side-by-side bar graph for each of the two environmental club membership categories (no and yes).

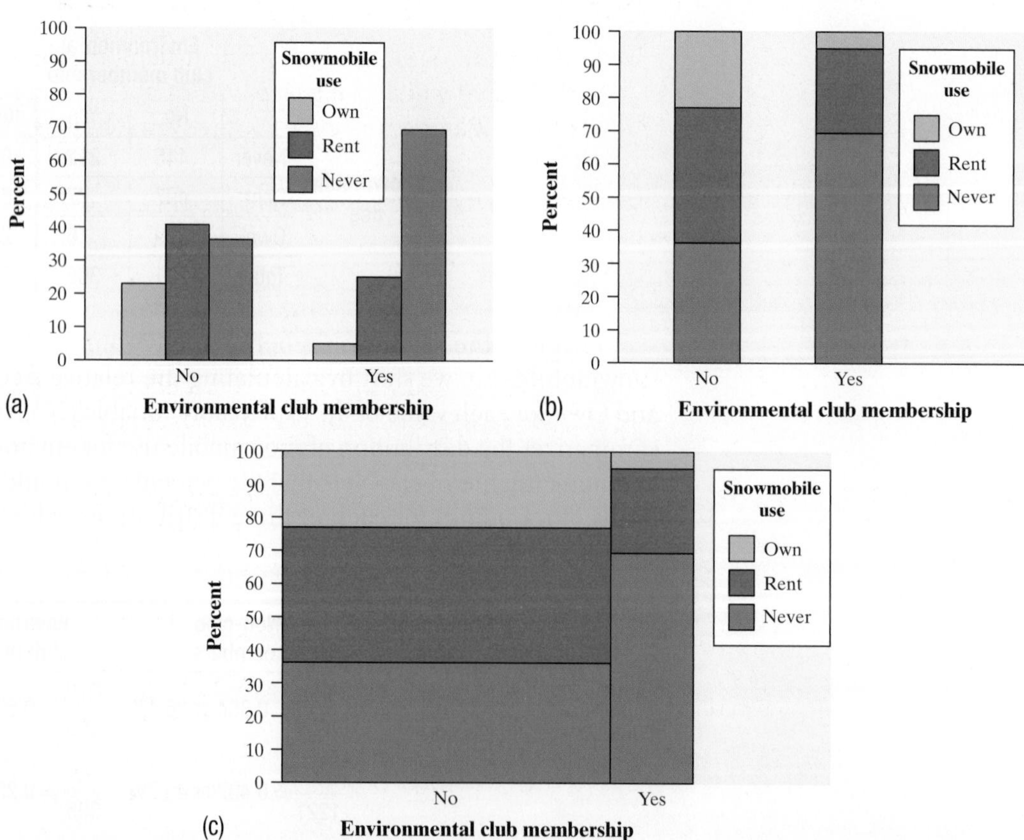

FIGURE 2.1 (a) Side-by-side bar graph, (b) segmented bar graph, and (c) mosaic plot displaying the distribution of snowmobile use among environmental club members and among non-club members for the 1526 randomly selected winter visitors to Yellowstone National Park.

DEFINITION Segmented bar graph

A **segmented bar graph** displays the distribution of a categorical variable as segments of a bar, with the area of each segment proportional to the number of individuals in the corresponding category.

Figure 2.1(c) shows a **mosaic plot,** which is a variation of a segmented bar graph that provides additional information about the distribution of the explanatory variable. By making the bar width proportional to the number of survey respondents who are (305) and are not (1221) environmental club members, we can see that the number of non-members in the survey was roughly 4 times larger than the number of members.

DEFINITION Mosaic plot

A **mosaic plot** is a modified segmented bar graph in which the width of each bar is proportional to the number of individuals in the corresponding category.

When you know how to make a side-by-side bar graph, it isn't hard to make a segmented bar graph.

HOW TO MAKE A SEGMENTED BAR GRAPH

1. **Identify the variables.** Determine which variable is the explanatory variable and which is the response variable. If there is no explanatory/response relationship, choose either variable as "explanatory" and follow the remaining steps.
2. **Draw and label the axes.** Put the name of the explanatory variable under the horizontal axis. To the left of the vertical axis, indicate whether the graph shows the percentage (or proportion) of individuals in each category of the response variable.
3. **Scale the axes.** Write the names of the categories of the explanatory variable at equally spaced intervals under the horizontal axis. On the vertical axis, start at 0% (or 0) and place tick marks at equal intervals until you reach 100% (or 1).
4. **Draw "100%" bars** above each of the category names for the explanatory variable on the horizontal axis so that each bar ends at the top of the graph. Make the bars equal in width and leave gaps between them.
5. **Segment each of the bars.** For each category of the explanatory variable, calculate the relative frequency for each category of the response variable. Then divide the corresponding bar so that the area of each segment corresponds to the proportion of individuals in each category of the response variable.
6. **Include a key** that identifies the different categories of the response variable.

To make a mosaic plot, follow the same instructions but make the widths of the bars proportional to the number of individuals in each category of the explanatory variable.

EXAMPLE

A *Titanic* disaster, part 2

Skill 2.B

Displaying the relationship between two categorical variables

PROBLEM: In 1912, the luxury liner *Titanic*, on its first voyage across the Atlantic, struck an iceberg and sank. Some passengers made it off the ship in lifeboats, but many died. The two-way table gives information about adult passengers who survived and who died, by class of travel. Construct a segmented bar graph to display the relationship between survival status and class of travel.

Matt Cardy/Getty Images

		Class of travel			
		First	Second	Third	Total
Survival status	Survived	197	94	151	442
	Died	122	167	476	765
	Total	319	261	627	1207

SOLUTION:

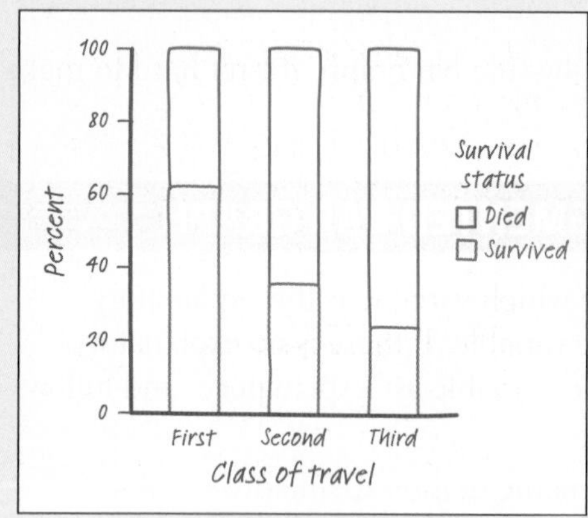

1. **Identify the variables.** Use class of travel as the explanatory variable, because class might help predict whether a passenger survived.
2. **Draw and label the axes.**
3. **Scale the axes.**
4. **Draw "100%" bars.**
5. **Segment each of the bars.**
   - In first class, 197/319 = 61.8% survived and 122/319 = 38.2% died.
   - In second class, 94/261 = 36.0% survived and 167/261 = 64.0% died.
   - In third class, 151/627 = 24.1% survived and 476/627 = 75.9% died.
6. **Include a key.**

FOR PRACTICE, TRY EXERCISE 7

Here is a mosaic plot for the *Titanic* data with a horizontal scale added along the top of the graph. Unlike the segmented bar graph in the preceding example, the mosaic plot offers an easy way to compare the percentages of people in first, second, and third class. We can now see that a little more than half of the passengers were in third class, about 25% were in first class, and the rest were in second class.

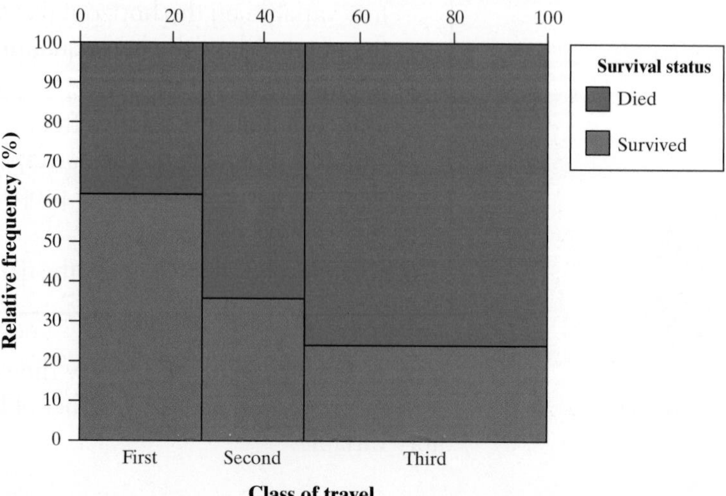

Describing the Relationship Between Two Categorical Variables

Once we calculate relative frequencies and make a graph to display the relationship between two categorical variables, the final step is to describe the nature of any **association** between the variables.

DEFINITION Association

There is an **association** between two variables if knowing the value of one variable helps us predict the value of the other. If knowing the value of one variable does not help us predict the value of the other, there is no association between the variables.

The following segmented bar graph from Figure 2.1 shows a clear association between environmental club membership and snowmobile use in the random sample of 1526 winter visitors to Yellowstone National Park. The environmental club members were much less likely to rent (about 25% versus 41%) or own (about 5% versus 23%) snowmobiles than non-club members and were more likely to have never used a snowmobile (about 70% versus 36%). Knowing whether a person in the sample is an environmental club member helps us predict that individual's snowmobile use.

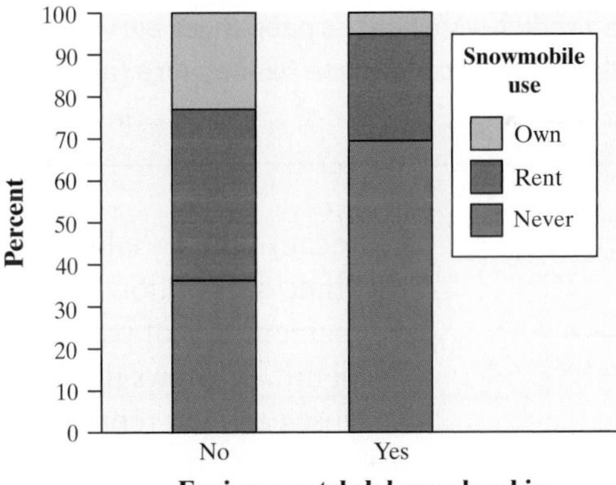

Is it reasonable to say that there is an association between environmental club membership and snowmobile use in the *population* of all winter visitors to Yellowstone National Park? Making this determination requires formal inference, which will have to wait until Unit 8.

What would the segmented bar graph look like if there were *no association* between environmental club membership and snowmobile use in the sample? The bottom segments would be the same height for both the "Yes" and "No" groups, as would the middle segments and the top segments, as shown in the graph at left. In that case, knowing whether a survey respondent is an environmental club member would *not* help us predict their snowmobile use.

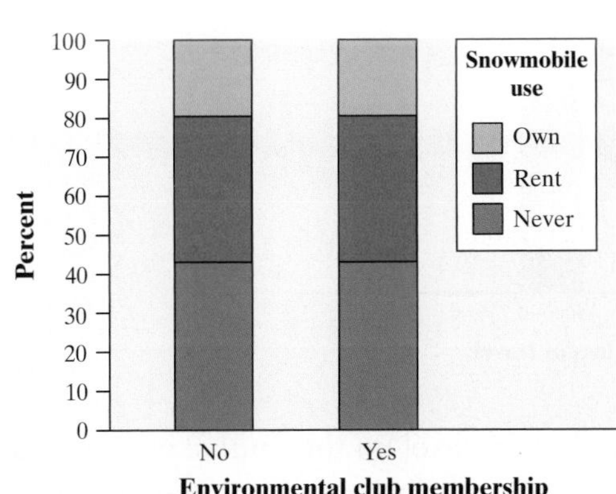

EXAMPLE

A *Titanic* disaster, part 3
Describing the relationship between two categorical variables

Skill 2.D

PROBLEM: In 1912, the luxury liner *Titanic*, on its first voyage across the Atlantic, struck an iceberg and sank. Some passengers made it off the ship in lifeboats, but many died. Describe what the segmented bar graph reveals about the relationship between class of travel and survival status for adult passengers on the *Titanic*.

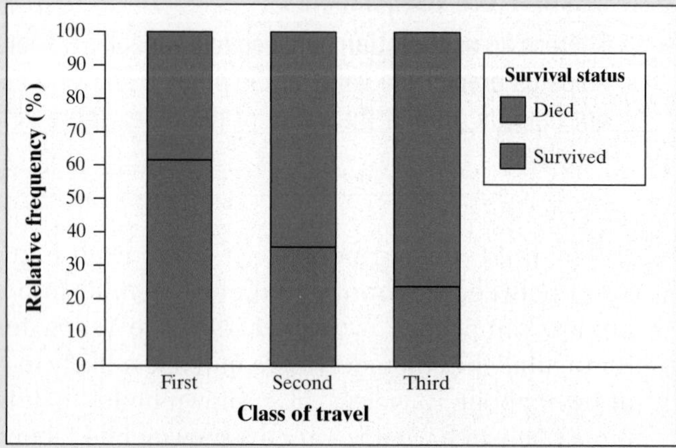

SOLUTION:

There is a clear association between survival status and class of travel on the *Titanic*. Knowing a passenger's class of travel helps us predict whether the passenger survived. Passengers in first class were the most likely to survive (about 62%), followed by second-class passengers (about 36%), and then third-class passengers (about 24%).

FOR PRACTICE, TRY EXERCISE 11

Because the variable "Survival status" has only two possible values, comparing the three distributions displayed in the segmented bar graph amounts to comparing the percentages of passengers in each class of travel who survived. The bar graph in Figure 2.2 shows this comparison. Note that the bar heights do *not* add to 100%, because each bar represents a different group of passengers on the *Titanic*.

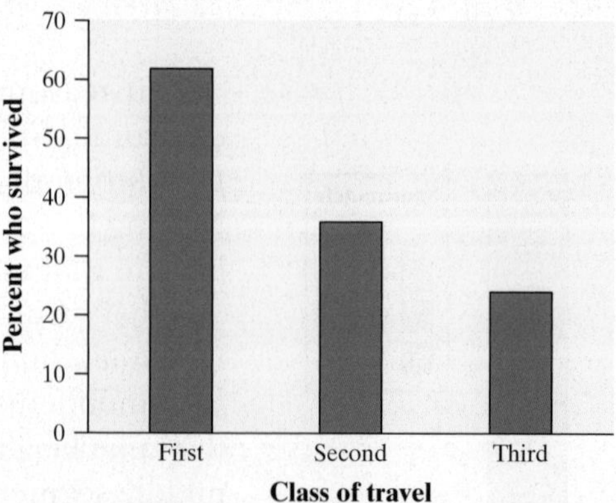

FIGURE 2.2 Bar graph comparing the percentages of passengers who survived in each of the three classes of travel on the *Titanic*.

It may be true that being in a higher class of travel on the *Titanic* increased a passenger's chance of survival. However, there isn't always a cause-and-effect relationship between two variables, even if they are clearly associated. For example, a recent study proclaimed that people who are overweight are less likely to die within a few years compared to other people. Does this mean that gaining weight will *cause* someone to live longer? Not at all. The study included smokers, who tend to be thinner as well as more likely to die in a given period than nonsmokers. Smokers increased the death rate among non-overweight people, making it look like extra pounds translated into a longer life span.[3] **Association does not imply causation!**

CHECK YOUR UNDERSTANDING

An article in the *Journal of the American Medical Association* reports the results of a study designed to see if the herb St. John's wort is effective in treating moderately severe cases of depression.[4] The 338 subjects were randomly assigned to receive one of three treatments: St. John's wort, Zoloft (a prescription drug), or placebo (an inactive treatment) for an 8-week period. The two-way table summarizes the data from the experiment.

		Change in depression			
		Full response	Partial response	No response	Total
Treatment	St. John's wort	27	16	70	113
	Zoloft	27	26	56	109
	Placebo	37	13	66	116
	Total	91	55	192	338

1. Identify the explanatory and response variables. Explain your reasoning.
2. What percentage of subjects had a full response? A partial response?
3. Make a segmented bar graph to display the relationship between treatment and change in depression.
4. Describe what the graph in Question 3 reveals about the relationship between treatment and change in depression for these subjects.

SECTION 2A Summary

- A **response variable** measures an outcome of a study. An **explanatory variable** may help predict or explain changes in a response variable.
- A **two-way table** is a table of counts or relative frequencies that summarizes data on the relationship between two categorical variables for some group of individuals.
- You can use a two-way table to calculate three types of relative frequencies:
  - A **marginal relative frequency** gives the percentage or proportion of individuals that have a specific value for one categorical variable. A marginal relative frequency is calculated by dividing a row or column total by the total for the entire two-way table.
  - A **joint relative frequency** gives the percentage or proportion of individuals that have a specific value for one categorical variable and a specific value

for another categorical variable. A joint relative frequency is calculated by dividing the value in one cell by the total for the entire two-way table.
  - A **conditional relative frequency** gives the percentage or proportion of individuals that have a specific value for one categorical variable among a group of individuals who share the same value of another categorical variable (the condition). A conditional relative frequency is calculated by dividing the value in one cell of a two-way table by the total for the appropriate row or column.
- Use a **side-by-side bar graph**, a **segmented bar graph**, or a **mosaic plot** to display the relationship between two categorical variables or compare the distribution of a categorical variable for two or more groups.
- There is an **association** between two variables if knowing the value of one variable helps us predict the value of the other. If knowing the value of one variable does not help us predict the value of the other, then there is no association between the variables.

AP® EXAM TIP
AP® Daily Videos

Review the content of this section and get extra help by watching the AP® Daily Videos for Topics 2.1, 2.2, and 2.3, which are available in AP® Classroom.

SECTION 2A Exercises

Explanatory and Response Variables

1. **Coral reefs and smartphones** Identify the explanatory variable and the response variable for the following relationships, if possible. Explain your reasoning. pg 143

(a) The weight gain of corals in aquariums where the water temperature is controlled at different levels

(b) The brand of smartphone and the model of car driven by a sample of recent college graduates

2. **Teenagers and corn yield** Identify the explanatory variable and the response variable for the following relationships, if possible. Explain your reasoning.

(a) The height and arm span in a sample of 50 teenagers

(b) The yield of corn in bushels per acre and the amount of rain in the growing season for farms in Iowa

Summarizing Data on Two Categorical Variables

3. **A smash or a hit?** Two researchers asked 150 people to recall the details of a car accident they watched on video. They selected 50 of these people at random and asked, "About how fast were the cars going when they smashed into each other?" For another 50 randomly selected people, they replaced the words "smashed into" with "hit." They did not ask the remaining 50 people—the control group—to estimate speed at all. A week later, the researchers asked all 150 participants if they saw any broken glass at the accident (there wasn't any). The two-way table summarizes each group's response to the broken glass question.[5] pg 146

		Response		
		Yes	No	Total
Question wording	"Smashed into"	16	34	50
	"Hit"	7	43	50
	Control	6	44	50
	Total	29	121	150

(a) What proportion of people in the experiment responded yes?

(b) What proportion of people in the experiment were in the control group and responded yes?

(c) What proportion of the people in the control group responded yes?

4. **Python eggs** How does the temperature of a water python nest influence egg hatching? Researchers assigned newly laid eggs to one of three water temperatures: hot, neutral, or cold. Hot duplicates the extra warmth provided by the mother python, and cold duplicates the absence of the mother. The two-way table summarizes the data on nest temperature and hatching status.

		Hatching status		
		Hatched	Didn't hatch	Total
Nest temperature	Cold	16	11	27
	Neutral	38	18	56
	Hot	75	29	104
	Total	129	58	187

(a) What proportion of the eggs hatched?

(b) What proportion of the eggs were assigned to the cold nest and hatched?

(c) What proportion of the eggs in the cold nests hatched?

5. **Social media** Pew Research Center surveyed a random sample of 1310 U.S. teens about their use of social media.[6] The teens were asked about the amount of time they spend on social media, as well as other demographic questions. The two-way table summarizes the relationship between social media use and the location where students live.

		Location			
		Urban	Suburban	Rural	Total
Social media use	Too much	121	239	116	476
	About right	126	407	183	716
	Too little	40	56	22	118
	Total	287	702	321	1310

(a) Create a two-way table of relative frequencies by dividing the value in each cell by the table total of 1310.

(b) What proportion of students live in an urban location and said too much or about right?

(c) What percentage of urban students said they use social media too much? Suburban students? Rural students?

6. **Squirrel!** Do adult and juvenile Eastern gray squirrels in New York's Central Park exhibit different behaviors toward humans? That is one of many questions investigated by 323 volunteer squirrel sighters.[7] The two-way table summarizes the data for 2898 squirrel sightings in the park.

		Age		
		Juvenile	Adult	Total
Behavior toward humans	Approach	111	756	867
	Indifferent	138	1241	1379
	Run away	81	571	652
	Total	330	2568	2898

(a) Create a two-way table of relative frequencies by dividing the value in each cell by the table total of 2898.

(b) In what proportion of the squirrel sightings was the squirrel a juvenile and ran away or was indifferent?

(c) What percentage of juvenile squirrels approached the human observer? Adult squirrels?

Displaying the Relationship Between Two Categorical Variables

7. **A smash or a hit?** Refer to Exercise 3. Construct a segmented bar graph to display the relationship between question wording and response.
pg 150

8. **Python eggs** Refer to Exercise 4. Construct a segmented bar graph to display the relationship between nest temperature and hatching status.

9. **Social media** Refer to Exercise 5. Construct a mosaic plot to display the relationship between location and social media use.

10. **Squirrel!** Refer to Exercise 6. Construct a mosaic plot to display the relationship between age and behavior toward humans.

Describing the Relationship Between Two Categorical Variables

11. **A smash or a hit?** Refer to Exercises 3 and 7. Describe what the segmented bar graph from Exercise 7 reveals about the relationship between question wording and response.
pg 152

12. **Python eggs** Refer to Exercises 4 and 8. Describe what the segmented bar graph from Exercise 8 reveals about the relationship between nest temperature and hatching status.

13. **Social media** Refer to Exercises 5 and 9. Use the mosaic plot from Exercise 9 to describe the association between location and social media use for these teens. What information is provided by the mosaic plot that wouldn't be provided by a segmented bar graph of the same data?

14. **Squirrel!** Refer to Exercises 6 and 10. Use the mosaic plot from Exercise 10 to describe the association between age and behavior toward humans for these squirrels. What information is provided by the mosaic plot that wouldn't be provided by a segmented bar graph of the same data?

15. **Napping and heart disease** In a long-term study of 3462 randomly selected adults from Lausanne, Switzerland, researchers investigated the relationship between weekly napping frequency and whether a person experienced a major cardiovascular disease (CVD) event, such as a heart attack or stroke.[8] The bar graph summarizes the data.

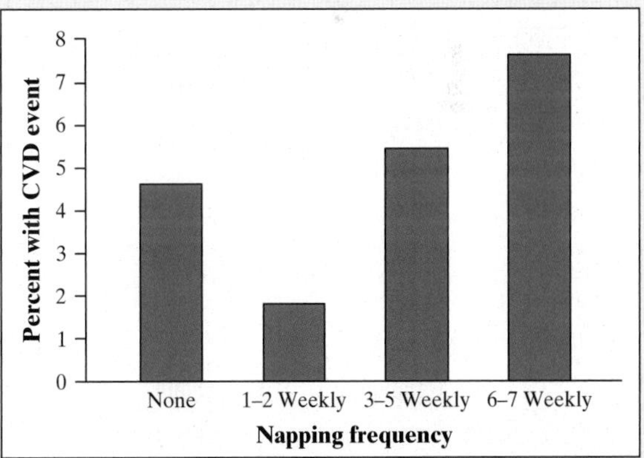

(a) Describe the association between napping frequency and whether or not a person had a CVD event.

(b) Based on the association, will taking 6–7 naps per week increase a person's risk of having a CVD event? Explain your reasoning.

16. **Turn off the nightlight?** Researchers at The Ohio State University wanted to know if there is an association between using a nightlight and myopia (nearsightedness) in children. They surveyed the parents of 1220 randomly selected children and recorded the lighting conditions in which the children slept during their first two years of life and whether they had myopia at age 10.[9] The bar graph summarizes the data.

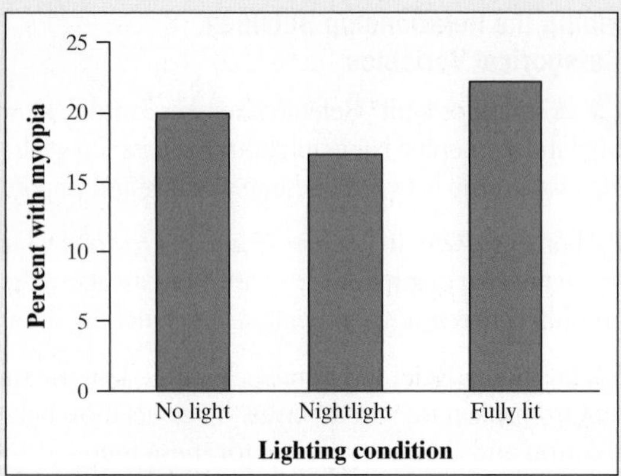

(a) Describe the association between lighting condition and whether or not a child developed myopia.

(b) Based on the association, will having a young child sleep in a fully lit room increase their chance of developing myopia? Explain your reasoning.

17. **Food preferences** Do elementary school students tend to have different food preferences than middle and high school students? A school district surveyed 32 students and recorded their school level and their preferred food choice of pizza, hot dogs, chicken nuggets, or hamburgers. Use the following data to create a two-way table summarizing the results of the study.

School level	Food choice	School level	Food choice
Elementary	Nuggets	Middle	Nuggets
Middle	Hamburger	Elementary	Nuggets
High	Pizza	Elementary	Hamburger
Elementary	Nuggets	High	Hot dogs
Elementary	Hot dogs	High	Pizza
High	Pizza	High	Hamburger
High	Nuggets	Middle	Pizza
Elementary	Nuggets	High	Nuggets
High	Hamburger	Elementary	Hamburger
Middle	Pizza	Elementary	Pizza
High	Pizza	Elementary	Nuggets
Elementary	Nuggets	Elementary	Hot dogs
High	Hamburger	High	Pizza
Elementary	Nuggets	Elementary	Pizza
Elementary	Pizza	Elementary	Hamburger
Elementary	Hot dogs	Middle	Hot dogs

For Investigation *Apply the skills from the section in a new context or nonroutine way.*

18. **Simpson's paradox** Accident victims are sometimes taken by helicopter from the accident scene to a hospital. Helicopters save time—but do they also save lives? The two-way table summarizes data from a sample of patients who were transported to the hospital by helicopter or by ambulance.[10]

		Method of transport		
		Helicopter	Ambulance	Total
Survival status	Died	64	260	324
	Survived	136	840	976
	Total	200	1100	1300

(a) What percentage of patients died with each method of transport?

Here are the same data, by severity of accident:

Serious accidents

		Method of transport		
		Helicopter	Ambulance	Total
Survival status	Died	48	60	108
	Survived	52	40	92
	Total	100	100	200

Less serious accidents

		Method of transport		
		Helicopter	Ambulance	Total
Survival status	Died	16	200	216
	Survived	84	800	884
	Total	100	1000	1100

(b) Calculate the percentage of patients who died with each method of transport for the serious accidents. Then calculate the percentage of patients who died with each method of transport for the less serious accidents. What do you notice?

(c) See if you can explain how the result in part (a) is possible given the result in part (b).

Note: This is an example of *Simpson's paradox*, which states that an association between two variables that holds for each value of a third variable can be changed or even reversed when the data for all values of the third variable are combined.

19. **Three-way table?** Does taking the drug dexamethasone help prevent death in patients who are hospitalized with Covid-19? Does the outcome depend on how much respiratory support (e.g., being on a ventilator) a patient is receiving when they start taking the drug? Researchers in the United Kingdom designed an experiment to find out.[11] They randomly assigned 2104 patients to receive dexamethasone along with the usual care and 4321 patients to receive only the usual care. They also noted the level of respiratory support each patient was receiving at the beginning of the experiment and whether the patient died in the following 28 days. The three-way table summarizes the data, with

the values in each cell showing the number of deaths and the number of patients. For example, the values in the upper-left cell indicate that 324 of the patients with invasive mechanical ventilation were assigned to dexamethasone, and 95 of these 324 patients died.

		Treatment		
		Dexamethasone	Usual care	Total
Initial respiratory support	Invasive mechanical ventilation	95/324	283/683	378/1007
	Oxygen only	298/1279	682/2604	980/3883
	No support	89/501	145/1034	234/1535
	Total	482/2104	1110/4321	1592/6425

(a) Is there evidence that dexamethasone helped? Compare the percentage of patients in each treatment group who died.

(b) Is there a relationship between initial respiratory support and outcome? Compare the percentage of patients at all three levels of initial respiratory support who died.

(c) Did dexamethasone work equally well for all three levels of initial respiratory support? Justify your answer.

Multiple Choice *Select the best answer for each question.*

Exercises 20–22 refer to the following setting: A Quinnipiac University poll asked a random sample of U.S. adults, "Would you support or oppose major new spending by the federal government that would help undergraduates pay tuition at public colleges without needing loans?" The two-way table shows the responses, grouped by age.[12]

		Age				
		18–34	35–49	50–64	65+	Total
Response	Support	91	161	272	332	856
	Oppose	25	74	211	255	565
	Don't know	4	13	20	51	88
	Total	120	248	503	638	1509

20. What proportion of the people in the sample support the proposal?

(A) 0.856
(B) 0.758
(C) 0.567
(D) 0.520
(E) 0.433

21. What percentage of the 18- to 34-year-olds responded that they opposed the proposal or didn't know?

(A) 1.9%
(B) 4.4%
(C) 24.2%
(D) 29.0%
(E) 43.3%

22. Which of the following is the best evidence of an association between age and response?

(A) The majority of people indicated they support the proposal.

(B) A greater number of people in the 65+ category said they support the proposal than in the 18–34 category.

(C) The number of people in each age category gets bigger as the ages increase.

(D) The proportions who support the proposal in each age category are different.

(E) In all age categories, the proportion who support the proposal is larger than the proportion who oppose or don't know.

23. The following partially completed two-way table shows the marginal distributions of grade level and handedness for a sample of 100 high school statistics students.

		Grade level		
		Junior	Senior	Total
Handedness	Right	x		90
	Left			10
	Total	40	60	100

If there is no association between grade level and handedness for the members of the sample, which of the following is the correct value of x?

(A) 20
(B) 30
(C) 36
(D) 45
(E) Impossible to determine without more information.

Recycle and Review *Practice what you learned in previous sections.*

24. **Hockey goals (1D, 2A)** In the National Hockey League, skaters (non-goalies) are classified as defensive players, centers, or wings. The boxplots summarize the distribution of goals scored for skaters in each of these position groups in the 2022 season (minimum 41 games played).[13]

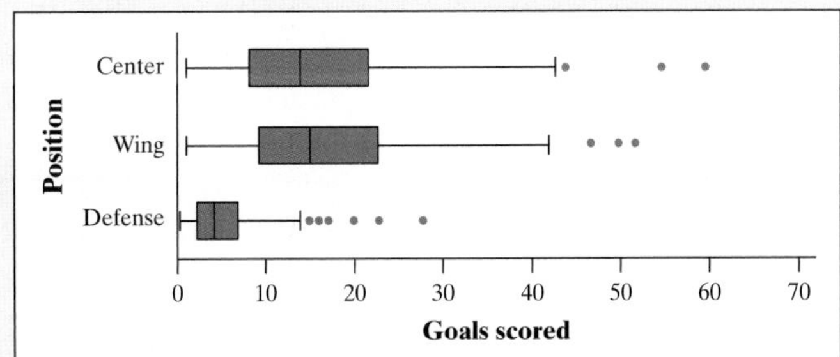

(a) Compare these distributions.

(b) Based on your answer to part (a), explain why it is reasonable to say that there is an association between position group (a categorical variable) and goals scored (a quantitative variable).

SECTION 2B Relationships Between Two Quantitative Variables

LEARNING TARGETS *By the end of the section, you should be able to:*

- Make a scatterplot to display the relationship between two quantitative variables.
- Describe the direction, form, and strength of a relationship displayed in a scatterplot and identify unusual features.
- Interpret the correlation for a linear relationship between two quantitative variables.
- Distinguish correlation from causation.
- Calculate the correlation for a linear relationship between two quantitative variables.

In Unit 1, we analyzed univariate data, starting with categorical data and moving to quantitative data. In this unit, we follow a similar path with bivariate data. In Section 2A, we analyzed the relationship between two categorical variables. For the rest of the unit, we'll focus on analyzing the relationship between two *quantitative* variables. Let's get started with a tasty activity.

ACTIVITY Candy grab

Kylie McManis

In this activity, you will investigate if students with a larger hand span can grab more candy than students with a smaller hand span.[14]

1. Measure the span of your dominant hand to the nearest half-centimeter (cm). Hand span is the distance from the tip of the thumb to the tip of the pinkie finger on your fully stretched-out hand.
2. One student at a time, go to the front of the class and use your dominant hand to grab as many candies as possible from the container. You must grab the candies with your fingers pointing down (no scooping!) and hold the candies for 2 seconds before counting them. After counting, put the candy back into the container.
3. On the board, record your hand span and the number of candies in a table with the following headings:

   Hand span (cm)   Number of candies

4. While other students record their values on the board, copy the table onto a piece of paper and make a graph. Begin by constructing a set of coordinate axes. Label the horizontal axis "Hand span (cm)" and the vertical axis "Number of candies." Choose an appropriate scale for each axis and plot each point from your class data table as accurately as you can on the graph.
5. What does the graph tell you about the relationship between hand span and number of candies? Summarize your observations in a sentence or two.

Displaying the Relationship Between Two Quantitative Variables

Although there are many ways to display the distribution of a single quantitative variable, a **scatterplot** is the best way to display the relationship between two quantitative variables.

> **DEFINITION Scatterplot**
>
> A **scatterplot** shows the relationship between two quantitative variables measured for the same individuals. The values of one variable appear on the horizontal axis, and the values of the other variable appear on the vertical axis. Each individual in the data set appears as a point in the graph.

Figure 2.3 is a scatterplot that displays the relationship between hand span (cm) and number of Starburst™ candies grabbed for the 24 students in Mr. Tyson's class who did the Candy Grab activity. As you can see, students with larger hand spans were typically able to grab more candies.

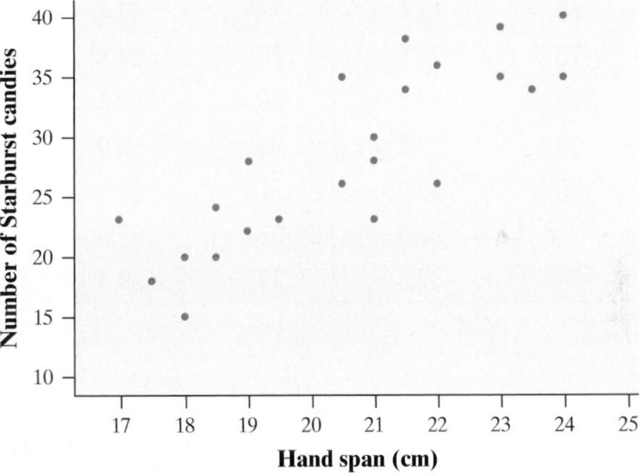

FIGURE 2.3 Scatterplot of hand span (cm) and number of Starburst candies grabbed by 24 students. Only 23 points appear because two students had hand spans of 19 cm and grabbed 28 Starburst candies.

After collecting bivariate quantitative data, it is easy to make a scatterplot.

HOW TO MAKE A SCATTERPLOT

1. **Label the axes.** Put the name of the explanatory variable under the horizontal axis and the name of the response variable to the left of the vertical axis. If there is no explanatory variable, either variable can go on the horizontal axis.
2. **Scale the axes.** Place equally spaced tick marks along each axis, beginning at a number equal to or less than the smallest value of the variable and continuing until you equal or exceed the largest value.
3. **Plot individual data values.** For each individual, plot a point directly above that individual's value for the variable on the horizontal axis and directly to the right of that individual's value for the variable on the vertical axis.

The following example illustrates the process of constructing a scatterplot.

EXAMPLE

Height and free-throws
Displaying the relationship between two quantitative variables

Skill 2.B

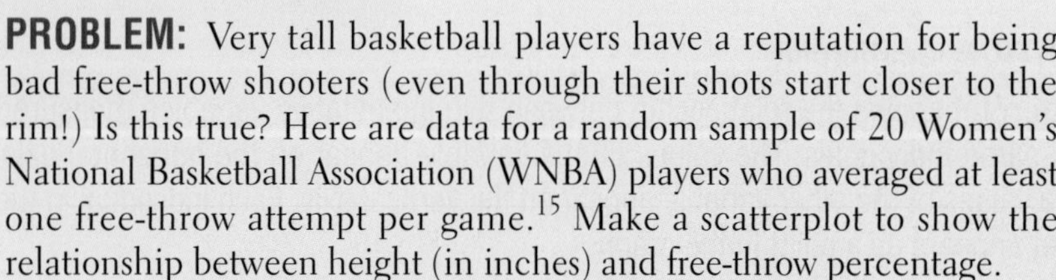

PROBLEM: Very tall basketball players have a reputation for being bad free-throw shooters (even through their shots start closer to the rim!) Is this true? Here are data for a random sample of 20 Women's National Basketball Association (WNBA) players who averaged at least one free-throw attempt per game.[15] Make a scatterplot to show the relationship between height (in inches) and free-throw percentage.

Mike Kirschbaum/NBAE/Getty Images

Height (in.)	Free-throw percentage	Height (in.)	Free-throw percentage
75	58.1	76	88.7
69	85.0	72	85.9
70	95.0	68	66.7
69	82.4	77	62.4
73	76.5	76	64.5
78	74.4	69	92.3
76	73.6	78	66.0
70	89.3	71	68.1
73	87.5	67	75.6
74	82.6	72	89.4

SOLUTION:

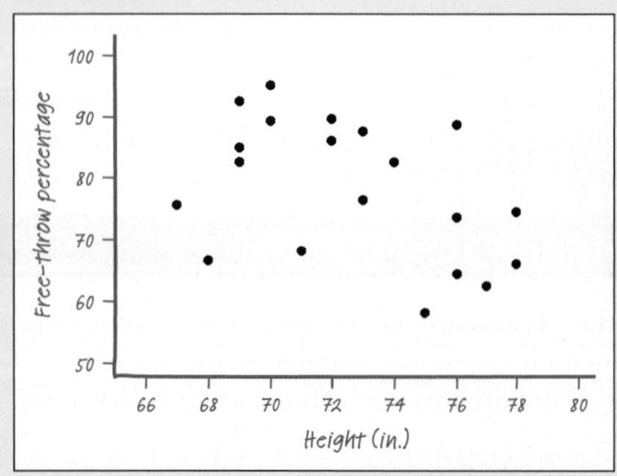

1. **Label the axes.** The explanatory variable is height because we think it might help explain the player's free-throw percentage.
2. **Scale the axes.**
3. **Plot individual data values.**

FOR PRACTICE, TRY EXERCISE 1

Making scatterplots with technology is much easier than constructing them by hand.

6. Tech Corner MAKING SCATTERPLOTS

TI-Nspire and other technology instructions are on the book's website at bfwpub.com/tps7e.

We'll use the WNBA data from page 160 to show how to construct a scatterplot on a TI-83/84.

1. Press STAT, choose Edit..., and enter the heights in L1 and the free-throw percentages in L2.

NORMAL FLOAT AUTO REAL RADIAN MP

L1	L2	L3	L4	L5
75	58.1	--------	--------	--------
69	85			
70	95			
69	82.4			
73	76.5			
78	74.4			
76	73.6			
70	89.3			
73	87.5			
74	82.6			
76	88.7			

L3(1)=

2. Press 2nd Y= (STAT PLOT) and press ENTER or 1 to go into Plot 1. Then adjust the settings as shown.

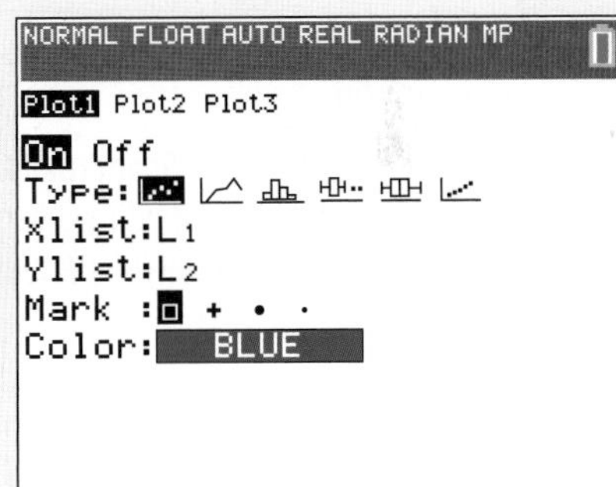

3. Press ZOOM and choose ZoomStat to let the calculator choose an appropriate window. Pressing TRACE will allow you to move around the plot and see the coordinates of each point.

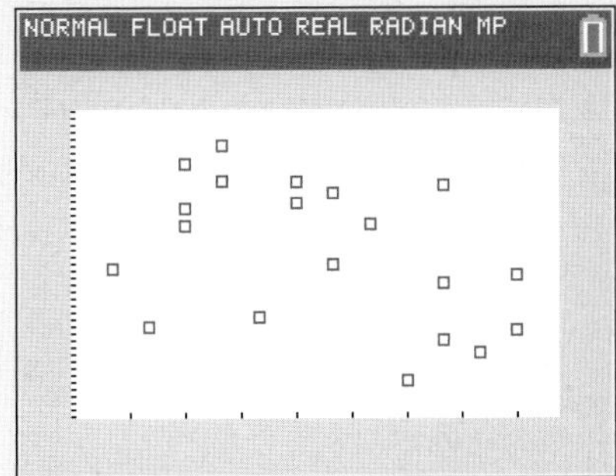

AP® EXAM TIP

If you are asked to make a scatterplot, be sure to label and scale both axes. *Don't* just copy an unlabeled calculator graph onto your paper.

Describing a Scatterplot

To describe a scatterplot, follow the basic strategy of data analysis from Unit 1: look for patterns and important departures from those patterns. Recall from Section 2A that two variables have an *association* if knowing the value of one variable helps us predict the value of the other variable.

The scatterplot in Figure 2.4(a) shows a **positive association** between hand span and number of candies a student can grab. That is, students with larger hands can typically grab more candy. Other scatterplots, such as the one in Figure 2.4(b), show a **negative association.** WNBA players who are taller tend to be worse free-throw shooters.

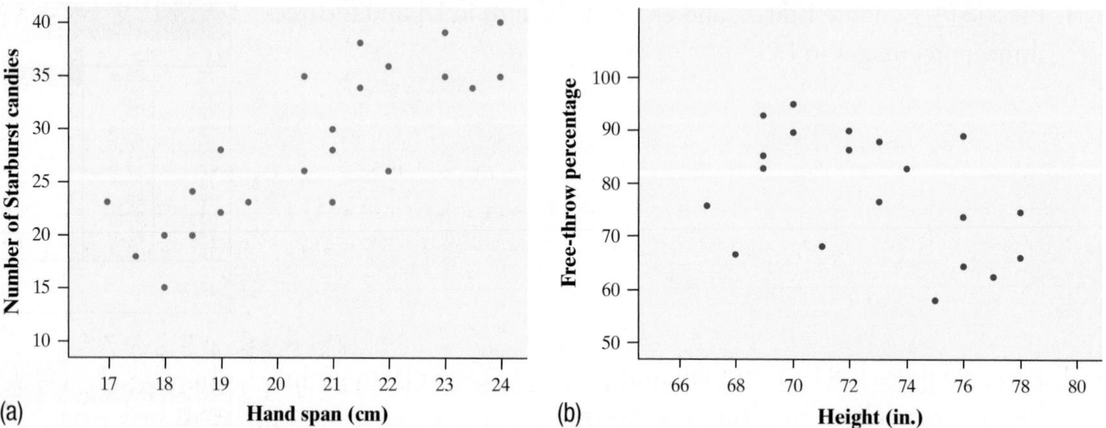

FIGURE 2.4 Scatterplots showing (a) a positive association between hand span (cm) and number of Starburst candies grabbed by a sample of students and (b) a negative association between height (in.) and free-throw percentage for WNBA players.

In some cases, there is **no association** between two variables. For example, the following scatterplot shows the relationship between height (in centimeters) and the typical amount of sleep on a nonschool night (in hours) for a sample of students.[16] Knowing the height of a student doesn't help predict how much they like to sleep on Saturday night!

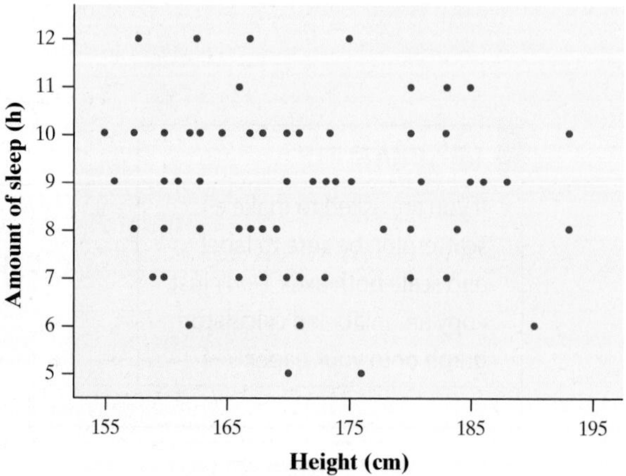

DEFINITION Positive association, Negative association, No association

Two variables have a **positive association** when the values of one variable tend to increase as the values of the other variable increase.

Two variables have a **negative association** when the values of one variable tend to decrease as the values of the other variable increase.

There is **no association** between two variables if knowing the value of one variable does not help us predict the value of the other variable.

Identifying the direction of an association in a scatterplot is a good start, but we also need to address several other characteristics when describing a scatterplot.

AP® EXAM TIP

When you are asked to *describe* the association shown in a scatterplot, you are expected to discuss the direction, form, and strength of the association, along with any unusual features, *in the context of the problem*. This means that you need to use both variable names in your description.

HOW TO DESCRIBE A SCATTERPLOT

To describe a scatterplot, make sure to address the following four characteristics in the context of the data:

- **Direction:** A scatterplot can show a positive association, a negative association, or no association.
- **Form:** A scatterplot can show a linear form or a nonlinear form. The form is linear if the overall pattern follows a straight line. Otherwise, the form is nonlinear.
- **Strength:** A scatterplot can show a weak, moderate, or strong association. An association is strong if the points closely follow a specific form. An association is weak if the points deviate quite a bit from the form identified.
- **Unusual features:** A scatterplot can show individual points that fall outside the overall pattern and distinct clusters of points.

Even though they have opposite directions, both associations in Figure 2.4 on page 162 have a linear form. The association between hand span and number of candies grabbed is stronger than the relationship between height and free-throw percentage because the points in Figure 2.4(a) follow the linear form more closely than the points in Figure 2.4(b) do. Neither scatterplot shows any unusual points or clusters of points.

Even when there is a clear relationship between two variables in a scatterplot, the direction of the association describes only the overall trend—not an absolute relationship. For example, even though taller WNBA players typically have lower free-throw percentages, there are plenty of exceptions. The 75-inch-tall player who made only 58.1% of her free throws was outshot by 6 taller players.

EXAMPLE

Old Faithful and fertility rate
Describing a scatterplot

Skill 2.A

PROBLEM: Describe the relationship in each of the following contexts.

(a) The scatterplot on the left shows the relationship between the duration (in minutes) of an eruption and the wait time until the next eruption (in minutes) of the Old Faithful geyser in Yellowstone National Park during a particular month.

f11photo/Shutterstock

(b) The scatterplot on the right shows the relationship between the average income (gross domestic product per person, in dollars) and fertility rate (number of children per woman) in 187 countries.[17]

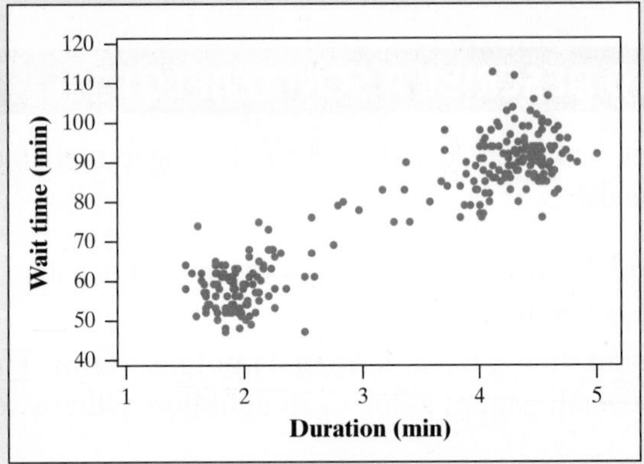

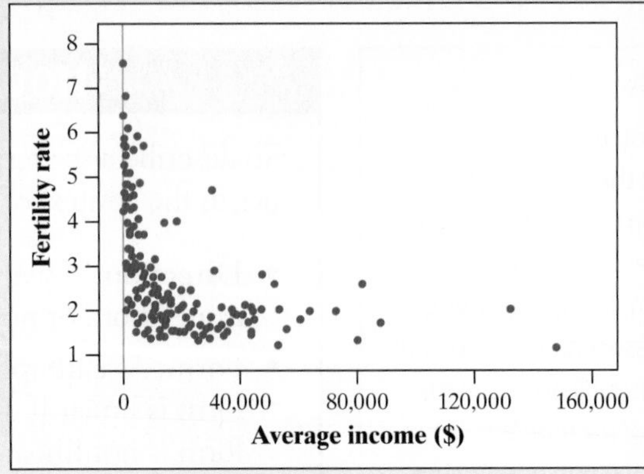

SOLUTION:

(a) There is a strong, positive linear relationship between the duration of an eruption and the wait time until the next eruption. There are two main clusters of points: one cluster has durations around 2 minutes with wait times around 55 minutes, and the other cluster has durations around 4.5 minutes with wait times around 90 minutes.

Even with the clusters, the overall direction is still positive. In some cases, however, the points in a cluster go in the opposite direction of the overall association.

(b) There is a moderately strong, negative nonlinear relationship between average income and fertility rate in these countries. There is a country outside this pattern with an average income around $30,000 and a fertility rate around 4.7 children per woman.

The association is called "nonlinear" because the pattern of points is clearly curved.

FOR PRACTICE, TRY EXERCISE 3

CHECK YOUR UNDERSTANDING

Is there a relationship between the amount of sugar (in grams) and the number of calories in movie-theater candy? Here are the data from a sample of 12 types of candy.[18]

Name	Sugar (g)	Calories	Name	Sugar (g)	Calories
Butterfinger Minis	45	450	Reese's Pieces	61	580
Junior Mints	107	570	Skittles	87	450
M&M'S®	62	480	Sour Patch Kids	92	490
Milk Duds	44	370	SweeTarts	136	680
Peanut M&M'S®	79	790	Twizzlers	59	460
Raisinets	60	420	Whoppers	48	350

1. Identify the explanatory and response variables. Explain your reasoning.
2. Make a scatterplot to display the relationship between amount of sugar and the number of calories in movie-theater candy.
3. Describe the relationship shown in the scatterplot.

Interpreting Correlation

A scatterplot displays the direction, form, and strength of a relationship between two quantitative variables. It is usually fairly easy to identify the direction and form of an association, but evaluating the strength of an association can be more difficult. When the association between two quantitative variables is linear, we can use the **correlation *r*** to help describe the strength and direction of the association. You may also see the term "correlation coefficient" used to refer to *r*.

DEFINITION Correlation *r*

The **correlation *r*** gives the direction and measures the strength of the linear association between two quantitative variables.

Here are some important properties of the correlation *r*:

- The correlation *r* is a value between −1 and 1 ($-1 \leq r \leq 1$).
- The correlation *r* indicates the direction of a linear relationship by its sign: $r > 0$ for a positive association and $r < 0$ for a negative association.
- The extreme values $r = -1$ and $r = 1$ occur *only* in the case of a perfect linear relationship, when the points lie exactly along a straight line.
- If the linear relationship is strong (very little scatter from the linear form), the correlation *r* will be close to 1 or −1. If the linear relationship is weak, the correlation *r* will be close to 0.

Use the correlation only to describe strength and direction for a linear association. That's why the word *linear* kept appearing in the preceding list!

Figure 2.5 shows six scatterplots that correspond to various values of *r*. To make the meaning of *r* clearer, the horizontal and vertical scales are the same. The correlation *r* describes the direction and strength of the linear relationship in each scatterplot.

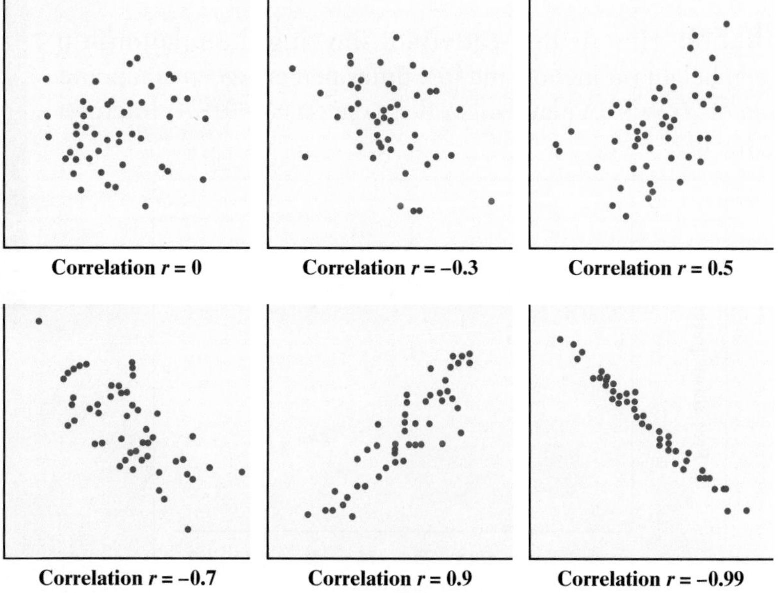

FIGURE 2.5 How correlation measures the strength and direction of a linear relationship. When the points are tightly packed around a line, the correlation will be close to 1 or −1.

ACTIVITY Guess the correlation

In this activity, we will have a class competition to see who can best guess the correlation.

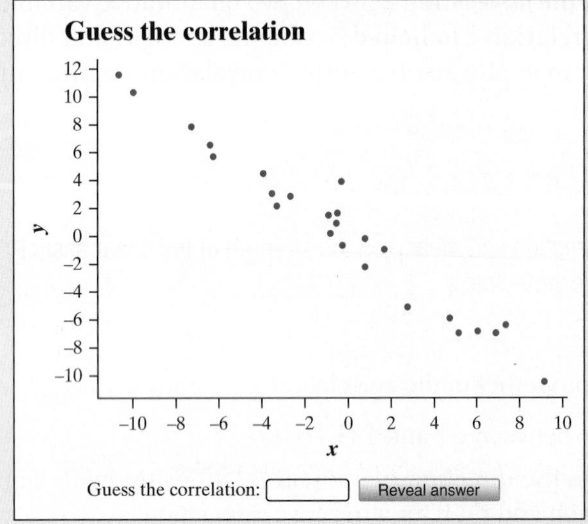

1. Go to www.stapplet.com and launch the *Guess the Correlation* applet (under Activities).
2. As a class, try to guess the correlation shown in the scatterplot. Enter your estimate and click the "Reveal answer" button. The applet will calculate the difference between your guess and the actual correlation. How close were you? Click the "Do it again!" button several times to get more practice. For the competition, there will be two rounds.
3. Starting on one side of the classroom and moving in order to the other side, the teacher will give each student *one* new sample and have them guess the correlation. The teacher will then record how far off the guess was from the true correlation.
4. When every student has guessed, the teacher will give each student a second sample. This time, the students will go in reverse order so that the student who went first in Round 1 will go last in Round 2. The student who has the closest guess in either round wins a prize!

The following example illustrates how to interpret the correlation.

EXAMPLE Height and free-throws, part 2
Interpreting correlation

Skill 4.B

PROBLEM: Here is the scatterplot showing the relationship between height (in inches) and free-throw percentage for a random sample of 20 WNBA players. For these data, $r = -0.427$. Interpret the value of r.

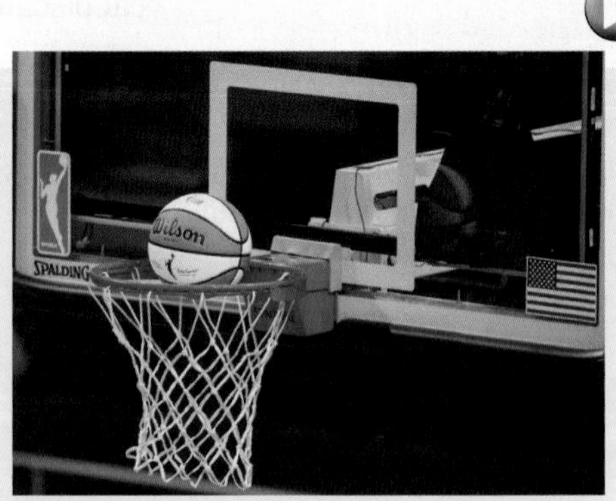

Scott Taetsch/Getty Images

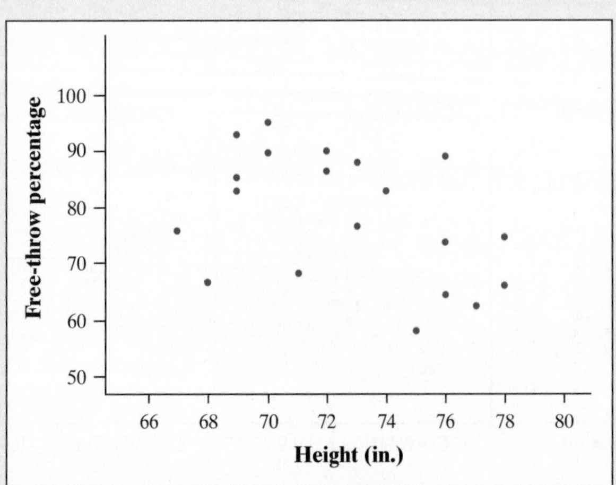

SOLUTION:

The correlation of $r = -0.427$ indicates that the linear association between height and free-throw percentage is negative and fairly weak.

AP® EXAM TIP

Remember to include context by using the variable names in your answer. Otherwise, you'll lose credit on the AP® Statistics exam.

FOR PRACTICE, TRY EXERCISE 11

Cautions about Correlation

Although the correlation is a good way to measure the strength and direction of a linear relationship, it has several limitations.

Correlation doesn't imply causation. In many cases, two variables might have a strong correlation, but changes in one variable are highly unlikely to cause changes in the other variable. Consider the following scatterplot showing total revenue generated by skiing facilities in the United States and the number of people who died by becoming tangled in their bedsheets in 10 recent years.[19] The correlation for these data is $r = 0.97$. Does an increase in skiing revenue *cause* more people to die by becoming tangled in their bedsheets? We doubt it!

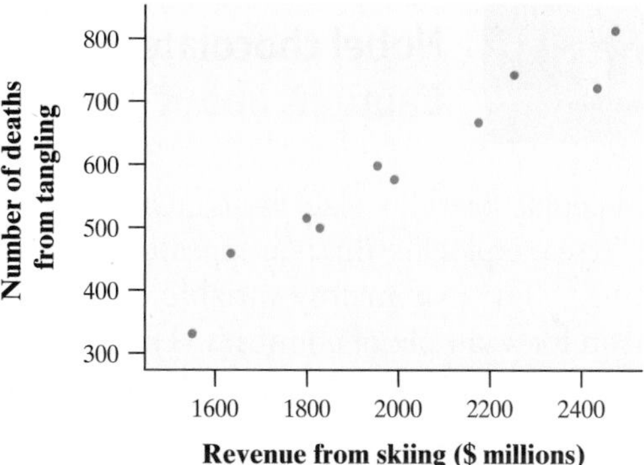

Even though we shouldn't automatically conclude that there is a cause-and-effect relationship between two variables when they have an association, in some cases there might actually be a cause-and-effect relationship. You will learn how to distinguish these cases in Unit 3.

A correlation close to 1 or −1 doesn't imply that an association is linear. Here is a scatterplot showing the speed (in miles per hour) and the distance (in feet) needed to come to a complete stop when a motorcycle's brake was applied.[20] The association is clearly curved, but the correlation is quite large: $r = 0.98$. In fact, the correlation for this *nonlinear* association is much stronger than the correlation of $r = -0.427$ for the *linear* relationship between height and free-throw percentage for WNBA players.

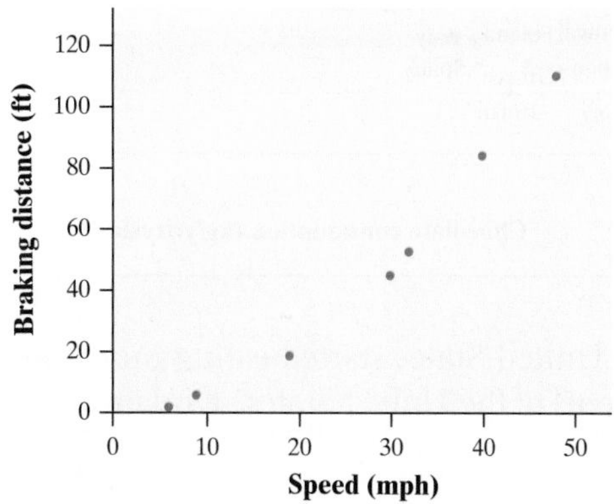

Correlation should be used only to describe linear relationships. The association displayed in the following scatterplot is extremely strong, but the correlation is $r = 0$. This isn't a contradiction because correlation doesn't measure the strength of nonlinear relationships.

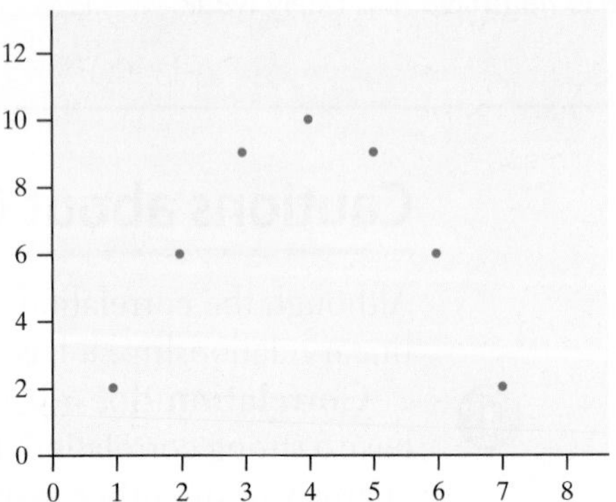

Correlation alone doesn't provide any information about form. To determine the form of an association, you must look at a scatterplot.

EXAMPLE

Nobel chocolate
Cautions about correlation

Skill 4.B

PROBLEM: Most people love chocolate for its great taste. But does it also make you smarter? A scatterplot like this one appeared in the *New England Journal of Medicine.*[21] The explanatory variable is the chocolate consumption per person for a sample of countries. The response variable is the number of Nobel Prizes per 10 million residents of that country.

amphotora/Getty Images

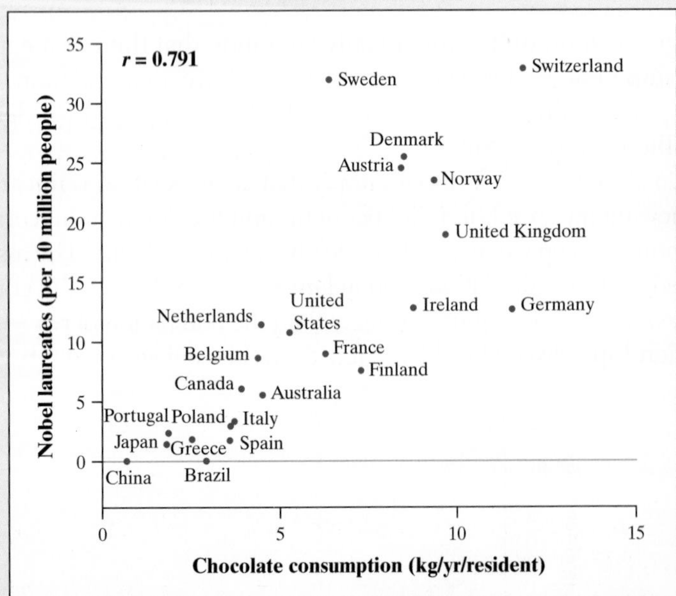

If people in the United States started eating more chocolate, should we expect more Nobel Prizes to be awarded to residents of the United States? Explain your answer.

SOLUTION:

Not necessarily. Even though there is a strong correlation between chocolate consumption and the number of Nobel laureates in a country, you should not infer causation. It is possible that both variables are changing due to another variable, such as income per person. Maybe countries with higher income per person tend to have greater chocolate consumption and a higher rate of Nobel laureates.

Not all questions about cause and effect include the word *cause*. Make sure to read questions—and reports in the media—very carefully.

FOR PRACTICE, TRY EXERCISE 15

Calculating Correlation

Now that you understand the meaning and limitations of the correlation, let's look at how it's calculated.

HOW TO CALCULATE THE CORRELATION *r*

Suppose we have data on variables x and y for n individuals. The values for the first individual are x_1 and y_1, the values for the second individual are x_2 and y_2, and so on. The correlation r between x and y is

$$r = \frac{1}{n-1}\sum\left(\frac{x_i - \bar{x}}{s_x}\right)\left(\frac{y_i - \bar{y}}{s_y}\right) = \frac{\sum z_{x_i} z_{y_i}}{n-1}$$

To calculate this quantity:

1. Find the mean $\bar{x}$ and the standard deviation s_x of the explanatory variable. Then calculate the corresponding z-score for each individual: $z_{x_i} = \dfrac{x_i - \bar{x}}{s_x}$.
2. Find the mean $\bar{y}$ and the standard deviation s_y of the response variable. Then calculate the corresponding z-score for each individual: $z_{y_i} = \dfrac{y_i - \bar{y}}{s_y}$.
3. Multiply the two z-scores for each individual: $z_{x_i} z_{y_i}$.
4. Add the z-score products for all the individuals and divide by $n - 1$.

AP® EXAM TIP
Formula sheet

The formula for the correlation *r* is included on the formula sheet provided on both sections of the AP® Statistics exam.

Although the formula for the correlation r is complex, it helps us understand some properties of correlation. In practice, you should use your calculator or software to calculate r. See the Tech Corner at the end of this section.

EXAMPLE

Grab that candy
Calculating correlation

Skill 2.C

PROBLEM: Mr. Tyson's class did the Candy Grab activity described at the beginning of the section. All 24 students measured their hand span (cm) and how many Starburst candies they could grab with their dominant hand. Here are their data, along with a scatterplot. Calculate the correlation.

Hand span (cm)	Number of candies	Hand span (cm)	Number of candies
19	22	21	23
21	28	23.5	34
18	15	17.5	18
24	40	19	28
18.5	24	18.5	20
22	36	21.5	38
23	35	23	39
19.5	23	20.5	26
21	30	20.5	35
18	20	24	35
19	28	17	23
22	26	21.5	34

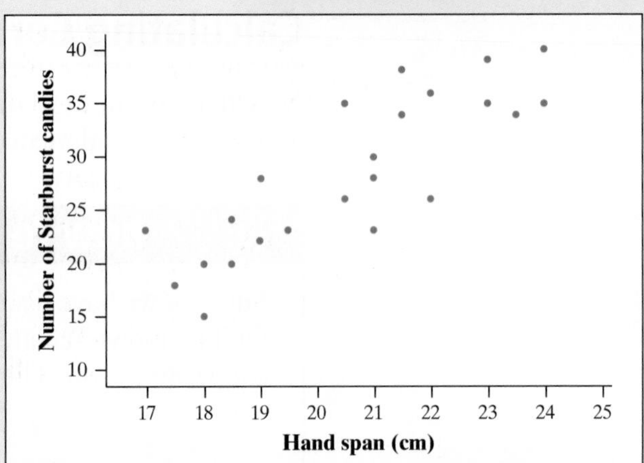

SOLUTION:

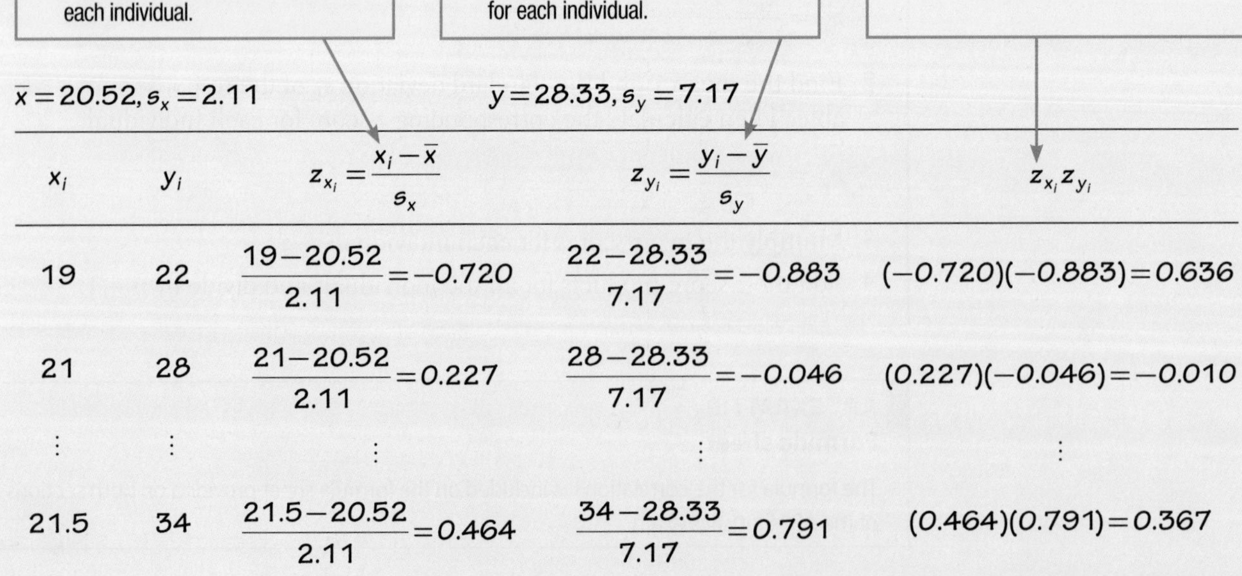

1. Find the mean $\bar{x}$ and the standard deviation s_x. Calculate the z-score for each individual.

2. Find the mean $\bar{y}$ and the standard deviation s_y. Calculate the z-score for each individual.

3. Multiply the two z-scores.

$\bar{x} = 20.52, s_x = 2.11$ $\qquad$ $\bar{y} = 28.33, s_y = 7.17$

x_i	y_i	$z_{x_i} = \frac{x_i - \bar{x}}{s_x}$	$z_{y_i} = \frac{y_i - \bar{y}}{s_y}$	$z_{x_i} z_{y_i}$
19	22	$\frac{19-20.52}{2.11} = -0.720$	$\frac{22-28.33}{7.17} = -0.883$	$(-0.720)(-0.883) = 0.636$
21	28	$\frac{21-20.52}{2.11} = 0.227$	$\frac{28-28.33}{7.17} = -0.046$	$(0.227)(-0.046) = -0.010$
⋮	⋮	⋮	⋮	⋮
21.5	34	$\frac{21.5-20.52}{2.11} = 0.464$	$\frac{34-28.33}{7.17} = 0.791$	$(0.464)(0.791) = 0.367$

$$r = \frac{0.636 + (-0.010) + \ldots + 0.367}{24 - 1} = 0.836$$

4. Add the z-score products for all individuals and divide the sum by $n - 1$. *Note that technology provides a value of $r = 0.835$. The difference is a result of rounding in the by-hand calculations.*

FOR PRACTICE, TRY EXERCISE 19

ADDITIONAL PROPERTIES OF THE CORRELATION

Now that you have seen how the correlation is calculated, here are some additional facts about correlation.

1. *Correlation requires that both variables be quantitative*, so that it makes sense to do the arithmetic indicated by the formula for r. We cannot calculate a correlation between the incomes of a group of people and what city they live in because city is a categorical variable. When one or both of the variables are categorical, use the term *association* rather than *correlation*.
2. *Correlation makes no distinction between explanatory and response variables.* When calculating the correlation, it makes no difference which variable you call x and which you call y. The correlation r will be the same in either case. As you can see in the formula, reversing the roles of x and y would change only the order of the multiplication, not the product.

$$r = \frac{1}{n-1}\sum\left(\frac{x_i - \bar{x}}{s_x}\right)\left(\frac{y_i - \bar{y}}{s_y}\right)$$

3. Because r uses the standardized values of the observations, *r does not change when we change the units of measurement of x, y, or both.* As you can see in the following scatterplot, the correlation between hand span and number of candies grabbed is the same whether we measure hand span in centimeters or in inches.

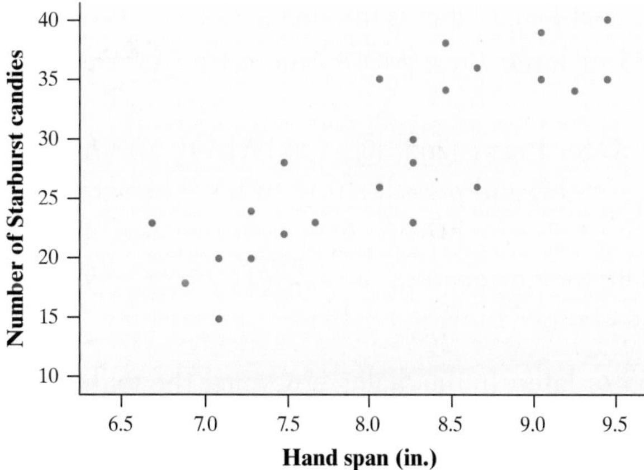

4. *The correlation r has no unit of measurement* because we are using standardized values (z-scores) in the calculation and standardized values have no units.

CHECK YOUR UNDERSTANDING

Here again are the data for a sample of 12 types of movie-theater candy.[22]

Name	Sugar (g)	Calories	Name	Sugar (g)	Calories
Butterfinger Minis	45	450	Reese's Pieces	61	580
Junior Mints	107	570	Skittles	87	450
M&M'S®	62	480	Sour Patch Kids	92	490
Milk Duds	44	370	SweeTarts	136	680
Peanut M&M'S®	79	790	Twizzlers	59	460
Raisinets	60	420	Whoppers	48	350

1. Calculate the correlation r using technology.
2. Interpret the correlation.
3. What happens to the correlation if you measure sugar in milligrams instead of grams? (*Note:* There are 1000 milligrams in 1 gram.)
4. What happens to the correlation if you use calories as the explanatory variable and sugar as the response variable?

7. Tech Corner CALCULATING CORRELATION

TI-Nspire and other technology instructions are on the book's website at bfwpub.com/tps7e.

Calculating the correlation r with technology is much easier than calculating it by hand. We'll use the WNBA data from page 160 to show how to calculate r on a TI-83/84. Note that the TI-83/84 cannot calculate correlation directly from two lists of data, except as part of calculating the equation of a regression line.

1. Press STAT, choose Edit..., and enter the heights in L1 and the free-throw percentages in L2.
2. Make sure Stat Diagnostics is turned on.
   - **OS 2.55 or later:** Press MODE and set STAT DIAGNOSTICS to ON.
   - **Older OS:** Press 2nd 0 (CATALOG), scroll down to DiagnosticOn, and press ENTER twice. The screen should say "Done."
3. To calculate the correlation, press STAT, arrow over to the CALC menu, and choose "LinReg(a + bx)."
   - **OS 2.55 or later:** In the dialog box, enter the following: Xlist:L1, Ylist:L2, FreqList(leave blank), Store RegEQ (leave blank), and choose Calculate.
   - **Older OS:** Finish the command to read LinReg(a + bx) L1,L2 and press ENTER.

Note: The correlation is labeled as r in the output. You will learn about a, b, and r^2 in Section 2C.

NORMAL FLOAT AUTO REAL RADIAN MP

L1	L2	L3	L4	L5
75	58.1	--------	--------	--------
69	85			
70	95			
69	82.4			
73	76.5			
78	74.4			
76	73.6			
70	89.3			
73	87.5			
74	82.6			
76	88.7			

L3(1)=

SECTION 2B Summary

- A **scatterplot** shows the relationship between two quantitative variables measured for the same individuals. Each individual in the data set appears as a point in the graph. The values of the explanatory variable appear on the horizontal axis, and the values of the response variable appear on the vertical axis. If there is no explanatory variable, either variable can appear on the horizontal axis.
- When describing a scatterplot, address the overall pattern (direction, form, strength) and departures from the pattern (unusual features) in context using the names of the variables.
  - **Direction:** A relationship has a **positive association** when the values of one variable tend to increase as the values of the other variable increase, a **negative association** when the values of one variable tend to decrease as the values of the other variable increase, and **no association** when knowing the value of one variable doesn't help predict the value of the other variable.
  - **Form:** A scatterplot can show a linear form or a nonlinear form. The form is linear if the overall pattern follows a straight line. Otherwise, the form is nonlinear.
  - **Strength:** A scatterplot can show a weak, moderate, or strong association. An association is strong if the points closely follow a specific form. An association is weak if the points deviate quite a bit from the form identified.
  - **Unusual features:** Look for individual points that fall outside the pattern and distinct clusters of points.
- The **correlation *r*** gives the direction and measures the strength of the linear association between two quantitative variables.
  - The value of the correlation is always between -1 and 1 ($-1 \leq r \leq 1$).
  - Correlation indicates the direction of a linear relationship by its sign: $r > 0$ for a positive association and $r < 0$ for a negative association.
  - Stronger linear associations have values of r closer to 1 or -1. Correlations of $r = 1$ and $r = -1$ occur only when the points on a scatterplot lie exactly on a straight line.
  - Correlation doesn't measure form and should be used only to describe linear relationships.
- Correlation does not imply causation.
- Calculate r using technology or the following formula:

$$r = \frac{1}{n-1}\sum\left(\frac{x_i - \bar{x}}{s_x}\right)\left(\frac{y_i - \bar{y}}{s_y}\right) = \frac{\sum z_{x_i} z_{y_i}}{n-1}$$

- The correlation isn't affected by reversing the variables or changing the units of either variable. The correlation has no units.

AP® EXAM TIP
AP® Daily Videos

Review the content of this section and get extra help by watching the AP® Daily Videos for Topics 2.4 and 2.5, which are available in AP® Classroom.

2B Tech Corner

TI-Nspire and other technology instructions are on the book's website at bfwpub.com/tps7e.

6. Making scatterplots Page 161
7. Calculating correlation Page 172

SECTION 2B Exercises

Displaying the Relationship Between Two Quantitative Variables

1. **Heavy backpacks** Ninth-grade students at a residential school go on a backpacking trip each fall. Students are divided into hiking groups of size 8 by selecting names from a hat. Before leaving, both the students and their backpacks are weighed. The data shown here are from one hiking group. Make a scatterplot by hand that shows how backpack weight relates to body weight.
pg 160

Body weight (lb)	120	187	109	103	131	165	158	116
Backpack weight (lb)	26	30	26	24	29	35	31	28

2. **Putting success** How well do professional golfers putt from different distances to the hole? The data show various distances to the hole (in feet) and the percentage of putts made at each distance for a sample of golfers.[23] Make a scatterplot by hand that shows how the percentage of putts made relates to the distance of the putt.

Distance (ft)	Percent made	Distance (ft)	Percent made
2	93.3	12	25.7
3	83.1	13	24.0
4	74.1	14	31.0
5	58.9	15	16.8
6	54.8	16	13.4
7	53.1	17	15.9
8	46.3	18	17.3
9	31.8	19	13.6
10	33.5	20	15.8
11	31.6		

Describing a Scatterplot

3. **Butterflies** Greenland is home to the *Boloria chariclea* butterfly. Scientists studying this species of butterfly investigated some of the factors that might affect wing size, including temperature during the previous summer when larvae were growing. Here is a scatterplot that shows the relationship between y = wing length (mm) and x = average temperature in the previous summer (°C). Describe the relationship shown in the scatterplot.
pg 163

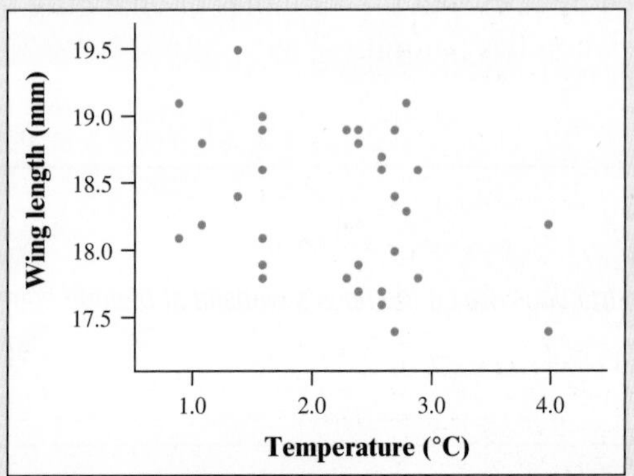

4. **Starbucks** The scatterplot shows the relationship between the amount of fat (in grams) and the number of calories in products sold at Starbucks.[24] Describe the relationship shown in the scatterplot.

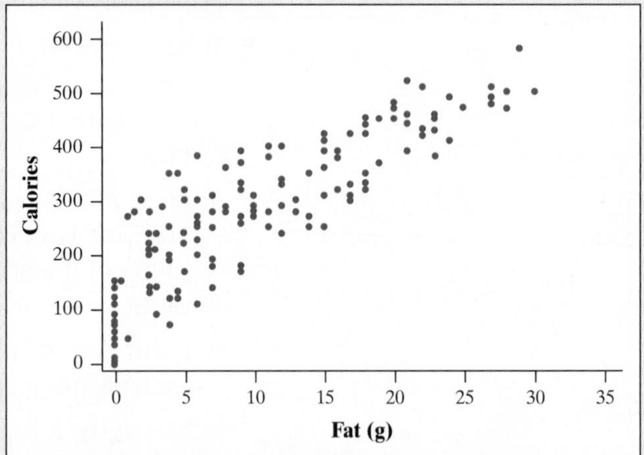

5. **More heavy backpacks** Refer to your graph from Exercise 1. Describe the relationship between body weight and backpack weight for this group of hikers.

6. **More putting success** Refer to your graph from Exercise 2. Describe the relationship between distance from hole and percentage of putts made for the sample of professional golfers.

7. **More butterflies** Refer to Exercise 3. Scientists studying the *Boloria chariclea* butterfly investigated some of the factors that might affect wing size, including temperature during the previous summer and the sex of the butterfly. The scatterplot shown here enhances the scatterplot from Exercise 3 by plotting female butterflies with blue squares. How are the relationships between temperature and wing length the same for male and female butterflies of this species? How are the relationships different?

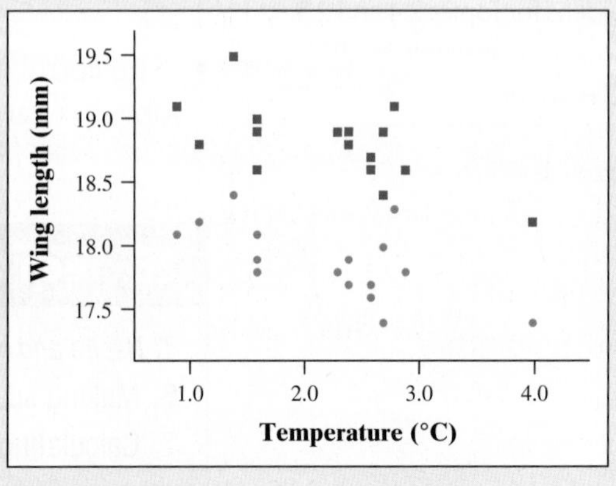

8. **More Starbucks** Refer to Exercise 4. How do the nutritional characteristics of food products differ from those of drink products at Starbucks? The scatterplot shown here enhances the scatterplot from Exercise 4 by plotting the food products with blue squares. How are the relationships between fat and calories the same for the two types of products? How are the relationships different?

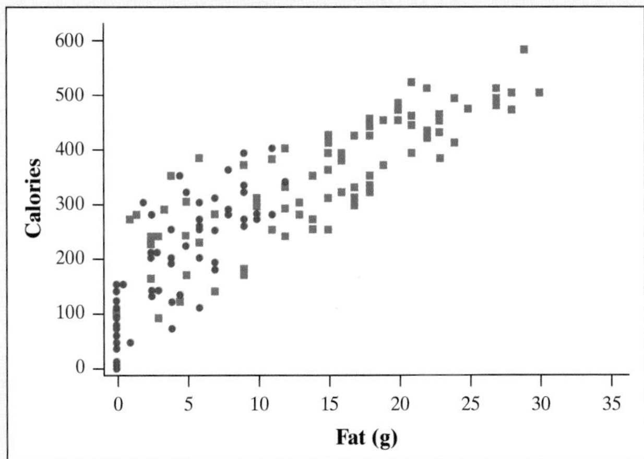

Interpreting Correlation

9. **Whales** Thanks to restrictions on whaling, populations of North Atlantic right whales are making a comeback. One way to measure the health of a whale is by its length. After all, we'd expect healthy whales to grow longer than whales that are not as healthy. Here is a scatterplot showing the relationship between $x =$ age (years) and $y =$ length (m) for a sample of 145 North Atlantic right whales ages 2–20 years that were observed from 2000 to 2019.[25] Is $r > 0$ or $r < 0$? Closer to $r = 0$ or $r = \pm 1$? Explain your reasoning.

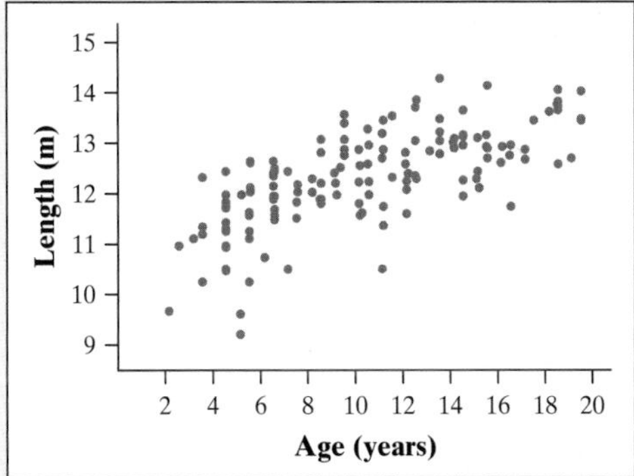

10. **Windy city** Is it possible to use temperature to predict wind speed? Here is a scatterplot showing the average temperature (in degrees Fahrenheit) and average wind speed (in miles per hour) for 365 consecutive days at O'Hare International Airport in Chicago.[26] Is $r > 0$ or $r < 0$? Closer to $r = 0$ or $r = \pm 1$? Explain your reasoning.

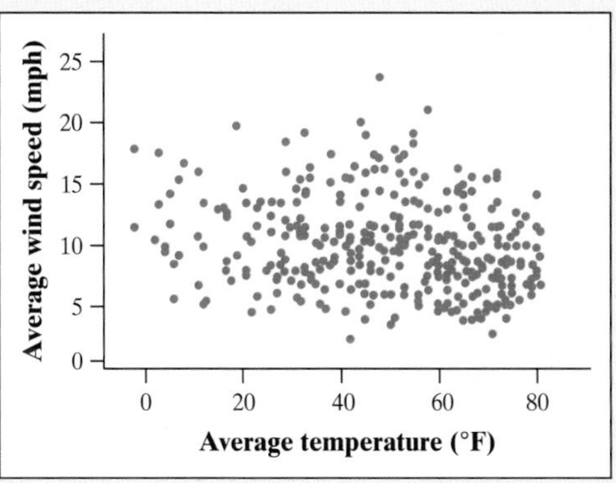

11. **Midterm elections** Many political scientists believe that midterm elections (elections held two years after a presidential election) often go against the party of the current U.S. president. Here is a scatterplot showing the net approval rating (Approve – Disapprove) for each president from Harry Truman in 1946 to Joe Biden in 2022, along with the net change in seats in the U.S. House of Representatives for the president's party in the midterm election.[27] Note that in all but two midterm elections, the president's party lost seats. The correlation for these data is $r = 0.52$. Interpret the correlation.

pg 166

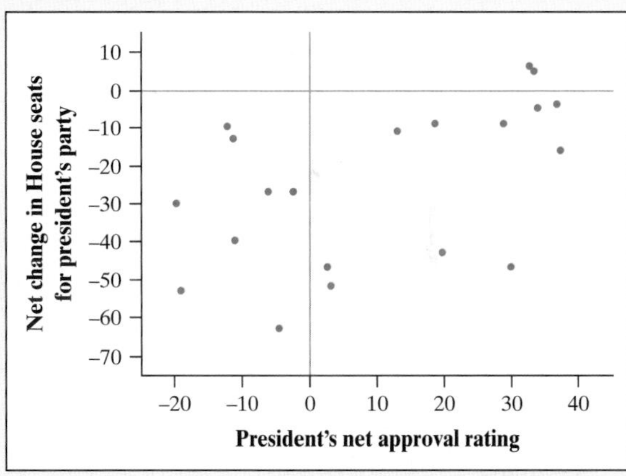

12. **Points and turnovers** Here is a scatterplot showing the relationship between the number of turnovers and the number of points scored for players in a recent National Basketball Association (NBA) season.[28] The correlation for these data is $r = 0.92$. Interpret the correlation.

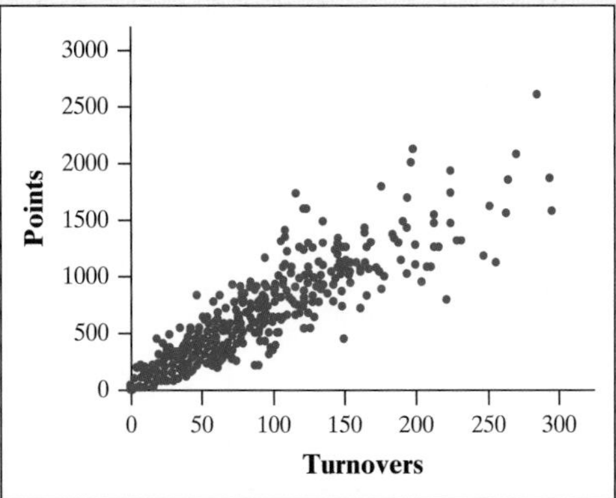

13. **Match the correlations** Suppose that a physical education (PE) teacher collected data about the students in a large class. The variables for which data were collected included number of pull-ups in 1 minute, number of push-ups in 1 minute, number of sit-ups in 1 minute, and weight. The teacher then calculated the correlations for the following relationships:
    - Number of pull-ups and number of push-ups
    - Number of pull-ups and number of sit-ups
    - Number of pull-ups and weight

    The correlations for these relationships are $r = 0.9$, $r = -0.5$, and $r = 0.3$. Which correlation goes with which relationship? Explain your reasoning.

14. **Rank the correlations** Each month for a year, an economist records data on the mean price of oil, the mean price of gasoline, the mean price of airline tickets, and the mean number of miles driven by residents of a city. The economist then calculates the correlations for the following relationships:
    - Mean price of oil and mean number of miles driven
    - Mean price of oil and mean price of gasoline
    - Mean price of oil and mean price of airline tickets

    The correlations for these relationships are $r = 0.95$, $r = -0.25$, and $r = 0.35$. Which correlation goes with which relationship? Explain your reasoning.

Cautions About Correlation

15. **More midterm elections** Refer to Exercise 11. Does the fact that $r = 0.52$ show that a popular president is guaranteed to improve the president's party's chances in the midterm election? Explain your reasoning.

pg 168

16. **More turnovers?** Refer to Exercise 12. Does the fact that $r = 0.92$ suggest that an increase in turnovers will cause NBA players to score more points? Explain your reasoning.

17. **Limitations of correlation** Companies work hard to have their website listed at the top of an internet search. Is there a relationship between a website's position in the results of an internet search (1 = top position, 2 = second position, etc.) and the percentage of people who click on the link for the website? Here are click-through rates for the top 10 positions in searches on a mobile device.[29]

Position	Click-through rate (%)	Position	Click-through rate (%)
1	23.53	6	3.80
2	14.94	7	2.79
3	11.19	8	2.11
4	7.47	9	1.57
5	5.29	10	1.18

(a) Make a scatterplot to show the relationship between position and click-through rate.

(b) The correlation for these data is $r = -0.90$. Explain how this demonstrates that correlation doesn't measure the form of a relationship between two quantitative variables.

18. **More limitations of correlation** A carpenter sells handmade wooden benches at a craft fair every week. Over the past year, the carpenter has varied the price of the bench from \$80 to \$120 and recorded the average weekly profit he made at each selling price. The prices of the bench and the corresponding average profits are shown in the table.

Price	\$80	\$90	\$100	\$110	\$120
Average profit	\$2400	\$2800	\$3000	\$2800	\$2400

(a) Make a scatterplot to show the relationship between price and profit.

(b) The correlation for these data is $r = 0$. Explain how this can be true even though there is a strong relationship between price and average profit.

Calculating Correlation

19. **Dem bones** *Archaeopteryx* is an extinct beast that had feathers like a bird but teeth and a long bony tail like a reptile. Only six fossil specimens are known to exist today. Because these specimens differ greatly in size, some scientists think they are different species rather than individuals from the same species. If the specimens belong to the same species and differ in size because some are younger than others, there should be a positive linear relationship between the lengths of a pair of bones from all individuals. A point outside the pattern would suggest a different species. Here are data on the lengths (in centimeters) of the femur (a leg bone) and the humerus (a bone in the upper arm) for the five specimens in which both bones are preserved, along with a scatterplot.[30] Calculate the correlation using the formula on page 169.

pg 170

Femur (x)	38	56	59	64	74
Humerus (y)	41	63	70	72	84

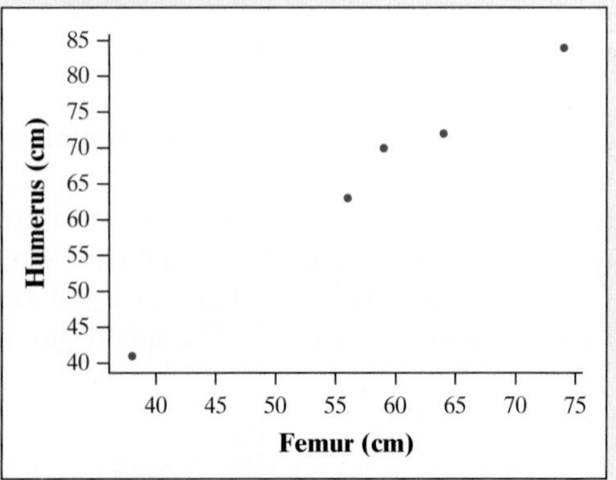

20. **Time and sweetness** People who drink diet sodas may notice a decrease in the sweetness of the soda if it has been sitting around for a few months—particularly in a warm environment. To determine how fast the sweetener aspartame degrades over time, researchers obtained diet sodas with three different acidity levels (pH). They kept some at 20°C, some at 30°C, and the rest at 40°C. Here are data on the time (in months) and the remaining amount of aspartame (%) in a soda with pH = 2.75 stored at 20°C, along with a scatterplot.[31] Calculate the correlation using the formula on page 169.

Time (months)	0	1	2	3	4	5
Aspartame (%)	100	90	85	81	74	65

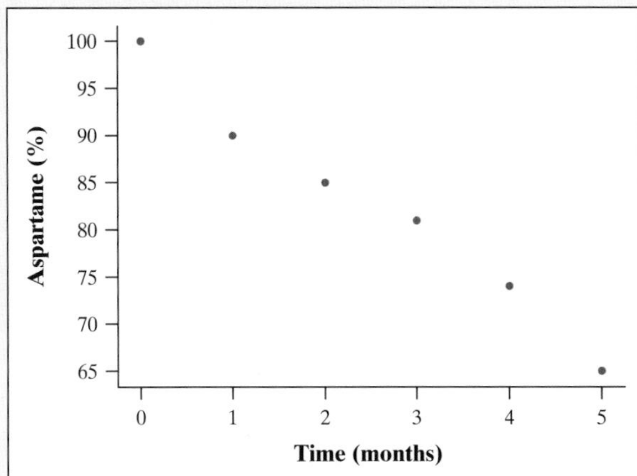

21. **Roller coasters** Many people like to ride roller coasters. Amusement parks try to attract visitors by offering roller coasters that have a variety of speeds and elevations. Here are data on the height (in feet) and maximum speed (in miles per hour) for nine roller coasters, along with a scatterplot.[32] Use technology to calculate the correlation.

Coaster	Height (ft)	Max speed (mph)
Apocalypse	100	55
Bullet	196	83
Corkscrew	70	55
Flying Turns	50	24
Goliath	192	66
Hidden Anaconda	152	65
Iron Shark	100	52
Stinger	131	50
Wild Eagle	210	61

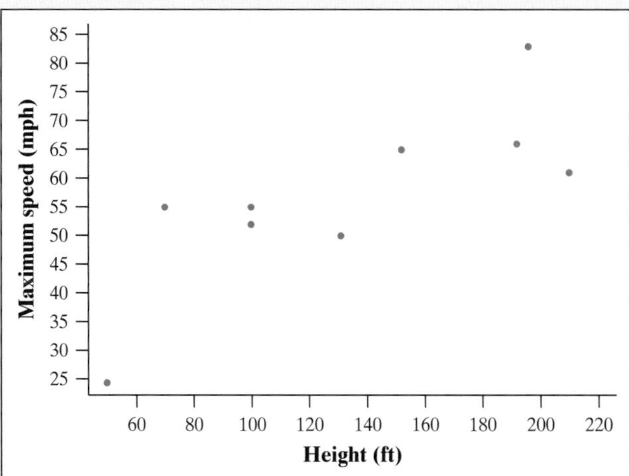

22. **Temperature and elevation** Here are data on the elevation (in feet) and average January temperature (in degrees Fahrenheit) for 10 cities and towns in Colorado, along with a scatterplot.[33] Use technology to calculate the correlation.

City	Elevation (ft)	Average January temperature (°F)
Limon	5452	27
Denver	5232	31
Golden	6408	29
Flagler	5002	29
Eagle	6595	21
Vail	8220	18
Glenwood Springs	7183	25
Rifle	5386	26
Grand Junction	4591	29
Dillon	9049	16

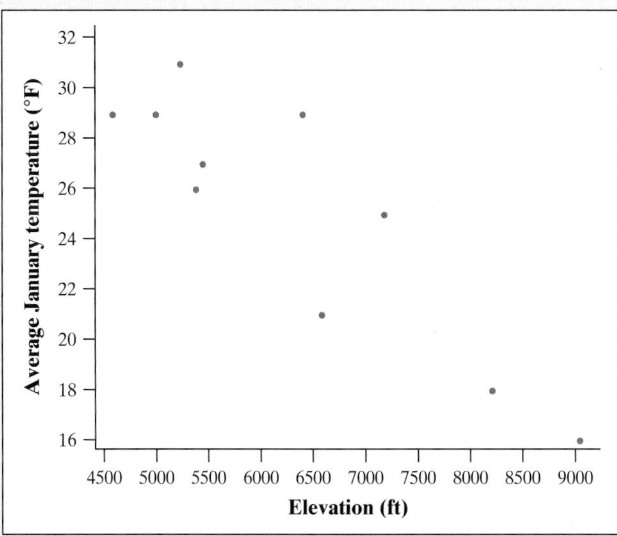

23. **Dem bones again** Refer to Exercise 19.

(a) What would happen to the correlation if the variables were reversed?

(b) What would happen to the correlation if the bone lengths were measured in inches?

(c) What are the units for r in this context?

24. **More time and sweetness** Refer to Exercise 20.

(a) What would happen to the correlation if the variables were reversed?

(b) What would happen to the correlation if time was measured in days?

(c) What are the units for r in this context?

25. **More roller coasters** Refer to Exercise 21.

(a) Use technology to calculate the value of the correlation using $x =$ maximum speed and $y =$ height.

(b) Use technology to convert each of the heights from feet to meters by dividing them by 3.28. Then calculate the correlation between $x =$ height (in meters) and $y =$ maximum speed.

(c) How do your answers to parts (a) and (b) compare with your answer to Exercise 21?

26. **More temperature and elevation** Refer to Exercise 22.

(a) Use technology to calculate the value of the correlation using $x =$ average January temperature and $y =$ elevation.

(b) Use technology to convert each of the elevations from feet to meters by dividing them by 3.28. Then calculate the correlation between $x =$ elevation (in meters) and $y =$ average January temperature.

(c) How do your answers to parts (a) and (b) compare with your answer to Exercise 22?

For Investigation *Apply the skills from the section in a new context or nonroutine way.*

27. **Bubble charts** Most scatterplots display only two variables. However, by using different colors and dot sizes, we can include additional variables. Examine the scatterplot of data from Gapminder (at the bottom of the page).[34] The individuals in this data set are all the world's nations for which data were available in 2020. The variable on the horizontal axis is income per person, a measure of how rich a country is. The variable on the vertical axis is life expectancy at birth. The color of the dot indicates the region where the country is located and the size of the dot is proportional to the size of the country's population.

(a) Describe the relationship between income per person and life expectancy.

(b) Describe the relationship between region and life expectancy.

(c) Describe the relationship between population size and income per person.

28. **Timeplots** Timeplots show how a single quantitative variable changes over time, by making a scatterplot with time on the horizontal axis and the quantitative variable on the vertical axis. Consecutive points are often connected with line segments to make trends easier to see. Here is a timeplot that shows the number of monthly visitors to Grand Canyon National Park from January 2014 to December 2019:

(Graph for Exercise 27)

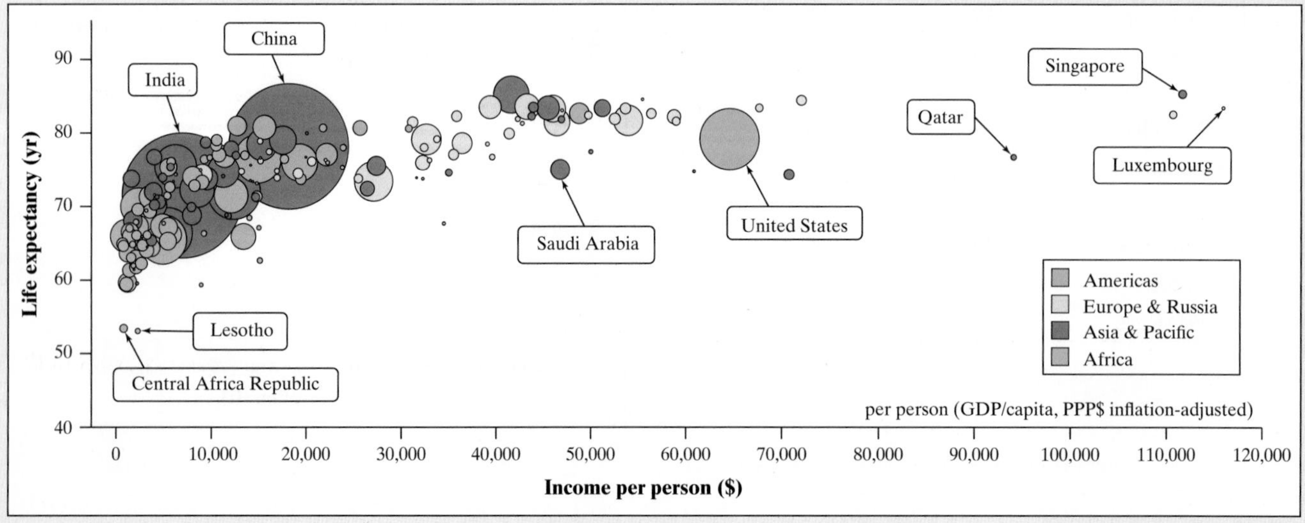

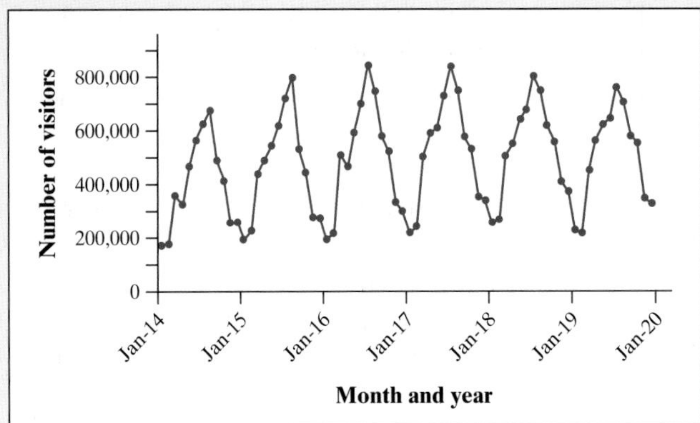

(a) Describe any trends that you see in the timeplot.

(b) Based on the timeplot, how many visitors would you expect in June 2020?

(c) In June 2020, there were just over 300,000 visitors to the Grand Canyon National Park. What does this suggest about using timeplots to predict the future?

29. **What affects correlation?** Here is a scatterplot showing the relationship between two variables with a correlation of $r = 0.81$:

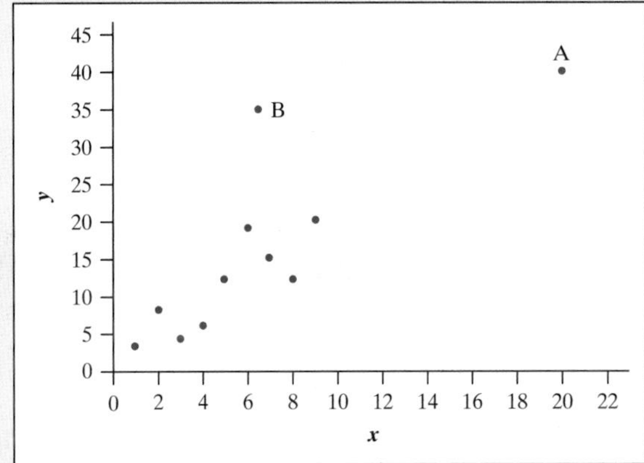

(a) Interpret the correlation.

Here is a scatterplot showing the *z*-scores of the *x* variable and the *z*-scores of the *y* variable:

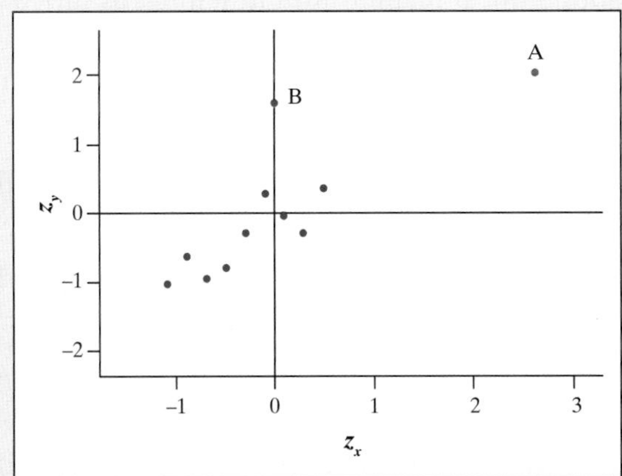

(b) Based on the formula for the correlation, what effect is point A having on the correlation? Explain your answer.

(c) Based on the formula for the correlation, what effect is point B having on the correlation? Explain your answer.

Multiple Choice *Select the best answer for each question.*

30. Which of the following is not important to address when describing the relationship shown in a scatterplot?

(A) Direction

(B) Center

(C) Unusual features

(D) Strength

(E) Form

31. Investment reports often include correlations. Commenting on a report that includes a table of correlations among mutual funds, a reporter says, "For every increase of \$1 in the price of Fund A, the price of Fund B increases by \$2. For example, if the price of Fund A increases by \$1.75, the price of Fund B increases by \$3.50." If this is true, what is the correlation between the price of Fund A and the price of Fund B?

(A) $r = 1$

(B) $r = 0.5$

(C) $r = 0$

(D) $r = -0.5$

(E) $r = -1$

32. In a sample of 42 brands of cookies, the correlation between y = number of calories and x = amount of saturated fat (in milligrams) is $r = 0.55$.[35] If you were to convert the saturated fat values to grams (1 gram = 1000 milligrams) and reverse the variables so that y = amount of saturated fat (in grams) and x = number of calories, what would the correlation be?

(A) $r = 0.55$

(B) $r = 0.00055$

(C) $r = -0.00055$

(D) $r = -0.55$

(E) The correlation cannot be determined without the raw data.

Recycle and Review *Practice what you learned in previous sections.*

33. **Big diamonds (1C)** Here are the weights (in milligrams) of 58 diamonds from a nodule carried up to the earth's surface in surrounding rock. These data represent a population of diamonds formed in a single event deep in the earth.[36]

13.8	3.7	33.8	11.8	27.0	18.9	19.3	20.8	25.4	23.1	7.8
10.9	9.0	9.0	14.4	6.5	7.3	5.6	18.5	1.1	11.2	7.0
7.6	9.0	9.5	7.7	7.6	3.2	6.5	5.4	7.2	7.8	3.5
5.4	5.1	5.3	3.8	2.1	2.1	4.7	3.7	3.8	4.9	2.4
1.4	0.1	4.7	1.5	2.0	0.1	0.1	1.6	3.5	3.7	2.6
4.0	2.3	4.5								

Make a histogram to display the distribution of weight. Describe the distribution.

34. **Fruit fly thorax lengths (1F)** Fruit flies are used frequently in genetic research because of their short reproductive cycle. The length of the thorax (in millimeters) for male fruit flies is approximately normally distributed with a mean of 0.80 mm and a standard deviation of 0.08 mm.[37]

(a) What proportion of male fruit flies have a thorax length greater than 1 mm?

(b) What is the 30th percentile for male fruit fly thorax lengths?

SECTION 2C Linear Regression Models

LEARNING TARGETS *By the end of the section, you should be able to:*

- Make predictions using a regression line, keeping in mind the dangers of extrapolation.
- Calculate and interpret a residual.
- Interpret the slope and *y* intercept of a regression line.
- Determine the equation of a least-squares regression line using formulas.
- Determine the equation of a least-squares regression line using technology.
- Construct and interpret residual plots to assess whether a regression model is appropriate.
- Interpret the coefficient of determination r^2 and the standard deviation of the residuals *s*.

In the preceding section, we found linear relationships in settings as varied as the WNBA, the Old Faithful geyser, and Nobel Prizes. When a scatterplot shows a linear relationship, we can summarize the overall pattern with a **regression line** that models the relationship between the two variables.

DEFINITION Regression line

A **regression line** is a line that models how a response variable *y* changes as an explanatory variable *x* changes. Regression lines are expressed in the form $\hat{y} = a + bx$, where $\hat{y}$ (pronounced "*y*-hat") is the predicted value of *y* for a given value of *x*.

AP® EXAM TIP
Formula sheet

The regression line is defined as $\hat{y} = a + bx$ on the formula sheet provided on both sections of the AP® Statistics exam.

We could also express a regression line in the form $y = mx + b$, as we do in algebra. However, statisticians prefer the reordered format $\hat{y} = a + bx$ because it works better when they use several explanatory variables in a *multiple* linear regression model. In this course, we focus on *simple* linear regression models with a single explanatory variable. Regardless of which format you use for a simple linear regression model, the slope is always the coefficient of the x variable.

It is common knowledge that cars and trucks lose value the more they are driven. Can we predict the price of a used Ford F-150 SuperCrew 4 × 4 truck if we know how many miles it has on the odometer? A random sample of 16 used Ford F-150 SuperCrew 4 × 4s was selected from among those listed for sale at autotrader.com. The number of miles driven and the price (in dollars) were recorded for each of the trucks.[38] Here are the data:

Tim Graham/Alamy Stock Photo

Miles driven	70,583	129,484	29,932	29,953	24,495	75,678	8359	4447
Price ($)	21,994	9500	29,875	41,995	41,995	28,986	31,891	37,991
Miles driven	34,077	58,023	44,447	68,474	144,162	140,776	29,397	131,385
Price ($)	34,995	29,988	22,896	33,961	16,883	20,897	27,495	13,997

Figure 2.6 is a scatterplot of these data. The plot shows a moderately strong, negative linear association between miles driven and price. There are two distinct clusters of trucks: a group of 12 trucks with between 0 and 80,000 miles driven and a group of 4 trucks with between 120,000 and 160,000 miles driven. The correlation is $r = -0.815$. The line on the plot is a regression line for predicting price from miles driven.

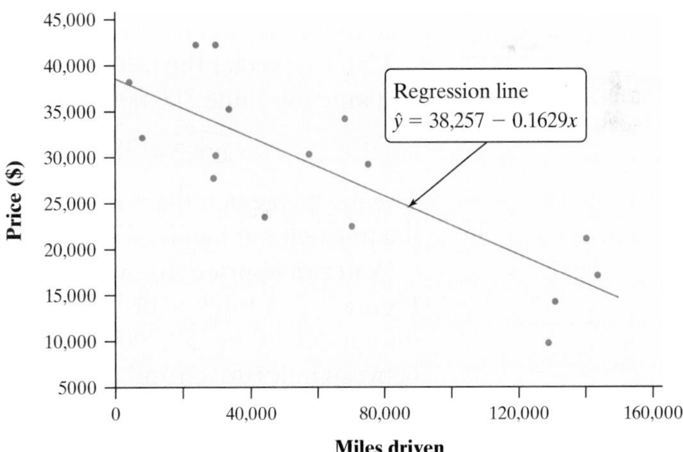

FIGURE 2.6 Scatterplot showing the price and miles driven for used Ford F-150s, along with a regression line.

Prediction

We can use a regression line to predict the value of the response variable for a specific value of the explanatory variable. For the Ford F-150 data, the equation of the regression line is

$$\widehat{\text{price}} = 38{,}257 - 0.1629(\text{miles driven})$$

where $\widehat{\text{price}}$ refers to the predicted price of a used Ford F-150.

If a used Ford F-150 has 100,000 miles driven, substitute $x = 100{,}000$ in the equation. The predicted price is

$$\widehat{\text{price}} = 38{,}257 - 0.1629(100{,}000) = \$21{,}967$$

This prediction is illustrated in Figure 2.7.

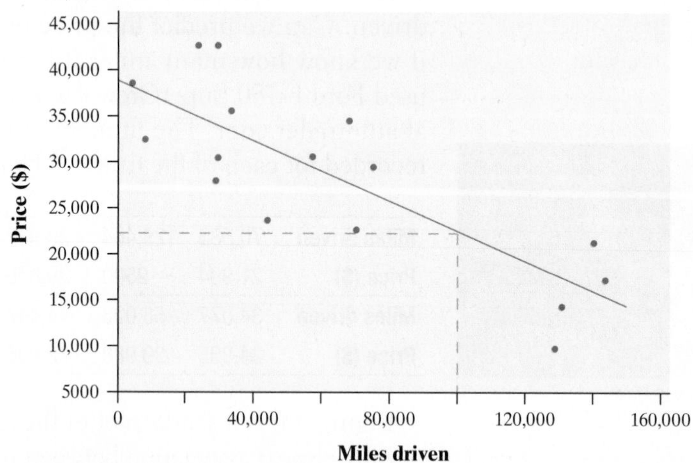

FIGURE 2.7 Using the regression line to predict price for a Ford F-150 with 100,000 miles driven.

Even though the value $\hat{y} = \$21{,}967$ is unlikely to be the actual price of a truck that has been driven 100,000 miles, it's our best guess based on the linear model using $x =$ miles driven. We can also think of $\hat{y} = \$21{,}967$ as the average price for a sample of trucks that have each been driven 100,000 miles.

Can we predict the price of a Ford F-150 with 300,000 miles driven? We can certainly substitute 300,000 into the equation of the line. The prediction is

$$\widehat{\text{price}} = 38{,}257 - 0.1629(300{,}000) = -\$10{,}613$$

The model predicts that we would need to *pay* \$10,613 just to have someone take the truck off our hands!

A negative price doesn't make much sense in this context. Look again at Figure 2.7. A truck with 300,000 miles driven is far outside the set of x-values for our data (0 to 150,000 miles driven). We can't say whether the relationship between miles driven and price remains linear at such extreme values. Predicting the price for a truck with 300,000 miles driven is an **extrapolation** of the relationship beyond what the data show.

DEFINITION Extrapolation

Extrapolation is the use of a regression line for prediction outside the interval of x-values used to obtain the line. The further we extrapolate, the less reliable the predictions become.

Few relationships are linear for all values of the explanatory variable. **Don't make predictions using values of x that are much larger or much smaller than those that actually appear in your data.**

EXAMPLE

How much candy can you grab?
Prediction

Skill 2.C

PROBLEM: The scatterplot shows the hand span (in cm) and the number of Starburst™ candies grabbed by each student when Mr. Tyson's class did the "Candy grab" activity. The regression line $\hat{y} = -29.8 + 2.83x$ has been added to the scatterplot.

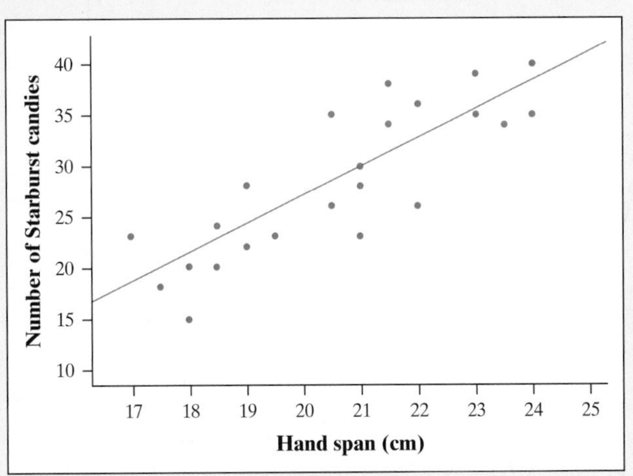

Josh Tabor

(a) Andres has a hand span of 22 cm. Predict the number of Starburst candies he can grab.
(b) Mr. Tyson's young daughter McKayla has a hand span of 12 cm. Predict the number of Starburst candies she can grab.
(c) How confident are you in each of these predictions? Explain your answer.

SOLUTION:

(a) $\hat{y} = -29.8 + 2.83(22)$
$= 32.46$ Starburst candies

(b) $\hat{y} = -29.8 + 2.83(12)$
$= 4.16$ Starburst candies

Don't worry that the predicted number of Starburst candies isn't an integer. Think of 32.46 as the average number of Starburst candies that a group of students, each with a hand span of 22 cm, could grab.

(c) The prediction for Andres is believable because $x = 22$ is within the interval of x-values used to create the model. However, the prediction for McKayla is not trustworthy because $x = 12$ is far outside of the set of x-values used to create the regression line. The linear form may not extend to hand spans this small.

FOR PRACTICE, TRY EXERCISE 1

Residuals

In most cases, no line will pass exactly through all the points in a scatterplot. Because we use the line to predict y from x, the prediction errors we make are errors in y, the vertical direction in the scatterplot.

Figure 2.8 shows a scatterplot of the Ford F-150 data with a regression line added. The prediction errors are marked as bold vertical segments in the graph. These vertical deviations represent "leftover" variation in the response variable after fitting the regression line. For that reason, they are called **residuals.**

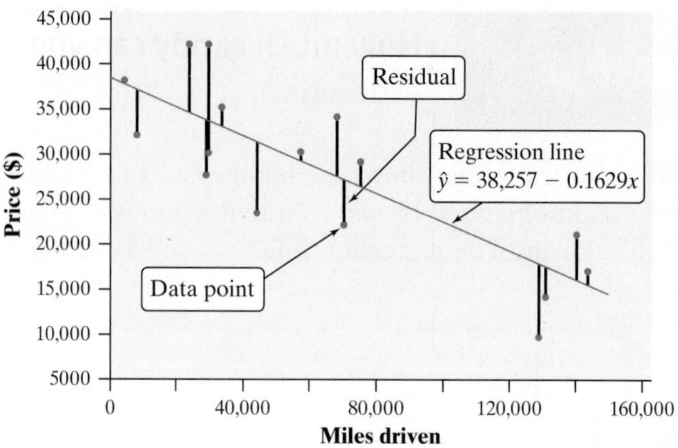

FIGURE 2.8 Scatterplot of the Ford F-150 data with a regression line added. A good regression line should make the residuals (shown as bold vertical line segments) as small as possible.

DEFINITION Residual

A **residual** is the difference between the actual value of *y* and the value of *y* predicted by the regression line. That is,

$$\begin{aligned}\text{residual} &= \text{actual } y - \text{predicted } y \\ &= y - \hat{y}\end{aligned}$$

To make sure you subtract in the correct order when calculating a residual, remember the acronym AP (Actual y − Predicted y).

In Figure 2.8, the highlighted data point represents a Ford F-150 that had 70,583 miles driven and a price of \$21,994. The regression line predicts a price of

$$\widehat{\text{price}} = 38{,}257 - 0.1629(70{,}583) = \$26{,}759$$

for this truck, but its actual price was \$21,994. This truck's residual is

$$\begin{aligned}\text{residual} &= \text{actual } y - \text{predicted } y \\ &= y - \hat{y} \\ &= 21{,}994 - 26{,}759 = -\$4765\end{aligned}$$

The actual price of this truck is \$4765 *less* than the cost predicted by the regression line with $x =$ miles driven. Why is the actual price less than predicted? There are many possible reasons. Perhaps the truck needs body work, has mechanical issues, or has been in an accident.

EXAMPLE

Can you grab more than expected?
Residuals

Skill 2.C

PROBLEM: Here again is the scatterplot showing the hand span (in cm) and the number of Starburst candies grabbed by each student in Mr. Tyson's class. The regression line is $\hat{y} = -29.8 + 2.83x$.

Josh Tabor

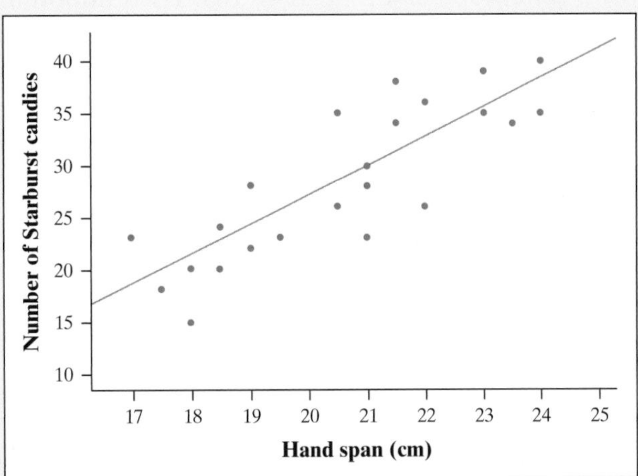

Find and interpret the residual for Andres, who has a hand span of 22 cm and grabbed 36 Starburst candies.

SOLUTION:

$\hat{y} = -29.8 + 2.83(22) = 32.46$ Starburst candies

residual $= 36 - 32.46 = 3.54$ Starburst candies

residual = actual y − predicted y
$= y - \hat{y}$

Andres grabbed 3.54 more Starburst candies than the number predicted by the regression line with x = hand span.

FOR PRACTICE, TRY EXERCISE 3

Interpreting the Slope and y Intercept

A regression line is a *model* for the data. The y **intercept** and **slope** of the regression line describe what the model tells us about the relationship between the response variable y and the explanatory variable x.

DEFINITION *y* intercept, Slope

In the regression equation $\hat{y} = a + bx$:

- a is the ***y* intercept**, the predicted value of y when $x = 0$.
- b is the **slope**, the amount by which the predicted value of y changes when x increases by 1 unit.

AP® EXAM TIP

When asked to interpret the slope or y intercept, it is very important to include the word *predicted* (or equivalent) in your response. Otherwise, it might appear that you believe the regression equation provides actual values of y.

The data used to calculate a regression line typically come from a sample. The statistics a and b in the sample regression model estimate the y intercept and slope parameters of the population regression model. You'll learn more about how this works in Unit 9.

Let's return to the Ford F-150 data. The equation of the regression line for these data is $\hat{y} = 38{,}257 - 0.1629x$, where x = miles driven and y = price. The slope $b = -0.1629$ tells us that the *predicted* price of a used Ford F-150 decreases by \$0.1629 (16.29 cents) for each additional mile that the truck has been driven. The y intercept $a = 38{,}257$ is the *predicted* price (in dollars) of a Ford F-150 that has been driven 0 miles.

The slope of a regression line is an important numerical description of the relationship between the two variables. Although we need the value of the y intercept to draw the line, it is statistically meaningful only when the explanatory variable can actually take values close to 0, as in the Ford F-150 data. In other cases, using the regression line to make a prediction for $x = 0$ is an extrapolation.

EXAMPLE

Grabbing more candy
Interpreting the slope and y intercept

Skill 4.B

PROBLEM: The scatterplot shows the hand span (in cm) and the number of Starburst candies grabbed by each student in Mr. Tyson's class, along with the regression line $\hat{y} = -29.8 + 2.83x$.

Felix Choo/Alamy Stock Photo

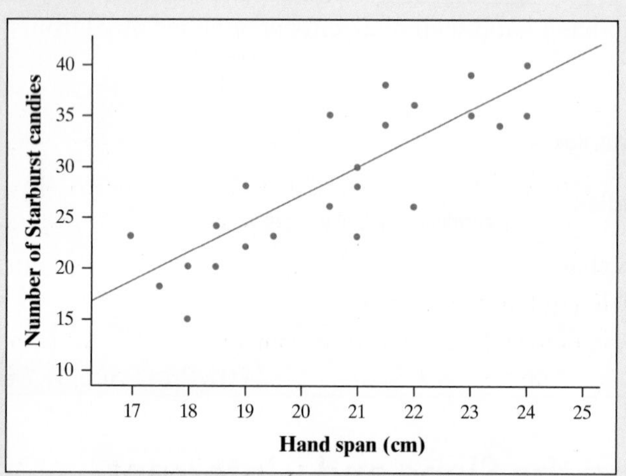

(a) Interpret the slope of the regression line.

(b) Does the value of the y intercept have meaning in this context? If so, interpret the y intercept. If not, explain why.

SOLUTION:

(a) The predicted number of Starburst candies grabbed increases by 2.83 for each increase of 1 cm in hand span.

Remember that the slope describes how the *predicted* value of y changes, not the actual value of y.

(b) The y intercept does not have meaning in this case, as it is impossible to have a hand span of 0 cm.

Predicting the number of Starburst candies grabbed when $x = 0$ is an extrapolation—and results in an unrealistic prediction of −29.8.

FOR PRACTICE, TRY EXERCISE 7

For the Ford F-150 data, the slope $b = -0.1629$ is very close to 0. This does *not* mean that change in miles driven has little effect on price. The size of the slope depends on the units in which we measure the two variables. In this example, the slope is the predicted change in price (in dollars) when the distance driven increases by 1 mile. There are 100 cents in a dollar. If we measured price in cents instead of dollars, the slope would be 100 times steeper, $b = -16.29$. *You can't say how strong a relationship is by looking at the slope of the regression line.*

CHECK YOUR UNDERSTANDING

In Section 2B, we investigated the relationship between height (in inches) and free-throw percentage for a random sample of WNBA players. Here is a scatterplot of the data, along with the regression line $\hat{y} = 177.4 - 1.366x$.

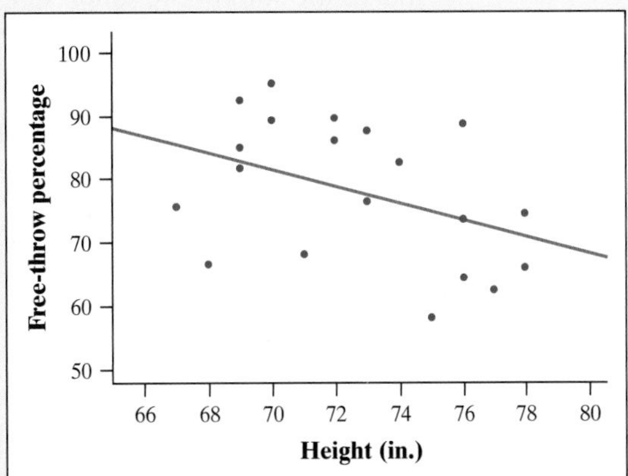

1. Predict the free-throw percentage for a player who is 75 inches tall.
2. Calculate and interpret the residual for the 75-inch-tall player who made 58.1% of her free throws.
3. Interpret the slope of the regression line.
4. Does the value of the y intercept have meaning in this context? If so, interpret the y intercept. If not, explain why.

The Least-Squares Regression Line

We could use many different lines to model the linear association in a particular scatterplot. A *good* regression line makes the residuals as small as possible.

The scatterplots in Figure 2.9 show the relationship between the price of a Ford F-150 and the number of miles it has been driven, along with the least-squares regression line. Figure 2.9(a) shows the residuals as vertical line segments. To determine how well the line works, we could add the residuals and hope for a small total. If the line is a good fit, some of the residuals will be positive and others negative.

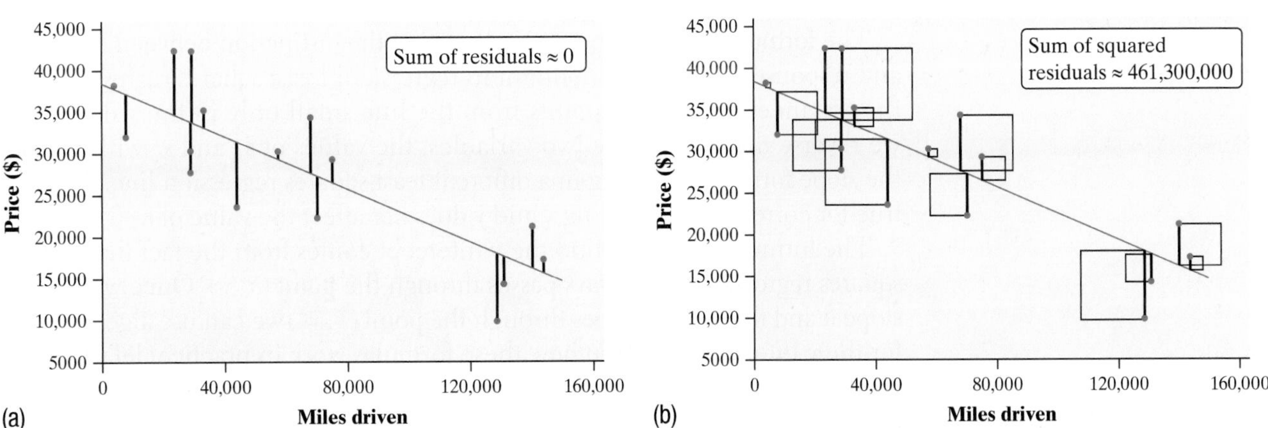

FIGURE 2.9 Scatterplots of the Ford F-150 data with the regression line added. (a) A good regression line will have a sum of residuals of approximately 0. (b) The least-squares regression line makes the sum of squared residuals as small as possible.

If we add these residuals, the positive and negative residuals will cancel each other out and add to approximately 0, as in Figure 2.9(a). Unfortunately, the residuals of some worse-fitting lines (such as a horizontal line at $y = \$27{,}834$) will also add to 0, making the sum of residuals a flawed measure of quality. To avoid this problem, we square the residuals before adding them, as shown in Figure 2.9(b). Of all the possible lines we could use, the **least-squares regression line** is the one that makes the sum of the *squares* of the residuals the *least*.

DEFINITION Least-squares regression line

The **least-squares regression line** is the line that makes the sum of the squared residuals as small as possible.

It is possible to calculate the equation of the least-squares regression line using only the means and standard deviations of the two variables and their correlation. Exploring this method will highlight an important relationship between the correlation and the slope of a least-squares regression line—and reveal why we include the word *regression* in the expression *least-squares regression line*.

AP® EXAM TIP

Formula sheet

The formulas to calculate the slope and *y* intercept of the least-squares regression line are included on the formula sheet provided on both sections of the AP® Statistics exam.

HOW TO CALCULATE THE LEAST-SQUARES REGRESSION LINE USING SUMMARY STATISTICS

Suppose we have data for an explanatory variable x and a response variable y for n individuals and want to calculate the least-squares regression line $\hat{y} = a + bx$. From the data, calculate the means $\bar{x}$ and $\bar{y}$, the standard deviations s_x and s_y, and correlation r. The slope is

$$b = r\frac{s_y}{s_x}$$

Because the point $(\bar{x}, \bar{y})$ is always on the least-squares regression line, solve for the y intercept a using this equation:

$$\bar{y} = a + b\bar{x}$$

The formula for the slope reminds us that the distinction between explanatory and response variables is important in regression. Least-squares regression makes the distances of the data points from the line small only in the y direction. If we reverse the roles of the two variables, the values of s_x and s_y will reverse in the slope formula, resulting in a different least-squares regression line. This is *not* true for correlation: switching x and y does *not* affect the value of r.

The formula we use to find the y intercept comes from the fact that the least-squares regression line always passes through the point $(\bar{x}, \bar{y})$. Once we know the slope b and that the line goes through the point $(\bar{x}, \bar{y})$, we can use algebra to solve for the y intercept a. To see how these formulas work in practice, let's look at an example.

EXAMPLE

Foot length and height
The least-squares regression line

PhotoAlto sas/Alamy Stock Photo

Skill 2.C

PROBLEM: A random sample of 15 high school students was selected to investigate the relationship between foot length (in centimeters) and height (in centimeters). The mean of the foot lengths is 24.76 with standard deviation 2.71. The mean of the heights is 171.43 with standard deviation 10.69. The correlation between foot length and height is $r = 0.697$. Calculate the equation of the least-squares regression line for predicting height from foot length using these data.

SOLUTION:

$b = 0.697\frac{10.69}{2.71} = 2.75$

> $b = r\frac{s_y}{s_x}$
> Because we are predicting height from foot length, y = height and x = foot length.

$171.43 = a + 2.75(24.76)$

> $\bar{y} = a + b\bar{x}$

$a = 103.34$

The equation of the least-squares regression line is $\hat{y} = 103.34 + 2.75x$.

FOR PRACTICE, TRY EXERCISE 13

There is a close connection between the correlation and the slope of the least-squares regression line. The slope is

$$b = r\frac{s_y}{s_x} = \frac{rs_y}{s_x}$$

This equation says that along the regression line, a change of 1 standard deviation in x corresponds to a change of r standard deviations in y. When the variables are perfectly correlated ($r = 1$ or $r = -1$), the change in the predicted response $\hat{y}$ is the same (in standard deviation units) as the change in x. For example, if $r = 1$ and x is 2 standard deviations above $\bar{x}$, then the corresponding value of $\hat{y}$ will be 2 standard deviations above $\bar{y}$.

If the variables are not perfectly correlated ($-1 < r < 1$), the change in $\hat{y}$ is *less than* the change in x, when measured in standard deviation units. For example, for each increase of 1 standard deviation in the value of the explanatory variable x, the least-squares regression line predicts an increase of *only* r standard deviations in the response variable y. This tendency for values of the explanatory variable to be paired with less extreme values of the response variable is called *regression to the mean*. In fact, we use r as the symbol for correlation because it indicates how much values tend to *regress* to the mean.

Regression to the mean occurs in many different contexts. The scatterplot shows the heights of fathers and their adult sons, along with the line $y = x$. Points on this line indicate that a father and son have the same height. Notice that tall fathers tend to have sons who are tall, but not quite as tall, on average. Likewise, short fathers tend to have sons who are short, but not quite as short, on average.

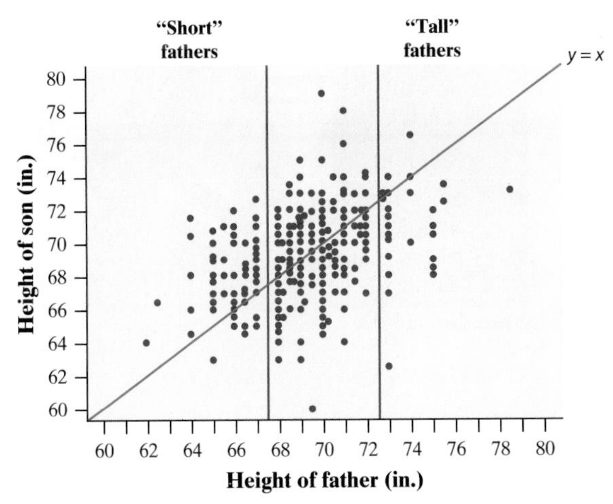

Using Technology to Calculate a Least-Squares Regression Line

Your calculator or statistical software will give the equation of the least-squares line for the data that you enter. As an example, Figure 2.10 displays the basic regression output for the Ford F-150 data from two statistical software packages: Minitab and JMP. Other software produces very similar output. Each output records the slope and y intercept of the least-squares line. The software also provides information that we don't need right now, although we will use much of it later. *Once you understand the statistical ideas, you can read and work with almost any software output.*

Minitab

(Slope → -0.16292; *y* intercept → 38257)

Predictor	Coef	SE Coef	T	P
Constant	38257	2446	15.64	0.000
Miles Driven	-0.16292	0.03096	-5.26	0.000

S = 5740.13   R-Sq = 66.4%   R-Sq(adj) = 64.0%

JMP

Summary of Fit

RSquare	0.664248
RSquare Adj	0.640266
Root Mean Square Error	5740.131
Mean of Response	27833.69
Observations (or Sum Wgts)	16

Parameter Estimates

Term	Estimate	Std Error	t Ratio	Prob>|t|
Intercept	38257.135	2445.813	15.64	<.0001
Miles Driven	-0.162919	0.030956	-5.26	0.0001

(*y* intercept → 38257.135; Slope → -0.162919)

FIGURE 2.10 Least-squares regression output for the Ford F-150 data from Minitab and JMP statistical software. Other software produces similar output.

AP® EXAM TIP

Be prepared to get information from computer output on the AP® Statistics exam, especially in settings involving the relationship between two quantitative variables. When displaying the slope and intercept of a least-squares regression line, technology will often report the slope and intercept with much more precision than we need. There is no firm rule for how many decimal places to show for answers on the AP® Statistics exam. Our advice: decide how much to round based on the context of the problem you are working on.

EXAMPLE

Foot length and height, part 2
Using technology to calculate a least-squares regression line

Skill 2.C

© Fancy/Alamy Stock Photo

PROBLEM: A random sample of 15 high school students was selected to investigate the relationship between foot length (in centimeters) and height (in centimeters). Use the statistical software output to determine the equation of the least-squares regression line for predicting height from foot length.

Predictor	Coef	SE Coef	T	P
Constant	103.4100	19.5000	5.30	0.000
Foot length	2.7469	0.7833	3.51	0.004

S = 7.95126   R-Sq = 48.6%   R-Sq(adj) = 44.7%

SOLUTION:

The equation of the least-squares regression line is
$\widehat{\text{height}} = 103.41 + 2.75\ \text{footlength}$.

Because of rounding in the preceding example, the answer here doesn't match exactly.

AP® EXAM TIP

To avoid losing credit on the AP® Statistics exam, make sure to define the variables when stating the equation of a least-squares regression line. Do this by including the names of the variables in the equation or by explicitly defining *x* and *y*.

FOR PRACTICE, TRY EXERCISE 17

Finally, here are the steps for calculating the equation of the least-squares regression line using a graphing calculator.

8. Tech Corner — CALCULATING LEAST-SQUARES REGRESSION LINES

TI-Nspire and other technology instructions are on the book's website at bfwpub.com/tps7e.

Let's use the Ford F-150 data to show how to find the equation of the least-squares regression line on the TI-83/84. Here are the data again:

Miles driven	70,583	129,484	29,932	29,953	24,495	75,678	8359	4447
Price ($)	21,994	9500	29,875	41,995	41,995	28,986	31,891	37,991
Miles driven	34,077	58,023	44,447	68,474	144,162	140,776	29,397	131,385
Price ($)	34,995	29,988	22,896	33,961	16,883	20,897	27,495	13,997

1. Enter the miles driven data into L1 and the price data into L2.
2. To determine the least-squares regression line, press STAT, choose CALC, and then choose LinReg(a + bx).
   - **OS 2.55 or later:** In the dialog box, enter the following: Xlist:L1, Ylist:L2, FreqList (leave blank), Store RegEQ (leave blank), and choose Calculate.
   - **Older OS:** Finish the command to read LinReg(a + bx) L1,L2 and press ENTER.

Note: If r^2 and r do not appear on the TI-83/84 screen, do this one-time series of keystrokes:

- **OS 2.55 or later:** Press MODE and set STAT DIAGNOSTICS to ON. Then redo Step 2 to calculate the least-squares line. The values of r^2 and r should now appear.
- **Older OS:** Press 2nd 0 (CATALOG), scroll down to DiagnosticOn, and press ENTER twice. The screen should say "Done." Then redo Step 2 to calculate the least-squares line. The value of r^2 and r should now appear.

To graph the least-squares regression line on the scatterplot:

1. Set up a scatterplot (see the Tech Corner on page 161).
2. Press Y= and enter the equation of the least-squares regression line in Y1.
3. Press ZOOM and choose ZoomStat to see the scatterplot with the least-squares regression line.

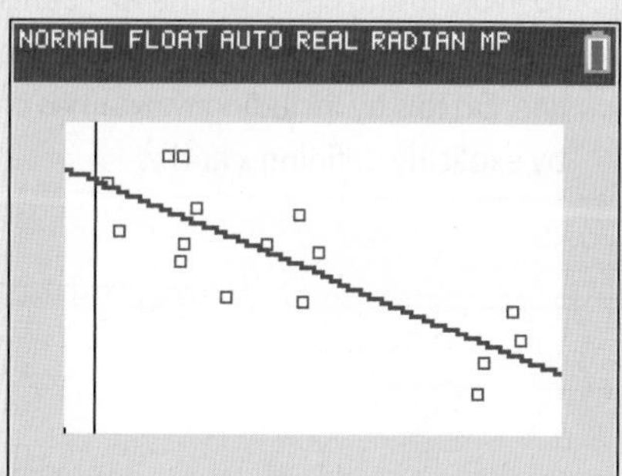

Determining Whether a Linear Model Is Appropriate: Residual Plots

One of the first principles of data analysis is to look for an overall pattern and for striking departures from the pattern. A regression line describes the overall pattern of a linear relationship between an explanatory variable and a response variable. We can identify departures from this pattern by looking at a **residual plot.**

DEFINITION Residual plot

A **residual plot** is a scatterplot that displays the residuals on the vertical axis and the values of the explanatory variable (or the predicted values) on the horizontal axis.

Although most residual plots you'll encounter in AP® Statistics show the values of the explanatory variable on the *x* axis, some software packages make residual plots with the predicted *y*-values on the *x* axis. The basic shape of the two plots is the same because $\hat{y}$ is linearly related to *x*.

Residual plots help us assess whether a regression model is appropriate. In Figure 2.11(a), the scatterplot shows the relationship between the average income (gross domestic product per person, in dollars) and fertility rate (number of children per woman) in 187 countries, along with the least-squares regression line. The residual plot in Figure 2.11(b) shows the average income for each country and the corresponding residual.

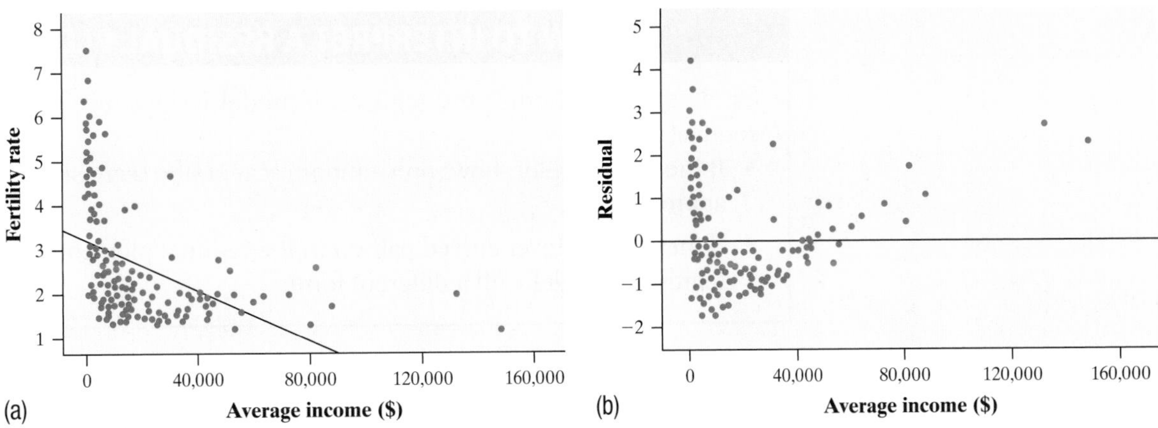

FIGURE 2.11 The (a) scatterplot and (b) residual plot for the linear model relating fertility rate to average income for a sample of countries.

The least-squares regression line clearly doesn't fit this association very well! For most countries with average incomes less than \$5000, the actual fertility rates are greater than predicted, resulting in positive residuals. For countries with average incomes between \$5000 and \$60,000, the actual fertility rates tend to be smaller than predicted, resulting in negative residuals. Countries with average incomes greater than \$60,000 all have fertility rates greater than predicted, again resulting in positive residuals. This U-shaped pattern in the residual plot indicates that the linear form of our model doesn't match the form of the association. A curved model might be better in this case.

In Figure 2.12(a), the scatterplot shows the Ford F-150 data, along with the least-squares regression line. The corresponding residual plot is shown in Figure 2.12(b).

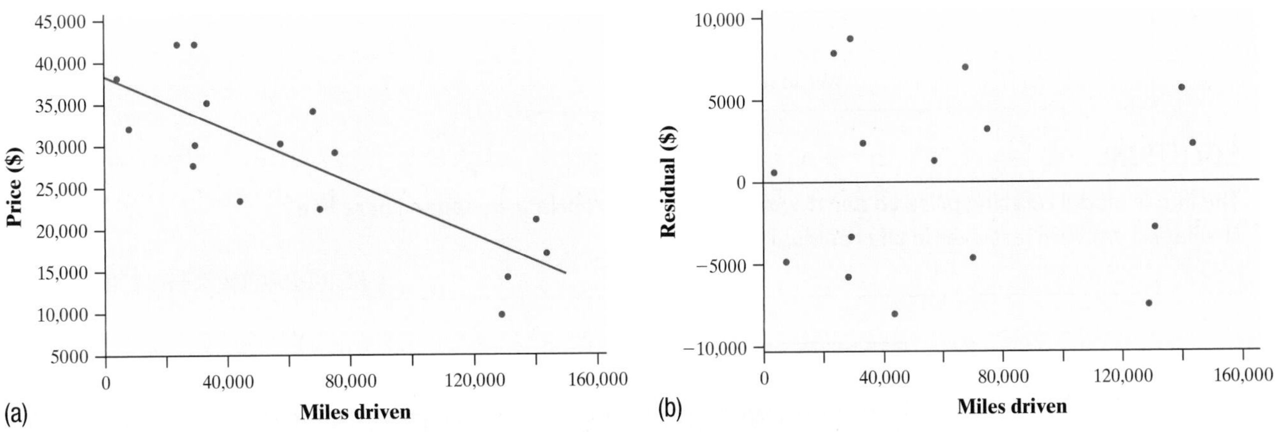

FIGURE 2.12 The (a) scatterplot and (b) residual plot for the linear model relating price to miles driven for Ford F-150s.

Looking at the scatterplot, the least-squares regression line seems to be a good fit for this relationship. You can "see" that the line is appropriate by the lack of a leftover curved pattern in the residual plot. In fact, the residuals look randomly scattered around the residual = 0 line.

HOW TO INTERPRET A RESIDUAL PLOT

To determine whether the regression model is appropriate, look at the residual plot.

- If the residual plot shows only random scatter, the regression model is appropriate.
- If there is a leftover curved pattern in the residual plot, consider using a regression model with a different form.

EXAMPLE

Pricing diamonds
Residual plots

Skill 2.A

PROBLEM: Is a linear model appropriate to describe the relationship between the weight (in carats) and the price (in dollars) of round, clear, internally flawless diamonds with excellent cuts? We calculated a least-squares regression line using x = weight and y = price and made the corresponding residual plot shown here.[39] Use the residual plot to determine if the linear model is appropriate.

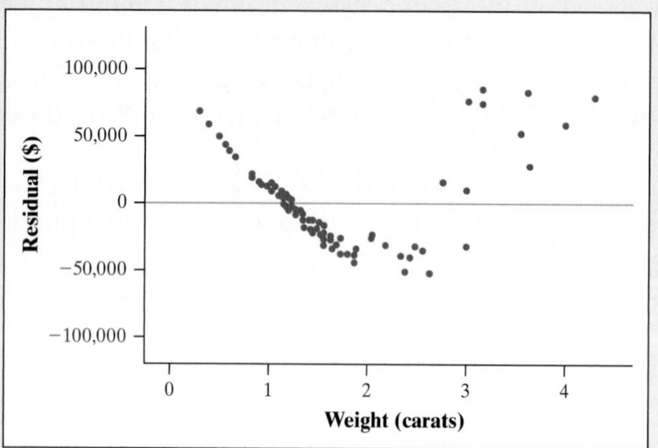

JGI/Getty Images

SOLUTION:

The linear model relating price to carat weight is not appropriate because there is a U-shaped pattern left over in the residual plot.

FOR PRACTICE, TRY EXERCISE 21

Think About It

WHY DO WE LOOK FOR PATTERNS IN RESIDUAL PLOTS? The word *residual* comes from the Latin word *residuum,* meaning "left over." When we calculate a residual, we are calculating what is left over after subtracting the predicted value from the actual value:

$$\text{residual} = \text{actual } y - \text{predicted } y$$

Likewise, when we look at the form of a residual plot, we are looking at the form that is left over after subtracting the form of the model from the form of the association:

$$\text{form of residual plot} = \text{form of association} - \text{form of model}$$

When there is a leftover form in the residual plot, the form of the association and the form of the model are not the same. However, if the form of the association and the form of the model are the *same*, the residual plot should have no form, other than random scatter.

If a residual plot isn't provided and you have the raw data, you can use technology to construct a residual plot.

9. Tech Corner MAKING RESIDUAL PLOTS

TI-Nspire and other technology instructions are on the book's website at bfwpub.com/tps7e.

Let's continue the analysis of the Ford F-150 miles driven and price data from the Tech Corner on page 191. You should have already made a scatterplot, calculated the equation of the least-squares regression line, and graphed the line on the scatterplot. Now, we want to calculate the residuals and make a residual plot. Fortunately, your calculator has already done most of the work. Each time the calculator computes a regression line, it finds the residuals and stores them in a list named RESID. *Note: If you have performed regression calculations on a different data set since the previous Tech Corner, you must recalculate the equation of the least-squares regression line using the Ford F-150 data for the calculator to store the correct residuals in the list menu.*

1. Press [2nd] [Y=] (STAT PLOT) and press [ENTER] or [1] go into Plot1. Then adjust the settings as shown. The RESID list is found in the List menu by pressing [2nd] [STAT].

2. Press [ZOOM] and choose 9: ZoomStat.

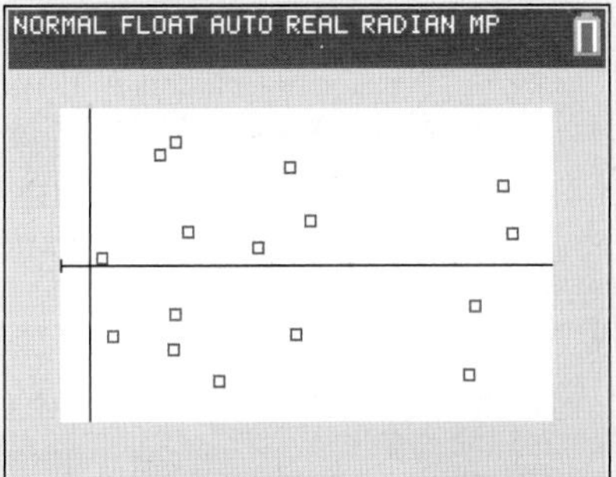

Note: If you want to see the values of the residuals, you can press [TRACE] or have the calculator store them in L3 (or any list). In the list editor, highlight the heading of L3, choose the RESID list from the LIST menu, and press [ENTER].

How Well the Line Fits the Data: Interpreting r^2 and s

We use a residual plot to determine if a least-squares regression line is an appropriate model for the relationship between two variables. If we determine that a least-squares regression line is appropriate, it makes sense to ask a follow-up question: How well does the line work? That is, if we use the least-squares regression line to make predictions, how good will these predictions be?

We already know that a residual measures how far an actual y-value is from its corresponding predicted value $\hat{y}$. Earlier in this section, we calculated the residual for the Ford F-150 with 70,583 miles driven and price \$21,994. As shown in Figure 2.13, the residual was −\$4765, meaning that the actual price was \$4765 *less* than we predicted.

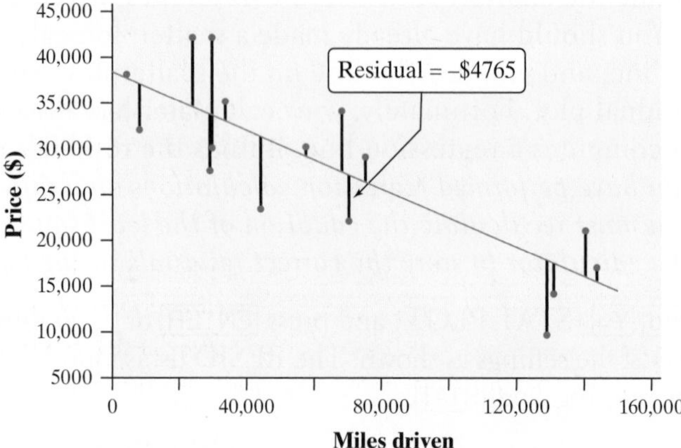

FIGURE 2.13 Scatterplot of the Ford F-150 data with a regression line added. Residuals for each truck are shown with vertical line segments.

To assess how well the line fits *all* the data, we need to consider the residuals for all of the trucks, not just one. Here are the residuals for all 16 trucks:

−4765	−7664	−3506	8617	7728	3057	−5004	458
2289	1183	−8121	6858	2110	5572	−5973	−2857

David Paul Morris/Bloomberg/Getty Images

If we add these residuals, we will get a sum ≈ 0 because the positive and negative residuals balance each other out. To avoid this problem, we'll square the residuals before adding them, just as we did in Section 1D when calculating the standard deviation:

$$(-4765)^2 + (-7664)^2 + \ldots + (-2857)^2 \approx 461,300,000$$

We can then use the *sum of squared residuals* in two different ways to answer the question of how well the line works.

THE COEFFICIENT OF DETERMINATION r^2

Figure 2.14 shows two different models for predicting the price of a Ford F-150, along with the sum of squared residuals for each model. Figure 2.14(a) shows the model $\hat{y} = \bar{y} = 27,834$. We'd use this model to predict the price of a used Ford F-150 if we had no idea how many miles it had been driven. That is, if we don't know how many miles a used Ford F-150 has been driven, our best guess for the price would be the mean price of other used Ford F-150s (\$27,834). The sum of squared residuals for this model is approximately 1,374,000,000.

Figure 2.14(b) shows the least-squares regression line using $x =$ number of miles driven. As stated earlier, the sum of squared residuals for this model is approximately 461,300,000.

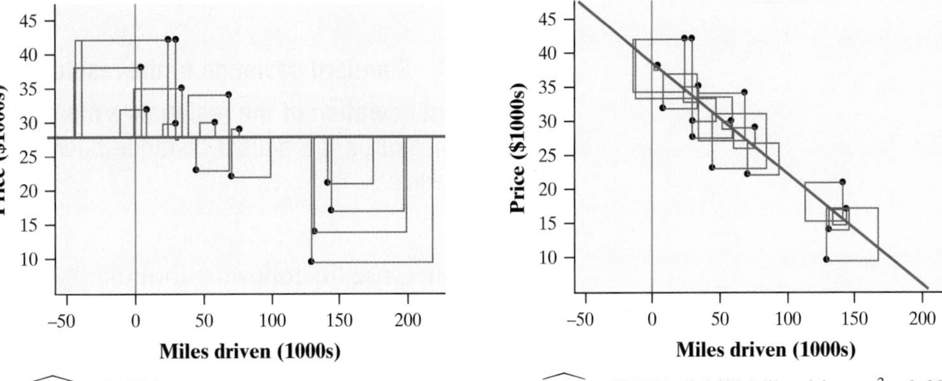

(a) $\widehat{\text{Price}} = 27{,}834$
Sum of squares ≈ 1,374,000,000

(b) $\widehat{\text{Price}} = 38{,}257 - 0.1629$ Miles driven; $r^2 = 0.66$
Sum of squares ≈ 461,300,000

FIGURE 2.14 (a) The sum of squared residuals is about 1,374,000,000 if we use the mean price as our prediction for all 16 trucks. (b) The sum of squared residuals from the least-squares regression line is about 461,300,000.

When looking at the squared residuals, you can see that the sum of squared residuals is smaller when we use the least-squares regression line. How much smaller? The answer is the **coefficient of determination** r^2, which measures the percent reduction in the sum of squared residuals:

$$r^2 = \frac{1{,}374{,}000{,}000 - 461{,}300{,}000}{1{,}374{,}000{,}000} = \frac{912{,}700{,}000}{1{,}374{,}000{,}000} = 0.66$$

Interpretation: About 66% of the variation in the price of a Ford F-150 is accounted for by the least-squares regression line with $x =$ miles driven. The remaining 34% is due to other factors, including age, color, and condition.

DEFINITION The coefficient of determination r^2

The **coefficient of determination r^2** measures the percent reduction in the sum of squared residuals when using the least-squares regression line to make predictions, rather than the mean value of y. In other words, r^2 measures the proportion (or percentage) of the variation in the response variable that is accounted for (or explained) by the explanatory variable in the linear model.

In the worst-case scenario, the least-squares regression line does no better at predicting y than the model using just the average y-value ($\hat{y} = \bar{y}$) does. Then the two sums of squared residuals are the same and $r^2 = 0$. In contrast, if all the points fall directly on the least-squares line, the sum of squared residuals is 0 and $r^2 = 1$. For example, if a taco stand always charges \$3 per taco, a scatterplot of $x =$ daily number of tacos sold and $y =$ daily revenue from tacos would be perfectly linear with $r^2 = 1$. That is, 100% of the variation in the daily revenue from tacos is explained by the linear relationship with the daily number of tacos sold.

THE STANDARD DEVIATION OF THE RESIDUALS *s*

Another way to measure how well the line works is to estimate the "typical" prediction error when using the least-squares regression line. To do this, we calculate the **standard deviation of the residuals** s.

DEFINITION Standard deviation of the residuals *s*

The **standard deviation of the residuals *s*** measures the size of a typical residual. That is, s measures the typical distance between the actual y-values and the predicted y-values.

To calculate s, use the following formula:

$$s = \sqrt{\frac{\text{sum of squared residuals}}{n-2}} = \sqrt{\frac{\sum (y_i - \hat{y}_i)^2}{n-2}}$$

For the Ford F-150 data, the standard deviation of the residuals is

$$s = \sqrt{\frac{(-4765)^2 + (-7664)^2 + \ldots + (-2857)^2}{16-2}} = \sqrt{\frac{461{,}300{,}000}{14}} = \$5740$$

Interpretation: The actual price of a Ford F-150 is typically about \$5740 away from the price predicted by the least-squares regression line with x = miles driven. If we look at the residual plot in Figure 2.12, this seems like a reasonable value. Although some of the residuals are close to 0, others are close to \$10,000 or −\$10,000.

Think About It

DOES THE FORMULA FOR *s* LOOK SLIGHTLY FAMILIAR? It should. In Unit 1, we defined the standard deviation for one quantitative variable as

$$s_x = \sqrt{\frac{\sum (x_i - \bar{x})^2}{n-1}}$$

We interpreted the resulting value as the "typical" distance of the data values from the mean. In the case of two quantitative variables, we're interested in the typical (vertical) distance of the data points from the regression line. We find this value in much the same way: first add up the squared deviations, then average them (again, in a funny way), and take the square root to get back to the original units of measurement.

Does $s = \$5740$ indicate that the line works well? That depends on what our typical prediction error would be if we had to make our predictions without knowing the number of miles driven. In that case, we'd use the mean price ($\bar{y} = \$27{,}834$) as our prediction and use the standard deviation of price ($s_y = \$9570$) as the typical prediction error. Using the least-squares regression line with x = number of miles driven reduces the typical prediction error by almost \$4000.

What's the relationship between s and r^2? Both are calculated from the sum of squared residuals, and both measure how well the line fits the data. The standard deviation of the residuals reports the size of a typical prediction error, in the same units as the response variable. In the Ford F-150 example, s = 5740 *dollars*.

The value of r^2 does not have units and is usually expressed as a percentage between 0% and 100%, such as $r^2 = 66\%$. Because these values assess how well the line fits the data in different ways, we recommend that you follow the example of most statistical software and report both.

EXAMPLE

Grabbing candy, again
Interpreting r^2 and s

Skill 4.B

Michael Neelon/Alamy Stock Photo

PROBLEM: Here is computer output summarizing the relationship between hand span (in centimeters) and the number of Starburst candies grabbed by the 24 students in Mr. Tyson's class.

```
Predictor        Coef   SE Coef        T        P
Constant     -29.7567    8.2088   -3.625   0.0015
Hand span      2.8308    0.3980    7.112   0.0000
S = 4.03422       R-Sq = 69.7%     R-Sq(adj) = 68.3%
```

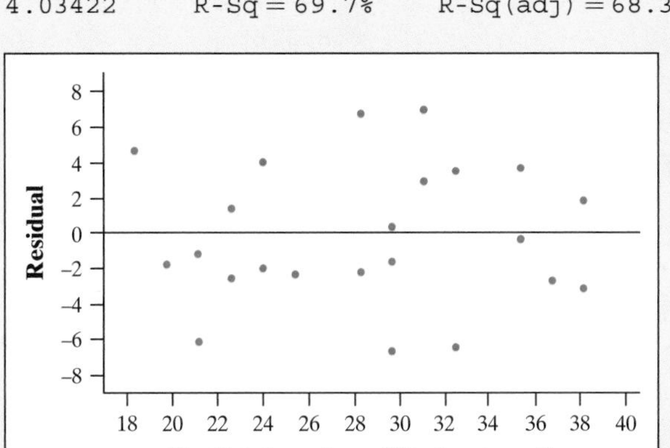

(a) What does the residual plot indicate about the linear model?
(b) Interpret the coefficient of determination.
(c) Interpret the standard deviation of the residuals.

SOLUTION:

(a) Because there is nothing but random scatter in the residual plot, the least-squares regression line is an appropriate model for the relationship between hand span and number of candies grabbed.

> On several previous AP® Statistics exam questions, a residual plot was presented with the predicted y-values on the horizontal axis, as in this example. However, the interpretation of the plot is the same—if there are no leftover patterns, the model is appropriate.

(b) About 69.7% of the variation in number of candies grabbed is accounted for by the least-squares regression line with $x =$ hand span.

> The coefficient of determination is displayed in the middle of the bottom row of this computer output as "R-sq." Ignore the value of R-sq (adj).

(c) The actual number of candies grabbed is typically about 4.03 away from the number predicted by the least-squares regression line with $x =$ hand span.

> The standard deviation of the residuals is displayed in the lower-left corner of this computer output as "S." In some types of computer output, the standard deviation of the residuals is called "root mean square error."

FOR PRACTICE, TRY EXERCISE 27

Although the value of the correlation isn't provided on most computer output, it is relatively easy to calculate r using the coefficient of determination r^2 and the sign (positive or negative) of the slope of the least-squares regression line: $r = \pm\sqrt{r^2}$. In the preceding example, $r^2 = 0.697$ and the slope is positive, so $r = +\sqrt{0.697} = 0.835$.

CHECK YOUR UNDERSTANDING

In Section 2B, you read about the Old Faithful geyser in Yellowstone National Park. The computer output shows the results of a linear regression analysis of x = duration (in minutes) of an eruption and y = wait time (in minutes) until the next eruption for each eruption of Old Faithful in a particular month.

Summary of Fit

RSquare	0.853725
RSquare Adj	0.853165
Root Mean Square Error	6.493357
Mean of Response	77.543730
Observations (or Sum Wgts)	263.000000

Parameter Estimates

Term	Estimate	Std Error	t Ratio	Prob>|t|
Intercept	33.347442	1.201081	27.76	<.0001*
Duration	13.285406	0.340393	39.03	<.0001*

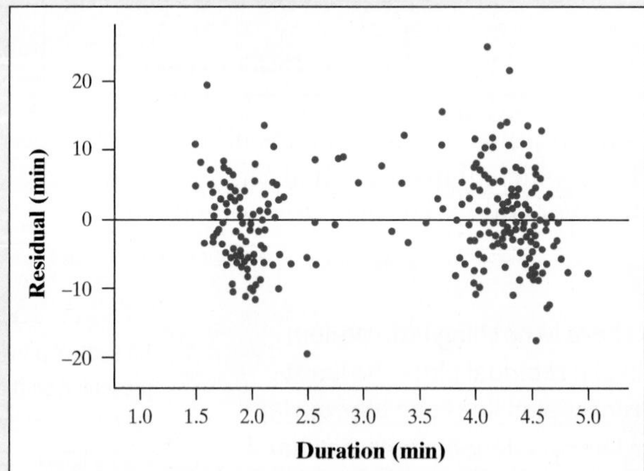

1. What is the equation of the least-squares regression line that models the relationship between wait time and duration? Define any variables that you use.
2. What information does the residual plot provide about the linear model?
3. What percentage of the variation in wait time is accounted for by the least-squares regression line with x = duration?
4. The standard deviation of the residuals is labeled as "root mean square error" in the output. Interpret this value.
5. Calculate the correlation.

SECTION 2C Summary

- The **regression line** $\hat{y} = a + bx$ models how a response variable y changes as an explanatory variable x changes. The symbol $\hat{y}$ indicates that the line gives the predicted value of y for a particular value of x, not the actual value of y.
- Avoid **extrapolation,** which is using a regression line to make predictions with values of the explanatory variable outside the interval of values used to calculate the line.
- The **slope** b of a regression line $\hat{y} = a + bx$ describes how the predicted value of y changes for each increase of one unit in x.
- The y **intercept** a of a regression line $\hat{y} = a + bx$ is the predicted value of y when the explanatory variable x equals 0. This prediction does not have a logical interpretation unless x can actually take values near 0.
- The **least-squares regression line** is the line that minimizes the sum of the squares of the vertical distances of the observed points from the line. Calculate the equation of the least-squares regression line using technology or with formulas.
- You can examine the fit of a regression model by analyzing the **residuals,** which are the differences between the actual values of y and the predicted values of y: residual $= y - \hat{y}$.
- A **residual plot** is a scatterplot that plots the residuals on the vertical axis and the values of the explanatory variable (or the predicted y-values) on the horizontal axis.
    - Random scatter in a residual plot indicates that the regression model used to calculate the residuals is appropriate.
    - A leftover curved pattern in the residual plot indicates that the model isn't appropriate.
- The **coefficient of determination** r^2 is the percentage of the variation in the response variable that is accounted for by the least-squares regression line using a particular explanatory variable.
- The **standard deviation of the residuals** s measures the typical size of a residual when using the least-squares regression line.

AP® EXAM TIP
AP® Daily Videos

Review the content of this section and get extra help by watching the AP® Daily Videos for Topics 2.6, 2.7, and 2.8, which are available in AP® Classroom.

2C Tech Corners

TI-Nspire and other technology instructions are on the book's website at bfwpub.com/tps7e.

8. Calculating least-squares regression lines — Page 191
9. Making residual plots — Page 195

SECTION 2C Exercises

Prediction

1. **Predicting CO_2** The scatterplot shows the relationship
pg 183 between x = engine size (in liters) and y = CO_2 emissions (in grams per mile) for a random sample of 176 gas-powered cars and trucks produced in 2021.[40] Also included is the least-squares regression line $\hat{y} = 244.7 + 54.9x$.

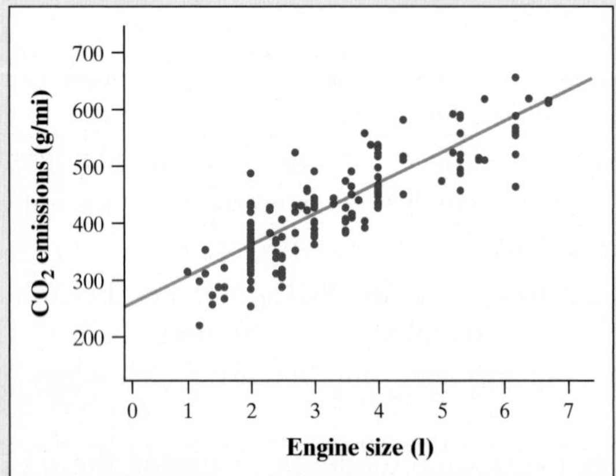

(a) Predict the CO_2 emissions for a gas-powered car with a 4-liter engine.

(b) Predict the CO_2 emissions for a gas-powered car with a 10-liter engine.

(c) How confident are you in each of these predictions? Explain your reasoning.

2. **How much gas?** Joan is concerned about the amount of energy she uses to heat her home. The scatterplot shows the relationship between x = mean temperature in a particular month (in °F) and y = mean amount of natural gas used per day (in cubic feet) in that month, along with the regression line $\hat{y} = 1425 - 19.87x$.

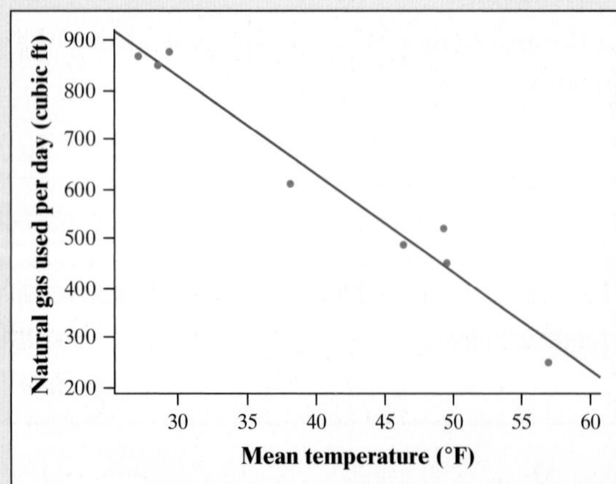

(a) Predict the mean amount of natural gas Joan will use per day in a month with a mean temperature of 30°F.

(b) Predict the mean amount of natural gas Joan will use per day in a month with a mean temperature of 65°F.

(c) How confident are you in each of these predictions? Explain your reasoning.

Residuals

3. **Residual CO_2** Refer to Exercise 1. The 2021 Toyota
pg 184 Sequoia 4WD has an engine size of 5.7 L and CO_2 emissions of 618 g/mi. Calculate and interpret the residual for this vehicle.

4. **Residual gas** Refer to Exercise 2. During March, the average temperature was 46.4°F and Joan used an average of 490 cubic feet of natural gas per day. Calculate and interpret the residual for this month.

5. **LPGA golfers** Here is a scatterplot showing the relationship between x = average driving distance (yards) and y = scoring average for 164 Ladies Professional Golf Association (LPGA) golfers in 2021.[41] Lower scores are better in golf, so the scatterplot shows that players who hit the ball farther typically have better (lower) scores. Also included is the regression line $\hat{y} = 81.9886 - 0.0392x$. Calculate and interpret the residual for Inbee Park, who had a scoring average of 69.3 and an average driving distance of 241.6 yards.

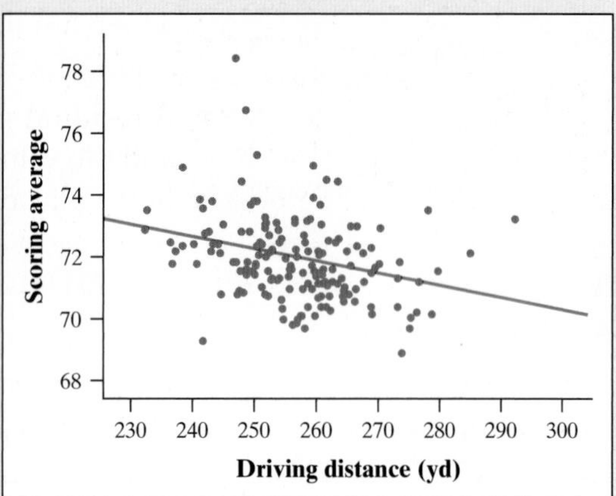

6. **Crickets chirping** The scatterplot shows the relationship between x = temperature in degrees Fahrenheit and y = chirps per minute for the striped ground cricket, along with the regression line $\hat{y} = -0.31 + 0.212x$.[42] Calculate and interpret the residual for the cricket who chirped 20 times per minute when the temperature was 88.6°F.

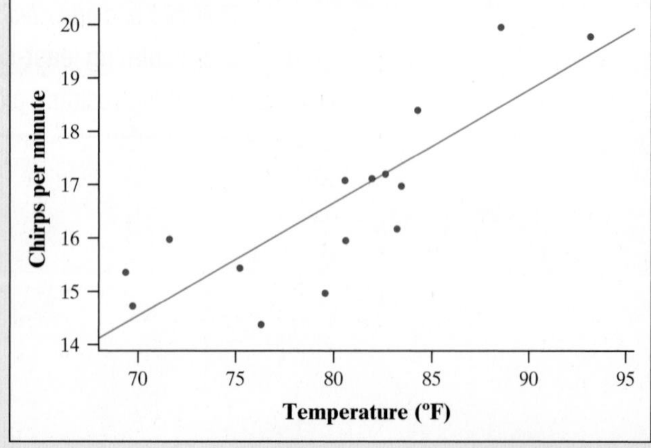

Interpreting the Slope and y Intercept

7. **More CO_2** Refer to Exercise 1.
pg 186
(a) Interpret the slope of the regression line.
(b) Does the value of the *y* intercept have meaning in this context? If so, interpret the *y* intercept. If not, explain why.

8. **Less gas?** Refer to Exercise 2.
(a) Interpret the slope of the regression line.
(b) Does the value of the *y* intercept have meaning in this context? If so, interpret the *y* intercept. If not, explain why.

9. **More distance?** Refer to Exercise 5. About how much lower would you predict an LPGA golfer's scoring average to be if she increased her average driving distance by 20 yards?

10. **More chirps?** Refer to Exercise 6. About how many additional chirps per minute do you predict a cricket to make if the temperature increases by 10°F?

11. **More butterflies** In Exercises 3 and 7 from Section 2B, you described the relationship between y = wing length (mm) and x = average temperature in the previous summer (°C) for the *Boloria chariclea* butterfly. The scatterplot shows this relationship, along with two regression lines. The regression line for female butterflies (blue squares) is $\hat{y} = 19.34 - 0.239x$ and the regression line for male butterflies (orange dots) is $\hat{y} = 18.40 - 0.231x$.

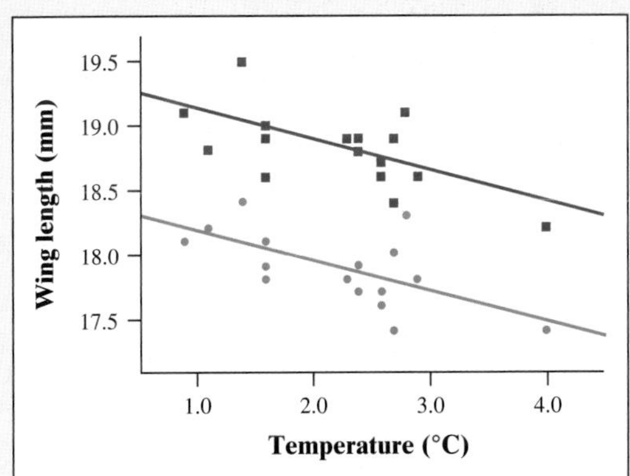

(a) How do the regression lines compare?
(b) How much longer do you expect the wing length of a female butterfly of this variety to be than the wing length of a male butterfly of this variety if the average summer temperature was 3°C?

12. **More Starbucks** In Exercises 4 and 8 from Section 2B, you described the relationship between fat (in grams) and the number of calories in products sold at Starbucks. The scatterplot shows this relationship, along with two regression lines. The regression line for the food products (blue squares) is $\hat{y} = 170 + 11.8x$. The regression line for the drink products (green dots) is $\hat{y} = 88 + 24.5x$.

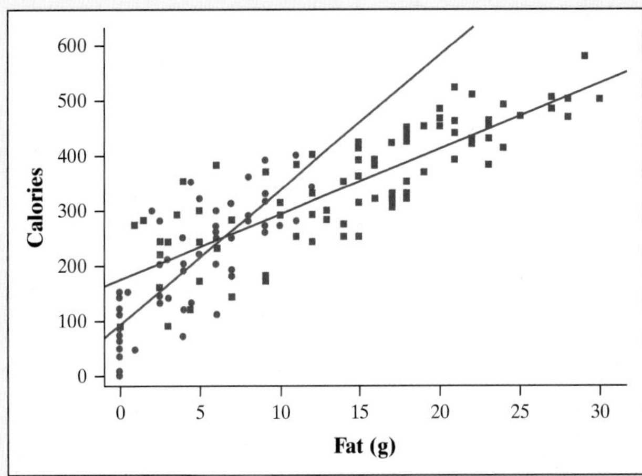

(a) How do the regression lines compare?
(b) How many more calories do you expect to find in a food item with 5 grams of fat compared to a drink item with 5 grams of fat?

The Least-Squares Regression Line

13. **Possums** Can you predict the total length of a possum from the length of its footprint? In a sample of 104 mountain brushtail possums from Australia, the mean total length is 87.1 cm with a standard deviation of 4.3 cm. The mean foot length is 6.8 cm with a standard deviation of 0.4 cm. The correlation between total length and foot length is $r = 0.445$.[43] Calculate the equation of the least-squares regression line for predicting total length from foot length.
pg 189

14. **The stock market** Some people think that the behavior of the stock market in January predicts its behavior for the rest of the year. Calculations from an 18-year period show that the mean change in stock prices in January is 1.75% with a standard deviation of 5.36%. For the full year, the mean change in stock prices is 9.07% with a standard deviation of 15.35%. The correlation between January change and full-year change is $r = 0.5$. Find the equation of the least-squares line for predicting full-year change from January change.

15. **Will I crush the final?** We expect that students who do well on the midterm exam in a course will usually also do well on the final exam. Gary Smith of Pomona College looked at the exam scores of all 346 students who took his statistics class over a 10-year period.[44] Assume that both the midterm and final exam were scored out of 100 points.
(a) State the equation of the least-squares regression line if each student scored the same on the midterm and the final.
(b) The actual least-squares line for predicting final-exam score *y* from midterm-exam score *x* was $\hat{y} = 46.6 + 0.41x$. Predict the final-exam score of a student who scored 50 on the midterm and a student who scored 100 on the midterm.
(c) Explain how your answers to part (b) illustrate regression to the mean.

16. **It's still early** We expect that a baseball player who has a high batting average in the first month of the season will also have a high batting average in the rest of the season. Using a sample of full-time Major League Baseball (MLB) players from a recent season,[45] a least-squares regression line was calculated to predict rest-of-season batting average y from first-month batting average x. *Note:* A player's batting average is the proportion of times at bat that he gets a hit. A batting average of more than 0.300 is considered very good in MLB.

(a) State the equation of the least-squares regression line if each MLB player had the same batting average the rest of the season as he did in the first month of the season.

(b) The actual equation of the least-squares regression line is $\hat{y} = 0.245 + 0.109x$. Predict the rest-of-season batting average for a player who had a 0.200 batting average in the first month of the season and for a player who had a 0.400 batting average in the first month of the season.

(c) Explain how your answers to part (b) illustrate regression to the mean.

Using Technology to Calculate a Least-Squares Regression Line

17. **More mess?** When Mentos are dropped into a newly opened bottle of Diet Coke, carbon dioxide is released from the Diet Coke very rapidly, causing the Diet Coke to be expelled from the bottle. To see if using more Mentos causes more Diet Coke to be expelled, Brittany and Allie obtained twenty-four 2-cup bottles of Diet Coke and randomly assigned each bottle to receive either 2, 3, 4, or 5 Mentos. After waiting for the fizzing to stop, they measured the amount of liquid expelled (in cups) by subtracting the amount remaining from the original amount in the bottle.[46] Here are a scatterplot and computer output from a linear regression of y = amount expelled on x = number of Mentos. Note that only 14 dots appear because there were multiple results with the same values. Use the output to determine the equation of the least-squares regression line for predicting amount expelled from number of Mentos.

pg 190

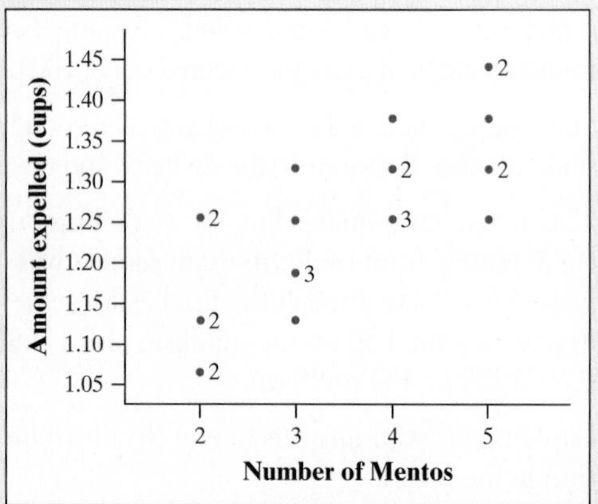

Term	Coef	SE Coef	T-Value	P-Value
Constant	1.0021	0.0451	22.21	0.000
Mentos	0.0708	0.0123	5.77	0.000

S = 0.06724   R-Sq = 60.21%   R-Sq(adj) = 58.40%

18. **Less mess?** Kerry and Danielle wanted to investigate whether tapping on a can of soda would reduce the amount of soda expelled after the can has been shaken. For their experiment, they vigorously shook 40 cans of soda and randomly assigned each can to be tapped for 0 seconds, 4 seconds, 8 seconds, or 12 seconds. After opening the cans and waiting for the fizzing to stop, they measured the amount of liquid expelled (in milliliters) by subtracting the amount remaining from the original amount in the can.[47] Here are a scatterplot and computer output from a linear regression of y = amount expelled on x = tapping time. Note that only 28 dots appear because there were multiple results with the same values. Use the output to determine the equation of the least-squares regression line for predicting amount expelled from tapping time.

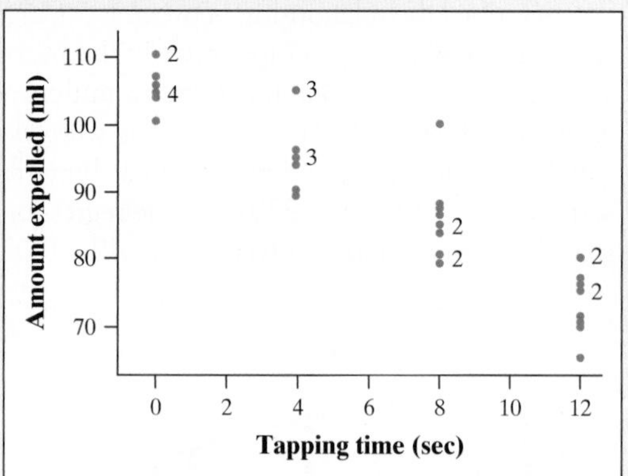

Term	Coef	SE Coef	T-Value	P-Value
Constant	106.360	1.320	80.34	0.000
Tapping time	−2.635	0.177	−14.90	0.000

S = 5.00347   R-Sq = 85.38%   R-Sq(adj) = 84.99%

19. **Movie candy** Is there a relationship between the amount of sugar (in grams) and the number of calories in movie-theater candy? Here are the data for a sample of 12 types of candy. Use technology to calculate the equation of the least-squares regression line for predicting the number of calories based on the amount of sugar.

Name	Sugar (g)	Calories	Name	Sugar (g)	Calories
Butterfinger Minis	45	450	Reese's Pieces	61	580
Junior Mints	107	570	Skittles	87	450
M&M'S®	62	480	Sour Patch Kids	92	490
Milk Duds	44	370	SweeTarts	136	680
Peanut M&M'S®	79	790	Twizzlers	59	460
Raisinets	60	420	Whoppers	48	350

20. **Long jumps** Here are the 40-yard-dash times (in seconds) and long-jump distances (in inches) for a small class of 12 students. Use technology to calculate the equation of the least-squares regression line for predicting the long-jump distance based on the dash time.

Dash time (sec)	5.41	5.05	7.01	7.17	6.73	5.68
Long-jump distance (in.)	171	184	90	65	78	130
Dash time (sec)	5.78	6.31	6.44	6.50	6.80	7.25
Long-jump distance (in.)	173	143	92	139	120	110

Residual Plots

21. **Infant weights in Nahya** A study of nutrition in developing countries collected data from the Egyptian village of Nahya. Researchers recorded the mean weight (in kilograms) for 170 infants in Nahya each month during their first year of life. The least-squares regression line $\widehat{\text{weight}} = 4.88 + 0.267(\text{age})$ results in the following residual plot. Use the residual plot to determine if the linear model is appropriate.

pg 194

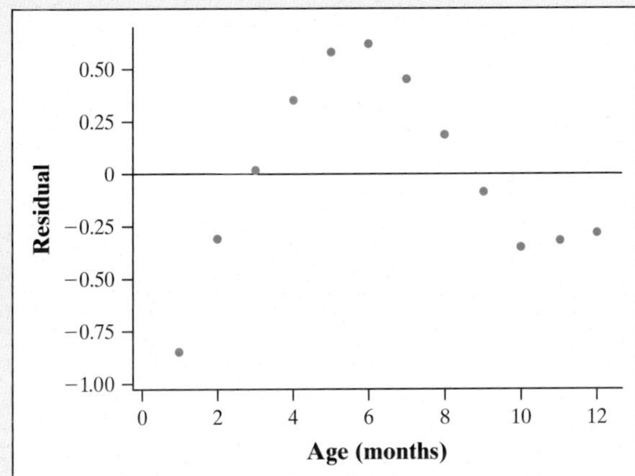

22. **Driving speed and fuel consumption** Researchers measured the fuel consumption y of a car at various speeds x. Fuel consumption is measured in liters of gasoline per 100 kilometers driven, and speed is measured in kilometers per hour. The least-squares regression line $\widehat{\text{fuel}} = 11.058 - 0.01466(\text{speed})$ results in the following residual plot. Use the residual plot to determine if the linear model is appropriate.

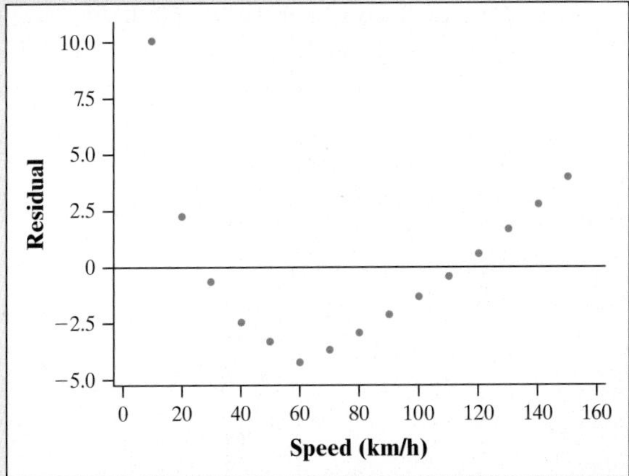

23. **Actual infant weight** Refer to Exercise 21. Use the equation of the least-squares regression line and the residual plot to estimate the *actual* mean weight of the infants when they were 1 month old.

24. **Actual fuel consumption** Refer to Exercise 22. Use the equation of the least-squares regression line and the residual plot to estimate the *actual* fuel consumption of the car when driving 20 kilometers per hour.

25. **More candy** Refer to Exercise 19. Use technology to create a residual plot. Sketch the residual plot and explain what information it provides.

26. **More long jumps** Refer to Exercise 20. Use technology to create a residual plot. Sketch the residual plot and explain what information it provides.

Interpreting r^2 and s

27. **Ponderosa pines** The U.S. Forest Service randomly selected ponderosa pine trees in western Montana to investigate the relationship between diameter at breast height (DBH), height, and volume of usable lumber.[48] Here is computer output summarizing the relationship between x = DBH (in inches) and y = height (feet) for these trees.

pg 199

```
Predictor     Coef     SE Coef     T        P
Constant     43.4740   5.2214    8.326   0.0000
DBH           2.6181   0.2042   12.820   0.0000
S = 9.42788   R-Sq = 81.2%   R-Sq(adj) = 80.7%
```

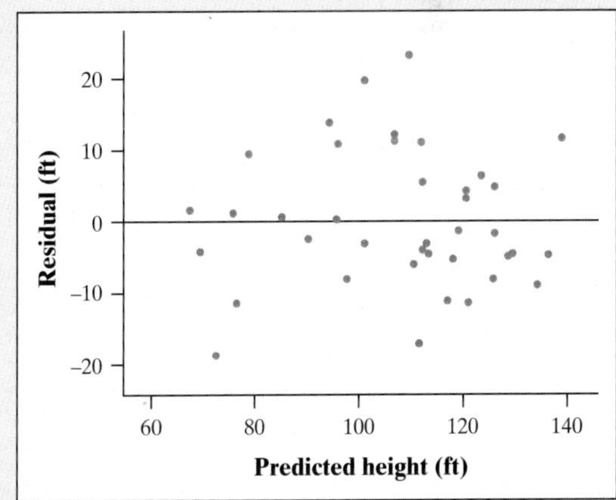

(a) What does the residual plot indicate about the linear model?

(b) Interpret the coefficient of determination.

(c) Interpret the standard deviation of the residuals.

28. **Hand size** One of the many factors National Football League (NFL) teams consider when drafting a quarterback is hand size. Does having larger hands help prevent the quarterback from throwing interceptions? Here is computer output summarizing the relationship between x = hand size (in inches) and y = number of interceptions.[49]

Predictor	Coef	SE Coef	T	P
Constant	28.4803	14.9086	1.910	0.0660
Hand size	−1.7995	1.5536	−1.158	0.2562
S = 3.27494	R-Sq = 4.4%		R-Sq(adj) = 1.1%	

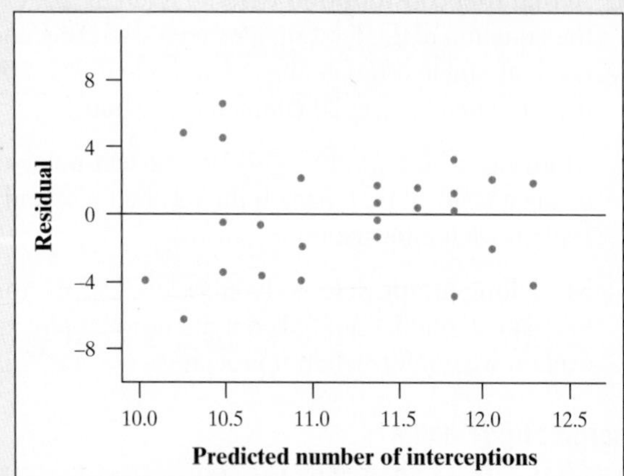

(a) What does the residual plot indicate about the linear model?

(b) Interpret the coefficient of determination.

(c) Interpret the standard deviation of the residuals.

29. **Even more mess** Refer to Exercise 17. Here again is the computer output for the least-squares regression line relating y = amount expelled (cups) and x = number of Mentos.

Term	Coef	SE Coef	T-Value	P-Value
Constant	1.0021	0.0451	22.21	0.000
Mentos	0.0708	0.0123	5.77	0.000
S = 0.06724	R-Sq = 60.21%		R-Sq(adj) = 58.40%	

(a) Interpret the slope of the least-squares regression line.

(b) Interpret r^2.

(c) Interpret s.

(d) How would the values of r^2 and s change if the amount of liquid expelled was measured in ounces rather than cups? (1 cup = 8 ounces)

30. **Even less mess** Refer to Exercise 18. Here again is the computer output for the least-squares regression line relating y = amount expelled (milliliters) and x = tapping time (sec).

Term	Coef	SE Coef	T-Value	P-Value
Constant	106.360	1.320	80.34	0.000
Tapping time	−2.635	0.177	−14.90	0.000
S = 5.00347	R-Sq = 85.38%		R-Sq(adj) = 84.99%	

(a) Interpret the slope of the least-squares regression line.

(b) Interpret r^2.

(c) Interpret s.

(d) How would the values of r^2 and s change if the amount of liquid expelled was measured in liters rather than milliliters? (1 liter = 1000 milliliters)

31. **Dissolved oxygen** Dissolved oxygen is important for the survival of aquatic life. Researchers measured the concentration of dissolved oxygen in Lake Champlain at various times and depths.[50] They also recorded the temperature of the water for each measurement. Here is computer output for a regression of y = dissolved oxygen concentration (mg/L) on x = temperature (°C).

Summary of Fit

RSquare	0.428475
RSquare Adj	0.42844
Root Mean Square Error	1.15233
Mean of Response	10.2577
Observations (or Sum Wgts)	16141

Parameter Estimates

Term	Estimate	Std Error	t Ratio	Prob>|t|
Intercept	12.1526	0.0195	624.205	<.0001*
Temperature	−0.1935	0.0018	−109.998	<.0001*

(a) Calculate the correlation.

(b) Calculate and interpret the residual for the measurement when the temperature was 12.84°C and the dissolved oxygen concentration was 10.33 mg/L.

(c) The standard deviation of the residuals is labeled as "root mean square error" in the output. Interpret this value.

(d) What percentage of the variation in dissolved oxygen concentration is accounted for by the least-squares regression line with x = temperature?

32. **Beetles and beavers** Do beavers benefit beetles? Researchers laid out 23 circular plots, each 4 meters in diameter, in an area where beavers were cutting down cottonwood trees. In each plot, they counted the number of stumps from trees cut by beavers and the number of clusters of beetle larvae. Ecologists believe that the new sprouts from stumps are more tender than other cottonwood growth, so beetles prefer them. If so, more stumps should produce more beetle larvae.[51] Here is computer output for a regression of y = number of beetle larvae on x = number of stumps.

Summary of Fit

RSquare	0.839144
RSquare Adj	0.831484
Root Mean Square Error	6.419386
Mean of Response	25.086960
Observations (or Sum Wgts)	23

Parameter Estimates

Term	Estimate	Std Error	t Ratio	Prob>|t|
Intercept	−1.286104	2.853182	−0.45	0.6568
Number of stumps	11.893733	1.136343	10.47	<.0001*

(a) Calculate the correlation.

(b) Calculate and interpret the residual for the plot that had 2 stumps and 30 beetle larvae.

(c) The standard deviation of the residuals is labeled as "root mean square error" in the output. Interpret this value.

(d) What percentage of the variation in number of larvae is accounted for by the least-squares regression line with x = number of stumps?

33. **More crickets** In Exercise 6, you considered the association between x = temperature and y = chirps per minute for a sample of 15 crickets. For these data, the standard deviation of chirps per minute is $s_y = 1.702$ and the standard deviation of the residuals is $s = 0.9715$.

(a) Use the value of s and the formula for the standard deviation of the residuals to calculate the sum of the squared residuals from the least-squares regression line.

(b) Use the value of s_y and the formula for the standard deviation of a single quantitative variable to calculate the sum of the squared residuals from the mean chirps per minute.

(c) Use the values you calculated in parts (a) and (b) to calculate the coefficient of determination r^2.

For Investigation *Apply the skills from the section in a new context or nonroutine way.*

34. **More LPGA golfers** In Exercise 5, you considered the association between x = average driving distance (yards) and y = scoring average for LPGA golfers in 2021. The least-squares regression line for this association is $\hat{y} = 81.9886 - 0.0392x$, with $r^2 = 0.087$ and $s = 1.295$. Of course, many factors affect a golfer's scoring average besides average driving distance. Here is the multiple regression model relating y = scoring average to x_1 = average driving distance, x_2 = driving accuracy (%), x_3 = putting average, and x_4 = sand saves (%). Remember that lower scores are better in golf!

$$\hat{y} = 60.6344 - 0.0556x_1 - 0.1024x_2 + 18.4071x_3 - 0.0197x_4$$

(a) Calculate and interpret the residual for Inbee Park, who had a scoring average of 69.3, an average driving distance of 241.6 yards, a driving accuracy of 80.3%, a putting average of 1.72, and a sand save percentage of 53.33.

(b) Here is the residual plot for the multiple regression model. We use the predicted y-values on the horizontal axis because the model includes four different explanatory variables and the predicted values are a combination of all four. What does the residual plot indicate about the appropriateness of the multiple regression model?

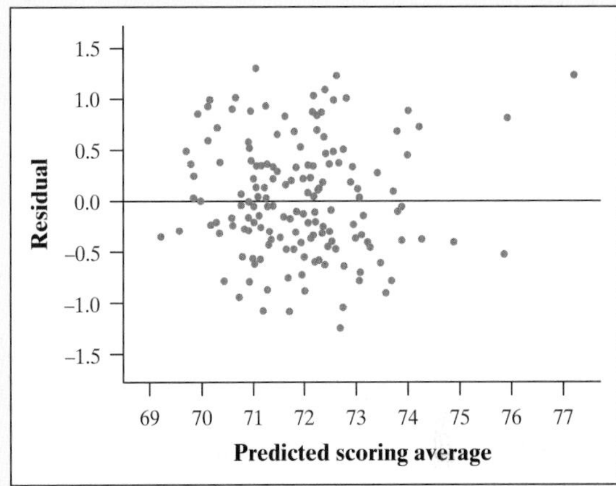

(c) For the multiple regression model, $R^2 = 0.833$ and $s = 0.560$. Interpret these values. *Note:* In multiple regression, it is common to use uppercase R^2 for the coefficient of determination.

(d) Compare R^2 and s for the multiple regression model to r^2 and s for the simple regression model. Does the multiple regression model do a better job of predicting scoring average than the simple regression model with only one explanatory variable (average driving distance)? Explain your reasoning.

35. **More butterflies** In Exercise 11, we fit separate least-squares regression lines to model the relationship between wing length (mm) and average temperature in the previous summer (°C) for male and female *Boloria chariclea* butterflies. It is also possible to create a single model for these data that uses an indicator variable for sex. Here is a model relating y = wing length (mm) to x_1 = average temperature and x_2 = sex (1 = female, 0 = male):

$$\hat{y} = 18.4042 - 0.2350x_1 + 0.9312x_2$$

(a) Calculate and interpret the residual for the female butterfly with a wing length of 19.1 mm and an average temperature of 0.9 °C.

(b) Letting $x_2 = 0$, find a simplified model for male butterflies.

(c) Letting $x_2 = 1$, find a simplified model for female butterflies.

(d) What is the same about the simplified models for males and females? What is different?

36. **Regression through the origin** The scatterplot shows the lean body mass and metabolic rate for a sample of 5 adults. For each person, the lean body mass is the subject's total weight in kilograms less any weight due to fat. The metabolic rate is the number of calories burned in a 24-hour period.

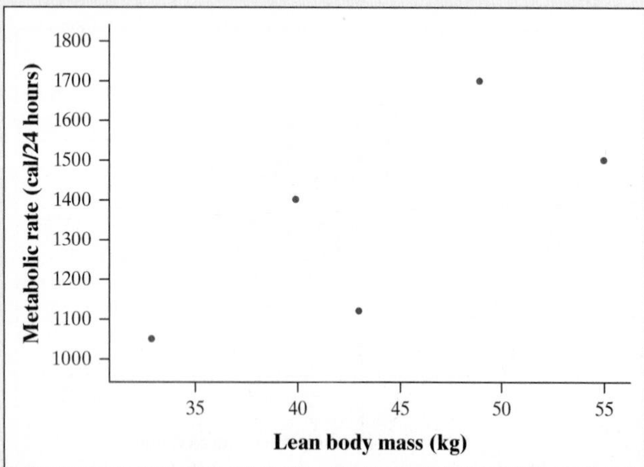

Because a person with no lean body mass should burn no calories, it makes sense to model the relationship with a function that goes through (0, 0) in the form $y = kx$. Models were tried using different values of k ($k = 25$, $k = 26$, etc.) and the sum of squared residuals was calculated for each value of k. Here is a scatterplot showing the relationship between sum of squared residuals and k. According to the scatterplot, what is the ideal value of k to use for predicting metabolic rate? Explain your reasoning.

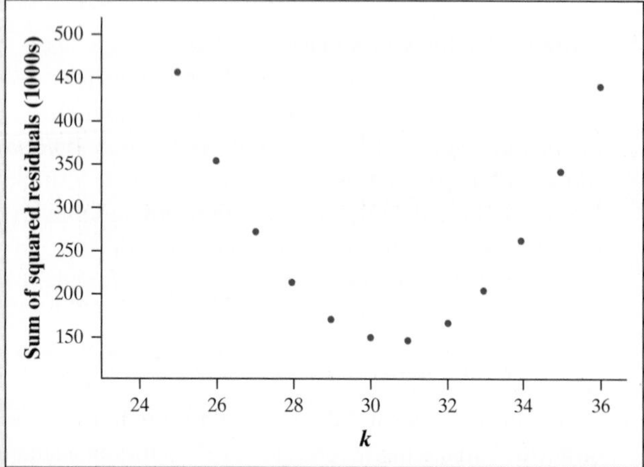

Multiple Choice *Select the best answer for each question.*

37. Which of the following is *not* a characteristic of the least-squares regression line?

(A) The slope of the least-squares regression line is always between −1 and 1.

(B) The least-squares regression line always goes through the point $(\bar{x}, \bar{y})$.

(C) The least-squares regression line minimizes the sum of squared residuals.

(D) The slope of the least-squares regression line will always have the same sign as the correlation.

(E) The least-squares regression line is appropriate if the corresponding residual plot shows random scatter.

38. Each year, students in an elementary school take a standardized math test at the end of the school year. For a class of fourth-graders, the average score was 55.1 with a standard deviation of 12.3. In the third grade, these same students had an average score of 61.7 with a standard deviation of 14.0. The correlation between the two sets of scores is $r = 0.95$. Calculate the equation of the least-squares regression line for predicting a fourth-grade score from a third-grade score.

(A) $\hat{y} = 3.58 + 0.835x$

(B) $\hat{y} = 15.69 + 0.835x$

(C) $\hat{y} = 2.19 + 1.08x$

(D) $\hat{y} = -11.54 + 1.08x$

(E) Cannot be calculated without the data

39. Using data from a sample of 42 brands of cookies, a regression analysis was performed using x = added sugar (grams) and y = calories.[52] Use the output from the regression analysis to determine the equation of the least-squares regression line.

```
Term          Coef     SE Coef  T-Value  P-Value
Constant      77.664   16.257   4.777    0.000
Added sugar   32.480    6.690   4.855    0.000
S=45.6103       R-Sq=37.1%      R-Sq(adj)=35.5%
```

(A) $\hat{y} = 77.664 + 45.6103x$

(B) $\hat{y} = 77.664 + 16.257x$

(C) $\hat{y} = 77.664 + 32.480x$

(D) $\hat{y} = 32.480 + 77.664x$

(E) $\hat{y} = 32.480 + 45.6103x$

Exercises 40–42 refer to the following setting. Measurements on young children in Mumbai, India, found this least-squares line for predicting y = height (in cm) from x = arm span (in cm):[53]

$$\hat{y} = 6.4 + 0.93x$$

40. In addition to the regression line, the report on the Mumbai measurements says that $r^2 = 0.95$. Based on this value, which of the following is the best conclusion?

(A) Although arm span and height are correlated, arm span does not predict height very accurately.

(B) Height increases by $\sqrt{0.95} = 0.97$ cm for each additional centimeter of arm span.

(C) 95% of the relationship between height and arm span is accounted for by the regression line.

(D) 95% of the variation in height is accounted for by the regression line with x = arm span.

(E) 95% of the height measurements are accounted for by the regression line with x = arm span.

41. One child in the Mumbai study had a height of 59 cm and an arm span of 60 cm. What is this child's residual?

(A) −3.2 cm
(B) −2.2 cm
(C) −1.3 cm
(D) 3.2 cm
(E) 62.2 cm

42. Suppose that the measurements of arm span and height were converted from centimeters to meters by dividing each measurement by 100. How will this conversion affect the values of r^2 and s?

(A) r^2 will increase; s will increase.

(B) r^2 will increase; s will stay the same.

(C) r^2 will increase; s will decrease.

(D) r^2 will stay the same; s will stay the same.

(E) r^2 will stay the same; s will decrease.

43. The relationship between x = number of wins and y = average attendance per game for Major League Baseball teams has a correlation of $r = 0.56$. If the mean attendance is 28,196 with a standard deviation of 9402, what is the predicted attendance for a team whose number of wins is 1 standard deviation above average?

(A) 25,192
(B) 28,196
(C) 33,461
(D) 37,598
(E) There is not enough information to predict the value of y.

Recycle and Review *Practice what you learned in previous sections.*

44. **The longest prime (2B)** Prime numbers are positive integers that have no integer factors other than 1 and the number itself. For example, 2, 3, 5, 7, and 11 are prime numbers. Since computers were developed, researchers have been using them to identify new prime numbers. The scatterplot shows the year and number of digits when a new largest prime number was discovered. In 1951, the longest known prime number had 79 digits; by 2018, the longest prime number had more than 24 million digits!

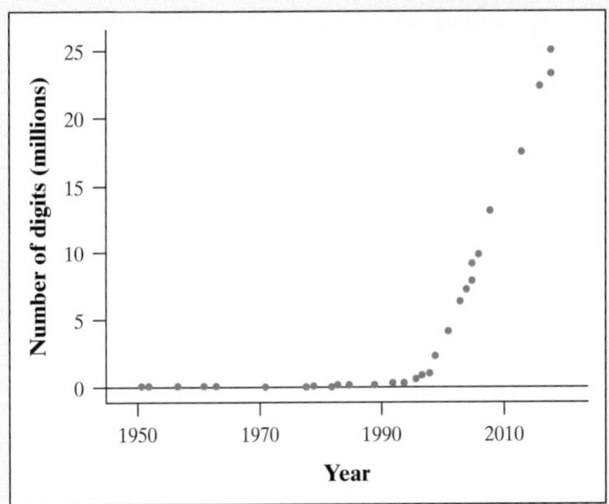

(a) Describe the association shown in the scatterplot.

(b) The correlation for these data is $r = 0.71$. Explain why the correlation is relatively small considering how strong the association is between year and number of digits.

45. **School satisfaction (2A)** How satisfied are parents of K–12 students with the overall quality of education their children are receiving at school? Does it depend on what type of school (elementary, middle, high) their children attend? Here are the results of Pew Research Center's survey of a random sample of 3238 U.S. parents.[54]

		Type of school			
		Elementary	Middle	High	Total
Level of satisfaction	Extremely/Very	929	357	595	1881
	Somewhat	427	232	397	1056
	Not too/Not at all	118	73	110	301
	Total	1474	662	1102	3238

(a) Make a graph that displays the relationship between level of satisfaction and type of school.

(b) Laura claims that parents of elementary school students are the most likely to respond "Not too" or "Not at all" satisfied because a greater number of parents of elementary students gave this response than parents of middle school or high school students. Explain why the data do not support Laura's claim.

SECTION 2D Analyzing Departures from Linearity

LEARNING TARGETS *By the end of the section, you should be able to:*

- Describe how unusual points influence the least-squares regression line, the correlation, r^2, and the standard deviation of the residuals.
- Calculate predicted values from linear models using variables that have been transformed with logarithms or powers.
- Determine which of several models does a better job of describing the relationship between two quantitative variables.

It would be nice if every association between two quantitative variables was linear with no unusual points. Unfortunately, this is often not the case. In this section, you'll learn how unusual points can influence the equation of the least-squares regression line and summary statistics such as the correlation. You'll also learn how transformations can be used to linearize associations that are curved.

Influential Points

In Unit 1, you learned that the mean and standard deviation are not resistant to outliers and skewness. Is the correlation resistant to unusual values? What about the equation of the least-squares regression line? The following activity will help you answer these questions.

ACTIVITY Investigating influential points

Applet

In this activity, you will use an applet to explore how unusual points influence correlation and regression calculations.

1. Launch the *Two Quantitative Variables* applet from www.stapplet.com.
2. Enter the following values for the explanatory and response variables. Leave the boxes for the variable names empty. Then click the "Begin analysis" button.

Explanatory	1	2	3	4	5	6	7	8	9
Response	6	10	13	9	7	13	10	9	14

3. Click the "Calculate correlation" button and confirm that $r = 0.447$.
4. Click the "Calculate least-squares regression line" button and confirm the following calculations: $\hat{y} = 7.8611 + 0.45x$, $s = 2.639$, and $r^2 = 0.20$.
5. What would happen if you add a point with an unusually large x-value that is on the original least-squares regression line $\hat{y} = 7.8611 + 0.45x$?
   (a) Make a guess about how each of the following values will be affected: r, slope, y intercept, s, and r^2.

(b) Click the "Edit inputs" button and add the point (20, 16.8611) to the data set. Then click the "Begin analysis" button to see the scatterplot with the new point added.

(c) Click the buttons to calculate the correlation and the least-squares regression line. How were each of the statistics in part (a) affected? How accurate were your guesses?

6. What would happen if you add a point with an unusually large x-value that is far below the original least-squares regression line $\hat{y} = 7.8611 + 0.45x$?

(a) Make a guess about how each of the following values will be affected: r, slope, y intercept, s, and r^2.

(b) Click the "Edit inputs" button and change the point (20, 16.8611) to (20, 0) in the data set. Then click the "Begin analysis" button to see the scatterplot with the new point added.

(c) Click the buttons to calculate the correlation and the least-squares regression line. How were each of the statistics in part (a) affected? How accurate were your guesses?

7. What would happen if you add a point with an x-value near $\bar{x}$ that is far above the original least-squares regression line $\hat{y} = 7.8611 + 0.45x$?

(a) Make a guess about how each of the following values will be affected: r, slope, y intercept, s, and r^2.

(b) Click the "Edit inputs" button and change the point (20, 0) to (5, 30) in the data set. Then click the "Begin analysis" button to see the scatterplot with the new point added.

(c) Click the buttons to calculate the correlation and the least-squares regression line. How were each of the statistics in part (a) affected? How accurate were your guesses?

8. Briefly summarize how unusual points influence r, slope, y intercept, s, and r^2.

As you learned in the activity, unusual points may or may not have an influence on the correlation, the least-squares regression line, r^2, and the standard deviation of the residuals. Here are four scatterplots that summarize the possibilities. In all four scatterplots, the eight points in the lower left are the same.

Case 1: No unusual points.

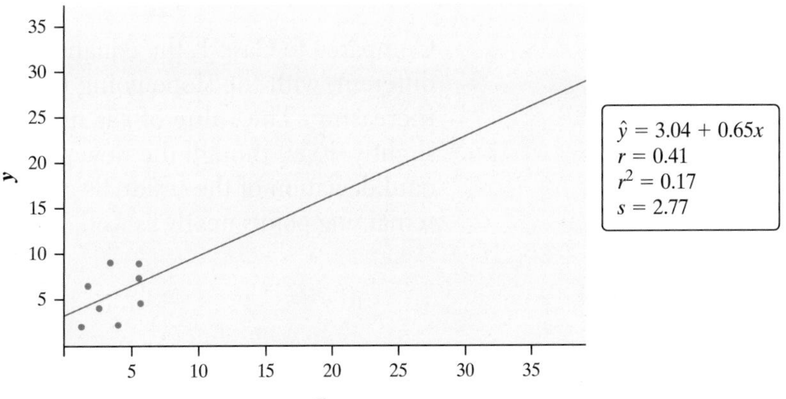

Case 2: A point that is far from the other points in the *x* direction, but in the same pattern.

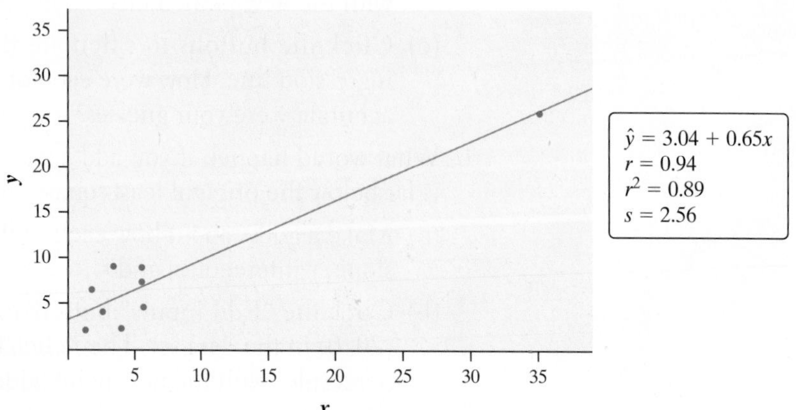

Compared to Case 1, the equation of the least-squares regression line remained the same, but the values of r and r^2 greatly increased. The standard deviation of the residuals got a bit smaller because the additional point has a very small residual.

Case 3: A point that is far from the other points in the *x* direction, and not in the same pattern.

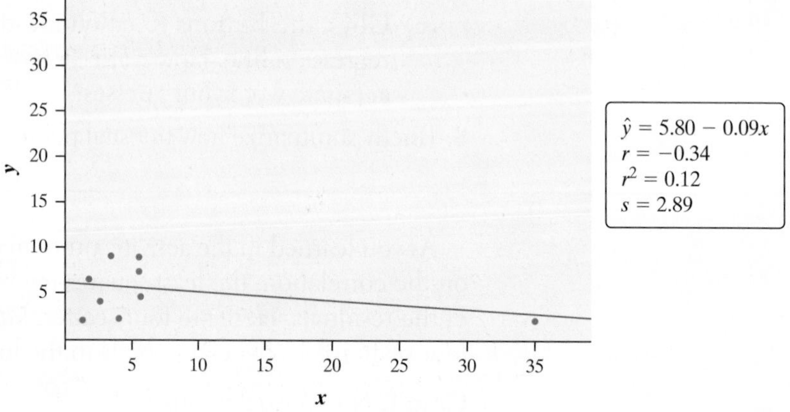

Compared to Case 1, the equation of the least-squares regression line is much different, with the slope going from positive to negative and the y intercept increasing. The value of r is now negative while the value of r^2 decreased slightly. Even though the new point has a relatively small residual, the standard deviation of the residuals got a bit larger because the line doesn't fit the remaining points nearly as well.

Case 4: A point that is far from the other points in the *y* direction, and not in the same pattern.

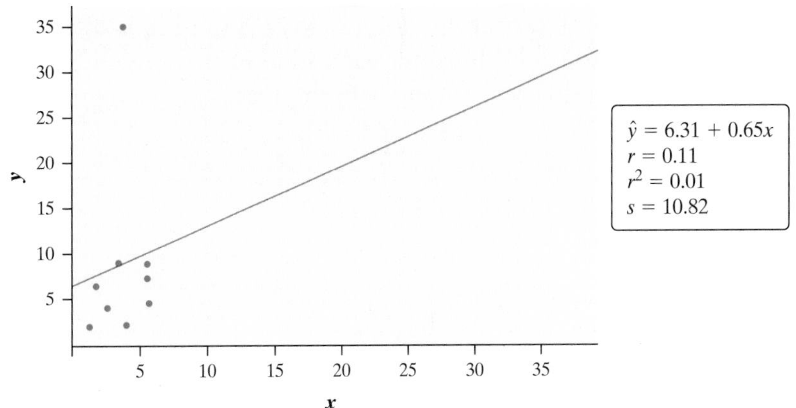

Compared to Case 1, the slope of the least-squares regression line is the same, but the *y* intercept is a little larger, as the line appears to have shifted up slightly. Because the new point has such a large residual, the values of r and r^2 are much smaller and the standard deviation of the residuals is much larger.

In Cases 2 and 3, the unusual point had a much bigger *x*-value than the other points. Points whose *x*-values are much smaller or much larger than the other points in a scatterplot have **high leverage.** In Case 4, the unusual point had a very large residual. Points with large residuals are called **outliers.** All of these unusual points are considered **influential points** because adding them to the scatterplot substantially changes either the equation of the least-squares regression line or one or more of the other summary statistics (r, r^2, s).

DEFINITION High leverage, Outlier, Influential point

Points with **high leverage** in regression have much larger or much smaller *x*-values than the other points in the data set.

An **outlier** in regression is a point that does not follow the pattern of the data and has a large residual.

An **influential point** in regression is any point that, if removed, substantially changes the slope, *y* intercept, correlation, coefficient of determination, or standard deviation of the residuals.

Outliers and high-leverage points are often influential in regression calculations! The best way to investigate the influence of such points is to perform regression calculations with and without them to see how much the results differ, as we did in the activity and follow-up discussion.

EXAMPLE

Forests and trees
Influential points

Skill 2.A

Greg Vaughn/Alamy Stock Photo

PROBLEM: Here is a scatterplot of forest area (in millions of acres) and number of trees (in billions) for each state in the United States, excluding Alaska and Hawaii, along with the least-squares regression line.[55]

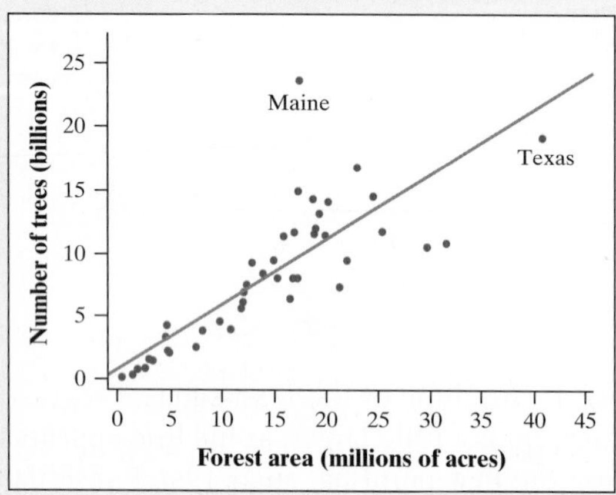

(a) Describe how the point representing Maine (17.6, 23.6) affects the equation of the least-squares regression line, r, r^2 and s. Is this point an outlier or a high-leverage point? Explain your reasoning.

(b) Describe how the point representing Texas (41.0, 19.0) affects the equation of the least-squares regression line, r, r^2, and s. Is this point an outlier or a high-leverage point? Explain your reasoning.

SOLUTION:

(a) Because the point for Maine is far above the regression line and slightly to the right of the mean forest area, it increases the y intercept and the slope. Because it has a large residual, the point for Maine is an outlier, making the values of r and r^2 closer to 0 and increasing the standard deviation of the residuals.

> The point for Maine is not a high-leverage point because its forest area is not unusually small or large.

(b) Because the point for Texas has an above-average forest area and is below the least-squares regression line, it is pulling the line toward itself, making the slope of the line closer to 0 and the y intercept greater. Because it has a very large forest area relative to other states, it is a high-leverage point. As it is in the linear pattern of the other points, the point for Texas makes the values of r and r^2 closer to 1. And because its residual appears fairly typical in size, the point for Texas doesn't affect the standard deviation of the residuals very much.

> The point for Texas is *not* considered an outlier because it is in the pattern of the rest of the points and doesn't have a large residual.

FOR PRACTICE, TRY EXERCISE 1

CHECK YOUR UNDERSTANDING

Does the age at which a child begins to talk predict their later score on a test of mental ability? A study of the development of young children recorded the age in months at which each of 24 children spoke their first word and their Gesell Adaptive

Score, the result of an aptitude test taken much later.[56] Here is a scatterplot of the data, with two points, Child 18 and Child 19, labeled on the graph.

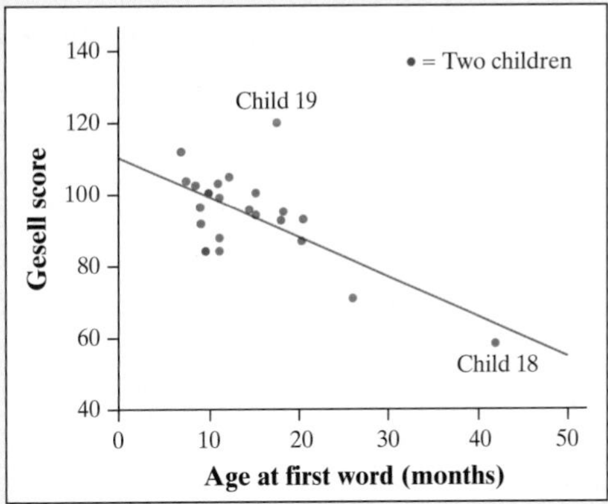

1. Describe how the point representing Child 18 affects the equation of the least-squares regression line, r, r^2, and s. Is this point an outlier or a high-leverage point? Explain your reasoning.
2. Describe how the point representing Child 19 affects the equation of the least-squares regression line, r, r^2, and s. Is this point an outlier or a high-leverage point? Explain your reasoning.

Transformations to Achieve Linearity

In Section 2C, we learned how to analyze relationships between two quantitative variables that showed a linear pattern. When bivariate data show a curved relationship, we can often transform one or both of the variables to create a linear association. Once the data have been transformed to achieve linearity, we can use least-squares regression to generate a useful model for making predictions.

The Gapminder website (www.gapminder.org) provides lots of data on the health and well-being of the world's inhabitants. Figure 2.15 shows a simplified version of the scatterplot from Exercise 27 in Section 2B. The individuals are all the world's nations for which data were available in 2020. The explanatory variable, income per person, is a measure of how rich a country is. The response variable is life expectancy at birth.

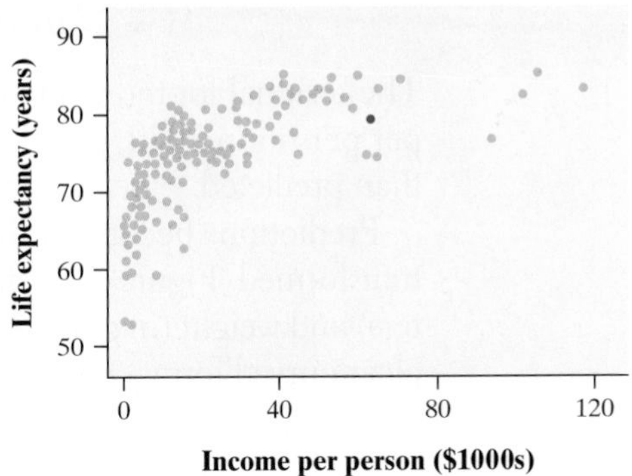

FIGURE 2.15 Scatterplot of the life expectancy of people in many nations versus each nation's income per person. The U.S. data is marked in purple.

We expect people in richer countries to live longer because they have better access to medical care and typically lead healthier lives. The overall pattern of the scatterplot does show this, but the relationship is not linear. Life expectancy rises quickly as income per person increases, but then levels off. People in very rich countries such as the United States (marked in purple) live no longer than people in poorer but not extremely poor nations. In some less-wealthy countries, people live longer than in the United States.

When the form of an association is nonlinear, as in Figure 2.15, we can often straighten things out by *transforming the data* using powers or logarithms. You learned a similar lesson in Exercise 28 from Section 1E, where you used square roots and logarithms to transform a distribution from skewed to roughly symmetric. Figure 2.16 plots the life expectancy against the *logarithm* of income per person for these same countries. The effect is remarkable, as the resulting graph has a clear, linear form.

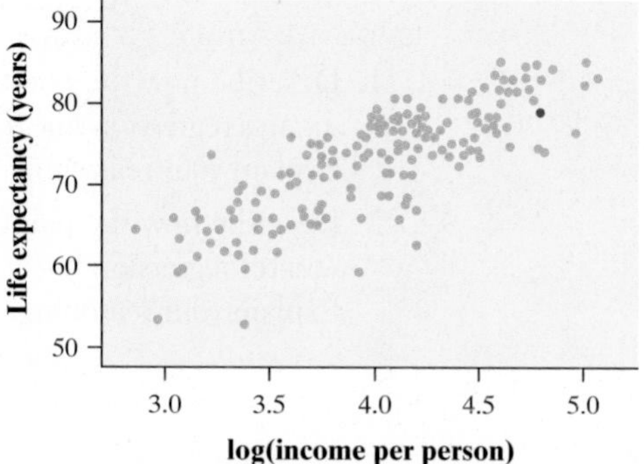

FIGURE 2.16 Scatterplot of life expectancy against log(income per person) for many nations. The U.S. data is marked in purple.

If one or more transformations produce a linear association, we can use least-squares regression lines to make predictions and calculate residuals. Here is computer output for a linear model of the association shown in Figure 2.16. Note that the explanatory variable is log(income per person), not income per person!

Predictor	Coef	SE Coef	T	P
Constant	27.9745	2.2492	12.437	0.0000
Log(income)	11.2003	0.5518	20.299	0.0000
S = 3.8432	R-Sq = 69.1%		R-Sq(adj) = 69.0%	

In 2020, the United States had an income per person of \$63,000 and a life expectancy of 79 years. According to the model, the predicted life expectancy for the United States is

$$\hat{y} = 27.9745 + 11.2003\log(63,000) = 81.73 \text{ years}$$

The residual for the United States is $79 - 81.73 = -2.73$ years. Based on income per person in 2020, people in the United States live, on average, 2.73 years less than predicted.

Predictions become a little more complicated when the response variable is transformed. Figure 2.17 shows the relationship between the length (in centimeters) and weight (in grams) for Atlantic Ocean rockfish of several sizes.[57] Note the clear curved form.

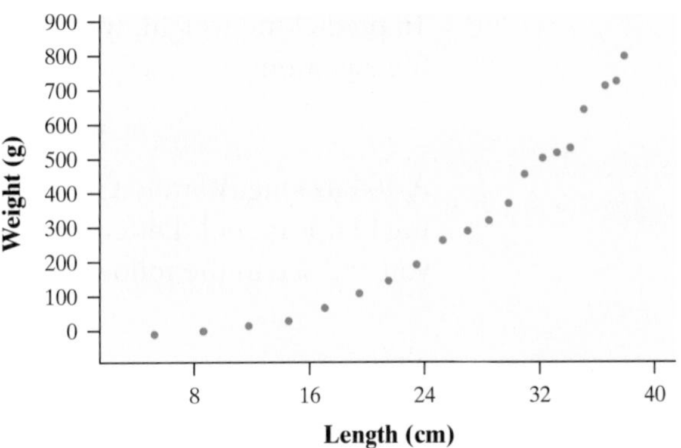

FIGURE 2.17 The scatterplot of weight versus length for Atlantic Ocean rockfish is clearly nonlinear.

Because length is one-dimensional and weight (like volume) is three-dimensional, taking the cube root of the weight values might produce a linear association. Figure 2.18 shows that the relationship between length and $\sqrt[3]{\text{weight}}$ is roughly linear for Atlantic Ocean rockfish.

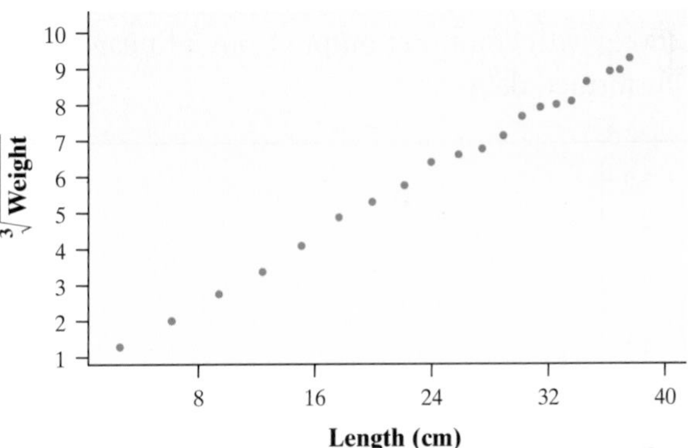

FIGURE 2.18 The scatterplot of the cube root of weight versus length is roughly linear for Atlantic Ocean rockfish.

Here is computer output from a linear regression analysis of $y = \sqrt[3]{\text{weight}}$ versus $x = $ length:

```
Predictor        Coef       SE Coef        T        P
Constant      −0.02204    0.07762      −0.28    0.780
Length         0.246616   0.002868     86.00    0.000
S = 0.124161        R-Sq = 99.8%       R-Sq(adj) = 99.7%
```

Based on this output, the equation of the least-squares regression line is

$$\widehat{\sqrt[3]{\text{weight}}} = -0.02204 + 0.246616(\text{length})$$

Note that the response variable is $\sqrt[3]{\text{weight}}$, not weight! If you write the equation as

$$\widehat{\sqrt[3]{y}} = -0.02204 + 0.246616x$$

you should make sure to define $y = $ weight and $x = $ length. In either form, make sure the hat is placed over the entire cube root, as the model was created using the cube root of the y-values, not the y-values themselves.

What is the predicted weight of a 36-cm-long Atlantic Ocean rockfish? Using this model, we can predict the *cube root* of the weight:

$$\widehat{\sqrt[3]{\text{weight}}} = -0.02204 + 0.246616(36) = 8.856$$

To predict the weight, we need to reverse the cube root by cubing the result from the equation:

$$\widehat{\text{weight}} = (8.856)^3 = 694.6 \text{ grams}$$

A 36-cm-long Atlantic Ocean rockfish is predicted to weigh 694.6 grams, according to this model. Different models will give slightly different predicted values, as you will see in the following example.

EXAMPLE

Go fish!
Transformations to achieve linearity

Skill 2.C

PROBLEM: Another way to linearize the Atlantic Ocean rockfish data is to transform the length and weight values using logarithms. Here is a scatterplot showing the relationship between log(weight) and log(length), along with computer output from a linear regression analysis of the transformed data.

Doug Wilson/Alamy Stock Photo

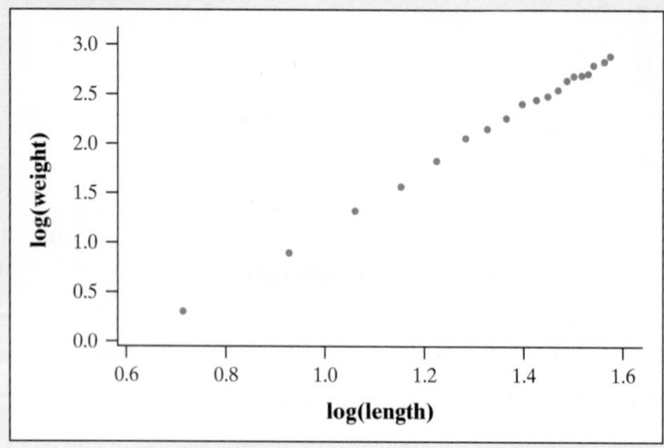

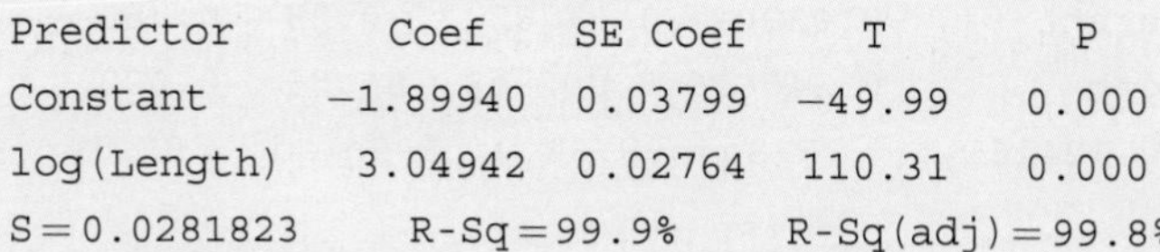

Predictor	Coef	SE Coef	T	P
Constant	−1.89940	0.03799	−49.99	0.000
log(Length)	3.04942	0.02764	110.31	0.000
S = 0.0281823	R-Sq = 99.9%		R-Sq(adj) = 99.8%	

Predict the weight of a 36-cm-long Atlantic Ocean rockfish using this model.

SOLUTION:

$\widehat{\log(\text{weight})} = -1.89940 + 3.04942 \log(36) = 2.8464$

$\widehat{\text{weight}} = 10^{2.8464} = 702.1$ grams

The least-squares regression line gives the predicted value of the base-10 logarithm of weight. To get the predicted weight, undo the logarithm by raising 10 to the 2.8464 power.

FOR PRACTICE, TRY EXERCISE 7

On the TI-83/84 calculator, you can "undo" the logarithm by using the [2nd] function keys. To solve $\widehat{\log(\text{weight})} = 2.8464$, press [2nd] [LOG] 2.8464 [ENTER].

In addition to base-10 logarithms, you can use natural (base-e) logarithms to transform the variables. Using the same Atlantic Ocean rockfish data, here is a scatterplot of ln(weight) versus ln(length).

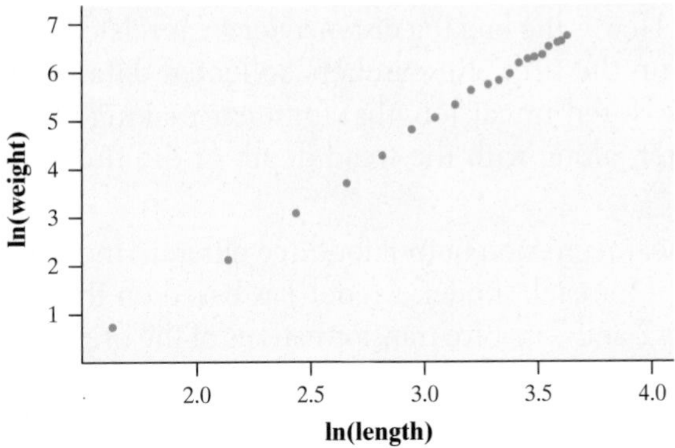

The least-squares regression line for these data is

$$\widehat{\ln(\text{weight})} = -4.3735 + 3.04942 \ln(\text{length})$$

To predict the weight of an Atlantic Ocean rockfish that is 36 cm long, we start by substituting 36 for length.

$$\widehat{\ln(\text{weight})} = -4.3735 + 3.04942 \ln(36) = 6.55415$$

To get the predicted weight, we then undo the natural logarithm by raising e to the 6.55415 power.

$$\widehat{\text{weight}} = e^{6.55415} = 702.2 \text{ grams}$$

On the TI-83/84 calculator, you can "undo" the natural logarithm using the [2nd] function keys. To solve $\widehat{\ln(\text{weight})} = 6.55415$, press [2nd] [ln] 6.55415 [ENTER].

Choosing the Most Appropriate Model

When a scatterplot shows a curved relationship between two quantitative variables, our strategy is to transform one or both variables to create a linear association. In some cases, more than one type of transformation can produce linearity. In that case, how do we decide which model is best?

HOW TO CHOOSE A MODEL

When choosing between different models to describe a relationship between two quantitative variables:

- Choose the model whose residual plot has the most random scatter.
- If more than one model produces a randomly scattered residual plot, choose the model with the largest coefficient of determination, r^2.

It is not advisable to use the standard deviation of the residuals s to help choose a model, as the y-values for the different models might be on different scales. The following example illustrates the process of choosing the most appropriate model for a curved relationship.

EXAMPLE

Stop that car!

Choosing the most appropriate model

Skill 2.D

PROBLEM: How is the braking distance for a car related to the amount of tread left on the tires? Researchers collected data on the braking distance (measured in car lengths) for a car making a panic stop in standing water, along with the tread depth of the tires (in 1/32-inch increments).[58]

Nawin Kitpipatphinyo/Shutterstock

Here is linear regression output for three different models, along with a residual plot for each model. Model 1 is based on the original data, while Models 2 and 3 involve transformations of the original data.

Model 1. Braking distance vs. Tread depth

Predictor	Coef	SE Coef	T	P
Constant	16.4873	0.7648	21.557	0.0000
Tread depth	−0.7282	0.1125	−6.457	0.0001
S=1.1827	R-Sq=0.822		R-sq(adj)=0.803	

Model 2. ln(braking distance) vs. ln(tread depth)

Predictor	Coef	SE Coef	T	P
Constant	2.9034	0.0051	566.34	0.0000
ln(tread depth)	−0.2690	0.0029	−91.449	0.0000
S=0.007	R-sq=0.999		R-sq(adj)=0.999	

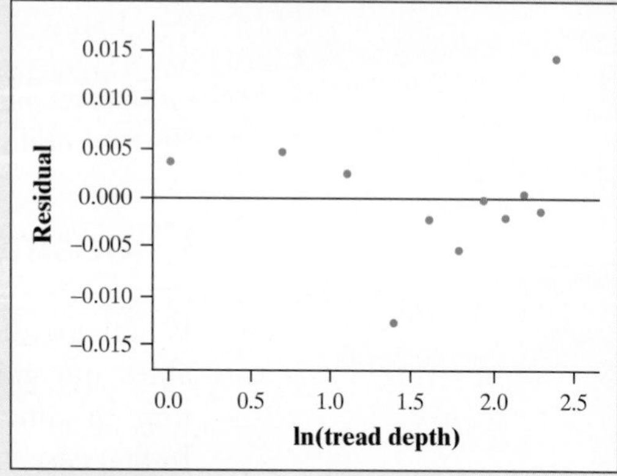

Model 3. ln(braking distance) vs. Tread depth

Predictor	Coef	SE Coef	T	P
Constant	2.8169	0.0461	61.077	0.0000
Tread depth	−0.0569	0.0068	−8.372	0.0000
S=0.071	R-sq=0.886		R-sq (adj)=0.874	

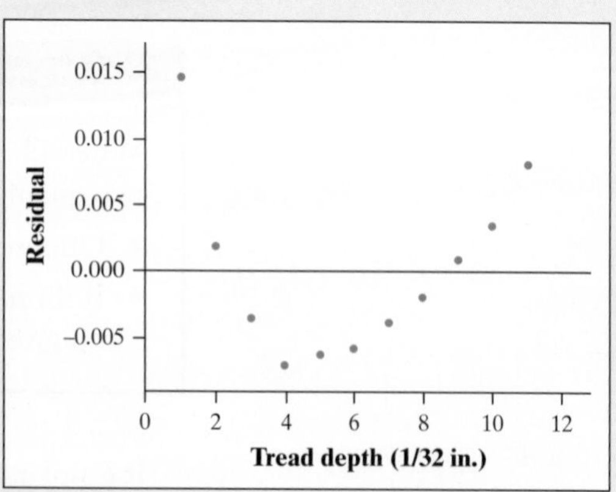

(a) Which model does the best job of summarizing the relationship between tread depth and braking distance? Explain your reasoning.

(b) Use the model chosen in part (a) to calculate and interpret the residual for the trial when the tread depth was 3/32 inch and the stopping distance was 13.6 car lengths.

SOLUTION:

(a) Because the model that uses $x = \ln(\text{tread depth})$ and $y = \ln(\text{braking distance})$ produced the most randomly scattered residual plot with no leftover curved pattern, it is the most appropriate model.

Note that the value of r^2 is also closest to 1 for the model that uses $x = \ln(\text{tread depth})$ and $y = \ln(\text{braking distance})$.

(b) $\widehat{\ln(\text{braking distance})} = 2.9034 - 0.2690\ln(3) = 2.608$

$\widehat{\text{braking distance}} = e^{2.608} = 13.57$ car lengths

residual $= 13.6 - 13.57 = 0.03$ car lengths

The residual calculated here is on the original scale (car lengths), while the residuals shown in the residual plot for this model are on a logarithmic scale.

When the tread depth was 3/32 inch, the car traveled 0.03 car length farther than the distance predicted by the model using $x = \ln(\text{tread depth})$ and $y = \ln(\text{braking distance})$.

FOR PRACTICE, TRY EXERCISE 13

We have used statistical software to do all the transformations and linear regression analysis in this section so far. You can also use your graphing calculator to transform variables to achieve linearity when you are provided with the raw data.

CHECK YOUR UNDERSTANDING

Many different types of life insurance policies are available. Some provide coverage throughout an individual's life (whole life), while others last only for a specified number of years (term life). The policyholder makes regular payments (premiums) to the insurance company in return for the coverage. When the insured person dies, a payment is made to designated family members or other beneficiaries.

How do insurance companies decide how much to charge for life insurance? They rely on a staff of highly trained actuaries—people with expertise in probability, statistics, and advanced mathematics—to establish premiums. For an individual who wants to buy life insurance, the premium will depend on the type and amount of the policy as well as personal characteristics such as age, sex, and health status. The table shows monthly premiums for a 10-year term-life insurance policy worth $1,000,000.[59]

Age (years)	Monthly premium
40	$29
45	$46
50	$68
55	$106
60	$157
65	$257

The computer output that follows shows three possible models for predicting monthly premium from age. Option 1 is based on the original data, while Options 2 and 3 involve transformations of the original data. Each set of output includes a scatterplot with a least-squares regression line added and a residual plot.

Option 1:

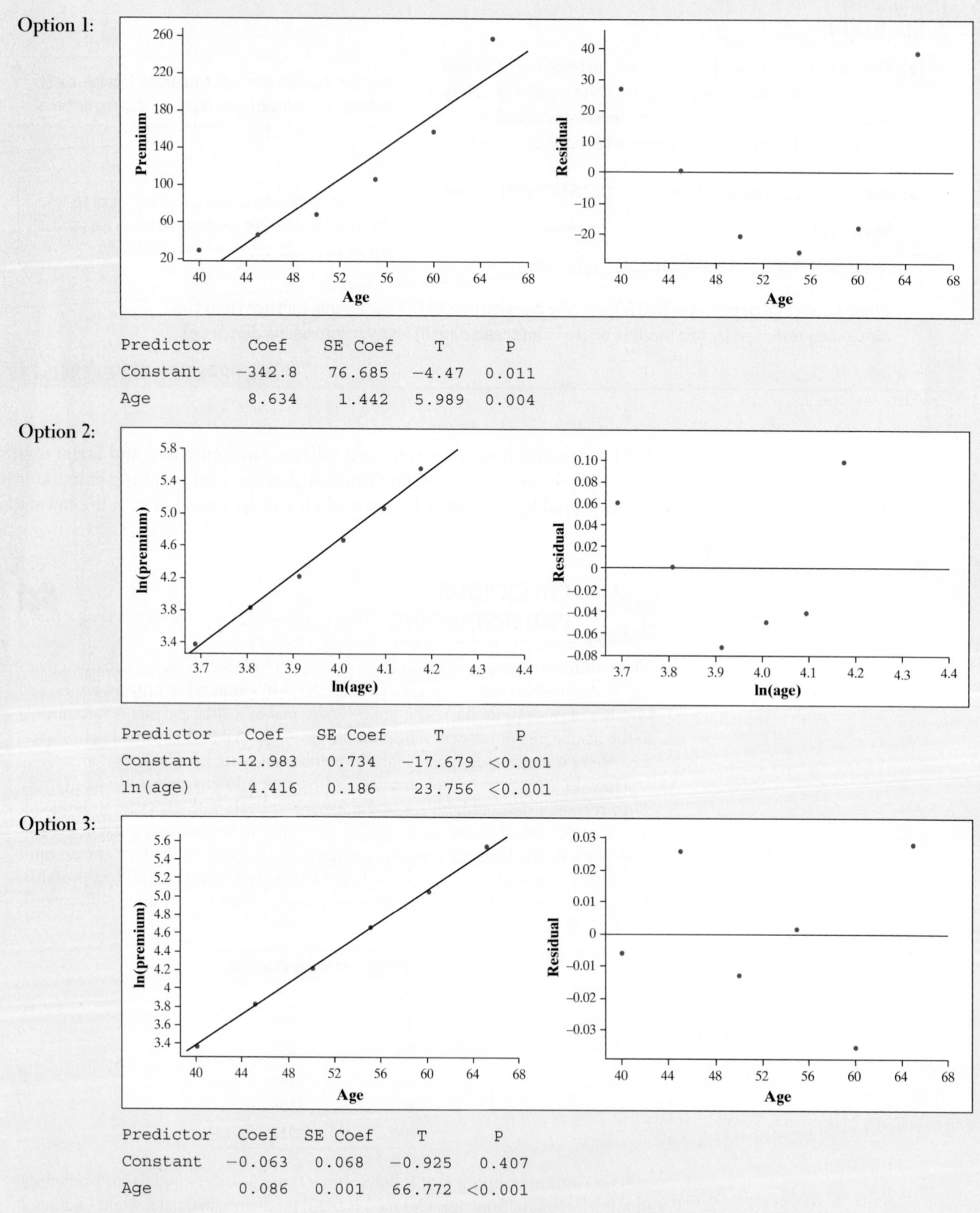

Predictor	Coef	SE Coef	T	P
Constant	−342.8	76.685	−4.47	0.011
Age	8.634	1.442	5.989	0.004

Option 2:

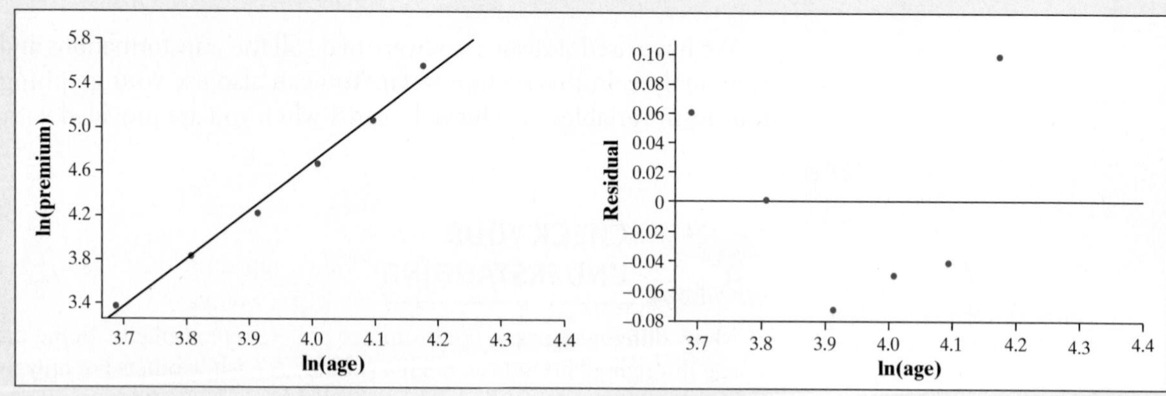

Predictor	Coef	SE Coef	T	P
Constant	−12.983	0.734	−17.679	<0.001
ln(age)	4.416	0.186	23.756	<0.001

Option 3:

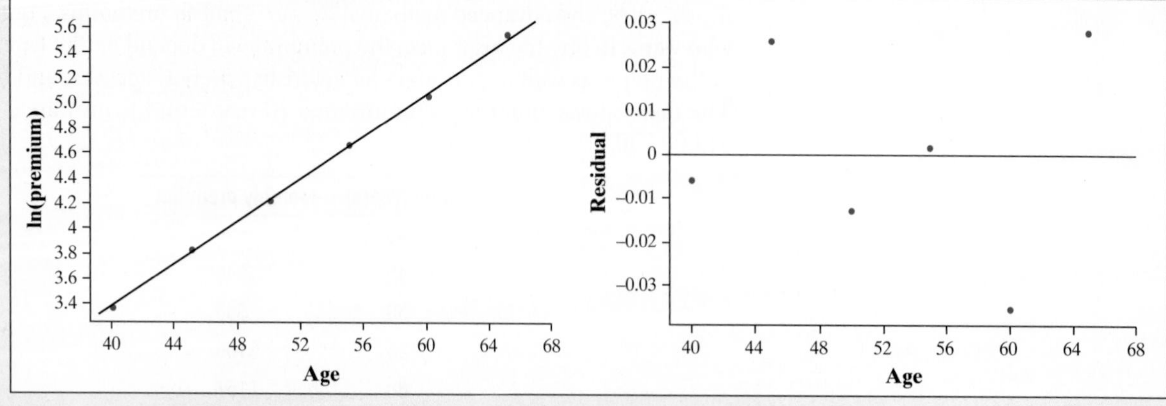

Predictor	Coef	SE Coef	T	P
Constant	−0.063	0.068	−0.925	0.407
Age	0.086	0.001	66.772	<0.001

1. Use each model to predict how much a 58-year-old person would pay for such a policy.
2. Which model does the best job of summarizing the relationship between age and monthly premium? Explain your answer.

SECTION 2D Summary

- **Influential points** can greatly affect the equation of the least-squares regression line and other summary statistics such as the correlation.
  - Points with x-values far from $\bar{x}$ have **high leverage** and can be very influential.
  - Points with large residuals are called **outliers** and can also affect correlation and regression calculations.
- Nonlinear associations between two quantitative variables can sometimes be changed into linear associations by transforming one or both variables. Once we transform the data to achieve linearity, we can fit a least-squares regression line to the transformed data and use this linear model to make predictions.
- To decide between competing models, choose the model with the most randomly scattered residual plot. If it is difficult to determine which residual plot is the most randomly scattered, choose the model with the largest value of r^2.

AP® EXAM TIP
AP® Daily Videos

Review the content of this section and get extra help by watching the AP® Daily Videos for Topic 2.9, which are available in AP® Classroom.

SECTION 2D Exercises

Influential Points

1. **All brawn?** The following scatterplot shows the average brain weight (in grams) versus average body weight (in kilograms) for 96 species of mammals.[60] There are many small mammals whose points overlap at the lower left.

pg 214

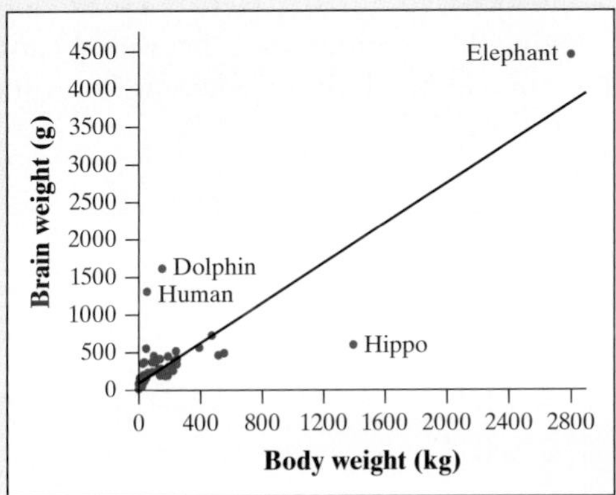

(a) Describe how the point representing the elephant affects the equation of the least-squares regression line, r, r^2, and s. Is this point an outlier or a high-leverage point? Explain your reasoning.

(b) Describe how the point representing the human affects the equation of the least-squares regression line, r, r^2, and s. Is this point an outlier or a high-leverage point? Explain your reasoning.

2. **Managing diabetes** People with diabetes measure their fasting plasma glucose (FPG, measured in milligrams per milliliter) after fasting for at least 8 hours. Another measurement, made at regular medical checkups, is called HbA1c. This is roughly the percentage of red blood cells that have a glucose molecule attached. It measures average exposure to glucose over a period of several months. The scatterplot shows the relationship between HbA1c and FPG for 18 people with diabetes five months after they had completed a diabetes education class.[61]

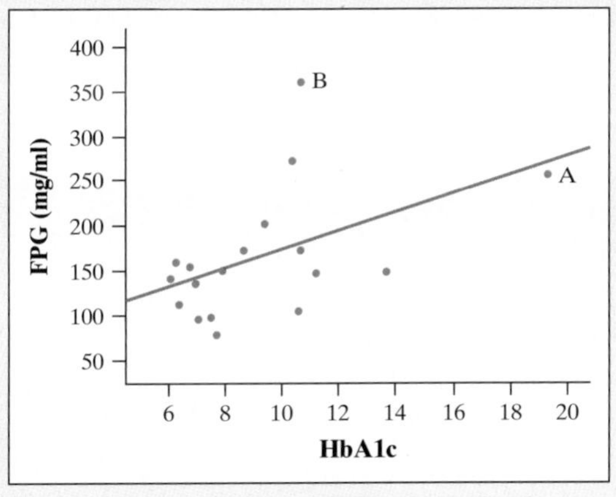

(a) Describe how point A affects the equation of the least-squares regression line, r, r^2, and s. Is this point an outlier or a high-leverage point? Explain your reasoning.

(b) Describe how point B affects the equation of the least-squares regression line, r, r^2, and s. Is this point an outlier or a high-leverage point? Explain your reasoning.

3. **Who's got hops?** Haley, Jeff, and Nathan measured the height (in inches) and vertical jump (in inches) of 74 students at their school.[62] Here is a scatterplot of the data, along with the least-squares regression line. Jacob (highlighted in red) had a vertical jump of nearly 3 feet! What would happen to the equation of the least-squares regression line if the point for Jacob was removed? What would happen to the values of r, r^2, and s if the point for Jacob was removed?

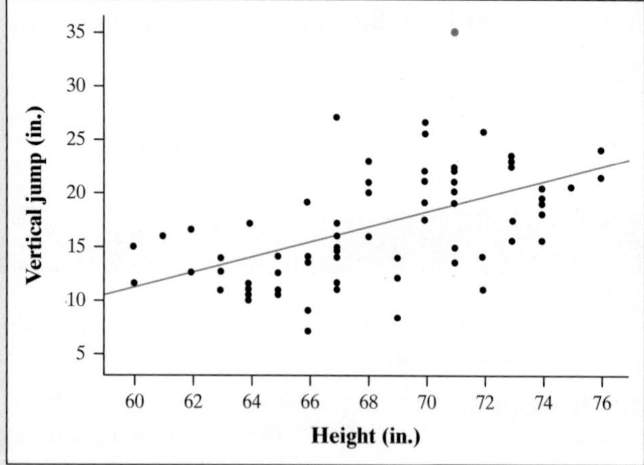

4. **Stand mixers** The scatterplot shows the weight (in pounds) and cost (in dollars) of 11 stand mixers.[63] The mixer from Walmart (highlighted in red) was much lighter—and cheaper—than the other mixers. What would happen to the equation of the least-squares regression line if the point for the Walmart mixer was removed? What would happen to the values of r, r^2, and s if the point for the Walmart mixer was removed?

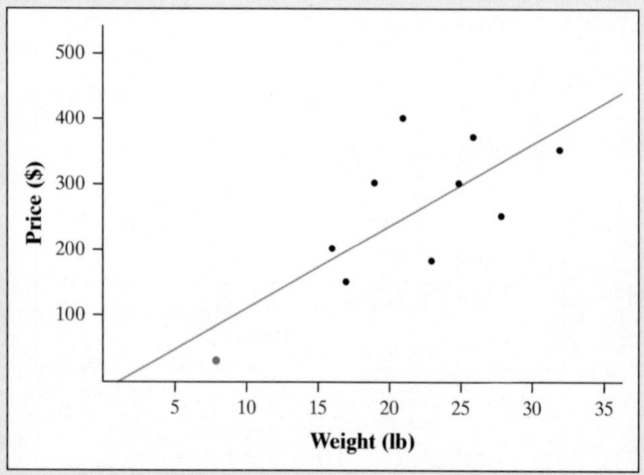

5. **High-protein beans** Beans and other legumes are an excellent source of protein. The scatterplot shows the relationship between protein (in grams) and carbohydrates (in grams) for a one-half cup portion of cooked beans for 12 different varieties.[64] The point for soybeans is highlighted in purple. What would happen to the values of r and s if the point for soybeans was removed?

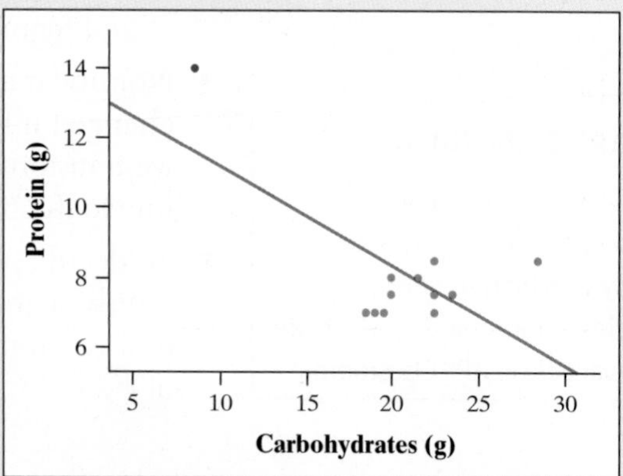

6. **Taxes and championships** Do states with lower income tax rates attract better professional athletes? Using data from 2000–2019, Ty Schalter at www.fivethirtyeight.com investigated the relationship between x = average state income tax rate and y = "championship points" per team-year.[65] Three championship points were awarded for winning the title and one point for being runner-up. For example, Massachusetts (highlighted in purple) had 41 championship points and 79 team-years, giving it a y-value of $41/79 = 0.519$. Here is a scatterplot summarizing this relationship. What would happen to the values of r and s if the point for Massachusetts was removed?

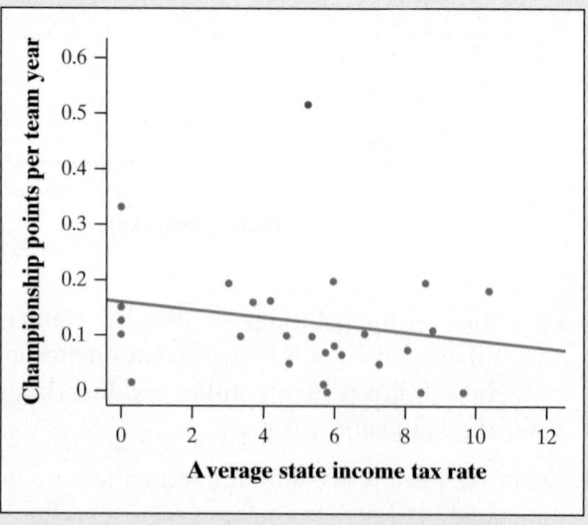

Transformations to Achieve Linearity

7. **The swinging pendulum** Mrs. Crosby's precalculus class collected data on the length (in centimeters) of a pendulum and the time (in seconds) the pendulum took to complete one back-and-forth swing (called its period). The theoretical relationship between a pendulum's length and its period is pg 218

$$\text{period} = \frac{2\pi}{\sqrt{g}}\sqrt{\text{length}}$$

where g is a constant representing the acceleration due to gravity (in this case, $g = 980 \text{ cm/s}^2$). Here is a graph of period versus $\sqrt{\text{length}}$, along with output from a linear regression analysis using these variables. Predict the period of a pendulum with length 80 cm.

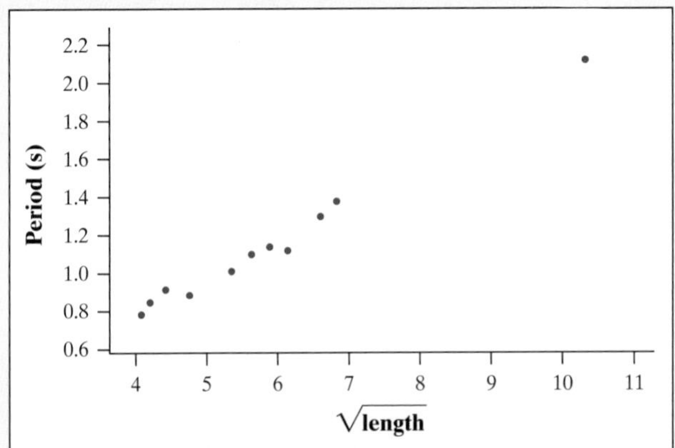

Predictor	Coef	SE Coef	T	P
Constant	−0.08594	0.05046	−1.70	0.123
sqrt(length)	0.209999	0.008322	25.23	0.000

S = 0.0464223 R-Sq = 98.6% R-Sq (adj) = 98.5%

8. **The swinging pendulum** Refer to Exercise 7. Here is a graph of period^2 versus length, along with output from a linear regression analysis using these variables. Predict the period of a pendulum with length 80 cm.

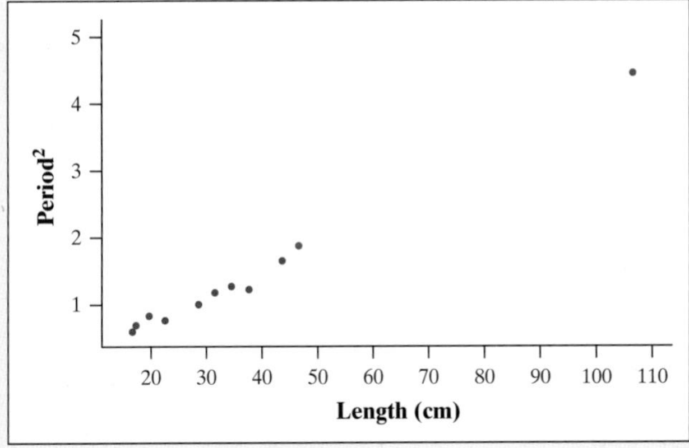

Predictor	Coef	SE Coef	T	P
Constant	−0.15465	0.05802	−2.67	0.026
Length	0.042836	0.001320	32.46	0.000

S = 0.105469 R-Sq = 99.2% R-Sq(adj) = 99.1%

9. **Brawn versus brain** Refer to Exercise 1. Researchers collected data on the brain weight (in grams) and body weight (in kilograms) for 96 species of mammals. The following figure is a scatterplot of the logarithm of brain weight versus the logarithm of body weight for all 96 species. The least-squares regression line for the transformed data is

$$\widehat{\log y} = 1.01 + 0.72 \log x$$

Based on footprints and some other sketchy evidence, some people believe that a large ape-like animal, called Sasquatch or Bigfoot, lives in the Pacific Northwest. Bigfoot's weight is estimated to be about 127 kilograms. How big do you expect Bigfoot's brain to be?

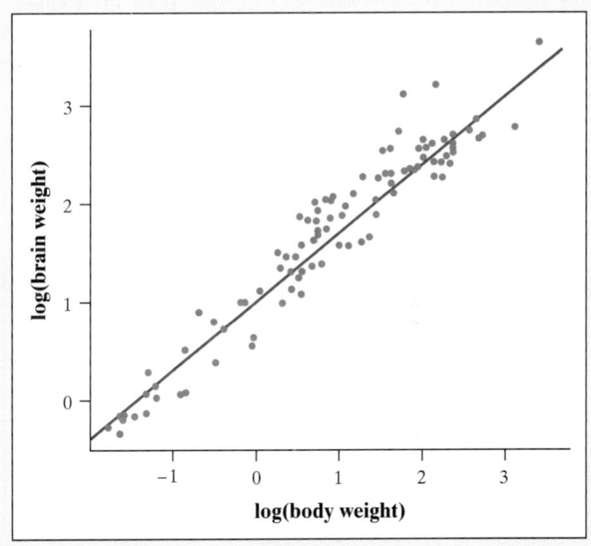

10. **Determining tree biomass** It is easy to measure the diameter at breast height (in centimeters) of a tree. It's hard to measure the total above-ground biomass (in kilograms) of a tree, because to do this, you must cut and weigh the tree. The following figure is a scatterplot of the natural logarithm of biomass versus the natural logarithm of diameter at breast height (DBH) for 378 trees in tropical rain forests.[66] The least-squares regression line for the transformed data is

$$\widehat{\ln y} = -2.00 + 2.42 \ln x$$

Use this model to estimate the biomass of a tropical tree 30 cm in diameter.

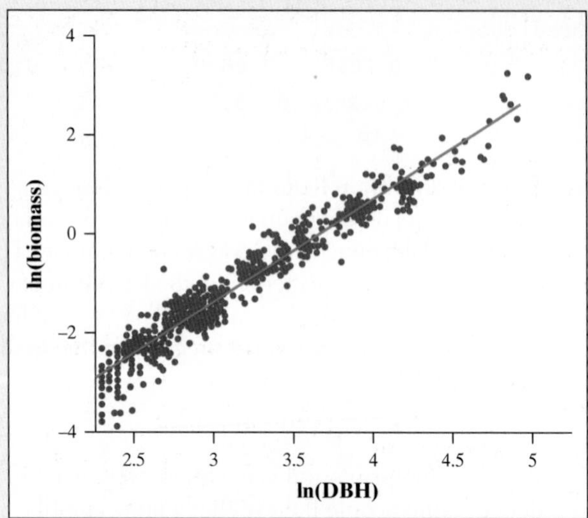

11. **Killing bacteria** A researcher exposed marine bacteria to X-rays for time periods from 1 to 15 minutes. Here is a scatterplot showing the natural logarithm of the number of surviving bacteria on a culture plate after each exposure time, along with output from a linear regression analysis.[67] Predict the number of surviving bacteria after 17 minutes.

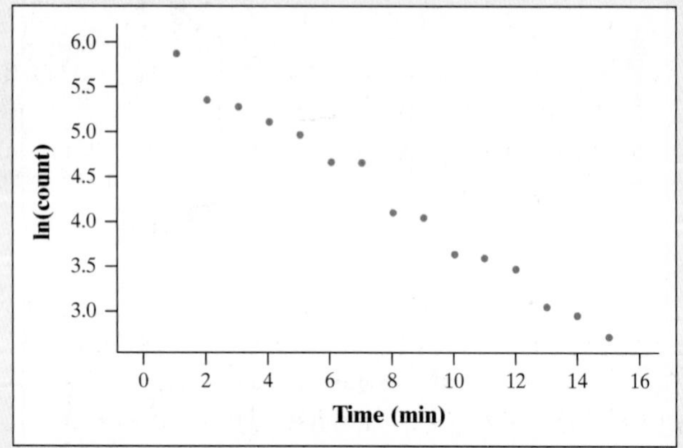

Predictor	Coef	SE Coef	T	P
Constant	5.97316	0.05978	99.92	0.000
Time	−0.218425	0.006575	−33.22	0.000

S = 0.110016 R-Sq = 98.8% R-Sq(adj) = 98.7%

12. **Click-through rates** Companies work hard to have their website listed at the top of an internet search. Is there a relationship between a website's position in the results of an internet search (1 = top position, 2 = second position, etc.) and the percentage of people who click on the link for the website? Here is a scatterplot showing the logarithm of the click-through rate for each of the top 10 positions in searches on a mobile device.[68] Predict the click-through rate for a website in the 11th position.

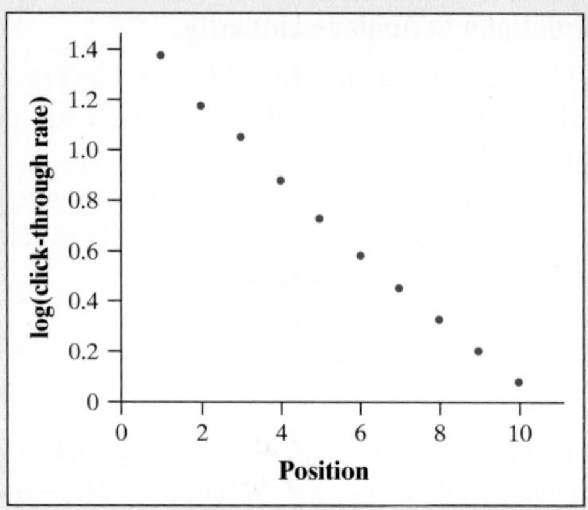

Predictor	Coef	SE Coef	T	P
Constant	1.4674	0.0194	75.634	0.000
Position	−0.1430	0.0031	−45.734	0.000

S = 0.0284 R-Sq = 99.6% R-Sq(adj) = 99.6

Choosing the Most Appropriate Model

13. **Putting success** How well do professional golfers putt from different distances? Researchers collected data on the percentage of putts made for various distances to the hole (in feet).[69] pg 220

Here is linear regression output for three different models, along with a residual plot for each model. Model 1 is based on the original data, while Models 2 and 3 involve transformations of the original data.

Model 1. Percent made vs. Distance

Predictor	Coef	SE Coef	T	P
Constant	83.6081	4.7206	17.711	0.0000
Distance	−4.0888	0.3842	−10.64	0.0000

S = 9.17 R-sq = 0.870 R-Sq(adj) = 0.862

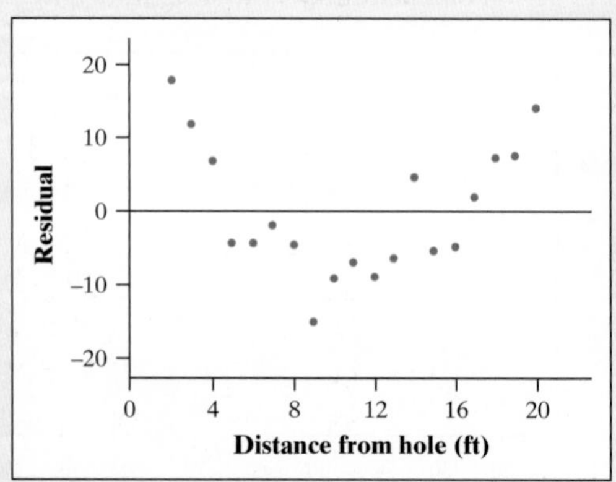

Model 2. ln(percent made) vs. ln(distance)

Predictor	Coef	SE Coef	T	P
Constant	5.5047	0.1628	33.821	0.0000
ln(distance)	−0.9154	0.0702	−13.04	0.0000
S = 0.196	R-sq = 0.909		R-sq(adj) = 0.904	

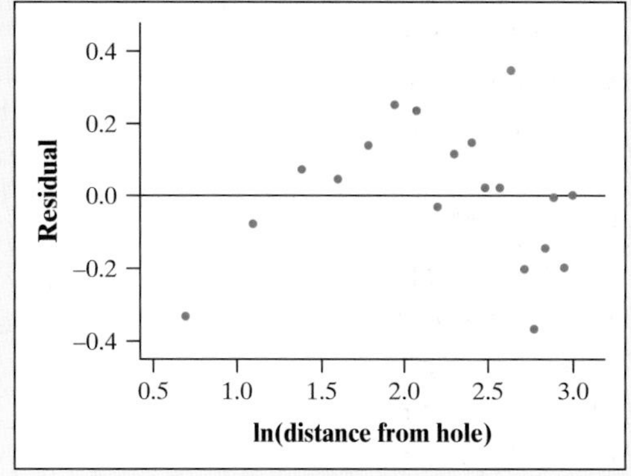

Model 3. ln(percent made) vs. Distance

Predictor	Coef	SE Coef	T	P
Constant	4.6649	0.0825	56.511	0.0000
Distance	−0.1091	0.0067	−16.24	0.0000
S = 0.160	R-sq = 0.939		R-sq(adj) = 0.936	

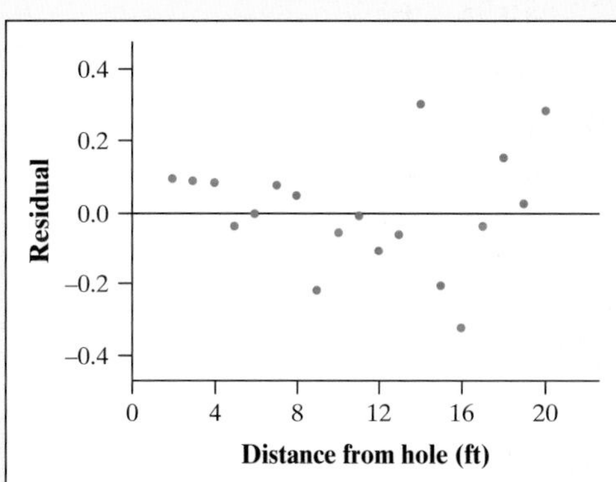

(a) Which model does the best job of summarizing the relationship between distance and percent made? Explain your reasoning.

(b) Using the model chosen in part (a), calculate and interpret the residual for the point where the golfers made 31% of putts from 14 feet away.

14. **Counting carnivores** Ecologists look at data to learn about nature's patterns. One pattern they have identified relates the size of a carnivore (body mass in kilograms) to how many of those carnivores exist in an area. A good measure of "how many" (abundance) is to count carnivores per 10,000 kg of their prey in the area. Researchers collected data on the abundance and body mass for 25 carnivore species.[70]

Here is linear regression output for three different models, along with a residual plot for each model. Model 1 is based on the original data, while Models 2 and 3 involve transformations of the original data.

Model 1. Abundance vs. Body mass

Predictor	Coef	SE Coef	T	P
Constant	158.3094	81.2586	1.948	0.0637
Body mass	−1.1140	0.9972	−1.007	0.3245
S = 345.5	R-sq = 0.042		R-sq(adj) = 0.001	

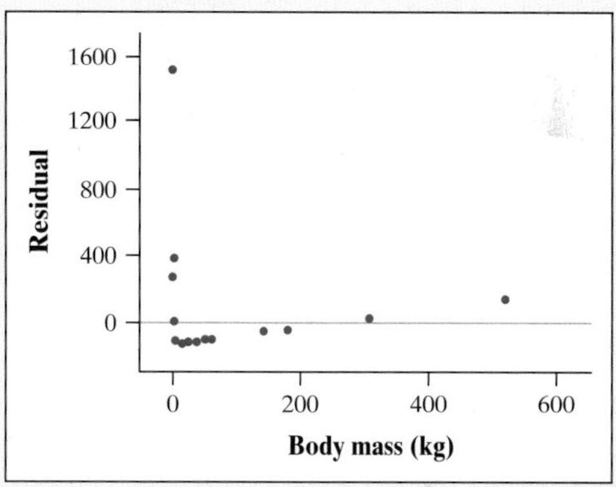

Model 2. ln(abundance) vs. ln(body mass)

Predictor	Coef	SE Coef	T	P
Constant	4.4907	0.3091	14.531	0.0000
ln(body mass)	−1.0481	0.0980	−10.693	0.0000
S = 0.975	R-sq = 0.833		R-sq(adj) = 0.825	

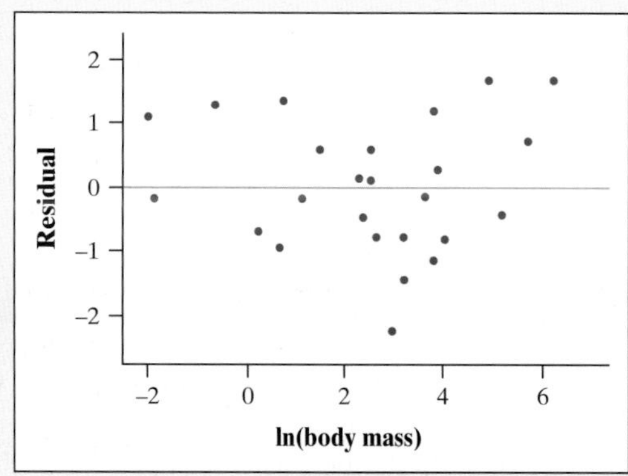

```
Model 3. ln(abundance) vs. Body mass
Predictor       Coef  SE Coef       T       P
Constant      2.6375   0.4843   5.447  0.0000
Body mass    -0.0166   0.0059  -2.791  0.0104
S=2.059       R-sq=0.253    R-sq(adj)=0.220
```

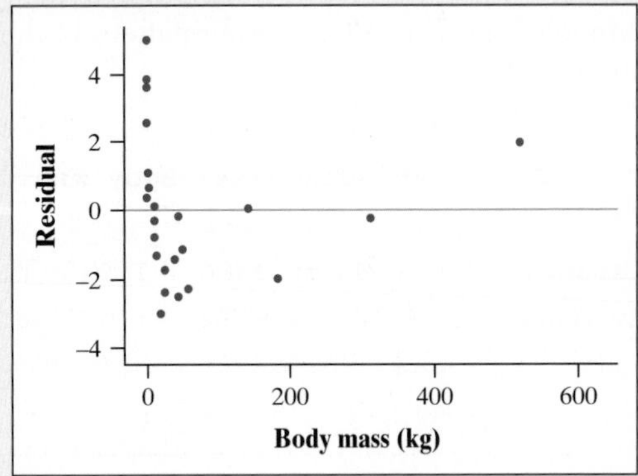

(a) Which model does the best job of summarizing the relationship between body mass and abundance? Explain your reasoning.

(b) Using the model chosen in part (a), calculate and interpret the residual for the coyote, which has a body mass of 13.0 kg and an abundance of 11.65.

15. **Heart weights of mammals** Is there a relationship between the length of the cavity of the left ventricle of a mammal's heart (in centimeters) and the heart's weight (in grams)? Here are the data for various mammals:[71]

Mammal	Length (cm)	Weight (g)
Mouse	0.55	0.13
Rat	1.00	0.64
Rabbit	2.20	5.80
Dog	4.00	102.00
Sheep	6.50	210.00
Ox	12.00	2030.00
Horse	16.00	3900.00

(a) Make a scatterplot using heart weight as the response variable. Describe what you see.

(b) Use technology to find the natural logarithm of the heart weights. Then make a scatterplot of ln(weight) versus length. Describe what you see.

(c) Use technology to find the natural logarithm of the lengths. Then make a scatterplot of ln(weight) versus ln(length). Describe what you see.

(d) If you were to calculate a least-squares regression line for each scatterplot in parts (a)–(c), which one would have the best fit? Use technology to calculate this line.

16. **Transforming back** In Exercise 10, you used the equation $\widehat{\ln y} = -2.00 + 2.42 \ln x$ to predict a tree's biomass based on its diameter at breast height. Use properties of exponents and logarithms to produce a model for these data in the form $\hat{y} = ax^b$.

For Investigation *Apply the skills from the section in a new context or nonroutine way.*

17. **Quadratic models** In addition to transforming to achieve linearity, we can fit models to nonlinear associations using other methods. Here is a scatterplot showing x = carapace (shell) length (cm) and y = clutch size (number of eggs) for a random sample of female gopher tortoises in Okeeheelee County Park, Florida.[72] Also included is a graph of the quadratic model $\hat{y} = -1018.95 + 66.181x - 1.0636x^2$.

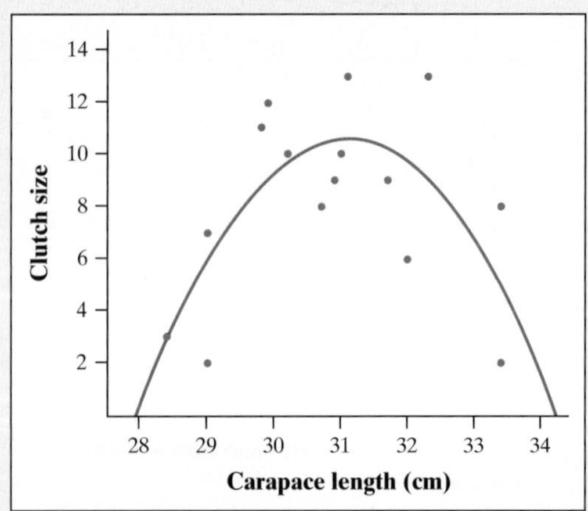

(a) Calculate and interpret the residual for the tortoise with a carapace length of 29 cm and a clutch size of 2.

(b) Sketch a residual plot for this model by estimating the remaining residuals from the graph.

Multiple Choice *Select the best answer for each question.*

18. Students in Mr. Handford's class dropped a kickball beneath a motion detector. The detector recorded the height of the ball (in feet) as it bounced up and down several times. Here is computer output from a linear regression analysis of the transformed data of log(height) versus bounce number. Predict the highest point the ball reaches on its seventh bounce.

Predictor	Coef	SE Coef	T	P
Constant	0.45374	0.01385	32.76	0.000
Bounce	−0.117160	0.004176	−28.06	0.000
S = 0.0132043	R-Sq = 99.6%		R-Sq(adj) = 99.5%	

(A) 0.35 ft

(B) 0.37 ft

(C) 0.43 ft

(D) 2.26 ft

(E) 2.32 ft

19. A scatterplot of y versus x shows a positive, nonlinear association. Two different transformations are attempted to try to linearize the association: using the logarithm of the y-values and using the square root of the y-values. Two least-squares regression lines are calculated, one that uses x to predict $\log(y)$ and the other that uses x to predict $\sqrt{y}$. Which of the following would be the best reason to prefer the least-squares regression line that uses x to predict $\log(y)$?

(A) The value of r^2 is smaller.

(B) The standard deviation of the residuals is smaller.

(C) The slope is greater.

(D) The residual plot has more random scatter.

(E) The distribution of residuals is more normal.

Recycle and Review *Practice what you learned in previous sections.*

20. **Shower time (1D, 1F)** Marcella takes a shower every morning when she gets up. Her time in the shower varies according to a normal distribution with mean 4.5 minutes and standard deviation 0.9 minute.

(a) How often will Marcella's shower last between 3 and 6 minutes?

(b) If Marcella took a 7-minute shower, would it be classified as an outlier by the $1.5 \times IQR$ rule? Justify your answer.

21. **NFL weights (1C, 1D)** Players in the National Football League (NFL) are bigger and stronger than ever before. And they are heavier, too.[73]

(a) Here is a boxplot showing the distribution of weight for NFL players in a recent season. Describe the distribution.

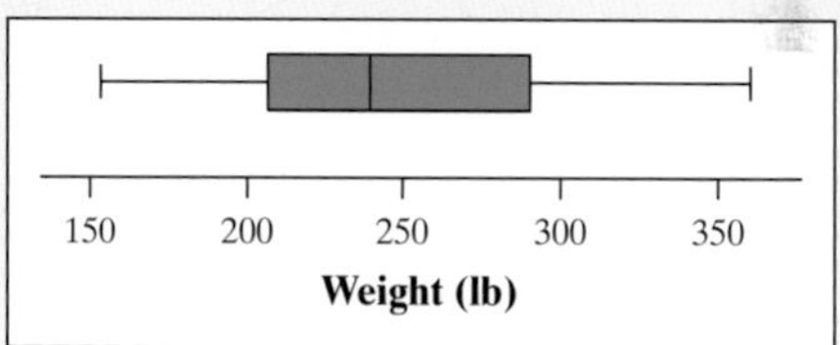

(b) Here is a dotplot of the same distribution. What feature of the distribution does the dotplot reveal that wasn't revealed by the boxplot?

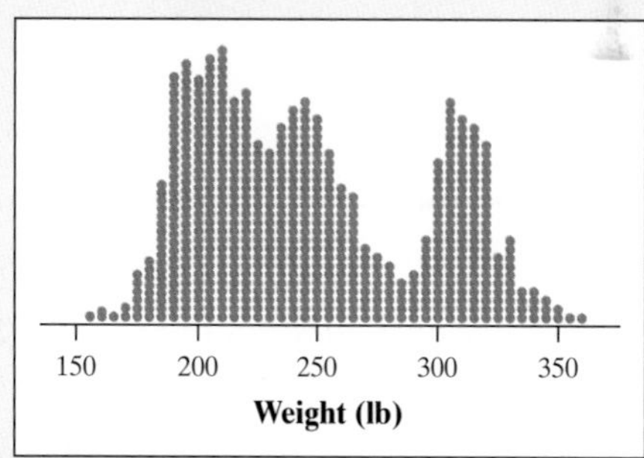

UNIT 2 Wrap-Up

FRAPPY! Free Response AP® Problem, Yay!

Directions: Show all your work. Indicate clearly the methods you use, because you will be scored on the correctness of your methods as well as on the accuracy and completeness of your results and explanations.

Two statistics students went to a flower shop and randomly selected 12 carnations. When they got home, the students prepared 12 identical vases with exactly the same amount of water in each vase. They put one tablespoon of sugar in 3 vases, two tablespoons of sugar in 3 vases, and three tablespoons of sugar in 3 vases. In the remaining 3 vases, they put no sugar. After the vases were prepared, the students randomly assigned 1 carnation to each vase and observed how many hours each flower continued to look fresh. A scatterplot of the data is shown here.

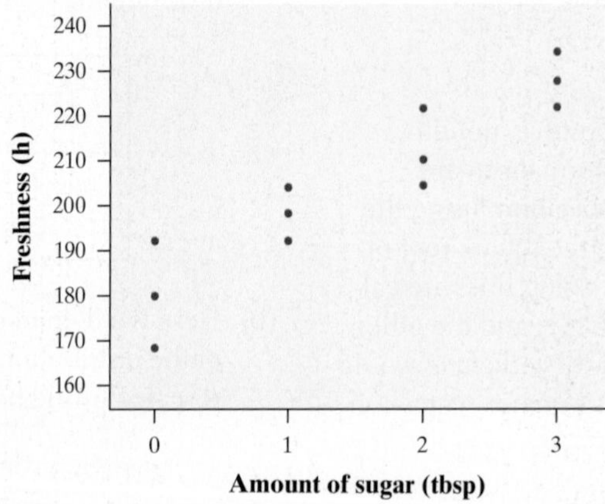

(a) Briefly describe the association shown in the scatterplot.

(b) The equation of the least-squares regression line for these data is $\hat{y} = 180.8 + 15.8x$. Interpret the slope of the line in the context of the study.

(c) Calculate and interpret the residual for the flower that had 2 tablespoons of sugar and looked fresh for 204 hours.

(d) Suppose that another group of students conducted a similar experiment using 12 flowers, but included different varieties in addition to carnations. Would you expect the value of r^2 for the second group's data to be greater than, less than, or about the same as the value of r^2 for the first group's data? Explain.

After you finish the FRAPPY!, you can view two example solutions on the book's website (**bfwpub.com/tps7e**). Determine whether you think each solution is "complete," "substantial," "developing," or "minimal." If the solution is not complete, what improvements would you suggest to the student who wrote it? Finally, your teacher will provide you with a scoring rubric. Score your response and note what, if anything, you would do differently to improve your own score.

UNIT 2 REVIEW

SECTION 2A Relationships Between Two Categorical Variables

In this section, you learned how to distinguish an **explanatory variable** from a **response variable.** Then you learned how to investigate the relationship between two categorical variables. Using a **two-way table,** you learned how to calculate and display **marginal relative frequencies** and **joint relative frequencies.** Calculating **conditional relative frequencies** and constructing **segmented bar graphs** or **mosaic plots** allow you to look for an **association** between the variables. If there is no association between the two variables, the distribution of the response variable will be the same for each value of the explanatory variable. However, if there are differences in the corresponding conditional relative frequencies, there is an association between the two variables. In other words, knowing the value of one variable helps you predict the value of the other variable.

SECTION 2B Relationships Between Two Quantitative Variables

In this section, you learned how to explore the relationship between two quantitative variables. As with univariate data, the first step when working with bivariate data is to make a graph. A **scatterplot** is the appropriate type of graph to investigate relationships between two quantitative variables. To describe a scatterplot, be sure to discuss four characteristics: direction, form, strength, and unusual features. The **direction** of a relationship might be described as a **positive association,** a **negative association,** or **no association.** The **form** of a relationship can be linear or nonlinear. The **strength** of a relationship is strong if it closely follows a specific form. Finally, **unusual features** include distinct clusters of points and points that clearly fall outside the pattern of the rest of the data.

The **correlation** r is a numerical summary for linear relationships that describes the direction and strength of the association. When $r > 0$, the association is positive; when $r < 0$, the association is negative. The correlation will always take values between -1 and 1, with $r = -1$ and $r = 1$ indicating a perfectly linear relationship. Strong linear relationships have correlations near 1 or -1, while weak linear relationships have correlations near 0. It isn't possible to determine the form of a relationship by using the correlation. Strong nonlinear relationships can have a correlation close to 1 or a correlation close to 0. You also learned that **correlation does not imply causation.** That is, we can't assume that changes in one variable cause changes in the other variable, just because the variables have a correlation close to 1 or -1. Finally, you should calculate r with technology whenever possible, but there is a formula just in case.

SECTION 2C Linear Regression Models

In this section, you learned how to use **regression lines** as models for relationships between two quantitative variables that have a linear association. To emphasize that the model provides only predicted values, regression lines are always expressed in terms of $\hat{y}$ instead of y. Likewise, when you are interpreting the y intercept or slope of a least-squares regression line, make sure you are describing the *predicted* value of y. The **slope** describes how the predicted value of y changes for each one-unit increase in x. The **y intercept** is the predicted value of y when $x = 0$.

The difference between the actual value of y and the predicted value of y is called a **residual.** Residuals are the key to understanding almost everything in this section. To find the equation of the **least-squares regression line,** find the line that minimizes the sum of the squared residuals. To see if a linear model is appropriate, make a **residual plot.** If there is no leftover curved pattern in the residual plot, you know the model is appropriate. To assess how well a line fits the data, use two values based on the sum of squared residuals. The **coefficient of determination r^2** measures the percentage of the variation in the response variable that is accounted for by the least-squares regression line that uses the explanatory variable. The **standard deviation of the residuals** s estimates the size of a typical prediction error.

It is best to get the equation of a least-squares regression line from technology. Make sure you can obtain the slope and y intercept from statistical software output. You can also find the equation with your calculator if you have raw data or with formulas that use the means and standard deviations of the two variables and their correlation.

SECTION 2D Analyzing Departures from Linearity

Unusual points can greatly influence the equation of the least-squares regression line and other summary statistics such as the correlation r, the coefficient of determination r^2, and the standard deviation of the residuals s. **Outliers** are points with large residuals. **High-leverage points** have x-values that are far from $\bar{x}$ relative to other points.

When the association between two variables is nonlinear, **transforming** one or both variables can result in a linear association. Once you have achieved linearity, calculate the equation of the least-squares regression line using the transformed data. Remember to include the transformed variables when you are writing the equation of the line. Likewise, when using the line to make predictions, make sure that the prediction is in the original units of y. If you transformed the y variable, you will need to undo the transformation after using the least-squares regression line.

To decide which of two or more models is most appropriate, choose the one that produces the most linear association and whose residual plot has the most random scatter. If more than one residual plot is randomly scattered, choose the model with the value of r^2 closest to 1.

What Did You Learn?

Learning Target	Section	Related Example on Page(s)	Relevant Unit Review Exercise(s)
Identify the explanatory and response variables in a given setting.	2A	143	R1
Calculate statistics for two categorical variables.	2A	146	R1
Display the relationship between two categorical variables.	2A	150	R1
Describe the relationship between two categorical variables.	2A	152	R1
Make a scatterplot to display the relationship between two quantitative variables.	2B	160	R6
Describe the direction, form, and strength of a relationship displayed in a scatterplot and identify unusual features.	2B	163	R6
Interpret the correlation for a linear relationship between two quantitative variables.	2B	166	R2
Distinguish correlation from causation.	2B	168	R3
Calculate the correlation for a linear relationship between two quantitative variables.	2B	170	R6
Make predictions using a regression line, keeping in mind the dangers of extrapolation.	2C	183	R4
Calculate and interpret a residual.	2C	184	R4
Interpret the slope and *y* intercept of a regression line.	2C	186	R4
Determine the equation of a least-squares regression line using formulas.	2C	189	R5
Determine the equation of a least-squares regression line using technology.	2C	190	R6, R9
Construct and interpret residual plots to assess whether a regression model is appropriate.	2C	194	R7
Interpret the coefficient of determination r^2 and the standard deviation of the residuals s.	2C	199	R7
Describe how unusual points influence the least-squares regression line, the correlation, r^2, and the standard deviation of the residuals.	2D	214	R8
Calculate predicted values from linear models using variables that have been transformed with logarithms or powers.	2D	218	R9
Determine which of several models does a better job of describing the relationship between two quantitative variables.	2D	220	R9

UNIT 2 REVIEW EXERCISES

These exercises are designed to help you review the important concepts and skills in the unit.

R1. Medieval bone fractures (2A) To study the challenges of living in medieval times, archaeologists in Cambridge, England, examined samples of skeletons in three different medieval cemeteries for bone fractures. The cemetery of the All Saints by the Castle parish was a place of burial for the majority of people in the area. The cemetery at the Hospital of St. John the Evangelist was a place of burial for those who were impoverished, chronically ill, or both. The cemetery at the Augustinian friary was used by friars and wealthier townspeople. The archeologists hypothesized that wealthier people would have fewer fractures due to several factors, including performing less physical labor.[74] The two-way table summarizes the relationship between cemetery and number of bone fractures.

		Cemetery			
		All Saints parish	Hospital of St. John	Augustinian friary	Total
Number of fractures	None	47	113	19	179
	1	21	22	5	48
	2 or more	16	20	4	40
	Total	84	155	28	267

(a) Identify the explanatory and response variables. Explain your reasoning.

(b) What percentage of the individuals in the sample had no fractures?

(c) What percentage of the individuals in the sample were buried in All Saints parish and had at least one fracture?

(d) Construct a segmented bar graph to display the association between cemetery and number of fractures.

(e) Describe the association shown in the segmented bar graph from part (d).

R2 **Warm body, fast heart? (2B)** Do people with warmer body temperatures have faster resting pulse rates? A researcher recorded x = body temperature (in degrees Fahrenheit) and y = pulse rate (in beats per minute) for a sample of 20 people.[75] The correlation for the relationship shown in the scatterplot is $r = 0.376$. Interpret this value.

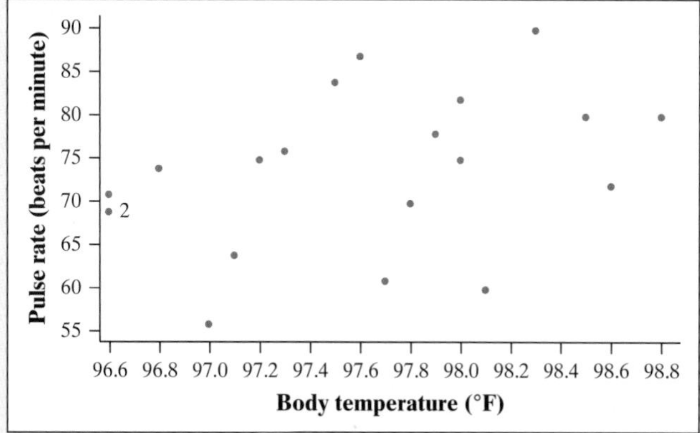

R3 **Calculating achievement (2B)** The principal of a high school read a study that reported a high correlation between the number of calculators owned by high school students and their math achievement. Based on this study, the principal decides to buy each student at the school two calculators, hoping to improve their math achievement. Explain the flaw in the principal's reasoning.

R4 **Late bloomers? (2C)** Japanese cherry trees tend to blossom early when spring weather is warm and later when spring weather is cool. Here is a scatterplot showing the average March temperature (in degrees Celsius) and the day in April when the first cherry blossom appeared over a 24-year period. The least-squares regression line $\hat{y} = 33.12 - 4.69x$.[76]

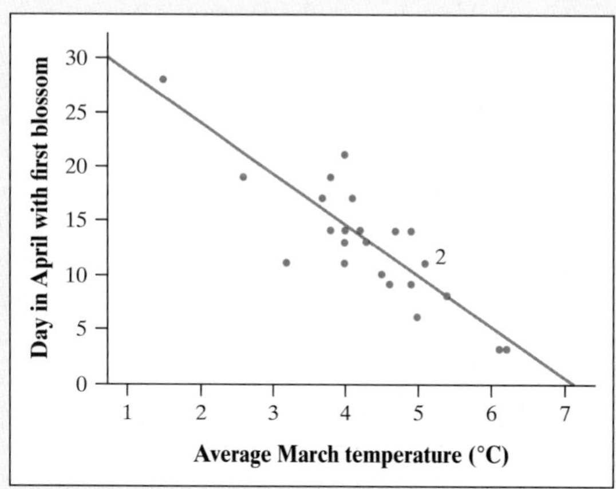

(a) Interpret the slope and y intercept of the line in this setting.

(b) Suppose that the average March temperature this year was 8.2°C. Would you be willing to use the least-squares regression line to predict the date of first blossom? Explain your reasoning.

(c) Calculate and interpret the residual for the year when the average March temperature was 4.5°C and the date of first blossom was April 10.

R5 **What's my grade? (2C)** In Professor Friedman's economics course, the correlation between the students' total scores prior to the final examination and their final exam scores is $r = 0.6$. The pre-exam totals for all students in the course have mean 280 and standard deviation 30. The final exam scores have mean 75 and standard deviation 8. Calculate the equation of the least-squares regression line for predicting final exam grade from the pre-exam totals.

R6 **Tennis serves (2B, 2C)** Are taller tennis players able to serve faster? Physics would suggest this is the case, as taller players have longer arms and can generate more racket speed. The table shows the height (in inches) and average first serve speed (in miles per hour) for 16 male professional tennis players.[77]

Height	Speed	Height	Speed
75	108	77	118
78	122	80	128
76	113	73	116
78	123	72	106
77	127	78	124
73	123	75	119
77	113	73	125
74	120	71	110

(a) Make a scatterplot to show the relationship between height and average serve speed.

(b) Describe the association shown in the scatterplot.

(c) Use technology to calculate the correlation and the equation of the least-squares regression line.

R7 **Hungry squirrels (2C)** A long-term study of red squirrels in Canada investigated the relationship between food availability and population size.[78] Red squirrels in the Yukon have a diet made up mostly of seeds from white spruce trees. They gather these seeds in the fall, store them underground, and survive on them for the following year. The following residual plot was created from a linear model relating x = cone index in autumn (a measure of food availability) to y = squirrel density the following spring (number of squirrels per hectare) for plots observed over a number of years.

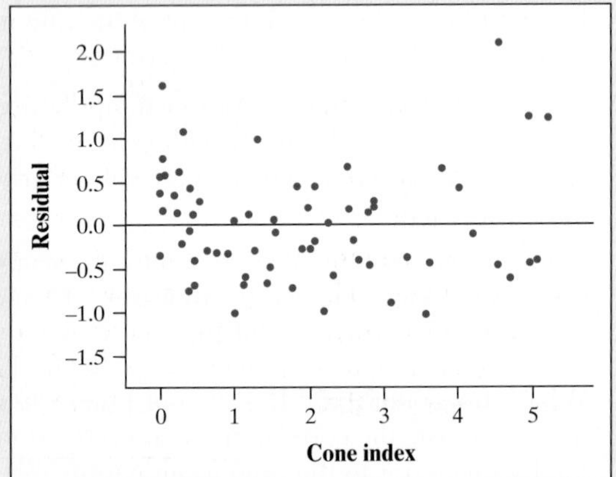

(a) What does the residual plot indicate about the linear model?

(b) For this model, $r^2 = 0.159$. Interpret this value.

(c) For this model, $s = 0.644$. Interpret this value.

R8 **Born to be old? (2D)** Is there a relationship between the gestational period (time from conception to birth) of an animal and its average life span? The figure shows a scatterplot of the gestational period and average life span for 43 species of animals.[79]

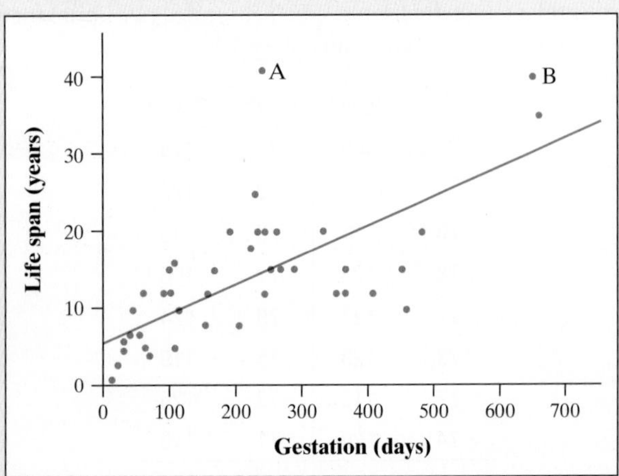

(a) Point A is the hippopotamus. What effect does this point have on the correlation, the equation of the least-squares regression line, and the standard deviation of the residuals? Is this point an outlier or a high-leverage point? Explain your reasoning.

(b) Point B is the Asian elephant. What effect does this point have on the correlation, the equation of the least-squares regression line, and the standard deviation of the residuals? Is this point an outlier or a high-leverage point? Explain your reasoning.

R9 **Diamonds! (2D)** Diamonds are expensive, especially big ones. To create a model to predict price from size, the weight (in carats) and price (in dollars) were recorded for each of 94 round, clear, internally flawless diamonds with excellent cuts.[80]

Here is linear regression output for three different models, along with a residual plot for each model. Model 1 is based on the original data, while Models 2 and 3 involve transformations of the original data.

Model 1. Price vs. Weight

Predictor	Coef	SE Coef	T	P
Constant	−98666	7594.1	−12.992	0.0000
Weight	105932	4219.5	25.105	0.0000
S = 34073	R-sq = 0.873		R-sq(adj) = 0.871	

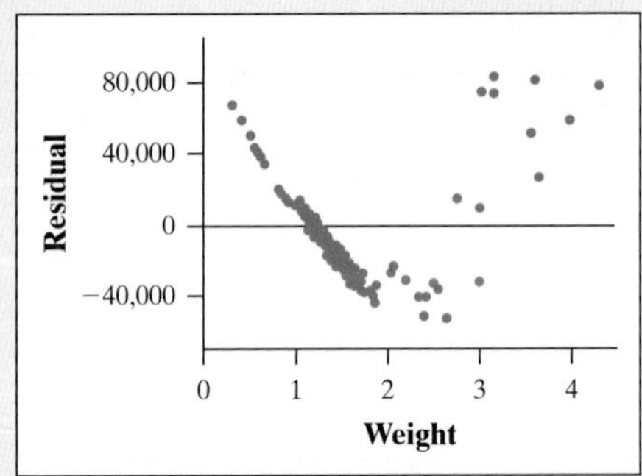

Model 2. ln(price) vs. ln(weight)

Predictor	Coef	SE Coef	T	P
Constant	9.7062	0.0209	465.102	0.0000
ln(weight)	2.2913	0.0332	68.915	0.0000
S = 0.171	R-sq = 0.981		R-sq(adj) = 0.981	

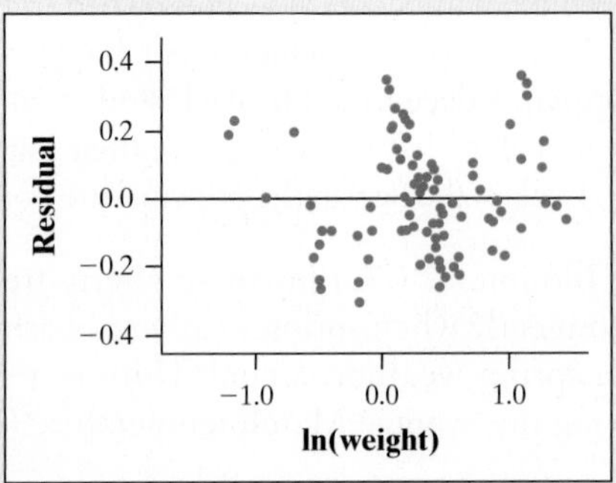

Model 3. ln(price) vs. Weight

Predictor	Coef	SE Coef	T	P
Constant	8.2709	0.0988	83.716	0.0000
Weight	1.3791	0.0549	25.123	0.0000

S = 0.443 R-sq = 0.873 R-sq(adj) = 0.871

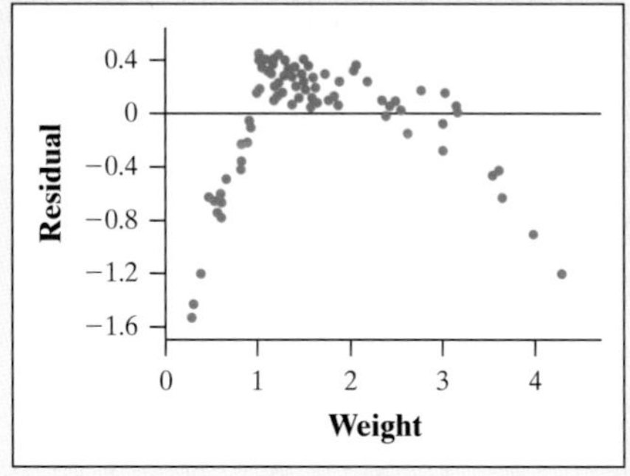

(a) Use each of the three models to predict the price of a diamond of this type that weighs 2 carats.

(b) Which model does the best job of summarizing the relationship between weight and price? Explain your reasoning.

UNIT 2 AP® STATISTICS PRACTICE TEST

Section I: Multiple Choice *Select the best answer for each question.*

Questions T1–T2 refer to the following setting. Is there an association between birth order and employment status for high school seniors? Each person in a random sample of 100 high school seniors was asked for their place in the birth order in their families and whether or not they had a part-time job.[81] The two-way table displays the results of the survey.

		Birth order				
		Oldest	Middle	Youngest	Only child	Total
Employment status	Employed	18	10	14	8	50
	Not employed	12	8	24	6	50
	Total	30	18	38	14	100

T1 What proportion of students in the sample were employed and either an only child or the oldest child in their family?

(A) 0.50
(B) 0.44
(C) 0.26
(D) 0.18
(E) 0.08

T2 Of the students who were the youngest in their family, what percentage were not employed?

(A) 63%
(B) 50%
(C) 48%
(D) 38%
(E) 24%

Questions T3–T5 refer to the following setting. Scientists examined the activity level of 7 fish at different temperatures. Fish activity was rated on a scale of 0 (no activity) to 100 (maximal activity). The temperature was measured in degrees Celsius. Computer output for a regression analysis and a residual plot are provided. Notice that the horizontal axis on the residual plot shows the fitted (predicted) values.

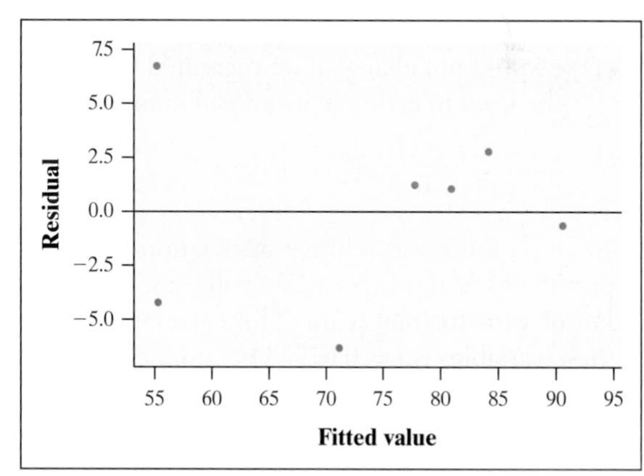

Predictor	Coef	SE Coef	T	P
Constant	148.62	10.71	13.88	0.000
Temperature	−3.2167	0.4533	−7.10	0.001

S = 4.78505 R-Sq = 91.0% R-Sq(adj) = 89.2%

T3 What is the correlation between temperature and fish activity?

(A) 0.95
(B) 0.91
(C) 0.45
(D) −0.91
(E) −0.95

T4 What was the actual activity level rating for the fish at a temperature of 20°C?

(A) 87
(B) 84
(C) 81
(D) 66
(E) 3

T5 Which of the following gives a correct interpretation of s in this setting?

(A) For every 1°C increase in temperature, fish activity is predicted to increase by 4.785 units.
(B) The typical distance of the temperature readings from the mean temperature is about 4.785°C.
(C) The typical distance of the activity level ratings from the least-squares line using $x =$ temperature is about 4.785 units.
(D) The typical distance of the activity level readings from the mean activity level is about 4.785 units.
(E) At a temperature of 0°C, this model predicts an activity level of 4.785 units.

T6 Which of the following statements about the correlation r between the lengths (in inches) and weights (in pounds) of a sample of brook trout is *not* true?

(A) r must take a value between -1 and 1.
(B) r is measured in pounds per inch.
(C) If longer trout tend to also be heavier, then $r > 0$.
(D) r would not change if we measured the lengths of the trout in centimeters instead of inches.
(E) r would not change if we reversed the variables in the calculation.

T7 In 2022, there was a linear association between the payroll of a Major League Baseball team and the number of wins for that team. The correlation between these variables is $r = 0.629$. The number of wins had a mean of 81 and standard deviation of 14.7, and the payroll had a mean of \$171 million and a standard deviation of \$62 million.[82] Calculate the equation of the least-squares regression line for predicting number of wins from payroll (in millions of dollars).

(A) $\hat{y} = 55.5 + 0.149x$
(B) $\hat{y} = 158.9 + 0.149x$
(C) $\hat{y} = -372.2 + 2.65x$
(D) $\hat{y} = -44.6 + 2.65x$
(E) The equation cannot be calculated without the raw data.

T8 Many people enjoy building with Lego™ blocks. To investigate the relationship between the number of pieces in a set and its cost, researchers gathered data on more than 1000 sets for sale on amazon.com. The relationship between $x =$ number of pieces and $y =$ cost (\$) is linear with a least-squares regression line of $\hat{y} = 18.92 + 0.0912x$. How much does the predicted price of a set of Legos increase for each additional 100 pieces?

(A) \$28.04
(B) \$19.01
(C) \$18.92
(D) \$9.12
(E) \$0.09

T9 The scatterplot shows the relationship between the number of people per television set and the number of people per physician for 40 countries, along with the least-squares regression line. In Ethiopia, there were 503 people per TV and 36,660 people per doctor. Which of the following statements is correct?

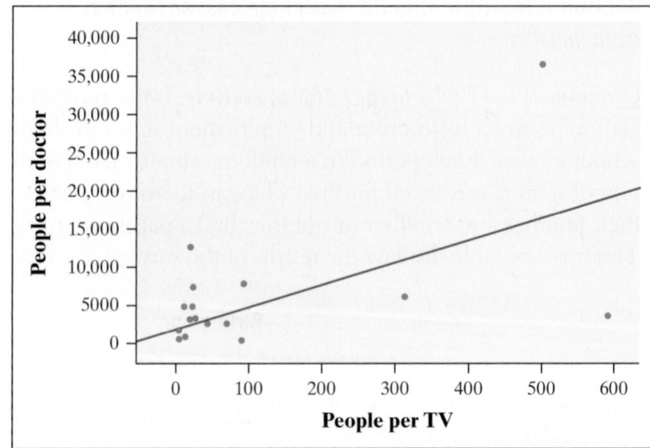

(A) Increasing the number of TVs in a country will attract more doctors.
(B) The slope of the least-squares regression line is less than 1.
(C) The correlation is greater than 1.
(D) The point for Ethiopia is decreasing the slope of the least-squares regression line.
(E) Ethiopia has more people per doctor than expected, based on how many people it has per TV.

T10 We recorded data on the population of a particular country from 1960 to 2010. A scatterplot reveals a clear curved relationship between population and year. However, a different scatterplot reveals a strong linear relationship between the logarithm (base 10) of the population and the year. The least-squares regression line for the transformed data is

$$\widehat{\log(\text{population})} = -13.5 + 0.01(\text{year})$$

Based on this equation, which of the following is the best estimate for the population of the country in the year 2020?

(A) 6.7
(B) 812
(C) 5,000,000
(D) 6,700,000
(E) 8,120,000

Section II: Free Response *Show all your work. Indicate clearly the methods you use, because you will be graded on the correctness of your methods as well as on the accuracy and completeness of your results and explanations.*

T11 Malaria is one of the leading causes of death worldwide. An experiment in Burkina Faso randomly assigned children aged 5–17 months to one of three treatment groups. Group 1 received a low-dose malaria vaccine, Group 2 received a high-dose malaria vaccine, and Group 3 received a control (rabies) vaccine. At the end of 12 months, researchers recorded whether or not each child had at least one episode of clinical malaria.[83] The two-way table summarizes the results of the experiment.

		Malaria outcome		
		None	At least once	Total
Type of vaccine	Control (rabies) vaccine	41	106	147
	Low-dose vaccine	96	50	146
	High-dose vaccine	107	39	146
	Total	244	195	439

(a) Explain why type of vaccine should be used as the explanatory variable in this context.

(b) Construct a segmented bar graph to display the relationship between type of vaccine and malaria outcome.

(c) Describe the association shown in the segmented bar graph from part (b).

T12 Long-term records from the Serengeti National Park in Tanzania show interesting ecological relationships. When wildebeest are more abundant, they graze the grass more heavily, so there are fewer fires and more trees grow. Lions feed more successfully when there are more trees, so the lion population increases. Researchers collected data on one part of this cycle, x = wildebeest abundance (in thousands of animals) and y = percentage of the grass area burned in the same year. Here are a residual plot and output from a least-squares regression analysis.[84]

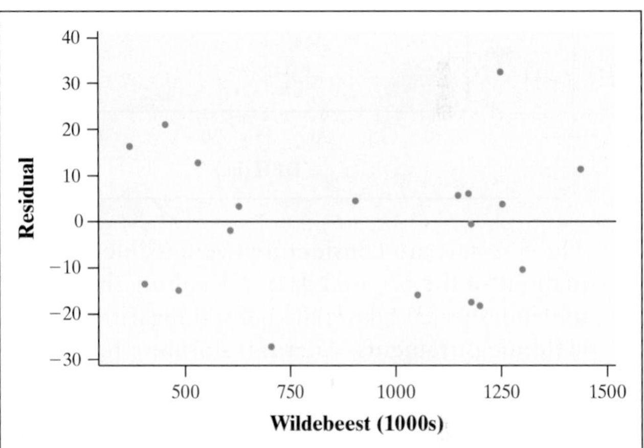


```
Predictor          Coef    SE Coef      T       P
Constant          92.29    10.06      9.17    0.000
Wildebeest     −0.05762  0.01035     −5.56    0.001
(1000s)
S = 15.9880     R-Sq = 64.6%     R-Sq(adj) = 62.5%
```

(a) Is a linear model appropriate for describing the relationship between wildebeest abundance and percentage of grass area burned? Explain your answer.

(b) Give the equation of the least-squares regression line. Be sure to define any variables you use.

(c) Interpret the slope. Does the value of the y intercept have meaning in this context? If so, interpret the y intercept. If not, explain why.

(d) Interpret the value of r^2.

T13 Foresters are interested in predicting the amount of usable lumber they can harvest from various tree species. They collect data on the diameter at breast height (DBH) in inches and the yield of usable lumber (in board feet) of a random sample of 40 ponderosa pine trees in Montana.[85] (Note that a board foot is defined as a piece of lumber 12 inches by 12 inches by 1 inch.) Here is a scatterplot of the data:

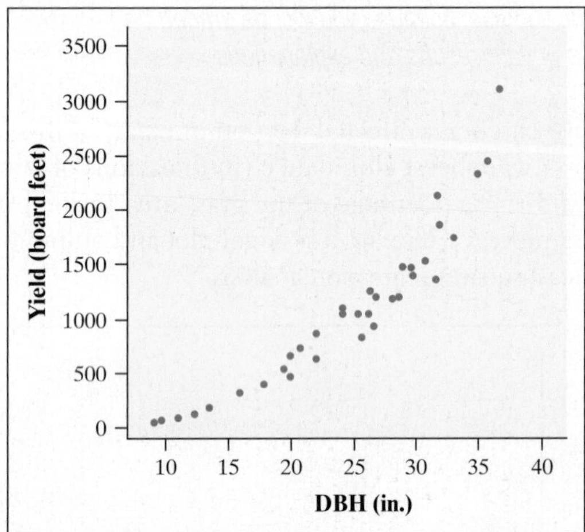

The foresters are considering two possible transformations of the original data: (1) cubing the diameter values or (2) taking the natural logarithm of the yield measurements. After transforming the data, a least-squares regression analysis is performed. Here is computer output and a residual plot for each of the two possible regression models:

Model 1: Cubing the diameter values

Predictor	Coef	SE Coef	T	P
Constant	92.177	44.6176	2.066	0.0457
DBH^3	0.0531	0.0020	26.656	0.0000
S=160.576	R-Sq=94.9%	R-Sq(adj)=94.8%		

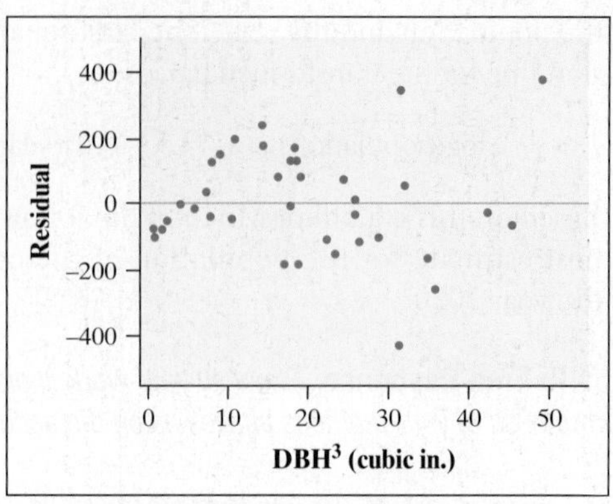

Model 2: Taking natural logarithm of yield measurements

Predictor	Coef	SE Coef	T	P
Constant	3.2469	0.1809	17.978	0.0000
DBH	0.1375	0.0071	19.436	0.0000
S=0.32665	R-Sq=90.9%	R-Sq(adj)=90.6%		

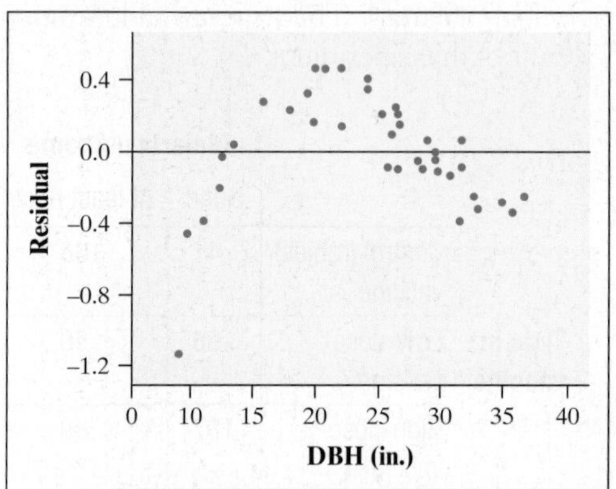

(a) Use both models to predict the amount of usable lumber from a ponderosa pine with diameter 30 inches.

(b) Which of the predictions in part (b) seems more reliable? Give appropriate evidence to support your choice.

UNIT 3 Collecting Data

Bob Krist/The Image Bank/Getty Images

Introduction

In Unit 1, you learned how to analyze distributions of categorical and quantitative data. In Unit 2, you continued to develop your data analysis skills by studying relationships between two categorical variables and relationships between two quantitative variables. In Unit 3, we focus on two different AP® Statistics course skill categories: selecting statistical methods and statistical argumentation. You'll learn how to describe appropriate methods for collecting data, along with the advantages and disadvantages of each method. You'll also learn how the method of data collection determines the types of conclusions we can draw from the data.

SECTION 3A Introduction to Data Collection

LEARNING TARGETS *By the end of the section, you should be able to:*

- Identify the population and sample in a statistical study.
- Distinguish between an observational study and an experiment.
- Determine what inferences are appropriate from an observational study.

Researchers who conduct statistical studies usually want to make an *inference* (i.e., conclusion) that goes beyond the data collected. Here are two examples:

- The U.S. Census Bureau carries out a monthly Current Population Survey of about 60,000 households. One goal is to use data from these randomly selected households to estimate the percentage of unemployed individuals in the population of all U.S. households.[1]
- Scientists performed an experiment involving 21 volunteers between the ages of 18 and 25. Each subject was randomly assigned to one of two treatments: sleep deprivation for one night or unrestricted sleep. The scientists hoped to show that sleep deprivation causes a decrease in performance 2 days later for volunteers like these.[2]

The type of inference we can make from a study depends on how the data were collected.

Populations and Samples

Suppose we want to find out what percentage of young drivers in the United States text while driving. To answer this question, we will survey 16- to 20-year-old drivers who live in the United States. Ideally, we would ask them all by conducting a **census.** Of course, contacting every driver in this age group wouldn't be practical. It would take too much time and cost too much money. Instead, we pose the question to a **sample** chosen to represent the entire **population** of young drivers.

DEFINITION Population, Census, Sample

The **population** in a statistical study is the entire group of individuals we want information about.

A **census** collects data from every individual in the population.

A **sample** is a subset of individuals in the population from which we collect data.

Recall from Unit 1 that individuals don't have to be humans. Individuals can also be animals or things, such as computer monitors.

EXAMPLE

Sampling monitors and voters
Populations and samples

Skill 1.C

Vladimir Vladimirov/E+/Getty Images

PROBLEM: Identify the population and the sample in each of the following settings.

(a) The quality control manager at a factory selects 10 computer monitors from the monitors produced during a particular hour and inspects each monitor for defects in construction and performance.

(b) Prior to an election, a news organization surveys 1000 registered voters to predict which candidate will be elected as president.

SOLUTION:

(a) The population is all the computer monitors produced in this factory during that hour. The sample is the 10 monitors selected and inspected for defects.

(b) The population is all registered voters. The sample is the 1000 registered voters surveyed.

> To identify the population, consider which individuals could have been selected for the sample. In part (a), the sample was only from computer monitors produced during that hour, so the population is limited to all monitors produced in this factory during that hour.

FOR PRACTICE, TRY EXERCISE 1

We often draw conclusions about a population based on a sample. Have you ever been given a sample of ice cream and ordered a cone because the sample tastes good? Because ice cream is fairly uniform, the single taste represents the whole. But choosing a representative sample from a large and varied population, like all young U.S. drivers, is not so easy. The first step in planning a **sample survey** is to decide what population we want to describe. The second step is to decide what we want to measure.

DEFINITION Sample survey

A **sample survey** is a study that collects data from a sample to learn about the population from which the sample was selected.

Some people use the terms *survey* or *sample survey* to refer *only* to studies in which people are asked questions, like the news organization's survey of registered voters in the preceding example. We'll avoid this more restrictive definition.

The final step in planning a sample survey is to decide how to choose a sample from the population. Here is an activity that illustrates the process of conducting a sample survey.

ACTIVITY Who wrote the Federalist Papers?

Bragin Alexey/Shutterstock.com

The Federalist Papers are a series of 85 essays supporting the ratification of the U.S. Constitution. At the time they were published, the identity of the authors was a secret known to only a few people. Over time, however, the authors were identified as Alexander Hamilton, James Madison, and John Jay. The authorship of 73 of the essays is fairly certain, leaving 12 in dispute. However, thanks in some part to statistical analysis, most scholars now believe that the 12 disputed essays were written by Madison alone or in collaboration with Hamilton.[3]

There are several ways to use statistics to help determine the authorship of a disputed text. One method is to estimate the average word length in a disputed text and compare it to the average word lengths of works where the authorship is not in dispute.

The following passage is the opening paragraph of Federalist Paper No. 51, one of the disputed essays.[4] The theme of this essay is the separation of powers between the three branches of government.

> To what expedient, then, shall we finally resort, for maintaining in practice the necessary partition of power among the several departments, as laid down in the Constitution? The only answer that can be given is, that as all these exterior provisions are found to be inadequate, the defect must be supplied, by so contriving the interior structure of the government as that its several constituent parts may, by their mutual relations, be the means of keeping each other in their proper places. Without presuming to undertake a full development of this important idea, I will hazard a few general observations, which may perhaps place it in a clearer light, and enable us to form a more correct judgment of the principles and structure of the government planned by the convention.

1. Choose 5 words from this passage. Count the number of letters in each of the words you selected, and find the average word length.
2. Your teacher will draw and label a horizontal axis for a class dotplot. Plot the average word length you obtained in Step 1 on the graph.
3. Your teacher will show you how to use a random number generator to select a random sample of 5 words from the 130 words in the opening passage. Count the number of letters in each of the selected words, and find the average word length.

4. Your teacher will draw and label another horizontal axis with the same scale for a comparative dotplot. Plot the average word length you obtained in Step 3 on the graph.
5. How do the dotplots compare? Can you think of any reasons why they might be different? Discuss with your classmates.

As you discovered in the Federalist Papers activity, using **random sampling** greatly improves our chances of getting a representative sample.

DEFINITION Random sampling

Random sampling involves using a chance process to determine which members of a population are chosen for the sample.

When the members of a sample are selected at random from a population, we can use the sample results to *infer* things about the population. That is, we can make *inferences* about the population from which the sample was randomly selected. You'll learn much more about sampling and surveys in Section 3B.

Observational Studies and Experiments

As seen in the Federalist Papers activity, a sample survey aims to gather information about a population without disturbing the population in the process. Sample surveys are one kind of **observational study.** Other examples of observational studies include recording the behavior of animals in the wild to discover their food preferences and tracking the medical history of volunteers to look for associations between variables such as type of diet, amount of exercise, and blood pressure. Observational studies can be either **retrospective** or **prospective.**

In contrast to observational studies, **experiments** don't involve just observing individuals or asking them questions. Instead, experimenters actively impose a *treatment* (condition) on *experimental units* (individuals) to measure their response. Experiments can answer questions like "Does aspirin reduce the chance of a heart attack?" and "Do plants grow better when classical music is playing?" You'll learn much more about experiments in Section 3C.

DEFINITION Observational study, Retrospective, Prospective, Experiment

An **observational study** observes individuals and measures variables of interest, but does not attempt to influence the responses.

Observational studies that examine existing data for a sample of individuals are called **retrospective.**

Observational studies that track individuals into the future are called **prospective.**

An **experiment** deliberately imposes treatments on experimental units to measure their responses.

The goal of an observational study can be to describe some group or situation, to compare groups, or to examine relationships between variables. The purpose of an experiment is to determine whether the treatment *causes* a change in the response. For reasons you'll soon learn, it is very difficult to make inferences about cause and effect from observational studies.

EXAMPLE

Family dinners and background music
Observational studies and experiments

Skill 1.C

Jasmin Merdan/Getty Images

PROBLEM: Determine whether each of the following settings describes an observational study or an experiment. Explain your answer.

(a) Researchers at Columbia University randomly selected 1000 teenagers in the United States for a survey. According to an ABC News story about the research, "Teenagers who eat with their families at least five times a week are more likely to get better grades in school."[5]

(b) Does the pace of background music affect how fast people eat? Researchers at Aarhus University in Denmark asked volunteers to eat a piece of chocolate and rate it for taste and other features. They randomly assigned each volunteer to listen to either a slow-paced version or a fast-paced version of the same composition while eating. The researchers secretly recorded the amount of time it took the volunteers to finish the chocolate. The average eating time was shorter for the group who listened to the faster-paced music.[6]

SOLUTION:

(a) This is an observational study because teenagers weren't told to eat with their families a certain number of times per week.

> This was a retrospective observational study because it used existing data and didn't follow the teenagers into the future.

(b) This is an experiment because some volunteers were assigned to eat with fast-paced music playing and others were assigned to eat with slow-paced music playing.

> An experiment imposes treatments (eating with fast music, eating with slow music) on experimental units (the volunteers).

FOR PRACTICE, TRY EXERCISE 5

Inference from Observational Studies

In the observational study about teenagers eating dinner with their families from the preceding example, the teenagers surveyed were randomly selected from the population of all U.S. teenagers. Due to the random selection, we can generalize the results of the study to the population of all teenagers in the United States. If this study had not used a random sample of teenagers, we could apply the results only to teenagers like those in the study.

Because this was an observational study and not an experiment, we can't conclude that eating more often with their family *causes* a teenager's grades to be better. It is possible that some teenagers have part-time jobs, which prevent them from eating with their families very often. And because of those part-time jobs, they may also have less time to study, so their grades are worse as a result. Observational studies like this one can't show that changes in one variable cause changes in another variable, even if they involve random sampling.

When the goal is to understand cause and effect, experiments are the only source of fully convincing data. For this reason, *the distinction between an observational study and an experiment is one of the most important ideas in statistics.*

EXAMPLE

A little something sweet
Inference from observational studies

Skill 4.A

Robin James/Image Source/Getty Images

PROBLEM: The National Health and Nutrition Examination Survey is a long-term research program that studies health and nutrition in adults and children in the United States. One random sample of 2437 young adults (aged 18–25 years) found that those who reported consuming food with added sugars, such as drinking soda or eating brownies, at least 5 times per week had a 73% higher risk of developing periodontal disease than did those who reported eating no added sugars.[7]

(a) Can we generalize this result to all U.S. adults? Explain your answer.

(b) What is the largest population to which we can generalize this result?

(c) Should we conclude that eating foods with added sugars causes periodontal disease based on this study? Explain your answer.

SOLUTION:

(a) No. Adults older than age 25 weren't part of the population from which the sample was selected.

(b) All young adults (aged 18–25 years) in the United States.

(c) No, because this is an observational study and not an experiment. Researchers didn't assign young adults to eat a certain number of foods with added sugars.

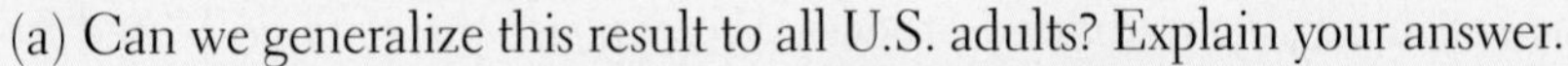

FOR PRACTICE, TRY EXERCISE 9

Although we can't conclude that eating foods with added sugars causes periodontal disease based on this study, it is still possible there is a cause-and-effect relationship between consuming food with added sugar and periodontal disease. But we'd need a well-designed experiment to find out.

CHECK YOUR UNDERSTANDING

1. The health office at a university randomly selects 100 students at the school and asks if they have received a flu vaccine. Identify the population and sample in this setting.
2. Does reducing screen brightness increase battery life in laptop computers? To find out, researchers obtained 30 new laptops of the same brand. They chose 15 of the computers at random and adjusted their screens to the brightest setting.

The other 15 laptop screens were left at the default setting — moderate brightness. Researchers then measured how long each machine's battery lasted. Was this an observational study or an experiment? Justify your answer.

3. Can drinking black tea extend your life? Researchers in the United Kingdom performed a prospective observational study using more than 500,000 volunteers. Those volunteers who drank more than 2 cups of black tea per day were significantly more likely to be alive 10 years after the beginning of the study. Can you apply these results to all people in the United Kingdom? Can you conclude that drinking black tea causes people to live longer? Explain your answers.

SECTION 3A Summary

- A **census** collects data from every individual in the **population.**
- A **sample survey** selects a **sample** from the population of all individuals about which we desire information. The goal of a sample survey is to draw conclusions about the population based on data from the sample.
- A **random sample** consists of individuals from the population who are selected for the sample using a chance process. We can use the data collected from a random sample to make inferences about the population from which the sample was selected.
- An **observational study** observes individuals and measures variables of interest but does not attempt to influence the responses. Observational studies that examine existing data for a sample of individuals are called **retrospective.** Observational studies that track individuals into the future are called **prospective.**
- An **experiment** deliberately imposes treatments (conditions) on experimental units (individuals) to measure their responses.
- Cause-and-effect relationships are very difficult to establish from observational studies. However, well-designed experiments allow for inference about cause and effect.

AP® EXAM TIP
AP® Daily Videos

Review the content of this section and get extra help by watching the AP® Daily Videos for Topics 3.1 and 3.2, which are available in AP® Classroom.

SECTION 3A Exercises

Populations and Samples

1. **Sampling stuffed envelopes** A large retailer prepares its customers' monthly credit card bills using an automatic machine that folds the bills, stuffs them into envelopes, and seals the envelopes for mailing. Are the envelopes completely sealed? Inspectors choose 40 envelopes at random from the 1000 stuffed each hour for visual inspection. Identify the population and the sample.
pg 241

2. **Student archaeologists** An archaeological dig turns up large numbers of pottery shards, broken stone tools, and other artifacts. Students working on the project classify each artifact and assign a number to it. The counts in different categories are important for understanding the site, so the project director chooses 2% of the artifacts at random and checks the students' work. Identify the population and the sample.

3. **Students as customers** A high school's student newspaper plans to survey local businesses about the importance of students as customers. From an alphabetical list of all local businesses, the newspaper staff chooses 150 businesses at random. Of these, 73 return the questionnaire mailed by the staff. Identify the population and the sample.

4. **Customer satisfaction** A department store mails a customer satisfaction survey to people who make credit card purchases at the store. This month, 45,000 people made credit card purchases. Surveys are mailed to 1000 of these people, chosen at random, and 137 people return the survey. Identify the population and the sample.

Observational Studies and Experiments

5. pg 244 **Social media and mood** How does social media use affect the mood and well-being of college students? A total of 143 students at the University of Pennsylvania helped answer this question. At the beginning of the study, they all took a survey that measured their mood and well-being. Then, half were chosen at random to continue their normal social media practices. The other half were limited to 10 minutes per day for each of Facebook, Snapchat, and Instagram. Social media use was verified by screen shots of battery use. The group that had limited social media use had significant decreases in loneliness and depression compared to the other group.[8] Is this an observational study or an experiment? Explain your answer.

6. **Indulgent veggies** Can using indulgent names make college students more likely to eat their veggies? Each day, a cafeteria at Stanford University labeled the featured vegetable dish in one of four randomly selected ways: basic ("beets"), healthy restrictive ("lighter-choice beets with no added sugar"), healthy positive ("high-antioxidant beets"), and indulgent ("tangy lime-seasoned beets"). The number of students choosing the veggie was highest when it had the indulgent label, followed by basic, healthy positive, and healthy restrictive.[9] Is this an observational study or an experiment? Explain your answer.

7. **Child care and aggression** A study of child care enrolled 1364 infants and followed them through their sixth year in school. Later, the researchers published an article in which they stated that "the more time children spent in child care from birth to age 4½, the more adults tended to rate them, both at age 4½ and at kindergarten, as less likely to get along with others, as more assertive, as disobedient, and as aggressive."[10] Is this a prospective observational study, a retrospective observational study, or an experiment? Justify your answer.

8. **Chocolate and happy babies** A University of Helsinki (Finland) study wanted to determine if chocolate consumption during pregnancy had an effect on infant temperament at age 6 months. Researchers began by asking 305 healthy pregnant women to report their chocolate consumption. Six months after the women gave birth, the researchers asked the new mothers to rate their infants' temperament using the traits of smiling, laughter, and fear. The babies born to women who had been eating chocolate daily during pregnancy were found to be more active and "positively reactive" — a measure that the investigators said encompasses traits such as smiling and laughter.[11] Is this a prospective observational study, a retrospective observational study, or an experiment? Justify your answer.

Inference from Observational Studies

9. pg 245 **Naps** A long-term study analyzed data from 3462 randomly selected adults in Lausanne, Switzerland. In the sample, 10.7% reported that they typically take 6 or 7 naps per week.[12] Furthermore, a significantly greater proportion of adults who reported taking 6 or 7 naps weekly had cardiovascular events (e.g., heart attacks) than did adults who reported taking fewer naps.

(a) Can we generalize this result to all adults? Explain your answer.

(b) What is the largest population to which we can generalize this result?

(c) Should we conclude that taking 6 or 7 naps per week increases the risk of a cardiovascular event based on this study? Explain your reasoning.

10. **Social media and sleep** A study on teen social media use analyzed data from 11,872 randomly selected adolescents (aged 13–15 years) in the United Kingdom. In the sample, 20.8% reported using social media 5 or more hours per day, on average.[13] Furthermore, those with the highest reported social media use had significantly more difficulty falling asleep and staying asleep than the teens who reported less social media use.

(a) Can we generalize this result to all teens? Explain your answer.

(b) What is the largest population to which we can generalize this result?

(c) Should we conclude that increased use of social media decreases the ability to fall asleep and stay asleep based on this study? Explain your reasoning.

11. **More child care** Refer to Exercise 7.

(a) Does this study show that child care makes children more aggressive? Explain your reasoning.

(b) What is the largest population to which we can generalize the results of this study?

12. **More chocolate** Refer to Exercise 8.

(a) Does this study show that eating chocolate regularly during pregnancy helps produce infants with a pleasant temperament? Explain your reasoning.

(b) What is the largest population to which we can generalize the results of this study?

For Investigation *Apply the skills from the section in a new context or nonroutine way.*

13. **Sleep habits and colds** In a study involving 153 healthy adults, researchers asked participants to record their sleep duration and efficiency for a period of 14 days. Then, all of the participants were given a nasal spray that contained a rhinovirus and monitored for symptoms of a common cold. Participants who averaged less than 7 hours of sleep were nearly 3 times more likely to develop a cold than those who averaged 8 or more hours of sleep.[14] Even though researchers imposed the virus on each participant, explain why we cannot conclude that reduced sleep causes an increase in the risk of developing a cold, based on this study.

Multiple Choice *Select the best answer for each question.*

14. Some news organizations maintain a database of customers who have volunteered to share their opinions on a variety of issues. Suppose that one of these databases includes 9000 registered voters in California. To measure the amount of support for a controversial ballot issue, 1000 registered voters in California are randomly selected from the database and asked their opinion. Which of the following is the largest population to which the results of this survey should be generalized?

(A) The 1000 people in the sample

(B) The 9000 registered voters from California in the database

(C) All registered voters in California

(D) All California residents

(E) All registered voters in the United States

15. Do parents who "clean" their babies' pacifiers by sucking on them help protect their children from allergies later in life? A cohort of 184 infants was followed for 36 months, with allergy tests being conducted at 18 and 36 months. Pacifier use and parental cleaning methods were recorded at 6 months in interviews with parents. Pacifier-using children whose parents cleaned their pacifiers by sucking on them were significantly less likely to have eczema at both 18 and 36 months than were pacifier-using children whose parents didn't use this method.[15] Based on this study, should we conclude that having parents clean their infant's pacifiers in this way helps prevent eczema?

(A) Yes, because this was an experiment.

(B) Yes, because the results were the same at both 18 months and 36 months.

(C) No, because the infants weren't randomly selected.

(D) No, because this was an observational study.

(E) No, because there were only 184 infants in the study.

Recycle and Review *Practice what you learned in previous sections.*

16. **Veteran leadership (2B)** Do basketball teams with older players win more games? The scatterplot shows the relationship between the average age of starters and the number of wins for the 30 National Basketball Association (NBA) teams in a recent season.

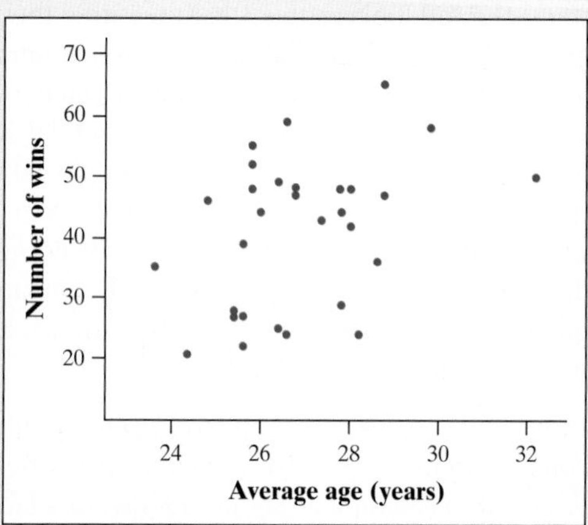

(a) Describe the association between average age and number of wins.

(b) Estimate the correlation for this association.

(c) Based on these data, would starting a 60-year-old player cause a team to win more games?

17. **Cold showers (2A)** A cold shower can help wake you up. Can it also keep you healthy? Researchers in the Netherlands randomly assigned 2426 participants to finish their shower with either 0, 30, 60, or 90 seconds of cold water for a month and recorded whether each participant reported a sickness in the next 90 days. The two-way table summarizes the results of this study.

		Cold shower time (sec)				
		0	30	60	90	Total
90-day outcome	Sick	379	437	387	384	1587
	Not sick	168	236	224	211	839
	Total	547	673	611	595	2426

(a) Construct a graph to display the relationship between cold shower time and 90-day outcome.

(b) Based on your graph in part (a), is there an association between cold shower time and 90-day outcome? Explain your answer.

SECTION 3B Sampling and Surveys

LEARNING TARGETS *By the end of the section, you should be able to:*

- Describe how to select a simple random sample.
- Describe other random sampling methods: stratified, cluster, systematic; explain the advantages and disadvantages of each method.
- Identify voluntary response sampling and convenience sampling, and explain how these sampling methods can lead to bias.
- Explain how undercoverage, nonresponse, question wording, and other aspects of a sample survey can lead to bias.

In Section 3A, you learned that random sampling is essential to make valid inferences about a population. We start this section by describing four different random sampling methods, along with the pros and cons of each method. To understand why random sampling is so important, we'll then consider some commonly used, nonrandom sampling methods. Finally, we'll discuss some additional problems that can arise when sampling.

Simple Random Samples

At the beginning of each class session, Mrs. Alvarez randomly selects 5 students to present homework problems on the whiteboard. To make the selection, she uses 30 equal-sized slips of paper and writes a different student's name on each slip. Then she puts the slips in a hat, mixes the slips, and selects 5 different slips. The students whose names are on the selected slips are the ones chosen to present the homework problems. This type of sample is called a **simple random sample,** or **SRS** for short.

DEFINITION Simple random sample (SRS)

A **simple random sample (SRS)** of size *n* is a sample chosen in such a way that every group of *n* individuals in the population has an equal chance to be selected as the sample.

Simple random sampling gives *every possible sample* of the desired size an equal chance to be chosen. In Mrs. Alvarez's random sampling method, any set of 5 students in the class has the same chance as any other set of 5 students to be selected for the sample. This also means that each individual student in the class has the same chance to be chosen as any other individual. However, giving each individual the same chance to be selected is *not* enough to guarantee that a sample is an SRS. Some other random sampling methods give each member of the population an equal chance to be selected, but not each possible sample. We'll look at some of these methods later.

Using slips of paper won't work well if the population is large. Imagine trying to select an SRS of 1000 registered voters in the United States using a slip of paper for each member of the population! In practice, most people use random numbers generated by technology to choose samples.

HOW TO CHOOSE AN SRS WITH TECHNOLOGY

1. **Label.** Give each individual in the population a distinct numerical label from 1 to N, where N is the number of individuals in the population.
2. **Randomize.** Use a random number generator to obtain *n different* integers from 1 to N, where *n* is the number of individuals in the sample.
3. **Select.** Choose the individuals that correspond to the randomly selected integers.

When choosing an SRS, we **sample without replacement.** That is, once an individual is selected for a sample, that individual cannot be selected again. Many random number generators **sample with replacement,** so it is important to explain that repeated numbers should be ignored when using technology to select an SRS.

DEFINITION Sampling without replacement, Sampling with replacement

When **sampling without replacement,** an individual from a population can be selected only once.

When **sampling with replacement,** an individual from a population can be selected more than once.

You can use a graphing calculator to generate random numbers, as shown in the following Tech Corner. There are also many random number generators available on the internet, including those at www.random.org.

10. Tech Corner CHOOSING AN SRS

TI-Nspire and other technology instructions are on the book's website at bfwpub.com/tps7e.

Let's use a graphing calculator to select an SRS of 5 students from a population of students numbered 1 to 30.

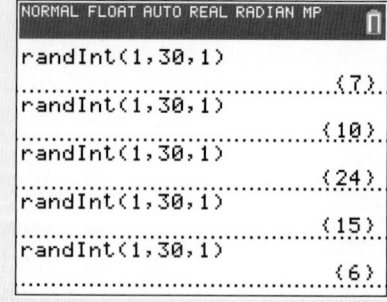

```
NORMAL FLOAT AUTO REAL RADIAN MP
randInt(1,30,1)
                    {7}
randInt(1,30,1)
                    {10}
randInt(1,30,1)
                    {24}
randInt(1,30,1)
                    {15}
randInt(1,30,1)
                    {6}
```

1. Press MATH, then select PROB (PRB) and choose randInt(.
   - **Newer OS:** In the dialog box, enter these values: lower: 1, upper: 30, n: 1, choose Paste, and press ENTER.
   - **Older OS:** Complete the command randInt(1,30) and press ENTER.
2. Keep pressing ENTER until you have chosen 5 *different* labels from 1 to 30.

Notes:

- If you have a TI-84 Plus CE, use the command RandIntNoRep(1,30,5) to get 5 distinct integers from 1 to 30.
- If you have a TI-84 with OS 2.55 or later, use the command RandIntNoRep(1,30) to sort the numbers from 1 to 30 in random order. The first 5 numbers listed give the labels of the chosen students.

If you don't have technology handy, you can use a table of random digits to choose an SRS. We have provided a table of random digits at the back of the book (Table D). Here is an excerpt:

Table D Random digits

LINE								
101	19223	95034	05756	28713	96409	12531	42544	82853
102	73676	47150	99400	01927	27754	42648	82425	36290
103	45467	71709	77558	00095	32863	29485	82226	90056

You can think of this table as the result of someone putting slips of paper numbered 0–9 in a hat, mixing, drawing one, replacing it, mixing again, drawing another, and so on. The digits have been arranged in groups of five within numbered rows to make the table easier to read. The groups and rows have no special meaning — Table D is just a long list of randomly chosen digits. As with technology, there are three steps in using Table D to choose a simple random sample.

HOW TO CHOOSE AN SRS USING TABLE D

1. **Label.** Give each member of the population a distinct numerical label with the *same number of digits*. Use as few digits as possible.
2. **Randomize.** Read consecutive groups of digits of the appropriate length from left to right across a line in Table D. Ignore spaces and any group of digits that wasn't used as a label or that duplicates a label already in the sample. Stop when you have chosen n different labels.
3. **Select.** Choose the individuals that correspond to the randomly selected labels.

Always use the shortest labels that will cover your population. For instance, you can label up to 100 individuals with two digits: 01, 02, ..., 99, 00. As standard practice, we recommend that you begin with label 1 (or 01 or 001 or 0001, as needed). Reading groups of digits from the table gives all individuals the same chance to be chosen because all labels of the same length have the same chance to be found in the table. For example, any pair of digits in the table is equally likely to be any of the 100 possible labels 01, 02, ..., 99, 00. Here's an example that shows how this process works.

EXAMPLE

Attendance audit
Simple random samples

Skill 1.C

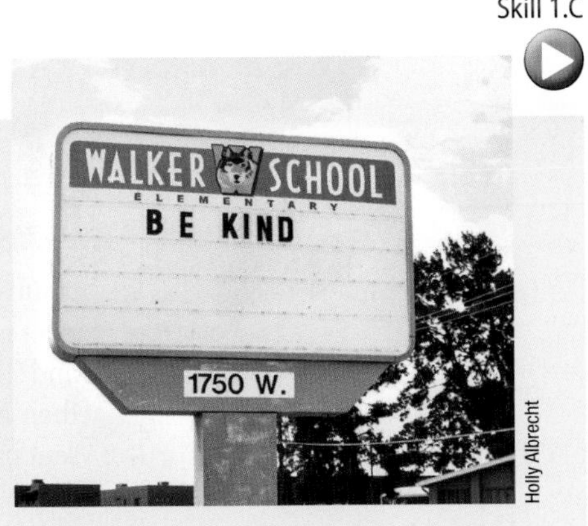

PROBLEM: Each year, the state Department of Education randomly selects three schools from each district and conducts a detailed audit of their attendance records.

(a) Describe how to use a table of random digits to select a simple random sample (SRS) of three schools from this list of 19 schools.

Amphitheater High School	Cross Middle School	Keeling Elementary School	Prince Elementary School
Amphitheater Middle School	Donaldson Elementary School	La Cima Middle School	Rio Vista Elementary School
Canyon del Oro High School	Harelson Elementary School	Mesa Verde Elementary School	Walker Elementary School
Copper Creek Elementary School	Holaway Elementary School	Nash Elementary School	Wilson K–8 School
Coronado K–8 School	Ironwood Ridge High School	Painted Sky Elementary School	

(b) Use the random digits here to choose the sample.

62081 64816 87374 09517 84534 06489 87201 97245

SOLUTION:

(a) Label the schools from 01 to 19 in alphabetical order. Move along a line of random digits from left to right, reading two-digit numbers, until three different numbers from 01 to 19 have been selected (ignoring repeated numbers and the numbers 20–99, 00). Audit the three schools that correspond with the numbers selected.

> Remember to include all three steps:
> 1. **Label.** *Remember to use the same number of digits for each label.*
> 2. **Randomize.** *Remember to address the possibility of repeated numbers.*
> 3. **Select.**

(b) 62—skip, 08—select, 16—select, 48—skip, 16—repeat, 87—skip, 37—skip, 40—skip, 95—skip, 17—select.

The three schools are 08: Harelson Elementary School, 16: Prince Elementary School, and 17: Rio Vista Elementary School.

FOR PRACTICE, TRY EXERCISE 1

When using a table of random digits, we often have to skip over unused labels. Because random number generators allow you to specify which labels they choose from, using technology is usually more efficient than using a table of random digits.

AP® EXAM TIP

Make sure to use statistical vocabulary correctly. In everyday life, some people use the word random to mean "haphazard," as in "That's so random." In statistics, random means "using chance." Don't say that a sample was chosen at random if a chance process wasn't used to select the individuals. Likewise, the terms "skewed" and "normal" are used in everyday life in ways that don't match their statistical meaning.

CHECK YOUR UNDERSTANDING

A furniture maker buys hardwood in batches that each contain 1000 pieces. The supplier is supposed to dry the wood before shipping (wood that isn't dry won't hold its size and shape). The furniture maker chooses five pieces of wood from each batch and tests their moisture content. If any piece exceeds 12% moisture content, the entire batch is sent back. Describe how to select a simple random sample of 5 pieces of wood.

Other Random Sampling Methods

One of the most common alternatives to a simple random sample is called a **stratified random sample.** Stratified random sampling involves dividing the population into non-overlapping groups (**strata**) of individuals that are expected to have similar responses, sampling from each of these groups, and combining these "subsamples" to form the overall sample.

DEFINITION Strata, Stratified random sample

Strata are groups of individuals in a population that share characteristics thought to be associated with the variables being measured in a study. The singular form of *strata* is *stratum.*

A **stratified random sample** is a sample selected by choosing an SRS from each stratum and combining the SRSs into one overall sample.

Stratified random sampling works best when the individuals within each stratum are similar (homogeneous) with respect to what is being measured and when there are large differences between strata. For example, in a study about the sleep habits of students at a high school, the population might be divided into four strata by grade level: 9th, 10th, 11th, and 12th grade. After all, it is reasonable to think that 9th-graders have different sleep habits than 12th-graders do.

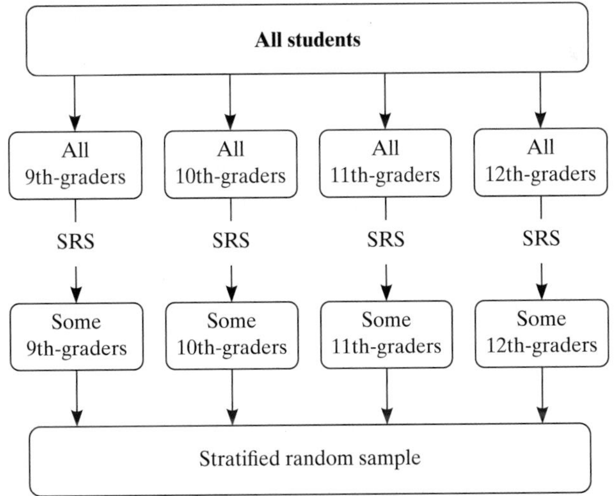

How many individuals should be selected from each stratum? It depends on several factors. The most straightforward method is to keep the sample size within each stratum roughly proportional to the size of the strata in the population. For example, if 20% of the students at the high school are 12th-graders and we want a stratified random sample of 250 students, we would randomly select an SRS of $(0.20)(250) = 50$ 12th-graders.

The following activity illustrates the benefit of choosing appropriate strata.

ACTIVITY Sampling sunflowers

Li Ding/Alamy Stock Photo

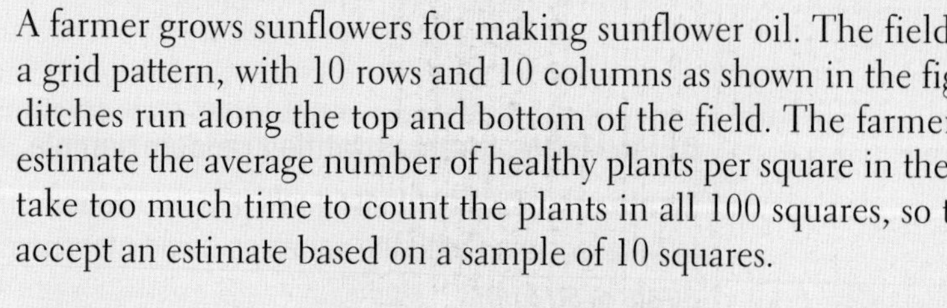

A farmer grows sunflowers for making sunflower oil. The field is arranged in a grid pattern, with 10 rows and 10 columns as shown in the figure. Irrigation ditches run along the top and bottom of the field. The farmer would like to estimate the average number of healthy plants per square in the field. It would take too much time to count the plants in all 100 squares, so the farmer will accept an estimate based on a sample of 10 squares.

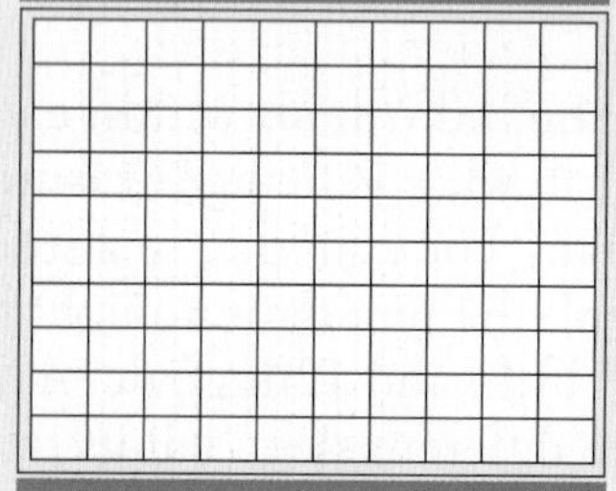

1. Go to www.stapplet.com and launch the *Sampling Sunflowers* applet. Then click the button to "Enter an existing class code" and input the code provided by your teacher.
2. Under Method 1, click the button to generate a sample. This will produce a simple random sample of 10 plots from the field. The plots chosen are marked with an X.
3. Under Method 2, click the button to generate a sample. This will produce a stratified random sample of 10 plots from the field, with one plot chosen at random from each *row*.
4. Under Method 3, click the button to generate a sample. This will produce a stratified random sample of 10 plots from the field, with one plot chosen at random from each *column*.
5. Which of these three methods do you think will work the best? That is, which method is most likely to give an estimate close to the true mean number of healthy plants in the field? Discuss this with your classmates before moving on.
6. Click the "Reveal results" button below the "If we did a census . . ." heading. This will reveal the number of healthy plants in each of the squares and the true mean number of healthy plants. It will also calculate the sample means for each of your three samples and plot each value on a separate dotplot. Which of your three samples gave a value closest to the true mean?
7. As your classmates complete the activity, their dots will be added to the dotplots. How do the dotplots compare? Which method seems best? Does this match your answer from Step 5?

The following dotplots show the mean number of healthy plants in 100 samples using each of the three sampling methods in the activity: simple random sampling, stratified random sampling with rows of the field as strata, and stratified random sampling with columns of the field as strata.

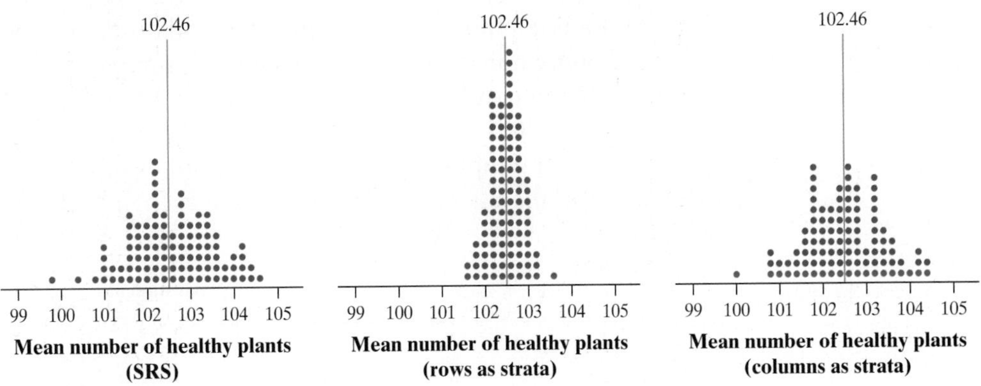

All three dotplots show *sampling variability* — the fact that different samples of the same size from the same population will produce different estimates. And all three distributions are centered at about 102.46, the true mean number of healthy plants in all squares of the field. That makes sense because all three methods use random sampling. Even with sampling variability, none of the methods *consistently* overestimated the true mean or *consistently* underestimated the true mean.

One key difference stands out in the graphs: there is much less variability in the estimates when we use the rows as strata — even though the sample sizes were the same with all three methods. *When strata are chosen wisely, the result is an estimate that is more precise than in a simple random sample of the same size.* In this case, stratifying by row worked best because the squares in a particular row are all the same distance from the irrigation ditches — and distance from the water source is likely to be associated with the health of the plants. With the other two methods, it is possible to get a majority of plots near the ditches (and a large mean) or to get a majority of plots far from the ditches (and a small mean).

CLUSTER SAMPLING

Both simple random sampling and stratified random sampling are hard to use when populations are large and spread out over a wide area. In that situation, we might prefer to use a **cluster sample.** Cluster sampling involves dividing the population into non-overlapping groups (**clusters**) of individuals that are "near" one another, then randomly selecting whole clusters to form the overall sample. Some people call this method *cluster random sampling*.

DEFINITION Cluster, Cluster sample

A **cluster** is a group of individuals in the population that are located near each other.

A **cluster sample** is a sample selected by randomly choosing clusters and including each member of the selected clusters in the sample.

Cluster samples are often used for practical reasons, such as saving time and money. Although the primary criterion for forming clusters is location, it is better when the individuals within each cluster are heterogeneous — mirroring the population — and when the clusters are similar to each other in their composition. Imagine a large high school that assigns students to homerooms alphabetically by last name, in groups of 25. Administrators want to survey 100 randomly selected students about a proposed schedule change. It would be difficult to track down an SRS of 100 students, so the administration opts for a cluster sample of homerooms. The principal selects an SRS of 4 homerooms and gives the survey to all 25 students in each of the selected homerooms.

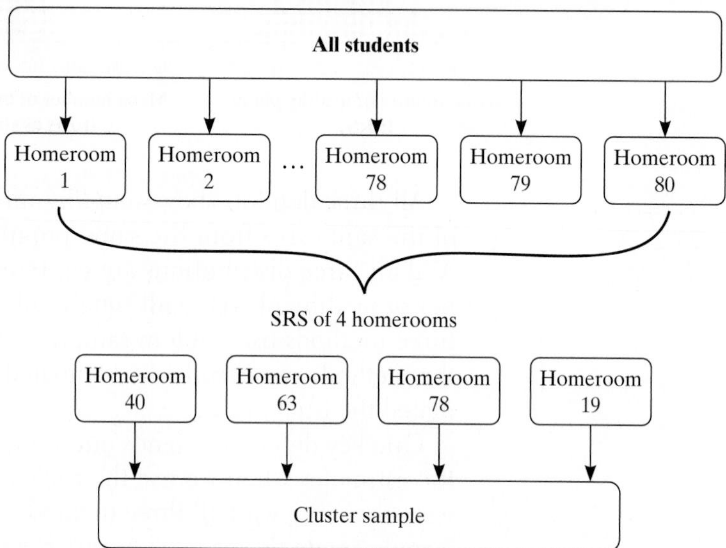

Be sure you understand the difference between strata and clusters. We want each stratum to contain similar individuals and for large differences to exist between strata. For a cluster sample, we'd *like* each cluster to look just like the population, but on a smaller scale. Unfortunately, even though the clusters are selected at random, cluster samples don't offer the statistical advantage of better information about the population that stratified random samples do.

SYSTEMATIC RANDOM SAMPLING

Many news organizations conduct exit polls of voters on election day. That is, they survey voters as they leave the polling place, asking questions about how they voted and why they voted that way, along with other demographic questions. To avoid favoring certain types of voters, pollsters should select the sample at random. But it is impossible to select a traditional SRS in this context, as this would require knowing which voters will show up, numbering them all, and then being able to identify the randomly selected voters as they leave the polling place. As an alternative, news organizations often use a **systematic random sample.**

DEFINITION Systematic random sample

A **systematic random sample** is a sample selected from an ordered arrangement of the population by randomly selecting one of the first *k* individuals and choosing every *k*th individual thereafter.

For example, at each polling place the interviewer might be instructed to interview every 20th voter as they leave. To choose a starting point, the interviewer would generate a random number from 1 to 20. If the random number was 6, the 6th voter would be interviewed, followed by the 26th voter, the 46th voter, and so on.

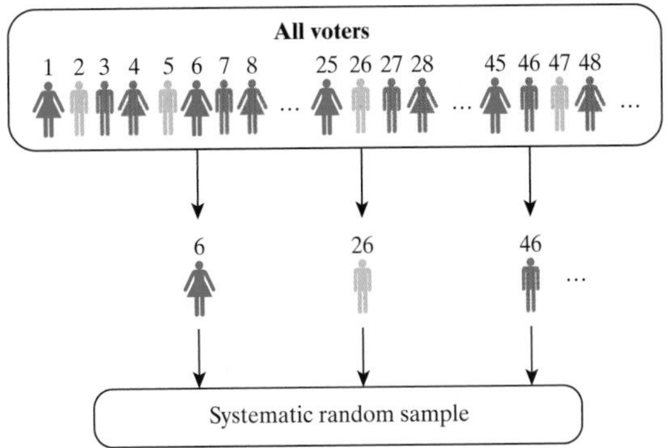

Here's an example that compares stratified random sampling, cluster sampling, and systematic random sampling.

EXAMPLE

Sampling at a school assembly
Other random sampling methods

Skill 1.C

PROBLEM: The student council wants to conduct a survey about use of the school library during the first 5 minutes of an all-school assembly in the auditorium. There are 800 students present at the assembly. Here is a map of the auditorium. Note that students are seated by grade level and that the seats are numbered from 1 to 800.

Marmaduke St. John/Alamy Stock Photo

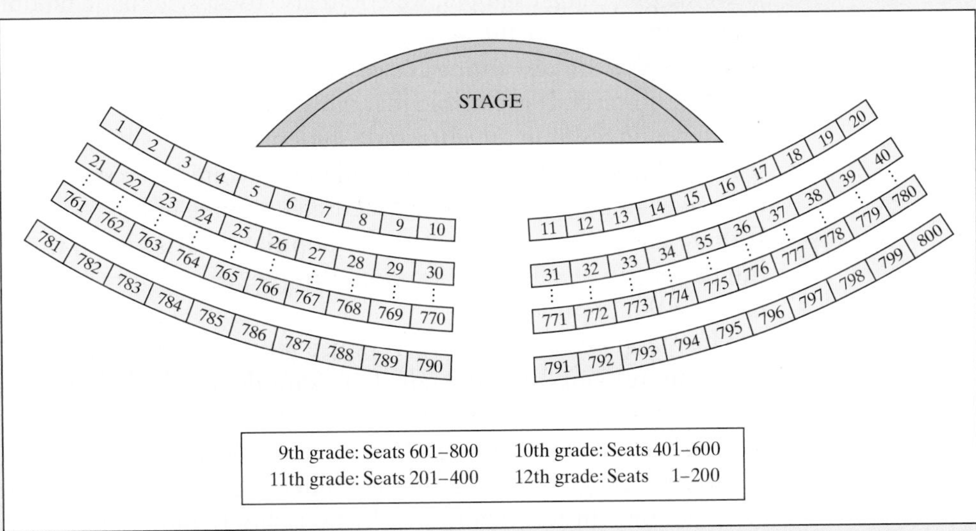

(a) Describe how to obtain a random sample of 80 students using stratified sampling. Explain your choice of strata and why this method might be preferred to simple random sampling.

(b) Describe how to obtain a random sample of 80 students using cluster sampling. Explain your choice of clusters and why this method might be preferred to simple random sampling.
(c) Describe how to obtain a random sample of 80 students using systematic sampling. Explain why this method might be preferred to simple random sampling.

SOLUTION:

(a) Because students' library use might be similar within grade levels but different across grade levels, use the grade-level seating areas as strata. For the ninth grade, use a random number generator to generate 20 different random integers from 601 to 800 and give the survey to the students in those seats. Do the same for 10th-, 11th-, and 12th-graders using their corresponding seat numbers. Stratification by grade level should result in more precise estimates of student opinion than an SRS of the same size.

(b) Each column of seats from the stage to the back of the auditorium could be used as a cluster because it would be relatively easy to hand out the surveys to an entire column. Label the columns from 1 to 20 starting at the left side of the stage, use a random number generator to generate two different integers from 1 to 20, and give the survey to the 80 students sitting in these two columns. Cluster sampling is much more efficient than finding 80 seats scattered around the auditorium, as required by simple random sampling.

Note that each cluster contains students from all four grade levels, so each should represent the population fairly well. Randomly selecting four rows as clusters would also be easy, but may over- or under-represent one grade level.

(c) Because we want a sample of 80 from a population of 800, we should select every $800/80 = 10$th student as they enter the auditorium. Use a random number generator to generate a random integer from 1 to 10 to determine which of the first 10 students to survey and then survey every 10th student after that. This would be much easier than simple random sampling, because you could survey students as they enter instead of having to find 80 seats scattered around the auditorium.

FOR PRACTICE, TRY EXERCISE 13

In the preceding example, we could also use a systematic random sample with every 10th seat inside the auditorium. That is, we could survey the student sitting in seat 8, the student sitting in seat 18, and so on. There is a drawback to this method, however. It is possible that our starting number could have been 1 or 10, which would result in selecting only students who are sitting in aisle seats. If these seats are filled primarily by tardy students who are less responsible, the results of the survey might not provide an accurate estimate of library use for all students.

If there are patterns in the way the population is ordered that coincide with the pattern in a systematic random sample, the sample may not be representative of the population.

Most large-scale sample surveys use *multistage sampling*, which combines two or more sampling methods. For example, the U.S. Census Bureau carries out a monthly Current Population Survey (CPS) of about 60,000 households. Researchers start by choosing a stratified random sample of neighborhoods in 756 of the 2007 geographic areas in the United States. Then they divide each neighborhood into clusters of four nearby households and randomly select clusters of households to interview.

Analyzing data from sampling methods other than simple random sampling takes us beyond introductory statistics. But the SRS is the building block of more elaborate methods, and the principles of analysis remain much the same for these other methods.

CHECK YOUR UNDERSTANDING

A factory runs 24 hours a day, producing 15,000 wood pencils per day over three 8-hour shifts — day, evening, and overnight. In the last stage of manufacturing, the pencils are packaged in boxes of 10 pencils each. Each day a sample of 300 pencils is selected and inspected for quality.

1. Describe how to select a stratified random sample of 300 pencils. Explain your choice of strata.
2. Describe how to select a cluster sample of 300 pencils. Explain your choice of clusters.
3. Describe how to select a systematic random sample of 300 pencils.
4. Explain a benefit of each of these three methods in this context.

Sampling Poorly: Convenience and Voluntary Response Samples

Suppose we want to know how long students at a large high school spent doing homework last week. We might go to the school library and ask the first 30 students we see about their homework time. The sample we get is called a **convenience sample.**

DEFINITION Convenience sample

A **convenience sample** consists of individuals from the population who are easy to reach.

Convenience sampling tends to produce unreliable conclusions because the members of the sample often differ from the population in ways that affect their responses. For example, students who are in the library probably spend more time doing homework than a typical student does. If we use this group as our sample, our estimate for the average amount of time spent doing homework will be too high. In fact, if we were to repeat this sampling process again and again, we would almost always overestimate the average time spent doing homework in the population of all students. This predictable overestimation is due to **bias,** which is created by using a method that systematically favors certain outcomes over others.

DEFINITION Bias

The design of a statistical study shows **bias** if it is very likely to underestimate or very likely to overestimate the value you want to know.

AP® EXAM TIP

If you're asked to describe how the design of a sample survey leads to bias, you're expected to do two things: (1) describe how the members of the sample might respond differently than the rest of the population, and (2) explain how this difference would lead to an underestimate or overestimate of the value you want to know. Suppose you were asked to explain how bias could result from using your statistics class as a sample to estimate the proportion of all students at your school who own a graphing calculator. You might respond, "This is a convenience sample. Because a graphing calculator is required for the statistics class, the sample would probably include a much higher proportion of students with a graphing calculator than in the school population. This method would probably lead to an overestimate of the actual population proportion."

Bias is not just bad luck in one sample. It's the result of a bad study design that will consistently miss the truth about the population in the same way. Convenience samples will almost always result in bias. So will **voluntary response samples,** such as those used for internet polls or call-in shows like *American Idol.*

DEFINITION Voluntary response sample

A **voluntary response sample** consists of people who choose to be in the sample by responding to a general invitation. Voluntary response samples are sometimes called *self-selected* samples.

Voluntary response sampling leads to *voluntary response bias* when the members of the sample differ from the population in ways that affect their responses. An advice columnist once asked her followers, "If you had it to do over again, would you have children?" She received nearly 10,000 responses from parents, with almost 70% saying "NO!"[16] Can it be true that 70% of parents regret having children? Not at all. People who feel strongly about an issue, particularly people with strong negative feelings, are more likely to take the trouble to respond. These results are misleading — the percentage of parents who said "No" in the columnist's sample is much higher than the proportion who would say "No" in the population of all parents. In fact, when *Newsday* asked the same question to a random sample of parents, only 9% said they wouldn't have kids again.[17]

EXAMPLE Boaty McBoatface
Convenience and voluntary response samples

Skill 1.C

NERC/Splash News/Newscom

PROBLEM: In 2016, Britain's Natural Environment Research Council invited the public to name its $300 million polar research ship. To vote on the name, people simply needed to visit a website and record their choice. Ignoring names suggested by the council, more than 124,000 people voted for "Boaty McBoatface," which ended up having more than 3 times as many votes as the second-place finisher.[18] What type of sample did the council use in their poll? Explain how bias in this sampling method could have affected the poll results.

SOLUTION:

The council used a voluntary response sample. People chose to go online and respond. The people who chose to be in the sample were probably less serious about science than the British population as a whole—and more likely to prefer a funny name. The proportion of people in the sample who prefer the name Boaty McBoatface is likely to be greater than the proportion of all British residents who would choose this name.

> Remember to describe how the responses from the members of the sample might differ from the responses from the rest of the population *and* how this difference will affect the estimate.

FOR PRACTICE, TRY EXERCISE 17

Other Sources of Bias in Surveys

As you have learned, the use of bad sampling methods (convenience or voluntary response) often leads to bias. Random sampling can help to prevent the problems created by these methods. Other problems in conducting sample surveys are more difficult to avoid.

Some samples are selected by randomly calling telephone numbers. Unfortunately, this means people who don't have a phone cannot be included in the sample. And other people have multiple phone numbers, which makes people with only one phone number less likely to be chosen than people with multiple numbers. Both of these problems result in **undercoverage.**

DEFINITION Undercoverage

Undercoverage occurs when some members of the population are less likely to be chosen or cannot be chosen in a sample.

Undercoverage bias occurs when the underrepresented individuals differ from the population in ways that affect their responses. Suppose that a dentist surveys a random sample of 100 patients to estimate the proportion of all U.S. adults who regularly floss their teeth. Because this sample includes only dental patients, people who don't go to the dentist are undercovered. And if the people who don't go to the dentist floss less often than people who do go to the dentist, the dentist's estimate will be too high.

To avoid the bias caused by undercoverage, every member of the population should be given the same chance to be selected. Ideally, samples would be chosen from a list that includes every member of the population, which is sometimes called a *sampling frame*. If the list doesn't contain every member of the population, undercoverage will occur.

Even if every member of the population is equally likely to be selected for a sample, not all members of the population are equally likely to provide a response. Some people are rarely at home and cannot be reached by pollsters on the phone or in person. Other people see an unfamiliar phone number on their caller ID and never pick up the phone or quickly hang up when they don't recognize the voice of the caller. These are examples of **nonresponse,** another major source of bias in surveys.

DEFINITION Nonresponse

Nonresponse occurs when an individual chosen for the sample can't be contacted or refuses to participate.

Nonresponse bias occurs when the individuals who can't be contacted or who refuse to participate differ from the population in ways that affect their responses. Consider a study that uses a mail-in survey to estimate the average number of hours that people work in a typical week. Because people who work very long hours are less likely to respond than people who work fewer hours, the estimated mean from the sample is likely to be less than the mean for the entire population.

How bad is nonresponse? According to the Pew Research Group, the response rate for phone surveys declined from 36% in 1997 to only 6% in 2018.[19] In contrast, the Census Bureau's American Community Survey (ACS) has a much higher response rate of approximately 85%.[20] Of course, it helps that responding to the ACS is required by law! Other polling groups resort to small payments and repeated requests to decrease nonresponse rates. To help account for nonresponse, polling groups use weighting based on population characteristics. For example, if 40% of residents in a state are Republicans and only 20% in the sample are Republicans, they count each Republican in the sample as 2 people.[21] But weighting works as intended only if the people who do respond would answer the same way as those who don't respond, which may or may not be the case.

Some students misuse the term *voluntary response* to explain why certain individuals don't respond in a sample survey. Their belief is that participation in the survey is optional (voluntary), so anyone can refuse to take part. What the students are actually describing is *nonresponse*. Think about it this way: nonresponse can occur only after a sample has been selected. In a voluntary response sample, every individual has opted to take part, so there won't be any nonresponse.

The wording of questions may strongly influence the answers given to a sample survey. Confusing or leading questions can introduce *question wording bias*. The Roper polling organization once asked a random sample of U.S. adults: "Does it seem possible or does it seem impossible to you that the Nazi extermination of the Jews never happened?" Only 65% said it was impossible that the Holocaust never happened, making it seem as if about one-third of U.S. adults doubt the Holocaust's reality. But the wording of the question included a double negative, making it difficult to understand. When the question was revised to remove the double negative, the percentage who were certain that the Holocaust occurred increased dramatically.[22]

The characteristics and behavior of the interviewer can also affect people's responses, especially if the survey is not anonymous. People may lie about their age, income, or drug use. They may misremember how many hours they spent on the internet last week. Or they might make up an answer to a question that they don't understand. All these issues can lead to **response bias.**

DEFINITION Response bias

Response bias occurs when there is a consistent pattern of inaccurate responses to a survey question.

Pilot testing a survey with a small sample is a good way to detect problems with question wording while there is still time to make revisions. And allowing people to respond anonymously can increase the chances of getting an honest response.

EXAMPLE

Wash your hands!
Other sources of bias in surveys

Skill 1.C

STAFF MUST WASH
HANDS BEFORE
RETURNING TO WORK.
YOU SHOULD TOO,
BUT WE'RE NOT
YOUR MOTHER.
Corrina Santos

PROBLEM: What percentage of Americans wash their hands after using the bathroom? The answer to this question depends on how you collect the data. In a telephone survey of 1006 U.S. adults, 96% said they always wash their hands after using a public restroom. An observational study of 6028 adults in public restrooms told a different story: only 85% of those observed washed their hands after using the restroom. Explain why the results of the two studies are so different.[23]

SOLUTION:

When asked by a live person, many people may lie about always washing their hands because they want to appear as if they have good hygiene. When people are only observed and not asked directly, the percentage who wash their hands will be smaller — and much closer to the truth.

The 85% figure was obtained by stationing an "observer" in the bathroom who was brushing teeth or putting on make-up. Imagine how much lower the percentage of hand washers might be if no one else was present in the bathroom to encourage good behavior!

FOR PRACTICE, TRY EXERCISE 27

Even the order in which questions are asked is important. For example, suppose we ask a sample of college students these two questions:

- "How happy are you with your life in general?" (Answer on a scale of 1 to 5.)
- "How many dates did you have last month?"

There is almost no association between responses to the two questions when asked in this order. Reverse the order of the questions, however, and a much stronger association appears: college students who say they had more dates tend to give higher ratings of happiness about life.

CHECK YOUR UNDERSTANDING

1. A farmer brings a juice company several crates of oranges each week. A company inspector looks at 10 oranges from the top of each crate before deciding whether to buy all the oranges. Identify the sampling method used and explain how bias in the sampling method could affect the estimate of the proportion of damaged oranges.
2. A survey paid for by makers of disposable diapers found that only 16% of the sample thought it would be fair to ban disposable diapers. Here is the actual question:

   It is estimated that disposable diapers account for less than 2% of the trash in today's landfills. In contrast, beverage containers, third-class mail, and yard wastes are estimated to account for about 21% of the trash in landfills. Given this, in your opinion, would it be fair to ban disposable diapers?[24]

   Do you think the estimate of 16% is less than, greater than, or about equal to the percentage of all people in the population who think it is fair to ban disposable diapers? Explain your reasoning.

SECTION 3B Summary

- A **simple random sample (SRS)** gives every possible sample of a given size the same chance to be chosen. Choose an SRS by labeling the members of the population and using slips of paper, a table of random digits, or technology to select the sample. Make sure to use **sampling without replacement** when selecting an SRS.
- To select a **stratified random sample,** divide the population into nonoverlapping groups of individuals (**strata**) that are similar in some way that might affect their responses. Then choose a separate SRS from each stratum and combine these SRSs to form the sample. When strata are "similar (homogeneous) within but different between," stratified random sampling tends to give more precise estimates of unknown population values than does simple random sampling.
- To select a **cluster sample,** divide the population into non-overlapping groups of individuals that are located near each other, called **clusters.** Randomly select some of these clusters. All the individuals in the chosen clusters are included in the sample. Ideally, clusters are "different (heterogeneous) within but similar between." Cluster sampling saves time and money by collecting data from entire groups of individuals that are close together.
- To select a **systematic random sample,** select a value of k based on the population size and desired sample size, randomly select a value from 1 to k to identify the first individual in the sample, and choose every kth individual thereafter. If there are patterns in the way the population is ordered that coincide with the value of k, the sample may not be representative of the population. Otherwise, systematic random sampling can be easier to conduct than other sampling methods.
- The design of a statistical study shows **bias** if it is very likely to underestimate or very likely to overestimate the value you want to know.
  - The members of a **convenience sample** are those individuals who are easiest to reach. The members of a **voluntary response sample** are those individuals who choose to join in response to an open invitation. Convenience sampling and voluntary response sampling often lead to bias because the members of the sample are not representative of the population.
  - **Undercoverage** occurs when some members of the population are less likely to be chosen or can't be chosen for the sample. Sampling methods that suffer from undercoverage can show bias if the individuals less likely to be included in the sample differ in relevant ways from the other members of the population.
  - **Nonresponse** occurs when an individual chosen for the sample can't be contacted or refuses to participate. Sampling methods that suffer from nonresponse can show bias if the individuals who don't respond differ in relevant ways from the other members of the population.
  - **Response bias** occurs when there is a consistent pattern of inaccurate responses to a survey question. This kind of bias can be caused by the wording of questions, characteristics of the interviewer, lack of anonymity, and other factors.

AP® EXAM TIP

AP® Daily Videos

Review the content of this section and get extra help by watching the AP® Daily Videos for Topics 3.3 and 3.4, which are available in AP® Classroom.

3B Tech Corner

TI-Nspire and other technology instructions are on the book's website at bfwpub.com/tps7e.

10. Choosing an SRS Page 250

SECTION 3B Exercises

Simple Random Samples

1. **Do you trust the internet?** You want to ask a sample of high school students this question: "How much do you trust information about health that you find on the internet — a great deal, somewhat, not much, or not at all?" You decide to try out this and other questions on a pilot group of 5 students chosen from a class of 40 students. pg 251

(a) Describe how to use a random digits table to choose a simple random sample (SRS) of 5 students from the following class list:

Anderson	Drasin	Kim	Rider
Arroyo	Eckstein	Molina	Rodriguez
Batista	Fernandez	Morgan	Samuels
Bell	Fullmer	Murphy	Shen
Burke	Gandhi	Nguyen	Tse
Cabrera	Garcia	Palmiero	Velasco
Calloway	Glaus	Percival	Wallace
Delluci	Helling	Prince	Washburn
Deng	Husain	Puri	Zabidi
De Ramos	Johnson	Richards	Zhao

(b) Use the random digits shown here to select the sample:

82739	57890	20807	47511	81676
55300	94383	14893	60940	72024

2. **Apartment living** You are planning a report on apartment living in a college town. You decide to select three apartment complexes at random at which to conduct in-depth interviews with residents.

(a) Describe how to use a random digits table to choose a simple random sample (SRS) of 3 complexes from the following list:

Ashley Oaks	Country View	Mayfair Village	Brandon Place	Fairington	Pheasant Run
Bay Pointe	Country Villa	Nobb Hill	Briarwood	Fairway Knolls	Richfield
Beau Jardin	Crestview	Pemberly Courts	Brownstone	Fowler	Sagamore Ridge
Bluffs	Del-Lynn	Peppermill	Burberry	Franklin Park	Salem Courthouse
			Cambridge	Georgetown	Village Manor
			Chauncey Village	Greenacres	Waterford Court
			Country Squire	Lahr House	Williamsburg

(b) Use the random digits shown here to select the sample:

38167	98532	62183	70632	23417
26185	41448	75532	73190	32533

3. **The landfill is full** When landfills run out of room, they are sometimes converted into parks. Does the presence of a landfill underneath affect the trees that grow in the park? One such park has 283 trees, and researchers want to closely inspect an SRS of 20 trees.

(a) Explain how you would use a random number generator to choose the SRS.

(b) Use your method from part (a) to choose the first 3 trees.

4. **Sampling gravestones** The local genealogical society in Coles County, Illinois, has compiled records on all 55,914 gravestones in cemeteries in the county for the years 1825 to 1985. Historians plan to use these records to learn about African Americans in Coles County's history. They first choose an SRS of 395 records to check their accuracy by visiting the actual gravestones.[25]

(a) Explain how you would use a random number generator to choose the SRS.

(b) Use your method from part (a) to choose the first 3 gravestones.

Other Random Sampling Methods

5. **No tipping** The owner of a large restaurant is considering a new "no tipping" policy and wants to survey a sample of employees using a stratified random sample.

The policy would add 20% to the price of food and beverages, and the additional revenue would be distributed equally among servers and kitchen staff.

(a) What would be a good variable to use when forming strata? Explain your answer.

(b) What would *not* be a good variable to use when forming strata? Explain your answer.

(c) Using the variable identified in part (a), describe how to select a stratified random sample of approximately 30 employees.

6. **Parking on campus** The director of student life at a university wants to use a stratified random sample to estimate the proportion of undergraduate students who regularly park a car on campus.

(a) What would be a good variable to use when forming strata? Explain your answer.

(b) What would *not* be a good variable to use when forming strata? Explain your answer.

(c) Using the variable identified in part (a), describe how to select a stratified random sample of approximately 100 students.

7. **SRS of employees?** A corporation employs 2000 assembly-line workers and 500 engineers. A stratified random sample of 200 assembly-line workers and 50 engineers gives every individual in the population the same chance to be chosen for the sample. Is this sample an SRS? Explain your answer.

8. **SRS of students?** At a party, there are 30 students older than age 21 and 20 students younger than age 21. You randomly select 3 of those older than 21 and then randomly select 2 of those younger than 21 to interview about their attitudes toward alcohol. You have given every student at the party the same chance to be interviewed. Is your sample an SRS? Explain your answer.

9. **High-speed internet** Laying fiber-optic cable is expensive. Cable companies want to make sure that if they extend their lines to less dense suburban or rural areas, there will be sufficient demand for their services so the work will be cost-effective. They decide to conduct a survey to determine the proportion of households in a rural subdivision that would buy the service. They select a simple random sample of 5 blocks in the subdivision and survey each family that lives on the selected blocks.

(a) What is the name for this kind of sampling method?

(b) Give a possible reason why the cable company chose this method.

10. **Orientation** Each year, new students at a small college are assigned to a 15-person orientation group. These groups participate in a variety of activities on campus before the rest of the students arrive. On the last day of orientation week, each group meets at a different place on campus for a picnic. To evaluate the orientation program, college administrators randomly select 6 orientation groups and survey each of the students in these 6 groups during the picnic.

(a) What is the name for this kind of sampling method?

(b) Give a possible reason why the administrators chose this method.

11. **Dead trees** In Rocky Mountain National Park, many mature pine trees along Highway 34 are dying due to infestation by pine beetles. Scientists would like to use a sample of size 200 to estimate the proportion of the approximately 5000 pine trees along the highway that have been infested.

(a) Explain why it wouldn't be practical for scientists to obtain an SRS in this setting.

(b) A possible alternative would be to use the first 200 pine trees along the highway as you enter the park. Why isn't this a good idea?

(c) Describe how to select a systematic random sample of 200 pine trees along Highway 34.

12. **iPhones** Suppose 1000 iPhones are produced at a factory today. Management would like to ensure that the phones' display screens meet the company's quality standards before shipping them to retail stores. Because it takes about 10 minutes to inspect an individual phone's display screen, managers decide to inspect a sample of 20 phones from the day's production.

(a) Explain why it would be difficult for managers to inspect an SRS of 20 iPhones that were produced today.

(b) An eager employee suggests that it would be easy to inspect the last 20 iPhones that were produced today. Why isn't this a good idea?

(c) Describe how to select a systematic random sample of 20 phones from the day's production.

13. pg 257 **How is your room?** A hotel has 30 floors with 40 rooms per floor. Half the rooms on each floor face the water and the remaining half face the golf course. There is an extra charge for the rooms with a water view. The hotel manager wants to select 120 rooms and survey the registered guest in each of the selected rooms about their overall satisfaction with the property.

(a) Describe how to obtain a random sample of 120 rooms using stratified sampling. Explain your choice of strata and why this method might be preferred to simple random sampling.

(b) Describe how to obtain a random sample of 120 rooms using cluster sampling. Explain your choice of clusters and why this method might be preferred to simple random sampling.

(c) Describe how to obtain a random sample of 120 rooms using systematic sampling. Explain why this method might be preferred to simple random sampling.

14. **Go Blue!** Michigan Stadium, also known as "The Big House," seats more than 100,000 fans for a football game. The University of Michigan Athletic Department wants to survey roughly 10% of fans in attendance about concessions that are sold during games. Tickets are most expensive for seats on the sidelines. The cheapest seats are in the end zones (where one of the authors sat as a student). A map of the stadium is shown.

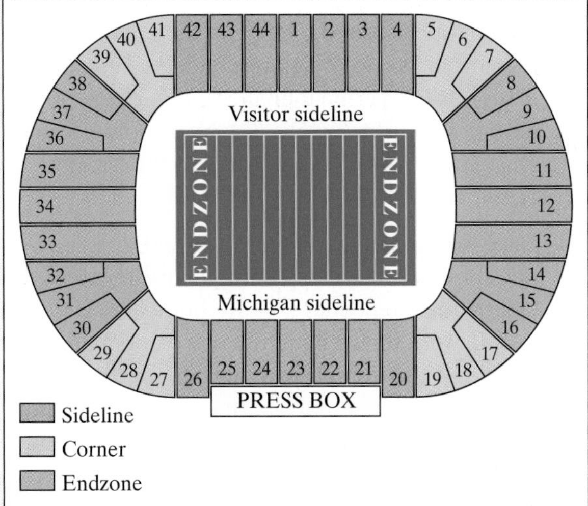

(a) Describe how to obtain a random sample of approximately 10,000 fans using stratified sampling. Explain your choice of strata and why this method might be preferred to simple random sampling.

(b) Describe how to obtain a random sample of approximately 10,000 fans using cluster sampling. Explain your choice of clusters and why this method might be preferred to simple random sampling.

(c) Describe how to obtain a random sample of approximately 10,000 fans using systematic sampling. Explain why this method might be preferred to simple random sampling.

15. **More sunflowers** Refer to the Sampling Sunflowers activity in this section.

(a) If you were to use a systematic random sample to select 10 squares, would it be better to start in the upper left and move across the rows or start in the upper left and move down the columns? Explain your reasoning.

(b) If you were to use a cluster sample to select 20 squares, would it be better to use rows as clusters or columns as clusters? Explain your reasoning.

16. **More Blue!** Refer to Exercise 14. Explain how you could select a random sample that incorporates both stratified sampling and cluster sampling. How is this combined method better than either method individually?

Convenience and Voluntary Response Samples

17. **Sleepless nights** How much sleep do high school students get on a typical school night? A counselor designed a survey to find out. To make data collection easier, the counselor surveyed the first 100 students to arrive at school on a particular morning. These students reported, on average, 7.2 hours of sleep on the previous night. What type of sample did the counselor obtain? Explain how bias in this sampling method could have affected the survey results.

pg 260

18. **Online polls** *Parade* magazine posed the following question: "Should drivers be banned from using all cell phones?" Readers were encouraged to vote online at parade.com. The subsequent issue of *Parade* reported the results: 2407 (85%) said "Yes" and 410 (15%) said "No." What type of sample did the *Parade* survey obtain? Explain how bias in this sampling method could have affected the survey results.

19. **Online reviews** Many websites include customer reviews of products, restaurants, hotels, and so on. The manager of a hotel was upset to see that 26% of reviewers on a travel website gave the hotel "one star" — the lowest possible rating. Explain how bias in the sampling method could affect the estimate.

20. **Funding for fine arts** The band director at a high school wants to estimate the percentage of parents who support a decrease in the budget for fine arts. Because many parents attend the school's annual musical production, the director surveys the first 30 parents who arrive at the show. Explain how bias in the sampling method could affect the estimate.

21. **Explain it to the senator** You are on the staff of a member of the U.S. Senate who is considering a bill that would provide government-sponsored insurance for nursing-home care. You report that 1128 comments have been received on the issue, of which 871 oppose the legislation. "I'm surprised that most of my constituents oppose the bill. I thought it would be quite popular," says the senator. Are you convinced that a majority of the constituents oppose the bill? How would you explain the statistical issue to the senator?

22. **Sleeping with dogs** Do most U.S. adult females sleep with a dog? In a survey of U.S. adult females, 55% of respondents said they sleep with a dog.[26] This was based on a study from researchers at Canisius College, whose report included the following description of their sampling method:[27]

Anyone over 18 years of age was eligible to participate in our online survey, but the results in this paper are restricted to female participants residing in the United States who were between 18 and 69 years old. We emailed personalized invitations to 1229 individuals who had previously participated in studies conducted by the Canisius Canine Research Team (CCRT) . . . Initial and reminder emails included a link to the survey that we asked recipients to share with their friends via email and social media sites.

Explain why the results of this study shouldn't be trusted.

Other Sources of Bias in Surveys

23. **Eating on campus** The facilities manager at a university wants to know what percentage of students eat regularly on campus. To find out, the manager selects an SRS of 300 students who live in the dorms.

(a) Describe how undercoverage might lead to bias in this study. Explain the likely direction of the bias.

(b) Would increasing the sample size address the bias described in part (a)? Explain your answer.

24. **Immigration reform** A news organization wants to know what percentage of U.S. residents support a "pathway to citizenship" for people who came to the United States illegally. The news organization randomly selects registered voters for the survey.

(a) Describe how undercoverage might lead to bias in this study. Explain the likely direction of the bias.

(b) Would increasing the sample size address the bias described in part (a)? Explain your answer.

25. **Driving distance** Researchers began a survey of drivers by randomly dialing telephone numbers in the United States. Of 45,956 calls to these numbers, 5029 were completed. The goal of the survey was to estimate how far people drive, on average, per day.[28] Describe how nonresponse might lead to bias in this study. Explain the likely direction of the bias.

26. **Reporting weight loss** A total of 300 people participated in a free 12-week weight-loss course at a community health clinic. After one year, administrators emailed each of the 300 participants to see how much weight they had lost since the end of the course. Only 56 participants responded to the survey. The mean weight loss for this sample was 13.6 pounds. Describe how nonresponse might lead to bias in this study. Explain the likely direction of the bias.

27. **Weight? Wait, what?** Marcos asked a random sample of 50 mall shoppers for their weight. Twenty-five of the shoppers were asked directly and the other 25 were asked anonymously by means of a "secret ballot." The mean reported weight was 13 pounds heavier for the anonymous group. Explain why the two means are different.[29]

pg 263

28. **Seat belt use** A study in El Paso, Texas, looked at seat belt use by drivers. Drivers were observed at randomly chosen convenience stores. After drivers left their cars, they were asked questions, including some about seat belt use. In all, 75% said they always used seat belts, yet only 61.5% were wearing seat belts when they pulled into the store parking lots.[30] Explain why the two percentages are different.

29. **Phone versus online** To investigate the effects of asking survey questions over the phone versus through an online poll, the Pew Research Center called a random sample of 1778 U.S. adults and recruited a different random sample of 2066 U.S. adults to fill out a survey online. Other than the mode of the survey, everything else was the same. When respondents were asked to describe their personal finances, 14% of the phone sample said "poor" while 20% of the online sample said "poor."[31] Give one reason why the two percentages are different.

30. **Care for homeless people** A researcher asked 80 randomly selected people if free health care should be provided to people who are homeless. Half of the people were shown a picture of a woman and small child who are homeless. When shown the picture, 67.5% agreed that free health care should be provided to people experiencing homelessness. When the picture was not shown, only 45% agreed with this statement.[32] Give one reason why the two percentages are different.

31. **National health insurance** Researchers select a random sample of 1200 U.S. adults and ask each person the following question:

In light of the huge national deficit, should the government at this time spend additional money to establish a national system of health insurance?

(a) Explain how the wording of this question could result in bias. Is the percentage from the sample who say yes likely to be greater than or less than the true percentage of U.S. adults who would say they favor establishing a national system of health insurance?

(b) Create two new questions about establishing a national system of health insurance. Write one that is unbiased and one that is biased in the opposite direction as the original question.

32. **Citizens United** Researchers select a random sample of 1200 U.S. adults and ask each person the following question:

Do you approve or disapprove of the U.S. Supreme Court decision in Citizens United that let corporations and wealthy individuals spend unlimited amounts on political campaigns?

(a) Explain how the wording of this question could result in bias. Is the percentage from the sample who say they approve likely to be greater than or less than the true percentage of U.S. adults who would say they approve of the *Citizens United* decision?

(b) Create two new questions about the *Citizens United* ruling. Write one that is unbiased and one that is biased in the opposite direction as the original question.

33. ***Literary Digest*** One of the most famous flops in survey history occurred in 1936. To predict the outcome of the U.S. presidential election between Republican Alf Landon and Democrat Franklin D. Roosevelt, the magazine *Literary Digest* sent more than 10 million "ballots" to its subscribers. It also sent "ballots" to registered owners of an automobile or telephone. Approximately 2.4 million of the ballots were returned, with a large majority (57%) favoring Landon. The election did turn out to be a landslide — but for Roosevelt (61%) instead of Landon.[33]

(a) Explain how undercoverage might have led to bias in this survey.

(b) Explain how nonresponse might have led to bias in this survey.

(c) If the magazine followed up with people who didn't return their ballots and was able to obtain responses, would this eliminate the bias described in part (a) or (b)? Explain your reasoning.

For Investigation *Apply the skills from the section in a new context or nonroutine way.*

34. **Randomized response** When asked sensitive questions, many people give untruthful responses — especially if the survey is not anonymous. To encourage honest answers, researchers developed the "randomized response" method. For example, suppose you want to ask students if they have cheated on an exam this year. Before they answer, have each student privately flip a coin. If the coin lands on heads, the student must answer truthfully. If it lands on tails, the student must answer "Yes."

(a) Explain why this method might encourage students to answer the question honestly.

(b) Suppose that 100 students used this method to answer the question about cheating on an exam this year. Estimate the proportion of students who have cheated if 63 students said "Yes" and 37 students said "No."

Multiple Choice *Select the best answer for each question.*

35. The website of a city newspaper places opinion poll questions next to many of its news stories. Simply click your response to join the sample. One of the questions was "Do you plan to diet this year?" More than 30,000 people responded, with 88% saying "Yes." Which of the following is true?

(A) About 88% of all residents of the city plan to diet this year.

(B) The poll used a convenience sample and the proportion of all city residents who plan to diet this year is likely to be less than 88%.

(C) The poll used a convenience sample and the proportion of all city residents who plan to diet this year is likely to be greater than 88%.

(D) The poll used a voluntary response sample and the proportion of all city residents who plan to diet this year is likely to be less than 88%.

(E) The poll used a voluntary response sample and the proportion of all city residents who plan to diet this year is likely to be greater than 88%.

36. To gather information about the validity of a new standardized test for 10th-grade students in a particular state, a random sample of 15 high schools was selected from the state. The new test was administered to every 10th-grade student in the selected high schools. What kind of sample is this?

(A) A simple random sample

(B) A stratified random sample

(C) A cluster random sample

(D) A systematic random sample

(E) A voluntary response sample

37. Suppose that 35% of the voters in a state are registered as Republicans, 40% as Democrats, and 25% as Independents. A newspaper wants to select a sample of 1000 registered voters to predict the outcome of the next election. If it randomly selects 350 Republicans, randomly selects 400 Democrats, and randomly selects 250 Independents, did this sampling procedure result in a simple random sample of registered voters from this state?

(A) Yes, because each registered voter had the same chance of being chosen.

(B) Yes, because random chance was involved.

(C) No, because not all registered voters had the same chance of being chosen.

(D) No, because a different number of registered voters was selected from each party.

(E) No, because not all possible groups of 1000 registered voters had the same chance of being chosen.

38. Professor Lee has a large collection of pennies that she uses for activities with her statistics students. To estimate the average age of the pennies in her collection, she asks Luke to select a random sample of 20 pennies. To avoid getting his hands dirty, Luke selects 20 of the shiniest pennies from the collection. Which of the following best describes the type of bias present and how the estimated average age compares to the actual average age of pennies in the collection?

(A) Undercoverage; the estimate will be too small

(B) Undercoverage; the estimate will be too large

(C) Nonresponse; the estimate will be too small

(D) Nonresponse; the estimate will be too large

(E) Response bias; the estimate will be too large

Recycle and Review *Practice what you learned in previous sections.*

39. **Coffee and longevity (3A)** A long-term study of more than 400,000 older people in eight states recorded whether participants drank coffee, whether they were alive at the end of the study, and their age when they died (if applicable). The researchers found that coffee drinkers tend to live longer than non-coffee drinkers.[34]

(a) Is this an observational study or an experiment? Explain your answer.

(b) What is the largest population to which we can generalize the results of this study?

(c) Should we conclude that drinking coffee causes people to live longer based on this study? Explain your answer.

40. **Internet charges (1E)** Some internet service providers (ISPs) charge companies based on how much bandwidth they use in a month. One method that ISPs use to calculate bandwidth is to find the 95th percentile of a company's usage based on samples of hundreds of 5-minute intervals during a month.

(a) Explain what "95th percentile" means in this setting.

(b) Explain what additional information would be necessary to determine the z-score that corresponds to the 95th percentile.

SECTION 3C Experiments

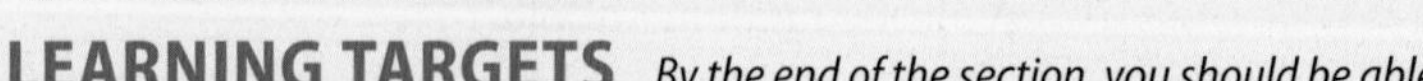

LEARNING TARGETS *By the end of the section, you should be able to:*

- Explain the concept of confounding and how it limits the ability to make cause-and-effect conclusions.
- Identify the experimental units and treatments in an experiment.
- Explain the purpose of a control group in an experiment.
- Describe the placebo effect and explain the purpose of blinding in an experiment.
- Describe how to randomly assign treatments in an experiment and explain the purpose of random assignment.
- Explain the purpose of controlling other variables in an experiment.
- Describe a randomized block design for an experiment and explain the benefits of blocking in an experiment.
- Describe a matched pairs design for an experiment.
- Explain the meaning of statistically significant in the context of an experiment.
- Identify when it is appropriate to make an inference about a population and when it is appropriate to make an inference about cause and effect.

In Section 3A, you learned how to distinguish observational studies from experiments. You also learned that experiments allow for cause-and-effect conclusions, unlike observational studies. In this section, you'll learn more about why it is difficult to determine causation from an observational study and how to design an experiment so that cause-and-effect conclusions are possible.

Confounding

Is taking a vitamin D supplement good for you? In one retrospective observational study, researchers found that people with vitamin D deficiencies were at greater risk for contracting Covid-19.[35] In a prospective observational study, researchers kept in contact with a sample of teenage girls and found those with higher vitamin D intakes were less likely to suffer broken bones.[36] Other observational studies have shown that people with a higher vitamin D concentration have less cardiovascular disease, better cognitive function, and less risk of diabetes than people with a lower concentration of vitamin D.[37]

In the observational studies involving vitamin D and diabetes, the **explanatory variable** is vitamin D concentration in the blood and the **response variable** is diabetes status — whether or not the person developed diabetes.

DEFINITION Response variable, Explanatory variable

A **response variable** measures an outcome of a study.

An **explanatory variable** may help explain or predict changes in a response variable.

Unfortunately, it is very difficult to show that taking vitamin D *causes* a lower risk of diabetes with an observational study. As shown in the table, there are many possible differences between the group of people with a high vitamin D concentration and the group of people with a low vitamin D concentration. Any of these differences could be causing the difference in diabetes risk between the two groups of people.

Variable	Group 1	Group 2
Vitamin D concentration (explanatory)	**High vitamin D concentration**	**Low vitamin D concentration**
Quality of diet	Better diet	Worse diet
Amount of exercise	More exercise	Less exercise
⋮	⋮	⋮
Amount of vitamin supplementation	More likely to take other vitamins	Less likely to take other vitamins
Diabetes status (response)	**Less likely to have diabetes**	**More likely to have diabetes**

For example, it is possible that people who have healthier diets eat lots of foods that are rich in vitamin D. It is also possible that people with healthier diets are less likely to develop diabetes. If both of these statements are true, we would see an association between vitamin D concentration and diabetes status, even if vitamin D has no effect on diabetes status. In this case, we say there is **confounding** between diet and vitamin D concentration because we cannot tell which variable is causing the change in diabetes status.

DEFINITION Confounding

Confounding occurs when two variables are associated in such a way that their effects on a response variable cannot be distinguished from each other.

AP® EXAM TIP

If you are asked to identify a possible confounding variable in a given setting, you are expected to explain how the variable you choose (1) is associated with the explanatory variable and (2) is associated with the response variable.

Likewise, because sun exposure increases vitamin D concentration, it is possible that people who exercise outside a lot have higher concentrations of vitamin D. If people who exercise a lot are also less likely to get diabetes, then amount of outdoor exercise and vitamin D concentration are confounded — we can't say which variable is the cause of the smaller diabetes risk. In other words, amount of outdoor exercise is a *confounding variable* because it is related to both vitamin D concentration and diabetes status.

Here is an example that identifies a potential confounding variable in an observational study.

EXAMPLE

Smoking and ADHD
Confounding

Skill 1.C

Highwaystarz-Photography/iStock/Getty Images

PROBLEM: In a study involving more than 4700 children, researchers from Cincinnati Children's Hospital Medical Center found that those children whose mothers smoked during pregnancy were more than twice as likely to develop attention-deficit/hyperactivity disorder (ADHD) as children whose mothers had not smoked.[38] Explain why confounding makes it unreasonable to conclude that a mother's smoking during pregnancy causes an increase in the risk of ADHD in her children based on this study.

SOLUTION:

It is possible that women who smoked during pregnancy had less healthy diets and that eating poorly during pregnancy increases the risk of ADHD in children. If both of these are true, then we would see a relationship between smoking and ADHD even if smoking has no effect on ADHD.

Notice that the solution describes how diet might be associated with the explanatory variable (smoking status) and how diet might be associated with the response variable (ADHD status). You need to describe both relationships to justify confounding.

FOR PRACTICE, TRY EXERCISE 1

The easiest way to identify confounding in an observational study is to think about other variables that are associated with the explanatory variable that might cause a change in the response variable. In the smoking and ADHD study, there are many potential differences between the group of mothers who smoked during pregnancy and the group of mothers who didn't. Any of these differences could be the cause of the change in the response variable.

The Language of Experiments

To determine whether taking vitamin D actually causes a reduction in diabetes risk, researchers in Norway performed an experiment. The researchers randomly assigned 500 **experimental units (subjects)** with prediabetes to one of two **treatments.** Half were given a high dose of vitamin D and half were given a **placebo** — a pill that looked exactly like the vitamin D supplement but contained no active ingredient.

DEFINITION Treatment, Experimental unit, Subjects, Placebo

A **treatment** is a specific condition applied to the individuals in an experiment. If an experiment has several explanatory variables, a treatment is a combination of specific values of these variables.

An **experimental unit** is the object to which a treatment is randomly assigned. When the experimental units are human beings, they are often called **subjects.**

A **placebo** is a treatment that has no active ingredient but is otherwise like other treatments.

The experiment in Norway avoided confounding by randomly assigning which participants took vitamin D and which didn't. That way, people with healthier diets were split about evenly between the two treatment groups. So were people who exercised a lot and people who took other vitamins. After 5 years, approximately 40% of the people in each treatment group were diagnosed with diabetes.[39] In other words, the association between vitamin D concentration and diabetes status disappeared when comparing two groups that were roughly the same to begin with.

In the vitamin D experiment, there was one **factor** (explanatory variable) with two **levels:** a dose of 20,000 IU of vitamin D per week and a dose of 0 IU of vitamin D per week. When there is only one factor, the levels are equivalent to the treatments. In an experiment with multiple factors, the treatments are formed using combinations of the levels of each of the factors.

DEFINITION Factor, Levels

In an experiment, a **factor** is an explanatory variable that is manipulated and may cause a change in the response variable.

The different values of a factor are called **levels.**

Here's an example of a multifactor experiment.

EXAMPLE

The five-second rule
The language of experiments

Skill 1.C

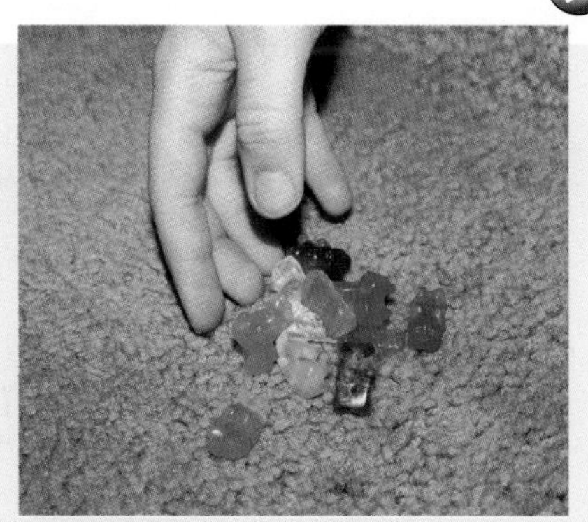

Zaia Snively

PROBLEM: Have you ever dropped a tasty piece of food on the ground, then quickly picked it up and eaten it? If so, you probably thought about the "five-second rule," which states that a piece of food is safe to eat if it has been on the floor less than 5 seconds. The rule is based on the belief that bacteria need time to transfer from the floor to the food. But does it work?

Researchers from Rutgers University put the five-second rule to the test. They dropped pieces of food onto four different surfaces — stainless steel, ceramic tile, wood, and carpet — and waited for four different lengths of time — less than 1 second, 5 seconds, 30 seconds, and

300 seconds. Finally, they used bacteria prepared two different ways — in a tryptic soy broth or a peptone buffer. Once the bacteria were ready, the researchers spread them out on the different surfaces and started dropping the pieces of food. After waiting the designated amount of time, each piece of food was removed, blended, and plated on tryptic soy agar to allow bacteria to grow.[40]

(a) What are the experimental units in this experiment?
(b) List the factors in this experiment and the number of levels for each factor.
(c) If the researchers used every possible combination of levels to form the treatments, how many treatments were included in the experiment?
(d) List two of the treatments.

SOLUTION:

(a) The experimental units are the pieces of food being dropped.
(b) Type of surface (4 levels), amount of time (4 levels), and method of bacterial preparation (2 levels).
(c) $4 \times 4 \times 2 = 32$ different treatments
(d) Stainless steel/less than 1 second/tryptic soy broth; wood/300 seconds/peptone buffer

FOR PRACTICE, TRY EXERCISE 5

The researchers in this study observed greater bacterial transfer when the food remained longer on the surface, although some transfer happened almost instantaneously. Of the four surfaces, they found that food dropped on carpet had the least bacterial transfer. Overall, the researchers concluded that the type of food and the type of surface were at least as important as the amount of time the food remained on the surface.

CHECK YOUR UNDERSTANDING

1. In a study conducted by researchers at the University of Texas, people were asked about their social media use and satisfaction with their marriage. Of the heavy social media users, 32% had thought seriously about leaving their spouse. Only 16% of non-social media users had thought seriously about leaving their spouse.[41] Based on this study, is it reasonable to conclude that using social media makes people more likely to consider leaving their spouse? Explain your reasoning.
2. Does the color of a restaurant server's lipstick affect how customers tip? To find out, Lexie wrote "Red" on 30 slips of paper and "Neutral" on 30 slips of paper, and mixed them well. Each time she got ready for work, she randomly selected one piece of paper to determine what lipstick color she would wear that day. The mean daily tip amount was \$23 greater when she was wearing red lipstick.[42] Identify the explanatory and response variables, the experimental units, and the treatments.

Comparison and Control Groups

Does caffeine affect pulse rate? Many students regularly consume caffeine to help them stay alert. So it seems plausible that taking caffeine might increase an individual's pulse rate. Is this true? One way to investigate this claim is to ask

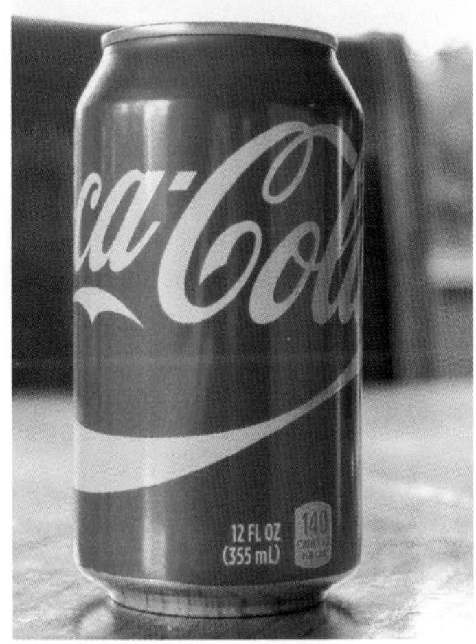
Zaia Snively

volunteers to measure their pulse rates, drink some cola with caffeine, measure their pulse rates again after 10 minutes, and calculate the increase in pulse rate.

This experiment has a very simple design. A group of subjects (the students) was exposed to a treatment (the cola with caffeine), and the response variable (change in pulse rate) was observed. Here is the design:

Students → Cola with caffeine → Change in pulse rate

Unfortunately, even if the pulse rate of every student went up, we couldn't attribute the increase to caffeine. Perhaps the excitement of being in an experiment made their pulse rates increase. Maybe it was the sugar in the cola, rather than the caffeine. Perhaps their teacher told them a funny joke during the 10-minute waiting period and made everyone laugh. In other words, many other variables are potentially confounded with caffeine intake.

The remedy for the confounding in the caffeine example is to do a comparative experiment with two groups: one group that receives caffeine and a **control group** that does not receive caffeine.

DEFINITION Control group

In an experiment, a **control group** is used to provide a baseline for comparing the effects of other treatments. Depending on the purpose of the experiment, a control group may be given an inactive treatment (placebo), an active treatment, or no treatment at all.

In all other ways, these two groups should be treated exactly the same, so the only consistent difference between the groups is the caffeine. This means that one group could get regular cola with caffeine, while the control group gets caffeine-free cola. Both groups would get the same amount of sugar, so sugar consumption would no longer be confounded with caffeine intake. Likewise, both groups would experience the same events during the experiment, so what happens during the experiment won't be confounded with caffeine intake either. *Using a design that compares two or more treatments is the first step in designing a good experiment.*

EXAMPLE Preventing malaria: Comparison and control groups

Skill 1.B

PROBLEM: Malaria causes hundreds of thousands of deaths each year, with many of the victims being children. Will regularly screening children for the malaria parasite and treating those who test positive reduce the proportion of children who develop the disease? Researchers worked with children in 101 schools in Kenya, randomly assigning half of the schools to provide regular screenings and follow-up treatments to their students and the remaining schools to provide no regular screening. Children at all 101 schools were tested for malaria at the end of the study.[43] Explain why it was necessary to include a control group of schools that didn't provide regular screenings.

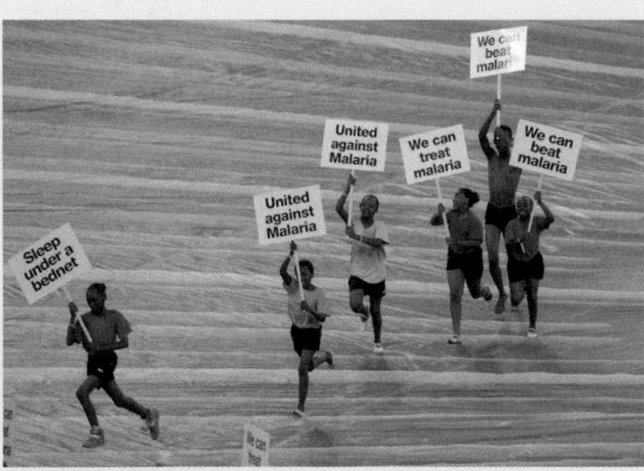

ALEXANDER JOE/AFP/Getty Images

SOLUTION:

The purpose of the control group is to provide a baseline for comparing the effect of the regular screenings and follow-up treatments. Otherwise, researchers wouldn't be able to determine if a decrease in malaria rates was due to the treatment or some other change that occurred during the experiment (such as a drought that killed off mosquitos, slowing the spread of malaria).

Note that the experimental units are the schools, not the students. The decision about who to screen was made school by school, not student by student. All students at the same school received the same treatment.

FOR PRACTICE, TRY EXERCISE 9

A control group was essential in the malaria experiment to determine whether the screening was effective. However, *not all experiments include a control group* — as long as comparison takes place. In the experiment about the five-second rule, a total of 32 different treatments were being compared — but there was no control group.

Blinding and the Placebo Effect

In the caffeine experiment, we used a control group to help prevent confounding. But even when there is comparison, confounding is still possible. If the subjects in the experiment know what type of soda they are receiving, the expectations of the two groups will be different. The knowledge that a subject is receiving caffeine may increase their pulse rate, apart from the caffeine itself. This is an example of the **placebo effect.**

DEFINITION Placebo effect

The **placebo effect** describes the fact that some subjects in an experiment will respond favorably to any treatment, even an inactive treatment.

In one study, researchers zapped the wrists of 24 test subjects with a painful jolt of electricity. Then they rubbed a cream with no active medicine on subjects' wrists and told them the cream should help soothe the pain. When researchers shocked them again, 8 subjects said they experienced significantly less pain.[44] When the ailment is psychological, such as depression, some experts think that the placebo effect accounts for about three-quarters of the effect of the most widely used drugs.[45]

Because of the placebo effect, it is important that subjects don't know which treatment they are receiving. It is also better if the people interacting with the subjects and measuring the response variable don't know which subjects are receiving which treatment. When neither group knows who is receiving which treatment, the experiment is **double-blind.** Other experiments are **single-blind.**

DEFINITION Double-blind, Single-blind

In a **double-blind** experiment, neither the subjects nor those who interact with them and measure the response variable know which treatment a subject is receiving.

In a **single-blind** experiment, either the subjects or the people who interact with them and measure the response variable don't know which treatment a subject is receiving.

The idea of a double-blind design is simple. Until the experiment ends and the results are in, only the study's statistician knows for sure which treatment a subject is receiving. However, some experiments cannot be carried out in a double-blind manner. For example, if researchers are comparing the effects of exercise and dieting on weight loss, then subjects will know which treatment they are receiving. Such an experiment can still be single-blind if the individuals who are interacting with the subjects and measuring the response variable don't know who is dieting and who is exercising. In other single-blind experiments, the subjects are unaware of which treatment they are receiving, but the people interacting with them and measuring the response variable do know.

EXAMPLE

Do magnets repel pain?
Blinding and the placebo effect

Skill 1.C

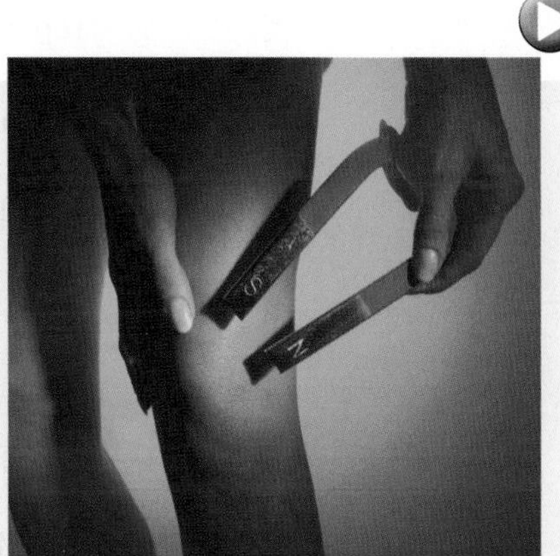

Eric O'Connell/Getty Images

PROBLEM: Early research showed that magnetic fields affected living tissue in humans. Some doctors have begun to use magnets to treat patients with chronic pain. Scientists wondered if this type of therapy really worked. They designed a double-blind experiment to find out. A total of 50 patients with chronic pain were recruited for the study. A doctor identified a painful site on each patient and asked the patient to rate the pain on a scale from 0 (mild pain) to 10 (severe pain). Then the doctor selected a sealed envelope containing a magnet at random from a box with a mixture of active and inactive magnets. The chosen magnet was applied to the site of the pain for 45 minutes. After being treated, each patient was again asked to rate the level of pain from 0 to 10.[46]

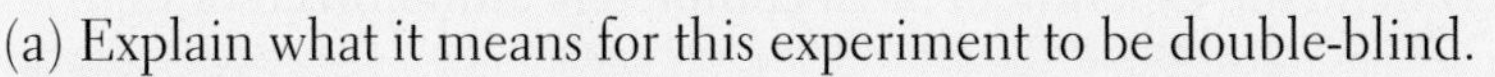

(a) Explain what it means for this experiment to be double-blind.

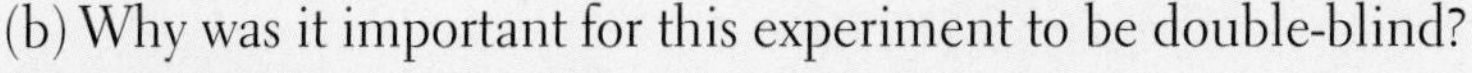

(b) Why was it important for this experiment to be double-blind?

SOLUTION:

(a) Neither the subjects nor the doctors applying the magnets and recording the pain ratings knew which subjects had the active magnets and which had the inactive magnets.

(b) If subjects knew they were receiving an active treatment, researchers wouldn't know if any improvement was due to the magnets or to the expectation of getting better (the placebo effect). If the doctors knew which subjects received which treatments, they might treat one group of subjects differently from the other group. This would make it difficult to know if the magnets were the cause of any improvement.

FOR PRACTICE, TRY EXERCISE 13

Random Assignment

Comparison alone isn't enough to produce results we can trust. Many other variables affect pulse rates besides caffeine. Some of these variables, such as caffeine tolerance, weight, and recent caffeine consumption, describe characteristics of the subjects. Other variables, such as sugar content, amount of soda consumed, and temperature, describe characteristics of the experimental process. To conduct a well-designed experiment, we want to treat every subject the same way and to create groups that are roughly equivalent at the beginning of an experiment. To create these groups, we use **random assignment** to determine which experimental units get which treatment.

DEFINITION Random assignment

In an experiment, **random assignment** means that treatments are assigned to experimental units (or experimental units are assigned to treatments) using a chance process.

Suppose there are 20 subjects in the caffeine experiment. To randomly assign treatments, we could number the students from 1 to 20 and randomly generate 10 *different* integers from 1 to 20. The subjects with these labels will receive the cola with caffeine. The remaining 10 subjects will receive the cola without caffeine. We could also write the letter "A" on 10 slips of paper, write the letter "B" on 10 slips of paper, shuffle the slips, and hand one to each subject without replacement. Subjects who get an "A" slip will receive the cola with caffeine and students who get a "B" slip will receive the cola without caffeine.

Random assignment should distribute the students who already have a caffeine tolerance in roughly equal numbers to each group. It should also roughly balance out the number of students who have consumed caffeine recently and make the average weight of subjects in each group about the same. In general, random assignment should create two groups that are roughly equivalent with respect to *every* variable that might affect the response, whether or not you identified the variable in advance. Because the only systematic difference between the groups is the treatment assigned, experiments with random assignment allow for cause-and-effect conclusions.

When treatments are randomly assigned to experimental units and there are no restrictions on which units can be assigned to which treatment, as in the caffeine experiment, the experiment has a **completely randomized design.**

DEFINITION Completely randomized design

In a **completely randomized design,** the experimental units are assigned to the treatments completely at random.

Figure 3.1 diagrams the details of the caffeine experiment: random assignment, the sizes of the groups and which treatment they receive, and the response variable.

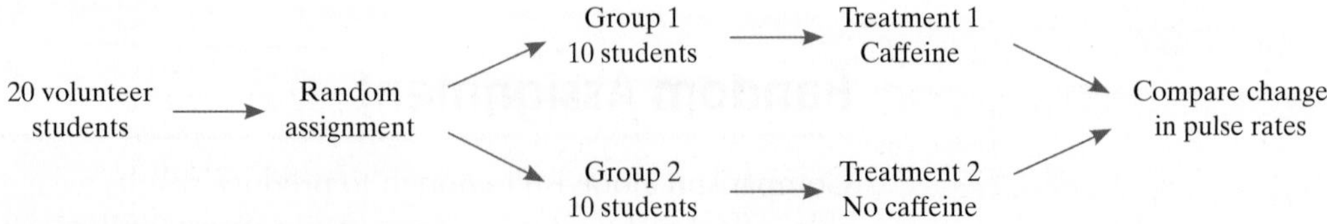

FIGURE 3.1 Diagram of a completely randomized design to compare caffeine and no caffeine.

Although there are good statistical reasons for using treatment groups that are roughly equal in size, the definition of a completely randomized design does not require that each treatment be assigned to an equal number of experimental units. It does specify that the assignment of treatments must occur completely at random.

AP® EXAM TIP

If you are asked to describe a completely randomized design, stay away from flipping coins. For example, suppose we ask each student in the caffeine experiment to toss a coin. If it's heads, then the student will drink the cola with caffeine. If it's tails, then the student will drink the caffeine-free cola. As long as all 20 students toss a coin, this is still a completely randomized design. Of course, the groups are unlikely to contain exactly 10 students each because it is unlikely that 20 coin tosses will result in a perfect 50-50 split between heads and tails.

The problem arises if we try to force the two groups to have equal sizes. Suppose we continue to have students toss coins until one of the groups has 10 students and then place the remaining students in the other group. In this case, the last two students in line are very likely to end up in the same group. However, in a completely randomized design, the last two subjects should have only a 50% chance of ending up in the same group.

EXAMPLE

Is the vaccine effective?
Random assignment

Skill 1.C

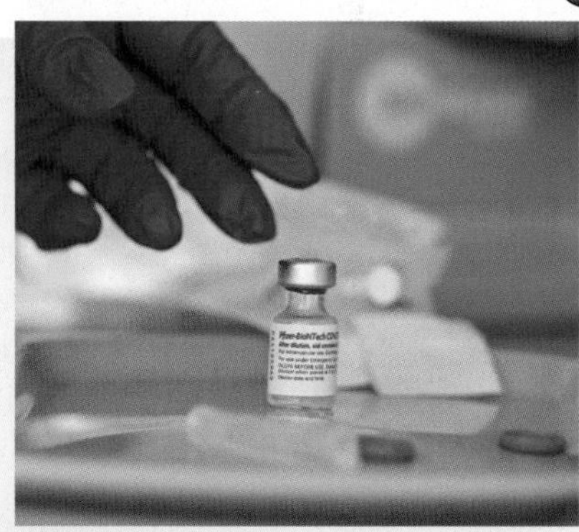

ROBYN BECK/AFP/Getty Images

PROBLEM: Before making a vaccine publicly available, pharmaceutical companies must conduct well-designed experiments (called clinical trials) to demonstrate a vaccine's effectiveness and safety. Some of the most important clinical trials in recent years were for the various Covid-19 vaccines. In Pfizer's clinical trial of its Covid-19 vaccine, half of the 43,548 volunteer subjects were randomly assigned to receive two shots of the vaccine 21 days apart. The other half of the volunteers were randomly assigned to receive two shots of a saline placebo 21 days apart. After several months, Pfizer determined that its vaccine was 95% effective in preventing Covid-19.[47]

(a) Describe how Pfizer could have randomly assigned the 43,548 subjects to the two treatments in a completely randomized design.

(b) Explain the purpose of random assignment in this experiment.

SOLUTION:

(a) Number the subjects from 1 to 43,548. Then, use a random number generator to generate 21,774 different integers from 1 to 43,548. Assign the subjects who correspond to these integers to the vaccine. Assign the remaining 21,774 subjects to the placebo.

We could have also used a table of random digits or 43,548 slips of paper to assign the treatments, but using a random number generator is the most efficient method with such a large number of subjects.

(b) The purpose of the random assignment was to create two groups of subjects who were roughly equivalent in terms of Covid-19 susceptibility before treatments were assigned. This allows the researchers to make cause-and-effect conclusions.

FOR PRACTICE, TRY EXERCISE 19

Random assignment works best when the experiment includes many subjects, as in the Pfizer clinical trial. If only 2 subjects participated in the clinical trial, the two treatment "groups" are guaranteed to be different, even if the two subjects were randomly assigned to treatments. In this case, confounding would prevent us from making any conclusions about the effectiveness of the vaccine.

In the actual trial, a total of 43,548 subjects were randomly assigned to the two treatments. With such a large number of subjects, age, health, occupation, and other variables that affect how likely a subject is to develop Covid-19 should be about the same, on average, in the two groups. The idea that we should use enough subjects to create roughly equivalent groups is called **replication.**

DEFINITION Replication

In an experiment, **replication** means giving each treatment to enough experimental units so that a difference in the effects of the treatments can be distinguished from chance variation due to the random assignment.

In statistics, replication means "use enough subjects." In other fields, the term *replication* has a different meaning. In these fields, replication means conducting an experiment in one setting and then having other investigators conduct a similar experiment in a different setting. That is, replication means repeatability.

Controlling Other Variables

Although random assignment should create two groups of experimental units that are roughly equivalent at the beginning of an experiment, other variables might also have an impact on the response variable. In our caffeine experiment, the sugar content, amount of soda consumed, and temperature of the room during the experiment will likely affect pulse rates. It is important that we **control** variables like these by keeping them the same for all treatment groups.

DEFINITION Control

In an experiment, **control** means keeping other variables constant for all experimental units.

Because sugar almost certainly affects pulse rates, it is important that each subject receive the same amount of sugar. If one group received regular cola with caffeine and the other group received caffeine-free *diet* cola with no sugar, then sugar consumption and caffeine intake would be confounded. *One reason to keep other variables the same for each subject is to prevent confounding, making it easier to determine if one treatment is more effective than another.*

We should also make sure that each student has the same amount of cola to drink. If each student was able to drink as much or as little as they wanted, the changes in pulse rates would be more variable than they would be otherwise. This increase in variability makes it harder to find convincing evidence that caffeine affects pulse rates.

The dotplots in Figure 3.2(a) show the results of an experiment in which the amount of cola consumed was the same for all participating students. Because there is very little overlap in the dotplots, it seems clear that caffeine does increase pulse rates. The dotplots in Figure 3.2(b) show the results of an experiment in which the students were able to choose the amount of soda they drank. Notice that the centers of the distributions haven't changed, but the distributions are more variable. The increased overlap in the dotplots makes the evidence supporting the effect of caffeine less convincing. *The second reason we keep other variables the same is to reduce the variability in the response variable, making it easier to determine if one treatment is more effective than another.*

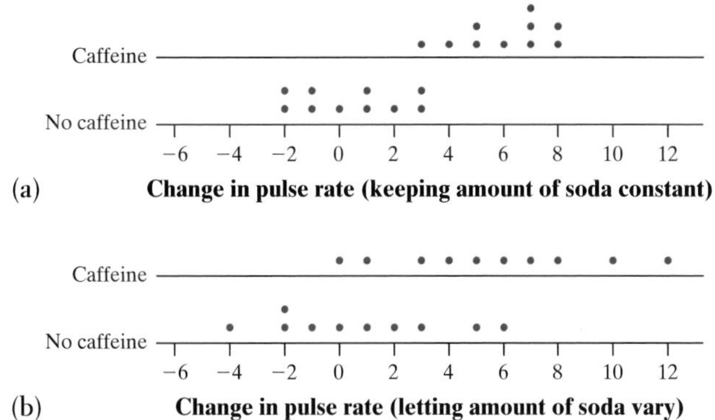

FIGURE 3.2 Dotplots showing the results of the caffeine experiment when (a) the amount of soda was kept the same and (b) the amount of soda was allowed to vary.

EXAMPLE

Multitasking
Controlling other variables

Skill 1.B

PROBLEM: Researchers in Canada performed an experiment with university students to examine the effects of multitasking on student learning. The 40 participants in the study were asked to attend the same lecture and take notes with their laptops. Half of the participants were randomly assigned to complete other online tasks not related to the lecture during that time. These tasks were meant to imitate typical student web browsing during classes. The remaining students simply took notes with their laptops. At the end of the lecture, students were given the same test about the content of the lecture.[48]

Ariel Skelley/Getty Images

(a) Identify two variables that the researchers kept the same during the experiment.
(b) Describe a benefit of keeping these variables the same for all subjects.

SOLUTION:

(a) All the students attended the same lecture and were given the same test.

(b) If students attended different lectures or took different tests, the test scores would likely be more variable than if they kept the lecture and test the same. This increase in test-score variability would make it harder to find convincing evidence that multitasking is harmful.

> Keeping these variables the same also helps to avoid confounding. For example, if the multitaskers attended one lecture and the non-multitaskers attended a different lecture, we wouldn't know whether the difference in multitasking or the difference in lectures caused a difference in test scores.

FOR PRACTICE, TRY EXERCISE 25

The following box summarizes the four key principles of experimental design: comparison, random assignment, replication, and control. Note that the principle of control isn't the same thing as including a control group, which is an aspect of comparison.

PRINCIPLES OF EXPERIMENTAL DESIGN

1. **Comparison.** Use a design that compares two or more treatments. Some experiments include a control group to establish a baseline for measuring the effects of other treatments.
2. **Random assignment.** Use chance to assign experimental units to treatments (or treatments to experimental units). Doing so helps create roughly equivalent groups of experimental units by balancing the effects of other variables among the treatment groups.
3. **Replication.** Give each treatment to enough experimental units so that any differences in the effects of the treatments can be distinguished from chance differences between the groups.
4. **Control.** Keep other variables the same for all groups, especially variables that are likely to affect the response variable. Control helps avoid confounding and reduces variability in the response variable.

Completely randomized designs are the simplest statistical designs for experiments. They illustrate clearly the principles of comparison, random assignment, replication, and control. But just as with sampling, sometimes the simplest method doesn't yield the most precise results. When there are sources of variability in the experimental units that can't be controlled, we can use more complicated experimental designs that account for these sources of variability.

CHECK YOUR UNDERSTANDING

Many utility companies have introduced programs to encourage energy conservation among their customers. An electric company considers placing small digital displays in households to show current electricity use and what the cost would be if this use continued for a month. Will the displays reduce electricity use? One cheaper approach is to give customers a chart and information about monitoring their electricity use from their outside meter. Would this method work almost as well? The company decides to conduct an experiment using 60 households from a small city during the month of April to compare these two approaches (display, chart) with a group of customers who receive information about energy consumption but no help in monitoring electricity use.

1. Explain the importance of a control group that doesn't get a display or chart.
2. Describe how to randomly assign the treatments to the 60 households.
3. What is the purpose of randomly assigning treatments in this context?
4. Identify one variable the researchers are keeping the same for all experimental units. Describe the benefit of keeping this variable constant.

Randomized Block Designs

When a population consists of groups of individuals that are "similar within but different between," a stratified random sample gives a more precise estimate than a simple random sample. We can apply a similar strategy when groups of experimental units are similar in ways that could affect the response variable.

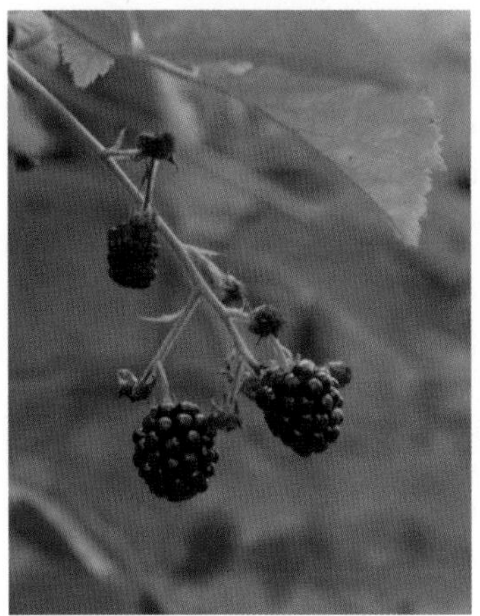
Gabriel Preda/Alamy Stock Photo

A university's school of agriculture has been developing a new variety of blackberry plant that is designed to have a greater yield than the current blackberry plant. To put the new variety to the test, agriculture students perform an experiment using two different fields. The smaller field is more fertile and has space for four plants. The larger field is less fertile and has space for six plants. In this context, the treatments are the new variety and current variety plants and the experimental units are the 10 available spaces where the blackberries will be planted.

Because of the difference in fertility between the two fields, the blackberry yields are likely to be more variable than if all the plants were grown in equally fertile soil. This additional variability may make it difficult to detect a difference in the average yield of the two varieties. To account for this source of variability, we form two **blocks** of experimental units: one block contains the six less-fertile spaces and the other block contains the four more-fertile spaces.

DEFINITION Block

A **block** is a group of experimental units that are known before the experiment to be similar in some way that is expected to affect the response to the treatments.

In general, blocking works best when blocks are formed using variables that are the most strongly associated with the response variable(s) in the experiment. In the blackberry experiment, it is reasonable to think that the fertility of the plots is more strongly associated with yield than with other variables in the plots such as sun exposure or proximity to other plants.

After forming blocks, we randomly assign treatments within each block. The result is a **randomized block design.**

DEFINITION Randomized block design

In a **randomized block design,** the random assignment of experimental units to treatments is carried out separately within each block.

For the blackberry experiment, the agriculture students will randomly assign half of the more-fertile spaces to receive the new variety and half of the more-fertile spaces to receive the current variety. Similarly, half of the less-fertile spaces will receive the new variety and half will receive the current variety, as shown in Figure 3.3.

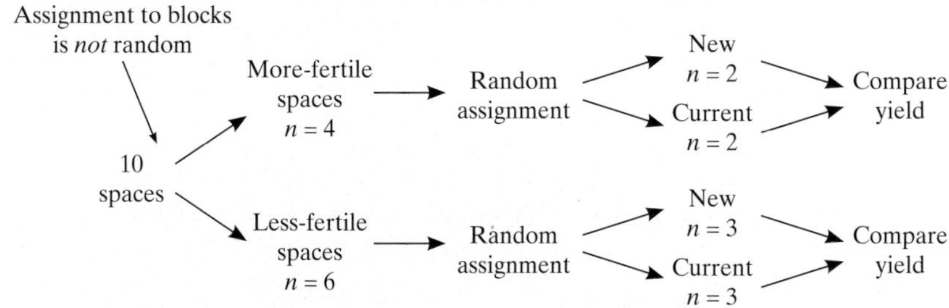

FIGURE 3.3 Diagram of a randomized block design to compare new and current varieties of blackberry plants.

Using a randomized block design allows us to account for the variation in the response that is due to the blocking variable of soil fertility. This makes it easier to determine if one variety of blackberry plant is really more productive than the other. To find out why, let's explore the results of the blackberry experiment. The dotplots show the yield (in pounds) for each of the five new-variety plants and the five current-variety plants.

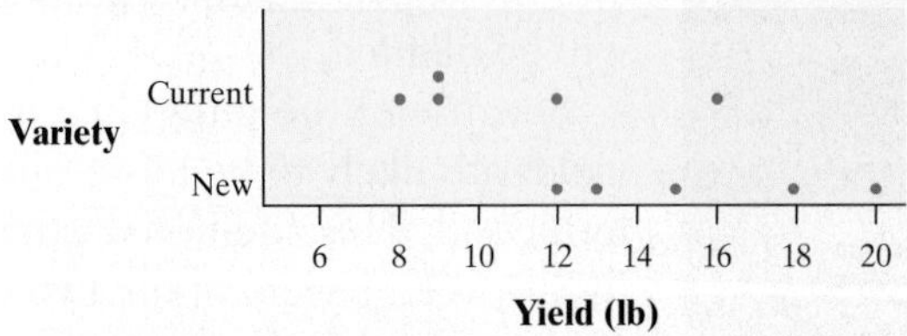

Because the center of the distribution of yield is slightly higher for the new variety of blackberry plants, the researchers have some evidence that the new variety results in higher yields. However, because the distributions overlap quite a bit, the evidence isn't especially convincing. It is possible that the difference in centers might simply be due to the chance variation in the random assignment.

If we compare the results for the two varieties of blackberry plants within each block, however, a different story emerges. Among the four plants in the more-fertile spaces (indicated by the blue dots), the new variety was the clear winner. Likewise, among the six plants in the less-fertile spaces (indicated by the red squares), the new variety was the clear winner.

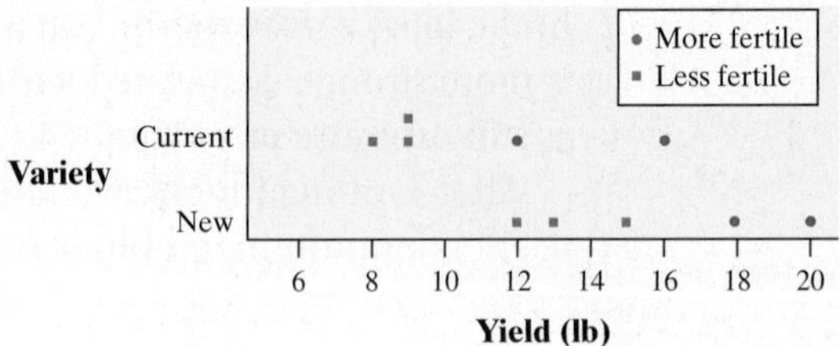

The overlap in the first set of dotplots was due almost entirely to the variation in fertility — the blackberry plants in the more-fertile spaces had a higher average yield than the plants in the less-fertile spaces, regardless of which variety was planted. In fact, the average yield was 5.5 pounds greater in the more-fertile spaces $(16.5 - 11 = 5.5)$. To account for the variation created by the difference in fertility, let's subtract 5.5 from the yields in the more-fertile spaces to "even the playing field." Here is a graph of the adjusted yields (in pounds):

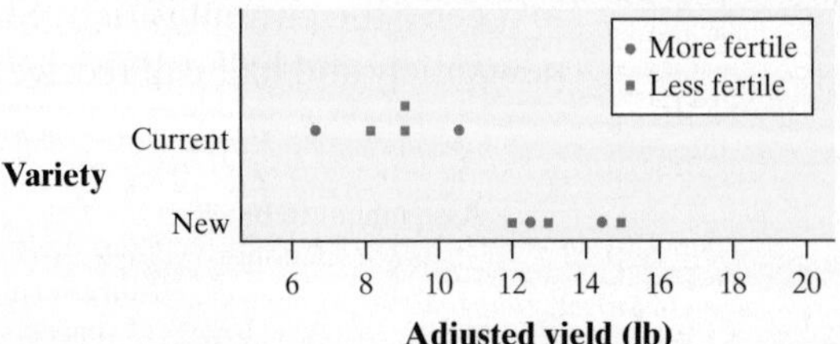

Because we accounted for the variation in yield due to the difference in fertility, the variation in each of the distributions has been reduced and there is no longer any overlap between the two distributions. *When blocks are formed wisely, it is easier to find convincing evidence that one treatment is more effective than another.*

EXAMPLE

Should I use the popcorn button?
Randomized block designs

Skill 1.C

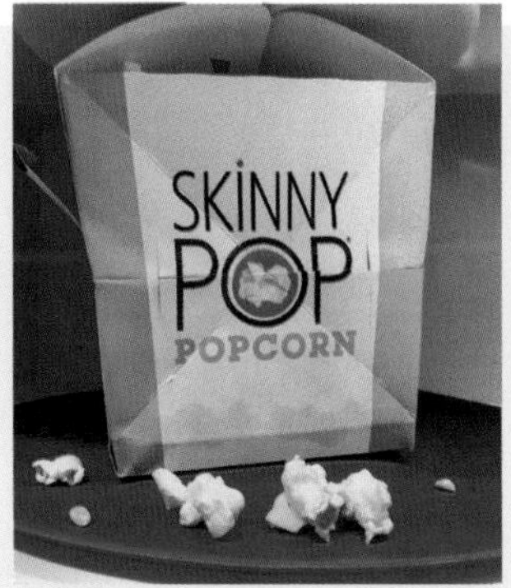

Ann Heath

PROBLEM: Some popcorn lovers want to determine if it is better to use the "popcorn button" on their microwave oven or to use the amount of time recommended on the bag of popcorn. To measure how well each method works, they will count the number of unpopped kernels remaining after popping. To obtain the experimental units, they buy 10 bags each of 4 different varieties of microwave popcorn (butter, cheese, natural, and kettle corn).

(a) Describe a randomized block design for this experiment. Justify your choice of blocks.

(b) Explain why a randomized block design might be preferable to a completely randomized design for this experiment.

SOLUTION:

(a) Form blocks based on variety, because the number of unpopped kernels is likely to differ by variety. Randomly assign 5 bags of each variety to the popcorn button treatment and 5 bags to the timed treatment by placing all 10 bags of a particular variety in a large box. Shake the box, pick 5 bags without replacement, and assign them to be popped using the popcorn button. The remaining 5 bags will be popped using the instructions on the bags. Repeat this process for the remaining 3 varieties. After popping each of the 40 bags in random order, count the number of unpopped kernels in each bag. Then compare the results for bags popped using the popcorn button and bags popped using the instructions on the bag for the 4 varieties.

> It is important to pop the bags in random order so that changes over time (e.g., temperature, humidity) aren't confounded with the explanatory variable. For example, if the 20 "popcorn button" bags are popped last, when the room temperature is greater, we wouldn't know if using the popcorn button or the warmer temperature was the cause of a difference in the number of unpopped kernels.

(b) A randomized block design accounts for the variability in the number of unpopped kernels created by the different varieties of popcorn (butter, cheese, natural, kettle). This makes it easier to determine if using the microwave button is more effective for reducing the number of unpopped kernels.

FOR PRACTICE, TRY EXERCISE 31

> **AP® EXAM TIP**
>
> Don't mix the language of experiments and the language of sample surveys or other observational studies. You will lose credit for saying things like "use a randomized block design to select the sample for this survey" or "randomly assign the treatments using three different strata."

Another way to address the variability created by using different varieties is to use only one variety of popcorn in the experiment. Because "variety of popcorn" is no longer a variable, it will not be a source of variability in the number of unpopped kernels. Of course, this means that the results of the experiment apply to only that one variety of popcorn.

Does blocking help with confounding? Yes, by helping ensure that the treatment groups in an experiment are as similar as possible at the beginning of the experiment. In the blackberry experiment, a completely randomized design could assign the new variety to all four of the more-fertile spaces just by chance. If the new variety has a greater average yield at the end of the experiment, we wouldn't know if it was due to the fertility or the variety. Because blocking ensures that both varieties are assigned to the more-fertile spaces in equal numbers, fertility won't be confounded with variety.

Blocking is an important additional principle of experimental design. A wise experimenter will form blocks based on the most important unavoidable sources of variation among the experimental units. Random assignment will then average out the effects of the remaining variables and allow for a fair comparison of the treatments. The moral of the story: *control what you can, block on what you can't control, and randomize to create comparable groups.*

Matched Pairs Designs

A common type of randomized block design for comparing two treatments is a **matched pairs design.** The idea is to create blocks by matching pairs of similar experimental units. The random assignment of treatments to experimental units is done within each matched pair. Just as with other forms of blocking, matching helps account for the variation in the response variable due to the variable(s) used to form the pairs.

DEFINITION Matched pairs design

A **matched pairs design** is a common experimental design for comparing two treatments that uses blocks of size 2. In some matched pairs designs, two very similar experimental units are paired and the two treatments are randomly assigned within each pair. In others, each experimental unit receives both treatments in a random order.

J-Elgaard/Getty Images

Suppose we want to investigate whether listening to classical music while taking a math test affects performance. A total of 30 students in a math class volunteer to take part in the experiment. The difference in mathematical ability among the volunteers is likely to create additional variation in the test scores, making it harder to see the effect of classical music. To account for this variation, we could pair the students by their grade in the class — the two students with the highest grades are paired together, the two students with the next highest grades are paired together, and so on. Within each pair, one student is randomly assigned to take a math test while listening to classical music and the other member of the pair is assigned to take the math test in silence.

Sometimes, each "pair" in a matched pairs design consists of just one experimental unit that gets both treatments in random order. In the classical music experiment, we could have each student take a math test in both conditions. To decide the order, we flip a coin for each student. If the coin lands on heads, the student takes a math test with classical music playing today and a similar math test without music playing tomorrow. If it lands on tails, the student does the opposite — no music today and classical music tomorrow.

Randomizing the order of treatments is important to avoid confounding. Suppose everyone did the classical music treatment on the first day and the no-music treatment on the second day, but the air conditioner wasn't working on the second day. We wouldn't know if any difference in mean test score was due to the difference in treatment or the difference in room temperature.

EXAMPLE

Will an additive improve my mileage?
Matched pairs design

Skill 1.C

Matthew Richardson/Alamy Stock Photo

PROBLEM: A consumer organization wants to know if using a certain fuel additive increases the fuel efficiency (in miles per gallon, or mpg) of cars. A total of 20 cars of different types are available for testing. Describe an experiment that uses a matched pairs design to investigate this question. Explain your method of pairing and how you'll randomly assign the treatments.

SOLUTION:

Give each car both treatments in random order. It is reasonable to think that some cars are more fuel efficient than others, so using each car as its own "pair" accounts for the variation in fuel efficiency in the experimental units. For each car, flip a coin. If the coin lands on heads, that car will use the additive first and no additive second. If the coin lands on tails, that car will use no additive first and then the additive second. For each car, record the fuel efficiency (mpg) after using each treatment. Then compare the fuel efficiency with and without the additive for the 20 cars.

In matched pairs experiments where each experimental unit receives both treatments in random order, the experimenters will often have a "wash-out" period between the treatments so the first treatment has no effect on the second. In this case, each car should be filled with additive-free gas and driven until the tank is empty between the treatments. This will help eliminate any residual effects of the additive for cars that got the additive first.

FOR PRACTICE, TRY EXERCISE 35

In the preceding example, it is also possible to form matched pairs of two similar cars. For instance, we could pair together the two most fuel-efficient cars, the next two most fuel-efficient cars, and so on. This is less ideal, however, because there will still be some differences between the members of each pair that may increase variation in the results. Using the same car twice creates perfectly matched "pairs," and also doubles the number of pairs used in the experiment. Both these features make it easier to find convincing evidence that the gas additive is effective, if it really is effective.

CHECK YOUR UNDERSTANDING

Researchers would like to design an experiment to compare the effectiveness of three different advertisements for a new television series featuring the works of Jane Austen. There are 300 volunteers available for the experiment.

1. Describe a completely randomized design to compare the effectiveness of the three advertisements.
2. Describe a randomized block design for this experiment. Justify your choice of blocks.
3. Why might a randomized block design be preferable in this context?

Inference for Experiments

Well-designed experiments allow for inference about cause and effect. But we should conclude that changes in the explanatory variable cause changes in the response variable only if the results of an experiment are **statistically significant.**

> **DEFINITION Statistically significant**
>
> When an observed difference in responses between the groups in an experiment is so large that it is unlikely to be explained by chance variation in the random assignment, the results are called **statistically significant.**

Mr. Wilcox and his students decided to perform the caffeine and pulse rate experiment described earlier in this section. In their experiment, 10 student volunteers were randomly assigned to drink cola with caffeine and the remaining 10 students were assigned to drink caffeine-free cola. The table and graph show the change in pulse rate for each student (Final pulse rate − Initial pulse rate), along with the mean change for each group.

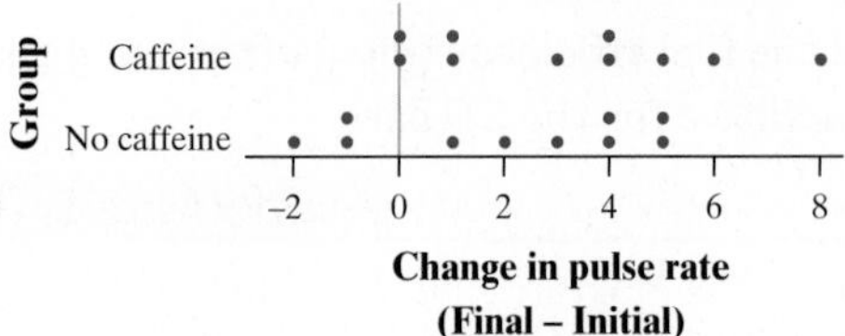

	Change in pulse rate (Final pulse rate – Initial pulse rate)										Mean change
Caffeine	8	3	5	1	4	0	6	1	4	0	3.2
No caffeine	3	−2	4	−1	5	5	1	2	−1	4	2.0

The dotplots provide some evidence that caffeine has an effect on pulse rates. The mean change for the 10 students who drank cola with caffeine was 3.2, which is 1.2 greater than for the group who drank caffeine-free cola. But are the results statistically significant?

Recall that the purpose of random assignment in this experiment was to create two groups that were roughly equivalent at the beginning of the experiment. Subjects with high caffeine tolerance should be split up in about equal numbers, subjects with high metabolism should be split up in about equal numbers, and so on.

Of course, the random assignment is unlikely to produce groups that are exactly equivalent. One group might get more "favorable" subjects just by chance. That is, the caffeine group might end up with a few extra subjects who were likely to have a pulse rate increase, just due to chance variation in the random assignment.

There are two ways to explain why the mean change in pulse rate was 1.2 greater for the caffeine group:

1. Caffeine does *not* have an effect on pulse rates, and the difference of 1.2 happened because of chance variation in the random assignment.
2. Caffeine increases pulse rates.

If it is somewhat likely to get a difference of 1.2 or more simply due to the chance variation in random assignment, the results of the experiment are not statistically significant. In that case, it is plausible that caffeine has no effect on pulse rates. But if it is unlikely to get a difference of 1.2 or more by chance alone, we rule out Explanation 1 and say the results are statistically significant.

How can we determine if a difference of 1.2 is statistically significant? You'll find out in the following activity.

ACTIVITY

Analyzing the caffeine experiment

Zaia Snively

In the experiment performed by Mr. Wilcox's class, the mean change in pulse rate for the caffeine group was 1.2 greater than the mean change for the no-caffeine group. This provides some evidence that caffeine increases pulse rates. But is the difference statistically significant? Or is it plausible that a difference of 1.2 would arise just due to chance variation in the random assignment? In this activity, we'll investigate by seeing what differences typically occur just by chance, assuming caffeine doesn't affect pulse rates. That is, we'll assume that the change in pulse rate for a particular student would be the same regardless of what treatment they were assigned.

	Change in pulse rate (Final pulse rate − Initial pulse rate)										Mean change
Caffeine	8	3	5	1	4	0	6	1	4	0	3.2
No caffeine	3	−2	4	−1	5	5	1	2	−1	4	2.0

1. Use 20 index cards to represent the 20 students in this experiment. On each card, write one of the 20 outcomes listed in the table. For example, write "8" on the first card, "3" on the second card, and so on.
2. Shuffle the cards and deal two piles of 10 cards each. This represents randomly assigning the 20 students to the two treatments, *assuming that the treatment received doesn't affect the change in pulse rate.* The first pile of 10 cards represents the caffeine group, and the second pile of 10 cards represents the no-caffeine group.
3. Find the mean change for each group and subtract the means (Caffeine – No caffeine). *Note:* It is possible to get a negative difference.
4. Your teacher will draw and label an axis for a class dotplot. Plot the difference you got in Step 3 on the graph.
5. In Mr. Wilcox's class, the observed difference in means was 1.2. Is a difference of 1.2 statistically significant? Discuss with your classmates.

We used technology to perform 100 trials of the simulation described in the activity. The dotplot in Figure 3.4 shows that getting a difference of 1.2 isn't that unusual. In 19 of the 100 trials, we obtained a difference of 1.2 or more simply due to chance variation in the random assignment.

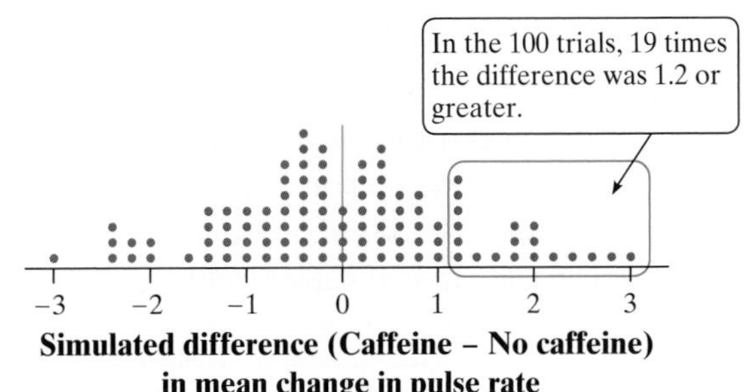

FIGURE 3.4 Dotplot showing the differences in means that occurred in 100 simulated random assignments, assuming that caffeine has no effect on pulse rates.

Because a difference of 1.2 or greater is somewhat likely to occur by chance alone, the results of Mr. Wilcox's class experiment aren't statistically significant. Based on this experiment, there isn't convincing evidence that caffeine increases pulse rates.

EXAMPLE

Distracted driving
Inference for experiments

Skill 4.B

Sean Locke Photography/Shutterstock

PROBLEM: Is talking on a cell phone while driving more distracting than talking to a passenger? David Strayer and his colleagues at the University of Utah designed an experiment to help answer this question. They used 48 undergraduate students as subjects. The researchers randomly assigned half of the subjects to drive in a simulator while talking on a cell phone, and the other half to drive in the simulator while talking to a passenger. One response variable was whether or not the driver stopped at a rest area that was specified by researchers before the simulation started. The table shows the results.[49]

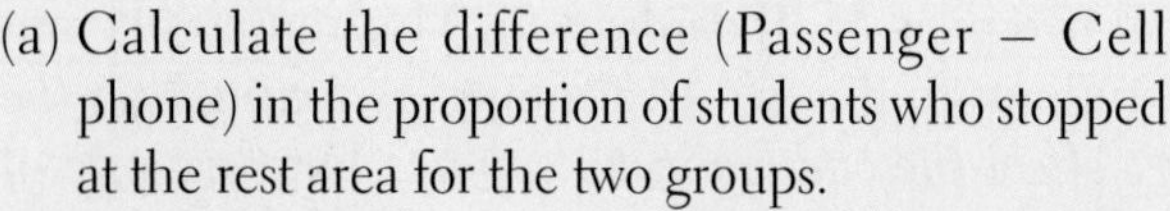

(a) Calculate the difference (Passenger – Cell phone) in the proportion of students who stopped at the rest area for the two groups.

		Treatment		
		Cell phone	Passenger	Total
Response	Stopped at rest area	12	21	33
	Didn't stop	12	3	15
	Total	24	24	48

One hundred trials of a simulation were performed to see what differences in proportions would occur due only to chance variation in the random assignment, assuming that the type of distraction did not affect whether a subject stopped at the rest area. That is, 33 "stoppers" and 15 "non-stoppers" were randomly assigned to two groups of 24. The dotplot displays the results of the simulation.

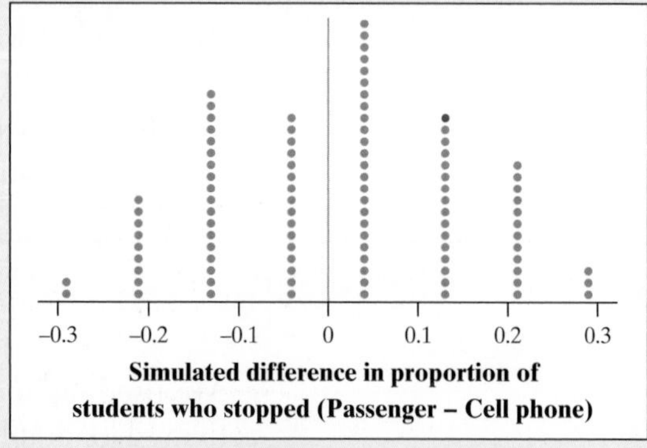

(b) There is a blue dot at a difference of 0.125. Explain what this dot represents.

(c) Use the results of the simulation to determine if the difference in proportions from part (a) is statistically significant. Explain your reasoning.

SOLUTION:

(a) Difference in proportions $= 21/24 - 12/24 = 0.875 - 0.500 = 0.375$

(b) One simulated random assignment resulted in a difference in proportions of students who stopped (Passenger — Cell phone) of 0.125.

(c) Because a difference of 0.375 or greater never occurred in the simulation, the difference is statistically significant. It is extremely unlikely to get a difference this big simply due to chance variation in the random assignment.

Because the difference is statistically significant, there is convincing evidence that talking on a cell phone is more distracting than talking with a passenger—at least for subjects like those in the experiment.

FOR PRACTICE, TRY EXERCISE 39

In the caffeine example, we said that a difference in means of 1.2 *was not unusual* because a difference that big or bigger occurred in 19% of the simulated random assignments by chance alone. In the distracted-drivers example, we said that a difference in proportions of 0.375 *was unusual* because a difference this big or bigger occurred in 0% of the simulated random assignments by chance alone. So the boundary between "not unusual" and "unusual" must be somewhere between 0% and 19%. For now, we recommend using a boundary of 5%, meaning that differences that would occur in less than 5% of simulated random assignments by chance alone are considered statistically significant. We will revisit this issue in Unit 6.

Putting It All Together: The Scope of Inference

In Sections 3A and 3B, you learned that random selection of individuals from a population allows us to make inferences about the population from which the sample was selected. In this section, you learned that random assignment of treatments to experimental units in an experiment allows us to make inferences about cause and effect, provided that the results of the experiment are statistically significant.

THE SCOPE OF INFERENCE

- Random selection of individuals justifies inference about the population from which the individuals were chosen.
- Random assignment of individuals to groups in an experiment with statistically significant results justifies inference about cause and effect.

The following chart summarizes the possibilities.[50]

		Were individuals randomly assigned to groups?	
		Yes	No
Were individuals randomly selected?	Yes	Inference about the population: YES Inference about cause and effect: YES	Inference about the population: YES Inference about cause and effect: NO
	No	Inference about the population: NO Inference about cause and effect: YES	Inference about the population: NO Inference about cause and effect: NO

Well-designed experiments randomly assign individuals to treatment groups. However, most experiments use volunteers rather than selecting experimental units at random from a larger population. This limits such experiments to inference about cause and effect *for individuals like those who received the treatments*. Determining whether a person is "like" those in an experiment isn't easy, but doctors and other researchers have to make this decision all the time in the real world.

EXAMPLE

When will I ever use this stuff?
The scope of inference

Skill 4.B

Frederic Cirou/PhotoAlto/Getty Images

PROBLEM: Researchers at the University of North Carolina were concerned about the increasing dropout rate in the state's high schools, especially for low-income students. Surveys of recent dropouts revealed that many of these students had started to lose interest during middle school. They said they saw little connection between what they were studying in school and their future plans. To change this perception, researchers developed a program called CareerStart. The central idea of the program is that teachers show students how the topics they're learning about can be applied to specific careers.

To test the effectiveness of CareerStart, the researchers recruited 14 middle schools in Forsyth County to participate in an experiment. Seven of the schools, determined at random, used CareerStart along with the district's standard curriculum. The other 7 schools just followed the standard curriculum. Researchers followed both groups of students for several years, collecting data on students' attendance, behavior, standardized test scores, level of engagement in school, and whether or not the students graduated from high school.

Results: Students at schools that used CareerStart had statistically significantly better attendance and fewer discipline problems, earned higher test scores, reported greater engagement in their classes, and were more likely to graduate.[51]

(a) Is it reasonable to make a cause-and-effect conclusion based on this study? Explain your answer.

(b) What is the largest population to which we can generalize the results of this study? Explain your answer.

SOLUTION:

(a) *Because treatments were randomly assigned to schools and the results were statistically significant, there is convincing evidence that using the CareerStart curriculum caused better attendance, fewer discipline problems, higher test scores, greater engagement, and increased graduation rates.*

(b) *Because the schools were not randomly selected from any population, the results of this study can be applied only to schools like these 14 from Forsyth County.*

FOR PRACTICE, TRY EXERCISE 43

CHECK YOUR UNDERSTANDING

When an athlete suffers a sports-related concussion, does it help to remove the athlete from play immediately? Researchers recruited 95 athletes seeking care for a sports-related concussion at a medical clinic and followed their progress during recovery. Researchers found statistically significant evidence that athletes who were removed from play immediately recovered more quickly, on average, than athletes who continued to play.[52]

1. Is it reasonable to make a cause-and-effect conclusion based on this study? Explain your answer.
2. What is the largest population to which we can generalize the results of this study? Explain your answer.

DATA ETHICS*

Although randomized experiments are the best way to make an inference about cause and effect, in some cases it isn't ethical to perform an experiment. For example, does texting while driving increase the risk of having an accident? Although a well-designed experiment would help answer this question, it would be unethical to randomly assign an individual to text while driving!

The most complicated ethical issues arise when we collect data from people. New medical treatments, for example, can do harm as well as good to subjects who participate in clinical trials. Likewise, ethical issues must be considered when administering a sample survey, even though no treatments are imposed.

Some basic standards of data ethics must be applied in all studies that gather data from human subjects, whether they are observational studies or experiments. These standards are summarized here.

BASIC DATA ETHICS

- All planned studies must be reviewed in advance by an *institutional review board* (*IRB*) charged with protecting the safety and well-being of the subjects.
- All individuals who are subjects in a study must give their *informed consent* before data are collected.
- All individual data must be kept *confidential*. Only statistical summaries for groups of subjects may be made public.

The law requires that studies carried out or funded by the U.S. federal government obey these principles.[53] The purpose of an *institutional review board* (*IRB*) is not to decide whether a proposed study will produce valuable information or if it is statistically sound. Instead, the board's purpose is, in the words of one university's board, "to protect the rights and welfare of human subjects (including patients) recruited to participate in research activities." An IRB would certainly reject an experiment that required subjects to text while driving.

Both words in the phrase "informed consent" are important. Subjects must be *informed* in advance about the nature of a study and any risk of harm it may bring. People who are asked to answer survey questions should be told what kinds of questions the survey will ask and approximately how much of their time it will take. Experimenters must tell subjects the nature and purpose of the study and outline possible risks. Subjects must then *consent*, or agree, to participate in writing.

It is important to protect individuals' privacy by keeping all data about them *confidential*. The report of an opinion poll may say what percentage of the 1200 respondents believed that legal immigration should be increased. It may not report what *you* said about this or any other issue. Confidentiality is not the same as *anonymity*. Anonymity means that the names of individuals are not known, even to the director of the study. Unfortunately, anonymity prevents any follow-up to reduce nonresponse or inform individuals of the study results.

*This is an important topic, but it is not required for the AP® Statistics exam.

SECTION 3C Summary

- A **response variable** measures an outcome of a study. An **explanatory variable** may help explain or predict changes in a response variable.
- Two variables are **confounded** when their effects on a response variable can't be distinguished from each other. It is difficult to make cause-and-effect conclusions from observational studies because of confounding.
- In an experiment, we impose one or more **treatments** on a group of **experimental units** (sometimes called **subjects** if they are human). Each treatment is a combination of the **levels** of the **factors** (explanatory variables).
- The **placebo effect** describes the fact that some subjects in an experiment will respond favorably to any treatment, even an inactive treatment. A **placebo** is a treatment that has no active ingredient but is otherwise like other treatments.
- In a **double-blind** experiment, neither the subjects nor those who interact with them and measure the response variable know which treatment a subject received. In a **single-blind** experiment, either the subjects or the people who interact with them and measure the response variable don't know which treatment a subject is receiving.
- In an experiment, a **control group** is used to provide a baseline for comparing the effects of other treatments. Depending on the purpose of the experiment, a control group may be given a placebo, an active treatment, or no treatment at all.
- The basic principles of experimental design are:
  - **Comparison:** Use a design that compares two or more treatments.
  - **Random assignment:** Use a chance process to assign treatments to experimental units (or experimental units to treatments). This helps create roughly equivalent groups before treatments are imposed.
  - **Replication:** Use each treatment with enough experimental units so that the effects of the treatments can be distinguished from chance differences between the groups.
  - **Control:** Keep other variables the same for all groups. Control helps avoid confounding and reduces the variation in the response variable, making it easier to decide if a treatment is effective.
- In a **completely randomized design,** the experimental units are assigned to the treatments completely at random.
- A **randomized block design** forms groups (**blocks**) of experimental units that are similar with respect to a variable that is expected to affect the response. Treatments are assigned at random within each block. Responses are then compared within each block and combined with the responses of other blocks after accounting for the differences between the blocks. When blocks are chosen wisely, it is easier to determine if one treatment is more effective than another.
- A **matched pairs design** is a common form of randomized block design for comparing two treatments. In some matched pairs designs, each subject receives both treatments in a random order. In others, two very similar subjects are paired, and the two treatments are randomly assigned within each pair.

- When an observed difference in responses between the groups in an experiment is so large that it is unlikely to be explained by chance variation in the random assignment, the results are called **statistically significant.**
- **The scope of inference** for a study describes the types of conclusions we can make based on how the data were collected.
  - **Inference about a population** requires that individuals are randomly selected from the population.
  - **Inference about cause and effect** requires a well-designed experiment with random assignment of treatments and statistically significant results.
- (Not required for the AP® Statistics exam) Studies involving humans must be screened in advance by an **institutional review board.** All participants must give their **informed consent** before taking part. Any information about the individuals in the study must be kept **confidential.**

AP® EXAM TIP
AP® Daily Videos

Review the content of this section and get extra help by watching the AP® Daily Videos for Topics 3.5, 3.6, and 3.7, which are available in AP® Classroom.

SECTION 3C Exercises

Confounding

1. **Good for the gut?** Is eating fish good for the gut? Researchers tracked 22,000 male physicians for 22 years. Those who reported eating seafood of any kind at least 5 times per week had a 40% lower risk of colon cancer than those who said they ate seafood less than once a week.[54] Explain how confounding makes it unreasonable to conclude that eating seafood causes a reduction in the risk of colon cancer based on this study.

pg 272

2. **Vegetables and eye health** In a long-term study, researchers collected information about the diets of participants and whether or not they developed an eye condition called age-related macular degeneration (AMD). People who ate lots of green leafy vegetables, which are high in lutein and zeaxanthin, had a lower risk of AMD.[55] Explain how confounding makes it unreasonable to conclude that eating green leafy vegetables causes a reduction in the risk of AMD based on this study.

3. **Breakfast** In a study of 294 British high school students, each student was asked to complete a 7-day food journal. Researchers used these journals to classify the students as rare, occasional, or frequent breakfast eaters. The mean score on an important national exam was significantly lower for those students who rarely ate breakfast compared to frequent eaters.[56] Based on this study, is it reasonable to conclude that eating breakfast more often causes higher test scores? Explain your reasoning.

4. **Screen time** In an investigation of early childhood development, researchers studied a sample of 2441 mothers and children. When the children were 24 and 36 months old, their mothers reported how much screen time their children got in a typical day. Then, at ages 36 months and 60 months, the children were assessed on various developmental milestones. The study found that higher amounts of screen time were associated with lower scores on the developmental test.[57] Based on this study, is it reasonable to conclude that more screen time causes impaired development in young children? Explain your reasoning.

The Language of Experiments

5. **Increasing response rates** Nonresponse can be a major source of bias in surveys, so researchers would like to investigate strategies for increasing response rates. In a study of households in Atlanta, researchers sent surveys to more than 4000 randomly selected households. They varied the length of the survey (11, 26, or 55 questions), the survey incentive ($2 cash or gift card), the number of follow-ups (0, 2, or 3), and the incentive for participating in a 5-night in-home sleep study ($100, $150, or $200).[58]

pg 273

(a) What are the experimental units in this experiment?

(b) List the factors in this experiment and the number of levels for each factor.

(c) If the researchers used every possible combination of levels to form the treatments, how many treatments were included in the experiment?

(d) List two of the treatments.

6. **Fabric science** A maker of fabric for clothing is setting up a new line to "finish" the raw fabric. The line will use either metal rollers or natural-bristle rollers to raise the surface of the fabric; a dyeing-cycle time of either

30 or 40 minutes; and a temperature of either 150°C or 175°C. Three specimens of fabric will be subjected to each treatment and scored for quality.

(a) What are the experimental units in this experiment?

(b) List the factors in this experiment and the number of levels for each factor.

(c) If the researchers used every possible combination of levels to form the treatments, how many treatments were included in the experiment?

(d) List two of the treatments.

7. **Growing in the shade** The ability to grow in shade may help pine trees found in the dry forests of Arizona to resist drought. How well do these pines grow in shade? Investigators planted pine seedlings in a greenhouse in either full light, light reduced to 25% of normal by shade cloth, or light reduced to 5% of normal. At the end of the study, they dried the young trees and weighed them. Identify the explanatory and response variables, the experimental units, and the treatments.

8. **Want a snack?** Can snacking on fruit rather than candy reduce later food consumption? Researchers randomly assigned 12 people to eat either 65 calories of berries or 65 calories of candy. Two hours later, all 12 people were given an unlimited amount of pasta to eat. The people who ate the berries consumed 133 fewer calories of pasta, on average. Identify the explanatory and response variables, the experimental units, and the treatments.

Comparison and Control Groups

9. **Shark repellents** Many people like to surf, but there are dangers involved, including the possibility of a shark attack. In response, companies have produced shark repellents in various forms such as magnetic bracelets and scented wax. But do they work? Researchers tested 5 different products in a clever experiment. They built 5 surfboards, each utilizing one of the deterrents. A sixth board had no deterrent but was otherwise identical to the other boards. Then they attached raw tuna about 30 centimeters below each board, where a surfer's foot might be, and recorded the behavior of sharks that came near the board. The shark ate the bait in 96% of the trials with the sixth board, but only 40% of the time with the most effective deterrent.[59] Explain why it was necessary to include the sixth board with no deterrent.

pg 275

10. **Dead jellyfish** Research by Andrew Sweetman of Norway's International Research Institute of Stavanger focused on whether deep-sea scavengers consume dead jellyfish. His team lowered platforms piled with dead jellyfish and other platforms piled with dead mackerel more than 4000 feet into Norway's largest fjord; they found that hagfish, crabs, and other scavengers consumed the jellyfish in a few hours — faster than they consumed the mackerel.[60] Explain why it was necessary to include platforms containing mackerel in the experiment, even though the study was focused on the consumption of jellyfish.

11. **Cocoa and blood flow** A study of blood flow involved 27 healthy people aged 18 to 72. Each subject consumed a cocoa beverage containing 900 milligrams of flavonols daily for 5 days. Using a finger cuff, the subjects' blood flow was measured on the first and fifth days of the study. After 5 days, researchers measured what they called "significant improvement" in blood flow and the function of the cells that line the blood vessels.[61] What flaw in the design of this experiment makes it impossible to say if the cocoa really caused the improved blood flow? Explain your answer.

12. **Reducing unemployment** Will cash bonuses speed the return to work of unemployed people? A state department of labor notes that last year 68% of people who filed claims for unemployment insurance found a new job within 15 weeks. As an experiment, this year the state offers $500 to people filing unemployment claims if they find a job within 15 weeks. The percentage who do so increases to 77%. What flaw in the design of this experiment makes it impossible to say if the bonus really caused the increase? Explain your answer.

Blinding and the Placebo Effect

13. **Oils and inflammation** Extracts from avocado and soybean oils have been shown to slow cell inflammation in test tubes. But will taking avocado and soybean unsaponifiables (called ASU) help relieve pain for subjects with joint stiffness due to arthritis? In an experiment, 345 people were randomly assigned to receive either 300 milligrams of ASU daily for three years or a placebo daily for three years.[62]

pg 277

(a) Could blinding be used in this experiment? Explain your reasoning.

(b) Why is blinding an important consideration in this context?

14. **Supplements for testosterone** As men age, their testosterone levels gradually decrease. This may cause a reduction in energy, an increase in fat, and other undesirable changes. Do testosterone supplements reverse some of these effects? A study in the Netherlands assigned 237 men aged 60 to 80 with low or low-normal testosterone levels to either a testosterone supplement or a placebo.[63]

(a) Could blinding be used in this experiment? Explain your reasoning.

(b) Why is blinding an important consideration in this context?

15. **A more expensive placebo** In a recent study, researchers had volunteers rate the pain of an electric shock before and after taking a new medication. Half of the subjects were told the medication cost $2.50 per dose, while the other half were told the medication cost $0.10 per dose. In reality, both medications were placebos. Of the "cheap" placebo users, 61% experienced pain relief, while 85% of the "expensive" placebo users experienced pain relief.[64] Explain how the results of this study support the idea of a placebo effect.

16. **Doctors and the placebo effect** In an experiment about how the placebo effect works, researchers assigned college students to the role of either doctor or patient. The "patients" were given a heat stimulus on their arm after having applied one of two different creams. The creams were administered by the "doctors," who were told that one was a pain reliever and one was a placebo. In reality, both were placebos. Patients who were treated by doctors who thought they were administering the real pain reliever felt less pain than the patients whose doctors thought they were administering the placebo.[65] Explain how this supports the value of double-blind experiments.

17. **Meditation for anxiety** An experiment that claimed to show that meditation lowers anxiety proceeded as follows. The experimenter interviewed the subjects and rated their level of anxiety. Then the subjects were randomly assigned to two groups. The experimenter taught one group how to meditate; this group then meditated daily for a month. The other group was simply told to relax more. At the end of the month, the experimenter interviewed all the subjects again and rated their anxiety level. The meditation group now had less anxiety. Psychologists said that the results were suspect because the ratings were not blind. Explain what this means and how lack of blindness could affect the reported results.

18. **Side effects** Even if an experiment is double-blind, the blinding might be compromised if side effects of the treatments differ. For example, suppose researchers at a skin-care company are comparing their new acne treatment against a treatment offered by the leading competitor. Fifty subjects are assigned at random to each treatment, and the company's researchers will rate the improvement for each of the 100 subjects. The researchers aren't told which subjects received which treatments, but they know that their new acne treatment causes a slight reddening of the skin. How might this knowledge compromise the blinding? Explain why this is an important consideration in the experiment.

Random Assignment

19. **Pizza prices** The cost of a meal might affect how customers evaluate and appreciate food. To investigate, researchers worked with an Italian all-you-can-eat buffet to perform an experiment. A total of 139 customers were randomly assigned to pay either $4 or $8 for the buffet and then asked to rate the quality of the pizza on a 9-point scale.[66]

pg 279

(a) Describe how the researchers could have randomly assigned the customers to the two treatments in a completely randomized design.

(b) Explain the purpose of random assignment in this experiment.

20. **Power and speech** Recent research suggests that people's speech patterns are influenced by how much power they think they have in a particular negotiation. In one study, researchers randomly assigned 161 college students to two groups. The students in one group were told they had higher status or inside information, giving them the sense that they had more power than those they were negotiating with. Students in the other group were told something that lowered their perceived power. All students were given the same statement to read and researchers analyzed the speech patterns of each student.[67]

(a) Describe how the researchers could have randomly assigned the students to the two treatments in a completely randomized design.

(b) Explain the purpose of random assignment in this experiment.

21. **Layoffs and "survivor guilt"** Workers who survive a layoff of other employees at their location may suffer from "survivor guilt." A study of survivor guilt and its effects used as subjects 120 students who were offered an opportunity to earn extra course credit by doing proofreading. Each subject worked in the same cubicle as another student, who was an accomplice of the experimenters. At a break midway through the work, one of three things happened:

Treatment 1: The accomplice was told to leave; it was explained that this was because the accomplice performed poorly.

Treatment 2: It was explained that unforeseen circumstances meant there was only enough work for one person. By "chance," the accomplice was chosen to be laid off.

Treatment 3: Both students continued to work after the break.

The subjects' work performance after the break was compared with their performance before the break. Overall, subjects worked harder when told the other student's dismissal was random.[68] Describe how the three treatments can be randomly assigned to the subjects so that each treatment has the same number of subjects.

22. **Reversing cognitive decline** As people age, they are at an increased risk of cognitive decline, including dementia. Can cognitive decline be slowed down or even reversed? In a recent experiment involving 160 older adults, half were randomly assigned to a heart-healthy DASH (Dietary Approaches to Stop Hypertension) diet, with the other half receiving no prescribed diet. Likewise, half of the adults in each diet group were randomly assigned to follow a specific exercise regimen and the other half were not told to exercise. After 6 months, the improvement in cognitive function was recorded for the subjects in each of the four groups. Subjects who were assigned the DASH diet and exercise regimen did best.[69] Describe how the four treatments can be randomly assigned to the subjects so that each treatment has the same number of subjects.

23. **Comparing treatments** A large study used records from Canada's national health care system to compare the effectiveness of two ways to treat prostate disease. The two treatments were traditional surgery and a new method that does not require surgery. The records described many cases where doctors had chosen the method of treatment for their patients. The study found that patients treated by the new method were significantly more likely to die within 8 years.[70] What flaw in this study makes it impossible to determine if the new method is more dangerous? Explain your reasoning.

24. **Stronger players** A football coach hears that a new exercise program will increase upper-body strength better than lifting weights. He is eager to test this new program in the off-season with the players on his high school team. The coach decides to let his players choose which of the two treatments they will undergo for 3 weeks — the new exercise program or weight lifting. He will use the number of push-ups a player can do at the end of the experiment as the response variable. What flaw in this study makes it impossible to determine if the new exercise program is better for increasing upper-body strength? Explain your reasoning.

Controlling Other Variables

25. **Pay to play?** Are politicians more available to people who donate money to their campaigns? Researchers had members of a political organization attempt to schedule an appointment with members of Congress to discuss the banning of a particular chemical. Callers were identified as either a "local campaign donor" or a "local constituent," with the identification determined at random for each member of Congress contacted. Otherwise, the protocol for the calls was identical, including the details of the meeting request, which were delivered by email. *Results*: "Donors" were more successful in obtaining meetings and were given better access to higher-level staff, including the member of Congress.[71]

pg 281

(a) Identify one variable that researchers kept the same during the experiment.

(b) Describe a benefit of keeping this variable the same for all experimental units.

26. **Apple calories** In an experiment to determine how calories in solid foods affect appetite, researchers randomly assigned volunteers to one of four treatments: 150 calories of apple slices, 150 calories of applesauce, 150 calories of apple juice, or 150 calories of apple juice with added fiber. Fifteen minutes later, researchers provided each participant with a large bowl of pasta and measured how many calories each person consumed. *Results:* The apple slice group ate 190 fewer calories, on average, compared to the other three groups.[72]

(a) Identify one variable that researchers kept the same during the experiment.

(b) Describe a benefit of keeping this variable the same for all experimental units.

27. **Boosting preemies** Do blood-building drugs help brain development in babies born prematurely? Researchers randomly assigned 53 babies, born more than a month premature and weighing less than 3 pounds, to one of three groups. Babies received either injections of erythropoietin (EPO) three times a week, darbepoetin once a week for several weeks, or no treatment. The results: the babies who got the medicines scored much better by age 4 on measures of intelligence, language, and memory than the babies who received no treatment.[73]

(a) Identify one characteristic of the subjects that was kept the same in this experiment.

(b) Describe a benefit and a drawback of keeping this characteristic the same for all subjects.

28. **The effects of day care** Does preschool help children stay in school and hold good jobs later in life? The Carolina Abecedarian Project (the name suggests the ABCs) has followed a group of 111 children for more than 40 years. Back then, these individuals were all healthy, low-income, Black infants in Chapel Hill, North Carolina. All the infants received nutritional supplements and help from social workers. Half were also assigned at random to an intensive preschool program. The results: children who were assigned to the preschool program had higher standardized test scores and were less likely to repeat a grade in school.[74]

(a) Identify one characteristic of the subjects that was kept the same in this experiment.

(b) Describe a benefit and a drawback of keeping this characteristic the same for all subjects.

29. **Precise offers** People often use round prices as first offers in a negotiation. But would a more precise number suggest that the offer was more reasoned and informed? In an experiment, 238 adults played the role of a person selling a used car. Each adult received one of three initial offers: \$2000, \$1865 (a precise under-offer), and \$2135 (a precise over-offer). After hearing the initial offer, each subject made a counter-offer.[75] Describe an experiment that uses a completely randomized design to determine if the counter-offers deviate more from the \$2000 initial offer than from the other offers, on average.

30. **Sealing your teeth** Many children have their molars sealed to help prevent cavities. In an experiment, 120 children aged 6–8 were randomly assigned to a control group, a group in which sealant was applied and reapplied periodically for 36 months, or a group in which fluoride varnish was applied and reapplied periodically for 42 months.[76] Describe an experiment that uses a completely randomized design to determine if sealing molars results in fewer cavities, on average.

Randomized Block Designs

31. **A fruitful experiment** A citrus farmer wants to know which of three fertilizers (A, B, and C) is most effective for increasing the number of oranges on orange trees. The farmer is willing to use 30 older orange trees and 60 younger orange trees in the experiment. pg 284

(a) Describe a randomized block design for this experiment. Justify your choice of blocks.

(b) Explain why a randomized block design might be preferable to a completely randomized design for this experiment.

32. **In the cornfield** An agriculture researcher wants to compare the yield of 5 corn varieties: A, B, C, D, and E. The field in which the experiment will be carried out increases in fertility from north to south. The researcher therefore divides the field into 25 plots of equal size, arranged in 5 east-west rows of 5 plots each, as shown in the diagram.

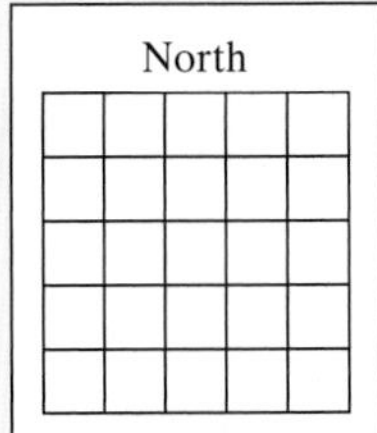

(a) Describe a randomized block design for this experiment. Justify your choice of blocks.

(b) Explain why a randomized block design might be preferable to a completely randomized design for this experiment.

33. **Doctors and nurses** Nurse-practitioners are nurses with advanced qualifications who often act much like primary-care physicians. Are they as effective as doctors at treating patients with chronic conditions? An experiment was conducted with 1316 patients who had been diagnosed with asthma, diabetes, or high blood pressure. Within each condition, patients were randomly assigned to either a doctor or a nurse-practitioner. The response variables included measures of the patients' health and of their satisfaction with their medical care after 6 months.[77]

(a) Which are the blocks in this experiment: the different diagnoses (asthma, diabetes, or high blood pressure) or the care provider (nurse or doctor)? Explain your answer.

(b) Explain why a randomized block design is preferable to a completely randomized design in this context.

34. **Comparing cancer treatments** Researchers want to design an experiment to compare three therapies for two common types of cancer. They recruit 600 subjects with the first type of cancer and randomly assign them to one of the three therapies. Researchers do the same with 300 subjects who have the second type of cancer.

(a) Which are the blocks in this experiment: the three cancer therapies or the two types of cancer? Explain your answer.

(b) Explain why a randomized block design is preferable to a completely randomized design in this context.

Matched Pairs Designs

35. **Valve surgery** Medical researchers want to compare the success rate of a new noninvasive method for replacing heart valves using a cardiac catheter with the success rate of traditional open-heart surgery. They have 40 male patients, ranging in age from 55 to 75, who need valve replacement. One of several response variables will be the percentage of blood that flows backward — in the wrong direction — through the valve on each heartbeat. Describe an experiment that uses a matched pairs design to compare these methods. Explain your method of pairing and how you'll randomly assign the treatments. pg 286

36. **Fresh bread** Who doesn't love the smell of freshly baked bread? Does the smell of fresh bread also encourage people to eat more? Researchers recruit 30 volunteers and ask them to fill out a survey in a room that either has no smell or is infused with the smell of freshly baked bread. After filling out the survey, volunteers will be given as much vegetable soup as they want and researchers will record the amount of soup consumed.[78] Describe an experiment that uses a matched pairs design to compare these methods. Explain your method of pairing and how you'll randomly assign the treatments.

37. **Chocolate gets my heart pumping** Cardiologists at Athens Medical School in Greece wanted to test if chocolate affects blood vessel function. The researchers recruited 17 healthy young volunteers, who were each given a 3.5-ounce bar of dark chocolate, either bittersweet or fake chocolate. On another day, the volunteers received the other treatment. The order in which subjects received the bittersweet and fake chocolate was determined at random. The subjects had no chocolate outside the study, and investigators didn't know if a subject had eaten the real or the fake chocolate. An ultrasound was taken of each volunteer's upper arm to observe the functioning of the cells in the walls of the main artery. The researchers found that blood vessel function was improved when the subjects ate the bittersweet chocolate, and that there were no such changes when they ate the placebo (fake chocolate).[79]

(a) What type of design did the researchers use in their study?

(b) Explain why the researchers chose this design instead of a completely randomized design.

(c) Why is it important to randomly assign the order of the treatments for the subjects?

38. **Bats in the forest** Although we typically think of bats as living in caves, bats play an important role in the health of a forest according to a recent study. At the beginning of summer, researchers identified 20 large, sub-canopy forest plots and built two large structures in each plot. One of the structures was covered every night with a mesh netting that prevented bats from entering (but had big enough holes for insects to move in and out). The other structure was left uncovered. At the end of the summer, the insect density on seedlings in the plot was 3 times greater in the bat-excluded areas and seedling defoliation was 5 times greater in the bat-excluded areas.[80]

(a) What type of design did the researchers use in their study?

(b) Explain why the researchers chose this design instead of a completely randomized design.

(c) Describe how the researchers could use random assignment in their study.

Inference for Experiments

39. **I work out a lot** Are people influenced by what others say? Michael conducted an experiment in front of a popular gym. As people entered, he asked them how many days they typically work out per week. As he asked the question, he showed the subjects one of two clipboards, determined at random. Clipboard A had the question and many responses written down, where the majority of responses were 6 or 7 days per week. Clipboard B was the same, except most of the responses were 1 or 2 days per week. The mean response for those shown Clipboard A was 4.68 and the mean response for those shown Clipboard B was 4.21.[81]

pg 290

(a) Calculate the difference (Clipboard A – Clipboard B) in the mean number of days for the two groups.

One hundred trials of a simulation were performed to see what differences in means would occur due only to chance variation in the random assignment, assuming that the responses on the clipboard don't matter. The results are shown in the dotplot.

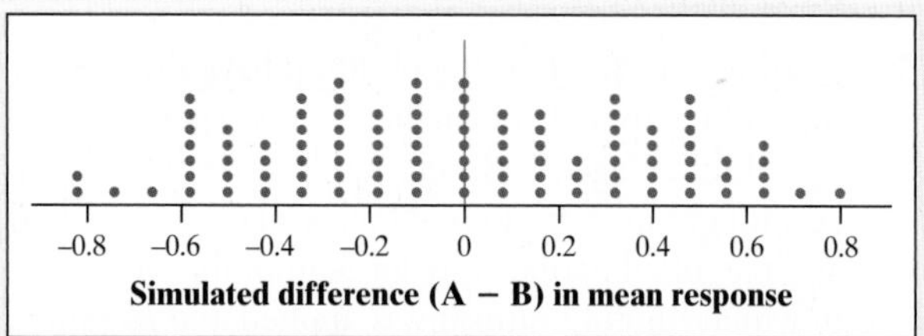

(b) There is one dot at 0.72. Explain what this dot means in this context.

(c) Use the results of the simulation to determine if the difference in means from part (a) is statistically significant. Explain your reasoning.

40. **A louse-y situation** A study published in the *New England Journal of Medicine* compared two medicines to treat head lice: an oral medication called ivermectin and a topical lotion containing malathion. Researchers studied 812 people in 376 households in seven areas around the world. Of the 185 households randomly assigned to ivermectin, 171 were free from head lice after 2 weeks, compared with only 151 of the 191 households randomly assigned to malathion.[82]

(a) Calculate the difference (Ivermectin – Malathion) in the proportion of households that were free from head lice in the two groups.

One hundred trials of a simulation were performed to see what differences in proportions would occur due only to chance variation in the random assignment, assuming that the type of medication doesn't matter. The results are shown in the dotplot.

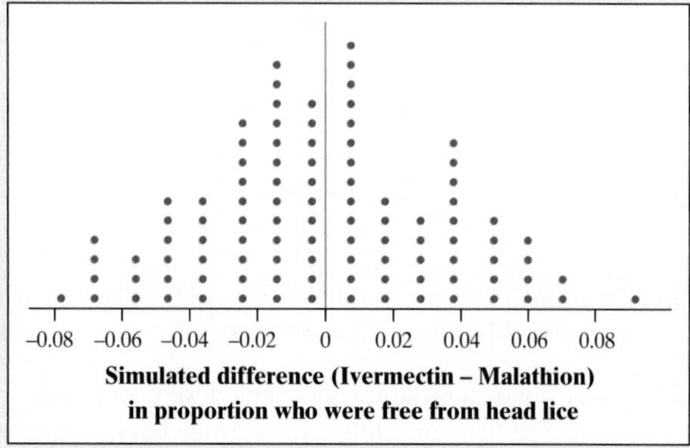

(b) There is one dot at 0.09. Explain what this dot means in this context.

(c) Use the results of the simulation to determine if the difference in proportions from part (a) is statistically significant. Explain your reasoning.

41. **The Physicians' Health Study** Does regularly taking aspirin help protect people against heart attacks? Does regularly taking beta-carotene help prevent cancer? The Physicians' Health Study I was a medical experiment that helped answer these questions. The subjects in this experiment were 21,996 male physicians. Each of the physicians was assigned to one of the following treatment groups: Group 1: aspirin and beta-carotene, Group 2: aspirin and placebo, Group 3: placebo and beta-carotene, or Group 4: placebo and placebo. The physicians were followed for several years and researchers recorded whether each of the physicians had a heart attack or developed cancer.[83]

(a) Why did researchers randomly assign the subjects to the four treatments?

(b) The difference (Group 2 − Group 4) in the proportion of subjects who had heart attacks was statistically significant. Explain what it means that this difference was statistically significant.

(c) The difference (Group 3 − Group 4) in the proportion of subjects who developed cancer was *not* statistically significant. Explain what it means that this difference was not statistically significant.

42. **Noninferiority study** In some situations, researchers hope to find results that are not statistically significant. People with anxiety disorders are often helped by taking medication. But can nondrug treatments produce equally good results? This would be helpful to know, especially if the commonly used drugs are expensive or have side effects. An experiment was conducted to compare mindfulness-based stress reduction (MBSR) with a commonly prescribed drug for anxiety, escitalopram. The 208 subjects were randomly assigned to the two treatments, and a blinded clinical interviewer assessed each subject at the beginning and the end of the study using the Clinical Global Impression of Severity (CGI-S) scale.[84]

(a) Why did the researchers randomly assign the subjects to the two treatments?

(b) Explain why it was important that the evaluator was blind.

(c) The difference (MBSR − Escitalopram) in mean improvement on the CGI-S scale was *not* statistically significant. Explain what it means that this difference was not statistically significant.

The Scope of Inference

43. **Foster care versus orphanages** Do abandoned children placed in foster homes do better than similar children placed in an institution? The Bucharest Early Intervention Project found statistically significant evidence that they do. The subjects were 136 young children abandoned at birth and living in orphanages in Bucharest, Romania. Half of the children, chosen at random, were placed in foster homes. The other half remained in the orphanages.[85] Foster care was not readily available in Romania at the time and so was paid for by the study.

pg 292

(a) It is reasonable to make a cause-and-effect conclusion based on this study? Explain your answer.

(b) What is the largest population to which we can generalize the results of this study? Explain your answer.

44. **Exercise and memory** To study strength training and memory, researchers randomly assigned 46 young adult volunteers to two groups. After both groups were shown 90 pictures, one group had to bend and extend one leg against heavy resistance 60 times. The other group stayed relaxed, while the researchers used the same exercise machine to bend and extend their legs with no resistance. Two days later, each subject was shown 180 pictures — the original 90 pictures plus 90 new pictures and asked to identify which pictures were shown 2 days earlier. The heavy resistance group was statistically significantly more successful in identifying these pictures than was the relax group.[86]

(a) It is reasonable to make a cause-and-effect conclusion based on this study? Explain your answer.

(b) What is the largest population to which we can generalize the results of this study? Explain your answer.

45. **Reflected glory** In a classic study, researchers investigated the tendency of sports fans to "bask in reflected glory" by associating themselves with winning teams. In the study, researchers called randomly selected students at a major university with a highly ranked football team. They randomly assigned half the students to answer questions about a recent game the team lost, and asked the other half about a recent game the team won. If the students were able to correctly identify the winner (showing they were fans of the team), they were asked to describe the game. Students were statistically significantly more likely to identify themselves with the team by their use of the word *we* in the description ("We won the game" versus "They won the game") when describing a win.[87] What conclusions can we draw from this study? Explain your answer.

46. **Berry good** Eating blueberries and strawberries might improve heart health, according to a long-term study of 93,600 women who volunteered to take part. These

berries are high in anthocyanins due to their pigment. Women who reported consuming the most anthocyanins had a statistically significantly smaller risk of heart attack compared to the women who reported consuming the least. What conclusion can we draw from this study? Explain your reasoning.[88]

Data Ethics*

47. ***Tuskegee syphilis study** In 1932, the U.S. Public Health Service and the Centers for Disease Control and Prevention recruited 600 male Black sharecroppers from Macon County, Alabama, to participate in a medical study. The goal of the study was to observe the natural progression of untreated syphilis, which roughly two-thirds of the men in the study had in latent form. The men in the study were told they were being given free health care and that the study would last 6 months. Instead, the study lasted 40 years, the men were never told about their diagnosis, and they were given placebo treatments, even though effective treatments for syphilis had been developed by 1947.[89] Which principle of data ethics did this study violate? Explain your answer. *Note: This terribly unethical study led to the creation of the U.S. government's Office for Human Research Protections (OHRP) and laws requiring institutional review boards to prevent such studies from ever occurring again.*

48. ***Facebook emotions** In cooperation with researchers from Cornell University, Facebook randomly selected almost 700,000 users for an experiment in "emotional contagion." Users' news feeds were manipulated (without their knowledge) to selectively show postings from their friends that were either more positive or more negative in tone, and the emotional tone of their own subsequent postings was measured. The researchers found evidence that people who read emotionally negative postings were more likely to post messages with a negative tone, whereas those who read positive messages were more likely to post messages with a positive tone.[90] Which principle of data ethics did this study violate? Explain your answer.

49. ***The Willowbrook hepatitis studies** In the 1960s, children entering the Willowbrook State School, an institution for intellectually disabled persons located on Staten Island in New York, were deliberately infected with hepatitis. The researchers argued that almost all children in the institution quickly became infected anyway. The studies showed for the first time that two strains of hepatitis existed. This finding contributed to the development of effective vaccines.[91] Despite these valuable results, the Willowbrook studies are now considered an example of unethical research. Explain why, according to current ethical standards, useful results are not enough to allow a study.

*Exercises 47–50: This is an important topic, but it is not required for the AP® Statistics exam.

50. ***Unequal benefits** Researchers on aging proposed to investigate the effect of supplemental health services on the quality of life of older people. Eligible patients on the rolls of a large medical clinic were to be randomly assigned to treatment and control groups. The treatment group would be offered hearing aids, dentures, transportation, and other services not available without charge to the control group. The IRB believed that providing these services to some, but not other, persons in the same institution raised ethical questions. Do you agree?

For Investigation *Apply the skills from the section in a new context or nonroutine way.*

51. **Magnet treatment** Refer to the example on page 277. Here are the improvements in pain for the group of patients who received the active magnets and the group of patients who received the inactive magnets.[92]

Active	0	0	0	0	1	1	1	4	4	5	5	5	5	6	6
	6	6	6	6	7	7	7	8	8	8	10	10	10	10	
Inactive	0	0	0	0	0	0	0	0	0	0	0	1	1	1	1
	1	2	3	4	4	5									

(a) Launch the *One Quantitative Variable, Multiple Groups* applet at www.stapplet.com. Then enter the variable name, group names, and data. Click the "Begin analysis" button and calculate the difference in the mean improvement (Active − Inactive).

(b) In the "Perform Inference" section of the applet, choose the option to simulate the difference in means and add at least 100 samples. What percentage of the simulated random assignments gave a difference at least as large as the one you calculated in part (a)?

(c) Based on your answer to part (b), is the difference from part (a) statistically significant? Explain your reasoning.

52. **Matched quadruplets** A total of 20 people have agreed to participate in a study of the effectiveness of four weight-loss treatments (A, B, C, and D). The researcher first calculates how overweight each subject is by comparing the subject's current weight with their "ideal" weight. These values are shown in the following table:

Birnbaum	35	Hernandez	25	Moses	25	Smith	29
Brown	34	Jackson	33	Nevesky	39	Stall	33
Brunk	30	Kendall	28	Obrach	30	Tran	35
Cruz	34	Loren	32	Rodriguez	30	Wilansky	42
Deng	24	Mann	28	Santiago	27	Williams	22

(a) Design an experiment that uses blocks of size 4 to investigate which of these treatments works the best. Explain your choice of blocks.

(b) Carry out the random assignment for your design in part (a).

Multiple Choice *Select the best answer for each question.*

53. Can a vegetarian or low-salt diet reduce blood pressure? Subjects with high blood pressure are assigned at random to one of four diets: (1) normal diet with unrestricted salt; (2) vegetarian diet with unrestricted salt; (3) normal diet with restricted salt; and (4) vegetarian diet with restricted salt. This experiment has

(A) one factor — the type of diet.

(B) two factors — high blood pressure and type of diet.

(C) two factors — normal/vegetarian diet and unrestricted/restricted salt.

(D) three factors — people, high blood pressure, and type of diet.

(E) four factors — the four diets being compared.

54. In the experiment described in the preceding exercise, the subjects were randomly assigned to the different treatments. What is the most important reason for this random assignment?

(A) Random assignment eliminates the effects of other variables such as stress and body weight.

(B) Random assignment balances the effects of other variables such as stress and body weight among the four treatment groups.

(C) Random assignment makes it possible to make a conclusion about all people.

(D) Random assignment reduces the amount of variation in blood pressure.

(E) Random assignment prevents the placebo effect from ruining the results of the study.

55. To investigate if standing up while studying affects performance in an algebra class, a teacher assigns half of the 30 students in the class to stand up while studying and assigns the other half to not stand up while studying. To determine who receives which treatment, the teacher identifies the two students who did best on the last exam and randomly assigns one to stand and one to not stand. The teacher does the same for the next two highest-scoring students and continues in this manner until each student is assigned a treatment. Which of the following best describes this plan?

(A) This is an observational study.

(B) This is a completely randomized experiment.

(C) This is an experiment with blocking, but not matched pairs.

(D) This is an experiment with blocking that uses matched pairs.

(E) This is a stratified random sample.

56. A gardener wants to try different combinations of fertilizer (none, 1 cup, 2 cups) and mulch (none, wood chips, pine needles, plastic) to determine which combination produces the highest yield for a variety of green beans. He has 60 green-bean plants to use in the experiment. If he wants an equal number of plants to be assigned to each treatment, how many plants will be assigned to each treatment?

(A) 1

(B) 3

(C) 4

(D) 5

(E) 12

57. Corn variety 1 yielded 140 bushels per acre last year at a research farm. This year, corn variety 2, planted in the same location, yielded only 110 bushels per acre. Based on these results, is it reasonable to conclude that corn variety 1 is more productive than corn variety 2?

(A) Yes, because 140 bushels per acre is greater than 110 bushels per acre.

(B) Yes, because the study was done at the same research farm.

(C) No, because there may be other differences between the two years besides the corn variety.

(D) No, because there was no use of a placebo in the experiment.

(E) No, because the experiment wasn't double-blind.

58. A report in a medical journal notes that the risk of developing Alzheimer's disease among subjects who regularly opted to take the drug ibuprofen was about half the risk of those who did not. Is this good evidence that ibuprofen is effective in preventing Alzheimer's disease?

(A) Yes, because the study was a randomized, comparative experiment.

(B) No, because the effect of ibuprofen is confounded with the placebo effect.

(C) Yes, because the results were published in a reputable professional journal.

(D) No, because this is an observational study.

(E) Yes, because a 50% reduction can't happen just by chance alone.

59. A farmer is conducting an experiment to determine which variety of apple tree, Fuji or Gala, will produce more fruit in the farmer's orchard. The orchard is divided into 20 equally sized square plots. There are 10 trees of each variety and the farmer randomly

assigns each tree to a separate plot in the orchard. What are the experimental unit(s) in this study?

(A) The trees

(B) The plots

(C) The apples

(D) The farmer

(E) The orchard

60. Two essential features of all statistically designed experiments are

(A) comparing two or more treatments; using the double-blind method.

(B) comparing two or more treatments; using chance to assign subjects to treatments.

(C) always having a placebo group; using the double-blind method.

(D) using a block design; using chance to assign subjects to treatments.

(E) using enough subjects; always having a control group.

61. Do product labels influence customer perceptions? To find out, researchers recruited more than 500 adults and asked them to estimate the number of calories, amount of added sugar, and amount of fat in a variety of food products. Half of the subjects were randomly assigned to evaluate products with the word "Natural" on the label, while the other half were assigned to evaluate the same products without the "Natural" label. On average, the products with the "Natural" label were judged to have significantly fewer calories. Based on this study, is it reasonable to conclude that including the word "Natural" on the label causes a reduction in estimated calories?

(A) No, because the adults weren't randomly selected from the population of all adults.

(B) No, because there wasn't a control group for comparison.

(C) No, because association doesn't imply causation.

(D) Yes, because the adults were randomly assigned to the treatments.

(E) Yes, because there were a large number of adults involved in the study.

Recycle and Review *Practice what you learned in previous sections.*

62. **Seed weights (1F)** Biological measurements on the same species often follow a normal distribution quite closely. The weights of seeds of a variety of winged bean are approximately normal with mean 525 milligrams (mg) and standard deviation 110 mg.

(a) What percentage of seeds weigh more than 500 mg?

(b) If we discard the lightest 10% of these seeds, what is the smallest weight among the remaining seeds?

63. **Comparing rainfall (1D)** The boxplots summarize the distributions of average monthly rainfall (in inches) for Tucson, Arizona, and Princeton, New Jersey.[93] Compare these distributions.

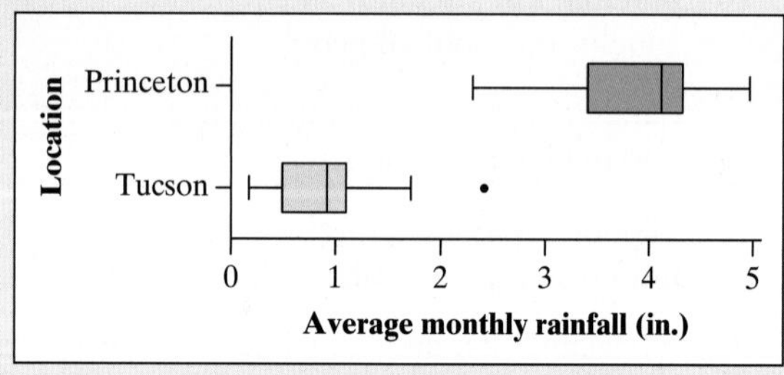

UNIT 3 Wrap-Up

FRAPPY! Free Response AP® Problem, Yay!

Directions: Show all your work. Indicate clearly the methods you use, because you will be scored on the correctness of your methods as well as on the accuracy and completeness of your results and explanations.

In a recent study, 166 adults from the St. Louis area were recruited and randomly assigned to receive one of two treatments for a sinus infection. Half of the subjects received an antibiotic (amoxicillin) and the other half received a placebo.[94]

(a) Describe how the researchers could have assigned treatments to subjects if they wanted to use a completely randomized design.

(b) All the subjects in the experiment had moderate, severe, or very severe symptoms at the beginning of the study. Describe one statistical benefit and one statistical drawback for using subjects with moderate, severe, or very severe symptoms instead of just using subjects with very severe symptoms.

(c) At different stages during the next month, all subjects took the sino-nasal outcome test. After 10 days, the difference in average test scores was not statistically significant. In this context, explain what it means for the difference to be not statistically significant.

(d) One possible way that researchers could have improved the study is to use a randomized block design. Explain how the researchers could have incorporated blocking in their design.

After you finish the FRAPPY!, you can view two example solutions on the book's website (**bfwpub.com/tps7e**). Determine whether you think each solution is "complete," "substantial," "developing," or "minimal." If the solution is not complete, what improvements would you suggest to the student who wrote it? Finally, your teacher will provide you with a scoring rubric. Score your response and note what, if anything, you would do differently to improve your own score.

UNIT 3 REVIEW

SECTION 3A Introduction to Data Collection

In this section, you learned that a **population** is the group of all individuals that we want information about. A **sample** is the subset of the population that we use to gather this information. The goal of most **sample surveys** is to use information from the sample to draw conclusions about the population. **Inference about a population** is justified when the sample is selected at random from that population.

You also learned about the difference between observational studies and experiments. **Experiments** deliberately impose treatments on experimental units to see if there is a cause-and-effect relationship between two variables. **Observational studies** look at relationships between two variables by using existing data (**retrospective**) or by following individuals into the future (**prospective**). **Inference about cause and effect** is very challenging with observational studies.

SECTION 3B Sampling and Surveys

Selecting people for a sample because they are easy to locate and letting people choose whether to be in the sample are poor ways to select a sample. Because **convenience sampling** and **voluntary response sampling** will produce estimates that are likely to underestimate or likely to overestimate the value you want to know, these methods of choosing a sample are **biased.**

To avoid bias in the way the sample is formed, the members of the sample should be chosen at random. One way to do this is with a **simple random sample (SRS),** which is equivalent to selecting well-mixed slips of paper from a hat without replacement. It is often more convenient to select an SRS using technology or a table of random digits.

Three other random sampling methods are stratified sampling, cluster sampling, and systematic sampling. To obtain a **stratified random sample,** divide the population into non-overlapping groups (**strata**) of individuals that are likely to have similar responses, select an SRS from each stratum, and combine the chosen individuals to form the sample. Stratified random samples can produce estimates with much greater precision than is possible with simple random samples. To obtain a **cluster sample,** divide the population into non-overlapping groups (**clusters**) of individuals that are in similar locations, randomly select clusters, and use every individual in the chosen clusters. Cluster samples are easier to obtain than simple random samples or stratified random samples, but they may not produce very precise estimates. To obtain a **systematic random sample,** choose a value of *k* based on the population size and desired sample size. Randomly select a number from 1 to *k* to determine which member of the population to survey first, and then survey every *k*th member thereafter. Systematic random samples can be easier to obtain than other types of random samples.

Several additional sources of bias can affect estimates from a sample. **Undercoverage** occurs when the sampling method systematically underrepresents one part of the population. **Nonresponse** describes when answers cannot be obtained from some people who were chosen to be in the sample. Bias can also occur when some people in the sample don't give accurate responses due to question wording, interviewer characteristics, or other factors (**response bias**).

SECTION 3C Experiments

It is difficult to justify a cause-and-effect relationship between an **explanatory variable** and a **response variable** in an observational study because of confounding. Variables are **confounded** when it is impossible to determine which of the variables is causing a change in the response variable. A well-designed experiment can help avoid confounding.

A common type of comparative experiment uses a **completely randomized design.** In this type of design, the experimental units are assigned to the treatments completely at random. With **random assignment,** the treatment groups should be roughly equivalent at the beginning of the experiment. **Replication** means giving each treatment to as many experimental units as possible. This makes it easier to see the effects of the treatments because the effects of other variables are more likely to be balanced among the treatment groups.

During an experiment, it is important that other variables be **controlled** (kept the same) for each experimental unit. Doing so helps avoid confounding and removes a possible source of variation in the response variable. Also, beware of the **placebo effect** — the tendency for people to improve because they expect to, not because of the treatment they are receiving. One way to make sure that all experimental units have the same expectations is to make them **blind** — unaware of which treatment they are receiving. When the people interacting with the subjects and measuring the response variable are also blind, the experiment is called **double-blind.** Many experiments include a **control group** to provide a baseline for measuring the effects of the other treatments.

Blocking in experiments is similar to stratifying in sampling. To form **blocks,** group together experimental units that are similar with respect to a variable that is associated with the response. Then randomly assign the treatments within each block. A **randomized block design** that uses blocks with two experimental units is called a matched pairs design. Blocking helps us estimate the effects of the treatments more precisely because we can account for the variability introduced by the variables used to form the blocks.

The results of an experiment are **statistically significant** if they are too unusual to occur by chance alone. When results from an experiment with random assignment of treatments are statistically significant, we can justify **inference about cause and effect.** In some cases, making a cause-and-effect conclusion is difficult because it is impossible or unethical to perform certain types of experiments. Good data ethics requires that studies are approved by an institutional review board, subjects give informed consent, and individual data are kept confidential.

What Did You Learn?

Learning Target	Section	Related Example on Page(s)	Relevant Chapter Review Exercise(s)
Identify the population and sample in a statistical study.	3A	241	R1
Distinguish between an observational study and an experiment.	3A	244	R1
Determine what inferences are appropriate from an observational study.	3A	245	R1

Learning Target	Section	Related Example on Page(s)	Relevant Chapter Review Exercise(s)
Describe how to select a simple random sample.	3B	251	R2
Describe other random sampling methods: stratified, cluster, systematic; explain the advantages and disadvantages of each method.	3B	257	R2, R3
Identify voluntary response sampling and convenience sampling, and explain how these sampling methods can lead to bias.	3B	260	R2
Explain how undercoverage, nonresponse, question wording, and other aspects of a sample survey can lead to bias.	3B	263	R4
Explain the concept of confounding and how it limits the ability to make cause-and-effect conclusions.	3C	272	R1
Identify the experimental units and treatments in an experiment.	3C	273	R5
Explain the purpose of a control group in an experiment.	3C	275	R6
Describe the placebo effect and explain the purpose of blinding in an experiment.	3C	277	R6
Describe how to randomly assign treatments in an experiment and explain the purpose of random assignment.	3C	279	R5, R6
Explain the purpose of controlling other variables in an experiment.	3C	281	R5
Describe a randomized block design for an experiment and explain the benefits of blocking in an experiment.	3C	284	R5
Describe a matched pairs design for an experiment.	3C	286	R8
Explain the meaning of statistically significant in the context of an experiment.	3C	290	R6
Identify when it is appropriate to make an inference about a population and when it is appropriate to make an inference about cause and effect.	3C	292	R7

UNIT 3 REVIEW EXERCISES

These exercises are designed to help you review the important concepts and skills in the unit.

R1 **Orange juice** (3A, 3C) A recent headline declared that "Orange Juice Protects Cognitive Function." The article went on to describe a study that followed 27,000 men for 26 years and found a substantially lower risk of cognitive troubles later in life for those men who drank the most orange juice.[95]

(a) Was this an observational study or an experiment? Explain your answer.

(b) What is the largest population to which we can generalize the results of this study?

(c) Explain how confounding makes it unreasonable to conclude that drinking more orange juice reduces the risk of cognitive troubles later in life, based on this study.

R2 **Parking problems** (3B) The administration at a high school with 1800 students wants to estimate the proportion of students who are satisfied with the amount of parking spaces available. It isn't practical to contact all students.

(a) Give an example of a way to obtain a voluntary response sample of students. Explain how this method could lead to bias.

(b) Give an example of a way to obtain a convenience sample of students. Explain how this method could lead to bias.

(c) Describe how to select an SRS of 50 students from the school. Explain how using an SRS helps avoid the biases you described in parts (a) and (b).

(d) Describe how to select a systematic random sample of 50 students from the school. What advantage does this method have over an SRS?

R3 **Surveying NBA fans (3B)** The manager of a sports arena wants to learn more about the financial status of the people who are attending a National Basketball Association (NBA) basketball game. The manager would like to give a survey to a representative sample consisting of approximately 10% of the fans in attendance. Ticket prices for the game vary a great deal: seats near the court cost more than $200 each, while seats in the top rows of the arena cost $50 each. The arena is divided into 50 numbered sections, from 101 to 150. Each section has rows of seats labeled with letters from A (nearest the court) to ZZ (top row of the arena).

(a) Explain why it might be difficult to give the survey to an SRS of fans.

(b) Explain why it would be better to select a stratified random sample using the lettered rows rather than the numbered sections as strata. What is the benefit of using a stratified sample in this context?

(c) Explain how to select a cluster sample of fans. What is the benefit of using a cluster sample in this context?

R4 **Been to the movies? (3B)** An opinion poll calls 2000 randomly selected telephone numbers, then asks to speak with an adult member of the household. The interviewer asks, "Box office revenues are at an all-time high. How many movies have you watched in a movie theater in the past 12 months?" In all, 1131 people responded. The researchers used the responses to estimate the mean number of movies adults had watched in a movie theater over the past 12 months.

(a) Describe a potential source of bias related to the wording of the question. Suggest a change that would help fix this problem.

(b) Describe how undercoverage might lead to bias and how this will affect the estimate.

(c) Describe how nonresponse might lead to bias and how this will affect the estimate.

R5 **Ugly fries (3C)** Few people want to eat discolored french fries. To prevent spoiling and to preserve flavor, potatoes are kept refrigerated before being cut for french fries. But immediate processing of cold potatoes causes discoloring due to complex chemical reactions. The potatoes must therefore be brought to room temperature before processing. Researchers want to design an experiment in which tasters will rate the color and flavor of french fries. The potatoes will be freshly picked, stored for a month at room temperature, or stored for a month refrigerated. Once retrieved, the potatoes will then be sliced and cooked immediately or after an hour at room temperature.

(a) Identify the experimental units, the factors, the number of levels for each factor, the treatments, and the response variables.

(b) Describe a completely randomized design for this experiment using 300 potatoes.

(c) A single supplier has made 300 potatoes available to the researchers. Describe a statistical benefit and a statistical drawback of using potatoes from only one supplier.

(d) The researchers decided to do a follow-up experiment using potatoes from several different suppliers. Describe how they should change the design of the experiment to account for the addition of other suppliers.

R6 **An herb for depression? (3C)** Does the herb St. John's wort relieve major depression? Here is an excerpt from the report of one study of this issue: "Design: Randomized, double-blind, placebo-controlled clinical trial."[96] The study concluded that the difference in effectiveness of St. John's wort and a placebo was not statistically significant.

(a) Explain the purpose of the control group in this experiment.

(b) Explain the purpose of random assignment in this experiment.

(c) Why is a double-blind design a good idea in this setting?

(d) Explain what "not statistically significant" means in this context.

R7 **Don't catch a cold! (3C)** A recent study of 1000 students at the University of Michigan investigated how to prevent catching the common cold. The students were randomly assigned to three different cold prevention methods for 6 weeks. Some wore masks, some wore masks and used hand sanitizer, and others took no precautions. The two groups who used masks reported 10% to 50% fewer cold symptoms than those who did not wear a mask.[97]

(a) Does this study allow for inference about a population? Explain your answer.

(b) Does this study allow for inference about cause and effect? Explain your answer.

R8 **Stressful puzzles (3C)** Do people get stressed out when other people watch them work? To find out, researchers recruited 30 volunteers to take part in an experiment.[98] Participants were timed while they completed a word search puzzle, with the researchers standing close by and taking notes or standing at a distance.

(a) Describe a matched pairs design for this experiment, including how treatments will be assigned to experimental units.

(b) Explain a statistical benefit of using a matched pairs design instead of a completely randomized design in this context.

UNIT 3 AP® STATISTICS PRACTICE TEST

Section I: Multiple Choice *Select the best answer for each question.*

T1 When we take a census, we attempt to collect data from

(A) a stratified random sample.

(B) every individual chosen in a simple random sample.

(C) every individual in the population.

(D) a voluntary response sample.

(E) a convenience sample.

T2 You want to select a simple random sample (SRS) of 50 of the 816 students who live in a dormitory on campus. You label the students 001 to 816 in alphabetical order. In the table of random digits, you read the following entries:

95592 94007 69769 33547 72450 16632 81194 14873

The first three students in your sample have labels

(A) 400, 769, 335.

(B) 400, 769, 769.

(C) 559, 294, 007.

(D) 929, 400, 769.

(E) 955, 929, 400.

T3 A study of treatments for angina (pain due to low blood supply to the heart) compared bypass surgery, angioplasty, and use of medications. The study looked at the medical records of thousands of patients with angina whose doctors had chosen one of these treatments. It found that the average survival time of patients given the medications was the highest. What do you conclude?

(A) This study proves that the medications prolong life and should be the treatment of choice.

(B) We can conclude that the medications prolong life because the study was a comparative experiment.

(C) We can't conclude that the medications prolong life because the patients were volunteers.

(D) We can't conclude that the medications prolong life because the groups might differ in ways besides the treatment.

(E) We can't conclude that the medications prolong life because no placebo was used.

T4 How much vitamin C does orange juice contain? A nutrition magazine measures the amount of vitamin C in 50 randomly selected half-gallon containers of a popular brand of orange juice from 10 different grocery stores and concludes that the containers produced by this company do not have as much vitamin C as advertised. Identify the population in this setting.

(A) All grocery stores

(B) All half-gallon containers of orange juice

(C) All half-gallon containers of this brand of orange juice

(D) All half-gallon containers of this brand of orange juice at these 10 stores

(E) All 50 of the half-gallon containers selected for this study

T5 Consider an experiment to investigate the effectiveness of different insecticides in controlling pests and their impact on the productivity of tomato plants. What is the best reason for randomly assigning treatments (Brand A, Brand B, Brand C) to the experimental units (farms)?

(A) Random assignment eliminates the effects of other variables, such as soil fertility.

(B) Random assignment eliminates chance variation in the responses.

(C) Random assignment allows researchers to generalize conclusions about the effectiveness of the insecticides to all farms.

(D) Random assignment will tend to average out all other uncontrolled factors such as soil fertility so that they are not confounded with the treatment effects.

(E) Random assignment helps avoid bias due to the placebo effect.

T6 Researchers randomly selected 1700 people from Canada and rated the happiness of each person. Ten years later, the researchers followed up with each person and found that people who were initially rated as happy were less likely to have a heart problem.[99] Which of the following is the most appropriate conclusion based on this study?

(A) Happier people in Canada are less likely to have heart problems.

(B) Happier people in the study were less likely to have heart problems.

(C) Happiness causes better heart health for all people.

(D) Happiness causes better heart health for Canadians.

(E) Happiness caused better heart health for the 1700 people in the study.

T7 The sales force for a publishing company is constantly on the road trying to sell books. As a result, each salesperson accumulates many travel-related expenses that they charge to a company-issued credit card. To prevent fraud, management hires an outside company to audit a sample of these expenses. For each salesperson, the auditor prints out the credit card statements for the entire year, randomly chooses one of the first 20 expenses to examine, and then examines every 20th expense from that point on. Which type of sampling method is the auditor using for each salesperson?

(A) Convenience sampling

(B) Simple random sampling

(C) Stratified random sampling

(D) Cluster sampling

(E) Systematic random sampling

T8 Bias in a sampling method is

(A) any difference between the estimate from a sample and the truth about the population.

(B) the difference between the estimate from a sample and the truth about the population due to using chance to select a sample.

(C) any difference between the estimate from a sample and the truth about the population due to practical difficulties such as contacting the subjects selected.

(D) any difference between the estimate from a sample and the truth about the population that tends to occur in the same direction whenever you use this sampling method.

(E) any difference between the estimate from an initial sample and the estimate from a second sample from the same population.

T9 You wonder if TV ads are more effective when they are longer or repeated more often, or both. So you design an experiment. You prepare 30-second and 60-second ads for a camera. Your subjects all watch the same TV program, but you assign them at random to four groups. One group sees the 30-second ad once during the program; another sees it three times; the third group sees the 60-second ad once; and the last group sees the 60-second ad three times. You ask all subjects how likely they are to buy the camera. Which of the following best describes the design of this experiment?

(A) This is a completely randomized design with one explanatory variable (factor).

(B) This is a completely randomized design with two explanatory variables (factors).

(C) This is a completely randomized design with four explanatory variables (factors).

(D) This is a randomized block design, but not a matched pairs design.

(E) This is a matched pairs design.

T10 Can texting make you healthier? Researchers randomly assigned 700 Australian adults to receive either usual health care or usual heath care plus automated text messages with positive messages, such as "Walking is cheap. It can be done almost anywhere. All you need is comfortable shoes and clothing." The group that received the text messages showed a statistically significant increase in physical activity.[100] What is the meaning of "statistically significant" in this context?

(A) The results of this study are very important.

(B) The results of this study should be generalized to all people.

(C) The difference in physical activity for the two groups is greater than 0.

(D) The difference in physical activity for the two groups is very large.

(E) The difference in physical activity for the two groups is larger than the difference that could be expected to happen by chance alone.

Section II: Free Response *Show all your work. Indicate clearly the methods you use, because you will be graded on the correctness of your methods as well as on the accuracy and completeness of your results and explanations.*

T11 Elephants sometimes damage trees in Africa. It turns out that elephants dislike bees. They recognize beehives in areas where they are common and avoid them. Can this information be used to keep elephants away from trees? Researchers want to design an experiment to answer these questions using 72 acacia trees and three treatments: active hives, empty hives, and no hives.[101]

(a) Identify the experimental units in this experiment.

(b) Explain why it is beneficial to include some trees that have no hives.

(c) Describe how the researchers could carry out a completely randomized design for this experiment. Include a description of how the treatments should be assigned.

T12 Many people start their day with a jolt of caffeine from coffee or a soft drink. Most experts agree that people who consume large amounts of caffeine each day may suffer from physical withdrawal symptoms if they stop ingesting their usual amounts of caffeine. Researchers recruited 11 volunteers who were caffeine dependent and who were willing to take part in a caffeine withdrawal experiment. The experiment was conducted on two 2-day periods that occurred one week apart. During one of the 2-day periods, each subject was given a capsule containing the amount of caffeine normally ingested by that subject in one day. During the other study period, the subjects were given placebos. The order in which each subject received the two types of capsules was randomized. The subjects' diets were restricted during each of the study periods. At the end of each 2-day study period, subjects were evaluated using a tapping task in which they were instructed to press a button 200 times as fast as they could.[102]

(a) Describe a statistical benefit of controlling the diets of subjects during the experiment.

(b) How was blocking used in the design of this experiment? What is the benefit of blocking in this context?

(c) Researchers randomized the order of the treatments to avoid confounding. Explain how confounding might occur if the researchers gave all subjects the placebo first and the caffeine second. In this context, what problem does confounding cause?

(d) Could this experiment have been carried out in a double-blind manner? Explain your answer.

T13 The counseling department at a large high school wants to estimate the proportion of last year's graduating class who are enrolled in a four-year college or university. To do this, they hope to survey 50 of the 800 graduates from last year.

(a) Describe how to select a stratified random sample of 50 students. Explain your choice of strata.

(b) What is the advantage of using a stratified random sample instead of a simple random sample in this context?

(c) When the counselors send an email to the 50 students selected for the sample, only 21 students respond. Explain how nonresponse could lead to bias and what effect the bias will have on the estimated proportion of students who are enrolled in a four-year college or university.

UNIT 3 PROJECT

Response Bias

In this project, your team will design and conduct an experiment to investigate the effects of response bias in surveys.[103] You may choose the topic for your surveys, but you must design your experiment so that it can answer at least one of the following questions.

- Can the wording of a question create response bias?
- Does providing additional information create response bias?
- Do the characteristics of the interviewer create response bias?
- Does anonymity change the responses to sensitive questions?
- Does manipulating the answer choices/order of answer choices change the response?
- Can revealing other individuals' answers to a question create response bias?

---

1. Write a proposal describing the design of your experiment. Be sure to include the following items:

(a) Your chosen topic and which of the bulleted questions you'll try to answer.

(b) A detailed description of how you will obtain your subjects (minimum of 50). Your plan must be practical!

(c) An explanation of the treatments in your experiment and how you will determine which subjects get which treatment.

(d) A clear explanation of how you will incorporate the principles of a good experiment and avoid potentially confounding variables.

(e) Precautions you will take to collect data ethically.

Here are two examples of successful student projects.

- "Cheerleading," by Hailey K.:

  Version A: "Is cheerleading a sport?" (Questioner wearing cheerleading uniform: 80% answered "Yes.")

  Version B: "Is cheerleading a sport?" (Questioner wearing regular clothes: 28% answered "Yes.")

- "Cartoons" by Sean W. and Brian H.:

  Version A: "Do you watch cartoons?" (90% answered "Yes.")

  Version B: "Do you *still* watch cartoons?" (60% answered "Yes.")

2. Once your teacher has approved your design, carry out the experiment. Record your data in a table.

3. Prepare a report that includes a brief introduction, a summary of your data collection process, the data you collected, graphs and summary statistics, the answer to your question of interest, and a discussion of any problems you encountered and how you dealt with them.

Cumulative AP® Practice Test 1

Section I: Multiple Choice *Choose the best answer for each question.*

AP1 You look at real estate ads for houses in Sarasota, Florida. Many houses have prices from \$200,000 to \$400,000. The few houses on the water, however, have prices up to \$15 million. Which of the following statements best describes the distribution of home prices in Sarasota?

(A) The distribution is most likely skewed to the left, and the mean is greater than the median.

(B) The distribution is most likely skewed to the left, and the mean is less than the median.

(C) The distribution is roughly symmetric with a few high outliers, and the mean is approximately equal to the median.

(D) The distribution is most likely skewed to the right, and the mean is greater than the median.

(E) The distribution is most likely skewed to the right, and the mean is less than the median.

AP2 The distribution of test scores for a large statistics class is summarized in the boxplot. If a score of at least 70 is considered passing, about what percentage of students in the class passed the test?

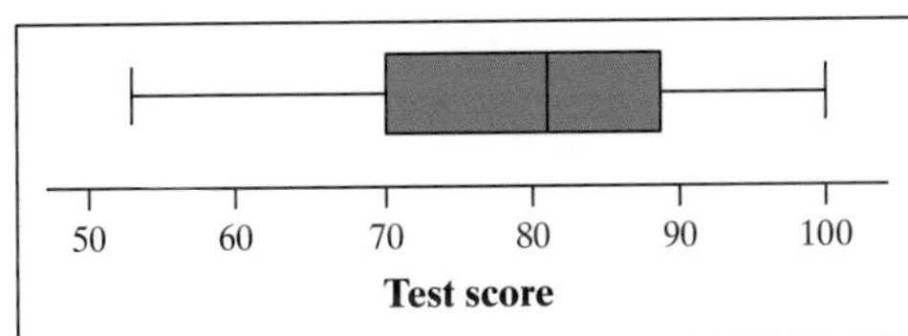

(A) 25
(B) 50
(C) 70
(D) 75
(E) Cannot be determined without the raw data.

AP3 Here are the times, in seconds, to run 400 meters for 10 students in a physical education class:

145 139 126 122 124 130 96 110 118 118

For these data, what is the lower outlier boundary according to the $1.5 \times IQR$ rule?

(A) 18
(B) 96
(C) 100
(D) 105
(E) 110

AP4 For a biology project, you measure the weight in grams (g) and the tail length in millimeters (mm) of a group of mice. The equation of the least-squares line for predicting tail length from weight is

$$\text{predicted tail length} = 20 + 3 \times \text{weight}$$

Which of the following is *not* correct?

(A) A mouse's predicted tail length increases by about 3 mm for each additional gram of weight.

(B) The predicted tail length of a mouse that weighs 38 g is 134 mm.

(C) The correlation between weight and tail length is positive.

(D) If you had measured the tail length in centimeters instead of millimeters, the slope of the regression line would have been $3/10 = 0.3$.

(E) Mice that have a weight of 0 g will have a tail of length 20 mm.

AP5 The weights of bananas from a certain tree are approximately normally distributed with a mean of 156 grams and a standard deviation of 11 grams. Approximately what proportion of bananas from this tree weigh between 150 and 160 grams?

(A) 0.10
(B) 0.35
(C) 0.65
(D) 0.68
(E) 0.95

AP6 In retail stores, there is a lot of competition for shelf space. To investigate the relationship between shelf space and sales, the amount of space allocated to a certain type of product is randomly varied between 3 and 6 linear feet over the next 12 weeks, and weekly sales revenue (in dollars) for that type of product is recorded. Here is some computer output from the study:

```
Predictor        Coef    SE Coef      T      P
Constant       317.940    31.32   10.15  0.000
Shelf length   152.680    6.445   23.69  0.000
S=22.9212          R-Sq=98.2%     R-Sq(adj)=98.1
```

Calculate the residual for the week where the shelf length was 3.5 feet and the sales were \$825.

(A) −\$427
(B) −\$27
(C) \$27
(D) \$427
(E) \$852

AP7 A large set of test scores has mean 60 and standard deviation 18. If each score is doubled, and then 5 is subtracted from the result, what are the mean and standard deviation of the new scores?

(A) mean = 115 and standard deviation = 31

(B) mean = 115 and standard deviation = 36

(C) mean = 120 and standard deviation = 6

(D) mean = 120 and standard deviation = 31

(E) mean = 120 and standard deviation = 36

AP8 Researchers investigating two different drugs to treat migraines in children conducted an experiment. They randomly assigned 328 children ages 9–17 who suffer from migraines to receive either amitriptyline, topiramate, or placebo. The primary outcome was a reduction of at least 50% in the number of headache days. The table summarizes the results.[104] What proportion of the children in the experiment were assigned to amitriptyline or topiramate and had at least a 50% reduction in the number of headache days?

		Drug			
		Amitriptyline	Topiramate	Placebo	Total
Outcome	At least 50% reduction	69	72	40	181
	Less than 50% reduction	63	58	26	147
	Total	132	130	66	328

(A) 0.43

(B) 0.54

(C) 0.55

(D) 0.78

(E) 0.80

AP9 The General Social Survey (GSS), conducted by the National Opinion Research Center at the University of Chicago, is a major source of data on social attitudes in the United States. Once each year, 1500 adults are interviewed in their homes all across the country. The subjects are asked their opinions about sex and marriage; attitudes toward women, welfare, foreign policy; and many other issues. The GSS begins by selecting a sample of counties from the 3000 counties in the country. The counties are divided into urban, rural, and suburban; a separate sample of counties is chosen at random from each group. The counties selected form which type of sample?

(A) A simple random sample

(B) A systematic random sample

(C) A cluster random sample

(D) A stratified random sample

(E) A voluntary response sample

AP10 The distribution of birth weight for infants born at a certain hospital last month had a mean of 128 ounces with a standard deviation of 10 ounces. Which of the following is the best interpretation of the standard deviation?

(A) All the infants born at this hospital last month weighed between 118 and 138 ounces.

(B) All the infants born at this hospital last month had weights that were 10 ounces from the mean of 128 ounces.

(C) About half of the infants born at this hospital last month weighed between 118 and 138 ounces.

(D) The weights of infants born at this hospital last month typically varied by about 10 ounces from the mean of 128 ounces.

(E) About 95% of the infants had weights between 108 and 148 ounces.

AP11 The scatterplot shows the relationship between the amount of saturated fat (g) and the number of calories for 17 different granola bars.[105] How does the point in the upper right corner affect the correlation and the slope of the least-squares regression line? Explain your answer.

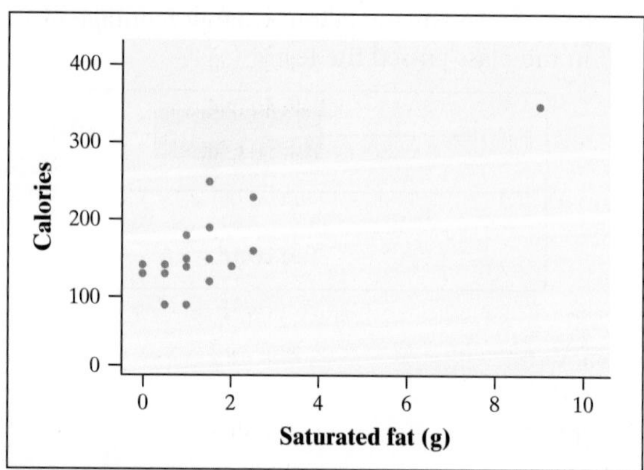

(A) Increases the correlation, decreases the slope

(B) Decreases the correlation, decreases the slope

(C) Increases the correlation, increases the slope

(D) Decreases the correlation, increases the slope

(E) This point doesn't affect the correlation or the slope.

AP12 How long does it take students to get to school in the morning? The table displays summary statistics, in minutes, for a random sample of 100 students.

n	Mean	StDev	Min	Q_1	Med	Q_3	Max
100	17.9	10.5	5	11	14.5	19.5	60

One student's travel time has a standardized score of $z = -0.56$. Approximately how long is this student's travel time?

(A) 10 minutes
(B) 12 minutes
(C) 15 minutes
(D) 17 minutes
(E) 24 minutes

AP13 For an experiment comparing two treatments (A and B), the experimental units are eight rats, of which four are female (F1, F2, F3, F4) and four are male (M1, M2, M3, M4). If a randomized block design is used, with the experimental units blocked by sex, which of the following is a valid assignment of treatments?

(A) A: (F1, F2, M3, F4), B: (M1, F3, M2, M4)
(B) A: (F2, M1, M4, M2), B: (F4, F3, M3, F1)
(C) A: (M2, F1, M3, F3), B: (M1, F4, F2, M4)
(D) A: (F1, F2, F3, F4), B: (M1, M2, M3, M4)
(E) A: (M4, M1, F1, M3), B: (F4, F2, M2, F3)

AP14 The frequency table summarizes the distribution of time that 140 patients at the emergency room of a small-city hospital waited to receive medical attention during the last month.

Waiting time	Frequency
Less than 10 minutes	5
At least 10 but less than 20 minutes	24
At least 20 but less than 30 minutes	45
At least 30 but less than 40 minutes	38
At least 40 but less than 50 minutes	19
At least 50 but less than 60 minutes	7
At least 60 but less than 70 minutes	2

Which of the following represents possible values for the median and *IQR* of waiting times for the emergency room last month?

(A) median = 27 minutes and *IQR* = 15 minutes
(B) median = 28 minutes and *IQR* = 25 minutes
(C) median = 31 minutes and *IQR* = 35 minutes
(D) median = 35 minutes and *IQR* = 45 minutes
(E) median = 45 minutes and *IQR* = 55 minutes

AP15 A child is 40 inches tall, which places the child at the 90th percentile of all children of similar age. The heights for children of this age form an approximately normal distribution with a mean of 38 inches. Based on this information, what is the standard deviation of the heights of all children of this age?

(A) 0.20 inch
(B) 0.31 inch
(C) 0.65 inch
(D) 1.21 inches
(E) 1.56 inches

Section II: Part A *Show all your work. Indicate clearly the methods you use, because you will be graded on the correctness of your methods as well as on the accuracy and completeness of your results and explanations.*

AP16 The manufacturer of exercise machines for fitness centers has designed two new elliptical machines that are meant to increase cardiovascular fitness. The two machines are being tested on 30 volunteers at a fitness center near the company's headquarters. The volunteers are randomly assigned to one of the machines and use it daily for two months. A measure of cardiovascular fitness is administered at the start of the experiment and again at the end. The following stemplot contains the differences (After – Before) in the two scores for subjects using the two machines. Note that greater differences indicate larger gains in fitness.

Machine A		Machine B
	0	2
54	1	0
876320	2	159
97411	3	2489
61	4	257
	5	359

Key: 2 | 1 represents a difference (After – Before) of 21 in fitness scores.

(a) Write a few sentences comparing the distributions of cardiovascular fitness gains from the two elliptical machines.

(b) Which machine should be chosen if the company wants to advertise it as achieving the highest overall gain in cardiovascular fitness? Explain your reasoning.

(c) Which machine should be chosen if the company wants to advertise it as achieving the most consistent gain in cardiovascular fitness? Explain your reasoning.

(d) What is the largest population to which we can apply the results of this study?

AP17 A national tire company wants to compare two different types of incentives for increasing sales: awarding cash bonuses or awarding noncash prizes such as vacations. The company has 60 retail sales districts of various sizes across the country and will randomly assign 30 districts to be given the opportunity to earn cash bonuses and assign the remaining 30 districts to be given the opportunity to earn non-cash prizes. The company will record the change in sales volume for each of the 60 retail sales districts over the following 3 months.

(a) Identify the treatments, experimental units, and response variable.

(b) Explain how to randomly assign treatments to experimental units for a completely randomized design.

(c) If the mean increase in sales is statistically significantly greater for the districts that received the cash bonus incentive, should the company conclude that the cash bonus incentive is more effective than the noncash prize incentive? Explain your answer.

Section II: Part B (Investigative Task) *Show all your work. Indicate clearly the methods you use, because you will be graded on the correctness of your methods as well as on the accuracy and completeness of your results and explanations.*

AP18 The age (years), number of car washes, and repair costs ($) were recorded during a 1-year period for a sample of 5 cars. The scatterplots show the relationship between age and number of car washes and the relationship between age and repair cost.

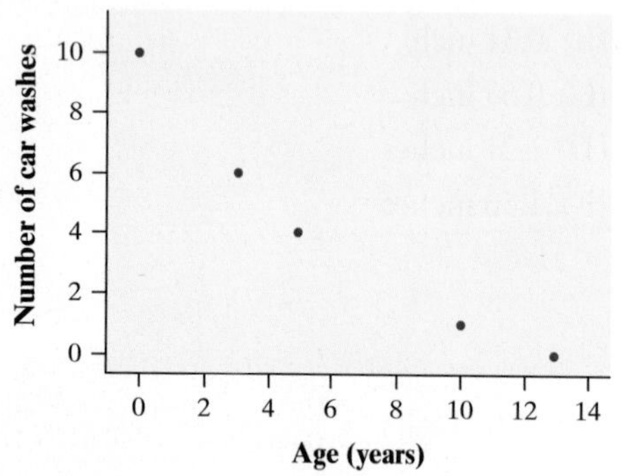

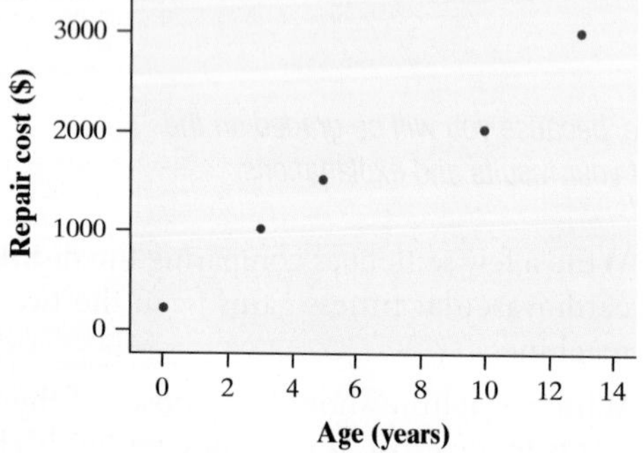

(a) Describe the relationship in each of the scatterplots.

(b) Using estimated values from the scatterplots, create a new scatterplot showing the relationship between x = number of car washes and y = repair cost.

(c) Describe the relationship between number of car washes and repair cost.

(d) Use your answers to parts (a) and (c) to explain the concept of confounding in this context.

UNIT 4

Probability, Random Variables, and Probability Distributions

PART I
Probability

Introduction

Chance is all around us. You and a friend play rock–paper–scissors to determine who gets the last slice of pizza. A coin toss decides which team gets to receive the ball first in a football game. Many adults regularly play the lottery, hoping to win a big jackpot with a few lucky numbers. People of all ages play games of chance involving cards, dice, or spinners. Chance also plays a large role in the genetic traits that children inherit from their birth parents, such as hair and eye color, blood type, handedness, dimples, or whether or not they can roll their tongues.

The mathematics of chance behavior is called *probability*. Probability is the topic of Part I of this unit. Here is an activity that gives you some idea of what lies ahead.

ACTIVITY Hiring discrimination — it just won't fly!

Digital Vision/Getty Images

In the late 2000s, an airline trained 25 pilots — 15 male and 10 female — to become captains. Unfortunately, only 8 captain positions were available at the end of the training. Airline managers announced that they would use a lottery to determine which pilots would fill the available positions. The names of all 25 pilots were written on identical slips of paper, placed in a hat, mixed thoroughly, and drawn out one at a time until all 8 positions were filled. A day later, managers announced the results of the lottery. Of the 8 captains chosen, 5 were female and 3 were male. Some of the male pilots who weren't selected suspected that the lottery was not carried out fairly. They wondered if there were grounds to file a grievance with the pilots' union.[1]

The key question in this possible discrimination case seems to be: *Is it plausible (believable) that 5 or more females were selected by chance alone?* To find out, you and your classmates will simulate the lottery process that airline managers said they used.

1. Your teacher will give you a bag with 25 beads (15 of one color and 10 of another color) or 25 slips of paper (15 labeled "M" and 10 labeled "F") to represent the 25 pilots. Mix the beads/slips thoroughly. Without looking, remove 8 beads/slips from the bag. Count the number of female pilots selected. Then return the beads/slips to the bag.
2. Your teacher will draw and label a number line for a class dotplot. On the graph, plot the number of females you got in Step 1.
3. Repeat Steps 1 and 2 if needed to get a total of at least 40 simulated lottery results in the class.
4. Based on the results of the simulation, is it plausible that 5 or more females were selected by chance alone? Or do these results provide convincing evidence that the lottery was unfair?

As the activity shows, *simulation* is a powerful method for modeling random behavior. Section 4A begins by examining the idea of probability and then illustrates how simulation can be used to estimate probabilities. In Sections 4B and 4C, we develop the basic rules and techniques of probability. This unit focuses on developing your skills in using probability and simulation, as well as interpreting statistical calculations to assign meaning or assess a claim.

Probability calculations are the basis for statistical inference (Units 6–9). When data are collected using random sampling or randomized experiments, the laws of probability answer the question, "What would happen if we repeated the random sampling or random assignment process many times?" Many of the examples, exercises, and activities in this unit focus on the connection between probability and inference.

SECTION 4A Randomness, Probability, and Simulation

LEARNING TARGETS *By the end of the section, you should be able to:*

- Interpret probability as a long-run relative frequency.
- Estimate probabilities using simulation.

Imagine flipping a coin 10 times. How likely are you to get a run of 3 or more consecutive heads? An airline knows that a certain percentage of customers who purchase tickets will not show up for a flight. If the airline overbooks a particular flight, what are the chances that it will have enough seats for the passengers who show up? To answer these questions, you need to understand how random behavior operates.

The Idea of Probability

Why do the rules of football require a coin toss to determine which team gets the ball first? Many people would agree that tossing a coin seems a fair way to decide. But what exactly does "fair" mean in this context? The following activity should help shed some light on this question.

ACTIVITY

What is probability?

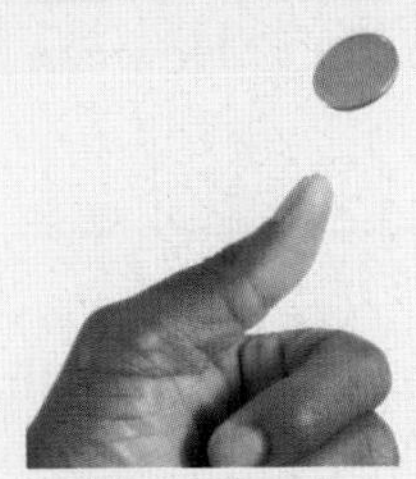
Gerville/E+/Getty Images

If you toss a fair coin, what's the probability that it shows heads? It's 1/2, or 0.5, right? But what does a probability of 1/2 really mean? In this activity, you will investigate by flipping a coin several times.

1. Flip your coin once. Record whether you get heads or tails in a table like the one that follows.
2. Flip your coin a second time. Record whether you get heads or tails. What proportion of your first two flips is heads?

3. Flip your coin 8 more times so that you have 10 flips in all. Record whether you get heads or tails on each flip.

Flip	1	2	3	4	5	6	7	8	9	10
Result (H or T)										
Proportion of heads										

4. Calculate the overall proportion of heads after each flip and record these values in the bottom row of the table. For instance, suppose you got tails on the first flip and heads on the second flip. Then your overall proportion of heads would be $0/1 = 0.00$ after the first flip and $1/2 = 0.50$ after the second flip.
5. Let's use technology to speed things up. Go to www.stapplet.com and launch *The Idea of Probability* applet (found under Concepts).
6. Keep the probability of heads as 0.5. Set the number of flips to 10 and click "Flip!" What proportion of the flips were heads?

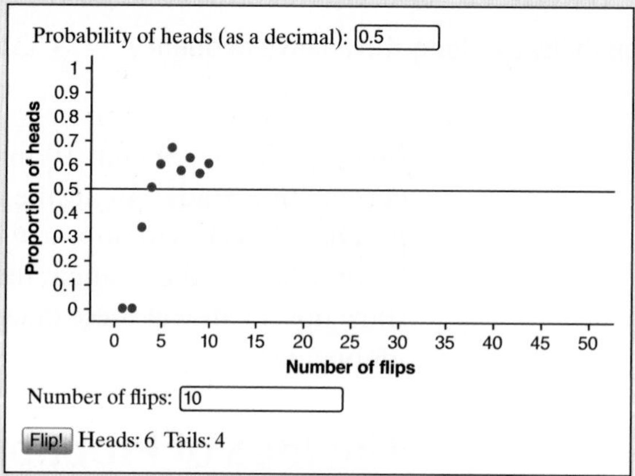

7. Keep clicking "Flip!" until you have a total of 100 flips. Is the proportion of heads exactly equal to 0.5? Close to 0.5?
8. Change the number of flips to 100 and click "Flip!" repeatedly. What happens to the proportion of heads?
9. Based on this exploration, write a sentence that explains what the following statement means: "If you flip a fair coin, the probability of heads is 0.5."

Extension: If you flip a coin, it can land heads or tails. If you "toss" a thumbtack, it can land with the point sticking up or with the point down. Does that mean the probability of a tossed thumbtack landing point up is 0.5? How can you find out? Discuss with your classmates.

Figure 4.1 shows some results from the activity. The proportion of flips that land heads varies from 0.30 to 1.00 in the first 10 flips. As we toss the coin more and more times, however, the proportion of heads gets closer to 0.5 and stays there.

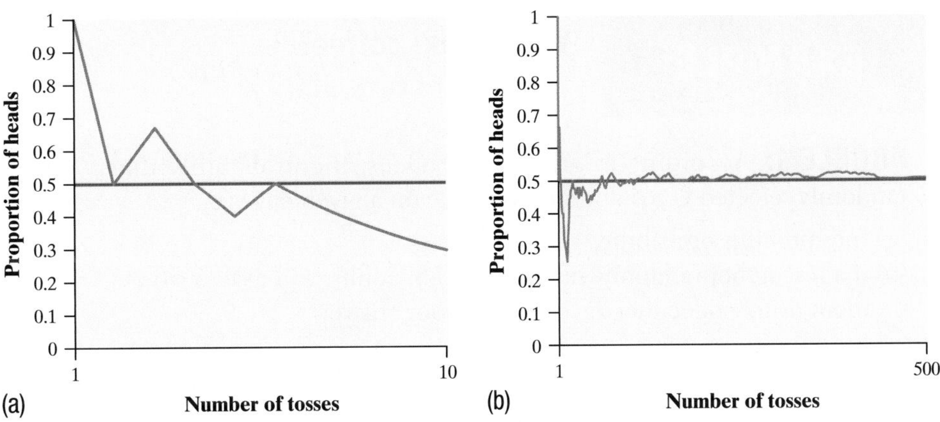

FIGURE 4.1 (a) The proportion of heads in the first 10 tosses of a coin. (b) The proportion of heads in the first 500 tosses of a coin.

When we watch coin tosses or the results of random sampling and random assignment closely, a remarkable fact emerges: A ***random process*** *is unpredictable in the short run but has a regular and predictable pattern in the long run.* This is the basis for the idea of **probability.**

DEFINITION Random process, Probability

A **random process** generates outcomes that are determined purely by chance.

The **probability** of any outcome of a random process is a number between 0 and 1 that describes the proportion of times the outcome would occur in a very long series of trials.

Outcomes that never occur have probability 0. An outcome that happens on every trial of a random process has probability 1. An outcome that happens half the time in a very long series of trials has probability 0.5.

The fact that the proportion of heads in many tosses eventually closes in on 0.5 is guaranteed by the **law of large numbers**. You can see this in Figure 4.1(b). The horizontal line represents the probability, and the proportion of heads in the simulation approaches this value as the number of trials becomes large. *Note:* Some people distinguish between *empirical probability*, which is based on actually tossing a coin several times to estimate the probability of a head, and *simulated probability*, which uses technology to imitate this random process.

DEFINITION Law of large numbers

The **law of large numbers** says that if we observe more and more trials of any random process, the proportion of times that a specific outcome occurs approaches its probability.

Life-insurance companies, casinos, and other businesses that make decisions based on probability rely on the long-run predictability of random processes. The law of large numbers helps ensure that these businesses make predictable profits from a high volume of transactions.

EXAMPLE

Who drinks coffee?
The idea of probability

Skill 4.B

Yagi Studio/DigitalVision/Getty Images

PROBLEM: According to *The Book of Odds*, the probability that a randomly selected U.S. adult drinks coffee on a given day is 0.56.

(a) Interpret this probability.

(b) If a researcher randomly selects 100 U.S. adults, will exactly 56 of them drink coffee that day? Explain your answer.

SOLUTION:

(a) If you take a very large random sample of U.S. adults, approximately 56% of them will drink coffee that day.

One trial of the random process consists of randomly selecting a U.S. adult and recording whether the person drinks coffee that day.

(b) Probably not. With only 100 randomly selected adults, the number who drink coffee that day is likely to differ from 56.

Probability describes what happens in many, many trials (way more than 100) of a random process.

FOR PRACTICE, TRY EXERCISE 1

MYTHS ABOUT RANDOMNESS

The idea of probability is that randomness is predictable *in the long run*. Unfortunately, our intuition leads us to think that random behavior should also be predictable in the short run. Suppose you toss a coin 6 times and get tails each time (TTTTTT). Is it logical to think that the next toss is more likely to be heads than tails? No. While it's true that in the long run, heads will appear half the time, it is a myth that future outcomes must make up for an imbalance like six straight tails. Some people use the phrase *law of averages* to refer to the misguided belief that the results of a random process must even out in the *short run*.

Coins and dice have no memories. A coin doesn't know that the first 6 outcomes were tails, and it can't try to get a head on the next toss to even things out. Of course, things do even out in the long run. That's the law of large numbers in action. After 10,000 tosses, the results of the first six tosses don't matter. They are overwhelmed by the results of the next 9994 tosses.

When asked to predict the sex — male (M) or female (F) — of the next seven puppies born in a large litter, most people will guess something like M-F-M-F-M-F-F. Few people would say F-F-F-M-M-M-F because this sequence of outcomes doesn't "look random." In reality, these two sequences of births are equally likely. "Runs" consisting of several of the same outcome in a row are surprisingly common in a random process.

CHECK YOUR UNDERSTANDING

1. Pedro drives the same route to work on Monday through Friday. The route includes one traffic light. According to the local traffic department, there is a 55% probability that the light will be red when Pedro reaches it. Interpret this probability.

2. Probability is a measure of how likely an outcome is to occur. Match one of the probabilities that follow with each statement. Be prepared to defend your answer.

   0 0.001 0.3 0.6 0.99 1

   (a) This outcome is impossible. It can never occur.
   (b) This outcome is certain. It will occur on every trial.
   (c) This outcome is very unlikely, but it will occur once in a while in a long sequence of trials.
   (d) This outcome will occur more often than not.

3. With each purchase, a local fast-food restaurant gives the customer a scratch-off card with a 1-in-6 chance of winning a prize. A group of 6 people goes to the restaurant together, and each person makes a separate purchase. The first 5 people in the group do not win prizes from their scratch-off cards. Lucky Louie, the sixth person in the group, says to the others, "That means I'm sure to be a winner!" Explain to Louie why he may not be so lucky.

Estimating Probabilities Using Simulation

We can model random behavior and estimate probabilities with a **simulation.**

DEFINITION Simulation

A **simulation** imitates a random process in such a way that simulated outcomes are consistent with real-world outcomes.

You already have some experience with simulations. In the Hiring Discrimination activity that opened this chapter, you drew beads or slips of paper to imitate a random lottery that chose which pilots would become captains. The Analyzing the Caffeine Experiment activity in Unit 3 asked you to determine whether the results of an experiment are statistically significant by shuffling and dealing piles of index cards to mimic the random assignment of subjects to treatments.

These simulations used different methods to imitate a random process. Even so, the same basic strategy was followed in each simulation.

HOW TO PERFORM A SIMULATION

1. Describe how to set up and use a random process to perform one trial of the simulation. Identify what you will record at the end of each trial.
2. Perform many trials.
3. Use the results of your simulation to answer the question of interest.

For the Hiring Discrimination activity, we wanted to estimate the probability of 5 or more female pilots being selected to become captains in a fair lottery.

1. We used a bag with 25 beads — 15 red and 10 white — to represent the 25 pilots, with red = male and white = female (you might have used beads of different colors in your own activity). After mixing the beads thoroughly, we removed 8 beads without looking and recorded the number of female pilots selected.
2. We performed 40 trials of the simulation. The dotplot shows the number of female pilots selected in each trial.

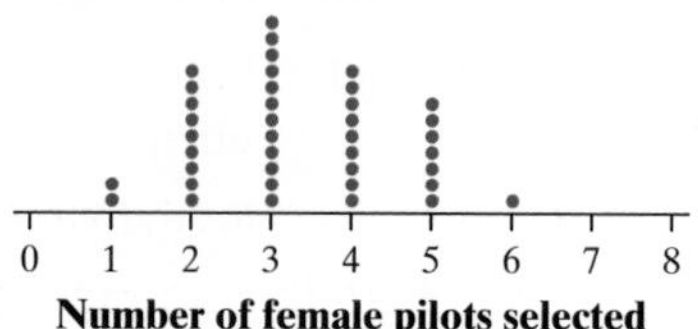

3. In 8 of the 40 trials, 5 or more female pilots were selected. So our estimate of the probability is 8/40 = 0.20 = 20%. Based on these results, it would not be unusual for 5 or more female pilots to be selected by chance alone, so we do not have convincing evidence that the airline managers conducted an unfair lottery.

EXAMPLE

Superhero comics and cereal boxes
Estimating probabilities using simulation

Skills 3.A, 4.B

PROBLEM: In an attempt to increase sales, a breakfast cereal company decides to offer a promotion. Each box of cereal will contain a collectible comic book featuring a popular superhero: The Flash, Batman, Superman, Wonder Woman, or Green Lantern. The company claims that the 5 comic books are equally likely to appear in each box of cereal.

A superfan decides to keep buying boxes of the cereal until they have all 5 superhero comic books. The fan is surprised when it takes 23 boxes to get the full set of comics. Does this outcome provide convincing evidence against the company's claim that the 5 comic books are equally likely to appear in each box of cereal? To help answer this question, we want to perform a simulation to estimate the probability that it will take 23 or more boxes to get a full set of the superhero comic books, assuming that the company's claim is true.

(a) Describe how to use a random number generator to perform one trial of the simulation.

We carried out 100 trials of the simulation and noted the results. The dotplot shows the number of cereal boxes it took to get all 5 superhero comic books in each trial.

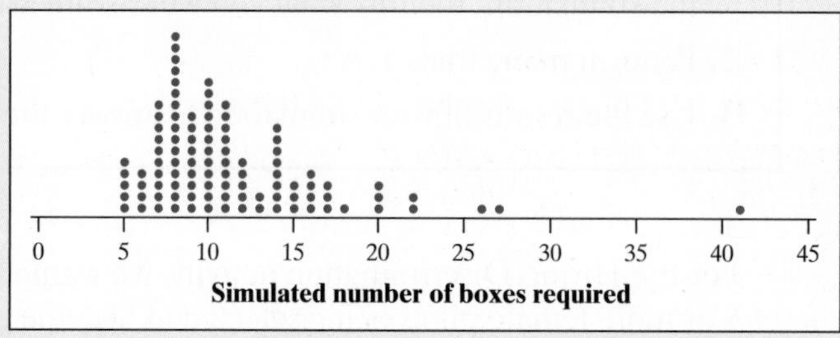

(b) Explain what the dot at 18 represents.
(c) Use the results of the simulation to estimate the probability that it will take 23 or more boxes to get a full set of comic books.
(d) Based on the actual result of 23 boxes and your answer to part (c), is there convincing evidence that the 5 superhero comic books are not equally likely to appear in each box? Explain your reasoning.

SOLUTION:

(a) Let 1 = The Flash; 2 = Batman; 3 = Superman; 4 = Wonder Woman; 5 = Green Lantern. Generate a random integer from 1 to 5 to simulate buying one box of cereal and looking at which comic book is inside. Keep generating random integers from 1 to 5 until all 5 labels from 1 to 5 appear. Record the number of boxes it takes to get all 5 superhero comic books.

> 1. Describe how to set up and use a random process to perform one trial of the simulation. Identify what you will record at the end of each trial.

(b) One trial where it took 18 boxes to get all 5 superhero comic books.

> 2. Perform many trials.

(c) Because 3 of the 100 dots are 23 or greater, the probability $\approx 3/100 = 0.03$.

> The simulation gives an estimate of the probability, so be sure to use an approximately equal sign.

(d) Because it is unlikely (probability < 0.05) that it would take 23 or more boxes to get a full set by chance alone when the superhero comic books are equally likely to be included, this result provides convincing evidence that the superhero comic books are not equally likely to appear in each box of cereal.

> 3. Use the results of your simulation to answer the question of interest.

FOR PRACTICE, TRY EXERCISE 9

It took our superfan 23 boxes to complete the set of 5 superhero comic books. Does that mean the company lied about how the comic books were distributed? Not necessarily. Our simulation says that it's unlikely for someone to have to buy at least 23 boxes to get a full set of superhero comic books *if* the 5 comic books are equally likely to appear in each box of cereal. However, it is still possible that the company was telling the truth and the superfan was just very unlucky.

EXAMPLE

Golden ticket parking lottery
Estimating probabilities using simulation

Skills 3.A, 4.B

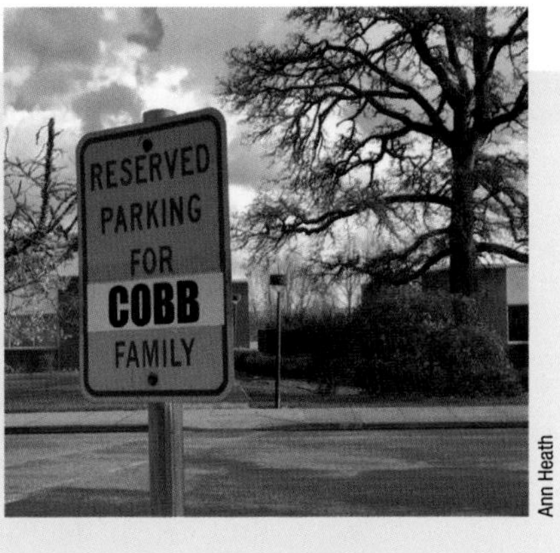

Ann Heath

PROBLEM: At a local suburban high school, 95 students have permission to park on campus. Each month, the student council holds a "golden ticket parking lottery" at a school assembly. The two lucky winners are given reserved parking spots next to the school's main entrance. Last month, the winning tickets were drawn by a student council member from the AP® Statistics class. When both golden tickets went to members of that same class, some people thought the lottery had been rigged. There are 28 students in the AP® Statistics class, all of whom are eligible to park on campus. We want to perform a simulation to estimate the probability that a fair lottery would result in two winners from the AP® Statistics class.

(a) Describe how you would use a table of random digits to carry out this simulation.

(b) Perform 3 trials of the simulation using the following row of random digits. Make your procedure clear so that someone can follow what you did.

70708 41098 55181 94904 43563 56934 48394 51719

(c) In 9 of the 100 trials of the simulation, both golden tickets were won by members of the AP® Statistics class. Do these results give convincing evidence that the lottery was not carried out fairly? Explain your reasoning.

SOLUTION:

(a) Label the students in the AP® Statistics class from 01 to 28, and label the remaining students from 29 to 95. Numbers from 96 to 99 and 00 will be skipped. Moving from left to right across a row, look at pairs of digits until we come across two *different* labels from 01 to 95. The two students with these labels win the reserved parking spots. Record whether both winners come from the AP® Statistics class. Perform many simulated lotteries. See what percentage of the time both winners come from this statistics class.

(b)

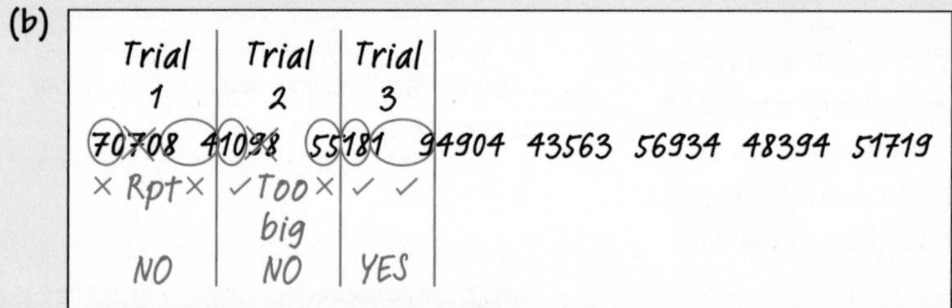

There was one trial out of the first 3 in which both golden parking tickets went to members of the AP® Statistics class.

(c) No; there's about a 9% chance of getting both winners from the AP® Statistics class in a fair lottery. Because this probability isn't small (probability > 0.05), we don't have convincing evidence that the lottery was unfair.

Does that mean the lottery *was* conducted fairly? Not necessarily.

FOR PRACTICE, TRY EXERCISE 11

In the golden ticket lottery example, we ignored repeated numbers from 01 to 95 within a given trial. That's because the random process involved sampling students *without* replacement. In the cereal box example, we allowed repeated numbers from 1 to 5 in a given trial. That's because the company claimed that each superhero comic book is equally likely to appear in a box of cereal. So, the probability of getting, say, a Batman comic book in any box of cereal is 1/5.

AP® EXAM TIP

On the AP® Statistics exam, you may be asked to describe how to perform a simulation using rows of random digits. If so, provide a clear enough description of your process for the reader to get the same results from *only* your written explanation. Remember that every label needs to be the same length. In the golden ticket lottery example, the labels should be 01 to 95 (all two digits), not 1 to 95. When sampling without replacement, be sure to mention that repeated numbers should be ignored or that *different (unique)* numbers should be selected.

CHECK YOUR UNDERSTANDING

A basketball announcer suggests that a certain player is a streaky shooter. That is, the announcer believes that if the player makes a shot, the player is more likely to make the next shot. As evidence, the announcer points to a recent game in which the player took 30 shots and had a streak of 10 made shots in a row. Is this convincing evidence of streaky shooting by the player? Assume that this player makes 50% of their shots and that the result of each shot doesn't depend on previous shots.

1. Describe how you would carry out a simulation to estimate the probability that a 50% shooter who takes 30 shots in a game will have a streak of 10 or more made shots.

The dotplot displays the results of 50 simulated games in which this player took 30 shots.

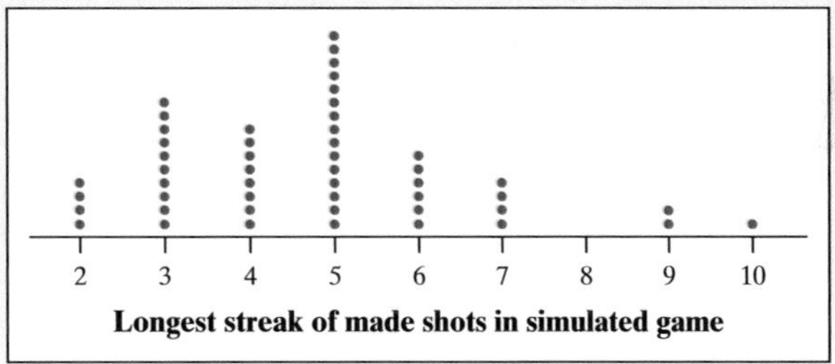

2. Explain what the two dots above 9 indicate.
3. Use the simulation results to estimate the probability that a 50% shooter will have a streak of 10 or more made shots in a game when the player takes 30 shots.
4. What conclusion would you make about whether this player was streaky? Explain your reasoning.

SECTION 4A Summary

- A **random process** generates outcomes that are determined purely by chance. Random behavior is unpredictable in the short run but shows a regular and predictable pattern in the long run.
- The **probability** of an outcome is the long-run relative frequency of the outcome after many trials of a random process. A probability is a number between 0 (never occurs) and 1 (always occurs).
- The **law of large numbers** says that in many trials of the same random process, the proportion of times that a particular outcome occurs will approach its probability.
- **Simulation** can be used to imitate a random process and to estimate probabilities. To perform a simulation:
  1. Describe how to set up and use a random process to perform one trial of the simulation. Identify what you will record at the end of each trial.
  2. Perform many trials.
  3. Use the results of your simulation to answer the question of interest.

AP® EXAM TIP

AP® Daily Videos

Review the content of this section and get extra help by watching the AP® Daily Videos for Topics 4.1 and 4.2, which are available in AP® Classroom.

SECTION 4A Exercises

The Idea of Probability

1. **Another commercial** If Lucretia launches her favorite streaming radio channel at a randomly selected time, there is a 0.20 probability that a commercial will be playing. pg 322

(a) Interpret this probability.

(b) If Lucretia launches this channel at 5 randomly selected times, will there be exactly 1 time when a commercial is playing? Explain your answer.

2. **Cystic fibrosis** It is not uncommon for two birth parents to both carry the gene for cystic fibrosis without having the disease themselves. Suppose we select one of these sets of birth parents at random. According to the laws of genetics, there is a 0.25 probability that any child they have will develop cystic fibrosis.

(a) Interpret this probability.

(b) If this set of birth parents has 4 children, is one child guaranteed to develop cystic fibrosis? Explain your answer.

3. **Mammograms** Many women choose to have annual mammograms to screen for breast cancer after age 40. A mammogram isn't foolproof. Sometimes the test suggests that a woman has breast cancer when she really doesn't (a "false-positive"). Other times, the test says that a woman doesn't have breast cancer when she actually does (a "false-negative"). Suppose the false-negative rate for a mammogram is 0.10.

(a) Explain what this probability means.

(b) Which is a more serious error in this case: a false-positive or a false-negative? Justify your answer.

4. **Liar, liar!** Sometimes police use a lie detector test to help determine whether a suspect is telling the truth. A lie detector test isn't foolproof — sometimes it suggests that a person is lying when they are actually telling the truth (a "false-positive"). Other times, the test says that the suspect is being truthful when they are actually lying (a "false-negative"). For one brand of lie detector, the probability of a false-positive is 0.08.

(a) Explain what this probability means.

(b) Which is a more serious error in this case: a false-positive or a false-negative? Justify your answer.

5. **Three pointers** The figure shows the results of a basketball player attempting many 3-point shots. Explain what this graph tells you about random behavior in the short run and the long run.

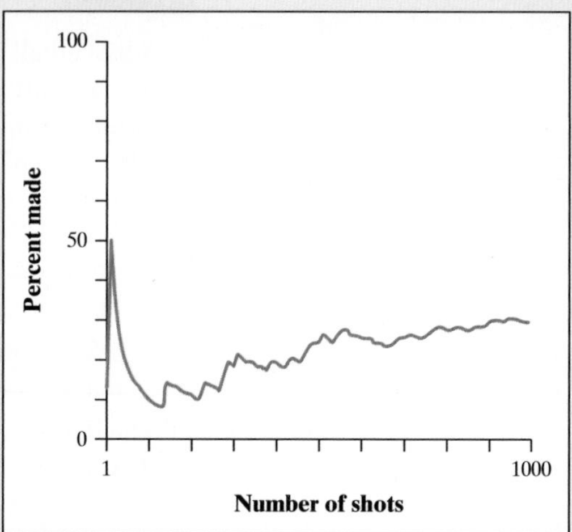

6. **Keep on tossing** The figure shows the results of two different sets of 5000 coin tosses. Explain what this graph tells you about random behavior in the short run and the long run. (Note that the horizontal axis is on a *logarithmic scale*.)

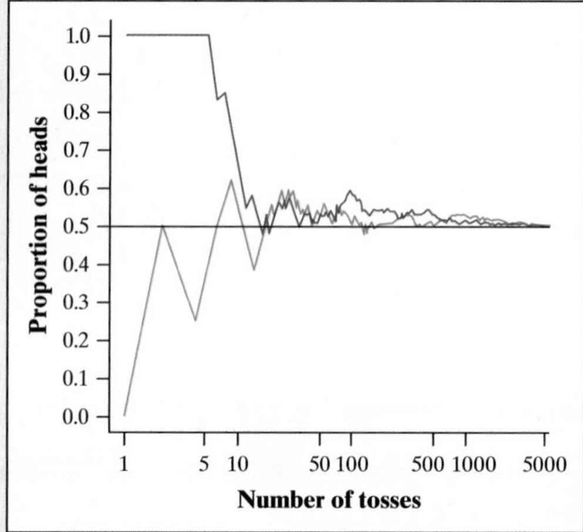

7. **Life insurance** Jake is an insurance salesperson who is able to complete a sale in about 15% of calls involving life insurance policies. One day, Jake fails to make a sale on the first 10 such calls. A colleague tells Jake not to worry because he is "due for a sale." Explain why the colleague is wrong.

8. **Softball stats** A very good professional softball player gets a hit about 35% of the time over an entire season. After the player failed to hit safely in six straight at-bats, a TV commentator said, "She is due for a hit." Explain why the commentator is wrong.

Estimating Probabilities Using Simulation

9. **Train arrivals** New Jersey Transit claims that its 8:00 A.M. train from Princeton to New York City has pg 324

probability 0.9 of arriving on time on a randomly selected day. Assume for now that this claim is true. Ariella takes the 8:00 A.M. train to work 20 days in a certain month, and is surprised when the train arrives late in New York on 4 of the 20 days. Should Ariella be surprised? To help answer this question, we want to carry out a simulation to estimate the probability that the train would arrive late on 4 or more of 20 days if New Jersey Transit's claim is true.

(a) Describe how to use a random number generator to perform one trial of the simulation.

The dotplot shows the number of days on which the train arrived late in 100 trials of the simulation.

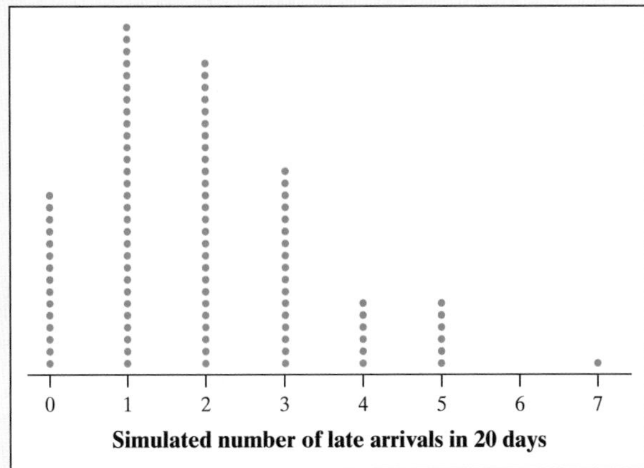

(b) Explain what the dot at 7 represents.

(c) Use the results of the simulation to estimate the probability that the train will arrive late on 4 or more of 20 days.

(d) Based on the actual result of 4 late arrivals in 20 days and your answer to part (c), is there convincing evidence that New Jersey Transit's claim is false? Explain your reasoning.

10. **Double-fault!** In tennis, a player gets two attempts to serve the ball in play. If a player misses the first serve, they are often more careful with the second serve to avoid a double-fault. A professional tennis player claims to make 90% of their second serves. In a recent match, the player missed 5 of 20 second serves. Is this a surprising result if the player's claim is true? Assume that the player has a 0.10 probability of missing a second serve. We want to carry out a simulation to estimate the probability that this player would miss 5 or more of 20 second serves.

(a) Describe how to use a random number generator to perform one trial of the simulation.

The dotplot displays the number of second serves missed by the player out of 20 second serves in 100 simulated matches.

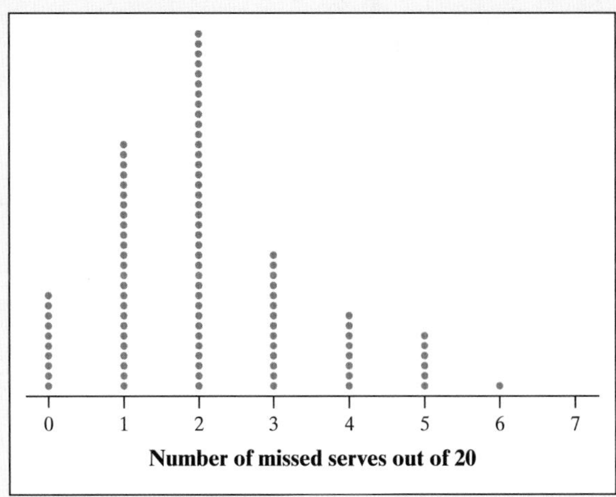

(b) Explain what the dot at 6 represents.

(c) Use the results of the simulation to estimate the probability that the player would miss 5 or more of 20 second serves in a match.

(d) Based on the actual result of 5 missed second serves and your answer to part (c), is there convincing evidence that the player's claim is false? Explain your reasoning.

11. pg 325 **Airport security** At a certain airport, security officials claim that they randomly select passengers for extra screening at the gate before boarding some flights. One such flight had 76 passengers — 16 in first class and 60 in economy class. Some passengers were surprised when none of the 14 passengers chosen for screening was seated in first class. Should they be surprised? We want to perform a simulation to estimate the probability that no first-class passengers would be chosen in a truly random selection.

(a) Describe how you would use a table of random digits to carry out this simulation.

(b) Perform one trial of the simulation using the random digits that follow. Copy the digits onto your paper and mark directly on or above them so that someone can follow what you did.

71487 09984 29077 14863 61683 47052 62224 51025

(c) We performed 1000 trials of the simulation. In 25 trials, none of the 14 passengers chosen was seated in first class. Does this result provide convincing evidence that the security officials did not carry out a truly random selection? Explain your reasoning.

12. **Scrabble** In the game of Scrabble, the first player draws 7 letter tiles at random from a bag containing 100 tiles. There are 42 vowels, 56 consonants, and 2 blank tiles in the bag. Anise draws first and is surprised to discover that all 7 tiles are vowels. Should Anise be surprised? We want to perform a simulation to estimate the

probability that a player will randomly select 7 vowels from the bag.

(a) Describe how you would use a table of random digits to carry out this simulation.

(b) Perform one trial of the simulation using the random digits that follow. Copy the digits onto your paper and mark directly on or above them so that someone can follow what you did.

```
00694 05977 19664 65441 20903 62371 22725 53340
```

(c) We performed 1000 trials of the simulation. In 2 trials, all 7 tiles were vowels. Does this result give convincing evidence that the bag of tiles was not well mixed when Anise selected the 7 tiles? Explain your reasoning.

13. **Bull's-eye!** In an archery competition, each player continues to shoot until they miss the center of the target twice. Quincy is one of the archers. Based on past experience, Quincy has a 0.60 probability of hitting the center of the target on each shot. Should we be surprised if Quincy stays in the competition for at least 10 shots?

(a) Describe how you would design a simulation to help answer this question.

(b) Carry out one trial of the simulation you described in part (a).

(c) In 3 out of 100 trials of the simulation, Quincy stayed in the competition for at least 10 shots. Based on this result, how would you answer the question of interest? Explain your reasoning.

14. **Color blindness** Approximately 7% of men in the United States have some form of red–green color blindness. Suppose we randomly select one U.S. adult male at a time until we find one who is red–green color-blind. Should we be surprised if it takes us 20 or more men?

(a) Describe how you would design a simulation to help answer this question.

(b) Carry out one trial of the simulation you described in part (a).

(c) In 24 out of 100 trials of the simulation, it took 20 or more randomly selected men to find one who is red–green color blind. Based on this result, how would you answer the question of interest?

15. **Notebook check** Every 9 weeks, Ms. Millar collects students' notebooks and checks their homework. She randomly selects 4 different assignments to inspect for each of the students. Marino, one of the students in Ms. Millar's class, completed 20 homework assignments and did not complete 10 assignments. Among the 4 assignments selected by Ms. Millar, Marino completed only 1 of them. Should Marino be surprised by this outcome? To find out, we want to design a simulation to estimate the probability that Ms. Millar will select 1 or fewer of the homework assignments that Marino completed in a random sample of 4 assignments. Determine whether the following simulation process is valid.

Get 30 identical slips of paper. Write "N" on 10 of the slips and "C" on the remaining 20 slips. Put the slips into a hat and mix well. Draw 1 slip without looking to represent the first randomly selected homework assignment, and record whether Marino completed it. Put the slip back into the hat, mix again, and draw another slip representing the second randomly selected assignment. Record whether Marino completed this assignment. Repeat this process two more times for the third and fourth randomly selected homework assignments. Record the number out of the 4 randomly selected homework assignments that Marino completed in this trial of the simulation. Perform many trials. Find the proportion of trials in which Ms. Millar randomly selects 1 or fewer of the homework assignments that Marino completed.

16. **Lefties** A website claims that 10% of U.S. adults are left-handed. A researcher believes that this figure is too low. She decides to test this claim by selecting a random sample of 20 U.S. adults and recording how many are left-handed. Four of the adults in the sample are left-handed. Does this result give convincing evidence that the website's 10% claim is too low? To find out, we want to perform a simulation to estimate the probability of getting 4 or more left-handed people in a random sample of size 20 from a very large population in which 10% of the people are left-handed. Determine whether the following simulation process is valid.

Let 00 to 09 indicate left-handed and 10 to 99 represent right-handed. Move from left to right across a row in a table of random digits. Each pair of digits represents one person. Keep going until you get 20 different pairs of digits. Record how many people in the simulated sample are left-handed. Repeat this process many, many times. Find the proportion of trials in which 4 or more people in the simulated sample were left-handed.

17. **Recycling** Researchers want to investigate whether a majority of students at the local community college regularly recycle. To find out, they survey a simple random sample (SRS) of 100 students at the college about their recycling habits. Suppose that 55 students in the sample say that they regularly recycle. Is this convincing evidence that more than half of the students at the college would say they regularly recycle? The dotplot

shows the results of taking 200 SRSs of 100 students from a population in which the true proportion who regularly recycle is 0.50.

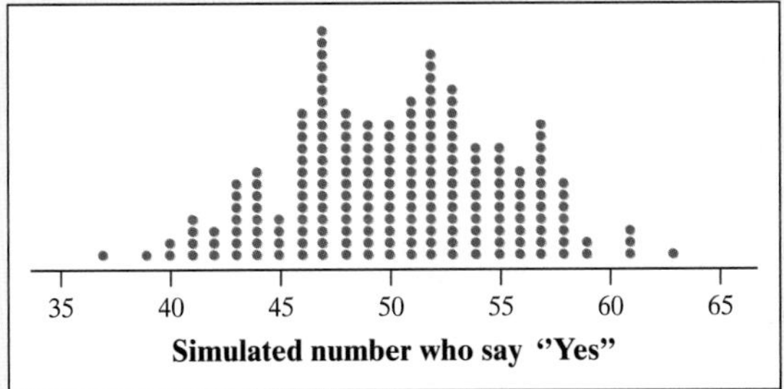

(a) Explain why the sample result (55 out of 100 said "Yes") does not give convincing evidence that a majority of the college's students would say that they regularly recycle.

(b) Suppose instead that 63 students in the researcher's sample had said "Yes." Explain why this result would give convincing evidence that a majority of the college's students would say that they regularly recycle.

18. **Conserving water** A recent study reported that two-thirds of young adults turn off the water while brushing their teeth.[2] Researchers suspect that the true proportion is lower at their local university. To find out, they ask an SRS of 60 students at the university if they usually brush with the water off. Suppose that 36 students in the sample say "Yes." The dotplot shows the results of taking 200 SRSs of 60 students from a population in which the true proportion who brush with the water off is two-thirds.

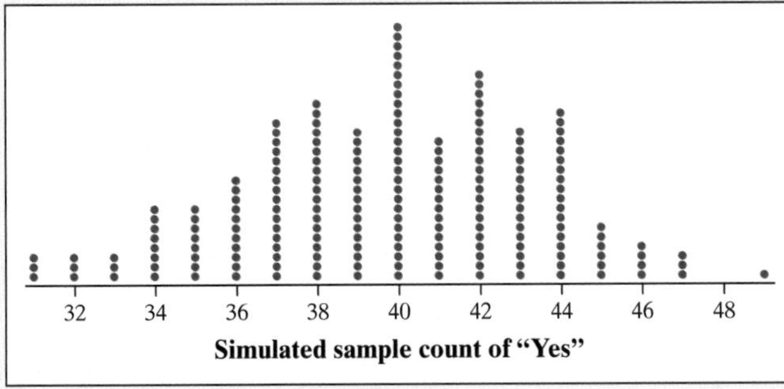

(a) Explain why the sample result (36 out of 60 said "Yes") does not give convincing evidence that fewer than two-thirds of the university's students would say that they brush their teeth with the water off.

(b) Suppose instead that 32 of the 60 students in the researchers' sample had said "Yes." Explain why this result would give convincing evidence that fewer than two-thirds of the university's students would say that they brush their teeth with the water off.

For Investigation *Apply the skills from the section in a new context or nonroutine way.*

19. **Tipping** A cashier at Starbucks has noticed that tips seem to come in streaks. That is, there are times when several customers in a row will leave tips and other times when several customers in a row won't leave tips. The cashier speculates that customers are influenced by the behavior of the customer in front of them and tip (or don't tip) accordingly. Is this true? Here are the results for 50 consecutive customers, where "T" represents one of the 25 customers who left a tip and "N" represents one of the 25 customers who didn't leave a tip:

N TTTT NN T NNNN TTT N T NNNN TTTT
NNN TT N T NNN TT NN TTT NN T N TT N T

(a) A run consists of one or more consecutive occurrences of the same outcome. How many runs did the cashier observe in the tipping behavior of these 50 customers?

(b) Given that there will be 25 T's and 25 N's, what is the smallest possible number of runs? Would an outcome with this number of runs suggest that customers are influenced by the previous customer? Explain your answer.

We used technology to simulate the number of runs in 100 different sets of 50 customers — half of whom left a tip and half of whom did not — assuming that one customer's behavior did not influence another's. The dotplot shows the results.

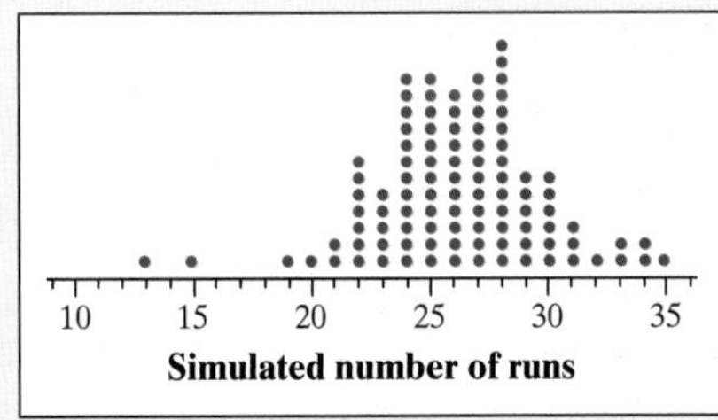

(c) Use the results of the simulation to estimate the probability that the cashier would observe a number of runs as small as or smaller than your answer to part (a) if one customer's tipping behavior does not influence another's.

(d) Based on the actual result observed by the cashier and your answer to part (c), is there convincing evidence that customers are influenced by the behavior of the customer in front of them and tip (or don't tip) accordingly? Explain your reasoning.

Multiple Choice *Select the best answer for each question.*

20. If I toss a fair coin 5 times and the outcomes are TTTTT, then the probability that tails appears on the next toss is

(A) 0.5.
(B) less than 0.5.
(C) greater than 0.5.
(D) 0.
(E) 1.

21. You read in a book about bridge that the probability that each of the four players is dealt exactly one ace is approximately 0.11. This means that

(A) in every 100 bridge deals, each player has one ace exactly 11 times.

(B) in 1 million bridge deals, the number of deals on which each player has one ace will be exactly 110,000.

(C) in a very large number of bridge deals, the percentage of deals on which each player has one ace will be very close to 11%.

(D) in a very large number of bridge deals, the average number of aces in a hand will be very close to 0.11.

(E) if each player gets an ace in only 2 of the first 50 deals, then each player should get an ace in more than 11% of the next 50 deals.

Exercises 22 and 23 refer to the following setting. A media report claims that 46% of U.S. teens are almost constantly online.[3] We want to simulate selecting a random sample of 10 U.S. teens and asking each of them if they are online almost constantly, assuming the report's claim is true.

22. To simulate the number of teens who say "Yes, I am online almost constantly" in a random sample of size 10, you would perform the simulation as follows:

(A) Use 46 random one-digit numbers, where 0–4 are Yes and 5–9 are No.

(B) Use 10 random two-digit numbers, where 01–46 are Yes and 47–99 are No.

(C) Use 10 random two-digit numbers, where 00–46 are Yes and 47–99 are No.

(D) Use 10 random two-digit numbers, where 01–46 are Yes and 47–00 are No.

(E) Use 10 random two-digit numbers, where 01–46 are Yes, 47–99 are No, and 00 is Unsure.

23. A total of 25 trials of the simulation were performed. The number of teens who said "Yes" in each simulated random sample of size 10 was recorded on the following dotplot. What is the approximate probability that 6 or more teens would say "Yes" in a random sample of size 10 if the report's claim is true?

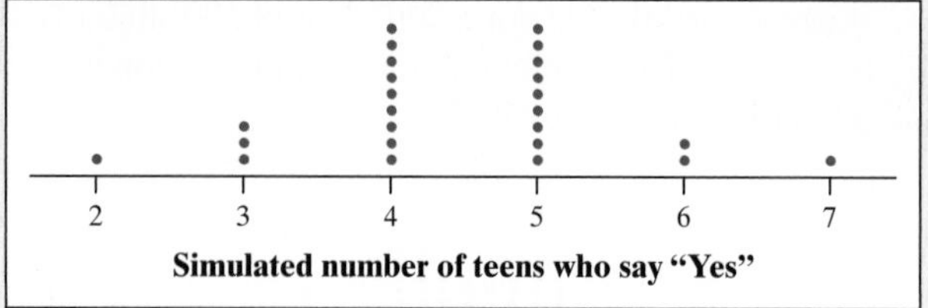

(A) 2/10
(B) 3/10
(C) 2/25
(D) 3/25
(E) 2/100

Recycle and Review *Practice what you learned in previous sections.*

24. **Waiting to park (1D, 3C)** Do drivers take longer to leave their parking spaces when someone is waiting? Researchers hung out in a parking lot and collected some data. The graphs and numerical summaries compare how long it took drivers to exit their spaces when someone was or was not waiting.[4]

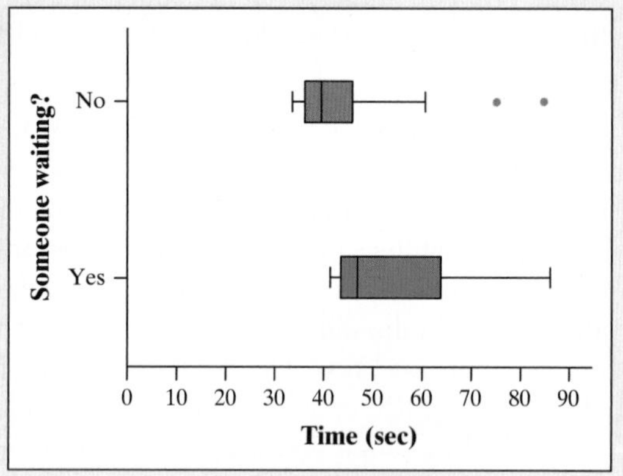

Descriptive Statistics: Time

Waiting	n	Mean	SD	Min	Q_1	Median	Q_3	Max
No	20	44.42	14.10	33.76	35.61	39.56	48.48	84.92
Yes	20	54.11	14.39	41.61	43.41	47.14	66.44	85.97

(a) Describe similarities and differences in the two distributions.

(b) Can we conclude that having someone waiting causes drivers to leave their spaces more slowly? Why or why not?

25. **AARP and Medicare (3B)** To find out what proportion of U.S. adults support proposed Medicare legislation to help pay medical costs, AARP conducted a survey of its members (people older than age 50 who pay membership dues). One of the questions was "Even if this plan won't affect you personally either way, do you think it should be passed so that people with low incomes or people with high drug costs can be helped?" Of the respondents, 75% answered "Yes."[5]

(a) Describe how undercoverage might lead to bias in this study. Explain the likely direction of the bias.

(b) Describe how the wording of the question might lead to bias in this study. Explain the likely direction of the bias.

SECTION 4B Probability Rules

LEARNING TARGETS *By the end of the section, you should be able to:*

- Give a probability model for a random process with equally likely outcomes and use it to find the probability of an event.
- Calculate probabilities using the complement rule.
- Use the addition rule for mutually exclusive events to find probabilities.
- Use a two-way table to find probabilities.
- Calculate probabilities with the general addition rule.

In Section 4A, we used simulation to imitate random behavior. Do we always have to repeat a random process — flipping coins, rolling dice, drawing slips from a hat — many times to determine the probability of a particular outcome? Fortunately, the answer is no.

Probability Models

Dorling Kindersley ltd./Alamy Stock Photo

Many board games involve rolling dice. Imagine rolling two fair, 6-sided dice — one that's red and one that's blue. How do we develop a **probability model** for this random process? Figure 4.2 displays the **sample space** of the 36 possible outcomes. Because the dice are fair, each of these outcomes will be equally likely and have probability 1/36.

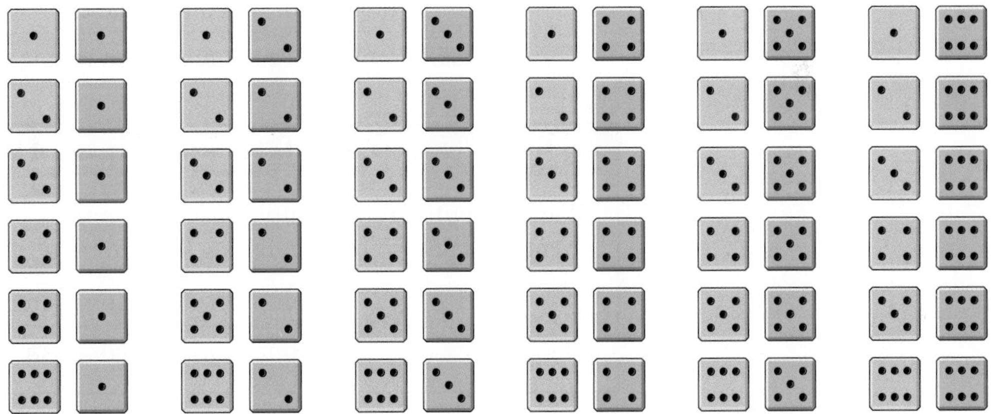

FIGURE 4.2 The 36 possible outcomes from rolling two 6-sided dice, one red and one blue. Each of these equally likely outcomes has probability 1/36.

DEFINITION Probability model, Sample space

A **probability model** is a description of a random process that consists of two parts: a list of all possible outcomes and the probability of each outcome.

The list of all possible outcomes is called the **sample space.**

What if the two dice were actually the same color? Figure 4.2 (with the colors of the dice adjusted) would still show the sample space for the random process of rolling two fair, 6-sided dice. Note that 1 on the first die and 6 on the second die is a different outcome than 6 on the first die and 1 on the second die, even if we can't tell the two dice apart.

A sample space can be very simple or very complex. If we toss a coin once, there are only two possible outcomes in the sample space, heads and tails. When Gallup takes a random sample of 1523 U.S. adults and asks a survey question, the sample space consists of all possible sets of responses from 1523 of the more than 250 million adults in the country.

A probability model does more than just assign a probability to each outcome — it allows us to find the probability of an **event.**

DEFINITION Event

An **event** is a subset of the possible outcomes from the sample space of a random process.

Events are usually denoted by capital letters, like A, B, C, and so on. For rolling two 6-sided dice, we can define event A as getting a sum of 5. We write the probability of event A as $P(\text{A})$ or $P(\text{sum is 5})$.

It is fairly straightforward to find the probability of an event in the case of equally likely outcomes. There are 4 outcomes in event A:

Because each of these outcomes is equally likely, the probability that event A occurs is

$$P(\text{A}) = \frac{\text{number of outcomes with sum of 5}}{\text{total number of outcomes when rolling two dice}} = \frac{4}{36} = 0.111$$

FINDING PROBABILITIES: EQUALLY LIKELY OUTCOMES

If all outcomes in the sample space are equally likely, the probability that event A occurs can be found using the formula

$$P(\text{A}) = \frac{\text{number of outcomes in event A}}{\text{total number of outcomes in sample space}}$$

EXAMPLE

Spin the spinner
Probability models

Skill 3.A

PROBLEM: A spinner has three equal sections: red, blue, and yellow. Suppose you spin the spinner two times.

(a) Give a probability model for this random process.

(b) Define event A as spinning blue at least once. Find $P(\text{A})$.

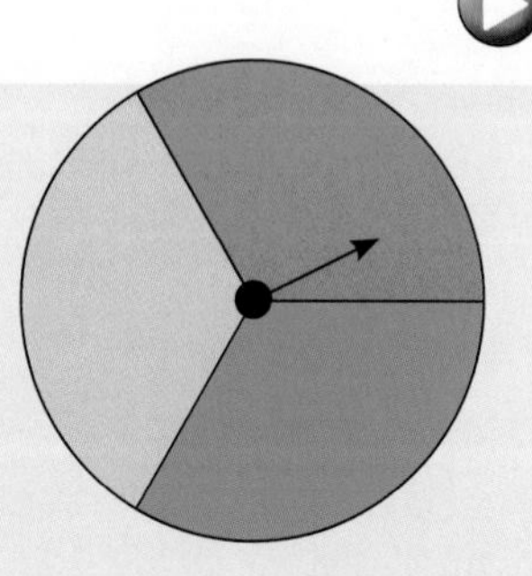

SOLUTION:

(a) Sample space: RR RB RY BR BB BY YR YB YY. Because the spinner has equal sections, each of these outcomes will be equally likely and have probability 1/9.

(b) There are 5 outcomes with at least one blue spin: RB BR BB BY YB. So $P(A) = \frac{5}{9} = 0.556$.

Remember: A probability model consists of a list of all possible outcomes and the probability of each outcome.

$$P(A) = \frac{\text{number of outcomes in event A}}{\text{total number of outcomes in sample space}}$$

FOR PRACTICE, TRY EXERCISE 1

Basic Probability Rules

AP® EXAM TIP

You will lose any chance for partial credit on probability free-response questions on the AP® Statistics exam if you state that a probability is less than 0 or greater than 1. Be extra careful if your calculator reports an answer in scientific notation, such as 2.32445E−4.

Our work so far suggests that a valid probability model must obey two common-sense rules:

- **The probability of any event is a number between 0 and 1.** This rule follows from the definition of probability in Section 4A: the proportion of times the event would occur in many trials of the random process. A proportion is a number between 0 and 1, so a probability is also a number between 0 and 1.
- **All possible outcomes together must have probabilities that add up to 1.** Anytime we observe a random process, some outcome must occur.

Here's one more rule that follows from the previous two:

- **The probability that an event does *not* occur is 1 minus the probability that the event does occur.** Earlier, we found that the probability of getting a sum of 5 when rolling two fair, 6-sided dice is 4/36. What's the probability that the sum is *not* 5?

$$P(\text{sum is not 5}) = 1 - P(\text{sum is 5}) = 1 - \frac{4}{36} = \frac{32}{36} = 0.889$$

We refer to the event "not A" as the **complement** of A and denote it by A^C. For that reason, this handy result is known as the **complement rule.** Using the complement rule in this setting is much easier than counting all 32 possible ways to get a sum that isn't 5.

DEFINITION Complement rule, Complement

The **complement rule** says that $P(A^C) = 1 - P(A)$, where A^C is the **complement** of event A; that is, the event that A does not occur.

Another commonly used notation for the complement of event A is A′. Here's an example that illustrates use of the complement rule when the outcomes of a random process are not equally likely.

EXAMPLE

Avoiding blue M&M'S®
Basic probability rules

Skill 3.A

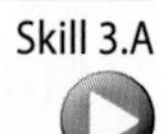

Niels Poulsen std/Alamy Stock Photo

PROBLEM: Suppose you tear open the corner of a bag of M&M'S® Milk Chocolate Candies, pour one candy into your hand, and observe the color. According to Mars, Incorporated, the maker of M&M'S, the probability model for a bag manufactured at its Cleveland, Tennessee, factory is:[6]

Color	Blue	Orange	Green	Yellow	Red	Brown
Probability	0.207	0.205	0.198	0.135	0.131	0.124

(a) Explain why this is a valid probability model.
(b) Find the probability that you don't get a blue M&M.

SOLUTION:

(a) The probability of each outcome is a number between 0 and 1, and $0.207 + 0.205 + 0.198 + 0.135 + 0.131 + 0.124 = 1$.

(b) $P(\text{not blue}) = 1 - P(\text{blue}) = 1 - 0.207 = 0.793$

$P(A^C) = 1 - P(A)$

FOR PRACTICE, TRY EXERCISE 5

Mutually Exclusive Events

Let's return to the random process of rolling two fair, 6-sided dice. Earlier, we found that $P(\text{sum is } 5) = 4/36$. Now consider the event "getting a sum of 6." The outcomes in this event are

  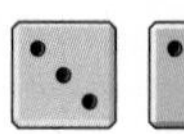   

So $P(\text{sum is } 6) = \frac{5}{36}$. What's the probability that we get a sum of 5 *or* a sum of 6?

$$P(\text{sum is 5 or sum is 6}) = P(\text{sum is } 5) + P(\text{sum is } 6) = \frac{4}{36} + \frac{5}{36} = \frac{9}{36} = 0.25$$

Why does this formula work? Because the events "getting a sum of 5" and "getting a sum of 6" have no outcomes in common — that is, they can't both happen at the same time. We say that these two events are **mutually exclusive** (sometimes referred to as *disjoint*). As a result, this intuitive formula is known as the **addition rule for mutually exclusive events.**

DEFINITION Mutually exclusive, Addition rule for mutually exclusive events

Two events A and B are **mutually exclusive** if they have no outcomes in common and so can never occur together—that is, if $P(\text{A and B}) = 0$.

The **addition rule for mutually exclusive events** A and B says that

$$P(\text{A or B}) = P(\text{A}) + P(\text{B})$$

Note that this rule works only for mutually exclusive events. We will soon develop a more general rule for finding $P(\text{A or B})$ that works for *any* two events.

EXAMPLE

More M&M'S®
Mutually exclusive events

Skill 3.A

PROBLEM: Having torn open a bag of M&M'S® Milk Chocolate Candies, you pour a candy into your hand, and observe the color. Recall that the probability model for a bag from Mars, Incorporated's Cleveland, Tennessee, factory is:[7]

Color	Blue	Orange	Green	Yellow	Red	Brown
Probability	0.207	0.205	0.198	0.135	0.131	0.124

Eric Carr/Alamy Stock Photo

Find the probability that you get an orange M&M or a brown M&M.

SOLUTION:

$$P(\text{orange or brown}) = P(\text{orange}) + P(\text{brown}) = 0.205 + 0.124 = 0.329$$

It is OK to add the two probabilities because the events "orange M&M" and "brown M&M" are mutually exclusive (disjoint)—there are no M&Ms that are both orange and brown.

FOR PRACTICE, TRY EXERCISE 7

AP® EXAM TIP

Many probability problems involve simple arithmetic that you can do in your head. It may be tempting to just write down your final answer without showing the supporting work. Don't do it! A "naked answer," even if it's correct, will usually be penalized on a free-response question.

What's the probability that you don't get an orange M&M or a brown M&M? We could use an expanded version of the addition rule for mutually exclusive events to calculate the probability:

$$\begin{aligned} P(\text{not "orange or brown"}) &= P(\text{blue or green or yellow or red}) \\ &= P(\text{blue}) + P(\text{green}) + P(\text{yellow}) + P(\text{red}) \\ &= 0.207 + 0.198 + 0.135 + 0.131 \\ &= 0.671 \end{aligned}$$

We could also use the complement rule:

$$P(\text{not "orange or brown"}) = 1 - P(\text{orange or brown}) = 1 - 0.329 = 0.671$$

Using the complement rule is much simpler than adding 4 probabilities together!

CHECK YOUR UNDERSTANDING

Suppose we choose a U.S. adult at random. Define two events:

A = the person has high cholesterol: 240 milligrams per deciliter of blood (mg/dl) or greater

B = the person has borderline high cholesterol level: 200 to <240 mg/dl

According to the American Heart Association, $P(\text{A}) = 0.16$ and $P(\text{B}) = 0.29$.

1. Explain why events A and B are mutually exclusive.
2. Say in plain language what the event "A or B" is. Then find $P(\text{A or B})$.
3. Let C be the event that the person chosen has normal cholesterol: less than 200 mg/dl. Find $P(\text{C})$.

Two-Way Tables and Probability

So far, you have learned how to model random behavior and seen some basic rules for finding the probability of an event. When you're trying to find probabilities involving two events, a two-way table can display the sample space in a way that makes probability calculations easier.

EXAMPLE

Who can roll their tongue or raise one eyebrow?
Two-way tables and probability

Skill 3.A

Vyacheslav Dumchev/Alamy Stock Photo

PROBLEM: Students in a college statistics class wanted to find out how common it is for young adults to be able to roll their tongues or to raise one eyebrow. They recorded data on whether each of the 200 people in the class could do either of these two things. The two-way table summarizes the data.

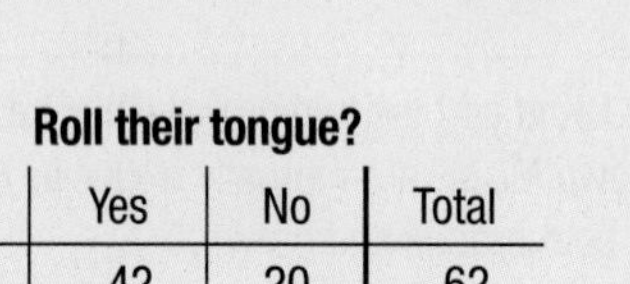

		Roll their tongue?		
		Yes	No	Total
Raise one eyebrow?	Yes	42	20	62
	No	107	31	138
	Total	149	51	200

Suppose we choose a student from the class at random. Define event A as "student can roll their tongue" and event B as "student can raise one eyebrow."

(a) Find $P(B)$. Describe this probability in words.
(b) Find $P(\text{roll tongue and raise one eyebrow})$.

SOLUTION:

(a) $P(B) = P(\text{raise one eyebrow}) = \frac{62}{200} = 0.31$

There is a 31% chance that a randomly selected student from this class can raise one eyebrow.

> Each student in the class is equally likely to be chosen, so
> $P(B) = \frac{\text{number of students who can raise one eyebrow}}{\text{total number of students in the class}}$

(b) $P(\text{roll tongue and raise one eyebrow}) = \frac{42}{200} = 0.21$

> The number of students who can roll their tongue and raise one eyebrow is in the "Yes"/"Yes" cell of the two-way table.

FOR PRACTICE, TRY EXERCISE 13

The probability $P(\text{A and B}) = 42/200$ that events A and B both occur is called a *joint probability* because it is found using the value where a row and a column come together. That's consistent with our description of 42/200 as a *joint relative frequency* in Unit 2. If the joint probability is 0, the events are disjoint (mutually exclusive).

When we found $P(\text{roll tongue and raise one eyebrow}) = 42/200$ in part (b) of the example, we could have described this as either $P(\text{A and B})$ or $P(\text{B and A})$. Why? Because "A and B" describes the same event as "B and A." Likewise, $P(\text{A or B})$ is the same as $P(\text{B or A})$. Don't get so caught up in the notation that you lose sight of what's really happening!

General Addition Rule

Bill Boch/Photodisc/Getty Images

There are two different uses of the word "or" in everyday life. In a restaurant, if you are asked if you want "soup or salad," the server expects you to choose one or the other, but not both. However, if you order coffee and are asked if you want "cream or sugar," it's OK to ask for one or the other or both. The same issue arises in statistics.

Mutually exclusive events A and B cannot both happen at the same time. For such events, "A or B" means that only event A happens or only event B happens. You can find *P*(A or B) with the addition rule for mutually exclusive events:

$$P(\text{A or B}) = P(\text{A}) + P(\text{B})$$

But how can we find *P*(A or B) when the two events are *not* mutually exclusive? Now we have to deal with the fact that in statistics, "A or B" means one or the other *or both*.

In the preceding example, *P*(A or B) = *P*(roll tongue or raise one eyebrow). We can't use the addition rule for mutually exclusive events in this case because events A and B share 42 outcomes — there are 42 students who can roll their tongue and raise one eyebrow. If we simply added the probabilities of A and B, we'd get 149/200 + 62/200 = 211/200. This is clearly wrong, because the probability is greater than 1. As Figure 4.3 illustrates, outcomes common to both events are counted twice when we add the probabilities of these two events.

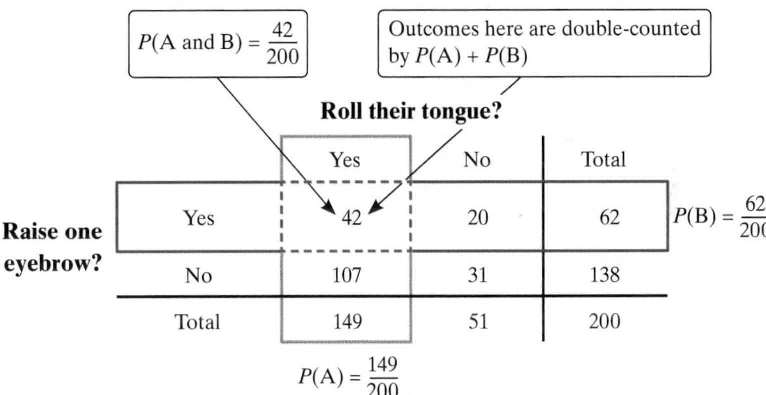

Raise one eyebrow?	Roll their tongue? Yes	No	Total
Yes	42	20	62
No	107	31	138
Total	149	51	200

FIGURE 4.3 Two-way table showing events A and B from the "roll tongue" and "raise one eyebrow" example. These events are *not* mutually exclusive, so we can't find *P*(A or B) by just adding the probabilities of the two events.

We can fix the double-counting problem illustrated in the two-way table by subtracting the probability *P*(roll tongue and raise one eyebrow) from the sum. That is,

$$\begin{aligned} &P(\text{roll tongue or raise eyebrow}) \\ &= P(\text{roll tongue}) + P(\text{raise one eyebrow}) - P(\text{roll tongue and raise one eyebrow}) \\ &= 149/200 + 62/200 - 42/200 \\ &= 169/200 \end{aligned}$$

This result is known as the **general addition rule.**

DEFINITION General addition rule

If A and B are any two events resulting from the same random process, the **general addition rule** says that

$$P(\text{A or B}) = P(\text{A}) + P(\text{B}) - P(\text{A and B})$$

There are other ways to find P(roll tongue or raise one eyebrow) from the two-way table without using the general addition rule. For instance, because each student in the class is equally likely to be selected:

$$P(\text{roll tongue or raise one eyebrow}) = \frac{\text{number of students who can roll their tongue or raise one eyebrow}}{\text{total number of students in the class}}$$
$$= \frac{42+107+20}{200}$$
$$= \frac{169}{200}$$

Sometimes it's easier to label events with letters that relate to the context, as the following example shows.

EXAMPLE

Facebook versus Instagram
General addition rule

Skill 3.A

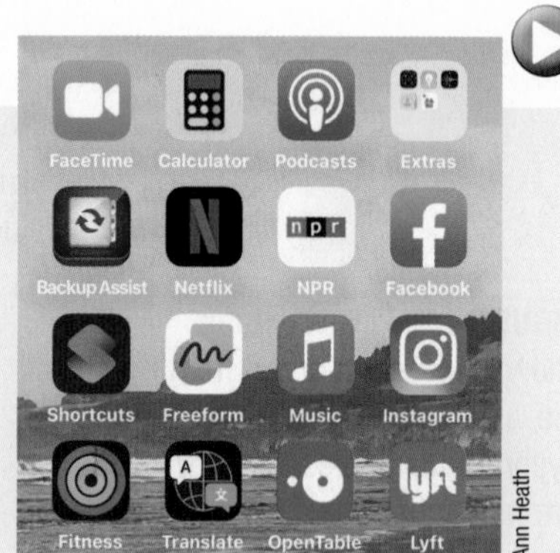

Ann Heath

PROBLEM: A survey about social media use suggests that 69% of U.S. adults use Facebook, 40% use Instagram, and 32% use both.[8] Suppose we select a U.S. adult at random. What's the probability that the person uses Facebook or Instagram?

SOLUTION:

Let F = uses Facebook and I = uses Instagram.

$$P(\text{F or I}) = P(\text{F}) + P(\text{I}) - P(\text{F and I})$$
$$= 0.69 + 0.40 - 0.32$$
$$= 0.77$$

$P(\text{A or B}) = P(\text{A}) + P(\text{B}) - P(\text{A and B})$

FOR PRACTICE, TRY EXERCISE 15

What's the probability that a randomly selected U.S. adult uses *neither* Facebook nor Instagram? The event "uses neither Facebook nor Instagram" is the complement of the event "uses at least one of Facebook or Instagram." Using the complement rule:

$P(\text{neither Facebook nor Instagram}) = 1 - P(\text{Facebook or Instagram}) = 1 - 0.77 = 0.23$

As you'll see in Section 5C, the fact that "none" is the opposite of "at least 1" comes in handy for a variety of probability questions.

What happens if we use the general addition rule for two mutually exclusive events A and B? In that case, $P(\text{A and B}) = 0$, and the formula reduces to

$$P(\text{A or B}) = P(\text{A}) + P(\text{B}) - P(\text{A and B}) = P(\text{A}) + P(\text{B}) - 0 = P(\text{A}) + P(\text{B})$$

In other words, the addition rule for mutually exclusive events is just a special case of the general addition rule.

VENN DIAGRAMS, INTERSECTIONS, AND UNIONS

We have seen that two-way tables can be used to illustrate the sample space of a random process involving two events. So can **Venn diagrams,** like the one shown in Figure 4.4.

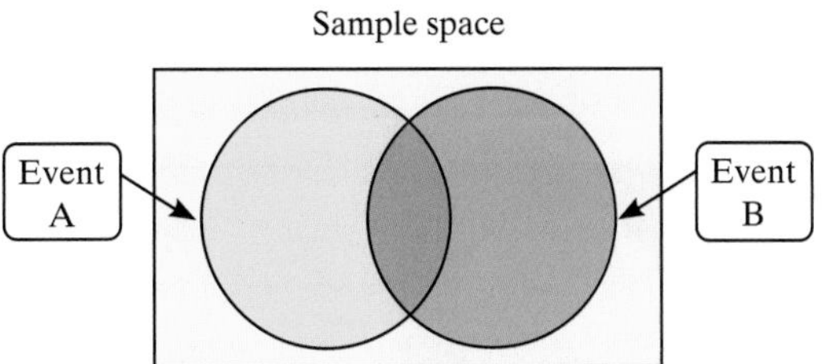

FIGURE 4.4 A typical Venn diagram that shows the sample space and the relationship between two events A and B.

DEFINITION Venn diagram

A **Venn diagram** consists of one or more circles surrounded by a rectangle. Each circle represents an event. The region inside the rectangle represents the sample space of the random process.

In an earlier example, we looked at data about who in a large class of college students can roll their tongue or raise one eyebrow. The chance process was selecting a student in the class at random. Our events of interest were A: student can roll their tongue and B: student can raise one eyebrow. Here is the two-way table that summarizes the data.

		Roll their tongue?		
		Yes	No	Total
Raise one eyebrow?	Yes	42	20	62
	No	107	31	138
	Total	149	51	200

The Venn diagram in Figure 4.5 displays the sample space in a slightly different way. There are four distinct regions in the Venn diagram, which correspond to the four cells in the two-way table.

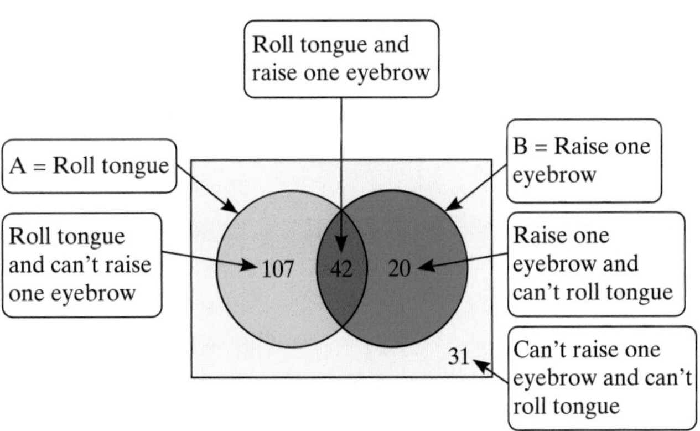

FIGURE 4.5 Venn diagram for the large class of college students in the earlier example. The circles represent the two events A = roll tongue and B = raise one eyebrow.

Statisticians have developed some standard vocabulary and notation to make our work with Venn diagrams a bit easier.

- We introduced the *complement* of an event earlier. The green area in Figure 4.6(a) shows the complement A^C, which contains the outcomes that are not in A.
- Figure 4.6(b) shows the event "A and B" in green. You can see why this event is also called the **intersection** of A and B. The corresponding notation is $A \cap B$.
- The event "A or B" is shown in green in Figure 4.6(c). This event is also known as the **union** of A and B. The corresponding notation is $A \cup B$.

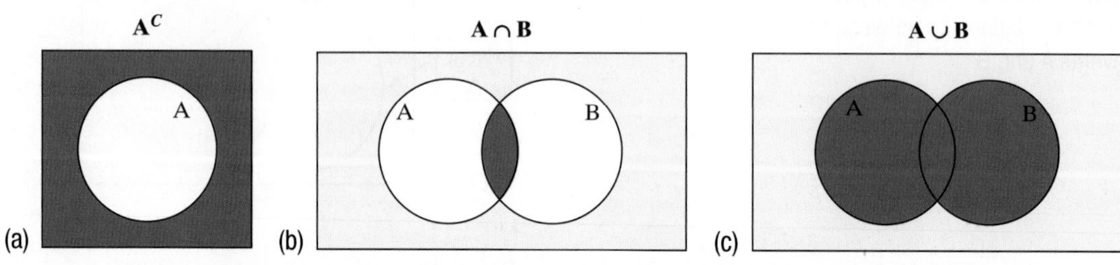

FIGURE 4.6 The green shaded region in each Venn diagram shows (a) the *complement* A^C of event A, (b) the *intersection* of events A and B, and (c) the *union* of events A and B.

DEFINITION Intersection, Union

The event "A and B" is called the **intersection** of events A and B. It consists of all outcomes that are common to both events, and is denoted $A \cap B$.

The event "A or B" is called the **union** of events A and B. It consists of all outcomes that are in event A, event B, or both, and is denoted $A \cup B$.

AP® EXAM TIP
Formula sheet

The formula for $P(A \cup B)$ is included on the formula sheet provided on both sections of the AP® Statistics exam.

Here's a way to keep the symbols straight: ∪ for **u**nion; ∩ for **int**ersection. With this new notation, we can rewrite the general addition rule in symbols as follows:

$$P(A \cup B) = P(A) + P(B) - P(A \cap B)$$

CHECK YOUR UNDERSTANDING

Yellowstone National Park staff surveyed a random sample of 1526 winter visitors to the park. They asked each person whether they belonged to an environmental club (like the Sierra Club). Respondents were also asked whether they owned, rented, or had never used a snowmobile. The two-way table summarizes the survey responses.[9]

		Environmental club membership	
		No	Yes
Snowmobile use	Never	445	212
	Rent	497	77
	Own	279	16

Suppose we choose one of the survey respondents at random.

1. What's the probability that the person is an environmental club member?
2. Find *P*(not a snowmobile renter). Describe this probability in words.
3. What's *P*(environmental club member and not a snowmobile renter)?
4. Find the probability that the person is not an environmental club member or is a snowmobile renter.

What if we focus on probabilities instead of numbers of students? Notice that

$$\frac{P(\text{roll tongue and raise one eyebrow})}{P(\text{raise one eyebrow})} = \frac{\frac{42}{200}}{\frac{62}{200}}$$

$$= \frac{42}{62} = P(\text{roll tongue} \mid \text{raise one eyebrow})$$

This observation leads to a general formula for calculating a conditional probability.

> **AP® EXAM TIP**
> **Formula sheet**
>
> The formula for $P(A \mid B)$ is included on the formula sheet provided on both sections of the AP® Statistics exam.

CONDITIONAL PROBABILITY FORMULA

To find the conditional probability $P(A \mid B)$, use the formula

$$P(A \mid B) = \frac{P(\text{A and B})}{P(\text{B})} = \frac{P(A \cap B)}{P(B)} = \frac{P(\text{both events occur})}{P(\text{given event occurs})}$$

By the same reasoning,

$$P(B \mid A) = \frac{P(\text{B and A})}{P(\text{A})} = \frac{P(B \cap A)}{P(A)}$$

EXAMPLE

Facebook versus Instagram
Conditional probability

Skill 3.A

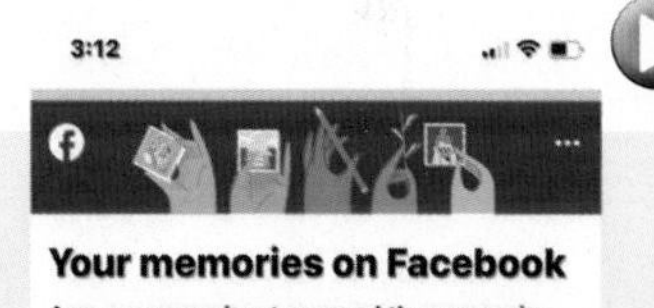
3:12
Your memories on Facebook
Ann, we care about you and the memories you share here. We thought you'd like to look back on this post from 2 years ago.

2 Years Ago

Ann Heath

PROBLEM: A survey about social media use suggests that 69% of U.S. adults use Facebook, 40% use Instagram, and 32% use both.[23] Suppose we select a U.S. adult at random. Given that the person uses Facebook, what's the probability that they also use Instagram?

SOLUTION:

$P(\text{Instagram} \mid \text{Facebook}) = \frac{0.32}{0.69} = 0.464$

$P(A \mid B) = \frac{P(A \cap B)}{P(B)}$

FOR PRACTICE, TRY EXERCISE 5

> **AP® EXAM TIP**
>
> You can write statements like $P(A \mid B)$ if events A and B are clearly defined in a problem. Otherwise, it's probably easier to use contextual labels, like $P(\text{Instagram} \mid \text{Facebook})$ or $P(I \mid F)$ in the preceding example.

Conditional Probability and Independence

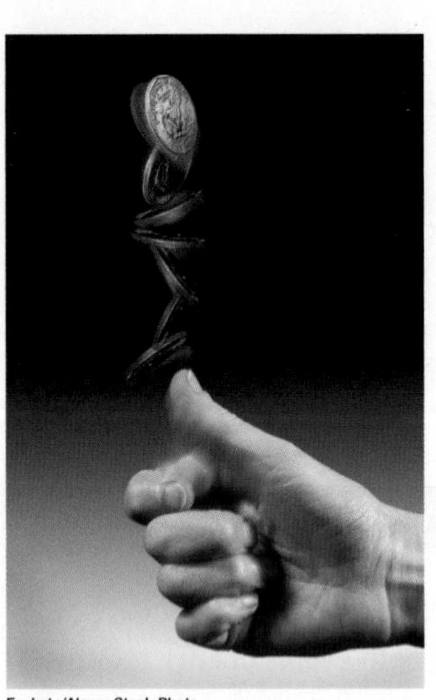

Eyebyte/Alamy Stock Photo

Suppose you toss a fair coin twice. Define events A: first toss is a head and B: second toss is a head. We know that $P(A) = 1/2$ and $P(B) = 1/2$.

- What's $P(B \mid A)$? It's the conditional probability that the second toss is a head given that the first toss was a head. The coin has no memory, so $P(B \mid A) = 1/2$.
- What's $P(B \mid A^C)$? It's the conditional probability that the second toss is a head given that the first toss was not a head. Getting a tail on the first toss does not change the probability of getting a head on the second toss, so $P(B \mid A^C) = 1/2$.

In this case, $P(B \mid A) = P(B \mid A^C) = P(B)$. Knowing whether or not the first toss was a head does not change the probability that the second toss is a head. We say that A and B are **independent events.**

DEFINITION Independent events

A and B are **independent events** if knowing whether or not one event has occurred does not change the probability that the other event will happen. In other words, events A and B are independent if

$$P(A \mid B) = P(A \mid B^C) = P(A)$$

Alternatively, events A and B are independent if

$$P(B \mid A) = P(B \mid A^C) = P(B)$$

Let's contrast the coin-toss scenario with our earlier tongue rolling/eyebrow raising example. In that case, the random process involved randomly selecting a student from a college statistics class. The events of interest were A: can roll their tongue and B: can raise one eyebrow. Are these two events independent?

		Roll their tongue?		
		Yes	No	Total
Raise one eyebrow?	Yes	42	20	62
	No	107	31	138
	Total	149	51	200

- Suppose that the chosen student can roll their tongue. We can see from the two-way table that $P(\text{raise one eyebrow} \mid \text{roll tongue}) = P(B \mid A) = 42/149 = 0.282$.
- Suppose that the chosen student cannot roll their tongue. From the two-way table, we see that $P(\text{raise one eyebrow} \mid \text{not roll tongue}) = P(B \mid A^C) = 20/51 = 0.392$.

Knowing that the chosen student can roll their tongue changes (reduces) the probability that the student can raise one eyebrow. So, these two events are not independent.

Another way to determine whether two events A and B are independent is to compare $P(B \mid A)$ to $P(B)$. For the tongue rolling/eyebrow raising scenario,

$$P(\text{raise one eyebrow} \mid \text{roll tongue}) = P(B \mid A) = 42/149 = 0.282$$

The unconditional probability that the chosen student can raise one eyebrow is

$$P(\text{raise one eyebrow}) = P(\text{B}) = 62/200 = 0.31$$

Again, knowing that the chosen student can roll their tongue changes (reduces) the probability that this person can raise one eyebrow. So, these two events are not independent.

EXAMPLE

Are more math teachers left-handed?
Conditional probability and independence

Skill 3.A

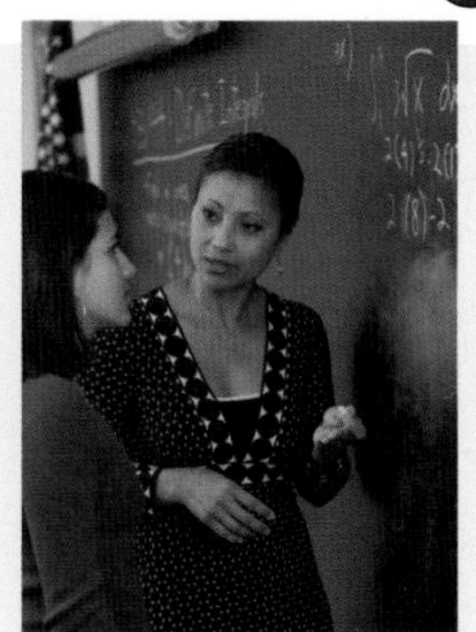
Ariel Skelley/DigitalVision/Getty Images

PROBLEM: Is there a relationship between subject taught and handedness? To find out, researchers chose an SRS of 100 teachers at an education conference. The two-way table summarizes the relationship between the subject taught and the dominant hand of each teacher. Suppose we choose one of the teachers in the sample at random. Are the events "teaches math" and "left-handed" independent? Justify your answer.

		Subject taught		
		Math	Not math	Total
Dominant hand	Right	39	51	90
	Left	7	3	10
	Total	46	54	100

SOLUTION:

$P(\text{left-handed} \mid \text{teaches math}) = 7/46 = 0.152$

$P(\text{left-handed}) = 10/100 = 0.10$

Does knowing whether the teacher teaches math change the probability of left-handedness?

Because these probabilities are not equal, the events "teaches math" and "left-handed" are not independent. Knowing that the teacher's subject is math increases the probability that the teacher is left-handed.

FOR PRACTICE, TRY EXERCISE 13

In the example, we could have also determined that the two events are not independent by showing that

$$P(\text{left-handed} \mid \text{teaches math}) = 7/46 = 0.152 \neq P(\text{left-handed} \mid \text{doesn't teach math}) = 3/54 = 0.056$$

Alternatively, we could have focused on whether knowing that the chosen teacher is left-handed changes the probability that they teach math. Because

$$P(\text{teaches math} \mid \text{left-handed}) = 7/10 = 0.70 \neq P(\text{teaches math}) = 46/100 = 0.46$$

the events "teaches math" and "left-handed" are not independent. Just remember: to assess independence, you need to compare the probability of the same event occurring under different conditions.

Think About It

IS THERE A CONNECTION BETWEEN INDEPENDENCE OF EVENTS AND ASSOCIATION BETWEEN TWO VARIABLES? Yes! In the preceding example, we found that the events "teaches math" and "left-handed" were not independent for the sample of 100 teachers. Knowing a teacher's subject area helped us predict their dominant hand. Applying what you learned in Section 2A, there is an association between subject taught and handedness for the teachers in the sample. The segmented bar chart shows this association in picture form.

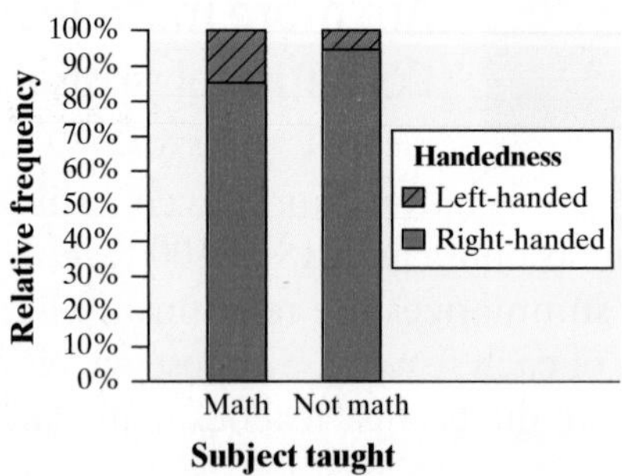

Does that mean there is an association between subject taught and handedness in the population of teachers at the conference? Maybe — or maybe not. Even if there is no association between the variables in the population, it would be surprising to choose a random sample of 100 teachers for which *P*(left-handed | teaches math), *P*(left-handed | doesn't teach math), and *P*(left-handed) were *exactly* equal. But these probabilities should be close to equal if there's no association between the variables in the population. How close is close? We'll discuss this issue further in Unit 8.

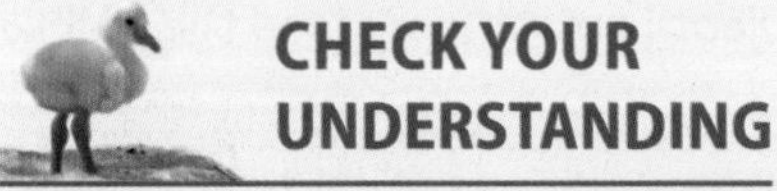

CHECK YOUR UNDERSTANDING

Yellowstone National Park staff surveyed a random sample of 1526 winter visitors to the park.[24] They asked each person whether they belonged to an environmental club (like the Sierra Club). Respondents were also asked whether they owned, rented, or had never used a snowmobile. The two-way table summarizes the survey responses.

		Environmental club membership		
		No	Yes	Total
Snowmobile use	Never	445	212	657
	Rent	497	77	574
	Own	279	16	295
	Total	1221	305	1526

Suppose we randomly select one of the survey respondents. Define events E: environmental club member, S: snowmobile owner, and N: never used.

1. Find $P(N \mid E)$. Describe this probability in words.
2. Given that the chosen person is not a snowmobile owner, what's the probability that they are an environmental club member? Write your answer as a probability statement using correct notation for the events.
3. Are the events "snowmobile owner" and "environmental club member" independent? Justify your answer.

The General Multiplication Rule

Suppose that A and B are two events resulting from the same random process. We can find the probability $P(\text{A or B})$ with the general addition rule:

$$P(\text{A or B}) = P(\text{A}) + P(\text{B}) - P(\text{A and B})$$

How do we find the probability that both events happen, $P(\text{A and B})$?

Consider this situation: approximately 55% of high school students participate on a school athletic team at some level. Roughly 6% of these athletes go on to play on a college team in the National Collegiate Athletic Association (NCAA).[25] What percentage of high school students play a sport in high school *and* go on to play on an NCAA team? Approximately 6% of 55%, or roughly 3.3%.

Let's restate the situation in probability language. Suppose we select a high school student at random. What's the probability that the student plays a sport in high school and goes on to play on an NCAA team? The given information suggests that

$$P(\text{high school sport}) = 0.55 \text{ and } P(\text{NCAA team} \mid \text{high school sport}) = 0.06$$

By the logic just stated,

$$\begin{aligned} &P(\text{high school sport and NCAA team}) \\ &\quad = P(\text{high school sport}) \cdot P(\text{NCAA team} \mid \text{high school sport}) \\ &\quad = (0.55)(0.06) \\ &\quad = 0.033 \end{aligned}$$

This is an example of the **general multiplication rule.**

DEFINITION General multiplication rule

For any random process, the probability that events A and B both occur can be found using the **general multiplication rule:**

$$P(\text{A and B}) = P(\text{A} \cap \text{B}) = P(\text{A}) \cdot P(\text{B} \mid \text{A})$$

The general multiplication rule says that for both of two events to occur, first one event must occur. Then, given that the first event has occurred, the second must occur. To confirm that this result is correct, start with the conditional probability formula

$$P(\text{B} \mid \text{A}) = \frac{P(\text{B} \cap \text{A})}{P(\text{A})}$$

The numerator gives the probability we want because $P(\text{B} \cap \text{A})$ is the same as $P(\text{A} \cap \text{B})$. Multiply both sides of this equation by $P(\text{A})$ to get

$$P(\text{A}) \cdot P(\text{B} \mid \text{A}) = P(\text{A} \cap \text{B})$$

EXAMPLE

Losing your marbles?
The general multiplication rule

Skill 3.A

Bombaert Patrick/Alamy Stock Photo

PROBLEM: A bag contains 5 red marbles and 8 yellow marbles. Suppose you randomly select 2 marbles from the bag. What's the probability that both are red?

SOLUTION:

P(both marbles red)
= *P*(first marble red and second marble red)
= *P*(first marble red) · *P*(second marble red | first marble red)
= (5/13)(4/12)
= 0.128

$P(A \text{ and } B) = P(A) \cdot P(B \mid A)$

FOR PRACTICE, TRY EXERCISE 19

In the example, it is easier to think about the random process of selecting 2 marbles from the bag in stages: choose the first marble at random, then (without replacing the first marble) choose the second marble at random. Doing so allows us to use the general multiplication rule to solve the problem. Keep this strategy in mind when you encounter questions involving random sampling—with or without replacement.

We can extend the general multiplication rule to more than two events. For instance, if you randomly selected 3 marbles from the bag in the preceding example,

P(all 3 marbles are red)
= *P*(1st marble red) · *P*(2nd marble red | 1st marble red) ·
P(3rd marble red | 1st two marbles red)
= (5/13)(4/12)(3/11)
= 0.035

Tree Diagrams and Probability

Shannon hits the snooze button on her alarm on 60% of school days. If she hits snooze, there is a 0.70 probability that she makes it to her first class on time. If she doesn't hit snooze and gets up right away, there is a 0.90 probability that she makes it to class on time. Suppose we select a school day at random and record whether Shannon hits the snooze button and whether she arrives in class on time. Figure 4.7 shows a **tree diagram** for this random process.

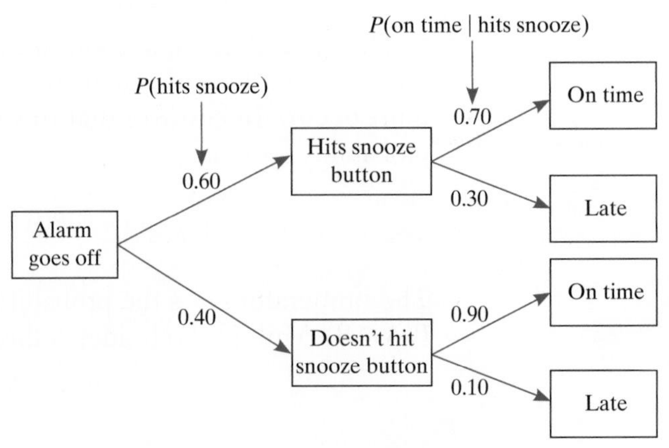

FIGURE 4.7 A tree diagram displaying the sample space of randomly choosing a school day and noting whether Shannon hits the snooze button and whether she gets to her first class on time.

DEFINITION Tree diagram

A **tree diagram** shows the sample space of a random process involving multiple stages. The probability of each outcome is shown on the corresponding branch of the tree. All probabilities after the first stage are conditional probabilities.

There are only two possible outcomes at the first "stage" of this random process: Shannon hits the snooze button or she doesn't. The first set of branches in the tree diagram displays these outcomes with their probabilities. The second set of branches shows the two possible results at the next "stage" of the process — Shannon gets to her first class either on time or late — and the probability of each result based on whether she hit the snooze button. Note that the probabilities on the second set of branches are *conditional* probabilities: $P(\text{on time} \mid \text{hits snooze}) = 0.70$ and $P(\text{on time} \mid \text{doesn't hit snooze}) = 0.90$.

We can ask some interesting questions related to the tree diagram:

- **What is the probability that Shannon hits the snooze button *and* is late for class on a randomly selected school day?** The general multiplication rule provides the answer:

$$\begin{aligned} P(\text{hits snooze and late}) &= P(\text{hits snooze}) \cdot P(\text{late} \mid \text{hits snooze}) \\ &= (0.60)(0.30) \\ &= 0.18 \end{aligned}$$

  There is an 18% chance that Shannon hits the snooze button and is late for class. Note that this calculation amounts to multiplying probabilities along the branches of the tree diagram.

- **What's the probability that Shannon is late to class on a randomly selected school day?** Figure 4.8 illustrates two ways this can happen: Shannon hits the snooze button and is late *or* she doesn't hit snooze and is late.

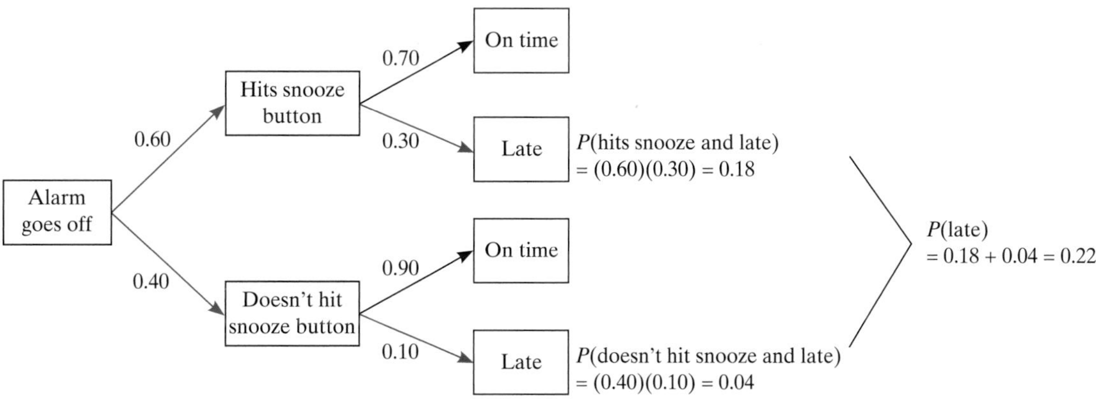

FIGURE 4.8 Tree diagram with blue arrows showing the two possible ways that Shannon can be late to class on a randomly selected day.

We already found that $P(\text{hits snooze and late}) = 0.18$. The general multiplication rule tells us that

$$\begin{aligned} P(\text{doesn't hit snooze and late}) &= P(\text{doesn't hit snooze}) \cdot P(\text{late} \mid \text{doesn't hit snooze}) \\ &= (0.40)(0.10) \\ &= 0.04 \end{aligned}$$

Because these outcomes are mutually exclusive,

$$\begin{aligned} P(\text{late}) &= P(\text{hits snooze and late}) + P(\text{doesn't hits snooze and late}) \\ &= 0.18 + 0.04 \\ &= 0.22 \end{aligned}$$

There is a 22% chance that Shannon will be late to class.

EXAMPLE Do people read more ebooks or print books? Tree diagrams and probability

Skill 3.A

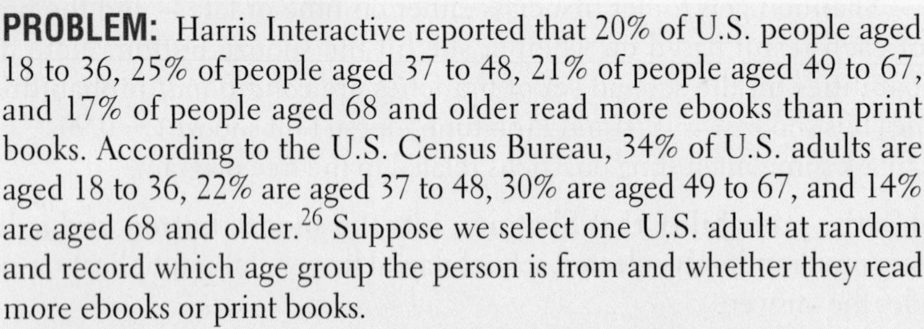

PROBLEM: Harris Interactive reported that 20% of U.S. people aged 18 to 36, 25% of people aged 37 to 48, 21% of people aged 49 to 67, and 17% of people aged 68 and older read more ebooks than print books. According to the U.S. Census Bureau, 34% of U.S. adults are aged 18 to 36, 22% are aged 37 to 48, 30% are aged 49 to 67, and 14% are aged 68 and older.[26] Suppose we select one U.S. adult at random and record which age group the person is from and whether they read more ebooks or print books.

Burlingham/Shutterstock

(a) Draw a tree diagram to model this random process.

(b) Find the probability that the person reads more ebooks than print books.

SOLUTION:

(a)

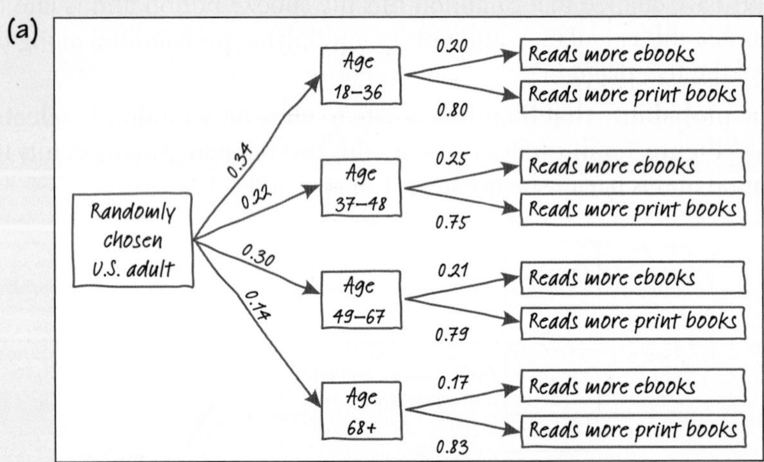

The age categories go on the first set of branches because the given relative frequencies are *not* conditional probabilities. Because the relative frequencies for reading more ebooks and reading more print books are conditional on the age categories, the reading tendencies go on the second set of branches.

(b)
$$\begin{aligned} P(\text{reads more ebooks}) &= (0.34)(0.20) + (0.22)(0.25) + (0.30)(0.21) + (0.14)(0.17) \\ &= 0.0680 + 0.0550 + 0.0630 + 0.0238 \\ &= 0.2098 \end{aligned}$$

FOR PRACTICE, TRY EXERCISE 21

WORKING BACKWARD ON A TREE DIAGRAM

Some interesting conditional probability questions involve "going in reverse" on a tree diagram. For instance, suppose that Shannon is late for class on a randomly chosen school day. What is the probability that she hit the snooze button that morning? To find this probability, we start with the given information that

Shannon is late, which is displayed on the second set of branches in Figure 4.8, and ask whether she hit the snooze button, which is shown on the first set of branches. We can use the information from the tree diagram and the conditional probability formula to do the required calculation:

$$P(\text{hit snooze button} \mid \text{late}) = \frac{P(\text{hit snooze button and late})}{P(\text{late})} = \frac{0.18}{0.22} = 0.818$$

Given that Shannon is late for school on a randomly selected day, there is a 0.818 probability that she hit the snooze button.

To answer questions that involve going backward on a tree diagram, we just use the conditional probability formula and plug in the appropriate values.

EXAMPLE

How reliable are mammograms?
Tree diagrams and probability

Skill 3.A

PROBLEM: Many women choose to have annual mammograms to screen for breast cancer starting at age 40. Unfortunately, a mammogram isn't foolproof. Sometimes the test suggests that a woman has breast cancer when she really doesn't (a "false-positive"). Other times, the test says that a woman doesn't have breast cancer when she actually does (a "false-negative").

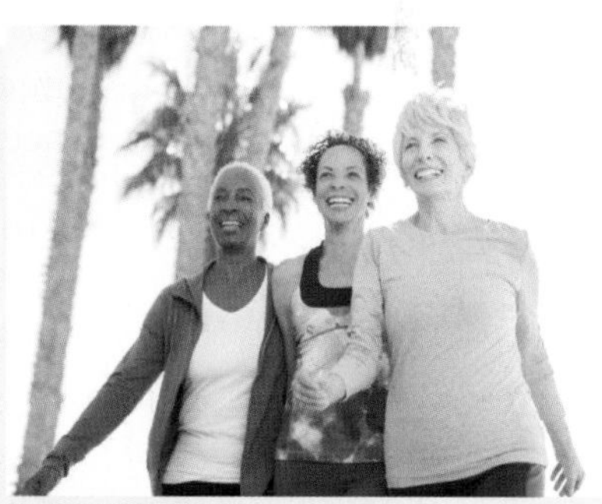
Sam Edwards/Getty Images

Suppose that we know the following information about breast cancer and mammograms in a particular population:

- One percent of the women aged 40 or older in this population have breast cancer.
- For women who have breast cancer, the probability of a negative mammogram is 0.03.
- For women who don't have breast cancer, the probability of a positive mammogram is 0.06.

A randomly selected woman aged 40 or older from this population tests positive for breast cancer in a mammogram. Find the probability that she actually has breast cancer.

SOLUTION:

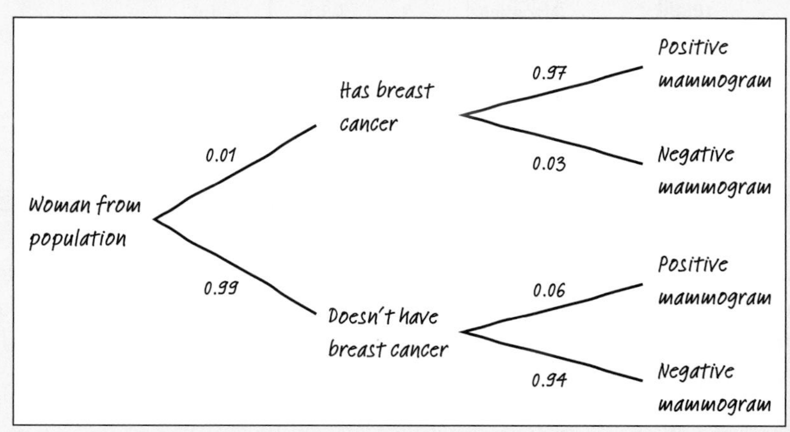

Start by making a tree diagram to summarize the possible outcomes.

- Because 1% of women in this population have breast cancer, 99% don't have breast cancer.
- Of those women who do have breast cancer, 3% would test negative on a mammogram. The remaining 97% would (correctly) test positive.
- Of the women who don't have breast cancer, 6% would test positive on a mammogram. The remaining 94% would (correctly) test negative.

$$P(\text{breast cancer} \mid \text{positive mammogram}) = \frac{P(\text{breast cancer and positive mammogram})}{P(\text{positive mammogram})}$$

$$= \frac{(0.01)(0.97)}{(0.01)(0.97)+(0.99)(0.06)}$$

$$= \frac{0.0097}{0.0691}$$

$$= 0.14$$

FOR PRACTICE, TRY EXERCISE 25

Given that a randomly selected woman from the population has a positive mammogram, there is only about a 14% chance that she has breast cancer. Are you surprised by this result? Most people are. Sometimes, a two-way table that includes counts is more convincing.[27]

To make the calculations easier, let's suppose that there are exactly 10,000 women aged 40 or older in this population, and that exactly 100 have breast cancer (that's 1% of the women in this population).

- How many of those 100 women would have a positive mammogram? That would be 97% of 100, or 97 of them. That leaves 3 who would test negative.
- How many of the 9900 women who don't have breast cancer would have a positive mammogram? That would be 6% of them, or $(9900)(0.06) = 594$ women. The remaining $9900 - 594 = 9306$ would test negative.
- In total, $97 + 594 = 691$ women would have positive mammograms and $3 + 9306 = 9309$ women would have negative mammograms.

This information is summarized in the two-way table.

		Has breast cancer?		
		Yes	No	Total
Mammogram result	Positive	97	594	691
	Negative	3	9306	9309
	Total	100	9900	10,000

Given that a randomly selected woman has a positive mammogram, the two-way table shows that the conditional probability

$$P(\text{breast cancer} \mid \text{positive mammogram}) = 97/691 = 0.14$$

This example illustrates an important fact when considering proposals for widespread screening for serious diseases or illegal drug use: if the condition being tested is uncommon in the population, many positive screening results will be false-positives. The best remedy is to retest any individual who tests positive.

CHECK YOUR UNDERSTANDING

A computer company makes desktop, laptop, and tablet computers at factories in two states, California and Texas. The California factory produces 40% of the company's computers and the Texas factory makes the rest. Of the computers made in California, 25% are desktops, 30% are laptops, and the rest are tablets. Of those made in Texas, 10% are desktops, 20% are laptops, and the rest are tablets. All computers are first shipped to a distribution center in Missouri before being sent out to stores. Suppose we select a computer at random from the distribution center and observe where it was made and whether it is a desktop, laptop, or tablet.[28]

1. Construct a tree diagram to model this random process.
2. Find the probability that the computer is a tablet.
3. Given that a tablet computer is selected, what is the probability that it was made in California?

The Multiplication Rule for Independent Events

What happens to the general multiplication rule in the special case when events A and B are independent? In that case, $P(\text{B} \mid \text{A}) = P(\text{B})$ because knowing that event A occurred doesn't change the probability that event B occurs. We can simplify the general multiplication rule as follows:

$$P(\text{A and B}) = P(\text{A} \cap \text{B}) = P(\text{A}) \cdot P(\text{B} \mid \text{A}) = P(\text{A}) \cdot P(\text{B})$$

This result is known as the **multiplication rule for independent events.**

DEFINITION Multiplication rule for independent events

The **multiplication rule for independent events** says that if A and B are independent events, the probability that A and B both occur is

$$P(\text{A and B}) = P(\text{A} \cap \text{B}) = P(\text{A}) \cdot P(\text{B})$$

Note that this rule applies *only* to independent events.

Suppose that Pedro drives the same route to work on Monday through Friday. Pedro's route includes one traffic light. The probability that the light will be green is 0.42, that it will be yellow is 0.03, and that it will be red is 0.55.

1. **What's the probability that the light is green on Monday and red on Tuesday?** Let event A be a green light on Monday and event B be a red light on Tuesday. These two events are independent because knowing whether or not the light was green on Monday doesn't help us predict the color of the light on Tuesday. By the multiplication rule for independent events,

$$\begin{aligned} P(\text{green on Monday and red on Tuesday}) &= P(\text{A and B}) \\ &= P(\text{A}) \cdot P(\text{B}) \\ &= (0.42)(0.55) \\ &= 0.231 \end{aligned}$$

There's about a 23% chance that the light will be green on Monday and red on Tuesday.

2. **What's the probability that Pedro finds the light red on Monday through Friday?** We can extend the multiplication rule for independent events to more than two events:

$$
\begin{aligned}
&P(\text{red Monday } \textit{and} \text{ red Tuesday } \textit{and} \text{ red Wednesday } \textit{and} \text{ red Thursday } \textit{and} \text{ red Friday}) \\
&= P(\text{red Monday}) \cdot P(\text{red Tuesday}) \cdot P(\text{red Wednesday}) \cdot P(\text{red Thursday}) \cdot P(\text{red Friday}) \\
&= (0.55)(0.55)(0.55)(0.55)(0.55) \\
&= (0.55)^5 \\
&= 0.0503
\end{aligned}
$$

There is about a 5% chance that Pedro will encounter a red light on all five days in a work week.

EXAMPLE

The *Challenger* disaster
Multiplication rule for independent events

Skill 3.A

PROBLEM: On January 28, 1986, Space Shuttle *Challenger* exploded on takeoff, killing all seven crew members aboard. Afterward, scientists and statisticians helped analyze what went wrong. They determined that the failure of O-ring joints in the shuttle's booster rockets caused the explosion. Experts estimated that the probability that an individual O-ring joint would function properly under the cold conditions that day was 0.977. But there were six O-ring joints, and all six had to function properly for the shuttle to launch safely. Assuming that O-ring joints succeed or fail independently, find the probability that the shuttle would launch safely under similar conditions.

Thom Baur/AP Photo

SOLUTION:

$$
\begin{aligned}
&P(\text{O-ring 1 works and O-ring 2 works and} \ldots \text{and O-ring 6 works}) \\
&= P(\text{O-ring 1 works}) \cdot P(\text{O-ring 2 works}) \cdot \ldots \cdot P(\text{O-ring 6 works}) \\
&= (0.977)(0.977)(0.977)(0.977)(0.977)(0.977) \\
&= (0.977)^6 \\
&= 0.87
\end{aligned}
$$

For the shuttle to launch safely, all six O-ring joints must function properly.

FOR PRACTICE, TRY EXERCISE 31

The multiplication rule for independent events can also be used to help find $P(\text{at least one})$. In the preceding example, the shuttle would *not* launch safely under similar conditions if one, two, three, four, five, or all six O-ring joints failed — that is, if *at least one* O-ring failed. In other words, the only possible number of O-ring failures excluded is 0. So, the events "at least one O-ring joint fails" and "no O-ring joints fail" are complementary events. By the complement rule,

$$
\begin{aligned}
P(\text{at least one O-ring fails}) &= 1 - P(\text{no O-ring fails}) \\
&= 1 - 0.87 \\
&= 0.13
\end{aligned}
$$

That's a very high chance of failure! As a result of this analysis following the *Challenger* disaster, NASA made important safety changes to the design of the shuttle's booster rockets.

Use Probability Rules Wisely

To find the probability of "A or B," we can use the general addition rule:

$$P(\text{A or B}) = P(\text{A} \cup \text{B}) = P(\text{A}) + P(\text{B}) - P(\text{A and B})$$

In the special case when A and B are *mutually exclusive* (have no outcomes in common), the addition rule simplifies to

$$P(\text{A or B}) = P(\text{A} \cup \text{B}) = P(\text{A}) + P(\text{B})$$

To find the probability of "A and B," we can use the general multiplication rule:

$$P(\text{A and B}) = P(\text{A} \cap \text{B}) = P(\text{A}) \cdot P(\text{B} \mid \text{A})$$

In the special case when A and B are *independent*, the multiplication rule simplifies to

$$P(\text{A and B}) = P(\text{A} \cap \text{B}) = P(\text{A}) \cdot P(\text{B})$$

caution !

Resist the temptation to use the simpler rules when the conditions that justify them are not met.

Hagar the Horrible

Is there a connection between mutually exclusive and independent? Let's start with a new random process. Imagine choosing a U.S. resident at random. Define event A: the person is younger than age 10 and event B: the person has a driver's license. It's fairly clear that these two events are mutually exclusive because they can't happen together! Are they also independent?

If you know that event A has occurred, does this change the probability that event B happens? Of course! If we know the person is younger than age 10, the probability that the person has a driver's license is 0. Because $P(\text{B} \mid \text{A}) \neq P(\text{B})$, the two events are not independent.

Two mutually exclusive events with nonzero probabilities can *never* be independent, because if one event happens, the other event is guaranteed not to happen.

EXAMPLE

Watch the weather!
Use probability rules wisely

Skill 3.A

PROBLEM: Hacienda Heights and La Puente are two neighboring suburbs in the Los Angeles area. According to the local newspaper, there is a 50% chance of rain tomorrow in Hacienda Heights and a 50% chance of rain in La Puente. Does this mean that there is a $(0.50)(0.50) = 0.25$ probability that it will rain in both cities tomorrow?

SOLUTION:

No; it is not appropriate to multiply the two probabilities, because "raining tomorrow in Hacienda Heights" and "raining tomorrow in La Puente" are not independent events. If it is raining in one of these locations, there is a greater than 50% chance that it is raining in the other location because they are geographically close to each other.

FOR PRACTICE, TRY EXERCISE 35

The multiplication rule $P(\text{A and B}) = P(\text{A}) \cdot P(\text{B})$ gives us another way to determine whether two events are independent. Let's return to the tongue rolling and eyebrow raising example from earlier in the section. The following two-way table summarizes data from a college statistics class.

		Roll their tongue?		
		Yes	No	Total
Raise one eyebrow?	Yes	42	20	62
	No	107	31	138
	Total	149	51	200

Our events of interest were A: can roll their tongue and B: can raise one eyebrow. Are these two events independent? No, because

$$P(\text{A and B}) = P(\text{roll tongue and raise one eyebrow}) = \frac{42}{200} = 0.21$$

is not equal to

$$P(\text{A}) \cdot P(\text{B}) = P(\text{roll tongue}) \cdot P(\text{raise one eyebrow}) = \frac{149}{200} \cdot \frac{62}{200} = 0.231$$

If the ability to roll one's tongue and the ability to raise one eyebrow were independent, then about 23.1% of the students would be able to do both. But only 21% of the students could do both — less than expected if these events were independent.

DETERMINING INDEPENDENCE WITH THE MULTIPLICATION RULE

Events A and B are independent if and only if $P(\text{A} \cap \text{B}) = P(\text{A}) \cdot P(\text{B})$.

CHECK YOUR UNDERSTANDING

Questions 1 and 2 refer to the following setting. New Jersey Transit claims that its 8:00 A.M. train from Princeton to New York City has probability 0.9 of arriving on time on a randomly selected day. Assume for now that this claim is true.

1. Find the probability that the train arrives late on Monday but on time on Tuesday.
2. What's the probability that the train arrives late at least once in a 5-day week?
3. Government data show that 5.4% of adults are full-time college students and that 37.8% of adults are age 55 or older.[29] If we randomly select an adult, is it true that

$$P(\text{full-time college student and age 55 or older}) = (0.054)(0.378) = 0.02?$$

Why or why not?

SECTION 4C Summary

- A **conditional probability** describes the probability that one event happens given that another event is already known to have happened.
- One way to calculate a conditional probability is to use the formula

$$P(\text{A} \mid \text{B}) = \frac{P(\text{A and B})}{P(\text{B})} = \frac{P(\text{A} \cap \text{B})}{P(\text{B})} = \frac{P(\text{both events occur})}{P(\text{given event occurs})}$$

- When knowing whether or not one event has occurred does not change the probability that another event happens, we say that the two events are **independent.** Events A and B are independent if

$$P(\text{A} \mid \text{B}) = P(\text{A} \mid \text{B}^C) = P(\text{A})$$

or, alternatively, if

$$P(\text{B} \mid \text{A}) = P(\text{B} \mid \text{A}^C) = P(\text{B})$$

- Use the **general multiplication rule** to calculate the probability that events A and B both occur:

$$P(\text{A and B}) = P(\text{A} \cap \text{B}) = P(\text{A}) \cdot P(\text{B} \mid \text{A})$$

- When a random process involves multiple stages, a **tree diagram** can be used to display the sample space and to help calculate probabilities.
- In the special case of independent events, the multiplication rule becomes

$$P(\text{A and B}) = P(\text{A} \cap \text{B}) = P(\text{A}) \cdot P(\text{B})$$

This formula gives us another way to determine whether events A and B are independent.

AP® EXAM TIP

AP® Daily Videos

Review the content of this section and get extra help by watching the AP® Daily Videos for Topic 4.5 and Topic 4.6 (Videos 1 and 3), which are available in AP® Classroom.

SECTION 4C Exercises

Conditional Probability

1. **Superpowers** Researchers took separate random samples of children from England and the United States. Each student's country was recorded, along with which superpower they would most like to have: the ability to fly, the ability to freeze time, invisibility, superstrength, or telepathy (the ability to read minds). The data are summarized in the two-way table.

pg 349

		Country		
		England	U.S.	Total
Superpower	Fly	54	45	99
	Freeze time	52	44	96
	Invisibility	30	37	67
	Superstrength	20	23	43
	Telepathy	44	66	110
	Total	200	215	415

Suppose we randomly select one of these students. Define events E: England, T: telepathy, and S: superstrength.

(a) Find $P(T \mid E)$. Describe this probability in words.

(b) Given that the child did not choose superstrength, what's the probability that they are from England? Write your answer as a probability statement using correct notation for the events.

2. **Squirrels and humans** Do adult and juvenile Eastern gray squirrels in New York's Central Park exhibit different behaviors toward humans? That is one of many questions investigated by 323 volunteer squirrel sighters.[30] The two-way table summarizes the data for 2898 squirrel sightings in the park.

		Age		
		Juvenile	Adult	Total
Behavior toward humans	Approach	111	756	867
	Indifferent	138	1241	1379
	Run away	81	571	652
	Total	330	2568	2898

Suppose we select one of these squirrel sightings at random. Define events J: juvenile and R: run away.

(a) Find $P(R \mid J)$. Describe this probability in words.

(b) Given that the squirrel spotted by the volunteer did not run away, find the probability that it was an adult. Write your answer as a probability statement using correct notation for the events.

3. **Teens on social media** Pew Research Center surveyed a random sample of 1310 U.S. teens about their use of social media.[31] The teens were asked about the amount of time they spend on social media, as well as other demographic questions. The two-way table summarizes the relationship between social media use and the location where students live.

		Social media use		
		Too much	About right	Too little
Location	Urban	121	126	40
	Suburban	239	407	56
	Rural	116	183	22

Suppose we select one of the survey respondents at random.

(a) Given that the chosen teen lived in a rural location, what is the probability that they admitted spending too much time on social media?

(b) If the chosen student said that the amount of time they spend on social media is about right, what is the probability that they did not live in a suburban location?

4. **Pythons hatching** How does the temperature of a water python nest influence egg hatching? Researchers randomly assigned newly laid eggs to one of three water temperatures: hot, neutral, or cold. The two-way table summarizes the data on nest temperature and hatching status.[32]

		Hatching status	
		Hatched	Didn't hatch
Nest temperature	Cold	16	11
	Neutral	38	18
	Hot	75	29

Suppose we select one of the eggs at random.

(a) Given that the chosen egg was assigned to hot water, what is the probability that it hatched?

(b) If the chosen egg hatched, what is the probability that it was not assigned to hot water?

5. **Household pets** In one large city, 40% of all households own a dog, 32% own a cat, and 18% own both. Suppose we randomly select a household that owns a cat. Find the probability that the household also owns a dog.

pg 351

6. **Streaming apps again** According to recent survey data, Pandora and Spotify are the music-streaming apps most widely used by young adults. The responses reveal that 26% use Pandora, 48% use Spotify, and 14% use both.[33] Suppose we randomly select a young adult who uses Pandora. Find the probability that the person also uses Spotify.

7. **Foreign-language study** Researchers selected students in grades 9 to 12 at random and asked if they were studying a language other than English. Here is the distribution of results. Given that a student is studying some language other than English, what is the probability that they are studying Spanish?

Language	Spanish	French	German	All others	None
Probability	0.26	0.09	0.03	0.03	0.59

8. **Income tax returns** Here is the distribution of the adjusted gross income (in thousands of dollars) reported on individual U.S. federal income tax returns in a certain year. Given that a randomly selected return shows an income of at least \$50,000, what is the probability that the income is at least \$100,000?

Income	<25	25–49	50–99	100–499	≥500
Probability	0.431	0.248	0.215	0.100	0.006

9. **Tall people and basketball players** Suppose we select an adult at random. Define events T: person is more than 6 feet tall and B: person is a professional basketball player. Rank the following probabilities from smallest to largest. Explain your reasoning.

$$P(T) \qquad P(B) \qquad P(T \mid B) \qquad P(B \mid T)$$

10. **Teachers and college degrees** Suppose we select an adult at random. Define events D: person has earned a college degree and T: person's career is teaching. Rank the following probabilities from smallest to largest. Explain your reasoning.

$$P(D) \qquad P(T) \qquad P(D \mid T) \qquad P(T \mid D)$$

Conditional Probability and Independence

11. **Big Papi** Hall-of-Fame baseball star David Ortiz — nicknamed "Big Papi" — was known for his ability to deliver hits in high-pressure situations. Here is a two-way table of his hits, walks, and outs in all of his regular-season and post-season plate appearances from 1997 through 2014.[34]

		At-bat			
		Hit	Walk	Out	Total
Season	Regular	2023	1474	5034	8531
	Post	87	57	208	352
	Total	2110	1531	5242	8883

Suppose we choose a plate appearance at random.

(a) Find $P(\text{hit} \mid \text{post-season})$.

(b) Use your answer from part (a) to help determine if the events "hit" and "post-season" are independent.

12. **Middle school values** Researchers carried out a survey of fourth-, fifth-, and sixth-grade students in Michigan. Students were asked whether getting good grades, having athletic ability, or being popular was most important to them. The two-way table summarizes the data.[35]

		Grade level			
		4th	5th	6th	Total
Most important	Grades	49	50	69	168
	Athletic	24	36	38	98
	Popular	19	22	28	69
	Total	92	108	135	335

Suppose we select one of these students at random.

(a) Find $P(\text{athletic} \mid \text{5th grade})$.

(b) Use your answer from part (a) to help determine if the events "5th grade" and "athletic" are independent.

13. **Who owns a home?** What is the relationship between educational achievement and home ownership? A random sample of 500 U.S. adults was selected. Each member of the sample was identified as having attended college (or not) and as being a homeowner (or not). The two-way table summarizes the data.

pg 353

		College attendee		
		Yes	No	Total
Homeowner	Yes	221	119	340
	No	89	71	160
	Total	310	190	500

Suppose we choose one member of the sample at random. Are the events "homeowner" and "college attendee" independent? Justify your answer.

14. **Is this your card?** A standard deck of playing cards (with jokers removed) consists of 52 cards in four suits — clubs, diamonds, hearts, and spades. Each suit has 13 cards, with denominations ace, 2, 3, 4, 5, 6, 7, 8, 9, 10, jack, queen, and king. The jack, queen, and king are referred to as "face cards." Imagine that we shuffle the deck thoroughly and deal one card. The two-way table summarizes the sample space for this random process based on whether or not the card is a face card and whether or not the card is a heart.

		Type of card		
		Face card	Nonface card	Total
Suit	Heart	3	10	13
	Nonheart	9	30	39
	Total	12	40	52

Are the events "heart" and "face card" independent? Justify your answer.

15. **Rolling dice** Suppose you roll two fair, 6-sided dice — one red and one blue. Are the events "sum is 7" and "blue die shows a 4" independent? Justify your answer.

(See Figure 4.2 on page 333 for the sample space of this random process.)

16. **Rolling dice again** Suppose you roll two fair, 6-sided dice — one red and one blue. Are the events "sum is 8" and "blue die shows a 4" independent? Justify your answer. (See Figure 4.2 on page 333 for the sample space of this random process.)

The General Multiplication Rule

17. **Coffee with cream** Employees at a local coffee shop recorded the drink orders of all the customers on a Saturday. They found that 64% of customers ordered a hot drink, and 80% of these customers added cream to their drink. Find the probability that a randomly selected Saturday customer ordered a hot drink and added cream to the drink.

18. **At the gym** Suppose that 10% of adults belong to health clubs, and 40% of these health club members go to the club at least twice a week. Find the probability that a randomly selected adult belongs to a health club and goes there at least twice a week.

19. **Box of chocolates** A candy maker offers a special box of 20 chocolate candies that look alike. In reality, 14 of the candies have soft centers and 6 have hard centers. Suppose you choose 3 of the candies from a special box at random. Find the probability that all 3 candies have soft centers.

pg 356

20. **Senior statisticians** A statistics class with 30 students includes 10 juniors and 20 seniors. Suppose you choose 4 of the students in the class at random. Find the probability that all 4 are seniors.

Tree Diagrams and Probability

21. **Fill 'er up!** In a recent month, 88% of automobile drivers filled their vehicles with regular gasoline, 2% purchased midgrade gas, and 10% bought premium gas.[36] Of those who bought regular gas, 28% paid with a credit card. Of the customers who bought midgrade and premium gas, 34% and 42%, respectively, paid with a credit card. Suppose we select a customer at random.

pg 358

(a) Draw a tree diagram to model this random process.

(b) Find the probability that the customer paid with a credit card.

22. **Lactose intolerance** Lactose intolerance means that a person has difficulty digesting dairy products that contain lactose (milk sugar). This condition is particularly common among people of African and Asian ancestry. In the United States, 80% of the population identify as White alone, 14% identify as Black alone, and 6% identify as Asian alone (these percentages do not include other groups and people who consider themselves to belong to more than one race). Moreover, according to the National Institutes of Health (NIH), 15% of White people, 70% of Black people, and 90% of Asian people are lactose intolerant.[37] Suppose we select a person from these populations at random.

(a) Draw a tree diagram to model this chance process.

(b) Find the probability that the person is lactose intolerant.

23. **Credit cards and gas** Refer to Exercise 21. Given that the customer paid with a credit card, find the probability that they bought premium gas.

24. **No lactose** Refer to Exercise 22. Given that the chosen person is lactose intolerant, what is the probability that they are of Asian heritage?

25. **Sensitivity and specificity** The *sensitivity* of a diagnostic test is its probability of correctly diagnosing an individual with the tested-for condition. The *specificity* of a diagnostic test is its probability of correctly diagnosing an individual without the tested-for condition. An at-home screening test for colorectal cancer has a sensitivity of 92.3% and a specificity of 89.8%.[38] Assume that 4% of a certain population has colorectal cancer. If an individual from this population gets a positive result on the at-home screening test, what is the probability that they have colorectal cancer?

pg 359

26. **HIV testing** Enzyme immunoassay (EIA) tests are used to screen blood specimens for the presence of antibodies to human immunodeficiency virus (HIV), the virus that causes acquired immunodeficiency syndrome (AIDS). Antibodies indicate the presence of the virus. The test is quite accurate but is not always correct. A false-positive occurs when the test gives a positive result but no HIV antibodies are actually present in the blood. A false-negative occurs when the test gives a negative result but HIV antibodies are actually present in the blood. Here are the approximate probabilities of positive and negative EIA outcomes when the blood tested does and does not actually contain antibodies to HIV:[39]

		Test result	
		+	−
Truth	Antibodies present	0.9985	0.0015
	Antibodies absent	0.0060	0.9940

Suppose that 1% of a large population carries antibodies to HIV in their blood. Imagine choosing a person from this population at random. If the person's EIA test is positive, what's the probability that the person has the HIV antibody?

27. **First serves** Tennis great Serena Williams made 59% of her first serves in a certain season. When Williams

made her first serve, she won 75% of the points. When Williams missed her first serve and had to serve again, she won only 49% of the points.[40] While watching video of a game from that season in which Williams is serving, you get distracted and miss seeing her serve, but look up in time to see Williams win the point. What's the probability that she missed her first serve?

28. **Metal detector** A prospector uses a homemade metal detector to look for valuable metal objects on a beach. The machine isn't perfect — it gives a signal for 98% of the metal objects over which it passes, and it gives a signal for 4% of the nonmetallic objects over which it passes. Suppose that 25% of the objects that the machine passes over are metal. If the machine gives a signal when it passes over an object, find the probability that the prospector has found a metal object.

29. **Fundraising by telephone** Tree diagrams can organize problems having more than two stages. The following figure shows probabilities for a charity calling potential donors by telephone.[41] Each person called is either a recent donor, a past donor, or a new prospect. At the next stage, the person called either does or does not pledge to contribute, with conditional probabilities that depend on the donor class to which the person belongs. Finally, those who make a pledge either do or do not actually contribute. Suppose we randomly select a person who is called by the charity.

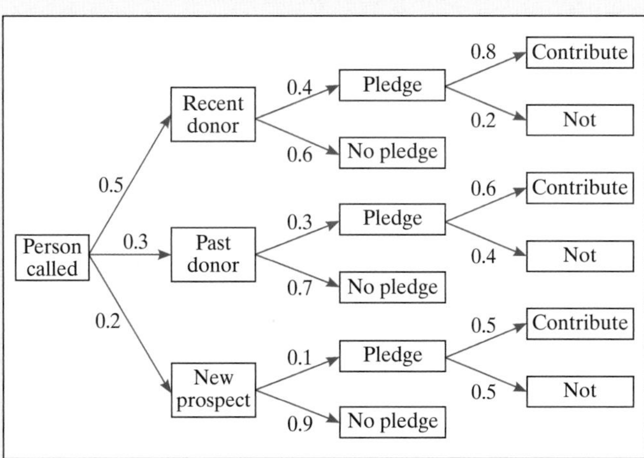

(a) What is the probability that the person contributed to the charity?

(b) Given that the person contributed, find the probability that they are a recent donor.

30. **HIV and confirmation testing** Refer to Exercise 26. Many of the positive results from EIA tests are false-positives. It is therefore common practice to perform a second EIA test on another blood sample from a person whose initial specimen tests positive. Assume that the false-positive and false-negative rates remain the same for a person's second test. Find the probability that a person who gets a positive result on both EIA tests actually has HIV antibodies. (*Hint:* Start by making a tree diagram.)

Multiplication Rule for Independent Events

31. **Holiday lights** A string of holiday lights contains 20 lights. The lights are wired in series, so that if any light fails, the whole string will go dark. Each light has probability 0.98 of working for a 3-year period. The lights fail independently of each other. Find the probability that the string of lights will remain bright for 3 years.

pg 362

32. **Get rid of the penny** Harris Interactive reported that 29% of all U.S. adults favor getting rid of the penny.[42] Assuming that responses from different individuals are independent, what is the probability of randomly selecting 3 U.S. adults who all say that they favor eliminating the penny?

33. **Is the package late?** A shipping company claims that 90% of its shipments arrive on time. Suppose this claim is true. If we take a random sample of 20 shipments made by the company, what's the probability that at least 1 of them arrives late?

34. **On a roll** Suppose that you roll a fair, six-sided die 10 times. What's the probability that you get at least one 6?

Use Probability Rules Wisely

35. **Late shows** Some TV shows begin after their scheduled times when earlier programs run late. According to a network's records, approximately 3% of its shows start late. To find the probability that 3 consecutive shows on this network start on time, can we multiply (0.97)(0.97)(0.97)? Why or why not?

pg 364

36. **Late flights** An airline reports that 85% of its flights into New York City's LaGuardia Airport arrive on time. To find the probability that its next 4 flights into LaGuardia Airport all arrive on time, can we multiply (0.85)(0.85)(0.85)(0.85)? Why or why not?

37. **Checking independence** Suppose A and B are two events such that $P(A) = 0.3$, $P(B) = 0.4$, and $P(A \cap B) = 0.12$. Are events A and B independent? Justify your answer.

38. **Checking independence again** Suppose C and D are two events such that $P(C) = 0.6$, $P(D) = 0.45$, and $P(C \cap D) = 0.3$. Are events C and D independent? Justify your answer.

39. **Mutually exclusive vs. independent** The two-way table summarizes data on the price paid and method of ticket purchase for the passengers on a trolley tour. Imagine choosing a passenger at random. Define

event A: paid full price and event B: purchased online.[43]

		Amount paid		
		Full price	Reduced price	Total
Purchase method	Online			10
	In person			40
	Total	20	30	50

(a) Copy and complete the two-way table so that events A and B are mutually exclusive.

(b) Copy and complete the two-way table so that events A and B are independent.

(c) Copy and complete the two-way table so that events A and B are not mutually exclusive and not independent.

40. **Independence and association** The two-way table summarizes data from an experiment comparing the effectiveness of three different diets (A, B, and C) for weight loss. Researchers randomly assigned 300 volunteer subjects to the three diets. The response variable was whether each subject lost weight over a 1-year period.

		Diet			
		A	B	C	Total
Lost weight?	Yes		60		180
	No		40		120
	Total	90	100	110	300

(a) Suppose we randomly select one of the subjects from the experiment. Show that the events "Diet B" and "Lost weight" are independent.

(b) Copy and complete the table so that there is no association between type of diet and whether a subject lost weight.

(c) Copy and complete the table so that there is an association between type of diet and whether a subject lost weight.

41. **Matching suits** A standard deck of playing cards consists of 52 cards, with 13 cards in each of four suits: spades, diamonds, clubs, and hearts. Suppose you shuffle the deck thoroughly and deal 5 cards face-up onto a table.

(a) What is the probability of dealing five spades in a row?

(b) Find the probability that all 5 cards on the table have the same suit.

42. **A perfect game** In baseball, a perfect game is one in which a pitcher doesn't allow any hitters to reach base in all 9 innings. Historically, Major League Baseball (MLB) pitchers throw a perfect inning — an inning where no hitters reach base — approximately 40% of the time.[44] So to throw a perfect game, a pitcher must have 9 perfect innings in a row.

(a) Find the probability that an MLB pitcher throws 9 perfect innings in a row, assuming the pitcher's performance in any one inning is independent of his performance in other innings.

In the previous 22 seasons, there have been 7 perfect MLB games. In each of these seasons, there were 30 teams, each of which played 162 games. Suppose we randomly select a game from the previous 22 seasons.

(b) What is the approximate probability of a perfect game, assuming that each game lasted a full 9 innings — that is, each team batted 9 times?

(c) How do your answers to parts (a) and (b) compare? What does this imply about the assumption of independence that you made in part (a)?

For Investigation *Apply the skills from the section in a new context or nonroutine way.*

43. **The geometric distributions** Suppose that you are tossing a pair of fair, six-sided dice in a board game. Tosses are independent. You land in a danger zone that requires you to roll doubles (both faces showing the same number of spots) before you are allowed to play again.

(a) What is the probability of rolling doubles on a single toss of the dice?

(b) What is the probability that you do not roll doubles on the first toss, but you do on the second toss?

(c) What is the probability that the first two tosses are not doubles and the third toss is doubles? This is the probability that the first doubles occurs on the third toss.

(d) Do you see the pattern? What is the probability that the first doubles occurs on the 100th toss? The kth toss?

Note: This type of problem, which involves performing repeated, independent trials of the same random process until a "success" occurs, is known as *geometric probability*.

44. **BMI** A person's body mass index (BMI) is their weight in kilograms divided by the square of their height in meters. One population of individuals is divided into Group A and Group B. The BMI of people in Group A is approximately normally distributed with mean 26.8 and standard deviation 7.4. The BMI of people in Group B is approximately normally distributed with mean 28.2 and standard deviation 8.1. In this population, 23% of people are in Group A and the remainder are in Group B. People with a BMI less than 18.5 are often classified as "underweight."

(a) What percentage of people in Group A are underweight?

(b) What proportion of people in this population are underweight?

(c) Suppose we select a person from this population at random and find that they are underweight. What's the probability that they are in Group A?

Multiple Choice *Select the best answer for each question.*

45. An athlete suspected of using steroids is given two tests that operate independently of each other. Test A has probability 0.9 of being positive if steroids have been used. Test B has probability 0.8 of being positive if steroids have been used. What is the probability that neither test is positive if the athlete has used steroids?

(A) 0.08
(B) 0.28
(C) 0.02
(D) 0.38
(E) 0.72

46. In an effort to find the source of an outbreak of food poisoning at a conference, a team of medical detectives carried out a study. They examined all 50 people who had food poisoning and a random sample of 200 people attending the conference who didn't get food poisoning. The detectives found that 40% of the people with food poisoning went to a cocktail party on the second night of the conference, while only 10% of the people in the random sample attended the same party. Which of the following statements is appropriate for describing the 40% of people who went to the party? (Let F = got food poisoning and A = attended party.)

(A) $P(F \mid A) = 0.40$
(B) $P(A \mid F^C) = 0.40$
(C) $P(F \mid A^C) = 0.40$
(D) $P(A^C \mid F) = 0.40$
(E) $P(A \mid F) = 0.40$

47. Suppose a loaded die has the following probability model:

Outcome	1	2	3	4	5	6
Probability	0.3	0.1	0.1	0.1	0.1	0.3

If this die is thrown and the top face shows an odd number, what is the probability that the die shows a 1?

(A) 0.10
(B) 0.17
(C) 0.30
(D) 0.50
(E) 0.60

48. If $P(A) = 0.24$, $P(B) = 0.52$, and A and B are independent events, what is $P(A \text{ or } B)$?

(A) 0.1248
(B) 0.2800
(C) 0.6352
(D) 0.7600
(E) The answer cannot be determined from the information given.

49. Which of the following could be compared to $P(A \mid B)$ to see if events A and B are independent?

I. $P(B)$ II. $P(A)$ III. $P(B \mid A)$ IV. $P(A \mid B^C)$ V. $P(A \cap B)$

(A) I only
(B) II only
(C) III only
(D) I and III only
(E) II and IV only

Recycle and Review *Practice what you learned in previous sections.*

50. **Clothing matters (2A, 3C)** Two young researchers suspect that people are more likely to agree to participate in a survey if the interviewers are dressed up. To test this, the researchers went to the local grocery store on two consecutive Saturday mornings at 10 A.M. On the first Saturday, they wore casual clothing (tank tops and jeans). On the second Saturday, they dressed in button-down shirts and nicer slacks. Each day, they asked every fifth person who walked into the store to participate in a survey. Their response variable was whether or not the person agreed to participate. Here are their results.[45]

		Clothing	
		Casual	Nice
Participation	Yes	14	27
	No	36	23

(a) Calculate the difference in the proportion of subjects who agreed to participate in the survey in the two groups (Casual − Nice).

(b) Assume the study design is equivalent to randomly assigning shoppers to the "casual" or "nice" groups. A total of 100 trials of a simulation were performed to see what differences in proportions would occur due only to chance variation in this random assignment. Use the results of the simulation in the following dotplot to determine whether the difference in proportions from part (a) is statistically significant. Explain your reasoning.

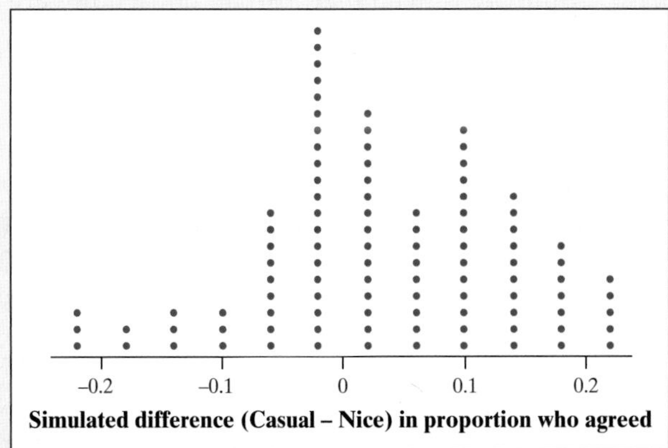

(c) What flaw in the design of this experiment would prevent the researchers from drawing a cause-and-effect conclusion about the impact of an interviewers' attire on nonresponse in a survey?

UNIT 4 Wrap-Up

PART I

FRAPPY! Free Response AP® Problem, Yay!

Directions: Show all your work. Indicate clearly the methods you use, because you will be scored on the correctness of your methods as well as on the accuracy and completeness of your results and explanations.

Prior to the lockdowns and social distancing associated with the Covid-19 pandemic, approximately 44% of all reported injuries in U.S. households occurred at home. With many more people working from home during the pandemic, researchers surveyed 2009 U.S. households in June 2020, roughly 3 months after lockdowns began.[46] Researchers asked whether anyone in the household had an injury that occurred at home or ingested something that could make them sick while at home. They also classified the households as urban, suburban, or rural. The two-way table summarizes the results.

		Type of incident			
		Both	Injury or ingestion, but not both	Neither	Total
Type of household	Urban	171	83	404	658
	Suburban	80	137	750	967
	Rural	18	81	285	384
	Total	269	301	1439	2009

Suppose that one of the 2009 households surveyed is randomly selected.

(a) What is the probability that the selected household is classified as urban or had both types of incidents?

(b) Given that the selected household is classified as urban, what is the probability that the household had both types of incidents?

(c) Are the events "selecting an urban household" and "selecting a household with both types of incidents" independent? Justify your answer.

Suppose instead that 3 urban households are randomly selected from the original sample.

(d) What is the probability that at least one of the households had both types of incidents?

After you finish the FRAPPY!, you can view two example solutions on the book's website (**bfwpub.com/tps7e**). Determine whether you think each solution is "complete," "substantial," "developing," or "minimal." If the solution is not complete, what improvements would you suggest to the student who wrote it? Finally, your teacher will provide a scoring rubric. Score your response and note what, if anything, you would do differently to improve your own score.

UNIT 4, PART I REVIEW

SECTION 4A Randomness, Probability, and Simulation

In this section, you learned about the idea of *probability*. The **law of large numbers** says that when you repeat a **random process** many, many times, the relative frequency of an outcome will approach a single number. This single number is called the **probability** of the outcome — how often we expect the outcome to occur in a very large number of trials of the random process. Be sure to remember the "large" part of the law of large numbers. Although clear patterns emerge in a large number of trials, we shouldn't expect such regularity in a small number of trials.

Simulation is a powerful tool that we can use to imitate a random process and estimate a probability. To conduct a simulation, describe how to set up and use a random process to perform one trial of the simulation. Identify what you will record at the end of each trial. Then perform many trials, and use the results of your simulation to answer the question of interest. If you are using random numbers to perform your simulation, be sure to consider whether numbers can be repeated within each trial.

SECTION 4B Probability Rules

In this section, you learned that random behavior can be described by a **probability model.** Probability models have two parts, a list of possible outcomes (the **sample space**) and a probability for each outcome. The probability of each outcome in a probability model must be between 0 and 1 (inclusive), and the probabilities of all the outcomes in the sample space must add to 1.

An **event** is a subset of possible outcomes from the sample space. The **complement rule** says the probability that an event doesn't occur is 1 minus the probability that the event does occur. In symbols, the complement rule says that $P(A^C) = 1 - P(A)$. Given two events A and B from the same random process, use the **general addition rule** to find the probability that event A or event B occurs:

$$P(\text{A or B}) = P(A \cup B) = P(A) + P(B) - P(A \cap B)$$

If the events A and B have no outcomes in common, use the addition rule for **mutually exclusive** events: $P(A \cup B) = P(A) + P(B)$.

Finally, you learned how to use two-way tables to display the sample space for a random process involving two events. Using a two-way table (or a **Venn diagram**) is a helpful way to organize information and calculate probabilities, including those involving the **union** $(A \cup B)$ and the **intersection** $(A \cap B)$ of two events.

SECTION 4C Conditional Probability and Independent Events

In this section, you learned that a **conditional probability** describes the probability of an event occurring given that another event is known to have already occurred. To calculate the probability that event A occurs given that event B has occurred, use the conditional probability formula:

$$P(A \mid B) = \frac{P(\text{A and B})}{P(B)} = \frac{P(A \cap B)}{P(B)} = \frac{P(\text{both events occur})}{P(\text{given event occurs})}$$

Two-way tables and **tree diagrams** are useful ways to organize the information provided in a conditional probability problem. Two-way tables are best when the problem describes the number or proportion of cases with certain characteristics. Tree diagrams are best when the problem describes a random process with multiple stages.

Use the **general multiplication rule** for calculating the probability that event A and event B both occur:

$$P(\text{A and B}) = P(A \cap B) = P(A) \cdot P(B \mid A)$$

If knowing whether or not event B occurs doesn't change the probability that event A occurs, then events A and B are **independent**. That is, events A and B are independent if $P(A \mid B) = P(A \mid B^C) = P(A)$. If events A and B are independent, use the **multiplication rule for independent events** to find the probability that events A and B both occur: $P(A \cap B) = P(A) \cdot P(B)$.

What Did You Learn?

Learning Target	Section	Related Example on Page(s)	Relevant Chapter Review Exercise(s)
Interpret probability as a long-run relative frequency.	4A	322	R1
Estimate probabilities using simulation.	4A	324, 325	R2
Give a probability model for a random process with equally likely outcomes and use it to find the probability of an event.	4B	334	R3
Calculate probabilities using the complement rule.	4B	336	R4
Use the addition rule for mutually exclusive events to find probabilities.	4B	337	R4
Use a two-way table to find probabilities.	4B	338	R5
Calculate probabilities with the general addition rule.	4B	340	R5
Calculate conditional probabilities.	4C	349, 351	R4, R5, R7

Learning Target	Section	Related Example on Page(s)	Relevant Chapter Review Exercise(s)
Determine whether two events are independent.	4C	353	R6
Use the general multiplication rule to calculate probabilities.	4C	356	R6, R7
Use a tree diagram to model a random process involving multiple stages and to find probabilities.	4C	358, 359	R7
Calculate probabilities using the multiplication rule for independent events.	4C	362	R8
Determine if it is appropriate to use the multiplication rule for independent events in a given setting.	4C	364	R8

UNIT 4, PART I REVIEW EXERCISES

These exercises are designed to help you review the important concepts and skills in Part I of this unit.

R1 **Butter side down (4A)** Researchers at Manchester Metropolitan University in England determined that if a piece of toast is dropped from a 2.5-foot-high table, the probability that it lands butter side down is 0.81.[47]

(a) Explain what this probability means.

(b) Suppose that the researchers dropped 4 pieces of toast, and all of them landed butter side down. Does that make it more likely that the next piece of toast will land with the butter side up? Explain your answer.

R2 **Tumbling toast (4A)** Refer to Exercise R1. Mariana decides to test this probability and drops 10 pieces of toast from a 2.5-foot table. Only 4 of them land butter side down. Mariana wants to perform a simulation to estimate the probability that 4 or fewer pieces of toast out of 10 would land butter side down if the researchers' 0.81 probability is correct.

(a) Describe how you would use a random number generator to perform one trial of the simulation.

The dotplot displays the results of 50 simulated trials of dropping 10 pieces of toast and recording the number that land butter side down.

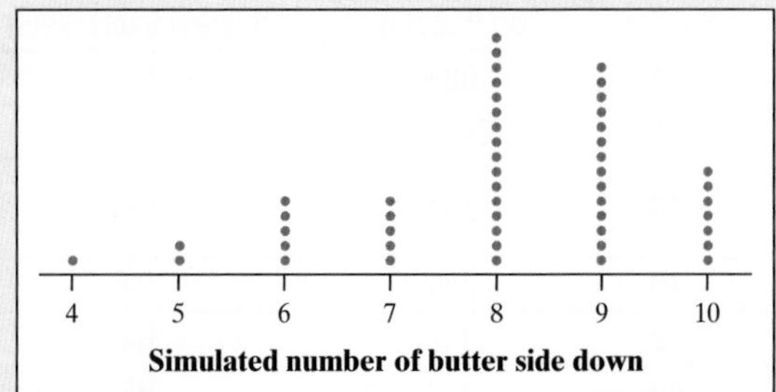

(b) Use the simulation results to estimate the probability that 4 or fewer pieces of toast out of 10 would land butter side down if the researchers' 0.81 probability is correct.

(c) Based on your answer to part (b) and the results of Mariana's experiment, is there convincing evidence that the researchers' 0.81 probability value is incorrect? Explain your answer.

R3 **Rock–paper–scissors (4B)** You're likely familiar with the game "rock–paper–scissors." Two players face each other and, at the count of 3, choose to make a fist (rock), extend a hand with an open palm (paper), or make a "V" with their index and middle fingers (scissors). The winner is determined by these rules: rock breaks scissors; paper covers rock; and scissors cut paper. If both players choose the same hand shape, then the game is a tie. Suppose that Player 1 and Player 2 are both equally likely to choose rock, paper, or scissors.

(a) Give a probability model for this random process.

(b) Find the probability that Player 1 wins on a single play of the game.

R4 **What kind of vehicle? (4B, 4C)** Suppose we randomly select a new vehicle sold in the United States in a recent year. The probability model for the type of vehicle chosen is given here.[48]

Vehicle type	Passenger car	Pickup truck	SUV	Crossover	Minivan
Probability	0.28	0.18	0.08	?	0.05

(a) What is the probability that the vehicle is a crossover? How do you know?

(b) Find the probability that the vehicle is not an SUV or a minivan.

(c) Given that the vehicle is not a passenger car, what is the probability that it is a pickup truck?

R5 **Astrology (4B, 4C)** The General Social Survey (GSS) asked a random sample of adults their opinion about whether astrology is very scientific, sort of scientific, or not at all scientific. Here is a two-way table that summarizes respondents' opinions by level of education.[49]

		Degree held			
		Associate's	Bachelor's	Master's	Total
Opinion about astrology	Not at all scientific	169	256	114	539
	Very or sort of scientific	65	65	18	148
	Total	234	321	132	687

Suppose one person who completed the survey is selected at random.

(a) Find the probability that the person thinks astrology is not at all scientific.

(b) Find the probability that the person has an associate's degree or thinks astrology is not at all scientific.

(c) Find the probability that the person has an associate's degree, given that the person thinks astrology is not at all scientific.

R6 **Mike's pizza (4B, 4C)** The chef at a local pizza shop gives you the following information about the pizzas currently in the oven: 6 of the 9 are thick-crust pizzas, and 2 of the 6 thick-crust pizzas have mushrooms. Of the remaining 3 pizzas, 1 has mushrooms. Suppose that a pizza is selected at random from the oven.

(a) Are the events "thick-crust pizza" and "pizza with mushrooms" mutually exclusive? Justify your answer.

(b) Are the events "thick-crust pizza" and "pizza with mushrooms" independent? Justify your answer.

Suppose instead that 2 pizzas are randomly selected from the oven.

(c) Find the probability that neither has mushrooms.

R7 **Does the new hire use drugs? (4C)** Many employers require prospective employees to take a drug test. A positive result on this test suggests that the prospective employee uses illegal drugs. However, not all people who test positive use illegal drugs. The test result could be a false-positive. A negative test result could be a false-negative if the person really does use illegal drugs. Suppose that 4% of prospective employees use illegal drugs, and that the drug test has a false-positive rate of 5% and a false-negative rate of 10%.[50] Imagine choosing a prospective employee at random.

(a) Draw a tree diagram to model this random process.

(b) Find the probability that the drug test result is positive.

(c) Given that the drug test result is positive, find the probability that the prospective employee actually uses illegal drugs.

R8 **Fire or medical? (4C)** Many fire stations handle more emergency calls for medical help than for fires. At one fire station, 77% of incoming calls are for medical help. Suppose we choose 4 incoming calls to the station at random.

(a) Find the probability that none of the 4 calls is for medical help.

(b) Explain why the calculation in part (a) may not be valid if we choose 4 consecutive calls to the station.

(c) What's the probability that at least 1 of the randomly selected calls is not for medical help?

UNIT 4, PART I AP® STATISTICS PRACTICE TEST

Section I: Multiple Choice *Select the best answer for each question.*

Questions T1–T3 refer to the following setting. A sample of 125 truck owners were asked what brand of truck they owned and whether or not the truck has four-wheel drive. The results are summarized in the two-way table. Suppose we randomly select one of these truck owners.

		Four-wheel drive?	
		Yes	No
Brand of truck	Ford	28	17
	Chevy	32	18
	Dodge	20	10

T1 What is the probability that the person owns a Dodge or has four-wheel drive?

(A) 20/80
(B) 20/125
(C) 80/125
(D) 90/125
(E) 110/125

T2 What is the probability that the person owns a Chevy, given that the truck has four-wheel drive?

(A) 32/50
(B) 32/80
(C) 32/125
(D) 50/125
(E) 80/125

T3 Which one of the following statements is true about the events "Owner's truck is a Chevy" and "Owner's truck has four-wheel drive"?

(A) These two events are mutually exclusive and independent.
(B) These two events are mutually exclusive, but not independent.
(C) These two events are not mutually exclusive, but they are independent.
(D) These two events are neither mutually exclusive nor independent.
(E) These two events are mutually exclusive, but we do not have enough information to determine if they are independent.

T4 A spinner has three equally sized regions: blue, red, and green. Jonelle spins the spinner 3 times and gets 3 blues in a row. Which of the following statements is correct?

(A) The probability that the next spin is blue is less than 1/3.
(B) The spinner is not working properly because 3 blues in a row should not happen.
(C) In the next 57 spins, Jonelle will get exactly 17 blues.
(D) In the next 57 spins, Jonelle will get about 17 blues.
(E) In the next 57 spins, Jonelle will get about 19 blues.

Questions T5 and T6 refer to the following setting. Wilt is a fine basketball player, but his free-throw shooting could use some work. For the past three seasons, Wilt has made only 56% of his free throws. His coach sends Wilt to a summer clinic to work on his shot, and when he returns, his coach has Wilt step to the free-throw line and take 50 shots. Wilt makes 34 shots. Is this result convincing evidence that Wilt's free-throw shooting has improved? We want to perform a simulation to estimate the probability that a 56% free-throw shooter would make 34 or more in a sample of 50 shots.

T5 Which of the following is a correct way to perform the simulation?

(A) Let integers from 1 to 34 represent making a free throw and from 35 to 50 represent missing a free throw. Generate 50 random integers from 1 to 50. Count the number of made free throws. Repeat this process many times.
(B) Let integers from 1 to 34 represent making a free throw and from 35 to 50 represent missing a free throw. Generate 50 random integers from 1 to 50 with no repeats allowed. Count the number of made free throws. Repeat this process many times.
(C) Let integers from 1 to 56 represent making a free throw and from 57 to 100 represent missing a free throw. Generate 50 random integers from 1 to 100. Count the number of made free throws. Repeat this process many times.
(D) Let integers from 1 to 56 represent making a free throw and from 57 to 100 represent missing a free throw. Generate 50 random integers from 1 to 100 with no repeats allowed. Count the number of made free throws. Repeat this process many times.
(E) None of the above is correct.

T6 The dotplot displays the number of made shots in 100 simulated sets of 50 free throws by someone with probability 0.56 of making a free throw.

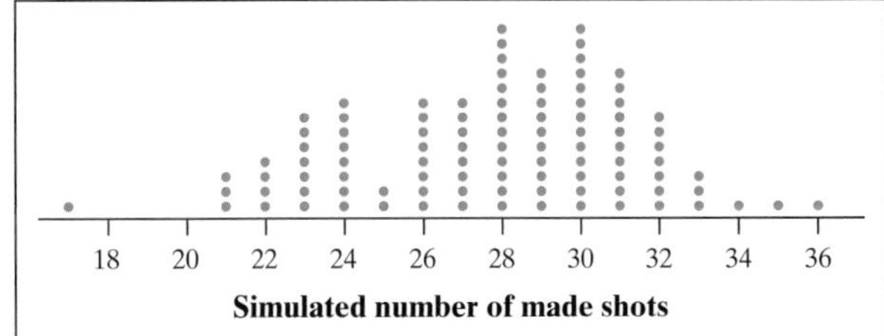

Which of the following is an appropriate statement about Wilt's free-throw shooting based on this dotplot?

(A) If Wilt is still only a 56% shooter, the probability that he would make at least 34 of his shots is about 0.03.

(B) If Wilt is still only a 56% shooter, the probability that he would make at least 34 of his shots is about 0.98.

(C) If Wilt is now shooting better than 56%, the probability that he would make at least 34 of his shots is about 0.03.

(D) If Wilt is now shooting better than 56%, the probability that he would make at least 34 of his shots is about 0.98.

(E) If Wilt is still only a 56% shooter, the probability that he would make at least 34 of his shots is about 0.02.

T7 This partially complete table shows the distribution of scores on the AP® Statistics exam for a class of students.

Score	1	2	3	4	5
Probability	0.10	0.20	???	0.25	0.15

Suppose we select a student from this class at random. If the student earned a score of 3 or higher on the AP® Statistics exam, what is the probability that the student scored a 5?

(A) 0.150

(B) 0.214

(C) 0.300

(D) 0.428

(E) 0.700

T8 A bowl holds 18 butterscotch candies and 14 peppermint candies. If two candies are randomly selected from the bowl, what is the probability that both are the same flavor?

(A) 0.49

(B) 0.50

(C) 0.51

(D) 0.52

(E) 0.53

T9 Suppose that a student is randomly selected from a large high school. The probability that the student is a senior is 0.22. The probability that the student has a driver's license is 0.30. If the probability that the student is a senior or has a driver's license is 0.36, what is the probability that the student is a senior and has a driver's license?

(A) 0.060

(B) 0.066

(C) 0.080

(D) 0.140

(E) 0.160

T10 The security system in a house has two units that set off an alarm when motion is detected. Neither one is entirely reliable, but one or both *always* go off when there is motion anywhere in the house. Suppose that for motion in a certain location, the probability that detector A goes off and detector B does not go off is 0.25, and the probability that detector A does not go off is 0.35. What is the probability that detector B goes off?

(A) 0.10

(B) 0.35

(C) 0.40

(D) 0.65

(E) 0.75

Section II: Free Response *Show all your work. Indicate clearly the methods you use, because you will be graded on the correctness of your methods as well as on the accuracy and completeness of your results and explanations.*

T11 The two-way table summarizes data on whether students at a certain high school eat regularly in the school cafeteria by grade level.

		Grade				
		9th	10th	11th	12th	Total
Eat in cafeteria?	Yes	130	175	122	68	495
	No	18	34	88	170	310
	Total	148	209	210	238	805

(a) If you choose a student at random, what is the probability that the student eats regularly in the cafeteria and is not a 10th-grader?

(b) If you choose a student at random who eats regularly in the cafeteria, what is the probability that the student is a 10th-grader?

(c) Are the events "10th-grader" and "eats regularly in the cafeteria" independent? Justify your answer.

T12 Three machines — A, B, and C — are used to produce a large quantity of identical parts at a factory. Machine A produces 60% of the parts, while Machines B and C produce 30% and 10% of the parts, respectively. Historical records indicate that 10% of the parts produced by Machine A are defective, compared with 30% for Machine B and 40% for Machine C. Suppose we randomly select a part produced at the factory.

(a) Find the probability that the part is defective.

(b) If the part is inspected and found to be defective, what's the probability that it was produced by Machine B?

T13 At Dicey Dave's Diner, the dinner buffet usually costs \$12.99. Once a month, Dave sponsors "lucky buffet" night. On that night, each patron can either pay the usual price or roll two fair, 6-sided dice and pay a number of dollars equal to the product of the numbers showing on the two faces. The table shows the sample space of this random process.

Second die \ First die	1	2	3	4	5	6
1	1	2	3	4	5	6
2	2	4	6	8	10	12
3	3	6	9	12	15	18
4	4	8	12	16	20	24
5	5	10	15	20	25	30
6	6	12	18	24	30	36

(a) A customer decides to play Dave's "lucky buffet" game. Find the probability that the customer will pay less than the usual cost of the buffet.

(b) A group of 4 friends comes to Dicey Dave's Diner to play the "lucky buffet" game. Find the probability that all 4 of these friends end up paying less than the usual cost of the buffet.

(c) Find the probability that at least 1 of the 4 friends ends up paying *more* than the usual cost of the buffet.

UNIT 4

Probability, Random Variables, and Probability Distributions

PART II Random Variables and Probability Distributions

Introduction

In Part I of Unit 4, you learned several methods for calculating probabilities. The following activity gives you a preview of what you'll learn about in Part II.

ACTIVITY The "1 in 6 wins" game

NIMA Stock/Alamy Stock Photo

As a special promotion for its 20-ounce bottles of soda, a soft drink company printed a message on the inside of each bottle cap. Some of the caps said, "Please try again!" while others said, "You're a winner!" The company advertised the promotion with the slogan "1 in 6 wins a prize." The prize is a free 20-ounce bottle of soda.

Grayson's statistics class wonders if the company's claim holds true for the bottles at a nearby convenience store. To find out, all 30 students in the class go to the store and each student buys one randomly selected 20-ounce bottle of the soda. Three of them get caps that say, "You're a winner!" Does this result give convincing evidence that the company's 1-in-6 claim is inaccurate?

For now, let's assume that the company is telling the truth, and that every 20-ounce bottle of soda it fills has a 1-in-6 chance of getting a cap that says, "You're a winner!" We can model the status of an individual bottle with a six-sided die: let 1 through 5 represent "Please try again!" and 6 represent "You're a winner!"

1. Roll your die 30 times to imitate the process of the students in Grayson's statistics class buying their sodas. How many 6's (prize winners) did you get?
2. Your teacher will draw and label a number line for a class dotplot. Plot the number of prize winners you got in Step 1 on the graph.
3. Repeat Steps 1 and 2, if needed, to get a total of at least 40 trials of the simulation for your class.
4. Discuss the results with your classmates. What percentage of the time did the simulation yield 3 or fewer prize winners in a class of 30 students just by chance? Does it seem plausible (believable) that the company is telling the truth, but that the class just got unlucky? Or is there convincing evidence that the 1-in-6 claim is wrong? Explain your reasoning.

If the company is telling the truth about its "1 in 6 wins" game, we'd expect about one-sixth of the 30 students in Grayson's statistics class (i.e., about 5 students) to win. How likely is it that 3 or fewer of the 30 students will win if the company's 1-in-6 claim is true? To answer this question without a simulation, we need a different kind of probability model from the ones we saw in Part I of this unit.

Section 4D introduces the concept of a *random variable*, a numerical outcome of some random process (like the number of winners out of 30 in the "1 in 6 wins" game). Each random variable has a *probability distribution* that gives us information about the likelihood that a specific event happens (like 3 or fewer winners out of 30) and about what's expected to happen if the random process is repeated many times. Section 4E examines the effect of transforming and combining random variables on their probability distributions. In Section 4F, we'll look at two types of random variables that are used frequently enough to have their own names — *binomial* and *geometric*.

SECTION 4D Introduction to Discrete Random Variables

LEARNING TARGETS *By the end of the section, you should be able to:*

- Calculate probabilities involving a discrete random variable.
- Display the probability distribution of a discrete random variable with a histogram and describe its shape.
- Calculate and interpret the mean, or expected value, of a discrete random variable.
- Calculate and interpret the standard deviation of a discrete random variable.

A probability model describes the possible outcomes of a random process and the likelihood that those outcomes will occur. For example, suppose you toss a fair coin 3 times. The sample space for this random process is

HHH HHT HTH THH HTT THT TTH TTT

Because there are 8 equally likely outcomes, the probability is 1/8 for each possible outcome.

Define the **random variable** X = the number of heads obtained in 3 tosses. The value of X will vary from one set of tosses to another, but it will always be one of the numbers 0, 1, 2, or 3. How likely is X to take each of those values? It will be easier to answer this question if we group the possible outcomes by the number of heads obtained:

$X = 0$: TTT $X = 1$: HTT THT TTH $X = 2$: HHT HTH THH $X = 3$: HHH

We can summarize the **probability distribution** of X in a table:

Number of heads	0	1	2	3
Probability	1/8	3/8	3/8	1/8

DEFINITION Random variable, Probability distribution

A **random variable** takes numerical values that describe the outcomes of a random process.

The **probability distribution** of a random variable gives its possible values and their probabilities.

Note that we use capital, italic letters (like X or Y) to represent random variables.

Recall from Unit 1 that there are two types of quantitative variables. Discrete variables typically result from counting something, while continuous variables typically result from measuring something. We make the same distinction with random variables and probability distributions. The random variable X in the coin-tossing setting is a **discrete random variable.**

DEFINITION Discrete random variable

A **discrete random variable** takes a countable set of possible values with gaps between them on a number line.

We can list the four possible values of X = the number of heads in 3 tosses of a fair coin as 0, 1, 2, 3. Note that there are gaps between these values on a number line. For instance, a gap exists between $X = 1$ and $X = 2$ because X cannot take values such as 1.2 or 1.84.

A countable set of possible values can be finite or infinite, as long we can number the possible values using whole numbers. For instance, let Y = the number of times a person plays the lottery until they win. The possible values of Y are 1, 2, 3, ...

The rest of this unit focuses on analyzing discrete random variables. We will discuss *continuous random variables* in Section 5A.

Discrete Random Variables and Probability

The probability distribution for the discrete random variable X = the number of heads in 3 tosses of a fair coin is

Number of heads	0	1	2	3
Probability	1/8	3/8	3/8	1/8

This probability distribution is valid because all the probabilities are between 0 and 1, and their sum is 1:

$$1/8 + 3/8 + 3/8 + 1/8 = 8/8 = 1$$

PROBABILITY DISTRIBUTION OF A DISCRETE RANDOM VARIABLE

The probability distribution of a discrete random variable X lists the values x_i and their probabilities $P(x_i)$:

Value	x_1	x_2	x_3	...
Probability	$P(x_1)$	$P(x_2)$	$P(x_3)$	...

For the probability distribution to be valid, the probabilities $P(x_i)$ must satisfy two requirements:

1. Every probability $P(x_i)$ is a number between 0 and 1.
2. The sum of the probabilities is 1: $P(x_1) + P(x_2) + P(x_3) + \ldots = 1$.

We can use the probability distribution of a discrete random variable to find the probability of an event. For instance, what's the probability that we get at least one head in three tosses of the coin? In symbols, we want to find $P(X \geq 1)$. We know that

$$P(X \geq 1) = P(X = 1 \text{ or } X = 2 \text{ or } X = 3)$$

Because the events $X = 1$, $X = 2$, and $X = 3$ are mutually exclusive, we can add their probabilities to get the answer:

$$\begin{aligned} P(X \geq 1) &= P(X = 1) + P(X = 2) + P(X = 3) \\ &= 3/8 + 3/8 + 1/8 \\ &= 7/8 \end{aligned}$$

Or we could use the complement rule from Section 4B:

$$\begin{aligned} P(X \geq 1) &= 1 - P(X < 1) \\ &= 1 - P(X = 0) \\ &= 1 - 1/8 \\ &= 7/8 \end{aligned}$$

EXAMPLE

Apgar scores: Babies' health at birth
Discrete random variables and probability

Skills 2.B, 3.A

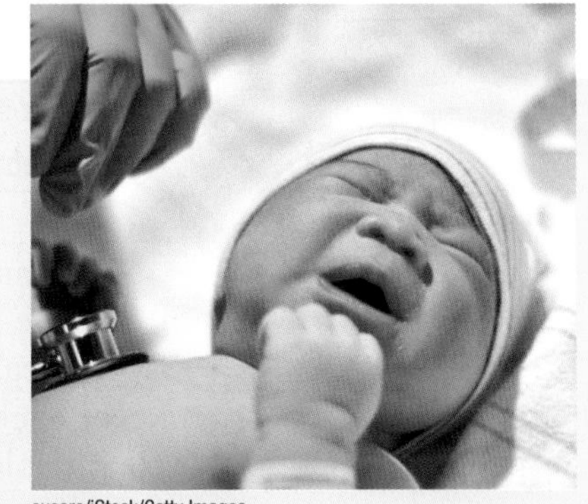
susaro/iStock/Getty Images

PROBLEM: In 1952, Dr. Virginia Apgar suggested five criteria for measuring a baby's health at birth: skin color, heart rate, muscle tone, breathing, and response when stimulated. She developed a 0–1–2 scale to rate a newborn on each of the five criteria. A baby's Apgar score is the sum of the ratings on each of the five scales, which gives a whole-number value from 0 to 10. Apgar scores are still used today to evaluate the health of newborns. Although this procedure was later named for Dr. Apgar, the acronym APGAR also represents the five scales: Appearance, Pulse, Grimace, Activity, and Respiration.

Which Apgar scores are typical? To find out, researchers recorded the Apgar scores of more than 2 million newborn babies in a single year.[51] Imagine selecting a newborn baby at random. Define the random variable X = Apgar score of a randomly selected newborn baby. The table gives the probability distribution of X.

Apgar score	0	1	2	3	4	5	6	7	8	9	10
Probability	???	0.006	0.007	0.008	0.012	0.020	0.038	0.099	0.319	0.437	0.053

(a) Write the event "the baby has an Apgar score of 0" in terms of X. Then find its probability.

(b) Doctors decided that Apgar scores of 7 or higher indicate a healthy baby. What's the probability that a randomly selected newborn is considered healthy?

SOLUTION:

(a) $$\begin{aligned} P(X = 0) &= 1 - (0.006 + 0.007 + \ldots + 0.053) \\ &= 1 - 0.999 \\ &= 0.001 \end{aligned}$$

Use the fact that the sum of the probabilities for all possible values of the random variable must be 1.

(b) $$\begin{aligned} P(X \geq 7) &= 0.099 + 0.319 + 0.437 + 0.053 \\ &= 0.908 \end{aligned}$$

The probability of choosing a healthy baby is $P(X \geq 7) = P(X = 7) + P(X = 8) + P(X = 9) + P(X = 10)$.

FOR PRACTICE, TRY EXERCISE 1

Note that the probability of randomly selecting a newborn whose Apgar score is *at least* 7 is not the same as the probability that the baby's Apgar score is *greater than* 7. The latter probability is

$$\begin{aligned} P(X > 7) &= P(X = 8) + P(X = 9) + P(X = 10) \\ &= 0.319 + 0.437 + 0.053 \\ &= 0.809 \end{aligned}$$

The outcome $X = 7$ is included in "at least 7" but is not included in "greater than 7." **Be sure to consider whether you need to include the boundary value in your calculations when dealing with discrete random variables.**

Displaying Discrete Probability Distributions

When we analyzed distributions of quantitative data in Unit 1, we made it a point to discuss their shape, center, and variability. We'll do the same with probability distributions of random variables.

For the discrete random variable X = Apgar score of a randomly selected newborn baby, the probability distribution is

Apgar score	0	1	2	3	4	5	6	7	8	9	10
Probability	0.001	0.006	0.007	0.008	0.012	0.020	0.038	0.099	0.319	0.437	0.053

We can display this probability distribution graphically using a histogram. Values of the random variable go on the horizontal axis and probabilities go on the vertical axis. There is one bar in the histogram for each value of X. The height of each bar gives the probability for the corresponding value of the random variable.

Figure 4.9 shows a histogram of the probability distribution of X. This distribution is skewed to the left and unimodal, with a single peak at an Apgar score of 9. You will learn how to describe its center and variability shortly.

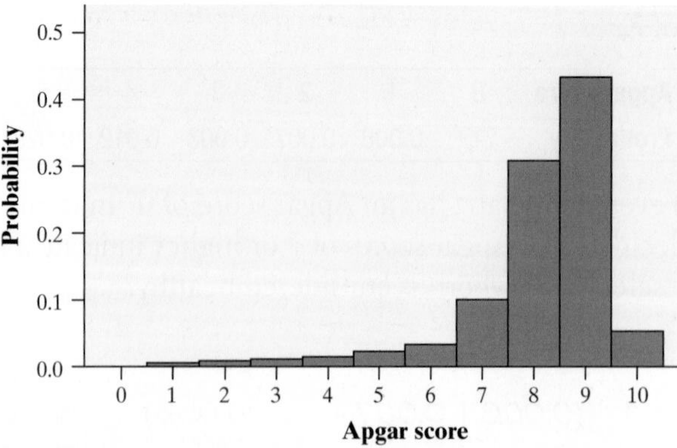

FIGURE 4.9 Histogram of the probability distribution of the random variable X = Apgar score of a randomly selected newborn.

There's another way to think about the graph displayed in Figure 4.9. The probability distribution of the random variable X models the *population distribution* of the quantitative variable "Apgar score of a newborn." So we can interpret our earlier result, $P(X > 7) = 0.809$, as saying that about 81% of all newborns have Apgar scores greater than 7. We also know that the shape of the population distribution is left-skewed with a single peak at 9.

EXAMPLE

Pete's Jeep Tours
Displaying discrete probability distributions

Skills 2.B, 4.B

Judy M. Starnes

PROBLEM: Pete's Jeep Tours offers a popular day trip in a tourist area. There must be at least 2 passengers for the trip to run, and the vehicle will hold up to 6 passengers. Pete charges \$150 per passenger. Let C = the total amount of money that Pete collects on a randomly selected trip. The probability distribution of C is given in the table.

Total collected ($)	300	450	600	750	900
Probability	0.15	0.25	0.35	0.20	0.05

Make a histogram of the probability distribution. Describe its shape.

SOLUTION:

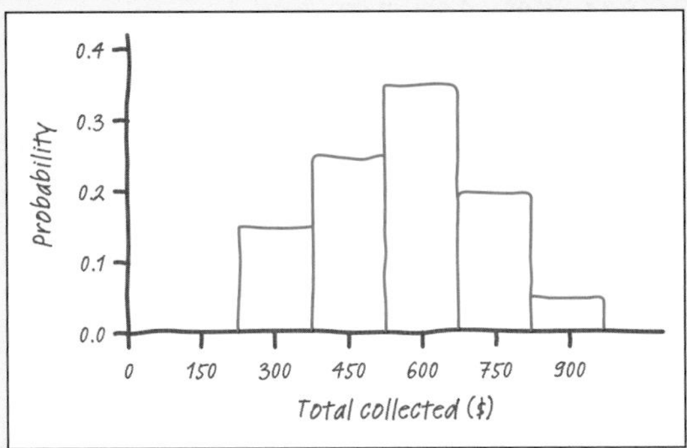

Remember: Values of the random variable go on the horizontal axis and probabilities go on the vertical axis. Don't forget to properly label and scale each axis!

The graph is roughly symmetric and has a single peak at \$600.

FOR PRACTICE, TRY EXERCISE 5

Notice the use of the label C (collected) for the random variable in the example. Sometimes we prefer contextual labels like this to the more generic X and Y.

CHECK YOUR UNDERSTANDING

High school students in Michigan sometimes get "snow days" in the winter when the roads are so bad that school is canceled for the day. Define the random variable S = the number of snow days at a certain high school in Michigan for a randomly selected school year. The table gives the probability distribution of S.

Number of snow days, s	0	1	2	3	4	5	6	7	8	9	10
Probability, $P(S = s)$	???	0.19	0.14	0.10	0.07	0.05	0.04	0.04	0.02	0.01	0.01

1. Find $P(S = 0)$. Describe this probability in words.
2. If there are more than 6 snow days in a given year, the school year will be extended. Find the probability that this happens.
3. Make a histogram of the probability distribution. Describe its shape.

Measuring Center: The Mean, or Expected Value, of a Discrete Random Variable

In Unit 1, you learned about summarizing the center of a distribution of quantitative data with either the mean or the median. For random variables, the mean is typically used to summarize the center of a probability distribution. Because a probability distribution can model the population distribution of a quantitative variable, we label the mean of a random variable X as μ_X. Like any parameter, μ_X is a single, fixed value.

To find the mean of a quantitative data set, we compute the sum of the individual observations and divide by the total number of data values. How do we find the *mean of a discrete random variable?*

Consider the random variable C = the total amount of money that Pete collects on a randomly selected jeep tour from the previous example. The probability distribution of C is given in the table.

Total collected ($)	300	450	600	750	900
Probability	0.15	0.25	0.35	0.20	0.05

What's the average amount of money that Pete collects on his jeep tours?

Imagine a hypothetical 100 trips. According to the probability distribution, Pete should collect \$300 on 15 of these trips, \$450 on 25 trips, \$600 on 35 trips, \$750 on 20 trips, and \$900 on 5 trips. Pete's average amount collected for these trips is

$$\begin{aligned}\mu_C &= \frac{300 \cdot 15 + 450 \cdot 25 + 600 \cdot 35 + 750 \cdot 20 + 900 \cdot 5}{100} \\ &= \frac{300 \cdot 15}{100} + \frac{450 \cdot 25}{100} + \frac{600 \cdot 35}{100} + \frac{750 \cdot 20}{100} + \frac{900 \cdot 5}{100} \\ &= 300(0.15) + 450(0.25) + 600(0.35) + 750(0.20) + 900(0.05) \\ &= \$562.50\end{aligned}$$

That is, the **mean of the discrete random variable** C is $\mu_C = \$562.50$. This is also known as the **expected value** of C, denoted by $E(C)$.

In the preceding calculation, the third line is just the values of the random variable C times their corresponding probabilities. The mean, or expected value, of any discrete random variable is found in a similar way. It is an average of the possible outcomes, but a weighted average in which each outcome is weighted by its probability.

AP® EXAM TIP
Formula sheet

The formula for the mean μ_X, or expected value $E(X)$, of a discrete random variable is included on the formula sheet provided on both sections of the AP® Statistics exam.

DEFINITION Mean, or expected value, of a discrete random variable

The **mean,** or **expected value, of a discrete random variable** is its average value over many, many trials of the same random process.

Suppose that *X* is a discrete random variable with probability distribution

Value	x_1	x_2	x_3	...
Probability	$P(x_1)$	$P(x_2)$	$P(x_3)$	...

To find the mean, or expected value, of *X*, multiply each possible value of *X* by its probability, then add all the products:

$$\mu_X = E(X) = x_1P(x_1) + x_2P(x_2) + x_3P(x_3) + \ldots = \sum x_iP(x_i)$$

Recall that the mean is the balance point of a distribution. For Pete's distribution of money collected on a randomly selected jeep tour, the histogram balances at $\mu_C = 562.50$. How do we interpret this parameter? If we randomly select many, many jeep tours, Pete will collect about \$562.50 per trip, on average.

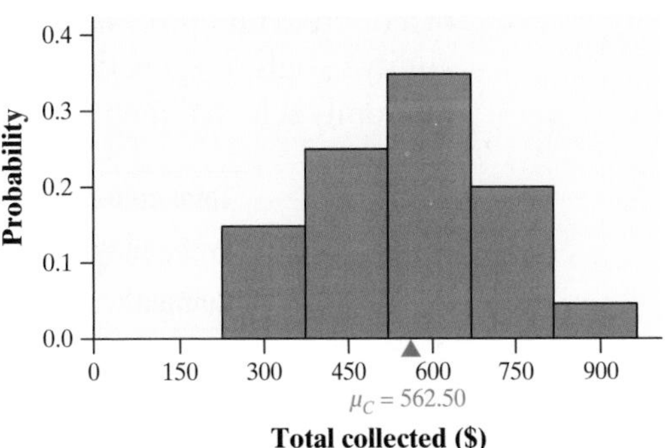

EXAMPLE

Skills 3.B, 4.B

Apgar scores: what's typical?
The mean, or expected value, of a discrete random variable

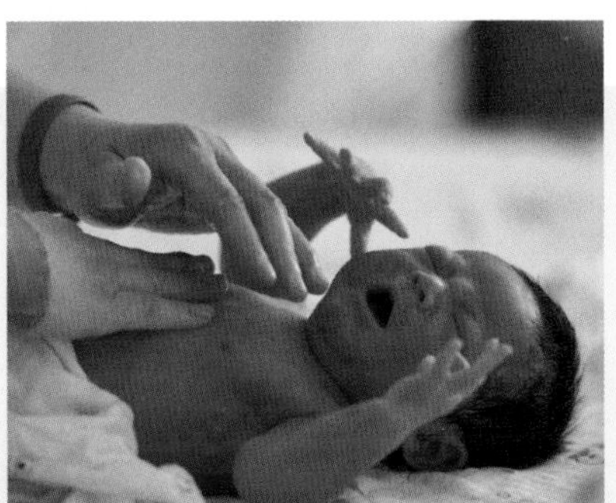
AN MING/ Feature China/Future Publishing/Getty Images

PROBLEM: Earlier, we defined the random variable X to be the Apgar score of a randomly selected newborn. The table gives the probability distribution of X once again.

Apgar score	0	1	2	3	4	5
Probability	0.001	0.006	0.007	0.008	0.012	0.020
Apgar score	6	7	8	9	10	
Probability	0.038	0.099	0.319	0.437	0.053	

Calculate and interpret the expected value of X.

SOLUTION:

$E(X) = \mu_X = (0)(0.001) + (1)(0.006) + \ldots + (10)(0.053)$
$= 8.128$

Remember that "expected value" is the same as "mean": $E(X) = \mu_X = \sum x_i P(x_i)$

If many, many newborns are randomly selected, their average Apgar score will be about 8.128.

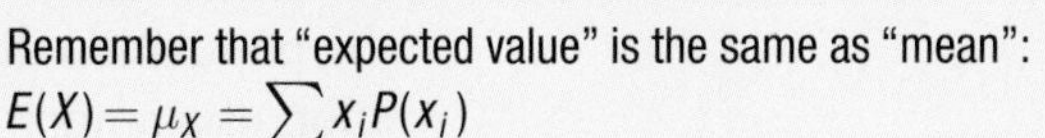
FOR PRACTICE, TRY EXERCISE 7

AP® EXAM TIP

If the mean of a random variable has a non-integer value but you report it as an integer, your answer will not get full credit.

Notice that the mean Apgar score, 8.128, is not a possible value of the random variable X because it is not a whole number between 0 and 10. The decimal value of the mean shouldn't bother you if you think of the expected value as a long-run average over many, many trials of the random process.

THE MEDIAN OF A DISCRETE RANDOM VARIABLE

How can we find the *median* of a discrete random variable? In Unit 1, we defined the median as "the midpoint of a distribution, the number such that about half the observations are smaller and about half are larger." The median of a discrete random variable is the 50th percentile of its probability distribution.

We can find the median from a *cumulative probability distribution*, like the one shown here for the random variable C = total amount of money Pete collects on a randomly selected jeep tour. We see from the table that $P(C \leq 300) = 0.15$ and that $P(C \leq 450) = 0.40$. So \$300 is the 15th percentile and \$450 is the 40th percentile of the probability distribution of C. You can think of the median of a discrete random variable as the smallest value for which the cumulative probability equals or exceeds 0.5. So the median amount of money Pete collects on a randomly selected jeep tour is \$600.

Total collected (\$)	300	450	600	750	900
Probability	0.15	0.25	0.35	0.20	0.05
Cumulative probability	0.15	0.40	0.75	0.95	1.00

Measuring Variability: The Standard Deviation of a Discrete Random Variable

Because we're using the mean as our measure of center for a discrete random variable, it shouldn't surprise you that we'll use the standard deviation as our measure of variability. In Section 1D, we defined the standard deviation s_x of a distribution of quantitative data as the typical distance of the values in the data set from the mean. To get the standard deviation, we started by "averaging" the squared deviations from the mean and then took the square root:

$$s_x = \sqrt{\frac{(x_1 - \bar{x})^2 + (x_2 - \bar{x})^2 + \ldots + (x_n - \bar{x})^2}{n-1}}$$

We can modify this approach to calculate the **standard deviation of a discrete random variable** X. Start by finding a weighted average of the squared deviations $(x_i - \mu_X)^2$ of the values of the variable X from its mean μ_X. The probability distribution gives the appropriate weight for each squared deviation. We call this weighted average of squared deviations the *variance* of X. Then take the square root to get the standard deviation. Because a probability distribution can model the population distribution of a quantitative variable, we label the standard deviation of a random variable X as σ_X (and the variance as σ_X^2). Like any parameter, σ_X is a single, fixed value.

DEFINITION Standard deviation of a discrete random variable

The **standard deviation of a discrete random variable** measures how much the values of the variable typically vary from the mean in many, many trials of the random process.

Suppose that X is a discrete random variable with probability distribution

Value	x_1	x_2	x_3	...
Probability	$P(x_1)$	$P(x_2)$	$P(x_3)$	...

and that μ_X is the mean of X. Then the standard deviation of X is

$$\sigma_X = \sqrt{(x_1 - \mu_X)^2 P(x_1) + (x_2 - \mu_X)^2 P(x_2) + (x_3 - \mu_X)^2 P(x_3) + \ldots}$$
$$= \sqrt{\sum (x_i - \mu_X)^2 P(x_i)}$$

AP® EXAM TIP
Formula sheet

The formula for the standard deviation σ_X of a discrete random variable is included on the formula sheet provided on both sections of the AP® Statistics exam.

Let's return to the random variable C = the total amount of money that Pete collects on a randomly selected jeep tour. Here is its probability distribution once again. Recall that the mean of C is $\mu_C = 562.50$.

Total collected ($)	300	450	600	750	900
Probability	0.15	0.25	0.35	0.20	0.05

We can use the formula from the definition box to calculate the standard deviation of C:

$$\sigma_C = \sqrt{(300-562.50)^2(0.15)+(450-562.50)^2(0.25)+\ldots+(900-562.50)^2(0.05)}$$
$$= \sqrt{26{,}718.75}$$
$$= \$163.46$$

The value obtained before taking the square root in the standard deviation calculation is the variance: $\sigma_C^2 = 26{,}718.75$ squared dollars. Because variance is measured in squared units, it is not a very helpful way to describe the variability of a distribution.

How do we interpret $\sigma_C = \$163.46$? If many, many jeep tours are randomly selected, the amount of money that Pete collects typically varies from the mean of $562.50 by about $163.46.

EXAMPLE

How much do Apgar scores vary?
The standard deviation of a discrete random variable

Skills 3.B, 4.B

PROBLEM: Earlier, we defined the random variable X to be the Apgar score of a randomly selected newborn. The table gives the probability distribution of X once again. In the preceding example, we calculated the mean Apgar score of a randomly chosen newborn to be $\mu_X = 8.128$.

Apgar score	0	1	2	3	4	5	6	7	8	9	10
Probability	0.001	0.006	0.007	0.008	0.012	0.020	0.038	0.099	0.319	0.437	0.053

Calculate and interpret the standard deviation of X.

Ben Edwards/Stockbyte/Getty Images

SOLUTION:

$$\sigma_X = \sqrt{(0-8.128)^2(0.001)+(1-8.128)^2(0.006)+\ldots+(10-8.128)^2(0.053)}$$
$$= \sqrt{2.066}$$
$$= 1.437$$

$$\sigma_X = \sqrt{\sum (x_i - \mu_X)^2 P(x_i)}$$

If many, many newborns are randomly selected, the babies' Apgar scores will typically vary from the mean of 8.128 by about 1.437 units.

FOR PRACTICE, TRY EXERCISE 13

Technology can be a big help when you are analyzing discrete random variables — but be sure to read the AP® Exam Tip following the Tech Corner.

11. Tech Corner ANALYZING DISCRETE RANDOM VARIABLES

TI-Nspire and other technology instructions are on the book's website at bfwpub.com/tps7e.

You can use technology to graph the probability distribution of a discrete random variable and to calculate its mean and standard deviation. We'll illustrate using the random variable X = Apgar score of a randomly selected newborn, whose probability distribution is shown in the preceding example.

1. Enter the values of the random variable in list L1 and the corresponding probabilities in list L2.

NORMAL FLOAT AUTO REAL RADIAN MP

L1	L2	L3	L4	L5
0	0.001	-------	-------	-------
1	0.006			
2	0.007			
3	0.008			
4	0.012			
5	0.02			
6	0.038			
7	0.099			
8	0.319			
9	0.437			
10	0.053			

L1(1)=0

2. To graph a histogram of the probability distribution:
   - Set up a statistics plot to be a histogram with Xlist: L1 and Freq: L2.
   - Adjust your window settings as follows:
     Xmin = −0.5, Xmax = 10.5, Xscl = 1, Ymin = −0.1, Ymax = 0.5, Yscl = 0.1
   - Press GRAPH.

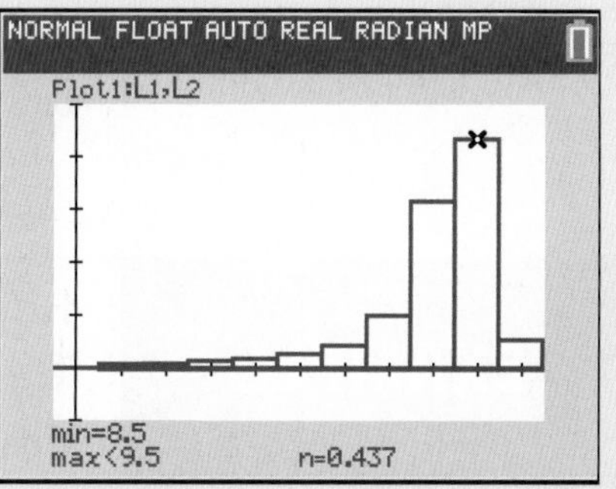

3. To calculate the mean and standard deviation of the random variable:
   - Press STAT, arrow over to the CALC menu, and choose 1-Var Stats.

     OS 2.55 or later: In the dialog box, specify List: L1 and FreqList: L2. Then choose Calculate.

     Older OS: Execute the command 1-Var Stats L1,L2.

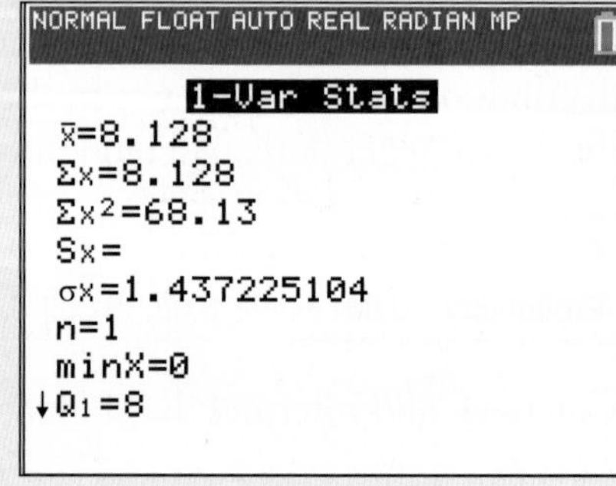

Note: Be sure to clear the FreqList before trying to calculate summary statistics for one-variable quantitative data!

Notice that the calculator's notation for the mean of the random variable X is incorrect. We should write $\mu_X = 8.128$. Fortunately, the notation for the standard deviation is correct: $\sigma_X = 1.437$.

AP® EXAM TIP

If you are asked to calculate the mean or standard deviation of a discrete random variable on a free response question, you must show numerical values substituted into the appropriate formula, as in the previous two examples. Feel free to use an ellipsis (. . .) if the summation includes a large number of terms, as we did. You may then use the method described in the Tech Corner to perform the calculation with 1-Var Stats. Writing only 1-Var Stats L1, L2 and then giving the correct values of the mean and standard deviation will *not* earn credit for showing work. Also, be sure to avoid incorrect notation when labeling these parameters.

CHECK YOUR UNDERSTANDING

High school students in Michigan sometimes get "snow days" in the winter when the roads are so bad that school is canceled for the day. Define the random variable S = the number of snow days at a certain high school in Michigan for a randomly selected school year. The table gives the probability distribution of S.

Number of snow days, s	0	1	2	3	4	5	6	7	8	9	10
Probability, $P(S = s)$	0.33	0.19	0.14	0.10	0.07	0.05	0.04	0.04	0.02	0.01	0.01

1. Calculate the mean of S. Interpret this parameter.
2. Find the median. Show your method clearly.
3. Compare the mean and the median. Explain why this relationship makes sense based on the probability distribution.
4. Calculate the standard deviation of S. Interpret this parameter.

SECTION 4D Summary

- A **random variable** takes numerical values determined by the outcome of a random process. There are two types of random variables: *discrete* and *continuous*.
- A **discrete random variable** has a countable set of possible values with gaps between them on a number line.
- The **probability distribution** of a discrete random variable gives its possible values and their probabilities.
    - A valid probability distribution assigns each of its values a probability between 0 and 1 such that the sum of all the probabilities is exactly 1.
    - The probability of any event is the sum of the probabilities of all the values that make up the event.
    - We can display the probability distribution as a histogram, with the values of the random variable on the horizontal axis and the probabilities on the vertical axis.
- We can describe the *shape* of a probability distribution's graph in the same way as we did a distribution of quantitative data — by identifying symmetry or skewness and any clear peaks.

- Use the mean to summarize the *center* of a probability distribution. The **mean, or expected value, of a random variable** is the balance point of the probability distribution's graph.
  - The mean, or expected value, is the long-run average value of the variable after many, many trials of the random process. It is denoted by μ_X or $E(X)$.
  - If X is a discrete random variable, the mean (expected value) is the average of the values of X, each weighted by its probability:

$$\mu_X = E(X) = \sum x_i P(x_i) = x_1 P(x_1) + x_2 P(x_2) + x_3 P(x_3) + \ldots$$

- Use the standard deviation to summarize the *variability* of a probability distribution. The **standard deviation of a random variable** σ_X measures how much the values of the variable typically vary from the mean in many, many trials of the random process.
  - If X is a discrete random variable, the standard deviation of X is

$$\sigma_X = \sqrt{\sum (x_i - \mu_X)^2 P(x_i)} = \sqrt{(x_1 - \mu_X)^2 P(x_1) + (x_2 - \mu_X)^2 P(x_2) + (x_3 - \mu_X)^2 P(x_3) + \ldots}$$

  - The value obtained before taking the square root is the *variance* σ_X^2.

AP® EXAM TIP

AP® Daily Videos

Review the content of this section and get extra help by watching the AP® Daily Videos for Topics 4.7 and 4.8, which are available in AP® Classroom.

4D Tech Corner

TI-Nspire and other technology instructions are on the book's website at bfwpub.com/tps7e.

11. Analyzing discrete random variables Page 390

SECTION 4D Exercises

Discrete Random Variables and Probability

1. **Languages spoken** Imagine selecting a U.S. high
pg 383 school student at random. Define the random variable Y = number of languages spoken by the student. The table gives the probability distribution of Y.[52]

Number of languages	1	2	3	4	5
Probability	0.630	???	0.065	0.008	0.002

(a) Write the event "student speaks 2 languages" in terms of Y. Then find its probability.

(b) What's the probability that a randomly selected student speaks at least 3 languages?

2. **Kids and toys** In an experiment on the behavior of young children, each subject is placed in an area with 5 toys. The probability distribution of the number X of toys played with by a randomly selected subject is as follows:

Number of toys	0	1	2	3	4	5
Probability	0.03	0.16	0.30	0.23	0.17	???

(a) Write the event "child plays with 5 toys" in terms of X. Then find its probability.

(b) What's the probability that a randomly selected subject plays with at most 3 toys?

3. **Get on the boat!** A small ferry runs every half-hour from one side of a large river to the other. The ferry can hold a maximum of 5 cars, and the charge is \$5 per car. The probability distribution for the random variable Y = money collected (in dollars) on a randomly selected ferry trip is shown here:

Money collected (\$), *y*	0	5	10	15	20	25
Probability, *P*(*Y* = *y*)	0.02	0.05	0.08	0.16	0.27	0.42

(a) Find $P(Y > 10)$. Describe this probability in words.

(b) Write the event "at least \$10 is collected" in terms of Y. What is the probability of this event?

4. **Skee Ball** Ana is a dedicated Skee Ball player (see photo) who always rolls for the 50-point slot. The probability distribution of Ana's score X on a randomly selected roll of the ball is shown here.

Score, x	10	20	30	40	50
Probability, $P(X = x)$	0.32	0.27	0.19	0.15	0.07

StanRohrer/Getty Images

(a) Find $P(X < 20)$. Describe this probability in words.

(b) Write the event "Ana scores at most 20" in terms of X. What is the probability of this event?

Displaying Discrete Probability Distributions

5. **More ferry fares** Refer to Exercise 3. Make a histogram of the probability distribution. Describe its shape. pg 385

6. **More Skee Ball** Refer to Exercise 4. Make a histogram of the probability distribution. Describe its shape.

Mean, or Expected Value, of a Discrete Random Variable

7. **Mean ferry trip** Refer to Exercise 3. Here again is the probability distribution for Y = money collected (in dollars) on a randomly selected ferry trip. pg 387

Money collected (\$), y	0	5	10	15	20	25
Probability, $P(Y = y)$	0.02	0.05	0.08	0.16	0.27	0.42

Calculate and interpret the expected value of Y.

8. **Mean Skee Baller** Refer to Exercise 4. Here again is the probability distribution of Ana's score X on a randomly selected roll of the ball.

Score, x	10	20	30	40	50
Probability, $P(X = x)$	0.32	0.27	0.19	0.15	0.07

Calculate and interpret the expected value of X.

9. **Benford's law** Faked numbers in tax returns, invoices, or expense account claims often display patterns that aren't present in legitimate records. Some patterns, like too many round numbers, are obvious and easily avoided by a clever crook. Others are more subtle. It is a striking fact that the first digits of numbers in legitimate records often follow a model known as Benford's law.[53] Call the first digit of a randomly chosen legitimate record X for short. Here is the probability distribution of X according to Benford's law (note that a first digit can't be 0).

First digit	1	2	3	4	5	6	7	8	9
Probability	0.301	0.176	0.125	0.097	0.079	0.067	0.058	0.051	0.046

(a) A histogram of the probability distribution is shown. Describe its shape.

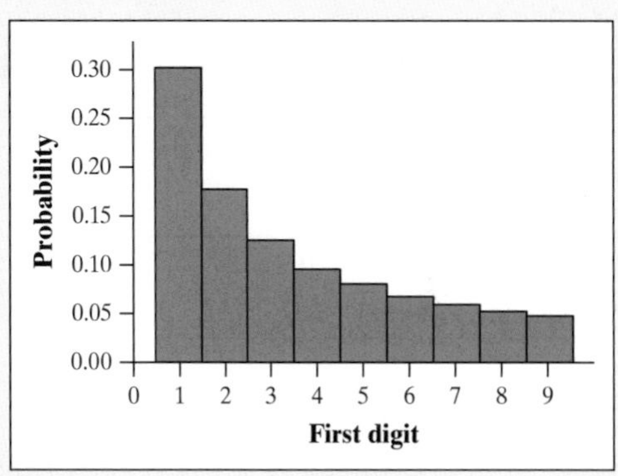

(b) Calculate the mean of X. Interpret this parameter.

10. **Working out** Choose a person aged 19 to 25 years at random and ask, "In the past seven days, how many times did you go to an exercise or fitness center or work out?" Call the response Y for short. Based on a large sample survey, here is the probability distribution of Y:[54]

Days	0	1	2	3	4	5	6	7
Probability	0.68	0.05	0.07	0.08	0.05	0.04	0.01	0.02

(a) A histogram of the probability distribution is shown. Describe its shape.

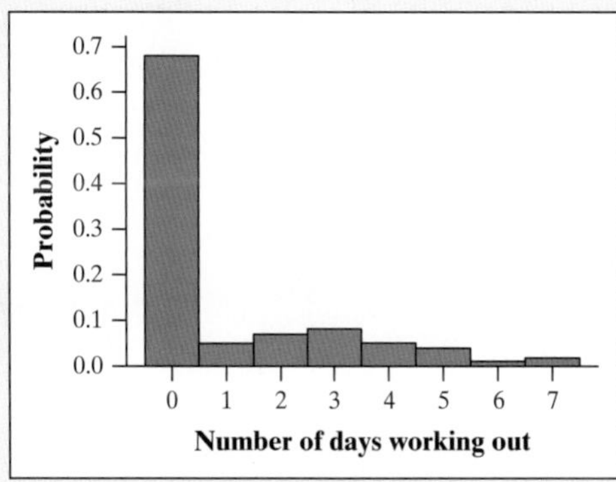

(b) Calculate the mean of Y. Interpret this parameter.

11. **Median ferry trip** A small ferry runs every half-hour from one side of a large river to the other. The probability distribution for the random variable Y = money

collected on a randomly selected ferry trip is shown here. From Exercise 7, $E(Y) = \$19.35$.

Money collected (\$), y	0	5	10	15	20	25
Probability, $P(Y = y)$	0.02	0.05	0.08	0.16	0.27	0.42

(a) Construct the cumulative probability distribution of Y.

(b) Find the median of Y.

(c) Compare the mean and the median. Explain why this relationship makes sense based on the probability distribution.

12. **Median Skee Baller** Ana is a dedicated Skee Ball player (see photo in Exercise 4) who always rolls for the 50-point slot. The probability distribution of Ana's score X on a randomly selected roll of the ball is shown here. From Exercise 8, $E(X) = 23.8$.

Score, x	10	20	30	40	50
Probability, $P(X = x)$	0.32	0.27	0.19	0.15	0.07

(a) Construct the cumulative probability distribution of X.

(b) Find the median of X.

(c) Compare the mean and the median. Explain why this relationship makes sense based on the probability distribution.

Standard Deviation of a Discrete Random Variable

13. **Ferry trips vary** Refer to Exercise 11. Calculate and interpret the standard deviation of Y.

pg 389

14. **Scores vary** Refer to Exercise 12. Calculate and interpret the standard deviation of X.

15. **More Benford's law** Exercise 9 described how the first digits of numbers in legitimate records often follow a model known as Benford's law. Call the first digit of a randomly chosen legitimate record X for short. The probability distribution for X = the first digit of a randomly chosen legitimate record is shown again here (note that a first digit can't be 0). From Exercise 9, $\mu_X = 3.441$.

First digit	1	2	3	4	5	6	7	8	9
Probability	0.301	0.176	0.125	0.097	0.079	0.067	0.058	0.051	0.046

(a) Find the variance of X.

(b) Find σ_X. Interpret this parameter.

16. **More working out** Exercise 10 described a large sample survey that asked a sample of people aged 19 to 25 years, "In the past seven days, how many times did you go to an exercise or fitness center or work out?" The response Y for a randomly selected survey respondent has the probability distribution shown here. From Exercise 10, $\mu_Y = 1.03$.

Days	0	1	2	3	4	5	6	7
Probability	0.68	0.05	0.07	0.08	0.05	0.04	0.01	0.02

(a) Find the variance of Y.

(b) Find σ_Y. Interpret this parameter.

17. **Life insurance** A life insurance company sells a term insurance policy to 21-year-olds that pays \$100,000 if the insured dies within the next 5 years. The probability that a randomly chosen 21-year-old will die each year can be found in mortality tables. The company collects a premium of \$250 each year as payment for the insurance. The amount Y that the company earns on a randomly selected policy of this type is \$250 per year, less the \$100,000 that it must pay if the insured dies. Here is the probability distribution of Y:

Age at death	Profit	Probability
21	–\$99,750	0.00183
22	–\$99,500	0.00186
23	–\$99,250	0.00189
24	–\$99,000	0.00191
25	–\$98,750	0.00193
26 or older	\$1,250	0.99058

(a) Explain why the company suffers a loss of \$98,750 on such a policy if a customer dies at age 25.

(b) Calculate the expected value of Y. Explain what this result means for the insurance company.

(c) Calculate the standard deviation of Y. Explain what this result means for the insurance company.

18. **Fire insurance** Suppose a homeowner spends \$300 for a home insurance policy that will pay out \$200,000 if the home is destroyed by fire. Let T = the profit made by the company on a randomly selected policy. Based on previous data, the probability that a home in this area will be destroyed by fire is 0.0002. Here is the probability distribution of T:

Profit	Probability
–\$199,700	0.0002
\$300	0.9998

(a) Explain why the company loses \$199,700 if the policyholder's home is destroyed by fire.

(b) Calculate the expected value of T. Explain what this result means for the insurance company.

(c) Calculate the standard deviation of T. Explain what this result means for the insurance company.

Exercises 19 and 20 examine how Benford's law (Exercise 9) can be used to detect fraud.

19. **Back to Benford's law** A not-so-clever employee decided to fake a monthly expense report. This employee believed that the first digits of expense amounts should be equally likely to be any of the numbers from 1 to 9. In that case, the first digit Y of a randomly selected expense amount would have the probability distribution shown in the histogram.

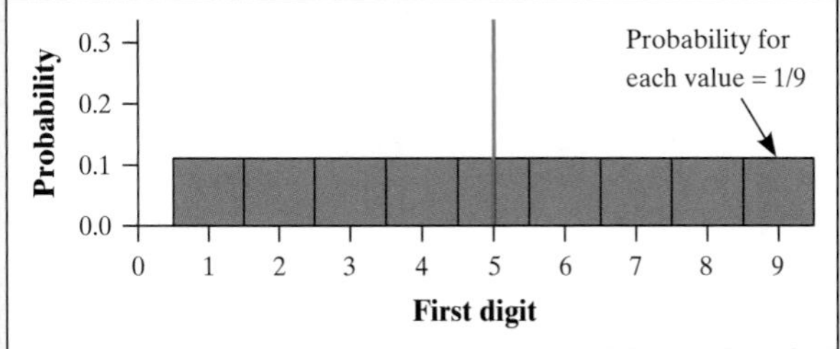

(a) According to the histogram, what is $P(Y > 6)$? According to Benford's law (see Exercise 9), what proportion of first digits in the employee's expense amounts should be greater than 6? How could this information be used to detect a fake expense report?

(b) Explain why the mean of the random variable Y is located at the solid red line in the figure.

(c) According to Benford's law, the expected value of the first digit is $\mu_X = 3.441$. Explain how this information could be used to detect a fake expense report.

20. **Benford's law finale**

(a) Using the histogram from Exercise 19, calculate the standard deviation σ_Y. It gives us an idea of how much variation we would expect in the employee's expense records if the employee assumed that first digits from 1 to 9 were equally likely.

(b) The standard deviation of first digits of randomly selected expense amounts that follow Benford's law is $\sigma_X = 2.46$. Would using standard deviations be a good way to detect fraud? Explain your answer.

For Investigation *Apply the skills from the section in a new context or nonroutine way.*

21. **Batch testing** Suppose that 12 people need to be given a blood test for a certain disease. Assume that each person has a 6% probability of having the disease, with this chance occurring independently for each person. Consider two different plans for conducting the tests:[55]

Plan A: Give an individual blood test to each person.
Plan B: Combine blood samples from all 12 people into one batch and test that batch.

- If at least one person has the disease, then the batch test result will be positive, and all 12 people must be tested individually.
- If no one has the disease, then the batch test result will be negative, and no additional tests will be needed.

Let X = the total number of tests needed with plan B (batch testing).

(a) Determine the probability distribution of X.

(b) If you implement plan B once, what is the probability that the number of tests needed will be smaller than are needed with plan A?

(c) Calculate and interpret the expected value of X.

(d) If thousands of groups of 12 people need to be tested, which plan — A or B — would be better? Justify your answer.

22. **Play ball!** When a Major League Baseball team has a runner on first base with no outs, its expected number of runs scored is 0.8666.[56] If a runner can successfully steal second base, then the expected runs scored increases to 1.0857, but if the runner is caught, the expected runs scored decreases to 0.2643.

(a) Interpret the value 0.8666.

(b) A particular base runner has an 80% chance of successfully stealing second base. If X = expected runs scored when attempting to steal second base, display the probability distribution of X for this base runner.

(c) Calculate the expected value of X. Based on your calculation, should this base runner attempt to steal second base? Explain your reasoning.

(d) If a different base runner is in the same situation, what minimum probability of success should the runner have to make attempting to steal second base a good strategy? Justify your answer.

Multiple Choice *Select the best answer for each question.*

Exercises 23–25 refer to the following setting. Suppose we choose a U.S. household at random and let the random variable X be the number of cars (including SUVs and light trucks) they own. Here is the probability distribution if we ignore the few households that own more than 5 cars:

Number of cars, *x*	0	1	2	3	4	5
Probability, *P*(*x*)	0.09	0.36	0.35	0.13	0.05	0.02

23. What's the expected number of cars in a randomly selected U.S. household?

(A) 1.00
(B) 1.75
(C) 1.84
(D) 2.00
(E) 2.50

24. The standard deviation of X is $\sigma_X = 1.08$. If many, many U.S. households were selected at random, which of the following is the best interpretation of the value 1.08?

(A) The mean number of cars for those households would be about 1.08.

(B) The number of cars would typically be about 1.08 from the mean.

(C) The number of cars would be at most 1.08 from the mean.

(D) The number of cars would be at least 1.08 from the mean.

(E) The mean number of cars would be about 1.08 from the expected value.

25. What is the approximate probability that a randomly selected household has a number of cars that is within 2 standard deviations of the mean?

(A) 68%
(B) 71%
(C) 93%
(D) 95%
(E) 98%

26. A deck of cards contains 52 cards, of which 4 are aces. You are offered the following wager: Draw one card at random from the deck. You win \$10 if the card drawn is an ace. Otherwise, you lose \$1. If you make this wager very many times, what will be the mean amount you win?

(A) About −\$1, because you will lose most of the time.

(B) About \$9, because you win \$10 but lose only \$1.

(C) About −\$0.15; that is, on average, you lose about 15 cents.

(D) About \$0.77; that is, on average, you win about 77 cents.

(E) About \$0, because the random draw gives you a fair bet.

Recycle and Review *Practice what you learned in previous sections.*

Exercises 27 and 28 refer to the following setting. Many chess masters and chess advocates believe that chess play develops general intelligence, analytical skills, and the ability to concentrate. According to such beliefs, improved reading skills should result from studying to improve chess-playing skills. To investigate this belief, researchers conducted a study. All the subjects in the study participated in a comprehensive chess program, and their reading performances were measured before and after the program. The graphs and numerical summaries that follow provide information on the subjects' pretest scores, their posttest scores, and the difference (Post − Pre) between these two scores.[57]

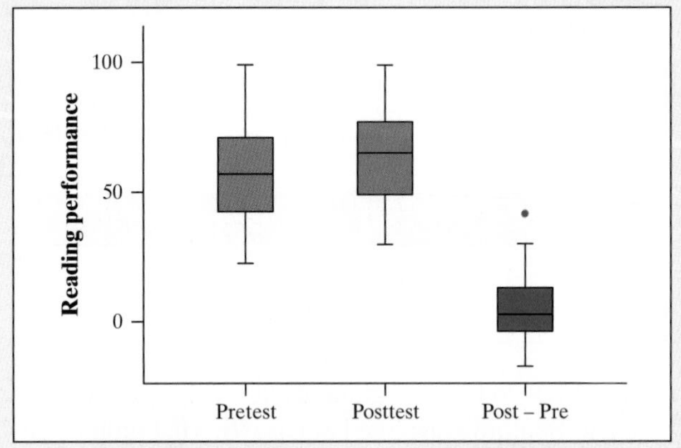

Descriptive Statistics: Pretest, Posttest, Post − Pre

Variable	n	Mean	Median	SD	Min	Max	Q_1	Q_3
Pretest	53	57.70	58.00	17.84	23.00	99.00	44.50	70.50
Posttest	53	63.08	64.00	18.70	28.00	99.00	48.00	76.00
Post − Pre	53	5.38	3.00	13.02	−19.00	42.00	−3.50	14.00

27. **Better readers? (1D, 3C)**

(a) Did students tend to have higher reading scores after participating in the chess program? Justify your answer.

(b) If the study found a statistically significant improvement in the average reading score, could you conclude that playing chess causes an increase in reading skills? Justify your answer.

Some graphical and numerical information about the relationship between pretest and posttest scores is provided here.

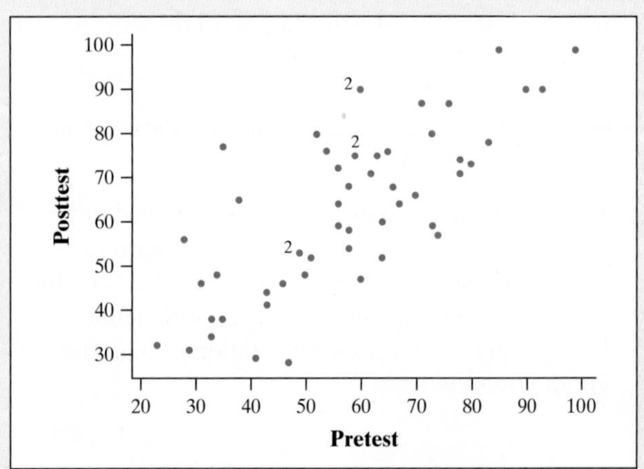

Regression Analysis: Posttest Versus Pretest

Predictor	Coef	SE Coef	T	P
Constant	17.897	5.889	3.04	0.004
Pretest	0.78301	0.09758	8.02	0.000

S = 12.55 R-Sq = 55.8% R-Sq(adj) = 54.9%

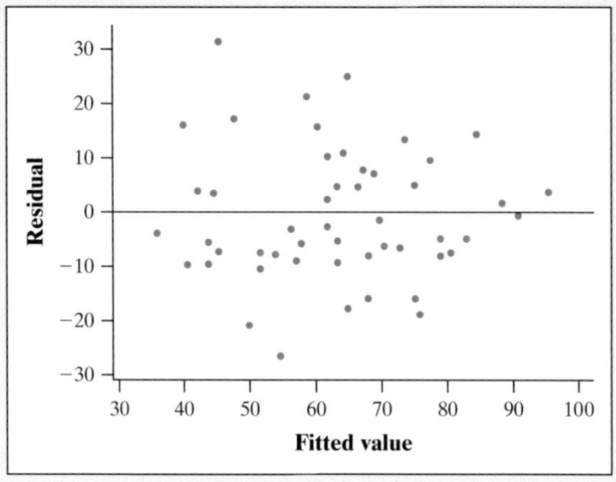

28. **Predicting posttest scores (2C)**

(a) What is the equation of the least-squares regression line relating posttest and pretest scores? Define any variables used.

(b) Is a linear model appropriate for describing this relationship? Justify your answer.

(c) If we use the least-squares regression line to predict students' posttest scores from their pretest scores, how far off will our predictions typically be?

SECTION 4E Transforming and Combining Random Variables

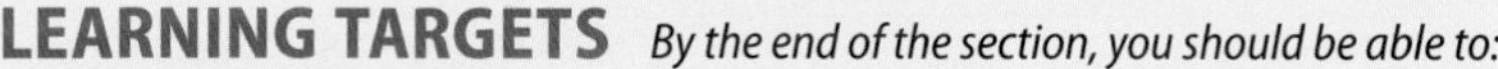

LEARNING TARGETS *By the end of the section, you should be able to:*

- Describe the effect of a linear transformation—adding or subtracting a constant and/or multiplying or dividing by a constant—on the probability distribution of a random variable.
- Calculate the mean of a sum, difference, or other linear combination of random variables.
- If appropriate, calculate the standard deviation of a sum, difference, or other linear combination of random variables.

In Section 4D, we looked at several examples of discrete random variables and their probability distributions. A histogram of the probability distribution enabled us to describe its shape. We also saw that the parameters μ_X and σ_X give us important information about the center and variability of a random variable's probability distribution. In this section, we start by examining the effect of adding, subtracting, multiplying by, or dividing by a constant on the probability distribution of a random variable. Then, we consider how to calculate the mean and standard deviation of a sum or difference of random variables.

Transforming Random Variables

In Section 1E, we studied the effects of linear transformations on the shape, center, and variability of a distribution of quantitative data. Here's what we discovered:

1. *Adding or subtracting a constant:* Adding the same positive number a to or subtracting a from each data value:
   - Adds a to or subtracts a from measures of center (mean, median).
   - Does not change measures of variability (range, IQR, standard deviation).
   - Does not change the shape of the distribution.
2. *Multiplying or dividing by a constant:* Multiplying or dividing each data value by the same positive number b:
   - Multiplies or divides measures of center (mean, median) by b.
   - Multiplies or divides measures of variability (range, IQR, standard deviation) by b.
   - Does not change the shape of the distribution.

How are the probability distributions of random variables affected by similar transformations?

EFFECT OF ADDING OR SUBTRACTING A CONSTANT

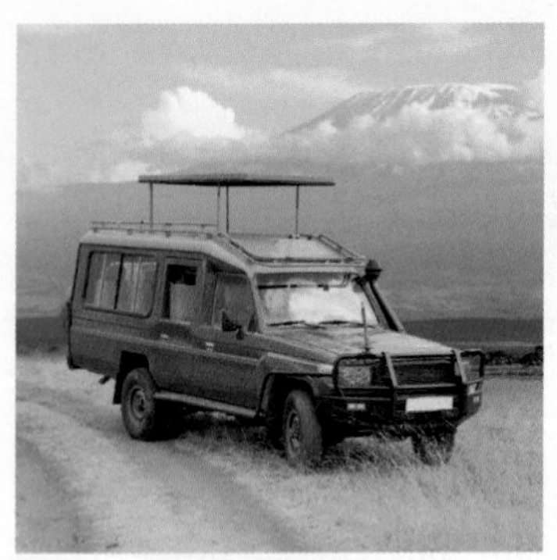
Volodymyr Burdiak/Shutterstock

Let's return to a familiar setting from Section 4D. Pete's Jeep Tours offers a popular day trip in a tourist area. There must be at least 2 passengers for the trip to run, and the vehicle will hold up to 6 passengers. Pete charges \$150 per passenger. Let C = the total amount of money that Pete collects on a randomly selected trip. The probability distribution of C is shown in the table and the histogram.

Total collected (\$)	300	450	600	750	900
Probability	0.15	0.25	0.35	0.20	0.05

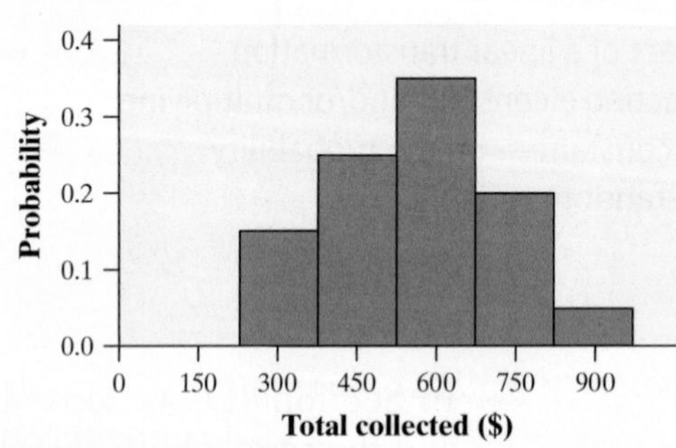

Earlier, we calculated the mean of C as $\mu_C = \$562.50$ and the standard deviation of C as $\sigma_C = \$163.46$. We can describe the probability distribution of C as follows:

Shape: Roughly symmetric with a single peak

Center: $\mu_C = \$562.50$

Variability: $\sigma_C = \$163.46$

It costs Pete \$100 to buy permits, gas, and a ferry pass for each day trip. The amount of profit V that Pete makes on a randomly selected trip is the total amount of money C that he collects from passengers minus \$100. That is, $V = C - 100$. The probability distribution of V is

Profit (\$)	200	350	500	650	800
Probability	0.15	0.25	0.35	0.20	0.05

A histogram of this probability distribution is shown here.

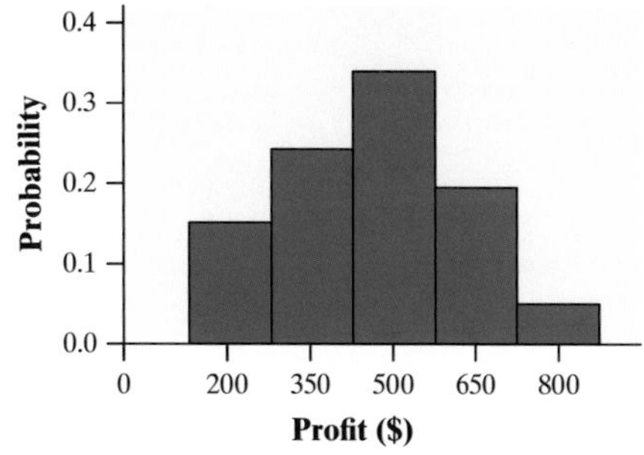

We can see that the probability distribution of V has the same shape as the probability distribution of C. The mean of V is

$$\begin{aligned}\mu_V &= (200)(0.15) + (350)(0.25) + (500)(0.35) + (650)(0.20) + (800)(0.05)\\ &= \$462.50\end{aligned}$$

On average, Pete will make a profit of \$462.50 from the trip. That's \$100 less than μ_C, his mean amount of money collected per trip. The standard deviation of V is

$$\begin{aligned}\sigma_V &= \sqrt{(200 - 462.50)^2(0.15) + (350 - 462.50)^2(0.25) + \ldots + (800 - 462.50)^2(0.05)}\\ &= \$163.46\end{aligned}$$

That's the same as the standard deviation of C.

It's fairly clear that subtracting 100 from the values of the random variable C just shifts the probability distribution to the left by 100. This transformation decreases the mean by 100 — from \$562.50 to \$462.50 — but it doesn't change the standard deviation, \$163.46, or the shape. Adding a positive constant to the values of a random variable would just shift the probability distribution to the right by that constant. These results can be generalized for any random variable.

THE EFFECT OF ADDING OR SUBTRACTING A CONSTANT ON A PROBABILITY DISTRIBUTION

Adding the same positive number a to or subtracting a from each value of a random variable:

- Adds a to or subtracts a from measures of center (mean, median).
- Does not change measures of variability (range, *IQR*, standard deviation).
- Does not change the shape of the probability distribution.

Note that adding or subtracting a constant affects the distribution of a quantitative variable and the probability distribution of a random variable in exactly the same way.

EFFECT OF MULTIPLYING OR DIVIDING BY A CONSTANT

Pete is curious about the amount of profit he makes per hour from his jeep tours. He works 10 hours on a tour day. The random variable $H = \frac{V}{10}$ describes Pete's hourly profit on a randomly selected trip. Here is the probability distribution of H, along with a histogram.

Hourly profit ($)	20	35	50	65	80
Probability	0.15	0.25	0.35	0.20	0.05

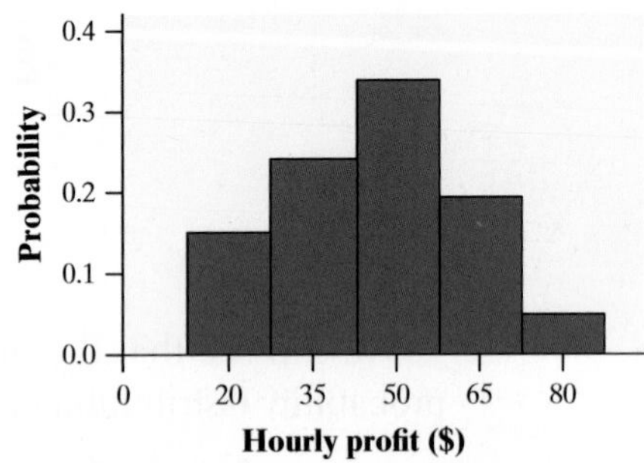

The probability distribution of H is the same shape as the probability distribution of V. However, the mean of H is

$$\mu_H = (20)(0.15) + (35)(0.25) + (50)(0.35) + (65)(0.20) + (80)(0.05)$$
$$= \$46.25$$

That's the mean amount of profit Pete expects to make per day, \$462.50, divided by 10 hours: $\mu_H = \frac{\mu_V}{10}$. The standard deviation of H is

$$\sigma_H = \sqrt{(20 - 46.25)^2(0.15) + (35 - 46.25)^2(0.25) + \ldots + (80 - 46.25)^2(0.05)}$$
$$= 16.346$$

or about \$16.35. That's just the standard deviation of V, \$163.46, divided by 10: $\sigma_H = \frac{\sigma_V}{10}$.

To get the mean and standard deviation of H, we simply divide the mean and standard deviation of V by 10. But the shape of the two probability distributions is the same. Multiplying or dividing all the values of a random variable by a positive constant stretches or compresses the probability distribution by that factor.

THE EFFECT OF MULTIPLYING OR DIVIDING BY A CONSTANT ON A PROBABILITY DISTRIBUTION

Multiplying or dividing each value of a random variable by the same positive number b:

- Multiplies or divides measures of center (mean, median) by b.
- Multiplies or divides measures of variability (range, *IQR*, standard deviation) by b.
- Does not change the shape of the distribution.

It is not common to multiply or divide a random variable by a negative number b. Doing so would multiply or divide the measures of variability by $|b|$. Multiplying or dividing by a negative number would also affect the shape of the probability distribution, as all values would be reflected over the y axis.

Once again, multiplying or dividing by a constant has the same effect on the probability distribution of a random variable as it does on a distribution of quantitative data.

EXAMPLE How much does college cost? Transforming random variables

Skill 3.C

FG Trade/E+/Getty Images

PROBLEM: El Dorado Community College considers a student to be attending full-time if they are taking between 12 and 18 units. The number of units X that a randomly selected El Dorado Community College full-time student is taking in the fall semester has the following distribution:

Number of units, x	12	13	14	15	16	17	18
Probability, $P(X = x)$	0.25	0.10	0.05	0.30	0.10	0.05	0.15

Here is a histogram of the probability distribution. The mean is $\mu_X = 14.65$ and the standard deviation is $\sigma_X = 2.056$.

At El Dorado Community College, the tuition for full-time students is \$50 per unit. Let T = the tuition charge for a randomly selected student.

(a) What shape does the probability distribution of T have?
(b) Find the mean of T.
(c) Calculate the standard deviation of T.

SOLUTION:

$T = 50X$

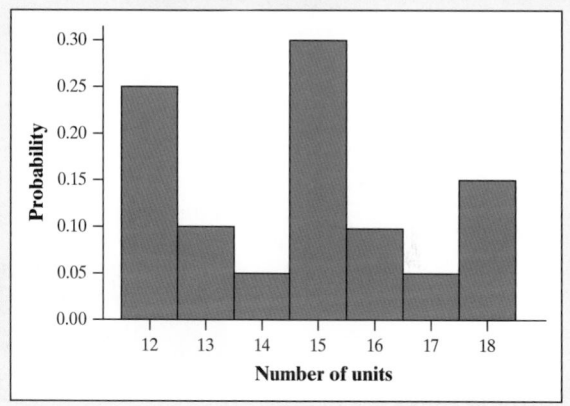

(a) The same shape as the probability distribution of X: roughly symmetric with three peaks.

Multiplying by a constant doesn't change the shape.

(b) $\mu_T = 50(14.65) = \$732.50$

$\mu_T = 50\mu_X$

(c) $\sigma_T = 50(2.056) = \$102.80$

$\sigma_T = 50\sigma_X$

FOR PRACTICE, TRY EXERCISE 3

Think About It

HOW DOES MULTIPLYING BY A CONSTANT AFFECT THE VARIANCE? For El Dorado Community College, the variance of the number of units that a randomly selected full-time student takes is $\sigma_X^2 = 4.2275$. The variance of the tuition charge for such a student is $\sigma_T^2 = 10{,}568.75$. That's $(2500)(4.2275)$.

So $\sigma_T^2 = 2500\sigma_X^2$. Where did 2500 come from? It's just $(50)^2$. In other words, because $\sigma_T = (50)\sigma_X$, $\sigma_T^2 = (50)^2\sigma_X^2$. Multiplying a random variable by a constant b multiplies the variance by b^2.

PUTTING IT ALL TOGETHER: LINEAR TRANSFORMATIONS

What happens if we transform a random variable by both adding or subtracting a constant and multiplying or dividing by a constant? Let's return to the preceding example.

El Dorado Community College charges each student a $100 fee per semester in addition to tuition charges. We can calculate a randomly selected full-time student's total charges Y for the fall semester directly from the number of units X the student is taking, using the equation $Y = 50X + 100$ or, equivalently, $Y = 100 + 50X$. This is called a *linear transformation* of the random variable X. Why? Because the equation describing the sequence of transformations has the form $Y = a + bX$, which you should recognize as a linear equation.

The linear transformation $Y = 100 + 50X$ includes two different transformations: (1) multiplying by 50 and (2) adding 100. Because neither of these transformations affects the distribution's shape, the probability distribution of Y will have the same shape as the probability distribution of X. To get the mean of Y, we multiply the mean of X by 50, then add 100:

$$\mu_Y = 100 + 50\mu_X = 100 + 50(14.65) = \$832.50$$

To get the standard deviation of Y, we multiply the standard deviation of X by 50 (adding 100 doesn't affect the standard deviation):

$$\sigma_Y = 50\sigma_X = 50(2.056) = \$102.80$$

This logic generalizes to any linear transformation.

THE EFFECT OF A LINEAR TRANSFORMATION ON A RANDOM VARIABLE

If $Y = a + bX$ is a linear transformation of the random variable X,

- The probability distribution of Y has the same shape as the probability distribution of X if $b > 0$.
- $\mu_Y = a + b\mu_X$.
- $\sigma_Y = |b|\sigma_X$ (because b could be a negative number).

AP® EXAM TIP
Formula sheet

Unfortunately, *no* formulas for linear transformations are included on the formula sheet provided on both sections of the AP® Statistics exam.

Note that these results apply to both discrete and continuous random variables.

Combining Random Variables: The Mean

So far, we have looked at scenarios that involved a single random variable. Many interesting statistics problems require us to combine two or more random variables using addition or subtraction. Unfortunately, there is no simple way to determine the shape of the resulting probability distribution when we add or subtract discrete random variables, except for one special case that you'll learn about in Section 5A. So, we will focus our attention for now on calculating the mean and standard deviation.

Frank Herholdt/Stone/Getty Images

Let's return to the familiar setting of Pete's Jeep Tours. Earlier, we focused on the amount of money C that Pete collects on a randomly selected day trip. This time we'll consider a different but related random variable: X = the number of passengers on a randomly selected trip. Here is its probability distribution:

Pete's Jeep Tours					
Number of passengers	2	3	4	5	6
Probability	0.15	0.25	0.35	0.20	0.05

You can use what you learned earlier to confirm that $\mu_X = 3.75$ passengers and $\sigma_X = 1.0897$ passengers.

Pete's sister Erin runs jeep tours in another part of the country on the same days as Pete in her slightly smaller vehicle, under the name Erin's Adventures. The number of passengers Y on a randomly selected trip has the following probability distribution. You can confirm that $\mu_Y = 3.10$ passengers and $\sigma_Y = 0.943$ passengers.

Erin's Adventures				
Number of passengers	2	3	4	5
Probability	0.3	0.4	0.2	0.1

How many total passengers $S = X + Y$ can Pete and Erin expect to have on their tours on a randomly selected day? Because Pete averages $\mu_X = 3.75$ passengers per day trip and Erin averages $\mu_Y = 3.10$ passengers per trip, they will average a total of $\mu_S = 3.75 + 3.10 = 6.85$ passengers per day.

What's the mean of the difference $D = X - Y$ in the number of passengers that Pete and Erin have on their tours on a randomly selected day? Because Pete averages $\mu_X = 3.75$ passengers per trip and Erin averages $\mu_Y = 3.10$ passengers per trip, the mean difference is $\mu_D = 3.75 - 3.10 = 0.65$ passenger. That is, Pete averages 0.65 more passengers per day than Erin does.

We can generalize these results for any two random variables.

MEAN OR EXPECTED VALUE OF A SUM OR DIFFERENCE OF TWO RANDOM VARIABLES

For any two random variables X and Y,

- If $S = X + Y$, the mean (expected value) of S is

$$\mu_S = \mu_{X+Y} = \mu_X + \mu_Y$$

  In other words, the mean of the sum of two random variables is equal to the sum of their means.

- If $D = X - Y$, the mean (expected value) of D is

$$\mu_D = \mu_{X-Y} = \mu_X - \mu_Y$$

  In other words, the mean of the difference of two random variables is equal to the difference of their means.

The order of subtraction is important. If we had defined $D = Y - X$, then $\mu_D = \mu_Y - \mu_X = 3.10 - 3.75 = -0.65$. In other words, Erin averages 0.65 fewer passengers than Pete does on a randomly chosen day.

EXAMPLE

Comparing college costs
Combining random variables: The mean

Skill 3.B

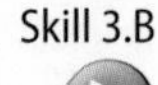

PROBLEM: El Dorado Community College has a main campus and a rural campus. The amount spent on tuition by a randomly selected full-time student at the main campus, M, follows a distribution with mean \$732.50 and standard deviation \$102.80. The amount spent on tuition by a randomly selected full-time student at the rural campus, R, follows a distribution with mean \$825 and standard deviation \$126.50. Let $D = M - R$. Calculate and interpret the mean of D.

Edwin Tan /E+/Getty Images

SOLUTION:

$\mu_D = \$732.50 - \$825 = -\$92.50$ $\mu_{M-R} = \mu_M - \mu_R$

If you were to repeat the process of randomly selecting one student from each campus and finding the difference (Main — Rural) in their tuition charges many, many times, the average difference would be about −\$92.50.

FOR PRACTICE, TRY EXERCISE 9

LINEAR COMBINATIONS OF RANDOM VARIABLES: THE MEAN

Earlier, we defined X = the number of passengers that Pete has and Y = the number of passengers that Erin has on a randomly selected day trip. Recall that $\mu_X = 3.75$ and $\mu_Y = 3.10$. Pete charges \$150 per passenger and Erin charges \$175 per passenger. The total amount collected by Pete and Erin on a randomly selected day is given by $T = 150X + 175Y$. How can we find the expected value of T?

Let $C = 150X$ represent the amount that Pete collects and $E = 175Y$ represent the amount that Erin collects, so that $T = C + E$. Using the earlier results for linear transformations,

$$\mu_C = 150\mu_X = 150(3.75) = 562.50 \text{ and } \mu_E = 175\mu_Y = 175(3.10) = 542.50$$

Now using the rule for the mean of a sum of two random variables,

$$\mu_T = \mu_C + \mu_E = 562.50 + 542.50 = 1105$$

Interpretation: Over many, many randomly selected days, Pete and Erin collect a total of \$1105 per day, on average.

The preceding discussion shows that $\mu_T = 150\mu_X + 175\mu_Y$. More generally, if $S = aX + bY$, then $\mu_S = a\mu_X + b\mu_Y$. The expression $aX + bY$ is called a *linear combination* of the random variables X and Y.

Combining Random Variables: The Standard Deviation

How much variation is there in the total number of passengers $S = X + Y$ who go on Pete's and Erin's tours on a randomly chosen day? Here are the probability distributions of X and Y once again. Let's think about the possible values of S.

The number of passengers X on Pete's tour is between 2 and 6, and the number of passengers Y on Erin's tour is between 2 and 5. So the total number of passengers S is between 4 and 11. That is, there's more variability in the values of S than in the values of X or Y alone. This makes sense, because the variation in X and the variation in Y both contribute to the variation in S.

Pete's Jeep Tours					
Number of passengers	2	3	4	5	6
Probability	0.15	0.25	0.35	0.20	0.05

$\mu_X = 3.75 \quad \sigma_X = 1.0897$

Erin's Adventures				
Number of passengers	2	3	4	5
Probability	0.3	0.4	0.2	0.1

$\mu_Y = 3.10 \quad \sigma_Y = 0.943$

What's the standard deviation of $S = X + Y$? If we had the probability distribution of S, then we could calculate σ_S. Let's try to construct this probability distribution starting with the smallest possible value, $S = 4$. The only way to get a total of 4 passengers is if Pete has $X = 2$ passengers and Erin has $Y = 2$ passengers. We know that $P(X = 2) = 0.15$ and that $P(Y = 2) = 0.3$. If the events $X = 2$ and $Y = 2$ are *independent*, we can use the multiplication rule for independent events to find $P(X = 2 \text{ and } Y = 2)$. Otherwise, we're stuck. In fact, we can't calculate the probability for any value of S unless X and Y are **independent random variables.**

DEFINITION Independent random variables

If knowing the value of *X* does not help us predict the value of *Y*, then *X* and *Y* are **independent random variables.** In other words, two random variables are independent if knowing the value of one variable does not change the probability distribution of the other variable.

It's reasonable to treat the random variables X = number of passengers on Pete's trip and Y = number of passengers on Erin's trip on a randomly chosen day as independent, because the siblings operate their trips in different parts of the country. Because X and Y are independent,

$$P(S = 4) = P(X = 2 \text{ and } Y = 2) = P(X = 2) \times P(Y = 2) = (0.15)(0.3) = 0.045$$

There are two ways to get a total of $S = 5$ passengers on a randomly selected day: $X = 3, Y = 2$ or $X = 2, Y = 3$. So

$$\begin{aligned} P(S = 5) &= P(X = 2 \text{ and } Y = 3) + P(X = 3 \text{ and } Y = 2) \\ &= (0.15)(0.4) + (0.25)(0.3) \\ &= 0.06 + 0.075 \\ &= 0.135 \end{aligned}$$

We can construct the probability distribution of S by listing all combinations of X and Y that yield each possible value of S and adding the corresponding probabilities. Here is the result:

Sum	4	5	6	7	8	9	10	11
Probability	0.045	0.135	0.235	0.265	0.190	0.095	0.030	0.005

Using what you learned in Section 4D, you can check that the mean of S is $\mu_S = 6.85$ passengers and the standard deviation of S is $\sigma_S = \sqrt{2.0775} = 1.441$

passengers. As expected, $\mu_S = \mu_X + \mu_Y = 3.75 + 3.10 = 6.85$. Is there a similarly intuitive method for calculating σ_S? Let's see what we can figure out from our previous calculations.

- When we add two independent random variables, their **standard deviations do not add.**

$$\sigma_X + \sigma_Y = 1.0897 + 0.943 = 2.0327 \text{ does not equal } \sigma_S = 1.441$$

- The *variance* of the sum of two independent random variables is the sum of their *variances*.

$$\sigma_X^2 = 1.1875, \sigma_Y^2 = 0.89, \text{ and } \sigma_X^2 + \sigma_Y^2 = 1.1875 + 0.89 = 2.0775 = \sigma_S^2$$

- To find the standard deviation of S, take the square root of the variance:

$$\sigma_S = \sqrt{2.0775} = 1.441$$

Interpretation: Over many, many randomly selected days, the total number of passengers on Pete's and Erin's trips typically varies from the mean of 6.85 passengers by about 1.441 passengers.

The formula $\sigma_S^2 = \sigma_X^2 + \sigma_Y^2$ is sometimes referred to as the "Pythagorean theorem of statistics." It certainly looks similar to $c^2 = a^2 + b^2$! Just as the real Pythagorean theorem applies only to right triangles, the formula $\sigma_S^2 = \sigma_X^2 + \sigma_Y^2$ applies only if X and Y are independent random variables.

Can you guess what the variance of the *difference* of two independent random variables will be? If you were thinking something like "the difference of their variances," think again! Here are the probability distributions of X and Y from the jeep tours scenario once again:

Pete's Jeep Tours

Number of passengers	2	3	4	5	6
Probability	0.15	0.25	0.35	0.20	0.05

$\mu_X = 3.75 \quad \sigma_X = 1.0897$

Erin's Adventures

Number of passengers	2	3	4	5
Probability	0.3	0.4	0.2	0.1

$\mu_Y = 3.10 \quad \sigma_Y = 0.943$

By following the process we used earlier with the random variable $S = X + Y$, you can build the probability distribution of $D = X - Y$.

Difference	−3	−2	−1	0	1	2	3	4
Probability	0.015	0.055	0.145	0.235	0.260	0.195	0.080	0.015

You can use the probability distribution to confirm that

1. $\mu_D = \mu_X - \mu_Y = 3.75 - 3.10 = 0.65$
2. $\sigma_D^2 = \sigma_X^2 + \sigma_Y^2 = 1.1875 + 0.89 = 2.0775$
3. $\sigma_D = \sqrt{2.0775} = 1.441$

Result 2 shows that, just as with addition, when we subtract two independent random variables, variances add. There's more variability in the values of the difference D than in the values of X or Y alone. This should make sense, because the variation in X and the variation in Y both contribute to the variation in D.

STANDARD DEVIATION OF THE SUM OR DIFFERENCE OF TWO INDEPENDENT RANDOM VARIABLES

For any two *independent* random variables X and Y,

- If $S = X + Y$, the variance of S is
$$\sigma_S^2 = \sigma_{X+Y}^2 = \sigma_X^2 + \sigma_Y^2$$
The variance of the sum of two independent random variables is the sum of their variances.
- If $D = X - Y$, the variance of D is
$$\sigma_D^2 = \sigma_{X-Y}^2 = \sigma_X^2 + \sigma_Y^2$$
The variance of the difference of two independent random variables is the sum of their variances.

To get the standard deviation in either case, take the square root of the variance:
$$\sigma_S = \sigma_D = \sqrt{\sigma_X^2 + \sigma_Y^2}$$

You might be wondering whether there's a formula for computing the variance or standard deviation of the sum or difference of two random variables that are *not* independent. There is, but it's beyond the scope of this course.

EXAMPLE

Comparing college costs
Combining random variables: The standard deviation

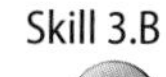
Skill 3.B

PROBLEM: El Dorado Community College has a main campus and a rural campus. The amount spent on tuition by a randomly selected full-time student at the main campus, M, follows a distribution with mean \$732.50 and standard deviation \$102.80. The amount spent on tuition by a randomly selected full-time student at the rural campus, R, follows a distribution with mean \$825 and standard deviation \$126.50. Let $D = M - R$. In the preceding example, we found that $\mu_D = -\$92.50$.

(a) Explain why it is reasonable to view M and R as independent random variables.

(b) Calculate and interpret the standard deviation of D.

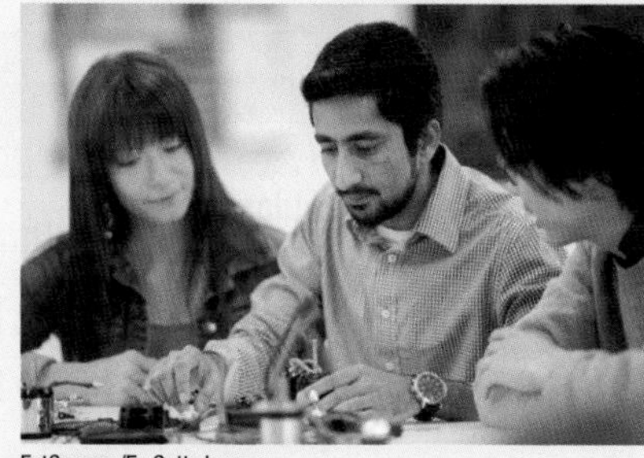
FatCamera/E+Getty Images

SOLUTION:

(a) Knowing the amount M spent on tuition by a randomly selected full-time student at El Dorado Community College's main campus does not help us predict the amount R spent on tuition by a randomly selected full-time student at the rural campus.

(b) Because D is the difference of two independent random variables,

$$\sigma_D^2 = (102.80)^2 + (126.50)^2 = 26{,}570.09$$
$$\sigma_D = \sqrt{26{,}570.09} = \$163$$

> $\sigma_D^2 = \sigma_X^2 + \sigma_Y^2$ if X and Y are independent random variables.

In many, many random selections of a full-time student from each campus, the difference in the amount spent on tuition (Main − Rural) by the two students typically varies from the mean difference of −\$92.50 by about \$163.

FOR PRACTICE, TRY EXERCISE 15

LINEAR COMBINATIONS OF RANDOM VARIABLES: THE STANDARD DEVIATION

Earlier, we defined X = the number of passengers that Pete has and Y = the number of passengers that Erin has on a randomly selected day trip. Recall that $\mu_X = 3.75$, $\sigma_X = 1.0897$, $\mu_Y = 3.10$, and $\sigma_Y = 0.943$. Pete charges \$150 per passenger and Erin charges \$175 per passenger. The total amount collected by Pete and Erin on a randomly selected day is given by $T = 150X + 175Y$. Earlier, we found that the mean of T is $\mu_T = \$1105$. How can we calculate the standard deviation of T?

Let $C = 150X$ represent the amount that Pete collects and $E = 175Y$ represent the amount that Erin collects, so that $T = C + E$. Using the earlier results for linear transformations,

$$\sigma_C = 150\sigma_X = 150(1.0897) = \$163.46 \text{ and } \sigma_E = 175\sigma_Y = 175(0.943) = \$165.03$$

Because C and E are independent random variables,

$$\sigma_T^2 = \sigma_C^2 + \sigma_E^2 = (163.46)^2 + (165.03)^2 = 53{,}954.07$$

$$\sigma_T = \sqrt{53{,}954.07} = \$232.28$$

Interpretation: Over many, many randomly selected days, the total (Pete + Erin) amount collected on their jeep tours typically varies from the mean total of \$1105 by about \$232.28.

Recall that $T = 150X + 175Y$ is a *linear combination* of the random variables X and Y. The discussion here shows that

$$\sigma_T^2 = (150\sigma_X)^2 + (175\sigma_Y)^2 = 150^2\sigma_X^2 + 175^2\sigma_Y^2$$

More generally, if $S = aX + bY$, then $\sigma_S^2 = a^2\sigma_X^2 + b^2\sigma_Y^2$ for independent random variables X and Y.

AP® EXAM TIP
Formula sheet

Unfortunately, *no* formulas for linear combinations are included on the formula sheet provided on both sections of the AP® Statistics exam.

MEAN AND STANDARD DEVIATION OF A LINEAR COMBINATION OF RANDOM VARIABLES

If $aX + bY$ is a linear combination of the random variables X and Y,

- Its mean is $a\mu_X + b\mu_Y$.
- Its standard deviation is $\sqrt{a^2\sigma_X^2 + b^2\sigma_Y^2}$ if X and Y are independent.

Note that these results apply to both discrete and continuous random variables.

COMBINING VERSUS TRANSFORMING RANDOM VARIABLES

In algebra, $x + x = 2x$. Does a similar result hold in statistics? That is, if X is a random variable, does X + X = 2X? Let's investigate with a simple game of rolling dice.

Imagine rolling a fair, six-sided die. Let X = the outcome of the roll. The probability distribution of X is shown in the table. You can check that $\mu_X = 3.5$ and $\sigma_X = 1.708$.

Outcome	1	2	3	4	5	6
Probability	1/6	1/6	1/6	1/6	1/6	1/6

- Suppose that you roll *two* fair, 6-sided dice. The sum of the outcomes on the two dice can be represented by the random variable $T = X_1 + X_2$. Here is a histogram of its probability distribution.

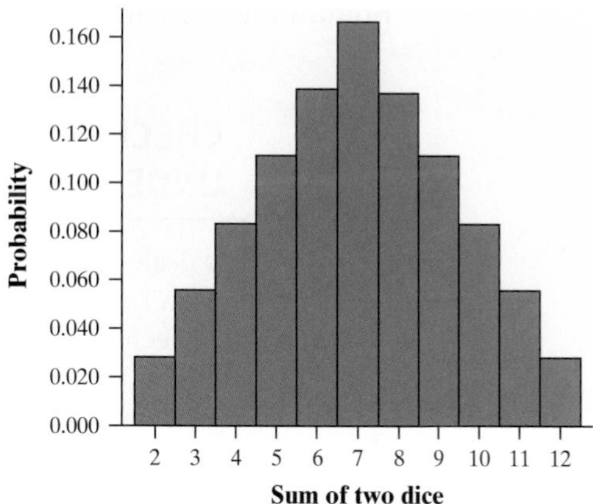

$T = X_1 + X_2$ is a *linear combination* of two random variables. Both X_1 and X_2 have the same probability distribution as X and, therefore, the same mean (3.5) and standard deviation (1.708). The mean of T is

$$\mu_T = \mu_{X_1} + \mu_{X_2} = 3.5 + 3.5 = 7$$

Because knowing the outcome for one die tells us nothing about the outcome for the other die, X_1 and X_2 are independent random variables. As a result,

$$\sigma_T = \sqrt{\sigma_{X_1}^2 + \sigma_{X_2}^2} = \sqrt{(1.708)^2 + (1.708)^2} = 2.415$$

- Now suppose that you roll one fair, 6-sided die and double the outcome. We can represent the result with the random variable $D = 2X$. Here is a histogram of its probability distribution.

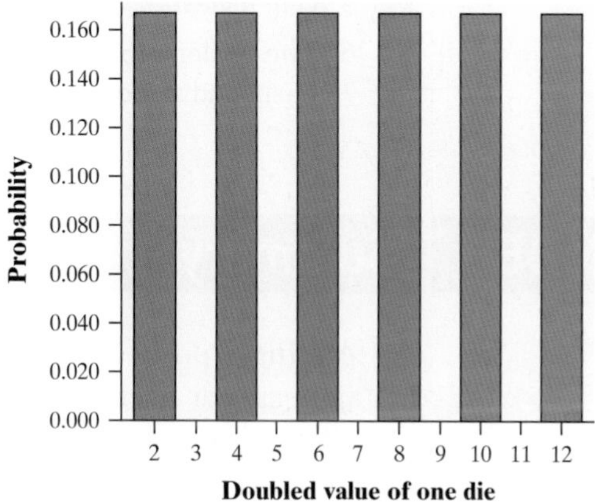

$D = 2X$ is a *linear transformation* of X because we are multiplying the values of the random variable by 2. The mean of D is

$$\mu_D = 2\mu_X = 2(3.5) = 7$$

Notice that this is the same as the mean of $T = X_1 + X_2$. The standard deviation of Y is

$$\sigma_D = 2\sigma_X = 2(1.708) = 3.416$$

This is *larger* than the standard deviation of $T = X_1 + X_2$. Looking at the probability distributions of T and D helps explain why: the values of T tend to be closer to the mean of 7, on average, than do the values of D.

Bottom line: in statistics, $X_1 + X_2$ is not the same as 2X.

CHECK YOUR UNDERSTANDING

A large auto dealership keeps track of sales and lease agreements made during each hour of the day. Let X = the number of cars sold during the first hour of business on a randomly selected Friday. Based on previous records, the probability distribution of X is as follows:

Cars sold, *x*	0	1	2	3
Probability, $P(X = x)$	0.3	0.4	0.2	0.1

$$\mu_X = 1.1 \quad \sigma_X = 0.943$$

Suppose the dealership's manager receives a \$500 bonus for each car sold. Let C = the bonus received from car sales during the first hour on a randomly selected Friday.

1. Write a formula that expresses the random variable C in terms of X. Explain why the probability distributions of C and X have the same shape.
2. Find the mean of C.
3. Calculate and interpret the standard deviation of C.

Let Y = the number of cars leased during the first hour of business on a randomly selected Friday. Based on previous records, the probability distribution of Y has mean 0.7 and standard deviation 0.64. It is safe to assume that X and Y are independent. Define $T = X + Y$.

4. Find and interpret the expected value of T.
5. Calculate σ_T.
6. The dealership's manager receives a \$300 bonus for each car leased. Find the mean and standard deviation of the manager's total bonus B.

SECTION 4E Summary

- Adding a positive constant *a* to or subtracting *a* from a random variable increases or decreases measures of center (mean, median) by *a*, but does not affect measures of variability (range, *IQR*, standard deviation) or the shape of its probability distribution.
- Multiplying or dividing a random variable by a positive constant *b* multiplies or divides measures of center (mean, median) by *b* and multiplies or divides measures of variability (range, *IQR*, standard deviation) by *b*, but does not change the shape of its probability distribution.
- If $Y = a + bX$ is a *linear transformation* of the random variable X with $b > 0$,
  - The probability distribution of Y has the same shape as the probability distribution of X.
  - $\mu_Y = a + b\mu_X$
  - $\sigma_Y = b\sigma_X$

- If X and Y are *any* two random variables,
  - $\mu_{X+Y} = \mu_X + \mu_Y$: the mean of the sum of two random variables is the sum of their means.
  - $\mu_{X-Y} = \mu_X - \mu_Y$: the mean of the difference of two random variables is the difference of their means.
- If X and Y are **independent random variables,** then knowing the value of one variable does not change the probability distribution of the other variable. In that case, variances add:
  - $\sigma^2_{X+Y} = \sigma^2_X + \sigma^2_Y$: the variance of the sum of two independent random variables is the sum of their variances.
  - $\sigma^2_{X-Y} = \sigma^2_X + \sigma^2_Y$: the variance of the difference of two independent random variables is the sum of their variances.
- To get the standard deviation of the sum or difference of two independent random variables, calculate the variance and then take the square root:

$$\sigma_{X+Y} = \sigma_{X-Y} = \sqrt{\sigma^2_X + \sigma^2_Y}$$

- If $aX + bY$ is a *linear combination* of the random variables X and Y,
  - Its mean is $a\mu_X + b\mu_Y$.
  - Its standard deviation is $\sqrt{a^2\sigma^2_X + b^2\sigma^2_Y}$ if X and Y are independent.

AP® EXAM TIP
AP® Daily Videos

Review the content of this section and get extra help by watching the AP® Daily Videos for Topic 4.9, which are available in AP® Classroom.

SECTION 4E Exercises

Transforming Random Variables

1. **Airline overbooking** Airlines typically accept more reservations for a flight than the number of seats on the plane. Suppose that for a certain route, an airline accepts 40 reservations on a plane that carries 38 passengers. Based on experience, the probability distribution of Y = the number of passengers who actually show up for a randomly selected flight with 40 reservations is given in the following table. You can check that $\mu_Y = 37.4$ and $\sigma_Y = 1.24$.

Number of passengers, *y*	35	36	37	38	39	40
Probability, *P*(*y*)	0.10	0.10	0.30	0.35	0.10	0.05

There is also a crew of two flight attendants and two pilots on each flight. Let *T* = the total number of people (passengers plus crew) on a randomly selected flight.

(a) Write a formula that expresses the random variable *T* in terms of Y. Explain why the probability distributions of *T* and Y have the same shape.

(b) Calculate μ_T.

(c) Find and interpret σ_T.

2. **City parking** Victoria parks her car at the same garage every time she goes to work. Because she stays at work for different lengths of time each day, the fee the parking garage charges on a randomly selected day is a random variable, G. The table gives the probability distribution of G. You can check that $\mu_G = \$14$ and $\sigma_G = \$2.74$.

Garage fee, *g*	\$10	\$13	\$15	\$20
Probability, *P*(*g*)	0.20	0.25	0.45	0.10

In addition to the garage's fee, the city charges a \$3 use tax each time Victoria parks her car. Let *M* = the total amount of money she pays on a randomly selected day.

(a) Write a formula that expresses the random variable *M* in terms of G. Explain why the probability distributions of *M* and G have the same shape.

(b) Calculate μ_M.

(c) Find and interpret σ_M.

3. **Back on the boat!** A small ferry runs every half-hour from one side of a large river to the other. The ferry can hold a maximum of 5 cars, and the charge is \$5 per car. The random variable Y = money collected

pg 401

on a randomly selected ferry trip has the probability distribution shown here, with mean $\mu_Y = \$19.35$ and $\sigma_Y = \$6.43$.

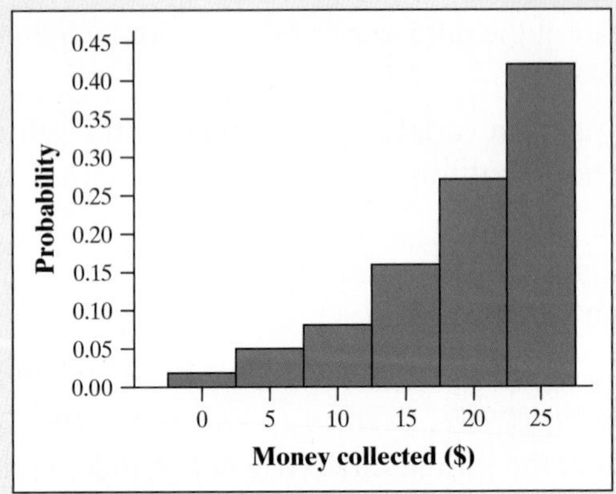

Define X = the number of cars on a randomly selected ferry trip.

(a) What shape does the probability distribution of X have?

(b) Find the mean of X.

(c) Calculate the standard deviation of X.

4. **Skee Ball encore** Ana is a dedicated Skee Ball player who always rolls for the 50-point slot. Ana's score X on a randomly selected roll of the ball has the probability distribution shown here, with mean $\mu_X = 23.8$ and standard deviation $\sigma_X = 12.63$.

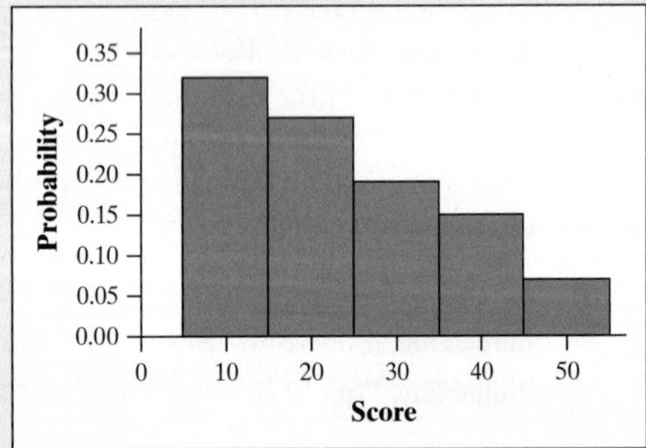

A player receives one ticket from the game for every 10 points scored. Define T = number of tickets Ana gets on a randomly selected roll.

(a) What shape does the probability distribution of T have?

(b) Find the mean of T.

(c) Calculate the standard deviation of T.

Exercises 5 and 6 refer to the following setting. Ms. Hall gave her class a 10-question multiple-choice quiz. Let X = the number of questions that a randomly selected student in the class answered correctly. The computer output gives information about the probability distribution of X. To determine each student's grade on the quiz (out of 100), Ms. Hall will multiply their number of correct answers by 5 and then add 50. Let G = the grade of a randomly chosen student in the class.

Mean	Median	SD	Min	Max	Q_1	Q_3
7.6	8.5	1.32	4	10	8	9

5. **Easy quiz** Find the median and the interquartile range (*IQR*) of G.

6. **More easy quiz** Find the expected value and the range of G.

Combining Random Variables: The Mean

7. **Time and motion** A time-and-motion study measures the time required for an assembly-line worker to perform a repetitive task. The data show that the time X required to bring a part from a bin to its position on an automobile chassis follows a distribution with mean 11 seconds and standard deviation 2 seconds. The time Y required to attach the part to the chassis follows a distribution with mean 20 seconds and standard deviation 4 seconds. Let $T = X + Y$.

(a) Explain what the random variable T represents in this context.

(b) Calculate the expected value of T.

8. **Essay errors** Typographical and spelling errors can be either "nonword errors" or "word errors." A nonword error is not a real word, as when "the" is typed as "teh." A word error is a real word, but not the right word, as when "lose" is typed as "loose." When students are asked to write a 250-word essay on a computer with the spell-check feature turned off, the number of nonword errors X in a randomly selected essay has mean 2.1 and standard deviation 1.136. The number of word errors Y in the essay has mean 1.0 and standard deviation 1.0. Let $S = X + Y$.

(a) Explain what the random variable S represents in this context.

(b) Calculate the expected value of S.

9. **Study habits** pg 404 The Survey of Study Habits and Attitudes (SSHA) is a psychological test that measures academic motivation and study habits. The SSHA score F of a randomly selected first-year student at a large university follows a distribution with mean 105 and standard deviation 35, and the SSHA score S of a randomly selected second-year student at the university follows a distribution with mean 120 and standard deviation 28. Let $D = S - F$. Calculate and interpret the mean of D.

10. **Commuting to work** Sulé's job is just a few bus stops away from his city apartment. While it can be faster to take the bus to work than to walk in the city's underground pedestrian tunnels, the travel time is more

variable due to traffic. The commute time B if Sulé takes the bus to work on a randomly selected day follows a distribution with mean 12 minutes and standard deviation 4 minutes. The commute time W if Sulé walks underground to work on a randomly selected day follows a distribution with mean 16 minutes and standard deviation 1 minute. Let $D = B - W$. Calculate and interpret the mean of D.

Combining Random Variables: The Standard Deviation

11. **Rainy days** Imagine that we randomly select a day from the past 10 years. Let X be the recorded rainfall on this date at the airport in Orlando, Florida, and let Y be the recorded rainfall on this date at Disney World, which is located just outside Orlando. Suppose that you know the means μ_X and μ_Y and the variances σ_X^2 and σ_Y^2 of both variables.

(a) Can we calculate the mean of the total rainfall $X + Y$ to be $\mu_X + \mu_Y$? Explain your answer.

(b) Can we calculate the variance of the total rainfall to be $\sigma_X^2 + \sigma_Y^2$? Explain your answer.

12. **Partner earnings** Researchers randomly select a family in which both parents are employed. Let X be the income of the first parent and Y be the income of the second parent. Suppose that you know the means μ_X and μ_Y and the variances σ_X^2 and σ_Y^2 of both variables.

(a) Can we calculate the mean of the total income $X + Y$ to be $\mu_X + \mu_Y$? Explain your answer.

(b) Can we calculate the variance of the total income to be $\sigma_X^2 + \sigma_Y^2$? Explain your answer.

13. **More time and motion** Refer to Exercise 7. The study finds that the times required for the two steps are independent. Find the standard deviation of T.

14. **More essay errors** Refer to Exercise 8. Assume that the number of word and nonword errors in a randomly selected essay are independent. Find the standard deviation of S.

15. **More study habits** Refer to Exercise 9.

pg 407

(a) Explain why it is reasonable to view the SSHA scores of a randomly selected first-year student and a randomly selected second-year student at the university as independent random variables.

(b) Calculate and interpret the standard deviation of the probability distribution of D.

16. **More commuting** Refer to Exercise 10.

(a) Explain why it is reasonable to view Sulé's commute times to work by bus and by walking underground on a randomly selected day as independent random variables.

(b) Calculate and interpret the standard deviation of the probability distribution of D.

17. **Linear combination** The random variable X has mean 20 and standard deviation 4. The random variable Y has mean 15 and standard deviation 3. If X and Y are independent random variables, find the mean and standard deviation of $2X + 3Y$.

18. **Another linear combination** The random variable X has mean 50 and standard deviation 12. The random variable Y has mean 15 and standard deviation 5. If X and Y are independent random variables, find the mean and standard deviation of $3X - 4Y$.

19. **Swim team** Hanover High School has the best women's swimming team in the region. The 400-meter freestyle relay team is undefeated this year. In the 400-meter freestyle relay, each swimmer swims 100 meters. The times, in seconds, for the four swimmers this season have the means and standard deviations shown in the table. Assume that the four swimmers' individual times in a given race are independent.

Swimmer	Mean	SD
Wendy	55.2	2.8
Jill	58.0	3.0
Carmen	56.3	2.6
Latrice	54.7	2.7

Find the mean and standard deviation of the total team time in the 400-meter freestyle relay for a randomly selected race.

20. **Toothpaste** Ken is traveling for his business. He has a new 0.85-ounce tube of toothpaste that's supposed to last him the whole trip. The amount of toothpaste Ken squeezes out of the tube each time he brushes can be modeled by a distribution with mean 0.13 ounce and standard deviation 0.02 ounce. If Ken brushes his teeth six times on a randomly selected trip and the amounts of toothpaste used each time are independent, find the mean and standard deviation of the total amount of toothpaste used.

21. **Gift box** A company offers gift boxes containing 4 pears and 5 chocolate truffles with an advertised weight of 800 grams. The weights of individual pears can be modeled by a distribution with mean 188 grams and standard deviation 6 grams. The weights of individual chocolate truffles can be modeled by a distribution with mean 12.5 grams and standard deviation 0.7 gram. The weight of a box can be modeled by a distribution with mean 20 grams and standard deviation 0.9 gram. Assume that the weights of individual pears, chocolate truffles, and boxes are independent. Calculate the mean and standard deviation of the total weight of a gift box.

For Investigation *Apply the skills from the section in a new context or nonroutine way.*

22. **Golf balls** A manufacturer offers a package of golf balls that consists of 1 randomly selected empty box and 15 randomly selected golf balls. The weights of such packages can be modeled by a distribution with a mean of 30 ounces and a standard deviation of 0.25 ounce. The weights of the empty boxes have a mean of 5.85 ounces and a standard deviation of 0.2 ounce.

(a) Explain why it is reasonable to assume that the weights of the individual golf balls are independent, and that the weights of the empty boxes and the weights of the golf balls are independent.

Let the random variable X be the weight of a single randomly selected golf ball.

(b) What is the mean of X?

(c) What is the standard deviation of X?

23. **Life insurance** A life insurance company sells a term insurance policy to 21-year-olds that pays $100,000 if the insured dies within the next 5 years. The probability distribution of the random variable X = the amount the company earns on a randomly chosen 5-year term life policy has mean $\mu_X = \$303.35$ and $\sigma_X = \$9707.57$.

(a) Suppose that we randomly select two insured 21-year-olds, and that their ages at death are independent. If X_1 and X_2 are the insurer's income from the two insurance policies, the insurer's average income $\bar{x}$ on the two policies is $\bar{x} = \dfrac{X_1 + X_2}{2}$. Find the mean and standard deviation of $\bar{x}$.

(b) If we randomly select four insured 21-year-olds, the insurer's average income is $\bar{x} = \dfrac{X_1 + X_2 + X_3 + X_4}{4}$ where X_i is the income from insuring one 21-year-old. Assuming that the amount of income earned on individual policies is independent, find the mean and standard deviation of $\bar{x}$.

(c) Use your results from parts (a) and (b) to explain why averaging over more insured individuals reduces risk.

(d) If we randomly select n insured 21-year-olds, the insurer's average income is

$$\bar{x} = \frac{X_1 + X_2 + \ldots + X_n}{n}$$

where X_i is the income from insuring one 21-year-old. Assuming that the amount of income earned on individual policies is independent, find the mean and standard deviation of $\bar{x}$.

Multiple Choice *Select the best answer for each question.*

Exercises 24 and 25 refer to the following setting. The number of calories in 1 serving of a certain breakfast cereal is a random variable with mean 110 and standard deviation 10. The number of calories in a cup of whole milk is a random variable with mean 140 and standard deviation 12. For breakfast, you eat 1 serving of the cereal with ½ cup of whole milk. Let T be the random variable that represents the total number of calories in this breakfast.

24. The mean of T is

(A) 110.
(B) 140.
(C) 180.
(D) 195.
(E) 250.

25. The standard deviation of T is

(A) 22.
(B) 16.
(C) 15.62.
(D) 11.66.
(E) 4.

Recycle and Review *Practice what you learned in previous sections.*

26. **Fluoride varnish (3C)** In an experiment to measure the effect of fluoride "varnish" on the incidence of tooth cavities, thirty-four 10-year-old girls whose parents volunteered them for the study were randomly assigned to two groups. One group was given fluoride varnish annually for 4 years, along with standard dental hygiene; the other group followed only the standard dental hygiene regimen. The mean number of cavities in the two groups was compared at the end of the 4 years.

(a) Are the participants in this experiment subject to the placebo effect? Explain your answer.

(b) Describe how you could alter this experiment to make it double-blind.

(c) Explain the purpose of the random assignment in this experiment.

27. **Buying stock (4C, 4D)** You purchase a hot stock for $1000. The stock either gains 30% or loses 25% each day, each with probability 0.5. Its returns on consecutive days are independent of each other. You plan to sell the stock after two days.

(a) What are the possible values of the stock after two days, and what is the probability for each value?

(b) What is the probability that the stock is worth more after two days than the $1000 you paid for it?

(c) What is the mean value of the stock after two days?

Note: The two criteria in parts (b) and (c) give different answers to the question, "Should I invest?"

SECTION 4F Binomial and Geometric Random Variables

LEARNING TARGETS *By the end of the section, you should be able to:*

- Determine whether a random variable has a binomial distribution.
- Calculate and interpret probabilities involving binomial random variables.
- Find the mean and standard deviation of a binomial distribution. Interpret these values.
- Calculate and interpret probabilities involving geometric random variables.
- Find the mean and standard deviation of a geometric distribution. Interpret these values.

When the same random process is repeated several times, we are often interested in whether a particular outcome does or doesn't happen on each trial. Here are two examples:

- A shipping company claims that 90% of its shipments arrive on time. To test this claim, we select a random sample of 100 shipments made by the company last month and see how many arrived on time. *Random process*: randomly select a shipment and check when it arrived. *Outcome of interest*: arrived on time. *Random variable*: Y = number of on-time shipments.
- In the game of Pass the Pigs, a player rolls a pair of pig-shaped dice. On each roll, the player earns points according to how the pigs land. If the player gets a "pig out," in which the two pigs land on opposite sides, they lose all points earned in that round and must pass the pigs to the next player. A player can choose to stop rolling at any point during their turn and to keep the points they have earned before passing the pigs. *Random process*: roll the pig dice. *Outcome of interest*: pig out. *Random variable*: T = number of rolls it takes the player to pig out.

Alex Clark/Alamy Stock Photo

Some random variables, like Y in the first example, count the number of times the outcome of interest occurs in a fixed number of trials. They are called *binomial random variables*. Other random variables, like T in the Pass the Pigs scenario, count the number of trials of the random process it takes for the outcome of interest to occur. They are known as *geometric random variables*. These two special types of discrete random variables are the focus of this section.

Binomial Random Variables

Let's start with an activity that involves repeating a random process several times.

ACTIVITY

Pop quiz!

It's time for a pop quiz! We hope you are ready. The quiz consists of 10 multiple-choice questions. Each question has five answer choices, labeled A through E. Now for the bad news: you will not get to see the questions. You just have to guess the answer for each one!

1. Get out a blank sheet of paper. Write your name at the top. Number your paper from 1 to 10. Then guess the answer to each question: A, B, C, D, or E. Do not look at anyone else's paper! You have 2 minutes.
2. Now it's time to grade the quizzes. Exchange papers with a classmate. Your teacher will display the answer key. The correct answer for each of the 10 questions was determined randomly so that A, B, C, D, or E was equally likely to be chosen.
3. How did you do on your quiz? Make a class dotplot that shows the number of correct answers for each student in your class. As a class, describe what you see.

In the Pop Quiz activity, each student is performing repeated *trials* of the same random process: guessing the answer to a multiple-choice question with a randomly generated correct answer. We're interested in the number of times that a specific event occurs: getting a correct answer (which we'll call a "success"). Knowing the outcome of one question (right or wrong answer) tells us nothing about the outcome of any other question. That is, the trials are independent. The number of trials is fixed in advance: $n = 10$. Also, a student's probability of getting a "success" is the same on each trial: $p = 1/5 = 0.2$. When these conditions are met, we have a **binomial setting.**

DEFINITION Binomial setting

A **binomial setting** arises when we perform n independent trials of the same random process and count the number of times that a particular outcome (called a "success") occurs.

The four conditions for a binomial setting are:

- **B**inary? The possible outcomes of each trial can be classified as "success" or "failure."
- **I**ndependent? Trials must be independent. That is, knowing the outcome of one trial must not tell us anything about the outcome of any other trial.
- **N**umber? The number of trials n of the random process must be fixed in advance.
- **S**ame probability? There is the same probability of success p on each trial.

The boldface letters in the definition box give you a helpful way to remember the conditions for a binomial setting: just check the BINS!

When checking the binary condition, note that there can be more than two possible outcomes per trial. In the Pop Quiz activity, each question (trial) had five possible answer choices: A, B, C, D, and E. If we define "success" as guessing the correct answer to a question, then "failure" occurs when the student guesses any of the four incorrect answer choices.

If we let X = the number of correct answers that a student gets on the pop quiz, then X is a **binomial random variable** with $n = 10$ and $p = 0.2$. The probability distribution of X is called a **binomial distribution.**

DEFINITION Binomial random variable, Binomial distribution

The count of successes X in a binomial setting is a **binomial random variable.** The possible values of X are 0, 1, 2, ..., n.

The probability distribution of X is a **binomial distribution.** Any binomial distribution is completely specified by two numbers: the number of trials n of the random process and the probability p of success on each trial.

Because X can take only whole-number values from 0 to n, binomial random variables are discrete random variables. It is important for you to be able to determine whether or not a given random variable has a binomial distribution.

EXAMPLE

From blood types to aces
Binomial random variables

Skill 3.A

PROBLEM: Determine whether the given random variable is a binomial random variable. If so, state its probability distribution.

tam_odin/Shutterstock

(a) Genetics says that the genes children receive from their birth parents are independent from one child to another. Each child of a particular set of birth parents has probability 0.25 of having type O blood. Suppose these parents have 5 children. Let X = the number of children with type O blood.

(b) Start with a well-shuffled deck of 52 playing cards. Imagine that you turn over the first 10 cards, one at a time. Let Y = the number of aces you observe.

(c) Start with a well-shuffled deck of 52 playing cards. Imagine that you turn over the top card, look at it, put the card back in the deck, and shuffle again. Repeat this process until you get an ace. Let W = the number of cards you had to turn over.

SOLUTION:

(a)
- Binary? "Success" = has type O blood. "Failure" = doesn't have type O blood.
- Independent? Knowing one child's blood type tells you nothing about another child's blood type because they inherit genes independently from their birth parents.
- Number? $n = 5$
- Same probability? $p = 0.25$

This is a binomial setting, and X is counting the number of successes (children with type O blood). So X is a binomial random variable with $n = 5$ and $p = 0.25$.

> Check the BINS! A trial consists of observing the blood type for one of these parents' children.

(b)
- Binary? "Success" = get an ace. "Failure" = don't get an ace.
- Independent? No. If the first card you turn over is an ace, then the next card is less likely to be an ace because you're not replacing the top card in the deck. If the first card isn't an ace, the second card is more likely to be an ace.

This is not a binomial setting, so Y is not a binomial random variable.

> Check the BINS! A trial consists of turning over a card from the deck and observing what's on the card.

> To check for independence, you could also write
> P(2nd card ace | 1st card ace) = 3/51 and
> P(2nd card ace | 1st card not ace) = 4/51
> Because the two probabilities are not equal, the trials are not independent.

(c) • Binary? "Success" = get an ace. "Failure" = don't get an ace.
- Independent? Yes. Because you are replacing the card in the deck and shuffling each time, the result of one trial doesn't tell you anything about the outcome of any other trial.
- Number? No. The number of trials is not fixed in advance.

Because there is no fixed number of trials, this is not a binomial setting. Also, you are counting the number of trials until you get a success, not the number of successes. So W is not a binomial random variable.

W is actually a *geometric* random variable. You will learn more about geometric random variables shortly.

FOR PRACTICE, TRY EXERCISE 1

What's the difference between the Independent condition and the Same probability condition? The Independent condition involves *conditional* probabilities. In part (b) of the example,

$$P(\text{2nd card ace} \mid \text{1st card ace}) = 3/51 \neq P(\text{2nd card ace} \mid \text{1st card not ace}) = 4/51$$

so the trials are not independent. The Same probability of success condition is about *unconditional* probabilities. Because

$$P(x\text{th card in a shuffled deck is an ace}) = 4/52$$

this condition is met in part (b) of the example. Be sure you understand the difference between these two conditions. When sampling is done without replacement, the Independent condition is violated. The Same probability condition would be violated if someone added a few aces to the deck between two of the trials.

We can use the simulation methods from Section 4A to estimate probabilities involving binomial random variables. For example, what's the probability that a student gets 4 or more correct answers on the 10 multiple-choice questions in the Pop Quiz activity? The dotplot shows the simulated distribution of the number of correct answers on the pop quiz by a group of 100 students. Because there are 11 students who got 4 or more correct answers, the estimated probability is $P(X \geq 4) = 11/100 = 0.11$.

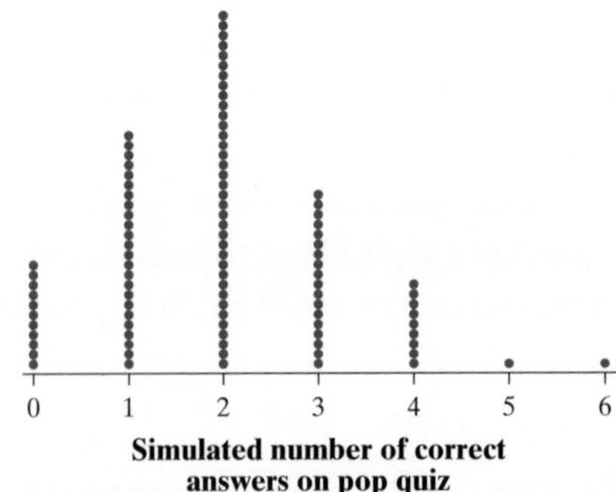

We will soon develop other ways to obtain more accurate binomial probabilities.

Calculating Binomial Probabilities

How can we calculate probabilities involving binomial random variables? Let's return to the scenario from part (a) of the preceding example:

> Genetics says that the genes children receive from their birth parents are independent from one child to another. Each child of a particular set of birth parents has probability 0.25 of having type O blood. Suppose these parents have 5 children. Let X = the number of children with type O blood.

In this binomial setting, a child with type O blood is a "success" (S) and a child with another blood type is a "failure" (F). The count X of children with type O blood is a binomial random variable with $n = 5$ trials and probability of success $p = 0.25$ on each trial.

- What's $P(X = 0)$? That is, what's the probability that *none* of the 5 children has type O blood? The probability that any one of this couple's children doesn't have type O blood is $1 - 0.25 = 0.75$ (complement rule). By the multiplication rule for independent events (Section 4C),

$$P(X = 0) = P(\text{FFFFF}) = (0.75)(0.75)(0.75)(0.75)(0.75) = (0.75)^5 = 0.23730$$

- How about $P(X = 1)$? There are several different ways in which exactly 1 of the 5 children could have type O blood. For instance, the first child born might have type O blood, while the remaining 4 children don't have type O blood. The probability that this happens is

$$P(\text{SFFFF}) = (0.25)(0.75)(0.75)(0.75)(0.75) = (0.25)^1(0.75)^4$$

Alternatively, Child 2 could be the one who has type O blood. The corresponding probability is

$$P(\text{FSFFF}) = (0.75)(0.25)(0.75)(0.75)(0.75) = (0.25)^1(0.75)^4$$

There are three more possibilities to consider — the situations in which Child 3, Child 4, and Child 5 are the only ones to inherit type O blood. Of course, the probability will be the same for each of those cases. In all, there are five different ways in which exactly 1 child would have type O blood, each with the same probability of occurring. As a result,

$$\begin{aligned} P(X = 1) &= P(\text{exactly 1 child with type O blood}) \\ &= 5(0.25)^1\,(0.75)^4 = 0.39551 \end{aligned}$$

Number of ways to get 1 child out of 5 with type O blood

1 child with type O blood

4 children don't have type O blood

The pattern of this calculation works for any binomial probability:

$$P(X = x) = (\text{number of ways to get } x \text{ successes in } n \text{ trials})(\text{success probability})^x(\text{failure probability})^{n-x}$$

To use this formula, we must count the number of arrangements of x successes in n trials. This number is called the **binomial coefficient.** We use the following fact to do the counting without actually listing all the arrangements.

DEFINITION Binomial coefficient

The number of ways to arrange x successes among n trials is given by the **binomial coefficient**

$$\binom{n}{x} = \frac{n!}{x!(n-x)!}$$

for $x = 0, 1, 2, \ldots, n$, where $n!$ (read as "n factorial") is given by

$$n! = n(n-1)(n-2)\cdot\ldots\cdot(3)(2)(1)$$

and $0! = 1$.

The larger of the two factorials in the denominator of a binomial coefficient will cancel much of the $n!$ in the numerator. For example, the binomial coefficient we need to find the probability that exactly 2 of the couple's 5 children inherit type O blood is

$$\binom{5}{2} = \frac{5!}{2!3!} = \frac{(5)(4)(\cancel{3})(\cancel{2})(\cancel{1})}{(2)(1)(\cancel{3})(\cancel{2})(\cancel{1})} = \frac{(5)(4)}{(2)(1)} = 10$$

The binomial coefficient $\binom{5}{2}$ is not related to the fraction $\frac{5}{2}$. A helpful way to remember its meaning is to read it as "5 choose 2" — as in, how many ways are there to choose which 2 children have type O blood in a family with 5 children? Binomial coefficients have many uses, but we are interested in them only as an aid to finding binomial probabilities. If you need to compute a binomial coefficient, use your calculator. *Note:* Some people and some technology tools use the notation ${}_5C_2$ instead of $\binom{5}{2}$ for the binomial coefficient.

12. Tech Corner CALCULATING BINOMIAL COEFFICIENTS

TI-Nspire and other technology instructions are on the book's website at bfwpub.com/tps7e.

To calculate a binomial coefficient like $\binom{5}{2}$ on the TI-83/84, proceed as follows:

- Type 5, press MATH, arrow over to PROB, choose nCr, and press ENTER. Then type 2 and press ENTER again to execute the command 5 nCr 2 (which displays as ${}_5C_2$ on devices with math print).

```
NORMAL FLOAT AUTO REAL RADIAN MP
5C2
                                10
```

The binomial coefficient $\binom{n}{x}$ counts the number of different ways in which x successes can be arranged among n trials. The binomial probability $P(X = x)$ is this count multiplied by the probability of any one specific arrangement of the x successes.

BINOMIAL PROBABILITY FORMULA

Suppose that X is a binomial random variable with n trials and probability p of success on each trial. The probability of getting exactly x successes in n trials ($x = 0, 1, 2, \ldots, n$) is

$$P(X = x) = \binom{n}{x} p^x (1-p)^{n-x}$$

where

$$\binom{n}{x} = \frac{n!}{x!(n-x)!}$$

AP® EXAM TIP
Formula sheet

The formula for a binomial probability $P(X = x)$ is included on the formula sheet provided on both sections of the AP® Statistics exam.

A table, a graph, or a function (like the binomial probability formula) can be used to specify the probability distribution of a random variable. With our formula in hand, we can now calculate a binomial probability.

EXAMPLE

How many correct on the pop quiz?
Calculating binomial probabilities

Skill 3.A

PROBLEM: To introduce a statistics class to binomial distributions, Mr. Desai does the Pop Quiz activity, in which each student in the class guesses an answer from A through E on each of 10 multiple-choice questions. Mr. Desai determined the "correct" answer for each of the 10 questions randomly so that A, B, C, D, or E was equally likely to be chosen. We are interested in the probability that a student gets exactly 4 correct answers on the pop quiz.

Juan Moyano/Moment/Getty Images

(a) Define the random variable of interest, state its probability distribution, and identify the value(s) of interest.

(b) Determine the probability that a student gets exactly 4 correct answers on the pop quiz.

SOLUTION:

(a) Let $X =$ the number of correct answers a student gets on the pop quiz.
X is a binomial random variable with $n = 10$ and $p = 1/5 = 0.2$.
We want to find $P(X = 4)$.

(b) $$P(X=4)=\binom{10}{4}(0.2)^4(0.8)^6$$
$$=210(0.2)^4(0.8)^6$$
$$=0.0881$$

$$P(X = x)=\binom{n}{x}p^x(1-p)^{n-x}$$

The probability that a student gets exactly 4 correct answers on the pop quiz is 0.0881.

FOR PRACTICE, TRY EXERCISE 5

Did you notice that the two parts of the preceding example were similar to the two-step process for performing normal distribution calculations in Section 1F? Here is a revised version of the process that you should follow when calculating probabilities involving binomial, geometric, and normal distributions.

HOW TO CALCULATE PROBABILITIES INVOLVING BINOMIAL, GEOMETRIC, AND NORMAL DISTRIBUTIONS

Step 1: Define the random variable of interest, state how it is distributed, and identify the value(s) of interest.

Step 2: Perform calculations — show your work!

Be sure to answer the question that was asked.

Sometimes we want to calculate a probability involving more than one value of a binomial random variable. As the following example illustrates, we can just use the binomial probability formula several times.

EXAMPLE

Did Hannah cheat on the pop quiz?
Calculating binomial probabilities

Skills 3.A, 4.B

Klaus Vedfelt/DigitalVision/Getty Images

PROBLEM: Refer to the preceding example. Hannah is one of the students in Mr. Desai's class. After taking the pop quiz, Hannah is surprised because she got 4 correct answers. Find the probability that a student would get 4 or more correct answers on the quiz purely by chance. Should Hannah be surprised?

SOLUTION:

Let X = the number of correct answers a student gets on the pop quiz. X is a binomial random variable with $n = 10$ and $p = 1/5 = 0.2$. We want to find $P(X \geq 4)$.

> **Step 1:** Define the random variable of interest, state how it is distributed, and identify the value(s) of interest.

$$P(X \geq 4) = 1 - P(X \leq 3)$$
$$= 1 - [P(X=0) + P(X=1) + P(X=2) + P(X=3)]$$
$$= 1 - \left[\binom{10}{0}(0.2)^0(0.8)^{10} + \binom{10}{1}(0.2)^1(0.8)^9 + \binom{10}{2}(0.2)^2(0.8)^8 + \binom{10}{3}(0.2)^3(0.8)^7\right]$$
$$= 1 - [0.1074 + 0.2684 + 0.3020 + 0.2013]$$
$$= 1 - 0.8791$$
$$= 0.1209$$

> **Step 2:** Perform calculations—show your work!

> You could calculate $P(X \geq 4) = P(X = 4) + P(X = 5) + \ldots + P(X = 10)$. But that would require 7 separate uses of the binomial formula! It is a lot easier to use the complement rule. $P(X \geq 4) = 1 - P(X < 4) = 1 - P(X \leq 3)$. Then you have to use the binomial formula only four times, for $X = 0, 1, 2$, and 3.

Because the probability of getting 4 or more correct answers by chance alone is 0.1209, which is not small (not less than 5%), Hannah should not be surprised by her quiz result.

> Be sure to answer the question that was asked.

FOR PRACTICE, TRY EXERCISE 9

We can also use technology to perform binomial probability calculations. The following Tech Corner shows how to do it. But be sure to read the AP® Exam Tip at the end of the Tech Corner.

13. Tech Corner CALCULATING BINOMIAL PROBABILITIES

TI-Nspire and other technology instructions are on the book's website at bfwpub.com/tps7e.

There are two handy commands on the TI-83/84 for finding binomial probabilities: binompdf and binomcdf. The inputs for both commands are the number of trials n, the success probability p, and the value(s) of interest for the binomial random variable X.

binompdf(n, p, x) computes $P(X = x)$
binomcdf(n, p, x) computes $P(X \leq x)$

Let's use these commands to confirm our answers in the previous two examples.

1. To find $P(X = 4)$:
   - Press 2nd VARS (DISTR) and choose binompdf(.

   OS 2.55 or later: In the dialog box, enter these values: trials: 10, p: 0.2, x value: 4. Choose Paste, and then press ENTER.

   Older OS: Complete the command binompdf(10,0.2,4) and press ENTER.

```
NORMAL FLOAT AUTO REAL RADIAN MP
binompdf(10,.2,4)
                 0.088080384
```

These results agree with our previous answer using the binomial probability formula: 0.0881.

2. To find $P(X \geq 4)$, use the complement rule:

$$P(X \geq 4) = 1 - P(X \leq 3) = 1 - \text{binomcdf(trials: 10, p: 0.2, x value: 3)}$$

   - Press 2nd VARS (DISTR) and choose binomcdf(.

   OS 2.55 or later: In the dialog box, enter these values: trials: 10, p: 0.2, x value: 3. Choose Paste, and then press ENTER. Subtract this result from 1 to get the answer.

   Older OS: Complete the command binomcdf(10,0.2,3) and press ENTER. Subtract this result from 1 to get the answer.

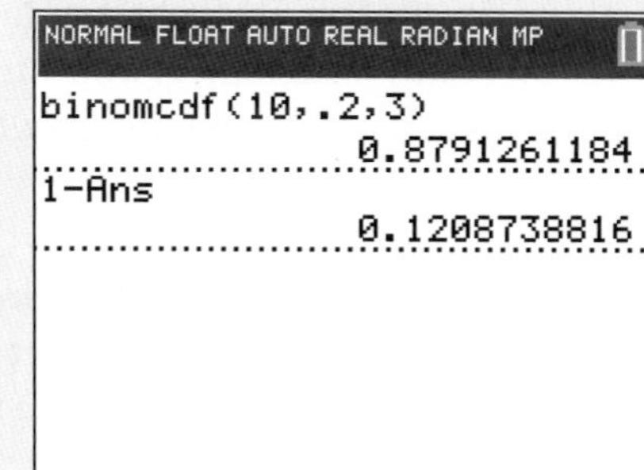

This result agrees with our previous answer using the binomial probability formula: 0.1209.

AP® EXAM TIP

Don't rely on unlabeled "calculator speak" when showing your work on free response questions. Writing only binompdf(10,0.2,4) = 0.088 will *not* earn you full credit for a binomial probability calculation. You can earn some credit by indicating what each of those calculator inputs represents. For example, "binompdf (trials: 10, p: 0.2, x value: 4) = 0.088." To give yourself the best chance to earn full credit, follow the two-step process!

Note the use of the complement rule to find $P(X \geq 4)$ in the preceding Tech Corner: $P(X \geq 4) = 1 - P(X \leq 3)$. This is necessary because the calculator's binomcdf(n,p,x) command computes the probability of getting *x or fewer* successes in *n* trials. Remember that a *cumulative probability distribution* typically gives $P(X \leq x)$.

Students often have trouble identifying the correct third input for the binomcdf command when a question asks them to find the probability of getting less than, more than, or at least so many successes. Here's a helpful tip to avoid making such a mistake: write out the possible values of the variable, circle the ones you want to find the probability of, and cross out the rest. In the preceding example, X can take values from 0 to 10 and we want to find $P(X \geq 4)$:

$$\not{0}\ \not{1}\ \not{2}\ \not{3}\ \boxed{4\ 5\ 6\ 7\ 8\ 9\ 10}$$

Crossing out the values from 0 to 3 shows why the correct calculation is $1 - P(X \leq 3)$.

EXAMPLE

Free lunch?
Calculating binomial probabilities

Skills 3.A, 4.B

PROBLEM: A local fast-food restaurant is running a "Draw a three, get it free" lunch promotion. After each customer orders, a touchscreen display shows the message "Press here to win a free lunch." A computer program then simulates one card being drawn from a standard deck. If the chosen card is a 3, the customer's order is free. Otherwise, the customer must pay the bill. On the first day of the promotion, 250 customers place lunch orders.

(a) Find the probability that fewer than 10 of the 250 customers win a free lunch.

(b) In fact, only 9 customers won a free lunch. Does this result give convincing evidence that the computer program is flawed? Explain your reasoning.

eldadcarin/Getty Images

SOLUTION:

(a) Let Y = the number of customers who win a free lunch. Y has a binomial distribution with $n = 250$ and $p = 4/52$. We want to find $P(Y < 10)$.

> **Step 1:** Define the random variable of interest, state how it is distributed, and identify the value(s) of interest.

$P(Y < 10) = P(Y \leq 9)$

$= \text{binomcdf(trials: 250, p: 4/52, x value: 9)}$

$= 0.00613$

> **Step 2:** Perform calculations—show your work!
> The values of Y that interest us are
>
> $\boxed{0\ 1\ 2\ 3\ 4\ 5\ 6\ 7\ 8\ 9}$ 10 11 12 . . . 250
>
> To use the binomial probability formula, we would have to add the probabilities for $Y = 0, 1, \ldots, 9$. That's too much work!

(b) There is only about a 0.006 probability that fewer than 10 customers would win a free lunch if the computer program is working properly. Because only 9 customers won a free lunch on this day, we have convincing evidence that the computer program is flawed.

> Be sure to answer the question that was asked.

FOR PRACTICE, TRY EXERCISE 13

CHECK YOUR UNDERSTANDING

As a special promotion for its 20-ounce bottles of soda, a soft drink company printed a message on the inside of each bottle cap. Some of the caps said, "Please try again!" while others said, "You're a winner!" The company advertised the promotion with the slogan "1 in 6 wins a prize." Grayson's statistics class wonders if the company's claim holds true at a nearby convenience store. To find out, all 30 students in the class go to the store and each buys one randomly selected 20-ounce bottle of the soda. Let X = the number of prize winners out of 30 if the company's claim is true.

1. Confirm that X is a binomial random variable, and state its probability distribution.
2. Find $P(X = 3)$. Describe this probability in words.
3. Find the probability that three or fewer students would win a prize if the company's claim is true.
4. Three of the students in Grayson's class got caps that say, "You're a winner!" Does this result give convincing evidence that the company's 1-in-6 claim is false? Explain your reasoning.

Describing a Binomial Distribution: Shape, Center, and Variability

What does the probability distribution of a binomial random variable look like? Let's return to the birth parents with 5 children. Recall that X = the number of children with type O blood. We determined that X is a binomial random variable with $n = 5$ and $p = 0.25$. Its probability distribution is shown in the table.

Number of children with type O blood, *x*	0	1	2	3	4	5
Probability, *P*(*x*)	0.23730	0.39551	0.26367	0.08789	0.01465	0.00098

Figure 4.10 shows a histogram of the probability distribution. The binomial distribution with $n = 5$ and $p = 0.25$ has a clear right-skewed shape. Why? Because the probability that any one of the couple's children inherits type O blood is 0.25, it's quite likely that 0, 1, or 2 of the children will have type O blood. Larger values of X are much less likely.

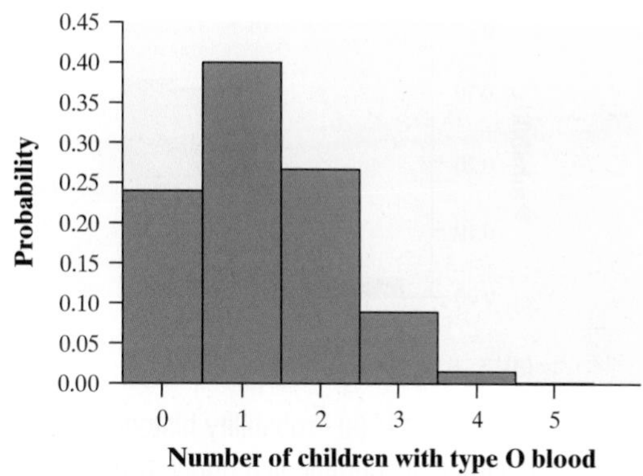

FIGURE 4.10 Histogram showing the probability distribution of the binomial random variable X = number of children with type O blood in a family with 5 children from the same birth parents.

You can use technology to graph a binomial probability distribution like the one shown in Figure 4.10.

14. Tech Corner GRAPHING BINOMIAL PROBABILITY DISTRIBUTIONS

TI-Nspire and other technology instructions are on the book's website at bfwpub.com/tps7e.

To graph the binomial probability distribution with $n = 5$ and $p = 0.25$:

- Type the possible values of the random variable X into list L1: 0, 1, 2, 3, 4, 5.
- Highlight L2 with your cursor. Enter the command binompdf(5,0.25) and press ENTER.
- Make a histogram of the probability distribution using the method shown in Tech Corner 11 (page 390).

NORMAL FLOAT AUTO REAL RADIAN MP

L1	L2	L3	L4	L5
0	0.2373	-------	-------	-------
1	0.3955			
2	0.2637			
3	0.0879			
4	0.0146			
5	9.8E-4			
-------	-------			

L2=binompdf(5,0.25,L1)

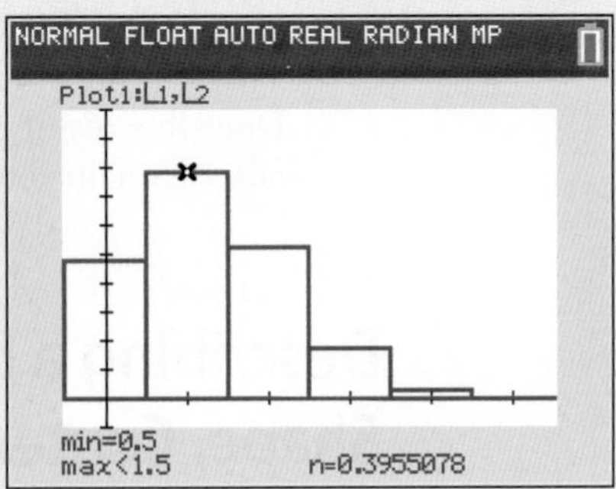

The binomial distribution with $n = 5$ and $p = 0.25$ is skewed to the right. Figure 4.11 shows two more binomial distributions with different shapes. The binomial distribution with $n = 5$ and $p = 0.51$ in Figure 4.11(a) is roughly symmetric. The binomial distribution with $n = 5$ and $p = 0.8$ in Figure 4.11(b) is skewed to the left. In general, when n is small, the probability distribution of a binomial random variable will be roughly symmetric if p is close to 0.5, right-skewed if p is much less than 0.5, and left-skewed if p is much greater than 0.5.

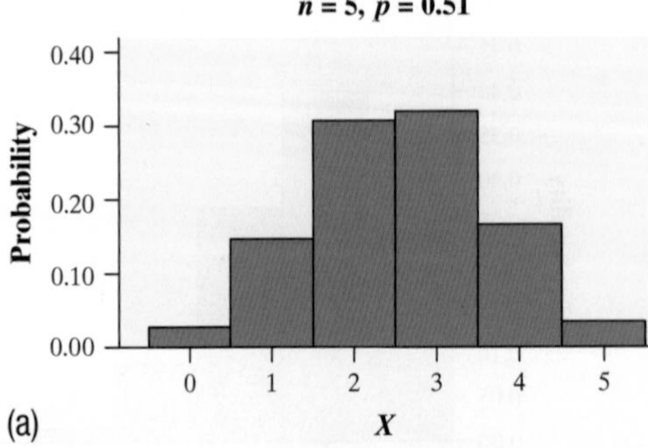

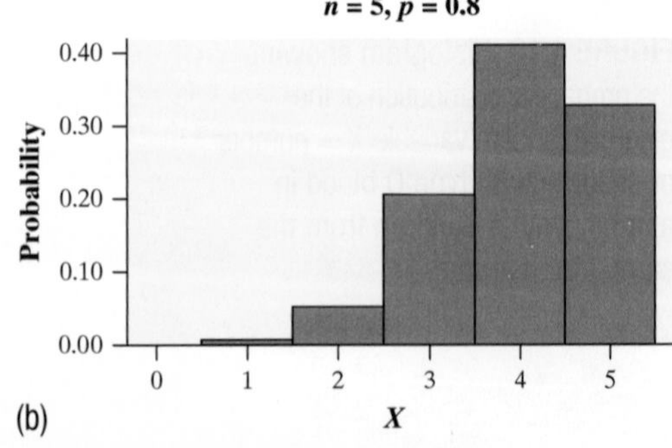

FIGURE 4.11 (a) Probability histogram for the binomial random variable X with $n = 5$ and $p = 0.51$. This binomial distribution is roughly symmetric. (b) Probability histogram for the binomial random variable X with $n = 5$ and $p = 0.8$. This binomial distribution has a left-skewed shape.

MEAN AND STANDARD DEVIATION OF A BINOMIAL RANDOM VARIABLE

Let's return to the birth parents with 5 children one more time. The random variable X = the number of children with type O blood has a binomial distribution with $n = 5$ and $p = 0.25$. Its probability distribution is shown in the table.

Number of children with type O blood, *x*	0	1	2	3	4	5
Probability, *P*(*x*)	0.23730	0.39551	0.26367	0.08789	0.01465	0.00098

Because X is a discrete random variable, we can calculate its mean using the formula

$$\mu_X = E(X) = \sum x_i P(x_i) = x_1 P(x_1) + x_2 P(x_2) + x_3 P(x_3) + \ldots$$

from Section 4D. We get

$$\mu_X = (0)(0.23730) + (1)(0.39551) + \ldots + (5)(0.00098) = 1.25$$

Interpretation: If we randomly select many, many families like this one with 5 children, the average number of children per family with type O blood will be about 1.25.

Did you think about why the mean of X is 1.25? Because each child has a 0.25 probability of inheriting type O blood, we'd expect 25% of the 5 children to have this blood type. In other words,

$$\mu_X = 5(0.25) = 1.25$$

This method can be used to find the mean of any binomial random variable.

AP® EXAM TIP
Formula sheet

The formula for the mean μ_X of a binomial random variable is included on the formula sheet provided on both sections of the AP® Statistics exam.

MEAN OF A BINOMIAL RANDOM VARIABLE

If a count X of successes has a binomial distribution with number of trials n and probability of success p, the mean of X is

$$\mu_X = np$$

To calculate the standard deviation of X, we use the formula

$$\sigma_X = \sqrt{\sum (x_i - \mu_X)^2 P(x_i)} = \sqrt{(x_1 - \mu_X)^2 P(x_1) + (x_2 - \mu_X)^2 P(x_2) + (x_3 - \mu_X)^2 P(x_3) + \ldots}$$

from Section 4A. So the standard deviation of X is

$$\begin{aligned}\sigma_X &= \sqrt{(0-1.25)^2(0.23730) + (1-1.25)^2(0.39551) + \ldots + (5-1.25)^2(0.00098)} \\ &= \sqrt{0.9375} \\ &= 0.968\end{aligned}$$

Interpretation: If we randomly select many, many families like this one with 5 children, the number of children per family with type O blood will typically vary from the mean of 1.25 by about 0.968 children.

There is a simple formula for the standard deviation of a binomial random variable, but it isn't easy to explain. For our family with $n = 5$ children and $p = 0.25$ of type O blood, the *variance* of X is

$$5(0.25)(0.75) = 0.9375$$

To get the standard deviation, we just take the square root:

$$\sigma_X = \sqrt{5(0.25)(0.75)} = \sqrt{0.9375} = 0.968$$

This method works for any binomial random variable.

AP® EXAM TIP
Formula sheet

The formula for the standard deviation σ_X of a binomial random variable is included on the formula sheet provided on both sections of the AP® Statistics exam.

STANDARD DEVIATION OF A BINOMIAL RANDOM VARIABLE

If a count X of successes has a binomial distribution with number of trials n and probability of success p, the standard deviation of X is

$$\sigma_X = \sqrt{np(1-p)}$$

Remember that these formulas for the mean and standard deviation work *only* for binomial distributions. The interpretation of μ_X and σ_X is the same as for any discrete random variable.

EXAMPLE

Return of the pop quiz
Describing a binomial distribution

Skills 3.B, 4.B

PROBLEM: To introduce a statistics class to binomial distributions, Mr. Desai does the Pop Quiz activity, in which each student in the class guesses an answer from A through E on each of 10 multiple-choice questions. Mr. Desai determined the "correct" answer for each of the 10 questions randomly so that A, B, C, D, or E was equally likely to be chosen. Let X = the number of questions that a specific student answers correctly. Earlier, we showed that X is a binomial random variable with $n = 10$ and $p = 0.2$. A histogram of its probability distribution is shown here.

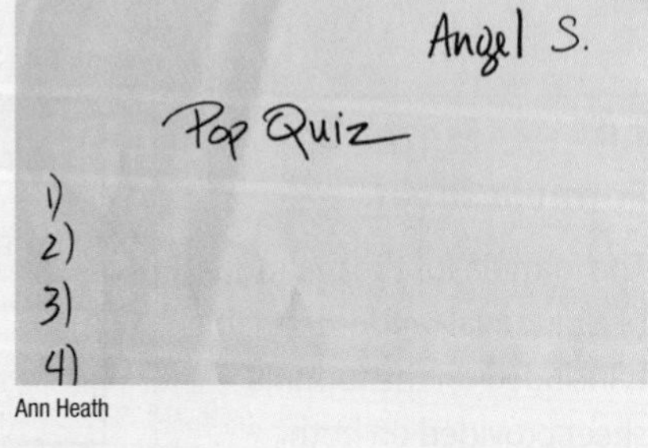

Ann Heath

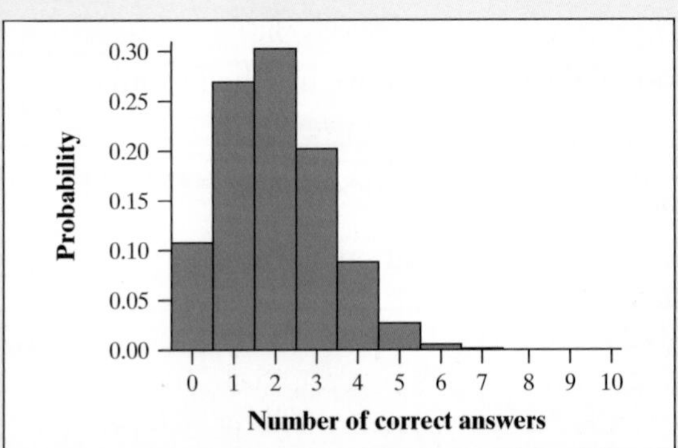

(a) Describe the shape of the probability distribution.
(b) Calculate and interpret the mean of X.
(c) Calculate and interpret the standard deviation of X.

SOLUTION:

(a) The probability distribution of X is skewed to the right with a single peak at $X = 2$.

(b) X has a binomial distribution with $n = 10$ and $p = 0.2$.

$\mu_X = 10(0.2) = 2$

If many, many students took the pop quiz, we'd expect them to get about 2 answers correct, on average.

The mean of a binomial random variable is $\mu_X = np$.

(c) $\sigma_X = \sqrt{10(0.2)(0.8)} = 1.265$

If many, many students took the pop quiz, their scores would typically vary from the mean of 2 by about 1.265 correct answers.

The standard deviation of a binomial random variable is $\sigma_X = \sqrt{np(1-p)}$.

FOR PRACTICE, TRY EXERCISE 17

BINOMIAL DISTRIBUTIONS IN STATISTICAL SAMPLING

The binomial distributions are important in statistics when we want to make inferences about the proportion p of successes in a population. For instance, suppose that a supplier inspects a random sample of 10 flash drives from a shipment of 10,000 flash drives in which 200 are defective (bad). Let X = the number of bad flash drives in the sample.

This is not quite a binomial setting. Because we are sampling without replacement, the Independent condition is violated. The conditional probability that the second flash drive chosen is bad changes when we know whether the first flash drive chosen is good or bad:

$$P(\text{second is bad} \mid \text{first is good}) = 200/9999 = 0.0200$$

but

$$P(\text{second is bad} \mid \text{first is bad}) = 199/9999 = 0.0199$$

These probabilities are very close because removing 1 flash drive from a shipment of 10,000 changes the makeup of the remaining 9999 flash drives very little. The probability distribution of X is close to a binomial distribution with $n = 10$ and $p = 0.02$.

To illustrate this, let's compute the probability that none of the 10 flash drives is defective. Using the binomial distribution, it's

$$P(X = 0) = \binom{10}{0}(0.02)^0(0.98)^{10} = 0.8171$$

The actual probability of getting no defective flash drives is

$$P(\text{no defectives}) = \frac{9800}{10{,}000} \times \frac{9799}{9999} \times \frac{9798}{9998} \times \ldots \times \frac{9791}{9991} = 0.8170$$

Those two probabilities are quite close!

Almost all real-world sampling, such as taking an SRS from a population of interest, is done without replacement. As the flash drive context illustrates, the random variable X = number of successes will have approximately a binomial distribution whenever the sample size n is relatively small compared to the population size N. What counts as relatively small? In practice, the binomial distribution gives a good approximation as long as we sample less than 10% of the population. We refer to this as the **10% condition.**

DEFINITION 10% condition

The **10% condition** states that when selecting a random sample of size n from a population of size N, we can treat individual observations as independent when performing calculations as long as $n < 0.10N$.

CHECK YOUR UNDERSTANDING

As a special promotion for its 20-ounce bottles of soda, a soft drink company printed a message on the inside of each bottle cap. Some of the caps said, "Please try again!" while others said, "You're a winner!" The company advertised the promotion with the slogan "1 in 6 wins a prize." Grayson's statistics class wonders if the company's claim holds true at a nearby convenience store. To find out, all 30 students in the class go to the store and each buys one randomly selected 20-ounce bottle of the soda. Let $X =$ the number of prize winners out of 30 if the company's claim is true. From the preceding Check Your Understanding, X is a binomial random variable with $n = 30$ and $p = 1/6$. Here is a graph of the probability distribution.

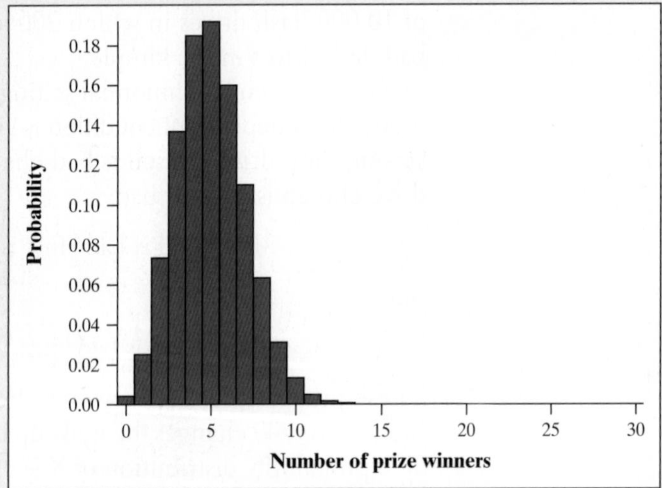

1. Describe the shape of the probability distribution.
2. Find the expected value of X. Explain why this result makes sense.
3. Calculate and interpret the standard deviation of X.

Geometric Random Variables and Probability

In a binomial setting, the number of trials n is fixed in advance, and the binomial random variable X counts the number of successes. The possible values of X are $0, 1, 2, \ldots, n$. In other situations, the goal is to repeat a random process *until a success occurs:*

- Roll a pair of dice until you get doubles.
- In basketball, attempt a 3-point shot until you make one.
- Keep trying to guess a random integer from 1 to 10 until you get it right.

These are all examples of a **geometric setting.**

DEFINITION Geometric setting

A **geometric setting** arises when we perform independent trials of the same random process and record the number of trials it takes to get one success. On each trial, the probability *p* of success must be the same.

In a geometric setting, if we define the random variable X to be the number of trials needed to get the first success, then X is called a **geometric random variable.** The probability distribution of X is a **geometric distribution.**

DEFINITION Geometric random variable, Geometric distribution

The number of trials *X* that it takes to get a success in a geometric setting is a **geometric random variable.**

The probability distribution of *X* is a **geometric distribution** with probability of success *p* on any trial. The possible values of *X* are 1, 2, 3, ...

Here's an activity your class can try that involves a geometric random variable.

ACTIVITY Is this your lucky day?

Your teacher is planning to give you 10 problems for homework. As an alternative, you can agree to play the Lucky Day game. Here's how it works. A student will be selected at random from your class and asked to pick a day of the week (e.g., Thursday). Then your teacher will use technology to randomly choose a day of the week as the "lucky day." If the student picks the correct day, the class will have only one homework problem. If the student picks the wrong day, your teacher will once again select a student from the class at random. The chosen student will pick a day of the week and your teacher will use technology to choose a "lucky day." If this student gets it right, the class will have two homework problems. The game continues until a randomly selected student correctly picks the lucky day. Your teacher will assign a number of homework problems that is equal to the total number of picks made by members of your class. Are you ready to play the Lucky Day game?

1. Decide as a class whether to "gamble" on the number of homework problems you will receive. You have 30 seconds.
2. Whatever decision the class makes, play the Lucky Day game and see what happens!

As with binomial random variables, it's important to be able to distinguish situations in which a geometric distribution does and doesn't apply. Let's consider the Lucky Day game. The random variable of interest in this game is X = the number of picks it takes to correctly match the lucky day. Each pick is one trial of the random process. Knowing the result of one pick tells us nothing about the result of any other pick. On each trial, the probability of a correct pick is 1/7. This is a geometric setting. Because X counts the number of trials to get the first success, it is a geometric random variable with $p = 1/7$.

What is the probability that the first student picks correctly and wins the Lucky Day game? It's $P(X = 1) = 1/7$. That's also the class's chance of having only one homework problem assigned. For the class to have two homework problems assigned, the first student selected must pick an incorrect day of the week and the second student selected must pick the lucky day correctly. The probability that this happens is

$$P(X = 2) = (6/7)(1/7) = 0.1224$$

Likewise,

$$P(X = 3) = (6/7)(6/7)(1/7) = (6/7)^2(1/7) = 0.1050$$

and

$$P(X = 4) = (6/7)(6/7)(6/7)(1/7) = (6/7)^3(1/7) = 0.0900$$

In general, the probability that the first correct pick comes on the xth trial is

$$P(X = x) = (6/7)^{x-1}(1/7)$$

Let's summarize what we've learned about calculating a geometric probability.

AP® EXAM TIP
Formula sheet

The formula for a geometric probability $P(X = x)$ is included on the formula sheet provided on both sections of the AP® Statistics exam.

GEOMETRIC PROBABILITY FORMULA

If X has the geometric distribution with probability of success p on each trial, the possible values of X are 1, 2, 3, ... If x is any one of these values,

$$P(X = x) = (1 - p)^{x-1}p$$

With the geometric probability formula in hand, we can calculate probabilities involving geometric random variables. Follow the two-step process we introduced earlier when calculating binomial probabilities.

EXAMPLE

The Lucky Day game
Geometric random variables and probability

Skill 3.A, 4.B

PROBLEM: After a quick discussion, Mr. Lochel's class decides not to play the Lucky Day game because they don't want to risk getting more than 10 homework problems. Did the class make a good decision?

(a) Find the probability that the class receives exactly 10 homework problems as a result of playing the Lucky Day game.

(b) Find $P(X < 10)$. Describe this probability in words.

Burazin/The Image Bank/Getty Images

SOLUTION:

(a) Let X = the number of homework problems received when playing the Lucky Day game. X has a geometric distribution with $p = 1/7$. We want to find $P(X = 10)$.

Step 1: Define the random variable of interest, state how it is distributed, and identify the values of interest.

$P(X = 10) = (6/7)^9(1/7) = 0.0357$

Step 2: Perform calculations—show your work!

(b) Let X = the number of homework problems received when playing the Lucky Day game. X has a geometric distribution with $p = 1/7$. We want to find $P(X < 10)$.

Step 1: Define the random variable of interest, state how it is distributed, and identify the values of interest.

$$P(X < 10) = 1/7 + (6/7)(1/7) + (6/7)^2(1/7) + \ldots + (6/7)^8(1/7) = 0.7503$$

Step 2: Perform calculations—show your work! The values of X that interest us are 0 1 2 3 4 5 6 7 8 9 | 10 11 12...

There's about a 75% probability that the class will get fewer than 10 homework problems by playing the Lucky Day game.

Be sure to answer the question that was asked.

FOR PRACTICE, TRY EXERCISE 27

There's a clever alternative approach to finding the probability in part (b) of the example. By the complement rule, $P(X < 10) = 1 - P(X \geq 10)$. What's the probability that it will take at least 10 picks for Mr. Lochel's class to win the Lucky Day game? It's the chance that the first 9 picks are all incorrect: $\left(\frac{6}{7}\right)^9 = 0.2497$. So the probability that the class will win the Lucky Day game in fewer than 10 picks (and therefore have fewer than 10 homework problems assigned) is

$$P(X < 10) = 1 - P(X \geq 10) = 1 - 0.2497 = 0.7503$$

Maybe Mr. Lochel's class should have taken the risk and played the game!

As you probably guessed, we can use technology to calculate geometric probabilities. The following Tech Corner shows how to do it.

15. Tech Corner CALCULATING GEOMETRIC PROBABILITIES

TI-Nspire and other technology instructions are on the book's website at bfwpub.com/tps7e.

There are two handy commands on the TI-83/84 for finding geometric probabilities: geometpdf and geometcdf. The inputs for both commands are the success probability p and the value(s) of interest for the geometric random variable X.

geometpdf(p, x) computes $P(X = x)$
geometcdf(p, x) computes $P(X \leq x)$

Let's use these commands to confirm our answers in the preceding example.

(a) Find the probability that the class receives exactly 10 homework problems as a result of playing the Lucky Day game.

- Press 2nd VARS (DISTR) and choose geometpdf(.

  OS 2.55 or later: In the dialog box, enter these values: p: 1/7, *x* value: 10. Choose Paste, and then press ENTER.

  Older OS: Complete the command geometpdf(1/7,10) and press ENTER.

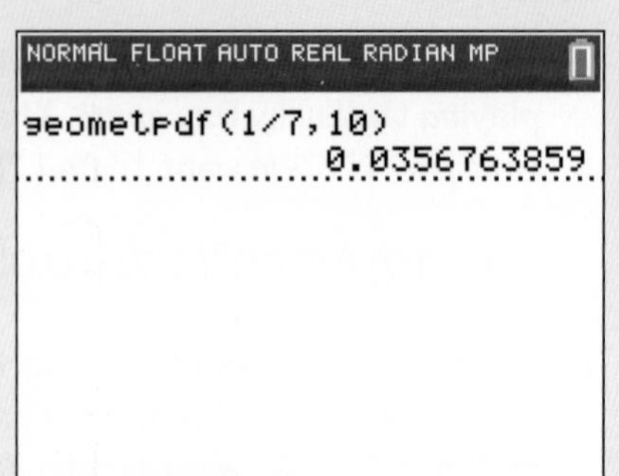

These results agree with our previous answer using the geometric probability formula: 0.0357.

(b) To find P(X < 10), use the geometcdf command:

$$P(X < 10) = P(X \leq 9) = \text{geometcdf(p:1/7, x value:9)}$$

- Press 2nd VARS (DISTR) and choose geometcdf.

  OS 2.55 or later: In the dialog box, enter these values: p: 1/7, x value: 9. Choose Paste, and then press ENTER.

  Older OS: Complete the command geometcdf(1/7,9) and press ENTER.

These results agree with our previous answer using the geometric probability formula: 0.7503.

Describing a Geometric Distribution: Shape, Center, and Variability

The table shows part of the probability distribution of X = the number of picks it takes to match the lucky day. We can't show the entire distribution because the number of trials it takes to get the first success could be a very large number.

Number of picks	1	2	3	4	5	6	7	8	9	...
Probability	0.143	0.122	0.105	0.090	0.077	0.066	0.057	0.049	0.042	...

Figure 4.12 is a histogram of the probability distribution for values of X from 1 to 26. Let's describe what we see.

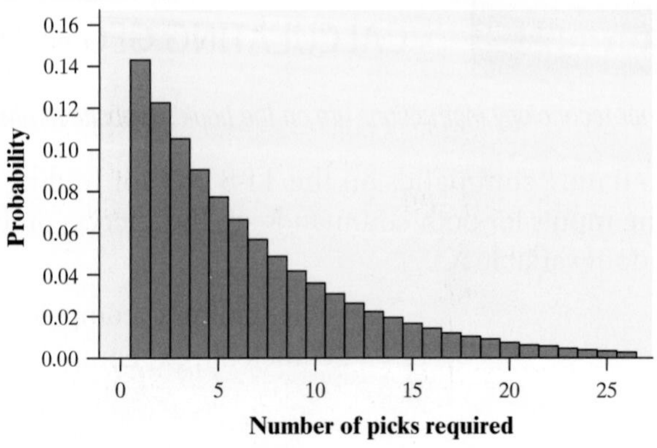

FIGURE 4.12 Histogram showing the probability distribution of the geometric random variable X = number of picks required for students to win in the Lucky Day game.

Shape: Skewed to the right. Every geometric distribution has this shape. That's because the most likely value of any geometric random variable is 1. The probability of each successive value decreases by a factor of $(1 - p)$.

Center: The mean (expected value) of X is $\mu_X = 7$. If the class played the Lucky Day game many, many times, they would receive an average of 7 homework problems. It's no coincidence that $p = 1/7$ and $\mu_X = 7$. With probability of success 1/7 on each trial, we'd expect it to take an average of 7 trials to get the first success. That is, $\mu_X = 1/(1/7) = 7$.

Variability: The standard deviation of X is $\sigma_X = 6.48$. If the class played the Lucky Day game many, many times, the number of homework problems they receive would typically vary from the mean of 7 by about 6.48 problems. That could mean a lot of homework! There is a simple formula for the standard deviation of a geometric random variable, but it isn't easy to explain. For the Lucky Day game,

$$\sigma_X = \frac{\sqrt{1 - 1/7}}{1/7} = 6.48$$

We can generalize these results for the mean and standard deviation of any geometric random variable.

AP® EXAM TIP
Formula sheet

The formulas for the mean μ_X and the standard deviation σ_X of a geometric random variable are included on the formula sheet provided on both sections of the AP® Statistics exam.

MEAN AND STANDARD DEVIATION OF A GEOMETRIC RANDOM VARIABLE

If X is a geometric random variable with probability of success p on each trial, then its mean (expected value) is $\mu_X = \frac{1}{p}$ and its standard deviation is $\sigma_X = \frac{\sqrt{1-p}}{p}$.

We interpret the parameters μ_X and σ_X in the same way as for any discrete random variable.

EXAMPLE

Waiting for a free lunch
Describing a geometric distribution

Skills 3.B, 4.B

PROBLEM: A local fast-food restaurant is running a "Draw a three, get it free" lunch promotion. After each customer orders, a touchscreen display shows the message "Press here to win a free lunch." A computer program then simulates one card being drawn from a standard deck. If the chosen card is a 3, the customer's order is free. Otherwise, the customer must pay the bill. Let X = the number of customers it takes to get the first free order on a given day.

(a) Calculate and interpret the mean of X.
(b) Calculate and interpret the standard deviation of X.

SOLUTION:

X has a geometric distribution with $p = 4/52$.

(a) $\mu_X = \frac{1}{4/52} = 13$

$\mu_X = \frac{1}{p}$

If the restaurant runs this lunch promotion on many, many days, the average number of customers it takes to get the first free order would be about 13.

(b) $\sigma_X = \frac{\sqrt{1-4/52}}{4/52} = 12.49$

$\sigma_X = \frac{\sqrt{1-p}}{p}$

If the restaurant runs this lunch promotion on many, many days, the number of customers it takes to get the first free order would typically vary from the mean of 13 by about 12.5.

FOR PRACTICE, TRY EXERCISE 31

CHECK YOUR UNDERSTANDING

Suppose you roll a pair of fair, six-sided dice until you get doubles. Let T = the number of rolls it takes to get doubles. Note that the probability of getting doubles on any roll is $6/36 = 1/6$.

1. Show that T is a geometric random variable.
2. Find $P(T = 3)$. Describe this probability in words.
3. In the game of Monopoly, a player can get out of jail free by rolling doubles within 3 turns. Find the probability that this happens.
4. Calculate the mean and standard deviation of T. Interpret these values.

SECTION 4F Summary

- A **binomial setting** arises when we perform n independent trials of the same random process and count the number of times that a particular outcome (a "success") occurs. The following conditions must be met to have a binomial setting:
  - Binary? The possible outcomes of each trial can be classified as "success" or "failure."
  - Independent? Trials must be independent. That is, knowing the result of one trial must not tell us anything about the result of any other trial.
  - Number? The number of trials n of the random process must be fixed in advance.
  - Same probability? There is the same probability of success p on each trial.

  Remember to check the BINS!

- The count X of successes in a binomial setting is a special type of discrete random variable known as a **binomial random variable.** Its probability distribution is a **binomial distribution.** Any binomial distribution is completely specified by two numbers: the number of trials n of the random process and the probability of success p on any trial. The possible values of X are the whole numbers 0, 1, 2, . . . , n.
- Use the binomial probability formula to calculate the probability of getting exactly x successes in n trials:

$$P(X = x) = \binom{n}{x} p^x (1-p)^{n-x}$$

  - The **binomial coefficient**

$$\binom{n}{x} = \frac{n!}{x!(n-x)!}$$

    counts the number of ways x successes can be arranged among n trials.
  - The factorial of n is

$$n! = n(n-1)(n-2)\cdot \ldots \cdot (3)(2)(1)$$

    for positive whole numbers n, where $0! = 1$.
- You can also use technology to calculate binomial probabilities. The TI-83/84 command binompdf(n,p,x) computes $P(X = x)$. The TI-83/84 command binomcdf(n,p,x) computes the cumulative probability $P(X \leq x)$.
- A binomial distribution can have a shape that is roughly symmetric, skewed to the right, or skewed to the left depending on the values of n and p.
- The mean and standard deviation of a binomial random variable X are

$$\mu_X = np \text{ and } \sigma_X = \sqrt{np(1-p)}$$

- The binomial distribution with n trials and probability of success p gives a good approximation of the count of successes in a random sample of size n selected without replacement from a large population containing proportion p of successes. This is true as long as the sample size n is less than 10% of the population size N. We refer to this as the **10% condition.** When the 10% condition is met, we can view individual observations as independent.
- A **geometric setting** consists of repeated trials of the same random process in which the probability p of success is the same for each trial, and the goal is to count the number of trials it takes to get one success. If X = the number of trials required to obtain the first success, then X is a **geometric random variable.** Its probability distribution is called a **geometric distribution.**
- If X has the geometric distribution with probability of success p, the possible values of X are the positive integers 1, 2, 3, ... The probability that it takes exactly x trials to get the first success is given by

$$P(X = x) = (1-p)^{x-1}p$$

- You can also use technology to calculate geometric probabilities. The TI-83/84 command geometpdf(p,x) computes $P(X = x)$. The TI-83/84 command geometcdf(p,x) computes the cumulative probability $P(X \leq x)$.

- The mean and standard deviation of a geometric random variable X are

$$\mu_X = \frac{1}{p} \quad \text{and} \quad \sigma_X = \frac{\sqrt{1-p}}{p}$$

- When calculating probabilities involving binomial or geometric random variables, follow this two-step process:

  Step 1: Define the random variable of interest, state how it is distributed, and identify the values of interest.

  Step 2: Perform calculations—show your work! If you use technology, be sure to label the inputs you used for your calculator command.

  Be sure to answer the question that was asked.

AP® EXAM TIP
AP® Daily Videos

Review the content of this section and get extra help by watching the AP® Daily Videos for Topics 4.10–4.12, which are available in AP® Classroom.

4F Tech Corners

TI-Nspire and other technology instructions are on the book's website at bfwpub.com/tps7e.

SECTION 4F Exercises

Binomial Random Variables

In Exercises 1–4, determine whether the given random variable is a binomial random variable. If so, state its probability distribution.

1. **Baby elk** Biologists estimate that a randomly selected baby elk has a 44% chance of surviving to adulthood. Assume this estimate is correct. Suppose researchers choose 7 baby elk at random to monitor. Let X = the number that survive to adulthood. pg 417

2. **Long or short?** Imagine putting the names of all the students in your statistics class in a hat. Suppose you mix up the names, and draw four without looking, one after the other. Let X = the number whose last names have more than six letters.

3. **Bull's-eye!** Lawrence likes to shoot a bow and arrow in his free time. On any shot, he has about a 10% chance of hitting the bull's-eye. As a challenge one day, Lawrence decides to keep shooting until he gets a bull's-eye. Let Y = the number of shots he takes.

4. **Taking the train** According to New Jersey Transit, the 8:00 A.M. weekday train from Princeton to New York City has a 90% chance of arriving on time on a randomly selected day. Suppose this claim is true. Imagine choosing 6 days at random. Let Y = the number of days on which the train arrives on time.

Calculating Binomial Probabilities

5. **More baby elk** Refer to Exercise 1. We are interested in the probability that exactly 4 of the 7 randomly selected baby elk survive to adulthood. pg 421

   (a) Define the random variable of interest, state its probability distribution, and identify the value(s) of interest.

   (b) Determine the probability that exactly 4 of the 7 randomly selected baby elk survive to adulthood.

6. **Taking the train again** Refer to Exercise 4. We are interested in the probability that the train arrives on time on exactly 4 of the 6 randomly selected days.

   (a) Define the random variable of interest, state its probability distribution, and identify the value(s) of interest.

   (b) Determine the probability that the train arrives on time on exactly 4 of the 6 randomly selected days.

7. **Take a spin** An online spinner has two colored regions — blue and yellow. According to the website, the probability that the spinner lands in the blue region on any spin is 0.80. Assume for now that this claim is correct. Suppose we spin the spinner 12 times and let X = the number of times it lands in the blue region.

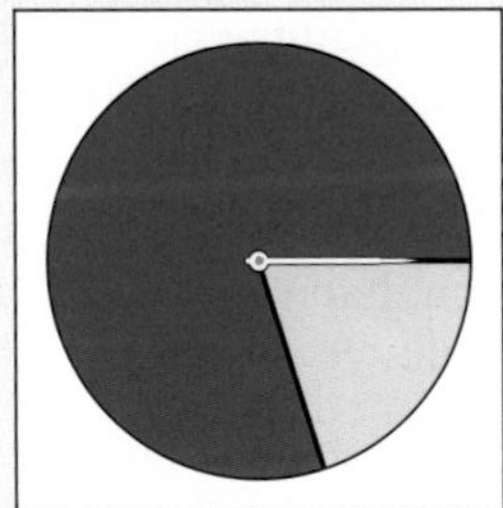

(a) Explain why X is a binomial random variable.

(b) Find the probability that exactly 8 spins land in the blue region.

8. **Red light!** Pedro drives the same route to work on Monday through Friday. The route includes one traffic light. According to the local traffic department, there is a 55% chance that the light will be red on a randomly selected work day. Suppose we choose 10 of Pedro's work days at random and let Y = the number of times that the light is red.

(a) Explain why Y is a binomial random variable.

(b) Find the probability that the light is red on exactly 7 days.

9. **Adult elk?** Refer to Exercise 1. Find the probability that more than 4 elk in the sample survive to adulthood. Should researchers be surprised if this happens?

pg 422

10. **Taking the train** Refer to Exercise 4. Find the probability that the train arrives on time on fewer than 4 of the randomly selected days. Should passengers be surprised if this happens?

11. **Take another spin** Refer to Exercise 7. Calculate $P(X \leq 7)$. Describe this probability in words.

12. **More red lights** Refer to Exercise 8. Calculate $P(Y \geq 7)$. Describe this probability in words.

13. **The last Kiss** Do people show a preference for the last thing they taste? Researchers at the University of Michigan designed a study to find out. The researchers gave 22 students five different Hershey's Kisses (milk chocolate, dark chocolate, crème, caramel, and almond) in random order and asked the student to rate each one. Participants were not told how many Kisses they would be tasting. However, when the fifth and final Kiss was presented, participants were told that it would be their last one.[58] Assume that the participants in the study don't have a special preference for the last thing they taste — that is, the probability of preferring the last Kiss tasted is 0.20.

pg 424

(a) Find the probability that 14 or more of the students would prefer the last Kiss tasted.

(b) Of the 22 students, 14 gave the final Kiss the highest rating. Does this result give convincing evidence that the participants show a preference for the last thing they taste? Explain your reasoning.

14. **Smelling Parkinson's disease** The Smelling Parkinson's activity at the beginning of Unit 1 described an experiment to test whether Joy Milne could smell Parkinson's disease. Joy was presented with 12 shirts, each worn by a different person, some of whom had Parkinson's disease and some of whom did not. The shirts were given to Joy in random order, and she had to decide whether or not each shirt was worn by a patient with Parkinson's disease. If we assume that Joy was just guessing, her probability of correctly identifying each shirt was 1/2.

(a) Determine the probability that Joy would identify at least 11 shirts correctly by guessing.

(b) Joy identified 11 of the 12 shirts correctly. Does this give convincing evidence that Joy really can smell Parkinson's disease? Explain your reasoning.

Note: The researchers later discovered that Joy had correctly identified all 12 shirts. Her one "mistake" was a person who was diagnosed with Parkinson's disease a few months later.

15. **Bag check** Thousands of travelers pass through the airport in Guadalajara, Mexico, each day. Mexican customs agents want to be sure that passengers are not bringing in illegal items, but do not have time to search every traveler's luggage. Instead, customs requires each person to press a button. Either a red bulb or a green bulb lights up. If the light is red, the passenger will be searched by customs agents. Green means "go ahead." Customs agents claim that the light has probability 0.30 of showing red on any push of the button. Assume for now that this claim is true.

Suppose we watch 20 passengers press the button. Let R = the number who get a red light.

(a) What probability distribution does R have? Justify your answer.

(b) Find the probability that at most 3 people out of 20 would get a red light if the agents' claim is true.

(c) Suppose that only 3 of the 20 passengers get a red light after pressing the button. Does this give convincing evidence that the customs agents' claimed value of $p = 0.3$ is too high? Explain your reasoning.

16. **Easy-start mower** A company has developed an "easy-start" mower that cranks the engine with the push of a button. The company claims that the probability the mower will start on any push of the button is 0.9. Assume for now that this claim is true.

On the next 30 uses of the mower, let T = the number of times it starts on the first push of the button.

(a) What probability distribution does T have? Justify your answer.

(b) Find the probability of at most 24 starts in 30 attempts if the company's claim is true.

(c) Suppose that the mower starts on only 24 of the 30 attempts. Does this give convincing evidence that the company's claim is exaggerated? Explain your reasoning.

Describing a Binomial Distribution

17. **Bag check distribution** Refer to Exercise 15. Here is a histogram of the probability distribution of R.

pg 428

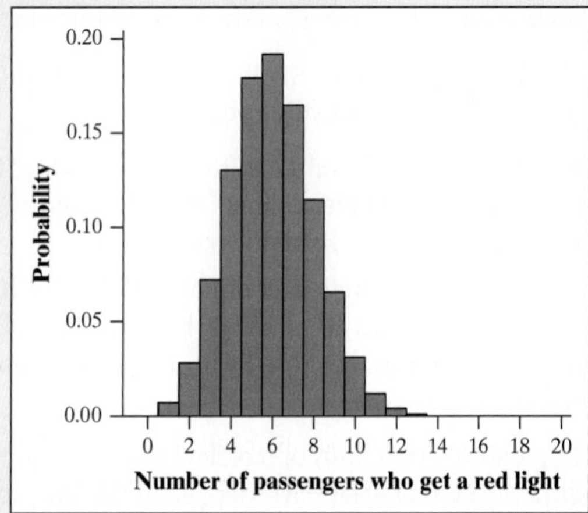

(a) Describe the shape of the probability distribution.

(b) Calculate and interpret the mean of R.

(c) Calculate and interpret the standard deviation of R.

18. **Mower distribution** Refer to Exercise 16. Here is a histogram of the probability distribution of T.

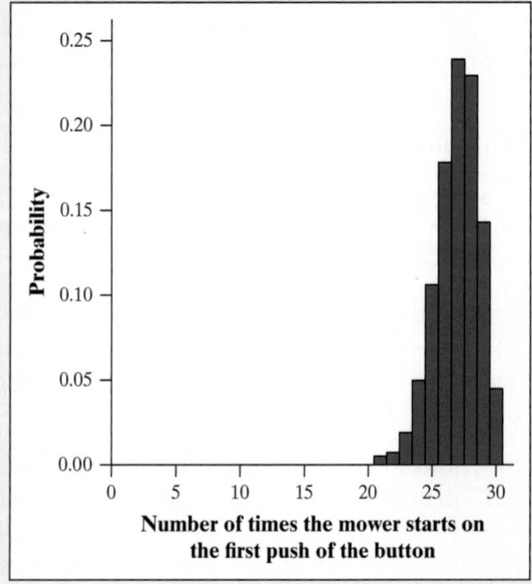

(a) Describe the shape of the probability distribution.

(b) Calculate and interpret the mean of T.

(c) Calculate and interpret the standard deviation of T.

19. **Take another spin** An online spinner has two colored regions — blue and yellow. According to the website, the probability that the spinner lands in the blue region on any spin is 0.80. Assume for now that this claim is correct. Suppose we spin the spinner 12 times and let X = the number of times it lands in the blue region. Exercise 7 shows that X is a binomial random variable.

(a) Make a graph of the probability distribution of X. Describe its shape.

(b) Calculate μ_X and σ_X. Interpret these values.

20. **Another red light!** Pedro drives the same route to work on Monday through Friday. The route includes one traffic light. According to the local traffic department, there is a 55% chance that the light will be red on a randomly selected work day. Suppose we choose 10 of Pedro's work days at random and let Y = the number of times that the light is red. Exercise 8 shows that Y is a binomial random variable.

(a) Make a graph of the probability distribution of Y. Describe its shape.

(b) Calculate μ_Y and σ_Y. Interpret these values.

21. **Airport security** At a certain airport, security officials claim that they randomly select passengers for extra screening at the gate before boarding some flights. One such flight had 76 passengers — 16 in first class and 60 in economy class. Some passengers were surprised when none of the 14 passengers chosen for screening was seated in first class. Let X = the number of first class passengers in a random sample of size 14 from this flight. We would like to find $P(X = 0)$.

(a) Explain why the Independent condition for a binomial setting is not met.

(b) Is the 10% condition met? Justify your answer.

22. **Scrabble** In the game of Scrabble, the first player draws 7 letter tiles at random from a bag containing 100 tiles. There are 42 vowels, 56 consonants, and 2 blank tiles in the bag. Anise draws first and is surprised to discover that all 7 tiles are vowels. Let Y = the number of vowels in a random sample of 7 Scrabble tiles. We would like to find $P(Y = 7)$.

(a) Explain why the Independent condition for a binomial setting is not met.

(b) Is the 10% condition met? Justify your answer.

23. **Lefties** Eleven percent of students at a large high school are left-handed. A statistics teacher selects a random sample of 100 students and records L = the number of left-handed students in the sample.

(a) Explain why L can be modeled by a binomial distribution even though the sample was selected without replacement.

(b) Use a binomial distribution to estimate the probability that 15 or more students in the sample are left-handed.

24. **In debt?** According to financial records, 24% of U.S. adults have more debt on their credit cards than they have money in their savings accounts. Suppose that we take a random sample of 100 U.S. adults. Let D = the number of adults in the sample with more debt than savings.

(a) Explain why D can be modeled by a binomial distribution even though the sample was selected without replacement.

(b) Use a binomial distribution to estimate the probability that 30 or more adults in the sample have more debt than savings.

Geometric Random Variables and Probability

25. **Geometric or not?** Determine whether each of the following scenarios describes a geometric setting. If so, define the random variable of interest and state how it is distributed.

(a) A popular brand of cereal puts a card bearing the image of 1 of 5 famous NASCAR drivers in each box. There is a 1/5 chance that any particular driver's card ends up in any box of cereal. Suppose you buy boxes of the cereal until you have all 5 drivers' cards.

(b) A puzzle company offers a set of 6 different animal puzzles — lions, tigers, bears, elephants, flamingoes, and peacocks — suitable for young children. However, the contents of each puzzle remain a mystery until the bag is opened. According to the company, each puzzle is equally likely to feature any one of the six animals. Lola decides to keep buying one puzzle at a time until she gets a flamingo puzzle.

26. **Is it geometric?** Determine whether each of the following scenarios describes a geometric setting. If so, define the random variable of interest and state how it is distributed.

(a) Suppose you shuffle a standard deck of 52 playing cards well, then turn over one card at a time from the top of the deck until you get an ace.

(b) Billy likes to play cornhole in his free time. On any toss, he has about a 20% chance of getting a bag into the hole. As a challenge one day, Billy decides to keep tossing bags until he gets one in the hole.

27. **Cranky mower** To start the family's old lawn mower, Rita has to pull a cord and hope for some luck. On any particular pull, the mower has a 20% chance of starting. Let X = the number of pulls it takes Rita to start the mower.

pg 432

(a) Find the probability that it takes Rita exactly 3 pulls to start the mower.

(b) Find $P(X < 6)$. Describe this probability in words.

28. **1-in-6 wins** As a special promotion for its 20-ounce bottles of soda, a soft drink company printed a message on the inside of each bottle cap. Some of the caps said, "Please try again!" while others said, "You're a winner!" The company advertised the promotion with the slogan "1 in 6 wins a prize." To investigate the company's claim, Alex decides to keep buying one 20-ounce bottle of the soda at a time until he gets a winner. Let Y = the number of bottles Alex has to buy if the company's claim is true.

(a) Find the probability that Alex buys exactly 5 bottles.

(b) Find $P(Y \leq 6)$. Describe this probability in words.

29. **Using Benford's law** According to Benford's law (Section 4D, Exercise 9), the probability that the first digit of the amount of a randomly chosen invoice is an 8 or a 9 is 0.097. Suppose that a forensic accountant randomly selects invoices from a vendor until they find one whose amount begins with an 8 or a 9. Would you be surprised if it took the accountant 40 or more invoices to find the first one with an amount that starts with an 8 or 9? Calculate an appropriate probability to support your answer.

30. **Lucky spin** A carnival game that costs $1 to play starts with the player picking a whole number from 1 to 38. The game's operator then drops a ball into a spinning wheel with slots numbered from 1 to 38. If the ball lands in the slot with the number that the player picked, the player wins a giant stuffed teddy bear. Marti decides to keep playing the game until she wins, picking the number 15 each time. On any spin, there's a 1-in-38 chance that the ball will land in the 15 slot. Would you be surprised if Marti won in 3 or fewer spins? Compute an appropriate probability to support your answer.

Describing a Geometric Distribution

31. **More cranky mower** Refer to Exercise 27.

pg 435

(a) Calculate and interpret the mean of X.

(b) Calculate and interpret the standard deviation of X.

32. **More 1-in-6 wins** Refer to Exercise 28.

(a) Calculate and interpret the mean of Y.

(b) Calculate and interpret the standard deviation of Y.

33. **Back to Benford's law** Consider the probability distribution in Exercise 29.

(a) What shape does it have? Justify your answer.

(b) Find its expected value.

(c) Calculate its standard deviation.

34. **Another lucky spin** Consider the probability distribution in Exercise 30.

(a) What shape does it have? Justify your answer.

(b) Find its expected value.

(c) Calculate its standard deviation.

For Investigation *Apply the skills from the section in a new context or nonroutine way.*

35. **Diet cola** The makers of a diet cola claim that its taste is indistinguishable from the taste of the full-calorie version of the same cola. To investigate this claim, a researcher prepared small samples of each type of soda in identical cups. Then the researcher had 30 volunteers taste each cola in random order and try to identify which was the diet cola and which was the regular cola.[59] If we assume that the volunteers couldn't tell the difference, each one was guessing with a 1/2 chance of being correct. Let $X =$ the number of volunteers who correctly identify the colas.

(a) Calculate the mean and standard deviation of X.

(b) Of the 30 volunteers, 23 made correct identifications. Find the standardized score (*z*-score) corresponding to $X = 23$.

(c) Based on your answer to part (b), explain why the observed result of 23 correct identifications is surprising if the 30 volunteers are just guessing.

36. **Bernoulli random variables** Consider a single trial of some random process with probability p of a success occurring. The number of successes B is a *Bernoulli random variable* with the following probability distribution:

Number of successes, b	0	1
Probability, $P(B = b)$	$1 - p$	p

(a) Calculate the mean and standard deviation of B.

Let the random variable $X = B_1 + B_2 + \ldots + B_n$. Note that X is the sum of n independent Bernoulli random variables.

(b) Using rules for combining random variables,

i. Find the mean of X.

ii. Find the standard deviation of X.

(c) What probability distribution does the random variable X have? Justify your answer.

Multiple Choice *Select the best answer for each question.*

37. Joe reads that 1 out of 4 eggs contains *Salmonella* bacteria. As a result, Joe never uses more than 3 eggs in cooking. If eggs do or don't contain *Salmonella* independently of each other, the number of contaminated eggs when Joe uses 3 eggs chosen at random has the following probability distribution:

(A) binomial with $n = 4$ and $p = 1/4$

(B) binomial with $n = 3$ and $p = 1/4$

(C) binomial with $n = 3$ and $p = 1/3$

(D) geometric with $p = 1/4$

(E) geometric with $p = 1/3$

Exercises 38 and 39 refer to the following setting. A fast-food restaurant runs a promotion in which certain food items come with game pieces. According to the restaurant, 1 in 4 game pieces is a winner.

38. If Jeff gets 4 game pieces, what is the probability that he wins exactly 1 prize?

(A) 0.25

(B) 1.00

(C) $\binom{4}{1}(0.25)^1(0.75)^3$

(D) $\binom{4}{1}(0.25)^3(0.75)^1$

(E) $(0.75)^3(0.25)^1$

39. If Jeff keeps playing until he wins a prize, what is the probability that he has to play the game exactly 5 times?

(A) $(0.25)^5$

(B) $(0.75)^4$

(C) $(0.75)^5$

(D) $(0.75)^4(0.25)$

(E) $\binom{5}{1}(0.75)^4(0.25)$

40. Each entry in a table of random digits like Table D has probability 0.1 of being a 0, and the digits are independent of one another. Each line of Table D contains 40 random digits. The mean and standard deviation of the number of 0s in a randomly selected line are approximately

(A) mean = 0.1, standard deviation = 0.05.

(B) mean = 0.1, standard deviation = 0.1.

(C) mean = 4, standard deviation = 0.05.

(D) mean = 4, standard deviation = 1.90.

(E) mean = 4, standard deviation = 3.60.

Recycle and Review *Practice what you learned in previous sections.*

41. **Spoofing (1A, 3A)** To collect information such as passwords, online criminals use "spoofing" to direct internet users to fraudulent websites. In one study of internet fraud, students were warned about spoofing and then asked to log into their university account starting from the university's home page. In some cases, the login link led to the genuine dialog box. In others, the box looked genuine but actually linked to a different site that recorded the ID and password the student entered. The box that appeared for each student was determined at random. An alert student could detect the fraud by looking at the true internet address displayed in the browser status bar, but most just entered their ID and password.

(a) Is this an observational study or an experiment? Justify your answer.

(b) What are the explanatory and response variables? Identify each variable as categorical or quantitative.

42. **Hay, hay! (1F)** On a certain farm, the baling machine produces small hay bales whose weights can be modeled by a normal distribution with mean 100 pounds and standard deviation 6 pounds.

(a) About what proportion of the small hay bales weigh between 90 and 110 pounds?

(b) What is the 99th percentile of the distribution of weight for these small hay bales?

UNIT 4 Wrap-Up

PART II

FRAPPY! Free Response AP® Problem, Yay!

Directions: Show all your work. Indicate clearly the methods you use, because you will be scored on the correctness of your methods as well as on the accuracy and completeness of your results and explanations.

Buckley Farms makes homemade potato chips that it sells in bags labeled "16 ounces." With the current production process, there is a 0.08 probability that a randomly selected bag is underweight. Buckley Farms puts 10 randomly selected bags of chips in each box for shipping to customers.

(a) A quality control supervisor is interested in the probability that more than 2 of the bags in a box of chips will be underweight. Assume that the weights of individual bags are independent of each other.

(i) Define the random variable of interest and state how the random variable is distributed.

(ii) Determine the probability that more than 2 of the bags in a randomly selected box of chips will be underweight.

(b) As a check on the production process, the quality control supervisor decides to inspect individual boxes until one is found with more than 2 bags of chips that are underweight. Should the supervisor be surprised if it takes 25 or more boxes to find one that meets this criterion? Give appropriate evidence to justify your answer.

(c) The empty boxes have a mean weight of 12 ounces and a standard deviation of 0.25 ounce. The bags of chips have a mean weight of 16.15 ounces and a standard deviation of 0.12 ounce. Consider the total weight of a box containing 10 bags of chips.

(i) Calculate the mean total weight.

(ii) Calculate the standard deviation of the total weight.

After you finish the FRAPPY!, you can view two example solutions on the book's website (**bfwpub.com/tps7e**). Determine whether you think each solution is "complete," "substantial," "developing," or "minimal." If the solution is not complete, what improvements would you suggest to the student who wrote it? Finally, your teacher will provide you with a scoring rubric. Score your response and note what, if anything, you would do differently to improve your own score.

UNIT 4, PART II REVIEW

SECTION 4D Introduction to Discrete Random Variables

A **random variable** assigns numerical values to the outcomes of a random process. The **probability distribution** of a random variable describes its possible values and their probabilities. There are two types of random variables: discrete and continuous. **Discrete random variables** take a countable (finite or infinite) set of possible values with gaps between them on the number line.

As with the distributions of quantitative data described in Unit 1, we are often interested in the shape, center, and variability of a probability distribution. The shape of a discrete probability distribution can be identified by graphing a probability histogram, with the height of each bar representing the probability of a single value. The center is usually identified by the **mean, or expected value, of the random variable,** which is the average value of the random variable if the random process is repeated many, many times. The variability of a probability distribution is usually identified

2. Each member of the class should then select an SRS of five pennies from the population and note the year on each penny.
   - Record the average year of these five pennies (rounded to the nearest year) with an "$\bar{x}$" on a new class dotplot. For example, if the average year is 2014.4, put an $\bar{x}$ above 2014. Make sure this dotplot is on the same scale as the dotplot in Step 1.
   - Record the proportion of pennies with a shield on the back with a "$\hat{p}$" on a different dotplot provided by your teacher. For example, if two of the pennies have a shield on the back, put a $\hat{p}$ above $2/5 = 0.4$.
   - Return the pennies to the population. Repeat this process until there are at least 100 $\bar{x}$'s and 100 $\hat{p}$'s.
3. Repeat Step 2 with SRSs of size $n = 20$. Make sure these dotplots are on the same scale as the corresponding dotplots from Step 2.
4. Compare the distribution of X (year of penny) with the two distributions of $\bar{x}$ (sample mean year). How are the distributions similar? How are they different? What effect does sample size seem to have on the shape, center, and variability of the distribution of $\bar{x}$?
5. Compare the two distributions of $\hat{p}$. How are the distributions similar? How are they different? What effect does sample size seem to have on the shape, center, and variability of the distribution of $\hat{p}$?

Many of the AP® Statistics course topics and skill categories from Units 1–4 come together in this unit, including ways of describing distributions, normal distribution calculations, random sampling and bias, and rules for random variables. Furthermore, the content in this unit forms the foundation for Units 6–9, so it is very important that you have a solid understanding of the topics in this unit. Section 5A introduces continuous random variables and revisits normal distribution calculations. Section 5B presents the big ideas of sampling distributions. Section 5C focuses on sampling distributions involving proportions, while Section 5D investigates sampling distributions involving means.

SECTION 5A Normal Distributions, Revisited

LEARNING TARGETS *By the end of the section, you should be able to:*

- Calculate probabilities and percentiles involving normal random variables.
- Find probabilities involving a sum, a difference, or another linear combination of independent, normal random variables.

In Unit 4, you learned how to analyze discrete random variables and their probability distributions. Recall that discrete random variables often result from *counting* something, such as the number of heads in five coin tosses or the number of Democrats in a random sample of 100 registered voters.

This section focuses on **continuous random variables,** with an emphasis on random variables that can be modeled by a normal distribution. Continuous random variables often result from *measuring* something, such as the distance a car can drive on a single tank of gas or the weight of a randomly selected newborn baby. The rules for transforming and combining random variables that we developed in Section 4E apply to both discrete and continuous random variables.

DEFINITION Continuous random variable

A **continuous random variable** can take any value in a specified interval on the number line.

Blaine Harrington/agefotostock/Newscom

Let's look at an example of a continuous random variable. Selena works at a bookstore in the Denver International Airport. To get to work, she can either take the airport train from the main terminal or use a moving walkway that would allow her to get from the main terminal to the bookstore in 4 minutes. Selena wonders if it will be faster to walk or take the train to work. Let the random variable X = journey time (in minutes) using the airport train.

In Section 4D, you learned that we can use a histogram to display the probability distribution of a discrete random variable. When we want to display the probability distribution of a continuous random variable, like X = journey time, we use a **density curve** similar to the one shown here.

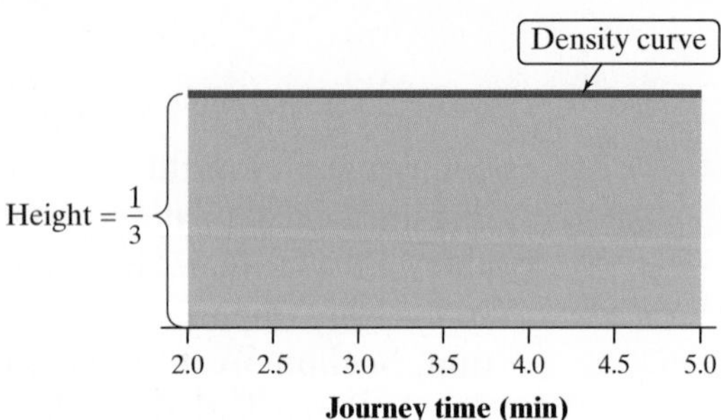

You might wonder why this density curve is drawn at a height of 1/3. Recall that in a probability histogram for a discrete random variable, the heights of the bars add to 1, or 100%. Likewise, we want the rectangular area under the density curve between 2 minutes and 5 minutes to equal 1, or 100%:

$$\text{area} = \text{base} \times \text{height} = 3 \times 1/3 = 1.00 = 100\%$$

This type of density curve is called a *uniform density curve* because it has constant height.

DEFINITION Density curve

A **density curve** models the probability distribution of a continuous random variable and has two characteristics:

- Is always on or above the horizontal axis
- Has an area of exactly 1 underneath it

The area under the density curve and above any specified interval of values on the horizontal axis gives the probability that the random variable falls within that interval.

To find the probability that Selena takes less than 4 minutes to get to work when using the airport train, we calculate the area under the density curve between 2 and 4 minutes (shown in red).

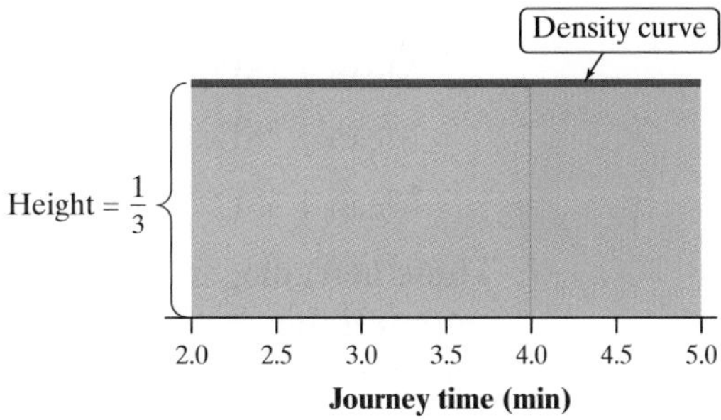

$$P(X < 4) = \text{base} \times \text{height} = 2 \times 1/3 = 2/3 = 0.667$$

The probability that Selena gets to work in less than 4 minutes when using the airport train is 0.667. Because the train will be faster on about 2/3 of the days, she should plan to take the train instead of walking.

What's the probability $P(X = 4)$ that it takes Selena *exactly* 4 minutes to get to work using the airport train? The probability of this event is the area under the density curve that's directly above the point 4.0000 . . . on the horizontal axis. But this vertical line segment has no width, so the area is 0. In fact, all continuous probability distributions assign probability 0 to every individual outcome. For that reason,

$$P(X < 4) = P(X \leq 4) = 0.667$$

Remember: Unlike for discrete random variables, the probability distribution of a continuous random variable assigns probabilities to *intervals* of values rather than to individual values.

Normal Random Variables

The density curves that are most familiar are the normal distributions from Section 1F. Normal curves can be probability distributions as well as models for data.

The histogram in Figure 5.1 shows the distribution of total length (in centimeters, cm) for a sample of 104 mountain brushtail possums.[2] The histogram is roughly symmetric and "bell-shaped," with both tails falling off from a single center peak. Is this distribution approximately normal?

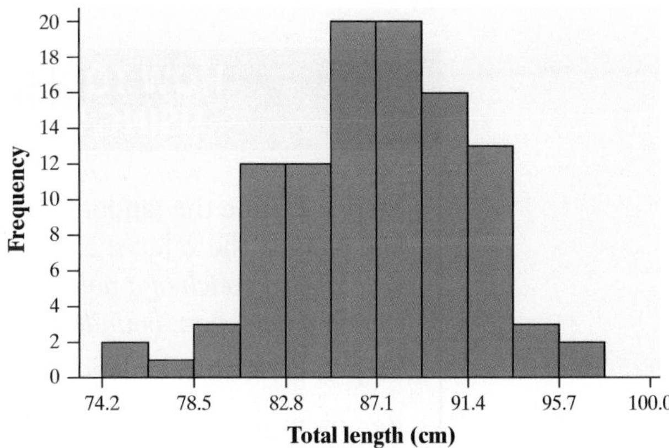

FIGURE 5.1 Histogram showing the distribution of total length (cm) for a random sample of 104 mountain brushtail possums.

As you learned in Section 1F, the empirical rule can give additional evidence in favor of or against normality. For these data, the mean is 87.1 cm and the standard deviation is 4.3 cm. Let's count the number of data values within 1, 2, and 3 standard deviations of the mean:

$$\text{Mean} \pm 1\text{SD:} \quad 87.1 \pm 1(4.3) \quad 82.8 \text{ to } 91.4 \quad 68 \text{ out of } 104 = 65.4\%$$

$$\text{Mean} \pm 2\text{SD:} \quad 87.1 \pm 2(4.3) \quad 78.5 \text{ to } 95.7 \quad 99 \text{ out of } 104 = 95.2\%$$

$$\text{Mean} \pm 3\text{SD:} \quad 87.1 \pm 3(4.3) \quad 74.2 \text{ to } 100.0 \quad 104 \text{ out of } 104 = 100\%$$

These percentages are quite close to the 68%, 95%, and 99.7% specified by the empirical rule. So the distribution of total length for mountain brushtail possums is approximately normal.

Recall from Section 4D that a probability distribution can be used to model the population distribution of a quantitative variable. If we define $X =$ the total length (cm) of a randomly selected mountain brushtail possum, then X is a continuous random variable. Its probability distribution is the normal density curve with mean $\mu_X = 87.1$ cm and standard deviation $\sigma_X = 4.3$ cm shown in Figure 5.2. We refer to X as a **normal random variable.**

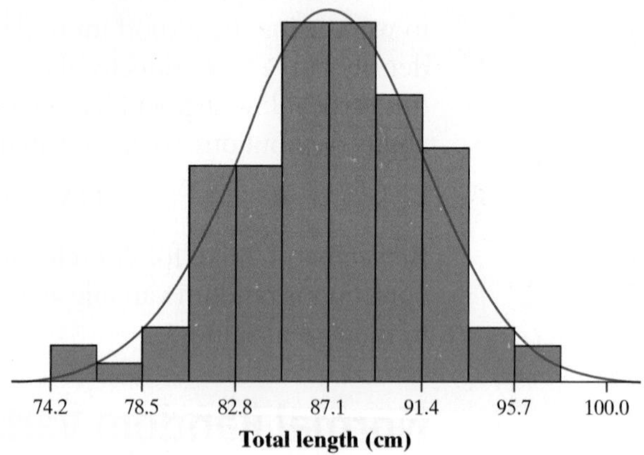

FIGURE 5.2 A normal density curve models the probability distribution of the continuous random variable $X =$ the total length of a randomly selected mountain brushtail possum.

DEFINITION Normal random variable

A **normal random variable** is a continuous random variable whose probability distribution is described by a normal curve.

The methods for calculating probabilities and percentiles involving normal random variables should seem familiar from Section 1F. When doing normal probability calculations, follow the two-step process we used for binomial and geometric probability calculations.

HOW TO PERFORM CALCULATIONS USING NORMAL DISTRIBUTIONS

Step 1: Define the random variable of interest, state how it is distributed, and identify the value(s) of interest. *This can be done verbally or with a well-labeled sketch of a normal curve that includes the variable name, mean, standard deviation, boundary value(s), and area(s) of interest.*

Step 2: Perform calculations — show your work!

Be sure to answer the question that was asked.

EXAMPLE

The baby and the bathwater
Normal random variables

Skills 3.A, 3.B

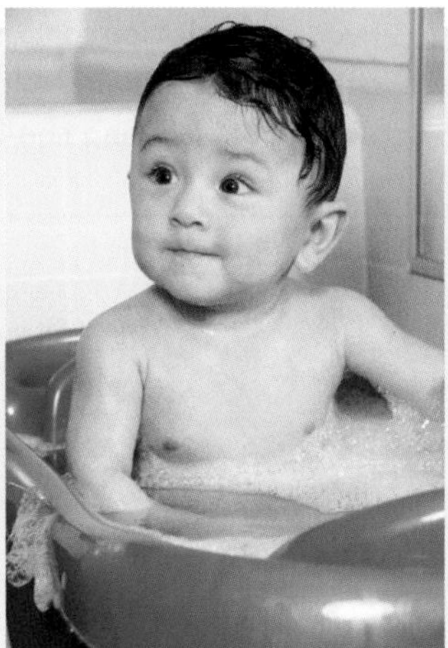

MichaelDeLeon/Getty Images

PROBLEM: One brand of baby bathtub comes with a dial to set the water temperature. When the "baby-safe" setting is selected and the tub is filled, the temperature X of the water in a randomly selected bath can be modeled by a normal probability distribution with a mean of 34°C and a standard deviation of 2°C. Let Y be the water temperature in degrees Fahrenheit for the randomly selected bath. Recall that $°F = 32 + \frac{9}{5}°C$.

According to one source, the temperature of a baby's bathwater should be between 90°F and 100°F. We are interested in finding the probability that this happens on a randomly selected day when the "baby-safe" setting is used.

(a) Find the mean and standard deviation of Y.

(b) Determine the probability that the bathwater has a temperature between 90°F and 100°F on a randomly selected day when the "baby-safe" setting is used.

(c) Based on your answer to part (b), should a parent feel comfortable using the "baby-safe" setting with this brand of baby bathtub? Explain your answer.

SOLUTION:

(a) $\mu_Y = 32 + \frac{9}{5}(34) = 93.2°F$

$\sigma_Y = \frac{9}{5}(2) = 3.6°F$

> Recall from Section 4E that for a linear transformation of the form $Y = a + bX$, the mean of Y is $\mu_Y = a + b\mu_X$ and the standard deviation of Y is $\sigma_Y = |b|\sigma_X$.

(b) Let Y = the temperature (in °F) of the bathwater on a randomly selected day when the "baby-safe" setting is used. Y has a normal distribution with mean 93.2 °F and standard deviation 3.6 °F. We want to find $P(90 < Y < 100)$.

> **Step 1:** Define the random variable of interest, state how it is distributed, and identify the value(s) of interest. *This can be done verbally or with a well-labeled sketch of a normal curve that includes the variable name, mean, standard deviation, boundary value(s), and area(s) of interest.*

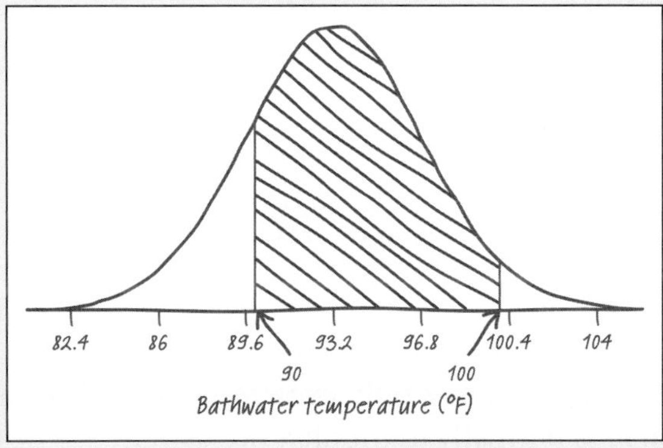

(i) $z = \frac{90 - 93.2}{3.6} = -0.889$ and $z = \frac{100 - 93.2}{3.6} = 1.889$

Using technology: normalcdf(lower: −0.889, upper: 1.889, mean: 0, SD: 1) = 0.7836

Using Table A: 0.9706 − 0.1867 = 0.7839

(ii) normalcdf(lower: 90, upper: 100, mean: 93.2, SD: 3.6) = 0.7835

> **Step 2:** Perform calculations — show your work!
> **(i)** Standardize the boundary value(s) and use technology or Table A to find the desired probability; or
> **(ii)** Use technology to find the desired area without standardizing. Label the calculator inputs.

(c) When using the baby-safe setting, there's about a 78% chance that the water temperature meets the recommendation for a randomly selected bath. There is about a 22% chance that the bathwater will be either too cold or too hot, so a parent should not be very comfortable using the "baby-safe" setting with this brand of bathtub.

Be sure to answer the question that was asked.

FOR PRACTICE, TRY EXERCISE 7

You can also use a normal probability distribution to find percentiles. Just follow the process you learned in Section 1F. For instance, what is the 99th percentile of the bathwater temperature (in °F) when the baby-safe setting is used?

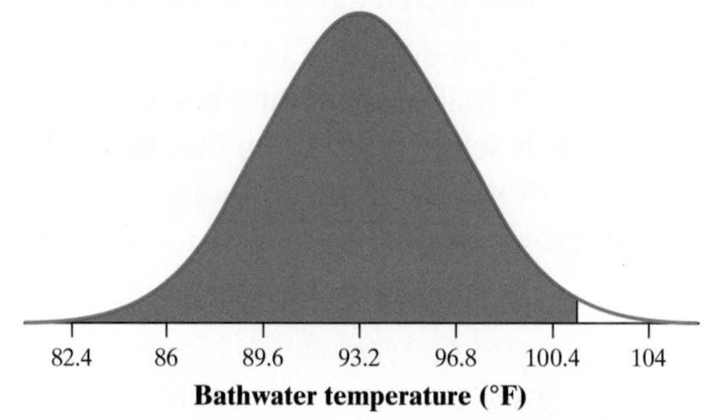

(i) *Using technology*: invNorm(area: 0.99, mean: 0, SD: 1) = 2.326

Using Table A: 0.99 area to the left → $z = 2.33$

$$2.326 = \frac{y - 93.2}{3.6}$$

$$2.326(3.6) + 93.2 = y$$

$$101.6 = y$$

(ii) invNorm(area: 0.99, mean: 93.2, SD: 3.6) = 101.6

The 99th percentile of the bathwater temperature on a randomly selected day when the baby-safe setting is used is 101.6°F.

Combining Normal Random Variables

In Section 4E, we developed formulas for calculating means and variances of a sum, a difference, or another linear combination of two random variables. Recall that the formulas for calculating variances only work for *independent* random variables. If a random variable is normally distributed, we can use its mean and standard deviation to compute probabilities. What happens if we combine two independent, normal random variables?

We used software to simulate separate random samples of size 1000 for each of two independent, normally distributed random variables, X and Y. Their means and standard deviations are as follows:

$$\mu_X = 2.5, \sigma_X = 0.9 \quad \mu_Y = 1.5, \sigma_Y = 1.2$$

What do we know about the sum and the difference of these two random variables? The histograms in Figure 5.3 show the results of (a) adding and (b) subtracting the corresponding values of X and Y for the 1000 randomly generated observations from each probability distribution.

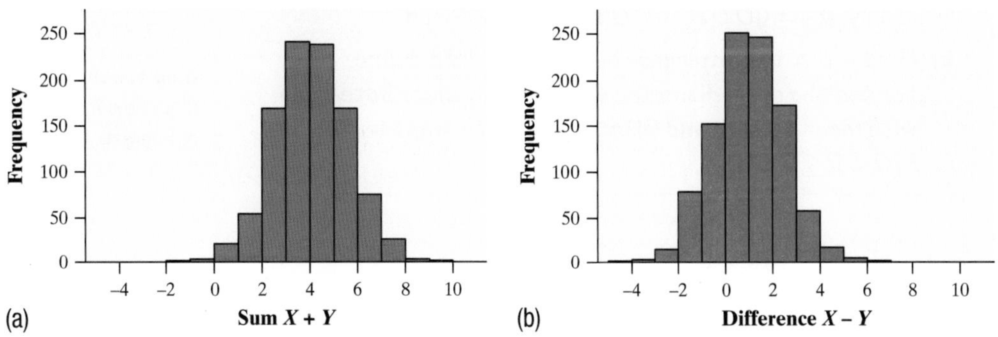

FIGURE 5.3 Histograms showing the results of randomly selecting 1000 values of two independent, normal random variables X and Y and finding (a) the sum $X + Y$ and (b) the difference $X - Y$ of the corresponding values.

As the simulation illustrates, *any sum or difference of independent, normal random variables is also normally distributed.* In fact, any *linear combination* of independent normal random variables will be normally distributed. The mean and standard deviation of the resulting normal distributions can be found using the appropriate rules for means and standard deviations from Section 4E:

	Sum $X + Y$	Difference $X - Y$
Mean	$\mu_{X+Y} = \mu_X + \mu_Y = 2.5 + 1.5 = 4$	$\mu_{X-Y} = \mu_X - \mu_Y = 2.5 - 1.5 = 1$
SD	$\sigma^2_{X+Y} = \sigma^2_X + \sigma^2_Y = 0.9^2 + 1.2^2 = 2.25$ $\sigma_{X+Y} = \sqrt{2.25} = 1.5$	$\sigma^2_{X-Y} = \sigma^2_X + \sigma^2_Y = 0.9^2 + 1.2^2 = 2.25$ $\sigma_{X-Y} = \sqrt{2.25} = 1.5$

EXAMPLE

Will the lid fit?
Combining normal random variables

Skills 3.A, 3.B

karandaev/Getty Images

PROBLEM: The diameter C of the top of a randomly selected large drink cup at a fast-food restaurant can be modeled by a normal distribution with a mean of 3.96 inches and a standard deviation of 0.01 inch. The diameter L of a randomly selected large lid at this restaurant can be modeled by a normal distribution with mean 3.98 inches and standard deviation 0.02 inch. Assume that L and C are independent random variables. For a lid to fit on a cup, the value of L has to be bigger than the value of C, but not by more than 0.06 inch. Let the random variable $D = L - C$ be the difference between the lid's diameter and the cup's diameter.

(a) Describe the probability distribution of D.

(b) What is the probability that a randomly selected lid will fit on a randomly selected cup? Describe this probability in words.

SOLUTION:

(a) Shape: Normal because D is the difference of independent, normal random variables.

Center: $\mu_D = 3.98 - 3.96 = 0.02$ inch

Variability: $\sigma_D = \sqrt{0.02^2 + 0.01^2} = 0.0224$ inch

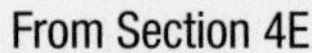

From Section 4E,

- $\mu_D = \mu_L - \mu_C$
- $\sigma_D = \sqrt{\sigma_L^2 + \sigma_C^2}$

(b) $D = L - C$ = the difference between the lid's diameter and the cup's diameter. D is normally distributed with mean $= 0.02$ and SD $= 0.0224$. We want to find $P(0 < D \leq 0.06)$.

Step 1: Define the random variable of interest, state how it is distributed, and identify the values of interest.

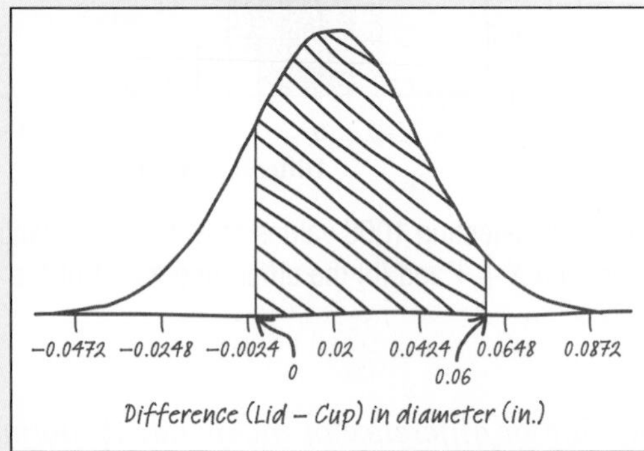

(i) $z = \dfrac{0 - 0.02}{0.0224} = -0.893$ and $z = \dfrac{0.06 - 0.02}{0.0224} = 1.786$

Step 2: Perform calculations—show your work!

Using technology: normalcdf(lower: −0.893, upper: 1.786, mean: 0, SD: 1) = 0.7770

Using Table A: 0.9633 − 0.1867 = 0.7766

(ii) normalcdf(lower: 0, upper: 0.06, mean: 0.02, SD: 0.0224) = 0.7770

There's about a 78% chance that a randomly selected lid will fit on a randomly selected cup.

Be sure to answer the question that was asked.

FOR PRACTICE, TRY EXERCISE 9

domin/Getty Images

We can extend what we have learned about combining independent normal random variables to settings that involve repeated observations from the same probability distribution. Consider this scenario. Mr. Starnes likes sugar in his hot tea — but not too much sugar! From experience, he knows that more than 9 grams of sugar in a cup makes the tea taste too sweet. While making his tea one morning, Mr. Starnes adds four randomly selected packets of sugar. Suppose the amount of sugar in these packets follows a normal distribution with mean 2.17 grams and standard deviation 0.08 gram. What's the probability that Mr. Starnes's tea tastes too sweet?

Let X = amount of sugar in a randomly selected packet. Then X_1 = amount of sugar in Packet 1, X_2 = amount of sugar in Packet 2, X_3 = amount of sugar in Packet 3, and X_4 = amount of sugar in Packet 4. Each of these random variables has a normal distribution with mean 2.17 grams and standard deviation 0.08 gram. We're interested in the total amount of sugar that Mr. Starnes puts in his tea: $T = X_1 + X_2 + X_3 + X_4$.

The random variable T is a sum of four independent normal random variables. So T follows a normal distribution with mean

$$\mu_T = \mu_{X_1} + \mu_{X_2} + \mu_{X_3} + \mu_{X_4} = 2.17 + 2.17 + 2.17 + 2.17 = 8.68 \text{ grams}$$

and variance

$$\sigma_T^2 = \sigma_{X_1}^2 + \sigma_{X_2}^2 + \sigma_{X_3}^2 + \sigma_{X_4}^2 = 0.08^2 + 0.08^2 + 0.08^2 + 0.08^2 = 0.0256 \text{ gram}^2$$

The standard deviation of T is

$$\sigma_T = \sqrt{0.0256} = 0.16 \text{ gram}$$

We want to find the probability $P(T > 9)$ that the total amount of sugar in Mr. Starnes's tea is greater than 9 grams. Figure 5.4 shows this probability as the area under a normal curve.

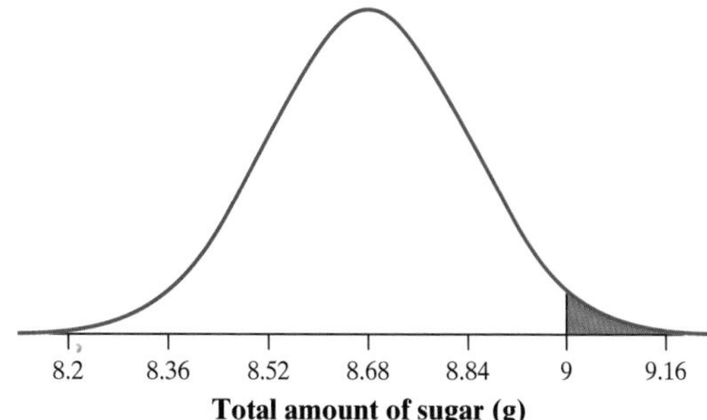

FIGURE 5.4 Normal distribution of the total amount of sugar in Mr. Starnes's tea.

To find this area, we can use either of our two familiar methods:

(i) Standardize the boundary values and use technology or Table A to find the area:

$$z = \frac{9 - 8.68}{0.16} = 2.000$$

Using technology: normalcdf(lower: 2.000, upper: 1000, mean: 0, SD: 1) = 0.0228
Using Table A: 1 − 0.9772 = 0.0228

(ii) Use technology to find the desired area without standardizing:

normalcdf(lower: 9, upper: 1000, mean: 8.68, SD: 0.16) = 0.0228

There's about a 2.28% chance that Mr. Starnes's tea will taste too sweet.

CHECK YOUR UNDERSTANDING

The amount of liquid soap X dispensed by an automatic soap dispenser each time it is activated can be modeled by a normal probability distribution with mean 0.9 milliliter (ml) and standard deviation 0.042 ml.

1. Calculate the probability that the amount of soap dispensed is less than 0.8 ml when the dispenser is activated once.
2. Find the 95th percentile of X. Describe this value in words.

Some experts suggest that at least 1.6 ml of liquid soap should be used by someone washing their hands. Suppose that the amount of soap dispensed on one activation is independent of the amount dispensed on another activation.

3. Determine the probability that a person will get at least 1.6 ml of soap from the dispenser if they activate it twice.

SECTION 5A Summary

- A **continuous random variable** can take any value in a specified interval on the number line.
- The probability distribution of a continuous random variable is described by a **density curve** that is always on or above the horizontal axis and has total area 1 underneath it.
- The area under the density curve and above any specified interval of values on the horizontal axis gives the probability that the random variable falls within that interval.
- A **normal random variable** is a continuous random variable whose probability distribution is described by a normal curve.
- When calculating probabilities involving normal random variables, follow this process:

  Step 1: Define the random variable of interest, state how it is distributed, and identify the values of interest. *This can be done verbally or with a well-labeled sketch of a normal curve that includes the variable name, mean, standard deviation, boundary value(s), and area(s) of interest.*

  Step 2: Perform calculations — show your work!

  Be sure to answer the question that was asked.
- A linear combination of independent, normal random variables is a normal random variable.

AP® EXAM TIP
AP® Daily Videos

Review the content of this section and get extra help by watching the AP® Daily Videos for Topics 5.1 and 5.2, which are available in AP® Classroom.

SECTION 5A Exercises

Introduction

1. **Reaction time** An internet reaction time test asks subjects to click their mouse as soon as a light flashes on the screen. The light is programmed to illuminate at a randomly selected time from 1 to 5 seconds after the subject clicks "Start." The density curve models the probability distribution of the amount of time the subject has to wait for the light to flash.

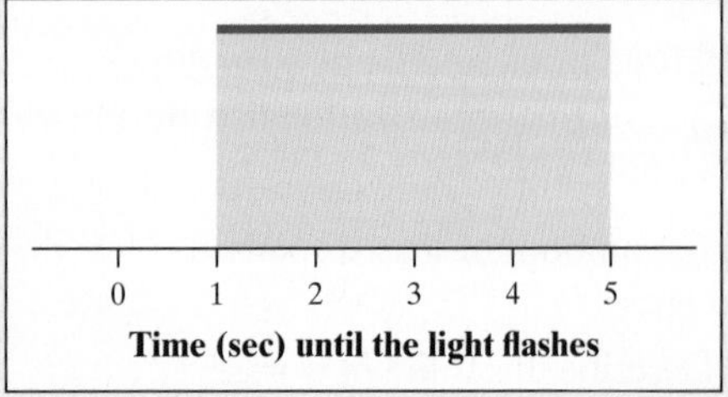

(a) What height must the density curve have? Justify your answer.

(b) Calculate the probability that the light flashes between 2.5 and 4 seconds after the subject clicks "Start."

2. **Class dismissal** Mr. Shrager does not always let his statistics class out at the scheduled time. In fact, he seems to end class according to his own "internal clock." The density curve models the probability distribution of the amount of time after the scheduled end of class (in minutes) when Mr. Shrager dismisses the class on a randomly selected day. (A negative value indicates he dismissed his class early.)

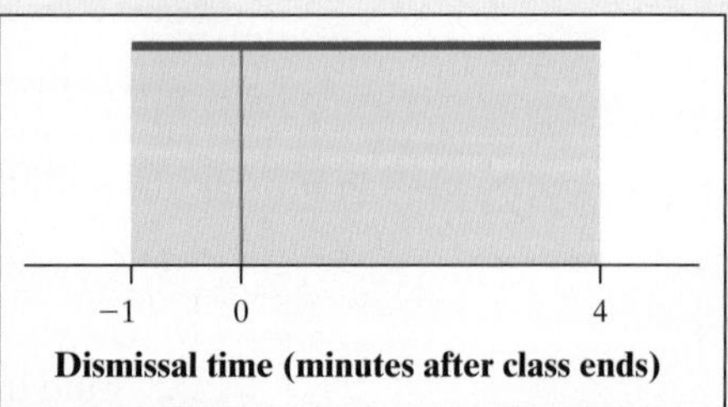

(a) What height must the density curve have? Justify your answer.

(b) Calculate the probability that Mr. Shrager ends class within 1 minute (before or after) of the scheduled end of class on a randomly selected day.

Normal Random Variables

3. **Normal aptitude?** The stemplot displays the scores of 60 fifth-grade students on an aptitude test. For these aptitude test scores, the mean is 115 and the standard deviation is 15. Use the stemplot and the empirical rule to determine whether it would be appropriate to use a normal curve to model the probability distribution of the random variable X = aptitude test score of a randomly selected fifth-grade student.

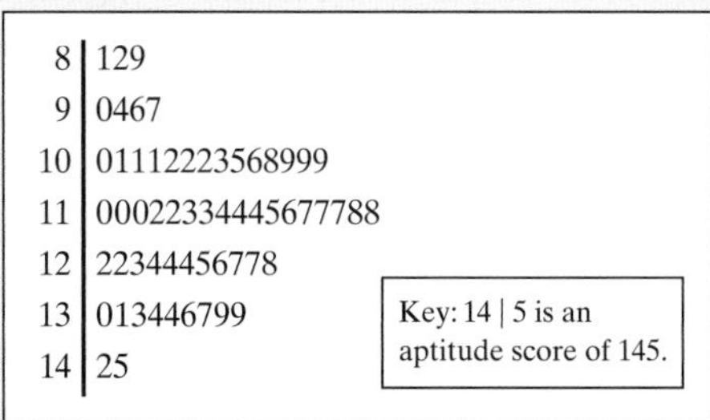

8 | 129
9 | 0467
10 | 01112223568999
11 | 000223344456777788
12 | 22344456778
13 | 013446799
14 | 25
Key: 14 | 5 is an aptitude score of 145.

4. **Normal to be older?** Here is a histogram of the percentage of residents aged 65 and older in the 50 U.S. states.[3] For these data, the mean is 16.5% and the standard deviation is 2%. Use the histogram and the empirical rule to determine whether it would be appropriate to use a normal curve to model the probability distribution of the random variable Y = percentage of residents aged 65 or older in a randomly selected U.S. state.

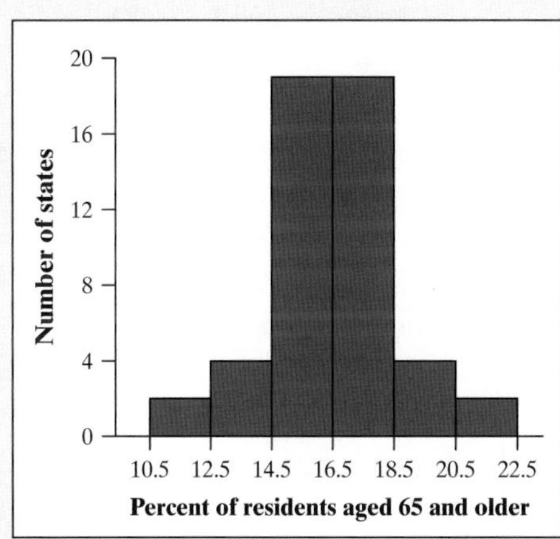

5. **Running a mile** A study of 12,000 students at the University of Illinois found that their times for the mile run were approximately normally distributed with mean 7.11 minutes and standard deviation 0.74 minute.[4] Suppose we choose a student at random from this group.

(a) What's the probability that the student's mile run time is between 6 and 7 minutes? Describe this probability in words.

(b) Find the 10th percentile of the probability distribution.

6. **Horse pregnancies** Bigger animals tend to carry their young longer before giving birth. The length of horse pregnancies from conception to birth varies according to a roughly normal distribution with mean 336 days and standard deviation 6 days. Suppose we randomly select a horse pregnancy.

(a) What's the probability that the pregnancy lasts between 325 and 345 days? Describe this probability in words.

(b) Find the 80th percentile of the probability distribution.

7. pg 457 **Too cool at the cabin?** During the winter months, the temperatures at the Starnes family cabin in Colorado can stay well below freezing (32°F or 0°C) for weeks at a time. To prevent the pipes from freezing, Mrs. Starnes sets the thermostat at 50°F. She also buys a digital thermometer that records the indoor temperature each night at midnight. Unfortunately, the thermometer is programmed to measure the temperature in degrees Celsius. Based on several years' worth of data, the temperature *T* in the cabin at midnight on a randomly selected night can be modeled by a normal distribution with mean 8.5°C and standard deviation 2.25°C. Let Y = the temperature in the cabin at midnight on a randomly selected night in degrees Fahrenheit. Recall that $°F = 32 + \frac{9}{5}°C$.

(a) Find the mean and standard deviation of Y.

(b) Determine the probability that the midnight temperature in the cabin is less than 40°F.

(c) Based on your answer to part (b), can Mrs. Starnes be fairly confident that the cabin temperature will be at least 40°F at midnight? Explain your answer.

8. **How much cereal?** A company's single-serving cereal boxes advertise that they contain 1.63 ounces of cereal. In fact, the amount of cereal X in a randomly selected box can be modeled by a normal distribution with a mean of 1.70 ounces and a standard deviation of 0.03 ounce. Let Y = the *excess* amount of cereal beyond what's advertised in a randomly selected box, measured in grams (1 ounce = 28.35 grams).

(a) Find the mean and standard of Y.

(b) Determine the probability of getting at least 1 gram more cereal than advertised.

(c) Based on your answer to part (b), can a customer who buys a single-serving box of this company's cereal be fairly confident that they will get at least 1 gram more cereal than advertised? Explain your answer.

Combining Normal Random Variables

9. pg 459 **Yard work** Zoe and Chloe run a two-person lawn-care service. They have been caring for one customer's very large lawn for several years, and they have found that the time Z it takes Zoe to mow the lawn on a randomly selected day can be modeled by a normal distribution with a mean of 105 minutes and a standard

deviation of 10 minutes. The time C it takes Chloe to use the edger and string trimmer on a randomly selected day can be modeled by a normal distribution with a mean of 98 minutes and a standard deviation of 15 minutes. Assume that Z and C are independent random variables. Let the random variable $D = Z - C$ be the difference in the time it takes Zoe and Chloe to finish their work on this lawn on a randomly selected day.

(a) Describe the probability distribution of D.

(b) What is the probability that Zoe and Chloe will finish their jobs within 5 minutes of each other on a randomly selected day?

10. **Hit the track** Andrea and Barsha are middle-distance runners for their school's track team. Andrea's time A in the 400-meter race on a randomly selected day can be modeled by a normal distribution with a mean of 62 seconds and a standard deviation of 0.8 second. Barsha's time B in the 400-meter race on a randomly selected day can be modeled by a normal distribution with a mean of 62.8 seconds and a standard deviation of 1 second. Assume that A and B are independent random variables. Let the random variable $D = A - B$ be the difference in the two runners' 400-meter race times on a randomly selected day.

(a) Describe the probability distribution of D.

(b) What is the probability that Barsha beats Andrea in the 400-meter race on a randomly selected day?

11. **Hay, weight!** On a certain farm, the baling machine produces hay bales with weights that are approximately normally distributed with mean 100 pounds and standard deviation 6 pounds. Assume that the weights of individual bales are independent of each other. A hay wagon that is designed to hold 10 bales of hay has a weight limit of 1050 pounds. Find the probability that 10 randomly selected hay bales from this farm will have a total weight that exceeds this limit.

12. **Toothpaste revisited** Ken is traveling for business. He has a new 0.85-ounce tube of toothpaste that's supposed to last him the whole trip. The amount of toothpaste Ken squeezes out of the tube each time he brushes is approximately normally distributed with mean 0.13 ounce and standard deviation 0.02 ounce. If Ken brushes his teeth six times on a randomly selected trip and the amounts of toothpaste used each time are independent, what's the probability that he will use all of the toothpaste in the tube?

13. **Gift box revisited** A company offers gift boxes containing 4 pears and 5 chocolate truffles with an advertised weight of 800 grams. The weights of individual pears can be modeled by a normal distribution with mean 188 grams and standard deviation 6 grams. The weights of individual chocolate truffles can be modeled by a normal distribution with mean 12.5 grams and standard deviation 0.7 gram. The weight of a box can be modeled by a normal distribution with mean 20 grams and standard deviation 0.9 gram. Assume that the weights of individual pears, chocolate truffles, and boxes are independent. Find the probability that a gift box consisting of a randomly selected box, 4 randomly selected pears, and 5 randomly selected chocolate truffles weighs less than the advertised 800 grams.

14. **Swim team revisited** Hanover High School has the best women's swimming team in the region. The 400-meter freestyle relay team is undefeated this year. In the 400-meter freestyle relay, each swimmer swims 100 meters. The times, in seconds, for the four swimmers this season are approximately normally distributed with the means and standard deviations as shown. Assume that all four swimmers' individual times in a race are independent. Find the probability that the total team time in the 400-meter freestyle for a randomly selected race is less than 220 seconds.

Swimmer	Mean	StDev
Wendy	55.2	2.8
Jill	58.0	3.0
Carmen	56.3	2.6
Latrice	54.7	2.7

For Investigation *Apply the skills from the section in a new context or nonroutine way.*

15. **Unusual density curve** The figure shows a density curve that models the probability distribution of a continuous random variable X.

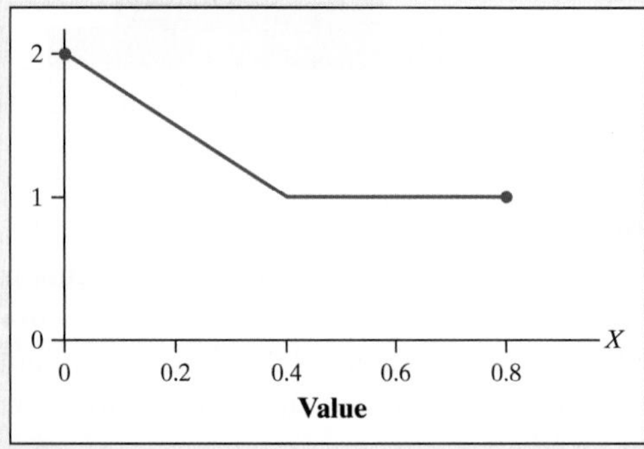

(a) Show that this is a valid density curve.

(b) What is the probability that X takes a value between 0 and 0.2?

(c) The median of X is between 0.2 and 0.4. Explain why.

(d) Is the mean of X less than, equal to, or greater than the median of the density curve? Justify your answer.

16. **Uniform distributions** What percentage of values in a uniform probability distribution are within 1 standard deviation of the mean? Within 2 standard deviations? For any uniform probability distribution, the standard deviation is $\sigma = \sqrt{\frac{(b-a)^2}{12}}$, where a is the minimum value in the distribution and b is the maximum value. Suppose that X is a uniform random variable that takes values from 0 to 10.

(a) What is the mean of X? Explain your reasoning.

(b) Use the formula to calculate the standard deviation of X.

(c) What percentage of the values of X are within 1 standard deviation of the mean? How does this compare with the empirical rule?

(d) What percentage of the values of X are within 2 standard deviations of the mean? How does this compare with the empirical rule?

Multiple Choice *Select the best answer for each question.*

17. A grinding machine in an auto parts plant prepares axles with a target diameter $\mu = 40.125$ millimeters (mm). The machine has some variability, so the diameters are approximately normally distributed with a standard deviation of $\sigma = 0.002$ mm. Assume the machine is working properly. What is the probability that a randomly selected axle has a diameter between 40.12 and 40.13 mm?

(A) 0.002
(B) 0.004
(C) 0.197
(D) 0.950
(E) 0.988

18. An ecologist studying starfish populations records values for each of the following variables from randomly selected plots on a rocky coastline.

X = The number of starfish in the plot
Y = The total weight of starfish in the plot
W = The percentage of area in the plot that is covered by barnacles (a popular food for starfish)

How many of these are continuous random variables and how many are discrete random variables?

(A) Three continuous
(B) Two continuous, one discrete
(C) One continuous, two discrete
(D) Three discrete
(E) One continuous, one discrete, one neither

19. The density curve shown models the probability distribution of a continuous random variable that is equally likely to take any value in the interval from 0 to 2. What is the probability of randomly selecting a value between 0.5 and 1.2?

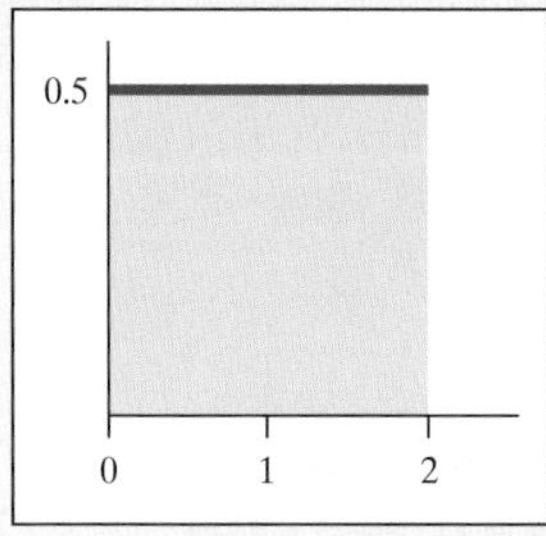

(A) 0.25
(B) 0.35
(C) 0.50
(D) 0.70
(E) 1.40

20. The weight of bananas chosen at random from a bin at the supermarket follows a normal distribution with mean $\mu = 4$ ounces and standard deviation $\sigma = 0.3$ ounce. Suppose we pick four bananas at random from the bin and find their total weight T. The random variable T is

(A) normal, with mean 4 ounces and standard deviation 0.3 ounce.
(B) normal, with mean 16 ounces and standard deviation 1.2 ounces.
(C) normal, with mean 16 ounces and standard deviation 0.60 ounce.
(D) binomial, with mean 16 ounces and standard deviation 1.2 ounces.
(E) binomial, with mean 16 ounces and standard deviation 0.60 ounce.

Recycle and Review *Practice what you learned in previous sections.*

21. **Denver quarters (4F)** In 2020, 49.5% of quarters were minted by the U.S. Mint in Denver, Colorado; the rest were produced in Philadelphia.[5] Suppose we select a random sample of 50 quarters produced in 2020. Let D = the number of quarters in the sample that were minted in Denver. (You can identify Denver as the mint by the small "D" to the lower right of George Washington.)

(a) Explain why D can be modeled by a binomial distribution even though the sample was selected without replacement.

(b) Calculate and interpret the mean and standard deviation of D.

22. **Colorado quarters (4F)** Refer to Exercise 21. Are quarters that were minted in Denver more likely to be found in Colorado? A Colorado resident collects 50 2020 quarters from local businesses and finds that 30 of them were minted in Denver. Assume that this can be considered a random sample of all 2020 quarters in this resident's town.

(a) If 49.5% of all 2020 quarters in the town were minted in Denver, find the probability that 30 or more of the quarters in the sample were minted in Denver.

(b) Based on your answer to part (a), is there convincing evidence that more than 49.5% of all 2020 quarters in this town were minted in Denver?

SECTION 5B What Is a Sampling Distribution?

LEARNING TARGETS *By the end of the section, you should be able to:*

- Distinguish between a parameter and a statistic.
- Create a sampling distribution using all possible samples from a small population.
- Use the sampling distribution of a statistic to evaluate a claim about a parameter.
- Determine if a statistic is an unbiased estimator of a population parameter.
- Describe the relationship between sample size and the variability of an estimator.

What is the mean income of U.S. residents with a college degree? Each March, the government's Current Population Survey (CPS) selects a random sample of U.S. residents and asks detailed questions about income. In 2021, the mean income for the *sample* of college graduates was $\bar{x} = \$91{,}892$.[6] How close is this estimate to $\mu =$ the mean income for the *population* of all college graduates in the United States? To find out how much an estimate varies from sample to sample, you need to understand *sampling distributions*.

Parameters and Statistics

As we begin to use sample data to draw conclusions about a larger population, we must be clear about whether a number describes a sample or a population. For the sample of college graduates contacted by the CPS, the mean income was $\bar{x} = \$91{,}892$. The number $91,892 is a **statistic** because it describes this one CPS sample. The population that the poll wants to draw conclusions about is the approximately 100 million U.S. residents with a college degree. In this case, the **parameter** of interest is the mean income μ of all these college graduates. We typically don't know the value of a parameter, but we can estimate it using data from the sample.

DEFINITION Statistic, Parameter

A **statistic** is a number that describes some characteristic of a sample.

A **parameter** is a number that describes some characteristic of a population.

A sample statistic is sometimes called a *point estimator* of the corresponding population parameter because the estimate — \$91,892 in this case — is a single point on the number line. You'll learn about *interval estimates* in Units 6–9.

Recall our hint from Unit 1 about **s** and **p**: statistics come from samples, and parameters come from populations. As long as we were doing descriptive statistics, the distinction between a population and a sample rarely came up. Now that we are focusing on statistical inference, however, this distinction is essential. The table shows three commonly used statistics and their corresponding parameters.

Sample statistic		Population parameter
$\bar{x}$ (the sample mean)	estimates	μ (the population mean)
$\hat{p}$ (the sample proportion)	estimates	p (the population proportion)
s_x (the sample SD)	estimates	σ (the population SD)

AP® EXAM TIP
Formula sheet

The formula sheet provided on both sections of the AP® Statistics exam uses both $\bar{x}$ and $\bar{X}$ to represent the sample mean.

Although we typically use capital letters to represent random variables, it is common practice to use lowercase letters for sample statistics. For example, we use s_x rather than S_X to represent the sample standard deviation, even though s_x is a random variable.

EXAMPLE

From ghosts to cold cabins
Parameters and statistics

Skill 3.B

PROBLEM: Identify the population, the parameter, the sample, and the statistic in each of the following scenarios.

(a) A Gallup poll asked 515 randomly selected U.S. adults if they believe in ghosts. Of the respondents, 160 said "Yes."[7]

(b) During the winter months, the temperatures outside the Starnes family cabin in Colorado can stay well below freezing for weeks at a time. To prevent the pipes from freezing, Mrs. Starnes sets the thermostat at 50°F. She wants to know how low the temperature actually gets in the cabin. A digital thermometer records the indoor temperature at 10 randomly chosen times during a given day. The minimum reading is 40°F.

Ryerson Clark/Getty Images

SOLUTION:

(a) Population: all U.S. adults. Parameter: p = the proportion of all U.S. adults who believe in ghosts. Sample: the 515 U.S. adults who were surveyed. Statistic: $\hat{p}$ = the proportion in the sample who say they believe in ghosts = 160/515 = 0.31.

(b) Population: all times during the day in question. Parameter: the true minimum temperature in the cabin at all times that day. Sample: the 10 randomly selected times. Statistic: the sample minimum temperature, 40°F.

Some parameters and statistics don't have their own symbols, such as the population minimum and sample minimum in part (b). To distinguish parameters and statistics in these cases, use descriptors like "all" and "sample."

FOR PRACTICE, TRY EXERCISE 1

AP® EXAM TIP

Many students lose credit on the AP® Statistics exam when defining parameters because their description refers to the sample instead of the population or because the description isn't clear about which group of individuals the parameter is describing. We strongly suggest including the word *population, all,* or *true* in your description of a parameter and the word *sample* in your description of a statistic.

The Idea of a Sampling Distribution

The students in Mr. Simonetti's class did a shorter version of the Penny for Your Thoughts activity described at the beginning of this unit. Figure 5.5 shows their "dotplot" of the sample proportion of pennies with a shield on the back for 50 samples of size $n = 20$.

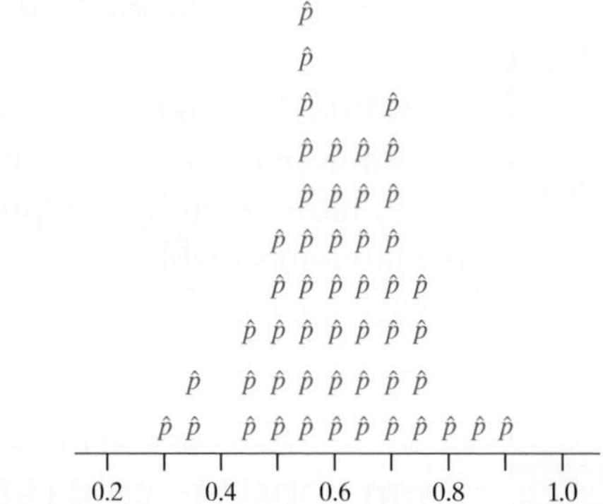

FIGURE 5.5 Distribution of $\hat{p}$ = sample proportion of pennies with a shield on the back for 50 samples of size $n = 20$ from Mr. Simonetti's population of pennies.

It shouldn't be surprising that the statistic $\hat{p}$ is a random variable. After all, different random samples of $n = 20$ pennies from the same population will produce different proportions. This basic fact is called **sampling variability.**

DEFINITION Sampling variability

Sampling variability refers to the fact that different random samples of the same size from the same population produce different values for a statistic.

It is because of sampling variability that the distinction between parameters and statistics is so important. After all, the value of a statistic will rarely be equal to the value of the parameter it is trying to estimate. Fortunately, we can estimate how much the value of the statistic typically varies from the parameter if we know the **sampling distribution** of the statistic.

DEFINITION Sampling distribution

The **sampling distribution** of a statistic is the distribution of values taken by the statistic in all possible samples of the same size from the same population.

For large populations, such as Mr. Simonetti's collection of pennies, it is too difficult to select all possible samples of size n to obtain the exact sampling distribution of a statistic. Instead, we can approximate a sampling distribution by taking many samples, calculating the value of the statistic for each of these samples, and graphing the results. Because the students in Mr. Simonetti's class didn't select all possible samples of 20 pennies, their dotplot of $\hat{p}$'s in Figure 5.5 is called an *approximate sampling distribution*.

The following example demonstrates how to construct a complete sampling distribution using a small population.

EXAMPLE

Sampling heights
The idea of a sampling distribution

Skill 3.C

PROBLEM: John and Carol have four grown sons. Their heights (in inches) are 71, 75, 72, and 68. List all 6 possible simple random samples (SRSs) of size $n = 2$ from this population of 4 sons, calculate the mean height for each sample, and display the sampling distribution of the sample mean on a dotplot.

Elizabeth Miller

SOLUTION:

Sample	Sample mean
71, 75	$\bar{x} = 73.0$
71, 72	$\bar{x} = 71.5$
71, 68	$\bar{x} = 69.5$
75, 72	$\bar{x} = 73.5$
75, 68	$\bar{x} = 71.5$
72, 68	$\bar{x} = 70.0$

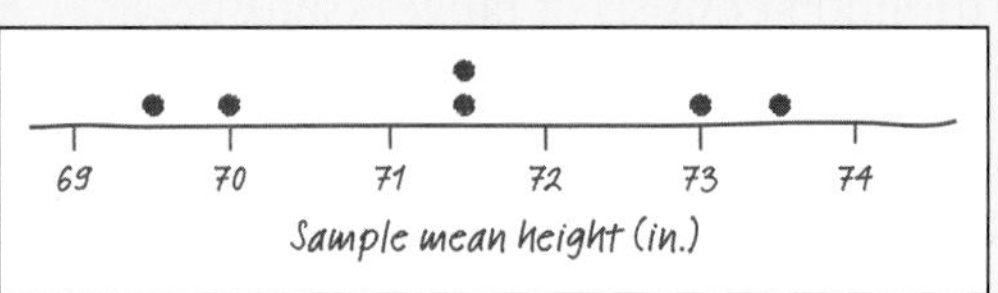

FOR PRACTICE, TRY EXERCISE 7

Every statistic has its own sampling distribution. For example, Figure 5.6 shows the sampling distribution of the sample range of height for SRSs of size $n = 2$ from John and Carol's four sons.

Sample 1: 71, 75; sample range = 4	Sample 4: 75, 72; sample range = 3
Sample 2: 71, 72; sample range = 1	Sample 5: 75, 68; sample range = 7
Sample 3: 71, 68; sample range = 3	Sample 6: 72, 68; sample range = 4

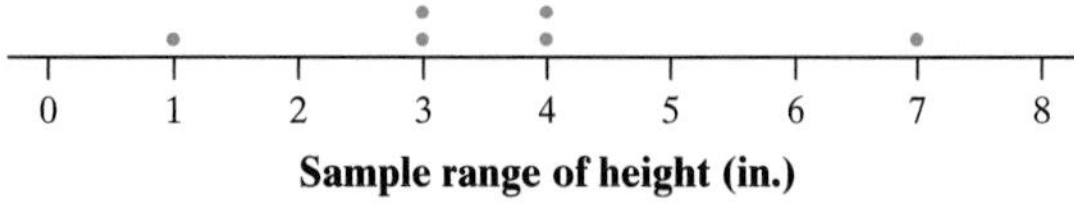

FIGURE 5.6 Dotplot showing the sampling distribution of the sample range of height for SRSs of size $n = 2$.

Be specific when you use the word *distribution*. There are three different types of distributions in this setting:

1. The distribution of height in the population (the four heights):

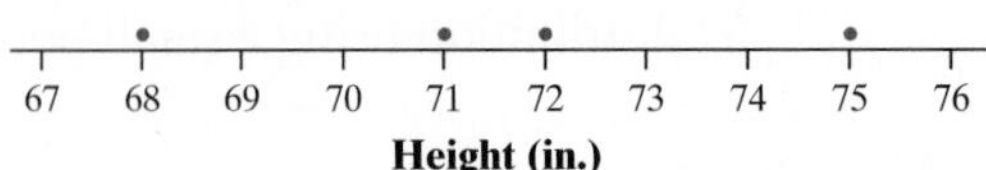

2. The distribution of height in a particular sample (two of the heights):

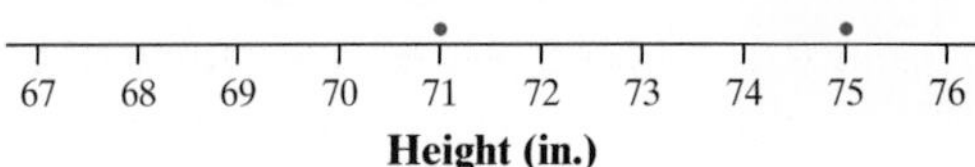

3. The sampling distribution of the sample range of height for all possible samples (the six sample ranges):

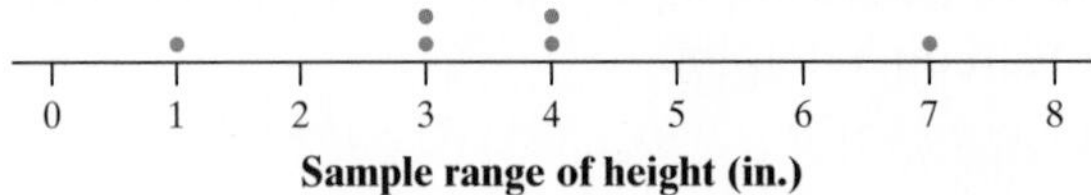

AP® EXAM TIP

Terminology matters. Never just say "the distribution." Always say "the distribution of [blank]," being careful to distinguish the distribution of the population, the distribution of sample data, and the sampling distribution of a statistic. Likewise, don't use ambiguous terms like "sample distribution," which could refer to the distribution of sample data or to the sampling distribution of a statistic. You will lose credit on free response questions for misusing statistical terms.

Notice that the first two distributions consist of heights (data values), while the third distribution consists of ranges (statistics).

Using Sampling Distributions to Evaluate Claims

Being able to construct (or approximate) the sampling distribution of a statistic allows us to determine the values of the statistic that are likely to occur by chance alone — and the values that should be considered unusual. The next example shows how we can use a *simulated* sampling distribution to evaluate a claim. *Note:* When we simulate a sampling distribution by using assumed values for the parameters, the resulting distribution is sometimes called a *randomization distribution.*

EXAMPLE

Reaching for chips
Using sampling distributions to evaluate claims

Skill 4.B

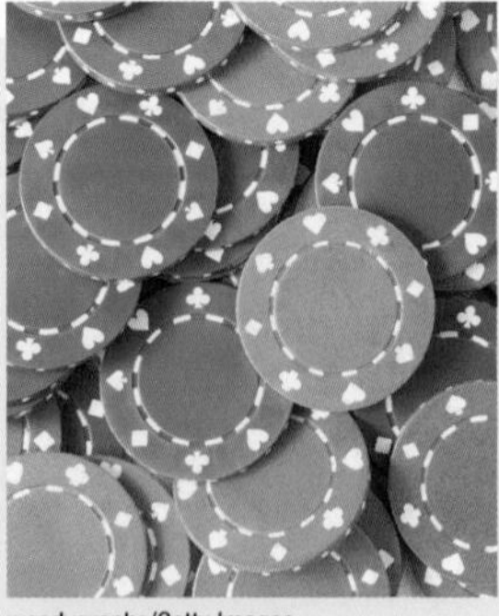
woodygraphs/Getty Images

PROBLEM: To determine how much homework time students will get in class, Mrs. Lin has a student select an SRS of 20 chips from a large bag. The number of red chips in the sample determines the number of minutes in class students get to work on homework. Mrs. Lin claims that there are 200 chips in the bag and that 100 of them are red. When Jenna selected a random sample of 20 chips from the bag (without looking), she got 7 red chips. Does this provide convincing evidence that less than half of the chips in the bag are red?

(a) What is the evidence that less than half of the chips in the bag are red?

(b) Provide two explanations for the evidence described in part (a).

We used technology to simulate choosing 100 SRSs of size $n = 20$ from a population of 200 chips, 100 red and 100 blue. The dotplot shows $\hat{p}$ = the sample proportion of red chips for each of the 100 simulated samples.

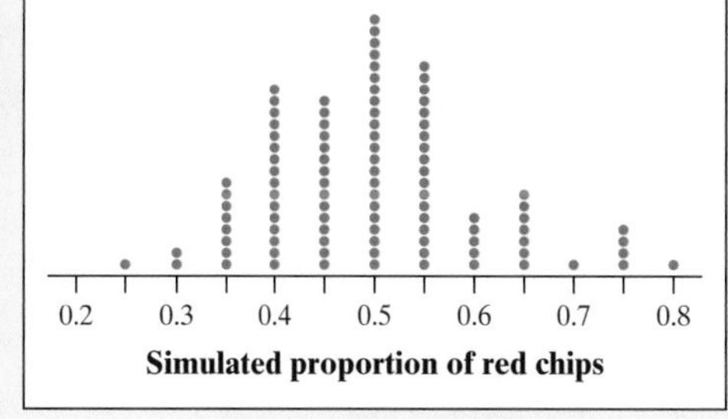

(c) There is one dot on the graph at 0.80. Explain what this value represents.

(d) Would it be surprising to get a sample proportion of $\hat{p} = 7/20 = 0.35$ or smaller in an SRS of size 20 when $p = 0.5$? Justify your answer.

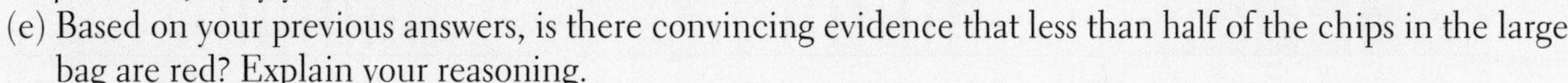

(e) Based on your previous answers, is there convincing evidence that less than half of the chips in the large bag are red? Explain your reasoning.

SOLUTION:

(a) Jenna's sample proportion was $\hat{p} = 7/20 = 0.35$, which is less than 0.50.

(b) It is possible that Mrs. Lin is telling the truth and Jenna got a $\hat{p}$ less than 0.50 because of sampling variability. It is also possible that Mrs. Lin is lying and less than half of the chips in the bag are red.

(c) In one simulated SRS of 20 chips, there were 16 red chips. So $\hat{p} = 16/20 = 0.80$ for this sample.

(d) No; there were 11 simulated samples (out of 100) that had $\hat{p}$ values less than or equal to 0.35.

Recall from Unit 3 that events with probabilities less than 0.05 are considered unusual.

(e) Because it isn't surprising to get a $\hat{p}$ less than or equal to 0.35 by chance alone when $p = 0.50$, there isn't convincing evidence that less than half of the chips in the bag are red.

FOR PRACTICE, TRY EXERCISE 13

Suppose that Jenna's sample included only 3 red chips, giving $\hat{p} = 3/20 = 0.15$. Would this provide convincing evidence that less than half of the chips in the bag are red? Yes. According to the simulated sampling distribution in the example, it would be very unusual to get a $\hat{p}$ value this small when $p = 0.50$. Therefore, sampling variability would not be a plausible explanation for the outcome of Jenna's sample. The only plausible explanation for a $\hat{p}$ value of 0.15 is that less than half of the chips in the bag are red.

CHECK YOUR UNDERSTANDING

Mars,® Inc. says that the mix of colors in its M&M'S® Milk Chocolate Candies from its Hackettstown, New Jersey, factory is 25% blue, 25% orange, 12.5% green, 12.5% yellow, 12.5% red, and 12.5% brown. Assume that the company's claim is true and that you will calculate $\hat{p}$ = the proportion of orange M&M'S in a random sample of 50 M&M'S.

1. Identify the population, the parameter, the sample, and the statistic in this setting.
2. Graph the population distribution.

3. Imagine taking a random sample of 50 M&M'S Milk Chocolate Candies. Make a graph showing a possible distribution of the sample data. Give the value of the statistic for this sample.
4. Which of these three graphs could be the approximate sampling distribution of the statistic? Explain your choice.

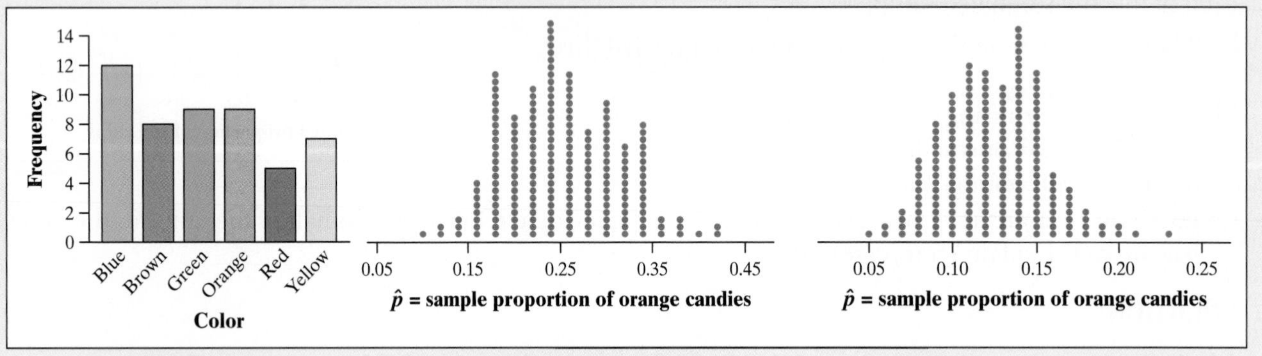

Biased and Unbiased Estimators

The fact that statistics from random samples have definite sampling distributions allows us to answer the question, "How trustworthy is a statistic as an estimate of a parameter?" To get a complete answer, we will consider the center, variability, and shape of the sampling distribution. For reasons that will be clear later, we'll save our discussion of shape until Sections 5C and 5D.

Here is an activity to get you thinking about the center and variability of a sampling distribution.

ACTIVITY

The craft stick problem

Stephen Orsillo/Alamy Stock Photo

In this activity, you will create a statistic for estimating the total number of craft sticks in a bag (N). The sticks are numbered 1, 2, 3, . . ., N. Near the end of the activity, your teacher will select a random sample of $n = 7$ sticks and read the number on each stick to the class. The team that has the best estimate for the total number of sticks will win a prize.

Let's start by investigating three different statistics as a class: twice the sample median, the sample mean + 3 sample standard deviations, and the sample maximum. For now, we'll assume that there are $N = 100$ sticks in the bag and we'll be selecting a random sample of $n = 7$ sticks.

1. Using your TI-83/84 calculator, press STAT and choose Edit... to get into the list editor. Move your cursor so that the heading of L1 is highlighted and clear the list (if necessary). While L1 is highlighted, press the MATH button,

scroll to the Prob menu, and choose RandInt. Enter the values (lower: 1, upper: 100, n: 7), highlight "Paste," and press ENTER twice.

- **Older OS:** While L1 is highlighted, use the command RandInt(1, 100, 7).

If there are repeated values, redo the process until there are no repeats in the sample.

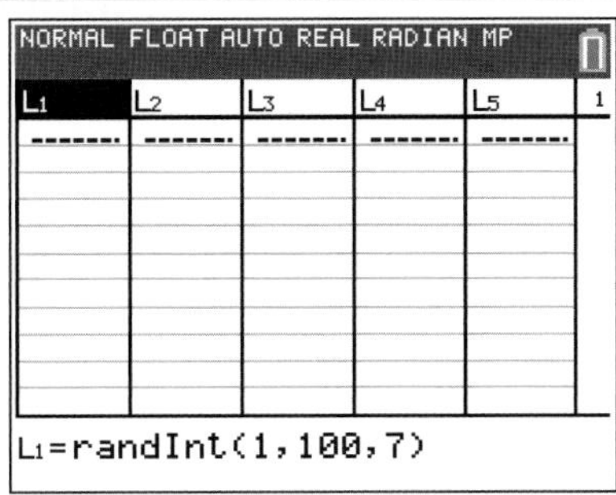

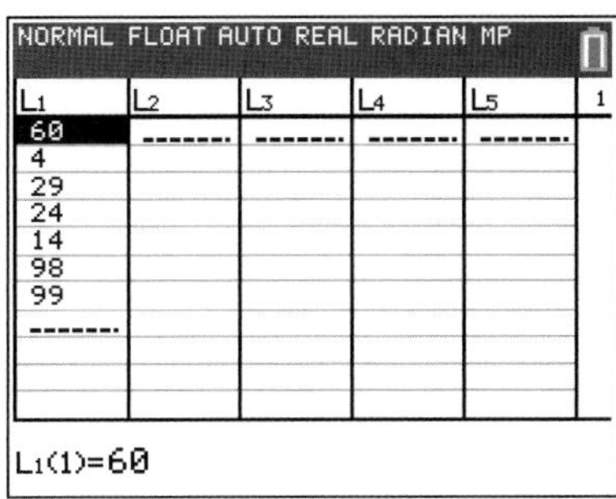

2. Press STAT, scroll to the Calc menu, and choose 1-Var Stats. Enter L1 for the list and press calculate.
3. Use the output to calculate the value of each of the following statistics:
   (a) $2 \times$ sample median
   (b) $\bar{x} + 3s_x$
   (c) sample maximum
4. Add each of these values to the class dotplots provided by your teacher.
5. As a class, discuss the quality of each of these statistics. Do any of them consistently overestimate or consistently underestimate the truth? Are some of the statistics more variable than others?
6. Form teams of three or four students. In your group, spend about 10–15 minutes brainstorming how to improve one of the original three statistics — or create your own statistic. Then create a simulated sampling distribution for each of your statistics to determine which one you will use for the competition.
7. Your teacher will now select a random sample of 7 sticks from the bag and read out the stick numbers. On a sheet of paper, write the names of your group, the statistic you think is best (a formula), and the value of the statistic calculated from the sample provided by the teacher. The closest estimate wins!

In the craft sticks activity, the goal was to estimate the maximum value in a population, with the assumption that the members of the population are numbered 1, 2, . . . , N. Two possible statistics that might be used to estimate N are the sample maximum (max) and twice the sample median ($2 \times$ median).

Assuming that the population has $N = 100$ members and we use SRSs of size $n = 7$, Figure 5.7 shows simulated sampling distributions of the sample maximum and twice the sample median, along with a red line at $N = 100$.

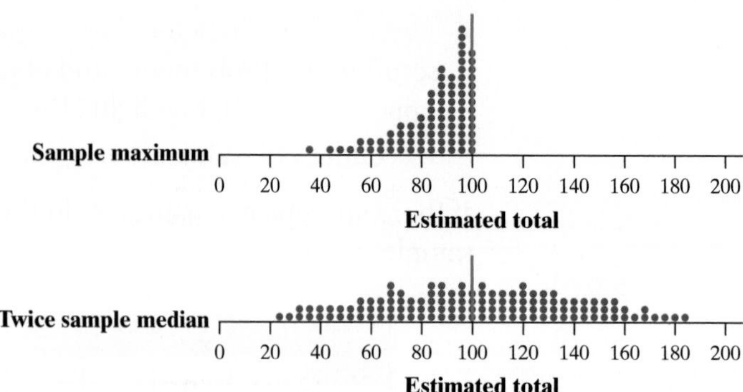

FIGURE 5.7 Simulated sampling distributions of the sample maximum and twice the sample median for samples of size $n = 7$ from a population with $N = 100$.

These simulated sampling distributions look quite different. The values of the sample maximum are consistently less than the population maximum $N = 100$. However, the values of twice the sample median aren't consistently less than or consistently greater than the population maximum $N = 100$. It appears that twice the sample median might be an **unbiased estimator** of the population maximum, while the sample maximum is clearly biased.

DEFINITION Unbiased estimator

A statistic used to estimate a parameter is an **unbiased estimator** if the mean of its sampling distribution is equal to the value of the parameter being estimated. The mean of the sampling distribution is also known as the *expected value of the estimator.*

In a particular sample, the value of an unbiased estimator might be greater than the value of the parameter or it might be less than the value of the parameter. However, because the sampling distribution of the statistic is centered at the true value, the statistic will not consistently overestimate or consistently underestimate the parameter. This fits with our definition of bias from Unit 3. The design of a statistical study shows bias if it is very likely to underestimate or very likely to overestimate the value we want to know, as in the Federalist Papers activity from Section 3A.

In Section 5C, we will confirm mathematically that the sample proportion $\hat{p}$ is an unbiased estimator of the population proportion p. This is a very helpful result if we're dealing with a categorical variable like color, for example. With quantitative variables, we might be interested in estimating the population mean, median, minimum, maximum, Q_1, Q_3, variance, standard deviation, IQR, or range. Which (if any) of these have unbiased estimators?

Let's revisit the "Sampling heights" example with John and Carol's four sons to investigate one of these statistics. Recall that the heights of the four sons are 71, 75, 72, and 68 inches. Here again is the sampling distribution of the sample mean $\bar{x}$ for samples of size 2:

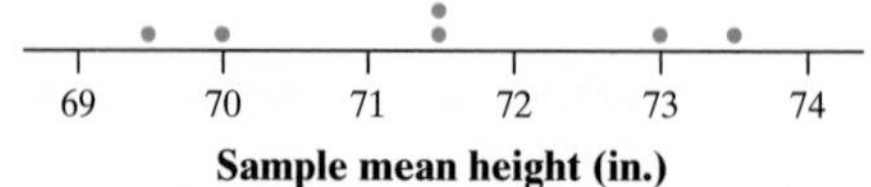

To determine if the sample mean is an unbiased estimator of the population mean, we need to compare the mean of the sampling distribution to the value we are trying to estimate — the mean of the population μ.

- The mean of the sampling distribution of $\bar{x}$ is

$$\mu_{\bar{x}} = \frac{69.5 + 70 + 71.5 + 71.5 + 73 + 73.5}{6} = 71.5$$

- The mean of the population distribution is

$$\mu = \frac{71+75+72+68}{4} = 71.5$$

Because these values are equal, this example suggests that the sample mean $\overline{x}$ is an unbiased estimator of the population mean μ. We will confirm this fact in Section 5D.

EXAMPLE

Estimating the range
Biased and unbiased estimators

Skill 4.B

PROBLEM: In the "Sampling heights" example, we created the sampling distribution of the sample range for samples of size $n = 2$ from the population of John and Carol's four sons with heights of 71, 75, 72, and 68 inches tall. Is the sample range an unbiased estimator of the population range? Explain your answer.

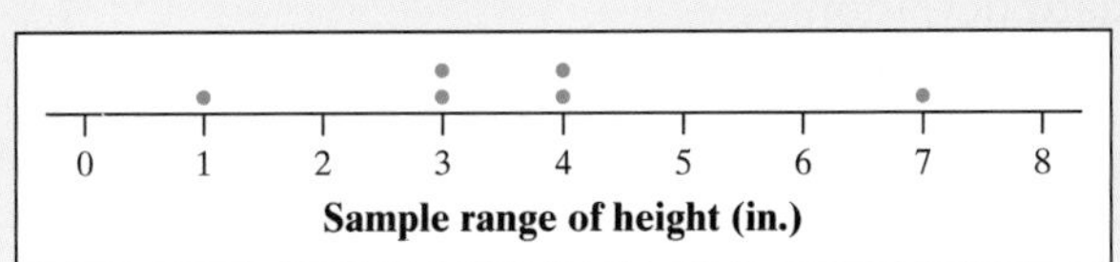

SOLUTION:

The mean of the sampling distribution of the sample range is

$$\frac{1+3+3+4+4+7}{6} = 3.67$$

The range of the population distribution is

Population range $= 75 - 68 = 7$

Because the mean of the sampling distribution of the sample range (3.67) is not equal to the value it is trying to estimate (7), the sample range is not an unbiased estimator of the population range.

It makes sense that the sample range is a biased estimator of the population range. After all, the sample range can't be larger than the population range, but it can be (and usually is) smaller.

FOR PRACTICE, TRY EXERCISE 19

Because the sample range is consistently smaller than the population range, the sample range is a *biased estimator* of the population range.

Think About It

WHY DO WE DIVIDE BY $n - 1$ WHEN CALCULATING THE SAMPLE STANDARD DEVIATION? Now that we know about sampling distributions and unbiased estimators, we can finally answer this question. In Unit 1, you learned that the formula for the sample standard deviation is

$$s_x = \sqrt{\frac{\sum(x_i - \overline{x})^2}{n-1}}$$

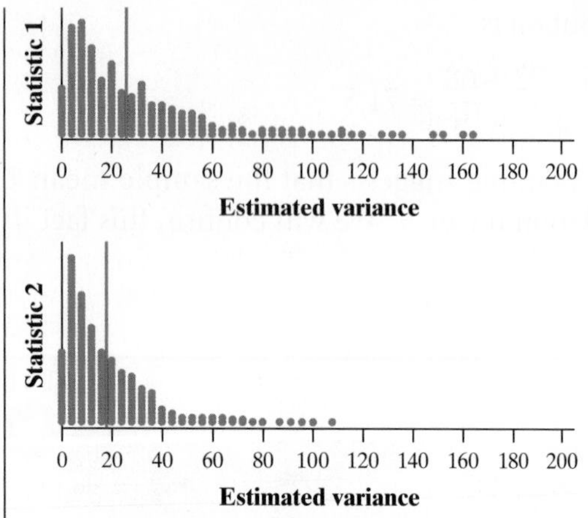

What if you divided by n instead of $n-1$? Let's simulate the sampling distributions of two statistics that can be used to estimate the variance of a population, where variance is the square of the standard deviation (variance $= s_x^2$):

$$\text{Statistic 1: } \frac{\sum(x_i - \bar{x})^2}{n-1} \qquad \text{Statistic 2: } \frac{\sum(x_i - \bar{x})^2}{n}$$

The simulated sampling distributions shown in the dotplots are based on 1000 SRSs of size $n = 3$ from a population with variance $= 25$. The mean of each distribution is indicated by a blue line segment.

We can see that Statistic 2 is a *biased* estimator of the population variance because the mean of its simulated sampling distribution is clearly less than the value of the population variance (25). Dividing by n instead of $n-1$ gives estimates that are consistently too small. Statistic 1, however, is unbiased because the mean of its simulated sampling distribution is equal to the population variance (25). That's why we divide by $n-1$ when calculating the sample variance — and when calculating the sample standard deviation.

Variability of an Estimator

To get a trustworthy estimate of an unknown population parameter, start by using a statistic that's an unbiased estimator. This ensures that you won't consistently overestimate or consistently underestimate the parameter. Unfortunately, using an unbiased estimator doesn't guarantee that the value of your statistic will be close to the actual parameter value.

Figure 5.8(a) shows the approximate sampling distribution of $\bar{x}$ = sample mean year of penny for 50 random samples of size $n = 5$ from Mr. Simonetti's collection of pennies. Notice that the values of $\bar{x}$ vary from 1993 to 2019, with a standard deviation of about 6.2 years. Figure 5.8(b) shows the results of increasing the sample size to $n = 20$. In this case, the values of $\bar{x}$ vary from 1999 to 2013, with a standard deviation of 3.3 years. Increasing the sample size reduced the variability of the sampling distribution of $\bar{x}$ by quite a bit.

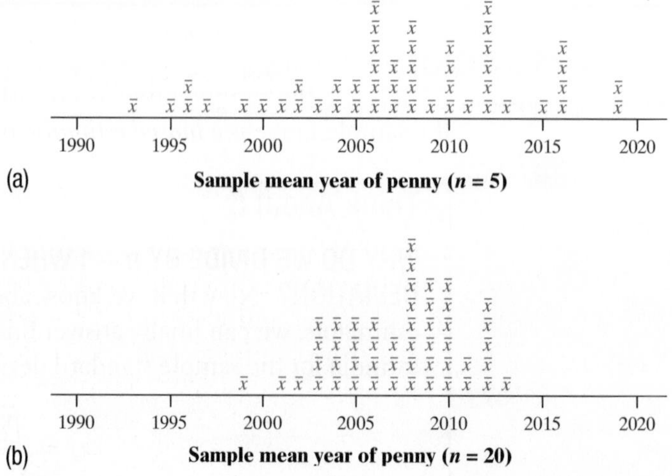

FIGURE 5.8 Approximate sampling distributions of the sample mean year of penny in 50 random samples of (a) size $n = 5$ and (b) size $n = 20$ from a population of pennies.

DECREASING SAMPLING VARIABILITY

The sampling distribution of any estimator (statistic) will have less variability when the sample size is larger.

EXAMPLE

Battery life
Variability of an estimator

Skill 3.B

PROBLEM: For quality control purposes, supervisors at a battery factory regularly select random samples of batteries to estimate the mean lifetime of the batteries they produce. Here is a simulated sampling distribution of $\bar{x}$, the sample mean lifetime (in hours) for 1000 random samples of size $n = 100$ from a population of AAA batteries.

Chanook Photography/Shutterstock

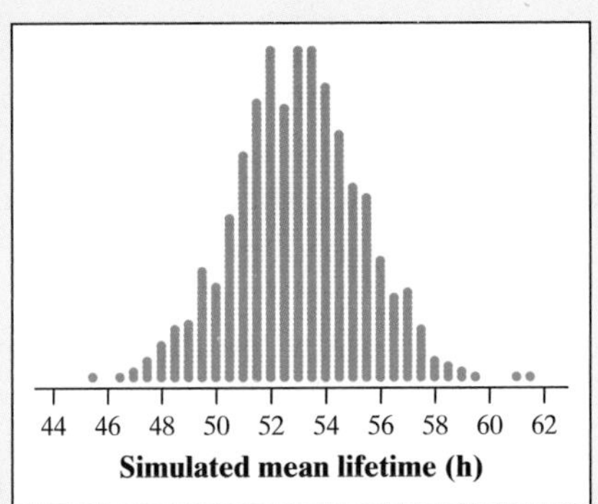

(a) What would happen to the sampling distribution of the sample mean $\bar{x}$ if the sample size were $n = 50$ instead? Explain your answer.
(b) How will this change in sample size affect the estimate?

SOLUTION

(a) The sampling distribution of the sample mean $\bar{x}$ will be more variable because the sample size is smaller.

(b) The estimated mean lifetime is less likely to be close to the population mean lifetime. In other words, the estimate will be less precise.

> Changing the sample size affects the variability of an estimator, but not bias. As long as the sample is randomly selected, the sample mean will be an unbiased estimator of the population mean, no matter the sample size.

FOR PRACTICE, TRY EXERCISE 23

Look back at Figure 5.8. In addition to decreasing the variability in the sampling distribution of $\bar{x}$, increasing the sample size changed the *shape* of the distribution. Both distributions are skewed to the left, but when $n = 20$, the skewness is less strong.

Here is an approximate sampling distribution for random samples of $n = 50$ pennies.

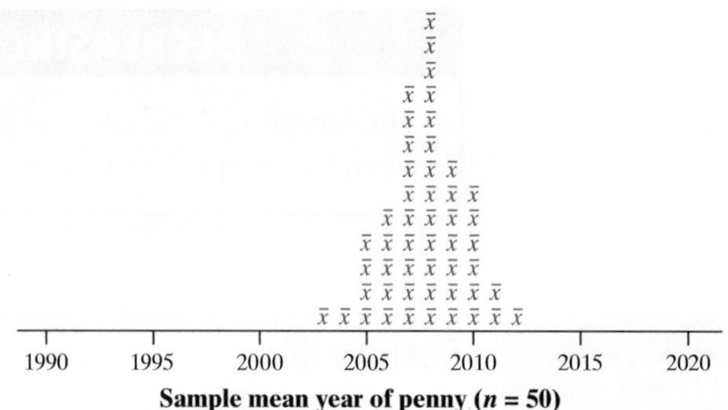

The sampling distribution of $\bar{x}$ now looks approximately normal. In fact, the *central limit theorem* states that when the sample size is sufficiently large, the sampling distribution of the mean will be approximately normally distributed. You'll learn more about the central limit theorem in Section 5D.

CHOOSING AN ESTIMATOR

In many cases, it is obvious which statistic should be used as an estimator of a population parameter. If we want to estimate a population mean μ, use the sample mean $\bar{x}$. If we want to estimate a population proportion p, use the sample proportion $\hat{p}$. However, in other cases, there isn't an obvious best choice. When trying to estimate the population maximum in the Craft Stick activity, for example, we could have used many different estimators.

To decide which estimator to use when there are several choices, consider both bias and variability. We can think of the true value of the population parameter as the bull's-eye on a target and of the sample statistic as an arrow fired at the target. Both bias and variability describe what happens when we take many shots at the target.

- *Bias* means that our aim is off and we consistently miss the bull's-eye in the same direction. That is, our sample statistics do not center on the population parameter.
- *High variability* means that repeated shots are widely scattered on the target. In other words, repeated samples do not give very similar results.

Figure 5.9 shows this target illustration of bias and variability. Notice that low variability (shots are close together) can accompany high bias (shots are consistently away from the bull's-eye in one direction). Likewise, low or no bias (shots center on the bull's-eye) can accompany high variability (shots are widely scattered). Ideally, we'd like our estimates to be *accurate* (unbiased) and *precise* (have low variability).

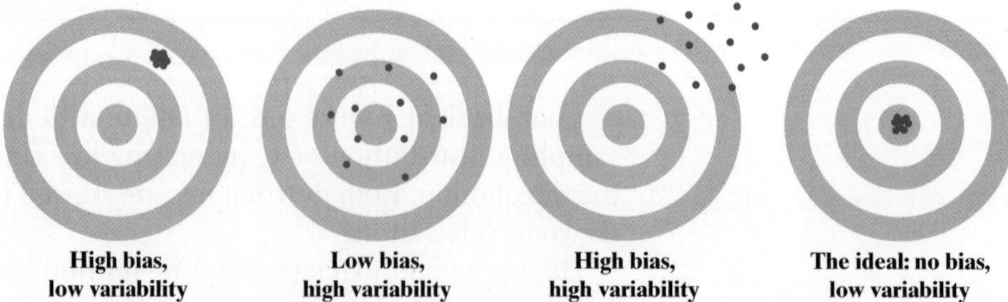

FIGURE 5.9 A visual representation of bias and variability. The center of the target represents the value of the parameter and the dots represent possible values of the statistic.

AP® EXAM TIP

Make sure you understand the difference between accuracy and precision when writing your responses on the AP® Statistics exam. Many students use the word *accurate* when they really mean *precise*. For example, a response that says "increasing the sample size will make an estimate more *accurate*" is incorrect. It should say that "increasing the sample size will make an estimate more *precise*." If you can't remember which term to use, don't use either of them. Instead, explain what you mean without using statistical vocabulary.

CHECK YOUR UNDERSTANDING

The histogram on the left shows the wait time (in minutes) between eruptions of the Old Faithful geyser for all 263 recorded eruptions during a particular month. For this population, the median is 83 minutes. We used technology to select 500 SRSs of size 15 from the population. The 500 values of the sample median are displayed in the histogram on the right. The mean of these 500 values is 80.5.

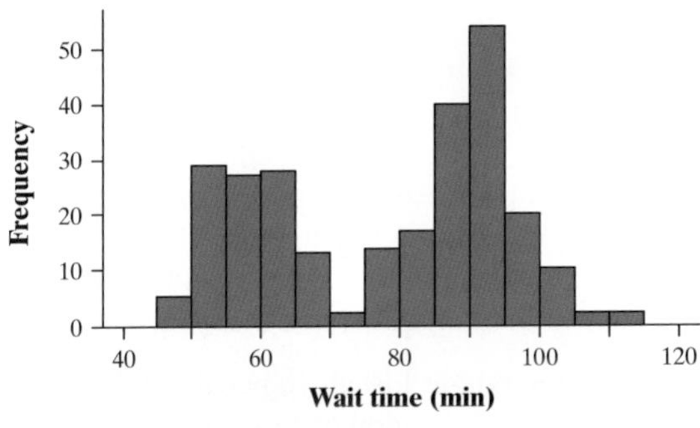

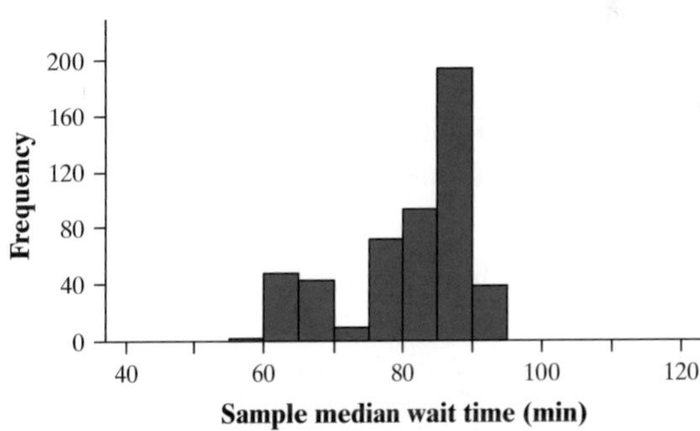

1. Does the simulation provide evidence that the sample median is a biased estimator of the population median? Justify your answer.
2. Suppose we had selected samples of size 30 instead of size 15. Would the variability of the sampling distribution of the sample median be larger, smaller, or about the same? Explain your answer.

SECTION 5B Summary

- A **parameter** is a number that describes some characteristic of a population. A **statistic** is a number that describes some characteristic of a sample. We use statistics to estimate parameters.
- The **sampling distribution** of a statistic describes the values of the statistic in all possible samples of the same size from the same population.
- To determine a sampling distribution, list all possible samples of a particular size, calculate the value of the statistic for each sample, and graph the distribution of the statistic. If there are many possible samples, approximate the sampling distribution: repeatedly select (or simulate) random samples of a particular size, calculate the value of the statistic for each sample, and graph the distribution of the statistic.
- We can use sampling distributions to evaluate a claim by determining which values of a statistic are likely to happen by chance alone.

> **AP® EXAM TIP**
> **AP® Daily Videos**
> Review the content of this section and get extra help by watching the AP® Daily Videos for Topics 5.3 and 5.4, which are available in AP® Classroom.

- A statistic used to estimate a parameter is an **unbiased estimator** if the mean of its sampling distribution is equal to the value of the parameter being estimated. That is, the statistic doesn't consistently overestimate or consistently underestimate the value of the parameter.
- The sampling distribution of any statistic will have less variability when the sample size is larger. That is, the statistic will be a more precise estimator of the parameter with larger sample sizes.
- When estimating a parameter, choose a statistic with low or no bias and minimum variability.

SECTION 5B Exercises

Parameters and Statistics

For Exercises 1–6, identify the population, the parameter, the sample, and the statistic in each setting.

1. **Instagram** A Pew Research Center poll asked 1502 adults in the United States if they use a variety of social media sites. Of the respondents, 601 said they use Instagram.[8] pg 467

2. **Unemployment** Each month, the Current Population Survey interviews about 60,000 randomly selected U.S. adults. One of its goals is to estimate the national unemployment rate. In November 2022, 3.7% of those interviewed were unemployed.[9]

3. **Fillings** How much do prices vary for filling a cavity? To find out, an insurance company randomly selects 10 dental practices in California and asks for the cash (non-insurance) price for this procedure at each practice. The interquartile range is $74.

4. **Warm turkey** Tom is roasting a large turkey breast for a holiday meal. He wants to be sure that the turkey is safe to eat, which requires a minimum internal temperature of 165°F. Tom uses a thermometer to measure the temperature of the turkey meat at four randomly chosen locations. The minimum reading is 170°F.

5. **Iced tea** On Tuesday, the bottles of Arizona Iced Tea filled in a plant were supposed to contain an average of 20 ounces of iced tea. Quality control inspectors selected 50 bottles at random from the day's production. These bottles contained an average of 19.6 ounces of iced tea.

6. **Bearings** A production run of ball bearings is supposed to have a mean diameter of 2.5000 centimeters (cm). An inspector chooses 100 bearings at random from the run. These bearings have mean diameter 2.5009 cm.

The Idea of a Sampling Distribution

Exercises 7–10 refer to the following scenario. An investor has five stocks in a portfolio. Here are the net returns for each of the five stocks during the previous 12 months:

8% 12% −5% −20% 25%

7. **Sample means** List all 10 possible SRSs of size $n = 2$ from this population of 5 stocks, calculate the mean return for each sample, and display the sampling distribution of the sample mean on a dotplot. pg 469

8. **Sample ranges** List all 10 possible SRSs of size $n = 3$ from this population of 5 stocks, calculate the range of return for each sample, and display the sampling distribution of the sample range on a dotplot.

9. **Sample proportions** List all 10 possible SRSs of size $n = 2$ from this population of 5 stocks, calculate the proportion with a positive return for each sample, and display the sampling distribution of the sample proportion on a dotplot.

10. **Sample medians** List all 10 possible SRSs of size $n = 3$ from this population of 5 stocks, calculate the median return for each sample, and display the sampling distribution of the sample median on a dotplot.

Using Sampling Distributions to Evaluate Claims

11. **Doing homework** A school newspaper article claims that 60% of the students at a large high school completed their assigned homework last week. Assume that this claim is true for the 2000 students at the school.

(a) Make a bar graph of the population distribution.

(b) Imagine one possible SRS of size 100 from this population. Sketch a bar graph of the distribution of sample data.

12. **Giraffe weights** The weights of male giraffes in the wild can be modeled by a normal distribution with a mean of 2628 pounds and a standard deviation of 541 pounds.[10]

(a) Make a graph of the population distribution.

(b) Imagine one possible SRS of size 20 from this population. Sketch a dotplot of the distribution of sample data.

13. **More homework** Some skeptical AP® Statistics students want to investigate the newspaper's claim from Exercise 11, so they choose an SRS of 100 students from the school to interview. In their sample, 45 students completed their homework last week. Does this provide convincing evidence that less than 60% of all students at the school completed their assigned homework last week?

pg 470

(a) What is the evidence that less than 60% of all students completed their assigned homework last week?

(b) Provide two explanations for the evidence described in part (a).

We used technology to simulate choosing 250 SRSs of size $n = 100$ from a population of 2000 students where 60% completed their assigned homework last week. The dotplot shows $\hat{p}$ = the sample proportion of students who completed their assigned homework last week for each of the 250 simulated samples.

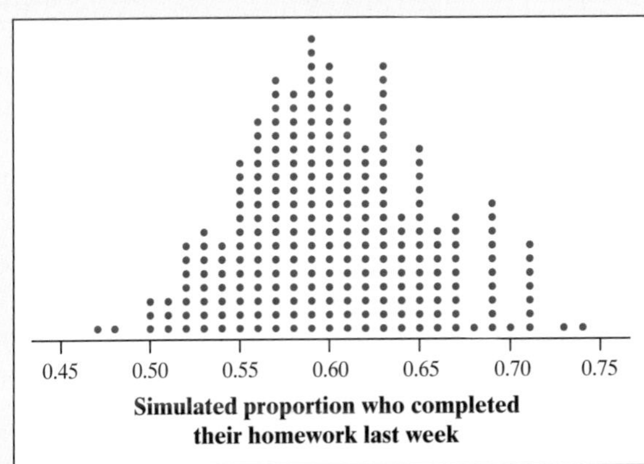

(c) There is one dot on the graph at 0.73. Explain what this value represents.

(d) Would it be surprising to get a sample proportion of $\hat{p} = 0.45$ or smaller in an SRS of size 100 when $p = 0.60$? Justify your answer.

(e) Based on your previous answers, is there convincing evidence that less than 60% of all students at the school completed their assigned homework last week? Explain your reasoning.

14. **Heavy giraffes** Refer to Exercise 12. Do giraffes that live in captivity weigh more than giraffes in the wild? A random sample of $n = 10$ male giraffes that live in zoos gives a sample mean of 3004 pounds. Does this provide convincing evidence that male giraffes that live in zoos weigh more than 2628 pounds, on average?

(a) What is the evidence that the average weight of all male giraffes that live in zoos is more than 2628 pounds?

(b) Provide two explanations for the evidence described in part (a).

We used technology to simulate choosing 100 SRSs of size $n = 10$ from a normal distribution with mean $\mu = 2628$ pounds and standard deviation $\sigma = 541$ pounds. The dotplot shows $\bar{x}$ = the sample mean weight for each of the 100 simulated samples.

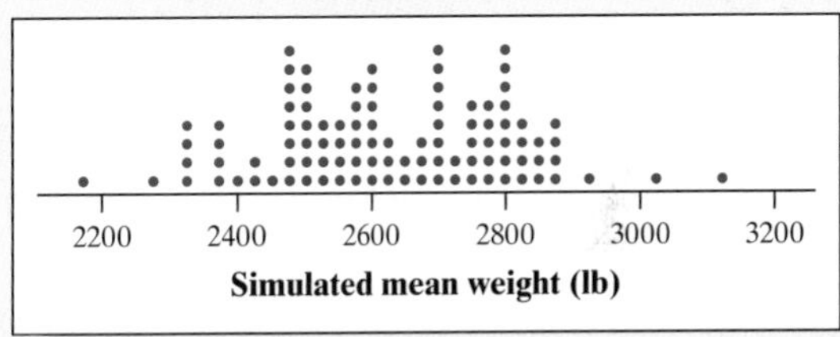

(c) There is one dot on the graph at 3025. Explain what this value represents.

(d) Would it be surprising to get a sample mean of 3004 or larger in an SRS of size 10 when $\mu = 2628$ and $\sigma = 541$? Justify your answer.

(e) Based on your previous answers, is there convincing evidence that the average weight of all male giraffes that live in zoos is more than 2628 pounds? Explain your reasoning.

15. **Even more homework** Refer to Exercises 11 and 13. Suppose that the sample proportion of students who did all their assigned homework last week is $\hat{p} = 57/100 = 0.57$. Would this sample proportion provide convincing evidence that less than 60% of all students at the school completed all their assigned homework last week? Explain your reasoning.

16. **Even more giraffes** Refer to Exercises 12 and 14. Suppose that the sample mean weight of the 10 male giraffes in zoos was 2775 pounds. Would this sample mean provide convincing evidence that the average weight of all male giraffes that live in zoos is more than 2628 pounds? Explain your reasoning.

Exercises 17 and 18 refer to the following scenario. During the winter months, outside temperatures at the Starnes family cabin in Colorado can stay well below freezing (32°F or 0°C) for weeks at a time. To prevent the pipes from freezing, Mrs. Starnes sets the thermostat at 50°F. The manufacturer claims that the thermostat allows variation in home temperature that follows a normal distribution with $\mu = 50$°F and $\sigma = 3$°F.

17. **Cold cabin?** The dotplot shows the results of taking 300 SRSs of 10 temperature readings from a normal distribution with $\mu = 50$ and $\sigma = 3$ and recording the sample standard deviation s_x each time. Suppose that the standard deviation from an actual sample is $s_x = 5$°F. What would you conclude about the thermostat manufacturer's claim? Explain your reasoning.

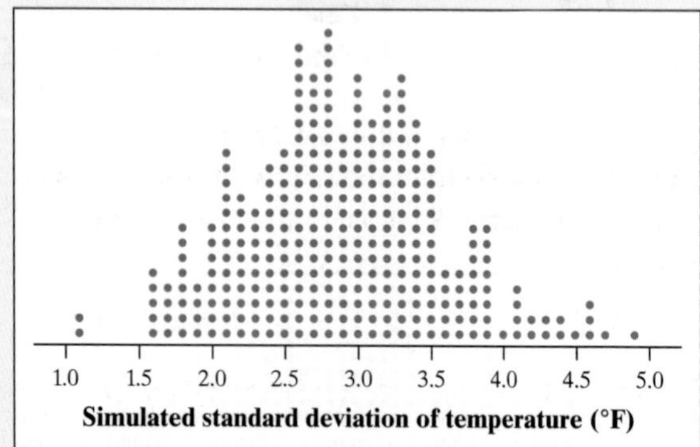

18. **Really cold cabin** The dotplot shows the results of taking 300 SRSs of 10 temperature readings from a normal distribution with $\mu = 50$ and $\sigma = 3$ and recording the sample minimum each time. Suppose that the minimum of an actual sample is 40°F. What would you conclude about the thermostat manufacturer's claim? Explain your reasoning.

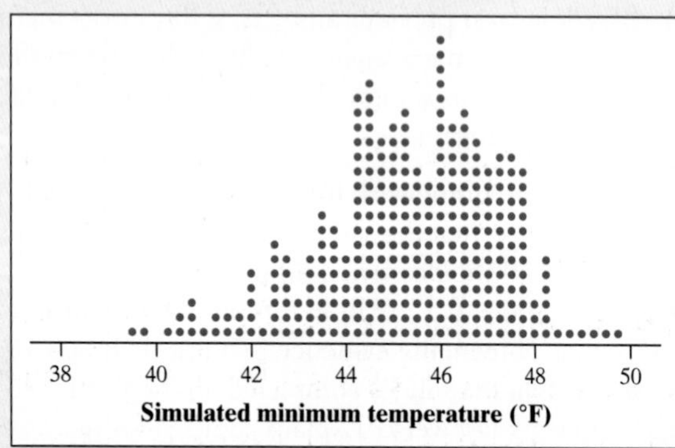

Biased and Unbiased Estimators

Exercises 19 and 20 refer to the small population of 4 cars listed in the table.

Color	Age (years)
Red	1
White	5
Silver	8
Red	20

19. **Red cars** The dotplot shows the sampling distribution of $\hat{p}$ = sample proportion of red cars for all 6 possible SRSs of size $n = 2$ from this population. Is the sample proportion an unbiased estimator of the population proportion? Explain your answer.

pg 475

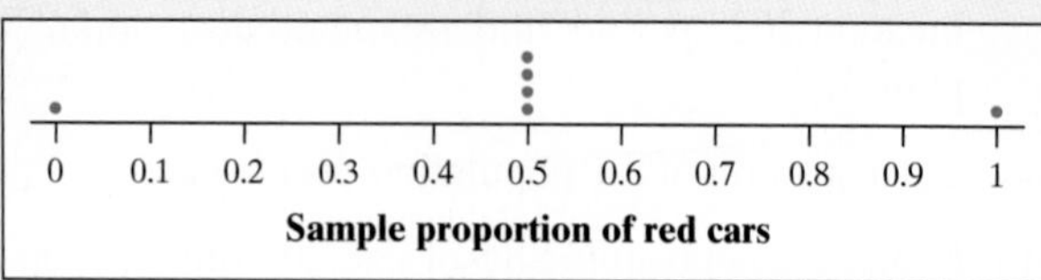

20. **Old cars** The dotplot shows the sampling distribution of the sample minimum age for all possible SRSs of size $n = 2$ from this population. Is the sample minimum an unbiased estimator of the population minimum? Explain your answer.

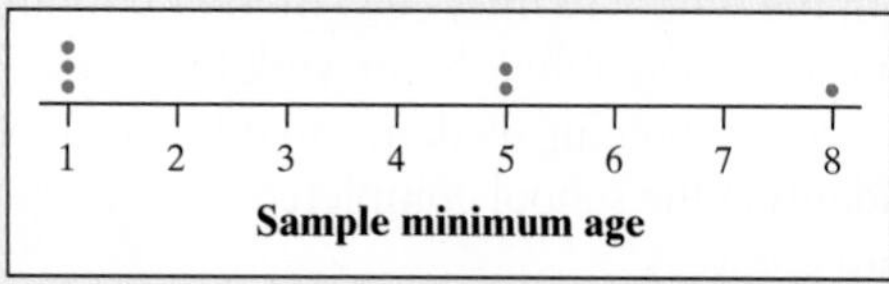

21. **A sample of teens** A study of the health of teenagers plans to measure the blood cholesterol levels of an SRS of 13- to 16-year-olds. The researchers will report the mean $\bar{x}$ from their sample as an estimate of the mean cholesterol level μ in this population. Explain to someone who knows little about statistics what it means to say that $\bar{x}$ is an unbiased estimator of μ.

22. **Predict the election** A polling organization plans to ask a random sample of likely voters who they plan to vote for in an upcoming election. The researchers will report the sample proportion $\hat{p}$ that favors the incumbent as an estimate of the population proportion p that favors the incumbent. Explain to someone who knows little about statistics what it means to say that $\hat{p}$ is an unbiased estimator of p.

Variability of an Estimator

23. **More red cars** In Exercise 19, you analyzed the sampling distribution of $\hat{p}$ = sample proportion of red cars for samples of size $n = 2$ from a small population of 4 cars.

pg 477

(a) What would happen to the sampling distribution of the sample proportion $\hat{p}$ if the sample size were $n = 3$ instead? Explain your answer.

(b) How will this change in sample size affect the estimate?

24. **More old cars** In Exercise 20, you analyzed the sampling distribution of the sample minimum age of car for samples of size $n = 2$ from a small population of 4 cars.

(a) What would happen to the sampling distribution of the sample minimum age if the sample size were $n = 3$ instead? Explain your answer.

(b) How will this change in sample size affect the estimate?

25. **Housing prices** In a residential neighborhood, the distribution of house values is unimodal and skewed to the right with a median of \$200,000 and an *IQR* of \$100,000. For which of the following sample sizes, $n = 10$ or $n = 100$, is the sample median more likely to be greater than \$250,000? Explain your reasoning.

26. **Basements** In a particular city, 74% of houses have basements. For which of the following sample sizes, $n = 25$ or $n = 150$, is the sample proportion of houses with a basement more likely to be less than 0.70? Explain your reasoning.

27. **Bias and variability** The figure shows approximate sampling distributions of four different statistics intended to estimate the same parameter.

(a) Which statistics appear to be unbiased estimators? Justify your answer.

(b) Which statistic does the best job of estimating the parameter? Explain your answer.

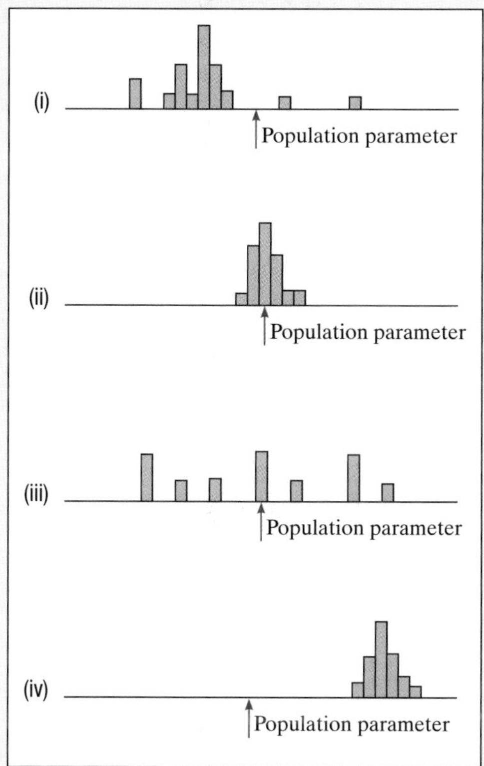

For Investigation *Apply the skills from the section in a new context or nonroutine way.*

28. **Internet equity** A school superintendent is concerned about equity within the district. In particular, the superintendent worries that the proportion of South High School students with internet access at home is less than the proportion of North High School students with internet access at home. To investigate this issue, the superintendent selects SRSs of size $n = 50$ from each school and finds $\hat{p}_S = 36/50 = 0.72$ and $\hat{p}_N = 46/50 = 0.92$.

To determine if a difference in proportions of $\hat{p}_S - \hat{p}_N = -0.20$ provides convincing evidence that South High School has a smaller proportion of students with internet access at home, we simulated two random samples of size $n = 50$ from populations having the same proportion of students with internet access. Then, we subtracted the simulated sample proportions. Here are the results from repeating this process 100 times. Based on the actual samples and the results of the simulation, is there convincing evidence that South High School has a smaller proportion of students with internet access at home? Explain your reasoning.

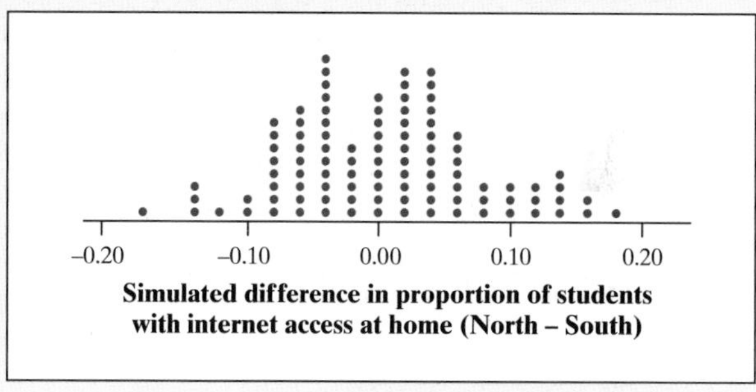

Multiple Choice *Select the best answer for each question.*

29. At a particular college, 78% of all students are receiving some kind of financial aid. The school newspaper selects a random sample of 100 students and 72% of the respondents say they are receiving some sort of financial aid. Which of the following is true?

(A) 78% is a population and 72% is a sample.

(B) 72% is a population and 78% is a sample.

(C) 78% is a parameter and 72% is a statistic.

(D) 72% is a parameter and 78% is a statistic.

(E) 72% is a parameter and 100 is a statistic.

30. A statistic is an unbiased estimator of a parameter when

(A) the statistic is calculated from a random sample.

(B) in a single sample, the value of the statistic is equal to the value of the parameter.

(C) in all possible samples, the values of the statistic are very close to the value of the parameter.

(D) in all possible samples, the mean of the values of the statistic equals the value of the parameter.

(E) in all possible samples, the distribution of the statistic has a shape that is approximately normal.

31. Increasing the sample size of an opinion poll will reduce the
    (A) bias of the estimates made from the data collected in the poll.
    (B) variability of the estimates made from the data collected in the poll.
    (C) effect of nonresponse on the poll.
    (D) variability of opinions in the sample.
    (E) variability of opinions in the population.

32. The math department at a small school has 5 teachers. The ages of these teachers are 23, 34, 37, 42, and 58. Suppose you select a random sample of 4 teachers and calculate the sample minimum age. Which of the following shows the sampling distribution of the sample minimum age?

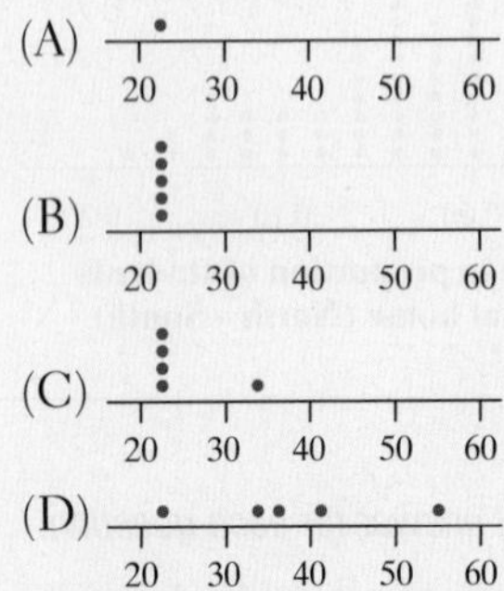

    (E) None of these

Recycle and Review *Practice what you learned in previous sections.*

33. **Light and plant growth (3C)** Meadowfoam seed oil is used in making various skin care products. Researchers interested in maximizing the productivity of meadowfoam plants designed an experiment to investigate the effect of different light intensities on plant growth. The researchers planted 120 meadowfoam seedlings in individual pots, randomly assigned 10 pots to each of 12 trays, and put all the trays into a controlled enclosure. Two trays were randomly assigned to each light intensity level (in micromoles per square meter per second): 150, 300, 450, 600, 750, and 900. The number of flowers produced by each plant was recorded and the average number of flowers was calculated for each tray.[11]
    (a) What are the explanatory and response variables?
    (b) What are the experimental units in this context?
    (c) Describe how to randomly assign the treatments to the experimental units.
    (d) How did the experiment incorporate replication?

34. **More light and plant growth (2C)** Refer to Exercise 33. Here is computer output for the least-squares regression line relating y = average number of flowers to x = light intensity.

Term	Coef	SE Coef	T-Value	P-Value
Constant	71.623	4.688	15.278	0.000
Intensity	−0.041	0.008	−5.118	0.000
S = 7.122	R-Sq = 72.4%		R-Sq(adj) = 69.6%	

    (a) What is the equation of the least-squares regression line?
    (b) How does the predicted average number of flowers change for each increase in light intensity of 150 micromoles per square meter per second?
    (c) Interpret the value of r^2.

SECTION 5C Sample Proportions

LEARNING TARGETS *By the end of the section, you should be able to:*

- Calculate and interpret the mean and standard deviation of the sampling distribution of a sample proportion $\hat{p}$.
- Determine if the sampling distribution of $\hat{p}$ is approximately normal.
- If appropriate, use a normal distribution to calculate probabilities involving sample proportions.
- Describe the shape, center, and variability of the sampling distribution of a difference in sample proportions $\hat{p}_1 - \hat{p}_2$.

What proportion of all U.S. adults support a proposal to have the federal government forgive all student loan debt? A Harris poll found that 558 of 1015 randomly selected U.S. adults supported this proposal.[12] The sample proportion

$\hat{p} = 558/1015 = 0.55$ is the statistic that we use to estimate the unknown population proportion p. Because a random sample of 1015 U.S. adults is unlikely to perfectly represent all U.S. adults, we can only say that "about" 55% of all U.S. adults support the student debt proposal.

To determine if the sample proportion of 0.55 provides convincing evidence that more than half of all U.S. adults would support the proposal, we need to know about the **sampling distribution of the sample proportion $\hat{p}$**.

DEFINITION Sampling distribution of the sample proportion $\hat{p}$

The **sampling distribution of the sample proportion $\hat{p}$** describes the distribution of values taken by the sample proportion $\hat{p}$ in all possible samples of the same size from the same population.

The dotplot shows a simulated sampling distribution of $\hat{p}$ = the sample proportion of U.S. adults who support the proposal in 100 random samples of size $n = 1015$, assuming that $p = 0.50$. That is, our simulation assumes that exactly 50% of all U.S. adults support the proposal.

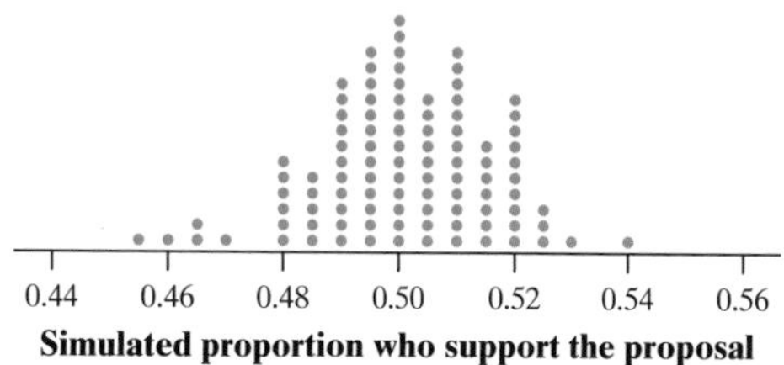

The distribution is approximately normal with a mean of about 0.50 and a standard deviation of about 0.015. By the end of this section, you should be able to anticipate the center, variability, and shape of sampling distributions like this one without having to perform a simulation.

The Sampling Distribution of $\hat{p}$: Center and Variability

When we select random samples of size n from a population with proportion of successes p, the value of $\hat{p}$ will vary from sample to sample. Figure 5.10 shows simulated sampling distributions of $\hat{p}$ for SRSs of size $n = 25$ from three different populations.

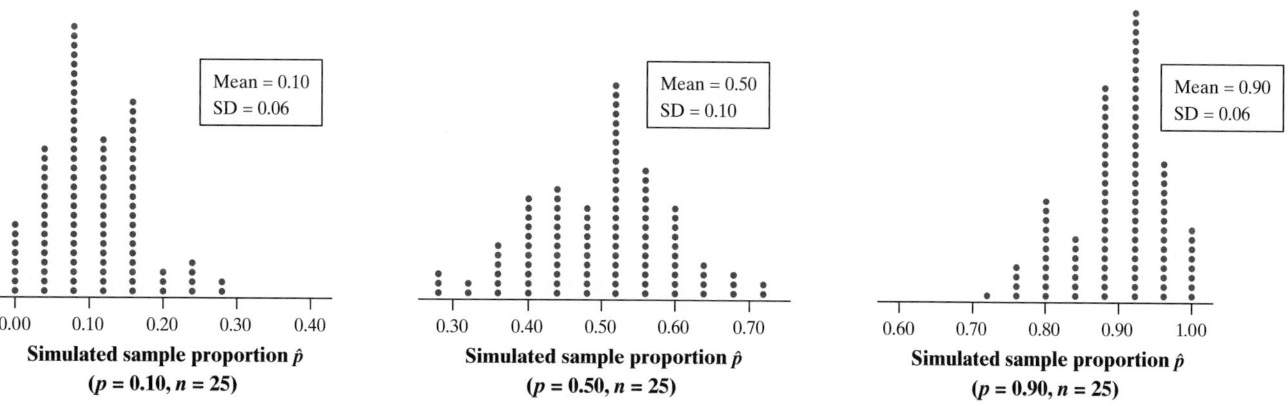

FIGURE 5.10 Simulated sampling distributions of $\hat{p}$ for 100 SRSs of size $n = 25$ from populations with $p = 0.10$, $p = 0.50$, or $p = 0.90$.

As you can see in Figure 5.10, the mean of each sampling distribution is very close to the population proportion. This isn't a coincidence! Also notice that the variability of the first and third distributions is roughly the same, while the variability is a little larger for the second distribution. Finally, note that the shape of the sampling distribution of $\hat{p}$ can be skewed to the right, roughly symmetric, or skewed to the left. We'll investigate shape more thoroughly in the next subsection. For now, here are the formulas that describe the center and variability of the sampling distribution of $\hat{p}$.

HOW TO CALCULATE $\mu_{\hat{p}}$ AND $\sigma_{\hat{p}}$

Let $\hat{p}$ be the sample proportion of successes in an SRS of size n from a population of size N with proportion p of successes. Then:

- The mean of the sampling distribution of $\hat{p}$ is $\mu_{\hat{p}} = p$.
- The standard deviation of the sampling distribution of $\hat{p}$ is approximately

  $\sigma_{\hat{p}} = \sqrt{\frac{p(1-p)}{n}}$ as long as the 10% condition is met: $n < 0.10N$.

AP® EXAM TIP
Formula sheet

The formulas for $\mu_{\hat{p}}$ and $\sigma_{\hat{p}}$ are included on the formula sheet provided on both sections of the AP® Statistics exam.

Here are some important facts about the mean and standard deviation of the sampling distribution of the sample proportion $\hat{p}$:

- The value $\mu_{\hat{p}}$ gives the average value of $\hat{p}$ in all possible samples of a given size from a population. Because $\mu_{\hat{p}} = p$, we know that the sample proportion $\hat{p}$ is an unbiased estimator of the population proportion p.
- The value $\sigma_{\hat{p}}$ measures the typical distance between a sample proportion $\hat{p}$ and the population proportion p in all possible samples of a given size from a population. Because the sample size appears in the denominator of the formula for $\sigma_{\hat{p}}$, we know that the sample proportion $\hat{p}$ is less variable when the sample size is larger. *Specifically, multiplying the sample size by 4 cuts the standard deviation in half.*
- When we sample *with* replacement, the standard deviation of the sampling distribution of $\hat{p}$ is exactly $\sigma_{\hat{p}} = \sqrt{\frac{p(1-p)}{n}}$. When we sample *without* replacement, the observations are not independent, and the actual standard deviation of the sampling distribution of $\hat{p}$ is smaller than the value given by the formula. However, if the sample size is less than 10% of the population size (the 10% condition), the value given by the formula is nearly correct.

Because larger random samples give better information, it sometimes makes sense to sample more than 10% of a population. In such a case, we can adjust the formula for $\sigma_{\hat{p}}$ in a way that correctly reduces the standard deviation. This adjustment is called a *finite population correction (FPC)*. We'll avoid situations that require the FPC in this text.

EXAMPLE

Backing the pack
Center and variability

Skills 3.B, 4.B

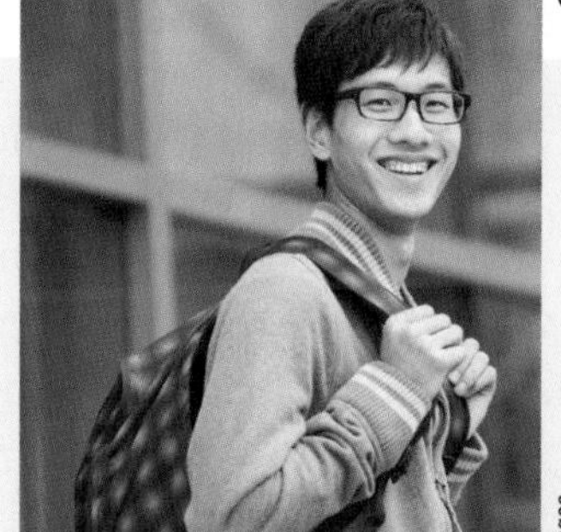
bo1982/Getty Images

PROBLEM: At a large high school, 84% of students regularly use a backpack to carry their books from class to class. Imagine taking an SRS of 100 students and calculating $\hat{p}$ = the proportion of students in the sample who regularly use a backpack.

(a) Calculate and interpret the mean of the sampling distribution of $\hat{p}$.
(b) Verify that the 10% condition is met. Then calculate and interpret the standard deviation of the sampling distribution of $\hat{p}$.

SOLUTION:

(a) $\mu_{\hat{p}} = 0.84$. If you selected all possible samples of 100 students from the school and calculated the sample proportion who regularly use a backpack for each sample, the sample proportions would have an average value of 0.84.

$\mu_{\hat{p}} = p$

(b) It is safe to assume that $n = 100$ students is less than 10% of the students in a large high school.

When the 10% condition is met, $\sigma_{\hat{p}} = \sqrt{\frac{p(1-p)}{n}}$.

$$\sigma_{\hat{p}} = \sqrt{\frac{0.84(1-0.84)}{100}} = 0.0367$$

If you selected all possible samples of 100 students from the school and calculated the sample proportion who regularly use a backpack for each sample, the sample proportions would typically vary from the population proportion of 0.84 by about 0.0367.

FOR PRACTICE, TRY EXERCISE 1

The Sampling Distribution of $\hat{p}$: Shape

Both the sample size n and the proportion of successes in the population p affect the shape of the sampling distribution of the sample proportion $\hat{p}$. The following activity will help you investigate the effects of these two factors.

ACTIVITY

Shape of the sampling distribution of $\hat{p}$

In this activity, you will use an applet to investigate the shape of the sampling distribution of the sample proportion $\hat{p}$ for different sample sizes (n) and different proportions of success in the population (p).

1. Go to www.stapplet.com and launch the *Simulating Sampling Distributions* applet.
2. In the box labeled "Population," choose "Categorical" from the drop-down menu. Keep the true proportion of successes as 0.5.

3. In the box labeled "Sample," change the sample size to $n = 20$ and click the "Select sample" button. The applet will graph the distribution of the sample and display the value of the sample proportion of successes $\hat{p}$. Was your sample proportion of successes close to the true proportion of successes, $p = 0.5$?
4. In the box labeled "Sampling Distribution," notice that the value of $\hat{p}$ from Step 3 is displayed on a dotplot. Click the "Select samples" button 9 more times, so that you have a total of 10 sample proportions. Look at the dotplot of your $\hat{p}$ values. Does the distribution have a recognizable shape?
5. To better see the shape of the sampling distribution of $\hat{p}$, enter 990 in the box for quickly selecting random samples and click the "Select samples" button. You have now selected a total of 1000 random samples of size $n = 20$ from a population with $p = 0.5$. Describe the shape of the simulated sampling distribution of $\hat{p}$ shown in the dotplot. Click the button to show the corresponding normal curve. How well does it fit?

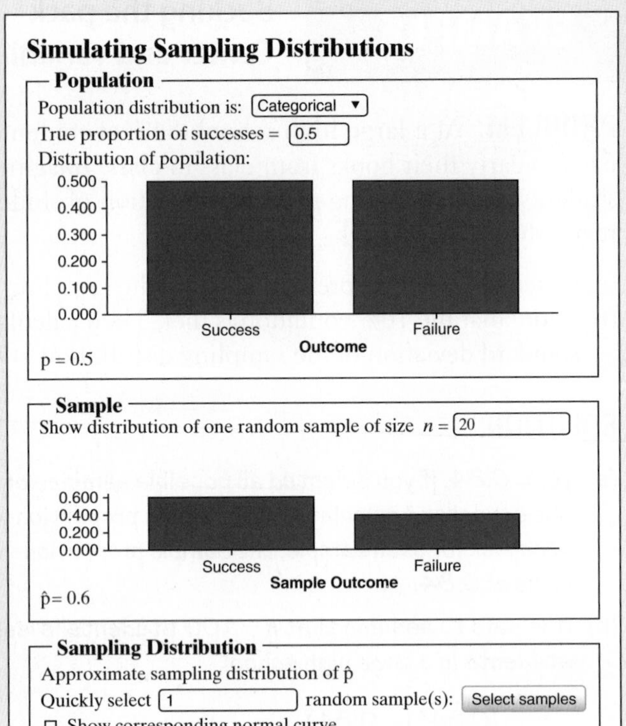

6. How does the shape of the sampling distribution of $\hat{p}$ change if the true proportion of successes changes?
   (a) Click the "Reset everything" button. In the "Population" box, change the true proportion to $p = 0.1$. Keeping $n = 20$ in the "Sample" box, quickly select 1000 random samples in the "Sampling Distribution" box. Describe the shape of the sampling distribution of $\hat{p}$.
   (b) Now change the true proportion to $p = 0.9$ and repeat Step 6(a).
   (c) Describe how the value of p affects the shape of the sampling distribution of $\hat{p}$.
7. How does the shape of the sampling distribution of $\hat{p}$ change if the sample size increases?
   (a) Click the "Reset everything" button. Keeping the true proportion as $p = 0.9$ in the "Population" box, increase the sample size to $n = 50$ in the "Sample" box. Then quickly select 1000 random samples in the "Sampling Distribution" box. Describe the shape of the sampling distribution of $\hat{p}$.
   (b) Repeat Step 7(a) with sample sizes of $n = 100$ and $n = 500$.
   (c) Describe how the value of n affects the shape of the sampling distribution of $\hat{p}$.

As you learned in the activity, the shape of the sampling distribution of $\hat{p}$ will be closer to normal when the value of p is closer to 0.5 and when the sample size n is larger. Figure 5.11 shows three simulated sampling distributions of $\hat{p}$ from a population where $p = 0.9$. Figure 5.11(a) shows the distribution of $\hat{p}$ for 100 random samples of size $n = 20$, Figure 5.11(b) shows the distribution of $\hat{p}$ for 100 random samples of size $n = 50$, and Figure 5.11(c) shows the distribution of $\hat{p}$ for 100 random samples of size $n = 100$.

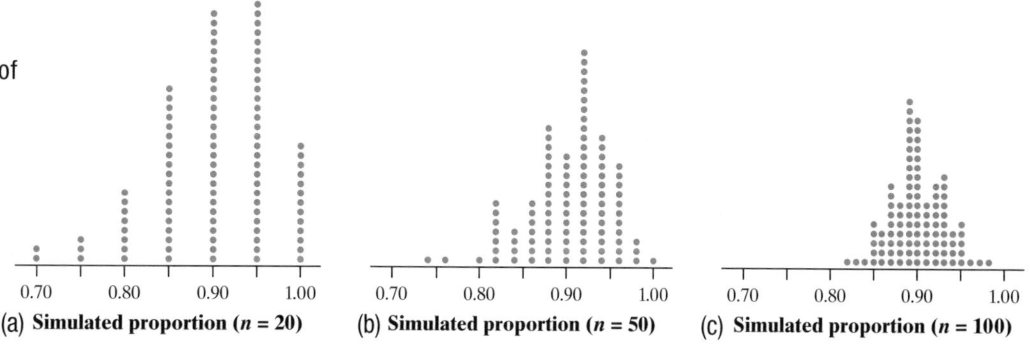

FIGURE 5.11 Simulated sampling distribution of $\hat{p}$ from 100 random samples of size (a) $n = 20$, (b) $n = 50$, and (c) $n = 100$ from a population where $p = 0.9$.

Notice how the simulated sampling distributions of $\hat{p}$ become less skewed and more approximately normal as the sample size increases. In fact, we can use a normal approximation to the sampling distribution of $\hat{p}$ whenever the **Large Counts condition** is met.

DEFINITION Large Counts condition

Let $\hat{p}$ be the proportion of successes in a random sample of size n from a population with proportion of successes p. The **Large Counts condition** says that the sampling distribution of $\hat{p}$ will be approximately normal when $np \geq 10$ and $n(1-p) \geq 10$.

We call it the "Large Counts" condition because np is the expected *count* of successes in the sample and $n(1-p)$ is the expected *count* of failures in the sample.

EXAMPLE

Backing the pack
Shape

Skill 3.C

PROBLEM: Let's return to the large high school where 84% of students regularly use a backpack to carry their books from class to class. Imagine taking an SRS of 100 students and calculating $\hat{p}$ = the proportion of students in the sample who regularly use a backpack. Describe the shape of the sampling distribution of $\hat{p}$. Justify your answer.

SOLUTION:

Because $100(0.84) = 84 \geq 10$ and $100(1-0.84) = 16 \geq 10$, the sampling distribution of $\hat{p}$ is approximately normal.

Justify that the distribution of $\hat{p}$ is approximately normal using the Large Counts condition: $np \geq 10$ and $n(1-p) \geq 10$.

FOR PRACTICE, TRY EXERCISE 9

Using the Normal Approximation for $\hat{p}$

When the Large Counts condition is met, we can use a normal distribution to calculate probabilities involving $\hat{p}$ = the proportion of successes in a random sample of size n. Here is an example.

EXAMPLE

Going to college
Using the normal approximation for $\hat{p}$

Skills 3.A, 3.B, 3.C

Ivan Vukovic/Shutterstock

PROBLEM: Suppose that 35% of all first-year students attend college within 50 miles of home. A polling organization asks an SRS of 1500 first-year college students how far away their home is. Find the probability that the random sample of 1500 students will give a result within 2 percentage points of the population percentage.

SOLUTION:

Let $\hat{p}$ = sample proportion of first-year college students who attend college within 50 miles of home.

$\mu_{\hat{p}} = 0.35$

$\sigma_{\hat{p}} = \sqrt{\frac{0.35(0.65)}{1500}} = 0.0123$ because $1500 < 10\%$ of all first-year college students.

Step 1: Define the random variable of interest, state how it is distributed, and identify the values of interest.

$$\mu_{\hat{p}} = p \quad \sigma_{\hat{p}} = \sqrt{\frac{p(1-p)}{n}}$$

Because $1500(0.35) = 525 \geq 10$ and $1500(0.65) = 975 \geq 10$, the distribution of $\hat{p}$ is approximately normal.

Justify that the distribution of $\hat{p}$ is approximately normal using the Large Counts condition: $np \geq 10$ and $n(1-p) \geq 10$.

We want to find $P(0.33 < \hat{p} < 0.37)$.

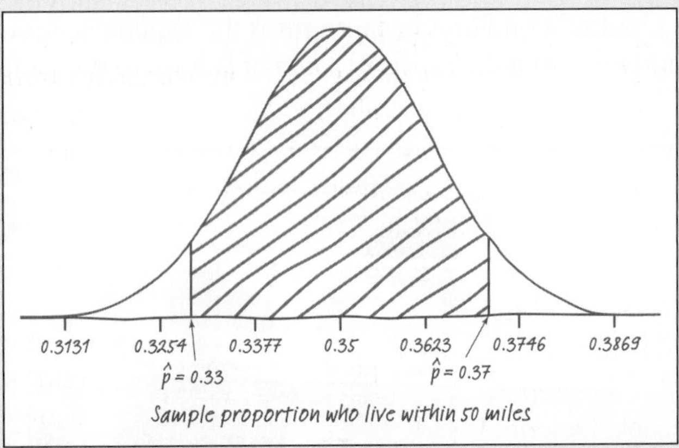

(i) $z = \frac{0.33 - 0.35}{0.0123} = -1.626$ and $z = \frac{0.37 - 0.35}{0.0123} = 1.626$

Step 2: Perform calculations—show your work!

Using technology: normalcdf(lower: −1.626, upper: 1.626, mean: 0, SD: 1) = 0.8961.

Using Table A: 0.9484 − 0.0516 = 0.8968

(ii) normalcdf(lower: 0.33, upper: 0.37, mean: 0.35, SD: 0.0123) = 0.8961

FOR PRACTICE, TRY EXERCISE 15

In the preceding example, about 90% of all SRSs of size 1500 from this population will give a result within 2 percentage points of the truth about the population. This result also suggests that in about 90% of all SRSs of size 1500 from this population, the population percentage will be within 2 percentage points of the sample percentage. This fact will become very important in Unit 6 when we use sample data to create an interval of plausible values for a population parameter.

THE NORMAL APPROXIMATION TO THE BINOMIAL DISTRIBUTION

In Section 4F, you learned about the binomial random variable X = the *number* of successes in a fixed number of trials n. As you have been learning about the sampling distribution of $\hat{p}$ = the *proportion* of success in a sample of size n, you might have wondered if there is a connection between these two random variables. There is! To calculate $\hat{p}$, we divide the number of successes X (a binomial random variable) by the sample size:

$$\hat{p} = \frac{\text{number of successes}}{\text{sample size}} = \frac{X}{n}$$

There is a close connection between the formulas for the mean and standard deviation of a binomial distribution and the formulas for the mean and standard deviation of the sampling distribution of $\hat{p}$. Because $\hat{p} = \frac{1}{n}X$, we can use the rules about linear transformations from Section 4E to determine the relationship between the formulas for the mean and standard deviation in these two contexts.

$$\mu_{\hat{p}} = \frac{1}{n}\mu_X = \frac{1}{n}(np) = p$$

$$\sigma_{\hat{p}} = \frac{1}{n}\sigma_X = \frac{1}{n}\sqrt{np(1-p)} = \sqrt{\frac{np(1-p)}{n^2}} = \sqrt{\frac{p(1-p)}{n}}$$

Furthermore, because the distribution of X is just a rescaled version of the distribution of $\hat{p}$, the same rules for shape apply: If the Large Counts condition is met, the distribution of X will be approximately normal.

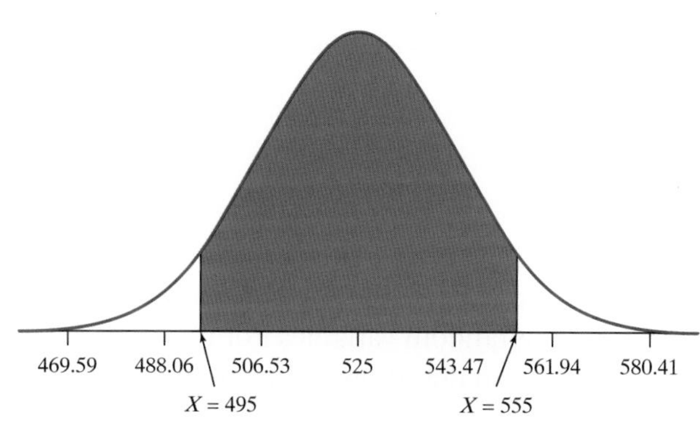

Number who live within 50 miles

In the preceding example, $P(0.33 < \hat{p} < 0.37)$ is equivalent to $P(495 < X < 555)$ where $495 = 1500(0.33)$ and $555 = 1500(0.37)$. Because the Large Counts condition is met, we can calculate this probability using the normal distribution with $\mu_X = 1500(0.35) = 525$ and $\sigma_X = \sqrt{1500(0.35)(0.65)} = 18.47$.

Not surprisingly, we get the same answer using the normal approximation to the binomial distribution. About 90% of all random samples of 1500 first-year college students will include between 495 and 555 students who live within 50 miles of campus.

CHECK YOUR UNDERSTANDING

Suppose that 75% of young-adult internet users (ages 18 to 29) watch online videos. A polling organization contacts an SRS of 1000 young-adult internet users and calculates the proportion $\hat{p}$ in the sample who watch online videos.

1. Calculate and interpret the mean of the sampling distribution of $\hat{p}$.
2. Calculate and interpret the standard deviation of the sampling distribution of $\hat{p}$. Verify that the 10% condition is met.
3. Is the sampling distribution of $\hat{p}$ approximately normal? Check that the Large Counts condition is met.
4. What is the probability that more than 80% of the members of the sample watch online videos?

The Sampling Distribution of a Difference Between Two Proportions

Are Democrats or Republicans more likely to use social media? Many statistical questions involve comparing the proportion of individuals with a certain characteristic in two populations. Let's call these parameters of interest p_1 and p_2. The preferred strategy is to select a separate random sample from each population and use the difference $\hat{p}_1 - \hat{p}_2$ in sample proportions as our point estimate. To make inferences about $p_1 - p_2$, you need to understand the sampling distribution of $\hat{p}_1 - \hat{p}_2$.

Suppose that there are two large high schools in a suburban community — each with more than 2000 students. In School 1, 70% of students ride the bus to school. In School 2, only 55% of students ride the bus to school. Imagine taking SRSs of size $n = 100$ from each school, calculating the sample proportion of bus riders from each school, and subtracting the two sample proportions. Figure 5.12 shows the simulated sampling distribution of $\hat{p}_1 - \hat{p}_2$.

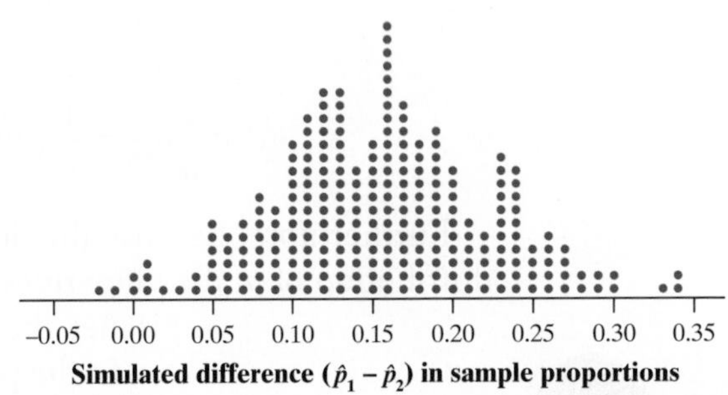

FIGURE 5.12 Simulated sampling distribution of $\hat{p}_1 - \hat{p}_2$, where $p_1 = 0.70$, $n_1 = 100$, $p_2 = 0.55$, and $n_2 = 100$.

It's no surprise that the sampling distribution of $\hat{p}_1 - \hat{p}_2$ is centered at $p_1 - p_2 = 0.70 - 0.55 = 0.15$. The standard deviation is harder to predict, however. Fortunately, we can use the rules for combining random variables from Section 4E to derive the formulas for the mean and standard deviation of the sampling distribution of $\hat{p}_1 - \hat{p}_2$.

Also note that the sampling distribution of $\hat{p}_1 - \hat{p}_2$ appears approximately normal. When the Large Counts condition is met for each sample, we can use the fact that the difference of two independent, normal random variables is also normal to determine that the sampling distribution of $\hat{p}_1 - \hat{p}_2$ is approximately normal.

AP® EXAM TIP

Formula sheet

The formulas for $\mu_{\hat{p}_1-\hat{p}_2}$ and $\sigma_{\hat{p}_1-\hat{p}_2}$ are included on the formula sheet provided on both sections of the AP® Statistics exam.

THE SAMPLING DISTRIBUTION OF $\hat{p}_1 - \hat{p}_2$

Let $\hat{p}_1$ be the sample proportion of successes in an SRS of size n_1 from Population 1 of size N_1 with proportion of successes p_1 and let $\hat{p}_2$ be the sample proportion of successes in an independent SRS of size n_2 from Population 2 of size N_2 with proportion of successes p_2. Then:

- The mean of the sampling distribution of $\hat{p}_1 - \hat{p}_2$ is $\mu_{\hat{p}_1-\hat{p}_2} = p_1 - p_2$.

- The standard deviation of the sampling distribution of $\hat{p}_1 - \hat{p}_2$ is approximately

$$\sigma_{\hat{p}_1-\hat{p}_2} = \sqrt{\frac{p_1(1-p_1)}{n_1} + \frac{p_2(1-p_2)}{n_2}}$$

as long as the samples are independent and the 10% condition is met for both samples: $n_1 < 0.10N_1$ and $n_2 < 0.10N_2$.
- The sampling distribution of $\hat{p}_1 - \hat{p}_2$ is approximately normal if the Large Counts condition is met for both samples: n_1p_1, $n_1(1-p_1)$, n_2p_2, and $n_2(1-p_2)$ are all at least 10.

Note that the formula for the standard deviation is exactly correct only when we have two types of independence:

- Independent samples, so that we can add the variances of $\hat{p}_1$ and $\hat{p}_2$.
- Independent observations within each sample. When sampling without replacement, the actual value of the standard deviation is smaller than the formula suggests. However, if the 10% condition is met for both samples, the difference is negligible.

EXAMPLE

Yummy goldfish! The sampling distribution of $\hat{p}_1 - \hat{p}_2$

Skills 3.B, 3.C, 4.B

PROBLEM: Suppose that your teacher brings two bags of colored goldfish crackers to class. Bag 1 has 25% red crackers and Bag 2 has 35% red crackers. Each bag contains approximately 1000 crackers. Using a paper cup, your teacher selects an SRS of 50 crackers from Bag 1 and an independent SRS of 40 crackers from Bag 2. Let $\hat{p}_1 - \hat{p}_2$ be the difference in the sample proportions of red crackers.

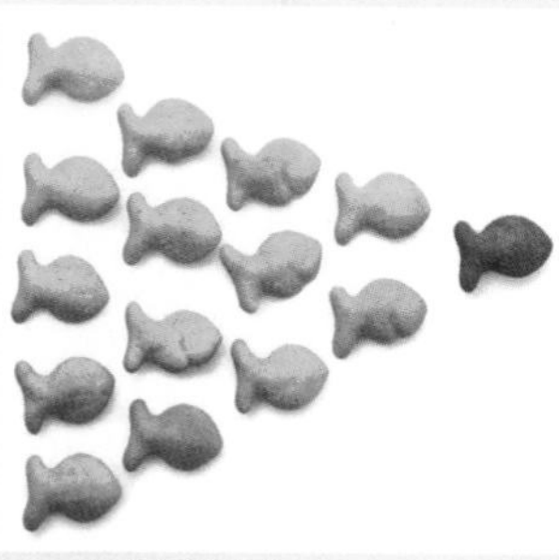

Juanmonino/iStock/Getty Images

(a) Is the shape of the sampling distribution of $\hat{p}_1 - \hat{p}_2$ approximately normal? Justify your answer.

(b) Calculate and interpret the mean of the sampling distribution.

(c) Verify both types of independence in this setting. Then, calculate and interpret the standard deviation of the sampling distribution.

SOLUTION:

(a) Yes, because $50(0.25)=12.5, 50(0.75)=37.5,$ $40(0.35)=14$, and $40(0.65)=26$ are all ≥ 10.

Check the Large Counts condition for both samples: n_1p_1, $n_1(1-p_1)$, n_2p_2, and $n_2(1-p_2)$ are all ≥ 10.

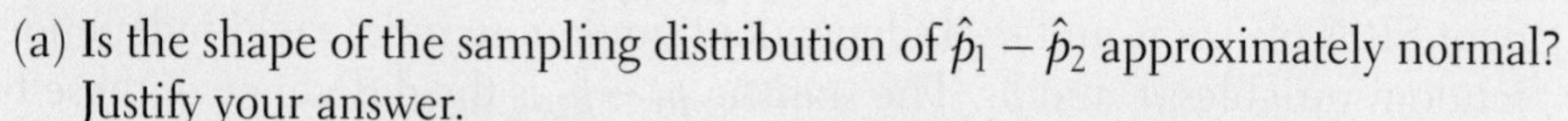

(b) $\mu_{\hat{p}_1-\hat{p}_2} = 0.25 - 0.35 = -0.10$

$\mu_{\hat{p}_1-\hat{p}_2} = p_1 - p_2$

In all possible independent random samples of 50 crackers from Bag 1 and 40 crackers from Bag 2, the resulting differences in the sample proportions $(\hat{p}_1 - \hat{p}_2)$ of red crackers have an average of -0.10.

(c) Because 50 < 10% of the 1000 crackers in Bag 1, 40 < 10% of the 1000 crackers in Bag 2, and the two samples are independent,

$$\sigma_{\hat{p}_1-\hat{p}_2}=\sqrt{\frac{0.25(0.75)}{50}+\frac{0.35(0.65)}{40}}=0.0971$$

$\sigma_{\hat{p}_1-\hat{p}_2}=\sqrt{\frac{p_1(1-p_1)}{n_1}+\frac{p_2(1-p_2)}{n_2}}$ when the samples are independent and the 10% condition is met for both samples.

In all possible independent random samples of 50 crackers from Bag 1 and 40 crackers from Bag 2, the resulting differences in the sample proportions ($\hat{p}_1-\hat{p}_2$) of red crackers typically vary from the difference in population proportions of −0.10 by 0.0971.

FOR PRACTICE, TRY EXERCISE 19

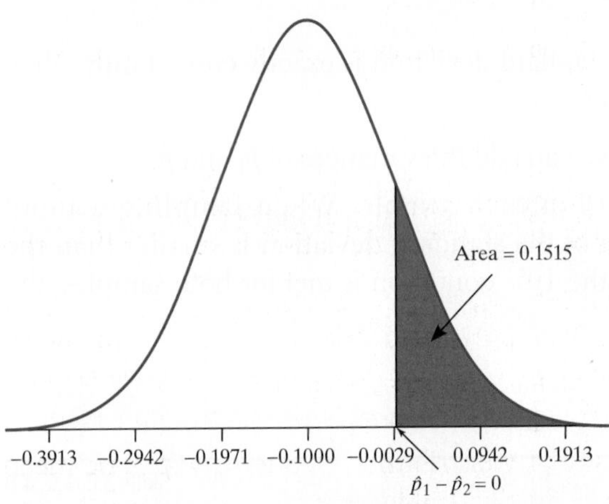

Difference in sample proportion ($\hat{p}_1-\hat{p}_2$) of red crackers from Bag 1 and Bag 2

Once we know that the sampling distribution of $\hat{p}_1-\hat{p}_2$ can be modeled by a normal distribution, we can use a normal curve to do probability calculations. To calculate the probability that the proportion of red crackers in the sample from Bag 1 is greater than the proportion of red crackers in the sample from Bag 2, we want to find $P(\hat{p}_1>\hat{p}_2)$, which is equivalent to $P(\hat{p}_1-\hat{p}_2>0)$.

Using technology or Table A, you can verify that the probability is approximately 0.1515. That is, there is about a 15% chance that a random sample of 50 crackers from Bag 1 has a greater proportion of red crackers than a random sample of 40 crackers from Bag 2.

Think About It

WHERE DO THE FORMULAS FOR THE MEAN AND STANDARD DEVIATION OF THE SAMPLING DISTRIBUTION OF $\hat{p}_1-\hat{p}_2$ COME FROM? Both $\hat{p}_1$ and $\hat{p}_2$ are random variables. That is, their values would vary in repeated independent SRSs of size n_1 and n_2. Independent random samples yield independent random variables $\hat{p}_1$ and $\hat{p}_2$. The statistic $\hat{p}_1-\hat{p}_2$ is the difference of these two independent random variables.

In Section 4E, we learned that for any two random variables X and Y,

$$\mu_{X-Y}=\mu_X-\mu_y$$

For the random variables $\hat{p}_1$ and $\hat{p}_2$, we have

$$\mu_{\hat{p}_1-\hat{p}_2}=\mu_{\hat{p}_1}-\mu_{\hat{p}_2}=p_1-p_2$$

We also learned in Section 4E that for *independent* random variables X and Y,

$$\sigma^2_{X-Y}=\sigma^2_X+\sigma^2_Y$$

For the independent random variables $\hat{p}_1$ and $\hat{p}_2$, we have

$$\sigma_{\hat{p}_1-\hat{p}_2}=\sqrt{\sigma^2_{\hat{p}_1}+\sigma^2_{\hat{p}_2}}=\sqrt{\left(\sqrt{\frac{p_1(1-p_1)}{n_1}}\right)^2+\left(\sqrt{\frac{p_2(1-p_2)}{n_2}}\right)^2}=\sqrt{\frac{p_1(1-p_1)}{n_1}+\frac{p_2(1-p_2)}{n_2}}$$

CHECK YOUR UNDERSTANDING

Among the 50,000 fans at a soccer game in Stadium A, 61% are wearing clothing that represents the home team (e.g., a jersey of one of the home team players). On the other side of the country, 65% of the 70,000 fans at Stadium B are wearing clothing that represents the home team. A marketing intern for each team selects a random sample of 200 fans from their home stadium and records the sample proportion of fans who are wearing clothing that represents the home team.

1. What is the shape of the sampling distribution of $\hat{p}_A - \hat{p}_B$? Justify your answer.
2. Calculate and interpret the mean of the sampling distribution.
3. Verify both types of independence in this setting. Then, calculate and interpret the standard deviation of the sampling distribution.
4. Calculate the probability that the sample proportions are within 0.05 of each other.

SECTION 5C Summary

- When we want information about the population proportion p of successes, we often select an SRS and use the sample proportion $\hat{p}$ to estimate the unknown parameter p. The **sampling distribution of the sample proportion $\hat{p}$** describes how the statistic $\hat{p}$ varies in all possible samples of the same size from the population.
  - **Center:** The mean of the sampling distribution of $\hat{p}$ is $\mu_{\hat{p}} = p$. The mean describes the average value of $\hat{p}$ in all possible samples of a certain size from a population.
  - **Variability:** The standard deviation of the sampling distribution of $\hat{p}$ is approximately $\sigma_{\hat{p}} = \sqrt{\frac{p(1-p)}{n}}$ when the 10% condition is met: $n < 0.10N$. The standard deviation measures how far the values of $\hat{p}$ typically vary from p in all possible samples of a certain size from a population.
  - **Shape:** The sampling distribution of $\hat{p}$ is approximately normal when the Large Counts condition is met: np and $n(1-p)$ are both at least 10.
- Choose independent SRSs of size n_1 from Population 1 of size N_1 with proportion of successes p_1 and of size n_2 from Population 2 of size N_2 with proportion of successes p_2. The **sampling distribution of the difference in sample proportions $\hat{p}_1 - \hat{p}_2$** has the following properties:
  - **Center:** The mean of the sampling distribution is $\mu_{\hat{p}_1-\hat{p}_2} = p_1 - p_2$.
  - **Variability:** The standard deviation of the sampling distribution is approximately $\sigma_{\hat{p}_1-\hat{p}_2} = \sqrt{\frac{p_1(1-p_1)}{n_1} + \frac{p_2(1-p_2)}{n_2}}$ as long as the samples are independent and the 10% condition is met for both samples: $n_1 < 0.10N_1$ and $n_2 < 0.10N_2$.
  - **Shape:** Approximately normal if the Large Counts condition is met for both samples: n_1p_1, $n_1(1-p_1)$, n_2p_2, and $n_2(1-p_2)$ are all at least 10.
- When the Large Counts condition is met, you can use a normal distribution to calculate probabilities involving the sampling distribution of $\hat{p}$ or the sampling distribution of $\hat{p}_1 - \hat{p}_2$.

AP® EXAM TIP
AP® Daily Videos

Review the content of this section and get extra help by watching the AP® Daily Videos for Topics 5.5 and 5.6, which are available in AP® Classroom.

SECTION 5C Exercises

Center and Variability

1. **Registered voters** In a congressional district, 55% of registered voters are Democrats. A polling organization selects a random sample of 500 registered voters from this district. Let $\hat{p}$ = the proportion of Democrats in the sample.

pg 487

(a) Calculate and interpret the mean of the sampling distribution of $\hat{p}$.

(b) Verify that the 10% condition is met. Then calculate and interpret the standard deviation of the sampling distribution of $\hat{p}$.

2. **Woodpeckers** In a forest, 63% of the pine trees have visible woodpecker damage. A researcher selects a random sample of 30 pine trees and visually inspects each tree for woodpecker damage. Let $\hat{p}$ = the proportion of pine trees with visible woodpecker damage in the sample.

(a) Calculate and interpret the mean of the sampling distribution of $\hat{p}$.

(b) Verify that the 10% condition is met. Then calculate and interpret the standard deviation of the sampling distribution of $\hat{p}$.

3. **Orange Skittles®** The makers of Skittles claim that 20% of Skittles candies are orange. Suppose this claim is true and you select a random sample of 30 Skittles from a large bag. Let $\hat{p}$ = the proportion of orange Skittles in the sample.

(a) Calculate the mean and standard deviation of the sampling distribution of $\hat{p}$.

(b) Interpret the mean and standard deviation from part (a).

4. **Assembly-line workers** A factory employs 3000 unionized workers, 90% of whom work on an assembly line. A random sample of 15 workers is selected for a survey about worker satisfaction. Let $\hat{p}$ = the proportion of assembly-line workers in the sample.

(a) Calculate the mean and standard deviation of the sampling distribution of $\hat{p}$.

(b) Interpret the mean and standard deviation from part (a).

5. **Less variable Skittles®** Refer to Exercise 3. What sample size would be required to reduce the standard deviation of the sampling distribution to one-half the value you found in Exercise 3(b)? Justify your answer.

6. **Less variable workers** Refer to Exercise 4. What sample size would be required to reduce the standard deviation of the sampling distribution to one-third the value you found in Exercise 4(b)? Justify your answer.

7. **Airport security** At a certain airport, security officials claim that they randomly select passengers for extra screening at the gate before boarding some flights. One such flight had 76 passengers — 16 in first class and 60 in economy class. Some passengers were surprised when none of the 14 passengers chosen for screening were seated in first class. Let $\hat{p}$ be the proportion of first-class passengers in the sample.

(a) Show that the 10% condition is not met in this case.

(b) What effect does violating the 10% condition have on the standard deviation of the sampling distribution of $\hat{p}$?

8. **Don't pick me!** Instead of collecting homework from all of her students, Ms. Friedman randomly selects 5 of her 30 students and collects homework from only those students. Let $\hat{p}$ be the proportion of students in the sample who completed their homework.

(a) Show that the 10% condition is not met in this case.

(b) What effect does violating the 10% condition have on the standard deviation of the sampling distribution of $\hat{p}$?

Shape

9. **More registered voters** Refer to Exercise 1. Describe the shape of the sampling distribution of $\hat{p}$ = the proportion of Democrats in the sample. Justify your answer.

pg 489

10. **More woodpeckers** Refer to Exercise 2. Describe the shape of the sampling distribution of $\hat{p}$ = the proportion of pine trees with visible woodpecker damage in the sample. Justify your answer.

11. **More Skittles®** Refer to Exercise 3. Is the shape of the sampling distribution of $\hat{p}$ approximately normal? Justify your answer.

12. **More workers** Refer to Exercise 4. Is the shape of the sampling distribution of $\hat{p}$ approximately normal? Justify your answer.

Using the Normal Approximation for $\hat{p}$

13. **Do you drink the cereal milk?** A *USA Today* poll asked a random sample of 1012 U.S. adults what they do with the milk in the bowl after they have eaten the cereal. Let $\hat{p}$ be the proportion of people in the sample who drink the cereal milk. A spokesperson for the dairy industry claims that 70% of all U.S. adults drink the cereal milk. Suppose this claim is true.

(a) What is the mean of the sampling distribution of $\hat{p}$?

(b) Find the standard deviation of the sampling distribution $\hat{p}$. Verify that the 10% condition is met.

(c) Justify that the sampling distribution of $\hat{p}$ is approximately normal.

(d) Of the poll respondents, 67% said that they drink the cereal milk. Find the probability of obtaining a random sample of 1012 adults in which 67% or fewer say they drink the cereal milk, assuming the milk industry spokesperson's claim is true.

(e) Does this poll give convincing evidence against the spokesperson's claim? Explain your reasoning.

14. **Jury duty** What proportion of U.S. residents receive a jury summons each year? A polling organization plans to survey a random sample of 500 U.S. residents to find out. Let $\hat{p}$ be the proportion of residents in the sample who received a jury summons in the previous 12 months. According to the National Center for State Courts (NCSC), 15% of U.S. residents receive a jury summons each year.[13] Suppose that this claim is true.

(a) What is the mean of the sampling distribution of $\hat{p}$?

(b) Find the standard deviation of the sampling distribution $\hat{p}$. Verify that the 10% condition is met.

(c) Justify that the sampling distribution of $\hat{p}$ is approximately normal.

(d) Of the poll respondents, 13% said that they received a jury summons in the previous 12 months. Find the probability of obtaining a random sample of 500 adults in which 13% or fewer say they got a jury summons in the previous 12 months, assuming the NCSC's claim is true.

(e) Does this poll give convincing evidence against the NCSC's claim? Explain your reasoning.

15. **Second jobs** According to the National Center for Education Statistics, 18% of teachers work at non-school second jobs during the school year.[14] You plan to select an SRS of 200 teachers. Find the probability that more than 20% of the teachers in the sample work a non-school second job during the school year. Describe this probability in words.

pg 490

16. **Motorcycle ownership** In the United States, 8% of households own a motorcycle.[15] You plan to send surveys to an SRS of 500 households. Find the probability that more than 10% of the households in the sample own a motorcycle. Describe this probability in words.

17. **On-time shipping** A mail-order company advertises that it ships 90% of its orders within 3 working days. You select an SRS of 100 of the 5000 orders received in the past week for an audit. Let X = the number of orders in the sample that are shipped on time. Assume that the company's claim is true.

(a) Verify that X is a binomial random variable and state n and p.

(b) Explain why X can be approximated by a normal distribution.

(c) Use an appropriate normal distribution to calculate the probability of getting 86 or fewer orders that were shipped on time.

(d) The audit reveals that 86 of these orders were shipped on time. Based on your answer to part (c), is there convincing evidence that less than 90% of all orders from this company are shipped within 3 working days? Explain your reasoning.

18. **Wait times** A hospital claims that 75% of people who come to its emergency room are seen by a doctor within half an hour of checking in. To verify this claim, an auditor inspects the medical records of 50 randomly selected patients who checked into the emergency room during the last year. Let Y = the number of patients in the sample who were seen by a doctor within half an hour of checking in. Assume that the hospital's claim is true.

(a) Verify that Y is a binomial random variable and state n and p.

(b) Explain why Y can be approximated by a normal distribution.

(c) Use an appropriate normal distribution to calculate the probability of getting 30 or fewer patients who are seen by a doctor within half an hour.

(d) In the actual sample, only 30 of these patients were seen by a doctor within half an hour of checking in. Based on your answer to part (c), is there convincing evidence that less than 75% of all patients are seen by a doctor within half an hour? Explain your reasoning.

The Sampling Distribution of $\hat{p}_1 - \hat{p}_2$

19. **AP® enrollment** Suppose that 30% of seniors and 25% of juniors at a large high school are enrolled in an AP® class. Independent random samples of 20 seniors and 20 juniors are selected and are asked if they are enrolled in an AP® class. Let $\hat{p}_S$ represent the sample proportion of seniors enrolled in an AP® class, and let $\hat{p}_J$ represent the sample proportion of juniors enrolled in an AP® class.

pg 493

(a) Is the shape of the sampling distribution of $\hat{p}_S - \hat{p}_J$ approximately normal? Justify your answer.

(b) Calculate and interpret the mean of the sampling distribution.

(c) Verify both types of independence in this setting. Then, calculate and interpret the standard deviation of the sampling distribution.

20. **Athletic participation** Suppose that 20% of students at High School A and 18% of students at High School B participate on a school athletic team. Independent random samples of 30 students from each school are selected and are asked if they participate on a school athletic team. Let $\hat{p}_A$ represent the sample proportion of students at High School A who participate on a school athletic team, and let $\hat{p}_B$ represent the sample proportion of students at High School B who participate on a school athletic team.

(a) Is the shape of the sampling distribution of $\hat{p}_A - \hat{p}_B$ approximately normal? Justify your answer.

(b) Calculate and interpret the mean of the sampling distribution.

(c) Verify both types of independence in this setting. Then, calculate and interpret the standard deviation of the sampling distribution.

21. **I want red!** A candy maker offers Child and Adult bags of jelly beans with different color mixes. The company claims that the Child mix has 30% red jelly beans, while the Adult mix contains 15% red jelly beans. Assume that the candy maker's claim is true. Suppose we select a random sample of 50 jelly beans from the Child mix and an independent random sample of 100 jelly beans from the Adult mix. Let $\hat{p}_C$ and $\hat{p}_A$ be the sample proportions of red jelly beans from the Child and Adult mixes, respectively.

(a) What is the shape of the sampling distribution of $\hat{p}_C - \hat{p}_A$? Justify your answer.

(b) Calculate the mean and standard deviation of the sampling distribution.

(c) Interpret the values from part (b).

22. **Literacy** A researcher reports that 80% of adults who graduated from high school, but only 40% of adults who didn't graduate from high school, would pass a basic literacy test.[16] Assume that the researcher's claim is true. Suppose we give a basic literacy test to a random sample of 60 adults who graduated and an independent random sample of 75 adults who didn't graduate. Let $\hat{p}_G$ and $\hat{p}_D$ be the sample proportions of adults who graduated and who didn't graduate, respectively, who pass the test.

(a) What is the shape of the sampling distribution of $\hat{p}_G - \hat{p}_D$? Justify your answer.

(b) Calculate the mean and standard deviation of the sampling distribution.

(c) Interpret the values from part (b).

23. **I want more red!** Refer to Exercise 21.

(a) Find the probability that the proportion of red jelly beans in the Child sample is less than or equal to the proportion of red jelly beans in the Adult sample, assuming that the company's claim is true.

(b) Suppose that the Child and Adult samples contain an equal proportion of red jelly beans. Based on your result in part (a), would this give you reason to doubt the company's claim? Explain your reasoning.

24. **More literacy** Refer to Exercise 22.

(a) Find the probability that the proportion of adults who graduated who pass the test is at most 0.20 higher than the proportion of adults who didn't graduate who pass the test, assuming that the researcher's report is correct.

(b) Suppose that the difference (Graduate – Didn't graduate) in the sample proportions who pass the test is exactly 0.20. Based on your result in part (a), would this give you reason to doubt the researcher's claim? Explain your reasoning.

For Investigation *Apply the skills from the section in a new context or nonroutine way.*

25. **10% condition for proportions** In this section, you learned that the formula for the standard deviation of the sampling distribution of $\hat{p}$ is valid when the sample size is less than 10% of the population size. What if the sample size is larger than 10% of the population size? In this case, we use the *finite population correction factor* in the formula for the standard deviation, where N is the population size:

$$\sigma_{\hat{p}} = \sqrt{\frac{p(1-p)}{n}}\sqrt{\frac{N-n}{N-1}}$$

Imagine that you are choosing an SRS from a population of size $N = 1000$ where $p = 0.60$.

(a) Calculate the standard deviation of the sampling distribution of $\hat{p}$ for samples of size $n = 10$, with and without the finite population correction factor. How do these values compare?

(b) Calculate the standard deviation of the sampling distribution of $\hat{p}$ for samples of size $n = 500$, with and without the finite population correction factor. How do these values compare?

(c) Based on your answers to parts (a) and (b), explain why we can use the formula $\sigma_{\hat{p}} = \sqrt{\frac{p(1-p)}{n}}$ when the sample size is less than 10% of the population size.

(d) Calculate the standard deviation of the sampling distribution of $\hat{p}$ for samples of size $n = 1000$ using the finite population correction factor. Why does the value of the standard deviation make sense in this situation?

26. **Continuity correction** A hospital claims that 75% of people who come to its emergency room are seen by a doctor within half an hour of checking in. To verify this claim, an auditor inspects the medical records of 50 randomly selected patients who checked into the emergency room during the last year. In Exercise 18, you verified that Y = the number of patients in the sample who were seen by a doctor within half an hour of checking in is a binomial random variable with $n = 50$ and $p = 0.75$.

(a) Use the binomial distribution to calculate $P(Y \leq 30)$. How does this answer compare to your answer in Exercise 18(c)?

When using a normal distribution to approximate the binomial distribution (or any discrete probability distribution), we can get more accurate results by using a *continuity correction*. Because any values between 29.5 and 30.5 in the normal distribution correspond to $Y = 30$ patients, we can get a better estimate of $P(Y \leq 30)$ by calculating $P(Y \leq 30.5)$.

(b) Use an appropriate normal distribution to calculate $P(Y \leq 30.5)$. How does this answer compare to the exact binomial probability from part (a)?

(c) If you wanted to use the continuity correction to calculate the probability that at least 35 to at most 40 patients are seen within half an hour, state the values of interest for the appropriate normal distribution calculation. Do not perform the calculation.

Multiple Choice *Select the best answer for each question.*

Exercises 27–29 refer to the following scenario. The magazine *Sports Illustrated* asked a random sample of 750 Division I college athletes, "Do you believe performance-enhancing drugs are a problem in college sports?" Suppose that 30% of all Division I athletes think that these drugs are a problem. Let $\hat{p}$ be the sample proportion who say that these drugs are a problem.

27. Which of the following are the mean and standard deviation of the sampling distribution of the sample proportion $\hat{p}$?

(A) Mean = 0.30, SD = 0.017

(B) Mean = 0.30, SD = 0.55

(C) Mean = 0.30, SD = 0.0003

(D) Mean = 225, SD = 12.5

(E) Mean = 225, SD = 157.5

28. Decreasing the sample size from 750 to 375 would multiply the standard deviation by which of the following?

(A) 4

(B) 2

(C) $\sqrt{2}$

(D) 1/2

(E) $1/\sqrt{2}$

29. Which of the following is the smallest sample size that would allow use of the normal distribution to approximate the sampling distribution of $\hat{p}$ in this setting?

(A) 10

(B) 20

(C) 30

(D) 40

(E) 50

30. In a large company, 55% of the employees work full-time. Which of the following is equivalent to the probability of getting less than 50% full-time employees in a random sample of 100 employees from this company?

(A) $P\left(z < \dfrac{0.50 - 0.55}{100}\right)$

(B) $P\left(z < \dfrac{0.50 - 0.55}{\sqrt{\dfrac{0.55(0.45)}{100}}}\right)$

(C) $P\left(z < \dfrac{0.55 - 0.50}{\sqrt{\dfrac{0.55(0.45)}{100}}}\right)$

(D) $P\left(z < \dfrac{0.50 - 0.55}{\sqrt{100(0.55)(0.45)}}\right)$

(E) $P\left(z < \dfrac{0.55 - 0.50}{\sqrt{100(0.55)(0.45)}}\right)$

Recycle and Review *Practice what you learned in previous sections.*

31. **Dem bones (1F)** Osteoporosis is a condition in which the bones become brittle due to loss of minerals. To diagnose osteoporosis, an elaborate apparatus measures bone mineral density (BMD). BMD is usually reported in standardized form, with the standardization being based on a population of healthy young adults. The World Health Organization (WHO) criterion for osteoporosis is a BMD score that is 2.5 standard deviations below the mean for young adults. BMD measurements in a population of people similar in age roughly follow a normal distribution.

(a) What percentage of healthy young adults have osteoporosis by the WHO criterion?

(b) Women aged 70 to 79 are, of course, not young adults. The mean BMD in this age group is about

−2 on the standard scale for young adults. Suppose that the standard deviation is the same as for young adults. What percentage of this older population has osteoporosis?

32. **Whole grains** (3C) A series of observational studies revealed that people who typically consume 3 servings of whole grains per day have about a 20% lower risk of dying from heart disease and about a 15% lower risk of dying from stroke or cancer than those who consume no whole grains.[17]

(a) Explain how confounding makes it difficult to establish a cause-and-effect relationship between whole grains consumption and risk of dying from heart disease, stroke, or cancer, based on these studies.

(b) Explain how researchers could establish a cause-and-effect relationship in this context.

SECTION 5D Sample Means

LEARNING TARGETS *By the end of the section, you should be able to:*

- Calculate and interpret the mean and standard deviation of the sampling distribution of a sample mean $\bar{x}$.
- Determine if the sampling distribution of $\bar{x}$ is approximately normal.
- If appropriate, use a normal distribution to calculate probabilities involving sample means.
- Describe the shape, center, and variability of the sampling distribution of a difference in sample means $\bar{x}_1 - \bar{x}_2$.

When sample data are categorical, we often use the proportion of successes in the sample $\hat{p}$ to make an inference about the population proportion p. What if we collect data on a quantitative variable, such as commute time? In these kinds of cases, we typically use the sample mean $\bar{x}$ to estimate the population mean μ. But how close will our estimate be to the truth? To understand how much $\bar{x}$ typically varies from μ, you need to understand the **sampling distribution of the sample mean $\bar{x}$**.

DEFINITION Sampling distribution of the sample mean $\bar{x}$

The **sampling distribution of the sample mean $\bar{x}$** describes the distribution of values taken by the sample mean $\bar{x}$ in all possible samples of the same size from the same population.

For example, here is a simulated sampling distribution of $\bar{x}$ = sample mean commute time (in minutes) for 300 random samples of size $n = 10$ from a population of employed adults who don't work at home.

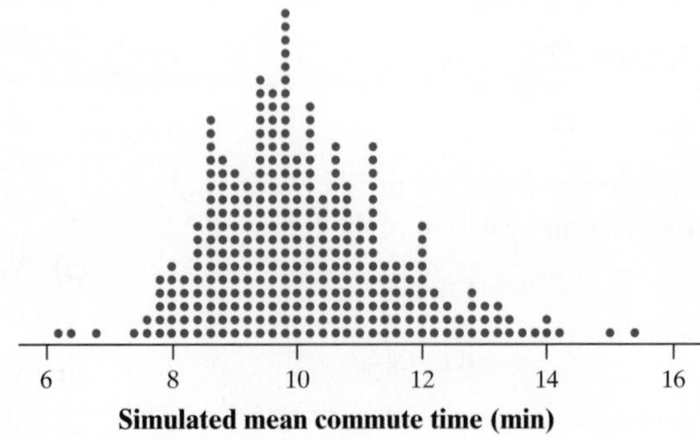

This distribution is slightly skewed to the right with a mean of 9.96 minutes and a standard deviation of 1.48 minutes. By the end of this section, you should be able to anticipate the center, variability, and shape of sampling distributions like this one without having to perform a simulation.

The Sampling Distribution of $\bar{x}$: Center and Variability

Figure 5.13(a) shows a histogram of a population distribution with mean $\mu = 100$ and $\sigma = 2.8$. Notice that the population distribution is clearly skewed to the left. Figure 5.13(b) shows a simulated sampling distribution of $\bar{x}$ with 200 SRSs of size $n = 4$ from this population. Figure 5.13(c) shows a simulated sampling distribution of $\bar{x}$ with 200 SRSs of size $n = 10$ from this population.

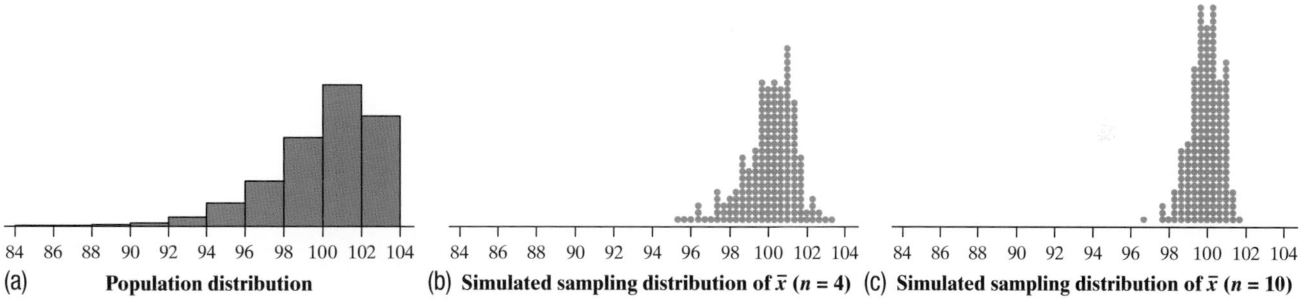

FIGURE 5.13 (a) Population distribution and simulated sampling distributions of $\bar{x}$ for SRSs of size (b) $n = 4$ and (c) $n = 10$.

Notice that both of the simulated sampling distributions are centered at the population mean of 100. Also note two important changes as the sample size increases from $n = 4$ to $n = 10$: the sampling distribution becomes less variable and it becomes less skewed. As with the sampling distribution of $\hat{p}$, there are some simple rules that describe the mean and standard deviation of the sampling distribution of $\bar{x}$. Describing the shape of the sampling distribution of $\bar{x}$ is more complicated, so we'll save that for later.

HOW TO CALCULATE $\mu_{\bar{x}}$ AND $\sigma_{\bar{x}}$

Let $\bar{x}$ be the sample mean in an SRS of size n from a population of size N with mean μ and standard deviation σ. Then:

- The mean of the sampling distribution of $\bar{x}$ is $\mu_{\bar{x}} = \mu$.
- The standard deviation of the sampling distribution of $\bar{x}$ is approximately $\sigma_{\bar{x}} = \dfrac{\sigma}{\sqrt{n}}$ as long as the 10% condition is met: $n < 0.10N$.

AP® EXAM TIP
Formula sheet

The formulas for $\mu_{\bar{x}}$ and $\sigma_{\bar{x}}$ are included on the formula sheet provided on both sections of the AP® Statistics exam.

The behavior of $\bar{x}$ in repeated random samples is much like that of the sample proportion $\hat{p}$:

- The value $\mu_{\bar{x}}$ gives the average value of $\bar{x}$ in all possible samples of a given size from a population. Because $\mu_{\bar{x}} = \mu$, we know that the sample mean $\bar{x}$ is an unbiased estimator of the population mean μ.

- The value $\sigma_{\bar{x}}$ measures the typical distance between a sample mean $\bar{x}$ and the population mean μ in all possible samples of a given size from a population. Because the sample size appears in the denominator of the formula for $\sigma_{\bar{x}}$, we know that the sample mean $\bar{x}$ is less variable when the sample size is larger. *Specifically, multiplying the sample size by 4 cuts the standard deviation in half.*
- When we sample *with* replacement, the standard deviation of the sampling distribution of $\bar{x}$ is exactly $\sigma_{\bar{x}} = \frac{\sigma}{\sqrt{n}}$. When we sample *without* replacement, the observations are not independent and the actual standard deviation of the sampling distribution of $\bar{x}$ is smaller than the value given by the formula. However, if the sample size is less than 10% of the population size (the 10% condition), the value given by the formula is nearly correct.

AP® EXAM TIP

Notation matters. The symbols $\hat{p}$, $\bar{x}$, n, p, μ, σ, $\mu_{\hat{p}}$, $\sigma_{\hat{p}}$, $\mu_{\bar{x}}$, and $\sigma_{\bar{x}}$ all have specific and different meanings. Either use notation correctly—or don't use it at all. You can expect to lose credit if you use incorrect notation on the AP® Statistics exam.

These facts about the mean and standard deviation of $\bar{x}$ are true *no matter what shape the population distribution has.*

EXAMPLE

Been to the movies recently?
Center and variability

Skills 3.B, 4.B

PROBLEM: The distribution of the number of movies viewed in the last year by students at a large high school is skewed to the right with a mean of 25.3 movies and a standard deviation of 15.8 movies. Suppose we select an SRS of 100 students from this school and calculate $\bar{x}$ = the mean number of movies viewed by the members of the sample.

(a) Calculate and interpret the mean of the sampling distribution of $\bar{x}$.

(b) Verify that the 10% condition is met. Then calculate and interpret the standard deviation of the sampling distribution of $\bar{x}$.

Antonio Gravante/EyeEm/Getty Images

SOLUTION:

(a) $\mu_{\bar{x}} = 25.3$ movies

If you selected all possible samples of 100 students from the school and calculated the sample mean number of movies for each sample, the sample means would have an average value of 25.3 movies.

$\mu_{\bar{x}} = \mu$

(b) Because we can assume that $n = 100$ is less than 10% of students at a large high school,

$$\sigma_{\bar{x}} = \frac{15.8}{\sqrt{100}} = 1.58 \text{ movies}$$

When $n < 0.10N$, $\sigma_{\bar{x}} = \frac{\sigma}{\sqrt{n}}$.

If you selected all possible samples of 100 students from the school and calculated the sample mean number of movies for each sample, the sample means would typically vary from the population mean of 25.3 movies by about 1.58 movies.

FOR PRACTICE, TRY EXERCISE 1

Think About It

WHERE DO THE FORMULAS FOR THE MEAN AND STANDARD DEVIATION OF $\bar{x}$ COME FROM? Suppose we select an SRS of size n from a population and record the value X of a quantitative variable for each member of the sample. Call the individual observations $X_1, X_2, \ldots, X_n$. If the sample size is less than 10% of the population size, we can think of these X_i's as independent random variables, each with mean μ and standard deviation σ. Because

$$\bar{x} = \frac{X_1 + X_2 + \ldots + X_n}{n} = \frac{1}{n}(X_1 + X_2 + \ldots + X_n)$$

we can use the rules for random variables from Section 4E to find the mean and standard deviation of $\bar{x}$.

Using the addition rules for means and variances and the rules for multiplying and dividing by a constant, we get

$$\mu_{\frac{1}{n}(X_1+X_2+\ldots+X_n)} = \frac{1}{n}(\mu_{X_1} + \mu_{X_2} + \ldots + \mu_{X_n}) = \frac{1}{n}(\mu + \mu + \ldots + \mu) = \frac{1}{n}(n\mu) = \mu$$

$$\sigma_{\frac{1}{n}(X_1+X_2+\ldots+X_n)} = \frac{1}{n}\sqrt{\sigma^2_{X_1} + \sigma^2_{X_2} + \ldots + \sigma^2_{X_n}} = \frac{1}{n}\sqrt{\sigma^2 + \sigma^2 + \ldots + \sigma^2} = \frac{1}{n}\sqrt{n\sigma^2}$$

$$= \sqrt{\frac{n\sigma^2}{n^2}} = \sqrt{\frac{\sigma^2}{n}} = \frac{\sigma}{\sqrt{n}}$$

The Sampling Distribution of $\bar{x}$: Shape

We have described the mean and standard deviation of the sampling distribution of a sample mean $\bar{x}$ but not its shape. That's because the shape of the sampling distribution of $\bar{x}$ depends on the shape of the population distribution. In the next activity, you will explore what happens when you sample from a normal population and when you sample from a non-normal population.

ACTIVITY Shape of the sampling distribution of $\bar{x}$

In this activity, you will use an applet to investigate the shape of the sampling distribution of the sample mean $\bar{x}$ for different sample sizes (n) and different population distribution shapes.

1. Go to www.stapplet.com and launch the *Simulating Sampling Distributions* applet.

Sampling from a normal population

2. In the box labeled "Population," there are four choices for the shape of the population distribution when the data are quantitative: normal, uniform, skewed, and bimodal. Select normal.
3. In the box labeled "Sample," change the sample size to $n = 2$ and click the "Select sample" button. The applet will graph the distribution of the sample and display the value of the sample mean $\bar{x}$. Was your sample mean close to the population mean $\mu = 10$?

4. In the box labeled "Sampling Distribution," notice that the value of $\bar{x}$ from Step 3 is displayed on a dotplot. Click the "Select samples" button 9 more times, so that you have a total of 10 sample means. Look at the dotplot of your $\bar{x}$ values. Does the distribution have a recognizable shape?
5. To better see the shape of the sampling distribution of $\bar{x}$, enter 990 in the box for quickly selecting random samples and click the "Select samples" button. You have now selected a total of 1000 random samples of size $n = 2$. Describe the shape of the simulated sampling distribution of $\bar{x}$ shown in the dotplot. Click the button to show the corresponding normal curve. How well does it fit?
6. Change the sample size to $n = 10$ in the "Sample" box and quickly select 1000 random samples in the "Sampling Distribution" box. Describe the shape of the sampling distribution of $\bar{x}$.
7. Repeat Step 6 with samples of size $n = 30$ and $n = 100$.
8. What have you learned about the shape of the sampling distribution of $\bar{x}$ when the population distribution has a normal shape?

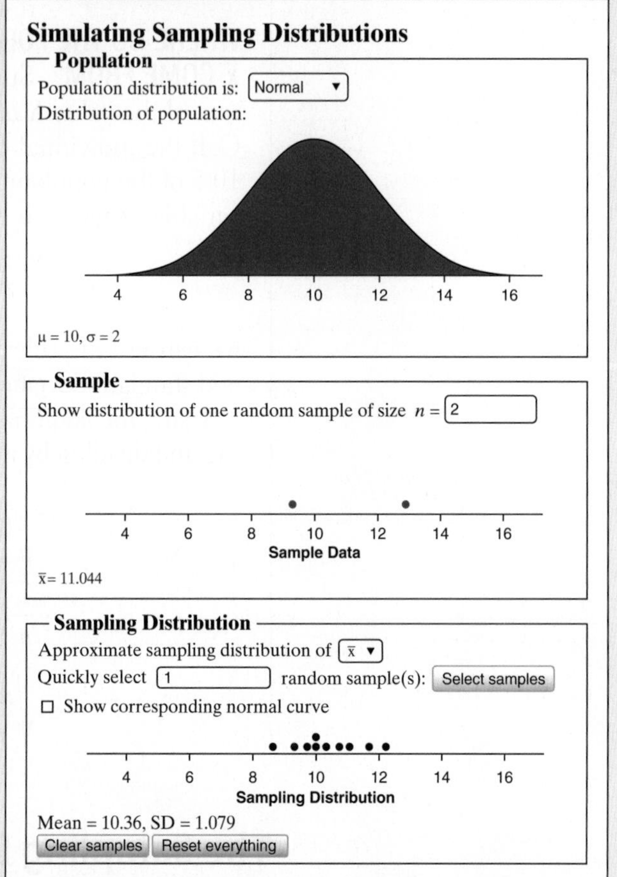

Sampling from a non-normal population

9. In the "Population" box, change the population shape to uniform and repeat Steps 3–7.
10. In the "Population" box, change the population shape to skewed and repeat Steps 3–7.
11. In the "Population" box, change the population shape to bimodal and repeat Steps 3–7.
12. What have you learned about the shape of the sampling distribution of $\bar{x}$ when the population distribution has a non-normal shape?

As the preceding activity demonstrates, if the population distribution is normal, so is the sampling distribution of $\bar{x}$. *This is true regardless of the sample size n.* Think back to what you learned in Section 5A to help make sense of this important idea. First, note that

$$\bar{x} = \frac{X_1 + X_2 + \ldots + X_n}{n} = \frac{1}{n}(X_1 + X_2 + \ldots + X_n)$$

The random variable $\bar{x}$ is a linear combination of the n independent, normally distributed random variables $X_1, X_2, \ldots, X_n$, so the sampling distribution of $\bar{x}$ is normal.

Of course, in the real world, no population distribution is exactly normal. Fortunately, the result of the activity still holds for *approximately* normal population distributions. If the shape of the population distribution is approximately normal, the shape of the sampling distribution of $\bar{x}$ will also be approximately normal for any sample size n.

What about non-normal populations? Figure 5.14 shows screen shots of simulated sampling distributions of $\overline{x}$ for random samples of size $n = 2$ and $n = 30$ from each of three different populations.

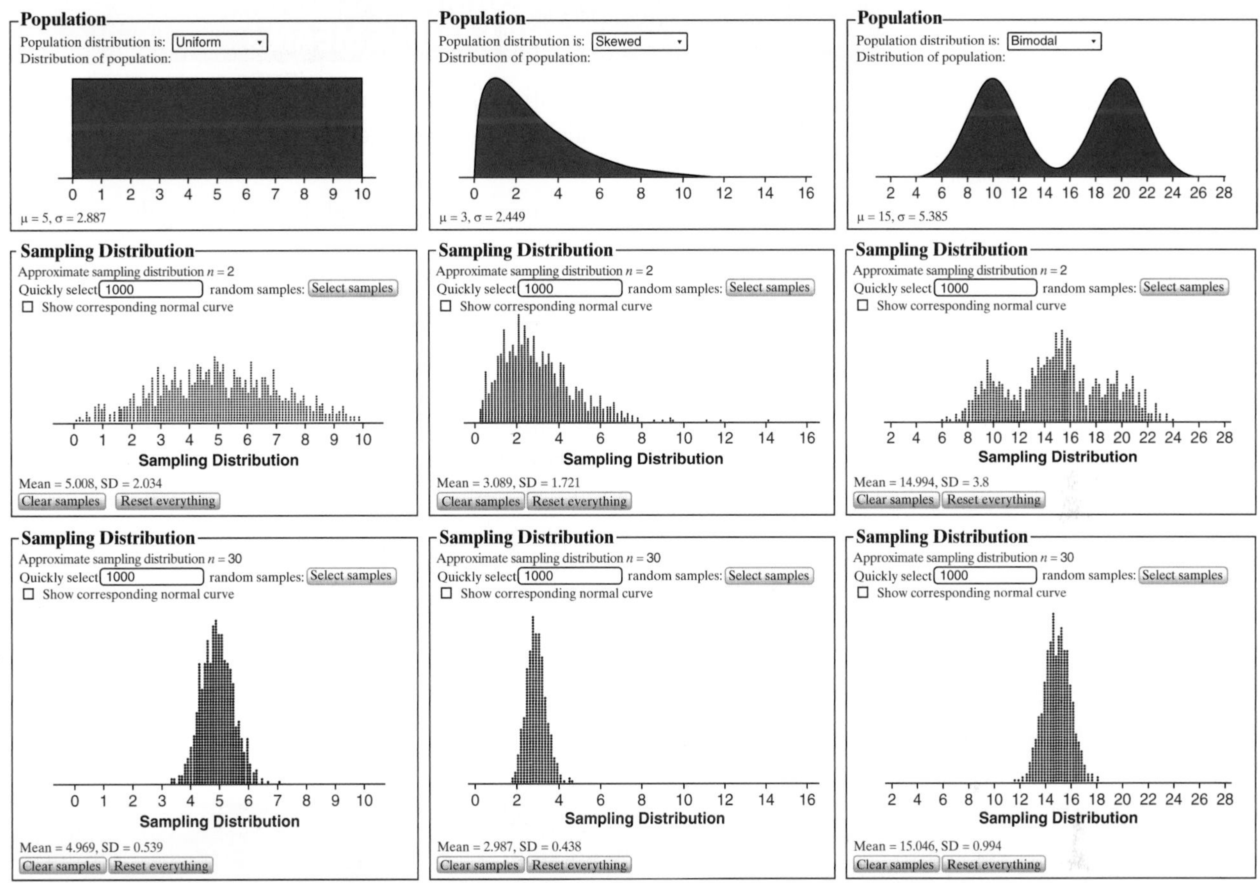

FIGURE 5.14 Three different populations (top row), sampling distributions of $\overline{x}$ for random samples of size $n = 2$ (middle row), and sampling distributions of $\overline{x}$ for random samples of size $n = 30$ (bottom row).

It is a remarkable fact that when the population distribution is non-normal, the sampling distribution of $\overline{x}$ looks more like a normal distribution as the sample size increases. This is true no matter what shape the population distribution has, as long as the population has a finite mean μ and standard deviation σ, and the observations in the sample are independent. This famous fact of probability theory is called the **central limit theorem** (CLT).

DEFINITION Central limit theorem (CLT)

Suppose we select an SRS of size n from any population with mean μ and standard deviation σ. The **central limit theorem (CLT)** says that when n is sufficiently large, the sampling distribution of the sample mean $\overline{x}$ is approximately normal.

The sample size needed for the sampling distribution of $\overline{x}$ to be approximately normal, as guaranteed by the central limit theorem, depends on the shape of the population distribution. For some non-normal population distributions, the sampling distribution of $\overline{x}$ may be approximately normal with samples as small as 5 or 10.

However, a much larger sample size is required if the population distribution is extremely non-normal. *To be safe, we'll require that n be at least 30 to invoke the CLT.*

SHAPE OF THE SAMPLING DISTRIBUTION OF THE SAMPLE MEAN $\bar{x}$

- If the population distribution is approximately normal, the sampling distribution of $\bar{x}$ will also be approximately normal, no matter what the sample size n is.
- If the population distribution is non-normal, the sampling distribution of $\bar{x}$ will be approximately normal when the sample size is sufficiently large ($n \geq 30$ in most cases).

What if the shape of the population distribution is non-normal and the sample size is small? With population distributions that are clearly skewed, such as the second population in Figure 5.14, the sampling distribution of $\bar{x}$ will be less skewed than the population distribution, but still not approximately normal. For other population shapes, such as the first and third populations in Figure 5.14, the shape of the sampling distribution of $\bar{x}$ is harder to determine when the sample size is small.

EXAMPLE

Going back to the movies
Shape

Skill 3.C

PROBLEM: The distribution of the number of movies viewed in the last year by students at a large high school is skewed to the right with a mean of 25.3 movies and a standard deviation of 15.8 movies, as shown in the histogram.

Klaus Vedfelt/DigitalVision/Getty Images

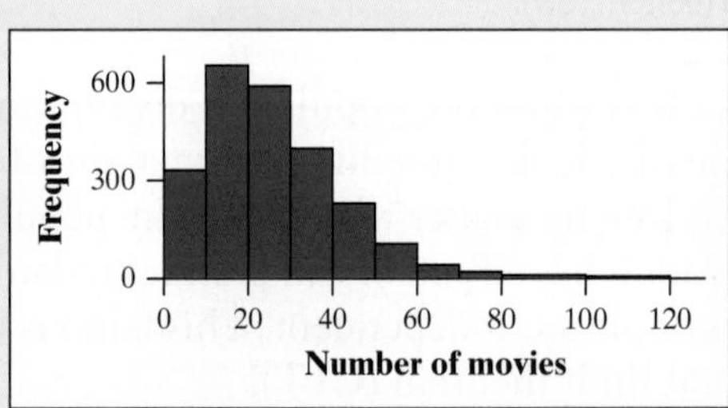

(a) Describe the shape of the sampling distribution of $\bar{x}$ for SRSs of size $n = 5$ from this population.
(b) Describe the shape of the sampling distribution of $\bar{x}$ for SRSs of size $n = 100$ from this population.

SOLUTION:

(a) Because $n = 5 < 30$, the sampling distribution of $\bar{x}$ will be skewed to the right, but not as strongly as the population distribution is.

(b) Because $n = 100 \geq 30$, the sampling distribution of $\bar{x}$ will be approximately normal.

Thanks to the central limit theorem!

FOR PRACTICE, TRY EXERCISE 7

The dotplots in Figure 5.15 show simulated sampling distributions of $\bar{x}$ = mean number of movies watched for (a) 200 SRSs of size $n = 5$ and (b) 200 SRSs of size $n = 100$ from the population of students at this school.

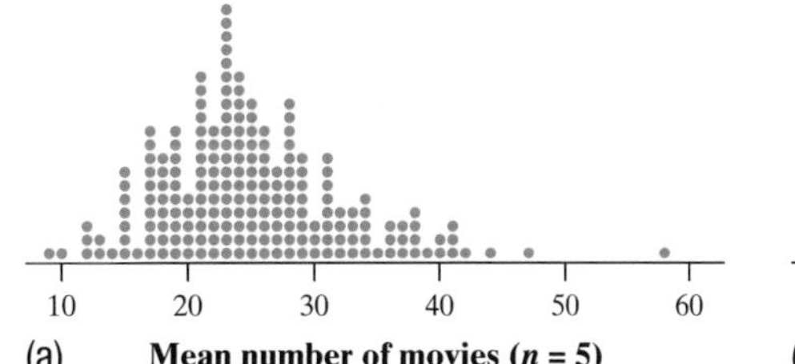

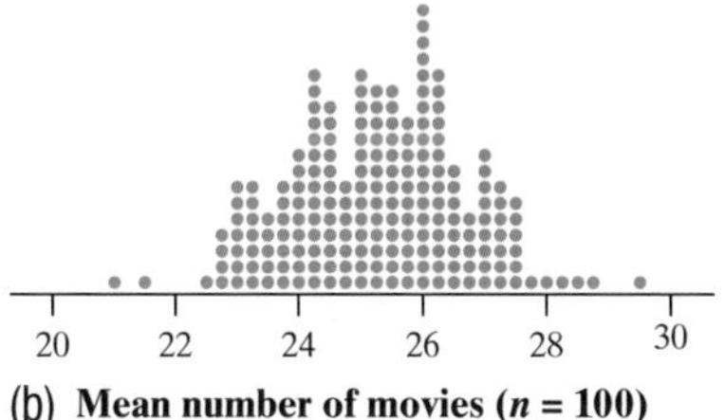

FIGURE 5.15 Simulated sampling distributions of the sample mean number of movies for (a) 200 SRSs of size $n = 5$ and (b) 200 SRSs of size $n = 100$ from a population of high school students.

As expected, the simulated sampling distribution of $\bar{x}$ for SRSs of size $n = 5$ is skewed to the right, but not as strongly as the population distribution is. The simulated sampling distribution of $\bar{x}$ for SRSs of size $n = 100$ is approximately normal — thanks to the central limit theorem.

Probabilities Involving $\bar{x}$

We can do probability calculations involving $\bar{x}$ if the shape of the population distribution is approximately normal or if the sample size is large ($n \geq 30$).

EXAMPLE

Free oil changes
Probabilities involving $\bar{x}$

Skills 3.A, 3.B, 3.C

Keith is the manager of an auto-care center. Based on service records from the past year, the time (in hours) that a technician requires to complete a standard oil change and inspection follows a right-skewed distribution with $\mu = 30$ minutes and $\sigma = 13.4$ minutes. For a promotion, Keith randomly selects 40 current customers and offers them a free oil change and inspection if they redeem the offer during the next month. Keith budgets an average of 35 minutes per customer for a technician to complete the work. Will this be enough? Calculate the probability that the average time it takes to complete the work exceeds 35 minutes.

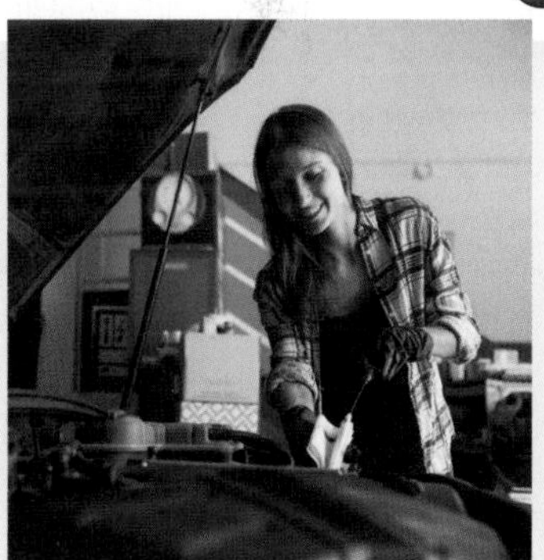

Westend61/Getty Images

SOLUTION:

Let $\bar{x}$ = sample mean time to complete work (in minutes).
The distribution of $\bar{x}$ is approximately normal because $n = 40 \geq 30$.
$\mu_{\bar{x}} = 30$ minutes

$\sigma_{\bar{x}} = \dfrac{13.4}{\sqrt{40}} = 2.119$ minutes because it is reasonable to assume $40 < 10\%$ of customers.

We want to find $P(\bar{x} > 35)$.

Step 1: Define the random variable of interest, state how it is distributed, and identify the values of interest.

$$\mu_{\bar{x}} = \mu$$

$$\sigma_{\bar{x}} = \frac{\sigma}{\sqrt{n}}$$

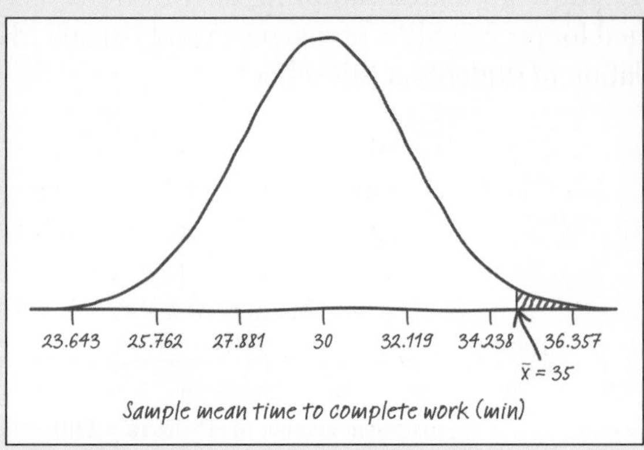

Step 2: Perform calculations—show your work!

(i) $z = \dfrac{35 - 30}{2.119} = 2.360$

Using technology: normalcdf(lower: 2.360, upper: 1000, mean: 0, SD: 1) = 0.0091
Using Table A: 1 − 0.9909 = 0.0091

(ii) normalcdf(lower: 35, upper: 1000, mean: 30, SD: 2.119) = 0.0091

Because there is such a small chance of exceeding a mean of 35 minutes, Keith has likely budgeted enough time.

Don't forget to answer the question!

FOR PRACTICE, TRY EXERCISE 13

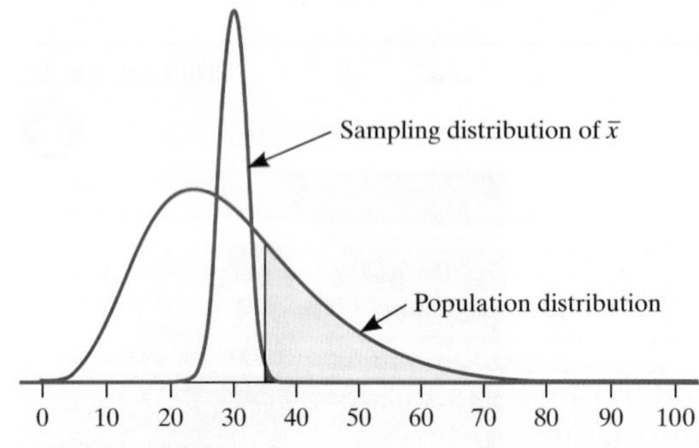

Averages are less variable than individuals, so randomly selecting a *single* oil change that takes longer than 35 minutes is more likely than randomly selecting a sample of 40 oil changes with a *mean* time larger than 35 minutes. As you can see in the figure, the area under the blue curve to the right of 35 is much larger than the area under the green curve to the right of 35.

The fact that averages of several observations are less variable than individual observations is important in many settings. For example, it is common practice in science and medicine to repeat a measurement several times and report the average of the results.

AP® EXAM TIP

Many students lose credit on probability calculations involving $\bar{x}$ because they forget to divide the population standard deviation by $\sqrt{n}$. Remember that averages are less variable than individual observations!

CHECK YOUR UNDERSTANDING

A snack-foods company uses a machine to place dry-roasted, shelled peanuts in jars labeled "16 ounces." The distribution of weight in the jars is approximately normal with a mean of 16.1 ounces and a standard deviation of 0.15 ounce.

1. Find the probability that a randomly selected jar has a weight less than the advertised weight of 16 ounces.
2. Find the probability that the mean weight of 10 randomly selected jars is less than the advertised weight of 16 ounces.

The Sampling Distribution of a Difference Between Two Means

In Section 5C, we introduced the sampling distribution of a difference between two proportions. What if we want to compare the mean of some quantitative variable for the individuals in Population 1 and Population 2? Our parameters of interest are the population means μ_1 and μ_2. Once again, the best approach is to select independent random samples from each population and to use $\bar{x}_1 - \bar{x}_2$ as our point estimate. To make inferences about $\mu_1 - \mu_2$, you need to understand the sampling distribution of $\bar{x}_1 - \bar{x}_2$.

Suppose that in City 1, the mean household income is \$65,000 with a standard deviation of \$35,000 and in City 2, the mean household income is \$50,000 with a standard deviation of \$30,000. Imagine taking SRSs of size $n = 50$ from each city, calculating the sample mean household income for each city, and subtracting the two sample means. Figure 5.16 shows the simulated sampling distribution of $\bar{x}_1 - \bar{x}_2$.

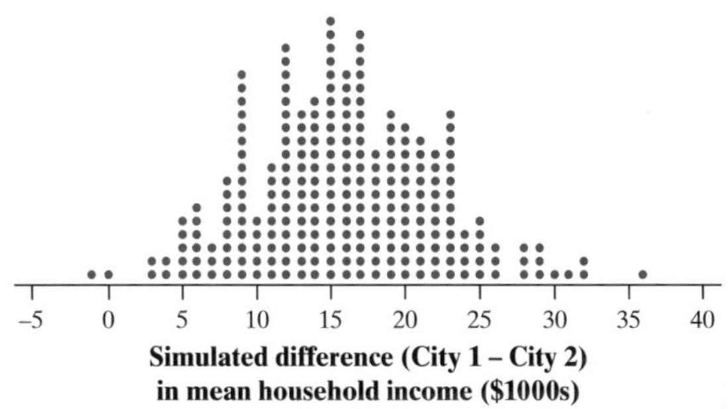

FIGURE 5.16 Simulated sampling distribution of $\bar{x}_1 - \bar{x}_2$, where $\mu_1 = \$65{,}000$, $\sigma_1 = \$35{,}000$, $n_1 = 50$, $\mu_2 = \$50{,}000$, $\sigma_2 = \$30{,}000$, and $n_2 = 50$.

It shouldn't come as a surprise that the distribution is centered at the difference in population means of \$65,000 – \$50,000 = \$15,000. The value of the standard deviation is not as obvious. Fortunately, we can use the rules for combining random variables from Section 4E to derive the formulas for the mean and standard deviation of the sampling distribution of $\bar{x}_1 - \bar{x}_2$. And using the fact that the difference of two independent, normal random variables is also normal, we can determine when the sampling distribution of $\bar{x}_1 - \bar{x}_2$ will be approximately normal.

THE SAMPLING DISTRIBUTION OF $\bar{x}_1 - \bar{x}_2$

Let $\bar{x}_1$ be the sample mean in an SRS of size n_1 from Population 1 of size N_1 with mean μ_1 and standard deviation σ_1, and let $\bar{x}_2$ be the sample mean in an independent SRS of size n_2 from Population 2 of size N_2 with mean μ_2 and standard deviation σ_2. Then:

- The mean of the sampling distribution of $\bar{x}_1 - \bar{x}_2$ is $\mu_{\bar{x}_1 - \bar{x}_2} = \mu_1 - \mu_2$.

- The standard deviation of the sampling distribution of $\bar{x}_1 - \bar{x}_2$ is approximately

$$\sigma_{\bar{x}_1-\bar{x}_2} = \sqrt{\frac{\sigma_1^2}{n_1} + \frac{\sigma_2^2}{n_2}}$$

as long as the samples are independent and the 10% condition is met for both samples: $n_1 < 0.10N_1$ and $n_2 < 0.10N_2$.

- The sampling distribution of $\bar{x}_1 - \bar{x}_2$ is approximately normal if both population distributions are approximately normal or if both sample sizes are large ($n_1 \geq 30$ and $n_2 \geq 30$).

AP® EXAM TIP
Formula sheet
The formulas for $\mu_{\bar{x}_1-\bar{x}_2}$ and $\sigma_{\bar{x}_1-\bar{x}_2}$ are included on the formula sheet provided on both sections of the AP® Statistics exam.

The formula for the standard deviation is exactly correct only when we have two types of independence:

- Independent samples, so that we can add the variances of $\bar{x}_1$ and $\bar{x}_2$.
- Independent observations within each sample. When sampling without replacement, the actual value of the standard deviation is smaller than the formula suggests. However, if the 10% condition is met for both samples, the difference is negligible.

Also note that the condition for normality is met when one sample size is large and the other sample size is small, as long as the small sample comes from an approximately normally distributed population.

EXAMPLE

Medium or large drink?
The sampling distribution of $\bar{x}_1 - \bar{x}_2$

Skills 3.B, 3.C, 4.B

Jackyenjoyphotography/Moment/Getty Images

PROBLEM: A fast-food restaurant uses an automated filling machine to pour its soft drinks. The machine has different settings for small, medium, and large drink cups. According to the machine's manufacturer, when the large setting is chosen, the amount of liquid L dispensed by the machine follows a normal distribution with mean 27 ounces and standard deviation 0.8 ounce. When the medium setting is chosen, the amount of liquid M dispensed follows a normal distribution with mean 17 ounces and standard deviation 0.5 ounce. To test this claim, the manager selects independent random samples of 20 cups filled using the large setting and 25 cups filled using the medium setting during one week. Let $\bar{x}_L - \bar{x}_M$ be the difference in the sample mean amount of liquid under the two settings. Assume the manufacturer's claim is true.

(a) Is the shape of the sampling distribution of $\bar{x}_L - \bar{x}_M$ approximately normal? Justify your answer.
(b) Calculate and interpret the mean of the sampling distribution.
(c) Verify both types of independence in this setting. Then, calculate and interpret the standard deviation of the sampling distribution.

SOLUTION:

(a) The shape of the sampling distribution of $\bar{x}_L - \bar{x}_M$ is approximately normal, because both population distributions are approximately normal.

(b) $\mu_{\bar{x}_L - \bar{x}_M} = 27 - 17 = 10$ ounces

In all possible independent random samples of 20 cups filled using the large setting and 25 cups filled using the medium setting, the resulting differences in sample mean volumes $(\bar{x}_L - \bar{x}_M)$ have an average of 10 ounces.

$$\mu_{\bar{x}_1 - \bar{x}_2} = \mu_1 - \mu_2$$

(c) Because the samples are independent, and because we can assume that $20 < 10\%$ of all cups filled using the large setting and $25 < 10\%$ of all cups filled using the medium setting that week at the restaurant,

$$\sigma_{\bar{x}_1 - \bar{x}_2} = \sqrt{\frac{\sigma_1^2}{n_1} + \frac{\sigma_2^2}{n_2}}$$

$$\sigma_{\bar{x}_L - \bar{x}_M} = \sqrt{\frac{0.80^2}{20} + \frac{0.50^2}{25}} = 0.205 \text{ ounce}$$

In all possible independent random samples of 20 cups filled using the large setting and 25 cups filled using the medium setting, the resulting differences in sample mean volume $(\bar{x}_L - \bar{x}_M)$ typically vary from the difference in population means of 10 ounces by about 0.205 ounce.

FOR PRACTICE, TRY EXERCISE 19

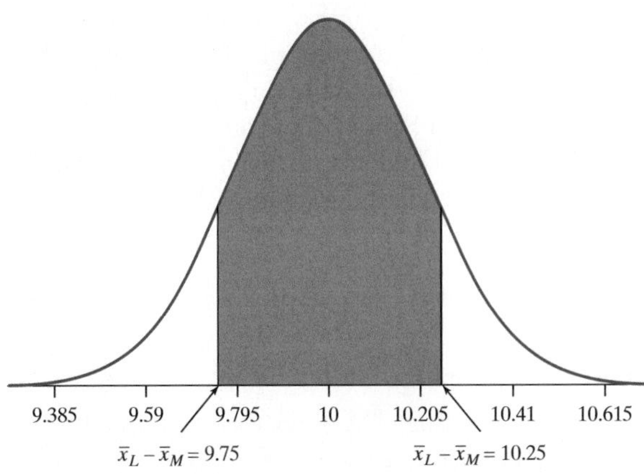

Difference in sample mean volume $(\bar{x}_L - \bar{x}_M)$ for cups filled using the large and medium settings (oz)

When the distribution of $\bar{x}_1 - \bar{x}_2$ is approximately normal, we can use a normal distribution to do probability calculations. For example, what is the probability that the difference in sample mean volume is within 0.25 ounce of the difference in population means? We want to find $P(9.75 \le \bar{x}_L - \bar{x}_M \le 10.25)$.

Using technology or Table A, you can verify that the probability is about 0.78. That is, there is about a 78% chance that the difference in sample mean volumes will be within 0.25 ounce of the difference in population mean volumes of 10 ounces.

In this unit, we have focused on *sampling* distributions. That is, we've limited our discussion to the distributions of statistics that arise when randomly sampling from one or two populations. As you learned in Section 3B, random selection allows us to make inferences about the population from which the sample was selected.

Inference about cause and effect is possible in the context of randomized experiments. Fortunately for us, the sampling distributions of $\hat{p}_1 - \hat{p}_2$ and $\bar{x}_1 - \bar{x}_2$ follow the same rules for shape, center, and variability when the data come from completely randomized experiments as they do with independent random samples. In fact, when we simulated the re-randomization of treatments in Section 3C to investigate statistical significance, we were using *randomization distributions*. You'll learn more about analyzing the results of experiments in Units 6–9.

Think About It

WHERE DO THE FORMULAS FOR THE MEAN AND STANDARD DEVIATION OF THE SAMPLING DISTRIBUTION OF $\bar{x}_1 - \bar{x}_2$ COME FROM? Both $\bar{x}_1$ and $\bar{x}_2$ are random variables. That is, their values would vary in repeated independent SRSs of size n_1 and n_2. Independent random samples yield independent random variables $\bar{x}_1$ and $\bar{x}_2$. The statistic $\bar{x}_1 - \bar{x}_2$ is the difference of these two independent random variables.

In Section 4E, we learned that for any two random variables X and Y,

$$\mu_{X-Y} = \mu_X - \mu_Y$$

For the random variables $\bar{x}_1$ and $\bar{x}_2$, we have

$$\mu_{\bar{x}_1-\bar{x}_2} = \mu_{\bar{x}_1} - \mu_{\bar{x}_2} = \mu_1 - \mu_2$$

We also learned in Section 4E that for *independent* random variables X and Y,

$$\sigma^2_{X-Y} = \sigma^2_X + \sigma^2_Y$$

For the independent random variables $\bar{x}_1$ and $\bar{x}_2$, we have

$$\sigma_{\bar{x}_1-\bar{x}_2} = \sqrt{\sigma^2_{\bar{x}_1} + \sigma^2_{\bar{x}_2}} = \sqrt{\left(\frac{\sigma_1}{\sqrt{n_1}}\right)^2 + \left(\frac{\sigma_2}{\sqrt{n_2}}\right)^2} = \sqrt{\frac{\sigma_1^2}{n_1} + \frac{\sigma_2^2}{n_2}}$$

CHECK YOUR UNDERSTANDING

At East High School, there are 2100 students. The distribution of the number of AP® classes that students are taking is strongly skewed to the right with a mean of 0.4 and a standard deviation of 0.8. At West High School, there are 2500 students. The distribution of the number of AP® classes that students are taking is also strongly skewed to the right with a mean of 0.5 and a standard deviation of 0.7. A random sample of 200 students is selected from each school.

1. Is the shape of the sampling distribution of $\bar{x}_E - \bar{x}_W$ approximately normal? Justify your answer.
2. Calculate and interpret the mean of the sampling distribution.
3. Verify both types of independence in this setting. Then, calculate and interpret the standard deviation of the sampling distribution.
4. What is the probability that the sample mean number of AP® classes at East High School is less than that at West High School?

SECTION 5D Summary

- When we want information about the population mean μ for some quantitative variable, we often select an SRS and use the sample mean $\bar{x}$ to estimate the unknown parameter μ. The **sampling distribution of the sample mean $\bar{x}$** describes how the statistic $\bar{x}$ varies in all possible samples of the same size from the population.
- **Center:** The mean of the sampling distribution of $\bar{x}$ is $\mu_{\bar{x}} = \mu$, where μ is the mean of the population. The mean describes the average value of $\bar{x}$ in all possible samples of a certain size from a population.
- **Variability:** The standard deviation of the sampling distribution of $\bar{x}$ is approximately $\sigma_{\bar{x}} = \frac{\sigma}{\sqrt{n}}$ for an SRS of size n if the population has standard deviation σ. This formula can be used when the 10% condition is met: $n < 0.10N$.

The standard deviation measures how far the values of $\bar{x}$ typically vary from μ in all possible samples of a certain size from a population.

- **Shape:** If the population distribution is approximately normal, then so is the sampling distribution of the sample mean $\bar{x}$. If the population distribution is non-normal, the **central limit theorem** (CLT) states that when n is sufficiently large, the sampling distribution of $\bar{x}$ is approximately normal. For most non-normal population distributions, it is safe to use a normal distribution to calculate probabilities involving $\bar{x}$ when $n \geq 30$.
- Suppose we select independent SRSs of size n_1 from Population 1 of size N_1 with mean μ_1 and standard derivation σ_1 and of size n_2 from Population 2 of size N_2 with mean μ_2 and standard derivation σ_2. The sampling distribution of $\bar{x}_1 - \bar{x}_2$ has the following properties:
  - **Center:** The mean of the sampling distribution is $\mu_{\bar{x}_1-\bar{x}_2} = \mu_1 - \mu_2$.
  - **Variability:** The standard deviation of the sampling distribution is approximately $\sigma_{\bar{x}_1-\bar{x}_2} = \sqrt{\frac{\sigma_1^2}{n_1} + \frac{\sigma_2^2}{n_2}}$ as long as the samples are independent and the 10% condition is met for both samples: $n_1 < 0.10N_1$ and $n_2 < 0.10N_2$.
  - **Shape:** Approximately normal if both population distributions are approximately normal or if both sample sizes are large ($n_1 \geq 30$ and $n_2 \geq 30$).
- When the population distribution(s) is/are approximately normal or the sample size(s) is/are large ($n \geq 30$), you can use a normal distribution to calculate probabilities involving $\bar{x}$ or $\bar{x}_1 - \bar{x}_2$.

AP® EXAM TIP

AP® Daily Videos

Review the content of this section and get extra help by watching the AP® Daily Videos for Topics 5.7 and 5.8, which are available in AP® Classroom.

SECTION 5D Exercises

Center and Variability

1. **Songs on a playlist** David's playlist has about 1000 songs. The distribution of the play times for these songs is heavily skewed to the right with a mean of 225 seconds and a standard deviation of 60 seconds. Suppose we choose an SRS of 10 songs from this population and calculate the mean play time $\bar{x}$ of these songs.

pg 502

(a) Calculate and interpret the mean of the sampling distribution of $\bar{x}$.

(b) Verify that the 10% condition is met. Then calculate and interpret the standard deviation of the sampling distribution of $\bar{x}$.

2. **High school GPAs** The distribution of grade point average for students at a large high school is skewed to the left with a mean of 3.53 and a standard deviation of 1.02. Suppose we choose an SRS of 4 students from this population and calculate the mean GPA $\bar{x}$ for these students.

(a) Calculate and interpret the mean of the sampling distribution of $\bar{x}$.

(b) Verify that the 10% condition is met. Then calculate and interpret the standard deviation of the sampling distribution of $\bar{x}$.

3. **More songs on playlist** Refer to Exercise 1. How many songs would you need to sample if you wanted the standard deviation of the sampling distribution of $\bar{x}$ to be 10 seconds? Justify your answer.

4. **More GPAs** Refer to Exercise 2. How many students would you need to sample if you wanted the standard deviation of the sampling distribution of $\bar{x}$ to be 0.10? Justify your answer.

5. **Screen time** Administrators at a small school with 200 students want to estimate the average amount of time students spend looking at a screen (phone, computer, television, and so on) per day. The administrators select a random sample of 50 students from the school to ask.

(a) Show that the 10% condition is not met in this case.

(b) What effect does violating the 10% condition have on the standard deviation of the sampling distribution of $\bar{x}$?

6. **Beautiful trees** As part of a school beautification project, students planted 35 trees along a road next to their school. Each month, students in the environmental science class randomly select 5 trees and use them to estimate the average height of all the trees planted for the project.
   (a) Show that the 10% condition is not met in this case.
   (b) What effect does violating the 10% condition have on the standard deviation of the sampling distribution of $\bar{x}$?

Shape

7. **Sampling songs on a playlist** David's playlist has about 1000 songs. The distribution of the play times for these songs is heavily skewed to the right with a mean of 225 seconds and a standard deviation of 60 seconds.
pg 506
   (a) Describe the shape of the sampling distribution of $\bar{x}$ for SRSs of size $n = 5$ from the population of songs on David's playlist. Justify your answer.
   (b) Describe the shape of the sampling distribution of $\bar{x}$ for SRSs of size $n = 100$ from the population of songs on David's playlist. Justify your answer.

8. **Sampling high school GPAs** The distribution of grade point average for students at a large high school is skewed to the left with a mean of 3.53 and a standard deviation of 1.02.
   (a) Describe the shape of the sampling distribution of $\bar{x}$ for SRSs of size $n = 4$ from the population of students at this high school. Justify your answer.
   (b) Describe the shape of the sampling distribution of $\bar{x}$ for SRSs of size $n = 50$ from the population of students at this high school. Justify your answer.

9. **Cola** A bottling company uses a filling machine to fill plastic bottles with cola. The bottles are supposed to contain 300 milliliters (ml). In fact, the contents vary according to an approximately normal distribution with mean $\mu = 298$ ml and standard deviation $\sigma = 3$ ml.
   (a) Describe the shape of the sampling distribution of $\bar{x}$ for random samples of 5 bottles from this population. Justify your answer.
   (b) Describe the shape of the sampling distribution of $\bar{x}$ for random samples of 40 bottles from this population. Justify your answer.

10. **Cereal** A company's cereal boxes advertise that each box contains 9.65 ounces of cereal. In fact, the amount of cereal in a randomly selected box follows an approximately normal distribution with mean $\mu = 9.70$ ounces and standard deviation $\sigma = 0.03$ ounce.
   (a) Describe the shape of the sampling distribution of $\bar{x}$ for random samples of 3 boxes from this population. Justify your answer.
   (b) Describe the shape of the sampling distribution of $\bar{x}$ for random samples of 35 boxes from this population. Justify your answer.

11. **What does the CLT say?** Asked what the central limit theorem says, a student replies, "As you take larger and larger samples from a population, the histogram of the sample values looks more and more normal." Is the student right? Explain your answer.

12. **Is this what the CLT says?** Asked what the central limit theorem says, a student replies, "As you take larger and larger samples from a population, the variability of the sampling distribution of the sample mean decreases." Is the student right? Explain your answer.

Probabilities Involving $\bar{x}$

13. **Bad yarn** The number of flaws in a ball of a certain variety of yarn varies with mean 1.6 flaws per ball and standard deviation 1.2 flaws per ball. Calculate the probability of getting a mean of 2 or more flaws in a random sample of 50 balls of this variety of yarn.
pg 507

14. **How many people in a car?** A study of rush-hour traffic in San Francisco counts the number of people in each car entering a freeway at a suburban interchange. Suppose the number of people in all cars entering this interchange during rush hour has a mean of 1.6 people and standard deviation 0.75 people. Calculate the probability of getting a mean greater than 2 people in a random sample of 35 cars entering this interchange.

15. **Dead battery?** A car company claims that the lifetime of its batteries is normally distributed with mean $\mu = 48$ months and standard deviation $\sigma = 8.2$ months. A consumer organization selects an SRS of $n = 8$ batteries, tests them under normal driving conditions, and calculates $\bar{x} = 42.2$ months.
   (a) Find the probability that the sample mean lifetime is 42.2 months or less if the company's claim is true.
   (b) Based on your answer to part (a), is there convincing evidence that the company is overstating the average lifetime of its batteries?

16. **Foiled again?** The manufacturer of a certain brand of aluminum foil claims that the amount of foil on each roll is normally distributed with a mean of 250 square feet (ft^2) and a standard deviation of 2 ft^2. To test this claim, a restaurant randomly selects 10 rolls of this aluminum foil and carefully measures the mean area to be $\bar{x} = 249.6$ ft^2.
   (a) Find the probability that the sample mean area is 249.6 ft^2 or less if the manufacturer's claim is true.
   (b) Based on your answer to part (a), is there convincing evidence that the company is overstating the average area of its aluminum foil rolls?

17. **Life insurance** A life insurance company sells a term insurance policy to 21-year-olds that pays $100,000 if the insured dies within the next 5 years. The company collects a premium of $250 each year as payment for the insurance. The amount Y that the company earns on a randomly selected policy of this type is $250 per year, less the $100,000 that it must pay if the insured dies. The probability distribution of Y is very strongly skewed to the left with a mean of $\mu_Y = \$303.35$ and a standard deviation of $\sigma_Y = \$9707.57$.

(a) Describe the sampling distribution of $\bar{y}$ for SRSs of $n = 10{,}000$ of these policies.

(b) Calculate the probability that the company will make an average profit greater than $0 in a random sample of 10,000 of these policies.

18. **Fire insurance** A very large insurance company sells a $300 home insurance policy that will pay out $200,000 if the home is destroyed by fire. The amount X the company earns on a randomly selected policy has a probability distribution that is very strongly skewed to the left with a mean of $\mu_X = \$260$ and a standard deviation of $\sigma_X = \$2828.14$.

(a) Describe the sampling distribution of $\bar{x}$ for SRSs of $n = 25{,}000$ of these policies.

(b) Calculate the probability that the company will make an average profit of at least $275 per policy in a random sample of 25,000 of these policies.

The Sampling Distribution of $\bar{x}_1 - \bar{x}_2$

19. **House prices** In the northern part of a large city, the distribution of home values is skewed to the right with a mean of $410,000 and a standard deviation of $250,000. In the southern part of the city, the distribution of home values is skewed to the right with a mean of $375,000 and a standard deviation of $240,000. Independent random samples of 10 houses in each part of the city are selected. Let $\bar{x}_N$ represent the sample mean value of homes in the northern part, and let $\bar{x}_S$ represent the sample mean value of homes in the southern part.

pg 510

(a) Is the shape of the sampling distribution of $\bar{x}_N - \bar{x}_S$ approximately normal? Justify your answer.

(b) Calculate and interpret the mean of the sampling distribution.

(c) Verify both types of independence in this setting. Then calculate and interpret the standard deviation of the sampling distribution.

20. **Young players** In the National Football League (NFL), the distribution of age is skewed to the right with a mean of 26.2 years and a standard deviation of 3.24 years. In the National Basketball Association (NBA), the distribution of age is skewed to the right with a mean of 25.8 years and a standard deviation of 4.24 years. Independent random samples of 20 players in each league are selected. Let $\bar{x}_{NFL}$ represent the sample mean age of NFL players, and let $\bar{x}_{NBA}$ represent the sample mean age of NBA players.

(a) Is the shape of the sampling distribution of $\bar{x}_{NFL} - \bar{x}_{NBA}$ approximately normal? Justify your answer.

(b) Calculate and interpret the mean of the sampling distribution.

(c) Verify both types of independence in this setting. Then calculate and interpret the standard deviation of the sampling distribution.

21. **Cholesterol** The level of cholesterol in the blood for young adults can be modeled by a normal distribution with mean $\mu_{YA} = 188$ milligrams per deciliter (mg/dl) and standard deviation $\sigma_{YA} = 41$ mg/dl. For teenagers, blood cholesterol levels can be modeled by a normal distribution with mean $\mu_T = 170$ mg/dl and standard deviation $\sigma_T = 30$ mg/dl. Suppose we select independent SRSs of 25 young adults and 36 teenagers and calculate the sample mean cholesterol levels $\bar{x}_{YA}$ and $\bar{x}_T$.

(a) What is the shape of the sampling distribution of $\bar{x}_{YA} - \bar{x}_T$? Justify your answer.

(b) Calculate the mean and standard deviation of the sampling distribution.

(c) Find the probability of getting a difference in sample means $\bar{x}_{YA} - \bar{x}_T$ less than 0 mg/dl.

22. **Heavy lions** The weights of adult male lions can be modeled by a normal distribution with mean $\mu_M = 450$ pounds and standard deviation $\sigma_M = 40$ pounds. The weights of adult female lions can be modeled by a normal distribution with mean $\mu_F = 335$ pounds and standard deviation $\sigma_F = 22$ pounds.[18] Suppose we select independent SRSs of 5 male and 5 female lions and calculate the sample mean weights $\bar{x}_M$ and $\bar{x}_F$.

(a) What is the shape of the sampling distribution of $\bar{x}_M - \bar{x}_F$? Justify your answer.

(b) Calculate the mean and standard deviation of the sampling distribution.

(c) Find the probability of getting a difference in sample means $\bar{x}_M - \bar{x}_F$ of at least 100 pounds.

23. **More cola** A bottling company uses a filling machine to fill plastic bottles with cola. The bottles are supposed to contain 300 milliliters (ml). In fact, the contents vary according to an approximately normal

distribution with mean $\mu = 298$ ml and standard deviation $\sigma = 3$ ml. A supervisor selects a random sample of 10 bottles each hour and calculates the sample mean volume. Assuming that the filling machine is working as intended, what is the probability that the sample means from two different random samples are more than 1 ml apart?

24. **More cereal** A company's cereal boxes advertise that each box contains 9.65 ounces of cereal. In fact, the amount of cereal in a randomly selected box follows an approximately normal distribution with mean $\mu = 9.70$ ounces and standard deviation $\sigma = 0.03$ ounce. A supervisor selects a random sample of 5 boxes each hour and calculates the sample mean weight. Assuming that the process is working as intended, what is the probability that the sample means from two different random samples are within 0.05 ounce of each other?

For Investigation *Apply the skills from the section in a new context or nonroutine way.*

25. **Lightning strikes** The number of lightning strikes on a square kilometer of open ground in a year has mean 6 and standard deviation 2.4. The National Lightning Detection Network (NLDN) uses automatic sensors to watch for lightning in 1-square-kilometer plots of land. A researcher randomly selects 50 1-square-kilometer plots of open ground and records the number of lightning strikes in each plot for 1 year.

(a) Describe the sampling distribution of $\bar{x}$ = sample mean number of strikes for random samples of $n = 50$ plots.

(b) Let T = total number of strikes in 50 randomly selected plots. Because $\bar{x} = T/50$, you can use the linear transformation $T = 50\bar{x}$ to determine the distribution of T. Use this transformation and your answer to part (a) to describe the distribution of T.

(c) Calculate the probability that there are fewer than 250 total strikes in 50 randomly selected plots.

You can also calculate the mean and standard deviation of T using the rules for the mean and standard deviation of a sum of random variables: $T = X_1 + X_2 + \ldots + X_{50}$, where X_i is the number of strikes in a randomly selected plot.

(d) Show that these rules give the same values for the mean and standard deviation of T as in part (b).

(e) Based on the relationship between T and $\bar{x}$, describe when it would be reasonable to conclude that $X_1 + X_2 + \ldots + X_n$ is approximately normal.

26. **10% condition for means** In this section, you learned that the formula for the standard deviation of the sampling distribution of $\bar{x}$ is valid when the sample size is less than 10% of the population size. But what if the sample size is larger than 10% of the population size? In this case, we use the *finite population correction factor* in the formula for the standard deviation, where N is the population size:

$$\sigma_{\bar{x}} = \frac{\sigma}{\sqrt{n}}\sqrt{\frac{N-n}{N-1}}$$

Imagine that you are choosing an SRS from a population of size $N = 1000$ where $\mu = 60$ and $\sigma = 5$.

(a) Calculate the standard deviation of the sampling distribution of $\bar{x}$ for samples of size $n = 10$, with and without the finite population correction factor. How do these values compare?

(b) Calculate the standard deviation of the sampling distribution of $\bar{x}$ for samples of size $n = 500$, with and without the finite population correction factor. How do these values compare?

(c) Based on your answers to parts (a) and (b), explain why we can use the formula $\sigma_{\bar{x}} = \frac{\sigma}{\sqrt{n}}$ when the sample size is less than 10% of the population size.

(d) Calculate the standard deviation of the sampling distribution of $\bar{x}$ for samples of size $n = 1000$ using the finite population correction factor. Why does the value of the standard deviation make sense in this situation?

Multiple Choice *Select the best answer for each question.*

27. The sampling distribution of the sample mean score on the mathematics part of the SAT exam in a recent year for SRSs of size 100 was approximately normal with mean 515 and standard deviation 11.4. Which of the following gives the mean and standard deviation of the population distribution of scores on the mathematics part of the SAT in this year?

(A) Mean = 515, SD = 11.4

(B) Mean = 515, SD = $\frac{11.4}{\sqrt{100}}$

(C) Mean = 515, SD = $11.4\sqrt{100}$

(D) Mean = 515/100, SD = $\frac{11.4}{\sqrt{100}}$

(E) Mean = 515/100, SD = $11.4\sqrt{100}$

28. Why is it important to check the 10% condition before calculating probabilities involving $\bar{x}$?

(A) To reduce the variability of the sampling distribution of $\bar{x}$

(B) To ensure that the distribution of $\bar{x}$ is approximately normal

(C) To ensure that we can generalize the results to a larger population

(D) To ensure that $\bar{x}$ will be an unbiased estimator of μ

(E) To ensure that the observations in the sample are close to independent

29. The number of hours a lightbulb burns before failing varies from bulb to bulb. The population distribution of burnout times is strongly skewed to the right. The central limit theorem says which of the following?

(A) As we look at more and more bulbs, their average burnout time gets ever closer to the mean μ for all bulbs of this type.

(B) The average burnout time of a large number of bulbs has a sampling distribution with the same shape (strongly skewed) as the population distribution.

(C) The average burnout time of a large number of bulbs has a sampling distribution with a similar shape but not as extreme (skewed, but not as strongly) as the population distribution.

(D) The average burnout time of a large number of bulbs has a sampling distribution that is close to normal.

(E) The average burnout time of a large number of bulbs has a sampling distribution that is exactly normal.

30. In City A, the distribution of the number of people in a household is skewed to the right with a mean of 2.4 people and a variance of 9.1 people^2. In City B, the distribution of the number of people in a household is skewed to the right with a mean of 2.6 people and a variance of 9.8 people^2. Suppose that independent random samples of 10 households are selected from each city and the difference (A – B) in the sample mean number of people in a household is calculated. Which of the following give the mean and standard deviation of the sampling distribution of this difference?

(A) Mean $= -0.2$, SD $= \sqrt{\frac{9.1}{10} + \frac{9.8}{10}}$

(B) Mean $= -0.2$, SD $= \sqrt{\frac{9.1^2}{10} + \frac{9.8^2}{10}}$

(C) Mean $= -0.2$, SD $= \sqrt{\frac{9.1}{10} - \frac{9.8}{10}}$

(D) Mean $= -0.2$, SD $= \sqrt{\frac{9.1^2}{10} - \frac{9.8^2}{10}}$

(E) Mean $= -0.2$, SD $= \sqrt{\frac{9.1}{10}} + \sqrt{\frac{9.8}{10}}$

Recycle and Review *Practice what you learned in previous sections.*

Exercises 31 and 32 refer to the following scenario. In the language of government statistics, you are "in the labor force" if you are available for work and either working or actively seeking work. The unemployment rate is the proportion of the labor force (not the entire population) that is unemployed. Here are estimates from the Current Population Survey for the civilian population aged 25 years and older in December 2022.[19] The table entries are counts in thousands of people.

Highest education	Total population	Labor force	Employed
Didn't finish high school	19,484	8885	8443
High school but no college	63,354	35,605	34,339
Less than bachelor's degree	57,080	35,789	34,735
College graduate	86,864	63,150	61,947

31. **Unemployment (2A)** Find the unemployment rate for people with each level of education. Is there an association between unemployment rate and education? Explain your answer.

32. **Unemployment (4B, 4C)** Suppose that you randomly select one person 25 years of age or older.

(a) What is the probability that this person is in the labor force?

(b) If you know that the person is a college graduate, what is the probability they are in the labor force?

(c) Are the events "in the labor force" and "college graduate" independent? Justify your answer.

UNIT 5 Wrap-Up

FRAPPY! Free Response AP® Problem, Yay!

Directions: Show all your work. Indicate clearly the methods you use, because you will be scored on the correctness of your methods as well as on the accuracy and completeness of your results and explanations.

The principal of a large high school is concerned about the number of absences for students at his school. To investigate, he prints a list showing the number of absences during the last month for each of the 2500 students at the school. For this population of students, the distribution of absences last month is skewed to the right with a mean of $\mu = 1.1$ and a standard deviation of $\sigma = 1.4$.

Suppose that a random sample of 50 students is selected from the list printed by the principal and the sample mean number of absences is calculated.

(a) What is the shape of the sampling distribution of the sample mean? Explain.

(b) What are the mean and standard deviation of the sampling distribution of the sample mean?

(c) What is the probability that the mean number of absences in a random sample of 50 students is less than 1?

(d) Because the population distribution is skewed, the principal is considering using the median number of absences last month instead of the mean number of absences to summarize the distribution. Describe how the principal could use a simulation to estimate the standard deviation of the sampling distribution of the sample median for samples of size 50.

After you finish the FRAPPY!, you can view two example solutions on the book's website (**bfwpub.com/tps7e**). Determine whether you think each solution is "complete," "substantial," "developing," or "minimal." If the solution is not complete, what improvements would you suggest to the student who wrote it? Finally, your teacher will provide you with a scoring rubric. Score your response and note what, if anything, you would do differently to improve your own score.

UNIT 5 REVIEW

SECTION 5A Normal Distributions, Revisited

In this section, you learned that a **continuous random variable** can take any value in a specified interval on the number line (unlike discrete random variables, which have gaps between possible values). Continuous random variables often result from measuring something, such as time or weight. The probability distribution of a continuous random variable is described by a **density curve** that is always on or above the horizontal axis and has total area 1 underneath it. The area under the density curve and above any specified interval of values on the horizontal axis gives the probability that the random variable falls within that interval. A **normal random variable** is a continuous random variable whose probability distribution is described by a normal curve. Any **linear combination** of independent, normal random variables is also a normal random variable.

SECTION 5B What Is a Sampling Distribution?

In this section, you learned the "big ideas" about sampling distributions. The first big idea is the difference between a statistic and a parameter. A **parameter** is a number that describes some characteristic of a population. A **statistic** estimates the value of a parameter using a sample from the population. Making the distinction between a statistic and a parameter will be crucial throughout the rest of the course.

The second big idea is that statistics vary and have distributions. For example, the mean weight for a sample of high school students is a variable that will change from sample to sample. The **sampling distribution of a statistic** gives the distribution of the statistic in all possible samples of the same size from the same population. Knowing the sampling distribution of a statistic tells us how far we can expect a statistic to vary from the parameter value and what values of the statistic should be considered unusual.

The third big idea is how to describe a sampling distribution. As in Unit 1, you need to address shape, center, and variability. If the center (mean) of the sampling distribution is the same as the value of the parameter being estimated, then the statistic is called an **unbiased estimator.** That is, an estimator is unbiased if it doesn't consistently underestimate or consistently overestimate the parameter. Ideally, the variability of a sampling distribution will be very small, meaning that the statistic provides precise estimates of the parameter. Larger sample sizes result in sampling distributions with less variability.

Finally, be very careful with your language. There is an important difference between the distribution of a population, the distribution of a sample, and the sampling distribution of a statistic. When you are writing your answers, be sure to indicate which distribution you are referring to. Don't make ambiguous statements like "the distribution will become less variable."

SECTION 5C Sample Proportions

In this section, you learned about the shape, center, and variability of the **sampling distribution of a sample proportion $\hat{p}$.** The mean of the sampling distribution of $\hat{p}$ is $\mu_{\hat{p}} = p$, the population proportion. As a result, the sample proportion $\hat{p}$ is an unbiased estimator of the population proportion p. When the sample size is less than 10% of the population size (the 10% condition), the standard deviation of the sampling distribution of the sample proportion is approximately $\sigma_{\hat{p}} = \sqrt{\frac{p(1-p)}{n}}$. The standard deviation measures how far the sample proportion $\hat{p}$ typically varies from the population proportion p. When $np \geq 10$ and $n(1-p) \geq 10$ (the Large Counts condition), the shape of the sampling distribution of $\hat{p}$ will be approximately normal. When the Large Counts condition is met, you can use normal distributions to calculate probabilities involving $\hat{p}$.

You also learned about the **sampling distribution of a difference in sample proportions $\hat{p}_1 - \hat{p}_2$.** The mean of the sampling distribution of $\hat{p}_1 - \hat{p}_2$ is $\mu_{\hat{p}_1-\hat{p}_2} = p_1 - p_2$. The standard deviation of the sampling distribution is approximately $\sigma_{\hat{p}_1-\hat{p}_2} = \sqrt{\frac{p_1(1-p_1)}{n_1} + \frac{p_2(1-p_2)}{n_2}}$ when the 10% condition is met for both samples and the samples are independent. The shape of the sampling distribution of $\hat{p}_1 - \hat{p}_2$ will be approximately normal when the Large Counts condition is met for both samples. When the Large Counts condition is met, you can use normal distributions to calculate probabilities involving $\hat{p}_1 - \hat{p}_2$.

SECTION 5D Sample Means

In this section, you learned about the shape, center, and variability of the **sampling distribution of a sample mean $\bar{x}$.** The mean of the sampling distribution of $\bar{x}$ is $\mu_{\bar{x}} = \mu$, the population mean. As a result, the sample mean $\bar{x}$ is an unbiased estimator of the population mean μ. When the sample size is less than 10% of the population size (the 10% condition), the standard deviation of the sampling distribution of the sample mean is approximately $\sigma_{\bar{x}} = \frac{\sigma}{\sqrt{n}}$. The standard deviation measures how far the sample mean $\bar{x}$ typically varies from the population mean μ.

When the population is approximately normally distributed, the shape of the sampling distribution of $\bar{x}$ will also be approximately normal for any sample size. When the population distribution is non-normal, the **central limit theorem (CLT)** says that the sampling distribution of $\bar{x}$ will become approximately normal when the sample size is sufficiently large. You can use a normal distribution to calculate probabilities involving the sampling distribution of $\bar{x}$ if the population distribution is approximately normal or the sample size is at least 30.

You also learned about the **sampling distribution of a difference in sample means $\bar{x}_1 - \bar{x}_2$.** The mean of the sampling distribution of $\bar{x}_1 - \bar{x}_2$ is $\mu_{\bar{x}_1-\bar{x}_2} = \mu_1 - \mu_2$. The standard deviation of the sampling distribution is approximately $\sigma_{\bar{x}_1-\bar{x}_2} = \sqrt{\frac{\sigma_1^2}{n_1} + \frac{\sigma_2^2}{n_2}}$ when the 10% condition is met for both samples and the samples are independent. The shape of the sampling distribution of $\bar{x}_1 - \bar{x}_2$ will be approximately normal when both population distributions are approximately normal or both sample sizes are at least 30.

Comparing Sampling Distributions

	Sampling distribution of $\hat{p}$	Sampling distribution of $\bar{x}$
Center	$\mu_{\hat{p}} = p$	$\mu_{\bar{x}} = \mu$
Variability	$\sigma_{\hat{p}} = \sqrt{\frac{p(1-p)}{n}}$ when the 10% condition is met: $n < 0.10N$	$\sigma_{\bar{x}} = \frac{\sigma}{\sqrt{n}}$ when the 10% condition is met: $n < 0.10N$
Shape	Approximately normal when the Large Counts condition is met: $np \geq 10$ and $n(1-p) \geq 10$	Approximately normal when the population distribution is approximately normal or the sample size is large ($n \geq 30$)
	Sampling distribution of $\hat{p}_1 - \hat{p}_2$	**Sampling distribution of $\bar{x}_1 - \bar{x}_2$**
Center	$\mu_{\hat{p}_1-\hat{p}_2} = p_1 - p_2$	$\mu_{\bar{x}_1-\bar{x}_2} = \mu_1 - \mu_2$
Variability	$\sigma_{\hat{p}_1-\hat{p}_2} = \sqrt{\frac{p_1(1-p_1)}{n_1} + \frac{p_2(1-p_2)}{n_2}}$ when the samples are independent and the 10% condition is met for both samples: $n_1 < 0.10N_1$ and $n_2 < 0.10N_2$	$\sigma_{\bar{x}_1-\bar{x}_2} = \sqrt{\frac{\sigma_1^2}{n_1} + \frac{\sigma_2^2}{n_2}}$ when the samples are independent and the 10% condition is met for both samples: $n_1 < 0.10N_1$ and $n_2 < 0.10N_2$
Shape	Approximately normal when the Large Counts condition is met for both samples: n_1p_1, $n_1(1-p_1)$, n_2p_2, and $n_2(1-p_2)$ all ≥ 10	Approximately normal when both population distributions are approximately normal or both sample sizes are at least 30

What Did You Learn?

Learning Target	Section	Related Example on Page(s)	Relevant Unit Review Exercise(s)
Calculate probabilities and percentiles involving normal random variables.	5A	457	R1
Find probabilities involving a sum, a difference, or another linear combination of independent, normal random variables.	5A	459	R2
Distinguish between a parameter and a statistic.	5B	467	R3
Create a sampling distribution using all possible samples from a small population.	5B	469	R4
Use the sampling distribution of a statistic to evaluate a claim about a parameter.	5B	470	R7
Determine if a statistic is an unbiased estimator of a population parameter.	5B	475	R4
Describe the relationship between sample size and the variability of an estimator.	5B	477	R4
Calculate and interpret the mean and standard deviation of the sampling distribution of a sample proportion $\hat{p}$.	5C	487	R5
Determine if the sampling distribution of $\hat{p}$ is approximately normal.	5C	489	R5
If appropriate, use a normal distribution to calculate probabilities involving sample proportions.	5C	490	R5, R6
Describe the shape, center, and variability of the sampling distribution of a difference in sample proportions $\hat{p}_1 - \hat{p}_2$.	5C	493	R6
Calculate and interpret the mean and standard deviation of the sampling distribution of a sample mean $\bar{x}$.	5D	502	R7
Determine if the sampling distribution of $\bar{x}$ is approximately normal.	5D	506	R7
If appropriate, use a normal distribution to calculate probabilities involving sample means.	5D	507	R7
Describe the shape, center, and variability of the sampling distribution of a difference in sample means $\bar{x}_1 - \bar{x}_2$.	5D	510	R8

UNIT 5 REVIEW EXERCISES

These exercises are designed to help you review the important concepts and skills of the unit.

R1 Birth weights (5A) Researchers in Norway analyzed data on the birth weights of 400,000 newborns over a 6-year period. The distribution of birth weights is approximately normal with a mean of 3668 grams and a standard deviation of 511 grams.[20]

(a) What is the probability that a randomly selected newborn weighs more than 3000 grams?

(b) What is the 90th percentile of the distribution of birth weight?

R2 Ohm-my! (5A) The design of an electronic circuit for a toaster calls for a 100-ohm resistor and a 250-ohm resistor to be connected in series so that their resistances add. The resistance X of a 100-ohm resistor in a randomly selected toaster follows a normal distribution with mean 100 ohms and standard deviation 2.5 ohms. The resistance Y of a 250-ohm resistor in a randomly selected toaster follows a normal distribution with mean 250 ohms and standard deviation 2.8 ohms. The resistances X and Y are independent. Find the probability that the total resistance for a randomly selected toaster lies between 345 and 355 ohms.

R3 Bad eggs (5B) Selling eggs that are contaminated with *Salmonella* bacteria can cause food poisoning in consumers. A large egg producer randomly selects 200 eggs from all the eggs shipped in one day. The laboratory reports that 9 of these eggs had *Salmonella* contamination. Identify the population, the parameter, the sample, and the statistic.

R4 Five books (5B) An author has written 5 children's books. The number of pages in each of these books is 64, 66, 71, 73, and 76, respectively.

(a) List all 10 possible SRSs of size $n = 3$, calculate the median number of pages for each sample, and display the sampling distribution of the sample median on a dotplot.

(b) Based on the sampling distribution in part (a), is the sample median an unbiased estimator of the population median?

(c) Describe how the variability of the sampling distribution of the sample median would change if the sample size was increased to $n = 4$.

R5 Do you jog? (5C) A Gallup poll asked a random sample of 1540 adults, "Do you regularly jog?" Suppose that the population proportion of all adults who jog is $p = 0.15$.

(a) Calculate and interpret the mean of the sampling distribution of $\hat{p}$.

(b) Verify that the 10% condition is met. Then calculate and interpret the standard deviation of the sampling distribution of $\hat{p}$.

(c) Is the sampling distribution of $\hat{p}$ approximately normal? Justify your answer.

(d) Find the probability that between 13% and 17% of people jog in a random sample of 1540 adults.

R6 American-made cars (5C) Eliav and Hyeyoung both work for the Department of Motor Vehicles (DMV), but they live in different states. In Eliav's state, 80% of the registered cars are made by American manufacturers. In Hyeyoung's state, only 60% of the registered cars are made by American manufacturers. Eliav selects a random sample of 100 cars in his state and Hyeyoung selects a random sample of 70 cars in her state. What is the probability that the proportion of American-made cars from Hyeyoung's sample exceeds the proportion of American-made cars from Eliav's sample?

R7 Detecting gypsy moths (5B, 5D) The gypsy moth poses a serious threat to oak and aspen trees. A state agriculture department places traps throughout the state to detect the moths. Each month, an SRS of 50 traps is inspected, the number of moths in each trap is recorded, and the mean number of moths is calculated. Based on years of data, the distribution of moth counts is discrete and strongly skewed with a mean of 0.5 and a standard deviation of 0.7.

(a) Estimate the probability that the mean number of moths in an SRS of 50 traps is greater than or equal to 0.6.

(b) In a recent month, the mean number of moths in an SRS of 50 traps was $\bar{x} = 0.6$. Based on this result, is there convincing evidence that the moth population is getting larger in this state? Explain your reasoning.

R8 Candles (5D) A company produces candles. Machine 1 makes candles with a mean length of 15 centimeters and a standard deviation of 0.15 centimeter. Machine 2 makes candles with a mean length of 15 centimeters and a standard deviation of 0.10 centimeter. A random sample of 49 candles is taken from each machine. Let $\bar{x}_1 - \bar{x}_2$ be the difference (Machine 1 − Machine 2) in the sample mean length of candles.

(a) Is the shape of the sampling distribution of $\bar{x}_1 - \bar{x}_2$ approximately normal? Justify your answer.

(b) Calculate and interpret the mean of the sampling distribution.

(c) Verify both types of independence in this setting. Then calculate and interpret the standard deviation of the sampling distribution.

UNIT 5 AP® STATISTICS PRACTICE TEST

Section I: Multiple Choice *Select the best answer for each question.*

T1 Researchers conducting a study of voting chose 663 registered voters at random shortly after an election. Of these people, 72% said they had voted in the election. Election records show that only 56% of registered voters voted in the election. Which of the following statements is true?

(A) 72% is a sample; 56% is a population.

(B) 72% and 56% are both statistics.

(C) 72% is a statistic and 56% is a parameter.

(D) 72% is a parameter and 56% is a statistic.

(E) 72% and 56% are both parameters.

T2 The Gallup organization, which conducts polls, has decided to increase the size of its random sample of voters from about 1500 people to about 4000 people right before an election. Its latest poll is designed to estimate the proportion of voters who favor a new law banning smoking within 100 feet of public buildings. What is the goal of this increase in sample size?

(A) To increase the bias of the estimate

(B) To reduce the bias of the estimate

(C) To increase the variability of the estimate

(D) To reduce the variability of the estimate

(E) To reduce the bias and variability of the estimate

T3 Suppose we select an SRS of size $n = 100$ from a large population having proportion p of successes. Let $\hat{p}$ be the proportion of successes in the sample. For which value of p would it be safe to use a normal distribution to approximate the sampling distribution of $\hat{p}$?

(A) 0.01

(B) 0.09

(C) 0.85

(D) 0.975

(E) 0.999

T4 The central limit theorem is important in statistics because it allows us to use a normal distribution to find probabilities involving the sample mean in which of the following settings?

(A) When the sample size is sufficiently large (for any population)

(B) When the population is approximately normally distributed (for any sample size)

(C) When the population is approximately normally distributed and the sample size is reasonably large

(D) When the population is approximately normally distributed and the population standard deviation is known (for any sample size)

(E) When the population size is reasonably large (whether the shape of the population distribution is known or not)

T5 The total time Nikhil has to wait to catch the bus, including the time it takes him to walk to the bus stop, follows a uniform distribution from 3 to 11 minutes. What is the height of the density curve that models this distribution?

(A) 1

(B) 1/3

(C) 1/8

(D) 1/11

(E) 1/14

T6 One dimension of bird beaks is "depth" — the height of the beak where it arises from the bird's head. During a research study on one island in the Galápagos archipelago, the beak depth of all Medium Ground Finches on the island was found to be approximately normally distributed with mean $\mu = 9.5$ millimeters (mm) and standard deviation $\sigma = 1.0$ mm.[21] What is the 35th percentile of the distribution of beak depth for this type of finch?

(A) 6.00 mm

(B) 7.69 mm

(C) 9.11 mm

(D) 9.85 mm

(E) 9.89 mm

T7 The student newspaper at a large university plans to ask an SRS of 250 undergraduates, "Do you favor eliminating the carnival from the term-end celebration?" Suppose that 55% of all undergraduates favor eliminating the carnival. How far do you expect the sample proportion who favor the proposal to typically vary from the population proportion?

(A) 0.001

(B) 0.031

(C) 0.035

(D) 0.063

(E) 0.248

T8 At a weekly street fair, a vendor sells hot dogs for \$4 and corn dogs for \$3. The number of hot dogs sold is approximately normally distributed with a mean of 42 and a standard deviation of 12. The number of corn dogs sold is approximately normally distributed with a mean of 17 and a standard deviation of 8. In a randomly selected week, what is the probability that the vendor makes less than \$200?

(A) 0.25
(B) 0.36
(C) 0.40
(D) 0.64
(E) 0.75

T9 A newborn baby has extremely low birth weight (ELBW) if it weighs less than 1000 grams. A study of the health of such children in later years examined a random sample of 219 children. Their mean weight at birth was $\bar{x} = 810$ grams. What does it mean to claim that the sample mean is an *unbiased estimator* of the mean weight μ in the population of all ELBW babies?

(A) As we select larger and larger samples from this population, $\bar{x}$ will get closer and closer to 810.
(B) As we select larger and larger samples from this population, $\bar{x}$ will get closer and closer to μ.
(C) In all possible samples of size 219 from this population, the values of $\bar{x}$ will have a distribution that is close to normal.
(D) In all possible samples of size 219 from this population, the mean of the values of $\bar{x}$ will equal 810.
(E) In all possible samples of size 219 from this population, the mean of the values of $\bar{x}$ will equal μ.

T10 An SRS of size 100 is selected from Population A with proportion 0.8 of successes. An independent SRS of size 400 is selected from Population B with proportion 0.5 of successes. The sampling distribution of the difference (A – B) in sample proportions has what mean and standard deviation?

(A) mean = 0.3; standard deviation = 1.3
(B) mean = 0.3; standard deviation = 0.40
(C) mean = 0.3; standard deviation = 0.047
(D) mean = 0.3; standard deviation = 0.0022
(E) mean = 0.3; standard deviation = 0.0002

Section II: Free Response *Show all your work. Indicate clearly the methods you use, because you will be graded on the correctness of your methods as well as on the accuracy and completeness of your results and explanations.*

T11 In a large city, the mean monthly fee for internet service is \$50 and the standard deviation is \$20. The distribution is skewed to the right: many households pay a low rate as part of a bundle with phone or television service, but some pay much more for internet service only or for faster connections.[22] A pollster surveys an SRS of 10 households with internet access and calculates $\bar{x}$ = sample mean fee for internet service.

(a) Calculate and interpret the mean of the sampling distribution of $\bar{x}$.

(b) Verify that the 10% condition is met. Then calculate and interpret the standard deviation of the sampling distribution of $\bar{x}$.

(c) What is the shape of the sampling distribution of $\bar{x}$? Justify your answer.

T12 According to government data, 22% of U.S. children younger than age 6 live in households with incomes less than the official poverty level. A study of learning in early childhood chooses an SRS of 300 children from one state and finds that $\hat{p} = 0.29$.

(a) Find the probability that at least 29% of the children in the sample are from poverty-level households, assuming that 22% of all children younger than age 6 in this state live in poverty-level households.

(b) Based on your answer to part (a), is there convincing evidence that the percentage of children younger than age 6 living in households with incomes less than the official poverty level in this state is greater than the national value of 22%? Explain your reasoning.

T13 In a children's book, the mean word length is 3.7 letters with a standard deviation of 2.1 letters. In a novel aimed at teenagers, the mean word length is 4.3 letters with a standard deviation of 2.5 letters. Both distributions of word length are unimodal and skewed to the right. Independent random samples of 35 words are selected from each book. What is the probability that the sample mean word length is greater in the sample from the children's book than in the sample from the teen novel?

Cumulative AP® Practice Test 2

Section I: Multiple Choice *Select the best answer for each question.*

AP1 The five-number summary for a data set is given by min = 5, $Q_1 = 18$, median = 20, $Q_3 = 40$, max = 75. If you wanted to construct a boxplot for the data set that would show outliers, if any existed, what would be the maximum possible length of the right-side "whisker"?

(A) 33
(B) 35
(C) 45
(D) 53
(E) 55

AP2 Here is the probability distribution of X = the number of heads in four tosses of a fair coin.

Number of heads	0	1	2	3	4
Probability	0.0625	0.2500	0.3750	0.2500	0.0625

What is the probability of getting at least one head in four tosses of a fair coin?

(A) 0.2500
(B) 0.3125
(C) 0.6875
(D) 0.9375
(E) 0.0625

AP3 In the population of applicants to a selective college, the distribution of high school GPA is strongly left-skewed with a mean of 4.1 and a standard deviation of 0.5. Suppose 200 applicants are randomly selected from this population. For SRSs of size 200, the distribution of sample mean GPA is

(A) left-skewed with mean 4.1 and standard deviation 0.035.
(B) exactly normal with mean 4.1 and standard deviation 0.5.
(C) exactly normal with mean 4.1 and standard deviation 0.035.
(D) approximately normal with mean 4.1 and standard deviation 0.5.
(E) approximately normal with mean 4.1 and standard deviation 0.035.

AP4 A 10-question multiple-choice exam offers 5 choices for each question. Jason uses a random number generator to answer each question, so he has probability 1/5 of getting each answer correct. You want to simulate the number of correct answers that Jason gets. Which of the following is a proper way to use a table of random digits to do the simulation?

(A) One digit from the random digits table simulates one answer, with 5 = correct and all other digits = incorrect. Ten digits from the table simulate 10 answers.
(B) One digit from the random digits table simulates one answer, with 0 or 1 = correct and all other digits = incorrect. Ten digits from the table simulate 10 answers.
(C) One digit from the random digits table simulates one answer, with odd = correct and even = incorrect. Ten digits from the table simulate 10 answers.
(D) One digit from the random digits table simulates one answer, with 0 or 1 = correct and all other digits = incorrect, ignoring repeats. Ten digits from the table simulate 10 answers.
(E) Two digits from the random digits table simulate one answer, with 00 to 20 = correct and 21 to 99 = incorrect. Ten pairs of digits from the table simulate 10 answers.

AP5 Suppose we roll a fair die four times. What is the probability that a 6 occurs on exactly one of the rolls?

(A) $4\left(\frac{1}{6}\right)^3\left(\frac{5}{6}\right)^1$
(B) $\left(\frac{1}{6}\right)^3\left(\frac{5}{6}\right)^1$
(C) $4\left(\frac{1}{6}\right)^1\left(\frac{5}{6}\right)^3$
(D) $\left(\frac{1}{6}\right)^1\left(\frac{5}{6}\right)^3$
(E) $6\left(\frac{1}{6}\right)^1\left(\frac{5}{6}\right)^3$

AP6 In one episode of a podcast, the host encouraged listeners to visit a website and vote in a poll about proposed tax increases. Of the 4821 people who voted, 4277 were against the proposed increases. To which of the following populations should the results of this poll be generalized?

(A) All people who have ever listened to this podcast
(B) All people who listened to this episode of the podcast
(C) All people who visited the podcast host's website
(D) All people who voted in the poll
(E) All people who voted against the proposed increases

AP7 The number of unbroken charcoal briquets in a 20-pound bag filled at the factory follows an approximately normal distribution with a mean of 450 briquets and a standard deviation of 20 briquets. The company expects that a certain number of the bags will be underfilled, so the company will replace for free the 5% of bags that contain too few briquets. What is the minimum number of unbroken briquets the bag would have to contain for the company to avoid having to replace the bag for free?

(A) 404
(B) 411
(C) 418
(D) 425
(E) 448

AP8 Suppose that you have torn a tendon and are facing surgery to repair it. The orthopedic surgeon explains the risks to you. Infection occurs in 3% of such operations, the repair fails in 14%, and both infection and failure occur together 1% of the time. What is the probability that the operation is successful for someone who has an operation that is free from infection?

(A) 0.8342
(B) 0.8400
(C) 0.8600
(D) 0.8660
(E) 0.9900

AP9 Social scientists are interested in the association between the high school graduation rate (HSGR, measured in percent) and the percentage of U.S. families living in poverty (POV). Data were collected from all 50 states and the District of Columbia, and a regression analysis was conducted. The resulting least-squares regression line is $\widehat{POV} = 59.2 - 0.620(HSGR)$ with $r^2 = 0.802$. Based on the information, which of the following is the best interpretation of the slope of the least-squares regression line?

(A) For each 1% increase in the graduation rate, the percentage of families living in poverty is predicted to decrease by approximately 0.896.
(B) For each 1% increase in the graduation rate, the percentage of families living in poverty is predicted to decrease by approximately 0.802.
(C) For each 1% increase in the graduation rate, the percentage of families living in poverty is predicted to decrease by approximately 0.620.
(D) For each 1% increase in the percentage of families living in poverty, the graduation rate is predicted to decrease by approximately 0.802.
(E) For each 1% increase in the percentage of families living in poverty, the graduation rate is predicted to decrease by approximately 0.620.

Questions AP10 and AP11 refer to the following graph. Here is a dotplot of the adult literacy rates in 177 countries in a recent year, according to the United Nations. For example, the lowest literacy rate was 23.6%, in the African country of Burkina Faso. Mali had the next lowest literacy rate at 24.0%.

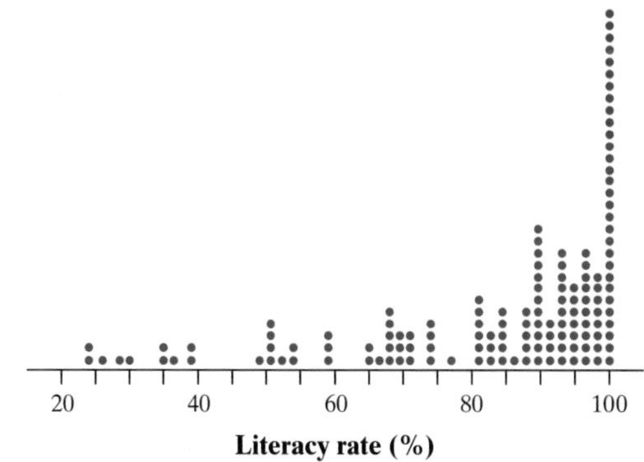

AP10 Which of the following is the best description of the shape of this distribution?

(A) Skewed to the right and unimodal
(B) Skewed to the left and unimodal
(C) Roughly symmetric and unimodal
(D) Approximately uniform
(E) Bimodal

AP11 The country with a literacy rate of 49% is closest to which of the following percentiles?

(A) 6th
(B) 11th
(C) 28th
(D) 49th
(E) 95th

AP12 The correlation between two quantitative variables is $r = 0.90$. Which of the following must be true?

I. There is a positive relationship between the two variables.
II. There is a linear relationship between the two variables.
III. Changes in one variable cause changes in the other variable.

(A) I only
(B) II only
(C) III only
(D) I and II only
(E) I, II, and III

AP13 An agronomist wants to test the effects of three different types of fertilizer (A, B, and C) on the yield of a new variety of wheat. The yield will be measured in bushels per acre. Six 1-acre plots of land are randomly assigned to each of the three fertilizers. Which of the following lists, in order, a treatment, an experimental unit, and the response variable?

(A) A specific fertilizer, bushels per acre, a plot of land
(B) Variety of wheat, bushels per acre, a specific fertilizer
(C) Variety of wheat, a plot of land, wheat yield
(D) A specific fertilizer, a plot of land, wheat yield
(E) A specific fertilizer, the agronomist, wheat yield

AP14 Records from a dairy farm yielded the following information on the number of male and female calves born at various times of the day.

		Time of day			
		Day	Evening	Night	Total
Sex of calf	Male	129	15	117	261
	Female	118	18	116	252
	Total	247	33	233	513

What is the probability that a randomly selected calf was born in the night or was a female?

(A) $\frac{369}{513}$ (D) $\frac{116}{252}$

(B) $\frac{485}{513}$ (E) $\frac{116}{233}$

(C) $\frac{116}{513}$

AP15 A grocery chain runs a game by giving each customer a scratch-off ticket that may win a prize. Printed on the ticket is a dollar value ($500, $100, $25) or the statement "This ticket is not a winner." Monetary prizes can be redeemed for groceries at the store. Here is the probability distribution of Y = the amount won on a randomly selected ticket:

Amount won, y	$500	$100	$25	$0
Probability, $P(Y = y)$	0.01	0.05	0.20	0.74

Which of the following are the mean and standard deviation of Y?

(A) $15.00, $2900.00
(B) $15.00, $53.85
(C) $15.00, $26.93
(D) $156.25, $53.85
(E) $156.25, $26.93

Section II: Part A begins on the next page.

Section II: Part A *Show all your work. Indicate clearly the methods you use, because you will be graded on the correctness of your methods as well as on the accuracy and completeness of your results and explanations.*

AP16 A researcher is interested in determining if omega-3 fish oil can help reduce cholesterol levels in adults. The researcher obtains permission to examine the health records of 200 people in a large medical clinic and classifies them according to whether or not they take omega-3 fish oil. The researcher also obtains their latest cholesterol readings and finds that the mean cholesterol reading for those who are taking omega-3 fish oil is 18 points less than the mean for the group not taking omega-3 fish oil.

(a) Is this an observational study or an experiment? Justify your answer.

(b) Explain the concept of confounding in the context of this study, and give one example of a variable that could be confounded with whether or not people take omega-3 fish oil.

(c) The researcher finds that the 18-point difference in the mean cholesterol readings of the two groups is statistically significant. Can they conclude that omega-3 fish oil is the cause? Why or why not?

AP17 Do you think colleges should pay their athletes? This question was asked to a random sample of 1264 U.S. adults. The two-way table displays the results for adults younger than age 45 and for adults age 45 or older.[23]

		Age group		
		Younger than 45	45 or older	Total
Should athletes be paid?	Yes	366	219	585
	No	203	410	613
	Unsure	12	54	66
	Total	581	683	1264

(a) Make a segmented bar graph to display the relationship between age and response to this question.

(b) Based on your graph in part (a), is there an association between age and response to this question? Justify your answer.

Section II: Part B begins on the next page.

Section II: Part B (Investigative Task) *Show all your work. Indicate clearly the methods you use, because you will be graded on the correctness of your methods as well as on the accuracy and completeness of your results and explanations.*

AP18 Administrators at a university have proposed an increase of \$200 in annual student fees to help pay for additional student parking on the campus. To see if students are in favor of this proposal, the administrators will select a random sample of 1000 students for a survey.

(a) Describe how to select a simple random sample of 1000 students from the population of 25,000 students at the university.

(b) Suppose that 41% of all students at the university support the proposal. Describe the sampling distribution of $\hat{p}$ = the proportion of students in the sample who support the proposal for simple random samples of size 1000.

One administrator suggests that a stratified random sample might be better, as students who live on campus might feel differently than students who live in nearby apartments or who live at home with their parents. Among all students at the university, 50% live on campus, 30% live in nearby apartments, and 20% live at home with their parents. To calculate the overall proportion of students who support the proposal in a stratified random sample, the administrators will use the following formula:

$$\hat{p}_{\text{overall}} = (0.50)\hat{p}_{\text{on-campus}} + (0.30)\hat{p}_{\text{apartment}} + (0.20)\hat{p}_{\text{home}}$$

If 10% of those who live on campus favor the proposal, 60% of those who live in nearby apartments favor the proposal, and 90% of those who live at home with their parents favor the proposal, the mean of the sampling distribution of $\hat{p}_{\text{overall}}$ is

$$\mu_{\hat{p}_{\text{overall}}} = (0.50)(0.10) + (0.30)(0.60) + (0.20)(0.90) = 0.41$$

(c) Assuming that the administrators randomly select 500 students who live on campus, 300 students who live in nearby apartments, and 200 students who live at home with their parents, show that the standard deviation of the sampling distribution of $\hat{p}_{\text{overall}}$ is 0.0116.

(d) Based on parts (b) and (c), explain the benefit of using a stratified random sample in this context.

UNIT

6

Inference for Categorical Data: Proportions

PART I
Inference for One Sample

Introduction

How long does a battery last on the newest iPhone, on average? What proportion of college undergraduates attended all of their classes last week? Do a majority of U.S. adults keep their New Year's resolutions? These are examples of the types of questions we can answer using the statistical inference methods you'll learn about in this chapter.

It isn't practical to determine the lifetime of *every* iPhone battery, to ask *all* undergraduates about their attendance, or to survey the *entire* population of U.S. adults. Instead, we choose a random sample of individuals (batteries, undergraduates, U.S. adults) to represent the population and collect data from those individuals. Using the information learned in Unit 3, we can generalize our results to the population of interest if we randomly select the sample. However, we cannot be certain that our conclusions are correct — a different sample would likely yield a different estimate. Fortunately, we can use what we learned about sampling distributions in Unit 5 to help account for the chance variation due to random selection.

In this unit, we begin the formal study of statistical inference — using information from a sample to draw conclusions about a population parameter such as p or μ. This is an important transition from Unit 5, where you were given information about a population and asked questions about the distribution of a sample statistic, such as the sample proportion $\hat{p}$ or the sample mean $\bar{x}$. This unit also combines many of the skills you've gained in the previous units — selecting statistical methods, using probability and simulation, and statistical argumentation.

The following activity sets the stage for what lies ahead.

ACTIVITY The beads

billnoll/Getty Images

Before class, your teacher prepared a large population of different-colored beads and put them into a container. In this activity, you and your team will create an interval of *plausible* (believable) values for p = the proportion of all beads in the container that are a particular color (e.g., red).

1. As a class, discuss how to use the cup provided to select a simple random sample of beads from the container.
2. Have one student select an SRS of beads. Separate the beads into two groups: those that are red and those that are not red. Count the number of beads in each group.
3. Calculate $\hat{p}$ = the sample proportion of beads that are red. Do you think this value is equal to the population proportion of red beads in the container? Explain your answer.
4. How much does the sample proportion typically vary from the population proportion in samples of the same size as the SRS selected? *Hint*: You'll need to use what you learned in Unit 5.

5. In teams of three or four students, use your answer from Step 4 to create an interval of plausible values for the population proportion of red beads in the container. How confident are you that your interval captures the population proportion?
6. Compare your results with those of the other teams in the class. Discuss any problems you encountered and how you dealt with them.

In this unit, we will introduce the two most common methods of formal statistical inference. Section 6A presents the fundamental ideas of *confidence intervals*, while Section 6C introduces the important concepts about *significance tests*. Sections 6B and 6D provide the details for constructing a confidence interval and performing a significance test for a population proportion. Finally, in Unit 6, Part II, Sections 6E and 6F present the details for constructing confidence intervals and performing significance tests for a difference in population proportions. Fortunately, both types of inference are based on the sampling distributions you studied in Unit 5.

SECTION 6A Confidence Intervals: The Basics

LEARNING TARGETS *By the end of the section, you should be able to:*

- Interpret a confidence interval in context.
- Use a confidence interval to make a decision about the value of a parameter.
- Interpret a confidence level in context.

Mr. Buckley's class did the Beads activity from the Introduction. In their sample of 251 beads, they selected 107 red beads and 144 other beads. If we had to give a single number to estimate $p =$ the population proportion of beads in the container that are red, what would it be? Because the sample proportion $\hat{p}$ is an unbiased estimator of the population proportion p, we use the statistic $\hat{p}$ as a **point estimator** of the parameter p. As you learned in Unit 5, the ideal point estimator will have no bias and little variability.

The best guess for the value of p is $\hat{p} = 107/251 = 0.426$. This value is known as a **point estimate** because it represents a single point on the number line.

DEFINITION Point estimator, Point estimate

A **point estimator** is a statistic that provides an estimate of a population parameter. The value of a point estimator from a sample is called a **point estimate.**

Even when we use an unbiased point estimator, the value of our point estimate is rarely correct. After all, it is unlikely that the proportion of red beads in the entire container is exactly equal to $\hat{p} = 0.426$. To increase our confidence, we can use an *interval estimate* instead.

Interpreting a Confidence Interval

When Mr. Buckley's class did the Beads activity, they obtained a sample proportion of $\hat{p} = 107/251 = 0.426$. To estimate how far sample proportions typically vary from the population proportion in SRSs of size $n = 251$, they estimated the standard deviation of the sampling distribution of $\hat{p}$ using the value of $\hat{p}$ from the sample:

$$\sigma_{\hat{p}} \approx \sqrt{\frac{0.426(1-0.426)}{251}} = 0.031$$

To account for sampling variability, they added and subtracted $2(0.031) = 0.062$ from the point estimate of $\hat{p} = 0.426$. This resulted in an interval of plausible values for the proportion of red beads in the container of 0.364 to 0.488.

When the estimate of a population parameter is reported as an interval of plausible values, it is called a **confidence interval.** A confidence interval is sometimes referred to as an **interval estimate** because it represents an interval of values on a number line, rather than a single point.

DEFINITION Confidence interval, Interval estimate

A **confidence interval,** also known as an **interval estimate,** gives a set of plausible values for a parameter based on sample data.

"Plausible" means that something is believable. Do not confuse this term with "probable," which means likely, or "possible," which means not impossible. In this case, "plausible" means that we shouldn't be surprised if any one of the values in the interval is equal to the value of the parameter. Mr. Buckley's class shouldn't be surprised if he revealed that the population proportion of red beads in the container is one of the values from 0.364 to 0.488. However, they should be surprised if the population proportion of red beads in the container is less than 0.364 or greater than 0.488.

We use an interval of plausible values rather than a point estimate to account for sampling variability and increase our confidence that we have a correct value for the parameter. Of course, as the cartoon illustrates, there is a trade-off between the amount of confidence we have that our estimate is correct and how much information the interval provides.

Confidence intervals are constructed so that we know how much confidence we should have in the interval. The most commonly used *confidence level* is 95%. You will learn how to interpret confidence levels shortly. For now, we'll focus on how to interpret a confidence interval.

The Pew Research Center regularly conducts surveys to learn about people in the United States and around the world. Each year, it conducts the National Public Opinion Reference Survey of randomly selected U.S. adults. One of the questions asks whether the respondent spent any time in the previous 12 months volunteering for any organization or association. In the most recent survey, 45.1% of the 4027 people who answered the question said yes.[1] A 95% confidence interval for the proportion of all U.S. adults who would say they have volunteered in the previous 12 months is 0.436 to 0.466. That is, we are 95% confident that the interval from 0.436 to 0.466 captures the proportion of all U.S. adults who would say they have volunteered for an organization or association in the previous 12 months.

HOW TO INTERPRET A CONFIDENCE INTERVAL

To interpret a *C*% confidence interval for an unknown parameter, say, "We are *C*% confident that the interval from ________ to ________ captures the [parameter in context]."

Some people like to describe the parameter before providing the endpoints. For example, they might say, "We are 95% confident that the proportion of all U.S. adults who would say they have volunteered for an organization or association in the previous 12 months is between 0.436 and 0.466." Either format is fine on the AP® Statistics exam, as long as the interpretation includes a statement about the confidence level, the endpoints of the interval, and a description of the parameter in context.

AP® EXAM TIP

When interpreting a confidence interval, make sure that you are describing the parameter and not the statistic. It's wrong to say that we are 95% confident the interval from 0.436 to 0.466 captures the proportion of U.S. adults who *said* they volunteered in the previous 12 months. The "proportion who *said* they volunteered in the previous 12 months" is the sample proportion, which is known to be $\hat{p} = 0.451$. The interval gives plausible values for the proportion of all U.S. adults who *would say* they have volunteered in the previous 12 months.

EXAMPLE

What's your blood pressure?
Interpreting a confidence interval

Skill 4.B

PROBLEM: The National Health and Nutrition Examination Survey (NHANES) is an ongoing research program conducted by the National Center for Health Statistics. Based on a random sample of 500 U.S. residents age 8 and older, a 95% confidence interval for the mean systolic blood pressure, the upper number in a blood pressure measurement, is 119.6 to 123.6 millimeters of mercury (mmHg).[2] Interpret this confidence interval.

Simon Marcus Taplin/The Image Bank/Getty Images

SOLUTION:

We are 95% confident that the interval from 119.6 to 123.6 mmHg captures the mean systolic blood pressure of all U.S. residents age 8 and older.

FOR PRACTICE, TRY EXERCISE 3

Using Confidence Intervals to Justify a Claim

In addition to estimating the value of a population parameter, we can use confidence intervals to evaluate claims about a parameter. For example, is there convincing evidence that fewer than half of all U.S. adults would say they have volunteered in the previous 12 months? Because every value in the interval of plausible values from 0.436 to 0.466 is less than 0.5, there is convincing evidence that fewer than half of all U.S. adults would say they have volunteered in the previous 12 months.

EXAMPLE

Is your blood pressure too high?

Using confidence intervals to justify a claim

Skill 4.D

AsiaVision/E+/Getty Images

PROBLEM: In the preceding example, you learned that a 95% confidence interval for the mean systolic blood pressure of all U.S. residents age 8 and older is 119.6 to 123.6 mmHg. According to the Centers for Disease Control and Prevention, "normal" systolic blood pressure is less than or equal to 120 mmHg.[3] Based on this interval, is there convincing evidence that the mean systolic blood pressure of all U.S. residents age 8 and older is greater than 120 mmHg? Explain your reasoning.

SOLUTION:

Because there are values in the interval from 119.6 to 123.6 that are less than or equal to 120, there is not convincing evidence that the mean systolic blood pressure of all U.S. residents age 8 and older is greater than 120 mmHg.

Note that the confidence interval gives plausible values for the *mean* systolic blood pressure, not the blood pressure of individual residents. In fact, fewer than 10% of individuals in the sample had a systolic blood pressure between 119.6 and 123.6 mmHg.

FOR PRACTICE, TRY EXERCISE 7

Interpreting Confidence Level

What does it mean to be 95% confident? The following activity gives you a chance to explore the meaning of the confidence level.

ACTIVITY

Investigating confidence level

In this activity, you will use an applet to learn what it means to say that we are "95% confident" that our confidence interval captures the parameter value.

1. Go to www.stapplet.com and launch the *Simulating Confidence Intervals* applet.
2. In the box labeled "Population," choose "Categorical" and change the true proportion of successes to 0.6.
3. In the box labeled "Sample," change the sample size to $n = 200$ and keep the default confidence level of 95%. Click the "Go!" button. The applet will display the distribution of the sample and calculate the sample

proportion and the endpoints of the 95% confidence interval. Does the interval include the population proportion, $p = 0.6$?

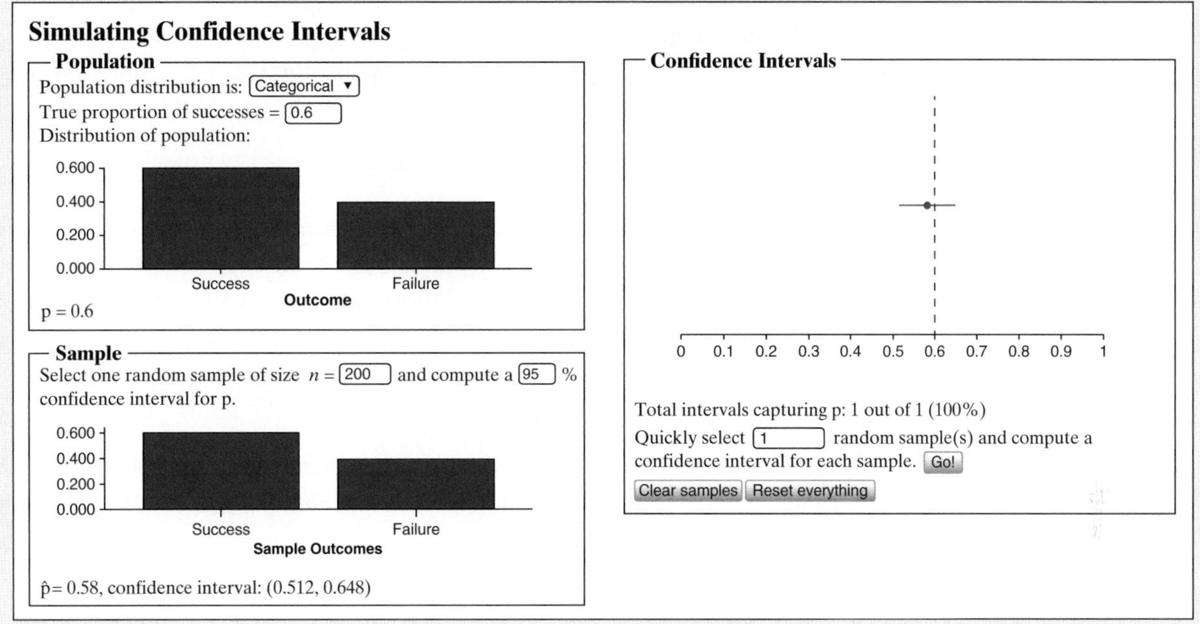

4. In the box labeled "Confidence Intervals," notice that the confidence interval is displayed as a horizontal line segment, along with a dashed vertical line segment at $p = 0.6$. If the interval includes $p = 0.6$, the interval is green. Otherwise, it is red. (You can hover your mouse over the interval to see the endpoints.) Click the "Go!" button 9 times to generate 9 additional random samples and their corresponding 95% confidence intervals. Look in the lower-left corner to determine how many of the intervals captured $p = 0.6$.
5. Change the number of intervals to 50 and click the "Go!" button many times to generate at least 1000 confidence intervals. What do you notice about the total percentage of the intervals that capture p?

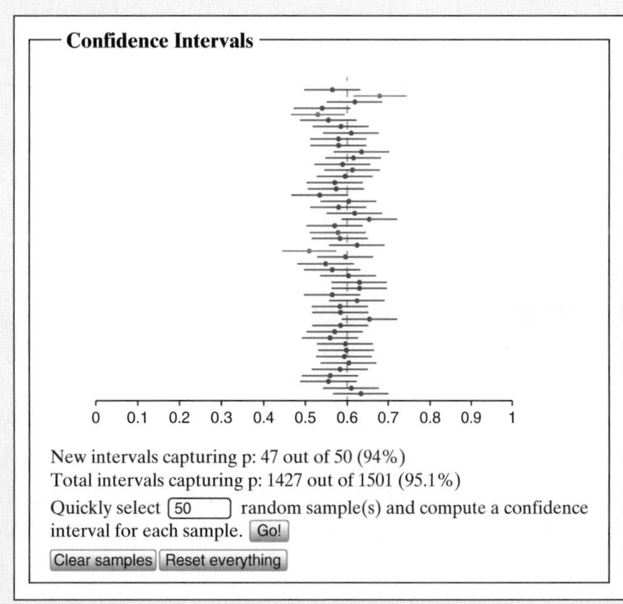

6. Click the button to "Clear samples." Then repeat Step 5 using a 99% confidence level.
7. Click the button to "Clear samples." Then repeat Step 5 using an 80% confidence level.
8. Summarize what you have learned about the relationship between the confidence level and the capture rate after taking many samples.

As the activity confirms, when many, many samples of the same size are randomly selected and a confidence interval is built from each sample, the percentage of intervals that contain the population parameter will be very close to the stated **confidence level C.**

DEFINITION Confidence level

The **confidence level *C*** gives the approximate percentage of confidence intervals that will capture the population parameter in repeated random sampling with the same sample size.

Recall that the 95% confidence interval for the proportion of all U.S. adults who would say they have volunteered for an organization or association in the previous 12 months is 0.436 to 0.466. If the Pew Research Center were to select many, many random samples of 4027 U.S. adults and construct a 95% confidence interval from each sample, about 95% of those intervals would capture the proportion of all U.S. adults who would say they volunteered in the previous 12 months.

AP® EXAM TIP

On a given problem, you may be asked to interpret the confidence interval, the confidence level, or both. Be sure you understand the difference: the confidence interval gives a set of plausible values for the parameter and the confidence level describes the overall capture rate of the method.

HOW TO INTERPRET A CONFIDENCE LEVEL

To interpret a confidence level C, say, "If we were to select many random samples of the same size from the same population and construct a C% confidence interval using each sample, about C% of the intervals would capture the [parameter in context]."

Note that the interpretation of the confidence level doesn't include the endpoints of any particular confidence interval. Rather, it refers to a collection of many, many confidence intervals, as you saw in the activity.

EXAMPLE

More blood pressure
Interpreting confidence level

Skill 4.B

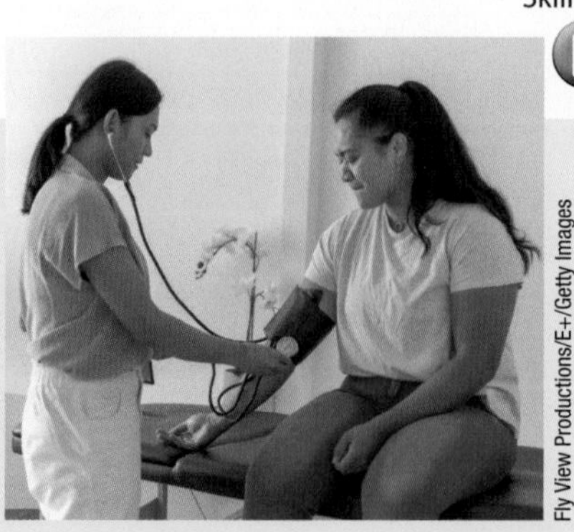
Fly View Productions/E+/Getty Images

PROBLEM: In the preceding examples, you learned that the 95% confidence interval for the mean systolic blood pressure of all U.S. residents age 8 and older is 119.6 to 123.6 mmHg. Interpret the confidence level.

SOLUTION:

If we were to select many random samples of 500 U.S. residents age 8 and older and construct a 95% confidence interval using each sample, about 95% of the intervals would capture the mean systolic blood pressure of all U.S. residents age 8 and older.

Remember that interpretations of the confidence level are about the method used to construct the interval—not one particular interval. In fact, we can interpret confidence levels before data are collected!

FOR PRACTICE, TRY EXERCISE 11

The confidence level does *not* tell us the probability that a particular confidence interval captures the population parameter. Once a particular confidence interval is calculated, its endpoints are fixed. And because the value of a parameter is also a constant, a particular confidence interval either includes the parameter (probability = 1) or doesn't include the parameter (probability = 0). To illustrate, we simulated 50 random samples of size $n = 200$ from a population where $p = 0.6$ and calculated a 95% confidence interval for p with each sample. As seen in Figure 6.1, no individual 95% confidence interval has a 95% probability of capturing the true parameter value.

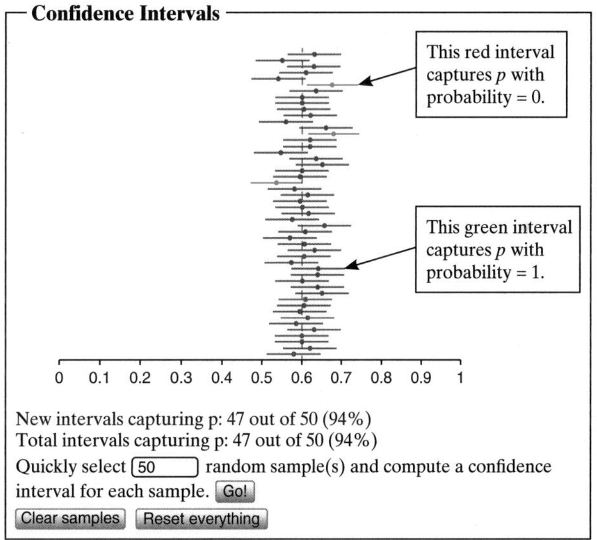

FIGURE 6.1 Confidence intervals from 50 simulated random samples showing that the probability a particular 95% confidence interval captures the true parameter value is either 0 (red intervals) or 1 (green intervals), and not 0.95.

CHECK YOUR UNDERSTANDING

The Pew Research Center and *Smithsonian* magazine recently quizzed a random sample of 1006 U.S. adults on their knowledge of science.[4] One of the questions asked, "Which gas makes up most of the Earth's atmosphere: hydrogen, nitrogen, carbon dioxide, or oxygen?" A 95% confidence interval for the proportion who would correctly answer nitrogen is 0.175 to 0.225.

1. Interpret the confidence interval.
2. Interpret the confidence level.
3. If people guess one of the four choices at random, about 25% should get the answer correct. Does this interval provide convincing evidence that fewer than 25% of all U.S. adults would answer this question correctly? Explain your reasoning.

SECTION 6A Summary

- To estimate an unknown population parameter, start with a statistic that will provide a reasonable guess. The chosen statistic is a **point estimator** for the parameter. The specific value of the point estimator that we calculate from sample data gives a **point estimate** for the parameter.
- A **confidence interval,** or **interval estimate,** gives a set of plausible values for an unknown population parameter based on sample data.
- To interpret a C% confidence interval, say, "We are C% confident that the interval from ____________ to ____________ captures the [parameter in context]." Be sure that your interpretation describes a parameter and not a statistic.
- If a proposed value for a population parameter is not included in a confidence interval for that parameter, there is convincing evidence that the proposed value is incorrect.
- The **confidence level C** is the long-run capture rate (success rate) of the method that produces the interval.
- To interpret a confidence level C, say, "If we were to select many random samples of the same size from the same population and construct a C% confidence interval using each sample, about C% of the intervals would capture the [parameter in context]."

AP® EXAM TIP
AP® Daily Videos

Review the content of this section and get extra help by watching the AP® Daily Video for Topic 6.1, which is available in AP® Classroom.

SECTION 6A Exercises

Introduction

1. **Pesticides** The U.S. Department of Agriculture (USDA) regularly tests produce products labeled as "organic" to estimate the proportion with chemical residue. In a recent random sample of 409 items labeled as organic, 87 contained some chemical residue.[5] Identify the point estimator the USDA should use to estimate the parameter. Then give the value of the point estimate.

2. **Reporting cheating** What proportion of students are willing to report cheating by other students? A student researcher asked this question to an SRS of 172 undergraduates at a large university: "You witness two students cheating on a quiz. Do you go to the professor?" Only 19 answered "Yes."[6] Identify the point estimator the student researcher should use to estimate the parameter. Then give the value of the point estimate.

Interpreting a Confidence Interval

3. **New Year's resolutions** At the end of a recent year, Morning Consult surveyed a random sample of U.S. adults about their New Year's resolutions. Among the 655 people in the sample who made a resolution, 402 kept their resolution. The 95% confidence interval for the proportion of all resolution-making U.S. adults who kept their resolution is 0.577 to 0.651.[7] Interpret the confidence interval.

pg 533

4. **International travel** A recent survey from Morning Consult asked a random sample of U.S. adults if they had ever traveled internationally. Based on the sample, the 95% confidence interval for the population proportion of U.S. adults who have traveled internationally is 0.436 to 0.478.[8] Interpret the confidence interval.

5. **Flint water** In Unit 1, you read about the dangerous lead levels in the drinking water in Flint, Michigan. Using the data collected by the Michigan Department of Environmental Quality in 2015, the 99% confidence interval for the mean lead level in Flint tap water in that year is 2.8 to 11.8 parts per billion (ppb).[9] Interpret the interval estimate.

6. **Exercise estimates** How accurately do people estimate the amount of exercise they get? Researchers randomly selected 3806 adults in New York City and asked them to estimate the number of minutes per week they spend doing moderate exercise. Based on the sample, a 90% confidence interval for the mean estimated amount of time doing moderate exercise per week is 517 minutes to 542 minutes.[10] Interpret the interval estimate.

Using Confidence Intervals to Justify a Claim

7. **More resolutions** Refer to Exercise 3. Is there convincing evidence that a majority of all U.S. resolution makers kept their resolutions? Explain your reasoning.

pg 534

8. **More travel** Refer to Exercise 4. Is there convincing evidence that fewer than half of all U.S. adults have traveled internationally? Explain your reasoning.

9. **Bottling cola** A particular type of diet cola advertises that each can contains 12 ounces of the beverage. Each hour, a supervisor selects 10 cans at random, measures their contents, and computes a 95% confidence interval for the true mean volume. For one particular hour, the 95% confidence interval is 11.97 ounces to 12.05 ounces.

(a) Does the confidence interval provide convincing evidence that the true mean volume is different than 12 ounces? Explain your answer.

(b) Does the confidence interval provide convincing evidence that the true mean volume is 12 ounces? Explain your answer.

10. **Fun size candy** A candy bar manufacturer sells a "fun size" version that is advertised to weigh 17 grams. A hungry teacher selected a random sample of 44 fun-size bars and found a 95% confidence interval for the true mean weight to be 16.945 to 17.395 grams.

(a) Does the confidence interval provide convincing evidence that the true mean weight is different than 17 grams? Explain your answer.

(b) Does the confidence interval provide convincing evidence that the true mean weight is 17 grams? Explain your answer.

Interpreting Confidence Level

11. **Resolutions kept** Refer to Exercise 3. Interpret the confidence level. pg 536

12. **Travel abroad** Refer to Exercise 4. Interpret the confidence level.

13. **How confident?** The figure shows the result of taking 25 SRSs from a normal population distribution and constructing a confidence interval for the population mean using each sample. Which confidence level — 80%, 90%, 95%, or 99% — do you think was used? Explain your reasoning.

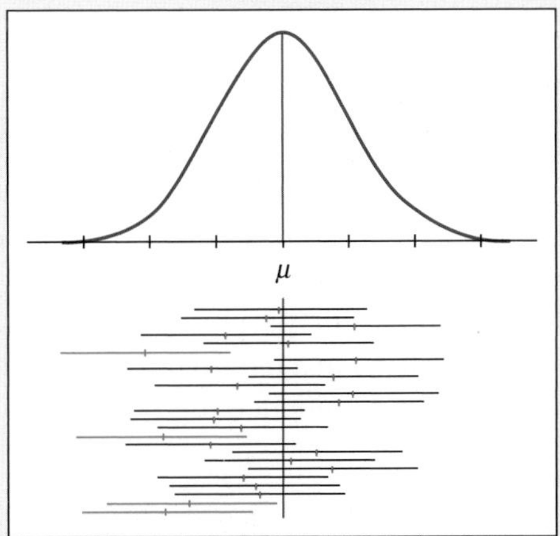

14. **How much confidence?** The figure shows the result of taking 25 SRSs from a normal population distribution and constructing a confidence interval for the population mean using each sample. Which confidence level — 80%, 90%, 95%, or 99% — do you think was used? Explain your reasoning.

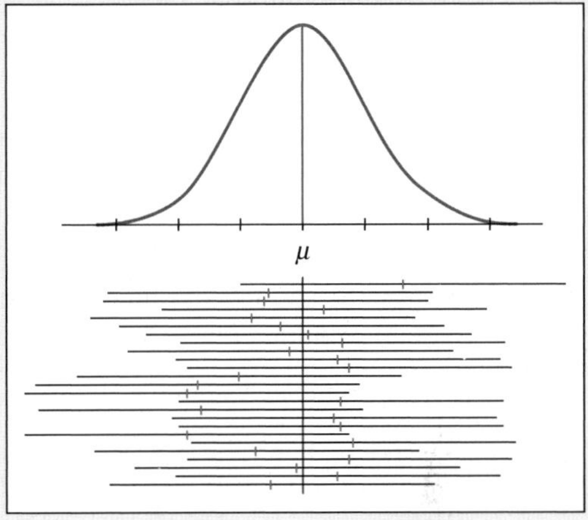

For Investigation *Apply the skills from the section in a new context or nonroutine way.*

15. **Crabby crabs** Biologists worried about the effect of noise pollution on crabs selected a sample of 34 shore crabs and randomly divided them into two groups. They exposed one group to 7.5 minutes of ship noise and exposed the other group to 7.5 minutes of ambient harbor noise. Because animals under stress typically increase their oxygen consumption, the researchers measured the amount of oxygen consumption for these two groups of crabs. A 95% confidence interval for the difference in mean oxygen consumption (Ship noise – Ambient noise) is 33.3 to 101.8 μmoles per hour.[11]

(a) Interpret the confidence interval.

(b) Does the confidence interval give convincing evidence that crabs have a greater mean oxygen consumption in the presence of ship noise? Explain your reasoning.

Multiple Choice *Select the best answer for each question.*

16. People love living in California for many reasons, but traffic isn't one of them. Based on a random sample of 572 employed California adults, a 90% confidence interval for the average travel time to work for all employed California adults is 23 to 26 minutes.[12] Which of the following statements is true?

(A) 90% of the sample have a travel time between 23 and 26 minutes.

(B) 90% of the population have a travel time between 23 and 26 minutes.

(C) There is a 90% probability that the population mean travel time is between 23 and 26 minutes.

(D) If the procedure were repeated many times, 90% of the resulting confidence intervals would contain the population mean travel time.

(E) If the procedure were repeated many times, 90% of the resulting confidence intervals would contain the sample mean travel time.

17. Based on the interval from Exercise 16, is there convincing evidence that the mean travel time for all employed California adults is less than 25 minutes?

(A) Yes, because 25 minutes is included in the interval.

(B) Yes, because the majority of values in the interval are less than 25 minutes.

(C) Yes, because there are values less than 25 in the interval.

(D) No, because there are values greater than 25 in the interval.

(E) No, because there are values less than 25 in the interval.

Recycle and Review *Practice what you learned in previous sections.*

18. **Oranges (5A, 5D)** A home gardener likes to grow various kinds of citrus fruit. One of the mandarin orange trees produces oranges whose circumferences can be modeled by a normal distribution with mean 21.1 cm and standard deviation 1.8 cm.

(a) What is the probability that a randomly selected orange from this tree has a circumference greater than 22 cm?

(b) What is the probability that a random sample of 20 oranges from this tree has a mean circumference greater than 22 cm?

19. **More oranges (1C, 1F)** The gardener in Exercise 18 randomly selects 20 mandarin oranges from the tree and counts the number of seeds in each orange. Here are the data:

3 4 6 6 9 11 11 12 13 13 14 14 16 17 22 23 23 24 28 30

(a) Graph the data using a dotplot.

(b) Based on your graph, is it plausible that the number of seeds from oranges on this tree follows a distribution that is approximately normal? Explain your answer.

SECTION 6B Confidence Intervals for a Population Proportion

LEARNING TARGETS *By the end of the section, you should be able to:*

- Check the conditions for calculating a confidence interval for a population proportion.
- Calculate a confidence interval for a population proportion.
- Construct and interpret a one-sample *z* interval for a proportion.
- Describe how the sample size and confidence level affect the margin of error.
- Determine the sample size required to obtain a confidence interval for a population proportion with a specified margin of error.

In Section 6A, you learned that a confidence interval can be used to estimate an unknown population parameter, such as a population mean μ or a population proportion p. In this section, you'll learn the details of calculating confidence intervals for a population proportion, such as the proportion of U.S. adults who are unemployed or the proportion of pine trees in a national park that are infested with beetles. We'll present the details about confidence intervals for a population mean in Unit 7.

Checking Conditions for a Confidence Interval for p

Before we calculate a confidence interval for a population proportion, we need to verify that the observations in the sample can be viewed as independent and that the sampling distribution of $\hat{p}$ is approximately normal. We do this by checking three conditions. Let's discuss them one at a time.

> **AP® EXAM TIP**
>
> The methods we use to calculate confidence intervals in AP® Statistics assume that the data come from an SRS from the population of interest. Other types of random samples (e.g., stratified or cluster) might be preferable to an SRS in a given setting, but they require more complex calculations than the ones you learn in AP® Statistics. When an example, exercise, or AP® Statistics exam item refers to a "random sample" without saying "stratified," "cluster," or "systematic," you can assume the sample is an SRS.

1. **The Random Condition** When our data come from a random sample, we can make an inference about the population from which the sample was selected. If the data come from a convenience sample or voluntary response sample, we should have no confidence that the resulting value of $\hat{p}$ is a good estimate of p. Random sampling also helps ensure that individual observations in the sample can be viewed as independent. Finally, random sampling introduces chance into the data-production process so we can use facts about sampling distributions from Unit 5 to build the confidence interval.
2. **The 10% Condition** As you learned in Unit 5, the formula for the standard deviation of the sampling distribution of $\hat{p}$ assumes that the individual observations are independent. Sampling without replacement violates this independence. However, when the sample size is less than 10% of the population size, we can view the observations in a sample as independent, even when sampling without replacement.
3. **The Large Counts Condition** The method we use to calculate a confidence interval for p requires that the sampling distribution of $\hat{p}$ be approximately normal. In Section 5C, you learned that this will be true whenever np and $n(1-p)$ are both at least 10. Because we don't know the value of p, we use $\hat{p}$ when checking the Large Counts condition.

CONDITIONS FOR CALCULATING A CONFIDENCE INTERVAL FOR A POPULATION PROPORTION

- **Random:** The data come from a random sample from the population of interest.
  - **10%:** When sampling without replacement, $n < 0.10N$.
- **Large Counts:** Both $n\hat{p}$ and $n(1-\hat{p})$ are at least 10.

Let's verify that the conditions were met for the interval calculated by Mr. Buckley's class.

EXAMPLE

The beads
Checking conditions for a confidence interval for p

Skill 4.C

Studioshots/Alamy Stock Photo

PROBLEM: Mr. Buckley's class wants to construct a confidence interval for p = the proportion of red beads in his container, which includes 3000 beads. Recall that the class's random sample of 251 beads had 107 red beads and 144 other beads. Check whether the conditions for constructing a confidence interval for p are met.

SOLUTION:

- Random: The class selected a random sample of 251 beads from the container. ✓
  - 10%: 251 beads is less than $0.10(3000) = 300$. ✓
- Large Counts:

To check the Large Counts condition, make sure both $n\hat{p}$ and $n(1-\hat{p})$ are at least 10.

$$251\left(\frac{107}{251}\right) = 107 \geq 10 \text{ and}$$

$$251\left(1-\frac{107}{251}\right) = 251\left(\frac{144}{251}\right) = 144 \geq 10 \checkmark$$

FOR PRACTICE, TRY EXERCISE 1

Notice that $n\hat{p}$ and $n(1-\hat{p})$ are the number of successes and failures in the sample. In the preceding example, we could address the Large Counts condition simply by saying, "The numbers of successes (107) and failures (144) in the sample are both at least 10."

WHAT HAPPENS IF ONE OF THE CONDITIONS IS VIOLATED?

If the data come from a convenience sample or if other sources of bias are present in the data collection process, there's no reason to calculate a confidence interval for p. You should have *no* confidence in an interval when the Random condition is violated.

To explore violations of the 10% condition, we performed a simulation with $p = 0.3$, $n = 500$, and $N = 1000$. This violates the 10% condition, as the sample size is 50% of the population size! Figure 6.2(a) shows 100 "95%" confidence intervals from the simulation. All 100 of the intervals captured the population proportion! In general, the actual capture rate is almost always *greater* than the reported confidence level when the 10% condition is violated.

To explore violations of the Large Counts condition, we performed a simulation with $p = 0.3$, $n = 10$, and $N = 1000$. This violates the Large Counts condition, as $np = 10(0.3) = 3$, which is clearly less than 10. Figure 6.2(b) shows 100 "95%" confidence intervals from the simulation. Only 81 of them captured the population proportion — much less than the advertised capture rate of 95%. In general, the actual capture rate will almost always be *smaller* than the stated confidence level when the Large Counts condition is violated.

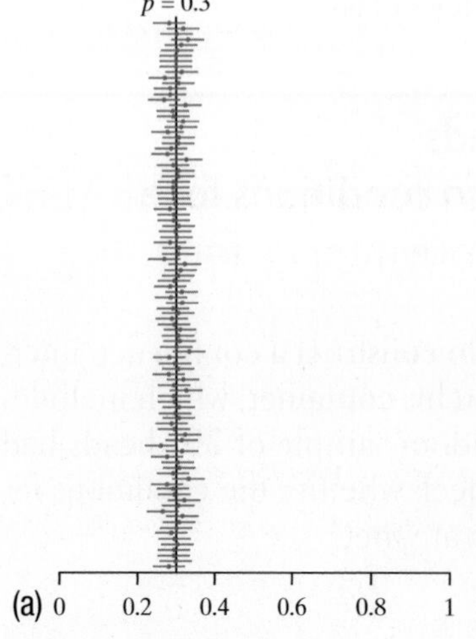

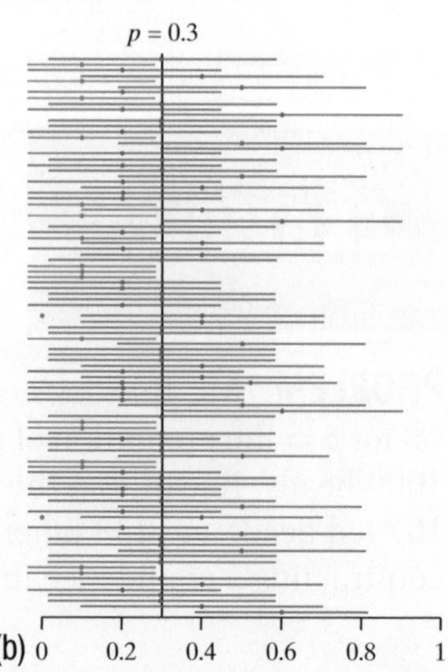

FIGURE 6.2 Results of simulating 100 "95%" confidence intervals when (a) the 10% condition is violated and (b) the Large Counts condition is violated.

There are advanced methods for calculating confidence intervals for population proportions that work well, even when the 10% condition or Large Counts condition is violated. However, these methods are beyond the scope of this course.

Calculating a Confidence Interval for *p*

To create a confidence interval for a population parameter based on data from a sample, we need two components: a point estimate to use as the midpoint of the interval and a **margin of error** to account for sampling variability. The structure of a confidence interval is

$$\text{point estimate} \pm \text{margin of error}$$

We can visualize a $C\%$ confidence interval like this:

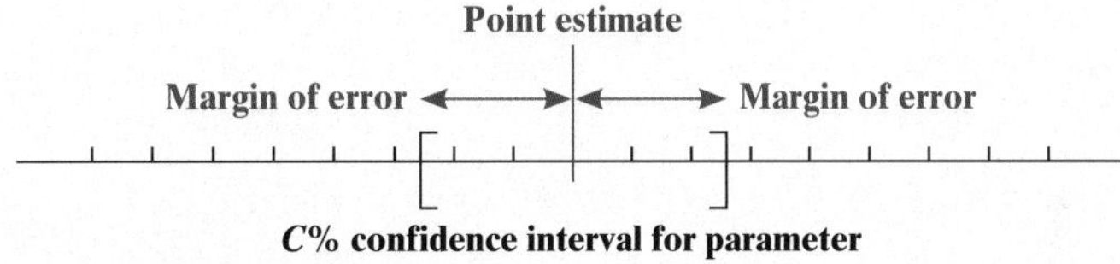

DEFINITION Margin of error

The **margin of error** of an estimate describes how far, at most, we expect the point estimate to vary from the population parameter.

In Section 6A, you learned that the 95% confidence interval for the proportion of all U.S. adults who have volunteered in the previous 12 months is 0.436 to 0.466. This interval could also be expressed as

$$0.451 \pm 0.015$$

where 0.451 is the midpoint of the interval (the point estimate) and 0.015 is the distance from the point estimate to either endpoint (the margin of error). The media often present confidence intervals in a similar way, but with the point estimate in the headline and the margin of error in the fine print.

The margin of error is calculated using two factors: the *critical value* and the *standard error* of the statistic used to estimate the parameter. Here is the formula for a confidence interval that shows these two components:

$$\text{statistic} \pm (\text{critical value})(\text{standard error of statistic})$$

AP® EXAM TIP
Formula sheet

This formula for a confidence interval is included on the formula sheet provided on both sections of the AP® Statistics exam.

CRITICAL VALUES

When Mr. Buckley's class did the Beads activity, they created the following confidence interval for the proportion of red beads in the container:

$$0.426 \pm 2(0.031)$$

In this calculation, the **critical value** is 2.

DEFINITION Critical value

The **critical value** is a multiplier that makes the interval wide enough to have the stated capture rate.

Mr. Buckley's class used a critical value of 2 based on the empirical rule and the fact that the sampling distribution of $\hat{p}$ is approximately normal when the Large Counts condition is met. Knowing that the value of $\hat{p}$ will be within 2 standard deviations of p in approximately 95% of samples, they reasoned that they could be 95% confident that p will be within 2 standard deviations of $\hat{p} = 0.426$. Following this logic, they would be 68% confident if they used a critical value of 1 and 99.7% confident if they used a critical value of 3.

Of course, the empirical rule gives approximate values — and only 3 of them. To get a more precise critical value for any confidence level, we use the standard normal distribution with technology or Table A. As Figure 6.3 shows, the central 95% of the standard normal distribution is marked off by two points, $-z^*$ and z^*. We use the * to remind you that this is a critical value, not a standardized score that has been calculated from data.

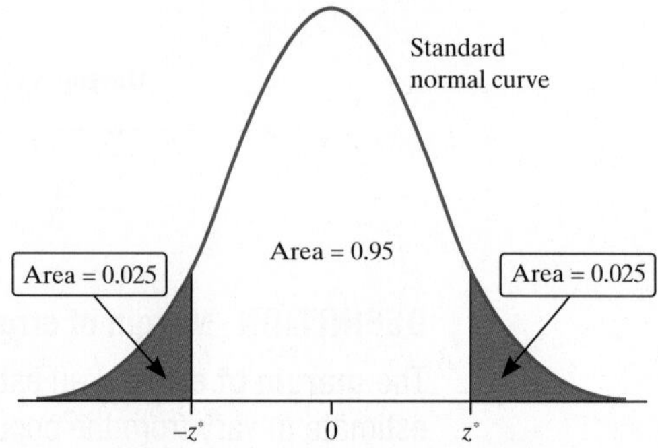

FIGURE 6.3 Finding the critical value z^* for a 95% confidence interval starts by labeling the middle 95% under a standard normal curve and calculating the area in each tail.

Due to the symmetry of the normal curve, the area in each tail is $0.05/2 = 0.025$. Once you know the tail areas, there are two ways to calculate the value of z^*:

- *Using technology:* The command invNorm(area: 0.025, mean: 0, SD: 1) gives $z = -1.960$, so $z^* = 1.960$.
- *Using Table A:* Search the body of Table A to find the point $-z^*$ with area 0.025 to its left. The entry $z = -1.96$ is what we are looking for, so $z^* = 1.96$.

z	.05	.06	.07
−2.0	.0202	.0197	.0192
−1.9	.0256	.0250	.0244
−1.8	.0322	.0314	.0307

Remember that critical values are always reported as positive numbers. If you get a negative value of z, such as -1.960, just drop the negative sign and report $z^* = 1.960$.

STANDARD ERROR

As you learned in Section 5C, when the 10% condition is met, the standard deviation of the sampling distribution of $\hat{p}$ is approximately

$$\sigma_{\hat{p}} = \sqrt{\frac{p(1-p)}{n}}$$

In practice, we don't know the value of p. If we did, we wouldn't need to construct a confidence interval for it! So we replace p with $\hat{p}$ in the formula for the standard deviation to get the **standard error of $\hat{p}$.**

DEFINITION Standard error of $\hat{p}$

The **standard error of $\hat{p}$** is an estimate of the standard deviation of the sampling distribution of $\hat{p}$:

$$s_{\hat{p}} = \sqrt{\frac{\hat{p}(1-\hat{p})}{n}}$$

The standard error describes how much the sample proportion $\hat{p}$ typically varies from the population proportion p in repeated random samples of size n.

Note that some people use the symbol $\text{SE}_{\hat{p}}$ for the standard error of $\hat{p}$ rather than $s_{\hat{p}}$. For the sample from Mr. Buckley's class,

$$s_{\hat{p}} = \sqrt{\frac{0.426(1-0.426)}{251}} = 0.031$$

In random samples of 251 beads from Mr. Buckley's container, the sample proportion of red beads typically varies from the population proportion by about 0.031.

CONFIDENCE INTERVALS FOR p

To complete the margin of error calculation, multiply the critical value and the standard error. Then add and subtract the margin of error from the point estimate to get the confidence interval:

$$\text{statistic} \pm (\text{critical value})(\text{standard error of statistic})$$

AP® EXAM TIP
Formula sheet

The specific formula for a confidence interval for a population proportion is *not* included on the formula sheet provided on both sections of the AP® Statistics exam. However, the formula sheet does include the general formula for the confidence interval:

$$\text{statistic} \pm (\text{critical value})(\text{standard error of statistic})$$

and the formula for the standard error of the sample proportion:

$$s_{\hat{p}} = \sqrt{\frac{\hat{p}(1-\hat{p})}{n}}$$

CALCULATING A CONFIDENCE INTERVAL FOR A POPULATION PROPORTION

When the conditions are met, a $C\%$ confidence interval for the unknown population proportion p is

$$\hat{p} \pm z^* \sqrt{\frac{\hat{p}(1-\hat{p})}{n}}$$

where z^* is the critical value for the standard normal curve with $C\%$ of its area between $-z^*$ and z^*.

We are now ready to calculate an "official" 95% confidence interval using the data from Mr. Buckley's class.

$$\begin{aligned}\hat{p} \pm z^* \sqrt{\frac{\hat{p}(1-\hat{p})}{n}} &= 0.426 \pm 1.960\sqrt{\frac{0.426(1-0.426)}{251}} \\ &= 0.426 \pm 0.061 \\ &= (0.365, 0.487)\end{aligned}$$

Notice that the margin of error is slightly smaller for this interval than when the class used 2 for the critical value. Mr. Buckley's class is 95% confident that the interval from 0.365 to 0.487 captures the population proportion of red beads in his container of 3000 beads. They can also be 95% confident that the interval from $3000(0.365) = 1095$ to $3000(0.487) = 1461$ captures the *number* of red beads in his container.

EXAMPLE

Read any good books lately?

Calculating a confidence interval for p

Skill 3.D

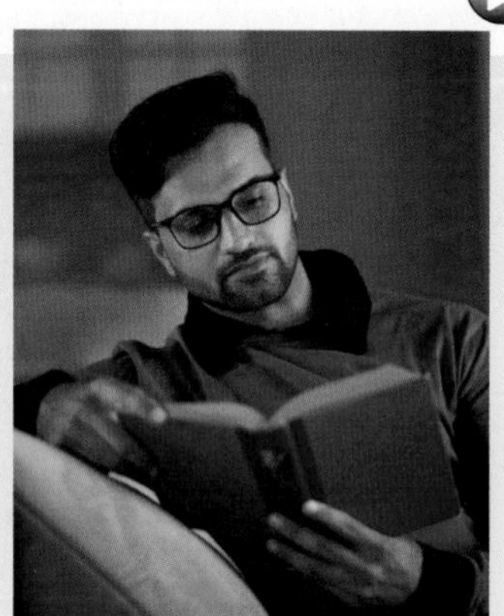
triloks/E+/Getty Images

PROBLEM: According to a 2022 Pew Research Center report, 75% of U.S. adults have read a book in the previous 12 months. This estimate was based on a random sample of 1502 U.S. adults.[13] Note that the conditions for inference are met.

(a) Determine the critical value z^* for a 99% confidence interval for a population proportion.

(b) Calculate the standard error of $\hat{p}$.

(c) Construct a 99% confidence interval for the proportion of all U.S. adults who have read a book in the previous 12 months.

SOLUTION:

(a)

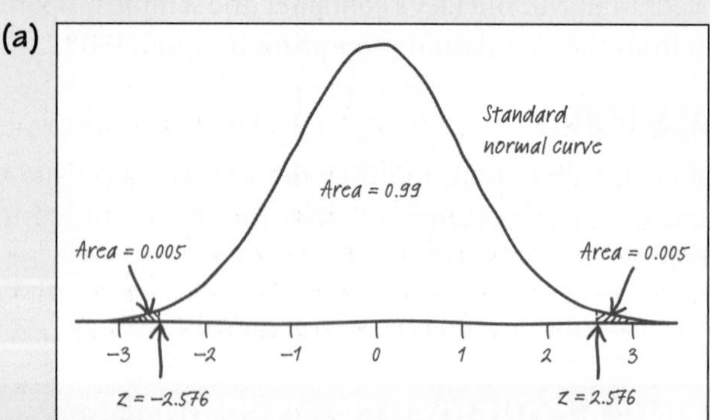

Using technology: invNorm(area: 0.005, mean: 0, SD: 1) = −2.576, so $z^* = 2.576$.

Using Table A: $z^* = 2.58$

(b) $s_{\hat{p}} = \sqrt{\dfrac{0.75(1-0.75)}{1502}} = 0.011$

$$s_{\hat{p}} = \sqrt{\frac{\hat{p}(1-\hat{p})}{n}}$$

(c) $0.75 \pm (2.576)(0.011)$
$= 0.75 \pm 0.028$
$= (0.722, 0.778)$

$$\hat{p} \pm z^* \sqrt{\frac{\hat{p}(1-\hat{p})}{n}}$$

FOR PRACTICE, TRY EXERCISE 13

We are 99% confident that the interval from 0.722 to 0.778 captures the proportion of all U.S. adults who have read a book in the previous 12 months. Because all of these values are greater than 0.5, there is convincing evidence that a majority of U.S. adults have read a book in the previous 12 months.

You can also use your calculator to compute a confidence interval for a population proportion, as the following Tech Corner illustrates.

16. Tech Corner CONFIDENCE INTERVALS FOR A PROPORTION

TI-Nspire and other technology instructions are on the book's website at bfwpub.com/tps7e.

The TI-83/84 calculator can be used to construct a confidence interval for an unknown population proportion. We'll demonstrate using data from the preceding example about reading books. Recall that 75% of the 1502 randomly selected U.S. adults said they read a book in the previous 12 months. To construct a confidence interval:

1. Press STAT, then choose TESTS and 1-PropZInt.

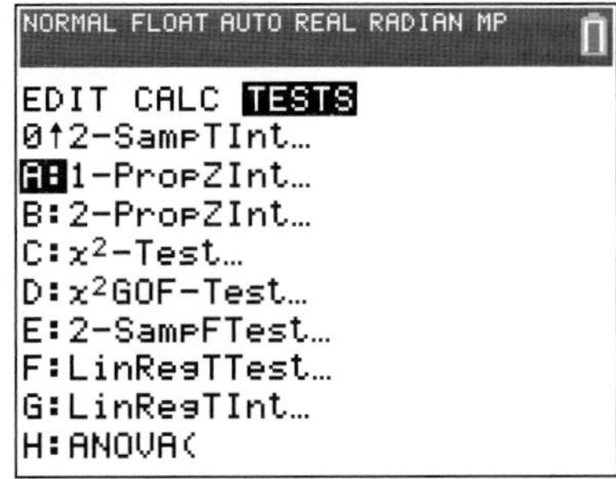

2. When the 1-PropZInt screen appears, enter x = 1127, n = 1502, and C-Level = 0.99. *Note*: Here, x is the *number* of successes and n is the number of trials. Both must be whole numbers or the calculator will give an error! In this case, we used $1502(0.75) = 1126.5 \approx 1127$ as the value of x.

3. Highlight "Calculate" and press ENTER. The 99% confidence interval for p is reported, along with the sample proportion $\hat{p}$ and the sample size n.

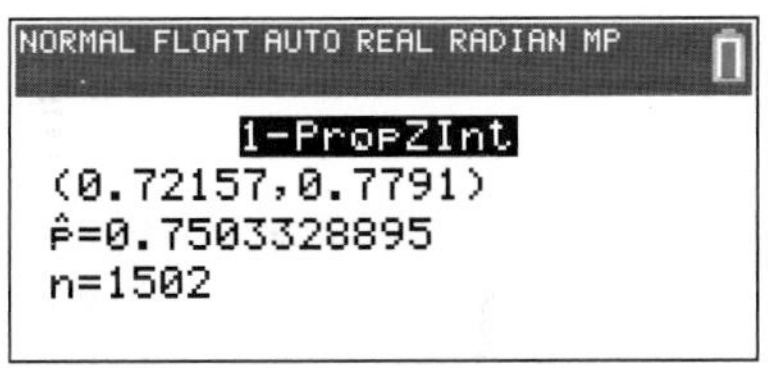

Confidence intervals calculated using the "full technology" approach described in the Tech Corner are generally more accurate than the values we obtain when we round while doing our calculations. For that reason, we will use the values from the full technology approach when reporting results "Using technology" in the rest of this section.

Putting It All Together: One-Sample *z* Interval for *p*

Whenever you are asked to construct and interpret a confidence interval in an exercise or on the AP® Statistics exam, use the following four-step process.

CONFIDENCE INTERVALS: A FOUR-STEP PROCESS

State: State the parameter you want to estimate and the confidence level.

Plan: Identify the appropriate inference method and check the conditions.

Do: If the conditions are met, perform calculations.

Conclude: Interpret your interval in the context of the problem.

The appropriate confidence interval method for estimating p is called a **one-sample z interval for a proportion.**

DEFINITION One-sample z interval for a proportion

A **one-sample z interval for a proportion** is a confidence interval used to estimate a population proportion p.

The next example illustrates the four-step process in action.

EXAMPLE

Distracted walking
One-sample z interval for p

Skills 1.D, 3.D, 4.B, 4.C

mimagephotography/Shutterstock

PROBLEM: A recent poll of 738 randomly selected cell-phone users found that 170 of the respondents admitted to walking into something or someone while talking on their cell phone.[14]

(a) Construct and interpret a 95% confidence interval for the proportion of all cell-phone users who would admit to walking into something or someone while talking on their cell phone.

(b) Interpret the confidence level.

SOLUTION:

(a)

Remember to follow the four-step process!

STATE: 95% CI for p = the proportion of all cell-phone users who would admit to walking into something or someone while talking on their cell phone.

State: State the parameter you want to estimate and the confidence level.

PLAN: One-sample z interval for a proportion

- Random: Random sample of 738 cell-phone users. ✓
  - 10%: It is reasonable to assume that 738 is less than 10% of all cell-phone users. ✓
- Large Counts: The number of successes (170) and the number of failures (738−170=568) are both at least 10. ✓

Plan: Identify the appropriate inference method and check the conditions.

Remember that $n\hat{p}$ is the number of successes and $n(1-\hat{p})$ is the number of failures in the sample:
$n\hat{p} = 738(170/738) = 170$
$n(1-\hat{p}) = 738(568/738) = 568$

DO: $\hat{p} = 170/738 = 0.230$

$$0.230 \pm 1.960\sqrt{\frac{0.230(1-0.230)}{738}}$$

$= 0.230 \pm 0.030$

$= (0.200, 0.260)$

Using technology: 0.200 to 0.261

Do: If the conditions are met, perform calculations.

$$\hat{p} \pm z^* \sqrt{\frac{\hat{p}(1-\hat{p})}{n}}$$

CONCLUDE: We are 95% confident that the interval from 0.200 to 0.261 captures the proportion of all cell-phone users who would admit to walking into something or someone while talking on their cell phone.

Conclude: Interpret your interval in the context of the problem. Make sure your conclusion is about the population (users *who would admit*) and not the sample (those *who admitted*).

(b) If we were to select many random samples of 738 cell-phone users and construct a 95% confidence interval using each sample, about 95% of the intervals would capture the proportion of all cell-phone users who would admit to walking into something or someone while talking on their cell phone.

FOR PRACTICE, TRY EXERCISE 15

AP® EXAM TIP

You may use your calculator to compute a confidence interval on the AP® Statistics exam. But there's a risk involved. If you just give the calculator answer with no work, you'll get either full credit for the "Do" step (if the interval is correct) or no credit (if it's wrong). If you opt for the calculator-only method, be sure to complete the other three steps (State, Plan, Conclude) and give the interval in the Do step (e.g., 0.200 to 0.261).

CHECK YOUR UNDERSTANDING

Sleep Awareness Week begins in the spring with the release of the National Sleep Foundation's annual poll of U.S. sleep habits and ends with the beginning of daylight saving time, when most people lose an hour of sleep.[15] In the foundation's random sample of 1029 U.S. adults, 48% reported that they "often or always" got enough sleep during the past 7 nights.

1. Construct and interpret a 90% confidence interval for the proportion of all U.S. adults who would report often or always getting enough sleep during the past 7 nights.
2. Does the interval in Question 1 provide convincing evidence that fewer than half of all U.S. adults would report they often or always got enough sleep during the past 7 nights? Justify your answer.

Factors That Affect the Margin of Error

In general, we prefer narrow confidence intervals — that is, confidence intervals with a small margin of error. To reduce the margin of error in a confidence interval, we can change two factors: the sample size and the confidence level. In the following activity, you will investigate how each of these factors affects the margin of error.

ACTIVITY Exploring margin of error with an applet

In this activity, you will use an applet to explore the relationship between the confidence level, the sample size, and the margin of error.

1. Go to www.stapplet.com and launch the *Simulating Confidence Intervals* applet.
2. In the box labeled "Population," choose "Categorical" for the population distribution. Change the true proportion of successes to $p = 0.6$.

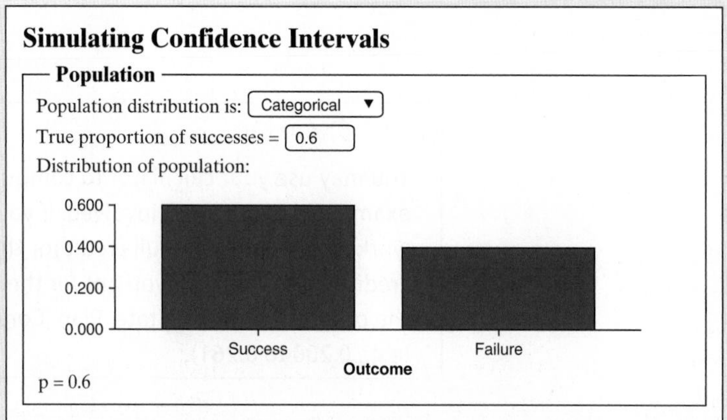

Part 1: Adjusting the Confidence Level

3. In the box labeled "Sample," change the sample size to 200 and keep the default confidence level of 95%. Click the "Go!" button to have the applet select an SRS of size $n = 200$, calculate $\hat{p}$, and construct a 95% confidence interval for p.

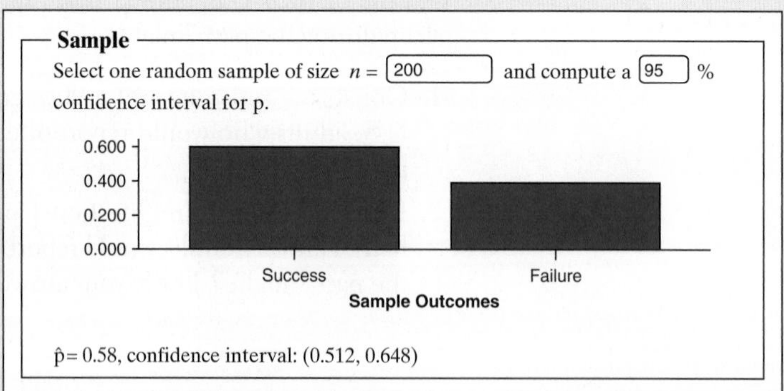

4. In the box labeled "Confidence Intervals," change the number of intervals to 50 and click the "Go!" button. Note the width of the intervals.

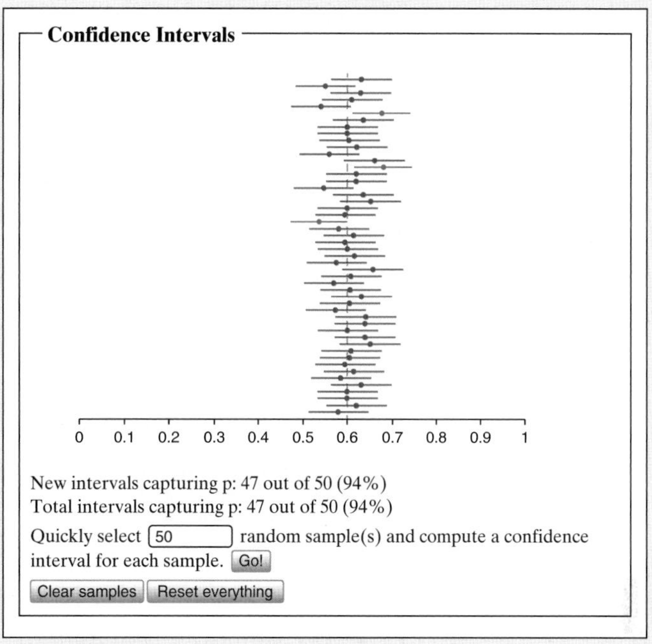

5. In the box labeled "Sample," gradually decrease the confidence level by clicking on the down arrow until you get to 80%. What happened to the width of the intervals? To the capture rate?
6. Now gradually increase the confidence level until you get to 99%. What happened to the width of the intervals? To the capture rate?
7. Summarize what you learned about the relationship between the confidence level and the margin of error for a fixed sample size.

Part 2: Adjusting the Sample Size

8. Click the "Reset everything" button, change the confidence level back to 95% in the box labeled "Sample," and click the "Go!" button. Then quickly generate 50 confidence intervals in the box labeled "Confidence Intervals." Note the width of the intervals.
9. In the box labeled "Sample," decrease the sample size to $n = 100$. Then quickly generate 50 confidence intervals in the box labeled "Confidence intervals." What happens to the width of the intervals?
10. Repeat Step 9 with samples of size $n = 1000$.
11. Summarize what you learned about the relationship between the sample size and the margin of error for a fixed confidence level.

As the activity illustrates, the price we pay for greater confidence is a wider interval. If we're satisfied with 90% confidence, then our interval of plausible values for the parameter will be narrower than if we insist on 95% or 99% confidence. For example, here is a 90% confidence interval and a 99% confidence interval for the proportion of red beads in Mr. Buckley's container based on the class's sample data.

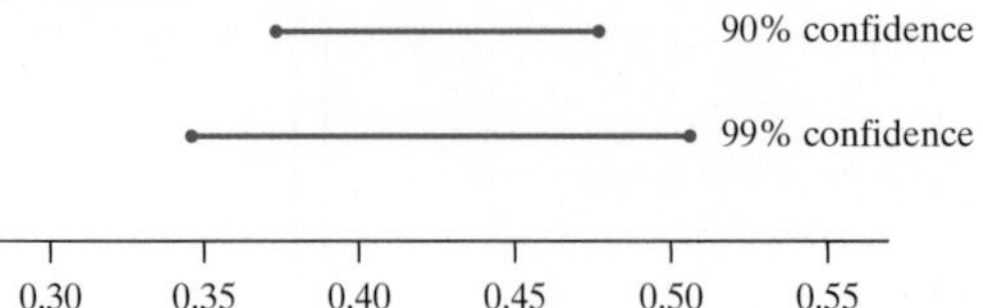

However, intervals constructed at a 90% confidence level will capture the true value of the parameter less often than intervals that use a 99% confidence level.

The activity also shows that we can get a more precise estimate of a parameter by increasing the sample size. Larger samples generally yield narrower confidence intervals at any confidence level. However, larger samples take more time and money to obtain.

DECREASING THE MARGIN OF ERROR

In general, we prefer an estimate with a small margin of error. The margin of error gets smaller when:

- *The confidence level decreases.* To obtain a smaller margin of error from the same data, you must be willing to accept a smaller capture rate.
- *The sample size n increases.* In general, increasing the sample size n reduces the margin of error for any fixed confidence level.

You can also determine the effect of changing the confidence level and sample size on the margin of error by considering its formula:

$$\text{margin of error} = z^* \sqrt{\frac{\hat{p}(1-\hat{p})}{n}}$$

Decreasing the confidence level means that there will be less area between $-z^*$ and z^*, forcing z^* to decrease. In addition, making z^* smaller decreases the margin of error, according to the formula. Because the sample size n is in the denominator of the standard error formula, increasing n will decrease the margin of error. In fact, *the width of a confidence interval for a proportion is proportional to $1/\sqrt{n}$, so quadrupling the sample size cuts the margin of error in half.*

EXAMPLE

Distracted walking
Factors that affect the margin of error

Skill 4.A

PROBLEM: In the preceding example, we were 95% confident that the interval from 0.200 to 0.261 captures p = the proportion of all cell-phone users who would admit to walking into something or someone while talking on their cell phone. This interval was based on a random sample of 738 cell-phone users.

(a) Explain what would happen to the width of the interval if the confidence level were increased to 99%.

(b) How would the width of a 95% confidence interval based on a sample of size 1500 compare to the original 95% confidence interval, assuming the sample proportion remained the same?

Maskot/Getty Images

SOLUTION:

(a) The confidence interval would be wider because increasing the confidence level from 95% to 99% increases the margin of error.

> To be more confident that our interval captures the population proportion, we need a wider interval, which requires a larger critical value.

(b) The confidence interval would be narrower because increasing the sample size from 738 to 1500 decreases the margin of error when everything else stays the same.

> Increasing the sample size decreases the standard error of the sampling distribution of $\hat{p}$ (assuming the sample proportion doesn't change).

FOR PRACTICE, TRY EXERCISE 21

When we calculate a confidence interval, we include the margin of error because we expect the value of the point estimate to vary somewhat from the parameter. However, the margin of error accounts for *only* the variability we expect from random sampling. It does not account for practical difficulties, such as undercoverage and nonresponse in a sample survey. These problems can produce estimates that are much farther from the parameter than the margin of error would suggest. Remember this unpleasant fact when reading the results of an opinion poll or other sample survey.  **The margin of error does not account for any sources of bias in the data collection process.**

Determining the Sample Size

When planning a study, we may want to choose a sample size that allows us to estimate a population proportion within a given margin of error. The formula for the margin of error (ME) in the confidence interval for p is

$$ME = z^* \sqrt{\frac{\hat{p}(1-\hat{p})}{n}}$$

To calculate the sample size, substitute values for ME, z^*, and $\hat{p}$, and solve for n. Unfortunately, we won't know the value $\hat{p}$ until *after* the study has been conducted. This means we have to guess the value of $\hat{p}$ when choosing n. Here are two ways to do this:

1. Use a guess for $\hat{p}$ based on a pilot (preliminary) study or past experience with similar studies.
2. Use $\hat{p} = 0.5$ as the guess. The margin of error ME is largest when $\hat{p} = 0.5$, so this guess yields an upper bound for the sample size that will result in a given margin of error. If we get a $\hat{p}$ other than 0.5 when we do our study, the margin of error will be smaller than planned.

Once you have a guess for $\hat{p}$, you can solve an inequality using the margin of error formula to determine the required sample size n.

CALCULATING THE SAMPLE SIZE FOR A DESIRED MARGIN OF ERROR WHEN ESTIMATING p

To determine the sample size n that will yield a $C\%$ confidence interval for a population proportion p with a maximum margin of error ME, solve the following inequality for n:

$$z^* \sqrt{\frac{\hat{p}(1-\hat{p})}{n}} \leq ME$$

where $\hat{p}$ is a guessed value for the sample proportion. The margin of error will always be less than or equal to ME if you use $\hat{p} = 0.5$.

Here's an example that shows you how to determine the sample size.

EXAMPLE

Customer satisfaction
Determining the sample size

Skill 3.D

PROBLEM: A company has received complaints about its customer service. The managers intend to hire a consultant to carry out a survey of customers. Before contacting the consultant, the company president wants some idea of the sample size that will be required. One value of interest is the proportion p of customers who are satisfied with the company's customer service. The company president wants the estimate to be within 3 percentage points (0.03) at a 95% confidence level. How large a sample is needed?

wbritten/Getty Images

SOLUTION:

$$1.96\sqrt{\frac{0.5(1-0.5)}{n}} \leq 0.03$$

We have no idea about the proportion p of satisfied customers, so we use $\hat{p} = 0.5$ as our guess to be safe.

$$\sqrt{\frac{0.5(1-0.5)}{n}} \leq \frac{0.03}{1.96}$$

Divide both sides by 1.96.

$$\frac{0.5(1-0.5)}{n} \leq \left(\frac{0.03}{1.96}\right)^2$$

Square both sides.

$$0.5(1-0.5) \leq n\left(\frac{0.03}{1.96}\right)^2$$

Multiply both sides by n.

$$\frac{0.5(1-0.5)}{\left(\frac{0.03}{1.96}\right)^2} \leq n$$

Divide both sides by $\left(\frac{0.03}{1.96}\right)^2$.

$$1067.11 \leq n$$

The sample needs to include at least 1068 customers.

Make sure to follow the inequality when rounding your answer.

FOR PRACTICE, TRY EXERCISE 23

Why not round to the nearest whole number — in this case, 1067? Because a smaller sample size will result in a larger margin of error, possibly more than the desired 3 percentage points for the survey. In general, we round to the next highest integer when solving for sample size to make sure the margin of error is less than or equal to the desired value.

CHECK YOUR UNDERSTANDING

Sleep Awareness Week begins in the spring with the release of the National Sleep Foundation's annual poll of U.S. sleep habits and ends with the beginning of daylight saving time, when most people lose an hour of sleep. In the foundation's random sample of 1029 U.S. adults, 48% reported that they "often or always" got enough sleep during the past 7 nights.

1. Describe two ways that the National Sleep Foundation could reduce the margin of error in its estimate of the proportion of all U.S. adults who would report they often or always got enough sleep during the past 7 nights. What are the drawbacks of these actions?
2. What sample size would be required to have a margin of error of at most 1% with 99% confidence? Use the value of $\hat{p}$ from the original sample in your calculations.

SECTION 6B Summary

- Confidence intervals for the proportion p of successes in a population use the sample proportion $\hat{p}$ as the point estimate.
- When constructing a confidence interval for a population proportion p, we need to ensure that the observations in the sample can be viewed as independent and that the sampling distribution of $\hat{p}$ is approximately normal. The required conditions are:
  - Random: The data come from a random sample from the population of interest.
    - 10%: When sampling without replacement, $n < 0.10N$.
  - Large Counts: Both $n\hat{p}$ and $n(1-\hat{p})$ are at least 10. That is, the number of successes and the number of failures in the sample are both at least 10.
- The general formula for a confidence interval is

  $$\text{point estimate} \pm \text{margin of error}$$

  where the **margin of error** of an estimate describes how far, at most, we expect the point estimate to vary from the population parameter.
- When calculating a confidence interval, it is common practice to use the form

  $$\text{statistic} \pm (\text{critical value})(\text{standard error of statistic})$$

  where the **critical value** is a multiplier that makes the interval wide enough to have the stated capture rate and the **standard error** of the statistic is an estimate of the standard deviation of the sampling distribution of the statistic used to estimate the parameter.

- When the conditions are met, the specific formula for a $C\%$ **one-sample z interval for a proportion** is

$$\hat{p} \pm z^* \sqrt{\frac{\hat{p}(1-\hat{p})}{n}}$$

where z^* is the critical value for the standard normal curve with $C\%$ of its area between $-z^*$ and z^*.

- When asked to construct and interpret a confidence interval, follow the **four-step process:**

State: State the parameter you want to estimate and the confidence level.
Plan: Identify the appropriate inference method and check the conditions.
Do: If the conditions are met, perform calculations.
Conclude: Interpret your interval in the context of the problem.

- Other things being equal, the margin of error of a confidence interval gets smaller as:
  - The confidence level C decreases.
  - The sample size n increases.
- The sample size needed to obtain a confidence interval with a maximum margin of error ME for a population proportion involves solving

$$z^* \sqrt{\frac{\hat{p}(1-\hat{p})}{n}} \leq ME$$

for n, where $\hat{p}$ is a guessed value for the sample proportion, and z^* is the critical value for the confidence level you want. Use $\hat{p} = 0.5$ if you don't have a good idea about the value of $\hat{p}$.

AP® EXAM TIP

AP® Daily Videos

Review the content of this section and get extra help by watching the AP® Daily Videos for Topics 6.2 and 6.3, which are available in AP® Classroom.

6B Tech Corner

TI-Nspire and other technology instructions are on the book's website at bfwpub.com/tps7e.

16. Confidence intervals for a proportion Page 547

SECTION 6B Exercises

Checking Conditions for a Confidence Interval for *p*

For Exercises 1–4, check whether each of the conditions is met for calculating a confidence interval for the population proportion p.

1. **Going to the prom** Tonya wants to estimate what proportion of her school's seniors plan to attend the prom. She interviews an SRS of 50 of the 750 seniors in her school and finds that 36 plan to go to the prom. pg 541

2. **Student government** The student body president of a high school claims to know the names of at least 1000 of the 1800 students who attend the school. To test this claim, the student government advisor randomly selects 100 students and asks the president to identify each by name. The president successfully names only 46 of the students.

3. **Salty chips** A quality control inspector selects a random sample of 25 bags of potato chips from the thousands of bags filled in an hour. Of the bags selected, 3 had too much salt.

4. **Whelks and mussels** The small round holes you often see in seashells were drilled by other sea creatures, who ate the former dwellers of the shells. Whelks often drill into mussels, but this behavior appears to be more

or less common in different locations. Researchers collected whelk eggs from the coast of Oregon, raised the whelks in the laboratory, then put each whelk in a container with some delicious mussels. Only 9 of 98 whelks drilled into a mussel.[16]

5. **The 10% condition** When constructing a confidence interval for a population proportion, we check that the sample size is less than 10% of the population size.

(a) Why is it necessary to check this condition?

(b) What happens to the capture rate if this condition is violated?

6. **The Large Counts condition** When constructing a confidence interval for a population proportion, we check that both $n\hat{p}$ and $n(1-\hat{p})$ are at least 10.

(a) Why is it necessary to check this condition?

(b) What happens to the capture rate if this condition is violated?

Calculating a Confidence Interval for *p*

7. **Keeping resolutions** At the end of a recent year, Morning Consult surveyed a random sample of U.S. adults about their New Year's resolutions. The 95% confidence interval for the proportion of all resolution-making U.S. adults who kept their resolution is 0.577 to 0.651.[17] Calculate the point estimate and margin of error used for this interval estimate.

8. **Traveling internationally** A recent survey from Morning Consult asked a random sample of U.S. adults if they had ever traveled internationally. Based on the sample, the 95% confidence interval for the population proportion of U.S. adults who have traveled internationally is 0.436 to 0.478.[18] Calculate the point estimate and margin of error used for this interval estimate.

9. **Finding z^*** Find the critical value z^* for a 98% confidence interval. Assume the Large Counts condition is met.

10. **Calculating z^*** Calculate the critical value z^* for a 96% confidence interval. Assume the Large Counts condition is met.

11. **Selling online** According to a recent Pew Research Center report, many U.S. adults have made money by selling something online. In a random sample of 4579 U.S. adults, 914 reported that they earned money by selling something online in the previous year.[19] Note that the conditions for inference are met. Calculate and interpret the standard error of $\hat{p}$ for these data.

12. **Pineapple pizza** To investigate which toppings people like on their pizza, YouGov surveyed 6168 randomly selected U.S. adults. In their sample, 1604 said they like pineapple on their pizza.[20] Note that the conditions for inference are met. Calculate and interpret the standard error of $\hat{p}$ for these data. (*Note:* In the sample, 2159 people said they *dislike* pineapple on their pizza.)

13. **Going to the prom** Tonya wants to estimate what proportion of her school's seniors plan to attend the prom. She interviews an SRS of 50 of the 750 seniors in her school and finds that 36 plan to go to the prom. Note that the conditions for inference are met.

pg 546

(a) Determine the critical value z^* for a 90% confidence interval for a proportion.

(b) Calculate the standard error of $\hat{p}$.

(c) Construct a 90% confidence interval for the proportion of seniors at her school who plan to go to the prom.

14. **Student government** The student body president of a high school claims to know the names of at least 1000 of the 1800 students who attend the school. To test this claim, the student government advisor randomly selects 100 students and asks the president to identify each by name. The president successfully names only 46 of the students. Note that the conditions for inference are met.

(a) Determine the critical value z^* for a 90% confidence interval for a proportion.

(b) Calculate the standard error of $\hat{p}$.

(c) Construct a 90% confidence interval for the proportion of all students at the school that the president can identify by name.

One-Sample *z* Interval for *p*

15. **Unknown callers** What do people do when an unknown number calls their cell phone? In a Pew Research Center survey, only 19% of the 10,211 randomly selected U.S. adults would answer the phone to see who it is.[21]

pg 548

(a) Construct and interpret a 95% confidence interval for the proportion of all U.S. adults who would answer their cell phone when an unknown number calls.

(b) Interpret the confidence level.

16. **Intelligent life** According to a Pew Research Center survey of 10,417 randomly selected U.S. adults, 65% said they believe that intelligent life exists on other planets.[22]

(a) Construct and interpret a 95% confidence interval for the proportion of all U.S. adults who believe that intelligent life exists on other planets.

(b) Interpret the confidence level.

17. **Three branches** According to a recent study by the Annenberg Foundation, only 47% of adults in the United States can name all three branches of government. This was based on a survey given to a random sample of 1113 U.S. adults.[23]

(a) Construct and interpret a 99% confidence interval for the proportion of all U.S. adults who can name all three branches of government.

(b) Does the interval from part (a) provide convincing evidence that fewer than half of all U.S. adults can name all three branches of government? Explain your answer.

18. **Food fight** A survey of 1480 randomly selected U.S. adults found that 45% of respondents agreed with the following statement: "Organic produce is better for health than conventionally grown produce."[24]

(a) Construct and interpret a 99% confidence interval for the proportion of all U.S. adults who think that organic produce is better for health than conventionally grown produce.

(b) Does the interval from part (a) provide convincing evidence that fewer than half of all U.S. adults think that organic produce is better for health? Explain your answer.

19. **Prom totals** Use your interval from Exercise 13 to construct and interpret a 90% confidence interval for the total number of seniors planning to go to the prom.

20. **Student body totals** Use your interval from Exercise 14 to construct and interpret a 90% confidence interval for the total number of students at the school that the student body president can identify by name. Then use your interval to evaluate the president's claim.

Factors that Affect the Margin of Error

21. **More unknown callers** Refer to Exercise 15.
pg 552

(a) Explain what would happen to the width of the interval if the confidence level were increased to 99%.

(b) How would the width of a 95% confidence interval based on a random sample of size 1000 compare to the original 95% confidence interval, assuming the sample proportion remained the same?

22. **More intelligent life** Refer to Exercise 16.

(a) Explain what would happen to the width of the interval if the confidence level were decreased to 90%.

(b) How would the width of a 95% confidence interval based on a random sample of size 1000 compare to the original 95% confidence interval, assuming the sample proportion remained the same?

Determining the Sample Size

23. **Starting a nightclub** A college student organization wants to start a nightclub for students younger than age 21. To assess support for this proposal, they will select an SRS of students and ask each respondent if they would patronize this type of establishment. What sample size is required to obtain a 90% confidence interval with a margin of error of at most 0.04?
pg 554

24. **Election polling** Gloria Chavez and Ronald Flynn are the candidates for mayor in a large city. We want to estimate the proportion p of all registered voters in the city who plan to vote for Chavez with 95% confidence and a margin of error no greater than 0.03. How large a random sample do we need?

25. **School vouchers** A small pilot study estimated that 44% of all U.S. adults agree that parents should be given vouchers that are good for education at any public or private school of their choice.

(a) How large a random sample is required to obtain an interval estimate with a margin of error of at most 0.03 and 99% confidence? Answer this question using the pilot study's result as the guessed value for $\hat{p}$.

(b) Answer the question in part (a) again, but this time use the conservative guess $\hat{p} = 0.5$. By how much do the two sample sizes differ?

26. **Can you taste PTC?** PTC is a substance that has a strong bitter taste for some people and is tasteless for others. The ability to taste PTC is inherited. About 75% of Italians can taste PTC, for example. You want to estimate the proportion of U.S. adults who have at least one Italian grandparent and who can taste PTC.

(a) How large a sample must you test to obtain an interval estimate with a margin of error of at most 0.04 and 90% confidence? Answer this question using the 75% estimate as the guessed value for $\hat{p}$.

(b) Answer the question in part (a) again, but this time use the conservative $\hat{p} = 0.5$. By how much do the two sample sizes differ?

27. **Teens and their devices** According to a Common Sense Media report, 69% of kids age 12 to 18 take their mobile device to bed with them. Here is part of the footnote to this report:

> Administered by Lake Research Partners, the survey consisted of telephone or online interviews with 1000 parents and children between 12 and 18 years old. The interviews were conducted between Feb. 2 and March 1 of this year, and the margin of error was ±4.4%.[25]

(a) Use what you have learned in this section to estimate the confidence level, assuming that Lake Research Partners used a simple random sample.

(b) Give an example of response bias that could affect the results of this survey. Is this bias accounted for by the margin of error?

28. **More organic food** Refer to Exercise 18. The study also estimated that 54% of adults age 18 to 29 would agree with the statement about organic foods, but only 39% of adults age 65 and older would agree.

(a) Explain why you do not have enough information to give confidence intervals for these two age groups separately.

(b) Do you think a 95% confidence interval for adults age 18 to 29 would have a larger or smaller margin of error than the estimate from Exercise 18? Explain your reasoning.

29. **Sample size and confidence** Does increasing the sample size make you more confident? To investigate this question, go to www.stapplet.com and launch the *Simulating Confidence Intervals* applet.

(a) In the box labeled "Population," choose "Categorical." Leave the true proportion of successes $= 0.5$. In the box labeled "Sample," change the sample size to $n = 125$ and leave the confidence level as 95%. Click the "Go!" button to generate one confidence interval. In the box labeled "Confidence Intervals," change the number of samples to 100 and click the "Go!" button 10 times. What percentage of the intervals captured the population proportion $p = 0.5$?

(b) Keeping the 95% confidence level, increase the sample size to $n = 200$ and repeat the process from part (a). What percentage of the intervals captured the population proportion $p = 0.5$?

(c) Keeping the 95% confidence level, increase the sample size to $n = 500$ and repeat the process from part (a). What percentage of the intervals captured the population proportion $p = 0.5$?

(d) Based on your answers, does increasing the sample size increase your confidence that a confidence interval will capture the parameter? Explain your reasoning.

For Investigation *Apply the skills from the section in a new context or nonroutine way.*

30. **The 10% condition, revisited** The principal of a school with 1000 students wants to use a 95% confidence interval to estimate the proportion of students who have a computer at home. Suppose this is true for 800 of the 1000 students at the school.

(a) Explain what it means to be 95% confident in this context.

(b) The principal selects a random sample of 400 students from the school and finds that 327 of these students have a computer at home. Show that the 10% condition isn't met.

(c) To explore the impact of violating the 10% condition, we simulated 2000 random samples of size $n = 400$ from a population of $N = 1000$ where $p = 0.8$. For each sample, we calculated a 95% confidence interval and found that 1979 of the 2000 intervals (98.95%) captured the population proportion of 0.80. Calculate a 95% confidence interval for the true capture rate of "95%" confidence intervals in this context. Assume the conditions for inference are met.

(d) Interpret your confidence interval from part (c). What does this indicate about violating the 10% condition?

(e) To avoid the problem created by violating the 10% condition, the principal can use the finite population correction factor in the confidence interval for the proportion of students who have a computer at home. The formula for a 95% confidence interval with this factor is

$$\hat{p} \pm z^* \sqrt{\frac{\hat{p}(1-\hat{p})}{n}} \sqrt{1 - \frac{n}{N}}$$

where N is the population size. Explain how including the finite population correction factor affects the lengths of the intervals and how this addresses the problem described in part (d).

Multiple Choice *Select the best answer for each question.*

31. A Gallup poll found that only 28% of American adults expect to inherit money or valuable possessions from a relative. The poll's margin of error was ±3 percentage points at a 95% confidence level. This means that

(A) the poll used a method that gets an answer within 3% of the truth about the population 95% of the time.

(B) the percentage of all adults who expect an inheritance must be between 25% and 31%.

(C) if Gallup takes another poll on this issue, the results of the second poll will lie between 25% and 31%.

(D) there's a 95% chance that the percentage of all adults who expect an inheritance is between 25% and 31%.

(E) Gallup can be 95% confident that between 25% and 31% of the sample expect an inheritance.

32. Refer to Exercise 31. Suppose that Gallup wanted to cut the margin of error in half from 3 percentage points to 1.5 percentage points. How should it adjust the sample size?

(A) Multiply the sample size by 4.

(B) Multiply the sample size by 2.

(C) Multiply the sample size by 1/2.

(D) Multiply the sample size by 1/4.

(E) There is not enough information to answer this question.

33. Most people can roll their tongues, but some can't. Suppose we are interested in determining what proportion of students can roll their tongues. We test a simple random sample of 400 students and find that 317 can roll their tongues. The margin of error for a 95% confidence interval for the population proportion of tongue rollers among students is closest to which of the following?

(A) 0.0008

(B) 0.02

(C) 0.03

(D) 0.04

(E) 0.05

34. A newspaper reporter asked an SRS of 100 residents in a large city for their opinion about the mayor's job performance. Using the results from the sample, the C% confidence interval for the proportion of all residents in the city who approve of the mayor's job performance is 0.565 to 0.695. What is the value of C?

(A) 82

(B) 86

(C) 90

(D) 95

(E) 99

Exercises 35 and 36 refer to the following scenario. A researcher plans to use a random sample of houses to estimate the proportion of all houses that have a basement.

35. The researcher is deciding between a 95% confidence level and a 99% confidence level. Compared with a 95% confidence interval, a 99% confidence interval will be

(A) narrower and would involve a larger risk of being incorrect.

(B) narrower and would involve a smaller risk of being incorrect.

(C) wider and would involve a smaller risk of being incorrect.

(D) wider and would involve a larger risk of being incorrect.

(E) wider and would have the same risk of being incorrect.

36. After deciding on a 95% confidence level, the researcher is deciding between a sample of size $n = 500$ and a sample of size $n = 1000$. Compared with using a sample size of $n = 500$, a confidence interval based on a sample size of $n = 1000$ will be

(A) narrower and would involve a larger risk of being incorrect.

(B) narrower and would involve a smaller risk of being incorrect.

(C) wider and would involve a smaller risk of being incorrect.

(D) wider and would involve a larger risk of being incorrect.

(E) narrower and would have the same risk of being incorrect.

37. In a poll conducted by phone,

I. Some people refused to answer questions.

II. People without telephones could not be in the sample.

III. Some people never answered the phone in several calls.

Which of these possible sources of bias is included in the ±2% margin of error announced for the poll?

(A) I only

(B) II only

(C) III only

(D) I, II, and III

(E) None of these

Recycle and Review *Practice what you learned in previous sections.*

38. **Instant winners (5B)** A fast-food restaurant promotes certain food items by giving a game piece with each item. Advertisements proclaim that "25% of the game pieces are Instant Winners!" To test this claim, a frequent diner collects 20 game pieces and gets only 1 instant winner.

(a) Identify the population, the parameter, the sample, and the statistic in this context.

Suppose the advertisements are correct and $p = 0.25$. The dotplot shows the distribution of the sample proportion of instant winners in 100 simulated SRSs of size $n = 20$.

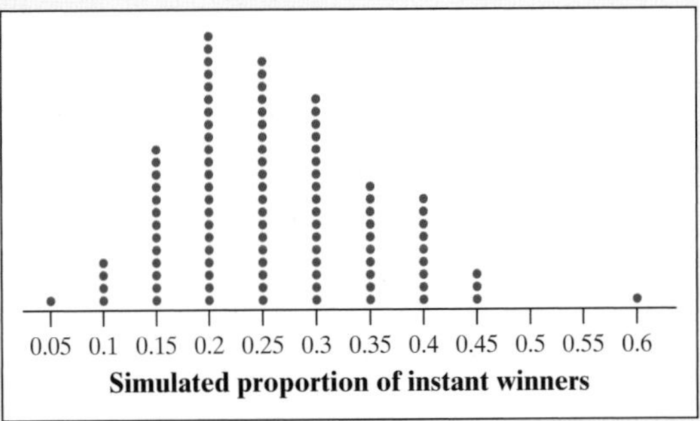

(b) Use the simulation to estimate the probability of getting a sample proportion of $\hat{p} = 1/20 = 0.05$ or less in a sample of size $n = 20$ when $p = 0.25$.

(c) Based on the actual sample and your answer to part (b), is there convincing evidence that fewer than 25% of all game pieces are instant winners? Explain your reasoning.

39. **More instant winners (4F)** Use the appropriate binomial distribution to calculate the probability described in Exercise 38(b). How does it compare with the estimate from the simulation?

SECTION 6C Significance Tests: The Basics

LEARNING TARGETS *By the end of the section, you should be able to:*

- State appropriate hypotheses for a significance test about a population parameter.
- Interpret a *P*-value in context.
- Make an appropriate conclusion for a significance test.

Confidence intervals are one of the two most commonly used methods of statistical inference. You can use a confidence interval to estimate parameters, such as the proportion p of all U.S. adults who exercise regularly.

What if we want to test a claim about a parameter? For instance, the U.S. Bureau of Labor Statistics claims that the national unemployment rate in December 2022 was 3.5%. A citizens' group suspects that the actual rate was higher. The second common method of inference, called a **significance test,** allows us to weigh the evidence in favor of or against a particular claim (hypothesis). A significance test is sometimes referred to as a *test of significance*, a *hypothesis test*, or a *test of hypotheses.*

DEFINITION Significance test

A **significance test** is a formal procedure for using observed data to decide between two competing claims (called hypotheses). The claims are usually statements about population parameters.

Here is an activity that illustrates the reasoning of a significance test.

ACTIVITY I'm a great free-throw shooter!

In this activity, you and your classmates will perform a simulation to test a claim about a population proportion.

A basketball player claims to be an 80% free-throw shooter. That is, the player claims that $p = 0.80$, where p is the proportion of free throws the

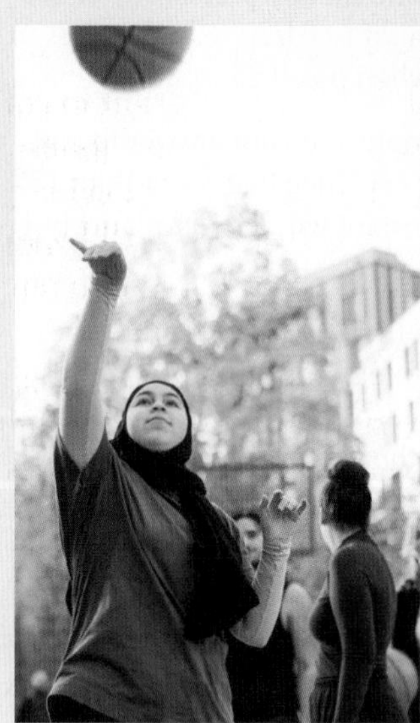
Portra/E+/Getty Images

player will make in the long run. We suspect that the player is exaggerating and that $p < 0.80$.

Suppose the player shoots 50 free throws and makes 32 of them. The sample proportion of made shots is $\hat{p} = 32/50 = 0.64$. This result gives *some* evidence that the player makes less than 80% of their free throws in the long run because $0.64 < 0.80$. But does it give *convincing* evidence that $p < 0.80$? Or is it plausible (believable) that an 80% shooter can have a performance this poor by chance alone? You can use a simulation to find out.

1. Go to www.stapplet.com and launch the *Logic of Significance Testing* applet.
2. Click the "Shoot" button 50 times to simulate a sample of 50 shots by an 80% free-throw shooter. The sample proportion $\hat{p}$ of made shots will be displayed on a dotplot in the "Simulation Results" section. Plot this value on the class dotplot drawn by your teacher.
3. Repeat Step 2 as needed to get at least 40 trials of the simulation for your class.
4. Based on the class's simulation results, how likely is it for an 80% shooter to make 64% or less of their shots when shooting 50 free throws?
5. Based on your answer in Step 4, does the observed $\hat{p} = 0.64$ result give convincing evidence that the player is exaggerating? Or is it plausible (believable) that an 80% shooter can have a performance this poor by chance alone?

In the activity, the shooter made only 32 of 50 free-throw attempts ($\hat{p} = 32/50 = 0.64$). There are two possible explanations for why the shooter made only 64% of their shots:

1. The player's claim is true ($p = 0.80$). That is, the player really is an 80% shooter and the poor performance happened by chance alone.
2. The player's claim is false ($p < 0.80$). That is, the player really makes less than 80% of their free throws in the long run.

If Explanation 1 is plausible, then we don't have convincing evidence that the shooter is exaggerating — the player's poor performance could have occurred purely by chance. However, if it is unlikely for an 80% shooter to get a proportion of 0.64 or less in 50 attempts, then we can rule out Explanation 1.

We used the applet to simulate 400 sets of 50 shots, assuming that the player is really an 80% shooter. Figure 6.4 shows a dotplot of the results. Each dot on the graph represents the simulated proportion $\hat{p}$ of made shots in one sample of 50 free throws.

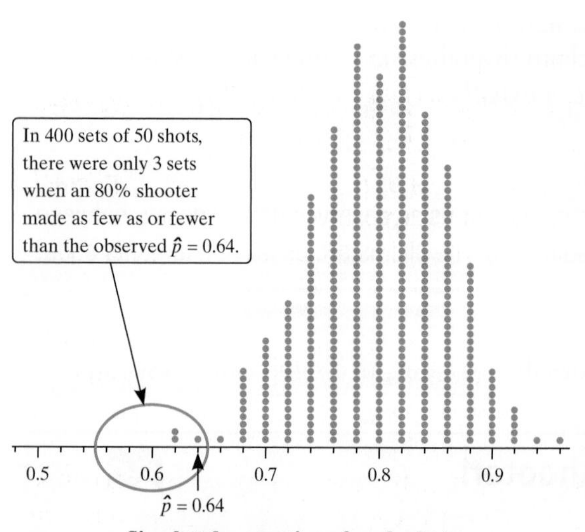
In 400 sets of 50 shots, there were only 3 sets when an 80% shooter made as few as or fewer than the observed $\hat{p}$ = 0.64.
0.5
0.6
0.7
0.8
0.9
$\hat{p}$ = 0.64
Simulated proportion of made shots

FIGURE 6.4 Dotplot of the simulated sampling distribution of $\hat{p}$ = the proportion of free throws made by an 80% shooter in a sample of 50 shots.

The simulation shows that it would be very unlikely for an 80% free-throw shooter to make 32 or fewer free throws in 50 attempts just by chance. This small probability ($\approx 3/400 = 0.0075$) leads us to rule out Explanation 1. The observed result gives us convincing evidence that Explanation 2 is correct: the player makes less than 80% of their free throws in the long run.

Stating Hypotheses

A significance test starts with a careful statement of the claims we want to compare. In the free-throw shooter activity, the player claims that their long-run proportion of made free throws is $p = 0.80$. This is the claim we seek evidence *against*. We call it the **null hypothesis**, abbreviated H_0. Usually, the null hypothesis is a statement of "no difference." For the free-throw shooter, no difference from the player's claim gives H_0: $p = 0.80$.

The claim we hope or suspect to be true instead of the null hypothesis is called the **alternative hypothesis.** We abbreviate the alternative hypothesis as H_a. In this case, we suspect the player might be exaggerating, so our alternative hypothesis is H_a: $p < 0.80$.

DEFINITION Null hypothesis H_0, Alternative hypothesis H_a

The claim that we weigh evidence against in a significance test is called the **null hypothesis (H_0).**

The claim that we are trying to find evidence for is the **alternative hypothesis (H_a).**

In the free-throw shooter context, our hypotheses are

$$H_0: p = 0.80$$
$$H_a: p < 0.80$$

where p is the proportion of free throws the player will make in the long run. The alternative hypothesis is **one-sided** because we suspect the player makes *less* than 80% of their free throws ($p < 0.80$). If you suspect that the true value of a parameter is *different from* the null value, use a **two-sided** alternative hypothesis. It is common to refer to a significance test with a one-sided alternative hypothesis as a *one-sided test* or *one-tailed test* and to a significance test with a two-sided alternative hypothesis as a *two-sided test* or *two-tailed test*.

DEFINITION One-sided, Two-sided

The alternative hypothesis is **one-sided** if it states that a parameter is *greater than* the null value or if it states that the parameter is *less than* the null value.

The alternative hypothesis is **two-sided** if it states that the parameter is *different from* the null value (it could be either greater than or less than).

The null hypothesis has the form H_0: parameter = null value. A one-sided alternative hypothesis has one of the forms H_a: parameter < null value or H_a: parameter > null value. A two-sided alternative hypothesis has the form H_a: parameter $\neq$ null value. To determine the correct form of H_a, read the problem carefully.

Note: Some people insist that all three possibilities — greater than, less than, and equal to — should be accounted for in the hypotheses. For the free-throw shooter example, because the alternative hypothesis is H_a: $p < 0.80$, they would write the null hypothesis as H_0: $p \geq 0.80$. Despite the mathematical appeal of covering all three cases, we use the claimed value $p = 0.80$ when carrying out the test. So we'll use a null hypothesis of the form H_0: $p = 0.80$ in this book.

EXAMPLE

Juicy pineapples
Stating hypotheses

Skill 1.F

Daisuke Kishi/Moment/Getty Images

PROBLEM: James is a manager at the Hawaii Pineapple Company and is interested in the size of the pineapples grown in the company's fields. Last year, the mean weight of the pineapples harvested from one large field was 31 ounces. A different irrigation system was installed in this field after the growing season. James wonders if this change will affect the mean weight of pineapples grown in the field this year. State appropriate hypotheses for performing a significance test. Be sure to define the parameter of interest.

SOLUTION:

H_0: $\mu = 31$

H_a: $\mu \neq 31$

where μ = the mean weight (in ounces) of all pineapples grown in the field this year.

Because James wonders if the mean weight of this year's pineapples will be affected (positively or negatively) by the new irrigation system, the alternative hypothesis is two-sided.

FOR PRACTICE, TRY EXERCISE 1

The hypotheses should express the belief or suspicion we have before we see the data. It is cheating to look at the data first and then frame the alternative hypothesis to fit what the data show. For example, the data for the pineapple study showed that $\bar{x} = 31.935$ ounces for a random sample of 50 pineapples grown in the field this year. You should *not* change the alternative hypothesis to H_a: $\mu > 31$ after looking at the data.

AP® EXAM TIP

Hypotheses always refer to a population, not to a sample. Be sure to state H_0 and H_a in terms of population parameters. It is *never* correct to write a hypothesis about a sample statistic, such as $H_0: \hat{p} = 0.80$ or $H_a: \bar{x} \neq 31$. Likewise, make sure to include the word *population, true,* or *all* when defining the parameter(s).

Interpreting *P*-Values

mgkaya/E+/Getty Images

It might seem strange to you that we state a null hypothesis and then try to find evidence *against* it. Maybe it would help to think about how a criminal trial works in the United States. The defendant is "innocent until proven guilty." That is, the null hypothesis is innocence and the prosecution must offer convincing evidence against this hypothesis and in favor of the alternative hypothesis: guilt. That's exactly how significance tests work, although in statistics we deal with evidence provided by data and use a probability to say how strong the evidence is.

In the Free-Throw Shooter activity at the beginning of the section, a player who claimed to be an 80% free-throw shooter made only $\hat{p} = 32/50 = 0.64$ of shots in a random sample of 50 free throws. This is evidence *against* the null hypothesis that $p = 0.80$ and *in favor of* the alternative hypothesis $p < 0.80$. But is the evidence convincing? To answer this question, we have to know how likely it is for an 80% shooter to make 64% or less of their free throws by chance alone in a random sample of 50 attempts. This probability is called a **P-value.**

DEFINITION *P*-value

The ***P*-value** of a test is the probability of getting evidence for the alternative hypothesis H_a as strong or stronger than the observed evidence when the null hypothesis H_0 is true.

We used simulation to estimate the *P*-value for our free-throw shooter: $3/400 = 0.0075$. How do we interpret this *P*-value? Assuming that the player makes 80% of their free throws in the long run, there is about a 0.0075 probability of getting a sample proportion of 0.64 or less just by chance in a random sample of 50 shots.

Small *P*-values give convincing evidence for H_a because they say that the observed result is unlikely to occur when H_0 is true. Large *P*-values fail to give convincing evidence for H_a because they say that the observed result is likely to occur by chance alone when H_0 is true.

We'll show you how to calculate *P*-values later. For now, let's focus on interpreting them.

EXAMPLE

Healthy bones
Interpreting *P*-values

Skill 4.B

PROBLEM: Calcium is a vital nutrient for healthy bones and teeth. The National Institutes of Health (NIH) recommends a calcium intake of 1300 milligrams (mg) per day for teenagers. The NIH is concerned that teenagers aren't getting enough calcium, on average. Is this true? Researchers decide to perform a test of

$$H_0: \mu = 1300$$
$$H_a: \mu < 1300$$

where μ is the population mean daily calcium intake for teenagers. They ask a random sample of 20 teens to record their food and drink consumption for one day. The researchers then compute the calcium intake for each teen. Data analysis reveals that $\bar{x} = 1198$ mg and $s_x = 411$ mg. Researchers performed a significance test and obtained a *P*-value of 0.1405.

(a) Explain what it would mean for the null hypothesis to be true in this setting.
(b) State the evidence for the alternative hypothesis.
(c) Interpret the *P*-value.

Martin Shields/Alamy Stock Photo

SOLUTION:

(a) If $H_0: \mu = 1300$ is true, then the mean daily calcium intake in the population of teenagers is 1300 mg.

(b) The evidence for H_a is: $\bar{x} = 1198 < 1300$.

(c) Assuming that the population mean daily calcium intake for teenagers is 1300 mg, there is a 0.1405 probability of getting a sample mean of 1198 mg or less by chance alone in a random sample of 20 teens.

When interpreting *P*-values, many students lose credit on the AP® Statistics exam because they don't answer in context or because they leave out the statement that the null hypothesis is assumed to be true when calculating the *P*-value.

FOR PRACTICE, TRY EXERCISE 9

The *P*-value is the probability of getting evidence for H_a as strong as or stronger than the observed result, given that H_0 is true. In other words, the *P*-value is a conditional probability. For the "Healthy bones" example, *P*-value $= P(\bar{x} \leq 1198 \mid \mu = 1300) = 0.1405$, as shown in the figure.

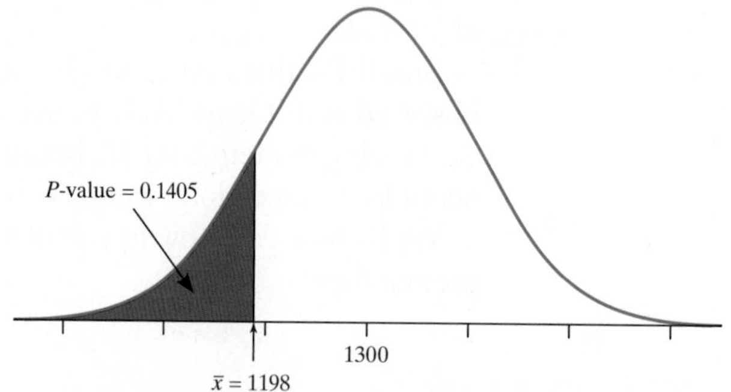

Sample mean daily calcium intake (mg) when $\mu = 1300$

When H_a is two-sided (parameter ≠ null value), values of the sample statistic less than or greater than the null value both count as evidence for H_a. In the "Juicy pineapples" example, we were testing H_0: $\mu = 31$ versus H_a: $\mu \neq 31$ and got $\bar{x} = 31.935$. This result gives some evidence for H_a: $\mu \neq 31$ because $31.935 \neq 31$. Because this is a two-sided test, evidence for H_a as strong as or stronger than the observed result includes any value of $\bar{x}$ greater than or equal to 31.935, as well as any value of $\bar{x}$ less than or equal to 30.065. Why? Because $\bar{x} = 30.065$ is just as different from the null value of $\mu = 31$ as $\bar{x} = 31.935$ is. For this scenario, the *P*-value is equal to the conditional probability $P(\bar{x} \leq 30.065 \text{ or } \bar{x} \geq 31.935 \mid \mu = 31)$, as shown in the following figure.

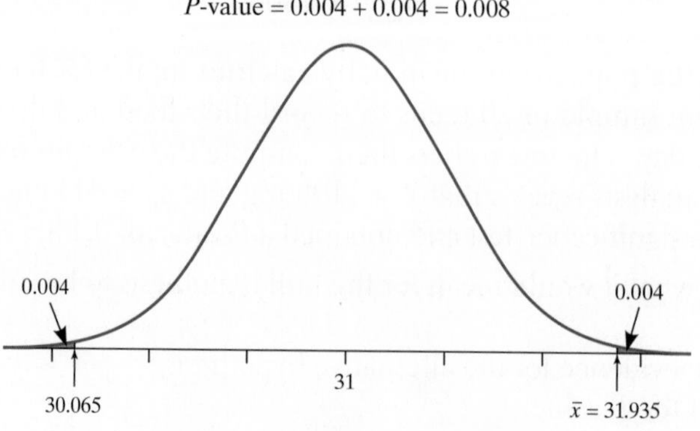

Sample mean weight (oz) when $\mu = 31$

Making Conclusions

The final step in performing a significance test is to make a conclusion about the competing claims being tested. We make this conclusion based on the strength of the evidence in favor of the alternative hypothesis and against the null hypothesis as measured by the *P*-value.

- If the *P*-value is small, we reject H_0 because the observed result is unlikely to occur when H_0 is true. In this case, there is convincing evidence for H_a.
- If the *P*-value is not small, we fail to reject H_0 because the observed result is at least somewhat likely to occur when H_0 is true. In this case, there is not convincing evidence for H_a.

Don Farrall/Photodisc/Getty Images

Saying we "reject H_0" or "fail to reject H_0" may seem unusual at first, but it's consistent with what happens in a criminal trial. Once the jury has weighed the evidence against the null hypothesis of innocence, they return one of two verdicts: "guilty" (reject H_0) or "not guilty" (fail to reject H_0). A not-guilty verdict doesn't guarantee that the defendant is innocent. It only says that there's not convincing evidence of guilt. Likewise, a fail-to-reject H_0 decision in a significance test doesn't guarantee that H_0 is true.

In the Free-Throw Shooter activity, the estimated *P*-value was 0.0075. Because this *P*-value is small, we reject H_0: $p = 0.80$. We have convincing evidence that the shooter makes fewer than 80% of their free throws in the long run.

How small does a *P*-value have to be for us to reject H_0? In Unit 3, we suggested that you use a boundary of 5% when determining whether a result is statistically significant. Choosing this boundary value means we require evidence for H_a so strong that it would happen at most 5% of the time just by chance when H_0 is true.

Sometimes it may be preferable to use a different boundary value — like 0.01 or 0.10 — when drawing a conclusion in a significance test. We will explain why at the end of Section 6D. The chosen boundary value is called the **significance level α**, where α is the Greek letter alpha.

DEFINITION Significance level α

The **significance level α** is the value that we use as a boundary to decide if an observed result is unlikely to happen by chance alone when the null hypothesis is true.

When we use a fixed significance level α to draw a conclusion in a significance test, there are only two possible conclusions.

HOW TO MAKE A CONCLUSION IN A SIGNIFICANCE TEST

- *P*-value $\leq \alpha$: "Because the *P*-value of ________ $\leq \alpha =$ ________, we reject H_0. There is convincing evidence for H_a (in context)."
- *P*-value $> \alpha$: "Because the *P*-value of ________ $> \alpha =$ ________, we fail to reject H_0. There is not convincing evidence for H_a (in context)."

Significance at the $\alpha = 0.05$ level is often expressed by the statement "The results were significant at the 5% level ($P < 0.05$)." Here, P stands for the P-value. *The P-value is more informative than a statement about significance* because it describes the strength of evidence for the alternative hypothesis. For example, both an observed result with P-value $= 0.03$ and an observed result with P-value $= 0.0003$ are significant at the 5% level. But the P-value of 0.0003 gives much stronger evidence against H_0 and in favor of H_a than the P-value of 0.03. This is why we always include the P-value in our conclusions, and not just a statement about significance.

EXAMPLE

More healthy bones
Making conclusions

Skill 4.E

PROBLEM: In the preceding example, researchers collected data on the daily calcium intake for a random sample of 20 teens. Data analysis revealed that $\bar{x} = 1198$ mg and $s_x = 411$ mg. The researchers used these data to perform a test of

$$H_0: \mu = 1300$$
$$H_a: \mu < 1300$$

where μ is the mean daily calcium intake in the population of teenagers. The resulting P-value is 0.1405. What conclusion would you make at the $\alpha = 0.05$ level?

Jack Andersen/Stone/Getty Images

SOLUTION:

Because the P-value of $0.1405 > \alpha = 0.05$, we fail to reject H_0. We don't have convincing evidence that teens are getting less than 1300 mg of calcium per day, on average.

FOR PRACTICE, TRY EXERCISE 13

Be careful how you write conclusions when the P-value isn't small. Don't conclude that the null hypothesis is true just because we didn't find convincing evidence for the alternative hypothesis. For example, it would be incorrect to conclude that teens *are* getting 1300 mg of calcium per day, on average. We found *some* evidence that the teens weren't getting enough calcium ($\bar{x} = 1198 < 1300$), but the evidence wasn't convincing enough to reject H_0.

Never "accept H_0" or conclude that H_0 is true! In fact, the 90% confidence interval for μ = the mean daily calcium intake in the population of teenagers is 1039.1 to 1356.9 milligrams. You can see that 1300 is just one of many plausible values for μ based on the sample data.

AP® EXAM TIP

On the AP® Statistics exam, if you are asked to make a conclusion for a significance test and no significance level is provided, use $\alpha = 0.05$.

When a researcher plans to draw a conclusion based on a significance level, **α should be stated *before* the data are produced.** Otherwise, a deceptive user of statistics might choose α *after* the data have been analyzed in an attempt to manipulate the conclusion. This is just as inappropriate as choosing an alternative hypothesis after looking at the data.

CHECK YOUR UNDERSTANDING

The manager of a fast-food restaurant wants to reduce the proportion of drive-thru customers who have to wait longer than 2 minutes to receive their food after placing an order. Based on store records from the past year, the proportion of customers who had to wait longer than 2 minutes was $p = 0.63$. To reduce this proportion, the manager assigns an additional employee to help with drive-thru orders. During the next month, the manager collects a random sample of 250 drive-thru times and finds that $\hat{p} = \frac{144}{250} = 0.576$. Does this provide convincing evidence that the proportion of all drive-through customers who have to wait longer than 2 minutes has been reduced?

1. State the hypotheses we are interested in testing. Make sure to define the parameter of interest.
2. The *P*-value of the manager's test is 0.0385. Interpret the *P*-value.
3. What conclusion should the manager make?

SECTION 6C Summary

- A **significance test** is a procedure for using observed data to decide between two competing claims, called hypotheses. The hypotheses are often statements about a parameter, like the population proportion p or the population mean μ.
- The claim that we weigh evidence *against* in a significance test is called the **null hypothesis (H_0)**. The null hypothesis is usually a statement of no change or no difference and has the form H_0: parameter = null value.
- The claim about the population that we are trying to find evidence *for* is the **alternative hypothesis (H_a)**.
  - A **one-sided** alternative hypothesis has the form H_a: parameter < null value or H_a: parameter > null value.
  - A **two-sided** alternative hypothesis has the form H_a: parameter ≠ null value.
- The **P-value** of a test is the probability of getting evidence for the alternative hypothesis H_a that is as strong as or stronger than the observed evidence when the null hypothesis H_0 is true.
- Small *P*-values are evidence against the null hypothesis and for the alternative hypothesis because they say that the observed result is unlikely to occur when H_0 is true. To determine if a *P*-value should be considered small, we compare it to the **significance level,** such as $\alpha = 0.05$.
- We make a conclusion in a significance test based on the *P*-value.
  - If *P*-value $\leq \alpha$: Reject H_0 and conclude there is convincing evidence for H_a (in context).
  - If *P*-value $> \alpha$: Fail to reject H_0 and conclude there is not convincing evidence for H_a (in context).

AP® EXAM TIP

AP® Daily Videos

Review the content of this section and get extra help by watching the AP® Daily Video for Topic 6.4 (Video 1), which is available in AP® Classroom.

SECTION 6C Exercises

Stating Hypotheses

In Exercises 1–6, state appropriate hypotheses for performing a significance test. Be sure to define the parameter of interest.

1. **No sleep?** According to the U.S. Centers for Disease Control and Prevention, 22.1% of high school students get at least 8 hours of sleep per night on school nights.[26] A school counselor worries that fewer than 22.1% of students at their school get this much sleep. To investigate, the counselor surveys a random sample of 50 students at the school.

pg 564

2. **Don't argue!** A Gallup poll report revealed that 72% of teens said they seldom or never argue with their friends. Yvonne wonders whether this result holds true in her large high school, so she surveys a random sample of 150 students at her school.

3. **How much juice?** One company's bottles of grapefruit juice are filled by a machine that is set to dispense an average of 180 milliliters (ml) of liquid. A quality-control inspector must check that the machine is working properly. The inspector selects a random sample of 40 bottles and measures the volume of liquid in each bottle.

4. **Attitudes** The Survey of Study Habits and Attitudes (SSHA) is a psychological test that measures students' attitudes toward school and study habits. Scores range from 0 to 200, with higher scores indicating more positive attitudes and better study habits. The mean score for U.S. college students is 115. A researcher suspects that older students have better attitudes toward school, on average. The researcher gives the SSHA to an SRS of 45 of the more than 1000 students at their college who are at least 30 years of age.

5. **Cold cabin?** During the winter months, outside temperatures at the Starnes family cabin in Colorado can stay well below freezing (32°F or 0°C) for weeks at a time. To prevent the pipes from freezing, Mrs. Starnes sets the thermostat at 50°F. The manufacturer claims that the thermostat allows variation in home temperature that follows an approximately normal distribution with $\sigma = 3$°F. Mrs. Starnes suspects that the manufacturer is overstating the consistency of the thermostat. To investigate, she programs a digital thermometer to take an SRS of $n = 10$ readings during a 24-hour period.

6. **New golf club** An avid golfer would like to improve their game. Based on years of experience, the golfer has established that the distance balls travel when hit with their current 3-iron follows an approximately normal distribution with standard deviation $\sigma = 15$ yards. The golfer is hoping that a new 3-iron will make their shots more consistent (less variable). To find out, the golfer hits 50 shots on a driving range with the new 3-iron.

7. **Parking changes** A change is made that should improve student satisfaction with the parking situation at a local high school. Before the change, 37% of students approve of the parking that's provided. Explain what's wrong with the following hypotheses for performing a significance test. Then give the correct hypotheses.

$$H_0: p > 0.37$$
$$H_a: p = 0.37$$

8. **Light babies** A researcher suspects that the mean birth weights of babies whose mothers did not see a doctor before delivery is less than 3000 grams. Explain what's wrong with the following hypotheses for performing a significance test. Then give the correct hypotheses.

$$H_0: \bar{x} = 3000$$
$$H_a: \bar{x} < 3000$$

Interpreting *P*-Values

9. **No sleep?** Refer to Exercise 1. The counselor finds that only $8/50 = 16\%$ of the students in the sample get at least 8 hours of sleep on school nights. A significance test yields a *P*-value of 0.1493.

pg 565

(a) Explain what it would mean for the null hypothesis to be true in this setting.

(b) State the evidence for the alternative hypothesis.

(c) Interpret the *P*-value.

10. **Attitudes** Refer to Exercise 4. In the study of older students' attitudes, the sample mean SSHA score was 125.7 and the sample standard deviation was 29.8. A significance test yields a *P*-value of 0.0101.

(a) Explain what it would mean for the null hypothesis to be true in this setting.

(b) State the evidence for the alternative hypothesis.

(c) Interpret the *P*-value.

11. **More juice** Refer to Exercise 3. The mean amount of liquid in the bottles is 179.6 ml and the standard deviation is 1.3 ml. A significance test yields a *P*-value of 0.0589. Interpret the *P*-value.

12. **More arguing** Refer to Exercise 2. Yvonne finds that 96 of the 150 students (64%) say they rarely or never argue with friends. A significance test yields a *P*-value of 0.0291. Interpret the *P*-value.

Making Conclusions

13. No sleep? Refer to Exercises 1 and 9. What conclusion would you make at the $\alpha = 0.05$ level? pg 568

14. **Attitudes** Refer to Exercises 4 and 10. What conclusion would you make at the $\alpha = 0.05$ level?

15. **Juice, again** Refer to Exercises 3 and 11.

(a) What conclusion would you make at the 10% significance level?

(b) Would your conclusion from part (a) change if a 5% significance level was used instead? Explain your reasoning.

16. **Arguing, again** Refer to Exercises 2 and 12.

(a) What conclusion would you make at the 1% significance level?

(b) Would your conclusion from part (a) change if a 5% significance level was used instead? Explain your reasoning.

For Investigation *Apply the skills from the section in a new context or nonroutine way.*

17. **More lefties?** In the population of people in the United States, about 10% are left-handed. After encountering some left-handed students at lunch, Simon wondered if more than 10% of students at his school are left-handed. To investigate, he selected an SRS of 50 students and found 8 lefties ($\hat{p} = 8/50 = 0.16$).

To determine if these data provide convincing evidence that more than 10% of the students at Simon's school are left-handed, 200 trials of a simulation were conducted. Each dot in the graph shows the proportion of students who are left-handed in a random sample of 50 students, assuming that each student has a 10% chance of being left-handed.

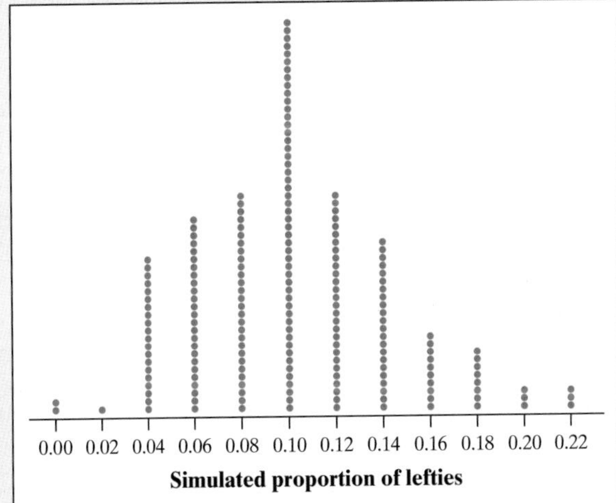

(a) State appropriate hypotheses for performing a significance test. Be sure to define the parameter of interest.

(b) Use the simulation results to estimate the *P*-value of the test in part (a). Interpret the *P*-value.

(c) What conclusion would you make?

Multiple Choice *Select the best answer for each question.*

18. Experiments on learning in animals sometimes measure how long it takes mice to find their way through a maze. The mean time is 18 seconds for one particular maze. A researcher thinks that a loud noise will cause the mice to complete the maze faster. The researcher measures how long each of 10 mice takes with a loud noise as stimulus. Which of the following are the appropriate hypotheses for the significance test?

(A) H_0: $\mu = 18$; H_a: $\mu \neq 18$

(B) H_0: $\mu = 18$; H_a: $\mu > 18$

(C) H_0: $\mu < 18$; H_a: $\mu = 18$

(D) H_0: $\mu = 18$; H_a: $\mu < 18$

(E) H_0: $\bar{x} = 18$; H_a: $\bar{x} < 18$

19. Members of the city council want to know if a majority of city residents support a 1% increase in the sales tax to fund road repairs. To investigate, they survey a random sample of 300 city residents and use the results to test the following hypotheses:

$$H_0: p = 0.50$$
$$H_a: p > 0.50$$

where p is the proportion of all city residents who support a 1% increase in the sales tax to fund road repairs. In the sample, $\hat{p} = 158/300 = 0.527$. The resulting *P*-value is 0.18. What is the correct interpretation of this *P*-value?

(A) Only 18% of the city residents support the tax increase.

(B) There is an 18% chance that the majority of residents support the tax increase.

(C) Assuming that 50% of residents support the tax increase, there is an 18% probability that the sample proportion would be 0.527 or greater by chance alone.

(D) Assuming that more than 50% of residents support the tax increase, there is an 18% probability that the sample proportion would be 0.527 or greater by chance alone.

(E) Assuming that 50% of residents support the tax increase, there is an 18% chance that the null hypothesis is true by chance alone.

20. Based on the *P*-value in Exercise 19, which of the following would be the most appropriate conclusion?

(A) Because the *P*-value is large, we reject H_0. There is convincing statistical evidence that more than 50% of city residents support the tax increase.

(B) Because the *P*-value is large, we fail to reject H_0. There is convincing statistical evidence that more than 50% of city residents support the tax increase.

(C) Because the *P*-value is large, we reject H_0. There is convincing statistical evidence that at most 50% of city residents support the tax increase.

(D) Because the *P*-value is large, we fail to reject H_0. There is convincing statistical evidence that at most 50% of city residents support the tax increase.

(E) Because the *P*-value is large, we fail to reject H_0. There is not convincing statistical evidence that more than 50% of city residents support the tax increase.

Recycle and Review *Practice what you learned in previous sections.*

21. **Kickstarter (5C)** The fundraising site Kickstarter regularly tracks the success rate of projects that seek funding on its site. Recently, the percentage of projects that were successfully funded was 37.5%.[27] You plan to select a random sample of 50 Kickstarter projects. Let $\hat{p}$ be the proportion of projects in the sample that were successfully funded.

(a) Describe the shape, center, and variability of the sampling distribution of $\hat{p}$.

(b) What is the probability that fewer than 30% of the projects were successfully funded?

22. **Explaining confidence (6A)** Here is an explanation from a newspaper concerning one of its opinion polls. Explain what is wrong with the following statement.

For a poll of 1600 adults, the variation due to sampling error is no more than 3 percentage points either way. The error margin is said to be valid at the 95% confidence level. This means that, if the same questions were repeated in 20 polls, the results of at least 19 surveys would be within 3 percentage points of the results of this survey.

SECTION 6D Significance Tests for a Population Proportion

LEARNING TARGETS *By the end of the section, you should be able to:*

- Check the conditions for performing a test about a population proportion.
- Calculate the standardized test statistic and *P*-value for a test about a population proportion.
- Perform a one-sample *z* test for a proportion.
- Interpret a Type I error and a Type II error in context and give a consequence of each type of error.
- Interpret the power of a significance test and describe which factors affect the power of a test.

In Section 6C, we met a basketball player who claimed to be an 80% free-throw shooter. But in a random sample of 50 free throws, the player made only 32/50 = 0.64 of the shots — much lower than the percentage that the free-throw shooter claimed to make. Does this poor performance provide *convincing* evidence against the player's claim? To find out, we need to perform a significance test of

$$H_0: p = 0.80$$
$$H_a: p < 0.80$$

where p = the proportion of free throws that the shooter makes in the long run.

In this section, you'll learn how to do a complete significance test for a population proportion. You'll also learn about the two types of errors we can make in a significance test, and how to reduce their likelihood.

Checking Conditions for a Test about p

In Section 6B, we introduced three conditions that should be met before we construct a confidence interval for a population proportion p. We called them the Random, 10%, and Large Counts conditions. These same conditions must be verified before carrying out a significance test. Recall that the purpose of these conditions is to ensure that the observations in the sample can be viewed as independent and that the sampling distribution of $\hat{p}$ is approximately normal.

In Unit 5, you learned that the Large Counts condition for proportions requires that both np and $n(1-p)$ are at least 10. When constructing a confidence interval for p in Section 6B, we used the sample proportion $\hat{p}$ in place of the unknown p to check the Large Counts condition. In this section, however, we use the parameter value specified by the null hypothesis (denoted p_0) when checking the Large Counts condition. We use p_0 instead of $\hat{p}$ because we are assuming the null hypothesis is true when performing the test.

CONDITIONS FOR PERFORMING A SIGNIFICANCE TEST ABOUT A POPULATION PROPORTION

- **Random:** The data come from a random sample from the population of interest.
  - **10%:** When sampling without replacement, $n < 0.10N$.
- **Large Counts:** Both np_0 and $n(1-p_0)$ are at least 10.

If the data come from a convenience sample or a voluntary response sample, there's no point in carrying out a significance test for p. The results will be invalid no matter what the *P*-value is. The same is true if there are other sources of bias during data collection. If the Large Counts condition is violated, a *P*-value calculated from a normal distribution will not be accurate. And if the 10% condition is violated, our estimate of the standard deviation of the sampling distribution of $\hat{p}$ will be too large, resulting in larger *P*-values.

Let's verify that the conditions are met for performing a significance test of the basketball player's claim that $p = 0.80$:

- Random: The 50 shots can be viewed as a random sample from the population of all possible shots that the shooter takes. ✓
  - 10%: We're not sampling without replacement from a finite population (because the player can keep on shooting), so we don't need to check the 10% condition.
- Large Counts: Assuming H_0 is true, $p = 0.80$ Then $np_0 = (50)(0.80) = 40$ and $n(1-p_0) = (50)(0.20) = 10$ are both at least 10. ✓

In Section 6B, the values of $n\hat{p}$ and $n(1-\hat{p})$ represent the *observed* numbers of successes and failures. In this section, the values $np_0 = 40$ and $n(1-p_0) = 10$ represent the *expected* numbers of successes and failures, assuming the null hypothesis is true.

EXAMPLE

Get a job!
Checking conditions for a test about *p*

Skill 4.C

ArtWell/Shutterstock

PROBLEM: According to the U.S. Census Bureau, the proportion of high school students who have a part-time job is 0.25. An administrator at a large high school suspects that the proportion of students at the school who have a part-time job is less than the national figure and would like to carry out a test at the $\alpha = 0.05$ significance level of

$$H_0: p = 0.25$$
$$H_a: p < 0.25$$

where $p =$ the population proportion of students at the school who have a part-time job. The administrator selects a random sample of 200 students from the school and finds that 43 of them have a part-time job. Check if the conditions for performing the significance test are met.

SOLUTION:

- Random: Random sample of 200 students from the school. ✓
  - 10%: It is reasonable to assume that 200 is less than 10% of all students at a large high school. ✓
- Large Counts: $np_0 = 200(0.25) = 50 \geq 10$ and $n(1 - p_0) = 200(1 - 0.25) = 150 \geq 10$ ✓

Be sure to use p_0 — not $\hat{p}$ — when checking the Large Counts condition!

FOR PRACTICE, TRY EXERCISE 1

Calculating the Standardized Test Statistic and *P*-Value for a Test about *p*

Viorika/Getty Images

In the free-throw shooter example context, the sample proportion of made shots is $\hat{p} = 32/50 = 0.64$. Because this result is less than 0.80, there is *some* evidence against $H_0: p = 0.80$ and in favor of $H_a: p < 0.80$. But do we have *convincing* evidence that the player is exaggerating? To answer this question, we have to know how likely it is to get a sample proportion of 0.64 or less by chance alone when the null hypothesis is true. In other words, we are looking for a *P*-value.

Suppose for now that the null hypothesis $H_0: p = 0.80$ is true. Consider the sample proportion $\hat{p}$ of made free throws in a random sample of size $n = 50$. You learned in Section 5C that the sampling distribution of $\hat{p}$ will have mean

$$\mu_{\hat{p}} = p = 0.80$$

and standard deviation

$$\sigma_{\hat{p}} = \sqrt{\frac{p(1-p)}{n}} = \sqrt{\frac{0.80(0.20)}{50}} = 0.0566$$

Because the Large Counts condition is met, the sampling distribution of $\hat{p}$ will be approximately normal. Figure 6.5 displays this distribution. We have added the player's sample result, $\hat{p} = 32/50 = 0.64$.

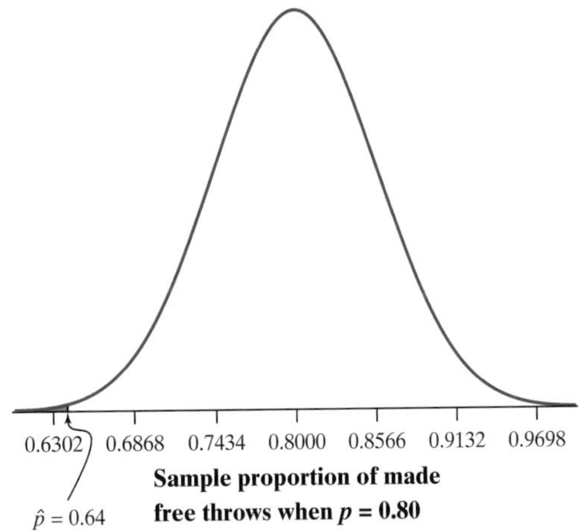

FIGURE 6.5 Normal distribution that models the sampling distribution of the sample proportion of made shots in random samples of 50 free throws by an 80% shooter.

To assess how far the statistic ($\hat{p} = 0.64$) is from the null value of the parameter ($p_0 = 0.80$), we standardize the statistic:

$$z = \frac{\hat{p} - p_0}{\sqrt{\frac{p_0(1-p_0)}{n}}} = \frac{0.64 - 0.80}{0.0566} = -2.827$$

The resulting value is called the **standardized test statistic.**

DEFINITION Standardized test statistic

A **standardized test statistic** measures how far a sample statistic is from what we would expect if the null hypothesis H_0 were true, in standard deviation units. That is,

$$\text{standardized test statistic} = \frac{\text{statistic} - \text{parameter}}{\text{standard error of statistic}}$$

AP® EXAM TIP
Formula sheet

The formula for the standardized test statistic is included on the formula sheet provided on both sections of the AP® Statistics exam. In most cases, we'll be using sample data to estimate the standard deviation of the relevant sampling distribution, which is why the formula sheet uses "standard error" instead of "standard deviation" in the denominator. The only exception is in a test about a population proportion, where we calculate the standard deviation using the hypothesized value of the proportion p_0, not the sample proportion $\hat{p}$. A footnote on the second page of the formula sheet addresses this special case. As you learned in Unit 5, the formula for the standard deviation of $\hat{p}$ is on the formula sheet as well.

The standardized test statistic says how far the sample result is from the null value, and in which direction, on a standardized scale. In this case, the sample proportion $\hat{p} = 0.64$ of made free throws is 2.827 standard deviations less than the null value of $p = 0.80$.

You can use the standardized test statistic to find the *P*-value for a significance test. In this case, the *P*-value is the probability of getting a sample proportion less than or equal to $\hat{p} = 0.64$ by chance alone when H_0: $p = 0.80$ is true. The shaded area in Figure 6.6(a) shows this probability. Figure 6.6(b) shows the corresponding area to the left of $z = -2.827$ in the standard normal distribution.

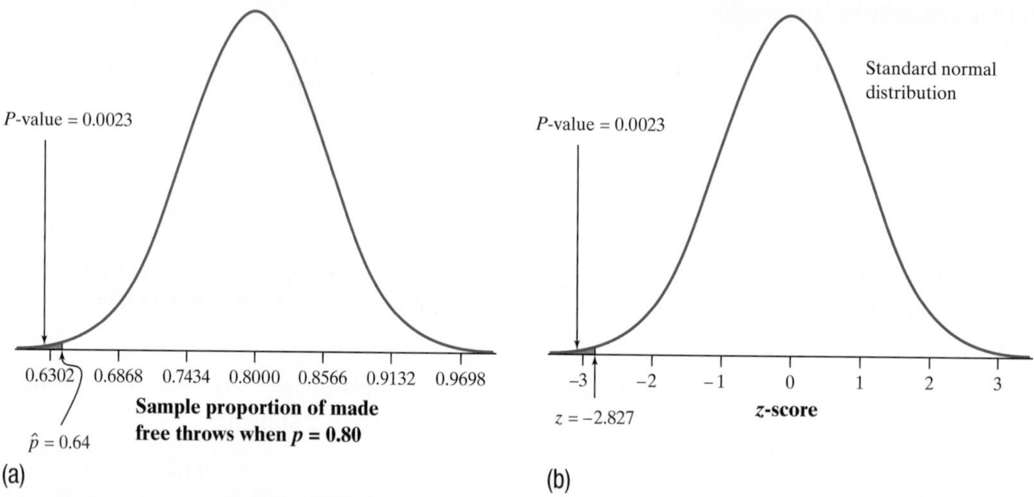

FIGURE 6.6 The shaded area shows the *P*-value for the player's sample proportion of made free throws (a) on the normal distribution that models the sampling distribution of $\hat{p}$ from Figure 6.5 and (b) on the standard normal curve.

We can calculate the *P*-value using technology or Table A. The TI-83/84 command normalcdf(lower: -1000, upper: -2.827, mean: 0, SD: 1) gives a *P*-value of 0.0023. Table A gives $P(z \leq -2.83) = 0.0023$. Remember that *P*-value calculations are valid only when our probability model is true — that is, when the conditions for inference are met.

If H_0 is true and the player makes 80% of their free throws in the long run, there's only about a 0.0023 probability that the player would make 32 or fewer of 50 shots by chance alone. This is even smaller than our estimated *P*-value of 0.0075 from the simulation in Section 6C. This small probability confirms our earlier decision to reject H_0 and gives convincing evidence that the player is exaggerating.

CALCULATING THE STANDARDIZED TEST STATISTIC AND *P*-VALUE IN A TEST ABOUT A POPULATION PROPORTION

Suppose the conditions are met. To perform a test of H_0: $p = p_0$, calculate the standardized test statistic

$$z = \frac{\hat{p} - p_0}{\sqrt{\dfrac{p_0(1-p_0)}{n}}}$$

Find the *P*-value by calculating the probability of getting a z statistic this large or larger in the direction specified by the alternative hypothesis H_a in the standard normal distribution.

Because some types of significance tests don't use a *standardized* test statistic, some people use the more general term "test statistic" to describe the value of z in a significance test about a population proportion.

EXAMPLE

Part-time jobs
Calculating the standardized test statistic and *P*-value for a test about *p*

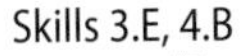
Skills 3.E, 4.B

PROBLEM: In the preceding example, an administrator at a large high school decided to perform a test at the $\alpha = 0.05$ significance level of

$$H_0: p = 0.25$$
$$H_a: p < 0.25$$

where p = the population proportion of students at the school who have a part-time job. The administrator selects a random sample of 200 students from the school and finds that 43 of them have a part-time job. We already confirmed that the conditions for performing a significance test are met.

Andersen Ross/Exactostock-1555/Superstock

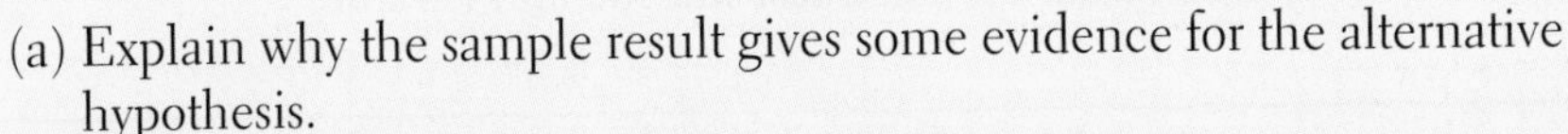
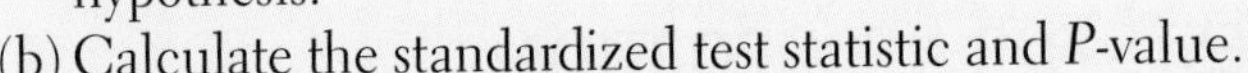

(a) Explain why the sample result gives some evidence for the alternative hypothesis.
(b) Calculate the standardized test statistic and *P*-value.
(c) Interpret the *P*-value.

SOLUTION:

(a) The sample proportion of students at this school with a part-time job is $\hat{p} = 43/200 = 0.215$, which is less than the national proportion of $p = 0.25$ (as suggested by H_a).

(b) $$z = \frac{0.215 - 0.25}{\sqrt{\frac{0.25(1-0.25)}{200}}} = -1.143$$

$$\text{standardized test statistic} = \frac{\text{statistic} - \text{parameter}}{\text{standard error of statistic}}$$

$$z = \frac{\hat{p} - p_0}{\sqrt{\frac{p_0(1-p_0)}{n}}}$$

P-value:

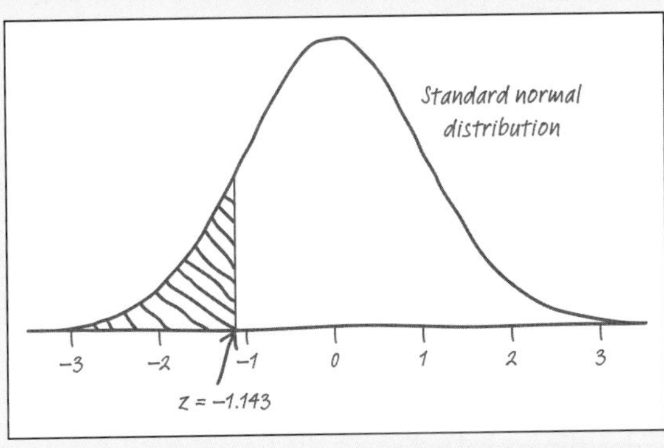

Using technology: normalcdf(lower: −1000, upper: −1.143, mean: 0, SD: 1) = 0.1265
Using Table A: 0.1271

(c) Assuming that 25% of all students at the school have a part-time job, there is a 0.1265 probability of getting a sample proportion of 0.215 or less by chance alone.

FOR PRACTICE, TRY EXERCISE 5

Remember that there are two possible explanations for why the sample proportion of students who have part-time jobs ($\hat{p} = 43/200 = 0.215$) in the example is less than $p = 0.25$. The first explanation is that the population proportion of students at the school who have part-time jobs is 0.25 and that we got a sample proportion this small due to sampling variability. The second explanation is that the population proportion of students at the school with part-time jobs is less than 0.25. We cannot rule out the first explanation because the *P*-value of 0.1265 isn't less than $\alpha = 0.05$. The administrator does not have convincing evidence that the population proportion of all students at the school with a part-time job is less than the national proportion of $p = 0.25$.

AP® EXAM TIP

Notice that we did not include an option (ii) to "Use technology to find the desired area without standardizing" when performing the normal calculation in part (b) of the example. That's because you are *always* required to give the standardized test statistic along with the *P*-value when performing a significance test on the AP® Statistics exam.

As with a confidence interval for a population proportion, you can use your calculator to compute the standardized test statistic and *P*-value.

17. Tech Corner SIGNIFICANCE TESTS FOR A PROPORTION

TI-Nspire and other technology instructions are on the book's website at bfwpub.com/tps7e.

You can use a TI-83/84 calculator to perform the calculations for a significance test about a population proportion. We'll demonstrate using the preceding example. In a random sample of size $n = 200$, the administrator found 43 students with a part-time job. To perform a significance test:

1. Press STAT, then choose TESTS and 1-PropZTest.
2. On the 1-PropZTest screen, enter the values $p_0 = 0.25$, x = 43, and n = 200. Specify the alternative hypothesis as "prop < p_0." *Note:* Here, x is the *number* of successes and n is the sample size. Both must be whole numbers or the calculator will give an error!

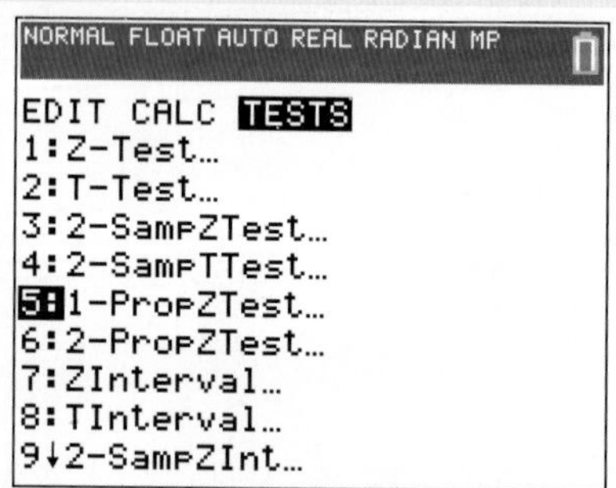

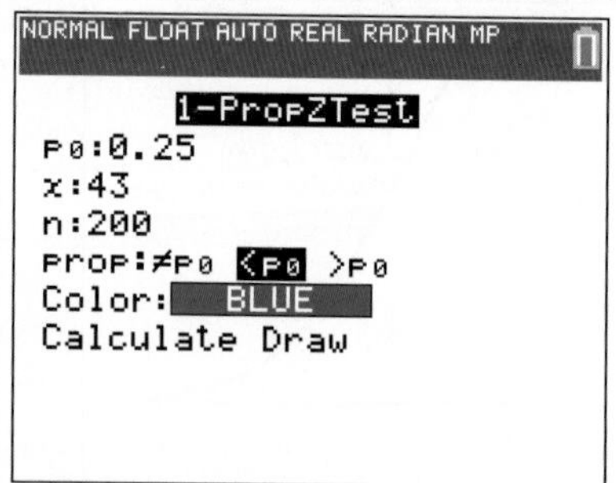

3. Highlight "Calculate" and press ENTER. The output includes the standardized test statistic and *P*-value, along with the sample proportion and sample size.

Note: If you select the "Draw" option, you will see a picture of the standard normal distribution with the area of interest shaded, the value of the standardized test statistic, and the *P*-value.

Test statistics and *P*-values calculated using the "full technology" approach described in the Tech Corner are generally more accurate than the values we obtain when we round while doing our calculations. For that reason, we will use the values from the full technology approach when reporting results "Using technology" in the rest of this section.

TWO-SIDED TESTS

The free-throw shooter and part-time job examples involved one-sided tests. The *P*-value in a one-sided test about a population proportion is the area in one tail of a standard normal distribution — the tail specified by H_a. In a two-sided test, the alternative hypothesis has the form H_a: $p \neq p_0$. The *P*-value in such a test is the probability of getting a sample proportion as far as or farther from p_0 *in either direction* than the observed value of $\hat{p}$. As a result, you have to find the area in both tails of a standard normal distribution to get the *P*-value. For this reason, a two-sided test is sometimes called a *two-tailed* test.

According to the National Center for Health Statistics (NCHS), 36.6% of U.S. adults eat fast food on a given day. A public health director wants to know if this figure holds true for adult residents of a particular county. A random sample of 150 adult residents finds that 68 of residents ate fast food on the previous day.

Because the sample proportion of $\hat{p} = 68/150 = 0.453$ is different from 0.366, there is evidence for the alternative hypothesis. To see how likely it is to get a sample proportion this different or more different than 0.366 by chance alone, we need to find $P(\hat{p} \geq 0.453) + P(\hat{p} \leq 0.279)$. Where did the 0.279 come from? Because 0.453 is 0.087 greater than 0.366, we also need to find the area less than or equal to $0.366 - 0.087 = 0.279$. The shaded area in Figure 6.7(a) shows this probability.

A test of H_0: $p = 0.366$ versus H_a: $p \neq 0.366$ with $\hat{p} = 68/150 = 0.453$ gives a standardized test statistic of

$$z = \frac{0.453 - 0.366}{\sqrt{\dfrac{0.366(1 - 0.366)}{150}}} = 2.212$$

The shaded area in Figure 6.7(b) is the *P*-value we are seeking.

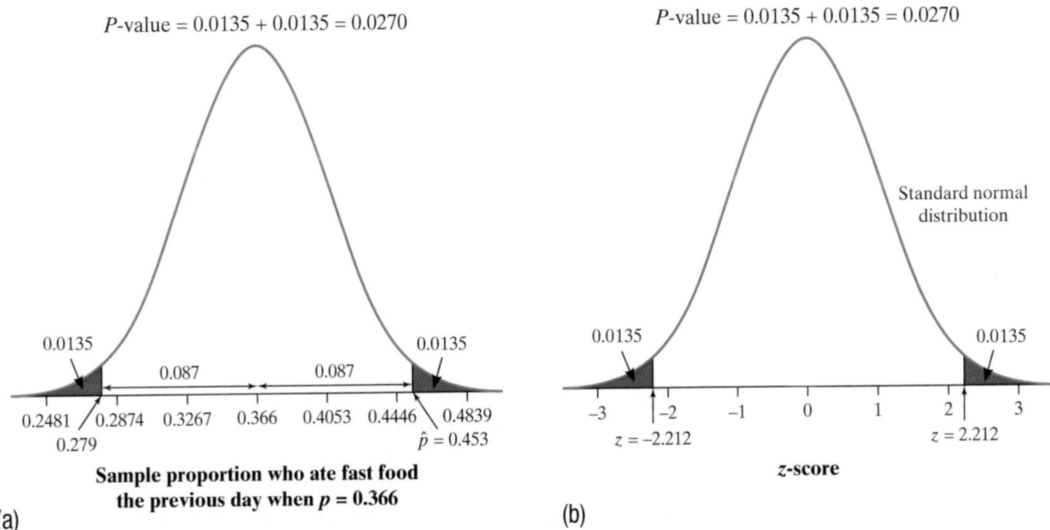

FIGURE 6.7 The shaded area shows the *P*-value for a two-sided test about a population proportion (a) on a normal curve that models the sampling distribution of $\hat{p}$ and (b) on the standard normal distribution.

Because of the symmetry of the standard normal curve, we can find the area in one tail and multiply it by 2 to get the *P*-value. Using technology, $P(z \geq 2.212) = 0.0135$, so the *P*-value is $2(0.0135) = 0.0270$. Assuming 36.6% of all adult residents in the county eat fast food on a given day, there is a 0.0270 probability of getting a sample proportion of at least 0.453 or at most 0.279 by chance alone. Because this *P*-value is less than $\alpha = 0.05$, we reject H_0. There is convincing evidence that the proportion of all adult county residents who eat fast food on a given day is different from the national proportion of 0.366.

CONFIDENCE INTERVALS GIVE MORE INFORMATION Knowing that the proportion of all adult county residents who eat fast food on a given day is different from 0.366 is interesting, but it doesn't give much information about the population proportion to the county health director. A confidence interval can help in situations like this. Using the method from Section 6B, we find that a 95% confidence interval for the proportion of all adult county residents who eat fast food on a given day is

$$0.453 \pm 1.960\sqrt{\frac{0.453(1-0.453)}{150}} = 0.453 \pm 0.080 = (0.373, 0.533)$$

The county health director should not be surprised if the proportion of all adult county residents who eat fast food on a given day is any value between 0.373 and 0.533. It would be surprising if the population proportion were any value outside this interval.

There is a link between confidence intervals and *two-sided* tests. The 95% confidence interval (0.373, 0.533) gives an approximate set of p_0's that should not be rejected by a two-sided test at the $\alpha = 0.05$ significance level. Any value of p outside this interval, including $p = 0.366$, should be rejected as implausible.

With proportions, the link isn't perfect because the standard error used for the confidence interval is based on the sample proportion $\hat{p}$ while the denominator of the standardized test statistic is based on the value p_0 from the null hypothesis.

$$\text{Standardized test statistic: } z = \frac{\hat{p} - p_0}{\sqrt{\frac{p_0(1-p_0)}{n}}}$$

$$\text{Confidence interval: } \hat{p} \pm z^*\sqrt{\frac{\hat{p}(1-\hat{p})}{n}}$$

But the big idea is still worth considering: a two-sided test at significance level α and a $100(1-\alpha)\%$ confidence interval (e.g., a 95% confidence interval if $\alpha = 0.05$) give similar information about the population parameter. There is a connection between *one*-sided tests and confidence intervals, but it is beyond the scope of this course.

CHECK YOUR UNDERSTANDING

According to the National Institute for Occupational Safety and Health, job stress poses a major threat to the health of workers. A news report claims that 75% of restaurant employees feel that work stress has a negative impact on their personal lives.[28]

The board of directors of a restaurant chain with thousands of employees wonder whether this claim is true for their employees. A random sample of 100 employees finds that 68 answer "Yes" when asked, "Does work stress have a negative impact on your personal life?"

1. State the hypotheses the board is interested in testing.
2. Check the conditions for a test about the population proportion.
3. Calculate the standardized test statistic and *P*-value.
4. What conclusion should the board make?

Putting It All Together: One-Sample *z* Test for *p*

To perform a significance test, we state hypotheses, check conditions, calculate a standardized test statistic and *P*-value, and make a conclusion in the context of the problem. The four-step process is ideal for organizing our work.

SIGNIFICANCE TESTS: A FOUR-STEP PROCESS

State: State the hypotheses, parameter(s), and significance level.

Plan: Identify the appropriate inference method and check the conditions.

Do: If the conditions are met, perform calculations.

- Calculate the test statistic.
- Find the *P*-value.

Conclude: Make a conclusion about the hypotheses in the context of the problem.

The appropriate inference method for testing a claim about p is called a **one-sample z test for a proportion**.

DEFINITION One-sample *z* test for a proportion

A **one-sample *z* test for a proportion** is a significance test of the null hypothesis that a population proportion p is equal to a specified value.

AP® EXAM TIP

You can use your calculator to carry out the mechanics of a significance test on the AP® Statistics exam. But there's a risk involved. If you give just the calculator answer with no work, and one or more of your values are incorrect, you will probably get no credit for the "Do" step. If you opt for the calculator-only method, be sure to complete the other three steps (State, Plan, Conclude) and report the standardized test statistic and *P*-value in the Do step.

Here is an example of the one-sample z test for a population proportion.

EXAMPLE One potato, two potato
One-sample z test for p

Skills 1.E, 1.F, 3.E, 4.C, 4.E

imagestock/Getty Images

PROBLEM: A potato chip producer and its main supplier agree that each shipment of potatoes must meet certain quality standards. If the producer finds convincing evidence that more than 8% of the potatoes in the shipment have "blemishes," the truck will be sent away to get another load of potatoes from the supplier. Otherwise, the entire truckload will be used to make potato chips.

The potato chip producer has just received a truckload of potatoes from the supplier. A supervisor selects a random sample of 500 potatoes from the truck. An inspection reveals that 47 of the potatoes have blemishes. Is there convincing evidence at the $\alpha = 0.10$ level that more than 8% of the potatoes in the shipment have blemishes?

SOLUTION:

Follow the four-step process!

STATE: We want to test

$H_0: p = 0.08$

$H_a: p > 0.08$

where $p =$ the proportion of all potatoes in this shipment with blemishes, using $\alpha = 0.10$.

State: State the hypotheses, parameters, and significance level.

PLAN: One-sample z test for a proportion

- Random: Random sample of 500 potatoes from the shipment. ✓
  - 10%: It's reasonable to assume that $500 < 10\%$ of all potatoes in the shipment. ✓
- Large Counts: $500(0.08) = 40 \geq 10$ and $500(0.92) = 460 \geq 10$ ✓

Plan: Identify the appropriate inference method and check the conditions.

Remember to use the null value p_0 when checking the Large Counts condition, not $\hat{p}$.

DO: $\hat{p} = 47/500 = 0.094$

- $z = \dfrac{0.094 - 0.08}{\sqrt{\dfrac{0.08(0.92)}{500}}} = 1.154$
- P-value:

Do: If the conditions are met, perform calculations.
- Calculate the test statistic.
- Find the P-value.

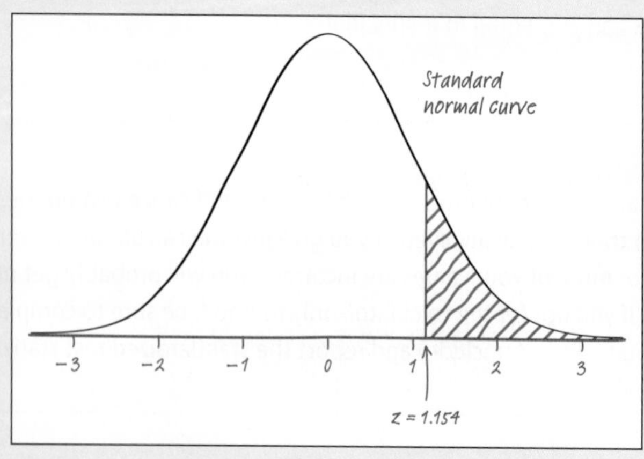

The sample result gives *some* evidence for H_a because $\hat{p} = 47/500 = 0.094 > 0.08$.

Using technology: $z = 1.154$, *P*-value $= 0.1243$

Using Table A: $1 - 0.8749 = 0.1251$

CONCLUDE: Because the *P*-value of $0.1243 > \alpha = 0.10$, we fail to reject H_0. There is not convincing evidence that the proportion of all potatoes in this shipment with blemishes is greater than 0.08.

Conclude: Make a conclusion about the hypotheses in the context of the problem.

FOR PRACTICE, TRY EXERCISE 11

AP® EXAM TIP

When a significance test leads to a fail-to-reject-H_0 decision, as in the preceding example, be sure to interpret the results as "We don't have convincing evidence for H_a." Saying anything that sounds like you believe H_0 is (or might be) true will lead to a loss of credit. For instance, it would be *wrong* to conclude, "There is convincing evidence that the population proportion of blemished potatoes is 0.08."

The preceding example reminds us why significance tests are important. The sample proportion of blemished potatoes was $\hat{p} = 47/500 = 0.094$. This result gave some evidence against H_0 and in favor of H_a. To see whether such an outcome is unlikely to occur by chance alone when H_0 is true, we had to carry out a significance test. The *P*-value told us that a sample proportion this large or larger would occur in about 12% of all random samples of 500 potatoes when H_0 is true. So we can't rule out sampling variability as a plausible explanation for getting a sample proportion of $\hat{p} = 0.094$.

WHAT HAPPENS WHEN THE DATA DON'T SUPPORT H_a?

Suppose the supervisor had inspected a random sample of 500 potatoes from the shipment and found 33 with blemishes. This yields a sample proportion of $\hat{p} = 33/500 = 0.066$. This sample doesn't give *any* evidence to support the alternative hypothesis H_a: $p > 0.08$! Don't continue with the significance test. The conclusion is clear: we should fail to reject H_0: $p = 0.08$. This truckload of potatoes will be used by the potato chip producer.

If you weren't paying attention, you might end up carrying out the test. Let's see what would happen. The corresponding standardized test statistic is

$$z = \frac{\hat{p} - p_0}{\sqrt{\frac{p_0(1-p_0)}{n}}} = \frac{0.066 - 0.08}{\sqrt{\frac{0.08(0.92)}{500}}} = -1.154$$

What's the *P*-value? It's the probability of getting a z statistic this large or larger in the direction specified by H_a, $P(z \geq -1.154)$. Figure 6.8 shows this *P*-value as an area under the standard normal curve. Using technology, the *P*-value is 0.8757. There's about an 88% chance of getting a sample proportion as large as or larger than $\hat{p} = 0.066$ if $p = 0.08$. As a result, we would fail to reject H_0. Same conclusion, but with lots of unnecessary work!

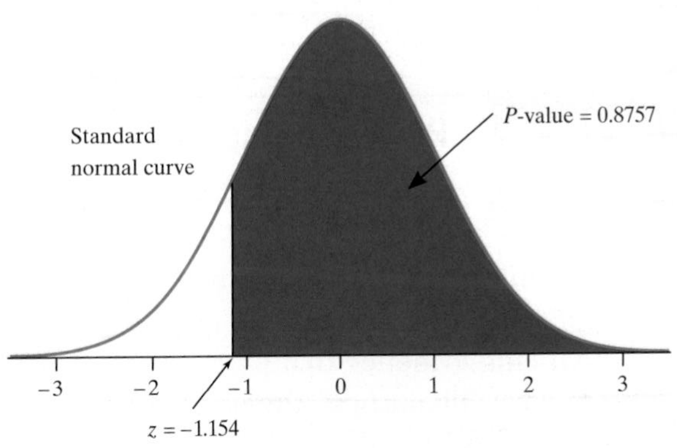

FIGURE 6.8 The *P*-value for the one-sided test of H_0: $p = 0.08$ versus H_a: $p > 0.08$ with $\hat{p} = 0.066$.

Always check to see whether the data give evidence against H_0 in the direction specified by H_a before you do calculations.

Type I and Type II Errors

When we make a conclusion from a significance test based on sample data, we hope our conclusion will be correct. But sometimes the data may lead us to an incorrect conclusion. There are two types of mistakes we can make: a **Type I error** or a **Type II error.**

DEFINITION Type I error, Type II error

A **Type I error** occurs if we reject H_0 when H_0 is true. That is, the data give convincing evidence that H_a is true when it really isn't.

A **Type II error** occurs if we fail to reject H_0 when H_a is true. That is, the data do not give convincing evidence that H_a is true when it really is.

The truth about the population is that either H_0 is true or H_a is true. It must be one or the other!

- If H_0 is true:
  - Our conclusion is correct if we don't find convincing evidence that H_a is true.
  - We make a Type I error if we find convincing evidence that H_a is true.
- If H_a is true:
  - Our conclusion is correct if we find convincing evidence that H_a is true.
  - We make a Type II error if we do not find convincing evidence that H_a is true.

Only one error is possible at a time, depending on the conclusion we make. To keep the two types of errors straight, remember that "fail to" goes with Type II. The possible outcomes of a significance test are summarized in Figure 6.9.

FIGURE 6.9 The two types of errors in significance tests.

		Truth about the population	
		H_0 true	H_a true
Conclusion based on sample	Find convincing evidence for H_a	Type I error	Correct conclusion
	Do not find convincing evidence for H_a	Correct conclusion	Type II error

It is important to be able to describe Type I and Type II errors in the context of a problem. Considering the consequences of each of these types of error is also important, as the following example shows.

EXAMPLE

Perfect potatoes
Type I and Type II errors

Skills 1.B, 4.B

Steve Cukrov/Alamy Stock Photo

PROBLEM: A potato chip producer and its main supplier agree that each shipment of potatoes must meet certain quality standards. If the producer finds convincing evidence that more than 8% of the potatoes in the shipment have "blemishes," the truck will be sent away to get another load of potatoes from the supplier. Otherwise, the entire truckload will be used to make potato chips. To make the decision, a supervisor will inspect a random sample of 500 potatoes from the shipment. The producer will then perform a test at the $\alpha = 0.05$ significance level of

$$H_0: p = 0.08$$
$$H_a: p > 0.08$$

where $p =$ the proportion of all potatoes in the shipment with blemishes. Describe a Type I error and a Type II error in this setting, and give a possible consequence of each.

SOLUTION:

Type I error: The supervisor finds convincing evidence that more than 8% of the potatoes in the shipment have blemishes, when the population proportion is really 0.08.

Consequence: The potato chip producer sends away the truckload of acceptable potatoes, wasting time and depriving the supplier of money.

Type II error: The supervisor does not find convincing evidence that more than 8% of the potatoes in the shipment have blemishes, when the population proportion is greater than 0.08.

Consequence: More potato chips are made with blemished potatoes, which may upset customers and lead to decreased sales.

FOR PRACTICE, TRY EXERCISE 19

Which is more serious: a Type I error or a Type II error? That depends on the situation. For the potato chip producer, a Type II error seems more serious because it may lead to lower-quality potato chips and decreased sales.

The most common significance levels are $\alpha = 0.05$, $\alpha = 0.01$, and $\alpha = 0.10$. Which is the best choice for a given significance test? That depends on whether a Type I error or a Type II error is more serious.

In the Perfect Potatoes example, a Type I error occurs if the population proportion of blemished potatoes in a shipment is $p = 0.08$, but we get a value of the sample proportion $\hat{p}$ large enough to yield a *P*-value less than $\alpha = 0.05$. When H_0 is true, this will happen 5% of the time just by chance. In other words, $P(\text{Type I error}) = \alpha$.

TYPE I ERROR PROBABILITY

The probability of making a Type I error in a significance test is equal to the significance level α.

We can decrease the probability of making a Type I error in a significance test by using a smaller significance level. For instance, the potato chip producer could use $\alpha = 0.01$ instead of $\alpha = 0.05$. But there is a trade-off between $P(\text{Type I error})$ and $P(\text{Type II error})$: as one increases, the other decreases, assuming everything else remains the same. If we make it more difficult to reject H_0 by decreasing α, we increase the probability that we will not find convincing evidence for H_a when it is really true. That's why it is important to consider the possible consequences of each type of error before choosing a significance level.

CHECK YOUR UNDERSTANDING

A June 2020 report says that 78% of U.S. students in grades K–12 have reliable internet access at home.[29] Researchers believe the proportion is smaller in their large, rural school district. To investigate, they choose a random sample of 120 K–12 students in the school district and find that 85 have reliable internet access at home.

1. Do these data provide convincing evidence that the proportion of students in this school district who have reliable internet access at home is less than the national figure?
2. Describe a Type I error and a Type II error in this context, along with a consequence of each error.

The Power of a Test

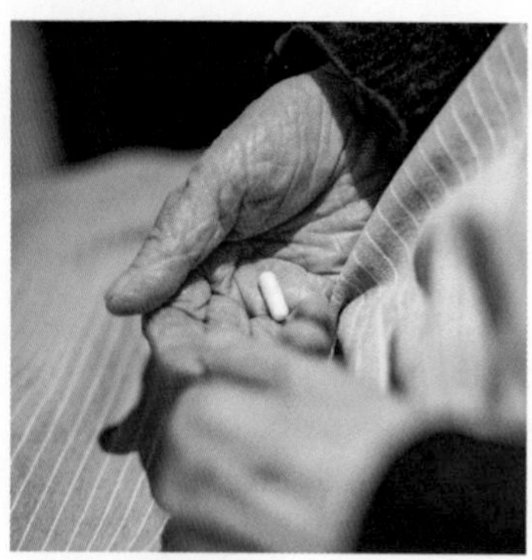

Ma yichao/Moment/Getty Images

Researchers often perform a significance test in hopes of finding convincing evidence *for* the alternative hypothesis. Why? Because H_a states the claim about the population parameter that they believe, suspect, worry, or hope is true. For instance, a drug manufacturer claims that fewer than 10% of patients who take its new drug for treating Alzheimer's disease will experience nausea. To test this claim, researchers want to carry out a test of

$$H_0: p = 0.10$$
$$H_a: p < 0.10$$

where $p =$ the true proportion of patients like the ones in the study who would experience nausea when taking the new Alzheimer's drug. They plan to give the new drug to a random sample of patients with Alzheimer's disease whose families have given informed consent for the patients to participate in the study.

Suppose that the true proportion of patients like the ones in the Alzheimer's drug study who would experience nausea after taking the new drug is $p = 0.08$. This means that the alternative hypothesis H_a: $p < 0.10$ is true. Researchers would make a correct conclusion if they found convincing evidence that $p < 0.10$. But if they failed to find convincing evidence that $p < 0.10$, they would make a Type II error. How likely is the significance test to *avoid* a Type II error in this case? We refer to this probability as the **power** of the test.

DEFINITION Power

The **power** of a test is the probability that the test will find convincing evidence for H_a when a specific alternative value of the parameter is true.

Similar to a *P*-value, the power of a test is a *conditional* probability: power $=$ $P(\text{reject } H_0 \mid \text{parameter} = \text{some specific alternative value})$. In other words, power is the probability that we find convincing evidence the alternative hypothesis is true, given that the alternative hypothesis really is true. To interpret the power of a test in a given setting, just interpret the relevant conditional probability. Start by assuming that a specific alternative parameter value is true. That's quite different from interpreting a *P*-value, when we start by assuming that the null hypothesis is true.

Let's return to the Alzheimer's drug study. Suppose the researchers decide to perform a test of H_0: $p = 0.10$ versus H_a: $p < 0.10$ at the $\alpha = 0.05$ significance level based on data from a random sample of 300 patients with Alzheimer's disease. Advanced calculations reveal that the power of this test to detect $p = 0.08$ is 0.29. *Interpretation:* If the true proportion of patients with Alzheimer's disease like these who would experience nausea when taking the new drug is $p = 0.08$, there is a 0.29 probability that the researchers will find convincing evidence for H_a: $p < 0.10$.

The researchers in this setting have a poor chance (power = 0.29) of rejecting H_0 if 8% of patients with Alzheimer's disease like the ones in this study would experience nausea — that is, $P(\text{reject } H_0 \mid p = 0.08) = 0.29$. What's the probability that the company makes a Type II error in this case? It's $P(\text{fail to reject } H_0 \mid p = 0.08) = 1 - 0.29 = 0.71$. We can generalize this relationship between the power of a significance test and the probability of a Type II error.

RELATING POWER AND TYPE II ERROR

The power of a test to detect a specific alternative parameter value is related to the probability of a Type II error for that alternative:

$$\text{power} = 1 - P(\text{Type II error}) \text{ and } P(\text{Type II error}) = 1 - \text{power}$$

What can researchers do to decrease the probability of making a Type II error and increase the power of the test? Here is an activity that will help you answer this question.

ACTIVITY Reducing nausea

In this activity, we will use an applet to investigate the factors that affect the power of a test. A drug manufacturer claims that fewer than 10% of patients who take its new drug for treating Alzheimer's disease will experience nausea. To test this claim, researchers want to carry out a test of

$$H_0: p = 0.10$$
$$H_a: p < 0.10$$

where p = the true proportion of patients like the ones in the study who would experience nausea when taking the new Alzheimer's drug. They plan to give the new drug to a random sample of 300 patients with Alzheimer's disease whose families have given informed consent for the patients to participate in the study.

1. Go to www.stapplet.com and launch the *Power* applet.
2. Select proportions for the type of test. Enter 0.10 for the null hypothesis value p_0; enter 0.08 for the true proportion p (indicating that 8% of all patients with Alzheimer's disease like the ones in this study would experience nausea when taking the drug); enter $p < p_0$ for the alternative hypothesis; enter 300 for the sample size n; and leave the significance level as $\alpha = 0.05$. Choose to calculate and plot α, the Type I error probability,

and then click "Calculate." Once the graphs appear, click the box to "Show rejection region."

- The dashed curve on the right shows the sampling distribution of the sample proportion $\hat{p}$ for random samples of size $n = 300$ when H_0: $p = 0.10$ is true. We refer to this as the *null distribution*. A value of $\hat{p}$ that falls along the horizontal axis within the green-shaded region (the rejection region) is far enough below 0.10 that we should reject H_0: $p = 0.10$. (You can hover your mouse over the green region to see that values of $\hat{p} < 0.072$ would lead to a "reject H_0" conclusion.) The corresponding area under the null distribution curve (the blue-shaded region) is equal to the significance level, $\alpha = 0.05$. Note that this is also the probability of making a Type I error — that is, rejecting H_0 when H_0 is true.
- The curve on the left shows the true sampling distribution of the sample proportion $\hat{p}$ for random samples of size $n = 300$ when $p = 0.08$. We refer to this as the *alternative distribution*. A value of $\hat{p}$ that falls along the horizontal axis within the green region is far enough below 0.10 that it would lead to a correct rejection of H_0: $p = 0.10$. In other words, the corresponding area under the alternative distribution curve represents the power of the test. A value of $\hat{p}$ that falls along the horizontal axis outside the green region in the alternative distribution would lead to a Type II error because it is not far enough below 0.10 to reject H_0: $p = 0.10$.

3. Use the applet to calculate and plot β, the probability of making a Type II error, using the same settings as in Step 2. The blue-shaded area represents this probability.
4. Use the applet to calculate and plot the power of the test using the same settings as in Step 2. Again, the blue-shaded area represents this probability.
5. How does increasing sample size affect power? Gradually increase n from 300 to 350 by clicking the up arrow in the "Sample size" box. How does the power change? Now type in $n = 100$ and click the "Calculate" button. Did the power increase or decrease? Explain why this makes sense.
6. How does the significance level affect power? Reset the sample size to $n = 300$ and click the "Calculate" button. Gradually decrease the significance level to $\alpha = 0.01$ by clicking the down arrow in the α box. How does this affect the power? Now change α to 0.10 and click the "Calculate" button. Did the power increase or decrease? Explain why this makes sense.
7. How does the difference between the null and alternative parameter value affect power? Reset the sample size to $n = 300$ and the significance level to $\alpha = 0.05$. Gradually decrease the true (alternative) proportion from 0.08 to 0.01 by clicking the down arrow in the "True proportion" box. How does this affect the power? Now change the true proportion to 0.09 and click the "Calculate" button. Did the power increase or decrease? Explain why this makes sense.
8. Summarize how the sample size, significance level, and difference between the null and alternative parameter values affect power.

As Step 5 of the activity confirms, we get better information about the true proportion of patients with Alzheimer's disease like these who would experience nausea when we increase the sample size. The power of the test increases from 0.29 to 0.33 when we increase the sample size from $n = 300$ to $n = 350$. Of course, increasing the sample size isn't free! There are real-world costs in terms of time and money for increasing the sample size.

In Step 6 of the activity, you learned that power increases when α increases. This makes sense, as it is easier to reject H_0 when α is larger, because the *P*-value doesn't need to be as small. The power increases from 0.29 to 0.44 when α increases from 0.05 to 0.10. Of course, increasing α in this way also increases the probability of a Type I error from 0.05 to 0.10. If a Type I error has serious consequences, this might not be a good plan.

Step 7 of the activity shows that it is easier to detect large differences between the null and alternative parameter values than to detect small differences. This difference is often referred to as the *effect size.* When $n = 300$ and $\alpha = 0.05$, the power of the test to detect $p = 0.08$ (0.02 from the null value of 0.10) is 0.29, whereas the power of the test to detect $p = 0.09$ (0.01 from the null value of 0.10) is only 0.13. Of course, researchers have little control over this difference. Presumably, they have already done the best job they could making the new drug nausea-free.

Figure 6.10 illustrates the connection between Type I and Type II error probabilities and the power of a test. The dashed curve shows the null distribution. Because $\alpha = 0.05$, the smallest 5% of the null distribution is shaded red to represent the probability of a Type I error. The value of α also determines the green-shaded rejection region. In this case, we would reject H_0 with any $\hat{p} < 0.072$.

The solid curve shows the alternative distribution. Any values under this curve in the rejection region leads to a correct rejection of H_0. Thus, the blue-shaded area represents power. Similarly, a value of $\hat{p}$ under this curve outside of the rejection region leads to a Type II error. This is represented by the purple area.

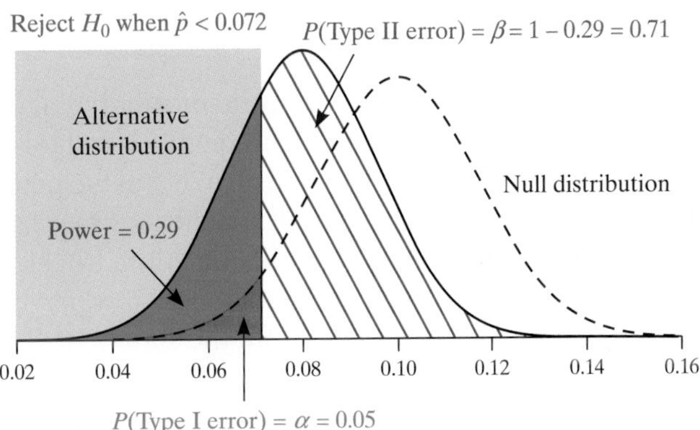

FIGURE 6.10 The connection between the rejection region (green area), Type I error (red area), Type II error (purple area), and power (blue area).

Assuming everything else stays constant:

- Increasing the Type I error probability α (red area) makes the boundary of the rejection region (green area) shift to the right, increasing the power (blue area) and decreasing the Type II error probability (purple area).
- Increasing the sample size makes both the null and alternative distributions less variable. Because α remains at 0.05, the boundary of the rejection region (green area) shifts to the right, increasing the power (blue area).

Simultaneously, the right tail of the alternative distribution doesn't extend as far to the right, also increasing the power. The probability of a Type II error (purple area) decreases.

- Increasing the distance between the null and alternative parameter values makes the center of the curves farther apart. This shifts the boundary of the rejection region (green area) to the right and increases the power (blue area). The probability of a Type II error (purple area) decreases.

In addition to manipulating these three factors, we can increase the power by decreasing the standard error of our test statistic. This is best accomplished by making wise choices when collecting data. In an experiment, using control to keep other variables constant will increase the power because there will be fewer sources of variability. Likewise, using blocking in an experiment or stratified random sampling can account for sources of variability, which increases the power.

INCREASING THE POWER OF A SIGNIFICANCE TEST

When H_0 is false and H_a is true, a significance test based on a random sample of size n and significance level α will have greater power and a smaller probability of making a Type II error when:

- The sample size n is larger.
- The significance level α is larger.
- The null and alternative parameter values are farther apart.
- The standard error of the statistic is smaller.

EXAMPLE

Powerful potatoes
The power of a test

Skills 4.A, 4.B

PROBLEM: A potato chip producer and its main supplier agree that each shipment of potatoes must meet certain quality standards. If the producer finds convincing evidence that more than 8% of the potatoes in the shipment have "blemishes," the truck will be sent away to get another load of potatoes from the supplier. Otherwise, the entire truckload will be used to make potato chips. To make the decision, a supervisor will inspect a random sample of 500 potatoes from the shipment. The producer will then perform a test at the $\alpha = 0.05$ significance level of

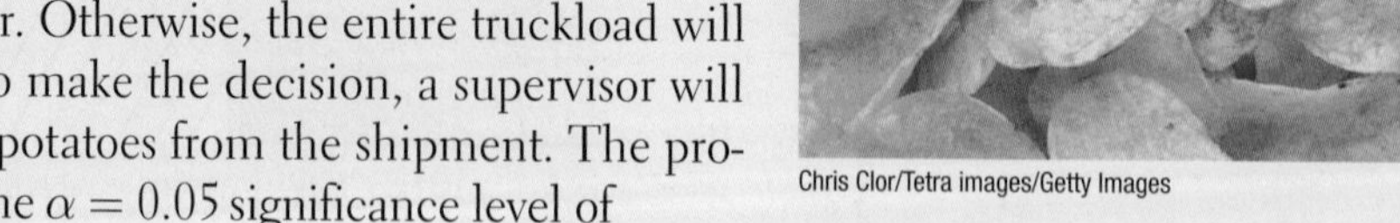

Chris Clor/Tetra images/Getty Images

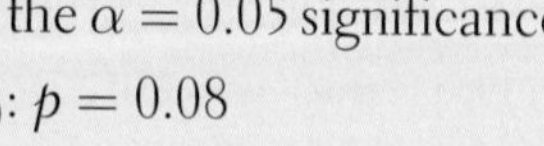

$$H_0: p = 0.08$$
$$H_a: p > 0.08$$

where p = the proportion of all potatoes in the shipment with blemishes.

(a) In one shipment, 11% of the potatoes have blemishes. In this case, the power of the producer's significance test is 0.764. Interpret this value.

(b) How will the power change if the producer uses $\alpha = 0.01$ instead of $\alpha = 0.05$? Explain your answer.

(c) How will the power change if the producer uses $n = 1000$ instead of $n = 500$? Explain your answer.

(d) In another shipment, only 10% of the potatoes have blemishes. Assuming the producer selects a random sample of $n = 500$ and uses $\alpha = 0.05$, how will the power compare to the shipment with 11% blemished potatoes? Explain your reasoning.

SOLUTION:

(a) If 11% of the potatoes in this shipment have blemishes, there is a 0.764 probability that the producer will find convincing evidence for H_a: $p > 0.08$.

(b) Decrease; decreasing the significance level from 0.05 to 0.01 makes it harder to reject H_0 when H_a is true.

(c) Increase; increasing the sample size from 500 to 1000 gives more information about the population proportion p.

(d) Decrease; it is harder to detect a smaller difference (0.02 versus 0.03) between the null and alternative parameter values.

FOR PRACTICE, TRY EXERCISE 25

In Section 6B, you learned how to calculate the sample size needed for a desired margin of error in a confidence interval for p. When researchers plan to use a significance test to analyze their results, they will often calculate the sample size needed for a desired power. For a specific study design, the sample size required depends on three factors:

1. *Significance level.* How much risk of a Type I error — rejecting the null hypothesis when H_0 is actually true — are we willing to accept? If a Type I error has serious consequences, we might opt for $\alpha = 0.01$. Otherwise, we should choose $\alpha = 0.05$ or $\alpha = 0.10$. Recall that using a higher significance level would decrease the Type II error probability and increase the power.
2. *Effect size.* How large a difference between the null parameter value and the actual parameter value is important for us to detect?
3. *Power.* What probability do we want our study to have to detect a difference of the size we think is important? Most researchers insist on a power of at least 0.80 for their significance tests.

Sometimes budget constraints get in the way of achieving high power. It can be expensive to collect data from a large enough sample of individuals to give a significance test the power that researchers desire.

CHECK YOUR UNDERSTANDING

Can a six-month exercise program increase the total body bone mineral content (TBBMC) of young adults? A team of researchers is planning a study to examine this question. The researchers would like to perform a test of

$$H_0: \mu = 0$$
$$H_a: \mu > 0$$

where μ is the true mean percent change in TBBMC during the exercise program for young adults like the ones in this study and $\alpha = 0.05$.

1. Describe a Type I error in this context. What is the probability of a Type I error?
2. The power of the test to detect a mean increase in TBBMC of 1% using $\alpha = 0.05$ and $n = 25$ subjects is 0.80. Interpret this value.
3. Describe a Type II error in this context. What is the probability of a Type II error?
4. Describe two ways that researchers could increase the power of the test.

SECTION 6D Summary

- Tests about the proportion p of successes in a population are based on the sample proportion $\hat{p}$.
- To perform a **one-sample z test for a proportion,** we need to verify that the observations in the sample can be viewed as independent and that the sampling distribution of $\hat{p}$ is approximately normal. The required conditions are:
  - Random: The data come from a random sample from the population of interest.
    - 10%: When sampling without replacement, $n < 0.10N$.
  - Large Counts: Both np_0 and $n(1 - p_0)$ are at least 10, where p_0 is the proportion specified by the null hypothesis.
- In general, the formula for the **standardized test statistic** is

$$\text{standardized test statistic} = \frac{\text{statistic} - \text{parameter}}{\text{standard error of statistic}}$$

- For a one-sample z test for a proportion specifically, the standardized test statistic is

$$z = \frac{\hat{p} - p_0}{\sqrt{\dfrac{p_0(1 - p_0)}{n}}}$$

- When the Large Counts condition is met, the standardized test statistic has approximately a standard normal distribution. You can use technology or Table A to find the P-value.
- Follow the **four-step process** when you perform a significance test:

  State: State the hypotheses, parameter(s), and significance level.

  Plan: Identify the appropriate inference method and check the conditions.

  Do: If the conditions are met, perform calculations:
    - Calculate the test statistic.
    - Find the P-value.

  Conclude: Make a conclusion about the hypotheses in the context of the problem.
- Confidence intervals provide additional information that significance tests do not — namely, a set of plausible values for the population proportion p. A two-sided test of H_0: $p = p_0$ at significance level α usually gives the same conclusion as a $100(1 - \alpha)\%$ confidence interval.
- When we make a conclusion in a significance test, there are two kinds of mistakes we can make.
  - A **Type I error** occurs if we reject H_0 when H_0 is true. In other words, the data give convincing evidence for H_a, but H_a isn't true.
  - A **Type II error** occurs if we fail to reject H_0 when H_a is true. In other words, the data don't give convincing evidence for H_a, but H_a is true.
- The probability of making a Type I error is equal to the significance level α. There is a trade-off between P(Type I error) and P(Type II error): as one increases, the other decreases, assuming other factors remain the same. It is important to consider the possible consequences of each type of error before choosing a significance level.

AP® EXAM TIP
AP® Daily Videos

Review the content of this section and get extra help by watching the AP® Daily Videos for Topics 6.4 (Video 2), 6.5, 6.6, and 6.7, which are available in AP® Classroom.

- The **power** of a test is the probability that the test will find convincing evidence for H_a when a specific alternative value of the parameter is true. In other words, the power of a test is the probability of avoiding a Type II error. For a specific alternative, power = 1 − P(Type II error).
- We can increase the power of a significance test (decrease the probability of a Type II error) by increasing the sample size, increasing the significance level, or increasing the difference that is important to detect between the null and alternative parameter values (known as the *effect size*). Reducing the standard error with better data collection methods can also increase power.

6D Tech Corner

TI-Nspire and other technology instructions are on the book's website at bfwpub.com/tps7e.

17. Significance tests for a proportion Page 578

SECTION 6D Exercises

Checking Conditions for a Test about *p*

1. **Online dating** Have a majority of U.S. young adults used an online dating app or website? The Pew Research Center surveyed a random sample of 929 U.S. adults age 18 to 29 and found that 492 had tried online dating.[30] We want to carry out a test at the $\alpha = 0.05$ significance level of H_0: $p = 0.50$ versus H_a: $p > 0.50$, where p = the population proportion of U.S. adults age 18 to 29 who have used an online dating app or website. Check if the conditions for performing the significance test are met. pg 574

2. **Walking to school** A recent report claimed that 13% of students typically walk to school.[31] DeAnna thinks that the proportion is higher than 0.13 at her large elementary school. She surveys a random sample of 100 students and finds that 17 typically walk to school. DeAnna would like to carry out a test at the $\alpha = 0.05$ significance level of H_0: $p = 0.13$ versus H_a: $p > 0.13$, where p = the proportion of all students at her elementary school who typically walk to school. Check if the conditions for performing the significance test are met.

3. **The chips project** Zenon decided to investigate whether students at his school prefer name-brand potato chips to generic potato chips. He randomly selected 50 of the 400 students at his school and asked each student to try both types of chips, in a random order. Overall, 41 of the 50 students preferred the name-brand chips. Zenon wants to perform a test of H_0: $p = 0.5$ versus H_a: $p > 0.5$, where p = the proportion of all students at his school who prefer name-brand chips.

(a) Can the observations in the sample be viewed as independent? Justify your answer.

(b) Is the sampling distribution of $\hat{p}$ approximately normal? Explain your reasoning.

4. **Better to be last?** On TV shows that feature singing competitions, contestants often wonder if there is an advantage in performing last. To investigate, researchers selected a random sample of 100 students from their large university and showed each student the audition video of 12 different singers with similar vocal skills. Each student viewed the videos in a random order. We would expect approximately 1/12 of the students to prefer the last singer seen, assuming order doesn't matter. In this study, 11 of the 100 students preferred the last singer they viewed. The researchers want to perform a test of H_0: $p = 1/12$ versus H_a: $p > 1/12$, where p = the population proportion of students at this university who prefer the singer they last see.

(a) Can the observations in the sample be viewed as independent? Justify your answer.

(b) Is the sampling distribution of $\hat{p}$ approximately normal? Explain your reasoning.

Calculating the Standardized Test Statistic and *P*-Value for a Test about *p*

5. **More online dating** Refer to Exercise 1. pg 577

(a) Explain why the sample result gives some evidence for the alternative hypothesis.

(b) Calculate the standardized test statistic and *P*-value.

(c) Interpret the *P*-value.

6. **More walking to school** Refer to Exercise 2.

(a) Explain why the sample result gives some evidence for the alternative hypothesis.

(b) Calculate the standardized test statistic and *P*-value.

(c) Interpret the *P*-value.

7. **Calculating P-values** A test of $H_0: p = 0.5$ versus $H_a: p < 0.5$ based on a sample of size 200 yields the standardized test statistic $z = -2.194$. Assume that the conditions for performing inference are met.

(a) Calculate the *P*-value.

(b) How would the *P*-value change if the alternative hypothesis was $H_a: p \neq 0.5$?

8. **P-value calculations** A test of $H_0: p = 0.65$ against $H_a: p < 0.65$ based on a sample of size 400 yields the standardized test statistic $z = -1.787$. Assume that the conditions for performing inference are met.

(a) Calculate the *P*-value.

(b) How would the *P*-value change if the alternative hypothesis was $H_a: p \neq 0.65$?

9. **Mendel and the peas** Gregor Mendel (1822–1884), an Austrian monk, is considered the father of genetics. Mendel studied the inheritance of various traits in pea plants. One such trait is whether the pea is smooth or wrinkled. Mendel predicted a ratio of 3 smooth peas for every 1 wrinkled pea. In one experiment, he observed 423 smooth peas and 133 wrinkled peas. Assume that the conditions for inference are met.

(a) State appropriate hypotheses for testing Mendel's claim about the true proportion of smooth peas.

(b) Calculate the standardized test statistic and *P*-value.

(c) What conclusion would you make? Use $\alpha = 0.05$.

10. **Spinning heads?** When a fair coin is flipped, we all know that the probability the coin lands on heads is 0.50. Is the probability also 0.50 when the coin is spun? According to the article "Euro Coin Accused of Unfair Flipping" in the *New Scientist*, two Polish math professors and their students spun a Belgian euro coin 250 times. It landed heads 140 times. One of the professors concluded that the coin was minted asymmetrically. A representative from the Belgian mint indicated the result was just chance. Assume that the conditions for inference are met.

(a) State appropriate hypotheses for testing these competing claims about the true proportion of spins that will land on heads.

(b) Calculate the standardized test statistic and *P*-value.

(c) What conclusion would you make? Use $\alpha = 0.05$.

One-Sample *z* Test for *p*

11. **Extreme poverty** "In the past 20 years, has the proportion of the world's population living in extreme poverty almost halved, remained more or less the same, or almost doubled?" When a random sample of 1000 U.S. adults were asked this question in the Gapminder Misconception Study, only 5% got it right — the proportion living in extreme poverty has been nearly cut in half.[32] As Gapminder points out, if people had randomly guessed, about 1/3 would have answered correctly. Do these data provide convincing evidence at the $\alpha = 0.01$ significance level that fewer than 1/3 of all U.S. adults would answer this question correctly?

pg 582

12. ***Survivor*** Advertisers want to know about the viewership of various reality shows before investing their marketing dollars. According to Nielsen ratings, *Survivor* was one of the most-watched shows in the United States during every week that it aired. An avid *Survivor* fan (your textbook author, Mr. Starnes) claims that 35% of all U.S. adults have watched *Survivor*. A skeptical editor believes this figure is too high. The editor asks a random sample of 200 U.S. adults if they have watched *Survivor*; 60 answer "Yes." Is there convincing evidence at the $\alpha = 0.10$ significance level to confirm the editor's belief?

13. **Teen drivers** A state's Division of Motor Vehicles (DMV) claims that 60% of all teens pass their driving test on the first attempt. An investigative reporter examines an SRS of the DMV records for 125 teens in this state; 86 of them passed the test on their first try. Is there convincing evidence at the 5% significance level that the DMV's claim is incorrect?

14. **We want to be rich** In a recent year, 73% of first-year college students responding to a national survey identified "being very well-off financially" as an important personal goal. A state university finds that 132 of an SRS of 200 of its first-year students say that this goal is important. Is there convincing evidence at the 5% significance level that the proportion of all first-year students at this university who think being very well-off is important differs from the national value of 73%?

15. **Teen drivers, part 2** Refer to Exercise 13.

(a) Construct and interpret a 95% confidence interval for the proportion p of all teens in the state who passed their driving test on the first attempt. Assume that the conditions for inference are met.

(b) Explain why the interval in part (a) provides more information than the test in Exercise 13.

16. **We want to be rich, part 2** Refer to Exercise 14.

(a) Construct and interpret a 95% confidence interval for the proportion p of all first-year students at the university who would identify being very well-off as an important personal goal. Assume that the conditions for inference are met.

(b) Explain why the interval in part (a) provides more information than the test in Exercise 14.

17. **Cell-phone passwords** A consumer organization suspects that fewer than half of all parents know their child's cell-phone password. The Pew Research Center asked a random sample of parents if they knew their child's cell-phone password. Of the 1060 parents surveyed, 551 reported that they knew the password. Explain why it isn't necessary to carry out a significance test in this setting.

18. **Proposition X** A political organization wants to determine if there is convincing evidence that a majority of registered voters in a large city favor Proposition X. In an SRS of 1000 registered voters, 482 favor the proposition. Explain why it isn't necessary to carry out a significance test in this setting.

Type I and Type II Errors

19. **Opening a restaurant** You are thinking about opening a restaurant and are searching for a good location. From research you have done, you know that the mean income of the people living near the restaurant must be more than \$85,000 to support the type of upscale restaurant you wish to open. You decide to take a simple random sample of 50 people living near one potential location. Based on the mean income of this sample, you will perform a test of

pg 584

$$H_0: \mu = \$85{,}000$$
$$H_a: \mu > \$85{,}000$$

where μ is the mean income (\$) in the population of people who live near the restaurant.[33] Describe a Type I error and a Type II error in this setting, and give a possible consequence of each.

20. **Clean water** The Environmental Protection Agency (EPA) has determined that safe drinking water should contain at most 1.3 mg/liter of copper, on average. A water supply company is testing water from a new source and will collect water in small bottles at each of 30 randomly selected locations. The company will perform a test of

$$H_0: \mu = 1.3$$
$$H_a: \mu > 1.3$$

where μ is the true mean copper content (mg/l) of the water from the new source. Describe a Type I error and a Type II error in this setting, and give a possible consequence of each.

21. **Risky opening** Refer to Exercise 19.

(a) Which error is worse for you as the potential restaurant owner? Explain your answer.

(b) Based on your answer to part (a), would you prefer to use $\alpha = 0.01$ or $\alpha = 0.10$? Explain your reasoning.

22. **Risky water** Refer to Exercise 20.

(a) Which error is worse for customers of this water company? Explain your answer.

(b) Based on your answer to part (a), should customers prefer to use $\alpha = 0.01$ or $\alpha = 0.10$? Explain your reasoning.

23. **Green tea** For their final project, two students decided to investigate whether green labeling makes consumers believe that a product is natural. They selected a random sample of 40 students at their school and served each person two cups of tea — one in a green cup and one in a clear cup. (Although the participants did not know it, both cups were filled with the same type of tea.) The researchers asked each person, "Which cup of tea do you believe has a more natural flavor?" Twenty-nine of the 40 people said the one in the green cup.[34]

(a) Is there convincing evidence that a majority of students at this school would identify the green cup of tea as having the more natural flavor?

(b) Based on the conclusion in part (a), could the students have made a Type I error or a Type II error? Explain your reasoning.

24. **Working students** According to the National Center for Education Statistics, 49% of full-time students in 2-year colleges are employed.[35] An administrator at a large, rural 2-year college suspects that the proportion of full-time students at this college who are employed is less than the national figure. The administrator selects a random sample of 200 full-time students from the college and finds that 91 of them are employed.

(a) Is there convincing evidence that the proportion of full-time students at this college who are employed is less than the national figure?

(b) Based on the conclusion in part (a), could the administrator have made a Type I error or a Type II error? Explain your reasoning.

The Power of a Test

25. **Upscale restaurant** You are thinking about opening a restaurant and are searching for a good location. From the research you have done, you know that the mean income of the people living near the restaurant must be more than \$85,000 to support the type of upscale restaurant you wish to open. You decide to take a simple random sample of 50 people living near one potential site. Based on the mean income of this sample, you will perform a test at the $\alpha = 0.05$ significance level of H_0: $\mu = \$85{,}000$ versus H_a: $\mu > \$85{,}000$,

pg 590

where μ is the true mean income in the population of people who live near the restaurant.

(a) The power of the test to detect that $\mu = \$86,000$ is 0.64. Interpret this value.

(b) How will the power change if you use a random sample of 30 people instead of 50 people? Explain your answer.

(c) How will the power change if you try to detect that $\mu = \$85,500$ instead of $\mu = \$86,000$? Explain your answer.

(d) How will the power change if you increase the significance level to $\alpha = 0.10$? Explain your answer.

26. **Finding clean water** The Environmental Protection Agency (EPA) has determined that safe drinking water should contain at most 1.3 mg/liter of copper, on average. A water supply company is testing water from a new source and will collect water in small bottles at each of 30 randomly selected locations. The company will perform a test at the $\alpha = 0.05$ significance level of H_0: $\mu = 1.3$ versus H_a: $\mu > 1.3$, where μ is the true mean copper content of the water from the new source.

(a) The power of the test to detect that $\mu = 1.4$ is 0.39. Interpret this value.

(b) How will the power change if the company collects water samples from 60 randomly selected locations instead of 30? Explain your answer.

(c) How will the power change if the company tries to detect that $\mu = 1.5$ instead of $\mu = 1.4$? Explain your answer.

(d) How will the power change if the company decreases the significance level to $\alpha = 0.01$? Explain your answer.

27. **Restaurant power problems** Refer to Exercise 25.

(a) Explain one disadvantage of using $\alpha = 0.10$ instead of $\alpha = 0.05$ when performing the test.

(b) Explain one advantage of selecting a random sample of 30 people instead of 50 people.

28. **Water power problems** Refer to Exercise 26.

(a) Explain one advantage of using $\alpha = 0.01$ instead of $\alpha = 0.05$ when performing the test.

(b) Explain one disadvantage of collecting water from 60 locations instead of 30.

29. **Better parking** A large high school makes a change that should improve student satisfaction with the parking situation. Before the change, 37% of the school's students approved of the parking that was provided. After the change, the principal plans to survey an SRS of 200 students at the school. The principal will then perform a test at the 5% significance level of H_0: $p = 0.37$ versus H_a: $p > 0.37$, where p is the proportion of all students at school who are satisfied with the parking situation after the change.

(a) Describe a Type I error in this context. What is the probability of a Type I error?

(b) The power of the test to detect that $p = 0.45$ based on a random sample of 200 students and a significance level of $\alpha = 0.05$ is 0.75. Interpret this value.

(c) Describe a Type II error in this context. What is the probability of a Type II error?

(d) Describe two ways to decrease the probability of a Type II error in this context.

30. **Awful accidents** Slow response times by paramedics, firefighters, and police officers can have serious consequences for accident victims. In the case of life-threatening injuries, victims generally need medical attention within 8 minutes of the accident. Several cities have begun to monitor emergency response times. In one such city, emergency personnel took more than 8 minutes to arrive on 22% of all calls involving life-threatening injuries last year. The city manager shares this information and encourages these first responders to "do better." After 6 months, the city manager will select an SRS of 400 calls involving life-threatening injuries and examine the response times. The manager will then perform a test at the 5% significance level of H_0: $p = 0.22$ versus H_a: $p < 0.22$, where p is the proportion of all calls involving life-threatening injuries during this 6-month period for which emergency personnel took more than 8 minutes to arrive.

(a) Describe a Type I error in this context. What is the probability of a Type I error?

(b) The power of the test to detect that $p = 0.20$ based on a random sample of 400 calls and a significance level of $\alpha = 0.05$ is 0.24. Interpret this value.

(c) Describe a Type II error in this context. What is the probability of a Type II error?

(d) Describe two ways to decrease the probability of a Type II error in this context.

31. **Error probabilities and power** You read that a significance test at the $\alpha = 0.01$ significance level has probability 0.14 of making a Type II error when a proposed alternative is true.

(a) What's the power of the test against this alternative?

(b) What's the probability of making a Type I error?

32. **Power and error** A scientist calculates that a test at the $\alpha = 0.05$ significance level has probability 0.23 of making a Type II error when a proposed alternative is true.

(a) What's the power of the test against this alternative?

(b) What's the probability of making a Type I error?

For Investigation *Apply the skills from the section in a new context or nonroutine way.*

33. **Cranky mower** A company has developed an "easy-start" mower that cranks the engine with the push of a button. The company claims that the probability this mower model will start on any push of the button is 0.9. A consumer testing agency suspects that this claim is exaggerated. To test the claim, an agency researcher selects a random sample of 20 of these mowers and attempts to start each one by pushing the button once. Only 15 of the mowers start.

(a) State appropriate hypotheses for performing a significance test. Be sure to define the parameter of interest.

(b) Show that the Large Counts condition is not met.

(c) Assuming that the null hypothesis from part (a) is true, use a binomial distribution to calculate the probability that 15 or fewer of the 20 randomly selected mowers would start.

(d) Based on your result in part (c), what conclusion would you make?

34. **Power curves** As mentioned at the end of the section, researchers should consider the power of a test to detect an alternative parameter value when planning a study. Most statistical software can produce *power curves* to help researchers determine the required sample size for various alternative parameter values or effect sizes. Suppose a researcher wants to perform a test at the $\alpha = 0.05$ significance level of H_0: $p = 0.60$ versus H_a: $p > 0.60$. The graph shows power curves for different alternative parameter values and sample sizes.

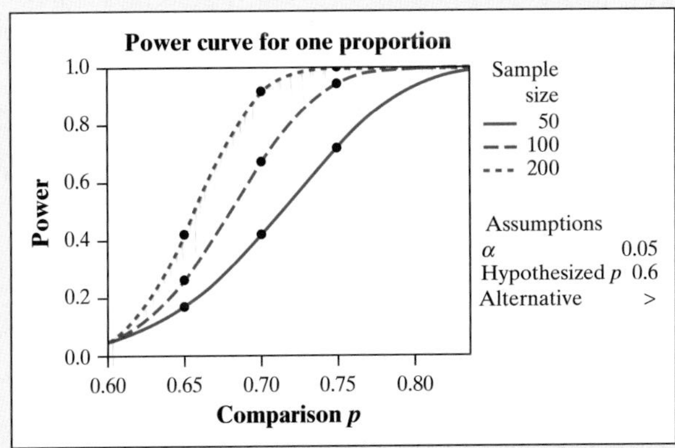

(a) Estimate the power of the test to detect that $p = 0.65$ using a sample size of $n = 100$.

(b) If the researcher wants to have a good chance to detect if $p = 0.70$, what sample size would you recommend? Explain your answer.

(c) Estimate the power of the test to detect that $p = 0.75$ for $n = 50$, 100, and 200.

(d) If the researcher believes that $p = 0.75$, which sample size would you recommend that the researcher use for the study, taking into account both power and cost?

35. **Power calculation: Parking** Refer to Exercise 29.

(a) Suppose that H_0: $p = 0.37$ is true. Describe the shape, center, and variability of the sampling distribution of $\hat{p}$ in random samples of size 200.

(b) Use the sampling distribution from part (a) to find the value of $\hat{p}$ with an area of 0.05 to the right of it. If the principal obtains a random sample of 200 students with a sample proportion of satisfied students greater than this value of $\hat{p}$, H_0: $p = 0.37$ will be rejected at the $\alpha = 0.05$ significance level.

(c) Now suppose that $p = 0.40$. Describe the shape, center, and variability of the sampling distribution of $\hat{p}$ in random samples of size 200.

(d) Use the sampling distribution from part (c) to find the probability of getting a sample proportion greater than the value you found in part (b). This result is the power of the test to detect $p = 0.40$.

Multiple Choice *Select the best answer for each question.*

36. After a college once again lost a football game to its archrival, the alumni association conducted a survey to see if alumni were in favor of firing the coach. An SRS of 100 alumni from the population of all living alumni was taken, and 64 of the alumni in the sample were in favor of firing the coach. Suppose you wish to see if a majority of all living alumni are in favor of firing the coach. Which of the following is the appropriate standardized test statistic?

(A) $z = \dfrac{0.64 - 0.5}{\sqrt{\dfrac{0.64(0.36)}{100}}}$

(B) $z = \dfrac{0.5 - 0.64}{\sqrt{\dfrac{0.64(0.36)}{100}}}$

(C) $z = \dfrac{0.64 - 0.5}{\sqrt{\dfrac{0.5(0.5)}{100}}}$

(D) $z = \dfrac{0.64 - 0.5}{\sqrt{\dfrac{0.64(0.36)}{64}}}$

(E) $z = \dfrac{0.5 - 0.64}{\sqrt{\dfrac{0.5(0.5)}{100}}}$

37. Which of choices (A) through (D) is *not* a condition for performing a significance test about a population proportion p?

(A) The data should come from a random sample from the population of interest.

(B) Both np_0 and $n(1-p_0)$ should be at least 10.

(C) If you are sampling without replacement from a finite population, then you should sample less than 10% of the population.

(D) The population distribution should be approximately normal, unless the sample size is at least 30.

(E) All of the above are conditions for performing a significance test about a population proportion.

38. The standardized test statistic for a test of H_0: $p = 0.4$ versus H_a: $p \neq 0.4$ is $z = 2.43$. This test is

(A) not significant at either $\alpha = 0.05$ or $\alpha = 0.01$.

(B) significant at $\alpha = 0.05$, but not at $\alpha = 0.01$.

(C) significant at $\alpha = 0.01$, but not at $\alpha = 0.05$.

(D) significant at both $\alpha = 0.05$ and $\alpha = 0.01$.

(E) inconclusive because we don't know the value of $\hat{p}$.

39. Which of the following 95% confidence intervals would lead us to reject H_0: $p = 0.30$ in favor of H_a: $p \neq 0.30$ at the 5% significance level?

(A) (0.19, 0.27)

(B) (0.24, 0.30)

(C) (0.27, 0.31)

(D) (0.29, 0.38)

(E) None of these

40. A political candidate funds a survey of randomly selected registered voters in the candidate's district. After the survey, the candidate will test H_0: $p = 0.50$ versus H_a: $p > 0.50$, where p = the proportion of all registered voters in the district who plan to vote for the candidate. Which of the following describes a Type I error in this context?

(A) Finding convincing evidence that a majority of all registered voters in the district plan to vote for the candidate, when a majority plan to vote for the candidate.

(B) Finding convincing evidence that a majority of all registered voters in the district plan to vote for the candidate, when a majority do not plan to vote for the candidate.

(C) Not finding convincing evidence that a majority of all registered voters in the district plan to vote for the candidate, when a majority plan to vote for the candidate.

(D) Not finding convincing evidence that a majority of all registered voters in the district plan to vote for the candidate, when a majority do not plan to vote for the candidate.

(E) Finding convincing evidence that a minority of all registered voters in the district plan to vote for the candidate, when a majority plan to vote for the candidate.

41. A researcher plans to conduct a significance test at the $\alpha = 0.01$ significance level. The researcher designs the study to have a power of 0.90 at a particular alternative value of the parameter of interest. Which of the following is the probability that the researcher will commit a Type II error for the particular alternative value of the parameter used?

(A) 0.01

(B) 0.10

(C) 0.89

(D) 0.90

(E) 0.99

Recycle and Review *Practice what you learned in previous sections.*

42. **Packaging vinyl (5A)** A manufacturer of vinyl records wants to be sure that the records will fit inside the square cardboard sleeve used as packaging. According to the supplier, the diameters of the records can be modeled by a normal distribution with mean = 12 inches and standard deviation = 0.05 inch. The width of the open side of the cardboard sleeves can be modeled by a normal distribution with mean = 12.2 inches and standard deviation = 0.1 inch.

(a) Let R = the diameter of a randomly selected record and W = the width of the opening of a randomly selected cardboard sleeve. Describe the shape, center, and variability of the distribution of the random variable $W - R$.

(b) Calculate the probability that a randomly selected record will fit inside a randomly selected cardboard sleeve.

43. **Cash to find work? (3C)** Will cash bonuses speed the return to work of unemployed people? The Illinois Department of Employment Security designed an experiment to find out. The subjects were 10,065 people age 20 to 54 who were filing claims for unemployment insurance. Some were offered \$500 if they found a job within 11 weeks and held it for at least 4 months. Others could tell potential employers that the state would pay the employer \$500 for hiring them. A control group got neither kind of bonus.[36]

(a) Describe a completely randomized design for this experiment. Explain how you would use a random number generator to assign the treatments.

(b) What is the purpose of the control group in this setting?

UNIT 6 Wrap-Up

PART I

FRAPPY! Free Response AP® Problem, Yay!

Directions: Show all your work. Indicate clearly the methods you use, because you will be scored on the correctness of your methods as well as on the accuracy and completeness of your results and explanations.

Members at a popular fitness club currently pay a \$40 per month membership fee. The owner of the club wants to raise the fee to \$50 but is concerned that some members will leave the gym if the fee increases. To investigate, the owner plans to survey a random sample of the club members and construct a 95% confidence interval for the proportion of all members who would quit if the fee was raised to \$50.

(a) Explain the meaning of "95% confidence" in the context of the study.

(b) After the owner conducted the survey, he calculated the confidence interval to be 0.18 ± 0.075. Interpret this interval in the context of the study.

(c) According to the club's accountant, the fee increase will be worthwhile if fewer than 20% of the members quit. According to the interval from part (b), can the owner be confident that the fee increase will be worthwhile? Explain.

(d) One of the conditions for calculating the confidence interval in part (b) is that $n\hat{p} \geq 10$ and $n(1-\hat{p}) \geq 10$. Explain why it is necessary to check this condition.

After you finish the FRAPPY!, you can view two example solutions on the book's website (**bfwpub.com/tps7e**). Determine whether you think each solution is "complete," "substantial," "developing," or "minimal." If the solution is not complete, what improvements would you suggest to the student who wrote it? Finally, your teacher will provide you with a scoring rubric. Score your response and note what, if anything, you would do differently to improve your own score.

UNIT 6, PART I REVIEW

SECTION 6A Confidence Intervals: The Basics

In this section, you learned that a **point estimate** is the single best guess for the value of a population parameter. You also learned that a **confidence interval,** also known as an **interval estimate,** provides an interval of plausible values for a parameter based on sample data. To interpret a confidence interval, say, "We are C% confident that the interval from ____________ to ____________ captures the [parameter in context]," where C is the confidence level of the interval. You can use a confidence interval to evaluate a claim about the value of a population parameter.

The **confidence level** C describes the percentage of confidence intervals that we expect to capture the value of the parameter in repeated sampling. To interpret a C% confidence level, say, "If we took many samples of the same size from the same population and used them to construct C% confidence intervals, about C% of those intervals would capture the [parameter in context]."

SECTION 6B Confidence Intervals for a Population Proportion

In this section, you learned how to construct and interpret confidence intervals for a population proportion. Three conditions must be met to ensure that the observations in the sample are independent and that the sampling distribution of $\hat{p}$ is approximately normal. First, the data used to calculate the interval must come from a random sample from the population of interest (the Random condition). When the sample is selected without replacement from the population, the sample size should be less than 10% of the population size (the 10%

condition). Finally, the observed number of successes $n\hat{p}$ and observed number of failures $n(1-\hat{p})$ must both be at least 10 (the Large Counts condition).

Confidence intervals are formed by adding and subtracting the **margin of error** from the point estimate (value of the statistic):

$$\text{CI} = \text{point estimate} \pm \text{margin of error}$$

The margin of error has two components: (1) the **standard error** of the statistic, which is an estimate of the standard deviation of the sampling distribution of the statistic, and (2) the **critical value,** which is a multiplier based on the confidence level of the interval.

$$\text{CI} = \text{statistic} \pm (\text{critical value})(\text{standard error of statistic})$$

The specific formula for calculating a confidence interval for a population proportion is

$$\hat{p} \pm z^* \sqrt{\frac{\hat{p}(1-\hat{p})}{n}}$$

where $\hat{p}$ is the sample proportion, z^* is the critical value, and n is the sample size. To find z^*, use technology or Table A to determine the values of z^* and $-z^*$ that capture the middle C% of the standard normal distribution, where C is the confidence level.

The **four-step process** (State, Plan, Do, Conclude) is perfectly suited for problems that ask you to construct and interpret a confidence interval:

State: State the parameter you want to estimate and the confidence level.
Plan: Identify the appropriate inference method and check the conditions.
Do: If the conditions are met, perform calculations.
Conclude: Interpret your interval in the context of the problem.

The size of the margin of error is determined by several factors, including the confidence level C and the sample size n. Increasing the sample size n makes the standard error of our statistic smaller, decreasing the margin of error. Increasing the confidence level C makes the margin of error larger, to ensure that the capture rate of the interval increases to C%. Remember that the margin of error only accounts for sampling variability — it does not account for any bias in the data collection process.

Finally, an important part of planning a study is determining the size of the sample to be selected. The necessary sample size is based on the confidence level, the proportion of successes, and the desired margin of error. To calculate the minimum sample size, solve the following inequality for n, where $\hat{p}$ is a guessed value for the sample proportion:

$$z^* \sqrt{\frac{\hat{p}(1-\hat{p})}{n}} \leq ME$$

If you do not have an approximate value of $\hat{p}$ from a previous study or a pilot study, use $\hat{p} = 0.5$ to determine the sample size that is guaranteed to yield a value less than or equal to the desired margin of error.

SECTION 6C Significance Tests: The Basics

In this section, you learned the basic ideas of **significance testing.** Start by stating the hypotheses that you want to test. The **null hypothesis** (H_0) is typically a statement of "no difference" and the **alternative hypothesis** (H_a) describes what we suspect is true. Remember that hypotheses are always about population parameters, not sample statistics.

When sample data provide evidence for the alternative hypothesis, there are two possible explanations: (1) the null hypothesis is true, and data supporting the alternative hypothesis occurred just by chance, or (2) the alternative hypothesis is true, and the data are consistent with an alternative value of the parameter. In a significance test, we evaluate Explanation 1 by assuming the null hypothesis is true and calculating the probability of getting evidence for the alternative hypothesis as strong as or stronger than the observed data. This probability is called a ***P*-value.**

To determine if the *P*-value is small enough to reject H_0, compare it to a predetermined **significance level** such as $\alpha = 0.05$. If the *P*-value $\leq \alpha$, reject H_0 — there is convincing evidence that the alternative hypothesis is true. However, if the *P*-value $> \alpha$, fail to reject H_0 — there is not convincing evidence that the alternative hypothesis is true.

SECTION 6D Significance Tests for a Population Proportion

In this section, you learned the details of performing a significance test about a population proportion p. After stating hypotheses and before doing calculations, check the conditions to verify that the observations in the sample are independent and that the sampling distribution of $\hat{p}$ is approximately normal. The Random condition requires that the sample be randomly selected from the population. The 10% condition requires that the sample size be less than 10% of the population size when sampling without replacement from the population. Finally, the Large Counts condition says that both np_0 and $n(1-p_0)$ must be at least 10, where p_0 is the value of p in the null hypothesis.

The **standardized test statistic** measures how far the observed value of $\hat{p}$ is from the null hypothesis value, in standard error units.

$$\text{standardized test statistic} = \frac{\text{statistic} - \text{parameter}}{\text{standard error of statistic}}$$

$$z = \frac{\hat{p} - p_0}{\sqrt{\frac{p_0(1-p_0)}{n}}}$$

Find the *P*-value by calculating the probability of getting a *z* statistic this large or larger in the direction specified by the alternative hypothesis H_a in the standard normal distribution. If you are performing a **two-sided test,** make sure to find the area in both tails of the standard normal distribution.

Whenever you are asked if there is convincing evidence for a claim about a population parameter, you are expected to respond using the familiar four-step process.

State: State the hypotheses, parameter(s), and significance level.

Plan: Identify the appropriate inference method and check the conditions.

Do: If the conditions are met, perform calculations.

- Calculate the test statistic.
- Find the *P*-value.

Conclude: Make a conclusion about the hypotheses in the context of the problem.

You can also use a confidence interval to make a conclusion for a two-sided test. If the null parameter value is one of the plausible values in the interval, there isn't convincing evidence that the alternative hypothesis is true. However, if the interval contains only values consistent with the alternative hypothesis, there is convincing evidence that the alternative hypothesis is true. Besides helping you draw a conclusion, the interval tells you which alternative parameter values are plausible.

Because conclusions are based on sample data, there is a possibility that the conclusion to a significance test will be incorrect. You can make two types of errors: A **Type I error** occurs if you find convincing evidence for the alternative hypothesis when, in reality, the null hypothesis is true. A **Type II error** occurs when you don't find convincing evidence that the alternative hypothesis is true when, in reality, the alternative hypothesis is true. The probability of making a Type I error is equal to the significance level (α) of the test. Decreasing the probability of a Type I error increases the probability of a Type II error, and increasing the probability of a Type I error decreases the probability of a Type II error.

The probability that you avoid making a Type II error when an alternative value of the parameter is true is called the **power** of the test. Power is good — if the alternative hypothesis is true, you want to maximize the probability of finding convincing evidence that it is true. We can increase the power of a significance test by increasing the sample size, by increasing the significance level, and by reducing the standard error with wise data collection methods. The power of a test will also be greater when the alternative value of the parameter is farther away from the null hypothesis value.

Inference for a Population Proportion

	Confidence Interval for *p*	**Significance Test for *p***
Name (TI-83/84)	One-sample *z* interval for *p* (1-PropZInt)	One-sample *z* test for *p* (1-PropZTest)
Null Hypothesis	*Not applicable.*	$H_0: p = p_0$
Conditions	• **Random:** The data come from a random sample from the population of interest. ○ **10%:** When sampling without replacement, $n < 0.10N$. • **Large Counts:** Both $n\hat{p}$ and $n(1-\hat{p})$ are at least 10. That is, the number of successes and the number of failures in the sample are both at least 10.	• **Random:** The data come from a random sample from the population of interest. ○ **10%:** When sampling without replacement, $n < 0.10N$. • **Large Counts:** Both np_0 and $n(1-p_0)$ are at least 10, where p_0 is the proportion specified by the null hypothesis.
Formula	$\hat{p} \pm z^* \sqrt{\frac{\hat{p}(1-\hat{p})}{n}}$ Critical value z^* from the standard normal distribution.	$z = \frac{\hat{p} - p_0}{\sqrt{\frac{p_0(1-p_0)}{n}}}$ *P*-value from the standard normal distribution.

What Did You Learn?

Learning Target	Section	Related Example on Page(s)	Relevant Unit Review Exercise(s)
Interpret a confidence interval in context.	6A	533	R1, R2
Use a confidence interval to make a decision about the value of a parameter.	6A	534	R1, R5
Interpret a confidence level in context.	6A	536	R1

Learning Target	Section	Related Example on Page(s)	Relevant Unit Review Exercise(s)
Check the conditions for calculating a confidence interval for a population proportion.	6B	541	R2
Calculate a confidence interval for a population proportion.	6B	546	R2
Construct and interpret a one-sample *z* interval for a proportion.	6B	548	R2
Describe how the sample size and confidence level affect the margin of error.	6B	552	R2
Determine the sample size required to obtain a confidence interval for a population proportion with a specified margin of error.	6B	554	R3
State appropriate hypotheses for a significance test about a population parameter.	6C	564	R4, R5, R6
Interpret a *P*-value in context.	6C	565	R4
Make an appropriate conclusion for a significance test.	6C	568	R4, R5
Check the conditions for performing a test about a population proportion.	6D	574	R5
Calculate the standardized test statistic and *P*-value for a test about a population proportion.	6D	577	R5
Perform a one-sample *z* test for a proportion.	6D	582	R5
Interpret a Type I error and a Type II error in context and give a consequence of each type of error.	6D	584	R6
Interpret the power of a significance test and describe which factors affect the power of a test.	6D	590	R6

UNIT 6, PART I REVIEW EXERCISES

These exercises are designed to help you review the important concepts and skills of the unit.

R1 **Sports fans (6A)** Are you a sports fan? That's the question the Gallup polling organization asked a random sample of 1527 U.S. adults.[37] Gallup reported that a 95% confidence interval for the proportion of all U.S. adults who are sports fans is 0.565 to 0.615.

(a) Interpret the confidence interval.

(b) Interpret the confidence level.

(c) Based on the interval, is there convincing evidence that a majority of U.S. adults are sports fans? Explain your answer.

R2 **Running red lights (6B)** A random sample of 880 U.S. drivers were asked, "Recalling the last 10 traffic lights you drove through, how many of them were red when you entered the intersection?" Of the 880 respondents, 171 admitted that at least one light had been red.[38]

(a) Construct and interpret a 95% confidence interval for the population proportion.

(b) Explain two ways you could reduce the margin of error of this confidence interval. What are the drawbacks to these actions?

R3 **Do you go to church? (6B)** The Gallup polling organization plans to ask a random sample of adults whether they attended a religious service in the past 7 days. How large a sample would be required to obtain a margin of error of at most 0.01 in a 99% confidence interval for the population proportion who would say that they attended a religious service in the past 7 days?

R4 **Signature verification (6C)** When a petition is submitted to government officials to put a political candidate's name on a ballot, a certain number of valid voters' signatures are required. Rather than check the validity of all the signatures, officials often randomly select a sample of signatures for verification and perform a significance test to see if the true proportion of signatures is less than the required value. Suppose a petition has 30,000 signatures and 18,000 valid signatures are required for a candidate to be on the ballot — which means at least 60% of the signatures on this petition must be valid. The officials select a random sample of 300 signatures and find that 171 are valid. Do these data provide convincing evidence that the proportion of all signatures that are valid is less than 0.6?

(a) State the hypotheses we are interested in testing.

(b) The *P*-value for this test is 0.1444. Interpret this value.

(c) What conclusion should you make?

R5 **Vaping in college (6D)** According to the University of Michigan's *Monitoring the Future* study, 39.6% of college students used e-cigarettes (vaped) in a recent year.[39] Public health researchers in a particular state wonder if this national result applies to students at their state's two- and four-year colleges and universities. To investigate, the researchers collected data from a random sample of 750 students attending these local institutions and found 331 of them vaped that year.

(a) Is there convincing evidence at the 5% level of significance that the proportion of all college students enrolled in this state who vaped that year differs from the national result?

(b) The 95% confidence interval for the population proportion of college students who vaped that year is (0.4058, 0.4769). Explain how this interval is consistent with your decision from part (a).

R6 **Flu vaccine (6D)** A drug company has developed a new vaccine for preventing the flu. The company claims that fewer than 5% of adults who use its vaccine will get the flu. To test the claim, researchers give the vaccine to a random sample of 1000 adults.

(a) State appropriate hypotheses for testing the company's claim. Be sure to define your parameter.

(b) Describe a Type I error and a Type II error in this setting, and give the consequences of each.

(c) Would you recommend a significance level of 0.01, 0.05, or 0.10 for this test? Justify your choice.

(d) The power of the test to detect the fact that only 3% of adults who use this vaccine would develop flu using $\alpha = 0.05$ is 0.9437. Interpret this value.

(e) Explain two ways that you could increase the power of the test from part (d).

UNIT 6, PART I AP® STATISTICS PRACTICE TEST

Section I: Multiple Choice *Select the best answer for each question.*

T1 A confidence interval for a proportion is 0.272 to 0.314. What are the point estimate and the margin of error for this interval?

(A) 0.293, 0.021
(B) 0.293, 0.042
(C) 0.293, 0.084
(D) 0.586, 0.021
(E) 0.586, 0.042

T2 Many television viewers express doubts about the validity of certain commercials. In an attempt to answer its critics, Timex Group USA wants to create an interval estimate for the proportion p of all consumers who believe what is shown in Timex television commercials. Which of the following is the smallest number of consumers that Timex can survey to guarantee a margin of error of 0.05 or less at a 99% confidence level?

(A) 550
(B) 600
(C) 650
(D) 700
(E) 750

T3 A random sample of 100 likely voters in a small city produced 59 voters in favor of Candidate A. The value of the standardized test statistic for performing a test of H_0: $p = 0.5$ versus H_a: $p > 0.5$ is which of the following?

(A) $z = \dfrac{0.59 - 0.5}{\sqrt{\dfrac{0.59(0.41)}{100}}}$

(B) $z = \dfrac{0.59 - 0.5}{\sqrt{\dfrac{0.5(0.5)}{100}}}$

(C) $z = \dfrac{0.5 - 0.59}{\sqrt{\dfrac{0.59(0.41)}{100}}}$

(D) $z = \dfrac{0.5 - 0.59}{\sqrt{\dfrac{0.5(0.5)}{100}}}$

(E) $z = \dfrac{0.59 - 0.5}{\sqrt{100}}$

T4 In a test of H_0: $p = 0.4$ against H_a: $p \neq 0.4$, a random sample of size 100 yields a standardized test statistic of $z = 1.28$. Which of the following is closest to the P-value for this test?

(A) 0.05
(B) 0.10
(C) 0.20
(D) 0.40
(E) 0.90

T5 Bags of a certain brand of tortilla chips are labeled as having a net weight of 14 ounces. A representative of a consumer advocacy group wishes to see if there is convincing evidence that the mean net weight is less than advertised and so intends to test H_0: $\mu = 14$ versus H_a: $\mu < 14$. Which of the following describes a Type II error?

(A) Finding convincing evidence the bags are being underfilled when they really aren't.
(B) Finding convincing evidence the bags are being underfilled when they really are.
(C) Not finding convincing evidence the bags are being underfilled when they really are.
(D) Not finding convincing evidence the bags are being underfilled when they really aren't.
(E) Finding convincing evidence the bags are being overfilled when they are really underfilled.

T6 A 95% confidence interval for the proportion of viewers of a certain reality television show who are older than age 30 is (0.26, 0.35). Suppose the show's producers want to test the hypothesis H_0: $p = 0.25$ against H_a: $p \neq 0.25$. Which of the following is an appropriate conclusion for them to draw at the $\alpha = 0.05$ significance level?

(A) Fail to reject H_0; there is convincing statistical evidence that the true proportion of viewers of this reality TV show who are older than age 30 equals 0.25.
(B) Fail to reject H_0; there is not convincing statistical evidence that the true proportion of viewers of this reality TV show who are older than age 30 differs from 0.25.
(C) Reject H_0; there is not convincing statistical evidence that the true proportion of viewers of this reality TV show who are older than age 30 differs from 0.25.
(D) Reject H_0; there is convincing statistical evidence that the true proportion of viewers of this reality TV show who are older than age 30 is greater than 0.25.
(E) Reject H_0; there is convincing statistical evidence that the true proportion of viewers of this reality TV show who are older than age 30 differs from 0.25.

T7 A marketing assistant for a technology firm plans to randomly select 1000 customers to estimate the proportion who are satisfied with the firm's performance. Based on the results of the survey, the assistant will construct a 95% confidence interval for the proportion of all customers who are satisfied. The marketing manager, however, says that the firm can afford to survey only 250 customers. How will this decrease in sample size affect the margin of error?

(A) The margin of error will be about 4 times larger.

(B) The margin of error will be about 2 times larger.

(C) The margin of error will be about the same size.

(D) The margin of error will be about half as large.

(E) The margin of error will be about one-fourth as large.

T8 In a random sample of 100 students from a large high school, 37 regularly bring a reusable water bottle from home. Which of the following gives the correct value and interpretation of the standard error of the sample proportion?

(A) In random samples of size 100 from this school, the sample proportion of students who bring a reusable water bottle from home will be at most 0.095 from the population proportion.

(B) In random samples of size 100 from this school, the sample proportion of students who bring a reusable water bottle from home will be at most 0.048 from the population proportion.

(C) In random samples of size 100 from this school, the sample proportion of students who bring a reusable water bottle from home typically varies by about 0.095 from the population proportion.

(D) In random samples of size 100 from this school, the sample proportion of students who bring a reusable water bottle from home typically varies by about 0.048 from the population proportion.

(E) There is not enough information to calculate the standard error.

T9 A radio talk show host with a large audience wants to estimate the proportion p of adults in his listening area who think the drinking age should be lowered to 18. To do so, he poses the following question to his listeners: "Do you think that the drinking age should be reduced to 18 in light of the fact that 18-year-olds are eligible for military service?" He asks listeners to go to his website and vote "Yes" if they agree the drinking age should be lowered and "No" if not. Of the 100 people who voted, 70 answered "Yes." Which of the following conditions for constructing a 95% confidence interval for p are violated?

I. Random

II. 10%

III. Large Counts

(A) I only

(B) II only

(C) III only

(D) I and II only

(E) I, II, and III

T10 The germination rate of seeds is defined as the proportion of seeds that sprout and grow when properly planted and watered. A certain variety of grass seed usually has a germination rate of 0.80. A company wants to see if spraying the seeds with a chemical that is known to increase germination rates in other species will increase the germination rate of this variety of grass. The company researchers spray a random sample of 400 grass seeds with the chemical, record the proportion that germinate, and test the hypotheses H_0: $p = 0.80$ versus H_a: $p > 0.80$. Which of the following steps will reduce the probability of making a Type II error?

(A) Only spraying half of the 400 seeds with the chemical.

(B) Using different varieties of grass seed instead of just one type.

(C) Using a larger value for the null hypothesis.

(D) Using $\alpha = 0.05$ instead of $\alpha = 0.01$.

(E) Using a two-sided test instead of a one-sided test.

Section II: Free Response *Show all your work. Indicate clearly the methods you use, because you will be graded on the correctness of your methods as well as on the accuracy and completeness of your results and explanations.*

T11 The U.S. Forest Service is considering additional restrictions on the number of vehicles allowed to enter Yellowstone National Park. To assess public reaction, the service asks a random sample of 150 visitors if they favor the proposal. Of these, 89 say "Yes."

(a) Construct and interpret a 99% confidence interval for the proportion of all visitors to Yellowstone who favor the restrictions.

(b) Interpret the confidence level.

T12 According to the Pew Research Center, "More than half of Americans live within an hour of extended family." In this study, "extended family" was defined as children, parents, grandparents, grandchildren, brothers, sisters, cousins, aunts, uncles, and in-laws, and "within an hour" was the amount of time it would take to drive to visit the extended family. Pew's claim was based on a random sample of 9676 U.S. adults in which 55% said they lived within an hour of extended family.

(a) Do these data provide convincing evidence for the Pew Research Center's claim?

(b) Which type of error, Type I or Type II, could you have made in part (a)? Explain your answer.

T13 A mattress manufacturer has designed a new production process that is intended to reduce the proportion of mattresses with defects in the stitching. Before the change, 15% of the mattresses had this type of defect. After the new production process is implemented, 100 mattresses will be randomly selected and inspected for stitching defects. Then the manufacturer will test H_0: $p = 0.15$ versus H_a: $p < 0.15$.

(a) The company statistician calculates that the power of the test to detect $p = 0.10$ using $\alpha = 0.05$ is 0.39. Interpret this value.

(b) After inspecting 100 randomly selected mattresses, 11 were found to have defects in the stitching, resulting in a *P*-value of 0.13. Interpret this value.

UNIT 6

Inference for Categorical Data: Proportions

PART II
Inference for Two Samples

Nils Hastrup/500px/Getty Images

Introduction

In Part I of Unit 6, you learned about confidence intervals and significance tests for a population proportion p. Many interesting statistical questions involve *comparing* the proportion of successes in two populations. What is the difference between the proportion of Democrats and the proportion of Republicans who favor the death penalty? Has there been a change in the proportion of U.S. adults who watch the Super Bowl in the last 5 years? In both of these cases, we are interested in the difference $p_1 - p_2$, where p_1 and p_2 are the proportions of success in Population 1 and Population 2.

Other statistical questions involve comparing the effectiveness of two treatments in an experiment. For example, is a new medication more effective than a current medication for relieving headaches? What is the difference in the survival rate for two cancer treatments? In these cases, we are also interested in the difference $p_1 - p_2$. But in these settings, p_1 and p_2 are the true proportions of success for individuals like the ones in the experiment who receive Treatment 1 or Treatment 2.

For their response bias project (page 312), Sarah and Miranda investigated whether the characteristics of the interviewer can affect the response to a survey question. At the Tucson Mall, they asked 60 shoppers the question, "Do you like tattoos?" When interviewing 30 of the shoppers, Sarah and Miranda wore long-sleeved shirts that covered their tattoos. For the remaining 30 shoppers, the interviewers wore tank tops that revealed their tattoos. The choice of long sleeve or tank top was determined at random for each subject. Sarah and Miranda suspected that more people would answer "Yes" when their tattoos were visible.[40]

What happened in the experiment? The two-way table summarizes the results:

		Clothing worn		
		Tank top	Long sleeves	Total
Like tattoos?	Yes	18	14	32
	No	12	16	28
	Total	30	30	60

The difference (Tank top − Long sleeves) in the proportions of people who said they like tattoos is $18/30 - 14/30 = 0.600 - 0.467 = 0.133$. Does this difference provide convincing evidence that the appearance of the interviewer has the intended effect on the response, or could the difference be due to chance variation in the random assignment? You'll find out in the following activity.

ACTIVITY Who likes tattoos?

In this activity, your class will investigate whether the results of Sarah and Miranda's experiment are statistically significant using index cards (or technology). Let's see what would happen just by chance if we randomly reassign the 60 people in this experiment to the two treatments (tank top and long sleeves) many times, *assuming the treatment received doesn't affect whether or not a person says they like tattoos.*

staticnak1983/E+/Getty Images

1. Using 60 index cards or equally sized pieces of paper, write "Yes" on 32 and "No" on 28.
2. Shuffle the cards and divide them at random into two piles of 30 — one for the tank top treatment group and one for the long sleeves treatment group. Be sure to determine which pile will represent each group before you deal.
3. Calculate the difference (Tank top − Long sleeves) in the proportions of "Yes" responses for the two groups.
4. Your teacher will draw and label axes for a class dotplot. Plot your result from Step 3 on the graph.
5. Repeat Steps 2–4 if needed to get a total of at least 40 trials of the simulation for your class.
6. How often did a difference in proportions of 0.133 or greater occur due only to the chance involved in the random assignment? What conclusion would you make about the effect of the interviewer's appearance?

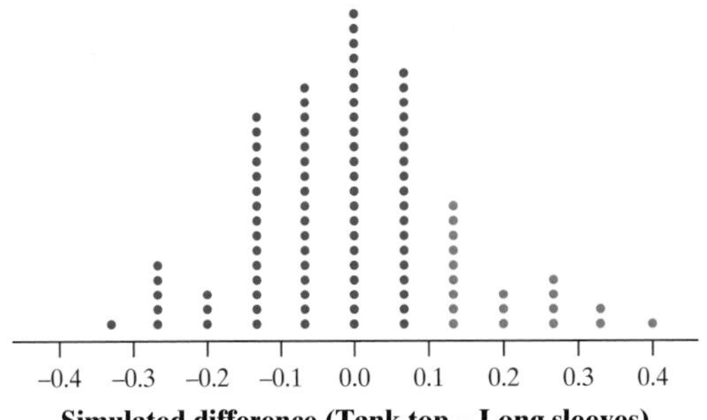

We used the *Two Categorical Variables* applet at www.stapplet.com to perform 100 trials of the simulation described in the activity. In each trial, the applet randomly reassigned the 60 subjects in the experiment to the two treatments, assuming that the treatment received would not affect whether each person says that they like tattoos. Each dot in the graph represents the difference (Tank top − Long sleeves) in the proportions of respondents in the two groups who say they like tattoos for a particular trial.

In this simulation, 19 of the 100 trials (in red) produced a difference in proportions of at least 0.133, so the approximate *P*-value is 0.19. A difference this large could easily occur just due to the chance variation in random assignment! Sarah and Miranda's data do not provide convincing evidence that the appearance of the interviewer causes a greater proportion of people like the ones in the study to say they like tattoos when the interviewer's tattoos are visible.

You'll learn how to conduct a formal significance test for data like these in Section 6F. We start this unit, however, with the details about constructing confidence intervals for a difference between two proportions.

SECTION 6E Confidence Intervals for a Difference in Population Proportions

LEARNING TARGETS *By the end of the section, you should be able to:*

- Check the conditions for calculating a confidence interval about a difference between two population proportions.
- Calculate a confidence interval for a difference between two population proportions.
- Construct and interpret a two-sample *z* interval for a difference in proportions.

In Section 6B, you learned how to construct and interpret a confidence interval for a population proportion p. In this section, you'll learn how to construct and interpret a confidence interval for a difference between two population proportions.

Checking Conditions for a Confidence Interval for $p_1 - p_2$

When data come from two independent random samples or two groups in a randomized experiment, the statistic $\hat{p}_1 - \hat{p}_2$ is our best guess for the value of $p_1 - p_2$. The method we use to calculate a confidence interval for $p_1 - p_2$ requires independence in the data collection and that the sampling distribution of $\hat{p}_1 - \hat{p}_2$ be approximately normal. To verify these things, we check slightly modified versions of the Random, 10%, and Large Counts conditions when calculating a confidence interval for a difference between two population proportions.

CONDITIONS FOR CONSTRUCTING A CONFIDENCE INTERVAL FOR A DIFFERENCE BETWEEN TWO POPULATION PROPORTIONS

- **Random:** The data come from two independent random samples or from two groups in a randomized experiment.
  - **10%:** When sampling without replacement, $n_1 < 0.10N_1$ and $n_2 < 0.10N_2$.
- **Large Counts:** The counts of "successes" and "failures" in each sample or group — $n_1\hat{p}_1$, $n_1(1-\hat{p}_1)$, $n_2\hat{p}_2$, and $n_2(1-\hat{p}_2)$ — are all at least 10.

There are now two ways to satisfy the Random condition. First, using independent random samples allows us to generalize our results to the populations of interest. Second, using random assignment in an experiment permits us to draw cause-and-effect conclusions. Although it would be ideal, it is rare that studies in the real world use *both* random selection and random assignment.

When sampling without replacement from two populations, the 10% condition allows us to view the observations within each sample as independent. As you learned in Section 5C, the formula for the standard deviation of $\hat{p}_1 - \hat{p}_2$ requires this type of independence. The formula also requires that the samples themselves are independent, which is why we require *independent* random samples in the Random condition when the goal is to generalize our results to the two populations.

The method we use to calculate a confidence interval for $p_1 - p_2$ requires that the sampling distribution of $\hat{p}_1 - \hat{p}_2$ be approximately normal. In Section 5C, you learned that this will be true whenever n_1p_1, $n_1(1-p_1)$, n_2p_2, and $n_2(1-p_2)$ are all at least 10. Because we don't know the value of p_1 or p_2 when we are estimating $p_1 - p_2$, we use $\hat{p}_1$ and $\hat{p}_2$ when checking the Large Counts condition.

EXAMPLE

Watching at home?

Checking conditions for a confidence interval for $p_1 - p_2$

Skill 4.C

Klaus Vedfelt/DigitalVision/Getty Images

PROBLEM: With more in-home viewing options available, do people see movies in a movie theater less often now than in the past? Gallup polled 800 randomly selected U.S. adults in 2001 and found that 66% had seen at least one movie in a movie theater during the previous year. In 2021, Gallup surveyed 811 U.S. adults and found that only 39% had been to the movie theater in the previous year.[41] Let p_{2021} = the proportion of all U.S. adults in 2021 who would say they saw at least one movie in a movie theater during the previous year and p_{2001} = the proportion of all U.S. adults in 2001 who would say they saw at least one movie in a movie theater during the previous year. Check that the conditions for calculating a confidence interval for $p_{2021} - p_{2001}$ are met.

SOLUTION:

- Random: Gallup selected independent random samples of U.S. adults in 2001 and 2021. ✓
  - 10%: 800 < 10% of all U.S. adults in 2001 and 811 < 10% of all U.S. adults in 2021. ✓
- Large Counts: 800(0.66) = 528, 800(1 − 0.66) = 272, 811(0.39) = 316.29 ≈ 316, and 811(1 − 0.39) = 494.71 ≈ 495 are all at least 10. ✓

Be sure to mention *independent* random samples (*plural*) from the populations of interest when checking the Random condition for an observational study.

Be sure to check that the counts of "successes" and "failures" in each sample or group—$n_1\hat{p}_1$, $n_1(1-\hat{p}_1)$, $n_2\hat{p}_2$, and $n_2(1-\hat{p}_2)$—are *all* at least 10. Because these are the observed counts of successes and failures, they should be rounded to the nearest integer.

FOR PRACTICE, TRY EXERCISE 1

Calculating a Confidence Interval for $p_1 - p_2$

When the conditions are met, we can use our familiar formula to calculate a confidence interval for $p_1 - p_2$:

$$\text{point estimate} \pm \text{margin of error}$$

Or, in slightly expanded form:

$$\text{statistic} \pm (\text{critical value})(\text{standard error of statistic})$$

The observed difference in sample proportions $\hat{p}_1 - \hat{p}_2$ is our point estimate for $p_1 - p_2$. We can find the critical value z^* for the given confidence level using either technology or Table A, as in Section 6B.

In Section 5C, you learned that the standard deviation of the sampling distribution of $\hat{p}_1 - \hat{p}_2$ is

$$\sigma_{\hat{p}_1-\hat{p}_2} = \sqrt{\frac{p_1(1-p_1)}{n_1} + \frac{p_2(1-p_2)}{n_2}}$$

when we have independent random samples and independent observations within each sample. Because we don't know the values of the parameters p_1 and p_2, we replace them in the standard deviation formula with the sample proportions $\hat{p}_1$ and $\hat{p}_2$. The result is the *standard error* of $\hat{p}_1 - \hat{p}_2$:

$$s_{\hat{p}_1-\hat{p}_2} = \sqrt{\frac{\hat{p}_1(1-\hat{p}_1)}{n_1} + \frac{\hat{p}_2(1-\hat{p}_2)}{n_2}}$$

This value estimates how much the difference in the sample proportions will typically vary from the difference in the population proportions if we repeat the random sampling or random assignment many times. Note that some people use the notation $\text{SE}_{\hat{p}_1-\hat{p}_2}$ rather than $s_{\hat{p}_1-\hat{p}_2}$.

CALCULATING A CONFIDENCE INTERVAL FOR A DIFFERENCE BETWEEN TWO POPULATION PROPORTIONS

When the conditions are met, a $C\%$ confidence interval for $p_1 - p_2$ is

$$(\hat{p}_1 - \hat{p}_2) \pm z^* \sqrt{\frac{\hat{p}_1(1-\hat{p}_1)}{n_1} + \frac{\hat{p}_2(1-\hat{p}_2)}{n_2}}$$

where z^* is the critical value for the standard normal curve with $C\%$ of the area between $-z^*$ and z^*.

AP® EXAM TIP
Formula sheet

The specific formula for a confidence interval for a difference between two population proportions is *not* included on the formula sheet provided on both sections of the AP® Statistics exam. However, the formula sheet does include the general formula for the confidence interval:

$$\text{statistic} \pm (\text{critical value})(\text{standard error of statistic})$$

and the formula for the standard error of the difference in sample proportions:

$$s_{\hat{p}_1-\hat{p}_2} = \sqrt{\frac{\hat{p}_1(1-\hat{p}_1)}{n_1} + \frac{\hat{p}_2(1-\hat{p}_2)}{n_2}}$$

EXAMPLE

Avoiding the theater
Calculating a confidence interval for $p_1 - p_2$

Skill 3.D

PROBLEM: Recall from the preceding example that Gallup polled 800 randomly selected U.S. adults in 2001 and found that 66% had seen at least one movie in a movie theater during the previous year. In 2021, Gallup surveyed 811 randomly selected U.S. adults and found that only 39% had been to the movie theater in the previous year. Calculate a 95% confidence interval for $p_{2021} - p_{2001}$. Note that we verified the conditions for inference were met in the preceding example.

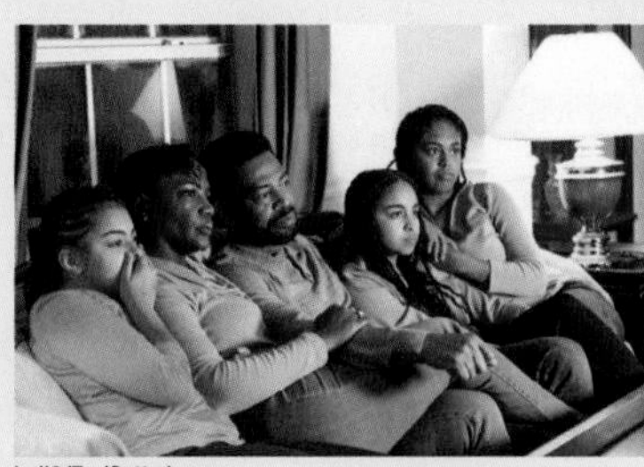

kali9/E+/Getty Images

SOLUTION:

$$(0.39-0.66)\pm 1.960\sqrt{\frac{0.39(1-0.39)}{811}+\frac{0.66(1-0.66)}{800}}$$

$=-0.27\pm 0.047$

$=(-0.317,-0.223)$

$$(\hat{p}_1-\hat{p}_2)\pm z^*\sqrt{\frac{\hat{p}_1(1-\hat{p}_1)}{n_1}+\frac{\hat{p}_2(1-\hat{p}_2)}{n_2}}$$

FOR PRACTICE, TRY EXERCISE 5

The way we interpret confidence intervals is unchanged from Section 6A: we are 95% confident that the interval from −0.317 to −0.223 captures the difference (2021 − 2001) in the proportions of all U.S. adults who saw a movie in a movie theater during the previous year.

Note that the confidence interval does not include 0 (no difference) as a plausible value for $p_{2021} - p_{2001}$, so we have convincing evidence of a difference between the population proportions. In fact, it is believable that $p_{2021} - p_{2001}$ has any value between −0.317 and −0.223. We can restate this in context as follows: the interval suggests that the percentage of all U.S. adults who saw a movie in a movie theater during the previous year decreased between 22.3 and 31.7 percentage points from 2001 to 2021.

Don't say "percent" when you mean "percentage points." A decrease of between 22.3 to 31.7 *percentage points* is not the same as a decrease of between 22.3 and 31.7 *percent*. In fact, the sample proportion decreased by (66 − 39)/66 = 0.41 = 41 percent, but only by 66 − 39 = 27 percentage points.

In the movie theater context, we used independent random samples of U.S. adults in two different years. In other contexts, it's common to select one random sample that includes individuals from both populations of interest and then to separate the chosen individuals into two groups. For instance, a polling company may randomly select 1000 U.S. adults, then separate the Republicans from the Democrats to estimate the difference in the proportion of all people in each party who favor the death penalty. The two-sample z procedures for comparing proportions are still valid in such situations, provided that the two groups can be viewed as independent samples from their respective populations of interest.

You can use technology to perform the calculations in the Do step of the four-step process. Remember that this comes with potential benefits and risks on the AP® Statistics exam.

18. Tech Corner

CONFIDENCE INTERVALS FOR A DIFFERENCE IN PROPORTIONS

TI-Nspire and other technology instructions are on the book's website at bfwpub.com/tps7e.

The TI-83/84 calculator can be used to construct a confidence interval for $p_1 - p_2$. We'll demonstrate using the preceding example. In 2001, 66% of the 800 randomly selected U.S. adults had seen at least one movie in a movie theater during the previous year. In 2021, it was 39% of the 811 randomly selected U.S. adults. To construct a confidence interval:

1. Press STAT, then choose TESTS and 2-PropZInt.

2. When the 2-PropZInt screen appears, enter xl = 316, nl = 811, x2 = 528, n2 = 800, and 0.95 for the confidence level, as shown in the screen shot. Note that these values must be integers, so we used 811(0.39) = 316.29 ≈ 316 for xl and 800(0.66) = 528 for x2.

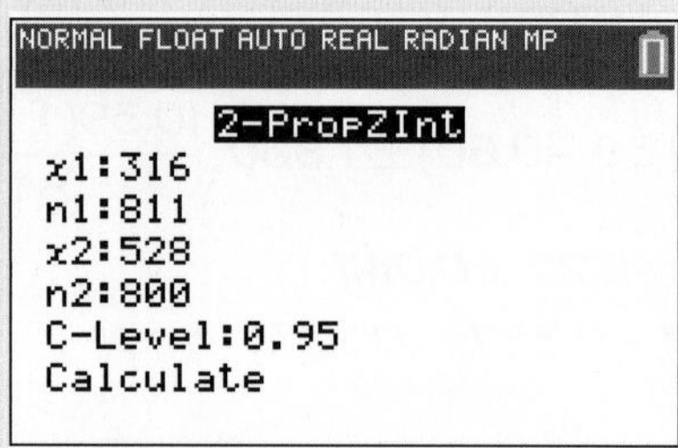

3. Highlight "Calculate" and press ENTER.

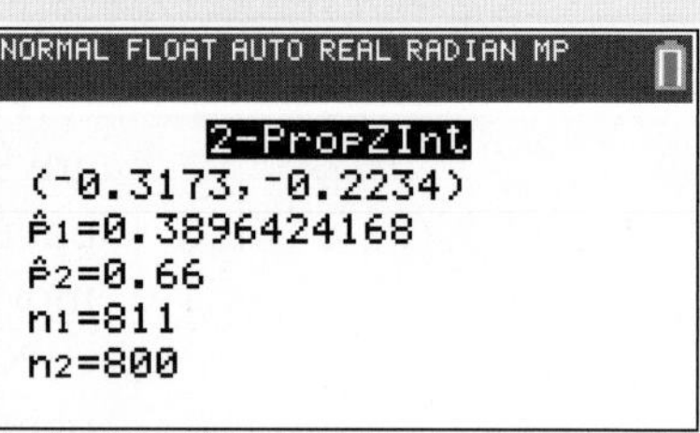

AP® EXAM TIP

The formula for the two-sample z interval for $p_1 - p_2$ often leads to calculation errors by students. As a result, your teacher may recommend using the calculator's 2-PropZInt feature when you are asked to construct and interpret this type of interval on the AP® Statistics exam. If you go this route, make sure to include the remaining three steps (State, Plan, Conclude) and provide the interval (−0.3173, −0.2234) in the Do step.

Confidence intervals calculated using the "full technology" approach described in the Tech Corner are generally more accurate than the values we obtain when we round while doing our calculations. For that reason, we will use the values from the full technology approach when reporting results "Using technology" in the rest of this section.

INFERENCE FOR EXPERIMENTS

So far, we have focused on doing inference using data that were produced by random sampling. However, many important statistical results come from randomized comparative experiments. Fortunately, the formula for calculating a confidence interval for a difference between two proportions is the same whether we have two independent random samples or two randomly assigned groups in an experiment. However, there are differences in the way we define parameters and check conditions.

AP® EXAM TIP

Many students lose credit when defining the parameters in an experiment because they describe the sample proportion rather than the true proportion. For example, "the true proportion of the men in the experiment who *had* surgery and survived 20 years" describes $\hat{p}_S$, not p_S.

In an experiment to compare treatments for prostate cancer, 731 men with localized prostate cancer were randomly assigned either to have surgery or to be observed only. After 20 years, $\hat{p}_S = 141/364 = 0.387$ of the men assigned to surgery were still alive and $\hat{p}_O = 122/367 = 0.332$ of the men assigned to observation were still alive.[42] The parameters in this setting are

p_S = the true proportion of men like the ones in the experiment who would survive 20 years when getting surgery

p_O = the true proportion of men like the ones in the experiment who would survive 20 years when being observed only

Most experiments on people use recruited volunteers as subjects. When subjects are not randomly selected, researchers cannot generalize the results of an experiment to some larger populations of interest. But researchers can draw cause-and-effect conclusions that apply to people like those who took part in the experiment, as long as the treatments were randomly assigned (the Random condition). This same logic applies to experiments on animals or things.

In addition to the difference in the Random condition, there is a change to the 10% condition when analyzing an experiment. Unless the experimental units are randomly selected without replacement from some population, we should not check it. Fortunately, there is no change to the Large Counts condition.

Putting It All Together: Two-Sample *z* Interval for $p_1 - p_2$

We are now ready to use the four-step process to construct and interpret a **two-sample *z* interval for a difference in proportions.**

DEFINITION Two-sample *z* interval for a difference in proportions

A **two-sample *z* interval for a difference in proportions** is a confidence interval used to estimate a difference in the proportions of successes for two populations or treatments.

As with any inference procedure, we follow the four-step process when calculating a confidence interval for a difference between two proportions.

EXAMPLE

Treating lower back pain
Two-sample *z* interval for $p_1 - p_2$

Skills 1.D, 3.D, 4.B, 4.C, 4.D

Syda Productions/Shutterstock

PROBLEM: Patients with lower back pain are often given nonsteroidal anti-inflammatory drugs (NSAIDs) like naproxen to help ease their pain. Researchers wondered if taking Valium along with the naproxen would affect pain relief. To find out, they recruited 112 patients with severe lower back pain and randomly assigned them to one of two treatments: naproxen and Valium or naproxen and placebo. After 1 week, 39 of the 57 subjects who took naproxen and Valium reported reduced lower back pain, compared with 43 of the 55 subjects in the naproxen and placebo group.[43]

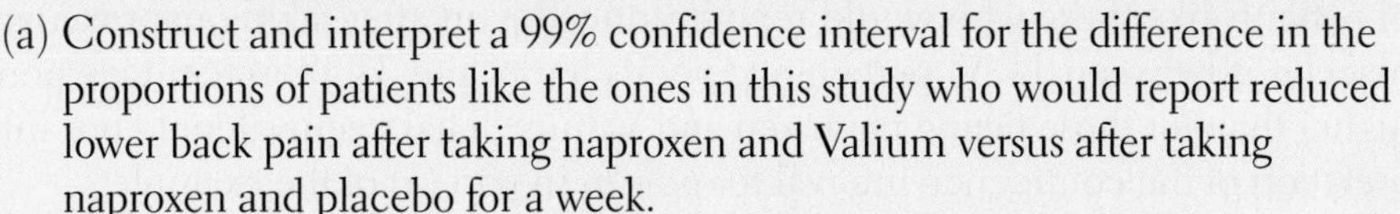

(a) Construct and interpret a 99% confidence interval for the difference in the proportions of patients like the ones in this study who would report reduced lower back pain after taking naproxen and Valium versus after taking naproxen and placebo for a week.

(b) Based on the confidence interval in part (a), what conclusion would you make about whether taking Valium along with naproxen affects pain relief? Justify your answer.

SOLUTION:

(a)

STATE: 99% CI for $p_{NV} - p_{NP}$, where p_{NV} = true proportion of patients like the ones in this study who would report reduced lower back pain after taking naproxen and Valium for a week and p_{NP} = true proportion of patients like the ones in this study who would report reduced lower back pain after taking naproxen and placebo for a week.

Be sure to indicate the order of subtraction when defining the parameter.

PLAN: Two-sample *z* interval for a difference in proportions

- Random: Randomly assigned patients to take naproxen and Valium or naproxen and placebo. ✓
- Large Counts: $39, 57-39=18, 43,$ and $55-43=12$ are all ≥ 10. ✓

It is not appropriate to check the 10% condition because researchers did not sample patients without replacement from a larger population.

DO: $\hat{p}_{NV} = \frac{39}{57} = 0.684$, $\hat{p}_{NP} = \frac{43}{55} = 0.782$

$$(\hat{p}_1 - \hat{p}_2) \pm z^* \sqrt{\frac{\hat{p}_1(1-\hat{p}_1)}{n_1} + \frac{\hat{p}_2(1-\hat{p}_2)}{n_2}}$$

$$(0.684 - 0.782) \pm 2.576\sqrt{\frac{0.684(0.316)}{57} + \frac{0.782(0.218)}{55}}$$

$= -0.098 \pm 0.214$

$= (-0.312, 0.116)$

Using technology: $(-0.3114, 0.1162)$

CONCLUDE: We are 99% confident that the interval from -0.3114 to 0.1162 captures the difference (Valium – Placebo) in the true proportions of patients like the ones in this study who would report reduced pain after taking naproxen and Valium versus after taking naproxen and a placebo for a week.

When the calculations from technology differ from the by-hand calculations, we'll use the values from technology in the conclusion.

(b) Because the interval from -0.3114 to 0.1162 includes 0 as a plausible value for $p_{NV} - p_{NP}$, we don't have convincing evidence that taking Valium along with naproxen affects pain relief for patients like these.

The interval suggests that the true proportion of patients like these who would report reduced pain after taking naproxen and Valium is between 31.14 percentage points lower and 11.62 percentage points higher than for those taking naproxen and placebo.

FOR PRACTICE, TRY EXERCISE 9

We could have subtracted the proportions in the opposite order in part (a) of the example. The resulting 99% confidence interval for $p_{NP} - p_{NV}$ is -0.1162 to 0.3114. Notice that the endpoints of the interval have the same values but opposite signs to the ones in the example. This interval suggests that the true proportion of patients like these who would report reduced pain after taking naproxen and placebo is between 11.62 percentage points lower and 31.14 percentage points higher than for those taking naproxen and Valium. That's equivalent to our interpretation of the confidence interval for $p_1 - p_2$ in part (a) of the example.

The fact that 0 is included in a confidence interval for $p_1 - p_2$ means that we don't have convincing evidence of a difference between the true proportions. Keep in mind that 0 is just one of many plausible values for $p_1 - p_2$ based on the sample data. **Never suggest that you believe the difference between the true proportions *is* 0 just because 0 is in the interval!**

What does it mean to be 99% confident in this context? If we were to repeat the random assignment process many, many times and compute a 99% confidence interval for $p_{NV} - p_{NP}$ each time, about 99% of these intervals would capture the difference (Naproxen and Valium − Naproxen and placebo) in the true proportions of patients like the ones in this study who would report reduced pain after taking naproxen and Valium versus after taking naproxen and a placebo for a week.

CHECK YOUR UNDERSTANDING

Do people use Instagram more now than in years past? A 2021 Pew Research Center survey selected a random sample of 1502 U.S. adults and found that 40% of the sample used Instagram. A similar survey in 2016 selected a random sample of 1520 U.S. adults and found that 32% of the sample used Instagram.[44] Calculate a 90% confidence interval for the difference (2021 − 2016) in the proportions of all U.S. adults who used Instagram in these two years.

SECTION 6E Summary

- Confidence intervals for the difference $p_1 - p_2$ between the proportions of successes in two populations or treatments use the difference $\hat{p}_1 - \hat{p}_2$ between the sample proportions as the point estimate.
- When constructing a confidence interval for a difference in population proportions, we must check for independence in the data collection process and that the sampling distribution of $\hat{p}_1 - \hat{p}_2$ is approximately normal. The required conditions are
  - Random: The data come from two independent random samples or from two groups in a randomized experiment.
    - 10%: When sampling without replacement, $n_1 < 0.10N_1$ and $n_2 < 0.10N_2$.
  - Large Counts: The counts of "successes" and "failures" in each sample or group — $n_1\hat{p}_1$, $n_1(1-\hat{p}_1)$, $n_2\hat{p}_2$, and $n_2(1-\hat{p}_2)$ — are all at least 10.
- When conditions are met, a C% confidence interval for $p_1 - p_2$ is

$$(\hat{p}_1 - \hat{p}_2) \pm z^* \sqrt{\frac{\hat{p}_1(1-\hat{p}_1)}{n_1} + \frac{\hat{p}_2(1-\hat{p}_2)}{n_2}}$$

  where z^* is the critical value for the standard normal curve with C% of its area between $-z^*$ and z^*. This is called a **two-sample z interval for a difference in proportions.**
- When performing inference for experiments, make sure to include "for individuals like the ones in the study" and avoid talking about the sample proportions when you are defining parameters and making your conclusion.
- Be sure to follow the four-step process whenever you construct and interpret a confidence interval for the difference between two proportions.

AP® EXAM TIP
AP® Daily Videos
Review the content of this section and get extra help by watching the AP® Daily Videos for Topics 6.8 and 6.9, which are available in AP® Classroom.

- You can use a confidence interval for a difference in proportions to determine if a claimed value is plausible. For example, if 0 is included in a confidence interval for $p_1 - p_2$, it is plausible that there is no difference between the population proportions. In other words, when 0 is in the confidence interval for $p_1 - p_2$, there is not convincing evidence of a difference between p_1 and p_2.

6E Tech Corner

TI-Nspire and other technology instructions are on the book's website at bfwpub.com/tps7e.

18. Confidence intervals for a difference in proportions Page 613

SECTION 6E Exercises

Checking Conditions for a Confidence Interval for $p_1 - p_2$

1. **Don't drink the water!** The movie *A Civil Action* tells the story of a major legal battle that took place in the small town of Woburn, Massachusetts. A town well that supplied water to east Woburn residents was contaminated by industrial chemicals. During the period that residents drank water from this well, 16 of 414 babies born had birth defects. On the west side of Woburn, 3 of 228 babies born during the same time period had birth defects. Let p_1 = the population proportion of babies born with birth defects in west Woburn and p_2 = the population proportion of babies born with birth defects in east Woburn. Check if the conditions for calculating a confidence interval for $p_1 - p_2$ are met. pg 611

2. **Broken crackers** We don't like to find broken crackers when we open the package. How can cracker makers reduce breaking? One idea is to microwave the crackers for 30 seconds right after baking them. Randomly assign 65 newly baked crackers to the microwave and another 65 to a control group that is not microwaved. After 1 day, none of the microwave group were broken and 16 of the control group were broken.[45] Let p_1 = the proportion of all crackers like the ones in this study that would break if baked in the microwave and p_2 = the proportion of all crackers like the ones in this study that would break if not microwaved. Check if the conditions for calculating a confidence interval for $p_1 - p_2$ are met.

3. **Fish on the moon** French researchers set an ambitious goal to farm fish on the moon. They performed an experiment to determine whether certain varieties of fish can be safely transported on a rocket into space. The researchers randomly assigned 400 European seabass eggs into two groups of 200 eggs each. They placed all of the eggs in both groups in a dish filled with seawater. They then placed the first group of eggs in a vibration chamber designed to simulate a typical takeoff. They kept the second group of eggs in similar environmental conditions, but with no vibrations. In the vibration group, 76% of the eggs went on to hatch, compared to 82% of the eggs in the control group.[46] Let p_V = the proportion of all eggs like the ones in this study that would hatch after being in a vibration chamber and p_{NV} = the proportion of all eggs like the ones in this study that would hatch if not vibrated. Check if the conditions for calculating a confidence interval for $p_V - p_{NV}$ are met.

4. **Name-brand clothes** A Harris Interactive survey asked random samples of adults from the United States and Germany about the importance of brand names when buying clothes. Of the 2309 U.S. adults surveyed, 26% said brand names were important, compared with 22% of the 1058 German adults surveyed.[47] Let p_U = the proportion of all U.S. adults who think brand names are important when buying clothes and p_G = the proportion of all German adults who think brand names are important when buying clothes. Check if the conditions for calculating a confidence interval for $p_U - p_G$ are met.

Calculating a Confidence Interval for $p_1 - p_2$

5. **More fish on the moon** Refer to Exercise 3. Calculate a 90% confidence interval for $p_V - p_{NV}$. pg 612

6. **More name-brand clothes** Refer to Exercise 4. Calculate a 99% confidence interval for $p_U - p_G$.

7. **Confident fish** Interpret the confidence level for the interval in Exercise 5.

8. **Confident clothes** Interpret the confidence level for the interval in Exercise 6.

Two-Sample *z* Interval for $p_1 - p_2$

9. **I want candy!** In an experiment carried out at Cambridge University, researchers wanted to determine if moving candy closer to people in a waiting room would increase the proportion of subjects who ate the candy. They randomly assigned subjects to a waiting room with a bowl of candy placed near the seating location (20 centimeters, cm) or far from the seating location (70 cm). Of the 61 subjects assigned to sit near the bowl, 39 of them ate the candy, while only 24 of the 61 subjects assigned to sit far from the bowl ate the candy.[48]

pg 615

(a) Construct and interpret a 90% confidence interval for the difference in the proportion of subjects like the ones in this study who would eat the candy when it is placed nearby and the proportion who would eat the candy when it is placed farther away.

(b) Based on the confidence interval in part (a), what conclusion would you make about whether the distance from a bowl of candy affects whether people eat from it? Justify your answer.

10. **Supportive texts** What can help young adults quit vaping? Researchers recruited more than 2500 young (18–24 years old) U.S. adults who had vaped recently and were interested in quitting. All subjects in the experiment were given monthly assessments via text message about e-cigarette use. The control group of 1284 subjects received no other intervention. The treatment group of 1304 subjects also received encouraging text messages that delivered social support and cognitive and behavioral coping skills training. At the end of the study, 239 members of the control group and 314 members of the treatment group had abstained from vaping for at least the previous 30 days.[49]

(a) Construct and interpret a 90% confidence interval for the difference in the proportion of subjects like the ones in this study who would abstain from vaping for at least 30 days when receiving encouraging texts and the proportion who would abstain when only receiving monthly assessments.

(b) Based on the confidence interval in part (a), what conclusion would you make about whether the encouraging texts make a difference? Justify your answer.

11. **Christmas trees** An association of Christmas tree growers in Indiana wants to know if there is a difference in preference for natural trees between urban and rural households. To investigate, the association sponsored a survey of a random sample of Indiana households that had a Christmas tree last year. Of the 160 rural households surveyed, 64 had a natural tree. Of the 261 urban households surveyed, 89 had a natural tree.[50] Even though the association selected a single random sample, it is reasonable to consider these as independent random samples of rural and urban households with Christmas trees in Indiana. Construct and interpret a 95% confidence interval for the difference (Rural – Urban) in the proportions of all rural and urban Indiana households with Christmas trees that had a natural tree last year.

12. **Partisan politics** The Harvard Youth Poll asked a random sample of young adults (ages 18 to 29) in the United States, "Do you agree or disagree with this statement: Politics has become too partisan?" Of the 1083 respondents who identified as Democrats, 57% said "Agree." Of the 564 respondents who identified as Republicans, 59% said "Agree."[51] Even though the Harvard Youth Poll selected a single random sample, it is reasonable to consider these as independent random samples from the populations of 18- to 29-year-old Democrats and Republicans in the United States. Construct and interpret a 95% confidence interval for the difference (Democrat – Republican) in the proportions of all U.S. young adult Democrats and Republicans who would say that politics has become too partisan.

13. **Vitamin D and depression** In some observational studies, researchers have found that people with higher concentrations of vitamin D in their blood are less likely to be diagnosed with depression. Can taking a vitamin D supplement help prevent depression? In an experiment, 9181 people were randomly assigned to take a vitamin D supplement and 9172 were randomly assigned to take a placebo. After 5 years, 6.6% of the vitamin D group and 6.8% of the placebo group were diagnosed with depression.[52] With 95% confidence, the interval estimate for the difference (Vitamin D – Placebo) in the proportions of people like the ones in this study who would be diagnosed with depression after 5 years is (−0.0093, 0.0052).

(a) Does the confidence interval provide convincing evidence of a difference in the effectiveness of the two treatments? Explain your answer.

(b) Explain what is misleading about the following headline: "Study shows vitamin D supplementation has no effect on depression."

14. **Aspirin and dementia** In an experiment that investigated the effectiveness of low-dose aspirin in preventing dementia, researchers recruited 19,114 participants aged 70 or older who were free from cardiovascular disease, physical disability, and dementia. Participants were randomly assigned to take either low-dose aspirin or a placebo daily. Of the 9525 participants who took low-dose aspirin, 283 developed dementia within

5 years. Of the 9589 participants who took placebo, 292 developed dementia within 5 years.[53] With 95% confidence, the interval estimate for the difference (Aspirin – Placebo) in the proportions of people like the ones in this study who would develop dementia after 5 years is $(-0.0056, 0.0041)$.

(a) Does the confidence interval provide convincing evidence of a difference in the effectiveness of the two treatments? Explain your answer.

(b) Explain what is misleading about the following headline: "Study shows aspirin is not effective in preventing dementia."

For Investigation *Apply the skills from the section in a new context or nonroutine way.*

15. **Text reminders** Will sending previous customers a text message reminder encourage people to get their annual flu vaccine? Researchers will conduct an experiment with Walmart pharmacy patients to help answer this question. Some pharmacy patients will be randomly assigned to "business as usual" — no text reminders. Others will be randomly assigned to receive a text message reminder to get a flu vaccine.[54]

(a) After the data are collected, the researchers will calculate a 95% confidence interval for the difference (Text – No text) in the proportions of Walmart pharmacy patients who would get a flu vaccine. Assuming that the text message has no effect, what is the probability that the interval will exclude 0?

(b) Which type of error, Type I or Type II, is being described in part (a)? Explain your answer.

(c) How can the researchers reduce the probability of the error described in part (b)?

(d) The researchers actually plan to compare each of 22 different text message reminders against getting no text reminder. If none of the 22 text messages is effective, how many of the resulting 95% confidence intervals would you expect to exclude 0?

(e) After the experiment, researchers calculated a 95% confidence interval for the difference (Text – No text) in the proportions of Walmart pharmacy patients who would get the flu vaccine for each of the 22 different text messages. If one of the intervals excluded 0, should we conclude that the corresponding text message was effective? Explain your reasoning. *Note:* In the actual study, each of the 22 confidence intervals excluded 0 and had only positive values.

16. **Partial doses** Is the occurrence of side effects from a flu vaccine the same whether you get a full dose or a half dose? Or does a half dose lead to fewer side effects? Researchers set out to study this question by randomly allocating 1259 patients to receive either the full dose (628 patients) or a half dose (631 patients) of the vaccine. Afterward, they measured how many patients in each group experienced a headache. Of those receiving a full dose, 37 reported having had a headache; of those receiving a half dose, 28 reported having a headache.[55]

(a) Construct and interpret a 95% confidence interval for the difference (Full – Half) in the proportions of subjects like the ones in this study who would experience a headache when taking a full dose of the vaccine versus a half dose.

(b) What if patients received a three-fourths (75%) dose? One way to estimate the proportion of patients like these who would experience a headache when taking a three-fourths dose is by averaging the proportion who experienced a headache with a full dose and the proportion who experienced a headache with a half dose. Compute this estimate, along with the standard error of this estimate.

(c) Consider a scatterplot showing the relationship between x = dosage of the vaccine and y = the proportion of patients who experience a headache. What must be true about this relationship for the average calculated in part (b) to be an accurate estimate of the proportion of patients like these who would experience a headache when taking a three-fourths dose?

Multiple Choice *Select the best answer for each question.*

17. Earlier in this section, you read about an experiment comparing surgery and observation as treatments for men with prostate cancer. After 20 years, $\hat{p}_S = 141/364 = 0.387$ of the men who were assigned to surgery were still alive and $\hat{p}_O = 122/367 = 0.332$ of the men who were assigned to observation were still alive. Which of the following is the 95% confidence interval for $p_S - p_O$?

(A) $(141-122) \pm 1.96\sqrt{\frac{141 \cdot 223}{364} + \frac{122 \cdot 245}{367}}$

(B) $(141-122) \pm 1.96\left(\sqrt{\frac{141 \cdot 223}{364}} + \sqrt{\frac{122 \cdot 245}{367}}\right)$

(C) $(0.387-0.332) \pm 1.96\sqrt{\frac{0.387 \cdot 0.613}{364} + \frac{0.332 \cdot 0.668}{367}}$

(D) $(0.387-0.332) \pm 1.96\left(\sqrt{\frac{0.387 \cdot 0.613}{364}} + \sqrt{\frac{0.332 \cdot 0.668}{367}}\right)$

(E) $(0.387-0.332) \pm 1.96\sqrt{\frac{0.387 \cdot 0.613}{364} - \frac{0.332 \cdot 0.668}{367}}$

Calculating the Standardized Test Statistic and *P*-Value for a Test about $p_1 - p_2$

If the conditions are met, we can proceed with calculations. To do a test of $H_0\colon p_1 - p_2 = 0$, standardize $\hat{p}_1 - \hat{p}_2$ to get a z statistic.

$$\text{standardized test statistic} = \frac{\text{statistic} - \text{parameter}}{\text{standard error of statistic}}$$

To calculate the standard error, we use $\hat{p}_C$ instead of $\hat{p}_1$ and $\hat{p}_2$ for the same reason we did when checking the Large Counts condition — because we are doing our calculations assuming the null hypothesis of $p_1 = p_2$ is true.

$$s_{\hat{p}_1-\hat{p}_2} = \sqrt{\frac{\hat{p}_C(1-\hat{p}_C)}{n_1} + \frac{\hat{p}_C(1-\hat{p}_C)}{n_2}} = \sqrt{\hat{p}_C(1-\hat{p}_C)\left(\frac{1}{n_1} + \frac{1}{n_2}\right)}$$

When the Large Counts condition is met, the z statistic

$$z = \frac{(\hat{p}_1 - \hat{p}_2) - 0}{\sqrt{\hat{p}_C(1-\hat{p}_C)\left(\frac{1}{n_1} + \frac{1}{n_2}\right)}}$$

will have approximately the standard normal distribution. We can find the appropriate *P*-value using technology or Table A.

CALCULATING THE STANDARDIZED TEST STATISTIC AND *P*-VALUE IN A TEST ABOUT THE DIFFERENCE BETWEEN TWO POPULATION PROPORTIONS

Suppose the conditions are met. To perform a test of $H_0\colon p_1 - p_2 = 0$, calculate the standardized test statistic

$$z = \frac{(\hat{p}_1 - \hat{p}_2) - 0}{\sqrt{\hat{p}_C(1-\hat{p}_C)\left(\frac{1}{n_1} + \frac{1}{n_2}\right)}} \quad \text{where } \hat{p}_C = \frac{X_1 + X_2}{n_1 + n_2}$$

Find the *P*-value by calculating the probability of getting a z statistic this large or larger in the direction specified by the alternative hypothesis H_a in the standard normal distribution.

AP® EXAM TIP

Formula sheet

The specific formula for the standardized test statistic for a difference in proportions is *not* included on the formula sheet provided on both sections of the AP® Statistics exam. However, the formula sheet does include the general formula for the standardized test statistic:

$$\text{standardized test statistic} = \frac{\text{statistic} - \text{parameter}}{\text{standard error of statistic}}$$

and the formula for the standard error of the difference in sample proportions when $p_1 = p_2$ is assumed, along with the formula for $\hat{p}_C$:

$$\sqrt{\hat{p}_C(1-\hat{p}_C)\left(\frac{1}{n_1}+\frac{1}{n_2}\right)} \quad \text{where } \hat{p}_C = \frac{X_1+X_2}{n_1+n_2}$$

EXAMPLE

Accurate orders

Calculating the standardized test statistic and P-value for a test about $p_1 - p_2$

Skills 3.E, 4.E

jessicaphoto/E+/Getty Images

PROBLEM: Refer to the previous two examples. In the 159 drive-thru orders at McDonald's, 145 were accurate. At Wendy's, 139 of the 163 drive-thru orders were accurate. Is there convincing evidence of a difference in the population proportion of accurate drive-thru orders at McDonald's and Wendy's?

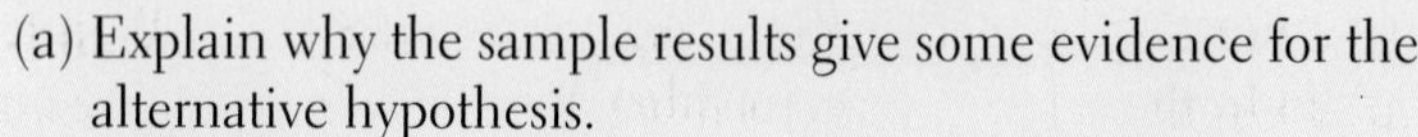

(a) Explain why the sample results give some evidence for the alternative hypothesis.

(b) Calculate the standardized test statistic and P-value.

(c) What conclusion would you make using $\alpha = 0.05$?

SOLUTION:

(a) The observed difference in the sample proportions is $\hat{p}_M - \hat{p}_W = \frac{145}{159} - \frac{139}{163} = 0.912 - 0.853 = 0.059$, which gives some evidence in favor of H_a: $p_M - p_W \neq 0$ because $0.059 \neq 0$.

(b) $\hat{p}_C = \frac{145+139}{159+163} = 0.882$

$$\hat{p}_C = \frac{X_1+X_2}{n_1+n_2}$$

- $z = \frac{(0.912-0.853)-0}{\sqrt{0.882(1-0.882)\left(\frac{1}{159}+\frac{1}{163}\right)}} = 1.641$

$$z = \frac{(\hat{p}_1-\hat{p}_2)-0}{\sqrt{\hat{p}_C(1-\hat{p}_C)\left(\frac{1}{n_1}+\frac{1}{n_2}\right)}}$$

- P-value:

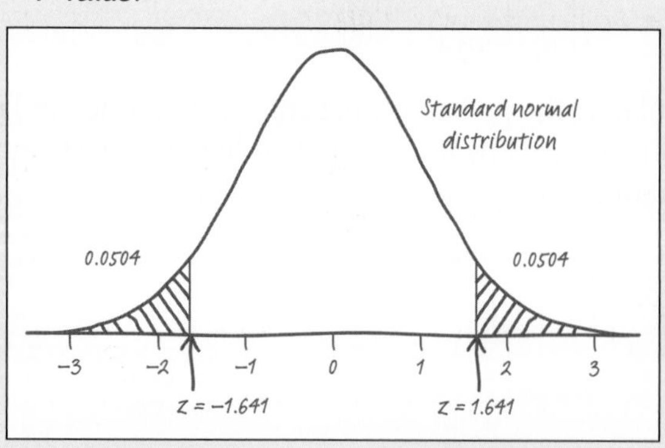

Using technology: 2 × normalcdf(lower: 1.641, upper: 1000, mean: 0, SD: 1) = 2(0.0504) = 0.1008

Using Table A: 2(1 − 0.9495) = 0.1010

(c) Because the P-value of $0.1008 > \alpha = 0.05$, we fail to reject H_0. There is not convincing evidence of a difference in the population proportions of accurate drive-thru orders at McDonald's and Wendy's.

FOR PRACTICE, TRY EXERCISE 9

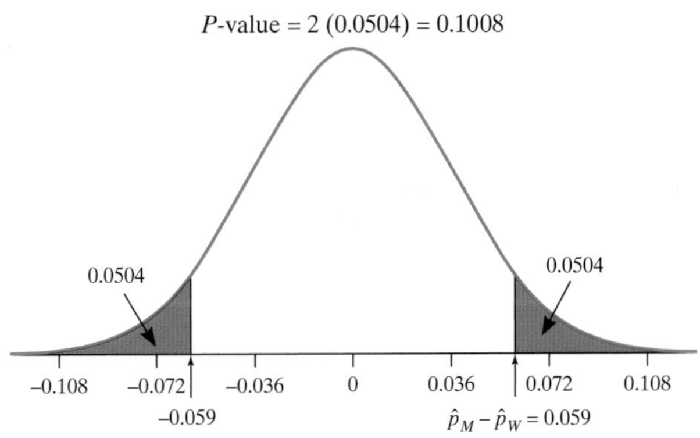

Difference (McDonald's – Wendy's) in sample proportion of accurate orders when $p_M - p_W = 0$

What does the *P*-value in the example tell us? If there is no difference in the population proportions of accurate drive-thru orders at McDonald's and Wendy's and we repeated the random sampling process many times, we'd get a difference in sample proportions as large as or larger than 0.059 in either direction about 10% of the time.

Because the probability of getting a result like this just by chance when the null hypothesis is true isn't that small, we don't have enough evidence to reject H_0. Of course, it is possible we made a Type II error — failing to find convincing evidence that the population proportions are different when they really are different.

We can get additional information about the difference between the population proportions of accurate drive-thru orders at McDonald's and Wendy's with a confidence interval. Technology gives the 95% confidence interval for $p_M - p_W$ as -0.011 to 0.129. That is, we are 95% confident that the population proportion of accurate drive-thru orders at McDonald's is between 1.1 percentage points lower and 12.9 percentage points higher than at Wendy's. Because 0 is included in the interval of plausible values for $p_M - p_W$, the confidence interval also leads to a "fail to reject H_0" conclusion. However, the two-sample *z* test and two-sample *z* interval for the difference between two proportions don't always give consistent results. That's because the test and confidence interval use slightly different standard errors.

19. Tech Corner — SIGNIFICANCE TESTS FOR A DIFFERENCE IN PROPORTIONS

TI-Nspire and other technology instructions are on the book's website at bfwpub.com/tps7e.

The TI-83/84 calculator can be used to perform significance tests for comparing two proportions when the null hypothesis is no difference. Here, we use the data from the "Accurate orders" example. To perform a test of H_0: $p_1 - p_2 = 0$ versus H_a: $p_1 - p_2 \neq 0$:

1. Press STAT, then choose TESTS and 2-PropZTest.
2. When the 2-PropZTest screen appears, enter x1 = 145, n1 = 159, x2 = 139, n2 = 163, and p1 ≠ p2 for the alternative hypothesis, as shown in the screen shot. Note that these values must be integers!

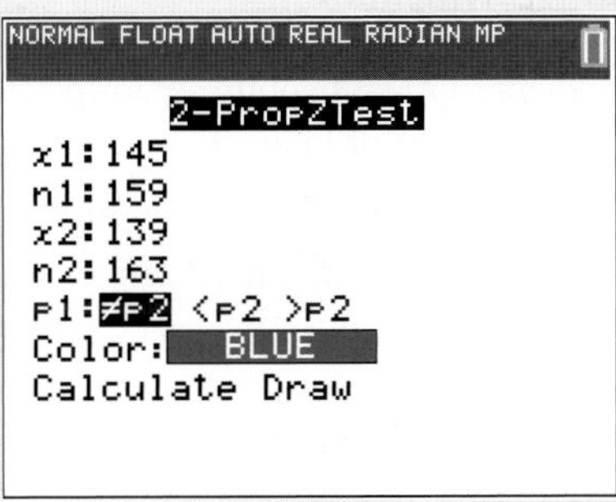

3. Highlight "Calculate" and press ENTER. The output includes the standardized test statistic and P-value, along with the sample proportions and sample sizes.

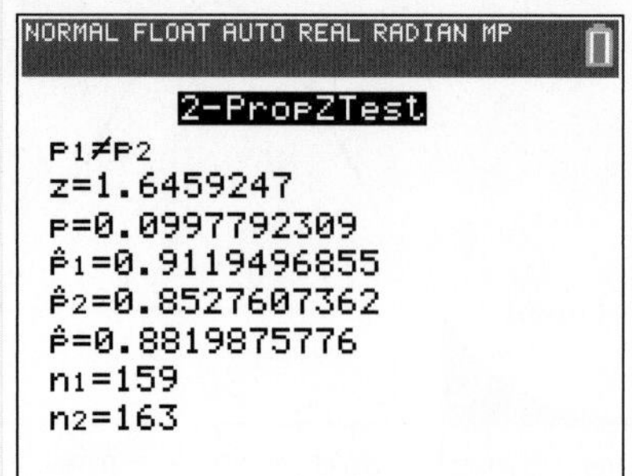

Notes:

- The value of the test statistic and P-value from technology might differ slightly from the values calculated by hand because of rounding.
- The calculator also gives the value of the combined sample proportion, but it uses the symbol $\hat{p}$ instead of $\hat{p}_C$.
- If you select the "Draw" option, you will see a picture of the standard normal distribution with the area of interest shaded, the value of the standardized test statistic, and the P-value.

AP® EXAM TIP

The formula for the two-sample *z* test for $H_0: p_1 - p_2 = 0$ often leads to calculation errors by students. As a result, your teacher may recommend using the calculator's 2-PropZTest feature to perform calculations on the AP® Statistics exam. If you go this route, make sure to include the remaining three steps (State, Plan, Conclude) and report the standardized test statistic ($z = 1.646$) and P-value (0.0998) in the Do step.

Test statistics and P-values calculated using the "full technology" approach described in the Tech Corner are generally more accurate than the values we obtain when we round while doing our calculations. For that reason, we will use the values from the full technology approach when reporting results "Using technology" in the rest of this section.

CHECK YOUR UNDERSTANDING

Restless legs syndrome (RLS) causes a powerful urge to move your legs — so much so that it becomes uncomfortable to sit or lie down. Sleep is difficult. Researchers conducted an experiment to determine if the drug pramipexole is effective in treating RLS. They randomly assigned patients to one of two groups: one group was treated with pramipexole, the other with a placebo. Of the 193 subjects in the pramipexole group, 158 reported "much improved" symptoms. In comparison, 50 of the 92 subjects in the placebo group reported "much improved" symptoms.[58]

1. State appropriate hypotheses for performing a significance test. Be sure to define the parameters of interest.
2. Check if the conditions for performing the test are met.
3. Calculate the standardized test statistic and P-value.
4. What conclusion would you make using $\alpha = 0.05$?

Putting It All Together: Two-Sample z Test for $p_1 - p_2$

We are now ready to use the four-step process to perform a **two-sample z test for a difference in proportions.**

> **DEFINITION Two-sample z test for a difference in proportions**
>
> A **two-sample z test for a difference in proportions** is a significance test of the null hypothesis that the difference in the proportions of successes for two populations or treatments is equal to 0.

As with any test, be sure to follow the four-step process.

EXAMPLE Cholesterol and heart attacks

Two-sample z test for $p_1 - p_2$

Skills 1.E, 1.F, 3.E, 4.B, 4.C, 4.E

jjubaphoto/E+/Getty Images

PROBLEM: The Helsinki Heart Study recruited middle-aged men with high cholesterol levels but no history of other serious medical problems to investigate whether a cholesterol-reducing drug could lower the risk of heart attacks. The volunteer subjects were assigned at random to one of two treatments: 2051 men took the drug gemfibrozil to reduce their cholesterol levels, and a control group of 2030 men took a placebo. During the next five years, 56 men in the gemfibrozil group and 84 men in the placebo group had heart attacks.[59]

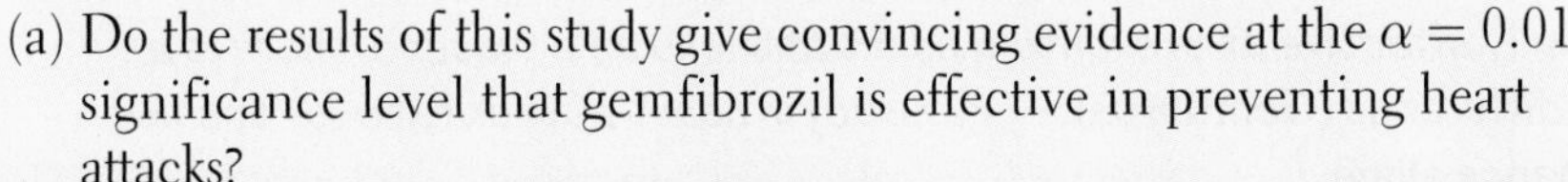

(a) Do the results of this study give convincing evidence at the $\alpha = 0.01$ significance level that gemfibrozil is effective in preventing heart attacks?

(b) Interpret the P-value from part (a).

SOLUTION:

(a)

STATE: We want to test

$H_0: p_G - p_{PL} = 0$

$H_a: p_G - p_{PL} < 0$

where p_G = the true heart attack rate for middle-aged men like the ones in this study who take gemfibrozil and p_{PL} = the true heart attack rate for middle-aged men like the ones in this study who take a placebo.

Use $\alpha = 0.01$.

> We could have subtracted in the opposite order when stating hypotheses:
>
> $H_0: p_{PL} - p_G = 0$
>
> $H_a: p_{PL} - p_G > 0$
>
> Make sure the direction of the alternative hypothesis corresponds to the order you select.

PLAN: Two-sample z test for a difference in proportions

- Random: Volunteer subjects were randomly assigned to gemfibrozil or placebo. ✓
- Large Counts: $\hat{p}_C = \dfrac{56+84}{2051+2030} = 0.0343$

  $2051(0.0343) = 70.35, 2051(1-0.0343) = 1980.65,$
  $2030(0.0343) = 69.63,$ and $2030(1-0.0343) = 1960.37$
  are all ≥ 10. ✓

> It is inappropriate to check the 10% condition because the subjects in the experiment were not sampled without replacement from some larger population.

> Check that $n_1\hat{p}_C$, $n_1(1-\hat{p}_C)$, $n_2\hat{p}_C$, $n_2(1-\hat{p}_C) \geq 10$ where $\hat{p}_C = \dfrac{X_1 + X_2}{n_1 + n_2}$.

DO: $\hat{p}_G = \dfrac{56}{2051} = 0.0273$, $\hat{p}_{PL} = \dfrac{84}{2030} = 0.0414$

> The sample result gives *some* evidence in favor of H_a because $0.0273 - 0.0414 = -0.0141 < 0$.

- $z = \dfrac{(0.0273 - 0.0414) - 0}{\sqrt{0.0343(1-0.0343)\left(\dfrac{1}{2051} + \dfrac{1}{2030}\right)}} = -2.475$

$$z = \frac{(\hat{p}_1 - \hat{p}_2) - 0}{\sqrt{\hat{p}_C(1-\hat{p}_C)\left(\frac{1}{n_1} + \frac{1}{n_2}\right)}}$$

- *P*-value:

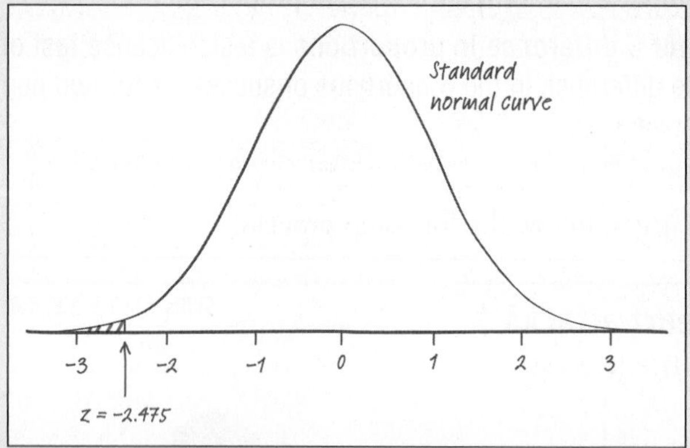

Using technology: $z = -2.470$, *P*-value $= 0.0068$

Using Table A: 0.0066

The *P*-value given for Table A uses $z = -2.48$. If the *z*-score is rounded to -2.47, the *P*-value would be 0.0068.

CONCLUDE: Because our *P*-value of $0.0068 < \alpha = 0.01$, we reject H_0. There is convincing evidence of a lower heart attack rate for middle-aged men like these who take gemfibrozil than for those who take only a placebo.

(b) Assuming that gemfibrozil is equally effective at preventing heart attacks as a placebo, there is a 0.0068 probability of getting a difference (Gemfibrozil − Placebo) in heart attack rate for the two groups of −0.0141 or less by chance alone.

FOR PRACTICE, TRY EXERCISE 13

We chose $\alpha = 0.01$ in the example to reduce the chance of making a Type I error — finding convincing evidence that gemfibrozil reduces heart attack risk when it really doesn't. This error could have serious consequences if an ineffective drug was given to lots of middle-aged men with high cholesterol levels! If we wanted to reduce the probability of a Type II error — and increase the power — we could have used $\alpha = 0.10$. This would make it more likely to find convincing evidence that gemfibrozil helps reduce heart attack risk if it really does help. And if the drug helps, the company and the potential patients would want to know.

Because the results were statistically significant and the subjects were randomly assigned to treatments, the researchers can say that gemfibrozil reduces the rate of heart attacks for middle-aged men like those who took part in the experiment. Because the subjects were not randomly selected from a larger population, researchers should not generalize the findings of this study any further.

Think About It

WHY DO THE INFERENCE METHODS FOR RANDOM SAMPLING WORK FOR RANDOMIZED EXPERIMENTS? Confidence intervals and tests for $p_1 - p_2$ are based on the sampling distribution of $\hat{p}_1 - \hat{p}_2$. But in most experiments, researchers don't select subjects at random from any larger populations. They do randomly assign subjects to treatments. We can think about what would happen if the random assignment were repeated many times under the assumption that H_0: $p_1 - p_2 = 0$ is true. That is, we assume that the specific treatment received doesn't affect an individual subject's response.

Let's see what would happen just by chance if we randomly reassign the 4081 subjects in the Helsinki Heart Study to the two groups many times, assuming the drug received *doesn't affect* whether or not each individual has a heart attack. We used software to redo the random assignment 500 times. The dotplot shows the value of $\hat{p}_G - \hat{p}_{PL}$ in the 500 simulated trials.

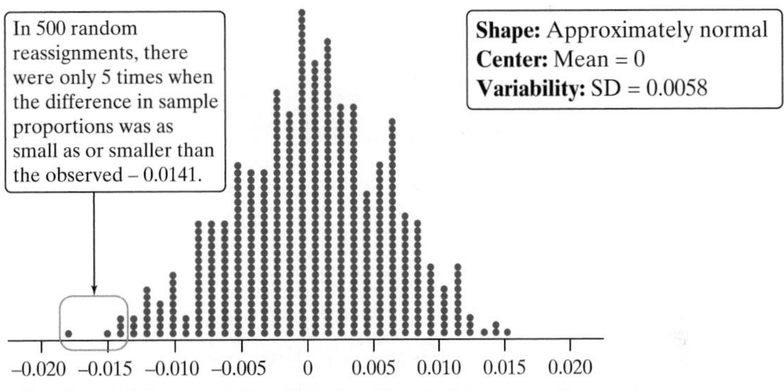

This distribution (sometimes referred to as the *randomization distribution* of $\hat{p}_G - \hat{p}_{PL}$) has an approximately normal shape with mean 0 and standard deviation 0.0058. This matches well with the distribution we used to perform calculations in the example. Because the Large Counts condition was met and we assumed that H_0: $p_G - p_{PL} = 0$ is true, we used a standard error of

$$\sqrt{0.0343(1-0.0343)\left(\frac{1}{2051}+\frac{1}{2030}\right)} = 0.0057$$

This is very close to the value of 0.0058 from the simulation!

In the Helsinki Heart Study, the difference in the proportions of subjects who had a heart attack in the gemfibrozil and placebo groups was $0.0273 - 0.0414 = -0.0141$. How likely is it that a difference this large or larger would happen just by chance when H_0 is true? The dotplot provides a rough answer: 5 of the 500 random reassignments yielded a difference in proportions less than or equal to -0.0141. That is, our estimate of the *P*-value is 0.01. This is quite close to the *P*-value of 0.0068 that we calculated in the preceding example, again suggesting that it's OK to use inference methods for random sampling to analyze randomized experiments.

CHECK YOUR UNDERSTANDING

To study the long-term effects of preschool programs for children living in poverty, researchers designed an experiment. They recruited 123 children who had never attended preschool from low-income families in Michigan. The researchers randomly assigned 62 of the children to attend preschool (paid for by the study budget) and the other 61 to serve as a control group who would not go to preschool. One response variable of interest was the need for social services as adults. Over a 10-year period, 38 children in the preschool group and 49 in the control group have needed social services.[60]

1. Do these data provide convincing evidence that preschool reduces the later need for social services for children like the ones in this study? Justify your answer.
2. Based on your conclusion to Question 1, could you have made a Type I error or a Type II error? Explain your reasoning.
3. Should you generalize the result in Question 1 to all children from low-income families who have never attended preschool? Why or why not?

SECTION 6F Summary

- Tests for the difference $p_1 - p_2$ between the proportions of successes in two populations or treatments are based on the difference $\hat{p}_1 - \hat{p}_2$ between the sample proportions.
- The usual null hypothesis for a significance test about the difference between two population proportions is H_0: $p_1 - p_2 = 0$ or H_0: $p_1 = p_2$. The alternative hypothesis says what kind of difference we expect.
- When testing a claim about a difference in proportions, we must check for independence in the data collection process and that the sampling distribution of $\hat{p}_1 - \hat{p}_2$ is approximately normal. The required conditions are
  - Random: The data come from two independent random samples or from two groups in a randomized experiment.
    - 10%: When sampling without replacement, $n_1 < 0.10N_1$ and $n_2 < 0.10N_2$.
  - Large Counts: The expected counts of successes and failures in each sample or group — $n_1\hat{p}_C$, $n_1(1-\hat{p}_C)$, $n_2\hat{p}_C$, and $n_2(1-\hat{p}_C)$ — are all at least 10, where $\hat{p}_C$ is the combined (pooled) sample proportion. To calculate $\hat{p}_C$, combine the two samples and divide the total number of successes by the total sample size:

$$\hat{p}_C = \frac{X_1 + X_2}{n_1 + n_2}$$

- When conditions are met, the **two-sample *z* test for a difference in proportions** uses the standardized test statistic

$$z = \frac{(\hat{p}_1 - \hat{p}_2) - 0}{\sqrt{\hat{p}_C(1-\hat{p}_C)\left(\frac{1}{n_1} + \frac{1}{n_2}\right)}}$$

with *P*-values calculated from the standard normal distribution.

- Be sure to follow the four-step process whenever you perform a significance test about a difference in proportions.

AP® EXAM TIP
AP® Daily Videos

Review the content of this section and get extra help by watching the AP® Daily Videos for Topics 6.10 and 6.11, which are available in AP® Classroom.

6F Tech Corner

TI-Nspire and other technology instructions are on the book's website at bfwpub.com/tps7e.

19. Significance tests for a difference in proportions Page 627

SECTION 6F Exercises

Stating Hypotheses for a Test about $p_1 - p_2$

1. **Psychological ownership** Does encouraging people pg 622 to take ownership for common areas make them more likely to care for those areas? At a public lake, researchers randomly assigned 54 kayak renters to think of and write down a nickname for the lake. The remaining 81 kayak renters were not asked any questions. Unknown to the kayakers, a research assistant anchored 4 pieces of trash (e.g., flip flop, water bottle) at fixed locations in the lake and observed whether or not each kayaker attempted to retrieve the trash. Of the 54 who were asked to give the lake a nickname, 22 attempted to pick up at least one of the pieces of trash. Only 6 of the 81 other kayakers attempted to pick up at least one of the pieces of trash.[61] State appropriate hypotheses for performing a significance test. Be sure to define the parameters of interest.

2. **Children make choices** Many new products introduced into the market are targeted toward children. The choice behavior of children with regard to new products is of particular interest to companies that design marketing strategies for these products. As part of one study, randomly selected children in different age groups were compared on their ability to sort new products into the correct product category (milk or juice).[62] Here are some of the data:

Age group	n	Number who sorted correctly
4- to 5-year-olds	50	10
6- to 7-year-olds	53	28

Researchers want to know if a greater proportion of 6- to 7-year-olds can sort correctly than 4- to 5-year-olds. State appropriate hypotheses for performing a significance test. Be sure to define the parameters of interest.

3. **Shrubs and fire** Fire is a serious threat to shrubs in dry climates. Some shrubs can resprout from their roots after their tops are destroyed. Researchers wondered if fire would help with resprouting. One study of resprouting took place in a dry area of Mexico.[63] The researchers randomly assigned shrubs to treatment and control groups. They clipped the tops of all the shrubs. They then applied a propane torch to the stumps of the treatment group to simulate a fire. All 12 of the shrubs in the treatment group resprouted. Only 8 of the 12 shrubs in the control group resprouted. State appropriate hypotheses for performing a significance test. Be sure to define the parameters of interest.

4. **Botox benefits?** You may have heard that Botox (botulinum toxin type A) is used for cosmetic surgery, but could it have other beneficial uses? A total of 31 patients who suffered chronic low back pain were randomly assigned to receive 200 units of either Botox or saline solution through 5 injections at 5 different locations in their backs. The saline injection was not expected to reduce pain but was given as a placebo treatment. Of the 15 people assigned to Botox, 10 had at least moderate pain relief. Only 3 of the 16 people assigned to the saline solution had at least moderate pain relief. State appropriate hypotheses for performing a significance test. Be sure to define the parameters of interest.

Checking Conditions for a Test about $p_1 - p_2$

5. **More psychological ownership** Refer to Exercise 1. pg 624 Check if the conditions for performing the test are met.

6. **More children and choices** Refer to Exercise 2. Check if the conditions for performing the test are met.

7. **More shrubs and fire** Refer to Exercise 3. Is it reasonable to use a normal distribution to model the sampling distribution of the difference in sample proportions? Explain your reasoning.

8. **More Botox benefits** Refer to Exercise 4. Is it reasonable to use a normal distribution to model the sampling distribution of the difference in sample proportions? Explain your reasoning.

Calculating the Standardized Test Statistic and *P*-Value for a Test about $p_1 - p_2$

9. **Final psychological ownership** Refer to Exercises pg 626 1 and 5.

(a) Explain why the sample results give some evidence for the alternative hypothesis.

(b) Calculate the standardized test statistic and *P*-value.

(c) What conclusion would you make at the 1% significance level?

10. **Final children and choices** Refer to Exercises 2 and 6.

(a) Explain why the sample results give some evidence for the alternative hypothesis.

(b) Calculate the standardized test statistic and *P*-value.

(c) What conclusion would you make at the 1% significance level?

11. **Simulating shrubs and fire** Refer to Exercises 3 and 7. We can use simulation to test the hypotheses

from Exercise 3. The dotplot shows the results of 50 trials of a simulation to see what differences (Fire − No fire) in the proportions of shrubs that resprout would occur due only to chance variation in the random assignment, assuming that the fire has no effect. What is the estimated *P*-value? What conclusion would you draw?

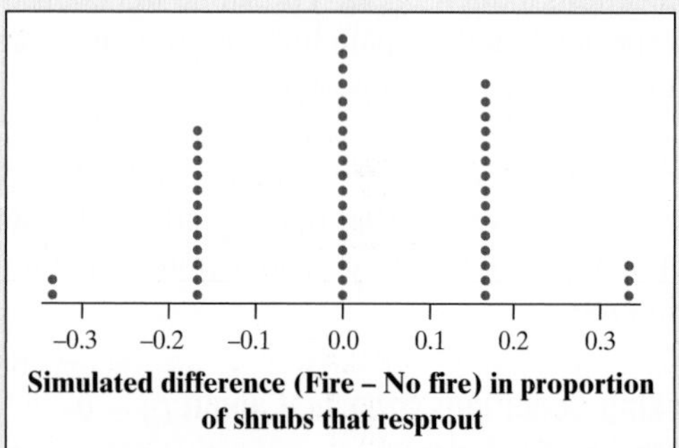

12. **Simulating Botox benefits** Refer to Exercises 4 and 8. We can use simulation to test the hypotheses from Exercise 4. The dotplot shows the results of 100 trials of a simulation to see what differences (Botox − Saline) in the proportions of patients who have at least moderate pain relief would occur due only to chance variation in the random assignment, assuming that the type of injection doesn't matter. What is the estimated *P*-value? What conclusion would you draw?

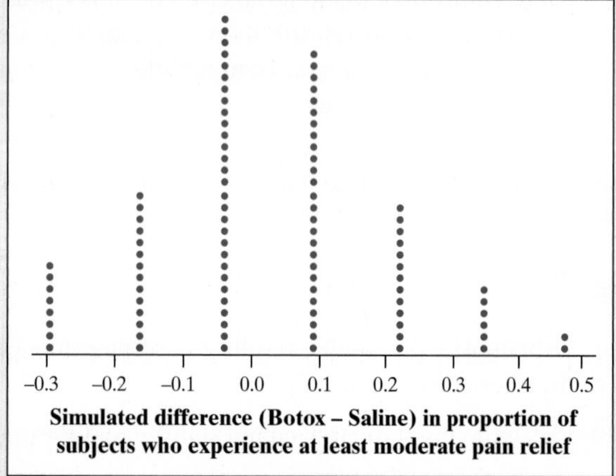

Two-Sample *z* Test for $p_1 - p_2$

13. **Low-birth-weight babies** Babies born weighing less than 1500 grams (about 3.3 pounds) are classified as very low birth weight (VLBW). A long-term study followed 242 randomly selected VLBW babies born in Cleveland to age 20 years, along with a group of 233 randomly selected babies born in Cleveland who had normal birth weight. At age 20 years, 179 of the VLBW group and 193 of the normal-birth-weight group had graduated from high school.[64] pg 629

(a) Do these data provide convincing evidence at the 1% significance level that the graduation rate among VLBW babies is less than that for normal-birth-weight babies in Cleveland?

(b) Interpret the *P*-value from part (a).

14. **City trees** The growth of a tree and its root system can damage nearby structures, including sidewalks. In New York City, are trees located on curbs more likely to have sidewalk damage nearby than trees offset from the curb? In a random sample of 7529 trees on the curb, 2181 had sidewalk damage nearby. In a random sample of 329 trees offset from the curb, 40 had sidewalk damage nearby.[65]

(a) Do these data provide convincing evidence at the 1% significance level that the proportion of trees with damaged sidewalks nearby is greater for trees on the curb than for trees that are offset from the curb in New York City?

(b) Interpret the *P*-value from part (a).

15. **Preventing peanut allergies** A recent study of peanut allergies — the LEAP trial — explored whether early exposure to peanuts helps or hurts subsequent development of an allergy to peanuts. Infants (4 to 11 months old) who had shown evidence of other kinds of allergies were randomly assigned to one of two groups. Group 1 consumed a baby-food form of peanut butter. Group 2 avoided peanut butter. At 5 years old, 10 of 307 children in the peanut-consumption group were allergic to peanuts, and 55 of 321 children in the peanut-avoidance group were allergic to peanuts.[66]

(a) Does this study provide convincing evidence at the $\alpha = 0.05$ significance level of a difference in the development of peanut allergies in infants like the ones in this study who consume and those who avoid peanut butter?

(b) Based on your conclusion in part (a), which mistake — a Type I error or a Type II error — could you have made? Explain your answer.

(c) Should you generalize the result in part (a) to all infants? Why or why not?

(d) A 95% confidence interval for $p_1 - p_2$ is $(-0.185, -0.093)$. Explain how the confidence interval provides more information than the test in part (a).

16. **Lowering bad cholesterol** Which of two widely prescribed drugs — Pravachol or Lipitor — helps lower "bad cholesterol" more? In an experiment, called the PROVE-IT Study, researchers recruited about 4000 people with heart disease as subjects. These

volunteers were randomly assigned to one of two treatment groups: Pravachol or Lipitor. At the end of the study, researchers compared the proportion of subjects in each group who died, had a heart attack, or suffered other serious consequences within two years. For the 2063 subjects using Pravachol, the proportion was 0.263. For the 2099 subjects using Lipitor, the proportion was 0.224.[67]

(a) Does this study provide convincing evidence at the $\alpha = 0.05$ significance level of a difference in the effectiveness of Pravachol and Lipitor for people like the ones in this study?

(b) Based on your conclusion in part (a), which mistake — a Type I error or a Type II error — could you have made? Explain your answer.

(c) Should you generalize the result in part (a) to all people with heart disease? Why or why not?

(d) A 95% confidence interval for $p_{PR} - p_L$ is (0.013, 0.065). Explain how the confidence interval provides more information than the test in part (a).

17. **More peanut allergies** Refer to Exercise 15. Explain how each of the following changes to the design of the experiment would affect the power of the test. Then give a drawback of making that change.

(a) Researchers recruit twice as many infants for the LEAP trial.

(b) Researchers use $\alpha = 0.10$ instead of $\alpha = 0.05$.

(c) Researchers use 628 male infants but no female infants in the study.

18. **More bad cholesterol** Refer to Exercise 16. Explain how each of the following changes to the design of the experiment would affect the power of the test. Then give a drawback of making that change.

(a) Researchers recruit twice as many subjects for the PROVE-IT Study.

(b) Researchers use $\alpha = 0.10$ instead of $\alpha = 0.05$.

(c) Researchers use 4162 subjects younger than age 60 but no older subjects in the study.

19. **Reflected glory** In a classic study, researchers investigated the tendency of sports fans to "bask in reflected glory" by associating themselves with winning teams. The researchers called randomly selected students at a major university with a highly ranked football team. About half of the students were randomly assigned to answer questions about a recent game the team had lost, and the remaining students were asked about a recent game the team had won. If the students were able to correctly identify the winner (showing they were fans of the team), they were asked to describe the game. Researchers recorded if the students identified themselves with the team by their use of the word *we* in the description ("We won the game" versus "They won the game").[68] The table shows the results of the experiment.

		Outcome of game		
		Win	Loss	Total
Association with team	"We"	27	15	42
	"They"	58	68	126
	Total	85	83	168

(a) Note that the researchers used both random selection and random assignment in their data collection method. What benefit does this provide?

(b) Is there convincing evidence that football fans at this university are more likely to associate themselves with the team after a win?

For Investigation *Apply the skills from the section in a new context or nonroutine way.*

20. **Financial incentives and smoking** In an effort to reduce health care costs, General Motors sponsored a study to help its employees stop smoking. In the study, half of the subjects (439) were randomly assigned to receive \$750 if they agreed to quit smoking for a year, while the other half (439) were simply encouraged to use traditional methods to stop smoking. None of the 878 volunteers knew that there was a financial incentive when they first signed up. At the end of one year, 15% of those in the financial rewards group had quit smoking, while only 5% in the traditional group had quit smoking.[69] In the media, it is common to report the results of a study like this one with a ratio of proportions, rather than a difference in proportions.

(a) Calculate the ratio of the sample proportions, $\hat{p}_F / \hat{p}_T$, where $\hat{p}_F$ = the proportion in the financial incentives group who quit smoking and $\hat{p}_T$ = the proportion in the traditional methods group who quit smoking. Interpret this value.

(b) The researchers suspect that the financial incentive will be more effective than the traditional method. State the appropriate hypotheses for a test of the ratio of the true proportions.

(c) To estimate the P-value, 1000 trials of a simulation were conducted assuming that the two methods are equally effective. For each trial, the ratio $\hat{p}_F / \hat{p}_T$

was calculated. Based on the results of the simulation shown in the histogram, what is the approximate P-value? Explain how you obtained your answer.

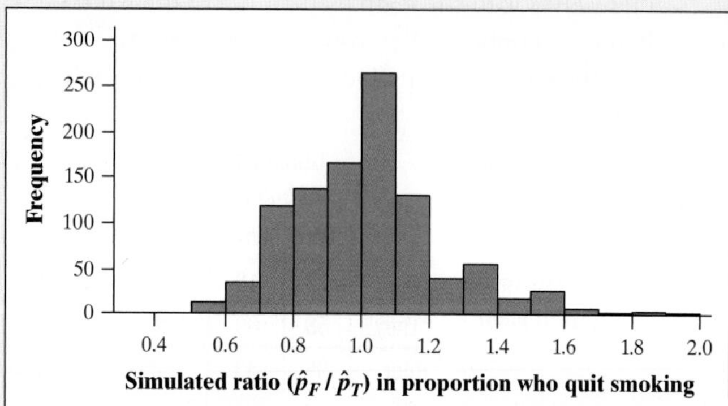

(d) Do the results of this study give convincing statistical evidence at the $\alpha = 0.01$ significance level that a financial incentive helps employees like the ones in this study stop smoking? Explain your reasoning.

21. **One-sided confidence intervals** Refer to Exercise 14. In general, 99% confidence intervals in the form

point estimate ± margin of error

give the same conclusions as two-sided significance tests with $\alpha = 0.01$. To be equivalent to a one-sided test of $H_0: p_1 - p_2 = 0$ versus $H_a: p_1 - p_2 > 0$ with $\alpha = 0.01$, we'd need a 99% confidence interval in the form

(lower boundary, ∞)

(a) Calculate the value of z^* for a 99% confidence interval in the form (lower boundary, ∞).

(b) Use the value of z^* from part (a) to calculate a 99% one-sided confidence interval for the difference (On curb − Offset from curb) in the proportions of trees on New York City streets that have damaged sidewalks nearby. Note that the conditions for inference are met.

(c) Does your interval provide convincing evidence at the 1% significance level that the proportion of trees with damaged sidewalks nearby is greater for trees on the curb than for trees that are offset from the curb in New York City? Explain your answer.

(d) What additional information does the interval provide that the test does not?

Multiple Choice *Select the best answer for each question.*

Exercises 22–24 refer to the following scenario. For patients needing an aortic-valve replacement, there are two commonly used approaches — traditional surgery and a method called transcatheter aortic-valve replacement (TAVR). Researchers wanted to know if there is a difference in the effectiveness of the two approaches, so they randomly assigned patients to the two approaches and then recorded the number of patients who had negative outcomes (death, stroke, or rehospitalization). Of the 454 patients assigned to undergo traditional surgery, 68 had negative outcomes. Of the 496 patients assigned to the TAVR method, only 42 had negative outcomes.[70]

22. Let p_S and p_T be the proportions of patients like the ones in this study who would have a negative outcome after surgery or TAVR. Which of the following are the correct hypotheses?

(A) $H_0: p_S - p_T = 0$ versus $H_a: p_S - p_T \neq 0$

(B) $H_0: p_S - p_T = 0$ versus $H_a: p_S - p_T > 0$

(C) $H_0: p_S - p_T = 0$ versus $H_a: p_S - p_T < 0$

(D) $H_0: p_S - p_T > 0$ versus $H_a: p_S - p_T = 0$

(E) $H_0: p_S - p_T \neq 0$ versus $H_a: p_S - p_T = 0$

23. The researchers report that the results were statistically significant at the 1% level. Which of the following is the most appropriate conclusion?

(A) Because the P-value is less than 1%, fail to reject H_0. There is not convincing evidence of a difference in the effectiveness of the two approaches for the patients who were in this study.

(B) Because the P-value is less than 1%, fail to reject H_0. There is not convincing evidence of a difference in the effectiveness of the two approaches for patients like those in this study.

(C) Because the P-value is less than 1%, reject H_0. There is convincing evidence of no difference in the effectiveness of the two approaches for patients like those in this study.

(D) Because the P-value is less than 1%, reject H_0. There is convincing evidence of a difference in the effectiveness of the two approaches for the patients who were in this study.

(E) Because the P-value is less than 1%, reject H_0. There is convincing evidence of a difference in the effectiveness of the two approaches for patients like those in this study.

24. Which of the following is the correct standard error for a test of the hypotheses in Exercise 22?

(A) $\sqrt{\dfrac{0.116(0.884)}{950}}$

(B) $\sqrt{\dfrac{0.116(0.884)}{454} + \dfrac{0.116(0.884)}{496}}$

(C) $\sqrt{\dfrac{0.150(0.850)}{454} + \dfrac{0.085(0.915)}{496}}$

(D) $\sqrt{\frac{0.116(0.884)}{950}+\frac{0.116(0.884)}{950}}$

(E) $\sqrt{\frac{0.150(0.850)}{950}+\frac{0.085(0.915)}{950}}$

25. Does providing additional information affect responses to a survey question? Two statistics students decided to investigate this issue by asking different versions of a question about texting and driving. Fifty mall shoppers were divided into two groups of 25 at random. The first group was asked version A and the other half were asked version B. The students believed that version A would result in more "Yes" answers. Here are the actual questions:

- *Version A:* A lot of people text and drive. Are you one of them?
- *Version B:* About 6000 deaths occur per year due to texting and driving. Knowing the potential consequences, do you text and drive?

Of the 25 shoppers assigned to version A, 18 admitted to texting and driving. Of the 25 shoppers assigned to version B, only 14 admitted to texting and driving. The two-sample z test for the difference in proportions

(A) gives $z = 1.179, P = 0.1193$.

(B) gives $z = 1.179, P = 0.2386$.

(C) gives $z = 1.195, P = 0.1160$.

(D) should not be used because the Random condition is violated.

(E) should not be used because the Large Counts condition is violated.

Recycle and Review *Practice what you learned in previous sections.*

Exercises 26 and 27 refer to the following scenario. Thirty randomly selected seniors at Council High School were asked to report the age (in years) and mileage of their main vehicles. Here is a scatterplot of the data:

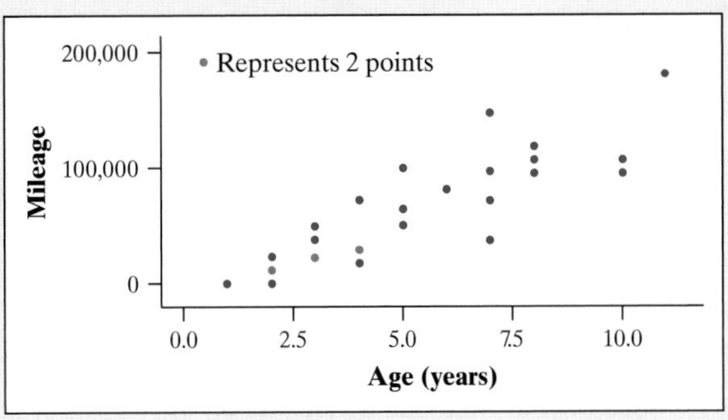

We used software to perform a least-squares regression analysis for these data. Part of the computer output from this regression is shown here.

Predictor	Coef	Stdev	t-ratio	P
Constant	−13832	8773	−1.58	0.126
Age	14954	1546	9.67	0.000
S = 22723	R-sq = 77.0%		R-sq(adj) = 76.1%	

26. **Drive my car (2C)**

(a) What is the equation of the least-squares regression line? Be sure to define any symbols you use.

(b) Interpret the slope of the least-squares line.

(c) One student reported that their 10-year-old car had 110,000 miles on it. Find and interpret the residual for this data point.

27. **More driving (2C, 3A)**

(a) Interpret the value of r^2.

(b) The mean age of the students' cars in the sample was $\bar{x} = 5$ years. Find the mean mileage of the cars in the sample.

(c) Interpret the value of s.

(d) Would it be reasonable to use the least-squares line to predict a car's mileage from its age for a Council High School teacher? Justify your answer.

UNIT 6 Wrap-Up

PART II

FRAPPY! Free Response AP® Problem, Yay!

Directions: Show all your work. Indicate clearly the methods you use, because you will be scored on the correctness of your methods as well as on the accuracy and completeness of your results and explanations.

Do employer-sponsored wellness programs work? About 5000 employees at the University of Illinois at Urbana-Champaign volunteered for a study to find out. Researchers randomly assigned 3300 volunteers to a treatment group and the remaining 1534 to a control group. Employees in the treatment group were invited to take paid time off to participate in a wellness program. Those in the control group were not allowed to participate. One measure of the program's effectiveness was employee attrition. In the 2-year period following the start of the study, 356 of the people in the treatment group left their job for any reason, compared to 184 of the people in the control group.

Do these results provide convincing evidence at the $\alpha = 0.05$ level that offering employees paid time off to participate in a wellness program reduces the proportion who leave their job within 2 years for people similar to the ones in this study?

After you finish the FRAPPY!, you can view two example solutions on the book's website (**bfwpub.com/tps7e**). Determine whether you think each solution is "complete," "substantial," "developing," or "minimal." If the solution is not complete, what improvements would you suggest to the student who wrote it? Finally, your teacher will provide you with a scoring rubric. Score your response and note what, if anything, you would do differently to improve your own score.

UNIT 6, PART II REVIEW

SECTION 6E Confidence Intervals for a Difference in Population Proportions

In this section, you learned how to construct confidence intervals for a difference between two population proportions. To verify independence in data collection and that the sampling distribution of $\hat{p}_1 - \hat{p}_2$ is approximately normal, we check three conditions. The Random condition says that the data must be from two independent random samples or two groups in a randomized experiment. The 10% condition says that each sample size should be less than 10% of the corresponding population size when sampling without replacement. The Large Counts condition says that the number of successes and the number of failures from each sample should be at least 10 — that is, $n_1\hat{p}_1$, $n_1(1-\hat{p}_1)$, $n_2\hat{p}_2$, and $n_2(1-\hat{p}_2)$ are ≥ 10.

A confidence interval for a difference between two proportions provides an interval of plausible values for the difference in the population proportions. The formula is

$$(\hat{p}_1 - \hat{p}_2) \pm z^* \sqrt{\frac{\hat{p}_1(1-\hat{p}_1)}{n_1} + \frac{\hat{p}_2(1-\hat{p}_2)}{n_2}}$$

The logic of confidence intervals, including how to interpret the confidence interval and the confidence level, is the same as when estimating a single population proportion. Likewise, you can use a confidence interval for a difference in proportions to evaluate claims about the population proportions. For example, if 0 is not included in a confidence interval for $p_1 - p_2$, there is convincing evidence that the population proportions are different.

SECTION 6F Significance Tests for a Difference in Population Proportions

In this section, you learned how to perform significance tests for a difference between two proportions. As in any test, you start by stating hypotheses. In AP® Statistics, we focus on the null hypothesis of H_0: $p_1 - p_2 = 0$ (or, equivalently, H_0: $p_1 = p_2$). The alternative hypothesis can be one-sided (<, >) or two-sided ($\neq$).

As with confidence intervals for $p_1 - p_2$, we check the conditions to verify independence in the data collection process and that the sampling distribution of $\hat{p}_1 - \hat{p}_2$ is approximately normal. The Random condition says that the data must be from two independent random samples or two groups in a randomized experiment. The 10% condition says that each sample size should be less than 10% of the corresponding population size when sampling without replacement. The Large Counts condition says that the *expected* numbers of successes and failures in each sample should be at least 10. For a test of $H_0: p_1 - p_2 = 0$, we estimate the common value p of the parameters p_1 and p_2 using the combined (pooled) proportion of successes in the two samples: $\hat{p}_C = \frac{X_1 + X_2}{n_1 + n_2}$. Consequently, the Large Counts condition requires us to check that $n_1\hat{p}_C$, $n_1(1-\hat{p}_C)$, $n_2\hat{p}_C$, and $n_2(1-\hat{p}_C)$ are all at least 10.

For a test of $H_0: p_1 - p_2 = 0$, the standardized test statistic is

$$z = \frac{(\hat{p}_1 - \hat{p}_2) - 0}{\sqrt{\hat{p}_C(1-\hat{p}_C)\left(\frac{1}{n_1} + \frac{1}{n_2}\right)}}$$

When conditions are met, *P*-values can be obtained from the standard normal distribution.

A significance test for a difference between two proportions uses the same logic as the significance test for one population proportion from Section 6D. Likewise, the way we interpret *P*-values and power and the way we describe Type I and Type II errors is very similar to what you learned in Section 6D.

Inference for a Difference in Proportions

	Confidence Interval for $p_1 - p_2$	Significance Test for $p_1 - p_2$
Name (TI-83/84)	Two-sample *z* interval for $p_1 - p_2$ (2-PropZInt)	Two-sample *z* test for $p_1 - p_2$ (2-PropZTest)
Null Hypothesis	*Not applicable.*	$H_0: p_1 - p_2 = 0$
Conditions	• **Random:** The data come from two independent random samples or from two groups in a randomized experiment. ○ **10%:** When sampling without replacement, $n_1 < 0.10N_1$ and $n_2 < 0.10N_2$. • **Large Counts:** The counts of "successes" and "failures" in each sample or group—$n_1\hat{p}_1$, $n_1(1-\hat{p}_1)$, $n_2\hat{p}_2$, and $n_2(1-\hat{p}_2)$—are all at least 10.	• **Random:** The data come from two independent random samples or from two groups in a randomized experiment. ○ **10%:** When sampling without replacement, $n_1 < 0.10N_1$ and $n_2 < 0.10N_2$. • **Large Counts:** The expected counts of "successes" and "failures" in each sample or group—$n_1\hat{p}_C$, $n_1(1-\hat{p}_C)$, $n_2\hat{p}_C$, and $n_2(1-\hat{p}_C)$—are all at least 10, where $\hat{p}_C = \frac{X_1 + X_2}{n_1 + n_2}$.
Formula	$(\hat{p}_1 - \hat{p}_2) \pm z^* \sqrt{\frac{\hat{p}_1(1-\hat{p}_1)}{n_1} + \frac{\hat{p}_2(1-\hat{p}_2)}{n_2}}$ Critical value z^* from the standard normal distribution.	$z = \frac{(\hat{p}_1 - \hat{p}_2) - 0}{\sqrt{\hat{p}_C(1-\hat{p}_C)\left(\frac{1}{n_1} + \frac{1}{n_2}\right)}}$ *P*-value from the standard normal distribution.

What Did You Learn?

Learning Target	Section	Related Example on Page(s)	Relevant Unit Review Exercise(s)
Check the conditions for calculating a confidence interval about a difference between two population proportions.	6E	611	R1
Calculate a confidence interval for a difference between two population proportions.	6E	612	R1
Construct and interpret a two-sample *z* interval for a difference in proportions.	6E	615	R1
State appropriate hypotheses for performing a test about a difference between two population proportions.	6F	622	R2, R3
Check the conditions for performing a test about a difference between two population proportions.	6F	624	R2
Calculate the standardized test statistic and *P*-value for a test about a difference between two population proportions.	6F	626	R2
Perform a two-sample *z* test for a difference in proportions.	6F	629	R2

UNIT 6, PART II REVIEW EXERCISES

These exercises are designed to help you review the important concepts and skills of the unit.

R1 **Student employment** (6E) Researchers want to estimate the difference in the proportions of high school students and college students who are employed. A sample survey is given to independent random samples of 500 high school students and 550 college students. In all, 20.4% of the high school students and 45.1% of the college students surveyed are employed.[71]

(a) Calculate and interpret a 99% confidence interval for the difference in the population proportions of high school students and college students who are employed.

(b) Based on the interval from part (a), is there convincing evidence of a difference in the proportions of high school students and college students who are employed? Explain your reasoning.

R2 **Treating AIDS** (6F) AZT was the first drug that seemed to be effective in delaying the onset of acquired immunodeficiency syndrome (AIDS). Evidence of AZT's effectiveness came from a large randomized comparative experiment. The subjects were 870 volunteers who were infected with human immunodeficiency virus (HIV, the virus that causes AIDS), but who did not yet have AIDS. The study assigned 435 of the subjects at random to take 500 milligrams of AZT each day and another 435 subjects to take a placebo. At the end of the study, 38 of the placebo subjects and 17 of the AZT subjects had developed AIDS.[72]

(a) If the results of the study are statistically significant, is it reasonable to conclude that AZT is the cause of the decrease in the proportion of people like these who will develop AIDS? Explain your answer.

(b) Do the data provide convincing evidence at the $\alpha = 0.05$ level that taking AZT decreases the proportion of infected people like the ones in this study who will develop AIDS in a given period of time?

R3 **Q collars** (6F) Concussions are a major concern for high school football players. Can wearing a device called a Q collar help prevent brain damage? Researchers randomly assigned 284 high school football players to either wear or not wear a Q collar during the season. Each player had magnetic resonance imaging (MRI) of their brain performed before and after the season to identify whether there was any damage in the white matter region of the brain.[73]

(a) State the hypotheses the researchers should test.

(b) Describe a Type I error and a Type II error in this context, along with a consequence of each type of error.

(c) Based on your answer to part (b), would you suggest using $\alpha = 0.01$ or $\alpha = 0.10$? Explain your reasoning.

(d) After collecting the data and performing the test of the hypotheses from part (a), the researchers got a *P*-value of approximately 0. Interpret this value.

UNIT 6, PART II AP® STATISTICS PRACTICE TEST

Section I: Multiple Choice *Select the best answer for each question*

T1 Thirty-five people from a random sample of 125 workers from Company A admitted to using sick leave when they weren't really ill. Seventeen employees from a random sample of 68 workers from Company B admitted that they had used sick leave when they weren't ill. Which of the following is a 95% confidence interval for the difference in the proportions of workers at the two companies who would admit to using sick leave when they weren't ill?

(A) $0.03 \pm \sqrt{\frac{(0.28)(0.72)}{125} + \frac{(0.25)(0.75)}{68}}$

(B) $0.03 \pm 1.96\sqrt{\frac{(0.28)(0.72)}{125} + \frac{(0.25)(0.75)}{68}}$

(C) $0.03 \pm 1.96\sqrt{\frac{(0.27)(0.73)}{125} + \frac{(0.27)(0.73)}{68}}$

(D) $18 \pm 1.96\sqrt{\frac{(0.28)(0.72)}{125} + \frac{(0.25)(0.75)}{68}}$

(E) $18 \pm 1.96\sqrt{\frac{(0.27)(0.73)}{125} + \frac{(0.27)(0.73)}{68}}$

T2 The power takeoff driveline on tractors used in agriculture can be a serious hazard to operators of farm equipment. The driveline is covered by a shield in new tractors, but the shield is often missing on older tractors. Two types of shields are the bolt-on and the flip-up. It was believed that the bolt-on shield was perceived as a nuisance by the operators and deliberately removed, but the flip-up shield is easily lifted for inspection and maintenance and may be left in place. In a study by the U.S. National Safety Council, random samples of older tractors with both types of shields were taken to see what proportion of shields were removed. Of 183 tractors designed to have bolt-on shields, 35 had been removed. Of the 136 tractors with flip-up shields, 15 were removed. We wish to perform a test of H_0: $p_B = p_F$ versus H_a: $p_B > p_F$, where p_B and p_F are the proportions of all tractors with the bolt-on and flip-up shields removed, respectively. Which of the following is *not* a condition for performing the significance test?

(A) Both populations are approximately normally distributed.

(B) The data come from two independent samples.

(C) Both samples were chosen at random.

(D) The expected counts of successes and failures are large enough to use normal calculations.

(E) Both populations are more than 10 times the corresponding sample sizes.

T3 A significance test allows you to reject a null hypothesis H_0 in favor of an alternative hypothesis H_a at the 5% significance level. What can you say about significance at the 1% level?

(A) H_0 can be rejected at the 1% significance level.

(B) There is insufficient evidence to reject H_0 at the 1% significance level.

(C) There is sufficient evidence to accept H_0 at the 1% significance level.

(D) H_a can be rejected at the 1% significance level.

(E) The answer can't be determined from the information given.

T4 Conference organizers wondered whether posting a sign that says "Please take only one cookie" would reduce the proportion of conference attendees who take multiple cookies from the snack table during a break. To find out, the organizers randomly assigned 212 attendees to take their break in a room where the snack table had the sign posted, and 189 attendees to take their break in a room where the snack table did not have a sign posted. In the room with the sign posted, 17.0% of attendees took multiple cookies. In the room without the sign posted, 24.3% of attendees took multiple cookies. Is this decrease in proportions statistically significant at the $\alpha = 0.05$ level? Note that the conditions for inference are met.

(A) No. The *P*-value is 0.034.

(B) No. The *P*-value is 0.068.

(C) Yes. The *P*-value is 0.034.

(D) Yes. The *P*-value is 0.068.

(E) Cannot be determined from the information given.

T5 At a baseball game, 42 of 65 randomly selected people own an iPhone. At a rock concert occurring at the same time across town, 34 of 52 randomly selected people own an iPhone. A researcher wants to test the claim that the proportion of iPhone owners at the two venues is different. A 90% confidence interval for the difference (Game − Concert) in population proportions is $(-0.154, 0.138)$. Which of the following gives the correct outcome of the researcher's test of the claim if $\alpha = 0.10$?

(A) Because the interval includes 0, the researcher can conclude that the proportion of iPhone owners at the two venues is the same.

(B) Because the center of the interval is -0.008, the researcher can conclude that a higher proportion of people at the rock concert own iPhones than at the baseball game.

(C) Because the interval includes 0, the researcher cannot conclude that the proportion of iPhone owners at the two venues is different.

(D) Because the interval includes -0.008, the researcher cannot conclude that the proportion of iPhone owners at the two venues is different.

(E) Because the interval includes more negative than positive values, the researcher can conclude that a higher proportion of people at the rock concert own iPhones than at the baseball game.

Section II: Free Response *Show all your work. Indicate clearly the methods you use, because you will be graded on the correctness of your methods as well as on the accuracy and completeness of your results and explanations.*

T6 Do "props" make a difference when researchers interact with their subjects? Emily and Madi asked 100 people if they thought buying coffee at Starbucks was a waste of money.[74] Half of the subjects were asked while Emily and Madi were holding cups from Starbucks, and the other half of the subjects were asked when the girls were empty handed. The choice of holding or not holding the cups was determined at random for each subject. When they were holding the cups, 19 of 50 subjects agreed that buying coffee at Starbucks was a waste of money. When they weren't holding the cups, 23 of 50 subjects said it was a waste of money. Calculate and interpret a 90% confidence interval for the difference in the proportions of people like the ones in this experiment who would say that buying coffee from Starbucks is a waste of money when asked by interviewers holding or not holding a cup from Starbucks.

T7 A random sample of 100 of last year's model of a certain popular car found that 20 had a specific minor defect in the brakes. The automaker adjusted the production process to try to reduce the proportion of cars with the brake problem. A random sample of 350 of this year's model found that 50 had the minor brake defect.

(a) Was the company's adjustment successful? Carry out an appropriate test using $\alpha = 0.05$ to support your answer.

(b) Suppose that the proportion of cars with the defect was reduced by 0.10 from last year to this year. The power of the test in part (a) to detect this decrease is 0.72. Interpret this value.

(c) Other than increasing the sample sizes, identify one way of increasing the power of the test.

UNIT 7

Inference for Quantitative Data: Means

Carlos Gil/Getty Images

Introduction

In Unit 6, you learned the basics of confidence intervals and significance tests, along with the details of estimating and testing claims about population proportions. In Unit 7, we revisit the big ideas — including possible errors that can occur when performing statistical inference — but focus on estimating and testing claims about population means. There's one key difference between these two units: inference about proportions involves *categorical* data; inference about means involves *quantitative* data. However, one thing remains the same: you will need to apply the skills of selecting statistical methods, using probability and simulation, and statistical argumentation to perform inference successfully.

SECTION 7A Confidence Intervals for a Population Mean or Mean Difference

LEARNING TARGETS *By the end of the section, you should be able to:*

- Determine the critical value t^* for calculating a confidence interval for a population mean.
- Check the conditions for calculating a confidence interval for a population mean.
- Calculate a confidence interval for a population mean.
- Construct and interpret a one-sample *t* interval for a mean.
- In the special case of paired data, construct and interpret a confidence interval for a population mean difference.

Inference about a population proportion becomes a possibility when we study categorical variables. We learned how to construct and interpret confidence intervals for a population proportion p in Section 6B. To estimate a population mean, we start by recording values of a quantitative variable for a sample of individuals. Consequently, it would — for example — make sense to try to estimate the mean amount of sleep μ that all students at a large high school got last night but not their mean eye color, because eye color is a categorical variable. In this section, we'll examine confidence intervals for a population mean μ.

In Section 6B, we presented the formula for a confidence interval (interval estimate) for a population proportion:

$$\hat{p} \pm z^* \sqrt{\frac{\hat{p}(1-\hat{p})}{n}}$$

In more general terms, this interval estimate is of the form

$$\begin{aligned} &\text{point estimate} \pm \text{margin of error} \\ &= \text{statistic} \pm (\text{critical value})(\text{standard error of statistic}) \end{aligned}$$

A confidence interval for a population mean has a formula with the same structure. Using the sample mean $\bar{x}$ as the point estimate for the population mean μ and $\sigma/\sqrt{n}$ as the standard deviation of the sampling distribution of $\bar{x}$ gives

$$\bar{x} \pm z^* \frac{\sigma}{\sqrt{n}}$$

Unfortunately, if we don't know the true value of μ, we rarely know the true value of σ, either. We can use s_x as an estimate for σ, but things don't work out as nicely as we might like.

ACTIVITY Confidence interval BINGO!

Adrian Burke/Getty Images

In this activity, you will investigate the problem caused by replacing σ with s_x when calculating a confidence interval for a population mean μ, and learn how to fix it.

A homesteading family wants to estimate the mean weight of all tomatoes grown on their farm. To do so, they select a random sample of 4 tomatoes, calculate the mean weight, and use the sample mean $\bar{x}$ to create a 99% confidence interval for the population mean μ. Suppose that the weights of all tomatoes on the farm are approximately normally distributed, with a mean of 10 ounces and a standard deviation of 2 ounces.

Let's use an applet to simulate taking an SRS of $n = 4$ tomatoes and calculating a 99% confidence interval for μ using three different methods.

Method 1 (assuming σ is known)

$$\bar{x} \pm z^* \frac{\sigma}{\sqrt{n}} = \bar{x} \pm 2.576 \frac{2}{\sqrt{4}}$$

1. Go to www.stapplet.com and launch the *Simulating Confidence Intervals* applet.
2. In the box labeled "Population," keep the default setting for the normal population distribution.
3. In the box labeled "Sample," change the sample size to $n = 4$, change the confidence level to 99%, and change the method to "z distribution with σ." Click the "Go!" button to select one random sample and calculate the corresponding 99% confidence interval.
4. In the box labeled "Confidence Intervals," the confidence interval is displayed as a horizontal line segment, along with a dashed vertical line segment at $\mu = 10$. If the interval includes $\mu = 10$, the interval is green. Otherwise, it is red.
5. In the box labeled "Confidence Intervals," quickly generate 1 confidence interval at a time by clicking the "Go!" button, shouting out "BINGO!" whenever you get an interval that misses $\mu = 10$ (i.e., a red interval). Stop when your teacher calls time.
6. How well did Method 1 work? Compare the running total in the lower-left corner with the stated confidence level of 99%.

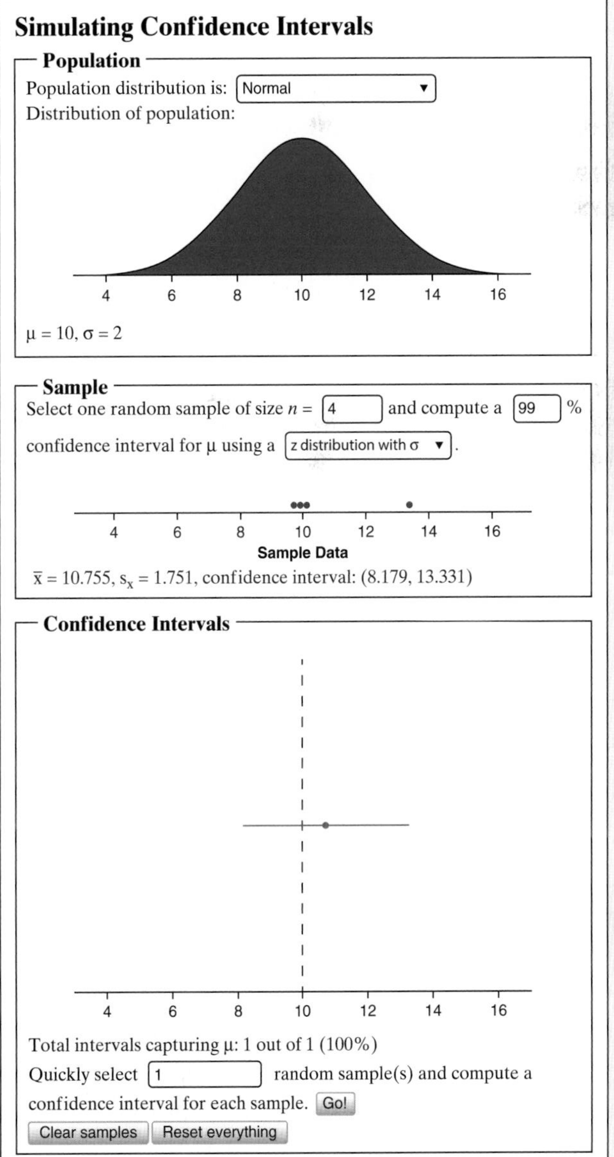

Method 2 (using s_x as an estimate for σ)

$$\bar{x} \pm z^* \frac{s_x}{\sqrt{n}} = \bar{x} \pm 2.576 \frac{s_x}{\sqrt{4}}$$

1. Click the "Clear samples" button at the bottom of the applet. Then, in the box labeled "Sample," change the method to "z distribution with sx." Keep everything else the same.
2. In the box labeled "Confidence Intervals," quickly generate 1 confidence interval at a time by clicking the "Go!" button, shouting out "BINGO!" whenever you get an interval that misses $\mu = 10$ (i.e., a red interval). Stop when your teacher calls time.
3. Did Method 2 work as well as Method 1? Discuss with your classmates.
4. Now, change the number of intervals to 50 and click the "Go!" button until you have more than 1000 intervals. Compare the running total in the lower-left corner with the stated confidence level of 99%. What do you notice about the width of the intervals that missed?

To increase the capture rate of the intervals to 99%, we need to make the intervals wider. We can do this by using a different critical value, called a t^* critical value. You'll learn how to find this number soon.

Method 3 (using s_x as an estimate for σ and a t^* critical value instead of a z^* critical value)

$$\bar{x} \pm t^* \frac{s_x}{\sqrt{n}} = \bar{x} \pm ??? \frac{s_x}{\sqrt{4}}$$

1. Click the "Clear samples" button at the bottom of the applet. Then, in the box labeled "Sample," change the method to "t distribution." Keep everything else the same.
2. In the box labeled "Confidence Intervals," reset the number of samples to 1 and quickly generate 1 confidence interval at a time, shouting out "BINGO!" whenever you get an interval that misses $\mu = 10$ (i.e., a red interval). Stop when your teacher calls time.
3. Did Method 3 work better than Method 2? How does it compare to Method 1? Discuss with your classmates.
4. Now, change the number of intervals to 50 and click the "Go!" button until you have more than 1000 intervals. Compare the running total in the lower-left corner with the stated confidence level of 99%. What do you notice about the width of the intervals compared to Method 2?

Figure 7.1 shows the results of repeatedly constructing confidence intervals using a z^* critical value and the sample standard deviation s_x, as described in Method 2 of the BINGO! activity. Of the 1000 intervals constructed, only 917 captured the population mean. That's far below our desired 99% confidence level!

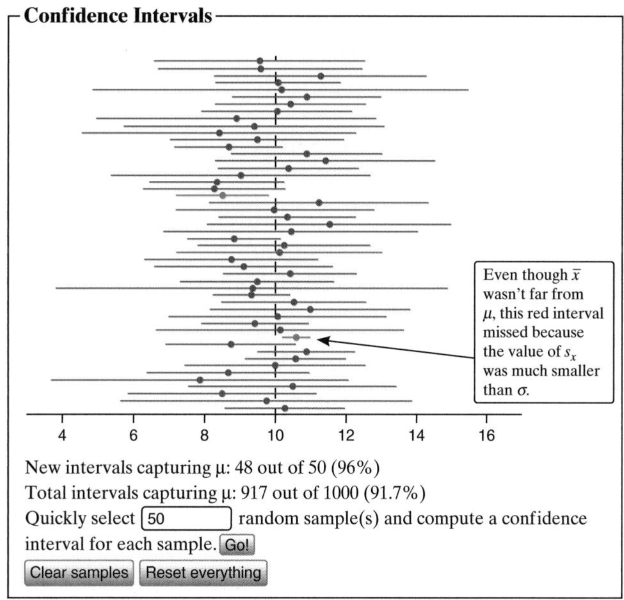

FIGURE 7.1 One thousand "99%" confidence intervals for μ calculated using a z^* critical value and the sample standard deviation s_x. The success rate for this method is less than 99%.

What went wrong? The intervals that missed (those in red) came from samples with a small standard deviation s_x and from samples in which the sample mean $\bar{x}$ was far from the population mean μ. In those cases, using a critical value of $z^* = 2.576$ didn't produce large enough intervals to reach $\mu = 10$. To achieve a 99% capture rate, we need to multiply by a larger critical value. But what critical value should we use?

Determining t^* Critical Values

When calculating a confidence interval for a population mean, we use a t^* critical value rather than a z^* critical value whenever we use s_x to estimate σ. The critical value is denoted t^* because it comes from a **t distribution,** not the standard normal distribution. There is a different t distribution for each sample size. We specify a particular t distribution by giving its *degrees of freedom* (df). When we perform inference about a population mean μ using a t distribution, the appropriate degrees of freedom are found by subtracting 1 from the sample size n, making df $= n - 1$.

DEFINITION *t* distribution

A ***t* distribution** is described by a symmetric, single-peaked, bell-shaped density curve centered at 0. Any t distribution is completely specified by its *degrees of freedom* (df) and has more area in its tails than the standard normal distribution.

Figure 7.2 compares the density curves of the standard normal distribution and the t distributions with 2 and 9 degrees of freedom. The figure illustrates these facts about the t distributions:

- The t distributions are similar in shape to the standard normal distribution. They are symmetric about 0, single-peaked, and bell-shaped.
- The t distributions have more variability — that is, more area in the tails — than the standard normal distribution. We are more likely to get an extremely large value of t (say, greater than 3) than an extremely large value of z.
- As the degrees of freedom increase, the area in the tails decreases and the t distributions approach the standard normal distribution. This makes sense because the value of s_x will typically be closer to σ as the sample size increases.

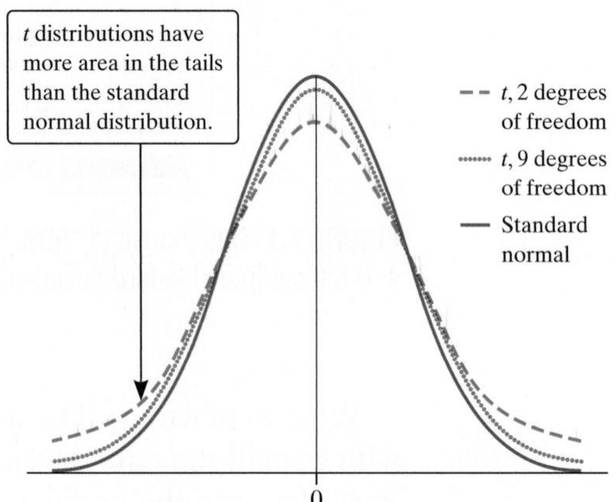

FIGURE 7.2 Density curves for the t distributions with 2 and 9 degrees of freedom and the standard normal distribution. All are symmetric with center 0. The t distributions have more area in the tails than the standard normal distribution does.

Because the t distributions have more area in the tails than the standard normal distribution does, t^* critical values will always be larger than z^* critical values for a specified confidence level. As you learned in the BINGO! activity, we need to use a critical value larger than z^* to compensate for the variability introduced by using the sample standard deviation s_x as an estimate for the population standard deviation σ. The critical value t^* has the same interpretation as z^*: it measures how many standard errors we need to extend from the point estimate to get the desired level of confidence.

How can we find the critical value t^* for a given confidence level and df? In the BINGO! activity, we calculated 99% confidence intervals based on a random sample of $n = 4$ tomatoes, so df $= 4 - 1 = 3$. As the figure illustrates, we need to determine the boundary values $-t^*$ and t^* that cut off the middle 99% of the t distribution with 3 degrees of freedom.

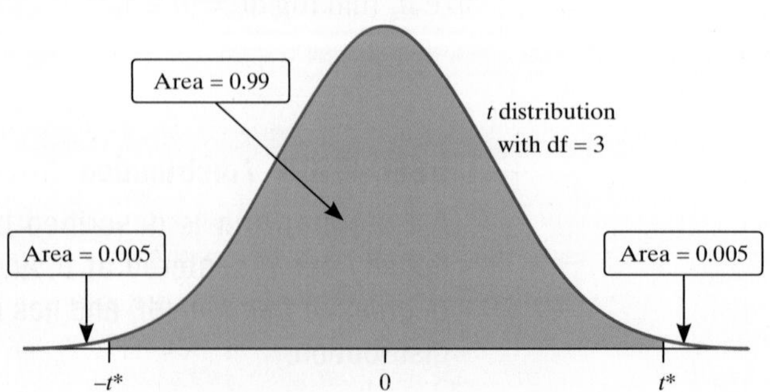

As when you determined z^* in Unit 6, you can use a table or technology to find the critical value t^* when calculating a confidence interval for a population mean. Table B in the back of the book gives critical values for many t distributions.

HOW TO FIND t^* USING TABLE B

1. Using Table B in the back of the book, find the correct confidence level at the bottom of the table.
2. On the left side of the table, find the correct number of *degrees of freedom* (df). For this type of confidence interval, df $= n - 1$.
3. If the correct df isn't listed, use the greatest df available that is *less than* the correct df.
4. In the body of the table, find the value of t^* that corresponds to the confidence level and df.

Here is an excerpt from Table B that shows how to find t^* for the BINGO! activity, using a 99% confidence level and df = 3.

	Tail probability *p*			
df	**.02**	**.01**	**.005**	**.0025**
1	15.89	31.82	63.66	127.3
2	4.849	6.965	9.925	14.09
3	3.482	4.541	5.841	7.453
⋮	⋮	⋮	⋮	⋮
∞	2.054	2.326	2.576	2.807
	96%	**98%**	**99%**	**99.5%**
	Confidence level *C*			

For 99% confidence and 3 degrees of freedom, $t^* = 5.841$. That is, the interval should extend 5.841 standard errors on both sides of the point estimate to have a capture rate of 99%. This t^* critical value is more than twice as large as the z^* critical value for 99% confidence ($z^* = 2.576$)!

The df $= \infty$ row at the bottom row of Table B gives z^* critical values for specified confidence levels, like $z^* = 2.576$ for 99% confidence. That's because the t distributions approach the standard normal distribution as the degrees of freedom approach infinity.

Unfortunately, Table B does not include all possible values for the confidence level or for the degrees of freedom (df). If the correct df isn't listed, be sure to use the greatest df available that is *less than* the correct df. "Rounding up" to a larger df will result in confidence intervals that are too narrow. The intervals won't be wide enough to capture the true population value as often as suggested by the confidence level. Due to these limitations of Table B, technology is often a better option for finding t^* critical values.

20. Tech Corner DETERMINING t^* CRITICAL VALUES

TI-Nspire and other technology instructions are on the book's website at bfwpub.com/tps7e.

You can use the TI-84 calculator to determine t^* critical values. We'll illustrate using the setting of the BINGO! activity with $n = 4$ and a 99% confidence level.

1. Press 2nd VARS (DISTR) and choose invT(.
2. Because we want a central area of 0.99, each tail has an area of 0.005.
   - **OS 2.55 or later:** In the dialog box, enter area: 0.005 and df: 3. Choose Paste, and then press ENTER.
   - **Older OS:** Complete the command invT(0.005,3) and press ENTER.

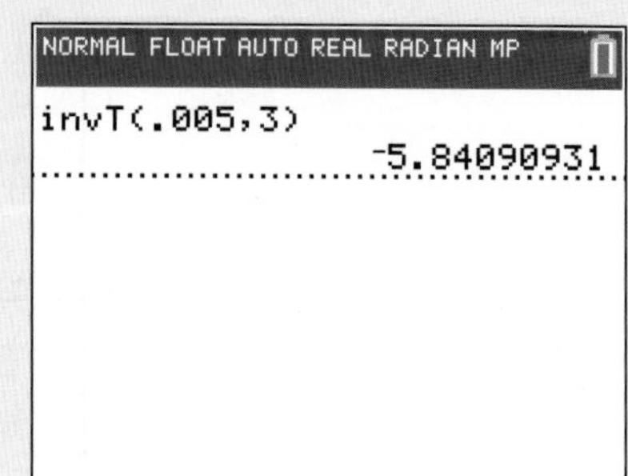

Note: In the inverse t command, the area entered always refers to the *area to the left*. The t^* critical value is positive 5.841. You could find t^* directly using invT(area: 0.995, df: 3).

EXAMPLE

How do you find t^*?
Determining t^* critical values

Skill 3.D

PROBLEM: What critical value t^* should be used in constructing a confidence interval for the population mean in each of the following settings?

(a) A 95% confidence interval based on an SRS of size $n = 12$

(b) A 90% confidence interval from a random sample of 48 observations

SOLUTION:

(a) df $= 12 - 1 = 11$

Using technology: invT(area:0.025, df:11) $= -2.201$, so $t^* = 2.201$

Using Table B: $t^* = 2.201$

In Table B, use the column for 95% confidence and the row corresponding to df $= 12 - 1 = 11$.

(b) df $= 48 - 1 = 47$

Using technology: invT(area:0.05, df:47) $= -1.678$, so $t^* = 1.678$

Using Table B: With df $= 40$, $t^* = 1.684$

There is no df $= 47$ row in Table B, so we use the more conservative df $= 40$. Technology gives the more accurate critical value using df $= 47$. The smaller critical value from technology will result in a smaller margin of error and a narrower confidence interval.

FOR PRACTICE, TRY EXERCISE 3

Now that you know how to calculate a t^* critical value, it's time to make a simple observation. Inference for *proportions* uses z; inference for *means* uses t. That's one reason why distinguishing categorical data from quantitative data is so important.

The t distributions and the t inference procedures were developed by William S. Gosset (1876–1937). Gosset worked for the Guinness brewery, and his role was to help the company make better beer. He used his new t procedures to find the best varieties of barley and hops. Because Gosset published under the pen name "Student," you will often see the t distribution called "Student's t" in his honor.

Checking Conditions for a Confidence Interval for μ

Before constructing a confidence interval for a population mean μ, you must check that the observations in the sample can be viewed as independent and that the sampling distribution of $\bar{x}$ is approximately normal. As with proportions, we check for independence using the Random condition and the 10% condition. If both of these conditions are met, our formula for the standard error will be approximately correct. However, if the data don't come from a random sample, you can't draw conclusions about a larger population. If the 10% condition is violated, our formula will overestimate the standard error, resulting in confidence intervals that have a capture rate *greater* than the stated confidence level.

As we learned in Section 6B, when calculating a confidence interval for a population proportion, we check the Large Counts condition to ensure that the sampling distribution of $\hat{p}$ is approximately normal. This allows us to use the standard normal distribution to calculate the z^* critical value. When calculating a confidence interval for a population mean, we check the *Normal/Large Sample condition* to ensure that the sampling distribution of $\bar{x}$ is approximately normal and that we are able to use a t distribution with df $= n - 1$ to calculate the t^* critical value. Violating this condition usually results in confidence intervals that have a capture rate *less than* the stated confidence level.

To meet the Normal/Large Sample condition, the population distribution must be approximately normal or the sample size must be large ($n \geq 30$). In either case, the sampling distribution of the sample mean $\bar{x}$ will be approximately normal, as you learned in Unit 5. More importantly, using a t^* critical value will produce confidence intervals with a capture rate that is approximately equal to the stated confidence level.

What if the population distribution has unknown shape and the sample size is small ($n < 30$)? We should graph the sample data and ask, "Is it plausible (believable) that these data came from an approximately normally distributed population?" If we do not see any strong skewness or outliers in the data, then the answer is "Yes." Remember to include a graph of the sample data when checking the Normal/Large Sample condition in this way.

Caia Image/Getty Images

Figure 7.3 is a boxplot of the SAT Math scores for a random sample of 20 students from a large high school. Although the sample size is small, the Normal/Large Sample condition is met because the boxplot is not strongly skewed and there are no outliers. It is plausible that this sample came from an approximately normally distributed population.

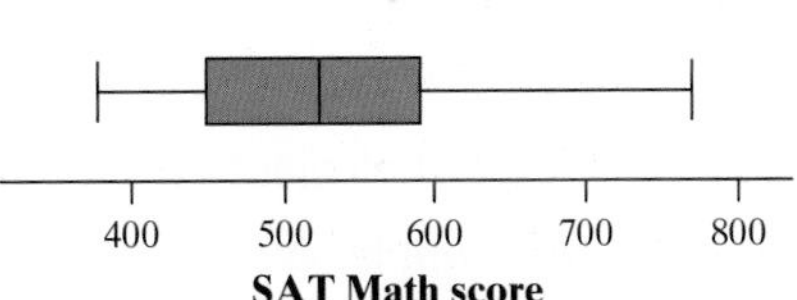

FIGURE 7.3 Boxplot of SAT Math scores for a random sample of 20 students at a large high school. Because the graph does not show strong skewness and has no outliers, it is believable that the population distribution of SAT Math scores at this school is approximately normal.

CONDITIONS FOR CALCULATING A CONFIDENCE INTERVAL FOR A POPULATION MEAN

- **Random:** The data come from a random sample from the population of interest.
    - **10%:** When sampling without replacement, $n < 0.10N$.
- **Normal/Large Sample:** The population distribution is approximately normal or the sample size is large ($n \geq 30$). If the population distribution has unknown shape and $n < 30$, a graph of the sample data shows no strong skewness or outliers.

EXAMPLE

Engagement rings and pulling wood apart
Checking conditions for a confidence interval for μ

Skill 4.C

PROBLEM: Check if the conditions for calculating a confidence interval for the population mean μ are met in each setting.

(a) When getting engaged, some people choose to buy a ring for their partner. How much do people spend on engagement rings, on average? To find out, *The New York Times* and Morning Consult surveyed a random sample of 1640 U.S. adults who bought an engagement ring.[1]

(b) How much force does it take to pull wood apart, on average? The stemplot shows the force (in pounds) required to pull apart a random sample of 20 pieces of Douglas fir.

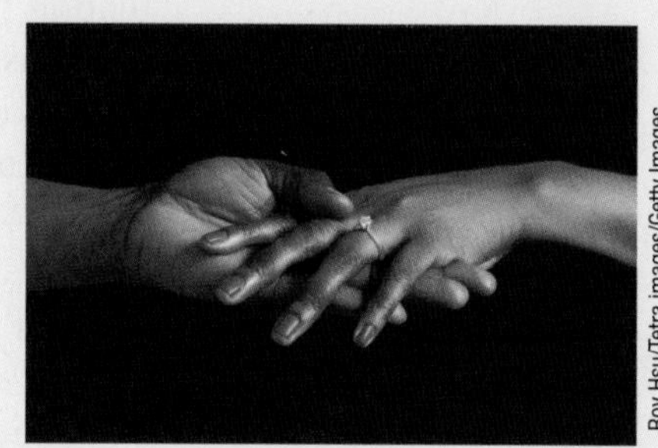

Roy Hsu/Tetra images/Getty Images

Stem	Leaf
23	0
24	0
25	
26	5
27	
28	7
29	
30	259
31	399
32	033677
33	0236

Key: 31|3 = 313 pounds of force required to pull a piece of Douglas fir apart

SOLUTION:

(a)
- Random: Random sample of U.S. adults who have bought an engagement ring. ✓
    - 10%: 1640 is less than 10% of all U.S. adults who have bought an engagement ring. ✓
- Normal/Large Sample: $n = 1640 \geq 30$. ✓

(b)
- Random: Random sample of 20 pieces of Douglas fir. ✓
    - 10%: 20 is less than 10% of all pieces of Douglas fir. ✓
- Normal/Large Sample: The sample size is small ($n = 20 < 30$) and the stemplot is strongly skewed to the left with several possible low outliers. ✗

It is *not* plausible that this sample came from an approximately normal population distribution due to the strong skewness and outliers.

FOR PRACTICE, TRY EXERCISE 5

You can use a dotplot, stemplot, histogram, or boxplot to check the Normal/Large Sample condition when $n < 30$ and the population distribution has unknown shape. Remember that boxplots give an incomplete picture of the shape of a distribution, hiding features like modes, gaps, and clusters. However, because boxplots clearly show skewness and outliers, they can be helpful for identifying important departures from normality.

AP® EXAM TIP

If a free-response question on the AP® Statistics exam asks you to calculate a confidence interval, all the conditions should be met. However, you are still required to state the conditions and show evidence that they are met—including a graph if the sample size is small and the data are provided.

CHECK YOUR UNDERSTANDING

1. Find the critical value t^* that you would use to calculate a confidence interval for a population mean μ in each of the following cases.
   (a) A 96% confidence interval based on a random sample of 22 observations
   (b) A 99% confidence interval from an SRS of 71 observations
2. To estimate the average GPA of students at a local community college, researchers randomly select 50 students and record their GPAs. Check if the conditions for calculating a confidence interval for the population mean μ are met.

Calculating a Confidence Interval for μ

The general form of a confidence interval is

$$\text{point estimate} \pm \text{margin of error}$$

When constructing a confidence interval for a population mean μ, we use the sample mean $\bar{x}$ as the point estimate. How do we calculate the margin of error?

Because we almost never know the population standard deviation σ, we have to use the sample standard deviation s_x as an estimate for σ. This means that we must estimate the standard deviation of the sampling distribution of $\bar{x}$, $\sigma_{\bar{x}} = \frac{\sigma}{\sqrt{n}}$, with the **standard error of the sample mean $\bar{x}$.**

DEFINITION Standard error of the sample mean $\bar{x}$

The **standard error of the sample mean $\bar{x}$** is an estimate of the standard deviation of the sampling distribution of $\bar{x}$.

$$s_{\bar{x}} = \frac{s_x}{\sqrt{n}}$$

The standard error of $\bar{x}$ estimates how much the sample mean $\bar{x}$ typically varies from μ in repeated random samples of size n.

Remember that standard error is sometimes abbreviated as SE.

When the Random, 10%, and Normal/Large Sample conditions are met, we can safely calculate a confidence interval for a population mean μ using the following formula:

$$\text{statistic} \pm (\text{critical value})(\text{standard error of statistic})$$

$$= \bar{x} \pm t^* \frac{s_x}{\sqrt{n}}$$

You already learned how to find the critical value t^* from a t distribution with $n - 1$ degrees of freedom.

CALCULATING A CONFIDENCE INTERVAL FOR A POPULATION MEAN

When the conditions are met, a $C\%$ confidence interval for the unknown population mean μ is

$$\bar{x} \pm t^* \frac{s_x}{\sqrt{n}}$$

where t^* is the critical value for a t distribution with df $= n - 1$ and $C\%$ of its area between $-t^*$ and t^*.

AP® EXAM TIP
Formula sheet

The specific formula for a confidence interval for a population mean is *not* included on the formula sheet provided on both sections of the AP® Statistics exam. However, the formula sheet does include the general formula for the confidence interval:

$$\text{statistic} \pm (\text{critical value})(\text{standard error of statistic})$$

and the formula for the standard error of the sample mean:

$$s_{\bar{x}} = \frac{s_x}{\sqrt{n}}$$

EXAMPLE

Put a ring on it!
Calculating a confidence interval for μ

Skill 3.D

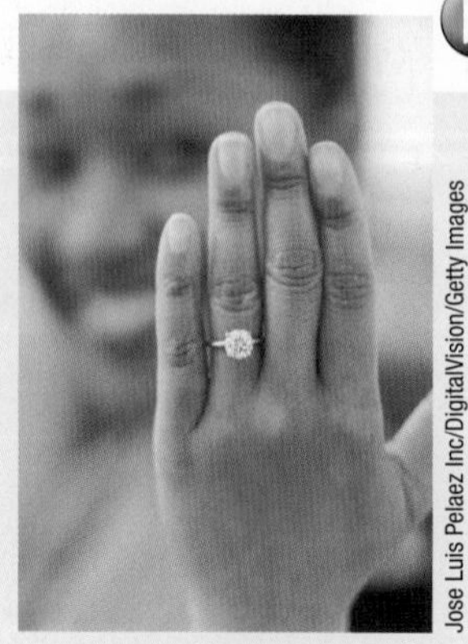
Jose Luis Pelaez Inc/DigitalVision/Getty Images

PROBLEM: In the preceding example, we verified that the conditions were met for calculating a confidence interval for μ = the population mean cost of an engagement ring for all U.S. adults who have bought one. After adjusting for inflation, the mean cost of the engagement rings in the random sample of 1640 adults was \$3329 with a standard deviation of \$4739.[2] Calculate a 99% confidence interval for μ.

SOLUTION:

df = 1640 − 1 = 1639

Using technology: invT(area: 0.005, df: 1639) = −2.579, so $t^* = 2.579$

In Table B, there is no row for df = 1639, so we would use the more conservative df = 1000 to get $t^* = 2.581$.

$$3329 \pm 2.579\frac{4739}{\sqrt{1640}}$$

$$\bar{x} \pm t^*\frac{s_x}{\sqrt{n}}$$

$= 3329 \pm 301.80$

$= (3027.20, 3630.80)$

FOR PRACTICE, TRY EXERCISE 11

How should we interpret the confidence interval in the example? We are 99% confident that the interval from \$3027.20 to \$3630.80 captures the population mean cost of an engagement ring for all U.S. adults who have bought one.

As you probably guessed, your calculator will compute a confidence interval for a population mean directly from sample data or summary statistics.

21. Tech Corner CONFIDENCE INTERVALS FOR A MEAN

TI-Nspire and other technology instructions are on the book's website at bfwpub.com/tps7e.

You can use a TI-83/84 calculator to construct a confidence interval for a population mean. We'll illustrate using the engagement ring scenario. Recall that the mean was \$3329, the standard deviation was \$4739, and the sample size was 1640.

1. Press STAT, arrow over to TESTS, and choose TInterval.
2. On the TInterval screen, choose Stats as the input method. Then enter the mean, standard deviation, and sample size, along with 0.99 as the confidence level.
3. Highlight Calculate and press ENTER. The 99% confidence interval for μ is reported, along with the sample mean $\bar{x}$, sample standard deviation s_x, and sample size n.

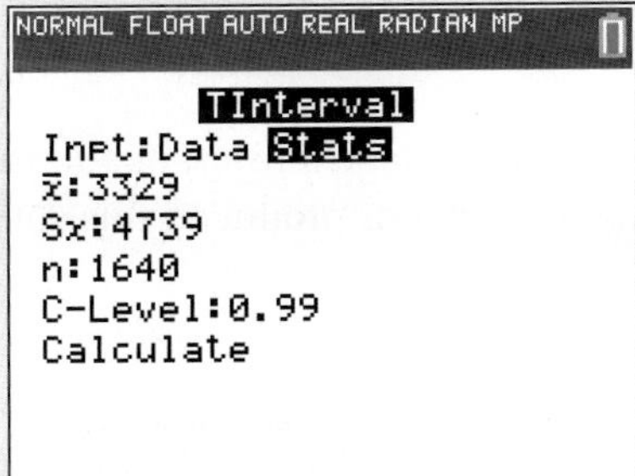

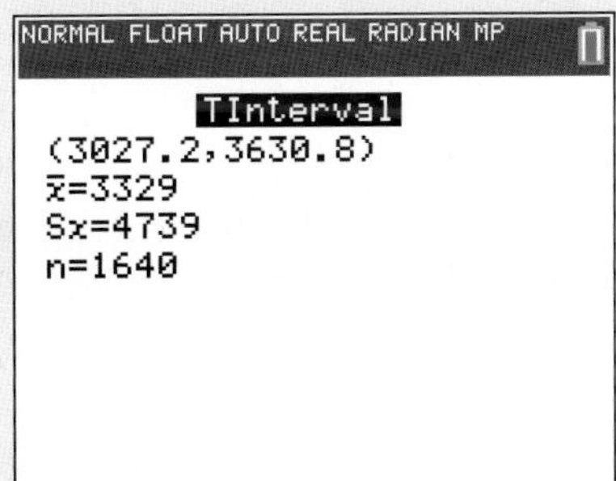

Note: If you have raw data, start by entering the data into a list. In Step 2, change the input method to Data and enter the name of the list where the data are stored (e.g., L1).

Confidence intervals calculated using the "full technology" approach described in the Tech Corner are generally more accurate than the values we obtain when we round while doing our calculations. For that reason, we will use the values from the full technology approach when reporting results "Using technology" in the rest of this section.

Putting It All Together: One-Sample *t* Interval for μ

The appropriate confidence interval procedure for estimating μ is called a **one-sample *t* interval for a mean.**

DEFINITION One-sample *t* interval for a mean

A **one-sample *t* interval for a mean** is a confidence interval used to estimate a population mean μ when the population standard deviation σ is unknown.

Whenever you construct and interpret a confidence interval for a population mean μ, be sure to follow the four-step process from Unit 6: State, Plan, Do, and Conclude.

EXAMPLE

Video screen tension
One-sample *t* interval for μ

Skills 1.D, 3.D, 4.B, 4.C, 4.D

PROBLEM: A manufacturer of high-resolution video screens must control the tension on the mesh of fine wires that lies behind the surface of the screen. Too much tension will tear the mesh, and too little will allow wrinkles. The tension is measured by an electrical device with output readings in millivolts (mV). Some variation is inherent in the production process. Here are the tension readings from a random sample of 20 screens from a single day's production:

269.5	297.0	269.6	283.3	304.8	280.4	233.5	257.4	317.5	327.4
264.7	307.7	310.0	343.3	328.1	342.6	338.8	340.1	374.6	336.1

John M Lund Photography Inc./DigitalVision/Getty Images

(a) Construct and interpret a 90% confidence interval for the mean tension μ of all the screens produced on this day.

(b) The manufacturer's goal is to produce video screens with an average tension of 300 mV. Based on the interval from part (a), is there convincing evidence that the screens produced this day don't meet the manufacturer's goal? Explain your reasoning.

SOLUTION:

Remember to follow the four-step process!

(a)

STATE: 90% CI for $\mu =$ the mean tension of all the video screens produced this day.

PLAN: One-sample *t* interval for a mean

- Random: Random sample of 20 video screens produced that day. ✓
  - 10%: Assume that 20 is less than 10% of all video screens produced that day. ✓

- **Normal/Large Sample: The sample size is small ($n = 20 < 30$), but the dotplot does not show strong skewness or outliers. ✓**

Because there is no strong skewness or outliers in the sample, it is plausible that the population distribution of video screen tension is approximately normal.

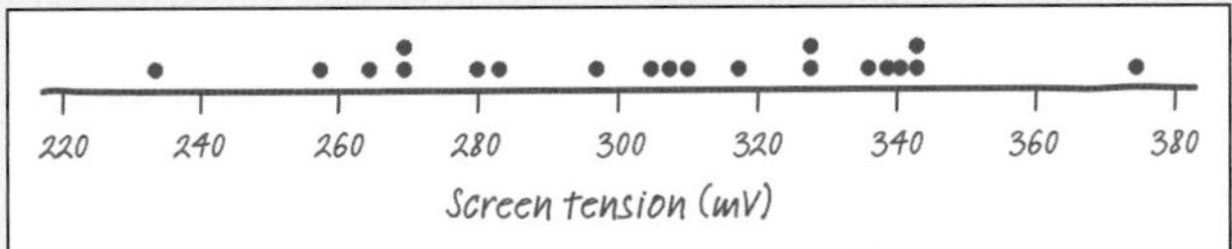

DO: $\bar{x} = 306.32$ mV, $s_x = 36.21$ mV, $n = 20$

With 90% confidence and df $= 20 - 1 = 19$, $t^* = 1.729$

Use your calculator to find the mean and standard deviation of the sample data.

$$306.32 \pm 1.729\frac{36.21}{\sqrt{20}}$$

$$\bar{x} \pm t^*\frac{s_x}{\sqrt{n}}$$

$= 306.32 \pm 14.00$

$= (292.32, 320.32)$

Using technology: (292.32, 320.32) with df $= 19$

CONCLUDE: We are 90% confident that the interval from 292.32 mV to 320.32 mV captures the mean tension of all the screens produced on this day.

Make sure that your conclusion is in context, is about a population mean, and includes units (mV).

(b) No. Because 300 mV is within the interval of plausible values for μ given by the confidence interval (292.32, 320.32), there is not convincing evidence that the video screens produced this day don't meet the manufacturer's goal.

FOR PRACTICE, TRY EXERCISE 13

> **AP® EXAM TIP**
>
> You may use your calculator to compute a confidence interval on the AP® Statistics exam — but there's a risk involved. If you just give the calculator answer with no work, you'll get either full credit for the "Do" step (if the interval is correct) or no credit (if it's wrong). If you opt for the calculator-only method, be sure to complete the other three steps (State, Plan, Conclude) and give the interval in the Do step.

How should we interpret the 90% confidence level in the example? If the manufacturer took many, many different random samples of 20 video screens produced on the given day, and constructed a 90% confidence interval for the population mean tension μ based on each sample, about 90% of those intervals would capture μ.

MORE ABOUT THE MARGIN OF ERROR

In Unit 6, you learned how different factors affect the margin of error when calculating a confidence interval for a population proportion. The lessons are the same in this section:

- The margin of error $t^*\frac{s_x}{\sqrt{n}}$ in a confidence interval for a population mean decreases as the sample size increases, assuming the confidence level and sample standard deviation s_x remain the same.
- The margin of error is proportional to $1/\sqrt{n}$, so quadrupling the sample size will cut the margin of error in half, assuming everything else remains the same.
- The margin of error will be larger for higher confidence levels. This makes sense, as wider intervals will have a greater capture rate than narrower intervals.
- The margin of error doesn't account for bias in the data collection process, only sampling variability.

In Section 6B, you learned how to calculate the sample size needed to ensure a certain margin of error when you're estimating a population proportion. Determining the required sample size is more difficult when you're estimating a population mean.

Think About It

IS IT POSSIBLE TO DETERMINE THE SAMPLE SIZE NEEDED TO ACHIEVE A SPECIFIED MARGIN OF ERROR WHEN ESTIMATING A POPULATION MEAN? Yes, but it's complicated. The margin of error (ME) in the confidence interval for μ is

$$ME = t^* \frac{s_x}{\sqrt{n}}$$

Unfortunately, there are two problems with using this formula to determine the sample size.

1. We don't know the sample standard deviation s_x because we haven't produced the data yet.
2. The critical value t^* depends on the sample size n that we choose.

The second problem is more serious. To get the correct value of t^*, we have to know the sample size. But that's precisely what we're trying to find!

One alternative is to come up with a reasonable estimate for the *population* standard deviation σ from a similar study that was done in the past or from a small-scale preliminary study. By pretending that σ is known, we can use a z^* critical value rather than a t^* critical value.

CHECK YOUR UNDERSTANDING

Biologists studying the healing of skin wounds measured the rate at which new cells closed a cut made in the skin of an anesthetized newt. Here are data from a random sample of 18 newts, measured in micrometers (millionths of a meter) per hour:[3]

29	27	34	40	22	28	14	35	26	35	12	30	23	18	11	22	23	33

Calculate and interpret a 95% confidence interval for the mean healing rate μ.

Paired Data: Confidence Intervals for a Population Mean Difference

Sean_Warren/E+/Getty Images

What if we want to construct a confidence interval for a population mean in a setting that involves measuring a quantitative variable twice for the same individual or for two very similar individuals? For instance, trace metals found in wells affect the taste of drinking water, and high concentrations can pose a health risk. Researchers measured the concentration of zinc (in milligrams per liter, mg/L) near the top and the bottom of 10 randomly selected wells in a large region. The data are provided in the table.[4]

Well	1	2	3	4	5	6	7	8	9	10
Top	0.415	0.238	0.390	0.410	0.605	0.609	0.632	0.523	0.411	0.612
Bottom	0.430	0.266	0.567	0.531	0.707	0.716	0.651	0.589	0.469	0.723

Notice that these two groups of zinc concentrations did *not* come from independent random samples of measurements made at the top of some wells and at the bottom of other wells. Instead, the data were obtained from *pairs* of measurements — one made at the top of a well and the other made at the bottom of the same well. This set of zinc concentrations is an example of **paired data.** Note that paired data can come from an observational study such as this one about zinc in wells, or from a *matched pairs* experiment in which two treatments are imposed on the same experimental unit or two very similar experimental units.

DEFINITION Paired data

Paired data result from recording two values of the same quantitative variable for each individual or for each pair of similar individuals.

When data are paired, the proper analysis uses the *differences* in each pair. Let's look at the difference in zinc concentrations at the top and bottom of each well.

Well	1	2	3	4	5	6	7	8	9	10
Top	0.415	0.238	0.390	0.410	0.605	0.609	0.632	0.523	0.411	0.612
Bottom	0.430	0.266	0.567	0.531	0.707	0.716	0.651	0.589	0.469	0.723
Difference (Bottom − Top)	0.015	0.028	0.177	0.121	0.102	0.107	0.019	0.066	0.058	0.111

The dotplot in Figure 7.4 displays these differences.

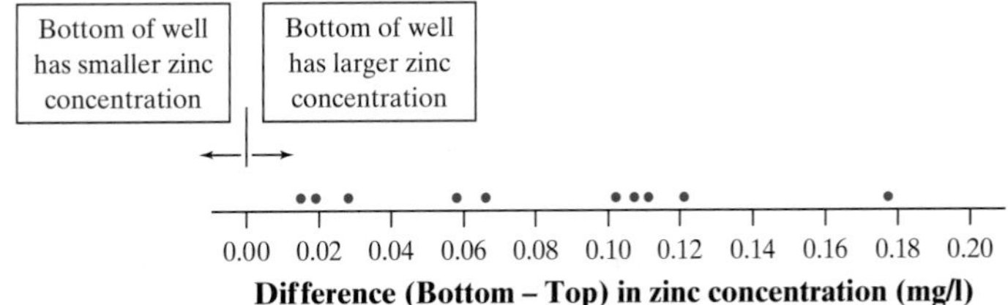

FIGURE 7.4 Dotplot of the difference in zinc concentrations at the top and bottom of each well.

For all 10 wells in the sample, the zinc concentration is greater at the bottom of the well than at the top of the well. This graph provides *some* evidence that zinc concentration tends to be larger at the bottom than at the top of all wells in this region.

Now that we have looked at a graph of the differences for paired data, it's time to calculate numerical summaries. The *mean difference* for these data is

$$\bar{x}_{\text{Diff}} = \bar{x}_{B-T} = \frac{0.015 + 0.028 + \ldots + 0.111}{10} = \frac{0.804}{10} = 0.0804 \text{ mg/L}$$

The standard deviation of the differences is

$$s_{\text{Diff}} = s_{B-T} = \sqrt{\frac{(0.015 - 0.0804)^2 + (0.028 - 0.0804)^2 + \ldots + (0.111 - 0.0804)^2}{10 - 1}}$$
$$= 0.0523 \text{ mg/L}$$

PUTTING IT ALL TOGETHER: PAIRED *t* INTERVAL FOR μ_{Diff}

When paired data come from a random sample or a randomized experiment, the statistic $\bar{x}_{\text{Diff}}$ is a point estimate for the population mean difference μ_{Diff}. Before constructing a confidence interval for μ_{Diff}, we have to check that the conditions are met. These conditions have been modified from the ones for calculating a one-sample *t* interval for a population mean to account for the fact that we are dealing with paired data. The purpose of these conditions is to ensure the independence of the individual differences and to verify that the sampling distribution of $\bar{x}_{\text{Diff}}$ is approximately normal. *Note:* The two measurements within each pair are generally *not* independent, because both are made on the same individual or two very similar individuals.

CONDITIONS FOR CALCULATING A CONFIDENCE INTERVAL FOR A POPULATION MEAN DIFFERENCE

- **Random:** Paired data come from a random sample from the population of interest or from a randomized experiment.
  - 10%: When sampling without replacement, $n_{\text{Diff}} < 0.10N_{\text{Diff}}$.
- **Normal/Large Sample:** The population distribution of differences is approximately normal or the sample size is large ($n_{\text{Diff}} \geq 30$). If the population distribution of differences has unknown shape and $n_{\text{Diff}} < 30$, a graph of the sample differences shows no strong skewness or outliers.

If these conditions are met, we can safely calculate a confidence interval for a population mean difference. The formula is the same one we used earlier, but substituting $\bar{x}_{\text{Diff}}$ for the sample mean, s_{Diff} for the sample standard deviation, and n_{Diff} for the sample size.

CALCULATING A CONFIDENCE INTERVAL FOR A POPULATION MEAN DIFFERENCE

When the conditions are met, a C% confidence interval for the population mean difference μ_{Diff} is

$$\bar{x}_{\text{Diff}} \pm t^* \frac{s_{\text{Diff}}}{\sqrt{n_{\text{Diff}}}}$$

where t^* is the critical value for a *t* distribution with df $= n_{\text{Diff}} - 1$ and C% of its area between $-t^*$ and t^*.

This inference procedure is often referred to as a **paired *t* interval for a mean difference.** It is also called a *one-sample t interval for a mean difference,* or when paired data come from a randomized experiment, a *matched-pairs t interval for a mean difference.*

DEFINITION Paired *t* interval for a mean difference

A **paired *t* interval for a mean difference** is a confidence interval used with paired data to estimate a population mean difference.

As with any inference method, follow the four-step process.

EXAMPLE

Does zinc sink?

Paired data: Confidence intervals for a population mean difference

Skills 1.D, 3.D, 4.B, 4.C

Sean_Warren/E+/Getty Images

PROBLEM: The data on zinc concentration (in mg/L) at the top and bottom of 10 randomly selected wells in a large region are shown again in the table, along with the differences we calculated earlier.

Well	1	2	3	4	5	6	7	8	9	10
Top	0.415	0.238	0.390	0.410	0.605	0.609	0.632	0.523	0.411	0.612
Bottom	0.430	0.266	0.567	0.531	0.707	0.716	0.651	0.589	0.469	0.723
Difference (Bottom − Top)	0.015	0.028	0.177	0.121	0.102	0.107	0.019	0.066	0.058	0.111

Construct and interpret a 95% confidence interval for the mean difference in zinc concentration at the top and bottom of all wells in this region.

SOLUTION:

STATE: 95% CI for μ_{Diff} = mean difference (Bottom − Top) in zinc concentration at the top and bottom of all wells in this large region.

Be sure to indicate the order of subtraction when defining the parameter.

PLAN: Paired *t* interval for a mean difference

- Random: Paired data come from zinc concentration measurements at the top and bottom of a random sample of 10 wells in the region. ✓
  - 10%: Assume $10 < 10\%$ of all wells in the region. ✓
- Normal/Large Sample: The sample size is small ($n_{Diff} = 10 < 30$), but the dotplot of differences doesn't show any strong skewness or outliers. ✓

0.00 0.02 0.04 0.06 0.08 0.10 0.12 0.14 0.16 0.18 0.20
Difference (Bottom – Top) in zinc concentration (mg/l)

DO: $\bar{x}_{Diff} = 0.0804, s_{Diff} = 0.0523, n_{Diff} = 10$

With 95% confidence and df $= 10 - 1 = 9$, $t^* = 2.262$

$$0.0804 \pm 2.262 \frac{0.0523}{\sqrt{10}}$$

$$= 0.0804 \pm 0.0374$$

$$= (0.0430, 0.1178)$$

Using technology: (0.04301, 0.11779) with df = 9

$$\bar{x}_{Diff} \pm t^* \frac{s_{Diff}}{\sqrt{n_{Diff}}}$$

CONCLUDE: We are 95% confident that the interval from 0.04301 to 0.11779 mg/L captures the mean difference (Bottom − Top) in the zinc concentrations at the top and bottom of all wells in this large region.

FOR PRACTICE, TRY EXERCISE 19

The 95% confidence interval in the example suggests that the zinc concentration is between 0.04301 and 0.11779 mg/L greater, on average, at the bottom than at the top of wells in this region. Because all the plausible values of μ_{Diff}

are positive, the interval gives convincing evidence that zinc concentrations are greater at the bottom than at the top of wells in this large region, on average.

CHECK YOUR UNDERSTANDING

Does music help or hinder performance in math? Researchers designed an experiment using 30 student volunteers to investigate. Each subject completed a 50-question, single-digit arithmetic test with and without music playing. For each subject, the order of the music and no music treatments was randomly assigned, and the time to complete the test (in seconds) was recorded for each treatment. Here are the data:[5]

Student	1	2	3	4	5	6	7	8	9	10	11	12	13	14	15
Time with music (sec)	83	119	77	75	64	106	70	69	60	76	47	97	68	77	48
Time without music (sec)	70	106	71	67	59	112	83	69	65	83	38	90	76	68	50
Student	16	17	18	19	20	21	22	23	24	25	26	27	28	29	30
Time with music (sec)	78	113	71	77	37	50	58	52	47	71	146	44	53	57	39
Time without music (sec)	73	93	59	70	39	52	60	54	51	60	141	40	56	53	37

1. Make a dotplot of the difference (Music − Without music) in time for each subject to complete the test. Describe what the graph reveals about whether music helps or hinders math performance.
2. Construct and interpret a 90% confidence interval for the population mean difference among all students like the ones in the study.
3. What does the interval in Question 2 suggest about whether music helps or hinders math performance? Explain your answer.

SECTION 7A Summary

- Confidence intervals for the population mean μ use the sample mean $\bar{x}$ as the point estimate.
- Because the population standard deviation σ is usually unknown, we use the sample standard deviation s_x to estimate σ when constructing a confidence interval for a population mean. Doing so requires use of a ***t* distribution** with $n-1$ *degrees of freedom* (df) to calculate a t^* critical value rather than the standard normal distribution and a z^* critical value.
- A t distribution is described by a symmetric, single-peaked, bell-shaped density curve centered at 0. Any t distribution is completely specified by its degrees of freedom (df) and has more area in its tails than the standard normal distribution.
- When constructing a confidence interval for a population mean μ, we need to ensure that the observations in the sample can be viewed as independent and that the sampling distribution of $\bar{x}$ is approximately normal. The required conditions are:
  - Random: The data come from a random sample from the population of interest.
    - 10%: When sampling without replacement, $n < 0.10N$.

  - Normal/Large Sample: The population distribution is approximately normal or the sample size is large ($n \geq 30$). If the population distribution has unknown shape and $n < 30$, a graph of the sample data shows no strong skewness or outliers.
- The **standard error of the sample mean** $\bar{x}$ is an estimate of the standard deviation of the sampling distribution of $\bar{x}$:

$$s_{\bar{x}} = \frac{s_x}{\sqrt{n}}$$

This value estimates how much the sample mean $\bar{x}$ typically varies from the population mean μ in random samples of size n.
- When conditions are met, a $C\%$ confidence interval for the population mean μ is given by

$$\bar{x} \pm t^* \frac{s_x}{\sqrt{n}}$$

where t^* is the critical value for a t distribution with df $= n - 1$ and $C\%$ of its area between $-t^*$ and t^*. This is known as a **one-sample *t* interval for a mean.**
- When estimating a population mean, the margin of error will be smaller when the sample size increases and larger when the confidence level increases, assuming that everything else remains the same.
- **Paired data** result from recording two values of the same quantitative variable for each individual or for each pair of similar individuals.
- To analyze paired data, start by computing the difference for each pair. Then make a graph of the differences. Use the mean difference $\bar{x}_{\text{Diff}}$ and the standard deviation of the differences s_{Diff} as summary statistics.
- Confidence intervals for a population mean difference μ_{Diff} use the sample mean difference $\bar{x}_{\text{Diff}}$ as the point estimate.
- Before estimating μ_{Diff}, we need to check for independence of the individual differences and confirm that the sampling distribution of $\bar{x}_{\text{Diff}}$ is approximately normal. The required conditions are:
  - Random: Paired data come from a random sample from the population of interest or from a randomized experiment.
    - 10%: When sampling without replacement, $n_{\text{Diff}} < 0.10N_{\text{Diff}}$.
  - Normal/Large Sample: The population distribution of differences is approximately normal or the sample size is large ($n_{\text{Diff}} \geq 30$). If the population distribution of differences has unknown shape and the number of differences in the sample is less than 30, a graph of the sample differences shows no strong skewness or outliers.
- When the conditions are met, a $C\%$ confidence interval for the population mean difference μ_{diff} is

$$\bar{x}_{\text{Diff}} \pm t^* \frac{s_{\text{Diff}}}{\sqrt{n_{\text{Diff}}}}$$

where t^* is the critical value for a t distribution with df $= n_{\text{Diff}} - 1$ and $C\%$ of its area between $-t^*$ and t^*. This is called a **paired *t* interval for a mean difference** (or a one-sample t interval for a mean difference).
- Follow the four-step process — State, Plan, Do, Conclude — whenever you are asked to construct and interpret a confidence interval for a population mean μ or a population mean difference μ_{Diff}.

AP® EXAM TIP

AP® Daily Videos

Review the content of this section and get extra help by watching the AP® Daily Videos for Topics 7.1, 7.2 and 7.3, which are available in AP® Classroom.

7A Tech Corners

TI-Nspire and other technology instructions are on the book's website at bfwpub.com/tps7e.

SECTION 7A Exercises

Determining t^* Critical Values

1. **Why t distributions?** The figure shows a t distribution with 5 degrees of freedom and a standard normal distribution.

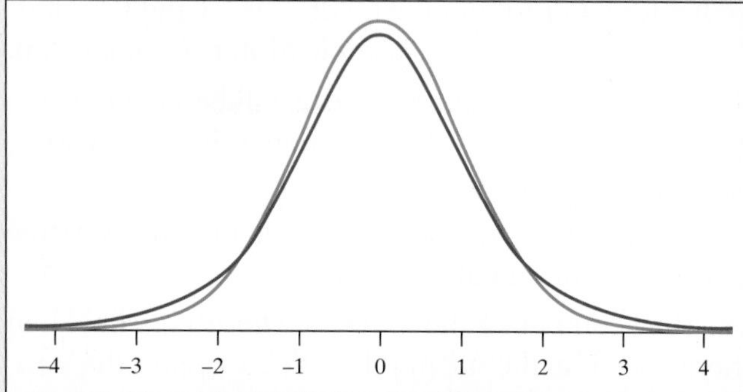

(a) Which density curve is the t distribution with df = 5? Explain your reasoning.

(b) When calculating a 99% confidence interval for a population mean based on a random sample of size $n = 6$, why should we use a t distribution with df = 5 rather than a standard normal distribution to determine the critical value?

2. **Why t and not z?** The figure shows a t distribution with 8 degrees of freedom and a standard normal distribution.

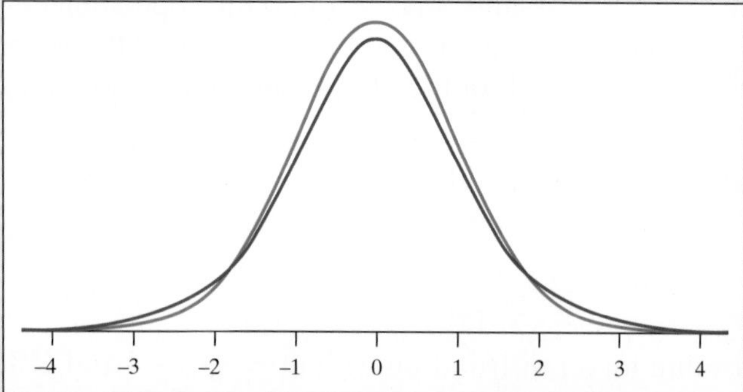

(a) Which density curve is the t distribution with df = 8? Explain your reasoning.

(b) When calculating a 95% confidence interval for a population mean based on a random sample of size $n = 9$, why should we use a t distribution with df = 8 rather than a standard normal distribution to determine the critical value?

3. pg 650 **Finding t^*** What critical value t^* should be used in constructing a confidence interval for the population mean in each of the following cases?

(a) A 99% confidence interval based on an SRS of size $n = 20$

(b) A 90% interval estimate from a random sample of 77 observations

4. **Finding t^* again** What critical value t^* should be used in constructing a confidence interval for the population mean in each of the following cases?

(a) A 95% confidence interval based on an SRS of 30 individuals

(b) A 99% interval estimate from a random sample of size 58

Checking Conditions for a Confidence Interval for μ

5. pg 652 **Velvetleaf and live events** Check if the conditions for calculating a confidence interval for the population mean μ are met in each scenario.

(a) The invasive weed known as velvetleaf is often found in U.S. cornfields, where it produces lots of seeds. How many seeds do velvetleaf plants produce, on average? The histogram shows the counts from a random sample of 28 plants that came up in a cornfield when no herbicide was used.[6]

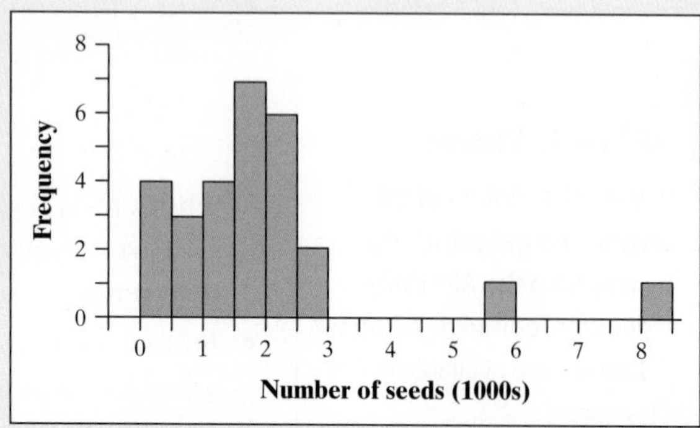

(b) How often do U.S. adults attend live music or theater events in a given year, on average? Gallup asked a random sample of 1025 U.S. adults about this and other leisure activities.[7]

6. **Salaries and possums** Check if the conditions for calculating a confidence interval for the population mean μ are met in each scenario.

(a) What is the mean base salary for full-time employees of community colleges in New York City? Researchers selected a random sample of 20 full-time NYC community college employees and recorded the base salary for each employee.[8] The results are summarized in the boxplot.

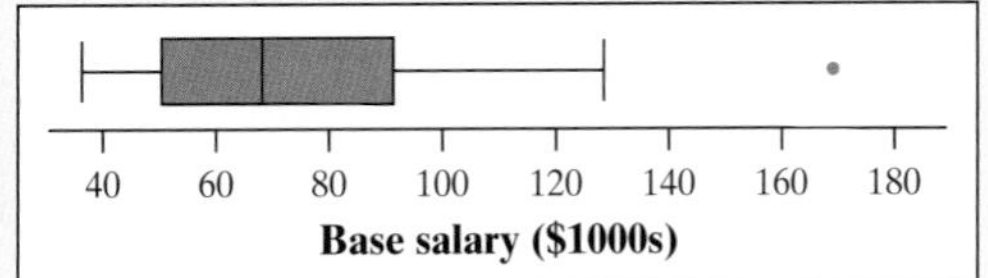

(b) How long are Australian possums, on average? Zoologists in Australia captured a random sample of 104 possums and recorded the values of 9 different variables, including total length (in centimeters).[9]

7. **Social media and home prices** Check if the conditions are met for constructing a confidence interval for the population mean in each scenario.

(a) How much time do students at a large high school spend on social media per day? Researchers collect data from the 32 members of the AP® Statistics class at the school and calculate the mean amount of time that these students spent on social media yesterday.

(b) Is the real estate market heating up? To estimate the mean sales price, a realtor in a large city randomly selects 100 home sales from the previous 6 months. These sales prices are displayed in the boxplot.

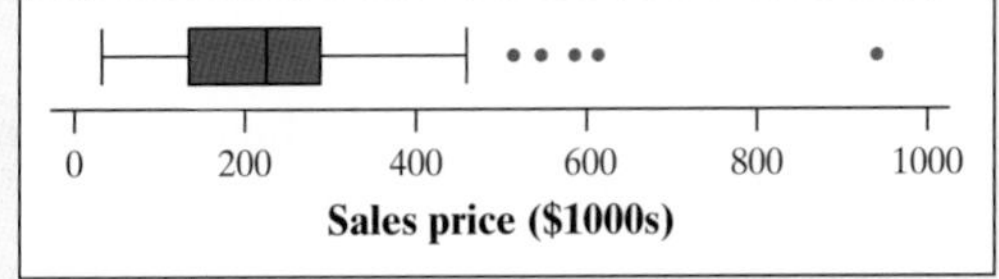

8. **Presidential life spans and medical journals** Check if the conditions are met for constructing a confidence interval for the population mean in each scenario.

(a) We want to estimate the average age at which U.S. presidents have died. To do so, we obtain a list of all U.S. presidents who have died and their ages at death.

(b) Judy is interested in estimating the reading level of a medical journal using average word length. She records the word lengths in a random sample of 100 words. The histogram displays the data.

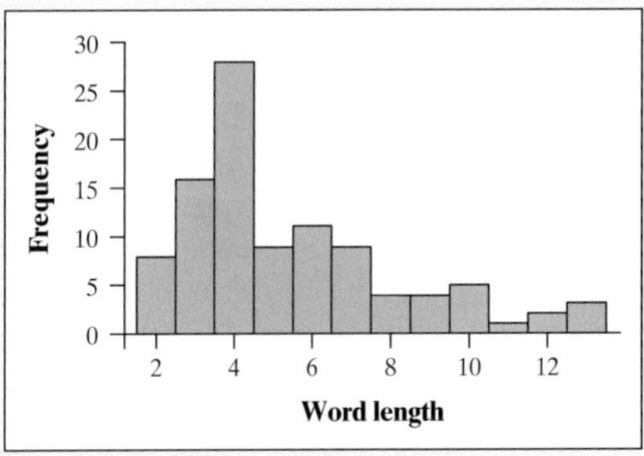

Calculating a Confidence Interval for μ

9. **Blood pressure** Researchers performing a medical study randomly selected 27 adults and measured the seated systolic blood pressure (in millimeters of mercury, mmHg) of each adult. In the sample, $\bar{x} = 114.9$ mmHg and $s_x = 9.3$ mmHg. What is the standard error of the mean? Interpret this value in context.

10. **Travel time to work** A study of commuting times reports the travel times to work of a random sample of 20 employed adults in New York State. The mean is $\bar{x} = 31.25$ minutes and the standard deviation is $s_x = 21.88$ minutes. What is the standard error of the mean? Interpret this value in context.

11. **Live events** Refer to Exercise 5(b). In the sample of 1025 adults, the mean number of live music and theater events attended was 3.8 with a standard deviation of 6.95. Note that the conditions are met. Calculate a 95% confidence interval for the mean number of live music and theater events U.S. adults attend in a given year.
pg 654

12. **Possums** Refer to Exercise 6(b). In the sample of 104 possums, the mean total length was 87.1 cm with a standard deviation of 4.3 cm. Note that the conditions are met. Calculate a 95% confidence interval for the mean total length of Australian possums.

One-Sample *t* Interval for μ

13. **Pepperoni pizza** Many people love pepperoni pizza, but sometimes they are disappointed with the small number of pepperonis on their pizza. To investigate, student researchers went to their favorite pizza restaurant at 10 random times during the week and ordered a large pepperoni pizza.[10] Here are the number of pepperonis on each pizza:
pg 656

47 36 25 37 46 36 49 32 32 34

(a) Construct and interpret a 90% confidence interval for the true mean number of pepperonis on a large pizza at this restaurant.

(b) According to the manager of the restaurant, there should be an average of 40 pepperonis on a large pizza.

Based on the interval from part (a), is there convincing evidence that the average number of pepperonis is less than 40? Explain your answer.

14. **Elephant size** How tall are adult male African elephants that lived through droughts during their first two years of life? Researchers measured the shoulder height (in centimeters) of a random sample of 14 mature (age 12 or older) male African elephants that lived through droughts during their first two years of life.[11] Here are the data:

200.00	272.91	217.57	294.15	296.84	212.00	257.00
251.39	266.75	237.19	265.85	212.00	220.00	225.00

(a) Construct and interpret a 90% confidence interval for the mean shoulder height of all mature male African elephants that lived through droughts during their first two years of life.

(b) The mean shoulder height of all mature male African elephants is 360 cm.[12] Based on the interval from part (a), is there convincing evidence that the mean shoulder height is less for mature male African elephants that went through a drought during their first two years of life? Explain your reasoning.

15. **A plethora of pepperoni?** Refer to Exercise 13.

(a) Explain why it was necessary to inspect a graph of the sample data when checking the Normal/Large Sample condition.

(b) Interpret the confidence level.

(c) Explain two ways that the student researchers could reduce the margin of error of their interval estimate. Why might they object to these changes?

16. **Elephants in drought** Refer to Exercise 14.

(a) Explain why it was necessary to inspect a graph of the sample data when checking the Normal/Large Sample condition.

(b) Interpret the confidence level.

(c) Explain two ways that the researchers could reduce the margin of error of their interval estimate. Why might they object to these changes?

Paired Data: Confidence Intervals for a Population Mean Difference

17. **Groovy tires** Researchers were interested in comparing two methods for estimating tire wear. The first method used the amount of weight lost by a tire. The second method used the amount of wear in the grooves of the tire. A random sample of 16 tires was obtained. Both methods were used to estimate the total distance traveled by each tire. The table provides the two estimates (in thousands of miles) for each tire.[13]

Tire	Weight	Groove	Tire	Weight	Groove
1	45.9	35.7	9	30.4	23.1
2	41.9	39.2	10	27.3	23.7
3	37.5	31.1	11	20.4	20.9
4	33.4	28.1	12	24.5	16.1
5	31.0	24.0	13	20.9	19.9
6	30.5	28.7	14	18.9	15.2
7	30.9	25.9	15	13.7	11.5
8	31.9	23.3	16	11.4	11.2

(a) Make a dotplot of the difference (Weight – Groove) in the estimate of wear for each tire using the two methods.

(b) Describe what the graph reveals about whether the two methods give similar estimates of tire wear.

(c) Calculate the mean and standard deviation of the differences.

18. **Internet speed** Ramon has found that his computer's internet connection is slower when he is farther from his wireless modem, which is located in the living room. To examine this difference, he randomly selects 14 times during the day and uses an online "speed test" to determine the download speeds to his computer in his bedroom and living room at each time (choosing which location to test first by flipping a coin). Here are the data, with download speeds in megabits per second (Mbps):[14]

Time	1	2	3	4	5	6	7
Bedroom	13.5	15.5	18.4	14.8	14.9	12.1	9.8
Living room	16.6	24.1	25.0	20.4	29.7	12.5	22.2
Time	**8**	**9**	**10**	**11**	**12**	**13**	**14**
Bedroom	16.0	11.1	14.3	15.6	10.5	15.6	11.3
Living room	17.6	26.7	18.5	28.7	15.7	22.8	27.0

(a) Make a dotplot of the difference (Living room – Bedroom) in the download speed for each room using the two methods.

(b) Describe what the graph reveals about the difference in download speeds in these two rooms.

(c) Calculate the mean and standard deviation of the differences.

19. **More groovy tires** Refer to Exercise 17. Construct and interpret a 95% confidence interval for the true mean difference in the estimates of tire wear using these two methods in the population of tires. pg 661

20. **More internet speed** Refer to Exercise 18. Construct and interpret a 95% confidence interval for the true mean difference in download speeds in these two rooms.

21. **Bone loss by nursing mothers** Breast-feeding mothers secrete calcium into their milk. Some of the calcium may come from their bones, so mothers may lose bone mineral. Researchers measured the percent change in bone mineral content (BMC) of the spines of 47 randomly selected mothers during 3 months of breast-feeding. The mean change in BMC was −3.587% and the standard deviation was 2.506%.[15]

(a) Construct and interpret a 99% confidence interval to estimate the mean percent change in BMC in the population of breast-feeding mothers.

(b) Based on your interval from part (a), do these data give convincing evidence that breastfeeding causes nursing mothers to lose bone mineral, on average? Explain your answer.

22. **Does playing the piano make kids smarter?** Do piano lessons improve the spatial-temporal reasoning of preschool children? A study designed to investigate this question measured the spatial-temporal reasoning of a random sample of 34 preschool children before and after 6 months of piano lessons. The difference (After − Before) in the reasoning scores for each student has mean 3.618 and standard deviation 3.055.[16]

(a) Construct and interpret a 90% confidence interval for the population mean difference in reasoning score.

(b) Based on your interval from part (a), can you conclude that taking 6 months of piano lessons would cause an increase in preschool students' average reasoning scores? Why or why not?

23. **Chewing gum** Researchers designed an experiment to investigate whether students can improve their short-term memory by chewing the same flavor of gum while studying for and taking an exam.[17] After recruiting 30 volunteers, they randomly assigned 15 to chew gum while studying a list of 40 words for 90 seconds. Immediately after the 90-second study period — and while chewing the same gum — the participants wrote down as many words as they could remember. The remaining 15 volunteers followed the same procedure without chewing gum. Two weeks later, each of the 30 participants did the opposite treatment. Researchers recorded the number of words each volunteer correctly remembered for each test.

(a) Explain why it was important for researchers to randomly assign the order in which each participant did the task with and without chewing gum.

(b) Verify that the conditions for constructing a matched pairs t interval for a mean difference are satisfied.

(c) The 95% confidence interval for the true mean difference (Gum − No gum) in number of words remembered is −0.67 to 1.54. Interpret the confidence interval and the confidence level.

(d) Based on the interval, is there convincing evidence that chewing gum helps subjects like these with short-term memory? Explain your answer.

24. **Stressful puzzles** Do people get stressed out when other people watch them work? To find out, researchers recruited 30 volunteers to take part in an experiment.[18] They randomly assigned 15 of the participants to complete a word search puzzle while researchers stood close by and visibly took notes. The remaining 15 were assigned to complete a word search puzzle while researchers stood at a distance. After each participant completed the word search, they completed a second word search under the opposite treatment. Researchers recorded the amount of time each volunteer required to complete each puzzle.

(a) Explain why it was important for researchers to randomly assign the order in which each participant did the task with researchers close by and standing at a distance.

(b) Verify that the conditions for constructing a matched pairs t interval for a mean difference are satisfied.

(c) The 95% confidence interval for the true mean difference (Close by − At a distance) in amount of time needed to complete the puzzle is −12.7 seconds to 119.4 seconds. Interpret the confidence interval and the confidence level.

(d) Based on the interval, is there convincing evidence that standing close by causes subjects like the ones in this study to take longer to complete a word search? Explain your answer.

For Investigation *Apply the skills from the section in a new context or nonroutine way.*

25. **Estimating BMI** The body mass index (BMI) of all individuals in a certain population is approximately normally distributed with a standard deviation of $\sigma = 7.5$. How large a sample would be needed to estimate the mean BMI in this population to within ±1 with 99% confidence? (Note that the population standard deviation is known in this case.)

26. **House appraisals** The mayor of a small town wants to estimate the average property value for the houses built there in the last year. She randomly selects 15 houses and pays an appraiser to determine the value of each house. The mean value of these houses is $183,100 with a standard deviation of $29,200. Assume the conditions are met.

(a) Calculate a 90% confidence interval for the population mean value of new houses in this town.

(b) There are a total of 300 new houses in the mayor's town and the town levies a 1% property tax on each house. Use your confidence interval from part (a) to calculate a 90% confidence interval for the total amount of tax revenue generated by the new houses.

Multiple Choice *Select the best answer for each question.*

27. One reason for using a t distribution instead of the standard normal distribution to find critical values when calculating a confidence interval for a population mean is that

(A) z can be used only for large samples.

(B) z requires that you know the population standard deviation σ.

(C) z requires that you can regard your data as a random sample from the population.

(D) z requires that the sample size is less than 10% of the population size.

(E) a z^* critical value will lead to a wider interval than a t^* critical value.

28. Suppose that you have an SRS of 23 observations from a large population. The distribution of sample data is roughly symmetric with no outliers. What critical value would you use to obtain a 98% confidence interval for the mean of the population?

(A) 2.177
(B) 2.183
(C) 2.326
(D) 2.500
(E) 2.508

29. A quality control inspector will measure the salt content (in milligrams) in a random sample of bags of potato chips from an hour of production. Which of the following would result in the smallest margin of error in estimating the mean salt content μ?

(A) 90% confidence; $n = 25$

(B) 90% confidence; $n = 50$

(C) 95% confidence; $n = 25$

(D) 95% confidence; $n = 50$

(E) $n = 100$ at any confidence level

30. Scientists collect data on the blood cholesterol levels (milligrams per deciliter of blood) of a random sample of 24 laboratory rats. A 95% confidence interval for the population mean blood cholesterol level μ is 80.2 to 89.8. Which of the following would cause the most worry about the validity of this interval?

(A) There is a clear outlier in the data.

(B) A stemplot of the data shows mild right skewness.

(C) The population standard deviation σ is unknown.

(D) The population distribution is not exactly normal.

(E) None of these is a problem when using a t interval.

Exercises 31 and 32 refer to the following scenario. Researchers suspect that Variety A tomato plants have a different average yield than Variety B tomato plants. To find out, researchers randomly select 10 Variety A and 10 Variety B tomato plants. Then the researchers divide in half each of 10 small plots of land in different locations. For each plot, a coin toss determines which half of the plot gets a Variety A plant; a Variety B plant goes in the other half. After harvest, the researchers compare the yield in pounds for the plants at each location. The 10 differences (Variety A – Variety B) in yield are recorded. A graph of the differences looks roughly symmetric and unimodal with no outliers. The mean difference is $\bar{x}_{A-B} = 0.34$ pound and the standard deviation of the differences is $s_{A-B} = 0.83$ pound. Let μ_{A-B} = the true mean difference (Variety A – Variety B) in yield for tomato plants of these two varieties (in pounds).

31. Which of the following is the best reason to use a paired t interval for a mean difference to calculate a confidence interval based on these data?

(A) The number of plots is the same for the Variety A and Variety B plants.

(B) The response variable, yield of tomatoes, is quantitative.

(C) This is an experiment with randomly assigned treatments.

(D) Each plot is given both varieties of tomato plant.

(E) The sample size is less than 30 for both treatments.

32. A 95% confidence interval for μ_{A-B} is given by

(A) $0.34 \pm 1.96(0.83)$
(B) $0.34 \pm 1.96\left(\frac{0.83}{\sqrt{10}}\right)$
(C) $0.34 \pm 2.228\left(\frac{0.83}{\sqrt{10}}\right)$
(D) $0.34 \pm 2.262(0.83)$
(E) $0.34 \pm 2.262\left(\frac{0.83}{\sqrt{10}}\right)$

Recycle and Review *Practice what you learned in previous sections.*

33. **Watching TV (4D, 5D)** Suppose we choose a young person (aged 19 to 25) at random and ask, "In the past seven days, how many days did you watch television?" Call the response X for short. Here is the probability distribution of X:[19]

Days, x	0	1	2	3	4	5	6	7
Probability, $P(X = x)$	0.04	0.03	0.06	0.08	0.09	0.08	0.05	???

(a) What is the probability that $X = 7$? Justify your answer.

(b) Calculate the mean and standard deviation of the random variable X.

(c) Suppose that you asked 100 randomly selected young people (aged 19 to 25) to respond to the question and found the mean $\bar{x}$ of their responses. Describe the sampling distribution of $\bar{x}$.

(d) Would a sample mean of $\bar{x} = 4.96$ be surprising? Explain your reasoning.

34. **Price cuts (3C)** Stores advertise price reductions to attract customers. What type of price cut is most attractive? Experiments with more than one factor allow insight into interactions between the factors. A study of the attractiveness of advertised price discounts focused on two factors: the percentage of all foods on sale (25%, 50%, 75%, or 100%) and whether the discount was stated precisely (e.g., "60% off") or as a range (e.g., "40% to 70% off"). Subjects rated the attractiveness of the sale on a scale of 1 to 7.

(a) List the treatments for this experiment, assuming researchers will use all combinations of the two factors.

(b) Describe how you would randomly assign 200 volunteer subjects to treatments.

(c) Explain the purpose of the random assignment in part (b).

(d) The figure shows the mean ratings for the eight treatments formed from the two factors.[20] Based on these results, write a careful description of how percentage on sale and precise discount versus range of discounts influence the attractiveness of a sale.

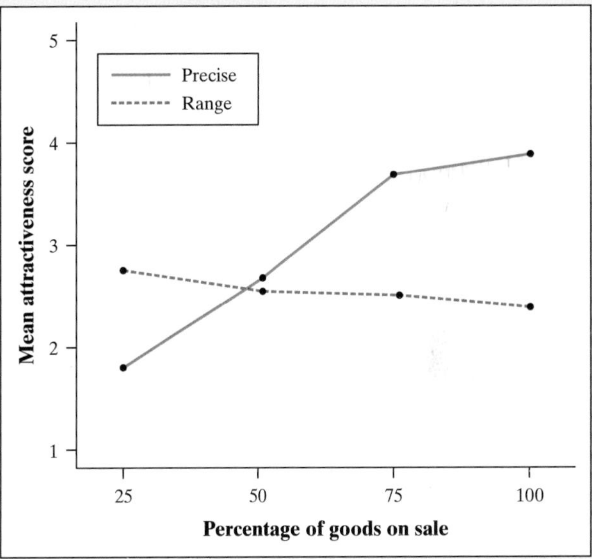

SECTION 7B Significance Tests for a Population Mean or Mean Difference

LEARNING TARGETS *By the end of the section, you should be able to:*

- State appropriate hypotheses and check the conditions for performing a test about a population mean.
- Calculate the standardized test statistic and *P*-value for a test about a population mean.
- Perform a one-sample *t* test for a mean.
- In the special case of paired data, perform a significance test about a population mean difference.

In Section 7A, you learned how to construct a confidence interval for a population mean μ and a confidence interval for a population mean difference μ_{Diff} in the special case of paired data. Now we'll examine the details of testing a claim about a population mean or a population mean difference.

Recall from Unit 6 that the first step in a significance test is stating hypotheses. When the conditions are met, we can calculate the *P*-value using an appropriate probability distribution. The *P*-value helps us make a conclusion about the competing claims being tested based on the strength of evidence in favor of the alternative hypothesis H_a and against the null hypothesis H_0.

Stating Hypotheses and Checking Conditions for a Test about μ

A "classic rock" radio station claims to play an average of 50 minutes of music every hour. However, it seems that every time you start listening to this station, a commercial is playing. To investigate the station's claim, you randomly select 12 different hours during the next week and record how many minutes of music the station plays in each of those hours. Here are the data:

44 49 45 51 49 53 49 44 47 50 46 48

You would like to perform a test at the $\alpha = 0.05$ significance level of

$$H_0: \mu = 50$$
$$H_a: \mu < 50$$

where μ = the true mean amount of music played (in minutes) during each hour by this station.

In Section 7A, we introduced the conditions that should be met before we construct a confidence interval for a population mean μ. We called them Random, 10%, and Normal/Large Sample. These same conditions must be verified before performing a significance test about a population mean. Recall that the purpose of checking these conditions is to ensure that the observations in the sample can be viewed as independent and that the sampling distribution of $\bar{x}$ is approximately normal.

CONDITIONS FOR PERFORMING A SIGNIFICANCE TEST ABOUT A POPULATION MEAN

- **Random:** The data come from a random sample from the population of interest.
    - **10%:** When sampling without replacement, $n < 0.10N$.
- **Normal/Large Sample:** The population distribution is approximately normal or the sample size is large ($n \geq 30$). If the population distribution has unknown shape and $n < 30$, a graph of the sample data shows no strong skewness or outliers.

Let's check the conditions for the radio station data.

- Random: Random sample of 12 different hours. ✓
    - 10%: 12 < 10% of all the hours of music played by the station in a week. ✓
- Normal/Large Sample: The sample size is small ($n = 12 < 30$), but the dotplot doesn't show any outliers or strong skewness. ✓

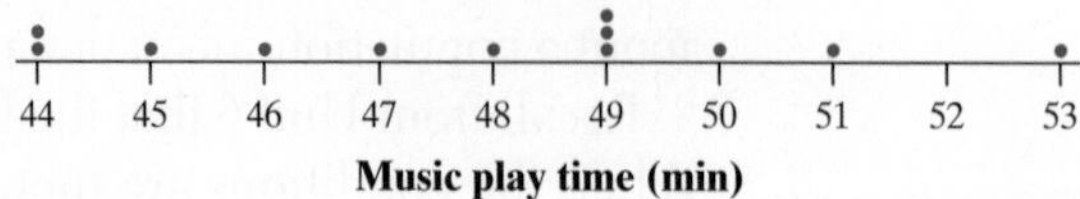

All the conditions are met, so we can safely proceed with a significance test about μ = the true mean amount of music played (in minutes) during each hour by this station.

EXAMPLE

Golden hamsters
Stating hypotheses and checking conditions for a test about μ

Skills 1.F, 4.C

Paul Starosta/Stone/Getty Images

PROBLEM: According to the Animal Diversity Web, adult golden hamsters have an average weight of 4 ounces.[21] An animal researcher wants to test this claim. The researcher selects and weighs a random sample of 30 golden hamsters.

(a) State appropriate hypotheses for the researcher's test. Be sure to define the parameter of interest.

(b) Check if the conditions for performing the test are met.

SOLUTION:

(a) $H_0: \mu = 4$

$H_a: \mu \neq 4$

where μ = the mean weight (in ounces) in the population of adult golden hamsters.

(b)
- Random: Random sample of 30 golden hamsters. ✓
  - 10%: Assume $30 < 10\%$ of all golden hamsters. ✓
- Normal/Large Sample: $n = 30 \geq 30$. ✓

FOR PRACTICE, TRY EXERCISE 1

When the conditions are met, we can proceed to calculations.

Calculating the Standardized Test Statistic and *P*-Value for a Test about μ

In the classic rock radio station example, you can check that the sample mean amount of music played during 12 randomly selected hours is $\bar{x} = 47.917$ minutes. Because this result is less than 50 minutes, there is *some* evidence against H_0: $\mu = 50$ and in favor of H_a: $\mu < 50$. But do we have *convincing* evidence that the true mean amount of music played during each hour by this station is less than 50 minutes? To answer this question, we want to know if it is unlikely to get a sample mean of 47.917 minutes or less by chance alone when the null hypothesis is true. To assess how far the sample mean $\bar{x}$ is from the null hypothesis value μ_0, we standardize the statistic:

$$\text{standardized test statistic} = \frac{\text{statistic} - \text{parameter}}{\text{standard error of statistic}}$$

Suppose for now that the null hypothesis H_0: $\mu = 50$ is true. Consider the sample mean amount of music played $\bar{x}$ (in minutes) for a random sample of $n = 12$ hours. You learned in Section 5D that the sampling distribution of $\bar{x}$ will have mean

$$\mu_{\bar{x}} = \mu = 50 \text{ minutes}$$

and standard deviation

$$\sigma_{\bar{x}} = \frac{\sigma}{\sqrt{n}}$$

In an ideal world, our standardized test statistic would be

$$z = \frac{\bar{x} - \mu_0}{\frac{\sigma}{\sqrt{n}}}$$

When the Normal/Large Sample condition is met, the standardized test statistic z can be modeled by the standard normal distribution. We could then use this distribution to find the *P*-value. Unfortunately, there are very few real-world situations in which we might know the population standard deviation σ when we don't know the population mean μ!

Because the population standard deviation σ is usually unknown, we use the sample standard deviation s_x in its place. For the radio station data, $s_x = 2.811$. Our resulting estimate of $\sigma_{\bar{x}}$ is the standard error of the mean:

$$s_{\bar{x}} = \frac{s_x}{\sqrt{n}} = \frac{2.811}{\sqrt{12}} = 0.8115$$

When we use the sample standard deviation s_x to estimate the unknown population standard deviation σ, the standardized test statistic is denoted by t (you can probably guess why from what you learned in Section 7A). So the formula becomes

$$t = \frac{\bar{x} - \mu_0}{\frac{s_x}{\sqrt{n}}}$$

The standardized test statistic for the radio station data is

$$t = \frac{47.917 - 50}{\frac{2.811}{\sqrt{12}}} = \frac{47.917 - 50}{0.8115} = -2.567$$

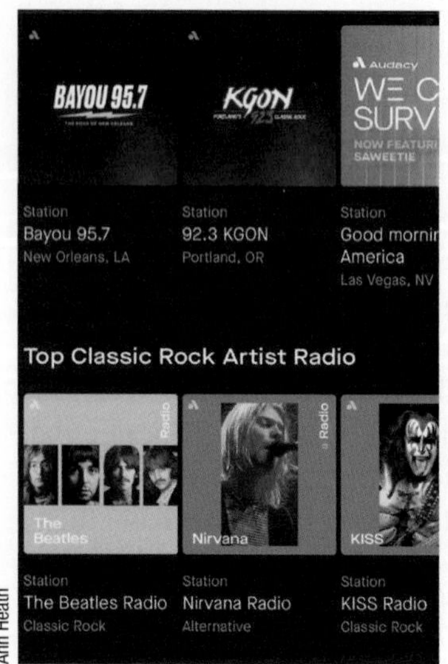

Ann Heath

Recall that the standardized test statistic tells us how far the sample result is from the null value, and in which direction, on a standardized scale. In this case, the mean amount of classic rock music played in the random sample of 12 hours, $\bar{x} = 47.197$ minutes, is 2.567 standard errors less than the null value of $\mu = 50$ minutes.

When the Normal/Large Sample condition is met and the null hypothesis H_0 is true, the standardized test statistic

$$t = \frac{\bar{x} - \mu_0}{\frac{s_x}{\sqrt{n}}}$$

can be modeled by a t distribution with degrees of freedom df $= n - 1$. We can use technology or Table B to find the *P*-value. In the radio station example, we planned to carry out a test at the $\alpha = 0.05$ significance level of

$$H_0: \mu = 50$$
$$H_a: \mu < 50$$

where $\mu =$ the true mean amount of music played (in minutes) during each hour by this station. In $n = 12$ randomly selected hours, the radio station played an average of $\bar{x} = 47.917$ minutes of music. The *P*-value is the probability of getting a result this small or smaller by chance alone when H_0: $\mu = 50$ is true. Figure 7.5(a) shows this probability as a shaded area in the sampling distribution of $\bar{x}$. Earlier, we calculated the standardized test statistic to be $t = -2.567$. So we calculate the *P*-value by finding $P(t \leq -2.567)$ in a t distribution with df $= 12 - 1 = 11$. The shaded area in Figure 7.5(b) shows this probability.

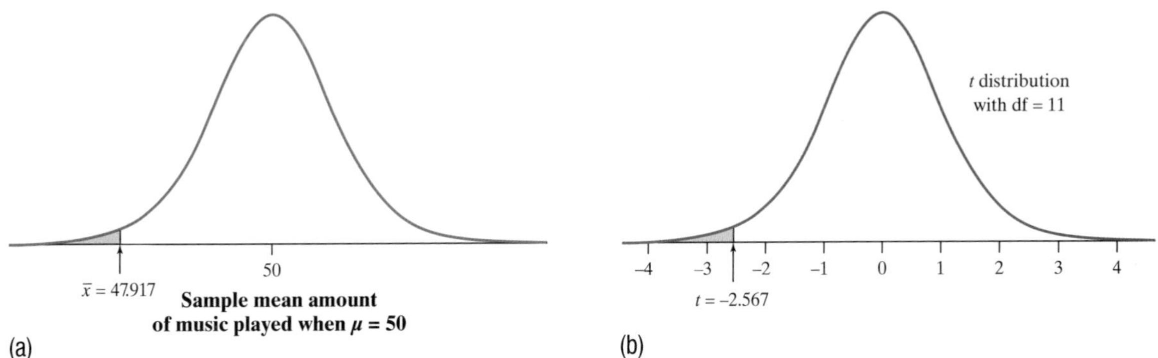

FIGURE 7.5 (a) The *P*-value for the radio station example is the area to the left of $\bar{x} = 47.917$ in the sampling distribution of the sample mean, assuming that H_0: $\mu = 50$ is true. (b) We calculate the *P*-value as the area to the left of $t = -2.567$ in a t distribution with 11 degrees of freedom.

We can find this *P*-value using Table B. Go to the df $= 11$ row. Unfortunately, Table B shows only ***positive*** *t*-values and areas in the *right* tail of the t distributions. But the t distributions are symmetric around their center of 0, so $P(t \leq -2.567) = P(t \geq 2.567)$. The t statistic falls between the values 2.328 and 2.718. Now look at the top of the corresponding columns in Table B. You see that the "Upper-tail probability p" is between 0.02 and 0.01. (See the excerpt from Table B.) Therefore, the *P*-value for this test is between 0.01 and 0.02.

	Upper-tail probability *p*		
df	**0.02**	**0.01**	**0.005**
10	2.359	2.764	3.169
11	2.328	2.718	3.106
12	2.303	2.681	3.055

As you can see, Table B gives an interval of possible *P*-values for a significance test. We can still make a conclusion from the test in the same way as if we had a single probability — by comparing the interval of possible *P*-values to the chosen significance level α.

Table B has another important limitation that we mentioned in Section 7A: it includes only t distributions with degrees of freedom from 1 to 30 and then skips to df $= 40, 50, 60, 80, 100$, and 1000. *Remember:* If the df you need isn't provided in Table B, use the next lower df that is available. It's not fair "rounding up" to a larger df, which is like pretending that your sample size is larger than it really is. Doing so would give you a smaller *P*-value than is true and would make it more likely that you would incorrectly reject H_0 when it's true (i.e., make a Type I error). Of course, "rounding down" to a smaller df will give you a larger *P*-value than is true, which makes it more likely that you will commit a Type II error!

Given the limitations of Table B, our advice is to use technology to find *P*-values when carrying out a significance test about a population mean.

22. Tech Corner CALCULATING *P*-VALUES FROM *t* DISTRIBUTIONS

TI-Nspire and other technology instructions are on the book's website at bfwpub.com/tps7e.

You can use the tcdf command on the TI-83/84 to calculate areas under a *t* distribution curve. The syntax is tcdf(lower bound, upper bound, df). Let's use the tcdf command to compute the *P*-values for the radio station example, where $t = -2.567$ and df $= 11$.

1. Press 2nd VARS (DISTR) and choose tcdf(.
2. We want to find the area to the left of $t = -2.567$ under the *t* distribution curve with df $= 11$.

OS 2.55 or later: In the dialog box, enter these values: lower: −1000, upper: −2.567, df: 11. Choose Paste, and then press ENTER.

Older OS: Complete the command tcdf(−1000, −2.567, 11) and press ENTER.

```
NORMAL FLOAT AUTO REAL RADIAN MP
tcdf(-1000,-2.567,11)
                    0.0130949981
```

Let's make a conclusion for the radio station example. Because the *P*-value of 0.0131 is less than $\alpha = 0.05$, we reject H_0: $\mu = 50$. We have convincing evidence that the classic rock radio station is playing fewer than 50 minutes of music per hour, on average.

CALCULATING THE STANDARDIZED TEST STATISTIC AND *P*-VALUE IN A TEST ABOUT A POPULATION MEAN

Suppose the conditions are met. To perform a test of H_0: $\mu = \mu_0$, calculate the standardized test statistic

$$t = \frac{\bar{x} - \mu_0}{\frac{s_x}{\sqrt{n}}}$$

Find the *P*-value by calculating the probability of getting a *t* statistic this large or larger in the direction specified by the alternative hypothesis H_a in a *t* distribution with df $= n - 1$.

AP® EXAM TIP

Formula sheet

The specific formula for the standardized test statistic in a test about a population mean is *not* included on the formula sheet provided on both sections of the AP® Statistics exam. However, the formula sheet does include the general formula for the standardized test statistic:

$$\text{standardized test statistic} = \frac{\text{statistic} - \text{parameter}}{\text{standard error of statistic}}$$

and the formula for the standard error of the sample mean:

$$s_{\bar{x}} = \frac{s_x}{\sqrt{n}}$$

EXAMPLE

How heavy are golden hamsters?

Calculating the standardized test statistic and *P*-value for a test about μ

Skills 3.E, 4.B, 4.E

PROBLEM: According to the Animal Diversity Web, adult golden hamsters have an average weight of 4 ounces.[22] To test this claim, an animal researcher selects and weighs a random sample of 30 golden hamsters. The mean weight is $\bar{x} = 4.033$ ounces with standard deviation $s_x = 0.199$ ounce. The researcher wants to perform a test at the 10% significance level of

$$H_0: \mu = 4$$
$$H_a: \mu \neq 4$$

Joe Blossom/Alamy Stock Photo

where μ = the mean weight (in ounces) in the population of adult golden hamsters. We verified that the conditions were met in the preceding example.

(a) Explain why the sample result gives some evidence for the alternative hypothesis.
(b) Calculate the standardized test statistic and *P*-value.
(c) Interpret the *P*-value.

SOLUTION:

(a) The sample mean is $\bar{x} = 4.033$, which is not equal to 4 (as suggested by H_a).

(b) $t = \dfrac{4.033 - 4}{\frac{0.199}{\sqrt{30}}} = 0.908$

P-value: df = 30 − 1 = 29

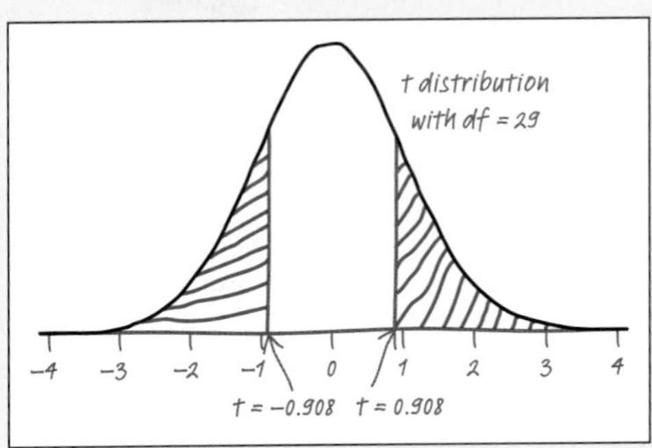

standardized test statistic $= \dfrac{\text{statistic} - \text{parameter}}{\text{standard error of statistic}}$

$$t = \frac{\bar{x} - \mu_0}{\frac{s_x}{\sqrt{n}}}$$

Using technology: tcdf(lower: 0.908, upper: 1000, df: 29) × 2 = 0.3714
Using Table B: The *P*-value is between 2(0.15) = 0.30 and 2(0.20) = 0.40.

(c) If the mean weight in the population of adult golden hamsters is 4 ounces, there is about a 0.3714 probability of getting a sample mean weight $\bar{x} \geq 4.033$ ounces or $\bar{x} \leq 3.967$ ounces by chance alone.

Because the alternative hypothesis H_a: $\mu \neq 4$ is two-sided, we need to consider values of the sample mean at least as surprising as $\bar{x} = 4.033$ *in either direction* from $\mu = 4$ when interpreting the *P*-value.

FOR PRACTICE, TRY EXERCISE 7

What conclusion should the animal researcher in the example make? Because 0.3714 is greater than $\alpha = 0.10$, we fail to reject H_0. We do not have convincing evidence that μ = the mean weight in the population of adult golden hamsters differs from 4 ounces.

As you probably guessed, your calculator will compute the standardized test statistic and *P*-value in a test for a population mean directly from sample data or summary statistics.

23. Tech Corner SIGNIFICANCE TESTS FOR A MEAN

TI-Nspire and other technology instructions are on the book's website at bfwpub.com/tps7e.

You can use the TI-83/84 calculator to perform all the calculations required for a significance test about a population mean. We'll illustrate this process using the radio station data from earlier in the section. Here are the data again:

44 49 45 51 49 53 49 47 50 46 48

1. Enter the data values into list L1.
2. Press STAT, then choose TESTS and T-Test.
3. When the T-Test screen appears, choose Data as the input method. Then enter $\mu_0 = 50$, List: L1, Freq: 1, and choose $\mu < \mu_0$.

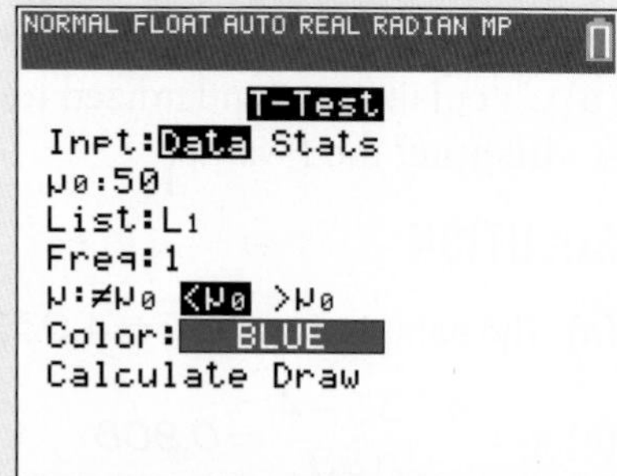

4. Highlight "Calculate" and press ENTER.

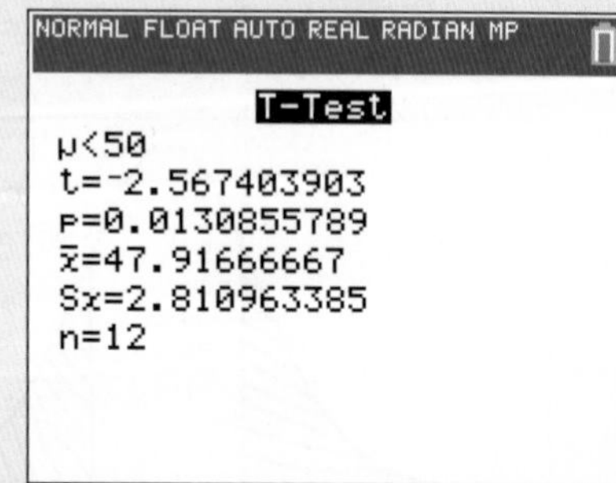

Notes:

- If you select the Draw option, you will get a picture of the appropriate *t* distribution with the area of interest shaded and the standardized test statistic and *P*-value labeled.
- If you have summary statistics instead of raw data, change the input method to Stats and enter the appropriate values.

The standardized test statistic and *P*-value calculated using the "full technology" approach described in the Tech Corner are generally more accurate than the values we obtain when we round while doing our calculations. For that reason, we will use the values from the full technology approach when reporting results "Using technology" in the rest of this section.

TWO-SIDED TESTS AND CONFIDENCE INTERVALS

You learned in Unit 6 that a confidence interval gives more information than a significance test because it provides the entire set of plausible values for the parameter based on the sample data. The connection between two-sided tests and confidence intervals is even stronger for means than it was for proportions because both inference methods for means use the standard error of $\bar{x}$ in the calculations:

$$\text{standardized test statistic: } t = \frac{\bar{x} - \mu_0}{\frac{s_x}{\sqrt{n}}} \qquad \text{confidence interval: } \bar{x} \pm t^* \frac{s_x}{\sqrt{n}}$$

This link between two-sided tests and confidence intervals for a population mean allows us to make a conclusion about H_0 directly from a confidence interval.

- If a 95% confidence interval for μ does not capture the null value μ_0, we can reject H_0: $\mu = \mu_0$ in a two-sided test at the 5% significance level ($\alpha = 0.05$).
- If a 95% confidence interval for μ captures the null value μ_0, then we should fail to reject H_0: $\mu = \mu_0$ in a two-sided test at the 5% significance level.

The same logic applies for other confidence levels, but *only* for a two-sided test.

For instance, the 90% confidence interval for the mean weight in the population of adult golden hamsters from the preceding example is

$$\bar{x} \pm t^* \frac{s_x}{\sqrt{n}} = 4.033 \pm 1.699 \frac{0.199}{\sqrt{30}} = 4.033 \pm 0.062 = (3.971, 4.095)$$

The null value $\mu = 4$ is contained in the 90% confidence interval, so we cannot reject H_0: $\mu = 4$ in favor of the two-sided alternative hypothesis H_a: $\mu \neq 4$ at the 10% significance level. This matches our conclusion in the example.

CHECK YOUR UNDERSTANDING

The makers of Aspro brand aspirin want to be sure that their tablets contain the right amount of active ingredient (acetylsalicylic acid), so they inspect a random sample of 30 tablets from a batch in production. When the production process is working properly, Aspro tablets contain an average of $\mu = 320$ milligrams (mg) of active ingredient. The amount of active ingredient in the 30 selected tablets has mean 319 mg and standard deviation 3 mg.

1. State appropriate hypotheses for a significance test in this setting.
2. Check that the conditions are met for carrying out the test.
3. Calculate the standardized test statistic and *P*-value.
4. What conclusion would you make at the $\alpha = 0.05$ significance level?
5. A 95% confidence interval for the mean amount of active ingredient (mg) in this production batch of Aspro tablets is (317.88, 320.12). Explain how this interval is consistent with, but gives more information than, your conclusion in Question 4.

Putting It All Together: One-Sample *t* Test for μ

The appropriate inference procedure for testing a claim about μ is called a **one-sample *t* test for a mean.**

DEFINITION One-sample *t* test for a mean

A **one-sample *t* test for a mean** is a significance test of the null hypothesis that a population mean μ is equal to a specified value, when the population standard deviation σ is unknown.

Whenever you perform a significance test about a population mean μ, be sure to follow the four-step process from Unit 6: State, Plan, Do, and Conclude.

EXAMPLE

Healthy streams
One-sample t test for μ

Skills 1.E, 1.F, 3.E, 4.C, 4.E

blackestockphoto/Getty Images

PROBLEM: The level of dissolved oxygen (DO) in a stream or river is an important indicator of the water's ability to support aquatic life. A researcher measures the DO level at 15 randomly chosen locations along a stream. Here are the results in milligrams per liter (mg/l):

4.53	5.04	3.29	5.23	4.13	5.50	4.83	4.40
5.42	6.38	4.01	4.66	2.87	5.73	5.55	

An average dissolved oxygen level less than 5 mg/l puts aquatic life at risk.

(a) Do the data provide convincing evidence at the $\alpha = 0.05$ significance level that aquatic life in this stream is at risk?

(b) Given your conclusion in part (a), which kind of error could you have made — a Type I error or a Type II error? Explain what this error would mean in context.

SOLUTION:

(a)

STATE: We want to test

$H_0: \mu = 5$

$H_a: \mu < 5$

where μ = the true mean dissolved oxygen (DO) level in the stream, using $\alpha = 0.05$.

> Follow the four-step process!

PLAN: One-sample t test for a mean

- Random: The researcher measured the DO level at 15 randomly chosen locations. ✓
  - 10%: 15 < 10% of the infinitely many possible locations along the stream. ✓
- Normal/Large Sample: The sample size is small ($n = 15 < 30$), but the histogram looks roughly symmetric and shows no outliers. ✓

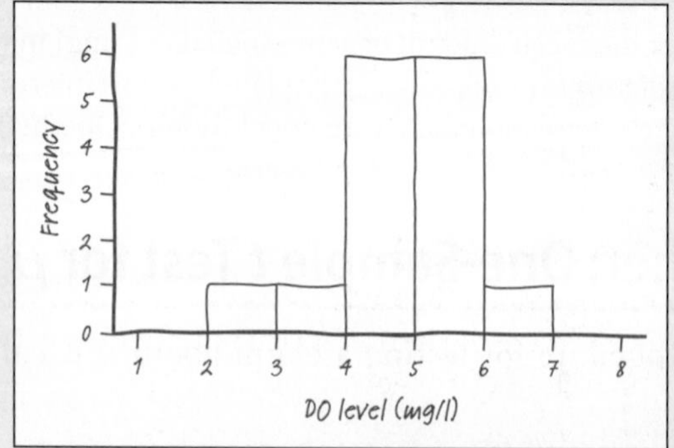

> Because the histogram shows no strong skewness or outliers in the sample, it is plausible that the population distribution of dissolved oxygen levels in the stream is approximately normal.

> Enter the data into your calculator to make a graph, and to calculate the sample mean and standard deviation for the Do step.

DO: $\bar{x} = 4.771$, $s_x = 0.9396$, $n = 15$

> The sample result gives *some* evidence in favor of H_a because $\bar{x} = 4.771 < 5$.

- $t = \dfrac{4.771 - 5}{\dfrac{0.9396}{\sqrt{15}}} = -0.944$

- *P*-value: df = 15 − 1 = 14

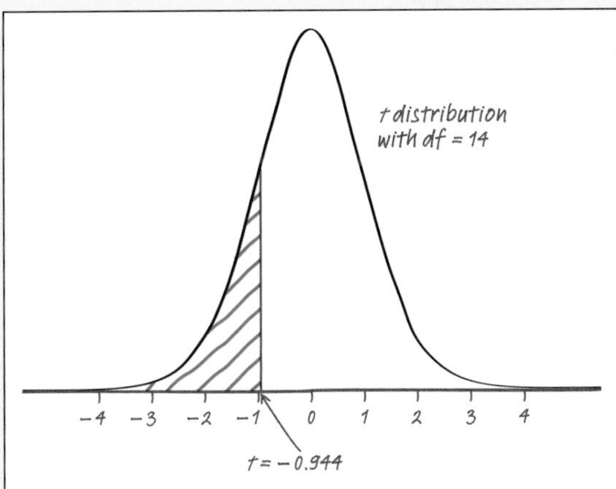

Using technology: $t = -0.9426$, *P*-value = 0.1809 with df = 14

Using Table B: *P*-value is between 0.15 and 0.20.

CONCLUDE: Because the *P*-value of $0.1809 > \alpha = 0.05$, we fail to reject H_0. We don't have convincing evidence that the true mean DO level in the stream is less than 5 mg/l.

(b) Because we failed to reject H_0 in part (a), we could have made a Type II error (failing to reject H_0 when H_a is true). If we did, then the true mean dissolved oxygen level μ in the stream is less than 5 mg/l, but we didn't find convincing evidence with our significance test. That would imply aquatic life in this stream is at risk, but we weren't able to detect that fact.

FOR PRACTICE, TRY EXERCISE 15

To reduce the chance of making a Type II error, the researcher in the preceding example could have taken a larger random sample of stream locations. Doing so would yield a more precise estimate of the stream's true mean dissolved oxygen level μ. That is, a larger sample size increases the power of the test to detect H_a: $\mu < 5$ is true. Another way to increase the power of the test would be to use a higher significance level, such as $\alpha = 0.10$. This change makes it easier to reject H_0: $\mu = 5$ when the average DO level in the stream is less than 5 mg/l. But increasing the significance level also increases the probability of making a Type I error — finding convincing evidence that the stream is unhealthy when it really isn't — from 0.05 to 0.10.

AP® EXAM TIP

You may use your calculator to carry out the mechanics of a significance test on the AP® Statistics exam — but there's a risk involved. If you give the calculator answer but don't show your work, and one or more of your values are incorrect, you will likely get no credit for the Do step. Make sure to include the test statistic, df, and *P*-value in the Do step, along with all the elements of the remaining three steps (State, Plan, Conclude).

CHECK YOUR UNDERSTANDING

Two young researchers noticed that the lengths of the "6-inch" sub sandwiches they get at their favorite restaurant seemed shorter than advertised. To investigate, they randomly selected 24 different times during the next month and ordered a "6-inch" sub. Here are the actual lengths (in inches) of each of the 24 sandwiches:[23]

4.50	4.75	4.75	5.00	5.00	5.00	5.50	5.50	5.50	5.50	5.50	5.50
5.75	5.75	5.75	6.00	6.00	6.00	6.00	6.00	6.50	6.75	6.75	7.00

Do these data give convincing evidence at the $\alpha = 0.10$ level that the "6-inch" sandwiches at this restaurant are shorter than advertised, on average?

Paired Data: Significance Tests about a Population Mean Difference

What if we want to compare means in a setting that involves *paired data*? Recall from Section 7A that paired data result from recording two values of the same quantitative variable for each individual or for each pair of similar individuals. To analyze paired data, start by finding the difference within each pair. Be sure to indicate the order of subtraction. Then calculate $\bar{x}_{\text{Diff}}$ and s_{Diff}.

When paired data come from a random sample or a randomized experiment, we may want to perform a significance test about the population mean difference μ_{Diff}. The null hypothesis has the general form

$$H_0: \mu_{\text{Diff}} = \text{hypothesized value}$$

We'll focus on situations where the hypothesized value is 0. Then the null hypothesis says that the population mean difference is 0:

$$H_0: \mu_{\text{Diff}} = 0$$

The alternative hypothesis says what kind of difference we expect.

The conditions for performing a significance test about μ_{Diff} are the same as the ones for constructing a confidence interval for a mean difference.

CONDITIONS FOR PERFORMING A SIGNIFICANCE TEST ABOUT A POPULATION MEAN DIFFERENCE

- **Random:** Paired data come from a random sample from the population of interest or from a randomized experiment.
  - **10%:** When sampling without replacement, $n_{\text{Diff}} < 0.10N_{\text{Diff}}$.
- **Normal/Large Sample:** The population distribution of differences is approximately normal or the sample size is large ($n_{\text{Diff}} \geq 30$). If the population distribution of differences has unknown shape and $n_{\text{Diff}} < 30$, a graph of the sample differences shows no strong skewness or outliers.

When these conditions are met, we can calculate the standardized test statistic and *P*-value. The formula for the standardized test statistic is the same one we used earlier, but substituting $\bar{x}_{\text{Diff}}$ for the sample mean, s_{Diff} for the sample standard deviation, and n_{Diff} for the sample size.

CALCULATING THE STANDARDIZED TEST STATISTIC AND *P*-VALUE IN A TEST ABOUT A POPULATION MEAN DIFFERENCE

Suppose the conditions are met. To test the hypothesis $H_0: \mu_{\text{Diff}} = 0$, compute the standardized test statistic

$$t = \frac{\bar{x}_{\text{Diff}} - 0}{\dfrac{s_{\text{Diff}}}{\sqrt{n_{\text{Diff}}}}}$$

Find the *P*-value by calculating the probability of getting a t statistic this large or larger in the direction specified by the alternative hypothesis H_a in a t distribution with $\text{df} = n_{\text{Diff}} - 1$.

This inference procedure is often referred to as a **paired *t* test for a mean difference.** It is also called a *one-sample t test for a mean difference*, or when paired data come from a randomized experiment, a *matched-pairs t test for a mean difference.*

DEFINITION Paired *t* test for a mean difference

A **paired *t* test for a mean difference** is a significance test of the null hypothesis that a population mean difference is equal to a specified value, usually 0.

As with any inference method, be sure to follow the four-step process.

EXAMPLE

Skills 1.E, 1.F, 3.E, 4.C, 4.E

Is caffeine dependence real?

Paired data: Significance tests about a population mean difference

MrPants/Wirelmage/Getty Images

PROBLEM: Researchers designed an experiment to study the effects of caffeine withdrawal. They recruited 11 volunteers who were diagnosed as being caffeine dependent to serve as subjects. Each subject was barred from consuming coffee, colas, and other substances with caffeine for the duration of the experiment. During one 2-day period, subjects took capsules containing their normal caffeine intake. During another 2-day period, they took placebo capsules. The order in which subjects took caffeine and the placebo was randomized. At the end of each 2-day period, a test for depression was given to all 11 subjects. Researchers wanted to know whether being deprived of caffeine would lead to an increase in depression.[24]

The table displays data on the subjects' depression test scores. Higher scores correspond to more symptoms of depression.

Subject	1	2	3	4	5	6	7	8	9	10	11
Depression (caffeine)	5	5	4	3	8	5	0	0	2	11	1
Depression (placebo)	16	23	5	7	14	24	6	3	15	12	0

Do these data provide convincing evidence that caffeine withdrawal increases the depression score, on average, for people like the ones in this study?

SOLUTION:

Subject	1	2	3	4	5	6	7	8	9	10	11
Difference (Placebo − Caffeine)	11	18	1	4	6	19	6	3	13	1	−1

Start by calculating the difference in depression test scores for each subject. We chose the order Placebo − Caffeine for subtraction so that most values would be positive.

STATE: $H_0: \mu_{\text{Diff}} = 0$

$H_a: \mu_{\text{Diff}} > 0$

where μ_{Diff} = the true mean difference (Placebo − Caffeine) in depression test score for caffeine-dependent people like the ones in this study. Because no significance level is given, we'll use $\alpha = 0.05$.

PLAN: Paired *t* test for a mean difference

- Random: Researchers randomly assigned the treatments—placebo then caffeine, caffeine then placebo—to the subjects. ✓
- Normal/Large Sample: The sample size is small ($n = 11 < 30$), but the boxplot of differences doesn't show any outliers or strong skewness. ✓

Note that we should not check the 10% condition because we are not sampling without replacement from a finite population.

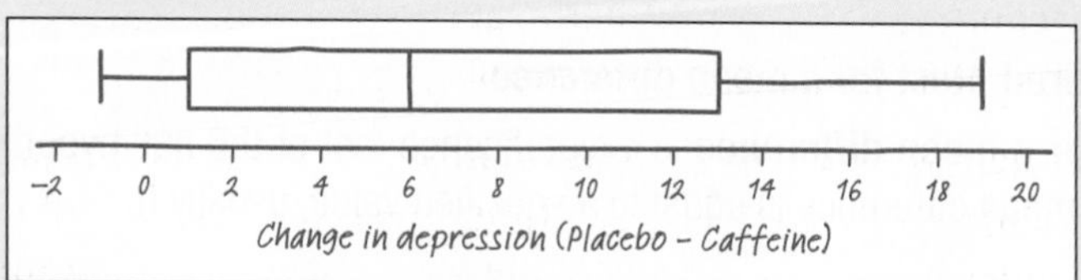

Because there is no strong skewness or outliers, it is plausible that the true distribution of difference (Placebo − Caffeine) in depression test scores for people like the ones in this study is approximately normal.

DO: $\bar{x}_{Diff} = 7.364$, $s_{Diff} = 6.918$, $n_{Diff} = 11$

The sample result gives some evidence in favor of H_a: $\mu_{Diff} > 0$ because $\bar{x}_{diff} = 7.364 > 0$.

- $t = \dfrac{7.364 - 0}{\dfrac{6.918}{\sqrt{11}}} = 3.530$

$$t = \frac{\bar{x}_{Diff} - 0}{\dfrac{s_{Diff}}{\sqrt{n_{Diff}}}}$$

- *P*-value: df = 11 − 1 = 10

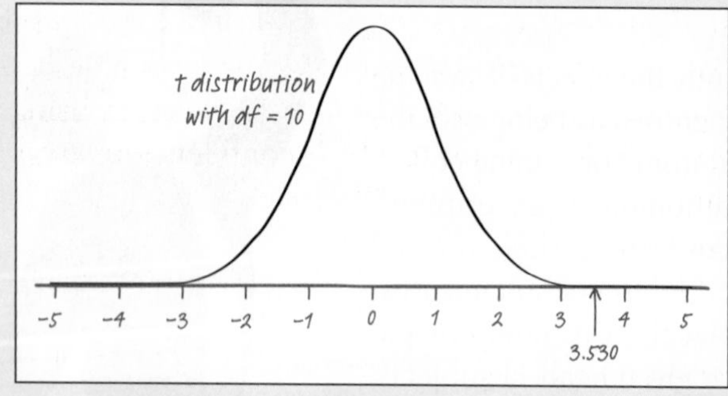

Using technology: $t = 3.530$ and *P*-value = 0.0027 with df = 10

Using Table B: *P*-value is between 0.0025 and 0.005

CONCLUDE: Because the *P*-value of $0.0027 < \alpha = 0.05$, we reject H_0. We have convincing evidence that caffeine withdrawal increases the depression test score, on average, for caffeine-dependent people like the ones in this study.

Be sure to include "on average" (or equivalent) in the conclusion to indicate that the test is about the population *mean* difference.

FOR PRACTICE, TRY EXERCISE 21

Why did the researchers randomly assign the order in which subjects received the placebo and caffeine in the example? The researchers want to be able to conclude that any statistically significant change in the depression score is due to the treatments themselves and not to some other variable. One obvious concern is the order of the treatments. Suppose that caffeine capsules were given to all the subjects during the first 2-day period. What if the weather were nicer on these 2 days than during the second 2-day period when all subjects were given a placebo? The researchers wouldn't be able to tell if a large increase in the mean depression score is due to the difference in weather or due to the treatments. Random assignment of the caffeine and placebo to the two time periods in the experiment should help ensure that no other variable (such as the weather) is systematically affecting subjects' responses.

Because researchers randomly assigned the treatments, they can make an inference about cause and effect. The data from this experiment provide convincing evidence that depriving caffeine-dependent individuals like these of caffeine causes an increase in their depression scores, on average.

In general, there are two ways that an experiment with two treatments can yield paired data:

1. Each experimental unit can be given both treatments in a random order.
2. The researcher can form pairs of similar experimental units and randomly assign each treatment to exactly one member of every pair.

As you learned in Section 3C, both of these approaches are known as *matched pairs* designs. The caffeine dependence experiment had the first type of design. The Check Your Understanding at the end of the section provides an example of the other type of matched pairs design.

Think About It

IS THERE A CONNECTION BETWEEN *ONE*-SIDED TESTS AND CONFIDENCE INTERVALS FOR A POPULATION MEAN OR MEAN DIFFERENCE? The significance test in the caffeine dependence example led to a simple decision: reject H_0. But we know that a confidence interval gives more information than a test — it provides the entire set of plausible values for the parameter based on the data. A *two-sided* significance test about a population mean using a 5% significance level gives equivalent results to a 95% confidence interval for μ. For the *one-sided* test in the example with $\alpha = 0.05$, a 90% confidence interval gives equivalent information.

The 90% confidence interval for μ_{diff} is

$$\bar{x}_{\text{Diff}} \pm t^* \frac{s_{\text{Diff}}}{\sqrt{n_{\text{Diff}}}} = 7.364 \pm 1.812 \frac{6.918}{\sqrt{11}} = 7.364 \pm 3.780 = (3.584, 11.144)$$

We are 90% confident that the interval from 3.584 to 11.144 captures the true mean difference (Placebo − Caffeine) in depression test scores for caffeine-dependent individuals like the ones in this study. The interval suggests that caffeine deprivation results in an average increase in depression test score of between 3.584 and 11.144 points for people like these. Because the null value of 0 is not contained in the interval, we would reject H_0, which is consistent with the decision made when performing the significance test.

CHECK YOUR UNDERSTANDING

Consumers Union designed an experiment to test whether nitrogen-filled tires would maintain their pressure better than air-filled tires. Its researchers obtained two tires from each of several brands and then randomly assigned one tire in each pair to be filled with air and the other to be filled with nitrogen. All tires were inflated to the same pressure and placed outside for a year. At the end of the year, the researchers measured the pressure in each tire. The pressure loss (in pounds per square inch, psi) during the year for the tires of each brand is shown in the table.[25]

Brand	Air	Nitrogen	Brand	Air	Nitrogen
BF Goodrich Traction T/A HR	7.6	7.2	Pirelli P6 Four Seasons	4.4	4.2
Bridgestone HP50 (Sears)	3.8	2.5	Sumitomo HTR H4	1.4	2.1
Bridgestone Potenza G009	3.7	1.6	Yokohama Avid H4S	4.3	3.0
Bridgestone Potenza RE950	4.7	1.5	BF Goodrich Traction T/A V	5.5	3.4
Bridgestone Turanza EL400	2.1	1.0	Bridgestone Potenza RE950_P195	4.1	2.8
Continental Premier Contact H	4.9	3.1	Continental ContiExtreme Contact	5.0	3.4
Cooper Lifeliner Touring SLE	5.2	3.5	Continental ContiPro Contact	4.8	3.3
Dayton Daytona HR	3.4	3.2	Cooper Lifeliner Touring SLE_T	3.2	2.5
Falken Ziex ZE-512	4.1	3.3	General Exclaim UHP	6.8	2.7
Fuzion Hrl	2.7	2.2	Hankook Ventus V4 H105	3.1	1.4
General Exclaim	3.1	3.4	Michelin Energy MXV4 Plus_S8	2.5	1.5
Goodyear Assurance Tripletred	3.8	3.2	Michelin Pilot Exalto A/S	6.6	2.2
Hankook Optimo H418	3.0	0.9	Michelin Pilot HX MXM4	2.2	2.0
Kumho Solus KH16	6.2	3.4	Pirelli P6 Four Seasons_Plus	2.5	2.7
Michelin Energy MXV4 Plus	2.0	1.8	Sumitomo HTR+	4.4	3.7
Michelin Pilot XGT H4	1.1	0.7			

Do these data give convincing evidence at the $\alpha = 0.05$ significance level that air-filled tires lose more pressure, on average, than nitrogen-filled tires for brands like the ones in this study?

SECTION 7B Summary

- Tests about a population mean μ are based on the sample mean $\bar{x}$.
- A significance test about a population mean starts with stating hypotheses. The null hypothesis is H_0: $\mu = \mu_0$ and the alternative hypothesis can be one-sided (H_a: $\mu < \mu_0$ or H_a: $\mu > \mu_0$) or two-sided (H_a: $\mu \neq \mu_0$).
- To perform a test of H_0: $\mu = \mu_0$, we need to ensure that the observations in the sample can be viewed as independent and that the sampling distribution of $\bar{x}$ is approximately normal. The required conditions are:
  - Random: The data come from a random sample from the population of interest.
    - 10%: When sampling without replacement, $n < 0.10N$.
  - Normal/Large Sample: The population distribution is approximately normal or the sample size is large ($n \geq 30$). If the population distribution has unknown shape and $n < 30$, a graph of the sample data shows no strong skewness or outliers.
- The standardized test statistic for a **one-sample *t* test for a mean** is

$$t = \frac{\bar{x} - \mu_0}{\frac{s_x}{\sqrt{n}}}$$

When the conditions are met, find the *P*-value by calculating the probability of getting a *t* statistic this large or larger in the direction specified by the alternative hypothesis H_a in a *t* distribution with df $= n - 1$.

- Confidence intervals provide additional information that significance tests do not — namely, a set of plausible values for the parameter μ. A 95% confidence interval for μ gives consistent results with a two-sided test of H_0: $\mu = \mu_0$ at the $\alpha = 0.05$ significance level.
- In the special case of paired data resulting from an observational study or a matched pairs experiment, the null hypothesis is usually H_0: $\mu_{\text{Diff}} = 0$. Be sure to specify the order of subtraction when defining the parameter.
- Significance tests about a population mean difference μ_{Diff} are based on the sample mean difference $\bar{x}_{\text{Diff}}$.
- Before testing a claim about μ_{Diff}, we need to check for independence of the differences and that the sampling distribution of $\bar{x}_{\text{Diff}}$ is approximately normal. The required conditions are:
  - Random: Paired data come from a random sample from the population of interest or from a randomized experiment.
    - 10%: When sampling without replacement, $n_{\text{Diff}} < 0.10N_{\text{Diff}}$.
  - Normal/Large Sample: The population distribution of differences is approximately normal or the sample size is large ($n_{\text{Diff}} \geq 30$). If the population distribution of differences has unknown shape and $n_{\text{Diff}} < 30$, a graph of the sample differences shows no strong skewness or outliers.
- The standardized test statistic in a **paired *t* test for a mean difference** (also called a one-sample *t* test for a mean difference) is

$$t = \frac{\bar{x}_{\text{Diff}} - 0}{\frac{s_{\text{Diff}}}{\sqrt{n_{\text{Diff}}}}}$$

When the conditions are met, find the *P*-value using the *t* distribution with $\text{df} = n_{\text{Diff}} - 1$.

- Follow the four-step process — State, Plan, Do, Conclude — whenever you are asked to perform a significance test about a population mean μ or a population mean difference μ_{Diff}.

> **AP® EXAM TIP**
> **AP® Daily Videos**
> Review the content of this section and get extra help by watching the AP® Daily Videos for Topics 7.4 and 7.5, which are available in AP® Classroom.

7B Tech Corners

TI-Nspire and other technology instructions are on the book's website at bfwpub.com/tps7e.

SECTION 7B Exercises

Stating Hypotheses and Checking Conditions for a Test about μ

1. **Candy!** A machine is supposed to fill bags with an average of 19.2 ounces of candy. The manager of the candy factory wants to ensure that the machine does not underfill or overfill the bags, on average. So, the manager plans to perform a significance test. The manager selects a random sample of 75 bags of candy produced that day and weighs each bag.

pg 671

(a) State appropriate hypotheses for the manager's test. Be sure to define the parameters of interest.

(b) Check if the conditions for performing the test are met.

2. **Water intake** A blogger claims that U.S. adults drink, on average, 40 ounces of water per day. Researchers suspect that the blogger's claim overstates the actual amount of water consumed, and want to perform a significance test. The researchers ask a random sample of 100 U.S. adults to report their daily water intake.

(a) State appropriate hypotheses for the researcher's test. Be sure to define the parameter of interest.

(b) Check if the conditions for performing the test are met.

3. **Battery life** A tablet computer manufacturer claims that its batteries last an average of 11.5 hours when the tablet is used to play videos. The quality-control department randomly selects 20 tablets from each day's production and tests the fully charged batteries by playing a video repeatedly until the battery dies. The quality-control department will discard the batteries from that day's production run if they find convincing evidence that the mean battery life is less than 11.5 hours. Here is a dotplot of the data from one day:

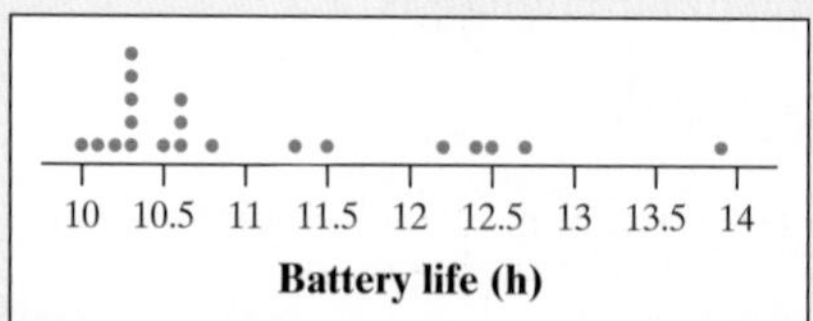

(a) State appropriate hypotheses for the quality-control department to test. Be sure to define the parameter.

(b) Check if the conditions for performing the test in part (a) are met.

4. **Paying high prices?** A retailer made an agreement with a supplier who guaranteed to provide all products at competitive prices. To be sure the supplier honored the terms of the agreement, the retailer had an audit performed on a random sample of 25 invoices. The auditor recorded the percentage of purchases on each invoice for which an alternative supplier offered a lower price than the original supplier.[26] For example, a data value of 38 means that the price would be lower from a different supplier for 38% of the items on the invoice. The retailer wants to determine if there is convincing evidence that the mean percentage of purchases for which an alternative supplier offered lower prices is greater than 50% in the population of this company's invoices. A histogram of the data is shown here. (Note that there were 14 invoices with 100% of the purchases having a lower price from an alternative supplier!)

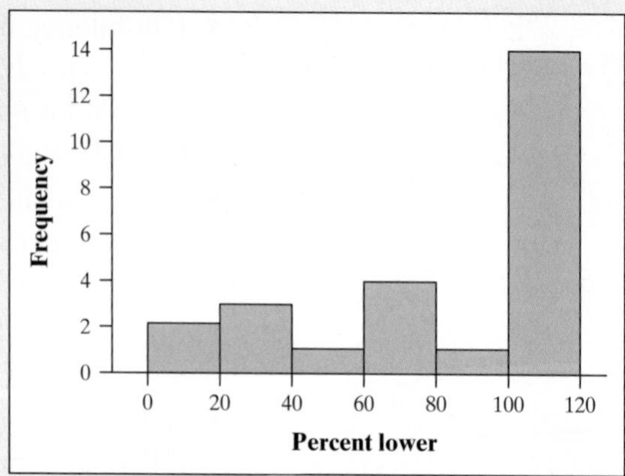

(a) State appropriate hypotheses for the retailer's test. Be sure to define the parameter.

(b) Check if the conditions for performing the test in part (a) are met.

5. **Salmon fillets** As part of a study on salmon health, researchers measured the pH of a random sample of 25 salmon fillets at a fish processing plant.[27] Here are the data:

6.34	6.39	6.53	6.36	6.39	6.25	6.45	6.38	6.33
6.26	6.24	6.37	6.32	6.31	6.48	6.26	6.42	6.43
6.36	6.44	6.22	6.52	6.32	6.32	6.48		

One concern is that the process of filleting salmon may result in fillets that are too acidic and unpleasant to eat. Plant managers will have to modify their process if the processed salmon fillets have an average pH less than 6.40. The researchers want to perform a test at the $\alpha = 0.05$ significance level of H_0: $\mu = 6.40$ versus H_a: $\mu < 6.40$.

(a) Define the parameter of interest.

(b) Check if the conditions are met for performing the significance test.

6. **Cooling reactions** A chemical production process uses water-cooling to carefully control the temperature of the system so that the correct products are obtained from the chemical reaction. The system must be maintained at a temperature of 120°F. A technician suspects that the average temperature is different from 120°F and checks the process by taking temperature readings at 12 randomly selected times over a 4-hour period. Here are the data:

120.1	124.2	122.4	124.4	120.8	121.4
121.8	119.6	120.2	121.5	118.7	122.0

The technician would like to use these data to perform a test of H_0: $\mu = 120$ versus H_a: $\mu \neq 120$ at the $\alpha = 0.05$ significance level.

(a) Define the parameter of interest.

(b) Check if the conditions are met for performing the significance test.

Calculating the Standardized Test Statistic and P-Value for a Test about μ

7. More candy In the study of the candy machine from Exercise 1, the sample mean weight was 19.28 ounces and the sample standard deviation was 0.81 ounce.
pg 675

(a) Explain why the sample result gives some evidence for the alternative hypothesis.

(b) Calculate the standardized test statistic and P-value.

(c) Interpret the P-value.

8. **More water intake** In the study of daily water consumption from Exercise 2, the sample mean amount of water consumed was 37.74 ounces and the sample standard deviation was 14.93 ounces.

(a) Explain why the sample result gives some evidence for the alternative hypothesis.

(b) Calculate the standardized test statistic and P-value.

(c) Interpret the P-value.

9. **Calculating P-values** Suppose you want to perform a test of H_0: $\mu = 64$ versus H_a: $\mu > 64$. A random sample of size $n = 25$ from the population of interest yields the standardized test statistic $t = 1.12$. Assume that the conditions for carrying out the test are met.

(a) Calculate the P-value.

(b) How would the P-value change if the alternative hypothesis was H_a: $\mu \neq 64$?

10. **P-value calculations** Suppose you want to perform a test of H_0: $\mu = 5$ versus H_a: $\mu < 5$ at the $\alpha = 0.05$ significance level. A random sample of size $n = 20$ from the population of interest yields $t = -1.81$. Assume that the conditions for carrying out the test are met.

(a) Calculate the P-value.

(b) How would the P-value change if the alternative hypothesis was H_a: $\mu \neq 5$?

11. **More salmon** Refer to Exercise 5.

(a) Calculate the standardized test statistic.

(b) Find the P-value.

(c) What conclusion would you make?

12. **More cooling reactions** Refer to Exercise 6.

(a) Calculate the standardized test statistic.

(b) Find the P-value.

(c) What conclusion would you make?

13. **Two-sided tests and confidence intervals** The P-value for a two-sided test of the null hypothesis H_0: $\mu = 10$ is 0.06.

(a) Does the 95% confidence interval for μ include 10? Why or why not?

(b) Does the 90% confidence interval for μ include 10? Why or why not?

14. **More two-sided tests and confidence intervals** The P-value for a two-sided test of the null hypothesis H_0: $\mu = 15$ is 0.03.

(a) Does the 99% confidence interval for μ include 15? Why or why not?

(b) Does the 95% confidence interval for μ include 15? Why or why not?

One-Sample *t* Test for μ

15. **Construction zones** Every road has one at some point—construction zones that have much lower speed limits. To see if drivers obey these lower speed limits, a police officer uses a radar gun to measure the speed (in miles per hour, mph) of a random sample of 10 drivers in a 25-mph construction zone. Here are the data:
pg 678

27 33 32 21 30 30 29 25 27 34

(a) Do these data provide convincing evidence at the $\alpha = 0.01$ significance level that the average speed of drivers in this construction zone is greater than the posted speed limit?

(b) Given your conclusion in part (a), which kind of error could you have made—a Type I error or a Type II error? Explain what this error would mean in context.

16. **Better batteries** A company has developed a new deluxe AAA battery that is supposed to last longer than its regular AAA battery. However, these new batteries are more expensive to produce, so the company would like to be convinced that they really do last longer. Based on years of experience, the company knows that its regular AAA batteries last for 30 hours of continuous use, on average. The company selects a random sample of 15 deluxe AAA batteries and uses them continuously until they are completely drained. Here are the battery lifetimes (in hours):

17 32 22 45 30 36 51 27 37 47 35 33 44 22 31

(a) Do these data provide convincing evidence at the $\alpha = 0.05$ significance level that the company's deluxe AAA batteries last longer than 30 hours, on average?

(b) Given your conclusion in part (a), which kind of error could you have made—a Type I error or a Type II error? Explain what this error would mean in context.

17. **Women and calcium** The recommended daily allowance (RDA) of calcium for women between the ages of 18 and 24 years is 1200 milligrams (mg). Researchers who were involved in a large-scale study of women's bone health suspected that their participants had significantly lower calcium intakes than the RDA. To test this suspicion, the researchers measured the daily calcium intake of a random sample of 36 women from the study who fell within the desired age range. The sample mean was 856.2 mg and the standard deviation was 306.7 mg. Do these data give convincing evidence that the researchers' suspicion is correct?

18. **Reading level** A school librarian purchases a novel for the library. The publisher claims that the book is written at a fifth-grade reading level, but the librarian suspects that the reading level is higher than that. The librarian selects a random sample of 40 pages and uses a standard readability test to assess the reading level of each page. The mean reading level of these pages is 5.4 with a standard deviation of 0.8. Do these data give convincing evidence that the average reading level of this novel is greater than 5?

19. **Pressing pills** A drug manufacturer forms tablets by compressing a granular material that contains the active ingredient and various fillers. The hardness of a sample from each batch of tablets produced is measured to control the compression process. The target value for the hardness is $\mu = 11.5$ newtons. Researchers select a random sample of 20 tablets and measure their hardness. Here is a boxplot of the data.

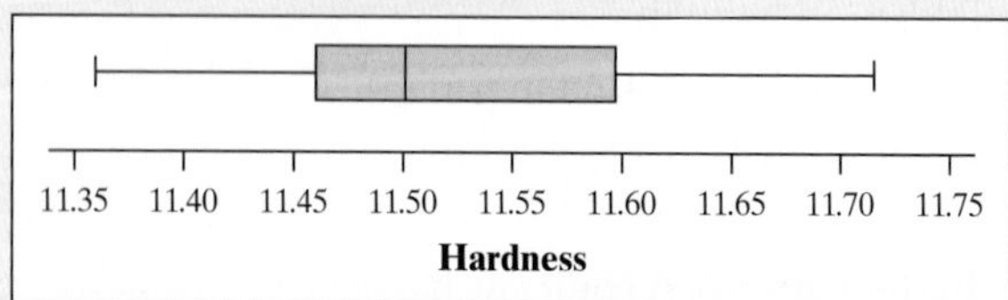

(a) State appropriate hypotheses for a test at the 5% significance level to determine whether the mean hardness of the tablets in this batch differs from the target value.

(b) Due to the small sample size ($n = 20 < 30$), one requirement for carrying out the test in part (a) is that the population distribution is approximately normal. Explain how this requirement is satisfied in this case.

(c) The power of the test to detect that $\mu = 11.55$ is 0.61. Interpret this value.

(d) Find the probability of a Type II error if $\mu = 11.55$.

(e) Describe two ways to decrease the probability in part (d).

20. **Jump around** Student researchers saw an article on the internet claiming that the average vertical jump for teens was 15 inches. They wondered if the average vertical jump of students at their school differed from 15 inches, so they obtained a list of student names and selected a random sample of 20 students. After contacting these students several times, they finally convinced them to allow their vertical jumps to be measured. Here is a histogram of the data (in inches):

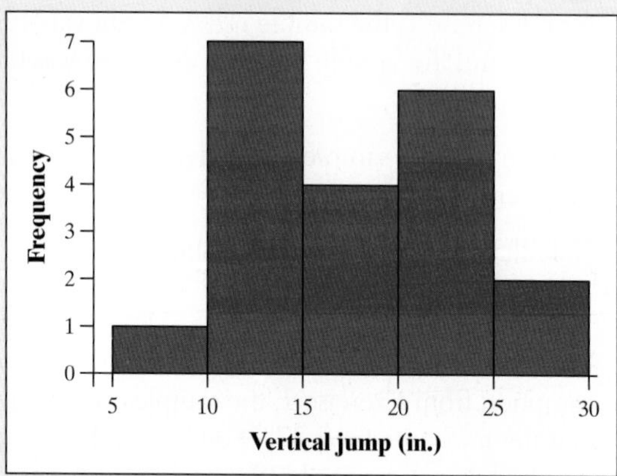

(a) State appropriate hypotheses for a test at the 10% significance level to determine whether the average vertical jump of students at this school differs from 15 inches.

(b) Due to the small sample size ($n = 20 < 30$), one requirement for carrying out the test in part (a) is that the population distribution is approximately normal. Explain how this requirement is satisfied in this case.

(c) The power of the test to detect that $\mu = 17$ inches is 0.49. Interpret this value.

(d) Find the probability of a Type II error if $\mu = 17$.

(e) Describe two ways to decrease the probability in part (d).

Significance Tests about a Population Mean Difference

21. **Drive-thru or go inside?** pg 681 Many people think it's faster to order at the drive-thru than to order inside at fast-food restaurants. Two student researchers used a random number generator to select 10 times over a 2-week period to visit a local Dunkin' restaurant. At each of these times, one student ordered an iced coffee at the drive-thru and the other ordered an iced coffee at the counter inside. A coin flip determined who went inside and who went to the drive-thru. The table shows the times, in seconds, that it took for each student to receive the iced coffee after placing the order.[28]

Visit	Inside time	Drive-thru time
1	62	55
2	63	50
3	325	321
4	105	110
5	135	124
6	55	54
7	92	90
8	75	69
9	203	200
10	100	103

Do these data provide convincing evidence at the $\alpha = 0.05$ level that it is faster to order at the drive-thru than to order inside, on average, at this Dunkin' restaurant?

22. **Better barley** Does drying barley seeds in a kiln increase the yield of barley? A famous experiment by William S. Gosset (who developed the t distributions) investigated this question. Eleven pairs of adjacent plots were marked out in a large field. For each pair, regular barley seeds were planted in one plot and kiln-dried seeds were planted in the other. A coin flip was used to determine which plot in each pair got the regular barley seed and which got the kiln-dried seed. The following table displays the data on barley yield (pound per acre) for each plot.[29]

Plot	Regular	Kiln
1	1903	2009
2	1935	1915
3	1910	2011
4	2496	2463
5	2108	2180
6	1961	1925
7	2060	2122
8	1444	1482
9	1612	1542
10	1316	1443
11	1511	1535

Do these data provide convincing evidence at the $\alpha = 0.05$ level that drying barley seeds in a kiln increases the yield of barley, on average?

23. **Friday the 13th** Does Friday the 13th have an effect on people's shopping behavior? Researchers collected data on the number of shoppers at a random sample of 45 grocery stores on Friday the 6th and Friday the 13th in the same month. A dotplot of the difference (Friday the 6th − Friday the 13th) in the number of shoppers at each store on these 2 days is shown here. The mean difference is −46.5 and the standard deviation of the differences is 178.0.[30]

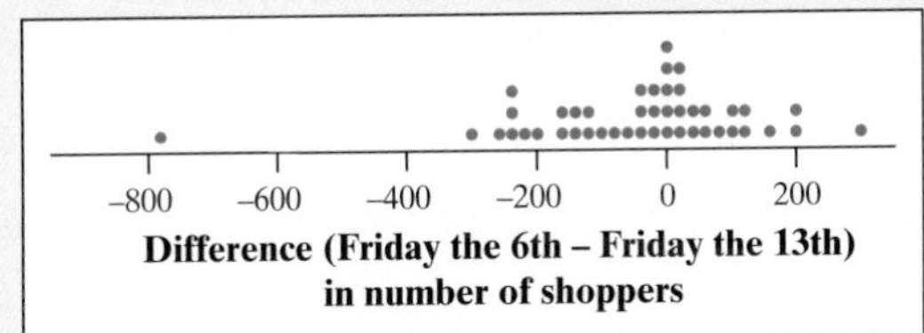

(a) Explain how you know that these are paired data.

(b) If the result of this study is statistically significant, can you conclude that the difference in shopping behavior is due to the effect of Friday the 13th on people's behavior? Why or why not?

(c) Do these data provide convincing evidence at the 10% significance level that the number of shoppers at grocery stores on these 2 days differs, on average?

24. **Coastal home sales** How do the list price and the sales price of homes along the coast compare? One of the authors recorded data from a random sample of 68 homes sold along the South Carolina coast during a 3-month period. A boxplot of the difference (List − Sales) in price is shown here. The mean difference is \$26,653.63 and the standard deviation of the differences is \$51,329.22.[31]

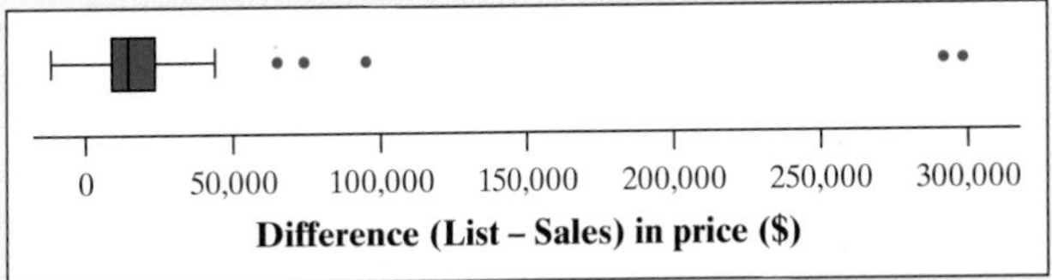

(a) Explain how you know that these are paired data.

(b) If the result of this study is statistically significant, can you conclude that the list prices of homes sold in the United States during this 3-month period differed from their sales prices, on average? Why or why not?

(c) Do these data provide convincing evidence at the 1% significance level of a nonzero mean difference in the list price and the sales price in the population of South Carolina coastal dwellings sold during this 3-month period?

25. **Friday the 13th, part II** Refer to Exercise 23.

(a) Construct and interpret a 90% confidence interval for the population mean difference. If you already defined the parameter and checked conditions, you don't need to do so again here.

(b) Explain how the confidence interval provides more information than the test in Exercise 23.

26. **Coastal home sales, again** Refer to Exercise 24.

(a) Construct and interpret a 99% confidence interval for the population mean difference. If you already defined the parameter and checked conditions, you don't need to do so again here.

(b) Explain how the confidence interval provides more information than the test in Exercise 24.

For Investigation *Apply the skills from the section in a new context or nonroutine way.*

27. **One-sided tests and confidence intervals** There is a connection between a one-sided test and a confidence interval for a population mean — but it's a little complicated. Consider a one-sided test at the $\alpha = 0.05$

significance level of H_0: $\mu = 10$ versus H_a: $\mu > 10$ based on an SRS of $n = 20$ observations.

(a) Complete this statement: We will reject H_0 if the standardized test statistic $t >$ ____________. In other words, we will reject H_0 if the sample mean is more than ____________ standard errors greater than $\mu = 10$.

(b) Find the t^* critical value for a 90% confidence level in this setting. The resulting interval is $\bar{x} \pm t^* (s_{\bar{x}})$.

(c) Suppose that the sample mean $\bar{x}$ leads us to reject H_0. Use parts (a) and (b) to explain why the 90% confidence interval *cannot* contain $\mu = 10$.

28. **Have a ball!** Can students throw a baseball farther than a softball? To find out, researchers conducted a study involving 24 randomly selected students from a large high school. After warming up, each student threw a baseball as far as they could and threw a softball as far as they could, in a random order. The distance in yards for each throw was recorded. Here are the data, along with the difference (Baseball − Softball) in distance thrown, for each student:

Student	Baseball	Softball	Difference (Baseball − Softball)
1	65	57	8
2	90	58	32
3	75	66	9
4	73	61	12
5	79	65	14
6	68	56	12
7	58	53	5
8	41	41	0
9	56	44	12
10	70	65	5
11	64	57	7
12	62	60	2
13	73	55	18
14	50	53	−3
15	63	54	9
16	48	42	6
17	34	32	2
18	49	48	1
19	48	45	3
20	68	67	1
21	30	27	3
22	26	25	1
23	28	25	3
24	26	31	−5

(a) Explain why these are paired data.

(b) A boxplot of the differences is shown. Explain how the graph gives *some* evidence that students like these can throw a baseball farther than a softball.

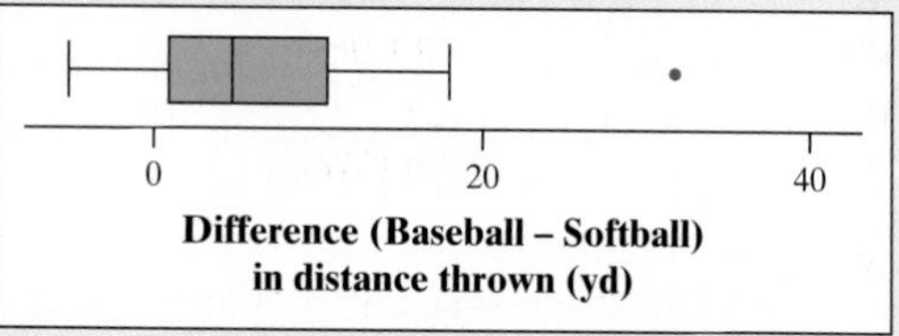

(c) State appropriate hypotheses for performing a test about the population mean difference. Be sure to define any parameter(s) you use.

(d) Explain why the Normal/Large Sample condition is not met in this case.

The mean difference (Baseball − Softball) in distance thrown for these 24 students is $\bar{x}_{diff} = 6.54$ yards. Is this a surprisingly large result if the null hypothesis is true? To find out, we can perform a simulation assuming that students have the same ability to throw a baseball and a softball. For each student, write the two distances thrown on different note cards. Shuffle the two cards and randomly designate one distance as that for the baseball and one distance as that for the softball. Then subtract the two distances (Baseball − Softball). Do this for all the students and find the simulated mean difference. Repeat many times. Here are the results of 100 trials of this simulation:

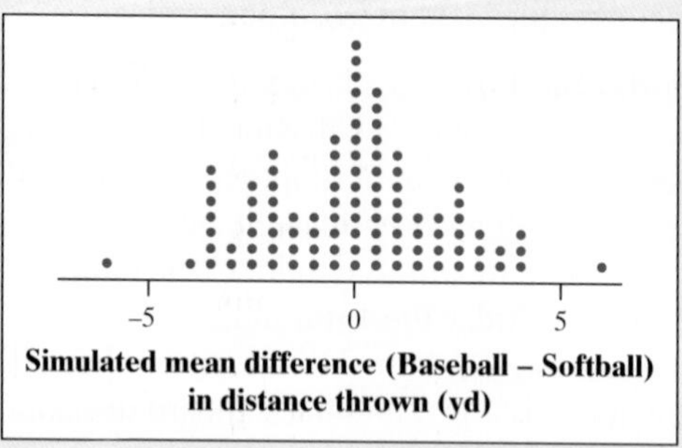

(e) Use the results of the simulation to estimate the P-value. What conclusion would you draw?

Multiple Choice *Select the best answer for each question.*

29. One condition that is always required to conduct a significance test about a population mean is that

(A) the data come from a random sample or a randomized experiment.

(B) the population distribution is approximately normal.

(C) the data contain no outliers.

(D) a graph of the data is roughly symmetric.

(E) the sample size is at least 30.

30. You are testing H_0: $\mu = 75$ versus H_a: $\mu < 75$ based on an SRS of 20 observations from an approximately normally distributed population. The standardized test statistic is $t = -2.25$. The *P*-value is

(A) less than 0.01.

(B) between 0.01 and 0.02.

(C) between 0.02 and 0.05.

(D) between 0.05 and 0.10.

(E) greater than 0.10.

31. A study of the impact of caffeine consumption on reaction time was designed to correct for the impact of subjects' prior sleep deprivation by dividing the 24 subjects into 12 sets of two on the basis of the average hours of sleep they had on the previous 5 nights. That is, the two subjects with the highest average sleep were a set, then the two with the next highest average sleep, and so on. One randomly assigned member of each set drank 2 cups of caffeinated coffee, and the other drank 2 cups of decaf. Each subject's performance on a standard reaction-time test was recorded. Assuming that the conditions for inference are met, which of the following procedures should be used to perform a significance test in this setting?

(A) one-sample *z* test for a population proportion

(B) one-sample *z* test for a population mean

(C) one-sample *z* interval for a population proportion

(D) one-sample *t* interval for a population mean

(E) paired *t* test for a population mean difference

32. Vigorous exercise helps people live several years longer, on average. Whether mild-intensity activities like slow walking extend life is not clear. Suppose that the added life expectancy from regular slow walking is just 2 months. A significance test is more likely to find a significant increase in mean life expectancy with regular slow walking if

(A) it is based on a very large random sample and a 5% significance level is used.

(B) it is based on a very large random sample and a 1% significance level is used.

(C) it is based on a very small random sample and a 5% significance level is used.

(D) it is based on a very small random sample and a 1% significance level is used.

(E) the size of the sample doesn't have any effect on the significance of the test.

33. After checking that the conditions are met, you perform a significance test of H_0: $\mu = 1$ versus H_a: $\mu \neq 1$. You obtain a *P*-value of 0.022. Which of the following must be true?

(A) A 95% confidence interval for μ will include the value 1.

(B) A 95% confidence interval for μ will include the value 0.022.

(C) A 99% confidence interval for μ will include the value 1.

(D) A 99% confidence interval for μ will include the value 0.022.

(E) None of these is necessarily true.

Recycle and Review *Practice what you learned in previous sections.*

34. **Is your food safe? (6A)** "Do you feel confident or not confident that the food available at most grocery stores is safe to eat?" When a Gallup poll asked this question, 87% of the sample said they were confident. Gallup announced the poll's margin of error for 95% confidence as ±3 percentage points.[32] Which of the following sources of error are included in this margin of error? Explain your answer.

(i) Gallup dialed cell phone numbers at random, so it missed all people without cell phones, including people whose only phone is a landline phone.

(ii) Some people whose numbers were chosen never answered the phone in several calls, or answered but refused to participate in the poll.

(iii) Other samples will provide different estimates due to chance variation in the random selection of telephone numbers.

35. **Spinning for apples (4C, 4F)** In the "Ask Marilyn" column of *Parade* magazine, a reader posed this question: "Say that a slot machine has five wheels, and each wheel has five symbols: an apple, a grape, a peach, a pear, and a plum. I pull the lever five times. What are the chances that I'll get at least one apple?" Suppose that the wheels spin independently and that the five symbols are equally likely to appear on each wheel in a given spin.

(a) Find the probability that the slot player gets at least one apple in one pull of the lever.

(b) Now answer the reader's question.

SECTION 7C Confidence Intervals for a Difference in Population Means

LEARNING TARGETS *By the end of the section, you should be able to:*

- Check the conditions for calculating a confidence interval for a difference between two population means.
- Calculate a confidence interval for a difference between two population means.
- Construct and interpret a two-sample *t* interval for a difference in means.

In Section 7A, you learned how to construct and interpret a confidence interval for a population mean μ or a population mean difference μ_{Diff} in the special case of paired data. What if we want to provide an interval estimate for the difference between two population means? For instance, maybe we want to estimate the difference $\mu_1 - \mu_2$ in the average annual income of all U.S. adults with college degrees and all U.S. adults who attended college but did not earn a degree. The ideal strategy is to take a separate random sample from each population and to use the difference $\bar{x}_1 - \bar{x}_2$ between the sample means as our point estimate. Or perhaps we want to use data from a randomized experiment to estimate the difference $\mu_{\text{Beta}} - \mu_{\text{Placebo}}$ in the average pulse rate during surgery for patients like the ones in the study who receive a beta blocker and patients like the ones in the study who receive a placebo. In this case, we use the difference $\bar{x}_{\text{Beta}} - \bar{x}_{\text{Placebo}}$ between the mean pulse rates of the subjects in the two groups as our point estimate.

This section focuses on constructing confidence intervals for a difference between two means. You will learn how to perform significance tests about $\mu_1 - \mu_2$ in Section 7D.

Checking Conditions for a Confidence Interval for $\mu_1 - \mu_2$

Before constructing a confidence interval for a difference in population means, we must check for independence in the data collection process and confirm that the sampling distribution of $\bar{x}_1 - \bar{x}_2$ is approximately normal. First, we need to modify the Random, 10%, and Normal/Large Sample conditions to account for the fact that we are comparing two means.

To ensure independence, the data should come from two independent random samples from the populations of interest or from two groups in a randomized experiment (the Random condition). Also, when sampling without replacement, the 10% condition should be met for each sample. Remember that it is not appropriate to check the 10% condition in experiments with volunteer subjects!

For the Normal/Large Sample condition to be met, both population distributions must be approximately normal, both sample sizes must be large ($n_1 \geq 30$ and $n_2 \geq 30$), or one population distribution must be approximately normal and the other sample size must be large. In any of these cases, the sampling distribution of $\bar{x}_1 - \bar{x}_2$ will be approximately normal, as you learned in Unit 5.

CONDITIONS FOR CONSTRUCTING A CONFIDENCE INTERVAL FOR A DIFFERENCE BETWEEN TWO POPULATION MEANS

- **Random:** The data come from two independent random samples or from two groups in a randomized experiment.
  - **10%:** When sampling without replacement, $n_1 < 0.10N_1$ and $n_2 < 0.10N_2$.
- **Normal/Large Sample:** For each sample, the data come from an approximately normally distributed population or the sample size is large ($n \geq 30$). For each sample, if the population distribution has unknown shape and $n < 30$, a graph of the sample data shows no strong skewness or outliers.

Recall from Unit 3 that the Random condition is important for determining the scope of inference. Random sampling allows us to generalize our results to the populations of interest; random assignment in an experiment permits us to draw cause-and-effect conclusions.

EXAMPLE

Windy City or Big Apple?

Checking conditions for a confidence interval for $\mu_1 - \mu_2$

Skill 4.C

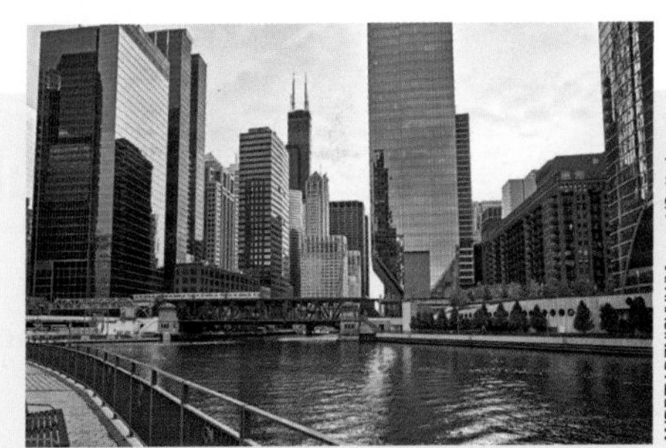
ANDREY DENISYUK/Moment/Getty Images

PROBLEM: A recent college graduate is considering a job offer that would allow them to live in either New York City or Chicago. To help make the decision, they want to estimate the typical cost of living in each city. Here are data on the monthly rents (in dollars) for independent random samples of 10 one-bedroom apartments in each city:[33]

New York City	2800	1450	2250	2075	2400	2400	1700	1900	1475	1800
Chicago	1695	1300	2200	1495	2200	1600	1450	1380	1260	1300

Let μ_1 = the population mean monthly rent (\$) of one-bedroom apartments in New York City and μ_2 = the population mean monthly rent (\$) of one-bedroom apartments in Chicago. Check if the conditions for constructing a 99% confidence interval for $\mu_1 - \mu_2$ are met.

SOLUTION:

- Random: Independent random samples of 10 one-bedroom apartments in New York City and 10 one-bedroom apartments in Chicago. ✓
  - 10%: We can assume that $10 < 10\%$ of all one-bedroom apartments in New York City and that $10 < 10\%$ of all one-bedroom apartments in Chicago.

- **Normal/Large Sample: Both sample sizes are small ($n = 10 < 30$), and both boxplots are skewed to the right — New York City slightly and Chicago moderately. But neither boxplot shows any strong skewness or outliers. ✓**

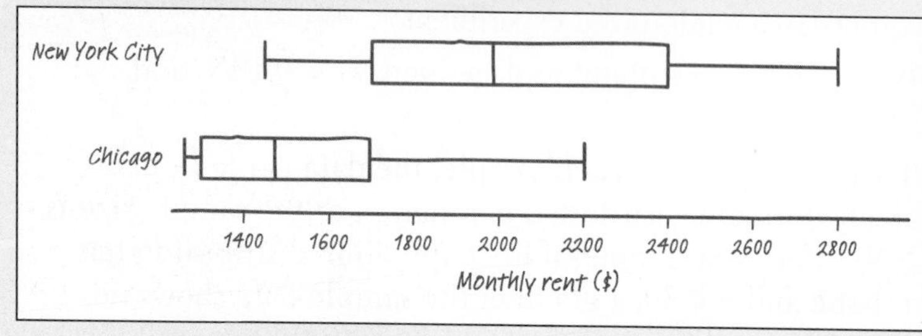

Because neither sample shows strong skewness nor has outliers, it is plausible that the population distributions of monthly rent for one-bedroom apartments in New York City and Chicago are approximately normal.

FOR PRACTICE, TRY EXERCISE 1

Note that although there are 10 apartments each from New York City and Chicago in the example, these are *not* paired data. There is nothing about the first apartment in the New York City sample that is related to the first apartment in the Chicago sample. In other words, the samples are independent.

We used boxplots to check the Normal/Large Sample condition in the example because they clearly display any outliers or strong skewness. Remember that you can also use dotplots, stemplots, or histograms to assess normality. Just be sure to include graphs of the sample data when the sample size is small ($n < 30$) and you are checking the Normal/Large Sample condition.

Calculating a Confidence Interval for $\mu_1 - \mu_2$

When conditions are met, we can calculate a confidence interval for $\mu_1 - \mu_2$ using the familiar formula for an interval estimate:

$$\text{point estimate} \pm \text{margin of error}$$

or, in slightly expanded form:

$$\text{statistic} \pm (\text{critical value})(\text{standard error of statistic})$$

The observed difference in sample means $\bar{x}_1 - \bar{x}_2$ is our point estimate for $\mu_1 - \mu_2$. In Section 5D, you learned that the standard deviation of the sampling distribution of $\bar{x}_1 - \bar{x}_2$ is

$$\sigma_{\bar{x}_1 - \bar{x}_2} = \sqrt{\frac{\sigma_1^2}{n_1} + \frac{\sigma_2^2}{n_2}}$$

when we have two types of independence:

- Independent samples, so we can add the variances of $\bar{x}_1$ and $\bar{x}_2$. This is why we reminded you to mention *independent* random samples when checking the Random condition in the preceding example.

- Independent observations within each sample. When sampling without replacement, the actual value of the standard deviation is smaller than the formula suggests. However, if the 10% condition is met for both samples, the given formula is approximately correct.

Because we usually don't know the values of σ_1 and σ_2, we replace them with the sample standard deviations s_1 and s_2, respectively. The result is the *standard error* of $\bar{x}_1 - \bar{x}_2$:

$$s_{\bar{x}_1-\bar{x}_2} = \sqrt{\frac{s_1^2}{n_1} + \frac{s_2^2}{n_2}}$$

Fortunately, this formula for the standard error is the same whether we have two independent samples or two groups in a randomized experiment. The value of $s_{\bar{x}_1-\bar{x}_2}$ estimates how much the difference $\bar{x}_1 - \bar{x}_2$ in sample means will typically vary from the difference $\mu_1 - \mu_2$ in the population means if we repeat the random sampling or random assignment many times. Note that some people use the notation $SE_{\bar{x}_1-\bar{x}_2}$ rather than $s_{\bar{x}_1-\bar{x}_2}$.

As in Section 7A, we use a critical value from a t distribution rather than from the standard normal distribution because the population standard deviations are unknown. When the Normal/Large Sample condition is met, the critical value t^* for a given confidence level is obtained by technology using a t distribution with degrees of freedom given by the following formula:

AP® EXAM TIP
Formula sheet

This formula for degrees of freedom is *not* included on the formula sheet provided on both sections of the AP® Statistics exam. However, you can use technology to obtain the df given by this formula.

$$\text{df} = \frac{\left(\frac{s_1^2}{n_1} + \frac{s_2^2}{n_2}\right)^2}{\frac{1}{n_1-1}\left(\frac{s_1^2}{n_1}\right)^2 + \frac{1}{n_2-1}\left(\frac{s_2^2}{n_2}\right)^2}$$

Note that the df given by this formula is usually *not* a whole number!

Our confidence interval for $\mu_1 - \mu_2$ is therefore

$$\text{statistic} \pm (\text{critical value})(\text{standard error of statistic})$$

$$= (\bar{x}_1 - \bar{x}_2) \pm t^* \sqrt{\frac{s_1^2}{n_1} + \frac{s_2^2}{n_2}}$$

CALCULATING A CONFIDENCE INTERVAL FOR A DIFFERENCE BETWEEN TWO POPULATION MEANS

When the conditions are met, a C% confidence interval for the difference $\mu_1 - \mu_2$ between two population means is

$$(\bar{x}_1 - \bar{x}_2) \pm t^* \sqrt{\frac{s_1^2}{n_1} + \frac{s_2^2}{n_2}}$$

where t^* is the critical value with C% of its area between $-t^*$ and t^* in the t distribution with degrees of freedom given by technology.

AP® EXAM TIP
Formula sheet

The specific formula for a confidence interval for a difference in population means is *not* included on the formula sheet provided on both sections of the AP® Statistics exam. However, the formula sheet does include the general formula for the confidence interval:

statistic ± (critical value)(standard error of statistic)

and the formula for the standard error of the difference in sample means:

$$s_{\bar{x}_1-\bar{x}_2} = \sqrt{\frac{s_1^2}{n_1} + \frac{s_2^2}{n_2}}$$

Let's return to the apartment-rent example. Recall that the recent college graduate recorded the monthly rent ($) for independent random samples of one-bedroom apartments in New York City and Chicago. Here are summary statistics calculated from the data:

New York City	$n_1 = 10$	$\bar{x}_1 = 2025.00$	$s_1 = 439.70$
Chicago	$n_2 = 10$	$\bar{x}_2 = 1588.00$	$s_2 = 350.41$

We already confirmed that the conditions are met. The 99% confidence interval for $\mu_1 - \mu_2$ is

$$(2025.00 - 1588.00) \pm t^* \sqrt{\frac{439.70^2}{10} + \frac{350.41^2}{10}}$$

At this point, it's time to let technology do the work!

24. Tech Corner CONFIDENCE INTERVALS FOR A DIFFERENCE IN MEANS

TI-Nspire and other technology instructions are on the book's website at bfwpub.com/tps7e.

You can use a TI-83/84 calculator to construct a confidence interval for $\mu_1 - \mu_2$. We'll illustrate using the data from the preceding example, which dealt with monthly rents of one-bedroom apartments in New York City and Chicago.

1. Enter the data for New York City in list L1 and the data for Chicago in list L2.
2. Press STAT, then choose TESTS and 2-SampTInt.
3. On the 2-SampTInt screen, choose Data as the input method and enter the values shown. Be sure to use 0.99 for the confidence level and choose "No" for Pooled. We'll discuss pooling at the end of this section.

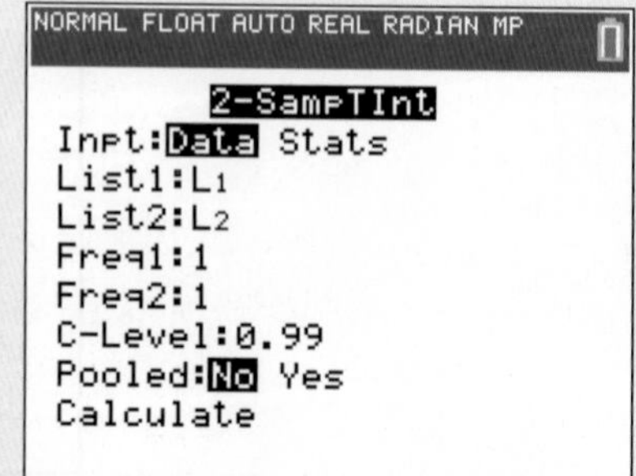

4. Highlight "Calculate" and press ENTER.

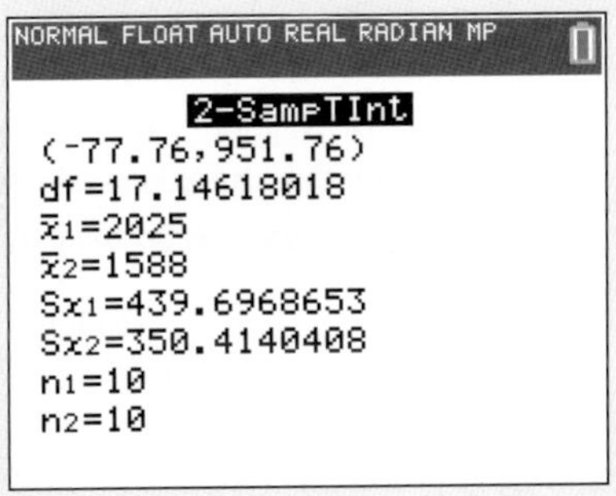

Note: To calculate a confidence interval for a difference in means from summary statistics, choose Stats as the input method in Step 3.

As the Tech Corner shows, the critical value t^* comes from a t distribution with df $= 17.146$. The resulting 99% confidence interval is $(-77.76,\ 951.76)$.

We are 99% confident that the interval from −\$77.76 to \$951.76 captures the difference in the population mean monthly rents of one-bedroom apartments in New York City and in Chicago. This interval suggests that the mean monthly rent of all one-bedroom apartments in New York City is between \$77.76 less and \$951.76 more than the mean monthly rent of all one-bedroom apartments in Chicago.

Skills 3.D, 4.D

EXAMPLE

Happy customers?
Calculating a confidence interval for $\mu_1 - \mu_2$

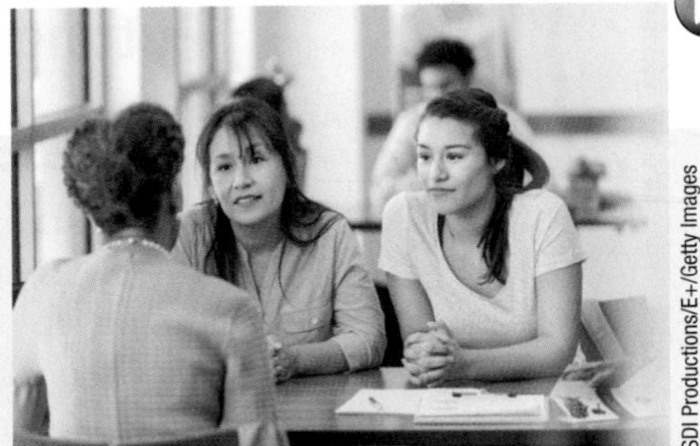
SDI Productions/E+/Getty Images

PROBLEM: As the Spanish-speaking population in the United States has grown, businesses have increased their focus on providing both English and Spanish options for customer service. One study interviewed a random sample of customers leaving a bank. Customers were asked if they spoke English or Spanish when talking to bank representatives. Each customer rated the importance of several aspects of bank service on a 10-point scale. Here are summary results for the importance of "reliability" (the accuracy of account records, etc.):[34]

Group	n	Mean	SD
English	92	6.37	0.60
Spanish	86	5.91	0.93

(a) Calculate a 95% confidence interval for the difference in the population mean reliability ratings of customers who speak English and customers who speak Spanish when talking with this bank's representatives. Assume the conditions for inference are met.

(b) Does the interval in part (a) provide convincing evidence of a difference in population mean reliability ratings? Justify your answer.

SOLUTION:

(a) 95% CI for $\mu_{English} - \mu_{Spanish}$, where $\mu_{English}$ = population mean reliability rating for customers who speak English to this bank's representatives and $\mu_{Spanish}$ = population mean reliability rating for customers who speak Spanish to this bank's representatives.

$$(6.37-5.91) \pm t^* \sqrt{\frac{0.60^2}{92}+\frac{0.93^2}{86}}$$

$$(\bar{x}_1 - \bar{x}_2) \pm t^* \sqrt{\frac{s_1^2}{n_1}+\frac{s_2^2}{n_2}}$$

Using technology: (0.226, 0.694) with df = 143.69

(b) Yes, the confidence interval does provide convincing evidence of a difference in the population mean reliability ratings because 0 is not included in the interval of plausible values (0.226, 0.694) for $\mu_{English} - \mu_{Spanish}$.

FOR PRACTICE, TRY EXERCISE 5

In the apartment-rent context, we used independent random samples of one-bedroom apartments in two different cities. In the bank reliability example, researchers selected one random sample that included individuals from both populations of interest — those who speak English and those who speak Spanish to this bank's representatives — and then separated the chosen individuals into two groups accordingly. Inference procedures for estimating or testing claims about a difference in population means are still valid in such situations, provided that the two groups can be viewed as independent samples from their respective populations of interest.

Putting It All Together: Two-Sample *t* Interval for $\mu_1 - \mu_2$

The apartment-rent and bank reliability examples involved inference about $\mu_1 - \mu_2$ using data that were produced by random sampling. In such cases, the parameters μ_1 and μ_2 are the means of the corresponding populations. However, many important statistical results involve comparing two means based on data from randomized experiments. Then the parameters μ_1 and μ_2 are the true mean responses for all individuals like the ones in the experiment who receive Treatment 1 or Treatment 2.

The appropriate confidence interval procedure for estimating $\mu_1 - \mu_2$ in either setting is called a **two-sample *t* interval for a difference in means.** As with any inference method, be sure to follow the four-step process.

DEFINITION Two-sample *t* interval for a difference in means

A **two-sample *t* interval for a difference in means** is a confidence interval used to estimate a difference in the means of two populations or treatments with unknown standard deviations.

EXAMPLE

Do portion sizes affect food consumption?
Two-sample t interval for $\mu_1 - \mu_2$

Skills 1.D, 3.D, 4.B, 4.C

VisualField/E+/Getty Images

PROBLEM: In a study published in the *American Journal of Clinical Nutrition*, researchers wanted to know if changing food portion sizes would influence subjects' food consumption one day later. The volunteer subjects were randomly assigned to two groups, with one group being served small portions of food and the other group being served large portions of food. The next day, all of the subjects were allowed to serve themselves, and the researchers recorded how much food they consumed (in grams). Here is a summary of the results:[35]

Treatment	n	Mean	SD
Small portion	38	144.66	72.36
Large portion	37	189.91	55.62

Construct and interpret a 95% confidence interval for the difference in the mean amount of food that would be consumed the next day by people like the ones in this study who are given large portions and those who are given small portions.

SOLUTION:

STATE: 95% CI for $\mu_{Large} - \mu_{Small}$, where μ_{Large} = true mean amount of food that would be consumed (grams) the next day by people like the ones in this study who are given large portions and μ_{Small} = true mean amount of food that would be consumed (grams) the next day by people like the ones in this study who are given small portions.

Be sure to indicate the order of subtraction when defining parameters.

PLAN: Two-sample t interval for a difference in means

- Random: Volunteers randomly assigned to receive large portions or small portions of food. ✓
- Normal/Large Sample: $n_{Large} = 37 \geq 30$ and $n_{Small} = 38 \geq 30$. ✓

It is *not* appropriate to check the 10% condition in this setting because the experiment used volunteer subjects.

DO: $(189.91 - 144.66) \pm t^* \sqrt{\dfrac{55.62^2}{37} + \dfrac{72.36^2}{38}}$

$$(\bar{x}_1 - \bar{x}_2) \pm t^* \sqrt{\frac{s_1^2}{n_1} + \frac{s_2^2}{n_2}}$$

Using technology: (15.57, 74.93) with df = 69.30

CONCLUDE: We are 95% confident that the interval from 15.57 grams to 74.93 grams captures the difference (Large − Small) in the mean amount of food that would be consumed the next day by all people like the ones in this study who are given large portions and those who are given small portions.

FOR PRACTICE, TRY EXERCISE 7

The 95% confidence interval in the example does not include 0. This gives convincing evidence that the difference in the true mean amount of food that would be consumed the next day by people like the ones in this study who are given large portions and those who are given small portions isn't 0. However, the confidence interval provides more information than a simple reject or fail to reject H_0 conclusion: it gives a set of plausible values for $\mu_{Large} - \mu_{Small}$.

The interval suggests that the true mean amount of food that would be consumed the next day by people like these is between 15.57 grams and 74.93 grams higher for those given large portions than for those given small portions. Also, because this was a randomized experiment, the researchers were able to conclude that changing food portion sizes influences food consumption one day later for people like the volunteers in the study.

We chose the order of subtraction in the example to estimate $\mu_{\text{Large}} - \mu_{\text{Small}}$. What if we had reversed the order and estimated $\mu_{\text{Small}} - \mu_{\text{Large}}$ instead? The resulting 95% confidence interval from technology is −74.93 to −15.57. This interval suggests that the true mean amount of food that would be consumed the next day by people like the ones in this study is between 15.57 grams and 74.93 grams less for those given small portions than for those given large portions. Note that this is equivalent to the conclusion in the preceding paragraph.

OTHER OPTIONS FOR A TWO-SAMPLE *t* INTERVAL

1. **The Conservative Approach** In the days when a four-function calculator and Table B were the main computation tools, many people used a "conservative approach" when performing inference about a difference in population means $\mu_1 - \mu_2$. Instead of using the complicated formula to find the appropriate degrees of freedom for two-sample t procedures, they just used df = the *smaller* of $n_1 - 1$ and $n_2 - 1$. Let's see what effect the conservative approach would have on the 99% confidence interval in the apartment-rent example.

   Recall that a recent college graduate recorded the monthly rent ($) for independent random samples of one-bedroom apartments in New York City and in Chicago. Here are the summary statistics once again:

New York City	$n_1 = 10$	$\bar{x}_1 = 2025.00$	$s_1 = 439.70$
Chicago	$n_2 = 10$	$\bar{x}_2 = 1588.00$	$s_2 = 350.41$

   We already showed that the 99% confidence interval for $\mu_1 - \mu_2$ is

$$(2025.00 - 1588.00) \pm t^* \sqrt{\frac{439.70^2}{10} + \frac{350.41^2}{10}}$$

By the conservative approach, df = smaller of $10 - 1$ and $10 - 1 = 9$. From Table B or using the invT command on the TI-83/84, the critical value for a 99% confidence level in a t distribution with 9 degrees of freedom is $t^* = 3.250$. Substituting this value into the confidence interval formula gives

$$\begin{aligned}(2025.00 - 1588.00) &\pm 3.250\sqrt{\frac{439.70^2}{10} + \frac{350.41^2}{10}}\\ &= 437.00 \pm 577.85\\ &= (-140.85, 1014.85)\end{aligned}$$

This interval is considerably wider than the 99% confidence interval calculated earlier (−77.76, 951.76) using df = 17.146. That's because the critical value for a 99% confidence level using df = 17.146 is only $t^* = 2.895$.

Calculating the degrees of freedom using the smaller of $n_1 - 1$ and $n_2 - 1$ is called the conservative approach because it will *always* result in a smaller df and a larger t^* value than when using the df given by technology, *making the interval wider than needed* for the given level of confidence. For that reason, we recommend using technology to get a more precise estimate of $\mu_1 - \mu_2$.

2. The Pooled Two-Sample *t* Procedures

The other option is a special version of the two-sample t procedures that assumes the two population distributions have equal variances. (Recall that the variance is the square of the standard deviation.) This procedure combines (*pools*) the two sample variances to estimate the common population variance σ^2. The formula for the pooled estimate of σ^2 is

$$s_p^2 = \frac{(n_1 - 1)s_1^2 + (n_2 - 1)s_2^2}{(n_1 - 1) + (n_2 - 1)}$$

If the two population variances are equal and both population distributions are close to normal, we can use a t distribution with $\text{df} = (n_1 - 1) + (n_2 - 1) = n_1 + n_2 - 2$ and s_p^2 to perform calculations. This method offers more degrees of freedom than the technology approach ($\text{df} = 10 + 10 - 2 = 18$ when pooling versus 17.146 with technology in the apartment-rent example context), which leads to narrower confidence intervals and smaller P-values. However, an additional condition must be met for these calculations to be valid: the population variances must be equal.

The pooled two-sample t procedures were widely used before technology made it easy to carry out the technology-based two-sample t procedures we present in this book. In practice, population variances are rarely equal, so our two-sample t procedures are almost always more accurate than the pooled procedures — and they work almost as well when the population variances *are* equal. Our advice: *don't use the pooled two-sample t procedures unless a statistician says it is OK to do so.*

CHECK YOUR UNDERSTANDING

Mr. Wilcox's class performed an experiment to investigate whether drinking a caffeinated beverage would affect pulse rates. Twenty students in the class volunteered to take part in the experiment. All of the students measured their initial pulse rates (in beats per minute). Then Mr. Wilcox randomly assigned the students into two groups of 10. Each student in the first group drank 12 ounces of cola with caffeine. Each student in the second group drank 12 ounces of caffeine-free cola. All students then measured their pulse rates again. The table displays the change in pulse rate for the students in both groups.

	Change in pulse rate (Final pulse rate − Initial pulse rate)										Mean change
Caffeine	8	3	5	1	4	0	6	1	4	0	3.2
No caffeine	3	−2	4	−1	5	5	1	2	−1	4	2.0

1. Construct and interpret a 95% confidence interval for the difference in the true mean change in pulse rate for people like the ones in this study who drink caffeine and those who drink no caffeine.
2. What does the interval in Question 1 suggest about whether drinking caffeine affects the average pulse rate of people like the ones in this study? Justify your answer.

SECTION 7C Summary

- Confidence intervals for the difference $\mu_1 - \mu_2$ between the means of two populations or the true mean responses to two treatments use the difference $\bar{x}_1 - \bar{x}_2$ between the sample means as the point estimate.
- Before constructing a confidence interval for $\mu_1 - \mu_2$, we need to check for independence and confirm that the sampling distribution of $\bar{x}_1 - \bar{x}_2$ is approximately normal. The required conditions are:
  - Random: The data come from two independent random samples or from two groups in a randomized experiment.
    - 10%: When sampling without replacement, $n_1 < 0.10N_1$ and $n_2 < 0.10N_2$.
  - Normal/Large Sample: For each sample, the data come from an approximately normally distributed population or the sample size is large ($n \geq 30$). For each sample, if the population distribution has unknown shape and $n < 30$, a graph of the sample data shows no strong skewness or outliers.
- When the conditions are met, a $C\%$ confidence interval for $\mu_1 - \mu_2$ is

$$(\bar{x}_1 - \bar{x}_2) \pm t^* \sqrt{\frac{s_1^2}{n_1} + \frac{s_2^2}{n_2}}$$

where t^* is the critical value with $C\%$ of its area between $-t^*$ and t^* for the t distribution with degrees of freedom given by technology. This is called a **two-sample t interval for a difference in means.**

AP® EXAM TIP

AP® Daily Videos

Review the content of this section and get extra help by watching the AP® Daily Videos for Topics 7.6, and 7.7, which are available in AP® Classroom.

7C Tech Corner

TI-Nspire and other technology instructions are on the book's website at bfwpub.com/tps7e.

24. Confidence Intervals for a difference in means Page 696

SECTION 7C Exercises

Checking Conditions for a Confidence Interval for $\mu_1 - \mu_2$

1. **Encouragement effect** A researcher wondered if people would perform better in the barbell curl if they were encouraged by a coach. The researcher recruited 31 people to participate in an experiment and then split them into two groups using random assignment. Group 1 received positive encouragement during the exercise, while Group 2 served as a control group that received no encouragement. The researcher recorded the number of barbell curl repetitions each participant was able to complete before setting the bar down. Here are the data:[37]

pg 693

With encouragement	50	11	60	5	40	45	11	30
	8	50	11	15	9	40	75	
No encouragement	43	19	33	1	0	2	6	61
	21	11	40	15	15	18	37	22

Let μ_1 = the true mean number of barbell curls that people like the ones in this study can do with encouragement and μ_2 = the true mean number of barbell curls that people like the ones in this study can do with no encouragement. Check if the conditions for constructing a 90% confidence interval for $\mu_1 - \mu_2$ are met.

2. **Polyphenols and heart attacks** Some studies suggest that polyphenols may reduce the risk of heart attacks. Researchers decided to investigate whether drinking red wine would increase polyphenol levels more, on average, than drinking white wine. The researchers randomly assigned healthy adults to drink half a bottle of either red

or white wine each day for 2 weeks. They measured the level of polyphenols in the participants' blood before and after the 2-week period. Here are the percent changes in polyphenol levels for the subjects in each group:[36]

Red wine	3.5	8.1	7.4	4.0	0.7	4.9	8.4	7.0	5.5
White wine	3.1	0.5	−3.8	4.1	−0.6	2.7	1.9	−5.9	0.1

Let μ_1 = the true mean percent change in polyphenol levels for adults like the ones in this study after drinking red wine daily for 2 weeks, and μ_2 = the true mean percent change in polyphenol levels for adults like the ones in this study after drinking white wine daily for 2 weeks. Check if the conditions for constructing a 90% confidence interval for $\mu_1 - \mu_2$ are met.

3. **Word length** Sanderson is interested in comparing the mean word length in articles from a medical journal and from an airline's in-flight magazine. Sanderson counts the number of letters in each of the first 400 words of an article in the medical journal and in each of the first 100 words of an article in the airline magazine, and then uses technology to produce the histograms shown here. Sanderson would like to construct a confidence interval for the difference in the population mean word length of articles in the medical journal and articles in the airline magazine.

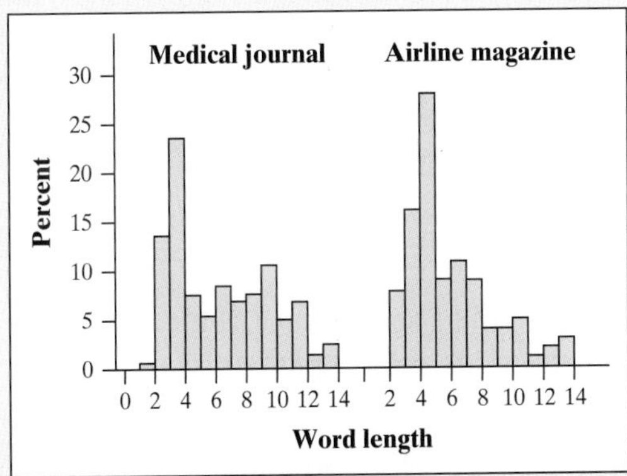

(a) Define the parameters of interest.

(b) Check if the conditions for calculating a confidence interval are met.

4. **Overthinking** Athletes often comment that they try not to "overthink it" when competing in their sport. Is it possible to "overthink"? To investigate, researchers put some golfers to the test. They recruited 40 experienced golfers and allowed them some time to practice their putting. After practicing, they randomly assigned the golfers in equal numbers to two groups. Golfers in one group had to write a detailed description of their putting technique (which could lead to "overthinking it"). Golfers in the other group had to do an unrelated verbal task for the same amount of time. After completing their tasks, each golfer was asked to attempt putts from a fixed distance until they made 3 putts in a row. The boxplots summarize the distribution of the number of putts required for the golfers in each group to make 3 putts in a row.[38] The researchers would like to construct a confidence interval for the difference in the true mean number of putts required for players to make three in a row after describing their putting and the true mean number of putts required for players to make three in a row after doing a verbal task for all golfers like the ones in this study.

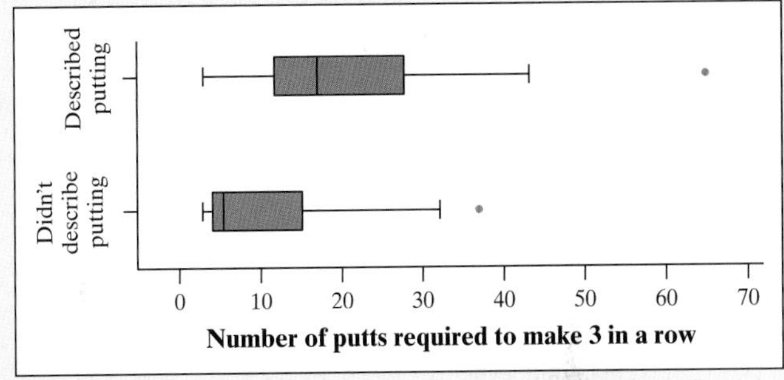

(a) Define the parameters of interest.

(b) Check if the conditions for calculating a confidence interval are met.

Calculating a Confidence Interval for $\mu_1 - \mu_2$

5. **More encouragement** Refer to Exercise 1. pg 697

(a) Calculate a 90% confidence interval for $\mu_1 - \mu_2$.

(b) Does the interval in part (a) provide convincing evidence of a difference in the true mean number of barbell curls that people like the ones in this study who receive encouragement and those who receive no encouragement can do? Justify your answer.

6. **More polyphenols** Refer to Exercise 2.

(a) Calculate a 90% confidence interval for $\mu_1 - \mu_2$.

(b) Does the interval in part (a) provide convincing evidence of a difference in the true mean percent change in polyphenol levels for adults like the ones in the study who drink red wine and those who drink white wine? Justify your answer.

Two-Sample *t* Interval for $\mu_1 - \mu_2$

7. **Value of a college degree** Is it true that students who earn an associate's degree or a bachelor's degree make more money than students who attend college but do not earn a degree? To find out, researchers selected a random sample of 500 U.S. residents aged 18 and older who had attended college.[39] They recorded the educational attainment and annual income of each person. Due to the sampling method used in this survey, it is reasonable to consider these as independent random samples of students who earned degrees and those who did not. Here pg 699

are relative frequency histograms and summary statistics of the income data for the two groups.

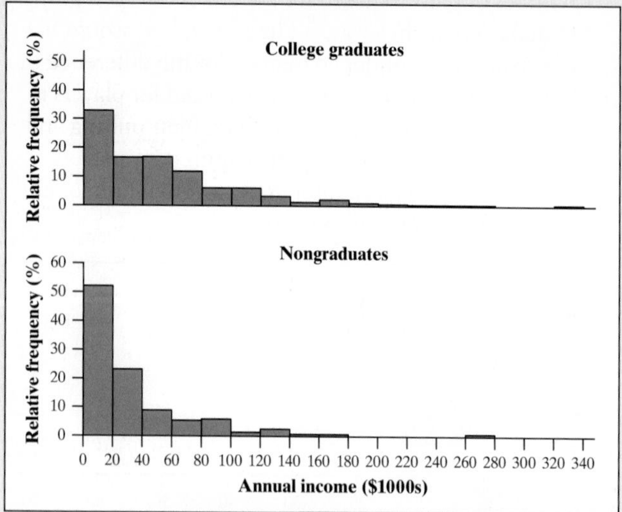

Education level	n	Mean	SD
College graduates	327	49,454.80	51,257.10
Nongraduates	173	29,299.20	38,298.00

Construct and interpret a 95% confidence interval for the difference (Graduates − Nongraduates) in the population mean annual income for U.S. adults with these two education levels.

8. **Household size** How do the numbers of people living in households in the United Kingdom and South Africa compare? To help answer this question, we used an online random data selector to choose separate random samples of 36 students from South Africa and 31 students from the United Kingdom. Here are dotplots and summary statistics of the household sizes reported by the students in the survey.

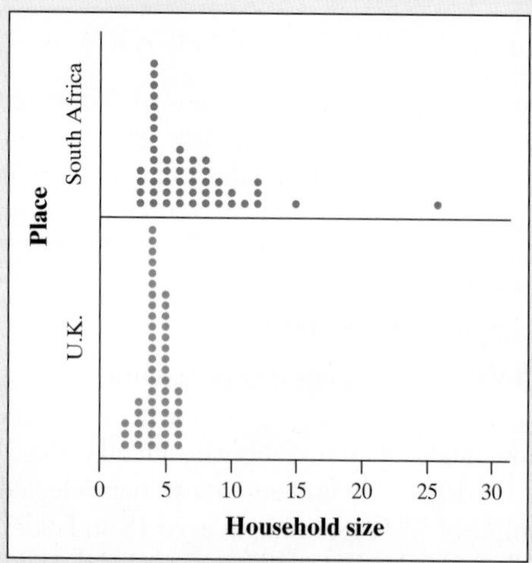

Country	n	Mean	SD
South Africa	36	7.06	4.38
United Kingdom	31	4.26	1.18

Construct and interpret a 99% confidence interval for the difference (South Africa − United Kingdom) in the population mean household size for students in the two countries.

9. **Beta blockers** In a study of heart surgery, one concern was the effect of drugs called beta blockers on the pulse rate of patients during surgery. The available subjects were randomly split into two groups of 30 patients each. One group received a beta blocker; the other group received a placebo. The pulse rate of each patient at a critical point during the operation was recorded. The treatment group had a mean pulse rate of 65.2 beats per minute (bpm) and a standard deviation of 7.8 bpm. For the control group, the mean pulse rate was 70.3 bpm and the standard deviation was 8.3 bpm.

(a) Calculate and interpret a 99% confidence interval for the difference in mean pulse rate for all patients like the ones in this study who receive a beta blocker and those who receive a placebo.

(b) Interpret the 99% confidence level.

10. **Tropical flowers** Different varieties of the tropical flower *Heliconia* are fertilized by different species of hummingbirds. Researchers believe that over time, the lengths of the flowers and the forms of the hummingbirds' beaks have adapted to match each other. Here are data on the lengths (in millimeters) for random samples of two color varieties of the same species of flower on the island of Dominica:[40]

H. caribaea red							
41.90	42.01	41.93	43.09	41.17	41.69	39.78	40.57
39.63	42.18	40.66	37.87	39.16	37.40	38.20	38.07
38.10	37.97	38.79	38.23	38.87	37.78	38.01	
H. caribaea yellow							
36.78	37.02	36.52	36.11	36.03	35.45	38.13	37.10
35.17	36.82	36.66	35.68	36.03	34.57	34.63	

Parallel boxplots of the data are shown, along with summary statistics.

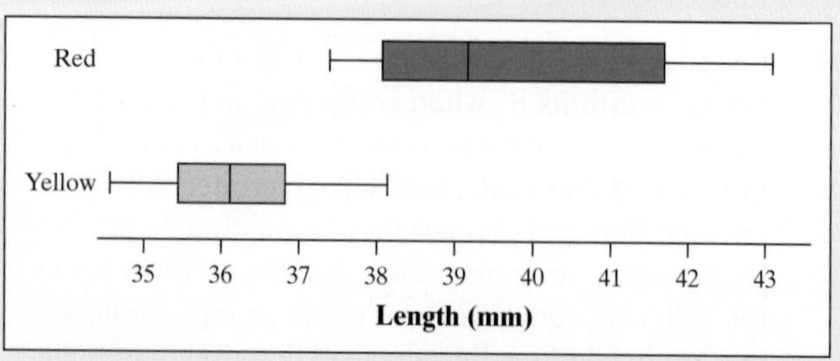

Group	n	Mean	SD	Min	Q_1	Med	Q_3	Max
Red	23	39.698	1.786	37.40	38.07	39.16	41.69	43.09
Yellow	15	36.180	0.975	34.57	35.45	36.11	36.82	38.13

(a) Calculate and interpret a 95% confidence interval for the difference in the population mean length of these two varieties of flowers on the island of Dominica.

(b) Interpret the 95% confidence level.

11. **Reaction times** Researchers wanted to know if student athletes (students on at least one varsity team) have faster reaction times than non-athletes. They took separate random samples of 33 athletes and 30 non-athletes from their school and tested their reaction times using an online reaction test, which measured the time (in seconds) between when a green light went on and the subject pressed a key on the computer keyboard. A 95% confidence interval for the difference (Non-athlete − Athlete) in the mean reaction times was 0.018 ± 0.034 second.

(a) Does the interval provide convincing evidence of a difference in the population mean reaction times of athletes and non-athletes at this school? Explain your reasoning.

(b) Does the interval provide convincing evidence that the population mean reaction time of athletes and the population mean reaction time of non-athletes at this school are the same? Explain your reasoning.

(c) Identify two ways the researchers could have reduced the width of their interval. Describe any drawbacks to these actions.

12. **Bird eggs** Researchers want to see if birds that build larger nests lay larger eggs. The researchers select two random samples of nests: one of small nests and the other of large nests. They then weigh one egg (chosen at random if there is more than one egg) from each nest. A 95% confidence interval for the difference (Large − Small) between the mean mass (in grams) of eggs in small and large nests is 1.6 ± 2.0.

(a) Does the interval provide convincing evidence of a difference in the population mean egg mass of birds with small nests and birds with large nests? Explain your answer.

(b) Does the interval provide convincing evidence that the population mean egg mass of birds with small nests and birds with large nests is the same? Explain your answer.

(c) Identify two ways the researchers could have reduced the width of the interval. Describe any drawbacks to these actions.

13. **Options for df** Refer to Exercise 9.

(a) Use the conservative approach to find the critical value t^* for a 99% confidence level.

(b) Calculate the 99% confidence interval using the critical value from part (a).

(c) How does the confidence interval from part (b) compare with the one from Exercise 9? Explain why this difference makes sense.

For Investigation *Apply the skills from the section in a new context or nonroutine way.*

14. **SE versus ME for means** In a random sample of 68 workers from Company 1, the mean number of sick days taken in the previous year was 4.8 days with a standard deviation of 2.1 days. In a separate random sample of 34 employees from Company 2, the mean number of sick days taken in the previous year was 2.5 days with a standard deviation of 1.1 days. The difference in the sample mean number of sick days from the two companies is $\bar{x}_1 - \bar{x}_2 = 4.8 - 2.5 = 2.3$.

(a) Calculate the standard error of $\bar{x}_1 - \bar{x}_2$. Interpret this value.

(b) Using technology, the margin of error for a 95% confidence interval for $\mu_1 - \mu_2$ is 0.45. Interpret this value.

(c) Use parts (a) and (b) to find the critical value t^* used to calculate the margin of error.

(d) Estimate the degrees of freedom for the t distribution that was used to obtain the critical value in part (c).

15. **Pooled two-sample *t* interval** Refer to Exercises 9 and 13.

(a) Explain why it might be reasonable to use pooled two-sample t procedures in this case.

(b) Find the critical value t^* for a pooled two-sample t interval using a 99% confidence level and $df = (n_1 - 1) + (n_2 - 1) = n_1 + n_2 - 2$.

(c) (i) Use the formula $s_p^2 = \dfrac{(n_1 - 1)s_1^2 + (n_2 - 1)s_2^2}{(n_1 - 1) + (n_2 - 1)}$ to calculate the pooled estimate of the population variance.

(ii) Use the formula $(\bar{x}_1 - \bar{x}_2) \pm t^* s_p\sqrt{\dfrac{1}{n_1} + \dfrac{1}{n_2}}$ to calculate the confidence interval.

(d) Explain why the interval from part (c-ii) is narrower than the ones calculated in Exercises 9 and 13. What additional assumption was required to obtain this narrower confidence interval?

Multiple Choice *Select the best answer for each question.*

16. Which of the following is *not* a requirement for calculating a confidence interval for a difference in population means?

(A) When sampling without replacement, both sample sizes are less than 10% of their respective population sizes.

(B) For each sample, the corresponding population distribution is approximately normally distributed or the sample size is at least 30.

(C) The data come from two groups in a randomized experiment or from independent random samples from the two populations of interest.

(D) Regardless of sample size, graphs of the sample data do not reveal any skewness or outliers.

(E) The data are quantitative.

17. A random sample of 30 words from Jane Austen's *Pride and Prejudice* had a mean length of 4.08 letters with a standard deviation of 2.40. A random sample of 30 words from Henry James's *What Maisie Knew* had a mean length of 3.85 letters with a standard deviation of 2.26. Which of the following is a correct expression for a 95% confidence interval for the difference in the population mean word length for these two novels?

(A) $(4.08-3.85)\pm 1.96\left(\frac{2.40}{\sqrt{30}}+\frac{2.26}{\sqrt{30}}\right)$

(B) $(4.08-3.85)\pm 2.002\left(\frac{2.40}{\sqrt{30}}+\frac{2.26}{\sqrt{30}}\right)$

(C) $(4.08-3.85)\pm 1.96\sqrt{\frac{2.40^2}{30}+\frac{2.26^2}{30}}$

(D) $(4.08-3.85)\pm 2.002\sqrt{\frac{2.40^2}{30}+\frac{2.26^2}{30}}$

(E) $(4.08-3.85)\pm 2.002\left(\frac{2.40^2}{30}+\frac{2.26^2}{30}\right)$

18. A researcher wondered if the bean burritos at Restaurant A tend to be heavier than the bean burritos at Restaurant B. To investigate, the researcher visited each restaurant at 10 randomly selected times, ordered a bean burrito, and weighed the burrito. The 95% confidence interval for the difference (A – B) in the population mean weight of burrito is 0.06 ounce ± 0.20 ounce. Based on the confidence interval, which conclusion is most appropriate?

(A) Because 0 is included in the interval, there is convincing evidence that the population mean bean burrito weight is the same at both restaurants.

(B) Because 0 is included in the interval, there isn't convincing evidence that the population mean burrito weight is different at the two restaurants.

(C) Because 0.06 is included in the interval, there is convincing evidence that the population mean burrito weight is greater at Restaurant A than Restaurant B.

(D) Because 0.06 is included in the interval, there isn't convincing evidence that the population mean burrito weight is greater at Restaurant A than Restaurant B.

(E) Because there are more positive values in the interval than negative values, there is convincing evidence that the population mean burrito weight is greater at Restaurant A than Restaurant B.

Recycle and Review *Practice what you learned in previous sections.*

19. **Quality control (Units 4 and 5)** Many manufacturing companies use statistical techniques to ensure that the products they make meet certain standards. One common way to do this is to take a random sample of products at regular intervals throughout the production shift. Assuming that the process is working properly, the mean measurement $\bar{x}$ from a random sample can be modeled by a normal distribution with mean $\mu_{\bar{x}}$ and standard deviation $\sigma_{\bar{x}}$. For each question that follows, assume that the process is working properly.

(a) What's the probability that at least one of the next two sample means will fall more than $2\sigma_{\bar{x}}$ from the target mean $\mu_{\bar{x}}$?

(b) What's the probability that the first sample mean that is greater than $\mu_{\bar{x}}+2\sigma_{\bar{x}}$ is the one from the fourth sample taken?

Plant managers are trying to develop a criterion for determining when the process is not working properly. One idea they have is to look at the 5 most recent sample means. If at least 4 of the 5 fall outside the interval $(\mu_{\bar{x}}-\sigma_{\bar{x}}, \mu_{\bar{x}}+\sigma_{\bar{x}})$, they will conclude that the process isn't working.

(c) Find the probability that at least 4 of the 5 most recent sample means fall outside the interval, assuming the process is working properly. Is this a reasonable criterion? Explain your reasoning.

20. **Stop doing homework! (3C)** Researchers in Spain interviewed 7725 13-year-olds about their homework habits — how much time they spent per night on homework and whether they got help from their parents or not — and then had them take a test with 24 math questions and 24 science questions. They found that students who spent between 90 and 100 minutes on their homework did only a little better on the test than those who spent 60 to 70 minutes on their homework. Beyond 100 minutes, students who spent more time on their homework did worse than those who spent less time on their homework. The researchers concluded that 60 to 70 minutes per night is the optimal amount of time for students to spend on homework.[41] Is it appropriate to conclude that students who reduce their homework time from 120 minutes to 70 minutes will likely improve their performance on tests such as those used in this study? Why or why not?

SECTION 7D Significance Tests for a Difference in Population Means

LEARNING TARGETS *By the end of the section, you should be able to:*

- State appropriate hypotheses and check the conditions for performing a test about a difference between two population means.
- Calculate the standardized test statistic and *P*-value for a test about a difference between two population means.
- Perform a two-sample *t* test for a difference in means.
- Determine when it is appropriate to use paired *t* procedures versus two-sample *t* procedures.

In Section 7B, you learned how to perform a significance test about a population mean μ or a population mean difference μ_{Diff}. Many interesting statistical questions involve comparing means μ_1 and μ_2 for two populations or treatments. Is the mean number of hours worked per week in the United States different now than it was several decades ago? For people with high blood pressure, does a diet that includes fish oil help reduce blood pressure more, on average, than a diet that includes other oils?

An observed difference between two sample means $\bar{x}_1$ and $\bar{x}_2$ can reflect an actual difference in the parameters μ_1 and μ_2, or it may be due to chance variation in random sampling or random assignment. Significance tests help us decide which explanation makes more sense. This section shows you how to perform a significance test about a difference $\mu_1 - \mu_2$ between two population means.

Stating Hypotheses and Checking Conditions for a Test about $\mu_1 - \mu_2$

In a test comparing two population means, the null hypothesis has the general form

$$H_0: \mu_1 - \mu_2 = \text{hypothesized value}$$

In AP® Statistics, we focus on the case where the hypothesized difference is 0. Then the null hypothesis says that there is no difference between the two parameters:

$$H_0: \mu_1 - \mu_2 = 0$$

The null hypothesis can also be written in the equivalent form $H_0: \mu_1 = \mu_2$. The alternative hypothesis says what kind of difference we expect.

The conditions for performing a significance test about $\mu_1 - \mu_2$ are the same as those for constructing a confidence interval for a difference in means that you learned in Section 7C. Recall that the purpose of these conditions is to check for independence in the data collection process and to verify that the sampling distribution of $\bar{x}_1 - \bar{x}_2$ is approximately normal.

CONDITIONS FOR PERFORMING A SIGNIFICANCE TEST ABOUT A DIFFERENCE BETWEEN TWO POPULATION MEANS

- **Random:** The data come from two independent random samples or from two groups in a randomized experiment.
  - **10%:** When sampling without replacement, $n_1 < 0.10N_1$ and $n_2 < 0.10N_2$.
- **Normal/Large Sample:** For each sample, the data come from an approximately normally distributed population or the sample size is large ($n \geq 30$). For each sample, if the population distribution has unknown shape and $n < 30$, a graph of the sample data shows no strong skewness or outliers.

Here's an example that illustrates how to state hypotheses and check conditions.

EXAMPLE

Big trees, small trees, short trees, tall trees
Stating hypotheses and checking conditions for a test about $\mu_1 - \mu_2$

Skills 1.F, 4.C

Danita Delimont/Alamy Stock Photo

PROBLEM: The Wade Tract Preserve in Georgia is an old-growth forest of longleaf pines that has survived in a relatively undisturbed state for hundreds of years. One question of interest to foresters who study the area is "How do the sizes of the trees in the northern and southern halves of the forest compare?" To find out, researchers took a random sample of 30 trees from each half and measured the diameter at breast height (DBH) in centimeters.[42] Here are comparative boxplots and numerical summaries of the data.

Forest location	*n*	Mean	SD
North	30	23.70	17.50
South	30	34.53	14.26

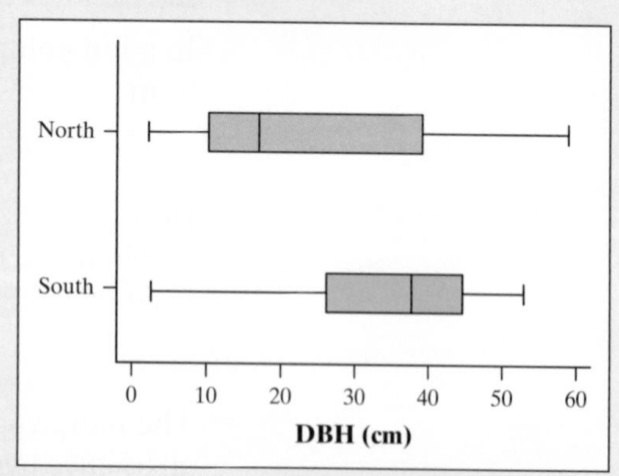

Foresters want to know whether these data provide convincing evidence of a difference in the mean DBH of trees in the northern and southern halves of the Wade Tract Preserve at the 5% significance level.

(a) State appropriate hypotheses for performing a significance test. Be sure to define the parameters of interest.
(b) Check if the conditions for performing the test are met.

SOLUTION:

(a) $H_0: \mu_{North} - \mu_{South} = 0$

$H_a: \mu_{North} - \mu_{South} \neq 0$

where μ_{North} = the mean DBH (cm) of all trees in the northern half of the forest and μ_{South} = the mean DBH (cm) of all trees in the southern half of the forest.

We could also state the hypotheses as

$H_0: \mu_{North} = \mu_{South}$

$H_a: \mu_{North} \neq \mu_{South}$

(b) • Random: Independent random samples of 30 trees each from the northern and southern halves of the forest. ✓

  ○ 10%: Assume 30 < 10% of all trees in the northern half of the forest and 30 < 10% of all trees in the southern half of the forest. ✓

• Normal/Large Sample: $n_{North} = 30 \geq 30$ and $n_{South} = 30 \geq 30$. ✓

Note: Due to the large sample sizes, we do not need to discuss the boxplots when addressing the Normal/Large sample condition.

FOR PRACTICE, TRY EXERCISE 1

Do the data in the example provide *some* evidence of a difference in the mean DBH of trees in the northern and southern halves of the Wade Tract Preserve? Yes, because $\bar{x}_{North} - \bar{x}_{South} = 23.70 - 34.53 = -10.83 \neq 0$. Also, the boxplots show that more than 75% of the southern trees have diameters that are greater than the northern sample's median. To determine whether there is *convincing* evidence for H_a: $\mu_{North} - \mu_{South} \neq 0$, we need to find the *P*-value.

Calculating the Standardized Test Statistic and *P*-Value for a Test about $\mu_1 - \mu_2$

If the conditions are met, we can proceed with calculations. To do a test of H_0: $\mu_1 - \mu_2 = 0$, start by standardizing $\bar{x}_1 - \bar{x}_2$:

$$\text{standardized test statistic} = \frac{\text{statistic} - \text{parameter}}{\text{standard error of statistic}}$$

Because we seldom know the population standard deviations σ_1 and σ_2, we use the standard error of $\bar{x}_1 - \bar{x}_2$ in the denominator of the standardized test statistic. If the hypothesized difference in means is 0, the standardized test statistic is

$$t = \frac{(\bar{x}_1 - \bar{x}_2) - 0}{\sqrt{\frac{s_1^2}{n_1} + \frac{s_2^2}{n_2}}}$$

When the Normal/Large Sample condition is met, we can use the *t* distribution with degrees of freedom calculated by technology to find the *P*-value. Recall from Section 7C that the df is usually not a whole number.

CALCULATING THE STANDARDIZED TEST STATISTIC AND *P*-VALUE IN A TEST FOR A DIFFERENCE BETWEEN TWO POPULATION MEANS

Suppose the conditions are met. To test the hypothesis H_0: $\mu_1 - \mu_2 = 0$, calculate the standardized test statistic

$$t = \frac{(\bar{x}_1 - \bar{x}_2) - 0}{\sqrt{\frac{s_1^2}{n_1} + \frac{s_2^2}{n_2}}}$$

Find the *P*-value by calculating the probability of getting a *t* statistic this large or larger in the direction specified by the alternative hypothesis H_a in a *t* distribution with df given by technology.

AP® EXAM TIP
Formula sheet

The specific formula for the standardized test statistic in a test about a difference in population means is *not* included on the formula sheet provided on both sections of the AP® Statistics exam. However, the formula sheet does include the general formula for the standardized test statistic:

$$\text{standardized test statistic} = \frac{\text{statistic} - \text{parameter}}{\text{standard error of statistic}}$$

and the formula for the standard error of the difference in sample means:

$$s_{\bar{x}_1 - \bar{x}_2} = \sqrt{\frac{s_1^2}{n_1} + \frac{s_2^2}{n_2}}$$

Here's a hypothetical example to illustrate the calculations. Suppose that we want to perform a test of $H_0: \mu_1 - \mu_2 = 0$ versus $H_a: \mu_1 - \mu_2 > 0$ based on a randomized experiment that yielded the following summary statistics for the two groups:

Group	*n*	Mean	SD
1	45	7.55	1.83
2	44	6.98	2.21

Note that the conditions for inference are met. The standardized test statistic is

$$t = \frac{(7.55 - 6.98) - 0}{\sqrt{\frac{1.83^2}{45} + \frac{2.21^2}{44}}} = \frac{0.57}{0.4306} = 1.324$$

The *P*-value is the probability of getting a difference in sample means greater than or equal to $\bar{x}_1 - \bar{x}_2 = 0.57$ by chance alone when $H_0: \mu_1 - \mu_2 = 0$ is true. Figure 7.6(a) shows this probability as a shaded area in the sampling distribution of $\bar{x}_1 - \bar{x}_2$. To find the *P*-value, we need to calculate the probability $P(t \geq 1.324)$ in a *t* distribution with df given by the complicated formula in Section 7C. The shaded area in Figure 7.6(b) shows this probability. We'll use technology to determine the df and *P*-value.

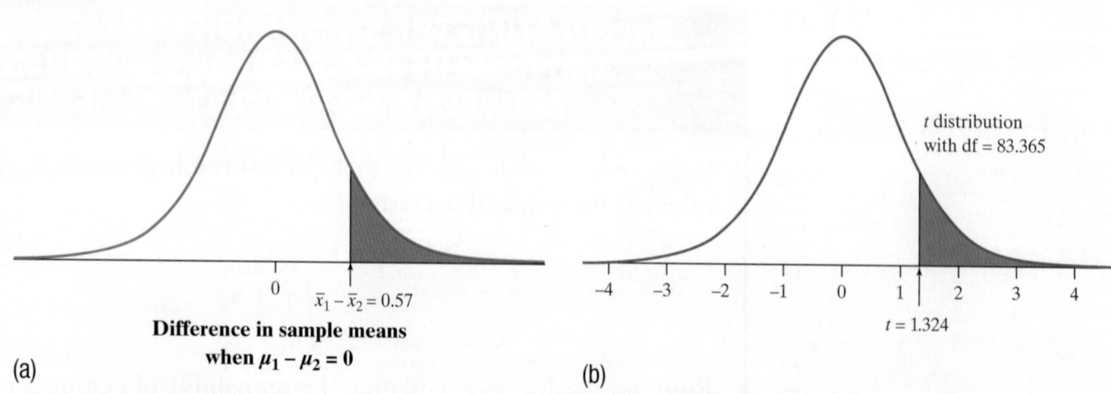

FIGURE 7.6 (a) The *P*-value is the area to the right of $\bar{x}_1 - \bar{x}_2 = 0.57$ in the sampling distribution of the difference in sample means, assuming that $H_0: \mu_1 - \mu_2 = 0$ is true. (b) We calculate the *P*-value as the area to the right of $t = 1.324$ in a *t* distribution with degrees of freedom obtained using technology.

convincing evidence for the alternative H_a: $\mu_C - \mu_P > 0$. We suspect that larger groups might show a similar difference in mean blood pressure reduction, which would indicate that calcium has a significant effect. If so, then the researchers in this experiment made a Type II error — they did not find convincing evidence for H_a when it is actually true.

In fact, later studies involving more subjects showed that an increase in calcium intake slightly reduces both systolic and diastolic blood pressures in healthy people.[45] *Sample size strongly affects the power of a test.* It is easier to detect a difference in the effectiveness of two treatments if both are applied to large numbers of subjects.

Think About It

WHY DO THE INFERENCE METHODS FOR RANDOM SAMPLING WORK FOR RANDOMIZED EXPERIMENTS? Confidence intervals and significance tests for $\mu_1 - \mu_2$ are based on the sampling distribution of $\bar{x}_1 - \bar{x}_2$. But in most experiments, researchers don't select subjects at random from any larger populations; instead, they randomly assign subjects to treatments. As in Section 6F, we can use simulation to see what would happen if the random assignment were repeated many times, assuming no treatment effect — that is, assuming that H_0: $\mu_1 - \mu_2 = 0$ is true.

We used technology to randomly reassign the 21 subjects in the calcium and blood pressure experiment to the two groups 1000 times, assuming the drug received (calcium or placebo) *doesn't affect* each individual's change in systolic blood pressure. The dotplot shows the value of $\bar{x}_C - \bar{x}_P$ in the 1000 simulated trials.

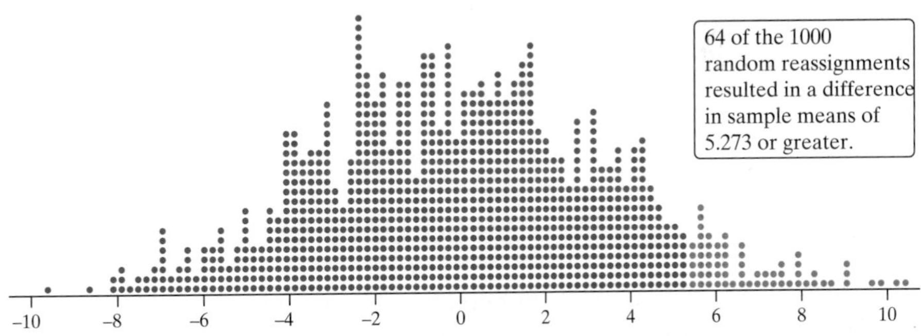

This distribution, sometimes referred to as a *randomization distribution* of $\bar{x}_C - \bar{x}_P$, has an approximately normal shape with mean 0 (no difference) and standard deviation 3.288. This matches well with the standard error we used when calculating the *t* statistic in the preceding example: 3.2878.

In the actual experiment, the difference between the mean decreases in systolic blood pressure in the calcium and placebo groups was $5.000 - (-0.273) = 5.273$. How likely is it that a difference this large or larger would happen purely by chance when H_0 is true? The dotplot provides a rough answer: 64 of the 1000 random reassignments (indicated by the red dots) yielded a difference in means greater than or equal to 5.273. That is, our estimate of the *P*-value is 0.064. This is quite close to the 0.0644 *P*-value that we calculated in the example, suggesting that it's OK to use inference methods for random sampling to analyze randomized experiments.

OTHER OPTIONS FOR A TWO-SAMPLE *t* TEST

In Section 7C, we discussed two alternative approaches for calculating a two-sample t interval for $\mu_1 - \mu_2$. These same options are available when carrying out a two-sample t test for a difference in means.

1. **The Conservative Approach** Instead of using technology to find the appropriate degrees of freedom for two-sample t procedures, this approach uses df = the *smaller* of $n_1 - 1$ and $n_2 - 1$. For the calcium and blood pressure experiment, the conservative df = smaller of $(10-1,\ 11-1) = 9$. The resulting P-value with $t = 1.604$ and df $= 9$ is 0.0716. In the example, the P-value calculated with technology using df $= 15.59$ is 0.0644. Notice that technology gives smaller, more accurate P-values for two-sample t tests than the conservative approach does. That's because calculators and software use the more complicated formula on page 695 to obtain a larger number of degrees of freedom.
2. **The Pooled Two-Sample *t* Procedures** This option assumes that the two population distributions have equal variances, and combines (*pools*) the two sample variances to estimate the common population variance σ^2. The formula for the pooled estimate of σ^2 is

$$s_p^2 = \frac{(n_1-1)s_1^2 + (n_2-1)s_2^2}{(n_1-1)+(n_2-1)}$$

   If the two population variances are equal and both population distributions are close to normal, we can use a t distribution with df $= (n_1 - 1) + (n_2 - 1) = n_1 + n_2 - 2$ and s_p^2 to perform our calculations. This method offers more degrees of freedom than the technology approach (df $= 10 + 11 - 2 = 19$ when pooling versus 15.59 with technology in the calcium and blood pressure example), which would lead to a smaller P-value. However, it may not be safe to assume that the population variances are equal because the two sample standard deviations are somewhat different: $s_C = 8.743$ versus $s_P = 5.901$. Our advice is the same as in Section 7C: *don't use pooled two-sample t procedures unless a statistician says it is OK to do so.*

Paired Data or Two Samples?

In Sections 7C and 7D, we used two-sample t procedures to perform inference about the difference $\mu_1 - \mu_2$ between two population means. These methods require data that come from *independent* random samples from the populations of interest or from two groups in a completely randomized experiment.

In Sections 7A and 7B, we use paired t procedures to perform inference about the population mean difference μ_{Diff}. Recall that paired data come from recording the same quantitative variable twice for each individual or from recording the same quantitative variable for each of two similar individuals. Sometimes paired data are called "dependent samples" to distinguish them from "independent samples."

AP® EXAM TIP

If you are presented with paired data on the AP® Statistics exam and you analyze them as if they were independent samples, you will lose credit. Because the choice between paired t procedures and two-sample t procedures depends on how the data were produced, make sure to read the question carefully! Here are some additional tips:

1. If the data sets have unequal sample sizes, the data can't be paired.
2. If the data table has a third row or column with labels for values in the other rows or columns, the data are likely paired. See pages 661 and 681 for examples.
3. Questions about paired data typically use the phrase "mean difference," whereas questions about two independent samples typically use the phrase "difference in means."

EXAMPLE

Are you all wet?
Paired data or two samples?

Skills 1.D, 1.E

Tammy616/E+/Getty Images

PROBLEM: In each of the following scenarios, decide whether you should use paired *t* procedures to perform inference about a mean difference or two-sample *t* procedures to perform inference about a difference in means. Explain your choice.

(a) Before exiting the water, scuba divers remove their fins. A maker of scuba equipment advertises a new style of fins that is supposed to be faster to remove. A consumer advocacy group suspects that the time to remove the new fins may be no different than the time required to remove the old fins, on average. Twenty experienced scuba divers are recruited to test the new fins. Each diver flips a coin to determine if they will wear the new fin on the left foot and the old fin on the right foot, or vice versa. The time to remove each type of fin is recorded for every diver.

(b) To study the health of aquatic life, scientists gathered a random sample of 60 White Piranha fish from a tributary of the Amazon River during one year. The average length of these fish was compared to the average length of a random sample of 82 White Piranha from the same tributary a decade ago.

(c) Can a wetsuit deter shark attacks? A researcher has designed a new wetsuit with color variations that are suspected to deter shark attacks. To test this idea, the researcher fills two identical drums with bait and covers one in the standard black neoprene wetsuit and the other in the new suit. Over a period of one week, the researcher selects 16 two-hour time periods and randomly assigns 8 of them to the drum in the black wetsuit. The other 8 are assigned to the drum with the new suit. During each time period, the appropriate drum is submerged in waters that sharks frequent, and the number of times a shark bites the drum is recorded.

SOLUTION:

(a) Paired *t* procedures. The data come from two measurements of the same variable (time to remove the fin) for each diver.

(b) Two-sample *t* procedures. The data come from independent random samples of White Piranha in two different years.

If the sample sizes are different, the data can't be paired.

(c) Two-sample *t* procedures. The data come from two groups in a randomized experiment, with each group consisting of 8 time periods in which a drum with a specific wetsuit (standard or new) was randomly assigned to be submerged.

If two drums—one with the standard suit and one with the new suit—were submerged at the same time, the data would be paired.

FOR PRACTICE, TRY EXERCISE 19

When designing an experiment to compare two treatments, a completely randomized design may not be the best option. A matched pairs design might be a better choice, as the following activity shows.

ACTIVITY Get your heart beating!

Are standing pulse rates higher, on average, than sitting pulse rates? In this activity, you will perform two experiments to try to answer this question.

Experiment 1: Completely randomized design

1. Your teacher will randomly assign half of the students in your class to stand and the other half to sit. Once the two groups have been formed, students should stand or sit as required. Then each student should measure their pulse and share their data anonymously.
2. Analyze the data for the completely randomized design. Make parallel dotplots and calculate the mean pulse rate for each group. Is there *some* evidence that standing pulse rates are higher, on average? Explain your answer.

Experiment 2: Matched pairs design

3. To produce paired data in this setting, each student should receive both treatments in a random order. Because each participant already sat or stood in Step 1, they just need to do the opposite now. Everyone should measure their pulse again in the new position. Then each person should calculate the difference (Standing − Sitting) in their pulse rate and share this value anonymously.
4. Analyze the data for the matched pairs design. Make a dotplot of these differences and calculate their mean. Is there *some* evidence that standing pulse rates are higher, on average? Explain your answer.
5. Which design provides more convincing evidence that standing pulse rates are higher, on average, than sitting pulse rates? Justify your answer.

A statistics class with 24 students performed the Get Your Heart Beating activity. Figure 7.7 shows a dotplot of the pulse rates for the class's completely randomized design. The mean pulse rate for the standing group is $\bar{x}_1 = 74.83$ bpm; the mean for the sitting group is $\bar{x}_2 = 68.33$ bpm. So the average pulse rate is 6.5 bpm higher in the standing group. However, the variability in pulse rates for the two groups creates a lot of overlap in the dotplots. A two-sample t test of H_0: $\mu_1 - \mu_2 = 0$ versus H_a: $\mu_1 - \mu_2 > 0$ yields $t = 1.42$ and a P-value of 0.09. These data do not provide convincing evidence that standing pulse rates are higher, on average, than sitting pulse rates for people like the students in this class.

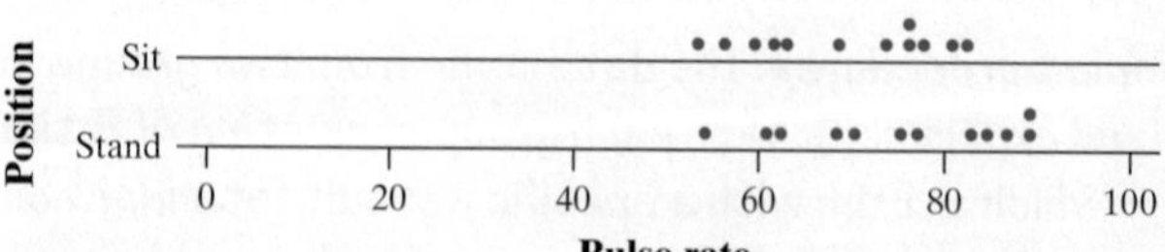

FIGURE 7.7 Parallel dotplots of the pulse rates for the standing and sitting groups in a statistics class's completely randomized experiment.

What about the class's matched pairs design? Figure 7.8 shows a dotplot of the difference (Standing − Sitting) in pulse rate for each of the 24 students. We can see that 21 of the 24 students recorded a positive difference, indicating that their standing pulse rate was higher. The mean difference is $\bar{x}_{\text{Diff}} = 6.83$ bpm. A paired t test of H_0: $\mu_{\text{Diff}} = 0$ versus H_a: $\mu_{\text{Diff}} > 0$ gives $t = 6.483$ and a P-value of approximately 0. These data provide *very* convincing evidence that standing pulse rates are higher, on average, than sitting pulse rates for people like the students in this class.

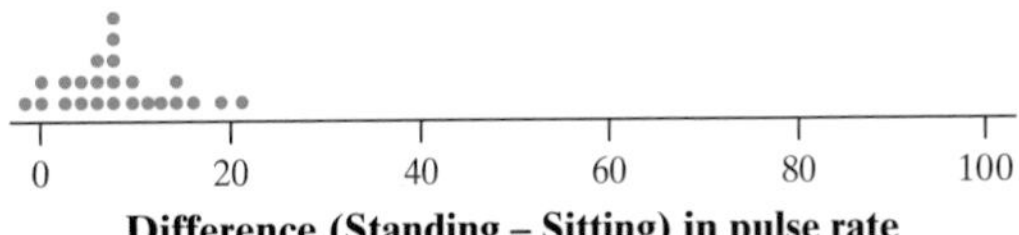

FIGURE 7.8 Dotplot of the difference in pulse rate (Standing – Sitting) for each student in a statistics class's matched pairs design.

Let's take one more look at Figures 7.7 and 7.8. Notice that we used the same scale for both graphs. The matched pairs design reduced the variability in the response variable by accounting for a big source of variability — the differences between individual students. That made it easier to detect the fact that standing causes an increase in the average pulse rate. In other words, using a paired (blocked) design resulted in more power. With the large amount of variability in pulse rates in the completely randomized design, we could not make such a conclusion.

USING TESTS WISELY

Significance tests are widely used in reporting the results of research in many fields. New drugs require convincing evidence of their effectiveness and safety before their manufacturers are allowed to market them. Courts ask about statistical significance when hearing discrimination cases. Marketers want to know whether a new ad campaign significantly outperforms the old one, and medical researchers want to know whether a new therapy performs significantly better than the existing treatments. In all these situations, statistical significance is valued because it points to an effect that is unlikely to occur by chance alone.

Carrying out a significance test is often straightforward, especially if you use technology. Using tests wisely is not so simple. Here are two points to keep in mind when using or interpreting significance tests.

1. Statistical significance is not the same thing as practical importance.

When a null hypothesis of no effect or no difference can be rejected at the usual significance levels ($\alpha = 0.10$ or $\alpha = 0.05$ or $\alpha = 0.01$), there is convincing evidence of a difference. But that difference may be very small. When large samples are used, even tiny deviations from the null hypothesis will be significant. Here's an illustrative example.

Researchers want to test a new antibacterial cream, "Formulation NS," to see how it affects the rate of healing in small cuts. Previous research shows that with no medication, the mean healing time, defined as the time for the scab to fall off, is 7.6 days. The researchers want to determine if Formulation NS speeds healing. They make minor cuts on a random sample of 250 college students who have given informed consent to participate in the study, and apply Formulation NS to the wounds. The researchers want to perform a test at the $\alpha = 0.05$ significance level of

$$H_0: \mu = 7.6$$
$$H_a: \mu < 7.6$$

where μ = the mean healing time (in days) in the population of college students whose cuts are treated with Formulation NS.

The mean healing time for these participants is $\bar{x} = 7.5$ days and the standard deviation is $s_x = 0.9$ day. After confirming that the conditions for inference are met, the researchers carry out a one-sample t test for μ, and find that $t = -1.757$ and P-value $= 0.0401$ with df $= 249$. Because 0.04 is less than $\alpha = 0.05$, we reject H_0. The researchers have convincing evidence that Formulation NS reduces the average healing time to less than 7.6 days in the population of college students whose cuts are treated with Formulation NS. In other words, the result is statistically significant. However, this result is not practically important because the observed sample mean was only $\bar{x} = 7.5$ days. Having scabs fall off one-tenth of a day sooner, on average, is no big deal!

The remedy for attaching too much importance to statistical significance is to give a confidence interval for the parameter in which you are interested. A confidence interval provides a set of plausible values for the parameter rather than simply asking if the observed result is unlikely to occur by chance alone when H_0 is true. For instance, the 90% confidence interval for the mean healing time in the population of college students whose cuts are treated with Formulation NS is 7.406 days to 7.594 days. This interval suggests that the population mean healing time μ is only slightly less than 7.6 days.

Confidence intervals are not used as often as they should be, whereas significance tests are perhaps overused. The remedy for attaching too much importance to statistical significance is to pay attention to the data as well as to the P-value. When planning a study, researchers should use a large enough sample size so the test will have adequate power to detect a *meaningful* difference from the null value.

2. Beware of multiple analyses!

Statistical significance should mean that you have found convincing evidence of a difference that you were looking for. The reasoning behind statistical significance works well if you decide what difference you are seeking, design a study to search for it, and use a significance test to weigh the evidence you get. In other settings, statistical significance may have little meaning. Here's one such example.

Might the radiation from cell phones be harmful to users? Many studies have found little or no connection between using cell phones and various illnesses. Here is part of a news account of one study:

> A hospital study that compared brain cancer patients and a similar group without brain cancer found no statistically significant difference in cell phone use for the two groups. But when 20 distinct types of brain cancer were considered separately, a significant difference in cell phone use was found for one rare type. Puzzlingly, however, this risk appeared to decrease rather than increase with greater mobile phone use.[46]

Think for a moment. Suppose that the 20 null hypotheses for these 20 significance tests are all true. Then each test has a 5% chance of making a Type I error — that is, of showing statistical significance at the $\alpha = 0.05$ level. That's what $\alpha = 0.05$ means: results this extreme occur only 5% of the time by chance alone when the null hypothesis is true. Therefore, we expect about 1 of 20 tests to give a significant result just by chance. Running one test and reaching the $\alpha = 0.05$ level is reasonably good evidence that you have found something; running 20 tests and reaching that level only once is not.

Searching data for patterns is a legitimate pursuit. Performing every conceivable significance test on a data set with many variables until you obtain a statistically significant result is not. This unfortunate practice is known by many names, including data dredging and *P*-hacking. To learn more about the pitfalls of multiple analyses, check out the XKCD comic about jelly beans causing acne (xkcd.com/882).

CHECK YOUR UNDERSTANDING

In a recent study at UCLA, researchers wanted to compare the amount of information recorded by students who write notes with pencil and paper and those who type notes on a laptop. They conducted an experiment in which 109 volunteer subjects were randomly assigned to one of these two strategies for taking notes. One of the variables they measured was the number of words written. The 55 students assigned to writing notes with pencil and paper had a mean of 390.7 words and a standard deviation of 143.9 words. The 54 students assigned to typing notes on a laptop had a mean of 548.7 words and a standard deviation of 252.7 words.[47] Researchers want to know if these data provide convincing evidence of a difference in the mean number of words that students like the ones in this study would hand write and the mean number of words that students like the ones in this study would type in their notes.

1. Should you use a paired t procedure or a two-sample t procedure to perform inference in this setting? Justify your answer.
2. Carry out the appropriate inference procedure to answer the researchers' question.

SECTION 7D Summary

- Tests for the difference $\mu_1 - \mu_2$ between the means of two populations or treatments are based on the difference $\bar{x}_1 - \bar{x}_2$ between the sample means.
- Before testing a claim about $\mu_1 - \mu_2$, we need to check for independence in the data collection process and verify that the sampling distribution of $\bar{x}_1 - \bar{x}_2$ is approximately normal. The required conditions are:
  - Random: The data come from two independent random samples or from two groups in a randomized experiment.
    - 10%: When sampling without replacement, $n_1 < 0.10N_1$ and $n_2 < 0.10N_2$.
  - Normal/Large Sample: For each sample, the data come from an approximately normally distributed population or the sample size is large ($n \geq 30$). For each sample, if the population distribution has unknown shape and $n < 30$, a graph of the sample data shows no strong skewness or outliers.
- To test H_0: $\mu_1 - \mu_2 = 0$, use a **two-sample t test for a difference in means.** The standardized test statistic is

$$t = \frac{(\bar{x}_1 - \bar{x}_2) - 0}{\sqrt{\dfrac{s_1^2}{n_1} + \dfrac{s_2^2}{n_2}}}$$

  P-values are calculated using the t distribution with degrees of freedom obtained by technology.

- Be sure to follow the four-step process whenever you perform a significance test for a difference between two population means.
- The proper inference method depends on how the data were produced. For paired data involving one quantitative variable, use paired t procedures to perform inference about μ_{Diff}. For quantitative data that come from independent random samples from two populations of interest or from two groups in a randomized experiment, use two-sample t procedures to perform inference about $\mu_1 - \mu_2$.
- Very small deviations from the null hypothesis can be highly significant (small P-value) when a test is based on a large sample. A statistically significant result may not be practically important.
- Many tests run at once will likely produce some significant results by chance alone, even if all the null hypotheses are true. Beware of P-hacking.

AP® EXAM TIP
AP® Daily Videos

Review the content of this section and get extra help by watching the AP® Daily Videos for Topics 7.8, 7.9, and 7.10, which are available in AP® Classroom.

7D Tech Corner

TI-Nspire and other technology instructions are on the book's website at bfwpub.com/tps7e.

25. Significance tests for a difference in means Page 711

SECTION 7D Exercises

Stating Hypotheses and Checking Conditions for a Test about $\mu_1 - \mu_2$

1. **Work hours** Has the mean number of hours worked in a week changed in the United States over time? One of the questions on the General Social Survey (GSS) asks respondents how many hours they work each week. Responses from random samples of U.S. employees in 1978 and 2018 are summarized in the table.[48]
pg 708

Year	n	Mean	SD
2018	1381	41.28	6.59
1978	855	40.81	8.40

Researchers want to know if there is a difference in the mean number of hours worked per week for all U.S. employees in 1978 and 2018.

(a) State appropriate hypotheses for performing a significance test. Be sure to define the parameters of interest.

(b) Check if the conditions for performing the test are met.

2. **Family income** How do incomes compare in Indiana and New Jersey? Here are dotplots and summary statistics of the total family income in separate random samples of 38 Indiana residents and 44 New Jersey residents.

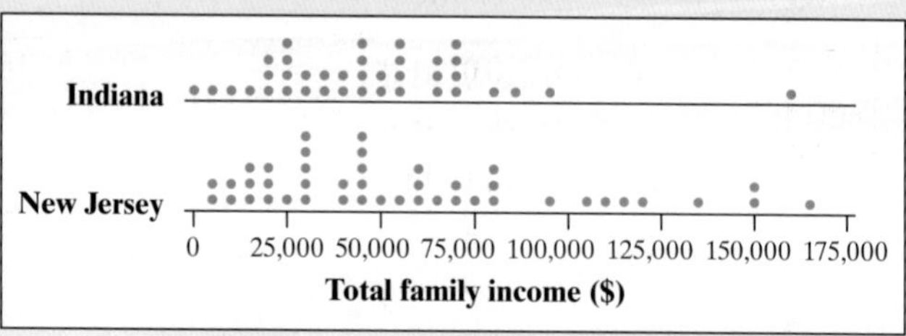

State	n	Mean	SD
Indiana	38	\$47,400	\$29,400
New Jersey	44	\$58,100	\$41,900

We want to know if there is a difference in the mean total family income of all Indiana residents and all New Jersey residents.

(a) State appropriate hypotheses for performing a significance test. Be sure to define the parameters of interest.

(b) Check if the conditions for performing the test are met.

3. **Fish oil** To see if fish oil can help reduce blood pressure, researchers recruited 14 men with high blood pressure and performed an experiment. They randomly assigned 7 of the men to a 4-week diet that included fish oil and the other 7 men to a 4-week diet that included a regular mixture of oils that approximated the types of fat in a typical diet. At the end of the 4 weeks, the researchers measured each volunteer's blood pressure (in mmHg) again and recorded the

reduction in diastolic blood pressure. These differences are shown in the table. Note that a negative value means that the subject's blood pressure increased.[49]

Fish oil	8	12	10	14	2	0	0
Regular oil	−6	0	1	2	−3	−4	2

Do these data provide convincing evidence at the 5% significance level that fish oil helps reduce blood pressure more than regular oil, on average, for men like the ones in this study?

(a) State appropriate hypotheses for the researchers' test. Be sure to define the parameters.

(b) Show that the conditions for performing the test in part (a) are met.

4. **Increasing tips** Does providing personalized service during a phone order for food increase tips, on average, compared to providing standard service? A researcher investigated this question with 48 customers who placed a food order on the phone at a local restaurant. The researcher randomly assigned 25 of the customers to receive personalized service at the end of the call, and the other 23 customers to receive the restaurant's standard service. When each customer came in to pick up the order, the researcher recorded the customer's tip as a percentage of the order's cost. Here are the percent tips for the customers in the two groups:[50]

Personalized service	9.3	12.9	0.0	19.3	14.6	0.0	18.4	0.0	10.3
	7.4	0.0	26.7	12.3	9.2	21.4	16.3	8.4	0.0
	18.3	26.3	10.1	19.3	12.3	9.2	21.4		
Standard service	3.3	10.7	11.5	10.2	7.0	0.0	0.0	10.2	0.0
	8.0	16.1	0.0	3.5	10.2	0.0	0.0	5.2	7.5
	11.8	10.2	7.0	0.0	0.0				

(a) State appropriate hypotheses for the researchers' test. Be sure to define the parameters.

(b) Show that the conditions for performing the test in part (a) are met.

Calculating the Standardized Test Statistic and *P*-Value for a Test about $\mu_1 - \mu_2$

5. **Changing work hours?** Refer to Exercise 1.

pg 711

(a) Calculate the standardized test statistic.

(b) Find the *P*-value.

(c) What conclusion would you make?

6. **More income?** Refer to Exercise 2.

(a) Calculate the standardized test statistic.

(b) Find the *P*-value.

(c) What conclusion would you make?

7. **More fish oil** Refer to Exercise 3.

(a) Explain why the sample results give some evidence for the alternative hypothesis.

(b) Calculate the standardized test statistic and *P*-value.

(c) What conclusion would you make?

8. **More tips** Refer to Exercise 4.

(a) Explain why the sample results give some evidence for the alternative hypothesis.

(b) Calculate the standardized test statistic and *P*-value.

(c) What conclusion would you make?

9. **Does breast-feeding weaken bones?** Breast-feeding mothers secrete calcium into their milk. Some of the calcium may come from their bones, so mothers may lose bone mineral. Researchers compared a random sample of 47 breast-feeding women with a random sample of 22 women of similar age who were neither pregnant nor lactating. They measured the percent change in the bone mineral content (BMC) of the women's spines over 3 months. Note that a negative percent change indicates a decrease in bone mineral content. Here are comparative boxplots of the data:[51]

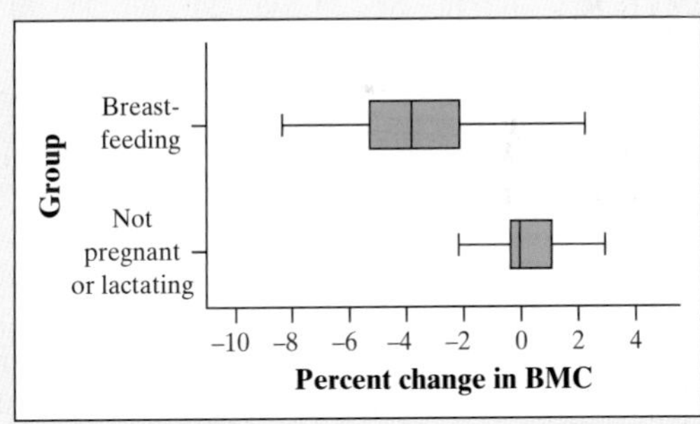

(a) Write a few sentences comparing the percent changes in BMC for the two groups.

After checking that the conditions for inference are met, the researchers perform a test of $H_0: \mu_{BF} - \mu_{NP} = 0$ versus $H_a: \mu_{BF} - \mu_{NP} < 0$, where μ_{BF} = the mean percent change in BMC for all breast-feeding women and μ_{NP} = the mean percent change in BMC for all women who are not pregnant or lactating. Computer output from the test is shown.

```
Two-sample T for BMC change
Group          N    Mean   StDev   SE Mean
Breastfeed    47   -3.59    2.51      0.37
Notpregnant   22    0.31    1.30      0.28
T-Value = -8.50    P-Value = 0.000    DF = 66
```

(b) What conclusion should the researchers make at the $\alpha = 0.05$ significance level?

(c) Can we conclude that breast-feeding causes a decrease in bone mineral content, on average? Why or why not?

(d) Based on your conclusion in part (b), which type of error — a Type I error or a Type II error — could you have made? Explain your answer.

10. **Teaching reading** An educator believes that new reading activities in the classroom will help elementary school students improve their reading ability. The educator recruits 44 third-grade students and randomly assigns them into two groups. One group of 21 students does these new activities for an 8-week period. A control group of 23 third-graders follows the same curriculum without the activities. At the end of the 8 weeks, all students are given the Degree of Reading Power (DRP) test, which measures the aspects of reading ability that the activities are designed to improve. Here are parallel boxplots of the DRP score data.[52]

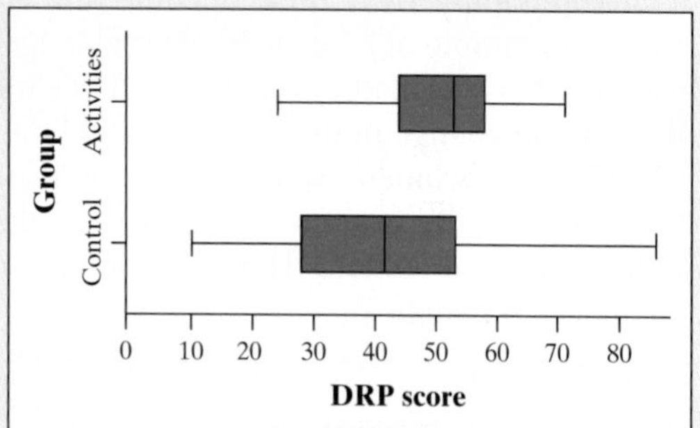

(a) Write a few sentences comparing the DRP scores for the two groups.

After checking that the conditions for inference are met, the educator performs a test of H_0: $\mu_A - \mu_C = 0$ versus H_a: $\mu_A - \mu_C > 0$, where μ_A = the true mean DRP score of third-graders like the ones in the study who do the new reading activities and μ_C = the true mean DRP score of third-graders like the ones in the study who follow the same curriculum without the activities. Computer output from the test is shown.

```
Two-sample T for DRP score
Group        N    Mean    StDev    SE Mean
Activities   21   51.5    11.0     2.4
Control      23   41.5    17.1     3.6

T-Value = 2.31  P-Value = 0.013  DF = 37
```

(b) What conclusion should the educator make at the $\alpha = 0.05$ significance level?

(c) Can the educator conclude that the new reading activities caused an increase in the mean DRP score? Explain your answer.

(d) Based on the conclusion in part (b), which type of error — a Type I error or a Type II error — could the educator have made? Explain your answer.

Two-Sample *t* Test for $\mu_1 - \mu_2$

11. pg 713 **Possum length** Does the size of Australian possums differ by sex? Zoologists in Australia captured a random sample of 104 possums and recorded the values of 9 different variables, including total length (in centimeters).[53] Due to the sampling method used in this survey, it is reasonable to view the male and female possums in this data set as independent random samples from the corresponding populations. The 43 female possums have a mean total length of 87.907 cm and a standard deviation of 4.182 cm. The 61 male possums have a mean total length of 86.511 cm and a standard deviation of 4.340 cm.

(a) Do the data provide convincing evidence at the $\alpha = 0.05$ significance level of a difference in the mean total lengths of all male and female Australian possums?

(b) Interpret the *P*-value you got in part (a).

12. **Gray squirrel** In many parts of the northern United States, two color variants of the Eastern Gray Squirrel — gray and black — are found in the same habitats. A scientist wonders if there is a difference in the sizes of the two color variants. The scientist collects separate random samples of 40 squirrels of each color from a large forest and weighs them. The 40 black squirrels have a mean weight of 20.3 ounces and a standard deviation of 2.1 ounces. The 40 gray squirrels have a mean weight of 19.2 ounces and a standard deviation of 1.9 ounces.

(a) Do these data provide convincing evidence at the $\alpha = 0.01$ significance level of a difference in the mean weights of all gray and black Eastern Gray Squirrels in this forest?

(b) Interpret the *P*-value you got in part (a).

13. **Rewards and creativity** Do external rewards — things like money, praise, fame, and grades — promote creativity? Researcher Teresa Amabile suspected that the answer is no, and that internal motivation enhances creativity. To find out, she recruited 47 experienced creative writers who were college students and divided them at random into two groups. The students in one group were given a list of statements about external reasons (E) for writing, such as public recognition, making money, or pleasing their parents. Students in the other group were given a list of statements about internal reasons (I) for writing, such as expressing yourself and enjoying playing with words. Both groups were then instructed to write a poem about laughter. Each student's poem was rated separately by 12 different poets using a creativity scale.[54] These ratings were averaged to obtain an overall creativity score for each

poem. Parallel dotplots and numerical summaries of the two groups' creativity scores are shown here.

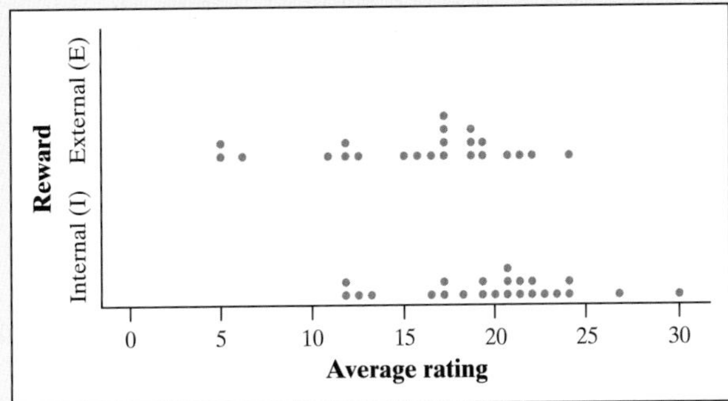

Group	n	Mean	SD
External	23	15.739	5.253
Internal	24	19.883	4.440

Do these data provide convincing evidence that giving internal reasons for writing increases the true mean creativity score compared to giving external reasons for writing among students like the ones in this study?

14. **Sleep deprivation** Does sleep deprivation linger for more than a day? Researchers designed a study using 21 volunteer subjects between the ages of 18 and 25. All 21 participants took a computer-based visual discrimination test at the start of the study. Then the subjects were randomly assigned into two groups. The 11 subjects in one group were deprived of sleep for an entire night in a laboratory setting. The 10 subjects in the other group were allowed unrestricted sleep for the night. Both groups were allowed as much sleep as they wanted for the next two nights. On day 4, all the subjects took the same visual discrimination test on the computer. Researchers recorded the improvement in time (measured in milliseconds) from day 1 to day 4 on each subject's tests. Here are the data:[55]

Sleep deprivation	−14.7	−10.7	−10.7	2.2	2.4	4.5
Unrestricted sleep	−7.0	11.6	12.1	12.6	14.5	18.6
Sleep deprivation	7.2	9.6	10.0	21.3	21.8	
Unrestricted sleep	25.2	30.5	34.5	45.6		

Do the data provide convincing evidence that sleep deprivation decreases the true mean improvement time on the visual discrimination task compared to unrestricted sleep for people like the volunteers in this study?

15. **More possum lengths** Refer to Exercise 11.

(a) Construct and interpret a 95% confidence interval for the difference in the population means. If you already defined parameters and checked conditions in Exercise 11, you don't need to do so again here.

(b) Explain how the confidence interval provides more information than the test in Exercise 11.

16. **More squirrels** Refer to Exercise 12.

(a) Construct and interpret a 99% confidence interval for the difference in the population means. If you already defined parameters and checked conditions in Exercise 12, you don't need to do so again here.

(b) Explain how the confidence interval provides more information than the test in Exercise 12.

17. **A better drug?** In a pilot study, a company's new cholesterol-reducing drug outperforms the currently available drug. If the data provide convincing evidence that the mean cholesterol reduction with the new drug is more than 10 milligrams per deciliter of blood (mg/dl) greater than with the current drug, the company will begin the expensive process of mass-producing the new drug. For the 14 subjects who were assigned at random to the current drug, the mean cholesterol reduction was 54.1 mg/dl with a standard deviation of 11.93 mg/dl. For the 15 subjects who were randomly assigned to the new drug, the mean cholesterol reduction was 68.7 mg/dl with a standard deviation of 13.3 mg/dl. Graphs of the data reveal no outliers or strong skewness.

Researchers want to perform a test of H_0: $\mu_{New} - \mu_{Cur} = 10$ versus H_a: $\mu_{New} - \mu_{Cur} > 10$.

(a) Calculate the standardized test statistic.

(b) Using df = 26.96, the *P*-value is 0.1675. Interpret the *P*-value. What conclusion would you make?

(c) Describe two ways to increase the power of the test.

18. **Down the toilet** A company that makes hotel toilets claims that its new pressure-assisted toilet reduces the average amount of water used by more than 0.5 gallon (gal) per flush when compared to its current model. To test this claim, the company randomly selects 30 toilets of each type and measures the amount of water that is used when each toilet is flushed once. For the current-model toilets, the mean amount of water used is 1.64 gal with a standard deviation of 0.29 gal. For the new toilets, the mean amount of water used is 1.09 gal with a standard deviation of 0.18 gal.

Researchers want to perform a test of H_0: $\mu_{Cur} - \mu_{New} = 0.5$ versus H_a: $\mu_{Cur} - \mu_{New} > 0.5$.

(a) Calculate the standardized test statistic.

(b) Using df = 48.46, the *P*-value is 0.2131. Interpret the *P*-value. What conclusion would you make?

(c) Describe two ways to increase the power of the test.

Paired Data or Two Samples?

19. **Tires, workers, and romance** In each of the following scenarios, decide whether you should use paired *t* procedures to perform inference about a mean difference or two-sample *t* procedures to

pg 717

perform inference about a difference in means. Explain your choice.[56]

(a) To test the wear characteristics of two tire brands, A and B, researchers randomly assign either four Brand A tires or four Brand B tires to each of 50 cars of the same make and model.

(b) To test the effect of background music on productivity, researchers observe factory workers. For one month, each subject works without music. For another month, each subject works while listening to music on an MP3 player. The month in which each subject listens to music is determined by a coin toss.

(c) How do young adults look back on adolescent romance? Investigators interviewed a random sample of 40 couples in their mid-20s. They interviewed each partner separately. They asked each participant about a romantic relationship that lasted at least 2 months when they were aged 15 or 16. One response variable was a measure on a numerical scale of how much the attractiveness of the adolescent partner mattered. Researchers want to find out how much the older and younger partner in a couple differ on this measure.

20. **Pigs, salaries, and fertilizer** In each of the following scenarios, decide whether you should use paired t procedures to perform inference about a mean difference or two-sample t procedures to perform inference about a difference in means. Explain your choice.[57]

(a) To compare the average weight gain of pigs fed two different diets, researchers used nine sets of pigs. The two pigs in each set were littermates. A coin toss was used to decide which pig in each pair got Diet A and which got Diet B.

(b) Researchers select separate random samples of professors at two-year colleges and four-year colleges. They compare the average salaries of professors at the two types of institutions.

(c) To test the effects of a new fertilizer, researchers treat 100 plots with the new fertilizer, and treat another 100 plots with another fertilizer. They use a computer's random number generator to determine which plots get which fertilizer.

Exercises 21 and 22 refer to the following scenario. Coaching companies claim that their courses can raise the SAT scores of high school students. Of course, the scores of students who retake the SAT without paying for coaching generally improve. A random sample of students who took the SAT twice included 427 who were coached and 2733 who were uncoached.[58] Here are summary statistics about their Verbal scores on the first and second tries:

		Try 1		Try 2		Gain	
	n	$\bar{x}$	s_x	$\bar{x}$	s_x	$\bar{x}$	s_x
Coached	427	500	92	529	97	29	59
Uncoached	2733	506	101	527	101	21	52

21. **Coaching and SAT scores** Do the scores of students who are coached increase significantly?

(a) You could use the information in the Coached row to carry out either a two-sample t test comparing Try 1 with Try 2 or a paired t test using Gain. Which is the correct test? Why?

(b) Carry out the proper test. What do you conclude?

22. **Coaching and SAT scores, again** What we really want to know is whether coached students improve more than uncoached students do, on average, and whether any advantage is large enough to be worth paying for.

(a) Carry out an appropriate test at the $\alpha = 0.05$ significance level to determine if there is convincing evidence that the SAT scores of coached students increase more, on average, than those of uncoached students.

(b) A 90% confidence interval for $\mu_{\text{Coached}} - \mu_{\text{Uncoached}}$ is (3.02, 12.98). Explain how this interval gives more information than the test in part (a).

(c) Based on your work, do you think SAT coaching courses are worth paying for? Why or why not?

For Investigation *Apply the skills from the section in a new context or nonroutine way.*

23. **Magnets and pain** Research has shown that magnetic fields can affect living tissue in humans. To investigate if magnets can be used to help patients with chronic pain, researchers conducted an experiment. A doctor identified a painful site on each patient and asked them to rate the pain on a scale from 0 (mild pain) to 10 (severe pain). Then, the doctor selected a sealed envelope containing a magnet at random from a box with a mixture of active and inactive magnets. The chosen magnet was applied to the site of the pain for 45 minutes. After being treated, each patient was again asked to rate the level of pain from 0 to 10 and the improvement in pain level was recorded.[59] Here are comparative boxplots of the data for the two groups.

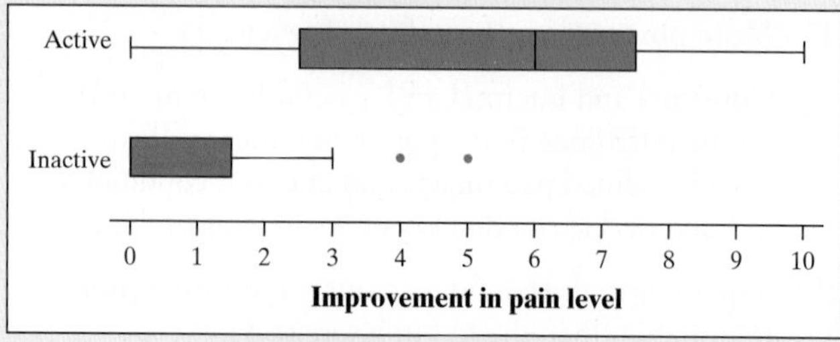

The mean improvement for 29 patients treated with active magnets is $\bar{x}_A = 5.24$ and the mean improvement for 21 patients treated with inactive magnets is $\bar{x}_I = 1.10$.

Researchers want to determine if these data provide convincing evidence that active magnets lead to a larger average improvement in pain level than inactive magnets for people with chronic pain like the ones in this study.

(a) State appropriate hypotheses for performing a significance test.

(b) Explain why it is not appropriate to perform a two-sample *t* test in this setting.

(c) Calculate the difference (Active − Inactive) in the mean improvement in pain for the two groups in the experiment.

We performed 500 trials of a simulation to see what differences in means would occur due only to chance variation in the random assignment, assuming that there is no treatment effect of using active magnets versus inactive magnets to treat chronic pain. The results are shown in the dotplot.

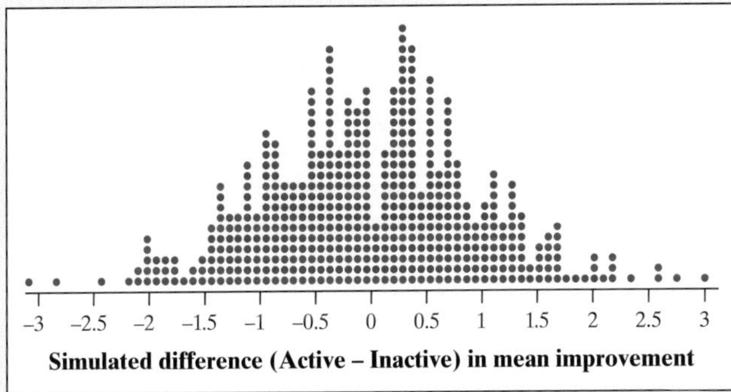

(d) Use the results of the simulation to make a conclusion about the hypotheses in part (a). Explain your reasoning.

Multiple Choice *Select the best answer for each question.*

24. Two methods are commonly used to measure the concentration of a pollutant in fish tissue. Do the two methods differ, on average? You apply both methods to each fish in a random sample of 18 carp. Assuming that the conditions for inference are met, which of the following significance tests should you perform?

(A) A paired *t* test for a population mean difference

(B) A one-sample *z* test for a population proportion

(C) A two-sample *t* test for a difference between two population means

(D) A two-sample *z* test for a difference between two population proportions

(E) None of these

Exercises 25–27 refer to the following scenario. A study of road rage asked random samples of 596 young adults (aged 25 or younger) and 523 senior citizens (aged 65 or older) about their behavior while driving. Based on their answers, each person was assigned a road rage score on a scale of 0 to 20. The participants were chosen by random digit dialing of phone numbers. The researchers performed a test of the following hypotheses: H_0: $\mu_{YA} = \mu_{SC}$ versus H_a: $\mu_{YA} \neq \mu_{SC}$.

25. Which of the following describes a Type II error in the context of this study?

(A) Finding convincing evidence that the population mean road rage scores are different for young adults and senior citizens, when in reality the population means are the same

(B) Finding convincing evidence that the population mean road rage scores are different for young adults and senior citizens, when in reality the population means are different

(C) Not finding convincing evidence that the population mean road rage scores are different for young adults and senior citizens, when in reality the population means are the same

(D) Not finding convincing evidence that the population mean road rage scores are different for young adults and senior citizens, when in reality the population means are different

(E) Not finding convincing evidence that the population mean road rage scores are different for young adults and senior citizens, when in reality there is convincing evidence that the population means are different

26. The *P*-value for the stated hypotheses is 0.002. Interpret this value in the context of this study.

(A) Assuming that the population mean road rage score is the same for young adults and senior citizens, there is a 0.002 probability of getting a difference in sample means equal to the one observed in this study.

(B) Assuming that the population mean road rage score is the same for young adults and senior citizens, there is a 0.002 probability of getting a difference in sample means at least as large in either direction as the one observed in this study.

(C) Assuming that the population mean road rage score is different for young adults and senior citizens, there is a 0.002 probability of getting a difference in sample means at least as large in either direction as the one observed in this study.

(D) Assuming that the population mean road rage score is the same for young adults and senior citizens, there is a 0.002 probability of incorrectly rejecting the null hypothesis.

(E) Assuming that the population mean road rage score is the same for young adults and senior citizens, there is a 0.002 probability that the sample means in the study will differ.

27. Based on the *P*-value in Exercise 26, which of the following must be true?

(A) A 90% confidence interval for $\mu_{YA} - \mu_{SC}$ will contain 0.

(B) A 95% confidence interval for $\mu_{YA} - \mu_{SC}$ will contain 0.

(C) A 99% confidence interval for $\mu_{YA} - \mu_{SC}$ will contain 0.

(D) A 99.9% confidence interval for $\mu_{YA} - \mu_{SC}$ will contain 0.

(E) It is impossible to determine whether any of these statements is true based only on the *P*-value.

Recycle and Review *Practice what you learned in previous sections.*

In each part of Exercises 28 and 29, state which inference procedure from Unit 6 or Unit 7 you would use. Be specific — for example, you might say, "Two-sample z test for a difference in proportions." You do not have to carry out any procedures.

28. **Which inference method?**

(a) Drowning in bathtubs is a major cause of death in children younger than age 5 years. Researchers asked a random sample of parents many questions related to bathtub safety. Overall, 85% of the sample said they used baby bathtubs for infants. What percentage of all parents of infants use baby bathtubs?

(b) How seriously do people view speeding in comparison with other annoying behaviors? A large random sample of adults was asked to rate a number of behaviors on a scale of 1 (no problem at all) to 5 (very severe problem). Do speeding drivers get a higher average rating than noisy neighbors?

(c) How large is the difference in the percentage of students from rural backgrounds among college graduates and those who attend college but do not graduate? To investigate this question, researchers ask a random sample of students who graduated from college within 7 years and a separate random sample of students who failed to graduate from college within 7 years if they had a rural background.

(d) Do experienced video game players earn higher scores, on average, when they play with someone present to cheer them on or when they play alone? Fifty teenagers with experience playing a particular video game have volunteered for a study. We randomly assign 25 of them to play the game alone and the other 25 to play the game with a supporter present. Each player's score is recorded.

29. **Which inference method, again?**

(a) A city planner wants to determine if there is convincing evidence of a difference in the average number of cars passing through two different intersections. The city planner randomly selects 12 times between 6:00 A.M. and 10:00 P.M. Volunteers record the number of cars passing through each of the two intersections during the 10-minute interval that begins at that time.

(b) Are more than 75% of Toyota owners generally satisfied with their vehicles? Let's design a study to find out. We'll select a random sample of 400 Toyota owners. Then we'll ask each individual in the sample, "Would you say that you are generally satisfied with your Toyota vehicle?"

(c) Are first-year college students more likely to binge drink than college students who have completed at least one year? The Harvard School of Public Health surveys random samples of first-year college students and college students who have completed at least one year about whether they have engaged in binge drinking.

(d) A bank wants to know which of two incentive plans, A or B, is most likely to increase the use of its credit cards and by how much. The bank offers incentive A to one group of current credit card customers and incentive B to another group of current credit card customers, with the offers determined at random. The amount charged by customers in the two groups during the following 6 months is compared.

UNIT 7 Wrap-Up

FRAPPY! Free Response AP® Problem, Yay!

Directions Show all your work. Indicate clearly the methods you use, because you will be scored on the correctness of your methods as well as on the accuracy and completeness of your results and explanations.

Will using name-brand microwave popcorn result in a greater percentage of popped kernels than using store-brand microwave popcorn? To find out, Briana and Maggie randomly selected 10 bags of name-brand microwave popcorn and 10 bags of store-brand microwave popcorn. The chosen bags were arranged in a random order. Then each bag was popped for 3.5 minutes, and the percentage of popped kernels was calculated. The results are displayed in the following table.

Name-brand	95	88	84	94	81	90	97	93	91	86
Store-brand	91	89	82	82	77	78	84	86	86	90

Do the data provide convincing evidence that the mean percentage of popped kernels is greater for all bags of name-brand microwave popcorn than for all bags of store-brand microwave popcorn?

After you finish the FRAPPY!, you can view two example solutions on the book's website (**bfwpub.com/tps7e**). Determine whether you think each solution is "complete," "substantial," "developing," or "minimal." If the solution is not complete, what improvements would you suggest to the student who wrote it? Finally, your teacher will provide you with a scoring rubric. Score your response and note what, if anything, you would do differently to improve your own score.

UNIT 7 REVIEW

SECTION 7A Confidence Intervals for a Population Mean or Mean Difference

In this section, you learned how to construct and interpret confidence intervals for a population mean μ. To verify independence in data collection and that the sampling distribution of $\bar{x}$ is approximately normal, we check three conditions. The Random condition says that the data come from a random sample from the population of interest. The 10% condition says that the sample size must be less than 10% of the population size when sampling without replacement. The Normal/Large Sample condition says that the population is approximately normally distributed or the sample size is at least 30. If the population distribution's shape is unknown and the sample size is less than 30, graph the sample data and check for strong skewness or outliers. If there is no strong skewness or outliers, it is reasonable to assume that the population distribution is approximately normal.

Because the population standard deviation σ is usually unknown, we use the sample standard deviation s_x to estimate σ when constructing a confidence interval for a population mean μ. Doing so requires use of a **t distribution** with $n-1$ *degrees of freedom* (df) to calculate a t^* critical value rather than the standard normal distribution. A t distribution is described by a symmetric, single-peaked, bell-shaped density curve centered at 0. Any t distribution is completely specified by its degrees of freedom (df) and has more area in its tails than the standard normal distribution.

The formula for calculating a **one-sample t interval for a mean** is

$$\bar{x} \pm t^* \frac{s_x}{\sqrt{n}}$$

where $\bar{x}$ is the sample mean, t^* is the critical value, s_x is the sample standard deviation, and n is the sample size. To find the critical value, use technology or Table B to determine the values of $-t^*$ and t^* that capture the middle C% of a t distribution with df $= n - 1$, where C is the confidence level.

In this section, you also learned how to analyze **paired data,** which result from measuring the same quantitative variable twice for each individual or once for each of two very similar individuals. Start by finding the difference between the values in each pair. Then make a graph of the differences. Use the mean difference $\bar{x}_{\text{Diff}}$ and the standard deviation of the differences s_{Diff} as summary statistics.

Three conditions need to be verified before calculating a confidence interval for a population mean difference μ_{Diff}. The Random condition says that paired data must come from a random sample from the population of interest or from a randomized experiment. The 10% condition says that the sample size n_{Diff} should be less than 10% of the size of the corresponding population of differences N_{Diff} when sampling without replacement. The Normal/Large Sample condition says that the population distribution of differences is approximately normal or the sample size is large ($n_{\text{Diff}} \geq 30$). If the number of differences is small and the population distribution's shape is unknown, graph the difference values to make sure there is no strong skewness or outliers.

Once you have computed the differences, calculations proceed as in a one-sample t interval for a population mean. Find the critical value t^* using a t distribution with df $= n_{\text{Diff}} - 1$. The formula for the **paired *t* interval for a mean difference** is

$$\bar{x}_{\text{Diff}} \pm t^* \frac{s_{\text{Diff}}}{\sqrt{n_{\text{Diff}}}}$$

Always use the four-step process whenever you are asked to construct and interpret a confidence interval for a population mean μ or a population mean difference μ_{Diff}.

The logic of confidence intervals, including how to interpret the confidence interval and the confidence level, is the same as in Unit 6. Likewise, you can use a confidence interval for μ or μ_{Diff} to evaluate a claim about the population mean or mean difference. As with other intervals, increasing the confidence level leads to a larger margin of error and increasing the sample size leads to a smaller margin of error, assuming other things remain the same.

SECTION 7B Significance Tests for a Population Mean or Mean Difference

In this section, you learned how to perform a significance test about a population mean μ. As in any test, you start by stating hypotheses. The null hypothesis is H_0: $\mu = \mu_0$. The alternative hypothesis can be one-sided (< or >) or two-sided ($\neq$).

Next, we check three conditions to verify independence in data collection and confirm that the sampling distribution of $\bar{x}$ is approximately normal. The Random, 10%, and Normal/Large Sample conditions for significance tests about μ are the same as the ones for confidence intervals for a population mean.

For a test of H_0: $\mu = \mu_0$ when the population standard deviation σ is unknown, the standardized test statistic is

$$t = \frac{\bar{x} - \mu_0}{\frac{s_x}{\sqrt{n}}}$$

When the conditions are met, the *P*-value can be obtained from a t distribution with df $= n - 1$. Use the *P*-value to make an appropriate conclusion about the hypotheses, in context. This inference method is called a **one-sample *t* test for a mean.**

In this section, you also learned how to perform a significance test about a population mean difference μ_{Diff} in the special case of paired data. It is very similar to the one-sample t test for a population mean except that you use the difference in each pair as the sample data. The null hypothesis is usually H_0: $\mu_{\text{Diff}} = 0$. The Random, 10%, and Normal/Large Sample conditions are the same for significance tests about μ_{Diff} as for confidence intervals about a population mean difference.

For a test of H_0: $\mu_{\text{Diff}} = 0$, the standardized test statistic is

$$t = \frac{\bar{x}_{\text{Diff}} - 0}{\frac{s_{\text{Diff}}}{\sqrt{n_{\text{Diff}}}}}$$

When the conditions are met, the *P*-value can be obtained from a t distribution with df $= n_{\text{Diff}} - 1$. This inference method is called a **paired *t* test for a mean difference.**

Whenever you are asked if there is convincing evidence for a claim about a population mean or mean difference, be sure to follow the four-step process.

A significance test for a population mean or mean difference uses the same logic as the significance tests for a population proportion from Section 6D. Likewise, the way we interpret *P*-values and power and the way we describe Type I and Type II errors are very similar to what you learned in Unit 6.

SECTION 7C Confidence Intervals for a Difference in Population Means

In this section, you learned how to construct confidence intervals for the difference $\mu_1 - \mu_2$ between the means of two populations or treatments. To verify independence in data collection and confirm that the sampling distribution of $\bar{x}_1 - \bar{x}_2$ is approximately normal, we check three conditions. The Random condition says that the data must be from two independent random samples or two groups in a randomized experiment. The 10% condition says that each sample size should be less than 10% of the corresponding population size when sampling without replacement. The Normal/Large Sample condition says that for each sample, the data come from an approximately normally distributed population or the sample size is large ($n \geq 30$). For each sample, if the population distribution has unknown shape and $n < 30$, confirm that a graph of the sample data shows no strong skewness or outliers.

Because we rarely know the two population standard deviations, we use a t distribution with degrees of freedom calculated by technology to determine the critical value t^* for a given confidence level C. Be sure to choose the technology's unpooled option when constructing a confidence interval for a difference between two means. Also, note that the df is usually not a whole number. The formula for the **two-sample *t* interval for a difference in means** is

$$(\bar{x}_1 - \bar{x}_2) \pm t^* \sqrt{\frac{s_1^2}{n_1} + \frac{s_2^2}{n_2}}$$

The logic of confidence intervals, including how to interpret the confidence interval and the confidence level, is the same as when estimating a single population mean. Likewise, you can use a confidence interval for a difference in means to evaluate claims about the population means. For example, if 0 is not included in a confidence interval for $\mu_1 - \mu_2$, there is convincing evidence that the population means are different.

SECTION 7D Significance Tests for a Difference in Population Means

In this section, you learned how to perform significance tests for the difference between the means of two populations or treatments. As in any test, you start by stating hypotheses. The null hypothesis is usually $H_0: \mu_1 - \mu_2 = 0$ (or equivalently, $H_0: \mu_1 = \mu_2$). The alternative hypothesis can be one-sided ($<$ or $>$) or two-sided ($\neq$).

Next, we need to verify independence in data collection and confirm that the sampling distribution of $\bar{x}_1 - \bar{x}_2$ is approximately normal. The Random, 10%, and Normal/Large Sample conditions for significance tests about a difference in means $\mu_1 - \mu_2$ are the same as the conditions for confidence intervals about a difference in means.

For a test of $H_0: \mu_1 - \mu_2 = 0$, the standardized test statistic is

$$t = \frac{(\bar{x}_1 - \bar{x}_2) - 0}{\sqrt{\frac{s_1^2}{n_1} + \frac{s_2^2}{n_2}}}$$

When the conditions are met, *P*-values can be obtained using a t distribution with degrees of freedom calculated by technology. This inference method is called a **two-sample *t* test for a difference in means.** Be sure to choose the technology's unpooled option when performing a significance test for a difference between two population means.

To decide whether a paired t procedure for a population mean difference μ_{Diff} or a two-sample t procedure for a difference between two means $\mu_1 - \mu_2$ is appropriate in a given setting, consider how the data were produced.

Remember to use significance tests wisely. Also, remember that statistically significant results aren't always practically important. Finally, be aware that the probability of making at least one Type I error goes up dramatically when conducting multiple tests.

Inference for Means

	Population mean	Population mean difference	Difference in population means
Parameter	μ	μ_{Diff}	$\mu_1 - \mu_2$
Conditions	• **Random:** The data come from a random sample from the population of interest. ◦ **10%:** When sampling without replacement, $n < 0.10N$. • **Normal/Large Sample:** The population distribution is approximately normal or the sample size is large ($n \geq 30$). If the population distribution has unknown shape and $n < 30$, a graph of the sample data shows no strong skewness or outliers.	• **Random:** Paired data come from a random sample from the population of interest or from a randomized experiment. ◦ **10%:** When sampling without replacement, $n_{\text{Diff}} < 0.10N_{\text{Diff}}$. • **Normal/Large Sample:** The population distribution of differences is approximately normal or the sample size is large ($n_{\text{Diff}} \geq 30$). If the population distribution of differences has unknown shape and $n_{\text{Diff}} < 30$, a graph of the sample differences shows no strong skewness or outliers.	• **Random:** The data come from two independent random samples or from two groups in a randomized experiment. ◦ **10%:** When sampling without replacement, $n_1 < 0.10N_1$ and $n_2 < 0.10N_2$. • **Normal/Large Sample:** For each sample, the data come from an approximately normally distributed population or the sample size is large ($n \geq 30$). For each sample, if the population distribution has unknown shape and $n < 30$, a graph of the sample data shows no strong skewness or outliers.

Inference for Means

Confidence Intervals

Parameter	μ	μ_{Diff}	$\mu_1 - \mu_2$
Name (TI-83/84)	One-sample *t* interval for a mean (TInterval)	Paired *t* interval for a mean difference (TInterval)	Two-sample *t* interval for a difference in means (2-SampTInt)
Formula	$\bar{x} \pm t^* \frac{s_x}{\sqrt{n}}$ Critical value from *t* distribution with df $= n - 1$	$\bar{x}_{Diff} \pm t^* \frac{s_{Diff}}{\sqrt{n_{Diff}}}$ Critical value from *t* distribution with df $= n_{Diff} - 1$	$(\bar{x}_1 - \bar{x}_2) \pm t^* \sqrt{\frac{s_1^2}{n_1} + \frac{s_2^2}{n_2}}$ Critical value from *t* distribution with df from technology

Significance Tests

Parameter	μ	μ_{Diff}	$\mu_1 - \mu_2$
Name (TI-83/84)	One-sample *t* test for a mean (T-Test)	Paired *t* test for a mean difference (T-Test)	Two-sample *t* test for a difference in means (2-SampTTest)
Null hypothesis	H_0: $\mu = \mu_0$	H_0: $\mu_{Diff} = 0$	H_0: $\mu_1 - \mu_2 = 0$
Formula	$t = \frac{\bar{x} - \mu_0}{\frac{s_x}{\sqrt{n}}}$ *P*-value from *t* distribution with df $= n - 1$	$t = \frac{\bar{x}_{Diff} - 0}{\frac{s_{Diff}}{\sqrt{n_{Diff}}}}$ *P*-value from *t* distribution with df $= n_{Diff} - 1$	$t = \frac{(\bar{x}_1 - \bar{x}_2) - 0}{\sqrt{\frac{s_1^2}{n_1} + \frac{s_2^2}{n_2}}}$ *P*-value from *t* distribution with df from technology

What Did You Learn?

Learning Target	Section	Related Example on Page(s)	Relevant Chapter Review Exercise(s)
Determine the critical value t^* for calculating a confidence interval for a population mean.	7A	650	R1
Check the conditions for calculating a confidence interval for a population mean.	7A	652	R1
Calculate a confidence interval for a population mean.	7A	654	R1
Construct and interpret a one-sample *t* interval for a mean.	7A	656	R1
In the special case of paired data, construct and interpret a confidence interval for a population mean difference.	7A	661	R2
State appropriate hypotheses and check the conditions for performing a significance test about a population mean.	7B	671	R3
Calculate the standardized test statistic and *P*-value for a test about a population mean.	7B	675	R3
Perform a one-sample *t* test for a population mean.	7B	678	R3
In the special case of paired data, perform a significance test about a population mean difference.	7B	681	R4
Check the conditions for calculating a confidence interval for a difference between two population means.	7C	693	R5
Calculate a confidence interval for a difference between two population means.	7C	697	R5

Learning Target	Section	Related Example on Page(s)	Relevant Chapter Review Exercise(s)
Construct and interpret a two-sample *t* interval for a difference in means.	7C	699	R5
State appropriate hypotheses and check the conditions for performing a significance test about a difference between two population means.	7D	708	R6
Calculate the standardized test statistic and *P*-value for a test about a difference between two population means.	7D	711	R6
Perform a two-sample *t* test for a difference in means.	7D	713	R6
Determine when it is appropriate to use paired *t* procedures versus two-sample *t* procedures.	7D	717	R2, R6

UNIT 7 REVIEW EXERCISES

These exercises are designed to help you review the important concepts and skills of the unit.

R1 **Engine parts (7A)** A random sample of 16 of the more than 200 auto engine crankshafts produced in one day was selected. Here are measurements of the length (in millimeters) of a critical component on these crankshafts, along with a boxplot and numerical summaries of the data:

224.120	224.001	224.017	223.982	223.989	223.961
223.960	224.089	223.987	223.976	223.902	223.980
224.098	224.057	223.913	223.999		

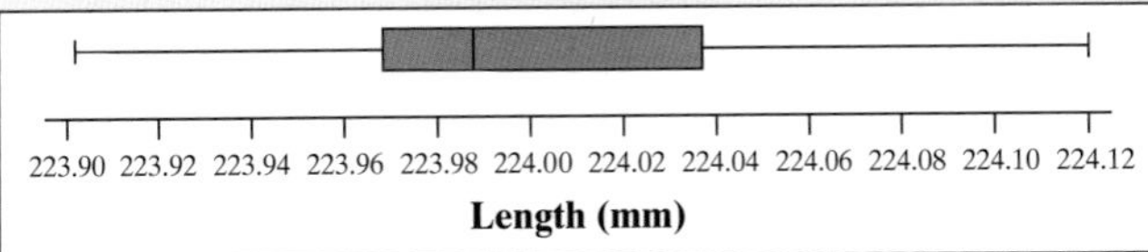

n	Mean	SD
16	224.0019	0.0618

(a) Construct and interpret a 95% confidence interval for the mean length of this component on all the crankshafts produced on that day.

(b) The mean length is supposed to be $\mu = 224$ mm but can drift away from this target during production. Does your interval from part (a) provide convincing evidence that the mean has drifted from 224 mm? Explain your answer.

R2 **No annual fee? (7A)** A bank wonders whether eliminating the annual credit card fee for customers who charge at least $2400 in a year will increase the amount charged on its credit cards. The bank makes this change for a random sample of 200 of its credit card customers. It then compares these customers' charges for this year with their charges for last year. The mean increase in the sample is $332, and the standard deviation is $108. Bank managers would like to use these data to calculate a 99% confidence interval.

(a) Define the parameter of interest.

(b) Explain why you should use paired *t* procedures to calculate a confidence interval for the parameter in part (a).

(c) Check that the conditions for inference are met.

(d) The 99% confidence interval is (312.14, 351.86). Find the critical value t^*.

(e) Based on the confidence interval, can bank managers conclude that eliminating the annual fee would cause an increase in the average amount spent by this bank's credit card customers? Why or why not?

R3 **Icebreaker? (7B)** In the children's game Don't Break the Ice, small plastic ice cubes are squeezed into a square frame. Each child takes turns tapping out a cube of "ice" with a plastic hammer, hoping that the remaining cubes don't collapse. For the game to work correctly, the cubes must be big enough so that they hold each other in place in the plastic frame, but not so big that they are too difficult to tap out. The machine that produces the plastic cubes is designed to make cubes that are 30 millimeters (mm) wide, but the width varies a little. To ensure that the machine is working well, a supervisor inspects a random sample of 50 cubes from each hour of production and measures their width. If the sample provides convincing evidence at the $\alpha = 0.05$ significance level that the mean width μ of all the cubes produced that hour differs from 30 mm, the supervisor will discard all of the cubes.

For the random sample of 50 cubes from the most recent hour of production, the sample mean width is 30.02 mm and the standard deviation is $s_x = 0.08$ mm.

(a) Carry out an appropriate test to determine whether the supervisor should discard all of the cubes produced in this hour.

(b) Describe a Type II error in this setting, and give a possible consequence of making this error.

(c) Identify one benefit and one drawback of changing the significance level to $\alpha = 0.10$.

(d) A 95% confidence interval for the mean width of the cubes produced in the last hour is 29.997 mm to 30.043 mm. Explain how this interval gives more information than the test in part (a).

R4 **On your mark (7B)** In track events, sprinters typically use starting blocks because they think it will help them run a faster race. To test this belief, an experiment was designed where each sprinter on a track team ran a 50-meter dash two times, once using starting blocks and once with a standing start. The order of the two different types of starts was determined at random for each sprinter. The times (in seconds) for 8 different sprinters are shown in the table.

Sprinter	With blocks	Standing start
1	6.12	6.38
2	6.42	6.52
3	5.98	6.09
4	6.80	6.72
5	5.73	5.98
6	6.04	6.27
7	6.55	6.71
8	6.78	6.80

(a) Do these data provide convincing evidence that sprinters like these run a faster race when using starting blocks, on average?

(b) Interpret the *P*-value from part (a) in context.

R5 **Water Fleas (7C)** *Daphnia pulicaria* is a water flea — a small crustacean that lives in lakes and is a major food supply for many species of fish. When fish are present in the lake water, they release chemicals called kairomones that induce water fleas to grow long tail spines that make them more difficult for the fish to eat. One study of this phenomenon compared the relative length of tail spines in two populations of *D. pulicaria*: one grown when kairomones were present and one grown when they were not. Here are summary statistics on the relative tail spine lengths, measured as a percentage of the entire length of the water flea, for separate random samples from the two populations:[60]

Group	*n*	Mean	SD
Fish kairomone present	214	37.26	4.68
Fish kairomone absent	152	30.67	4.19

(a) Construct and interpret a 99% confidence interval for the difference in the population mean relative tail spine lengths of *D. pulicaria* with fish kairomone present and *D. pulicaria* with fish kairomone absent.

(b) Interpret the 99% confidence level.

R6 **Subliminal messages (7D)** A "subliminal" message is below our threshold of awareness but may nonetheless influence us. Can subliminal messages help students learn math? A group of 18 students who had failed the mathematics part of the City University of New York Skills Assessment Test agreed to participate in a study to find out. All received a daily subliminal message, flashed on a screen too rapidly to be consciously read. The treatment group of 10 students (assigned at random) was exposed to a positive message, "Each day I am getting better in math." The control group of 8 students was exposed to a neutral message, "People are walking on the street." All 18 students participated in a summer program designed to improve their math skills, and all took the assessment test again at the end of the program. The following table gives data on the subjects' scores before and after the program, along with the differences (After – Before).[61]

Treatment group			Control group		
Pretest	Posttest	Difference	Pretest	Posttest	Difference
18	24	6	18	29	11
18	25	7	24	29	5
21	33	12	20	24	4
18	29	11	18	26	8
18	33	15	24	38	14
20	36	16	22	27	5
23	34	11	15	22	7
23	36	13	19	31	12
21	34	13			
17	27	10			

(a) Explain why a two-sample *t* test and not a paired *t* test is the appropriate inference procedure in this setting.

Here are comparative dotplots and summary statistics of the differences in pretest and posttest scores for the students in the control (C) and treatment (T) groups.

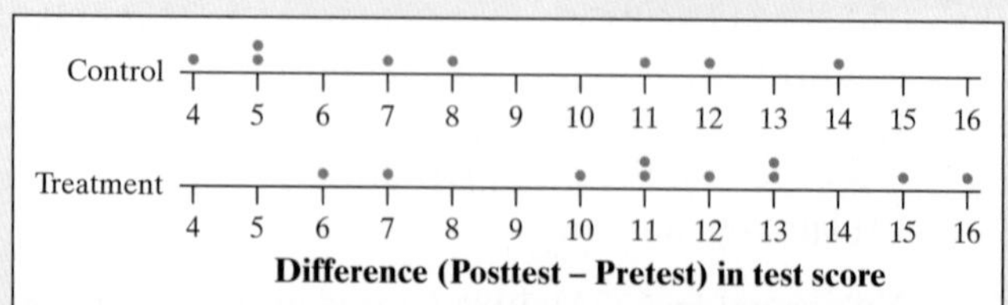

Group	*n*	Mean	SD	Min	Q_1	Med	Q_3	Max
Control	8	8.25	3.6936	4	5	7.5	11.5	14
Treatment	10	11.4	3.1693	6	10	11.5	13	16

(b) Explain why these data give some evidence that positive subliminal messages help students like the ones in this study learn math.

(c) Do these data provide convincing evidence at the $\alpha = 0.01$ significance level that positive subliminal messages help students like the ones in this study learn math, on average?

(d) Can we generalize the results of this study to the population of all students who failed the mathematics part of the City University of New York Skills Assessment Test? Why or why not?

UNIT 7 AP® STATISTICS PRACTICE TEST

Section I: Multiple Choice *Select the best answer for each question.*

T1 The weights (in pounds) of three adult goats are 160, 215, and 195. What is the standard error of the mean for these data?

(A) 13.13
(B) 16.07
(C) 22.73
(D) 27.84
(E) 190

T2 We want to construct a one-sample t interval for a population mean using data from a population distribution with unknown shape. In which of the following circumstances would it be inappropriate to construct the interval based on an SRS of size 14 from the population?

(A) A stemplot of the data is slightly skewed to the left and unimodal.
(B) A histogram of the data is approximately uniform.
(C) A boxplot of the data is moderately skewed to the right.
(D) A dotplot of the data is roughly bell-shaped, except for one high outlier.
(E) The population consists of 200 individuals.

T3 A researcher wished to compare the average amount of time spent on extracurricular activities by high school students in a large suburban school district with the average time spent on extracurricular activities by high school students in a large city school district. The researcher obtained a random sample of 60 high school students in the suburban school district and found the mean time spent on extracurricular activities per week to be 6 hours with a standard deviation of 3 hours. The researcher also obtained a random sample of 40 high school students in the city school district and found the mean time spent on extracurricular activities per week to be 5 hours with a standard deviation of 2 hours. Suppose that the researcher decides to carry out a significance test of H_0: $\mu_{\text{suburban}} = \mu_{\text{city}}$ versus a two-sided alternative. Which of the following is the correct standardized test statistic?

(A) $z = \dfrac{(6-5)-0}{\sqrt{\dfrac{3}{60}+\dfrac{2}{40}}}$

(B) $z = \dfrac{(6-5)-0}{\sqrt{\dfrac{3^2}{60}+\dfrac{2^2}{40}}}$

(C) $t = \dfrac{(6-5)-0}{\dfrac{3}{\sqrt{60}}+\dfrac{2}{\sqrt{40}}}$

(D) $t = \dfrac{(6-5)-0}{\sqrt{\dfrac{3}{60}+\dfrac{2}{40}}}$

(E) $t = \dfrac{(6-5)-0}{\sqrt{\dfrac{3^2}{60}+\dfrac{2^2}{40}}}$

T4 A significance test was performed to test H_0: $\mu = 2$ versus the alternative H_a: $\mu \neq 2$. A sample of size 28 produced a standardized test statistic of $t = -2.047$. Assuming all conditions for inference were met, which of the following intervals contains the P-value for this test?

(A) $0.01 < P\text{-value} < 0.02$
(B) $0.02 < P\text{-value} < 0.025$
(C) $0.025 < P\text{-value} < 0.05$
(D) $0.05 < P\text{-value} < 0.10$
(E) $P\text{-value} > 0.10$

T5 Which of the following has the smallest probability?

(A) $P(t > 2)$ in a t distribution with 5 degrees of freedom
(B) $P(t > 2)$ in a t distribution with 3 degrees of freedom
(C) $P(z > 2)$ in the standard normal distribution
(D) $P(t < 2)$ in a t distribution with 5 degrees of freedom
(E) $P(z < 2)$ in the standard normal distribution

T6 A Census Bureau report says that, with 90% confidence, the median income of all U.S. households in a recent year was \$70,784 with a margin of error of \$605. Which of the following statements is correct?

(A) 90% of all U.S. households had incomes in the interval $\$70,784 \pm \$605$.

(B) We can be sure that the median income for all households in the country lies in the interval $\$70,784 \pm \$605$.

(C) 90% of the households in the sample interviewed by the Census Bureau had incomes in the interval $\$70,784 \pm \$605$.

(D) The Census Bureau got the result $\$70,784 \pm \$605$ using a method that will capture the population median household income in 90% of all possible samples.

(E) 90% of all possible samples of this same size would result in a population median household income that falls within \$605 of \$70,784.

T7 Anne wants to determine how much better a store-brand fertilizer works as a soil enhancement rather than homemade compost when growing tomatoes. To investigate, she plants two tomato plants in each of five planters. One plant in each planter is grown in soil with store-brand fertilizer; the other plant is grown in soil with homemade compost, with the choice of soil determined at random. In three months, she will harvest and weigh the tomatoes from each plant. Assuming that the conditions for inference are met, which of the following is the correct confidence interval procedure that Anne should use for these data?

(A) Two-sample z interval for a difference between two population means

(B) Paired z interval for a population mean difference

(C) Two-sample t interval for a difference between two population means

(D) Paired t interval for a population mean difference

(E) Two-sample z interval for a difference between two population proportions

T8 A 95% confidence interval for μ based on a random sample from an approximately normal population distribution is $(-0.73, 1.92)$. If we use this confidence interval to perform a test of H_0: $\mu = 0$ against H_a: $\mu \neq 0$, which of the following is the most appropriate conclusion?

(A) Reject H_0 at the $\alpha = 0.05$ level of significance

(B) Fail to reject H_0 at the $\alpha = 0.05$ level of significance

(C) Reject H_0 at the $\alpha = 0.10$ level of significance

(D) Fail to reject H_0 at the $\alpha = 0.10$ level of significance

(E) We cannot perform the required test because we do not know the value of the standardized test statistic.

T9 Researchers want to evaluate the effect of a natural product on reducing blood pressure. They plan to carry out a randomized experiment to compare the mean reduction in blood pressure in a treatment (natural product) group and a placebo group. Then they will use the data to perform a test of H_0: $\mu_T - \mu_P = 0$ versus H_a: $\mu_T - \mu_P > 0$, where $\mu_T =$ the true mean reduction in blood pressure when taking the natural product and $\mu_P =$ the true mean reduction in blood pressure when taking a placebo for people like the ones in the experiment. The researchers would like to detect whether the natural product reduces blood pressure by at least 7 points more, on average, than the placebo. If groups of size 50 are used in the experiment, a two-sample t test using $\alpha = 0.01$ will have a power of 80% to detect a 7-point difference in mean blood pressure reduction. If the researchers want to be able to detect a 5-point difference instead, then the power of the test

(A) would be less than 80%.

(B) would be greater than 80%.

(C) would still be 80%.

(D) could be either less than or greater than 80%.

(E) would vary depending values of μ_T and μ_P.

T10 A 90% confidence interval for the mean μ of a population is computed from a random sample and is found to be 90 ± 30. Which of the following could be the 95% confidence interval based on the same data?

(A) 90 ± 21

(B) 90 ± 30

(C) 90 ± 39

(D) 90 ± 70

(E) Without knowing the sample size, any of the above answers could be the 95% confidence interval.

Section II: Free Response *Show all your work. Indicate clearly the methods you use, because you will be graded on the correctness of your methods as well as on the accuracy and completeness of your results and explanations.*

T11 A government report says that the average amount of money spent per U.S. household per week on food is \$158. A random sample of 50 households in a small city is selected, and their weekly spending on food is recorded. The sample data have a mean of \$165 and a standard deviation of \$32. Is there convincing evidence that the mean weekly spending on food in this city differs from the national figure of \$158?

(a) State appropriate hypotheses for performing a significance test in this setting. Be sure to define the parameter of interest.

(b) The distribution of household spending in this small city is heavily skewed to the right. Explain why the Normal/Large Sample condition is met in this case.

(c) The *P*-value of the test is 0.128. Interpret this value.

(d) What conclusion would you make?

T12 Researchers wondered whether maintaining a patient's body temperature close to normal by warming the patient during surgery would affect rates of infection of wounds. Patients were assigned at random to two groups: the normothermic group (core temperatures were maintained at near normal, 36.5°C, using heating blankets) and the hypothermic group (core temperatures were allowed to decrease to about 34.5°C). If keeping patients warm during surgery alters the chance of infection, patients in the two groups should show a difference in the average length of their hospital stays. Here are summary statistics on hospital stay (in number of days) for the two groups:[62]

Group	n	$\bar{x}$	s_x
Normothermic	104	12.1	4.4
Hypothermic	96	14.7	6.5

(a) Construct and interpret a 95% confidence interval for the difference in the true mean length of hospital stay for normothermic and hypothermic patients like the ones in this study.

(b) Does the interval in part (a) provide convincing evidence that keeping patients like the ones in this study warm during surgery affects the average length of their hospital stays? Justify your answer.

T13 Researchers investigated which line was faster in the supermarket: the express lane or the regular lane. They randomly selected 15 times during a week, went to the same store, and bought the same item. One of the researchers used the express lane and the other used the closest regular lane. To decide which lane each of them would use, they flipped a coin. They entered their lanes at the same time, paid in the same way, and recorded the time (in seconds) it took them to complete the transaction. The table displays the original data, along with the difference in times for each visit, and some summary statistics.[63]

Visit	Express lane	Regular lane	Difference (Regular − Express)
1	337	342	5
2	226	472	246
3	502	456	−46
4	408	529	121
5	151	181	30
6	284	339	55
7	150	229	79
8	357	263	−94
9	349	332	−17
10	257	352	95
11	321	341	20
12	383	397	14
13	565	694	129
14	363	324	−39
15	85	127	42
Mean	315.867	358.533	42.667
SD	129.764	141.266	84.019

(a) Do these data provide convincing evidence at the $\alpha = 0.05$ significance level that it is faster to use the express lane at this store than a regular lane, on average?

(b) Which type of error — a Type I error or a Type II error — could you have made in part (a)? Explain your answer.

Cumulative AP® Practice Test 3

Section I: Multiple Choice *Select the best answer for each question.*

AP1 National Park rangers keep data on the bears that inhabit their park. Here is a histogram of the weights of 143 bears measured in a recent year:

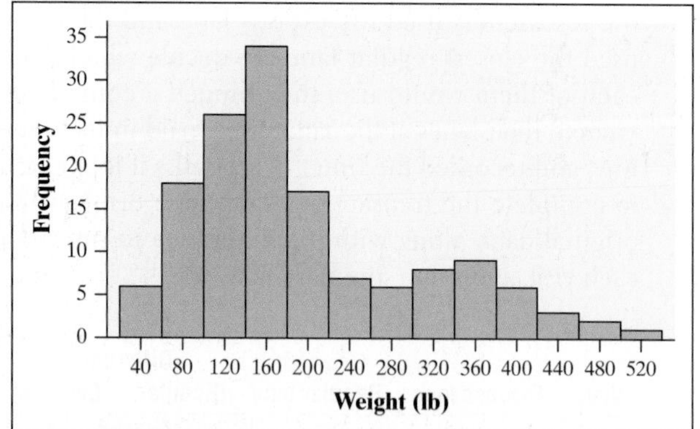

Which of the following statements could be correct?

(A) The median is between 140 and 180, and the mean is between 180 and 220.

(B) The median is between 140 and 180, and the mean is between 260 and 300.

(C) The median is between 100 and 140, and the mean is between 180 and 220.

(D) The median is between 260 and 300, and the mean is between 140 and 180.

(E) The median is between 180 and 220, and the mean is between 100 and 140.

AP2 The two-way table summarizes the responses of 120 people to a survey in which they were asked, "Do you exercise for at least 30 minutes 4 or more times per week?" and "What kind of vehicle do you drive?"

		Vehicle type		
		Sedan	SUV	Truck
Exercise?	Yes	25	15	12
	No	20	24	24

Suppose one person from this sample is randomly selected. What is the probability that the person exercises for at least 30 minutes 4 or more times per week given that they do not drive a truck?

(A) 0.298
(B) 0.333
(C) 0.476
(D) 0.700
(E) 0.769

AP3 A large university is considering establishing a schoolwide recycling program. To measure interest in the program by means of a questionnaire, the university administration takes separate random samples of undergraduate students, graduate students, faculty, and staff. This is an example of which type of sampling design?

(A) Simple random sample
(B) Stratified random sample
(C) Convenience sample
(D) Cluster random sample
(E) Systematic random sample

AP4 Which of the following would be a correct interpretation if your score on an exam taken by a large number of students corresponds to a z-score of $+2.0$?

(A) It means that you missed two questions on the exam.

(B) It means that you got twice as many questions correct as the average student.

(C) It means that your score was 2 points higher than the mean score on this exam.

(D) It means that you scored higher than about 97.5% of students on this exam.

(E) It means that your score is 2 standard deviations above the mean score on this exam.

AP5 A random sample of 200 New York State adults included 88 who said they love broccoli, while a random sample of 300 California adults included 141 who said they love broccoli. Which of the following represents the 95% confidence interval for the difference in the population proportion of adults in New York State and California who would say they love broccoli?

(A) $(0.44-0.47)\pm 1.96\left(\dfrac{(0.44)(0.56)+(0.47)(0.53)}{\sqrt{200+300}}\right)$

(B) $(0.44-0.47)\pm 1.96\left(\dfrac{(0.44)(0.56)}{\sqrt{200}}+\dfrac{(0.47)(0.53)}{\sqrt{300}}\right)$

(C) $(0.44-0.47)\pm 1.96\sqrt{\dfrac{(0.44)(0.56)}{200}+\dfrac{(0.47)(0.53)}{300}}$

(D) $(0.44-0.47)\pm 1.96\sqrt{\dfrac{(0.44)(0.56)+(0.47)(0.53)}{200+300}}$

(E) $(0.44-0.47)\pm 1.96\sqrt{(0.458)(0.542)\left(\dfrac{1}{200}+\dfrac{1}{300}\right)}$

AP6 Which of the following is *not* a property of a binomial setting?

(A) Outcomes of different trials are independent.

(B) The random process consists of a fixed number of trials, n.

(C) The probability of success is the same for each trial.

(D) Trials are repeated until a success occurs.

(E) Each trial can result in either a success or a failure.

AP7 An agricultural station is testing the yields for six different varieties of seed corn. The station has four large fields available, located in four distinctly different parts of the county. The agricultural researchers consider the climatic and soil conditions in the four parts of the county as being quite different, but are reasonably confident that the conditions within each field are fairly similar throughout. The researchers divide each field into six sections and then randomly assign a different variety of corn seed to each section in that field. At the end of the growing season, the corn will be harvested, and the yield (measured in tons per acre) will be compared. Which of the following statements about the study design is correct?

(A) This is an observational study because the researchers are watching the corn grow.

(B) This a randomized block design with fields as blocks and seed types as treatments.

(C) This is a randomized block design with seed types as blocks and fields as treatments.

(D) This is a completely randomized design because the six seed types were randomly assigned to the four fields.

(E) This is a completely randomized design with 24 treatments — 6 seed types and 4 fields.

AP8 According to sleep researchers, if you are between the ages of 12 and 18 years, you need 9 hours of sleep to function well. A simple random sample of 28 students was chosen from a large high school, and these students were asked how much sleep they got the previous night. The mean of the responses was 7.9 hours with a standard deviation of 2.1 hours. If we are interested in determining whether these data give convincing evidence that students at this high school are getting too little sleep, on average, which of the following is the standardized test statistic for the appropriate significance test?

(A) $t = \dfrac{7.9-9}{\frac{2.1}{\sqrt{28}}}$

(B) $t = \dfrac{9-7.9}{\frac{2.1}{\sqrt{28}}}$

(C) $t = \dfrac{7.9-9}{\sqrt{\frac{2.1}{28}}}$

(D) $t = \dfrac{7.9-9}{\frac{2.1}{\sqrt{27}}}$

(E) $t = \dfrac{9-7.9}{\frac{2.1}{\sqrt{27}}}$

AP9 A 96% confidence interval for the proportion of the labor force that is unemployed in a certain city is (0.07, 0.10). Which of the following statements is true?

(A) The probability is 0.96 that between 7% and 10% of the labor force in this city is unemployed.

(B) About 96% of the intervals constructed by this method will contain the population proportion of the labor force that is unemployed in the city.

(C) In repeated samples of the same size, there is a 96% chance that the sample proportion will fall between 0.07 and 0.10.

(D) The true rate of unemployment in the labor force lies within this interval 96% of the time.

(E) About 96% of the intervals constructed by this method will contain the sample proportion of the labor force that is unemployed in the city.

AP10 The following dotplots show the average high temperatures (in degrees Celsius) for a sample of tourist cities from around the world. Both the January and July average high temperatures for each city are shown. What is one statement that can be made with certainty from an analysis of the graphical display?

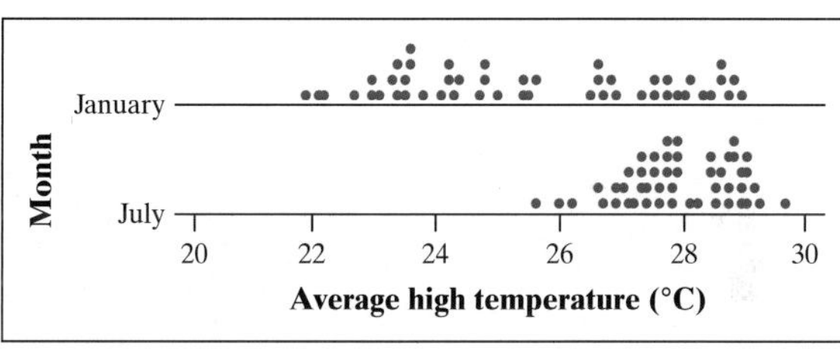

(A) Every city in the sample has a larger average high temperature in July than in January.

(B) The distribution of temperatures in July is skewed right, while the distribution of temperatures in January is skewed left.

(C) The median average high temperature for January is greater than the median average high temperature for July.

(D) There appear to be outliers in the average high temperatures for January and July.

(E) There is more variability in average high temperatures in January than in July.

AP11 According to the U.S. Census Bureau, the proportion of adults in a certain county who own their own home is 0.71. A random sample of 100 adults in a certain section of the county found that 65 own their home. Which of the following represents the approximate probability of obtaining a random sample of 100 adults in which 65 or fewer own their home, assuming that this section of the county has the same overall proportion of adults who own their home as does the entire county?

(A) $\binom{100}{65}(0.71)^{65}(0.29)^{35}$

(B) $\binom{100}{65}(0.29)^{65}(0.71)^{35}$

(C) $P\left(z \le \frac{0.65-0.71}{\sqrt{\frac{(0.71)(0.29)}{100}}}\right)$

(D) $P\left(z \le \frac{0.65-0.71}{\sqrt{\frac{(0.65)(0.35)}{100}}}\right)$

(E) $P\left(z \le \frac{0.65-0.71}{\frac{(0.71)(0.29)}{\sqrt{100}}}\right)$

AP12 The U.S. Environmental Protection Agency (EPA) is charged with monitoring industrial emissions that pollute the atmosphere and water. So long as emission levels stay within specified guidelines, the EPA does not take action against the polluter. If the polluter violates regulations, the offender can be fined, forced to clean up the problem, or possibly closed. Suppose that for a particular industry the acceptable emission level has been set at no more than 5 parts per million (5 ppm). The null and alternative hypotheses are H_0: $\mu = 5$ versus H_a: $\mu > 5$. Which of the following describes a Type II error?

(A) The EPA fails to find convincing evidence that emissions exceed acceptable limits when, in fact, they are within acceptable limits.

(B) The EPA finds convincing evidence that emissions exceed acceptable limits when, in fact, they are within acceptable limits.

(C) The EPA fails to find convincing evidence that emissions exceed acceptable limits when, in fact, they do exceed acceptable limits.

(D) The EPA finds convincing evidence that emissions exceed acceptable limits when, in fact, they do exceed acceptable limits.

(E) The EPA fails to find convincing evidence that emissions exceed acceptable limits when, in fact, there is convincing evidence.

AP13 A large company is interested in improving the efficiency of its customer service and decides to examine the length of the business phone calls made to clients by its sales staff. Here is a cumulative relative frequency graph from data collected over the past year. According to the graph, the shortest 80% of calls took how long to complete?

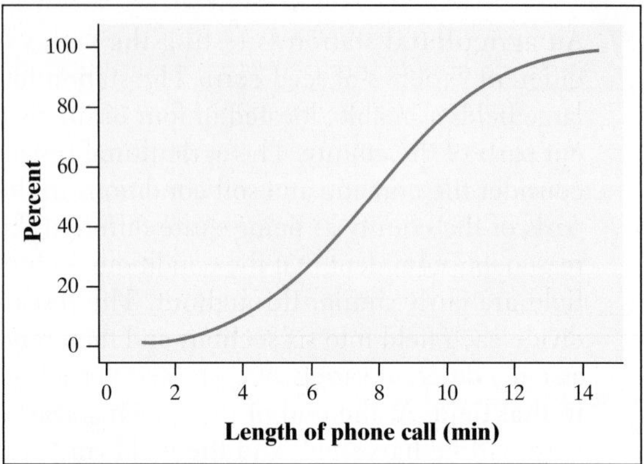

(A) 10 minutes or less

(B) More than 10 minutes

(C) Exactly 10 minutes

(D) More than 5.5 minutes

(E) 5.5 minutes or less

AP14 Are TV commercials louder than the programs they precede? To find out, researchers collected data on 50 randomly selected commercials in a given week. With the television's volume at a fixed setting, they determined whether or not the maximum loudness of each commercial exceeded the maximum loudness in the first 30 seconds of regular programming that followed. Assuming conditions for inference are met, the most appropriate method for answering the question of interest is a

(A) one-sample z test for a proportion.

(B) two-sample z test for a difference in proportions.

(C) one-sample t test for a mean.

(D) paired t test for a mean difference.

(E) two-sample t test for a difference in means.

AP15 A certain variety of candy has different wrappers for various holidays. During Holiday 1, the candy wrappers are 30% silver, 30% red, and 40% pink. During Holiday 2, the wrappers are 50% silver and 50% blue. In separate random samples of 40 candies on Holiday 1 and 40 candies on Holiday 2, what are the mean and standard deviation of the total number of silver wrappers?

(A) 32, 18.4

(B) 32, 6.06

(C) 32, 4.29

(D) 80, 18.4

(E) 80, 4.29

Section II: Part A *Show all your work. Indicate clearly the methods you use, because you will be graded on the correctness of your methods as well as on the accuracy and completeness of your results and explanations.*

AP16 A nuclear power plant releases water into a nearby lake every afternoon at 4:51 P.M. Environmental researchers are concerned that fish are being driven away from the area around the plant. They believe that the temperature of the water discharged may be a factor. The researchers collect data on the temperature of the water (in degrees Celsius) released by the plant and the measured distance (in meters) from the outflow pipe of the plant to the nearest fish found in the water on eight randomly chosen afternoons. Computer output from a least-squares regression analysis of these data and a residual plot are shown.

```
Predictor      Coef     SE Coef     T        P
Constant      -73.64    15.48     -4.76    0.003
Temperature   5.7188    0.5612    10.19    0.000
S = 11.4175   R-Sq = 94.5%    R-Sq (adj) = 93.6%
```

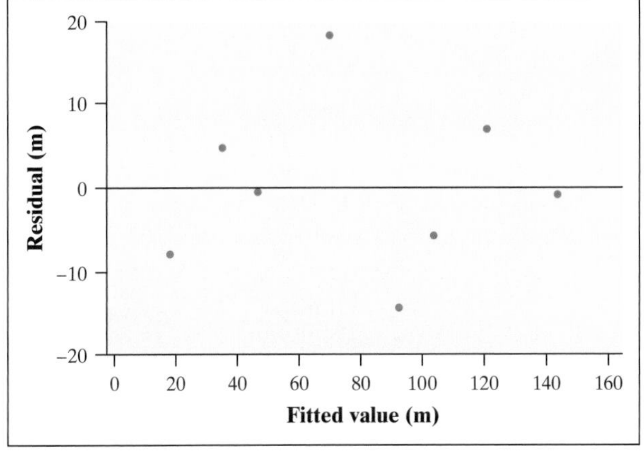

(a) Explain why a linear model is appropriate for describing the relationship between temperature and distance to the nearest fish.

(b) Write the equation of the least-squares regression line. Define any variables you use.

(c) Interpret the slope of the regression line.

(d) Calculate and interpret the residual for the point (29, 78).

AP17 The length of commercially raised copper rockfish can be modeled by a normal distribution with mean $\mu = 16.4$ inches and standard deviation $\sigma = 2.1$ inches.[64]

(a) What proportion of commercially raised copper rockfish are between 15 and 20 inches long?

Suppose that a biologist selects a random sample of 10 commercially raised copper rockfish and measures the length of each fish.

(b) The biologist is interested in finding the probability that fewer than half of the fish in the sample have lengths between 15 and 20 inches.

(i) Define the random variable of interest and state its probability distribution.

(ii) Calculate the probability that fewer than half of the fish in the sample have lengths between 15 and 20 inches.

(c) Let $\bar{x}$ = the mean length of the 10 randomly selected rockfish.

(i) Find the mean and standard deviation of the sampling distribution of $\bar{x}$.

(ii) Explain why the sampling distribution of $\bar{x}$ can be modeled with a normal distribution.

Section II: Part B (Investigative Task) *Show all your work. Indicate clearly the methods you use, because you will be graded on the correctness of your methods as well as on the accuracy and completeness of your results and explanations.*

AP18 An investor is comparing two stocks, A and B. The investor wants to know if over the long run, there is a significant difference in the average daily return on investment as measured by the annualized percent increase or decrease in the price of the stock. To find out, the investor selects a random sample of 50 annualized daily returns over the past 5 years for each stock. The data are summarized in the table.

Stock	Mean return	Standard deviation
A	11.8%	12.9%
B	7.1%	9.6%

(a) The investor uses the data to perform a two-sample *t* test of $H_0: \mu_A - \mu_B = 0$ versus $H_a: \mu_A - \mu_B \neq 0$, where μ_A = the true mean annualized daily return for Stock A and μ_B = the true mean annualized daily return for Stock B. The resulting *P*-value is 0.042. Interpret this value in context. What conclusion would you make?

(b) The investor believes that although the return on investment for Stock A usually exceeds that of Stock B, Stock A represents a riskier investment, where the risk is measured by the price volatility of the stock. The sample variance s_x^2 is a statistical measure of the price volatility and indicates how much an investment's actual performance during a specified period varies from its average performance over a longer period. Do the price fluctuations in Stock A significantly exceed those of Stock B, as measured by their variances? State an appropriate set of hypotheses about the *ratio* of the variances for Stock A and Stock B, $\frac{\sigma_A^2}{\sigma_B^2}$, that the investor is interested in testing.

(c) To test the hypotheses in part (b), we will construct a test statistic defined as

$$F = \frac{\text{larger sample variance}}{\text{smaller sample variance}}$$

(i) Calculate the value of the F statistic using the information given in the table.

(ii) Explain how the value of the statistic provides some evidence for the alternative hypothesis you stated in part (b).

(d) Two hundred simulated values of the F test statistic were calculated assuming that the two stocks have the same variance in daily price. The results of the simulation are displayed in the following dotplot.

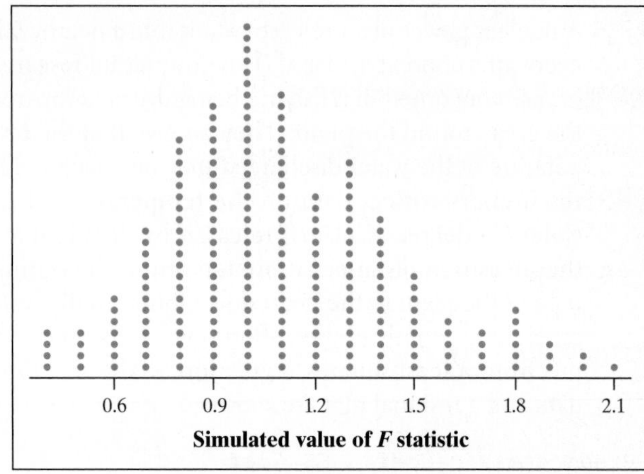

Use these simulated values to determine whether the observed data provide convincing evidence that Stock A is a riskier investment than Stock B. Explain your reasoning.

UNIT
8

Inference for Categorical Data: Chi-Square

Introduction

In Section 6C, you learned how to perform tests for the proportion of successes in a single population. These tests were based on a single categorical variable with values that were divided into two categories: success and failure. Sometimes we want to perform a test for the distribution of a categorical variable with two *or more* categories. The *chi-square test for goodness of fit* allows us to determine whether a hypothesized distribution seems valid. This test is useful in fields such as genetics, where the laws of probability give the expected proportion of outcomes in each category.

In Section 2A, you learned how to analyze the relationship between two categorical variables. For instance, we used two-way tables and segmented bar graphs to show there was an association between environmental club membership and snowmobile use for a random sample of winter visitors to Yellowstone National Park. Do these data provide convincing evidence of an association between these two variables in the population of winter visitors to Yellowstone? We can answer this question with a *chi-square test for independence.*

Tests for independence use data that can be summarized in a two-way table. It is also possible to use the information in a two-way table to compare the distribution of a categorical variable for two *or more* populations or treatments, where the variable can have two *or more* categories. We can decide whether the distribution of a categorical variable differs for two or more populations or treatments using a *chi-square test for homogeneity.* This test will help us answer questions like "Does background music influence customer purchases?"

Here's an activity that gives you a taste (pun intended) of what lies ahead.

ACTIVITY The candy man can

Ramón Rivera Moret

Mars®, Inc., is famous for its milk chocolate candies. Here's what the company's Consumer Affairs Department says about the distribution of color for M&M'S® Milk Chocolate Candies produced at its Hackettstown, New Jersey, factory:

Brown: 12.5%	Red: 12.5%	Yellow: 12.5%
Green: 12.5%	Orange: 25%	Blue: 25%

The purpose of this activity is to investigate if the distribution of color in a large bag of M&M'S Milk Chocolate Candies differs from the distribution of color claimed by the Hackettstown factory.

1. Your class will select a random sample of 60 M&M'S Milk Chocolate Candies from a large bag and count the number of candies of each color. Make a table on the board that summarizes these *observed counts.* Does the distribution of color in the sample match the claimed distribution? Which colors are closest to the number expected? Farthest from the number expected?
2. How can you tell if the sample data give convincing evidence against the company's claim? Each team of three or four students should discuss this

question and devise a formula for a test statistic that measures the difference between the observed and expected color distributions. The test statistic should yield a single number when the observed and expected values are plugged in. Also, larger differences between the observed and expected distributions should result in a larger value for the statistic.

3. Each team will share its proposed test statistic with the class. Your teacher will then reveal how the *chi-square test statistic* χ^2 is calculated. We'll discuss it in more detail shortly.
4. Discuss as a class: If your sample is consistent with the company's claim, will the value of χ^2 be large or small? If your sample is not consistent with the company's claim, will the value of χ^2 be large or small?
5. Compute the value of the χ^2 test statistic for the class's data.

We can use simulation to determine if your class's chi-square test statistic is large enough to provide convincing evidence that the distribution of color in the large bag of M&M'S® Milk Chocolate Candies differs from the company's claim.

6. Go to www.stapplet.com and launch the *M&M's/Skittles/Froot Loops* applet. Keep the color distribution you are comparing to as M&M's Milk Chocolate. Enter the class's observed counts in the table and click the "Begin analysis" button. The applet will display the distribution of color with a bar chart and calculate the value of the chi-square statistic for these data.
7. To see which values of the chi-square statistic are likely to happen by chance alone when taking samples of the same size from the claimed population, enter 1 for the number of samples to simulate and click the "Simulate" button. A graph of the simulated sample will appear, along with the chi-square test statistic for that sample. Was the simulated chi-square statistic smaller or larger than the chi-square statistic from the class's sample?
8. Keep clicking the "Simulate" button to generate more simulated chi-square statistics. To speed up the process, change the number of samples to 100 and click the "Simulate" button again and again. Describe the shape, center, and variability of the sampling distribution of the simulated chi-square test statistic.
9. Click the box to plot the approximate χ^2 density curve. Does it fit well?
10. To estimate the *P*-value, use the counter at the bottom of the applet to count the percentage of dots greater than or equal to the class's chi-square statistic from Step 5. Based on the *P*-value, what conclusion would you make about the distribution of color in the large bag?

Here is one possible dotplot from Step 8 in the activity. It shows the value of the chi-square test statistic for each of 100 random samples from the company's claimed M&M'S® Milk Chocolate Candies color distribution.

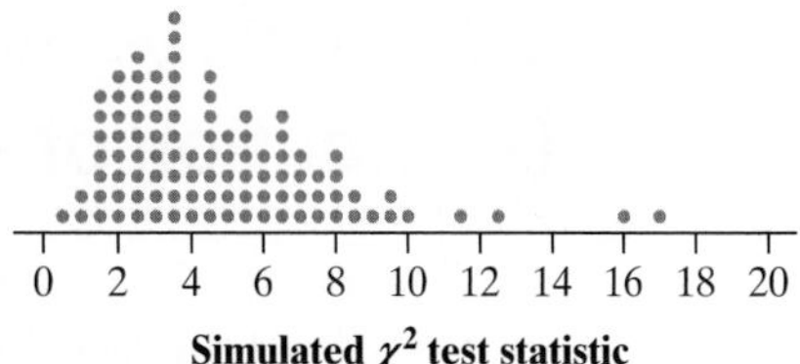

You may have noticed that the shape of the distribution *isn't* approximately normal. Will it always look like this? You will learn more about the sampling distribution of the chi-square test statistic shortly.

SECTION 8A Chi-Square Tests for Goodness of Fit

LEARNING TARGETS *By the end of the section, you should be able to:*

- State appropriate hypotheses for a test about the distribution of a categorical variable.
- Calculate expected counts for a test about the distribution of a categorical variable.
- Check the conditions for a test about the distribution of a categorical variable.
- Calculate the test statistic and *P*-value for a test about the distribution of a categorical variable.
- Perform a chi-square test for goodness of fit.

Milan's class did the Candy Man Can activity using a bag from the Hackettstown factory. The following one-way table summarizes the data from the class's sample of M&M'S® Milk Chocolate Candies:

Color	Brown	Red	Yellow	Green	Orange	Blue	Total
Count	12	3	7	9	9	20	60

The sample proportion of brown candies is $\hat{p} = \frac{12}{60} = 0.20$. Because the company claims that 12.5% of M&M'S Milk Chocolate Candies are brown, the class might believe that something fishy is going on. They could use the one-sample z test for a proportion from Section 6D to test the hypotheses

$$H_0: p = 0.125$$
$$H_a: p \neq 0.125$$

where p is the population proportion of brown M&M'S Milk Chocolate Candies in the large bag. They could then perform additional significance tests for each of the remaining colors.

Besides being fairly inefficient, this method would also lead to the problem of multiple tests, which we discussed in Section 7D. More importantly, this approach wouldn't tell us how likely it is to get a random sample of 60 candies with a *distribution* of color that differs as much from the one claimed by the company as the class's sample does, taking all the colors into consideration at one time. To make that determination, we need to use a significance test about the distribution of a categorical variable.

Stating Hypotheses for Tests about the Distribution of a Categorical Variable

As with any significance test, we begin by stating hypotheses. The null hypothesis in a test about the distribution of a categorical variable should state a claim about the distribution of that variable in the population of interest. In the Candy Man

Can activity, the categorical variable we measured was color and the population of interest was all the M&M'S® Milk Chocolate Candies in the large bag. The appropriate null hypothesis is

H_0: The distribution of color in the large bag of M&M'S Milk Chocolate Candies is the same as the claimed distribution.

The alternative hypothesis in a test about the distribution of a categorical variable is that the categorical variable does *not* have the specified distribution. For the M&M'S, our alternative hypothesis is

H_a: The distribution of color in the large bag of M&M'S Milk Chocolate Candies is *not* the same as the claimed distribution.

Although we usually write the hypotheses in words for this test, we can also write the hypotheses in symbols. For example, here are the hypotheses for the Candy Man Can activity:

$$H_0: p_{\text{brown}} = 0.125, p_{\text{red}} = 0.125, p_{\text{yellow}} = 0.125, p_{\text{green}} = 0.125, p_{\text{orange}} = 0.25, p_{\text{blue}} = 0.25$$

H_a: At least two of these proportions differ from the values stated by the null hypothesis.

where p_{color} = the population proportion of M&M'S Chocolate Candies of the specified color in the large bag.

Why don't we write the alternative hypothesis as "H_a: At least one of these proportions differs from the values stated by the null hypothesis" instead? If the stated proportion in one category is wrong, then the stated proportion in at least one other category must be wrong because the sum of the proportions is always equal to 1.

Don't state the alternative hypothesis in a way that suggests that all the proportions in the null hypothesis are wrong. For instance, it would be *incorrect* to write the alternative hypothesis as

$$H_a: p_{\text{brown}} \neq 0.125, p_{\text{red}} \neq 0.125, p_{\text{yellow}} \neq 0.125, p_{\text{green}} \neq 0.125, p_{\text{orange}} \neq 0.25, p_{\text{blue}} \neq 0.25$$

EXAMPLE

Are more NHL players born earlier in the year?

Stating hypotheses for tests about the distribution of a categorical variable

Skill 1.F

PROBLEM: In his book *Outliers,* Malcolm Gladwell introduces the idea of the relative-age effect with an example from hockey. Many National Hockey League (NHL) players come from Canada. Because January 1 is the cutoff birth date for youth hockey leagues in Canada, children born earlier in the year compete against players who may be as much as 12 months younger. Gladwell argues that these "relatively old" players tend to be bigger, stronger, and more coordinated and hence get more playing time and more coaching, and have a better chance of being successful.

Canva Pty Ltd./Alamy Stock Photo

To see if the birth dates of NHL players are uniformly distributed across the four quarters of the year, we selected a random sample of 80 NHL players and recorded their birthdays.[1] The table summarizes the distribution of birthday for these 80 players by quarter.

Birthday	Jan–Mar	Apr–Jun	Jul–Sep	Oct–Dec
Number of players	32	20	16	12

Do these data provide convincing evidence that the population proportions of NHL players born in each quarter are not all the same?

(a) State appropriate hypotheses for a test that addresses this question.
(b) Calculate the proportion of players in the sample who were born in each quarter.
(c) Explain how the proportions in part (b) give some evidence for the alternative hypothesis.

SOLUTION:

(a) H_0: The population proportions of NHL players born in each quarter are all the same.

H_a: The population proportions of NHL players born in each quarter are not all the same.

(b) Jan–Mar: $32/80 = 0.40$

Apr–Jun: $20/80 = 0.25$

Jul–Sep: $16/80 = 0.20$

Oct–Dec: $12/80 = 0.15$

(c) There is some evidence for H_a because the sample proportions are not all the same.

> The hypotheses could also be stated symbolically:
> H_0: $p_{\text{Jan–Mar}} = p_{\text{Apr–Jun}} = p_{\text{Jul–Sep}} = p_{\text{Oct–Dec}} = 0.25$
> H_a: At least 2 of the proportions $\neq 0.25$
> where p_{quarter} = the population proportion of NHL players born in the specified quarter.

FOR PRACTICE, TRY EXERCISE 1

Calculating Expected Counts for Tests about the Distribution of a Categorical Variable

Milan's class did the Candy Man Can activity. The following table summarizes the data from the class's sample of M&M'S® Milk Chocolate Candies:

Color	Brown	Red	Yellow	Green	Orange	Blue	Total
Count	12	3	7	9	9	20	60

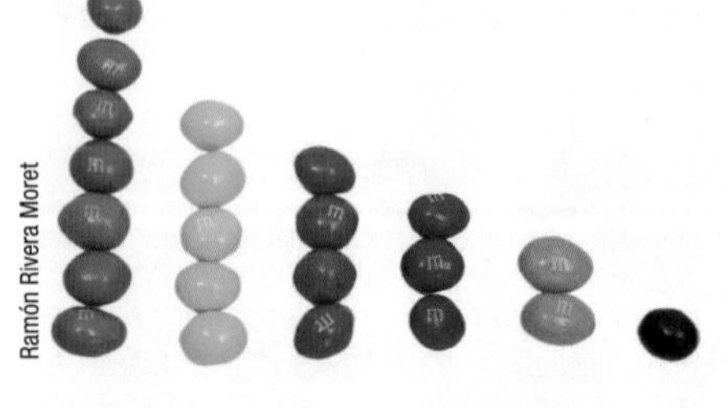
Ramón Rivera Moret

To begin their analysis, Milan's class compared the observed counts from their sample with the counts that would be expected if the manufacturer's claim is true. To calculate the expected counts, they started with the manufacturer's claim:

Brown: 12.5%	Red: 12.5%	Yellow: 12.5%
Green: 12.5%	Orange: 25%	Blue: 25%

Assuming that the claimed distribution is true, 12.5% of all M&M'S Milk Chocolate Candies in the large bag should be brown. For random samples of 60 candies, the expected count of brown M&M'S is $(60)(0.125) = 7.5$.

CALCULATING EXPECTED COUNTS IN A TEST ABOUT THE DISTRIBUTION OF A CATEGORICAL VARIABLE

The expected count for category i in the distribution of a categorical variable is

$$np_i$$

where n is the overall sample size and p_i is the proportion for category i specified by the null hypothesis.

Using this same method, we find the expected counts for the other color categories:

$$\text{Red: } (60)(0.125) = 7.5$$
$$\text{Yellow: } (60)(0.125) = 7.5$$
$$\text{Green: } (60)(0.125) = 7.5$$
$$\text{Orange: } (60)(0.25) = 15$$
$$\text{Blue: } (60)(0.25) = 15$$

Did you notice that the "expected count" sounds a lot like the "expected value" of a random variable from Unit 4? That's no coincidence. The number of M&M'S Milk Chocolate Candies of a specific color in a random sample of 60 candies is a binomial random variable. Its expected value is np, the average number of candies of this color in many samples of 60 M&M'S Milk Chocolate Candies. *Expected counts are often not whole numbers and shouldn't be rounded to a whole number.*

EXAMPLE

Are more NHL players born earlier in the year?

Calculating expected counts for tests about the distribution of a categorical variable

Skill 3.A

PROBLEM: A random sample of 80 NHL players was selected to determine if their birthdays are uniformly distributed across the four quarters of the year. The table summarizes birthday data for these 80 players.

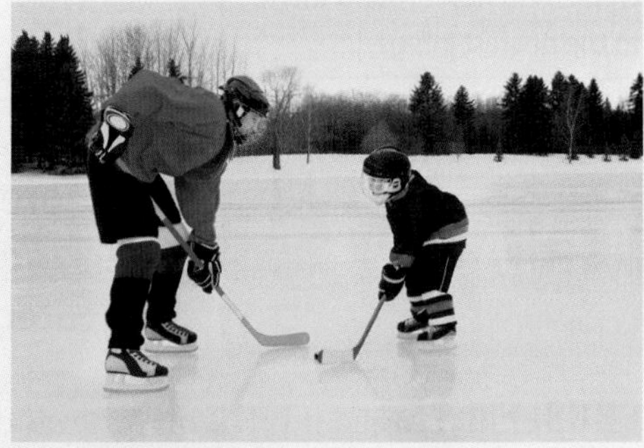

Design Pics Inc/Alamy Stock Photo

Birthday	Jan–Mar	Apr–Jun	Jul–Sep	Oct–Dec
Number of players	32	20	16	12

Calculate the expected counts for a test of the null hypothesis that the population proportions of NHL players born in each quarter are the same.

SOLUTION:

Jan–Mar: $(80)(0.25) = 20$

Apr–Jun: $(80)(0.25) = 20$

Jul–Sep: $(80)(0.25) = 20$

Oct–Dec: $(80)(0.25) = 20$

If the proportion of players born in each of the four quarters is the same, then 100%/4 = 25% = 0.25 of NHL players should be born in each quarter.

FOR PRACTICE, TRY EXERCISE 5

Checking Conditions for Tests about the Distribution of a Categorical Variable

As with the other significance tests you have studied, three conditions need to be met when performing a test about the distribution of a categorical variable. The Random and 10% conditions are the same as for the other one-sample tests from Units 6 and 7. The Large Counts condition has the same name as other tests about categorical data (one- and two-sample z tests for proportions), but we check it in a slightly different way. Together, these conditions ensure that the observations in the sample can be viewed as independent and that the sampling distribution of the test statistic can be modeled by the appropriate distribution. (You'll learn more about this shortly.)

CONDITIONS FOR PERFORMING A TEST ABOUT THE DISTRIBUTION OF A CATEGORICAL VARIABLE

- **Random:** The data come from a random sample from the population of interest.
  - **10%:** When sampling without replacement, $n < 0.10N$.
- **Large Counts:** All *expected* counts are at least 5.

AP® EXAM TIP

When checking the Large Counts condition, be sure to address the *expected* counts, not the observed counts. Make sure to write the expected counts on your paper and label them as expected counts, or you won't receive credit. And never round them to the nearest integer!

For the data collected by Milan's class, these three conditions are met:

- Random: Milan's class selected a random sample of M&M'S® Milk Chocolate Candies from the large bag. ✓
  - 10%: The sample of 60 M&M'S is less than 10% of all M&M'S in the large bag. ✓
- Large Counts: All the expected counts are at least 5. ✓

Color	Brown	Red	Yellow	Green	Orange	Blue	Total
Expected count	7.5	7.5	7.5	7.5	15	15	60

EXAMPLE

Are more NHL players born earlier in the year?

Checking conditions for tests about the distribution of a categorical variable

Skill 4.C

PROBLEM: To see if NHL player birthdays are uniformly distributed across the four quarters of the year, a random sample of 80 NHL players was selected and their birthdays recorded. The table shows the observed and expected counts for a test of the null hypothesis that the population proportions of NHL players born in each quarter are the same.

Hero Images/iStock/Getty Images

Birthday	Jan–Mar	Apr–Jun	Jul–Sep	Oct–Dec
Observed count	32	20	16	12
Expected count	20	20	20	20

Check whether the conditions are met for performing a test about the distribution of a categorical variable.

SOLUTION:

- Random: Random sample of 80 NHL players. ✓
  - 10%: We must assume that 80 is less than 10% of all NHL players. ✓
- Large Counts: All the expected counts (20, 20, 20, 20) are at least 5. ✓

Make sure to check the values of the *expected* counts when checking the Large Counts condition.

FOR PRACTICE, TRY EXERCISE 9

CHECK YOUR UNDERSTANDING

Mars, Inc., reports that the M&M'S® Peanut Chocolate Candies from its Cleveland, Tennessee, factory have the following distribution of color: 20% blue, 20% orange, 20% green, 20% yellow, 10% red, and 10% brown. Joey bought a large bag of M&M'S Peanut Chocolate Candies and selected a random sample of 65 candies. He found 14 blue, 9 orange, 15 green, 14 yellow, 5 red, and 8 brown.

1. State appropriate hypotheses for testing the company's claim about the color distribution of M&M'S Peanut Chocolate Candies in Joey's large bag.
2. Calculate the expected count for each color.
3. Verify that the conditions are met for a test about the distribution of a categorical variable.

Calculating the Test Statistic and *P*-Value for Tests about the Distribution of a Categorical Variable

Figure 8.1 is a graph comparing the observed and expected counts of M&M'S® Milk Chocolate Candies for Milan's class.

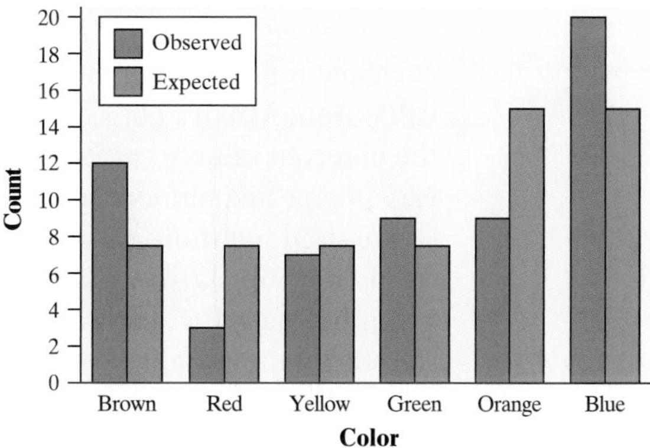

FIGURE 8.1 Bar graph comparing observed and expected counts for Milan's class sample of 60 M&M'S Milk Chocolate Candies.

Because the observed counts are different from the expected counts, there is some evidence that the claimed distribution is not correct. However, perhaps the claimed distribution is correct, and the differences found by Milan's class were simply due to sampling variability.

To assess if these differences are larger than what is likely to happen by chance alone when H_0 is true, we calculate a test statistic and *P*-value. For a test about the distribution of a categorical variable, we use the **chi-square test statistic** χ^2. (The symbol χ is the lowercase Greek letter chi, pronounced "kye" like "rye.")

AP® EXAM TIP
Formula sheet

The formula for the chi-square test statistic is included on the formula sheet provided on both sections of the AP® Statistics exam.

DEFINITION Chi-square test statistic χ^2

The **chi-square test statistic χ^2** is a measure of how different the observed counts are from the expected counts, relative to the expected counts. The formula for the statistic is

$$\chi^2 = \sum \frac{(\text{Observed} - \text{Expected})^2}{\text{Expected}}$$

where the sum is over all possible values of the categorical variable.

The following table shows the observed and expected counts for Milan's class data:

Color	Brown	Red	Yellow	Green	Orange	Blue	Total
Observed count	12	3	7	9	9	20	60
Expected count	7.5	7.5	7.5	7.5	15	15	60

For these data, the chi-square test statistic is

$$\begin{aligned}\chi^2 &= \sum \frac{(\text{Observed} - \text{Expected})^2}{\text{Expected}} \\ &= \frac{(12-7.5)^2}{7.5} + \frac{(3-7.5)^2}{7.5} + \frac{(7-7.5)^2}{7.5} + \frac{(9-7.5)^2}{7.5} + \frac{(9-15)^2}{15} + \frac{(20-15)^2}{15} \\ &= 2.7 + 2.7 + 0.03 + 0.30 + 2.4 + 1.67 \\ &= 9.8\end{aligned}$$

Make sure to use the observed and expected *counts*, not the observed and expected *proportions*, when calculating the chi-square test statistic. In addition to giving an incorrect value for the test statistic, using proportions instead of counts removes valuable information about the sample size from the calculation.

Large values of χ^2 arise when the differences between the observed counts and the expected counts are larger. Thus, the larger the value of χ^2, the stronger the evidence is for the alternative hypothesis (and against the null hypothesis). Is the value from Milan's class, $\chi^2 = 9.8$, a large value? You know the drill: compare the observed value $\chi^2 = 9.8$ to the sampling distribution that shows how χ^2 would vary in repeated random sampling if the null hypothesis were true.

We used technology to simulate selecting 1000 random samples of size 60 from the population distribution of M&M'S® Milk Chocolate Candies given by Mars, Inc. Figure 8.2 shows a dotplot of the values of the chi-square test statistic for these 1000 samples.

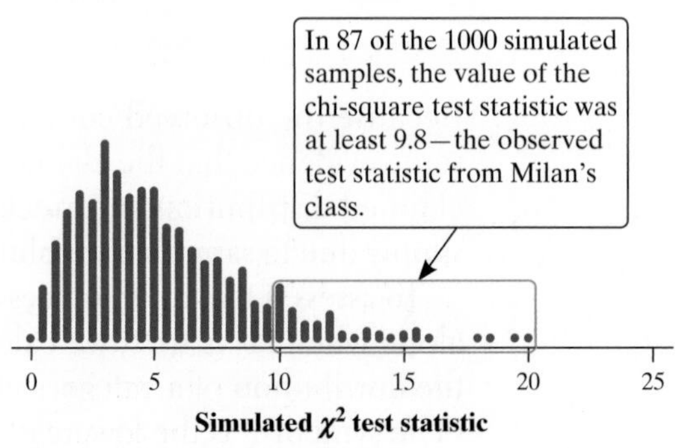
In 87 of the 1000 simulated samples, the value of the chi-square test statistic was at least 9.8—the observed test statistic from Milan's class.
0
5
10
15
20
25
Simulated χ^2 test statistic

FIGURE 8.2 Dotplot showing values of the chi-square test statistic in 1000 simulated samples of size $n = 60$ from the population distribution of M&M'S Milk Chocolate Candies stated by the company. Some people call a simulated distribution like this one a *randomization distribution.*

Recall that larger values of χ^2 give more convincing evidence against H_0 and in favor of H_a. According to the dotplot, 87 of the 1000 simulated samples resulted in a chi-square test statistic of 9.8 or higher. Thus, the *P*-value is approximately $87/1000 = 0.087$. There is about a 0.087 probability of getting an observed distribution of color this different or more different from the expected distribution of color by chance alone, assuming that the distribution of color in the bag is the same as the company claims.

As Figure 8.2 suggests, the sampling distribution of the chi-square test statistic is *not* a normal distribution. It is a right-skewed distribution that allows only non-negative values because χ^2 can never be negative. When the Large Counts condition is met, the sampling distribution of the χ^2 test statistic is modeled well by a **chi-square distribution** with degrees of freedom (df) equal to the number of categories minus 1. As with the *t* distributions, there is a different chi square distribution for each possible df value. Unlike for *t* tests and intervals, however, the degrees of freedom for a test about the distribution of a categorical variable are based on the number of categories, not the sample size.

DEFINITION Chi-square distribution

A **chi-square distribution** is described by a density curve that takes only non-negative values and is skewed to the right. A particular chi-square distribution is specified by its degrees of freedom.

Figure 8.3 shows the density curves for three members of the chi-square family of distributions. As the degrees of freedom (df) increase, the density curves become less skewed, and larger values of χ^2 become more likely.

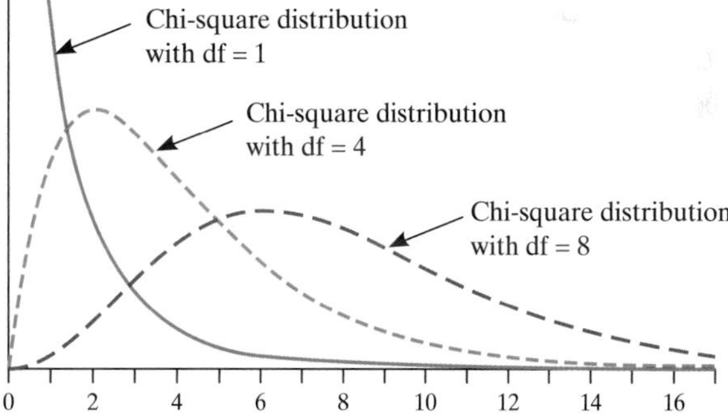

FIGURE 8.3 The density curves for three members of the chi-square family of distributions.

Here are two other interesting facts about the chi-square distributions:

- The mean of a particular chi-square distribution is equal to its degrees of freedom.
- For df > 2, the mode (peak) of the chi-square density curve is at df − 2.

For example, when df = 8, the chi-square distribution has a mean of 8 and a mode of 6.

To get *P*-values from a chi-square distribution, we can use technology or Table C. For the data collected by Milan's class, the conditions are met and $\chi^2 = 9.8$. Because there are 6 different categories for color, df = 6 − 1 = 5. The *P*-value is the probability of getting a value of χ^2 as large as or larger than 9.8 when H_0 is true. Figure 8.4 shows this probability as an area under the chi-square density curve with 5 degrees of freedom.

FIGURE 8.4 The *P*-value for a test about the distribution of a categorical variable using Milan's M&M'S® Milk Chocolate Candies class data.

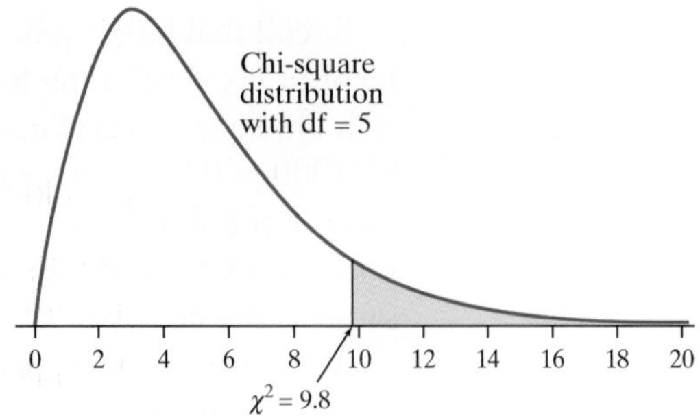

	Tail probability *p*		
df	**.15**	**.10**	**.05**
4	6.74	7.78	9.49
5	8.12	9.24	11.07
6	9.45	10.64	12.59

To find the *P*-value using Table C, look in the df = 5 row. The value $\chi^2 = 9.8$ falls between the critical values 9.24 and 11.07. Looking at the top of the corresponding columns, we find that the right tail area of the chi-square distribution with 5 degrees of freedom is between 0.10 and 0.05. So, the *P*-value for a test based on Milan's class data is between 0.05 and 0.10. This is consistent with the value of 0.087 we got from the simulation earlier.

Table C gives an interval of possible *P*-values for a significance test. We can still make a conclusion from the test in the same way as if we had a single probability — by comparing the interval of possible *P*-values to the chosen significance level α. However, we recommend that you use technology to calculate a more precise *P*-value, as shown in the following Tech Corner.

26. Tech Corner — CALCULATING *P*-VALUES FROM χ^2 DISTRIBUTIONS

TI-Nspire and other technology instructions are on the book's website at bfwpub.com/tps7e.

You can use the χ^2cdf command on the TI-83/84 calculator to calculate areas under a χ^2 density curve. The syntax is χ^2cdf(lower bound, upper bound, df). Let's use the χ^2cdf command to compute the *P*-value for the M&M'S® example, where $\chi^2 = 9.8$ and df = 5.

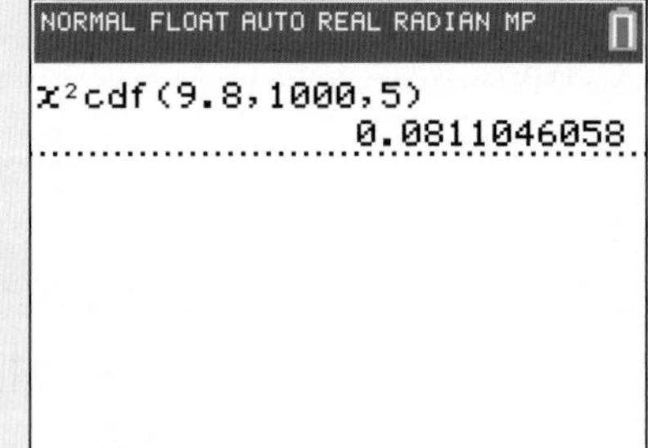

1. Press 2nd VARS (DISTR) and choose χ^2cdf(.
2. We want to find the area to the right of $\chi^2 = 9.8$ under the χ^2 density curve with df = 5.

   OS 2.55 or later: In the dialog box, enter these values: lower: 9.8, upper: 1000, df: 5. Choose Paste, and then press ENTER.

   Older OS: Complete the command χ^2cdf(9.8, 1000, 5) and press ENTER.

Because our *P*-value of 0.081 is greater than $\alpha = 0.05$, we fail to reject H_0. We don't have convincing evidence that the distribution of color in the large bag of M&M'S® Milk Chocolate Candies differs from the claimed distribution. This is consistent with the results of the simulation from the activity.

Failing to reject H_0 does not mean that the null hypothesis is true! We can't conclude that the large bag *has* the distribution of color claimed by Mars®, Inc. — all we can say is that the sample data did not provide convincing evidence to reject H_0.

CALCULATING THE TEST STATISTIC AND *P*-VALUE IN A TEST ABOUT THE DISTRIBUTION OF A CATEGORICAL VARIABLE

Suppose the conditions are met. To perform a test of the null hypothesis

H_0: The stated distribution of a categorical variable in the population of interest is correct

compute the chi-square test statistic:

$$\chi^2 = \sum \frac{(\text{Observed} - \text{Expected})^2}{\text{Expected}}$$

where the sum is over all categories. Find the *P*-value by calculating the probability of getting a χ^2 statistic this large or larger in a chi-square distribution with df = number of categories − 1.

EXAMPLE

Are more NHL players born earlier in the year?

Calculating the test statistic and *P*-value for tests about the distribution of a categorical variable

Skill 3.E

PROBLEM: To see if NHL player birthdays are uniformly distributed across the four quarters of the year, we recorded the birthdays of a random sample of 80 NHL players. The table shows the observed and expected counts for a test of the null hypothesis that the population proportions of NHL players born in each quarter are the same.

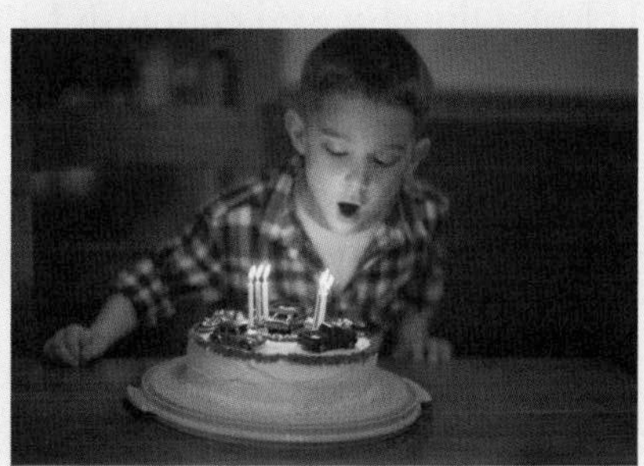

Rebecca Nelson/The Image Bank/Getty Images

Birthday	Jan–Mar	Apr–Jun	Jul–Sep	Oct–Dec
Observed count	32	20	16	12
Expected count	20	20	20	20

(a) Calculate the chi-square test statistic.
(b) Find the *P*-value.
(c) Interpret the *P*-value.

SOLUTION:

(a) $\chi^2 = \frac{(32-20)^2}{20} + \frac{(20-20)^2}{20} + \frac{(16-20)^2}{20} + \frac{(12-20)^2}{20}$

$= 7.2 + 0 + 0.8 + 3.2$

$= 11.2$

$$\chi^2 = \sum \frac{(\text{Observed} - \text{Expected})^2}{\text{Expected}}$$

(b) df = 4 − 1 = 3

df = number of categories − 1

Using technology:
χ^2cdf(lower: 11.2, upper: 1000, df: 3) = 0.0107

Using Table C:
0.01 < *P*-value < 0.025

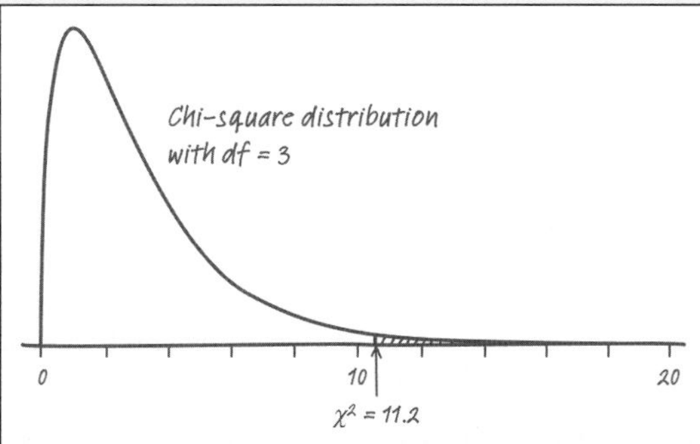

(c) **Assuming that the population proportions of NHL players born in each quarter are all the same, there is a 0.0107 probability of getting a distribution of birthdays in a random sample of size 80 this different or more different than the hypothesized distribution by chance alone.**

FOR PRACTICE, TRY EXERCISE 17

Because the P-value of 0.0107 is less than $\alpha = 0.05$, we should reject H_0. There is convincing evidence that the birthdays of NHL hockey players are not uniformly distributed across the four quarters of the year.

As you probably guessed, your calculator will compute the chi-square test statistic and P-value in a test about the distribution of a categorical variable directly from the observed and expected counts.

27. Tech Corner

SIGNIFICANCE TESTS FOR THE DISTRIBUTION OF A CATEGORICAL VARIABLE

TI-Nspire and other technology instructions are on the book's website at bfwpub.com/tps7e.

You can use the TI-84 to perform the calculations for a test about the distribution of a categorical variable. We'll use the data from the hockey and birthdays example to illustrate the steps.

1. Enter the observed counts in L1 and the expected counts in L2.

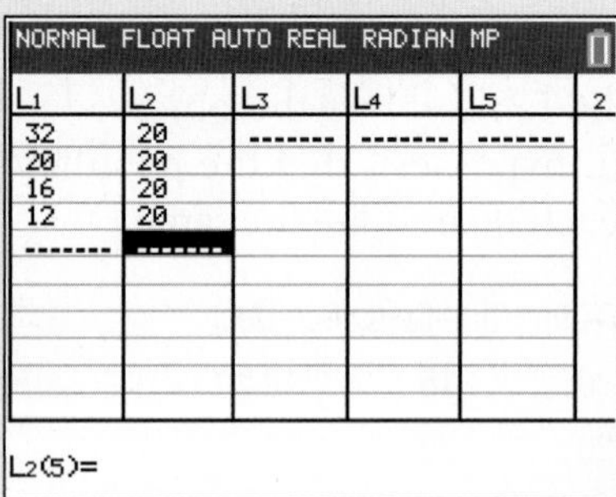

2. Press STAT, arrow over to TESTS, and choose χ^2 GOF–Test.... (*Note*: "GOF" stands for *goodness of fit.*)
3. When the χ^2 GOF–Test screen appears, enter L1 for Observed, L2 for Expected, 3 for df, select Calculate, and press the ENTER button.

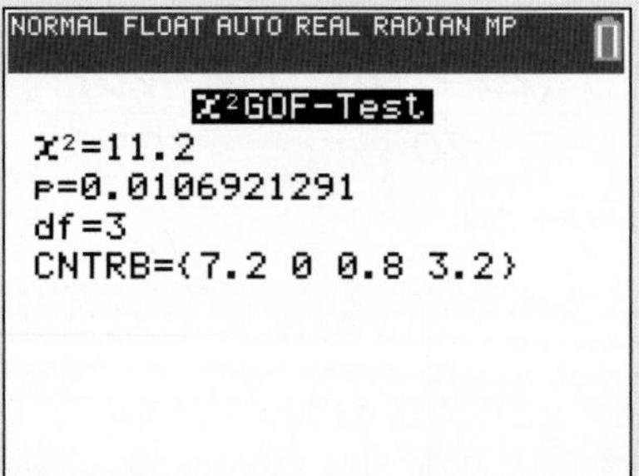

Notes:

- TI-83s and some older TI-84s don't have this test. TI-84 users can get this functionality by upgrading their operating systems.
- If you select the Draw option, you will get a picture of the appropriate χ^2 distribution with the area of interest shaded and the χ^2 test statistic and P-value labeled.
- CNTRB stands for "contributions to the χ^2 statistic." We'll discuss these shortly.

AP® EXAM TIP

You can use your calculator to carry out the mechanics of a significance test on the AP® Statistics exam, but there's a risk involved. If you give the calculator answer but don't show your work, and one or more of your values are incorrect, you will likely get no credit for the Do step. Make sure to include the χ^2 test statistic, df, and *P*-value in the Do step, along with all the elements of the remaining three steps (State, Plan, Conclude).

The χ^2 test statistic and *P*-value calculated using the "full technology" approach described in the Tech Corner are generally more accurate than the values we obtain when we round while doing our calculations. For that reason, we will use the values from the full technology approach when reporting results "Using technology" in the rest of this section.

Think About It

WHY DO WE DIVIDE BY THE EXPECTED COUNT WHEN CALCULATING THE CHI-SQUARE TEST STATISTIC? When Milan's class collected their sample, they got $12 - 7.5 = 4.5$ more brown candies than expected and $20 - 15 = 5$ more blue candies than expected. Which of these results is more surprising? In both cases, the number of M&M'S® Milk Chocolate Candies in the sample exceeds the expected count by about the same amount. But it's much more surprising to be off by 4.5 out of an expected 7.5 brown candies (a 60% discrepancy) than to be off by 5 out of an expected 15 blue candies (a 33% discrepancy). For that reason, we want the category with a larger *relative* difference to contribute more heavily to the evidence in favor of H_a measured by the χ^2 test statistic.

If we computed (Observed count − Expected count)^2 for each category instead, the contributions of these two color categories would be about the same (with the contribution from blue being slightly larger):

$$\text{Brown: } (12 - 7.5)^2 = 20.25 \quad \text{Blue: } (20 - 15)^2 = 25$$

By using (Observed count − Expected count)^2/Expected count, we guarantee that the color category with the larger relative difference will contribute more heavily to the total:

$$\text{Brown: } \frac{(12 - 7.5)^2}{7.5} = 2.7 \quad \text{Blue: } \frac{(20 - 15)^2}{15} = 1.67$$

Putting It All Together: The Chi-Square Test for Goodness of Fit

The appropriate inference procedure for testing a claim about the distribution of a categorical variable is called a **chi-square test for goodness of fit.**

DEFINITION Chi-square test for goodness of fit

A **chi-square test for goodness of fit** is a significance test of the null hypothesis that a categorical variable has a specified distribution in the population of interest.

Whenever you perform a chi-square test for goodness of fit, be sure to follow the four-step process.

EXAMPLE

Trees of New York
The chi-square test for goodness of fit

Skills 1.E, 1.F, 3.A, 3.E, 4.C, 4.E

PROBLEM: Many people love tree-lined streets, especially in big cities where tall buildings can dominate the landscape. Are trees on New York City streets proportionally distributed across the five boroughs? That is, if a borough has 30% of the land area, does it also have 30% of the street trees? The table shows the land area for each borough, along with each borough's percentage of total area.

Alexander Spatari/Moment/Getty Images

Borough	Area (square miles)	Area (percentage)
Bronx	42.2	14.04
Brooklyn	69.4	23.09
Manhattan	22.7	7.55
Queens	108.7	36.17
Staten Island	57.5	19.13

Here are the observed number of trees in each borough from a random sample of 1000 trees on New York City streets.[2]

Borough	Number of trees
Bronx	123
Brooklyn	249
Manhattan	84
Queens	392
Staten Island	152
Total	1000

(a) Do these data provide convincing evidence that trees on New York City streets are not proportionally distributed across the five boroughs?
(b) Which type of error, Type I or Type II, could you have made in part (a)? Explain your answer.

SOLUTION:

Follow the four-step process!

(a)

STATE:

H_0: Trees on New York City streets are proportionally distributed across the five boroughs.

H_a: Trees on New York City streets are not proportionally distributed across the five boroughs.

We could write the hypotheses in symbols as
H_0: $p_{\text{Bronx}} = 0.1404, p_{\text{Brooklyn}} = 0.2309, p_{\text{Manhattan}} = 0.0755, p_{\text{Queens}} = 0.3617, p_{\text{StatenIsland}} = 0.1913$
H_a: At least two of these proportions differ from the values stated by the null hypothesis.
where p_{borough} = the proportion of all trees on New York City streets that are in the indicated borough.

We'll use $\alpha = 0.05$.

If a significance level isn't provided, use $\alpha = 0.05$.

PLAN: Chi-square test for goodness of fit

- Random: The data came from a random sample of trees on New York City streets. ✓
  - 10%: We must assume that 1000 is less than 10% of all trees on New York City streets. ✓
- Large Counts: All expected counts (140.4, 230.9, 75.5, 361.7, 191.3) ≥ 5. ✓

To calculate the expected counts, multiply the sample size (1000) times the expected proportion for each borough.

There is some evidence for H_a because the observed counts are different than the expected counts.

DO:

- $$\chi^2 = \frac{(123-140.4)^2}{140.4} + \frac{(249-230.9)^2}{230.9} + \frac{(84-75.5)^2}{75.5} + \frac{(392-361.7)^2}{361.7} + \frac{(152-191.3)^2}{191.3}$$
  $$= 2.16 + 1.42 + 0.96 + 2.54 + 8.07$$
  $$= 15.15$$

$$\chi^2 = \sum \frac{(\text{Observed} - \text{Expected})^2}{\text{Expected}}$$

If the χ^2 statistic includes many terms, you can save some time by writing out the first two terms and then using an ellipsis (...).

- *P*-value: df $= 5 - 1 = 4$

  Using technology: $\chi^2 = 15.144$, *P*-value $= 0.0044$, df $= 4$

  Using Table C: The *P*-value is between 0.0025 and 0.005.

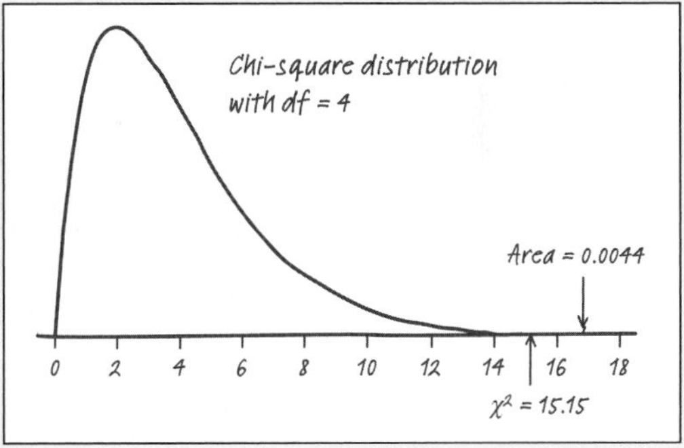

CONCLUDE: Because the *P*-value of $0.0044 < \alpha = 0.05$, we reject H_0. We have convincing evidence that trees on New York City streets are not proportionally distributed across the five boroughs.

(b) Type I error. We found convincing evidence that trees on New York City streets are not proportionally distributed across the five boroughs, but in reality, they could be proportionally distributed.

FOR PRACTICE, TRY EXERCISE 19

FOLLOW-UP ANALYSIS

In a chi-square test for goodness of fit, we test the null hypothesis that a categorical variable has a specified distribution in the population of interest. If the sample data lead to a statistically significant result, we can conclude that our variable has a distribution different from the one stated. To investigate *how* the distribution is different, start by identifying the categories that contribute the most to the chi-square statistic. Then describe how the observed and expected counts differ in those categories, noting the direction of the difference.

Let's return to the New York City street trees example. The following table shows the observed counts, expected counts, differences between these counts, and the contributions to the chi-square statistic.

Borough	Observed	Expected	O – E	$(O-E)^2/E$
Bronx	123	140.4	–17.4	2.16
Brooklyn	249	230.9	18.1	1.42
Manhattan	84	75.5	8.5	0.96
Queens	392	361.7	30.3	2.54
Staten Island	152	191.3	–39.3	8.07

The two biggest contributions to the chi-square statistic came from Queens (2.54) and Staten Island (8.07). In Queens, there were 30.3 more trees than expected. In Staten Island, there were 39.3 fewer trees than expected.

Note: When we ran the chi-square test for goodness of fit on the calculator, a list of these individual components was produced and stored in the list menu. On the TI-84, the list is called CNTRB (for contribution).

CHECK YOUR UNDERSTANDING

Does the warm, sunny weather in Arizona affect a driver's choice of car color? Cass thinks that Arizona drivers might opt for a lighter color with the hope that it will reflect some of the heat from the sun. To see if the distribution of car colors in Oro Valley, near Tucson, is different from the distribution of car colors across North America, she selected a random sample of 300 cars in Oro Valley. The table shows the distribution of car color for Cass's sample in Oro Valley and the distribution of car color in North America, according to www.ppg.com.[3]

Color	White	Black	Gray	Silver	Red	Blue	Green	Other	Total
Oro Valley sample	84	38	31	46	27	29	6	39	300
North America	23%	18%	16%	15%	10%	9%	2%	7%	100%

1. Do these data provide convincing evidence that the distribution of car color in Oro Valley differs from the North American distribution?
2. If there is convincing evidence of a difference in the distribution of car color, perform a follow-up analysis.

SECTION 8A Summary

- The **chi-square test for goodness of fit** tests the null hypothesis that a categorical variable has a specified distribution in the population of interest. The alternative hypothesis is that the variable does not have the specified distribution in the population of interest.
- This test compares the **observed count** in each category with the counts that would be expected if H_0 were true. The **expected count** for any category is found by multiplying the sample size by the proportion specified by the null hypothesis for that category.
- To ensure that the observations in the sample can be viewed as independent and that the sampling distribution of the χ^2 test statistic can be modeled by a chi-square distribution, the following conditions must be met:
  - Random: The data come from a random sample from the population of interest.
    - 10%: When sampling without replacement, $n < 0.10N$.
  - Large Counts: All expected counts are at least 5.

- The **chi-square test statistic** χ^2 is

$$\chi^2 = \sum \frac{(\text{Observed} - \text{Expected})^2}{\text{Expected}}$$

where Observed is the observed count, Expected is the expected count, and the sum is over all possible categories.

- Large values of χ^2 are evidence against H_0 and in favor of H_a. The P-value is the area to the right of χ^2 under the chi-square density curve with degrees of freedom df = number of categories − 1.
- If the test finds a statistically significant result, consider doing a *follow-up analysis* that identifies the largest contributions to the chi-square test statistic and compares the observed and expected counts in the identified categories.

AP® EXAM TIP

AP® Daily Videos

Review the content of this section and get extra help by watching the AP® Daily Videos for Topics 8.1, 8.2, and 8.3, which are available in AP® Classroom.

8A Tech Corners

TI-Nspire and other technology instructions are on the book's website at bfwpub.com/tps7e.

SECTION 8A Exercises

Stating Hypotheses for Tests about the Distribution of a Categorical Variable

1. pg 747 **Covid-19 and blood type** Are people with certain blood types more susceptible to developing Covid-19? A study in Denmark used medical records to identify the blood types of 7422 randomly selected individuals who tested positive for Covid-19 between March 2020 and July 2020.[4] The table shows the distribution of blood type for these individuals and the distribution of blood type in the population of Denmark. Do these data provide convincing evidence that the distribution of blood type for Danish individuals who test positive for Covid-19 differs from the distribution of blood type in the population of Denmark?

Blood type	0	A	B	AB	Total
Number in Covid-19–positive sample	2851	3296	897	378	7422
Population percentage	41.7	42.4	11.4	4.5	100.0

(a) State appropriate hypotheses for a test that addresses this question.

(b) Calculate the proportion of people in the sample with each blood type.

(c) Explain how the proportions in part (b) give some evidence for the alternative hypothesis.

2. **Fruit flies** Biologists wish to mate pairs of randomly selected fruit flies having genetic makeup RrSs, indicating that each has one dominant gene (R) and one recessive gene (r) for eye color, along with one dominant (S) and one recessive (s) gene for wing type. Each offspring will receive one gene for each of the two traits from each parent, so the biologists predict that the following phenotypes should occur in a ratio of 9:3:3:1. Do these data provide convincing evidence that the predicted 9:3:3:1 ratio is incorrect?

Phenotype	Red eyes and straight wings	Red eyes and curly wings	White eyes and straight wings	White eyes and curly wings
Frequency	99	42	49	10

(a) State appropriate hypotheses for a test that addresses the biologists' prediction.

(b) Calculate the proportion of fruit flies with each phenotype.

(c) Explain how the proportions in part (b) give some evidence for the alternative hypothesis.

3. **Animal crackers** Are the different animals equally likely to be present in a box of animal crackers? To investigate, a student opened a box and recorded the number of each type of animal in the box. The

table shows the distribution of animal in the student's sample. State the appropriate hypotheses for testing whether the animals are equally likely.

Animal	Cow	Horse	Buffalo	Moose	Elephant
Frequency	1	5	4	3	5
Animal	Camel	Goat	Polar bear	Donkey	Cat (tail)
Frequency	6	4	3	6	2
Animal	Lion	Cat (no tail)	Wombat		Total
Frequency	3	3	5		50

4. **A fair die?** As a project in her ceramics class, Carrie made a six-sided die. To investigate if each side was equally likely to show up, she rolled it 20 times and recorded the outcome of each roll. The table shows the results of her 20 rolls. State the appropriate hypotheses for testing the fairness of Carrie's die.

Side	1	2	3	4	5	6	Total
Number of rolls	3	2	5	1	7	2	20

Calculating Expected Counts for Tests about the Distribution of a Categorical Variable

5. **Expected blood types** Calculate the expected counts for the hypotheses stated in Exercise 1(a). pg 749

6. **Expected fruit flies** Calculate the expected counts for the hypotheses stated in Exercise 2(a).

7. **Expected animals** Refer to Exercise 3.

(a) Explain what it would mean for the null hypothesis to be true in this setting.

(b) Calculate the expected number of animals of each type if the null hypothesis is true.

8. **Expected rolls** Refer to Exercise 4.

(a) Explain what it would mean for the null hypothesis to be true in this setting.

(b) Calculate the expected number of rolls for each side if the null hypothesis is true.

Checking Conditions for Tests about the Distribution of a Categorical Variable

9. **Blood type conditions** Refer to Exercises 1 and 5. Check whether the conditions are met for performing a test about the distribution of a categorical variable. pg 750

10. **Fruit fly conditions** Refer to Exercises 2 and 6. Check whether the conditions are met for performing a test about the distribution of a categorical variable.

11. **Animal cracker conditions** Refer to Exercises 3 and 7. Explain why the Large Counts condition isn't met.

12. **Die roll conditions** Refer to Exercises 4 and 8. Explain why the Large Counts condition isn't met.

Calculating the Test Statistic and *P*-Value for Tests about the Distribution of a Categorical Variable

13. **χ^2 distributions** The figure shows a χ^2 distribution with df = 3 and a χ^2 distribution with df = 7.

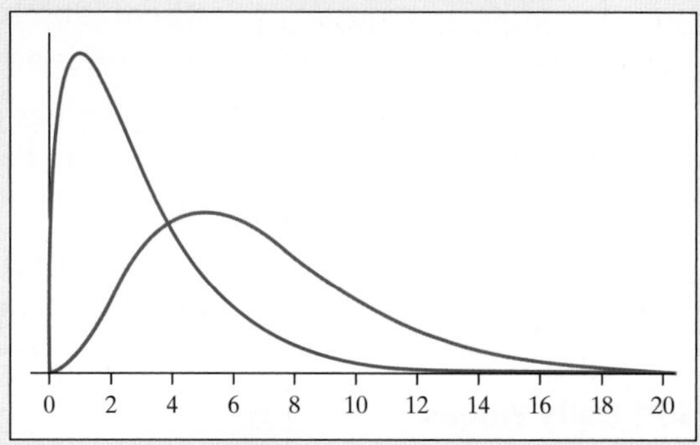

(a) Which density curve is the χ^2 distribution with df = 3? Explain your reasoning.

(b) In which χ^2 distribution is it more likely to get a χ^2 value greater than 10?

14. **More χ^2 distributions** The figure shows a χ^2 distribution with df = 2 and a χ^2 distribution with df = 5.

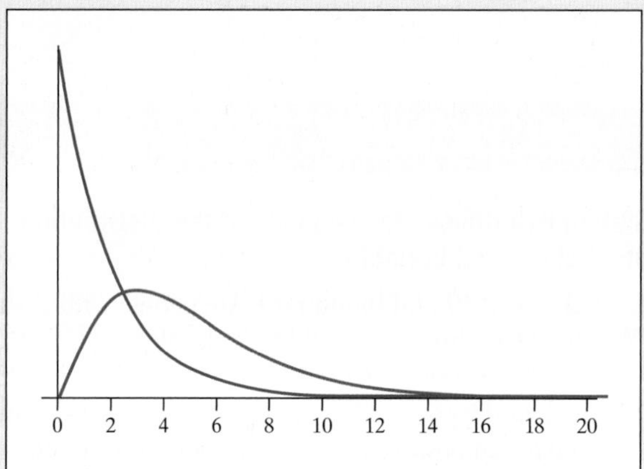

(a) Which density curve is the χ^2 distribution with df = 5? Explain your reasoning.

(b) In which χ^2 distribution is it more likely to get a χ^2 value greater than 10?

15. ***P*-values** For each of the following, estimate the *P*-value.

(a) $\chi^2 = 19.03$, df = 11 (b) $\chi^2 = 19.03$, df = 3

16. **More *P*-values** For each of the following, estimate the *P*-value.

(a) $\chi^2 = 4.49$, df = 5 (b) $\chi^2 = 4.49$, df = 1

17. **Testing blood types** In Exercises 1, 5, and 9, you read about a study that compared the distribution of blood type in Danish individuals who test positive for Covid-19 to the distribution of blood type in the population of Denmark. pg 755

(a) Calculate the chi-square test statistic.

(b) Find the *P*-value.

(c) Interpret the *P*-value.

18. Testing fruit flies In Exercises 2, 6, and 10, you read about a study that compared the distribution of phenotype for fruit fly offspring to the distribution predicted by genetic theory.

(a) Calculate the chi-square test statistic.

(b) Find the *P*-value.

(c) Interpret the *P*-value.

The Chi-Square Test for Goodness of Fit

19. **Chicken nuggets** According to McDonald's, Chicken
pg 758 McNuggets® come in four different shapes: Bone, Bell, Boot, and Ball. Are these shapes equally likely? To find out, Carlos and Nathaniel randomly selected 200 nuggets and identified the shape of each one.[5]

Shape	Bone	Bell	Boot	Ball
Count	50	40	59	51

(a) Do these data provide convincing evidence at the 5% significance level that the four McNuggets shapes are not all equally likely?

(b) Which type of error, Type I or Type II, could you have made in part (a)? Explain your answer.

20. Munching Froot Loops Kellogg's Froot Loops cereal comes in six colors: orange, yellow, purple, red, blue, and green. Charise randomly selected 120 loops and noted the color of each.

Color	Orange	Yellow	Purple	Red	Blue	Green
Count	28	21	16	25	14	16

(a) Do these data provide convincing evidence at the 5% significance level that Kellogg's Froot Loops do not contain an equal proportion of each color?

(b) Which type of error, Type I or Type II, could you have made in part (a)? Explain your answer.

21. Birds in the trees Researchers studied the behavior of birds that were searching for seeds and insects in an Oregon forest. In this forest, 54% of the trees are Douglas firs, 40% are ponderosa pines, and 6% are other types of trees. At a randomly selected time during the day, the researchers observed 156 red-breasted nuthatches: 70 were seen in Douglas firs, 79 in ponderosa pines, and 7 in other types of trees.[6]

(a) Do these data provide convincing evidence that nuthatches prefer particular types of trees when they're searching for seeds and insects?

(b) Relative to the proportion of each tree type in the forest, which type of trees do the nuthatches seem to prefer the most? The least?

22. Seagulls by the seashore Do seagulls show a preference for where they land? To answer this question, biologists conducted a study in an enclosed outdoor space with a piece of shore whose area was made up of 56% sand, 29% mud, and 15% rocks. The biologists chose 200 seagulls at random. Each seagull was released into the outdoor space on its own and observed until it landed somewhere on the piece of shore. In all, 128 seagulls landed on the sand, 61 landed in the mud, and 11 landed on the rocks.

(a) Do these data provide convincing evidence that seagulls show a preference for where they land?

(b) Relative to the proportion of each ground type on the shore, which type of ground do the seagulls seem to prefer the most? The least?

23. Mendel and the peas Gregor Mendel (1822–1884), an Austrian monk, is considered the father of genetics. Mendel studied the inheritance of various traits in pea plants. One such trait is whether the pea is smooth or wrinkled. Mendel predicted a ratio of 3 smooth peas for every 1 wrinkled pea. In one experiment, he observed 423 smooth and 133 wrinkled peas. Assume that the conditions for inference are met.

(a) Carry out a chi-square test for goodness of fit for the genetic model that Mendel predicted.

(b) In Section 6D, Exercise 9, you tested Mendel's prediction using a one-sample z test for a proportion. The hypotheses were H_0: $p = 0.75$ and H_a: $p \neq 0.75$, where $p =$ the true proportion of smooth peas. Calculate the z statistic and *P*-value for this test.

(c) How do the *P*-values from parts (a) and (b) compare? Calculate the value of z^2 and compare it to the value of χ^2. What do you notice?

24. No chi-square A school's principal selects a random sample of 50 students to learn about students' homework habits.

(a) The following table displays the average amount of time (in minutes) students reported spending on homework per night. Explain carefully why it would not be appropriate to perform a chi-square test for goodness of fit using these data.

Night	Sunday	Monday	Tuesday	Wednesday
Average time	130	108	115	104
Night	Thursday	Friday	Saturday	
Average time	99	37	62	

(b) The following table shows the number of students out of 50 who did all the assigned homework from each school day. Explain carefully why it would not be appropriate to perform a chi-square test for goodness of fit using these data.

School day	Monday	Tuesday	Wednesday	Thursday	Friday
Frequency	34	29	32	28	19

For Investigation *Apply the skills from the section in a new context or nonroutine way.*

25. **Carrot weights** A family claims that the weights of carrots from their garden are approximately normally distributed with a mean of 60 grams and a standard deviation of 10 grams. To test this claim, you select a random sample of 45 carrots from their garden and weigh them. The table summarizes the distribution of weight for the carrots in the sample.

Weight	Less than 50 g	50 to < 60 g	60 to < 70 g
Frequency	10	16	13
Weight	**At least 70 g**	**Total**	
Frequency	6	45	

(a) If the gardeners' claim is true, what proportion of the carrots should be in each of the weight categories?

(b) Calculate the expected number of carrots in each of the weight categories if the gardeners' claim is true.

(c) Calculate the value of the chi-square test statistic for these data, along with the *P*-value.

(d) What conclusion would you make?

26. **Fishing line** A certain brand of fishing line is advertised as having a breaking strength of 6 pounds. In addition to making sure that the mean breaking strength meets or exceeds the advertised weight, it is important that the breaking strength is consistent. It would certainly be a problem if some pieces of line could hold 10 pounds, but other pieces could hold only 2 pounds. Consistency of breaking strength implies a small standard deviation. An angler randomly selects 30 pieces of fishing line, attaches each to a bucket, and fills the bucket with water until the line breaks.[7] For this sample, $\bar{x} = 6.44$ pounds and $s_x = 0.75$ pound. Is there convincing evidence at the $\alpha = 0.05$ significance level that the true standard deviation of the breaking strength is greater than 0.5 pound?

(a) State appropriate hypotheses for this test. Be sure to define the parameter of interest.

(b) What is the evidence for the alternative hypothesis?

When the conditions for inference are met, the test statistic $\frac{(n-1)s_x^2}{\sigma^2}$ follows a χ^2 distribution with df $= n - 1$.

(c) Calculate the value of the test statistic and the *P*-value. Assume the conditions are met.

(d) What conclusion should you make?

27. **Is my random number generator working?** A student generated 200 random digits from 0 to 9 to test whether the random number generator on a calculator was working correctly. The table shows the distribution of digits produced by the calculator.

Digit	Frequency	Digit	Frequency
0	18	5	21
1	22	6	17
2	23	7	14
3	21	8	21
4	21	9	22

(a) Carry out a test at the $\alpha = 0.05$ significance level.

(b) Assuming that a student's calculator is working properly, what is the probability that the student will make a Type I error in part (a)?

(c) Suppose that the student tested 25 different calculators and that all of the calculators are working properly. If the student carries out a significance test for each calculator as in part (a), what is the probability that at least one of the tests results in a Type I error?

Multiple Choice *Select the best answer for each question.*

Exercises 28–30 refer to the following scenario. The manager of a high school cafeteria is planning to offer several new types of food for student lunches next school year. To find out if each type of food will be equally popular, the manager selects a random sample of 100 students and asks them, "Which type of food do you prefer: ramen, tacos, pizza, or burgers?" Here are the data:

Type of food	Ramen	Tacos	Pizza	Burgers
Count	18	22	39	21

28. Which of the following is an appropriate null hypothesis to test whether the food choices are equally popular?

(A) H_0: $\mu = 25$, where $\mu =$ the mean number of students who prefer each type of food.

(B) H_0: $p = 0.25$, where $p =$ the proportion of all students who prefer ramen.

(C) H_0: $n_R = n_T = n_P = n_H = 25$ where $n_R =$ the number of students who would choose ramen, and so on.

(D) H_0: $p_R = p_T = p_P = p_H = 0.25$ where $p_R =$ the proportion of students who would choose ramen, and so on.

(E) H_0: $\hat{p}_R = \hat{p}_T = \hat{p}_P = \hat{p}_H = 0.25$ where $\hat{p}_R =$ the proportion of students in the sample who chose ramen, and so on.

29. Which of the following is the value of the chi-square test statistic?

(A) $\frac{(18-25)^2}{25}+\frac{(22-25)^2}{25}+\frac{(39-25)^2}{25}+\frac{(21-25)^2}{25}$

(B) $\frac{(25-18)^2}{18}+\frac{(25-22)^2}{22}+\frac{(25-39)^2}{39}+\frac{(25-21)^2}{21}$

(C) $\frac{(18-25)}{25}+\frac{(22-25)}{25}+\frac{(39-25)}{25}+\frac{(21-25)}{25}$

(D) $\frac{(18-25)^2}{100}+\frac{(22-25)^2}{100}+\frac{(39-25)^2}{100}+\frac{(21-25)^2}{100}$

(E) $\frac{(0.18-0.25)^2}{0.25}+\frac{(0.22-0.25)^2}{0.25}+\frac{(0.39-0.25)^2}{0.25}+\frac{(0.21+0.25)^2}{0.25}$

30. The *P*-value for a chi-square test for goodness of fit is 0.0129. Which of the following is the most appropriate conclusion at a significance level of 0.05?

(A) Because 0.0129 is less than $\alpha = 0.05$, reject H_0. There is convincing evidence that the food choices are equally popular.

(B) Because 0.0129 is less than $\alpha = 0.05$, reject H_0. There is not convincing evidence that the food choices are equally popular.

(C) Because 0.0129 is less than $\alpha = 0.05$, reject H_0. There is convincing evidence that the food choices are not equally popular.

(D) Because 0.0129 is less than $\alpha = 0.05$, fail to reject H_0. There is not convincing evidence that the food choices are equally popular.

(E) Because 0.0129 is less than $\alpha = 0.05$, fail to reject H_0. There is convincing evidence that the food choices are equally popular.

31. Which of the following is *false*?

(A) A chi-square distribution with k degrees of freedom is more right-skewed than a chi-square distribution with $k+1$ degrees of freedom.

(B) A chi-square distribution never takes negative values.

(C) The degrees of freedom for a chi-square test for goodness of fit are determined by the sample size.

(D) $P(\chi^2 > 10)$ is greater when df $= k+1$ than when df $= k$.

(E) The total area under a chi-square density curve is always equal to 1.

32. When geneticists cross tall cut-leaf tomatoes with dwarf potato-leaf tomatoes, they expect the resulting plants will have phenotypes in the ratio 9:3:3:1. What is the minimum number of plants the geneticists should produce so that the Large Counts condition will be met?

(A) 1
(B) 5
(C) 30
(D) 80
(E) 256

33. A spinner has three colored segments (blue, red, yellow), each of different size. To test the spinner, you spin it 50 times and record the color the spinner lands on each time. Given the following chi-square statistic, which could be the null hypothesis for this test?

$$\chi^2 = \frac{(35-30)^2}{30}+\frac{(11-15)^2}{15}+\frac{(4-5)^2}{5}$$

(A) H_0: $p_{Blue} = 0.30$, $p_{Red} = 0.15$, $p_{Yellow} = 0.05$

(B) H_0: $p_{Blue} = 0.35$, $p_{Red} = 0.11$, $p_{Yellow} = 0.04$

(C) H_0: $p_{Blue} = 0.60$, $p_{Red} = 0.30$, $p_{Yellow} = 0.10$

(D) H_0: $p_{Blue} = 0.70$, $p_{Red} = 0.22$, $p_{Yellow} = 0.08$

(E) H_0: $p_{Blue} = 0.33$, $p_{Red} = 0.33$, $p_{Yellow} = 0.33$

Recycle and Review *Practice what you learned in previous sections.*

34. **Mountain sickness (2A, 3C)** Acute mountain sickness can be a problem for people in the mountains once they get above 2000 m (6500 feet). Himalayan trekkers often go far above that height, so they are especially susceptible to this illness. Researchers wondered whether gingko biloba, acetazolamide, or both would help trekkers to avoid acute mountain sickness. They devised a randomized, double-blind, placebo-controlled experiment to evaluate whether any of these treatments would help. The subjects were 487 Western trekkers in the Himalayas.[8] The two-way table summarizes the results of the experiment:

		Treatment				
		Placebo	Acetazolamide	Gingko biloba	Both	Total
Acute mountain sickness	Yes	40	14	43	18	115
	No	79	104	81	108	372
	Total	119	118	124	126	487

(a) Explain what the terms *randomized*, *double-blind*, and *placebo-controlled* mean in this context.

(b) Construct a segmented bar graph to display the relationship between the treatment and whether subjects had acute mountain sickness.

(c) Describe the association shown in the segmented bar graph from part (b).

35. **Reading for fun (1D, 7C)** Do students who read more books for pleasure tend to earn higher grades in

English? The boxplots and statistics summarize data from a simple random sample of 79 students at a large high school. Students were classified as light readers if they read fewer than 3 books for pleasure per year. Otherwise, they were classified as heavy readers. Each student's average English grade for the previous two marking periods was converted to a GPA scale, where A = 4.0, A− = 3.7, B+ = 3.3 and so on.

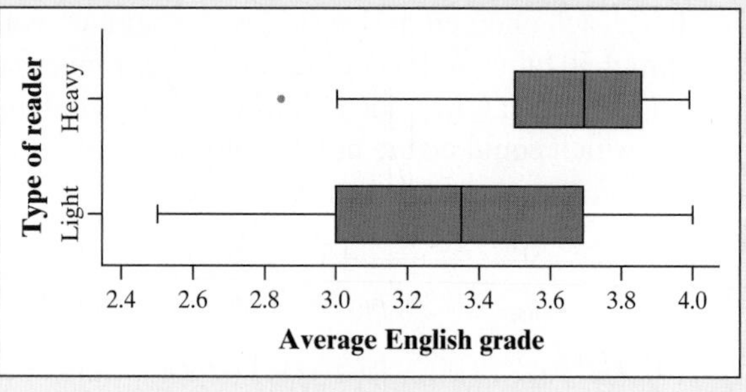

Type of reader	n	Mean	StDev	SE Mean
Heavy	47	3.640	0.324	0.047
Light	32	3.356	0.380	0.067

(a) Write a few sentences comparing the distributions of average English grade for light and heavy readers.

(b) Construct and interpret a 95% confidence interval for the difference in the mean English grade for light and heavy readers. Note that it is reasonable to treat these two groups as independent random samples even though researchers selected one random sample and split it into two groups.

(c) Does the interval in part (b) provide convincing evidence that reading more causes a difference in students' English grades? Justify your answer.

SECTION 8B Chi-Square Tests for Independence or Homogeneity

LEARNING TARGETS *By the end of the section, you should be able to:*

- State appropriate hypotheses for a test about the relationship between two categorical variables.
- Calculate expected counts for a test about the relationship between two categorical variables.
- Check the conditions for a test about the relationship between two categorical variables.
- Calculate the test statistic and *P*-value for a test about the relationship between two categorical variables.
- Perform a chi-square test for independence.
- Perform a chi-square test for homogeneity.
- Distinguish between a chi-square test for independence and a chi-square test for homogeneity.

In Section 2A, you learned how to analyze the relationship between two categorical variables, such as snowmobile use and environmental club membership for winter visitors to Yellowstone National Park. Here is a two-way table summarizing the data from a random sample of 1526 winter visitors to Yellowstone:

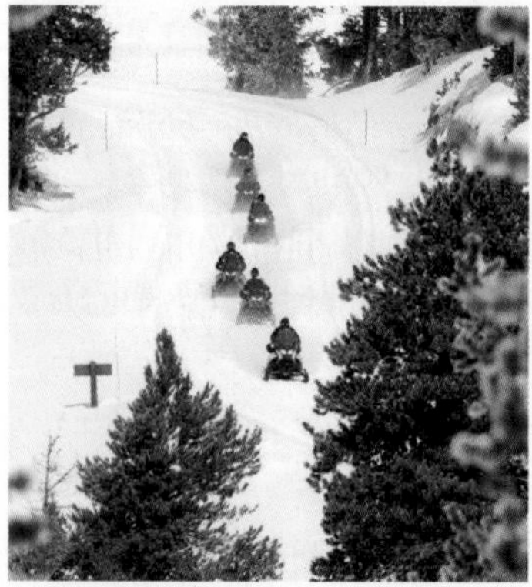
Grant Faint/Moment/Getty Images

		Environmental club membership		
		No	Yes	Total
Snowmobile use	Never	445	212	657
	Rent	497	77	574
	Own	279	16	295
	Total	1221	305	1526

To begin the analysis, make a side-by-side bar graph, a segmented bar graph, or a mosaic plot. Then look for an association. Recall that two variables have an *association* if knowing the value of one variable helps to predict the value of the other variable.

The segmented bar graph in Figure 8.5 shows an association between environmental club membership and snowmobile use for members of the Yellowstone sample. Environmental club members were much less likely to rent (about 25% versus 41%) or own (about 5% versus 23%) snowmobiles than non-club members, and were more likely to have never used a snowmobile (about 70% versus 36%). In other words, knowing whether a person in the sample is an environmental club member helps us predict that individual's snowmobile use.

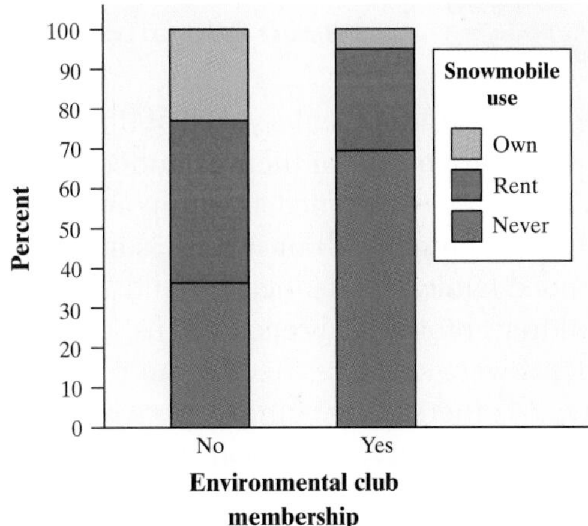

FIGURE 8.5 Segmented bar graph displaying the distribution of snowmobile use among environmental club members and among non-club members for 1526 randomly selected winter visitors to Yellowstone National Park.

Is it reasonable to say that there is an association between environmental club membership and snowmobile use in the *population* of all winter visitors to Yellowstone National Park? Or is it plausible that there is no association between these variables in the population and the association in the sample happened by chance alone? Making this determination requires a significance test about the relationship between two categorical variables.

Stating Hypotheses for Tests about the Relationship Between Two Categorical Variables

As with every other significance test, we start by stating hypotheses. In a test about the relationship between two categorical variables, the null hypothesis is that there is *no association* between the two variables in the population of interest. The alternative says that there *is* an association.

Here are the hypotheses for the example about winter visitors to Yellowstone National Park:

H_0: There is no association between snowmobile use and environmental club membership in the population of winter visitors to Yellowstone National Park.

H_a: There is an association between snowmobile use and environmental club membership in the population of winter visitors to Yellowstone National Park.

No association between two variables means that knowing the value of one variable does not help us predict the value of the other. That is, the variables are *independent*. An equivalent way to state the preceding hypotheses is as follows:

H_0: Snowmobile use and environmental club membership are independent in the population of winter visitors to Yellowstone National Park.

H_a: Snowmobile use and environmental club membership are not independent in the population of winter visitors to Yellowstone National Park.

AP® EXAM TIP

Many students lose credit on the AP® Statistics exam because they reverse the null and alternative hypotheses in a test about the relationship between two categorical variables. Remember that the *null* hypothesis is the *dull* hypothesis: there is *no association* between the two variables.

We could substitute the word *dependent* in place of *not independent* in the alternative hypothesis. We'll avoid this practice, however, because saying that two variables are dependent sounds too much like saying that changes in one variable cause changes in the other.

EXAMPLE

Living close to family

Stating hypotheses for tests about the relationship between two categorical variables

Skill 1.F

PROBLEM: The Pew Research Center asked 500 randomly selected U.S. adults about the proportion of their extended family members who live nearby. "Nearby" was defined as within an hour's drive, and the response options included "all or most," "some," "only a few," and "none." Extended family was defined as children, parents, grandparents, grandchildren, brothers, sisters, cousins, aunts, uncles, and in-laws. Respondents were also classified by the region of the country where they live.[9] Is there convincing evidence of an association between region of the country and proportion of extended family nearby for U.S. adults? The two-way table summarizes the data:

Lawrence Manning/Corbis/Getty Images

		Region				
		Northeast	South	Midwest	West	Total
Proportion of extended family nearby	None	13	45	21	28	107
	Only a few	22	50	29	22	123
	Some	31	47	45	29	152
	All or most	18	45	31	24	118
	Total	84	187	126	103	500

(a) State appropriate hypotheses for a test about the relationship between region of the country and proportion of extended family nearby.
(b) Calculate the proportion in each region who responded "none."
(c) Explain how your answer to part (b) gives some evidence for the alternative hypothesis.

SOLUTION:

(a) H_0: There is no association between region of the country and proportion of extended family nearby in the population of U.S. adults.

H_a: There is an association between region of the country and proportion of extended family nearby in the population of U.S. adults.

> You could also state the hypotheses as
> H_0: Region of the country and proportion of extended family nearby are independent in the population of U.S. adults.
> H_a: Region of the country and proportion of extended family nearby are not independent in the population of U.S. adults.

(b) Northeast: $13/84 = 0.155$, South: $45/187 = 0.241$, Midwest: $21/126 = 0.167$, West: $28/103 = 0.272$

(c) There is some evidence for H_a because the proportion of people with no extended family nearby isn't the same in each region of the country. People who live in the South and West are more likely to have no extended family nearby.

> Knowing a person's region of the country helps to predict if the person has extended family nearby.

FOR PRACTICE, TRY EXERCISE 1

Calculating Expected Counts for Tests about the Relationship Between Two Categorical Variables

The two-way table shows the observed counts for the random sample of 1526 winter visitors to Yellowstone National Park.

Observed counts

		Environmental club membership		
		No	Yes	Total
Snowmobile use	Never	445	212	657
	Rent	497	77	574
	Own	279	16	295
	Total	1221	305	1526

As with chi-square tests for goodness of fit, we compare the observed counts with the counts that we would expect when the null hypothesis is true.

The null hypothesis says that there is no association between snowmobile use and environmental club membership. That is, the proportion of visitors who never use a snowmobile should be the same for club members and for non-club members if the null hypothesis is true. In the entire sample, $657/1526 = 43.1\%$ of the visitors never use a snowmobile. Therefore, we'd expect 43.1% of the club members and 43.1% of the non-club members to never use a snowmobile if the null hypothesis is true. Because there were 1221 non-club members in the sample, the expected count of non-club members who never use a snowmobile is

$$(1221)(0.431) = 526.3$$

Note that this calculation is a group size multiplied by a proportion determined by the null hypothesis, which is very similar to the way we calculate expected counts for the chi-square test for goodness of fit.

This calculation can be rewritten as

$$(1221)(0.431) = 1221\left(\frac{657}{1526}\right) = \frac{(657)(1221)}{1526}$$

The fraction on the right is the total number of people who never use a snowmobile (row total) multiplied by the total number of people who are non-club members (column total), divided by the total number of people in the sample (table total). This leads to the general rule for computing expected counts in a test about the relationship between two categorical variables.

CALCULATING EXPECTED COUNTS IN A TEST ABOUT THE RELATIONSHIP BETWEEN TWO CATEGORICAL VARIABLES

When H_0 is true, the expected count for any cell in a two-way table is

$$\text{expected count} = \frac{(\text{row total})(\text{column total})}{\text{table total}}$$

We can complete the table of expected counts by using the formula in each cell or by using the formula in some cells and subtracting to find the remaining cells. For example, if we know the expected count of non-club members who never use a snowmobile, we can subtract that from the total number who never use snowmobiles to get the expected count of club members who never use a snowmobile. Here is the completed table of expected counts. The expected count for non-club members who never use a snowmobile is slightly different (and more accurate) because we didn't have to do any intermediate rounding.

Expected counts

		Environmental club membership		
		No	Yes	Total
Snowmobile use	Never	$\frac{(657)(1221)}{1526} = 525.7$	$657 - 525.7 = 131.3$	657
	Rent	$\frac{(574)(1221)}{1526} = 459.3$	$574 - 459.3 = 114.7$	574
	Own	$1221 - 525.7 - 459.3 = 236.0$	$295 - 236.0 = 59.0$	295
	Total	1221	305	1526

Note that we used subtraction to calculate the expected counts for the cells in the last row and the cells in the last column. In other words, we needed to use the formula for expected counts in only two cells. This observation will be useful when we discuss degrees of freedom and calculating *P*-values for tests about the relationship between two categorical variables.

EXAMPLE

Skill 3.A

Living close to family
Calculating expected counts for tests about the relationship between two categorical variables

PROBLEM: Here again is the two-way table summarizing the relationship between region of the country and proportion of extended family who live nearby for a random sample of 500 U.S. adults. Calculate the expected counts for a test of the null hypothesis that there is no association between region of the country and proportion of extended family who live nearby in the population of U.S. adults.

Jack Hollingsworth/Getty Images

		Region				
		Northeast	**South**	**Midwest**	**West**	**Total**
Proportion of extended family nearby	None	13	45	21	28	107
	Only a few	22	50	29	22	123
	Some	31	47	45	29	152
	All or most	18	45	31	24	118
	Total	84	187	126	103	500

SOLUTION:

$$\text{expected count} = \frac{(\text{row total})(\text{column total})}{\text{table total}}$$

		Region				
		Northeast	South	Midwest	West	Total
Proportion of extended family nearby	None	$\frac{(107)(84)}{500} = 18.0$	$\frac{(107)(187)}{500} = 40.0$	$\frac{(107)(126)}{500} = 27.0$	$107 - 18.0 - 40.0 - 27.0 = 22.0$	107
	Only a few	$\frac{(123)(84)}{500} = 20.7$	$\frac{(123)(187)}{500} = 46.0$	$\frac{(123)(126)}{500} = 31.0$	$123 - 20.7 - 46.0 - 31.0 = 25.3$	123
	Some	$\frac{(152)(84)}{500} = 25.5$	$\frac{(152)(187)}{500} = 56.8$	$\frac{(152)(126)}{500} = 38.3$	$152 - 25.5 - 56.8 - 38.3 = 31.4$	152
	All or most	$84 - 18.0 - 20.7 - 25.5 = 19.8$	$187 - 40.0 - 46.0 - 56.8 = 44.2$	$126 - 27.0 - 31.0 - 38.3 = 29.7$	$103 - 22.0 - 25.3 - 31.4 = 24.3$	118
	Total	84	187	126	103	500

FOR PRACTICE, TRY EXERCISE 3

Checking Conditions for Tests about the Relationship Between Two Categorical Variables

As with the chi-square test for goodness of fit, we need to check the Random, 10%, and Large Counts conditions before we perform a test about the relationship between two categorical variables. This ensures independence in the data collection process and verifies that a chi-square distribution will be a good model for the sampling distribution of the chi-square test statistic. Good news! The conditions are exactly the same as in the test for goodness of fit.

CONDITIONS FOR PERFORMING A TEST ABOUT THE RELATIONSHIP BETWEEN TWO CATEGORICAL VARIABLES

- **Random:** The data come from a random sample from the population of interest.
  - **10%:** When sampling without replacement, $n < 0.10N$.
- **Large Counts:** All *expected* counts are at least 5.

Expected counts

		Environmental club membership		
		No	Yes	Total
Snowmobile use	Never	525.7	131.3	657
	Rent	459.3	114.7	574
	Own	236.0	59.0	295
	Total	1221	305	1526

For the Yellowstone data, these three conditions are met:

- Random: The data came from a random sample of 1526 winter visitors to Yellowstone National Park. ✓
  - 10%: We assume 1526 is less than 10% of all winter visitors to Yellowstone National Park. ✓
- Large Counts: All the expected counts are at least 5, as shown in the table. ✓

EXAMPLE

Living close to family
Checking conditions for tests about the relationship between two categorical variables

Skill 4.C

fotostorm/E+/Getty Images

PROBLEM: Refer to the preceding examples. Check if the conditions are met for performing a test about the relationship between region of the country and proportion of extended family who live nearby in the population of U.S. adults.

SOLUTION:

- Random: Random sample of 500 U.S. adults. ✓
  - 10%: 500 < 10% of all U.S. adults. ✓
- Large Counts: All the expected counts (18.0, 40.0, 27.0, 22.0, 20.7, 46.0, 31.0, 25.3, 25.5, 56.8, 38.3, 31.4, 19.8, 44.2, 29.7, 24.3) are at least 5. ✓

Make sure to reference the values of the *expected* counts when checking the Large Counts condition.

FOR PRACTICE, TRY EXERCISE 5

Calculating the Test Statistic and *P*-Value for Tests about the Relationship Between Two Categorical Variables

In a test about the relationship between two categorical variables, we use the familiar chi-square test statistic to measure how much the observed counts differ from the expected counts, relative to the expected counts. This time, the sum is over all the cells in the two-way table, not including the totals.

AP® EXAM TIP
Formula sheet
The formula for the chi-square test statistic is included on the formula sheet provided on both sections of the AP® Statistics exam.

$$\chi^2 = \sum \frac{(\text{Observed} - \text{Expected})^2}{\text{Expected}}$$

The following computer output shows the observed and expected counts for the Yellowstone data. Note that it displays both counts in the same cell, with the expected counts appearing below the observed counts.

```
Chi-square test: No, Yes
Expected counts are printed
below observed counts
            No     Yes   Total
Never      445     212     657
         525.7   131.3
Rent       497      77     574
         459.3   114.7
Own        279      16     295
         236.0    59.0
Total     1221     305    1526
```

The chi-square test statistic is

$$\chi^2 = \frac{(445-525.7)^2}{525.7} + \frac{(212-131.3)^2}{131.3} + \ldots + \frac{(16-59.0)^2}{59.0} = 116.6$$

Once again, large values of χ^2 are evidence against H_0 and for H_a. The *P*-value measures the strength of this evidence. When the conditions for inference are met, we calculate *P*-values using a chi-square distribution with

$$\text{df} = (\text{number of rows} - 1)(\text{number of columns} - 1)$$

For the Yellowstone data, df $= (3-1)(2-1) = 2$. The *P*-value is the probability of getting a value of χ^2 as large as or larger than 116.6 when H_0 is true. Figure 8.6 shows this probability as an area under the chi-square density curve with 2 degrees of freedom.

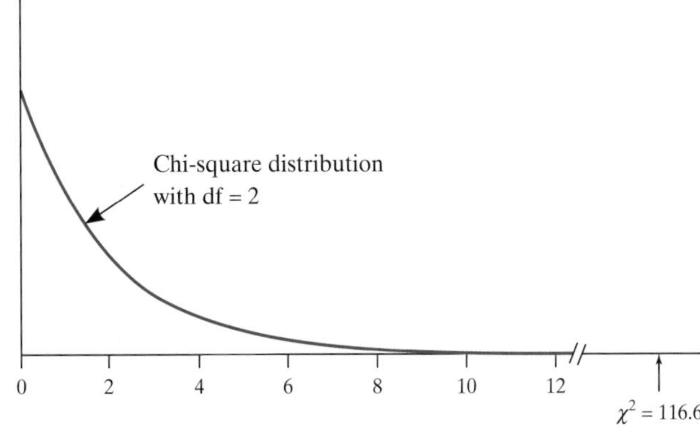

FIGURE 8.6 The *P*-value for a test about the relationship between snowmobile use and environmental club membership in the population of winter visitors to Yellowstone National Park.

Using Table C, look in the df = 2 row and notice that 116.6 is larger than any of the χ^2 critical values provided. Thus, the *P*-value < 0.0005. Using technology, we can get a more precise *P*-value: χ^2cdf(lower: 116.6, upper: 1000, df: 2) ≈ 0. Assuming there is no association between snowmobile use and environmental club membership in the population of winter visitors to Yellowstone National Park, there is about a 0 probability of getting an association as strong as or stronger than the association observed in the sample of 1526 winter visitors.

CALCULATING THE TEST STATISTIC AND *P*-VALUE IN A TEST ABOUT THE RELATIONSHIP BETWEEN TWO CATEGORICAL VARIABLES

Suppose the conditions are met. To perform a test of the null hypothesis

H_0: There is no association between two categorical variables in the population of interest

compute the chi-square test statistic:

$$\chi^2 = \sum \frac{(\text{Observed} - \text{Expected})^2}{\text{Expected}}$$

where the sum is over all cells in the two-way table (not including the totals). Find the *P*-value by calculating the probability of getting a χ^2 statistic this large or larger in a chi-square distribution with df = (number of rows – 1)(number of columns – 1).

EXAMPLE

Skills 3.E, 4.B, 4.E

Living close to family

Calculating the test statistic and *P*-value for tests about the relationship between two categorical variables

PROBLEM: Refer to the preceding examples. The computer output shows the observed and expected counts for a test of H_0: There is no association between region of the country and proportion of extended family who live nearby in the population of U.S. adults.

kate_sept2004/E+/Getty Images

Chi-square test: Northeast, South, Midwest, West

Expected counts are printed below observed counts

	Northeast	South	Midwest	West	Total
None	13	45	21	28	107
	18.0	40.0	27.0	22.0	
Only a few	22	50	29	22	123
	20.7	46.0	31.0	25.3	
Some	31	47	45	29	152
	25.5	56.8	38.3	31.4	
All or most	18	45	31	24	118
	19.8	44.2	29.7	24.3	
Total	84	187	126	103	500

(a) Calculate the test statistic and *P*-value.
(b) Interpret the *P*-value.
(c) What conclusion would you make at the 1% significance level?

SOLUTION:

(a) $$\chi^2 = \frac{(13-18.0)^2}{18.0} + \frac{(45-40.0)^2}{40.0} + \ldots + \frac{(24-24.3)^2}{24.3} = 10.44$$

$$\chi^2 = \sum \frac{(\text{Observed} - \text{Expected})^2}{\text{Expected}}$$

df $= (4-1)(4-1) = 9$

df = (number of rows − 1)(number of columns − 1)

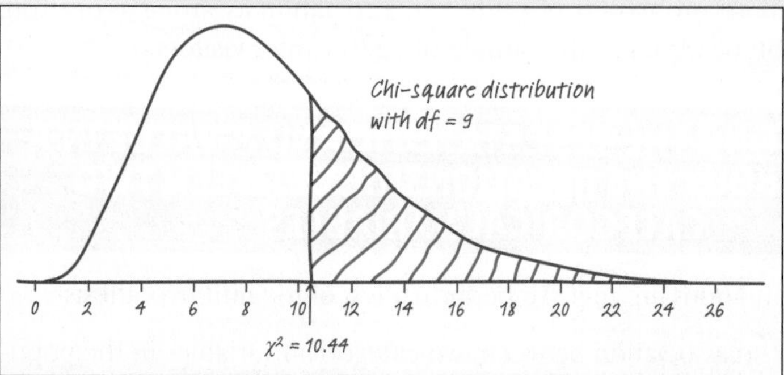

Using technology: χ^2cdf(lower: 10.44, upper: 1000, df: 9) = 0.3161

Using Table C: *P*-value > 0.25

(b) Assuming there is no association between region of the country and proportion of extended family who live nearby in the population of U.S. adults, there is a 0.3161 probability of getting an association as strong as or stronger than the association in the sample by chance alone.

(c) Because the *P*-value of 0.3161 > α = 0.01, we fail to reject H_0. There is not convincing evidence of an association between region of the country and proportion of extended family who live nearby in the population of U.S. adults.

FOR PRACTICE, TRY EXERCISE 9

As you might have suspected, your calculator will compute the chi-square test statistic and *P*-value in a test about the relationship between two categorical variables given only the observed counts.

28. Tech Corner

SIGNIFICANCE TESTS FOR THE RELATIONSHIP BETWEEN TWO CATEGORICAL VARIABLES

TI-Nspire and other technology instructions are on the book's website at bfwpub.com/tps7e.

You can use the TI-83/84 to perform the calculations for a test about the relationship between two categorical variables. We'll use the data from the Yellowstone example to illustrate the steps.

1. Enter the observed counts in matrix [A].
   - Press 2nd X⁻¹ (matrix), arrow to EDIT, and choose A.
   - Enter the dimensions of the matrix: 3 × 2.
   - Enter the observed counts from the two-way table in the same locations in the matrix.

   Don't enter any totals into the matrix.

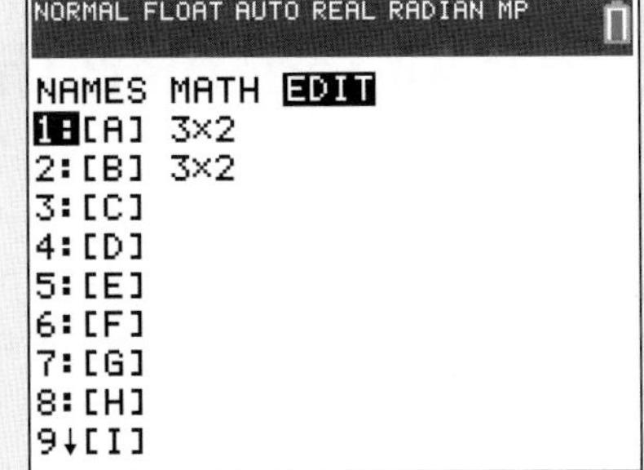

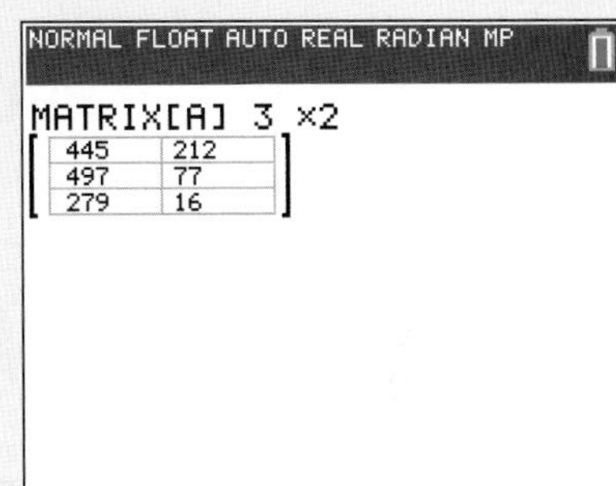

2. Press STAT, arrow over to TESTS, and choose χ^2 – Test…. Enter matrix [A] for the observed counts and matrix [B] for the expected counts, select Calculate, and press the ENTER button. You can find the matrix names in the matrix menu, if necessary. *Note:* You do not have to enter the expected counts in matrix [B]. Once you have run the test, the expected counts will be stored in matrix [B].

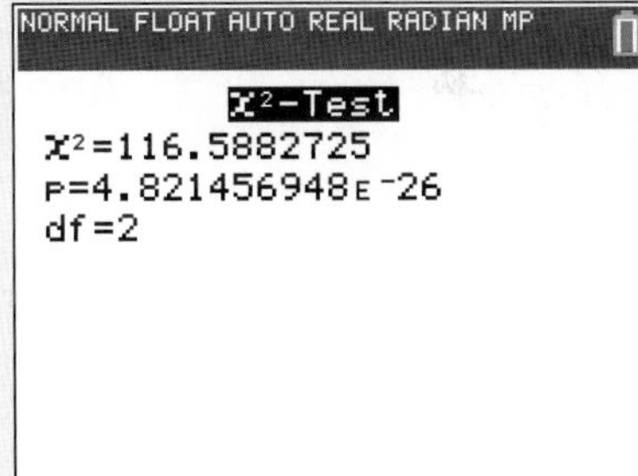

3. To see the expected counts, press 2nd X⁻¹ (matrix), arrow to EDIT, and choose [B].

Note: If you select the "Draw" option, you will get a picture of the appropriate χ^2 distribution with the area of interest shaded and the χ^2 test statistic and *P*-value labeled.

AP® EXAM TIP

You can use your calculator to carry out the mechanics of a significance test on the AP® Statistics exam — but there's a risk involved. If you give the calculator answer but don't show your work, and one or more of your values are incorrect, you will likely get no credit for the Do step. Make sure to include the χ^2 test statistic, df, and *P*-value in the Do step, along with all the elements of the remaining three steps (State, Plan, Conclude).

The χ^2 test statistic and *P*-value calculated using the "full technology" approach described in the Tech Corner are generally more accurate than the values we obtain when we round while doing our calculations. For that reason, we will use the values from the full technology approach when reporting results "Using technology" in the rest of this section.

Putting It All Together: The Chi-Square Test for Independence

The appropriate inference procedure for testing a claim about the relationship between two categorical variables is the **chi-square test for independence.**

DEFINITION Chi-square test for independence

A **chi-square test for independence** is a significance test of the null hypothesis that there is no association between two categorical variables in the population of interest.

Whenever you perform a chi-square test for independence, be sure to follow the four-step process.

EXAMPLE Anger and heart disease
The chi-square test for independence

Skills 1.E, 1.F, 3.A, 3.E, 4.C, 4.E

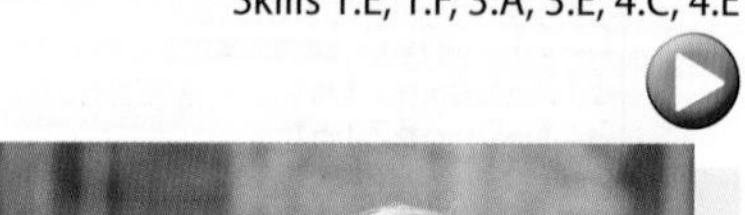

Francesco Carta fotografo/Getty Images

PROBLEM: Are people who are prone to sudden anger more likely to develop heart disease? A prospective observational study followed a random sample of 8474 people with normal blood pressure for about 4 years. All the individuals were free of heart disease at the beginning of the study. Each person took the Spielberger Trait Anger Scale test, which measures how prone a person is to sudden anger. The researchers then categorized each person's anger level as Low, Moderate, or High based on their test score. They also recorded whether each individual developed coronary heart disease. This classification includes people who had heart attacks and those who needed medical treatment for heart disease.[10] The two-way table summarizes the results of the study.

		Anger level			
		Low	Moderate	High	Total
Heart disease status	Yes	53	110	27	190
	No	3057	4621	606	8284
	Total	3110	4731	633	8474

Do the data provide convincing evidence at the 10% significance level of an association between anger level and heart disease status for people with normal blood pressure?

SOLUTION:

STATE:

H_0: There is no association between anger level and heart disease status in the population of people with normal blood pressure.

H_a: There is an association between anger level and heart disease status in the population of people with normal blood pressure.

Use $\alpha = 0.10$.

> Use the four-step process!

> You could also state the hypotheses as
> H_0: Anger level and heart disease status are independent in the population of people with normal blood pressure.
> H_a: Anger level and heart disease status are not independent in the population of people with normal blood pressure.

PLAN: Chi-square test for independence

- Random: Random sample of 8474 people with normal blood pressure. ✓
  - 10%: 8474 < 10% of people with normal blood pressure. ✓
- Large Counts: All the expected counts are at least 5 (see the following table). ✓

> Remember that expected counts should not be rounded to the nearest integer.

Expected counts

		Anger level			
		Low	Moderate	High	Total
Heart disease status	Yes	69.73	106.08	14.19	190
	No	3040.27	4624.92	618.81	8284
	Total	3110	4731	633	8474

DO:

- $\chi^2 = \frac{(53-69.73)^2}{69.73} + \frac{(110-106.08)^2}{106.08} + \ldots = 16.077$
- *P*-value: df $= (2-1)(3-1) = 2$

> Because the observed counts differ from the expected counts, there is *some* evidence for H_a.

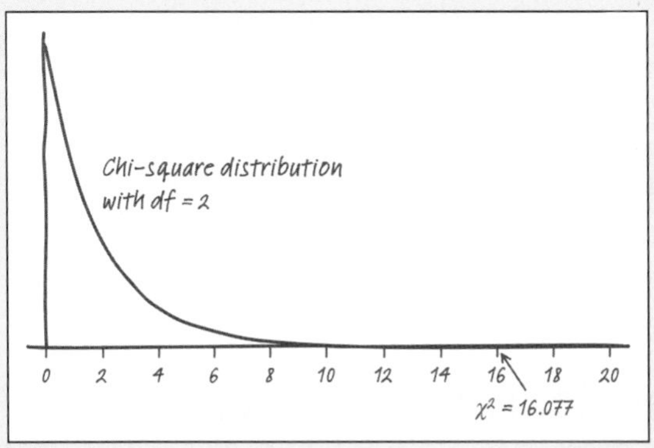

Using technology: $\chi^2 = 16.077$, *P*-value $= 0.0003$, df $= 2$

Using Table C: *P*-value < 0.0005

CONCLUDE: Because the *P*-value of $0.0003 < \alpha = 0.10$, we reject H_0. There is convincing evidence of an association between anger level and heart disease status in the population of people with normal blood pressure.

FOR PRACTICE, TRY EXERCISE 13

AP® EXAM TIP

When the *P*-value is very small, the calculator will report it using scientific notation. Remember that *P*-values are probabilities and must be between 0 and 1. If your calculator reports the *P*-value with a number that appears to be greater than 1, look to the right, and you will see that the *P*-value is being expressed in scientific notation. If you claim that the *P*-value is 3.228 instead of 0.0003228, you will certainly lose credit.

A follow-up analysis reveals that two cells contribute the most to the chi-square test statistic: Low anger, Yes (4.014) and High anger, Yes (11.564). Many fewer low-anger people developed heart disease than expected, and many more high-anger people got heart disease than expected.

Can we conclude that a propensity to quick anger *causes* heart disease? No. The anger and heart-disease study is an observational study, not an experiment. It isn't surprising that some other variables are confounded with anger level. For example, if people prone to sudden anger are more likely to drink alcohol and smoke and if people who drink alcohol and smoke are more likely to develop heart disease, then we would see an association between anger level and heart disease status even if anger level had no effect on heart disease.

To investigate the relationship between age and playing video games, the Pew Research Center asked randomly selected U.S. adults for their age and if they "ever play video games on a computer, TV, game console, or portable device like a cell phone."[11] Do the data summarized in the two-way table provide convincing evidence of an association between age and gaming status in the population of U.S. adults?

		Age group				
		18–29	30–49	50–64	65+	Total
Gaming status	Plays video games	887	1217	650	279	3033
	Doesn't play video games	429	872	958	840	3099
	Total	1316	2089	1608	1119	6132

The Chi-Square Test for Homogeneity

Two-way tables can summarize data from different types of studies. The examples involving the survey of winter visitors to Yellowstone National Park, the study of how close U.S. adults live to extended family, and the study of anger level and heart disease all analyzed the relationship between two categorical variables in a single population. We use the chi-square test for independence to perform inference in such settings.

Another common situation that leads to a two-way table is when we compare the distribution of a *single* categorical variable for *multiple* populations or treatments. We need a different chi-square test for these situations. Fortunately, the test that compares the distribution of a categorical variable for two or more

Ingram Publishing/Alamy Stock Photo

populations or treatments is very similar to the test about the relationship between two categorical variables.

Does background music influence what customers buy? One experiment in a European restaurant compared three randomly assigned treatments: no music, French accordion music, and Italian string music. Under each condition, the researchers recorded the number of customers who ordered French, Italian, and other entrées.[12] The two-way table summarizes the data from this experiment:

		Type of background music			
		None	French	Italian	Total
Entrée ordered	French	30	39	30	99
	Italian	11	1	19	31
	Other	43	35	35	113
	Total	84	75	84	243

STATING HYPOTHESES

The null hypothesis in this example is

H_0: There is no difference in the true distributions of entrées ordered at this restaurant when no music, French accordion music, or Italian string music is played.

In general, the null hypothesis in a test comparing the distribution of a categorical variable for two or more populations or treatments says that there is no difference in the distributions. As with the chi-square test for independence, the alternative hypothesis says that the null hypothesis isn't true:

H_a: There is a difference in the true distributions of entrées ordered at this restaurant when no music, French accordion music, or Italian string music is played.

The alternative hypothesis does not state that all of the distributions are different. Instead, the alternative hypothesis is true even if just one of the true distributions is different from the others. Consequently, any difference among the three observed distributions of entrées ordered is evidence against the null hypothesis and for the alternative hypothesis.

CHECKING CONDITIONS

As with the χ^2 test for independence, three conditions must be met to ensure independence in the data collection and verify that the sampling distribution of the χ^2 test statistic can be well modeled by a chi-square distribution. The Large Counts condition is the same as that for the test for independence, but the Random and 10% conditions are a little different in the test about the distribution of a categorical variable for two or more populations or treatments.

CONDITIONS FOR PERFORMING A TEST ABOUT THE DISTRIBUTION OF A CATEGORICAL VARIABLE FOR TWO OR MORE POPULATIONS OR TREATMENTS

- **Random:** The data come from independent random samples or from groups in a randomized experiment.
  - **10%:** When sampling without replacement, $n < 0.10N$ for each sample.
- **Large Counts:** All *expected* counts are at least 5.

Because a stratified random sample consists of a simple random sample from each stratum and these samples are independent, you could also use a test for homogeneity to compare the distribution of a categorical variable among the different strata.

In the background music experiment, the Random condition is met because researchers randomly assigned the three treatments. We don't need to check the 10% condition because the customers in the experiment weren't selected without replacement from a population. To check the Large Counts condition, we need to calculate the expected counts. Fortunately, the method for calculating expected counts is exactly the same as in a χ^2 test for independence.

CALCULATING EXPECTED COUNTS FOR TEST ABOUT THE DISTRIBUTION OF A CATEGORICAL VARIABLE FOR TWO OR MORE POPULATIONS OR TREATMENTS

When H_0 is true, the expected count for any cell in a two-way table is

$$\text{expected count} = \frac{(\text{row total})(\text{column total})}{\text{table total}}$$

The table shows the expected counts for the background music experiment. Because they are all at least 5, the Large Counts condition is met.

Expected counts

		Type of background music			
		None	French	Italian	Total
Entrée ordered	French	34.22	30.56	34.22	99
	Italian	10.72	9.57	10.72	31
	Other	39.06	34.88	39.06	113
	Total	84	75	84	243

CALCULATING THE TEST STATISTIC AND *P*-VALUE

As with the chi-square test for independence, we use the chi-square test statistic to measure how much the observed counts deviate from the expected counts, relative to the expected counts. To calculate the *P*-value, we use a chi-square distribution with df = (number of rows – 1)(number of columns – 1).

For the restaurant experiment,

$$\chi^2 = \frac{(30-34.22)^2}{34.22} + \frac{(39-30.56)^2}{30.56} + \ldots + \frac{(35-39.06)^2}{39.06} = 18.28$$

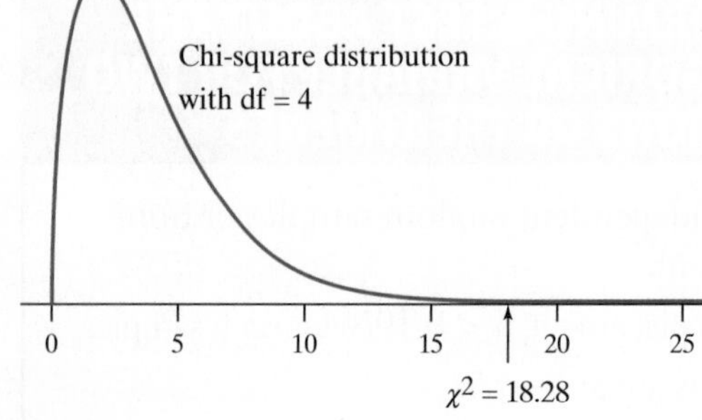

With df = (3 – 1)(3 – 1) = 4, the *P*-value is approximately 0.0011.

Assuming background music has no effect on which entrée a customer orders, there is a 0.0011 probability of getting distributions of entrée choice this different or more different by chance alone. Because the *P*-value of $0.0011 < \alpha = 0.05$, we reject H_0. There is convincing evidence of a difference in the true distributions of entrées ordered at this restaurant when no music, French accordion music, or Italian string music is played.

CALCULATING THE TEST STATISTIC AND *P*-VALUE IN A TEST ABOUT THE DISTRIBUTION OF A CATEGORICAL VARIABLE FOR TWO OR MORE POPULATIONS OR TREATMENTS

Suppose the conditions are met. To perform a test of the null hypothesis

H_0: There is no difference in the distributions of a categorical variable in the populations of interest or for the treatments in an experiment

compute the chi-square test statistic:

$$\chi^2 = \sum \frac{(\text{Observed} - \text{Expected})^2}{\text{Expected}}$$

where the sum is over all cells in the two-way table (not including the totals). Find the *P*-value by calculating the probability of getting a χ^2 statistic this large or larger in a chi-square distribution with df = (number of rows − 1)(number of columns − 1).

PUTTING IT ALL TOGETHER: THE CHI-SQUARE TEST FOR HOMOGENEITY

The appropriate inference method for testing a claim about the distribution of a categorical variable for two or more populations or treatments is the **chi-square test for homogeneity.**

DEFINITION Chi-square test for homogeneity

A **chi-square test for homogeneity** is a significance test of the null hypothesis that there is no difference in the distributions of a categorical variable in the populations of interest or for the treatments in an experiment.

Whenever you perform a chi-square test for homogeneity, be sure to follow the four-step process.

EXAMPLE

Speaking English
The chi-square test for homogeneity

Skills 1.E, 1.F, 3.A, 3.E, 4.C, 4.E

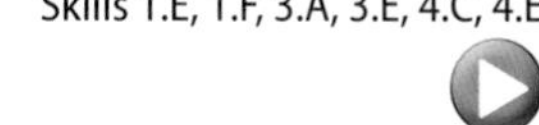

PROBLEM: The Pew Research Center conducts surveys about a variety of topics in many different countries. One such survey investigated how residents of different countries feel about the importance of speaking the national language.[13] Independent random samples of adult residents of Australia, the United Kingdom, and the United States were asked, "How important do you think it is to be able to speak English?" The two-way table summarizes the responses to this question.

Creative-Touch/DigitalVision Vectors/Getty Images

		Country			
		Australia	U.K.	U.S.	Total
Opinion about speaking English	Very important	690	1177	702	2569
	Somewhat important	250	242	221	713
	Not very important	40	28	50	118
	Not at all important	20	13	30	63
	Total	1000	1460	1003	3463

Do these data provide convincing evidence at the $\alpha = 0.05$ level that the distributions of opinion about speaking English differ for residents of Australia, the United Kingdom, and the United States?

SOLUTION:

STATE: H_0: There is no difference in the population distributions of opinion about speaking English for residents of Australia, the United Kingdom, and the United States.

Use the four-step process!

H_a: There is a difference in the population distributions of opinion about speaking English for residents of Australia, the United Kingdom, and the United States.

We'll use $\alpha = 0.05$.

PLAN: Chi-square test for homogeneity

- Random: Independent random samples of adult residents from the three countries. ✓
  - 10%: 1000 < 10% of all Australian residents, 1460 < 10% of all U.K. residents, and 1003 < 10% of all U.S. residents. ✓
- Large Counts: All expected counts ≥ 5 (see the following table). ✓

Expected counts

		Country			
		Australia	U.K.	U.S.	Total
Opinion about speaking English	Very important	741.8	1083.1	744.1	2569
	Somewhat important	205.9	300.6	206.5	713
	Not very important	34.1	49.7	34.2	118
	Not at all important	18.2	26.6	18.2	63
	Total	1000	1460	1003	3463

Because the observed counts differ from the expected counts, there is *some* evidence for H_a.

DO:

- Test statistic: $\chi^2 = \dfrac{(690-741.8)^2}{741.8} + \dfrac{(1177-1083.1)^2}{1083.1} + \ldots = 68.57$
- *P*-value: df = (4 − 1)(3 − 1) = 6

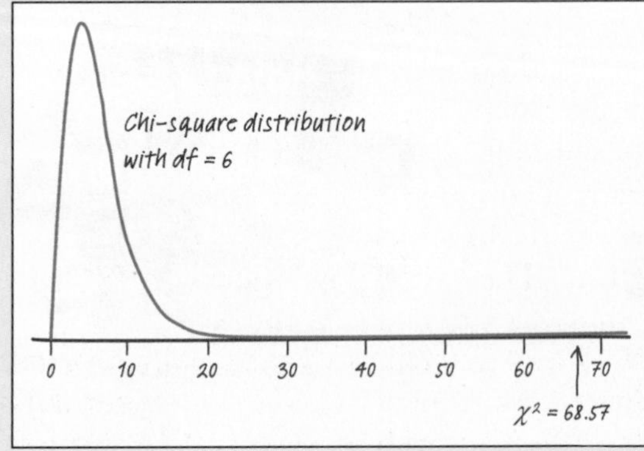

Using technology: $\chi^2 = 68.57$, P-value ≈ 0, df $= 6$

Using Table C: P-value < 0.0005

CONCLUDE: Because the P-value of approximately $0 < \alpha = 0.05$, we reject H_0. There is convincing evidence of a difference in the population distributions of opinion about speaking English for residents of Australia, the United Kingdom, and the United States.

If you want to know how the distributions differ, do a follow-up analysis.

FOR PRACTICE, TRY EXERCISE 17

AP® EXAM TIP

Many students lose credit on the AP® Statistics exam because they don't write down and label the expected counts in their response. It isn't enough to claim that all the expected counts are at least 5; you must provide clear evidence to support your claim.

Independence or Homogeneity?

Both the chi-square test for independence and the chi-square test for homogeneity start with a two-way table of observed counts. They even calculate the test statistic, degrees of freedom, and P-value in the same way. However, the questions answered by these two tests are different. The chi-square test for independence tests whether two categorical variables are associated in some population of interest. A chi-square test for homogeneity tests whether the distribution of a categorical variable is the same for each of several populations or treatments.

The key to distinguishing these two tests is to consider *how the data were produced.* If the data come from a single random sample, with the individuals classified according to two categorical variables, use a chi-square test for independence. In contrast, if the data come from two or more independent random samples or treatment groups in a randomized experiment, use a chi-square test for homogeneity.

Another way to help you distinguish these two tests is to consider the two sets of totals in the two-way table. In tests for homogeneity, one set of totals is known by the researchers *before the data are collected.* For example, in the observational study comparing opinions about speaking English, researchers knew in advance that they would survey 1000 people from Australia, 1460 from the United Kingdom, and 1003 from the United States.

Test for homogeneity

		Country			
		Australia	U.K.	U.S.	Total
Opinion about speaking English	Very important	?	?	?	?
	Somewhat important	?	?	?	?
	Not very important	?	?	?	?
	Not at all important	?	?	?	?
	Total	1000	1460	1003	3463

Likewise, in a randomized experiment, researchers know the number of individuals in each treatment group before the experiment begins. In both cases, only one set of totals was left to vary. This is consistent with the design of the study: select independent random samples (or randomly assign treatments) and compare the distribution of a single categorical variable.

However, in a test for independence, neither set of totals is known in advance. In the observational study about snowmobile use in Yellowstone, the researchers didn't know anything about either variable ahead of time — they only knew that they would survey 1526 winter visitors. This is consistent with the design of the study: select one sample and record the values of two variables for each member.

Test for independence

		Environmental club membership		
		No	Yes	Total
Snowmobile use	Never	?	?	?
	Rent	?	?	?
	Own	?	?	?
	Total	?	?	1526

EXAMPLE

A smash or a hit?
Independence or homogeneity?

Skills 1.E, 1.F, 4.A, 4.E

PROBLEM: Two researchers asked 150 volunteers to recall the details of a car accident they watched on video. Fifty of the volunteers were randomly assigned to respond to the question "About how fast were the cars going when they smashed into each other?" Another 50 volunteers were randomly assigned the same question, but with the words "smashed into" replaced by the word "hit." Researchers did not ask the remaining 50 volunteers — the control group — to estimate speed at all. A week later, the researchers asked all 150 volunteers if they saw any broken glass at the accident (there wasn't any). The two-way table summarizes each group's response to the broken glass question.[14]

Peter Stark/iStop/Getty Images

		Response		
		Yes	No	Total
Question wording	"Smashed into"	16	34	50
	"Hit"	7	43	50
	Control	6	44	50
	Total	29	121	150

(a) Which chi-square test should be used to analyze these data? Explain your reasoning.
(b) State the hypotheses for the test identified in part (a).
(c) The *P*-value for the appropriate test is 0.02. What conclusion would you make at the $\alpha = 0.01$ significance level?
(d) Name two ways you could increase the power of the test.

SOLUTION:

(a) *Because this was an experiment with randomly assigned treatments, a chi-square test for homogeneity should be used to analyze these data.*

Researchers determined in advance that there would be 50 volunteers in each group, so only one set of totals was free to vary. This indicates that a test of homogeneity is the correct choice.

(b) H_0: The true proportions of people like the ones in the study who would say there was broken glass are the same for all three treatments.

H_a: The true proportions of people like the ones in the study who would say there was broken glass are not the same for all three treatments.

When there are only two response categories, you can also state the hypotheses symbolically. In this case, H_0: $p_1 = p_2 = p_3$, where p_i is the true proportion of people like the ones in this experiment who would say there was broken glass for treatment i. The alternative hypothesis would be H_a: At least one of these proportions is different than the others.

(c) Because the P-value of $0.02 > \alpha = 0.01$, we fail to reject H_0. There is not convincing evidence that the proportions of people like the ones in the study who would say there was broken glass are not the same for all three treatments.

(d) To increase the power of the test, you could make the significance level greater than 0.01 or make the group sizes greater than 50.

FOR PRACTICE, TRY EXERCISE 25

Many studies involve comparing the proportion of successes for each of several populations or treatments. The two-sample z test from Section 6F allows us to test the null hypothesis H_0: $p_1 = p_2$, where p_1 and p_2 are the proportions of successes for the two populations or treatments. The chi-square test for homogeneity allows us to test H_0: $p_1 = p_2 = \ldots = p_k$. This null hypothesis says that there is no difference in the proportions of successes for the k populations or treatments. The alternative hypothesis is that at least one of the proportions is different.

In the preceding example, the null hypothesis is H_0: The true proportions of people like the ones in the study who would say there was broken glass are the same for all three treatments. The alternative hypothesis is that the proportions are not all the same. Many students *incorrectly state* H_a as "all the proportions are different." Think about it this way: the opposite of "all the proportions are the same" is "at least one of the proportions is not the same."

CHECK YOUR UNDERSTANDING

A random sample of 200 children from the United Kingdom and a random sample of 215 children from the United States were selected. For each child, researchers recorded the superpower the child would most like to have: the ability to fly, the ability to freeze time, invisibility, super strength, or telepathy (the ability to read minds). The data are summarized in the two-way table.[15] Is there convincing evidence that the distributions of superpower preference are different for children in the two countries?

		Country		
		U.K.	U.S.	Total
Superpower preference	Fly	54	45	99
	Freeze time	52	44	96
	Invisibility	30	37	67
	Super strength	20	23	43
	Telepathy	44	66	110
	Total	200	215	415

1. Explain how you know that a chi-square test for homogeneity is the appropriate test in this setting.
2. Perform the chi-square test for homogeneity using $\alpha = 0.05$.

SECTION 8B Summary

- We use the **chi-square test for independence** to test for an association between two categorical variables in a population of interest.
- The hypotheses for a test of independence are

  H_0: There is no association between two categorical variables in the population of interest.

  H_a: There is an association between two categorical variables in the population of interest.
- When performing a chi-square test for independence, we need to check that the observations in the sample can be viewed as independent and that the sampling distribution of the χ^2 test statistic can be modeled by a chi-square distribution. The required conditions are
  - Random: The data come from a random sample from the population of interest.
    - 10%: When sampling without replacement, $n < 0.10N$.
  - Large Counts: All expected counts must be at least 5.
- The **expected count** in any cell of a two-way table when H_0 is true is

$$\text{expected count} = \frac{(\text{row total})(\text{column total})}{\text{table total}}$$

- The chi-square test statistic is

$$\chi^2 = \sum \frac{(\text{Observed} - \text{Expected})^2}{\text{Expected}}$$

  where the sum is over all cells in the two-way table (not including the totals).
- Calculate the P-value by finding the area to the right of χ^2 in a chi-square distribution with df = (number of rows − 1)(number of columns − 1).
- We use the **chi-square test for homogeneity** to compare the distribution of a single categorical variable for each of several populations or treatments.
- The hypotheses for a test for homogeneity are

  H_0: There is no difference in the distributions of a categorical variable in the populations of interest or for the treatments in an experiment.

  H_a: There is a difference in the distributions of a categorical variable in the populations of interest or for the treatments in an experiment.
- When performing a chi-square test for homogeneity, we need to check for independence and verify that the sampling distribution of the χ^2 test statistic can be modeled by a chi-square distribution. The required conditions are
  - Random: The data come from independent random samples or groups in a randomized experiment.
    - 10%: When sampling without replacement, $n < 0.10N$ for each sample.
  - Large Counts: All expected counts must be at least 5.
- If a test for independence or a test for homogeneity finds a statistically significant result, consider doing a *follow-up analysis* that looks for the largest components of the chi-square test statistic and compares the observed and expected counts in the corresponding cells.

AP® EXAM TIP
AP® Daily Videos

Review the content of this section and get extra help by watching the AP® Daily Videos for Topics 8.4, 8.5, 8.6, and 8.7, which are available in AP® Classroom.

8B Tech Corner

TI-Nspire and other technology instructions are on the book's website at bfwpub.com/tps7e.

28. Significance tests for the relationship between two categorical variables Page 775

SECTION 8B Exercises

Stating Hypotheses for Tests about the Relationship Between Two Categorical Variables

1. **Relaxing in the sauna** Researchers followed a random sample of 2315 middle-aged men from eastern Finland for up to 30 years. They recorded how often each man went to a sauna and whether or not he experienced sudden cardiac death (SCD). The two-way table shows the data from the study.[16]

pg 768

		Weekly sauna frequency			
		1 or fewer	2–3	4 or more	Total
SCD Status	Yes	61	119	10	190
	No	540	1394	191	2125
	Total	601	1513	201	2315

(a) State appropriate hypotheses for a test about the relationship between weekly sauna frequency and SCD.

(b) Calculate the proportion of men in each weekly sauna frequency category who experienced SCD.

(c) Explain how your answer to part (b) gives some evidence for the alternative hypothesis.

2. **Napping and heart disease** In a long-term study of 3462 randomly selected adults from Lausanne, Switzerland, researchers investigated the relationship between weekly napping frequency and whether a person experienced a major cardiovascular disease (CVD) event, such as a heart attack or stroke. The two-way table shows the data from the study.[17]

		Napping frequency				
		None	1–2 weekly	3–5 weekly	6–7 weekly	Total
CVD status	Yes	93	12	22	28	155
	No	1921	655	389	342	3307
	Total	2014	667	411	370	3462

(a) State appropriate hypotheses for a test about the relationship between napping frequency and CVD.

(b) Calculate the proportion of people in each napping frequency category who experienced CVD.

(c) Explain how your answer to part (b) gives some evidence for the alternative hypothesis.

Calculating Expected Counts for Tests about the Relationship Between Two Categorical Variables

3. **Expected saunas** Calculate the expected counts for a test of the null hypothesis from Exercise 1.

pg 770

4. **Expected naps** Calculate the expected counts for a test of the null hypothesis from Exercise 2.

Checking Conditions for Tests about the Relationship Between Two Categorical Variables

5. **Sauna conditions** Refer to Exercises 1 and 3. Check if the conditions are met for performing a test about the relationship between weekly sauna frequency and SCD status in the population of middle-aged men from eastern Finland.

pg 772

6. **Nap conditions** Refer to Exercises 2 and 4. Check if the conditions are met for performing a test about the relationship between napping frequency and CVD status in the population of adults from Lausanne, Switzerland.

7. **Paying college athletes** Do sports fans think college athletes should be paid? Does their opinion depend on which region of the country the fan is from? The table summarizes the results of a survey of 707 randomly selected U.S. sports fans.[18]

		Should pay college athletes?			
		Yes	No	Unsure	Total
Region	Northeast	61	47	5	113
	Midwest	61	84	3	148
	South	146	124	6	276
	West	95	68	7	170
	Total	363	323	21	707

(a) Show that the Large Counts condition is not met.

(b) When the Large Counts condition isn't met, we can combine two or more columns (or two or more rows) so that the Large Counts condition will be satisfied. Combine two rows or two columns and show that the Large Counts condition is now met.

8. **Healthy trees** In Section 8A, one of the examples showed that trees on New York City streets were not proportionally distributed across the five boroughs of the city. What about the health of the trees? The table classifies 946 randomly selected trees on New York City streets by borough and health.[19]

		Health			
		Good	Fair	Poor	Total
Borough	Bronx	99	17	3	119
	Brooklyn	209	28	5	242
	Manhattan	56	12	7	75
	Queens	300	49	17	366
	Staten Island	122	15	7	144
	Total	786	121	39	946

(a) Show that the Large Counts condition is not met.

(b) When the Large Counts condition isn't met, we can combine two or more columns (or two or more rows) so that the Large Counts condition will be satisfied. Combine two rows or two columns and show that the Large Counts condition is now met.

Calculating the Test Statistic and *P*-Value for Tests about the Relationship Between Two Categorical Variables

9. **Sauna calculations** Refer to Exercises 1, 3, and 5 and the following computer output.

pg 774

```
Chi-square test: 1 or fewer, 2-3, 4 or more
Expected counts are printed below observed
counts
          1 or fewer     2-3   4 or more   Total
Yes               61     119          10     190
                49.3   124.2        16.5
No               540    1394         191    2125
               551.7  1388.8       184.5
Total            601    1513         201    2315
```

(a) Calculate the test statistic and *P*-value.

(b) Interpret the *P*-value.

(c) What conclusion would you make at the 5% significance level?

10. **Nap calculations** Refer to Exercises 2, 4, and 6 and the following computer output.

```
Chi-square test: none, 1-2 weekly,
3-5 weekly, 6-7 weekly
Expected counts are printed below observed
counts
           None     1-2      3-5      6-7   Total
                 weekly   weekly   weekly
Yes          93      12       22       28     155
           90.2    29.9     18.4     16.6
No         1921     655      389      342    3307
         1923.8   637.1    392.6    353.4
Total      2014     667      411      370    3462
```

(a) Calculate the test statistic and *P*-value.

(b) Interpret the *P*-value.

(c) What conclusion would you make at the 5% significance level?

11. **More saunas, better health?** Based on the evidence for H_a in Exercise 1 and your conclusion in Exercise 9, should people go to the sauna more often if they want to avoid sudden cardiac death? Explain your reasoning.

12. **Fewer naps, better health?** Based on the evidence for H_a in Exercise 2 and your conclusion in Exercise 10, should people aim for 1–2 naps per week if they want to avoid cardiovascular disease? Explain your reasoning.

The Chi-Square Test for Independence

13. **Birth order and employment** Each person in a random sample of 100 high school seniors was asked for their place in the birth order in their families and whether or not they had a part-time job.[20] The two-way table displays the results of the survey.

pg 776

		Birth order				
		Oldest	Middle	Youngest	Only child	Total
Employment status	Employed	18	10	14	8	50
	Not employed	12	8	24	6	50
	Total	30	18	38	14	100

Do these data provide convincing evidence of an association between birth order and employment status in the population of high school seniors? Use $\alpha = 0.05$.

14. **Nightlights and vision problems** Researchers at The Ohio State University wanted to know if there was an association between using a nightlight and myopia (nearsightedness) in U.S. children. They surveyed the parents of 1220 randomly selected children and recorded the lighting condition in which the children slept during their first two years of life and whether they had myopia at age 10.[21] The two-way table summarizes the data.

		Lighting condition			
		No light	Nightlight	Fully lit	Total
Myopia status	Yes	83	129	10	222
	No	334	629	35	998
	Total	417	758	45	1220

Do these data provide convincing evidence of an association between lighting condition and myopia status in the population of U.S. children? Use $\alpha = 0.05$.

15. **Tuition bills** A random sample of U.S. adults was recently asked, "Would you support or oppose major new spending by the federal government that would help undergraduates pay tuition at public colleges

without needing loans?" The computer output shows the results of a chi-square analysis of the relationship between opinion and age group.[22]

```
Chi-square test: 18-34, 35-49, 50-64, 65+
Expected counts are printed below observed
counts
Chi-square contributions are printed below
expected counts
```

	18-34	35-49	50-64	65+	Total
Support	91	161	272	332	856
	68.1	140.7	285.3	361.9	
	7.72	2.94	0.62	2.47	
Oppose	25	74	211	255	565
	44.9	92.9	188.3	238.9	
	8.84	3.83	2.73	1.09	
Don't know	4	13	20	51	88
	7.0	14.5	29.3	37.2	
	1.28	0.15	2.97	5.11	
Total	120	248	503	638	1509

Chi-Sq = 39.755 DF = 6 P-value < 0.001

(a) Do these data provide convincing evidence of an association between age group and opinion about loan-free tuition in the population of U.S. adults?

(b) Perform a follow-up analysis for the chi-square test in part (a).

16. **Online banking** A recent poll conducted by the Pew Research Center asked a random sample of 1846 U.S. adult internet users if they do any of their banking online. The table summarizes their responses by age.[23]

```
Chi-square test: 18-34, 35-49, 50-64, 65+
Expected counts are printed below observed
counts
Chi-square contributions are printed below
expected counts
```

	18-34	35-49	50-64	65+	Total
Yes	265	352	304	167	1088
	232.8	319.4	325.9	209.8	
	4.45	3.32	1.48	8.74	
No	130	190	249	189	758
	162.2	222.6	227.1	146.2	
	6.39	4.76	2.12	12.54	
Total	395	542	553	356	1846

Chi-Sq = 43.797 DF = 3 P-value < 0.001

(a) Do these data provide convincing evidence of an association between age group and use of online banking for U.S. adult internet users?

(b) Perform a follow-up analysis for the chi-square test in part (a).

The Chi-Square Test for Homogeneity

17. **Gummy bears** Courtney and Lexi wondered if the distribution of color was the same for name-brand gummy bears (Haribo Gold) and store-brand gummy bears (Great Value). To investigate, they randomly selected 6 bags of each type and counted the number of gummy bears of each color.[24] The two-way table summarizes the data.

pg 781

		Brand		
		Name	Store	Total
Color	Red	137	212	349
	Green	53	104	157
	Yellow	50	85	135
	Orange	81	127	208
	White	52	94	146
	Total	373	622	995

Do these data provide convincing evidence at the $\alpha = 0.05$ significance level that the distributions of color differ for name-brand gummy bears and store-brand gummy bears?

18. **Going to the movies** During the Covid-19 pandemic, attendance at movie theaters was down dramatically. But was attendance at movie theaters going down anyway, considering how many in-home viewing options are now available? Gallup polled 1002 randomly selected U.S. adults in 2001 and 1025 randomly selected U.S. adults in 2019 and asked how often these adults went to the movies.[25] The two-way table summarizes the results of the two surveys.

		Year		
		2001	2019	Total
Number of trips to the movie theater	None	261	277	538
	1	70	102	172
	2	110	123	233
	3–5	210	246	456
	6–9	90	82	172
	10+	261	195	456
	Total	1002	1025	2027

Do these data provide convincing evidence at the $\alpha = 0.05$ significance level that the distribution of the number of trips to the movie theater changed from 2001 to 2019?

19. **St. John's wort** An article in the *Journal of the American Medical Association* reports the results of a study that sought to determine if the herb St. John's wort is effective in treating moderately severe cases of depression.[26] The 338 subjects were randomly assigned to receive one of three treatments: St. John's wort, Zoloft (a prescription drug), or placebo (an inactive treatment) for

an 8-week period. The two-way table summarizes the data from the experiment.

		Change in depression			
		Full response	Partial response	No response	Total
Treatment	St. John's wort	27	16	70	113
	Zoloft	27	26	56	109
	Placebo	37	13	66	116
	Total	91	55	192	338

Do these data provide convincing evidence that the distribution of change in depression is different for people like the ones in the study who take St. John's wort, Zoloft, or placebo? Use $\alpha = 0.01$.

20. **Python eggs** How is the hatching of water python eggs influenced by the temperature of the snake's nest? Researchers randomly assigned newly laid eggs to one of three water temperatures: hot, neutral, or cold.[27] The two-way table summarizes the data from the experiment.

		Water temperature			
		Cold	Neutral	Hot	Total
Hatching status	Yes	16	38	75	129
	No	11	18	29	58
	Total	27	56	104	187

Do these data provide convincing evidence that the distribution of hatching status is different for python eggs like the ones in the study that are placed in cold, neutral, or hot water? Use $\alpha = 0.10$.

21. **How to quit smoking** It's hard for smokers to quit. Perhaps prescribing a drug to fight depression will work as well as the usual nicotine patch. Perhaps combining the patch and the drug will work better than either treatment alone. Here are data from a randomized, double-blind trial that compared four treatments.[28] A "success" means that the subject did not smoke for a year following the start of the study.

Group	Treatment	Subjects	Successes
1	Nicotine patch	244	40
2	Drug	244	74
3	Patch plus drug	245	87
4	Placebo	160	25

(a) Summarize these data in a two-way table.

(b) Do the data provide convincing evidence that the proportions of people like the ones in this experiment who would be able to quit smoking are not the same for the four treatments?

22. **Preventing strokes** Aspirin prevents blood from clotting and so helps prevent strokes. The Second European Stroke Prevention Study asked whether adding another anticlotting drug named dipyridamole would be more effective for patients who had already had a stroke. Here are the data on strokes during the two years of the study:[29]

Group	Treatment	Number of patients	Number who had a stroke
1	Placebo	1649	250
2	Aspirin	1649	206
3	Dipyridamole	1654	211
4	Both	1650	157

(a) Summarize these data in a two-way table.

(b) Do the data provide convincing evidence that the proportions of people like the ones in this experiment who would have a stroke are not the same for the four treatments?

Independence or Homogeneity?

23. **Which test?** Determine which chi-square test is appropriate in each of the following scenarios. Explain your reasoning.

(a) With many babies being delivered by planned cesarean section, Mrs. McDonald's statistics class hypothesized that there would be fewer younger people born on the weekend. To investigate, they selected a random sample of people born before 1980 and a separate random sample of people born after 1993. In addition to year of birth, they recorded the day of the week on which each person was born.

(b) Are younger people more likely to be vegan/vegetarian? To investigate, the Pew Research Center asked a random sample of 1480 U.S. adults and categorized respondents by age and whether or not they are vegan/vegetarian.

24. **What test?** Determine which chi-square test is appropriate in each of the following scenarios. Explain your reasoning.

(a) Does chocolate help heart-attack victims live longer? Researchers in Sweden randomly selected 1169 people who had suffered heart attacks and asked them about their consumption of chocolate in the previous year. The researchers then followed these people and categorized respondents by chocolate consumption and whether or not they had died within 8 years.[30]

(b) In the past, random-digit-dialing telephone surveys excluded cell-phone numbers. If the opinions of people who have only cell phones differ from those of people who have landline service, the poll results may not represent the entire adult population. The Pew Research Center interviewed separate random samples of cell-phone-only and landline telephone users who were younger than age 30 and asked them to describe their political party affiliation.[31]

25. **Cold showers** A cold shower can help wake pg 784 you up. Can it also keep you healthy? Researchers randomly assigned 2426 participants to finish their shower with either 0, 30, 60, or 90 seconds of cold water for a month and recorded whether each participant reported a sickness in the next 90 days.[32] The two-way table summarizes the results of this study.

		Cold shower time (sec)				
		0	30	60	90	Total
90-day outcome	Sick	379	437	387	384	1587
	Not sick	168	236	224	211	839
	Total	547	673	611	595	2426

(a) Which chi-square test should be used to analyze these data? Explain your reasoning.

(b) The *P*-value for the appropriate test is 0.17. What conclusion would you make at the $\alpha = 0.05$ significance level?

(c) Name two ways you could increase the power of the test.

26. **Kids and migraines** Researchers investigating two different drugs to treat migraines in children conducted an experiment. They randomly assigned 328 children ages 9–17 who suffered from migraines to receive either amitriptyline, topiramate, or placebo. The primary outcome was a reduction of at least 50% in the number of headache days. The table summarizes the results.[33]

		Drug			
		Amitriptyline	Topiramate	Placebo	Total
Outcome	At least 50% reduction	69	72	40	181
	Less than 50% reduction	63	58	26	147
	Total	132	130	66	328

(a) Which chi-square test should be used to analyze these data? Explain your reasoning.

(b) The *P*-value for the appropriate test is 0.54. What conclusion would you make at the $\alpha = 0.05$ significance level?

(c) Name two ways you could increase the power of the test.

27. **Treating ulcers** Gastric freezing was once a recommended treatment for ulcers in the upper intestine. Use of gastric freezing stopped after experiments showed it had no effect. One randomized comparative experiment found that 28 of the 82 patients who underwent the gastric-freezing procedure improved, while 30 of the 78 patients in the placebo group improved.[34] We can test for a difference in the effectiveness of the treatments in two ways: with a two-sample z test or with a chi-square test. Note that the conditions are met for both tests.

(a) Create a two-way table for these data.

(b) Calculate the value of the chi-square test statistic, along with the *P*-value.

(c) Calculate the value of the standardized test statistic for a two-sample z test, along with the *P*-value for the two-sided test.

(d) How do the *P*-values from parts (b) and (c) compare? How does the value of z^2 compare to the value of χ^2?

28. **Sorry, no chi-square** How do U.S. residents who travel overseas for leisure differ from those who travel for business? The following table provides a breakdown by occupation.[35]

Occupation	Leisure travelers (%)	Business travelers (%)
Professional/technical	36	39
Manager/executive	23	48
Retired	14	3
Student	7	3
Other	20	7
Total	**100**	**100**

Explain why we can't use a chi-square test to learn whether these two distributions differ significantly.

For Investigation *Apply the skills from the section in a new context or nonroutine way.*

29. **Apartment distributions** The general manager of a chain of apartment complexes wants to compare the number of residents per apartment in two of the complexes, Complex A and Complex B. The manager selects a random sample of 40 apartments from each complex and records the number of residents

in each of the selected apartments. The table shows the distribution of number of residents for each complex, along with the mean and standard deviation.

	Number of residents						
	1	2	3	4	n	Mean	SD
Complex A	11	9	8	12	40	2.525	1.198
Complex B	6	15	15	4	40	2.425	0.874

(a) Make graphs to compare these two distributions. Describe what you see.

(b) The *P*-value for a two-sample *t* test of H_0: $\mu_A - \mu_B = 0$ versus H_a: $\mu_A - \mu_B \neq 0$ is 0.671. What conclusion would you make using $\alpha = 0.05$?

(c) You can also compare these two complexes using a chi-square test for homogeneity.

(i) State the hypotheses for this test.

(ii) The *P*-value for this test is 0.028. What conclusion would you make using $\alpha = 0.05$?

(d) Based on your answers to parts (a)–(c), explain to the general manager how the distributions of number of residents compare in the two complexes.

30. **Distance from home** Randomly selected first-year college students at private and public universities in the United States were asked the following question: "How many miles is this university from your permanent home?" Here is a two-way table summarizing the responses:[36]

		Type of university		
		Public	Private	Total
Distance from home (miles)	5 or fewer	1951	1028	2979
	6 to 10	2688	1285	3973
	11 to 50	10,971	5527	16,498
	51 to 100	6765	2211	8976
	101 to 500	15,177	6195	21,372
	More than 500	5811	9486	15,297
	Total	43,363	25,732	69,095

(a) If you were to randomly select one public university student and one private university student from the sample, which student is more likely to live more than 500 miles from the university they attend? Justify your answer.

(b) If the goal of the study is to estimate the average distance that all first-year college students live from their university, under what circumstances would it be beneficial to use a stratified random sample, where the strata are the two types of universities, as compared to a simple random sample of first-year university students? What is the benefit?

(c) Assuming these data came from a stratified random sample, with the two types of universities as the strata, which chi-square test would be appropriate to analyze the data in the two-way table? Explain your reasoning.

Multiple Choice *Select the best answer for each question.*

Exercises 31–34 refer to the following scenario. Morning Consult surveyed a random sample of 2119 U.S. adults about their plans for the upcoming Christmas holiday. In addition to recording each person's plans, the researchers recorded the generation to which each person belonged.[37] Here is a two-way table summarizing the results:

		Christmas tree plans				
		Buy a real tree	Buy a new artificial tree	Reuse an artificial tree	Don't decorate for or celebrate Christmas	Total
Generation	Gen Z	47	50	119	34	250
	Millennial	129	132	256	57	574
	Gen X	119	70	267	119	575
	Baby boomer	86	18	485	131	720
	Total	381	270	1127	341	2119

31. Which of the following is the appropriate null hypothesis for performing a chi-square test?

(A) Equal proportions of each generation plan to buy a real tree.

(B) There is no difference between the distributions of Christmas tree plans for the four generations in the sample.

(C) There is no difference between the distributions of Christmas tree plans for the four generations in the population of U.S. adults.

(D) There is no association between generation and Christmas tree plans in the sample.

(E) There is no association between generation and Christmas tree plans in the population of U.S. adults.

32. Which of the following is the expected count of Gen Zers who plan to buy a real tree?

(A) 44.95
(B) 47.00
(C) 62.50
(D) 95.25
(E) 132.44

33. Which of the following is the correct number of degrees of freedom for the chi-square test using these data?

(A) 3
(B) 9
(C) 12
(D) 16
(E) 2118

34. For these data, $\chi^2 = 206.2$ with a *P*-value of approximately 0. Assuming that the researchers used a significance level of 0.05, which of the following is true?

(A) A Type I error is possible.
(B) A Type II error is possible.
(C) Both a Type I and a Type II error are possible.
(D) There is no chance of making a Type I or Type II error because the *P*-value is approximately 0.
(E) There is no chance of making a Type I or Type II error if the calculations are correct.

35. When analyzing survey results from a two-way table, which of the following is the main distinction between a test for independence and a test for homogeneity?

(A) How the degrees of freedom are calculated
(B) How the expected counts are calculated
(C) The number of samples obtained
(D) The number of rows in the two-way table
(E) The number of columns in the two-way table

36. Can labels on menus influence what customers order? Researchers randomly assigned 5049 volunteers to order food from one of three randomly assigned menus. The first menu had green low-climate-impact labels on chicken, fish, and vegetarian items, the second menu had red high-climate-impact labels on red meat items, and the third menu had no climate-related labels. Researchers recorded whether or not each volunteer chose an item with red meat.[38] Which of the following conditions must be satisfied to perform the appropriate chi-square test using the data from this study?

I. The population distribution is approximately normal.
II. The treatments were randomly assigned.
III. The observed counts are all at least 5.

(A) I only
(B) II only
(C) III only
(D) II and III
(E) I, II, and III

Recycle and Review *Practice what you learned in previous sections.*

37. **Whales (2C)** Thanks to restrictions on whaling, populations of North Atlantic right whales are making a comeback. One way to measure the health of a whale is by its length. After all, we'd expect healthy whales to grow longer than whales that are not as healthy. Here is a scatterplot showing the relationship between x = age (years) and y = length (m) for a sample of 145 North Atlantic right whales ages 2–20 years that were observed from 2000 to 2019, along with some computer output.[39]

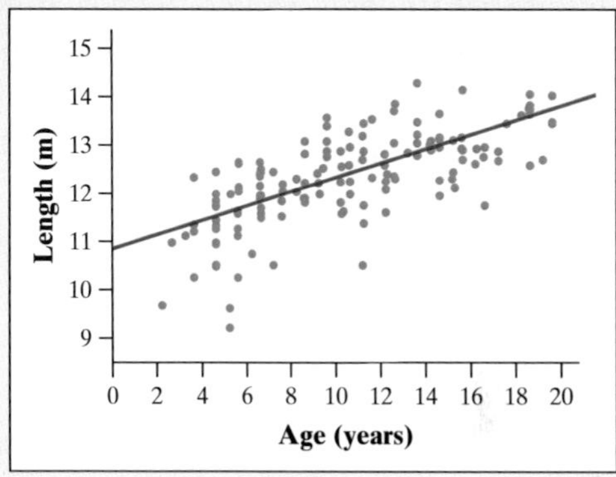

Term	Coef	SE Coef	T-Value	P-Value
Constant	10.7917	0.1410	76.512	0.000
Age	0.1476	0.0124	11.909	0.000
S = 0.68398	R-Sq = 49.8%		R-Sq(adj) = 49.4%	

(a) State the equation of the least-squares regression line, defining any variables used.

(b) Interpret the slope of the least-squares regression line.

(c) By about how much do the actual lengths of the whales vary from the length predicted by the least-squares regression line using x = age?

38. **Inference recap (Units 6–8)** In each of the following settings, state which inference procedure from Units 6–8 you would use. Be specific. For example, you might answer, "Two-sample z test for the difference between two proportions." You do not have to carry out any procedures.

(a) Do a majority of U.S. adults favor abolishing the Electoral College? A random sample of 1000 U.S. adults will be surveyed.

(b) We want to estimate how many more servings of meat high school students eat than servings of vegetables, on average. Each student in a random sample of 100 high school students will be asked for the number of servings of meat they eat per week and the number of servings of fruit they eat per week.

UNIT 8 Wrap-Up

FRAPPY! Free Response AP® Problem, Yay!

Directions: Show all your work. Indicate clearly the methods you use, because you will be scored on the correctness of your methods as well as on the accuracy and completeness of your results and explanations.

Two statistics students wanted to know if including additional information in a survey question would change the distribution of responses. To find out, they randomly selected 30 teenagers and asked them one of the following two questions. Fifteen of the teenagers were randomly assigned to answer Question A, and the other 15 students were assigned to answer Question B.

Question A: When choosing a college, how important is a good athletic program: very important, important, somewhat important, not that important, or not important at all?
Question B: It's sad that some people choose a college based on its athletic program. When choosing a college, how important is a good athletic program: very important, important, somewhat important, not that important, or not important at all?

The table below summarizes the responses to both questions. For these data, the chi-square test statistic is $\chi^2 = 6.12$.

		Question		
		A	B	Total
Importance of a good athletic program	Very important	7	2	9
	Important	4	3	7
	Somewhat important	2	3	5
	Not that important	1	2	3
	Not important at all	1	5	6
	Total	15	15	30

(a) State the hypotheses that the students are interested in testing.

(b) Describe a Type I error and a Type II error in the context of the hypotheses stated in part (a).

(c) For these data, explain why it would *not* be appropriate to use a chi-square distribution to calculate the *P*-value.

(d) To estimate the *P*-value, 100 trials of a simulation were conducted, assuming that the additional information didn't have an effect on the response to the question. In each trial of the simulation, the value of the chi-square test statistic was calculated. These simulated chi-square test statistics are displayed in the dotplot shown here.

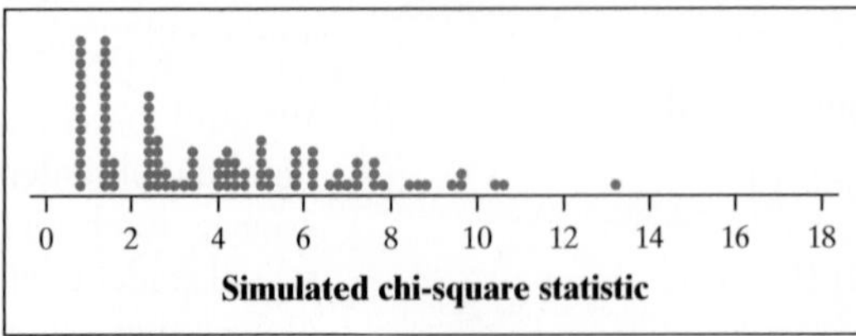

Based on the results of the simulation, what conclusion would you make about the hypotheses stated in part (a)?

After you finish the FRAPPY!, you can view two example solutions on the book's website (**bfwpub.com/tps7e**). Determine whether you think each solution is "complete," "substantial," "developing," or "minimal." If the solution is not complete, what improvements would you suggest to the student who wrote it? Finally, your teacher will provide you with a scoring rubric. Score your response and note what, if anything, you would do differently to improve your own score.

UNIT 8 REVIEW

SECTION 8A Chi-Square Tests for Goodness of Fit

In this section, you learned how to perform a **chi-square test for goodness of fit.** The null hypothesis is that a single categorical variable follows a specified distribution in a population of interest. The alternative hypothesis is that the variable does not follow the specified distribution in the population of interest.

The **chi-square test statistic** χ^2 measures the difference between the observed distribution of a categorical variable and its hypothesized distribution. To calculate the chi-square test statistic, use the following formula with the observed and expected counts:

$$\chi^2 = \sum \frac{(\text{Observed} - \text{Expected})^2}{\text{Expected}}$$

To calculate the expected counts, multiply the sample size by the proportion specified by the null hypothesis for each category. Larger values of the chi-square test statistic provide more convincing evidence that the distribution of the categorical variable differs from the hypothesized distribution in the population of interest.

To ensure that the observations in the sample can be viewed as independent and that the sampling distribution of the chi-square test statistic can be modeled by a chi-square distribution, we check three conditions. The Random condition says that the data are from a random sample from the population of interest. The 10% condition says that the sample size should be less than 10% of the population size when sampling without replacement. The Large Counts condition says that the *expected* counts for each category must be at least 5. In a test for goodness of fit, use a chi-square distribution with degrees of freedom = number of categories − 1.

When the results of a test for goodness of fit are significant, consider doing a *follow-up analysis*. Identify which categories of the variable made the largest contributions to the chi-square test statistic and whether the observed values in those categories were larger or smaller than expected.

A chi-square test for goodness of fit uses the same logic as the previously introduced significance tests. Likewise, the way we interpret *P*-values and power and the way we describe Type I and Type II errors is very similar to what you learned earlier.

SECTION 8B Chi-Square Tests for Independence or Homogeneity

In this section, you learned two different tests to analyze categorical data that are summarized in a two-way table. A **chi-square test for independence** looks for an association between two categorical variables in a single population. A **chi-square test for homogeneity** compares the distribution of a single categorical variable for two or more populations or treatments.

In a chi-square test for independence, the null hypothesis is that there is no association between two categorical variables in one population (or that the two variables are independent in the population). The alternative hypothesis is that there is an association between the two variables (or that the two variables are not independent).

As with other significance tests, we need to verify that the observations in the sample can be viewed as independent and that the sampling distribution of the chi-square test statistic can be modeled by a chi-square distribution. The conditions for a chi-square test for independence are exactly the same as the conditions for a chi-square test for goodness of fit. However, there is a new method for calculating the expected counts:

$$\text{expected count} = \frac{(\text{row total})(\text{column total})}{\text{table total}}$$

To calculate the *P*-value, compute the chi-square test statistic and use a chi-square distribution with degrees of freedom = (number of rows − 1)(number of columns − 1).

In a chi-square test for homogeneity, the null hypothesis is that there is no difference in the distribution of a categorical variable for two or more populations or treatments. The alternative hypothesis is that there is a difference in the distributions.

The conditions are slightly different, but the purpose for checking them is still the same: to ensure independence in the data collection process and verify that the sampling distribution of the chi-square test statistic can be modeled by a chi-square distribution. The Random condition is that the data come from independent random samples or from groups in a randomized experiment. The 10% condition applies for each sample when sampling without replacement, but not for experiments when there is no sampling without replacement from a population. Finally, the Large

Counts condition remains the same — the expected counts must be at least 5 in each cell of the two-way table. The methods for calculating expected counts, the chi-square test statistic, the degrees of freedom, and the *P*-value are exactly the same in a test for homogeneity as in the test for independence.

As with tests for goodness of fit, when the results of a test for homogeneity or a test for independence are significant, consider doing a follow-up analysis. Identify which cells in the two-way table made the largest contributions to the chi-square test statistic and whether the observed counts in those cells were larger or smaller than expected.

The chi-square tests for independence and homogeneity use the same logic as the previously discussed significance tests. Likewise, the way we interpret *P*-values and power and the way we describe Type I and Type II errors is very similar to what you learned in previous units.

Comparing the Three Chi-Square Tests

	Goodness of fit	Independence	Homogeneity
Number of samples/ treatments	1	1	2 or more
Null hypothesis	The stated distribution of a categorical variable in the population of interest is correct.	There is no association between two categorical variables in the population of interest.	There is no difference in the distribution of a categorical variable for several populations or treatments.
Random condition	The data come from a random sample from the population of interest.		The data come from independent random samples or groups in a randomized experiment.
10% condition	When sampling without replacement, $n < 0.10N$ for each sample.		
Large Counts condition	All expected counts ≥ 5.		
Expected counts	np_i where p_i is the proportion specified by the null hypothesis for a particular category	$\frac{\text{(row total)(column total)}}{\text{table total}}$	
Formula for test statistic	$\chi^2 = \sum \frac{(\text{Observed} - \text{Expected})^2}{\text{Expected}}$		
Degrees of freedom	# categories − 1	(# rows − 1)(# columns − 1)	
TI-83/84 name	χ^2GOF-test	χ^2-test	

What Did You Learn?

Learning Target	Section	Related Example on Page(s)	Relevant Chapter Review Exercise(s)
State appropriate hypotheses for a test about the distribution of a categorical variable.	8A	747	R1
Calculate expected counts for a test about the distribution of a categorical variable.	8A	749	R1
Check the conditions for a test about the distribution of a categorical variable.	8A	750	R1
Calculate the test statistic and *P*-value for a test about the distribution of a categorical variable.	8A	755	R1
Perform a chi-square test for goodness of fit.	8A	758	R1
State appropriate hypotheses for a test about the relationship between two categorical variables.	8B	768	R2
Calculate expected counts for a test about the relationship between two categorical variables.	8B	770	R2
Check the conditions for a test about the relationship between two categorical variables.	8B	772	R2
Calculate the test statistic and *P*-value for a test about the relationship between two categorical variables.	8B	774	R2
Perform a chi-square test for independence.	8B	776	R2
Perform a chi-square test for homogeneity.	8B	781	R3
Distinguish between a chi-square test for independence and a chi-square test for homogeneity.	8B	784	R2, R3

UNIT 8 REVIEW EXERCISES

These exercises are designed to help you review the important concepts and skills of the unit.

R1 Testing a genetic model (8A) Biologists wish to cross pairs of tobacco plants having genetic makeup Gg, indicating that each plant has one dominant gene (G) and one recessive gene (g) for color. Each offspring plant will receive one gene for color from each parent. Genetic theory suggests that the ratio of green (GG) to yellow-green (Gg) to albino (gg) tobacco plants should be 1:2:1. In other words, the biologists predict that 25% of the offspring will be green, 50% will be yellow-green, and 25% will be albino. To test their hypothesis about the distribution of offspring, the biologists mate 84 randomly selected pairs of yellow-green parent plants. Of the 84 offspring, 23 plants were green, 50 were yellow-green, and 11 were albino. Do these data provide convincing evidence at the $\alpha = 0.01$ level that the true distribution of offspring color is different from what the biologists predict?

R2 Social media (8B) Pew Research Center surveyed a random sample of 1310 U.S. teens about their use of social media.[40] The teens were asked about the amount of time they spend on social media, as well as other demographic questions. The two-way table summarizes the relationship between social media use and the location where students live. Is there convincing evidence at the 10% level of significance of an association between location and social media use for U.S. teens?

		Location			
		Urban	Suburban	Rural	Total
Social media use	Too much	121	239	116	476
	About right	126	407	183	716
	Too little	40	56	22	118
	Total	287	702	321	1310

R3 Stress and heart attacks (8B) You read a newspaper article that describes a study of whether stress management can help reduce heart attacks. The 107 subjects all had reduced blood flow to the heart and so were at risk of a heart attack. They were assigned at random to three groups. The article goes on to say:

> One group took a four-month stress management program, another underwent a four-month exercise program, and the third received usual heart care from their personal physicians. In the next three years, only 3 of the 33 people in the stress management group suffered "cardiac events," defined as a fatal or non-fatal heart attack or a surgical procedure such as a bypass or angioplasty. In the same period, 7 of the 34 people in the exercise group and 12 out of the 40 patients in usual care suffered such events.[41]

Do these data provide convincing evidence that the true proportions of patients like the ones in this study who would have cardiac events are not the same for the three treatments?

(a) Use the information in the news article to construct a two-way table that describes the study results.

(b) State the hypotheses for a significance test to analyze the results of this study.

(c) The *P*-value for the hypotheses in part (b) is 0.0889. What conclusion would you make at the $\alpha = 0.05$ significance level?

(d) Name two ways the researchers could reduce the probability of a Type II error. What are the drawbacks of these actions?

UNIT 8 AP® STATISTICS PRACTICE TEST

Section I: Multiple Choice *Select the best answer for each question.*

Exercises T1 and T2 refer to the following scenario. Recent revenue shortfalls in a midwestern state led to a reduction in the state budget for higher education. To offset the reduction, the largest state university proposed a 25% tuition increase. It was determined that such an increase was needed simply to compensate for the lost support from the state. Separate random samples of 50 freshmen, 50 sophomores, 50 juniors, and 50 seniors from the university were asked whether they were strongly opposed to the increase, given that it was the minimum increase necessary to maintain the university's budget at current levels. Here are the results:

		Year				
		First	Second	Third	Fourth	Total
Student opinion	Opposed	39	36	29	18	122
	Not opposed	11	14	21	32	78
	Total	50	50	50	50	200

T1 Which null hypothesis would be appropriate for performing a chi-square test?

(A) The closer students get to graduation, the less likely they are to be opposed to tuition increases.

(B) The observed counts are the same as the expected counts.

(C) The distribution of student opinion about the proposed tuition increase is the same for each of the 4 student-years at this university.

(D) Year in school and student opinion about the tuition increase are independent in the sample.

(E) There is an association between year in school and opinion about the tuition increase at this university.

T2 The conditions for carrying out the chi-square test in Exercise T1 are

I. Independent random samples from the populations of interest.

II. All sample sizes are less than 10% of the corresponding populations of interest.

III. All expected counts are at least 5.

Which of the conditions is (are) satisfied in this case?

(A) I only

(B) II only

(C) I and III only

(D) II and III only

(E) I, II, and III

Exercises T3–T4 refer to the following scenario. Faked numbers in tax returns, invoices, or expense account claims often display patterns that aren't present in legitimate records. Some patterns are obvious and easily avoided by a clever crook; others are more subtle. It is a striking fact that the first digits of numbers in legitimate records often follow a model known as Benford's law.[42] Here is the distribution of first digit for variables that follow Benford's law.

First digit	1	2	3	4	5
Proportion	0.301	0.176	0.125	0.097	0.079
First digit	6	7	8	9	Total
Proportion	0.067	0.058	0.051	0.046	1.000

A forensic accountant who is familiar with Benford's law inspects a random sample of 250 invoices from a company that is accused of committing fraud.

T3 Assuming H_0 is true, what is the expected number of invoices that start with the digit 1?

(A) 25.00

(B) 27.78

(C) 30.10

(D) 75.25

(E) 250.00

T4 Based on the random sample of 250 invoices, the value of the chi-square statistic is $\chi^2 = 21.56$. Assuming that the conditions for inference are met, which of the following intervals contains the *P*-value?

(A) P-value > 0.05

(B) $0.01 < P$-value < 0.05

(C) $0.005 < P$-value < 0.01

(D) $0.0005 < P$-value < 0.005

(E) P-value < 0.0005

T5 Which of the following statements about chi-square distributions is false?

(A) For all chi-square distributions, $P(\chi^2 \geq 0) = 1$.

(B) A chi-square distribution with fewer than 10 degrees of freedom is roughly symmetric.

(C) The more degrees of freedom a chi-square distribution has, the larger the mean of the distribution is.

(D) You are more likely to get a large value of χ^2 when the degrees of freedom are larger.

(E) The mean of a chi-square distribution is always greater than the median of the distribution.

Section II: Free Response *Show all your work. Indicate clearly the methods you use, because you will be graded on the correctness of your methods as well as on the accuracy and completeness of your results and explanations.*

T6 **Skittles** According to www.skittles.com, the five flavors in the Skittles® original blend are strawberry, grape, lemon, orange, and lime. To see if these flavors are equally likely, Grace bought a large bag of Skittles and randomly selected 162 candies from the bag.[43] Here are her results:

Flavor	Strawberry	Grape	Lemon	Orange	Lime	Total
Frequency	36	35	34	30	27	162

Do these data provide convincing evidence that the five flavors are not equally likely in Grace's large bag? Use $\alpha = 0.10$.

T7 **Cows and predators** Cows in Botswana are regularly attacked by predators, including lions. In response, farmers often attempt to kill the predators. However, if predators can be tricked into thinking that the cows have spotted them, attacks might decrease, saving the lives of both cows and predators. Would painting eyes on the rear ends of cows be enough to scare off predators? Researchers randomly assigned cows to one of three treatments: eyes painted on the rear end, two Xs painted on the rear end, or no marks on the rear end. After a month, the researchers counted the number of cows that had been killed by predators in each group.[44] None of the 683 cows with painted eyes was killed by a predator, 4 of the 543 cows with Xs were killed by a predator, and 15 of the 835 cows with no marks were killed by a predator.

(a) Is a chi-square test for independence or a chi-square test for homogeneity appropriate in this setting? Explain your answer.

(b) Conduct the test you identified in part (a) using $\alpha = 0.05$.

(c) Based on your conclusion in part (a), could you have made a Type I or a Type II error? What is a potential consequence of this error? Explain your answer.

UNIT
9

Inference for Quantitative Data: Slopes

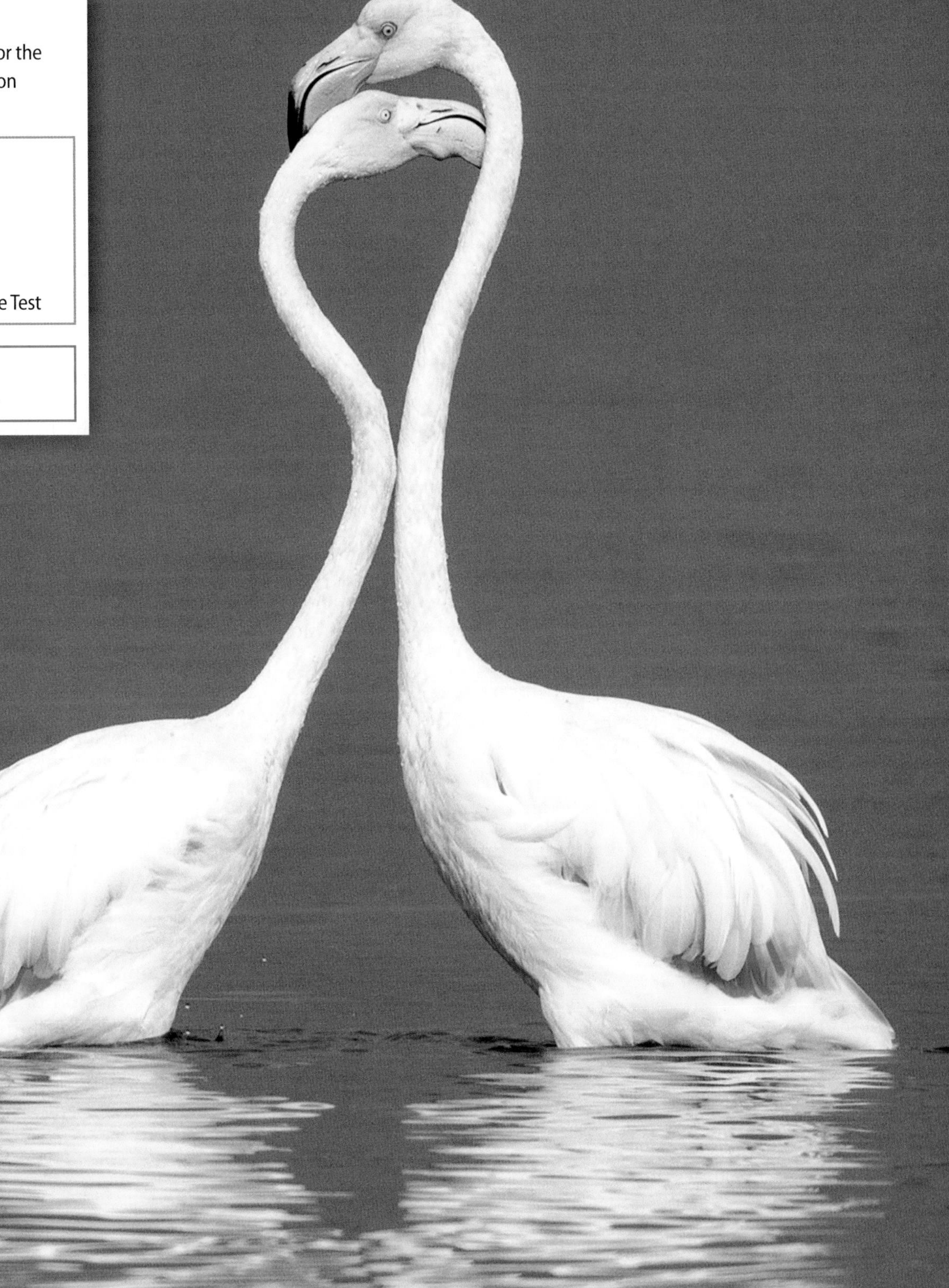

Introduction

In Unit 2, you learned how to display the relationship between two quantitative variables with a scatterplot and how to calculate and interpret the correlation r. You also learned how to summarize a linear relationship between x and y with the least-squares regression line $\hat{y} = a + bx$. Recall that the slope b gives us the predicted change in the response variable y for each one-unit increase in the explanatory variable x.

When the data come from a random sample or a randomized experiment, we can use the slope b of the *sample regression line* to estimate or test a claim about the slope β of the *population regression line*. To perform inference about the population slope β, you need to understand how the sample slope b varies in repeated random sampling. The following activity will get you started.

ACTIVITY Sampling from Old Faithful

Westend61/Getty Images

The Old Faithful geyser is one of the most popular attractions in Yellowstone National Park. As you saw in Unit 2, it is possible to use a least-squares regression line to predict y = the wait time until the next eruption (in minutes) from x = the duration of an eruption (in minutes). In one particular month, Old Faithful erupted 263 times.

In this activity, you will use an applet to repeatedly select a random sample of 15 eruptions from this population, calculate the least-squares regression line $\hat{y} = a + bx$, and plot the value of the sample slope b on a dotplot.

1. Go to www.stapplet.com and launch the *Old Faithful* applet.
2. In the box initially labeled "Population," the applet will display the population of 263 eruptions, along with the population regression line. Near the bottom of the box, keep the default sample size of $n = 15$ and click the "Go!" button. The applet will highlight the 15 randomly selected eruptions and display the sample least-squares regression line and the slope b of the sample regression line. Was your sample slope close to the population slope $\beta = 13.3$?
3. In the box labeled "Sampling Distribution," notice that the value of the sample slope from Step 2 is displayed on a dotplot. Click the "Select samples" button 9 times so that you have a total of 10 sample slopes. Look at the dotplot of sample slopes. Does the distribution have a recognizable shape?

4. To see the shape of the sampling distribution of the sample slope clearly, enter 990 in the box for quickly selecting random samples and click the "Select samples" button. You have now selected a total of 1000 random samples of size $n = 15$ and plotted 1000 sample slopes.
5. Describe the shape of the approximate sampling distribution of b shown in the dotplot. Click the button to show the corresponding normal curve. How well does it fit?
6. What is the mean of your approximate sampling distribution? How does it compare to the population slope $\beta = 13.3$?
7. What is the standard deviation of the simulated sampling distribution? Now change the sample size to $n = 50$ in the "Sample" box and quickly select 1000 random samples in the "Sampling Distribution" box. What happened to the value of the standard deviation? Explain why this makes sense.

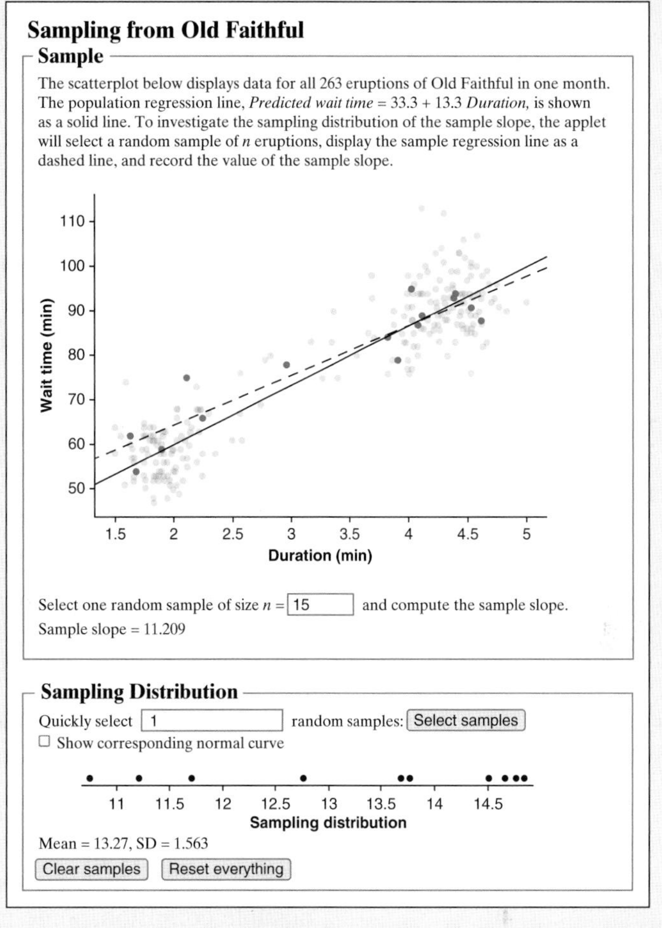
Sampling from Old Faithful
Sample
The scatterplot below displays data for all 263 eruptions of Old Faithful in one month. The population regression line, Predicted wait time = 33.3 + 13.3 Duration, is shown as a solid line. To investigate the sampling distribution of the sample slope, the applet will select a random sample of n eruptions, display the sample regression line as a dashed line, and record the value of the sample slope.
110
100
90
80
70
60
50
Wait time (min)
1.5
2
2.5
3
3.5
4
4.5
5
Duration (min)
Select one random sample of size n = 15 and compute the sample slope.
Sample slope = 11.209
Sampling Distribution
Quickly select 1 random samples: Select samples
Show corresponding normal curve
11
11.5
12
12.5
13
13.5
14
14.5
Sampling distribution
Mean = 13.27, SD = 1.563
Clear samples
Reset everything

Section 9A shows you how to construct and interpret confidence intervals for the slope of a population regression line. Section 9B focuses on performing significance tests about the slope.

SECTION 9A Confidence Intervals for the Slope of a Population Regression Line

LEARNING TARGETS *By the end of the section, you should be able to:*

- Describe the sampling distribution of the sample slope b.
- Check the conditions for calculating a confidence interval for the slope β of a population regression line.
- Calculate a confidence interval for the slope of a population regression line.
- Construct and interpret a t interval for the slope.

When two quantitative variables x and y have a linear relationship, we can model that relationship with the least-squares regression line $\hat{y} = a + bx$. If the data come from a random sample from the population of interest or from a randomized experiment, we may want to use the **sample regression line** to make an inference about the **population regression line.** In Unit 2, we interpreted $\hat{y}$ as the predicted value of y for a given value of x. That interpretation is still valid. We can also think of $\hat{y}$ as being an estimate for μ_y, the mean value of y for all individuals in the population with the given value of x. For that reason, we write the equation of the population regression line as $\mu_y = \alpha + \beta x$.

DEFINITION Population regression line, Sample regression line

A regression line calculated from every value in the population is called a **population regression line.** The equation of a population regression line is $\mu_y = \alpha + \beta x$, where

- μ_y is the mean y-value for a given value of x.
- α is the population y intercept.
- β is the population slope.

A regression line calculated from sample data is called a **sample regression line.** The equation of a sample regression line is $\hat{y} = a + bx$, where

- $\hat{y}$ is the predicted y-value or the estimated mean y-value for a given value of x.
- a is the sample y intercept.
- b is the sample slope.

The symbols α and β here refer to the y intercept and the slope of the population regression line, respectively. They are not related to the probabilities of Type I and Type II errors, which are also designated by the Greek letters α and β. To prevent possible confusion, some people use the form $\mu_y = \beta_0 + \beta_1 x$ for the population regression line.

How does the slope of the sample regression line b relate to the slope of the population regression line β? To find out, we'll explore the *sampling distribution of the sample slope b.*

Sampling Distribution of the Sample Slope *b*

Figure 9.1 shows the relationship between x = duration of an eruption (in minutes) and y = wait time until the next eruption (in minutes) for all 263 eruptions of the Old Faithful geyser during a particular month. Because the scatterplot includes all the eruptions in that month, the least-squares regression line shown is the population regression line $\mu_y = 33.3 + 13.3x$.

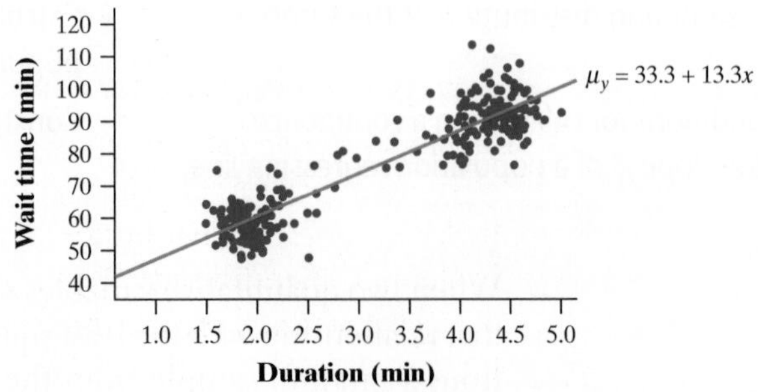

FIGURE 9.1 Scatterplot of the duration and wait time between eruptions of Old Faithful for all 263 eruptions in a single month. The population regression line is shown in orange.

Figure 9.2 shows the results of taking three different SRSs of 15 Old Faithful eruptions from the population described earlier. Each graph displays the selected points and the least-squares regression line for that sample (in purple). The population regression line $\mu_y = 33.3 + 13.3x$ is also shown (in orange).

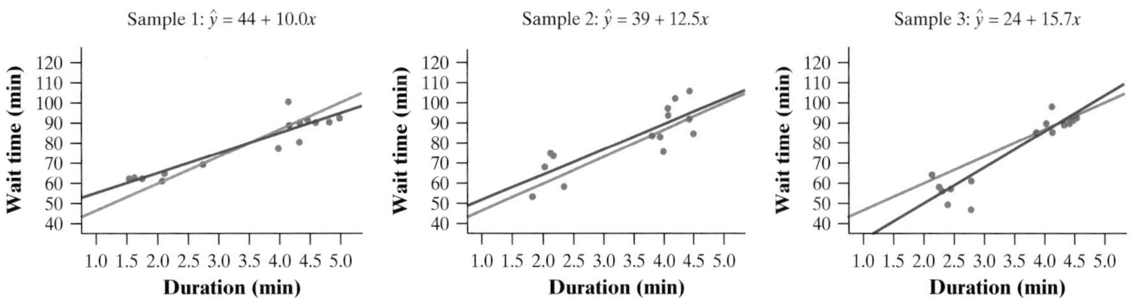

FIGURE 9.2 Scatterplots and least-squares regression lines (in purple) for three different SRSs of 15 Old Faithful eruptions, along with the population regression line (in orange).

Notice that the slopes of the sample regression lines ($b = 10.0$, $b = 12.5$, and $b = 15.7$) vary quite a bit from the slope of the population regression line, $\beta = 13.3$. The pattern of variation in the sample slope b is described by its sampling distribution.

To get a better picture of this variation, we used technology to select 1000 SRSs of $n = 15$ points from the Old Faithful population, each time calculating the sample regression line. Figure 9.3 displays the values of the slope b for the 1000 sample regression lines. We have added a vertical line (in orange) at 13.3 corresponding to the slope of the population regression line β.

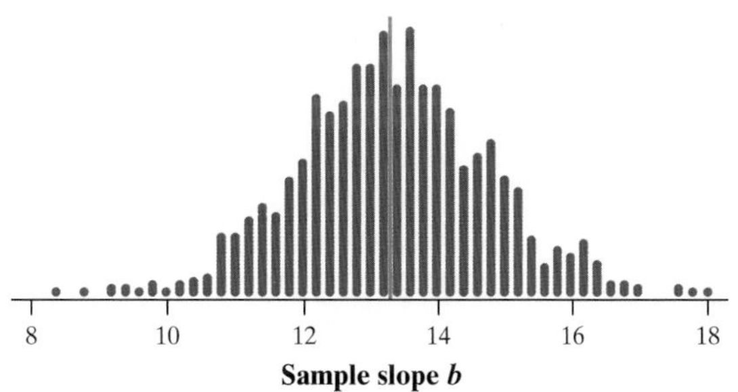

FIGURE 9.3 Dotplot of the slope b of the sample regression line for 1000 SRSs of $n = 15$ eruptions. The slope of the population regression line, $\beta = 13.3$, is marked with an orange vertical line.

The simulated sampling distribution of b is approximately normal, with a mean of about 13.3 (the population slope) and a standard deviation of about 1.42. If we take *all* possible SRSs of size $n = 15$ from this population, we get the actual sampling distribution of b.

Inference about the slope β of a population linear regression model requires that the sampling distribution of the sample slope b is approximately normal, as in the Old Faithful setting. We'll discuss the necessary conditions for inference shortly. For now, here are the formulas that describe the center and variability of the sampling distribution of b.

HOW TO CALCULATE μ_b AND σ_b

Let b be the slope of the sample regression line in an SRS of n observations (x, y) from a population of size N with regression line

$$\mu_y = \alpha + \beta x$$

Then

- The **mean** of the sampling distribution of b is $\mu_b = \beta$.
- When the 10% condition ($n < 0.10N$) is met, the **standard deviation** of the sampling distribution of b is approximately $\sigma_b = \dfrac{\sigma}{\sigma_x\sqrt{n}}$, where σ is the population standard deviation of the residuals and σ_x is the population standard deviation of the explanatory variable x.

AP® EXAM TIP
Formula sheet

The formulas for μ_b and σ_b are included on the formula sheet provided on both sections of the AP® Statistics exam.

Here are some important facts about the mean and standard deviation of the sampling distribution of the sample slope b:

- The value μ_b gives the average value of b for all possible samples of a given size from a population. Because $\mu_b = \beta$, we know that the slope b of the sample regression line is an unbiased estimator of the slope β of the population regression line.
- The value σ_b measures the typical distance between a sample slope b and the population slope β for all possible samples of a given size from a population. Because the sample size appears in the denominator of the formula for σ_b, we know that the sample slope b is less variable when the sample size is larger.
- When we sample *with* replacement, the standard deviation of the sampling distribution of b is exactly $\sigma_b = \dfrac{\sigma}{\sigma_x\sqrt{n}}$. When we sample *without* replacement, the observations are not independent and the actual standard deviation of the sampling distribution of b is smaller than the value given by the formula. However, if the sample size is less than 10% of the population size (the 10% condition), the value given by the formula is nearly correct.

EXAMPLE

Predicting Old Faithful
Sampling distribution of the sample slope b

Skills 3.B, 4.B

PROBLEM: For the population of 263 Old Faithful eruptions in a particular month, the population regression line is $\mu_y = 33.3 + 13.3x$, the standard deviation of the residuals is $\sigma = 6.47$ minutes, and the standard deviation of eruption duration is $\sigma_x = 1.18$ minutes. Imagine taking a random sample of 15 eruptions that occurred during this month and using the sample data to calculate b = the slope of the sample regression line for predicting y = wait time until the next eruption (in minutes) from x = duration of the previous eruption (in minutes).

Grant Faint/Moment/Getty Images

(a) Calculate and interpret the mean of the sampling distribution of b.

(b) Verify that the 10% condition is met. Then calculate and interpret the standard deviation of the sampling distribution of b.

SOLUTION:

$\mu_b = \beta$

(a) $\mu_b = 13.3$

If you selected all possible samples of 15 Old Faithful eruptions from this particular month and calculated the slope of the least-squares regression line for predicting wait time until the next eruption (in minutes) from the duration of the previous eruption (in minutes) for each sample, the sample slopes would have an average value of 13.3.

When $n < 0.10N$, $\sigma_b = \dfrac{\sigma}{\sigma_x\sqrt{n}}$.

(b) Because 15 is less than 10% of the 263 eruptions of Old Faithful in this particular month,

$$\sigma_b = \frac{6.47}{1.18\sqrt{15}} = 1.42$$

If you selected all possible samples of 15 Old Faithful eruptions from this particular month and calculated the slope of the least-squares regression line for predicting wait time until the next eruption (in minutes) from the duration of the previous eruption (in minutes) for each sample, the sample slopes would typically vary from the population slope of 13.3 by about 1.42.

FOR PRACTICE, TRY EXERCISE 1

The values we calculated in the example match quite well with the approximate sampling distribution of b in Figure 9.3.

Checking Conditions for a Confidence Interval for β

As with inference for proportions and means, certain conditions must be met to construct confidence intervals and perform significance tests about the slope β of a population regression line. Figure 9.4 shows the regression model in picture form *when the conditions are met.*

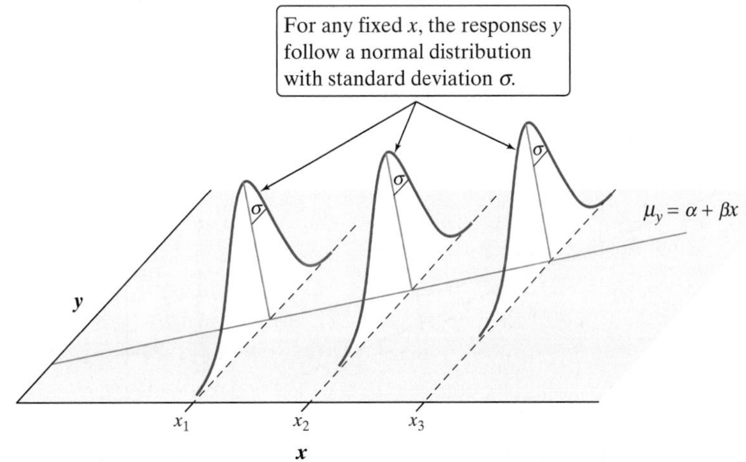

FIGURE 9.4 The population regression model when the conditions for inference are met. The line is the population regression line, which shows how the mean response μ_y changes as the explanatory variable x changes. For any fixed value of x, the observed values of the response variable y follow a normal distribution with mean μ_y and standard deviation σ.

Here are some key observations about Figure 9.4:

- **Shape:** For each value of the explanatory variable x, the distribution of the response variable y is normal.

- **Center:** For each value of x, the mean μ_y of the distribution of y falls on the population regression line.
- **Variability:** For each value of x, the distribution of y has the same standard deviation σ.

Before constructing a confidence interval for the population slope β, you need to check that the conditions for inference are met. These conditions are necessary for the population regression model in Figure 9.4 to be valid. They are also meant to ensure independence in data collection and that the sampling distribution of the sample slope b is approximately normal. You can use the acronym LNERI to help you remember them.

CONDITIONS FOR CONSTRUCTING A CONFIDENCE INTERVAL FOR THE SLOPE OF A POPULATION REGRESSION LINE

Before constructing a confidence interval for the slope β of a population regression line, check that the following conditions are met:

- **Linear:** The form of the relationship between x and y is linear. For each value of x, the mean value of y (denoted by μ_y) falls on the population regression line $\mu_y = \alpha + \beta x$.
- **Normal:** For each value of x, the distribution of y is approximately normal.
- **Equal SD:** For each value of x, the standard deviation of y (denoted by σ) is the same.
- **Random:** The data come from a random sample from the population of interest or a randomized experiment.
  - **10%:** When sampling without replacement, $n < 0.10N$.

Here's a summary of how to check the conditions one by one.

- **Linear:** Examine a scatterplot to see if the overall pattern is roughly linear. Make sure there are no leftover curved patterns in the residual plot.

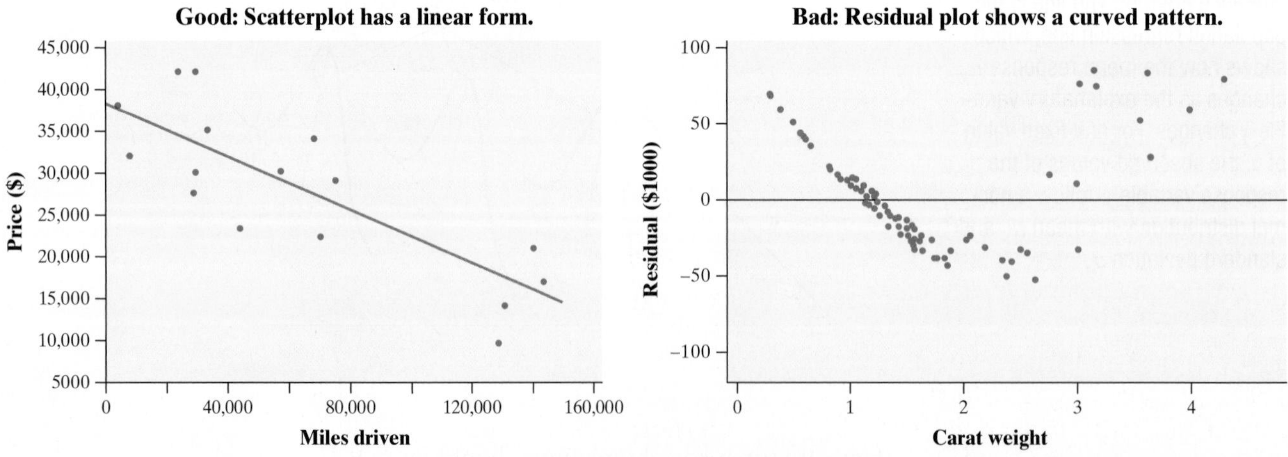

- **Normal:** Make a dotplot, stemplot, histogram, or boxplot of the residuals and check for strong skewness or outliers. Because we rarely have enough observations to check for normality at each x-value, we create one graph of all the residuals to check this condition. If that graph shows no strong skewness and no outliers, it is plausible that the distribution of y is approximately normal for each x. *Note:* Simulation studies suggest that inference about the slope will be reasonably accurate in some cases even when the distribution of residuals has strong skewness or outliers if the sample size is large ($n \geq 30$).

- **Equal SD:** Look at the scatter of the residuals above and below the "residual = 0" line in the residual plot. The variability of the residuals in the vertical direction should be roughly the same from the smallest to the largest x-value. Common violations of this condition include a $<$ pattern (residuals tend to grow in size as x increases) or a $>$ pattern (residuals tend to shrink in size as x increases).

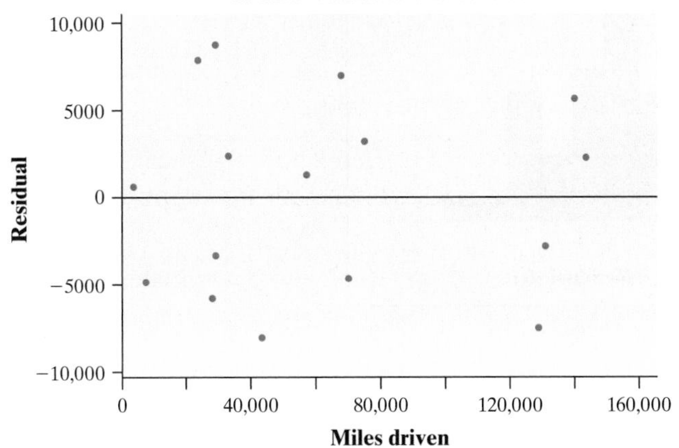

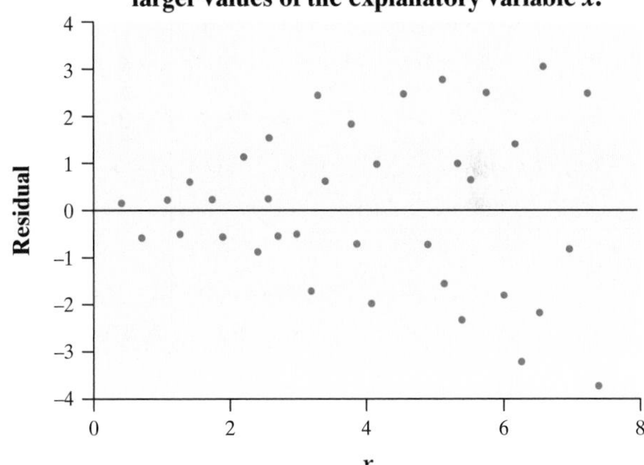

- **Random:** Ensure that the data came from a random sample from the population of interest or a randomized experiment. If not, we can't make inferences about a larger population or about cause and effect.
  - **10%:** If sampling is done without replacement, check that the sample size n is less than 10% of the population size N.

Let's look at an example that illustrates the process of checking conditions.

EXAMPLE Studying ponderosa pines
Checking conditions for a confidence interval for β

Skill 4.C

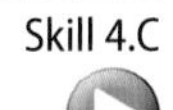

PROBLEM: The U.S. Forest Service randomly selected ponderosa pine trees in western Montana to investigate the relationships between diameter at breast height (DBH), height, and volume of usable lumber.[1] The scatterplot shows the relationship between x = DBH (in inches) and y = height (in feet) for a random sample of 40 ponderosa pines, along with the least-squares regression line and the corresponding residual plot and histogram of residuals. Check if the conditions for constructing a confidence interval for the slope of the population regression line are met.

Federica Grassi/Moment/Getty Images

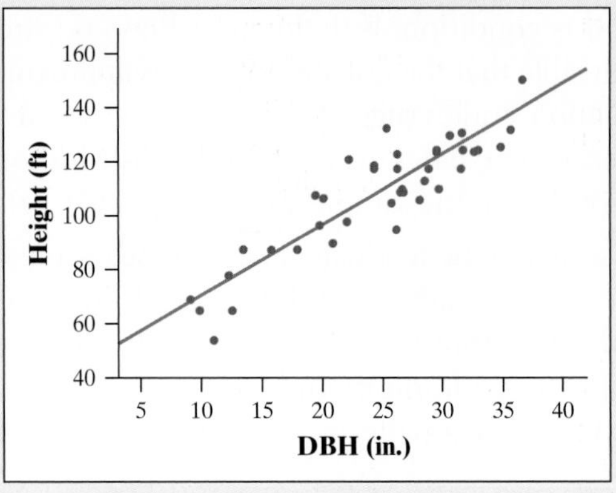

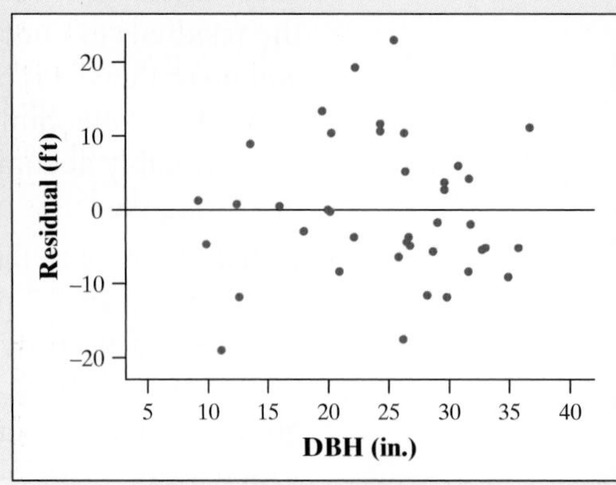

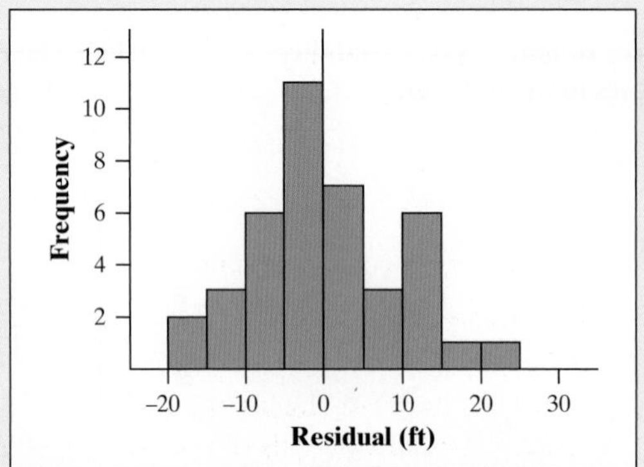

SOLUTION:

Use the LNER1 acronym.

- Linear: The scatterplot shows a linear relationship between DBH and height, and there is no leftover curved pattern in the residual plot. ✓
- Normal: The histogram of residuals shows no strong skewness or outliers. ✓
- Equal SD: In the residual plot, we do not see a clear < pattern or > pattern. ✓
- Random: Random sample of 40 ponderosa pines in western Montana. ✓
  - 10%: 40 is less than 10% of all ponderosa pine trees in western Montana. ✓

FOR PRACTICE, TRY EXERCISE 5

You will always see some irregularity when you look for normality and equal standard deviations in the residuals, especially when you have few observations. **Don't overreact to minor issues in the graphs when checking the Normal and Equal SD conditions.**

1. Here is a scatterplot showing the relationship between x = average driving distance (yards) and y = scoring average for all 164 Ladies Professional Golf Association (LPGA) golfers in 2021.[2] Lower scores are better in golf, so the scatterplot shows that players who hit the ball farther typically have better (lower) scores.

The population regression line is $\mu_y = 81.9886 - 0.0392x$, the standard deviation of the residuals is $\sigma = 1.29$, and the standard deviation of average driving distance is $\sigma_x = 10.17$ yards. Imagine taking a random sample of 12 LPGA players in 2021 and calculating $b =$ the slope of the sample regression line for predicting scoring average from average driving distance.

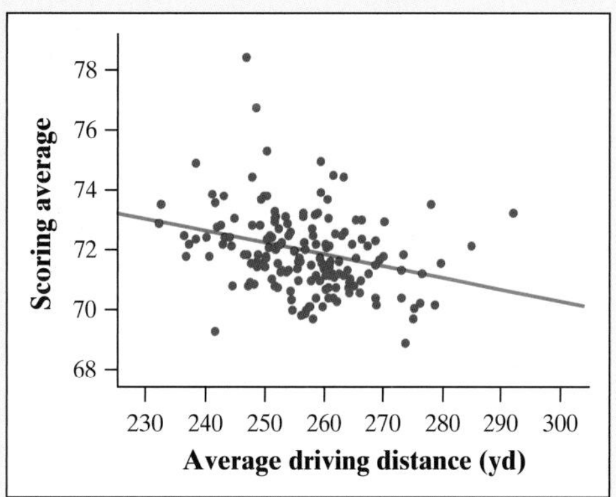

(a) Calculate and interpret the mean of the sampling distribution of b.

(b) Verify that the 10% condition is met. Then calculate and interpret the standard deviation of the sampling distribution of b.

2. A physics class did an experiment in which they dropped 70 paper helicopters from various heights. Each helicopter was assigned at random to a drop height. The class suspects that helicopters dropped from a greater height will take longer to land on the ground. Here are a scatterplot, residual plot, and histogram of residuals created from the least-squares regression line relating $y =$ flight time (sec) to $x =$ drop height (cm).[3]

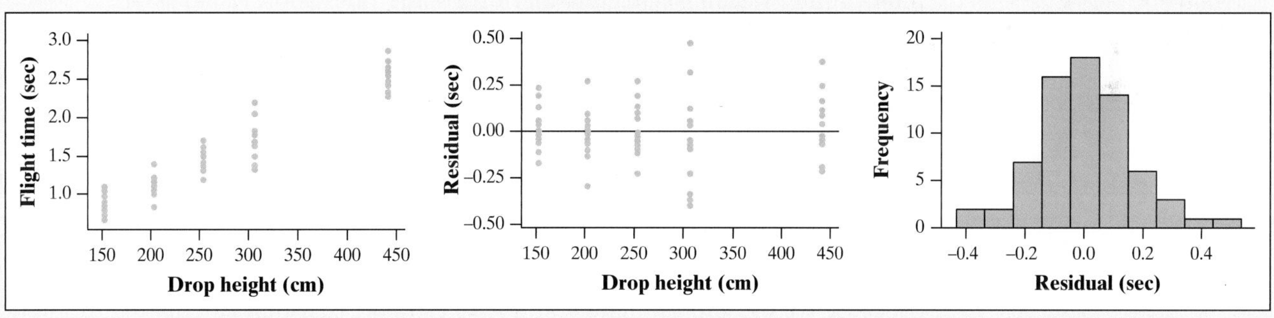

Check if the conditions for constructing a confidence interval for the slope of the true regression line are met.

Calculating a Confidence Interval for β

When the conditions are met, the sampling distribution of the sample slope b is approximately normal with mean $\mu_b = \beta$ and standard deviation

$$\sigma_b = \frac{\sigma}{\sigma_x \sqrt{n}}$$

Two issues typically keep us from using this standard deviation formula in practice.

1. **We don't know the standard deviation of the residuals σ for the population regression line.** So we estimate it with the standard deviation of the residuals s calculated from the sample regression line:

$$s = \sqrt{\frac{\sum \text{residuals}^2}{n-2}}$$

Recall from Unit 2 that s describes the size of a typical prediction error when using the regression line to predict y from x. *Note:* This formula uses $n - 2$ in the denominator instead of $n - 1$ because two parameters, α and β, from the population regression line $\mu_y = \alpha + \beta x$ must be estimated to obtain the predicted values from the sample regression line $\hat{y} = a + bx$.

2. **We don't know the standard deviation σ_x for the population of *x*-values.** So we estimate it with the standard deviation s_x for the sample of *x*-values:

$$s_x = \sqrt{\frac{\sum (x_i - \bar{x})^2}{n-1}}$$

Our resulting estimate for the standard deviation of the sampling distribution of b is the *standard error of the sample slope b*:

$$s_b = \frac{s}{s_x\sqrt{n-1}}$$

(The reason for the use of $\sqrt{n-1}$ rather than $\sqrt{n}$ in the denominator is beyond the scope of this course.)

We interpret s_b like any standard error: It measures how far the sample slope typically varies from the population slope if we repeat the data collection process many times. Recall that standard error is sometimes abbreviated as *SE*.

Although we give the formula for the standard error of b, you should rarely have to calculate it by hand. Computer output shows the standard error s_b immediately to the right of the sample slope b.

When the conditions for inference are met, the sample slope b is our point estimate for the slope β of the population regression line. The interval estimate for β has the familiar form

$$\text{statistic} \pm (\text{critical value})(\text{standard error of statistic})$$

Because we use s to estimate σ and s_x to estimate σ_x when calculating the standard error of the slope s_b, we must use a t^* critical value rather than a z^* critical value. The t^* critical value comes from a t distribution with $n - 2$ degrees of freedom. The resulting formula for the confidence interval is

$$b \pm t^* s_b$$

CALCULATING A CONFIDENCE INTERVAL FOR THE SLOPE OF A POPULATION REGRESSION LINE

When the conditions are met, a C% confidence interval for the slope β of the population regression line is

$$b \pm t^* s_b$$

where t^* is the critical value for the t distribution with $n - 2$ degrees of freedom and C% of the area between $-t^*$ and t^*.

AP® EXAM TIP
Formula sheet

The specific formula for a confidence interval for a population slope is *not* included on the formula sheet provided on both sections of the AP® Statistics exam. However, the formula sheet does include the general formula for the confidence interval:

$$\text{statistic} \pm (\text{critical value})(\text{standard error of statistic})$$

and the formula for the standard error of the sample slope:

$$s_b = \frac{s}{s_x\sqrt{n-1}}$$

EXAMPLE

Studying ponderosa pines
Calculating a confidence interval for β

Skills 3.D, 4.D

PROBLEM: In the preceding example about the height and diameter at breast height (DBH) of 40 randomly selected ponderosa pines in western Montana, we verified that the conditions for constructing a confidence interval for the slope of the population regression line are met. Here is computer output summarizing the relationship between x = DBH (in inches) and y = height (feet) for these trees.

Mint Images/Getty Images

Predictor	Coef	SE Coef	T	P
Constant	43.4740	5.2214	8.326	0.0000
DBH	2.6181	0.2042	12.820	0.0000
S = 9.42788	R-Sq = 81.2%		R-Sq(adj) = 80.7%	

(a) Calculate a 99% confidence interval for the slope of the population regression line.

(b) Does the confidence interval in part (a) provide convincing evidence of a linear association between DBH and height in the population of ponderosa pines? Explain your answer.

SOLUTION:

(a) With 99% confidence and df $= 40 - 2 = 38$, $t^* = 2.712$

$2.6181 \pm 2.712(0.2042)$

$= 2.6181 \pm 0.5538$

$= (2.0643, 3.1719)$

> $b \pm t^* s_b$; get the values of b and s_b from the computer output. The TI-83/84 command invT(area: 0.005, df: 38) gives -2.712, so $t^* = 2.712$. Using Table B with df $= 30$ gives $t^* = 2.750$. The corresponding interval is (2.057, 3.180).

(b) Yes. Because all of the plausible values for β in the interval (2.0643, 3.1719) are different from 0, there is convincing evidence of a linear association between DBH and height in the population of ponderosa pines.

FOR PRACTICE, TRY EXERCISE 9

How should we interpret the 99% confidence level in the example? If the U.S. Forest Service took many, many different random samples of 40 ponderosa pine trees in western Montana, and constructed a 99% confidence interval for the slope β of the population regression line for predicting height (in feet) from DBH (in inches) based on each sample, about 99% of those intervals would capture β.

The values of t given in the computer regression output are not the critical values for a confidence interval. They come from carrying out a significance test about the y intercept or slope of the population regression line. We'll discuss tests about the slope in Section 9B.

Putting It All Together: *t* Interval for the Slope

The appropriate confidence interval method for estimating the slope β of a population regression line is called a ***t* interval for the slope.**

DEFINITION *t* interval for the slope

A ***t* interval for the slope** is a confidence interval used to estimate the slope β of a population regression line.

As with any inference procedure, be sure to follow the four-step process.

EXAMPLE

How much is that truck worth?
t interval for the slope

Skills 1.D, 3.D, 4.B, 4.C

PROBLEM: Cars and trucks lose value the more they are driven. Can we predict the price of a used Ford F-150 SuperCrew 4 × 4 if we know how many miles it has on the odometer? A random sample of 16 used Ford F-150 SuperCrew 4 × 4s was selected from among those listed for sale on autotrader.com. The number of miles driven and price (in dollars) were recorded for each of the trucks. Here are the data:[4]

Oleksiy Maksymenko/imageBROKER/Newscom

Miles driven	70,583	129,484	29,932	29,953	24,495	75,678	8359	4447
Price ($)	21,994	9500	29,875	41,995	41,995	28,986	31,891	37,991
Miles driven	34,077	58,023	44,447	68,474	144,162	140,776	29,397	131,385
Price ($)	34,995	29,988	22,896	33,961	16,883	20,897	27,495	13,997

Some graphs and computer output from a least-squares regression analysis of these data are shown.

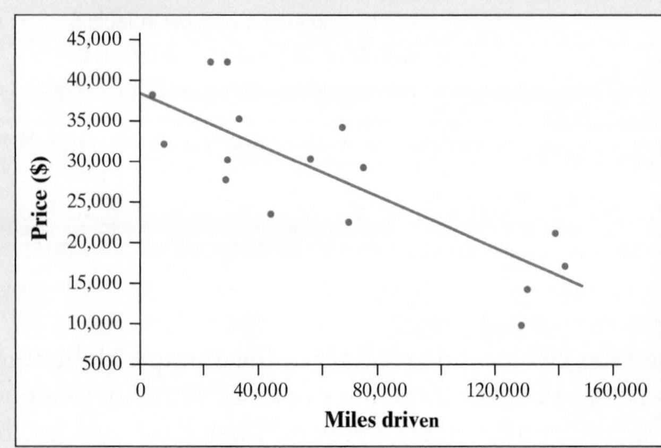

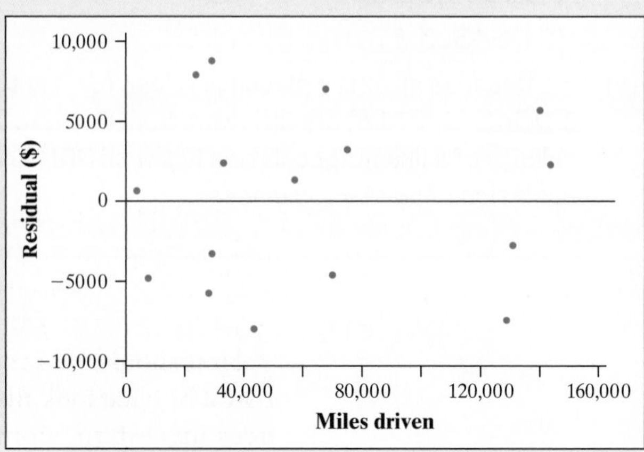

(a) Calculate and interpret the mean of the sampling distribution of b.

(b) Verify that the 10% condition is met. Then calculate and interpret the standard deviation of the sampling distribution of b.

2. **Height and free throws** Very tall basketball players have a reputation for being bad free-throw shooters, even through their shots start closer to the rim! Is this true? Here is a scatterplot of y = free-throw percentage versus x = height (in inches) for all 129 Women's National Basketball Association (WNBA) players in a recent season who averaged at least 10 minutes of playing time per game, with the population least-squares regression line added.[6] The population regression line is $\mu_y = 134.14 - 0.769x$, the standard deviation of the residuals is $\sigma = 10.67$, and the standard deviation of height is 3.61 inches. Imagine taking a random sample of 9 players from this season who averaged at least 10 minutes of playing time per game, and using the sample data to calculate b = the slope of the sample regression line for predicting free-throw percentage from height.

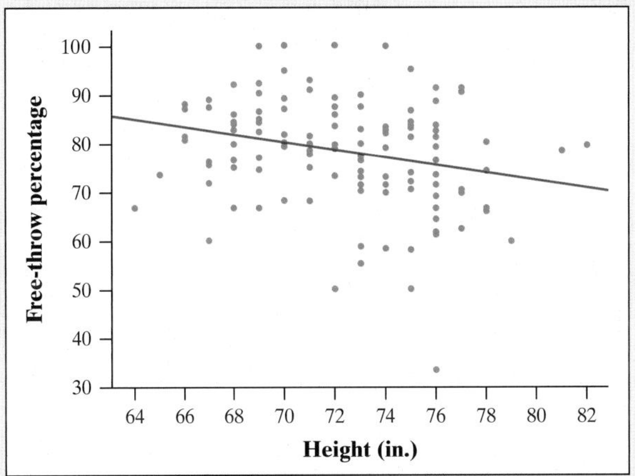

(a) Calculate and interpret the mean of the sampling distribution of b.

(b) Verify that the 10% condition is met. Then calculate and interpret the standard deviation of the sampling distribution of b.

Checking Conditions for a Confidence Interval for β

3. **Oil and residuals** Researchers examined data on the depth of small defects in the Trans-Alaska Oil Pipeline. The researchers compared the results of measurements made in the field on a random sample of 100 defects with measurements of the same defects made in the laboratory.[7] The figure shows a residual plot for the least-squares regression line relating y = field measurement to x = lab measurement. Explain why the conditions for calculating a confidence interval for the slope β of the population regression line are *not* met.

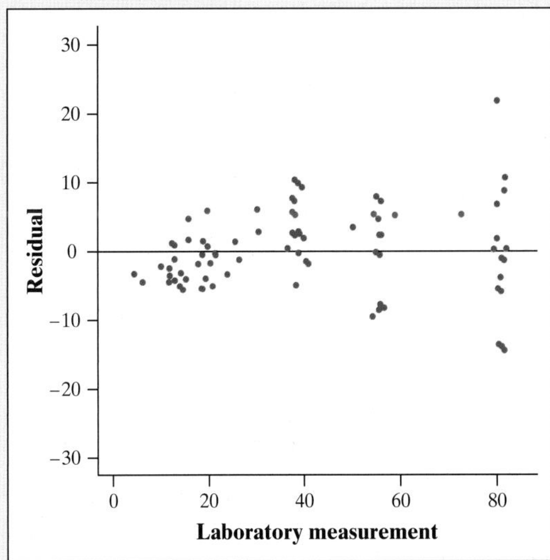

4. **Income and mortality** What does a country's income per person (in thousands of dollars) tell us about the mortality rate for children younger than 5 years of age (per 1000 live births) in that country? A random sample of 14 countries was selected to investigate. The figure shows a residual plot for the least-squares regression line relating y = mortality rate to x = income per person.[8] Explain why the conditions for constructing a confidence interval for the slope of the population regression line are *not* met.

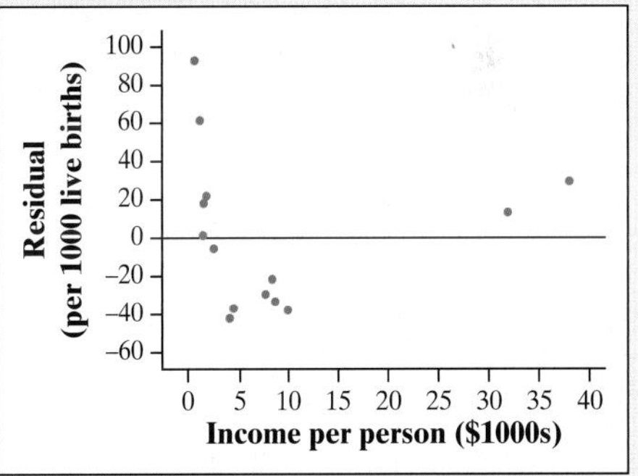

5. pg 809 **Light and plant growth** Meadowfoam seed oil is used in making various skin care products. Researchers interested in maximizing the productivity of meadowfoam plants designed an experiment to investigate the effect of different light intensities on plant growth. The researchers planted 120 meadowfoam seedlings in individual pots, randomly assigned 10 pots to each of 12 trays, and put all the trays into a controlled enclosure. Two trays were then randomly assigned to each light intensity level (micromoles per square meter per second): 150, 300, 450, 600, 750, and 900. The number of flowers produced by each plant was recorded and the average number of flowers was

calculated for each tray.[9] A linear regression analysis was performed for the 12 data points using x = light intensity and y = average number of flowers. Here is a residual plot and a dotplot of the residuals. Check if the conditions for constructing a confidence interval for the slope of the regression model $\mu_y = \alpha + \beta x$ are met.

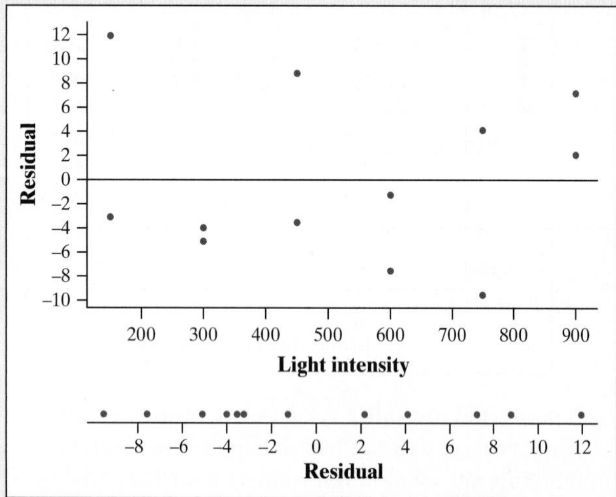

6. **Prey attracts predators** Does a larger concentration of fish attract more predators? One study looked at kelp perch and their common predator, the kelp bass. The researcher set up 16 large circular pens on sandy ocean bottoms off the coast of southern California, and randomly assigned young perch to the pens so that 4 pens had 10 perch each, 4 pens had 20 perch each, 4 pens had 40 perch each, and 4 pens had 60 perch each. The researcher then dropped the nets protecting the pens, allowing bass to swarm in, and counted the perch left after 2 hours.[10] A regression analysis was performed on the 16 data points using x = number of perch in the pen and y = proportion of perch killed. Here is a residual plot and a histogram of the residuals. Check if the conditions for constructing a confidence interval for the slope of the regression model $\mu_y = \alpha + \beta x$ are met.

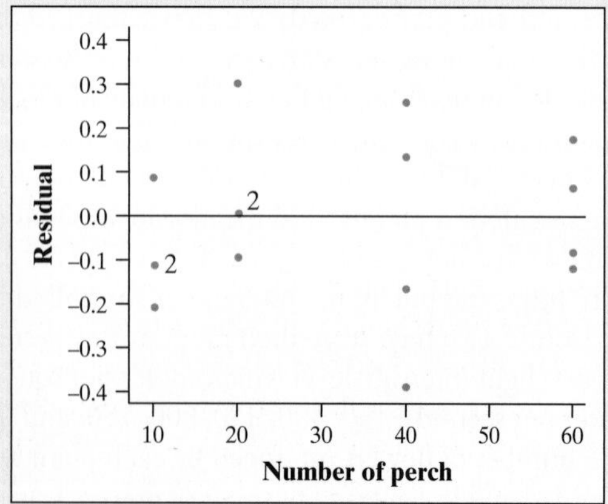

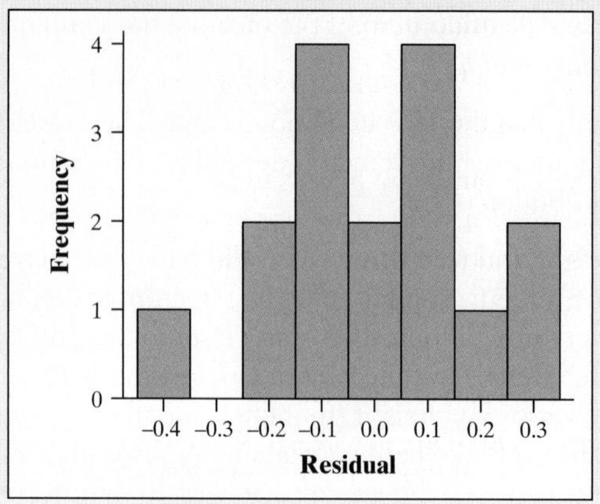

Calculating a Confidence Interval for β

7. **Predicting flowers** Refer to Exercise 5. Here is computer output from the least-squares regression analysis of the data on light intensity and the average number of meadowfoam flowers per plant.

Term	Coef	SE Coef	T-Value	P-Value
Constant	71.62	4.69	15.28	0.000
Light intensity	−0.04108	0.00803	−5.12	0.000
S = 7.12178		R-sq = 72.73%		R-sq(adj) = 69.61%

(a) What is the estimate for α? Interpret this value.

(b) What is the estimate for β? Interpret this value.

(c) What is the estimate for σ? Interpret this value.

(d) Give the standard error of the slope s_b. Interpret this value.

8. **Predicting predators** Refer to Exercise 6. Here is computer output from the least-squares regression analysis of the perch data.

Predictor	Coef	Stdev.	t-ratio	p
Constant	0.12049	0.09269	1.30	0.215
Perch	0.008569	0.002456	3.49	0.004
S = 0.1886	R-Sq = 46.5%		R-Sq(adj) = 42.7%	

(a) What is the estimate for α? Interpret this value.

(b) What is the estimate for β? Interpret this value.

(c) What is the estimate for σ? Interpret this value.

(d) Give the standard error of the slope s_b. Interpret this value.

9. **Meadowfoam inference** Refer to Exercises 5 and 7.
pg 813

(a) Calculate a 95% confidence interval for the slope of the true regression line.

(b) Does the confidence interval in part (a) provide convincing evidence of a linear association between light intensity and average number of meadowfoam flowers per plant? Explain your answer.

10. **Predator inference** Refer to Exercises 6 and 8.

(a) Calculate a 90% confidence interval for the slope of the true regression line.

(b) Does the confidence interval in part (a) provide convincing evidence of a linear association between number of perch in the pen and the proportion killed? Explain your answer.

11. **Temperature and elevation** Do higher elevations mean lower temperatures? To investigate, a researcher collected data on x = the elevation (in feet) and y = the average January temperature (in degrees Fahrenheit) for a random sample of 10 cities and towns in Colorado.[11] The equation of the sample regression model for these data is $\hat{y} = 44.59564 - 0.00309x$. The standard error of the slope is $s_b = 0.0006$.

(a) Calculate and interpret a 90% confidence interval for the slope of the population regression model. Assume the conditions for inference are met.

(b) Interpret the 90% confidence level.

(c) What is the largest population to which we can generalize the result in part (a)? Justify your answer.

12. **Flight costs** Do longer flights cost more money? A frequent flier recorded the distance from Philadelphia to a random sample of 6 cities and the cost of the cheapest flight to that city on a popular discount airline.[12] The equation of the sample regression model relating y = cost (dollars) to x = distance (miles) is $\hat{y} = 107.08 + 0.0416x$. The standard error of the slope is $s_b = 0.0106$.

(a) Calculate and interpret a 99% confidence interval for the slope of the population regression model. Assume the conditions for inference are met.

(b) Interpret the 99% confidence level.

(c) What is the largest population to which we can generalize the result in part (a)? Justify your answer.

t Interval for the Slope

13. **Less mess?** Kerry and Danielle wanted to investigate if tapping on a can of soda would reduce the amount of soda expelled after the can has been shaken. For their experiment, they vigorously shook 40 cans of soda and randomly assigned each can to be tapped for 0 seconds, 4 seconds, 8 seconds, or 12 seconds. After opening the cans and waiting for the fizzing to stop, they measured the amount expelled (in milliliters) by subtracting the amount remaining from the original amount in the can.[13] Here is some computer output from a least-squares regression analysis of y = amount expelled on x = tapping time.

Predictor	Coef	SE Coef	T	P
Constant	106.36	1.3238	80.345	0.000
Tapping time	−2.6350	0.1769	−14.895	0.000

S = 5.00347 R-Sq = 85.4% R-Sq(adj) = 85.0%

(a) Explain why the Random condition is met in this setting.

(b) Construct and interpret a 95% confidence interval for the slope of the true regression line. Assume the conditions for inference are met.

(c) What does the confidence interval in part (b) suggest about the relationship between tapping time on a can of soda and the amount of soda expelled? Explain your answer.

(d) Describe two ways the researchers could reduce the width of the confidence interval in part (b). Explain any drawbacks of these actions.

14. **Dissolved oxygen** Dissolved oxygen is important for the survival of aquatic life. Researchers measured the concentration of dissolved oxygen in Lake Champlain at various times and depths. They also recorded the temperature of the water for each measurement.[14] Here is computer output for a linear regression of y = dissolved oxygen concentration (mg/l) on x = temperature (°C) for a random sample of 500 measurements.

Summary of Fit

RSquare	0.4340
RSquare Adj	0.4329
Root Mean Square Error	1.12786
Mean of Response	10.275
Observations (or Sum Wgts)	500

Term	Estimate	Std Error	t Ratio	Prob > |t|
Intercept	12.165	0.109	111.52	< .0001*
Temperature	−0.18973	0.00971	−19.54	< .0001*

(a) Explain why the Random condition is met in this setting.

(b) Construct and interpret a 95% confidence interval for the slope of the population regression line. Assume the conditions for inference are met.

(c) What does the confidence interval in part (b) suggest about the association between dissolved oxygen concentration and temperature in Lake Champlain? Explain your answer.

(d) Describe two ways the researchers could reduce the width of the confidence interval in part (b). Explain any drawbacks of these actions.

15. pg 814 **Beavers and beetles** Do beavers benefit beetles? Researchers laid out 23 circular plots, each 4 meters in diameter, at randomly selected locations in a large area where beavers were cutting down cottonwood trees. In each plot, they counted the number of stumps from trees cut by beavers and the number of clusters of beetle larvae. Ecologists believe that the new sprouts from stumps are more tender than other cottonwood growth, so beetles prefer them. If so, more stumps should produce more beetle larvae. Here are the data:[15]

Stumps	Larvae	Stumps	Larvae
2	10	2	25
2	30	1	8
1	12	2	21
3	24	2	14
3	36	1	16
4	40	1	6
3	43	4	54
1	11	1	9
2	27	2	13
5	56	1	14
1	18	4	50
3	40		

Some graphs and computer output from a linear regression of $y =$ number of beetle larvae on $x =$ number of stumps are shown. Construct and interpret a 99% confidence interval for the slope of the population regression line.

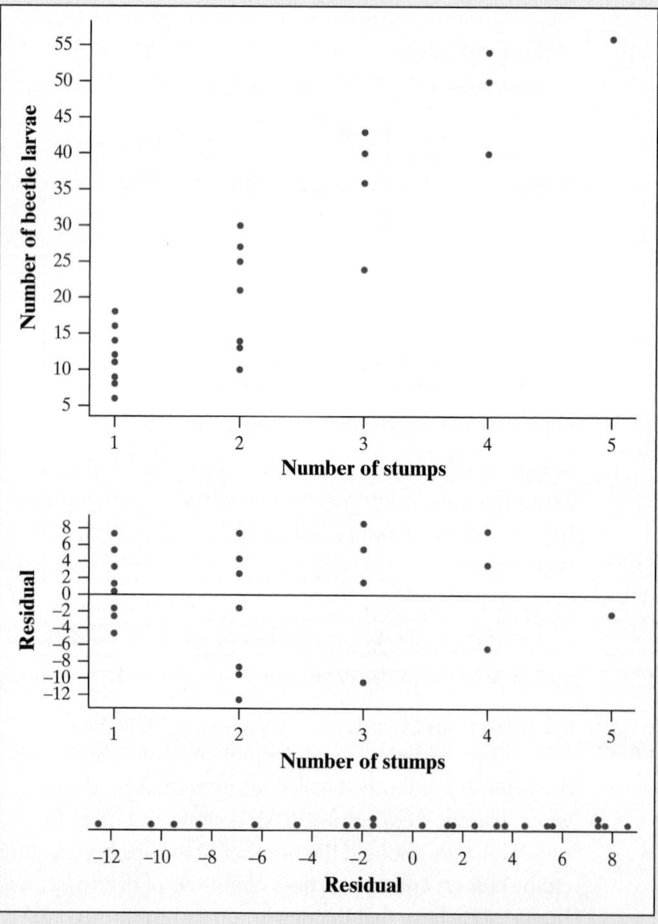

Summary of Fit

RSquare	0.839144
RSquare Adj	0.831484
Root Mean Square Error	6.419386
Mean of Response	25.086960
Observations (or Sum Wgts)	23

Parameter Estimates

Term	Estimate	Std Error	t Ratio	Prob >|t|
Intercept	−1.286104	2.853182	−0.45	0.6568
Number of stumps	11.893733	1.136343	10.47	<.0001*

16. **Beer and BAC** How well does the number of beers a person drinks predict their blood alcohol content (BAC)? Sixteen volunteers aged 21 or older with an initial BAC of 0 took part in a study to find out. Each volunteer drank a randomly assigned number of cans of beer and had their BAC measured 30 minutes later. Here are the data:[16]

Beers	BAC	Beers	BAC
5	0.10	3	0.02
2	0.03	5	0.05
9	0.19	4	0.07
8	0.12	6	0.10
3	0.04	5	0.085
7	0.095	7	0.09
3	0.07	1	0.01
5	0.06	4	0.05

Some graphs and computer output from a linear regression of $y =$ BAC on $x =$ number of beers are shown. Construct and interpret a 99% confidence interval for the slope of the population regression line.

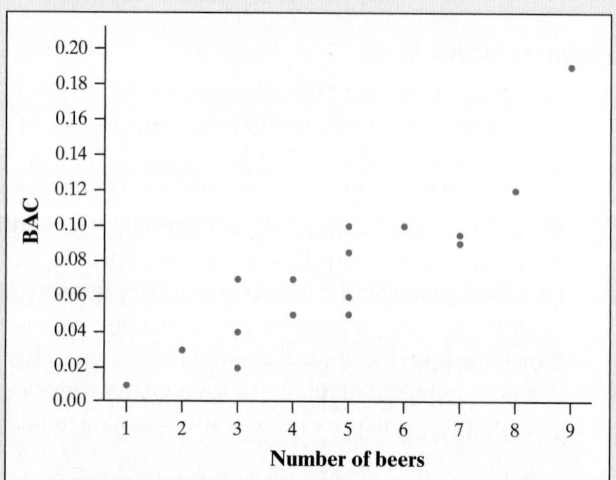

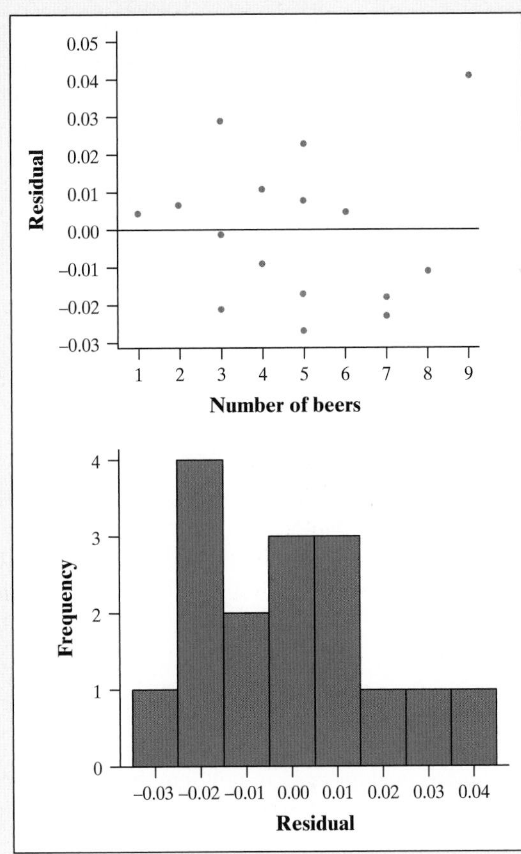

Term	Coef	SE Coef	T-Value	P-Value
Constant	−0.012701	0.0126	−1.00	0.3320
Beers	0.017964	0.0024	7.84	≤ 0.0001

S = 0.0204   R-sq = 80.0%   R-sq(adj) = 78.6%

For Investigation *Apply the skills from the section in a new context or nonroutine way.*

17. **Teenage height** Using the health records of every student at a high school, the school nurse created a scatterplot relating y = height (in centimeters) to x = age (in years). After verifying that the conditions for the regression model were met, the nurse calculated the equation of the population regression line to be $\mu_y = 105 + 4.2x$ with a standard deviation of the residuals of $\sigma = 7$ cm. The nurse also calculated the standard deviation of age to be $\sigma_x = 1.28$ years.

(a) According to the population regression line, what is the average height of 15-year-old students at this high school?

(b) About what percentage of 15-year-old students at this school are taller than 180 cm? (*Hint:* For each x-value, the distribution of y-values is approximately normally distributed with mean μ_y and standard deviation σ.)

Suppose the nurse selects a random sample of 10 students from the school and calculates the equation of the least-squares regression line for this sample.

(c) Calculate the mean and standard deviation of the sampling distribution of the sample slope for samples of size 10. What shape does the sampling distribution have?

(d) What is the probability that the slope of the least-squares regression line for this sample is less than 4?

18. **Transforming income and mortality** Refer to Exercise 4. To help meet the conditions for inference about the slope, we can often transform one or both variables using mathematical operations such as square roots and logarithms. Using technology, we calculated the base-10 logarithm of each mortality value and the base-10 logarithm of each of income value. Here is a scatterplot of the original data and a scatterplot of the transformed data, along with computer output from a least-squares regression analysis of y = log(mortality rate) on x = log(income).

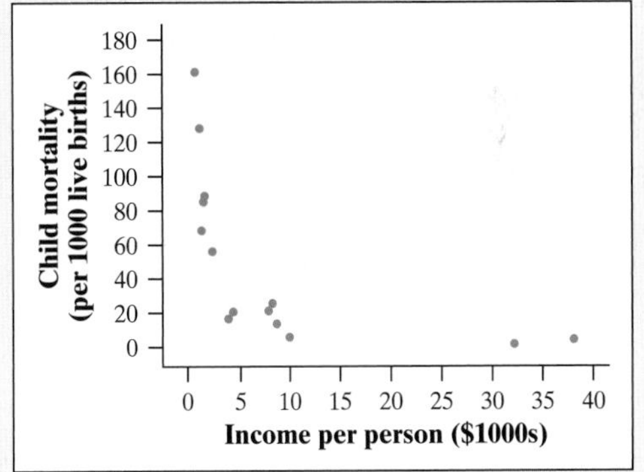

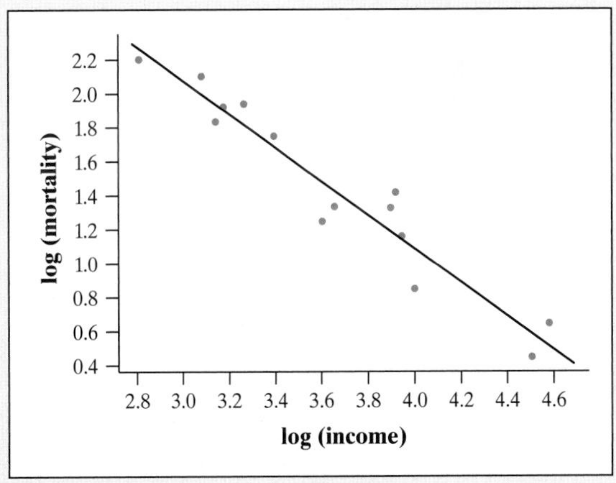

Predictor	Coef	SE Coef	T	P
Constant	5.0431	0.2941	17.1491	< 0.001
log(income)	−0.9892	0.08	−12.3644	< 0.001

(a) What is the equation of the sample least-squares regression line? Be sure to define any variables you use.

(b) One of the countries in the sample had an income per person of $10,005 and a mortality rate of 7.1 per 100,000 people. Use the regression line from part (a) to help calculate the predicted mortality rate for this country.

(c) Calculate and interpret a 95% confidence interval for the slope of the population regression line. Assume that the conditions for inference are met.

(d) Does the confidence interval in part (c) give convincing evidence of a negative association between income and mortality rate in the population of countries? Justify your answer.

Multiple Choice *Select the best answer for each question.*

Exercises 19–22 refer to the following scenario. Can you predict the total length of an Australian brushtail possum from the length of its footprint? To find out, researchers collected data from a sample of 104 mountain brushtail possums from Australia. For each possum, foot length (cm) and total length (cm) were recorded.[17] We checked that the conditions for constructing a confidence interval for the slope of the population regression line are met. Here is some computer output from a least-squares regression analysis using these data:

Predictor	Coef	SE Coef	T	P
Constant	57.2491	5.9981	9.5445	< 0.001
Foot length (cm)	0.4364	0.0874	4.9916	< 0.001
s = 3.8814	R-Sq = 19.8%			

19. Which of the following would have resulted in a violation of the conditions for inference?

(A) If the entire sample of possums was selected from one forest in eastern Australia

(B) If the sample size was 52 instead of 104

(C) If the residual plot showed a random scatter of points about the horizontal line at residual = 0

(D) If a histogram of the residuals is slightly skewed to the right

(E) If the residual plot showed a roughly equal amount of variation around the horizontal line at residual = 0 for all foot lengths

20. Which of the following is the best interpretation of the value 0.0874 in the computer output?

(A) For each increase of 1 cm in foot length, the average total length increases by about 0.0874 cm.

(B) When using this model to predict total length from foot length, the predictions will typically be off by about 0.0874 cm.

(C) The linear relationship between foot length and total length accounts for 8.74% of the variation in total length.

(D) The linear relationship between foot length and height is weak and positive.

(E) In repeated samples of size 104, the slope of the sample regression line for predicting total length from foot length will typically vary from the population slope by about 0.0874.

21. Which of the following is a 95% confidence interval for the population slope β?

(A) 0.4364 ± 0.0874

(B) 0.4364 ± 0.1713

(C) 0.4364 ± 0.1733

(D) 0.4364 ± 3.8814

(E) 0.4364 ± 7.6968

22. The slope β of the population regression line describes

(A) the exact increase in total length (cm) for Australian mountain brushtail possums when foot length increases by 1 cm.

(B) the average increase in total length (cm) for Australian mountain brushtail possums when foot length increases by 1 cm.

(C) the average increase in foot length (cm) for Australian mountain brushtail possums when total length increases by 1 cm.

(D) the average increase in total length (cm) for possums in the sample when foot length increases by 1 cm.

(E) the average increase in foot length (cm) for possums in the sample when total length increases by 1 cm.

Recycle and Review *Practice what you learned in previous sections.*

Exercises 23–25 refer to the following scenario. Does the color in which words are printed affect your ability to read them? Do the words themselves affect your ability to name the color in which they are printed? Mr. Starnes had his 16 students investigate these questions. Each student performed two tasks in a random order while a partner timed the activity: (1) Read 32 words aloud as quickly as possible, and (2) say the color in which each of 32 words is printed as quickly as possible. Try both tasks for yourself using the following word list:

BROWN	RED	BLUE	GREEN
RED	GREEN	BROWN	BROWN
GREEN	RED	BLUE	BLUE
BROWN	BLUE	GREEN	RED
BLUE	BROWN	RED	RED
RED	BLUE	BROWN	GREEN
BLUE	GREEN	GREEN	BLUE
GREEN	BROWN	RED	BROWN

23. **Color words** (3A, 3C) Let's review the design of the study.

(a) Explain why this was an experiment and not an observational study.

(b) Did Mr. Starnes use a completely randomized design or a randomized block design? Why do you think he chose this experimental design?

(c) Explain the purpose of the random assignment in the context of the study.

Here are the data from the experiment. For each student, the time to perform the two tasks is given to the nearest second.

Subject	Words	Colors	Subject	Words	Colors
1	13	20	9	10	16
2	10	21	10	9	13
3	15	22	11	11	11
4	12	25	12	17	26
5	13	17	13	15	20
6	11	13	14	15	15
7	14	32	15	12	18
8	16	21	16	10	18

24. **More color words** (1D, 7A) Now let's analyze the data.

(a) Calculate the difference in time for each student, make a boxplot of the differences, and describe what you see.

(b) Calculate and interpret the mean difference.

(c) Explain why it is not safe to use paired *t* procedures to do inference about the mean difference in the time to complete the two tasks.

25. **Color words finale** (2B, 2C, 9A) Can we use a student's word task time to predict their color task time?

(a) Make an appropriate scatterplot to help answer this question. Describe what you see.

(b) Use technology to find the equation of the least-squares regression line. Define any variables you use.

(c) Find and interpret the residual for the student who completed the word task in 9 seconds.

(d) Assume that the conditions for performing inference about the slope of the true regression line are met. Calculate and interpret a 95% confidence interval for the slope.

SECTION 9B Significance Tests for the Slope of a Population Regression Line

LEARNING TARGETS *By the end of the section, you should be able to:*

- State appropriate hypotheses and check the conditions for a test about the slope of a population regression line.
- Calculate the standardized test statistic and *P*-value for a test about the slope of a population regression line.
- Perform a *t* test for the slope.

In Section 8B, you learned how to perform a test about the relationship between two *categorical* variables, such as anger level and heart disease status. In this section, you will learn how to state hypotheses, calculate the standardized test statistic and *P*-value, and carry out a test about the relationship between two *quantitative* variables. The following activity gives you a preview of what's to come.

ACTIVITY Should you sit in front?

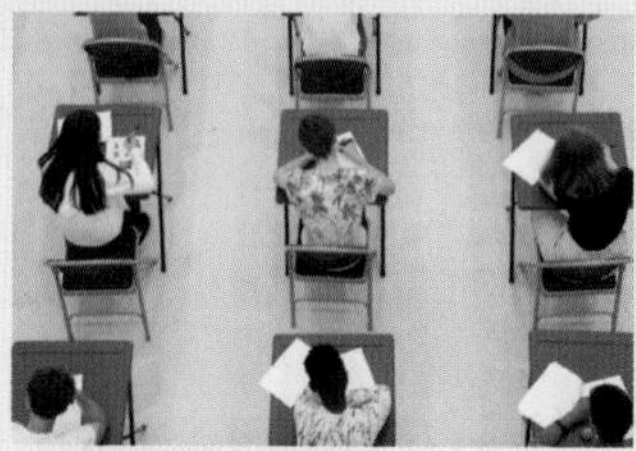
Caia Image/Getty Images

Many people believe that students learn better if they sit closer to the front of the classroom. Does sitting closer *cause* higher achievement, or do better students simply choose to sit in the front? To investigate, a statistics teacher randomly assigned students to seat locations in the classroom for a particular unit and recorded the test score for each student at the end of the unit.

Here are the data. (Note that one student earned a test score of 101 due to extra credit.)

Row	1	1	1	1	2	2	2	2	3	3	3	3	4	4	4
Score	76	77	94	99	83	85	74	79	90	88	68	78	94	72	101
Row	4	4	5	5	5	5	5	6	6	6	6	7	7	7	7
Score	70	79	76	65	90	67	96	88	79	90	83	79	76	77	63

In this activity, you and your classmates will use simulation to determine if these data provide convincing evidence that sitting closer to the front improves test scores.

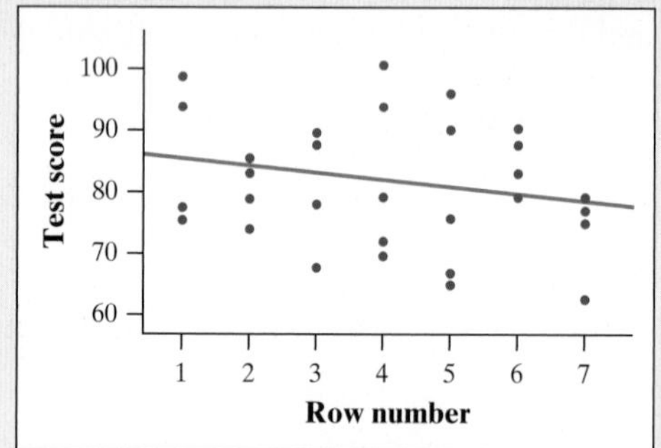

1. The scatterplot shows the relationship between x = row number (the rows are equally spaced, with row 1 closest to the front and row 7 farthest away) and y = test score, along with the least-squares regression line $\hat{y} = 85.71 - 1.12x$. Do these data provide *some* evidence that sitting closer to the front improves test scores? How do you know?

   Does the sample slope of $b = -1.12$ provide *convincing* evidence that sitting closer to the front improves test scores for students like these, or is it plausible that the association is due to the chance variation in the random assignment? Let's perform a simulation to help answer this question.

2. Go to www.stapplet.com and launch the *Two Quantitative Variables* applet. Enter the Row values as the explanatory variable and the Score values as the response variable. Then click the "Begin analysis" button to generate a scatterplot of the data.
3. Click the "Calculate least-squares regression line" button and confirm that the slope of the sample regression line for these data is $b = -1.12$.
4. In the Perform Inference section, choose "Simulate sample slope" as the inference procedure. Enter 1 for the number of samples to add and click "Add samples." This randomly reassigns the score values to the row values assuming there is no association between row and score, and calculates the resulting slope. The simulated slope is displayed on the dotplot. How does it compare to the slope of $b = -1.12$ from the actual experiment?
5. Keep clicking "Add samples" to perform more random reassignments (or increase the number of samples to add in the entry box). Is $b = -1.12$ unusual or something that is likely to occur by chance alone? What conclusion would you make based on the statistics teacher's experiment?

Stating Hypotheses and Checking Conditions for a Test about β

In Section 8B, you learned how to state hypotheses for a test about the relationship between two *categorical* variables. Stating hypotheses for a test about the relationship between two *quantitative* variables is very similar, except that we can now specify a direction for the relationship in the alternative hypothesis. For example, the null hypothesis usually says that there is no linear association between the explanatory and response variables. The alternative hypothesis can say that there is a positive association, a negative association, or an association (either positive or negative) between the two variables.

Although we can verbally state the hypotheses in a test about the relationship between two quantitative variables, we typically state them symbolically using the slope of the population least-squares regression line β. If there is no linear association between the x and y variables, the slope of the population least-squares regression line is 0. Consequently, our usual null hypothesis is H_0: $\beta = 0$. Put another way, H_0 says that *linear regression of y on x is no better for predicting y than the mean of the response variable $\bar{y}$*.

In the seating chart activity, we wanted to test if sitting closer to the front improves test scores for students like the ones in the study. That is, we wanted to know if smaller row numbers are associated with larger test scores (a negative association). Here is one way to state the hypotheses for this test:

H_0: There is no linear association between row number and test score for students like the ones in this study.

H_a: There is a negative linear association between row number and test score for students like the ones in this study.

We can also state the hypotheses symbolically using β, the slope of the true regression line:

H_0: $\beta = 0$
H_a: $\beta < 0$

where β is the slope of the true least-squares regression line relating y = test score to x = row number for students like the ones in this experiment.

As in tests for means and proportions, a two-sided alternative hypothesis is possible. For example, if we had no initial belief about the relationship between seat location and test scores, the alternative hypothesis would be H_a: $\beta \neq 0$ (there *is* a linear association between row number and test score for students like these). As always, the alternative hypothesis is formulated based on the statistical question being investigated — before the data are collected.

CHECKING CONDITIONS

The conditions for performing a test about the slope β of a population linear regression model are exactly the same as the ones for constructing a confidence interval for β.

CONDITIONS FOR PERFORMING A SIGNIFICANCE TEST ABOUT THE SLOPE OF A POPULATION REGRESSION LINE

- **Linear:** The form of the relationship between x and y is linear. For each value of x, the mean value of y (denoted by μ_y) falls on the population regression line $\mu_y = \alpha + \beta x$.
- **Normal:** For each value of x, the distribution of y is approximately normal.
- **Equal SD:** For each value of x, the standard deviation of y (denoted by σ) is the same.
- **Random:** The data come from a random sample from the population of interest or a randomized experiment.
  - **10%:** When sampling without replacement, $n < 0.10N$.

AP® EXAM TIP

Sometimes on an AP® Statistics exam free response question, you are told to assume that the conditions for inference are met. You should be pleased when that happens! Don't waste any time checking the conditions in this case—you won't earn any credit for doing so and you take a risk of saying something incorrect that might result in a deduction.

Let's confirm that the conditions are met in the seating chart context.

- Linear? There is a linear association between row number and test score in the scatterplot shown earlier, and there is no leftover curved pattern in the residual plot. ✓

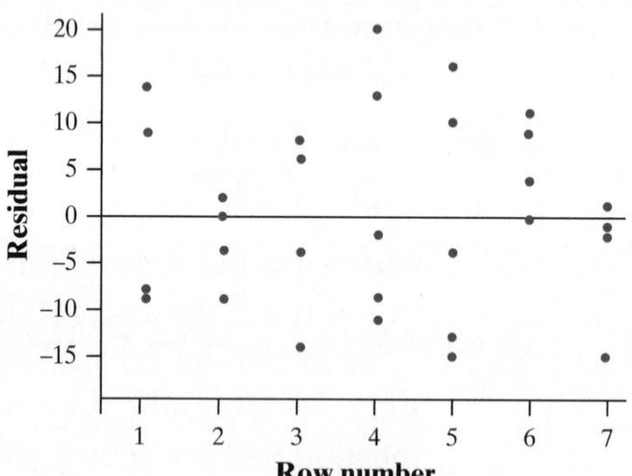

- Normal? A dotplot of the residuals shows no strong skewness or outliers. ✓

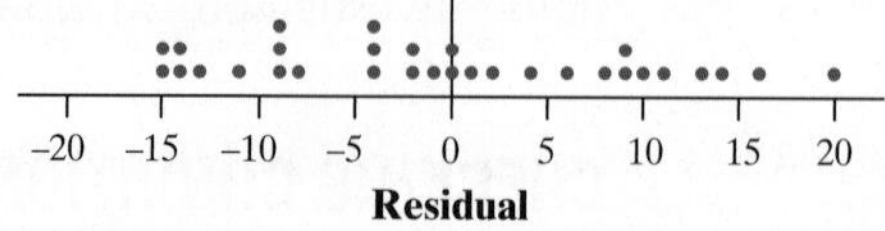

- Equal SD? The residual plot shows a similar amount of variation around the residual $= 0$ line for the different rows. ✓
- Random? The students were randomly assigned to seats for the experiment. ✓

Notice that we did not check the 10% condition in this case because the data were not obtained using sampling without replacement from a population.

EXAMPLE

Can foot length predict height?
Stating hypotheses and checking conditions for a test about β

Skills 1.F, 4.C

Frederic Cirou/PhotoAlto/Getty Images

PROBLEM: Are students with longer feet typically taller than their classmates with shorter feet? Fifteen high school students were selected at random and asked to measure their foot length and height, both in centimeters. Some computer output from a least-squares regression analysis of the data is shown here:

Predictor	Coef	SE Coef
Constant	103.4100	19.5000
Foot length	2.7469	0.7833
S = 7.95126	R-Sq = 48.6%	

(a) State appropriate hypotheses for a test about the slope of the regression model for predicting height from foot length in the population of high school students.
(b) Explain why the data provide some evidence for H_a.
(c) Is the Random condition met in this case? Justify your answer.

SOLUTION:

(a) $H_0: \beta = 0$
$H_a: \beta > 0$
where β is the slope of the population regression line relating y = height (cm) to x = foot length (cm) for high school students.

> The alternative hypothesis specifies a positive association because the question asks if students with longer feet are typically taller than those with shorter feet.

(b) There is some evidence for $H_a: \beta > 0$ because $b = 2.7469 > 0$.
(c) Yes, because the 15 high school students were randomly selected.

FOR PRACTICE, TRY EXERCISE 1

Calculating the Standardized Test Statistic and *P*-Value for a Test about β

The standardized test statistic in a test about the slope of a population regression line has the same form as the standardized test statistic for a test about a mean or a test about a proportion:

$$\text{standardized test statistic} = \frac{\text{statistic} - \text{parameter}}{\text{standard error of statistic}}$$

The statistic we use is b, the slope of the sample regression line. We use the standard error of the slope s_b in the denominator, which is almost always calculated with technology. Here is the formula:

$$t = \frac{b - \beta_0}{s_b}$$

In most cases, we want to test the null hypothesis that the slope of the population regression line is $\beta = 0$. However, sometimes we might use a null value different than 0. Regression output from statistical software gives the value of t for a test of H_0: $\beta = 0$ by default. If you want to test a null hypothesis other than H_0: $\beta = 0$, get the slope and standard error from the output and use the formula to calculate the t statistic.

When the conditions are met for a test about the slope of a population regression model, we use a t distribution with $n - 2$ degrees of freedom to calculate the P-value.

Here is computer output from a least-squares regression analysis of the data from the seating chart experiment:

Term	Coef	SE Coef	T-Value	P-Value
Constant	85.71	4.24	20.22	0.000
Row	−1.117	0.947	−1.18	0.248
S = 10.0673	R-sq = 4.73%		R-sq(adj) = 1.33%	

The slope of the least-squares regression line is −1.117 and the standard error of the slope is $s_b = 0.947$. For a test of H_0: $\beta = 0$ versus H_a: $\beta < 0$, the value of the standardized test statistic is

$$t = \frac{-1.117 - 0}{0.947} = -1.180$$

This value is conveniently displayed in the "T-value" column for the "Row" term in the computer output.

The P-value is the probability of getting a sample slope of −1.117 or less by chance alone when H_0: $\beta = 0$ is true. Figure 9.5(a) shows this probability as a shaded area in the sampling distribution of b. Earlier, we calculated the standardized test statistic to be $t = -1.180$. Because there were $n = 30$ students in the experiment, we calculate the P-value by finding $P(t \leq -1.180)$ in a t distribution with df $= 30 - 2 = 28$. The shaded area in Figure 9.5(b) shows this probability.

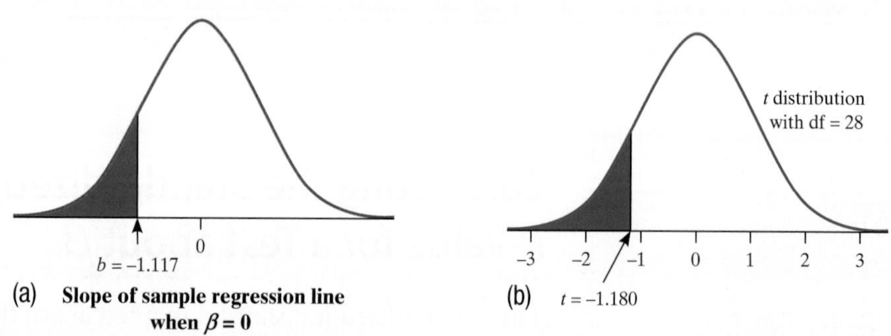

FIGURE 9.5 (a) The P-value for the seating chart experiment is the area to the left of $b = -1.117$ in the sampling distribution of the sample slope, assuming that H_0: $\beta = 0$ is true. (b) We calculate the P-value as the area to the left of $t = -1.180$ in a t distribution with 28 degrees of freedom.

Using Table B, the P-value is between 0.10 and 0.15. Using technology, we get a more precise P-value of 0.124. Because this P-value is larger than $\alpha = 0.05$, we fail to reject H_0. These data do not provide convincing evidence that the slope β of the true regression line is less than 0, so we do not have convincing evidence that sitting closer to the front of the classroom improves test scores for students like the ones in this experiment.

Notice that the P-value reported in the computer output, 0.248, is for a *two-sided test*. To get the correct P-value for a one-sided test when there is some evidence for H_a, we simply divide the given P-value by 2: $0.248/2 = 0.124$.

CALCULATING THE STANDARDIZED TEST STATISTIC AND *P*-VALUE IN A TEST ABOUT THE SLOPE OF A POPULATION REGRESSION LINE

Suppose the conditions are met. To perform a test of $H_0: \beta = \beta_0$, compute the standardized test statistic

$$t = \frac{b - \beta_0}{s_b}$$

Find the *P*-value by calculating the probability of getting a t statistic this large or larger in the direction specified by the alternative hypothesis H_a using a t distribution with $n - 2$ degrees of freedom.

In some contexts, it makes sense to fit a regression model that goes through the point (0, 0). Because there is only one parameter to estimate in this case (the slope), the degrees of freedom would be df $= n - 1$.

AP® EXAM TIP

Formula sheet

The specific formula for the standardized test statistic in a test about a population slope is *not* included on the formula sheet provided on both sections of the AP® Statistics exam. However, the formula sheet does include the general formula for the standardized test statistic:

$$\text{standardized test statistic} = \frac{\text{statistic} - \text{parameter}}{\text{standard error of statistic}}$$

and the formula for the standard error of the sample slope:

$$s_b = \frac{s}{s_x\sqrt{n-1}}$$

EXAMPLE

Can foot length predict height for high schoolers?

Calculating the standardized test statistic and *P*-value for a test about β

Skills 3.E, 4.B, 4.E

PROBLEM: In the preceding example, 15 randomly selected high school students measured their height (cm) and foot length (cm). Some computer output from a least-squares regression analysis of the data is again shown here:

Daniel de la Hoz/Moment/Getty Images

```
Predictor          Coef    SE Coef
Constant       103.4100    19.5000
Foot length      2.7469     0.7833
S = 7.95126       R-Sq = 48.6%
```

We want to use these data to perform a test of $H_0: \beta = 0$ versus $H_a: \beta > 0$, where β is the slope of the population least-squares regression line relating y = height to x = foot length. Assume the conditions for inference are met.

(a) Calculate the standardized test statistic and *P*-value.
(b) Interpret the *P*-value.
(c) What conclusion would you make?

SOLUTION:

(a) $t = \dfrac{2.7469 - 0}{0.7833} = 3.507$

$$t = \frac{b - \beta_0}{s_b}$$

P-value: df = 15 − 2 = 13

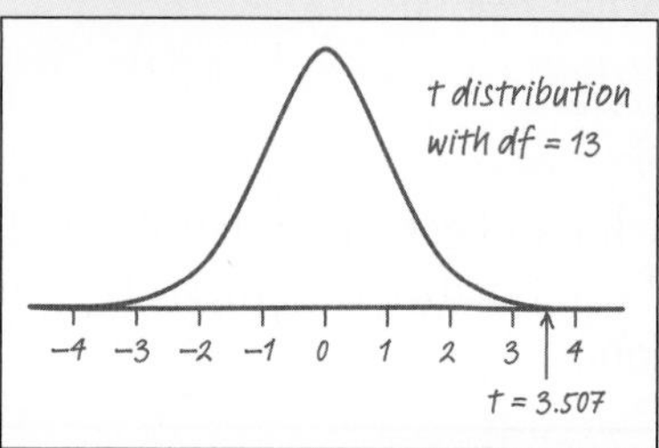

Using technology: tcdf(lower: 3.507, upper: 1000, df: 13) = 0.0019

Using Table B: The P-value is between 0.001 and 0.0025.

(b) Assuming there is no association between x = foot length and y = height in the population of high school students, there is a 0.0019 probability of getting a sample slope of 2.7469 or greater by chance alone.

(c) Because 0.0019 is less than $\alpha = 0.05$, we reject H_0. These data provide convincing evidence that the slope β of the population regression line is greater than 0, so we have convincing evidence of a positive, linear association between foot length (cm) and height (cm) in the population of high school students.

FOR PRACTICE, TRY EXERCISE 5

As we saw in Units 6 and 7, confidence intervals can provide more information than significance tests. You can check that the 95% confidence interval for the slope of the population regression line in this context is 1.055 to 4.439. Because all of the values in the interval are positive, the confidence interval gives the same conclusion as the test — plus an interval of plausible values for the slope.

Putting It All Together: *t* Test for the Slope

The appropriate inference method for testing a claim about β is called a ***t* test for the slope.**

DEFINITION *t* test for the slope

A ***t* test for the slope** is a significance test of the null hypothesis that the slope β of a population regression line is equal to a specified value, most commonly 0.

As with any inference procedure, be sure to follow the four-step process.

EXAMPLE

Can infant crying predict aptitude later in life?

t test for the slope

Skills 1.E, 1.F, 3.E, 4.C, 4.E

PROBLEM: Child development researchers explored the relationship between the crying of 38 randomly selected infants when they were 4 to 10 days old and their scores on an aptitude test later in life; they thought that infants who are more easily stimulated might cry more as an infant, but also develop faster. A snap of a rubber band on the sole of the foot caused the infants to cry. The researchers recorded the crying and measured its intensity by the number of peaks in the most active 20 seconds. Several years later, they recorded the aptitude score for each child.[18] Here are some graphs and computer output from a least-squares regression analysis of the data.

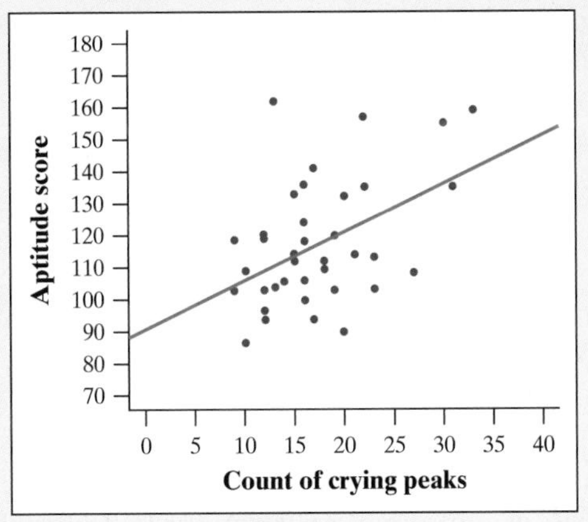

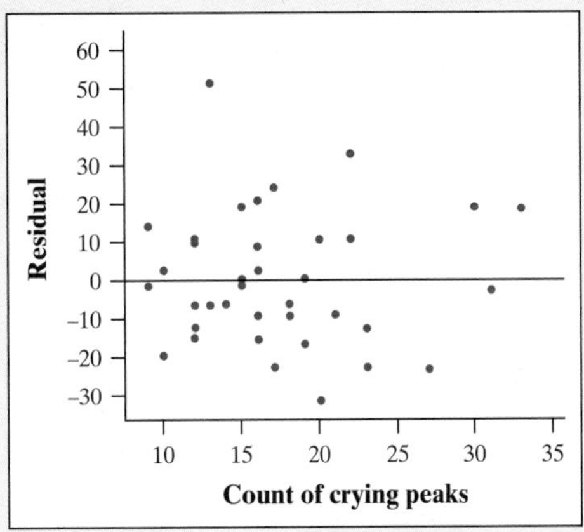

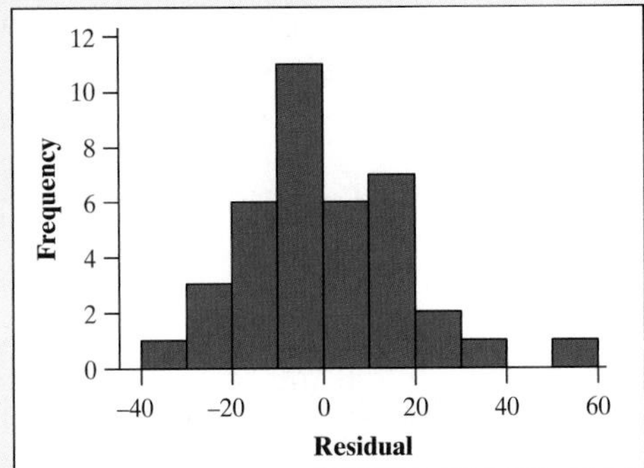

Predictor	Coef	SE Coef	T	P
Constant	91.268	8.934	10.22	0.000
Cry count	1.4929	0.4870	3.07	0.004

S = 17.50 R-Sq = 20.7% R-Sq(adj) = 18.5%

Do these data provide convincing evidence at the $\alpha = 0.05$ significance level of a positive linear relationship between cry count and later aptitude score in the population of infants?

SOLUTION:

Follow the four-step process!

STATE: We want to test

$H_0: \beta = 0$

$H_a: \beta > 0$

where β = the slope of the population regression line relating y = aptitude score to x = count of crying peaks in the population of infants. Use $\alpha = 0.05$.

PLAN: *t* test for the slope

- Linear: The scatterplot shows a linear relationship between cry count and aptitude score, and the residual plot shows a fairly random scatter about the residual = 0 line. ✓
- Normal: The histogram of residuals does not show strong skewness or obvious outliers. ✓
- Equal SD: The residual plot shows a fairly equal amount of scatter around the horizontal line at 0 for all x-values. ✓
- Random: Random sample of 38 infants. ✓
    - 10%: 38 is less than 10% of all infants. ✓

DO:

- $t = 3.07$
- P-value $= 0.004/2 = 0.002$ using df $= 38 - 2 = 36$

The sample data give some evidence for H_a because $b = 1.4929 > 0$.

The *t* statistic and *P*-value *for a two-sided test* are found in the row for "Cry count" under the corresponding headings. Because there is evidence for H_a, to get the *P*-value for this one-sided test, divide the *P*-value reported in the computer output by 2.

CONCLUDE: Because the P-value of $0.002 < \alpha = 0.05$, we reject H_0. These data provide convincing evidence that the slope β of the population regression line relating is greater than 0, so we have convincing evidence of a positive linear relationship between cry count and later aptitude score in the population of infants.

FOR PRACTICE, TRY EXERCISE 9

Based on the results of the crying study, should we ask doctors and parents to make infants cry more to improve their aptitude scores later in life? Hardly. This observational study gives statistically significant evidence of a positive linear relationship between the two variables — but we can't conclude that more intense crying as an infant *causes* an increase in aptitude. Maybe infants who cry more are born more alert and tend to score higher on aptitude tests.

30. Tech Corner SIGNIFICANCE TESTS FOR THE SLOPE

TI-Nspire and other technology instructions are on the book's website at bfwpub.com/tps7e.

You can use the TI-83/84 to perform a significance test for the slope of a population regression line when raw data are provided. We'll illustrate using the data from the 38 randomly selected infants in the crying and aptitude score study.[19]

Cry count	Score	Cry count	Score	Cry count	Score	Cry count	Score
10	87	20	90	17	94	12	94
12	97	16	100	19	103	12	103
9	103	23	103	13	104	14	106
16	106	27	108	18	109	10	109
18	109	15	112	18	112	23	113
15	114	21	114	16	118	9	119
12	119	12	120	19	120	16	124
20	132	15	133	22	135	31	135
16	136	17	141	30	155	22	157
33	159	13	162				

1. Enter the *x*-values (cry count) into L1 and the *y*-values (score) into L2.
2. Press STAT, then choose TESTS and LinRegTTest....
3. In the LinRegTTest screen, enter Xlist: L1, Ylist: L2, Freq: 1, and choose β & ρ: > 0.

4. Highlight "Calculate" and press ENTER. The linear regression t test results take two screens to present. We show only the first screen. The value of s is the standard deviation of the residuals, not the standard error of the slope.

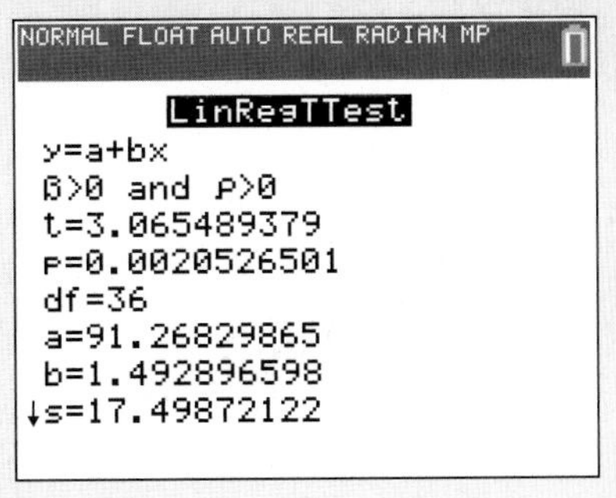

Think About It

WHAT'S WITH THAT $\rho > 0$ IN THE LinRegTTest SCREEN? The slope b of the sample least-squares regression line is closely related to the correlation r between the explanatory and response variables x and y (recall that $b = r\frac{s_y}{s_x}$). In the same way, the slope β of the population regression line is closely related to the correlation ρ (lowercase Greek letter rho) between x and y in the population. In particular, the slope is 0 when the correlation is 0.

Testing the null hypothesis H_0: $\beta = 0$ is exactly the same as testing that there is *no correlation* between x and y in the population from which we selected our data. You can use the test for zero slope to test the hypothesis H_0: $\rho = 0$ of zero correlation between any two quantitative variables. That's a useful trick. Because correlation also makes sense when there is no explanatory variable–response variable distinction, it is handy to be able to test correlation without doing regression.

CHECK YOUR UNDERSTANDING

Do countries with higher female literacy rates have lower birth rates? To investigate, a researcher collected data on x = female literacy rate (for women 15 years of age or older) and y = birth rate (births per 1000 population) for 20 randomly selected countries.[20] The equation of the least-squares regression line is $\hat{y} = 54.6392 - 0.3769x$ and the standard error of the slope is 0.087. The researcher verified that the conditions for performing inference about the slope of the population linear regression model are met.

1. State appropriate hypotheses for a significance test in this setting. Be sure to define the parameter of interest.
2. Calculate the standardized test statistic and the P-value.
3. What conclusion would you make?
4. Based on your conclusion in Question 3, which type of error could you have made — a Type I error or a Type II error? Describe this error in context.

SECTION 9B Summary

- When an association between two quantitative variables is linear, use a least-squares regression line to model the relationship between the explanatory variable x and the response variable y. To test a claim about the slope β of the population regression line $\mu_y = \alpha + \beta x$, use the slope b of the sample regression line $\hat{y} = a + bx$ as the statistic.
- A significance test about the slope of a population regression line starts with stating hypotheses. The null hypothesis is H_0: $\beta = \beta_0$ and the alternative hypothesis can be one-sided or two-sided.
  - The usual null hypothesis for this test is that there is no linear association between the two variables in the population of interest — that is, $\beta = 0$.
  - The alternative hypothesis is that there is some kind of linear association between the two variables in the population of interest. That is, $\beta > 0$ (positive linear association), $\beta < 0$ (negative linear association), or $\beta \neq 0$ (a linear association).
- Before performing a test of H_0: $\beta = \beta_0$, we need to ensure that the population regression model is valid, that there is independence in the data collection, and that the sampling distribution of the sample slope b is approximately normal. The required conditions are
  - Linear: The form of the relationship between x and y is linear. For each value of x, the mean value of y (denoted by μ_y) falls on the population regression line $\mu_y = \alpha + \beta x$.
  - Normal: For each value of x, the distribution of y is approximately normal.
  - Equal SD: For each value of x, the standard deviation of y (denoted by σ) is the same.
  - Random: The data come from a random sample from the population of interest or a randomized experiment.
    - 10%: When sampling without replacement, $n < 0.10N$.
- The standardized test statistic for a **t test for the slope** is

$$t = \frac{b - \beta_0}{s_b}$$

  When the conditions are met, find the P-value by calculating the probability of getting a t statistic this large or larger in the direction specified by the alternative hypothesis H_a in a t distribution with df $= n - 2$.
- Be sure to follow the four-step process whenever you perform a significance test about the slope β of a population regression line.

AP® EXAM TIP
AP® Daily Videos

Review the content of this section and get extra help by watching the AP® Daily Videos for Topics 9.4, 9.5, and 9.6, which are available in AP® Classroom.

9B Tech Corner

TI-Nspire and other technology instructions are on the book's website at bfwpub.com/tps7e.

30. Significance tests for the slope Page 834

SECTION 9B Exercises

Stating Hypotheses and Checking Conditions for a Test about β

1. **Wine and heart health** Is higher per capita wine consumption in a country associated with a lower death rate from heart disease? A researcher from the University of California, San Diego, collected data on average per capita wine consumption (in liters per year) and heart disease death rate (number of deaths per 100,000 people) in a random sample of 19 countries for which data were available.[21] Some computer output from a least-squares regression analysis of the data is shown here. Assume that the conditions for inference are met.

pg 829

Predictor	Coef	SE Coef
Constant	260.5634	13.8354
Wine consumption	−22.9688	3.5574
s = 37.88	R-sq = 71.0%	

(a) State hypotheses for a test about the slope of the regression model for predicting heart disease death rate from wine consumption in the population of countries.

(b) Explain why the data provide some evidence for H_a.

(c) Is the Random condition met in this case? Justify your answer.

2. **Swimming and pulse rate** Professor Moore recorded the time (in minutes) he took to swim 2000 yards and his pulse rate (in beats per minute) after swimming on a random sample of 23 days. Does his pulse rate tend to be higher on days that he swims faster (has smaller times)? Some computer output from a least-squares regression analysis of the data is shown here. Assume that the conditions for inference are met.

Predictor	Coef	SE Coef
Constant	479.9341	66.2278
Time	−9.6949	1.8887
s = 6.455	R-sq = 55.7%	

(a) State hypotheses for a test about the slope of the regression model for predicting pulse rate from swimming time on days when Professor Moore swims 2000 yards.

(b) Explain why the data provide some evidence for H_a.

(c) Is the Random condition met in this case? Justify your answer.

3. **Fouls and points** Is the number of points scored by a National Basketball Association (NBA) player related to the number of fouls he commits? A random sample of 50 of the 450 NBA players from a recent season was selected to investigate. Here are a scatterplot, histogram of residuals, and residual plot created from the least-squares regression line relating y = number of points to x = number of fouls.[22]

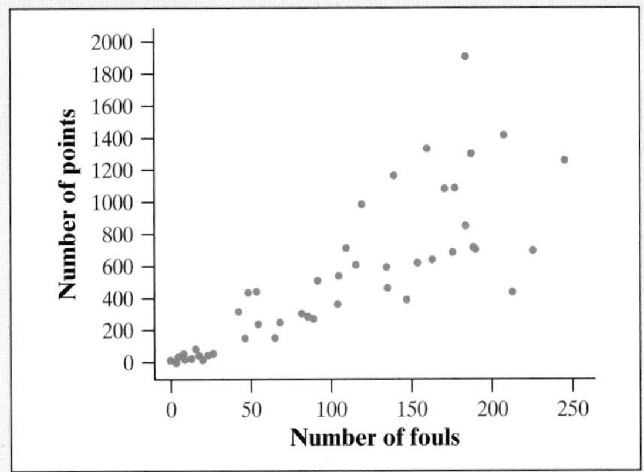

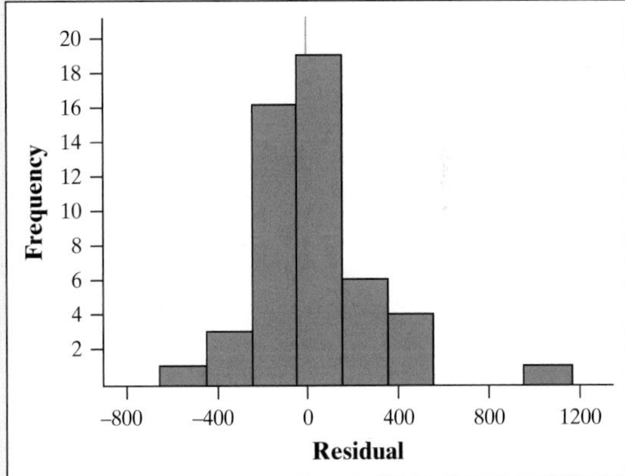

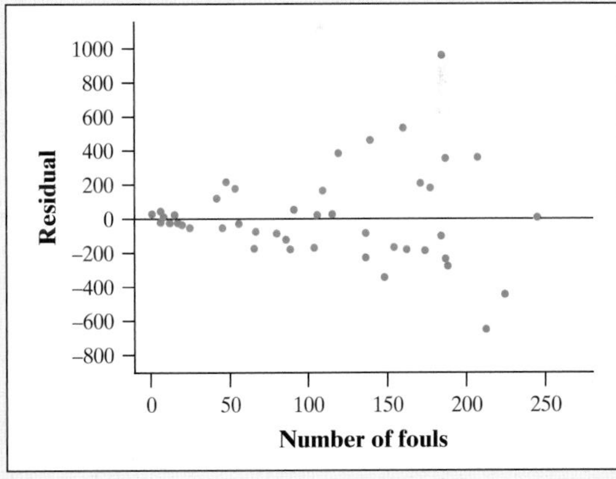

(a) State appropriate hypotheses for a significance test to answer the question of interest. Be sure to define any parameter you use.

(b) Check if the conditions for performing the test in part (a) are met.

4. **Tall saguaros** Saguaro National Park near Tucson, Arizona, is famous for its saguaro cactus. To track the health of saguaros in the park and estimate the total number of saguaros, researchers randomly select saguaros for inspection every 10 years.[23] Is there an association between the height of a saguaro and the elevation where it grows? Here are a scatterplot, residual plot, and dotplot of residuals created from the least-squares

regression line relating y = height (meters) to x = elevation (meters) for a random sample of 100 saguaro cactus from the approximately 2 million saguaros in the park.

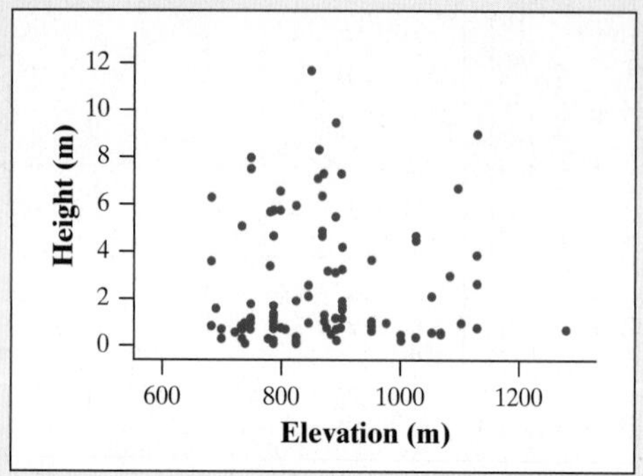

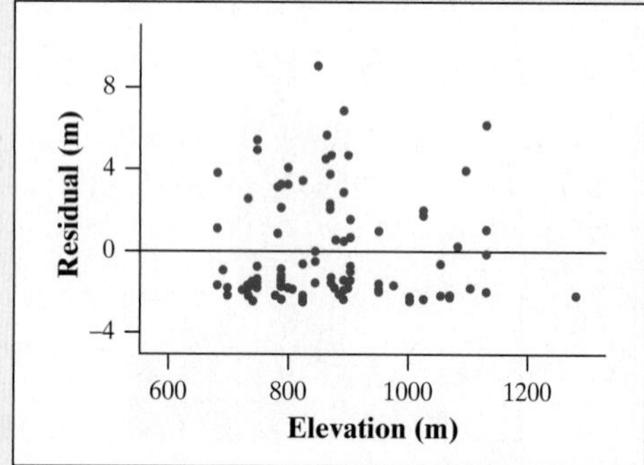

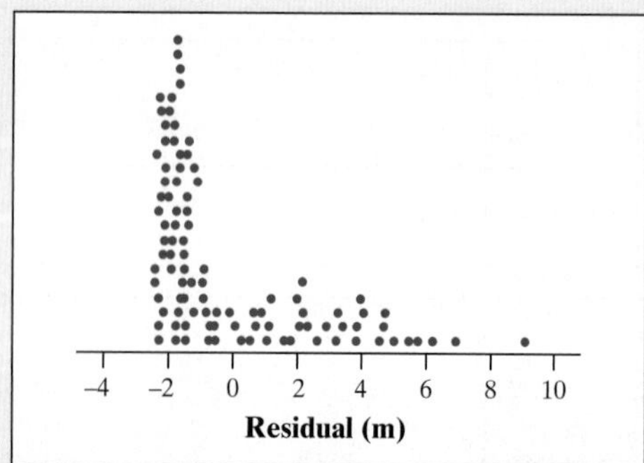

(a) State appropriate hypotheses for a significance test to answer the question of interest. Be sure to define any parameters you use.

(b) Check if the conditions for performing the test in part (a) are met.

Calculating the Standardized Test Statistic and *P*-Value for a Test about β

5. **More wine and heart health** Refer to Exercise 1.
pg 831

(a) Calculate the standardized test statistic and *P*-value.

(b) Interpret the *P*-value.

(c) What conclusion would you make?

6. **More swimming** Refer to Exercise 2.

(a) Calculate the standardized test statistic and *P*-value.

(b) Interpret the *P*-value.

(c) What conclusion would you make?

7. **Heartbeats** Is there a relationship between a person's body temperature and their resting pulse rate? A researcher collected data on x = body temperature (in °F) and y = pulse rate (in beats per minute) for each of 20 randomly selected students at a university.[24] The equation of the least-squares regression line relating these two variables is $\hat{y} = -408.904 + 4.944x$ and the standard error of the slope is 2.8712. Do these data provide convincing evidence at the 1% significance level of a linear relationship between body temperature and resting pulse rate for students at this university? Assume the conditions for inference are met.

(a) State appropriate hypotheses. Be sure to define any parameters you use.

(b) Interpret the standard error of the slope.

(c) Calculate the standardized test statistic and *P*-value.

(d) What conclusion would you make?

8. **More mess?** When Mentos are dropped into a newly opened bottle of Diet Coke, carbon dioxide is released from the Diet Coke very rapidly, causing the Diet Coke to be expelled from the bottle. To see if using more Mentos causes more Diet Coke to be expelled, Brittany and Allie obtained twenty-four 2-cup bottles of Diet Coke and randomly assigned each bottle to receive either 2, 3, 4, or 5 Mentos. After waiting for the fizzing to stop, they measured the amount expelled (in cups) by subtracting the amount remaining from the original amount in the bottle.[25] The equation of the least-squares regression line relating y = amount expelled (in cups) to x = number of Mentos is $\hat{y} = 1.0021 + 0.0708x$ and the standard error of the slope is 0.0123. Do these data provide convincing evidence at the 5% significance level of a positive linear relationship between number of Mentos added and amount of Diet Coke expelled? Assume the conditions for inference are met.

(a) State appropriate hypotheses. Be sure to define any parameters you use.

(b) Interpret the standard error of the slope.

(c) Calculate the standardized test statistic and *P*-value.

(d) What conclusion would you make?

t Test for the Slope

9. **Weeds among the corn** Lamb's-quarter is a common weed that interferes with the growth of corn. An agriculture researcher planted corn at the same rate in 16 small plots of ground and then weeded the plots by
pg 832

hand to allow a fixed number of lamb's-quarter plants to grow in each meter of the corn rows. The decision of how many of these plants to leave in each plot was made at random. No other weeds were allowed to grow. Here are the data on x = number of weeds per meter and y = corn yield (bushels per acre):[26]

Weeds per meter	Corn yield	Weeds per meter	Corn yield
0	166.7	3	158.6
0	172.2	3	176.4
0	165.0	3	153.1
0	176.9	3	156.0
1	166.2	9	162.8
1	157.3	9	142.4
1	166.7	9	162.8
1	161.1	9	162.4

Here are a scatterplot, residual plot, and dotplot of residuals, along with computer output from a linear regression analysis of the data.

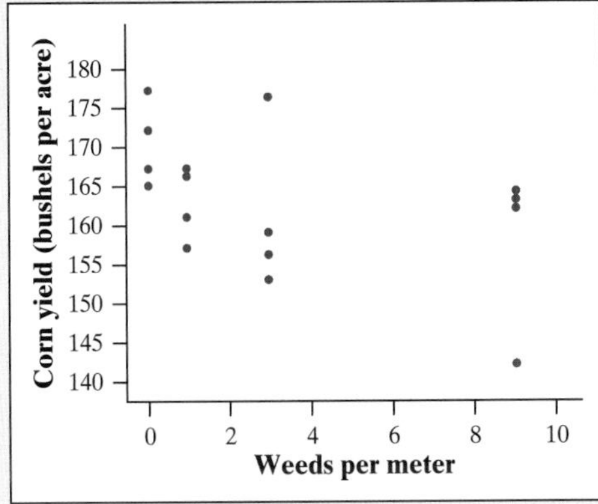

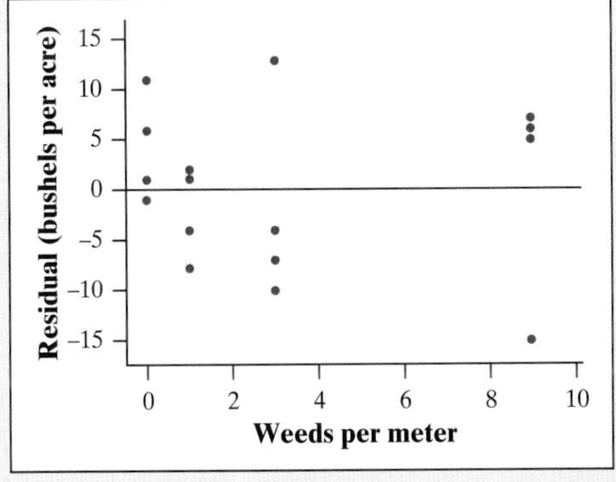

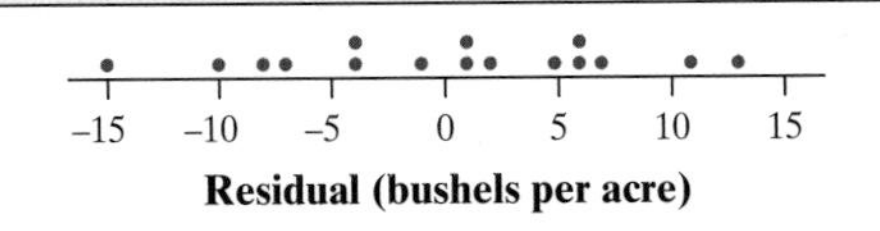


```
Predictor              Coef    SE Coef       T       P
Constant            166.483      2.725   61.11   0.000
Weeds per meter     -1.0987     0.5712   -1.92   0.075
S = 7.97665       R-Sq = 20.9%    R-Sq(adj) = 15.3%
```

Do these data provide convincing evidence at the $\alpha = 0.05$ significance level of a negative linear association between the number of lamb's-quarter plants and corn yield?

10. Time at the table Is there an association between the length of time a toddler sits at the lunch table and the number of calories consumed? Researchers collected data on a random sample of 20 toddlers observed over several months. Here are the data on x = the mean time a child spent at the lunch table (minutes) and y = the mean number of calories consumed:[27]

Time	Calories	Time	Calories	Time	Calories	Time	Calories
21.4	472	42.4	450	39.5	437	32.9	436
30.8	498	43.1	410	22.8	508	30.6	480
37.7	465	29.2	504	34.1	431	35.1	439
33.5	456	31.3	437	33.9	479	33.0	444
32.8	423	28.6	489	43.8	454	43.7	408

Here are a scatterplot, residual plot, and dotplot of residuals, along with computer output from a linear regression analysis of the data.

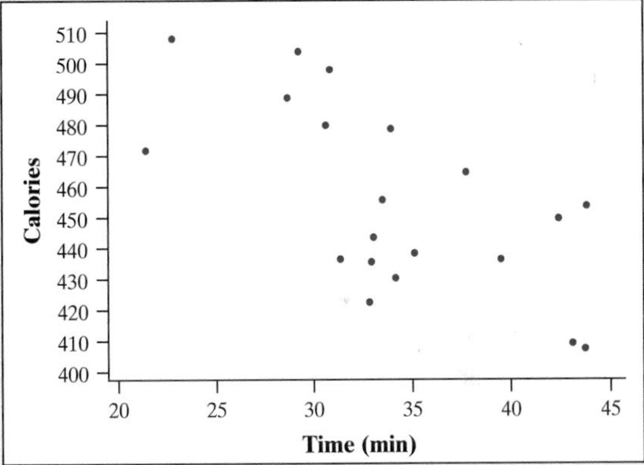

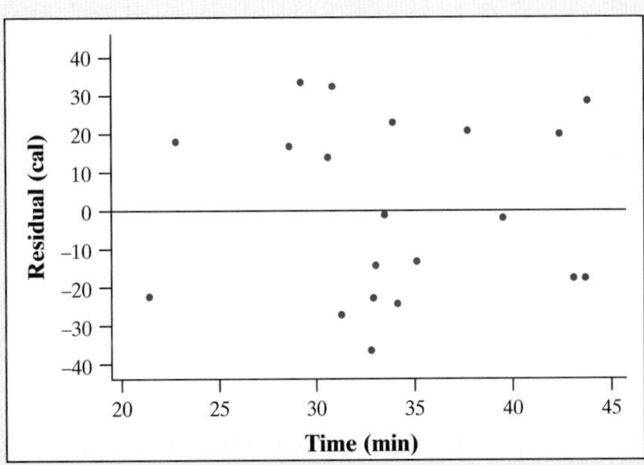

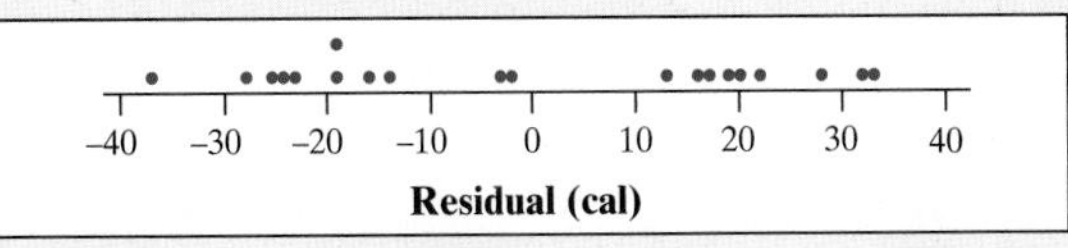


```
Predictor       Coef    SE Coef       T       P
Constant      560.65      29.37   19.09   0.000
Time         -3.0771     0.8498   -3.62   0.002
S = 23.3980    R-Sq = 42.1%    R-Sq(adj) = 38.9%
```

Do these data provide convincing evidence at the $\alpha = 0.01$ significance level of a linear relationship between time at the table and calories consumed in the population of toddlers?

11. **Loud music and tests** Two psychology students wanted to know if listening to music at a louder volume hurts test performance. To investigate, they recruited 30 volunteers and randomly assigned 10 to listen to music at 30 decibels, 10 to listen to music at 60 decibels, and 10 to listen to music at 90 decibels. While listening to the music, each student took a 10-question math test.[28] Here is computer output from a least-squares regression analysis using x = volume of music (in decibels) and y = score on the math test.

```
Predictor      Coef   SE Coef       T        P
Constant     9.9000   0.7525   13.156   0.0000
Volume      -0.0483   0.0116   -4.163   0.0003
S = 1.55781     R-Sq = 38.2%
```

(a) Explain why we don't need to check the 10% condition in this case.

(b) If these data provide convincing evidence of a negative association between music volume and math test score, can you conclude that listening to music at a louder volume causes a decrease in test performance for people like the volunteers in this study? Explain your reasoning.

(c) Carry out an appropriate significance test. Assume the conditions for inference are met.

(d) Which kind of error — a Type I error or a Type II error — could you have made in part (c)? Explain your answer.

12. **Sugary flowers** Does adding sugar to the water in a vase help flowers stay fresh? To find out, two researchers prepared 12 identical vases with exactly the same amount of water in each vase. They put 1 tablespoon of sugar in 3 vases, 2 tablespoons of sugar in 3 vases, and 3 tablespoons of sugar in 3 vases. In the remaining 3 vases, they put no sugar. After the vases were prepared, the researchers randomly assigned 1 carnation to each vase and observed how many hours each flower continued to look fresh. Here is computer output from a least-squares regression analysis using x = amount of sugar (in tablespoons) and y = freshness time (in hours).

```
Predictor          Coef   SE Coef        T        P
Constant          181.2   3.6354   49.8435  < 0.001
Amount of sugar    15.2   1.9432    7.8222  < 0.001
S = 7.526             R-Sq = 86.0%
```

(a) Explain why we don't need to check the 10% condition in this case.

(b) If these data provide convincing evidence of a positive association between amount of sugar and freshness time, can you conclude that adding sugar to the water in a vase helps flowers like the ones in this study stay fresh? Explain your reasoning.

(c) Carry out an appropriate significance test. Assume the conditions for inference are met.

(d) Which kind of error — a Type I error or a Type II error — could you have made in part (c)? Explain your answer.

13. **Stats teachers' cars** A random sample of 21 AP® Statistics teachers was asked to report the age (in years) and mileage of their primary vehicles. The following scatterplot displays the data.

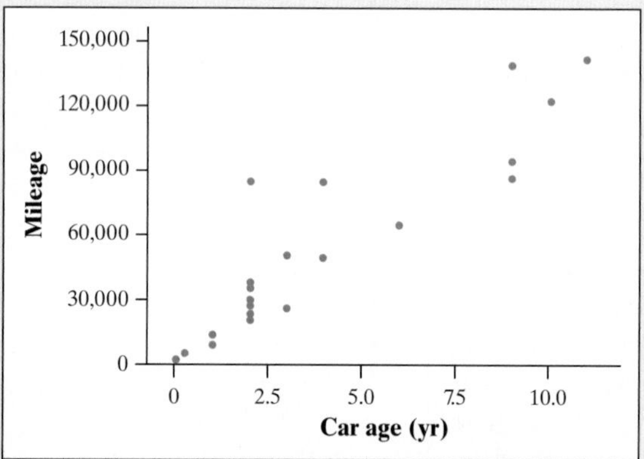

Here is some computer output from a least-squares regression analysis of these data. Assume that the conditions for inference about the slope of the population linear regression model are met.

```
Variable       Coef   SE Coef   t-ratio   prob
Constant    7288.54     6591      ****    ****
Car age     11630.6     1249      ****    ****
S = 19280     R-Sq = 82.0%     R-Sq(adj) = 81.1%
```

(a) A national automotive group claims that the typical driver puts 15,000 miles per year on their main vehicle. We want to test whether AP® Statistics teachers are typical drivers. Explain why an appropriate pair of hypotheses for this test is H_0: $\beta = 15{,}000$ versus H_a: $\beta \neq 15{,}000$.

(b) Compute the standardized test statistic and P-value for the test in part (a). What conclusion would you make at the $\alpha = 0.05$ significance level?

(c) Verify that the 95% confidence interval for the slope of the population regression line is (9016.4, 14,244.8).

(d) Explain how the confidence interval in part (c) is consistent with but gives more information than the test in part (b).

14. **Paired tires** Exercise 17 in Section 7A (page 666) compared two methods for estimating tire wear. The first method used the amount of weight lost by a tire, and the second method used the amount of wear in the grooves of the tire. A random sample of 16 tires was obtained. Both methods were used to estimate the total distance traveled by each tire. The following scatterplot displays the two estimates (in thousands of miles) for each tire.[29]

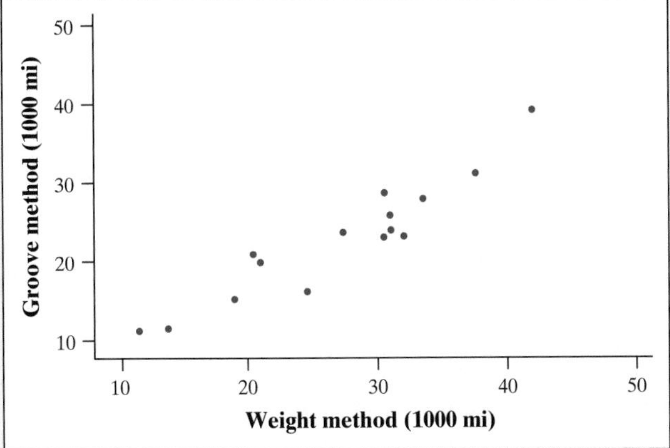

Here is some computer output from a least-squares regression analysis of these data. Assume that the conditions for inference about the slope of the population regression model are met.

```
Predictor       Coef   SE Coef       T        P
Constant       1.351     2.105    0.64    0.531
Weight       0.79021   0.07104   11.12    0.000
S = 2.62078    R-Sq = 89.8%    R-Sq(adj) = 89.1%
```

(a) Researchers want to test whether there is a difference in the results obtained with the two methods of estimating tire wear. Researchers plan to test H_0: $\beta = 1$ versus H_a: $\beta \neq 1$. Explain why the alternative hypothesis indicates that the methods won't give the same estimates of tire wear.

(b) Compute the standardized test statistic and *P*-value for the test in part (a). What conclusion would you make at the $\alpha = 0.01$ significance level?

(c) Verify that the 99% confidence interval for the slope of the population regression line is (0.5787, 1.0017).

(d) Explain how the confidence interval in part (c) is consistent with but gives more information than the test in part (b).

For Investigation *Apply the skills from the section in a new context or nonroutine way.*

15. **Car statistics** Refer to the car data in Exercise 13. In addition to performing tests about the slope of a least-squares regression line, it is possible to perform a test about the *y* intercept.

(a) If x = age (years) and y = miles driven, what should the value of the *y* intercept equal? Explain your answer.

(b) Based on your answer to part (a), state the hypotheses for a test about the *y* intercept of the population regression line (denoted by α).

(c) The equation of the least-squares regression line is $\hat{y} = 7288.54 + 11{,}630.6x$, where x = age (years) and y = miles driven. The standard error of the *y* intercept is $s_a = 6591$ and the sample size is $n = 21$. Calculate the *t* statistic and *P*-value for a test of the hypotheses in part (b), assuming the conditions for inference are met.

(d) Based on the *P*-value in part (c), what would you conclude?

16. **Infant crying and aptitude** In the final example of Section 9B (page 831), we used the slope to perform a test about the relationship between x = cry count and y = later aptitude score in the population of infants. It is also possible to classify the values for each variable into categories and use a chi-square test for independence. The following two-way table summarizes these data:

		Cry count		
		Less than 16	At least 16	Total
Aptitude score	Less than 115	10	12	22
	At least 115	5	11	16
	Total	15	23	38

(a) Do these data provide convincing evidence that cry count and later aptitude score are independent in the population of infants?

(b) Describe a disadvantage of using a chi-square test instead of a *t* test for slope in this context.

Multiple Choice *Select the best answer for each question.*

Exercises 17–20 refer to the following scenario. Josh wears an activity tracker to record the number of steps he takes each day, along with several other health-related measurements. Based on these variables, the activity tracker calculates an estimate of the number of calories he burns each day. Josh wonders if there is a positive, linear relationship between x = number of steps and y = calories burned. To investigate, he performs a least-squares regression analysis for these two variables using data from 10 randomly selected days. A graph of the residuals and some computer output are shown here.

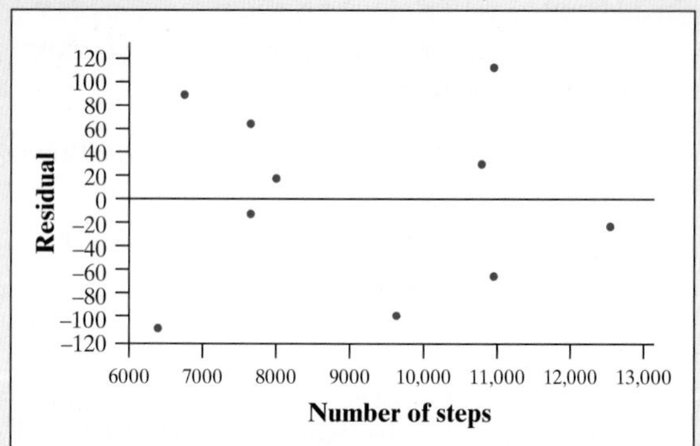

Predictor	Coef	SE Coef
Constant	2015.3123	118.6682
Number of steps	0.0671	0.0127
s = 81.7232	R-sq = 77.8%	

17. Is there convincing evidence at the 5% significance level of a positive, linear relationship between Josh's daily number of steps and calories burned? What hypotheses should be used for a significance test to answer this question?

(A) H_0: $\beta = 0$ versus H_a: $\beta \neq 0$

(B) H_0: $\beta = 0$ versus H_a: $\beta > 0$

(C) H_0: $\beta = 0$ versus H_a: $\beta < 0$

(D) H_0: $\beta > 0$ versus H_a: $\beta = 0$

(E) H_0: $\beta = 1$ versus H_a: $\beta > 1$

18. Which of the following is indicated by the residual plot?

(A) There is not a linear relationship between number of steps and calories burned.

(B) The Normal condition is met because the points appear to be randomly scattered about the residual = 0 line.

(C) The Equal SD condition is met because the points appear to be randomly scattered about the residual = 0 line.

(D) The Normal condition is met because the variability of the residuals appears to be similar for all x-values.

(E) The Equal SD condition is met because the variability of the residuals appears to be similar for all x-values.

19. Which of the following is the best interpretation of the value 81.7232 in the computer output?

(A) For every additional step that the activity tracker records in a day, Josh burns an additional 81.7232 calories, on average.

(B) About 81.7% of the variation in Josh's number of calories burned per day is accounted for by the linear model with x = number of steps recorded by the activity tracker.

(C) The predictions of Josh's number of calories burned in a day using this linear model with x = number of steps recorded by the activity tracker will typically differ from his actual number of calories burned that day by about 81.7.

(D) On a day when Josh's activity tracker records 0 steps, he is predicted to burn about 81.7 calories.

(E) The number of steps recorded each day by Josh's activity tracker typically varies from the mean number of steps recorded on these 10 days by about 81.7 steps.

20. Which of the following is closest to the P-value of this test?

(A) 0.05
(B) 0.01
(C) 0.005
(D) 0.0005
(E) 0.00005

Recycle and Review *Practice what you learned in previous sections.*

21. **Inference methods (Units 6–9)** In each of the following settings, state which inference method from Unit 6, 7, 8, or 9 you would use. For example, you might say "two-sample z test for the difference between two proportions." You do not have to carry out any procedures.[30]

(a) What is the average number of city council members for cities in a large state? A random sample of 38 cities was asked to report the number of city council members for their city.

(b) Are people who live in urban areas more likely to visit libraries than people who live in rural areas? Separate random samples of 300 urban residents and 300 rural residents were asked if they had been to a library in the previous year.

(c) Is there a relationship between attendance at religious services and alcohol consumption? A random sample of 1000 adults was asked if they regularly attend religious services and if they drink alcohol daily.

(d) Separate random samples of 75 undergraduate students and 75 graduate students were asked how much time, on average, they spend watching television each week. We want to estimate the difference in the average amount of TV watched by undergraduate and graduate students.

UNIT 9 Wrap-Up

FRAPPY! Free Response AP® Problem, Yay!

Directions: Show all your work. Indicate clearly the methods you use, because you will be scored on the correctness of your methods as well as on the accuracy and completeness of your results and explanations.

How much carbon dioxide (CO_2) do motor vehicles emit? The following scatterplot shows the relationship between x = engine size (in liters) and y = CO_2 emissions (in grams per mile) for a random sample of 176 gas-powered cars and trucks produced in 2021.[31]

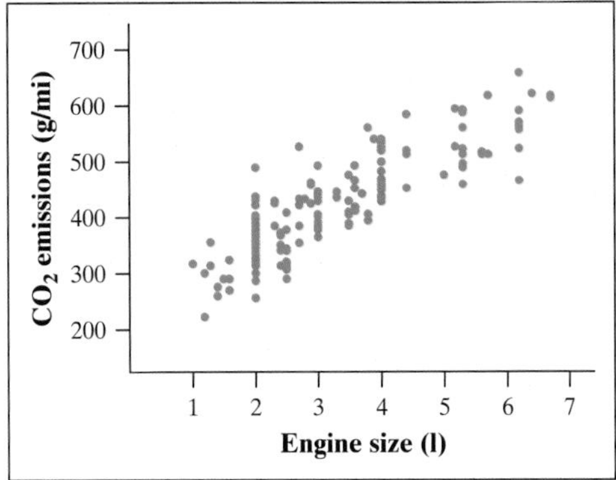

Here is some computer output from a linear regression analysis of the data.

Predictor	Coef	SE Coef	T	P
Constant	244.7181	8.7175	28.072	0.0000
Engine size	54.8730	2.4782	22.142	0.0000
S = 45.2329		R-Sq = 73.8%		R-Sq(adj) = 73.7%

(a) Identify and interpret the standard error of the slope.

(b) Calculate a 99% confidence interval for the slope of the population regression line. Assume that the conditions for inference are met.

(c) Interpret the confidence interval from part (b) in context.

(d) For the population of gas-powered cars and trucks produced in 2021, is it plausible that for every 1-liter increase in engine size, CO_2 emissions increase by an average of 50 grams per mile? Justify your answer.

After you finish the FRAPPY!, you can view two example solutions on the book's website (**bfwpub.com/tps7e**). Determine whether you think each solution is "complete," "substantial," "developing," or "minimal." If the solution is not complete, what improvements would you suggest to the student who wrote it? Finally, your teacher will provide you with a scoring rubric. Score your response and note what, if anything, you would do differently to improve your own score.

UNIT 9 REVIEW

SECTION 9A Confidence Intervals for the Slope of a Population Regression Line

When the relationship between two quantitative variables x and y is linear, and the data come from a random sample or a randomized experiment, we can use the **sample regression line** $\hat{y} = a + bx$ to perform inference about the **population regression line** $\mu_y = \alpha + \beta x$. The sampling distribution of the sample slope b is the foundation for doing inference about the population slope β. When the conditions are met, the sampling distribution of b is approximately normal with mean $\mu_b = \beta$ and standard deviation $\sigma_b = \dfrac{\sigma}{\sigma_x \sqrt{n}}$, where σ is the population standard deviation of the residuals and σ_x is the population standard deviation of x.

After learning about the sampling distribution of b, you learned how to construct and interpret confidence intervals for the slope β of a population regression line. Five conditions need to be met to calculate a confidence interval for β. The Linear condition says that the form of the relationship between x and y is linear and that the mean value of the response variable μ_y for each value of x falls on the population regression line $\mu_y = \alpha + \beta x$. The Normal condition says that the distribution of y is approximately normal for each value of x. The Equal SD condition says that for each value of x, the distribution of y should have the same standard deviation. The Random condition says that the data come from a random sample from the population of interest or from a randomized experiment. When sampling without replacement from a population, the 10% condition says that $n < 0.10N$. These conditions help ensure independence in data collection and verify that the sampling distribution of the sample slope b is approximately normal.

Because the values σ and σ_x in the formula for the standard deviation of the sampling distribution of b are usually unknown, we use the standard deviation of the residuals s from the sample regression line to estimate σ and the standard deviation s_x of the sample x-values to estimate σ_x. Then we use the *standard error of the slope* to estimate σ_b:

$$s_b = \frac{s}{s_x \sqrt{n-1}}$$

The standard error of the slope s_b describes how far the sample slope typically varies from the population slope in repeated random samples or random assignments. Fortunately, the standard error of the slope is typically provided with standard computer output for linear regression models.

The formula for calculating a ***t* interval for the slope** is

$$b \pm t^* s_b$$

where t^* is the critical value for a t distribution with df $= n - 2$ and C% of its area between $-t^*$ and t^*. We use a t^* critical value instead of a z^* critical value whenever we use s_b to estimate the unknown standard deviation σ_b.

Always use the four-step process whenever you are asked to construct and interpret a confidence interval for the slope β of a population regression line.

The logic of confidence intervals for the slope of a population regression line, including how to interpret the confidence interval and the confidence level, is the same as in Units 6 and 7. Similarly, you can use a confidence interval for β to evaluate a claim about the population slope. As with other intervals, increasing the confidence level leads to a larger margin of error and increasing the sample size leads to a smaller margin of error, assuming other things remain the same.

SECTION 9B Significance Tests for the Slope of a Population Regression Line

In this section, you learned how to perform a significance test about the slope β of a population regression line. As in any test, you start by stating hypotheses. The null hypothesis has the form H_0: $\beta = \beta_0$. The usual null hypothesis for this test is that there is no linear association between the two variables, H_0: $\beta = 0$. The alternative hypothesis can be one-sided ($<, >$) or two-sided ($\neq$).

Next, we check conditions to ensure independence in data collection and verify that the sampling distribution of b is approximately normal. The Linear, Normal, Equal SD, Random, and 10% conditions for significance tests about the population slope β are the same as the ones for confidence intervals for β.

For a test of H_0: $\beta = \beta_0$ when σ_b is unknown, the standardized test statistic is

$$t = \frac{b - \beta_0}{s_b}$$

When the conditions are met, find the *P*-value by calculating the probability of getting a t statistic this large or larger in the direction specified by the alternative hypothesis H_a in a t distribution with df $= n - 2$. The value of the standardized test statistic for a test of H_0: $\beta = 0$, along with a two-sided *P*-value, is typically provided with standard computer output for least-squares regression. Use the *P*-value to make an appropriate conclusion about the hypotheses, in context. This inference method is called a ***t* test for the slope.**

Whenever you are asked if there is convincing evidence for a claim about the slope β of a population regression line, be sure to follow the four-step process.

A significance test for a population slope uses the same logic as previous significance tests. Likewise, the way we interpret *P*-values and power and the way we describe Type I and Type II errors are very similar to what you learned earlier.

Inference for the Slope of a Population Regression line		
	Confidence Interval for β	**Significance Test for β**
Name (TI-83/84)	*t* interval for slope (LinRegTInt)	*t* test for slope (LinRegTTest)
Null Hypothesis	*Not applicable*	$H_0: \beta = \beta_0$
Conditions	• **Linear:** The form of the relationship between *x* and *y* is linear. For each value of *x*, the mean response μ_y falls on the population regression line $\mu_y = \alpha + \beta x$. • **Normal:** For each value of *x*, the distribution of *y* is approximately normal. • **Equal SD:** The standard deviation of *y* (call it σ) is the same for all values of *x*. • **Random:** The data come from a random sample from the population of interest or a randomized experiment. ○ **10%:** When sampling without replacement, $n < 0.10N$.	
Formula	$b \pm t^* s_b$ Critical value t^* from a *t* distribution with df $= n - 2$	$t = \frac{b - \beta_0}{s_b}$ *P*-value from a *t* distribution with df $= n - 2$

What Did You Learn?

Learning Target	Section	Related Example on Page(s)	Relevant Chapter Review Exercise(s)
Describe the sampling distribution of the sample slope *b*.	9A	806	R1
Check the conditions for calculating a confidence interval for the slope β of a population regression line.	9A	809	R2
Calculate a confidence interval for the slope of a population regression line.	9A	813	R2
Construct and interpret a *t* interval for the slope.	9A	814	R2
State appropriate hypotheses and check the conditions for a test about the slope of a population regression line.	9B	829	R3
Calculate the standardized test statistic and *P*-value for a test about the slope of a population regression line.	9B	831	R3
Perform a *t* test for the slope.	9B	832	R3

UNIT 9 REVIEW EXERCISES

These exercises are designed to help you review the important concepts and skills of the unit.

R1 **Sampling distribution of slope (9A)** Consider the population of all eruptions of the Old Faithful geyser in a given year. For each eruption, let *x* be the duration (in minutes) and *y* be the wait time (in minutes) until the next eruption. Suppose that the conditions for regression inference are met, the population regression line is $\mu_y = 34 + 13x$, the variability around the line is $\sigma = 6$ minutes, and the standard deviation of the *x*-values is $\sigma_x = 1.38$ minutes. Imagine that we select repeated random samples of 25 eruptions and calculate the equation of the least-squares regression line $\hat{y} = a + bx$ for each sample. Describe the shape, center (mean), and variability (standard deviation) of the sampling distribution of *b*.

R2 **Walking tall (9A)** Do taller students require fewer steps to walk a fixed distance? The scatterplot shows the relationship between *x* = height (in inches) and *y* = number of steps required to walk the length of a school hallway for a random sample of 36 students at a large high school.

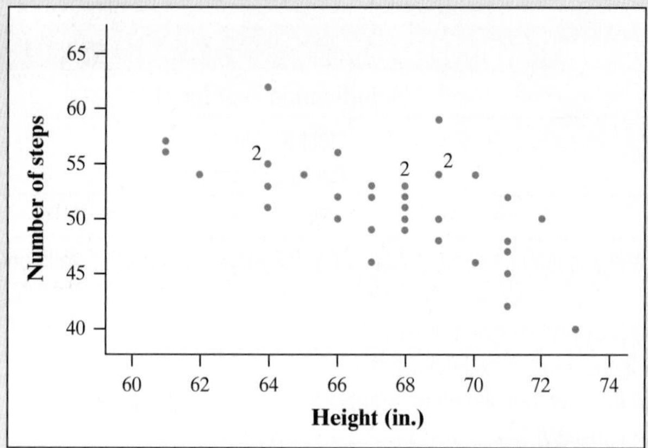

A least-squares regression analysis was performed on the data. Here are some graphs and computer output from the analysis:

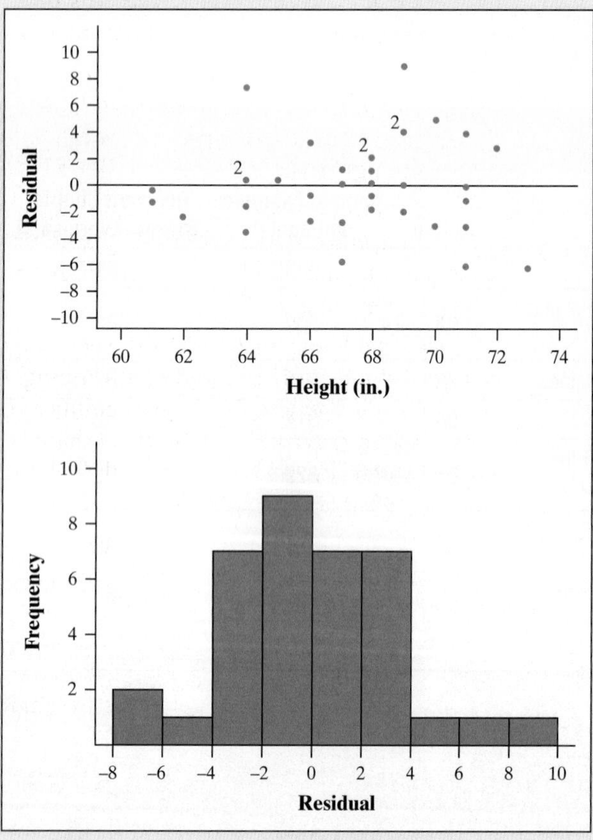

Predictor	Coef	SE Coef	T	P
Constant	113.57	13.085	8.679	0.000
Height	−0.9211	0.1938	−4.753	0.000

S = 3.50429 R-Sq = 39.9% R-Sq(adj) = 38.1%

(a) Calculate and interpret a 90% confidence interval for the slope.

(b) Does the confidence interval provide convincing evidence that taller students at this school tend to require fewer steps to walk a fixed distance? Justify your answer.

(c) Interpret the meaning of "90% confidence" in context.

R3 **Butterflies (9B)** Scientists were interested in the effect of changing climate on butterflies. They took a random sample of butterflies of the species *Boloria chariclea* in 32 different years and measured y = the wing length for the butterfly and x = the average temperature during the larval growing season in the previous summer.[32] The bigger the butterfly, the hardier it is considered to be. If the butterflies are getting smaller as the temperature increases with changing climate, this could be a problem for the species. Here is a scatterplot of the data along with the least-squares regression line $\hat{y} = 18.87 - 0.24x$. The standard error of the slope is 0.163.

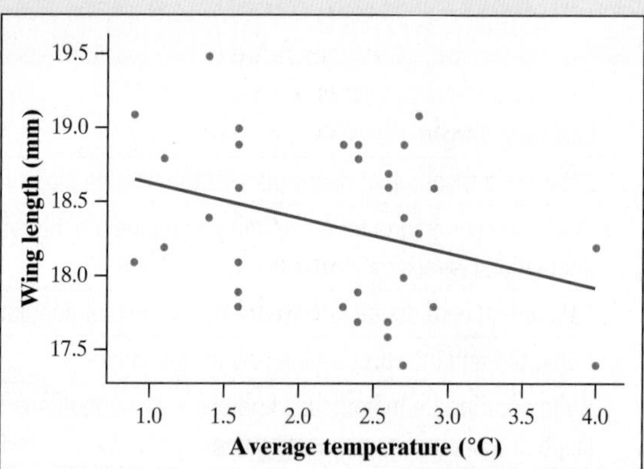

Do these data provide convincing evidence at the $\alpha = 0.05$ significance level of a negative linear relationship between average temperature during the larval growing season and wing length in the population of *Boloria chariclea* butterflies? Assume that the conditions for inference are met.

UNIT 9 AP® STATISTICS PRACTICE TEST

Section I: Multiple Choice *Select the best answer for each question.*

T1 Which of the following does *not* need to be true in order to perform inference about the slope of a population linear regression model?

(A) For each value of x, the distribution of y-values in the population follows an approximately normal distribution.

(B) For each value of x, the distribution of y-values in the population has the same standard deviation.

(C) The sample size — that is, the number of paired observations (x, y) — is at least 30.

(D) For each value of x, the mean of the distribution of y-values is on the population regression line.

(E) The data come from a random sample or a randomized experiment.

T2 Critical values for confidence intervals and P-values for significance tests about the slope β of a population regression line come from which of the following distributions?

(A) The t distribution with $n-1$ degrees of freedom

(B) The standard normal distribution

(C) The chi-square distribution with $n-1$ degrees of freedom

(D) The t distribution with $n-2$ degrees of freedom

(E) The chi-square distribution with $n-2$ degrees of freedom.

Exercises T3–T5 refer to the following scenario. An old saying in golf is "You drive for show and you putt for dough." The point is that good putting is more important than long driving for shooting lower (better) scores and hence winning money. To see if this is true, data from a random sample of 69 of the nearly 1000 players on the PGA Tour's world money list are examined. The average number of putts per hole (fewer is better) and the player's total winnings for the previous season are recorded, and a least-squares regression line is fitted to the data. Assume the conditions for inference about the slope are met. Here is computer output from the regression analysis:

```
Predictor         Coef   SE Coef       T        P
Constant       7897179   3023782    6.86    0.000
Avg. putts    -4139198   1698371    ****     ****
S = 281777         R-Sq = 8.1%     R-Sq(adj) = 7.8%
```

T3 Suppose that the researchers perform a significance test using the hypotheses H_0: $\beta = 0$ versus H_a: $\beta < 0$. Which of the following is the value of the standardized test statistic?

(A) −20.24

(B) −2.44

(C) 0.081

(D) 2.44

(E) 2.61

T4 The P-value for the test in Exercise T3 is 0.0087. Which of the following is a correct interpretation of this result?

(A) The probability there is no linear relationship between average number of putts per hole and total winnings for these 69 players is 0.0087.

(B) The probability there is no linear relationship between average number of putts per hole and total winnings for all players on the PGA Tour's world money list is 0.0087.

(C) If there is no linear relationship between average number of putts per hole and total winnings for the players in the sample, the probability of getting a random sample of 69 players that yields a least-squares regression line with a slope of −4,139,198 or less by chance alone is 0.0087.

(D) If there is no linear relationship between average number of putts per hole and total winnings for the players on the PGA Tour's world money list, the probability of getting a random sample of 69 players that yields a least-squares regression line with a slope of −4,139,198 or less by chance alone is 0.0087.

(E) The probability of making a Type I error is 0.0087.

T5 Which of the following is the 95% confidence interval for the slope of the population least-squares regression line relating winnings to putts per hole?

(A) $-4{,}139{,}198 \pm 2.437(1{,}698{,}371)$

(B) $-4{,}139{,}198 \pm 1.996(1{,}698{,}371)$

(C) $-4{,}139{,}198 \pm 1.960(1{,}698{,}371)$

(D) $-4{,}139{,}198 \pm 1.996(281{,}777)$

(E) $-4{,}139{,}198 \pm 1.960(281{,}777)$

Section II: Free Response *Show all your work. Indicate clearly the methods you use, because you will be graded on the correctness of your methods as well as on the accuracy and completeness of your results and explanations.*

T6 Growth hormones are often used to increase the weight gain of chickens. In an experiment using 15 chickens, researchers randomly assigned 3 chickens to each of 5 different doses of growth hormone (0, 0.2, 0.4, 0.8, and 1.0 milligrams). The subsequent weight gain (in ounces) was recorded for each chicken. The equation of the least-squares regression line for predicting weight gain from dose of growth hormone is $\hat{y} = 4.5459 + 4.8323x$. The standard error of the slope is 1.0164.

(a) Interpret the standard error of the slope.

(b) Calculate and interpret a 95% confidence interval for the slope of the true least-squares regression line. Assume the conditions for inference are met.

(c) What does the interval in part (b) suggest about whether the dose of growth hormone affects weight gain for chickens like the ones in this study?

T7 Companies that make processed foods often add some combination of sugar and fat to their products. Because the makers of protein bars want to keep the number of calories low, they will typically add either mostly sugar or mostly fat, but not both. Here are a scatterplot, residual plot, and some computer output for a linear regression analysis of x = amount of sugar (tsp) and y = amount of saturated fat (g) in 19 randomly selected types of protein bars.[33]

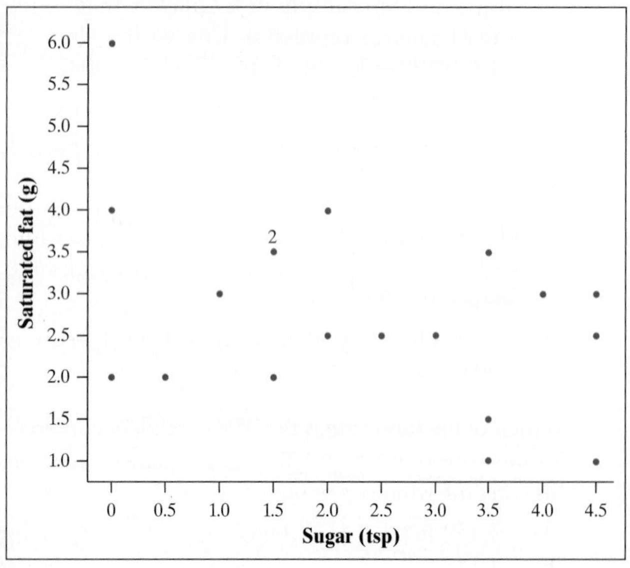

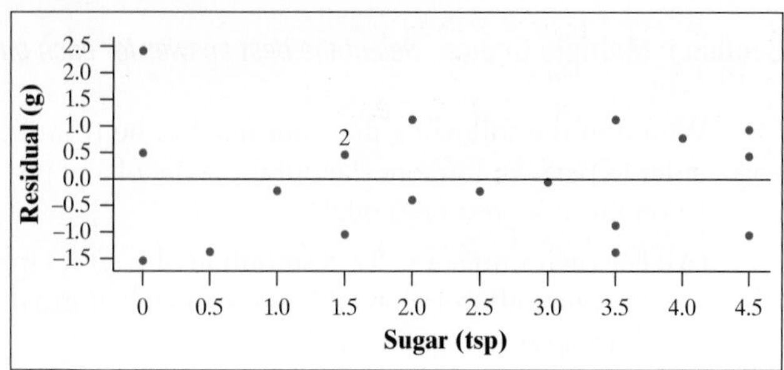

Predictor	Coef	SE Coef	T	P
Constant	3.542	0.4486	7.8953	< 0.001
Sugar (tsp)	−0.3287	0.1625	−2.0222	0.0592

S = 1.0922 R-Sq = 19.4%

(a) State appropriate hypotheses for a test to determine whether these data provide convincing evidence at the 5% significance level of a negative linear relationship between sugar and fat content in protein bars. Be sure to define any parameters you use.

(b) Identify both conditions for performing the test in part (a) that can be checked using the residual plot, and explain why those conditions are met.

(c) Assume that the remaining conditions for the test in part (a) are met. Based on the computer output, what conclusion would you make? Explain your answer.

(d) Which kind of error — a Type I error or a Type II error — could you have made in part (c)? Justify your answer.

Cumulative AP® Practice Test 4

Section I: Multiple Choice

- *Time limit:* 1 hour and 30 minutes
- *Number of questions:* 40
- *Percentage of AP® Statistics exam score:* 50

Directions: *Select the best answer for each question.*

AP1 Which of the following is a categorical variable?

(A) The weight of an automobile

(B) The time required to complete the Olympic marathon

(C) The fuel efficiency (in miles per gallon) of a hybrid car

(D) The brand of shampoo purchased by shoppers in a grocery store

(E) The closing price of a particular stock on the New York Stock Exchange

AP2 The school board in a certain school district obtained a random sample of 200 residents and asked if they were in favor of raising property taxes to fund the hiring of more teachers. The resulting confidence interval for the population proportion of residents in favor of raising taxes was (0.183, 0.257). Which of the following is the margin of error for this confidence interval?

(A) 0.037
(B) 0.074
(C) 0.183
(D) 0.220
(E) 0.257

AP3 A state agency wants to survey 1500 registered voters about an upcoming election. The agency randomly selects 50 registered voters from each of the state's 30 senatorial districts. Which type of sample was selected by the state agency?

(A) Simple random sample

(B) Stratified random sample

(C) Cluster sample

(D) Systematic random sample

(E) Convenience sample

AP4 The distribution of the number of cars X parked in a randomly selected residential driveway on any night is given by

Number of cars, x	0	1	2	3	4
Probability, $P(X = x)$	0.10	0.30	0.25	0.25	0.10

What is the expected value of X?

(A) 1.00
(B) 1.45
(C) 1.95
(D) 2.00
(E) 2.05

AP5 The following back-to-back stemplots compare the ages of players from two minor-league hockey teams, each with 25 players.

Team A		Team B
98777	1	788889
44333221	2	00123444
7766555	2	556679
521	3	023
86	3	55

1 | 7 = 17 years old

Which of the following statements *cannot* be justified from the stemplots?

(A) Team A has the same number of players in their 30s as does Team B.

(B) The median age of both teams is the same.

(C) Both age distributions are slightly skewed to the right.

(D) The range of age is greater for Team A.

(E) The value of Q_1 for Team B is greater than the value of Q_1 for Team A.

AP6 The distribution of grade point averages (GPAs) for a certain college is approximately normal with a mean of 2.5 and a standard deviation of 0.6. Any student with a GPA less than 1.0 is put on probation, while any student with a GPA of 3.5 or higher is added to the dean's list. About what percentage of students at the college are on probation or on the dean's list?

(A) 0.6
(B) 4.7
(C) 5.4
(D) 94.6
(E) 95.3

AP7 The dotplot shows the approximate sampling distribution of the sample proportion $\hat{p}$ for 100 random samples of size $n = 50$ from a population with $p = 0.3$. How would the approximate sampling distribution change if we selected 100 random samples of size $n = 500$ from the same population instead?

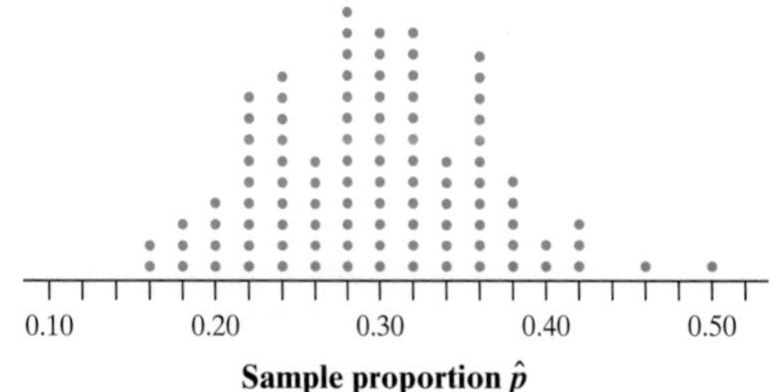

(A) The approximate sampling distribution would be more variable.

(B) The approximate sampling distribution would be less variable.

(C) The approximate sampling distribution would have more dots.

(D) The approximate sampling distribution would have fewer dots.

(E) The approximate sampling distribution would be centered at 0.5.

AP8 A survey of 100 high school students was conducted. Each member of the sample was asked two questions: "Are you in 12th grade?" and "Do you have a driver's license, a learner's permit, or neither?" Overall, 23 members of the sample were in 12th grade. Also, 33 members of the sample had a driver's license and 12 had a learner's permit. Assuming that there is no association between grade and driving status in the population of high school students, what is the expected number of 12th-graders with a driver's license?

(A) 7.59

(B) 15.41

(C) 25.41

(D) 56.00

(E) Impossible to determine without more information

AP9 Do hummingbirds prefer store-bought food made from concentrate or a simple mixture of sugar and water? To find out, a researcher obtains 10 identical hummingbird feeders and fills 5, chosen at random, with store-bought food from concentrate and the other 5 with a mixture of sugar and water. The feeders are then randomly assigned to 10 possible hanging locations in the researcher's yard. After one week, the researcher records the amount of mixture remaining for each feeder. Which inference procedure should you use to test whether hummingbirds show a preference for store-bought food based on amount consumed?

(A) A one-sample z test for a proportion

(B) A two-sample z test for a difference in proportions

(C) A chi-square test for independence

(D) A two-sample t test for a difference in means

(E) A matched pairs t test for a mean difference

AP10 A national newspaper in the United States asked the following question on its website: "Do you prefer watching first-run movies at a movie theater, or waiting until they are available to watch at home or on a digital device?" In all, 8896 people responded, with only 12% (1118 people) saying they preferred theaters. Which of the following is the best conclusion?

(A) People in the United States strongly prefer watching movies at home or on their digital devices.

(B) The high nonresponse rate prevents us from drawing a conclusion.

(C) The sample is too small to draw any conclusion.

(D) The poll uses a voluntary response sample, so the results tell us little about all people in the United States.

(E) People in the United States strongly prefer seeing movies at a movie theater.

AP11 In an experiment, volunteers were randomly assigned to complete one of two word search puzzles. One was very easy and the other very difficult. When the volunteer completed the word search, they were offered a choice of snacks (candy bar, apple, granola bar, cookie). Among the 50 volunteers who received the easy word search, 27 chose one of the unhealthy snacks (candy bar, cookie). Among the 50 volunteers who received the hard word search, 35 chose an unhealthy snack. Which of the following is a 95% confidence interval for the difference (Easy – Hard) in the proportions of people like the ones in this experiment who would choose an unhealthy snack after completing an easy or hard word search?

(A) $(0.27-0.35)\pm 1.960\sqrt{\frac{0.27(0.73)}{50}+\frac{0.35(0.65)}{50}}$

(B) $(0.54-0.70)\pm 1.960\sqrt{\frac{0.54(0.46)}{50}+\frac{0.70(0.30)}{50}}$

(C) $(0.54-0.70)\pm 1.960\sqrt{\frac{0.62(0.38)}{50}+\frac{0.62(0.38)}{50}}$

(D) $(0.54-0.70)\pm 2.010\sqrt{\frac{0.54(0.46)}{50}+\frac{0.70(0.30)}{50}}$

(E) $(0.54-0.70)\pm 2.010\sqrt{\frac{0.62(0.38)}{50}+\frac{0.62(0.38)}{50}}$

AP12 Suppose that the mean weight of a certain breed of pig is 280 pounds with a standard deviation of 80 pounds. The distribution of weight for these pigs is somewhat skewed to the right. A random sample of 100 pigs is selected. Which of the following statements about the sampling distribution of the sample mean weight $\bar{x}$ is true?

(A) It is normally distributed with a mean of 280 pounds and a standard deviation of 80 pounds.

(B) It is normally distributed with a mean of 280 pounds and a standard deviation of 8 pounds.

(C) It is approximately normally distributed with a mean of 280 pounds and a standard deviation of 80 pounds.

(D) It is approximately normally distributed with a mean of 280 pounds and a standard deviation of 8 pounds.

(E) It is somewhat skewed to the right with a mean of 280 pounds and a standard deviation of 8 pounds.

AP13 A survey asked a random sample of U.S. adults where they live and what their primary car is. The responses are summarized in the following two-way table.

		Location			
		Rural	Suburban	Urban	Total
Type of car	Sedan	79	134	101	314
	Crossover	169	163	84	416
	SUV	43	49	23	115
	Truck	69	58	26	153
	Total	360	404	234	998

Suppose we select one of the survey respondents at random. Which of the following probabilities is the largest?

(A) *P*(Suburban and Sedan)

(B) *P*(Rural or SUV)

(C) *P*(Rural | SUV)

(D) *P*(Crossover | Rural)

(E) *P*(Crossover)

AP14 A scatterplot and a least-squares regression line are shown in the figure. What effect does point *P* have on the slope of the regression line and the correlation?

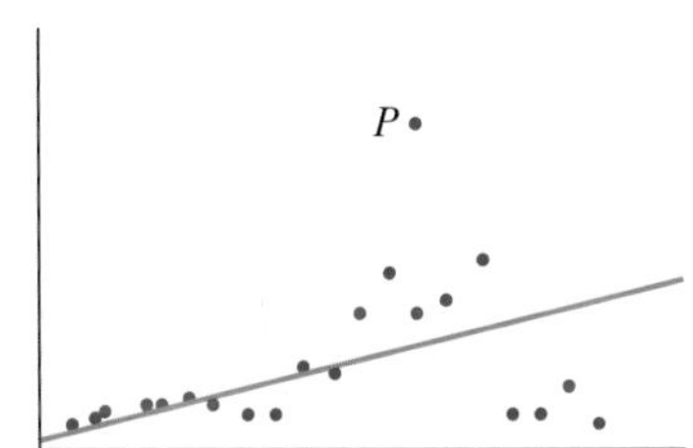

(A) Point *P* increases the slope and increases the correlation.

(B) Point *P* increases the slope and decreases the correlation.

(C) Point *P* decreases the slope and decreases the correlation.

(D) Point *P* decreases the slope and increases the correlation.

(E) Point *P* increases the slope but does not change the correlation.

AP15 The town council wants to estimate the proportion of all adults in their medium-sized town who favor a change in zoning to allow for more affordable housing. Which of the following sampling plans is most appropriate for estimating this proportion?

(A) A random sample of 150 adults who have registered to speak at town council meetings in the past

(B) A systematic random sample of 200 adults who enter the town shopping mall

(C) A sample consisting of 500 people from the city who take an online survey about the issue

(D) A random sample of 300 homeowners in the town

(E) A random sample of 100 people from an alphabetical list of all adults who live in the town

AP16 After a name-brand drug has been sold for several years, the Food and Drug Administration (FDA) will allow other companies to produce a generic equivalent. The FDA will permit the generic drug to be sold as long as there isn't convincing evidence that it is less effective than the name-brand drug. For a proposed generic drug intended to lower blood pressure, the following hypotheses will be used:

$$H_0: \mu_G = \mu_N \text{ versus } H_a: \mu_G < \mu_N$$

where μ_G = true mean reduction in blood pressure using the generic drug and μ_N = true mean reduction in blood pressure using the name-brand drug. In this context, which of the following describes a Type I error?

(A) The FDA finds convincing evidence that the generic drug is less effective, when in reality it is less effective.

(B) The FDA finds convincing evidence that the generic drug is less effective, when in reality it is not less effective.

(C) The FDA finds convincing evidence that the generic drug is equally effective, when in reality it is less effective.

(D) The FDA fails to find convincing evidence that the generic drug is less effective, when in reality it is less effective.

(E) The FDA fails to find convincing evidence that the generic drug is less effective, when in reality it is not less effective.

AP17 Every 17 years, swarms of cicadas emerge from the ground in the eastern United States, live for about 6 weeks, and then die. (There are several different "broods," so we experience cicada eruptions more often than every 17 years.) There are so many cicadas that their dead bodies can serve as fertilizer and increase plant growth. In a study, a researcher added 10 dead cicadas under randomly selected plants in a natural plot of American bellflowers on the forest floor, leaving other plants undisturbed. One of the response variables measured was the size of seeds produced by the plants. Here are the boxplots of seed mass (in milligrams) for the 39 cicada plants and the 33 undisturbed (control) plants:

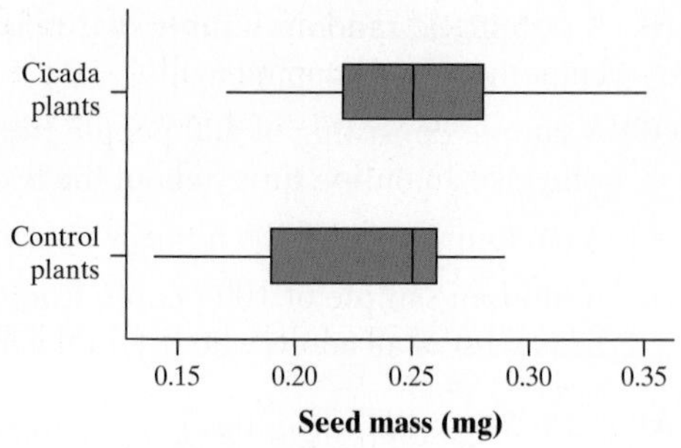

Which of the following *must* be true based on the boxplots?

(A) The means of the two distributions are the same.

(B) The medians of the two distributions are the same.

(C) The distribution for cicada plants is approximately normal and the distribution for control plants is skewed to the left.

(D) The range and interquartile range are larger for cicada plants than for control plants.

(E) Cicada plants are more likely to have outliers than control plants.

AP18 Sam has determined that the weights of unpeeled bananas from one local store have a mean of 116 grams with a standard deviation of 9 grams. Assuming that the distribution of weight is approximately normal, to the nearest gram, the heaviest 30% of these bananas weigh at least how much?

(A) 107 g
(B) 111 g
(C) 116 g
(D) 121 g
(E) 125 g

AP19 Are the five flavors of Skittles® candies equally likely to be found in a large bag of regular Skittles? A student selected a random sample of Skittles from the large bag, verified the conditions for inference were met, performed a chi-square test, and got a *P*-value of 0.023. Which of the following could have been the value of the test statistic?

(A) $\chi^2 = 13.03$
(B) $\chi^2 = 11.35$
(C) $\chi^2 = 9.53$
(D) $\chi^2 = 5.18$
(E) We cannot determine the value of the test statistic without the sample size.

AP20 The distribution of commute times for the employees at a large company is skewed to the right with a mean of 22 minutes and a standard deviation of 14 minutes. If the human resources director selects a random sample of 50 employees, which of the following is closest to the probability that the sample mean commute time is less than 20 minutes?

(A) $P\left(z < \frac{20-22}{14/\sqrt{50}}\right)$

(B) $P\left(z < \frac{20-22}{14}\right)$

(C) $P\left(t < \frac{20-22}{14/\sqrt{50}}\right)$

(D) $P\left(t < \frac{20-22}{14}\right)$

(E) $P\left(t < \frac{20-22}{14^2/50}\right)$

AP21 Park rangers are interested in estimating the weight of the bears that inhabit their state. The rangers have data on weight (in pounds) and neck girth (distance around the neck in inches) for 10 randomly selected bears. Here is some computer output from a linear regression analysis of these data:

Predictor	Coef	SE Coef	T	P
Constant	−241.70	38.57	−6.27	0.000
Neck girth	20.230	1.695	11.93	0.000

S = 26.7565 R-Sq = 94.7%

Which of the following represents a 95% confidence interval for the slope of the population least-squares regression line relating the weight of a bear and its neck girth, assuming that the conditions for inference are met?

(A) 20.230 ± 1.695
(B) 20.230 ± 3.83
(C) 20.230 ± 3.91
(D) 20.230 ± 20.22
(E) 26.7565 ± 3.83

AP22 A distribution of exam scores has mean 60 and standard deviation 18. If each score is doubled, and then 5 is subtracted from that result, what will the mean and standard deviation of the new scores be?

(A) mean = 115 and standard deviation = 31
(B) mean = 115 and standard deviation = 36
(C) mean = 120 and standard deviation = 6
(D) mean = 120 and standard deviation = 31
(E) mean = 120 and standard deviation = 36

AP23 Suppose the population proportion of people who use public transportation to get to work in the Washington, D.C., area is 0.45. In a simple random sample of 250 people who work in that city, about how far do you expect the sample proportion to be from the population proportion?

(A) 0.0010
(B) 0.0285
(C) 0.0315
(D) 0.4975
(E) 7.8661

AP24 Why is random assignment an important part of a well-designed comparative experiment?

(A) Because it eliminates chance variation in the results.
(B) Because it helps create roughly equivalent groups before treatments are imposed on the subjects.
(C) Because it allows researchers to generalize the results of their experiment to a larger population.
(D) Because it helps eliminate any possibility of bias in the experiment.
(E) Because it prevents the placebo effect from occurring.

AP25 A class of 20 students includes 14 seniors and 6 juniors. Each day, the teacher randomly selects an SRS of 4 students to present the answer to a homework question. Which of the following correctly describes one trial of a simulation to estimate the probability that all 4 of the chosen students are juniors?

(A) Let 1–70 = senior and 71–100 = junior. Generate 4 random integers with replacement from 1 to 100 and count how many integers are between 71 and 100.
(B) Let 1–70 = senior and 71–100 = junior. Generate 4 random integers without replacement from 1 to 100 and count how many integers are between 71 and 100.
(C) Let 1–14 = senior and 15–20 = junior. Generate 4 random integers with replacement from 1 to 20 and count how many integers are between 15 and 20.
(D) Let 1–14 = senior and 15–20 = junior. Generate 4 random integers without replacement from 1 to 20 and count how many integers are between 15 and 20.
(E) Let 1–14 = senior and 15–20 = junior. Generate 6 random integers without replacement from 1 to 20 and count how many integers are between 15 and 20.

AP26 When constructing a confidence interval for a population mean, which of the following is the best reason for using a t^* critical value rather than a z^* critical value?

(A) Because the population may not be normally distributed.
(B) Because the sample may not be normally distributed.
(C) Because we do not know the population mean.
(D) Because we do not know the sample standard deviation.
(E) Because we do not know the population standard deviation.

AP27 Two members of a gardening club use different fertilizers, and each claims that theirs is the best fertilizer to use when growing tomatoes. They agree to do a study using the weight of their tomatoes as the response variable. Each gardener planted the same varieties of tomatoes on the same day and fertilized the plants on the same schedule throughout the growing season. At harvest time, each gardener randomly selects 15 tomatoes from their garden and weighs them. A two-sample t test for the difference in mean weights of tomatoes from the two gardens gives $t = 5.24$ and P-value $= 0.0008$. Can the gardener with the larger mean claim that the fertilizer caused their tomatoes to be heavier?

(A) Yes, because a different fertilizer was used on each garden.
(B) Yes, because random samples were taken from each garden.
(C) Yes, because the P-value is so small.
(D) No, because the condition of the soil in the two gardens is a potential confounding variable.
(E) No, because $15 < 30$.

AP28 The probability that a softball player gets a hit in any single at-bat is 0.300. Assuming that the player's chance of getting a hit during a particular time at bat is independent of the player's other times at bat, what is the probability that the player will not get a hit until their fourth time at bat?

(A) $\binom{4}{3}(0.3)^1(0.7)^3$
(B) $\binom{4}{3}(0.3)^3(0.7)^1$
(C) $\binom{4}{1}(0.3)^3(0.7)^1$
(D) $(0.3)^3(0.7)^1$
(E) $(0.3)^1(0.7)^3$

AP29 In a clinical trial, 30 patients with a certain blood disease are randomly assigned to two groups. One group is then randomly assigned to take the currently marketed medicine, and the other group receives an experimental medicine. Every week, patients report to the clinic where blood tests are conducted. The clinic technician is unaware of the kind of medicine each patient is taking, and the patient is also unaware of which medicine they have been given. Which of the following is the best description of this design?

(A) A double-blind, completely randomized experiment, with the currently marketed medicine and the experimental medicine as the two treatments.

(B) A single-blind, completely randomized experiment, with the currently marketed medicine and the experimental medicine as the two treatments.

(C) A double-blind, matched pairs design, with the currently marketed medicine and the experimental medicine forming a pair.

(D) A double-blind, randomized block design that is not a matched pairs design, with the currently marketed medicine and the experimental medicine as the two blocks.

(E) A double-blind, randomized observational study.

AP30 A local investment club that meets monthly has 200 members ranging in age from 27 to 81 years. A cumulative relative frequency graph of age is shown. Approximately how many members of the club are older than age 60 years?

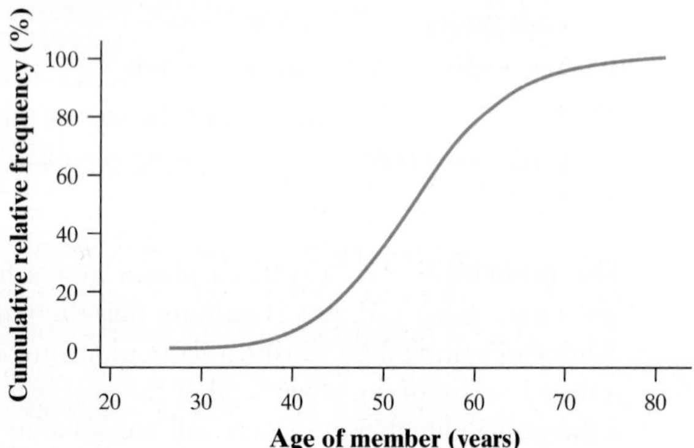

(A) 20

(B) 44

(C) 78

(D) 90

(E) 110

AP31 Do children's fear levels change over time and, if so, in what ways? Several years ago, two researchers surveyed a randomly selected group of 94 third- and fourth-grade children, asking them to rate their level of fearfulness about a variety of situations. Two years later, the children again completed the same survey. The researchers computed the overall fear rating for each child in both years and want to see if there is convincing statistical evidence of a positive relationship between the scores from the two years. Here is a scatterplot of the data, along with a residual plot created from a linear regression model:

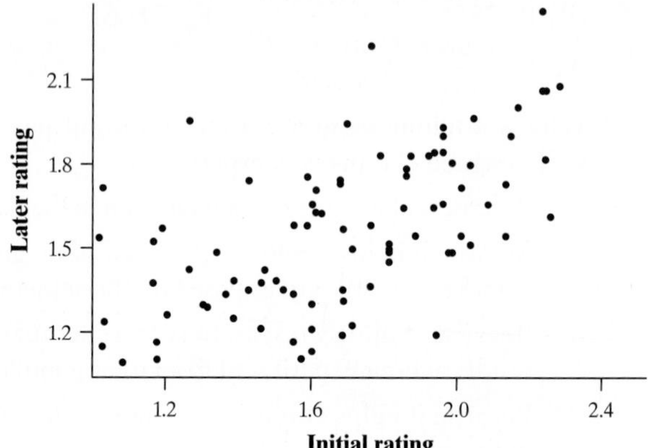

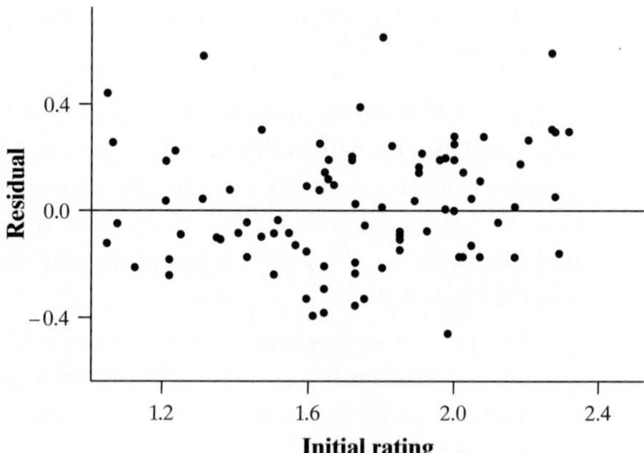

Which of the following statements is supported by these plots?

(A) The abundance of outliers and influential observations in the plots means that the conditions for inference are clearly violated.

(B) These plots contain strong evidence that the standard deviation of the response variable about the population regression line is not approximately the same for each x-value.

(C) These plots call into question the validity of the condition that the response variable varies normally about the least-squares regression line for each x-value.

(D) A linear model isn't appropriate here because the residual plot shows no association.

(E) There is no clear evidence that the conditions for inference are violated.

AP32 A large toy company introduces many new toys to its product line each year. The company wants to predict the demand as measured by y, first-year sales (in millions of dollars) using x, awareness of the product (as measured by the percentage of customers who had heard of the product by the end of the second month after its introduction). A random sample of 65 new products was taken, and a correlation of 0.96 between the values of x and y for these products was computed. Which of the following statements is true?

(A) The probability that the least-squares regression line accurately predicts first-year sales is 0.96.

(B) For each increase of 1% in awareness of the new product, the predicted sales will go up by 0.96 million dollars.

(C) About 92% of the time, the percentage of people who have heard of the product by the end of the second month will correctly predict first-year sales.

(D) About 92% of first-year sales can be accounted for by the percentage of people who have heard of the product by the end of the second month.

(E) About 92% of the variation in first-year sales can be accounted for by the least-squares regression line with the percentage of people who have heard of the product by the end of the second month as the explanatory variable.

AP33 Suppose the null and alternative hypotheses for a significance test are defined as H_0: $\mu = 40$ versus H_a: $\mu < 40$. Which of the following values of μ will result in the highest power for this test?

(A) $\mu = 38$

(B) $\mu = 39$

(C) $\mu = 41$

(D) $\mu = 42$

(E) $\mu = 43$

AP34 A Harris poll found that 54% of U.S. adults are concerned about the economy. The poll's margin of error for 95% confidence is 3%. Which of the following statements is true?

(A) There is a 95% probability that the percentage of all U.S. adults who are concerned about economy is between 51% and 57%.

(B) The poll used a method that provides an estimate within 0.03 of the proportion of all U.S. adults who are concerned about the economy in 95% of samples.

(C) If Harris conducts another poll using the same method, the percentage of U.S. adults in the sample who are concerned about the economy will be between 51% and 57%.

(D) There is a 3% chance that the interval from 0.51 to 0.57 misses the proportion of all U.S. adults who are concerned about the economy.

(E) The poll used a method that would result in an interval that contains 54% in 95% of all possible samples of the same size from the population of U.S. adults.

AP35 Western lowland gorillas, whose main habitat is in central Africa, have a mean weight of 275 pounds with a standard deviation of 40 pounds. Capuchin monkeys, whose main habitat is in Brazil and other parts of Latin America, have a mean weight of 6 pounds with a standard deviation of 1.1 pounds. Both distributions of weight are approximately normal. If a particular western lowland gorilla is known to weigh 345 pounds, approximately how much would a capuchin monkey have to weigh, in pounds, to have the same standardized weight as the gorilla?

(A) 4.075

(B) 7.750

(C) 7.925

(D) 8.200

(E) There is not enough information to determine the weight of a capuchin monkey.

AP36 Which of the following statements about the t distribution with k degrees of freedom is (are) true?

I. It is symmetric.

II. It has more variability than the t distribution with $k+1$ degrees of freedom.

III. It is centered at 0.

(A) I only

(B) II only

(C) III only

(D) I and III

(E) I, II, and III

AP37 A manufacturer of electronic components is testing the durability of a newly designed integrated circuit to determine whether its life span is longer than that of the current model, which has a mean life span of 58 months. The company selects a simple random sample of 120 new integrated circuits and simulates typical use until they stop working. The null and alternative hypotheses used for the significance test are H_0: $\mu = 58$ and H_a: $\mu > 58$. The sample mean is $\bar{x} = 61$ and the P-value for the resulting one-sample t test is 0.035. Which of the following best describes what the P-value measures?

(A) The probability that the new integrated circuit has the same life span as the current model is 0.035.

(B) The probability that the test correctly rejects the null hypothesis in favor of the alternative hypothesis is 0.035.

(C) The probability that a single new integrated circuit will not last as long as one of the current circuits is 0.035.

(D) The probability of getting a sample mean greater than or equal to 61 if there really is no difference between the new and the current circuits is 0.035.

(E) The probability of getting a sample mean greater than or equal to 58 if there really is no difference between the new and the current circuits is 0.035.

AP38 If $P(A) = 0.24$, $P(B) = 0.52$, and events A and B are independent, what is $P(A \text{ or } B)$?

(A) 0.1248

(B) 0.2800

(C) 0.6352

(D) 0.7600

(E) The answer cannot be determined from the given information.

AP39 Suppose that a test of $H_0: \mu_1 - \mu_2 = 0$ versus $H_a: \mu_1 - \mu_2 \neq 0$ resulted in a decision to reject the null hypothesis at a significance level of 0.05. Which of the following statements must be true?

(A) A 99% confidence interval for $\mu_1 - \mu_2$ will include 0.

(B) A 99% confidence interval for $\mu_1 - \mu_2$ will not include 0.

(C) A 90% confidence interval for $\mu_1 - \mu_2$ will include 0.

(D) A 90% confidence interval for $\mu_1 - \mu_2$ will not include 0.

(E) It is not possible to determine if any of these statements are true without more information.

AP40 A university president estimated the proportion of students who plan to take a summer school course with a 95% confidence interval. If the interval is (0.258, 0.342), which of the following is closest to the sample size the administrator used?

(A) 30

(B) 120

(C) 460

(D) 545

(E) There is not enough information to determine the sample size.

Section II: Free Response

- *Time limit:* 1 hour and 30 minutes
- *Number of questions:* 6
- *Percentage of AP® Statistics exam score:* 50

Part A

- *Suggested time:* 1 hour and 5 minutes
- *Number of questions:* 5
- *Percentage of AP® Statistics exam score:* 37.5

Directions: *Show all your work. Indicate clearly the methods you use, because you will be scored on the correctness of your methods as well as on the accuracy and completeness of your results and explanations.*

AP41 A random sample of 100 students from a large high school was surveyed about transportation to school. Students were asked to estimate the total amount of time (in minutes) it took them to get to school on a typical school day and their method of transportation. The histograms summarize the distributions of travel time for the 40 students who ride the bus and the 60 students with other forms of transportation.

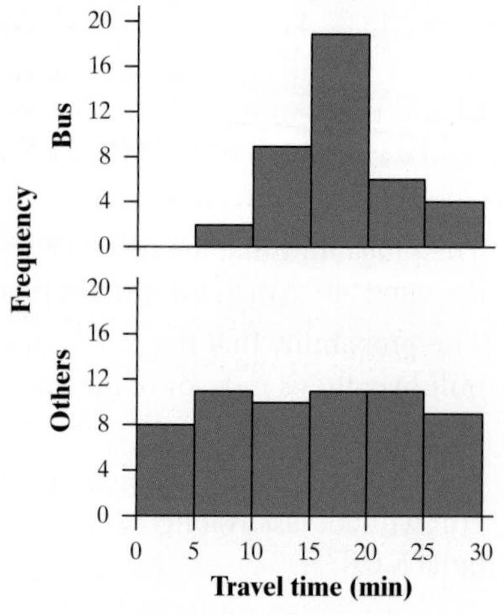

(a) Compare these distributions.

(b) Is it possible that the median travel time is the same for both groups? Justify your answer.

AP42 Will posting anti-littering signs at bus stops encourage people to be more mindful of their trash? Administrators in a large city identified 206 bus stops that were already equipped with trash bins. Half of these bus stops were randomly assigned to get a sign with the statement: "A neighbor helps you keep this stop clean. Please use the trash can." The remaining bus stops had no sign. After three weeks, researchers counted the number of pieces of trash at each bus stop, including cigarette butts and other pieces of trash.[34]

(a) Identify the experimental units, the treatments, and the response variable.

(b) Describe a method of randomly assigning the treatments in this experiment so that the group sizes will be the same.

(c) If the results of the experiment are statistically significant, would it be reasonable to conclude that the signs caused a change in behavior? Explain your answer.

AP43 The weights of chocolate truffles made by a candy shop are approximately normally distributed with a mean of 2 ounces and a standard deviation of 0.3 ounce.

(a) What is the probability that a randomly selected chocolate truffle from this candy shop will weigh less than 1.5 ounces?

(b) The candy shop assembles gift boxes that contain 8 randomly selected chocolate truffles. What is the probability that none of the truffles in a gift box weighs less than 1.5 ounces?

(c) The weights of gift boxes, including all packaging materials, are approximately normally distributed with a mean of 3 ounces and a standard deviation of 0.5 ounce. If 8 randomly selected chocolate truffles are placed into a randomly selected gift box, describe the distribution of the total weight of the gift box and truffles.

AP44 Each year, a news organization polls a random sample of adult residents in a particular state to ask about the governor's job performance. In one year, 529 of the 1002 randomly selected adult residents approved of the governor's job performance. The next year, 498 of the 1009 randomly selected adult residents approved of the governor's job performance. Do these data provide convincing evidence of a change in the proportion of all adult residents in the state who approved of the governor's job performance during these two years? Complete an appropriate significance test using $\alpha = 0.05$.

AP45 The amount of saturated fat (g) and the number of calories was recorded for 19 types of dairy-free frozen desserts. Here is a scatterplot of these data, along with the least-squares regression line $\hat{y} = 125.3 + 11.54x$.

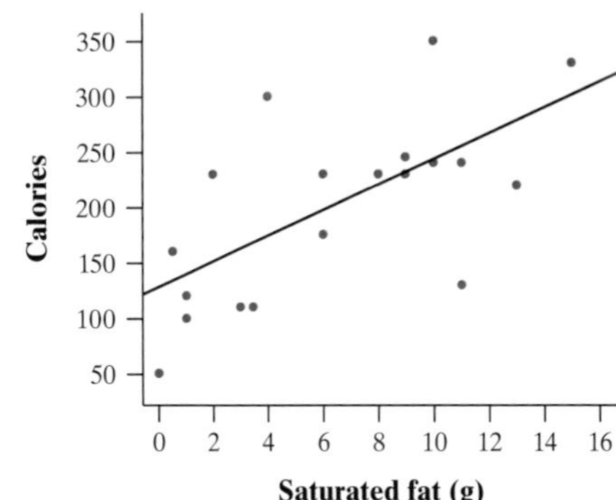

(a) Interpret the slope of the least-squares regression line.

(b) Calculate and interpret the residual for the dessert with 13 grams of saturated fat and 220 calories.

(c) For these data, $\bar{x} = 6.5$ grams and $\bar{y} = 200$ calories. If you randomly select one of these desserts that is above average in saturated fat, what is the probability that it is also above average in calories?

Part B

- *Suggested time:* 25 minutes
- *Number of questions:* 1
- *Percentage of AP® Statistics exam score:* 12.5

Directions: *Show all your work. Indicate clearly the methods you use, because you will be scored on the correctness of your methods as well as on the accuracy and completeness of your results and explanations.*

AP46 Do polyurethane swimsuits help swimmers swim faster, on average? To investigate this question, a swim coach recruited 8 members of the swim team to participate in an experiment. On two consecutive Saturdays, each swimmer performed the same warm-up routine and swam a 50-meter freestyle. On one Saturday, the swimmer wore a polyurethane suit and on the other Saturday, the swimmer wore a traditional suit, with the order determined at random for each swimmer. The times (in seconds) for each swimmer are shown in the table, along with the difference (Traditional − Polyurethane) in times for each swimmer.[35]

Traditional	Polyurethane	Difference
26.32	25.89	0.43
28.13	28.01	0.12
27.06	27.03	0.03
27.57	27.22	0.35
25.68	24.44	1.24
28.32	28.29	0.03
26.77	26.91	−0.14
27.26	27.01	0.25

(a) The coach wants to perform a matched pairs t test using the following hypotheses:

$$H_0: \mu_{diff} = 0 \text{ versus } H_a: \mu_{diff} > 0$$

where μ_{diff} = the mean difference (Traditional – Polyurethane) in 50-meter freestyle times for all swimmers like the ones in this study. Show that the conditions for performing this test are not met.

(b) Instead of a test about the mean difference, the coach decides to perform a test about the median difference using the following hypotheses:

$$H_0: \text{Median}_{diff} = 0 \text{ versus } H_a: \text{Median}_{diff} > 0$$

where Median_{diff} = the median difference (Traditional – Polyurethane) in 50-meter freestyle times for all swimmers like the ones in this study.

(i) If the null hypothesis is true, what proportion of the differences should be greater than 0?

(ii) In the coach's experiment, how many of the 8 differences were greater than 0?

(c) Use your answers to part (b) to calculate the P-value for the test about the median difference.

(d) Based on the P-value from part (c), what conclusion should the coach make?

Strategies for Success on the AP® Statistics Exam

ABOUT THE AP® STATISTICS EXAM

The AP® Statistics exam consists of two distinct sections: Multiple Choice and Free Response. Here are specific details about the structure and scoring of the exam, as of Fall 2023.

AP® STATISTICS EXAM STRUCTURE

Section I: Multiple Choice — **90 minutes** — **50% of exam score**

40 multiple choice questions, each with 5 answer choices

The composition of the Multiple Choice section is based on both content and skills. Here are the number of questions for each unit and skill category.

Unit	Number of questions
Unit 1: Exploring One-Variable Data	6–9
Unit 2: Exploring Two-Variable Data	2–3
Unit 3: Collecting Data	5–6
Unit 4: Probability, Random Variables, and Probability Distributions	4–8
Unit 5: Sampling Distributions	3–5
Unit 6: Inference for Categorical Data: Proportions	5–6
Unit 7: Inference for Quantitative Data: Means	4–7
Unit 8: Inference for Categorical Data: Chi-Square	1–2
Unit 9: Inference for Quantitative Data: Slopes	1–2

Skill Category	Number of questions
Skill 1: Selecting Statistical Methods	6–9
Skill 2: Data Analysis	6–9
Skill 3: Using Probability and Simulation	12–16
Skill 4: Statistical Argumentation	10–14

Section II: Free Response	**90 minutes**	**50% of exam score**
Part A: Questions 1–5	65 minutes	37.5% of exam score
Part B: Question 6 (Investigative Task)	25 minutes	12.5% of exam score

The composition of the Free Response section is based on skills. The first 5 free-response questions (Part A) include the following:

- One multi-part question focusing primarily on Exploring Data (Skill Category 2)
- One multi-part question focusing primarily on Collecting Data (Skill Category 1)
- One multi-part question focusing primarily on Probability and Sampling Distributions (Skill Category 3)
- One question focusing primarily on Inference (Skill Categories 1, 3, 4)
- One question focusing on two or more skill categories

The sixth free-response question (Part B) is the Investigative Task, which assesses multiple skill categories and content areas, focusing on applying the skills and content in new contexts or nonroutine ways.

Formulas, Tables, and Calculator Use

The Formula Sheet and Tables (Table A: Standard normal probabilities, Table B: *t* distribution critical values, Table C: χ^2 critical values) are provided on both sections of the exam. Find examples in AP® Classroom, on a recent AP® Statistics exam (2021 or later), or in the back of the book. The formulas are found at the beginning of the test booklets and the tables are found at the end, for both the Multiple Choice and Free Response sections. You may use your approved calculator throughout the exam.

AP® STATISTICS EXAM SCORING

Section I: Multiple Choice The score is based only on the number of questions answered correctly. So don't leave any questions unanswered!

$$\textit{Weighted Section I Score} = \textit{Number of correct answers} \times 1.25$$

Section II: Free Response Each free-response question is scored holistically on a 0 to 4 scale. The score categories represent different levels of quality in a student's response across two dimensions: statistical knowledge and communication.

$$\textit{Weighted Section II Score} = (\textit{Sum of scores on Questions } 1-5) \times 1.875 + \textit{Question } 6 \textit{ score} \times 3.125$$

$$\textit{Composite Score} = \textit{Weighted Section I Score} + \textit{Weighted Section II Score}$$

Composite scores (on a 100-point scale) are converted to AP® Scores (on a 1 to 5 scale) using cutoffs determined each year based on statistical analysis of overall student performance on the exam. Using the median cut scores from nine recent exams, it takes about 32 points to earn a 2, 43 points to earn a 3, 56 points to earn a 4, and 70 points to earn a 5.

STRATEGIES FOR SUCCESS

Before the Exam

1. Understand the structure of the exam and how it is scored.
2. Review the Formula Sheet and memorize the formulas that aren't included (e.g., z-score, finding the mean and standard deviation of a linear combination of random variables). Don't bother memorizing the formulas on the Formula Sheet, as you'll have access to the Formula Sheet on both sections of the exam. Specific formulas for confidence intervals and test statistics aren't on the Formula Sheet (other than the chi-square test statistic), but the general formulas are included on the bottom of page 1, along with the standard error of each statistic on page 2.
3. Practice using your graphing calculator and/or the tables that are provided on the AP® Statistics exam (Table A: Standard normal probabilities, Table B: t distribution critical values, Table C: χ^2 critical values).
4. Do lots of practice AP® Statistics exam questions, focusing on exams from recent years.
5. Take a full-length practice exam, using the official time limits to get a sense of pacing. Cumulative AP® Exam 4 (page 849) closely models a complete AP® Statistics exam.
6. Use the practice exam and other practice questions to identify weak spots. Then focus on those areas by watching the AP® Daily videos about those topics, rereading the relevant sections in the book, and doing some exercises.
7. Watch the AP® Daily review videos available in the Review tab of AP® Classroom.
8. Ask your teacher for the flash cards provided in the Teacher's Resource Materials. Go through the cards once or twice and sort them into three piles: "I know this," "I kind of know this," and "I don't know this." Review the first set of cards once a week, the second set of cards every other day, and the third set every day. Shift the cards to other piles as you master the concepts and skills on the cards.
9. Eat a typical meal before the test and get lots of sleep on the *two* nights before the exam.
10. *What to bring:* Pencils, a fully charged calculator, water, snack, and a non-internet-connected watch. You can also bring a second calculator and replacement batteries, just in case. Don't bring formula sheets or tables—these are provided on both sections of the exam. And don't bring any phones or other internet-connected devices into the testing room. No white-out, rulers, or highlighters are permitted, either.

During the Exam: Multiple Choice

1. The Formula Sheets are found in the front of the test booklet and the Tables are found in the back.
2. Pace yourself. There are 40 questions to complete in 90 minutes. Most students feel like they have plenty of time to answer every question and spend some time double-checking.
3. Underline key words (e.g., "and," "not," "must be") and circle key values as you encounter them.
4. Make sure the option you choose answers the question that was asked. Some multiple choice questions contain distractors with true statements that don't answer the question asked.
5. Eliminate obviously incorrect options. Cross out answer options once you can eliminate them.
6. Circle your answers in the test booklet in case you have a bubbling error and have to rapidly fix your bubbled answers. The bubble sheet is the only thing that is scored.
7. There is no penalty for wrong answers, so make sure to answer every question.
8. All questions are worth the same amount, but they typically get harder toward the end. Skip tough questions and return to them later.

During the Exam: Free Response

1. The Formula Sheets are found in the front of the test booklet and the Tables are found in the back.
2. Pace yourself. There are 6 questions to complete in 90 minutes. Students typically find that the Free Response section takes longer than the Multiple Choice section.
3. Know the structure of the Free Response section. Although not a guarantee, questions on most recent exams have been in the following order:
   Question 1: Data analysis (Units 1–2)
   Question 2: Collecting data (Unit 3)
   Question 3: Probability and sampling distributions (Units 4–5)
   Question 4: Inference (Units 6–9)
   Question 5: Multi-focus (Units 1–9)
   Question 6: Investigative Task: According to the Course and Exam Description, the investigative task assesses multiple skill categories and content areas, focusing on the application of skills and content in new contexts or in nonroutine ways. *Translation:* You'll be asked to do something new, so don't panic if parts of the question seem unfamiliar. The question is typically scaffolded to help you through the new parts, so let the question guide you. And don't give up—sometimes the final part is the easiest part.

   You don't have to answer the questions in order. The first 5 questions are each worth the same amount (15% of the Free Response section). Question 6 is worth 25% of the Free Response section, so don't save this question for the end when you are tired and rushed. One strategy is to start with Question 1, move to Question 6, and then work through the remaining questions in order from easiest to hardest. Another strategy is to do the three questions you feel most comfortable with, then attempt Question 6 before returning to the remaining questions. Spend no more than 25 minutes on Question 6, which leaves about 13 minutes for each of the other questions.
4. Show your work! Here are the directions for the Free Response section:

   Show all your work. Indicate clearly the methods you use, because you will be graded on the correctness of your methods as well as on the accuracy and completeness of your results and explanations.

   Responses with no work will not receive full credit, even if the answer is correct. This is often an issue on probability questions.
5. Remember that AP® Readers, the graders of the exam, will be scoring a black-and-white scan of your test. So don't use different colors on a graph and expect that AP® Readers will be able to distinguish them. And don't waste time erasing, as it is hard for AP® Readers to determine what is erased and what is written lightly on the scanned responses. Instead, cross out wrong answers and draw arrows to help AP® Readers follow your work.
6. Communicate clearly, correctly, completely, concisely, and in context.
   a. Read the question carefully and make sure to answer the question that was asked.
   b. Improper use of statistical vocabulary or symbols will hurt your score.
      i. Be cautious when including terms that are frequently misused (e.g., bias, confounding). It is better to clearly explain your answer than to rely on statistical vocabulary.
      ii. Don't include formulas with symbols, where it is easy to use the wrong symbol (e.g., σ instead of s_x or $\hat{p}$ instead of p). Instead, show your work by including formulas with numbers substituted in.
   c. Make sure to give a complete answer. AP® Readers aren't allowed to do any thinking for you, so make sure to give enough details—even if you think they are obvious.
   d. But don't say too much. Answer the question that was asked and stop. There is often more space provided than is necessary, so don't feel obliged to fill the space available. And don't give more than one solution—what AP® Readers call parallel solutions. The worst of the solutions will be the one that is graded.
   e. All explanations, interpretations, descriptions, and conclusions should be in context. The easiest way to include context is by including the variable name(s) and group names, if applicable.
7. **Data Analysis questions:**
   a. When asked to describe a distribution, remember to address shape, center, variability, and potential outliers, and to use the variable name (context).
   b. When asked to compare distributions (or discuss similarities and differences), make sure to use comparison phrases (e.g., "greater than," "approximately the same as") when addressing center and variability. Also, remember to describe the shape of each distribution and note whether it has potential outliers or other unusual features (e.g., gaps). Don't just list characteristics of each distribution separately. And make sure to include the variable name and not just the group names.
   c. When asked to interpret the slope or y intercept of a least-squares regression line, make sure to include nondeterministic language (e.g., "predicted").
   d. If you are asked to do two things (e.g., calculate and interpret), make sure to do both. Circle the word "and" in the question to help you remember.
   e. Never describe a distribution as "normal"—real distributions are "approximately normal" at best. And remember that boxplots don't reveal peaks or gaps, so never conclude that a distribution is approximately normal from a boxplot.
8. **Collecting Data questions:**
   a. Don't mix and match vocabulary for sampling and experiments. For example, stratifying is for sampling, and blocking is for experiments.
   b. If you are asked to choose between several options, give reasons for your choice and reasons why you did not choose the others.
   c. If you are asked to describe a method for random selection or random assignment, be thorough. Include details about whether you'll be sampling with or without replacement when using a random number generator, a table of random digits, or slips of paper. And if you use slips of paper, make sure to mix/shuffle them before selection.

9. **Probability questions:**
   a. Show your work, even if it seems obvious.
   b. For random variable questions (especially normal, binomial, and geometric random variables), remember to define the random variable and state its distribution. Then give a probability statement to identify the boundary value and direction. For normal distribution questions, make sure to draw a well-labeled picture.
   c. If you use technology, make sure to write down the calculator command *with the inputs labeled.* For example, write "normalcdf(lower: 90, upper: 100, mean: 98, SD: 2)" rather than "normalcdf(90, 100, 98, 2)."
   d. For sampling distribution questions, clearly identify which distribution you are discussing: the distribution of the population, the distribution of a sample, or the sampling distribution of a statistic.
   e. If you cannot get an answer for an early part of a question but need it for a later part, make up a reasonable value. Remember that probabilities have to be between 0 and 1. Don't give up—the last part might be easy!
10. **Inference questions:**
   a. Every AP® Statistics exam includes at least one opportunity to carry out a full inference procedure. Use the four-step process to organize your solution and include the State, Plan, Do, and Conclude labels in your response. You'll know that you're expected to do a full inference procedure when you are asked to "construct and interpret" a confidence interval or determine if data "provide convincing statistical evidence" for a claim.
   b. Practice identifying the correct inference procedure to use. There are some online applets that can help you with this, along with the Inference Summary in the back of the book.
   c. Don't type data into your calculator just because it's there. Items often include graphs, summary statistics, or computer output that you can use instead.
   d. In your conclusions, don't interpret the *P*-value or interpret the confidence level unless you are specifically asked to do this.
   e. Exception to the "show your work" rule: For inference questions, you can use the calculator for the Do step and report only the endpoints of the confidence interval or the test statistic, df, and *P*-value, as long as the inference procedure is identified by name. But, if you go this route and make a mistake entering values into your calculator, there is no way to earn partial credit for the calculations.
   f. Don't argue with the question writers—believe what they tell you (e.g., conditions are met).
   g. If you can't remember how to calculate the *P*-value or the endpoints of a confidence interval, make up a reasonable answer (e.g., a *P*-value between 0 and 1) and write your conclusion using the made-up value(s).
   h. Always use nondeterministic language in your conclusions (e.g., "We are 95% confident ...," "There is convincing evidence that ..."). Make sure your conclusion in a significance test justifies the conclusion by comparing the *P*-value to α and addresses the alternative hypothesis in context. Context includes a clear reference to the parameter(s), population(s), and response variable.

Final thoughts

1. The AP® Statistics exam is harder than a normal classroom test. Scoring at least 45% will almost guarantee a 3 or higher on the exam. DON'T PANIC if you cannot answer a question or two. If a question is hard, it will be hard for everyone and the cut scores will be adjusted for this fact.
2. Be confident. You've got this!

Solutions

UNIT 1, PART I

Section 1A

Answers to Check Your Understanding

page 7: 1. *Individuals:* the cars in the student parking lot. *Variables:* model, year, number of stickers, color, weight, whether or not it has a navigation system, highway gas mileage.
2. *Categorical:* model, year, color, navigation system. *Quantitative:* number of stickers, weight, highway gas mileage.
3.

Freq. Table

Num. correct	Freq.
2	1
3	2
4	2
5	4
6	8
7	15
8	14
9	4
Total	**50**

Rel. Freq. Table

Num. correct	Rel. Freq.
2	1/50 = 0.02
3	2/50 = 0.04
4	2/50 = 0.04
5	4/50 = 0.08
6	8/50 = 0.16
7	15/50 = 0.30
8	14/50 = 0.28
9	4/50 = 0.08
Total	**50/50 = 1**

4. Proportion who got less than 7 correct answers $= 0.02 + 0.04 + 0.04 + 0.08 + 0.16 = 0.34$. This does not support the professor's belief because only 34% of the students got less than 7 correct answers.

Answers to Odd-Numbered Section 1A Exercises

1. **(a)** *Individuals:* students who completed the questionnaire. *Variables:* grade level, dominant hand, GPA, number of children in family, time spent on homework last night (min), type of phone **(b)** *Categorical:* grade level, dominant hand, time spent on homework last night (min), type of phone. *Quantitative:* GPA, number of children in family
3. *Individuals:* movies. Variables: year, rating, time (min), genre, box office ($). *Categorical:* year, rating, genre; *Quantitative:* time (min), box office ($). *Note:* Year might be considered quantitative if we want to know the average year.
5. (a)

Freq. Table

Response	Freq.
Approach	17
Run	17
Indifferent	41
Total	**75**

Rel. Freq. Table

Response	Rel. Freq.
Approach	17/75 = 0.227
Run	17/75 = 0.227
Indifferent	41/75 = 0.547
Total	**75/75 = 1**

(b) The squirrels in the random sample were most likely to be indifferent (54.7%) and were equally likely to approach (22.7%) or run from the humans (22.7%).
7. **(a)** $50 \times 0.12 = 6$ communication missions. **(b)** Only 34% of the missions were imaging, which is not a majority (it's not greater than 50%).
9. B

Section 1B

Answers to Check Your Understanding

page 16: 1. Most of the adults (1217/1502 = 81.0%) who were surveyed own a smartphone. Only 212 adults (or 14.1%) own a cell phone that is not a smartphone, and even fewer adults (73/1502 or 4.9%) do not own a cell phone.

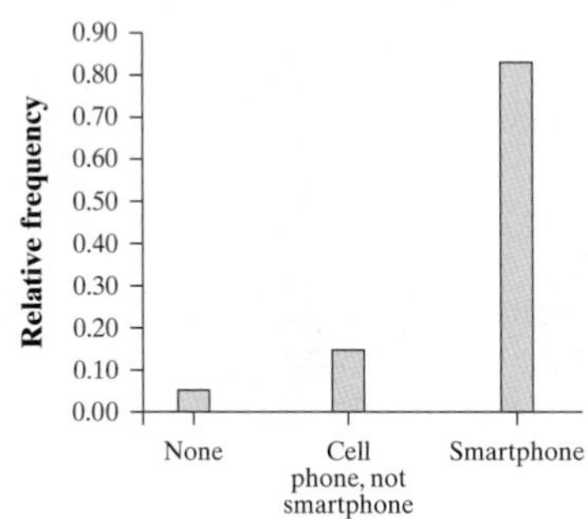

2. Younger adults are more likely to own a smartphone than older adults, less likely to own a cell phone that is not a smartphone, and less likely to own no phone at all. The older the age group, the smaller the percentage of adults who own a smart phone. Also, the older the age group, the greater the percentage of adults who own a cell phone, but not a smartphone. Lastly, the older the age group, the greater the percentage of adults who do not own a cell phone.

Answers to Odd-Numbered Section 1B Exercises

1. It appears that births occurred with similar frequencies on weekdays, but with noticeably smaller frequencies on the weekend days.

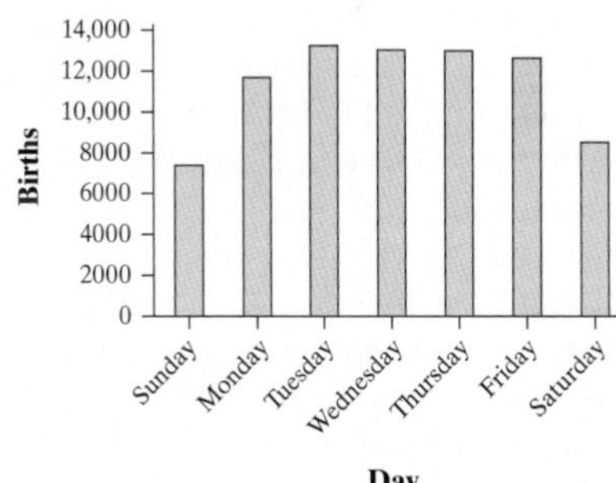

3. **(a)** $100 - 19 - 7 - 3 - 15 - 1 - 5 - 9 - 38 - 2 = 1\%$ **(b)** The most popular color of vehicles sold that year was white, followed by black, gray, and silver. It appears that a majority of car buyers that year preferred vehicles that were shades of black and white.

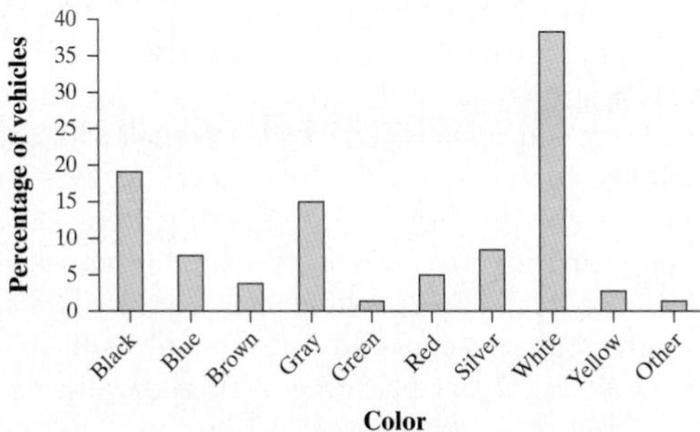

(c) Yes, because the numbers in the table refer to parts of a single whole.

5. Estimates may vary. About 58% Mexican and 10.5% Puerto Rican.

7. Regardless of whether a student went to a private or public college, most students chose a school that was at least 11 miles from home. Those who went to a public university were most likely to choose a school that was 11 to 50 miles from home (about 30%), while those who went to a private university were most likely to choose a school that was 101 to 500 miles from home (about 29%).

9. (a)

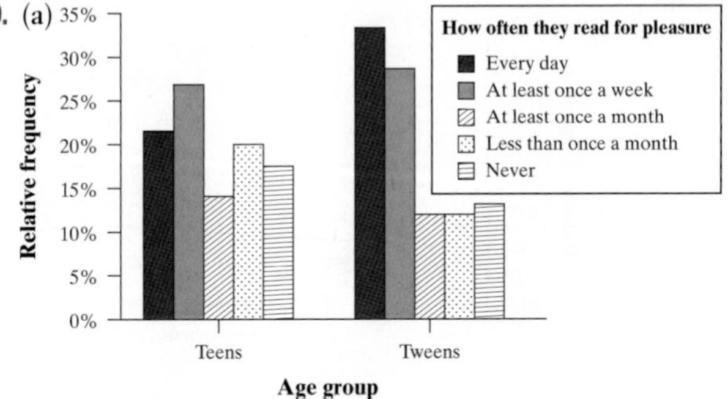

(b) Teens and tweens are similar in that for both groups, the two most frequent categories are reading every day or at least once a week for pleasure. However, teens are less likely than tweens to read every day for pleasure (21% to 34%) and are slightly less likely to read at least once a week (27% to 29%). They are slightly more likely than tweens to read at least once a month for pleasure (14% to 12%), and are much more likely to read less than once a month (20% to 12%) or never (18% to 13%).

11. The areas of the flags should be proportional to the percentage of adults they represent. As drawn, it appears that adults from the United States are more than twice as likely to dislike shopping for clothes than are adults in Germany and about 4 times as likely as adults in Spain—when, in fact, the percentages are approximately U.S. 30%, Germany 20%, and Spain 12%.

13. By starting the vertical scale at 12 instead of 0, it looks like the percentage of binge-watchers who think that 5 to 6 episodes is too many to watch in one viewing session is almost 20 times higher than the percentage of binge-watchers who think that 3 to 4 episodes is too many to watch in one viewing session. In truth, the percentage is less than 3 times higher (31% to 13%). Similar arguments can be made for the relative sizes of the other categories represented in the bar graph.

15. (a) The graph reveals that those who have a technical degree or some college and those who have a college degree are more likely to have purchased a lottery ticket (just over 50%) than are those who have a high school diploma or less (about 47%) or those who have a postgraduate education (about 45%). (b) No, because the data do not represent parts of a whole. The sum of all these percentages is greater than 100%. This is because we have the percentage of each education group who has bought a lottery ticket, rather than the percentage of all lottery ticket buyers who are in each education group.

17. The pie chart shows that 35% of methane emissions come from livestock production, 30% come from oil and gas production, 17% come from landfills, 8% come from coal mining, and 10% come from other sources. The graph further breaks down the sources of the methane emissions from livestock production: 20% come from beef cows, 11% from dairy cows, and 4% from pigs. The graph also provides the distribution of methane emission from the front and back end of beef cows, dairy cows, and pork. It is interesting to note that beef cows produce only 2.6% of their methane emissions out the back end, whereas dairy cows produce 43% of their emissions out the back end, and pigs produce 89% of their emissions out the back end. Life lesson: Don't stand downwind of the back of a pig. Also keep in mind the back end of a beef cow isn't nearly as toxic as you might think.

19. D

Section 1C

Answers to Check Your Understanding

page 27:

1.

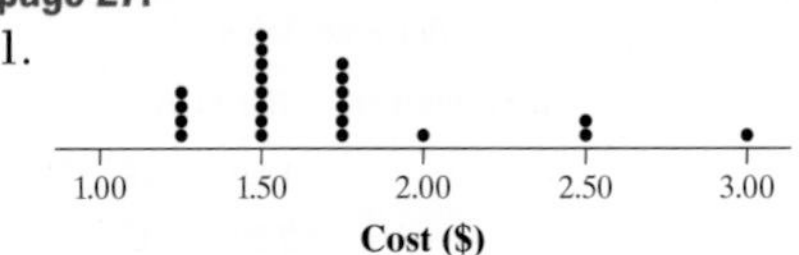

2. The distribution of cost is skewed to the right with a single peak at \$1.50. There are two small gaps at \$2.25 and \$2.75. The median cost is \$1.50 and the cost varies from \$1.25 to \$3.

page 31: *Similarities:* The distributions have similar shapes: Both are skewed to the right with a single peak (dairy yogurt at 130 calories and plant-based yogurt at 140 calories). Also, each distribution has at least one possible high outlier—Dairy: 280 and 290 calories; Plant-based: 260 calories. *Differences:* The number of calories in dairy yogurt tends to be less (median ≈ 130 calories) than for plant-based yogurt (median ≈ 145 calories). Also, the number of calories in dairy yogurt varies more (from about 70 calories to about 290 calories) than for plant-based yogurt (from about 110 calories to about 260 calories).

2. 20.6%

3. The split stemplot reveals that there are some gaps in the distribution. There is a gap between 11.8% and 12.6%, 12.6% and 13.9%, 14.3% and 15.4%, and also 19.9% and 20.5%.

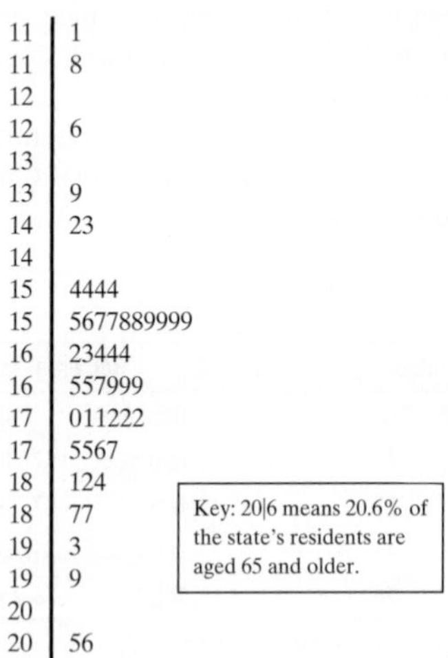

page 37: 1. Answers may vary but should be around 49.4%.

$$1-\frac{255+215+210+150}{1640}=0.494=49.4\%$$

2. Single-peaked and skewed to the right.

3. The median cost of an engagement ring is between \$1500 and \$1999 and the costs vary from as low as \$0 to as much as \$15,500.

Answers to Odd-Numbered Section 1C Exercises

1. *Categorical:* type of wood, paint color. *Quantitative (discrete):* number of blemishes, weathering time. *Quantitative (continuous):* paint thickness.

3. (a)

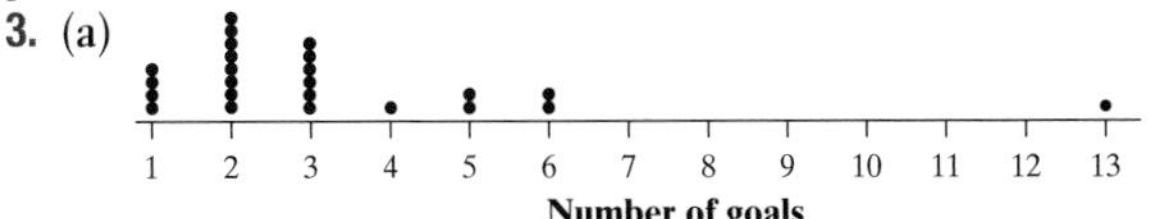

(b) $6/24 = 0.25$

5. (a) The dot above −2 represents one of the U.S. women's 2019 soccer games in which the difference in goals (U.S. team − Opponent) is −2. This means that the U.S. team lost that game by 2 goals. (b) The U.S. women's 2019 soccer team won 20 games, tied 3 games, and lost 1 game. Overall, they won $20/24 = 0.833$, or about 83.3% of the games. The team did very well in 2019!

7. (a) Left-skewed with a peak between 90 and 100 years. There is a small gap around 70 years. (b) Approximately uniform, meaning the frequency of each random integer is similar.

9. *Shape:* The distribution of amount of sleep is skewed to the left with a single peak around 7 hours. There is a gap from 0 to 3 hours. *Outliers:* The student who got 0 hours of sleep is an apparent outlier. *Center:* The median amount of sleep is 7 hours. *Variability:* The amount of sleep varies from 0 hours to 9 hours.

11. (a) The distribution of number of goals is skewed to the right with a single peak at 2 goals. There is a gap from 6 to 13 goals. There is an apparent outlier of 13 goals scored in a single game. (b) The team scored about 2 or 3 goals per game that season.

13. *Shape:* The distribution of calcium content for flavored yogurts is skewed to the left and single-peaked. The distribution of calcium content for plain yogurts is fairly symmetric and single-peaked. *Outliers:* Neither calcium content distribution appears to have any outliers. *Center:* Both types of yogurt have a median calcium content of 20% of the recommended daily value, so one type of yogurt does not tend to have more calcium than the other. *Variability:* The variability in the calcium content is greater among the plain yogurts (10% to 35%) than the flavored yogurts (10% to 25%).

15. (a) Both distributions have about the same amount of variability. The "external reward" distribution varies from 5 to about 24, and the "internal reward" distribution varies from about 12 to 30. (b) The center of the internal distribution (median ≈ 20.5) is greater than the center of the external distribution (median ≈ 17), indicating that external rewards do not promote creativity.

17. (a) The graph reveals that there was one Fun Size Snickers® bar that is "gigantic" compared to the others! It weighs 19.2 grams.

Stem	Leaves
15	9
16	055678
17	111344778
18	
19	2

Key: 19 | 2 = Snickers Fun Size bar that weighs 19.2 g

(b) Seven of the 17 candy bars in this sample, or about 41%, weigh less than advertised.

19. (a) The area of the largest county in South Carolina is 1220 square miles (rounded to the nearest 10 mi^2). (b) *Shape:* The distribution of the area for the 46 South Carolina counties is skewed to the right with a clear peak in the 500 mi^2 stem. *Outliers:* There are no clear outliers. *Center:* Counties in South Carolina have a median area of about 655 square miles. *Variability:* The area of the counties varies from about 390 square miles to 1220 square miles.

21. (a) If we had not split the stems, all of the data would appear on just 4 stems, making it hard to identify the shape of the distribution. (b) Key: 11 | 5 = One day in July on which the high temperature in Phoenix was 115°F. (c) The distribution of high temperature is skewed to the left with a single peak in the 105°F – 109°F stem. There is a gap between 84°F and 93°F. The day in July with a high temperature of 84°F is a possible outlier.

23. *Shape:* The distribution of acorn volume for the Atlantic Coast is skewed to the right. The distribution of acorn volume for California is roughly symmetric with one apparent high outlier. *Outliers:* The distribution of volume of acorn for the Atlantic Coast has several potential outliers, including 8.1, 9.1, and 10.5 cubic centimeters. The California distribution has 1 potential outlier: 17.1 cubic centimeters. *Center:* The typical acorn volume for Atlantic Coast oak tree species (median = 1.7 cubic centimeters) is less than the typical acorn volume for California oak tree species (median = 4.1 cubic centimeters). *Variability:* The Atlantic Coast distribution (with acorn volumes from 0.3 to 10.5 cubic centimeters) varies less than the California distribution (with acorn volumes from 0.4 to 17.1 cubic centimeters.)

25. (a)

(b) $6/48 = 0.125$

27.

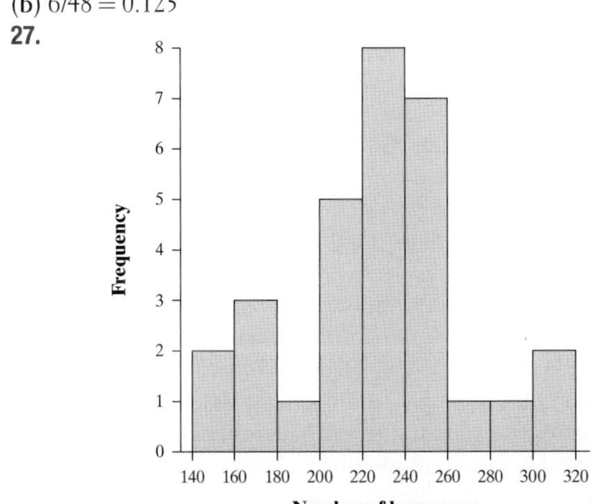

Shape: The distribution of number of home runs hit is fairly symmetric with a single peak in the 220 to <240 interval, and a minor peak on each end of the distribution. *Outliers:* There are no obvious outliers. *Center:* The median number of home runs hit is 223.5 home runs. *Variability:* The number of home runs hit varies from 146 to 307 home runs.

29. **(a)** About 37% of these months (102 out of 273) **(b)** Slightly skewed left with a single peak in the 0 to <2.5% interval; there may be one low outlier. **(c)** The median is between 0% and 2.5% return on common stocks. The returns vary from as low as −25% to as high as 12.5%.

31. **(a)** *Shape:* Both distributions of annual income are skewed to the right and single-peaked. *Outliers:* There are some possible high outliers in both distributions. *Center:* The center of the distribution is larger for college graduates ($40,000 to < $60,000 vs. $0 to < 20,000), indicating that college graduates typically have greater annual income than nongraduates in this sample. *Variability:* The annual income for college graduates varies a lot more (from as low as $0 to as much as $340,000) than the annual income for nongraduates (from as low as $0 to as much as $280,000). **(b)** No, because there were many more graduates surveyed (327) than nongraduates (173).

33. **(a)** *Similarities:* The shapes of the BMI distributions for semi-urban and rural woman in Ghana are similar. Both are single-peaked and skewed to the right. Also, neither distribution has any obvious outliers. *Differences:* The centers of the two distributions are different. The median for the semi-urban women (20 to <22) is greater than that of the rural women (18 to <20). The variabilities of the two distributions are also different. The BMI measurements vary more (as little as 12 to as much as 42) among semi-urban women than among the rural women (as little as 12 to as much as 34). **(b)** Semi-urban women: about $1+2+10=13\%$ of women are underweight. Rural women: about $2+7+24=33\%$ of women are underweight. A greater percentage of rural women are underweight than semi-urban women.

35. A bar graph because birth month is a categorical variable. A possible bar graph is given here.

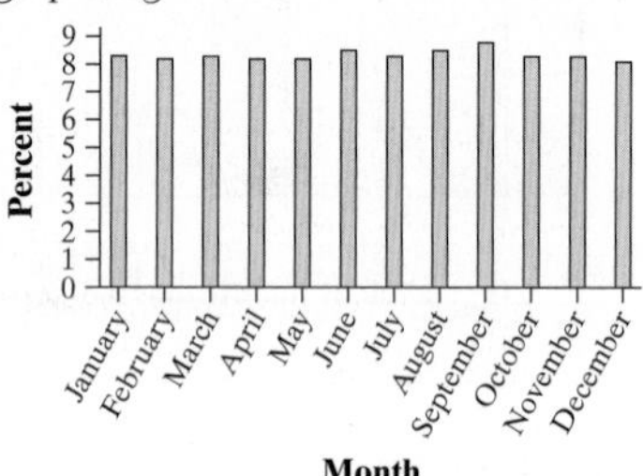

37. **(a)** 28/50 = 56% of the states **(b)** *Shape:* The shape of the distribution of seat belt usage is skewed to the left with a single peak in the 90–95% stem. **(c)** We use relative frequency rather than frequency because the groups are of different sizes. In this case, relative frequencies enable us to compare the distributions fairly. **(d)** *Shape:* The shape of the distribution of seat belt use is single-peaked and slightly skewed left for both primary enforcement states and secondary enforcement states. *Outliers:* There are two possible outliers among the primary enforcement states (the two states in the 80% to <82.5% interval) and one possible outlier among the secondary enforcement states (the state in the 70% to <72.5% interval). *Center:* The median of the percentage of observed people who were wearing seat belts in primary enforcement states was greater (median = 90% to <92.5%) than that of secondary enforcement states (median = 85% to <87.5%). *Variability:* In primary enforcement states, seat belt usage varied less (as little as 80% to as much as 97.5%) than in secondary enforcement states (as little as 75% to as much as 95%).

39. **(a)** This dotplot is oriented vertically rather than horizontally. The variable, cell length, is on the vertical axis rather than the horizontal axis. Also, the dots are "stacked" symmetrically on either side of the central axis for each dosage group. **(b)** The data do provide evidence that vitamin C helps teeth grow in guinea pigs similar to the ones used in this experiment because the median cell length increases as the dose of vitamin C increases. The median cell length for the guinea pigs given 0.5 mg/day is 10 micrometers, which is less than the median cell length of about 19 micrometers for guinea pigs given 1 mg/day, which is less than the median cell length of about 26 micrometers for guinea pigs given 2 mg/day.

41. A

43. D

45. D

Section 1D

Answers to Check Your Understanding

page 54: 1. First, we must put the weights in order: 2, 2.8, 3.4, 3.6, 4, 4, 5.4, 5.4, 6, 6, 6.1, 6.6, 9.6, 9.6, 11, 11.9, 12.4, 12.7, 13, 14, 15, 31, and 33. Because there are 23 weights, the median is the 12th weight in this ordered list. The median weight of the pumpkins is 6.6 pounds.

2. The mean weight of the pumpkins is

$$\bar{x}=\frac{3.6+4+9.6+\ldots+5.4+31+33}{23}=9.935 \text{ pounds.}$$

3. I knew the mean would be greater than the median because the distribution of pumpkin weights is skewed to the right with two possible upper outliers: 31 and 33 pounds.

page 63: 1. We cannot calculate the range exactly from the histogram because we cannot identify the exact value of the maximum or minimum from the histogram. Using the data, the range = max − min = 33 − 2 = 31 lb.

2. The standard deviation of 8.01 pounds tells us that the weight of these pumpkins typically varies from the mean by about 8.01 pounds.

3. In order: 2, 2.8, 3.4, 3.6, 4, 4, 5.4, 5.4, 6, 6, 6.1, 6.6, 9.6, 9.6, 11, 11.9, 12.4, 12.7, 13, 14, 15, 31, 33. There are 23 observations in the data set, so the median is the 12th observation, which is 6.6 pounds. There are 11 values in the lower half of the data set and 11 values in the upper half of the data set. The first quartile is the 6th observation from the bottom of the list, which is 4 pounds. The third quartile is the 6th observation from the top of the list, which is 12.7 pounds. The interquartile range of the distribution of weight is $IQR = Q_3 - Q_1 = 12.7 - 4 = 8.7$ pounds.

4. I would use the median and *IQR* to describe center and variability because there appears to be two high outliers. The *IQR* and median are resistant to outliers, but the mean, range, and standard deviation are not.

page 68: 1. 5-number summary: min = 2.0, $Q_1 = 4$, median = 6.6, $Q_3 = 12.7$, max = 33. $IQR = 12.7 - 4 = 8.7$ pounds.
Low outliers $< Q_1 - 1.5 \times IQR = 4 - 1.5(8.7) = -9.05$ pounds
High outliers $> Q_3 + 1.5 \times IQR = 12.7 + 1.5(8.7) = 25.75$ pounds
The pumpkins that weighed 31 and 33 pounds are outliers.

2.

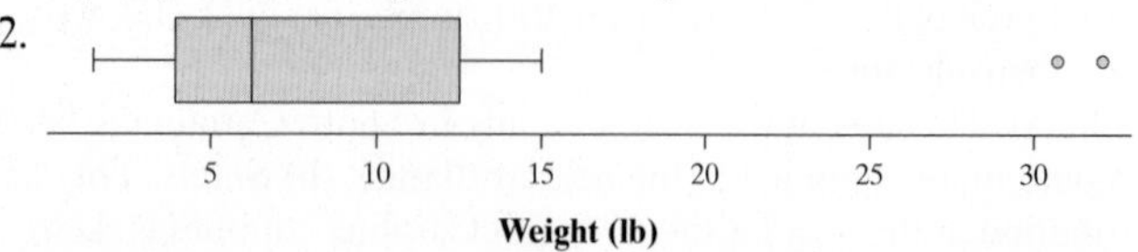

3. The boxplot does not completely display the shape of the distribution because it does not show that there is a single peak in the 5- to 10-pound weight interval.

Answers to Odd-Numbered Section 1D Exercises

1. The median is the average of the 12th and 13th value in the ordered list.

$$\text{median} = \frac{2+3}{2} = 2.5 \text{ goals}$$

3. **(a)** Mean = 3.208 goals per game **(b)** Removing the possible outlier, mean = 2.783 goals per game. By removing the outlier, the mean was reduced by 0.425 goal per game. The mean is nonresistant.

5. **(a)** The median is the 26th observation, or 8 electoral votes. About half of the states have fewer than 8 electoral votes and about half of the states have more than 8 electoral votes. **(b)** The distribution of electoral votes is skewed to the right and has several possible high outliers, so the mean will be greater than the median. **(c)** A parameter because it is a number that describes the entire population of all 50 states and the District of Columbia.

7. **(a)** $\text{mean} = \dfrac{\$1{,}710{,}000{,}000{,}000}{45{,}300{,}000} = \$37{,}748.34$ per borrower

(b) The median amount of student loan debt is probably less than the mean amount of student loan debt, because the distribution of student loan debt is most likely skewed to the right with several possible high outliers. Of the 45.3 million borrowers with loan debt, over 41 million of them had debt less than \$100,000, whereas more than 3 million had debt greater than \$100,000 and about 900,000 had debt greater than \$200,000.

9. Here are estimates of the frequencies of the bars (from left to right): 15, 11, 15, 11, 8, 5, 3, 3, and 3. Although the answers may vary slightly, the frequencies must sum to 74. The median is the average of the 37th and 38th values, or 2 servings of fruit per day. We can estimate the mean by adding 0 fifteen times, 1 eleven times, and so on. This gives us a sum of 194. The mean is then calculated by dividing by the number of responses:

$\bar{x} = \dfrac{194}{74} = 2.62$ servings of fruit per day.

11. **(a)** Range = max − min = 13 − 1 = 12 goals **(b)** Removing the possible outlier, Range = 6 − 1 = 5 goals. By removing the outlier, the range was reduced by 7 goals. The range is nonresistant.

13. The mean foot length is $\bar{x} = 24$ cm.

x	$x_i - \bar{x}$	$(x_i - \bar{x})^2$
25	25 − 24 = 1	$1^2 = 1$
22	22 − 24 = −2	$(-2)^2 = 4$
20	20 − 24 = −4	$(-4)^2 = 16$
25	25 − 24 = 1	$(1)^2 = 1$
24	24 − 24 = 0	$(0)^2 = 0$
24	24 − 24 = 0	$(0)^2 = 0$
28	28 − 24 = 4	$(4)^2 = 16$
	Sum = 0	**Sum = 38**

The sample variance is $s_x^2 = \dfrac{38}{7-1} = 6.33$. The sample standard deviation is $s_x = \sqrt{6.33} = 2.52$ cm. The foot lengths typically vary from the mean by about 2.52 cm.

15. **(a)** The file sizes typically vary by about 1.9 megabytes from the mean. **(b)** The mean would decrease slightly because 4 megabytes < 7.5 megabytes. The standard deviation would decrease as well because a file of size 4 megabytes will be closer to the new mean than the file of 7.5 megabytes was to the former mean.

17. Variable B has a smaller standard deviation because more of the observations have values closer to the mean than in Variable A's distribution. That is, the typical distance from the mean is smaller for Variable B than for Variable A.

19. $Q_1 = 2, Q_3 = 3.5, IQR = Q_3 - Q_1 = 3.5 - 2 = 1.5$ goals

21. **(a)** $Q_1 = 19.27$, $Q_3 = 45.40$, $IQR = 45.40 - 19.27 = \$26.13$. The range of the middle half of the amounts spent by these 50 grocery shoppers is \$26.13. **(b)** The distribution is likely skewed to the right because the mean is much larger than the median. Also, Q_3 is much farther from the median than is Q_1.

23. We should use the mean and standard deviation to summarize this distribution because the shape is fairly symmetric and there are no apparent outliers.

25. **(a)** Answers may vary. Correct answers will have 4 values that are all the same (e.g., 1, 1, 1, 1 or 5, 5, 5, 5). **(b)** 0, 0, 10, 10. This data set will give the largest possible deviations from the mean and therefore the highest standard deviation. **(c)** There are 11 possible correct answers for part (a): 0, 0, 0, 0 through 10, 10, 10, 10, all of which have a standard deviation of 0. There is only one possible answer for part (b).

27. Min = 1, $Q_1 = 2$, median = 2.5, $Q_3 = 3.5$, max = 13, $IQR = 3.5 - 2 = 1.5$ goals.
Low outliers $< Q_1 - 1.5 \times IQR = 2 - 1.5(1.5) = -0.25$ goals
High outliers $> Q_3 + 1.5 \times IQR = 3.5 + 1.5(1.5) = 5.75$ goals
The 3 games in which 6, 6, and 13 goals are scored are high outliers because they are greater than 5.75. There are no low outliers because the min of 1 is not less than −0.25.

29. **(a)** $IQR = 4.37 - 3.97 = 0.4\%$ butterfat
Low outliers $< Q_1 - 1.5 \times IQR = 3.97 - 1.5(0.4) = 3.37\%$ butterfat
High outliers $> Q_3 + 1.5 \times IQR = 4.37 + 1.5(0.4) = 4.97\%$ butterfat
There are no low outliers according to the $1.5 \times IQR$ rule because there are no values less than 3.37 (the minimum = 3.52% butterfat). There are no high outliers according to the $1.5 \times IQR$ rule because there are no values greater than 4.97 (the maximum = 4.91% butterfat).
(b) Low outliers $< \bar{x} - 2s_x = 4.173 - 2(0.291) = 3.591\%$ butterfat
High outliers $> \bar{x} + 2s_x = 4.173 + 2(0.291) = 4.755\%$ butterfat
There are two low outliers according to the 2 × SD rule because the histogram shows 2 values in the 3.3 to 3.591 interval, which are less than 3.591% butterfat. There are also two high outliers according to the 2 × SD rule because the histogram shows 2 values in the 4.755 to 5.046 interval, which are greater than 4.755% butterfat.

31. **(a)** Min = 0, $Q_1 = 3$, Med = 11.5, $Q_3 = 48$, Max = 268, $IQR = Q_3 - Q_1 = 48 - 3 = 45$ text messages
Low Outliers $< Q_1 - 1.5 \times IQR = 3 - 1.5(45) = -64.5$
High Outliers $> Q_3 + 1.5 \times IQR = 48 + 1.5(45) = 115.5$
Because 118 and 268 are greater than 115.5, they are considered outliers.

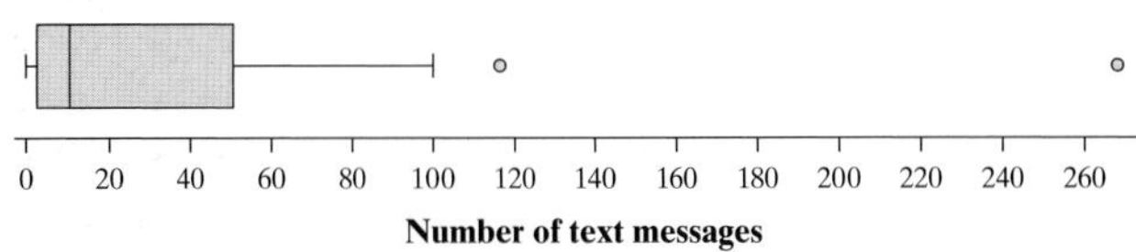

(b) Based on the boxplot, less than 25% of Dr. Williams's students sent more than 64 texts in the previous day because 64 is greater than the third quartile. The data do confirm Dr. Williams's suspicion because 75% of the students sent fewer than 48 text messages, which is far less than 64 texts.

33. **(a)** $IQR \approx 12 - 4 = 8$ electoral votes **(b)** The dotplot reveals that the distribution has a single peak at 3–4 electoral votes. This cannot be discerned from the boxplot.

35. *Shape:* The distribution of number of putts is slightly skewed right for those who described their putting technique and is strongly skewed to the right for those who did an unrelated task. *Outliers:* There is one outlier of 65 putts among those who described their putting technique, and there is one outlier of 37 putts among those who did an unrelated task. *Center:* Those who described their putting technique tended to require more putts to make 3 in a row (median = 17 putts) than those who did an unrelated task (median = 5.5 putts). *Variability:* There is more variation in number of putts required to make 3 in a row for those who described their putting technique ($IQR = 16.5$) than for those who did an unrelated task ($IQR = 11.5$).

37. **(a)** Approximately 25% of refrigerators with top freezers, almost all refrigerators with side freezers, and more than 75% of refrigerators with bottom freezers cost more than \$60 per year to operate. **(b)** *Shape:* The distribution of energy cost (in dollars) for refrigerators with top freezers and for refrigerators with side freezers is roughly symmetric. The distribution of energy cost (in dollars) for refrigerators with bottom freezers is skewed to the right. *Outliers:* There are no outliers for the refrigerators with top or side freezers. There are two refrigerators with bottom freezers that have unusually high energy costs (more than \$140 per year). *Center:* The typical energy cost for the refrigerators with side freezers (median ≈ \$75) is greater than the typical cost for the refrigerators with bottom freezers (median ≈ \$69), which is greater than the typical cost for the refrigerators with top freezers (median ≈ \$56). *Variability:* There is much more variability in the energy costs for refrigerators with bottom freezers ($IQR \approx \$20$) than for those with side freezers ($IQR \approx \$12$) or top freezers ($IQR \approx \$8$).

39. **(a)**

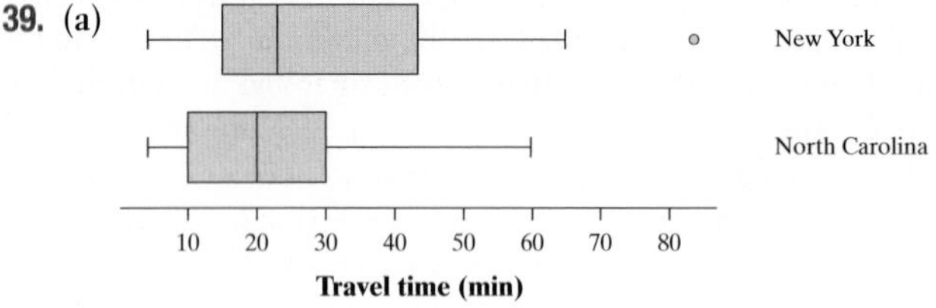

(b) No. The data do not provide strong evidence that the travel times to work differ for workers in these two states because the medians are similar and there is a substantial amount of overlap of the boxplots.

41. **(a)** Mean = 11.65 pairs of shoes per person **(b)** There are 20 values in the data set. 10% of 20 = 2, so we will remove the 2 largest and 2 smallest values from the data set and recalculate the mean. Trimmed data set: 5, 5, 6, 7, 7, 7, 7, 8, 10, 10, 10, 10, 11, 12, 14, 22. The 10% trimmed mean = 9.4375 pairs of shoes per person. **(c)** The trimmed mean provides a better summary of the center of this distribution than does the mean because it removes the 4 most extreme values in order to give a more representative measure of the center of the distribution.

43. D

45. E

47. Answers may vary. A boxplot, histogram, dotplot, or stemplot are appropriate graphs. A histogram and boxplot are given here.

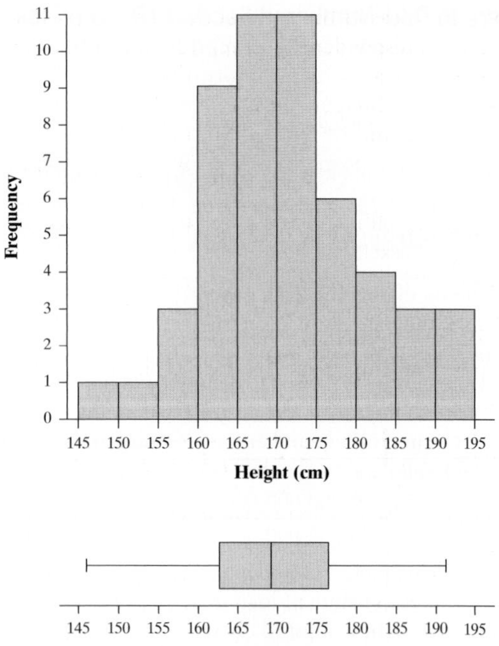

Shape: This distribution is roughly symmetric with a single peak at 170 cm. *Outliers:* There are no outliers. *Center:* The center of the distribution of heights can be described by the mean of 169.88 cm or the median of 169.5 cm. *Variability:* The heights vary from 145.5 cm to 191 cm, so the range is 45.5 cm. The standard deviation of heights is 9.687 cm and the IQR is $177 - 163 = 14$ cm.

Answers to Unit 1, Part I Review Exercises

R1. **(a)** *Individuals:* car buyers **(b)** *Variables:* zip code, whether they are a first-time buyer, distance from dealer, car model, model year, and price. *Categorical:* zip code, whether they are a first-time buyer, car model, model year. *Quantitative:* distance from dealer (mi), price (\$). *Note:* Model year may be considered quantitative if we want to know the average model year.

R2. **(a)**

Candy Selected	Rel. Freq.
Snickers®	8/30 = 0.267 = 26.7%
Milky Way®	3/30 = 0.100 = 10.0%
Butterfinger®	7/30 = 0.233 = 23.3%
Twix®	10/30 = 0.333 = 33.3%
3 Musketeers®	2/30 = 0.067 = 6.7%

(b) Students preferred Twix the most (≈33% of the students chose this candy), followed by Snickers (≈27%), Butterfinger (≈23%), Milky Way (10%), and lastly 3 Musketeers (≈7%).

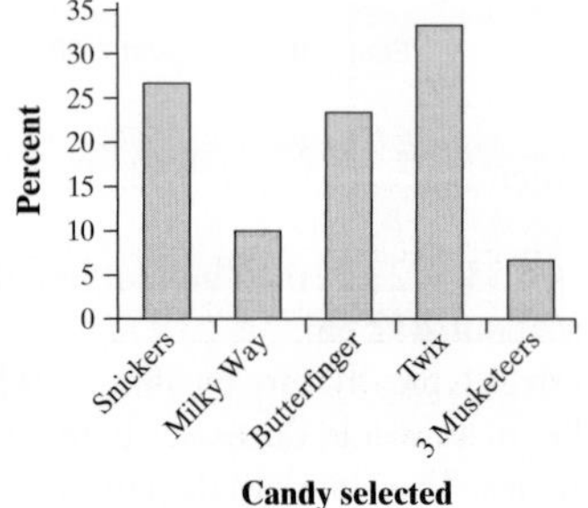

R3. (a) The areas of the icons should be proportional to the percentage of respondents they represent. As drawn, it appears that YouTube was selected more than 3 times as often as TikTok and that Snapchat was selected more than 2 times as often as TikTok and Instagram. But in reality, the percentages are: YouTube 32%, Snapchat 20%, TikTok 13%, and Instagram 13%.

R4. For all race/ethnicity categories, the percentage of students who frequently asked questions in class were similar for those who were first-generation and not first-generation college students. Black students were most likely to frequently ask questions in class, followed by White students, other students, students of two or more races, American Indian students, Hispanic students, and Asian students.

R5. (a)

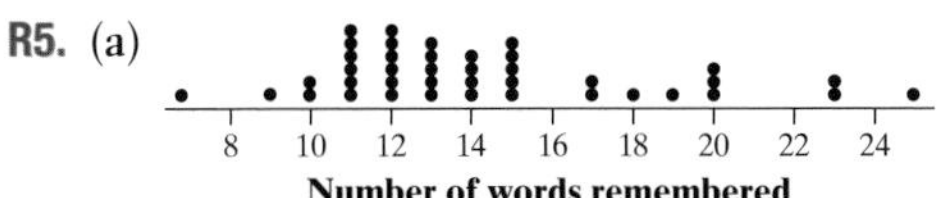

(b) $6/40 = 0.15$, or 15%, of this group of students remembered 20 or more words. (c) The median is the average of the 20th and 21st value in the ordered list. Median = 13 words. (d) The distribution of number of words remembered is skewed to the right, so the mean > median.

R6. (a)

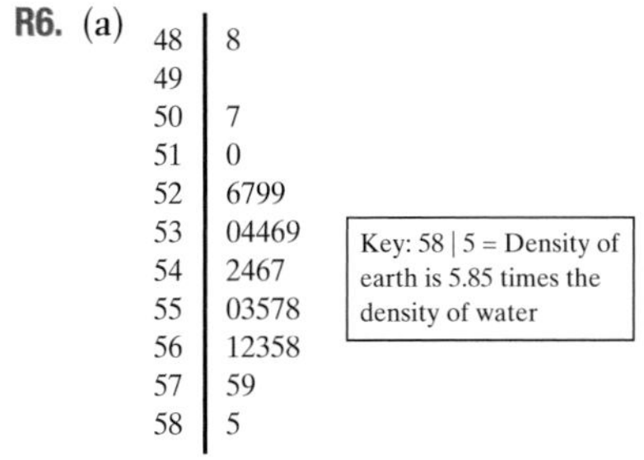

(b) *Shape:* The distribution of Cavendish's earth density measurements is roughly symmetric. *Outliers:* There is one possible outlier at 4.88. *Center:* The median of the distribution is 5.46 and the mean is 5.45. *Variability:* The density measurements vary from 4.88 to 5.85, $IQR = 5.615 - 5.295 = 0.32$, and the standard deviation is 0.22. (c) The mean of the distribution of Cavendish's 29 measurements is 5.45, which is fairly close to the currently accepted value for the density of the earth (5.51 times the density of water).

R7. (a)

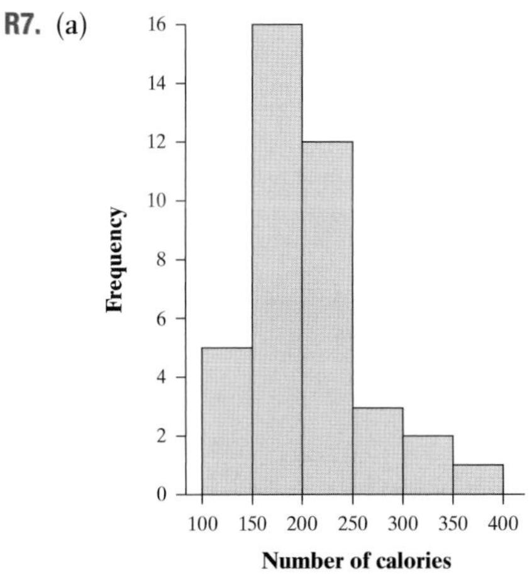

The histogram is skewed to the right with a peak between 150 and 200 calories.

(b) $IQR = Q_3 - Q_1 = 230 - 160 = 70$ calories

(c)

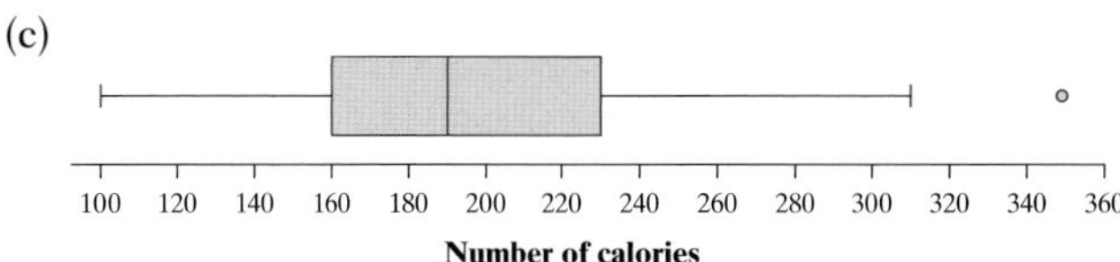

(d) The histogram shows that there is a single peak in the 150–200 calorie interval, which cannot be seen in the boxplot. The boxplot shows that the oatmeal with 350 calories is an outlier, which cannot be clearly seen in the histogram.

R8. (a) About 11% of low-income and 40% of high-income households consisted of four or more people. (b) *Similarities:* Both the low-income and high-income distributions include household sizes that vary from 1 to 7 people. Also, neither distribution shows any obvious outliers. *Differences:* The shapes of both distributions are skewed to the right; however, the skewness is much stronger in the distribution for low-income households. On average, household size is larger for high-income households. In fact, the majority of low-income households consist of only one person, while only about 7% of high-income households consist of one person.

R9. (a) Range $= 1.5 - 0.012 = 1.488$ ppm (b) The mercury concentration typically varies from the mean by about 0.3 ppm. (c) $IQR = 0.38 - 0.071 = 0.309$, so any point below $0.071 - 1.5(0.309) = -0.3925$ or above $0.38 + 1.5(0.309) = 0.8435$ would be considered an outlier. Because the smallest value of $0.012 > -0.3925$, there are no low outliers. According to the dotplot, there are values greater than 0.8435, so there are several high outliers. (d) The distribution of mercury concentration is skewed to the right and has several high outliers, so it would be best to use the median and *IQR* to summarize the center and variability of this distribution.

R10. *Shape:* The distribution of mercury concentration for light tuna is skewed to the right. The distribution of mercury concentration for albacore tuna is fairly symmetric. *Outliers:* The light tuna has several high outliers. The albacore tuna has just a couple high outliers. *Center:* The albacore tuna generally has more mercury (median = 0.4 ppm) than light tuna (median = 0.16 ppm). Albacore's minimum, first quartile, median, and third quartile are all greater than the respective values for light tuna. But some cans of light tuna have about twice as much mercury as the maximum for the cans of albacore tuna. *Variability:* Light tuna has much larger variation in mercury concentration ($IQR = 0.288$ ppm) than albacore tuna ($IQR = 0.167$ ppm).

Answers to Unit 1, Part I AP® Statistics Practice Test

T1. C
T2. C
T3. B
T4. B
T5. C
T6. E
T7. C
T8. D
T9. C
T10. D
T11. (a) Year 2. In year 2, about $24 + 34 = 58\%$ of the students said pop or hip hop. In year 1, about $28 + 21 = 49\%$ of students said pop or hip hop. (b) For the 250 students in year 1, the most

preferred music type was pop at around 28%, but in year 2 the top preference switched to hip hop at about 34%. In both years, the music type that was least preferred was jazz. From year 1 to year 2, hip hop had the largest increase, while rock and jazz saw the largest decreases.

T12. (a)

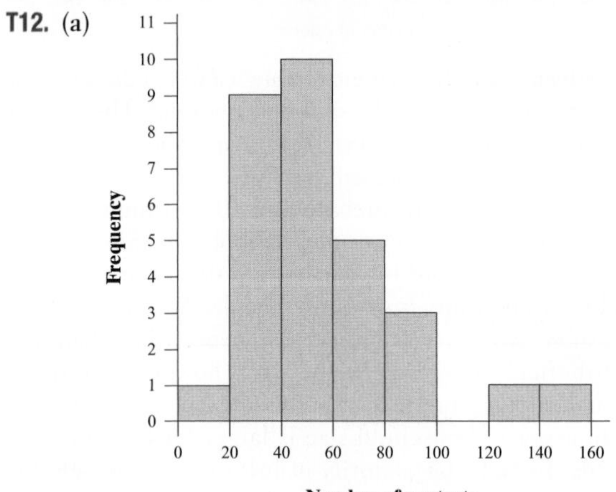

(b) $Q_1 = 30, Q_3 = 77, IQR = 77 - 30 = 47$ contacts.
Low outliers $< 30 - 1.5(47) = -40.5$
High outliers $> 77 + 1.5(47) = 147.5$
The individual who has 151 contacts is an outlier because $151 > 147.5$. There are no low outliers because min $= 7 > -40.5$ **(c)** It would be better to use the median and *IQR* to describe the center and variability because the distribution of number of contacts is skewed to the right and has a high outlier.

T13. *Shape:* The distribution of reaction time for the athletes is slightly skewed to the right. The distribution of reaction time for the non-athletes is roughly symmetric with two high outliers. *Outliers:* It appears that the athlete's distribution has one high outlier, whereas the non-athlete's distribution has two high outliers. *Center:* The reaction times for the students who are not athletes tended to be slower (median = 292 milliseconds) than for the athletes (median = 261 milliseconds). *Variability:* The distribution of reaction time for the non-athletes has more variability ($IQR = 70$ milliseconds) than the athletes' reaction times ($IQR = 64$ milliseconds).

UNIT 1, PART II

Section 1E

Answers to Check Your Understanding

page 93: 1. $44/50 = 0.88$. Ohio is at the 88th percentile in the distribution of number of representatives. 88% of the states have 15 representatives or fewer.
2. $50 \times 0.52 = 26$. There are 26 states that have the same number of representatives as South Carolina, or fewer representatives. Counting from the minimum, South Carolina has 6 representatives.
3. $z = \dfrac{15 - 8.7}{9.69} = 0.65$. *Interpretation:* The number of representatives that Ohio has is 0.65 standard deviation above the mean number of representatives for all 50 states.
4. Ohio's *z*-score for number of counties is $z = \dfrac{88 - 62.82}{46.421} = 0.542$. Ohio's relative position is more unusual among the distribution of number of representatives because its *z*-score within that distribution is farther from 0.

page 100: 1. About 65% of calls lasted less than or equal to 30 minutes. This means that about 35% of calls lasted more than 30 minutes.
2. The first quartile (25th percentile) is at about $Q_1 = 13$ minutes. The third quartile (75th percentile) is at about $Q_3 = 32$ minutes. $IQR \approx 32 - 13 = 19$ minutes.
3. Converting the cost of the rides from dollars to cents will not change the shape. However, it will multiply the mean and the standard deviation by 100. So the shape will still be skewed to the right with possible high outliers, the mean will be 171 cents, and the standard deviation will be 45 cents.
4. Adding 25 cents to the cost of each ride will not change the shape of the distribution, nor will it change the standard deviation. It will, however, add 25 cents to the mean.
5. Converting the costs to *z*-scores will not change the shape of the distribution. It will change the mean to 0 and the standard deviation to 1.

Answers to Odd-Numbered Section 1E Exercises

1. **(a)** $18/20 = 0.90$. Jackie is at the 90th percentile. *Interpretation:* 90% of the students own less than or equal to the number of pairs of shoes that Jackie owns. **(b)** $0.45(20) = 9$ students. Therefore, Raul's response is the 9th value in the ordered list. Raul owns 7 pairs of shoes.

3. **(a)** Because 10 of the 30 observations (33.3%) are less than or equal to Antawn's head circumference (22.3 inches), Antawn is at the 33rd percentile in the head circumference distribution. **(b)** $0.90(30) = 27$. The player at the 90th percentile will have a head circumference that is the 27th value in the ordered list. The player with a head circumference of 23.9 inches is at the 90th percentile of the distribution.

5. This means that the speed limit is set at such a speed that 85% of the vehicle speeds are less than or equal to the posted speed.

7. **(a)** $z = \dfrac{2.3 - 9.494}{6.214} = -1.158$. *Interpretation:* Montana's percentage of foreign-born residents is 1.158 standard deviations below the mean percentage of foreign-born residents for all states. **(b)** If we let *x* denote the percentage of foreign-born residents in New York at that time, then we can solve for x in the equation $2.08 = \dfrac{x - 9.494}{6.214}$. Thus, $x = 22.419\%$ foreign-born residents.

9. **(a)** The Washington Nationals number of wins in 2019 is 0.75 standard deviation above the mean of 81 wins. **(b)** $0.75 = \dfrac{93 - 81}{SD}$, $0.75(SD) = 12$, $SD = 16$. The standard deviation of the number of wins during the 2019 season is 16 wins.

11. SAT: $z = \dfrac{1280 - 1059}{210} = 1.052$

ACT: $z = \dfrac{27 - 20.7}{5.9} = 1.068$

Alejandra scored better on the ACT relative to her peers because her *z*-score on the ACT was greater than her *z*-score on the SAT. On the

SAT, she scored only 1.052 standard deviations above the mean; on the ACT, she scored 1.068 standard deviations above the mean.

13. This means that 48% of boys his age weigh less than or equal to his weight and 76% of boys his age are less than or equal to his height. Because he is as tall as or taller than 76% of boys, but only weighs as much as or more than 48% of boys, he is probably fairly thin.

15. **(a)** No. A sprint time of 8.05 seconds is not unusually slow. A student with an 8.05-second sprint is at about the 75th percentile, so about 25% of the students took longer than that (were slower than that). **(b)** The 20th percentile of the distribution is approximately 6.7 seconds.

17. **(a)** $Q_1 \approx 860$ hours and $Q_3 \approx 1050$ hours, so $IQR \approx 1050 - 860 = 190$ hours.

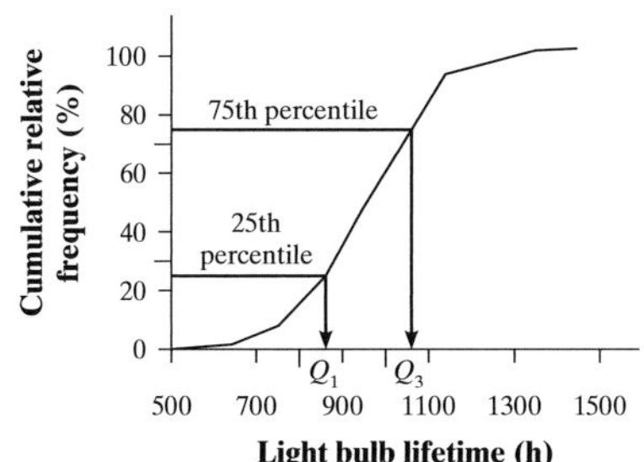

(b) The graph is less steep between 650 and 850 than it is between 850 and 1150 because there were fewer light bulbs with lifetimes in the interval of 650 to 850 than in the interval of 850 to 1150 hours.

19. **(a)** The shape of the distribution of corrected long-jump distance will be the same as the original distribution of long-jump distance: roughly symmetric with a single peak. **(b)** The median of the distribution of corrected long-jump distance is $577 - 20 = 557$ centimeters. **(c)** The IQR of the distribution of corrected long-jump distance is the same as the IQR of the distribution of long-jump distance as originally measured: $581.5 - 574.5 = 7$ centimeters.

21. **(a)** The shape of the resulting salary distribution will be the same as the original distribution of salaries. **(b)** The measures of center will increase by 5% because each value in the distribution is multiplied by 1.05. Multiplying each value in a distribution by a constant multiplies measures of center by the same amount. **(c)** The measures of variability will increase by 5% because each value in the distribution is multiplied by 1.05. Multiplying each value in a distribution by a constant multiplies the measures of variability by the same amount.

23. **(a)** The mean temperature reading in degrees Celsius

$$= \frac{5}{9}(77) - \frac{160}{9} = 25°\text{C}.$$

(b) The variance of the temperature reading in degrees Celsius

$$= \left(\frac{5}{9}\right)^2 (3^2) = 2.778(°\text{C})^2.$$

25. mean fare $= 2.85 + 2.7 \times$ mean miles

$15.45 = 2.85 + 2.7 \times$ mean miles

mean miles $= 4.667$

The mean length of the passenger's cab rides is 4.667 miles.

SD Fare $= 2.7 \times$ SD miles

$10.20 = 2.7 \times$ SD miles

SD miles $= 3.778$

The standard deviation of the lengths of their cab rides is 3.778 miles.

27. **(a)** The median nitrate concentration for Stony Brook is about 4.5%, which is less than the median nitrate concentration for Mill Brook of about 7.5%. **(b)** The Mill Brook stream has more variability in nitrate concentrations. The nitrate concentrations for the Mill Brook stream varies from 0 to about 20 mg/l, which is greater than the Stony brook stream, which varies from 0 to about 12 mg/l.

29. C

31. C

33. C

35. **(a)**

(b) Because $3/50 = 6\%$ of the sample was left-handed, our best estimate of the proportion of the population of Canadian high school students like these that is left-handed is 0.06.

Section 1F

Answers to Check Your Understanding

page 116:

1.

2.

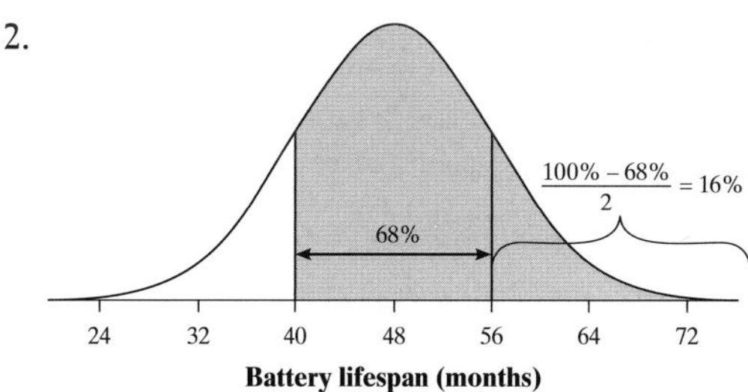

About $68\% + 16\% = 84\%$ of car batteries would last more than 40 months.

3. The histogram of these data is roughly symmetric and bell-shaped. The mean and standard deviation of these data are $\bar{x} = 56.70$ years and $s_x = 7.247$ years.

$\bar{x} \pm 1s_x = (49.453, 63.947)$ 31 of 46 = 67.4%

$\bar{x} \pm 2s_x = (42.206, 71.194)$ 45 of 46 = 97.8%

$\bar{x} \pm 3s_x = (34.959, 78.441)$ 46 of 46 = 100%

In a normal distribution, about 68% of the values fall within 1 standard deviation of the mean, about 95% within 2 standard deviations of the mean, and about 99.7% within 3 standard deviations of the mean. For the distribution of the presidents' ages, about 67.4% of the presidents were inaugurated at ages within 1 standard deviation of the mean, about 97.8% within 2, and 100% within 3 standard deviations of the mean. None of the observed percentages were substantially different than the estimated percentages. Thus, the distribution of age at inauguration is approximately normal.

page 123:

1. (i) $z = \dfrac{120 - 151.6}{25} = -1.264$. *Tech:* normalcdf(lower: -1000, upper: -1.264, mean: 0, SD: 1) $= 0.1031$; *Table* A: Area to the left of $z = -1.26$ is 0.1038.

(ii) normalcdf(lower: -1000, upper: 120, mean: 151.6, SD: 25) $=$ 0.1031. The proportion of teenagers who have blood cholesterol less than 120 mg/dl is about 0.1031.

2. (i) $z = \dfrac{200 - 151.6}{25} = 1.936$. *Tech:* normalcdf(lower: 1.936, upper: 1000, mean: 0, SD: 1) $= 0.0264$; *Table* A: Area to the right of $z = 1.94$ is $1 - 0.9738 = 0.0262$.
(ii) normalcdf(lower: 200, upper: 1000, mean: 151.6, SD: 25) $=$ 0.0264. About 2.64% of teenagers have blood cholesterol of 200 mg/dl or higher.

3. (i) $z = \dfrac{170 - 151.6}{25} = 0.736$ and $z = \dfrac{200 - 151.6}{25} = 1.936$.

Tech: normalcdf(lower: 0.736, upper: 1.936, mean: 0, SD: 1) $=$ 0.2044; *Table* A: Area between $z = 0.74$ and $z = 1.94$ is $0.9738 - 0.7704 = 0.2034$. (ii) normalcdf(lower: 170, upper: 200, mean: 151.6, SD: 25) $= 0.2044$. The proportion of teenagers who have borderline high blood cholesterol is about 0.2044.

page 128:

1. (i) *Tech:* invNorm(area: 0.20, mean: 0, SD: 1) $= -0.842$; *Table* A: 0.20 area to the left $\rightarrow z = -0.84$.

$-0.842 = \dfrac{x - 151.6}{25} \rightarrow x = (-0.842)(25) + 151.6 \rightarrow x = 130.559.$

(ii) invNorm(area: 0.20, mean: 151.6, SD: 25) $= 130.559$. About 20% of teenagers have cholesterol levels that are less than or equal to 130.559 mg/dl.

2. *Tech:* invNorm(area: 0.911, mean: 0, SD: 1) $= 1.347$; *Table* A: 0.911 area to the left $\rightarrow z = 1.35$.

$1.347 = \dfrac{200 - 157.5}{\sigma} \rightarrow 1.347\sigma = 42.5 \rightarrow \sigma = 31.552$. The standard deviation of the distribution of blood cholesterol for young adults is about 31.552 mg/dl.

Answers to Odd-Numbered Section 1F Exercises

1.

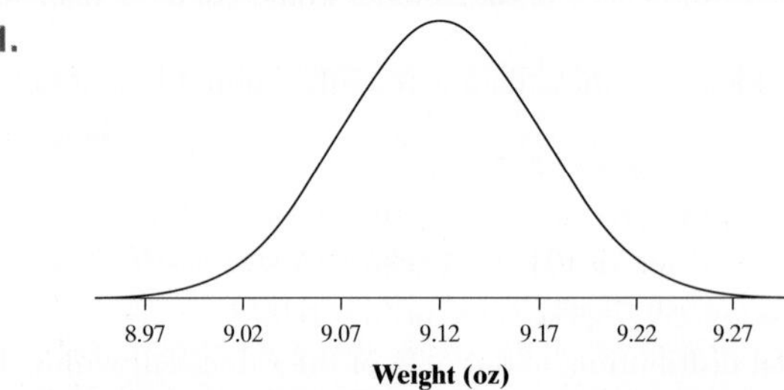

3. The mean is approximately 10. About 95% of the values are between 6 and 14, so the standard deviation is about 2. This value can also be obtained by finding the inflection point and estimating the horizontal distance between the inflection point and the mean.

5.

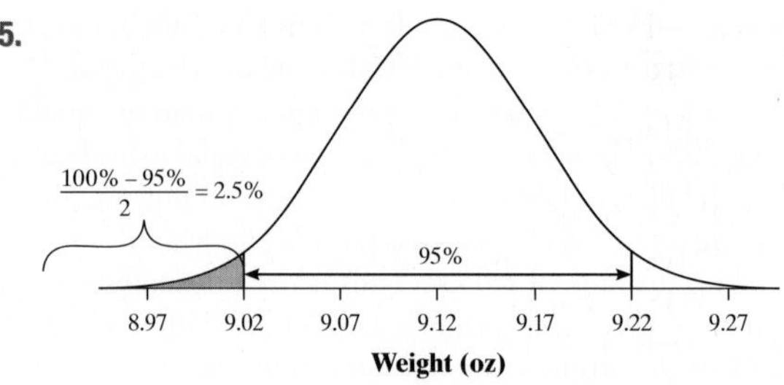

About 2.5% of all 9-oz bags of potato chips weigh less than 9.02 ounces.

7. The standard deviation is 40 pounds. First, notice that 130 is 120 pounds below the mean and 370 is 120 pounds above the mean. Because these values are symmetric about the mean and 99.7% of bears have weights between 130 and 370 pounds, 130 must be 3SD below the mean and 370 must be 3SD above the mean. The difference $370 - 250 = 120$ pounds is equal to 3SD, so $120 / 3 = 40$ pounds is 1SD.

9. The histogram of these data is roughly symmetric and bell-shaped. The mean and standard deviation of these data are $\bar{x} = 15.825$ cubic feet and $s_x = 1.217$ cubic feet.

$\bar{x} \pm 1s_x = (14.608, 17.042)$ 24 of 36 = 66.7%
$\bar{x} \pm 2s_x = (13.391, 18.259)$ 34 of 36 = 94.4%
$\bar{x} \pm 3s_x = (12.174, 19.476)$ 36 of 36 = 100%

These percentages are quite close to what we would expect based on the empirical rule of about 68%, 95% and 99.7%. Combined with the graph, this gives good evidence that this distribution is approximately normal.

11. The distribution of highway gas mileage is not approximately normal because the distribution is skewed to the right.

13. The distribution of tuitions in Michigan is not approximately normal. If it was normal, then the minimum value should be about 3 standard deviations below the mean. However, the actual minimum has a z-score of just $z = \dfrac{1873 - 10{,}614}{8049} = -1.086$. Also, if the distribution was normal, the minimum and maximum should be about the same distance from the mean. However, the mean is much farther from the maximum ($30{,}823 - 10{,}614 = 20{,}209$) than the minimum ($10{,}614 - 1873 = 8741$).

15. (i) $z = \dfrac{9 - 9.12}{0.05} = -2.400$. *Tech:* normalcdf(lower: -1000, upper: -2.4, mean: 0, SD: 1) $= 0.0082$; *Table* A: Area to the left of $z = -2.40$ is 0.0082.

(ii) normalcdf(lower: -1000, upper: 9, mean: 9.12, SD: 0.05) $=$ 0.0082. About 0.82% of 9-oz bags of this brand of potato chips weigh less than the advertised 9 ounces. This is not likely to pose a problem for the company that produces these chips because the percentage of bags that weigh less than the advertised amount is very small.

17. (i) $z = \dfrac{2400 - 1993}{593} = 0.686$. *Tech:* normalcdf(lower: 0.686, upper: 1000, mean: 0, SD: 1) $= 0.2464$; *Table* A: Area to the right of $z = 0.69$ is $1 - 0.7549 = 0.2451$.

(ii) normalcdf(lower: 2400, upper: 10000, mean: 1993, SD: 593) $=$ 0.2462. The proportion of meals ordered that exceed the recommended daily allowance of 2400 mg of sodium is about 0.2462.

19. (i) $z = \frac{1200 - 1993}{593} = -1.337$ and $z = \frac{1800 - 1993}{593} = -0.325$.

Tech: normalcdf(lower: −1.337, upper: −0.325, mean: 0, SD: 1) = 0.2820; *Table* A: Area between $z = -1.34$ and $z = -0.33$ is $0.3707 - 0.0901 = 0.2806$.
(ii) normalcdf(lower: 1200, upper: 1800, mean: 1993, SD: 593) = 0.2818. About 28.18% of meals ordered contained between 1200 mg and 1800 mg of sodium.
21. **(a)** $z = -1.66$. *Tech:* normalcdf(lower: −1.66, upper: 1000, mean: 0, SD: 1) = 0.9515; *Table* A: Area to the right of $z = -1.66$ is $1 - 0.0485 = 0.9515$. The proportion of observations in a standard normal distribution that satisfy $z > -1.66$ is 0.9515. **(b)** $z = -1.66$ and $z = 2.85$. *Tech:* normalcdf(lower: −1.66, upper: 2.85, mean: 0, SD: 1) = 0.9494; *Table* A: Area between $z = -1.66$ and $z = 2.85$ is $0.9978 - 0.0485 = 0.9493$. The proportion of observations in a standard normal distribution that satisfy $-1.66 < z < 2.85$ is 0.9494. **(c)** $z = 3.90$. *Tech:* normalcdf(lower: 3.90, upper: 1000, mean: 0, SD: 1) = 0.000048; *Table* A: Area to the right of $z = 3.90$ is 1 − (greater than 0.9998) = less than 0.0002. The proportion of observations in a standard normal distribution that satisfy $z > 3.90$ is 0.000048.

23. **(a)** (i) $z = \frac{3 - 5.3}{0.9} = -2.556$. *Tech:* normalcdf(lower: −1000, upper: −2.556, mean: 0, SD: 1) = 0.0053; *Table* A: Area to the left of $z = -2.56$ is 0.0052. **(ii)** normalcdf(lower: −1000, upper: 3, mean: 5.3, SD: 0.9) = 0.0053. Mrs. Starnes completes an easy Sudoku puzzle in less than 3 minutes about 0.0053, or about 0.53%, of the time. **(b)** (i) $z = \frac{6 - 5.3}{0.9} = 0.778$. *Tech:* normalcdf(lower: 0.778, upper: 1000, mean: 0, SD: 1) = 0.2183;

Table A: Area to the right of $z = 0.78$ is $1 - 0.7823 = 0.2177$.
(ii) normalcdf(lower: 6, upper: 1000, mean: 5.3, SD: 0.9) = 0.2183. About 21.83% of the time, it takes Mrs. Starnes more than 6 minutes to complete an easy puzzle.

(c) (i) $z = \frac{6 - 5.3}{0.9} = 0.778$ and $z = \frac{8 - 5.3}{0.9} = 3$. *Tech:* normalcdf(lower: 0.778, upper: 3, mean: 0, SD: 1) = 0.2169; *Table* A: Area between $z = 0.78$ and $z = 3.00$ is $0.9987 - 0.7823 = 0.2164$.
(ii) normalcdf(lower: 6, upper: 8, mean: 5.3, SD: 0.9) = 0.2170 About 21.7% of easy puzzles take Mrs. Starnes between 6 and 8 minutes to complete.
25. (i) *Tech:* invNorm(area: 0.99, mean: 0, SD: 1) = 2.326. *Table* A: 0.99 area to the left → $z = 2.33$.

$$2.326 = \frac{x - 22}{6.9} \rightarrow x = (2.326)(6.9) + 22 \rightarrow x = 38.05.$$

(ii) invNorm(area: 0.99, mean: 22, SD: 6.9) = 38.05 minutes. The longest 1% of response times take at least 38.05 minutes.
27. First Quartile: *Tech:* invNorm(area: 0.25, mean: 0, SD: 1) = −0.6745; *Table* A: 0.25 area to the left → $z = -0.67$.
Third Quartile: *Tech:* invNorm(area: 0.75, mean: 0, SD: 1) = 0.6745; *Table* A: 0.75 area to the left → $z = 0.67$. The first quartile of the standard normal distribution is $z = -0.6745$ and the third quartile of the standard normal distribution is $z = 0.6745$.

29. **(a)** (i) $z = \frac{2500 - 3668}{511} = -2.286$. *Tech:* normalcdf(lower: −1000, upper: −2.286, mean: 0, SD: 1) = 0.0111; *Table* A: Area to the left of $z = -2.29$ is 0.0110.

(ii) normalcdf(lower: −1000, upper: 2500, mean: 3668, SD: 511) = 0.0111. The proportion of newborns who would be identified as low birth weight is 0.0111. **(b)** (i) *Tech:* invNorm(area: 0.8, mean: 0, SD: 1) = 0.842; *Table* A: 0.8 area to the left → $z = 0.84$.

$$0.842 = \frac{x - 3668}{511} \rightarrow x = (0.842)(511) + 3668 \rightarrow x = 4098.26$$

(ii) invNorm(area: 0.8, mean: 3668, SD: 511) = 4098.07 grams. The 80th percentile of the distribution of birth weight is 4098.07 grams.
31. *Tech:* invNorm(area: 0.943, mean: 0, SD: 1) = 1.580. *Table* A: 0.943 area to the left → $z = 1.58$. $1.580 = \frac{84 - 78.4}{\text{SD}} \rightarrow$

$$\text{SD} = \frac{84 - 78.4}{1.580} \rightarrow \text{SD} = 3.544$$

The standard deviation of the distribution of height is approximately 3.544 inches.
33. *Tech:* invNorm(area: 0.2, mean: 0, SD: 1) = −0.842. *Table* A: 0.20 area to the left → $z = -0.84$.

$$-0.842 = \frac{11.7 - \mu}{0.2} \rightarrow \mu = 11.7 - (-0.842)(0.2) \rightarrow \mu = 11.868$$

The mean length of footlong sub sandwiches at this sub shop is about 11.87 inches.

35. **(a)** (i) $z = \frac{3.95 - 3.98}{0.02} = -1.500$. *Tech:* normalcdf(lower: −1000, upper: −1.5, mean: 0, SD: 1) = 0.0668; *Table* A: Area to the left of $z = -1.5$ is 0.0668.
(ii) normalcdf(lower: −1000, upper: 3.95, mean: 3.98, SD: 0.02) = 0.0668. About 6.68% of the large lids are too small to fit.

(b) (i) $z = \frac{4.05 - 3.98}{0.02} = 3.500$. *Tech:* normalcdf(lower: 3.5, upper: 1000, mean: 0, SD: 1) = 0.0002; *Table* A: Area to the right of $z = 3.50$ is less than 0.0002.
(ii) normalcdf(lower: 4.05, upper: 1000, mean: 3.98, SD: 0.02) = 0.0002. Approximately 0.02% of the large lids are too big to fit.
(c) It makes more sense to have a larger proportion of lids too small rather than too big. If lids are too small, customers will just try another lid. But if lids are too large, the customer may not notice and then spill the drink.
37. **(a)** First Quartile: (i) *Tech:* invNorm(area: 0.25, mean: 0, SD: 1) = −0.674; *Table* A: 0.25 area to the left gives $z = -0.67$.

$$-0.674 = \frac{x - 3668}{511} \rightarrow x = (-0.674)(511) + 3668 \rightarrow x = 3323.59$$

(ii) invNorm(area: 0.25, mean: 3668, SD: 511) = 3323.34 grams. Third Quartile: (i) *Tech:* invNorm(area: 0.75, mean: 0, SD: 1) = 0.674; *Table* A: 0.75 area to the left gives $z = 0.67$.

$$0.674 = \frac{x - 3668}{511} \rightarrow x = (0.674)(511) + 3668 \rightarrow x = 4012.41$$

(ii) invNorm(area: 0.75, mean: 3668, SD: 511) = 4012.66 grams The first quartile of the birth weight distribution is about 3323.34 grams, and the third quartile of the birth weight distribution is about 4012.66 grams. $IQR = 4012.66 - 3323.34 = 689.32$ grams
Low outlier $< Q_1 - 1.5 \times IQR = 3323.34 - 1.5(689.32) = 2289.36$
High outlier $> Q_3 + 1.5 \times IQR = 4012.66 + 1.5(689.32) = 5046.64$
According to the $1.5 \times IQR$ rule, any birth weight less than 2289.36 grams or greater than 5046.64 grams will be outliers.
(b) Proportion of low outliers: (i) $z = \frac{2289.36 - 3668}{511} = -2.698$.
Tech: normalcdf(lower: −1000, upper: −2.698, mean: 0, SD: 1) =

0.0035; *Table* A: Area to the left of $z=-2.70$ is 0.0035. (**ii**) normalcdf(lower: −10000, upper: 2289.36, mean: 3668, SD: 511) = 0.0035. Proportion of high outliers:
(**i**) $z=\frac{5046.64-3668}{511}=2.698$. *Tech:* normalcdf(lower: 2.698, upper: 1000, mean: 0, SD: 1) = 0.0035; *Table* A: Area to the right of $z=2.70$ is 0.0035; (**ii**) normalcdf(lower: 5046.64, upper: 10000, mean: 3668, SD: 511) = 0.0035. The proportion of Norwegian newborns who would have birth weights considered to be an outlier is 0.0035 + 0.0035 = 0.007.

39. (**a**) 5/500 = 0.01 = 1%. In 1% of the 500 simulated samples, the proportion of values within 1 standard deviation of the mean was greater than or equal to 0.818. (**b**) Yes! It is unlikely (probability ≈ 1%) that 77 values selected at random from a normally distributed population would have at least 81.8% of the values fall within 1 SD of the mean. (**c**) The distribution of calories in cereal is not approximately normally distributed because 81.8% is too far from the anticipated value of 68% of the data falling within 1SD of the mean.

41. C

43. B

45. *Indiana's percentile:* 37 of the 38 family incomes in Indiana are at or below \$95,000. Because 37/38 = 0.974, or 97.4%, this individual's family income is at the 97th percentile. *New Jersey's percentile:* 36 of the 44 family incomes in New Jersey are at or below \$95,000. Because 36/44 = 0.818, or 81.8%, this individual's family income is at the 81st percentile.

$$\text{Indiana: } z=\frac{95{,}000-47{,}400}{29{,}400}=1.619$$

$$\text{New Jersey: } z=\frac{95{,}000-58{,}100}{41{,}900}=0.881$$

A family income of \$95,000 in the state of Indiana is 1.619 standard deviations above the mean family income for that state, whereas a family income of \$95,000 in the state of New Jersey is only 0.881 standard deviation above the mean family income for that state. Individuals from Indiana with a family income of \$95,000 have a greater income relative to others in their state because they have a higher percentile (at about the 97th versus about the 81st) and have a higher *z*-score (1.619 versus 0.881) than individuals from New Jersey with a family income of \$95,000.

Answers to Unit 1, Part II Review Exercises

R1. (**a**) 27/40 = 0.675 → 67th percentile (**b**) $z=\frac{234{,}000-203{,}388}{87{,}609}=0.349$. *Interpretation:* The sale price for this home is 0.349 standard deviation above the average sale price for the homes in the sample.
(**c**) 5 years earlier: $z=\frac{212{,}500-191{,}223}{76{,}081}=0.280$. The home sold for more money recently, relatively speaking, because the *z*-score for the recent sale price is greater than the *z*-score of the sale price 5 years ago.

R2. (**a**) Reading up from 7 hours on the *x*-axis to the graphed line and then across to the *y*-axis, we see that 7 hours corresponds to about the 58th percentile. (**b**) To find Q_1, start at 25 on the *y*-axis, move across to the line and down to the *x*-axis. Q_1 is approximately 2.5 hours. To find Q_3, start at 75 on the *y*-axis, move across to the line and down to the *x*-axis. Q_3 is approximately 11 hours. Thus, $IQR\approx 11-2.5=8.5$ hours per week.

R3. (**a**) The shape of the distribution of adjusted salary is skewed to the right, just like the original distribution. Multiplying by a positive constant and adding a constant do not change the shape of the distribution. (**b**) The mean of the adjusted salary is (75,100)(1.02) + 500 = \$77,102. (**c**) The standard deviation of the adjusted salary is (36,554)(1.02) = \$37,285.08.

R4. (**a**)

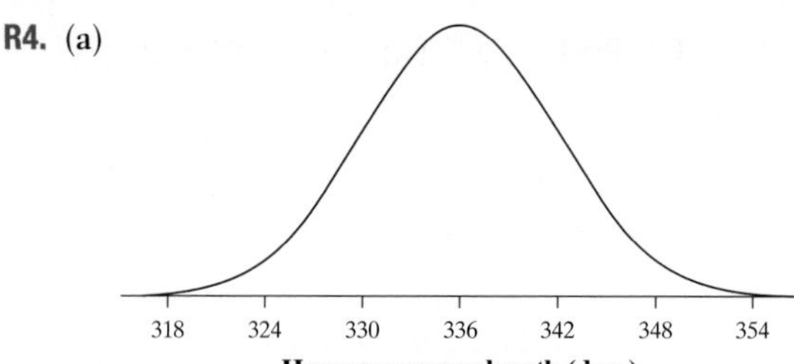

(**b**)

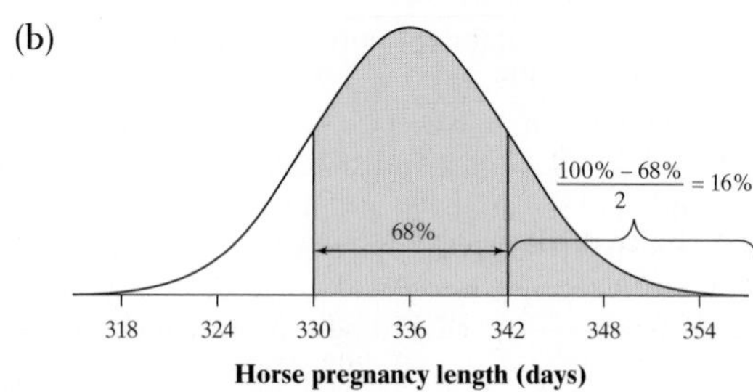

About 68% + 16% = 84% of horse pregnancies are longer than 330 days.

R5. The distribution of density of the earth is roughly symmetric and single-peaked. The mean and standard deviation of these data are $\bar{x}=5.448$ and $s_x=0.221$.

$\bar{x}\pm 1s_x=(5.227, 5.669)$	22 of 29 = 75.9%
$\bar{x}\pm 2s_x=(5.006, 5.890)$	28 of 29 = 96.6%
$\bar{x}\pm 3s_x=(4.785, 6.111)$	29 of 29 = 100%

Except for the small discrepancy with the percentage of values within 1 standard deviation of the mean, these percentages are fairly close to the 68%, 95%, and 99.7% that we would expect based on the empirical rule. Combined with the graph, this gives good evidence that this distribution of density measurements is approximately normal.

R6. (**a**) (**i**) $z=\frac{900-694}{112}=1.839$. *Tech:* normalcdf(lower: 1.839, upper: 1000, mean: 0, SD: 1) = 0.0330; *Table* A: Area to the right of $z=1.84$ is $1-0.9671=0.0329$.
(**ii**) normalcdf(lower: 900, upper: 10000, mean: 694, SD: 112) = 0.0329. About 3.29% of test-takers earn a score greater than or equal to 900 on the GRE Chemistry test. (**b**) (**i**) *Tech:* invNorm(area: 0.99, mean: 0, SD: 1) = 2.326. *Table* A: 0.99 area to the left → $z=2.33$. $2.326=\frac{x-694}{112}\rightarrow x=954.51$
(**ii**) invNorm(area: 0.99, mean: 694, SD: 112) = 954.551. The 99th percentile score on the GRE Chemistry test is about 954.551.

R7. (**a**) (**i**) $z=\frac{1-1.05}{0.08}=-0.625$ and $z=\frac{1.2-1.05}{0.08}=1.875$. *Tech:* normalcdf(lower: −0.625, upper: 1.875, mean: 0, SD: 1) = 0.7036; *Table* A: Area between $z=-0.63$ and $z=1.88$ is 0.9699 − 0.2643 = 0.7056. (**ii**) normalcdf(lower: 1, upper: 1.2, mean: 1.05, SD: 0.08) = 0.7036. About 70.36% of the time the dispenser will put between 1 and 1.2 ounces of ketchup on a burger. (**b**) Because the mean of 1.1 is in the middle of the interval from 1 to 1.2, we are looking for the middle 99% of the distribution. This leaves 0.5% in each tail. *Tech:* invNorm(area: 0.005, mean: 0, SD: 1) = −2.576; *Table* A: 0.005 area to the left → $z = -2.58$. $-2.576 = \frac{1-1.1}{\sigma} \rightarrow \sigma = 0.0388$

A standard deviation of at most 0.0388 ounce will result in at least 99% of burgers getting between 1 and 1.2 ounces of ketchup.

Answers to Unit 1, Part II AP® Statistics Practice Test

T1. E
T2. B
T3. A
T4. B
T5. A
T6. C
T7. C
T8. D
T9. C
T10. E

T11. (a) (i) $z = \frac{34,772-35,987}{607.5} = -2$ and $z = \frac{36,225-35,987}{607.5} = 0.392$. *Tech:* normalcdf(lower: −2, upper: 0.392, mean: 0, SD: 1) = 0.6297; *Table* A: Area between $z = -2$ and $z = 0.39$ is $0.6517 - 0.0228 = 0.6289$. (ii) normalcdf(lower: 34772, upper: 36225, mean: 35987, SD: 607.5) = 0.6296. About 62.96% of cars have "good" or "great" asking prices. (b) (i) *Tech:* invNorm(area: 0.98, mean: 0, SD: 1) = 2.054; *Table* A: 0.98 area to the left → $z = 2.05$.

$$2.054 = \frac{x-35,987}{607.5} \rightarrow x = (2.054)(607.5) + 35,987 \rightarrow x = \$37,234.81$$

(ii) invNorm(area: 0.98, mean: 35987, SD: 607.5) = \$37,234.65. About 98% of cars of this model have an MSRP of less than \$37,234.65, so a car with an MSRP of \$37,234.65 would be at the 98th percentile of the distribution.

T12. (a) $z = \frac{37.5-36.805}{0.407} = 1.708$. Mikaela's temperature is 1.708 standard deviations above the mean temperature. (b) 127/130 = 0.977. Mikaela's temperature is at the 97th percentile of this distribution. (c) Jin-Yu's *z*-score: $z = \frac{36.3-36.805}{0.407} = -1.241$. Mikaela's temperature is more unusual because her temperature is more standard deviations away from the mean than Jin-Yu's temperature.

T13. (a) The data do provide some evidence that normal body temperature is not 37.0°C because the mean body temperature of these 130 healthy individuals is 36.805°C, which is not quite 37.0°C. (b) The dotplot looks roughly symmetric, single-peaked, and somewhat bell-shaped. The percentage of values within 1, 2, and 3 standard deviations of the mean are as follows:

Mean ± 1SD: 36.805 ± 1(0.407) 36.398 to 37.212 97 out of 130 = 74.6%

Mean ± 2SD: 36.805 ± 2(0.407) 35.991 to 37.619 123 out of 130 = 94.6%

Mean ± 3SD: 36.805 ± 3(0.407) 35.584 to 38.026 129 out of 130 = 99.2%

Except for the small discrepancy with the percentage of values within 1 standard deviation of the mean, these percentages are fairly close to the 68%, 95%, and 99.7% that we would expect based on the empirical rule. Combined with the graph, this gives good evidence that this distribution of body temperatures is approximately normal.

UNIT 2

Section 2A

Answers to Check Your Understanding

page 153: 1. *Explanatory:* treatment. *Response:* change in depression. The type of treatment given helps explain the change in depression.
2. 91/338 = 0.269 or 26.9% of subjects had a full response; 55/338 = 0.163 or 16.3% of subjects had a partial response.
3.

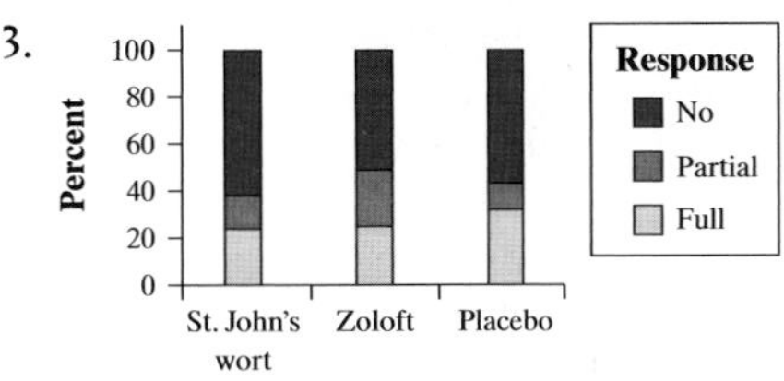

4. There does not appear to be a strong association between treatment and change in depression for these subjects because the distribution of response status is similar for the three different treatments. Although Zoloft had the highest percentage with some response, the treatment with the highest rate of "full response" was the placebo!

Answers to Odd-Numbered Section 2A Exercises

1. (a) *Explanatory:* water temperature. *Response:* weight gain. Water temperature helps explain changes in weight gain.
(b) Either variable could be the explanatory variable because each one could be used to predict the other.

3. (a) 29/150 = 0.193 (b) 6/150 = 0.04 (c) 6/50 = 0.12

5. (a)

		Location			
		Urban	Suburban	Rural	Total
Social media use	Too much	0.092	0.182	0.089	0.363
	About right	0.096	0.311	0.140	0.547
	Too little	0.031	0.043	0.017	0.090
	Total	0.219	0.536	0.245	1

(b) 0.092 + 0.096 = 0.188 (c) 121/287 = 0.422 or 42.2% of urban students, 239/702 = 0.340 or 34.0% of suburban students, and 116/321 = 0.361 or 36.1% of rural students said they use social media too much.

7.

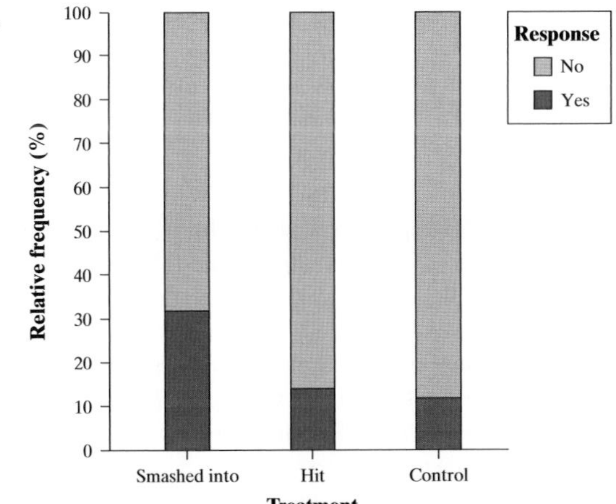

9.

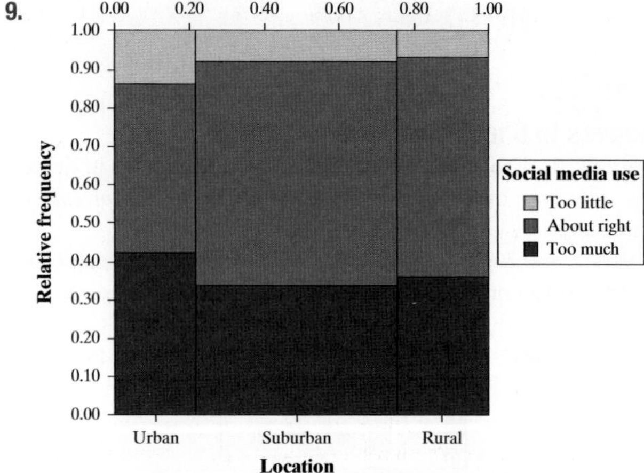

11. The subjects were most likely to recall seeing broken glass when they were told the cars "Smashed into" each other (32%), followed by 14% who recalled seeing broken glass when they were told the cars "Hit" each other, and 12% of the control group recalled seeing broken glass.

13. Urban teens were most likely to say "too much" (about 42%), followed by rural teens (about 36%), and suburban teens (about 34%). However, urban teens were also most likely to say "too little" (about 14%), followed by suburban teens (about 8%), and rural teens (about 7%). Unlike a segmented bar chart, the mosaic plot reveals that about 54% of the teens surveyed live in a suburban location, about 22% live in an urban location, and about 24% live in a rural location.

15. **(a)** Those who napped 1–2 times per week were the least likely to experience a CVD event (about 2%) and those who napped 6–7 times per week were the most likely to experience a CVD event (about 7.5%). Those who didn't nap at all or napped 3–5 times per week had a similar likelihood of experiencing a CVD event (about 4.5% and 5.5%, respectively). **(b)** No. There is an association between napping frequency and likelihood of experiencing a CVD event, but we cannot say that frequent napping may cause an increase in a person's risk of having a CVD event. It is possible that those with heart disease may feel fatigue and need to nap more. Association is not causation.

17.

		School level			
		Elementary	Middle	High	Total
Food choice	Nuggets	7	1	2	10
	Hamburger	3	1	3	7
	Pizza	3	2	5	10
	Hot dogs	3	1	1	5
	Total	16	5	11	32

19. **(a)** There is some evidence that dexamethasone helped. In the dexamethasone group, 22.9% died, which is less than the 25.7% who died while receiving the usual care. **(b)** Yes. The more extreme the initial respiratory support, the worse the outcome tends to be. Mechanical ventilation, 37.5% died; on oxygen only, 25.2% died; no respiratory support, 15.2% died. **(c)** No, dexamethasone appears to make the biggest impact among those who were on mechanical ventilation (29.3% died in the dexamethasone group, which is 12.1 percentage points less than the 41.4% who died in this group under usual care). Dexamethasone also appears to help those who were on oxygen only as the death rate decreased by 2.9 percentage points (usual care: 26.2%, dexamethasone: 23.3%). However, dexamethasone appears to have made things worse for those who were under no respiratory support because the death rate increased from 14% with usual care to 17.8% for those who received dexamethasone.

21. C

23. C

Section 2B

Answers to Check Your Understanding

page 164: 1. *Explanatory:* amount of sugar (in grams). *Response:* number of calories. The amount of sugar helps predict the number of calories in movie-theater candy.

2.

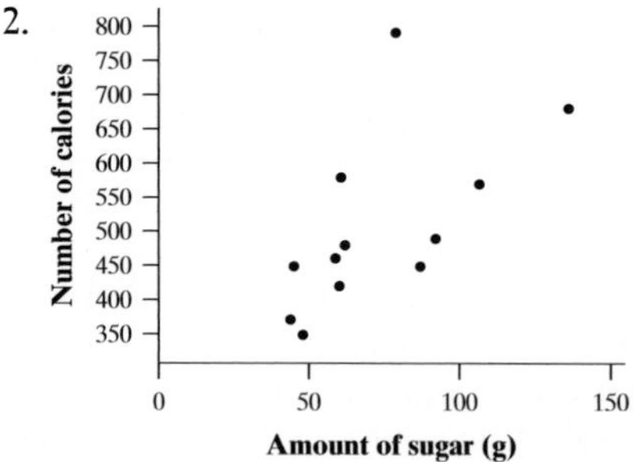

3. There is a moderately strong, positive, linear relationship between the amount of sugar contained in movie-theater candy and the number of calories in the candy. The point for Peanut M&M'S®, which has 79 grams of sugar and 790 calories, has an unusually high number of calories compared to other candies with similar amounts of sugar.

page 172: 1. Using technology, $r = 0.618$.

2. The correlation of $r = 0.618$ indicates that the linear association between amount of sugar and number of calories is positive and fairly strong.

3. The correlation would remain the same. Correlation is not affected by a change of units.

4. The correlation would remain the same. Correlation makes no distinction between explanatory and response variables. Reversing the explanatory and response variables does not affect the value of the correlation.

Answers to Odd-Numbered Section 2B Exercises

1.

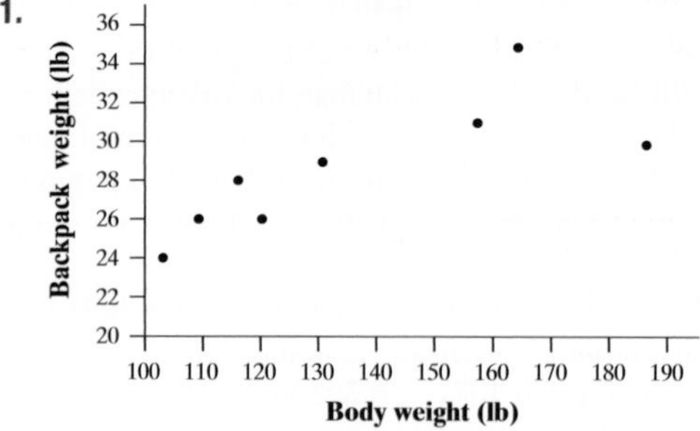

3. There is a weak, negative, linear relationship between average temperature in the previous summer and wing length. There are no obvious unusual values.

5. There is a moderately strong, positive, linear association between backpack weight and body weight for these students. There is an unusual point in the graph—the hiker with body weight 187 pounds and pack weight 30 pounds. This hiker makes the form appear to be nonlinear for weights above 140 pounds.

7. There is a moderately strong, negative, linear relationship between temperature and wing length for both the male and female butterflies. However, the female butterflies tend to have greater wing lengths than the male butterflies for any given temperature.

9. The relationship between age and length is positive, so $r > 0$. Also, r is closer to 1 than 0 because the relationship is fairly strong. There is not too much scatter from the linear pattern.

11. The correlation of 0.52 indicates that the linear relationship between the net approval rating and change in seats in the House of Representatives in midterm elections is somewhat weak and positive.

13. The correlation between number of pull-ups and the number of push-ups is $r = 0.9$. Both of these exercises require upper body strength, so we expect the relationship to be strong and positive. The correlation between the number of pull-ups and the number of sit-ups is $r = 0.3$. We expect these two variables to be positively associated, as they are both good indicators of overall fitness, but the relationship would be weaker than that of pull-ups and push-ups because pull-ups use upper body strength and sit-ups use core strength. The correlation between number of pull-ups and weight is $r = -0.5$. The smaller a person's weight, the easier it might be to do more pull-ups, so we expect a negative relationship here.

15. Not necessarily. Even though there is a moderate positive association between presidential approval and the change in the number of seats in the midterm election, it might be that both values are high or both values are low due to another variable, such as the state of the economy or being involved in a war.

17. (a)

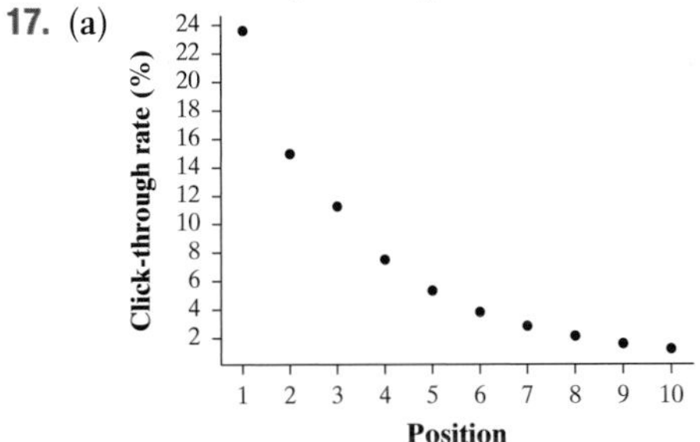

(b) When there is a strong linear association between two variables, the correlation will be close to 1 or −1. However, the converse isn't necessarily true. In this case, correlation is close to −1, but the association is clearly curved. In other words, we can't use the correlation to determine if the form of an association is linear.

19. The femur measurements have a mean of 58.2 cm and a standard deviation of 13.198 cm. The humerus measurements have a mean of 66 cm and a standard deviation of 15.89 cm.

Femur	Humerus	*z* femur	*z* humerus	Product
38	41	−1.531	−1.573	2.408
56	63	−0.167	−0.189	0.031
59	70	0.061	0.252	0.015
64	72	0.439	0.378	0.166
74	84	1.197	1.133	1.356

$$r = \frac{2.408 + 0.031 + 0.015 + 0.166 + 1.356}{5-1} = 0.994$$

21. $r = 0.783$

23. (a) The correlation would remain the same if the variables were reversed. Correlation makes no distinction between explanatory and response variables. (b) The correlation would remain the same if the bone lengths were measured in inches. The value of r does not change when we change the units of x, y, or both. (c) Correlation is unitless.

25. (a) $r = 0.783$ (b) $r = 0.783$ (c) The correlation of x = maximum speed and y = height is the same as the correlation of x = height and y = maximum speed. The correlation of x = height (in meters) and y = maximum speed is the same as the correlation of x = height (in feet) and y = maximum speed.

27. (a) There is a strong, positive, nonlinear relationship between income per person and life expectancy. (b) African countries tend to have the lowest life expectancy, followed by Asia/Pacific. The Americas and Europe/Russia tend to have the highest life expectancies and are fairly similar in this regard. (c) Small countries tend to span all income values, however the largest countries (e.g., India and China) tend to have less income per person, on average. The exception to this rule is the United States, which has a fairly large population size and also a fairly high income per person.

29. (a) The correlation of $r = 0.81$ indicates that the linear relationship between x and y is fairly strong and positive. (b) Because both z_x and z_y are large and positive for Point A, it greatly increases the positive correlation because the product of its z-scores is much greater than that of the other points in the scatterplot. (c) Because z_x is 0 for Point B, its product of z-scores is 0. Therefore, it contributes 0 to the numerator and adds 1 to the denominator. So, it makes the correlation slightly smaller.

31. A

33. The distribution of weight is skewed to the right and unimodal with several possible high outliers. The median weight is 5.4 mg and the *IQR* is 5.5 mg.

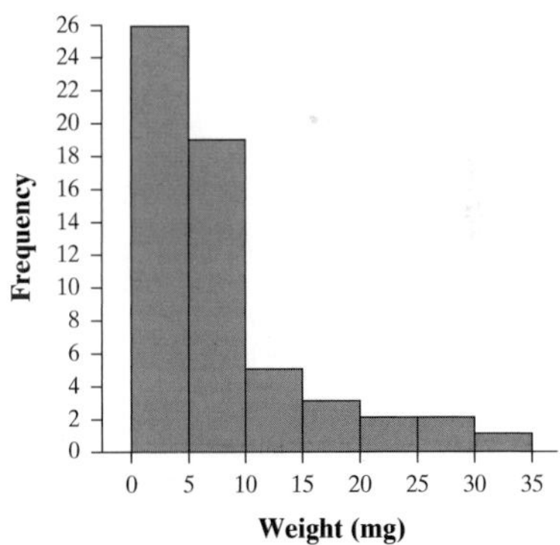

Section 2C

Answers to Check Your Understanding

page 187: 1. $\hat{y} = 177.4 - 1.366(75) = 74.95$

2. residual $= 58.1 - 74.95 = -16.85$. *Interpretation:* The free-throw percentage for the 75-inch-tall player was 16.85 percentage points below the free-throw percentage predicted by the regression line with $x = 75$.

3. The slope is −1.366. *Interpretation:* The predicted free-throw percentage decreases by 1.366 for each additional 1 inch of height.

4. The y intercept does not have meaning in this case, as it is impossible for a WNBA player to be 0 inches tall.

page 200: 1. predicted wait time = 33.347 + 13.285(duration)

2. Although there are two clusters in the residual plot, there is no leftover curved pattern, so the least-squares regression line is an appropriate model for the relationship between duration and wait time.

3. $r^2 = 85.4\%$ of the variation in wait time is accounted for by the least-squares regression line with x = duration.
4. $s = 6.493$ minutes. The actual wait time is typically about 6.493 minutes away from the wait time predicted by the least-squares regression line with x = duration (minutes).
5. $r = +\sqrt{0.854} = 0.924$. The relationship between duration and wait time is positive, so the correlation is positive 0.924.

Answers to Odd-Numbered Section 2C Exercises

1. **(a)** $\hat{y} = 244.7 + 54.9(4) = 464.3$ grams per mile **(b)** $\hat{y} = 244.7 + 54.9(10) = 793.7$ grams per mile **(c)** The prediction for the 4-liter engine is believable because 4 liters is within the interval of x-values used to create the model. However, the prediction for the 10-liter engine is not trustworthy because 10 liters is far outside of the x-values used to create the regression line. The linear form may not continue beyond 7 liters.
3. $\hat{y} = 244.7 + 54.9(5.7) = 557.63$ grams per mile. Residual $= 618 - 557.63 = 60.37$ grams per mile. *Interpretation:* The CO_2 emissions for the 2021 Toyota Sequoia was 60.37 grams per mile greater than the amount of CO_2 emissions predicted by the regression line with x = engine size.
5. $\hat{y} = 81.9886 - 0.0392(241.6) = 72.518$ strokes; residual $= 69.3 - 72.518 = -3.218$ strokes. *Interpretation:* Inbee Park's scoring average was 3.218 strokes fewer (better) than the scoring average predicted by the regression line with x = average driving distance.
7. **(a)** The slope is 54.9. *Interpretation:* The predicted CO_2 emissions increase by 54.9 grams per mile for each additional 1 liter in engine size. **(b)** The y intercept is 244.7. The y intercept does not have meaning in this case, as it is impossible to have an engine size of 0 liters.
9. I expect an LPGA golfer's scoring average to improve (decrease) by about $(0.0392)(20) = 0.784$ strokes if she increased her average driving distance by 20 yards.
11. **(a)** The slopes of the regression lines are nearly equal. The y intercept for the regression line for female butterflies is 0.94 greater than the y intercept for male butterflies. **(b)** We expect the wing length for a female butterfly to be about 0.94 mm greater than that of a male butterfly for any temperature because the regression lines are nearly parallel. To be precise, however, at 3°C, the female butterfly wing is predicted to be $(19.34 - 0.239(3)) - (18.40 - 0.231(3)) = 0.916$ mm greater.

13. *Explanatory:* x = footprint length. *Response:* y = total length. *Slope:* $b = 0.445\left(\frac{4.3}{0.4}\right) = 4.784$. *y intercept:* $87.1 = a + 4.784(6.8) \rightarrow a = 54.569$. *Regression equation:* $\hat{y} = 54.569 + 4.784x$.
15. **(a)** $\hat{y} = x$, where $\hat{y}$ = predicted grade on final and x = grade on midterm. **(b)** $x = 50$: $\hat{y} = 46.6 + 0.41(50) = 67.1$. $x = 100$: $\hat{y} = 46.6 + 0.41(100) = 87.6$ **(c)** The student who did poorly on the midterm (50) is predicted to do better on the final (closer to the mean) while the student who did very well on the midterm (100) is predicted to do worse on the final (closer to the mean).
17. $\hat{y} = 1.0021 + 0.0708x$, where x = the number of Mentos and y = the amount expelled.
19. predicted calories $= 300.040 + 2.829$(sugar)
21. The linear model relating age and weight is not appropriate because there is a negative-positive-negative ∩-shaped pattern left over in the residual plot.
23. Predicted mean weight: $\hat{y} = 4.88 + 0.267(1) = 5.147$ kg. From the residual plot, the mean weight of 1-month-old infants is about 0.85 kg less than predicted. Actual mean weight $= 5.147 - 0.85 = 4.297$ kg.

25.

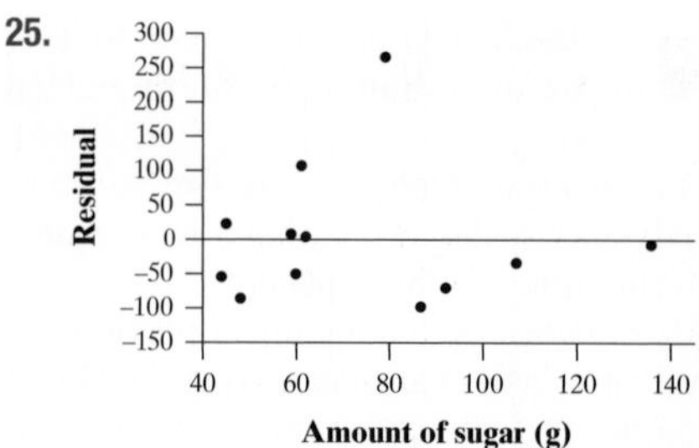

The linear model relating the amount of sugar to the number of calories is appropriate because there is nothing but random scatter in the residual plot.
27. **(a)** The linear model relating DBH to height is appropriate because there is nothing but random scatter in the residual plot. **(b)** $r^2 = 81.2\%$. About 81.2% of the variation in height is accounted for by the least-squares regression line with x = DBH. **(c)** $s = 9.42788$ feet. The actual height is typically about 9.42788 feet away from the height predicted by the least-squares regression line with x = DBH.
29. **(a)** $b = 0.0708$. The predicted amount expelled increases by 0.0708 cup for each additional Mento placed inside the bottle. **(b)** $r^2 = 60.21\%$. About 60.21% of the variation in amount expelled is accounted for by the least-squares regression line with x = number of Mentos. **(c)** $s = 0.06724$ cup. The actual amount expelled is typically about 0.067 cup away from the amount predicted by the least-squares regression line with x = number of Mentos. **(d)** The value of r^2 would not change because r^2 is not affected by the units of the explanatory and response variables. The value of s, in ounces, would be $0.06724(8) = 0.53792$ ounce.
31. **(a)** $r^2 = 0.428475$ and the slope is negative, so $r = -\sqrt{0.428475} = -0.655$. **(b)** $\hat{y} = 12.1526 - 0.1935(12.84) = 9.668$ mg/l; residual $= 10.33 - 9.668 = 0.662$ mg/l. *Interpretation:* The actual dissolved oxygen concentration was 0.662 mg/l greater than the amount predicted by the least-squares regression line with x = temperature. **(c)** $s = 1.15233$ mg/l. The actual dissolved oxygen concentration is typically about 1.15233 mg/l away from the amount predicted by the least-squares regression line with x = temperature. **(d)** $r^2 = 42.8475\%$ of the variation in dissolved oxygen concentration is accounted for by the least-squares regression line with x = temperature.

33. **(a)** Given that $s = 0.9715$ and $n = 15$, $0.9715 = \sqrt{\frac{\sum(y_i - \hat{y}_i)^2}{13}}$. Therefore, $12.2696 = \sum(y_i - \hat{y}_i)^2$. **(b)** Given that $s_y = 1.702$ and $n = 15$, $1.702 = \sqrt{\frac{\sum(y_i - \bar{y})^2}{14}}$. Therefore, $40.5553 = \sum(y_i - \bar{y})^2$.
(c) $r^2 = \frac{\sum(y_i - \bar{y})^2 - \sum(y_i - \hat{y}_i)^2}{\sum(y_i - \bar{y})^2} = \frac{40.5553 - 12.2696}{40.5553} = 0.697$
35. **(a)** Actual $y = 19.1$; predicted $y = 18.4042 - 0.2350(0.9) + 0.9312(1) = 19.124$; residual $= 19.1 - 19.124 = -0.024$. The actual wing length of this butterfly is 0.024 mm less than the wing length predicted by the multiple regression model with x_1 = average temperature and x_2 = sex. **(b)** For male butterflies: $\hat{y} = 18.4042 - 0.2350x$, where x = average temperature and y = wing length. **(c)** For female butterflies: $\hat{y} = 18.4042 - 0.2350x + 0.9312(1) = 19.3354 - 0.2350x$.
(d) Both models have the same slope, but they have different y intercepts. The females tend to have wing lengths that are about 0.93 mm longer than male butterflies.
37. A
39. C
41. A
43. C

company would have to visit individual homes all over the rural subdivision. With the cluster sampling method, the company would have to visit only 5 locations.

11. **(a)** To obtain an SRS, every tree would need to be identified and numbered. It is not practical to number every tree along the highway and then search for the trees that are selected. **(b)** This convenience sampling method is not a good idea because these trees are unlikely to be representative of the population. Trees closest to the entrance are more likely to be damaged by cars and people, and may be more susceptible to infestation. Another thing to consider is that beetles may be disturbed by cars and people. If so, then maybe the trees near the entrance will be less infested than other trees. **(c)** There are 5000 pine trees that are along Highway 34 and our desired sample size is 200, so we should select every $5000/200 = 25$th pine tree. For simplicity, we could walk down Highway 34 and select every 25th pine tree that we pass. To choose a starting point, we randomly select a number from 1 to 25. We select that pine tree and every 25th pine tree thereafter until 200 pine trees have been selected.

13. **(a)** We should stratify by floor and view because satisfaction with the property is likely to vary depending on the location of the room. For example, one stratum would be rooms on the first floor with a water view and another stratum would be rooms on the 30th floor with a golf course view. For each floor/view combination, give each room a distinct integer label from 1 to 20. Use a random number generator to select 2 different integers from 1 to 20 and survey the guest in the corresponding rooms. Using a stratified random sample would ensure that the manager got opinions from each type of room and would provide a more precise estimate of customer satisfaction than an SRS of the same size. **(b)** Give each floor a distinct integer label from 1 to 40 and use a random number generator to select 3 different integers from 1 to 40. The guests in each of the rooms on the corresponding floors should be surveyed. This would be a simpler option than selecting an SRS because the manager would need to survey guests on only three floors instead of surveying guests all over the hotel. **(c)** There are 1200 rooms, and our desired sample size is 120, so we should select every $1200/120 = 10$th room. To choose a starting point, we randomly select a number from 1 to 10. We select that room and every 10th room thereafter by room number until 120 rooms have been selected. This method is easier to implement than an SRS. This method also ensures that the sample is selected uniformly across all floors of the building.

15. **(a)** It would be better to start in the upper left and move across the rows. The end result of a systematic random sample would be that all squares in the same column will be selected. Each column serves as a small-scale replica of the population because it includes a square at each distance from the water. It is better to end up with all squares in a single column than to end up with all squares in a single row, which is what would happen by moving down the columns. **(b)** The columns. Each column serves as a small-scale replica of the population, whereas the rows are systematically different due to their proximity to the water. Therefore, the columns would be better to use as clusters than the rows.

17. This is a convenience sample. It is unlikely that the first 100 students to arrive at school are representative of the student population in general. The estimate of 7.2 hours is probably less than the true average because students who arrive first to school had to wake up earlier and may have gotten less sleep than those students who are able to sleep in.

19. This is a voluntary response sample. It is likely that those customers who volunteered to leave reviews feel strongly about the hotel, often due to a negative experience. Customers who had an average or high-quality experience are less likely to leave a review. As a result, the 26% from the sample is likely greater than the true percentage of all the hotel's customers who would give the hotel 1 star.

21. Comments to senators are an example of a voluntary response sample—the proportion of comments opposed to the bill should not be assumed to be a fair representation of the attitudes of all constituents. Only those who have very strong opinions will comment. It is likely that the true proportion of all constituents who oppose the bill is less than 871/1128.

23. **(a)** Undercoverage might lead to bias in this study because the students who do not live in the dorms cannot be part of the sample. Some of these students would live off campus, and would therefore be less likely to eat on campus than those students who live in the dorm—meaning that the director's estimate for the percentage of students who eat regularly on campus will likely be too high. **(b)** No, increasing the sample size while using the same flawed method would only result in a larger flawed sample.

25. Nonresponse might lead to bias in this study because the people who have long commutes are less likely to answer their phone (because they are driving), so they are less likely to be included in the sample. This will likely produce an estimate that is too small, as the people who are more likely to answer the phone may have shorter commutes (or none at all).

27. When asked in person, the mall shoppers might be embarrassed or ashamed of their weight and lie by giving a weight less than their true weight. Mall shoppers given an anonymous survey are more likely to be honest about their weight.

29. When asked about their finances over the phone, people may be inclined to lie and exaggerate their financial status because they are embarrassed or ashamed to admit to a live person that their financial status is poor. In the online survey, people may be more likely to be honest about their financial status.

31. **(a)** By pointing out the huge national deficit, U.S. adults might become concerned about the government spending additional money on social programs. The percentage from the sample who say "yes" is likely less than the true percentage of U.S. adults who favor spending money to establish a national system of health insurance. **(b)** Answers will vary. Unbiased: Should the government establish a national system of health insurance? Biased: A national system of health insurance would help the poor and would help many children get the life-saving healthcare they need. Do you support the establishment of a national system of health insurance?

33. **(a)** Undercoverage might have led to bias in this survey because people who did not have a subscription to *Literary Digest* and those who did not own an automobile or telephone were not sent a ballot. Those who were not sent ballots may have been more likely to support the Democratic candidate (Roosevelt), but were underrepresented in the sample, making the estimate of the proportion who supported Landon too high. **(b)** Nonresponse might have led to bias in this survey because only 2,400,000 of the ballots were returned, which means at least 7,600,000 ballots were not returned. Those who did not return their ballots may have been more likely to support the Democratic candidate (Roosevelt), but were underrepresented in the sample, making the estimate of the proportion who supported Landon too high. **(c)** Following up would have helped to eliminate the bias due to nonresponse, as a greater percentage of those selected for the sample would have to be contacted and surveyed. However, following up with people who didn't return their ballots would not have eliminated the bias due to undercoverage, as those people without an automobile, phone, or subscription to *Literary Digest* would still have been left out of the survey results.

35. D

37. E

39. **(a)** Observational study. There were no treatments imposed on the older people. In other words, the older people weren't assigned to drink a certain amount of coffee. **(b)** The largest population to which we can generalize the results of the study is all older people like the ones in the study from these eight states. **(c)** No. This is an observational study, not an experiment. The researchers did not assign the older people to drink coffee or not, so we cannot conclude that drinking coffee causes people to live longer.

Section 3C

Answers to Check Your Understanding

page 274: **1.** No, it is possible that people who are heavy social media users are more dissatisfied with their own lives and that people dissatisfied with their own lives are more likely to consider leaving their spouse. If both of these are true, then we would see a relationship between social media use and consideration of leaving a spouse, even if social media use has no effect on whether a person considers leaving their spouse. This also could be a case of "reverse causation"—those unhappy with their spouse might want to spend more time online.

2. *Explanatory variable:* lipstick color. *Response variable:* daily tips earned. *Experimental units:* work days. *Treatments:* red or neutral lipstick.

page 282: **1.** It is important to have a control group that did not get the display or chart for comparison purposes. A control group would show how much electricity customers would tend to use naturally. This would serve as a baseline to determine how much less electricity is used in each of the other treatment groups.

2. The random assignment can be done by using 60 identical slips of paper, using technology, or using Table D. The simplest is to use 60 identical slips of paper. Number the houses 1 to 60. Write the numbers 1 to 60 on the slips of paper. Shuffle well. Draw out 20 slips of paper (without replacement). Those households with the corresponding numbers will receive a display. Draw out another 20 slips of paper (without replacement). Those corresponding households will receive a chart. The remaining 20 households will receive only information about energy consumption.

3. The purpose of randomly assigning treatments is to create groups of households that are roughly equivalent at the beginning of the experiment in terms of typical electricity usage. This will ensure that the effects of other variables (the thrifty inclination of some households, usage changes due to work responsibilities, or other family changes that affect energy usage) are spread evenly among the three groups, allowing the researchers to make cause-and-effect conclusions.

4. One variable that researchers kept the same for all experimental units was that the customers were all from the same small city and all participated in the study during the same month (April). This is important because it helps to reduce variation in electricity use (the response variable).

page 287: **1.** Number the volunteers 1 to 300. Use a random number generator to produce 100 different random integers from 1 to 300 and show the first advertisement to the volunteers with those numbers. Use a random number generator to produce 100 additional different random integers from 1 to 300, skipping any numbers selected in the first set and show the second advertisement to the volunteers with those numbers. The remaining 100 volunteers will view the third advertisement. Measure the response for each person. Then compare the results for the three advertisements.

2. Answers may vary regarding suggested blocking variables. The advertisements may impact the volunteers differently, depending on whether or not they know who Jane Austen is. To control this influence, block by whether or not the individuals are familiar with the works of Jane Austen. Those who are not familiar with her works would form block 1 and those who are familiar with her works would form block 2. The volunteers in each block would be numbered and randomly assigned to one of three treatment groups: One-third of the volunteers in each block would view the first advertisement, one-third would view the second advertisement, and one-third would view the third advertisement. Measure the response for each person. Then compare the results of the three advertisements for the two blocks.

3. A randomized block design accounts for the variability in effectiveness that is due to subjects' familiarity with the works of Jane Austen. This makes it easier to determine the effectiveness of the three different advertisements.

page 292: **1.** This is an observational study, not an experiment. The individuals were not randomly assigned to continue to play or to stop playing, so we cannot make a cause-and-effect conclusion based on this study.

2. The individuals in this study were not randomly selected, so the results of this study can only be applied to athletes like the ones in this study.

Answers to Odd-Numbered Section 3C Exercises

1. It is possible that people who frequently eat seafood also exercise frequently and that exercising frequently reduces the risk of colon cancer. If both of these are true, then we would see a relationship between eating seafood frequently and a lower risk of colon cancer, even if eating seafood frequently has no effect on the risk of colon cancer.

3. No. It is possible that the students who ate breakfast in the morning were more conscientious and that conscientious students tend to score better on the national exam. If both of these are true, then we would see a relationship between eating breakfast and the score on the national exam even if eating breakfast has no effect on the national exam score.

5. **(a)** *Experimental units:* households in Atlanta. **(b)** *Factors:* length of survey (3 levels), survey incentive (2 levels), number of follow-ups (3 levels), and incentive for participating in a 5-night in-home sleep study (3 levels). **(c)** There are $3 \times 2 \times 3 \times 3 = 54$ possible treatment combinations. **(d)** *Possible treatment:* 11 questions, \$2 cash, 0 follow-ups, and \$100 sleep study incentive. *Another possible treatment:* 11 questions, \$2 cash, 0 follow-ups, and \$150 sleep study incentive.

7. *Explanatory variable:* light intensity. *Response variable:* weight. *Experimental units:* pine seedlings. *Treatments:* full light, 25% light and 5% light.

9. A control group would show what percentage of the time the shark would eat the bait from a surfboard with no deterrent. This would serve as a baseline to determine how much less frequently the shark ate the bait for each of the deterrents.

11. There was no control group for comparison purposes. We do not know if the improvement was due to the placebo effect or if the flavonols actually affected the blood flow. To make a cause-and-effect conclusion possible, we need to randomly assign some subjects to get flavonols and others to get a placebo.

13. **(a)** Yes. This experiment could be double-blind if the treatment (ASU or placebo) assigned to a subject was unknown to both the subject and to those responsible for assessing the effectiveness of that treatment. **(b)** It is important for the subjects to be blind because if they knew they were receiving the placebo, their expectations would differ from those who received the ASU. Then it would be impossible to know if an improvement in pain was due to the difference in expectations or the ASU. It is important for the experimenters to be blind so that they will be unbiased in the way that they interact and assess the subjects.

15. The placebo effect describes the fact that some subjects in an experiment will respond favorably to any treatment, even an inactive treatment. In this experiment, both treatment groups experienced pain relief, even though neither group was receiving an actual medication. Furthermore, because people generally expect that more expensive products work better, a greater proportion of people in the "expensive" group experienced pain relief.

17. The experimenter knew which subjects had learned the meditation techniques, so the experimenter is not blind. The experimenter may have some expectations about the outcome of the experiment. If the experimenter believes the meditation is beneficial, the experimenter may subconsciously rate subjects in the meditation group as being less anxious.

19. **(a)** Label each customer with a different integer from 1 to 139. Then, randomly generate 70 different integers from 1 to 139 using a random number generator. The customers with these labels will pay \$4 for the buffet. The remaining 69 customers will pay \$8 for the buffet. **(b)** The purpose of random assignment is to avoid confounding by creating two groups of customers that are roughly equivalent at the beginning of the experiment in terms of their taste in pizza. This allows the researchers to make cause-and-effect conclusions.

21. Label each subject with a different integer from 1 to 120. Then randomly generate 40 different integers from 1 to 120 using a random number generator. The subjects with these labels will be given Treatment 1. Next, randomly generate 40 additional different integers from 1 to 120, being careful to not choose subjects who have already been selected for Treatment 1. The subjects with these labels will be given Treatment 2. The remaining 40 subjects will be given Treatment 3.

23. There was no random assignment of treatments. When doctors were choosing the method of treatment for their patients, it is possible that the patients in the worst initial condition were more often given the new treatment. Patients in the worst initial condition are more likely to die within 8 years, so we wouldn't know if any difference in death rates was due to initial condition of the patient or the method of treatment.

25. **(a)** The details of the phone call to the Congress member were kept the same. **(b)** If one of the treatment groups was allowed to discuss different details during the phone call, we wouldn't know if the details of the phone call or the caller's designation (campaign donor or constituent) was the cause of a difference in success rate in obtaining meetings.

27. **(a)** All infants were born prematurely and weighed less than 3 pounds. **(b)** It is beneficial to control these variables, because otherwise they would provide additional sources of variability that would make it harder to determine the effectiveness of EPO and darbepoetin. A drawback is that the results of this study can only be applied to babies like those in the study, so the scope of inference is limited.

29. Using 238 identical slips of paper, write the name of each subject on one slip. Then, put them in a container and mix thoroughly. Pull out 80 slips of paper without replacement and assign the corresponding subjects to Treatment 1, \$2000 offer. Then, pull out 79 more slips of paper without replacement and assign the corresponding subjects to Treatment 2, \$1865 offer. The remaining 79 subjects are assigned to Treatment 3, \$2135 offer. Administer the treatments. Calculate the deviation from the initial offer for each person. Then compare the results for the three initial offers.

31. **(a)** Form blocks based on the age of the trees (older, younger) because it is reasonable to think that there is an association between the age of the tree and number of oranges produced. That is, younger trees are likely to have fewer oranges than older trees. Give each of the younger trees a distinct integer label from 1 to 60. Use a random number generator to select 20 different integers from 1 to 60 and assign the corresponding trees to Fertilizer A. Then use the random number generator to select an additional 20 different numbers from 1 to 60 and assign the corresponding trees to Fertilizer B. The remaining 20 trees get Fertilizer C. Do the same for the 30 older trees. Count the number of oranges on each tree. Then compare the results of Fertilizers A, B, and C for the two ages of trees. **(b)** In a completely randomized design, the differences in age will increase the variability in number of oranges for each treatment. However, a randomized block design would help us account for the variability in number of oranges that is due to the differences in age. This will make it easier to determine if one fertilizer is better than the others.

33. **(a)** The blocks are the different diagnoses (e.g., asthma) because the treatments (doctor or nurse-practitioner) were assigned to patients within each diagnosis. In general, blocks are formed by creating groups of experimental units that are similar to each other but different from the units in other blocks. **(b)** A randomized block design is preferable because it allows us to account for the variability in health and satisfaction due to differences in diagnosis by comparing the results within each block. For example, patients with asthma might be healthier initially than patients with diabetes and be healthier after 6 months as well. In a completely randomized design, the variability in health and satisfaction due to differences in diagnoses will be unaccounted for and will make it harder to determine if there is a difference in health and satisfaction due to the difference between doctors and nurse-practitioners.

35. Create pairs of similar patients based on their overall heart health (the two patients with the best overall heart health are paired together, the two patients with the next best overall heart health are paired together, etc.). Within each pair, one patient is assigned heads, the other tails. When a coin is tossed, the patient corresponding to the result is assigned to the noninvasive method and the other is assigned to open heart surgery. For each patient, record the percentage of blood that flows backward after receiving the treatment. Then compare the percentages with the noninvasive and surgical methods for the 20 pairs of patients.

37. **(a)** This was a matched pairs design because each volunteer was assigned both treatments. **(b)** In a completely randomized design, the differences between the volunteers will increase variability in blood vessel function, making it harder to detect if there is a difference caused by the chocolate. In a matched pairs design, each volunteer is compared with themselves, so the differences between volunteers are accounted for. **(c)** If everyone has the fake chocolate first and the bittersweet chocolate second, the type of chocolate will be confounded with other variables that vary between the two days. For example, if the weather is warmer on one of the days, we will not know if the difference in blood vessel function is due to the difference in chocolate or the difference in weather. By randomizing the order, some volunteers will have the bittersweet chocolate on the warmer day and some will have it on the colder day.

39. **(a)** Difference (A − B) in mean number of days $= 4.68 - 4.21 = 0.47$ days. **(b)** When we assumed that the type of clipboard does not matter, there was one simulated random assignment where the difference (A − B) in the mean number of days for the two groups was 0.72. **(c)** Because a difference of means of 0.47 or greater occurred 16 out of 100 times in the simulation, the difference is not statistically significant. It is plausible to get a difference this big simply due to chance variation in the random assignment.

41. **(a)** The researchers randomly assigned the subjects to the four treatments to avoid confounding by creating four groups of physicians that are roughly equivalent at the beginning of the experiment in terms of their risk of having a heart attack or developing cancer. **(b)** The researchers observed a difference in the proportions who had a heart attack among those who took aspirin and those who received a placebo that was so large that it is unlikely to be explained by chance variation in the random assignment. **(c)** The difference in the proportions who developed cancer among those who received beta-carotene and those who received the placebo could be plausibly attributed to chance alone.

43. **(a)** Yes. The children were randomly assigned to the treatments (foster care or institutional care) and the results were statistically significant, so there is convincing evidence that the difference between foster care or institutional care caused the difference in response. **(b)** The 136 children were not randomly selected, but rather were abandoned at birth and living in orphanages in Bucharest, Romania, so the results can be applied only to children like the ones in the study.

45. The students were randomly assigned to the treatment groups (win, loss) and the results were statistically significant, so there is convincing evidence that the fact that the team won or lost caused the students to identify themselves with the team (or not). The students were randomly selected from the population of all students who are fans of the team at the major university, so we can generalize the results to all students who are fans of the team at this university. However, these results may not apply to students at other universities or for other sports teams at this university.

47. Informed consent. The subjects in this study weren't informed about the true nature of the study, their diagnosis, and the planned use of placebo treatments. If they were informed of these facts, they most certainly would not have consented to be in the study.

49. In this case, the subjects were not able to give informed consent. They did not know what was happening to them and they were not old enough to understand the ramifications in any event.

51. **(a)** Difference (Active − Inactive) in mean improvement $= 5.2414 - 1.0952 = 4.1462$. **(b)** $0/100 = 0\%$ of simulated experiments gave a difference at least as large as 4.1462.

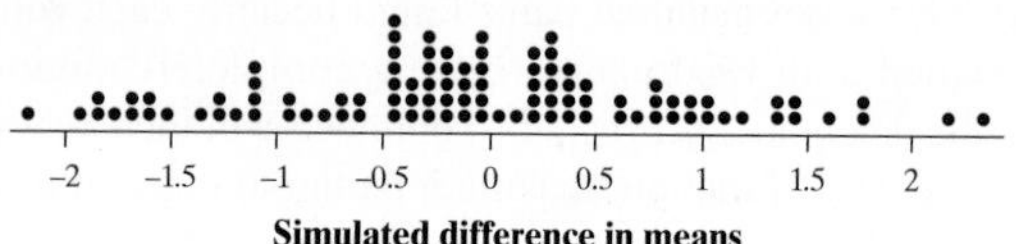

(c) Yes. A difference in means of 4.1462 or greater never occurred in the simulation ($0/100 = 0\%$), so the difference is statistically significant. It is unlikely to get a difference (Active − Inactive) in mean improvement this big simply due to chance variation in the random assignment.

53. C

55. D

57. C

59. B

61. D

63. The distribution of average monthly rainfall for Tucson is roughly symmetric with one high outlier. The distribution of average monthly rainfall for Princeton is skewed left with no outliers. The center of the distribution of average monthly rainfall for Tucson, Arizona, was considerably less (Median ≈ 0.95 inch) than for Princeton, New Jersey (Median ≈ 4.15 inches). The distribution of average monthly rainfall for Tucson has less variability ($IQR \approx 0.6$ inch) than the average monthly rainfall for Princeton ($IQR \approx 0.9$ inch).

Answers to Unit 3 Review Exercises

R1. **(a)** This was an observational study because the men weren't assigned to drink a specified amount of orange juice. **(b)** The men were not randomly selected, so the largest population to which we can generalize the results of this study is all men like the 27,000 in this study. **(c)** It is possible that men who drink more orange juice had also followed healthier diets, and that following a healthier diet leads to a lower risk of cognitive troubles later in life. If both of these are true, then we would see a relationship between orange juice consumption and lower risk of cognitive troubles later in life even if drinking orange juice has no effect on the risk of developing cognitive troubles later in life.

R2. **(a)** Answers will vary. One possible answer: Announce over the school speaker that there is a survey available in the main office for students who want to give their opinion about the amount of parking available. Because people who feel strongly (and often negatively) about an issue are more likely to respond to a voluntary response survey, the proportion in the sample who are satisfied with the amount of parking is likely less than the proportion of all students who are satisfied. **(b)** Answers will vary. One possible answer: Interview the first 25 students who park in the lot one day. Because the first students to park will have no trouble finding space to park, the proportion in the sample who are satisfied with the amount of parking is likely greater than the proportion of all students that are satisfied. **(c)** Label each student with a different integer from 1 to 1800. Randomly generate 50 different integers from 1 to 1800 using a random number generator. The students with these labels will form the SRS of 50 students from the school. An SRS reduces bias by selecting the sample in such a way that every group of 50 individuals in the population has an equal chance to be selected as the sample, making the sample selected much more likely to be representative of the population. **(d)** Because there are 1800 students at this high school and our desired sample size is 50, we would like to select every $1800/50 = 36$th student. Assuming that students enter the building at one location, we could select every 36th student that entered the school. To do this, we would select a random number between 1 and 36. That student would be selected for the sample along with every 36th student thereafter. The advantage of a systematic random sample over an SRS is that a systematic random sample selects the students on the spot as they enter the building, which is much easier. In an SRS, we would have to track down students all over campus to learn about student opinion regarding parking for students.

R3. (a) It might be difficult to give a survey to an SRS of fans because you would have to identify 10% of the seats, go to those seats in the arena, and find the people who are sitting there. This means going to many different locations throughout the arena, which would take time. (b) For a stratified sample, it is best to create strata where the people within a stratum are very similar to each other with regard to the variables being measured, but different from the people in other strata. In this case, it would be better to take the lettered rows as the strata because each lettered row is the same distance from the court and so would contain only seats with the same (or nearly the same) ticket price, and therefore people with similar financial status. (c) For a cluster sample, use the numbered sections as clusters because the people in a particular numbered section are in roughly the same location, making it easy to administer the survey. Furthermore, because each cluster contains seats of all different prices, the people in each cluster reflect the variability in financial status found in the population, which is ideal. To select the sample, use a random number generator to select 5 different integers from 101 to 150 and survey every fan in the corresponding 5 sections.

R4. (a) When the interviewer provides the additional information that "box-office revenues are at an all-time high," the listeners may believe that they contributed to this fact and be more likely to overestimate the number of movies they have seen in the past 12 months. A change that would fix the problem is to eliminate this sentence. (b) A sample that uses only phone numbers is likely to underrepresent low-income people who cannot afford a phone. Low-income people also likely go to movies less often, so leaving these people out of the sample will produce an estimated mean that is larger than that which would be produced by the population of all adults. (c) People who do not go to the movie theater very often may be less likely to respond because they are not interested in surveys about movies. If people who do not go to the movies are less likely to respond, the estimated mean will be larger than that which would be produced by the population of all adults.

R5. (a) The experimental units are the potatoes used in the experiment. The factors are the storage method (3 levels) and time from slicing until cooking (2 levels). There are six treatments: (1) fresh picked and cooked immediately, (2) fresh picked and cooked after an hour, (3) stored at room temperature and cooked immediately, (4) stored at room temperature and cooked after an hour, (5) stored in refrigerator and cooked immediately, (6) stored in refrigerator and cooked after an hour. The response variables are the rating of color and the rating of flavor. (b) Using 300 identical slips of paper, write "1" on 50 of them, write "2" on 50 of them, and so on. Put the papers in a hat and mix well. Then, select a potato and select a slip from the hat to determine which treatment that potato will receive. Repeat this process for the remaining 299 potatoes, making sure not to replace the slips of paper into the hat. After applying the treatments, measure the color and flavor of the french fries from each potato. Then compare the results for the 6 treatments. (c) A benefit to using 300 potatoes available from a single supplier is that the quality of the potatoes should be fairly consistent, reducing a source of variability. This makes it easier to estimate how the treatments affect color and flavor of the french fries. A drawback of using potatoes from only one supplier is that the results of the experiment could then be applied only to potatoes that come from that one supplier rather than to potatoes in general. (d) It would be best to use a randomized block design with the suppliers as the blocks. For each supplier, randomly assign the potatoes to the 6 treatments. After applying the treatments, measure the color and flavor of the french fries from each potato. Then compare the results of the 6 treatments for the different suppliers. Doing so would allow the researchers to account for the variability in color and flavor due to differences in the initial quality of the potatoes from different suppliers, making it easier to estimate how the treatments affect color and flavor of the french fries.

R6. (a) If all the patients received the St. John's wort, the researchers would not know if any improvement was due to the St. John's wort or to the expectations of the subjects (the placebo effect). By giving some patients a treatment that should have no effect at all (control group) but that looks, tastes, and feels like the St. John's wort, the researchers can account for the placebo effect by comparing the results for the two groups. (b) The purpose of random assignment is to create two groups of subjects that are roughly equivalent at the beginning of the experiment in terms of their depression and likelihood of improving. This allows the researchers to make cause-and-effect conclusions. (c) The subjects should not know which treatment they are getting, so they will all have the same expectations. Also, the researchers should be unaware of which subjects received which treatment, so they cannot consciously (or subconsciously) influence how the depression symptoms are measured. (d) In this context, "not statistically significant" means that the difference in improvement between the St. John's wort and placebo groups was not large enough to rule out random chance as the explanation. In other words, it is plausible that the difference was solely due to the variability caused by the random assignment to treatments.

R7. (a) No, the study does not allow for inference about a population. The 1000 students were not randomly selected from any larger population, so we should apply the results only to students like those in the study. (b) Yes, the study does allow for inference about cause and effect. The students were randomly assigned to the three treatments, so there is convincing evidence that the reduction in cold symptoms was caused by the masks, assuming that the results were statistically significant.

R8. (a) Give each person both treatments in random order. It is reasonable to think that some people are more focused than others, so using each person as their own "pair" accounts for the variation in completion time in the subjects. For each person, flip a coin. If the coin lands on heads, that person will complete a word search puzzle with the researchers standing close by first and will complete a word search puzzle with the researchers standing at a distance second. If the coin lands on tails, the treatment order will be reversed. For each person, record the time it took to complete the puzzle with each treatment. Then, compare the times with the researchers close by and the researchers farther away for the 30 people. (b) Matching helps account for a source of variability in completion time, such as skill at completing word searches. Comparing each volunteer to themselves makes it easier to tell if the researcher's proximity causes a difference in completion time.

Answers to Unit 3 AP® Statistics Practice Test

T1. C **T2.** A **T3.** D **T4.** D **T5.** D
T6. A **T7.** E **T8.** D **T9.** B **T10.** E

T11. (a) The experimental units are the acacia trees. (b) It is beneficial to include some trees that have no hives, as a control group, to provide a baseline for comparing the effects of other treatments. The control group allows the researchers to measure the effect of active hives and empty hives on tree damage compared to no hives at all. **(c)** Assign each tree a different integer from 1 to 72. Then randomly generate 24 different integers from 1 to 72 using a random number generator. The trees with these labels will get the active beehives. Next, randomly generate 24 additional different integers from 1 to 72, being careful to not choose trees that have already been selected. The trees with these labels will get the empty beehives. The remaining 24 trees will remain empty. Measure the damage caused to each tree. Then compare the amount of damage to the trees with active beehives, those with empty beehives, and those with no beehives.

T12. **(a)** By controlling the diets of the subjects during the experiment, the researchers reduce variability in the response variable, which is performance on the tapping test. This makes it easier to determine the effect of caffeine withdrawal. **(b)** Each of the 11 individuals will be a block in a matched pairs design. Each participant will take the caffeine tablets on one of the two-day sessions and the placebo on the other. The blocking was done to account for individual differences in dexterity. This makes it easier to determine the effect of caffeine withdrawal. **(c)** The order was randomized to control for any possible influence of the order in which the treatments were administered on the subject's tapping speed. For example, after the first trial, subjects might practice the tapping task and do better the second time. If all the subjects got caffeine the second time, the researchers would not know if the increase was due to the practice or the caffeine, preventing a cause-and-effect conclusion. **(d)** It is possible to carry out this experiment in a double-blind manner. This means that neither the subjects nor the people who come in contact with them during the experiment (including those who record the number of taps) have knowledge of the order in which the caffeine or placebo was administered.

T13. **(a)** The proportion of graduates who are enrolled in a four-year college or university may differ according to those who were in the top half of the graduating class versus the bottom half of the graduating class, so we will use class rank as strata. Label the top 400 students 1 to 400. Generate 25 different random integers from 1 to 400 using a random number generator and select those students. Label the bottom 400 students 401 to 800. Generate 25 different random integers from 401 to 800 using a random number generator and select those students. **(b)** A benefit of using a stratified random sample is that we are guaranteed to select 25 students from the top half and 25 from the bottom half of the class. This will lead to a more precise estimate of the proportion who are enrolled in a four-year college or university. **(c)** Nonresponse might lead to bias in this study because the people who are not enrolled in a four-year college or university may be less likely to respond to the survey. This will likely produce an estimated proportion that is too large, as the people who responded to the survey probably are more likely to be enrolled in a four-year college or university than those who did not respond.

Answers to Cumulative AP® Practice Test 1

AP1 D **AP2** D **AP3** C **AP4** E **AP5** B
AP6 B **AP7** B **AP8** A **AP9** D **AP10** D
AP11 A **AP12** B **AP13** C **AP14** A **AP15** E

AP16 **(a)** *Shape:* The distribution of gain for subjects using Machine A is roughly symmetric and unimodal, whereas the distribution of gain for subjects using Machine B is skewed to the left (toward the smaller values) and unimodal. *Outliers:* Neither distribution appears to contain any outliers. *Center:* The center of the distribution of gain for subjects using Machine B (median = 38) is greater than the center of the distribution of gain for subjects using Machine A (median = 28). *Variability:* The distribution of gain for subjects using Machine B (range = 57, *IQR* = 22) is more variable than the distribution of gain for subjects using Machine A (range = 32, *IQR* = 15). **(b)** The company should choose Machine B if they want to advertise it as achieving the highest overall gain in cardiovascular fitness. The median gain for Machine B (38) is greater than for Machine A (28), as is the mean ($\bar{x}_B = 35.4$ versus $\bar{x}_A = 28.9$). Overall, the cardiovascular fitness gains using Machine B tend to be higher than those for Machine A. **(c)** The company should choose Machine A if they want to advertise it as achieving the most consistent gains in cardiovascular fitness. Machine A exhibits less variation in gains than does Machine B. The *IQR* for Machine A (15) is less than the *IQR* for Machine B (22). Additionally, the standard deviation for Machine A (9.38) is less than the standard deviation for Machine B (16.19). **(d)** Volunteers were used for the experiment, so the largest population to which we can generalize the results are all people like the 30 volunteers in this study.

AP17 **(a)** *Treatments:* incentive type (cash bonuses or non–cash prizes). *Experimental units:* the retail sales districts. *Response variable:* change in sales volume. **(b)** Number the 60 retail sales districts with a two-digit number from 01 to 60. Using a table of random digits, read two-digit numbers until 30 unique numbers from 01 to 60 have been selected. The 30 districts corresponding to the numbers are assigned to the cash bonus incentives group and the remaining 30 to the non–cash prize incentives group. **(c)** Yes. This is an experiment, not an observational study. The retail sales districts were randomly assigned to the incentive types and the results were statistically significant, so there is convincing evidence that the cash bonus incentive is more effective than the noncash prize incentive.

AP18 **(a)** There is a strong, negative, nonlinear relationship between age and number of car washes. Generally, as the number of years increases, the number of car washes decreases. There is a strong, positive, linear relationship between age and repair cost. Generally, as the number of years increases, the repair cost increases. Neither scatterplot has any unusual points.

(b)

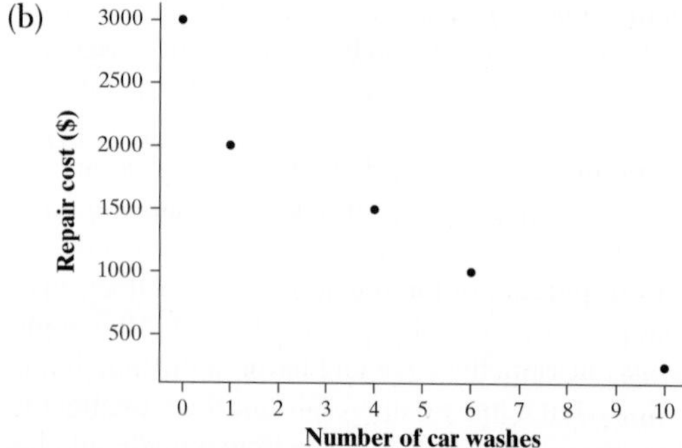

(c) There is a fairly strong, negative, nonlinear relationship between number of car washes and repair cost. Generally, as the number of car washes increases, the repair cost decreases. **(d)** Although the

(c) Answers may vary. Here is a two-way table where A and B are not mutually exclusive and not independent.

		Amount paid		
		Full price	Reduced price	Total
Purchase method	Online	5	5	10
	In person	15	25	40
	Total	20	30	50

The events "paid full price" and "purchased online" are *not* mutually exclusive because they can occur at the same time. P(full price and online) = 5/50 = 0.10 ≠ 0. The events "paid full price" and "purchased online" are *not* independent because P(paid full price) = 20/50 = 0.4 ≠ P(paid full price | purchased online) = 5/10 = 0.5.

41. (a) The probability of dealing a spade changes each time a card is dealt, so use the general multiplication rule. P(5 ♠ in a row) = $\frac{13}{52}\cdot\frac{12}{51}\cdot\frac{11}{50}\cdot\frac{10}{49}\cdot\frac{9}{48}=0.000495$ (b) There are 4 suits in a standard deck of cards, so P(5 cards with the same suit) = 0.000495 + 0.000495 + 0.000495 + 0.000495 = 0.00198.

43. (a) There are 6 ways to get doubles out of 36 possibilities so $P(\text{doubles})=\frac{6}{36}=\frac{1}{6}=0.167$. (b) Because the rolls are independent, we can use the multiplication rule for independent events. P(no doubles on first roll ∩ doubles on second roll) = P(no doubles first). $P(\text{doubles second})=\frac{5}{6}\left(\frac{1}{6}\right)=\frac{5}{36}=0.139$ (c) $P(\text{first doubles on third roll})=\frac{5}{6}\left(\frac{5}{6}\right)\left(\frac{1}{6}\right)=\frac{25}{216}=0.116$ (d) For the first doubles on the hundredth toss, the probability is $\left(\frac{5}{6}\right)^{99}\left(\frac{1}{6}\right)=2.41\cdot10^{-9}$. The probability that the first doubles are rolled on the kth toss is $\left(\frac{5}{6}\right)^{k-1}\left(\frac{1}{6}\right)$.

45. C

47. E

49. E

Answers to Unit 4, Part I Review Exercises

R1. (a) If you take many, many pieces of buttered toast and drop them from a 2.5-foot high table, about 81% of them will land butter side down. (b) No. If four dropped pieces of toast all landed butter side down, it does not make it more likely that the next piece of toast will land with the butter side up. Random behavior is unpredictable in the short run, but it has a regular and predictable pattern in the long run. The value 0.81 describes the proportion of times that toast will land butter side down in a very long series of trials.

R2. (a) Let 1 to 81 = butter side down and 82 to 100 = butter side up. Use a random number generator to generate 10 integers from 1 to 100 (with repeats allowed) to simulate dropping 10 pieces of toast. Record the number of pieces of toast out of 10 that are simulated to land butter side down. Determine if 4 or fewer pieces of toast out of 10 landed butter side down. (b) Assuming the probability that each piece of toast lands butter side down is 0.81, the estimate of the probability that dropping 10 pieces of toast yields 4 or less butter side down is 1/50 = 0.02. (c) Because it is unlikely (probability ≈ 0.02) that 4 or fewer pieces of toast out of 10 would land butter side down by chance alone if the researchers' 0.81 probability is correct, there is convincing evidence that the 0.81 claim is incorrect.

R3. (a) The sample space is: rock/rock, rock/paper, rock/scissors, paper/rock, paper/paper, paper/scissors, scissors/rock, scissors/paper, and scissors/scissors. Because each player is equally likely to choose any of the three hand shapes, each of these 9 outcomes will be equally likely and have probability 1/9. (b) There are three outcomes—rock/scissors, paper/rock, and scissors/paper—where player 1 wins on a single play of the game. P(Player 1 wins on a single play) = 3/9 = 0.333.

R4. (a) The probability that the vehicle is a crossover is 1 − 0.28 − 0.18 − 0.08 − 0.05 = 0.41. The sum of the probabilities must add to 1. (b) P(vehicle is not an SUV or a minivan) = 0.28 + 0.18 + 0.41 = 0.87. (c) $P(\text{pickup truck} \mid \text{not a passenger car})=\frac{0.18}{1-0.28}=\frac{0.18}{0.72}=0.25$.

R5. (a) P(Astrology is not at all scientific) = 539/687 = 0.785 (b) P(Associate's degree or thinks astrology is not at all scientific) = $\frac{234}{687}+\frac{539}{687}-\frac{169}{687}=0.879$ (c) P(Associate's degree | not at all scientific) = 169/539 = 0.314

R6. (a) The events "thick-crust pizza" and "pizza with mushrooms" are not mutually exclusive (disjoint). There are 2 pizzas that have thick-crust and have mushrooms. If mutually exclusive, there would be no pizzas with both thick–crust and mushrooms.
(b)

		Crust		
		Thick	Thin	Total
Mushrooms?	Yes	2	1	3
	No	4	2	6
	Total	6	3	9

P(mushrooms) = 3/9 = 0.333 and P(mushrooms | thick crust) = 2/6 = 0.333. P(mushrooms) = P(mushrooms | thick crust), so the events "mushrooms" and "thick crust" are independent.
(c) P(neither has mushrooms) = P(first no mushrooms) · P(second no mushrooms | first no mushrooms) = $\frac{6}{9}\cdot\frac{5}{8}=0.417$.

R7. (a)

Test result
New hire
0.04 Drugs: 0.90 +; 0.10 −
0.96 No drugs: 0.05 +; 0.95 −

(b) P(test result is positive) = (0.04)(0.90) + (0.96)(0.05) = 0.084
(c) P(uses illegal drugs | positive result)

$$=\frac{P(\text{uses illegal drugs} \cap \text{positive result})}{P(\text{positive result})}=\frac{(0.04)(0.90)}{0.084}=0.429$$

R8. (a) P(none of the 4 calls are for medical help)
$=(1-0.77)(1-0.77)(1-0.77)(1-0.77)$
$=(0.23)^4$
$=0.003$
(b) The calculation in part (a) might not be valid because the 4 consecutive calls being medical are not independent events. Knowing that the first call is medical might make it more likely that the next call is medical (e.g., several people might call for the same medical emergency).

(c) P(at least 1 of the calls is not for medical help)
$= 1 - P$(all 4 calls are for medical help)
$= 1 - (0.77)(0.77)(0.77)(0.77)$
$= 1 - (0.77)^4$
$= 0.648$

Answers to Unit 4, Part I AP® Statistics Practice Test

T1. D **T2.** B **T3.** C **T4.** E **T5.** C
T6. A **T7.** B **T8.** A **T9.** E **T10.** E

T11. (a) P(student eats regularly in the cafeteria and is not a 10th grader) $= \frac{130+122+68}{805} = \frac{320}{805} = 0.398$.

(b) P(10th grader | eats regularly in the cafeteria) $= \frac{175}{495} = 0.354$

(c) P(10th grader) $= \frac{209}{805} = 0.260$. The events "10th grader" and "eats regularly in the cafeteria" are not independent because P(10th grader | eats regularly in the cafeteria) $\neq P$(10th grader).

T12. Let's organize this information using a tree diagram.

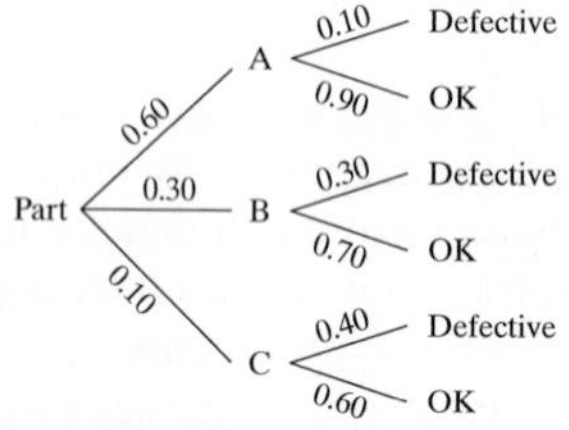

(a) P(defective) $= (0.60)(0.10) + (0.30)(0.30) + (0.10)(0.40) = 0.06 + 0.09 + 0.04 = 0.19$.

(b) P(Machine B | defective) $= \frac{(0.30)(0.30)}{0.19} = \frac{0.09}{0.19} = 0.474$

T13. (a) P(customer will pay less than the usual cost of the buffet) $= 23/36 = 0.639$ (b) P(all 4 friends end up paying less than the usual cost of the buffet) $= (23/36)^4 = 0.167$ (c) P(at least one of the four friends end up paying more than the usual cost of the buffet) $= 1 - 0.167 = 0.833$

Section 4D

Answers to Check Your Understanding

page 385: **1.** $P(S = 0) = 1 - (0.19 + 0.14 + 0.10 + 0.07 + 0.05 + 0.04 + 0.04 + 0.02 + 0.01 + 0.01) = 0.33$. There is a 0.33 probability that there are 0 snow days at this high school in Michigan for a randomly selected school year.
2. $P(S > 6) = 0.04 + 0.02 + 0.01 + 0.01 = 0.08$
3. The distribution of number of snow days is skewed to the right with a single peak at 0 snow days.

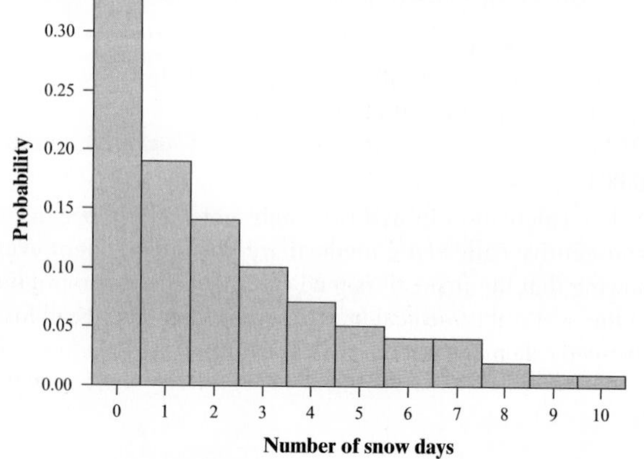

page 391: **1.** $\mu_S = 0(0.33) + 1(0.19) + \ldots + 10(0.01) = 2.17$ days. If many, many school years are randomly selected, the average number of snow days will be about 2.17 days.
2. The median number of snow days is 1 day because this is the smallest value for which the cumulative probability is at least 0.50.

Number of snow days	0	1	2	3	4	5
Probability	0.33	0.19	0.14	0.10	0.07	0.05
Cumulative Probability	0.33	0.52	0.66	0.76	0.83	0.88

Number of snow days	6	7	8	9	10
Probability	0.04	0.04	0.02	0.01	0.01
Cumulative Probability	0.92	0.96	0.98	0.99	1

3. The mean (2.17 snow days) is greater than the median (1 snow day). This relationship makes sense because the distribution of number of snow days is skewed to the right.
4. $\sigma_S = \sqrt{(0-2.17)^2(0.33) + (1-2.17)^2(0.19) + \ldots + (10-2.17)^2(0.01)} = \sqrt{5.801} = 2.409$ days. If many, many school years are randomly selected, the number of snow days will typically vary from the mean of 2.17 days by about 2.409 days.

Answers to Odd-Numbered Section 4D Exercises

1. (a) "Speaks 2 languages" is the event $Y = 2$; $P(Y = 2) = 1 - 0.630 - 0.065 - 0.008 - 0.002 = 0.295$. (b) $P(Y \geq 3) = P(Y = 3) + P(Y = 4) + P(Y = 5) = 0.065 + 0.008 + 0.002 = 0.075$
3. (a) $P(Y>10) = P(Y=15) + P(Y=20) + P(Y=25) = 0.16 + 0.27 + 0.42 = 0.85$. There is a 0.85 probability that more than \$10 is collected on a randomly selected ferry trip. (b) "At least \$10 is collected" is the event $Y \geq 10$. $P(Y \geq 10) = P(Y=10) + P(Y=15) + P(Y = 20) + P(Y = 25) = 0.08 + 0.16 + 0.27 + 0.42 = 0.93$.
5. The shape of the probability histogram is skewed to the left with a single peak at \$25 collected.

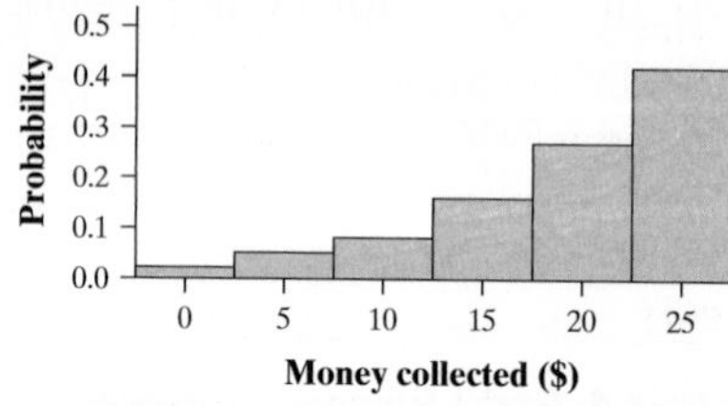

7. $\mu_Y = 0(0.02) + 5(0.05) + \ldots + 25(0.42) = \$19.35$. If many, many ferry trips are randomly selected, the average amount of money collected per trip would be about \$19.35.
9. (a) The histogram shows a right skewed distribution with a peak at 1. The most likely first digit is 1 and each subsequent digit is less likely than the previous digit. (b) $\mu_X = 1(0.301) + 2(0.176) + \ldots + 9(0.046) = 3.441$. If many, many legitimate records are randomly selected, the average of the first digit would be about 3.441.
11. (a)

Money collected ($)	0	5	10	15	20	25
Probability	0.02	0.05	0.08	0.16	0.27	0.42
Cumulative probability	0.02	0.07	0.15	0.31	0.58	1

(b) The smallest value for which the cumulative probability equals or exceeds 0.5 is 20. The median is \$20. (c) The mean of Y (\$19.35) is less than the median of Y (\$20). This relationship makes sense because the probability distribution is skewed left.
13. $\sigma_Y = \sqrt{(0-19.35)^2(0.02) + (5-19.35)^2(0.05) + \ldots + (25-19.35)^2(0.42)} = \sqrt{41.3275} = \$6.429$. If many, many ferry trips are randomly selected, the amount of money collected will typically vary from the mean of \$19.35 by about \$6.43.

15. (a) $\sigma_X^2 = (1-3.441)^2(0.301) + (2-3.441)^2(0.176) + \ldots + (9-3.441)^2(0.046) = 6.0605$ (b) $\sigma_X = \sqrt{6.0605} = 2.462$. If many, many records are randomly selected, the value of the first digit of the record will typically vary from the mean of 3.441 by about 2.462.

17. (a) The company has collected 5 payments of $250 each (for a total of $1250) and must pay out $100,000. This means the company earns $1250 – $100,000 = –$98,750. (b) $E(Y) = (-\$99,750)(0.00183) + (-\$99,500)(0.00186) + \ldots + (\$1250)(0.99058) = \$303.35$. If many, many 21-year-olds are insured by this company, the average amount the company would make on a policy would be about $303.35. (c) $\sigma_Y = \sqrt{(-99,750-303.35)^2(0.00183) + \ldots + (1250-303.35)^2(0.99058)} = \sqrt{94,236,826.64} = \$9707.57$. If many, many 21-year-olds are insured by this company, the amount that the company earns on a policy will typically vary by about $9707.57 from the mean of $303.35.

19. (a) $P(Y > 6) = 3/9 = 0.333$. According to Benford's law, the proportion of first digits in the employee's expense amounts that should be greater than 6 is $P(Y > 6) = P(Y = 7) + P(Y = 8) + P(Y = 9) = 0.058 + 0.051 + 0.046 = 0.155$. If an expense report contains expense amounts with a proportion of first digits greater than 6 that is closer to 0.333 than to 0.155, it may be a fake expense report. (b) The distribution is symmetric and 5 is located at the center. So, 5 is the balance point of the distribution. (c) To detect a fake expense report, compute the sample mean of the first digits and see if it is closer to 5 (suggesting a fake report) or 3.441 (consistent with a truthful report).

21. (a) Outcomes for Plan B: (1) No one has the disease or (2) At least one person has the disease. $P(\text{no one has the disease}) = (0.94)^{12} = 0.47592$. $P(\text{at least one person has the disease}) = 1 - (0.94)^{12} = 0.52408$. If no one has the disease, then X = 1 because only one test is needed: the batch test. If at least one person has the disease, then X = 13 because the batch test is needed as well as 12 individual tests. The probability distribution of X is:

X = Total number of tests with Plan B	1	13
Probability	0.47592	0.52408

(b) If you implement Plan B once, the probability that the number of tests needed will be smaller than with Plan A is 0.47592 because Plan A will always require 12 tests. (c) $E(X) = 1(0.47592) + 13(0.52408) = 7.289$. If many, many batches are tested, the average number of tests needed will be about 6.516 tests. (d) If thousands of groups of 12 people need to be tested, Plan B would be better because the expected number of tests is about 7.3 tests, whereas Plan A will always require 12 tests.

23. B

25. C

27. (a) Yes, in general, students did have higher scores after participating in the chess program. If we look at the differences (Post – Pre) in the scores, the mean difference was 5.38 and the median difference was 3. This means that at least half of the students (though less than three-quarters because Q_1 was negative) improved their reading scores. (b) No, we cannot conclude that chess causes an increase in reading scores. There was no control group (i.e., a group that did not participate in the chess program) for comparison. It may be that children of this age improved their reading scores for other reasons (e.g., regular school) and that the chess program had nothing to do with their improvement.

Section 4E

Answers to Check Your Understanding

page 410: 1. C = 500X. The probability distributions of C and X have the same shape because linear transformations do not change the shape of the distribution.
2. $\mu_C = 500(\mu_X) = 500(1.1) = \$550$
3. $\sigma_C = 500(0.943) = \$471.50$. If many, many Fridays are randomly selected, the bonus earned in the first hour of business will typically vary from the mean of $550 by about $471.50.
4. $\mu_T = \mu_X + \mu_Y = 1.1 + 0.7 = 1.8$ cars. If many, many Fridays are randomly selected, this dealership expects to sell or lease a total of 1.8 cars, on average, in the first hour of business.
5. Because X and Y are independent, $\sigma_T = \sqrt{\sigma_X^2 + \sigma_Y^2} = \sqrt{0.943^2 + 0.64^2} = 1.140$ cars.
6. The total bonus is $B = 500X + 300Y$. So $\mu_B = 500\mu_X + 300\mu_Y = 500(1.1) + 300(0.7) = \$760$. Because X and Y are independent, $\sigma_B = \sqrt{(500\sigma_X)^2 + (300\sigma_Y)^2} = \sqrt{(500 \cdot 0.943)^2 + (300 \cdot 0.64)^2} = \$509.09$.

Answers to Odd-Numbered Section 4E Exercises

1. (a) $T = Y + 4$. The probability distributions of T and Y have the same shape because adding a constant does not affect the shape of the probability distribution. (b) $\mu_T = \mu_Y + 4 = 37.4 + 4 = 41.4$ people. (c) $\sigma_T = \sigma_Y = 1.24$ people because adding a constant does not affect the standard deviation of the probability distribution. If many, many of these flights are randomly selected, the total number of people on the flight will typically vary from the mean of 41.4 people by about 1.24 people.

3. $X = \frac{1}{5}Y$ (a) The probability distribution of X has the same shape as the probability distribution of Y: skewed to the left and single-peaked. (b) $\mu_X = (1/5)\mu_Y = (1/5)(19.35) = \$3.87$ (c) $\sigma_X = (1/5)\sigma_Y = (1/5)(6.43) = \$1.29$

5. $G = 5X + 50$. So $\text{median}_G = 5\text{median}_X + 50 = 5(8.5) + 50 = 92.5$. $IQR_G = 5(IQR_X) = 5(9 - 8) = 5$.

7. (a) T = total time required to bring a part from a bin to its position on an automobile chassis and to attach the part to the chassis. (b) $E(T) = \mu_X + \mu_Y = 11 + 20 = 31$ seconds.

9. $\mu_D = \mu_S - \mu_F = 120 - 105 = 15$ points. If you were to repeat the process of randomly selecting a first-year student, randomly selecting a second-year student, and finding the difference in their scores (Second – First) many, many times, the average difference in SSHA scores would be about 15 points.

11. (a) Yes. The mean of a sum of two random variables is always equal to the sum of their means. (b) No. The variance of the sum is not equal to the sum of the variances, because it is not reasonable to assume that X and Y are independent. Orlando and Disney World are close in proximity, so it is likely that they will experience similar weather patterns.

13. $\sigma_T = \sqrt{\sigma_X^2 + \sigma_Y^2} = \sqrt{2^2 + 4^2} = 4.472$ seconds

15. (a) Knowing the value of one randomly selected student's score does not help us predict the value of the other randomly selected student's score, so it is reasonable to view the SSHA score of a randomly selected first-year and a randomly selected second-year student as independent random variables.
(b) $\sigma_D = \sigma_{S-F} = \sqrt{\sigma_S^2 + \sigma_F^2} = \sqrt{28^2 + 35^2} = \sqrt{2009} = 44.822$ points. If many, many pairs of first- and second-year college students are randomly selected, the difference (Second – First) of their SSHA scores will typically vary from the mean of 15 points by about 44.822 points.

17. $\mu_{2X+3Y} = 2\mu_X + 3\mu_Y = 2(20) + 3(15) = 85$

$\sigma_{2X+3Y} = \sqrt{2^2\sigma_X^2 + 3^2\sigma_Y^2} = \sqrt{4(4)^2 + 9(3)^2} = 12.042$

19. Let T = the total team swim time and X_1 be Wendy's time, X_2 be Jill's time, X_3 be Carmen's time, and X_4 be Latrice's time. Then $T = X_1 + X_2 + X_3 + X_4$ and the random variables are independent. $\mu_T = \mu_1 + \mu_2 + \mu_3 + \mu_4 = 55.2 + 58.0 + 56.3 + 54.7 = 224.2$ seconds

$\sigma_T = \sqrt{\sigma_{X_1}^2 + \sigma_{X_2}^2 + \sigma_{X_3}^2 + \sigma_{X_4}^2} = \sqrt{2.8^2 + 3.0^2 + 2.6^2 + 2.7^2} = 5.56$ seconds

21. Let T be the total weight of 4 pears (P), 5 chocolate truffles (C), and the box (B), so $T = (P_1 + P_2 + P_3 + P_4) + (C_1 + C_2 + C_3 + C_4 + C_5) + B$. $\mu_T = 188 + 188 + 188 + 188 + 12.5 + 12.5 + 12.5 + 12.5 + 12.5 + 20 = 834.5$ grams

$\sigma_T = \sqrt{6^2 + 6^2 + 6^2 + 6^2 + (0.7)^2 + (0.7)^2 + (0.7)^2 + (0.7)^2 + (0.7)^2 + (0.9)^2}$

$= 12.135$ grams

23. (a) $\mu_{\bar{x}} = \frac{1}{2}(\mu_{X_1} + \mu_{X_2}) = \frac{1}{2}(303.35 + 303.35) = \$303.35$

$\sigma_{\bar{x}} = \frac{1}{2}\sqrt{\sigma_{X_1}^2 + \sigma_{X_2}^2} = \frac{1}{2}\sqrt{9707.57^2 + 9707.57^2} = \$6864.29$

(b) $\mu_{\bar{x}} = \frac{1}{4}(\mu_{X_1} + \mu_{X_2} + \mu_{X_3} + \mu_{X_4}) = \frac{1}{4}(303.35 + 303.35 + 303.35 + 303.35) = \$303.35$

$\sigma_{\bar{x}} = \frac{1}{4}\sqrt{\sigma_{X_1}^2 + \sigma_{X_2}^2 + \sigma_{X_3}^2 + \sigma_{X_4}^2}$

$= \frac{1}{4}\sqrt{9707.57^2 + 9707.57^2 + 9707.57^2 + 9707.57^2}$

$= \$4853.79$

(c) As the number of insured individuals increases, the average income earned remains the same ($303.35 per person), but the standard deviation of the average income earned decreases, which reduces the variability in income earned, and therefore, the risk. (d) For n randomly selected individuals:

$\mu_{\bar{x}} = \frac{n\mu_X}{n} = \mu_X$ and $\sigma_{\bar{x}} = \frac{1}{n}\sqrt{n \cdot \sigma_X^2} = \frac{\sqrt{n}}{n}\sqrt{\sigma_X^2} = \frac{1}{\sqrt{n}}\sqrt{\sigma_X^2} = \frac{\sigma_X}{\sqrt{n}}$

25. D

27. (a) First, let's make a tree diagram.

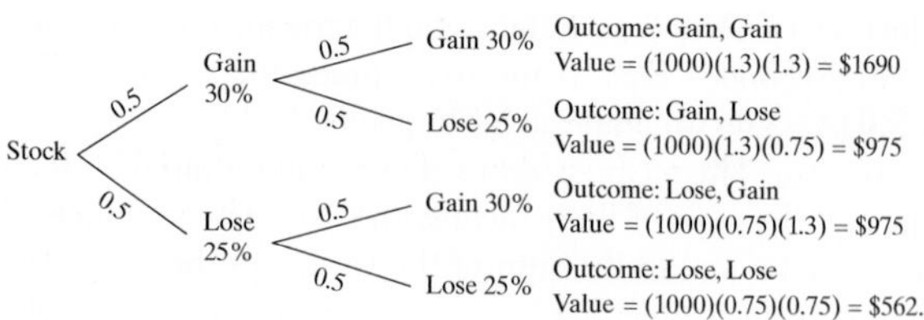

Because gain or loss is equally likely both days, the four outcomes each have probability 0.25. Note, however, there are two ways to get $975. The probability distribution is:

Amount	$562.50	$975	$1690
Probability	0.25	0.50	0.25

(b) The probability that the stock is worth more than the $1000 paid for it, after two days, is 0.25 (the probability of it being worth $1690). (c) $\mu = 562.50(0.25) + 975(0.5) + 1690(0.25) = \$1050.63$

Section 4F

Answers to Check Your Understanding

page 425: 1. *Binary?* "Success" = win a prize; "Failure" = don't win a prize. *Independent?* Knowing the outcome of one bottle does not tell us anything about the outcome of any other bottle. *Number?* $n = 30$. *Same probability?* $p = 1/6$. This is a binomial setting and X has a binomial distribution with $n = 30$ and $p = 1/6$.

2. $P(X = 3) = \binom{30}{3}\left(\frac{1}{6}\right)^3\left(\frac{5}{6}\right)^{27} = 0.137$. *Tech:* $P(X = 3) =$ binompdf(trials: 30, p: 1/6, x value: 3) = 0.137. There is a 0.137 probability that exactly 3 of the 30 students will win a prize.

3. $P(X \le 3) = P(X = 0) + P(X = 1) + P(X = 2) + P(X = 3)$

$$P(X \le 3) = \binom{30}{0}\left(\frac{1}{6}\right)^0\left(\frac{5}{6}\right)^{30} + \binom{30}{1}\left(\frac{1}{6}\right)^1\left(\frac{5}{6}\right)^{29} + \binom{30}{2}\left(\frac{1}{6}\right)^2\left(\frac{5}{6}\right)^{28} + \binom{30}{3}\left(\frac{1}{6}\right)^3\left(\frac{5}{6}\right)^{27} = 0.240.$$

Tech: $P(X \le 3) =$ binomcdf(trials: 30, p: 1/6, x value: 3) = 0.240

4. No. It is likely (probability = 0.24) that 3 or fewer out of 30 students would win a prize, by chance alone, when 1 in 6 wins overall. This result does not give convincing evidence that the company's claim is false.

page 430: 1. The probability distribution histogram is skewed to the right with a single peak at 5 successes. *Note:* There is a long tail to the right all the way to $X = 30$.

2. $E(X) = np = (30)(1/6) = 5$ winners. This makes sense because, in the long run, for every 30 sodas that are purchased, the average number of winners should be $(1/6)(30) = 5$ winners.

3. $\sigma_X = \sqrt{np(1-p)} = \sqrt{30(1/6)(5/6)} = 2.041$ winners. If many, many random samples of 30 sodas are purchased, the number of winners will typically vary from the mean of 5 winners by about 2.041 winners.

page 436: 1. Die rolls are independent, the probability of getting doubles is the same on each roll (1/6), we are repeating the random process until we get a success (doubles), and T is counting the number of trials it takes to get the first success. This is a geometric setting and T is a geometric random variable with $p = 1/6$.

2. $P(T = 3) = \left(\frac{5}{6}\right)^2\left(\frac{1}{6}\right) = 0.116$. *Tech:* geometpdf($p$: 1/6, x value: 3) = 0.116. There is an 11.6% probability that you will get the first set of doubles on the third roll of the dice.

3. $P(T \le 3) = P(T = 1) + P(T = 2) + P(T = 3) = \frac{1}{6} + \left(\frac{5}{6}\right)\left(\frac{1}{6}\right) + \left(\frac{5}{6}\right)^2\left(\frac{1}{6}\right) = 0.421$. *Tech:* geometcdf($p$: 1/6, x value: 3) = 0.421.

4. $\mu_T = \frac{1}{(1/6)} = 6$ rolls. If you repeat the process of rolling until you get doubles many, many times, the average number of rolls it would take to get doubles is about 6. $\sigma_T = \frac{\sqrt{1-(1/6)}}{1/6} = 5.477$ rolls. If you repeat the process of rolling until you get doubles many, many times, the number of rolls it would take to get doubles will typically vary by about 5.477 rolls from the mean of 6 rolls.

Answers to Odd-Numbered Section 4F Exercises

1. *Binary?* "Success" = survive to adulthood; "Failure" = does not survive to adulthood. *Independent?* Yes, because the baby elk were randomly selected, knowing the outcome of one elk doesn't tell us anything about the outcomes of other elk. *Number?* $n = 7$. *Same probability?* $p = 0.44$. This is a binomial setting and X is counting the number of baby elk that survive to adulthood. X is a binomial random variable with $n = 7$ and $p = 0.44$.

3. *Binary?* "Success" = hits bullseye; "Failure" = does not hit bullseye. *Independent?* Yes, the outcome of one shot does not tell us anything about the outcome of other shots. *Number?* No, there is not a fixed number of trials. So this is not a binomial setting and Y does not have a binomial distribution.

5. (a) X = number of elk who survive to adulthood. X is a binomial random variable with $n = 7$ and $p = 0.44$. We want to find $P(X = 4)$. (b) $P(X=4)=\binom{7}{4}(0.44)^4(0.56)^3=0.230$. *Tech:* $P(X=4)$ = binompdf(trials: 7, p: 0.44, x value: 4) = 0.230.

7. (a) *Binary?* "Success" = spinner lands in the blue region; "Failure" = spinner does not land in the blue region. *Independent?* Knowing whether or not one spin lands in the blue region tells you nothing about whether or not another spin lands in the blue region. *Number?* $n = 12$. *Same probability?* $p = 0.80$. This is a binomial setting and X is counting the number of times the spinner lands in the blue region. X is a binomial random variable with $n = 12$ and $p = 0.80$.

(b) $P(X = 8) = \binom{12}{8}(0.80)^8(0.20)^4 = 0.133$. *Tech:* $P(X = 8)$ = binompdf(trials: 12, p: 0.80, x value: 8) = 0.133.

9. X = number of elk who survive to adulthood. X is a binomial random variable with $n = 7$ and $p = 0.44$. We want to find $P(X > 4) = P(X = 5) + P(X = 6) + P(X = 7)$

$= \binom{7}{5}(0.44)^5(0.56)^2 + \binom{7}{6}(0.44)^6(0.56)^1 + \binom{7}{7}(0.44)^7(0.56)^0 = 0.140.$

Tech: $P(X>4)=1-P(X\leq 4)=1-$ binomcdf(trials: 7, p: 0.44, x value: 4) = 1 − 0.860 = 0.140. This probability is not small, so it is not surprising for more than 4 elk to survive to adulthood.

11. X = number of times the spinner lands on blue. X is a binomial random variable with $n = 12$ and $p = 0.80$. We want to find $P(X \leq 7) = P(X = 0) + P(X = 1) + \ldots + P(X = 7)$

$= \binom{12}{0}(0.8)^0(0.2)^{12} + \binom{12}{1}(0.8)^1(0.2)^{11} + \ldots + \binom{12}{7}(0.8)^7(0.2)^5$

= 0.073. *Tech:* $P(X\leq 7)$ = binomcdf(trials: 12, p: 0.80, x value: 7) = 0.073. Assuming the website's claim is true, there is about a 7.3% probability that there would be 7 or fewer spins landing in the blue region.

13. (a) X = number of students who prefer the last kiss. X is a binomial random variable with $n = 22$ and $p = 0.20$. *Tech:* $P(X\geq 14)=1-$ binomcdf(trials: 22, p: 0.20, x value: 13) = $0.00001 \approx 0$. (b) Yes. It is very unlikely (probability ≈ 0) for at least 14 out of 22 students to choose the last Kiss by chance alone, so we have convincing evidence that participants have a preference for the last thing they taste.

15. (a) R is a binomial random variable with $n = 20$ and $p = 0.30$ because the following conditions are met: *Binary?* "Success" = the light is red; "Failure" = the light is green. *Independent?* Knowing whether or not the light is red for one passenger tells you nothing about whether or not the light will be red for another passenger. *Number?* $n = 20$. *Same probability?* $p = 0.30$. (b) *Tech:* $P(R \leq 3)$ = binomcdf(trials: 20, p: 0.3, x value: 3) = 0.107. (c) No. It is somewhat likely (probability = 0.107) that 3 or fewer people out of 20 will get a red light by chance alone when $p = 0.30$, so we do not have convincing evidence against the custom agents' claim.

17. (a) The probability distribution of R is slightly skewed to the right with a single peak at $R = 6$. (b) R has a binomial distribution with $n = 20$ and $p = 0.3$. $\mu_R = np = 20(0.3) = 6$ passengers. If many, many groups of 20 passengers were selected, the average number of passengers who would get a red light would be about 6 passengers. (c) $\sigma_R = \sqrt{np(1-p)} = \sqrt{20(0.3)(0.7)} = 2.049$ passengers. If many, many groups of 20 passengers were selected, the number of passengers who would get a red light would typically vary from the mean of 6 passengers by about 2.049 passengers.

19. (a) The probability distribution of X is skewed to the left with a single peak at X = 10.

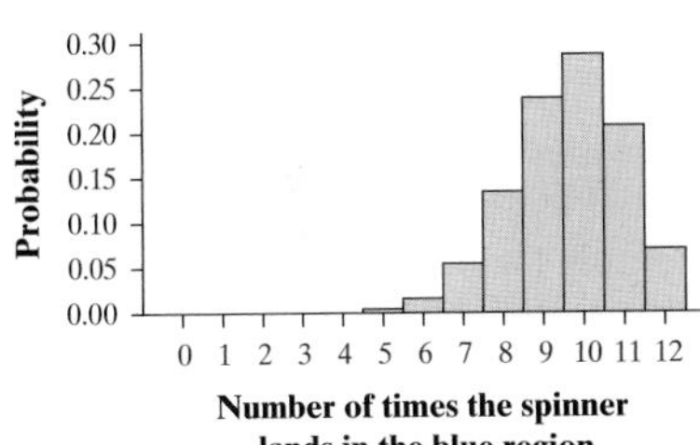

(b) X has a binomial distribution with $n = 12$ and $p = 0.8$. $\mu_X = np = 12(0.8) = 9.6$ times. If many, many rounds of 12 spins were completed, the average number of times the spinner would land in the blue region would be about 9.6 times. $\sigma_X = \sqrt{np(1-p)} = \sqrt{12(0.8)(0.2)} = 1.386$ times. If many, many rounds of 12 spins were completed, the number of times the spinner would land in the blue region would typically vary from the mean of 9.6 times by about 1.386 times.

21. (a) The Independent condition is not met because the security officials are sampling without replacement. (b) The sample size ($n = 10$) is not less than 10% of the population size ($N = 76$), so the 10% condition is not satisfied. We cannot view the observations as independent.

23. (a) *Binary?* "Success" = left-handed; "Failure" = not left-handed. *Independent?* The sample was selected without replacement, but this condition is met as long as $n = 100$ is less than 10% of the size of the population (the large high school). *Number?* $n = 100$. *Same probability?* $p = 0.11$. This is a binomial setting and L has a binomial distribution with $n = 100$ and $p = 0.11$. (b) *Tech:* $P(L\geq 15)\approx 1-$ binomcdf(trials: 100, p: 0.11, x value: 14) = 0.133.

25. (a) This is not a geometric setting because we can't classify the possible outcomes on each trial (card) as "success" or "failure" and we are not selecting cards until we get 1 success—we are selecting cards until we get one of each type. (b) This is a geometric setting with $p = 1/6$. The outcomes (animal featured) are independent, the probability of getting a flamingo is the same for each purchase ($p = 1/6$), and Lola continues purchasing puzzles until she gets a flamingo. The random variable F = the number of puzzles Lola purchases to get a flamingo puzzle has a geometric distribution with $p = 1/6$.

27. **(a)** X has a geometric distribution with $p=0.20$. $P(X=3)=(0.8)^2(0.2)=0.128$. *Tech:* $P(X=3)=$ geometpdf(p: 0.20, x value: 3) $=0.128$. **(b)** $P(X<6)=P(X=1)+\ldots+P(X=5)=0.2+(0.8)(0.2)+(0.8)^2(0.2)+(0.8)^3(0.2)+(0.8)^4(0.2)=0.672$. *Tech:* $P(X<6)=$ geometcdf(p: 0.20, x value: 5) $=0.672$. There is a 0.672 probability that it takes Rita less than 6 pulls to start the mower.

29. Let X = the number of invoices that must be examined to find one whose amount begins with an 8 or 9. X has a geometric distribution with $p=0.097$. *Tech:* $P(X\geq 40)=1-P(X\leq 39)=1-$ geometcdf(p: 0.097, x value: 39) $=1-0.981=0.019$. Yes, this would be surprising. It is unlikely (probability = 0.019) that the accountant would need to look at 40 or more invoices to find the first one with an amount that starts with an 8 or 9 if the amounts are truthful.

31. **(a)** X has a geometric distribution with $p=0.2$. $\mu_X=\dfrac{1}{0.2}=5$ pulls. If Rita repeats the process of starting the mower many, many times, the average number of pulls she would expect to need in order to start the mower is about 5 pulls. **(b)** $\sigma_X=\dfrac{\sqrt{1-0.2}}{0.2}=4.472$ pulls. If Rita repeats the process of starting the mower many, many times, the number of pulls it will take her to start the mower will typically vary from the mean of 5 pulls by about 4.472 pulls.

33. **(a)** Skewed to the right. Every geometric distribution has this shape. That's because the most likely value of a geometric random variable is 1. The probability of each successive value decreases by a factor of $(1-p)$. **(b)** Let X = the number of invoices that must be examined to find one whose amount begins with an 8 or 9. X has a geometric distribution with $p=0.097$. $E(X)=\dfrac{1}{0.097}=10.31$ invoices. **(c)** $\sigma_X=\dfrac{\sqrt{1-0.097}}{0.097}=9.797$ invoices.

35. **(a)** X is a binomial random variable with $n=30$ and $p=\dfrac{1}{2}=0.5$. $\mu_X=30(0.5)=15$ correct and $\sigma_X=\sqrt{30(0.5)(0.5)}=2.739$ correct. **(b)** $z=\dfrac{23-15}{2.739}=2.921$ **(c)** The observed result of 23 correct identifications is very surprising if the 30 volunteers are just guessing because this result falls 2.921 standard deviations above the mean, which is unlikely to occur by chance alone.

37. B

39. D

41. **(a)** This is an experiment because a treatment was deliberately imposed on the students, randomly assigned at the time they visited the site. **(b)** The explanatory variable is the type of login box (genuine or not), which is categorical, and the response variable is the student's action (logging in or not), which is also categorical.

Answers to Unit 4, Part II Review Exercises

R1. **(a)** $P(Y=5)=1-(0.1+0.2+0.3+0.3)=0.1$. There is a 0.1 probability that a randomly selected patient would rate their pain as a 5 on a scale of 1 to 5. **(b)** $P(Y\leq 2)=P(Y=1)+P(Y=2)=0.1+0.2=0.3$ **(c)** $E(X)=1(0.1)+2(0.2)+\ldots+5(0.1)=3.1$

$\sigma_X=\sqrt{(1-3.1)^2(0.1)+(2-3.1)^2(0.2)+\ldots+(5-3.1)^2(0.1)}=1.136$

R2. **(a)** The mean number of degrees off target is $\mu_D=550-550=0$°C. The standard deviation of the number of degrees off target stays the same: $\sigma_D=5.7$°C, because subtracting a constant does not change the variability. **(b)** $\mu_Y=\dfrac{9}{5}(550)+32=1022$°F and $\sigma_Y=\dfrac{9}{5}(5.7)=10.26$°F.

R3. **(a)** The distribution of amount won is skewed to the right with a single peak at \$0. The mean of the distribution is \$0.70 and the standard deviation is \$6.58. There is a large gap from \$3 won to \$120 won.

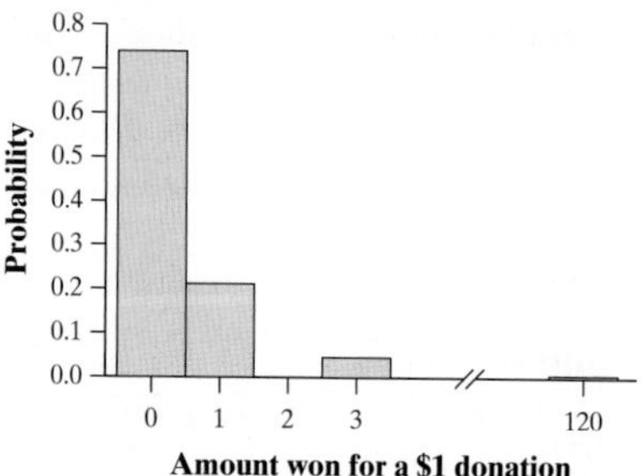

(b) If you were to play this game many, many times you would get, on average, about \$0.70 of your \$1 donation back (as the amount won) per game. If you were to play this game many, many times, the amount won would typically vary from the mean of \$0.70 by about \$6.58. **(c)** Let Y be the amount Jeree wins, so Y = 5X. $\mu_Y=5\mu_X=5(0.70)=\$3.50$ and $\sigma_Y=5\sigma_X=5(6.58)=\$32.90$. **(d)** Let W be the total amount won by Mario. Because he plays 5 separate \$1 games, $W=X_1+\ldots+X_5$, where all X's have the same distribution and are independent of each other.

$\mu_W=\mu_{X_1}+\ldots+\mu_{X_5}=0.70+0.70+0.70+0.70+0.70=\$3.50$

$\sigma_W=\sqrt{\sigma_{X_1}^2+\ldots+\sigma_{X_5}^2}=\sqrt{6.58^2+\ldots+6.58^2}=\$14.71$

R4. **(a)** It is reasonable to assume the cap strength and torque are independent because the machine that makes the caps and the machine that applies the torque are not the same. Knowing the cap strength does not change the probability distribution of machine torque, and vice versa. **(b)** $\mu_D=\mu_{C-T}=10-7=3$ inch-pounds $\sigma_D=\sigma_{C-T}=\sqrt{1.2^2+0.9^2}=1.5$ inch-pounds.

R5. **(a)** Because $n=8$ is less than 10% of the size of the population (all M&M'S® in the large bag), X can be modeled by a binomial distribution with $n=8$ and $p=0.205$ even though the sample was selected without replacement. **(b)** $P(X=3)=\binom{8}{3}(0.205)^3(0.795)^5=0.153$. *Tech:* $P(X=3)=$ binompdf(trials: 8, p: 0.205, x value: 3) $=0.153$. **(c)** *Tech:* $P(X\geq 4)=1-P(X<4)=1-P(X\leq 3)=1-$ binomcdf(trials: 8, p: 0.205, x value: 3) $=0.061$. There is a 6.1% chance that at least 4 of the 8 M&M'S will be orange. **(d)** No. It is not very unlikely (probability = 0.061) to receive 4 or more orange M&M'S in a sample of 8 M&M'S by chance alone when 20.5% of M&M'S produced are orange, so this result does not provide convincing evidence that Mars's claim is false for this bag.

R6. **(a)** $\mu_X=np=8(0.205)=1.64$. If we were to select many, many random samples of size 8, we would expect to get about 1.64 orange M&M'S per sample, on average. **(b)** $\sigma_X=\sqrt{np(1-p)}=\sqrt{8(0.205)(0.795)}=1.14$. If we were to select many, many random samples of size 8, the number of orange M&M'S would typically vary by about 1.14 orange candies from the mean of 1.64 orange candies.

R7. (a) Let Y = the number of spins it takes to get a wasabi bomb. Y has a geometric distribution with $p = 3/12 = 0.25$. (b) $P(Y \le 3) = 0.25 + (0.75)(0.25) + (0.75)^2(0.25) = 0.578$. *Tech*: $P(Y \le 3) =$ geometcdf(p: 0.25, x value: 3) = 0.578. (c) $\mu_Y = \frac{1}{0.25} = 4$. If a contestant repeats the process of spinning until they get a wasabi bomb many, many times, the average number of spins it would take to get the first wasabi bomb is 4 spins. $\sigma_Y = \frac{\sqrt{1-0.25}}{0.25} = 3.464$. If a contestant repeats the process of spinning until they get a wasabi bomb many, many times, the number of spins it would take to get the first wasabi bomb would typically vary from the mean of 4 spins by about 3.464 spins.

Answers to Unit 4, Part II AP® Statistics Practice Test

T1. B **T2.** A **T3.** D **T4.** E **T5.** D
T6. D **T7.** C **T8.** B **T9.** B **T10.** C

T11. (a) $\mu_Y = 0(0.78) + 1(0.11) + 2(0.07) + 3(0.03) + 4(0.01) = 0.38$ eggs. If many, many cartons of eggs were randomly selected, we would expect an average of about 0.38 eggs per carton to be broken. (b) $\sigma_Y = \sqrt{(0-0.38)^2(0.78) + (1-0.38)^2(0.11) + \ldots + (4-0.38)^2(0.01)} = 0.822$ eggs (c) $P(Y \ge 2) = P(Y = 2) + P(Y = 3) + P(Y = 4) = 0.07 + 0.03 + 0.01 = 0.11$. (d) Let X = the number of cartons inspected to get one carton with at least 2 broken eggs. X is a geometric random variable with $p = 0.11$. $P(X \le 3) = P(X = 1) + P(X = 2) + P(X = 3) = (0.11) + (0.89)(0.11) + (0.89)^2(0.11) = 0.295$. *Tech:* $P(X \le 3) =$ geometcdf(p: 0.11, x value: 3) = 0.295.

T12. (a) This is a binomial setting; X is a binomial random variable with $n = 12$ and $p = 0.66$. $P(X = 6) = \binom{12}{6}(0.66)^6(0.34)^6 = 0.118$.

Tech: $P(X = 6) =$ binompdf(trials:12, p: 0.66, x value: 6) = 0.118.

(b) $P(X \le 4) = \binom{12}{0}(0.66)^0(0.34)^{12} + \ldots + \binom{12}{4}(0.66)^4(0.34)^8 = 0.021$. *Tech:* $P(X \le 4) =$ binomcdf(trials: 12, p: 0.66, x value: 4) = 0.021. (c) It is unlikely (probability = 0.021) to have only 4 or fewer owners greet their dogs first by chance alone if 66% of all dog owners greet their dog first. This gives convincing evidence that the magazine's claim is incorrect.

T13. (a) It is reasonable to view A and E as independent random variables because knowing how much time one of these students spent on a randomly selected math assignment in their class does not change the probability distribution of the amount of time the other student spent on a randomly selected math assignment in a different class. (b) $\mu_{A-E} = 50 - 25 = 25$ minutes. (c) $\sigma_{A-E} = \sqrt{\sigma_A^2 + \sigma_E^2} = \sqrt{10^2 + 5^2} = 11.180$ minutes.

UNIT 5

Section 5A

Answers to Check Your Understanding

page 461: 1. Let X = the amount of liquid soap dispensed when the dispenser is activated once. The distribution of X is approximately normal with a mean of 0.9 ml and a standard deviation of 0.042 ml. We want to find $P(X < 0.8)$.

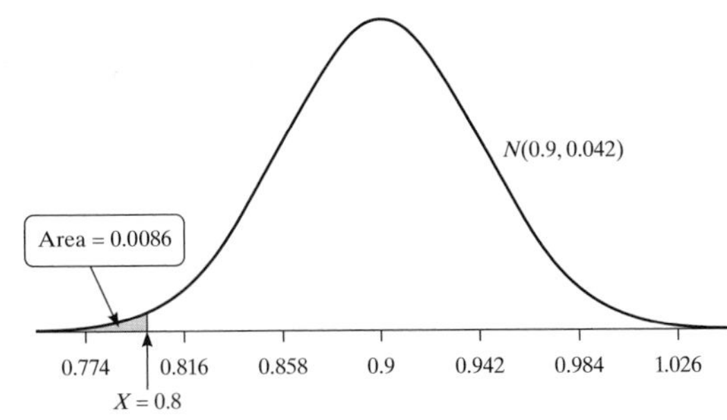

(i) $z = \frac{0.8 - 0.9}{0.042} = -2.381$; *Tech:* normalcdf(lower: −1000, upper: −2.381, mean: 0, SD: 1) = 0.0086; *Table* A: 0.0087. (ii) normalcdf (lower: −1000, upper: 0.8, mean: 0.9, SD: 0.042) = 0.0086.
2. We want to find the 95th percentile of the distribution of X.

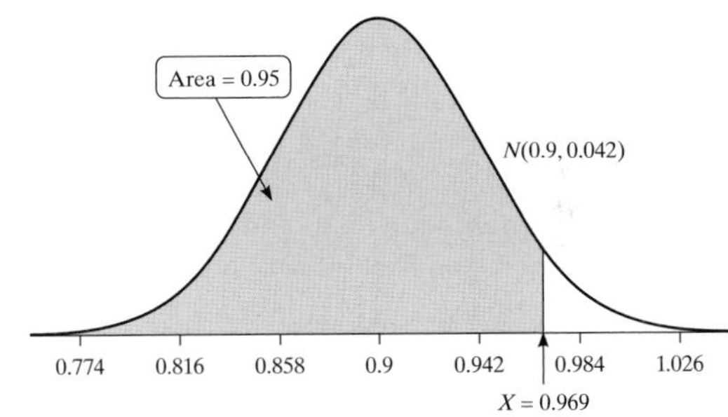

(i) *Tech:* invNorm(area: 0.95, mean: 0, SD: 1) = 1.645; *Table* A: $z = 1.64$ or 1.65

$$1.645 = \frac{x - 0.9}{0.042}$$
$$x = 0.969$$

(ii) invNorm(area: 0.95, mean: 0.9, SD: 0.042) = 0.969. The 95th percentile of the amount of soap dispensed when the dispenser is activated once is 0.969 ml. In other words, 95% of all activations will produce 0.969 ml or less soap.
3. Let T = the total amount of soap dispensed when the dispenser is activated twice. $\mu_T = \mu_X + \mu_X = 0.9 + 0.9 = 1.8$ ml; $\sigma_T = \sqrt{\sigma_X^2 + \sigma_X^2} = \sqrt{0.042^2 + 0.042^2} = 0.059$ ml. T is normally distributed with mean 1.8 ml and SD 0.059 ml. We want to find $P(T \ge 1.6)$.

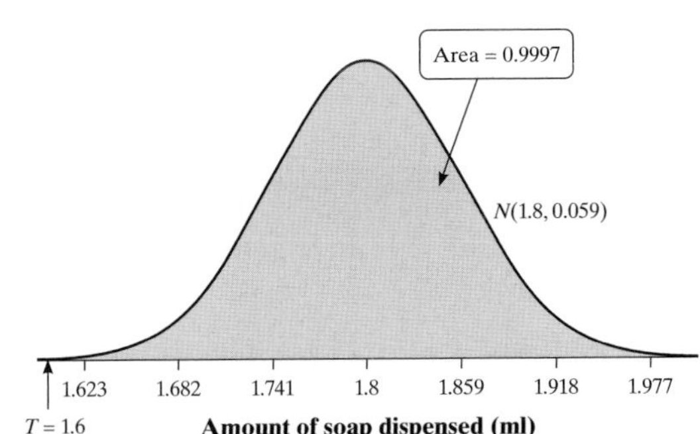

(i) $z = \frac{1.6-1.8}{0.059} = -3.390$; *Tech:* normalcdf(lower: −3.390, upper: 1000, mean: 0, SD: 1) = 0.9997; *Table A:* 1 − 0.0003 = 0.9997. (ii) normalcdf(lower: 1.6, upper: 1000, mean: 1.8, SD: 0.059) = 0.9997.

Answers to Odd-Numbered Section 5A Exercises

1. (a) The total area under the density curve between 1 second and 5 seconds equals 1; $A = b \times h \to 1 = 4 \times h \to h = \frac{1}{4}$.

(b) P(between 2.5 and 4 seconds) $= (4-2.5)\left(\frac{1}{4}\right) = 0.375$

3. The stemplot of these data is roughly symmetric and bell-shaped. The mean and standard deviation of these data are $\bar{x} = 115$ and $s_x = 15$. $\bar{x} \pm 1s_x = 100$ to $130 \to 43$ of $60 = 71.7\%$. $\bar{x} \pm 2s_x = 85$ to $145 \to 58$ of $60 = 96.7\%$. $\bar{x} \pm 3s_x = 70$ to $160 \to 60$ of $60 = 100\%$. In a normal distribution, about 68% of the values fall within 1 SD of the mean, about 95% fall within 2 SD of the mean, and about 99.7% fall within 3 SD of the mean. None of the observed percentages were substantially different from the estimated percentages. Thus, the distribution of aptitude test score is approximately normal.

5. (a) Let X = the 1-mile time for a randomly selected student from this group. The distribution of X is approximately normal with a mean of 7.11 minutes and SD of 0.74 minute. We want to find $P(6 < X < 7)$. (i) $z = \frac{6-7.11}{0.74} = -1.500$ and $z = \frac{7-7.11}{0.74} = -0.149$; *Tech:* normalcdf(lower: −1.5, upper: −0.149, mean: 0, SD: 1) = 0.3740; *Table A:* 0.4404 − 0.0668 = 0.3736 (ii) normalcdf(lower: 6, upper: 7, mean: 7.11, SD: 0.74) = 0.3741. There is about a 0.374 probability that a randomly selected student will run a mile between 6 and 7 minutes. **(b)** We want to find the 10th percentile of the probability distribution. (i) *Tech:* invNorm(area: 0.10, mean: 0, SD: 1) = −1.282; *Table A:* 0.10 area to the left $\to z = -1.28$; $-1.282 = \frac{x-7.11}{0.74}$ $x = 6.161$ minutes (ii) invNorm(area: 0.10, mean: 7.11, SD: 0.74) = 6.162 minutes

7. Note that $Y = 32 + \frac{9}{5}T$. **(a)** $\mu_Y = 32 + \frac{9}{5}\mu_T = 32 + \frac{9}{5}(8.5) = 47.3$°F; $\sigma_Y = \frac{9}{5}\sigma_T = \frac{9}{5}(2.25) = 4.05$°F **(b)** Y can be modeled by a normal distribution because it is a linear transformation of T, and T can be modeled by a normal distribution. We want to find $P(Y < 40)$. (i) $z = \frac{40-47.3}{4.05} = -1.802$; *Tech:* normalcdf(lower: −1000, upper: −1.802, mean: 0, SD: 1) = 0.0358; *Table A:* 0.0359 (ii) normalcdf(lower: −1000, upper: 40, mean: 47.3, SD: 4.05) = 0.0357 **(c)** Yes. It is unlikely (probability = 0.0357) that the cabin temperature at midnight will be below 40°F.

9. (a) Normal, because D is the difference of independent, normal random variables. $\mu_D = \mu_Z - \mu_C = 105 - 98 = 7$ minutes; $\sigma_D = \sqrt{\sigma_Z^2 + \sigma_C^2} = \sqrt{10^2 + 15^2} = 18.028$ minutes **(b)** We want to find $P(-5 < D < 5)$. (i) $z = \frac{-5-7}{18.028} = -0.666$ and $z = \frac{5-7}{18.028} = -0.111$; *Tech:* normalcdf(lower: −0.666, upper: −0.111, mean: 0, SD: 1) = 0.2031; *Table A:* 0.4562 − 0.2514 = 0.2048 (ii) normalcdf(lower: −5, upper: 5, mean: 7, SD: 18.028) = 0.2030

11. Let X = the weight of a randomly selected hay bale and T = the total weight of 10 randomly selected hay bales. T is a normally distributed random variable because T is the sum of independent, normal random variables. $\mu_T = 100 + 100 + \ldots + 100 = 1000$ pounds; $\sigma_T = \sqrt{6^2 + 6^2 + \ldots + 6^2} = 18.974$ pounds. We want to find $P(T > 1050)$. (i) $z = \frac{1050-1000}{18.974} = 2.635$; *Tech:* normalcdf (lower: 2.635, upper: 1000, mean: 0, SD: 1) = 0.0042; *Table A:* 1 − 0.9959 = 0.0041 (ii) normalcdf(lower: 1050, upper: 10000, mean: 1000, SD: 18.974) = 0.0042

13. Let T be the total weight of 4 pears (P), 5 chocolate truffles (C), and the box (B), so $T = P_1 + P_2 + P_3 + P_4 + C_1 + C_2 + C_3 + C_4 + C_5 + B$. T is a normally distributed random variable because T is the sum of independent, normal random variables. $\mu_T = 188 + 188 + 188 + 188 + 12.5 + 12.5 + 12.5 + 12.5 + 12.5 + 20 = 834.5$ grams; $\sigma_T = \sqrt{6^2 + 6^2 + 6^2 + 6^2 + 0.7^2 + 0.7^2 + 0.7^2 + 0.7^2 + 0.7^2 + 0.9^2} = 12.135$ grams. We want to find $P(T < 800)$. (i) $z = \frac{800-834.5}{12.135} = -2.843$; *Tech:* normalcdf(lower: −1000, upper: −2.843, mean: 0, SD: 1) = 0.0022; *Table A:* 0.0023 (ii) normalcdf (lower: −1000, upper: 800, mean: 834.5, SD: 12.135) = 0.0022

15. (a) This is a valid density curve because the density curve is entirely above the horizontal axis and the area under the density curve is 1. Split the area under the density curve into a trapezoid (from $X = 0$ to $X = 0.4$) and a rectangle (from $X = 0.4$ to $X = 0.8$). Total area = area of trapezoid + area of rectangle = $(1/2)(2+1)(0.4) + (0.8-0.4)(1) = 1$. **(b)** Area of trapezoid = $(1/2)(2+1.5)(0.2) = 0.35$. **(c)** The median is the equal-areas point. The area to the left of 0.2 is 0.35. The area to the left of 0.4 is 0.6 [from part (a)]. Therefore, we know somewhere between $X = 0.2$ and $X = 0.4$ is the equal-areas point with an area of 0.50 to the left. **(d)** The mean will be greater than the median because the shape of this distribution is skewed to the right.

17. E

19. B

21. (a) Because $n = 50$ is less than 10% of the size of the population (all quarters minted in Denver), X can be modeled by a binomial distribution with $n = 50$ and $p = 0.495$, even though the sample was selected without replacement. **(b)** $\mu_D = np = 50(0.495) = 24.75$ quarters. If we were to select many, many random samples of 50 quarters minted in 2020, we would expect to get about 24.75 Denver-minted quarters per sample, on average. $\sigma_D = \sqrt{np(1-p)} = \sqrt{50(0.495)(1-0.495)} = 3.535$ quarters. If we were to select many, many random samples of 50 quarters minted in 2020, the number of Denver-minted quarters would typically vary by about 3.535 quarters from the mean of 24.75 quarters.

Section 5B

Answers to Check Your Understanding

page 471: 1. The population is all M&M'S® Milk Chocolate Candies produced by the factory in Hackettstown, NJ. The parameter is p = the proportion of all M&M'S Milk Chocolate Candies produced by the factory in Hackettstown, NJ, that

are orange. The parameter is claimed to be $p = 0.25$. The sample is the 50 M&M'S Milk Chocolate Candies selected. The statistic is the proportion of the sample of 50 M&M'S that are orange, $\hat{p}$.
2. The graph shows the population distribution.

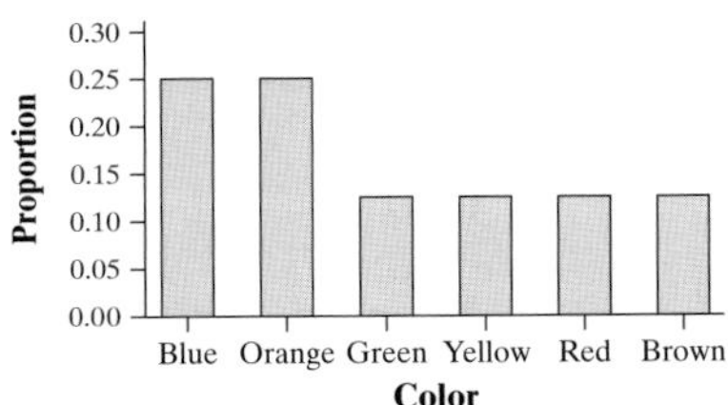

3. Answers will vary. The graph shows a possible distribution of sample data. For this sample, there are 11 orange M&M'S, so $\hat{p} = \frac{11}{50} = 0.22$.

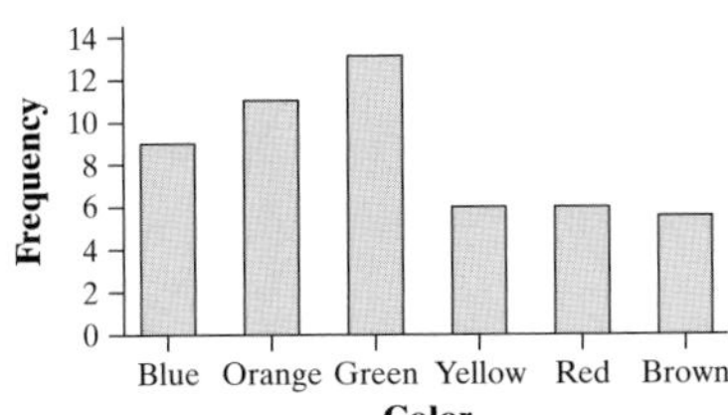

4. The middle graph is the approximate sampling distribution of $\hat{p}$. The statistic measures the proportion of orange candies in samples of 50 M&M'S. Assuming that the company is correct, 25% of the M&M'S are orange, so the center of the distribution of $\hat{p}$ should be at approximately 0.25. The first graph shows the distribution of the colors for one sample, rather than the distribution of $\hat{p}$ from many samples, and the third graph is centered at 0.125, rather than at 0.25.

page 479: 1. The simulation does provide evidence that the sample median is a biased estimator of the population median because the mean of the sampling distribution of the sample median wait time (80.5) is not equal to the value it is trying to estimate, the population median wait time (83).
2. Increasing the sample size from 15 to 30 will decrease the variability of the sampling distribution of the sample median wait time. Larger samples provide more precise estimates, because larger samples include more information about the population distribution.

Answers to Odd-Numbered Section 5B Exercises

1. *Population*: all adults in the United States. *Parameter*: $p =$ the proportion of all adults in the United States who use Instagram. *Sample*: the 1502 adults who were surveyed. *Statistic*: $\hat{p} =$ the proportion in the sample who use Instagram $= \frac{601}{1502} = 0.40$.

3. *Population*: all dental practices in California. *Parameter*: the interquartile range of the price to fill a cavity for all dental practices in California. *Sample*: the 10 randomly selected dental practices. *Statistic*: $IQR =$ the sample interquartile range of the price to fill a cavity for the 10 selected dental practices $= \$74$.

5. *Population:* all bottles of Arizona Iced Tea produced that day in the plant. *Parameter:* $\mu =$ the average number of ounces per bottle in all bottles of Arizona Iced Tea produced in the plant that day. *Sample*: the 50 bottles of tea selected at random from the day's production. *Statistic:* $\bar{x} =$ the average number of ounces of tea contained in the 50 bottles $= 19.6$ ounces.

7. #1: 8% and 12%, $\bar{x} = 10\%$. #2: 8% and −5%, $\bar{x} = 1.5\%$. #3: 8% and −20%, $\bar{x} = -6\%$. #4: 8% and 25%, $\bar{x} = 16.5\%$. #5: 12% and −5%, $\bar{x} = 3.5\%$; #6: 12% and −20%, $\bar{x} = -4\%$; #7: 12% and 25%, $\bar{x} = 18.5\%$. #8: −5% and −20%, $\bar{x} = -12.5\%$. #9: −5% and 25%, $\bar{x} = 10\%$. #10: −20% and 25%, $\bar{x} = 2.5\%$

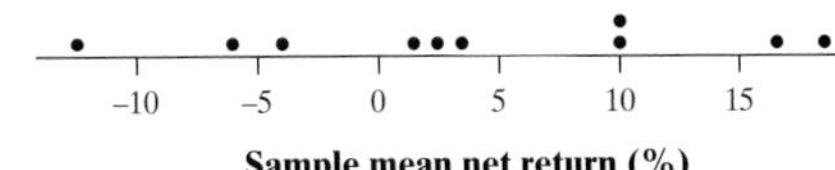

9. #1: 8% and 12%, $\hat{p} = 1$. #2: 8% and −5%, $\hat{p} = 0.5$. #3: 8% and −20%, $\hat{p} = 0.5$. #4: 8% and 25%, $\hat{p} = 1$. #5: 12% and −5%, $\hat{p} = 0.5$. #6: 12% and −20%, $\hat{p} = 0.5$. #7: 12% and 25%, $\hat{p} = 1$. #8: −5% and −20%, $\hat{p} = 0$. #9: −5% and 25%, $\hat{p} = 0.5$. #10: −20% and 25%, $\hat{p} = 0.5$.

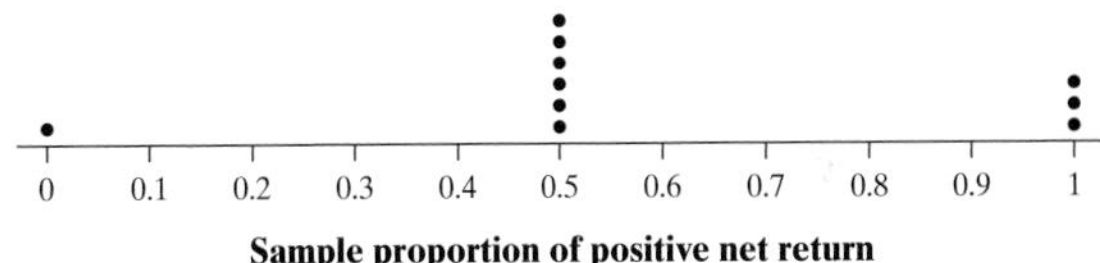

11. (a) Population distribution:

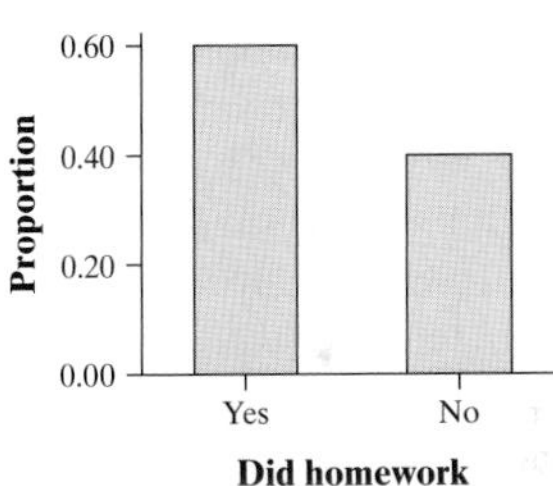

(b) Answers will vary. Distribution of sample data:

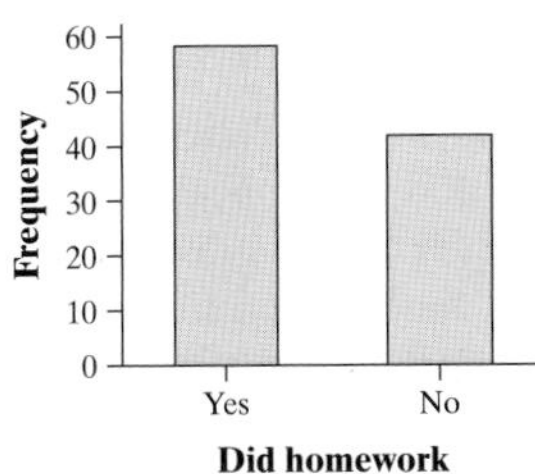

13. (a) $\hat{p} = 45/100 = 0.45$, which is less than 0.60. (b) It is possible that 60% of the students did their homework and that the students got a $\hat{p}$ less than 60% because of sampling variability. It is also possible that the sample proportion is less than 60% because less than 60% of all the students did their homework. (c) In one simulated SRS of 100 students, 73 students ($\hat{p} = 73\%$) did their assigned homework last week. (d) Yes. There were no values of $\hat{p}$ less than or equal to 0.45 in the simulation, so it would be very surprising to get a sample proportion of 0.45 or less by chance alone in an SRS of size 100 from a population in which $p = 0.60$. (e) It is unlikely (probability ≈ 0) to get a sample proportion of $\hat{p} = 0.45$ or less in an SRS of size 100 when $p = 0.60$, so we have convincing

evidence that less than 60% of all students at the school completed their assigned homework last week.

15. It is not unlikely (probability $\approx 78/250 = 31.2\%$) to get a sample proportion of $\hat{p} = 0.57$ or less in an SRS of size 100 from a population in which $p = 0.60$ by chance alone, so this would not provide convincing evidence that the population proportion of students who did all their assigned homework is less than $p = 60\%$.

17. A sample SD of 5°F is quite large compared with what we would expect to happen by chance alone, because none of the 300 simulated SRSs had a sample SD that large. A sample SD of 5°F provides convincing evidence that the manufacturer's claim is false and that the thermostat actually has more variability than claimed.

19. The population proportion of red cars is $p = 2/4 = 0.5$. The mean of the sampling distribution of the sample proportion is $\dfrac{0+0.5+0.5+0.5+0.5+1}{6} = 0.5$. The sample proportion is an unbiased estimator of the population proportion because the mean of the sampling distribution of the sample proportion of red cars is equal to the population proportion of red cars.

21. If we chose many SRSs of size n and calculated the sample mean cholesterol level $\bar{x}$ for each sample, the distribution of $\bar{x}$ would be centered at the value of the population mean cholesterol level, μ. In other words, when we use the sample mean cholesterol level to estimate the population mean cholesterol level, we will not consistently underestimate or consistently overestimate the population mean.

23. **(a)** The variability in the sampling distribution for samples of size $n = 3$ will be less than the variability in the sampling distribution for samples of size $n = 2$. Increasing the sample size decreases the variability. **(b)** By increasing the sample size, the estimated proportion is more likely to be close to the population proportion of red cars. In other words, the estimate will be more precise.

25. $n = 10$. Sample medians obtained from smaller sample sizes are more variable than those obtained from larger sample sizes, so the sample size of $n = 10$ is more likely to produce a sample median greater than \$250,000 (farther from \$200,000). The larger sample size ($n = 100$) is more likely to produce a sample median that is close to \$200,000, the population median.

27. **(a)** Statistics (ii) and (iii) appear to be unbiased estimators because the means of their sampling distributions appear to be equal to the corresponding population parameters. **(b)** Statistic (ii) does the best job at estimating the parameter. It is unbiased and has little variability.

29. C

31. B

33. **(a)** *Explanatory variable*: light intensity. *Response variable*: average number of flowers produced. **(b)** *Experimental units*: the 12 trays **(c)** Label the trays 1 to 12. Write 1 to 12 on slips of paper, put them in a container, and mix thoroughly. Pull out 2 different slips of paper and assign the corresponding trays to 150-level light intensity. Do not replace the slips of paper. Then, pull out 2 more different slips of paper and assign the corresponding trays to 300-level light intensity. Repeat this process for 450-, 600-, and 750-level light intensity. The final 2 trays will receive 900-level light intensity. **(d)** The experiment incorporated replication by using more than 1 tray in each treatment group.

Section 5C

Answers to Check Your Understanding

page 491: **1.** $\mu_{\hat{p}} = p = 0.75$. If you selected all possible samples of 1000 young adult internet users and calculated the sample proportion who watch online videos for each sample, the sample proportions would have an average value of 0.75.

2. It is safe to assume that $n = 1000$ is less than 10% of all young adult internet users. $\sigma_{\hat{p}} = \sqrt{\dfrac{p(1-p)}{n}} = \sqrt{\dfrac{0.75(1-0.75)}{1000}} = 0.0137$. If you selected all possible samples of 1000 young adult internet users and calculated the sample proportion who watch online videos for each sample, the sample proportions would typically vary from the population proportion of $p = 0.75$ by about 0.0137.

3. Because $np = 1000(0.75) = 750 \geq 10$ and $n(1-p) = 1000(1-0.75) = 250 \geq 10$, the sampling distribution of $\hat{p}$ is approximately normal.

4. We want to find $P(\hat{p} > 0.80)$.

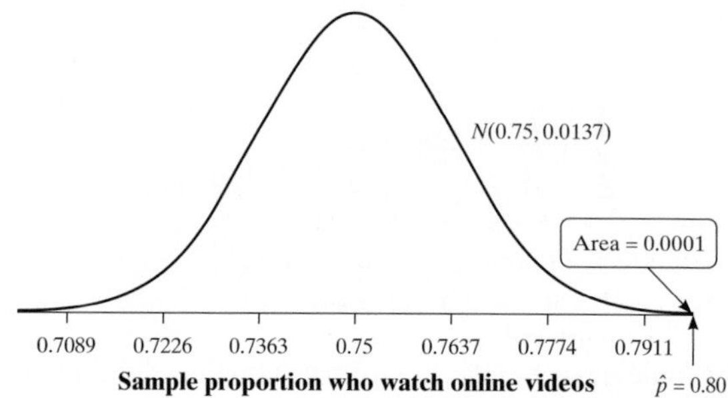

(i) $z = \dfrac{0.8-0.75}{0.0137} = 3.650$; *Tech:* normalcdf(lower: 3.650, upper: 1000, mean: 0, SD: 1) = 0.0001; *Table* A: less than $1-0.9998 = 0.0002$. (ii) normalcdf(lower: 0.8, upper: 1000, mean: 0.75, SD: 0.0137) = 0.0001.

page 495: **1.** Let $\hat{p}_A$ = the proportion of the sample of 200 fans from Stadium A who are wearing clothing that represents the home team and let $\hat{p}_B$ = the proportion of the sample of 200 fans from Stadium B who are wearing clothing that represents the home team. The shape of the sampling distribution of $\hat{p}_A - \hat{p}_B$ is approximately normal because $n_A p_A = 200(0.61) = 122$, $n_A(1-p_A) = 200(0.39) = 78$, $n_B p_B = 200(0.65) = 130$, and $n_B(1-p_B) = 200(0.35) = 70$ are all ≥ 10.

2. $\mu_{\hat{p}_A - \hat{p}_B} = p_A - p_B = 0.61 - 0.65 = -0.04$. In all possible independent random samples of 200 fans from Stadium A and 200 fans from Stadium B, the resulting differences in sample proportions $(\hat{p}_A - \hat{p}_B)$ of fans wearing clothing that represents the home team have an average of -0.04.

3. Because $200 < 10\%$ of the 50,000 fans at Stadium A and $200 < 10\%$ of the 70,000 fans at Stadium B, and the two samples are independent,

$$\sigma_{\hat{p}_A - \hat{p}_B} = \sqrt{\frac{p_A(1-p_A)}{n_A} + \frac{p_B(1-p_B)}{n_B}} = \sqrt{\frac{0.61(0.39)}{200} + \frac{0.65(0.35)}{200}} = 0.0482$$

In all possible independent random samples of 200 fans from Stadium A and 200 fans from Stadium B, the resulting differences in sample proportions $(\hat{p}_A - \hat{p}_B)$ of fans wearing clothing that represents the home team typically vary from the difference in population proportions of -0.04 by about 0.0482.

4. We want to find $P(-0.05 \le \hat{p}_A - \hat{p}_B \le 0.05)$.

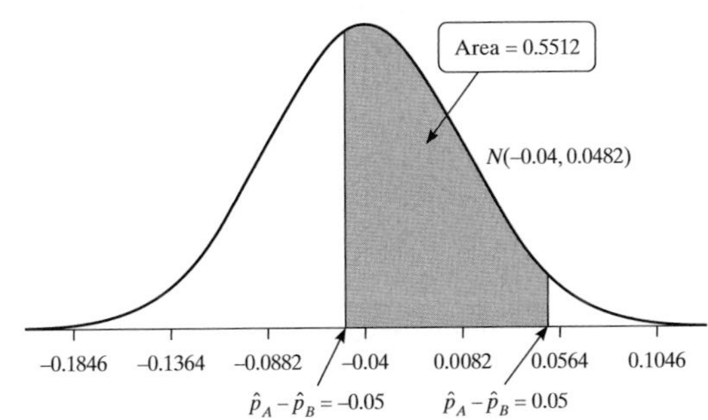

Difference in sample proportions of fans wearing clothing that represents the home team

(i) $z = \dfrac{-0.05-(-0.04)}{0.0482} = -0.207$ and $z = \dfrac{0.05-(-0.04)}{0.0482} = 1.867$; *Tech:* normalcdf (lower: −0.207, upper: 1.867, mean: 0, SD: 1) = 0.5510; *Table* A: 0.9693 − 0.4168 = 0.5525 (ii) normalcdf (lower: −0.05, upper: 0.05, mean: −0.04, SD: 0.0482) = 0.5512.

Answers to Odd-Numbered Section 5C Exercises

1. **(a)** $\mu_{\hat{p}} = p = 0.55$. If you selected all possible samples of 500 registered voters from this district and calculated the sample proportion who are Democrats for each sample, the sample proportions would have an average value of 0.55. **(b)** It is safe to assume that $n = 500$ registered voters is less than 10% of the population of all registered voters in this district.

$$\sigma_{\hat{p}} = \sqrt{\frac{p(1-p)}{n}} = \sqrt{\frac{0.55(1-0.55)}{500}} = 0.022$$

If you selected all possible samples of 500 registered voters from this district and calculated the sample proportion who are Democrats for each sample, the sample proportions would typically vary from the population proportion of 0.55 by about 0.022.

3. **(a)** $\mu_{\hat{p}} = p = 0.20$. It is safe to assume that $n = 30$ Skittles® is less than 10% of the population of all Skittles in the large bag, so

$$\sigma_{\hat{p}} = \sqrt{\frac{p(1-p)}{n}} = \sqrt{\frac{0.20(1-0.20)}{30}} = 0.073$$

(b) *Mean:* If you selected all possible samples of 30 Skittles from the large bag and calculated the sample proportion of orange Skittles for each sample, the sample proportions would have an average value of 0.20. *Standard deviation:* If you selected all possible samples of 30 Skittles from the large bag and calculated the sample proportion of orange Skittles for each sample, the sample proportions would typically vary from the population proportion of 0.20 by about 0.073.

5. Multiplying the sample size by 4, so $n = 4 \cdot 30 = 120$, would reduce the SD of the sampling distribution to one-half of the value found in Exercise 3(b).

$$\sigma_{\hat{p}} = \sqrt{\frac{p(1-p)}{n}} = \sqrt{\frac{0.20(1-0.20)}{120}} = 0.0365$$

7. **(a)** The 10% condition is not met because more than 10% of the population (14/76 = 18.4%) was selected. **(b)** Because the TSA officers sampled more than 10% of the population, the actual SD of the sampling distribution of $\hat{p}$ will be smaller than the value provided by the formula $\sigma_{\hat{p}} = \sqrt{\dfrac{p(1-p)}{n}}$.

9. Because $np = 500(0.55) = 275 \ge 10$ and $n(1-p) = 500(1-0.55) = 225 \ge 10$, the sampling distribution of $\hat{p}$ is approximately normal.

11. Because $np = 30(0.2) = 6 < 10$, the sampling distribution of $\hat{p}$ is not approximately normal. Because $p = 0.20$ is closer to 0 than to 1, the sampling distribution of $\hat{p}$ is skewed to the right.

13. **(a)** $\mu_{\hat{p}} = p = 0.70$ **(b)** $n = 1012$ is less than 10% of the population of all U.S. adults, so $\sigma_{\hat{p}} = \sqrt{\dfrac{0.7(1-0.3)}{1012}} = 0.0144$ **(c)** $np = 1012(0.70) = 708.4 \ge 10$ and $n(1-p) = 1012(1-0.70) = 303.6 \ge 10$, so the sampling distribution of $\hat{p}$ is approximately normal. **(d)** We want to find $P(\hat{p} \le 0.67)$. (i) $z = \dfrac{0.67-0.70}{0.0144} = -2.083$; *Tech:* normalcdf(lower: −1000, upper: −2.083, mean: 0, SD: 1) = 0.0186; *Table* A: 0.0188. (ii) normalcdf(lower: −1000, upper: 0.67, mean: 0.70, SD: 0.0144) = 0.0186 **(e)** Yes. It is unlikely (probability = 0.0186) to obtain a sample of size 1012 in which 67% or fewer drink the milk by chance alone when $p = 0.70$, so there is convincing evidence against the claim that $p = 0.70$.

15. We want to find $P(\hat{p} > 0.20)$, where $\hat{p}$ is the sample proportion of teachers who work second jobs. The sampling distribution of $\hat{p}$ is approximately normal because $np = (200)(0.18) = 36 \ge 10$ and $n(1-p) = (200)(1-0.18) = 164 \ge 10$.

$$\mu_{\hat{p}} = p = 0.18 \quad \sigma_{\hat{p}} = \sqrt{\frac{0.18(1-0.18)}{200}} = 0.027$$

(i) $z = \dfrac{0.20-0.18}{0.027} = 0.741$; *Tech:* normalcdf(lower: 0.741, upper: 1000, mean: 0, SD: 1) = 0.2293; *Table* A: 1 − 0.7704 = 0.2296 (ii) normalcdf(lower: 0.20, upper: 1000, mean: 0.18, SD: 0.027) = 0.2294. There is about a 0.2294 probability of selecting a random sample of 200 teachers and finding that more than 20% of the sample works a second job during the school year.

17. **(a)** X is a binomial random variable with $n = 100$ and $p = 0.90$ because the following conditions are met: *Binary?* "Success" = shipped on time; "Failure" = not shipped on time. *Independent?* Knowing whether one order is shipped on time tells you nothing about whether another order will be shipped on time, and $n = 100$ is less than 10% of N = 5000. *Number?* $n = 100$. *Same probability?* $p = 0.90$. **(b)** Because $np = 100(0.90) = 90 \ge 10$ and $n(1-p) = 100(1-0.90) = 10 \ge 10$, the sampling distribution of X can be approximated by a normal distribution. **(c)** $\mu_X = 100(0.90) = 90$ and $\sigma_X = \sqrt{100(0.9)(0.1)} = 3$. We want to find $P(X \le 86)$. (i) $z = \dfrac{86-90}{3} = -1.333$; *Tech:* normalcdf (lower: −1000, upper: −1.333, mean: 0, SD: 1) = 0.0913; *Table* A: 0.0918 (ii) normalcdf (lower: −1000, upper: 86, mean: 90, SD: 3) = 0.0912 **(d)** No. It isn't unlikely (probability = 0.0912) for 86 or fewer orders in an SRS of 100 to be shipped within 3 working days by chance alone when $p = 0.90$. Therefore, we do not have convincing evidence that less than 90% of all orders from this company are shipped within three working days.

19. **(a)** No, the shape of the sampling distribution is not approximately normal. $n_S p_S = 20(0.3) = 6$, $n_S(1-p_S) = 20(0.7) = 14$, $n_J p_J = 20(0.25) = 5$, $n_J(1-p_J) = 20(0.75) = 15$ are not all ≥ 10. **(b)** $\mu_{\hat{p}_S - \hat{p}_J} = 0.30 - 0.25 = 0.05$. In all possible independent random samples of 20 seniors and 20 juniors from this large high school, the resulting differences in sample proportions $(\hat{p}_S - \hat{p}_J)$ of students enrolled in an AP® class have an average of 0.05.

(c) Because 20 < 10% of all seniors at the large high school and 20 < 10% of all juniors at the large high school, and the two samples are independent, $\sigma_{\hat{p}_S-\hat{p}_J}=\sqrt{\frac{0.30(0.70)}{20}+\frac{0.25(0.75)}{20}}=0.141$. In all possible independent random samples of 20 seniors and 20 juniors from this large high school, the resulting differences in sample proportions $(\hat{p}_S-\hat{p}_J)$ of students enrolled in an AP® class typically vary from the difference in population proportions of 0.05 by about 0.141.

21. (a) The shape of the sampling distribution of $\hat{p}_C-\hat{p}_A$ is approximately normal because $n_C p_C=50(0.30)=15$, $n_C(1-p_C)=50(0.7)=35$, $n_A p_A=100(0.15)=15$, and $n_A(1-p_A)=100(0.85)=85$ are all ≥ 10. (b) $\mu_{\hat{p}_C-\hat{p}_A}=0.30-0.15=0.15$. Because 50 < 10% of the jelly beans in the Child mix and 100 < 10% of the jelly beans in the Adult mix, and the two samples are independent, $\sigma_{\hat{p}_C-\hat{p}_A}=\sqrt{\frac{0.3(0.7)}{50}+\frac{0.15(0.85)}{100}}=0.074$. (c) *Mean:* In all possible independent random samples of 50 jelly beans from the Child mix and 100 jelly beans from the Adult mix, the resulting differences in sample proportions $(\hat{p}_C-\hat{p}_A)$ of red jelly beans have an average of 0.15. *Standard deviation:* In all possible independent random samples of 50 jelly beans from the Child mix and 100 jelly beans from the Adult mix, the resulting differences in sample proportions $(\hat{p}_C-\hat{p}_A)$ of red jelly beans typically vary from the difference in population proportions of 0.15 by about 0.074.

23. (a) We want to find $P(\hat{p}_C \leq \hat{p}_A)$, which is equivalent to $P(\hat{p}_C-\hat{p}_A \leq 0)$ using a normal distribution with a mean of 0.15 and SD of 0.074. (i) $z=\frac{0-0.15}{0.074}=-2.027$; *Tech:* normalcdf(lower: −1000, upper: −2.027, mean: 0, SD: 1) = 0.0213; *Table* A: 0.0212 (ii) normalcdf (lower: −1000, upper: 0, mean: 0.15, SD: 0.0740) = 0.0213 (b) Yes, we should doubt the company's claim. There is only a 2% chance of getting a proportion of red jelly beans in the Child sample less than or equal to the proportion of red jelly beans in the Adult sample if the company's claim is true. This is not very likely.

25. (a) $\sigma_{\hat{p}}=\sqrt{\frac{0.6(1-0.6)}{10}}=0.1549$;

$$\sigma_{\hat{p}}=\sqrt{\frac{0.6(1-0.6)}{10}}\sqrt{\frac{1000-10}{1000-1}}=0.1542.$$

The SD of the sampling distribution of $\hat{p}$ is about the same whether the finite population correction factor is used or not.

(b) $\sigma_{\hat{p}}=\sqrt{\frac{0.6(1-0.6)}{500}}=0.0219$;

$$\sigma_{\hat{p}}=\sqrt{\frac{0.6(1-0.6)}{500}}\sqrt{\frac{1000-500}{1000-1}}=0.0155$$

The SD of the sampling distribution of $\hat{p}$ is much smaller when the finite population correction factor is used. (c) When the sample size is less than 10% of the population size [such as in part (a)], the finite population correction factor is not needed because the SD of the sampling distribution is about the same whether the finite population correction factor is used or not. However, when the sample size is at least 10% of the population size [such as part (b)], the finite population correction factor is needed to determine the SD of the sampling distribution of $\hat{p}$.

(d) $\sigma_{\hat{p}}=\sqrt{\frac{0.6(1-0.6)}{1000}}\sqrt{\frac{1000-1000}{1000-1}}=0$. In this case, all 1000 members of the population are selected for the sample. Because this is a census, the true value of the population proportion p will be known. There will be no variability in this value because it is a parameter.

27. A

29. D

31. (a) We are looking for $P(z \leq -2.5)$ in a normal distribution with mean 0 and SD 1. (i) $z=-2.5$; *Tech:* normalcdf(lower: −1000, upper: −2.5, mean: 0, SD: 1) = 0.0062; *Table* A: 0.0062. Less than 1% of healthy young adults have osteoporosis. (b) Let X be the BMD for women aged 70 to 79 on the standard scale. Then X has a normal distribution with a mean of −2 and SD of 1; we want to find $P(X \leq -2.5)$. (i) $z=\frac{-2.5-(-2)}{1}=-0.5$; *Tech:* normalcdf(lower: −1000, upper: −0.5, mean: 0, SD: 1) = 0.3085; *Table* A: 0.3085. (ii) normalcdf(lower: −1000, upper: −0.5, mean: −2, SD: 1) = 0.3085. About 30.85% of women aged 70–79 have osteoporosis.

Section 5D: Sample Means

Answers to Check Your Understanding

page 508: 1. Let X = weight (in ounces). The distribution of X is approximately normal with a mean of 16.1 ounces and a standard deviation of 0.15 ounce. We want to find $P(X < 16)$.

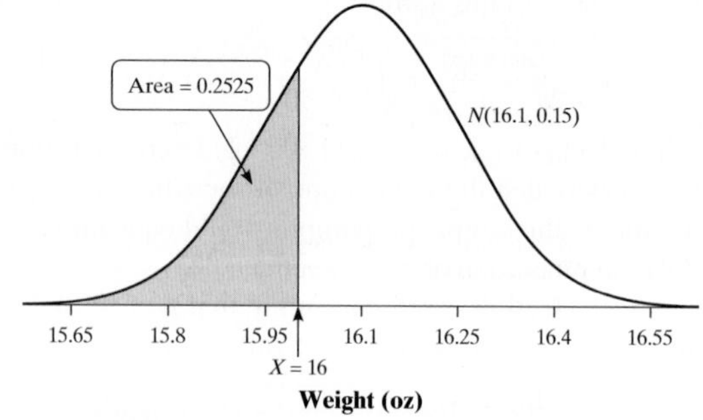

(i) $z=\frac{16-16.1}{0.15}=-0.667$; *Tech:* normalcdf(lower: −1000, upper: −0.667, mean: 0, SD: 1) = 0.2524; *Table* A: 0.2514 (ii) *Tech:* normalcdf(lower: −1000, upper: 16, mean: 16.1, SD: 0.15) = 0.2525. 2. Let $\bar{x}$ = the mean weight (in ounces) of 10 randomly selected jars. The distribution of $\bar{x}$ is approximately normal because the population distribution is approximately normal. $\mu_{\bar{x}}=\mu=16.1$ ounces; $\sigma_{\bar{x}}=\frac{\sigma}{\sqrt{n}}=\frac{0.15}{\sqrt{10}}=0.047$ ounce because it is reasonable to assume that 10 < 10% of all jars. We want to find $P(\bar{x} < 16)$.

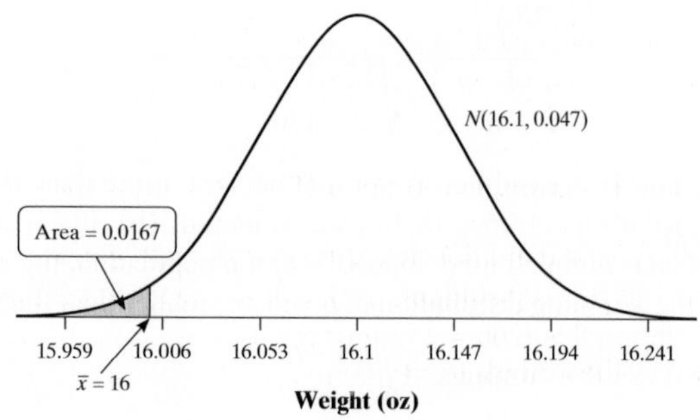

(i) $z = \dfrac{16-16.1}{0.047} = -2.128$; *Tech:* normalcdf(lower: -1000, upper: -2.128, mean: 0, SD: 1) $= 0.0167$; *Table* A: 0.0166 (ii) *Tech:* normalcdf(lower: -1000, upper: 16, mean: 16.1, SD: 0.047) $= 0.0167$.

page 512: **1.** The shape of the sampling distribution of $\bar{x}_E - \bar{x}_W$ is approximately normal because $n_E = 200 \geq 30$ and $n_W = 200 \geq 30$.
2. $\mu_{\bar{x}_E-\bar{x}_W} = 0.4 - 0.5 = -0.1$ class. In all possible independent random samples of 200 students from East High School and 200 students from West High School, the resulting differences in sample mean number of AP® classes that students are taking ($\bar{x}_E - \bar{x}_W$) have an average of -0.1 class.
3. Because the samples are independent, and because $200 < 10\%$ of the 2100 students at East High School and $200 < 10\%$ of the 2500 students at West High School, $\sigma_{\bar{x}_E-\bar{x}_W} = \sqrt{\dfrac{0.8^2}{200} + \dfrac{0.7^2}{200}} = 0.075$ class. In all possible independent random samples of 200 students from East High School and 200 students from West High School, the resulting differences in sample mean number of AP® classes that students are taking ($\bar{x}_E - \bar{x}_W$) typically vary from the difference in population means of -0.1 class by about 0.075 class.
4. We want to find $P(\bar{x}_E < \bar{x}_W)$, which is equivalent to $P(\bar{x}_E - \bar{x}_W < 0)$.

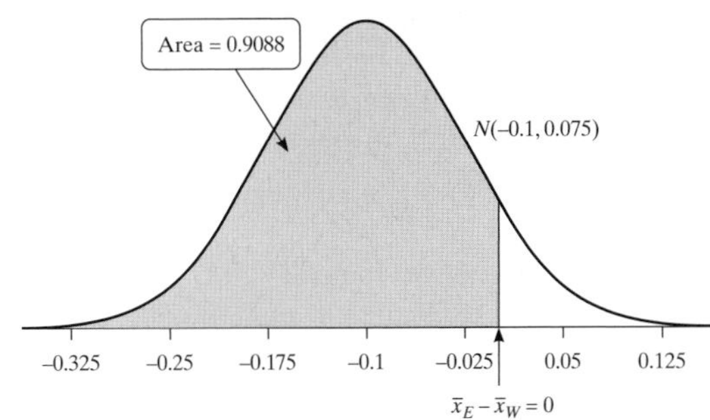

(i) $z = \dfrac{0-(-0.1)}{0.075} = 1.333$; *Tech:* normalcdf(lower: -1000, upper: 1.333, mean: 0, SD: 1) $= 0.9087$; *Table* A: 0.9082 (ii) *Tech:* normalcdf(lower: -1000, upper: 0, mean: -0.1, SD: 0.075) $= 0.9088$.

Answers to Odd-Numbered Section 5D Exercises

1. **(a)** $\mu_{\bar{x}} = \mu = 225$ seconds. If you selected all possible samples of 10 songs from the playlist and calculated the sample mean play time for each sample, the sample means would have an average value of 225 seconds. **(b)** Because $n = 10$ is less than 10% of the population of 1000 songs on David's playlist, $\sigma_{\bar{x}} = \dfrac{\sigma}{\sqrt{n}} = \dfrac{60}{\sqrt{10}} = 18.974$ seconds. If you selected all possible samples of 10 songs from the playlist and calculated the sample mean play time for each sample, the sample means would typically vary from the population mean play time of 225 seconds by about 18.974 seconds.

3. We need to sample $n = 36$ songs. $10 = \dfrac{60}{\sqrt{n}} \rightarrow n = 36$

5. **(a)** The 10% condition is not met because $n = 50$ is not less than 10% of the 200 students at the small school. The administrators sampled $50/200 = 25\%$ of the students from the school. **(b)** Because the administrators surveyed more than 10% of the population of 200 students, the actual SD of the sampling distribution of $\bar{x}$ will be smaller than the value provided by the formula $\sigma_{\bar{x}} = \dfrac{\sigma}{\sqrt{n}}$.

7. **(a)** Because $n = 5 < 30$, the sampling distribution of $\bar{x}$ will also be skewed to the right, but not quite as strongly as the population. **(b)** Because $n = 100 \geq 30$, the sampling distribution of $\bar{x}$ is approximately normal by the central limit theorem.

9. **(a)** Because the population distribution is approximately normal, the sampling distribution of $\bar{x}$ will also be approximately normal, regardless of the sample size. **(b)** Because the population distribution is approximately normal, the sampling distribution of $\bar{x}$ will also be approximately normal, regardless of the sample size.

11. No. The histogram of the sample values will look like the population distribution, whatever it might happen to be. The central limit theorem says that the histogram of the *sampling distribution of the sample mean* will look more and more normal as the sample size increases.

13. Let $\bar{x}$ = the mean number of flaws in 50 randomly selected balls of yarn. The distribution of $\bar{x}$ is approximately normal because $n = 50 \geq 30$. $\mu_{\bar{x}} = \mu = 1.6$ flaws, $\sigma_{\bar{x}} = \dfrac{\sigma}{\sqrt{n}} = \dfrac{1.2}{\sqrt{50}} = 0.170$ flaw because it is reasonable to assume that $50 < 10\%$ of all balls of this variety of yarn. We want to find $P(\bar{x} \geq 2)$. (i) $z = \dfrac{2-1.6}{0.17} = 2.353$; *Tech:* normalcdf(lower: 2.353, upper: 1000, mean: 0, SD: 1) $= 0.0093$; *Table* A: $1 - 0.9906 = 0.0094$. (ii) *Tech:* normalcdf(lower: 2, upper: 1000, mean: 1.6, SD: 0.17) $= 0.0093$

15. **(a)** Let $\bar{x}$ = the mean lifetime for a random sample of 8 batteries. The distribution of $\bar{x}$ is approximately normal because the population distribution is normally distributed. $\mu_{\bar{x}} = \mu = 48$ months, $\sigma_{\bar{x}} = \dfrac{\sigma}{\sqrt{n}} = \dfrac{8.2}{\sqrt{8}} = 2.899$ months because it is reasonable to assume that $8 < 10\%$ of all batteries of this type. We want to find $P(\bar{x} \leq 42.2)$. (i) $z = \dfrac{42.2-48}{2.899} = -2.001$; *Tech:* normalcdf(lower: -1000, upper: -2.001, mean: 0, SD: 1) $= 0.0227$; *Table* A: 0.0228. (ii) *Tech:* normalcdf(lower: -1000, upper: 42.2, mean: 48, SD: 2.899) $= 0.0227$ **(b)** Because this probability is very small, there is convincing evidence that the company is overstating the average lifetime of its batteries. It is not plausible to get a sample mean this small by chance alone.

17. **(a)** Let $\bar{y}$ = the mean profit for a random sample of 10,000 policies. The distribution of $\bar{y}$ is approximately normal because $n = 10{,}000 \geq 30$. $\mu_{\bar{y}} = \mu = \$303.35$, $\sigma_{\bar{y}} = \dfrac{\sigma}{\sqrt{n}} = \dfrac{9707.57}{\sqrt{10{,}000}} = \$97.076$ because it is reasonable to assume that $10{,}000 < 10\%$ of all insurance policies sold by this company. **(b)** We want to find $P(\bar{y} > 0)$.

(i) $z = \dfrac{0-303.35}{97.076} = -3.125$; *Tech:* normalcdf(lower: -3.125, upper: 1000, mean: 0, SD: 1) $= 0.9991$; *Table* A: $1 - 0.0009 = 0.9991$. (ii) *Tech:* normalcdf(lower: 0, upper: 10000, mean: 303.35, SD: 97.076) $= 0.9991$

19. **(a)** No, because the population distributions are skewed to the right and $n_N = 10 < 30$ and $n_S = 10 < 30$. **(b)** $\mu_{\bar{x}_N-\bar{x}_S} = 410{,}000 - 375{,}000 = \$35{,}000$. In all possible independent random samples of 10 homes in the northern part of the large city and 10 homes in the southern part of the large city, the resulting differences in sample mean home value ($\bar{x}_N - \bar{x}_S$) have an average of \$35,000. **(c)** Because the samples are independent, and because $10 < 10\%$ of all homes in the northern part of the city and $10 < 10\%$ of all homes in the southern part of the city, $\sigma_{\bar{x}_N - \bar{x}_S} = \sqrt{\frac{250{,}000^2}{10} + \frac{240{,}000^2}{10}} = \$109{,}590.15$. In all possible independent random samples of 10 homes in the northern part of the large city and 10 homes in the southern part of the large city, the resulting differences in sample mean home value ($\bar{x}_N - \bar{x}_S$) typically vary from the difference in population means of \$35,000 by about \$109,590.15.

21. **(a)** Both population distributions are approximately normal, so the distribution of $\bar{x}_{YA} - \bar{x}_T$ is approximately normal regardless of the individual sample sizes. **(b)** $\mu_{\bar{x}_{YA} - \bar{x}_T} = 188 - 170 = 18$ mg/dl. Because the samples are independent, and because $25 < 10\%$ of all young adults and $36 < 10\%$ of all teenagers, $\sigma_{\bar{x}_{YA} - \bar{x}_T} = \sqrt{\frac{41^2}{25} + \frac{30^2}{36}} = 9.604$ mg/dl. **(c)** We want to find $P(\bar{x}_{YA} - \bar{x}_T < 0)$. (i) $z = \frac{0 - 18}{9.604} = -1.874$; *Tech:* normalcdf(lower: −1000, upper: −1.874, mean: 0, SD: 1) = 0.0305; *Table* A: 0.0307 (ii) *Tech:* normalcdf(lower: −1000, upper: 0, mean: 18, SD: 9.604) = 0.0305

23. Let $\bar{x}_F$ = the mean volume of the first random sample of 10 bottles and $\bar{x}_S$ = the mean volume of the second random sample of 10 bottles. The sampling distribution of $\bar{x}_F - \bar{x}_S$ is approximately normal because both population distributions are approximately normal. $\mu_{\bar{x}_F - \bar{x}_S} = 298 - 298 = 0$ ml. Because the samples are independent, and because $10 < 10\%$ of all bottles filled the first hour and $10 < 10\%$ of all bottles filled the second hour, $\sigma_{\bar{x}_F - \bar{x}_S} = \sqrt{\frac{3^2}{10} + \frac{3^2}{10}} = 1.342$ ml. We want to find $P(\bar{x}_F - \bar{x}_S < -1$ or $\bar{x}_F - \bar{x}_S > 1)$. (i) $z = \frac{-1 - 0}{1.342} = -0.745$ and $z = \frac{1 - 0}{1.342} = 0.745$; *Tech:* normalcdf(lower: −1000, upper: −0.745, mean: 0, SD: 1) ×2 = 0.4563; *Table* A: 0.2266 × 2 = 0.4532 (ii) *Tech:* normalcdf (lower: −1000, upper: −1, mean: 0, SD: 1.342) × 2 = 0.4562

25. **(a)** The distribution of $\bar{x}$ is approximately normal because $n = 50 \geq 30$. $\mu_{\bar{x}} = \mu = 6$ strikes, $\sigma_{\bar{x}} = \frac{\sigma}{\sqrt{n}} = \frac{2.4}{\sqrt{50}} = 0.339$ strike because it is reasonable to assume that $50 < 10\%$ of all 1-square-kilometer plots of land. **(b)** Approximately normal, because the distribution of $\bar{x}$ is approximately normal and T is a linear transformation of $\bar{x}$. $\mu_T = 50(\mu_{\bar{x}}) = 50(6) = 300$ strikes, $\sigma_T = 50(\sigma_{\bar{x}}) = 50\left(\frac{2.4}{\sqrt{50}}\right) = 16.971$ strikes. **(c)** We want to find $P(T < 250)$. (i) $z = \frac{250 - 300}{16.971} = -2.946$; *Tech:* normalcdf(lower: −1000, upper: −2.946, mean: 0, SD: 1) = 0.0016; *Table* A: 0.0016 (ii) *Tech:* normalcdf(lower: −1000, upper: 250, mean: 300, SD: 16.971) = 0.0016 **(d)** $\mu_T = \mu_{X_1} + \mu_{X_2} + \ldots + \mu_{X_{50}} = 6 + 6 + \ldots + 6 = 50(6) = 300$ strikes. $\sigma_T = \sqrt{\sigma^2_{X_1} + \sigma^2_{X_2} + \ldots + \sigma^2_{X_{50}}} = \sqrt{2.4^2 + 2.4^2 + \ldots + 2.4^2} = \sqrt{50(2.4)^2}$ 16.971 strikes **(e)** The distribution of $X_1 + X_2 + \ldots + X_n$ would be approximately normal if X is approximately normally distributed or $n \geq 30$. The same rules apply as those for the sampling distribution of the sample mean.

27. C

29. D

31. Unemployment rate for each level of education:

Didn't finish high school: $\frac{(8885 - 8443)}{8885} = \frac{442}{8885} = 0.050$

High school, but no college: $\frac{(35{,}605 - 34{,}339)}{35{,}605} = \frac{1266}{35{,}605} = 0.036$

Less than a bachelor's degree: $\frac{(35{,}789 - 34{,}735)}{35{,}789} = \frac{1054}{35{,}789} = 0.029$

College graduate: $\frac{(63{,}150 - 61{,}947)}{63{,}150} = \frac{1203}{63{,}150} = 0.019$

There is an association between unemployment rate and education. The unemployment rate decreases with additional education.

Answers to Unit 5 Review Exercises

R1. **(a)** Let X = birth weight (in grams). The distribution of X is approximately normal with a mean of 3668 grams and SD of 511 grams. We want to find $P(X > 3000)$. (i) $z = \frac{3000 - 3668}{511} = -1.307$; *Tech:* normalcdf(lower: −1.307, upper: 1000, mean: 0, SD: 1) = 0.9044; *Table* A: 1 − 0.0951 = 0.9049 (ii) normalcdf(lower: 3000, upper: 10000, mean: 3668, SD: 511) = 0.9044. **(b)** We want to find the 90th percentile of the distribution of X. (i) *Tech:* invNorm(area: 0.90, mean: 0, SD: 1) = 1.282; *Table* A: $z = 1.28$, $1.282 = \frac{x - 3668}{511}$, $x = (1.282)(511) + 3668$, $x = 4323.102$ g (ii) invNorm(area: 0.90, mean: 3668, SD: 511) = 4322.873 g

R2. Let T = the total resistance of one 100-ohm resistor and one 250-ohm resistor. Then $T = X + Y$. The distribution of T is approximately normal because it is the sum of independent normally distributed random variables. $\mu_T = 100 + 250 = 350$ ohms; $\sigma_T = \sqrt{2.5^2 + 2.8^2} = 3.754$ ohms (i) $z = \frac{345 - 350}{3.754} = -1.332$ and $z = \frac{355 - 350}{3.754} = 1.332$; *Tech:* normalcdf(lower: −1.332, upper: 1.332, mean: 0, SD: 1) = 0.8171; *Table* A: 0.9082 − 0.0918 = 0.8164 (ii) normalcdf(lower: 345, upper: 355, mean: 350, SD: 3.754) = 0.8171

R3. *Population:* all eggs shipped in one day from a large egg producer. *Parameter:* p = the proportion of eggs shipped that day that had salmonella contamination. *Sample:* the 200 eggs examined. *Statistic:* $\hat{p}$ = the proportion of eggs in the sample that had salmonella contamination = 9/200 = 0.045.

R4. **(a)** The 10 possible SRSs of size $n = 3$ and the median score of each sample are: #1: 64, 66, 71, median = 66. #2: 64, 66, 73, median = 66. #3: 64, 66, 76, median = 66. #4: 64, 71, 73, median = 71. #5: 64, 71, 76, median = 71. #6: 64, 73, 76, median = 73. #7: 66, 71, 73, median = 71. #8: 66, 71, 76, median = 71. #9: 66, 73, 76, median = 73. #10: 71, 73, 76, median = 73.

66 67 68 69 70 71 72 73

Sample median number of pages

(b) Median of the population: 71 pages. Mean of the sampling distribution of the sample median: $\frac{66 + 66 + 66 + 71 + 71 + 71 + 71 + 73 + 73 + 73}{10} = 70.1$. Because the mean of the sampling distribution of the sample median number of pages is not equal to the median of the population, the sample median is not an unbiased estimator of the population median. **(c)** The variability in the sampling distribution of the sample median number of pages for samples of size $n = 4$ will be less than the variability in the sampling distribution for samples of size $n = 3$. Increasing the sample size decreases the variability of the sampling distribution.

R5. (a) Let $\hat{p}$ = the proportion of the 1540 adults who regularly jog. $\mu_{\hat{p}} = p = 0.15$. If you selected all possible samples of 1540 adults and calculated the sample proportion who regularly jog for each sample, the sample proportions would have an average value of 0.15. **(b)** Because $n = 1540$ is less than 10% of the population of all adults, $\sigma_{\hat{p}} = \sqrt{\frac{0.15(0.85)}{1540}} = 0.0091$. If you selected all possible samples of 1540 adults and calculated the sample proportion who regularly jog for each sample, the sample proportions would typically vary from the population proportion of $p = 0.15$ by about 0.0091. **(c)** Because $np = 1540(0.15) = 231 \geq 10$ and $n(1-p) = 1540(0.85) = 1309 \geq 10$, the sampling distribution of $\hat{p}$ is approximately normal. **(d)** We want to find $P(0.13 \leq \hat{p} \leq 0.17)$. (i) $z = \frac{0.13 - 0.15}{0.0091} = -2.198$ and $z = \frac{0.17 - 0.15}{0.0091} = 2.198$; *Tech:* normalcdf(lower: −2.198, upper: 2.198, mean: 0, SD: 1) = 0.9721; *Table A*: 0.9861 − 0.0139 = 0.9722 (ii) normalcdf(lower: 0.13, upper: 0.17, mean: 0.15, SD: 0.0091) = 0.9720

R6. Let $\hat{p}_H$ = the proportion of the 70 cars in Hyeyoung's sample that are made by American manufacturers and $\hat{p}_E$ = the proportion of the 100 cars in Eliav's sample that are made by American manufacturers. The shape of the sampling distribution of $\hat{p}_H - \hat{p}_E$ is approximately normal because $n_H p_H = 70(0.60) = 42$, $n_H(1-p_H) = 70(0.4) = 28$, $n_E p_E = 100(0.8) = 80$, and $n_E(1-p_E) = 100(0.2) = 20$ are all ≥ 10. $\mu_{\hat{p}_H - \hat{p}_E} = 0.60 - 0.80 = -0.20$. Because $70 < 10\%$ of all registered cars in Hyeyoung's state and $100 < 10\%$ of all registered cars in Eliav's state, and the two samples are independent, $\sigma_{\hat{p}_H - \hat{p}_E} = \sqrt{\frac{0.6(0.4)}{70} + \frac{0.8(0.2)}{100}} = 0.071$. We want to find $P(\hat{p}_H - \hat{p}_E > 0)$. (i) $z = \frac{0 - (-0.2)}{0.071} = 2.817$; *Tech:* normalcdf(lower: 2.817, upper: 10000, mean: 0, SD: 1) = 0.0024; *Table A*: 1 − 0.9976 = 0.0024 (ii) normalcdf (lower: 0, upper: 1000, mean: −0.2, SD: 0.071) = 0.0024.

R7. (a) Let $\bar{x}$ = the mean number of moths per trap in an SRS of 50 traps. The distribution of $\bar{x}$ is approximately normal because $n = 50 \geq 30$. $\mu_{\bar{x}} = \mu = 0.5$ moth, $\sigma_{\bar{x}} = \frac{\sigma}{\sqrt{n}} = \frac{0.7}{\sqrt{50}} = 0.099$ moth because it is reasonable to assume that $50 < 10\%$ of all traps throughout the state. We want to find $P(\bar{x} \geq 0.6)$. (i) $z = \frac{0.6 - 0.5}{0.099} = 1.010$; *Tech:* normalcdf(lower: 1.01, upper: 1000, mean: 0, SD: 1) = 0.1562; *Table A*: 1 − 0.8438 = 0.1562 (ii) *Tech:* normalcdf(lower: 0.6, upper: 1000, mean: 0.5, SD: 0.099) = 0.1562 **(b)** Because this probability is not small, there is not convincing evidence that the moth population is getting larger in this state. It is plausible to get a sample mean of 0.6 or more by chance alone when $\mu = 0.5$.

R8. (a) Yes, because $n_1 = 49 \geq 30$ and $n_2 = 49 \geq 30$. **(b)** $\mu_{\bar{x}_1 - \bar{x}_2} = 15 - 15 = 0$ cm. In all possible independent random samples of 49 candles made by Machine 1 and 49 candles made by Machine 2, the resulting differences in sample mean length $(\bar{x}_1 - \bar{x}_2)$ have an average of 0 cm. **(c)** Because the samples are independent, and because $49 < 10\%$ of all candles made by Machine 1 and $49 < 10\%$ of all candles made by Machine 2, $\sigma_{\bar{x}_1 - \bar{x}_2} = \sqrt{\frac{0.15^2}{49} + \frac{0.10^2}{49}} = 0.026$ cm. In all possible independent random samples of 49 candles made by Machine 1 and 49 candles made by Machine 2, the resulting differences in sample mean length $(\bar{x}_1 - \bar{x}_2)$ typically vary from the difference in population means of 0 cm by about 0.026 cm.

Answers to Unit 5 AP® Statistics Practice Test

T1. C **T2.** D **T3.** C **T4.** A **T5.** C
T6. C **T7.** B **T8.** B **T9.** E **T10.** C

T11. (a) $\mu_{\bar{x}} = \mu = \$50$. If you selected all possible samples of 10 households with internet access from the large city and calculated the sample mean fee for internet service for each sample, the sample means would have an average value of \$50. **(b)** Because $n = 10$ is less than 10% of the population of all households with internet access in this large city, $\sigma_{\bar{x}} = \frac{20}{\sqrt{10}} = \$6.32$. If you selected all possible samples of 10 households with internet access from the large city and calculated the sample mean fee for internet service for each sample, the sample means would typically vary from the population mean fee for internet service of \$50 by about \$6.32. **(c)** Because $n = 10 < 30$ and the shape of the population distribution is skewed to the right, the sampling distribution of $\bar{x}$ will be skewed to the right, but not as much as the population distribution.

T12. (a) Let $\hat{p}$ = the proportion of the 300 children in the sample who are from poverty-level households. The sampling distribution of $\hat{p}$ is approximately normal because $np = (300)(0.22) = 66 \geq 10$ and $n(1-p) = (300)(1-0.22) = 234 \geq 10$. $\mu_{\hat{p}} = p = 0.22$. Because $n = 300$ is less than 10% of the population of all children in this state, $\sigma_{\hat{p}} = \sqrt{\frac{0.22(0.78)}{300}} = 0.024$. We want to find $P(\hat{p} \geq 0.29)$. (i) $z = \frac{0.29 - 0.22}{0.024} = 2.917$; *Tech:* normalcdf(lower: 2.917, upper: 1000, mean: 0, SD: 1) = 0.0018; *Table A*: 1 − 0.9982 = 0.0018 (ii) normalcdf(lower: 0.29, upper: 1000, mean: 0.22, SD: 0.024) = 0.0018 **(b)** Because it is unlikely (probability = 0.0018) to get a sample proportion of 29% or greater by chance alone when $p = 0.22$, there is convincing evidence that the percentage of U.S. children under the age of six living in households with incomes less than the official poverty level in this state is greater than the national value of 22%.

T13. Let $\bar{x}_C$ = the sample mean word length in the children's book and $\bar{x}_T$ = the sample mean word length in the teen novel. The sampling distribution of $\bar{x}_C - \bar{x}_T$ is approximately normal because $n_C = 35 \geq 30$ and $n_T = 35 \geq 30$. $\mu_{\bar{x}_C - \bar{x}_T} = 3.7 - 4.3 = -0.6$ word. Because the samples are independent, and because $35 < 10\%$ of all words in the children's book and $35 < 10\%$ of all words in the teen novel, $\sigma_{\bar{x}_C - \bar{x}_T} = \sqrt{\frac{2.1^2}{35} + \frac{2.5^2}{35}} = 0.552$ word. We want to find $P(\bar{x}_C - \bar{x}_T > 0)$. (i) $z = \frac{0 - (-0.6)}{0.552} = 1.087$; *Tech:* normalcdf (lower: 1.087, upper: 1000, mean: 0, SD: 1) = 0.1385; *Table A*: 1 − 0.8621 = 0.1379. (ii) *Tech:* normalcdf(lower: 0, upper: 1000, mean: −0.6, SD: 0.552) = 0.1385

Answers to Cumulative AP® Practice Test 2

AP1 A **AP2** D **AP3** E **AP4** B **AP5** C
AP6 D **AP7** C **AP8** D **AP9** C **AP10** B
AP11 A **AP12** A **AP13** D **AP14** A **AP15** B

AP16 **(a)** This is an observational study. Subjects were not assigned to take (or not take) fish oil. **(b)** Two variables are confounded when their effects on the cholesterol level cannot be distinguished from one another. For example, people who take omega-3 fish oil might also be more health conscious in general and exercise more. If exercising more lowers cholesterol, researchers would not know whether it was the omega-3 fish oil or the exercise that lowered cholesterol. **(c)** No. Even though the difference was statistically significant, we cannot conclude that fish oil was the cause of the difference in mean cholesterol readings. This wasn't an experiment with random assignment and taking fish oil is possibly confounded with other good habits, such as healthful eating and exercise.

AP17 **(a)**

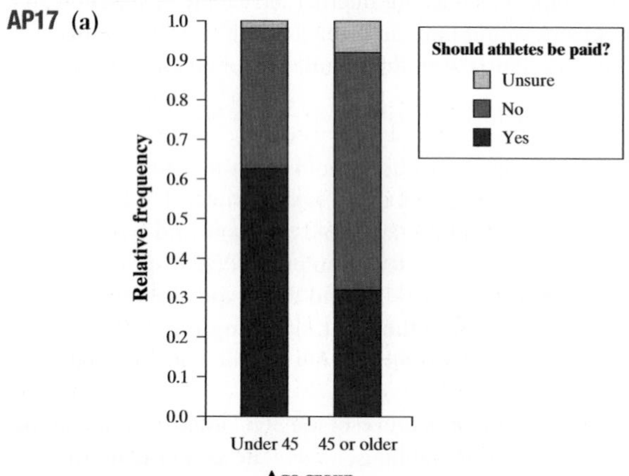

(b) There is an association between age group and opinion about whether athletes should be paid. Knowing the age group a person belongs to helps us predict whether they think athletes should be paid. Those under age 45 were more likely to say colleges should pay athletes ($366/581 = 0.63$) than were those who are age 45 or older ($219/683 = 0.32$). Those under age 45 were less likely to say colleges should not pay athletes ($203/581 = 0.35$) than were those age 45 and older ($410/683 = 0.60$). Finally, those who are under age 45 were less likely to be unsure ($12/581 = 0.02$) than were those age 45 or older ($54/683 = 0.08$).

AP18 **(a)** Give each student a distinct integer label from 1 to 25,000. Use a random number generator to select 1000 different integers from 1 to 25,000. Survey the 1000 students that correspond to the generated integers. **(b)** Approximately normal because $np = (1000)(0.41) = 410 \geq 10$ and $n(1-p) = (1000)(1-0.41) = 590 \geq 10$. $\mu_{\hat{p}} = p = 0.41, \sigma_{\hat{p}} = \sqrt{\frac{0.41(1-0.41)}{1000}} = 0.0156$

(c) $\sigma_{\hat{p}_{overall}} =$

$$\sqrt{\left(0.5\sqrt{\frac{(0.1)(0.9)}{500}}\right)^2 + \left(0.3\sqrt{\frac{(0.6)(0.4)}{300}}\right)^2 + \left(0.2\sqrt{\frac{(0.9)(0.1)}{200}}\right)^2} = 0.0116$$

(d) Both sampling methods are using a total sample size of 1000 students, but the variability of the sampling distribution of $\hat{p}_{\text{overall}}$ using the stratified random sampling method ($\sigma_{\hat{p}_{overall}} = 0.0116$) is much less than when a simple random sampling method is used ($\sigma_{\hat{p}} = 0.0156$). When strata are chosen wisely, stratified random samples produce less variable estimates of the population parameter than simple random samples of the same size.

UNIT 6

Section 6A

Answers to Check Your Understanding

page 537: **1.** We are 95% confident that the interval from 0.175 to 0.225 captures the proportion of all U.S. adults who would answer the question about Earth's atmosphere correctly.
2. If we were to select many random samples of 1006 U.S. adults and construct a 95% confidence interval using each sample, about 95% of the intervals would capture the proportion of all U.S. adults who would answer the question about Earth's atmosphere correctly.
3. Because all of the values in the interval from 0.175 to 0.225 are less than 0.25, there is convincing evidence that less than 25% of all U.S. adults would answer this question about Earth's atmosphere correctly.

Answers to Odd-Numbered Section 6A Exercises

1. *Point estimator:* The sample proportion; *point estimate:* $\hat{p} = 87/409 = 0.213$.
3. We are 95% confident that the interval from 0.577 to 0.651 captures the proportion of all resolution-making U.S. adults who kept their resolution.
5. We are 99% confident that the interval from 2.8 ppb to 11.8 ppb captures the mean lead level for all tap water in Flint, Michigan, that year.
7. Because all the values in the interval 0.577 to 0.651 are greater than 0.50, there is convincing evidence that a majority of all U.S. resolution makers kept their resolutions.
9. **(a)** Because 12 is one of the values in the interval from 11.97 to 12.05, there is not convincing evidence that the true mean volume is different from 12 oz. **(b)** No. Although 12 is a plausible value for the true mean volume of all cans of diet cola, there are many other values besides 12 in the confidence interval and any of these values could plausibly be the true mean.
11. If we were to select many random samples of 655 U.S. resolution-making adults and construct a 95% confidence interval using each sample, about 95% of the intervals would capture the proportion of all resolution-making U.S. adults who kept their resolution.
13. $21/25 = 84\%$ of the confidence intervals contain the population mean, which suggests that these are 80% or 90% confidence intervals.
15. **(a)** We are 95% confident that the interval from 33.3 μmoles per hour to 101.8 μmoles per hour captures the true difference in the mean oxygen consumption (Ship noise − Ambient noise) for shore crabs like the ones in this study. **(b)** Because all of the values in the interval from 33.3 to 101.8 are greater than 0, there is convincing evidence that crabs have a greater mean oxygen consumption in the presence of ship noise.
17. D
19. **(a)**

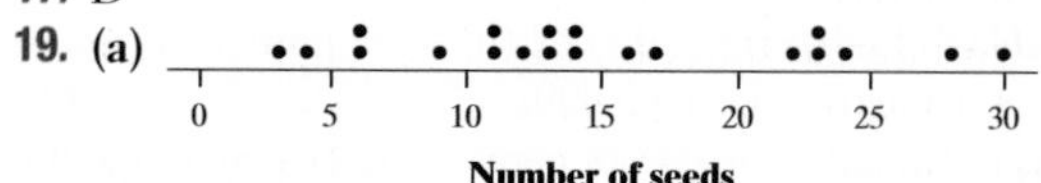

(b) Yes. The distribution of the number of seeds is fairly symmetric, there is no strong skewness, nor are there any outliers in the data—so it is plausible that the number of seeds from oranges on this tree follows a distribution that is approximately normal.

Section 6B

Answers to Check Your Understanding

page 549: 1. STATE: 90% CI for p = the proportion of all U.S. adults who would report they "often or always" got enough sleep during the past 7 nights. **PLAN:** One-sample z interval for a proportion. *Random:* Random sample of 1029 U.S. adults. ✓ *10%:* $1029 < 10\%$ of the population of all U.S. adults. ✓ *Large Counts:* $n\hat{p} = 1029(0.48) \approx 494 \geq 10$ and $n(1-\hat{p}) = 1029(1-0.48) \approx 535 \geq 10$. ✓ **DO:** $0.48 \pm 1.645\sqrt{\frac{0.48(1-0.48)}{1029}} = 0.48 \pm 0.026 = (0.454, 0.506)$. **CONCLUDE:** We are 90% confident that the interval from 0.454 to 0.506 captures the proportion of all U.S. adults who would report they "often or always" got enough sleep during the past 7 nights.
2. Because there are values in the interval from 0.454 to 0.506 that are greater than or equal to 0.5, there is not convincing evidence that less than half of all U.S. adults would report that they "often or always" got enough sleep during the past 7 nights.

page 555: 1. We could reduce the margin of error by decreasing the confidence level. *Drawback:* We won't be as confident that our interval will capture the population proportion. We could also reduce the margin of error by increasing the sample size. *Drawback:* Larger samples cost more time and money to obtain.

2. Solving $2.576\sqrt{\frac{0.48(1-0.48)}{n}} \leq 0.01$ for n gives $n \geq (0.48)(1-0.48)\left(\frac{2.576}{0.01}\right)^2 = 16{,}562.9$. This means that we should select a random sample of at least 16,563 U.S. adults.

Answers to Odd-Numbered Section 6B Exercises

1. *Random:* Tonya interviews an SRS of seniors in her school. ✓ *10%:* $50 < 10\%$ of all seniors in her school ($N = 750$). ✓ *Large Counts:* $36 \geq 10$ and $50 - 36 = 14 \geq 10$. ✓

3. *Random:* The inspector selects a random sample of bags of potato chips. ✓ *10%:* $25 < 10\%$ of the thousands of bags filled in an hour. ✓ *Large Counts:* This condition is not met because $n\hat{p} = 3$, which is not at least 10.

5. (a) To ensure that the observations can be viewed as independent. (b) The standard error of the sample proportion calculated using the formula $\sqrt{\frac{\hat{p}(1-\hat{p})}{n}}$ is larger than the actual standard deviation of the sampling distribution, which makes the margin of error larger than necessary. Therefore, confidence intervals that are calculated when the 10% condition is violated will capture the population parameter more often than the specified confidence level.

7. Point estimate $= \frac{0.577 + 0.651}{2} = 0.614$;
Margin of error $= 0.651 - 0.614 = 0.037$

9. *Tech:* invNorm(area: 0.01, mean: 0, SD: 1) $= -2.326$, so $z^* = 2.326$. *Table* A: 0.01 area to the left $\rightarrow z = -2.33$, so $z^* = 2.33$.

11. $\hat{p} = \frac{914}{4579} = 0.1996$. Standard error of $\hat{p} = \sqrt{\frac{0.1996(1-0.1996)}{4579}} = 0.0059$. In repeated SRSs of size 4579, the sample proportion of U.S. adults who have earned money by selling something online in the previous year typically varies from the population proportion by about 0.0059.

13. (a) *Tech:* invNorm(area: 0.05, mean: 0, SD: 1) $= -1.645$, so $z^* = 1.645$; *Table* A: 0.05 area to the left $\rightarrow$ Average of $z = -1.64$ and $z = -1.65$ is $z = -1.645$, so $z^* = 1.645$. (b) $\hat{p} = \frac{36}{50} = 0.72$; standard error of $\hat{p} = \sqrt{\frac{0.72(0.28)}{50}} = 0.0635$ (c) $0.72 \pm 1.645\sqrt{\frac{0.72(0.28)}{50}} = 0.72 \pm 0.104 = (0.616,\ 0.824)$

15. (a) **STATE:** 95% CI for p = the proportion of all U.S. adults who would answer their cell phone when an unknown number calls. **PLAN:** One-sample z interval for a proportion. *Random:* Random sample of 10,211 U.S. adults. ✓ *10%:* $10{,}211 < 10\%$ of all U.S. adults. ✓ *Large counts:* $10{,}211(0.19) \approx 1940 \geq 10$ and $10{,}211(1-0.19) \approx 8271 \geq 10$. ✓ **DO:** (0.182, 0.198). **CONCLUDE:** We are 95% confident that the interval from 0.182 to 0.198 captures the proportion of all U.S. adults who would answer their cell phone when an unknown number calls. (b) If the Pew Research Center selects many random samples of 10,211 U.S. adults and constructs a 95% confidence interval for each sample, about 95% of these intervals would capture the proportion of all U.S. adults who would answer their cell phone when an unknown number calls.

17. (a) **S:** 99% CI for p = the proportion of all U.S. adults who can name all three branches of government. **P:** One-sample z interval for a proportion. *Random:* Random sample of 1113 U.S. adults. ✓ *10%:* $1113 < 10\%$ of all U.S. adults. ✓ *Large counts:* $1113(0.47) \approx 523 \geq 10$ and $1113(1-0.47) \approx 590 \geq 10$. ✓ **D:** (0.431, 0.508). **C:** We are 99% confident that the interval from 0.431 to 0.508 captures the proportion of all U.S. adults who can name all three branches of government. (b) Because some of the values in the interval from 0.431 to 0.508 are greater than or equal to 0.5, there is not convincing evidence that fewer than half of all U.S. adults can name all three branches of government.

19. $750(0.616) = 462$ and $750(0.824) = 618$. We can be 90% confident that the interval from 462 to 618 captures the total number of seniors planning to go to the prom.

21. (a) The confidence interval would be wider because increasing the confidence level from 95% to 99% increases the margin of error. (b) The confidence interval would be wider because decreasing the sample size from 10,211 to 1000 increases the margin of error.

23. $1.645\sqrt{\frac{0.5(0.5)}{n}} \leq 0.04 \rightarrow n \geq 422.82$

Select a random sample of $n \geq 423$ students.

25. (a) $2.576\sqrt{\frac{0.44(0.56)}{n}} \leq 0.03 \rightarrow n \geq 1816.73$

Select a random sample of $n \geq 1817$ U.S. adults.

(b) $2.576\sqrt{\frac{0.5(0.5)}{n}} \leq 0.03 \rightarrow n \geq 1843.27$

Select a random sample of $n \geq 1844$ U.S. adults. The conservative approach requires 27 more people.

27. (a) $0.044 = z^*\sqrt{\frac{0.69(0.31)}{1000}} \rightarrow z^* = 3.008$

Tech: normalcdf(lower: -3.008, upper: 3.008, mean: 0, SD: 1) $= 0.9974$; *Table* A: Area between $z = -3.01$ and $z = 3.01$ is 0.9974. Confidence level $\approx 99.7\%$. (b) Kids (age 12 to 18) who take their mobile device to bed with them may not respond truthfully because they may be doing so without their parents' permission, making the estimate of 69% less than the true percentage. This bias is not accounted for by the margin of error, which only accounts for sampling variability.

29. **(a)** Answers may vary, but should be close to 95%. **(b)** Answers may vary, but should be close to 95%. **(c)** Answers may vary, but should be close to 95%. **(d)** No. A 95% confidence level will have a capture rate of about 95%, regardless of the sample size (when the conditions are met). As the sample size increases, the confidence intervals become narrower, but the capture rate (determined by the confidence level) stays about the same.

31. A

33. D

35. C

37. E

39. Let X = the number of game pieces that are instant winners. X has a binomial distribution with $n = 20$ and $p = 0.25$. Find $P(X \le 1)$. *Tech:* binomcdf(n: 20, p: 0.25, x value: 1) = 0.0243. This probability is similar to the probability of 0.01 estimated in Exercise 38(b).

Section 6C

Answers to Check Your Understanding

page 569: 1. H_0: $p = 0.63$ and H_a: $p < 0.63$, where p = the proportion of all drive-thru customers this month who had to wait longer than 2 minutes to receive their food.

2. Assuming that the proportion of all drive-thru customers this month who had to wait longer than 2 minutes to receive their food is 0.63, there is a 0.0385 probability of getting a sample proportion of 0.576 or less by chance alone in a random sample of 250 drive-thru customers.

3. Because the P-value of $0.0385 < \alpha = 0.05$, we reject H_0. We have convincing evidence that the proportion of all customers this month who had to wait longer than 2 minutes to receive their food is less than 0.63.

Answers to Odd-Numbered Section 6C Exercises

1. H_0: $p = 0.221$; H_a: $p < 0.221$, where p = the proportion of all students at this school who get at least 8 hours of sleep per night on school nights.

3. H_0: $\mu = 180$ ml; H_a: $\mu \ne 180$ ml, where μ = the mean volume of liquid dispensed by the machine in all bottles.

5. H_0: $\sigma = 3°F$; H_a: $\sigma > 3°F$, where σ = the true standard deviation of the temperature in the cabin during this 24-hour period.

7. The null hypothesis is always that there is "no difference" or "no change" and the alternative hypothesis is what we suspect is true. These ideas are reversed in the stated hypotheses. *Correct hypotheses:* H_0: $p = 0.37$ and H_a: $p > 0.37$.

9. **(a)** The proportion of all students at this school who get at least 8 hours of sleep on school nights is $p = 0.221$. **(b)** $\hat{p} = 0.16 < 0.221$ **(c)** Assuming that the proportion of all students at this school who get at least 8 hours of sleep on school nights is 0.221, there is a 0.1493 probability of getting a sample proportion of 0.16 or less by chance alone in a random sample of 50 students at the school.

11. Assuming the mean volume of liquid dispensed by the machine in all bottles is 180 ml, there is a 0.0589 probability of getting a sample mean less than or equal to 179.6 or greater than or equal to 180.4 by chance alone in a random sample of 40 bottles filled by the machine.

13. Because the P-value of $0.1493 > \alpha = 0.05$, we fail to reject H_0. We do not have convincing evidence that the proportion of all students at this school who get at least 8 hours of sleep on school nights is less than 0.221.

15. **(a)** Because the P-value of $0.0589 < \alpha = 0.10$, we reject H_0. We have convincing evidence that the mean volume of liquid dispensed by the machine in all bottles is different than 180 ml. **(b)** Yes. Because the P-value of $0.0589 > \alpha = 0.05$, we would fail to reject H_0. This would mean we do not have convincing evidence that the mean volume of liquid dispensed by the machine in all bottles is different than 180 ml.

17. **(a)** H_0: $p = 0.10$; H_a: $p > 0.10$, where p = the proportion of all students at Simon's school who are left-handed. **(b)** P-value $\approx 24/200 = 0.12$. Assuming the proportion of all students at Simon's school who are left-handed is 0.10, there is a 0.12 probability of getting a sample proportion of 0.16 or greater by chance alone in a random sample of 50 students. **(c)** Using $\alpha = 0.05$: Because the P-value of $0.12 > \alpha = 0.05$, we fail to reject H_0. We do not have convincing evidence that the proportion of all students at Simon's school who are left-handed is greater than 0.10.

19. C

21. **(a)** Approximately normal because $50(0.375) = 18.75 \ge 10$ and $50(0.625) = 31.25 \ge 10$. $\mu_{\hat{p}} = 0.375$. $\sigma_{\hat{p}} = \sqrt{\frac{0.375(1-0.375)}{50}} = 0.068$. **(b)** (i) $z = \frac{0.30 - 0.375}{0.068} = -1.103$. *Tech:* normalcdf(lower: −1000, upper: −1.103, mean: 0, SD: 1) = 0.1350; *Table* A: 0.1357. (ii) normalcdf(lower: −1000, upper: 0.30, mean: 0.375, SD: 0.068) = 0.1350.

Section 6D

Answers to Check Your Understanding

page 580: 1. H_0: $p = 0.75$; H_a: $p \ne 0.75$, where p = the proportion of all employees of this restaurant chain who would say that work stress has a negative impact on their personal life.

2. *Random:* We have a random sample of 100 employees from the restaurant chain. ✓ *10%:* 100 < 10% of the thousands of employees at this restaurant chain. ✓ *Large Counts:* $np_0 = 100(0.75) = 75 \ge 10$ and $n(1-p_0) = 100(1-0.75) = 25 \ge 10$. ✓

3. $\hat{p} = \frac{68}{100} = 0.68$; standardized test statistic: $z = \frac{0.68 - 0.75}{\sqrt{\frac{0.75(0.25)}{100}}} = -1.617$. P-value: *Table* A: 2(0.0526) = 0.1052; *Tech:* $z = -1.617$, P-value = 0.1060.

4. Because the P-value of $0.1060 > \alpha = 0.05$, we fail to reject H_0. We do not have convincing evidence that the proportion of all restaurant employees at this large restaurant chain who would say that work stress has a negative impact on their personal lives is different from 0.75.

page 586: 1. **STATE:** We want to test H_0: $p = 0.78$, H_a: $p < 0.78$, where p = the proportion of all students in this school district who have reliable internet access at home, using $\alpha = 0.05$. **PLAN:** One-sample z test for a proportion. *Random:* We have a random sample of 120 students from this school district. ✓ *10%:* It is safe to assume that 120 < 10% of all students in this large school district. ✓ *Large Counts:* $np_0 = 120(0.78) = 93.6 \ge 10$ and $n(1-p_0) = 120(1-0.78) = 26.4 \ge 10$. ✓ **DO:** $\hat{p} = \frac{85}{120} = 0.708$.

$z = \frac{0.708 - 0.78}{\sqrt{\frac{0.78(0.22)}{120}}} = -1.904$. P-value: *Table* A: 0.0287; *Tech*: $z = -1.895$, *P*-value $= 0.0290$. **CONCLUDE:** Because the *P*-value of $0.0290 < \alpha = 0.05$, we reject H_0. We have convincing evidence that the proportion of all students in this school district who have reliable internet access at home is less than the national figure of 0.78.

2. *Type I error:* The researcher finds convincing evidence that less than 78% of all students in this school district have reliable internet access at home, when the population proportion is really 0.78 or more. *Consequence:* The district may unnecessarily invest more money into technology to counteract the perceived lack of access at home. *Type II error:* The researcher fails to find convincing evidence that less than 78% of all students in this school district have reliable internet access at home, when the population proportion is less than 0.78. *Consequence:* The district may not invest more money into technology because they believe at-home access is sufficient, when in reality it is not.

page 591: **1.** *Type I error:* The researchers find convincing evidence that the true mean percent change in TBBMC during the exercise program is greater than 0 for young adults like the ones in this study, when the true mean percent change is really 0. The probability of a Type I error is $\alpha = 0.05$.

2. If the true mean percent change in TBBMC is $\mu = 1\%$, there is a 0.80 probability that the researchers will find convincing evidence for H_a: $\mu > 0$.

3. *Type II error:* The researchers fail to find convincing evidence that the true mean percent change in TBBMC during the exercise program is greater than 0 for young adults like the ones in this study, when the true mean percent change in TBBMC is really 1%. The probability of a Type II error $= 1 - \text{power} = 1 - 0.80 = 0.20$.

4. The researchers could increase the power of the test by increasing the sample size or using a larger significance level.

Answers to Odd-Numbered Section 6D Exercises

1. *Random:* We have a random sample of 929 U.S. adults age 18–29. ✓ *10%:* 929 < 10% of all U.S. adults age 18–29. ✓ *Large Counts:* $929(0.50) = 464.5 \geq 10$ and $929(1 - 0.5) = 464.5 \geq 10$. ✓

3. **(a)** No, because the sample size ($n = 50$) is more than 10% of the population size (N = 400). **(b)** Yes, because the Large Counts condition is satisfied. Assuming $p = 0.5$ is true, $50(0.5) = 25$ and $50(1 - 0.5) = 25$ are both ≥ 10.

5. **(a)**The sample proportion of U.S. adults age 18–29 who had tried online dating is $\hat{p} = \frac{492}{929} = 0.530$, which is greater than 0.50. **(b)** $z = \frac{0.530 - 0.50}{\sqrt{\frac{0.50(0.50)}{929}}} = 1.829$; *Tech*: *P*-value $= 0.0337$; *Table* A: *P*-value $= 1 - 0.9664 = 0.0336$ **(c)** Assuming that the population proportion of U.S. adults age 18–29 who have tried online dating is 0.50, there is a 0.0337 probability of getting a sample proportion of 0.530 or greater by chance alone in a random sample of 929 U.S. adults age 18–29.

7. **(a)** *Tech*: 0.0141; *Table* A: 0.0143 **(b)** The *P*-value is double the *P*-value in part (a). *Tech*: $0.0141 \times 2 = 0.0282$; *Table* A: $2(0.0143) = 0.0286$.

9. **(a)** H_0: $p = 0.75$, H_a: $p \neq 0.75$, where $p =$ the true proportion of smooth peas. **(b)** $\hat{p} = \frac{423}{423 + 133} = 0.761$; $z = \frac{0.761 - 0.75}{\sqrt{\frac{0.75(0.25)}{556}}} = 0.599$. *Tech:* *P*-value $= 0.2746 \times 2 = 0.5492$; *Table* A: *P*-value $= 2(0.2743) = 0.5486$ **(c)** Because the *P*-value of $0.5492 > \alpha = 0.05$, we fail to reject H_0. We do not have convincing evidence that the true proportion of peas that are smooth is different than 0.75.

11. **STATE:** H_0: $p = 1/3$, H_a: $p < 1/3$, where $p =$ the proportion of all U.S. adults who would answer this question about poverty correctly; $\alpha = 0.01$. **PLAN:** One-sample z test for a proportion. *Random:* We have a random sample of 1000 U.S. adults. ✓ *10%:* 1000 < 10% of all U.S. adults. ✓ *Large Counts:* $1000(1/3) = 333.\overline{3} \geq 10$ and $1000(1 - 1/3) = 666.\overline{6} \geq 10$. ✓ **DO:** $z = -19.007$, *P*-value ≈ 0. **CONCLUDE:** Because the *P*-value of approximately $0 < \alpha = 0.01$, we reject H_0. We have convincing evidence that less than 1/3 of all U.S. adults would answer this question about poverty correctly.

13. **S:** H_0: $p = 0.60$, H_a: $p \neq 0.60$, where $p =$ the proportion of all teens in this state who pass their driving test on the first attempt; $\alpha = 0.05$. **P:** One-sample z test for a proportion. *Random:* SRS of 125 teens from this state who attempt the driving test. ✓ *10%:* 125 < 10% of all teens in this state who attempt the driving test. ✓ *Large Counts:* $125(0.60) = 75 \geq 10$ and $125(1 - 0.60) = 50 \geq 10$. ✓ **D:** $z = 2.008$, *P*-value $= 0.0446$. **C:** Because the *P*-value of $0.0446 < \alpha = 0.05$, we reject H_0. There is convincing evidence that the proportion of all teens in this state who pass the driving test on their first attempt is different from 0.60.

15. **(a)** **S:** 95% confidence interval for $p =$ the proportion of all teens in this state who pass their driving test on the first attempt. **P:** One-sample z interval for a proportion. The problem states that the conditions have been met. **D:** (0.607, 0.769). **C:** We are 95% confident that the interval from 0.607 to 0.769 captures the proportion of all teens in this state who pass the driving test on the first attempt. **(b)** In addition to providing convincing evidence to reject $p = 0.60$ (because 0.60 is not in the interval from 0.607 to 0.769), the confidence interval uses sample data to provide plausible values of $p =$ the proportion of all teens in this state who pass the driving test on the first attempt.

17. Because $\hat{p} = \frac{551}{1060} = 0.520$ is not less than 0.50, there is no evidence for H_a. A lower-tailed test will never reject the null hypothesis when the sample proportion is greater than the hypothesized value.

19. *Type I error:* You find convincing evidence that the mean income of all residents near the restaurant exceeds $85,000 when in reality it does not. *Consequence:* You will open your restaurant in a location where the residents will not be able to support it, so your restaurant may go out of business. *Type II error:* You fail to find convincing evidence that the mean income of all residents near the restaurant exceeds $85,000 when in reality it does. *Consequence:* You will not open your restaurant in a location where the residents would have been able to support it and you lose potential income.

21. **(a)** A Type I error is worse for the restaurant owner because they will lose a lot of money by investing in a restaurant in a location that likely will not be profitable. A Type II error is a missed opportunity, which is not as bad as losing money from a failing restaurant. **(b)** To minimize the possibility of a Type I error,

I would prefer to use $\alpha = 0.01$ because reducing alpha reduces the probability of a Type I error.

23. **S:** H_0: $p = 0.5$, H_a: $p > 0.5$, where $p =$ the proportion of all students at this school who would identify the green cup of tea as having the more natural flavor; $\alpha = 0.05$. **P:** One-sample z test for a proportion. *Random:* We have a random sample of 40 students from this school. ✓ *10%:* It is reasonable to assume that $40 < 10\%$ of all students at this school. ✓ *Large Counts:* $40(0.5) = 20 \geq 10$ and $40(1-0.5) = 20 \geq 10$. **D:** $z = 2.846$, P-value $= 0.0022$. **C:** Because the P-value of $0.0022 < \alpha = 0.05$, we reject H_0. We have convincing evidence that a majority of students at this school would identify the green cup of tea as having the more natural flavor. **(b)** Because the null hypothesis is rejected, there is a possibility that the students could have made a Type I error: finding convincing evidence that a majority of students would identify the green cup of tea as having the more natural flavor, when in reality at most half of the students would.

25. **(a)** If the mean income in the population of people who live near the restaurant is $\mu = \$86{,}000$, there is a 0.64 probability that I will find convincing evidence for H_a: $\mu > \$85{,}000$. **(b)** Power will decrease. Reducing the sample size from 50 to 30 gives less information about the population mean μ. **(c)** Power will decrease. It is harder to detect a smaller difference (\$500 vs. \$1000) between the null and alternative parameter value. **(d)** Power will increase. Increasing the significance level from 0.05 to 0.10 makes it easier to reject H_0 when H_a is true.

27. **(a)** A disadvantage of using $\alpha = 0.10$ instead of $\alpha = 0.05$ is that the larger significance level will increase the probability of a Type I error. By increasing the probability of making a Type I error, the owner takes on a greater risk of opening a restaurant in an area that can't sustain it. **(b)** An advantage of selecting a random sample of 30 people instead of 50 people is that the smaller sample size would require less time and money.

29. **(a)** A Type I error would mean the principal finds convincing evidence that more than 37% of all students in this school are satisfied with the parking situation after the change, when the population proportion is really 0.37; $P(\text{Type I error}) = \alpha = 0.05$. **(b)** If the proportion of all students at the school who are satisfied with the parking situation after the change is $p = 0.45$, there is a 0.75 probability that the principal will find convincing evidence for H_a: $p > 0.37$. **(c)** The principal fails to find convincing evidence that more than 37% of all students in this school are satisfied with the parking situation after the change, when the population proportion is really 0.45. $P(\text{Type II error}) = 1 - 0.75 = 0.25$. **(d)** To decrease the probability of a Type II error, increase the sample size or use a larger significance level.

31. **(a)** Power $= 1 - 0.14 = 0.86$ **(b)** $P(\text{Type I error}) = \alpha = 0.01$

33. **(a)** The hypotheses are H_0: $p = 0.9$, H_a: $p < 0.9$, where $p =$ the proportion of all mowers of this model that will start with the first push of the button. **(b)** *Large counts:* $20(0.9) = 18 \geq 10$, but $20(1-0.9) = 2 < 10$. This condition is not met. **(c)** Let X = the number of mowers that start out of 20 mowers. X is a binomial random variable with $n = 20$ and $p = 0.9$; $P(X \leq 15) = 0.043$. **(d)** Because the P-value of $0.043 < \alpha = 0.05$, we reject H_0. We have convincing evidence that the proportion of all mowers of this model that will start with the first push of the button is less than the company's claim of 0.9.

35. **(a)** Approximately normal because $200(0.37) = 74 \geq 10$ and $200(1-0.37) = 126 \geq 10$. $\mu_{\hat{p}} = 0.37$. It is reasonable to assume that $200 < 10\%$ of the population of all students at this large school, so $\sigma_{\hat{p}} = \sqrt{\frac{0.37(0.63)}{200}} = 0.0341$. **(b)** *Tech:* $\hat{p} = 0.4261$; *Table* A: 0.95 area to the left $\rightarrow z = 1.645$. Solving $1.645 = \frac{\hat{p} - 0.37}{0.0341}$ gives $\hat{p} = 0.4261$. **(c)** Approximately normal because $200(0.40) = 80 \geq 10$ and $200(1-0.40) = 120 \geq 10$. $\mu_{\hat{p}} = 0.40$. It is reasonable to assume that $200 < 10\%$ of the population of all students at this school, so $\sigma_{\hat{p}} = \sqrt{\frac{0.40(0.60)}{200}} = 0.0346$. **(d)** (i) $z = \frac{0.4261 - 0.4}{0.0346} = 0.754$. *Tech:* 0.2254; *Table* A: $1 - 0.7734 = 0.2266$ (ii) normalcdf(lower: 0.4261, upper: 1000, mean: 0.4, SD: 0.0346) = 0.2253

37. D

39. A

41. B

43. **(a)** Label the people from 1 to 10,065 in alphabetical order. Use a random number generator to select 3355 different integers from 1 to 10,065. Identify the individuals with these labels and assign them to Treatment 1: \$500 bonus for the worker. Next, use a random number generator to select an additional 3355 different integers from 1 to 10,065, making sure to not include the integers selected the first time. Identify the individuals with these labels and assign them to Treatment 2: \$500 bonus for employer. All remaining individuals will be assigned to Treatment 3: no bonus. After the experiment, record the time to work for each individual. **(b)** The purpose of a control group is to give the researchers a baseline for comparison. This will allow them to see if a financial incentive is better than no incentive at all. If there were no control group, the researchers could only make a conclusion about which type of incentive (to the worker or to the employer) is more effective rather than whether an incentive is effective at all.

Answers to Unit 6, Part I Review Exercises

R1. **(a)** We are 95% confident that the interval from 0.565 to 0.615 captures the proportion of all U.S. adults who would say they are sports fans. **(b)** If we were to select many random samples of 1527 U.S. adults and construct a 95% confidence interval using each sample, about 95% of the intervals would capture the proportion of all U.S. adults who would say they are sports fans. **(c)** Because all the values in the interval from 0.565 to 0.615 are greater than 0.5, there is convincing evidence that a majority of U.S. adults are sports fans.

R2. **(a)** **S:** 95% confidence interval for $p =$ the proportion of all U.S. drivers who have run at least one red light in the last 10 traffic lights they drove through. **P:** One-sample z interval for a proportion. *Random:* We have a random sample of 880 U.S. drivers. ✓ *10%:* $880 < 10\%$ of all U.S. drivers. ✓ *Large Counts:* $171 \geq 10$ and $709 \geq 10$. ✓ **D:** (0.168, 0.220). **C:** We are 95% confident that the interval from 0.168 to 0.220 captures the proportion of all U.S. drivers who have run at least one red light in the last 10 traffic lights they drove through. **(b)** We could reduce the margin of error by decreasing the confidence level. The drawback is that we can't be as confident that our interval will capture the population proportion. We could also reduce the margin of error by increasing the sample size. The drawback is that larger samples cost more time and money to obtain.

R3. Using $\hat{p} = 0.5$, we want $2.576\sqrt{\frac{0.5(0.5)}{n}} \leq 0.01$, so $n \geq 16{,}589.44$. They should select an SRS of at least $n = 16{,}590$ adults.
R4. **(a)** $H_0: p = 0.60$, $H_a: p < 0.60$, where $p =$ the proportion of all signatures that are valid. **(b)** Assuming that the proportion of all signatures that are valid is 0.6, there is a 0.1444 probability of getting a sample proportion of $\hat{p} = 171/300 = 0.57$ or less by chance alone in a random sample of 300 signatures. **(c)** Because the *P*-value of $0.1444 > \alpha = 0.05$, we fail to reject H_0. We do not have convincing evidence that the proportion of all signatures on the petition that are valid is less than 0.60.
R5. **(a)** **S:** $H_0: p = 0.396, H_a: p \neq 0.396$, where $p =$ the proportion of all college students in this state who vaped that year; $\alpha = 0.05$. **P:** One-sample z test for a proportion. *Random:* We have a random sample of 750 college students from this state. ✓ *10%:* It is reasonable to assume that 750 < 10% of all college students in this state. ✓ *Large Counts:* $750(0.396) = 297 \geq 10$ and $750(1-0.396) = 453 \geq 10$. ✓ **D:** $z = 2.539$, *P*-value $= 0.0111$. **C:** Because the *P*-value of $0.0111 < \alpha = 0.05$, we reject H_0. We have convincing evidence that the proportion of all college students in this state who vaped that year is different than 0.396. **(b)** Because 0.396 is not in the interval from 0.4058 to 0.4769, we would reject $H_0: p = 0.396$ in favor of $H_a: p \neq 0.396$ at the $\alpha = 0.05$ significance level. This is the same decision that was reached by the significance test.
R6. **(a)** $H_0: p = 0.05, H_a: p < 0.05$, where $p =$ the proportion of all adults who will get the flu after using the vaccine. **(b)** *Type I:* Finding convincing evidence that fewer than 5% of vaccinated adults would get the flu, when in reality at least 5% would. *Consequence:* A higher proportion of adults than expected will get the flu and could potentially become very ill. *Type II:* Failing to find convincing evidence that fewer than 5% of vaccinated adults would get the flu, when in reality less than 5% would. *Consequence:* The new vaccine that would have been effective might not be used. **(c)** Answers will vary. For example, if a Type I error is considered worse, a 0.01 significance level should be used because 0.01 is less than the other options and significance level = *P*(Type I error). **(d)** If the proportion of the population of all adults who use the vaccine and get the flu is $p = 0.03$, there is a 0.9437 probability that the researchers will find convincing evidence for $H_a: p < 0.05$. **(e)** To increase the power, increase the sample size or use a larger significance level.

Answers to Unit 6, Part I AP® Practice Test

T1. A **T2.** D **T3.** B **T4.** C **T5.** C
T6. E **T7.** B **T8.** D **T9.** A **T10.** D
T11. **(a)** **S:** 99% confidence interval for $p =$ the proportion of all visitors to Yellowstone who would say they favor the restrictions. **P:** One-sample z interval for a proportion. *Random:* We have a random sample of visitors. ✓ *10%:* 150 < 10% of all visitors to Yellowstone National Park. ✓ *Large Counts:* $89 \geq 10$ and $61 \geq 10$. ✓ **D:** (0.490, 0.697). **C:** We are 99% confident that the interval from 0.490 to 0.697 captures the proportion of all visitors to Yellowstone who would say that they favor the restrictions. **(b)** If we were to select many random samples of 150 visitors to Yellowstone and construct a 99% confidence interval using each sample, about 99% of the intervals would capture the proportion of all visitors to Yellowstone who would say that they favor the restrictions.
T12. **S:** $H_0: p = 0.5, H_a: p > 0.5$, where $p =$ the proportion of all U.S. adults who live within an hour of extended family; $\alpha = 0.05$. **P:** One-sample z test for a proportion. *Random:* We have a random sample of 9676 U.S. adults. ✓ *10%:* 9676 < 10% of all U.S. adults. ✓ *Large Counts:* $9676(0.5) = 4838 \geq 10$ and $9676(1-0.5) = 4838 \geq 10$. ✓ **D:** $z = 9.841$, *P*-value ≈ 0. **C:** Because the *P*-value of approximately $0 < \alpha = 0.05$, we reject H_0. We have convincing evidence that more than half of all U.S. adults live within an hour of extended family. **(b)** Because the null hypothesis is rejected, there is a possibility that the Pew Research Center made a Type I error: finding convincing evidence that a majority of U.S. adults live within an hour of extended family, when in reality this isn't true.
T13. **(a)** If the proportion of all mattresses with defects in the stitching is $p = 0.10$, there is a 0.39 probability that the manufacturer will find convincing evidence for $H_a: p < 0.15$. **(b)** Assuming that the proportion of all mattresses with defects in the stitching is $p = 0.15$, there is a 0.13 probability of getting a sample proportion of 0.11 or less by chance alone in a random sample of 100 mattresses.

Section 6E

Answers to Check Your Understanding

page 617: **STATE:** 90% CI for $p_{2021} - p_{2016}$, where $p_{2021} =$ proportion of all U.S. adults who used Instagram in 2021 and $p_{2016} =$ proportion of all U.S. adults who used Instagram in 2016. **PLAN:** Two-sample z interval for a difference in proportions. *Random:* The data come from independent random samples of 1502 U.S. adults in 2021 and 1520 U.S. adults in 2016. ✓ *10%:* 1502 < 10% of all U.S. adults in 2021 and 1520 < 10% of all U.S. adults in 2016. ✓ *Large Counts:* $1502(0.40) = 600.8 \approx 601$, $1502(1-0.40) = 901.2 \approx 901$, $1520(0.32) = 486.4 \approx 486$, and $1520(1-0.32) = 1033.6 \approx 1034$ are all ≥ 10. ✓ **DO:** $(0.40 - 0.32) \pm 1.645\sqrt{\frac{0.40(0.60)}{1502} + \frac{0.32(0.68)}{1520}} = 0.08 \pm 0.029 = (0.051,\ 0.109)$; *Tech:* (0.052, 0.109). **CONCLUDE:** We are 90% confident that the interval from 0.052 to 0.109 captures the difference (2021 – 2016) in the proportions of all U.S. adults who used Instagram in 2021 and in 2016.

Answers to Odd-Numbered Section 6E Exercises

1. *Random:* The Random condition is not met because these data do not come from independent random samples or two groups in a randomized experiment. *10%:* The 10% condition is not met because both "samples" were 100% of the populations of babies born in each area. *Large Counts:* The Large Counts condition is not met because there were less than 10 successes (3) in the group from the west side of Woburn.
3. *Random:* The European seabass eggs were randomly assigned to one of the two treatment groups, vibration or no vibration. ✓ *Large Counts:* $200(0.76) = 152$; $200(1-0.76) = 48$; $200(0.82) = 164$; $200(1-0.82) = 36$ are all ≥ 10.✓
5. $(0.76 - 0.82) \pm 1.645\sqrt{\frac{0.76(0.24)}{200} + \frac{0.82(0.18)}{200}} = -0.06 \pm 0.067 = (-0.127, 0.007)$
7. If we repeated many random assignments of 200 eggs to vibration and 200 eggs to no vibration and constructed a 90% confidence interval for $p_V - p_{NV}$ each time, about 90% of the resulting intervals would capture the difference (Vibration – No vibration) in the true proportions that would hatch for eggs like the ones in this study.

9. **(a)** **S:** 90% CI for $p_N - p_F$, where p_N = the proportion of all subjects like the ones in this study who eat the candy when sitting near a bowl of candy and p_F = the proportion of all subjects like the ones in this study who eat the candy when sitting far from a bowl of candy. **P:** Two-sample z interval for a difference in proportions. *Random:* Subjects were randomly assigned to be close to or far from the bowl of candy. ✓ *Large Counts:* $39, 61-39=22, 24, 61-24=37$ are all at least 10. ✓ **D:** (0.102, 0.390). **C:** We are 90% confident that the interval from 0.102 to 0.390 captures the difference (Near – Far) between the proportions of subjects like the ones in this study who would eat the candy when sitting near or sitting far from a bowl of candy. **(b)** Because the interval from 0.102 to 0.390 does not include 0, we have convincing evidence that the distance from a bowl of candy affects whether people like those in this study will eat from it.

11. **S:** 95% CI for $p_{\text{rural}} - p_{\text{urban}}$, where p_{rural} = the proportion of all rural households in Indiana with Christmas trees that had a natural tree last year and p_{urban} = the proportion of all urban households in Indiana with Christmas trees that had a natural tree last year. **P:** Two-sample z interval for a difference in proportions. *Random:* It is reasonable to consider these as independent random samples of rural and urban households in Indiana with Christmas trees. ✓ *10%:* $160 <$ 10% of all rural households in Indiana with Christmas trees and $261 <$ 10% of all urban households in Indiana with Christmas trees. ✓ *Large Counts:* $64, 160-64=96, 89, 261-89=172$ are all at least 10. ✓ **D:** (−0.036, 0.154). **C:** We are 95% confident that the interval from −0.036 to 0.154 captures the difference (Rural – Urban) in the proportions of all rural and all urban Indiana households with Christmas trees that had a natural tree last year.

13. **(a)** Because the interval from −0.0093 to 0.0052 includes 0 as a plausible value for the difference (Vitamin D – Placebo) in the proportions of people like the ones in this study who would be diagnosed with depression after 5 years, we do not have convincing evidence of a difference in the effectiveness of the two treatments. **(b)** Even though 0 (no difference) is a plausible value for the difference (Vitamin D – Placebo) in the proportions of people like the ones in this study who would be diagnosed with depression after 5 years, all of the other plausible values in the interval suggest that vitamin D supplementation has an effect on depression.

15. **(a)** There is a 0.05 probability that the interval will exclude 0 because, assuming the text message has no effect, 95% of all confidence intervals will include 0, the true difference in the population proportions. **(b)** Type I error. There is a 0.05 probability that we will find convincing evidence that there is a difference in the population proportions when the true difference in the proportions of Walmart pharmacy patients who would get a flu vaccine is 0. **(c)** They can increase the confidence level in order to reduce the probability of the error described in part (b). **(d)** $22(0.05)=1.1\approx 1$. If none of the 22 text messages is effective, we would expect about 1 of the resulting 95% confidence intervals to exclude 0 by chance alone. **(e)** No, we would expect one of the twenty-two 95% confidence intervals to exclude 0 by chance alone.

17. C

19. D

21. **(a)** There is some evidence that it is harder for golfers like the ones in this study to make putts after being asked to describe their putting technique because the difference (Described putting – Unrelated task) in the median number of putts is greater than 0. This means that those who described their putting technique have a median number of putts that is greater than those who did an unrelated task. **(b)** Yes, there is convincing evidence that it is harder for golfers like the ones in this study to make putts after being asked to describe their putting technique. Of the 100 trials in the simulation, none of the simulated differences in the median number of putts required was 11.5 or greater by chance alone.

Section 6F

Answers to Check Your Understanding

page 628: 1. We want to test $H_0: p_1 - p_2 = 0$, $H_a: p_1 - p_2 > 0$, where p_1 = the true proportion of subjects like the ones in the study with RLS who would experience much improved symptoms when taking pramipexole and p_2 = the true proportion of subjects like the ones in the study with RLS who would experience much improved symptoms when taking a placebo.

2. *Random:* The subjects were randomly assigned to receive pramipexole or a placebo. ✓ *Large Counts:* $\hat{p}_C = \frac{158+50}{193+92} = \frac{208}{285} = 0.7298$; $n_1\hat{p}_C = 193(0.7298) = 140.85$, $n_1(1-\hat{p}_C) = 193(1-0.7298) = 52.15$, $n_2\hat{p}_C = 92(0.7298) = 67.14$, $n_2(1-\hat{p}_C) = 92(1-0.7298) = 24.86$ are all ≥ 10. ✓

3. $z = \dfrac{(0.8187-0.5435)-0}{\sqrt{0.7298(0.2702)\left(\frac{1}{193}+\frac{1}{92}\right)}} = 4.892$,

Table A: less than 0.0002; *Tech:* $z = 4.891$, *P*-value ≈ 0

4. Because the *P*-value of approximately $0 < \alpha = 0.05$, we reject H_0. There is convincing evidence that the true proportion of subjects like the ones in the study with RLS who would experience much improved symptoms when taking pramipexole is greater than the true proportion of subjects like the ones in the study with RLS who would experience much improved symptoms when taking a placebo. In other words, we have reason to believe that pramipexole is effective in treating RLS.

page 631: 1. **STATE:** We want test $H_0: p_1 - p_2 = 0$, $H_a: p_1 - p_2 < 0$, where p_1 = the true proportion of children like the ones in the study who attend preschool who use social services later and p_2 = the true proportion of children like the ones in the study who do not attend preschool who use social services later using $\alpha = 0.05$. **PLAN:** Two-sample z test for a difference in proportions. *Random:* The children were randomly assigned to attend or not attend preschool. ✓ *Large Counts:* $\hat{p}_C = \frac{38+49}{62+61} = \frac{87}{123} = 0.7073$, $n_1\hat{p}_C = 62(0.7073) = 43.85$, $n_1(1-\hat{p}_C) = 62(1-0.7073) = 18.15$, $n_2\hat{p}_C = 61(0.7073) = 43.15$, $n_2(1-\hat{p}_C) = 61(1-0.7073) = 17.85$ are all ≥ 10. ✓ **DO:** $z = -2.320$, *P*-value: 0.0102. **CONCLUDE:** Because the *P*-value of $0.0102 < \alpha = 0.05$, we reject H_0. There is convincing evidence that the true proportion of children like the ones in the study who do attend preschool who use social services later is less than the true proportion of children like the ones in the study who do not attend preschool who use social services later. In other words, children like those in this study who participate in preschool are less likely to use social services later in life.

2. Because we rejected H_0, it is possible that we made a Type I error: finding convincing evidence that the true proportion of children like the ones in the study who attend preschool who use social services later is less than the true proportion of children like the

ones in the study who do not attend preschool who use social services later, when in reality the true proportions are equal.
3. No. The children that participated in the study were recruited from low-income families in Michigan, not randomly selected, so the results cannot be generalized to all children from low-income families who have never attended preschool.

Answers to Odd-Numbered Section 6F Exercises

1. H_0: $p_O - p_N = 0$, H_a: $p_O - p_N > 0$, where p_O = the true proportion of kayakers like the ones in the study who would attempt to pick up at least one piece of trash after experiencing ownership of the lake by giving it a nickname and p_N = the true proportion of kayakers like the ones in the study who would attempt to pick up at least one piece of trash without experiencing ownership of the lake.
3. H_0: $p_1 - p_2 = 0$, H_a: $p_1 - p_2 > 0$, where p_1 = the true proportion of shrubs like the ones in this study that would resprout after being clipped and burned and p_2 = the true proportion of shrubs like the ones in this study that would resprout after being clipped only.
5. *Random:* The kayakers were randomly assigned to experience ownership of the lake by giving it a nickname or to not be prompted to give the lake a nickname. ✓ *Large Counts:* $\hat{p}_C = \dfrac{22+6}{54+81} = 0.2074$; $54(0.2074) = 11.20$, $54(1-0.2074) = 42.80$, $81(0.2074) = 16.80$, $81(1-0.2074) = 64.20$ are all ≥ 10. ✓
7. No, because the Large Counts condition is not met. *Large Counts:* $\hat{p}_C = \dfrac{12+8}{12+12} = 0.8333$. $12(0.8333) = 9.996$, $12(1-0.8333) = 2.004$, $12(0.8333) = 9.996$, $12(1-0.8333) = 2.004$ are not all ≥ 10.
9. **(a)** There is some evidence in favor of H_a: $p_O - p_N > 0$ because $\dfrac{22}{54} - \dfrac{6}{81} = 0.4074 - 0.0741 = 0.3333 > 0$.
(b) $z = \dfrac{(0.4074-0.0741)-0}{\sqrt{0.2074(0.7926)\left(\dfrac{1}{54}+\dfrac{1}{81}\right)}} = 4.679$, *Table A:* less than 0.0002; *Tech:* $z = 4.680$, *P*-value ≈ 0. **(c)** Because the *P*-value of approximately $0 < \alpha = 0.01$, we reject H_0. There is convincing evidence that the proportion of all kayakers like the ones in the study who would attempt to pick up at least one piece of trash after experiencing ownership of the lake is greater than the proportion of all kayakers like the ones in the study who would attempt to pick up at least one piece of trash without experiencing ownership of the lake.
11. $\hat{p}_1 - \hat{p}_2 = \dfrac{12}{12} - \dfrac{8}{12} = 1 - 0.667 = 0.333$. There are 3 trials of the simulation with a difference of proportions greater than or equal to 0.333. Estimated *P*-value = 3/50 = 0.06. Because the estimated *P*-value of $0.06 > \alpha = 0.05$, we fail to reject H_0. We do not have convincing evidence that the true proportion of shrubs like the ones in this study that would resprout after being clipped and burned is greater than the true proportion of shrubs like the ones in this study that would resprout after being clipped only.
13. **(a)** **S:** H_0: $p_1 - p_2 = 0$, H_a: $p_1 - p_2 < 0$, where p_1 = the proportion of all VLBW babies born in Cleveland who graduate from high school and p_2 = the proportion of all normal-birth-weight babies born in Cleveland who graduate from high school; $\alpha = 0.01$. **P:** Two-sample *z* test for a difference in proportions. *Random:* The data come from independent random samples of 242 VLBW babies born in Cleveland and 233 normal-birth-weight babies born in Cleveland. ✓ *10%:* We must assume 242 < 10% of all VLBW babies born in Cleveland and 233 < 10% of all normal-birth-weight babies born in Cleveland. ✓ *Large Counts:* $\hat{p}_C = \dfrac{179+193}{242+233} = 0.7832$, $242(0.7832) = 189.53$, $242(1-0.7832) = 52.47$, $233(0.7832) = 182.49$, $233(1-0.7832) = 50.51$ are all ≥ 10. ✓ **D:** $z = -2.344$, *P*-value = 0.0095. **C:** Because the *P*-value of $0.0095 < \alpha = 0.01$, we reject H_0. We have convincing evidence that the high school graduation rate among VLBW babies born in Cleveland is less than for normal-birth-weight babies born in Cleveland. **(b)** Assuming that the difference in the proportions of all VLBW babies born in Cleveland and normal-birth-weight babies born in Cleveland who graduate high school is 0, there is a 0.0095 probability of getting a difference in sample proportions of −0.0886 or less by chance alone in independent random samples of 242 VLBW and 233 normal-birth-weight babies born in Cleveland.
15. **(a)** **S:** H_0: $p_1 - p_2 = 0$, H_a: $p_1 - p_2 \neq 0$, where p_1 = the true proportion of children like the ones in this study who are exposed to peanut butter as infants who are allergic to peanuts at age 5 and p_2 = the true proportion of children like the ones in this study who are not exposed to peanut butter as infants who are allergic to peanuts at age 5; $\alpha = 0.05$. **P:** Two-sample *z* test for a difference in proportions. *Random:* Infants were randomly assigned to either consume baby food form peanut butter or avoid peanut butter. ✓ *Large Counts:* $\hat{p}_C = \dfrac{10+55}{307+321} = 0.1035$; $307(0.1035) = 31.77$, $307(1-0.1035) = 275.23$, $321(0.1035) = 33.22$, $321(1-0.1035) = 287.78$ are all ≥ 10. ✓ **D:** $z = -5.707$, *P*-value ≈ 0. **C:** Because the *P*-value of approximately $0 < \alpha = 0.05$, we reject H_0. There is convincing evidence that the true proportion of children like the ones in this study who are exposed to peanut butter as infants who are allergic to peanuts at age 5 is different than the true proportion of children like the ones in this study who are not exposed to peanut butter as infants who are allergic to peanuts at age 5. **(b)** Because we rejected H_0, it is possible that we made a Type I error: finding convincing evidence of a difference in the risk of developing peanut allergies in infants like the ones in this study who consume or avoid peanut butter when there is actually no difference in the true proportions. **(c)** No. This experiment likely involved parents who volunteered their infants as subjects. When subjects are not randomly selected, we should not generalize the results of an experiment to some larger population of interest. **(d)** The confidence interval from −0.185 to −0.093 does not include 0 as a plausible value for $p_1 - p_2$, which is consistent with the decision to reject H_0: $p_1 - p_2 = 0$ in part (a). The confidence interval also tells us that any value between −0.185 and −0.093 is plausible for $p_1 - p_2$ based on the sample data, but the significance test in part (a) didn't give us an estimate for $p_1 - p_2$.
17. **(a)** Increasing the sample size will increase the power of the test. Increasing the sample size gives more information about the difference in the true proportions. *Drawback*: An experiment that uses twice as many infants will be more expensive and will require a lot more work in following up with the parents of these infants when they are 5 years old. **(b)** Using $\alpha = 0.10$ instead of $\alpha = 0.05$

will increase the power of the test. When α is larger, it is easier to reject the null hypothesis because the *P*-value doesn't need to be as small. *Drawback*: Increasing α increases the probability of making a Type I error. **(c)** Using only male infants would increase the power of the test by eliminating a source of variability, making it easier to reject the null hypothesis when it is false. *Drawback*: This limits the scope of inference to only male infants like the ones in the study.

19. **(a)** By using both random selection and random assignment, the researchers can draw conclusions about cause and effect for all football fans at this major university. **(b)** **S:** $H_0: p_W - p_L = 0$, $H_a: p_W - p_L > 0$, where p_W = the proportion of all football fans at this university who would identify with the team when the team won and p_L = the proportion of all football fans at this university who would identify with the team when the team lost; $\alpha = 0.05$. **P:** Two-sample *z* test for a difference in proportions. *Random:* We have a random sample of 168 football fans from this major university, and the football fans were randomly assigned to answer questions about a recent game the team lost or a recent game the team won. ✓ *10%:* 168 < 10% of all football fans at this major university. ✓ *Large Counts:* $\hat{p}_C = \frac{42}{168} = 0.25$. $85(0.25) = 21.25$, $85(1 - 0.25) = 63.75$, $83(0.25) = 20.75$, $83(1-0.25) = 62.25$ are all ≥ 10. ✓ **D:** $z = 2.049$, *P*-value $= 0.0202$. **C:** Because the *P*-value of $0.0202 < \alpha = 0.05$, we reject H_0. There is convincing evidence that football fans at this university are more likely to associate themselves with the team after a win.

21. **(a)** The value of *z* for which an area of 0.99 is to the right of *z* (thus, 0.01 to the left) is −2.326. *Tech:* invNorm(area: 0.01, mean: 0, SD: 1) = −2.326.

(b) $\hat{p}_{Curb} = \frac{2181}{7529} = 0.2897$, $\hat{p}_{Off} = \frac{40}{329} = 0.1216$.

$$(0.2897 - 0.1216) - 2.326\sqrt{\frac{0.2897(1-0.2897)}{7529} + \frac{0.1216(1-0.1216)}{329}}$$

$= 0.1245 \rightarrow (0.1245, \infty)$. **(c)** Because all the values in the interval from 0.1245 to infinity are greater than 0, there is convincing evidence at the 1% significance level that the proportion of trees with damaged sidewalk nearby is greater for trees on the curb than for trees that are offset from the curb in New York City. **(d)** The confidence interval provides more information than the test because the test only provides a decision about $p_{Curb} - p_{Off} = 0$, whereas the confidence interval provides a set of plausible values for $p_{Curb} - p_{Off}$.

23. E

25. E

27. **(a)** 77% of the variation in mileage is accounted for by the least-squares regression line with *x* = age (years). **(b)** $\hat{y} = -13{,}832 + 14{,}954\bar{x} = -13{,}832 + 14{,}954(5) = 60{,}938$ miles **(c)** The actual mileage is typically about 22,723 miles away from the amount predicted by the least-squares regression line with *x* = age (years). **(d)** No, it would not be reasonable to use the least-squares line to predict a car's mileage from its age for a teacher. The least-squares line is based on a sample of cars owned and driven by students, not teachers.

Answers to Unit 6, Part II Review Exercises

R1. **(a)** **S:** 99% CI for $p_C - p_H$, where p_C = the proportion of all college students who are employed and p_H = the proportion of all high school students who are employed. **P:** Two-sample *z* interval for a difference in proportions. *Random:* Independent random samples of 550 college students and 500 high school students. ✓ *10%:* 550 < 10% of all college students and 500 < 10% of all high school students. ✓ *Large Counts:* $550(0.451) = 248.05 \rightarrow 248$, $550(1-0.451) = 301.95 \rightarrow 302$, $500(0.204) = 102$, $500(1-0.204) = 398$ are all ≥ 10. ✓ **D:** (0.175, 0.319). **C:** We are 99% confident that the interval from 0.175 to 0.319 captures the difference (College − High school) in the proportions of all college and high school students who are employed. **(b)** Because 0 is not in the interval from 0.175 to 0.319, there is convincing evidence of a difference in the proportions of high school students and college students who are employed.

R2. **(a)** Yes, because the subjects were randomly assigned to either AZT or placebo. **(b)** **S:** $H_0: p_1 - p_2 = 0$, $H_a: p_1 - p_2 < 0$, where p_1 = the true proportion of HIV-infected patients like the ones in this study who take AZT and develop AIDS and p_2 = the true proportion of HIV-infected patients like the ones in this study who take placebo and develop AIDS; $\alpha = 0.05$. **P:** Two-sample *z* test for $p_1 - p_2$. *Random:* The subjects were assigned at random to take AZT or a placebo. ✓ *Large Counts:* $\hat{p}_C = \frac{17+38}{435+435} = 0.0632$; $435(0.0632) = 27.492$, $435(1-0.0632) = 407.508$, $435(0.0632) = 27.492$, $435(1-0.0632) = 407.508$ are all ≥ 10. ✓ **D:** $z = -2.926$, *P*-value $= 0.0017$. **C:** Because the *P*-value of $0.0017 < \alpha = 0.05$, we reject H_0. We have convincing evidence that taking AZT lowers the proportion of HIV-infected people like the ones in this study who will develop AIDS compared to a placebo.

R3. **(a)** We want to test $H_0: p_Q - p_N = 0$, $H_a: p_Q - p_N < 0$, where p_Q = the true proportion of football players like the ones in this study who have brain damage after wearing a Q collar and p_N = the true proportion of football players like the ones in this study who have brain damage after not wearing a Q collar; $\alpha = 0.05$. **(b)** *Type I:* The researchers find convincing evidence that the Q collar helps prevent brain damage when in reality it does not. *Consequence:* Schools invest in purchasing Q collars for all football players and make the players wear them when they do not prevent brain damage. *Type II:* The researchers fail to find convincing evidence that the Q collar helps prevent brain damage when in reality it does. *Consequence:* Schools will not invest in purchasing Q collars for all football players, when doing so could help prevent brain damage. **(c)** Answers will vary. For example, if a Type II error is considered more serious, it would be better to use the larger value of α, which is $\alpha = 0.10$. The larger the significance level, which is the probability of a Type I error, the smaller the probability of making a Type II error. **(d)** Assuming that the difference (Q collar − No Q collar) in the true proportions of football players like the ones in this study who have brain damage is 0, there is approximately a 0 probability of getting a difference in sample proportions as small as or smaller than the observed difference in proportions by chance alone.

Answers to Unit 6, Part II AP® Practice Test

T1. B **T2.** A **T3.** E **T4.** C **T5.** C

T6. **S:** 90% CI for $p_1 - p_2$, where p_1 = the true proportion of people like the ones in this study who would say that buying coffee at Starbucks is a waste of money when the girls hold cups from Starbucks and p_2 = the true proportion of people like the ones in this study who would say that buying coffee at Starbucks is a waste of money when the girls were empty handed. **P:** Two-sample z-interval for a difference in proportions. *Random:* Subjects were randomly assigned to be asked while the girls held Starbucks cups or were empty handed. ✓ *Large Counts:* 19, $50-19=31$, 23, $50-23=27$ are all ≥ 10. ✓ **D:** $(-0.242,\ 0.082)$. **C:** We are 90% confident that the interval from -0.242 to 0.082 captures the difference (Holding Starbucks cups − Empty handed) in the true proportions of people like the ones in this study who would say that buying coffee at Starbucks is a waste of money.

T7. **(a)** **S:** $H_0: p_1 - p_2 = 0$, $H_a: p_1 - p_2 > 0$, where $p_1 =$ the population proportion of cars that have the brake defect in last year's model and $p_2 =$ the population proportion of cars that have the brake defect in this year's model; $\alpha = 0.05$. **P:** Two-sample z test for a difference in proportions. *Random:* We have independent random samples of 100 cars of last year's model and 350 cars of this year's model. ✓ *10%:* $100 < 10\%$ of all cars made of last year's model and $350 < 10\%$ of all cars made of this year's model. ✓ *Large Counts:* $\hat{p}_C = \dfrac{20+50}{100+350} = 0.1556$. $100(0.1556) = 15.56$, $100(1-0.1556) = 84.44$, $350(0.1556) = 54.46$, $350(1-0.1556) = 295.54$ are all ≥ 10. ✓ **D:** $z = 1.390$, P-value $= 0.0822$. **C:** Because the P-value of $0.0822 > \alpha = 0.05$, we fail to reject H_0. We do not have convincing evidence that the population proportion of brake defects is smaller in this year's model compared to last year's model. **(b)** If the population proportion of cars with the defect has decreased by 0.10 from last year to this year, there is a 0.72 probability that the automaker will find convincing evidence for $H_a: p_1 - p_2 > 0$. **(c)** Besides increasing the sample sizes, the power of the test can be increased by using a larger value for α, the significance level of the test.

UNIT 7

Section 7A

Answers to Check Your Understanding

page 653: **1.** **(a)** $df = 22 - 1 = 21$; *Tech:* invT(area: 0.02, df: 21) $= -2.189$, so $t^* = 2.189$; *Table B:* $t^* = 2.189$. **(b)** $df = 71 - 1 = 70$; *Tech:* invT(area: 0.005, df: 70) $= -2.648$, so $t^* = 2.648$; *Table B:* Using $df = 60$, $t^* = 2.660$.

2. *Random:* The students were randomly selected. ✓ *10%:* It is safe to assume that $n = 50 < 10\%$ of all students at this local community college. ✓ *Normal/Large Sample:* $n = 50 \geq 30$. ✓

page 658: **STATE:** 95% CI for $\mu =$ the true mean healing rate (in micrometers per hour) for all newts. **PLAN:** One-sample t interval for a mean. *Random:* Random sample of 18 newts. ✓ *10%:* $n = 18 < 10\%$ of the population of all newts. ✓ *Normal/Large Sample:* The sample size is small ($n = 18 < 30$), but the dotplot does not show strong skewness or outliers. ✓

15 20 25 30 35 40

Healing rate (μm/hr)

DO: $\bar{x} = 25.667$, $s_x = 8.324$, $n = 18$; with 95% confidence and $df = 18 - 1 = 17$, $t^* = 2.110$; $25.667 \pm 2.110\dfrac{8.324}{\sqrt{18}} = 25.667 \pm 4.140 = (21.527,\ 29.807)$; *Tech:* $(21.527,\ 29.806)$ with $df = 17$ **CONCLUDE:** We are 95% confident that the interval from 21.527 to 29.806 micrometers per hour captures the true mean healing rate for all newts.

page 662: **1.** First, compute the difference for each student (Music − Without music).

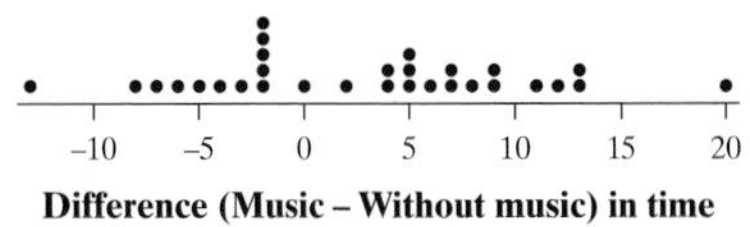

Difference (Music – Without music) in time to complete the test (sec)

A majority of the students (17/30 = 0.567 or 56.7%) took longer to complete the test when they were listening to music, so listening to music seems to hinder math performance.

2. **STATE:** 90% CI for $\mu_{Diff} =$ the true mean difference (Music − Without music) in time (in seconds) for students like the ones in this study to complete the arithmetic test. **PLAN:** Paired t interval for a mean difference. *Random:* The students were randomly assigned the order of the music and no music treatments. ✓ *Normal/Large Sample:* $n_{Diff} = 30 \geq 30$. ✓ **DO:** $\bar{x}_{Diff} = 2.8$, $s_{Diff} = 7.490$, $n_{Diff} = 30$; with 90% confidence and $df = 30 - 1 = 29$, $t^* = 1.699$; $2.8 \pm 1.699\dfrac{7.490}{\sqrt{30}} = 2.8 \pm 2.323 = (0.477,\ 5.123)$; *Tech:* $(0.477,\ 5.123)$ with $df = 29$. **CONCLUDE:** We are 90% confident that the interval from 0.477 second to 5.123 seconds captures the true mean difference (Music − Without music) in time for students like the ones in this study to complete the arithmetic test.

3. The 90% confidence interval in Question 2 suggests that students are between 0.477 and 5.123 seconds tslower, on average, when completing the arithmetic test while music is playing. Because all of the values in the interval 0.477 second to 5.123 seconds are positive, there is convincing evidence that music hinders math performance for students like the ones in this study.

Answers to Odd-Numbered Section 7A Exercises

1. **(a)** Purple, because a t distribution has more area in the tails than the standard normal distribution. **(b)** We need the critical value to be larger than z^* to compensate for the variability introduced by using the sample standard deviation as an estimate for the population standard deviation. The larger critical value will help to ensure that the capture rate is equal to the confidence level.

3. **(a)** *Tech:* invT(area: 0.005, df: 19) $= -2.861$, so $t^* = 2.861$. *Table B:* $df = 19$, $t^* = 2.861$. **(b)** *Tech:* invT(area: 0.05, df: 76) $= -1.665$, so $t^* = 1.665$. *Table B:* Using $df = 60$, $t^* = 1.671$.

5. **(a)** *Random:* Random sample of 28 velvetleaf plants in this cornfield. ✓ *10%:* $n = 28 < 10\%$ of all velvetleaf plants in this cornfield. ✓ *Normal/Large Sample:* The sample size is small ($n = 28 < 30$) and the histogram shows one or two apparent outliers, so it is not appropriate to use a t critical value to calculate the confidence interval. (Condition not met) **(b)** *Random:* Random sample of 1025 U.S. adults. ✓ *10%:* $n = 1025 < 10\%$ of all U.S. adults. ✓ *Normal/Large Sample:* $n = 1025 \geq 30$. ✓

7. **(a)** *Random:* Members of the AP® Statistics class are not a random sample of all students. (Condition not met) *10%:* Assume $n = 32 < 10\%$ of all students in the school. ✓ *Normal/Large Sample:* $n = 32 \geq 30$. ✓ **(b)** *Random:* Random sample of 100 home sales from the previous 6 months in the city. ✓ *10%:* It is safe to assume that $n = 100 < 10\%$ of all homes that were sold during the previous 6 months in the city. ✓ *Normal/Large Sample:* $n = 100 \geq 30$. ✓

9. $s_{\bar{x}} = \dfrac{9.3}{\sqrt{27}} = 1.7898$ mmHg. If we select many random samples of 27 adults, the sample mean blood pressure will typically vary from the population mean blood pressure by about 1.7898 mmHg.

11. df $= 1024$; $t^* = 1.962$; *Table B*: df $= 1000$, $t^* = 1.962$; $3.8 \pm 1.962\frac{6.95}{\sqrt{1025}} = 3.8 \pm 0.426 = (3.374,\ 4.226)$; *Tech*: (3.374, 4.226) with df $= 1024$.

13. **(a)** **S**: 90% CI for μ = the true mean number of pepperonis on a large pizza at this restaurant. **P**: One-sample t interval for a mean. *Random*: Random sample of 10 pizzas. ✓ *10%*: $n = 10 < 10\%$ of all pepperoni pizzas made at this restaurant. ✓ *Normal/Large Sample*: The sample size is small ($n = 10 < 30$), but the dotplot does not show strong skewness or outliers. ✓

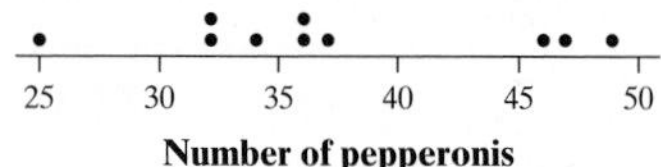

D: $\bar{x} = 37.4$, $s_x = 7.662$, $n = 10$; (32.958, 41.842) with df $= 9$. **C**: We are 90% confident that the interval from 32.958 to 41.842 captures the mean number of pepperonis for all large pizzas at this restaurant. **(b)** Because there are values in the interval 32.958 to 41.842 that are 40 or more, there is not convincing evidence that the average number of pepperonis is less than 40.

15. **(a)** It was necessary because the shape of the population distribution is unknown and the sample size was not large ($n = 10 < 30$). In this case, it must be plausible that the population is approximately normally distributed. If a graph of the sample data displays any strong skewness or any outliers, we shouldn't assume the population is approximately normally distributed. **(b)** If the student researchers select many random samples of 10 large pizzas from this restaurant and construct a 90% confidence interval for each sample, about 90% of these intervals would capture the mean number of pepperonis for all large pizzas at this restaurant. **(c)** To reduce the margin of error, the student researchers could increase the sample size or decrease the confidence level. They may not want to increase the sample size because it would be expensive to buy more pizzas. They may not want to decrease the confidence level because it is desirable to have a high degree of confidence that the population mean number of pepperonis will be captured by the interval.

17. **(a)** The calculated differences are: 10.2, 2.7, 6.4, 5.3, 7.0, 1.8, 5.0, 8.6, 7.3, 3.6, −0.5, 8.4, 1.0, 3.7, 2.2, 0.2

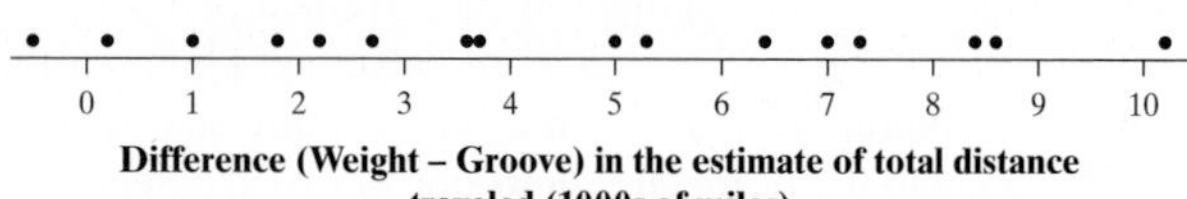

(b) The graph reveals that most (15/16 = 0.9375 or 93.75%) of the differences are positive, meaning that the estimates of distance traveled using weight tend to be greater than the estimates of distance traveled using the grooves of the tire. **(c)** $\bar{x}_{\text{Diff}} = 4.556$ thousand miles, $s_{\text{Diff}} = 3.226$ thousand miles.

19. **S**: 95% CI for μ_{Diff} = the true mean difference (Weight − Groove) in the estimates of tire wear (1000s of miles) using these two methods in the population of tires. **P**: Paired t interval for a mean difference. *Random*: A random sample of 16 tires was selected. ✓ *10%*: $n_{\text{Diff}} = 16 < 10\%$ of all tires. ✓ *Normal/Large Sample*: The number of differences is small ($n_{\text{Diff}} = 16 < 30$), but the dotplot in Exercise 17(a) does not show any strong skewness or outliers. ✓ **D**: (2.837, 6.275) with df $= 15$. **C**: We are 95% confident that the interval from 2.837 to 6.275 thousand miles captures the true mean difference (Weight − Groove) in the estimates using these two methods in the population of tires.

21. **(a)** **S**: 99% CI for μ_{Diff} = the true mean percent change in BMC for breast-feeding mothers. **P**: Paired t interval for a mean difference. *Random*: Random sample of 47 breast-feeding mothers.✓ *10%*: $n_{\text{Diff}} = 47 < 10\%$ of all breast-feeding mothers.✓ *Normal/Large Sample*: $n_{\text{Diff}} = 47 \geq 30$.✓ **D**: $(-4.569,\ -2.605)$ using df $= 46$. **C**: We are 99% confident that the interval from −4.569% to −2.605% captures the true mean percent change in BMC for breast-feeding mothers. **(b)** No. Because all the values in the confidence interval −4.569% to −2.605% are negative, these data provide convincing evidence that breast-feeding mothers lose bone mineral content, on average. However, because this was an observational study and not an experiment, we can't conclude that the bone mineral loss was *caused* by breast-feeding.

23. **(a)** If the volunteers were all given the same treatment order, the order of the treatment may be confounded with the number of words the volunteers remembered. For example, they may remember more words the second time around due to their experience, not the treatment. **(b)** *Random*: Paired data come from the number of words remembered with and without chewing gum generated by the random assignment of 30 volunteers to a treatment order. ✓ *Normal/Large Sample*: $n_{\text{Diff}} = 30 \geq 30$. ✓ **(c)** *Confidence interval*: We are 95% confident that the interval from −0.67 to 1.54 words captures the true mean difference (Gum − No gum) in number of words remembered for subjects like the ones in this study. *Confidence level*: If we repeated many random assignments of 30 volunteers to each of the treatments and each time constructed a 95% confidence interval in this same way, about 95% of the resulting intervals would capture the true mean difference (Gum − No gum) in the number of words recalled for volunteers like the ones in this study. **(d)** No. Zero is included in the confidence interval (−0.67, 1.54), so zero is a plausible value for the true mean difference. Therefore, we do not have convincing evidence that chewing gum helps subjects like the ones in this study with short-term memory.

25. For 99% confidence and $\sigma = 7.5$ known, $z^* = 2.576$; $2.576\frac{7.5}{\sqrt{n}} \leq 1 \rightarrow n \geq 373.26$. Select an SRS of at least 374 individuals.

27. B

29. B

31. D

33. **(a)** $P(X = 7) = 1 - (0.04 + 0.03 + \ldots + 0.05) = 0.57$
(b) $\mu_X = E(X) = 0(0.04) + 1(0.03) + \ldots + 7(0.57) = 5.44$

$\sigma_X = \sqrt{(0-5.44)^2(0.04) + (1-5.44)^2(0.03) + \ldots + (7-5.44)^2(0.57)}$ $= \sqrt{4.5664} = 2.137$

(c) *Mean*: $\mu_{\bar{x}} = 5.44$; *SD*: $\sigma_{\bar{x}} = \frac{2.137}{\sqrt{100}} = 0.2137$; *Shape*: Because $n = 100 \geq 30$, the sampling distribution of $\bar{x}$ is approximately normal. **(d)** (i) $z = \frac{4.96 - 5.44}{0.2137} = -2.246$; *Tech*: normalcdf(lower: −1000, upper: −2.246, mean: 0, SD: 1) = 0.0124; *Table A*: 0.0122. (ii) normalcdf(lower: −1000, upper: 4.96, mean: 5.44, SD: 0.2137) = 0.0123. Yes, because it is unlikely (probability ≈ 0.0123) to get a sample mean of 4.96 or less by chance alone.

Section 7B

Answers to Check Your Understanding

page 677: 1. H_0: $\mu = 320$, H_a: $\mu \neq 320$, where μ = the population mean amount of active ingredient (in mg) in Aspro tablets from this batch of production.

2. *Random:* Random sample of 30 tablets. ✓ *10%:* Assume $n = 30 < 10\%$ of all tablets in this batch. ✓ *Normal/Large Sample:* $n = 30 \geq 30$. ✓

3. $t = \dfrac{319 - 320}{\frac{3}{\sqrt{30}}} = -1.826$, df $= 30 - 1 = 29$, *P*-value = tcdf(lower: −1000, upper: −1.826, df: 29) ×2 = 0.0782; *Table B:* $2(0.025) = 0.05 < P\text{-value} < 2(0.05) = 0.10$

4. Because the *P*-value of $0.0782 > \alpha = 0.05$, we fail to reject H_0. There is not convincing evidence that the population mean amount of the active ingredient (mg) in Aspro tablets from this batch of production differs from 320 mg.

5. The test only allowed us to fail to reject H_0: $\mu = 320$. The confidence interval tells us that any value of μ between 317.88 mg and 320.12 mg is plausible based on the sample data. This is consistent with, but gives more information than, the test because the interval contains 320 as well as other values as plausible for μ.

page 679: STATE: H_0: $\mu = 6$, H_a: $\mu < 6$, where μ = the true mean length (in inches) of "6-inch" subs at this restaurant. Use $\alpha = 0.10$. **PLAN:** One-sample *t* test for a mean. *Random:* Random sample of 24 "6-inch" subs. ✓ *10%:* $n = 24 < 10\%$ of all "6-inch" subs made by this restaurant. ✓ *Normal/Large Sample:* The sample size is small ($n = 24 < 30$), but the dotplot doesn't show any strong skewness or outliers. ✓

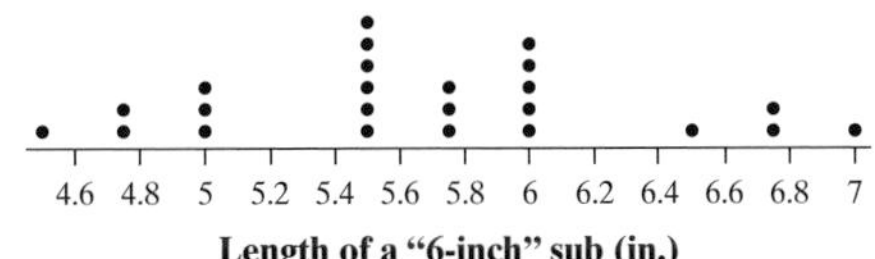

DO: $\bar{x} = 5.677$, $s_x = 0.657$, $n = 24$; $t = \dfrac{5.677 - 6}{\frac{0.657}{\sqrt{24}}} = -2.408$, *P*-value: df = 23; *Table B:* $0.01 < P\text{-value} < 0.02$; *Tech:* $t = -2.407$, *P*-value = 0.0123, df = 23. **CONCLUDE:** Because the *P*-value of $0.0123 < \alpha = 0.10$, we reject H_0. There is convincing evidence that the mean length of all "6-inch" subs at this restaurant is less than 6 inches.

page 683: STATE: H_0: $\mu_{\text{Diff}} = 0$, H_a: $\mu_{\text{Diff}} > 0$, where μ_{Diff} = the true mean difference (Air − Nitrogen) in pressure loss. Use $\alpha = 0.05$. **PLAN:** Paired *t* test for a mean difference. *Random:* Tires in each pair are randomly assigned to be filled with air or nitrogen. ✓ *Normal/Large Sample:* $n_{\text{Diff}} = 31 \geq 30$. ✓ **DO:** $\bar{x}_{\text{Diff}} = 1.252$, $s_{\text{Diff}} = 1.202$, and $n_{\text{Diff}} = 31$. $t = \dfrac{1.252 - 0}{\frac{1.202}{\sqrt{31}}} = 5.799$, *P*-value: df $= 31 - 1 = 30$; *Table B:* *P*-value < 0.0005; *Tech:* $t = 5.797$, *P*-value ≈ 0, df = 30. **CONCLUDE:** Because the *P*-value of approximately $0 < \alpha = 0.05$, we reject H_0. We have convincing evidence that air-filled tires lose more pressure, on average, than nitrogen-filled tires for brands like the ones in this study.

Answers to Odd-Numbered Section 7B Exercises

1. **(a)** H_0: $\mu = 19.2$, H_a: $\mu \neq 19.2$, where μ = the mean weight (in ounces) for all bags of candy filled by this machine that day. **(b)** *Random:* Random sample of 75 bags of candy that day. ✓ *10%:* Assume $n = 75 < 10\%$ of all bags of candy produced that day. ✓ *Normal/Large Sample:* $n = 75 \geq 30$. ✓

3. **(a)** H_0: $\mu = 11.5$, H_a: $\mu < 11.5$, where μ = the mean battery life (in hours) for all tablets produced that day. **(b)** *Random:* Random sample of 20 tablets produced that day. ✓ *10%:* Assume $n = 20 < 10\%$ of all tablets made by this manufacturer that day. ✓ *Normal/Large Sample:* The sample size is small ($n = 20 < 30$) and the dotplot of battery life is strongly skewed to the right with at least one potential high outlier. (Condition not met)

5. **(a)** μ = the mean pH for all salmon fillets at this fish processing plant. **(b)** *Random:* Random sample of 25 salmon fillets at this processing plant. ✓ *10%:* Assume $n = 25 < 10\%$ of all salmon fillets at this processing plant. ✓ *Normal/Large Sample:* The sample size is small ($n = 25 < 30$), but the dotplot doesn't show any outliers or strong skewness. ✓

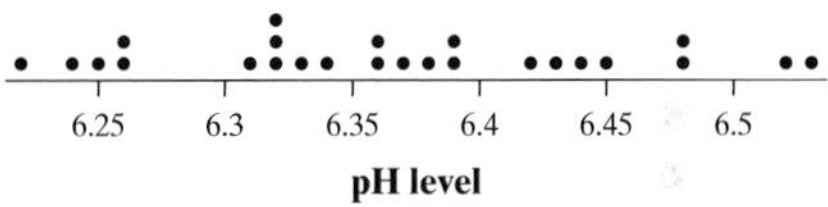

7. **(a)** Because $\bar{x} = 19.28$ ounces $\neq 19.2$ ounces. **(b)** $t = \dfrac{19.28 - 19.2}{\frac{0.81}{\sqrt{75}}} = 0.855$, *P*-value: df = 74; *Tech:* tcdf(lower: 0.855, upper: 1000, df: 74) × 2 = 0.3953; *Table B:* $2(0.15) = 0.30 < P\text{-value} < 2(0.20) = 0.40$ using df = 60. **(c)** Assuming that the true mean weight of candy bags filled by this machine is 19.2 ounces, there is a 0.3953 probability of getting a sample mean as different from 19.2 ounces in either direction as 19.28 ounces by chance alone.

9. **(a)** df = 24; *Tech:* *P*-value = 0.1369; *Table B:* $0.10 < P\text{-value} < 0.15$. **(b)** If the alternative hypothesis was H_a: $\mu \neq 64$, the *P*-value would be $2 \times 0.1369 = 0.2738$.

11. **(a)** $t = \dfrac{6.367 - 6.4}{\frac{0.087}{\sqrt{25}}} = -1.897$ **(b)** *P*-value: df = 24; *Tech:* tcdf (lower: −1000, upper: −1.897, df: 24) = 0.0350; *Table B:* $0.025 < P\text{-value} < 0.05$. **(c)** Because the *P*-value of $0.0350 < \alpha = 0.05$, we reject H_0. We have convincing evidence that the mean pH for all salmon fillets at this fish processing plant is less than 6.4.

13. **(a)** Yes. Because the *P*-value of $0.06 > \alpha = 0.05$, we fail to reject H_0: $\mu = 10$ at the 5% level of significance. Thus, the 95% confidence interval will include 10. **(b)** No. Because the *P*-value of $0.06 < \alpha = 0.10$, we reject H_0: $\mu = 10$ at the 10% level of significance. Thus, the 90% confidence interval would not include 10 as a plausible value.

15. **(a)** **S:** H_0: $\mu = 25$, H_a: $\mu > 25$, where μ = the mean speed of all drivers in the construction zone. Use $\alpha = 0.01$. **P:** One-sample *t* test for a mean. *Random:* Random sample of 10 drivers. ✓ *10%:* $n = 10 < 10\%$ of all drivers in this construction zone. ✓ *Normal/Large Sample:* The sample is small ($n = 10 < 30$), but the graph of the sample data shows no strong skewness or outliers. ✓

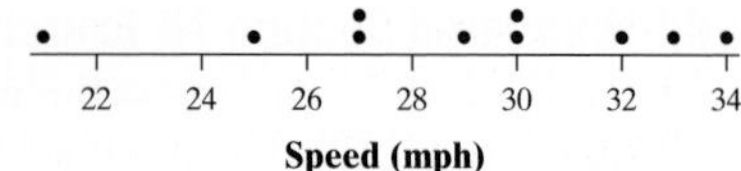

D: $t = 3.051$, P-value $= 0.0069$; df $= 9$. **C:** Because the P-value of $0.0069 < \alpha = 0.01$, we reject H_0. We have convincing evidence that the true mean speed of all drivers in the construction zone is greater than 25 mph. **(b)** Because we rejected H_0, it is possible we made a Type I error—finding convincing evidence that the true mean speed is greater than 25 mph when it really isn't.

17. **S:** H_0: $\mu = 1200$, H_a: $\mu < 1200$, where $\mu =$ the mean calcium intake for all women in the desired age group from this large-scale study. Use $\alpha = 0.05$. **P:** One-sample t test for a mean. *Random:* Random sample of 36 women in the desired age group from this study. ✓ *10%:* $n = 36 < 10\%$ of all women between ages 18 and 24 in this large-scale study. ✓ *Normal/Large Sample:* $n = 36 \geq 30$. ✓ **D:** $t = -6.726$; P-value ≈ 0, df $= 35$; *Table B:* P-value < 0.0005 using df $= 30$. **C:** Because the P-value of approximately $0 < \alpha = 0.05$, we reject H_0. There is convincing evidence that the mean calcium intake for all women in the desired age group from this large-scale study is less than 1200 mg.

19. **(a)** H_0: $\mu = 11.5$, H_a: $\mu \neq 11.5$, where $\mu =$ the true mean hardness of the tablets in this batch (in newtons). **(b)** Although the sample size is small ($n = 20 < 30$), the boxplot of the sample data shows no strong skewness or outliers, so it is plausible that the population distribution is approximately normal. **(c)** If the true mean hardness of the tablets is $\mu = 11.55$ newtons, there is a 0.61 probability that the drug manufacturer will find convincing evidence for H_a: $\mu \neq 11.5$ using a random sample of 20 tablets. **(d)** The probability of a Type II error if $\mu = 11.55$ is $1 - 0.61 = 0.39$. **(e)** Two ways to decrease the probability of a Type II error are to increase the sample size and use a larger significance level.

21. The differences (Inside – Drive-thru) in order are 7, 13, 4, −5, 11, 1, 2, 6, 3, −3. **S:** H_0: $\mu_{\text{Diff}} = 0$, H_a: $\mu_{\text{Diff}} > 0$, where $\mu_{\text{Diff}} =$ the true mean difference (Inside – Drive-thru) in time it would take to receive an iced coffee at this Dunkin' restaurant. Use $\alpha = 0.05$. **P:** Paired t test for a mean difference. *Random:* Random sample of 10 times to place an order. ✓ *10%:* $n = 10 < 10\%$ of all possible times the students could have placed the order. ✓ *Normal/Large Sample:* The number of differences is small ($n = 10 < 30$), but the dotplot does not show any strong skewness or outliers. ✓

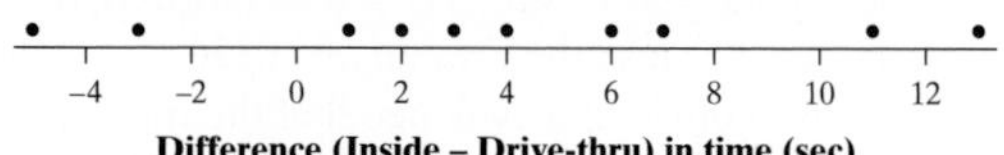

D: $t = 2.184$, P-value $= 0.0284$, df $= 9$. **C:** Because the P-value of $0.0284 < \alpha = 0.05$, we reject H_0. We have convincing evidence that it is faster to order at the drive-thru than inside, on average, at this Dunkin' restaurant.

23. **(a)** These are paired data because there are two measurements made for each grocery store: the number of shoppers at the store on Friday the 6th and the number of shoppers at the store on Friday the 13th. **(b)** No. This is an observational study, and not a controlled experiment, so even if the result of this study is statistically significant, we cannot conclude that the difference in shopping behavior is due to the effect of Friday the 13th. **(c)** **S:** H_0: $\mu_{\text{Diff}} = 0$, H_a: $\mu_{\text{Diff}} \neq 0$, where $\mu_{\text{Diff}} =$ the true mean difference (6th – 13th) in the number of shoppers at grocery stores. Use $\alpha = 0.10$. **P:** Paired t test for a mean difference. *Random:* Researchers collected paired data for a random sample of 45 grocery stores. ✓ *10%:* $n_{\text{Diff}} = 45 < 10\%$ of all grocery stores. ✓ *Normal/Large Sample:* $n_{\text{Diff}} = 45 \geq 30$. ✓ **D:** $t = -1.752$, P-value $= 0.0867$, df $= 44$. *Table B:* $2(0.025) = 0.05 <$ P-value $< 2(0.05) = 0.10$ using df $= 40$. **C:** Because the P-value of $0.0867 < \alpha = 0.10$, we reject H_0. We have convincing evidence that the number of shoppers at grocery stores on these 2 days differs, on average.

25. **(a)** See Exercise 23 for definition of parameters and check of conditions. **S:** 90% CI for μ_{Diff}. **P:** Paired t interval for a mean difference. **D:** (−91.080, −1.916) using df $= 44$. *Table B:* (−91.185, −1.815) using df $= 40$. **C:** We are 90% confident that the interval from −91.080 to −1.916 captures the true mean difference (6th – 13th) in the number of shoppers at grocery stores. **(b)** The two-sided test only allowed us to reject H_0: $\mu_{\text{Diff}} = 0$, but the confidence interval provides a set of plausible values (−91.080, −1.916) for the true mean difference (6th – 13th) in the number of shoppers at grocery stores on these two days.

27. **(a)** 1.729; 1.729 **(b)** Using df $= 20 - 1 = 19$, the t^* critical value for a 90% confidence level is 1.729. **(c)** If we reject H_0, we know from part (a) that the sample mean $\bar{x}$ is more than 1.729 standard errors greater than $\mu = 10$. If $\bar{x}$ is more than 1.729 standard errors away from the mean, we know from part (b) that the 90% confidence interval does not contain $\mu = 10$.

29. A

31. E

33. C

35. **(a)** P(apple on a given wheel) $= 0.20$, so P(no apple on a given wheel) $= 0.80$. P(at least one apple in one pull of the lever) $= 1 - P$(no apples in one pull of the lever) $= 1 - (0.80)^5 = 0.67232$ **(b)** Assume that pulls of the lever are independent. P(no apples on a given pull of the lever) $= (0.80)^5 = 0.32768$. P(no apples on any of 5 pulls) $= (0.32768)^5 = 0.00378$. P(at least one apple in 5 pulls of the lever) $= 1 - 0.00378 = 0.99622$.

Section 7C

Answers to Check Your Understanding

page 701: 1. STATE: 95% CI for $\mu_1 - \mu_2$, where $\mu_1 =$ the true mean change in pulse rate (bpm) for students like the ones in this study after drinking 12 ounces of cola with caffeine and $\mu_2 =$ the true mean change in pulse rate (bpm) for students like the ones in this study after drinking 12 ounces of caffeine-free cola. **PLAN:** Two-sample t interval for a difference in means. *Random:* The volunteers were randomly assigned to drink either cola with caffeine or caffeine-free cola. ✓ *Normal/Large Sample:* The sample sizes are small ($n_1 = 10 < 30$ and $n_2 = 10 < 30$), but the dotplots do not show any outliers or strong skewness. ✓

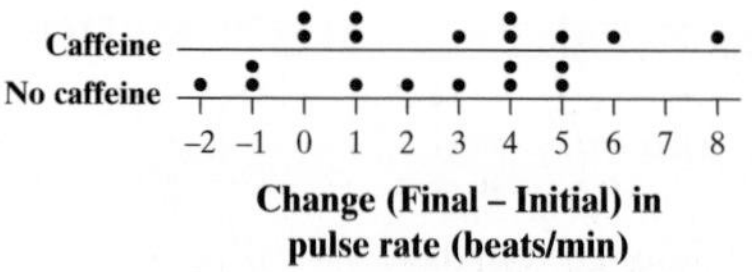

DO: $\bar{x}_1 = 3.2$, $s_1 = 2.70$, $n_1 = 10$, $\bar{x}_2 = 2$, $s_2 = 2.62$, $n_2 = 10$; $(3.2 - 2) \pm t^* \sqrt{\frac{2.70^2}{10} + \frac{2.62^2}{10}}$; *Tech:* (−1.302, 3.702) with df $=$ 17.986. **CONCLUDE:** We are 95% confident that the interval from −1.302 beats per minute to 3.702 beats per minute captures the difference (Caffeine – No caffeine) in the true mean change in pulse rate for all students like the ones in this study after drinking 12 ounces of cola with caffeine and for all students like the ones in this study after drinking 12 ounces of cola without caffeine. **2.** The interval in Question 1 does not suggest that caffeine affects the average pulse rate of subjects like the ones in this study. Because the interval contains 0, it is plausible that the true mean change in pulse rate for all students like the ones in this study after drinking 12 ounces of cola with caffeine and the true mean change in pulse rate after drinking 12 ounces of cola without caffeine equals zero, indicating no difference.

Answers to Odd-Numbered Section 7C Exercises

1. *Random:* The participants were randomly assigned to do curls either with or without encouragement. ✓ *Normal/Large Sample:* Both sample sizes are small ($n_1 = 15 < 30$ and $n_2 = 16 < 30$), but the graphs of the sample data show no strong skewness or outliers. ✓

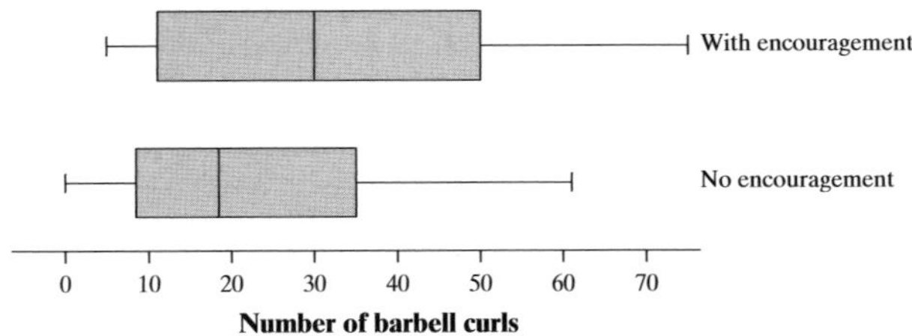

3. **(a)** μ_M = the mean word length for all articles in the medical journal and μ_A = the mean word length for all articles in the airline's in-flight magazine. **(b)** *Random:* The words chosen from the medical journal and the airline magazine were the first words in a single article, not random samples of words. (Condition not met.) *10%:* Assume $n_M = 400 < 10\%$ of the words in the medical journal and assume $n_A = 100 < 10\%$ of the words in the airline magazine. ✓ *Normal/Large Sample:* $n_M = 400 \geq 30$ and $n_A = 100 \geq 30$. ✓

5. **(a)** $(30.67 - 21.5) \pm t^* \sqrt{\frac{22.36^2}{15} + \frac{17.23^2}{16}}$; *Tech:* (−3.115, 21.449) with df = 26.314. **(b)** No, 0 is included in the interval (−3.115, 21.449), so there is not convincing evidence of a difference in the true mean number of barbell curls that people like the ones in this study who receive encouragement and those who receive no encouragement can do.

7. **S:** 95% CI for $\mu_C - \mu_N$, where μ_C = the mean annual income for all college graduates and μ_N = the mean annual income for all college attendees who did not graduate. **P:** Two-sample *t* interval for a difference in means. *Random:* Independent random samples of $n_C = 327$ college graduates and $n_N = 173$ nongraduates. ✓ *10%:* $n_C = 327 < 10\%$ of all college graduates and $n_N = 173 < 10\%$ of all nongraduates. ✓ *Normal/Large Sample:* $n_C = 327 \geq 30$ and $n_N = 173 \geq 30$. ✓ **D:** $(49{,}454.80 - 29{,}299.20) \pm t^* \sqrt{\frac{51{,}257.10^2}{327} + \frac{38{,}298^2}{173}}$; *Tech:* (12169, 28142) with df = 442.7.

C: We are 95% confident that the interval from \$12,169 to \$28,142 captures the difference (Graduates − Nongraduates) in the population mean annual income for U.S. adults with these two education levels.

9. **(a)** **S:** 99% CI for $\mu_1 - \mu_2$, where μ_1 = true mean pulse rate (bpm) of patients similar to those in the experiment who take beta-blockers and μ_2 = true mean pulse rate (bpm) of similar patients who take a placebo. **P:** Two-sample *t* interval for a difference in means. *Random:* The patients were randomly assigned to receive a beta blocker or a placebo. ✓ *Normal/Large Sample:* $n_1 = 30 \geq 30$ and $n_2 = 30 \geq 30$. ✓ **D:** (−10.64, 0.439) with df = 57.778. **C:** We are 99% confident that the interval from −10.64 bpm to 0.439 bpm captures the difference (Beta blocker − Placebo) in mean pulse rate for all patients like the ones in this study who receive a beta blocker and those who receive a placebo. **(b)** If we repeated many random assignments of 30 patients to each of the treatments and each time constructed a 99% confidence interval in this same way, about 99% of the resulting intervals would capture the true difference in mean pulse rate for patients like the ones in this study who receive a beta blocker and those who receive a placebo.

11. **(a)** Because the interval (−0.016, 0.052) includes a difference of 0 (no difference) as a plausible value, there is not convincing evidence of a difference in the population mean reaction times of athletes and non-athletes at this school. **(b)** No, we do not have convincing evidence that the reaction times of athletes and non-athletes are the same. Zero is a plausible value for the difference in means, but there are many other plausible values besides 0 in the confidence interval. **(c)** In order to reduce the width of their interval, the researchers could increase the sample sizes or decrease the confidence level. *Drawbacks:* Increasing the sample sizes would require more work to recruit more participants and take more time to test each of their reaction times. Decreasing the confidence level would reduce researchers' certainty that the confidence interval will capture the difference in the true mean reaction time of athletes and non-athletes.

13. **(a)** df = smaller of $(30-1, 30-1) = 29 \rightarrow t^* = 2.756$

(b) $(65.2 - 70.3) \pm 2.756\sqrt{\frac{7.8^2}{30} + \frac{8.3^2}{30}} = -5.1 \pm 5.731 = (-10.831,$ $0.631)$. **(c)** From Exercise 9: (−10.64, 0.439) with df = 57.778. The confidence interval from part (b) is wider than the one from Exercise 9 because in part (b) df = 29, which is less than df = 57.778 from Exercise 9. The smaller the degrees of freedom, the more area in the tails of the *t* distribution, the larger the value of t^* and the wider the interval.

15. **(a)** It might be reasonable to use the pooled two-sample *t* procedures in this case because the sample standard deviations are similar ($s_1 = 7.8$ and $s_2 = 8.3$), so it is plausible that the population standard deviations are equal. **(b)** df = (30 − 1) + (30 − 1) = 58; *Tech:* $t^* = 2.663$.

(c) **(i)** $s_p^2 = \frac{(30-1)(7.8^2) + (30-1)(8.3^2)}{(30-1) + (30-1)} = 64.865$;

(ii) $(65.2 - 70.3) \pm 2.663\sqrt{64.865}\left(\sqrt{\frac{1}{30} + \frac{1}{30}}\right) = -5.1 \pm 5.538 =$ (−10.638, 0.438). **(d)** Using conservative df = 29 → (−10.831, 0.631); using df = 57.778 → (−10.64, 0.439); pooled two-sample *t* procedure df = 58 → (−10.638, 0.438). The confidence interval obtained using the pooled two-sample *t* procedure is the narrowest of the three intervals because the degrees of freedom is the greatest for this procedure. However, the assumption of equal population variances was required to obtain this slightly narrower confidence interval.

17. D

19. **(a)** By the empirical rule, about 95% of all sample means fall within the interval $\mu_{\bar{x}} - 2\sigma_{\bar{x}}$ to $\mu_{\bar{x}} + 2\sigma_{\bar{x}}$. About 5% of all sample means will fall outside of this interval. *P*(at least one mean outside interval) = 1 − *P*(neither mean outside interval) = $1 - (0.95)^2 = 0.0975$. **(b)** By the empirical rule, about 95% of all sample means fall within the interval $\mu_{\bar{x}} - 2\sigma_{\bar{x}}$ to $\mu_{\bar{x}} + 2\sigma_{\bar{x}}$. About 2.5% ($5\%/2 = 2.5\%$) of all sample means will be greater than $\mu_{\bar{x}} + 2\sigma_{\bar{x}}$. Let Y = the number of samples that must be selected to observe one greater than $\mu_{\bar{x}} + 2\sigma_{\bar{x}}$. Y is a geometric random variable with $p = 0.025$. $P(Y = 4) = (1 - 0.025)^3(0.025) = 0.0232$. **(c)** By the empirical rule, the probability of any one sample mean falling within the interval $\mu_{\bar{x}} - \sigma_{\bar{x}}$ to $\mu_{\bar{x}} + \sigma_{\bar{x}}$ is about 0.68, and the probability of falling outside this interval is about 0.32. Let W = the number of sample means out of 5 that fall outside this interval. Assuming that the samples are independent, W is a binomial random variable with $n = 5$ and $p = 0.32$. $P(W \geq 4) = 1 -$ binomcdf(trials: 5, p: 0.32, x value: 3) $= 1 - 0.961 = 0.039$. There is a 0.039 probability that at least 4 of the 5 sample means fall outside of this interval by chance alone. This is a reasonable criterion, because when the process is under control, we would get a "false alarm" only about 3.9% of the time.

Section 7D

Answers to Check Your Understanding

page 712: 1. H_0: $\mu_1 - \mu_2 = 0$, H_a: $\mu_1 - \mu_2 \neq 0$, where μ_1 = the true mean growth (in millimeters) after 2 weeks for bean plants like the ones in this study that are exposed to heavy metal music each night and μ_2 = the true mean growth (in millimeters) after 2 weeks for bean plants like the ones in this study that are exposed to classical music each night.

2. *Random:* The bean seeds were randomly assigned to be exposed to either heavy metal or classical music. ✓ *Normal/Large Sample:* The sample sizes are small ($n_1 = 5 < 30$ and $n_2 = 5 < 30$), but the dotplots show no strong skewness and no outliers. ✓

Heavy metal music
Classical music
10 20 30 40 50 60 70 80 90 100 110 120
Growth (mm)

3. $\bar{x}_1 = 38.2$, $s_1 = 27.707$, $n_1 = 5$, $\bar{x}_2 = 85.8$, $s_2 = 42.494$, $n_2 = 5$;

$$t = \frac{(38.2 - 85.8) - 0}{\sqrt{\frac{27.707^2}{5} + \frac{42.494^2}{5}}} = -2.098$$

4. P-value $= 0.0748$, df $= 6.88$

5. Because the P-value of $0.0748 > \alpha = 0.05$, we fail to reject H_0. We do not have convincing evidence that the true mean growth of plants like the ones in this study that are exposed to heavy metal music at night differs from the true mean growth of plants like the ones in this study that are exposed to classical music at night.

page 721: 1. A two-sample t procedure should be used to perform inference because the 109 volunteers were randomly assigned to the two treatment groups. There is no pairing in this context.

2. **STATE:** H_0: $\mu_1 - \mu_2 = 0$, H_a: $\mu_1 - \mu_2 \neq 0$, where μ_1 = the true mean number of words hand written by all students like the ones in this study who take notes with pencil and paper and μ_2 = the true mean number of words written by all students like the ones in this study who take notes using a laptop. Use $\alpha = 0.05$. **PLAN:** Two-sample t test for a difference in means. *Random:* The subjects were randomly assigned to take notes with pencil and paper or a laptop. ✓ *Normal/Large Sample:* $n_1 = 55 \geq 30$ and $n_2 = 54 \geq 30$. ✓ **DO:** $t = -4.002$, P-value $= 0.000135$, df $= 83.78$. **CONCLUDE:** Because the P-value of $0.000135 < \alpha = 0.05$, we reject H_0. There is convincing evidence of a difference in the mean number of words that students like the ones in this study would hand write and the mean number of words that students would type in their notes.

Answers to Odd-Numbered Section 7D Exercises

1. (a) H_0: $\mu_1 - \mu_2 = 0$, H_a: $\mu_1 - \mu_2 \neq 0$, where μ_1 = the mean number of hours worked by all U.S. employees in 2018 and μ_2 = the mean number of hours worked by all U.S. employees in 1978. **(b)** *Random:* Independent random samples of 1381 U.S. employees in 2018 and 855 U.S. employees in 1978. ✓ *10%:* 1381 < 10% of all U.S. employees in 2018 and 855 < 10% of all U.S. employees in 1978. ✓ *Normal/Large Sample:* $n_1 = 1381 \geq 30$ and $n_2 = 855 \geq 30$. ✓

3. (a) H_0: $\mu_1 - \mu_2 = 0$, H_a: $\mu_1 - \mu_2 > 0$, where μ_1 = the mean change in blood pressure (mmHg) for all subjects like the ones in this study who follow a 4-week diet that includes fish oil and μ_2 = the mean change in blood pressure (mmHg) for all subjects like the ones in this study who follow a 4-week diet that includes a regular mixture of oils. **(b)** *Random:* The subjects were randomly assigned to a diet that includes fish oil or a regular mixture of oils. ✓ *Normal/Large Sample:* The sample sizes are small ($n_1 = 7 < 30$ and $n_2 = 7 < 30$), but the graphs of the sample data show no strong skewness and no outliers. ✓

Fish oil
Regular oil
–6 –4 –2 0 2 4 6 8 10 12 14
Change in blood pressure (mmHg)

5. (a) $t = \dfrac{(41.28 - 40.81) - 0}{\sqrt{\frac{6.59^2}{1381} + \frac{8.40^2}{855}}} = 1.392$ **(b)** P-value $= 0.1641$, df $= 1494.54$ **(c)** Because the P-value of $0.1641 > \alpha = 0.05$, we fail to reject H_0. There is not convincing evidence of a difference in the mean number of hours worked per week for all U.S. employees in 1978 and 2018.

7. (a) Because $\bar{x}_1 - \bar{x}_2 = 6.571 - (-1.143) = 7.714 > 0$ **(b)** $t = \dfrac{[6.571 - (-1.143)] - 0}{\sqrt{\frac{5.855^2}{7} + \frac{3.185^2}{7}}} = 3.062$; P-value $= 0.0065$, df $= 9.26$

(c) Because the P-value of $0.0065 < \alpha = 0.05$, we reject H_0. There is convincing evidence that a 4-week diet that includes fish oil helps reduce blood pressure more than a 4-week diet that includes a regular mixture of oils, on average, for men like the ones in this study.

9. (a) Both distributions of BMC change are slightly skewed to the right. Neither distribution has any outliers. The center of the BMC change distribution is much less for breast-feeding mothers (median $\approx -4\%$) than for women who are not pregnant or lactating (median $\approx 0\%$). Also, all of the women who were neither pregnant nor lactating had percent changes in BMC greater than the median for the women who were breast-feeding. The BMC changes are much more variable for breast-feeding mothers ($IQR \approx 3.5\%$) than for women who are not pregnant or lactating ($IQR \approx 1.75\%$). Overall, it appears that breast-feeding mothers do lose bone mineral because a more negative percent change represents a more substantial bone loss. **(b)** Because the P-value of approximately 0 is less than $\alpha = 0.05$, we reject H_0. We have convincing evidence that breast-feeding women have a greater mean percent bone mineral loss than women who are neither pregnant nor lactating. **(c)** Because this was not a randomized controlled experiment, we cannot conclude that breast-feeding is the cause of the bone mineral loss. The women who are breast-feeding may be doing other things that are causing the loss in bone mineral. **(d)** Because we rejected the null hypothesis, we could have committed a Type I error. It is possible that the difference in mean percent bone mineral loss is the same for women who are breast-feeding and women who are not pregnant or lactating, but we found convincing evidence that the mean percent bone mineral loss is greater among breast-feeding women.

11. **(a)** **S:** H_0: $\mu_F - \mu_M = 0$, H_a: $\mu_F - \mu_M \neq 0$, where μ_F = the mean length (cm) of all female Australian possums and μ_M = the mean length (cm) of all male Australian possums. Use $\alpha = 0.05$. **P:** Two-sample *t* test for a difference in means. *Random:* Independent random samples of 43 female Australian possums and 61 male Australian possums. ✓ *10%:* $n_F = 43 < 10\%$ of all female Australian possums and $n_M = 61 < 10\%$ of all male Australian possums. ✓ *Normal/Large Sample:* $n_F = 43 \geq 30$ and $n_M = 61 \geq 30$. ✓ **D:** $t = 1.650$, *P*-value = 0.1023, df = 92.61. **C:** Because the *P*-value of $0.1023 > \alpha = 0.05$, we fail to reject H_0. There is not convincing evidence of a difference in the mean total lengths of all male and female Australian possums. **(b)** If there is no difference in the population mean total length of all male and female Australian possums, there is a 0.1023 probability of getting a difference in sample means as large as or larger than 1.396 cm in either direction ($\bar{x}_F - \bar{x}_M \leq -1.396$ or $\bar{x}_F - \bar{x}_M \geq 1.396$) by chance alone.

13. **S:** H_0: $\mu_E - \mu_I = 0$, H_a: $\mu_E - \mu_I < 0$, where μ_E = the true mean creativity score for all students like the ones in this study when given external reasons for writing and μ_I = the true mean creativity score for all students like the ones in this study when given internal reasons for writing. Use $\alpha = 0.05$. **P:** Two-sample *t* test for a difference in means. *Random:* The subjects were randomly assigned to receive external or internal reasons for writing. ✓ *Normal/Large Sample:* The sample sizes are small ($n_E = 23 < 30$ and $n_I = 24 < 30$), but the graphs of the sample data show no strong skewness and no outliers. ✓ **D:** $t = -2.915$, *P*-value = 0.0028, df = 43.109. **C:** Because the *P*-value of $0.0028 < \alpha = 0.05$, we reject H_0. There is convincing evidence that giving internal reasons for writing increases the true mean creativity score compared to giving external reasons for writing among students like the ones in this study.

15. **(a)** See Exercise 11 for definition of parameters and check of conditions. **S:** 95% CI for $\mu_F - \mu_M$. **P:** Two-sample *t* interval for a difference in means. **D:** (−0.284, 3.076) with df = 92.61. **C:** We are 95% confident that the interval from −0.284 cm to 3.076 cm captures the difference in the mean total lengths of all male and female Australian possums. **(b)** The confidence interval from −0.284 to 3.076 includes 0 as a plausible value for $\mu_F - \mu_M$, which is consistent with the decision to fail to reject H_0: $\mu_F - \mu_M = 0$ in Exercise 11. The confidence interval also tells us that any value between −0.284 cm and 3.076 cm is plausible for $\mu_F - \mu_M$ based on the sample data, but the significance test in part (a) didn't give us an estimate for $\mu_F - \mu_M$.

17. **(a)** $t = \dfrac{(68.7 - 54.1) - 10}{\sqrt{\dfrac{13.3^2}{15} + \dfrac{11.93^2}{14}}} = 0.982$ **(b)** *Interpretation:* Assuming that the true mean cholesterol reduction for subjects like the ones in the study who take the new drug is 10 mg/dl more than for subjects like the ones in the study who take the old drug, there is a 0.1675 probability that we would observe a difference in sample means that is at least 4.6 mg/dl or greater beyond a difference of 10 mg/dl by chance alone. *Conclusion:* Because the *P*-value of $0.1675 > \alpha = 0.05$, we fail to reject H_0. We do not have convincing evidence that the true mean cholesterol reduction is more than 10 mg/dl greater for the new drug than for the current drug for people like the ones in this experiment. **(c)** To increase the power of this test, we could increase the sample sizes or we could increase the significance level.

19. **(a)** Two-sample *t* procedures. The data are being produced using two distinct groups of cars in a randomized experiment. **(b)** Paired *t* procedures. This is a matched pairs experimental design where both treatments are applied to each subject in a random order. **(c)** Paired *t* procedures. Two values are recorded for each pair (each of the 40 couples).

21. **(a)** The appropriate test is the paired *t* test because we have paired data (two scores for each student). **(b)** **S:** H_0: $\mu_{Diff} = 0$, H_a: $\mu_{Diff} > 0$, where μ_{Diff} = the true mean increase in SAT verbal scores of students who were coached using $\alpha = 0.05$. **P:** Paired *t* test for a mean difference. *Random:* Random sample of 427 students who are coached. ✓ *10%:* $n_{Diff} = 427 < 10\%$ of students who are coached for their second attempt on the SAT. ✓ *Normal/Large Sample:* $n_{Diff} = 427 \geq 30$. ✓ **D:** $t = 10.157$, *P*-value ≈ 0, df = 426. (*Table B: P*-value < 0.0005 using df = 100.) **C:** Because the *P*-value of approximately $0 < \alpha = 0.05$, we reject H_0. There is convincing evidence that students who are coached increase their scores on the SAT verbal test, on average.

23. **(a)** H_0: $\mu_A - \mu_I = 0$, H_a: $\mu_A - \mu_I > 0$, where μ_A = the true mean improvement in pain level for all people with chronic pain like the ones in this study who receive an active magnet and μ_I = the true mean improvement in pain level for all people with chronic pain like the ones in this study who receive an inactive magnet. **(b)** The sample sizes are small ($n_A = 29 < 30$ and $n_I = 21 < 30$) and the graph of the sample data for the participants who received an inactive magnet is strongly skewed to the right and has two high outliers. **(c)** $\bar{x}_A - \bar{x}_I = 5.24 - 1.10 = 4.14$ **(d)** None of the values in the simulation has a difference (Active – Inactive) in the mean improvement in pain of 4.14 or greater, so the estimated *P*-value = 0. Because the *P*-value $\approx 0 < \alpha = 0.05$, we reject H_0. We have convincing evidence that active magnets lead to a larger average improvement in pain level than inactive magnets for people with chronic pain like the ones in this study.

25. D

27. D

29. **(a)** Paired *t* test for the mean difference. We have one random sample of paired data (two intersections) and are asked to conduct a test ("a city planner wants to determine") for a mean difference. **(b)** One-sample *z* test for a proportion. We have one random sample and are asked to conduct a test ("are more than 75% of Toyota owners") for a population proportion. **(c)** Two-sample *z* test for a difference in proportions. We have two independent random samples (first-year college students and college students who have completed at least one year) and are asked to conduct a test ("are first-year college students more likely") for a difference in population proportions. **(d)** Two-sample *t* interval for a difference in means. We have two groups (random assignment to incentive A or B) and are asked to estimate ("how much") the difference in population means.

Answers to Unit 7 Review Exercises

R1. **(a)** **S:** 95% CI for μ = the mean length of this component (mm) for all engine crankshafts produced this day. **P:** One-sample *t* interval for a mean. *Random:* Random sample of 16 crankshafts produced this day. ✓ *10%:* $n = 16 < 10\%$ of the 200 crankshafts produced this day. ✓ *Normal/Large Sample:* The sample size is small ($n = 16 < 30$), but the boxplot does not show strong skewness or outliers. ✓ **D:** (223.97, 224.03) using df = 15. **C:** We are 95% confident that the interval from 223.97 mm to 224.03 mm captures μ = the mean length of the critical component for all the engine crankshafts produced on this day. **(b)** No, the value 224 mm is in the 95% confidence interval from 223.97 to 224.03, so it is a plausible value for the population mean length.

R2. **(a)** μ_{Diff} = the true mean increase in the amount this bank's credit card customers would spend with no annual fee. **(b)** We have two measurements (charges for this year and charges for last year) for each of the 200 randomly selected credit card customers. Because this is paired data, we should use paired *t* procedures to calculate a confidence interval for the true mean increase in the amount this bank's credit card customers would spend with no annual fee. **(c)** *Random:* Random sample of 200 credit card customers. ✓ *10%:* Assume $n_{Diff} = 200 < 10\%$ of all credit card customers at this bank. ✓ *Normal/Large Sample:* $n_{Diff} = 200 \geq 30$. ✓ **(d)** df = 199, $t^* = 2.601$ **(e)** No, even though the confidence interval (312.14, 351.86) does not include 0, we cannot conclude that dropping the annual fee would cause an increase in the average amount spent by the bank's credit card customers. There is no control group to compare results. It may be that the increase in spending is due to an improved economy and not the no-annual-fee credit card.

R3. **(a)** **S:** $H_0: \mu = 30$, $H_a: \mu \neq 30$, where μ = the mean width (mm) of all cubes produced that hour. Use $\alpha = 0.05$. **P:** One-sample *t* test for a mean. *Random:* Random sample of 50 cubes produced that hour. ✓ *10%:* $n = 50 < 10\%$ of all cubes produced that hour. ✓ *Normal/Large Sample:* $n = 50 \geq 30$. ✓ **D:** $t = 1.768$, *P*-value = 0.0833, df = 49. *Table B:* $2(0.025) = 0.05 <$ *P*-value $< 2(0.05) = 0.10$ using df = 40. **C:** Because the *P*-value of $0.0833 > \alpha = 0.05$, we fail to reject H_0. There is not convincing evidence that the mean width of all cubes produced this hour is different from 30 mm. **(b)** *Type II error:* Failing to find convincing evidence that the true mean width of the cubes differs from 30 mm, when the true mean width of the cubes differs from 30 mm. *Consequence:* The manager will keep a bad batch of cubes that will cause the game to not function properly. **(c)** *Benefit:* This will reduce the probability of making a Type II error. *Drawback:* We increase the probability of making a Type I error, which equates to an increased chance of rejecting batches of cubes that are of the appropriate width. **(d)** The test only allowed us to fail to reject $\mu = 30$ as a plausible value for the population mean width, but the interval gives a set of plausible values for the population mean width of all cubes produced that hour, which is any value between 29.997 mm and 30.043 mm.

R4. Here are the differences (Standing − Block) in order: 0.26, 0.10, 0.11, −0.08, 0.25, 0.23, 0.16, 0.02. **S:** $H_0: \mu_{Diff} = 0$, $H_a: \mu_{Diff} > 0$, where μ_{Diff} = the true mean difference (Standing − Blocks) in 50-meter run time (sec) for sprinters like the ones in this study. Use $\alpha = 0.05$. **P:** Paired *t* test for a mean difference. *Random:* Sprinters were randomly assigned a treatment order. ✓ *Normal/Large Sample:* The number of differences is small ($n_{Diff} = 8 < 30$), but the dotplot does not show any strong skewness or outliers. ✓

−0.1 −0.05 0 0.05 0.1 0.15 0.2 0.25 0.3

Difference (Standing − Blocks) in time (sec)

D: $t = 3.111$, *P*-value = 0.0085, df = 7. **C:** Because the *P*-value of $0.0085 < \alpha = 0.05$, we reject H_0. We have convincing evidence that sprinters like the ones in this study run a faster race when using starting blocks, on average. **(b)** Assuming $H_0: \mu_{Diff} = 0$ is true, there is a 0.0085 probability of getting a mean difference (Standing − Blocks) in 50-meter run time of 0.131 second or greater by chance alone.

R5. **(a)** **S:** 99% CI for $\mu_1 - \mu_2$, where μ_1 = the mean relative tail spine lengths of *D. pulicaria* with fish kairomone present and μ_2 = the mean relative tail spine lengths of *D. pulicaria* with fish kairomone absent. **P:** Two-sample *t* interval for a difference in means. *Random:* Independent random samples of $n_1 = 214$ *D. pulicaria* with fish kairomone present and $n_2 = 152$ *D. pulicaria* with fish kairomone absent. ✓ *10%:* Assume $n_1 = 214 < 10\%$ of all *D. pulicaria* with fish kairomone present and $n_2 = 152 < 10\%$ of all *D. pulicaria* with fish kairomone absent. ✓ *Normal/Large Sample:* $n_1 = 214 \geq 30$ and $n_2 = 152 \geq 30$. ✓ **D:** (5.381, 7.799) using df = 345.08. **C:** We are 99% confident that the interval from 5.381 to 7.799 captures the difference (Present − Absent) in the population mean relative tail spine lengths (measured as a percentage of the entire length of the water flea) of *D. pulicaria* with fish kairomone present and *D. pulicaria* with fish kairomone absent. **(b)** If we took many random samples of 214 *D. pulicaria* with fish kairomone present and 152 *D. pulicaria* with fish kairomone absent and each time constructed a 99% confidence interval in this same way, about 99% of the resulting intervals would capture the difference (Present − Absent) in the population mean relative tail spine lengths.

R6. **(a)** Even though each subject has two scores (before and after), we are comparing two groups in a randomized experiment. **(b)** Because the students in the treatment group had a greater mean improvement in score than those in the control group. $\bar{x}_{Treatment} - \bar{x}_{Control} = 11.4 - 8.25 = 3.15 > 0$. **(c)** **S:** $H_0: \mu_T - \mu_C = 0$, $H_a: \mu_T - \mu_C > 0$, where μ_T = the true mean improvement in test score for students like the ones in this study who get the treatment message and μ_C = the true mean improvement in test score for students like the ones in this study who get the neutral (control) message. Use $\alpha = 0.01$. **P:** Two-sample *t* test for a difference in means. *Random:* The students were randomly assigned to the treatment or control message. ✓ *Normal/Large Sample:* The sample sizes are small ($n_T = 10 < 30$ and $n_C = 8 < 30$), but the graphs of the sample data show no strong skewness and no outliers. ✓ **D:** $t = 1.914$, *P*-value = 0.0382, df = 13.919. **C:** Because the *P*-value of $0.0382 > \alpha = 0.01$, we fail to reject H_0. There is not convincing evidence that positive subliminal messages help students like the ones in this study learn math, on average. **(d)** We cannot generalize to all students who failed the test because our sample was not a random sample of all students who failed the test. It was a group of students who agreed to participate in the experiment.

Answers to Unit 7 AP® Statistics Practice Test

T1. B	**T2.** D	**T3.** E	**T4.** D	**T5.** C
T6. D	**T7.** D	**T8.** B	**T9.** A	**T10.** C

T11. **(a)** $H_0: \mu = \$158$, $H_a: \mu \neq \$158$, where μ = the true mean amount spent per week on food by households in this city. Use $\alpha = 0.05$. **(b)** The Normal/Large Sample condition is met because $n = 50 \geq 30$. **(c)** Assuming that the true mean amount of money spent per week on food by households in this city is \$158, there is a 0.128 probability of getting a sample mean as different as or more different than \$165 (in either direction) by chance alone. **(d)** Because the *P*-value of $0.128 > \alpha = 0.05$, we fail to reject H_0. We do not have convincing evidence that the true mean amount spent on food per household in this city is different from the national average of \$158.

T12. S: 95% CI for $\mu_1 - \mu_2$, where μ_1 = the true mean length of hospital stay (in days) for patients like the ones in this study who get heating blankets during surgery and μ_2 = the true mean length of hospital stay (in days) for patients like the ones in this study who are allowed to have core temperatures decrease during surgery. **P:** Two-sample *t* interval for a difference in means. *Random:* The patients were assigned at random to the normothermic group and the hypothermic group. ✓ *Normal/Large Sample:* $n_1 = 104 \geq 30$ and $n_2 = 96 \geq 30$. ✓ **D:** $(-4.162, -1.038)$ using df = 165.12. **C:** We are 99% confident that the interval from -4.162 days to -1.038 days captures the true difference (Normothermic – Hypothermic) in mean length of hospital stay for patients like the ones in this study who get heating blankets during surgery and those who are allowed to have their core temperatures decrease during surgery. **(b)** Yes. Because 0 is not in the interval from -4.162 to -1.038, we have convincing evidence that the true mean hospital stay for patients like the ones in this study who get heating blankets during surgery is different than the true mean hospital stay for patients like the ones in this study who are allowed to have their core temperatures decrease during surgery. And because this was a randomized experiment, these data provide convincing evidence that keeping patients like the ones in this study warm during surgery affects the average length of their hospital stays.

T13. S: H_0: $\mu_{\text{Diff}} = 0$, H_a: $\mu_{\text{Diff}} > 0$, where μ_{Diff} = the true mean difference (Regular – Express) in wait time for shoppers like the ones in this study at this store. Use $\alpha = 0.05$. **P:** Paired *t* test for a mean difference. *Random:* Random selection of 15 times to visit the store. ✓ *10%:* $n_{\text{Diff}} = 15 < 10\%$ of all possible times the store could be visited. ✓ *Normal/Large Sample:* The number of differences is small ($n_{Diff} = 15 < 30$), but the dotplot does not show any strong skewness or outliers. ✓

-100 -50 0 50 100 150 200 250

Difference (Regular – Express) in wait time (sec)

D: $t = 1.967$, *P*-value $= 0.0347$, df $= 14$. **C:** Because the *P*-value of $0.0347 < \alpha = 0.05$, we reject H_0. There is convincing evidence that it is faster to use the express lane at this store than a regular lane, on average. **(b)** Because we rejected H_0, it is possible we made a Type I error, which occurs when we find convincing evidence that it is faster to use the express lane at this store than a regular lane, on average, when it is actually not.

Answers to Cumulative AP® Statistics Practice Test 3

AP1. A	**AP2.** C	**AP3.** B	**AP4.** E	**AP5.** C
AP6. D	**AP7.** B	**AP8.** A	**AP9.** B	**AP10.** E
AP11. C	**AP12.** C	**AP13.** A	**AP14.** A	**AP15.** C

AP16. (a) A linear model is appropriate for describing the relationship between temperature and distance to the nearest fish because there is no leftover curved pattern in the residual plot. The residuals look randomly scattered around the residual = 0 line. **(b)** $\hat{y} = -73.64 + 5.7188x$, where y = distance (in meters) from the outflow pipe and x = temperature (°C). **(c)** The predicted distance from the nearest fish to the outflow pipe increases by 5.7188 meters for each additional 1°C increase in water temperature. **(d)** $\hat{y} = -73.64 + 5.7188(29) = 92.21$ meters. Residual $= 78 - 92.21 = -14.21$ meters. The actual distance from the nearest fish to the outflow pipe on this afternoon was 14.21 meters closer than the distance predicted by the regression line with $x = 29°$C.

AP17. (a) (i) $z = \dfrac{15 - 16.4}{2.1} = -0.667$ and $z = \dfrac{20 - 16.4}{2.1} = 1.714$. *Tech:* normalcdf(lower: -0.667, upper: 1.714, mean: 0, SD: 1) = 0.7043; *Table A:* Area between $z = -0.67$ and $z = 1.71$ is $0.9564 - 0.2514 = 0.7050$. (ii) normalcdf(lower: 15, upper: 20, mean: 16.4, SD: 2.1) = 0.7043. The proportion of commercially raised copper rockfish that are between 15 and 20 inches long is about 0.7043. **(b)** (i) Let Y = number of copper rockfish that are between 15 and 20 inches long. Y is a binomial random variable with $n = 10$ and $p = 0.7043$. (ii) $P(Y < 5)$ = binomcdf (trials: 10, p: 0.7043, x value: 4) = 0.0443. **(c)** (i) $\mu_{\bar{x}} = 16.4$ inches. $\sigma_{\bar{x}} = \dfrac{2.1}{\sqrt{10}} = 0.664$ inch because it is reasonable to assume that $10 < 10\%$ of all copper rockfish at this commercial fish plant. (ii) The distribution of $\bar{x}$ is approximately normal because the population distribution of length is approximately normal.

AP18. (a) *Interpretation:* Assuming that the true difference in the mean annualized daily return for Stock A and Stock B is zero, there is a 0.042 probability that we would observe a difference in sample means of 4.7 percentage points or greater by chance alone. *Conclusion:* Because the *P*-value of $0.042 < \alpha = 0.05$, we reject H_0. We have convincing evidence that the true mean annualized return for Stock A is different than the true mean annualized return for Stock B. **(b)** H_0: $\dfrac{\sigma_A^2}{\sigma_B^2} = 1$, H_a: $\dfrac{\sigma_A^2}{\sigma_B^2} > 1$, where σ_A^2 = the true variance of returns for Stock A and σ_B^2 = the true variance of returns for Stock B. **(c)** (i) $F = \dfrac{12.9^2}{9.6^2} = 1.806$ (ii) When the standard deviation of Stock A is greater than the standard deviation of Stock B, the variance of Stock A will be bigger than the variance of Stock B. Thus, values of F that are greater than 1 would indicate that the price volatility for Stock A is higher than that for Stock B. The value of the statistic provides some evidence for the alternative hypothesis because $1.806 > 1$. **(d)** In the simulation, a test statistic of 1.806 or greater occurred in only 6 out of the 200 trials. Thus, the approximate *P*-value is $6/200 = 0.03$. Because the approximate *P*-value of $0.03 < \alpha = 0.05$, we reject H_0. There is convincing evidence that the true variance of returns for Stock A is greater than the true variance of returns for Stock B. (i.e., that Stock A is a riskier investment than Stock B).

UNIT 8

Section 8A

Answers to Check Your Understanding

page 751: 1. H_0: The company's claimed color distribution in a large bag of Peanut M&M'S® is correct. H_a: The company's claimed color distribution in the large bag of Peanut M&M'S is not correct.

2. There were 65 candies in the sample from the bag. The expected counts of blue, orange, green, and yellow candies are $65(0.20) = 13$, and the expected counts of red and brown are $65(0.10) = 6.5$.

3. *Random:* Random sample of 65 M&M'S from the large bag. ✓ *10%:* $n = 65 < 10\%$ of all M&M'S in the large bag. ✓ *Large Counts:* All the expected counts (see #2) are at least 5. ✓

page 760: 1. **STATE:** H_0: The distribution of car color in Oro Valley is the same as the distribution of car color across North America. H_a: The distribution of car color in Oro Valley is not the same as the distribution of car color across North America. Use $\alpha = 0.05$. **PLAN:** Chi-square test for goodness of fit. *Random:* Random sample of 300 cars in Oro Valley. ✓ *10%:* $n = 300 < 10\%$ of all cars in Oro Valley. ✓ *Large Counts:* All expected counts [White: 300(0.23) = 69, Black: 300(0.18) = 54, Gray: 300(0.16) = 48, Silver: 300(0.15) = 45, Red: 300(0.10) = 30, Blue: 300(0.09) = 27, Green: 300(0.02) = 6, and Other: 300(0.07) = 21] are at least 5. ✓ **DO:** $\chi^2 = \frac{(84-69)^2}{69} + \frac{(38-54)^2}{54} + \ldots = 29.921$. *P*-value: df = $8 - 1 = 7$. *Table C:* *P*-value < 0.0005; *Tech:* $\chi^2 = 29.921$, *P*-value ≈ 0, df = 7. **CONCLUDE:** Because the *P*-value of approximately $0 < \alpha = 0.05$, we reject H_0. We have convincing evidence that the distribution of car color in Oro Valley is not the same as the distribution of car color across North America.

2. The table of chi-square contributions is shown here.

Color	Observed	Expected	0 − E	(0 − E)²/E
White	84	69	15	3.2609
Black	38	54	−16	4.7407
Gray	31	48	−17	6.0208
Silver	46	45	1	0.0222
Red	27	30	−3	0.3
Blue	29	27	2	0.1481
Green	6	6	0	0
Other	39	21	18	15.4286

The two biggest contributions to the chi-square statistic came from gray and other colored cars. There were fewer gray cars than expected and more other colored cars than expected. As for Cass's question: It does seem that drivers in Oro Valley prefer lighter colored cars as there were more white cars than expected and fewer black cars than expected, so it seems Cass might be onto something!

Answers to Odd-Numbered Section 8A Exercises

1. **(a)** H_0: The blood type distribution for Danish individuals who test positive for Covid-19 is the same as the blood type distribution in the Danish population. H_a: The blood type distribution for Danish individuals who test positive for Covid-19 is different than the blood type distribution in the Danish population. **(b)** $\hat{p}_O = \frac{2851}{7422} = 0.384, \hat{p}_A = \frac{3296}{7422} = 0.444, \hat{p}_B = \frac{897}{7422} = 0.121$, $\hat{p}_{AB} = \frac{378}{7422} = 0.051$. **(c)** Because the proportions in the sample (0.384, 0.444, 0.121, 0.051) don't all match the population percentages.

3. H_0: The animals in boxes of animal crackers are equally likely. H_a: The animals in boxes of animal crackers are not equally likely.

5. *Type O:* 7422(0.417) = 3094.974; *Type A:* 7422(0.424) = 3146.928; *Type B:* 7422(0.114) = 846.108; *Type AB:* 7422(0.045) = 333.99

7. **(a)** If the null hypothesis is true, then the animals in boxes of animal crackers are equally likely. **(b)** 50/13 = 3.846 for each type of animal

9. *Random:* Random sample of 7422 Danish individuals who test positive for Covid-19. ✓ *10%:* Assume $n = 7422 < 10\%$ of all Danish individuals who test positive for Covid-19. ✓ *Large Counts:* All expected counts (see Exercise 5) are at least 5. ✓

11. Not all of the expected counts are at least 5 (all expected counts = 3.846).

13. **(a)** The blue curve has df = 3. As the degrees of freedom (df) increase, the density curves become less skewed and have a greater mean. The blue curve (df = 3) is more skewed and has a smaller mean than the green curve (df = 7). **(b)** It is more likely to get a χ^2 value greater than 10 in the distribution with df = 7, which has a greater mean. Increasing the df causes larger values of χ^2 to become more likely.

15. **(a)** *Table C:* $0.05 <$ *P*-value < 0.10; *Tech:*χ^2cdf(lower:19.03, upper:10000, df:11) = 0.0606. **(b)** *Table C:* The *P*-value is less than 0.0005. *Tech:* χ^2cdf(lower:19.03, upper:10000, df:3) = 0.0003

17. **(a)** $\chi^2 = \frac{(2851-3094.974)^2}{3094.974} + \ldots + \frac{(378-333.99)^2}{333.99} = 35.154$ **(b)** *Tech:* $\chi^2 = 35.154$, *P*-value ≈ 0, df = 3. *Table C:* *P*-value < 0.0005. **(c)** Assuming that the population proportion of Danish individuals who test positive for Covid-19 of each blood type is the same as that of the population proportion of all Danish individuals of each blood type, there is about a 0% probability of getting a distribution of blood type in a random sample of size 7422 this different or more different than the hypothesized distribution by chance alone.

19. **(a)** **S:** H_0: The four Chicken McNugget shapes are equally likely. H_a: The four Chicken McNugget shapes are not equally likely. Use $\alpha = 0.05$. **P:** Chi-square test for goodness of fit. *Random:* Random sample of 200 Chicken McNuggets. ✓ *10%:* $n = 200 < 10\%$ of all Chicken McNuggets. ✓ *Large Counts:* Each expected count = $50 \geq 5$. ✓ **D:** $\chi^2 = 3.640$, *P*-value = 0.3031, df = 3. **C:** Because the *P*-value of $0.3031 > \alpha = 0.05$, we fail to reject H_0. We do not have convincing evidence that the four Chicken McNugget shapes are not equally likely. **(b)** Because we failed to reject H_0, it is possible to have made a Type II error. This error would occur if the population proportions of Chicken McNuggets shapes are not all the same, but we failed to find convincing evidence that the four Chicken McNugget shapes are not equally likely.

21. **(a)** **S:** H_0: Nuthatches do not prefer particular types of trees when searching for seeds and insects. H_a: Nuthatches do prefer particular types of trees when searching for seeds and insects. Use $\alpha = 0.05$. **P:** Chi-square test for goodness of fit. *Random:* Random sample of 156 red-breasted nuthatches. ✓ *10%:* $n = 156 < 10\%$ of all red-breasted nuthatches. ✓ *Large Counts:* Expected counts (84.24, 62.40, and 9.36) are all at least 5. ✓ **D:** $\chi^2 = 7.418$, *P*-value = 0.0245, df = 2. **C:** Because the *P*-value of $0.0245 < \alpha = 0.05$, we reject H_0. There is convincing evidence that nuthatches prefer particular types of trees when they are searching for seeds and insects. **(b)** The chi-square statistic is: $\chi^2 = 2.407 + 4.416 + 0.595$. We see that the largest contributors to the statistic are the Douglas firs and the ponderosa pines. There are fewer red-breasted nuthatches observed in Douglas firs (70 − 84.24 = −14.24) and more red-breasted nuthatches observed in ponderosa pines (79 − 62.4 = 16.6) than we would expect, so the nuthatches seem to prefer the ponderosa pines the most and the Douglas firs the least.

2. Because all the values in the interval (0.0053218, 0.0061270) are positive, there is convincing evidence to support the class's suspicion that helicopters dropped from a greater height will take longer to land on the ground.

Answers to Odd-Numbered Section 9A Exercises

1. **(a)** $\mu_b = 1.02$. *Interpretation:* If you selected all possible samples of 25 days from this year and calculated the slope of the least-squares regression line for predicting high temperature (°F) from low temperature (°F) for each sample, the sample slopes would have an average value of 1.02. **(b)** Because 25 < 10% of the 365 days this year, $\sigma_b = \frac{6.64}{19.65\sqrt{25}} = 0.068$. *Interpretation:* If you selected all possible samples of 25 days from this year and calculated the slope of the least-squares regression line for predicting high temperature (°F) from low temperature (°F) for each sample, the sample slopes would typically vary from the population slope of 1.02 by about 0.068.
3. Because the Equal SD condition is violated. The residuals increase in magnitude as the laboratory measurement (x) increases. There is a clear < pattern in the residual plot.
5. *Linear:* There is no leftover curved pattern in the residual plot. ✓ *Normal:* The dotplot of residuals shows no strong skewness or outliers. ✓ *Equal SD:* In the residual plot, we do not see a clear < pattern or > pattern. ✓ *Random:* The trays were randomly assigned a light intensity. ✓
7. **(a)** Estimate: $a = 71.62$. If a tray receives no light, the predicted average number of meadowfoam flowers per plant is 71.62. **(b)** Estimate: $b = -0.04108$. For each increase of 1 micromole per square meter per second in light intensity, the average number of meadowfoam flowers decreases by 0.04108, on average. **(c)** Estimate: $s = 7.12178$. The average number of meadowfoam flowers per plant typically varies by about 7.12178 from the number predicted with the least-squares regression line using x = light intensity. **(d)** Standard error: $s_b = 0.00803$. If we repeated the random assignment many times, and calculated the slope of the sample regression line each time, the slopes of the sample regression lines would typically vary by about 0.00803 from the slope of the true regression line for predicting the average number of meadowfoam flowers per plant from light intensity.
9. **(a)** With 95% confidence and df $= 12-2 = 10$, $t^* = 2.228$; $-0.04108 \pm 2.228(0.00803) = -0.04108 \pm 0.01789 = (-0.05897, -0.02319)$. **(b)** Because 0 is not one of the values in the interval $(-0.05897, -0.02319)$, there is convincing evidence of a linear association between light intensity and average number of meadowfoam flowers per plant.
11. **(a)** **S:** 90% CI for β = the slope of the population regression line relating y = average January temperature (in °F) to x = elevation (in feet) for all cities and towns in Colorado. **P:** t interval for the slope. The conditions for inference have been met. ✓ **D:** With 90% confidence and df $= 10 - 2 = 8$, $t^* = 1.860$; $-0.00309 \pm 1.860(0.0006) = -0.00309 \pm 0.00112 = (-0.00421, -0.00197)$. **C:** We are 90% confident that the interval from -0.00421 to -0.00197 captures the slope of the population regression line relating y = average January temperature (°F) to x = elevation (in feet) for all cities and towns in Colorado. **(b)** If the researcher selected many, many different random samples of 10 cities and towns in Colorado, and constructed a 90% confidence interval for the slope β of the population regression line for predicting average January temperature (in °F) from elevation (in feet) based on each sample, about 90% of those intervals would capture β. **(c)** Because the sample was randomly selected from the population of all cities and towns in Colorado, the largest population to which we can generalize the results in part (a) is all cities and towns in Colorado.
13. **(a)** Because the cans of soda were randomly assigned a tapping time. **(b)** **S:** 95% CI for β = the slope of the true regression line relating y = amount expelled (in milliliters) to x = tapping time (in seconds) for all cans of soda like the ones in this study. **P:** t interval for the slope. The conditions for inference have been met. ✓ **D:** With 95% confidence and df $= 40 - 2 = 38$, $t^* = 2.024$; $-2.6350 \pm 2.024(0.1769) = -2.6350 \pm 0.3580 = (-2.993, -2.277)$. **C:** We are 95% confident that the interval from -2.993 to -2.277 captures the slope of the true regression line relating y = amount expelled (in milliliters) to x = tapping time (in seconds) for all cans of soda like the ones in this study. **(c)** Because all of the values in the interval from -2.993 to -2.277 are negative and this was a randomized experiment, the interval suggests that increasing tapping time on a can of soda decreases the amount expelled. **(d)** Increase the sample size or decrease the confidence level. *Drawbacks:* Increasing the sample size would be more expensive and time consuming to conduct the experiment. Decreasing the confidence level would reduce researchers' confidence that they have captured the population slope β in the confidence interval.
15. **S:** 99% CI for β = the slope of the population regression line relating y = number of beetle larvae to x = number of stumps for all circular plots in the area where beavers were cutting down cottonwood trees. **P:** t interval for the slope. *Linear:* The scatterplot shows a clear linear pattern between number of stumps and number of beetle larvae, and the residual plot shows no leftover curved patterns. ✓ *Normal:* The dotplot of residuals shows no strong skewness or outliers. ✓ *Equal SD:* The variability of points around the residual = 0 line appears to be about the same at all x-values. ✓ *Random:* Random sample of 23 locations. ✓ *10%: Assume* $n = 23 < 10\%$ of all locations in this area where beavers were cutting down cottonwood trees. ✓ **D:** With 99% confidence and df $= 23 - 2 = 21$, $t^* = 2.831$; $11.893733 \pm 2.831(1.136343) = 11.893733 \pm 3.216987 = (8.676746, 15.110720)$. *Tech:* (8.6763, 15.1111) with df = 21. **C:** We are 99% confident that the interval from 8.6763 to 15.1111 captures the slope of the population regression line relating y = number of beetle larvae to x = number of stumps for all circular plots in the area where beavers were cutting down cottonwood trees.
17. **(a)** $\mu_y = 105 + 4.2(15) = 168$ cm **(b)** The distribution of height of 15-year-old students is approximately normal, with a mean of 168 cm and a standard deviation of 7 cm. (i) $z = \frac{180-168}{7} = 1.714$. *Tech:* normalcdf(lower: 1.714, upper: 1000, mean: 0, SD: 1) = 0.0433; *Table A:* $1 - 0.9564 = 0.0436$ (ii) normalcdf(lower: 180, upper: 1000, mean: 168, SD: 7) = 0.0432. So, about 4.32%. **(c)** $\mu_b = 4.2$. Assume that 10 < 10% of all 15-year-old students at this school, so $\sigma_b = \frac{7}{1.28\sqrt{10}} = 1.729$. The shape of the sampling distribution is approximately normal because the conditions for the regression model are met. **(d)** We want to find $P(b < 4)$. (i) $z = \frac{4-4.2}{1.729} = -0.116$. *Tech:* normalcdf(lower: −1000, upper: −0.116, mean: 0, SD: 1) = 0.4538; *Table A:* 0.4522 (ii) normalcdf(lower: −1000, upper: 4, mean: 4.2, SD: 1.729) = 0.4540.

19. A

21. C

23. **(a)** This was an experiment because the order of the two treatments (say the color of the printed word and read the word) were randomly assigned to the students. **(b)** He used a randomized block design where each student was a block. In other words, he used a matched pairs design because there were only two treatments per block. He did this to help account for the different abilities of individual students to read the words or to say the color they were printed in. **(c)** The random assignment was used to help average out the effects of the order in which people did the two treatments. For example, if every subject said the color of the printed word first and were frustrated by this task, the times for the second treatment (reading the word) might be worse. Then we wouldn't know the reason the times were longer for the second treatment — because of frustration or because the second method actually takes longer.

25. **(a)** There appears to be a moderately strong, positive linear association between the length of time to read the word and length of time to identify the color. There are no apparent outliers.

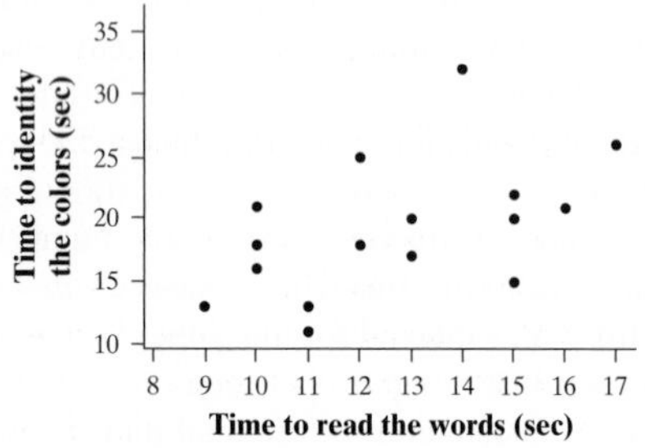

(b) $\hat{y} = 4.887 + 1.1321x$, where $y =$ the time to say the color and $x =$ the amount of time to read the words. **(c)** $\hat{y} = 4.887 + 1.1321(9) = 15.076$ seconds. Residual $= 13 - 15.076 = -2.076$ seconds. *Interpretation:* The actual time to complete the color task was 2.076 seconds less than the time predicted by the regression line with $x = 9$ seconds to complete the word task. **(d)** **S:** 95% CI for $\beta =$ the slope of the true regression line relating $y =$ color time (in sec) to $x =$ word time (in sec) for all subjects like the ones in this study. **P:** t interval for the slope. The conditions for inference have been met. ✓ **D:** *Tech:* (0.0404, 2.2237) using df $= 14$. **C:** We are 95% confident that the interval from 0.0404 to 2.2237 captures the slope of the true regression line relating $y =$ color time (in sec) to $x =$ word time (in sec) for all subjects like the ones in this study.

Section 9B

Answers to Check Your Understanding

page 835: 1. H_0: $\beta = 0$, H_a: $\beta < 0$, where $\beta =$ the slope of the population regression line relating $x =$ female literacy rate (for women 15 years of age or older) and $y =$ birth rate (births per 1000 population) in the population of countries.

2. $t = \dfrac{-0.3769 - 0}{0.087} = -4.332$; *P*-value: df $= 20 - 2 = 18$

Tech: tcdf(lower: -1000, upper: -4.332, df: 18) $= 0.0002$; *Table B:* *P*-value < 0.0005

3. Because the *P*-value $= 0.0002 < \alpha = 0.05$, we reject H_0. These data provide convincing evidence that the slope β of the true regression line is less than 0, so we have convincing evidence of a negative, linear association between $x =$ female literacy rate (for women 15 years of age or older) and $y =$ birth rate (births per 1000 population) in the population of countries.

4. Because we rejected the null hypothesis, it is possible that we made a Type I error, which occurs when we find convincing evidence of a negative, linear association between $x =$ female literacy rate (for women 15 years of age or older) and $y =$ birth rate (births per 1000 population) in the population of countries when there is actually no linear association.

Answers to Odd-Numbered Section 9B Exercises

1. **(a)** H_0: $\beta = 0$, H_a: $\beta < 0$, where $\beta =$ the slope of the population regression line relating $x =$ wine consumption (in liters per year) and $y =$ heart disease death rate (number of deaths per 100,000 people) in the population of countries for which data were available. **(b)** Because $b = -22.9688 < 0$. **(c)** Yes, because the 19 countries were randomly selected.

3. **(a)** H_0: $\beta = 0$, H_a: $\beta \neq 0$, where $\beta =$ the slope of the population regression line relating $x =$ number of fouls and $y =$ number of points for all NBA players. **(b)** *Linear:* The scatterplot shows a clear linear relationship between number of fouls and number of points, and the residual plot shows no leftover curved patterns. ✓ *Normal:* The histogram of residuals contains one potential outlier (condition not met). *Equal SD:* In the residual plot, we see a clear < pattern. As the number of fouls increases, the standard deviation of the residuals increases (condition not met). *Random:* Random sample of 50 NBA players. ✓ *10%:* 50 is not less than 10% of the 450 players in the NBA (condition not met).

5. **(a)** $t = \dfrac{-22.9688 - 0}{3.5574} = -6.457$; df $= 17$; *Tech:* *P*-value $=$ tcdf (lower: -1000, upper: -6.457, df: 17) $= 0.000003$; *Table B:* *P*-value < 0.0005. **(b)** If there is no linear relationship between $x =$ wine consumption and $y =$ heart disease death rate for all countries for which data were available, the probability of getting a random sample of 19 countries that yields a least-squares regression line with a slope of -22.9688 or less by chance alone is 0.000003. **(c)** Because the *P*-value $= 0.000003 < \alpha = 0.05$, we reject H_0. These data provide convincing evidence that the slope β of the population regression line is less than 0, so we have convincing evidence of a negative, linear association between $x =$ wine consumption (in liters per year) and $y =$ heart disease death rate (number of deaths per 100,000 people) in the population of countries for which data were available.

7. **(a)** H_0: $\beta = 0$, H_a: $\beta \neq 0$, where $\beta =$ the slope of the population regression line relating $x =$ body temperature (in °F) and $y =$ pulse rate (in beats per minute) for all students at this university. **(b)** If we select many random samples of 20 students at this university and calculate the regression line for predicting pulse rate from body temperature for each sample, the sample slopes will typically vary from the population slope by about 2.8712. **(c)** $t = \dfrac{4.944 - 0}{2.8712} = 1.722$; df $= 18$; *Tech:* *P*-value $=$ tcdf (lower: 1.722, upper: 1000, df: 18) $\times 2 = 0.1022$; *Table B:* $2(0.05) = 0.10 <$ *P*-value $< 2(0.10) = 0.20$. **(d)** Because the *P*-value $= 0.1022 > \alpha = 0.01$, we fail to reject H_0. These data do not provide convincing evidence that the slope β of the population regression line is different from 0, so we do not have convincing evidence of a linear relationship between body temperature and resting pulse rate for students at this university.

9. **S:** H_0: $\beta = 0$, H_a: $\beta < 0$, where $\beta =$ the slope of the true regression line relating $y =$ corn yield (in bushels per acre) to $x =$ number of weeds per meter. Use $\alpha = 0.05$. **P:** t test for the slope. *Linear:* The scatterplot shows a clear linear relationship between number of weeds per meter and corn yield, and the residual plot shows no leftover curved patterns. ✓ *Normal:* The dotplot of residuals shows no strong skewness or outliers. ✓ *Equal SD:* In the residual plot, we do not see a clear < pattern or > pattern. ✓ *Random:* The plots were randomly assigned a number of lamb's-quarter plants to remain. ✓ **D:** $t = -1.92$, P-value $= 0.075/2 = 0.0375$ using df $= 14$. **C:** Because the P-value of $0.0375 < \alpha = 0.05$, we reject H_0. These data provide convincing evidence that the slope β of the true regression line is less than 0, so we have convincing evidence of a negative linear association between the number of lamb's-quarter plants and corn yield.

11. **(a)** We do not need to check the 10% condition because the volunteers were not randomly selected, but rather randomly assigned to listen to music at either 30, 60, or 90 decibels. **(b)** Yes. Because the volunteers were randomly assigned to the music volume treatments. **(c)** **S:** H_0: $\beta = 0$, H_a: $\beta < 0$, where $\beta =$ the slope of the true regression line relating $x =$ volume of music (in decibels) and $y =$ score on the math test for people like the volunteers in this study. Use $\alpha = 0.05$. **P:** t test for the slope. Assume the conditions for inference are met. ✓ **D:** $t = -4.163$, P-value $= 0.0003/2 = 0.00015$ using df $= 28$. **C:** Because the P-value of $0.00015 < \alpha = 0.05$, we reject H_0. These data provide convincing evidence that the slope β of the true regression line is less than 0, so we have convincing evidence of a negative association between music volume and math test score for people like the volunteers in this study. **(d)** Because the null hypothesis is rejected, there is a possibility that the students could have made a Type I error. If the slope of the population regression line relating volume of music to score on the math test is 0, we will have made a Type I error because we found convincing evidence of a negative linear relationship.

13. **(a)** If the national automotive group's claim is correct, the slope β of the population regression line relating $x =$ age (in years) and $y =$ number of miles driven would be 15,000 miles per year of age of the car. The alternative hypothesis says that the claim is incorrect. **(b)** $t = \dfrac{11{,}630.6 - 15{,}000}{1249} = -2.698$; df $= 19$; *Tech:* P-value $=$ tcdf(lower: -1000, upper: -2.698, df: 19) $\times$ $2 = 0.0143$; *Table B:* $2(0.005) = 0.01 < P\text{-value} < 2(0.01) = 0.02$. Because the P-value of $0.0143 < \alpha = 0.05$, we reject H_0. These data provide convincing evidence that the slope β of the population regression line is not equal to 15,000, so we have convincing evidence that AP® Statistics teachers do not drive their main vehicle 15,000 miles per year, on average. **(c)** With 95% confidence and df $= 19$, $t^* = 2.093$; $11630.6 \pm 2.093(1249) = 11630.6 \pm 2614.157 =$ (9016.443, 14244.757), which rounds to (9016.4, 14,244.8). **(d)** Because 15,000 is not in the 95% confidence interval (9016.4, 14,244.8), we would reject H_0: $\beta = 15{,}000$ in favor of H_a: $\beta \neq 15{,}000$ at the $\alpha = 0.05$ significance level. This is the same decision that was reached by the significance test. The confidence interval provides more information because the interval (9016.4, 14,244.8) provides plausible values for the average number of miles driven by AP® Statistics teachers per year.

15. **(a)** The y intercept should equal 0 because a car that is 0 years old should have 0 miles driven. **(b)** H_0: $\alpha = 0$, H_a: $\alpha \neq 0$, where $\alpha =$ the y intercept of the population regression line relating $x =$ age (in years) and $y =$ number of miles driven for the primary vehicles of all AP® Statistics teachers. **(c)** $t = \dfrac{7288.54 - 0}{6591} = 1.106$; df $= 19$; *Tech:* P-value $=$ tcdf(lower: 1.106, upper: 1000, df: 19) $\times 2 = 0.2825$; *Table B:* $2(0.10) = 0.20 < P\text{-value} < 2(0.15) = 0.30$. **(d)** Because the P-value of $0.2825 > \alpha = 0.05$, we fail to reject H_0. These data do not provide convincing evidence that the y intercept α of the population regression line relating $x =$ age (in years) and $y =$ number of miles driven for the primary vehicles of all AP® Statistics teachers is different than 0.

17. B

19. C

21. **(a)** One-sample t interval for a mean. **(b)** Two-sample z test for a difference in proportions. **(c)** Chi-square test for independence. **(d)** Two-sample t interval for a difference in means.

Unit 9 Review Exercises

R1. *Shape:* Approximately normal. The conditions for inference are met, so we know that for each value of the explanatory variable x, the distribution of the response variable y is approximately normal. *Center:* $\mu_b = 13$. *Variability:* $\sigma_b = \dfrac{6}{1.38\sqrt{25}} = 0.8696$

R2. **(a)** **S:** 90% CI for $\beta =$ the slope of the population regression line relating $x =$ height (in inches) and $y =$ number of steps required to walk the length of a school hallway for all students at this high school. **P:** t interval for the slope. *Linear:* The scatterplot shows a linear relationship between height and number of steps, and there is no leftover curved pattern in the residual plot. ✓ *Normal:* The histogram of residuals shows no strong skewness or outliers. ✓ *Equal SD:* In the residual plot, we do not see a clear < pattern or > pattern. ✓ *Random:* Random sample of 36 students. ✓ *10%:* 36 < 10% of all students at this large high school. ✓ **D:** With 95% confidence and df $= 36 - 2 = 34$, $t^* = 1.691$; $-0.9211 \pm 1.691(0.1938) = -0.9211 \pm 0.3277 = (-1.2488, -0.5934)$. **C:** We are 90% confident that the interval from -1.2488 to -0.5934 captures the slope β of the population regression line relating $x =$ height (in inches) and $y =$ number of steps required to walk the length of a school hallway for all students at this high school. **(b)** Because all of the values in the interval (-1.2488, -0.5934) are negative, there is convincing evidence that the slope β of the population regression line is negative, so we have convincing evidence that taller students at this school tend to require fewer steps to walk a fixed distance. **(c)** If many, many different random samples of 36 students were selected from this high school, and we constructed a 90% confidence interval for the slope β of the population regression line for predicting number of steps from height (in inches) based on each sample, about 90% of those intervals would capture β.

R3. **S:** H_0: $\beta = 0$, H_a: $\beta < 0$, where $\beta =$ the slope of the population regression line relating $x =$ average temperature (in °C) and $y =$ wing length (in mm) in the population of *Boloria chariclea* butterflies. Use $\alpha = 0.05$. **P:** t test for the slope. Assume the conditions for inference are met. ✓ **D:** $t = \dfrac{-0.24 - 0}{0.163} = -1.472$; df $= 30$; *Tech:* P-value $=$ tcdf(lower: -1000, upper: -1.472, df: 30) $= 0.0757$; *Table B:* $0.05 < P\text{-value} < 0.10$. **C:** Because the P-value of $0.0757 > \alpha = 0.05$, we fail to reject H_0. These data do not provide convincing evidence that the slope β of the population regression line is negative, so we do not have convincing evidence of a negative linear relationship between average temperature during the larval growing season and wing length in the population of *Boloria chariclea* butterflies.

Unit 9 AP® Statistics Practice Test

T1. C
T2. D
T3. B
T4. D
T5. B

T6. **(a)** If we repeated the random assignment many times and calculated the least-squares regression line each time, the slope of the sample regression line would typically vary by about 1.0164 from the slope of the true regression line for predicting weight gain from dose of growth hormone. **(b)** **S:** 95% CI for β = the slope of the true regression line relating y = weight gain (in ounces) to x = dose of growth hormone (in mg) for all chickens like the ones in this study. **P:** t interval for the slope. Assume the conditions for inference have been met. ✓ **D:** With 95% confidence and df = 13, $t^* = 2.160$; $4.8323 \pm 2.160(1.0164) = 4.8323 \pm 2.1954 = (2.6369, 7.0277)$. **C:** We are 95% confident that the interval from 2.6369 to 7.0277 captures the slope of the true regression line relating y = weight gain (in ounces) to x = dose of growth hormone (in mg) for all chickens like the ones in this study. **(c)** Because all of the plausible values of the slope of the true regression line in the interval (2.6369, 7.0277) are positive and this was a randomized experiment, the interval suggests that chickens like the ones in this study that are given more growth hormones will gain more weight, on average.

T7. **(a)** H_0: $\beta = 0$, H_a: $\beta < 0$, where β = the slope of the population regression line relating x = amount of sugar (in tsp) and y = amount of saturated fat (in g) for all protein bars. Use $\alpha = 0.05$. **(b)** The Linear condition and the Equal SD condition can both be checked with the residual plot. *Linear:* There is no leftover curved pattern in the residual plot. ✓ *Equal SD:* In the residual plot, we do not see a clear < pattern or > pattern. ✓ **(c)** P-value $= 0.0592/2 = 0.0296$. Because the P-value of $0.0296 < \alpha = 0.05$, we reject H_0. These data provide convincing evidence that the slope β of the population regression line is negative, so we have convincing evidence of a negative linear relationship between sugar and fat in protein bars. **(d)** Because we rejected the null hypothesis, we could have made a Type I error. This would occur if we find convincing evidence of a negative linear relationship between amount of sugar and amount of fat in protein bars when there actually is not a negative relationship between amount of sugar and amount of fat among all protein bars.

Cumulative AP® Practice Test 4

AP1 D	**AP2** A	**AP3** B	**AP4** C	**AP5** E	**AP6** C
AP7 B	**AP8** A	**AP9** D	**AP10** D	**AP11** B	**AP12** D
AP13 D	**AP14** B	**AP15** E	**AP16** B	**AP17** B	**AP18** D
AP19 B	**AP20** A	**AP21** C	**AP22** B	**AP23** C	**AP24** B
AP25 D	**AP26** E	**AP27** D	**AP28** E	**AP29** A	**AP30** B
AP31 E	**AP32** E	**AP33** A	**AP34** B	**AP35** C	**AP36** E
AP37 D	**AP38** C	**AP39** D	**AP40** C		

AP41 **(a)** *Shape:* The distribution of travel time by bus is single-peaked and roughly symmetric, and the distribution of travel time by other methods is roughly symmetric and approximately uniform. *Outliers:* There are no clear outliers in either distribution. *Center:* The typical travel time by bus (median: 15 to < 20 minutes) is similar to the typical travel time by other methods (median: 15 to < 20 minutes). *Variability:* The distribution of travel time by bus is less variable than by other methods. Travel time by bus varies from as little as 5 minutes to as much as 30 minutes (range is as much as 25 minutes). Travel time by other methods varies from as little as 0 minutes to as much as 30 minutes (range is as much as 30 minutes). **(b)** Yes, it is possible that the median travel time is the same for both groups. For the students who ride the bus, the median is the average of the 20th and 21st values in the ordered data set. From the histogram, both of these values fall in the 15- to < 20-minute interval. For the students with other forms of transportation, the median is the average of the 30th and 31st values in the ordered data set. From the histogram, both of these values fall in the 15- to < 20-minute interval.

AP42 **(a)** *Experimental units:* bus stops. *Treatments:* sign or no sign. *Response variable:* Number of pieces of trash at each bus stop **(b)** Number the bus stops 1 to 206. Randomly generate 103 different integers between 1 and 206. The bus stops with those labels will receive a sign about using the trash can. The remaining 103 bus stops will not receive a sign. **(c)** Yes. If the results are statistically significant, it would be reasonable to conclude that the signs caused a change in behavior because the bus stops were randomly assigned to receive a sign or not receive a sign.

AP43 **(a)** (i) $z = \dfrac{1.5-2}{0.3} = -1.667$. *Tech:* normalcdf(lower: −1000, upper: −1.667, mean: 0, SD: 1) = 0.0478; *Table A:* Area to the left of $z = -1.67$ is 0.0475. (ii) normalcdf(lower: −1000, upper: 1.5, mean: 2, SD: 0.3) = 0.0478. **(b)** P(none weigh less than 1.5 ounces) $= (1-0.0478)^8 = 0.676$ **(c)** Let T = the total weight of the gift box and truffles. *Shape:* Approximately normal, because the distribution of truffle weight and the distribution of gift box weight are each approximately normal, and independent of each other. *Center:* $\mu_T = 3+2+2+2+2+2+2+2+2 = 19$ ounces. *Variability:* $\sigma_T = \sqrt{0.5^2+0.3^2+0.3^2+0.3^2+0.3^2+0.3^2+0.3^2+0.3^2+0.3^2}$ = 0.985 ounce

AP44 **S:** H_0: $p_1 - p_2 = 0$, H_a: $p_1 - p_2 \neq 0$, where p_1 = the proportion of all adult residents who approve of the governor's job performance in the first year and p_2 = the proportion of all adult residents who approve of the governor's job performance in the next year. Use $\alpha = 0.05$. **P:** Two-sample z test for a difference in proportions. *Random:* Independent random samples of 1002 adult residents in the first year and 1009 adult residents in the next year. ✓ *10%:* 1002 < 10% of all adult residents in the state in the first year and 1009 < 10% of all adult residents in the state in the next year. ✓ *Large Counts:* $\hat{p}_C = \dfrac{529+498}{1002+1009} = \dfrac{1027}{2011} = 0.5107$. $1002(0.5107) = 511.72$, $1002(1-0.5107) = 490.28$, $1009(0.5107) = 515.30$, $1009(1-0.5107) = 493.70$ are all ≥ 10. ✓ **D:** *Tech:* $z = 1.542$, P-value = 0.1230. **C:** Because the P-value of $0.1230 > \alpha = 0.05$, we fail to reject H_0. There is not convincing evidence of a change in the proportion of all adult residents in the state who approved of the governor's job performance during these two years.

AP45. **(a)** The predicted number of calories increases by 11.54 for each additional 1 gram of saturated fat for these dairy-free frozen desserts. **(b)** residual $= 220 - [125.3 + 11.54(13)] = -55.32$ calories. The number of calories for the dairy-free frozen dessert with 13 grams of saturated fat is 55.32 less than the number of calories predicted by the regression line with $x = 13$. **(c)** There are

9 frozen desserts that are above average in saturated fat, of which 8 are above average in calories. P(above average in calories | above average in saturated fat) $= \frac{8}{9} = 0.889$.

AP46 **(a)** The Normal/Large Sample condition is not met because the sample size is small and there is an outlier in the sample data.

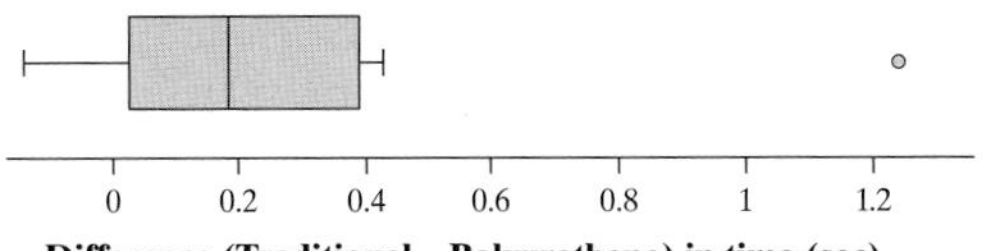

(b) (i) If the null hypothesis is true, half of the differences (a proportion of 0.5) should be greater than zero. (ii) Seven of the 8 differences in the coach's experiment are greater than zero. **(c)** P-value $=$ P(at least 7 out of 8 of the differences are greater than zero | the true proportion is 0.5) is equivalent to $P(X \geq 7)$, where $X =$ the number of differences that are greater than zero. X is a binomial random variable with $n = 8$ and $p = 0.5$. *Tech:* $P(X \geq 7) = 1-$ binomcdf(n: 8, p: 0.5, x value: 6) $= 0.0352$. **(d)** Because the P-value of $0.0352 < \alpha = 0.05$, we reject H_0. There is convincing evidence that the median difference (Traditional − Polyurethane) in 50-meter freestyle times for all swimmers like the ones in this study is greater than zero. The data provide convincing evidence that polyurethane swimsuits help swimmers like these swim faster.

Notes and Data Sources

Unit 1

1 Adapted from A. Bargagliotti, C. Franklin, et al., *Pre-K–12 Guidelines for Assessment and Instruction in Statistics Education II* (American Statistical Association and National Council of Teachers of Mathematics: 2020).

2 *AP Statistics: Course and Exam Description* (College Board: 2020).

3 This data set was obtained by taking a random sample of 3000 households from the American Community Survey's Public Use Microdata Sample files at www.census.gov/programs-surveys/acs/microdata.html.

4 Roller coaster data from www.rcdb.com.

5 Data obtained from The Numbers website: https://www.the-numbers.com/box-office-records/worldwide/all-movies/cumulative/all-time, accessed December 10, 2022.

6 Data obtained from skyscraperpage.com and Wikipedia.

7 Squirrel data set obtained from data.cityofnewyork.us/Environment/2018-Central-Park-Squirrel-Census-Squirrel-Data/vfnx-vebw/data.

8 Data on winners of the World Canine Disc Championship for 1975–1999 from https://en.wikipedia.org/wiki/Frisbee_Dog_World_Championship and for 2000–2019 from https://skyhoundz.com/world-champions/.

9 Data from the CubeSat database: https://sites.google.com/a/slu.edu/swartwout/cubesat-database, accessed October 28, 2022.

10 Data on U.S. radio station formats in 2021 obtained from statista.com.

11 V. Rideout, A. Peebles, S. Mann, and M. B. Robb, *Common Sense Census: Media Use by Tweens and Teens, 2021* (San Francisco, CA: Common Sense: 2022).

12 We got the idea for this example from David Lane's case study "Who Is Buying iMacs?", onlinestatbook.com/case_studies_rvls/.

13 Mobile Fact Sheet from www.pewresearch.org/internet/fact-sheet/mobile/, published April 7, 2021.

14 Centers for Disease Control and Prevention, National Vital Statistics Reports, www.cdc.gov.

15 Data from Oliver Roeder, "New York's Elevators Define the City," May 4, 2016, fivethirtyeight.com.

16 Axalta Coating Systems, *Global Automotive 2020 Color Popularity Report*.

17 Found at www.toptenreviews.com, which claims to have compiled data "from a number of different reputable sources."

18 U.S. Census Bureau, Current Population Survey, Annual Social and Economic Supplement, published April 2020.

19 Data for 2020 from the National Center for Education Statistics, www.nces.ed.gov.

20 K. Eagan, J. B. Lozano, S. Hurtado, and M. H. Case, *The American Freshman: National Norms, Fall 2013* (Los Angeles, CA: Higher Education Research Institute, UCLA, 2013).

21 Data from the CubeSat database: https://sites.google.com/a/slu.edu/swartwout/cubesat-database, accessed October 28, 2022.

22 See note 11.

23 See note 16.

24 Regina A. Corso, "Few Hate Shopping for Clothes, But Love of It Varies by Country," *Harris Poll*, June 24, 2011.

25 This exercise is based on information from www.minerandcostudio.com.

26 https://news.gallup.com/poll/193874/half-americans-play-state-lotteries.aspx.

27 ITU Facts and Figures, 2019.

28 Joanna Foster, "Moo-ving the Dial on Methane," *Solutions* 53, no. 3 (2022), Environmental Defense Fund.

29 flintwaterstudy.org.

30 *Nutrition Action Healthletter*, May 2021.

31 Recommendation from the American Heart Association website, www.heart.org.

32 *Nutrition Action Healthletter*, May 2021.

33 Data collected by Doug Tyson in 2020.

34 Data on household size from separate random samples of 50 students from South Africa and the United Kingdom who completed the Census at School survey in a recent year.

35 Data provided by the team manager for the youth football team of one of Mr. Starnes's grandsons.

36 *Nutrition Action Healthletter*, April 2016.

37 Data from the U.S. Census Bureau, Population Estimates Program, 2020.

38 Data on percentage of foreign-born residents in the states from www.statista.com, accessed January 28, 2021.

39 Histogram of engagement ring costs from www.nytimes.com/2020/02/06/learning/whats-going-on-in-this-graph-engagement-ring-costs.html.

40 Data on the U.S. Women's National Soccer Team from the U.S. Soccer Federation, www.ussoccer.com.

41 *Model Year 2022 Fuel Economy Guide*, www.fueleconomy.gov.

42 Frozen pizza data obtained from *Consumer Reports* magazine, January 2011.

43 *Nutrition Action Healthletter*, April 2016.

44 The original paper is T. M. Amabile, "Motivation and Creativity: Effects of Motivational Orientation on Creative Writers," *Journal of Personality and Social Psychology*, 48, no. 2 (February 1985): 393–399. The data for this exercise came from Fred L. Ramsey and Daniel W. Schafer, *The Statistical Sleuth*, 3rd ed. (Pacific Grove, CA: Brooks/Cole Cengage Learning, 2013).

45 The cereal data came from the Data and Story Library, https://dasl.datadescription.com/.

46 USDA National Nutrient Database for Standard Reference 26 Software v.1.4.

47 Data provided by Josh Tabor.

48 From the *Electronic Encyclopedia of Statistics Examples and Exercises* story, "Acorn Size and Oak Tree Range."

49 Finch data from Fred L. Ramsey and Daniel W. Schafer, *The Statistical Sleuth*, 3rd ed. (Pacific Grove, CA: Brooks/Cole Cengage Learning, 2013).

50 CO_2 emissions data from Our World in Data website: https://ourworldindata.org/share-co2-emissions, accessed January 28, 2021.

51 Data on travel time to work from U.S. Census Bureau, 2015–2019 American Community Survey 5-Year Estimates.

52 Monthly stock returns from the website of Professor Kenneth French of Dartmouth University: https://mba.tuck.dartmouth.edu/pages/faculty/ken.french/data_library.html. A fine point: the data are the "excess returns" on stocks, which are actual returns less the small monthly returns on Treasury bills.

53 See note 45.

54 Data on annual income for college graduates and nongraduates obtained from the March 2021 Annual Social and Economic Supplement, downloaded from www.census.gov. We took a random sample of 500 people who had attended some college but earned no degree, or who had earned an associate or bachelor's degree.

55 Fisher iris data from UCI Machine Learning Repository, https://archive.ics.uci.edu/ml/datasets/iris.

56 S. M. Kerry, F. B. Micah, J. Plange-Rhule, J. B. Eastwood, and F. P. Cappuccio, "Blood Pressure and Body Mass Index in Lean Rural and Semi-urban Subjects in West Africa," *Journal of Hypertension* 23, no. 9 (2005): 1645–1651.

57 Data from a student project in Josh Tabor's Introductory Statistics class.

58 National Center for Statistics and Analysis, *Seat Belt Use in 2019: Use Rates in the States and Territories, Traffic Safety Facts Crash Stats*. Report No. DOT HS 812 947 (Washington, DC: National Highway Traffic Safety Administration, April 2020).

59 Data on the population distributions by age group in Australia and Ethiopia from the U.S. Census Bureau's International Database: www.census.gov, accessed October 4, 2021.

60 Guinea pig data from rdocumentation.org. The original source is listed as E. W. Crampton, "The Growth of the Odontoblast of the Incisor Teeth as a Criterion of Vitamin C Intake of the Guinea Pig," *Journal of Nutrition* 33, no. 5 (1947): 491–504, doi: 10.1093/jn/33.5.491.

61 College tuition data from www.petersons.com.

62 Data from a student project by Rikki Schlott, Daniel Joseph, and Mandi Marom. Daniel was Interviewer 1 and Rikki was Interviewer 2.

63 Data for 2020 from U.S. Census Bureau, *Current Population Survey, 2020 Annual Social and Economic Supplement*. Data for 1980 were obtained from www.census.gov.

64 www.flintwaterstudy.org.

65 Population density data for 2020 from www.indexmundi.com, which cites its source as the *CIA World Factbook*, www.cia.gov/the-world-factbook/.

66 kitchencabinetkings.com.

67 Data on PVC pipe length from a related example in Minitab statistical software.

68 Modified based on data from http://pirate.shu.edu/~wachsmut/Teaching/MATH1101/Descriptives/variability.html.

69 See note 64.

70 NBA scoring data from www.basketball-reference.com.

71 Burger King nutrition data from https://www.bk.com/nutrition-explorer, accessed April 7, 2021.

72 Data from a student project provided by Josh Tabor.

73 Tablet ratings from www.consumerreports.org.

74 Data on the U.S. Women's National Soccer Team from the U.S. Soccer Federation, www.ussoccer.com.

75 *Model Year 2023 Fuel Economy Guide*, www.fueleconomy.gov.

76 Data on birth rates obtained from www.indexmundi.com, which cites its source as the *CIA World Factbook*, www.cia.gov/the-world-factbook/.

77 Zack Friedman, "Student Loan Debt Statistics in 2021: A Record $1.7 Trillion," *Forbes*, February 20, 2021, www.forbes.com.

78 Home price data from the Federal Reserve Bank of St. Louis, 2022.

79 Tom Lloyd et al., "Fruit Consumption, Fitness, and Cardiovascular Health in Female Adolescents: The Penn State Young Women's Health Study," *American Journal of Clinical Nutrition* 67 (1998): 624–630.

80 C. B. Williams, *Style and Vocabulary: Numerological Studies* (Griffin, 1970).

81 Data provided by Dr. Tim Brown.

82 Sidney Crosby data from www.hockey-reference.com.

83 These data are a random sample from the values given in R. Sokal and J. Rohlf, *Introduction to Biostatistics* (Mineola, NY: Dover, 2009).

84 Claim about number of texts per day from www.textrequest.com/blog/how-many-texts-people-send-per-day/.

85 Copper data from a Hubday news release on March 29, 2021, which was highlighted in Tony Davis, "Hubday Says It Found Copper That Could Result in Open Pit Mines on Santa Ritas' West Side," *Arizona Daily Star*, April 1, 2021.

86 K. E. Flegal and M. C. Anderson, "Overthinking Skilled Motor Performance: Or Why Those Who Teach Can't Do," *Psychonomic Bulletin & Review* 15, no. 5 (2008): 927–932, doi: 10.3758/PBR.15.5.927. Thanks to Kristin Flegal for sharing the data from this study.

87 See note 70.

88 Freezer data from *Consumer Reports*, May 2010.

89 Joshua E. Stubbs and Toby C. B. Stubbs, "How to Lose Weight Well, According to *How to Lose Weight Weight Well*," *Significance Magazine*, February 2021.

90 From the American Community Survey, at the Census Bureau website, www.census.gov. The data are a subsample of the individuals in the ACS New York sample who had travel times greater than zero.

91 Data from Andrew Hurst, "Most Americans Are Concerned about Climate Change, But That Doesn't Mean They're Prepared for It," October 15, 2019, www.valuepenguin.com.

92 Data set shared at statcrunch.com by Amanda Moulton, College of the Canyons.

93 See note 11.

94 2019 CIRP Freshman Survey; data from the Higher Education Research Institute, *The American Freshman: National Norms Fall 2019*, January 2020.

95 Data from a student project by Alex and Tempe in Mr. Starnes's Introductory Statistics class.

96 S. M. Stigler, "Do Robust Estimators Work with Real Data?" *Annals of Statistics* 5 (1977): 1055–1078.

97 *Nutrition Action Healthletter*, February 2019.

98 Data from Bureau of Labor Statistics, Annual Demographic Supplement, www.bls.gov.

99 Data from the report "Is Our Tuna Family-Safe?" prepared by Defenders of Wildlife, 2006.
100 Pew Research Center, "Teens, Social Media and Technology 2022," August 2022.
101 These data are a random sample from the values given in R. Sokal and J. Rohlf, *Introduction to Biostatistics* (Mineola, NY: Dover, 2009).
102 See note 64.
103 National percentile rank tables from Assessment Technologies Institute, Inc. and National League for Nursing.
104 Values for the mean and standard deviation obtained from Centers for Disease Control and Prevention's Growth Charts, www.cdc.gov.
105 Data collected by Doug Tyson in 2020.
106 Data from the U.S. Census Bureau, Population Estimates Program, 2020.
107 Data provided by the team manager for the youth football team of one of Mr. Starnes's grandsons.
108 Data on percentage of foreign-born residents in the states from www.statista.com, accessed January 28, 2021.
109 Median household incomes in 2021 from dqydj.com/average-income-by-state-median-top-percentiles/.
110 Data on number of wins for each MLB team from www.baseball-reference.com.
111 SAT summary statistics from https://reports.collegeboard.org/media/pdf/2019-total-group-sat-suite-assessments-annual-report.pdf. ACT summary statistics from www.act.org.
112 Gymnastics score data from http://www.olympedia.org/.
113 Published by Statista Research Department, September 30, 2022.
114 Sheldon Ross, *Introduction to Probability and Statistics for Engineers and Scientists*, 3rd ed. (Academic Press, 2004).
115 Information on GRE scores from www.ets.org/gre/score-users/scores.html.
116 Data from Gary Community School Corporation, courtesy of Celeste Foster, Department of Education, Purdue University.
117 See note 45.
118 Project data from Ryan Wells and Allie Quiroz, Canyon del Oro High School.
119 www.lpga.com/statistics/driving/average-driving-distance?year=2022.
120 E. J. Benjamin et al., on behalf of American Heart Association Council on Epidemiology and Prevention Statistics Committee and Stroke Statistics Subcommittee. "Heart Disease and Stroke Statistics—2019 Update: A Report from the American Heart Association," *Circulation* [published online ahead of print January 31, 2019]. www.ahajournals.org/doi/full/10.1161/CIR.0000000000000659.
121 www.cdc.gov/growthcharts/.
122 https://blog.rescuetime.com/screen-time-stats-2018.
123 See note 120.
124 Refrigerator data from *Consumer Reports*, May 2010.
125 Data provided by Chris Olsen, who found the information in *Scuba News* and *Skin Diver* magazines.
126 *Model Year 2022 Fuel Economy Guide*, www.fueleconomy.gov.
127 Data on percentage of foreign-born residents in the states from www.statista.com, accessed January 28, 2021.
128 https://wwwn.cdc.gov/nchs/nhanes/continuousnhanes/default.aspx?BeginYear=2017.
129 Kevin Quealy, Amanda Cox, and Josh Katz, "At Chipotle, How Many Calories Do People Really Eat?" *New York Times*, February 17, 2015.
130 We found the information on birth weights of Norwegian children on the National Institute of Environmental Health Sciences website: www.ncbi.nlm.nih.gov/pubmed/1536353.
131 stats.nba.com/players/bio.
132 See note 130.
133 Thanks to Jeff Eicher for sharing the idea that led to this exercise.
134 S. M. Stigler, "Do Robust Estimators Work with Real Data?" *Annals of Statistics* 5 (1977): 1055–1078.
135 Based on an exercise from Ignacio Bello, Anton Kaul, and Jack R. Britton, *Topics in Contemporary Mathematics*, 10th ed. (Boston, MA: Cengage Learning, 2013).
136 P. A. Mackowiak, S. S. Wasserman, and M. M. Levine, "A Critical Appraisal of 98.6 Degrees F, the Upper Limit of the Normal Body Temperature, and Other Legacies of Carl Reinhold August Wunderlich," *Journal of the American Medical Association* 268 (1992): 1578–1580. We produced this data set with similar properties as the original temperature readings.

Unit 2

1 Survey data from the Visitor Use Study at www.nps.gov/yell.
2 British Parliamentary Papers, *Shipping Casualties (Loss of the Steamship "Titanic")*, 1912, cmd. 6352, *Report of a Formal Investigation into the Circumstances Attending the Foundering on the 15th April, 1912, of the British Steamship "Titanic," of Liverpool, after Striking Ice in or near Latitude 41° 46' N., Longitude 50° 14' W., North Atlantic Ocean, Whereby Loss of Life Ensued* (London, UK: His Majesty's Stationery Office, 1912). See also www.anesi.com/titanic.htm.
3 *Nutrition Action*, "Weighing the Options: Do Extra Pounds Mean Extra Years?" (March 2013).
4 "Effect of *Hypericum perforatum* (St John's Wort) in Major Depressive Disorder," *Journal of the American Medical Association* 287, no. 14 (2002): 1807–1814.
5 Elizabeth F. Loftus and John C. Palmer, "Reconstruction of Automobile Destruction: An Example of the Interaction Between Language and Memory," *Journal of Verbal Learning and Verbal Behavior* 13 (1974): 585–589, www.researchgate.net/publication/222307973_Reconstruction_of_Automobile_Destruction_An_Example_of_the_Interaction_Between_Language_and_Memory.
6 Pew Research Center, "Teens, Social Media and Technology 2022," August 2022.
7 Squirrel data obtained from data.cityofnewyork.us/Environment/2018-Central-Park-Squirrel-Census-Squirrel-Data/vfnx-vebw/data.
8 N. Häusler, J. Haba-Rubio, R. Heinzer, et al., "Association of Napping with Incident Cardiovascular Events in a Prospective Cohort Study," *Heart* 105 (2019): 1793–1798.
9 Ohio State University. "Night Lights Don't Lead to Nearsightedness, Study Suggests," *ScienceDaily*, March 9, 2000.
10 Siem Oppe and Frank De Charro, "The Effect of Medical Care by a Helicopter Trauma Team on the Probability of Survival and the Quality of Life of Hospitalized Victims," *Accident Analysis and Prevention* 33 (2001): 129–138. The authors give the data in this example as a "theoretical example" to illustrate the need for

more elaborate analysis of actual data using severity scores for each victim.

11 RECOVERY Collaborative Group, "Dexamethasone in Hospitalized Patients with Covid-19," *New England Journal of Medicine* 384 (2021): 693–704.

12 poll.qu.edu/Poll-Release-Legacy?releaseid=2275

13 www.hockey-reference.com

14 Thanks to Paul Myers for sharing this idea.

15 stats.wnba.com/players

16 From the random sampler at ww2.amstat.org/CensusAtSchool/.

17 www.gapminder.org

18 *Nutrition Action*, December 2009.

19 www.tylervigen.com

20 Data from Aaron Waggoner.

21 www.nejm.org/doi/full/10.1056/NEJMon1211064

22 *Nutrition Action*, December 2009.

23 www.stat.columbia.edu/~gelman/research/published/golf.pdf

24 www.starbucks.com/promo/nutrition

25 Joshua D. Stewart et al., "Decreasing Body Lengths in North Atlantic Right Whales," *Current Biology* (2021), doi.org/10.1016/j.cub.2021.04.067. Whales younger than 2 years and older than 20 years were excluded to avoid curvature in the relationship between age and total length.

26 www.ncei.noaa.gov

27 fivethirtyeight.com/features/some-early-clues-about-how-the-midterms-will-go

28 www.basketball-reference.com

29 www.advancedwebranking.com/cloud/ctrstudy

30 M. A. Houck et al., "Allometric Scaling in the Earliest Fossil Bird, *Archaeopteryx lithographica*," *Science*, 247 (1990): 195–198. The authors conclude from a variety of evidence that all specimens represent the same species.

31 Tulin Yakici and Muhammet Arici, "Storage Stability of Aspartame in Orange Flavored Soft Drinks," *International Journal of Food Properties* 16, no. 3 (2013): 698–705, doi:10.1080/10942912.2011.565903.

32 Data from the roller coaster database: www.rcdb.com.

33 www.weatherbase.com

34 www.gapminder.org/tools/#$model$markers$bubble$encoding$x$scale$type=linear;;&frame$speed:100;;;;;&chart-type=bubbles&url=v1

35 *Nutrition Action*, April 2022.

36 E. Thomassot et al., "Methane-Related Diamond Crystallization in the Earth's Mantle: Stable Isotopes Evidence from a Single Diamond-Bearing Xenolith," *Earth and Planetary Science Letters*, 257 (2007), pp. 362–371, Table 1.

37 From a graph in L. Partridge and M. Farquhar, "Sexual Activity Reduces Lifespan of Male Fruit Flies," *Nature* 294 (1981): 580–582. Provided by Brigitte Baldi.

38 Data on used car prices from autotrader.com, September 8, 2012. We searched for F-150 4 × 4's on sale within 50 miles of College Station, Texas.

39 www.lumeradiamonds.com/diamonds/results?price=1082-1038045&carat=0.30-16.03&shapes=B%20&cut=EX&clarity=FL,IF&color=D,E,F#

40 www.epa.gov/greenvehicles/fast-facts-transportation-greenhouse-gas-emissions

41 stats.washingtonpost.com/golf/averages.asp?tour=LPGA&rank=05. Data through August 16, 2021.

42 Data from George W. Pierce, *The Songs of Insects* (Cambridge, MA: Harvard University Press, 1949), pp. 12–21.

43 D. B. Lindenmayer, K. L. Viggers, R. B. Cunningham, and C. F. Donnelly, "Morphological Variation among Columns of the Mountain Brushtail Possum, *Trichosurus caninus Ogilby* (Phalangeridae: Marsupiala)," *Australian Journal of Zoology* 43 (1995): 449–458.

44 Gary Smith, "Do Statistics Test Scores Regress Toward the Mean?" *Chance* 10, no. 4 (1997): 42–45.

45 www.baseball-reference.com

46 Data from Brittany Foley and Allie Dutson, Canyon del Oro High School.

47 Data from Kerry Lane and Danielle Neal, Canyon del Oro High School.

48 www.fs.usda.gov/pnw/pubs/pnw_rp283.pdf

49 www.pro-football-reference.com and www.espn.com/nfl/story/_/id/33108568/nfl-quarterback-hand-size-measurements-smallest-biggest-all-32-teams-taysom-hill-russell-wilson. Data for quarterbacks with at least 10 starts in the 2021 season.

50 www.nature.com/articles/s41586-021-03550-y and portal.edirepository.org/nis/mapbrowse?packageid=edi.698.2

51 From a graph in G. D. Martinsen, E. M. Driebe, and T. G. Whitham, "Indirect Interactions Mediated by Changing Plant Chemistry: Beaver Browsing Benefits Beetles," *Ecology*, 79 (1998): 192–200.

52 *Nutrition Action*, April 2022.

53 A. K. Yousafzai et al., "Comparison of Armspan, Arm Length and Tibia Length as Predictors of Actual Height of Disabled and Nondisabled Children in Dharavi, Mumbai, India," *European Journal of Clinical Nutrition* 57 (2003): 1230–1234.

54 www.pewresearch.org/social-trends/2022/10/26/parents-differ-sharply-by-party-over-what-their-k-12-children-should-learn-in-school

55 Based on an article by Matthew Russell, "Do Forests Have the Capacity for 1 Trillion Extra Trees?" *Significance Magazine* 17, no. 6 (2020): 8–9.

56 N. R. Draper and J. A. John, "Influential Observations and Outliers in Regression," *Technometrics* 23 (1981): 21–26.

57 Gordon L. Swartzman and Stephen P. Kaluzny, *Ecological Simulation Primer* (New York, NY: Macmillan, 1987), p. 98.

58 From Discount Tire advertisement (Tandy Engineering and Associates), November 28, 2015.

59 Sample Pennsylvania female rates provided by Life Quotes, Inc., in *USA Today*, December 20, 2004.

60 G. A. Sacher and E. F. Staffelt, "Relation of Gestation Time to Brain Weight for Placental Mammals: Implications for the Theory of Vertebrate Growth," *American Naturalist* 108 (1974): 593–613. We found the data in Fred L. Ramsey and Daniel W. Schafer, *The Statistical Sleuth: A Course in Methods of Data Analysis* (Duxbury Press, 1997), p. 228.

61 Debora L. Arsenau, "Comparison of Diet Management Instruction for Patients with Non–Insulin Dependent Diabetes Mellitus: Learning Activity Package vs. Group Instruction," MS thesis, Purdue University, 1993.

62 Data from Haley Vaughn, Nate Trona, and Jeff Green, Central York High School.

63 *Consumer Reports*, November 2005.

64 beaninstitute.com/bean-nutrition-overview

65 fivethirtyeight.com/features/are-states-with-lower-income-tax-rates-better-at-winning-championships

66 Jérôme Chave, Bernard Riéra, and Marc-A. Dubois, "Estimation of Biomass in a Neotropical Forest of French Guiana: Spatial and Temporal Variability," *Journal of Tropical Ecology* 17 (2001): 79–96.

67 S. Chatterjee and B. Price, *Regression Analysis by Example* (Hoboken, NJ: Wiley, 1977).

68 www.advancedwebranking.com/cloud/ctrstudy

69 www.stat.columbia.edu/~gelman/research/published/golf.pdf

70 Chris Carbone and John L. Gittleman, "A Common Rule for the Scaling of Carnivore Density," *Science* 295 (2002): 2273–2276.

71 Data originally from A. J. Clark, *Comparative Physiology of the Heart* (New York, NY: Macmillan, 1927), p. 84. Obtained from Frank R. Giordano and Maurice D. Weir, *A First Course in Mathematical Modeling* (Monterey, CA: Brooks/Cole, 1985), p. 56.

72 Data estimated from a graph in Kyle G. Ashton, Russell L. Burke, and James N. Layne, "Geographic Variation in Body and Clutch Size of Gopher Tortoises," *Copeia* 2 (2007): 355–363.

73 Thanks to Jeff Eicher for collecting and cleaning these data from www.espn.com.

74 Jenna M. Dittmar et al., "Medieval Injuries: Skeletal Trauma as an Indicator of Past Living Conditions and Hazard Risk in Cambridge, England," *American Journal of Physical Anthropology* 175 (2021): 626–645, doi:10.1002/ajpa.24225.

75 Data from James R. Jordan.

76 We found the data on cherry blossoms in the paper "Linear Equations and Data Analysis," which was posted on the North Carolina School of Science and Mathematics website, www.ncssm.edu.

77 www.wimbledon.com

78 Ben Dantzer et al., "Decoupling the Effects of Food and Density on Life-History Plasticity of Wild Animals Using Field Experiments: Insights from the Steward Who Sits in the Shadow of Its Tail, the North American Red Squirrel," *Journal of Animal Ecology*, doi:10.1111/1365-2656.13341.

79 The World Almanac and Book of Facts (2009).

80 www.lumeradiamonds.com/diamonds/results?price=1082-1038045&carat=0.30-16.03&shapes=B%20&cut=EX&clarity=FL,IF&color=D,E,F#

81 Hannah Salisbury and Joey Manas, Canyon del Oro High School.

82 www.mlb.com

83 Mehreen S. Datoo et al., "Efficacy of a Low-Dose Candidate Malaria Vaccine, R21 in Adjuvant Matrix-M, with Seasonal Administration to Children in Burkina Faso: A Randomised Controlled Trial," *Lancet* 397 (2021): 1809–1818, doi.org/10.1016/S0140-6736(21)00943-0.

84 From a graph in Craig Packer et al., "Ecological Change, Group Territoriality, and Population Dynamics in Serengeti Lions," *Science* 307 (2005): 390–393.

85 www.fs.usda.gov/pnw/pubs/pnw_rp283.pdf

Unit 3

1 www.bls.gov/cps

2 The sleep deprivation study is described in R. Stickgold, L. James, and J. Hobson, "Visual Discrimination Learning Requires Post-Training Sleep," *Nature Neuroscience* 3, no. 12 (2000): 1237–1238.

3 Frederick Mosteller and David L. Wallace, *Inference and Disputed Authorship: The Federalist* (Reading, MA: Addison-Wesley, 1964).

4 billofrightsinstitute.org/primary-sources/federalist-no-51

5 John Mackenzie, "Family Dinner Linked to Better Grades for Teens: Survey Finds Regular Meal Time Yields Additional Benefits," ABC News, *World News Tonight*, September 13, 2005.

6 Signe Lund Mathiesen, Line Ahm Mielby, Derek Victor Byrne, and Qian Janice Wang, "Music to Eat By: A Systematic Investigation of the Relative Importance of Tempo and Articulation on Eating Time," *Appetite* 155 (2020): 104801.

7 academic.oup.com/ajcn/article/100/4/1182/4576550#110598138

8 Melissa G. Hunt, Rachel Marx, Courtney Lipson, and Jordyn Young, "No More FOMO: Limiting Social Media Decreases Loneliness and Depression," *Journal of Social and Clinical Psychology* 37, no. 10 (2018): 751–768.

9 B. Turnwald, D. Boles, and A. Crum, "Association Between Indulgent Descriptions and Vegetable Consumption: Twisted Carrots and Dynamite Beets," *JAMA Internal Medicine* 177, no. 8 (2017): 1216–1218. doi:10.1001/jamainternmed.2017.1637.

10 National Institute of Child Health and Human Development (NICHD), Study of Early Child Care and Youth Development. The article appears in the July 2003 issue of *Child Development*. The quotation is from the summary on the NICHD website: www.nichd.nih.gov.

11 Katri Räikkönen, Anu-Katriina Pesonen, Anna-Liisa Järvenpää, and Timo E. Strandberg, "Sweet Babies: Chocolate Consumption During Pregnancy and Infant Temperament at Six Months," *Early Human Development* 76, no. 2 (February 2004): 139–145.

12 heart.bmj.com/content/105/23/1793

13 H. Scott et al., "Social Media Use and Adolescent Sleep Patterns: Cross-Sectional Findings from the UK Millennium Cohort Study," *BMJ Open* 9 (2019): e031161.

14 Sheldon Cohen, William J. Doyle, Cuneyt M. Alper, Denise Janicki-Deverts, and Ronald B. Turner, "Sleep Habits and Susceptibility to the Common Cold," *Archives of Internal Medicine* 169, no. 1 (January 2009): 62–67. doi: 10.1001/archinternmed.2008.505.

15 Bill Hesselmar, Fei Sjöberg, Robert Saalman, Nils Åberg, Ingegerd Adlerberth, and Agnes E. Wold, "Pacifier Cleaning Practices and Risk of Allergy Development," *Pediatrics* 131, no. 6 (2013): e1829–e1837, doi:10.1542/peds.2012-3345

16 The advice columnist is Ann Landers.

17 www.econlib.org/archives/2009/02/parents_and_buy.html

18 *Arizona Daily Star*, April 18, 2016.

19 www.pewresearch.org/fact-tank/2019/02/27/response-rates-in-telephone-surveys-have-resumed-their-decline

20 www.census.gov/acs/www/methodology/sample-size-and-data-quality/response-rates

21 www.nytimes.com/2022/10/12/upshot/midterms-polling-phone-calls.html

22 www.nytimes.com/1994/07/08/us/poll-on-doubt-of-holocaust-is-corrected.html

23 www.cleaninginstitute.org/assets/1/AssetManager/2010%20Hand%20Washing%20Findings.pdf

24 Cynthia Crossen, "Margin of Error: Studies Galore Support Products and Positions, But Are They Reliable?" *Wall Street Journal*, November 14, 1991.

25 Gary S. Foster and Craig M. Eckert, "Up from the Grave: A Socio-Historical Reconstruction of an African American

Community from Cemetery Data in the Rural Midwest," *Journal of Black Studies* 33 (2003): 468–489.

26 www.huffpost.com/entry/women-sleep-better-with-dogs-than-with-human-partners-study_n_5c002dede4b0864f4f6b706f

27 Christy L. Hoffman, Kaylee Stutz, and Terrie Vasilopoulos, "An Examination of Adult Women's Sleep Quality and Sleep Routines in Relation to Pet Ownership and Bedsharing," *Anthrozoös* 31, no. 6 (2018): 711–725. doi:10.1080/08927936.2018.1529354.

28 Bryan E. Porter and Thomas D. Berry, "A Nationwide Survey of Self-Reported Red Light Running: Measuring Prevalence, Predictors, and Perceived Consequences," *Accident Analysis and Prevention* 33 (2001): 735–741.

29 Data from Marcos Chavez-Martinez, Canyon del Oro High School.

30 Mario A. Parada et al., "The Validity of Self-Reported Seatbelt Use: Hispanic and Non-Hispanic Drivers in El Paso," *Accident Analysis and Prevention* 33 (2001): 139–143.

31 www.pewresearch.org/fact-tank/2017/08/04/personal-finance-questions-elicit-slightly-different-answers-in-phone-surveys-than-online

32 Data from Emma Merry, Canyon del Oro High School.

33 en.wikipedia.org/wiki/The_Literary_Digest

34 tucson.com/news/science/health-med-fit/coffee-buzz-study-finds-java-drinkers-live-longer/article_c8de3c15-67da-5c57-8efd-a120d86ca967.html

35 David O. Meltzer, MD, PhD; Thomas J. Best, PhD; Hui Zhang, PhD; et al. "Association of Vitamin D Status and Other Clinical Characteristics with COVID-19 Test Results," *JAMA Network Open* 3, no. 9 (2020): e2019722. doi:10.1001/jamanetworkopen.2020.19722.

36 "Study Links Vitamin D, Stronger Bones in Girls,"*Arizona Daily Star*, March 6, 2012.

37 fivethirtyeight.com/features/dont-take-your-vitamins

38 www.nbcwashington.com/news/health/ADHD_Linked_To_Lead_and_Mom_s_Smoking.html

39 *Journal of Clinical Endocrinology and Metabolism* 2016. doi:10.1210/jc.2015-4013. Cited in *Nutrition Action*, April 2016.

40 journals.asm.org/doi/10.1128/AEM.01838-16

41 www.sciencedirect.com/science/article/abs/pii/S0747563214001563

42 Data from Lexie Gardner and Erica Chauvet, Waynesburg University.

43 www.plosmedicine.org/article/info%3Adoi%2F10.1371%2Fjournal.pmed.1001595

44 The placebo effect examples are from Sandra Blakeslee, "Placebos Prove So Powerful Even Experts Are Surprised," *New York Times*, October 13, 1998.

45 The "three-quarters" estimate is cited by Martin Enserink, "Can the Placebo Be the Cure?" *Science*, 284 (1999): 238–240. An extended treatment is Anne Harrington, ed., *The Placebo Effect: An Interdisciplinary Exploration* (Cambridge, MA: Harvard University Press, 1997).

46 Carlos Vallbona et al., "Response of Pain to Static Magnetic Fields in Postpolio Patients: A Double Blind Pilot Study," *Archives of Physical Medicine and Rehabilitation* 78 (1997): 1200–1203.

47 Fernando P. Polack, et al., "Safety and Efficacy of the BNT162b2 mRNA Covid-19 Vaccine," *New England Journal of Medicine* 383, (December 10, 2020): 2603–2615. doi:10.1056/NEJMoa2034577.

48 www.sciencedirect.com/science/article/pii/S0360131512002254

49 David L. Strayer, Frank A. Drews, and William A. Johnston, "Cell Phone–Induced Failures of Visual Attention During Simulated Driving," *Journal of Experimental Psychology: Applied* 9 (2003): 23–32.

50 The idea for this chart came from Fred L. Ramsey and Daniel W. Schafer, *The Statistical Sleuth: A Course in Methods of Data Analysis*, 2nd ed. (Duxbury Press, 2002).

51 journals.sagepub.com/doi/pdf/10.3102/0002831213488818

52 pediatrics.aappublications.org/content/pediatrics/early/2016/08/25/peds.2016-0910.full.pdf

53 See the details on the website of the Office for Human Research Protections of the Department of Health and Human Services, hhs.gov/ohrp.

54 *Nutrition Action*, July/August 2008.

55 *Nutrition Action*, October 2013.

56 Katie Adolphus, Clare L. Lawton, and Louise Dye, "Associations Between Habitual School-Day Breakfast Consumption Frequency and Academic Performance in British Adolescents," *Frontiers in Public Health* 7 (2019): 283. doi:10.3389/fpubh.2019.00283.

57 Sheri Madigan, Dillon Browne, Nicole Racine, et al., "Association Between Screen Time and Children's Performance on a Developmental Screening Test," *JAMA Pediatrics* 173, no. 3 (2019): 244–250.

58 Michael G. Smith, Maryam Witte, Sarah Rocha, and Mathias Basner, "Effectiveness of Incentives and Follow-up on Increasing Survey Response Rates and Participation in Field Studies," *BMC Medical Research Methodology* 19 (2019): 230. doi:10.1186/s12874-019-0868-8.

59 Charlie Huveneers, Sasha Whitmarsh, Madeline Thiele, Lauren Meyer, Andrew Fox, and Corey J. A. Bradshaw, "Effectiveness of Five Personal Shark-Bite Deterrents for Surfers," *Peer Journal* 6 (2018): e5554. doi:10.7717/peerj.5554.

60 *New York Times*, November 11, 2014. www.nytimes.com/2014/11/11/science/dead-jellyfish-are-more-nutrition-than-nuisance.html?_r=0.

61 Naomi D. L. Fisher, Meghan Hughes, Marie Gerhard-Herman, and Norman K. Hollenberg, "Flavonol-Rich Cocoa Induces Nitricoxide-Dependent Vasodilation in Healthy Humans," *Journal of Hypertension*, 21, no. 12 (2003): 2281–2286.

62 *Nutrition Action*, October 2013. Originally published in *Archives of Physical Medicine and Rehabilitation* 93 (2012): 1269.

63 Marielle H. Emmelot-Vonk et al., "Effect of Testosterone Supplementation on Functional Mobility, Cognition, and Other Parameters in Older Men," *Journal of the American Medical Association* 299 (2008): 39–52.

64 www.nytimes.com/2008/03/05/health/research/05placebo.html?_r=0

65 P.-H. A. Chen, J. H. Cheong, E. Jolly, et al., "Socially Transmitted Placebo Effects," *Nature Human Behavior* 3 (2019): 1295–1305. doi:10.1038/s41562-019-0749-5.

66 archive.ciser.cornell.edu/reproduction-packages/2780/project-description

67 www.sciencedaily.com/releases/2014/11/141124081040.htm

68 Joel Brockner et al., "Layoffs, Equity Theory, and Work Performance: Further Evidence of the Impact of Survivor Guilt," *Academy of Management Journal* 29 (1986): 373–384.

69 www.cnn.com/2018/12/19/health/reverse-cognitive-aging-exercise-diet-study/index.html
70 Christopher Anderson, "Measuring What Works in Health Care," *Science* 263 (1994): 1080–1082.
71 www.jstor.org/stable/24877480
72 *Nutrition Action*, March 2009.
73 "Blood Boosters May Help Preemies Develop," *Arizona Daily Star*, February 15, 2016.
74 Details of the Carolina Abecedarian Project, including references to published work, can be found at abc.fpg.unc.edu.
75 M. F. Mason et al., "Precise Offers Are Potent Anchors: Conciliatory Counteroffers and Attributions of Knowledge in Negotiations," *Journal of Experimental Social Psychology* 49, no. 4 (2013): 759–763. doi:10.1016/j.jesp.2013.02.012
76 www.ncbi.nlm.nih.gov/pubmed/16304443
77 Mary O. Mundinger et al., "Primary Care Outcomes in Patients Treated by Nurse Practitioners or Physicians," *Journal of the American Medical Association* 238 (2000): 59–68.
78 C. Proserpio, C. Invitti, S. Boesveldt, L. Pasqualinotto, M. Laureati, C. Cattaneo, and E. Pagliarini, "Ambient Odor Exposure Affects Food Intake and Sensory Specific Appetite in Obese Women," *Frontiers in Psychology* 10 (2019): 7. doi:10.3389/fpsyg.2019.00007.
79 Study conducted by cardiologists at Athens Medical School, Greece, and announced at a European cardiology conference in February 2004.
80 aces.illinois.edu/news/bats-protect-young-trees-insect-damage-three-times-fewer-bugs and esajournals.onlinelibrary.wiley.com/doi/epdf/10.1002/ecy.3903
81 Data from Michael Khawam, Canyon del Oro High School.
82 www.nejm.org/doi/full/10.1056/NEJMoa0905471
83 Steering Committee of the Physicians' Health Study Research Group, "Final Report on the Aspirin Component of the Ongoing Physicians' Health Study," *New England Journal of Medicine* 321 (1989): 129–135.
84 E. A. Hoge, E. Bui, M. Mete, et al., "Mindfulness-Based Stress Reduction vs Escitalopram for the Treatment of Adults with Anxiety Disorders: A Randomized Clinical Trial," *JAMA Psychiatry* 80, no. 1 (2023): 13–21. doi:10.1001/jamapsychiatry.2022.3679.
85 Charles A. Nelson III et al., "Cognitive Recovery in Socially Deprived Young Children: The Bucharest Early Intervention Project," *Science* 318 (2007): 1937–1940.
86 pubmed.ncbi.nlm.nih.gov/25262058/
87 R. Cialdini et al., "Basking in Reflected Glory: Three (Football) Field Studies," *Journal of Personality and Social Psychology* 34, no. 3 (1976): 366–373.
88 *Nutrition Action*, March 2013 (*Circulation* 127, p. 188), www.ncbi.nlm.nih.gov/pubmed/23319811
89 en.wikipedia.org/wiki/Tuskegee_Syphilis_Study
90 www.pnas.org/content/111/24/8788.full
91 en.wikipedia.org/wiki/Willowbrook_State_School
92 Carlos Vallbona et al., "Response of Pain to Static Magnetic Fields in Postpolio Patients: A Double Blind Pilot Study," *Archives of Physical Medicine and Rehabilitation* 78 (1997): 1200–1203.
93 weatherspark.com/compare/y/145550~23814/Comparison-of-the-Average-Weather-at-Tucson-International-Airport-and-Princeton
94 medicine.wustl.edu/news/antibiotics-and-sinus-infections/#:~:text=In%20fact%2C%20a%20study%20from,get%20better%20on%20their%20own
95 blog.thealzheimerssite.greatergood.com/orange-juice-protects and n.neurology.org/content/92/1/e63.
96 R. C. Shelton et al., "Effectiveness of St. John's Wort in Major Depression," *Journal of the American Medical Association* 285 (2001): 1978–1986.
97 *Arizona Daily Star*, October 29, 2008.
98 Data from a study by Sean Leader and Shelby Zismann, Canyon del Oro High School.
99 "Scientists: Act Happy; It Might Help Your Heart," *Arizona Daily Star*, February 20, 2010.
100 health.usnews.com/health-news/news/articles/2015/09/22/txt-msgs-may-lead-to-broad-heart-linked-benefits-study-says
101 Based on "Bee Off with You," *The Economist*, November 2, 2002, p. 78.
102 From the Electronic Encyclopedia of Statistical Examples and Exercises (EESEE) case study "Is caffeine dependence real?"
103 This project is based on an activity suggested in Richard L. Schaeffer, Ann Watkins, Mrudulla Gnanadesikan, and Jeffrey A. Witmer, *Activity-Based Statistics* (Emeryville, CA: Springer, 1996).
104 Scott W. Powers, Christopher S. Coffey, Leigh A. Chamberlin, et al., "Trial of Amitriptyline, Topiramate, and Placebo for Pediatric Migraine," *New England Journal of Medicine* 376 (2017): 115–124. doi:10.1056/NEJMoa1610384.
105 *Nutrition Action*, September 2019.

Unit 4

1 This scenario is based on actual events reported by Mr. Starnes's youngest son, who at the time was one of the junior pilots. Details have been changed to protect confidential information, but the calculated probability is consistent with the original scenario.
2 Hoang Nguyen, "One in Three Leave the Tap Running While Brushing Their Teeth," June 4, 2018, yougov.com.
3 Pew Research Center, "Teens, Social Media and Technology 2022," August 2022.
4 R. B. Ruback and D. Juieng, "Territorial Defense in Parking Lots: Retaliation Against Waiting Drivers," *Journal of Applied Social Psychology* 27, no. 9 (1997): 821–834. doi.org/10.1111/j.1559-1816.1997.tb00661.x.
5 www.prweek.com/article/1247089/paul-holmes-aarp-fooling-itself-misrepresenting-facts-gop-medicare-bill-members
6 We obtained the color distribution for M&M'S Milk Chocolate Candies from blogs.sas.com/content/iml/2017/02/20/proportion-of-colors-mandms.html for bags packaged at Mars, Incorporated's factory in Cleveland, Tennessee.
7 See note 6.
8 Brooke Auxier and Monica Anderson, "Social Media Use in 2021," Pew Research Center, April 2021.
9 Survey data from Visitor Use Study, www.nps.gov/yell.
10 Distribution of blood types from American Red Cross, www.redcrossblood.org/donate-blood/blood-types.html.
11 Data from Statistics Canada, www.statcan.gc.ca.
12 U.S. Census Bureau, "Current Population Survey, March and Annual Social and Economic Supplements," released December 2020.

13 Data from National Household Travel Survey, nhts.ornl.gov.

14 Data on educational attainment in 2019 from nces.ed.gov/programs/coe/indicator_caa.asp.

15 Exercise modified based on data provided in Gail Burrill, *Two-Way Tables: Introducing Probability Using Real Data*, paper presented at the Mathematics Education into the Twenty-First Century Project, Czech Republic, September 2003. Burrill cites as her source H. Kranendonk, P. Hopfensperger, and R. Scheaffer, *Exploring Probability* (New York, NY: Dale Seymour Publications, 1999).

16 Pew Research Center, "Teens, Social Media and Technology 2022," August 2022.

17 Squirrel data obtained from data.cityofnewyork.us/Environment/2018-Central-Park-Squirrel-Census-Squirrel-Data/vfnx-vebw/data.

18 Results based on a survey reported at statista.com.

19 AP® Statistics exam score distribution obtained from apcentral.collegeboard.org/courses/ap-statistics/exam.

20 From the EESEE story, "Is It Tough to Crawl in March?"

21 Pierre J. Meunier et al., "The Effects of Strontium Ranelate on the Risk of Vertebral Fracture in Women with Postmenopausal Osteoporosis," *New England Journal of Medicine* 350 (2004): 459–468.

22 British Parliamentary Papers, *Shipping Casualties (Loss of the Steamship "Titanic")*, 1912, cmd. 6352, *Report of a Formal Investigation into the Circumstances Attending the Foundering on the 15th April, 1912, of the British Steamship "Titanic," of Liverpool, After Striking Ice in or near Latitude 41° 46' N., Longitude 50° 14' W., North Atlantic Ocean, Whereby Loss of Life Ensued* (London, UK: His Majesty's Stationery Office, 1912). See also www.anesi.com/titanic.htm.

23 See note 8.

24 See note 9.

25 www.ncaa.org/about/resources/research/probability-competing-beyond-high-school

26 www.infodocket.com/2014/04/17/new-findings-from-the-harris-poll-americans-who-read-more-electronically-read-more/

27 Thanks to Gerd Gigerenzer for suggesting this approach.

28 Thanks to Michael Legacy for suggesting the context of this problem.

29 Data on school enrollment from U.S. Census Bureau, "Current Population Survey, School Enrollment Supplement," October 2019. Data on population aged 55 and older from U.S. Census Bureau, "Current Population Survey, Annual Social and Economic Supplement," 2019.

30 See note 17.

31 See note 16.

32 R. Shine et al., "The Influence of Nest Temperatures and Maternal Brooding on Hatchling Phenotypes in Water Pythons," *Ecology* 78 (1997): 1713–1721.

33 See note 18.

34 Data on David Ortiz from www.baseball-reference.com.

35 From the EESEE story, "What Makes a Pre-teen Popular?"

36 We got these data from the Energy Information Administration website, www.eia.gov.

37 From National Institutes of Health's National Digestive Diseases Information Clearinghouse, www.niddk.nih.gov/health-information/digestive-diseases.

38 Thomas F. Imperiale et al., "Multitarget Stool DNA Testing for Colorectal-Cancer Screening," *New England Journal of Medicine* 370 (2014): 1287–1297.

39 Probabilities from trials with 2897 people known to be free of HIV antibodies and 673 people known to be infected are reported in J. Richard George, *Alternative Specimen Sources: Methods for Confirming Positives*, 1998 Conference on the Laboratory Science of HIV, found online at Centers for Disease Control and Prevention website, www.cdc.gov.

40 Data on Serena Williams's serve percentages from www.wtatennis.com.

41 The probabilities given are realistic, according to the fundraising firm SCM Associates, scmassoc.com.

42 www.prnewswire.com/news-releases/penny-for-your-thoughts-americans-oppose-abolishing-the-penny-300146827.html

43 Thanks to Corey Andreasen for suggesting the idea for this exercise.

44 The probabilities in this exercise are taken from Tommy Bennett, "Expanded Horizons: Perfection," June 8, 2010, www.baseballprospectus.com.

45 Research performed by Matthew O'Brien and Diego Bustos Arrieta, Canyon del Oro High School, under the supervision of Josh Tabor.

46 A. C. Gielen et al., "National Survey of Home Injuries During the Time of COVID-19: Who Is at Risk?" *Injury Epidemiology* 7 (2020): 63.

47 www.bbc.com/news/blogs-magazine-monitor-23957303

48 The probability distribution was based on data found at www.statista.com/statistics/276506/change-in-us-car-demand-by-vehicle-type.

49 The table in this exercise was constructed using the search function at the GSS archive, sda.berkeley.edu/archive.htm.

50 This exercise was inspired by the report at www.cbsnews.com/news/drug-tests-not-immune-from-false-positives.

51 The Apgar score data came from National Center for Health Statistics, *Monthly Vital Statistics Reports* 30, no. 1 (May 6, 1981): suppl.

52 Probability distribution based on a sample of students from the U.S. Census at School database, www.amstat.org/censusatschool.

53 You can find a mathematical explanation of Benford's law in Ted Hill, "The First-Digit Phenomenon," *American Scientist* 86 (1996): 358–363; and Ted Hill, "The Difficulty of Faking Data," *Chance* 12, no. 3 (1999): 27–31. Applications in fraud detection are discussed in the second paper by Hill and in Mark A. Nigrini, "I've Got Your Number," *Journal of Accountancy* (May 1999), www.journalofaccountancy.com/issues/1999/may/nigrini.

54 The National Longitudinal Study of Adolescent Health interviewed a stratified random sample of 27,000 adolescents, then reinterviewed many of the subjects 6 years later, when most were aged 19 to 25. These data are from the Wave III reinterviews in 2000 and 2001, found at the website of the Carolina Population Center, www.cpc.unc.edu.

55 These exercises are based on an activity shared by Allan Rossman in his Ask Good Questions blog, askgoodquestions.blog/2020/11/02/70-batch-testing-part-2.

56 legacy.baseballprospectus.com/sortable/index.php?cid=975409 (using data from 2010)

57 From the EESEE story, "Checkmating and Reading Skills."
58 Ed O'Brien and Phoebe C. Ellsworth, "Saving the Last for Best: A Positivity Bias for End Experiences," *Psychological Science* 23, no. 2 (2011): 163–165.
59 Data from Emily Clymer, Canyon del Oro High School.

Unit 5

1 This activity is based on a similar activity suggested in Richard L. Schaeffer, Ann Watkins, Mrudulla Gnanadesikan, and Jeffrey A. Witmer, *Activity-Based Statistics* (Springer, 1996).
2 D. B. Lindenmayer, K. L. Viggers, R. B. Cunningham, and C. F. Donnelly, "Morphological Variation Among Columns of the Mountain Brushtail Possum, *Trichosurus caninus Ogilby* (Phalangeridae: Marsupiala)," *Australian Journal of Zoology* 43 (1995): 449–458.
3 Data from the U.S. Census Bureau, Population Estimates Program, 2020.
4 Thomas K. Cureton et al., "Endurance of Young Men," *Monographs of the Society for Research in Child Development* 10, no. 1 (1945).
5 www.coinnews.net/2021/01/22/u-s-mint-produces-14-77-billion-coins-for-circulation-in-2020
6 www.census.gov/data/tables/time-series/demo/income-poverty/cps-pinc/pinc-01.html
7 This and similar results of Gallup polls are from the Gallup Organization website, www.gallup.com.
8 www.pewresearch.org/internet/2021/04/07/social-media-use-in-2021
9 www.bls.gov/cps
10 en.wikipedia.org/wiki/Giraffe
11 M. Seddigh and G. D. Jolliff, "Light Intensity Effects on Meadowfoam Growth and Flowering," *Crop Science* 34 (1994): 497–503.
12 www.insidehighered.com/views/2021/01/19/national-opinion-survey-shows-growing-public-support-helping-students-debt-opinion
13 www.pewresearch.org/fact-tank/2017/08/24/jury-duty-is-rare-but-most-americans-see-it-as-part-of-good-citizenship
14 www.pewresearch.org/fact-tank/2019/07/01/about-one-in-six-u-s-teachers-work-second-jobs-and-not-just-in-the-summer
15 www.ultimatemotorcycling.com/2019/02/07/motorcycle-statistics-in-america-demographics-change-for-2018
16 The idea for this exercise was inspired by an example in David M. Lane's *Hyperstat Online*, davidmlane.com/hyperstat.
17 *Nutrition Action Healthletter*, September 2016.
18 www.pbs.org/wnet/nature/blog/lion-fact-sheet
19 www.bls.gov/web/empsit/cpseea05.htm
20 We found the information on birth weights of Norwegian children on the National Institute of Environmental Health Sciences website: www.ncbi.nlm.nih.gov/pubmed/1536353.
21 Based on a figure in Peter R. Grant, *Ecology and Evolution of Darwin's Finches* (Princeton, NJ: Princeton University Press, 1986).
22 www.time.com/money/3712480/pay-less-internet-cut-cable
23 maristpoll.marist.edu/wp-content/uploads/2022/03/Marist-Poll_Center-for-Sports-Communication_USA-NOS-and-Banners_202203031619.pdf

Unit 6

1 www.pewresearch.org/methods/fact-sheet/national-public-opinion-reference-survey-npors
2 wwwn.cdc.gov/nchs/nhanes/continuousnhanes/default.aspx?BeginYear=2017. Data based on the second of three readings of systolic blood pressure.
3 www.cdc.gov/bloodpressure/about.htm
4 www.smithsonianmag.com/innovation/how-much-do-americans-know-about-science-27747364/
5 www.forbes.com/sites/stevensavage/2016/02/08/inconvenient-truth-there-are-pesticide-residues-on-organics
6 Michele L. Head, *Examining College Students' Ethical Values*, Consumer Science and Retailing honors project, Purdue University, 2003.
7 Morning Consult National Tracking Poll #191255, December 13–15, 2019.
8 Morning Consult National Tracking Poll #200158, January 23–24, 2020.
9 Robert Langkjaer-Bain, "The Murky Tale of Flint's Deceptive Water Data," *Significance Magazine*, April 2017. rss.onlinelibrary.wiley.com/doi/full/10.1111/j.1740-9713.2017.01016.x.
10 Sungwoo Lim, Brett Wyker, Katherine Bartley, and Donna Eisenhower, "Measurement Error in Self-Reported Physical Activity Levels in New York City: Assessment and Correction," *American Journal of Epidemiology* 181, no. 9 (2015): 648–655, doi.org/10.1093/aje/kwu470. Confidence interval was inferred based on summary statistics in the study. Results from a follow-up study that had respondents use a fitness tracker showed that respondents overestimated their activity.
11 M. A. Wale, S. D. Simpson, and A. N. Radford, "Size-Dependent Physiological Responses of Shore Crabs to Single and Repeated Playback of Ship Noise," *Biology Letters* 9 (2013): 20121194. dx.doi.org/10.1098/rsbl.2012.1194.
12 Data from 2013 Current Population Survey, found at www.eeps.com/zoo/acs/source/index.php.
13 www.pewresearch.org/fact-tank/2022/01/06/three-in-ten-americans-now-read-e-books
14 www.usatoday.com/picture-gallery/news/2015/04/07/usa-today-snapshots/6340793
15 sleepfoundation.org/sleep-polls-data/2015-sleep-and-pain
16 Eric Sanford et al., "Local Selection and Latitudinal Variation in a Marine Predator–Prey Interaction," *Science*, 300 (2003): 1135–1137.
17 Morning Consult National Tracking Poll #191255, December 13–15, 2019.
18 Morning Consult National Tracking Poll #200158, January 23–24, 2020.
19 Pew Research Center, November 2016, "Gig Work, Online Selling and Home Sharing."
20 today.yougov.com/topics/consumer/articles-reports/2021/02/08/most-liked-disliked-pizza-toppings-poll-data
21 www.pewresearch.org/wp-content/uploads/2020/12/Phones-and-scams-methods-and-topline.pdf
22 www.pewresearch.org/fact-tank/2021/06/30/most-americans-believe-in-intelligent-life-beyond-earth-few-see-ufos-as-a-major-national-security-threat
23 www.annenbergpublicpolicycenter.org/americans-civics-knowledge-drops-on-first-amendment-and-branches-of-government
24 www.pewresearch.org/fact-tank/2018/11/26/americans-are-divided-over-whether-eating-organic-foods-makes-for-better-health

25 www.commonsensemedia.org/articles/how-are-screens-affecting-my-kids-sleep#:~:text=And%20devices%20definitely%20play%20a,by%20notifications%20in%20the%20night

26 yrbs-explorer.services.cdc.gov/#/graphs?questionCode=H88&topicCode=C08&location=XX&year=2019

27 www.kickstarter.com/help/stats

28 National Institute for Occupational Safety and Health, "Stress at Work," 2000, www.cdc.gov/niosh/docs/99-101/default.html/. Results of this survey were reported in *Restaurant Business*, September 15, 1999, pp. 45–49.

29 Robin Lake and Alvin Makori, "The Digital Divide Among Students During COVID-19: Who Has Access? Who Doesn't?" *The Lens*, June 16, 2020.

30 www.pewresearch.org/short-reads/2023/02/02/key-findings-about-online-dating-in-the-u-s

31 Thanks to DeAnna McDonald for allowing us some creative license with her teaching assignment!

32 www.gapminder.org/ignorance/gms and www.ipsos.com/ipsos-mori/en-uk/mind-gap-ipsos-mori-survey-gapminder. Gapminder points out that one-third of chimpanzees would get this question correct when asked to choose from three bananas with the answer choices written on them.

33 The idea for this exercise was provided by Michael Legacy and Susan McGann.

34 Data from a project by Daniel Brown and Chad Porter, Canyon del Oro High School.

35 U.S. Department of Commerce, Census Bureau, Current Population Survey (CPS), October 2017.

36 Based on Stephen A. Woodbury and Robert G. Spiegelman, "Bonuses to workers and employers to reduce unemployment: randomized trials in Illinois," *American Economic Review*, 77 (1987), pp. 513–530.

37 news.gallup.com/poll/183689/industry-grows-percentage-sports-fans-steady.aspx

38 Bryan E. Porter and Thomas D. Berry, "A Nationwide Survey of Self-Reported Red Light Running: Measuring Prevalence, Predictors, and Perceived Consequences," *Accident Analysis and Prevention*, 33 (2001): 735–741.

39 J. E. Schulenberg, M. E. Patrick, L. D. Johnston, P. M. O'Malley, J. G. Bachman, and R. A Miech, *Monitoring the Future National Survey Results on Drug Use, 1975–2020: Volume II, College Students and Adults Ages 19–60* (Ann Arbor, MI: Institute for Social Research, University of Michigan, 2021).

40 Data from student project by Miranda Edwards and Sarah Juarez, Canyon del Oro High School.

41 news.gallup.com/poll/3166/State-Movies-Summer-2001.aspx and news.gallup.com/poll/388538/movie-theater-attendance-far-below-historical-norms.aspx

42 www.nejm.org/doi/full/10.1056/NEJMoa1615869

43 Benjamin W. Friedman et al., "Diazepam Is No Better Than Placebo When Added to Naproxen for Acute Low Back Pain," *Annals of Emergency Medicine*, February 7, 2017.

44 Instagram data from Pew Research Center, "Social Media Update 2016," November 2016, and Pew Research Center, "Social Media Use in 2021," April 2021.

45 Saiyad S. Ahmed, "Effects of Microwave Drying on Checking and Mechanical Strength of Low-Moisture Baked Products," MS thesis, Purdue University, 1994.

46 www.hakaimagazine.com/news/the-plan-to-rear-fish-on-the-moon

47 www.prnewswire.com/news-releases/few-hate-shopping-for-clothes-but-love-of-it-varies-by-country-124480498.html

48 www.ncbi.nlm.nih.gov/pubmed/29183701

49 Amanda L. Graham et al., "Effectiveness of a Vaping Cessation Text Message Program Among Young Adult e-Cigarette Users," *JAMA Internal Medicine* 181, no. 7 (2021): 923–930, doi:10.1001/jamainternmed.2021.1793.

50 Data on Christmas trees based on James Farmer et al., "Beauty Is in the Eye of the Tree Holder: Indiana Christmas Tree Consumer Survey," conducted by Indiana University.

51 iop.harvard.edu/youth-poll/41st-edition-spring-2021

52 Oliva I. Okereke et al., "Effect of Long-Term Vitamin D_3 Supplementation vs Placebo on Risk of Depression or Clinically Relevant Depressive Symptoms and on Change in Mood Scores," *Journal of the American Medical Association* 324, no. 5 (2020): 471–480, doi:10.1001/jama.2020.10224.

53 Joanne Ryan et al., "Randomized Placebo-Controlled Trial of the Effects of Aspirin on Dementia and Cognitive Decline," *Neurology* 95 (2020): e320–e331, doi:10.1212/WNL.0000000000009277.

54 Katherine L. Milkman et al., "A 680,000-Person Megastudy of Nudges to Encourage Vaccination in Pharmacies," Proceedings of the National Academy of Sciences 119, no. 6 (2022): e2115126119, www.pnas.org/doi/full/10.1073/pnas.2115126119.

55 R. J. M. Engler, M. R. Nelson, M. M. Klote, et al., "Half- vs Full-Dose Trivalent Inactivated Influenza Vaccine (2004–2005): Age, Dose, and Sex Effects on Immune Responses," *Archives of Internal Medicine* 168, no. 22 (2008): 2405–2414, doi:10.1001/archinternmed.2008.513.

56 K. E. Flegal and M. C., Anderson, "Overthinking Skilled Motor Performance: Or Why Those Who Teach Can't Do," *Psychonomic Bulletin & Review* 15, no. 5 (2008): 927–932, doi:10.3758/PBR.15.5.927. Thanks to Kristin Flegal for sharing the data from this study.

57 Based on the story "Drive-Thru Competition" in the *Electronic Encyclopedia of Statistical Examples and Exercises* (EESEE). Updated data were obtained from the *QSR* magazine website, www.qsrmagazine.com/reports/2021-qsr-magazine-drive-thru-study.

58 J. W. Winkelman et al., "Efficacy and Safety of Pramipexole in Restless Legs Syndrome," *Neurology* 67, no. 6 (2006): 1034–1039, doi:10.1212/01.wnl.0000231513.23919.a1.

59 M. Frick et al., "Helsinki Heart Study: Primary-Prevention Trial with Gemfibrozil in Middle-Aged Men with Dyslipidemia," *New England Journal of Medicine* 317 (1987): 1237–1245, doi:10.1056/NEJM198711123172001.

60 The study is reported in William Celis III, "Study Suggests Head Start Helps Beyond School," *New York Times*, April 20, 1993. See highscope.org.

61 Joann Peck et al., "Caring for the Commons: Using Psychological Ownership to Enhance Stewardship Behavior for Public Goods," *Journal of Marketing* 85, no. 2 (2021): 33–49.

62 Based on Deborah Roedder John and Ramnath Lakshmi-Ratan, "Age Differences in Children's Choice Behavior: The Impact of Available Alternatives," *Journal of Marketing Research*, 29 (1992): 216–226.

63 Francisco Lloret et al., "Fire and Resprouting in Mediterranean Ecosystems: Insights from an External Biogeographical Region, the Mexican Shrubland," *American Journal of Botany*, 88 (1999): 1655–1661.

64 Maureen Hack et al., "Outcomes in Young Adulthood for Very Low-Birth-Weight Infants," *New England Journal of Medicine* 346 (2002): 149–157. This exercise is simplified, in that the measures reported in this paper have been statistically adjusted for "sociodemographic status."

65 Data set randomly selected from full data set available at data.cityofnewyork.us/Environment/2015-Street-Tree-Census-Tree-Data/pi5s-9p35.

66 George Du Toit et al., "Randomized Trial of Peanut Consumption in Infants at Risk for Peanut Allergy," *New England Journal of Medicine* 372 (2015): 803–813.

67 C. P. Cannon et al., "Intensive Versus Moderate Lipid Lowering with Statins After Acute Coronary Syndromes," *New England Journal of Medicine* 350 (2004): 1495–1504.

68 Robert B. Cialdini et al., "Basking in Reflected Glory: Three (Football) Field Studies," *Journal of Personality and Social Psychology* 34, no. 3 (1976): 366–373.

69 *Arizona Daily Star,* February 11, 2009.

70 Michael J. Mack et al., "Transcatheter Aortic-Valve Replacement with a Balloon-Expandable Valve in Low-Risk Patients," *New England Journal of Medicine*, www.nejm.org/doi/full/10.1056/NEJMoa1814052.

71 Based on data on youth employment from the Bureau of Labor Statistics, www.bls.gov.

72 The original randomized clinical trial testing the effectiveness of AZT in treating patients with AIDS was conducted by Burroughs Wellcome. Data for this exercise came from the original "Against All Odds" video series.

73 www.fda.gov/news-events/press-announcements/fda-authorizes-marketing-novel-device-help-protect-athletes-brains-during-head-impacts

74 Emily Cohen and Madi McDole, Canyon del Oro High School.

Unit 7

1 www.nytimes.com/interactive/2019/02/13/upshot/engagement-rings-cost-two-weeks-pay.html

2 Mean and standard deviation estimated from graph at www.nytimes.com/interactive/2019/02/13/upshot/engagement-rings-cost-two-weeks-pay.html.

3 Data provided by Drina Iglesia, Purdue University. The data are part of a larger study reported in D. D. S. Iglesia, E. J. Cragoe, Jr., and J. W. Vanable, "Electric Field Strength and Epithelization in the Newt (*Notophthalmus viridescens*)," *Journal of Experimental Zoology* 274 (1996): 56–62.

4 Data from Pennsylvania State University Stat 500 Applied Statistics online course, online.stat.psu.edu/statprogram/stat500.

5 Data from a student project in Mr. Starnes's class at The Lawrenceville School.

6 Harry B. Meyers, *Investigations of the Life History of the Velvetleaf Seed Beetle*, Althaeus folkertsi *Kingsolver,* MS thesis, Purdue University, 1996.

7 news.gallup.com/poll/284009/library-visits-outpaced-trips-movies-2019.aspx

8 data.cityofnewyork.us/City-Government/Citywide-Payroll-Data-Fiscal-Year-/k397-673e

9 D. B. Lindenmayer, K. L. Viggers, R. B. Cunningham, and C. F. Donnelly, "Morphological Variation Among Columns of the Mountain Brushtail Possum, *Trichosurus caninus Ogilby* (Phalangeridae: Marsupiala)," *Australian Journal of Zoology* 43 (1995): 449–458.

10 Data from Melissa Silva and Madeline Dunlap, Canyon del Oro High School.

11 Data are from Phyllis Lee, Stirling University, and are related to P. Lee et al., "Enduring Consequences of Early Experiences: 40-Year Effects on Survival and Success Among African Elephants (*Loxodonta africana*)," *Biology Letters* 9 (2013): 20130011.

12 en.wikipedia.org/wiki/African_elephant#Size

13 R. D. Stichler, G. G. Richey, and J. Mandel, "Measurement of Treadware of Commercial Tires," *Rubber Age*, 73, no. 2 (May 1953).

14 Data provided by Ramon Olivier.

15 M. Ann Laskey et al., "Bone Changes After 3 Mo of Lactation: Influence of Calcium Intake, Breast-Milk Output, and Vitamin D–Receptor Genotype," *American Journal of Clinical Nutrition*, 67 (1998): 685–692.

16 F. H. Rauscher et al., "Music Training Causes Long-Term Enhancement of Preschool Children's Spatial-Temporal Reasoning," *Neurological Research* 19 (1997): 2–8.

17 Data on chewing gum and short-term memory from a study by Leila El-Ali and Valerie Pederson.

18 Data from a study by Sean Leader and Shelby Zismann.

19 Based on interviews by the National Longitudinal Study of Adolescent Health; found at the website of the Carolina Population Center, www.cpc.unc.edu.

20 Simplified from Sanjay K. Dhar, Claudia Gonzalez-Vallejo, and Dilip Soman, "Modeling the Effects of Advertised Price Claims: Tensile Versus Precise Pricing," *Marketing Science*, 18 (1999): 154–177.

21 Information about golden hamsters from animaldiversity.org.

22 See note 21.

23 Data from a project by Abigail Polsky and Raquel Quesada, Canyon del Oro High School.

24 E. C. Strain et al., "Caffeine Dependence Syndrome: Evidence from Case Histories and Experimental Evaluation," *Journal of the American Medical Association* 272 (1994): 1604–1607.

25 Tire pressure loss data from *Consumer Reports* website, www.consumerreports.org/tire-buying-maintenance/should-you-use-nitrogen-in-car-tires.

26 This exercise is based on events that are real. The data and details have been altered to protect the privacy of the individuals involved.

27 Data provided by Tim Brown.

28 Data from a student project by Patrick Baker and William Manheim, Canyon del Oro High School.

29 W. S. Gosset, "The Probable Error of a Mean," *Biometrika* 6 (1908): 1–25.

30 Data and Story Library, "Friday the 13th," dasl.datadescription.com.

31 Data provided by Judy Starnes, who compiled the data from a local realtor.

32 news.gallup.com/poll/4831/americans-confident-safety-nations-food.aspx

33 Data sourced from www.craigslist.com, February 18, 2023.

34 Gabriela S. Castellani, *The Effect of Cultural Values on Hispanics' Expectations About Service Quality*, MS thesis, Purdue University, 2000.

35 www.ncbi.nlm.nih.gov/pubmed/29635503

36 Shailija V. Nigdikar et al., "Consumption of Red Wine Polyphenols Reduces the Susceptibility of Low-Density Lipoproteins to Oxidation in Vivo," *American Journal of Clinical Nutrition* 68 (1998): 258–265.

37 Data from a student project by Daniel Flexas, Canyon del Oro High School.

38 Kristen E. Flegal and Michael C. Anderson, "Overthinking Skilled Motor Performance: Or Why Those Who Teach Can't Do," *Psychonomic Bulletin & Review* 15, no. 5 (2008): 927– 932, doi: 10.3758/PBR.15.5.927. Thanks to Kristin Flegal for sharing the data from this study.

39 Data on annual income for college graduates and non-graduates obtained from the March 2021 Annual Social and Economic Supplement, downloaded from www.census.gov. We took a random sample of 500 people who had attended some college but earned no degree, or who had earned an associate's or bachelor's degree.

40 Ethan J. Temeles and W. John Kress, "Adaptation in a Plant–Hummingbird Association," *Science*, 300 (2003): 630–633. We thank Ethan J. Temeles for providing the data.

41 blogs.edweek.org/edweek/curriculum/2015/03/homework_math_science_study.html, from this study: www.apa.org/pubs/journals/releases/edu-0000032.pdf.

42 Noel Cressie, *Statistics for Spatial Data* (Hoboken, NJ: Wiley, 1993).

43 Data from a final project in Daren Starnes's Introductory Statistics class.

44 This study is reported in Roseann M. Lyle et al., "Blood Pressure and Metabolic Effects of Calcium Supplementation in Normotensive White and Black Men," *Journal of the American Medical Association* 257 (1987): 1772–1776. The data were provided by Dr. Lyle.

45 www.cochrane.org/CD010037/HTN_extra-calcium-prevent-high-blood-pressure

46 Warren E. Leary, "Cell Phones: Questions But No Answers," *New York Times*, October 26, 1999.

47 cpb-us-w2.wpmucdn.com/sites.udel.edu/dist/6/132/files/2010/11/Psychological-Science-2014-Mueller-0956797614524581-1u0h0yu.pdf

48 General Social Survey results from gss.norc.org.

49 H. R. Knapp and G. A. FitzGerald, "The Antihypertensive Effects of Fish Oil: A Controlled Study of Polyunsaturated Fatty Acid Supplements in Essential Hypertension," *New England Journal of Medicine* 320 (1989): 1037–1043. Cited in Fred Ramsey and Daniel Schafer, *The Statistical Sleuth* (Belmont, CA: Duxbury Press, 2002), 23.

50 Data from a student project by Vanessa Claude, Canyon del Oro High School.

51 This exercise is based on M. Ann Laskey et al., "Bone Changes After 3 Months of Lactation: Influence of Calcium Intake, Breast-Milk Output, and Vitamin D–Receptor Genotype," *American Journal of Clinical Nutrition* 67 (1998): 685–692.

52 Adapted from Maribeth Cassidy Schmitt, *The Effects of an Elaborated Directed Reading Activity on the Metacomprehension Skills of Third Graders*, Ph.D. dissertation, Purdue University.

53 See note 9.

54 The original paper is T. M. Amabile, "Motivation and Creativity: Effects of Motivational Orientation on Creative Writers," *Journal of Personality and Social Psychology*, 48, no. 2 (1985): 393–399. The data for this exercise came from Fred L. Ramsey and Daniel W. Schafer, *The Statistical Sleuth*, 3rd ed. (Pacific Grove, CA: Brooks/Cole Cengage Learning, 2013).

55 The data for this exercise came from Allan Rossman, George Cobb, Beth Chance, and John Holcomb's National Science Foundation project shared at the Joint Mathematics Meeting (JMM) 2008 in San Diego. Their original source was Robert Stickgold, LaTanya James, and J. Allan Hobson, "Visual Discrimination Learning Requires Sleep After Training," *Nature Neuroscience* 3 (2000): 1237–1238.

56 The idea for this exercise was provided by Robert Hayden.

57 See note 56.

58 Wayne J. Camera and Donald Powers, "Coaching and the SAT I," *TIP* (July 1999), www.siop.org/tip.

59 Carlos Vallbona et al., "Response of Pain to Static Magnetic Fields in Postpolio Patients: A Double Blind Pilot Study," *Archives of Physical Medicine and Rehabilitation* 78 (1997): 1200–1203.

60 Piet Spaak and Maarten Boersma, "Tail Spine Length in the *Daphnia galeata* Complex: Costs and Benefits of Induction by Fish," *Aquatic Ecology* 31 (1997): 89–98.

61 Data provided by Warren Page, New York City Technical College, from a study done by John Hudesman.

62 *Electronic Encyclopedia of Statistical Examples and Exercises*, "Surgery in a Blanket."

63 Data from a student project by Libby Foulk and Kathryn Hilton.

64 www.dfw.state.or.us/MRP

Unit 8

1 Biographical data from www.nhl.com.

2 Random sample of trees selected from data set at data.cityofnewyork.us. Land area values from en.wikipedia.org/wiki/Boroughs_of_New_York_City.

3 Data from Cassandra Randal-Greene, Canyon del Oro High School, and corporate.ppg.com/Media/Newsroom/2014/PPG-data-shows-white-continues-to-dominate-as-most

4 ashpublications.org/bloodadvances/article/4/20/4990/463793/Reduced-prevalence-of-SARS-CoV-2-infection-in-ABO

5 McNuggets information from www.thehits.co.nz/the-latest/mcdonalds-reveals-the-truth-about-chicken-mcnugget-shapes. Data from Carlos Poblano and Nathaniel Benavidez, Canyon del Oro High School.

6 R. W. Mannan and E. C. Meslow, "Bird Populations and Vegetation Characteristics in Managed and Old-Growth Forests, Northwestern Oregon," *Journal of Wildlife Management* 48 (1984): 1219–1238.

7 Nick Blanchard, Canyon del Oro High School.

8 J. H. Gertsch, B. Basnyat, E. W. Johnson, J. Onopa, and P. S. Holck, "Randomised, Double Blind, Placebo Controlled Comparison of Ginkgo Biloba and Acetazolamide for Prevention of Acute Mountain Sickness Among Himalayan Trekkers: The Prevention of High Altitude Illness Trial (PHAIT)," *BMJ* 328 (2004): 797. Data downloaded from users.stat.ufl.edu/~winner/datasets.html.

9 www.pewresearch.org/fact-tank/2022/05/18/more-than-half-of-americans-live-within-an-hour-of-extended-family. We selected a random sample of 500 from the more than 9000 members of the Pew Research Center's sample.

10 Janice E. Williams et al., "Anger Proneness Predicts Coronary Heart Disease Risk," *Circulation* 101 (2000): 63–95.

11 www.pewinternet.org/2015/12/15/gaming-and-gamers

12 The context of this example was inspired by C. M. Ryan et al., "The Effect of In-Store Music on Consumer Choice of Wine," *Proceedings of the Nutrition Society* 57 (1998): 1069A.

13 Pew Research Center, "What It Takes to Truly Be 'One of Us,'" February 2017.

14 Elizabeth F. Loftus and John C. Palmer, "Reconstruction of Automobile Destruction: An Example of the Interaction Between Language and Memory," *Journal of Verbal Learning and Verbal Behavior* 13 (1974): 585–589, www.researchgate.net/publication/222307973_Reconstruction_of_Automobile_Destruction_An_Example_of_the_Interaction_Between_Language_and_Memory.

15 ww2.amstat.org/CensusAtSchool

16 Tanjaniina Laukkanen, Hassan Khan, Francesco Zaccardi, et al., "Association Between Sauna Bathing and Fatal Cardiovascular and All-Cause Mortality Events," *JAMA Internal Medicine* 175, no. 4 (2015): 542–548, doi:10.1001/jamainternmed.2014.8187.

17 N. Häusler, J. Haba-Rubio, R. Heinzer, et al., "Association of Napping with Incident Cardiovascular Events in a Prospective Cohort Study," *Heart* 105 (2019): 1793–1798.

18 maristpoll.marist.edu/wp-content/uploads/2022/03/Marist-Poll_Center-for-Sports-Communication_USA-NOS-and-Banners_202203031619.pdf

19 See note 2.

20 Hannah Salisbury and Joey Manas, Canyon del Oro High School.

21 Ohio State University, "Night Lights Don't Lead to Nearsightedness, Study Suggests," *ScienceDaily*, March 9, 2000.

22 poll.qu.edu/Poll-Release-Legacy?releaseid=2275

23 www.pewresearch.org/internet/wp-content/uploads/sites/9/media/Files/Reports/2013/PIP_OnlineBanking.pdf

24 Data from Lexi Epperson and Courtney Johnson, Canyon del Oro High School.

25 news.gallup.com/poll/284009/library-visits-outpaced-trips-movies-2019.aspx

26 Hypericum Depression Trial Study Group; J. R. Davidson, K. M. Gadde, J. A. Fairbank, et al., "Effect of *Hypericum perforatum* (St John's Wort) in Major Depressive Disorder," *Journal of the American Medical Association* 287, no. 14 (2002): 1807–1814.

27 R. Shine et al., "The Influence of Nest Temperatures and Maternal Brooding on Hatchling Phenotypes in Water Pythons," *Ecology* 78 (1997): 1713–1721.

28 Douglas E. Jorenby et al., "A Controlled Trial of Sustained Release Bupropion, a Nicotine Patch, or Both for Smoking Cessation," *New England Journal of Medicine* 340 (1990): 685–691.

29 Martin Enserink, "Fraud and Ethics Charges Hit Stroke Drug Trial," *Science* 274 (1996): 2004–2005.

30 I. Janszky, K. J. Mukamal, R. Ljung, et al., "Chocolate Consumption and Mortality Following a First Acute Myocardial Infarction: The Stockholm Heart Epidemiology Program," *Journal of Internal Medicine* 266 (2009): 248–257.

31 Pew Research Center for the People and the Press, "The Cell Phone Challenge to Survey Research" [press release], May 15, 2006, www.people-press.org.

32 G. A. Buijze, I. N. Sierevelt, B. C. J. M. van der Heijden, et al. "The Effect of Cold Showering on Health and Work: A Randomized Controlled Trial," *PLoS ONE* 11(9): e0161749, doi:10.1371/journal.pone.0161749.

33 Scott W. Powers, Christopher S. Coffey, Leigh A. Chamberlin, et al., "Trial of Amitriptyline, Topiramate, and Placebo for Pediatric Migraine," *New England Journal of Medicine* 376 (2017): 115–124, doi: 10.1056/NEJMoa1610384.

34 Lillian Lin Miao, "Gastric Freezing: An Example of the Evaluation of Medical Therapy by Randomized Clinical Trials," in John P. Bunker, Benjamin A. Barnes, and Frederick Mosteller (eds.), *Costs, Risks, and Benefits of Surgery* (Oxford, UK: Oxford University Press, 1977), pp. 198–211.

35 U.S. Department of Commerce, Office of Travel and Tourism Industries, in-flight survey, 2007.

36 K. Eagan, J. B. Lozano, S. Hurtado, and M. H. Case, *The American Freshman: National Norms, Fall 2013* (Los Angeles, CA: Higher Education Research Institute, UCLA, 2013).

37 morningconsult.com/2020/11/18/christmas-tree-real-artificial-preference

38 Julia A. Wolfson, Aviva A. Musicus, Cindy W. Leung, Ashley N. Gearhardt, and Jennifer Falbe, "Effect of Climate Change Impact Menu Labels on Fast Food Ordering Choices Among US Adults: A Randomized Clinical Trial," *JAMA Network Open* 5, no. 12 (2022): e2248320, doi:10.1001/jamanetworkopen.2022.48320.

39 Joshua D. Stewart et al., "Decreasing Body Lengths in North Atlantic Right Whales," *Current Biology* 31, no. 14 (2021): 3174–3179.e3, doi.org/10.1016/j.cub.2021.04.067. Whales younger than 2 years and older than 20 years were excluded to avoid curvature in the relationship between age and total length.

40 Pew Research Center, "Teens, Social Media and Technology 2022," August 2022.

41 Brenda C. Coleman, "Study: Heart Attack Risk Cut 74% by Stress Management," Associated Press dispatch appearing in *Lafayette (Ind.) Journal and Courier*, October 20, 1997.

42 You can find a mathematical explanation of Benford's law in Ted Hill, "The First-Digit Phenomenon," *American Scientist* 86 (1996): 358–363; and Ted Hill, "The Difficulty of Faking Data," *Chance* 12, no. 3 (1999): 27–31. Applications in fraud detection are discussed in the second paper by Hill and in Mark A. Nigrini, "Fraud Detection: I've Got Your Number," *Journal of Accountancy* 187, no. 5 (1999): 79–83, www.journalofaccountancy.com/issues/1999/may/nigrini.

43 Grace Gephardt, Canyon del Oro High School

44 Cameron Radford et al., "Artificial Eyespots on Cattle Reduce Predation by Large Carnivores," *Communications Biology* 3 (2020): 430. The study is also described here: www.npr.org/2020/08/23/905181717/study-finds-painting-eyes-on-cows-butts-can-save-their-lives.

Unit 9

1 www.fs.usda.gov/pnw/pubs/pnw_rp283.pdf

2 stats.washingtonpost.com/golf/averages.asp?tour=LPGA&rank=05. Data through August 16, 2021.

3 The idea for this exercise came from Gloria Barrett, Floyd Bullard, and Dan Teague at the North Carolina School of Science and Math.

4 Data on used car prices from autotrader.com. We searched for F-150 44's on sale within 50 miles of College Station, Texas.

5 www.ncei.noaa.gov

6 stats.wnba.com/players

7 Data from National Institute of Standards and Technology, *Engineering Statistics Handbook*, www.itl.nist.gov/div898/handbook. The analysis there does not comment on the bias of field measurements.

8 www.gapminder.org

9 M. Seddigh and G. D. Jolliff, "Light Intensity Effects on Meadowfoam Growth and Flowering," *Crop Science* 34 (1994): 497–503.

10 Todd W. Anderson, "Predator Responses, Prey Refuges, and Density-Dependent Mortality of a Marine Fish," *Ecology* 81 (2001): 245–257.

11 Data from www.weatherbase.com.

12 Cheapest "wanna-get-away" fare on Southwest Airlines as of August 8, 2014.

13 Data from Kerry Lane and Danielle Neal, Canyon del Oro High School.

14 www.nature.com/articles/s41586-021-03550-y and portal.edirepository.org/nis/mapbrowse?packageid=edi.698.2

15 From a graph in G. D. Martinsen, E. M. Driebe, and T. G. Whitham, "Indirect Interactions Mediated by Changing Plant Chemistry: Beaver Browsing Benefits Beetles," *Ecology*, 79 (1998): 192–200.

16 bcs.whfreeman.com/WebPub/Statistics/shared_resources/EESEE/BloodAlcoholContent/index.html

17 D. B. Lindenmayer, K. L. Viggers, R. B. Cunningham, and C. F. Donnelly, "Morphological Variation Among Columns of the Mountain Brushtail Possum, *Trichosurus caninus Ogilby* (Phalangeridae: Marsupiala)," *Australian Journal of Zoology* 43 (1995): 449–458.

18 Samuel Karelitz et al., "Relation of Crying Activity in Early Infancy to Speech and Intellectual Development at Age Three Years," *Child Development* 35 (1964): 769–777.

19 See note 18.

20 www.gapminder.org

21 M. H. Criqui, University of California, San Diego, reported in *The New York Times*, December 28, 1994.

22 www.basketball-reference.com

23 tucson.com/news/local/saguaro-census-shows-more-giants-low-reproduction-in-namesake-park/article_c659cb2d-ddef-5371-b5a3-abe351f886ad.html. Thanks to Don Swann of the National Park Service for sharing additional information.

24 Data from James R. Jordan, Lawrenceville School.

25 Data from Brittany Foley and Allie Dutson, Canyon del Oro High School.

26 Data from Samuel Phillips, Purdue University.

27 Based on Marion E. Dunshee, "A Study of Factors Affecting the Amount and Kind of Food Eaten by Nursery School Children," *Child Development* 2 (1931): 163–183. This article gives the means, standard deviations, and correlation for 37 children but does not give the actual data.

28 Data from Nicole Enos and Elena Tesluk, Canyon del Oro High School.

29 R. D. Stichler, G. G. Richey, and J. Mandel, "Measurement of Treadware of Commercial Tires," *Rubber Age* 73, no. 2 (May 1953).

30 Thanks to Larry Green, Lake Tahoe Community College, for giving us permission to use several of the contexts from his website at www.ltcconline.net/greenl/java/Statistics/catStatProb/categorizingStatProblemsJavaScript.html.

31 www.epa.gov/greenvehicles/fast-facts-transportation-greenhouse-gas-emissions

32 Data from J. Bowden et al., "High-Arctic Butterflies Become Smaller with Rising Temperatures," *Biology Letters* 11 (2015): 20150574.

33 *Nutrition Action*, September 2019.

34 www.route-fifty.com/management/2022/10/new-ideas-how-cities-can-stop-littering/378670

35 Data from Lauren Baker, Canyon del Oro High School.

Glossary/Glosario

English	Español
$1.5 \times IQR$ rule for outliers An observation is called an outlier if it falls more than $1.5 \times IQR$ above the third quartile or more than $1.5 \times IQR$ below the first quartile. (p. 64)	**regla $1.5 \times$ la gama entre cuartiles para valores atípicos** Se le dice valor atípico a una observación si cae a más de $1.5 \times$ la gama entre cuartiles por encima del tercer cuartil o a más de $1.5 \times$ la gama por debajo del primer cuartil. (p. 64)
$2 \times SD$ rule for outliers Any value more than 2 standard deviations from the mean of a distribution is sometimes classified as an outlier. (p. 64)	**regla $2 \times$ desvío estándar para valores atípicos** Todo valor mayor que 2 desvíos estándar de la media de una distribución en ocasiones se puede clasificar como valor atípico. (p. 64)
10% condition When selecting a random sample of size n (without replacement) from a population of size N, we can treat individual observations as independent when performing calculations as long as $n < 0.10N$. (pp. 430, 486, 493, 501, 510, 541, 573, 610, 624, 652, 660, 670, 680, 693, 708, 750, 771, 779, 806, 808, 828)	**condición del 10%** Cuando se selecciona una muestra aleatoria de tamaño n (sin reposición) de una población de tamaño N, es posible tratar las observaciones individuales de modo independiente al realizar cálculos en tanto $n < 0.10N$. (págs. 430, 486, 493, 501, 510, 541, 573, 610, 624, 652, 660, 670, 680, 693, 708, 750, 771, 779, 806, 808, 828)
A	
accurate An estimator is accurate if it is unbiased. (p. 478)	**preciso** Un estimador se considera preciso si no presenta sesgo. (p. 478)
addition rule for mutually exclusive events If A and B are mutually exclusive events, $P(\text{A or B}) = P(\text{A}) + P(\text{B})$. (p. 336)	**regla de suma para eventos que se excluyen mutuamente** Si A y B son eventos que se excluyen entre sí, $P(\text{A o B}) = P(\text{A}) + P(\text{B})$. (p. 336)
alternative hypothesis H_a The claim that we are trying to find evidence for in a significance test. (p. 563)	**hipótesis H_a alternativa** La proposición de que en una prueba de significancia estadística estamos tratando de hallar evidencia que esté a favor. (p. 563)
anonymity The names of individuals participating in a study are not known even to the director of the study. (p. 293)	**anonimato** Cuando se desconocen los nombres de las personas que participan en un estudio; incluso el director del estudio los ignora. (p. 293)
approximate sampling distribution The distribution of a statistic in many samples (but not all possible samples) of the same size from the same population. (p. 469)	**distribución aproximada del muestreo** Distribución de una estadística entre muchas muestras (aunque no entre todas las muestras posibles) del mismo tamaño de la misma población. (p. 469)
approximately uniform A distribution in which the frequency (relative frequency) of each possible value is about the same. (p. 25)	**aproximadamente uniforme** Distribución en la cual la frecuencia (frecuencia relativa) de cada valor posible es aproximadamente la misma (p. 25)
association A relationship between two variables in which knowing the value of one variable helps predict the value of the other. If knowing the value of one variable does not help predict the value of the other, there is no association between the variables. (p. 151)	**asociación** Relación entre dos variables en la cual saber el valor de una variable facilita la predicción del valor de la otra. Si saber el valor de una variable no facilita la predicción del valor de la otra, entonces no existe ninguna asociación entre las variables. (p. 151)
B	
bar graph (*also called* **bar chart**) Graph used to display the distribution of a categorical variable. The horizontal axis of a bar graph identifies the categories being compared. The heights of the bars show the frequency or relative frequency for each value of the categorical variable. The graph is drawn with blank spaces between the bars to separate the categories being compared. (p. 11)	**gráfico de barras** (*también llamado* **diagrama de barras**) Se usa para ilustrar la distribución de una variable categorizada. El eje horizontal del gráfico de barras identifica las categorías que se han de comparar. La altura de las barras muestra la frecuencia o la frecuencia relativa de cada valor de la variable categórica. Se puede dibujar con espacios en blanco entre las barras a fin de separar las diversas categorías que se desea comparar. (p. 11)

bias The design of a statistical study shows bias if it is very likely to underestimate or very likely to overestimate the value you want to know. (p. 259)	**sesgo** El diseño de un estudio estadístico refleja un sesgo si existe una alta probabilidad de subestimar o sobreestimar el valor que se busca. (p. 259)
biased estimator A statistic used to estimate a parameter is biased if the mean of its sampling distribution is not equal to the value of the parameter being estimated. (p. 475)	**calculador sesgado** La estadística que se usa para computar un parámetro está sesgada si la media de la distribución de su muestreo no equivale al valor del parámetro que se está computando. (p. 475)
bimodal A graph of quantitative data with two clear peaks. (p. 25)	**bimodal** Gráfico de datos cuantitativos con dos picos bien definidos. (p. 25)
binomial coefficient The number of ways to arrange x successes among n trials is given by the binomial coefficient $\binom{n}{x} = \frac{n!}{x!(n-x)!}$ for $x = 0, 1, 2, \ldots, n$ where $n! = n(n-1)(n-2) \cdot \ldots \cdot 3 \cdot 2 \cdot 1$ and $0! = 1$. (p. 419)	**coeficiente binomial** La cantidad de maneras de organizar x aciertos entre n ensayos se representa con el coeficiente binomial $\binom{n}{x} = \frac{n!}{x!(n-x)!}$ para $x = 0, 1, 2, \ldots, n$ en el que $n! = n(n-1)(n-2) \cdot \ldots \cdot 3 \cdot 2 \cdot 1$ y $0! = 1$. (p. 419)
binomial distribution In a binomial setting, suppose we let X = the number of successes. The probability distribution of X is a binomial distribution with parameters n and p, where n is the number of trials of the random process and p is the probability of a success on each trial. (p. 417)	**distribución binomial** En un entorno binomial, supongamos que se permite que X = la cantidad de aciertos. La distribución de la probabilidad de X es una distribución binomial con los parámetros n y p, en la que n es la cantidad de ensayos del proceso aleatorio y p es la probabilidad de un acierto en cualquiera de los ensayos. (p. 417)
binomial probability formula Suppose that X is a binomial random variable with n trials and probability p of success on each trial. The probability of getting exactly x successes in n trials ($x = 0, 1, 2, \ldots, n$) is $P(X = x) = \binom{n}{x} p^x (1-p)^{n-x}$. (p. 420)	**fórmula de probabilidad binomial** Supongamos que X es una variable aleatoria binomial con n ensayos y la probabilidad p de acierto en cada ensayo. La probabilidad de obtener exactamente x aciertos en n ensayos ($x = 0, 1, 2, \ldots, n$) es $P(X = x) = \binom{n}{x} p^x (1-p)^{n-x}$. (p. 420)
binomial random variable The count of successes X in a binomial setting. The possible values of X are 0, 1, 2, . . . , n. (p. 417)	**variable aleatoria binomial** La cuenta de aciertos X en un entorno binomial. Los valores posibles de X son 0, 1, 2, . . . , n. (p. 417)
binomial setting Arises when we perform n independent trials of the same random process and count the number of times that a particular outcome (called a "success") occurs. The four conditions for a binomial setting are: • Binary? The possible outcomes of each trial can be classified as "success" or "failure." • Independent? Trials must be independent. That is, knowing the outcome of one trial must not tell us anything about the outcome of any other trial. • Number? The number of trials n of the random process must be fixed in advance. • Same probability? There is the same probability p of success on each trial. (p. 416)	**entorno binomial** Surge cuando se realizan n ensayos independientes del mismo proceso aleatorio y se cuenta la cantidad de veces que se produce un resultado dado, denominado "acierto". Las cuatro condiciones que definen un entorno binomial son: • ¿Binario? Los resultados posibles de cada ensayo se pueden clasificar como "acierto" o "fracaso." • ¿Independiente? Los ensayos han de ser independientes. Es decir, saber el resultado de un ensayo no debe indicar nada acerca del resultado de otro ensayo. • ¿Número? El número de ensayos de n en el proceso aleatorio se tiene que fijar con anticipación. • ¿Misma probabilidad? Existe la misma probabilidad p de lograr un acierto en cada ensayo. (p. 416)
bivariate data A data set that describes the relationship between two variables. (p. 158)	**datos bivariados** Grupo de datos que describen la relación entre dos variables. (p. 158)
block Group of experimental units that are known before the experiment to be similar in some way that is expected to affect the response to the treatments. (p. 283)	**bloque** Grupo de unidades experimentales que desde antes del experimento se sabe son similares de alguna manera previsible que afecte la respuesta a los tratamientos. (p. 283)
boxplot A visual representation of the five-number summary of a distribution of quantitative data. The box spans the quartiles and shows the variability of the middle half of the distribution. The median is marked with a vertical line segment in the box. Lines extend from the ends of the box to the smallest and largest observations that are not outliers. Outliers are marked with a special symbol such as an asterisk (*). (p. 65)	**diagrama de caja y bigotes** Representación visual del resumen de cinco cifras de una distribución de datos cuantitativos. La caja abarca los cuartiles y muestra la variabilidad de la mitad central de la distribución. La media se marca con un segmento lineal en la caja. Las líneas se extienden a partir de los extremos de la caja a las observaciones más pequeña y más grande que no son valores atípicos. Los valores atípicos se marcan con un símbolo especial tal como un asterisco (*). (p. 65)

C	
categorical variable A variable that assigns labels that place each individual into a particular group, called a category. (p. 4)	**variable categorizada** Variable que asigna una etiqueta a cada individuo para colocarlo dentro un grupo particular, conocido como categoría. (p. 4)
census Study that collects data from every individual in the population. (p. 241)	**censo** Estudio en el que se recogen datos acerca de cada individuo en la población. (p. 241)
central limit theorem (CLT) In an SRS of size n from any population with mean μ and finite standard deviation σ, when n is sufficiently large, the sampling distribution of the sample mean $\bar{x}$ is approximately normal. (p. 505)	**teorema del límite central** En una muestra aleatoria sencilla de tamaño n a partir de una población con la media μ y una desviación estándar finita de σ, la distribución de muestreo de la media de la muestra $\bar{x}$ es aproximadamente normal. (p. 505)
chi-square distribution A distribution that is defined by a density curve that takes only non-negative values and is skewed to the right. A particular chi-square distribution is specified by giving its degrees of freedom. (p. 753)	**distribución de ji cuadrado** Distribución que se define por una curva de densidad que sólo acepta valores no negativos y está sesgada hacia la derecha. Se especifica una distribución de ji cuadrado dada citando sus grados de libertad. (p. 753)
chi-square test for goodness of fit A significance test of the null hypothesis that a categorical variable has a specified distribution in the population of interest. (p. 757) For more details, see the inference summary in the back of the book.	**prueba de ji cuadrado para confirmar la bondad de ajuste** Prueba de significancia de hipótesis nula en la cual la variable categórica tiene una distribución especificada en la población de interés. (p. 757) Para más información, ver el resumen de inferencia del final del libro.
chi-square test for homogeneity A significance test of the null hypothesis that there is no difference in the distributions of a categorical variable in the populations of interest or for the treatments in an experiment. (p. 781) For more details, see the inference summary in the back of the book.	**prueba de ji cuadrado de homogeneidad** Prueba de significancia de la hipótesis nula en la cual no hay diferencia en las distribuciones de una variable categórica en las poblaciones de interés o en los tratamientos de un experimento. (p. 781) Para más información, ver el resumen de inferencia del final del libro.
chi-square test for independence A significance test of the null hypothesis that there is no association between two categorical variables in the population of interest. (p. 776) For more details, see the inference summary in the back of the book.	**prueba de ji cuadrado de independencia** Prueba de significancia de la hipótesis nula en la cual no hay asociación entre dos variables categóricas dentro de la población de interés. (p. 776) Para mayor información, ver el resumen de inferencia del final del libro.
chi-square test statistic Measure of how far the observed counts are from the expected counts relative to the expected counts. The formula is $$\chi^2 = \sum \frac{(\text{Observed} - \text{Expected})^2}{\text{Expected}}$$ where the sum is over all possible values of the categorical variable or all cells in the two-way table. (pp. 752, 772)	**prueba estadística de ji cuadrado** Una medición de la distancia entre las cuentas observadas y las cuentas previstas en relación con las cuentas previstas. La fórmula es $$\chi^2 = \sum \frac{(\text{Observadas} - \text{Previstas})^2}{\text{Previstas}}$$ en la que la suma está sobre todos los valores posibles de la variable categorizada o sobre todas las celdas en la tabla de doble vía. (págs. 752, 772)
cluster A group of individuals in the population that are located near each other. (p. 255)	**clúster** Grupo de individuos de una población que se ubican cerca uno del otro. (p. 255)
cluster sample A sample selected by dividing the population into non-overlapping groups *(clusters)* of individuals that are located near each other, randomly choosing clusters, and including each member of the selected clusters in the sample. (p. 255)	**muestra de clúster** Muestra seleccionada al dividir a la población en grupos de individuos (clústers) no superpuestos, al seleccionar aleatoriamente clústers y al incluir a cada miembro de los clústers seleccionados en la muestra. (p. 255)
coefficient of determination r^2 A measure of the percent reduction in the sum of squared residuals when using the least-squares regression line to make predictions, rather than the mean value of y. In other words, r^2 measures the proportion or percentage of the variability in the response variable that is accounted for by the explanatory variable in the linear model. (p. 197)	**coeficiente de determinación r^2** Medida del porcentaje de reducción de la suma de cuadrados residuales cuando se hacen predicciones por medio de una línea de regresión de mínimos cuadrados, en lugar de usar el valor medio de y. Es decir, r^2 mide la proporción o el porcentaje de variabilidad en la variable de respuesta que se calcula a través de la variable explicativa del modelo lineal. (p. 197)

combined (pooled) sample proportion The overall proportion of successes in the two samples, which is given by $\hat{p}_C = \frac{\text{number of successes in both samples combined}}{\text{number of individuals in both samples combined}} = \frac{X_1 + X_2}{n_1 + n_2}$ (p. 623)	**proporción combinada (agrupada) de la muestra** La proporción total de aciertos en las dos muestras, que se calcula mediante $\hat{p}_C = \frac{\text{cuenta de aciertos en ambas muetras combinadas}}{\text{cuenta de individuos en ambas muestras combinadas}} = \frac{X_1 + X_2}{n_1 + n_2}$ (p. 623)
comparison Experimental design principle. Use a design that compares two or more treatments. (p. 282)	**comparación** Principio de diseño experimental. Se usa un diseño que compara dos o más tratamientos. (p. 282)
complement The complement of event A, written as A^C, is the event that A does not occur. (p. 335)	**complemento** El complemento del evento A, escrito como A^C, es el evento en el que A no ocurre. (p. 335)
complement rule The probability that an event does not occur is 1 minus the probability that the event does occur. In symbols, $P(A^C) = 1 - P(A)$. (p. 335)	**regla del complemento** La probabilidad de que no suceda un evento es 1 menos la probabilidad de que el evento sí suceda. En representación simbólica, $P(A^C) = 1 - P(A)$. (p. 335)
completely randomized design Design in which the experimental units are assigned to the treatments completely at random. (p. 278)	**diseño completamente aleatorizado** Cuando las unidades experimentales se les asignan a los tratamientos de manera completamente aleatoria. (p. 278)
conditional probability Probability that one event happens given that another event is already known to have happened. The probability that event A happens given that event B has happened is denoted by $P(A \mid B)$. To find the conditional probability $P(A \mid B)$, use the formula $P(A \mid B) = \frac{P(A \cap B)}{P(B)} = \frac{P(\text{both events occur})}{P(\text{given event occurs})}$ (pp. 349, 351)	**probabilidad condicional** La probabilidad de que un evento suceda a la luz de que se sabe que otro evento ya sucedió. La probabilidad de que el evento A suceda, dado que el evento B ya sucedió, se denota con $P(A \mid B)$. Para hallar la probabilidad condicional $P(A \mid B)$, se usa la fórmula $P(A \mid B) = \frac{P(A \cap B)}{P(B)} = \frac{P(\text{ambos eventos ocurren})}{P(\text{el evento dado ocurre})}$ (págs. 349, 351)
conditional relative frequency Gives the percentage or proportion of individuals that have a specific value for one categorical variable among a group of individuals that share the same value of another categorical variable (the condition). (p. 146)	**frecuencia relativa condicional** Ofrece el porcentaje o proporción de individuos que tienen un valor específico para una variable categórica entre un grupo de individuos que comparten el mismo valor de otra variable categórica (condición). (p. 146)
confidence interval Gives a set of plausible values for a parameter based on sample data. Confidence intervals have the form point estimate ± margin of error or, alternatively, statistic ± (critical value)(standard error of statistic) (pp. 532, 543)	**intervalo de confianza** Ofrece un intervalo de valores plausibles para un parámetro. El intervalo se computa a partir de los datos muestrales y tiene la forma estimado de punto ± margen de error o alternativamente, estadística ± (valor crítico)(desviación estándar de la estadística) (págs. 532, 543)
confidence level C Gives the approximate percentage of confidence intervals that will capture the population parameter in repeated random sampling with the same sample size. (p. 536)	**nivel de confianza C** Indica el porcentaje aproximado de intervalos de confianza que captarán el parámetro de población en muestras aleatorias reiteradas con el mismo tamaño de muestra. (p. 536)
confidential A basic principle of data ethics that requires that an individual's data be kept private. Only statistical summaries for groups of subjects may be made public. (p. 293)	**confidencial** Principio básico de la ética de la gestión de datos. Requiere que los datos de un individuo se mantengan en reserva. Solo pueden hacerse públicos los resúmenes estadísticos de grupos de individuos. (p. 293)
confounding When two variables are associated in such a way that their effects on a response variable cannot be distinguished from each other. (p. 271)	**confuso** Cuando dos variables se asocian de tal manera que sus efectos en una variable de respuesta no se pueden distinguir el uno del otro. (p. 271)

confounding variable A variable related to both the explanatory variable and the response variable in a study that may create the false impression of a cause-and-effect relationship between the explanatory and response variables. (p. 271)	**variable confusa** Variable relacionada tanto a la variable explicativa como a la variable de respuesta en un estudio que puede crear la falsa impresión de una relación de causa y efecto entre las variables explicativa y de respuesta. (p. 271)
contingency table *See* two-way table. (p. 144)	**tabla de contingencia** *Ver* tabla de doble vía. (p. 144)
continuous random variable Variable that can take any value in a specified interval on the number line. Continuous random variables are modeled by density curves. (p. 454)	**variable aleatoria continua** Emplea cualquier valor en un intervalo especificado de cifras. Las variables aleatorias continuas se modelan mediante curvas de densidad. (p. 454)
continuous variable A quantitative variable that can take any value in an interval on the number line (p. 21)	**variable continua** Variable cuantitativa que puede tomar cualquier valor en un intervalo de cifras. (p. 21)
contributions Individual terms that are added together to produce the chi-square test statistic. Also called *components:* $\text{contribution} = \frac{(\text{Observed} - \text{Expected})^2}{\text{Expected}}$ (p. 759)	**contribuciones** Los términos individuales que se suman para producir la estadística de prueba de ji cuadrado. También se denominan *componentes:* $\text{contribución} = \frac{(\text{Observadas} - \text{Previstas})^2}{\text{Previstas}}$ (p. 759)
control Experimental design principle. Keeping variables (other than the explanatory variable) the same for all groups, especially variables that are likely to affect the response variable. Helps avoid confounding and reduces variability in the response variable. (pp. 280, 282)	**control** Principio del diseño experimental. Se mantienen las mismas variables (con excepción de la variable explicativa) para todos los grupos, en especial las variables que podrían afectar la variable de respuesta. Permite evitar la confusión y reduce la variabilidad en la variable de respuesta. (págs. 280, 282)
control group Experimental group whose primary purpose is to provide a baseline for comparing the effects of the other treatments. Depending on the purpose of the experiment, a control group may be given an inactive treatment (placebo), an active treatment, or no treatment at all. (p. 275)	**grupo de control** Grupo experimental cuyo fin primario es establecer una línea base mediante la cual se comparan los efectos de otros tratamientos. Según el objeto del experimento, a un grupo de control se le puede administrar un tratamiento inactivo (placebo), un tratamiento activo, o ningún tratamiento. (p. 275)
convenience sample A sample that consists of individuals from the population who are easy to reach. Convenience sampling leads to bias when the members of the sample differ from the population in ways that affect their responses. (p. 259)	**muestra de conveniencia** Muestra formada por individuos de la población con quienes es fácil hacer contacto. Un muestreo de conveniencia genera sesgo cuando los miembros de la muestra difieren de la población en maneras que afecten sus respuestas. (p. 259)
correlation r (*also called* **correlation coefficient**) Gives the direction and measures the strength of the linear relationship between two quantitative variables. We can calculate r using technology or the formula $r = \frac{1}{n-1}\sum\left(\frac{x_i - \bar{x}}{s_x}\right)\left(\frac{y_i - \bar{y}}{s_y}\right)$. (pp. 165, 169)	**correlación r** (*también conocida como* **coeficiente de correlación**) Indica el sentido y mide la fuerza de la relación lineal entre dos variables cuantitativas. La r puede calcularse con tecnología o a través de la fórmula $r = \frac{1}{n-1}\sum\left(\frac{x_i - \bar{x}}{s_x}\right)\left(\frac{y_i - \bar{y}}{s_y}\right)$. (págs. 165, 169)
critical value Multiplier that makes a confidence interval wide enough to have the stated capture rate. The critical value depends on both the confidence level C and the sampling distribution of the statistic. (p. 543)	**valor crítico** Multiplicador que amplía el intervalo de confianza lo suficiente para retener la tasa de captación indicada. El valor crítico depende de tanto el nivel de confianza C como de la distribución de muestreo de la estadística. (p. 543)
cumulative probability distribution Gives $P(X \leq x)$, the percentile corresponding to each possible value of the variable X. (p. 388)	**distribución de probabilidad acumulativa** Indica $P(X \leq x)$, el percentil correspondiente a cada valor posible de la variable X. (p. 388)
cumulative relative frequency graph A cumulative relative frequency graph plots a point corresponding to the percentile of a given value in a distribution of quantitative data. Consecutive points are then connected with a line segment to form the graph. (p. 94)	**gráfico de la frecuencia relativa acumulada** El gráfico de frecuencia relativa acumulada traza un punto correspondiente al percentil de un valor dado en una distribución de datos cuantitativos. A partir de ahí, los puntos consecutivos se conectan con un segmento lineal para formar el gráfico. (p. 94)

D	
density curve Models the probability distribution of a continuous random variable with a curve that (a) is always on or above the horizontal axis and (b) has area exactly 1 underneath it. The area under the curve and above any specified interval of values on the horizontal axis gives the probability that the random variable falls within that interval. (p. 454)	**curva de densidad** Modela la distribución de la probabilidad de una variable aleatoria continua con una curva que (a) siempre está sobre o por encima del eje horizontal y (b) tiene 1 área exactamente debajo. El área debajo de la curva y por encima de todo intervalo especificado de valores en el eje horizontal estima la probabilidad de que la variable aleatoria caiga en dicho intervalo. (p. 454)
descriptive statistics Process of describing data using graphs and numerical summaries. (p. 10)	**estadística descriptiva** Proceso que describe los datos haciendo uso de gráficos y resúmenes numéricos. (p. 10)
discrete random variable Variable that takes a countable set of possible values with gaps between them on a number line. The probability of any event is the sum of the probabilities for the values of the variable that make up the event. (p. 382)	**variable aleatoria discreta** Variable que emplea un conjunto contable de valores posibles entre los cuales hay brechas a lo largo de una línea de cifras. La probabilidad de cualquier evento es la suma de las probabilidades de los valores de la variable que compone el evento. (p. 382)
discrete variable A quantitative variable that takes a countable set of possible values with gaps between them on the number line. (p. 21)	**variable discreta** Variable cuantitativa que toma un conjunto contable de valores posibles entre los cuales hay brechas a lo largo de una línea de cifras. (p. 21)
distribution Tells what values a variable takes and how often it takes each value. (p. 5)	**distribución** Indica qué valores adopta una variable y con qué frecuencia adopta cada valor. (p. 5)
distribution of sample data Tells what values a variable takes for all individuals in a particular sample. (p. 470)	**distribución de valores muestrales** Indica qué valores asume una variable para todos los individuos en una muestra particular. (p. 470)
dotplot A graph that displays the distribution of a quantitative variable by plotting each data value as a dot above its location on a number line. (p. 22)	**gráfico de puntos** Gráfico que muestra la distribución de una variable cuantitativa trazando el valor de cada dato encima de su ubicación a lo largo de una línea de cifras. (p. 22)
double-blind An experiment in which neither the subjects nor those who interact with them and measure the response variable know which treatment a subject is receiving. (p. 276)	**doble ciego** Experimento en el que ninguno de los sujetos ni aquellos que interactúan con los sujetos y que miden la variable de repuesta saben qué tratamiento recibe el sujeto. (p. 276)
E	
empirical probability Estimated probability of a specific outcome of a random process (like getting a head when tossing a fair coin) obtained by actually performing many trials of the random process. (p. 321)	**probabilidad empírica** Probabilidad estimada de obtener un resultado específico en un proceso aleatorio (como lanzar una moneda al aire y salga cara) luego de realizar muchos ensayos del proceso aleatorio. (p. 321)
empirical rule (*also known as the* **68–95–99.7 rule**) In a normal distribution with mean μ and standard deviation σ (a) about 68% of the values fall within 1σ of the mean μ (b) about 95% of the values fall within 2σ of μ and (c) about 99.7% of the values fall within 3σ values of μ (p. 112)	**regla empírica** (*también conocida como* **regla 68–95–99.7**) En una distribución normal, com media μ y desviación estándar σ (a) alrededor del 68% de los valores caen dentro de 1 σ de la media μ, (b) alrededor del 95% de los valores caen dentro de 2σ de la media μ y (c) alrededor del 99,7% de los valores caen dentro de 3σ de la media μ (p. 112)
event A subset of the possible outcomes from the sample space of a random process. Events are usually designated by capital letters, like A, B, C, and so on. (p. 334)	**evento** Subconjunto de los posibles resultados del espacio de la muestra de un proceso aleatorio. Los eventos generalmente se designan en letra mayúscula, como A, B, C, y así sucesivamente. (p. 334)
expected counts Numbers of individuals in the sample that would fall in each cell of the one-way or two-way table if H_0 were true. (pp. 749, 769, 780)	**cuentas previstas** Cantidades de individuos en la muestra que caerían en cada celda en la tabla, sea de una vía o de dos vías, si H_0 fuera verdad. (págs. 749, 769, 780)
experiment A study in which researchers deliberately impose treatments on experimental units to measure their responses. (p. 243)	**experimento** Estudio en el que los investigadores imponen deliberadamente tratamientos a unidades experimentales con el fin de medir sus respuestas. (p. 243)

experimental unit The object to which a treatment is randomly assigned. When the experimental units are human beings, they are often called *subjects*. (p. 273)	**unidad experimental** Objeto al cual se le asigna un tratamiento aleatorio. Cuando las unidades experimentales son seres humanos, generalmente se se refiere a ellos como *sujetos*. (p. 273)
explanatory variable Variable that may help predict or explain changes in a response variable. (pp. 143, 271)	**variable explicativa** Variable que puede ayudar a predecir o explicar cambios en una variable de respuesta. (págs. 143, 271)
extrapolation Use of a regression model for prediction outside the interval of x values used to obtain the model. The further we extrapolate, the less reliable the predictions become. (p. 182)	**extrapolación** Uso de un modelo de regresión para hacer predicciones por fuera del intervalo de valores x que se utiliza para obtener el modelo. Cuanto mayor sea la extrapolación, menos confiable serán las predicciones. (p. 182)
F	
factor Explanatory variable in an experiment that is manipulated and may cause a change in the response variable. (p. 273)	**factor** La variable explicativa en un experimento que se manipula y puede causar un cambio en la variable de respuesta. (p. 273)
factorial For any positive whole number n, its factorial $n!$ is $n! = n(n-1)(n-2)\cdot\ldots\cdot3\cdot2\cdot1$ In addition, we define $0! = 1$. (p. 419)	**factorial** Para cualquier número entero positivo n, su factorial n! es $n! = n(n-1)(n-2)\cdot\ldots\cdot3\cdot2\cdot1$ Además, definimos $0! = 1$. (p. 419)
fail to reject H_0 If the observed result is not unlikely to occur when the null hypothesis is true, we should fail to reject H_0 and say that we do not have convincing evidence for H_a. (p. 567)	**no rechazar H_0** Si no es improbable que el resultado observado suceda cuando es verdad la hipótesis nula, no se debe rechazar H_0 y se ha de indicar que no contamos con evidencia convincente de H_a. (p. 567)
first quartile Q_1 If the observations in a data set are arranged left to right from smallest to largest, the first quartile Q_1 is the median of the data values that are to the left of the median in the ordered list (p. 59)	**primer cuartil Q_1** Si las observaciones del conjunto de datos se disponen de izquierda a derecha de menor a mayor, el primer cuartil Q_1 es la media de los valores de los datos ubicados a la izquierda de la media en la lista ordenada. (p. 59)
five-number summary The minimum, first quartile Q_1, median, third quartile Q_3, and maximum of a distribution of quantitative data. (p. 65)	**resumen de cinco cifras** El mínimo, el primer cuartil Q_1, la media, el tercer cuartil Q_3, y el máximo de una distribución de datos cuantitativos. (p. 65)
frequency table Table that shows the number of individuals having each value (p. 5)	**tabla de frecuencias** Tabla que indica el número de individuos que tiene cada valor. (p. 5)
G	
general addition rule If A and B are any two events resulting from the same random process, then the probability that event A or event B (or both) occur is $P(\text{A or B}) = P(\text{A} \cup \text{B}) = P(\text{A}) + P(\text{B}) - P(\text{A} \cap \text{B})$. (pp. 339, 342)	**regla general de adición** Si A y B son dos eventos resultantes cualesquiera del mismo proceso aleatorio, la probabilidad de que el evento A o el evento B (o ambos) suceda es $P(\text{A o B}) = P(\text{A} \cup \text{B}) = P(\text{A}) + P(\text{B}) - P(\text{A} \cap \text{B})$. (págs. 339, 342)
general multiplication rule For any random process, the probability that events A and B both occur can be found using the formula $P(\text{A and B}) = P(\text{A} \cap \text{B}) = P(\text{A}) \cdot P(\text{B} \mid \text{A})$. (p. 355)	**regla general de multiplicación** En todo proceso aleatorio, la probabilidad de que sucedan los eventos A y B se puede determinar utilizando la fórmula $P(\text{A y B}) = P(\text{A} \cap \text{B}) = P(\text{A}) \cdot P(\text{B} \mid \text{A})$. (p. 355)
geometric distribution In a geometric setting, suppose we let X = the number of trials it takes to get a success. The probability distribution of X is a geometric distribution with parameter p, the probability of a success on any trial. (p. 431)	**distribución geométrica** En un entorno geométrico, supongamos que se permite que X = la cantidad de ensayos que se precisan para lograr un acierto. La distribución de la probabilidad de X es una distribución geométrica con el parámetro p, la probabilidad de lograr un acierto en cualquier ensayo. (p. 431)
geometric probability formula If X has the geometric distribution with probability p of success on each trial, the probability of getting the first success on the xth trial ($x = 1,2,3,\ldots$) is $P(X = x) = (1-p)^{x-1}p$. (p. 432)	**fórmula de probabilidad geométrica** Si X tiene una distribución geométrica con la probabilidad p de acierto en cada ensayo, la probabilidade de acertar por primera vez en el ensayo número x ($x = 1,2,3,\ldots$) es $P(X = x) = (1-p)^{x-1}p$. (p. 432)

geometric random variable The number of trials X that it takes to get a success in a geometric setting. The possible values of X are 1, 2, 3, ... (p. 431)	**variable aleatoria geométrica** La cantidad de ensayos X que se precisan para lograr un acierto en un entorno geométrico. Los valores posibles de X son 1, 2, 3, ... (p. 431)
geometric setting Arises when we perform independent trials of the same random process and record the number of trials it takes to get one success. On each trial, the probability p of success must be the same. (p. 431)	**entorno geométrico** Surge un entorno geométrico cuando se realizan ensayos independientes del mismo proceso aleatorio y se graban la cantidad de ensayos que se precisan para lograr un acierto. En cada ensayo, la probabilidad p de lograr un acierto tiene que ser la misma. (p. 431)
H	
heterogeneous When the individuals in a group differ considerably with respect to the variable of interest. Ideally, the individuals in a cluster are heterogeneous, and each cluster mirrors the variability in the population. (p. 255)	**heterogéneo** Condición que se presenta cuando los individuos de un grupo difieren considerablemente respecto de la variable de interés. Idealmente, los individuos de un grupo son heterogéneos y cada grupo refleja la variabilidad de la población. (p. 255)
high leverage Points that have much larger or much smaller x values than the other points in a bivariate quantitative data set. (p. 213)	**apalancamiento alto** Puntos con valores x mucho mayores o mucho menores que los otros puntos de un conjunto de datos cuantitativos bivariados. (p. 213)
histogram Graph that displays the distribution of a quantitative variable by showing each interval of values as a bar. The heights of the bars show the frequencies or relative frequencies of values in each interval. (p. 32)	**histograma** Muestra la distribución de una variable cuantitativa que expresa cada intervalo de valores en forma de barra. La altura de las barras muestra las frecuencias o las frecuencias relativas de los valores en cada intervalo. (p. 32)
homogeneous When the individuals in a group are quite similar with respect to the variable of interest. Ideally, the individuals in a stratum are homogeneous. (p. 253)	**homogéneo** Condición que se presenta cuando los individuos de un grupo son bastante parecidos respecto de la variable de interés. Idealmente, los individuos en un estrato son homogéneos. (p. 253)
I	
independent events Two events are independent if knowing whether or not one event has occurred does not change the probability that the other event will happen. In other words, events A and B are independent if $P(A \mid B) = P(A \mid B^C) = P(A)$ and $P(B \mid A) = P(B \mid A^C) = P(B)$ (p. 352)	**eventos independientes** Dos eventos son independientes cuando el hecho de saber o desconocer si un evento ha ocurrido no cambia la probabilidad de que el otro suceda. Es decir, los eventos A y B son independientes si $P(A \mid B) = P(A \mid B^C) = P(A)$ y $P(B \mid A) = P(B \mid A^C) = P(B)$ (p. 352)
independent random variables If knowing the value of X does not help us predict the value of Y, then X and Y are independent random variables. In other words, two random variables are independent if knowing the value of one variable does not change the probability distribution of the other variable. (p. 405)	**variables aleatorias independientes** Si conocer los valores de X no nos sirve para predecir el valor de Y, entonces X e Y son variables aleatorias independientes. Es decir, dos variables aleatorias son independientes si conocer el valor de una variable no cambia la probabilidad de distribución de la otra variable. (p. 405)
individual An object described by a set of data. Individuals can be people, animals, or things. (p. 4)	**individuo** Un objeto descrito por un conjunto de datos. Los individuos pueden ser personas, animales o cosas. (p. 4)
inference about a population Conclusion about the larger population based on sample data. Requires that the individuals taking part in a study be randomly selected from the population of interest. (p. 291)	**inferencia sobre una población** Conclusión sobre una población en general con base en datos muestrales. Se precisa que los participantes del estudio sean escogidos de manera aleatoria a partir de la población de interés. (p. 291)
inference about cause and effect Conclusion from the results of an experiment that the treatments caused the difference in responses. Requires a well-designed experiment in which the treatments are randomly assigned to the experimental units and the results are statistically significant. (p. 291)	**inferencia sobre causa y efecto** Uso de los resultados de un experimento para llegar a la conclusión de que los tratamientos son los que marcan la diferencia en las respuestas. Exige un experimento bien diseñado en el que los tratamientos se asignan de manera aleatoria a las unidades experimentales y los resultados son estadísticamente significativos. (p. 291)

inferential statistics Using sample statistics to make conclusions about population parameters. (p. 51)	**estadística inferencial** Usar estadísticas de muestras para llegar a conclusiones sobre parámetros de poblaciones. (p. 51)
influential point Any point that, if removed, substantially changes the slope, *y* intercept, correlation, coefficient of determination, or standard deviation of the residuals. (p. 213)	**punto influyente** Cualquier punto que, al quitarse, cambia sustancialmente la pendiente, la interceptación *y*, el coeficiente de determinación o la desviación estándar de las residuales. (p. 213)
informed consent Basic principle of data ethics that states that individuals must be informed in advance about the nature of a study and any risk of harm it may bring. Participating individuals must then consent in writing. (p. 293)	**consentimiento informado** Principio básico de ética en la gestión de los datos. A los individuos se les ha de informar, con antelación, de la naturaleza de un estudio y de los riesgos o perjuicios que podría conllevar. Entonces los participantes tendrán que dar autorización por escrito. (p. 293)
institutional review board (IRB) Board charged with protecting the safety and well-being of the participants in advance of a planned study and with monitoring the study itself. (p. 293)	**junta de revisión institucional (IRB,** *por su sigla en inglés***)** Junta encargada de salvaguardar la seguridad y el bienestar de los participantes, anticipándose a un estudio planeado, además de supervisar el estudio mismo. (p. 293)
interquartile range (*IQR*) The distance between the first and third quartiles of a distribution of quantitative data. In symbols, $IQR = Q_3 - Q_1$. (p. 59)	**gama entre cuartiles (*IQR*,** *por su sigla en inglés***)** Distancia entre el primer cuartil y el tercer cuartil de una distribución de datos cuantitativos. En símbolos, $IQR = Q_3 - Q_1$. (p. 59)
intersection The event "A and B" is called the intersection of events A and B. It consists of all outcomes that are common to both events, and is denoted by $A \cap B$. (p. 342)	**intersección** Al evento "A y B" se le conoce como la intersección de los eventos A y B. Consiste en todos los resultados que son comunes para ambos eventos, y se expresa $A \cap B$. (p. 342)
interval estimate *See* confidence interval. (p. 532)	**estimación de intervalo** *Ver* intervalo de confianza. (p. 532)
J	
joint probability The probability *P*(A and B) that events A and B both occur. (p. 338)	**probabilidad conjunta** Probabilidad *P*(A y B) de que ocurran ambos eventos A y B. (p. 338)
joint relative frequency Gives the percent or proportion of individuals in a two-way table that have a specific value for one categorical variable and a specific value for another categorical variable. (p. 144)	**frecuencia relativa conjunta** Ofrece el porcentaje o proporción de individuos que tienen un valor específico para una variable categórica y un valor específico para otra variable categórica. (p. 144)
L	
Large Counts condition A condition for using a normal distribution to perform calculations involving the sampling distribution of $\hat{p}$ or $\hat{p}_1 - \hat{p}_2$. When p is known, check that both np and $n(1-p)$ are at least 10. If both p_1 and p_2 are known, check that n_1p_1, $n_1(1-p_1)$, n_2p_2, and $n_2(1-p_2) \geq 10$. When estimating p with a confidence interval, check that $n\hat{p}$ and $n(1-\hat{p}) \geq 10$. When testing a claim about p, check that np_0 and $n(1-p_0) \geq 10$, where p_0 is the value of p in the null hypothesis. When estimating $p_1 - p_2$, check that $n_1\hat{p}_1$, $n_1(1-\hat{p}_1)$, $n_2\hat{p}_2$, and $n_2(1-\hat{p}_2) \geq 10$. When testing a claim about $p_1 - p_2$, check that $n_1\hat{p}_C$, $n_1(1-\hat{p}_C)$, $n_2\hat{p}_C$, and $n_2(1-\hat{p}_C) \geq 10$, where $\hat{p}_C = \frac{X_1 + X_2}{n_1 + n_2}$. (pp. 486, 493, 541, 573, 610, 624)	**condición de cuentas grandes** Condición para usar una distribución normal para hacer cálculos que impliquen el uso de la distribución del muestreo de $\hat{p}$ o $\hat{p}_1 - \hat{p}_2$. Cuando se conoce p, comprobar que tanto np como $n(1-p)$ sean al menos 10. Si se conocen tanto p_1 como p_2, comprobar que n_1p_1, $n_1(1-p_1)$, n_2p_2, y $n_2(1-p_2) \geq 10$. Al estimar p con un intervalo de confianza, comprobar que $n\hat{p}$ y $n(1-\hat{p}) \geq 10$. Al probar una afirmación sobre p, comprobar que np_0 y $n(1-p_0) \geq 10$, donde p_0 es el valor de p en la hipótesis nula. Al estimar $p_1 - p_2$, comprobar que $n_1\hat{p}_1$, $n_1(1-\hat{p}_1)$, $n_2\hat{p}_2$, y $n_2(1-\hat{p}_2) \geq 10$. Al probar una afirmación sobre $p_1 - p_2$, comprobar que $n_1\hat{p}_C$, $n_1(1-\hat{p}_C)$, $n_2\hat{p}_C$, y $n_2(1-\hat{p}_C) \geq 10$ donde $\hat{p}_C = \frac{X_1 + X_2}{n_1 + n_2}$. (págs. 486, 493, 541, 573, 610, 624).
Large Counts condition for a chi-square test It is safe to use a chi-square distribution to perform *P*-value calculations if all expected counts are at least 5. (pp. 750, 771, 779)	**condición de cuentas grandes en la prueba de ji cuadrado** Se puede utilizar sin problemas la distribución de ji cuadrado para realizar cómputos de valor *P* si todas las cuentas previstas son de al menos 5. (págs. 750, 771, 779)

law of large numbers If we observe more and more trials of any random process, the proportion of times that a specific outcome occurs approaches its probability. (p. 321)	**ley de las cifras grandes** Si se observan más y más ensayos en cualquier proceso aleatorio, la proporción de veces que se da un resultado específico se aproxima a su probabilidad. (p. 321)
least-squares regression line The line that makes the sum of the squared residuals as small as possible. (p. 188)	**línea de regresión de mínimos cuadrados** La línea que reduce al mínimo posible la suma de los cuadrados residuales. (p. 188)
level Specific value of an explanatory variable (factor) in an experiment. (p. 273)	**nivel** Valor específico de una variable explicativa (factor) en un experimento. (p. 273)
linear combination A linear combination of two random variables X and Y can be written in the form $aX + bY$, where a and b are constants. (p. 404)	**combinación lineal** La combinación lineal de dos variables aleatorias X e Y puede representarse con la fórmula $aX + bY$, en la cual a y b son constantes. (p. 404)
linear transformation A linear transformation of the random variable X involves multiplying (dividing) the values of the variable by a constant and/or adding (subtracting) a constant. It can be written in the form $Y = a + bX$. (p. 402)	**transformación lineal** La transformación lineal de la variable aleatoria X implica multiplicar (dividir) los valores de la variable por una constante o sumar (restar) una constante. Puede representarse con la fórmula $Y = a + bX$. (p. 402)
M	
margin of error Describes how far, at most, we expect the point estimate to vary from the population parameter. That is, the difference between the point estimate and the population parameter will be less than the margin of error in C% of all samples, where C is the confidence level. (p. 543)	**margen de error** Describe el máximo punto en el que puede esperarse una variación del valor estimado del punto respecto del parámetro de la población. Esto es, que la diferencia entre el estimado del punto y el parámetro de la población será menor que el margen de error en C% de todas las muestras, donde C es el nivel de confianza. (p. 543)
marginal relative frequency Gives the percentage or proportion of individuals in a two-way table that have a specific value for one categorical variable. (p. 144)	**frecuencia relativa marginal** Indica el porcentaje o proporción de individuos en una tabla de doble vía que tienen un valor específico para una variable categórica. (p. 144)
matched pairs design Common experimental design for comparing two treatments that uses blocks of size 2. In some matched pairs designs, each subject receives both treatments in a random order. In others, two very similar experimental units are paired and the two treatments are randomly assigned within each pair. (p. 286)	**diseño de pares coincidentes** Diseño experimental común que emplea bloques de tamaño 2 para comparar dos tratamientos. En algunos diseños de pares coincidentes, cada sujeto se somete a ambos tratamientos en un orden aleatorio. En otros, se emparejan dos unidades experimentales muy similares y los tratamientos se asignan aleatoriamente a los sujetos de cada par. (p. 286)
mean The average of all the individual data values in a distribution of quantitative data. To find the mean, add all the values and divide by the total number of data values. In symbols, the sample mean $\bar{x}$ is given by $$\bar{x} = \frac{\sum x_i}{n}$$ (Use μ to denote the population mean.) (p. 50)	**media** El promedio de todos los valores de datos individuales en una distribución de datos cuantitativos. Para obtener la media, se suman todos los valores y se divide el resultado entre el número total de observaciones. En símbolos, la media de la muestra $\bar{x}$ se determina de la siguiente manera: $$\bar{x} = \frac{\sum x_i}{n}$$ (Para referirse a la media de la población debe usarse μ.) (p. 50)
mean (expected value) of a discrete random variable Describes the variable's long-run average value over many, many trials of the same random process. To find the mean (expected value) of X, multiply each possible value by its probability, then add all the products: $$\mu_X = E(X) = x_1P(x_1) + x_2P(x_2) + x_3P(x_3) + \ldots = \sum x_iP(x_i)$$ (p. 386)	**media (valor previsto) de una variable aleatoria discreta** Describe el valor medio de la variable en el largo plazo a partir de muchos ensayos de un proceso aleatorio. Para hallar la media (un valor previsto) de X, se multiplica cada valor posible por su probabilidad y luego se suman todos los productos: $$\mu_X = E(X) = x_1P(x_1) + x_2P(x_2) + x_3P(x_3) + \ldots = \sum x_iP(x_i)$$ (p. 386)

median The midpoint of a distribution of quantitative data; the number such that about half the observations are smaller and about half are larger. To find the median, arrange the data values from smallest to largest. If the number n of data values is odd, the median is the middle value in the ordered list. If the number n of data values is even, use the average of the two middle values in the ordered list as the median. (p. 49)	**mediana** El punto intermedio de una distribución de datos cuantitativos, con una cifra tal que aproximadamente la mitad de las observaciones son más pequeñas y la mitad son más grandes. Para hallar la mediana, los datos deben organizarse en orden ascendente, es decir de menor a mayor valor. Si el número n valores es impar, la mediana es el valor del medio de la lista ordenada. Si el número n de valores es par, la mediana es el promedio de los dos valores del medio de la lista ordenada. (p. 49)
mode Value in a distribution having the greatest frequency. (p. 48)	**modo** En una distribución, el valor que tiene la mayor frecuencia. (p. 48)
mosaic plot A modified segmented bar graph in which the width of each bar is proportional to the number of individuals in the corresponding category. (p. 149)	**gráfico de mosaico** Variación del gráfico de barras segmentadas en el cual el ancho de cada barra es proporcional al número de individuos de la categoría correspondiente. (p. 149)
multiplication rule for independent events If A and B are independent events, then the probability that A and B both occur is $P(\text{A and B}) = P(\text{A} \cap \text{B}) = P(\text{A}) \cdot P(\text{B})$. (p. 361)	**regla de multiplicación de eventos independientes** Si A y B son eventos independientes, la probabilidad de que sucedan ambos, tanto A como B es $P(\text{A y B}) = P(\text{A} \cap \text{B}) = P(\text{A}) \cdot P(\text{B})$. (p. 361)
mutually exclusive (*also called* **disjoint**) Two events A and B that have no outcomes in common and so can never occur together. That is, $P(\text{A and B}) = 0$. (p. 336)	**exclusivos mutuamente** (*también llamado* **desencajamiento**) Dos eventos A y B que no tienen resultados en común y por lo tanto nunca pueden suceder a la vez. Es decir, $P(\text{A y B}) = 0$. (p. 336)
N	
negative association When values of one variable tend to decrease as the values of the other variable increase. (p. 162)	**asociación negativa** Se presenta cuando los valores de una varia-ble tienden a disminuir al tiempo que los de la otra variable aumentan. (p. 162)
no association A relationship between two variables where knowing the value of one variable does not help predict the value of the other variable. (pp. 151, 162)	**sin asociación** Relación entre dos variables en la que conocer el valor de una variable no sirve para anticipar el valor de la otra variable. (págs. 151, 162)
nonresponse Occurs when an individual chosen for the sample can't be contacted or refuses to participate. Nonresponse bias occurs when the individuals who can't be contacted or who refuse to participate differ from the population in ways that affect their responses. (p. 261)	**no respondió** Sucede cuando a un individuo escogido para la muestra no se le puede contactar o el sujeto se niega a participar. Se produce un "sesgo de no respondió" cuando los individuos que no se pueden contactar o que se niegan a participar difieren de la población en maneras que afecten sus respuestas. (p. 261)
normal approximation to a binomial distribution Suppose that a count X of successes has the binomial distribution with n trials and success probability p. The distribution of X has mean np and standard deviation $\sqrt{np(1-p)}$. The distribution of X is approximately normal when the Large Counts condition is met: $np \geq 10$ and $n(1-p) \geq 10$. (p. 491)	**aproximación normal hacia una distribución binomial** Supongamos que una cuenta X de aciertos tiene la distribución binomial con n ensayos y una probabilidad de acierto p. La distribución de X tiene media np y desviación estándar $\sqrt{np(1-p)}$. La distribución de X es aproximadamente normal cuando se cumple la condición de cuentas grandes: $np \geq 10$ y $n(1-p) \geq 10$. (p. 491)
normal curve Important kind of curve that is symmetric, single-peaked, and mound-shaped. (p. 108)	**curva normal** Tipo importante de curva que es simétrica, de un solo pico y con la forma de curva de campana. (p. 108)
normal distribution Distribution described by a normal curve. Any normal distribution is completely specified by two parameters, its mean μ and standard deviation σ. The mean of a normal distribution is at the center of the symmetric normal curve. The standard deviation is the distance from the center to the change-of-curvature points on either side. (p. 108)	**distribución normal** Según la describe una curva normal. Cualquier distribución normal se especifica completamente con dos parámetros, su media μ y la desviación estándar σ. La media de una distribución normal yace en el centro de la curva normal simétrica. La desviación estándar es la distancia del centro a los puntos a ambos lados en los que cambia la curva. (p. 108)

Normal/Large Sample condition A condition for performing inference about a mean, which requires that the population distribution is approximately normal or the sample size is large ($n \geq 30$). If the population distribution has unknown shape and $n < 30$, a graph of the sample data shows no strong skewness or outliers. When performing inference about a difference between two means, check that this condition is met for both samples. (pp. 652, 660, 670, 680, 693, 708)	**condición de muestra Normal/Grande** Condición para hacer una inferencia sobre una media, que requiere que la distribución de la población sea aproximadamente normal o que el tamaño de la muestra sea grande ($n \geq 30$) Si la distribución de la población tiene forma desconocida y $n < 30$, un gráfico de los datos de la muestra no muestra una fuerte asimetría ni valores atípicos. Al hacer una inferencia sobre la diferencia entre dos medias, revisa que ambas muestras cumplan con esta condición. (págs. 652, 660, 670, 680, 693, 708)
normal random variable A continuous random variable whose probability distribution is described by a normal curve. (p. 456)	**variable aleatoria normal** variable aleatoria continua cuya distribución de probabilidad está descrita por una curva normal. (p. 456)
null hypothesis H_0 Claim we weigh evidence against in a significance test. Often the null hypothesis is a statement of "no difference." (p. 563)	**hipótesis nula H_0** Contrapeso de la evidencia en una prueba de significancia. A menudo la hipótesis nula es una declaración de que "no hay diferencia." (p. 563)
O	
observational study Study that observes individuals and measures variables of interest but does not attempt to influence the responses. (p. 243)	**estudio de observación** Se observan los individuos y se miden las variables de interés, pero no se trata de influir en las respuestas. (p. 243)
observed counts Actual numbers of individuals in the sample that fall in each cell of the one-way or two-way table. (pp. 748, 769)	**cuentas observadas** Las cifras reales que corresponden a individuos en la muestra que caen en cada celda de la tabla de una vía o en la de dos vías. (págs. 748, 769)
one-sample t interval for a mean Confidence interval used to estimate a population mean μ when the population standard deviation σ is unknown. (p. 656) For more details, see the inference summary in the back of the book.	**intervalo t de una sola muestra para una media** Intervalo de confianza que sirve para estimar la media de una población μ cuando la desviación estándar de la población σ es desconocida. (p. 656) Para más información, ver el resumen de inferencia del final del libro.
one-sample t test for a mean A significance test of the null hypothesis that a population mean μ is equal to a specified value, when the population standard deviation σ is unknown. (p. 677) For more details, see the inference summary in the back of the book.	**prueba t de una sola muestra para una media** Prueba de significancia de la hipótesis nula en la cual la media μ de la población es igual a un valor específicado, cuando la desviación estándar de la población σ es desconocida. (p. 677) Para más información, ver el resumen de inferencia del final del libro.
one-sample z interval for a proportion Confidence interval used to estimate a population proportion p. (p. 548) For more details, see the inference summary in the back of the book.	**intervalo z de una sola muestra para una proporción** Intervalo de confianza que se usa para estimar la proporción de una población p. (p. 548) Para más información, ver el resumen de inferencia del final del libro.
one-sample z test for a proportion A significance test of the null hypothesis that a population proportion p is equal to a specified value. (p. 581) For more details, see the inference summary in the back of the book.	**prueba z de una sola muestra para una proporción** Prueba de significancia de la hipótesis nula en la cual la proporción p de una población es igual a un valor especificado. (p. 581) Para más información, ver el resumen de inferencia del final del libro.
one-sided alternative hypothesis An alternative hypothesis is one-sided if it states that a parameter is greater than the null value or if it states that the parameter is less than the null value. Tests with a one-sided alternative hypothesis are sometimes called *one-sided tests* or *one-tailed tests*. (p. 563)	**hipótesis alternativa unilateral** Una hipótesis alternativa es unilateral si indica que un parámetro es mayor que el valor nulo o si indica que el parámetro es más pequeño que el valor nulo. Las pruebas con una hipótesis alternativa de un solo lado a veces se llaman *pruebas de un solo lado o pruebas de una cola*. (p. 563)
one-way table Table used to display the distribution of a single categorical variable. (p. 746)	**tabla de una vía** Se usa para mostrar la distribución de una sola variable categorizada. (p. 746)
outlier Individual value that falls outside the overall pattern of a distribution of quantitative data. (p. 26)	**valor atípico** Valor individual que queda por fuera del patrón general de la distribución de datos cuantitativos. (p. 26)
outlier in regression A point that does not follow the pattern of the data and has a large residual. (p. 213)	**valor atípico en regresión** Punto que no sigue el patrón de los datos y cuenta con una residual grande. (p. 213)

P	
P-value The probability of getting evidence for the alternative hypothesis H_a as strong as or stronger than the observed evidence when the null hypothesis H_0 is true. The smaller the *P*-value, the stronger the evidence against H_0 and in favor of H_a provided by the data. (p. 565)	**valor *P*** La probabilidad de obtener evidencia para la hipótesis alternativa H_a es tan fuerte o más fuerte que la evidencia observada cuando la hipótesis H_0 es verdadera. Cuanto menor sea el valor *P*, más fuerte será la evidencia contra H_0 y en favor de H_a que proporcionan los datos. (p. 565)
paired data The result of recording two values of the same quantitative variable for each individual or for each pair of similar individuals. (p. 659)	**datos apareados** El resultado de registrar dos valores de la misma variable cuantitativa por cada individuo o por cada par de individuos similares. (p. 659)
paired *t* interval for a mean difference (*also called a* **one-sample *t* interval for a mean difference)** Confidence interval used with paired data to estimate a population mean difference. (p. 660) For more details, see the inference summary in the back of the book.	**intervalo *t* apareado para la diferencia media** (*también llamado* **intervalo *t* para una diferencia media)** Intervalo de confianza que sirve para estimar la diferencia media de una población con datos apareados. (p. 660) Para más información, ver el resumen de inferencia del final del libro.
paired *t* test for a mean difference (*also called a* **one-sample *t* test for a mean difference)** A significance test of the null hypothesis that a population mean difference is equal to a specified value, usually 0. (p. 681) For more details, see the inference summary in the back of the book.	**prueba *t* apareada para la diferencia media** (*también llamada* **prueba *t* de una sola muestra para la diferencia media)** Prueba de significancia de la hipótesis nula en la cual la diferencia media de una población es igual a un valor especificado, generalmente 0. (p. 681) Para más información, ver el resumen de inferencia del final del libro.
parameter A number that describes some characteristic of a population. (pp. 51, 466)	**parámetro** Número que describe ciertas características de una población. (págs. 51, 466)
percentile The *p*th percentile of a distribution is the value with *p*% of observations less than or equal to it. (p. 89)	**percentil** El percentil *p*th de una distribución es el valor cuyo porcentaje de las observaciones es menor o igual que la cifra. (p. 89)
pie chart Graph that shows the distribution of a categorical variable as a "pie" whose slices have areas proportional to the category frequencies or relative frequencies. A pie chart must include all the categories that make up a whole. (p. 12)	**gráfico circular** Gráfico que muestra la distribución de una variable categorizada con la forma de un círculo subdividido en porciones cuyo tamaño es proporcional a las frecuencias de categoría o frecuencias relativas. El gráfico circular tiene que incluir todas las categorías que componen la totalidad. (p. 12)
placebo A treatment that has no active ingredient but is otherwise like other treatments. (p. 273)	**placebo** Tratamiento que, más allá de no tener un ingrediente activo, es como todos los tratamientos. (p. 273)
placebo effect Describes the fact that some subjects in an experiment will respond favorably to any treatment, even an inactive treatment (placebo). (p. 276)	**efecto placebo** Describe el hecho de que algunos sujetos del experimento responden de manera favorable a cualquier tratamiento, incluso un tratamiento inactivo (con placebo). (p. 276)
point estimate Specific value of a point estimator from a sample. (p. 531)	**estimado de punto** El valor especifico de un estimador de punto tomado de una muestra. (p. 531)
point estimator Statistic that provides an estimate of a population parameter. (p. 531)	**estimador de punto** Estadística que nos da un estimado de un parámetro de la población. (p. 531)
pooled sample proportion *See* combined (pooled) sample proportion. (p. 623)	**proporción agrupada de la muestra** *Ver* proporción combinada de la muestra. (p. 623)
population In a statistical study, the entire group of individuals we want information about. (p. 241)	**población** En un estudio estadístico, la población es el grupo completo de individuos sobre el cual deseamos obtener información. (p. 241)
population distribution The distribution of a variable for all individuals in the population. (pp. 384, 470)	**distribución de la población** Distribución de una variable para todos los individuos de la población. (págs. 384, 470)
population regression line (*also called* **population regression model)** Regression line $\mu_y = \alpha + \beta x$ calculated from every value in the population. (p. 804)	**línea de regresión (real) de la población** (*también llamada* **modelo de regresión poblacional)** La línea de regresión $\mu_y = \alpha + \beta x$ calculada a partir de cada valor de la población. (p. 804)

positive association When values of one variable tend to increase as the values of the other variable increase. (p. 162)	**asociación positiva** Cuando los valores de una variable tienden a aumentar a medida que los otra variable aumentan. (p. 162)
power The probability that a test will find convincing evidence for H_a when a specific alternative value of the parameter is true. The power of a test against any alternative is 1 minus the probability of a Type II error for that alternative; that is, power $= 1 - P(\text{Type II error})$. (pp. 586–587)	**poder** La probabilidad de que en una prueba halle evidencia convincente de H_a cuando un valor alternativo especificado del parámetro es verdadero. El poder de una prueba con respecto a cualquier alternativa es 1 menos la probabilidad de un error Tipo II para dicha alternativa; es decir, poder $= 1 - P(\text{error Tipo II})$. (págs. 586–587)
precise An estimator is precise if it has low variability. (p. 478)	**preciso** Un estimador es preciso si tiene una variabilidad baja. (p. 478)
predicted value In regression, $\hat{y}$ (read "y hat") is the predicted value of the response variable y for a given value of the explanatory variable x. (p. 180)	**valor proyectado** $\hat{y}$ es el valor proyectado de la variable de respuesta y para un valor dado de la variable explicativa x. (p. 180)
probability A number between 0 and 1 that describes the proportion of times an outcome of a random process would occur in a very long series of trials. (p. 321)	**probabilidad** Cifra entre 0 y 1 que describe la proporción de veces que un resultado de un proceso aleatorio sucedería en una serie muy prolongada de repeticiones. (p. 321)
probability distribution Gives the possible values of a random variable and their probabilities. (p. 381)	**distribución de la probabilidad** Presenta los valores posibles de una variable aleatoria y sus probabilidades. (p. 381)
probability model Description of a random process that consists of two parts: a list of all possible outcomes and the probability of each outcome. (p. 333)	**modelo de probabilidad** Descripción de un proceso aleatorio que consta de dos partes: una lista de todos los resultados posibles y la probabilidad de cada resultado. (p. 333)
prospective observational study An observational study that tracks individuals into the future. (p. 243)	**estudio de observación prospectivo** Estudio de observación que sigue la trayectoria futura de los individuos. (p. 243)
Q	
quantitative variable Variable that takes number values that are quantities—counts or measurements. (p. 4)	**variable cuantitativa** Variable que toma valores numéricos que representan cantidades: cuentas o medidas. (p. 4)
quartiles The quartiles of a distribution divide an ordered data set into four groups having roughly the same number of values. (p. 59)	**cuartiles** Los cuartiles de una distribución dividen un conjunto de datos ordenados en cuatro grupos que tienen aproximadamente el mismo número de valores. (p. 59)
question wording bias Confusing or leading questions can introduce strong bias, and changes in wording can greatly change a survey's outcome. Even the order in which questions are asked matters. (p. 262)	**sesgo en la formulación de preguntas** Preguntas confusas o capciosas puede introducir un sesgo marcado, y los cambios de terminología pueden modificar por mucho los resultados de tal sondeo. Incluso el orden en que se hacen las preguntas tiene importancia. (p. 262)
R	
r^2 *See* coefficient of determination. (p. 197)	r^2 *Ver* coeficiente de determinación (p. 197)
random assignment Experimental design principle. Use of a chance process to assign experimental units to treatments (or treatments to experimental units). Doing so helps create roughly equivalent groups of experimental units by balancing the effects of other variables among the treatment groups, allowing for cause-and-effect conclusions. (pp. 278, 282)	**asignación aleatoria** Principio de diseño experimental. Uso del azar para asignar unidades experimentales a los tratamientos (o tratamientos a las unidades experimentales) de manera tal que se formen grupos de unidades experimentales más o menos equivalentes al equilibrar los efectos de otras variables entre los grupos de tratamiento, que permite conclusiones de causa y efecto. (págs. 278, 282)
random condition A condition for performing inference, which requires that the data come from a random sample from the population of interest or from a randomized experiment. When comparing two or more populations or treatments, check that the data come from independent random samples from the populations of interest or from groups in a randomized experiment. (pp. 541, 573, 610, 624, 652, 660, 670, 680, 693, 708, 750, 771, 779, 808, 828)	**condición aleatoria** Condición para hacer inferencias que requiere que los datos hayan sido tomados de una muestra aleatoria de la población de interés o de un experimento aleatorio. Al comparar dos o más poblaciones o tratamientos, verifica que los datos hayan sido tomados por muestras aleatorias independientes de la población de interés o de grupos de un experimento aleatorio. (págs. 541, 573, 610, 624, 652, 660, 670, 680, 693, 708, 750, 771, 779, 808, 828)

random process Generates outcomes that are determined purely by chance. (p. 321)	**proceso aleatorio** Proceso que genera resultados determinados puramente al azar. (p. 321)
random sampling Using a chance process to determine which members of a population are chosen for the sample. (p. 243)	**muestreo aleatorio** Uso de un proceso de probabilidad para determinar qué miembros de la población son elegidos para la muestra. (p. 243)
random variable Variable that takes numerical values that describe the outcomes of a random process. (p. 381)	**variable aleatoria** Toma valores numéricos que describen los resultados de un proceso aleatorio. (p. 381)
randomization distribution Distribution of a statistic (like $\hat{p}$ or $\bar{x}_1 - \bar{x}_2$) generated by simulation, assuming known values for the parameters, or in repeated random assignments of experimental units to treatment groups, assuming that the specific treatment received doesn't affect individual responses. (pp. 470, 631)	**distribución de la aleatoriedad** La distribución de una estadística (como $\hat{p}$ o $\bar{x}_1 - \bar{x}_2$) generados mediante una simulación asumiendo valores conocidos para los parámetros o en designaciones aleatorizadas reiteradas de unidades experimentales a grupos de tratamiento, asumiendo que el tratamiento específico no afecte las respuestas individuales. Cuando se cumplan las condiciones, los procedimientos de inferencia corrientes que se basan en la distribución del muestreo de la estadística serán aproximadamente correctos. (págs. 470, 631)
randomized block design Experimental design that forms groups (*blocks*) consisting of individuals that are similar in some way that is expected to affect the response to the treatments and then randomly assigns experimental units to treatments separately within each block. (p. 283)	**diseño de bloques aleatorios** Diseño experimental en el que se forman grupos (*bloques*) compuestos por individuos que se parecen en algo que supuestamente afectará la respuesta a los tratamientos y aleatoriamente asigna unidades experimentales a tratamientos de manera separada dentro de cada bloque. (p. 283)
range The distance between the minimum value and the maximum value of a distribution of quantitative data. That is, range = maximum − minimum. (p. 55)	**gama** La distancia entre el valor mínimo y el valor máximo de una distribución. Es decir, gama = máximo − mínimo. (p. 55)
regression line (*also called a* **simple linear regression model**) Line that models how a response variable y changes as an explanatory variable x changes. Regression lines are expressed in the form $\hat{y} = a + bx$, where $\hat{y}$ is the predicted value of y for a given value of x. (p. 180)	**línea de regresión** (*también llamada* **modelo de regresión lineal simple**) Línea que describe cómo una variable de respuesta y cambia a medida que cambia una variable explicativa x. Las líneas de regresión se expresan mediante la fórmula $\hat{y} = a + bx$, en la que $\hat{y}$ es el valor anticipado de y para un valor dado de x. (p. 180)
regression to the mean The tendency for values of the explanatory variable to be paired with less extreme values of the response variable. (p. 189)	**regresión a la media** tendencia de los valores de la variable explicativa a emparejarse con valores menos extremos de la variable de respuesta. (p. 189)
reject H_0 If the observed result is too unlikely to occur by chance alone when the null hypothesis is true, we can reject H_0 and say that there is convincing evidence for H_a. (p. 567)	**rechazar H_0** Si es demasiado improbable que el resultado observado suceda solo por simple casualidad cuando la hipótesis nula es verdad, se puede rechazar H_0 y decir que existe evidencia convincente a favor de H_a. (p. 567)
relative frequency table Table that shows the proportion or percentage of individuals having each value. (p. 5)	**tabla de frecuencia relativa** Tabla que muestra la proporción o porcentaje de individuos que tiene cada valor. (p. 5)
replication Experimental design principle. Giving each treatment to enough experimental units so that a difference in the effects of the treatments can be distinguished from chance variation due to the random assignment. (pp. 280, 282)	**replicación** Principio de diseño experimental. Cada tratamiento se administra a suficientes unidades experimentales con el propósito de que cualquier diferencia en los efectos de los tratamientos pueda distinguirse de una variación casual debida a la asignación aletaoria. (págs. 280, 282)
residual Difference between an actual value of y and the value of y predicted by the regression line: residual = actual y − predicted $y = y - \hat{y}$ (p. 184)	**residual** La diferencia entre un valor real de y el valor de y proyectado por la línea de regresión: residual = actual y − predicción de $y = y - \hat{y}$ (p. 184)
residual plot A scatterplot that displays the residuals on the vertical axis and the explanatory variable (or the predicted values) on the horizontal axis. Residual plots help us assess whether a regression model is appropriate. (p. 192)	**gráfico residual** Gráfico de dispersión que muestra los residuales sobre el eje vertical y la variable explicativa sobre el eje horizontal. Los trazados residuales nos permiten evaluar si el modelo de regresión es apropiado. (p. 192)

resistant A statistical measure that isn't affected much by extreme values. (p. 52)	**resistente** Medida estadística que no resulta demasiado afectada por valores extremos. (p. 52)
response bias Occurs when there is a consistent pattern of inaccurate responses to a survey question. Includes bias due to question wording. (p. 262)	**sesgo de la respuesta** Sucede cuando existe un patrón consistente de respuestas imprecisas a una pregunta de encuesta. Se incluye el sesgo dada la formulación de la pregunta. (p. 262)
response variable Variable that measures the outcome of a study. (pp. 143, 271)	**variable de respuesta** Variable que mide el resultado de un estudio. (págs. 143, 271)
retrospective observational study An observational study that uses existing data for a sample of individuals. (p. 243)	**estudio de observación retrospectiva** Estudio de observación que usa los datos existentes para las muestras de individuos. (p. 243)
roughly symmetric When the right side of a graph of quantitative data, which contains the half of the observations with the largest values, is approximately a mirror image of the left side. (p. 24)	**aproximadamente simétrico** Cuando el lado derecho de un gráfico de datos cuantitativos, que contiene la mitad de las observaciones con los valores más grandes, es casi como un espejo del lado izquierdo. (p. 24)
S	
sample Subset of individuals in the population from which we collect data. (p. 241)	**muestra** Subconjunto de individuos en la población a partir de la cual se recogen datos. (p. 241)
sample regression line Least-squares regression line $\hat{y} = a + bx$ computed from sample data. (p. 804)	**línea de regresión de la muestra** La línea de regresión de mínimos cuadrados $\hat{y} = a + bx$ computada a partir de datos de muestra. (p. 804)
sample space List of all possible outcomes of a random process. (p. 333)	**espacios de la muestra** Lista de todos los resultados posibles de un proceso aleatorio. (p. 333)
sample survey Study that collects data from a sample to learn about the population from which the sample was selected. (p. 241)	**valoración de la muestra** Estudio que reúne datos de una muestra para conocer a la población de la cual se tomó la muestra. (p. 241)
sampling distribution The distribution of values taken by a statistic in all possible samples of the same size from the same population. (p. 468)	**distribución del muestreo** La distribución de valores tomados por la estadística en todas las muestras posibles del mismo tamaño obtenidas de una misma población. (p. 468)
sampling distribution of a sample mean $\bar{x}$ The distribution of values taken by the sample mean $\bar{x}$ in all possible samples of the same size from the same population. (p. 500)	**distribución de muestreo de la media de una muestra $\bar{x}$** Distribución de valores tomados de la media muestral $\bar{x}$ en todas las posibles muestras del mismo tamaño obtenidas de la misma población. (p. 500)
sampling distribution of a sample proportion $\hat{p}$ The distribution of values taken by the sample proportion $\hat{p}$ in all possible samples of the same size from the same population. (p. 485)	**distribución del muestreo de una proporción de la muestra $\hat{p}$** Distribución de valores tomados de la proporción muestral $\hat{p}$ en todas las posibles muestras del mismo tamaño de la misma población. (p. 485)
sampling distribution of a sample slope b The distribution of values taken by the sample slope b in all possible samples of the same size from the same population. (pp. 804–807)	**distribución del muestreo de una pendiente de muestra b** Distribución de valores tomados de la pendiente muestral b en todas las posibles muestras del mismo tamaño de la misma población. (págs. 804–807)
sampling distribution of $\hat{p}_1 - \hat{p}_2$ The distribution of values taken by the statistic $\hat{p}_1 - \hat{p}_2$ in all possible samples of size n_1 from population 1 and all possible samples of size n_2 from population 2. (p. 492)	**distribución del muestreo $\hat{p}_1 - \hat{p}_2$** Distribución de valores tomados por la estadística $\hat{p}_1 - \hat{p}_2$ de todas las muestras posibles de tamaño n_1 de la población 1 y de todas las muestras posibles del tamaño n_2 de la población 2. (p. 492)
sampling distribution of $\bar{x}_1 - \bar{x}_2$ The distribution of values taken by the statistic $\bar{x}_1 - \bar{x}_2$ in all possible samples of size n_1 from population 1 and all possible samples of size n_2 from population 2. (p. 509)	**distribución de muestreo de $\bar{x}_1 - \bar{x}_2$** Distribución de valores tomados por la estadística $\bar{x}_1 - \bar{x}_2$ de todas las muestras posibles de tamaño n_1 de la población 1 y de todas las muestras posibles del tamaño n_2 de la población 2. (p. 509)
sampling frame A list of individuals in a population. (p. 261)	**marco de muestreo** Lista de individuos de una población. (p. 261)

sampling variability The fact that different random samples of the same size from the same population produce different values for an estimate (statistic). (pp. 255, 468)	**variabilidad del muestreo** El hecho de que muestras aleatorias de un mismo tamaño tomadas de la misma población producen valores distintos para una estimación (estadística). (págs. 255, 468)
sampling with replacement When an individual from a population can be selected more than once when choosing a sample. (p. 250)	**muestreo con reemplazo** Muestreo en el cual un individuo puede elegirse más de una vez cuando se selecciona una muestra. (p. 250)
sampling without replacement When an individual from a population can be selected only once when choosing a sample. (p. 250)	**muestreo sin reemplazo** Muestreo en el cual al seleccionar una muestra cada individuo de la población puede elegirse solo una vez. (p. 250)
scatterplot Graph that shows the relationship between two quantitative variables measured on the same individuals. The values of one variable appear on the horizontal axis, and the values of the other variable appear on the vertical axis. Each individual in the data appears as a point in the graph. (p. 159)	**gráfico de dispersión** Permite apreciar la relación entre dos variables cuantitativas midiendo los mismos individuos. Los valores de una variable figuran en el eje horizontal y los valores de la otra variable figuran en el eje vertical. Cada individuo en los datos figura como un punto en el gráfico. (p. 159)
segmented bar graph Graph that displays the distribution of a categorical variable as segments of a bar, with the area of each segment proportional to the number of individuals in the corresponding category. (p. 149)	**gráfico de barras segmentado** Gráfico que muestra la distribución de una variable categorizada como segmentos de una barra cuyo tamaño es proporcional al número de individuos de la categoría correspondiente. (p. 149)
side-by-side bar graph Graph used to compare the distribution of a categorical variable in each of two or more groups. For each value of the categorical variable, there is a bar corresponding to each group. The height of each bar is determined by the count or percentage of individuals in the group with that value. (p. 13)	**gráfico de barras contiguas** Se usa para comparar la distribución de una variable categorizada en cada uno de dos o más grupos. Para cada valor de la variable categorizada, hay una barra que corresponde a cada grupo. La altura de la barra la determina el conteo o el porcentaje de individuos en el grupo que tengan ese valor. (p. 13)
significance level α Value that we use as a boundary to decide if an observed result is unlikely to happen by chance alone when the null hypothesis is true. The significance level gives the probability of a Type I error. (pp. 567, 585)	**nivel de significancia** α Valor que utilizamos como límite para decidir si es poco probable que un resultado observado ocurra solo al azar cuando la hipótesis nula es verdad. El nivel de significancia nos da la probabilidad de un error Tip. 1. (págs. 567, 585)
significance test Formal procedure for using observed data to decide between two competing claims (the null hypothesis and the alternative hypothesis). The claims are usually statements about parameters. Also called a *test of significance*, a *hypothesis test*, or a *test of hypotheses*. (p. 561)	**prueba de significancia** Procedimiento formal en el que se usan datos observados para decidir entre dos opciones que compiten entre sí (la hipótesis nula y la hipótesis alternativa). Las opciones comúnmente son enunciados acerca de un parámetro. También llamada *prueba de significancia* o *prueba de hipótesis*. (p. 561)
simple random sample (SRS) Sample chosen in such a way that every group of n individuals in the population has an equal chance to be selected as the sample. (p. 249)	**muestra aleatoria sencilla** Muestra tomada de tal manera que cada grupo de n individuos en la población tiene la misma oportunidad de ser escogido como la muestra. (p. 249)
simulation Imitation of a random process in such a way that simulated outcomes are consistent with real-world outcomes. (p. 323)	**simulación** Imitación de un proceso aleatorio de manera que los resultados simulados sean consecuentes con los resultados reales. (p. 323)
single-blind An experiment in which either the subjects or the people who interact with them and measure the response variable don't know which treatment a subject is receiving. (p. 276)	**ciego sencillo** Experimento en el que ya sea los sujetos, o bien las personas que interactúan con los sujetos y miden la variable de respuesta, desconocen qué tratamiento recibe un sujeto. (p. 276)
skewed to the left When the left side of a graph of quantitative data, which contains the half of the observations with the smallest values, is much longer than the right side. (p. 24)	**sesgo hacia la izquierda** sesgo hacia la izquierda Cuando el lado izquierdo de un gráfico de datos cuantitativos, que contiene la mitad de las observaciones con los valores más pequeños, es mucho más largo que el lado derecho. (p. 24)
skewed to the right When the right side of a graph of quantitative data, which contains the half of the observations with the largest values, is much longer than the left side. (p. 24)	**sesgo hacia la derecha** cuando el lado derecho de un gráfico de datos cuantitativos, que contiene la mitad de las observaciones con los valores más grandes, es mucho más largo que el lado izquierdo. (p. 24)

slope In the regression equation $\hat{y} = a + bx$, the slope b is the amount by which the predicted value of y changes when x increases by 1 unit. (p. 185)	**pendiente** En la ecuación de regresión $\hat{y} = a + bx$, la pendiente b es la cantidad por la cual el valor anticipado de y cambia si x aumenta 1 unidad. (p. 185)
standard deviation Measures the typical distance of the values in a distribution from the mean. It is calculated by finding an "average" of the squared deviations of the individual data values from the mean, and then taking the square root. In symbols, the sample standard deviation s_x is given by $$s_x = \sqrt{\frac{\sum(x_i - \bar{x})^2}{n-1}}$$ (Use σ_x to denote the population standard deviation.) (p. 56)	**desviación estándar** Mide la distancia típica de los valores en una distribución a partir de la media. Se calcula hallando un "promedio" de las desviaciones al cuadrado de los datos individuales a las que luego se les computa la raíz cuadrada. En símbolos, la desviación estándar de la muestra s_x se representa de la siguiente manera: $$s_x = \sqrt{\frac{\sum(x_i - \bar{x})^2}{n-1}}$$ (Para referirse a la media de la población debe usarse σ_x.) (p. 56)
standard deviation of a discrete random variable Measures how much the values of the random variable typically vary from the mean in many, many trials of the random process. In symbols, $$\sigma_X = \sqrt{\sum(x_i - \mu_x)^2 P(x_i)}$$ (p. 388)	**desviación estándar de una variable aleatoria discreta** La raíz cuadrada de la variación de una variable aleatoria discreta σ_x^2. La desviación estándar mide cuánto varían normalmente los valores de la variable aleatoria respecto de la media repitiendo una gran cantidad de veces los ensayos del proceso aleatorio. En símbolos se representa de la siguiente manera: $$\sigma_X = \sqrt{\sum(x_i - \mu_x)^2 P(x_i)}$$ (p. 388)
standard deviation of the residuals (s) If we use a least-squares line to predict the values of a response variable y from an explanatory variable x, the standard deviation of the residuals (s) is given by $$s = \sqrt{\frac{\sum \text{residuals}^2}{n-2}} = \sqrt{\frac{\sum(y_i - \hat{y}_i)^2}{n-2}}$$ This value measures the size of a typical residual. That is, s measures the typical distance between the actual y values and the predicted y values. (p. 198)	**desviación estándar de las residuales (s)** Si se usa la línea de cuadrados mínimos para predecir los valores de una variable de respuesta y a partir de una variable explicativa x, la desviación estándar de las residuales (s) la da $$s = \sqrt{\frac{\sum \text{residuales}^2}{n-2}} = \sqrt{\frac{\sum(y_i - \hat{y}_i)^2}{n-2}}$$ Este valor mide el tamaño de una residual típica. Es decir, s mide la distancia típica entre los valores de y reales y los valores de y esperados. (p. 198)
standard error of $\hat{p}$ An estimate of the standard deviation of the sampling distribution of $\hat{p}$: $$s_{\hat{p}} = \sqrt{\frac{\hat{p}(1-\hat{p})}{n}}$$ The standard error describes how much the sample proportion $\hat{p}$ typically varies from the population proportion p in repeated random samples of size n. (p. 545)	**error estándar de $\hat{p}$** Estimación de la desviación estándar de la distribución de muestreo de $\hat{p}$: $$s_{\hat{p}} = \sqrt{\frac{\hat{p}(1-\hat{p})}{n}}$$ El error estándar describe cuánto suele variar la proporción de la muestra $\hat{p}$ con respecto a la proporción de la población p en muestras aleatorias repetidas de tamaño n. (p. 545)
standard error of $\bar{x}$ An estimate of the standard deviation of the sampling distribution of $\bar{x}$. $$s_{\bar{x}} = \frac{s_x}{\sqrt{n}}$$ The standard error of $\bar{x}$ estimates how much the sample mean $\bar{x}$ typically varies from the population mean μ in repeated random samples of size n. (p. 653)	**error estándar $\bar{x}$** Estimación de la desviación estándar de la distribución de muestreo de $\bar{x}$. $$s_{\bar{x}} = \frac{s_x}{\sqrt{n}}$$ El error estándar de $\bar{x}$ estima cuánto varía típicamente la media muestral $\bar{x}$ con respecto a la media poblacional μ en muestras aleatorias repetidas de tamaño n. (p. 653)
standard normal distribution Normal distribution with mean 0 and standard deviation 1. (p. 117)	**distribución normal estándar** La distribución normal con media de 0 y desviación estándar 1. (p. 117)

standardized score (z-score) For an individual value in a distribution, the standardized score (z-score) tells us how many standard deviations from the mean the value falls, and in what direction. To find the standardized score (z-score), compute $$z = \frac{\text{value} - \text{mean}}{\text{standard deviation}}$$ (p. 91)	**puntuación estandarizada (puntuación z)** Para un valor individual en una distribución, la puntuación estandarizada (puntuación z) expresa cuántas desviaciones estándar se producen respecto de la media y en qué dirección ocurren. Para determinar la puntuación estandarizada (puntuación z) debe aplicarse la siguiente fórmula: $$z = \frac{\text{valor} - \text{media}}{\text{desviación estándar}}$$ (p. 91)
standardized test statistic Value that measures how far a sample statistic is from what we would expect if the null hypothesis H_0 were true, in standard deviation units. That is, $$\text{standardized test statistic} = \frac{\text{statistic} - \text{parameter}}{\text{standard error of statistic}}$$ (p. 575)	**estadística de prueba estandarizada** Valor que mide la divergencia entre la estadística de muestra y lo que esperaríamos si la hipótesis nula H_0 fuera cierta, en unidades de desviación estándar. Esto es: $$\text{estadística de prueba estandarizada} = \frac{\text{estadística} - \text{parámetro}}{\text{error estándar de la estadística}}$$ (p. 575)
statistic Number that describes some characteristic of a sample. (pp. 51, 466)	**dato estadístico** Número que describe alguna característica de una muestra. (págs. 51, 466)
statistically significant (1) In an experiment, when the difference in responses between the groups is so large that it is unlikely to be explained by the chance variation in the random assignment, the results are called statistically significant. (p. 288) (2) If the P-value is less than α, we say that the results of a statistical study are statistically significant at level α. In that case, we reject the null hypothesis H_0 and conclude that there is convincing evidence for the alternative hypothesis H_a. (p. 567)	**estadísticamente significativo** (1) En un experimento, cuando la diferencia en las respuestas entre los grupos es tan grande que es poco probable que pueda explicarse por la variación aleatoria en la asignación al azar, los resultados se denominan estadísticamente significativos. (p. 288) (2) Si el valor P es menor que alfa, se dice que los resultados de un estudio estadístico son estadísticamente significativos al nivel α. En tal caso, se rechaza la hipótesis nula H_0 y se concluye que hay evidencia convincente para la hipótesis H_a alternativa. (p. 567)
statistics The science and art of collecting, analyzing, and drawing conclusions from data. (p. 2)	**estadística** La ciencia y arte de coleccionar, analizar y llegar a conclusiones a partir de un conjunto de datos. (p. 2)
stemplot (*also called* **stem-and-leaf plot**) Graph that displays the distribution of a quantitative variable while including the actual numerical values in the graph. Each data value is separated into two parts: a *stem*, which consists of the leftmost digits, and a *leaf*, consisting of the final digit. The stems are ordered from least to greatest and arranged in a vertical column. The leaves are arranged in increasing order out from the appropriate stems. (p. 29)	**gráfico de tallos al que** (*también se le dice* **gráfico de tallos y hojas**) Gráfico que muestra la distribución de una variable cuantitativa, al tiempo que incluyen los valores numéricos mismos en el gráfico. Cada valor se separa en dos partes: un *tallo*, compuesto los dígitos colocados más a la izquierda, y una *hoja*, que consiste del último dígito. Los tallos se ordenan de menor a mayor en una columna. Las hojas, en orden ascendente partiendo de los tallos aproximados. (p. 29)
strata Groups of individuals in a population that share characteristics thought to be associated with the variables being measured in a study. The singular form of *strata* is *stratum*. (p. 253)	**estratos** Grupos de individuos de una población que comparten características supuestamente asociadas con las medidas en un estudio. La forma singular de *estrato* es *estratos*. (p. 253)
stratified random sample A sample selected by dividing the population into non-overlapping groups *(strata)* of individuals that share characteristics thought to be associated with the variables being measured in a study, selecting an SRS from each stratum, and combining the SRSs into one overall sample. (p. 253)	**muestra aleatoria estratificada** Muestra seleccionada mediante la división de la población en grupos no superpuestos (*estratos*) de individuos que comparten características parecidas supuestamente asociadas con las variables medidas en el estudio, selecciona una muestra aleatoria sencilla de cada estrato y combina las muestras aleatorias sencillas en una muestra general. (p. 253)
subjects Experimental units that are human beings. (p. 273)	**sujetos** Unidades experimentales que son seres humanos. (p. 273)
systematic random sample A sample selected by choosing individuals from an ordered arrangement of the population by randomly selecting one of the first k individuals and choosing every kth individual thereafter. (p. 256)	**muestra sistemática aleatoria** Muestra seleccionada mediante la elección de individuos de un grupo ordenado de la población de manera aleatoria mediante la elección de uno de los primeros individuos k y cada individuo k-ésimo de ahí en adelante. (p. 256)

T	
***t* distribution** A distribution described by a symmetric, single-peaked, bell-shaped density curve centered at 0 that is completely specified by its degrees of freedom (df) and has more area in its tails than the standard normal distribution. When performing inference about a population mean μ based on a random sample of size n using the sample standard deviation s_x to estimate the population standard deviation σ, use a t distribution with df $= n - 1$. (p. 647)	**distribución *t*** Distribución descrita por una curva de densidad simétrica, con un solo vértice y forma de campana centrada en 0, completamente especificada por sus grados de libertad (df) y tiene más área en sus colas que la distribución normal estándar. Al realizar la inferenecia de una población μ basada en una muestra aleatoria de tamaño n usando la desviación estándar de la muestra s_x para estimar la desviación estándar de población σ se debe usar distribución t con df $= n - 1$. (p. 647)
***t* interval for the slope** Confidence interval used to estimate the slope β of a population regression line. (p. 814) For more details, see the inference summary in the back of the book.	**intervalo *t* para la pendiente** Intervalo de confianza que se usa para estimar la pendiente β de la línea de regresión de una población. (p. 814) Para más información, ver el resumen de inferencia del final del libro.
***t* test for the slope** A significance test of the null hypothesis that the slope β of a population regression line is equal to a specified value, most commonly 0. (p. 832) For more details, see the inference summary in the back of the book.	**prueba *t* para la pendiente** Prueba de significancia de la hipótesis nula en la cual a pendiente β de una línea de regresión de la población es igual a un valor especificado, más comúnmente 0. (p. 832) Para más información, ver el resumen de inferencia del final del libro.
third quartile Q_3 If the observations in a data set are arranged left to right from smallest to largest, the third quartile Q_3 is the median of the data values that are to the right of the median in the ordered list. (p. 59)	**tercer cuartil Q_3** Si las observaciones en un conjunto de datos se ordenan de izquierda a derecha, de menor a mayor, el tercer cuartil Q_3 es la media de los valores ubicados a la derecha de la media de la lista ordenada. (p. 59)
transforming data Changing the scale of measurement for a quantitative data set by adding a constant to (subtracting a constant from) every data value, and/or multiplying (dividing) every data value by a constant. (p. 96)	**transformación de datos** Conlleva cambiar la escala de medición de un conjunto de datos cuantitativos al agregar una constante a (restar una constante de) cada valor de datos y/o multiplicar (dividir) cada valor de datos por una constante. (p. 96)
transforming to achieve linearity Using powers or logarithms to transform one or both variables to create a linear association. (p. 216)	**transformar para lograr linealidad** Usar potencias o logaritmos para transformar una o ambas variables y crear una asociación lineal. (p. 216)
treatment Specific condition applied to the individuals in an experiment. If an experiment has several explanatory variables, a treatment is a combination of specific values of these variables. (p. 273)	**tratamiento** Una condición específica que se les aplica a los individuos en un experimento. Si un experimento tiene varias variables explicativas, el tratamiento es una combinación de los valores específicos de estas variables. (p. 273)
tree diagram A diagram that shows the sample space of a random process involving multiple stages. The probability of each outcome is shown on the corresponding branch of the tree. All probabilities after the first stage are conditional probabilities. (p. 357)	**diagrama de árbol** Diagrama que presenta el espacio de muestra de un proceso aleatorio de múltiples etapas. La probabilidad de cada resultado se muestra en la rama correspondiente del árbol. Todas las probabilidades que están después de la primera etapa son probabilidades condicionales. (p. 357)
trial One repetition of a random process. (p. 321)	**ensayo** Repetición de un proceso aleatorio. (p. 321)
two-sample *t* interval for a difference in means Confidence interval used to estimate a difference in the means of two populations or treatments, when both population standard deviations are unknown. (p. 698) For more details, see the inference summary in the back of the book.	**intervalo *t* de dos muestras para obtener una diferencia en las medias** Intervalo de confianza que se usa para estimar la diferencia entre las medias de dos poblaciones o tratamientos, cuando las desviaciones estándar de ambas poblaciones son desconocidas. (p. 698) Para más información, ver el resumen de inferencia del final del libro.
two-sample *t* test for a difference in means A significance test of the null hypothesis that the difference in the means of two populations or treatments is equal to a specified value (usually 0), when both population standard deviations are unknown. (p. 713) For more details, see the inference summary in the back of the book.	**prueba *t* de dos muestras para obtener una diferencia en las medias** Prueba de significancia de la hipótesis nula en la cual la diferencia entre las medias de dos poblaciones o tratamientos es igual a un valor especificado (normalmente 0), cuando las desviaciones estándar de ambas poblaciones son desconocidas. (p. 713) Para más información, ver el resumen de inferencia del final del libro.

two-sample z interval for a difference in proportions Confidence interval used to estimate a difference in the proportions of successes for two populations or treatments. (p. 615) For more details, see the inference summary in the back of the book.	**intervalo z de dos muestras para obtener una diferencia en las proporciones** Intervalo de confianza que se usa para estimar la diferencia entre las proporciones de éxitos de dos poblaciones o tratamientos. (p. 615) Para más información, ver el resumen de inferencia del final del libro.
two-sample z test for the difference in proportions A significance test of the null hypothesis that the difference in the proportions of successes for two populations or treatments is equal to a specified value (usually 0). (p. 629) For more details, see the inference summary in the back of the book.	**prueba z de dos muestras para obtener una diferencia en las proporciones** Prueba de la hipótesis nula en la cual diferencia en la proporción de aciertos de dos poblaciones o tratamientos es igual a un valor especificado (normalmente 0). (p. 629) Para más información, ver el resumen de inferencia del final del libro.
two-sided alternative hypothesis The alternative hypothesis is two-sided if it states that the parameter is different from the null value (it could be either greater than or less than). Tests with a two-sided alternative hypothesis are sometimes called *two-sided tests* or *two-tailed tests*. (p. 563)	**hipótesis alternativa bilateral** La hipótesis alternativa es bilateral si indica que el parámetro es diferente del valor nulo (podría ser mayor o menor que dicho valor). Las pruebas con una hipótesis alternativa lateral en ocasiones se conocen como *pruebas bilaterales* o *pruebas de dos colas*. (p. 563)
two-way table Table of counts or relative frequencies that summarizes data on the relationship between two categorical variables for some group of individuals. (p. 144)	**tabla de doble vía** Una tabla de cuentas o de frecuencias relativas que resume los datos sobre la relación entre dos variables categorizadas para un grupo de individuos. (p. 144)
Type I error An error that occurs if we reject H_0 when H_0 is true. That is, the data give convincing evidence that H_a is true when it really isn't. (p. 584)	**error Tipo I** Error que sucede si se rechaza H_0 aunque H_0 sea verdadera. Es decir, la prueba ofrece evidencia contundente de que H_a es verdadera cuando en realidad no lo es. (p. 584)
Type II error An error that occurs if we fail to reject H_0 when H_a is true. That is, the data do not give convincing evidence that H_a is true when it really is. (p. 584)	**error Tipo II** Error que sucede si no se rechaza H_0 incluso cuando H_a es verdadera. Es decir, los datos no ofrecen evidencia contundente de que H_a sea verdadera cuando sí lo es. (p. 584)
U	
unbiased estimator A statistic used for estimating a parameter is unbiased if the mean of its sampling distribution is equal to the value of the parameter being estimated. The mean of the sampling distribution is also known as the *expected value of the estimator*. (p. 474)	**estimador sin sesgo** La estadística que se usa para computar un parámetro es un estimador sin sesgo si la media de distribución de su muestreo equivale al valor del parámetro que se está computando. La media de distribución del muestreo también se denomina *valor esperado del estimador*. (p. 474)
undercoverage Occurs when some members of the population are less likely to be chosen or cannot be chosen in a sample. Undercoverage bias occurs when the underrepresented individuals differ from the population in ways that affect their responses. (p. 261)	**subcobertura** Sucede cuando algunos miembros de la población tienen menor probablidad de ser elegidos o no pueden ser elegidos para una muestra. Se produce un "sesgo de subcobertura" cuando los individuos subrepresentados difieren de la población en maneras que afecten sus respuestas. (p. 261)
unimodal A graph of quantitative data with a single peak. (p. 25)	**unimodal** Gráfico de datos cuantitativos con un solo pico. (p. 25)
union The event "A or B" is called the union of events A and B. It consists of all outcomes that are in event A, event B, or both, and is denoted by $A \cup B$. (p. 342)	**unión** El evento "A or B" "se denomina la unión de los eventos A y B. Abarca todos los resultados que hay en el evento A, en el evento B o en ambos y se expresa con $A \cup B$. (p. 342)
univariate data A data set with one variable. (p. 158)	**datos univariados** Conjunto de datos con una sola variable. (p. 158)
V	
variability of an estimator Describes the variation in the sampling distribution of an estimator (statistic). The variability of an estimator decreases as the sample size increases. (p. 477)	**variabilidad de un estimador** Describe la variabilidad de la distribución de muestreo de un estimador (estadístico). La variabilidad de un estimador disminuye a medida que aumenta el tamaño de la muestra. (p. 477)
variable Any characteristic of an individual. A variable can take different values for different individuals. (p. 4)	**variable** Cualquier característica de un individuo. Una variable puede tener distintos valores según cada individuo. (p. 4)

variance The square of the standard deviation. In symbols, the sample variance is given by s_x^2. (p. 56)	**variación** El cuadrado de la desviación estándar. En símbolos se expresa con s_x^2. (p. 56)
variance of a discrete random variable Weighted average of the squared deviations of the values of the random variable from the mean, denoted by σ_X^2. (p. 388)	**variación de una variable aleatoria discreta** Promedio sopesado de las desviaciones cuadráticas de los valores de la variable aleatoria a partir de su media, que se expresa con σ_X^2. (p. 388)
Venn diagrams A figure that consists of one or more circles surrounded by a rectangle. Each circle represents an event. The region inside the rectangle represents the sample space of the random process. (p. 341)	**diagrama de Venn** Figura que consiste en uno o más círculos colocados dentro de un rectángulo. Cada círculo representa un evento. El interior del rectángulo representa el espacio muestral del proceso aleatorio. (p. 341)
voluntary response sample A sample that consists of people who choose to be in the sample by responding to a general invitation. Voluntary response samples are sometimes called *self-selected* samples. Voluntary response sampling leads to voluntary response bias when the members of the sample differ from the population in ways that affect their responses. (p. 260)	**muestra de respuesta voluntaria** Muestra que consiste en personas que eligen ser parte de la muestra al aceptar una invitación general. Las muestras de respuesta voluntaria en ocasiones se denominan *muestras autoescogidas*. El muestreo de respuesta voluntaria conduce a sesgo de respuesta voluntaria cuando los miembros de la muestra difieren de la población en maneras que afecten sus respuestas. (p. 260)
Y	
***y* intercept** In the regression equation $\hat{y} = a + bx$, the y intercept a is the predicted value of y when $x = 0$. (p. 185)	**interceptación *y*** En la ecuación de regresión $\hat{y} = a + bx$ la interceptación y a es el valor anticipado de y cuando $x = 0$ (p. 185)
Z	
***z*-score** *See* standardized score.	**puntaje *z*** *Ver* puntajes estandarizados.

Index

Note: Page numbers followed by f indicate figures.

S

Formulas for the AP® Statistics Exam

Students are provided with the following formulas on both the multiple choice and free-response sections of the AP® Statistics exam.

I. Descriptive Statistics

$$\bar{x} = \frac{1}{n}\sum x_i = \frac{\sum x_i}{n}$$

$$s_x = \sqrt{\frac{1}{n-1}\sum (x_i - \bar{x})^2} = \sqrt{\frac{\sum (x_i - \bar{x})^2}{n-1}}$$

$$\hat{y} = a + bx$$

$$\bar{y} = a + b\bar{x}$$

$$r = \frac{1}{n-1}\sum\left(\frac{x_i - \bar{x}}{s_x}\right)\left(\frac{y_i - \bar{y}}{s_y}\right)$$

$$b = r\frac{s_y}{s_x}$$

II. Probability and Distributions

$$P(A \cup B) = P(A) + P(B) - P(A \cap B)$$

$$P(A \mid B) = \frac{P(A \cap B)}{P(B)}$$

Probability Distribution	Mean	Standard Deviation
Discrete random variable, X	$\mu_X = E(X) = \sum x_i P(x_i)$	$\sigma_X = \sqrt{\sum (x_i - \mu_X)^2 P(x_i)}$
If X has a **binomial** distribution with parameters n and p, then: $P(X = x) = \binom{n}{x} p^x (1-p)^{n-x}$ where $x = 0, 1, 2, 3, \ldots, n$	$\mu_X = np$	$\sigma_X = \sqrt{np(1-p)}$
If X has a **geometric** distribution with parameter p, then: $P(X = x) = (1-p)^{x-1}p$ where $x = 1, 2, 3, \ldots$	$\mu_X = \frac{1}{p}$	$\sigma_X = \frac{\sqrt{1-p}}{p}$

III. Sampling Distributions and Inferential Statistics

Standardized test statistic: $\dfrac{\text{statistic} - \text{parameter}}{\text{standard error of the statistic}}$

Confidence interval: statistic $\pm$ (critical value)(standard error of statistic)

Chi-square statistic: $\chi^2 = \sum \dfrac{(\text{observed} - \text{expected})^2}{\text{expected}}$

III. Sampling Distributions and Inferential Statistics *(continued)*

Sampling distributions for proportions

Random Variable	Parameters of Sampling Distribution		Standard Error* of Sample Statistic
For one population: $\hat{p}$	$\mu_{\hat{p}} = p$	$\sigma_{\hat{p}} = \sqrt{\dfrac{p(1-p)}{n}}$	$s_{\hat{p}} = \sqrt{\dfrac{\hat{p}(1-\hat{p})}{n}}$
For two populations: $\hat{p}_1 - \hat{p}_2$	$\mu_{\hat{p}_1-\hat{p}_2} = p_1 - p_2$	$\sigma_{\hat{p}_1-\hat{p}_2} = \sqrt{\dfrac{p_1(1-p_1)}{n_1} + \dfrac{p_2(1-p_2)}{n_2}}$	$s_{\hat{p}_1-\hat{p}_2} = \sqrt{\dfrac{\hat{p}_1(1-\hat{p}_1)}{n_1} + \dfrac{\hat{p}_2(1-\hat{p}_2)}{n_2}}$ When $p_1 = p_2$ is assumed: $s_{\hat{p}_1-\hat{p}_2} = \sqrt{\hat{p}_c(1-\hat{p}_c)\left(\dfrac{1}{n_1} + \dfrac{1}{n_2}\right)}$ where $\hat{p}_c = \dfrac{X_1 + X_2}{n_1 + n_2}$

Sampling distributions for means

Random Variable	Parameters of Sampling Distribution		Standard Error* of Sample Statistic
For one population: $\bar{x}$	$\mu_{\bar{x}} = \mu$	$\sigma_{\bar{x}} = \dfrac{\sigma}{\sqrt{n}}$	$s_{\bar{x}} = \dfrac{s}{\sqrt{n}}$
For two populations: $\bar{x}_1 - \bar{x}_2$	$\mu_{\bar{x}_1-\bar{x}_2} = \mu_1 - \mu_2$	$\sigma_{\bar{x}_1-\bar{x}_2} = \sqrt{\dfrac{\sigma_1^2}{n_1} + \dfrac{\sigma_2^2}{n_2}}$	$s_{\bar{x}_1-\bar{x}_2} = \sqrt{\dfrac{s_1^2}{n_1} + \dfrac{s_2^2}{n_2}}$

Sampling distributions for simple linear regression

Random Variable	Parameters of Sampling Distribution		Standard Error* of Sample Statistic
For slope: b	$\mu_b = \beta$	$\sigma_b = \dfrac{\sigma}{\sigma_x\sqrt{n}}$ where $\sigma_x = \sqrt{\dfrac{\sum(x_i-\mu)^2}{n}}$	$s_b = \dfrac{s}{s_x\sqrt{n-1}}$ where $s = \sqrt{\dfrac{\sum(y_i-\hat{y}_i)^2}{n-2}}$ and $s_x = \sqrt{\dfrac{\sum(x_i-\bar{x})^2}{n-1}}$

*Standard deviation is a measurement of variability from the theoretical population. Standard error is the estimate of the standard deviation. If the standard deviation of the statistic is assumed to be known, then the standard deviation should be used instead of the standard error.

Tables

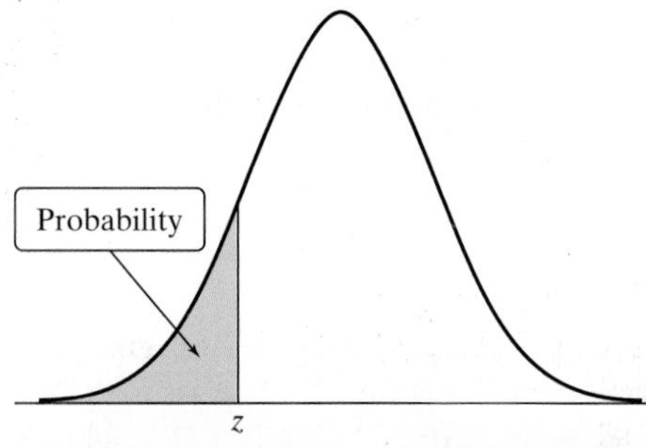

Table entry for z is the area under the standard normal curve to the left of z.

Table A Standard normal probabilities

z	.00	.01	.02	.03	.04	.05	.06	.07	.08	.09
−3.4	.0003	.0003	.0003	.0003	.0003	.0003	.0003	.0003	.0003	.0002
−3.3	.0005	.0005	.0005	.0004	.0004	.0004	.0004	.0004	.0004	.0003
−3.2	.0007	.0007	.0006	.0006	.0006	.0006	.0006	.0005	.0005	.0005
−3.1	.0010	.0009	.0009	.0009	.0008	.0008	.0008	.0008	.0007	.0007
−3.0	.0013	.0013	.0013	.0012	.0012	.0011	.0011	.0011	.0010	.0010
−2.9	.0019	.0018	.0018	.0017	.0016	.0016	.0015	.0015	.0014	.0014
−2.8	.0026	.0025	.0024	.0023	.0023	.0022	.0021	.0021	.0020	.0019
−2.7	.0035	.0034	.0033	.0032	.0031	.0030	.0029	.0028	.0027	.0026
−2.6	.0047	.0045	.0044	.0043	.0041	.0040	.0039	.0038	.0037	.0036
−2.5	.0062	.0060	.0059	.0057	.0055	.0054	.0052	.0051	.0049	.0048
−2.4	.0082	.0080	.0078	.0075	.0073	.0071	.0069	.0068	.0066	.0064
−2.3	.0107	.0104	.0102	.0099	.0096	.0094	.0091	.0089	.0087	.0084
−2.2	.0139	.0136	.0132	.0129	.0125	.0122	.0119	.0116	.0113	.0110
−2.1	.0179	.0174	.0170	.0166	.0162	.0158	.0154	.0150	.0146	.0143
−2.0	.0228	.0222	.0217	.0212	.0207	.0202	.0197	.0192	.0188	.0183
−1.9	.0287	.0281	.0274	.0268	.0262	.0256	.0250	.0244	.0239	.0233
−1.8	.0359	.0351	.0344	.0336	.0329	.0322	.0314	.0307	.0301	.0294
−1.7	.0446	.0436	.0427	.0418	.0409	.0401	.0392	.0384	.0375	.0367
−1.6	.0548	.0537	.0526	.0516	.0505	.0495	.0485	.0475	.0465	.0455
−1.5	.0668	.0655	.0643	.0630	.0618	.0606	.0594	.0582	.0571	.0559
−1.4	.0808	.0793	.0778	.0764	.0749	.0735	.0721	.0708	.0694	.0681
−1.3	.0968	.0951	.0934	.0918	.0901	.0885	.0869	.0853	.0838	.0823
−1.2	.1151	.1131	.1112	.1093	.1075	.1056	.1038	.1020	.1003	.0985
−1.1	.1357	.1335	.1314	.1292	.1271	.1251	.1230	.1210	.1190	.1170
−1.0	.1587	.1562	.1539	.1515	.1492	.1469	.1446	.1423	.1401	.1379
−0.9	.1841	.1814	.1788	.1762	.1736	.1711	.1685	.1660	.1635	.1611
−0.8	.2119	.2090	.2061	.2033	.2005	.1977	.1949	.1922	.1894	.1867
−0.7	.2420	.2389	.2358	.2327	.2296	.2266	.2236	.2206	.2177	.2148
−0.6	.2743	.2709	.2676	.2643	.2611	.2578	.2546	.2514	.2483	.2451
−0.5	.3085	.3050	.3015	.2981	.2946	.2912	.2877	.2843	.2810	.2776
−0.4	.3446	.3409	.3372	.3336	.3300	.3264	.3228	.3192	.3156	.3121
−0.3	.3821	.3783	.3745	.3707	.3669	.3632	.3594	.3557	.3520	.3483
−0.2	.4207	.4168	.4129	.4090	.4052	.4013	.3974	.3936	.3897	.3859
−0.1	.4602	.4562	.4522	.4483	.4443	.4404	.4364	.4325	.4286	.4247
−0.0	.5000	.4960	.4920	.4880	.4840	.4801	.4761	.4721	.4681	.4641

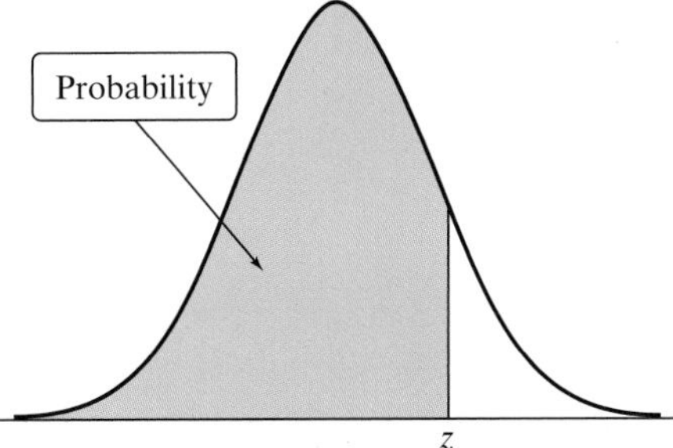

Table entry for z is the area under the standard normal curve to the left of z.

Table A Standard normal probabilities (continued)

z	.00	.01	.02	.03	.04	.05	.06	.07	.08	.09
0.0	.5000	.5040	.5080	.5120	.5160	.5199	.5239	.5279	.5319	.5359
0.1	.5398	.5438	.5478	.5517	.5557	.5596	.5636	.5675	.5714	.5753
0.2	.5793	.5832	.5871	.5910	.5948	.5987	.6026	.6064	.6103	.6141
0.3	.6179	.6217	.6255	.6293	.6331	.6368	.6406	.6443	.6480	.6517
0.4	.6554	.6591	.6628	.6664	.6700	.6736	.6772	.6808	.6844	.6879
0.5	.6915	.6950	.6985	.7019	.7054	.7088	.7123	.7157	.7190	.7224
0.6	.7257	.7291	.7324	.7357	.7389	.7422	.7454	.7486	.7517	.7549
0.7	.7580	.7611	.7642	.7673	.7704	.7734	.7764	.7794	.7823	.7852
0.8	.7881	.7910	.7939	.7967	.7995	.8023	.8051	.8078	.8106	.8133
0.9	.8159	.8186	.8212	.8238	.8264	.8289	.8315	.8340	.8365	.8389
1.0	.8413	.8438	.8461	.8485	.8508	.8531	.8554	.8577	.8599	.8621
1.1	.8643	.8665	.8686	.8708	.8729	.8749	.8770	.8790	.8810	.8830
1.2	.8849	.8869	.8888	.8907	.8925	.8944	.8962	.8980	.8997	.9015
1.3	.9032	.9049	.9066	.9082	.9099	.9115	.9131	.9147	.9162	.9177
1.4	.9192	.9207	.9222	.9236	.9251	.9265	.9279	.9292	.9306	.9319
1.5	.9332	.9345	.9357	.9370	.9382	.9394	.9406	.9418	.9429	.9441
1.6	.9452	.9463	.9474	.9484	.9495	.9505	.9515	.9525	.9535	.9545
1.7	.9554	.9564	.9573	.9582	.9591	.9599	.9608	.9616	.9625	.9633
1.8	.9641	.9649	.9656	.9664	.9671	.9678	.9686	.9693	.9699	.9706
1.9	.9713	.9719	.9726	.9732	.9738	.9744	.9750	.9756	.9761	.9767
2.0	.9772	.9778	.9783	.9788	.9793	.9798	.9803	.9808	.9812	.9817
2.1	.9821	.9826	.9830	.9834	.9838	.9842	.9846	.9850	.9854	.9857
2.2	.9861	.9864	.9868	.9871	.9875	.9878	.9881	.9884	.9887	.9890
2.3	.9893	.9896	.9898	.9901	.9904	.9906	.9909	.9911	.9913	.9916
2.4	.9918	.9920	.9922	.9925	.9927	.9929	.9931	.9932	.9934	.9936
2.5	.9938	.9940	.9941	.9943	.9945	.9946	.9948	.9949	.9951	.9952
2.6	.9953	.9955	.9956	.9957	.9959	.9960	.9961	.9962	.9963	.9964
2.7	.9965	.9966	.9967	.9968	.9969	.9970	.9971	.9972	.9973	.9974
2.8	.9974	.9975	.9976	.9977	.9977	.9978	.9979	.9979	.9980	.9981
2.9	.9981	.9982	.9982	.9983	.9984	.9984	.9985	.9985	.9986	.9986
3.0	.9987	.9987	.9987	.9988	.9988	.9989	.9989	.9989	.9990	.9990
3.1	.9990	.9991	.9991	.9991	.9992	.9992	.9992	.9992	.9993	.9993
3.2	.9993	.9993	.9994	.9994	.9994	.9994	.9994	.9995	.9995	.9995
3.3	.9995	.9995	.9995	.9996	.9996	.9996	.9996	.9996	.9996	.9997
3.4	.9997	.9997	.9997	.9997	.9997	.9997	.9997	.9997	.9997	.9998

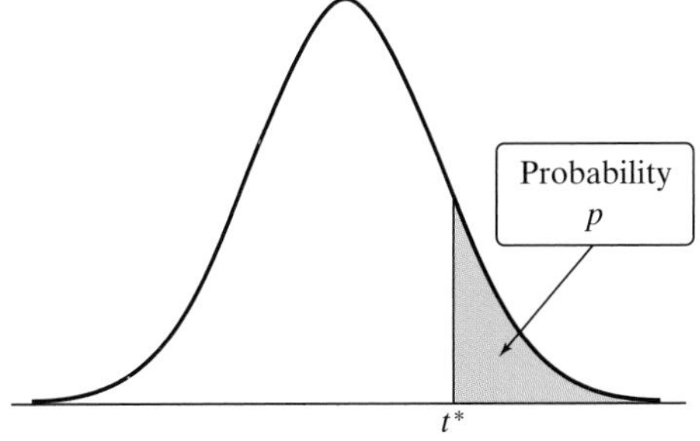

Table entry for p and C is the point t^* with probability p lying to its right and probability C lying between $-t^*$ and t^*.

Table B *t* distribution critical values

	Tail probability *p*											
df	**.25**	**.20**	**.15**	**.10**	**.05**	**.025**	**.02**	**.01**	**.005**	**.0025**	**.001**	**.0005**
1	1.000	1.376	1.963	3.078	6.314	12.71	15.89	31.82	63.66	127.3	318.3	636.6
2	0.816	1.061	1.386	1.886	2.920	4.303	4.849	6.965	9.925	14.09	22.33	31.60
3	0.765	0.978	1.250	1.638	2.353	3.182	3.482	4.541	5.841	7.453	10.21	12.92
4	0.741	0.941	1.190	1.533	2.132	2.776	2.999	3.747	4.604	5.598	7.173	8.610
5	0.727	0.920	1.156	1.476	2.015	2.571	2.757	3.365	4.032	4.773	5.893	6.869
6	0.718	0.906	1.134	1.440	1.943	2.447	2.612	3.143	3.707	4.317	5.208	5.959
7	0.711	0.896	1.119	1.415	1.895	2.365	2.517	2.998	3.499	4.029	4.785	5.408
8	0.706	0.889	1.108	1.397	1.860	2.306	2.449	2.896	3.355	3.833	4.501	5.041
9	0.703	0.883	1.100	1.383	1.833	2.262	2.398	2.821	3.250	3.690	4.297	4.781
10	0.700	0.879	1.093	1.372	1.812	2.228	2.359	2.764	3.169	3.581	4.144	4.587
11	0.697	0.876	1.088	1.363	1.796	2.201	2.328	2.718	3.106	3.497	4.025	4.437
12	0.695	0.873	1.083	1.356	1.782	2.179	2.303	2.681	3.055	3.428	3.930	4.318
13	0.694	0.870	1.079	1.350	1.771	2.160	2.282	2.650	3.012	3.372	3.852	4.221
14	0.692	0.868	1.076	1.345	1.761	2.145	2.264	2.624	2.977	3.326	3.787	4.140
15	0.691	0.866	1.074	1.341	1.753	2.131	2.249	2.602	2.947	3.286	3.733	4.073
16	0.690	0.865	1.071	1.337	1.746	2.120	2.235	2.583	2.921	3.252	3.686	4.015
17	0.689	0.863	1.069	1.333	1.740	2.110	2.224	2.567	2.898	3.222	3.646	3.965
18	0.688	0.862	1.067	1.330	1.734	2.101	2.214	2.552	2.878	3.197	3.611	3.922
19	0.688	0.861	1.066	1.328	1.729	2.093	2.205	2.539	2.861	3.174	3.579	3.883
20	0.687	0.860	1.064	1.325	1.725	2.086	2.197	2.528	2.845	3.153	3.552	3.850
21	0.686	0.859	1.063	1.323	1.721	2.080	2.189	2.518	2.831	3.135	3.527	3.819
22	0.686	0.858	1.061	1.321	1.717	2.074	2.183	2.508	2.819	3.119	3.505	3.792
23	0.685	0.858	1.060	1.319	1.714	2.069	2.177	2.500	2.807	3.104	3.485	3.768
24	0.685	0.857	1.059	1.318	1.711	2.064	2.172	2.492	2.797	3.091	3.467	3.745
25	0.684	0.856	1.058	1.316	1.708	2.060	2.167	2.485	2.787	3.078	3.450	3.725
26	0.684	0.856	1.058	1.315	1.706	2.056	2.162	2.479	2.779	3.067	3.435	3.707
27	0.684	0.855	1.057	1.314	1.703	2.052	2.158	2.473	2.771	3.057	3.421	3.690
28	0.683	0.855	1.056	1.313	1.701	2.048	2.154	2.467	2.763	3.047	3.408	3.674
29	0.683	0.854	1.055	1.311	1.699	2.045	2.150	2.462	2.756	3.038	3.396	3.659
30	0.683	0.854	1.055	1.310	1.697	2.042	2.147	2.457	2.750	3.030	3.385	3.646
40	0.681	0.851	1.050	1.303	1.684	2.021	2.123	2.423	2.704	2.971	3.307	3.551
50	0.679	0.849	1.047	1.299	1.676	2.009	2.109	2.403	2.678	2.937	3.261	3.496
60	0.679	0.848	1.045	1.296	1.671	2.000	2.099	2.390	2.660	2.915	3.232	3.460
80	0.678	0.846	1.043	1.292	1.664	1.990	2.088	2.374	2.639	2.887	3.195	3.416
100	0.677	0.845	1.042	1.290	1.660	1.984	2.081	2.364	2.626	2.871	3.174	3.390
1000	0.675	0.842	1.037	1.282	1.646	1.962	2.056	2.330	2.581	2.813	3.098	3.300
∞	0.674	0.841	1.036	1.282	1.645	1.960	2.054	2.326	2.576	2.807	3.091	3.291
	50%	**60%**	**70%**	**80%**	**90%**	**95%**	**96%**	**98%**	**99%**	**99.5%**	**99.8%**	**99.9%**
	Confidence level *C*											

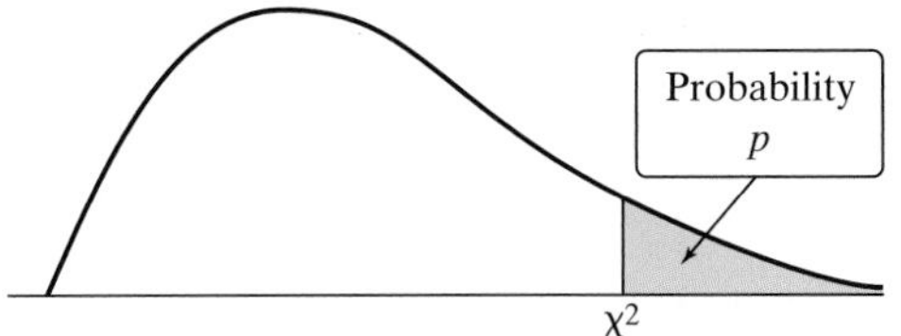

Table entry for p is the point χ^2 with probability p lying to its right.

Table C Chi-square distribution critical values

	Tail probability p											
df	.25	.20	.15	.10	.05	.025	.02	.01	.005	.0025	.001	.0005
1	1.32	1.64	2.07	2.71	3.84	5.02	5.41	6.63	7.88	9.14	10.83	12.12
2	2.77	3.22	3.79	4.61	5.99	7.38	7.82	9.21	10.60	11.98	13.82	15.20
3	4.11	4.64	5.32	6.25	7.81	9.35	9.84	11.34	12.84	14.32	16.27	17.73
4	5.39	5.99	6.74	7.78	9.49	11.14	11.67	13.28	14.86	16.42	18.47	20.00
5	6.63	7.29	8.12	9.24	11.07	12.83	13.39	15.09	16.75	18.39	20.51	22.11
6	7.84	8.56	9.45	10.64	12.59	14.45	15.03	16.81	18.55	20.25	22.46	24.10
7	9.04	9.80	10.75	12.02	14.07	16.01	16.62	18.48	20.28	22.04	24.32	26.02
8	10.22	11.03	12.03	13.36	15.51	17.53	18.17	20.09	21.95	23.77	26.12	27.87
9	11.39	12.24	13.29	14.68	16.92	19.02	19.68	21.67	23.59	25.46	27.88	29.67
10	12.55	13.44	14.53	15.99	18.31	20.48	21.16	23.21	25.19	27.11	29.59	31.42
11	13.70	14.63	15.77	17.28	19.68	21.92	22.62	24.72	26.76	28.73	31.26	33.14
12	14.85	15.81	16.99	18.55	21.03	23.34	24.05	26.22	28.30	30.32	32.91	34.82
13	15.98	16.98	18.20	19.81	22.36	24.74	25.47	27.69	29.82	31.88	34.53	36.48
14	17.12	18.15	19.41	21.06	23.68	26.12	26.87	29.14	31.32	33.43	36.12	38.11
15	18.25	19.31	20.60	22.31	25.00	27.49	28.26	30.58	32.80	34.95	37.70	39.72
16	19.37	20.47	21.79	23.54	26.30	28.85	29.63	32.00	34.27	36.46	39.25	41.31
17	20.49	21.61	22.98	24.77	27.59	30.19	31.00	33.41	35.72	37.95	40.79	42.88
18	21.60	22.76	24.16	25.99	28.87	31.53	32.35	34.81	37.16	39.42	42.31	44.43
19	22.72	23.90	25.33	27.20	30.14	32.85	33.69	36.19	38.58	40.88	43.82	45.97
20	23.83	25.04	26.50	28.41	31.41	34.17	35.02	37.57	40.00	42.34	45.31	47.50
21	24.93	26.17	27.66	29.62	32.67	35.48	36.34	38.93	41.40	43.78	46.80	49.01
22	26.04	27.30	28.82	30.81	33.92	36.78	37.66	40.29	42.80	45.20	48.27	50.51
23	27.14	28.43	29.98	32.01	35.17	38.08	38.97	41.64	44.18	46.62	49.73	52.00
24	28.24	29.55	31.13	33.20	36.42	39.36	40.27	42.98	45.56	48.03	51.18	53.48
25	29.34	30.68	32.28	34.38	37.65	40.65	41.57	44.31	46.93	49.44	52.62	54.95
26	30.43	31.79	33.43	35.56	38.89	41.92	42.86	45.64	48.29	50.83	54.05	56.41
27	31.53	32.91	34.57	36.74	40.11	43.19	44.14	46.96	49.64	52.22	55.48	57.86
28	32.62	34.03	35.71	37.92	41.34	44.46	45.42	48.28	50.99	53.59	56.89	59.30
29	33.71	35.14	36.85	39.09	42.56	45.72	46.69	49.59	52.34	54.97	58.30	60.73
30	34.80	36.25	37.99	40.26	43.77	46.98	47.96	50.89	53.67	56.33	59.70	62.16
40	45.62	47.27	49.24	51.81	55.76	59.34	60.44	63.69	66.77	69.70	73.40	76.09
50	56.33	58.16	60.35	63.17	67.50	71.42	72.61	76.15	79.49	82.66	86.66	89.56
60	66.98	68.97	71.34	74.40	79.08	83.30	84.58	88.38	91.95	95.34	99.61	102.7
80	88.13	90.41	93.11	96.58	101.9	106.6	108.1	112.3	116.3	120.1	124.8	128.3
100	109.1	111.7	114.7	118.5	124.3	129.6	131.1	135.8	140.2	144.3	149.4	153.2

Table D Random digits

Line								
101	19223	95034	05756	28713	96409	12531	42544	82853
102	73676	47150	99400	01927	27754	42648	82425	36290
103	45467	71709	77558	00095	32863	29485	82226	90056
104	52711	38889	93074	60227	40011	85848	48767	52573
105	95592	94007	69971	91481	60779	53791	17297	59335
106	68417	35013	15529	72765	85089	57067	50211	47487
107	82739	57890	20807	47511	81676	55300	94383	14893
108	60940	72024	17868	24943	61790	90656	87964	18883
109	36009	19365	15412	39638	85453	46816	83485	41979
110	38448	48789	18338	24697	39364	42006	76688	08708
111	81486	69487	60513	09297	00412	71238	27649	39950
112	59636	88804	04634	71197	19352	73089	84898	45785
113	62568	70206	40325	03699	71080	22553	11486	11776
114	45149	32992	75730	66280	03819	56202	02938	70915
115	61041	77684	94322	24709	73698	14526	31893	32592
116	14459	26056	31424	80371	65103	62253	50490	61181
117	38167	98532	62183	70632	23417	26185	41448	75532
118	73190	32533	04470	29669	84407	90785	65956	86382
119	95857	07118	87664	92099	58806	66979	98624	84826
120	35476	55972	39421	65850	04266	35435	43742	11937
121	71487	09984	29077	14863	61683	47052	62224	51025
122	13873	81598	95052	90908	73592	75186	87136	95761
123	54580	81507	27102	56027	55892	33063	41842	81868
124	71035	09001	43367	49497	72719	96758	27611	91596
125	96746	12149	37823	71868	18442	35119	62103	39244
126	96927	19931	36809	74192	77567	88741	48409	41903
127	43909	99477	25330	64359	40085	16925	85117	36071
128	15689	14227	06565	14374	13352	49367	81982	87209
129	36759	58984	68288	22913	18638	54303	00795	08727
130	69051	64817	87174	09517	84534	06489	87201	97245
131	05007	16632	81194	14873	04197	85576	45195	96565
132	68732	55259	84292	08796	43165	93739	31685	97150
133	45740	41807	65561	33302	07051	93623	18132	09547
134	27816	78416	18329	21337	35213	37741	04312	68508
135	66925	55658	39100	78458	11206	19876	87151	31260
136	08421	44753	77377	28744	75592	08563	79140	92454
137	53645	66812	61421	47836	12609	15373	98481	14592
138	66831	68908	40772	21558	47781	33586	79177	06928
139	55588	99404	70708	41098	43563	56934	48394	51719
140	12975	13258	13048	45144	72321	81940	00360	02428
141	96767	35964	23822	96012	94591	65194	50842	53372
142	72829	50232	97892	63408	77919	44575	24870	04178
143	88565	42628	17797	49376	61762	16953	88604	12724
144	62964	88145	83083	69453	46109	59505	69680	00900
145	19687	12633	57857	95806	09931	02150	43163	58636
146	37609	59057	66967	83401	60705	02384	90597	93600
147	54973	86278	88737	74351	47500	84552	19909	67181
148	00694	05977	19664	65441	20903	62371	22725	53340
149	71546	05233	53946	68743	72460	27601	45403	88692
150	07511	88915	41267	16853	84569	79367	32337	03316

Inference Summary

How to Organize an Inference Problem: The Four-Step Process

Confidence intervals	Significance tests
STATE: State the parameter you want to estimate and the confidence level.	**STATE:** State the hypotheses, parameter(s), and significance level.
PLAN: Identify the appropriate inference method and check conditions.	**PLAN:** Identify the appropriate inference method and check conditions.
DO: If the conditions are met, perform calculations.	**DO:** If the conditions are met, perform calculations. • Calculate the test statistic. • Find the *P*-value.
CONCLUDE: Interpret your interval in the context of the problem.	**CONCLUDE:** Make a conclusion about the hypotheses in the context of the problem.

confidence interval = statistic ± (critical value)(standard error of statistic)

$$\text{standardized test statistic} = \frac{\text{statistic} - \text{parameter}}{\text{standard error of statistic}}$$

Inference about	Number of samples/ groups	Interval or test (Section)	Name of procedure (TI-83/84 name) Formula	Conditions
Proportions	1	Interval **(6B)**	One-sample z interval for p (1-PropZInt) $\hat{p} \pm z^* \sqrt{\frac{\hat{p}(1-\hat{p})}{n}}$	**Random:** The data come from a random sample from the population of interest. ○ **10%:** When sampling without replacement, $n < 0.10N$. **Large Counts:** *Interval:* Both $n\hat{p}$ and $n(1-\hat{p}) \geq 10$. *Test:* Both np_0 and $n(1-p_0) \geq 10$, where p_0 is the value of p specified by the null hypothesis.
		Test **(6D)**	One-sample z test for p (1-PropZTest) $z = \frac{\hat{p} - p_0}{\sqrt{\frac{p_0(1-p_0)}{n}}}$	
	2	Interval **(6E)**	Two-sample z interval for $p_1 - p_2$ (2-PropZInt) $(\hat{p}_1 - \hat{p}_2) \pm z^* \sqrt{\frac{\hat{p}_1(1-\hat{p}_1)}{n_1} + \frac{\hat{p}_2(1-\hat{p}_2)}{n_2}}$	**Random:** The data come from two independent random samples or from two groups in a randomized experiment. ○ **10%:** When sampling without replacement, $n_1 < 0.10N_1$ and $n_2 < 0.10N_2$. **Large Counts:** *Interval:* The observed counts of successes and failures in each sample or group—$n_1\hat{p}_1$, $n_1(1-\hat{p}_1)$, $n_2\hat{p}_2$ and $n_2(1-\hat{p}_2)$—are all ≥ 10. *Test:* The expected counts of successes and failures in each sample or group—$n_1\hat{p}_C$, $n_1(1-\hat{p}_C)$, $n_2\hat{p}_C$, $n_2(1-\hat{p}_C)$—are all ≥ 10.
		Test **(6F)**	Two-sample z test for $p_1 - p_2$ (2-PropZTest) $z = \frac{(\hat{p}_1 - \hat{p}_2) - 0}{\sqrt{\hat{p}_C(1-\hat{p}_C)\left(\frac{1}{n_1} + \frac{1}{n_2}\right)}}$ where $\hat{p}_C = \frac{\text{total successes}}{\text{total sample size}} = \frac{X_1 + X_2}{n_1 + n_2}$	

Inference about	Number of samples/ groups	Interval or test (Section)	Name of procedure (TI-83/84 name) Formula	Conditions
Means	1	Interval **(7A)**	One-sample t interval for μ (TInterval) $\bar{x} \pm t^* \frac{s_x}{\sqrt{n}}$ $\quad df = n - 1$	**Random:** The data come from a random sample from the population of interest. ○ **10%:** When sampling without replacement, $n < 0.10N$. **Normal/Large Sample:** The population distribution is approximately normal or the sample size is large ($n \geq 30$). If the population distribution has unknown shape and $n < 30$, a graph of the sample data shows no strong skewness or outliers.
		Test **(7B)**	One-sample t test for μ (T-Test) $t = \frac{\bar{x} - \mu_0}{\frac{s_x}{\sqrt{n}}}$ $\quad df = n - 1$	
	Paired data	Interval **(7A)**	Paired t interval for μ_{Diff} (TInterval) $\bar{x}_{Diff} + t^* \frac{s_{Diff}}{\sqrt{n_{Diff}}}$ $\quad df = n_{Diff} - 1$	**Random:** Paired data come from a random sample from the population of interest or from a randomized experiment. ○ **10%:** When sampling without replacement, $n_{Diff} < 0.10N_{Diff}$. **Normal/Large Sample:** The population distribution of differences is approximately normal or the sample size is large ($n_{Diff} \geq 30$). If the population distribution of differences has unknown shape and $n_{Diff} < 30$, a graph of the sample differences shows no strong skewness or outliers.
		Test **(7B)**	Paired t test for μ_{Diff} (T-Test) $t = \frac{\bar{x}_{Diff} - \mu_0}{\frac{s_{Diff}}{\sqrt{n_{Diff}}}}$ $\quad df = n_{Diff} - 1$	
	2	Interval **(7C)**	Two-sample t interval for $\mu_1 - \mu_2$ (2-SampTInt) $(\bar{x}_1 - \bar{x}_2) \pm t^* \sqrt{\frac{s_1^2}{n_1} + \frac{s_2^2}{n_2}}$ df from technology	**Random:** The data come from two independent random samples or from two groups in a randomized experiment. ○ **10%:** When sampling without replacement, $n_1 < 0.10N_1$ and $n_2 < 0.10N_2$. **Normal/Large Sample:** For each sample, the data come from an approximately normally distributed population or the sample size is large ($n \geq 30$). For each sample, if the population distribution has unknown shape and $n < 30$, a graph of the sample data shows no strong skewness or outliers.
		Test **(7D)**	Two-sample t test for $\mu_1 - \mu_2$ (2-SampTTest) $t = \frac{(\bar{x}_1 - \bar{x}_2) - 0}{\sqrt{\frac{s_1^2}{n_1} + \frac{s_2^2}{n_2}}}$ df from technology	